Genetic Instabilities and Hereditary Neurological Diseases

Genetic Instabilities and Hereditary Neurological Diseases

Editors

ROBERT D. WELLS
Institute of Biosciences and Technology
Texas A&M University
The Texas Medical Center
Houston, Texas

STEPHEN T. WARREN
Howard Hughes Medical Institute
Emory University School of Medicine
Atlanta, Georgia

Associate Editor

MARION SARMIENTO
Institute of Biosciences and Technology
Texas A&M University
The Texas Medical Center
Houston, Texas

ACADEMIC PRESS
San Diego London Boston New York Sydney Tokyo Toronto

This book is printed on acid-free paper. 

Academic Press
a division of Harcourt Brace & Company
525 B Street, Suite 1900, San Diego, California 92101-4495, USA
http://www.apnet.com

Academic Press Limited
24-28 Oval Road, London NW1 7DX, UK
http://www.hbuk.co.uk/ap/

Library of Congress Catalog Card Number: 98-84947

International Standard Book Number: 0-12-742935-2

PRINTED IN THE UNITED STATES OF AMERICA
98 99 00 01 02 03 MM 9 8 7 6 5 4 3 2 1

Contents

PART III Other Fragile Sites

CHAPTER 5 FRAXE Mental Retardation and Other Folate-Sensitive Fragile Sites

DAVID L. NELSON

CHAPTER 6 Common Fragile Sites

THOMAS W. GLOVER

PART IV Spinal and Bulbar Muscular Atrophy

CHAPTER 7 Kennedy's Disease: Clinical Aspects

JEFFREY D. ZAJAC AND HELEN E. MACLEAN

CHAPTER 8 Genetics and Molecular Biology of the Androgen Receptor CAG Repeat

DIANE E. MERRY AND KENNETH H. FISCHBECK

PART V Myotonic Dystrophy

CHAPTER 9 Myotonic Dystrophy as a Trinucleotide Repeat Disorder—A Clinical Perspective

PETER S. HARPER

CHAPTER 10 Genetic Studies of the Myotonic Dystrophy CTG Repeat

JAMES D. WARING AND ROBERT G. KORNELUK

CHAPTER 11 Biochemical Studies of DM Protein Kinase (DMPK)

HENRY F. EPSTEIN AND SHIDA JIN

Chapter 12 Is Myotonic Dystrophy (DM) the Result of a Contiguous Gene Defect?

Catherine L. Winchester and Keith J. Johnson

Chapter 13 Murine Models for Myotonic Dystrophy

Darren G. Monckton, Tetsuo Ashizawa, and Michael J. Siciliano

Part VI Dentatorubral–Pallidoluysian Atrophy

Chapter 14 Clinical Aspects of DRPLA

Ikuko Kondo

Chapter 15 Molecular Genetics of Dentatorubral–Pallidoluysian Atrophy (DRPLA)

Shoji Tsuji

Chapter 16 The Allelic Variant: Haw River Syndrome

Jeffery M. Vance

Part VII Spinocerebellar Ataxia Type 1

Chapter 17 Clinical Aspects of Spinocerebellar Ataxia 1

S. H. Subramony and P. J. S. Vig

Chapter 18 Clues about the Pathogenesis of SCA1 from Biochemical and Molecular Studies of Ataxin-1

Beena T. Koshy, Antoni Matilla, and Huda Y. Zoghbi

Chapter 19 Murine Model of SCA1: Repeat Instability and Neurobiology

Michael D. Kaytor and Harry T. Orr

PART VIII Other Dominant Spinocerebellar Ataxias

PART IX Huntington's Disease

Chapter 25 Murine Models of Huntington's Disease

Gillian P. Bates, Erich E. Wanker, and Stephen W. Davies

Part X Friedreich's Ataxia

Chapter 26 Friedreich's Ataxia

Massimo Pandolfo and Michel Koenig

Part XI Candidate Disorders Revealing Anticipation

Chapter 27 Anticipation and Psychiatric Disorders

Melvin G. McInnis and Russell L. Margolis

Chapter 28 Genetic Anticipation in Neurological and Other Disorders

Andrew D. Paterson, David M. J. Naimark, John B. Vincent, James L. Kennedy, and Arturas Petronis

Part XII Approaches to Detect Unstable Trinucleotide Repeat Loci

Chapter 29 Detection of Unstable Trinucleotide Repeat Loci: Genome and cDNA Screening

Russell L. Margolis and Christopher A. Ross

Chapter 30 RED Technology

Martin Schalling, Kerstin Lindblad, Qiu-Ping Yuan, Catherine E. Burgess, Cecilia Zander, and Tom Hudson

Chapter 31 Selective Recognition of Proteins with Pathological Polyglutamine Tracts by a Monoclonal Antibody

Yvon Trottier, Gabrielle Zeder-Lutz, and Jean-Louis Mandel

CHAPTER 32 DIRECT Technologies

K. SANPEI AND S. TSUJI

PART XIII Systems for the Study of Genetic Instabilities

CHAPTER 33 Systems for the Study of Genetic Instabilities: *Escherichia coli*

ALBINO BACOLLA, RICHARD P. BOWATER, AND ROBERT D. WELLS

CHAPTER 34 Genetic Instabilities in Yeast

SUE JINKS-ROBERTSON, CHRISTOPHER GREENE, AND WENLIANG CHEN

CHAPTER 35 Systems for the Study of Triplet Repeat Instability: Cultured Mammalian Cells

PETER STEINBACH, DORIS WÖHRLE, DIETER GLÄSER, AND WALTHER VOGEL

CHAPTER 36 Genetic Instability in Transgenic Mice Deficient in DNA Mismatch Repair

PETER M. GLAZER, SUSAN E. ANDREW, FRANK R. JIRIK, AND LATHA NARAYANAN

CHAPTER 37 Human Germline Mutation Analysis by Single Genome PCR: Application to Dynamic Mutations

ESTHER P. LEEFLANG, SIMON TAVARÉ, PAUL MARJORAM, RAJI GREWAL, CAROLYN O. S. NEAL, AND NORMAN ARNHEIM

PART XIV Biophysical and Structural Studies on Triplet Repeat Sequences

CHAPTER 38 Biophysical and Structural Studies on Triplet Repeat Sequences: Duplex Triplet Repeat Structures

ROBERT GELLIBOLIAN AND ALBINO BACOLLA

CHAPTER 45 *In Vitro* DNA Synthesis of Triplet Repeat Sequences

KEIICHI OHSHIMA AND ROBERT D. WELLS

APPENDIX: Replication Fork Progression through CTG · CAG Triplet Repeats *in Vitro*

HIROSHI HIASA AND KENNETH J. MARIANS

PART XVI Oligoglutamine Effects

CHAPTER 46 Amino Acid Repeats in Proteins and the Neurological Diseases Produced by Polyglutamine

HOWARD GREEN AND PHILIPPE DJIAN

CHAPTER 47 Pathogenesis of Polyglutamine Neurodegenerative Diseases: Toward a Unifying Mechanism

CHRISTOPHER A. ROSS, RUSSELL L. MARGOLIS, MARK W. BECHER, JONATHAN D. WOOD, SIMONE ENGELENDER, AND ALAN H. SHARP

PART XVII Genome Instability Not Involving Triplet Repeats

CHAPTER 48 Expansion of a 12-mer Repeat in Progressive Myoclonus Epilepsy

STYLIANOS E. ANTONARAKIS

Contributors

Numbers in parentheses indicate the pages on which the author's contributions begin.

Nacer Abbas (273)
INSERM U289, Hôpital de la Salpêtrière, 75651 Paris Cedex 13, France

Yves Agid (273)
INSERM U289 and Fédération de Neurologie, Hôpital de la Salpêtrière, 75651 Paris Cedex 13, France

Susan Andrew (529)
Biomedical Research Center and Department of Medicine, University of British Columbia, 2222 Health Sciences Mall, Vancouver, British Columbia, Canada V6T 1Z3

Stylianos E. Antonarakis (779)
Laboratory of Human Molecular Genetics, Department of Genetics and Microbiology, University of Geneva Medical School, 1, rue Michel-Servet Cantonal Hospital of Geneva, Geneva 4 1211, Switzerland

Norman Arnheim (543)
Molecular Biology Program, University of Southern California, Stauffe Hall of Science 172, Los Angeles, California 90089-1340

Tetsuo Ashizawa (181)
Department of Neurology, Baylor College of Medicine, One Baylor Plaza, Houston, Texas 77030

Albino Bacolla (467, 561)
Center for Genome Research, Institute of Biosciences and Technology, Texas A&M University, 2121 West Holcombe Boulevard, Houston, Texas 77030-3303

Gillian P. Bates (355)
Medical and Molecular Genetics, United Medical and Dental Schools, Guy's Hospital, London, SE1 9RT United Kingdom

Mark W. Becher (761)
Department of Pathology, Johns Hopkins University School of Medicine, 720 Rutland Avenue, Ross Research Building 618, Baltimore, Maryland 21205-2196

Lisa A. Boardman (791)
Division of Experimental Pathology, Mayo Clinic, 200 First Street, Southwest, Rochester, Minnesota 55905

Richard P. Bowater (467)
Imperial Cancer Research Center, Clare Hall Laboratories, South Mimms, Hertfordshire EN6 3LD, United Kingdom

E. Morton Bradbury (647)
Department of Biology and Chemistry, School of Medicine, University of California, Building 1A, Room 4110, Davis, California 95616

Alexis Brice (273)
INSERM U289 and Fédération de Neurologie, Hôpital de la Salpêtrière, 75651 Paris Cedex 13, France

Lawrence J. Bugart (791)
Department of Pathology, Mayo Clinic, 200 First Street, Southwest, Rochester, Minnesota 55905

Catherine E. Burgess (439)
Center for Molecular Medicine, Karolinska Institute, Neurogenetics Unit, L8:00, S-171 76 Stockholm, Sweden

Géraldine Cancel (273)
INSERM U289, Hôpital de la Salpêtrière, 75651 Paris Cedex 13, France

Paolo Catasti (647)
Life Sciences Division, Los Alamos National Laboratory, LS-2, MS 880, Los Alamos, New Mexico 87545

Wenliang Chen (485)
Department of Biology, Graduate Program in Genetics and Molecular Biology, Emory University, 1510 Clifton Road, Atlanta, Georgia 30322

Xian Chen (647)
Life Sciences Division, Los Alamos National Laboratory, LS-2, MS 880, Los Alamos, New Mexico 87545

Julie M. Cunningham (791)
Division of Experimental Pathology, Mayo Clinic, 200 First Street, Southwest, Rochester, Minnesota 55905

Gilles David (273)
INSERM U289, Hôpital de la Salpêtrière, 75651 Paris Cedex 13, France

Stephen W. Davies (355)
Department of Anatomy and Developmental Biology, University College London, London, WC1E 6BT United Kingdom

Philippe Djian (739)
Moléculaire et le Développement, Centre de Recherche sur l'Endocrinologie, CNRS, 9, rue Jules Hetzel, 92190 Meudon, Bellevue, France

Alexandra Dürr (273)
INSERM U289 and Fédération de Neurologie, Hôpital de la Salpêtrière, 75651 Paris Cedex 13, France

Simone Engelender (761)
Department of Psychiatry, Johns Hopkins University School of Medicine, 720 Rutland Ave., Baltimore, Maryland 21205-2196

Henry F. Epstein (147)
Departments of Neurology, Biochemistry, Cell Biology, Neuroscience, and Molecular Physiology and Biophysics, Baylor College of Medicine, One Baylor Plaza, Houston, Texas 77030

Yue Feng (27)
Howard Hughes Medical Institute, Departments of Biochemistry, Pediatrics, and Genetics, Emory University School of Medicine, Atlanta, Georgia 30322

Kenneth H. Fischbeck (101)
Department of Neurology, University of Pennsylvania School of Medicine, 415 Curie Blvd., Philadelphia, Pennsylvania 19104

Xiaolian Gao (623)
Department of Chemistry, University of Houston, 4800 Calhoun, Houston, Texas 77204-5641

Robert Gellibolian (561)
Gene Expression Laboratories, Salk Institute for Biological Studies, 10010 North Torrey Pines Road, La Jolla, California 92037

Dieter Gläser (509)
Department of Medical Genetics, University of Ulm, 89070 Ulm, Germany

Peter Glazer (529)
Department of Therapeutic Radiology, Yale University School of Medicine, 295 Congress Avenue, Room 354 BCMM, New Haven, Connecticut 06536

Thomas W. Glover (75)
Departments of Pediatrics and Human Genetics, University of Michigan, Box 0618, 4708 Medical Sciences II, 1137 E. Catherine, Ann Arbor, Michigan 48109

Rona K. Graham (325)
Department of Medical Genetics, University of British Columbia, 416-2125 E. Mall, 2211 Wesbrook Mall, Vancouver, British Columbia, V6T 2B5 Canada

Howard Green (739)
Department of Cell Biology, Harvard Medical School, 240 Longwood Avenue, Boston, Massachusetts 02115-6092

Christopher Greene (485)
Department of Biology, Graduate Program in Genetics and Molecular Biology, Emory University, 1510 Clifton Road, Atlanta, Georgia 30322

Raji Grewal (543)
Molecular Biology Program, University of Southern California, Stauffer Hall of Science 172, Los Angeles, California 90089-1340

Jack D. Griffith (677)
Lineberger Cancer Center, University of North Carolina, Campus Box 7295, Mason Farm Rd., Room 119, Chapel Hill, North Carolina 27599-7295

Goutam Gupta (647)
Theoretical Biology and Biophysics, Los Alamos National Laboratory, T-10, MS-K710, Los Alamos, New Mexico 87545

Abigail Hackam (325)
Department of Medical Genetics, University of British Columbia, 416-2125 E. Mall, 2211 Wesbrook Mall, Vancouver, British Columbia, V6T 2B5 Canada

Randi J. Hagerman (15)
University of Colorado Health Science Center and Child Development Unit, Children's Hospital, 1056 East 19th Ave., Denver, Colorado 80218

Peter S. Harper (115)
Institute of Medical Genetics, University of Wales College of Medicine, Cardiff CF4 4XN, United Kingdom

Michael R. Hayden (325)
Department of Medical Genetics, University of British Columbia, 416-2125 E. Mall, 2211 Wesbrook Mall, Vancouver, British Columbia, Canada V6T 2B5

Hiroshi Hiasa (732)
Sloan-Kettering Cancer Center, Program in Molecular Biology, 1275 York Avenue, New York, New York 10021

Xuening Huang (623)
Department of Chemistry, University of Houston, 4800 Calhoun, Houston, Texas 77204-5641

Thomas Hudson (439)
Center for Genome Research, Whitehead Institute, Massachusetts Institute of Technology, Cambridge, Massachusetts 02139

A. H. M. Mahbubul Huq (325)
Department of Medical Genetics, University of British Columbia, 416-2125 E. Mall, 221 Wesbrook Mall, Vancouver, British Columbia, V6T 2B5 Canada

Hanako Ikeda (283)
The Fourth Department, Osaka Bioscience Institute, Osaka 565, Japan

Georges Imbert (27, 273)
Institut de Génétique et de Biologie Moléculaire et Cellulaire, CNRS/INSERM/ULP/HUS, 67404 Illkirch Cedex, Strasbourg, France

Shida Jin (147)
Departments of Neurology and Molecular and Human Genetics, Baylor College of Medicine, One Baylor Plaza, Houston, Texas 77030

Sue Jinks-Robertson (485)
Department of Biology, Graduate Program in Genetics and Molecular Biology, Emory University, 1510 Clifton Road, Atlanta, Georgia 30322

Frank R. Jirik (529)
Biomedical Research Center and Department of Medicine, University of British Columbia, 2222 Health Sciences Mall, Vancouver, British Columbia, Canada V6T 1Z3

Keith J. Johnson (169)
Division of Molecular Genetics, Institute of Biomedical and Life Sciences, University of Glasgow, Anderson College, Glasgow, G11 6NU, United Kingdom

Akira Kakizuka (283)
The Fourth Department, Osaka Bioscience Institute, Osaka 565, Japan

Michael D. Kaytor (249)
Institute of Human Genetics, Department of Laboratory Medicine and Pathology, University of Minnesota, Minneapolis, Minnesota 55455

James L. Kennedy (413)
Neurogenetics Section, Clarke Institute of Psychiatry, 250 College Street, Toronto, Ontario, M5T 1R8 Canada

Michel Koenig (373)
Institut de Génétique et de Biologie Moléculaire et Cellulaire, CNRS/INSERM/ULP/HUS, 67404 Illkirch, Strasbourg, France

Ikuko Kondo (197)
Department of Hygiene, Ehime University School of Medicine, Ehime 791-02, Japan

Robert G. Korneluk (131)
Department of Microbiology and Immunology, University of Ottawa, Ottawa, Ontario K1H 8M5, Canada

Beena T. Koshy (241)
Departments of Pediatrics and Molecular and Human Genetics, Howard Hughes Medical Institute, Baylor College of Medicine, One Baylor Plaza, Houston, Texas 77030

Lisa C. Kroutil (699)
Department of Chemistry, Beloit College, 700 College St., Beloit, Wisconsin 53511

Thomas A. Kunkel (699)
Laboratory of Molecular Genetics, E3-01, Building 101, Room E342B, National Institute of Environmental Health Sciences, 111 T. W. Alexander Drive, Research Triangle Park, North Carolina 27709-2233

Esther P. Leeflang (543)
Molecular Biology Program, University of Southern California, Stauffer Hall of Sciences 172, Los Angeles, California 90089-1340

Kerstin Lindblad (439)
Center for Molecular Medicine, Karolinska Institute, Neurogenetics Unit, L8:00, S-171 76 Stockholm, Sweden

Marcy E. MacDonald (301)
Molecular Neurogenetics Unit, Massachusetts General Hospital, 13th Street, Bldg. 149, Rm. 6115, Charlestown, Massachusetts 02129

Helen E. MacLean (87)
Centre for Hormone Research, Royal Children's Hospital, Parkville, Victoria 3052, Australia

Jean-Louis Mandel (27, 273, 447)
Institut de Génétique et de Biologie Moléculaire et Cellulaire, CNRS/INSERM/ULP/HUS, 67404 Illkirch Cedex, Strasbourg, France

Karen S. Marans (301)
VA Medical Center, Boston, Massachusetts 02130

Russell L. Margolis (401, 431, 761)
Laboratory of Genetic Neurobiology, Department of Psychiatry, Johns Hopkins University School of

Medicine, 720 Rutland Ave., Baltimore, Maryland 21287-7463

Kenneth J. Marians (732)
Sloan-Kettering Cancer Center, Program in Molecular Biology, 1275 York Avenue, New York, New York 10021

S. V. Santhana Mariappan (647)
Life Sciences Division, Los Alamos National Laboratory, LS-2, MS 880, Los Alamos, New Mexico 87545

Paul Marjoram (543)
Department of Mathematics, University of Southern California, Los Angeles, California 90089-1113

Antoni Matilla (241)
Laboratory of Human Neurogenetics, Rockefeller University, 1230 York Avenue, Box 313, New York, New York 10021

Melvin G. McInnis (401)
Department of Psychiatry and Behavioral Sciences, Johns Hopkins University School of Medicine, 720 Rutland Ave., Baltimore, Maryland 21287-7463

Diane E. Merry (101)
Department of Neurology, University of Pennsylvania School of Medicine, 415 Curie Blvd., Philadelphia, Pennsylvania 19104

Darren G. Monckton (181)
Division of Molecular Genetics, Institute for Biomedical and Life Sciences, University of Glasgow, Anderson College, Glasgow, G11 6NU, United Kingdom

Richard H. Myers (301)
Department of Neurology, Neurogenetics Laboratory, Boston University School of Medicine, 80 E. Concord St., Boston, Massachusetts 02114

David M. J. Naimark (413)
Neurogenetics Section, Clarke Institute of Psychiatry, 250 College Street, Toronto, Ontario, M5T 1R8 Canada

Mika Nakamoto (283)
The Fourth Department, Osaka Bioscience Institute, Osaka 565, Japan

Latha Narayanan (529)
Department of Therapeutic Radiology, Yale University School of Medicine, 295 Congress Avenue, Room 339 BCMM, New Haven, Connecticut 06530

Carolyn O. S. Neal (543)
Molecular Biology Program, University of Southern California, Stauffer Hall of Science 172, Los Angeles, California 90089-1340

David L. Nelson (27, 65)
Department of Molecular and Human Genetics, Baylor College of Medicine, One Baylor Plaza, Houston, Texas 77030

Keiichi Oshima (717)
Research Center Louis-Charles Simard, CHUM, Pavillon Notre-Dame, 1560 Sherbrooke Street, East, Montreal, Quebec H2L 4M1, Canada

Ben A. Oostra (55)
Department of Clinical Genetics, Erasmus University Rotterdam, Post Office Box 1738, 3000 DR Rotterdam, The Netherlands

Harry T. Orr (249)
Institute of Human Genetics, Department of Laboratory Medicine and Pathology and Department of Biochemistry, University of Minnesota, Box 206, Minneapolis, Minnesota 55455

Massimo Pandolfo (373)
Département de Médecine, Université de Montréal and Department of Neurology and Neurosurgery, McGill University, 1560 Sherbrooke Street, CHUM, Montreal, Quebec, H2L 4M1 Canada

Andrew D. Paterson (413)
Neurogenetics Section, Clarke Institute of Psychiatry, 250 College Street, Toronto, Ontario, M5T 1R8 Canada

Christopher E. Pearson (585)
Center for Genome Research, Institute of Biosciences and Technology, Department of Biochemistry and Biophysics, Texas A&M University, Texas Medical Center, 2121 W. Holcombe Blvd., Houston, Texas 77030-3303

Arturas Petronis (413)
Neurogenetics Section, Clarke Institute of Psychiatry, 250 College Street, Toronto, Ontario, M5T 1R8 Canada

Stefan-M. Pulst (265)
Division of Neurology, Cedars-Sinai Medical Center, UCLA School of Medicine, 8700 Beverly Blvd., Room 8911, Los Angeles, California 90048

Christopher A. Ross (431, 761)
Laboratory of Molecular Neurobiology, Departments of Psychiatry and Neuroscience, Program in Cellular and Molecular Medicine, Johns Hopkins University School of Medicine, 720 Rutland Ave., Baltimore, Maryland 21205-2196

K. Sanpei (455)
Department of Neurology, Brain Research Institute, Nigata University, Nigata 951, Japan

Frédéric Saudou (273)
Institut de Génétique et de Biologie Moléculaire et Cellulaire, CNRS/INSERM/ULP/HUS, 67404 Illkirch Cedex, Strasbourg, France

Martin Schalling (439)
Center for Molecular Medicine, Karolinska Institute, Neurogenetics Unit, L8:00, S-171 76 Stockholm, Sweden

Alan H. Sharp (761)
Department of Psychiatry, Johns Hopkins University School of Medicine, 720 Rutland Ave., Baltimore, Maryland 21205-2196

Michael J. Siciliano (181)
Department of Molecular Genetics, University of Texas M. D. Anderson Cancer Center, Texas Medical Center, 1515 Holcombe Blvd., Houston, Texas 77030-3303

Richard Sinden (585)
Center for Genome Research, Institute of Biosciences and Technology, Department of Biochemistry and Biophysics, Texas A&M University, Texas Medical Center, 2121 W. Holcombe Blvd., Houston, Texas 77030-3303

Rakesh K. Singhal (693)
Department of Pediatrics, Perinatology Center, New York Hospital, Cornell Medical Center, 515-E. 71 St., Rm. S709, New York, New York 10022

Kenneth G. Smith (623)
Department of Biochemistry and Biophysical Sciences, University of Houston, Houston, Texas 77204-5641

Peter Steinbach (509)
Department of Medical Genetics, University of Ulm, Ulm, 89070 Germany

Giovanni Stevanin (273)
INSERM U289, Hôpital de la Salpêtrière, 75651 Paris Cedex 13, France

S. H. Subramony (231)
Department of Neurology, University of Mississippi School of Medicine, 2500 North State Street, Jackson, Mississippi 39216-4505

Simon Tavaré (543)
Department of Mathematics and Molecular Biology Program, University of Southern California, Los Angeles, California 90089

Stephen N. Thibodeau (791)
Laboratory Genetics, Mayo Clinic, 200 First Street, Southwest, Rochester, Minnesota 55905

Yvon Trottier (447)
Institut de Génétique et de Biologie Moléculaire et Cellulaire, CNRS/INSERM/ULP/HUS, 67404 Illkirch Cedex, Strasbourg, France

Shoji Tsuji (213, 455)
Department of Neurology, Brain Research Institute, Nigata University, Nigata 951, Japan

Jeffery M. Vance (223)
Division of Neurology, Department of Medicine, Duke University Medical Center, Box 2900, Durham, North Carolina 27710-0001

Parminder J. S. Vig (231)
Department of Neurology, Neurology Research Division, University of Mississippi School of Medicine, 2500 North State Street, Jackson, Mississippi 39216-4505

John B. Vincent (413)
Neurogenetics Section, Clarke Institute of Psychiatry, 250 College Street, Toronto, Ontario, M5T 1R8 Canada

Walther Vogel (509)
Department of Medical Genetics, University of Ulm, 89070 Ulm, Germany

Yuh-Hwa Wang (677)
Lineberger Cancer Center, University of North Carolina, Campus Box 7295, Rm. 119, Mason Farm Rd., Chapel Hill, North Carolina 27599-7295

Erich E. Wanker (355)
Max Planck Institute for Molecular Genetics, Berlin, Germany

James D. Waring (131)
Solange Gauthier Karsh Laboratory, Research Institute, Children's Hospital of Eastern Ontario, Ottawa, Ontario K1H 8L1, Canada

Stephen T. Warren (3, 27, 811)
Howard Hughes Medical Institute, Departments of Biochemistry, Pediatrics, and Genetics, Emory University School of Medicine, Atlanta, Georgia 30322

Cheryl L. Wellington (325)
Department of Medical Genetics, University of British Columbia, 416-2125 E. Mall, 2211 Wesbrook Mall, Vancouver, British Columbia, V6T 2B5 Canada

Robert D. Wells (467, 717, 811)
Department of Biochemistry and Biophysics, Institute of Biosciences and Technology, Texas A&M University, Texas Medical Center, 2121 W. Holcombe Blvd., Houston, Texas 77030-3303

Patrick J. Willems (55)
Department of Medical Genetics, University of Antwerp, Antwerp, Belgium

George R. Wilmot (3)
Department of Neurology, Emory University School of Medicine, Atlanta, Georgia 30322

Samuel H. Wilson (693)
Department of Health and Human Services, NIH, NIEHS, Post Office Box 12233, Research Triangle Park, North Carolina 27709

Catherine L. Winchester (169)
Division of Molecular Genetics, Institute of Biomedical and Life Sciences, University of Glasgow, Anderson College, Glasgow G11 6NU, United Kingdom

Doris Wöhrle (509)
Department of Medical Genetics, University of Ulm, 89070 Ulm, Germany

Jonathan D. Wood (761)
Department of Psychiatry, Johns Hopkins University School of Medicine, 720 Rutland Ave., Baltimore, Maryland 21205-2196

Qiu-Ping Yuan (439)
Neurogenetics Unit L8:00, Center for Molecular Medicine, Karolinska Institute, S-171 76 Stockholm, Sweden

Gaël Yvert (273)
Institut de Génétique et de Biologie Moléculaire et Cellulaire, CNRS/INSERM/ULP/HUS, 67404 Illkirch Cedex, Strasbourg, France

Jeffrey D. Zajac (87)
Department of Medicine, University of Melbourne, Royal Melbourne Hospital, Victoria 3050, Australia

Cecilia Zander (439)
Neurogenetics Unit L8:00, Center for Molecular Medicine, Karolinska Institute, S-171 76 Stockholm, Sweden

Gabrielle Zeder-Lutz (447)
Institut de Biologie Moléculaire et Cellulaire, CNRS/ULP-9021, 67404 Illkirch Cedex, Strasbourg, France

Minxue Zheng (623)
Department of Chemistry, University of Houston, 4800 Calhoun, Houston, Texas 77204-5641

Barbara Z. Zmudzka (693)
Department of Health and Human Services, NIH, NIEHS, Post Office Box 12233, Research Triangle Park, North Carolina 27709

Huda Y. Zoghbi (241)
Departments of Pediatrics and Molecular and Human Genetics, Howard Hughes Medical Institute, Baylor College of Medicine, One Baylor Plaza, Houston, Texas 77030

Foreword

There are occasions in the history of scientific discovery when observations in two completely different research disciplines unexpectedly converge to explain previously unexplained phenomena and open new sets of challenges for the researcher. Such an occasion was the discovery in 1991 that fragile X syndrome, a cytogenetic phenomenon associated with mental retardation and with the puzzling and controversial genetic process of anticipation, was caused by the germline expansion of a simple DNA sequence (CGG). Earlier, the work of H. G. Khorana's group at the University of Wisconsin–Madison in the early 1960s had shown that short synthetic oligonucleotide duplexes, containing di-, tri-, and tetranucleotide repeats, were substrates for the enzyme DNA polymerase I from *Escherichia coli* and that the products were much longer double-strand DNAs containing the priming repeat motif. Biologically, these polynucleotides were important because they were key tools in Khorana's approach to deciphering the genetic code, an approach that resulted in a Nobel Prize in 1968. Otherwise, they were interesting targets for biophysical studies. The phenomenon of genetic anticipation, along with genetic imprinting, was one of a much older, confusing, and controversial origin and history (F. Vogel and A. G. Motulsky, *Human Genetics,* 3rd ed.). The discovery of the cause of fragile X syndrome, and the subsequent identification of a substantial number of other human diseases and cytogenetic anomalies associated with genomic nucleotide triplet expansions, was a genetic discovery of profound importance.

The editors, Robert D. Wells and Stephen T. Warren, have assembled a stellar group of authors whose comprehensive surveys of the clinical, genetic, and biochemical features of all the known diseases involving triplet and other simple repeats will be the primary source of information on this important family of diseases. The publication is timely. The gathering momentum of the Human Genome Project will result in an enormous expansion of our awareness and knowledge of repeat sequences that are bound to be found distributed throughout the three billion base pairs of the haploid human genome. That knowledge will stimulate an increasing interest in the possible role of such sequences in the human condition, diversity, and disease—an interest for which this volume will undoubtedly be the source book for clinicians, biologists, geneticists, molecular biologists, and biochemists.

MICHAEL SMITH
Nobel Laureate, Chemistry, 1993
Vancouver, Canada
February 1998

Preface

Seldom in the history of medicine has so much progress toward the understanding of hereditary diseases been made so quickly. The combined advances in the past 20 years in human genetics, cell biology, biochemistry, and biophysics have been brought to focus on the non-Mendelian expansion process and on their protein products. These advances have been realized due to the broad spectrum of researchers, including chemists, biochemists, geneticists, cell biologists, and clinical scientists, who have devoted their attention to these issues. The substantial discoveries since 1991 are due to the identity of disease genes that have expanded triplet repeat sequences as their responsible mutations. We hope that this book will serve as a comprehensive treatise covering many aspects of this broad-based attack on these neurological diseases. This book should be of interest to clinicians and scientists who wish to become better informed about these diseases and their etiology, as well as to practitioners in the field.

The purpose of this book is to provide the first authoritative review of all diseases that are related to repeat expansions. The diseases include fragile X syndrome, spino and bulbar muscular atrophy, myotonic dystrophy, spinocerebellar ataxia type 1 and type 7, dentatorubral–pallidoluysian atrophy, Huntington's disease, and Friedreich's ataxia. Furthermore, a number of chapters describe investigations of the underlying molecular mechanisms responsible for these syndromes. In addition, this book includes chapters on related diseases and on candidate disorders that reveal anticipation.

Despite the remarkable developments identified above, we are still a long way from developing effective therapeutic strategies for these diseases. This book is dedicated to those patients afflicted with these diseases and to their families, with the conviction that future generations will not experience their pain and suffering.

A large number of dedicated and talented scientists have willingly contributed chapters to this book. Their prompt and dedicated participation has made publication of this treatise possible in a timely fashion.

We thank Mrs. Lorrie A. Adams and Ms. Janelle Clark for their expert assistance and Mr. Craig Panner and Ms. Melanie Gross of Academic Press for their encouragement and participation.

Robert D. Wells

Stephen T. Warren

Marion Sarmiento

PART I

Introduction

CHAPTER 1

A New Mutational Basis for Disease

GEORGE R. WILMOT Department of Neurology, Emory University School of Medicine, Atlanta, Georgia 30322

STEPHEN T. WARREN Departments of Biochemistry, Pediatrics, and Genetics and Howard Hughes Medical Institute, Emory University School of Medicine, Atlanta, Georgia 30322

The concept of anticipation does not readily fit in the system of genetic facts. . . . Indeed, anticipation seems to be statistical rather than a biological phenomenon.

Curt Stern [1]

I. INTRODUCTION

Prior to this decade, the adage of "like-begets-like" fit nicely with the dogma of human genetics where the genetic material was considered stable upon transmission and, in the rare instance of mutation, the new variant itself was stably inherited. Although there had been clear instances when this did not hold, most notably in McClintock's study of maize, the human genetics community did not embrace the notion of "dynamic mutations." Indeed, clinical investigators had long suspected something was amiss when carefully examining families with dominant myotonic dystrophy where offspring of affected individuals often had a more severe form of the disease. This progressive increase in expressivity of the identical mutation over a number of generations was termed "anticipation." Since anticipation did not easily fit with the biological tenets of the genetics of that era, the concept of anticipation was summarily dismissed as an ascertainment bias. Many years later, the recognition of increasing penetrance through subsequent generations in fragile X syndrome, subsequently known as the Sherman paradox, invoked a resemblance to the genetic anticipation of myotonic dystrophy. While the observation in fragile X syndrome was treated with somewhat more acceptance by the scientific community, perhaps owing to its origin from quantitative geneticists rather than clinicians, the Sherman paradox, like anticipation, was not easily explained at a molecular level. Since nothing resembling anticipation had been noted in any genetically studied organism, the phenomenon, if true, represented one of the darkest of black boxes in molecular genetics.

In 1991, the genes responsible for fragile X syndrome and spinobulbar muscular atrophy were found to contain unstable, expanded trinucleotide repeats. In the

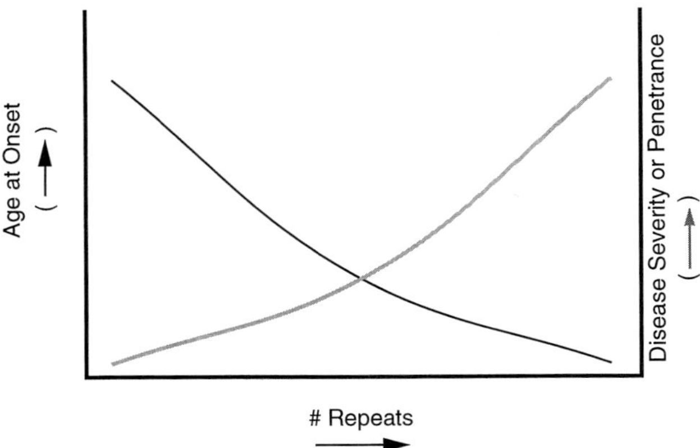

FIGURE 1-1 Genetic anticipation in diseases with dynamic mutations. In these disorders, anticipation (decreasing age at onset or increasing disease severity with subsequent generations) can be related to the size of the expansion mutation, given as the number of repeats . An inverse relationship exists between age at onset and size of the repeat expansion, and a direct relationship exists between expansion size and disease severity. In disorders with partial penetrance, such as fragile X syndrome, penetrance is also related to repeat size. The true shapes of these curves vary with each disorder; the figure is not meant to portray actual data.

following year, myotonic dystrophy was also found to be the result of an expanded trinucleotide repeat. These discoveries were soon followed by a remarkable number of neurological disorders, each sharing the mutational mechanism of the unstable expansion of a repeat, most often being a triplet in form. Soon, the behavior of these repeats in affected families clearly revealed a pattern where the increasing penetrance (i.e., Sherman paradox) or increasing expressivity (anticipation) through subsequent generations correlated with increasing lengths of the triplet repeat (Fig. 1-1). Thus a biological basis of anticipation, the dynamic mutation, was defined.

In this volume, the dozen neurological diseases due to dynamic mutations are reviewed, approaches to identify further examples are considered, and the mechanisms by which these repeats expand are analyzed. In this chapter, these comprehensive discussions will be condensed as a brief overview.

II. FEATURES OF DYNAMIC MUTATIONS

From the experience of many laboratories over the past several years, certain consistencies have emerged as typical for this class of mutation. Since pathologic trinucleotide repeats are found in the proximity of specific genes, the precise location within an affected gene serves as a useful subclass discriminator. As summarized in Fig. 1-2, such triplet repeats can be found in transcribed RNA destined to be untranslated (either 5′ or 3′ such as in fragile X syndrome or myotonic dystrophy, respectively), spliced out intronic sequence (such as in Friedreich's ataxia), or coding exonic sequence (such as the dominant ataxias). In general, the noncoding repeats are able to undergo massive expansions from a normal number of 6–40 triplets to an abnormal range of many hundreds or thousands of repeats. This leads to either transcriptional suppression, as in the case of fragile X syndrome, or abnormal RNA processing, limiting the amount of mature cytoplasmic message, as in the case of myotonic dystrophy. In contrast, the coding expansions undergo much more modest expansions from a normal range of approximately 10–35 repeats to an abnormal range of approximately 40 to 90 triplets. Since these are CAG repeats coding for polyglutamine tracts, constraints of the individual protein structures significantly modify this range. This is most apparent in SCA6 where the largest normal allele contains 18 repeats whereas the smallest abnormal allele contains 21 repeats. In all coding expansions, the mechanism(s), while still poorly understood, appear to reflect a gain or change of function of the abnormal protein, eventually leading to neurodegeneration.

The sequence of the repetitive element is no longer confined to CG-rich triplets as was assumed earlier, with the discovery of the GAA repeat involved in Friedreich's ataxia. More recently, repeats other than triplets have been shown to exhibit behavior consistent with dynamic mutations with the observation that an unstable dodecamer repeat in the upstream region of the cystatin B genes leads to progressive myoclonus epilepsy type 1 when expanded. Friedreich's ataxia, an autosomal recessive disease, is also the exception to the rule of the dominant mode of inheritance for dynamic mutations. From the exceptions, one is forced to conclude that the range of involvement of dynamic mutations in human disease is yet to be fully appreciated.

III. SPECIFIC DISORDERS

Trinucleotide repeat expansions are now known to cause 12 diseases (excluding allelic variants) and 3 additional cytogenetic abnormalities (fragile sites). The expansions have been found in every gene region (5′UTR, 3′UTR, intron and exon; Fig. 1-2) and in a wide variety of chromosomes (Fig. 1-3). Some of the features of these conditions are summarized in Table 1-1 and discussed briefly below.

A. Diseases Involving Noncoding Repeats

These tend to be multisystem disorders characterized by long expansions in symptomatic individuals as well

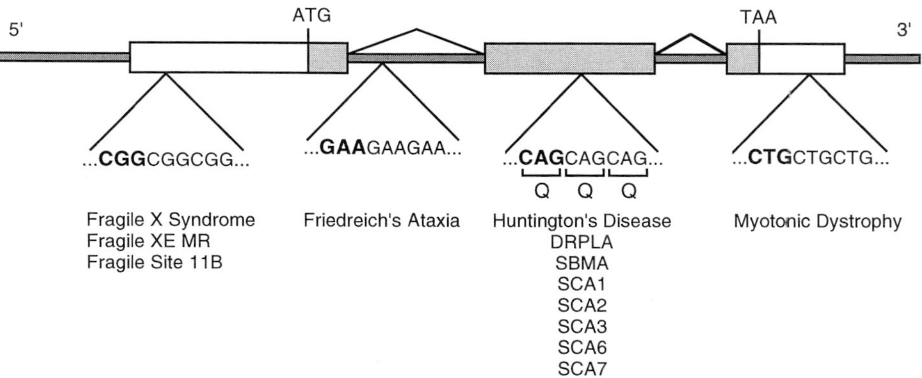

FIGURE 1-2 Location of trinucleotide expansions in humans. A summary of the locations of the known trinucleotide expansions is indicated within this factitious gene consisting of a coding sequence (gray) distributed between three exons with two intervening introns. The sequence of the involved repetitive triplet is indicated above the named diseases. CGG expansions occur in the 5′UTR in fragile X syndrome, fragile XE mental retardation, and fragile site 11B, which has been associated with disease only indirectly through its predisposition to Jacobsen syndrome, a chromosomal deletion disorder. FRAXF and FRA16A (not indicated) are also fragile sites associated with CGG expansions, but they have not been associated with a clinical phenotype and their relationship to nearby genes is currently unclear. Other diseases with repeat expansions occurring in noncoding regions include Friedreich's ataxia (an intronic GAA repeat) and myotonic dystrophy (CTG expansion in the 3′UTR). The expansion of CAG repeats in Huntington's disease, dentatorubral-pallidoluysian atrophy (DRPLA), spinobulbar muscular atrophy (SBMA), and the spinocerebellar atrophies (SCAs) results in proteins with elongated tracts of glutamine (Q) because of the expansion's location within a coding segment of the exon.

as intermediate expansions ("premutations") that are clinically silent but can evolve into the full mutation through germ-line transmission.

1. FRAGILE X SYNDROME

Fragile X syndrome is the most common inherited form of mental retardation [2]. As discussed in the next two chapters, it is inherited in an X-linked dominant fashion with reduced penetrance, and is clinically characterized by mild to severe mental retardation, macro orchidism, and mild facial dysmorphia (narrow faces with prominent jaw and forehead and large ears). The disorder was originally identified through its association with a folate-sensitive fragile site at Xq27.3 [3], and in 1991 the mutational basis of the disease was identified as a hypermethylated $(CGG)_n$ repeat expansion in the gene FMR1 [4]. Subsequent work has shown that FMRP, the protein product of the FMR1 gene, is an RNA binding protein which is shuttled between the nucleus and the cytoplasm and is associated with polysomes [5]. Delineation of the mechanisms by which loss of FMRP leads to the fragile X phenotype is ongoing and has benefited from a murine model of the disease (Chapter 4).

2. FRAGILE XE MENTAL RETARDATION

A small proportion of patients possessing a folate-sensitive fragile site at Xq28 did not test positive for the FMR1 $(CGG)_n$ expansion. This observation led to the discovery of a nearby fragile site and eventually to the identification of the FMR2 gene [6, 7]. Like FMR1, FMR2 contains a trinucleotide repeat which is polymorphic in normal individuals and when expanded into a full mutation becomes transcriptionally silent due to methylation. Fragile XE mental retardation and FMR2 are discussed in Chapter 5.

3. FRAGILE 11B WITH PREDISPOSITION TO JACOBSEN SYNDROME

Jacobsen syndrome, a chromosomal deletion disorder involving 11q23, is characterized by multiple dysmorphic features and severe mental retardation. In at least some patients, the deletion appears to be triggered via the heritable folate-sensitive fragile site FRA11B, which has been localized to a $(CGG)_n$ expansion in the 5′-untranslated region of the CBL2 protooncogene [8]. FRA11B and Jacobsen syndrome therefore demonstrate that trinucleotide expansions can result in chromosomal rearrangements and are discussed further in Chapter 5. Other fragile sites that demonstrate genetic instability but have not yet been linked with clinical phenotypes are then discussed in Chapters 6 and 7.

4. MYOTONIC DYSTROPHY

Myotonic dystrophy (DM) is generally regarded as the most common inherited muscle disease, affecting

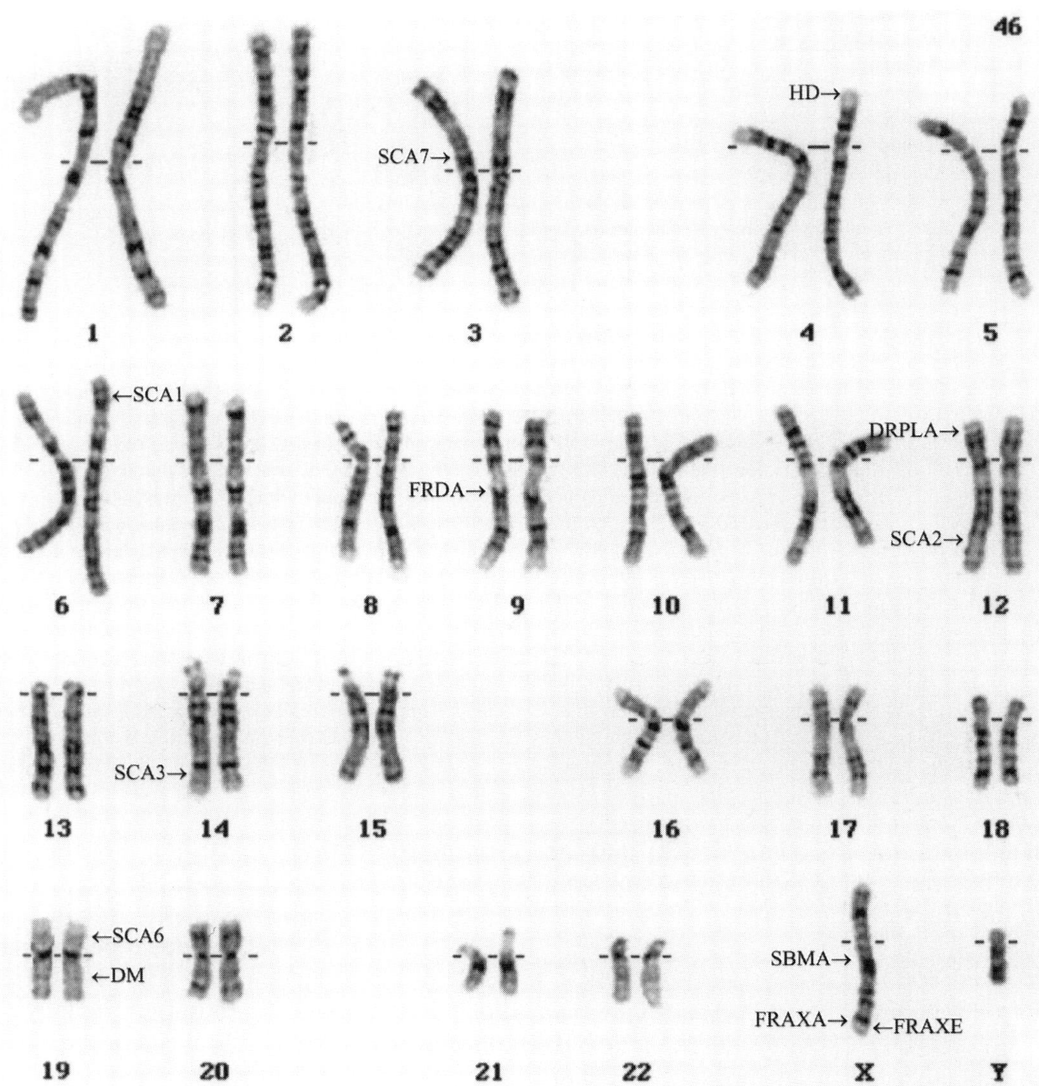

FIGURE 1-3 Chromosomal location of the disease-causing trinucleotide repeat expansions. Locations are indicated on the G-banded chromosomes representing a normal human 46,XY karyotype. DM, myotonic dystrophy; DRPLA, dentatorubral-pallidoluysian atrophy; FRAXA, fragile X syndrome; FRAXE, fragile XE mental retardation; FRDA, Friedreich's ataxia; HD, Huntington's disease; SBMA, spinobulbar muscular atrophy; SCA, spinocerebellar ataxia.

approximately 1 in 8000 individuals in North American and European populations [9]. A wide variation in DM frequency exists among global populations, which appears to be due to distinct frequencies of predisposed alleles with the various populations [10]. DM is a progressive multisystem disorder with highly variable manifestations which include myotonia, muscle weakness and atrophy, cardiac conduction system abnormalities, frontal balding, cataracts, endocrine abnormalities, and mild mental impairment. The disorder displays marked anticipation resulting in a wide range of severity, and in its most severe form occurs congenitally with hypotonia, facial diplegia, swallowing difficulties, and often severe mental retardation. The mutation causing DM is a $(CTG)_n$ expansion occurring in the 3′-UTR of a novel serine–threonine protein kinase termed DMPK (see Chapters 10 and 11). Like all of the noncoding trinucleotide repeat expansions, this expansion can reach a very large size, especially in the congenital form in which repeat lengths are most often greater than 1000 [11]. Congenital DM, which is phenotypically different enough from adult DM to cause some authors to wonder if its pathogenesis is somewhat different from adult DM (see Chapter 12), occurs almost exclusively through maternal transmission. A parental gender bias is a general feature of the trinucleotide repeat expansion diseases, and is displayed most prominently in fragile X syndrome, in which males are incapable of transmitting a

TABLE 1-1 Summary of Diseases Caused by Trinucleotide Repeat Expansions

Disorder	Inheritance	Gene/locus	Chromosomal localization	Protein product	Expansion size		Repeat location	Mutation type	Parental gender bias
					Normal	Mutant			
Fragile X syndrome	X-linked dominant	FMR1 (FRAXA)	Xq27.3	FMRP	$(CGG)_{6\text{-}52}$	$(CGG)_{60\text{-}200}$ (premutation) $(CGG)_{230\text{-}1000}$ (full)	5′-UTR	LOF Fragile site	Maternal
Fragile XE mental retardation	X-linked ?dominant	FMR2 (FRAXE)	Xq28	FMR2 protein	$(GCC)_{7\text{-}35}$	$(GCC)_{130\text{-}150}$ (premutation) $(GCC)_{230\text{-}750}$ (full)	5′-UTR	LOF Fragile site	ND
Friedreich's ataxia	Autosomal recessive	X25	9q13-21.1	Frataxin	$(GAA)_{6\text{-}34}$	$(GAA)_{80}$ (premutation) $(GAA)_{112\text{-}1700}$ (full)	Intron 1	LOF (partial)	Maternal
Myotonic dystrophy	Autosomal dominant	DMPK	19q13	Myotonic dystrophy protein kinase	$(CTG)_{5\text{-}37}$	$(CTG)_{50\text{-}3000}$	3′-UTR	?Dominant negative	Maternal
Spinobulbar muscular atrophy (Kennedy disease)	X-linked recessive	AR	Xq13-21	Androgen receptor	$(CAG)_{11\text{-}33}$	$(CAG)_{38\text{-}66}$	Coding	GOF LOF (partial)	ND
Huntington's disease	Autosomal dominant	IT15	4p16.3	Huntingtin	$(CAG)_{6\text{-}39}$	$(CAG)_{36\text{-}121}$	Coding	GOF	Paternal
Dentatorubral-pallidoluysian atrophy/Haw River syndrome	Autosomal dominant	DRPLA (B37)	12p13.31	Atrophin-1 (drplap)	$(CAG)_{6\text{-}35}$	$(CAG)_{51\text{-}88}$	Coding	GOF	Paternal
Spinocerebellar ataxia type 1	Autosomal dominant	SCA1	6p23	Ataxin-1	$(CAG)_{6\text{-}39}$	$(CAG)_{41\text{-}81}$	Coding	GOF	Paternal
Spinocerebellar ataxia type 2	Autosomal dominant	SCA2	12q24.1	Ataxin-2	$(CAG)_{14\text{-}31}$	$(CAG)_{35\text{-}64}$	Coding	GOF	Paternal
Spinocerebellar ataxia type 3/Machado-Joseph disease	Autosomal dominant	SCA3 (MJD1)	14q32.1	Ataxin-3	$(CAG)_{12\text{-}41}$	$(CAG)_{40\text{-}84}$	Coding	GOF	Paternal
Spinocerebellar ataxia type 6/Episodic ataxia type 2	Autosomal dominant	CACNA1A	19p13	α_{1A}-Voltage-dependent calcium channel subunit	$(CAG)_{7\text{-}18}$	$(CAG)_{20\text{-}23}$ (EA2) $(CAG)_{21\text{-}27}$ (SCA6)	Coding	ND	ND
Spinocerebellar ataxia type 7	Autosomal dominant	SCA7	3p12-13	Ataxin-7	$(CAG)_{7\text{-}17}$	$(CAG)_{38\text{-}130}$	Coding	GOF	Paternal

Note. GOF, gain of function; LOF, loss of function; ND, not determined.

full mutation to their offspring. As a general rule in these diseases, small expansions tend to be favored by paternal transmission, yet very large expansions are hindered by paternal transmission. Presumably, parental gender bias can be governed by both the meiotic instability of the repeat and the fitness of the resulting gametes, and is discussed in detail in Chapter 37, in Section VI of Chapter 10, and briefly in various chapters throughout this volume.

5. Friedreich's Ataxia

Friedreich's ataxia is the most common inherited form of ataxia, with an estimated frequency of 1–2 per 100,000 [12]. The disorder is in fact a multisystem degenerative disease predominantly affecting the spinal cord, peripheral nerves, and heart. Friedreich's ataxia is unique among trinucleotide expansion diseases in its inheritance pattern (autosomal recessive), in its trinucleotide sequence (GAA), and in the location of its $(GAA)_n$ expansion within an intron. Since the disease is autosomal recessive, a large number of mutant alleles exist in the general population in carrier individuals. In fact, the frequency of mutated alleles has been estimated to be between 1 in 140 and 1 in 240, making this expansion the most commonly occurring trinucleotide repeat mutation yet identified [13].

Campuzano *et al.* [13] identified a $(GAA)_n$ expansion in the first intron of the X25 gene as the disease mutation in the vast majority of patients. An inverse relationship exists between the age of onset/clinical severity and GAA repeat number [14], suggesting that the disease results from a partial, rather than a total, loss of function of the X25 gene product "frataxin." Recent work has confirmed that the mutation results in lowered levels of frataxin and has suggested that the amount of frataxin in affected individuals is largely determined by the $(GAA)_n$ expansion length. Studies of the normal function of frataxin appear to indicate that it is a mitochondrial protein which might be involved in cellular iron metabolism. These findings are discussed in detail in Chapter 26.

B. Diseases Involving Coding Repeats

Unlike the diseases noted above, these disorders tend to lack systemic involvement and are all caused by $(CAG)_n$ expansions that are much shorter (typically <100) than the expansions noted above.

1. Spinobulbar Muscular Atrophy

SBMA, also known as Kennedy's disease, is an uncommon disorder (there are no published prevalence estimates) characterized by X-linked recessive inheritance, mild androgen insensitivity, motor neuron degeneration with prominent bulbar involvement, and a mild sensory neuronopathy (see Chapter 7). SBMA was the first disorder found to be caused by a $(CAG)_n$ expansion within the coding region of a gene. The affected gene in SBMA is the androgen receptor, which together with calcium channel subunit gene affected in spinocerebellar atrophy type 6 (SCA6) are the only two genes in trinucleotide expansion disorders with predetermined cellular functions. Since CAG is the codon for glutamine, the $(CAG)_n$ expansion results in an enlarged polyglutamine domain within the androgen receptor. This causes a partial functional inhibition of the androgen receptor and therefore leads to the mild androgen insensitivity seen in the disease. As in all of the other $(CAG)_n$ repeat diseases discussed below, the neurodegenerative aspects of SBMA have not yet been explained mechanistically, but are presumed to arise from a toxic gain of function caused by the enlarged polyglutamine domain (see Chapters 8, 46, and 47).

2. Huntington's Disease

Huntington's disease (HD; discussed in Section IX of this volume) is a progressive autosomal dominant disorder characterized by abnormal involuntary movements, impaired motor control, psychiatric disturbance, and dementia. In Caucasian populations, the prevalence of HD has been estimated at 1 in 10,000 [15]. The disease is caused by a $(CAG)_n$ expansion within exon 1 of the IT15 gene which codes for the protein huntingtin. As with the other $(CAG)_n$ expansion disorders, the expansion is more unstable with paternal transmission, and paternal transmission can result in more highly expanded alleles and a more severe juvenile-onset form of the disease characterized by bradykinesia, rigidity, dystonia, and seizures. A number of lines of evidence suggest that HD results from a gain-of-function process. A possible clue to the nature of this process has recently been provided by experiments with mice transgenic for an N-terminal fragment of huntingtin with an expanded CAG repeat. These mice show some of the phenotypic features of HD [16] and further analysis of this model has suggested that the polyglutamine expansion forms insoluble protein aggregates which accumulate in neuronal nuclei [17]. Additional work assessing the validity of this HD model and the extent to which it can be applied to other $(CAG)_n$ expansion diseases will no doubt be forthcoming.

3. Dentatorubral-Pallidoluysian Atrophy

Although DRPLA occurs in other populations, this rare autosomal dominant disorder is most commonly seen in Japan, where its prevalence has been estimated to be 0.2–0.7 per 100,000 [18]. Affected individuals demonstrate varying degrees of myoclonus, seizures, choreoathetosis, and cognitive changes. The disease is caused by a $(CAG)_n$ expansion in the DRPLA gene (originally called E27) which codes for the protein atrophin-1. The disorder is detailed further in Chapters 14 and 15, and in Chapter 16 an allelic variant termed the Haw River syndrome is discussed.

4. The Spinocerebellar Ataxias (SCA1, SCA2, SCA3, SCA6, and SCA7)

The classification of the spinocerebellar ataxias has historically been confusing, largely because of the phenotypic overlap between the different disorders. In addition, a subset of clinicians and investigators have used the pathologically derived term "familial olivopontocerebellar atrophy (OPCA)" to refer to patients who may have had one of the autosomal dominant spinocerebellar ataxias. This situation has been dramatically improved through linkage analysis and the identification of five separate disease-associated genes. A comparison of the most current genotype-based classification scheme with Harding's [19] previously used phenotypic scheme appears in Table 1-2.

Although all of these disorders have some overlapping features (progressive ataxia, varying degrees of peripheral neuropathy, pyramidal, extrapyramidal, ocular, and cognitive involvement), the distinction between SCA1, SCA2, and SCA3 (also called Machado-Joseph disease) is the most difficult to make on clinical grounds and requires DNA testing to be definitive. The specific clinical features of these disorders are summarized briefly in Table 1-2 and are detailed in the literature and in subsequent chapters later in this volume.

The genes for SCAs 1,2,3,6, and 7 have been cloned, and each contains a coding $(CAG)_n$ repeat which is expanded in affected individuals. SCA6 is somewhat different than the other disorders in a variety of ways. First, the SCA6 gene (CACNA1A) codes for a membrane-bound voltage-dependent calcium channel subunit, whereas all of the other CAG repeat expansions occur in genes coding for soluble proteins. Second, the range of expansion lengths in disease-associated SCA6 alleles (21–27 repeats) is much smaller than the other disorders and in fact falls within the normal range of these conditions. This appears to limit the instability of the expansion and may prevent anticipation from occurring. Third, point mutations in the CACNA1A gene have been described, and these lead to other phenotypes (familial hemiplegic migraine and episodic ataxia EA2). Since these disorders have been associated with missense and nonsense mutations, respectively, their mutational mechanisms (i.e., loss of function vs gain of function vs dominant negative) may differ from each other and from SCA6. This distinction has recently been called into question with the description of 2 SCA6 families with small $(CAG)_n$ expansions occurring in

TABLE 1-2 Classification Schemes for the Autosomal Dominant Spinocerebellar Ataxias

Classification of Harding [19]		Current classification		
Disorder	Features in addition to ataxia	Disorder	Chromosomal localization	Prominent clinical features
ADCA I	Ophthalmoplegia, optic atrophy, dementia, pyramidal and extrapyramidal features	SCA 1	Gene at 6p22-p23	Gait ataxia, ophthalmoparesis, dysarthria, often hyperreflexia
		SCA 2	Gene at 12q23-24.1	Similar to SCA1 with more marked anticipation, often hyporeflexia
		SCA 3/MJD	Gene at 14q32.1	Extrapyramidal features often prominent; marked phenotypic variability
		SCA 4	Mapped to 14q32.1	Ataxia is slowly progressive; significant sensory axonal polyneuropathy
ADCA II	Pigmentary retinopathy +/− ophthalmoplegia and extrapyramidal features	SCA 7	Gene at 3p12-13	Severe retinopathy with macular degeneration and visual loss
ACDA III	"Pure" cerebellar ataxia of late onset	SCA 5	Mapped to cent. 11	Slowly progressive ataxia and dysarthria
		SCA 6	Gene at 19p13	Slowly progressive ataxia, dysarthria, and vibratory/proprioceptive sensory loss

Note. ACDA, autosomal dominant cerebellar ataxia; MJD, Machado-Joseph disease; SCA, spinocerebellar ataxia.

individuals with the EA2 phenotype [20]. Thus, the pathenogenesis of EA2 and SCA6 may be more related than previously thought.

C. Nontriplet Dynamic Mutations and Candidate Disorders

Other than the disorders with known trinucleotide repeat expansions, several additional diseases are considered to be candidates for similar mutations, largely because their inheritance patterns are suggestive of anticipation (see Tables 28-1 and 28-2). The establishment of true genetic anticipation in diseases in which the disease mutation is not known is very difficult, in part because clinical information is subject to a number of potential biases (see Table 9-7). Claims of anticipation must therefore be interpreted very cautiously. Candidate psychiatric diseases are discussed in more detail in Chapter 27.

Various strategies exist for identifying trinucleotide repeat expansion-containing genes, including library screening, "repeat expansion detection" (RED), "direct identification of repeat expansion and cloning technique" (DIRECT), and reactivity with a monoclonal antibody that recognizes expanded polyglutamine domains. Each of these approaches has inherent advantages and disadvantages, and all are discussed in detail within Section XII of this volume. Within the past year, application of these approaches has led to the identification of the SCA2, SCA6, and SCA7 genes. Which of the many candidate disorders will eventually be found to be caused by trinucleotide repeat expansions is difficult to say, but further application and refinement of these techniques appears certain to identify some.

Expansions of repetitive elements other than trinucleotides have also been linked to disease. Microsatellite instability has been associated with hereditary nonpolyposis colon cancer and is discussed in Chapter 49. As previously mentioned, an enlarged dodecamer repeat in the upstream region of the cystatin B gene has been shown to explain the majority of cases of the Unvericht-Lundborg variety of progressive myoclonus epilepsy (EPM1) ([21] Chapter 48).

In addition, enlarged homopolymers of polyalanine have been associated with synpolydactyly [22] and cleidocranial dysplasia [23]. In these cases, the mutations do not appear to represent true dynamic mutations, since they do not result in intergenerational instability of the repeat length. Unlike dynamic mutations, the enlarged polyalanine tracts in these cases appear to arise from single recombination events following unequal crossing-over within cryptic repeats of polyalanine codons [24]. Many genes with cryptic repetitive sequences encoding runs of a single amino acid have been described (discussed in Chapter 46), and these may likewise be prone to unequal crossing-over with maintenance of the reading frame. Thus, the lengthening of a tract of a single amino acid leading to a change-of-function protein may be a common mechanism of human genetic disease, even if it is not always caused by a dynamic mutation. It should also be kept in mind that true dynamic expansions of polyalanine (or other homopolymer) tracts may also occur. One would predict from what is currently known about the stabilizing influence of sequence interruptions that these mutations would occur in pure rather than cryptic repeats of variant codons for the same amino acid.

D. Mechanism of Repeat Expansion

Why are some repetitive elements stable, whereas others appear to have the capacity to expand? What aspects of DNA replication and repair are involved in the mechanism of expansion? Are multiple mechanisms involved, and if so, are they specific for repeat types? For specific loci? What role does the process of DNA packaging into nucleosomes play in repeat instability and gene expression in these disorders? These fundamental questions of DNA structure, replication, and repair are currently being addressed most actively in a number of model systems which are discussed in Section XIII of this volume.

One crucial question concerns the structural motifs that can be formed by stretches of repetitive DNA. For many years it has been known that DNA can adopt a variety of conformations in addition to the standard double-helical, right-handed B-form (see Table 38-1 and the discussion in Chapter 38). Given that alternate structures may be differentially processed during DNA replication and repair events, a great deal of current work is directed toward defining these conformations for the various possible trinucleotide repeat combinations. This work is detailed in Chapters 38–41. Although most of the current evidence seems point to some form of a hairpin structure for the G/C-rich repeats (e.g., CGG, CAG) and triplex formation for the GAA repeat, much of this evidence comes from the study of simplified systems and the direct applicability of this information to *in vivo* repeat instability awaits further experimental confirmation.

In addition to a detailed knowledge of the structural features of these dynamic mutations, our understanding of the mechanisms underlying repeat instability relies heavily upon how the cellular processes of DNA replication and repair are affected by these structures. Three general mechanisms can be proposed to account for repeat instability: slippage during DNA replication, misalignment with subsequent excision repair, and unequal crossover and recombination. Of these three, slippage during DNA replication has received the most attention, in part because of the observation that in unstable trinucleotide repeat expansions, the repeat size associated with the threshold for instability corresponds to the size of the Okazaki fragment (the nascent lagging strand at the replication fork). It should also be noted that these general mechanisms cannot be assumed to be mutually exclusive; excision repair of a hairpin structure may occur at the replication fork, for instance. These mechanisms of instability have been addressed in *Escherichia coli* (Chapter 33) and yeast (Chapter 34) and are also addressed in detail in Chapters 39, 44, and 45. In addition, our understanding of repeat instability is likely to benefit from the recent development of murine models of DM and HD that show intergenerational repeat instability ([25], Chapters 13 and 25).

Our understanding of dynamic mutations would be incomplete without consideration of the role that nucleosome assembly and chromatin structure may be playing in these diseases. There is clear evidence that nucleosome assembly is strongly affected by CTG and CCG repeats, and the chromatin structure of adjacent regions can be affected as well (reviewed in Chapter 42). To what degree the pathophysiology of these diseases is actually determined by altered chromatin structure remains an open question.

Finally, two recently proposed mechanisms have been suggested to participate in repeat expansion. Gordenin *et al.* [26] suggested that the FEN1 product, the flap endonuclease responsible for removal of the 5′ portion of Okazaki fragments which are displaced by the invading, more recent, Okazaki fragment, could play a role in repeat expansion. Specifically, it is suggested that certain triplet repeats could form a FEN1-resistant secondary structure which may impede movement of the replication fork. This, in turn, could lead to a double strand break within the repeat which could result in expansion via end joining. Indeed, Samadashwily *et al.* [27] recently provided evidence that trinucleotide repeats do lead to stalling of the replication fork. Moreover their data are very consistent with the biological data for triplet repeat behavior by observing that stalling is dependent upon the orientation, purity, and length of the triplet. It is important to note that these two recent models are not mutually exclusive and still require formation of an unusual structure of the triplet repeat. Hence, the many models of repeat expansion each may provide a valuable clue in understanding the mechanism of repeat expansion.

Although the pathophysiologies of each of the diseases discussed in this chapter certainly differ, a few generalizations can be made. First, some of these diseases result from a loss of function of their respective gene product. The pathophysiology of these disorders will therefore differ greatly and can be studied by determining the normal cellular function of the encoded protein. Second, at least some of the diseases caused by $(CAG)_n$ expansions encoding enlarged polyglutamine domains may have key pathophysiologic events in common. Multiple mechanisms have been proposed to explain how an enlarged polyglutamine domain may result in neurodegeneration, including altered associations with interacting proteins, abnormal proteolytic cleavage, and abnormal aggregation. These are discussed in detail in Chapters 46 and 47.

IV. CONCLUDING REMARKS

Without question, the examples of disease discussed in this volume share certain properties; indeed, it is partly for this reason that this new class of mutation has received such attention. Genetic anticipation, mosaicism, and marked phenotypic variability are prominent in these diseases and may be a direct result of their defining element: a dynamic, often-changing DNA mutation. But there are other qualities of these disorders that are less well grounded in the disease mechanism, and over time these properties might not hold up well as over all generalizations. This has already happened with the discovery of the GAA repeat in Friedreich's ataxia, which invalidated the prior observation that all trinucleotide repeat expansions occurred with C/G-rich triplets [28]. Furthermore, the discovery that this autosomal recessive disease is caused by an intronic expansion sets the stage for the additional recessive diseases to be linked to expansions. Can we assume that all of the future disease caused by dynamic expansions will have neurologic involvement, as each of the currently known ones do? There simply is no firm mechanistic reason to believe so. With every new example that is uncovered, our previous assumptions about dynamic mutations and disease are challenged. Will expansions of homopolymers other than polyglutamine, e.g., polyalanine, be found to cause disease via dynamic mutations? Will trinucleotide repeat mutations continue to dominate the landscape or will di-, tetra-, or other-sized nucleotide expansions become increasingly linked to disease? Surely, if the past is any indication of the future, a number of surprises are in store for us.

Acknowledgments

This work was supported in part by NIH Grants R37HD20521 and P01HD35576 to S.T.W. and K08NS01977 to G.R.W. S.T.W. is an investigator of the Howard Hughes Medical Institute.

References

1. Stern, C. (1973). "Principles of Human Genetics," 3rd ed., pp. 409–410. W. H. Freeman, San Francisco.
2. Webb, T. P., Bundey, S. E., Thake, A. I., and Todd, J. (1986). Population incidence and segregation ratios in the Martin-Bell syndrome. *Am. J. Med. Genet.* **23,** 573–580.
3. Sutherland, G. R. (1977). Fragile sites on human chromosomes: demonstration of their dependence on the type of tissue culture medium. *Science* **197,** 265–266.
4. Warren, S. T., and Ashley, C. T. (1995). Triplet repeat expansion mutations: the example of fragile X syndrome. *Annu. Rev. Neurosci.* **18,** 77–99.
5. Feng, Y., Gutekunst, C. A., Eberhart, D. E., Yi, H., Warren, S. T., and Hersch, S. M. (1997). Fragile X mental retardation protein: nucleocytoplasmic shuttling and association with somatodendritic ribosomes. *J. Neurosci.* **17,** 1539–1547.
6. Gecz, J., Gedeon, A. K., Sutherland, G. R., and Mulley, J. C. (1996). Identification of the gene FMR2, associated with FRAXE mental retardation. *Nature Genet.* **13,** 105–108.
7. Gu, Y., Shen, Y., Gibbs, R. A., and Nelson, D. L. (1996). Identification of FMR2, a novel gene associated with the FRAXE CCG repeat and CpG island. *Nature Genet.* **13,** 109–113.
8. Jones, C., Penny, L., Mattina, T., Yu, S., Baker, E., Voullaire, L., Langdon, W. Y., Sutherland, G. R., Richards, R. I., and Tunnacliffe, A. (1995). Association of a chromosome deletion syndrome with a fragile site within the proto-oncogene CBL2. *Nature* **376,** 145–149.
9. Harper, P. S. (1989). "Myotonic Dystrophy," 2nd ed. W. B. Saunders, Philadelphia.
10. Zerylnick, C., Torroni, A., Sherman, S. L., and Warren, S. T. (1995). Normal variation at the myotonic dystrophy locus in global human populations. *Am. J. Hum. Genet.* **56,** 123–130.
11. Tsilfidis, C., MacKenzie, A. E., Mettler, G., Barcelo, J., and Korneluk, R. G. (1992). Correlation between CTG trinucleotide repeat length and frequency of severe congenital myotonic dystrophy. *Nature Genet.* **1,** 192–195.
12. Romeo, G., Menozzi, P., and Ferlini, A. (1983). Incidence of Friedreich ataxia in Italy estimated from consanguineous marriages. *Am. J. Hum. Genet.* **35,** 523–529.
13. Campuzano, V., Montermini, L., Molto, M. D., Pianese, L., Cossee, M., Cavalcanti, F., Monros, E., Rodius, F., Duclos, F., Monticelli, A., Zara, F., Canizares, J., Koutnikova, H., Bidichandani, S. I., Gellera, C., Brice, A., Trouillas, P., De Michele, G., Filla, A., De Frutos, R., Palau, F., Patel, P. I., Di Donato, S., Mandel, J. L., Cocozza, S., Koenig, M., and Pandolfo, M. (1996). Friedreich's ataxia: autosomal recessive disease caused by an intronic GAA triplet repeat expansion. *Science* **271,** 1423–1427.
14. Filla, A., De Michele, G., Cavalcanti, F., Pianese, L., Monticelli, A., Campanella, G., and Cocozza, S. (1996). The relationship between trinucleotide (GAA) repeat length and clinical features in Freidreich ataxia. *Am. J. Hum. Genet.* **59,** 554–560.
15. Harper, P. S. (1991). The epidemiology of Huntington's disease. *In* "Huntington's Disease," (P. S. Harper, Ed.), pp. 251–280. Saunders, London.
16. Mangiarini, L., Sathasivam, K., Seller, M., Cozens, B., Harper, A., Hetherington, C., Lawton, M., Trottier, Y., Lehrach, H., Davies, S. W., and Bates, G. P. (1996). Exon 1 of the HD gene with an expanded CAG repeat is sufficient to cause a progressive neurological phenotype in transgenic mice. *Cell* **87,** 493–506.
17. Scherzinger, E., Lurz, R., Turmaine, M., Mangiarini, L., Hollenbach, B., Hasenbank, R., Bates, G. P., Davies, S. W., Lehrach, H., and Wanker, E. E. (1997). Huntingtin-encoded polyglutamine expansions form amyloid-like protein aggregates in vitro and in vivo. *Cell* **90,** 549–558.
18. Inazuki, G., Kumagai K., and Naito, H. (1990). Dentatorubral-pallidoluysian atrophy (DRPLA): its distribution in Japan and prevalence rate in Nigata. *Seishin Igaku* **32,** 1135–1138.
19. Harding, A. E. (1983). Classification of the hereditary ataxias and paraplegias. *Lancet* **I**(8334), 1151–1155.
20. Jodice, C, Mantuano, E., Veneziano, L., Trettel, F., Sabbadini, G., Calandriello, L., Francia, A., Spadaro, M., Pierelli, F., Salvi, F., Ophoff, R., Frants, R. R., and Frontali, M. (1997). Episodic ataxia type 2 (EA2) and spinocerebellar ataxia type 6 (SCA6) due to CAG repeat expansion in the CACNA1A gene on chromosome 19p. *Hum. Mol. Genet.* **6,** 1973–1978.
21. Lalioti, M. D., Scott, H. S., Buresi, C., Rossier, C., Bottani, A., Morris, M. A., Malafosse, A., and Antonarakis, S. E. (1997). Dodecamer repeat expansion in cystatin B gene in progressive myoclonus epilepsy. *Nature* **386,** 847–851.
22. Akarsu, A. N., Stoilov, I., Yilmaz, E., Sayli, B. S., and Sarfarazi, M. (1996). Genomic structure of HOXD13 gene: a nine polyalanine duplication causes synpolydactyly in two unrelated families. *Hum. Mol. Genet.* **5,** 945–952.
23. Mundlos, S., Otto, F., Mundlos, C., Mulliken, J. B., Aylsworth, A. S., Albright, S., Lindhout, D., Cole, W. G., Henn, W., Knoll, J. H., Owen, M. J., Mertelsmann, R., Zabel, B. U., and Olsen, B. R. (1997). Mutations involving the transcription factor CBFA1 cause cleidocranial dysplasia. *Cell* **89,** 773–779.
24. Warren, S. T. (1997). Polyalanine expansion in synpolydactyly might result from unequal crossing-over of HOXD13. *Science* **275,** 408–409.
25. Korneluk, R. G., and Narang, M. A. (1997). Anticipating anticipation. *Nature Genet.* **15,** 119–120.
26. Gordenin, D. A., Kunkel, T. A., and Resnick, M. A. (1997). Repeat expansion—all in a flap? *Nature Genet.* **16,** 116–118.
27. Samadashwily, G. M., Raca, G., and Mirkin, S. M. (1997). Trinucleotide repeats affect DNA replication in vivo. *Nature Genet.* **17,** 298–304.
28. Warren, S. T. (1996). The expanding world of trinucleotide repeats. *Science* **271,** 1374–1375.

PART II

Fragile X Syndrome

CHAPTER 2

Clinical and Diagnostic Aspects of Fragile X Syndrome

RANDI HAGERMAN Child Development Unit and the Fragile X Treatment & Research Center, The Children's Hospital, Denver, Colorado; The Department of Pediatrics and the Colorado University Affiliated Program, University of Colorado Health Sciences Center, Denver, Colorado

I. INTRODUCTION

In the subsequent chapters you will be informed regarding the latest advances in the molecular biology of the Fragile X Mental Retardation 1 gene (FMR1) mutation which causes fragile X syndrome (FXS). This chapter will focus on the clinical aspects including the physical, behavioral, and cognitive features of FXS. In addition, diagnostic issues including who should be screened or tested for FXS will be reviewed.

FXS is not only the most common inherited cause of mental retardation, representing approximately 30% of X-linked mental retardation [83], but it also causes a broad spectrum of learning and behavior problems in those who do not have mental retardation [28, 21]. Therefore, the clinician must have a high index of suspicion regarding FXS when evaluating children or adults with the symptoms and features described below. The prevalence of carriers in the general population is higher than previously thought. Rousseau *et al.* [73] screened over 10,000 women in Quebec who donated to a blood bank and found 1 in 259 with the premutation (50 to 200 CGG repeats). In a similar screening of males in the general population, approximately 1 in 700 carried the premutation [74]. The prevalence of the full mutation (greater than 230 CGG repeats) is uncertain because of limited screening studies. Cytogenetic studies found approximately 1 in 1250 males and 1 in 2500 females with mental retardation and FXS in England [98]. Subsequent molecular studies decreased this prevalence figure to 1 in 4000 since several cytogenetically positive individuals did not demonstrate the full mutation at FMR1. A similar prevalence rate of 1 in 4000 was found by Turner *et al.* [95] in screening mentally retarded males in New South Wales, Australia. However, the full spectrum of FXS, including nonretarded individuals, has not been assessed in screening with molecular studies. Although the prevalence of moderate to severe mental retardation from FXS may be 1 in 4000, the prevalence of the full spectrum of involvement may be substantially higher.

II. PHYSICAL AND BEHAVIORAL FEATURES IN MALES

A typical adult male with the full mutation of FXS usually has a long face, prominent ears, and large testicles. These features are considered the triad of clinical involvement in FXS. The long face is more common in adolescence and adulthood and sometimes the chin may become prominent. The cause of the long face which is often also associated with a high arched palate is not known. Because these features are seen in acromegaly, growth hormone abnormalities have been hypothesized but not documented in FXS. Some young patients may have evidence of excessive growth including height, weight, and head circumference measurements in childhood, but Loesch *et al.* [47] have shown that growth in puberty may not be normal such that adult height may be smaller than expected. DeVries *et al.* [19] have documented that a subgroup of individuals with FXS have a Soto syndrome-like phenotype with general overgrowth presumably secondary to endocrinological abnormalities. In addition, another subgroup may have a Prader-Willi-like phenotype including short extremities, hypotonia in early childhood, and excessive eating related to obsessive–compulsive behavior [19]. All of the growth studies in FXS suggest hypothalamic dysfunction as predicted by Fryns *et al.* [26].

Prominent ears are seen in the majority of patients with FXS, but at least 25% do not have prominent ears (Fig. 2-1) [28]. The ears are often soft and flexible and frequently the upper pinnae cups out with poor development of the antihelical fold. The prominent ears are thought to be secondary to an overall connective tissue abnormality in FXS [60, 30]. Other evidence of a connective tissue dysplasia in FXS includes soft, velvet-like skin, hyperextensible finger joints, double-jointed thumbs, flat feet, and the occasional occurrence of a joint dislocation or hernia. See Table 2-1 for the frequency of these physical findings in males with FXS. Very little research has taken place regarding the etiology or cause of the connective tissue abnormality. Early studies by Waldstein *et al.* [96, 97] documented abnormal elastin fibers in the skin and other tissues of males with FXS. The normal elastin network present between the dermis and the epidermis was absent in patients with FXS compared to controls. How the absence or a deficiency of the FMR1 protein (FMRP) causes these elastin abnormalities or other problems with the connective tissue in patients with FXS is not known and is worthy of further research.

FIGURE 2-1 Two young adult brothers with fragile X syndrome. Note long face in both and mildly prominent ears especially in brother on the right.

TABLE 2-1 Physical Features in Males with the Full Mutation[a]

	Children (%)	Adolescents and adults (%)
Long face	64	80
Prominent ears	78	66
High arched palate	51	63
Hyperextensible finger joints	81	49
Double jointed thumbs	58	48
Single palmar crease	26	22
Hand calluses	18	52
Flat feet	82	60
Heart murmur or click	16	29
Macroorchidism (>25 ml volume)	—	92

[a] Adapted from [28].

On occasion, patients with FXS may present with more remarkable malformations such as a club foot deformity or a cleft palate, both of which are seen in less that 5% of patients with FXS. These abnormalities, however, may be more common in patients with FXS compared to the normal population because of the connective tissue abnormalities seen in FXS. An occasional urological abnormality, such as hypospadius or dilation of the ureter associated with reflux, also appears to be slightly increased in individuals with FXS compared to the general population [28].

The third feature of the classical triad, large testicles or macroorchidism, is present in approximately 80 to 95% of adolescent and adult males. An enlargement of the testicles is only occasionally seen in early childhood; in most males with FXS, the enlargement begins between 8 and 9 years of age [41]. The testicles consistently increase in size throughout puberty and subsequently level off with a stable enlargement at age 16 to 18 years of age [10]. Most males with FXS have between two to three times the normal testicle size with a mean at 45 ml in volume [10]. The cause of the macroorchidism does not appear to be related to the connective tissue abnormality in FXS. Instead, hypothalamic dysfunction as first reported by Fryns *et al.* [26] is the hypothesized cause of the macroorchidism. Perhaps hypothalamic dysfunction causes an increase in gonadotropins in FXS. This may be related to a melatonin deficiency which has been reported in a small number of males with FXS by O'Hare *et al.* [59]. Since melatonin inhibits the release of gonadotropin from the pituitary [44], a deficiency of melatonin could lead to an enhanced level of gonadotropin release in males with FXS [37]. Although some researchers have found mild elevations in gonadotropin levels (FSH and LH) [52, 75, 93], others have not [5, 6, 11]. Clearly further studies are needed regarding these endocrine findings in patients with FXS.

Young boys with FXS usually do not have a long face, prominent jaw, or large testicles. However, other connective tissue abnormalities such as hyperextensible finger joints and flat feet are seen in the majority of children with FXS. Prominent ears are helpful when they are present and particularly when they are associated with cupping, but in approximately 25% of boys with FXS, the ears are not prominent (see Table 2-1). As more cases are diagnosed, this percentage appears to be increasing. Therefore, at first glance, many young children with FXS may not demonstrate the typical facial features that are associated with this syndrome. Therefore the behavioral features and the cognitive features should be considered at the time of presentation.

Young children with FXS commonly present with language delays and hyperactivity [25, 84]. They may also have temper tantrums and behavioral difficulties which lead the parent to seek treatment before the diagnosis of FXS is made. A number of autistic-like features are also associated with this syndrome including poor eye contact, tactile defensiveness or sensitivity to touch, hand flapping, hand biting, other hand stereotypies, and perseveration or repetition in speech. Usually children with FXS will ask the same question over and over again even after the answer is given. They may also be perseverative in their behavior, that is, watching the same video many times or following a similar routine in a compulsive fashion. Most of these features are described as autistic-like because they are commonly seen in children with autism or pervasive developmental disorder. The exact cause of these autistic-like features is not known. However, these features appear to be related to a hypersensitivity to sensory stimuli including visual, auditory, olfactory, and tactile stimuli [12, 28]. Electrophysiological studies have documented a lack of normal habituation to stimuli in patients with FXS compared to normal controls [56]. The lack of normal habituation may be the cause of hyperarousal and the frequent occurrence of tantrums or outburst behavior. The lack of normal habituation may also lead to an oversensitivity to stimuli and a subsequent adverse response even with minor stimulation such as direct eye contact or light touch. This hypersensitivity to stimuli may also be associated with difficulties such as turning away when shaking hands [102] and approach/avoidance behaviors [12].

In general, however, patients with FXS are usually interested in social interaction, although the sensory distortions may cause avoidance of social interaction at times. Usually patients with FXS are shy and socially anxious at first but then they connect emotionally with others such that the majority of children with FXS do not have autism [2]. Patients with FXS usually do not have the disinterest in social interaction which is typical of children with autism. Only approximately 15% of males with FXS will be diagnosed with autism and this usually represents individuals with severe cognitive deficits or severe problems with sensory integration (reviewed in [28]). Usually autistic behavior also improves with therapy over time in children with FXS compared to controls [2]. It is important, however, to do FMR1 testing in children with autism, mental retardation, or pervasive developmental disorder when the etiology for these problems is not known.

III. NEUROANATOMICAL CHANGES

Detailed quantitative MRI studies have been carried out on patients with FXS compared to controls. Reiss *et al.* [63, 64, 66–68] have demonstrated a smaller cerebellar vermis in both males and females affected with FXS in addition to a larger caudate, thalamus, and hippocampus. Schapiro *et al.* [76] have also demonstrated an overall larger brain in patients with FXS compared to controls. Reiss *et al.* [67] and Cohen [12] hypothesized a lack of the normal pruning process of neuronal connections early on in development when there is a lack of the FMR1 protein. Subsequent studies by Comery *et al.* [14] have demonstrated immature neuronal connections and an increase in dentritic branches in knockout mice who are missing the FMR1 gene. This further supports the hypothesis of a lack of normal pruning and maturation of connections in patients affected by FXS. Perhaps the neuronal overconnectedness also relates to the findings of hypersensitivity to sensory stimuli. The neuroanatomical studies document the CNS abnormalities which are consistent with the cognitive and behavioral phenotype in FXS.

IV. ADDITIONAL MEDICAL PROBLEMS

Approximately 30% of babies with FXS have problems with recurrent emesis in early childhood [27, 28]. In 15% this can lead to failure to thrive in infancy [28]. The recurrent emesis is secondary to gastroesophageal reflux. Hypotonia in addition to the connective tissue abnormalities could impact the lower esophageal sphincter and may cause reflux. In addition, approximately a third of boys with FXS suffer from strabismus or a weak eye muscle [39, 48]. This usually requires treatment which may include patching or the use of glasses to strengthen the weaker eye. A referral to an optometrist or an ophthalmologist is essential if this finding is present. Recurrent otitis media infections occur in over 80% of children with FXS [28]. This requires aggressive treatment because decreased hearing secondary to the otitis media can further damage language development in FXS. Just over 20% of patients also may have recurrent sinusitis, and this may be related to changes in the facial structure in addition to the connective tissue findings.

Seizures are seen in just over 20% of patients with FXS [57, 58]. Wisniewski [101] have found that the seizures are usually easily treated; the most common anticonvulsant used is carbamazepine.

V. COGNITIVE FEATURES IN MALES

Approximately 87% of males with FXS will have cognitive abilities in the mentally retarded range [33]. Young males, however, in the preschool age range often present with developmental or early cognitive testing which is in the normal or borderline range [25]. As the patient ages and further cognitive testing is carried out, there is a decrease in IQ scores which was first reported by Lachiewicz *et al.* [42]. Subsequent studies regarding IQ decline by Hodapp *et al.* [36] and Wright-Talamante *et al.* [103] suggest that approximately 30% of males with FXS experience significant IQ decline and that this decline is related to the amount of the FMR1 protein present [92]. Merenstein *et al.* [52a] documented different IQ outcomes in individuals with variant DNA patterns including the presence of mosaicism or a lack of methylation of the full mutation. Although the average adult IQ of males with a full mutation that is fully methylated is 41, the average IQ of a male with a mosaic pattern (some cells with the premutation and other cells with the full mutation) was 60, and the average IQ for a male with a full mutation but less that 50% methylation was 88. There have been other reports of high functioning or nonretarded males with FXS including the reports by Loesch *et al.* [46], Smeets *et al.* [85], McConkie-Rosell *et al.* [50], Merenstein *et al.* [53], Rousseau *et al.* [71], Milà *et al.* [55], Steyaert *et al.* [90], and Lachiewicz *et al.* [43]. These individuals demonstrated a complete or partial lack of methylation of the full mutation, and in some cases, FMRP production has been documented [33, 53, 85]. A recent report by Tassone *et al.* [92] has found a significant correlation between the percentage lack of methylation in partially unmethylated males or the percentage of the premutation in mosaic males and the IQ level. Documentation of FMRP levels utilizing the technique first described by Willemsen *et al.* [99, 100] has also shown a correlation between FMRP levels and IQ in both of these groups of males [92]. Cohen [13] also demonstrated better adaptive skills in mosaic males compared to those with the full mutation.

High functioning or nonretarded males with FXS usually demonstrate significant problems in the emotional and behavioral area in addition to having learning disabilities. Generally, strengths occur in the area of reading and spelling but math difficulties are usually a pervasive problem [28, 53]. The unusual behavioral manifestations that were previously termed autistic-like features may often be called schizotypal features in a nonretarded male with FXS. They include poor eye contact, odd mannerisms, difficulties in modulating speech, and poor pragmatic language. These behaviors are also typical of individuals with Asperger syndrome (high functioning individuals with autistic features) so some high functioning males with FXS may also be diagnosed with Asperger syndrome. Problems with shyness, social anxiety, obsessive–compulsive behavior, and sometimes attentional problems may be present. These difficulties without significant cognitive deficits are associated with a small deficit of FMRP levels [92].

Similar problems may also be seen in females with the full mutation who are not retarded.

VI. PHYSICAL AND BEHAVIORAL FEATURES IN FEMALES

Females with the full mutation are significantly less affected than males with the full mutation in physical, behavioral, and cognitive areas. This is because their second X chromosome is producing FMRP to a variable extent depending on the activation ratio, that is, the percentage of cells that have the normal X as the active X chromosome [1, 70]. This has been well studied in peripheral blood but the activation ratio in blood may not be reflective of the activation ratio in the central nervous system. There is evidence in animal studies by Tan *et al.* [91] of a layering of inactivation patterns throughout the CNS in females. Binstock [4] suggested that this may account for the variable phenotypic affects related to cognitive and behavioral functioning in females with FXS. Studies similar to the report of Tan *et al.* [91] have not been carried out in females with the full mutation nor in female knockout mice as of yet.

Previous studies carried out by Fryns [26], Loesch and Hay [45], and subsequently Cronister *et al.* [17] suggest that approximately 50% of females affected by FXS demonstrate physical features which are typical of males with FXS including a long face, prominent ears, mandibular prognanthism, hypermobility at the finger joints, and double-jointed thumbs. In these studies, molecular analysis of the FMR1 mutation was not available but the general finding was that women with more severe cognitive deficits were more likely to demonstrate typical physical features of FXS. The most common finding in young girls with FXS was prominent ears which helped to differentiate them from their normal female sisters [31, 38].

Riddle *et al.* [69] studied 41 women with the full mutation, 114 women with the premutation, and 126 controls and found that women with the full mutation had a longer face, more prominent ears, hand calluses, double-jointed thumbs, and poorer eye contact than controls or women with the premutation. In addition, women identified by history with the full mutation also had a greater incidence of hand flapping, special education help for math and reading, and grade retention than controls or women with the premutation. A compilation of these pertinent findings by history correlated negatively with the activation ratio in women with the full mutation. That is, the more involvement historically, the lower the activation ratio, or the fewer number of cells with the normal X as the active X.

Subsequent protein studies reported by Tassone *et al.* [92] found a significant correlation between the overall Physical Index score, which included the presence of long face, prominent ears, long ears, high arched palate, hyperextensible finger joints, double jointed thumbs, hand calluses, single palmar crease, flat feet, and cardiac murmur, and the level of FMRP in females with the full mutation.

In the behavioral area, there are two classical presentations for females with the full mutation. Freund *et al.* [24] demonstrated that shyness and social anxiety with a psychiatric diagnosis of avoidant personality disorder or avoidant disorder in childhood was present in the majority of females with the full mutation; these findings were significantly different then females matched on IQ but without the fragile X mutation. Lachiewicz and Dawson [40] found that 40% of girls with FXS over 9 years of age had an anxiety rating in the clinical range on behavioral questionnaires. Anxiety, shyness, and social phobia are frequently seen together in females with FXS. An occasional female has such severe problems in this area that selective mutism, that is the inability to speak in certain social circumstances, particularly school, can occasionally occur [35].

The second classical presentation of women with the full mutation includes hyperactivity and a short attention span. This problem is seen in approximately one-third of girls with the full mutation [24, 31]. Although the degree of hyperactivity is less severe than what is typically seen in males, the attention and concentration problems which are linked to the executive function deficits described below are often responsive to treatment including counseling and medication [29, 86].

Shyness and anxiety are such significant core features of FXS that they occasionally occur in women who carry only the premutation. A controlled study by Franke *et al.* [23] found anxiety disorders with three times the prevalence in women with the premutation compared to control women without the mutation who had children with autism. The findings of mild phenotypic involvement in women with the premutation have also been seen in other areas. For instance, Hull and Hagerman [38] demonstrated more prominent ears and a higher Physical Index in women with the premutation compared to controls. This finding was also replicated in the study by Riddle *et al.* [69]. Premature ovarian failure is a significant finding in over 20% of women with the premutation which is significantly different from women with the full mutation [61, 82]. Turner *et al.* [94] found a higher twinning rate in women with the premutation compared to controls. The twinning rate was three times that seen in controls. From a clinical perspective, the finding of excessive worrying combined with mood lability and on occasion obsessive–

compulsive behavior is not uncommon in women with the premutation. There is no evidence of cognitive deficits in women with the premutation [49, 65].

VII. COGNITIVE FEATURES IN FEMALES

Approximately 50% of females with the full mutation demonstrate cognitive deficits in the borderline or mentally retarded range [31, 72]. Abrams *et al.* [1] found a significant correlation between cognitive measures and both the activation ratio and the CGG repeat length in 31 females ranging in age from 4 to 27 years with the full mutation. Subsequently, Reiss *et al.* [68] studied both molecular and background genetic effects on IQ by analyzing the IQ of the parents of females with the full mutation. They found that the mean parental IQ predicted 26% of the variance in the IQ of their daughter with FXS whereas the activation ratio predicted 33% of the IQ variance. In controls, only half of the child's IQ was predicted by the mean parental IQ [68]. Subsequent studies regarding FMRP levels in females with the full mutation demonstrated a significant correlation between FMRP level and IQ but this was not nearly as strong a correlation as compared to the males [92]. This difference is most likely related to the variation in tissues, such that the activation ratio in blood may be different from the activation ratios seen in different parts of the brain.

A recent study by deVries *et al.* [20] evaluated 33 females with the full mutation. Their evaluation process included visiting the homes so that individuals who were too anxious to come into clinic were assessed and their cognitive findings documented. They found that 71% of females with the full mutation demonstrated a significant cognitive deficit in the borderline or mentally retarded range. In addition, the activation ratio correlated with the Performance IQ and the Full Scale IQ but not the Verbal IQ. In females without mental retardation, executive function deficits are seen in the majority [49]. These deficits includes problems with organization, attention, shifting problem-solving strategies, and maintaining topic in conversation. Sobesky *et al.* [87] recently found that the activation ratio correlated with executive function deficits. Therefore, the learning, attention, and organization problems even in nonretarded females with the full mutation are part of a continuum of involvement from the FMR1 mutation.

VIII. WHO TO TEST

There is ample documentation in the literature that those individuals with mental retardation, autism, or pervasive developmental disorder should be tested for FXS when the etiology is unknown. FMR1 DNA testing should be carried out using a Southern blot or screening with PCR followed by a Southern blot if the finding is abnormal [8]. Newer testing procedures using antibodies to the FMR1 protein (FMRP) developed by Willemsen *et al.* [99, 100] have not been utilized clinically and may not detect individuals who are higher functioning and may show only a mild deficit of FMRP. If there is a family history of mental retardation, there is an even greater imperative to carry out DNA testing for the FMR1 mutation because approximately 30% of individuals with X-linked mental retardation turn out to have FXS [83].

Higher functioning individuals with FXS may present only with learning disabilities, executive functioning deficits, ADHD, or language delays. Not all individuals with these problems should be tested for the FMR1 mutation. However, those with typical physical or behavioral features consistent with FXS or those who have a family history of mental retardation should be considered for FMR1 testing. Some specific medical diagnoses have been associated with FXS. These include Soto syndrome, Prader-Willi syndrome, Pierre Robin sequence, and motor tics or Tourette's syndrome associated with cognitive deficits. Individuals who present with these diagnoses should also be tested for the FMR1 mutation.

In the obstetric population, pregnant females who have a family history of mental retardation should also be tested for FXS. Approximately 1 in 40 pregnant women with this history are found to carry the FMR1 mutation [9]. Since Rousseau *et al.* [73] have demonstrated that 1 in 259 women in the general population carries the FMR1 premutation, perhaps all pregnant females should be considered for FMR1 testing [22]. Certainly additional subgroups in the OBGYN population should be considered for testing. For instance, women who present with premature menopause have a higher risk for the FMR1 mutation, particularly if the premature menopause is associated with a family history of premature menopause [15, 61].

In the psychiatric population, there are several diagnostic categories that should be considered for FMR1 testing besides those who present with autism. For instance, individuals with Asperger syndrome (an individual with a normal IQ and autistic-like features) should be considered for testing. Also schizotypy or a schizotypal personality disorder, which involves odd communication patterns and odd mannerisms in addition to social deficits, can occur in both males and females mildly affected with FXS, so therefore these diagnoses should warrant testing. Psychosis associated with mental retardation should be consistently screened for the FMR1 mutation. In addition, avoidant disorder, selective mutism, and severe anxiety disorders or social phobias should also be strongly considered for FMR1 testing. These disorders are seen frequently in females affected by FXS.

IX. TREATMENT

Although there is no cure at the present time for FXS, there is treatment available for all patients; this includes a variety of interventions. Both speech and language therapy and sensory integration/occupational therapy can be helpful for all children affected by FXS. These interventions can be incorporated even in the first year of life through an infant stimulation program. As the child becomes a toddler, individualized therapy can be helpful for improving both speech and language and motor development. Sensory integration therapy can also help to decrease hyperarousal from a variety of stimuli, and this has a secondary beneficial effect regarding behavior, particularly tantrums, outbursts, or aggression [77, 81]. Although speech and language therapy early on will focus on vocabulary development, the focus on the ability to generalize, higher linguistic skills, and abstract reasoning becomes important as the child ages [77]. The strong emphasis on abstract reasoning will be helpful in ameliorating the IQ decline which occurs in middle to late childhood related to the difficulties in problem solving and reasoning that are demanded of the older child [3].

In the learning arena, children with FXS demonstrate significant strengths including imitation abilities, a sense of humor, memory skills, and an interest and expertise in computer use, in addition to significant weakness including attentional problems, hyperactivity, and math deficits. The use of computer software programs to enhance academic development, particularly math, spelling, and reading, can be beneficial and can be started even in the preschool years [77–79]. Special education programs such as the Logo Reading, the Edmark Series, and the Lindamood–Bell [80] can enhance reading skills and additional programs can be helpful in math [54].

For individuals who experience significant behavior problems at home, including severe hyperactivity, tantrums, mood lability, or outburst behavior, the use of counseling for both the child and the parents in addition to other family members can be beneficial particularly with the addition of behavioral modification techniques [7, 86]. Individual counseling can be helpful for females who have severe shyness, social anxiety, poor self image, and/or depression. This type of intervention can be especially helpful around the time of adolescence when social pressures become more intense. Transition into adulthood, including the independence from family and the development of job expertise and confidence, can be aided by individual counseling. Group therapy with a focus on building social skills can also be helpful for the child and adolescent with FXS.

In general, in the educational setting, inclusion into the regular classroom with the help of an aid is helpful for children affected by FXS [77, 88]. The skills in imitation and the utilization of modeling can help to normalize social skills and facilitate social acceptance. Although children with FXS and mental retardation will not keep up with the regular classroom from an academic perspective, their work can be modified by an aid or the help of a special education teacher [81, 88].

Treatment of specific medical problems which are associated with FXS are essential to normalize vision, hearing, and cognitive awareness. For instance, seizures which occur in 20% of children with FXS should be treated with anticonvulsants such as carbamazepine or valproic acid since spike wave discharges with or without clinical seizures can interfere with cognitive processing [29, 57, 101]. Chronic otitis media infections can cause a conductive hearing loss for a prolonged period of time in childhood which in turn can interfere with auditory processing and the development of appropriate language skills. Therefore, chronic ear infections should be treated vigorously from a medical perspective including the use of antibiotics and PE tubes when necessary. Disturbances in vision including strabismus leading to amblyopia or significant refraction errors also require prompt treatment. Children who are diagnosed with FXS should be seen by an optometrist or an ophthalmologist within the first 4 years of life, and for more obvious problems, they should be seen as soon as the problems are noted.

On occasion, other medical complications such as orthopedic problems or mitral valve prolapse also require a referral to an appropriate specialist. For instance, if a cardiac murmur or click is heard, referral to a cardiologist for an ultrasound study will facilitate the diagnosis of mitral valve prolapse. Significant prolapse requires the use of antibiotic prophylaxis [29].

Lastly, the use of psychopharmacology to improve behavioral difficulties which can interfere with learning is often helpful as part of an overall treatment program. In the school-aged child, significant attentional problems or ADHD is usually treated successfully with stimulant medication such as methylphenidate, dextroamphetamine, or pemoline. The latter stimulant may often be better tolerated if a negative response is seen with methylphenidate or dextroamphetamine; however, pemoline requires the frequent monitoring of liver function studies since the FDA recently reported 13 cases of liver failure associated with pemoline's use over the last 20 years. I have not seen liver dysfunction in patients with FXS but these problems should still be monitored until further research is carried out to better understand which individuals are predisposed to liver dysfunction.

Often mood lability or aggression, even in the preschool period, can be treated with clonidine, which is a presynaptic α_2 agonist that has an overall calming effect in addition to improving attention. Although clonidine is not nearly as effective for improving distractibility and attention compared to the stimulants, clonidine is more effective in decreasing tics and calming outburst behavior [29, 34]. Because of recent reports of cardiac dysrhythmias on rare occasion with clonidine, the use of an EKG in follow-up is recommended [29].

Often anxiety or mood lability can be a problem for both males and females affected by FXS and this can respond well to a selective serotonin reuptake inhibitor (SSRI) [29]. SSRIs, in addition, have been used to treat aggression because aggression may stem from significant anxiety, particularly associated with transitions in the environment [32]. In general, SSRIs are relatively safe, without significant cardiac or liver toxicity. However, in approximately 15% of patients with FXS, an increase in hyperactivity, aggressive behavior, or manic symptoms may occur. SSRIs are helpful for individuals with the premutation who are suffering from significant anxiety, especially when this is associated with sleep disturbances. The SSRI agents such as fluoxetine, sertraline, paroxetine, or venlafaxine can decrease anxiety and also improve mood. Chronic anxiety can often lead to depression and an SSRI agent will treat both symptoms. Obsessive–compulsive behavior is also common in FXS and in general this symptom improves with an SSRI agent. A newer SSRI, fluvoxamine, has recently been approved for use in children down to 8 years of age for treatment of obsessive–compulsive disorder. It is effective for the majority of children with these symptoms. It is found to be anecdotally helpful in individuals with FXS who suffer from significant obsessive–compulsive behavior including obsessive eating, obsessive picking, and even perseverative behavior.

On occasion, psychosis or significant paranoia can be seen in both males and females with FXS and the use of an antipsychotic agent, such as risperidone, which is a newer antipsychotic with a lower risk of tardive dyskinesias compared to other antipsychotic agents, can be helpful. Risperidone can also be used to treat aggression which is not responsive to the other agents described above. Preferably a low dose of risperidone, even starting with just 0.5 mg at bedtime (which can be increased to a twice daily dosage), can be beneficial.

After the diagnosis of FXS is made, it is essential to carry out genetic counseling with the family. The counselor reviews the pedigree and explains the full spectrum of involvement in the syndrome and how it may impact other family members [16, 89]. The genetic counselor should be available to send out written information to extended family members explaining the diagnosis of FXS [51]. The counselor should also establish a long-term relationship with the family, such that when future pregnancies arise, the genetic counselor will be contacted by the family to facilitate prenatal diagnosis [89].

It is important to link newly diagnosed families with parent support groups where other families who are impacted by FXS gather to discuss their concerns and their solutions to daily problems. The National Fragile X Foundation (phone number (800) 688-8765 or (303) 333-6155) can be contacted for a national and international list of resource centers and parent support groups. The National Fragile X Foundation also has a variety of parent educational materials to send to families and professionals with additional information regarding FXS.

The diagnosis and treatment of children and adults who are affected by FXS are both challenging and rewarding. The advances in the molecular biology of FXS described in the subsequent chapters will eventually lead to more effective interventions including protein replacement therapy and gene therapy, although many obstacles must be overcome before this becomes a reality [62].

References

1. Abrams, M. T., Reiss, A. L., Freund, L. S., Baumgardner, T. L., Chase, G. A., and Denekla, M. B. (1994). Molecular-neurobehavioral associations in females with the fragile X full mutation. *Am. J. Med. Genet.* **51,** 317–327.
2. Baumgardner, T., Reiss, A. L., Freund, L. S., and Abrams, M. T. (1995). Specifications of the neurobehavioral associations in males with fragile X syndrome. *Pediatrics* **95,** 744–752.
3. Bennetto, L., and Pennington, B. F. (1996). The neuropsychology of fragile X syndrome. *In* "Fragile X Syndrome: Diagnosis, Treatment, & Research," 2nd edition (R. J. Hagerman and A. Cronister, Eds.), pp. 210–248. Johns Hopkins University Press, Baltimore.
4. Binstock, T. (1995). From FMR1 mutation to phenotype: more than one pathway? Poster presented at the 7th International Workshop on the Fragile X and X-Linked Mental Retardation. Aug. 2–5, 1995, Tromsø, Norway.
5. Bowen, P., Biederman, B., and Swallow, K. A. (1978). The X-linked syndrome of macroorchidism and mental retardation; Further observations *Am. J. Med. Genet.* **2,** 409–414.
6. Brøndum-Nielsen, K., Tommerup, N., Dyggve, V., and Schou, C. (1982). Macroorchidism and fragile X in mentally retarded males: clinical, cytogenetic and some hormonal investigations in mentally retarded males, including two with the fragile site at Xq28, fra(X) (q28). *Hum. Genet.* **61,** 113–117.
7. Brown, J., Braden, M., and Sobesky, W. (1991). The treatment of behavioral and emotional problems. *In* "Fragile X Syndrome:

Diagnosis, Treatment, & Research" (R. J. Hagerman and A. Cronister Silverman, Eds.), pp. 311–326. Johns Hopkins University Press, Baltimore.

8. Brown, W. T. (1996). The molecular biology of the fragile X mutation. *In* "Fragile X Syndrome: Diagnosis, Treatment, & Research," 2nd edition (R. J. Hagerman and A. Cronister, Eds.), pp. 88–113. Johns Hopkins University Press, Baltimore.
9. Brown, W. T., Houck, G. E., Jeziorowska, A., Levinson, F. N., Ding, X., Dobkin, C., Zhong, N., Henderson, J., Brooks, S. S., and Jenkins, E. C. (1993). Rapid fragile X carrier screening and prenatal diagnosis using a nonradioactive PCR test. *JAMA* **270**, 1569–1575.
10. Butler, M. G., Brunschwig, A., Miller, L. K., and Hagerman, R. J. (1992). Standards for selected anthropometric measurements in males with the fragile X syndrome. *Pediatrics* **89,** 1059–62.
11. Cantú, J. M., Scaglia, H. E., Medina, M., Gonzalez-Diddi, M., Morato, T., Moreno, M. E., and Perez-Palacios, G. (1976). Inherited congenital normofunctional testicular hyperplasia and mental deficiency. *Hum. Genet.* **33,** 23–33.
12. Cohen, I. L. (1995). Behavioral profiles of autistic and non autistic fragile X males. *Dev. Brain Dysfunct.* **8,** 252–269.
13. Cohen, I. L., Nolin, S. L., Sudhalter, V., Ding, X-H., Dobkin, S. C., and Brown, W. T. (1996). Mosaicism for the FMR1 gene influences adaptative skills development in fragile X-affected males. *Am. J. Med. Genet.* **64,** 365–369.
14. Comery, T. A., Harris, J. B., Willems, P. J., Oostra, B. A., Irwin, S. A., Weiler, I. J., and Greenough, W. T. (1997). Abnormal dendritic spines in fragile X knockout mouse: Maturation and pruning deficits. *Proc. Natl. Acad. Sci.* **94,** 5401–5404.
15. Conway, G. S., Hettiarachchi, S., Murray, A., and Jacobs, P. A. (1995). Fragile X premutations in familial premature ovarian failure. *Lancet* **346,** 309–310.
16. Cronister, A. C. (1996). Genetic counseling. *In* "Fragile X Syndrome: Diagnosis, Treatment, & Research," 2nd edition (R. J. Hagerman and A. Cronister, Eds.), pp. 251–282. Johns Hopkins University Press, Baltimore.
17. Cronister, A., Schreiner, R., Wittenberger, M., Amiri, K., Harris, K., and Hagerman, R. J. (1991). The heterozygous fragile X female: historical, physical, cognitive and cytogenetic features. *Am. J. Med. Genet.* **38,** 269–274.
18. deVries, B. B. A., Fryns, J-P., Butler, M. G., Canziani, F., Wesby-van Swaay, E., vanHemel, J. O., Oostra, B. A., Halley, D. J. J., and Niermeyer, M. F. (1993). Clinical and molecular studies in fragile X patients with a Prader-Willi-like phenotype. *J. Med. Genet.* **30,** 761–766.
19. deVries, B. B. A., Robinson, H., Stolte-Dijkstra, I., Gi, C. V. T. P., Dijkstra, D. F., vanDoorn, J., Halley, D. J. J., Oostra, B. A., Turner, G., and Niermeijer. M. F. (1995). General overgrowth in the fragile X syndrome: Variability in the phenotype expresion of the FMR1 gene mutation. *J. Med. Genet.* **32,** 764–769.
20. deVries, B. B. A., Wiegers, A. M., Smits, A. P. T., Mohkamsing, S., Duivenvoorden, H. J., Fryns, J-P., Curfs, L. M. G., Halley, D. J. J., Oostra, B. A., van den Ouweland, A. M. W., and Niermeijer, M. F. (1996). Mental status of females with an FMR1 gene full mutation. *Am. J. Human Genet.* **58,** 1025–1032.
21. Dykens, E. M., Hodapp, R. M., and Leckman, J. F. (1994). "Behavior and Development in Fragile X Syndrome." Sage, Thousand Oaks, CA.
22. Finucane, B. (1996). Should all pregnant women be offered carrier testing for fragile X syndrome? *Clin. Obstet. Gynecol.* **39,** 772–782.
23. Franke, P., Maier, W., Iwers, B., Hautzinger, M., and Froster, U. G. (1996). Fragile X carrier females: evidence for a distinct psychopathological phenotype? *Am. J. Med. Genet.* **64,** 334–339.
24. Freund, L. S., Reiss, A. L., and Abrams, M. (1993). Psychiatric disorders associated with fragile X in the young female. *Pediatrics* **91,** 321–329.
25. Freund, L. S., Peebles, C. D., Aylward, E., and Reiss, A. L. (1995). Preliminary report on cognitive and adaptive behaviors of preschool aged males with fragile X. *Dev. Brain. Dysfunct.* **8,** 242–251.
26. Fryns, J. P., Dereymaeker, A. M., Hoefnagels, M., Volcke, P., and Van den Berghe, H. (1986). Partial fra(X) phenotype with megalotestes in fra(X) negative patients with acquired lesions of the central nervous system. *Am. J. Med. Genet.* **23,** 213–219.
27. Goldson, E., and Hagerman, R. J. (1993). Fragile X syndrome and failure to thrive. *AJDC* **147,** 605–607.
28. Hagerman, R. J. (1996). Physical and behavioral phenotype. *In* "Fragile X Syndrome: Diagnosis, Treatment, & Research," 2nd edition (R. J. Hagerman and A. Cronister, Eds.), pp. 3–87. Johns Hopkins University Press, Baltimore.
29. Hagerman, R. J. (1996). Medical follow-up and psychopharmacology. *In* "Fragile X Syndrome: Diagnosis, Treatment, & Research," 2nd edition (R. J. Hagerman and A. Cronister, Eds.), pp. 283–331. Johns Hopkins University Press, Baltimore.
30. Hagerman, R. J., Van Housen, K., Smith, A. C. M., and McGavran, L. (1984). Consideration of connective tissue dysfunction in the fragile X syndrome. *Am. J. Med. Genet.* **17,** 111–122.
31. Hagerman, R. J., Jackson, C., Amiri, K., Silverman, A. C., O'Connor, R., and Sobesky, W. E. (1992). Fragile X girls: physical and neurocognitive status and outcome. *Pediatrics* **89,** 395–400.
32. Hagerman, R. J., Fulton, M. J., Leaman, A., Riddle, J., Hagerman, K., and Sobesky, W. (1994). Fluoxetine therapy in fragile X syndrome. *Dev. Brain. Dysfunct.* **7,** 155–164.
33. Hagerman, R. J., Hull, C. E., Safanda, J. F., Carpenter, I., Staley, L. W., O'Connor, R. O., Seydel, C., Mazzocco, M. M., Snow, K., Thibodeau, S., Kuhl, D., Nelson, D. L., Caskey, C. T., and Taylor, A. (1994). High functioning fragile X males: demonstration of an unmethylated, fully expanded FMR-1 mutation associated with protein expression. *Am. J. Med. Genet.* **51,** 298–308.
34. Hagerman, R. J., Riddle, J., Roberts, L. S., Breese, K., and Fulton, M. (1995). A survey of the efficacy of clonidine in fragile X syndrome. *Dev. Brain Dys.* **8,** 336–344.
35. Hagerman, R. J., Hills, J., Scharfenaker, S., and Lewis, H.(1997). Fragile X syndrome and selective mutism. In press. *Am. J. Med. Genetics.*
36. Hodapp, R. M., Dykens, E. M., Hagerman, R. J., Schreiner, R., Lachiewicz, A. M., and Leckman, J. F. (1990). Developmental implications of changing trajectories of IQ in males with fragile X syndrome. *J. Am. Acad. Child. Adolesc. Psych.* **9,** 214–219.
37. Holloway, S., Loesch, D., and Hagerman, R. J. (1997) Temporal sleep characteristics of young boys with fragile X syndrome. In press. *Am J. Mental Retard.*
38. Hull, C., and Hagerman, R. J. (1993). A study of the physical, behavioral, and medical phenotype, including anthropometric measures of females with fragile X syndrome. *Am. J. Dis. Child.* **147,** 1236–1241.
39. King, R. A., Hagerman, R. J., and Houghton, M. (1995). Ocular findings in fragile X syndrome. *Dev. Brain. Dys.* **8,** 223–229.
40. Lachiewicz, A. M., and Dawson, D. V. (1994). Behavioral problems of young girls with fragile X syndrome: factor scores on the Conner's parent questionnaire. *Am. J. Med. Genet.* **15,** 364–369.
41. Lachiewicz, A. M., and Dawson, D. V. (1994). Do young boys with fragile X syndrome have macroorchidism? *Pediatrics* **93,** 992–995.

42. Lachiewicz, A. M., Gullion, C., Spiridigliozzi, G., and Aylsworth, A. (1987). Declining IQs of young males with the fragile X syndrome. *Am. J. Ment. Retard.* **92,** 272–278.
43. Lachiewicz, A. M., Spiridigliozzi, G. A., McConkie-Rosell, A., Burgess, D., Feng, Y., Warren, S. T., and Tarleton, J. (1996). A fragile X male with a broad smear on southern blot analysis representing 100-500 CGG repeats and no methylation at the EagI site of the FMR1 gene. *Am. J. Med. Genet.* **64,** 278–282.
44. Lewin, D. (1996). Researchers cycle down the path to elucidating melatonin's rhythms. *J. NIH Res.* **8,** 45–50.
45. Loesch, D. Z., and Hay, D. A. (1988). Clinical features and reproductive patterns in fragile X female heterozygotes. *J. Med. Genet.* **25,** 407–414.
46. Loesch, D. Z., Huggins, R. M., and Chin, W. F. (1993). Effect of fragile X on physical and intellectual traits estimated by pedigree analysis. *Am. J. Med. Genet.* **46,** 415–422.
47. Loesch, D. Z., Huggins, R. M., and Hoang, N. H. (1995). Growth in stature in fragile X families: a mixed longitudinal study. *Am. J. Med. Genet.* **58,** 249–256.
48. Maino, D. M., Wesson, M., Schlange, D., Cibis, G., and Maino, J. H. (1991). Optometric findings in the fragile X syndrome. *Optometry Vis. Sci.* **68,** 634–640.
49. Mazzocco, M. M. M., Pennington, B. F., and Hagerman, R. J. (1993). The neurocognitive phenotype of female carriers of fragile X; further evidence for specificity. *J. Dev. Behav. Pediatr.* **14,** 328–335.
50. McConkie-Rosell, A., Lachewicz, A., Spiridigliozzi, G. A., Tarleton, J., Scheonwald, S., Phelan, M. C., Goonewardena, P., Ding, X., and Brown, W. T. (1993). Evidence that methylation of the FMR-1 locus is responsible for variable phenotypic expression of the fragile X syndrome. *Am. J. Hum. Genet.* **53,** 800–809.
51. McConkie-Rosell, A., Robinson, H., Wake, S., Staley, L. W., Heller, K., and Cronister, A. (1995). Dissemination of genetic risk information to relatives in the fragile X syndrome: Guidelines for genetic counselors. *Am. J. Med. Genet.* **59,** 426–430.
52. McDermott, A., Walters, R., Howell, R. T., and Gardner, A. (1983). Fragile X chromosome: clinical and cytogenetic studies on cases from seven families. *J. Med. Genet.* **20,** 169–178.
52a. Merenstein, S. A., Sobesky, W. E., Taylor, A. K., Riddle, J. E., Tran, H. X., and Hagerman, R. J. (1996). Molecular–clinical correlations in males with an expanded FMRI mutation. *Am. J. Med. Genet.* **64,** 389–394.
53. Merenstein, S. A., Shyu, V., Sobesky, W. E., Staley, L., Taylor, A., and Hagerman, R. J. (1994). Fragile X syndrome in a normal IQ male with learning and emotional problems. *J. Am. Acad. Child Adolesc. Psych.* **33,** 1316–1321.
54. Miezejeski, C. M., and Hinton, V. J. (1992). Fragile X learning disability; neurobehavioral research, diagnostic models and treatment options. *In* "The 1992 International Fragile X Conference Proceedings" (R. J. Hagerman and P. McKenzie, Eds.), pp. 85–98. The National Fragile X Foundation and Spectra Publishing, Dillion.
55. Milá, M., Kruyer, H., Glover, G., Sanchez, A., Carbonell, P., Castellvi-Bel, S., Volpini, V.,Rosell, J., Gabarrón, J, López, I., Villa, M., Ballestra, F., and Estivill, X. (1994). Molecular analysis of the $(CGG)_n$ expansion in the FMR-1 gene in 59 Spanish families. *Hum. Genet.* **94,** 395–400.
56. Miller, L., McIntosh, D., McGrath, J., Shyu, V., Lampe, M., Taylor, A., Tassone, F., Nietzel, K., Stackhouse, T., and Hagerman, R. (1998). Electrodermal responses to sensory stimuli in individuals with fragile X syndrome. In press. *Am. J. Med. Genetics.*
57. Musumeci, S. A., Ferri, R., Colognola, R. M., Neri, G., Sanfilippo, S., and Bergonzi, P. (1988). Prevalence of a novel epileptogenic EEG pattern in the Martin-Bell syndrome. *Am. J. Med. Genet.* **30,** 207–212.
58. Musumeci, S. A., Colognola, R. M., Ferri, R., Gigli, G. L., Petrella, M. A., Sanfilippo, S., Bergonzi, P., and Tassinari, C. A. (1988). Fragile-X syndrome: aparticular epileptogenic EEG pattern. *Epilepsia* **29,** 41–47.
59. O'Hare, J. P., O'Brien, I. A. D., Arendt, J., Astley, P., Ratcliffe, W., Andrews, H., Walters, R., and Corrall, R. J. M. (1986). Does melatonin deficiency cause the enlarged genitalia of the fragile X syndrome? *Clin. Endocrinol.* **24,** 327–333.
60. Opitz, J. M., Westphal, J. M., and Daniel, A. (1984). Discovery of a connective tissue dysplasia in the Martin-Bell syndrome. *Am. J. Med. Genet.* **17,** 101–109.
61. Partington, M. W., Moore, D. Y., and Turner, G. M. (1996). Confirmation of early menopause in fragile X carriers. *Am. J. Med. Genet.* **64,** 370–372.
62. Rattazzi, M. C., and Ioannou, Y. A. (1996). Molecular approaches to therapy. *In* "Fragile X Syndrome: Diagnosis, Treatment, & Research,' 2nd edition (R. J. Hagerman and A. Cronister, Eds.), pp. 412–452. Johns Hopkins University Press, Baltimore.
63. Reiss, A. L., Aylward, E., Freund, L. S., Joshi, P. K., and Bryan R. N. (1991). Neuroanatomy of the fragile X syndrome: the posterior fossa. *Ann. Neurol.* **29,** 26–32.
64. Reiss, A. L., Freund, L., Tseng, J. E., and Joshi, P. K. (1991). Neuroanatomy in fragile X females: the posterior fossa. *Am. J. Hum. Genet.* **49,** 279–288.
65. Reiss, A. L., Freund, L., Abrams, M. T., Boehm, C., and Kazazian, H. (1993). Neurobehavioral effects of the fragile X premutation in adult women; a controlled study. *Am. J. Hum. Genet.* **52,** 884–894.
66. Reiss, A. L., Lee, J., and Freund, L. (1994). Neuroanatomy of fragile X syndrome: the temporal lobe. *Neurology* **44,** 1317–1324.
67. Reiss, A., Abrams, M., Greenlaw, R., Freund, L., and Denckla, M. (1995). Neuro developmental effects of the FMR-1 full mutation in humans. *Nat. Med.* **1,** 159–167.
68. Reiss, A. L., Freund, L. S., Baumgardner, T. L., Abrams, M. T., and Denckla, M. B. (1995). Contribution of the FMR1 gene mutation to human intellectual dysfunction. *Nat. Genet.* **11,** 331–334.
69. Riddle, M., Riddle, J. E., Cheema, A., Sobesky, W. E., Gardner, S. C., Taylor, A. K., Pennington, B. F., and Hagerman, R. J. (1998). Phenotype in females with the FMRI gene mutation. In press. *Am. J. Ment. Retard.*
70. Rousseau, F., Heitz, D., Oberle, I., and Mandel, J-L. (1991). Selection in blood cells from female carriers responsible for variable phenotypic expression of the fragile X syndrome: inverse correlation between age and proportion of active X-carrying the full mutation. *J. Am. Genet.* **28,** 830–836.
71. Rousseau, F., Robb, L. J., Rouillard, P., and der Kaloustian, V. M. (1994). No mental retardation in a man with 40% abnormal methylation at the FMR1 locus and transmission of sperm cell mutations as premutations. *Hum. Mol. Genet.* **6,** 927–930.
72. Rousseau, F., Heitz, D., Tarleton, J., MacPherson, J., Malmgren, H., Dahl, N., Barnicost, A., Matthew, C., Mornet, E., Teuada, I., Maddalena, A., Spiegel, R., Schinzel, A., Marcos, J. A. G., Schorderet, D. F., Schaap, T., Maccioni, L., Russo, S., Jacobs, P. A., Schwartz, C., and Mandel, J. L. (1994). A multicenter study on genotype-phenotype correlations in fragile X syndrome, using direct diagnosis with probe StB12.3: the first 2253 cases. *Am. J. Hum. Genet.* **55,** 225–237.
73. Rousseau, F., Rouillard, P., Morel, M-L., Khandjian, E. W., and Morgan, K. (1995). Prevalence of carriers of premutation-sized alleles of the FMR1 gene-and implications for the population

genetics of the fragile X syndrome. *Am. J. Hum. Genet.* **57,** 1006–1018.

74. Rousseau, F., Morel, M-L., Rouillard, P., Khandjian, E. W., and Morgan, K. (1996). Suprisingly low prevalance of FMR1 premutation among males from the general population. *Am. J. Hum. Genet.* **59**(Suppl.), A188:1069.
75. Ruvalcaba, R. H., Myhre, S. A., Roosen-Runge, E. C., and Beckwith, J. B. (1977). X-linked mental deficiency megalotestes syndrome. *JAMA* **238,** 1646–1650
76. Schapiro, M., Murphy, G., Hagerman, R. J., Azari, N., Alexander, G., Miezejeski, C., Hinton, V., Hoiowitz, B., Haxby, J., Kumar, A., White, B., and Grady, C. (1995). Adult fragile X syndrome: neu opsychology; brain anatomy and metabolism. *Am. J. Med. Genet.* **60,** 480–493.
77. Scharfenaker, S., O'Connor, R., Stackhouse, T., Braden, M., Hickman, L., and Gray, K. (1996). An integrated approach to intervention. *In* "Fragile X Syndrome: Diagnosis, Treatment, & Research," 2nd edition (R. J. Hagerman and A. Cronister, Eds.), pp. 349–411. Johns Hopkins University Press, Baltimore.
78. Scharfenaker, S., O'Connor, R., Stackhouse, T., Braden, M., Hickman, L., and Gray, K. (1996). Computer software information. *In* "Fragile X Syndrome: Diagnosis, Treatment, & Research," 2nd edition (R. J. Hagerman and A. Cronister, Eds.), pp. 453–462. Johns Hopkins University Press, Baltimore.
79. Scharfenaker, S., O'Connor, R., Stackhouse, T., Braden, M., Hickman, L., and Gray, K. (1996). Computer technology resource centers. *In* "Fragile X Syndrome: Diagnosis, Treatment, & Research," 2nd edition (R. J. Hagerman and A. Cronister, Eds.), pp. 463–466. Johns Hopkins University Press, Baltimore.
80. Scharfenaker, S., O'Connor, R., Stackhouse, T., Braden, M., Hickman, L., and Gray, K. (1996). Suggested academic materials for use with fragile X children. *In* "Fragile X Syndrome: Diagnosis, Treatment, & Research," 2nd edition (R. J. Hagerman and A. Cronister, Eds.), pp. 467–469. Johns Hopkins University Press, Baltimore.
81. Schopmeyer, B. B., and Lowe, F. (Eds). (1992). "The Fragile X Child." Singular, San Diego, CA.
82. Schwartz, C. E., Dean, J., Howard-Peebles, P. N., Bugge, M., Mikkelsin, M., Tommerup, N., Hull, C. E., Hagerman, R. J., Holden, J. J. A., and Stevenson, R. E. (1994). Obstetrical and gynecological complication in fragile X carriers: a multicenter study. *Am. J. Med. Genet.* **51,** 400–402.
83. Sherman, S. (1996). Epidemiology. *In* "Fragile X Syndrome: Diagnosis, Treatment, & Research," 2nd edition (R. J. Hagerman and A. Cronister, Eds.), pp. 165–192. Johns Hopkins University Press, Baltimore.
84. Simko, A., Hornstein, L., Soukup, S., and Bagamery, N. (1989). Fragile X syndrome: recognition in young children. *Pediatrics* **83,** 547–552.
85. Smeets, H., Smits, A., Verheif, C. E., Theelen, J., Willemsen, R., van de Burgt, I., Beverstock, G., Hoogereen, A. T., Oosterwijk, J. C., and Oostra, B. A. (1995). Normal phenotype in two brothers with a full FMR1 mutation. *Hum. Mol. Genet.* **4,** 2103–2108.
86. Sobesky, W. E. (1996). The treatment of emotional and behavioral problems. *In* "Fragile X Syndrome: Diagnosis, Treatment, & Research," 2nd edition" (R. J. Hagerman and A. Cronister, Eds.), pp. 332–348. Johns Hopkins University Press, Baltimore.
87. Sobesky, W. E., Taylor, A. K., Pennington, B. F., Riddle, J. E., and Hagerman, R. J. (1996). Molecular/clinical correlations in females with fragile X. *Am. J. Med. Genet.* **64,** 340–345.
88. Spiridigliozzi, G. A., Lachiewicz, A. M., MacMurdo, C. S., Vizoso, A. D., O'Donnell, C. M., McConkie-Rosell, A., and Burgess, D. J. (1994). "Educating Boys with Fragile X Syndrome: A Guide for Parents and Professionals." Duke University Medical Center.
89. Staley-Gane, L. W., Flynn, L., Neitzal, K., Cronister, A., and Hagerman, R. J. (1996). Expanding the role of the genetic counselor. *Am. J. Med. Genet.* **64,** 382–387.
90. Steyaert, J., Tegius, E., and Fryns, J. (1996). Molecular-intelligence correlations in young fragile X males with a mild CGG repeat expansion in the FMR1 gene. *Am. J. Med. Genet.* **64,** 274–277.
91. Tan, S. S., Faulkner-Jones, B., Breen, S. J., Walsh, M., Betram, J. F., and Reese, B. E. (1995). Cell dispersion patterns in different cortical regions studied with an X-inactivation transgenic marker. *Development* **121,** 1029–1039.
92. Tassone, F., Hagerman, R. J., Ikle, D., Dyer, P. N., Lampe, M., Willemsen, R., Oostra, B. A., and Taylor, A. K. (1998). FMRP expression as a potential prognostic indicator in fragile X syndrome. [Submitted for publication]
93. Turner, G., Eastman, C., Casey, J., McLeay, A., Procopis, P., and Turner, B. (1975). X-linked mental retardation associated with macro-orchidism. *J. Med. Genet.* **12,** 367–371.
94. Turner, G., Robinson, H., Wake, S., and Martin, N. (1994). Dizygous twinning and premature menopause in fragile X syndrome. *Lancet* **344,** 1500.
95. Turner, G., Webb, T., Wake, S., and Robinson, H. (1996). Prevalence of the fragile X syndrome. *Am J. Med. Genet.* **64,** 196–197.
96. Waldstein, G., and Hagerman, R. J. (1988). Aortic hypoplasia and cardiac valvular abnormalities in a boy with fragile X syndrome. *Am. J. Med. Genet.* **30,** 83–98.
97. Waldstein, G., Mierau, G., Ahmad, R., Thibodeau, S. N., Hagerman, R. J., and Caldwell, S. (1987). Fragile X syndrome: skin elastin abnormalities. Birth Defects: *Original Article Series* **23,** 103–114.
98. Webb, T. P., Bundey, S., Thake, A., and Todd, J. (1986). The frequency of the fragile X chromosome among school children in Coventry. *J. Med. Genet.* **23,** 396–399.
99. Willemsen, R., Mohkemsing, S., deVries, B., Devys, D., Van den Ouweland, A., Mandel, J. L., Galjaard, H., and Oostra, B. A. (1995). Rapid antibody test for fragile X syndrome. *Lancet* **345,** 1147–1148.
100. Willemsen, R., Smits, A., Mohkamsing, S., van Beerendonk, H., de Haan, A., de Vries, B., van den Ouweland, A., Sistermans, E., Galjaard, H., and Oostra, B. (1997). Rapid antibody test for diagnosing fragile X syndrome: a validation of the technique. *Hum. Genet.* **99,** 308–311.
101. Wisniewski, K. E., Segan, S. M., Miezejeski, C. M., Sersen, E. A., and Rudelli, R. D. (1991). The fra(X) syndrome: neurological, electrophysiological, and neuropathological abnormalities. *Am. J. Med. Genet.* **38,** 476–480.
102. Wolff, P. H., Gardner, J., Paccia, J. J., and Lappen, J. (1989). The greeting behavior of fragile X males. *Am. J. Ment. Retard.* **93,** 406–411.
103. Wright-Talamante, C., Cheema, A., Riddle, J. E., Luckey, D. W., Taylor, A. K., and Hagerman, R. J. (1996). A controlled study of longitudinal IQ changes in females and males with fragile X syndrome. *Am. J. Med. Genet.* **64,** 350–355.

Chapter 3

FMR1 and Mutations in Fragile X Syndrome: Molecular Biology, Biochemistry, and Genetics

Georges Imbert Institut de Génétique et de Biologie Moléculaire et Cellulaire, CNRS/INSERM/ULP, 67404 Illkirch Cedex, C. U. de Strasbourg, France

Yue Feng Howard Hughes Medical Institute, Departments of Biochemistry, Pediatrics, and Genetics, Emory University School of Medicine, Atlanta, Georgia 30322

David L. Nelson Department of Molecular and Human Genetics. Baylor College of Medicine, Houston, Texas 77030

Stephen T. Warren Howard Hughes Medical Institute, Departments of Biochemistry, Pediatrics, and Genetics, Emory University School of Medicine, Atlanta, Georgia 30322

Jean-Louis Mandel Institut de Génétique et de Biologie Moléculaire et Cellulaire, CNRS/INSERM/ULP, 67404 Illkirch Cedex, C. U. de Strasbourg, France

I. INTRODUCTION

Much of our current understanding of the unstable trinucleotide repeats involved in human genetic disorders stems from analysis of fragile X syndrome. This is due in part to the fact that the unstable expanding repeat in this disorder was the first to be characterized in detail, but another significant factor has been the high frequency of the disease and mutation in the human population, allowing large numbers of patients and families to be studied. This chapter will provide a review of the current understanding of the nature and behavior of the CGG repeats responsible for the vast majority of fragile X syndrome cases, and of the gene (FMR1) whose function is disrupted by the repeat expansion. Discussion of the repeats and their behavior will include the presumed population history of these mutations, while that regarding the gene will summarize current knowledge of the function of the protein encoded by FMR1. While issues regarding the behavior of the CGG repeats in the FMR1 locus are central to the theme of this volume, and provide lessons of interest to those interested in all trinucleotide repeat disorders, the effect of the repeat expansion in this disorder is manifest through the loss of function of the FMR1 gene. The greatest benefit to patients and families with fragile X syndrome will accrue through a more complete understanding of this gene and its partners.

II. FMR1 AND RELATED GENES

FMR1 was identified by positional cloning in pursuit of the gene(s) involved in fragile X syndrome [188]; these efforts focused on identifying genes and mutations in a region marked by the presence of a cytogenetically visible, folate-sensitive chromosomal fragile site (FRAXA) [91, 171] near the distal boundary of Xq27.3. The ability to identify DNA clones and fragments suspected of carrying the fragile site was greatly enhanced by the development of a series of monochromosomal hybrids carrying fragments of the human fragile X chromosome in hamster cells. These cell lines had undergone fragmentation and translocation of the human chromosome after induction of fragility at the FRAXA site [195]. Breakpoints in a number of these cell lines were characterized cytogenetically, and found to be coincident with the location of the fragile site found in patients. The FMR1 gene was identified by using DNA fragments found to be at or near the chromosomal breakpoints as hybridization probes in cDNA libraries. Two cDNA clones were initially identified and together were found to carry the entire coding sequence of FMR1 [188]. The mRNA for FMR1 is 3.9 kilobase pairs (kb), with a long (1.8 kb) 3′-untranslated region. The CGG trinucleotide repeat responsible for mutations in the majority of fragile X syndrome cases was found to be present at the 5′ end of one of these cDNAs, and was later shown to be located within the 5′ untranslated region of the FMR1 transcript.

A. Gene Structure and Expression

The gene spans about 40 kb and is transcribed in a proximal to distal orientation (Fig. 3-1). There are 17 exons which encode the human mRNA [38], and the entire locus has been subjected to genomic sequence analysis; currently 185 kb of contiguous sequence is available starting 12 kb upstream of exon 1 and continuing for nearly 100 kb distal to the gene (GenBank Accession No. L29074). No additional genes have been identified within the sequenced region.

Polymorphic loci have been described in or near the FMR1 gene, in addition to the CGG repeats. Several of these have been utilized in studies aimed at tracking background chromosomes. Three microsatellites (DXS548 [141], FRAXAC1 and FRAXAC2 [136, 214]) have been characterized extensively, as have single nucleotide polymorphisms, FMRa and FMRb [88] and BanI [150]. The positions of these are shown in Fig. 3-1.

Several of the FMR1 exons are found to be alternatively spliced, leading to a number of potentially different mRNAs and protein isoforms (Fig. 3-2) [5, 189]. Numerous distinct isoforms of FMRP, including some with alternative reading frames resulting in shorter C-termini and strikingly altered biochemical properties, have been predicted (see below). Western blot analysis demonstrates several protein products ranging from 70 to 80 kDa [31, 187, 228]. Widespread expression of FMR1 is seen in both human and mouse embryos [1, 71]. The protein product of the FMR1 gene has been found present in many adult and fetal tissues, but is concentrated in neurons [31]. Similar concentration of FMRP is found in brain neurons throughout various brain regions [46], including hippocampus, whose importance in learning and memory is well recognized. High expression of FMR1 is also found in spermatogonia [7, 31]. The high-level expression of FMRP in brain and testes is consistent with the major phenotypic aspects of the disorder. The FMR1 promoter has recently been characterized by footprinting studies and by a transgenic experiment with a reporter gene [221, 224, 227]. Initial studies suggested a largely cytoplasmic location of FMRP, with occasional nuclear signals [31, 187]. A more detailed view of the intracellular localization and associations of the protein is described below.

B. Gene Conservation

The FMR1 coding sequence is highly conserved in vertebrate evolution; homologues have been character-

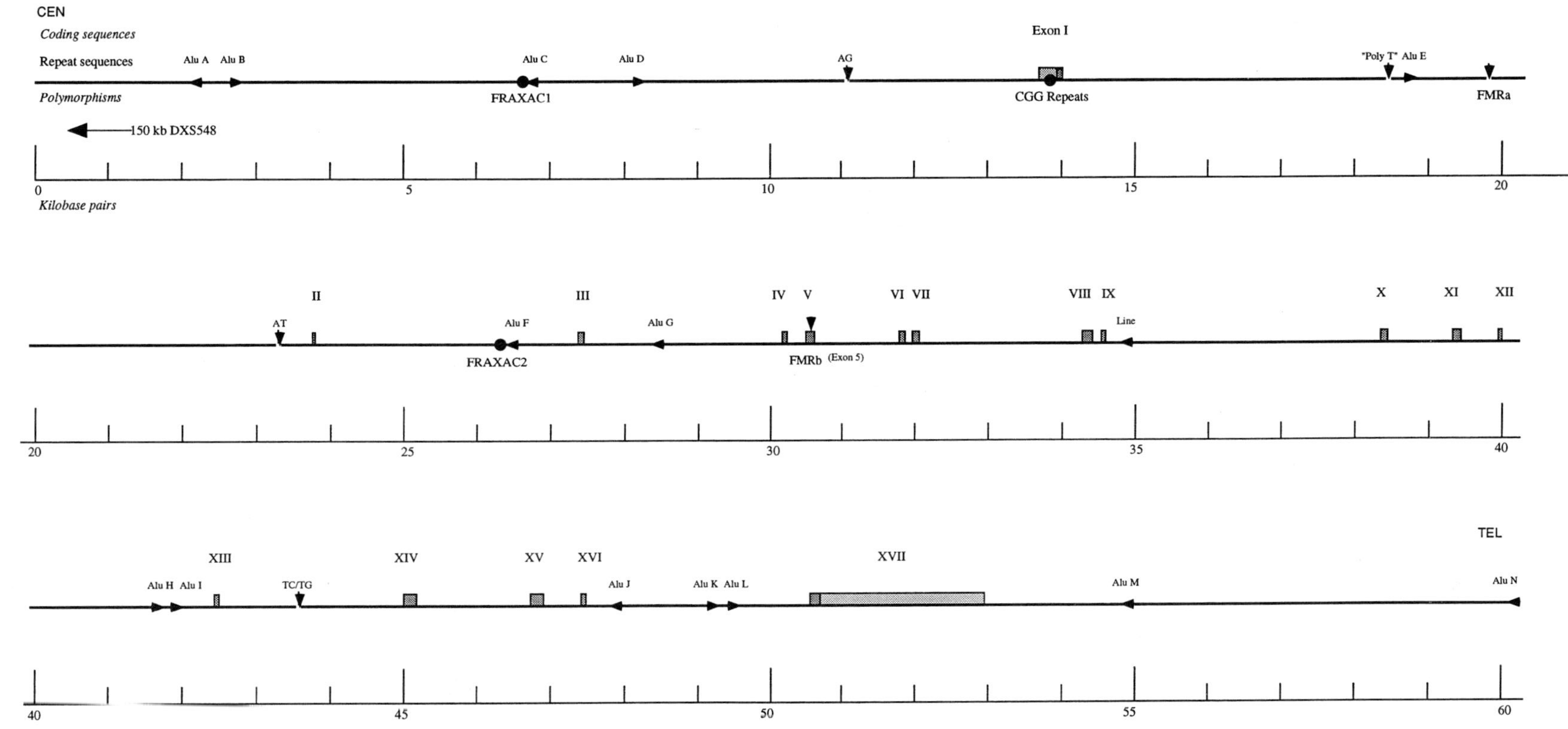

FIGURE 3-1 Sequence-based map of the human FMR1 locus. Sixty kilobases surrounding the FMR1 gene are depicted using the GenBank file L29074. Positions of coding sequences (exons) are given as shaded boxes labeled with roman numerals (lighter shading indicates untranslated regions of exons), and common repetitive elements are indicated with horizontal arrows. Alu repeats are lettered from A to N in proximal to distal order. Known polymorphic microsatellites are indicated with filled circles. Several other sequence features are indicated at their positions.

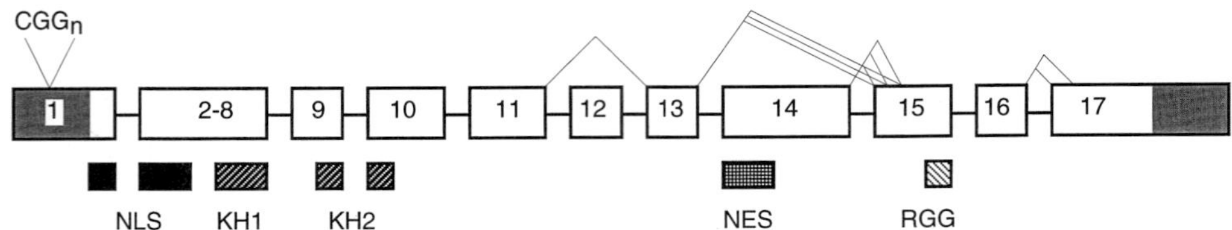

FIGURE 3-2 Schematic depiction of the structure of the human FMR1 gene. KH domains, RGG boxes, and nuclear localization (NLS) and export (NES) signals are shown below. Untranslated regions are indicated with shading. Alternative splices are given.

ized from the mouse [5], the chicken [131], and *Xenopus laevis* [161]. The level of similarity is striking; FMR1 is more highly conserved than many proteins with fundamental biochemical functions, such as glucose 6-phosphate dehydrogenase or α-globin [131]. Figure 3-3 shows an alignment of the four proteins with exon boundaries derived from the human and the positions of functional domains (see below). Among mammals, the FMR1 gene is highly conserved at the DNA level. The 5′-untranslated region has been found to carry CGG repeats in approximately the same location in all placental mammals as well as marsupials analyzed [39]. In the chicken, the region of the CGG repeats in mammals carries a CCT repeat [131].

C. Related Genes

FMR1 is a member of a family of genes identified by cross-hybridization [161]. Two additional genes termed FXR1 (FMR1L1 is the name sanctioned by the Human Genome Organization nomenclature committee) and FXR2 (FMR1L2 is the official name) have been described from both mammals (human and mouse) and amphibians (*Xenopus*). These are autosomally encoded in mammals, and no mutations have yet been described in either. In the human, FXR1 has been localized to 3q28, and FXR2 to 17p13.1 [24, 201]. Both related genes carry the functional domains described in FMR1, and show significant identity, particularly in their N-termini (Fig. 3-4). High-level expression of FXR1 is found in heart, skeletal muscle, and testis, although it appears to be present in all tissues tested by Northern and RT-PCR analyses [24, 161]. A recent study by Tamanini *et al.* [175] demonstrates that both FXR1 and FXR2 are expressed at significant levels in adult Purkinje cells, as well as in other neurons. The Purkinje cell staining was cytoplasmic, while that in other neurons appeared to show concentration in the proximal dendrites. Significantly, the expression of the FXR1 and FXR2 proteins appeared unchanged in the brain of a fragile X syndrome patient. A nuclear location was seen for FXR1 in fetal human brain. Overlap of expression of the three proteins was not as apparent in adult and fetal testis, suggesting that the normal roles of these proteins may not require coexpression. On the other hand, all three proteins in this family have been shown to form homo- and heteromers *in vitro* [213], and coimmunoprecipitate with their biological complexes formed in Hela cells [162]. In spite of their possible distinctive functions, perhaps during development and in different tissues, these proteins may share functional similarities. Thus, a question has been raised whether these proteins may partially complement each other's function, especially considering that mutations in FMR1 gene does not cause deficits in all the FMRP-expressing tissues of fragile X patients.

III. CHARACTERISTICS OF FMRP, THE PROTEIN PRODUCT OF FMR1

A. RNA Binding

Subsequent to the original isolation of FMR1, a class of proteins with RNA-binding activity was defined by sequence similarity; the hnRNP K Homology (KH) domain and RGG box were found shared among a growing family of proteins characterized from organisms ranging from archeabacteria to human [15, 104, 158, 159]. The presence of two KH domains and RGG boxes allowed the hypothesis that FMR1 was an RNA-binding protein to be put forward [56], while two studies demonstrated *in vitro* RNA-binding activity for FMR1 [6, 159]. Ashley *et al.* [6] also found that FMRP appears to have a selectivity for a fraction of mRNAs expressed in brain, including its own message. RNA binding has been disrupted by deletions of the C-terminus of the protein, which ablate the RGG sequences [159].

A patient carrying a missense mutation in FMR1 was identified with unusually severe mental retardation, remarkable macroorchidism, and a violent temperament [25]. The point mutation in this patient alters an Ile residue to Asn at position 304 of the

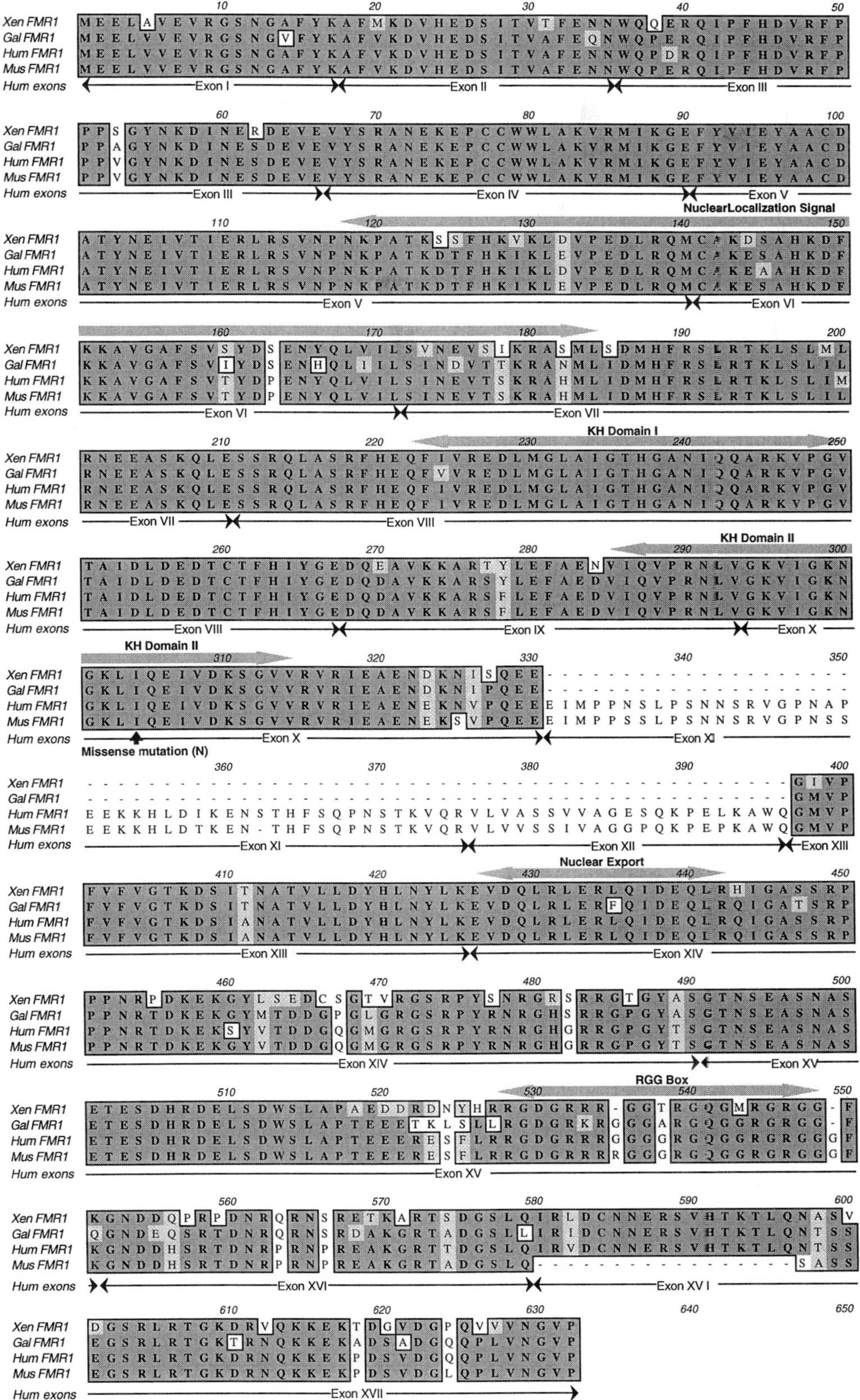

FIGURE 3-3 Alignment of FMR1 peptide sequences derived from four species with indication of presumptive functional domains. Species are indicated to the side: Xen is *Xenopus laevis,* Gal is *Gallus gallus,* Hum is human, and Mus is *Mus musculus.* Alignment was performed by Clustal W with default parameters using sequences available in GenBank (see text); the human sequence represents the longest known isoform. Lines above indicate positions of functional domains as labeled. Below are lines to indicate the positions of the exons composing the human FMR1 gene. The position of the only known point mutation in FMR1 is indicated with a vertical arrow.

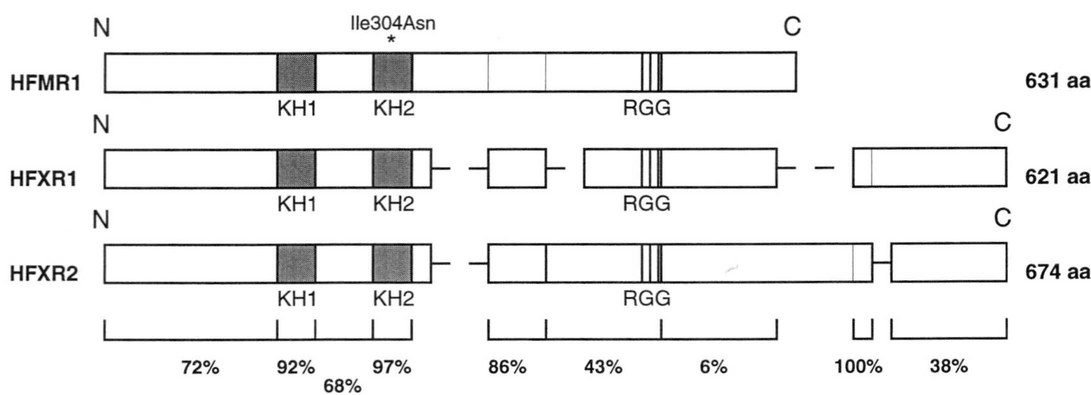

FIGURE 3-4 Graphical alignment of peptide sequences for FMR1 and homologues FXR1 and FXR2. The position of the only known point mutation in FMR1 is indicated with an asterisk. Note the high similarity through the KH domains.

common isoform. The Ile in this position is conserved in nearly every protein carrying a KH domain described to date, suggesting a critical role for this residue in the function of the domain. This mutant protein also shows reduced *in vitro* RNA binding activity [160, 174] while retaining its ability to form protein–protein interactions with FXR1 and FXR2 [162]. Recent evidence derived from structural analysis of another member of the KH family, human vigilin, suggests that this mutation destroys the ability of the protein to fold properly [118]. Interestingly, missense mutations within KH domains often cause more severe phenotypes as compared to those caused by null mutations. Besides the I304N mutation of FMRP found in the severe fragile X patient, more severe phenotypes have also been reported due to KH domain mutations in the *C. elegans* GLD-1 protein [78], and the *Drosophila* Bicaudal-C protein [98]. These more severe phenotypes cannot be explained simply by reduced RNA binding of the corresponding proteins. Since KH domain-containing proteins are usually incorporated into RNP complexes, it is conceivable that the altered structure of the KH domains may change the spectrum of interaction within the complex, and therefore resulting in a dominant-negative-like phenotype. Indeed, as discussed below, the FMRP missense mutation (I304N) is likely to involve such an interaction.

B. FMRP Is Associated with Ribosomes

Immunolocalization studies with antibodies directed against FMRP have revealed that FMRP is predominantly localized in the cytoplasm with a diffuse granular staining pattern [31, 187]. Recent reports have indicated that FMRP is associated with translating ribosomes [35, 81, 174]. FMRP–ribosome association is sensitive to low levels of RNase, which removes the mRNAs linking translating polyribosomes without disturbing ribosome assembly [23, 35, 174]. In addition, EDTA treatment, which dissociates ribosomes into subunits, releases FMRP into complexes with a similar size range to messenger ribonucleoprotein (mRNP) particles [35, 47]. Furthermore, nearly all the cytoplasmic FMRP can be cocaptured with poly-A RNA [23]. Taken as a whole, these data suggest that FMRP associates with ribosomes as a component of mRNP particles. Upon translation termination, FMRP seems to dissociate from ribosomes [47], ready to be recycled for the next round of translation. It remains to be elucidated how FMRP may enter ribosomes in relation to the selection of its bound mRNA for translation.

The association of FMRP with translating ribosomes has been observed in many cell and tissue types, including brain neurons. The localization of FMRP in neurons has been analyzed using immunogold electron microscopy [46, 234]. The highest concentration of immunogold particles were seen in the cell body in regions rich in ribosomes. In neuronal processes, immunogold particles were obvious in proximal dendrites. However, no FMRP was detected in axons where ribosomes are absent. In addition, FMRP–immunogold particles were frequently seen in dendritic spines at sites where ribosomes are customarily found. Neuronal FMRP was also found in synaptosome preparations in association with polyribosomes [46, 234], and its expression was found increased upon glutamate receptor stimulation [234]. The association of FMRP with somatodendritic polyribosomes is consistent with the hypothesis that fragile X mental retardation may be a result of translation abnormality due to the absence of FMRP [46, 234].

C. FMRP Shuttles between the Nucleus and Cytoplasm

Although FMRP is predominantly localized in the cytoplasm, nuclear staining has been reported [31, 187], especially for isoforms lacking exon 14 [228]. Since FMRP seemed to be a component of mRNP particles and most RNP particles are formed in the nucleus, and given the fact that other RNA-binding proteins can shuttle between the nucleus and cytoplasm, this suggested that FMRP might enter the nucleus as a part of its normal function. FMRP deletion and fusion constructs were used to delineate a nuclear localization signal (NLS) in the amino terminus of FMRP and a nuclear export signal (NES) encoded by exon 14 [35, 217, 222] (Fig. 3-3). The FMRP NES resembles the NES motifs recently described for the HIV-1 Rev protein [49] and protein kinase inhibitor a (PKI) (Fig. 3-5A) [200]. A 17- or 20-amino-acid peptide containing the FMRP NES can direct temperature-dependent nuclear export of a microinjected protein conjugate, while a mutant peptide cannot direct export [35, 222] (Fig. 3-5B). Immunogold electron microscopy provides further evidence that FMRP is capable of shuttling between the nucleus and the cytoplasm. While FMRP is predominantly in the cytoplasm of neurons, it was also detected in the nucleus and in transit through the nuclear pore [46].

The process of nuclear export is poorly understood, but the discovery of NESs has provided a great deal of insight into RNA export [54, 57]. It has been suggested that protein-based NESs may be a common feature of nuclear RNA export [54]. Rev binds and directs export of viral mRNAs containing the Rev response element binding site [49]. TFIIIA, which has been implicated in the export of 5S rRNA, contains a putative leucine-rich NES which can substitute for the Rev NES [57, 49]. hnRNP A1, which is associated with poly(A) RNA in both the nucleus and the cytoplasm, contains the M9 domain which allows shuttling between the two compartments [107]. Other proteins which resemble hnRNP

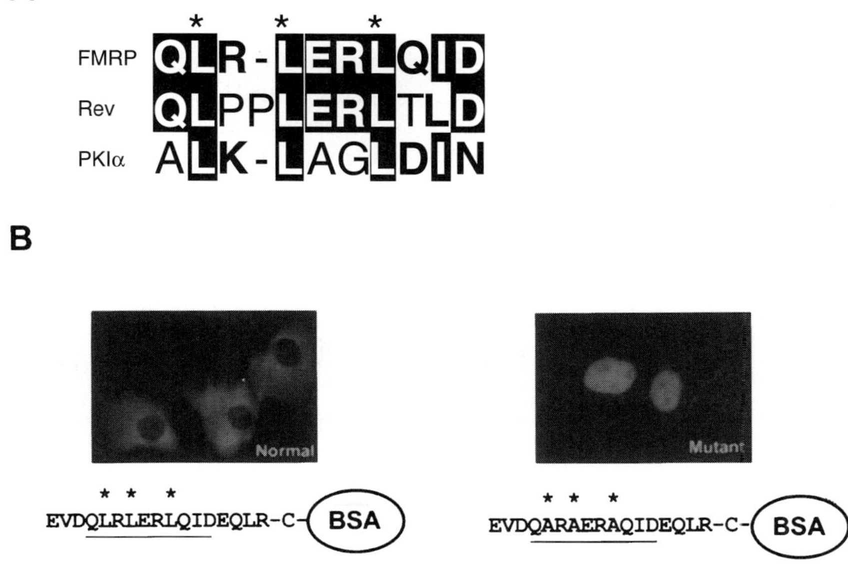

FIGURE 3-5 FMRP contains a functional NES. (A) Amino acid alignment of NES sequences of FMRP, Rev, and PKIa. Identical residuals are boxed, with each conserved functional leucine marked by an asterisk; similar residuals are in bold. (B) FMRP's NES directs nuclear export. The 17-amino-acid peptide containing either the normal FMRP's NES (underlined, with conserved leucine residuals marked by asterisks) or a mutant NES (underlined, with all leucine residuals substituted by alanines as marked by asterisks) was conjugated to BSA. FITC-labeled conjugates were nuclear injected into COS-7 cells. After incubation at 37°C for 1 h, direct fluorescent microscopy analysis was performed on injected cells. The normal NES–BSA conjugate was exported to the cytoplasm, and the mutant NES–BSA conjugate remained in the nucleus.

A1, such as the predominantly cytoplasmic hrp36 protein of *C. tentans* [191], have also been implicated in RNA export. While further analysis is necessary to determine the export characteristics of FMRP, the identification of domains which mediate nucleocytoplasmic shuttling suggests that FMRP may be involved in RNA export.

D. Possible Function of FMRP

The subcellular behavior of FMRP, including its selective mRNA binding, nucleocytoplasmic shuttling, and predominant association with cytoplasmic polyribosomes, has led to a hypothesis that FMRP may play a role in nuclear export of its target mRNAs and further present these mRNAs to the translation machinery [36]. The discovery of FMRP–polyribosome association in the neuronal dendritic spine further implies the possible involvement of FMRP in modulating localization and/or translation of its target mRNAs in the appropriate subcellular compartment(s) [46, 234].

Regulation of localized protein synthesis has been demonstrated to be functionally important in cell growth and polarity development (for review see [79, 143]). In neurons, the selective localization of translation machinery and a subclass of mRNAs to the postsynaptic sites [18, 108, 167, 168] has led to a hypothesis that proteins important for synaptic plasticity may be synthesized near their functional sites. Direct evidence has been found demonstrating that local protein synthesis is required in neurotrophin-induced hippocampal synaptic plasticity [80]. The elevated protein synthesis within the dendritic spine during synaptic development [167], especially upon activation of the metabotropic glutamate receptors [198], further reinforced the importance of regulated translation in neuronal function. The formation of abnormal dendritic spines due to the lack of FMRP as found in fragile X patients' brain [72, 151], as well as in the fmr1 knockout mice brain [21], is consistent with the view that fragile X mental retardation may result from protein synthesis abnormality during synapse development.

Support for the functional importance of FMRP–polyribosome association, in relation to translation, has been recently obtained [23, 47]. The I304N mutation in FMRP which causes severe fragile X syndrome abolished FMRP–polyribosome association (Fig. 3-6A). The impaired FMRP–polyribosome association does not seem to be a result of failure in RNA-binding as suggested by previous *in vitro* study [160], since the mutant FMRP can be cocaptured with cytoplasmic poly-A RNA similar to what was observed for normal FMRP [47]. Rather, the inability of the mutant FMRP to associate with polyribosomes most likely results from the formation of abnormally small mRNP particles which fail to be selected for translation. These data suggest that FMRP is important in forming translation-competent mRNP particles, and may further direct their translation in the appropriate subcellular compartment(s). In a typical fragile X patient who lacks FMRP, mRNAs that are normally directed to translation by FMRP may be handled by other mRNPs, possibly including FMRP-related proteins, resulting in partial translation, perhaps abnormally regulated and/or localized in cells such as brain neurons. The dominant formation of abnormal FMRP–mRNP particles that fail to associate with translating polyribosomes by the I304N mutation implies that FMRP-bound mRNAs may be sequestered from translation, thus resulting in the unusually severe fragile X phenotype. A model describing FMRP's involvement in mRNP formation and polyribosome association is postulated in Fig. 3-6B.

In addition to modulating mRNA translation and subcellular localization, polyribosmal mRNPs may also influence mRNA stability either by intrinsic endonucleolitic activity as demonstrated by the polysomal messenger ribonuclease, PMR1 [17], or by increasing the resistance of the bound mRNA to ribonuclease attack [77]. In addition, numerous nuclear RNPs are responsible for mRNA processing. Whether FMRP, especially particular FMRP isoforms that display differential subcellular localization, may also be involved in multiple aspects of mRNA metabolism still awaits elucidation. Furthermore, accumulating evidence has pointed to the possible regulation of mRNP complex formation and mRNA export by modification of specific mRNA-binding proteins through phosphorylation and/or methylation [16, 68]. Modification of FMRP and/or its related proteins under the control of growth- and neurotransmitter-mediated signals would be conceivable with the predicted function of FMRP in learning and memory. Finally, identification of FMRP's RNA targets should help to define the influence of FMRP in mRNA metabolism, and finally lead to the elucidation of pathogenesis in fragile X syndrome due to the lack of FMRP.

IV. CHARACTERISTICS OF THE FRAGILE X SYNDROME MUTATIONS

A. General Features

The fragile X syndrome, as defined by the association of mental retardation and a fragile site in Xq27.3 at the FRAXA locus, is caused by a single type of mutation, the unstable expansion of a CGG trinucleotide repeat

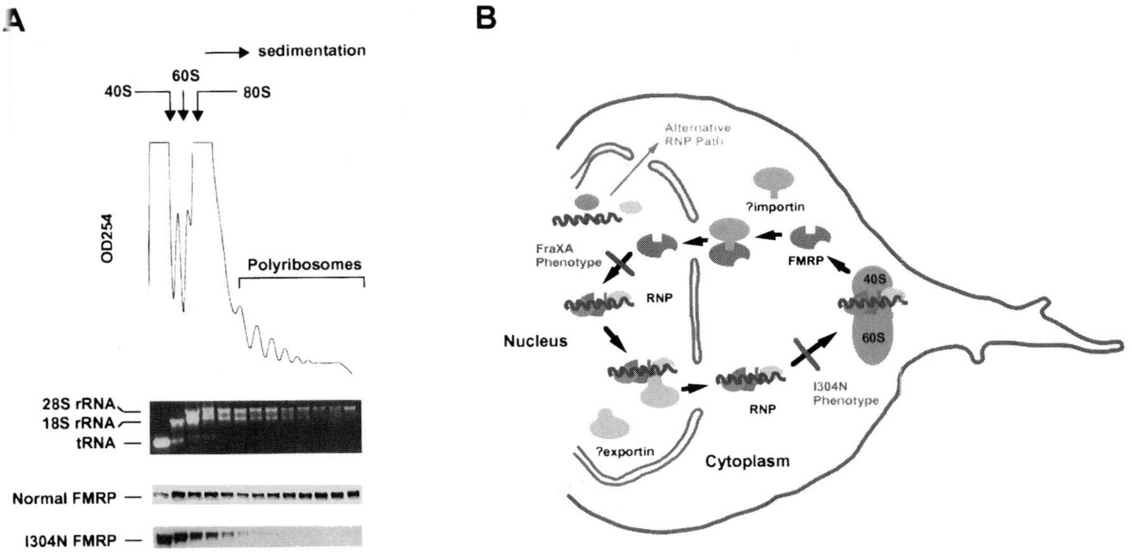

FIGURE 3-6 Prospective function of FMRP as a nucleocytoplasmic shuttling polyribosomal mRNP. (A) FMRP normally associates with translating polyribosomes, and the I304N mutation in FMRP abolishes such association. Cytoplasmic extracts from normal and the I304N lymphoblastoids were fractionated through parallel linear sucrose gradients, allowing the separation of translating inactive components including mRNP particles, ribosomal subunits, and monosomes from actively translating polyribosomes. The sedimentation of these components was monitored by the absorption at 254 nm and further confirmed by the distribution of ribosomal RNA through the gradients. Normal FMRP cofractionated with translating polyribosomes, whereas the I304N FMRP was restricted in the translation inactive fractions most likely as abnormally formed mRNP particles. (B) Possible function of FMRP in mRNP complex formation and predicted biochemical abnormality in fragile X syndrome. FMRP may enter the nucleus with the aid of importins to form nuclear RNP particles with its target mRNAs, and direct the nuclear export of these RNP particles perhaps mediated by exportins. In the cytoplasm, FMRP further presents these mRNP particles to translation machinery and modulates their translation on the polyribosomes. In typical fragile X patients who lack FMRP, these mRNAs may be directed by an alternative RNP path leading to partial translation, although it may be regulated abnormally in affected tissues such as brain neurons. In contrast, the I304N mutation sequesters FMRP-bound mRNAs from translation by forming translation-incompetent mRNP complexes, therefore causing an unusually severe fragile X phenotype.

sequence found in the 5′-untranslated region of the X-linked, fragile X mental retardation gene (FMR1) [51, 85, 124, 188]. Much rarer cases of male patients with the same clinical phenotype are caused by deletions, ranging from 1.6 kb to several megabases, that include part or all of the FMR1 gene [52, 58, 74, 106, 176, 181, 207]. A 9-megabase deletion encompassing the FMR1 gene and the Xq27 band was found in a male with mental retardation and atypical features (obesity and anal atresia) that may be due to deletion of other genes [132]. An even larger deletion of 13 Mb was reported [235]. The fact that nullisomy of such a large region is compatible with life indicates a low gene density proximal to FMR1. Three point mutations within the FMR1 gene (causing a frameshift or affecting splicing) were also reported in patients with a classic phenotype [92, 231]. As described above, a single missense point mutation has been described in a patient with an unusually severe phenotype, suggesting alteration of a critically important amino acid residue [25]. Taken together, these observations indicate that most or all of the fragile X phenotype is due to the deficiency in FMR1 expression. There is thus no indication that the expansion mutation may affect other neighboring genes.

There are essentially four forms of the CGG repeat region found in the population and they are referred to as normal, gray zone, premutation, and full mutation (Table 3-1).

The normal form is polymorphic and contains from 6 to about 50 CGG repeats, interspersed in most alleles with one to three single AGG triplets found approximately every 10 CGG repeats [39, 51, 88, 165, 166, 216]. The polymorphic variation in both number of repeats and position of interspersed AGGs is considerable: of 82 alleles sequenced by Kunst and Warren [88], 47 were unique. The most common alleles have 29–30 repeats and include 2 interspersed AGGs. They account for more than 40% of X chromosomes. Alleles with less than 18 or more than 40 triplets are rare and represent about 7% of the total. Very similar distributions of allele size were found in Japanese and Chinese populations [3, 138, 215]. An extensive analysis of the CGG repeat

TABLE 3-1 Allele Categories of the CGG Repeat at the FRAXA Locus

	Number of CGG triplets (including AGG interruptions)	Methylation on active X	Presence of FMR1 mRNA/protein	Transmission instability	Somatic instability	Clinical phenotype
Normal alleles Major alleles	6 to 45 (0 to 3 AGG) CGG_9 AGG CGG_9 AGG $CGG_{9\text{-}10}$	Unmethylated	+/+	No	No	Normal
Grey zone alleles Major alleles	45 to 60 (0 to 2 AGG) CGG_9 AGG CGG_x CGG_9 AGG CGG_9 AGG CGG_x	Unmethylated	+/+	Limited (depends on AGG interspersion)	No	Normal
Premutation alleles	≈60 to 200	Unmethylated	+/+	Yes, increasing with size	Limited	Normal
Full mutation alleles	≈230 to >1000	Methylated	−/−	Yes	Extensive	Affected
Mosaics	Presence of a full mutation and a premutation allele (or a normal allele or a small deletion allele in rarer cases)	Full mutation allele is methylated	±/±	Yes	Extensive	Affected
Methylation mosaics	≈230 to >1000	Partly unmethylated in the CGG_{230} to CGG_{350} range	±/±	Yes	Extensive	Affected

polymorphism in nine global populations from Asian, African, and Native American origins confirmed its extreme heterogeneity and suggested that the ancestral allele had 2 interrupting AGGs [89]. Analysis of the repeat in higher primates showed variable locations of AGGs, with a somewhat different interspersion pattern than in humans [40, 216]. Gorillas exhibit only 3 alleles, while chimpanzees are highly polymorphic [40].

Normally, the repeat region is stable when transmitted from parent to child. An increase in the number of perfect CGG repeats on the 3′ side or the loss of an interspersed AGG may cause the repeat region to become unstable, and to expand in size when transmitted from mother to child [39, 73, 88, 166, 216].

Grey zone alleles are alleles at the upper range of the normal ones. They show limited instability (of a few repeat units) upon transmission, but with a very low probability of transition to full mutation in one generation. They are characterized by an uninterrupted stretch of ~34–50 CGGs, corresponding to allele length in the range of 40–60 repeats [39, 117]. Such alleles are found at a frequency of about 1% in the normal population.

In fragile X families, two basic types of abnormal unstable alleles may be observed [51, 124, 144, 212]. Premutations are found in male or female individuals who do not exhibit any obvious clinical symptoms related to the mutation, and usually consist of about 60 to 200 repeats. They do not affect transcription of the FMR1 gene, stability of the mRNA, and synthesis of FMRP, at least in lymphoblastoid cells [44]. When transmitted by a female, premutations have a high risk (up to 100%) of changing to a disease causing full mutation.

The full mutation form of the FMR1 gene consists of over 230 CGG repeats (up to 1000 or more) and is abnormally hypermethylated [62, 75, 124, 144]. Consequently, no mRNA is produced and the lack of the encoded protein FMRP, an RNA-binding protein [6, 159], is responsible for the mental retardation [31, 105, 130, 170]. Due to X-linkage, males have more severe expression of the phenotype associated with the syndrome compared to females.

B. Instability of the Premutation

The increase in size of premutations is due to an expansion of pure CGG repeats at the 3′ side of the repeat array [39]. Upon maternal transmission, premutations have a strong tendency to increase in size, either to larger premutations or to full mutation. The risk of transition to full mutation is very strongly dependent on the size of the maternal premutation. It appears very low (probably <5%) for alleles with less than 60 repeats, as very few such events have been reported [48, 67] for unsequenced premutations in this size range. The risk is around 50% for alleles around 80 repeats, and is above 95% for alleles with more than 100 repeats [48, 51, 67, 123, 165, 193, 212]. Eichler *et al.* [39] analyzed AGG interspersion in premutation alleles: the smallest to undergo transition to full mutation had 66 triplets, with a single AGG [$(CGG)_9AGG(CGG)_{56}$], while an allele with 59 pure CGGs (and no AGG) expanded to full mutation within two generations.

In contrast, premutations show a much more limited instability upon male transmission, with no obvious trend toward preferential increase or decrease in size. Gain or loss of repeats is very rarely larger than 20 repeat units [48, 165, 193]. Nolin *et al.* [123] observed, however, that small length increases occur in a majority of daughters when the father carries ≤80 CGGs, while retractions affect predominantly paternal alleles with ≥80 CGGs. Transition from premutation to full mutation has never been observed (to our knowledge) upon male transmission.

The characteristics of premutation instability account for the complex inheritance of the fragile X syndrome that had been previously uncovered by segregation analysis in affected families [154, 155], and in particular for the fact that the penetrance (or risk of having mentally retarded children), is much lower for mothers of normal transmitting males than for their daughters. Female carriers with small premutations will tend to have offspring with larger premutations, including normal transmitting males. The latter will pass, in general with little change, this premutation to their daughters, who in turn will be at much greater risk to have affected offspring with a full mutation. It should be noted that very rare cases of reversion of a maternal premutation to an apparently normal allele in one offspring have been described: from 110 to 44 repeats in one case [192], and from 80 to 29 repeats [186]. In the latter case, analysis of flanking polymorphic markers suggested that the reversion was due to a very localized gene conversion event implicating the normal maternal allele.

Occasional alleles with sizes at the high range of normal alleles or the low range of premutations (from about 45 to 60 repeats) have been identified by population studies or as husbands or wives of fragile X families members [51, 67, 95, 133, 147]. These are called gray zone alleles, and some of them show limited instability when transmitted to offspring. The most thorough study, by Rousseau *et al.* [150], on 10,624 unselected women, uncovered 23 carriers of alleles between 55 and 63 repeats, and 18 with sizes from 66 to 101 repeats. Eichler *et al.* [39] showed that the AGG interspersion pattern was most important for the stability of such alleles. No instability was observed for 10 transmissions of alleles with 29–33 pure CGGs, while 6 of 12 transmissions of alleles with 34 to 40 CGGs resulted in a change in repeat length. A somewhat lower instability was reported by Murray *et al.* [117], affecting 2 of 13 transmissions of alleles with ≥37 pure CGGs, and by Nolin *et al.* [123]. One does not know how many generations would be needed for such gray zone alleles to become full mutations. An exceptional jump from a 29-repeat allele (with 2 AGGs) to 39 repeats in a paternal transmission was concomitant with a rare double recombination event between DXS548 and FRAXE on the maternal chromosome [117].

While the premutation is thought not to affect FMR1 expression, several studies showed that female carriers of a premutation may have increased risk of premature menopause [22, 129, 153, 190], although a claim of increased dizygotic twinning in such females [183] has not been confirmed [157]. The molecular basis of such effects of premutation on ovarian function is unknown.

Recently, a significant excess of gray zone alleles (at both FRAXA and FRAXE) in boys with learning difficulties was reported by Murray *et al.* [116], suggesting that these may play a direct or indirect (through linkage disequilibrium) role in mental impairment. It will be important to confirm these results in other populations.

C. The Full Mutation

The full mutation shows both germinal and somatic instability [124, 144, 211]. Most mutation analyses are performed by Southern blotting on leukocyte DNA, where the full mutation appears often as multiple discrete bands, sometimes as a smear, but a single apparently homogeneous band may also be observed in about 25% of patients [114]. Limited somatic instability is in fact also observed in large premutations. It may then appear as a fuzzy band [67]. Abnormal methylation was initially observed for methylation-sensitive restriction sites (*Eag*I, *Bss*HII, or *Sac*II) that lie within a few hundred base pairs of the CGG repeat (and that operationally define the FMR1 CpG island) [10, 66, 124]. It was later shown that the CGG repeat and adjacent CpG dinucleotides are also abnormally methylated [62, 75, 230]. The FMR1 CpG island is methylated on the inactive X chromosome in somatic cells of normal females, and of female fetuses by 6 weeks from conception (with the exception of chorionic villi, at least up to 12–14 weeks of gestation) [30, 93]. Methylation of CpG islands on the inactive X chromosome is part of the mechanism of maintenance of the stably inactive state. The abnormal methylation of this region is correlated with a locally altered chromatin conformation [37, 223, 232] and is responsible for the transcriptional shut-off of the FMR1 gene. Indeed, in chorionic villi from affected male fetuses with large expansions, methylation is often absent [30, 172], and in such cases, FMR1 mRNA and FMRP are present [170, 204]. Small full mutations (in the range of 230–350 repeats) may show partial methylation (as tested on the *Eag*I site), a pattern that can be called "methylation mosaic" [144]. A nuclear protein that binds unmethylated but not fully methylated CGG repeats has been recently purified and may be involved in the transcriptional silencing of FMR1 in fragile X patients [29, 220].

Almost invariably, transmission of a full mutation by a female resulted in offspring exhibiting the full mutation. However, very rare cases of contraction from a maternal full mutation allele to a premutation have been described [144]. It is impossible to know whether these latter cases correspond to a true reduction in size, or to undetected mosaicism for a premutation in the mother. In contrast, in all documented cases, daughters of affected males had premutations, even though the number of such males who reproduce is small [149, 202]. This regression (or absence of expansion, see below) appears to be specific to the male germ-line. Indeed in three males with full mutation and no apparent somatic mosaicism, only premutation size alleles were found in sperm [135].

D. Mutation Mosaicism

In about 15% of the cases, patients with a large methylated full mutation show also, in leukocyte DNA, the presence of a significant proportion of unmethylated fragment of premutation size [144]. This mosaic pattern may in fact be more frequent, as it was reported that under appropriate technical conditions allowing the detection of faint premutation bands, 41% of fragile X males appeared as mosaics [122]. In rarer cases, the mosaic pattern in males includes a band of apparently normal size [147, 165]. Indeed, in some of these, PCR amplification substantiated this finding ([122, 165, 185]; V. Biancalana, personal communication). However in other rare cases, mosaicism implicates both full mutations and small deletions in the region containing the CGG repeat. In four such cases, the 5′ breakpoint of the deletion was located 75–53 bp proximal to the CGG repeat [26]. A preliminary comparison of mutation patterns between leukocytes and buccal epithelial cells from the same individuals suggested that the expansion pattern (including mosaicism for full mutation and premutation) may be very different in different cell types [177]. Further similar cases were presented by Dobkin *et al.* [32] and Maddalena *et al.* [97].

E. Methylation: Exceptions to the Rule

A relatively small number of cases have been reported in which expansions of full mutation size show little (≤10 %) or no methylation (in leukocytes) (see [28] for a review; [90, 229, 233]). Somatic mosaicism for both repeat size and FMR1 methylation status in such patients is usually obvious. Male carriers of such mutations often show borderline or no mental retardation, while other typical somatic features of fragile X syndrome may be present [28, 90]. FMRP levels appeared diminished in nonneuronal biopsy tissues in a few cases analyzed [45, 90, 229]. Feng *et al.* [45] presented evidence that large expansions (≥200 repeats), when present on the mRNA, may inhibit translation in lymphoblastoids and fibroblast clones. Unfortunately, information regarding the methylation status of FMR1 locus and FMRP expression level in these patient brains are not available. Whether a reduced FMRP production may still "protect" these patients from being severely affected mentally or, alternatively, critical brain regions in these patients contain alleles with smaller repeat expansion and therefore express a nearly normal level of FMRP, still remains unclear. It is important to keep in mind that it is difficult to extrapolate from findings in leukocytes to the situation in the brain. It was recently shown that one can demethylate a full mutation in lymphoblasts and restore FMRP expression [219].

Another curious exception in a patient with a full mutation (present in many tissues) was reported [27]. This patient had a lung tumor with a homogeneous premutation of 60 repeats that was methylated at the *Eag*I/*Bss*HII diagnostic sites. However, immunohistochemical analysis showed high FMRP expression. Possible explanations are that the CpGs that are significant for expression were unmethylated, or that in this tumor, another promoter was being used.

V. TIMING OF EXPANSION

A most puzzling question regarding the fragile X mutation is the absolute requirement for female transmission for the transition from premutation (60–200 repeats, unmethylated) to the disease causing full mutation (~230 to ≥1000 repeats, methylated) and the timing of this transition (pre- or postzygotic). The observation that males with a full mutation appear to have a premutation in sperm [135, 202] (see above) led Reyniers *et al.* [135] to suggest that expansion from premutation to full mutation is a postzygotic event that spares the germ line, and occurs in somatic cells after separation from the germ cell lineage. It should thus occur independently in the trophectoderm, as the full mutation is present in chorionic villi [30]. The postzygotic model assumes, however, an imprinting mechanism that would differentiate between a zygote carrying a premutation on the paternal X chromosome (that would not lead to expansion) and a zygote carrying the premutation on the maternal X chromosome (where expansion to full mutation is an obligatory event if the maternal premutation is longer than 100 CGG repeats). Such an imprint could be either maternal or paternal. In favor of the latter alternative, one may invoke that the X chromosome in the male germ line is inactivated by a process likely to involve XIST gene expression, but not DNA methylation [94].

The alternative prezygotic expansion model necessitates postzygotic contraction events followed by a selection for premutation sized alleles in the male germ line, assuming for instance that the FMR1 protein (FMRP), which is present in early spermatogonia [8, 31], is important for spermatogenesis. Two observations seemed, however, to exclude that FMRP is required for spermatogenesis. The knock-out mice that are totally deficient in FMRP are fertile, and appear to have normal spermatogenesis [178]. A family has been described showing segregation of a 1.6-kb deletion encompassing the promoter and part of the first exon of the FMR1 gene, resulting in extinction of FMR1 expression and a clinical phenotype of fragile X syndrome in four males. The deletion, arising from an expanded allele, was inherited by four females from their deceased father (who was a likely carrier of a premutation, as he was described to have normal intelligence) [106]. This case shows that human spermatogonia can also function without expression of FMRP.

Finally, a more elaborate model was recently proposed by Ashley and Sherman [4], which assumes a moderate expansion during gametogenesis, followed by a postzygotic transition to full mutation.

Several observations recently put forward seem to tilt the advantage in favor of the prezygotic expansion model. First, a number of studies reported that in an affected fetus, various tissues present a very similar mutation pattern [30, 114, 208, 209]. In several cases, a major band was found common to chorionic villi and other embryonic tissues [30, 114]. This renders unlikely a transition to full mutation after the determination of the male germ line (at day 5) and suggests that expansion must have occurred before the separation of the trophectoderm and the inner cell mass.

Second, since small maternal premutations in the range of 60–90 give rise either to premutation or full mutation offspring, with a ratio that strongly depends on the size of the repeat, one would expect that such dependence reflects a probability of transition inferior to 100% in individual postzygotic cells. Another prediction of the postzygotic model would therefore be that mosaic children, who carry both premutations and full mutation-sized fragments, should arise preferentially from mothers with small premutations. This was confirmed by simulations using various postzygotic transition models [114]. Analysis of premutation size in a large series of 112 carrier mothers with 212 mutated offspring failed to show any dependence of mosaicism upon maternal premutation size. This can be taken as strong, albeit indirect, evidence against a postzygotic transition model [114].

A third piece of evidence indicates that selection for FMRP expression may indeed occur in spermatogonia. Malter *et al.* [101] showed that FMRP expression was absent in testes from a 13-week-old fetus carrying a full mutation (with no detectable premutation), while FMRP was abundantly expressed in testes from a normal fetus of the same age. A 17-week-old affected fetus showed both a faint premutation band and evidence of FMRP expression (not observed in neuronal tissues). The same authors also reported that oocytes from a full mutation female fetus (16 weeks) carried a full but unmethylated expansion in her oocytes. The critical experiment would have required analysis of oocytes from young premutation carrier females, but such samples may be difficult to obtain for obvious reasons.

Selection for premutation carrying cells during male gametogenesis could, however, also occur at the level of DNA replication, favoring smaller repeat sizes. It has been shown that the full mutation is responsible for a late replication of the FMR1 gene during the cell cycle [63], and the induction of the fragile site indicates that under certain conditions, this may lead to an abnormal chromatin structure and may thus be detrimental. The large number of replication cycles linked to spermatogenesis may enhance such a selection. It has indeed been suggested that the average size and/or heterogeneity of the CGG repeat of full mutations decreases with age in leukocytes [109, 145]. Alternatively, despite the observations suggesting that the FMRP protein is not absolutely required for spermatogenesis, it may still provide a selective advantage to spermatogonia in which it is expressed. Such a selective advantage conferred by FMRP expression has indeed been found in leukocytes. A slow selection operates in females carrying a full mutation, which results in adult carriers in a strongly biased X-inactivation pattern, favoring the presence of cells with the normal allele on the active X chromosome [145]. This selection occurs despite the absence of any detectable dysfunction of white blood cells in males carrying a full mutation.

The prezygotic model with selection against full mutation in the male germ line thus appears more likely. Mosaicism for premutation would thus result from a postzygotic retraction of full mutations, arising in a random (stochastic) manner. This would also account for the very good correlation between clinical status, likely to reflect the mutation pattern in the brain, and the mutation pattern observed in leukocytes. It would also fit with the observation that in premutation/full mutation mosaics, the premutation is usually found in a minority of cells. If expansion was postzygotic and mosaicism created by a lack of expansion in some cells, one would expect large differences between the number of mutated and premutated cells in various tissues, and a frequent lack of correlation between clinical status and mutation pattern in leukocytes.

It should be noted that, contrary to the case of myotonic dystrophy, the somatic instability of the full mutation, which is indicated by the very heterogeneous expansion pattern observed in leukocytes from many fragile X patients, or in fetal tissues from fetuses with a full mutation, is most likely to take place during early embryogenesis. This conclusion is derived from the observation of very similar expansions in leukocytes from five of six pairs of monozygotic twins studied [30, 87, 99, 134], or in different tissues of affected fetuses [30, 114, 208, 209], and by the clonal stability of the full mutation in cells in culture [208]. It was suggested by Wöhrle *et al.* [208, 209] that methylation may stabilize the expanded CGG repeat. The greater heterogeneity often observed for the unmethylated full mutation in chorionic villi is consistent with this hypothesis [114].

Additional light may be brought on the timing of expansion by an analysis of the transmission pattern for other CGG expansions linked to folate-sensitive fragile sites (see other chapters). Indeed, initial results for FRA16A suggest that retraction of large expansions does not occur in an obligatory fashion upon paternal transmission [120]. More direct but technically difficult approaches could involve single-cell analysis of expansion in oocytes from premutation carrier females (or female fetuses), and the construction of mouse models carrying various sizes of repeats. However, a pure FMR1 60 CGG repeat present in three different founder mice was recently shown to be stable in both male and female transmission [12] (see also [226]).

VI. LINKAGE DISEQUILIBRIUM AND FOUNDER EFFECTS AT THE FRAGILE X LOCUS

Founder effects, reflected by linkage disequilibrium between polymorphic markers and mutant alleles, are not expected in X-linked diseases that severely affect reproductive fitness in males. A high proportion of new mutations is expected to counter the loss of mutated alleles and maintain at equilibrium the incidence of such diseases. This is the case for the fragile X syndrome, as almost all males carrying the full mutation exhibit significant cognitive deficit, and the vast majority of such males do not reproduce; thus, their reproductive fitness is close to zero. However, a new mutation, defined as the transition from a normal FMR 1 allele to an unstable premutation or a full mutation, has never been observed in the fragile X syndrome. This could be due to the fact that ancestors in older generations, where such event could have occurred, are not available for study. However, an alternative explanation was proposed, i.e., that the fragile X mutations evolve by a multistep process over a large number of generations ([84, 111, 112]; reviewed in [19]). It had indeed been noticed that apparently unrelated affected individuals may sometimes have far removed common ancestors [34], and haplotype studies have since confirmed the hypothesis of a common mutation origin in such cases, and have shown that a premutation (or predisposing allele) can be carried silently over more than seven generations [100, 164]. The apparent discrepancy between the lack of fresh mutations and the high incidence of the disease, and the existence of a strong founder effect in myotonic dystrophy, another trinucleotide expansion disease, prompted an analysis of the origin of fragile X mutations using haplotype studies.

Indeed, the analysis of highly polymorphic markers located very close to the FRAXA locus has indicated the existence of linkage disequilibrium between some haplotypes and the fragile X mutation. The initial studies were carried in Caucasian populations, using either the FRAXAC1 and FRAXAC2 microsatellites that flank, and are within 10 kb of the CGG repeat [136], or FRAXAC2 and the DXS548 microsatellite at 150 kb proximal to the CGG repeat [126] (see Fig. 3-1). Richards *et al.* [136] found that the major haplotype in the normal population (50% of normal Xs) was only on 18% of fragile X chromosomes, while a haplotype found on 31% of the fragile X chromosomes was only present in 6% of the normal ones. Very similar results were obtained by Oudet *et al.* [126]; however, the different marker used uncovered an increased heterogeneity of fragile X-associated haplotypes (2 of them, accounting for 25% of fragile X chromosomes, were only found on 3% of the normals).

An even more striking linkage disequilibrium was found when the more homogeneous Finnish population was analyzed, since a single FRAXAC2/DXS548 haplotype was found on 75% of the fragile X chromosomes (versus 5% of normals). The major normal haplotype (that is the same as in the Caucasian population studied by Oudet *et al.* [126]), is only found on 1% of the fragile X chromosomes [59, 127]. The widely spread geographic origin of the earliest known fragile X carriers led Oudet *et al.* [127] to hypothesize that the predisposing haplotype was present in the initial settlers of Finland, 20 centuries ago, while Haataja *et al.* [59] argued that a predominance of the fragile X syndrome in the sparsely populated eastern and northern parts of the country supports a later founder effect, around the 16th century. The predisposing Finnish fragile X haplotype is also present in one-third of fragile X patients in Sweden [100], but appears extremely rare in the French population [125, 127]. An extreme linkage disequilibrium was also found in Tunisian Jews, who show an increased prevalence of fragile X syndrome compared to other populations in Israel. A single previously unreported haplotype accounts for all observed cases of the disease in Tunisian Jews [43].

Linkage disequilibrium has also been found in Japanese fragile X chromosomes, although the haplotypes involved are different than in the Caucasian population [3, 138]. This distinguishes fragile X from myotonic dystrophy, since in the latter disease, the DM mutation shows complete association to the same allele at an adjacent insertion–deletion polymorphism in Caucasians and in Japanese populations, suggesting a much earlier founder effect [65, 210].

Zhong *et al.* [214] confirmed in a U.S. fragile X population a linkage disequilibrium with FRAXAC1, but failed to find it with FRAXAC2. They remarked that the latter polymorphism was more complex than initially described, and exhibited both variation in length (by 1-bp increments rather than 2 bp expected for dinucleotide repeats), and also by internal sequence variation. They proposed that their observed absence of linkage disequilibrium with FRAXAC2 was due to increased instability of this microsatellite in fragile X maternally derived meioses [214]. Such instability was not observed in other laboratories, however [110, 139], and was in contradiction with the linkage disequilibrium observed in other populations, especially in Finland. Analyzing all three markers in 44 patients from southern England, Macpherson *et al.* [96] found a markedly different distribution of haplotypes between normal and fragile X chromosomes, with a greater haplotype diversity in the fragile X sample than in normals. This appeared in contradiction with the hypothesis of a founder effect, and suggested that two alternate mechanisms could create predisposing alleles: a gradual increase from large normal alleles accounting for founder haplotypes (see below), and rare events of jumps (or loss of interrupting AGGs) that might be linked to generalized microsatellite instability (as found in tumor cells with mismatch repair defects), thus generating the rare haplotypes. Indeed, Murray *et al.* [117] reported that instability of FRAXA or FRAXE normal or gray zone alleles was more often observed in conjunction with a second instability (mutation or recombination affecting very close microsatellites), suggesting that such rare events may occur in a context of generalized genomic instability.

Richards *et al.* [136] observed that large normal CGG repeats ($n \geq 39$) were more frequently associated with the fragile X preferential haplotype, and suggested that such large alleles are the founder fragile X chromosomes. The finding of polymorphic AGG interspersion and its effect on allele stability led to a reappraisal of this hypothesis. The sequencing of a large number of alleles [73, 88, 166] revealed that about half of the alleles have two interspersed AGGs on the model $(CGG)_{9\text{-}11}AGG(CGG)_{9\text{-}11}AGG(CGG)_n$, 25% have a single AGG, and less than 10% have no AGG. However, 10 of 13 premutation size alleles (>55 repeats) had 0 or 1 AGG [39]. Using *MnlI* restriction analysis to detect AGG interruption, Zhong *et al.* [216] reported that 63% of 54 premutations analyzed had no AGG, versus 1.5% in the control population.

Kunst and Warren [88] found that alleles with greater than 24 pure 3′ CGGs (2% of normal chromosomes) appear more frequently on haplotypes overrepresented among fragile X chromosomes, and may constitute a pool from which the majority of mutant alleles are derived. Similar results were reported by Snow *et al.* [166] and Hirst *et al.* [73] (although the similarity in marker association between long pure CGG alleles and fraX chromosomes was not as striking). The haplotype linked to fragile X syndrome in Tunisian Jews is found associated to large normal alleles lacking AGG interruption, which are found at high frequency in this population (5% of alleles have ≥35 pure CGGs) [43].

Both Kunst and Warren [88] and Snow *et al.* [166] remarked that some haplotypes associated with long pure CGGs are not overrepresented in fragile X chromosomes. A similar observation was made for premutation size alleles in the range of 54–64 repeats found in the Quebec population by Rousseau *et al.* [150]. Kunst and Warren [88] proposed that alleles linked to the FRAXAC1 allele A are of more recent origin (they carry a more regular AGG interspersion pattern than in the other haplotypes) and may not have reached equilibrium with mutant alleles (see also [89] and [41]). Alternately (but probably less likely), these observations could be due to putative stabilizing effects of some adjacent sequences carried by these haplotypes.

From these studies it was concluded that at-risk alleles can be derived either by increase in length of the 3′ pure CGG tract, or by loss of the most 3′ AGG [41]. Rare jumps may also be possible (for instance by unequal recombination in sister chromatid exchange). A jump from 28 to 38 repeat length has indeed been observed once [96] as well as a mutation from 29 to 21 repeats [3]. The threshold of instability at the population level would be around 24 pure CGGs [88] while the threshold of instability that can be directly observed in families is around 34 pure CGGs [39]. In parallel studies, it was shown that alleles with 20–30 CTGs are the reservoir for myotonic dystrophy mutations, and arose from one or a very small number of jumps from the $(CTG)_5$ frequent allele [76], accounting for the extreme linkage disequilibrium observed in this disease, while observed new mutations in Huntington's disease arise on alleles in the range of 30–35 pure CAGs [119]. These threshold values for instability are remarkably similar to those proposed for the FMR1 CGG repeat.

VII. POPULATION FREQUENCY OF FRAGILE X MUTATIONS

Most epidemiological studies on the frequency of the fragile X syndrome were based on cytogenetic analyses of mentally retarded patients, often with ascertainment biases (especially in females). These have been reviewed by Sherman [156] (see also [180]). The most thorough and systematic studies were conducted on children of school age in Coventry by Webb *et al.* [196], and by Turner *et al.* [182] in mentally retarded patients from New South Wales. The English study concluded initially a frequency of about 1 in 1040 school children [197], while the Australian study gave a lower approximate frequency of 1 in 2610 males and 1 in 4220 females. However, cytogenetic analysis may be subject to false negatives (cells from individuals with small full mutations or from mosaics may show very little fragile site expression) or to false positives (due to a high background of fragile site expression according to the particular technique used). There may also be confusion with the FRAXE or FRAXF fragile sites (although these appear much rarer than FRAXA; see below). Indeed molecular reanalysis of the cytogenetic positives in the Coventry study led to an overall prevalence figure of 1 in 2700 to 1 in 5700 [113], and of 1 in 4350 in males in the Australian study [184]. A systematic molecular study of children with special education needs in Wessex found only 5 children with a full mutation, resulting in an estimate of 1 in 5000 males [116]. Cytogenetic analysis was of course unable to document the frequency of premutations. In a systematic study of 10,624 females from the general population (from Quebec), Rousseau *et al.* [150] found 18 alleles of premutation size (≥66 repeats), a frequency of about 1 in 500 females, while an equivalent number of females (23) carried 1 allele in the gray zone area (55 to 63 repeats), albeit with a likely low risk of changing to full mutation in a single generation. Although the design of the study does not allow the fate of these premutation alleles to be followed, if their risk of expansion to full mutation is equivalent to that of premutations calculated in the fragile X families, a frequency of about 3.5 newborns carrying a full mutation per 10,000 births (male and females) would be expected in this Quebec population. Founder effect can result in higher frequency of the disease in some populations. For instance, Tunisian Jews show a 10-fold increased incidence compared to other Israeli populations, due to the presence of a high proportion of alleles lacking AGG interruptions. The high frequency of premutations and mutations (4 in 154 chromosomes analyzed) in this population might warrant implementation of a specific screening program [43].

In other trinucleotide expansion diseases, such as myotonic dystrophy, Huntington's disease, or dentatorubro-pallidoluysian atrophy, wide variations in disease frequency in various populations appear to be linked to variations in the frequency of large normal alleles. These alleles constitute a reservoir for recurring unstable mutations [76, 102, 119]. In the case of the FMR1 repeat, few studies of the normal allele distribution in various populations have been published. Although the distribution of the most common alleles observed in Japanese and Chinese populations appear quite similar to that of Caucasians, this does not exclude significant differences in the frequency of large normal alleles. Moreover, AGG interspersion, which plays an important role in the stability of such alleles, has not been studied at all in these populations. Based on the observation of linkage disequilibrium between flanking polymorphic markers and either large normal CGG repeat alleles or fragile X mutations, it was suggested that some large normal alleles are of relatively recent origin [88] (see above). They may therefore not have yet reached equilibrium with respect to transition to full mutation and allele loss due to decreased reproductive fitness. It was proposed that such alleles might be responsible for "a small but steady increase in the frequency of fragile X syndrome over many future generations" [88] (see discussion in [102]).

VIII. POSSIBLE MECHANISMS OF EXPANSION

One of the most fascinating aspects of the DNA instability at the fragile X locus is the apparently ineluctable tendency to size-increase when a premutation evolves into a full mutation. This strong tendency toward expansion is shared by the other unstable trinucleotide repeats (CGG, CAG, or CTG) linked to various diseases, for a comparable range of parental allele length (50–200 repeats), although the parental sex biases for such expansions are different for the CAG/CTG repeats [9]. It is in fact striking to note that the threshold for instability that is observed in families is at about 50 repeats for all these diseases. The process that causes the expansion is unknown, but several mutation mechanisms have been proposed. Two major types of mutation event may be distinguished [199]. Unequal homologous recombination is a well known mechanism of length variation of tandem repeats (of longer sequence units). However, unequal crossing over at meiosis is excluded, since exchange of very close flanking markers consecutive to the expansion has never been observed (even at a population level, as testified by the linkage disequilibrium observed in studies of patients, see above). A mechanism involving unequal sister chromatid exchange or gene conversion events is also unlikely since it should cause both increase and decrease of the number of repeats. In any case, a double-strand break–recombination event could only account for limited size expansion, by a factor of two (or possibly three) [199]. Several recombination events would thus be necessary to induce a transition from premutation to full mutation. Rare cases of reversion of the fragile X mutation have been described [186], and one such case appears to be due to a conversion-like event.

Thus, replication-based errors appear to be the favored mechanism. DNA polymerase slippage is a well known mechanism that accounts for the polymorphism of short tandem repeats (by increasing or decreasing the repeat length by a small number of units), but may only explain the small-scale variation of the short premutation alleles. Richards and Sutherland [140] and Eichler *et al.* [39] have proposed a simple mechanism to explain the mutation process. They remarked that the threshold for instability (about 50 repeats, or 150 bp), is similar to the average size of Okazaki replication fragments. They suggest that expansion is due to slippage and sliding of such fragments during replication. If an Okazaki fragment is anchored by a unique (nonrepetitive) flanking sequence, no expansion would occur. However, if the fragment is not anchored (because it is smaller than the parental repeat), extensive slippage during replication might result. Alternatively, the presence of an abnormal structure during replication (which might be stabilized by repeat binding proteins [137]) could induce DNA polymerase stalling, resulting in multiple slippage events. In fact, long CGGs appear very difficult to duplicate *in vitro* by *Taq*I DNA polymerase. Such replication problems, however, would manifest only under special conditions (for instance, limiting substrates or replication factors), as full mutations appear stable in fibroblasts or leukocytes [30, 208].

The CGG repeats may indeed form unusual DNA structures. *In vitro* studies allowed Fry and Loeb [50] to propose that, under physiological conditions, short $(CGG)_n$ oligonucleotides form tetrahelices that could obstruct replication and transcription. These structures are stabilized by methylation of the C residues but are not observed with the complementary $(CCG)_n$ oligonucleotides. Recently, several authors have proposed that single-strand CGG (and CAG) repeats may form hairpin structures. Kuryavyi and Jovin [86] suggested on the basis of modeling studies that CGG and CAG triplet repeats may form a triad-DNA structure, consisting of an anti-parallel double helix of base-triads (instead of base pairs in the conventional B-DNA). These authors suggested that tetrahelix structures of parallel strands cannot account for the genomic nature of the unstable triplet sequences. Physicochemical analysis of CAG and CTG repeats has suggested that these may form hairpins, and modelization studies predicted that similar structures would be formed by CGG/CCG repeats [103]. The stability of these hairpins would be similar for repeat length at the threshold of allele instability. The features of these hairpins might account for the length dependence of expansion and the stabilizing effect of AGG interspersion (i.e., they destabilize the hairpin and thus stabilize the alleles) [103, 218]. Smith *et al.* [163] observed that $(CGG)_{15}$ and $(CCG)_{15}$ oligonucleotides adopt unimolecular foldback conformations that are detected using nondenaturing gel electrophoresis. Furthermore, the $(CCG)_{15}$ foldbacks are exceptionally good substrates for the human methyltransferase, suggesting that methylation might be induced as a consequence of the CGG expansion.

The unstable triplet repeats that have been identified to date are CAG/CTG and CGG/CCG, with the single exception of the GAA repeat in Friedreich's ataxia (see Pandolfo and Koenig, this volume). It is possible that these three triplets are the only ones that have the ability to expand. But there is also a bias in the identification of these trinucleotides. The CGG expansions were identified because they are capable to generate fragile sites under appropriate cell growth conditions [173]. The CAG expansions causing neurodegenerative disorders share the common characteristic of being translated into glutamine residues in the corresponding proteins. It remains possible, therefore, that other triplet sequences elsewhere in the human genome have the same ability to expand, but that these have not been identified yet, due for example to the absence of associated clinical or cytogenetic phenotype.

IX. CONSEQUENCES OF THE CGG EXPANSION (OTHER THAN ON FMR1 EXPRESSION)

In addition to the hypermethylation of the FRAXA locus and the loss of transcription of the FMR1 gene, the fragile X full mutation has at least three other consequences at the molecular level.

Under culture conditions in a medium deficient in folic acid or thymidine (or that impairs dTMP synthesis), cells carrying the fragile X mutation express cytogenetically the fragile site. The level of expression ranges from 5 to 50% in lymphocytes from affected males. There is some correlation between fragile site expression and repeat length [212], and methylation may also play a role [232]. Short full mutations in the range $n = 250–400$ often show no fragile site expression. The apparent contradiction between the effect of a deficiency in dTTP for the expression of the fragile site and the absence of T bases in the repeats has not been solved and may be related to secondary effects on dCTP pools.

It has been shown that the full mutation alters the replication timing around the FRAXA locus [63, 64, 169, 179]. In normal lymphoblastoid cell lines, the FMR1 region replicates in the second half of the S phase. Premutation alleles behave identically. But full mutated FMR1 alleles replicate later, during the G2/M phase, although other X-linked loci (G6PD, F9, PGK1, X-alpha) in the same fragile X cells replicate normally. It was shown that the affected region extends at least 400 kb 5′ of FMR1. The distal border of delayed replication varies among different fragile X males, up to about 600 kb 3′ of FMR1, far beyond the repeats themselves [64]. It was therefore suggested that the delayed replication might be due to the inactivation of a nearby replication origin, or even that the CGG repeats may function as a replication origin [63, 179].

Several cases of fragile X patients have been described in which CGG expansions result in a small deletion of the region surrounding the CGG repeats. Meijer *et al.* [106] studied a family with a 1.6-kb deletion proximal to the repeat and ending in the CGG repeats. Based on haplotype and sequence analyses, they concluded that the deceased grandfather most likely carried a premutation allele in his somatic cells but may have been a mosaic for a premutation and a deletion allele in his germline. The deletion allele was transmitted through his four daughters, to six grandchildren. De Graaff *et al.* [26] studied four unrelated individuals carriers of both a full mutation and a deletion of the CGG repeat and its flanking sequences. The deletions varied in size from 150 to 600 bp and all had a 5′ breakpoint located between 75 and 53 bp proximal to the CGG repeat, resulting in the identification of a hotspot for deletions. The sequence flanking the deletion breakpoints did not show the methylation pattern of a full mutation. Another case of mosaicism for full mutation and a microdeletion was reported, with a 5′ breakpoint more upstream than the proposed hotspot [152].

X. DIAGNOSTIC APPLICATIONS

The ability to detect both premutations and full mutations at the DNA level has revolutionized the diagnosis of the disease (cytogenetic analysis was subject to both false-positive and false-negative results; see above), genetic counseling through reliable detection of female and male carriers, and prenatal diagnosis.

At present the main procedures are based on Southern blotting and on PCR amplification, and will be outlined here: technical aspects are more thoroughly discussed elsewhere [125, 146, 165]. Southern blot analysis using a single enzyme (*Eco*RI or *Hin*dIII) allows the detection of full mutations and large premutations, and is sufficient for the screening of mentally retarded probands (expected to carry a full mutation if their mental handicap is due to an expansion at the FRAXA locus). *Bcl*I or *Pst*I digests give smaller CGG-containing fragments, allowing a better detection and sizing of premutations (as an alternative to PCR). Combined detection of expansion and methylation (for instance with *Eco*RI and *Eag*I) is widely used. It offers a very good compromise in family studies to detect full mutations, mosaic patterns, and premutations [144, 146, 147] and may be performed using nonradioactive (digoxigenin-labeled) probes [42]. Accurate sizing of premutations, important in females for assessing the risk of having affected offspring and for distinguishing large normal alleles from premutations, can be done by PCR [51, 225]. Use of *Pfu*I polymerase or of a proprietary PCR system [69] instead of *Taq*I polymerase may facilitate analysis [20]. For gray zone alleles, the use of *Mnl*I digestion after PCR to analyze the AGG interspersion has been utilized [39]. This may be of diagnostic interest for gray zone alleles detected in systematic population analyses, where there is no *a priori* knowledge about their instability in the family.

Indirect diagnosis using very close microsatellite markers (FRAXAC1 and C2, DXS548, or even the FRAXE CGG repeat) is of value because of its rapidity in a prenatal diagnosis context, especially when there is little fetal DNA available. It should be kept in mind that very rare events of recombination between DXS548 and FRAXA, or of apparent gene conversion, have been reported that could lead to diagnostic errors if indirect diagnosis was based on a single close marker.

The Southern blot is a tedious and rather expensive technique, especially when applied to systematic screening of mentally retarded probands, in which a positive rate of 2–5% is observed if only minimal clinical preselection is used. More stringent preselection leads to a higher positive rate [55, 182], at the probable cost of missing cases who lack typical phenotypic features, especially very young children or affected females. Thus alternative techniques have been proposed. The combined PCR/Southern method of Brown *et al.* [13] is nonradioactive and more rapid than conventional Southern, but does not seem to have gained wide use (detection of full mutations may be tricky and not always reliable, especially in females [42], and the detection of premutation sized fragments would be favored in mosaics). A nonradioactive PCR exclusion diagnosis was proposed by Wang *et al.* [194] for males, under conditions in which only normal alleles are amplified. This is applicable only to males, as only females heterozygous for alleles of clearly different size would also be excluded. A similar test, also able to detect premutation sized alleles and with an internal control for PCR amplification, was described by Haddad *et al.* [60].

An interesting development of this scheme has been recently proposed in which both FRAXA and FRAXE CGG repeats are coamplified, one locus serving as an internal control for the other [194]. All individuals showing absence of one the two locus-specific bands are subsequently analyzed by Southern blotting. One possible problem is caused by the rare occurrence of mosaic individuals who carry a full mutation and an amplifiable normal size band ([122, 186]; Biancalana. personal communication), and who might be found normal in such PCR tests (although one could expect dosage difference if the normal sized fragment is a very minor one). PCR-based methods require very strict conditions to avoid contamination, which might lead to false-negative results. Adaptation of PCR tests for automated ABI sequencers appears able to detect most full mutations in addition to normal and premutation alleles, in both males and females and starting from a very small amount of DNA, avoiding the necessity of blood sampling. If fully validated, such a method would provide a most useful alternative to present tests (W. Block, personal communication).

Prevention of fragile X could in principle be attained by systematic screening for carriers in females of child bearing age. However, available techniques of PCR of premutations appear too tricky for population screening and would miss full mutation carriers, awaiting the adaptation to fluorescent automated sequencers. Scaling down the Southern blot may be an alternative, as shown by Rousseau *et al.* [148], who screened in such a way some 10,600 females [150]; the method used may also miss some carriers of very heterogeneous full mutations, and the pooling scheme used might not be acceptable in a diagnostic situation. The price of this scaled-down test has, however, been estimated to be less than that of the triple test for prediction of risk of trisomy 21 (the latter test having a lower diagnostic sensitivity and specificity than the test for a premutation) (Rousseau, personal communication). Cost–benefit studies should reveal whether this strategy may be economically sound. Inductive screening, i.e., starting from a fragile X proband and following segregation of the abnormal allele through the extended family, is at present a practical but not very efficient approach.

Detection of the FMR1 protein using the monoclonal antibodies from Devys *et al.* [31], can be performed on blood smears and may provide an inexpensive and very rapid test in mentally retarded males [203, 205]. Males with a full mutation were found to have less than 40% FMRP-positive lymphocytes (the latter probably being due to mosaicism for a premutation), and were well distinguished from controls. This test would have the advantage of detecting the rare male patients with deletions, and those with nonsense or frameshift mutations who are presently missed by standard DNA analysis. This test is much less specific for females with a full mutation, because of the overlap of the proportion of FMRP-positive cells between carriers and controls [205]. In particular, females with biased inactivation (the majority of adult female carriers of full mutations [144]) would show FMRP in the majority of their lymphocytes. The antibody test can be performed on smears, stored at room temperature up to 3 weeks [205]. Details concerning the technique are posted at the following site on the world wide web: http://www.eur.nl/FGG/CH1/fragx/index.html.

If an immunotest could be adapted to dried blood (Guthrie spots), it could then be used for presymptomatic testing in newborn males. Although this would be of great epidemiological interest, and of potential value in providing early genetic counseling, such population screening poses serious ethical questions, and would have to be carefully evaluated because of potentially harmful consequences for the affected children, for whom no specific treatment can be proposed. Indeed, neonatal screening schemes for Duchenne muscular dystrophy were shown to be of controversial value [70]. In any case, the protein-based test will not detect premutations, as these do not affect FMRP expression. A promising application of the immunotest may lie in prenatal diagnosis, where it may provide a very rapid test for full mutation in chorionic villi, at least in male fetuses. However, the full mutation is in general unmethylated in chorionic villi at 11 weeks, thus allowing expression of FMRP. Willemsen *et al.* [206] suggest that only villi taken at >12.5 weeks gestational age may be used. An alternative is to use uncultured amniocytes, allowing same-day results [206]. However, before these techniques are fully validated, they should be carried in parallel to DNA tests.

Preimplantation diagnosis would be particularly valuable for fragile X, given its semidominant nature, leading to a 50% rate of abortion, when prenatal diagnosis is performed and parents do not wish to take the risk of an affected female. The full mutation cannot, however, be amplified from a single cell. Such diagnoses would thus have to be based on the amplification of a normal allele (distinguishing the parental alleles for a female) or on indirect diagnosis (with linked microsatellites) [33]. However, preliminary studies suggest that subtle endocrine dysfunction in female carriers may render inefficient ovarian stimulation, resulting in increased failure rate of the *in vitro* fertilization procedure [11].

XI. SUMMARY

Remarkable progress has been made since the initial isolation of the fragile X site and FMR1 gene. Significant advances have been realized in understanding the history and behavior of the CGG repeats responsible for the fragile site and the loss of FMR1 function. Many of these lessons have been applied to other trinucleotide repeats more recently found expanded in human genetic disease and discussed in other chapters of this volume. Still, much remains to be learned. Many of the open questions have been discussed throughout this chapter. In the area of repeat instability, it remains unclear when and where the transition from premutation to full mutation occurs, although the bulk of the evidence is leaning toward the prezygotic. The mechanism that drives expansion of this and other repeats remains elusive. Models which fully recapitulate the events seen in human fragile X families (ideally in a mammalian system) will provide a much more facile approach to the study of these questions. Such models have not yet been developed, possibly due to fundamental differences between the DNA replication and/or repair machinery found in humans and other mammals. Analyses of the FMR1 gene product and of the consequences of its absence are beginning to demonstrate their utility. Hypotheses regarding the normal function of FMR1 are beginning to emerge and to be tested. The possibility of developing rational treatments for patients with fragile X syndrome, while still some distance ahead, is now within the grasp of the imagination.

References

1. Abitbol, M., Menini, C., Delezoide, A.-L., Rhyner, T., Vekemans, M., and Mallet, J. (1993). Nucleus basalis magnocellularis and hippocampus are the major sites of FMR-1 expression in the human fetal brain. *Nature Genet.* **4,** 147–153.
2. Allingham-Hawkins, D. J., and Ray, P. N. (1995). FRAXE expansion is not a common etiological factor among developmentally delayed males. *Am. J. Hum. Genet.* **56,** 72–76.
3. Arinami, T., Asano, M., Kobayashi, K., Yanagi, H., and Hamaguchi, H. (1993). Data on the CGG repeat at the fragile X site in the non-retarded Japanese population and family suggest the presence of a subgroup of normal alleles predisposing to mutate. *Hum. Genet.* **92,** 431–436.
4. Ashley, A. E., and Sherman, S. L. (1995). Population dynamics of a meiotic/mitotic expansion model for the fragile X syndrome. *Am. J. Hum. Genet.* **57,** 1414–1425.
5. Ashley, C. T., Sutcliffe, J. S., Kunst, C. B., Leiner, H. A., Eichler, E. E., Nelson, D. L., and Warren, S. T. (1993). Human and murine FMR-1: alternative splicing and translational initiation downstream of the CGG-repeat. *Nature Genet.* **4,** 244–251.
6. Ashley, C. T., Jr., Wilkinson, K. D., Reines, D., and Warren, S. T. (1993). FMR1 protein: conserved RNP family domains and selective RNA binding. *Science* **262,** 563–566.
7. Bächner, D., Steinbach, P., Wöhrle, D., Just, W., Vogel, W., Hameister, H., Manca, A., and Poustka, A. (1993). Enhanced Fmr-1 expression in testis. *Nature Genet.* **4,** 115–116.
8. Bachner, D., Manca, A., Steinbach, P., Wöhrle, D., Just, W., Vogel, W., Hameister, H., and Poustka, A. (1993). Enhanced expression of the murine FMR1 gene during germ cell proliferation suggests a special function in both the male and the female gonad. *Hum. Mol. Genet.* **2,** 2043–2050.
9. Bates, G., and Lehrach, H. (1994). Trinucleotide repeat expansions and human genetic disease. *Bioessays* **16,** 277–284.
10. Bell, M. V., Hirst, M. C., Nakahori, Y., MacKinnon, R. N., Roche, A., Flint, T. J., Jacobs, P. A., Tommerup, N., Tranebjaerg, L., Froster-Iskenius, U., Kerr, B., Turner, G., Lindenbaum, R. H., Winter, R., Pembrey, M., Thibodeau, S., and Davies, K. E. (1991). Physical mapping across the fragile X: hypermethylation and clinical expression of the fragile X syndrome. *Cell* **64,** 861–866.
11. Black, S. H., Levinson, G., Harton, G. L., Palmer, F. T., Sisson, M. E., Schoener, C., Nance, C., Fugger, E. F., and Fields, R. A. (1996). Preimplantation genetic testing (PGT) for fragile X (fraX). *Am. J. Hum. Genet.* **57,** A31. (Abstract)
12. Bontekoe, C. J. M., de Graaff, E., Nieuwenhuizen, I. M., Willemsen, R., and Oostra, B. A. (1997). FMR1 premutation is stable in mice. *Eur. J. Hum. Genet.* **5,** 293–298.
13. Brown, W. T., Houck, G. E., Jr., Jeziorowska, A., Levinson, F. N., Ding, X., and Dobkin, C., Zhong, N., Henderson, J., Brooks, S. S., and Jenkins, E. C. (1993). Rapid fragile X carrier screening and prenatal diagnosis using a nonradioactive PCR test. *JAMA* **270,** 1569–1575.
14. Brown, W. T., Zhong, N., Curley, D., Wang, D. W., Weina, J., and Dobkin, C. S. (1995). A general method for finding new polymorphisms: identification of an Alu SSCP polymorphism in the fragile X gene. 6th X chromosome workshop, Banff, Canada. [Abstract].
15. Burd, C. G., and Dreyfuss, G. (1994). Conserved structures and diversity of functions of RNA-binding proteins. *Science* **265,** 615–621.
16. Chen, T., Damaj, B. B., Herrera, C., Lasko, P., and Richard, S. (1997). Self-association of the single-KH-domain family members Sam68, GRP33, GLD-1, and Qk1: role of the KH domain. *Mol. Cell. Biol.* **17,** 5707–5718.
17. Chernokalskaya, E., Dompenciel, R., and Schoenberg, D. R. (1997). Cleavage properties of an estrogen-regulated polysomal ribonuclease involved in the destabilization of albumin mRNA. *Nucleic Acids Res.* **25,** 735–42.

18. Chicurel, M. E., Terrian, D. M., and Potter, H. (1993). mRNA at the synapse of a synaptpsomal preperation enriched in hippocampal dendritic spines. *J. Neurosci.* **13,** 4054–4063.
19. Chiurazzi, P., MacPherson, J., Sherman, S., and Neri, G. (1996). Significance of linkage disequilibrium between the fragile X locus and its flanking markers. *Am. J. Med. Genet.* **64,** 203–208. [Editorial]
20. Chong, S. S., Eichler, E. E., Hughes, M. R., and Nelson, D. L. (1994). Robust amplification of the fragile X syndrome CGG repeat using Pfu polymerase: ethidium bromide detection of normal and premutation alleles. *Am. J. Med. Genet.* **51,** 522–526.
21. Comery, T. A., Harris, J. B., Willems, P. J., Oostra, B. A., Irwin, S. A., Weiler, I. J., and Greenough, W. T. (1997). Abnormal dendritic spines in fragile X knockout mice: maturation and pruning deficits. *Proc. Natl. Acad. Sci. USA* **94,** 5401–5404.
22. Conway, G. S., Hettiarachchi, S., Murray, A., and Jacobs, P. A. (1995). Fragile X premutations in familial premature ovarian failure. *Lancet* **346,** 309–310. [Letter; comment]
23. Corbin, F., Bouillon, M., Fortin, A., Morin, S., Rousseau, F., and Khandjian, E. W. (1997). The fragile X mental retardation protein is associated with poly(A)+ mRNA in actively translating polyribosomes. *Hum. Mol. Genet.* **6,** 1465–1472
24. Coy, J. F., Sedlacek, Z., Bachner, D., Hameister, H., Joos, S., Lichter, P., Delius, H., and Poustka, A. (1995). Highly conserved 3′ UTR and expression pattern of FXR1 points to a divergent gene regulation of FXR1 and FMR1. *Hum. Mol. Genet.* **4,** 2209–2218.
25. De Boulle, K., Verkerk, A. J. M. H., Reyniers, E., Vits, L., Hendrikx, J., Van Roy, B., Van Den Bos, F., de Graaff, E., Oostra, B. A., and Willems, P. J. (1993). A point mutation in the FMR-1 gene associated with fragile X mental retardation. *Nature Genet.* **3,** 31–35.
26. de Graaff, E., Rouillard, P., Willems, P. J., Smits, A. P. T., Rousseau, F., and Oostra, B. A. (1995). Hotspot for deletions in the CGG repeat region of FMR1 in fragile X patients. *Hum. Mol. Genet.* **4,** 45–49.
27. de Graaff, E., Willemsen, R., Zhong, N., de Die-Smulders, C. E., Brown, W. T., Freling, G., and Oostra, B. (1995). Instability of the CGG repeat and expression of the FMR1 protein in a male fragile X patient with a lung tumor. *Am. J. Hum. Genet.* **57,** 609–618.
28. de Vries, B. B., Jansen, C. C., Duits, A. A., Verheij, C., Willemsen, R., Van Hemel, J. O., van den Ouweland, A. M., Niermeijer, M. F., Oostra, B. A., and Halley, D. J. (1996). Variable FMR1 gene methylation of large expansions leads to variable phenotype in three males from one fragile X family. *J. Med. Genet.* **33,** 1007–1010.
29. Deissler, H., Behn-Krappa, A., and Doerfler, W. (1996). Purification of nuclear proteins from human HeLa cells that bind specifically to the unstable tandem repeat (CGG)n in the human FMR1 gene. *J. Biol. Chem.* **271,** 4327–4334.
30. Devys, D., Biancalana, V., Rousseau, F., Boue, J., Mandel, J. L., and Oberle, I. (1992). Analysis of full fragile X mutations in fetal tissues and monozygotic twins indicate that abnormal methylation and somatic heterogeneity are established early in development. *Am. J. Med. Genet.* **43,** 208–216.
31. Devys, D., Lutz, Y., Rouyer, N., Bellocq, J. P., and Mandel, J. L. (1993). The FMR-1 protein is cytoplasmic, most abundant in neurons and appears normal in carriers of a fragile X premutation. *Nature Genet.* **4,** 335–340.
32. Dobkin, C. S., Nolin, S. L., Cohen, I., Sudhalter, V., Bialer, M. G., Ding, X. H., Jenkins, E. C., Zhong, N., and Brown, W. T. (1996). Tissue differences in fragile X mosaics: mosaicism in blood cells may differ greatly from skin. *Am. J. Med. Genet.* **64,** 296–301.
33. Dreesen, J. C., Geraedts, J. P., Dumoulin, J. C., Evers, J. L., and Pieters, M. H. (1995). RS46(DXS548) genotyping of reproductive cells: approaching preimplantation testing of the fragile-X syndrome. *Hum. Genet.* **96,** 323–329.
34. Drugge, U., Holmgren, G., Blomquist, H. K., Dahl, N., Gustavson, K. H., and Malmgren, H. (1992). Study of individuals possibly affected with the fragile X syndrome in a large Swedish family in the 18th to 20th centuries. *Am. J. Med. Genet.* **43,** 353–354. [Letter]
35. Eberhart, D. E., Malter, H. E., Feng, Y., and Warren, S. T. (1996). The fragile X mental retardation protein is a ribonucleoprotein containing both nuclear localization and nuclear export signals. *Hum. Mol. Genet.* **5,** 1083–1091.
36. Eberhart, D. E. and Warren, S. T. (1996). Molecular basis of fragile X syndrome. *In* "Cold Spring Harbor Symposia on Quantitative Biology," Vol. LXI, pp. 679–687. Cold Spring Harbor Laboratory, Cold Spring Harbor, NY.
37. Eberhart, D. E., and Warren, S. T. (1997). Nuclease sensitivity of permeabilized cells confirms altered chromatin formation at the fragile X locus. *Somat. Cell Mol. Genet.* **22,** 435–441.
38. Eichler, E. E., Richards, S., Gibbs, R. A., and Nelson, D. L. (1993). Fine structure of the human FMR1 gene. *Hum. Mol. Genet.* **2,** 1147–1153.
39. Eichler, E. E., Holden, J. J. A., Popovich, B. W., Reiss, A. L., Snow, K., Thibodeau, S. N., Richards, C. S., Ward, P. A., and Nelson, D. L. (1994). Length of uninterrupted CGG repeats determines instability in the FMR1 gene. *Nature Genet.* **8,** 88–94.
40. Eichler, E. E., Kunst, C. B., Lugenbeel, K. A., Ryder, O. A., Warren, S. T., and Nelson, D. L. (1995). Evolution of the cryptic FMR1 CGG repeat. *Nature Genet.* **11,** 301–307.
41. Eichler, E. E., Macpherson, J. N., Murray, A., Jacobs, P. A., Chakravarti, A., and Nelson, D. L. (1996). Haplotype and interspersion analysis of the FMR1 CGG repeat identifies two different mutational pathways for the origin of the fragile X syndrome. *Hum. Mol. Genet.* **5,** 319–330.
42. el-Aleem, A. A., Bohm, I., Temtamy, S., el-Awady, M., Awadalla, M., Schmidtke, J., and Stuhrmann, M. (1995). Direct molecular analysis of the fragile X syndrome in a sample of Egyptian and German patients using non-radioactive PCR and Southern blot followed by chemiluminescent detection. *Hum. Genet.* **96,** 577–584.
43. Falik-Zaccai, T. C., Shachak, E., Yalon, M., Lis, Z., Borochowitz, Z., Macpherson, J. N., Nelson, D. L., and Eichler, E. E. (1997). Predisposition to the fragile X syndrome in Jews of Tunisian descent is due to the absence of AGG interruptions on a rare Mediterranean haplotype. *Am. J. Hum. Genet.* **60,** 103–112.
44. Feng, Y., Lakkis, L., Devys, D., and Warren, S. T. (1995). Quantitative comparison of FMR1 gene expression in normal and premutation alleles. *Am. J. Hum. Genet.* **56,** 106–113.
45. Feng, Y., Zhang, F., Lokey, L. K., Chastain, J. L., Lakkis, L., Eberhart, D., and Warren, S. T. (1995). Translational supression by trinucleotide repeat expansion at FMR1. *Science* **268,** 731–734.
46. Feng, Y., Gutekunst, C. A., Eberhart, D. E., Yi, H., Warren, S. T., and Hersch, S. M. (1997). Fragile X mental retardation protein: nucleocytoplasmic shuttling and association with somatodendritic ribosomes. *J. Neurosci.* **17,** 1539–1547.
47. Feng, Y., Absher, D., Eberhart, D. E., Brown, V., Malter, H. E., and Warren, S. T. (1997). FMRP associates with polyribosomes as an mRNP and the I304N mutation of severe fragile X syndrome abolishes this association. *Mol. Cell.* **1,** 109–118.

48. Fisch, G. S., Snow, K., Thibodeau, S. N., Chalifaux, M., Holden, J. J. A., Nelson, D. L., Howard-Peebles, P. N., and Maddalena, A. (1995). The fragile X premutation in carriers and its effect on mutation size in offspring. *Am. J. Hum. Genet.* **56,** 1147–1155.
49. Fischer, U., Huber, J., Boelens, W. C., Mattaj, I. W., and Luhrmann, R. (1995). The HIV-1 Rev activation domain is a nuclear export signal that accesses an export pathway used by specific cellular RNAs. *Cell* **82,** 475–483.
50. Fry, M., and Loeb, L. A. (1994). The fragile X syndrome d(C-GG)n nucleotide repeats form a stable tetrahelical structure. *Proc. Natl. Acad. Sci. USA* **91,** 4950–4954.
51. Fu, Y. H., Kuhl, D. P. A., Pizzuti, A., Pieretti, M., Sutcliffe, J. S., Richards, S., Verkerk, A. J. M. H., Holden, J. J. A., Fenwick, R. G., Jr., Warren, S. T., Oostra, B. A., Nelson, D. L., and Caskey, C. T. (1991). Variation of the CGG repeat at the Fragile X site results in genetic instability: resolution of the Sherman paradox. *Cell* **67,** 1047–1058.
52. Gedeon, A. K., Baker, E., Robinson, H., Partington, M. W., Gross, B., Manca, A., Korn, B., Poustka, A., Yu, S., Sutherland, G. R., and Mulley, J. C. (1992). Fragile X syndrome without CCG amplification has an FMR1 deletion. *Nature Genet.* **1,** 341–344.
53. Gedeon, A. K., Keinänen, M., Adès, L. C., Kääriäinen, H., Gécz, J., Baker, E., Sutherland, G. R., and Mulley, J. C. (1995). Overlapping submicroscopic deletions in Xq28 in two unrelated boys with developmental disorders: identification of a gene near FRAXE. *Am. J. Hum. Genet.* **56,** 907–914.
54. Gerace, L. (1995). Nuclear export signals and the fast track to the cytoplasm. *Cell* **82,** 341–344.
55. Giangreco, C. A., Steele, M. W., Aston, C. E., Cummins, J. H., and Wenger, S. L. (1996). A simplified six-item checklist for screening for fragile X syndrome in the pediatric population. *J. Pediatr.* **129,** 611–614.
56. Gibson, T. J., Rice, P. M., Thompson, J. D., and Heringa, J. (1993). KH domains within the FMR1 sequence suggest that fragile X syndrome stems from a defect in RNA metabolism. *TIBS* **18,** 331–333.
57. Gorlich, D., and Mattaj, I. W. (1996). Nucleocytoplasmic transport. *Science* **271,** 1513–1518.
58. Gu, Y., Lugenbeel, K. A., Vockley, J. G., Grody, W. W., and Nelson, D. L. (1994). A de novo deletion in FMR1 in a patient with developmental delay. *Hum. Mol. Genet.* **3,** 1705–1706.
59. Haataja, R., Väisänen, M. L., Li, M., Ryynänen, M., and Leisti, J. (1994). The fragile X syndrome in Finland: demonstration of a founder effect by analysis of microsatellite haplotypes. *Hum. Genet.* **94,** 479–483.
60. Haddad, L. A., Mingroni-Netto, R. C., Vianna-Morgante, A. M., and Pena, S. D. (1996). A PCR-based test suitable for screening for fragile X syndrome among mentally retarded males. *Hum. Genet.* **97,** 808–812.
61. Hamel, B. C. J., Smits, A. P. T., de Graaff, E., Smeets, D. F. C. M., Schoute, F., Eussen, B. H. J., Knight, S. J. L., Davies, K. E., Assman-Hulsmans, C. F. C. H., and Oostra, B. A. (1994). Segregation of FRAXE in a large family: clinical, psychometric, cytogenetic, and molecular data. *Am. J. Hum. Genet.* **55,** 923–931.
62. Hansen, R. S., Gartler, S. M., Scott, C. R., Chen, S. H., and Laird, C. D. (1992). Methylation analysis of CGG sites in the CpG island of the human FMR1 gene. *Hum. Mol. Genet.* **1,** 571–578.
63. Hansen, R. S., Canfield, T. K., Lamb, M. M., Gartler, S. M., and Laird, C. D. (1993). Association of fragile X syndrome with delayed replication of the FMR1 gene. *Cell* **73,** 1403–1409.
64. Hansen, R. S., Canfield, T. K., Fjeld, A. D., Mumm, S., Laird, C. D., and Gartler, S. M. (1997). A variable domain of delayed replication in FRAXA fragile X chromosomes: X inactivation-like spread of late replication. *Proc. Natl. Acad. Sci. USA* **94,** 4587–4592.
65. Harley, H. G., Brook, J. D., Rundle, S. A., Crow, S., Reardon, W., Buckler, A. J., Harper, P. S., Housman, D. E., and Shaw, D. J. (1992). Expansion of an unstable DNA region and phenotypic variation in myotonic dystrophy. *Nature* **355,** 545–546.
66. Heitz, D., Rousseau, F., Devys, D., Saccone, S., Abderrahim, H., Le Paslier, D., Cohen, D., Vincent, A., Toniolo, D., Dellamonica, C., Johnson, S., Schlessinger, D., Oberl—, I., and Mandel, J. L. (1991). Isolation of sequences that span the fragile X and identifiation of a fragiles X-related CpG island. *Science* **251,** 1236–1239.
67. Heitz, D., Devys, D., Imbert, G., Kretz, C., and Mandel, J. L. (1992). Inheritance of the fragile X syndrome: size of the fragile X premutation is a major determinant to full mutation. *J. Med. Genet.* **29,** 794–801.
68. Henry, M. F., and Silver, P. A. (1996). A novel methyltransferase (Hmt1p) modifies poly(A)+-RNA-binding proteins. *Mol. Cell. Biol.* **16,** 3668–3678.
69. Hilbert, P., and Sabine, M. (1996). Variation of CGG repeats at the fragile X site: clear signal enhancement using Expand(TM) Long Template. *Biochemica* **4,** 29.
70. Hildes, E., Jacobs, H. K., Cameron, A., Seshia, S. S., Booth, F., Evans, J. A., Wrogemann, K., and Greenberg, C. R. (1993). Impact of genetic counselling after neonatal screening for Duchenne muscular dystrophy. *J. Med. Genet.* **30,** 670–674.
71. Hinds, H. L., Ashley, C. T., Sutcliffe, J. S., Nelson, D. L., Warren, S. T., Housman, D. E., and Schalling, M. (1993). Tissue specific expression of FMR-1 provides evidence for a functional role in fragile X syndrome. *Nature Genet.* **3,** 36–43.
72. Hinton, V. J., Brown, W. T., Wisniewski, K., and Rudelli, R. D. (1991). Analysis of neocortex in three males with the fragile X syndrome. *Am. J. Med. Genet.* **41,** 289–294
73. Hirst, M. C., Grewal, P. K., and Davies, K. E. (1994). Precursor arrays for triplet repeat expansion at the fragile X locus. *Hum. Mol. Genet.* **3,** 1553–1560.
74. Hirst, M., Grewal, P., Flannery, A., Slatter, R., Maher, R., Maher, E., Barton, D., Fryns, J. P., and Davies, K. (1995). Two new cases of FMR1 deletion associated with mental impairment. *Am. J. Hum. Genet.* **56,** 67–74.
75. Hornstra, I. K., Nelson, D. L., Warren, S. T., and Yang, T. P. (1993). High resolution methylation analysis of the FMR1 gene trinucleotide repeat region in fragile X syndrome. *Hum. Mol. Genet.* **2,** 1659–1665.
76. Imbert, G., Kretz, C., Johnson, K., and Mandel, J. L. (1993). Origin of the expansion mutation in myotonic dystrophy. *Nature Genet.* **4,** 72–76.
77. Jain, R. G., Andrews, L. G., McGowan, K. M., Pekala, P. H., and Keene, J. D. (1997). Ectopic expression of Hel-N1, an RNA-binding protein, increases glucose transporter (GLUT1) expression in 3T3-L1 adipocytes. *Mol. Cell. Biol.* **17,** 954–962.
78. Jones, A. J., and Schedl, T. (1995). Mutations in gld-1, a female germ cell-specific tumor suppresser gene in Caenorhabditis elegans, affect a conserved domain also found in Src-associated protein Sam68. *EMBO J.* **9,** 1491–1504.
79. Johnston, D. S. (1995). The intracellular localization of messenger RNAs. *Cell* **81,** 161–170.
80. Kang, H., and Schuman, E. M. (1996). A requirement for local protein synthesis in neurotrophin-induced hippocampal synaptic plasticity. *Science* **273,** 1042–1046.
81. Khandjian, E. W., Corbin, F., Woerly, S., and Rousseau, F. (1996). The fragile X mental retardation protein is associated with ribosomes. *Nature Genet.* **12,** 91–93

82. Knight, S. J., Flannery, A. V., Hirst, M. C., Campbell, L., Christodoulou, Z., Phelps, S. R., Pointon, J., Middleton-Price, H. R., Barnicoat, A., Pembrey, M. E., *et al.* (1993). Trinucleotide repeat amplification and hypermethylation of a CpG island in FRAXE mental retardation. *Cell* **74,** 127–134.
83. Knight, S. J., Voelckel, M. A., Hirst, M. C., Flannery, A. V., Moncla, A., and Davies, K. E. (1994). Triplet repeat expansion at the FRAXE locus and X-linked mild mental handicap. *Am. J. Hum. Genet.* **55,** 81–86.
84. Kolehmainen, K. (1994). Population genetics of fragile X: a multiple allele model with variable risk of CGG repeat expansion. *Am. J. Med. Genet.* **51,** 428–435.
85. Kremer, E. J., Pritchard, M., Lynch, M., Yu, S., Holman, K., Baker, E., Warren, S. T., Schlessinger, D., Sutherland, G. R., and Richards, R. I. (1991). Mapping of DNA instability at the Fragile X to a trinucleotide repeat sequence (pCCG)n. *Science* **252,** 1711–1714.
86. Kuryavyi, V. V., and Jovin, T. M. (1995). Triad-DNA: a model for trinucleotide repeats. *Nature Genet.* **9,** 339–341
87. Kruyer, H., Mila, M., Glover, G., Carbonell, P., Ballesta, F., and Estivill, X. (1994). Fragile X syndrome and the (CGG)n mutation: two families with discordant MZ twins. *Am. J. Hum. Genet.* **54,** 437–442.
88. Kunst, C. B., and Warren, S. T. (1994). Cryptic and polar variation of the fragile X repeat could result in predisposing normal alleles. *Cell* **77,** 853–861.
89. Kunst, C. B., Zerylnick, C., Karickhoff, L., Eichler, E., Bullard, J., Chalifoux, M., Holden, J. J., Torroni, A., Nelson, D. L., and Warren, S. T. (1996). FMR1 in global populations. *Am. J. Hum. Genet.* **58,** 513–522.
90. Lachiewicz, A. M., Spiridigliozzi, G. A., McConkie-Rosell, A., Burgess, D., Feng, Y., Warren, S. T., and Tarleton, J. (1996). A fragile X male with a broad smear on Southern blot analysis representing 100–500 CGG repeats and no methylation at the EagI site of the FMR-1 gene. *Am. J. Med. Genet.* **64,** 278–282.
91. Lubs, H. A. (1969). A marker X chromosome. *Am. J. Hum. Genet.* **21,** 231–244.
92. Lugenbeel, K. A., Peier, A. M., Carson, N. L., Chudley, A. E., and Nelson, D. L. (1995). Intragenic loss of function mutations demonstrate the primary role of FMR1 in fragile X syndrome. *Nature Genet.* **10,** 483–485.
93. Luo, S., Robinson, J. C., Reiss, A. L., and Migeon, B. R. (1993). DNA methylation of the fragile X locus in somatic and germ cells during fetal development: relevance to the fragile X syndrome and X inactivation. *Somat. Cell Mol. Genet.* **19,** 393–404.
94. Lyon, M. F. (1993). Controlling the X chromosome. *Curr. Biol.* **3,** 242–244.
95. Macpherson, J. N., Nelson, D. L., and Jacobs, P. A. (1992). Frequent small amplifications in the FMR-1 gene in fra(X) families: limits to the diagnosis of 'premutations'. *J. Med. Genet.* **29,** 802–806.
96. Macpherson, J. N., Bullman, H., Youings, S. A., and Jacobs, P. A. (1994). Insert size and flanking haplotype in fragile X and normal populations: possible multiple origins for the fragile X mutation. *Hum. Mol. Genet.* **3,** 399–405.
97. Maddalena, A., Yadvish, K. N., Spence, W. C., and Howard-Peebles, P. N. (1996). A fragile X mosaic male with a cryptic full mutation detected in epithelium but not in blood. *Am. J. Med. Genet.* **64,** 309–312.
98. Mahone, M., Saffman, E. E., and Lasko, P. F. (1995). Localized Bicaudal-C RNA encodes a protein containing a KH domain, the RNA binding motif of FMR1. *EMBO J.* **14,** 2043–2055.
99. Malmgren, H., Steen-Bondeson, M. L., Gustavson, K. H., Seemanova, E., Holmgren, G., Oberle, I., Mandel, J. L., Pettersson, U., and Dahl, N. (1992). Methylation and mutation patterns in the fragile X syndrome. *Am. J. Med. Genet.* **43,** 268–278.
100. Malmgren, H., Gustavson, K. H., Oudet, C., Holmgren, G., Pettersson, U., and Dahl, N. (1994). Strong founder effect for the fragile X syndrome in Sweden. *Eur. J. Hum. Genet.* **2,** 103–109.
101. Malter, H. E., Iber, J. C., Willemsen, R., de Graaff, E., Tarleton, J. C., Leisti, J., Warren, S. T., and Oostra, B. A. (1997). Characterization of the full fragile X syndrome mutation in fetal gametes. *Nature Genet.* **15,** 165–169.
102. Mandel, J.-L. (1994). Trinucleotide diseases on the rise. *Nature Genet.* **7,** 453–455.
103. Marquis Gacy, A., Goellner, G., Juranic, N., Macura, S., and McMurray, C. T. (1995). Trinucleotide repeats that expand in human disease form hairpin structures in vitro. *Cell* **81,** 533–540.
104. Mattaj, I. W. (1993). RNA recognition: a family matter? *Cell* **73,** 837–840.
105. McConkie-Rosell, A., Lachiewicz, A. M., Spiridigliozzi, G. A., Tarleton, J., Schoenwald, S., Phelan, M. C., Goonewardena, P., Ding, X., and Brown, W. T. (1993). Evidence that methylation of the FMR-I locus is responsible for variable phenotypic expression of the fragile X syndrome. *Am. J. Hum. Genet.* **53,** 800–809.
106. Meijer, H., de Graaff, E., Merckx, D. M., Jongbloed, R. J., de Die-Smulders, C. E., Engelen, J. J., Fryns, J. P., Curfs, P. M., and Oostra, B. A. (1994). A deletion of 1.6 kb proximal to the CGG repeat of the FMR1 gene causes the clinical phenotype of the fragile X syndrome. *Hum. Mol. Genet.* **3,** 615–620.
107. Michael, W. M., Choi, M., and Dreyfuss, G. (1995). A nuclear export signal in hnRNP A1: a signal-mediated, temperature-dependent nuclear protein export pathway. *Cell* **83,** 415–422.
108. Miyashiro, K., Dichter, M., and Eberwine, J. (1994). On the nature and differential distribution of mRNAs in hippocampal neurites: implications for neuronal functioning. *Proc. Natl. Acad. Sci. USA* **91,** 10800–10804.
109. Mornet, E., Jokic, M., Bogyo, A., Tejada, I., Deluchat, C., Boue, J., and Boue, A. (1993). Affected sibs with fragile X syndrome exhibit an age-dependent decrease in the size of the fragile X full mutation. *Clin. Genet.* **43,** 157–159.
110. Mornet, E., Chateau, C., Taillandier, A., Montagnon, M., Simon-Bouy, B., Serre, J. L., and Boue, A. (1994). FRAXAC2 instability. *Nature Genet.* **7,** 122–123. [Letter]
111. Morris, A., Morton, N. E., Collins, A., MacPherson, J., Nelson, D., and Sherman, S. (1995). An n-allele model for progressive amplification in the FMR1 locus. *Proc. Natl. Acad. Sci. USA* **92,** 4833–4837.
112. Morton, N. E., and Macpherson, J. N. (1992). Population genetics of the fragile-X syndrome: multiallelic model for the FMR1 locus. *Proc. Natl. Acad. Sci. USA* **89,** 4215–4217.
113. Morton, J. E., Bundey, S., Webb, T. P., MacDonald, F., Rindl, P. M., and Bullock, S. (1997). Fragile X syndrome is less common than previously estimated. *J. Med. Genet.* **34,** 1–5.
114. Moutou, C., Vincent, M.-C., Biancalana, V., and Mandel, J.-L. (1997). Transition from premutation to full mutation in fragile X syndrome is likely to be prezygotic. *Hum. Mol. Genet.* **6,** 971–979.
115. Mulley, J. C., Yu, S., Loesch, D. Z., Hay, D. A., Donnelly, A., Gedeon, A. K., Carbonell, P., Lopez, I., Glover, G., Gabarron, I., Yu, P. W. L., Baker, E., Haan, E. A., Hockey, A., Knight, S. J. L., Davies, K. E., Richards, R. I., and Sutherland, G. R. (1995). FRAXE and mental retardation. *J. Med. Genet.* **32,** 162–169.
116. Murray, A., Youings, S., Dennis, N., Latsky, L., Linehan, P., McKechnie, N., MacPherson, J., Pound, M., and Jacobs, P. (1996). Population screening at the FRAXA and FRAXE loci: molecular analyses of boys with learning difficulties and their mothers. *Hum. Mol. Genet.* **5,** 727–735.

117. Murray, A., Macpherson, J. N., Pound, M. C., Sharrock, A., Youings, S. A., Dennis, N. R., McKechnie, N., Linehan, P., Morton, N. E., and Jacobs, P. A. (1997). The role of size, sequence and haplotype in the stability of FRAXA and FRAXE alleles during transmission. *Hum. Mol. Genet.* **6,** 173–184.

118. Musco, G., Stier, G., Joseph, C., Morelli, M. A. C., Nilges, M., Gibson, T. J., and Pastore, A. (1996). Three-dimensional structure and stability of the KH domain: molecular insights into the fragile X syndrome. *Cell* **85,** 237–245.

119. Myers, R. H., Mac Donald, M. E., Koroshetz, W. J., Duyao, M. P., Ambrose, C. M., Taylor, S. A., Barnes, G., Srinidhi, J., Lin, C. S., Whaley, W. L., *et al.* (1993). De novo expansion of a (CAG)n repeat in sporadic Huntington's disease. *Nature Genet.* **5,** 168–173.

120. Nancarrow, J. K., Kremer, E., Holman, K., Eyre, H., Doggett, N. A., Le Paslier, D., Callen, D. F., Sutherland, G. R., and Richards, R. I. (1994). Implications of FRA16A structure for the mechanism of chromosomal fragile site genesis. *Science* **264,** 1938–1941.

121. Nancarrow, J. K., Holman, K., Mangelsdorf, M., Hori, T., Denton, M., Sutherland, G. R., and Richards, R. I. (1995). Molecular basis of p(CCG)n repeat instability at the FRA16A fragile site locus. *Hum. Mol. Genet.* **4,** 367–372.

122. Nolin, S. L., Glicksman, A., Houck, G. E., Brown, W. T., and Dobkin, C. S. (1994). Mosaicism in fragile X affected males. *Am. J. Med. Genet.* **51,** 509–512.

123. Nolin, S. L., Lewis, F. A., Ye, L. L., Houck, G. E., Jr., Glicksman, A. E., Limprasert, P., Li, S. Y., Zhong, N., Ashley, A. E., Feingold, E., Sherman, S. L., and Brown, W. T. (1996). Familial transmission of the FMR1 CGG repeat. *Am. J. Hum. Genet.* **59,** 1252–1261.

124. Oberlé, I., Rousseau, F., Heitz, D., Kretz, C., Devys, D., Hanauer, A., Boue, J., Bertheas, M. F., and Mandel, J. L. (1991). Instability of a 550-base pair DNA segment and abnormal methylation in fragile X syndrome. *Science* **252,** 1097–1102.

125. Oostra, B. A., Jacky, P. B., Brown, W. T., and Rousseau, F. (1993). Guidelines for the diagnosis of fragile X syndrome. National Fragile X Foundation. *J. Med. Genet.* **30,** 410–413.

126. Oudet, C., Mornet, E., Serre, J. L., Thomas, F., Lentes-Zingerling, S., Kretz, C., Deluchat, C., Tejada, I., Boué, J., Boué, A., and Mandel, J. L. (1993). Linkage disequilibrium between the fragile X mutation and two closely linked CA repeats suggests that fragile X chromosomes are derived from a small number of founder chromosomes. *Am. J. Hum. Genet.* **52,** 297–304.

127. Oudet, C., von Koskull, H., Nordstrom, A. M., Peippo, M., and Mandel, J. L. (1993). Striking founder effect for the fragile X syndrome in Finland. *Eur. J. Hum. Genet.* **1,** 181–189.

128. Parrish, J. E., Oostra, B. A., Verkerk, A. J. M. H., Richards, C. S., Reynolds, J., Spikes, A. S., Shaffer, L. G., and Nelson, D. L. (1994). Isolation of a GCC repeat showing expansion in FRAXF, a fragile site distal to FRAXA and FRAXE. *Nature Genet.* **8,** 229–235.

129. Partington, M. W., Moore, D. Y., and Turner, G. M. (1996). Confirmation of early menopause in fragile X carriers. *Am. J. Med. Genet.* **64,** 370–372.

130. Pieretti, M., Zhang, F., Fu, Y.-H., Warren, S. T., Oostra, B. A., Caskey, C. T., and Nelson, D. L. (1991). Absence of expression of the FMR1 gene in fragile X syndrome. *Cell* **66,** 817–822.

131. Price, D. K., Zhang, F., Ashley, C. T., Jr., and Warren, S. T. (1996). The chicken FMR1 gene is highly conserved with a CCT 5′ untranslated repeat and encodes an RNA-binding protein. *Genomics* **31,** 3–12.

132. Quan, F., Zonana, J., Gunter, K., Peterson, K. L., Magenis, R. E., and Popovich, B. W. (1995). An atypical case of fragile X syndrome caused by a deletion that includes the FMR1 gene. *Am. J. Hum. Genet.* **56,** 1042–1051.

133. Reiss, A. L., Kazazian, H. H., Jr., Krebs, C. M., McAughan, A., Boehm, C. D., Abrams, M. T., and Nelson, D. L. (1994). Frequency and stability of the fragile X premutation. *Hum. Mol. Genet.* **3,** 393–398.

134. Reiss, A. L., Abrams, M. T., Greenlaw, R., Freund, L., and Denckla, M. B. (1995). Neurodevelopmental effects of the FMR-1 full mutation in humans. *Nature Med.* **1,** 159–167.

135. Reyniers, E., Vits, L., De Boulle, K., Van Roy, B., Van Velzen, D., de Graaff, E., Verkerk, A. J., Jorens, H. Z., Darby, J. K., Oostra, B., and Willems, P. J. (1993). The full mutation in the FMR-1 gene of male fragile X patients is absent in their sperm. *Nature Genet.* **4,** 143–146.

136. Richards, R. I., Holdman, K., Friend, K., Kremer, E., Hillen, D., Staples, A., Brown, W. T., Goonewardena, P., Tarleton, J., Schwartz, C., and Sutherland, G. R. (1992). Evidence of founder chromosomes in fragile X syndrome. *Nature Genet.* **1,** 257–260.

137. Richards, R. I., Holman, K., Yu, S., and Sutherland, G. R. (1993). Fragile X syndrome unstable element, p(CCG)n, and other simple tandem repeat sequences are binding sites for specific nuclear proteins. *Hum. Mol. Genet.* **2,** 1429–1435.

138. Richards, R. I., Kondo, I., Holman, K., Yamauchi, M., Seki, N., Kishi, K., Staples, A., Sutherland, G. R., and Hori, T-A. (1994). Haplotype analysis at the FRAXA locus in the Japanese population. *Am. J. Med. Genet.* **51,** 412–416.

139. Richards, R. I., Holman, K., Friend, K., Staples, A., Sutherland, G. R., Oudet, C., Biancalana, V., and Mandel, J. L. (1994). FRAXAC2 instability. *Nature Genet.* **7,** 122. [Letter]

140. Richards, R. I., and Sutherland, G. R. (1994). Simple repeat DNA is not replicated simply. *Nature Genet.* **6,** 114–116.

141. Riggins, G. J., Sherman, S. L., Oostra, B. A., Sutcliffe, J. S., Feitell, D., Nelson, D. L., van Oost, B. A., Smits, A. P. T., Ramos, F. J., Pfendner, E., Kuhl, D. P. A., Caskey, C. T., and Warren, S. T. (1992). Characterization of a highly polymorphic dinucleotide repeat 150 kb proximal to the fragile X site. *Am. J. Med. Genet.* **43,** 237–243.

142. Ritchie, R. J., Knight, S. J. L., Hirst, M. C., Grewal, P. K., Bobrow, M., Cross, G. S., and Davies, K. E. (1994). The cloning of FRAXF: trinucleotide repeat expansion and methylation at a third fragile site in distal Xqter. *Hum. Mol. Genet.* **3,** 2115–2121.

143. Ross, A. F. Oleynikov, Y. Kislauskis, E. H. Taneja, K. L., and Singer, R. H. (1997). Characterization of a beta-actin mRNA zipcode-binding protein. *Mol. Cell. Biol.* **17,** 2158–65

144. Rousseau, F., Heitz, D., Biancalana, V., Blumenfeld, S., Kretz, C., Boué, J., Tommerup, N., Van der Hagen, C., DeLozier-Blanchet, C., Croquette, M.-F., Gilgenkrantz, S., Jalbert, P., Voelckel, M. A., Oberlé, I., and Mandel, J. L. (1991). Direct diagnosis by DNA analysis of the fragile X syndrome of mental retardation. *N. Engl. J. Med.* **325,** 1673–1681.

145. Rousseau, F., Heitz, D., Oberle, I., and Mandel, J. L. (1991). Selection in blood cells from female carriers of the fragile X syndrome: inverse correlation between age and proportion of active X chromosomes carrying the full mutation. *J. Med. Genet.* **28,** 830–836.

146. Rousseau, F., Heitz, D., Biancalana, V., Oberle, I., and Mandel, J. L. (1992). On some technical aspects of direct DNA diagnosis of the fragile X syndrome. *Am. J. Med. Genet.* **43,** 197–207.

147. Rousseau, F., Heitz, D., Tarleton, J., MacPherson, J., Malmgren, H., Dahl, N., Barnicoat, A., Mathew, C., Mornet, E., Tejada, I., *et al.* (1994). A multicenter study on genotype-phenotype correlations in the fragile X syndrome, using direct diagnosis with probe StB12.3: the first 2,253 cases. *Am. J. Hum. Genet.* **55,** 225–237.

148. Rousseau, F., Rehel, R., Rouillard, P., De Granpre, P., and Khandjian, E. W. (1994). High throughput and economical mutation detection and RFLP analysis using a minimethod for DNA preparation from whole blood and acrylamide gel electrophoresis. *Hum. Mutat.* **4,** 51–54.
149. Rousseau, F., Robb, L. J., Rouillard, P., and Der Kaloustian, V. M. (1994). No mental retardation in a man with 40% abnormal methylation at the FMR-1 locus and transmission of sperm cell mutations as premutations. *Hum. Mol. Genet.* **3,** 927–930.
150. Rousseau, F., Rouillard, P., Morel, M. L., Khandjian, E. W., and Morgan, K. (1995). Prevalence of carriers of premutation-size alleles of the FMR1 gene and implications for the population genetics of the fragile X syndrome. *Am. J. Hum. Genet.* **57,** 1006–1018.
151. Rudelli, R. D., Brown, W. T., Wisniewski, K., Jenkins, E. C., Laure-Kamionowska, M., and Connel, F. (1985). Adult Fragile X Syndrome. *Acta Neuropathol.* **67,** 289–295.
152. Schmucker, B., Ballhausen, W. G., and Pfeiffer, R. A. (1996). Mosaicism of a microdeletion of 486 bp involving the CGG repeat of the FMR1 gene due to misalignment of GTT tandem repeats at chi-like elements flanking both breakpoints and a full mutation. *Hum. Genet.* **98,** 409–414.
153. Schwartz, C. E., Dean, J., Howard-Peebles, P. N., Bugge, M., Mikkelsen, M., Tommerup, N., Hull, C., Hagerman, R., Holden, J. J., and Stevenson, R. E. (1994). Obstetrical and gynecological complications in fragile X carriers: a multicenter study. *Am. J. Med. Genet.* **51,** 400–402.
154. Sherman, S. L., Morton, N. E., Jacobs, P. A., and Turner, G. (1984). The marker (X) syndrome: a cytogenetic and genetic analysis. *Ann. Hum. Genet.* **48,** 21–37.
155. Sherman, S. L., Jacobs, P. A., Morton, N. E., Froster-Iskenius, U., Howard-Peebles, P. N., Brondum-Nielsen, K., Partington, M. W., Sutherland, G. R., Turner, G., and Watson, M. (1985). Further segregation analysis of the Fragile X syndrome with special reference to transmitting males. *Hum. Genet.* **69,** 289–299.
156. Sherman, S. L. (1991). Genetic epidemiology of the fragile X syndrome with special reference to genetic counseling. *Prog. Clin. Biol. Res.* **368,** 79–99.
157. Sherman, S. L., Meadows, K. L., and Ashley, A. E. (1996). Examination of factors that influence the expansion of the fragile X mutation in a sample of conceptuses from known carrier females. *Am. J. Med. Genet.* **64,** 256–260.
158. Siomi, H., Matunis, M. J., Michael, W. M., and Dreyfuss, G. (1993). The pre-mRNA binding K protein contains a novel evolutionarily conserved motif. *Nucleic Acids. Res.* **21,** 1193–1198.
159. Siomi, H., Siomi, M. C., Nussbaum, R. L., and Dreyfuss, G. (1993). The protein product of the fragile X gene, FMR1, has characteristics of an RNA-binding protein. *Cell* **74,** 291–298.
160. Siomi, H., Choi, M., Siomi, M., Nussbaum, R., and Dreyfuss, G. (1994). Essential role for KH domains in RNA binding: Impaired RNA binding by a mutation in the KH domains of FMR1 that causes Fragile X Syndrome. *Cell* **77,** 33–39.
161. Siomi, M. C., Siomi, H., Sauer, W. H., Srinivasan, S., Nussbaum, R. L., and Dreyfuss, G. (1995). FXR1, an autosomal homolog of the fragile X mental retardation gene. *EMBO J.* **14,** 2401–2408.
162. Siomi, M. C., Zhang, Y., Siomi, H., and Dreyfuss, G. (1996). Specific sequences in the fragile X syndrome protein FMR1 and the FXR proteins mediate their binding to 60S ribosomal subunits and the interactions among them. *Mol. Cell Biol.* **16,** 3825–3832.
163. Smith, S. S., Laayoun, A., Lingeman, R. G., Baker, D. J., and Riley, J. (1994). Hypermethylation of telomere-like foldbacks at codon 12 of the human c-Ha-ras gene and the trinucleotide repeat of the FMR-1 gene of fragile X. *J. Mol. Biol.* **243,** 143–151.
164. Smits, A. P., Dreesen, J. C., Post, J. G., Smeets, D. F., de Die-Smulders, C., Spaans-van der Bijl, T., Govaerts, L. C., Warren, S. T., Oostra, B. A., and Van Oost, B. A. (1993). The fragile X syndrome: no evidence for any recent mutations. *J. Med. Genet.* **30,** 94–96.
165. Snow, K., Doud, L. K., Hagerman, R., Pergolizzi, R. G., Erster, S. H., and Thibodeau, S. N. (1993). Analysis of a CGG sequence at the FMR-1 locus in fragile X families and in the general population. *Am. J. Hum. Genet.* **53,** 1217–1228.
166. Snow, K., Tester, D. J., Kruckeberg, K. E., Schaid, D. J., and Thibodeau, S. N. (1994). Sequence analysis of the fragile X trinucleotide repeat: implications for the origin of the fragile X mutation. *Hum. Mol. Genet.* **3,** 1543–1551.
167. Steward, O. (1994). Dendrites as compartments for macromolecular synthesis. *Proc. Natl. Acad. Sci. USA* **91,** 10766–10768.
168. Steward, O., and Reeves, T. (1988). Protein-synthetic machinery beneath postsynaptic sites on CNS neurons: association between polyribosomes and other organelles at the synaptic site. *J. Neurosci.* **8,** 176–184.
169. Subramanian, P. S., Nelson, D. L., and Chinault, A. C. (1996). Large domains of apparent delayed replication timing associated with triplet repeat expansion at FRAXA and FRAXE. *Am. J. Hum. Genet.* **59,** 407–416
170. Sutcliffe, J. S., Nelson, D. L., Zhang, F., Pieretti, M., Caskey, C. T., Saxe, D., and Warren, S. T. (1992). DNA methylation represses FMR-1 transcription in fragile X syndrome. *Hum. Mol. Genet.* **1,** 397–400.
171. Sutherland, G. R. (1977). Fragile sites on human chromosomes: demonstration of their dependence on the type of tissue culture medium. *Science* **197,** 265–266.
172. Sutherland, G. R., Gedeon, A., Kornman, L., Donnelly, A., Byard, R. W., Mulley, J. C., Kremer, E., Lynch, M., Pritchard, M., Yu, S., and Richards, R. I. (1991). Prenatal diagnosis of fragile X syndrome by direct detection of the unstable DNA sequence. *N. Engl. J. Med.* **325,** 1720–1722. [See comments]
173. Sutherland, G. R., and Richards, R. I. (1995). Simple tandem DNA repeats and human genetic disease. *Proc. Natl. Acad. Sci. USA* **92,** 3636–3641.
174. Tamanini, F., Meijer, N., Verheij, C., Willems, P. J., Galjaard, H., Oostra, B. A., and Hoogeveen, A. T. (1996). FMRP is associated to the ribosomes via RNA. *Hum. Mol. Genet.* **5,** 809–813.
175. Tamanini, F., Willemsen, R., van Unen, L., Bontekoe, C., Galjaard, H., Oostra, B. A., and Hoogeveen, A. T. (1997). Differential expression of FMR1, FXR1 and FXR2 proteins in human brain and testis. *Hum. Mol. Genet.* **6,** 1315–1322.
176. Tarleton, J., Richie, R., Schwartz, C., Rao, K., Aylsworth, A. S., and Lachiewicz, A. (1993). An extensive de novo deletion removing FMR1 in a patient with mental retardation and the fragile X syndrome phenotype. *Hum. Mol. Genet.* **2,** 1973–1974.
177. Taylor, A. K., Safanda, J. F., Fall, M. Z., Quince, C., Lang, K. A., Hull, C. E., Carpenter, I., Staley, L. W., and Hagerman, R. J. (1994). Molecular predictors of cognitive involvement in female carriers of fragile X syndrome [see comments]. *JAMA* **271,** 507–514.
178. The Dutch-Belgian fragile X consortium (1994). Fmr1 knockout mice: a model to study fragile X mental retardation. *Cell* **78,** 23–33.
179. Torchia, B. S., Call, L. M., and Migeon, B. R. (1994). DNA replication analysis of FMR1, XIST, and factor 8C loci by FISH shows nontranscribed X-linked genes replicate late. *Am. J. Hum. Genet.* **55,** 96–104.
180. Tranebjaerg, L., Hilling, S., Jessen, J., Lind, D., and Hansen, M. S. (1994). Prevalence of fra(X) in the county of Funen in Denmark is lower expected. *Am. J. Med. Genet.* **51,** 423–427.

181. Trottier, Y., Imbert, G., Poustka, A., Fryns, J. P., and Mandel, J. L. (1994). Male with typical fragile X phenotype is deleted for part of the FMR1 gene and for about 100kb of upstream region. *Am. J. Med. Genet.* **51,** 454–457.
182. Turner, G., Robinson, H., Laing, S., van den Berk, M., Colley, A., Goddard, A., Sherman, S., and Partington, M. (1992). Population screening for fragile X. *Lancet* **339,** 1210–1213.
183. Turner, G., Robinson, H., Wake, S., and Martin, N. (1994). Dizygous twinning and premature menopause in fragile X syndrome. *Lancet* **344,** 1500. [Letter; see comments]
184. Turner, G., Webb, T., Wake, S., and Robinson, H. (1996). Prevalence of fragile X syndrome. *Am. J. Med. Genet.* **64,** 196–197.
185. van den Ouweland, A. M. W., de Vries, B. B. A., Bakker, P. L. G., Deelen, W. H., de Graaff, E., Van Hemel, J. O., Oostra, B. A., Niermeijer, M. F., and Halley, D. J. J. (1994). DNA diagnosis of the fragile X syndrome in a series of 236 mentally retarded subjects and evidence for a reversal of mutation in the FMR-1 gene. *Am. J. Med. Genet.* **51,** 482–485.
186. van den Ouweland, A. M. W., Deelen, W. H., Kunst, C. B., Giovannucci Uzielli, M. L., Nelson, D. L., Warren, S. T., Oostra, B. A., and Halley, D. J. J. (1994). Loss of mutation at the FMR1 locus through multiple exchanges between maternal X chromosomes. *Hum. Mol. Genet.* **3,** 1823–1827.
187. Verheij, C., Bakker, C. E., de Graaff, E., Keulemans, J., Willemsen, R., Verkerk, A. J. M. H., Galjaard, H., Reuser, A. J. J., Hoogeveen, A. T., and Oostra, B. A. (1993). Characterization and localization of the FMR-1 gene product associated with the fragile X syndrome. *Nature* **363,** 722–724.
188. Verkerk, A. J. M. H., Pieretti, M., Sutcliffe, J. S., Fu, Y. H., Kuhl, D. P. A., Pizzuti, A., Reiner, O., Richards, S., Victoria, M. F., Zhang, F., Eussen, B. E., Van Ommen, G. J. B., Blonden, L. A. J., Riggins, G. J., Chastain, J. L., Kunst, C. B., Galjaard, H., Caskey, C. T., Nelson, D. L., Oostra, B. A., and Warren, S. T. (1991). Identification of a gene (FMR-1) containing a CGG repeat coincident with a breakpoint cluster region exhibiting length variation in fragile X syndrome. *Cell* **65,** 905–914.
189. Verkerk, A. M. J. H., de Graff, E., De Boulle, K., Eichler, E. E., Konecki, D. S., Reyniers, E., Manca, A., Poustka, A., Willems, P. J., Nelson, D. L., and Oostra, B. A. (1993). Alternative splicing in the fragile X gene FMR1. *Hum. Mol. Genet.* **2,** 399–404.
190. Vianna-Morgante, A. M., Costa, S. S., Pares, A. S., and Verreschi, I. T. (1996). FRAXA premutation associated with premature ovarian failure. *Am. J. Med. Genet.* **64,** 373–375.
191. Visa, N., Alzhanova-Ericsson, A. T., Sun, X., Kiseleva, E., Bjorkroth, B., Wurtz, T., and Daneholt, B. (1996). A pre-mRNA-binding protein accompanies the RNA from the gene through the nuclear pores and into polysomes. *Cell* **84,** 253–264.
192. Vits, L., De Boulle, K., Reyniers, E., Handig, I., Darby, J. K., Oostra, B., and Willems, P. J. (1994). Apparent regression of the CGG repeat in FMR1 to an allele of normal size. *Hum. Genet.* **94,** 523–526.
193. Väisänen, M. L., Kähkönen, M., and Leisti, J. (1994). Diagnosis of fragile X syndrome by direct mutation analysis. *Hum. Genet.* **93,** 143–147.
194. Wang, Q., Green, E., Bobrow, M., and Mathew, C. G. (1995). A rapid, non-radioactive screening test for fragile X mutations at the FRAXA and FRAXE loci. *J. Med. Genet.* **32,** 170–173.
195. Warren, S. T., Zhang, F., Licameli, G. R., and Peters, J. F. (1987). The fragile X site in somatic cell hybrids: an approach for molecular cloning of fragile sites. *Science* **237,** 420–423.
196. Webb, T. P., Bundey, S., Thake, A., and Todd, J. (1986). The frequency of the fragile X chromosome among schoolchildren in Coventry. *J. Med. Genet.* **23,** 396–399.
197. Webb, T., and Bundey, S. (1991). Prevalence of fragile X syndrome. *J. Med. Genet.* **28,** 358. [Letter; see comments]
198. Weiler, I. J., and Greenough, W. T. (1993). Metabotropic glutamate receptors trigger postsynaptic protein synthesis. *Neurobiology* **90,** 7168–7171.
199. Wells, R. D., and Sinden, R. R. (1993). Defined ordered sequence DNA, DNA structure, and DNA-directed mutation. *In* "Genome Analysis" (K. E. Davies and S. T. Warren, Eds.), pp. 107–137. Cold Spring Harbor Laboratory, Cold Spring Harbor, NY.
200. Wen, W., Meinkoth, J. L., Tsien, R. Y., and Taylor, S. S. (1995). Identification of a signal for rapid export of proteins from the nucleus. *Cell* **82,** 463–473.
201. Wilgenbus, K. K., Coy, J. F., Mincheva, A., Nicolai, H., Solomon, E., Lichter, P., and Poustka, A. (1996). Ordering of 66 STSs along the entire short arm of human chromosome 17 and chromosome assignment of a transcribed sequence (FMR1L2) homologous to FMR1. *Cytogenet. Cell Genet.* **73,** 240–243.
202. Willems, P. J., Van Roy, B., De Boulle, K., Vits, L., Reyniers, E., Beck, O., Dumon, J. E., Verkerk, A., and Oostra, B. (1992). Segregation of the fragile X mutation from an affected male to his normal daughter. *Hum. Mol. Genet.* **1,** 511–515.
203. Willemsen, R., Mohkamsing, S., De Vries, B., Devys, D., van den Ouweland, A., Mandel, J. L., Galjaard, H., and Oostra, B. (1995). Rapid antibody test for fragile X syndrome. *Lancet* **345,** 1147–1148.
204. Willemsen, R., Oosterwijk, J. C., Los, F. J., Galjaard, H., and Oostra, B. A. (1996). Prenatal diagnosis of fragile X syndrome. *Lancet* **348,** 967–968. [Letter]
205. Willemsen, R., Smits, A., Mohkamsing, S., van Beerendonk, H., de Haan, A., De Vries, B., van den Ouweland, A., Sistermans, E., Galjaard, H., and Oostra, B. A. (1997). Rapid antibody test for diagnosing fragile X syndrome: a validation of the technique. *Hum. Genet.* **99,** 308–311.
206. Willemsen, R., Los, F., Mohkamsing, S., van den Ouweland, A., Deelen, W., Galjaard, H., and Oostra, B. (1997). Rapid antibody test for prenatal diagnosis of fragile X syndrome on amniotic fluid cells: a new appraisal. *J. Med. Genet.* **34,** 250–251.
207. Wörhle, D., Kotzot, D., Hirst, M. C., Manca, A., Korn, B., Schmidt, A., Barbi, G., Rott, H.-D., Poustka, A., Davies, K. E., and Steinbach, P. (1992). A microdeletion of less than 250 kb, including the proximal part of the FMR-1 gene and the fragile-X site, in a male with the clinical phenotype of fragile-X syndrome. *Am. J. Hum. Genet.* **51,** 299–306.
208. Wöhrle, D., Hennig, I., Vogel, W., and Steinbach, P. (1993). Mitotic stability of fragile X mutations in differentiated cells indicates early post-conceptional trinucleotide repeat expansion. *Nature Genet.* **4,** 140–142. [See comments]
209. Wöhrle, D., Kennerknecht, I., Wolf, M., Enders, H., Schwemmle, S., and Steinbach, P. (1995). Heterogeneity of DM kinase repeat expansion in different fetal tissues and further expansion during cell proliferation in vitro: evidence for a causal involvement of methyl-directed DNA mismatch repair in triplet stability. *Hum. Mol. Genet.* **4,** 1147–1153.
210. Yamagata, H., Miki, T., Ogihara, T., Nakagawa, M., Higuchi, I., Osame, M., Shelbourne, P., Davies, J., and Johnson, K. (1992). Expansion of unstable DNA region in Japanese myotonic dystrophy patients. *Lancet* **339,** 692.
211. Yu, S., Pritchard, M., Kremer, E., Lynch, M., Nancarrow, J., Baker, E., Holman, K., Mulley, J. C., Warren, S. T., Schlessinger, D., Sutherland, G. R., and Richards, R. I. (1991). Fragile X

genotype characterized by an unstable region of DNA. *Science* **252,** 1179–1181.

212. Yu, S., Mulley, J., Loesch, D., Turner, G., Donelly, A., Gedeon, A., Hillen, D., Kremer, E., Lynch, M., Pritchard, M., Sutherland, G. R., and Richards, R. I. (1992). Fragile-X syndrome: unique genetics of the heritable unstable element. *Am. J. Hum. Genet.* **50,** 968–980.

213. Zhang, Y., O'Connor, P., Siomi, M., Srinivasan, S., Dutra, A., Nussbaum, R. L., and Dreyfuss, G. (1995). The fragile X mental retardation syndrome protein interacts with novel homologs FXR1 and FXR2. *EMBO J.* **14,** 5358–5366.

214. Zhong, N., Dobkin, C., and Brown, W. T. (1993). A complex mutable polymorphism located within the fragile X gene. *Nature Genet.* **5,** 248–253.

215. Zhong, N., Liu, X., Gou, S., Houck, G. E., Li, S., Dobkin, C., and Brown, W. T. (1994). Distribution of FMR-1 and associated microsatellite alleles in a normal Chinese population. *Am. J. Med. Genet.* **51,** 417–422.

216. Zhong, N., Yang, W., Dobkin, C., and Brown, W. T. (1995). Fragile X gene instability: anchoring AGGs and linked microsatellites. *Am. J. Hum. Genet.* **57,** 351–361. [See comments]

217. Bardoni, B., Sittler, A., Shen, Y., and Mandel, J.-L. (1997). Analysis of domains affecting intracellular localization of the FMRP protein. *Neurobiol. Disease* **4,** 329–336.

218. Chen, X., Mariappan, S. V., Catasti, P., Ratliff, R., Moyzis, R. K., Laayoun, A., Smith, S. S., Bradbury, E. M., and Gupta, G. (1995). Hairpins are formed by the single DNA strands of the fragile X triplet repeats: structure and biological implications. *Proc. Natl. Acad. Sci. USA* **92,** 5199–5203.

219. Chiurazzi, P., Pomponi, M. G., Willemsen, R., Oostra, B. A., and Neri, G. (1998). In vitro reactivation of the FMR1 gene involved in fragile X syndrome. *Hum. Mol. Genet.* **7,** 109–113.

220. Deissler, H., Wilm, M., Genc, B., Schmitz, B., Ternes, T., Naumann, F., Mann, M., and Doerfler, W. (1997). Rapid protein sequencing by tandem mass spectrometry and cDNA cloning of p20-CGGBP. A novel protein that binds to the unstable triplet repeat 5′-d(CGG)n-3′ in the human FMR1 gene. *J. Biol. Chem.* **272,** 16761–16768.

221. Drouin, R., Angers, M., Dallaire, N., Rose, T. M., Khandjian, E. W., and Rousseau, F. (1997). Structural and functional characterization of the human FMR1 promoter reveals similarities with the hnRNP-A2 promoter region. *Hum. Mol. Genet.* **6,** 2051–2060.

222. Fridell, R. A., Benson, R. E., Hua, J., Bogerd, H. P., and Cullen, B. R. (1996). A nuclear role for the fragile X mental retardation protein. *EMBO J.* **15,** 5408–5414.

223. Godde, J. S., Kass, S. U., Hirst, M. C., and Wolffe, A. P. (1996). Nucleosome assembly on methylated CGG triplet repeats in the fragile X mental retardation gene 1 promoter. *J. Biol. Chem.* **271,** 24325–24328.

224. Hegersberg, M., Matsuo, K., Gassmann, M., Schaffner, W., Lüscher, B., Rülicke, T., Aguzzi, A., *et al.* (1995). Tissue-specific expression of a FMR1/beta-galactosidase fusion gene in transgenic mice. *Hum. Mol. Genet.* **4,** 359–366.

225. Larsen, L. A., Gronskov, K., Norgaard-Pedersen, B., Brondum-Nielsen, K., Hasholt, L., and Vuust, J. (1997). High-throughput analysis of fragile X (CGG)n alleles in the normal and premutation range by PCR amplification and automated capillary electrophoresis. *Hum. Genet.* **100,** 564–568.

226. Lavedan, C. N., Garrett, L., and Nussbaum, R. L. (1997). Trinucleotide repeats (CGG)22TGG(CGG)43TGG(CGG)21 from the fragile X gene remain stable in transgenic mice. *Hum. Genet.* **100,** 407–414.

227. Schwemmle, S., de Graaff, E., Deissler, H., Glaser, D., Wöhrle, D., Kennerknecht, I., Just, W., Oostra, B. A., Dorfler, W., Vogel, W., and Steinbach, P. (1997). Characterization of FMR1 promoter elements by *in vivo*-footprinting analysis. *Am. J. Hum. Genet.* **60,** 1354–1362.

228. Sittler, A., Devys, D., Weber, C., and Mandel, J.-L. (1996). Alternative splicing of exon 14 determines nuclear or cytoplasmic localisation of fmr1 protein isoforms. *Hum. Mol. Genet.* **5,** 95–102.

229. Smeets, H. J., Smits, A. P., Verheij, C. E., Theelen, J. P., Willemsen, R., van de Burgt, I., Hoogeveen, A. T., Oosterwijk, J. C., and Oostra, B. A. (1995). Normal phenotype in two brothers with a full FMR1 mutation. *Hum. Mol. Genet.* **4,** 2103–2108.

230. Stöger, R., Kajimura, T. M., Brown, W. T., and Laird, C. D. (1997). Epigenetic variation illustrated by DNA methylation patterns of the fragile X gene FMR1. *Hum. Mol. Genet.* **11,** 1791–1801.

231. Wang, Y. C., Lin, M. L., Lin, S. J., Li, Y. C., and Li, S. Y. (1997). Novel point mutation within intron 10 of FMR-1 gene causing fragile X syndrome. *Hum. Mutat.* **10,** 393–399.

232. Wang, Y. H., and Griffith, J. (1996). Methylation of expanded CCG triplet repeat DNA from fragile X syndrome patients enhances nucleosome exclusion. *J. Biol. Chem.* **271,** 22937–22940.

233. Wang, Z., Taylor, A. K., and Bridge, J. A. (1996). FMR1 fully expanded mutation with minimal methylation in a high functioning fragile X male. *J. Med. Genet.* **33,** 376–378.

234. Weiler, I. J., Irwin, S. A., Klintsova, A. Y., Spencer, C. M., Brazelton, A. D., Miyashiro, K., Comery, T. A., Patel, B., Eberwine, J., and Greenough, W. T. (1997). Fragile X mental retardation protein is translated near synapses in response to neurotransmitter activation. *Proc. Natl. Acad. Sci. USA* **94,** 5395–5400.

235. Wolff, D. J., Gustashaw, K. M., Zurcher, V., Ko, L., White, W., Weiss, L., Van Dyke, D. L., Schwartz, S., and Willard, H. F. (1997). Deletions in Xq26.3-q27.3 including FMR1 result in a severe phenotype in a male and variable phenotypes in females depending upon the X inactivation pattern. *Hum. Genet.* **100,** 256–261.

CHAPTER 4

Murine Model of Fragile X Syndrome

BEN A. OOSTRA MGC Department of Clinical Genetics, Erasmus University, Rotterdam, The Netherlands
PATRICK J. WILLEMS Department of Medical Genetics, University of Antwerp, Antwerp, Belgium

I. ANIMAL MODEL FOR FRAGILE X SYNDROME

Clues about the mechanisms that cause the abnormalities observed in fragile X syndrome are missing. To gain more insight in the pathologic and physiologic processes, an animal model for the disease has been generated [1]. The development of an animal model has major advantages. First, the unlimited supply of tissues gives the opportunity to study the effects of the lack of FMRP expression on the morphological and molecular level. Second, the behavior of the knockout mice can be studied in order to understand the mechanisms involved in learning. Third, the knockout mice can be used in crossing experiments with transgenic mice for different mutations or with different expression patterns.

The FMR1 gene is highly conserved among species and the murine homolog Fmr1 shows 97% homology in amino acid sequence [2]. The expression patterns at the mRNA and protein level are very similar in humans and mice [3–7]. The function of the FMR1 gene is therefore thought to be similar in both species. The amplification of the CGG repeat causes silencing of the FMR1 gene in patients [8, 9]. Cloning an expanded CGG repeat in the full mutation range is extremely difficult because of technical restrictions and therefore an FMR1 knockout mouse was developed.

The knockout mice were generated by replacing the wild-type murine Fmr1 gene with a nonfunctional Fmr1 gene in which a neomycin resistance cassette was placed in exon 5, using homologous recombination in embryonic stem cells employing conventional transgenic ES technology (Fig. 4-1). These ES cells were injected into blastocysts and transferred to pseudopregnant females. Highly chimeric males were crossed with wild-type C57Bl6 females to give birth to females heterozygous for the knockout mutation. Breeding those females with wild-type males resulted in knockout males. As a result of the integration of the neo cassette in the Fmr1 gene the mutant mice are no longer able to make normal Fmr1 mRNA. Although the knockout mutation in the animal model is different from the mutation found in

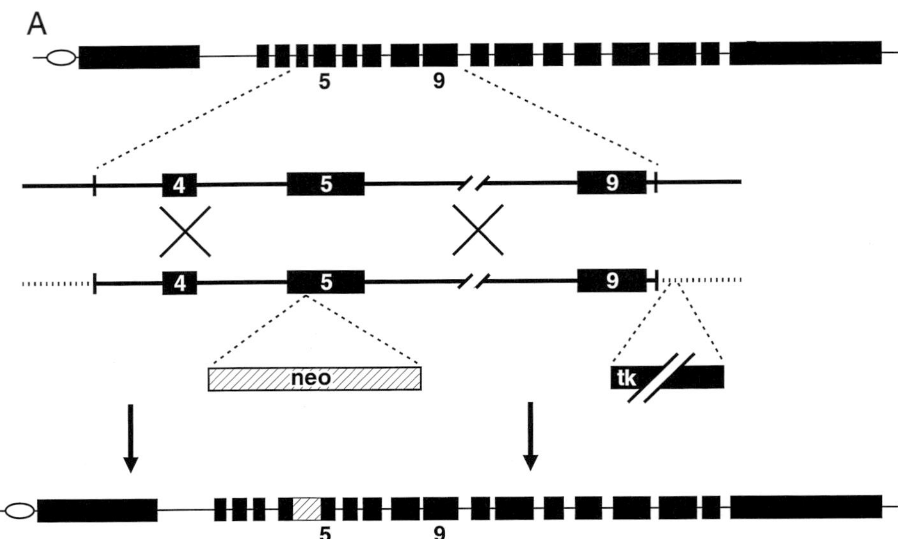

FIGURE 4-1 Schematic representation of the method used for the silencing of the endogenous Fmr1 gene in ES cells. (A) Homologous recombination of the target vector with the mouse Fmr1 gene leads to the integration of the neo cassette in the Fmr1 gene. Random integration of the target vector into the mouse genome leads to the integration of both the neo and tk cassettes. The addition of G418 and gangcyclovir to the culture medium eventually leads to the selection of clones where homologous recombination has occurred. (B) The ES cells in which the Fmr1 gene is silenced, through the introduction of the neo cassette, are introduced in blastocysts of normal mice. The blastocysts are transferred to foster mothers. Fmr1 knockout mice are obtained by crossing chimeric F1 mice with normal mice.

human fragile X patients, both mutations lead to an absence of the FMR1 protein (Fig. 4-2). Weight and light microscopic appearance of kidney, heart, spleen, liver, lung, or brain were not different between knockouts and normal littermates [1]. Other phenotypic characteristics such as the long face, prominent ears, high-arched palate, flat feet, hand calluses, and hyperextensible finger joints have not been found in fragile X mice [10]. No macroscopic or microscopic abnormalities could be detected in complete autopsies of knockout mice. The weights of several organs of mutant mice except testis did not significantly differ from those of control mice [11]. No significant statistical difference in weight was found between age-matched groups of control and mutant mice [11]. The absence of FMRP did not influence the reproduction or viability of the knockout mice. Also, human fragile X patients have a normal life span.

Male knockout mice do not express the Fmr1 protein in any of their cells; in contrast, in fragile X males the Fmr1 protein is absent in all cells except their germ cells [12]. It was therefore suggested that FMR1 is essential for gametogenesis. The observation that both male and female knockout mice without any protein expression are fertile and have the same size of progeny as controls indicates that Fmr1 is not necessary for spermatogenesis and oogenesis in mice.

A. Macroorchidism

One of the most obvious phenotypic characteristics of fragile X patients is macroorchidism; sometimes this is manifested in childhood, but it is present in almost all fragile X patients after puberty [13, 14]. Histological studies of human fragile X testes have revealed limited abnormalities: testicular enlargement seems to be due to interstitial edema or increased amount of interstitial tissue [15–17]. Microscopic examination of the testes of mutant mice revealed no structural differences as compared with controls, including a normal pattern of tubule size, a normal amount of interstitial mass, and a normal spermatogenesis [1]. The most advanced type of spermatogenic cells present at 15 days after birth varied considerably but ranged from early to mid-pachytene spermatocytes in both wild-type and control, indicating that the increase in testis size in knockout mice is not caused by differences in onset of spermatogenesis [18]. Testicular weight, however, was significantly higher in knockout mice than in normal littermates (Fig. 4-3). It was found that at 15 days of age FMR1 knockout mice already have larger testes than their wild-type littermates [18]. In adult mice the differences become much more pronounced (Fig. 4-3). Like in fragile X males, macroorchidism is present in >90% of adult knockout mice [10].

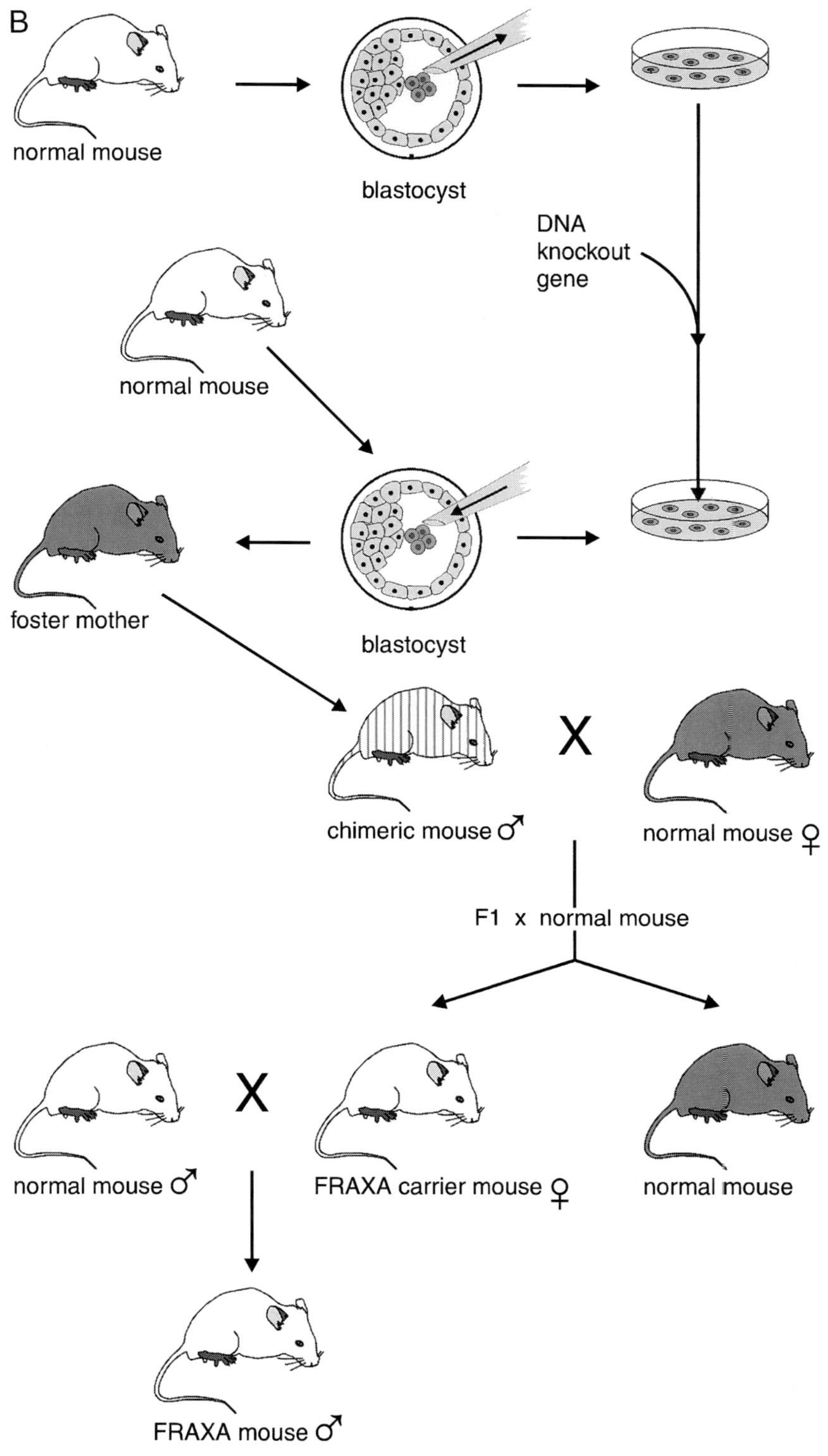

Figure 4-1 *(continued)*

In adult mice, testis size is mainly a reflection of the number of germ cells present. Since only a limited number of germ cells can be supported by one Sertoli cell [19], the maximal quantity of germ cells per testis is dependent on the number of Sertoli cells present. In male FMR1 knockout mice, Sertoli cell proliferation was increased during the period of Sertoli cell divisions: from embryonic day 12 (E12) to 15 days postnatally. The length of the period of Sertoli cell proliferation was not changed compared to the control males. Furthermore, a slightly increased testicular level of FSH receptor mRNA expression was found in the Fmr1 knockout

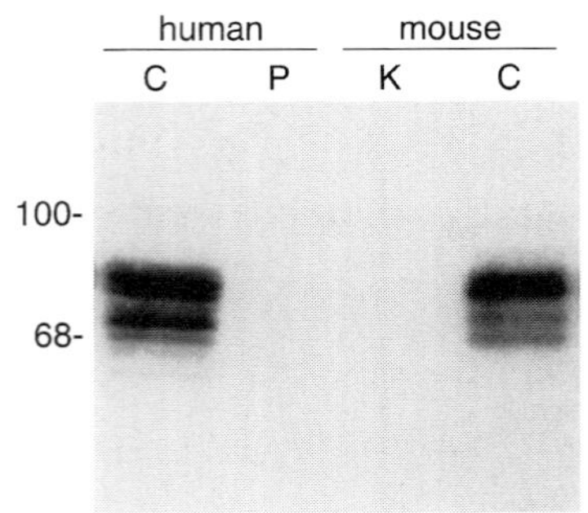

FIGURE 4-2 FMRP expression shown by Western blotting using an antibody against FMRP. Lane C, control; lane P, fragile X patient; lane K, knockout mouse.

mice on days 10 and 15 after birth. However, serum levels of FSH and short-term effects of FSH on Sertoli cell function as measured by down-regulation of FSH receptor mRNA were not changed [18]. It can be concluded that macroorchidism in Fmr1 knockout male mice is caused by an increased Sertoli cell proliferation, subsequently causing increased germ cell numbers and elevated testis weight.

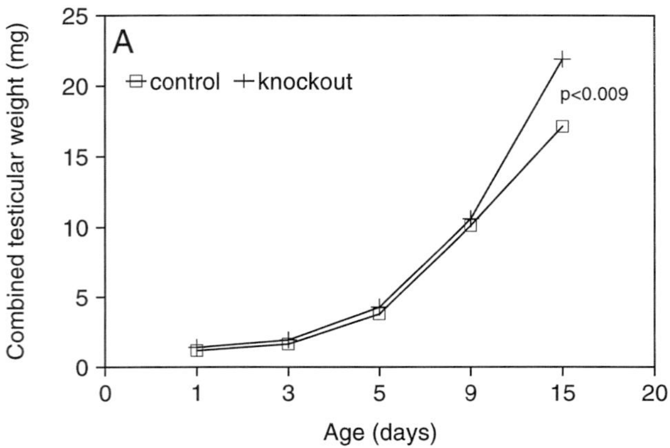

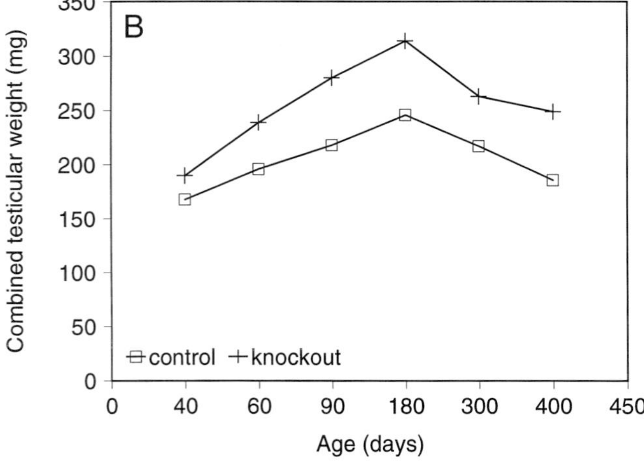

FIGURE 4-3 Combined testicular weight of knockout (■) and control mice (+) at different ages.

B. Brain Studies on Knockout Mice

Macroscopical and microscopical examination of different parts of the brain did not show any differences between knockout and normal mice [1]. Staining with a number of antibodies did not reveal differences except for the expression of the Fmr1 gene (Fig. 4-4). It was suggested that the sizes of the vermis of the cerebellum and the caudate nucleus are reduced in human fragile X patients [20, 21]. However, preliminary measurements of the vermis sizes showed that the vermis size is not significantly different between knockout and control mice (P. Cras, personal communication).

After neurotransmitter activation with the metabotrophic glutamate receptor an increased expression of Fmrp was observed [22]. It is suggested that rapid production of Fmrp near synapses in response to activation may be important for normal maturation of synaptic connections. Furthermore, it was noted that dendritic spines on apical dendrites in occipital cortex of knockout mice were longer and thinner than those in wild-type mice [23]. Also in fragile X patients immature, long tortuous spines are found on the dendrites in the neocortex [24]. This suggests a possible function of FMR1 protein in maturation and pruning of neurons.

Given the possible function of FMRP in maturation of dendritic spines and in synaptic transmission, and the malperformance in the Morris water maze test (see below) often associated with impaired long-term potentiating (LTP), it could be speculated that the LTP could be influenced in knockout mice. Electrophysiological studies have been performed on hippocampal slices of control and Fmr1 knockout mice [25] (Y. Ben-Ari, personal communication). No differences in LTP expression were found between knockouts and wild-type littermates. Also, short-term potentiation was similar in both types of mice.

C. Behavior

The behavior of the knockout mice was tested with the use of exploratory and motor activity tests as well as with the Morris water maze test [1, 10]. The tests revealed that the knockout mice had a higher motor activity than control and were impaired in the acquisition of novel spatial information.

In the Morris water maze test impairments in visual short-term memory and visual–spatial abilities can be

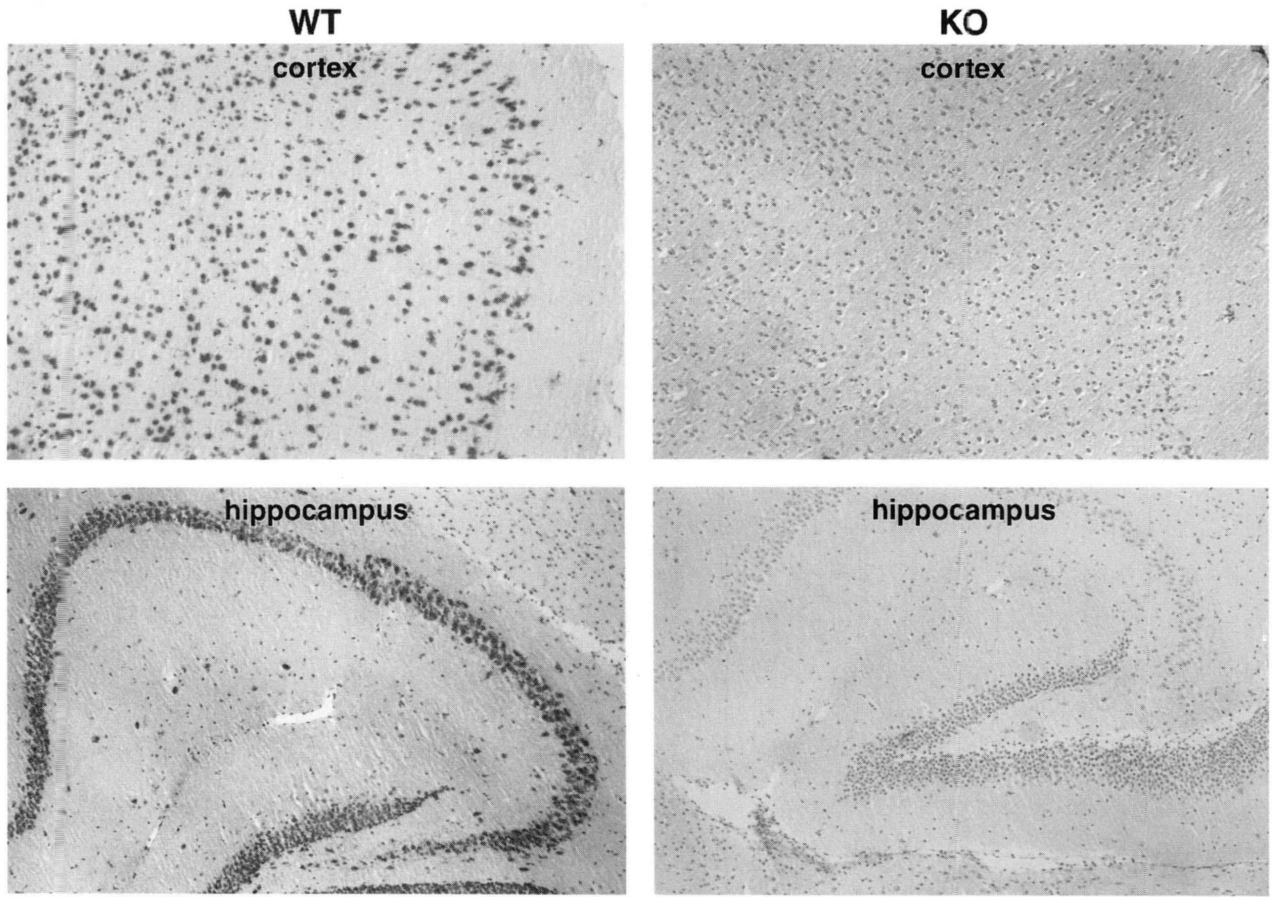

FIGURE 4-4 Immunohistochemical staining of brain sections of wild-type (WT) and knockout mice, stained with antibodies against FMRP.

measured in small rodents [26, 27]. In the water maze test the mice are placed in a large circular pool filled with opaque water and must swim to a platform that is visible or hidden. No gross impairment in swimming ability was observed in either the knockout or the control group. Knockouts and normal littermates were able to locate the hidden platform as a result of training. An example of a training trial of a normal littermate is given in Fig. 4-5. Using video tracking the successive swims of an animal were recorded. During successive swims the time and the path length are dramatically reduced. Similar results were obtained with the knockout animals. Thus, the hidden platform test showed that mutant mice reached levels of performance equal to those of controls (Fig. 4-5). Following 12 training trials, the position of the hidden platform was changed and four reversal trials were carried out. Now the knockout mice experienced more difficulty than normal littermates in learning the new position, which was apparent in both increased escape latency and path length. It appears to be due to a relative inability of knockouts to change a learned spatial strategy. Spatial memory abilities like those in the hidden platform learning test have been shown to be highly dependent upon hippocampal function. However, as is described earlier no differences in LTP could be detected between the mutant and control mice.

The original Morris water maze tests were carried out in a C57Bl6 background [1]. The knockout allele was bred into a background of FvB that were selected for proper visual abilities. These animals were tested in the Morris water maze test and similar results as described before were obtained (unpublished results).

D. Transgenic Mice

The question is whether the defect in knockout mice can be restored. To address this question the FMR1 gene was introduced into wild-type oocytes under the control of a CMV promoter. The transgene was subsequently bred into a knockout background (Bakker *et*

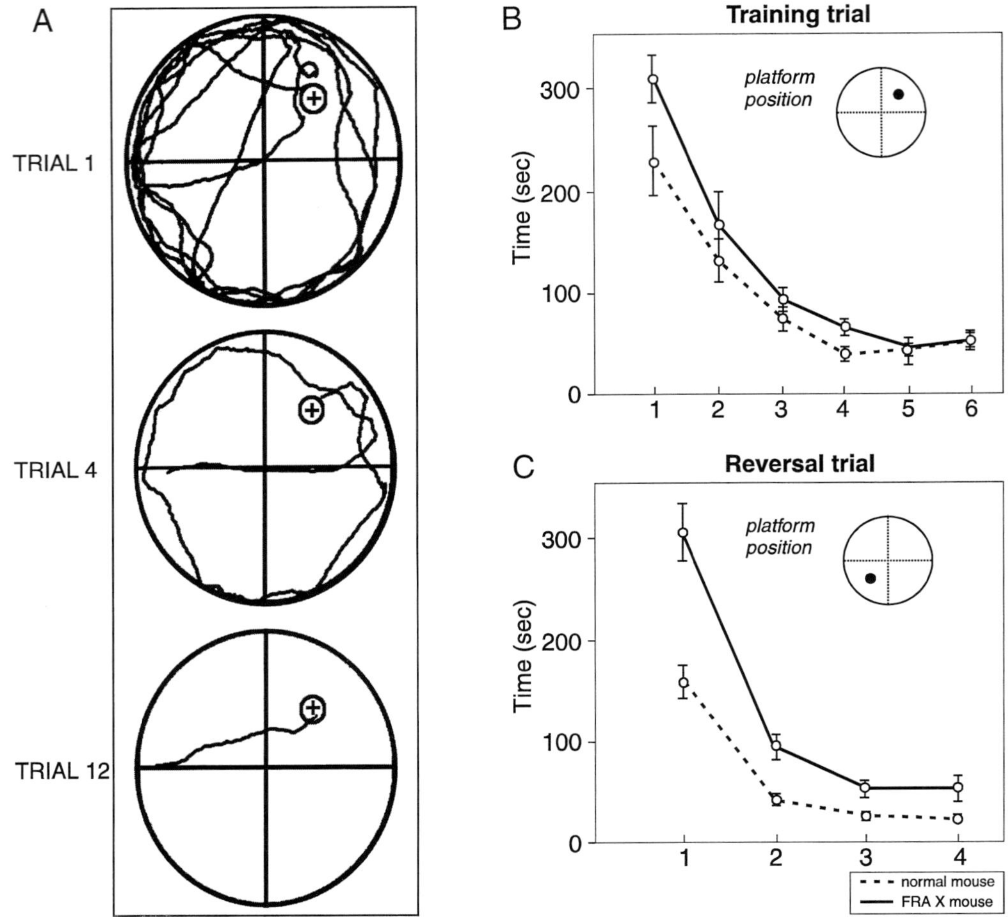

FIGURE 4-5 Morris water maze test performance of mutant and control mice. (A) Swimming trajectory of a control mouse during three training trials of a hidden-platform water maze training. (B) Escape latency to find a hidden platform: training trials. (C) Escape latency during reversal trials. Filled circles, knockout mice; open circles, control littermates.

al., unpublished results). The gene was expressed, although the level of expression per cell was low (around 5%). No restoration of normal testis size occurred. Preliminary experiments did not show a restoration of normal behavior in a Morris water maze test (unpublished results). It might mean that a higher expression of the FMR1 gene is needed for a normal phenotype. With this promoter, even with injecting much higher concentrations of DNA constructs, no higher expression was found which could suggest that a general expression of the FMR1 gene, especially during early embryogenesis, could be lethal.

Knockout mice with a transgene under the control of a promoter that is only active after birth might give an answer to the question of whether the expression of the FMR1 gene is necessary during embryonal development or whether the FMR1 protein is just necessary for a normal brain functioning.

II. INSTABILITY OF THE CGG REPEAT IN MICE

Very little is known about mechanism and timing of the CGG repeat amplification. Several models for the repeat amplification have been proposed; the most likely model, suggested by Wells, is that the expansion is the result of slippage during replication [28] (see Chapters 39 and 45). Moreover, somatic instability does also occur. Most patients show different lengths of full mutations often present as a smear on Southern blot analysis [29]. This shows that repeat length changes still occur in the early embryo. Analysis of intact oocytes of full mutation fetuses have shown that only full expansion alleles can be detected in oocytes [30]. On the other hand, in sperm of patients only premutations were found [12], suggesting contraction of full mutations in the immature testis. A model for the repeat expansion has been proposed by Malter *et al.*, but the exact timing of the repeat expansion and retraction is still not known [30]. To study the mechanism and timing of repeat instability an animal model is required. Only in an animal model will it be possible to study gametogenesis and early embryogenesis at specific time points.

In the murine Fmr1 CGG repeats of 9 to 12 repeat units have been identified in different mice strains. To study the behavior of a premutation allele upon subsequent generations, transgenic mice with a premutation allele, with a repeat of $(CGG)_{81}$, in the FMR1 promoter were generated [31]. Three independent lines (in total, 263 transgenic animals) were tested for repeat instability. In all meiosis and mitosis tested the repeat was inherited stably. In a similar type of experiment Lavedan *et al.* also found that a cloned 88-trinucleotide CGG repeat remained stable in mice [32]. A number of hypotheses can be put forward to explain these results.

First, the size of the repeat necessary to cause instability in mice is not known. It is not known whether CGG repeat instability in mice exists. It might be a human-specific phenomenon. We assumed that underlying mechanisms causing instability would be present in humans and mice. Second, the site of integration of the transgene might determine the stability of the repeat. Sequences in the interrupted chromosomal region might influence the behavior of the transgene. The FMR1 gene is normally present at the X chromosome, while the transgene was found to be at one of the autosomes. Since we do not know which mechanisms are involved in CGG expansions, the X chromosome might play an importance role. Third, flanking sequences of the FMR1 promoter might be important in generating instability. Fourth, in both described experiments impurities were present in the repeat. This might have provided stability to the repeat. In humans it was observed that AGG triplets interrupting the CGG tract provided stability to the repeat [33, 34].

Transgenic mice containing an expanded repeat in Huntington and myotonic dystrophy have shown that small changes in repeat length can occur [35–37]. This suggests that different factors might play a role in (in)-stability of the different repeats.

III. CONCLUSION

Very little is known about the function of the FMR1 protein, especially in transport and translation of mRNAs. Regulation of translation of specific mRNAs might be an important role of FMRP as a response of neurons to stimuli. Disturbance of translation through the lack of FMRP might have great effects on the normal functioning of the brain. The disturbance might lead in fragile X patients to learning and memory deficits. Animal models might help us to learn more about the function of the FMR1 gene and the effect that the lack of the protein has on brain functioning. Furthermore, animal models might help in studying the timing and mechanism of the repeat amplification.

Acknowledgments

We thank our colleagues in Rotterdam and Antwerp for their stimulating collaboration. This work was supported in part by grants from the FRAXA Foundation, the Netherlands Organization of Scientific Research (Telethon-Italy), and BIOMEDII.

References

1. Bakker, C. E., Verheij, C., Willemsen, R., Vanderhelm, R., Oerlemans, F., Vermey, M., Bygrave, A., Hoogeveen, A. T., Oostra, B. A., Reyniers, E., Deboulle, K., Dhooge, R., Cras, P., Van Velzen, D., Nagels, G., Martin, J. J., Dedeyn, P. P., Darby, J. K., and Willems, P. J. (1994). Fmr1 knockout mice: a model to study fragile X mental retardation. *Cell* **78,** 23–33.
2. Ashley, C. T., Sutcliffe, J. S., Kunst, C. B., Leiner, H. A., Eichler, E. E., Nelson, D. L., and Warren, S. T. (1993). Human and murine FMR-1: alternative splicing and translational initiation downstream of the CGG-repeat. *Nature Genet.* **4,** 244–251.
3. Hinds, H. L., Ashley, C. T., Sutcliffe, J. S., Nelson, D. L., Warren, S. T., Housman, D. E., and Schalling, M. (1993). Tissue specific expression of FMR-1 provides evidence for a functional role in fragile X syndrome. *Nature Genet.* **3,** 36–43.
4. Abitbol, M., Menini, C., Delezoide, A. L., Rhyner, T., Vekemans, M., and Mallet, J. (1993). Nucleus basalis magnocellularis and hippocampus are the major sites of FMR-1 expression in the human fetal brain. *Nature Genet.* **4,** 147–153.
5. Devys, D., Lutz, Y., Rouyer, N., Bellocq, J. P., and Mandel, J. L. (1993). The FMR-1 protein is cytoplasmic, most abundant in neurons and appears normal in carriers of a fragile X premutation. *Nature Genet.* **4,** 335–340.
6. Khandjian, E. W., Fortin, A., Thibodeau, A., Tremblay, S., Cote, F., Devys, D., Mandel, J. L., and Rousseau, F. (1995). A heterogeneous set of FMR1 proteins is widely distributed in mouse tissues and is modulated in cell culture. *Hum. Mol. Genet.* **4,** 783–790.
7. Verheij, C., De Graaff, E., Bakker, C. E., Willemsen, R., Willems, P. J., Meijer, N., Galjaard, H., Reuser, A. J. J., Oostra, B. A., and Hogeveen, A. T. (1995). Characterization of FMR1 proteins isolated from different tissues. *Hum. Mol. Genet.* **4,** 895–901.
8. Pieretti, M., Zhang, F. P., Fu, Y. H., Warren, S. T., Oostra, B. A., Caskey, C. T., and Nelson, D. L. (1991). Absence of expression of the FMR-1 gene in fragile X syndrome. *Cell* **66,** 817–822.
9. Sutcliffe, J. S., Nelson, D. L., Zhang, F., Pieretti, M., Caskey, C. T., Saxe, D., and Warren, S. T. (1992). DNA methylation represses FMR-1 transcription in fragile X syndrome. *Hum. Mol. Genet.* **1,** 397–400.
10. Kooy, R. F., Dhooge, R., Reyniers, E., Bakker, C. E., Nagels, G., Deboulle, K., Storm, K., Clincke, G., Dedeyn, P. P., Oostra, B. A., and Willems, P. J. (1996). Transgenic mouse model for the fragile X syndrome. *Am. J. Med. Genet.* **64,** 241–245.
11. Willems, P. J., Reyniers, E., and Oostra, B. A. (1995). An animal model for fragile X syndrome. *Mental Ret. Dev. Disab. Res. Rev.* **1,** 298–302.

12. Reyniers, E., Vits, L., De Boulle, K., Van Roy, B., Van Velzen, D., de Graaff, E., Verkerk, A. J. M. H., Jorens, H. Z., Darby, J. K., Oostra, B. A., and Willems, P. J. (1993). The full mutation in the FMR-1 gene of male fragile X patients is absent in their sperm. *Nature Genet.* **4,** 143–146.
13. Butler, M. G., Brunswig, A., Miller, L. K., and Hagerman, R. J. (1992). Standards for selected anthropogenetic measurements in males with fragile X syndrome. *Pediatrics* **89,** 1059–1062.
14. Hagerman, R. J. (1996). Physical and behavioral phenotype. *In* "Fragile X Syndrome: Diagnosis, Treatment and Research" (R. J. Hagerman and A. C. Silverman, Eds.), pp 3–87. Johns Hopkins University Press, Baltimore.
15. Turner, G., Eastman, C., Casey, J., McLeay, A., Procopis, P., and Turner, B. (1975). X-linked mental retardation associated with macro-orchidism. *J. Med. Genet.* **12,** 367–71.
16. De Graaff, E., Willemsen, R., Zhong, N., De Die-Smulders, C. E. M., Brown, W. T., Freling, G., and Oostra, B. A. (1995). Instability of the CGG repeat and expression of the *FMR1* protein in a male fragile X patient with a lung tumour. *Am. J. Hum. Genet.* **57,** 609–618.
17. Tamanini, F., Willemsen, R., van Unen, L., Bontekoe, C., Galjaard, H., Oostra, B. A., and Hoogeveen, A. T. (1997). Differential expression of FMR1, FXR1 and FXR2 proteins in human brain and testis. *Hum. Mol. Genet.* **6,** 1315–1322.
18. Slegtenhorst-Eegdeman, K. E., van de Kant, H. J. G., Post, M., Ruiz, A., Uilenbroek, J. T. J., Bakker, C. E., Oostra, B. A., Grootegoed, J. A., de Rooij, D. G., and Themmen, A. P. N. (1997). Macro-orchidism in *FMR1* knockout mice is caused by increased Sertoli cell proliferation during testis development. *Endocrinol.* **139,** 156–162.
19. Sharpe, R. M. (1993). Experimental evidence for Sertoli-germ cell and Sertoli-Leydig cell interactions. *In* "The Sertoli Cell" (L. G. D. Russul and M. D. Griswold, Eds.), pp. 391–418. Cache River Press, Clearwater, FL.
20. Reiss, A. L., Patel, S., Kumar, A. J., and Freund, L. (1988). Preliminary communication: neuronanatomical variations of the posterior fossa in men with the fragile X (Martin Bell) syndrome. *Am. J. Med. Genet.* **31,** 407–414.
21. Reiss, A. L., Abrams, M. T., Greenlaw, R., Freund, L., and Denckla, M. B. (1995). Neurodevelopmental effects of the FMR-1 full mutation in humans. *Nature Med.* **1,** 159–167.
22. Weiler, I. J., Irwin, S. A., Klintsova, A. Y., Spencer, C. M., Brazelton, A. D., Miyashiro, K., Comery, T. A., Patel, B., Eberwine, J., and Greenough, W. T. (1997). Fragile X mental retardation protein is translated near synapses in response to neurotransmitter activation. *Proc. Natl. Acad. Sci. USA* **94,** 5395–5400.
23. Comery, T. A., Harris, J. B., Willems, P. J., Oostra, B. A., Irwin, S. A., Weiler, I. J., and Greenough, W. T. (1997). Abnormal dendritic spines in fragile X knockout mice: maturation and pruning deficits. *Proc. Natl. Acad. Sci. USA* **94,** 5401–5404.
24. Hinton, V. J., Brown, W. T., Wisniewski, K., and Rudelli, R. D. (1991). Analysis of neocortex in three males with the fragile X syndrome. *Am. J. Med. Genet.* **41,** 289–294.
25. Godfraind, J. M., Reyniers, E., Deboulle, K., Dhooge, R., Dedeyn, P. P., Bakker, C. E., Oostra, B. A., Kooy, R. F., and Willems, P. J. (1996). Long-term potentiation in the hippocampus of fragile X knockout mice. *Am. J. Med. Genet.* **64,** 246–251.
26. Morris, R. G. M. (1981). Spatial localization does not require the presence of local clues. *Learn Motiv.* **12,** 239–260.
27. Morris, R. G., Garrud, P., Rawlins, J. N., and O'Keefe, J. (1982). Place navigation impaired in rats with hippocampal lesions. *Nature* **297,** 681–683.
28. Wells, R. D. (1996). Molecular basis of genetic instability of triplet repeats. *J. Biol. Chem.* **271,** 2875–2878.
29. Rousseau, F., Heitz, D., Biancalana, V., Blumenfeld, S., Kretz, C., Boue, J., Tommerup, N., Van Der Hagen, C., DeLozier-Blanchet, C., Croquette, M. F., Gilgenkranz, S., Jalbert, P., Voelckel, M. A., Oberlé, I., and Mandel, J. L. (1991). Direct diagnosis by DNA analysis of the fragile X syndrome of mental retardation. *N. Engl. J. Med.* **325,** 1673–1681.
30. Malter, H. E., Iber, J. C., Willemsen, R., De Graaff, E., Tarleton, J. C., Leisti, J., Warren, S. T., and Oostra, B. A. (1997). Characterization of the full fragile X syndrome mutation in fetal gametes. *Nature Genet.* **15,** 165–169.
31. Bontekoe, C. J. MC. J., de Graaff, E., Nieuwenhuizen, I. M., Willemsen, R., and Oostra, B. A. (1997). FMR1 premutation allele is stable in mice. *Eur. J. Hum. Genet.* **5,** 293–298.
32. Lavedan, C. N., Garrett, L., and Nussbaum, R. L. (1997). Trinucleotide repeats (CGG)22TGG(CGG)43TGG(CGG)21 from the fragile X gene remain stable in transgenic mice. *Hum. Genet.* **100,** 407–414.
33. Kunst, C. B., and Warren, S. T. (1994). Cryptic and polar variation of the fragile X repeat could result in predisposing normal alleles. *Cell* **77,** 853–861.
34. Eichler, E. E., Hammond, H. A., Macpherson, J. N., Ward, P. A., and Nelson, D. L. (1995). Population survey of the human FMR1 CGG repeat substructure suggests biased polarity for the loss of AGG interruption. *Hum. Mol. Genet.* **4,** 2199–2208.
35. Gourdon, G., Radvanyi, F., Lia, A. S., Duros, C., Blanche, M., Abitbol, M., Junien, C., and Hofmann-Radvanyi, H. (1997). Moderate intergenerational and somatic instability of a 55-CTG repeat in transgenic mice. *Nature Genet.* **15,** 190–192.
36. Monckton, D. G., Coolbaugh, M. I., Ashizawa, K. T., Siciliano, M. J., and Caskey, C. T. (1997). Hypermutable myotonic dystrophy CTG repeats in transgenic mice. *Nature Genet.* **15,** 193–196.
37. Mangiarini, L., Sathasivam, K., Mahal, A., Mott, R., Seller, M., and Bates, G. P. (1997). Instability of highly expanded CAG repeats in mice transgenic for the Huntington's disease mutation. *Nature Genet.* **15,** 197–200.

Part III

Other Fragile Sites

CHAPTER 5

FRAXE Mental Retardation and Other Folate-Sensitive Fragile Sites

DAVID L. NELSON Department of Molecular and Human Genetics, Baylor College of Medicine, Houston, Texas 77030

I. INTRODUCTION

Fragile sites were first characterized in detail in the mid-1980s by Sutherland and co-workers [1, 2]. Some 21 rare, folate-sensitive fragile sites have been described at the cytogenetic level to date [3]. This chapter reviews the current state of knowledge regarding the five folate-sensitive fragile sites that have been characterized at the molecular level, with the exception of FRAXA, the fragile site associated with fragile X syndrome, which is extensively discussed in preceding chapters.

Fragile sites are classified as either rare or common depending upon their frequency in the human population, with common sites being homozygous in nearly every person. Rare fragile sites are typically found in fewer than 1% of individuals. Fragile sites are visualized in metaphase preparations of chromosomes after "induction" or treatment of the cells with a variety of reagents that elicit the sites. In addition to folate, fragile sites can be induced by distamycin-A, bromodeoxyuridine (BrdU), aphidicolin, and 5-azacytidine. The folate-sensitive sites are the most numerous of the rare fragile sites, while aphidicolin-induced sites are plentiful among the common sites. There are 21 known folate-sensitive sites (Fig. 5-1), and these can be elicited by a variety of culture conditions [3]. Culture medium lacking folic acid and thymidine is the simplest method, although inhibitors of folate metabolism such as methotrexate can also be used. High thymidine can also induce these

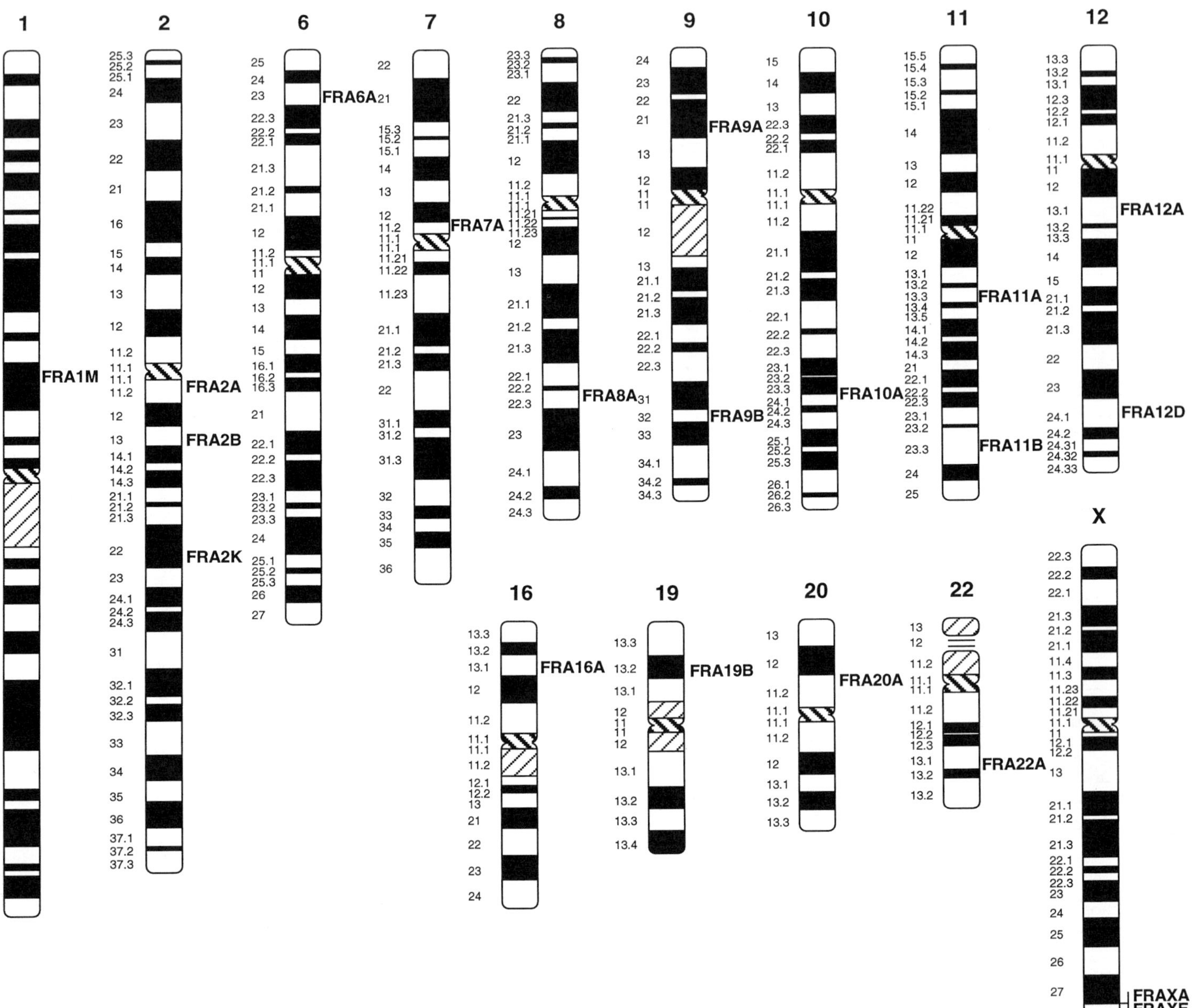

FIGURE 5-1 Partial karyotype of the human chromosome complement with associated positions of the rare folate-sensitive fragile sites. Each of the chromosomes which carries a folate-sensitive fragile site is shown to scale with its Giemsa staining pattern. The cytogenetic locations of each of the fragile sites are indicated by the name of the fragile site at the appropriate position.

fragile sites at concentrations below those that result in mitotic arrest. These treatments all impinge on the pyrimidine biosynthetic pathway, and result in diminished dTTP pools in the cell. The precise mechanism by which the depletion of dTTP results in a chromosomal abnormality is unknown.

The rare folate-sensitive fragile sites are rarely found present in all chromosomes prepared from an individual. In fact, it is unusual to find more than 40% of metaphase nuclei exhibiting a fragile site, and typically only 20% or fewer nuclei are found to carry the abnormality. This number has been used in fragile X syndrome studies in attempts to correlate with severity of disease; however, no consistent correlations have been found. Numbers below 1–2% induction can be due to false positive results; thus some caution is warranted in interpretation of cytogenetic fragile site studies.

Five folate-sensitive fragile sites have been identified at the molecular level; these include three in the distal part of the X chromosome (FRAXA, FRAXE, and FRAXF), and two on autosomes (FRA11B and FRA16A). In addition to FRAXA, the fragile site associated with fragile X syndrome, two of the other four folate-sensitive fragile sites have been found associated with human disease. FRAXE can lead to a mild learning disability and FRA11B predisposes to chromosome breakage which result in 11q- or Jacobsen's syndrome. All folate-sensitive fragile sites have been found to result from significant expansions of normally polymorphic CGG/CCG trinucleotide repeats. It is therefore likely that all remaining folate-sensitive fragile sites will be found to result from CGG repeat expansion. It is as yet unclear how folate-induced fragile sites carrying expanded regions of CGG triplets emanate from the reduction in dTTP pools elicited by the culture conditions described above. The precise molecular nature of the cytogenetically observed fragile site is also unknown.

Two common fragile sites have been characterized at the molecular level. These are the aphidicolin-induced FRA3B and the distamycin A-induced FRA16B. These clearly do not result from expansions of CGG repeats, and are discussed in more detail in the Chapter 6. It remains to be seen whether specific sequence elements will be found associated with common fragile sites.

II. HISTORICAL ASPECTS OF FRAGILE SITES IN Xq27.3-q28

When the FMR1 gene and FRAXA fragile site were identified in 1991 [4–7], it rapidly became apparent that several families and individuals in collections of fragile X syndrome patients did not exhibit the expected CGG repeat expansion in the FMR1 gene. After initial speculation that these patients had an unusual variant fragile site [8], cytogenetic analyses using fluorescence *in situ* hybridization (FISH) demonstrated that the fragile sites observed in these patients were more distal than FRAXA; this fragile site was termed FRAXE as it represented the fifth fragile site described on the X chromosome [9]. Additional studies showed methylation at a CpG island 600 kb distal to the FRAXA CpG island in FRAXE patients. This marked the segment of the chromosome to be targeted for DNA isolation, which was rapidly accomplished by Davies and co-workers [10]. This led to studies by several groups to identify and characterize the gene affected by FRAXE expansion. The gene was eventually isolated; however, those studies were significantly hampered by the size of the FMR2 gene and its early intervening sequences [11–13]. Additional details regarding FRAXE, the associated mental handicap, and the gene responsible (FMR2) are discussed below.

A third fragile site in Xq27.3-q28 has been characterized and termed FRAXF [14–16]. This fragile site has only been observed in a few families, in part due to its scarcity, but also since it apparently contributes no discernible phenotype to individuals carrying the expansion. Again, the realization that some patients or families with cytogenetic fragile sites apparent in Xq27.3 or Xq28 without the associated expansion of CGG repeats at FRAXA or FRAXE led to the appreciation that FRAXF was a separate site. It is located approximately 1.2 million base pairs distal to FRAXA. FRAXF is likewise discussed in greater detail below. A map of the Xq27.3 to Xq28 region is provided in Fig. 5-2.

III. FRAXE

A. Clinical Aspects of FRAXE-Associated Mental Retardation

As mentioned above, families carrying the FRAXE fragile site were initially ascertained due to mental retardation. Cytogenetic analyses of these patients revealed the presence of a fragile site in Xq27.3, and they were classified as fragile X syndrome patients. It was only with the identification of the FRAXA CGG repeat expansion and the FMR1 gene that it became apparent that the fragile site carried by these families was not that involved in fragile X syndrome. This allowed these families to be reevaluated clinically as a separate pathological entity, which both altered the description of fragile X syndrome, since these families no longer needed to be accounted for, and provided a clinical description of the FRAXE phenotype [17–25].

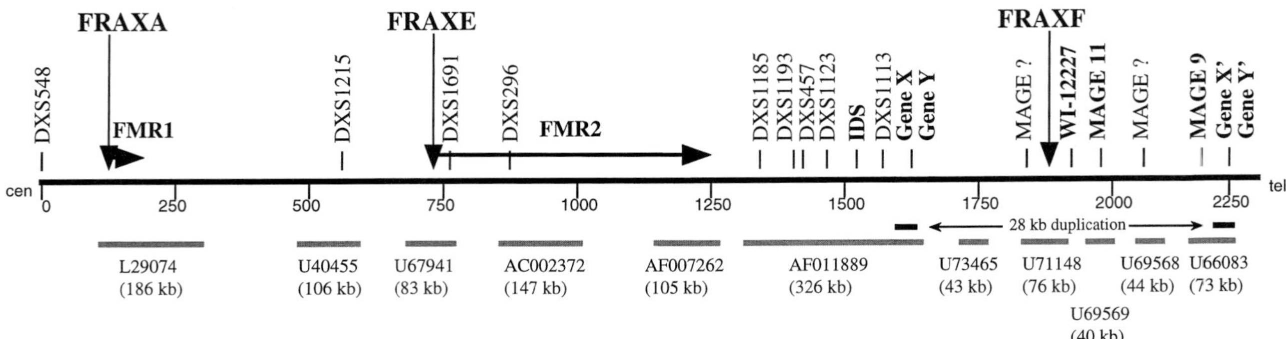

FIGURE 5-2 Map of 2.25 Mb of human Xq27.3-q28 showing the relative positions of the three fragile sites FRAXA, FRAXE, and FRAXF and associated genes and markers. Bona fide genes are in bold; other markers indicated are generally highly polymorphic microsatellites. Genes X, Y, X′, and Y′ are duplicated in a 28-kb interval that is highly similar between the two copies. MAGE genes with known transcripts are given numbers; those with question marks are found by similarity in genomic sequence, but have not been identified as transcripts. An indication of the current status of genomic sequence for this region, with accession numbers, is provided in the lower line. At present, over one-half of the interval is present in the sequence databases. Additional "in progress" data can be found at http://www.gc.bcm.tmc.edu:8088/home.html.

A number of descriptors have been used to characterize the mental status of male FRAXE patients. These range from mildly retarded to a mild mental handicap. In general, the mental functions of FRAXE patients are uniformly improved over fragile X syndrome patients. Total IQ scores range from the mid 60s to the normal range. Some individuals have been described as mentally normal (in family 1 of Knight *et al.* [26], and family 2 of Mulley *et al.* [26] with 20–40% fragile site expression (although these latter individuals were not assessed psychometrically). Although it was initially reported that females carrying a full mutation were not impaired intellectually, families studied by Hamel *et al.* [18] and Mulley *et al.* [20] showed an excess of females in the mild or borderline mental retardation range. There too, systematic psychometric testing would be required to evaluate the risk of mental retardation in FRAXE carrier females. No consistent physical phenotype has been described in FRAXE patients.

FRAXE expansion mutations appear to be quite rare, with a frequency of less than 1% of that of FRAXA expansions. Indeed, fewer than 20 families have been described to date. A number of studies have concluded that the utility of screening for this mutation in individuals with developmental delay is not warranted due to the low number of cases identified [27–29]. However, it remains to be determined what additional mutations in the FMR2 gene (see below) might exist in this population. While the FRAXE expansion mutation may be rare, the size of the gene (600 kb) may lead to a reasonably high frequency of loss of function mutations, which may be important in nonsyndromic X-linked mental retardation.

In this vein, two patients have been described with mutations other than repeat expansion mutations in the gene associated with FRAXE (FMR2—see below). These loss-of-function mutations resulted from deletions in the vicinity of the second and third exons of the gene, some 150 kb distal to the FRAXE site [30]. The two males were not significantly different in phenotype than those patients in families carrying the repeat expansion.

B. Characteristics of the FRAXE Repeat

In many respects, the FRAXE site behaves similarly with that at FRAXA. The fragile site causing mutation was shown to be an unstable expansion of a GCC (equivalent to CCG) repeat associated with abnormal methylation, with characteristics that appear very similar to those found in the fragile X (FRAXA) syndrome. Methylated large expansions (200–1000 GCCs), often heterogeneous (in leukocytes), were found linked to fragile site expression and constitute a "full mutation." Cytogenetic expression of the fragile site in females appears on average higher than in FRAXA, and in particular, very large expansions ($n = 2000$) were reported in one family associated to unusually high fragile site expression (60–75%) in two adult females [19]. More moderate ($n = 100$–130) expansions were found in a few carriers with no cytogenetic expression, resembling the FRAXA premutations. The threshold for methylation appears to be around 130 repeats [18, 20], lower than for FRAXA, where the threshold is between 200 and 230 repeats. In the $n = 130$–200 range, mutations will thus be partially or totally methylated, but are not associated with cytogenetic fragile site expression. In contrast, partial methylation was reported for a male with an expansion of 1100 bp (380 GCCs). Extent of methylation may depend on the restriction site used in the analysis. Inhibition of digestion at the *Eco*RI site just distal to the repeat, both on normal inactive Xs and in full FRAXE mutation alleles, is most likely due to an overlap of the GAATTC site with a CpG dinucleotide. This can be detected with the OxE18 probe, and allows the estimation of both expansion and methylation in a single *Eco*RI digest (i.e., the same digest that is generally used to test for FRAXA expansions [31]. Normal alleles range in length from 6 to 25, with 15 repeats being the most common allele in Caucasians [26].

The parental biases for expansion or contraction of FRAXE mutations appear to resemble those found for FRAXA. Maternal premutations were observed to expand to full mutation size in the offspring, and four males with full mutations had daughters with no fragile site expression and premutation sized expansions [19, 20]. However, in the family described by Hamel *et al.* [18], contraction was not as significant in male transmissions, as two daughters of males with full mutations had expansions in the $n = 220–280$ range and were mentally impaired (but did not exhibit a cytogenetic fragile site). As the number of premutation carriers (males or females) studied is still small, these trends require confirmation.

C. The FMR2 Gene

Mental retardation due to FRAXE expansion results from apparent reduction in activity of a gene at whose promoter the GCC repeats reside. Identification of this gene was hampered by the large size of the gene, and particularly of the large (150 kb) first intron. The fragile site and mutation were first identified in 1993 [10], yet it was not until 1996 that the complete FMR2 gene was characterized [11, 12]. Chakrabarti *et al.* had also identified a portion of the transcript [13]. Each group arrived at the FMR2 gene from different approaches. The Australian group [12] identified transcripts using evolutionarily conserved sequences near the marker VK21 (DXS296) that were involved in deletions in two unrelated males with speech delay [30]. The U.S. group [11] utilized prediction of gene coding content [32] to identify potential exons from the sequence of a cosmid carrying the DXS296 locus; this allowed isolation of cDNA clones which were found to carry sequences adjacent to the FRAXE site. The British group used direct hybridization of cDNA libraries with DNA fragments at near the fragile site. Each group was aided by the long-range sequencing efforts of Gibbs and co-workers at Baylor College of Medicine which have by now nearly completed the sequence of the FMR2 locus (Fig. 5-2).

The major FMR2 transcript is estimated to be between 8.7 and 9.5 kb, with alternate, smaller forms seen in some tissues [11–13, 33]. The coding sequence can encode isoforms as large as 1311 amino acids, and there is a long 3′ untranslated segment of the mRNA. Alternative splicing has been observed for several of the 22 exons (Fig. 5-3). Some exons exhibit alternative splice donors and/or acceptors as well [33]. The gene is estimated to span 600 kb of the chromosome (Fig. 5-3). Two introns (introns 1 and 3) are quite large, estimated to be 140 to 150 kb each.

FMR2 is not at all similar to FMR1 at either the DNA or protein sequence levels. The predicted protein is very rich in proline, serine, and threonine, and is similar to two other proteins described in humans, AF4 (MLLT2) [34] and LAF-4 [35]. Identity with each paralog is ~25% over the entire length of the protein, and is observed in regions outside of the low complexity sequences. AF-4 was identified as a fusion partner of the MLL gene (a homolog of the *Drosophila trithorax* gene) on chromosome 11 in mixed lymphocytic leukemias carrying a recurrent 4;11 translocation [36]. LAF-4 was recently identified from lymphoma cells [35]. Both proteins exhibit nuclear localization and features of transcriptional activators. Recent evidence suggests that FMR2 is likewise a nuclear protein and has characteristics consistent with transcriptional activation [33].

High-level expression of FMR2 is limited in adult tissues to brain and placenta [11–13, 33], although expression can be detected in whole blood, fibroblasts, hair roots, and cultures of amniocytes and chorionic villi [37]. Within the brain, structures with the highest levels of FMR2 expression are the amygdala and hippocampus, the latter structure likely significant for the gene's effect on learning and memory [13]. Since cells transfected with FMR2 constructs tagged with green fluorescent protein (GFP) show nuclear signals when one of two nuclear localizing signals is present [33], it is reasonable to conclude that FMR2 is a transcriptional activator involved in neuronal gene regulation which is disrupted by the FRAXE repeat expansion mutation. Much further effort is necessary to identify the precise function(s) of FMR2 and the consequences of FRAXE expansion in the developing and adult brain.

Expansion of the FRAXE repeat results in hypermethylation of the FMR2 promoter sequences. As in FMR1, this leads to transcriptional repression of FMR2 expression. FMR2 is not expressed to appreciable levels in lymphoblasts, however. Thus, fibroblasts from patients and normal individuals have been utilized to demonstrate reduced in expression in patients by reverse-transcriptase PCR [11, 12]. In a larger study involving 11 individuals from seven families, Gecz *et al.* found that expression in fibroblasts was eliminated

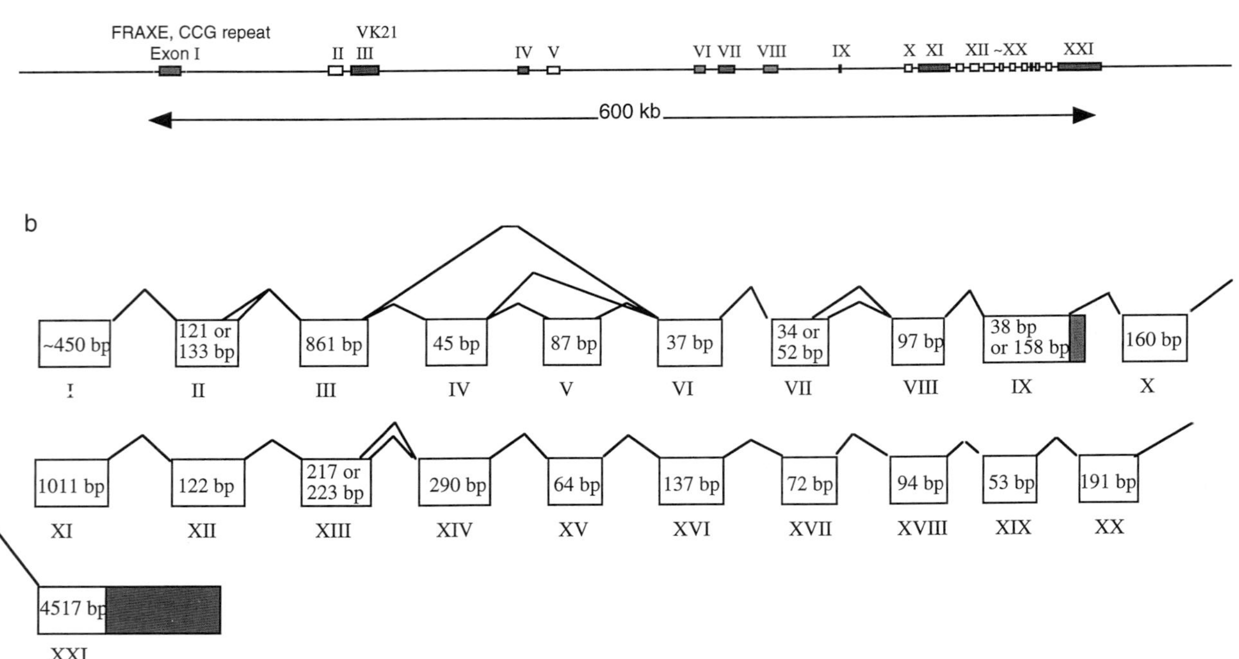

FIGURE 5-3 Diagrammatic representation of the FMR2 gene structure. (a) The organization of the 21 exons of the FMR2 gene is represented by boxes. The gene spans 600 kb of Xq27.3–Xq28. The FRAXE CCG repeat is indicated at the first exon. (b) Splicing patterns of FMR2 are indicated along with exon sizes. These data are from Gu and Nelson (unpublished) and are generally in agreement with data from Gecz and co-workers [33], who have determined that the last exon (exon 21 here) is interrupted by another intron to give 22 total exons.

when expansion mutations reached a threshold between 700 and 800 base pairs larger than normal (~250 repeats), and correlated with methylation [37]. One individual carried a 600-bp expansion that was not methylated, and his FMR2 expression was found to be normal. It is likely that the effect of FRAXE expansion is to down-regulate the FMR2 gene and is a loss of function mutation. This is consistent with the phenotypes characterized in the other loss of function mutations found in males with deletions in the region of exons 2 and 3 [30]. Identification of other mutations, particularly gain of function or dominant negative alleles, will be of interest to further characterize the function of this gene. Such mutations have been of significant interest in understanding the function of FMR1, for example [38].

FMR2 is conserved in mammals as evidenced by zoo blot studies [13]. Recent unpublished work by Gu and Nelson has demonstrated that the murine FMR2 homolog is well conserved (87% identity). Construction of a mouse model of the human disease may provide additional insight into the nature of the disorder in humans as well as allow additional experimental manipulation in order to develop therapies.

IV. FRAXF

The third fragile site in the Xq27.3-q28 region is FRAXF, which is located approximately one megabase distal to FRAXE. This position is distal to the iduronate sulfatase gene (IDS), defects in which result in Hunter syndrome (Fig. 5-2). This fragile site has been found in a very small number of families. Members of these families had been assessed cytogenetically for a variety of reasons, principally for suspicion of fragile X syndrome due to mental retardation or developmental delay. These cases were revisited upon identification of molecular markers for FRAXA and FRAXE and found to be negative for expansions at these sites. This allowed definition of a third fragile site in the region [39], which was termed FRAXF and estimated to lie more than one megabase distal to FRAXA. The family which was studied to define the fragile site was found to express the fragile site independently from the developmental delay seen in multiple affected females. Some individuals with the fragile site were found to be normal, while others with the phenotype were found not to express the fragile site in addition to some with both the fragile site and the phenotype. This finding was in accord with a study from New Zealand, where a family without phenotype was found to segregate the FRAXF site [40].

Upon molecular cloning of the FRAXF site [14, 15], it was found that expansions showed a large GCC expansion and methylation pattern in fragile site positive individuals similar to those seen in FRAXA and FRAXE. Although only four families have been (partially) studied, the initial observations suggest that the FRAXF fragile site is not causally associated with mental retardation in males or females. Further studies are needed to evaluate the frequency of this mutation, although it appears even rarer than FRAXE. It remains to be seen whether the pattern of transmission is similar to that found for FRAXA and FRAXE. Normal alleles were found to range in length from 6 to 38 triplets, with 14 being the most common [15, 41, 42].

FRAXA and FRAXE are found in the promoter regions of their respective genes. While FRAXF is likewise found in a CpG island, it does not appear to be associated with a gene promoter based on sequence analysis (Fig. 5-2 and Genbank Accession No. U71148) and efforts to define transcripts from the region [15]. However, a number of genes appear to be nearby (Fig. 5-2), as the FRAXF site is in proximal Xq28, a region that is much more gene dense than the Xq27.3 region where FRAXA and FRAXE are found. As yet, no evidence of diminished expression of nearby genes has been found in individuals expressing FRAXF.

The presence of three CGG/GCC loci susceptible to unstable expansions in a relatively small region of the genome is intriguing. Whether this finding is typical of the genome is not known. The presence of folate-sensitive fragile sites with frequencies as rare as those of FRAXE and FRAXF would be difficult to discover without the large catchment of cytogenetic tests for Xq27.3 fragile sites in the mentally retarded. Similarly rare autosomal sites may not be identified, or may be discarded as unimportant. Nonetheless, as the sequence of this region of the X chromosome is completed, it will be of interest to compare this region to the genome in general to attempt to understand the basis for the apparently high frequency of folate-sensitive fragile sites.

V. FRA16A

The FRA16A locus is the first folate-sensitive autosomal fragile site to be cloned [43]. It is characterized by very large (n = 1000 to 2000) methylated expansions of a CCG triplet. The similarity with the pattern seen for the X-linked fragile sites indicates that methylation is a consequence of expansion, and not a feature associated with X chromosome inactivation. While a tendency to greater instability was observed upon maternal transmission, paternal transmission of a large expansion alleles (n = 1000) did not result in contraction in the three children studied [43], at variance with the pattern observed for FRAXA and FRAXE. In normal alleles the FRA16A repeat is also interrupted by a CTG triplet, and the more variable alleles are associated with a loss of this CTG, confirming that unstable alleles are those with the longer perfect repeats [44]. The FRA16A locus does not appear associated with a specific pathology in the heterozygous state, but if it results in inactivation of an adjacent gene, a phenotype might only be present in homozygous individuals. Such individuals have not yet been identified, and given the rarity of the site, it is unlikely that they will be found unless there is a phenotype which leads to their discovery.

The CCG sequence at FRA16A is a complicated compound polymorphism, with four varying regions [44]. Of particular interest is the finding that the bulk of variant alleles identified in this study were found among Europeans, such that the vast majority of polymorphic variation is contained in this population. This has led to the suggestion that other *cis*-acting sequences are responsible for acceleration of the rate of variation, and that the identification of these might lead to a more comprehensive understanding of the factors which influence triplet repeat instability [45].

VI. FRA11B

FRA11B is a folate-sensitive fragile site located in 11q23.3. It is the only autosomal folate-sensitive fragile site to be linked to a human disease state to date. The genetic condition known as Jacobsen syndrome is a variable phenotype involving mental retardation and other dysmorphic features including growth retardation (for review, see [46]) which results from absence of sequences at the distal long arm of chromosome 11 (11q-syndrome). The possible association with chromosome instability resulting from the FRA11B fragile site led to the hypothesis that Jacobsen syndrome patients may have inherited deleted chromosomes which resulted from CGG expansion at FRA11B.

In 1994, Jones *et al.* reported an association between FRA11B and a deletion breakpoint identified in a Jacobsen syndrome patient [47]. In this study, the patient's normal brother and mother were found to carry the FRA11B fragile site (the patient did not, but was deleted for this region), suggesting that the fragile site may have led to the breakage. The breakpoint in the patient was found to be within 100 kb of the fragile site. In a following study in 1995, Jones and colleagues identified the CCG repeat responsible for the fragile site, and showed that it was present in the CBL2 proto-oncogene [48]. They were further able to demonstrate fragile sites or larger than normal repeat lengths in relatives of Jacobsen patients. This study also demonstrated methylation of a nearby CpG island upon repeat expansion. Breakpoints in patients were not coincident with the CCG repeat, but were found in one case to lie within 20 kb proximal to the repeat.

There is considerable variation in the extent of deletion in patients with the 11q- syndrome, and it is clear that not all patients result from CBL2/FRA11B expansions. However, this is an important demonstration of *in vivo* fragility induced by expanded CCG repeats associated with a folate-sensitive fragile site, since it is the first such example of *in vivo* breakage for this class of fragile site. The possibility that the CCG repeat can also be involved in somatic events leading to tumorigenesis is important to consider as well [48], since the embedded gene is potentially oncogenic. With the discovery of the FRA3B fragile site association with cancer [49, 50] (see Chapter 6), this provides further fuel for the speculation that fragile sites may play a role in chromosomal instability leading to oncogenesis.

VII. Conclusions

As stated above, each of the five folate-sensitive fragile sites identified to date results from expansion of a polymorphic CCG/CGG trinucleotide repeat to considerable lengths. Each is found to be methylated in its expanded state, and this methylation leads to pathogenesis through reduction in gene expression in FRAXA and FRAXE. Given the concordance of these characteristics among the five sites, it appears likely that the remaining known folate-sensitive fragile sites will behave in a similar fashion. It is anticipated that each will result from an expanded CCG/CGG repeat and that methylation will likely result. Since each appears benign, it is unlikely that the remaining sites will have a profound impact on the health of their carriers, although the limited sample sizes currently limit such predictions.

It is also possible that novel fragile sites may remain unknown. The human genome carries many polymorphic CCG/CGG repeat sequences [51, 52], and, based on the characteristics of the five known sites, it is impossible to rule any out as candidates for expansion mutation. It is possible that *cis*-acting sequences are important to the events that lead to expansions, and with additional study of each region, it is likely that these sequences will be defined.

The role(s) of additional folate-sensitive fragile sites in inherited or somatic genetic conditions can only be speculative at present. It is likely that loss of function mutations will be the norm due to methylation-associated transcriptional repression. However, the possibility that such fragile sites lead to chromosome instability is also worthy of consideration, particularly with respect to somatic events.

Acknowledgments

The author thanks his many colleagues for their advice and suggestions during the preparation of the manuscript, and thanks Yanghong Gu for unpublished data. This work was supported in part by grants from the US NIH (HD29256 and GM52982).

References

1. Sutherland, G. R. (1979). Heritable fragile sites on human chromosomes. I. Factors affecting expression in lymphocyte culture. *Am. J. Hum. Genet.* **31,** 125–135.
2. Sutherland, G. R. (1977). Fragile sites on human chromosomes: demonstration of their dependence on the type of tissue culture medium. *Science* **197,** 265–266.
3. Sutherland, G. R. (1991). Chromosomal fragile sites. *GATA* **8,** 161–166.
4. Kremer, E. J., Pritchard, M., Lynch, M., Yu, S., Holman, K., Baker, E., Warren, S. T., Schlessinger, D., Sutherland, G. R., and Richards, R. I. (1991). Mapping of DNA instability at the fragile X to a trinucleotide repeat sequence p(CCG)n. *Science* **252,** 1711–1714.
5. Oberlé, I., Rousseau, F., Heitz, D., Kretz, C., Devys, D., Hanauer, A., Boué, J., Bertheas, M. F., and Mandel, J. L. (1991). Instability of a 550-base pair DNA segment and abnormal methylation in fragile X syndrome. *Science* **252,** 1097–1102.
6. Yu, S., Pritchard, M., Kremer, E., Lynch, M., Nancarrow, J., Baker, E., Holman, K., Mulley, J. C., Warren, S. T., Schlessinger, D., Sutherland, G. R., and Richards, R. I. (1991). Fragile X genotype characterized by an unstable region of DNA. *Science* **252,** 1179–1181.
7. Verkerk, A. J. M. H., Pieretti, M., Sutcliffe, J. S., Fu, Y. H., Kuhl, D. P. A., Pizutti, A., Reiner, O., Richards, S., Victoria, M. F., Zhang, F., Eussen, B. E., van Ommen, G. J. B., Blonden, L. A. J., Riggins, G. J., Chastain, J. L., Kunst, C. B., Galjaard, H., Caskey, C. T., Nelson, D. L., Oostra, B. A., and Warren, S. T. (1991). Identification of a gene (FMR-1) containing a CGG repeat coincident with a breakpoint cluster region exhibiting length variation in fragile X syndrome. *Cell* **65,** 905–914.
8. Nakahori, Y., Knight, S. J. L., Holland, J., Schwartz, C., Roche, A., Tarleton, J., Wong, S., Flint, T. J., Froster-Iskenius, U., Bentley, D., Davies, K. E., and Hirst, M. C. (1991). Molecular heterogeneity of the fragile X syndrome. *Nucleic Acids Res.* **19,** 4355–4359.
9. Sutherland, G., and Baker, E. (1992). Characterisation of a new rare fragile site easily confused with the fragile X. *Hum. Mol. Genet.* **1,** 111–113.
10. Knight, S. J. L., Flannery, A. V., Hirst, M. C., Campbell, L., Christodoulou, Z., Phelps, S. R., Pointon, J., Middleton-Price, H. R., Barnicoat, A., Pembrey, M. E., Holland, J., Oostra, B. A., Bobrow, M., and Davies, K. E. (1993). Trinucleotide repeat amplification and hypermethylation of a CpG island in FRAXE mental retardation. *Cell* **74,** 127–134.
11. Gu, Y., Shen, Y., Gibbs, R. A., and Nelson, D. L. (1996). Identification of FMR2, a novel gene associated with the FRAXE CCG repeat and CpG island. *Nat. Genet.* **13,** 109–113.

12. Gecz, J., Gedeon, A., Sutherland, G., and Mulley, J. (1996). Identification of the gene FMR2, associated with FRAXE mental retardation. *Nat. Genet* **13,** 105–108.
13. Chakrabarti, L., Knight, S. J. L., Flannery, A. V. and Davies, K. E. (1996). A candidate gene for mild mental handicap at the FRAXE fragile site. *Hum. Mol. Genet.* **5,** 275–282.
14. Ritchie, R. J., Knight, S. J. L., Hirst, M. C., Grewal, P. K., Bobrow, M., Cross, G. S., and Davies, K. E. (1994). The cloning of FRAXF: trinucleotide repeat expansion and methylation at a third fragile site in distal Xqter. *Hum. Mol. Genet.* **3,** 2115–2121.
15. Parrish, J. E., Oostra, B. A., Verkerk, A. J. M. H., Richards, C. S. Reynolds, J., Spikes, A. S., Shaffer, L. G., and Nelson, D. L. (1994). Isolation of a GCC repeat showing expansion in FRAXF, a fragile site distal to FRAXA and FRAXE. *Nature Genet.* **8,** 229–235.
16. Hirst, M., Barnicoat, A., Flynn, G., Wang, Q., Daker, M., Buckle, V., Davies, K., and Bobrow, M. (1993). The identification of a third fragile site, FRAXF, in Xq27--q28 distal to both FRAXA and FRAXE. *Hum. Mol. Genet.* **2,** 197–200.
17. Flynn, G. A., Hirst, M. C., Knight, S. J. L., Macpherson, J. N., Barber, J. C. K., Flannery, A. V., Davies, K. E., and Buckle, V. J. (1993). Identification of the FRAXE fragile site in two families ascertained for X linked mental retardation. *J. Med. Genet.* **30,** 97–100.
18. Hamel, B. C. J., Smits, A. P. T., de Graaff, E., Smeets, D. F. C. M., Schoute, F., Eussen, B. H. J., Knight, S. J. L., Davies, K. E., Assman-Hulsmans, C. F. C. H., and Oostra, B. A. (1994). Segregation of FRAXE in a large family: Clinical, pyschometric, cytogenetic and molecular data. *Am. J. Hum. Genet.* **55,** 923–931.
19. Knight, S. J. L., Voelckel, M. A., Hirst, M. C., Flannery, A. V., Moncla, A., and Davies, K. E. (1994). Triplet repeat expansion at the FRAXE locus and X-linked mild mental handicap. *Am. J. Hum. Genet.* **55,** 81–86.
20. Mulley, J. C., Yu, S., Loesch, D. Z., Hay, D. A., Donnelly, A., Gedeon, A. K., Carbonell, P., Lopez, I., Glover, G., Gabarron, I., Yu, P. W. L., Baker, E., Haan, E. A., Hockey, A., Knight, S. J. L., Davies, K. E., Richards, R. I., and Sutherland, G. R. (1995). FRAXE and mental retardation. *J. Med. Genet.* **32,** 162–169.
21. Carbonell, P., Lopez, I., Gabarron, J., Bernabe, M., Lucas, J., Guitart, M., Gabau, E., and Glover, G. (1996). FRAXE mutation analysis in three Spanish families. *Am. J. Med. Genet.* **64,** 434–40.
22. Barnicoat, A., Wang, Q., Turk, J., Green, E., Mathew, C., Flynn, G., Buckle, V., Hirst, M., Davies, K., and Bobrow, M. (1997). Clinical, cytogenetic, and molecular analysis of three families with FRAXE. *J. Med. Genet.* **34,** 13–17.
23. Abrams, M., Doheny, K., Mazzocco, M., Knight, S., Baumgardner, T., Freund, L., Davies, K., and Reiss, A. (1997). Cognitive, behavioral, and neuroanatomical assessment of two unrelated male children expressing FRAXE. *Am. J. Med. Genet.* **74,** 73–81.
24. Murgia, A., Polli, R., Vinanzi, C., Salis, M., Drigo, P., Artifoni, L., and Zacchello, F. (1996). Amplification of the Xq28 FRAXE repeats: extreme phenotype variability? *Am. J. Med. Genet.* **64,** 441–444.
25. Knight, S. J. L., Ritchie, R. J., Chakrabarti, L., Cross, G., Taylor, G. R., Mueller, R. F., Hurst, J., Paterson, J., Yates, J. R. W., Dow, D. J., and Davies, K. E. (1996). A study of FRAXE in mentally retarded individuals referred for Fragile X syndrome (FRAXA) testing in the United Kingdom. *Am. J. Hum. Genet.* **58,** 906–913.
26. Knight, S., Voelckel, M., Hirst, M., Flannery, A., Moncla, A., and Davies, K. (1994). Triplet repeat expansion at the FRAXE locus and X-linked mild mental handicap. *Am. J. Hum. Genet.* **55,** 81–86.
27. Brown, W. T. (1996). The FRAXE syndrome: is it time for routine screening? *Am. J. Hum. Genet.* **58,** 903–905.
28. Allingham-Hawkins, D. J., and Ray, P. N. (1995). FRAXE expansion is not a common etiological factor among developmentally delayed males. *Am. J. Hum. Genet.* **56,** 72–76.
29. Murray, A., Youings, S., Dennis, N., Latsky, L., Linehan, P., McKechnie, N., Macpherson, J., Pound, M., and Jacobs, P. (1996). Population screening at the FRAXA and FRAXE loci: molecular analyses of boys with learning difficulties and their mothers. *Hum. Mol. Genet.* **5,** 727–735.
30. Gedeon, A. K., Keinanen, M., Ades, L. C., Kaariainen, H., Gecz, J., Baker, E., Sutherland, G. R., and Mulley, J. C. (1995). Overlapping submicroscopic deletions in Xq28 in two unrelated boys with developmental disorders: identification of a gene near FRAXE. *Am. J. Hum. Genet.* **56,** 907–914.
31. Biancalana, V., Taine, L., Bouix, J.-C., Finck, S., Chauvin, A., De Verneuil, H., Knight, S. J. L., Stoll, C., Lacombe, D., and Mandel, J.-L. (1996). Expansion and methylation status at FRAXE can be detected on EcoR1 blots used for FRAXA diagnosis: analysis of four FRAXE families with mild mental retardation in males. *Am. J. Hum. Genet.* **59,** 847–854.
32. Uberbacher, E. C., and Mural, R. J. (1991). Locating protein-coding regions in human DNA sequences by a multiple sensor-neural network approach. *Proc. Natl. Acad. Sci. USA* **88,** 11261–11265.
33. Gecz, J., Bielby, S., Sutherland, G., and Mulley, J. (1997). Gene structure and subcellular localization of FMR2, a member of a new family of putative transcription activators. *Genomics* **44,** 201–213.
34. Biondi, A., Rambaldi, A., Rossi, V., Elia, L., Caslini, C., Basso, G., Battista, R., Barbui, T., Mandelli, F., Masera, G., Croce, C., Canaani, E., and Cimino, G. (1993). Detection of ALL-1/AF4 fusion transcript by reverse transcription-polymerase chain reaction for diagnosis and monitoring of acute leukemias with the t(4;11) translocation. *Blood* **82,** 2943–2947.
35. Ma, C., and Staudt, L. (1996). LAF-4 encodes a lymphoid nuclear protein with transactivation potential that is homologous to AF-4, the gene fused to MLL in t(4;11) leukemias. *Blood* **87,** 734–745.
36. Gu, Y., Nakamura, T., Alder, T., Prasad, R., Canaani, O., Cimino, G., Croce, C. M., and Canaani, E. (1992). The t(4;11) chromosome translocation of human acute leukemias fuses the ALL-1 gene, related to drosophila trithorax, to the AF-4 gene. *Cell* **71,** 701–708.
37. Gecz, J., Oostra, B., Hockey, A., Carbonell, P., Turner, G., Haan, E., Sutherland, G., and Mulley, J. (1997). FMR2 expression in families with FRAXE mental retardation. *Hum. Mol. Genet.* **6,** 435–441.
38. De Boulle, K., Verkerk, A. J. M. H., Reyniers, E., Vits, L., Hendrickx, J., Van Roy, B., Van Den Bos, F., de Graaff, E., Oostra, B. A., and Willems, P. J. (1993). A point mutation in the FMR-1 gene associated with fragile X mental retardation. *Nature Genet.* **3,** 31–35.
39. Hirst, M. C., Barnicoat, A., Flynn, G., Wang, Q., Daker, M., Buckle, V. J., Davies, K. E., and Bobrow, M. (1993). The identification of a third fragile site, FRAXF in Xq27-28 distal to both FRAXA and FRAXE. *Hum. Mol. Genet.* **2,** 197–200.
40. Romain, D., and Chapman, C. (1992). Fragile site Xq27.3 in a family without mental retardation. *Clin. Genet.* **41,** 33–35.
41. Holden, J., Walker, M., Chalifoux, M., and White, B. (1996). Trinucleotide repeats at the FRAXF locus: frequency and distribution in the general population. *Am. J. Med. Genet.* **64,** 424–7.
42. Ritchie, R., Knight, S., Hirst, M., Grewal, P., Bobrow, M., Cross, G., and Davies, K. (1994). The cloning of FRAXF: trinucleotide repeat expansion and methylation at a third fragile site in distal Xqter. *Hum. Mol. Genet.* **3,** 2115–2121.
43. Nancarrow, J. K., Kremer, E., Holman, K., Eyre, H., Doggett, N. A., Le Paslier, D., Callen, D. F., Sutherland, G. R., and Rich-

ards, R. I. (1994). Implications of FRA16A structure for the mechanism of chromsomal fragile site genesis. *Science* **264,** 1938–1941.

44. Nancarrow, J. K., Holman, K., Mangelsdorf, M., Hori, T., Denton, M., Sutherland, G. R., and Richards, R. I. (1995). Molecular basis of p(CCG)n repeat instability at the FRA16A fragile site locus. *Hum. Mol. Genet.* **4,** 367–372.
45. Richards, R., Crawford, J., Narahara, K., Mangelsdorf, M., Friend, K., Staples, A., Denton, M., Easteal, S., Hori, T., Kondo, I., Jenkins, T., Goldman, A., Panich, V., Ferakova, E., and Sutherland, G. (1996). Dynamic mutation loci: allele distributions in different populations. *Ann. Hum. Genet.* **60** (Pt 5), 391–400.
46. Pivnick, E., Velagaleti, G., Wilroy, R., Smith, M., Rose, S., Tipton, R., and Tharapel, A. (1996). Jacobsen syndrome: report of a patient with severe eye anomalies, growth hormone deficiency, and hypothyroidism associated with deletion 11 (q23q25) and review of 52 cases. *J. Med. Genet.* **33,** 772–778.
47. Jones, C., Slijepcevic, P., Marsh, S., Baker, E., Langdon, W., Richards, R., and Tunnacliffe, A. (1994). Physical linkage of the fragile site FRA11B and a Jacobsen syndrome chromosome deletion breakpoint in 11q23.3. *Hum. Mol. Genet.* **3,** 2123–2130.
48. Jones, C., Penny, L., Mattina, T., Yu, S., Baker, E., Voullaire, L., Langdon, W., Sutherland, G., Richards, R., and Tunnacliffe, A. (1995). Association of a chromosome deletion syndrome with a fragile site within the proto-oncogene CBL2. *Nature* **376,** 145–149.
49. Pennisi, E. (1996). New gene forges link between fragile site and many cancers. *Science* **272,** 649.
50. Ohta, M., Inoue, H., Cotticelli, M. G., Kastury, K., Baffa, R., Palazzo, J., Siprashvili, Z., Mori, M., McCue, P., Druck, T., Croce, C. M., and Huebner, K. (1996). The FHIT gene, spanning the chromosome 3p14.2 fragile site and renal carcinoma-associate t(3;8) breakpoint, is abnormal in digestive tract cancers. *Cell* **84,** 587–597.
51. Eichler, E. E., Lu, F., Shen, Y., Antonacci, R., Jurecic, V., Doggett, N. A., Moyzis, R. K., Baldini, A., Gibbs, R. A., and Nelson, D. L. (1996). Duplication of a gene-rich cluster between 16p11.1 and Xq28: a novel pericentromeric-directed mechanism for paralogous genome evolution. *Hum. Mol. Genet.* **5,** 899–912.
52. Mazzarella, R., Pengue, G., Yoon, J., Jones, J., and Schlessinger, D. (1997). Differential expression of XAP5, a candidate disease gene. *Genomics* **45,** 216–219.

CHAPTER 6

Common Fragile Sites

THOMAS W. GLOVER Departments of Pediatrics and Human Genetics, University of Michigan, Ann Arbor, Michigan 48109

I. INTRODUCTION

The phenomenon of chromosomal fragile sites has intrigued investigators for over two decades. The study of fragile sites was most notably forwarded in the 1970s by Sutherland in his studies of inherited autosomal fragile sites, and gained notoriety with the description of the fragile X site (FRAXA) associated with the fragile X syndrome (reviewed in [1]). FRAXA provided a target for positional cloning, leading to the discovery of CCG trinucleotide repeat expansion in the associated FMR-1 gene and marking the unfolding of a new principal for mutagenesis in human genetic disorders [2–4]. Today, we know of over 100 fragile sites. Some, like FRAXA, are caused by expanded CCG repeats. The majority of fragile sites, however, are the so-called "common fragile sites" [5] about which much less is known.

Chromosomal fragile sites are loci that are especially sensitive to forming chromosome gaps or breaks on metaphase chromosomes. They are site specific in that they occur at cytogenetically identifiable points in the genome and do not generally appear on metaphase chromosomes unless the cells are cultured under conditions that inhibit DNA synthesis or repair. Fragile sites now are generally grouped into two classes based on their relative frequency of occurrence and means of induction. These are the "rare" fragile sites, such as FRAXA in the FMR1 locus, and "common" (or "constitutive") fragile sites, such as FRA3B at 3p14.2, which are found in many if not all chromosomes. Appearance of the rare fragile sites is a manifestation of mutation and genetic variation while common fragile sites apparently represent a constant feature of chromosome structure. A separate category of fragile sites, not included in this discussion, are associated with adenovirus infection of human cells [6].

Growth of cells under conditions of folate or thymidylate stress has led to the identification of 17 rare "folate sensitive" fragile sites, a class which includes FRAXA. Other rare fragile sites are induced by 5-bromodeoxyuridine (BrdU) or distamycin A [1]. Seven of the rare fragile sites have been cloned and 6, including the fragile X site, FRAXA, are associated with CCG trinucleotide repeats [7]. These fragile site loci therefore represent a subclass of the trinucleotide repeat expansion mutations involved in a number of human diseases. However, only the CCG repeat expansion disorders are known to be associated with chromosomal fragile sites.

Expanded CAG, CTG, or GAA repeats apparently do not give rise to fragile sites [8]. The associated fragile site contributed immensely to the positional cloning of the *FMR1* gene and the discovery that an expanded CCG repeat in the 5′ untranslated region of the *FMR1* gene leads to inactivation of the gene and associated chromosomal instability [2–4]. This expansion explains the unusual inheritance and genetic anticipation seen in the fragile X syndrome [9], and accounts for appearance of the associated fragile site on metaphase chromosomes from affected individuals and some carrier females.

In addition to FRAXA, expanded CCG repeats have also been found at the FRAXE and FRAXF loci in Xq27-28, FRA16A in 16p13, and FRA11B in 11q23 [10–13]. FRAXE is associated with mild mental retardation in males with expanded repeats. FRA11B, found in the *CBL2* gene, is believed to lead to terminal deletions of distal chromosome 11 in some cases of Jacobsen syndrome. FRA16A lies adjacent to a CpG island but is not associated with disease in heterozygotes. Expanded CCG repeats are associated with methylation of the repeat and adjacent CpG dinucleotides leading to inactivation of associated genes [12]. A related, but different, rare fragile site was recently cloned and characterized by Yu *et al.* [14]. This fragile site, FRA16B, maps to 16q22.1 and is associated with an expanded 33-bp AT-rich microsatellite repeat. The finding that this sequence contains long AT tracts was predicted based on the fact that it and a few other rare fragile sites are induced by AT-binding drugs such as distamycin A and netropsin.

In all of these cases, the instability observed at the nucleotide level as expanded repeats is translated through an apparent change in chromatin structure to a higher order instability observed at the level of the metaphase chromosome. These changes can give rise to genetic disease by modifying the expression of neighboring genes, as is the case with FRAXA and FRAXE, or by mediating chromosome deletions, as is seen in some cases of Jacobsen syndrome. In all cases, the changes give rise to fragile sites on metaphase chromosomes by mechanisms that likely involve formation of complex intramolecular structures that act as a block to DNA synthesis.

II. COMMON FRAGILE SITES

In the early investigations of the cytogenetics of the fragile X syndrome, the occurrence of site-specific chromosomal gaps and breaks was noted on chromosomes of all individuals, hence the term "common fragile sites" [5]. Certain sites such as 3p14.2, 16q23, and Xp22.3 repeatedly displayed gaps and breaks which appear on the cytological level just like the fragile X site. While common fragile sites are seen when cells are grown under those conditions of folate or thymidylate stress that also induce the rare (CCG)n-associated fragile sites, they are more efficiently seen when cells are treated with aphidicolin, an inhibitor of DNA polymerases α and γ [5]. However, the rare fragile sites are not induced by aphidicolin, suggesting related, but different, mechanisms for instability.

At present, 84 common fragile sites are listed in the Genome Database (GDB). The exact number of common fragile sites that exist is a matter of interpretation based on criteria for inclusion and statistical analyses of data. The greater the stress placed on DNA replication or G_2 repair, the more breaks and gaps are observed until replication ceases altogether. The most frequently observed common fragile sites occur at 3p14.2 (FRA3B), 16q23 (FRA16D), 6q26 (FRA6E), 7q32 (FRA7H), Xp22.3 (FRAXB), and 5–10 other sites [5] (Fig. 6-1).

Most common fragile sites occur in G-light bands, leading to suggestions that they may be associated with active gene regions [15, 16]. Variation among normal individuals in the proportion of treated cells expressing common fragile sites has been reported [25], and a study of expression in twins [17, 18] has suggested that the level of expression is under stringent genetic control. However, there is also very likely an environmental component to fragile site expression. Cigarette smokers, for example, consistently show a higher level of expression of specific common fragile sites [19, 20].

A. Significance of Common Fragile Sites

The biological significance of some rare fragile sites, especially FRAXA in the FMR1 locus, is well estab-

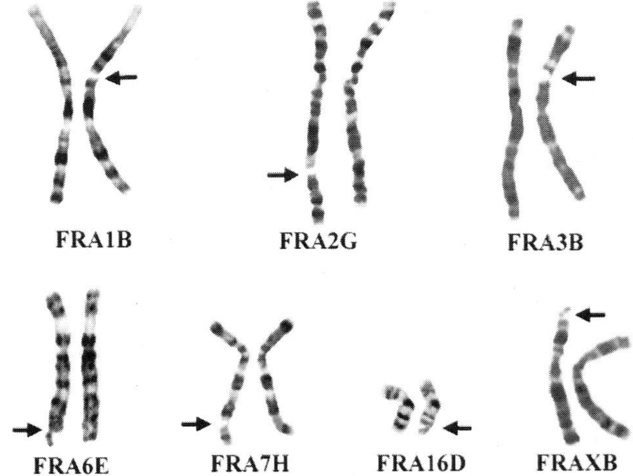

FIGURE 6-1 Examples of some of the most frequently expressed common fragile sites in cultured human lymphocytes.

lished due to their association with genetic diseases. What is the significance of common fragile sites? Even though these sites compose the vast majority of fragile sites, little is known about their structure or biological role. While most rare fragile sites arise via expansion of normally occurring CCG repeats, and represent aberrations in nucleotide and chromatin structure, the large number of common fragile sites are a constant feature of the genome. This constancy of common fragile sites suggests that their overall nucleotide and/or chromatin structure is also unvarying, and implies a conservation of function.

Common fragile sites have been of interest because they represent an unknown component of chromosome structure, and because of their instability and high recombination *in vitro*. Based on their behavior *in vitro* and their locations, suggestions have been made that fragile sites may play a mechanistic role in chromosome breakage and rearrangements involved in cancer, birth defects, and evolution [15, 21–23]. This association is based on the coincident location of a number of fragile sites and the location of breakpoints in characteristic chromosome aberrations in cancer. However, these studies have been based largely on statistical correlation of breakpoints and fragile sites with little direct evidence either to support or refute this suggestion. They make the assumption that instability at fragile sites can lead to chromosome breakage *in vivo,* as they do *in vitro,* and that these breaks result in chromosome rearrangements or deletions. Recent data made possible by molecular analyses of the FRA3B site appear to support this assumption.

In addition to forming gaps or breaks on metaphase chromosomes, common fragile sites have been shown to display a number of characteristics of unstable and highly recombinogenic DNA *in vitro.* All of these characteristics are consistent with the hypothesis that DNA strand breaks are associated with fragile site induction. Following induction, these sites are "hot spots" for increased sister chromatid exchange on metaphase chromosomes [24] and show a high rate of translocations and deletions in somatic cell hybrid systems [25, 26]. An interesting characteristic of common fragile sites that supports the hypothesis that they represent DNA lesions was reported by Rassool *et al.* [27] who showed that they are preferred sites of recombination, or integration, with pSV2neo-plasmid DNA transfected into cells pretreated for fragile site induction. Perhaps related to this characteristic are the findings of Wilke *et al.* [28] that FRA3B was the site of integration of HPV16 in a primary cervical carcinoma, and other reports of the coincidence of viral integration sites in tumors or tumor cell lines and fragile sites as studied by FISH [29–31].

Fragile sites have recently been implicated in intrachromosomal gene amplification events in cultured CHO cells. Kuo *et al.* [32] found a fragile site in CHO cells to be consistent with the site of three independently established P-glycoprotein gene amplification events, and that pretreatment of cells with aphidicolin greatly increased such amplification events. These results suggested that fragile sites may also play a pivotal role in some gene amplification by initiating a breakage–fusion bridge cycle. More recently, Coquelle *et al.* [33] have expanded on the findings that show a relationship between fragile sites and gene amplification. They showed that actinomycin D, a known clastogen with CG binding preference, and related compounds induce fragile sites in CHO cells, including fragile sites flanking the *mdr1* gene. Following exposure to the drugs, the *mdr1* gene is amplified. Other clastogens such as adriamycin that cause random breakage do not induce *mdr1* amplification. Methotrexate, an inhibitor of dihydrofolate reductase (which also induces human common fragile sites), had the same effect at the *dhfr* locus and at other loci near fragile sites. The *mdr1* and *dhfr* loci are flanked by fragile sites induced by the respective drugs. The authors hypothesize that DNA strand breaks preferentially at fragile sites initiate breakage–bridge–fusion cycles, triggering, and setting limits to, amplification of large DNA segments between the fragile sites. Additional studies of the connection between fragile sites and gene amplification are necessary to understand this process, and are important in fully understanding the significance of fragile site expression in cancer cells.

B. Molecular Analysis of the FRA3B Common Fragile Site

Only one common fragile site, FRA3B, has so far been extensively studied on the molecular level. This fragile site was targeted for study because it is the most frequently seen fragile site on human metaphase chromosomes and can be induced in the majority of treated cells. It maps to a region of 3p known to be associated with deletions in a number of solid tumors, including small cell lung cancer (SCCL) [34], hereditary [35] and sporadic [36] renal cell cancer (RCC), breast cancer, and cervical cancer [37]. In addition, FRA3B was found to lie very near the t(3;8) translocation breakpoint segregating in a family described with familial renal cell carcinoma [38].

Sequences at FRA3B have been identified by a number of related approaches. Wilke *et al.* [28, 39] used a positional cloning approach and FISH to localize the fragile site to a large region of >100 kb that is ~160 kb telomeric to the renal cell carcinoma t(3;8) breakpoint.

Sequence analysis of 9 kb at the "center" of the defined region revealed a site of integration of HPV 16 (Fig. 6-2) in a human cervical carcinoma, suggesting that fragile sites may be hot spots for integration of viral DNA *in vivo* as they are for transfected DNA *in vitro* [27]. Paradee *et al.* [40, 41] mapped a series of aphidicolin-induced chromosome 3 breakpoints to the same region, and identified two clusters of breaks flanking the region studied by Wilke *et al.* [28]. By cloning sequences surrounding pSV2neo-plasmid DNA integrated into FRA3B after being transfected into cells treated with aphidicolin, Rassool *et al.* [42] identified sequences within the fragile site region ~350 kb distal to the t(3;8) breakpoint. Boldog *et al.* [43] cloned a 300-kb region spanning an aphidicolin-induced chromosome 3 translocation breakpoint and have sequenced over 110 kb of this region. This sequence matched the HPV16 integration site, the aphidicolin-induced breakpoints, and the pSV2-neo integration sites in the FRA3B region. Finally, in an approach aimed at investigating regions of LOH in tumors, Ohta *et al.* [44] cloned the *FHIT* gene from the same region, and it is now known that *FHIT* spans FRA3B.

The results of these studies show that the common fragile sites, and FRA3B in particular, differ from FRAXA and other rare fragile sites in many ways. First, gaps, breaks, and instability occur over a very large genomic region relative to FRAXA. FRA3B extends from approximately the t(3;8) breakpoint to roughly 500 kb telomeric. Whether this represents a single unstable region or multiple unstable regions is unknown. Second, trinucleotide repeats have not been identified in the region [28, 41, 43]. Thus, while all fragile sites appear similar on the level of the metaphase chromosome, there are significant differences on the molecular level.

C. FRA3B Lies within the *FHIT* Gene—An Unstable Gene in Tumor Cells

Are common fragile sites associated with genes? Given their apparent instability, this might have been considered doubtful not long ago. However, FRA3B, so far the only common fragile site to be characterized at the molecular level, lies within a gene that is highly unstable in tumor cells.

The identification of this gene followed from searches for regions that might contain tumor suppressor genes in tumor cell lines. Based on representational difference analysis (RDA) of genomic DNA, Lisitsyn *et al.* [45] identified a region on 3p (BE 758-6, Fig. 6-2) which showed homozygous deletions in gastrointestinal tumor cell lines. Kastery *et al.* [46] defined homozygous deletions encompassing these sequences in a variety of epithelial tumor cell lines. Focusing on this deleted region, Otha *et al.* [44] used exon trapping to identify an exon of what is now known as the *FHIT* gene (for fragile histadine triad) from a cosmid contig spanning this region. The region of homozygous deletion, and the *FHIT* gene, were subsequently found to map exactly at FRA3B.

The *FHIT* gene spans an estimated 700–900 kb of 3p14.2, and spans both FRA3B and the t(3;8) familial renal cell carcinoma breakpoint (Fig. 6-2). It contains 10 exons, 5 of which are coding. In contrast to its large genomic size, it encodes only a 1.1-kb transcript that is ubiquitously expressed at relatively low levels [44].

The product of the *FHIT* gene is a 147-amino-acid (16.8 kDa) protein with identity to diadenosine 5′,5‴-P1,P4-tetraposphate hydrolase (Ap4A) from *Schizosaccharomyces pompe* which functions in the cleavage of diadenosine tetraphosphates [44]. Both proteins have sequence homology to the HIT family of proteins. These proteins are characterized by four conserved histidines, three of which compose a histidine triad (HxHxH), or HIT sequence which is encoded by exon 8 for the human protein. Barnes *et al.* [47] have recently demonstrated that the preferred substrate for the *FHIT* protein is a diadenosine triphosphate, rather than tetraphosphate; thus, *FHIT* functions as a 5′5‴-P1,P3-diadenosine triphosphate (Ap3A) hydrolase (EC 3.6.1.29).

The biological significance of the *FHIT* gene is not clearly understood. Diadenosine tri- and tetraphosphates are ubiquitous small molecules produced from ATP and consist of two adenosine molecules bridged by two or three phosphate groups. They have been implicated in cell cycle arrest and are synthesized in response to cellular stress such as starvation and heat shock conditions in prokaryotes, leading to their being

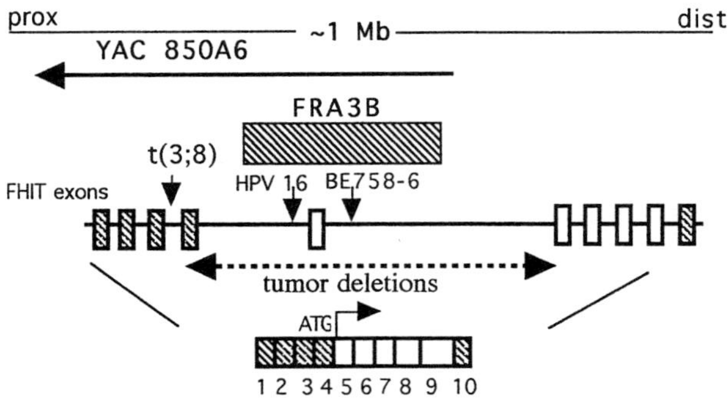

FIGURE 6-2 General diagram of maps of FRA3B and the *FHIT* gene. The relative position of FRA3B was determined from positioning of genomic clones used in FISH analysis of FRA3B and *FHIT* exons. The general position of deletions seen in tumor cell lines is indicated.

termed "alarmones" [48, 49]. This possible role is of particular interest when considering fragile sites given that they are induced under conditions of cellular stress or nucleotide starvation. Besides this, the significance of diadenosine phosphates as modulators of physiological processes such as hemostasis and neurotransmission have been the focus of intensive research interest for many years.

D. The *FHIT*/FRA3B Region Is Frequently Deleted in Tumor Cells

Because of its location in a region of frequent LOH in many tumors, and the fact that a homozygous deletion in an esophageal tumor provided a foothold into 3p14.2 for the identification of *FHIT,* Ohta *et al.* [44] and others from this group studied the *FHIT* gene in a number of tumors and tumor cell lines. Using RT-PCR to study RNA from cell lines as well as primary squamous cell esophageal, stomach, and colon carcinomas, aberrant *FHIT* transcripts, suggesting deletions, were identified in ~50% of the tumors and most of the cell lines. On the genomic level, the cell lines were also shown to have homozygous deletions within *FHIT,* and the FHIT protein was shown to be absent or reduced in some of these cell lines. Based on these findings, Ohta *et al.* [44] suggested that *FHIT* is a frequently mutated tumor suppressor gene.

Aberrant transcripts were subsequently identified by RT-PCR in numerous tumors or tumor cell lines including breast [50], lung [51], head and neck [52], cervical [53], and Barrett's esophageal tumors [54]. While the nature of the aberrant transcripts is not completely understood, genomic deletions have been identified in many of the cell lines and some of the primary tumors, including both premalignant metaplasia and adenonocarcinomas of Barrett's esophagus by our group [54]. In this study, deletions of exon 5 and within intron 5 of *FHIT,* both within FRA3B by FISH, were observed. The results of genomic deletion analysis coincided with the RT-PCR results in some but not all cases, suggesting that other factors such as aberrant splicing in tumor cells, a phenomenon also recently reported for the TS101 gene [55], may also contribute to the appearance of aberrant transcripts.

These and other studies have firmly shown a relationship between deletions and instability at the *FHIT*/FRA3B locus and cancer cells. Questions that remain include whether these deletions are a cause or an effect of cancer, and whether FRA3B is truly the cause of the instability. Arguments against the *FHIT* gene acting as a classical tumor suppressor gene include the fact that the known function of *FHIT* is not obviously related to tumor suppressor activity. In addition, some primary tumors and cancer cell lines harbor allelic and homozygous deletions only in intronic regions [43, 51, 56]. These deletions would presumably not affect *FHIT* function, although it is conceivable that they could affect normal splicing. Under certain conditions, "aberrant" *FHIT* transcripts have been seen in normal cells [43] and alternative splicing has been suggested to account for some of these findings [43, 57]. In addition, functional data to support the hypothesis that *FHIT* is a tumor suppressor has largely been lacking. Therefore, alternative hypothesis for the unusual findings with the *FHIT* gene include aberrant or alternative splicing, selection for another mutated tumor suppressor gene in the region, perhaps in the large *FHIT* intron 5, and genomic instability at FRA3B which lies at the center of the frequently deleted regions of *FHIT.*

Genomic instability is a frequent characteristic of tumor cells, and common fragile sites may be particularly sensitive to such instability. Thus, as now suggested by many investigators, *FHIT* deletions and aberrant transcripts may arise through FRA3B leading to genomic instability and possibly interfering with *FHIT* transcription and/or splicing without providing a selective advantage to the affected cell [43, 54, 56]. Whether the *FHIT* gene is causally involved in carcinogenesis or is an "innocent bystander," it represents a highly unstable gene in cancer. It contains a common fragile site which may cause the instability, although this has not been proven by direct experimentation. It is possible that similarly unstable genes could be associated with the 80 or so other common fragile sites, and such associations would lend support to this hypothesis.

E. Molecular Studies of Other Common Fragile Sites

Fragile sites are clearly recognizable targets for identification of YACs and other large genomic clones by FISH analysis. With the explosion of mapped resources now available, the precise mapping and identification of clones spanning additional common fragile sites is expected to proceed at a rapid rate for those who search. Huang *et al.* [58] have recently identified YACs spanning the common fragile site at 7q32.1 (FRA7G). This fragile site region also appears large, over 300 kb, and contains markers that are frequently lost in breast, prostate, and ovarian cancers. My laboratory has recently identified YAC clones spanning common fragile sites at Xp22.1 (FRAXB) and 16q23.1 (FRA16D) and has begun their analysis. FRAXB lies in a region of frequent deletions at Xp22, while the region in which FRA16D maps has been identified as a region of frequent LOH in prostate and breast cancer [59]. Whether these fragile sites are mechanistically related to the deletions noted in these regions is unknown at present. Comparison of sequences at these fragile sites could aid in identifying the mechanism of fragility, if a common sequence motif exists.

F. Fragile Sites in Evolution

Common fragile sites have been observed and characterized in a number of mammals, and have been suggested to play a role in chromosome rearrangements involved in speciation. These include the dog, cat, pig, horse, cow, Indian mole rat, deer mouse, and laboratory mouse [60–64]. In all cases, the fragile sites are induced by folate stress or aphidicolin, as in humans. For example, in the randomly bred ICR mouse strain, 19 fragile sites are repeatedly seen, including one on chromosome 14, band A2, in the region where we have mapped the mouse *FHIT* gene (Glover, unpublished). In the domestic dog, 16–20 fragile sites have been described, including 3 on the X chromosome at sites which appear to be homologous to human Xp22, Xq22, and Xq27, all of which contain human common fragile sites. Interesting data from wild populations of the Deer mouse [64], *Peromyscus,* in which two pericentric inversions are polymorphic in the population, support a role of fragile sites in chromosome evolution. The fragile sites are only seen in the wild-type chromosomes at the inversion breakpoints and disappear on the rearranged chromosomes.

Very little is known of variation of common fragile site expression in humans, and even less about fragile site expression in other mammals. Until now, molecular probes to specific fragile sites have not been available for studies of the evolution of common fragile sites. FISH analysis with probes spanning fragile sites could be used to unequivocally indicate if fragile sites are involved in chromosome breakage and rearrangements during evolution. Furthermore, such studies could help answer questions about the mechanisms and biological significance of fragile sites. For example, the absence of FRA3B or other common fragile sites in certain primates would provide a basis for comparison to investigate sequences responsible for fragile site expression.

G. Mechanism of Instability

The mechanism underlying instability at fragile sites, including the trinucleotide repeats, is not well understood. The mechanism for induction of all fragile sites appears to involve interruptions of normal DNA synthesis. DNA repair mechanisms may also play a role since caffeine, an inhibitor of G2 repair, increases the number of cells expressing fragile sites [15, 65]. Based on the fact that fragile sites are induced by agents that retard DNA replication or repair, and on the high frequency of sister chromatid exchanges and chromosome rearrangements, it was proposed many years ago that fragile sites were associated with unreplicated DNA or DNA strand breaks [5, 24, 25].

The same events that lead to instability at CCG repeats during meiosis or mitosis *in vivo* also likely lead to the manifestation of fragile sites on metaphase chromosomes. Richards and Sutherland have suggested that the massive expansion seen at trinucleotide repeats may be due to the occurrence of more than one single-strand DNA break causing slippage of unanchored Okazaki fragments during replication [66]. Instability in related tandem repeat sequences has been shown by Jeffreys *et al.* [67] to involve a complex process involving strand breaks, gene conversion or internal reduplication, and gap repair. Interestingly, mutations in these repeats tend to occur at one end of the repeat, i.e., they are polar, and this has led to the suggestion that sequences adjacent to the repeats may play a role in the instability. In recent years, investigators have demonstrated an association of trinucleotide repeats with exclusion of nucleosomes [68], DNA hairpin structures [69], and pausing of DNA synthesis [70].

Even less is known about the mechanism responsible for common fragile sites. No trinucleotide or other simple repeat sequence motif has yet been identified that could be responsible for the instability [39, 41, 43]. The only common fragile site region so far examined in detail, FRA3B, is very large and complete sequence is not yet available. Boldog *et al.* [43] have sequenced approximately 110 kb of the centromeric portion of FRA3B within intron 5 of the *FHIT* gene, and have found the region to be high in A-T content and in LINE and MER repeats, with few alu repeats. Interestingly, they identified a small polydispersed circular DNA (spcDNA) sequence in this region. Such sequences have been shown to be associated with clustered repeats such as beta-satellites and are elevated in Fanconi's anemia patients. They are also produced by DNA damaging agents including the fragile site inducer, aphidicolin.

If fragile sites are indeed regions of unreplicated DNA, this could arise from a number of reasons in addition to repeats leading to blocks to DNA replication, such as late replication or a paucity of replication origins in the region. Laird *et al.* [71] have proposed a model for fragile sites wherein they occur at alleles whose replication is delayed relative to nonfragile alleles. LeBeau *et al.* [72] have recently shown that FRA3B is a late-replicating region and that replication is further delayed, or inhibited, by aphidicolin used to induce expression of FRA3B. While late replication alone does not result in a fragile site, fragile site sequences may not be able to recover from a further delay in DNA synthesis. Such regions of unreplicated DNA persisting in G2 and M could explain why caffeine, an inhibitor of G2 repair, increases fragile site expression. A largely unexplored possibility is that common fragile site expression is related to transcriptional timing of associated late replicating genes, such as *FHIT* or other genes at FRA3B.

In addition to *cis*-acting sequences, there are undoubtedly other genetic and environmental factors that influence fragile site expression and associated genomic instability. As mentioned above, caffeine treatment *in vitro* increases fragile site expression [15, 65], and cigarette smokers show increased frequencies of fragile site expression in their cultured lymphocytes [19, 20]. Cigarette smokers have recently also been shown to have increased aberrant *FHIT* transcripts by PCR [73], and the two observations are likely linked to a common mechanism. How cigarette smoking can lead to increased expression of a fragile site, even after lymphocytes are cultured for 3–4 days, is presently unknown.

Candidates for genes influencing expression are those involved in DNA replication and repair, and those involved in cell cycle checkpoints such as p53. It is clear that one of the defining features of cancer cells is an increase in genome instability. Mutations in the p53 gene are known to result in polyploidy and widespread karyotypic instability, amplifications, and deletions [74, 75]. Loss of p53 function by mutation or inactivation of the protein leads to an inability of the cell to interrupt its cell division cycle in response to many types of DNA damage, perhaps including DNA damage at fragile sites. It is very possible that fragile site regions, and any associated genes (such as *FHIT* at FRA3B), may be especially sensitive to the effects of such instability. This could in part explain why the *FHIT* gene is so frequently deleted and rearranged in many tumors. Boldog *et al.* [43] have recently found a correlation between tumors with *FHIT* deletions and their p53 mutation status, supporting this suggestion. The recent findings that fragile sites are associated with intrachromosomal gene amplification in CHO cells may relate to their deficiency in p53 or related genes.

It is clear that, at present, there are more questions than answers regarding the mechanism of expression of common fragile sites and their biological significance. However, recent events such as cloning of the FRA3B site, the discovery of the *FHIT* gene, and the association of fragile sites with intrachromosomal gene amplification have focused new attention on these chromosomal phenomena. The near future should provide more of the answers.

Acknowledgments

I thank Ferenc Boldog, Harry Drabkin, Robert Gemmill, Michelle Lebeau, Diane Miller, Feyruz Rassool, and David Smith for reviewing this chapter and providing helpful comments. I am also grateful to Diane Miller for producing Fig. 6-1.

References

1. Sutherland, G. R., and Hecht, F. (1985). Fragile sites on human chromosomes. *In* "Oxford Monographs on Medical Genetics," Oxford University Press, Oxford.
2. Verkerk, A. J. M., Pieretti, M., Sutcliffe, J. S., Fu, Y. H., Kuhl, D. P., Pizzuti, A., Reiner, O., Richards, S., Victoria, M. F., and Zhang, F. P. (1991). Identification of a gene (*FMR-1*) containing a CGG repeat coincident with a breakpoint cluster region exhibiting length variation in fragile X syndrome. *Cell* **65,** 905–914.
3. Kremer, E. J., Pritchard, M., Lynch, M., Yu, S., Holman, K., Baker, E., Warren S. T., Schlessinger, D., Sutherland, G. R., and Richards, R. I. (1991). Mapping of DNA instability at the fragile X to a trinucleotide repeat sequence p(CCG)n. *Science* **252,** 1711–1718.
4. Oberle, I., Rousseau, F., Heitz, D., Kretz, C., Deveys, D., *et al.* (1991). Instability of a 550-base pair DNA segment and abnormal methylation in fragile X syndrome. *Science* **252,** 1097–1102.
5. Glover, T. W., Berger, C., Coyle, J., and Echo, B. (1984). DNA polymerase a inhibition by aphidicolin induces gaps and breaks at common fragile sites in human chromosomes. *Hum. Genet.* **67,** 136–142.
6. Bailey, A. D., Li, Z., Pavelitz, T., and Weiner, A. M. (1995). Adenovirus type 12-induced fragility of the human *RNU2* locus requires U2 small nuclear RNA transcriptional regulatory elements. *Mol. Cell. Biol.* **15,** 6246–55.
7. Warren, S. T. (1996). The expanding world of trinucleotide repeats. *Science* **271,** 1374–1375.
8. Jalal, S. M., Lindor, N. M., Michels, V. V., Buckley, D. D., Hoppe, D. A., Sarkar, G., and Dewald, G. W. (1993). Absence of chromosome fragility at 19q13.3 in patients with myotonic dystrophy. *Am. J. Med. Genet.* **46,** 441–439.
9. Fu, Y. H., Kuhl, D. P. A., Pizzuti, A., Pieretti, M., Sutcliffe, J. S., Richards, S., Verkerk, A. J., Holden, J. J., Fenwick, R. G., Jr., and Warren, S. T. (1991). Variation of the CGG repeat at the fragile X site results in genetic instability: resolution of the Sherman paradox. *Cell* **67,** 1047–1058.
10. Knight, S. J. L., Flannery, A. V., Hirst, A. C., Campbell, L., Christodoulou, Z., Phelps, S. R., Pointon, J., Middleton-Price, H. R., Barnicoat, A., and Pembrey, M. E. (1993). Trinucleotide repeat amplification and hypermethylation of a CpG island in FRAXE mental retardation. *Cell* **74,** 1–20.
11. Parrish, P., Oostra, B. A., Verkerk, A. J. M. H., Richards, C. S., *et al.* (1994). Isolation of a GCC repeat showing expansion in FRAXF, a fragile site distal to FRAXA and FRAXE. *Nature Genet.* **8,** 229–235.
12. Nancarrow, J. K., Kremer, E., Holman, K., Eyre, H., Dogget, N. A., Le Paslier, D., Callen, D. F., Sutherland, G. R., and Richards, R. I. (1994). Implications of FRA16A structure for the mechanism of chromosomal fragile site genesis. *Science* **264,** 1938–1941.

13. Jones, C., Penny, L., Mattina, T., Yu, S., Baker, E., Voullaire, L., Langdon, W. Y., Sutherland, G. R., Richards, R. I., and Tunnacliffe, A. (1995). Association of a chromosome deletion syndrome with a fragile site within the proto-oncogene CBL2. *Nature* **376,** 145–149.
14. Yu, S., Mangelsdorf, M., Hewett, D., Hobsen, L., Baker, E., Eyre, H. J., Lapsys, N., Le Paslier, D., Doggett, N. A., Sutherland, G. R., and Richards, R. I. (1997). Human chromosomal fragile site FRA16B is an amplified AT-rich minisatellitte repeat. *Cell* **88,** 367–374.
15. Yunis, J. J., and Soreng, A. L. (1984). Constitutive fragile sites and cancer. *Science* **226,** 1199–1204.
16. Hecht, F. (1988). Fragile sites, cancer chromosome breakpoints, and oncogenes all cluster in light G-bands. *Cancer Genet. Cytogenet.* **31,** 17–24.
17. Austin, M. J., Collins, J. M., Corey, L. A., Nance, W. E., Neale, M. C., Schieken, R. M., and Brown, J. A. (1992). Aphidicolin-inducible common fragile-site expression: results from a population survey of twins. *Am. J. Hum. Genet.* **50,** 76–83.
18. Tedeschi, B., Vernole, P., Sanna, M. L., and Nicoletti, B. (1992). Population cytogenetics of aphidicolin-induced fragile sites. *Hum. Genet.* **89,** 543–547.
19. Ban, S., Cologne, J. B., and Neriishi, K. (1995). Effect of radiation and cigarette smoking on expression of FUdR-inducible common fragile sites in human peripheral lymphocytes. *Mutat. Res.* **334,** 197–203.
20. Stein, C. K., Glover, T. W., Palmer, J. L., Naylor, S., and Glissen, B. S. Direct correlation between FRA3B expression and smoking. [Submitted for publication]
21. LeBeau, M. M., and Rowley, J. D. (1984). Heritable fragile sites and cancer. *Nature* **308,** 607–608.
22. Hecht, F., and Sutherland, G. R. (1984). Fragile sites and cancer breakpoints. *Cancer Genet. Cytogenet.* **12,** 179–181.
23. Hecht, F., and Glover, T. W. (1984). Cancer chromosome breakpoints and common fragile sites induced by aphidicolin. *Cancer Genet. Cytogenet.* **13,** 185–188.
24. Glover, T. W., and Stein, C. K. (1987). Induction of sister chromatid exchanges at common fragile sites. *Am. J. Hum. Genet.* **41,** 882–890.
25. Glover, T. W., and Stein, C. K. (1988). Chromosome breakage and recombination at fragile sites. *Am. J. Hum. Genet.* **43,** 265–273.
26. Wang, N. D., Testa, J. R., and Smith, D. I. (1993). Determination of the specificity of aphidicolin-induced breakage of the human 3p14.2 fragile site. *Genomics* **17,** 341–347.
27. Rassool, F. V., McKeithan, T. W., Neilly, M. E., van Melle, E., Espinosa, III R., and Lebeau, M. M. (1991). Preferential integration of marker DNA into the chromosomal fragile site at 3p14: An approach to cloning fragile sites. *Proc. Natl. Acad. Sci. USA* **88,** 6657–6661.
28. Wilke, C. M., Hall, B. K., Hoge, A., Pardee, W., Smith, D. I., and Glover, T. W. (1996). FRA3B extends over a broad region and contains a spontaneous HPV integration site: direct evidence for the coincidence of viral integration sites and fragiles sites. *Hum. Mol. Genet.* **5,** 187–195.
29. De Braekeleer, M., Sreekantaiah, C., and Haas, O. (1992). Herpes simplex virus and human papillomavirus sites correlate with chromosomal breakpoints in human cervical carcinoma. *Cancer Genet. Cytogenet.* **59,** 135–137.
30. Popescu, N. C., and DiPaolo, J. A. (1989). Preferential sites for viral integration on mammalian genome. *Cancer Genet. Cytogenet.* **42,** 157–171.
31. Smith, P. P., Friedman, C., Bryant, E. M., and McDougall, J. K. (1992). Viral integration and fragile sites in human papillomavirus-immortalized human keratinocyte cell lines. *Genes Chrom. Cancer* **5,** 150–157.
32. Kuo, M. T., Vyas, R. C., Jiang, L. X., and Hittelman, W. N. (1994). Chromosome breakage at a major fragile site associated with P-glycoprotein gene amplification in multidrug-resistant CHO cells. *Mol. Cell Biol.* **14,** 5202–11.
33. Coquelle, A., Pipiras, E., Toledo, F., Buttin, G., and Debatisse, M. (1997). Expression of fragile sites triggers intrachromosomal mammalian gene amplification and sets boundaries to early amplicons. *Cell* **89,** 215–225.
34. Hibi, K., Takahashi, T., Yamakawa, K., Ueda, R., Sekido, Y., Ariyoshi, Y., Suyama, M., Takagi, H., Nakamura, Y., and Takahashi, T. (1992). Three distinct regions involved in 3p deletion in human lung cancer. *Oncogene* **7,** 445–449.
35. Cohen, A. J., Li, F. P., Berg, S., Marchetto, D. J., Tsai, S., Jacobs, S. C., and Brown, R. S. (1979). Hereditary renal-cell carcinoma associated with a chromosomal translocation. *N. Engl. J. Med.* **301,** 592–595.
36. Zbar, B., Brauch, H., Talmadge, C., and Linehan, M. (1987). Loss of alleles of loci on the short arm of chromosome 3 in renal cell carcinoma. *Nature* **327,** 721–724.
37. Yamakawa, K., Morita, R., Takahashi, E., Hori, T., Ishikawa, J., and Nakamura, Y. (1991). A detailed deletion mapping of the short arm of chromosome 3 in sporadic renal cell carcinoma. *Cancer Res.* **51,** 4707–4711.
38. Glover, T. W., Coyle-Morris, J. F., Li, F. P., Brown, R. S., and Berger, C. S. (1988). Translocation t(3;8)(p14.2;q24.1) in renal cell carcinoma affects expression of the common fragile site at 3p14 (FRA3B) in Lymphocytes. *Cancer Genet. Cytogenet.* **31,** 69–73.
39. Wilke, C. M., Hall, B. K., Boldog, F., Gemmill, R. M., Chandrasekharappa, C., Drabkin, H., and Glover, T. W. (1994). Multicolor FISH mapping of YAC clonec in 3p14 and isolation of a YAC spanning both FRA3B and the t(3;8) translocation associated with heritable renal cell Carcinoma. *Genomics* **22,** 319–326.
40. Paradee, W., Wilke, C. M., Hoge, A., Glover, T. W., and Smith, D. I. (1996). A 350 Kb cosmid contig in 3p14.2 that crosses the t(3;8) hereditary renal cell carcinoma translocation breakpoint and 17 aphidicolin-induced FRA3B breakpoints. *Genomics* **35,** 87–93.
41. Paradee, W., Mullins, C., He, Z., Glover, T. W., Wilke, C., Opalka, B., Schutte, J., and Smith, D. I. (1995). Precise localization of aphidicolin-induced breakpoints on the short arm of human chromosome 3. *Genomics* **27,** 358–61.
42. Rassool, F. V., Le Beau, M. M., Shen, M. L., Neilly, M. E., Espinosa III, R., Ong, S. T., Boldog, F., Drabkin, H., McCarroll, R., and McKeithan, T. W. (1996). Direct cloning of DNA sequences from the common fragile site region at chromosome band 3p14.2. *Genomics* **35,** 109–17.
43. Boldog, F., Gemmill, R., West, J., Robinson, M., and Li, E. (1997). Chromosome 3p14 homozygous deletions and sequence analysis of FRA3B. *Hum. Mol. Genet.* **6,** 193–203.
44. Ohta, M., Inoue, H., Cotticelli, M. G., Kastury, K., Baffa, R., Palazzo, J., Siprashvili, Z., Mori, M., McCue, P., Druck, T., Croce, C. M., and Huebner, K. (1996). The *FHIT* gene, spanning the chromosome 3p14.2 fragile site and renal carcinoma-associated t(3;8) breakpoint, is abnormal in digestive tract cancers. *Cell* **84,** 587–597.
45. Lisitsyn, N. A., Lisitsina, N. M., Dalbagni, G., Barker, P., Sanchez, C. A., Gnarra, J., Linehan, W. M., Reid, B. J., and Wigler, M. H. (1995). Comparative genomic analysis of tumors: detection of DNA losses and amplification. *Proc. Natl. Acad. Sci. USA* **92,** 151–155.

46. Kastury, K., Baffa, R., Druck, T., Ohta, M., Cotticelli, M. G., Inoue, H., Negrini, M., Rugge, M., Huang, D., Croce, C. M., Palazzo, J., and Huebner, K. (1996). Potential gastrointestinal tumor suppressor locus at the 3p14.2 FRA3B site identified by homozygous deletions in tumor cell lines. *Cancer Res.* **56,** 978–983.
47. Barnes, L. D., Garrison, P. N., Siprashvili, Z., Guranowski, A., Robinson, A. K., Ingram, S. W., Croce, C. M., Ohta, M., and Huebner, K. (1996). *FHIT,* a putative tumor suppressor in humans, is a dinucleoside 5′,5′′′-P1,P3-triphosphate hydrolase. *Biochemistry* **35,** 11529–11535.
48. Segal, E., and Le Pecq, J. B. (1986). Relationship between cellular diadenosine 5′, 5″-P_1, P_4- tetraphosphate level, cell density, cell growth stimulation and toxic stresses. *Exp. Cell Res.* **167,** 119–126.
49. Garrison, P. N., Mathis, S. A., and Barnes, L. D. (1986). *In vivo* levels of diadenosine tetraphosphate and adenosine tetraphosphoguanosine in Physarum polycephalum during the cell cycle and oxidative stress. *Mol. Cell. Biol.* **6,** 1179–1186.
50. Negrini, M., Monaco, C., Vorechovsky, I., Ohta, M., Druck, T., Baffa, R., Huebner, K., and Croce, C. M. (1996). The *FHIT* gene at 3p14.2 is abnormal in breast carcinomas. *Cancer Res.* **56,** 3173–3179.
51. Sozzi G., Veronese, M. L., Negrini, M., Baffa, R., Cotticelli, M. G., Inoue H., Tornielli, S., Pilotti, S., De Gregorio, L., Pastorino, U., Pierotti, M. A., Ohta, M., Huebner, K., and Croce, C. M. (1996). The *FHIT* gene 3p14.2 is abnormal in lung cancer. *Cell* **85,** 7–26.
52. Virgilio, L., Shuster, M., Gollin, S. M., Veronese, M. L., Ohta, M., Huebner, K., and Croce, C. M. (1996). *FHIT* gene alterations in head and neck squamous cell carcinomas. *Proc. Natl. Acad. Sci. USA* **93,** 9770–9775.
53. Muller, C. Y., O'Boyle, J. D., Fong, K. M., Wistuba, I. I., Biesterveld, E., Ahmadian, M., Miller, D. S., Gazdar, A. F., and Minna, J. D. Abnormalities of *FHIT* genomic and cDNAs in human cervical cancer with HPV subtype correlation. *J. Natl. Cancer Inst.* [In press]
54. Michael, D., Beer, D. G., Wilke, C. M., Miller, D. M., and Glover, T. W. (1997). Frequent deletions of *FHIT* and FRA3B in Barrett's metaplasia and esophageal adenocarcinomas. *Oncogene* **15,** 1653–1659.
55. Feinberg, A., and Lee, M. P. (1997). Aberrant splicing but not mutations of TSG101 in human breast cancer. *Cancer Res.* **57,** 3131–3134.
56. Ong, S. T., Fong, K. M., Bader, S. A., Minna, J. D., LeBeau, M. M., Mc Keithan, T. W., and Rassool, F. V. (1997). Precise loacalization of the *FHIT* gene to the common fragile site at 3p14.2 (FRA3B), and characterization of homozygous deletions within FRA3B that affect transcription in tumor cell lines. *Genes Chrom. Cancer* **20,** 16–23.
57. Thiagalingam, S., Lisitsyn, N. A., Hamaguchi, M., Wigler, M. H., Willson, J. K., Markowitz, S. D., Leach, F. S., Kinzler, K. W., and Vogelstein, B. (1996). Evaluation of the *FHIT* gene in colorectal cancers. *Cancer Res.* **56,** 2936–2939.
58. Huang H., Qian, C., Jenkins, R., and Smith, D. (1998). FISH mapping of YAC clones at human chromsomal band 7q31.2: Identification of YACs spanning FRA7G within the common region of LOH in breast and prostate cancer. *Genes Chrom. Cancer* **21,** 152–159.
59. Cleton-Jansen, A. M., Moerland, E. W., Kuipers-Dijkshoorn, N. J., Callen, D. F., Sutherland, G. R., Hansen B., Devilee P., and Cornelisse C. J. (1994). At least two different regions are involved in allelic imbalance on chromosome arm 16q in breast cancer. *Genes Chrom. Cancer* **9,** 101–107.
60. Stone, D. M., Stephens, K. E., and Doles, J. (1993). Folate-sensitive and aphidicolin-inducible fragile sites are expressed in the genome of the domestic cat. *Cancer Genet. Cytogenet.* **65,** 130–134.
61. Stone, D. M., Jacky, P. B., Hancock, D. D., and Prieur, D. J. (1991). Chromosomal fragile site expression in dogs. *Am. J. Med. Genet.* **40,** 214–222.
62. Yang, M. Y., and Long, S. E. (1993). Folate sensitive common fragile sites in chromosomes of the domestic pig (Sus scrofa). *Res. Vet. Sci.* **55,** 231–235.
63. Elder, F. F. B., and Robinson, T. J. (1989). Rodent fragile sites: are they conserved? Evidence from mouse and rat. *Chromosoma* **97,** 459–464.
64. McAllister, B. F., and Greenbaum, I. F. How common are common fragile sites: variation of aphidicolin-induced chromosomal fragile sites in a population of the deer mouse (Peromyscus maniculatus). [In press]
65. Glover, T. W., Coyle-Morris, J., and Morgan, R. (1986). Fragile Sites: Overview, occurrence in acute nonlymphocytic leukemia-M4 and effects of caffeine on expression. *Cancer Genet. Cytogenet.* **19,** 141–150.
66. Richards, R. I., and Sutherland, G. R. (1994). Simple repeat DNA is not replicated simply. *Nat. Genet.* **6,** 114–116.
67. Jeffreys, A. J., Tamaki, K., MacLeod, A., Monckton, D. G., Neil, D. L., and Armour J. A. (1994). Complex gene conversion events in germline mutation at human satellites. *Nat. Genet.* **6,** 136–145.
68. Wang, Y. H., and Griffith, J. (1996). Methylation of expanded CCG triplet repeat DNA from fragile X syndrome patients enhances nucleosome exclusion. *J. Biol. Chem.* **271,** 22937–22940.
69. Gacy, A. M., Goellner, G., Juranic, N., Macura, S., and McMurray, C. T. (1995). Trinucleotide repeats that expand in human disease form hairpin structures *in vitro*. *Cell* **81,** 533–540.
70. Kang, S., Ohshima, K., Shimizu, M., *et al.* (1995). Pausing of DNA synthesis *in vitro* at specific loci in CGT and CGG triplet repeats from human hereditary disease genes. *J. Biol. Chem.* **270,** 27014–27021.
71. Laird, C., Jaffe, E., Karpsen, G., Lamb, M., and Nelson, R. (1987). Fragile sites in human chromosomes as regions of late-replicating DNA. *Trends Genet.* **3,** 274–281.
72. LeBeau, M. M., Rassool, R. V., Neilly, M. E., Espinosa, R., Glover, T. W., Smith, D. L., and McKeithan, T. W. Replication of a common fragile site, FRA3B, occurs late in S phase and is further delayed upon induction: implications for the mechanism of fragile site induction. *Hum. Mol. Genet.* [In press]
73. Sozzi, G., Sard, L., De Gregorio, L., Marchetti, A., Musso, K., *et al.* (1997). Association between cigarette smoking and *FHIT* gene alterations in lung cancer. *Cancer Res.* **57,** 2121–2123.
74. Donehower, L. A., Godley, L. A, Aldaz, C. M., Pyle, R., Shi, Y. P., Pinkel, D. Gray, J., Bradley, A., Medina, D., and Varmus, H. E. (1995). *Genes Dev.* **9,** 882–895.
75. Huang, L. C., Clarkin, K. C., and Wahl, G. M. (1996). Sensitivity and selectivity of the DNA damage sensor responsible for activating p53-dependent G1 arrest. *Proc. Natl. Acad. Sci. USA* **93,** 4827–4832.

PART IV

Spinal and Bulbar Muscular Atrophy

CHAPTER 7

Kennedy's Disease: Clinical Aspects

JEFFREY D. ZAJAC Department of Medicine, University of Melbourne, Royal Melbourne Hospital, Victoria 3050, Australia

HELEN E. MACLEAN Centre for Hormone Research, Royal Children's Hospital, Parkville, Victoria 3052, Australia

I. INTRODUCTION

Spinal and bulbar muscular atrophy (SBMA) or Kennedy's disease is a rare, X-linked, slowly progressive, adult-onset form of motor neuronopathy, caused by an expansion of the CAG repeat sequence in the first exon of the androgen receptor gene. This disease, with onset in the third to fifth decades, presents with proximal muscle and bulbar muscle weakness, atrophy, and fasciculation. Patients can have a postural tremor, particularly affecting the hands. Muscle cramps are an early feature. Almost all patients have areflexia of the lower limbs. Endocrine abnormalities, including gynecomastia, and testicular atrophy are common.

A. Historical Background

The first coherent description of this illness was given by Kennedy *et al.* [1]. Other cases, published in abstract form, were identified by Stefanis *et al.* [2]. Subsequent to Kennedy's original description, a series of case reports highlighted similar patients [2–12]. Arbizu and coworkers [9] first raised the possibility of hereditary androgen receptor failure involving mainly the spinal and bulbar motor neurons. Then in 1991, La Spada and colleagues [13] identified an expanded triplet DNA repeat in the first exon of the androgen receptor as causing this disease. This was the third disease identified which was caused by an expanded trinucleotide repeat.

II. CLINICAL FEATURES

A. Clinical Features

Spinal muscular atrophies are a group of clinically heterogeneous inherited neurological disorders involving the bulbar and spinal motor neurons with sparing of the corticospinal and sensory tracts. SBMA is an X-linked member of this group. Multiple clinical descriptions of this condition have now been published [2–12, 14–29].

1. Characteristic Features

The characteristic features of SBMA are as follows:

- SBMA is an X-linked recessive disorder.
- The age of onset of symptoms is usually between 30 and 50 years.
- Muscle cramps are almost universal and can precede the weakness by up to 20 years.
- Most patients develop weakness and atrophy of the shoulder and pelvic girdle muscles, with weakness usually first occurring in the lower limbs.
- Brain stem involvement is associated with weakness and wasting of the tongue and jaw muscles; fasciculation is particularly prominent on the tongue.
- Twitching of the muscles of the chin.
- Dysarthria.
- A small percentage of patients may develop aspiration pneumonia.
- Tremor, which is mainly postural.
- Deep tendon reflexes are decreased or absent, and there are no signs of upper motor neuron involvement.
- Sensation may be abnormal, although more commonly asymptomatic sensory impairment can be identified on more intensive testing.
- Gynecomastia is present in approximately 50% of affected individuals.
- Testicular volumes can be significantly reduced.

2. Case Report

The patient (Fig. 7-1) is a 68-year-old man with a relative with SBMA, related by a maternal great-grandmother. Cramps first occurred in this patient in his early 40s, and fasciculations in the late 40s. Cramps occurred mainly in the neck and tongue, and limbs. Fasciculation occurred mainly in his arms, chest, and facial muscles. Mild hand tremor was present. Weakness of the shoulder and pelvic muscles has progressed, but he is still able to walk 50 meters, after which he must rest.

His voice has hoarsened and become slurred and weak since his early 50s. Dysphagia occurs with certain solid foods. He occasionally aspirates food, but there has been no history of pneumonia. He was forced to retire from work prematurely at age 52 years because of his illness.

On examination, there was mild gynecomastia. The left breast was slightly tender. Wasting was more noticeable in the tongue than the limb muscles. Fasciculation

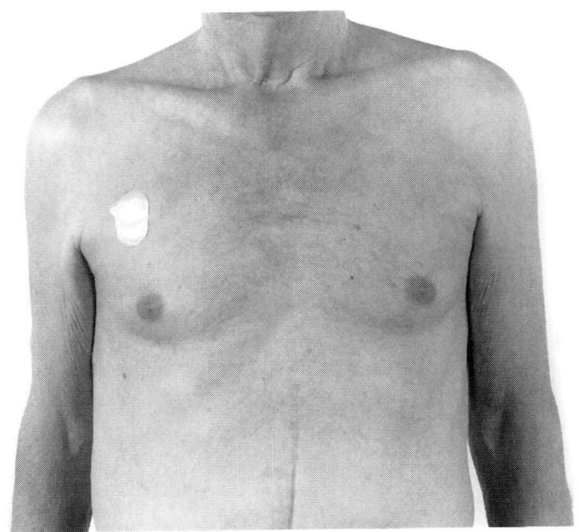

FIGURE 7-1 Subject with SBMA, aged 68 years, showing muscle atrophy in the arms and shoulder girdle and mild gynecomastia.

was clearly seen on the face and limbs. Bilaterally, facial muscles were weakened. Palatal movement was weak, and nasal escape was present. Muscle tone of the limbs was normal. Proximal shoulder and leg strength was 4/5. The patient developed muscle cramps during testing of muscle strength. Reflexes were absent despite reinforcement. Plantar responses were down-going. Decreased light touch sensation in a stocking distribution was present in the right lower leg. Coordination was normal. EMG analysis showed high-amplitude motor units with reduced interference pattern (Fig. 7-2), and nerve conduction studies showed very low-amplitude responses (Fig. 7-3).

3. Onset

The age at onset of symptoms of SBMA is usually between 30 and 50 [3], although cases as young as 12 and as old as 84 have been reported. We have identified a child dying in his teens, with onset of muscle weakness in infancy. Although DNA tests have not been performed, this child's maternal uncle has genetically confirmed SBMA (Zajac, unpublished data). In contrast, Doyu *et al.* [16] have identified an 85-year-old Japanese man with first symptoms at age 83. His neurological features were entirely consistent with SBMA. Cramps can occur 15 to 25 years before other symptoms [3, 4].

B. Neurological Features

1. Proximal Muscle Weakness

Proximal muscle weakness usually first occurs in the lower limbs and shoulder girdle. This is associated with muscular atrophy and significant fasciculation. Most pa-

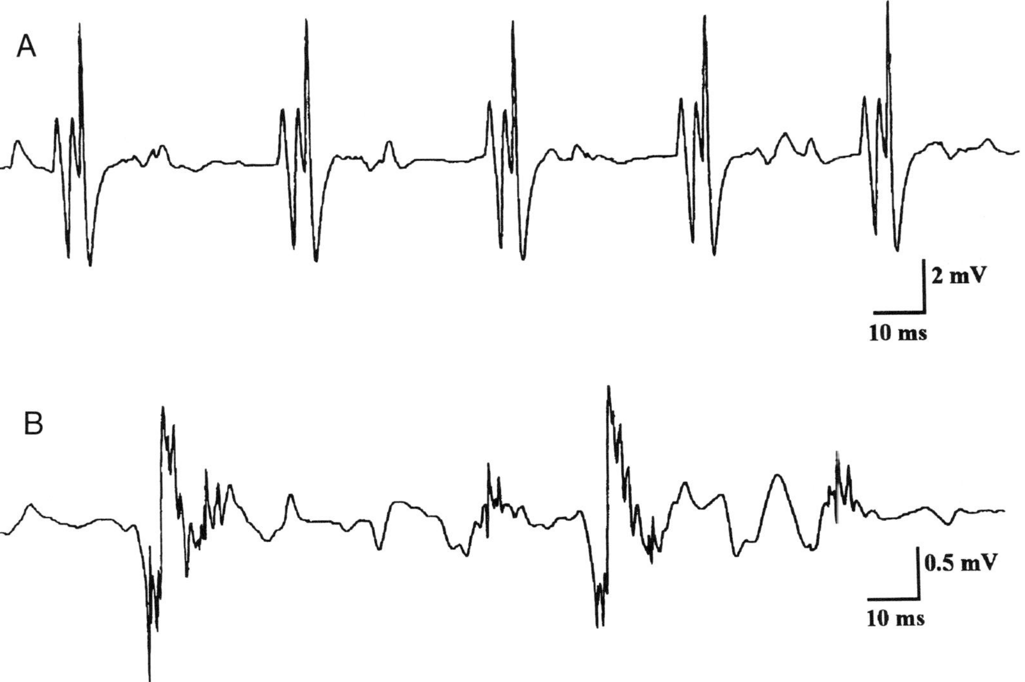

Figure 7-2 Concentric needle EMG recorded from (A) the right biceps muscle and (B) right vastus lateralis muscle, at maximum voluntary contraction in a patient with SBMA. High-amplitude and/or long-duration polyphasic motor units with reduced interference pattern are characteristic and are consistent with denervation.

tients develop weakness and atrophy of the shoulder and pelvic girdle muscles. This may result in the requirement of a cane for walking, or a total inability to walk. Harding *et al.* [3] report upper limb proximal weakness in all of the 10 patients they described, and distal weakness to a lesser degree in 9 of the 10 patients. Proximal lower limb weakness of significance was reported in all of the patients, and mild distal weakness in 6 out of 10 patients.

2. Brain Stem Involvement

Bulbar weakness affecting the muscles of the chin and oropharyngeal region, particularly the tongue, usually develops later. Brain stem involvement is associated with weakness and wasting of the tongue and jaw muscles [18], often with sagging of the lower lip, and wasting and fissuring of the tongue itself. Fasciculation is particularly prominent in the tongue. Brain stem involvement often results in dysarthria, making the patient's speech difficult to understand. Dysphagia is a late feature that usually does not interfere with adequate nutrition, although in a minority of patients this can be a significant problem. A small percentage of patients may develop aspiration pneumonia as a final event.

3. Muscle Cramps

Muscle cramps are almost universal in all the reported studies. In all these series, the cramps precede the weakness, in some cases by 20 years. The cramps are often induced by exercise. Schoenen *et al.* [46] reported a 28-year-old man who had gynecomastia and very significant cramps, in the absence of any other disease symptoms. Cramps are not uncommon in patients with motor neuron disease, but usually do not occur before symptoms of weakness.

4. Tremor

Tremor has been reported, particularly in the hands and particularly when outstretched. The incidence of tremor varied from 50 to 100%.

Tremor of the hands may occur before the onset of weakness. Tsukagoshi *et al.* [30] reported tremor in a boy at age 15, who did not develop limb weakness until 10 years later. Harding *et al.* [3] described one patient with extremely troublesome tremor which was the major symptom. In this case, tremor was improved by propranolol. Eighty percent of patients reported by Harding experienced tremor, which is similar to that in Kennedy's original series [1] and that of Stefanis *et al.* [4]. The tremor is mainly postural and is not made worse by purposeful movement.

5. Fasciculation

These are generalized and particularly noticeable around the face and chin. This has led to the description of a characteristic feature of this illness, which is a

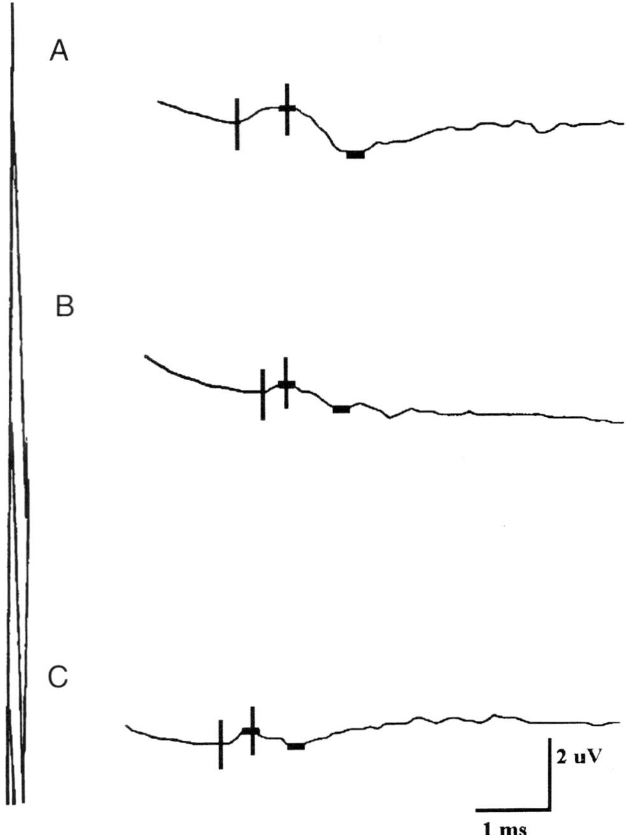

FIGURE 7-3 Orthodromic sensory nerve action potentials from the (A) right median (digit 2), (B) radial (digit 1), and (C) sural nerves, showing very low-amplitude responses. Each trace is the average of 50 stimuli.

twitching of the chin. In Harding's series [3], significant fasciculation of the face was present in all but one patient, present at rest, and more marked on muscle contraction. It was particularly prominent around the mouth and chin, and in this series was best elicited by asking patients to whistle or blow with the mouth closed. Two of their patients were aware of involuntary facial movement. Over 90% of studies report this twitching of the chin. This is not a common feature of any other inherited or acquired degeneration of motor nerve nuclei. Facial fasciculation in Harding's series was associated with weakness and wasting of the muscles supplied by the seventh and fifth motor nerve nuclei. Olney *et al.* [18] also reported this enhanced twitching of the face in five out five patients. In their series, weakness of the face was reported to be mild to moderate in all of the individuals reported.

6. Tendon Reflexes

Deep tendon reflexes are decreased or absent, and there are no signs of upper motor neuron involvement, clearly differentiating this illness from amyotrophic lateral sclerosis [3]. Complete absence of tendon reflexes is an almost universal finding in all the clinical cases reported. In a few cases, reflexes are present, but in these cases are significantly diminished.

7. Sensation

Sensation may be abnormal, although more commonly asymptomatic sensory impairment can be identified on more intensive testing.

Sensory symptoms and signs are usually absent in SBMA. However, one study reported mild distal decreased perception of vibration [18]. Barkhaus *et al.* [7] reported a patient with symmetrical distal diminution of sensation to pinprick and vibration sense in one case, and the diagnosis of mild peripheral neuropathy was diagnosed. There was no evidence of posterior column or pyramidal tract signs. In general, most of the studies have not reported any sensory abnormality. Those few studies that did report a mild distal sensory neuropathy identified this in a minority of patients.

8. Neuropsychological Features

Nonspecific neuropsychological disturbances and behavioral problems have been described [32, 33]. There is no consistent description, but long-term memory and frontal lobe function defects have been described. Guidetti *et al.* [33] report a number of neuropsychological studies on patients and heterozygote carriers. Abnormal test results were obtained in all three affected individuals as well as several carriers, but in only one out of five normal nonaffected relatives. Abnormal tests included the Stroop color word time, WAIS, prose memory, verbal judgments, phonemic fluency, prose memory, spatial supraspan, paired associate, and Benton. The neurological impairment was evident in the patients with symptoms; however, in asymptomatic males carrying the mutation, investigation either was normal, or showed only moderate abnormalities.

C. Endocrine Features

Gynecomastia is present in approximately 50% of affected individuals, with the occurrence of gynecomastia varying from 0 to 90% in the quoted studies. Testicular atrophy, progressive infertility, and impotence can occur [9]. In one patient a right supernumerary nipple has been reported [1]. In one typical patient reported by Arbizu *et al.* [9] there was bilateral gynecomastia without galactorrhea, progressive loss of libido, and inability to maintain an erection at age 30. This degree of hypogonadism is unusual in SBMA, as most studies have reported normal sexual functioning, although some decrease in fertility is present in most studies discussing this feature. Many studies quote the presence of testicular atrophy [3, 9, 50], but in only a small number of these studies is testicular size quoted [20, 28]. Testicular volumes measured by MacLean *et al.* [28] showed significantly reduced testicular size ranging from 6 to 11 ml (normal > 12 ml). A significant incidence of late impotence has been reported in many studies.

1. Fertility

Some degree of infertility has been reported in SBMA. In the study of Arbizu *et al.* [9], only four out of seven patients had children. In this study, testicular biopsy was performed, and spermatogenic arrest at the spermatid level was identified. There were normal, occasionally clustered Leydig cells present. Three out of seven patients in one study had small testicles, and six out of seven patients were impotent. In two patients in whom sperm density was measured, it was found to be less than five million per milliliter [34]. In most reports, many or all of the patients have fathered children, suggesting that progressive infertility may occur in some cases.

D. Other Characteristics

1. Ethnic Origin

MacLean *et al.* [28] describe patients from diverse ethnic groups, including Anglo-Australian, German, and Vietnamese backgrounds; and other groups have published reports of patients of Japanese, Chinese, Greek, Italian, and Spanish origin. Cases have been reported from Australia [20, 29], North America [1, 7, 13], Asia [40–42], and Europe [3, 4, 9, 21].

2. Progression, Prognosis, and Life Expectancy

The age at death of patients with SBMA has been identified in several papers [1, 3, 42, 45, 48, 50]. In the original paper by Kennedy *et al.* [1] postmortem examinations were performed on several patients. The age at death of these four patients ranged between 55 and 83 years. Two of the patients died of pneumonia, one of intraabdominal carcinoma, and two of unspecified or unknown causes. This pattern is typical of other publications. Harding *et al.* [3] reported two deaths at ages 61 and 65 years. Death in one patient was attributed to choking. In the study by Li *et al.* of seven autopsied patients, the age at death varied between 51 and 84 years, with the cause of death being aspiration pneumonia or bronchiectasis in five patients, silicosis and pneumonitis in one patient, and hepatic failure caused by liver metastases of colon cancer in one patient [42]. In this study, the date of onset of the disease was from the late 20s to age 72. In general, the rate of disease progression is slow and death is reported at an old age; however, there are clearly individuals dying from aspiration pneumonia directly caused by this disease. There have been a number of cases reported, typified by that described by Amato *et al.* [19], in which an individual presented and rapidly progressed from being healthy to wheelchair-bound within 1.5 years. Although this is not typical, there is a clear minority of cases in which the progression of disease can be rapid. There is no evidence that genetic anticipation (increasing severity of the disease over generations) occurs in SBMA.

In some other DNA triplet repeat diseases the severity of the clinical features and the age at onset correlate with the length of the repeat sequence (see other chapters). This remains a point of controversy with regard to SBMA. Several publications suggest such a correlation exists [40, 41, 45]. However, we and others [28, 19, 33] have not noted this finding, and we believe that there are inadequate data to come to a firm conclusion on this point. Significant intrafamily variation in onset and severity can occur [19, 45], and factors other than the size of the CAG repeat must modulate the severity of the disease.

E. Associated Illnesses

1. Neurological

A number of atypical presentations associated with an increased number of triplet repeats in the androgen receptor have been reported. These include one patient presenting with tremor [15]. In this case, the patient developed fine postural tremor of the fingers in his mid-50s. There was no relevant family history, and a diagnosis of essential tremor was made. At age 65, he developed fasciculation of the facial muscles and cramps, particularly at night. Deep tendon reflexes were normal. Investigation revealed 42 CAG repeats. Thus it may well be that essential tremor can be the first presenting sign of this illness.

Another patient initially presented with myasthenic symptoms [35]. This was a man who noted muscle cramps since the age of 30, with difficulty climbing stairs at age 47. After resting, he was able to walk further. He developed bilateral ptosis of the eyelids, atrophy and fasciculation of the tongue, and atrophy, weakness, and easy fatiguability of the proximal muscles of the shoulder and pelvic girdle. An EMG revealed high-amplitude motor unit potentials, with reduced interference patterns in all the muscles examined including the bulbar muscles, suggesting denervation. Administration of edrophonium was suggestive of the presence of myasthenia gravis. However, repeat testing with edrophonium showed no change in the compound muscle action potential. There was a definite decrement of the compound muscle action potential during repetitive nerve stimulation, very suggestive of myasthenia gravis. Thus, either this patient had two separate illnesses or this was SBMA. The number of CAG repeats in the androgen receptor gene was 45, well into the range for SBMA. The authors postulate that in a process of chronic denervation and renervation in SBMA, renervated motor endplates may be associated with abnormal neuromuscular transmission.

There has been an isolated case report of an atypical patient presenting with difficulty in masticating [36]. He had "to move his jaw manually to masticate when eating." There were severe muscular weakness and atrophy in the masseters, and fasciculation around the mouth. There was no weakness, atrophy, or fasciculation in the extremities, but deep tendon reflexes were absent. Fine postural tremor was present in the hands. There were no cranial nerve defects, and there was no gynecomastia. His brothers also complained of tremor. Analysis showed 41 CAG repeats. Nedelec *et al.* [37] describe a case in which weakness is confined to the lower limbs, and is strictly unilateral in two brothers.

2. Endocrine

In a number of reports, the presence of diabetes has been mentioned. In Kennedy's original report [1], three out of the nine affected patients had type II diabetes. Harding *et al.* [3] also noted type II diabetes in a significant number of her patients. There is however no clear evidence that diabetes is more common in SBMA than in the general population.

3. Other Associated Features

We have identified one patient with SBMA who developed carcinoma of the breast (Zajac unpublished). In our series of 18 unrelated men with SBMA, one had carcinoma of the breast. Hypertrophic cardiomyopathy has been reported in one patient with SBMA [38], but there is no clear evidence of a linkage. Type IV hyperlipoproteinemia [39] has also been described in one family. Again, there is no evidence that this is anything other than a coincidence. Hypobetalipoproteinemia has also been reported [34].

E. Summary of Clinical Features

Kennedy's original description remains accurate, with most patients initially experiencing muscle cramps, followed by generalized fasciculation and muscle weakness over a period of 10 to 20 years. Fasciculation persists throughout the course of the disease, and in some of Kennedy's original patients, fasciculations were so severe that they interrupted the patients' sleep. Proximal muscle weakness is the predominant finding, with some distal involvement at a later stage. Most of Kennedy's original patients, and all those in subsequent descriptions, had lordotic waddling gaits, and many of the more afflicted are unable to rise from a squatting position. All patients have bulbar involvement to some extent. It is important to note that the extraocular muscles have never been described as having been involved, and sphincter control is unimpaired. Kennedy's patients died from ages 55 to 83, with the average duration of illness being 26 years.

Table 7-1 summarizes most studies with more than six patients involved.

III. LABORATORY DATA

A. Laboratory Data

1. Biochemistry

Serum creatine kinase (CK) is elevated in the majority of patients with SBMA, often up to five times the upper limit of normal. This is no doubt related to the secondary myopathic wasting caused by denervation. CK can be normal [16, 17, 28]. In MacLean *et al.* [28] CK ranged from 195 to 1200 IU/liter, with the normal range <200. Other muscle enzymes, including aldose reductase, have been elevated. A lumbar puncture investigation was reported in Kennedy's original study, with 0.35 g of protein per liter (normal < 0.4 g/liter). In three other patients in that study in whom the cerebrospinal fluid was examined, the protein level was moderately elevated. Few other results of lumbar puncture are available in this condition. In those patients studied by Arbizu *et al.,* CSF protein was normal [9].

2. Endocrine Testing

Levels of testosterone have been reported as normal or slightly low, ranging, in the study of MacLean *et al.* [28], from 6.2 to 21.1 nmol/liter (normal range 10.5–35 nmol/liter). Other studies show similar results, with a majority of patients having serum testosterone levels within the normal range, or slightly depressed. Elevations of LH or FSH are uncommon, and tend to be mild when present. In Harding's series [3], two patients out of five had elevations in either FSH or LH. This is similar to the situation in the series of MacLean *et al.* [28]. Thus most patients have FSH and LH levels within the normal range, although there is clear evidence that a significant minority of these patients have elevations in these gonadotropins. The study of MacLean *et al.* [28] showed no elevation in serum estradiol levels, although other studies have shown a slight increase in estrogen levels.

Table 7-1 Summary of Clinical Features in Studies of SBMA

Study	No. of patients	Distinct kinships	Age of onset	Muscle groups	Course	Cramps	Fasciculations	Hand tremor	Reflexes	Gynecomastia
Kennedy [1]	11	2	35–83		Slow, >20 yrs	Most	Generalized, especially chin	Most	Decreased or absent	3/11
Harding [3]	10	8	30–50	Slow	Slow	Most	Face	8/10	Decreased or absent	9/10
Stefanis [4]	7	4	28–45	Proximal, distal, slight bulbar		5/7	Generalized, especially perioral	100%	Decreased or absent	6/7
Arbizu [9]	7	1	18–26	Proximal, bulbar		100%	Generalized	100%	Decreased or absent	6/6
Doyu [16, 40]	26	21	25–70	Slow		100%				
Amato [19]	17	7	33–60					9/17		9/17
Choi [29]	12	8	30–	Proximal, bulbar	Slow, >20 yrs	100%				
Tsukagoshi [30]	7	1	17–35	Proximal, distal, bulbar			Face, tongue, proximal limb	4/7		1/7
Igarashi [41]	19		10–54							11/19

For example in the study of Schoenen *et al.* [46] two patients had slight elevations of estradiol levels. All patients had normal levels of testosterone, and one patient had a significantly elevated peak LH level, after stimulation with LH-RH. Prolactin levels were normal. There does not appear to be any identifiable correlation between hormone levels and the development of gynecomastia.

3. Electrophysiological Studies

Electrophysiological studies have been reported by a number of groups [1, 3, 4, 7, 9, 12, 18, 19, 29, 33, 43, 44, 46]. The results of the study by Ferrante and Wilbourn [43] are typical. They report studies on 19 patients. Of note in this study is the very high percentage of patients with abnormalities of sensory nerve action potentials (SNAP). These patients were generally asymptomatic, and sensory loss was mostly clinically undetectable. In this study, 18 out of 19 clinically diagnosed patients demonstrated abnormal SNAPs. When those cases with genetically confirmed SBMA were separately considered, 87% of these patients had abnormal SNAPs. These abnormalities include low-amplitude potentials or potentials that were unelicitable.

Analysis of compound muscle actions potential (CMAP) responses revealed abnormalities in 7 out of 19 cases. All patients had low-amplitude potentials or these were unelicitable. Three out of 19 demonstrated mildly prolonged distal latency, and 3 had reduced conduction velocity. Thus in both types of study, the bulk of the abnormalities consisted of low-amplitude or unelicitable responses. The H reflex responses were abnormal in 17 out of 19 patients. The latency of the low-amplitude response was reported as normal. Finally, the needle electrode examination was abnormal in all patients examined, and showed evidence of both acute and chronic motor axon loss, with fibrillation potentials and fasciculations. These changes were widespread. Grouped repetitive discharges were found in the area of the chin in some but not all patients. Discussion of neurophysiological studies can be found in Refs. [33, 43, 44].

Guidetti *et al.* [33] report abnormalities of sensory nerve function in heterozygotes. Approximately 50% of heterozygote carriers in the family described showed mild neuromuscular abnormalities, and several heterozygote carriers had up to twofold elevations in CK level.

B. Pathology

1. Nerve

A number of studies [48–51] have investigated the pathological changes in patients with SBMA. Kennedy's original postmortem examinations revealed marked reduction in the number of anterior horn cells in all segments, while the remaining anterior horn cells were atrophic. Autopsies have been reported by a number of groups [3, 11, 48–52]. The principal pathological changes are spinal and bulbar motor neuron loss, but involvement of the primary sensory neurons has been demonstrated. Motor neuron loss in the spinal cord and brain stem nuclei is extremely severe in SBMA. These neurons are almost completely depleted. Oyanagi *et al.* [48], Li *et al.* [42], Terao *et al.* [51], Harding *et al.* [3], Sobue *et al.* [52], Wilde *et al.* [11], and Nagashima *et al.* [50] have also published autopsy studies on patients with SBMA. In one study, there was mild astrocytosis associated with the loss of nerve cells. In contrast to this, the nerve cells of Clarke's cell columns and the posterior horns were well-preserved. The white matter was well-conserved. The anterior spinal nerve roots contained a decreased number of fibers compared with the posterior nerve roots, and the peripheral nerves exhibited a mild loss of nerve fibers. Electron microscopic studies of the rough endoplasmic reticulum showed marked disaggregation of the polyribosomes surrounding the endoplasmic reticulum in many anterior horn cells [48].

Li *et al.* [42] performed quantitative assessment of sensory nerve involvement in seven autopsied cases from seven different families. The population of myelinated fibers in the fasciculus gracilis was significantly decreased, particularly in the rostral segment. Myelinated fibers in the fourth dorsal root, the most proximal portion of the central rami of the primary sensory neuron, were well-preserved, although there was a tendency to decrease in the number of large fibers. Segmental demyelination/remyelination was present. Analysis of the dorsal root ganglion neurons showed a decrease in large-diameter neurons 48.8 μm in diameter, and an increase in small-diameter neurons, smaller than 48.8 μm. Myelinated fibers had abnormally thick myelin, compared with the axonal areas, suggesting axonal atrophy. The large myelinated fibers in the sural nerve were markedly diminished. This study identified that primary sensory neurons are extensively involved in SBMA, with neuronal and axonal atrophy.

The density [51] and the distribution of large and small myelinated lateral corticospinal tract fibers were the same in SBMA patients and controls. Atrophy of the ventral horns was observed in SBMA through all spinal segments. Large and medium-sized neurons as well as small neurons in the intermediate zone of the ventral horn were significantly depleted. Simple atrophy of neurons was observed, but central chromatolysis, inclusion bodies, and neuronophagia were rarely seen. Reactive astrogliosis was minimal. The total number of neurons in the ventral horns at the L4 segment was decreased, being 14.5–44.1% of the corresponding controls. There was a significant loss of large neurons (2.4–17.9% of controls), small neurons (18.6–57.3%), and medium-sized neurons (10.0–43.7% of controls). There was a marked loss of large and medium-sized neurons in the lateral and medial nuclei, and also a high degree of small neuronal loss in the intermediate zone.

Few patients with upper motor neuron signs have been reported in the literature [53]. Some patients with hyperreflexia have been reported, mainly in the Japanese literature [51].

Nagashima *et al.* [50] reported two cases in which autopsy was performed. They reported degeneration of cranial and spinal motor nuclei, with marked loss of neurons in the motor nuclei of the 5th nerve, as well as the 7th, 11th, and 12th nerves, and the anterior horn of the entire spinal cord. Neurons at the motor cortex and the upper pyramidal tract were well-preserved, as were the cranial nuclei of nerves 3, 4, 6, 8, 10, and the spinal neurons of posterior, Clarke's, intermediolateral, and Onufrowicz's nuclei. The posterior columns and anterior white matter of the spinal cord were unremarkable, except for a slight pallor of the myelin sheaths

2. Muscle

Skeletal muscle biopsies show progressive neuropathic changes. These are atrophy of muscle fibers, either as isolated angular atrophic fibers or as small group atrophy among normal-sized fibers. Later large group atrophy occurs. Fiber atrophy will result in apparent increases in clumps of pyknotic nuclei, or be so severe that only nuclear "bags" remain. Muscle fiber hypertrophy is also common, with increased numbers of internal nuclei. Later some of these enlarged fibers undergo splitting and fragmentation. Scant endomysial connective tissue proliferation may follow, but fibrosis is rarely a feature. One patient reported by Ringel *et al.* [55] had a marked increase in connective tissue.

Myopathic changes, usually considered to be secondary changes, are uncommon, and usually consist of isolated fibers undergoing segmental degeneration, with necrosis and myophagia. Some may also show regenerative changes.

Enzyme histochemistry shows atrophy of all three fiber types (type 1, 2a, and 2b fibers). Later collateral reinnervation may occur, with numbers of the same fiber type forming enclosed fiber groups, leading to type grouping [1, 3, 4, 54, 55]. Internal rearrangements can occur, causing loss of ordered sarcomeres. These changes can be readily seen histochemically with mitochondrial stains, such as NADH-TR, as moth-eaten, core-targetoid, or target fibers [55], and by electron microscopy as Z-band streaming or core lesions. Loss of muscle fibers usually results in adipose tissue replacement.

Biopsies from heterozygous females have shown moderate type 2 fiber predominance and possible early type grouping, with an increase in spread of fiber diameters [46].

In our postmortem analysis of a 72-year-old man with SBMA (Zajac, manuscript in preparation), gross examination of the spinal cord showed wasting of anterior nerve rootlets, correlating with severe loss of anterior horn neurons in sections of the spinal cord (Fig. 7-4). Sections of the medulla did not show any significant loss of neurons from the dorsal vagal or hypoglossal nuclei. Sections of skeletal muscle showed muscle fibers ranging in diameter from <10 to 150 μm (Fig. 7-5). Small groups of denervated, angular, atrophic fibers of all three fiber types were still present, but there was a type 1 predominance. Internal nuclei were markedly elevated (87% compared to normal of $\leq$3%). Considerable compensatory hypertrophy had also occurred with many enlarged fibers, many of which contained core-targetoid fibers. Increased endomysial and perimysial adipose tissue was present. These changes can be seen in long-standing denervation, particularly in denervation due to loss or degeneration of anterior horn spinal neurons [56].

3. Androgen Receptor Binding

Three studies have investigated the ability of the androgen receptor (AR) to bind androgens in cultured genital skin fibroblasts from SBMA patients [28, 57, 58]. The largest study measured AR binding in cultured suprapubic skin fibroblasts from six unrelated men with SBMA [28]. Scatchard analysis was carried out to determine total number of binding sites (B_{max}) and the apparent binding affinity (K_d) of AR for ligand. This study found abnormal binding properties of the AR in patient cultured skin fibroblasts (Fig. 7-6), with five out of six patients exhibiting a decreased binding affinity for androgen. A significant correlation was found between the AR K_d and the severity of testicular atrophy and gynecomastia in the patients.

The progressive development of infertility and gynecomastia in subjects with SBMA could be related to the gradual decrease in bioavailable testosterone levels that occurs in aging males [59], although most individuals continue to have testosterone levels within the normal range throughout their life [34].

Warner *et al.* [57] measured AR binding in scrotal skin fibroblasts from eight SBMA men from four different families. These authors found a lowered total binding in three subjects from one family, but normal binding in the other families. Another group investigated AR binding in cultured scrotal skin fibroblasts from one

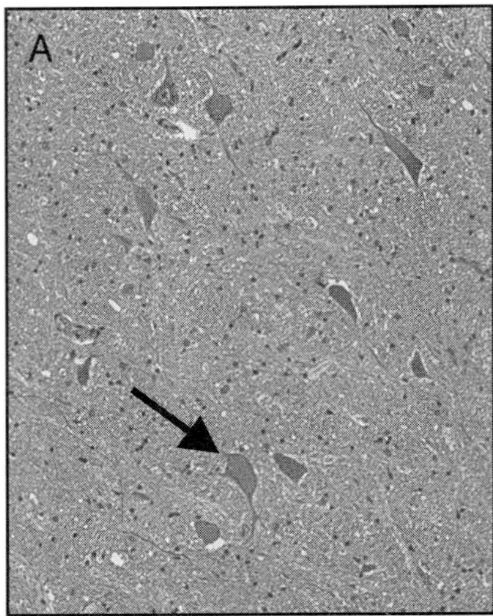

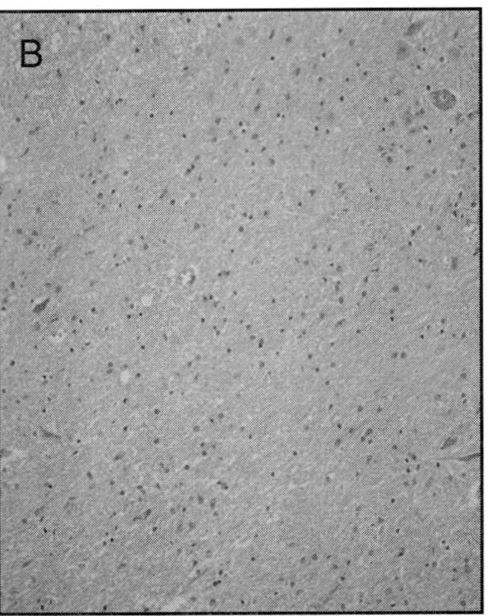

FIGURE 7-4 Anterior horn region of cervical spinal cord sections, stained with hematoxylin and eosin, 100× magnification. (A) Normal male, age-matched (arrow indicates typical motor neuron cell body). (B) SBMA subject.

patient, and found a marked decrease in total number of binding sites and an apparent binding affinity lower than the normal range, although this was described as being essentially normal [58]. Decreased total number of binding sites may indicate that the AR is unstable; however, these results remain to be confirmed by further studies.

Other evidence also suggests that AR levels are decreased in SBMA,although the limited numbers of patients and variability of techniques used in different studies make these data difficult to interpret. Nakamura *et al.* [26] found decreased AR mRNA and protein in the spinal cord from one patient with SBMA, compared to control subjects with amyotrophic lateral sclerosis (ALS). However, another brief report showed detectable immunoreactive AR in the spinal cords from two brothers with SBMA, similar to a control ALS spinal cord [60]. Another study showed no detectable staining in the upper prickle cells of the scrotum in three SBMA patients, compared to positive staining in the normal controls [49]. This lack of detectable AR could have arisen because of either absence of the receptor protein, or a change in the AR conformation causing the loss of the AR epitope.

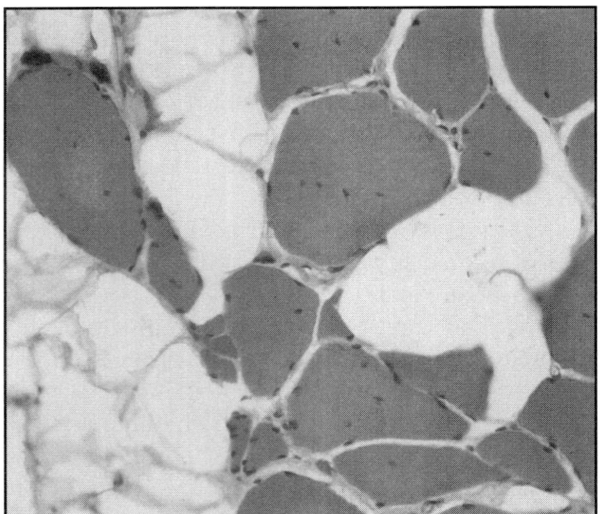

FIGURE 7-5 Quadriceps femoris muscle from SBMA subject, stained with hematoxylin and eosin, 200× original magnification.

IV. GENETIC TESTING

A. Genetic Testing

1. LABORATORY DIAGNOSIS

In most cases, the diagnosis of SBMA is a clinical one, immediately confirmed now by analysis of the triplet repeat sequence within the first exon of the androgen receptor (see Chapter 8). Biochemical testing is not particularly helpful in a diagnostic sense. Testosterone levels may be low or normal, with high or normal levels of LH and FSH. Creatine kinase may be elevated. Muscle biopsy is not indicated for diagnostic purposes, but the changes are described above. Nerve conduction studies also are not required for diagnostic purposes, and are mainly an investigational tool. Thus definitive DNA diagnosis can save significant expense and patient discomfort.

Approximately 30% of the cases are apparently sporadic, with no affected family member. The incidence

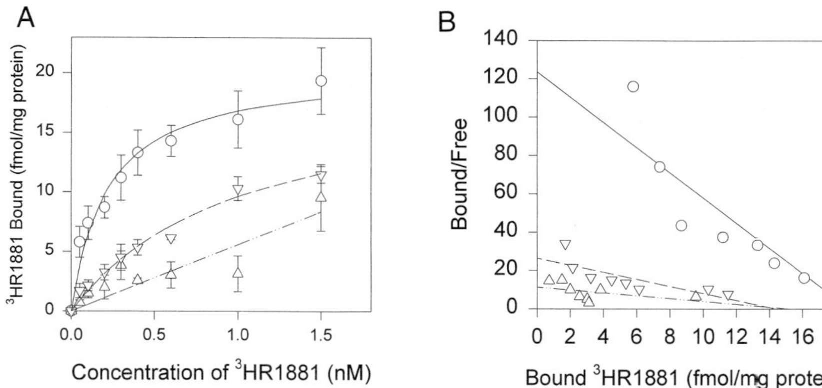

FIGURE 7-6 (A) Saturation binding curve and (B) Scatchard analysis of AR binding in cultured suprapubic skin fibroblasts from two SBMA subjects (triangles) and a normal male control (circles).

of sporadic cases may well be underestimated in this type of analysis, however, as the finding of one family member clearly leads to others. The frequency of SBMA in the population has not been reported. Prior to the identification of the androgen receptor gene mutation, the diagnosis of SBMA was often not made. Since the advent of the genetic test for SBMA, 38 different families in Australia (population 18,000,000) have been positively identified as carrying the disease (Zajac unpublished).

2. Prenatal Diagnosis

Prenatal diagnosis of SBMA has been reported [31]. A confirmed female carrier requested that chorionic villus sampling be performed on two pregnancies; in both cases the male fetuses were shown to carry the AR gene mutation, and the pregnancies were terminated. The severity of the disease in affected relatives will obviously influence whether the parents decide to terminate an affected male fetus.

3. Heterozygote Females

Female relatives of men with SBMA can be heterozygote carriers of the disease, and this can easily by confirmed by DNA analysis. There is little evidence that heterozygous females show clinical manifestations of SBMA despite the fact that cramps have been reported [1]; however, laboratory investigations may reveal subtle abnormalities of nerve function.

Sobue *et al.* [54] report analysis of eight heterozygous females. Four of these showed extensive high-amplitude motor unit potentials in the muscle examined. All the patients were neurologically normal. Biochemical measurement, including CK, myoglobin, myosin light chain, lactate and pyruvate, were all normal. Muscle biopsy showed a type 2 fiber predominance, and possible very mild type 2 fiber grouping in one heterozygous female. EMG of one heterozygous female has demonstrated mild chronic denervation in both upper and lower limb muscles [23]. Another study performed by Guidetti *et al.* [33] examined one extended family with SBMA, which included eight heterozygous females. Mild proximal weakness was observed in two carriers, and three heterozygotes had moderately elevated CK. Three of the carriers also showed neurogenic changes on EMG, and three other carriers had moderately slowed sensory conduction velocities of peripheral nerves.

B. Ethical Considerations

In any potentially fatal neurological disease with a genetic basis, consideration needs to be given to ethical factors relating to diagnosis and information supplied to the patients. Ethical issues which have proven to be difficult in our practice include the following:

1. What is the appropriate age for DNA testing? Although many parents wish to know the diagnosis in their children, it is not clear that it is ethically valid to test children merely to be able to document the genetic mutation. As with many ethical issues, there is no clear answer; in general, although the parents' initial impulse is to have their children tested, in our experience, parents on reconsideration often change their approach and opt not to test.

2. Testing of presymptomatic, potentially affected adults is also a delicate area. In general, we have found that all such adults immediately wish to be tested. Allowing a time delay before such testing has resulted in a significant percentage of such individuals opting not to have DNA testing while asymptomatic. This is clearly an issue in which there will be significant individual variation, but it does suggest that the instant reflex to test all such individuals should be resisted.

3. Genetic testing of one individual within a family will allow diagnosis in other individuals who may not wish to have this information. Patient confidentiality issues are therefore significantly broader than in nongenetic neurological disease.

V. THERAPY

Tremor was improved in one report by proprananol [3]. Standard symptomatic treatment of the increasing motor weakness, dysphagia, dysarthria, and recurrent aspiration is of course appropriate. However, there is no proven specific therapy resulting in effective amelioration of SBMA. Experimental approaches to the treatment of other forms of motor neuron disease may be tried, but reports of successful treatment have not been published.

Because of reports of abnormalities in androgen receptor function, and lowered levels of testosterone, some groups have used androgen replacement therapy in an attempt to ameliorate the features of this illness [24, 47, 58]. We have used replacement doses of androgens in several patients with no documented improvement (Zajac, unpublished data). Goldenberg and Bradley [47] report two brothers given high-dose oral testosterone for 18 months along with an exercise program. One patient made a significant improvement, although the other did not. The conclusion of that paper suggests that the testosterone had no significant effect, either positive or negative, on the long-term progression of SBMA. Two other publications have shown no beneficial effects of androgen replacement [24, 58].

The dominant hypothesis as to the cause of SBMA is of a positive effect of the mutant androgen receptor, either a nonspecific toxic effect or a gain of toxic function. It is necessary to explain the fact that individuals with complete deletion of the androgen receptor gene have no documented neurological disturbance, and the fact that heterozygote female carriers have no or only minor abnormalities. However, it is difficult to explain in terms of the toxic gene hypothesis how female heterozygotes, with one abnormal copy and one normal copy, do not develop the disease. One hypothesis which may explain this situation is that circulating androgens are required for activation of the toxic receptor effect. This would explain the absence of such an effect in females, who have low levels of androgens compared with males. Thus, one might expect anti-androgens, estrogens, or orchidectomy to have a beneficial result in this condition. On the other hand, complete withdrawal of androgens is associated with another set of clinical difficulties. This is a clearly testable hypothesis, but no data for this approach have yet appeared. However, it seems entirely plausible that anti-androgens may be a more beneficial therapy in SBMA than androgen agonists.

Although there is no currently available effective treatment for SBMA, this illness is an active area of research. A great deal is known about the function of the androgen receptor gene which causes this disease and this makes it likely that the basic mechanism of the disease will soon be understood. Continuing research in this area will allow design of rational therapeutic approaches to the treatment of this potentially severe disease.

VI. SUMMARY

In summary, SBMA is an X-linked form of motor neuronopathy caused by an expansion of the CAG repeat sequence in the androgen receptor gene. Late onset, slow progression, and characteristic clinical features differentiate SBMA from other forms of motor neuron disease. Genetic diagnosis is rapid and reliable. This may circumvent the need for multiple and uncomfortable investigations. There is no effective treatment, but the pathogenic mechanism is an active area of research. For other reviews in this area, see Refs. [14, 61–63].

Acknowledgments

We thank Dr. Michael Gonzales for performing the postmortem analysis, Dr. Lyn Kiers for performing the electrophysiological studies, and Dr. Trevor Kilpatrick and Dr. Xenia Dennett for their helpful comments on our manuscript.

References

1. Kennedy, W. R., Alter, M., and Sung, J. H. (1968). Progressive proximal spinal and bulbar muscular dystrophy. *Neurology* **18,** 671–680.
2. Stefanis, C., Papapetropoulos, T., and Scarpalezos, S. (1968). A peculiar form of X-linked hereditary degenerative motor neuron disease. Presented in The Hellenic Association of Neurology and Psychiatry, 26 November, 1968.
3. Harding, A. E., Thomas, P. K., Baraitser, M., Bradbury, P. G., Morgan-Hughes, J. A., and Ponsford, J. R. (1982). X-linked recessive bulbospinal neuronopathy: report of ten cases. *J. Neurol. Neurosurg. Psych.* **45,** 1012–1019.
4. Stefanis, C., Papapetropoulos, T., Scarpalezos, S., Lygidakis, G., and Panayiotopoulos, C. P. (1975). X-linked spinal and bulbar muscular atrophy of late onset: a separate type of motor neuron disease? *J. Neurol. Sci.* **24,** 493–503.
5. Pareyson, D., Castellotti, B., Botti, S., Defanti, C. A., Gellera, C., Taroni, F., and Sghirlanzoni, A. (1995). Kennedy's disease: clinical and molecular study of two Italian families. *Ital. J. Neurol. Sci.* **16,** 467–471.

6. Albers, J. W., and Bromberg, M. B. (1994). X-linked bulbospino-muscular atrophy (Kennedy's disease) masquerading as lead neuropathy. *Muscle Nerve* **17,** 419–423.
7. Barkhaus, P. E., Kennedy, W. R., Stern, L. Z., and Harrington, R. B. (1982). Hereditary proximal spinal and bulbar motor neuron disease of late onset. *Arch. Neurol.* **39,** 112–116.
8. Hausmanowa-Petrusewicz, I., Borkowska, J., and Janczewski, Z. (1983). X-linked adult form of spinal muscular atrophy. *J. Neurol.* **229,** 175–188
9. Arbizu, T., Santamaria, J., Gomez, J. M., Quilez, A., and Serra, J. P. (1983). A family with adult spinal and bulbar muscular atrophy, X linked inheritance and associated testicular failure. *J. Neurol. Sci.* **59,** 371–382.
10. Guidetti, D., Motti, L., Marcello, N., Vescovini, E., Marbini, A., Dotti, C., Lucci, B., and Solime, F. (1986). Kennedy's disease in an Italian kindred *Eur. Neurol.* **25,** 188–196.
11. Wilde, J., Moss, T., and Thrush, D. (1987). X-linked bulbo-spinal neuronopathy: a family study of three patients. *J. Neurol. Neurosurg. Psych.* **50,** 279–284.
12. Ringel, S. P., Lava, N. S., Treihaft, M. M., Lubs, M. L., and Lubs, H. A. (1978). Late onset X-linked recessive spinal and bulbar muscular atrophy. *Muscle Nerve* **1,** 297–307.
13. La Spada, A. R., Wilson, E. M., Lubahn, D. B., Harding, A. E., and Fischbeck, K. H. (1991). Androgen receptor gene mutations in X-linked spinal and bulbar muscular atrophy. *Nature* **352,** 77–79.
14. MacLean, H. E., Warne, G. L., and Zajac, J. D. (1996). Spinal and bulbar muscular atrophy: androgen receptor dysfunction caused by trinucleotide repeat expansion. *J. Neurol. Sci.* **135,** 149–157.
15. Kaneko, K., Igarashi, S., Miyatake, T., and Tsuji, S. (1993). 'Essential tremor' and CAG repeats in androgen receptor gene. *Neurology* **43,** 1618–1619.
16. Doyu, M., Sobue, G., Mitsuma, T., Uchida, M., Iwase, T., and Takahashi, A. (1993). Very late onset X-linked recessive bulbospinal neuronopathy: mild clinical features and a mild increase in the size of tandem CAG repeat in androgen receptor gene. *J. Neurol. Neurosurg. Psych.* **56**(7), 832.
17. Lumbroso, S., Lobaccaro, J.-M., Vial, C., Sassolas, G., Ollagnon, B., Beldon, C., Pouget, J., and Sultan, C. (1997). Molecular analysis of the androgen receptor gene in Kennedy's disease. *Horm. Res.* **47,** 23–29.
18. Olney, R. K., Aminoff, M. J., Yuen, T., and So, M. D. (1991). Clinical and electrodiagnostic features of X-linked recessive bulbospinal neuronopathy. *Neurology* **41,** 823–828.
19. Amato, A. A., Prior, T. W., Barohn, R. J., Snyder, B. S., Papp, B. S., and Mendell, J. R. (1993). Kennedy's disease: a clinicopathologic correlation with mutations in the androgen receptor gene. *Neurology* **43,** 791–794.
20. Faragher, M. W., Choi, W.-T., MacLean, H. E. Warne, G. L., and Zajac, J. D. (1993). Kennedy's disease: clinical presentation and laboratory diagnosis. *Clin. Exp. Neurol.***30,** 61–65.
21. Papapetropoulos, T., and Panayiotopoulos, C. P. (1981). X-linked spinal and bulbar muscular atrophy of late onset (Kennedy-Stefanis disease?). *Eur. Neurol.* **20,** 485–488.
22. Schanen, J., Mikol, J., Guiziou, C., Vital, C., Coquet, M., Lagueny, A., Julien, J., and Haguenau, M. (1984). Forme familiale liée au sexe d'amyotrophie spinale progressive de l'adulte. *Rev. Neurol. (Paris)* **140,** 720–727.
23. Belsham, D. D., Yee, W.-C., Greenberg, C. R., and Wrogemann, K. (1992). Analysis of CAG repeat region of the androgen receptor gene in a kindred with X-linked spinal and bulbar muscular atrophy. *J. Neurol. Sci.* **112,** 133–138.
24. Neuschmid-Kaspar, F., Gast, A., Peterziel, H., Schneikert, J., Muigg, A., Ransmayr, G., Klocker, H., Bartsch, G., and Cato, A. C. B. (1996). CAG-repeat expansion in androgen receptor in Kennedy's disease is not a loss of function mutation. *Mol. Cell. Endocrinol.* **117,** 149–156.
25. Takayama, H., Ichiki, K., Oishi, T., Suga, Y., Kuroda, S., and Ogawa, N. (1991). Case report: a sporadic case of bulbospinal muscular atrophy of late onset. *J. Med.* **22,** 123–130.
26. Nakamura, M., Mita, S., Murakami, T., Uchino, M., Watanabe, S., Tokunaga, M., Kumamoto, T., and Ando M. (1994). Exonic trinucleotide repeats and expression of androgen receptor gene in spinal cord from X-linked spinal and bulbar muscular atrophy. *J. Neurol. Sci.* **122,** 74–79.
27. Jöbsis, G. J., Louwerse, E. S., de Visser, M., Wolterman, R. A., Bolhuis, P. A., Busch, H. F. M., Brüggenwirth, H. T., Baas, F., Wiersinga, W. M., and Koelman, J. H. T. M. (1995). Differential diagnosis in spinal and bulbar muscular atrophy: clinical and molecular aspects. *J. Neurol. Sci.* **129,** (Suppl.) 56–57.
28. MacLean, H. E., Choi, W.-T., Rekaris, G., Warne, G. L., and Zajac, J. D. (1995). Abnormal androgen receptor binding affinity in subjects with Kennedy's disease (spinal and bulbar muscular atrophy). *J. Clin. Endocrinol. Metab.* **80,** 508–516.
29. Choi, W.-T., MacLean, H. E., Chu, S., Warne, G. L., and Zajac, J. D. (1993). Kennedy's disease: genetic diagnosis of an inherited form of motor neurone disease. *Aust. NZ J. Med.* **23,** 187–192.
30. Tsukagoshi, H., Shoji, H., and Furukawa, T. (1970). Proximal neurogenic muscular atrophy in adolescence and adulthood with X-linked recessive inheritance. *Neurology* **20,** 1188–1193.
31. Yapijakis, C., Kapaki, E., Boussiou, M., Vassilopoulos, D., and Papageorgiou, C. (1996). Prenatal diagnosis of X-linked spinal and bulbar muscular atrophy in a Greek family. *Prenatal Diagnosis* **16,** 262–265.
32. Ferlini, A., Patrosso, M. C., Guidetti, D., Merlini, L., Uncini, A., Ragno, M., Plasmati, R., Fini, S., Repetto, M., Vezzoni, P., and Forabosco, A. (1995). Androgen receptor gene (CAG)n repeat analysis in the differential diagnosis between Kennedy disease and other motoneuron disorders. *Am. J. Med. Genet.* **55,** 105–111.
33. Guidetti, D., Vescovini, E., Motti, L., Ghidoni, E., Gemignani, F., Marbini, A., Patrosso, M. C., Ferlini, A., and Sollime, F. (1996). X-linked bulbar ands spinal muscular atrophy, or Kennedy's disease: clinical, neurophysiological, neuropathological, neuropsychological and molecular study of a large family. *J. Neurol. Sci.* **135,** 140–148.
34. Warner, C. L., Servidei, S., Lange, D. J., Miller, E., Lovelace, R. E., and Rowland, L. P. (1990). X-linked spinal muscular atrophy (Kennedy's syndrome): a kindred with hypobetalipoproteinemia. *Arch. Neurol.* **47,** 1117–1120.
35. Yamada, M., Inaba, A., and Shiojiri, T. (1997). X-linked spinal and bulbar muscular atrophy with myasthenic symptoms. *J. Neurol. Sci.* **146,** 183–185.
36. Suzuki, T., Endo, K., Igarashi, S., Fukuda, M., and Tanaka, M. (1997). Isolated bilateral masseter atrophy in X-linked recessive bulbospinal neuropathy. *Neurology* **48,** 539–540.
37. Nedelec, C., Dubas, F., Truelle, J. L., Pouplard, F., Delestre, F., and Penisson-Besnier, I. (1987). Amyotrophie spinale progressive distale et asymetrique des membres inférieurs à caractère familial. *Rev. Neurol. (Paris)* **143,** 765–767.
38. Kaneko, K., Igarashi, S., Miyatake, T., and Tsuji, S. (1993). Hypertrophic cardiomyopathy and increased number of CAG repeats in the androgen receptor gene. *Am. Heart J.* **126,** 248–249.
39. Liu, C.-S., Chang, Y.-C., Chen, D.-F., Huang, C.-C., Pang, C.-Y., Lee, H.-C., Cheng, C.-C., Horng, C. J., and Wei, Y.-H. (1995). Type IV hyperlipoproteinemia and moderate instability of CAG triplet expansion in the androgen-receptor gene. *Acta Neurol. Scand.* **92,** 398–404.

40. Doyu, M., Sobue, G., Mukai, E., Kachi, T.,Yashuda, T., Mitsuma T., and Takahashi, A. (1992). Severity of X-linked recessive bublospinal neuropathy correlates with size of the tandem CAG repeat in androgen receptor gene. *Ann. Neurol.* **32,** 707–710.
41. Igarashi, S., Tanno, Y., Onodera, O., Yamazaki, M., Sato, S., Ishikawa, A., Miyatani, N., Nagashima, M., Ishikawa, Y., Sahashi, K., Ibi, T., Miyatake, T., and Tsuji, S. (1992). Strong correlation between the number of CAG repeats in androgen receptor genes and the clinical onset of features of spinal and bublar muscular atrophy. *Neurology* **42,** 2300–2302.
42. Li, M., Sobue, G., Doyu, M., Mukai, E., Hashizume, Y., and Mitsuma, T. (1995). Primary sensory neurons in X-linked recessive bublospinal neuropathy: histopathology and androgen receptor gene expression. *Muscle Nerve.* **18,** 301–308.
43. Ferrante, M. A., and Wilbourn, A. J. (1997). The characteristic electrodiagnostic features of Kennedy's disease. *Muscle Nerve.* **20,** 323–329.
44. Trojaborg, W., and Wulff, C. H. (1994). X-linked recessive bulbospinal neuropathy (Kennedy's syndrome); a neurophysiological study. *Acta Neurol. Scand.* **89,** 214–219.
45. Shimada, N., Sobue, G., Doyu, M., Yamamoto, K., Yasuda, T., Mukai, E., Kaci, T., and Mitsuma, T. (1995). X-linked recessive bublospinal neuropathy: clinical phenotypes and CAG repeat size in androgen receptor gene. *Muscle Nerve* **18,**1378–1384.
46. Schoenen, J., Delwaide, P. J., Legros, J. J., and Franchimont, P. (1979). Motoneuropathie héréditaire: la forme proximale de l'adulte liée au sexe (ou maladie de Kennedy). *J. Neurol. Sci.* **41,** 343–357.
47. Goldenberg, J. N., and Bradley, W. G. (1996). Testosterone therapy and the pathogenesis of Kennedy's disease (X-linked bulbopspinal muscular atrophy). *J. Neurol. Sci.* **135,** 158–161.
48. Oyanagi, K., Aoki, K., Morita, T., Igarashi, S., Inuzuka, T., and Horikawa, Y. (1996). Disaggregation of polyribosomes in the spinal anterior horn cells in a patient with X-linked spinal and bulbar muscular atrophy. *Acta Neuropathol.* **91,** 444–447.
49. Matsuura, T., Demura, T., Aimoto, Y., Mizuno, T., Moriwaka, F., and Tashiro, K. (1992). Androgen receptor abnormality in X-linked spinal and bulbar muscular atrophy. *Neurology* **42,** 1724–1726.
50. Nagashima, T., Seko, K., Hirose, K., Mannen, T., Yoshimura, S., Arima, R., Nagashima, K., and Morimatsu, Y. (1988). Familial bulbo-spinal muscular atrophy associated with testicular atrophy and sensory neuropathy (Kennedy-Alter-Sung syndrome) *J. Neurol. Sci.* **87,** 141–152.
51. Terao, S., Sobue, G., Li, M., Hashizume, Y., Tanaka, F., and Mitsuma, T. (1997). The lateral corticospinal tract and spinal ventral horn in X-linked recessive spinal and bublar muscular atrophy: a quantitative study. *Acta Neuropathol.* **93,** 1–6.
52. Sobue, G., Hashizume, Y., Mukai, E., Hirayama, M., Mitsuma, T., and Takahashi, A. (1989). X-linked recessive bulbospinal neuropathy. *Brain* **112,** 209–232.
53. Paulson, G. W., Liss, L., and Sweeney, P. J. (1980). Late onset spinal muscle atrophy - a sex linked variant of Kugelberg-Welander. *Acta Neurol. Scand.* **61,** 49–55.
54. Sobue, G., Doyu, M., Kaci, T., Yasuda, T., Mukai, E., Kumagai, T., and Mitsuma, T. (1993). Subclinical phenotypic expression in heterozygous females of X-linked recessive bulbospinal neuropathy. *J. Neurol. Sci.* **117,** 74–78.
55. Ringel, S. P., Lava, N. S., Treihaft, M. M., Lubs, M. L., and Lubs, H. A. (1978). Late onset X-linked recessive spinal and bulbar muscular atrophy. *Muscle Nerve* **1,** 297–307.
56. Drachman, D. B., Murphy, S. R., Nigam, M. P., and Hills, J. R. (1967). "Myopathic" changes in chronically denervated muscle. *Arch. Neurol.* **16,** 14–24.
57. Warner, C. L., Griffin, J. E., Wilson, J. D., Jacobs, L. D., Murray, K. R., Fischbeck, K. H., Dickoff, D., and Griggs, R. C. (1992). X-linked spinomuscular atrophy: a kindred with associated abnormal androgen receptor binding. *Neurology* **42,** 2181–2184.
58. Danek, L., Witt, T. N., Mann, K., Schweikert, H. U., Romalo, G., La Spada, A. R., and Fischbeck, K. H. (1994). Decrease in androgen binding and effect of androgen treatment in a case of X-linked bulbospinal neuronopathy. *Clin. Invest.* **72,** 892–897.
59. Nankin, H. R., and Calkins, J. H. (1986). Decreased bioavailable testosterone in aging normal and impotent men. *J. Clin. Endocrinol. Metab.* **63,** 1418–1420.
60. Ogata, A., Matsuura, T., Tashiro, K., Moriwaka, F., Demura, T., Koyanagi, T., and Nagashima, K. (1994). Expression of androgen receptor in X-linked spinal and bulbar muscular atrophy and amyotrophic lateral sclerosis. *J. Neurol. Neurosurg. Psych.* **57,** 1274–1275.
61. MacLean, H. E., Warne, G. L., and Zajac, J. D. (1995). Defects of androgen receptor function: from sex reversal to motor neurone disease. *Mol. Cell. Endocrinol.* **112,** 133–141.
62. Fischbeck, K. H. (1995). The expanded trinucleotide repeat in Kennedy's disease. *Proc. Assoc. Am. Physicians* **107,** 228–230.
63. Brooks, B. P., and Fischbeck, K. H. (1995). Spinal and bulbar muscular atrophy: a trinucleotide-repeat expansion neurodegenerative disease. *Trends Neurosci.* **18,** 459–461.

CHAPTER 8

Genetics and Molecular Biology of the Androgen Receptor CAG Repeat

DIANE E. MERRY AND KENNETH H. FISCHBECK
Department of Neurology, University of Pennsylvania School of Medicine, Philadelphia, Pennsylvania 19104

Spinal and bulbar muscular atrophy (SBMA), or Kennedy's disease, is a neurodegenerative disorder characterized by the adult onset of proximal muscle weakness, atrophy, and fasciculations, caused by the degeneration of motor neurons [1]. The disease is slowly progressive, and is complicated by the involvement of bulbar muscles, which may lead to recurrent aspiration. Patients often die from complications of this problem. Muscle weakness in SBMA patients is symmetrical, and there are no upper motor neuron manifestations, such as spasticity or hyperreflexia. SBMA patients also show sensory deficits, and there is degeneration of sensory neurons [2]. The disease may be, but is not always, fatal. Another feature of SBMA is that affected males show signs of androgen insensitivity [3], including gynecomastia, reduced fertility, and testicular atrophy, while serum testosterone levels are normal or increased. A clinical description of SBMA is provided in Chapter 7.

The X-linked inheritance of SBMA was described by Kennedy *et al.* [1]; subsequently, the disease gene was mapped to the proximal long arm of the X chromosome [4], specifically to a region where the AR gene was localized [5, 6]. This evidence, coupled with the finding of partial androgen insensitivity in SBMA patients, suggested the AR gene as a candidate gene in SBMA [7], and a common mutation, the expansion of a trinucleotide (CAG) repeat within the coding region of the AR gene, was identified [8].

Beyond the identification of the genetic mutation, the expansion of the CAG trinucleotide repeat, little is known about the molecular pathogenesis of SBMA. It is likely, however, that SBMA has a similar mechanism to other neurodegenerative diseases with the same kind of mutation (see below). Neuropathological evaluation in SBMA reveals the loss of anterior horn cells of the spinal cord, and motor neurons within the brainstem [2], consistent with the distribution of motor weakness in patients. Dorsal root ganglion neurons also degenerate. An electron microscopic study [9] revealed the disaggregation of polyribosomes within remaining anterior horn cells from an SBMA patient; whether this is a primary characteristic of SBMA neuronal degeneration or is secondary to intracellular pathology remains to be seen.

I. THE ANDROGEN RECEPTOR GENE

A. Structure

The human AR gene contains eight exons and a 2.7-kb open reading frame that encodes the 919-amino-acid AR protein and extends over 100 kb at Xq12 [10, 11]. A long 3′-untranslated region is encoded by exon 8 and is differentially spliced to give mRNAs of about 7 and 10 kb.

B. The Androgen Receptor CAG Expansion

The CAG repeat found within the first exon of the AR gene is polymorphic in normal human populations, and ranges in length from 10 to 36 repeat units, with a mean repeat length of 21 [12, 13]. SBMA patients show a CAG repeat length that is nonoverlapping with that of normal individuals and ranges from 40 to 62 . The mouse and rat AR genes also contain a CAG repeat that is polymorphic between strains, although in these species it is found approximately 150 bp 3′ to that of the human gene [14, 15]. Nonhuman primates show intraspecies as well as interspecies variation in CAG repeat length [16]; average repeat lengths range from 3 in the macaque to 20 in the chimpanzee [17].

As with other triplet repeat diseases, a correlation is seen between CAG repeat length and age of disease onset [18–20]; however, considerable variability in age of onset is seen between family members with similar CAG repeat lengths [18]. The correlation between CAG repeat length and disease severity is even weaker, albeit significant [20]. Thus, although significant correlations exist between repeat length and both age of onset and disease severity, other factors influence these clinical characteristics, negating the use of repeat length as a reliable predictor of clinical course.

While AR repeats in the normal range are transmitted with fidelity, expanded repeats in SBMA patients are transmitted much less stably, showing expansions or contractions in approximately 60% of transmissions [18, 21]. In addition, expansions occur more frequently than contractions, and mutations usually result in repeat length changes of only a few repeats, although occasional large changes in repeat length are seen. Also, as with the other polyglutamine repeat diseases, a parent-of-origin effect is seen; expansions, and most notably, large expansions, show predominantly paternal transmission. Studies of the AR CAG repeat mutation rate in single sperm from normal individuals and an SBMA patient corroborate the population findings [22, 23]. Of interest is the finding of higher mutation rates in normal individuals with high-normal CAG repeat lengths when compared with normal individuals with average-normal CAG repeat lengths. In particular, this class of alleles displayed an increase in the rate of contraction (2.9 vs 0.9% for average-normal alleles), while the rates of expansion for these two groups were the same. This may explain the lack of new mutations giving rise to SBMA. In addition, the finding of linkage disequilibrium in Japanese SBMA chromosomes suggests that a founder effect, rather than new mutation, accounts for the generation of SBMA chromosomes in Japan [24].

II. PROPOSED MECHANISMS OF POLYGLUTAMINE TOXICITY

A. Gain versus Loss of Function

Several lines of evidence support the hypothesis that CAG-polyglutamine expansions confer a gain of a toxic function on the associated gene product. First, in all diseases of this class, the mutant allele is both transcribed and translated, arguing against a simple loss of function as the pathogenic mechanism [25–34]. Second, all of these diseases are inherited as dominant diseases, except for SBMA, which follows an X-linked recessive pattern of inheritance. However, an X-linked dominant mode of inheritance cannot be ruled out, since most females may be protected by either lyonization or low androgen levels, and some females have been described with mild symptoms of SBMA [35]. Third, individuals with true loss of function mutations in the AR gene, including deletions, missense mutations, and nonsense mutations, exhibit the distinct phenotype of testicular feminization (Tfm), and show no signs of a motor neuronopathy [36]. Fourth, transgenic mice that overexpress expanded repeat ataxin-1 protein in Purkinje cells develop ataxia, while transgenic mice that overexpress the normal repeat ataxin-1 protein do not [37]. Likewise, transgenic mice that express an expanded repeat version of exon 1 of the HD gene develop symptoms of HD [38]. Lastly, loss of function mutations in the mouse huntingtin gene result in embryonic lethality, not an adult-onset neurodegenerative disease; heterozygotes appear to be normal [39–41].

While the evidence is strong for a gain of function mechanism of toxicity in CAG/polyglutamine diseases, additional effects due to the graded loss of function of these genes cannot be ruled out. For SBMA, the only disease in which the gene function is known, partial androgen insensitivity is often found in affected patients. This indicates either that the mutant AR protein is not completely normal in transducing the effects of its hormonal ligand, androgen, or that it is not present at normal levels, at least in tissues involved in secondary sex characteristics. In fact, immunoreactive AR protein is undetectable in scrotal skin fibroblasts of some SBMA patients [42]. Whether a partial loss of function of AR protein contributes to the *motor neuron* effects is unknown.

B. Protein or RNA?

The evidence for a toxic effect of CAG expansion at the protein level rests largely on the fact that all of these genes are expressed at the protein level, and that in every case, a polyglutamine tract, rather than a polyserine (AGC) or polyglutamate (GCA) tract, is encoded by the expanded sequence. A toxic effect of the CAG repeat expansion mediated by RNA cannot be ruled out, however. Investigators have identified an RNA-binding protein that binds *huntingtin* RNA in a repeat length-dependent manner [43], and propose that RNA may mediate the toxic effects of CAG expansion. Nonetheless, recent studies that reveal abnormal physical characteristics of expanded polyglutamine repeat-containing proteins support the notion that expanded CAG repeat toxicity is manifest at the level of the protein [44, 45]. In these studies, expanded repeat huntingtin protein was shown to form large ubiquitinated intranuclear inclusions. These results suggest that the nucleus is a major site of pathology, and confirm a role for the protein in the neurodegenerative process.

III. POLYGLUTAMINES

A. Polyglutamine Structure

What characteristics of polyglutamine tracts provide insight into a possible mechanism of toxicity of the expanded forms? Biophysical studies of polyglutamine tracts, both isolated and in the context of a larger protein [46, 47], support the idea of a toxic polyglutamine effect. In the first of these studies, Perutz *et al.* [46] proposed that polyglutamine tracts may form unusual structures and found, using the techniques of circular dichroism, X-ray diffraction, and electron microscopy, that synthetic polyglutamines form β-sheets that aggregate into 70- to 120-μm-thick particles, the peptides held together by hydrogen bonds between main chain and side chain amides. Perutz called this structure a "polar zipper," and went on to show [47] that the introduction of a polyglutamine tract into a small protein that normally occurs in monomeric form caused this protein to aggregate into dimers and trimers that were stable even at high temperatures. Thus, the expansion of polyglutamine tracts found in this family of neurodegenerative diseases may result in the acquisition of unusually high affinities of the expanded protein with itself and other proteins, perhaps with similar structures.

B. Polyglutamine Function

The function of polyglutamine tracts, if any, is unknown, although these tracts are found in a number of transcription factors and other regulatory proteins in organisms from yeast to man; their function in these proteins may provide insight for their behavior and function in other proteins. Polyglutamines are found in such transcription factors as TATA-binding protein, Sp1, AP1, and rat glucocorticoid receptor, *Antennapedia* of Drosophila, the Drosophila neurogenic genes *Notch* and *mastermind,* and the yeast transcription factor GAL11 (reviewed in [48]). Glutamine-rich domains have been shown to be essential for proper transcription factor interactions [49, 50], and can function themselves as transcriptional activators when in the size range of 10–30 amino acids [48]. These findings suggest that polyglutamine tracts may be important for mediating a variety of protein–protein interactions.

IV. ANDROGEN RECEPTOR PROTEIN

A complete understanding of the effect that expansion of the polyglutamine tract within the AR protein has on a given cell requires an understanding of the AR protein itself, and its normal function.

A. Androgen Receptor Domain Structure and Function

The androgen receptor, a 919-amino-acid transcriptional regulatory protein, is a member of the steroid/

thyroid hormone receptor family. Like other members of this family, it is a phosphoprotein [51–54], and is composed of several functional domains (Fig. 8-1) (reviewed in [55, 56]). The C-terminal ligand binding domain binds the two biologically active ligands, testosterone and dihydrotestosterone, with high affinity; it also binds other steroid hormones and antiandrogens with moderate affinity. The centrally located DNA binding domain is highly conserved among steroid hormone receptors and contains nine invariant cysteine residues; eight of these are proposed to complex zinc ions in two separate zinc finger structures. Such structures are known to bind DNA contact sites in hormone response elements (HRE). Members of the steroid hormone receptor family show a high degree of sequence homology within their DNA binding domains and can interact with common HREs. Specific androgen response elements (AREs) have been described and are contained within the androgen-regulated prostate genes prostatein C3 subunit and 20-kDa protein [57, 58], and within the pituitary-expressed glycoprotein hormone α subunit gene [59]. Whether AR makes direct contact with other proteins of the transcriptional apparatus is unclear. A nuclear localization signal has been defined that borders the DNA binding domain and extends into the hinge region [60, 61]. The amino terminus of the AR that contains the polyglutamine repeat is involved in transcriptional activation [62], although a specific role for the AR glutamine repeat is unknown.

The AR, after it is synthesized, is found in an unliganded "aporeceptor" complex, consisting of at least the heat shock proteins Hsp90, Hsp70, and Hsp56, as well as p23 and a dnaJ protein ([64, 65]; reviewed in [66]). Hsp90 association requires the C-terminus of the AR [67]. This aporeceptor complex represents a "poised" receptor conformation, ready to respond to ligand. AR mutants that lack the ligand binding domain and bind neither ligand nor Hsp90 are constitutively active [62, 68]. This suggests that the aporeceptor complex not only maintains the receptor in a form competent for ligand binding, but also maintains the receptor in a transcriptionally inactive state. Upon hormone binding, the AR dissociates from the aporeceptor complex, translocates to the nucleus, and binds as a dimer to DNA sequences through its zinc finger domain. The AR protein then functions to either stimulate or repress transcription. Dimerization of the AR, required for transcriptional activation, is thought to occur in an antiparallel manner, mediated by sequences in both the ligand-binding and amino-terminal domains [69]. While the AR is known to exist as a phosphoprotein, the role of phosphorylation in AR function is unclear [54].

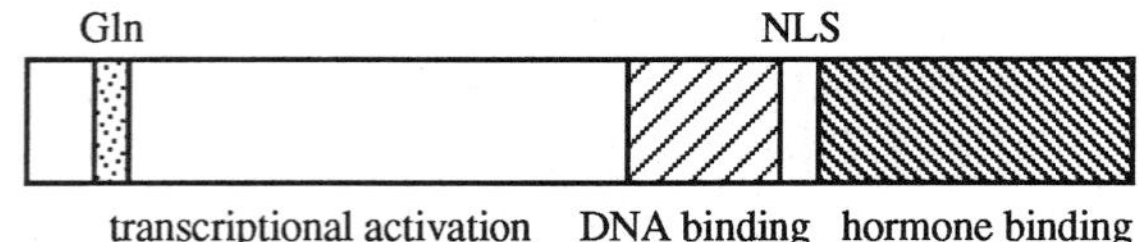

FIGURE 8-1 Schematic representation of the androgen receptor protein. Gln refers to the polymorphic glutamine repeat. NLS refers to the nuclear localization signal. While the complete N-terminal region of the protein has been implicated in transcriptional activation in some studies [63], a discrete 200-amino-acid activation domain has been defined C-terminal to the glutamine repeat by others [62].

B. Androgen Receptor Function in Neurons

Much is understood about the role of AR in gonadal tissues, but its role in the brain and spinal cord is less clear. The AR is expressed in many regions of the brain, in both sexually dimorphic and nondimorphic nuclei, including the somatostatin neurons of the periventricular nucleus of the hypothalamus [70–72], and neurons of the amygdala [70], midbrain, pons, medulla, cerebellum, and spinal cord [73]. Notably, high concentrations of radioactive ligand were found in motor neurons of the brain stem and spinal cord following injection of [^{3}H]dihydrotestosterone [73].

A role for the AR has been shown in the structure and function of central nervous system neurons [74–76]. A particularly well-studied example is the sexually dimorphic spinal nucleus of the bulbocavernosus in rodents (Onuf's nucleus in man) [77, 78]. These motor neurons accumulate particularly high levels of testosterone [77]; ligand insufficiency in females and loss of AR function (Tfm) in male animals lead to the death of neurons in this nucleus and the degeneration of the target muscle. Prenatal androgen administration rescues these cells from death [79–86]; however, recent studies suggest that this effect is largely target muscle-derived [87, 88]. Nonetheless, modulation of androgen levels in adult male rodents affects dendritic length of the same motor neuron population, suggesting that androgens and the AR can have an effect on the structural plasticity of adult neurons [89].

Testosterone has also been shown to attenuate the loss of non-sexually dimorphic motor neurons after injury, and to hasten axonal regeneration [90–94]; this process may be mediated in part by the androgenic regulation of tubulin gene expression [93]. Lastly, transgenic mice that overexpress AR in neurons show increased grip strength ([95]; unpublished observations), suggesting that AR expression in motor neurons may affect neuronal function.

To better understand the molecular mechanisms of androgen action in neurons, we created a model system by stably expressing the human AR in mouse motor neuron hybrid cells [96]. These hybrid cells express markers characteristic of anterior horn cells and can be differentiated into a neuronal phenotype. When differentiated in the presence of androgen, AR-expressing cells, but not control cells, exhibited a dose-dependent change in morphology: androgen-treated cells developed larger cell bodies and broader neuritic processes while continuing to express neuronal markers. In addition, androgen promoted the survival of AR-expressing cells, but not control cells, under low serum conditions. These results are consistent with a direct trophic effect of androgens on lower motor neurons, mediated by endogenous AR.

C. Androgen Receptor Protein Interactions

The AR normally interacts with a variety of heterologous proteins, including heat shock proteins and other members of the "aporeceptor" complex, as discussed above. In addition, it is phosphorylated at serine residues, binds to ligand, docks with and is transported by the nuclear transport machinery, associates with the nuclear matrix, and interacts with components of the transcriptional apparatus to activate or inhibit transcription [67].

In addition, the AR participates in important homologous intermolecular interactions during dimerization. Dimerization contact sites are located within the DNA binding domain, as well as within the amino-terminal and ligand-binding domains. Hormone-dependent receptor dimerization and DNA binding require an intact amino terminus; in addition, hormone-independent dimerization and DNA binding are seen in the absence of the amino terminus, suggesting that the amino terminus acts to inhibit dimerization and DNA binding in the absence of ligand [97]. Evidence suggests that the amino terminus interacts with the ligand-binding domain in a ligand-dependent manner [69], and the interaction is not affected by expansion of the repeat to 66 glutamines [69]. These data support a model in which the amino terminus of the AR directs a protein conformation in the absence of ligand (and in association with the aporeceptor complex) that inhibits dimerization and DNA binding. Upon ligand binding and release of the AR from the aporeceptor complex, the AR dimerizes in an antiparallel orientation with contacts between the amino-terminal and steroid-binding domains [69].

In addition to the protein interactions already discussed, the AR interacts with a number of different cellular proteins, including insulin degrading enzyme (also called receptor accessory factor, RAF) [98, 99], and glyceraldehyde 3-phosphate dehydrogenase (GAPDH) [100].

1. Polyglutamine–Protein Interactions

Polyglutamine tracts within the context of their associated proteins can be cross-linked by the enzyme transglutaminase, and Green [101] has hypothesized that long polyglutamine tracts may serve as better substrates for the enzyme and thereby become excessively cross-linked. The cross-linked proteins themselves or the isodipeptide produced by the transglutamination may then become toxic to neurons. Recent evidence indicates that glutamine residues within polyglutamine tracts can be cross-linked by transglutaminase; increasing repeat length increased this reactivity [102].

Other protein interactions that are modulated by polyglutamine repeat length include the huntingtin interactors HAP1 [103], for which polyglutamine expansion positively affects the interaction, and HIP1 [104, 105], for which polyglutamine expansion negatively affects interaction strength.

V. MODEL SYSTEMS OF SBMA

A. *In Vitro* Models

While many investigators have created cell culture systems in which the normal and expanded repeat AR have been transfected [96, 106–109], to date none of these systems reproduces the toxic effects of the expanded repeat AR. Nonetheless, substantial information has been gained from these experiments regarding the normal function of the AR.

1. Characteristics of Expanded Repeat AR

a. Ligand Binding. *In vitro* studies of AR ligand binding show that this intrinsic property of the protein is equivalent between normal and expanded repeat AR [96, 107, 108]. However, ligand binding capacity is diminished, in a manner that correlates with lower steady state mRNA and protein levels [96, 107]. These findings are consistent with SBMA patient studies, in which normal affinities but lower total ligand binding were found [33, 110]; one study, however, described a decreased affinity for ligand [111].

b. Steady-State Protein Levels. Consistent with decreased maximal ligand binding of expanded repeat AR, several investigators have noted a decrease in steady-state protein or mRNA levels, both *in vitro* and *in vivo.* In one study, no protein could be detected by immunohistochemistry in scrotal skin fibroblasts of three SBMA patients [42], while in another [112], decreased mRNA levels were seen in the spinal cord from an SBMA patient. However, immunoreactive AR was detected in spinal cord obtained from two SBMA patients at autopsy [113]. *In vitro,* decreased mRNA [107] or protein [96] levels were consistently seen in cells transfected with expanded repeat AR. This finding of decreased AR, with the secondary effect of decreased maximal ligand binding, may explain the symptoms of partial androgen insensitivity, such as gynecomastia, testicular atrophy, and reduced fertility seen in SBMA patients. In addition, a decrease in both circulating androgen levels [114] and normal AR protein [115] associated with aging may exacerbate the effects of low levels of the expanded repeat AR.

c. Transcriptional Competence. Because polyglutamines are found in a number of transcription factors, and may function to activate and/or modulate transcription, several investigators have addressed the question of transcriptional activation capability of the expanded repeat AR.

Several studies using transient transfection of nonneuronal cells [106, 108, 109, 116] showed decreased transcriptional activation by the expanded repeat AR; in addition, in one of these studies [109], the investigators also showed decreased transcriptional competence in transiently transfected neuronal cells. In another study [107], Choong and co-workers, using transient transfection of nonneuronal cells, showed no difference in transcriptional capability of expanded repeat AR. In concordance with this, Brooks *et al.* [96] used stably transfected neuronal cells and showed no effect of repeat length on transcriptional activation. Decreased steady state levels of AR mRNA [107] or protein [96] were noted in the latter two studies. It is likely that the decreased transcriptional competence seen by some is due in part to decreased steady-state AR levels; the cell type used for these experiments may also account for some of the differences seen. It is of interest, in addition, that a decrease in CAG repeat length has been positively associated with prostate cancer [117, 118]; this has been attributed to an increase in transcriptional activation competence of short repeat ARs [116], but may also be caused by increased steady-state levels of the AR protein.

d. Cellular Toxicity. In previous studies designed to address *in vitro* effects of AR CAG repeat expansion [96, 106–109], no cellular toxicity was observed, regardless of the cellular origin (i.e., neuronal or nonneuronal). This suggests that the intracellular milieu of a differentiated motor neuron (including the array of expressed proteins, and perhaps the functional state of neuronal interactions with other neurons and glia) is likely to be critical for the development of a degenerative phenotype. Specifically, cell-type-specific AR protein interactions might be required to generate the toxic effects of the repeat expansion. In keeping with this notion of cell-type-specific protein interactions, Ikeda *et al.* [119] have shown with the ataxin-3 protein that truncation of the expanded repeat protein produces toxicity in nontarget cell types; this suggests that a toxic derivative or conformation can be achieved more readily in the absence of full-length protein sequence. It may be that this holds for the AR, and for other expanded glutamine repeat-containing proteins as well. Indeed, we have found that truncated forms of the AR containing expanded repeats produce toxicity in Cos and neuronal cells [120a]. Thus, while toxicity may be mediated by the polyglutamine repeat itself, tissue specificity may be achieved through specific protein interactions with moieties outside of the glutamine repeat. In addition, truncated expanded repeat forms of the AR display repeat length-dependent proteolytic processing and protein aggregation [120]; the latter manifest as large intracellular inclusions.

B. Transgenic Models of SBMA

Efforts to create a mouse model of SBMA have fallen short of reproducing an SBMA-like phenotype. Bingham *et al.* [120] created transgenic mice with the neuron-specific enolase promoter and the inducible Mx promoter to drive expression of normal (24 CAG) and expanded (45 CAG) repeat AR. No phenotype was seen, but protein expression levels were low. Merry *et al.* ([95]; unpublished observations) also created transgenic mice using the neuron-specific enolase and neurofilament light chain promoters to drive expression of normal (24 CAG) and expanded (65 CAG) repeat AR. In these experiments, significant protein expression was achieved, yet no neurodegenerative phenotype arose, even in mice up to 2 years of age. In both transgenic experiments, CAG repeat length remained stable in 76 and 154 meioses, respectively.

C. Other Model Systems

Other systems have been created to model the repeat instability of the AR CAG repeat seen in SBMA patients. In a yeast system, Bingham *et al.* [121] found one contraction in 32 meioses in yeast containing AR YACs with 45 CAGs. AR YACs with 20 CAGs showed no instability. Repeat lengths of both sizes were stable through mitosis. In other experiments, La Spada *et al.* [122] found substantial mitotic instability. YACs containing 45 CAGs displayed instability in 5/90 (6%) colonies, and YACs containing 60 CAGs showed instability in 8/25 (32%) colonies; YACs containing 20 repeats showed no instability. Thus, increased instability was seen with an increase in CAG repeat length, similar to the repeat length instability in SBMA patients. Unlike the instability in humans, however, contractions outnumbered expansions, although the few expansions identified were large expansions (e.g., +10 CAGs, +14 CAGs). In yet another system, we have found substantial instability of highly expanded repeat containing AR clones in bacteria, with contractions exceeding expansions in number and size (unpublished observations). We have also observed the orientation-dependent instability in bacteria that has been observed by others [123].

VI. FUTURE STUDIES

Understanding the biochemical events that lead to neurodegeneration in SBMA will require suitable model systems in which the cell death can be reproduced. We have recently created longer CAG repeats within the AR and have found that truncated, amino-terminal AR proteins bearing repeats of approximately 100 or greater cause toxicity in Cos and neuronal cells [120a]. Whether these constructs will allow us to recapitulate the toxicity of SBMA in other systems, including mice, flies, nematodes, and yeast, remains to be seen. In addition, further studies to elucidate the molecular basis of the AR's trophic effect in neurons should provide a better understanding of the events that lead to neuronal demise in SBMA.

References

1. Kennedy, W. R., Alter, M., and Sung, J. H. (1968). Progressive proximal spinal and bulbar muscular atrophy of late onset: a sex-linked recessive trait. *Neurology* **18,** 671–680.
2. Sobue, G., Hashizume, Y., Mukai, E., Hirayama, M., Mitsuma, T., and Takahashi, A. (1989). X-linked recessive bulbospinal neuronopathy: a clinicopathological study. *Brain* **112,** 209–232.
3. Arbizu, T., Santamaria, J., Gomez, J. M., Quilez, A., and Serra, J. P. (1983). A family with adult spinal and bulbar muscular atrophy, X-linked inheritance and associated testicular failure. *J. Neurol. Sci.* **59,** 371–382.
4. Fischbeck, K. H., Ionasescu, V., Ritter, A. W., Ionasescu, R., Davies, K., Ball, S., Bosch, P., Burns, T., Hausmanowa-Petrusewicz, I., Borkowska, J., Ringel, S. P., and Stern, L. Z. (1986). Localization of the gene for X-linked spinal muscular atrophy. *Neurology* **36,** 1595–1598.
5. Migeon, B. R., Brown, T. R., Axelman, J., and Migeon, C. J. (1981). Studies of the locus for androgen receptor: localization on the human X chromosome and evidence for homology with the Tfm locus in the mouse. *Proc. Natl. Acad. Sci. USA* **78,** 6339–6343.
6. Brown, C. J., Goss, S. J., and Lubahn, D. B. (1989). Androgen receptor locus on the X chromosome: regional localization to Xq11-12 and description of a DNA polymorphism. *Am. J. Hum. Genet.* **44,** 264–269.
7. Fischbeck, K. H., Souders, D., and La Spada, A. R. (1991). A candidate gene for X-linked spinal muscular atrophy. *Adv. Neurol.* **56,** 209–213.
8. La Spada, A. R., Wilson, E. M, Lubahn, D. B., Harding, A. E., and Fischbeck, K. H. (1991). Androgen receptor gene mutations in X-linked spinal and bulbar muscular atrophy. *Nature* **353,** 77–79.
9. Oyanagi, K., Aoki, K., Morita, T., Igarashi, S., Inuzuka, T., and Horikawa, Y. (1996). Disaggregation of polyribosomes in the spinal anterior horn cells in a patient with X-linked spinal and bulbar muscular atrophy. *Acta Neuropathol.* **91,** 444–447.
10. Lubahn, D. B., Joseph, D. R., Sar, M., Tan, J., Higgs, H. N., Larson, R. E., French, F. S., and Wilson, E. M. (1988). The human androgen receptor: complementary deoxyribonucleic acid cloning, sequence analysis and gene expression in prostate. *Mol. Endocrinol.* **2,** 1265–1275.
11. Lubahn, D. B., Joseph, D. R., Sullivan, P. M., Willard, H. F., French, F. S., and Wilson, E. M. (1988). Cloning of human androgen receptor complementary DNA and localization to the X chromosome. *Science* **240,** 327–330.
12. Edwards, A., Hammond, H. A., Jin, L., Caskey, C. T., and Chakraborty, R. (1992). Genetic variation at five trimeric and tetrameric tandem repeat loci in four human population groups. *Genomics* **12,** 241–253.
13. Macke, J. P., Hu, N., Hu, S., Bailey, M., King, V. L., Brown, T., Hamer, D., and Nathans, J. (1993). Sequence variation in the androgen receptor gene is not a common determinant of male sexual orientation. *Am. J. Hum. Genet.* **53,** 844–852.
14. Faber, P. W., King, A., van Rooij, H. C. J., Brinkmann, A. O., de Both, N. J., and Trapman, J. (1991). The mouse androgen receptor: functional analysis of the protein and characterization of the gene. *Biochem. J.* **278,** 269–278.
15. Tan, J.-A., Jouseph, D. R., Quarmby, V. E., Lubahn, D. B., Sar, M., French, F. S., and Wilson, E. M. (1988). The rat androgen receptor: primary structure, autoregulation of its messenger ribonucleic acid, and immunocytochemical localization of the receptor protein. *Mol. Endocrinol.* **2,** 1276–1285.
16. Rubinsztein, D. C., Amos, W., Leggo, J., Goodburn, W. S., Jain, S., Li, S.-H., Margolis, R. L., Ross, C. A., and Ferguson-Smith, M. A. (1995). Microsatellite evolution—evidence for directionality and variation in rate between species. *Nature Genet.* **10,** 337–343.
17. Rubensztein, D. C., Leggo, J., Coetzee, G. A., Irvine, R. A., Buckley, M., and Ferguson-Smith, M. A. (1995). Sequence variation and size ranges of CAG repeats in the Machado-Joseph disease, spinocerebellar ataxia type 1 and androgen receptor genes. *Hum. Mol. Genet.* **4,** 1585–1590.
18. La Spada, A. R., Roling, D., Harding, A. E., Warner, C. L., and Speigel, R., Hausmanowa-Petrusewicz, I., Yee, W.-C., and Fischbeck, K. H. (1992). Meiotic stability and genotype-phenotype correlation of the expanded trinucleotide repeat in X-linked spinal and bulbar muscular atrophy. *Nature Genet.* **2,** 301–304.
19. Igarashi, S., Tanno, Y., Onodera, O., Yamazaki, M., Sato, S., Ishikawa, A., Mayatani, N., Nagashima, M., Ishikawa, Y., Sahashi, K., Ibi, T., Miyatake, T., and Tsuji, S. (1992). Strong correlation between the number of CAG repeats in androgen receptor genes and the clinical onset of features of spinal and bulbar muscular atrophy. *Neurology* **42,** 2300–2302.
20. Doyu, M., Sobue, G., Mukai, E., Kachi, T., Ysauda, T., Mitsuma, T., and Takahashi, A. (1992). Severity of X-linked recessive bulbosopinal neuronopathy correlates with size of the tandem CAG repeat in androgen receptor gene. *Ann. Neurol.* **32,** 707–710.

21. Biancalana, V., Serville, F., Pommier, J., Julien, J., Hanauer, A., and Mandel, J. L. (1992). Moderate instability of the trinucleotide repeat in spinobulbar muscular atrophy. *Hum. Mol. Genet.* **1,** 255–258.
22. Zhang, L., Leeflang, E. P., Yu, J., and Arnheim, N. (1994). Studying human mutations by sperm typing: instability of CAG trinucleotide repeats in the human androgen receptor gene. *Nature Genet.* **7,** 531–535.
23. Zhang, L., Fischbeck, K. H., and Arnheim, N. (1995). CAG repeat length variation in sperm from a patient with Kennedy's disease. *Hum. Mol. Genet.* **4,** 303–305.
24. Tanaka, F., Doyu, M., Ito, Y., Matsumoto, M., Mitsuma, T., Abe, K., Aoki, M., Itoyama, Y., Fischbeck, K. H., and Sobue, G. (1996). Founder effect in spinal and bulbar muscular atrophy (SBMA). *Hum. Mol. Genet.* **5,** 1253–1257.
25. Banfi, S., Servadio, A., Chung, M., Kwiatkowski, T. J., McCall, A. E., Duvick, L. A., Shen, Y., Roth, E. J., Orr, H. T., and Zoghbi, H. Y. (1994). Identification and characterization of the gene causing type 1 spinocerebellar ataxia. *Nature Genet.* **7,** 513–520.
26. Li, S. H., Schilling, G., Young, W. S., Li, X.-J., and Margolis, R. L., Stine, O. C., Wagster, M. V., Abbot, M. H., Franz, M. L., Ramen, N. G., Folstein, S. E., Hedreen, J. C., and Ross, C. A. (1993). Huntington's disease gene (IT15) is expressed widely in human and rat tissues. *Neuron* **11,** 985–993.
27. Nagafuchi, S., Yanagisawa, H., Ohsaki, E., Shirayama, T., and Tadokoro, K., Inoue, T., and Yamada, M. (1994). Structure and expression of the gene responsible for the triplet repeat disorder, dentatorubral and pallidoluysian atrophy (DRPLA). *Nature Genet.* **8,** 177–182.
28. Servadio, A., Koshy, B., Armstrong, L. D., Antalffy, B., Orr, H. T., and Zoghbi, H. Y. (1995). Expression analysis of the ataxin-1 protein in tissues from normal and spinocerebellar ataxia type 1 individuals. *Nature Genet.* **10,** 94–98.
29. Sharp, A. H., Loev, S. J., Schilling, G., Li, S.-H., and Li, X.-J., Bao, J., Wagster, M. V., Kotzuk, J. A., Steiner, J. P., Lo, A., Hedreen, J., Sisodia, S., Snyder, S. H., Dawson, T. M., Ryugo, D. K., and Ross, C. A. (1995). Widespread expression of Huntigton's disease gene (IT15) protein product. *Neuron* **14,** 1–20.
30. Strong, T. V., Tagle, D. A., Valdes, J. M., Elmer, L. W., and Boehm, K., Swaroop, M., Kaatz, K. W., Collngs, F. S., and Albin, R. L. (1993). Widespread expression of the human and rat Huntington's disease gene in brain and nonneural tissues. *Nature Genet.* **5,** 259–265.
31. Trottier, Y., Devys, D., Imbert, G., Saudou, F., An, I., Lutz, Y., Weber, C., Agid, Y., Hirsch, E. C., and Mandel, J.-L. (1995). Cellular localization of the Huntington's disease protein and discrimination of the normal and mutated form. *Nature Genet.* **10,** 104–110.
32. Yazawa, I., Nukina, N., Hashida, H., Goto, J., Yamada, M., and Kanazawa, I. (1995). Abnormal gene product identified in hereditary denatorubral-pallidoluysian atrophy (DRPLA) brain. *Nature Genet.* **10,** 99–103.
33. Warner, C. L., Griffen, J. E., Wilson, J. D., Jacobs, L. D., Murray, K. R., Fischbeck, K. H., Dickoff, D., and Griggs, R. C. (1992). X-linked spinomuscluar atrophy; a kindred with associated abnormal androgen binding. *Neurology* **42,** 2181–2184.
34. Trottier, Y., Lutz, Y., Stevanin, G., Imbert, G., Devys, D., Cancel, G., Saudou, F., Weber, C., David, G., Tora, L., Agid, Y., Brice, A., and Mandel, J.-L. (1995). Polyglutamine expansion as a pathological epitope in Huntington's disease and four dominant cerebellar ataxias. *Nature* **378,** 403–406.
35. Ferlini, A., Patrosso, M. C., Guidetti, D., Merlini, L., Uncini, A., Ragno, M., Plasmati, R., Fini, S., Repetto, M., Vezzoni, P., and Forabosco, A. (1995). Androgen receptor gene (CAG)n repeat analysis in the differential diagnosis between Kennedy Disease and other motoneuron disorders. *Am. J. Med. Genet.* **55,** 105–111.
36. Pinsky, L., Trifiro, M., Kaufman, M., Beitel, L. K., Mhatre, A., Kazemi-Esfarjani, P., Sabbaghian, N., Lumbroso, R., Alvarado, C., Vasiliou, M., and Gottlieb, B. (1992). Androgen resistance due to mutation of the androgen receptor. *Clin. Invest. Med.* **15,** 456–472.
37. Burright, E., Clark, H. B., Servadio, L. A., Mantilla, T., Feddersen, R. M., Yunis, W. S., Duvick, L. A., Zoghbi, H. Y., and Orr, H. T. (1995). SCA1 transgenic mice: a model for neurodegeneration caused by an expanded CAG trinucleotide repeat. *Cell* **82,** 937–948.
38. Mangiarini, L., Sathasivam, K., Seller, M., Cozens, B., Harper, A., Hetherington, C., Lawton, M., Trottier, Y., Lehrach, H., Davies, S. W., and Bates, G. P. (1996). Exon 1 of the HD gene with an expanded CAG repeat is sufficient to cause a progressive neurological phenotype in transgenic mice. *Cell* **87,** 493–506.
39. Zeitlin, S., Liu, J. P., Chapman, D. L., Papaioannou, V. E., and Efstratiadis, A. (1995). Increased apoptosis and early embryonic lethality in mice nullizygous for the Huntington's disease gene homologue. *Nature Genet.* **11,** 155–163.
40. Duyao, M. P., Auerbach, A. B., Ryan, A., Persichetti, F., Barnes, G. T., McNeil, S. M., Ge, P., Vonsattel, J.-P., Gusella, J. F., Joyner, A. L., and MacDonald, M. E. (1995). Inactivation of the mouse Huntington's disease gene homolog Hdh. *Science* **269,** 407–410.
41. Nasir, J., Floresco, S. B., O'Kusky, J. R., Diewert, V. M., Richman, J. M., Zeisler, J., Borowski, A., Marth, J. D., Phillips, A. G., and Hayden, M. R. (1995). Targeted disruption of the Huntington's Disease gene results in embryonic lethality and behavioral and morphological changes in heterozygotes. *Cell* **81,** 811–823.
42. Matsuura, T., Demura, T., Aimoto, Y., Mizuno, T., Moriwaka, F., and Tashiro, K. (1992). Androgen receptor abnormality in X-linked spinal and bulbar muscular atrophy. *Neurology* **42,** 1724–1726.
43. McLaughlin, B. A., Spencer, C., and Eberwine, J. (1996). CAG trinucleotide RNA repeats interact with mRNA-binding proteins. *Am. J. Hum. Genet.* **59,** 561–569.
44. Scherzinger, E., Lurz, R., Turmaine, M., Mangiarini, L., Hollenbach, B., Hasenbank, R., Bates, G. P., Davies, S. W., Lehrach, H., and Wanker, E. E. (1997). Huntingtin-encoded polyglutamine expansions form amyloid-like protein aggregates in vitro and in vivo. *Cell* **90,** 549–558.
45. Davies, S. W., Turmaine, M., Cozens, B. A., DiFiglia, M., Sharp, A. H., Ross, C. A., Scherzinger, E., Wanker, E. E., Mangiarini, L., and Bates, G. P. (1997). Formation of neuronal intranuclear inclusions underlies the neurological dysfunction in mice transgenic for the HD mutation. *Cell* **90,** 537–548.
46. Perutz, M. F., Johnson, T., Suzuki, M., and Finch, J. T. (1994). Glutamine repeats as polar zippers: their possible role in inherited neurodegenerative diseases. *Proc. Natl. Acad. Sci. USA* **91,** 5355–5358.
47. Stott, K., Backburn, J. M, Butler, P. J. G., and Perutz, M. (1995). Incorporation of glutamine repeats makes protein oligomerize: implications for neurodegenerative diseases. *Proc. Natl. Acad. Sci. USA* **92,** 6509–6513.
48. Gerber, H.-P., Seipel, K., Georgiev, O., Hofferer, M., Hug, M., Rusconi, S., and Schaffner, W. (1994). Transcriptional activation modulated by homopolymeric glutamine and proline stretches. *Science* **263,** 808–811.
49. Pascal, E., and Tjian, R. (1991). Different activation domains of Sp1 govern formation of multimers and mediate transcriptional synergism. *Genes Dev.* **5,** 1646–1656.

50. Bruhn, L., Hwang-Shum, J. J., and Sprague, G. F. (1992). The N-terminal 96 residues of MCM1, a regulator of cell type-specific genes in Saccharomyces cerevisiae, are sufficient for DNA binding, transcription activation and interaction with alpha 1. *Mol. Cell. Biol.* **12,** 3563–3572.
51. Kuiper, G. G. J. M., and Brinkmann, A. O. (1995). Phosphotryptic peptide analysis of the human androgen receptor: detection of a hormone-induced phosphopeptide. *Biochemistry* **34,** 1851–1857.
52. Kemppainen, J. A., Lane, M. V., Sar, M., and Wilson, E. M. (1992). Androgen receptor phosphorylation, turnover, nuclear transport and transcriptional activation: specificity for steroids and antihormones. *J. Biol. Chem.* **267,** 968–974.
53. van Laar, J. H., Berrevoets, C. A., Trapman, J., Zegers, N. D., and Brinkmann, A. O. (1991). Hormone-dependent androgen receptor phosphorylation is accompanied by receptor transformation in human lymph node carcinoma of the prostate cells. *J. Biol. Chem.* **266,** 3734–3738.
54. Zhou, Z.-x., Kemppainen, J. A., and Wilson, E. M. (1995). Identification of three proline-directed phosphorylation sites in the human androgen receptor. *Mol. Endocrinol.* **9,** 605–615.
55. Tsai, M.-J., and O'Malley, B. W. (1994). Molecular mechanisms of action of steroid/thryoid receptor superfamily members. *Annu. Rev. Biochem.* **63,** 451–486.
56. Zhou, Z.-x., Wong, C.-i., Sar, M., and Wilson, E. M. (1994). The androgen receptor: an overview. *Recent Prog. Horm. Res.* **49,** 249–274.
57. Tan, J.-a., Marschke, K. B., Ho, K.-C., Perry, S. T., Wilson, E. M., and French, F. S. (1992). Response elements of the androgen-regulated C3 gene. *J. Biol. Chem.* **267,** 4456–4466.
58. Ho K.-C., Marschke, K. B., Tan, J.-a., Power, S. G. A., Wilson, E. M., and French, F. S. (1993). A complex response element in intron 1 of the androgen-regulated 20-kDa protein gene displays cell type-dependent androgen receptor specificity. *J. Biol. Chem.* **268,** 27226–27235.
59. Clay, C. M., Keri, R. A., Finicle, A. B., Heckert, L. L., Hamernik, D. L., Marschke, K. M., Wilson, E. M. French, F. S., and Nilson, J. H. (1993). Transcriptional repression of the glycoprotein hormone α subunit gene by androgen may involve direct binding of androgen receptor to the proximal promoter. *J. Biol. Chem.* **268,** 13556–13564.
60. Jenster, G., Trapman, J., and Brinkmann, A. O. (1993). Nuclear import of the human androgen receptor. *Biochem. J.* **293,** 761–768.
61. Zhou, Z.-x., Sar, M., Simental, J. A., Lane, M. V., and Wilson, E. M. (1994). A ligand-dependent bipartite nuclear targeting signal in the human androgen receptor. *J. Biol. Chem.* **269,** 13115–13123.
62. Simental, J. A., Sar, M., Lane, M. V., French, F. S., and Wilson, E. M. (1991). Transcriptional activation and nuclear targeting signals of the human androgen receptor. *J. Biol. Chem.* **266,** 510–518.
63. Jenster, G., van der Korput, H. A. G. M., Trapman, J., and Brinkmann, A. O. (1995). Identification of two transcription activation units in the N-terminal domain of the human androgen receptor. *J. Biol. Chem.* **270,** 7341–7346.
64. Veldscholte, J., Berrevoets, C. A., Zegers, N. D., van der Kwast, T. H., Grootegoed, A., and Mulder, E. (1992). Hormone-induced dissociation of the androgen receptor-heat-shock protein complex: use of a new monoclonal antibody to distinguish transformed from nontransformed receptors. *Biochemistry* **31,** 7422–7430.
65. Caplan, A. J., Langley, E., Wilson, E. M., and Vidal, J. (1995). Hormone-dependent transactivation by the human androgen receptor is regulated by a dnaJ protein. *J. Biol. Chem.* **270,** 5251–5257.
66. Bohen, S. P., Kralli, A., and Yamamoto, K. R. (1995). Hold 'em and fold 'em: chaperones and signal transduction. *Science* **268,** 1303–1304.
67. Jenster, G. (1994) Functional domains of the human androgen receptor. Thesis, Erasmus University, Rotterdam.
68. Jenster, G., van der Korput, H. A. G. M., van Vroonhoven, C., van der Kwast, T. H., Trapman, J., and Brinkmann, A. O. (1991). Domains of the human androgen receptor involved in steroid binding, transcriptional activation, and subcellular localization. *Mol. Endocrinol.* **5,** 1396–1404.
69. Langley, E., Zhou, Z.-x., and Wilson, E. M. (1995). Evidence for an anti-parallel orientation of the ligand-activated human androgen receptor dimer. *J. Biol. Chem.* **270,** 29983–29990.
70. Bingaman, E., Baeckman, L. M., Yracheta, J. M., Handa, R. J., and Gray, T. S. (1994). Localization of androgen receptor within peptidergic nerons of the rat forebrain. *Brain Res. Bull.* **35,** 379–382.
71. Huang, X., and Harlan, R. E. (1994). Androgen receptor immunoreactivity in somatostatin neruons of the periventricular nucleus but not in the bed nucleus of the stria terminalis in the male rats. *Brain Res.* **652,** 291–296.
72. Zhou, L., Blaustein, J. D., and De Vries, G. J. (1994). Distribution of androgen receptor immunoreactivity in vasopressin- and oxytocin-immunoreactive neurons in the male rat brain. *Endocrinology* **134,** 2522–2627.
73. Sar, M., and Stumpf, W. E. (1977). Androgen concentration in motor neruons of cranial nerves and spinal cord. *Science* **19,** 77–79.
74. Breedlove, S. M. (1992). Sexual dimorphism in the vertebrate nervous system. *J. Neurosci.* **12,** 4133–4142.
75. Garcia-Segura, L. M., Chowen, J. A., Parducz, A., and Naftolin, F. (1994). Gonadal hormones as promoters of structural synaptic plasticity: cellular mechanisms. *Prog. Neurobiol.* **44,** 279–307.
76. Matsumoto, A., Arai, Y., Urano, A., and Hyodo, S. (1995). Molecular basis of neuronal plasticity to gonadal steroids. *Funct. Neurol.* **10,** 59–76.
77. Breedlove, S. M., and Arnold, A. P. (1980). Hormone accumulation in a sexually dimorphic nucleus in the rat spinal cord. *Science* **210,** 564–566.
78. Forger, N. G., Hodges, L. L., Roberts, S. L., and Breedlove, S. M. (1992). Regulation of motoneuron death in the spinal nucleus of the bulbocavernosus. *J. Neurobiol.* **23,** 1192–1203.
79. Breedlove, S. M., and Arnold, A. P. (1981). Sexually dimorphic motor nucleus in the rat spinal cord: response to adult hormone manipulation, absence in androgen-insensitive rats. *Brain Res.* **225,** 297–307.
80. Breedlove, S. M., and Arnold, A. P. (1983). Sex differences in the pattern of steroid accumulation by motoneurons in the rat lumbar spinal cord. *J. Comp. Neurol.* **215,** 211–216.
81. Breedlove, S. M., and Arnold, A. P. (1983). Hormonal control of a developing neuromuscular system. I. Complete demasculinization of the male rat spinal nucleus of the bulbocavernosus using the anti-androgen flutamide. *J. Neurosci.* **3,** 417–423.
82. Breedlove, S. M., and Arnold, A. P. (1983). Hormonal control of a developing neuromuscular system. II. Sensitive periods for the androgen-induced masculinization of the rat spinal nucleus of the bulbocavernosus. *J. Neurosci.* **3,** 424–432.
83. Jordan, C. L., Breedlove, S. M., and Arnold, A. P. (1982). Sexual dimorphism and the influence of neonatal androgen in the dorsolateral motor nucleus of the rat lumbar spinal cord. *Brain Res.* **249,** 309–314.

84. Nordeen, E. J., Nordeen, K. W., Sengelaub, D. R., and Arnold, A. P. (1985). Androgens prevent normally occurring cell death in a sexually dimorphic spinal nucleus. *Science* **229,** 671–673.
85. Sengelaub, D. R., Jordan, C. L., Kurz, E. M., and Arnold, A. P. (1989). Hormonal control of neuron number in sexually dimorphic spinal nuclei of the rat. III. Differential effects of the androgen dihydrotestosterone. *J. Comp. Neurol.* **280,** 637–644.
86. Sengelaub, D. R., and Arnold, A. P. (1989). Hormonal control of neuron number in sexually dimorphic spinal nuclei of the rat. I. Testosterone-regulated death in the dorsolateral nucleus. *J. Comp. Neurol.* **280,** 622–629.
87. Al-Shamma, H. A., and Arnold, A. P. (1995). Importance of target innervation in recovery from axotomy-induced loss of androgen receptor in rat perineal motoneurons. *J. Neurobiol.* **28,** 341–353.
88. Freeman, L. M. W. N. V., and Breedlove, S. M. (1996). Androgen spares androgen-insensitive motoneurons from apoptosis in the spinal nucleus of the bulbocavernosus in rats. *Horm. Behav.* **30,** 424–433.
89. Kurz, E. M., Sengelaub, D. R., and Arnold, A. P. (1986). Androgens regulate the dendritic length of mammalian motoneruons in adulthood. *Science* **232,** 395–398.
90. Kujawa, K. A., Emeric, E., and Jones, K. J. (1991). Testosterone differentially regulates the regenerative properties of injured hamster facial motor neurons. *J. Neurosci.* **11,** 3898–3908.
91. Kujawa, K. A., Jacob, J. M., and Jones, K. J. (1993). Testosterone regulation of the regenerative properties of injured rat sciatic motor neurons. *J. Neurosci. Res.* **35,** 268–273.
92. Yu, W. A. (1989). Administration of testoterone attenuates neuronal loss following axotomy in the brainstem motor nuclei of female rats. *J. Neurosci.* **9.**
93. Jones, K. J., and Oblinger, M. M. (1994). Androgenic regulation of tubulin gene expression in axotomized hamster facial motoneurons. *J. Neurosci.* **14,** 3620–3627.
94. Perez, J., and Kelley, D. B. (1996). Trophic effects of androgen: receptor expression and the survival of laryngeal motor neurons after axotomy. *J. Neurosci.* **16,** 6625–6633.
95. Merry, D. E., McCampbell, A., Taye, A. A., Winston, R. L., and Fischbeck, K. H. (1996). Toward a mouse model for spinal and bulbar muscular atrophy: effect of neuronal expression of androgen receptor in transgenic mice. *Am. J. Hum. Genet.* **59**(Suppl.), A271.
96. Brooks, B. P., Paulson, H. L., Merry, D. E., Salazar-Grueso, E. F., Brinkmann, A. O., Wilson, E. M., and Fischbeck, K. H. (1997). Characterization of an expanded glutamine repeat androgen receptor in a neuronal cell culture system. *Neurobiol. Disease* **4,** 313–323.
97. Wong, C.-i., Zhou, Z.-x., Sar, M., and Wilson, E. M. (1993). Steroid requirement for androgen receptor dimerization and DNA binding. *J. Biol. Chem.* **268,** 19004–19012.
98. Kupfer, S. R., Marschke, K. B., Wilson, E. M., and French, F. S. (1993). Receptor accessory factor enhances specific DNA binding of androgen and glucocorticoid recptors. *J. Biol. Chem.* **268,** 17519–17527.
99. Kupfer, S. R., Wilson, E. M., and French, F. S. (1994). Androgen and glucocorticoid receptors interact with insulin degrading enzyme. *J. Biol. Chem.* **269,** 20622–20628.
100. Koshy, B., Matilla, T., Burright, E. N., Merry D. E., Fischbeck, K. H., Orr, H. T., and Zoghbi, H. Y. (1996). Spinocerebellar ataxia type-1 and spinobulbar muscular atrophy gene products interact with glyceraldehyde-3-phosphate dehydrogenase. *Hum. Mol. Genet.* **5,** 1311–1318.
101. Green, H. (1993). Human genetic diseases due to codon reiteration: relationship to an evolutionary mechanism. *Cell* **74,** 955–956.
102. Kahlem, P., Terre, C., Green, H., and Djian, P. (1996). Peptides containing glutamine repeats as substrates for transglutaminase-catalyzed cross-linking: Relevance to diseases of the nervous system. *Proc. Natl. Acad. Sci. USA* **93,** 14580–14585.
103. Li, X.-J., Li, S.-H., Sharp, A. H., Nucifora, F. C., Jr., Schilling, G., Lanahan, A., Worley, P., Snyder, S. H., and Ross, C. A. (1995). A huntingtin-associated protein enriched in brain with implications for pathology. *Nature* **378,** 398–402.
104. Kalchman, M. A., Koide, H. B., McCutcheon, K., Graham, R. K., Nichol, K., Nishiyama, K., Kazemi-Esfarjani, P., Lynn, F. C., Wellington, C., Metzler, M., Goldberg, Y. P., Kanazawa, I., Gietz, R. D., and Hayden, M. R. (1997). HIP1, a human homologue of *S. Cerevisiae Sla2p,* interacts with membrane-associated huntingtin in the brain. *Nature Genet.* **16,** 44–53.
105. Wanker, E. E., Rovirra, C., Scherzinger, E., Hasenbank, R., Walter, S., Tait, D., Colicelli, J., and Lehrach, H. (1997). HIP-1: A huntingtin interacting protein isolated by the yeast two-hybrid system. *Hum. Mol. Genet.* **6,** 487–495.
106. Chamberlain, N. L., Driver, E. D., and Miesfeld, R. L. (1994). The length and location of CAG trinucleotide repeats in the androgen receptor N-terminal domain affect transactivation function. *Nucleic Acids Res.* **22,** 3181–3186.
107. Choong, C. S., Kamppainen, J. A., Zhou, Z. X., and Wilson, E. M. (1996). Reduced androgen receptor gene expression with first exon CAG repeat expansion. *Mol. Endocrinol.* **10,** 1527–1535.
108. Mhatre, A. N., Trifiro, M. A., Kaufman, M., Kazemi-Esfarjani, P., Figlewicz, D., Rouleau, G., and Pinsky, L. (1993). Reduced transcriptional regulatory competence of the androgen receptor in X-linked spinal and bulbar muscular atrophy. *Nature Genet.* **5,** 184–188.
109. Nakajima, H., Kimura, F., Nakagawa, T., Furutama, D., Shinoda, K., Shimizu, A., and Ohsawa, N. (1996). Transcriptional activation by the androgen receptor in X-linked spinal and bulbar muscular atrophy. *J. Neurol. Sci.* **142,** 12–16.
110. Danek, A., Witt, T. N., Mann, K., Schweikert, H. U., Romalo, G., LaSpada, A. R., and Fischbeck, K. H. (1994). Decrease in androgen binding and effect of androgen treatment in a case of X-linked bulbosopinal neuronopathy. *Clin. Invest.* **72,** 892–897.
111. Maclean, H. E., Choi, W.-T., Rekaris, G., Warne, G. L., and Zajac, J. D. (1995). *J. Clin. Endocrinol. Metab.* **80,** 508–516.
112. Nakamura, M., Mita, S., Murakami, T., Uchino, M., Watanabe, S., Tokunaga, M., Kumamoto, T., and Ando, M. (1994). Exonic trinucleotide repeats and expression of androgen receptor gene in spinal cord from X-linked spinal and bulbar muscular atrophy. *J. Neurol. Sci.* **122,** 74–79.
113. Ogata, A., Matsuura, T., Tashiro, K., Moriwaka, F., Demura, T., Koyanagi, T., and Nagashima, K. (1994). Expression of androgen receptor in X-linked spinal and bulbar muscular atrophy and amyotrophic lateral sclerosis. *J. Neurol. Neurosurg. Psych.* **57,** 1274–1275.
114. Nankin, H. R., and Calkins, J. H. (1986). Decreased bioavailable testosterone in aging normal and impotent men. *J. Clin. Endocrinol. Metab.* **63,** 1418–1420.
115. Roehrborn, C. G., Lange, J. L., George, F. W., and Wilson, J. D. (1987). Changes in amount and intracellular distribution of androgen receptor in human foreskin as a function of age. *J. Clin. Invest.* **79,** 44–47.
116. Kazemi-Esfarjani, P., Trifiro, M. A., and Pinsky, L. (1995). Evidence for a repressive function of the long polyglutamine tract in the human androgen receptor: possible pathogenetic relevance for the $(CAG)_n$-expanded neuronopathies. *Hum. Mol. Genet.* **4,** 523–527.

117. Irvine, R. A., Yu, M. C., Ross, R. K., and Coetzee, G. A. (1995). The CAG and GGC microsatellites of the androgen receptor gene are in linkage disequilibrium in men with prostate cancer. *Cancer Res.* **55,** 1937–1940.

118. Giovannucci, E., Stampfer, M. J., Krithivas, K., Brown, M., Brufsky, A., Talcott, J., Hennekens, C. H., and Kantoff, P. W. (1997). The CAG repeat within the androgen receptor gene and its relationship to prostate cancer. *Proc. Natl. Acad. Sci. USA* **94,** 3320–3323.

119. Ikeda, H., Yamaguchi, M., Sugai, S., Aze, Y., Naruiya, S., and Kakizuka, A. (1996). Expanded polyglutamine in the Machado-Joseph disease protein induces cell death in vitro and in vivo. *Nature Genet.* **13,** 196–202.

120a. Merry, D. E., Kobayashi, Y., Bailey, C. K., Taye, A. A., and Fischbeck, K. H. (1998). Cleavage, aggregation, and toxicity of the expanded androgen receptor in spinal and bulbar muscular atrophy. *Hum. Mol. Genet.* In press.

120. Bingham, P. M., Scott, M. O., Wang, S., McPhaul, M. J., Wilson, E. M., Garbern, J. Y., Merry, D. E., and Fischbeck, K. H. (1995). Stability of an expanded trinucleotide repeat in the androgen receptor gene in transgenic mice. *Nature Genet.* **9,** 191–196.

121. Bingham, P., Scott, M., Wang, S., Chen, K. L., den Dunnen, J., Steensma, Y., and Fischbeck, K. (1993). Stability of an expanded trinucleotide repeat in yeast artificial chromosomes and transgenic mice. *Am. J. Hum. Genet.* **53**(Suppl.), A234.

122. La Spada, A. R., Peterson, K., Jeng, G., Chen, K., Fischbeck, K., and McKnight, G. S. (1996). Instability of CAG repeat expansions introduced into yeast artificial chromosomes. *Am. J. Hum. Genet.* **59**(Suppl.), A153.

123. Kang, S., Jaworski, A., Ohshima, K., and Wells, R. D. (1995). Expansion and deletion of CTG repeats from human disease genes are determined by the direction of replication in *E. coli. Nature Genet.* **10,** 213–218.

Part V

Myotonic Dystrophy

CHAPTER 9

Myotonic Dystrophy as a Trinucleotide Repeat Disorder—A Clinical Perspective

PETER S. HARPER Institute of Medical Genetics, University of Wales College of Medicine, Heath Park, Cardiff CF4 4XN, Wales

In 1992 a series of reports [1–3] provided for the first time a clear molecular basis for our understanding of myotonic dystrophy. The discovery of an unstable mutation in affected patients followed shortly by the identification of a trinucleotide repeat expansion in a specific gene coding for a serine-threonine protein kinase [4–6] provided the key to resolving many of the previously puzzling genetic complexities. It also allowed the start of the much more difficult process of elucidating the pathological mechanisms involved in the disease process.

Other chapters in this book discuss in detail the molecular and cell biological studies that are contributing to our understanding, still extremely incomplete, of myotonic dystrophy. Here the clinical features are examined in the light of the recent advances to see how they stand in this new framework and what lessons they can contribute to our fuller understanding of this and other trinucleotide repeat disorders. More detailed clinical descriptions are available in the author's monograph [7] and in subsequent reviews [8, 9].

I. INTRODUCTION

Myotonic dystrophy has been recognised as a clinical entity since the early years of the 20th century. First separated from the nonprogressive disorder, myotonia congenita, by Steinert [10] and by Batten and Gibb [11] in 1909, the essential features were rapidly delineated, while its remarkable variability was also soon appreciated, in terms both of severity and of systems involved.

Pathological studies showed that skeletal muscle abnormality, rather than a defect in peripheral nerve or the central nervous system, was primarily responsible for the neurological features [12, 13], while subsequent electrophysiological studies also showed that the characteristic myotonia was a primary feature of muscle itself [14]. The familial pattern was soon accepted as an example of Mendelian autosomal dominant inheritance [15, 16].

Thus, 50 years ago, a consistent picture had already evolved of myotonic dystrophy as a genetically determined, primary degeneration of muscle, with clinical and pathological changes that clearly separated it from other muscle diseases, and which was also marked out by involvement of other body systems and by great clinical variability. Yet, even at this stage, puzzling features had been noted which posed serious questions not only about the nature of the pathological process and its relationship to the variable clinical picture, but also about its genetic basis, in particular the apparent progressive intergenerational trend in severity and age at onset known as anticipation.

Many of these puzzling features, almost entirely unexplained until the recognition of the mutational basis of the disorder in 1992, are now seen to be hallmarks of a trinucleotide repeat disorder, to such an extent that they are now carefully sought in other disorders where a comparable molecular basis is suspected. Not all of the puzzling features of myotonic dystrophy are readily explained on this basis, however, and it is necessary to look also at the specific function of the myotonic dystrophy gene and of neighbouring genes, along with the effects of the expanded trinucleotide repeat itself, in order to gain a fuller understanding of the processes that underlie the clinical disorder.

II. CLINICAL ASPECTS

A. The Pattern of Muscle Involvement

Despite its multisystem nature, myotonic dystrophy most frequently presents to the clinician as a progressive neuromuscular disorder with a characteristic pattern of muscle involvement. Figure 9-1 illustrates some of the typical features as seen in personally studied patients, while Table 9-1 summarises the principal muscle groups involved. Clearly, these features are of practical diagnostic importance—indeed their recognition commonly allows myotonic dystrophy to be diagnosed confidently without further investigation—but in the present context it is important to ask why certain muscles are particularly involved, such as those of the face and jaw, the sternomastoids, and the distal limb muscles, while others, such as the proximal weight-bearing muscles of the lower limbs, are relatively spared.

Are these differences principally related to the effects of the trinucleotide repeat on cellular processes, or are they due to the function and distribution of the myotonic dystrophy protein kinase, to the involvement of other genes immediately adjacent, or to inherent properties of the particular muscle groups? At present we do not know, but it is worth comparing the different forms of muscular dystrophy and related disorder (Table 9-2) as a starting point for a more detailed study. While the shared involvement of facial muscles in such disorders as facioscapulohumeral muscular dystrophy and myasthenia gravis may prove to be superficial in relation to the underlying mechanisms, such similarities, along with the marked differences between myotonic dystrophy and the large group of dystrophies mainly involving the limb girdles, may prove relevant once the detailed function of the myotonic dystrophy kinase and its relationship to the various subcellular protein components of muscle become clearer in the context of skeletal muscle function and development.

B. Myotonia

Myotonia, one of the key distinguishing features of myotonic dystrophy (Fig. 9-2), though not often a prominent complaint of those affected, has long been studied in terms of its electrophysiology, but has only more recently come to be understood at the molecular level (see Harper and Rudel [8] for a general review). Our model for pure myotonia has long been myotonia congenita, rather than myotonic dystrophy, with a well-studied animal homologue in the form of the myotonic goat [14, 17]. Here the primary defect was suspected from physiological studies to be one of chloride ion conductance across the muscle membrane [18, 19], and this has been confirmed in human myotonia congenita (both dominant and recessive forms) by the identification of mutations in the gene determining the skeletal muscle chloride ion channel [20]. Other disorders showing myotonia, such as the normokalaemic form of familial periodic paralysis, have proved to be potassium ion-channel defects [21], part of a rapidly increasing series of "channelopathies."

For myotonic dystrophy, the most relevant fact has proved to be that it is *not* a primary ion-channel defect. Here again the clinical features support the evidence from physiology and molecular analysis. It is a progressive disorder, unlike the primary ion-channel disorders which are essentially static in nature;

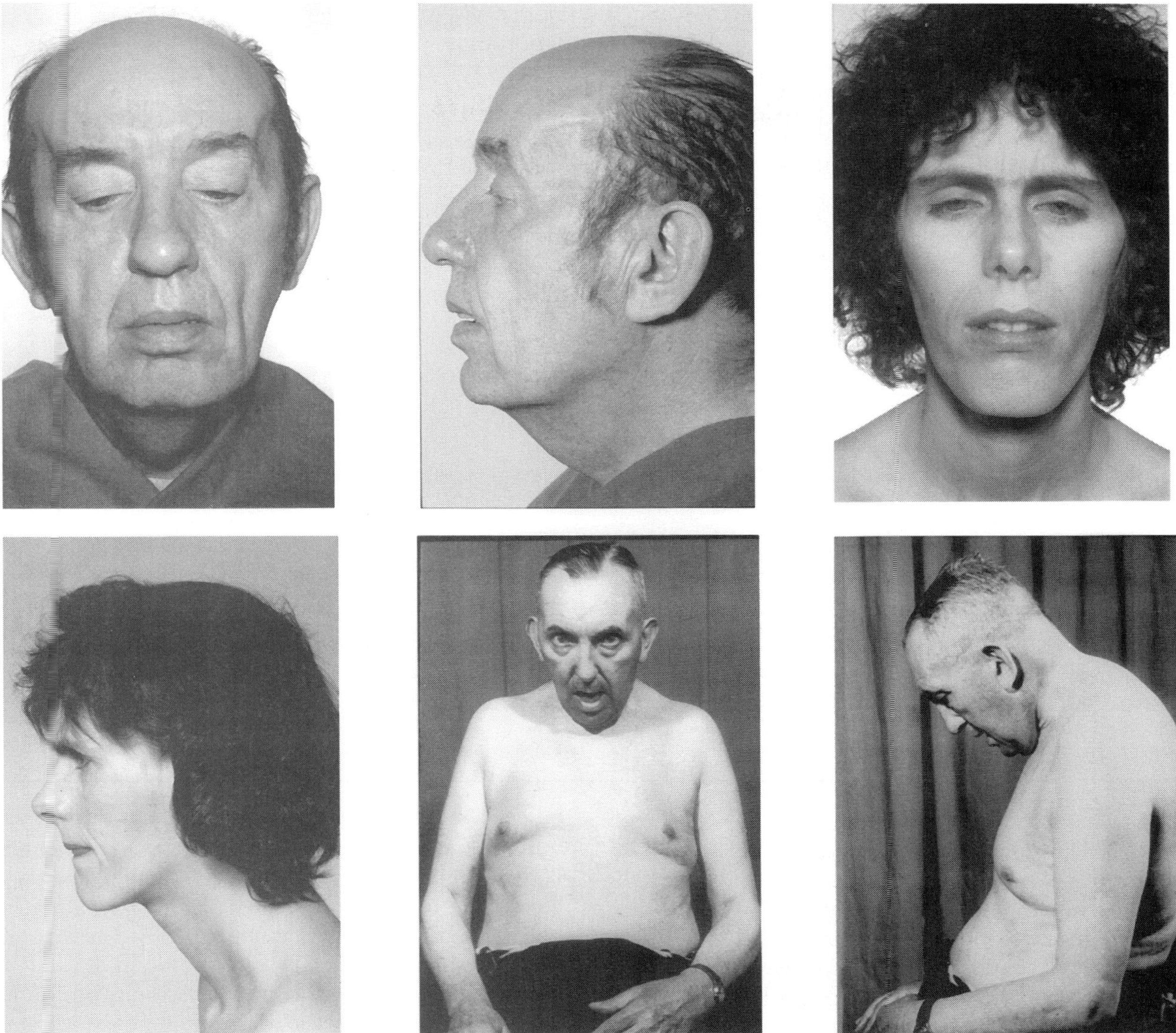

FIGURE 9-1 Clinical features of myotonic dystrophy. Adult patients, illustrating the facial and jaw weakness, ptosis, and sternomastoid muscle weakness and wasting. Reproduced in part from Harper [7].

TABLE 9-1 Myotonic Dystrophy—Distribution of Muscle Involvement

Muscles most prominently affected
Facial muscles
Jaw muscles
Sternomastoids
Distal limb muscles
Muscles frequently affected
Diaphragm and intercostals
Palatal and pharyngeal muscles
Small muscles of hands and feet
External ocular muscles
Muscles frequently spared
Pelvic girdle
Posterior thigh and calf muscles

its multisystem involvement extends beyond the likely distribution or function of a single ion channel, while none of the known ion-channel disorders show any of the genetic features that are suggestive of an unstable trinucleotide repeat disorder. The earlier physiological studies of muscle in myotonic dystrophy had also shown a mixed pattern of abnormalities inconsistent with a single ion-channel disturbance [22–24] and contrasting with those of human myotonia congenita or the myotonic goat.

Relevant questions, now that the specific molecular basis of myotonic dystrophy is known, include: what is the relationship of the myotonic dystrophy kinase to the different ion channels of skeletal muscle? Is the myotonia of myotonic dystrophy specifically related to disturbance of function of the kinase, or does it reflect a more general disruption of the membrane and its

TABLE 9-2 Patterns of Neuromuscular Involvement: Myotonic Dystrophy and Other Adult Muscle Disorders

Muscle groups/ system involved	Myotonic dystrophy	Proximal myotonic myopathy (PROMM)	Duchenne/Becker muscular dystrophy	Abnormal recessive limb girdle dystrophies	Facioscapulohumeral muscular dystrophy	Myasthenia gravis
Limb muscles	Predominantly distal	Mainly proximal	Limb girdles (especially pelvic)	Limb girdles (especially pelvic)	Limb girdles (especially scapular); also peroneal muscles	Mainly proximal
Facial muscles	Prominent	Minimal	Absent	Absent	Prominent	Prominent
Myotonia	Characteristic	Mild	Absent	Absent	Absent	Absent
Cardiac	Prominent (especially conducting tissue)	Usually absent	Variable; often general cardiomyopathy	Usually absent	Usually absent	Usually absent
Smooth muscle	Extensive, especially gastrointestinal	Usually absent	Unusual	Usually absent	Usually absent	Usually absent
CNS	Variable; especially prominent in congenital cases	Mild	Some IQ reduction frequent in DMD	Usually absent	Usually absent	Usually absent
Other systems	Widespread, variable (see Table 9-4)	Cataract	Rare	Absent	Absent	Absent

functions? Why is there often such a discrepancy in patients between the prominence of myotonia and the severity of muscle weakness? A striking clinical feature is that myotonia is often prominent in patients with minimal muscle weakness and very slow deterioration, while others, mostly infants with congenital myotonic dystrophy, may show profound weakness with a near absence of myotonia [25].

C. System Involvement

The recognition that myotonic dystrophy is not purely a muscle disease dates back to the original descriptions of Steinert, in 1909, noting the presence of testicular atrophy at autopsy in one of his cases [10]. The range of involvement is extremely wide (Table 9-3), and can broadly be divided into those effects that can be explained on the basis of muscle involvement, whether skeletal, smooth, or cardiac, and those which imply a primary defect in nonmuscular tissue.

Cardiac abnormalities in myotonic dystrophy are of obvious practical importance and underlie many of the early deaths from the disorder (see Phillips and Harper [26] for a general review). Correlation of cardiac abnormalities with the extent of trinucleotide repeat expansion has been shown in one study [27], though not in a second, smaller one [28], but such a correlation seems to be general for most aspects of the disease, rather than specific to any particular system [29, 30]. The myotonic dystrophy kinase gene was early noted as strongly expressed in heart muscle, with the possibility of different isoforms to skeletal muscle, while immunohistochemical localisation to the cardiac muscle intercalated disc regions is also characteristic [31]. It remains to be seen how these can be related to the conduction tissue degeneration which is the most prominent pathological feature of the cardiac involvement [32]. A primary metabolic basis resulting from failure of myocardial glucose utilisation is one possibility [33], while the finding that the genetic locus for progressive familial heart block is close to that for myotonic dystrophy has led to the suggestion that this locus might be involved [34], though whether the molecular distance will prove to make such an effect plausible remains to be seen.

The clinical features of heart disease in myotonic dystrophy also give some indications of what the pathogenesis is not likely to be. There is a notable absence of increased coronary heart disease [35], while the systemic blood pressure tends to be decreased [36]. There is not a high frequency of generalised cardiomyopathy, whether hypertrophic or dilated, such as is seen in primary defects involving cardiac forms of myosin or dystrophin.

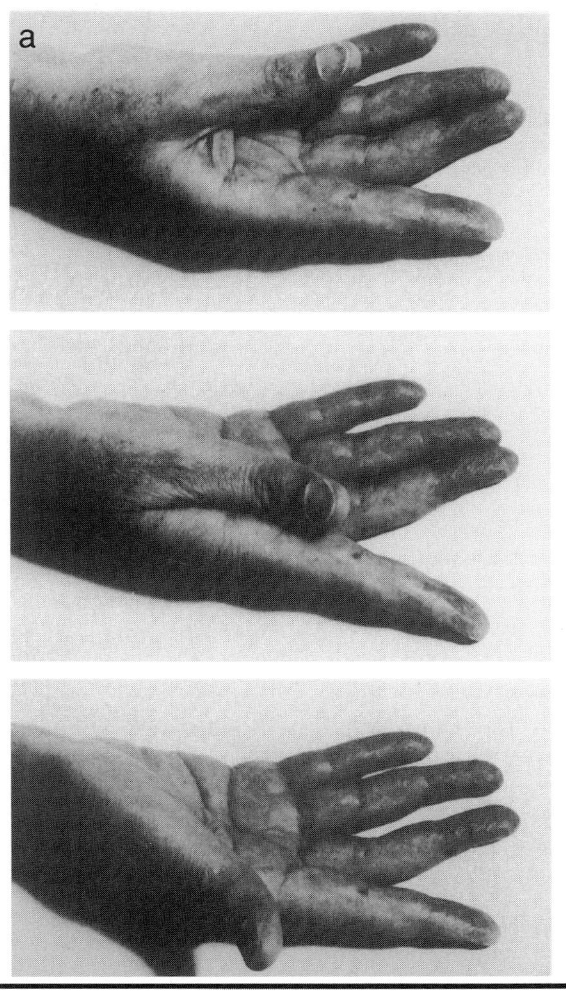

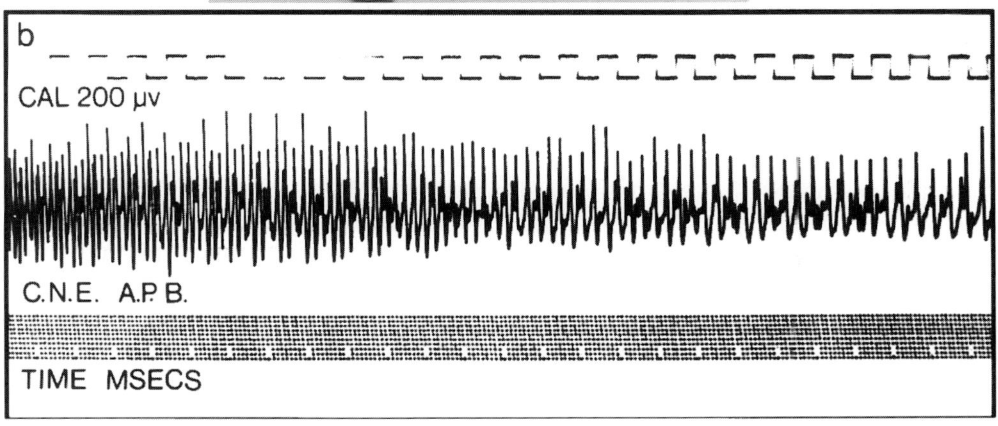

FIGURE 9-2 Myotonia, illustrating the clinical appearance (the patient is attempting to relax grip) and electrophysiology, showing the repetitive discharges on active muscle contraction. Reproduced from Harper [7].

Smooth muscle defects account for most of the widespread and important gastrointestinal features of myotonic dystrophy, ranging from loss of oesophageal motility [37] to anal sphincter disturbance [38]. Current lack of evidence relating these to any specific molecular process probably reflects a generally poor understanding of smooth muscle by comparison with skeletal muscle.

TABLE 9-3 Myotonic Dystrophy—Main Systemic Features

Smooth muscle
Oesophageal—failure of peristalsis, bronchial aspiration
Colon—abdominal pain, variable bowel habit, colonic dilatation
Gallbladder—increased gallbladder disease
Uterine—incoordinate labour
Anal sphincter—soiling, laxity (especially children)
Cardiac muscle
Cardiac conduction defects (especially atrioventricular)
Nonmuscular
Cataract
Retinopathy (rarely significant clinically)
Somnolence (partly central, partly respiratory)
Mental handicap (congenital cases)
Diabetes (some degree of insulin resistance usual)
Testicular atrophy (often male fertility only moderately reduced)
Early balding (often striking in males)
Multiple calcifying epitheliomas (benign, subcutaneous tumours)

From the viewpoint of understanding the underlying mechanisms, the greatest interest is provided by those systemic features of myotonic dystrophy that cannot be related to any type of muscle dysfunction, notably involvement of the brain, eye, and endocrine glands. Central nervous system involvement provides a particularly striking correlation with the degree of trinucleotide repeat expansion [30, 39, 40], being most prominent in cases of congenital onset with the largest expansions [30] (see below) and only partly, if at all, explicable in terms of perinatal anoxia and related problems [40]. At the other extreme, CNS involvement is essentially absent in those patients with minimal clinical features in later life. Although the mental handicap seen in congenital cases may relate to impaired cerebral development, the features in adults can undoubtedly have a progressive element, notably the increased daytime somnolence [41], but also some of the more subtle behavioural and personality changes [42]. To what extent these abnormalities reflect disturbance of the myotonic dystrophy kinase function, the differential involvement of different isoforms of the protein, other effects of the expanded repeat, or the function of adjacent genes, is still entirely uncertain.

The ocular lens and the cataract associated with myotonic dystrophy offer a particular challenge for the relating of clinical pathology to underlying mechanisms. The early lens opacities of myotonic dystrophy are extremely characteristic and their subcortical distribution shows that they are part of a progressive process, affecting those layers of the lens successively laid down under the lens capsule. This could offer the possibility of an experimental approach comparable to that used in dendrochronology, where recent climatic and nutritional influences are similarly expressed in the superficial layers. It is noticeable that congenital cataract is not a feature of congenital myotonic dystrophy [43], despite the severe involvement of other systems, while cataract is often the predominant, and even the only, clinical manifestation in later life of those with small expansions in the repeat sequence [43]. Can such changes be related to the myotonic dystrophy kinase and to possible damage to the membrane of functioning lens cells [44], or might the process relate more to dysfunction of other adjacent genes? It is difficult to see how the function of an adjacent gene could be disturbed by a very small expansion in the repeat without also causing some alteration of the functions of the kinase gene itself.

In the study of the widespread endocrine dysfunction of myotonic dystrophy, the mechanism of insulin resistance is of particular relevance. Recent studies suggesting a reduction of insulin receptor RNA levels [45], possibly as a general effect of the myotonic dystrophy protein kinase, may help to explain the frequent insulin resistance, and the less frequent clinical diabetes, seen in myotonic dystrophy [46]. They may also provide a starting point for clinical trials, a beneficial effect of insulin-like growth factor 1 having been shown in a preliminary therapeutic trial [47].

D. Congenital Myotonic Dystrophy

The existence of a clinical syndrome differing widely from the features of adult myotonic dystrophy [40, 48, 49] has provided a series of challenges to those attempting to relate it to the more usual form of the disease. Figure 9-3 shows some of the typical clinical features, which evolve over time (Fig. 9-4) and which are summarised in Table 9-4.

Some of the differences are secondary in nature and these are instructive in showing how a chain of events can lead to developmental defects. Congenital myotonic dystrophy can be regarded in many aspects as a dysmorphic syndrome, many of whose features can be considered as "deformations" rather than primary malformation. The striking facial features in part reflect the modelling influences that muscle function (and its lack) can have on bony structures of the face, jaw, and cranium. The joint contractures, ranging from talipes to general congenital contractures, can likewise be explained in terms of intrauterine muscle imbalance, while congenital heart defects such as patent ductus arteriosis and possibly septal defects could also result from impaired smooth and cardiac muscle function. Even such apparently unrelated findings as maternal uterine hydramnios have proved to be due to failure of intrauterine swallowing [50], while the fetal immobility can have consequences as superficially diverse as thin bones, in-

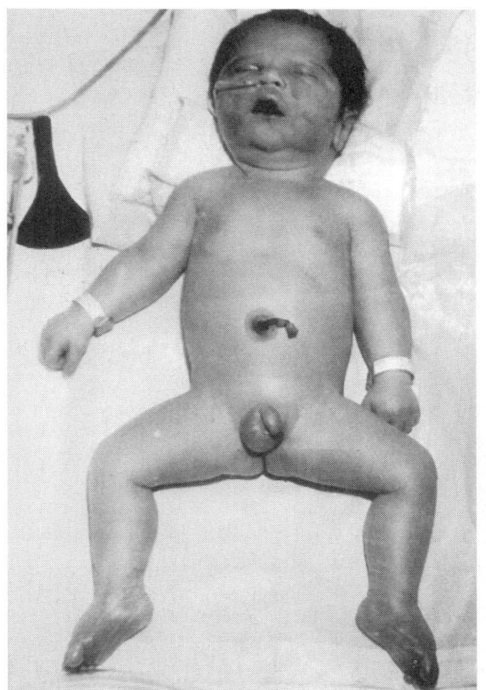
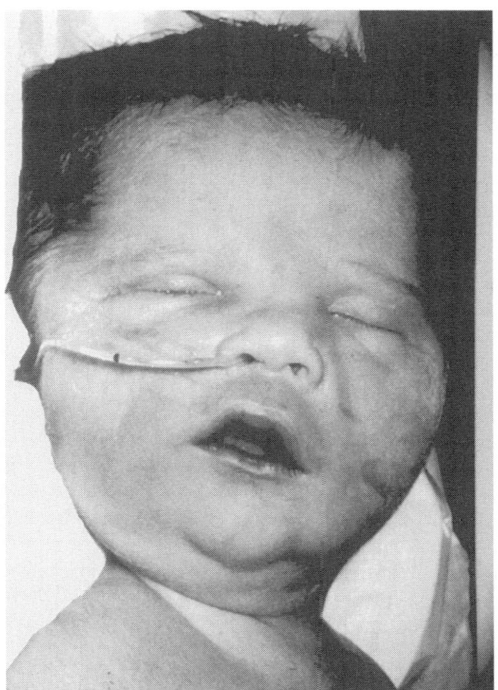
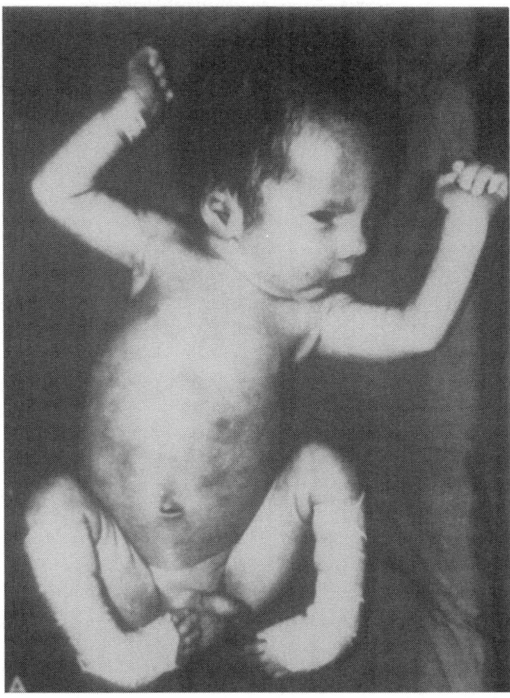
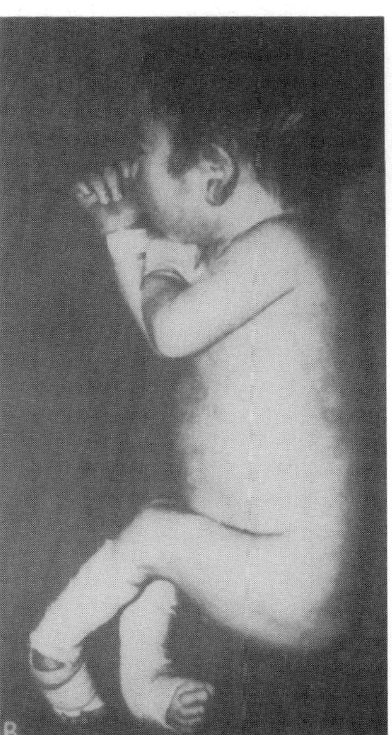

FIGURE 9-3 Congenital myotonic dystrophy, illustrating the facial diplegia, general immobility, and congenital contractures. Reproduced in part from Harper [7].

cluding ribs, and pulmonary hypoplasia, with consequent neonatal death from respiratory inadequacy [51].

If we pursue the underlying process back a stage, we find that a common feature of the above changes is a profound immaturity of fetal muscle [25], which lacks the degenerative and progressive changes of disorders such as Duchenne dystrophy and which differs markedly from the muscle changes of myotonic dystrophy in later life. The abundance of satellite cells and the central nuclei giving the appearance of myotubes [52] indicate that the process of muscle development has been arrested, rather than disturbed by a degenerative process.

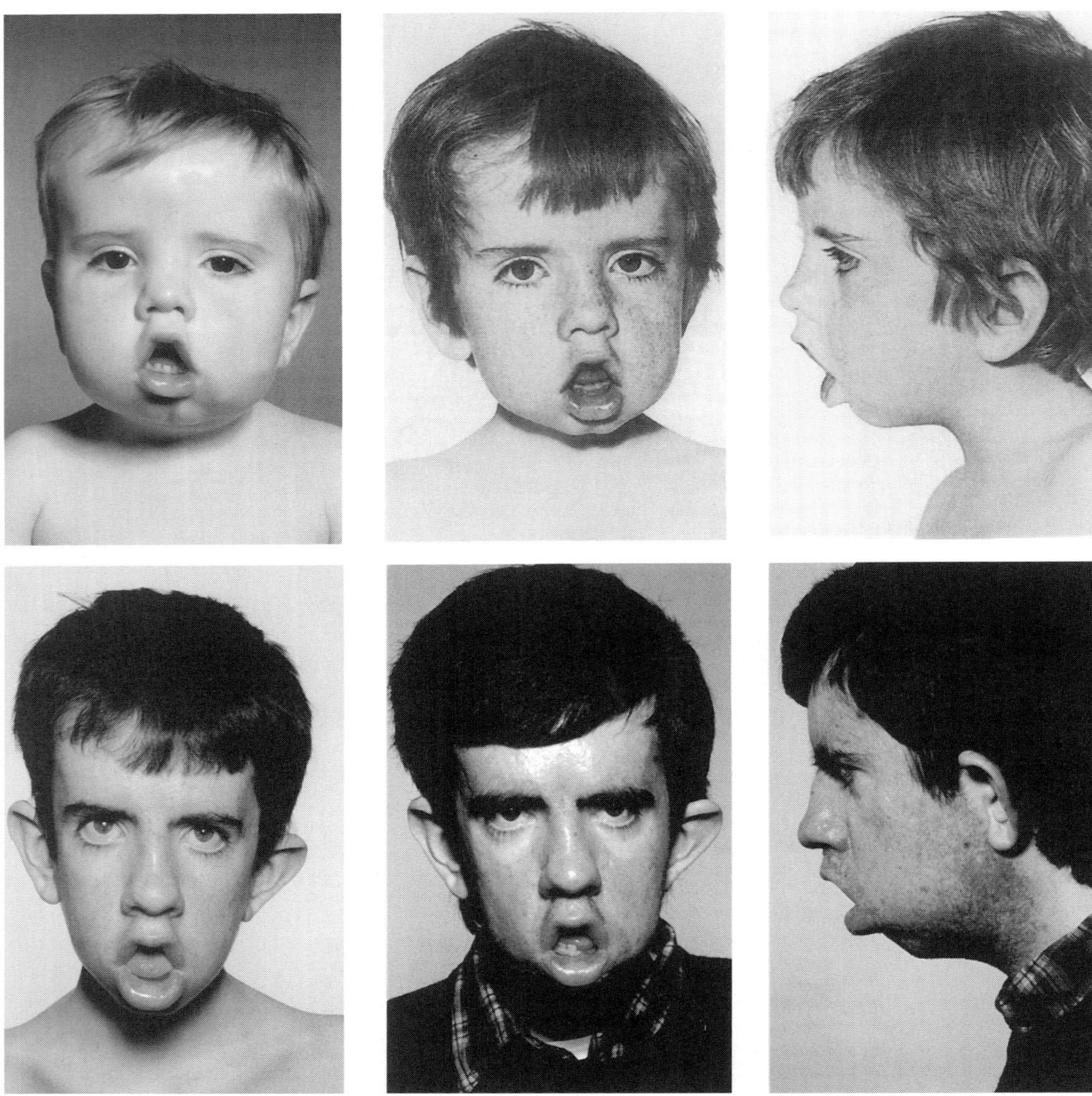

FIGURE 9-4 Congenital myotonic dystrophy. A sequence of photographs of the same patient illustrating changing features with increasing age. Reproduced in part from Harper [7].

The cerebral involvement in congenitally affected patients indicates that this process is not confined to muscle.

Although we still cannot explain those abnormalities directly at the molecular level, there are several points that are relevant: firstly, the intrauterine features of congenital myotonic dystrophy are only seen with the largest trinucleotide expansions, commonly those of over 1000 repeats [29, 30]. Secondly, some of the apparent discrepancies found between research studies in activity of the myotonic dystrophy kinase gene [53, 54] have depended on whether congenital or adult-onset patients have been studied, suggesting that the mechanism may indeed differ between the two forms. Finally, one suggestion can probably now be laid aside; the maternal transmission of the congenital form, combined with the unusual clinical features, led to the hypothesis of a maternal intrauterine factor being involved [55]. Not only has such a factor never been identified, but what we now know about differential mechanisms of repeat expansion adequately explains the parent of origin effect, so that repeat length alone is adequate to explain most of the correlation with severity, though not its mechanism, in congenital myotonic dystrophy.

III. CLINICAL GENETIC ASPECTS OF MYOTONIC DYSTROPHY

Long before myotonic dystrophy was recognised as a trinucleotide repeat disorders, a series of genetic ob-

TABLE 9-4 Congenital Myotonic Dystrophy—Characteristic Features

Clinical
Facial diplegia and jaw weakness (causing severe neonatal feeding problems)
Respiratory muscle weakness (causing severe neonatal respiratory problems)
General hypotonia and immobility
Congenital contractures (variable)
Intrauterine hydramnios
Mental handicap
High perinatal mortality rate
No significant myotonia
Genetic
Almost always maternally transmitted
Close concordance for affected sibs
Maternal disease variable but clinically significant
No evidence of genetic heterogeneity

servations, based on clinical and pedigree data, raised questions as to whether the condition was behaving as a straightforward autosomal dominant disorder. Some of the most important of these are summarised in Table 9-5. Their relationship with the trinucleotide repeat expansion itself is discussed in the following chapter, while the applied diagnostic aspects are reviewed elsewhere [56, 57]. However the general relationship of these genetic issues to the expression and transmission of the disorder requires separate consideration and the principal aspects are outlined here. Overwhelmingly the most important of these, in terms of its general implications as well as for myotonic dystrophy, is that of anticipation.

A. Anticipation in Myotonic Dystrophy

The study of the "rediscovery" of genetic anticipation is a fascinating one, intimately linked to the history of clinical and genetic research into myotonic dystrophy.

TABLE 9-5 Myotonic Dystrophy as a Trinucleotide Repeat Disorder: Genetic Issues

Anticipation—striking in comparison with other known and possible trinucleotide repeat disorders
Parent of origin effects (a) Maternal transmission of congenital cases (b) Predominant paternal origin of minimally affected
Segregation ratio distortion—existence and possible mechanisms (e.g., meiotic drive) still unresolved
Genetic fitness—variable between generations
Phenotype penetrance—variable between generations

It is a salutary story, too, with valuable lessons to be learned as to how clinical observations can lead to new biological insights—and how they can be ignored. In 1992 the author and his colleagues wrote a review on the topic [58], shortly after the discovery of the myotonic dystrophy mutation and before the recognition that trinucleotide repeat expansion was the mechanism underlying Huntington's disease and related central nervous system degenerations. Since then the attitude to anticipation has changed in an extraordinary way. From being a forgotten and neglected phenomena in 1992, ignored or dismissed by all the standard genetics textbooks, it has become accepted, even canonised, as a fundamental feature of the trinucleotide repeat disorders in general, and invoked, often on the basis of questionable evidence, in a much wider range of diseases where there would be no other reasons for postulating this type of mutational mechanism. The author's own view is that the attitude of the scientific community today is often as uncritical as it was before 1992—uncritical dismissal has too often given way to uncritical acceptance. Here the discussion of anticipation will largely be focused on myotonic dystrophy; helpful reviews have appeared since 1992 in relation to schizophrenia [59, 60] and to other disorders [61], while other chapters in this volume also cover the aspects involving psychiatric disorders.

The historical origins of anticipation lie well back in the past century [62], in the concept of progressive degeneration that was current in relation to many disease states, notably mental illness. This was taken up at the beginning of the present century [63, 64] and would thus have been part of current thinking at the time that myotonic dystrophy was first delineated and when its familial basis was being investigated. At this point it seems wise to define anticipation as: "The occurrence of a genetic disorder at progressively earlier age in successive generations." It should be noted that the definition does not specify progressively greater severity, though this is usually also the case (Fig. 9-5).

The first indication of anticipation in myotonic dystrophy comes from as far back as 1911, only 2 years after the first clear description of the disease. Greenfield [65], in documenting the association of cataract with myotonic dystrophy, noted the occurrence of cataract alone in previous generations (interestingly in the paternal line).

It was the 1918 paper of Fleischer [66] (Fig. 9-6) which clearly documented the phenomenon; he not only confirmed (as an ophthalmologist) the occurrence of cataract in individuals who had no muscle disease but whose offspring had full myotonic dystrophy, but was also able to link together different affected families through members who had cataract alone, as well as through individuals who were not known to have had

FIGURE 9-5 Anticipation in a four-generation family with myotonic dystrophy showing child with congenital myotonic dystrophy, moderately affected mother (lower left), age 23, grandfather with onset of mild features in mid-50s, and great-grandmother (in late 70s) with no neuromuscular abnormality but with cataract extractions in mid-60s.

any clinical abnormality. These perceptive observations thus established the key features of anticipation in myotonic dystrophy: the progression in successive generations, the occurrence of cataract as the only clinical feature in some patients, and the occurrence of gene carriers with no apparent abnormality at all.

In the light of these careful studies and in the broader framework of anticipation and degeneration postulated for mental illness, it is not surprising that anticipation should at that time have been readily accepted in myotonic dystrophy. That no biological explanation existed was of little consequence given the general level of ignorance concerning genetic mechanisms at this time. The evidence was reinforced almost 30 years later by the landmark study of Julia Bell [67], who not only collected the world literature on myotonic dystrophy, but undertook the first quantitative genetic analysis on the combined pedigree data. She found clear and statistically significant differences in age at onset between generations and observed a considerably stronger correlation of age at onset between sibs than between parents and offspring, leaving her in no doubt that anticipation, or antedating as she termed it, existed.

This consistency of clinical observations, genetic analysis, and conclusions over a 30-year period makes it all the more surprising that Penrose [68], working in the same Institute as Julia Bell and using her data, should have come to a radically different conclusion, one that would effectively banish the concept of anticipation from serious scientific thought and investigation for a further 40 years. It should not be thought, however, that Penrose was dismissive of anticipation as an observation; quite the contrary. He carefully analysed data of Julia Bell on a number of other Mendelian disorders, as well as data of his own on diabetes and on mental illness. Part of this analysis is reproduced in Table 9-6, and the data on other disorders are of new interest in the light of postulated trinucleotide repeat involvement in some of them. The original paper is well worth consulting.

It can be seen that Penrose's "index of anticipation," derived from the difference in mean values of onset in parent and offspring (D) and from the standard deviation (S), shows apparent evidence for anticipation in most of the disorders chosen. For myotonic dystrophy, however, the value (0.90) greatly exceeds that for any of the others. It is also noteworthy that the value for Huntington's disease (0.36) is comparable to that for glaucoma and less than that for diabetes and for mental illness (Penrose did not distinguish different forms of this).

Penrose's conclusions were that most of what appeared to be anticipation could be explained on the

FIGURE 9-6 Two central people in the history of anticipation. (left) Bruno Fleischer, German ophthalmologist. (right) Lionel Penrose, British human geneticist. Reproduced from [58].

basis of a variety of ascertainment biases, while the special case of myotonic dystrophy required an additional explanation. He suggested a common allele that influenced age at onset, which could never to passed to a child along with the disease gene, and which would thus explain both the exceptionally low parent–child

TABLE 9-6 Anticipation in Genetic Disorders (Based on Penrose [68])

Type of disease[a]	Number of parent–child pairs	Age of onset difference between parent and child (years) (D)	General standard deviation (S)	Index of anticipation (D/2S)	Parent–child correlation
Peroneal atrophy, dominant (Charcot-Marie-Tooth) disease)	86	4.94	13.59	0.18	0.76
Muscular dystrophy, dominant	90	6.44	13.08	0.25	0.62
Hereditary glaucoma	113	11.42	17.99	0.32	0.81
Huntington's chorea	153	8.82	12.38	0.36	0.59
Diabetes mellitus	216	17.22	15.26	0.56	0.44
Mental illness, all diagnoses	1728	16.30	16.80	0.48	0.38
Dystrophia myotonica (myotonic dystrophy)	51	23.24	12.92	0.90	0.32

[a]The terms used by Penrose are given first, with more familiar current names in parentheses.

correlation and the high index of anticipation. Penrose's biases (six in total) are summarised in Table 9-7.

Revisiting Penrose's data with the benefit of hindsight, I would agree with almost all of his conclusions. The apparent anticipation shown by most of the disorders in the table is likely to result, as Penrose suggested, from the various biases, and is unlikely to indicate a basis for postulating a trinucleotide repeat mechanism. This is equally the case for Huntington's disease, where the later documentation of true anticipation required a massive data set and was even then only significant for male transmissions. It is unlikely that the data analysed by Penrose could have detected the modest degree of anticipation shown by the relatively small expansion of Huntington's disease and related CAG repeats. For myotonic dystrophy, recognised as a special case by Penrose, we now have the advantage of a specific biological mechanism and no longer have to postulate an allelic modifier. We also expect a much greater degree of anticipation than in other dominantly inherited trinucleotide repeat disorders, owing to the extreme sizes of expansion seen, with consequent greater instability. Even now, though, it would not be surprising if Penrose were to be proved right regarding modifying genes, since factors other than the expanded repeat length are proving to be involved in the phenotype and age at onset of trinucleotide repeat disorders.

If any blame can be attributed for the 40-year neglect of anticipation following Penrose's 1948 paper, it is probably to the tendency of workers to accept the verdict of scientific authority rather than to base conclusions on raw data, whether these be clinical or genetic in nature. The complete omission of the topic in successive editions of the most respected genetics textbooks (now hastily rectified in the latest editions!) should stand as a warning.

TABLE 9-7 Potential Biases in Genetic Anticipation (after Penrose [68])

1. Selection of affected parents in whom the onset of the disease is late
   (early onset tends to diminish fertility)
2. Selection of affected offspring in whom the onset of the disease is early
   (early onset often correlates with severe disease, making recognition more likely)
3. Selection of cases with simultaneous onset in parents and offspring
   (Likely to occur when onset early in offspring and late in parent, but not when onset late in offspring and early in parent, since most studies done at a single point in time)
4. Weakness of real correlation of onset ages in parents and offspring
   (necessary if apparent anticipation is to be observed, since an exact correspondence would make it impossible)
5. General variability in age of onset
   (apparent anticipation will be a function of general total variability in onset age)

Finally, it should be noted that anticipation did not simply resurface after the discovery of its trinucleotide repeat expansion basis. The work of Höweler and coworkers [69, 70] showed convincingly that anticipation in myotonic dystrophy persisted even when most of the ascertainment biases were removed, while the recognition that the congenital form required transmission from an affected mother provided another piece of supporting evidence. The parallel with fragile X syndrome and the possibility of some premutational step in both disorders, already recognised by 1989, meant that what was required was a specific biological mechanism, a deficiency that was supplied by the genetic instability of expanded trinucleotide repeats.

B. Parent of Origin Effects

Pedigree studies of myotonic dystrophy have all agreed in finding an approximately equal sex ratio among affected individuals (once propositi have been removed), but the early series, mainly of adults [8, 67, 71], found an excess of males among affected parents, which was not satisfactorily explained. When childhood myotonic dystrophy became adequately recognised it was found, by contrast, that most patients had affected mothers, while this was almost universally the case for the severe congenital form [72]. When an effort was made to ascertain all age groups equally, it became clear that the parental sex ratio was extremely close (49.5%) to the expected 50% ratio for autosomal dominant inheritance, and that age-related biases were responsible for the differences (see Harper [8] p. 305). The question thus shifted from the primary parental ratio to why these age-related differences should exist, and has only been satisfactorily answered since the discovery of the trinucleotide repeat basis.

The most striking departure from the expected sex ratio is seen in the maternally transmitted congenital form, where only a handful of paternally transmitted cases have been documented [73, 74], in contrast to the many thousands of those that are maternally transmitted. The suggestion of a maternal intrauterine factor [55], acting only on those possessing the genetic mutation, has never been substantiated, as already noted; it now seems clear that for these cases, resulting from extreme repeat expansions up to several thousand repeats in length, some mechanism exists that does not permit male expansion beyond a limit—or at least does not allow it to achieve a viable pregnancy [75]. The absence of such a limit in female transmission results in the selectively maternal origin of the congenital cases, and it now seems unnecessary to postulate additional factors, though the mechanisms involved remain poorly understood.

Much less appreciated until recently is that there is an excess of male grandparental transmissions and in particular an excess of males among those first documented in a pedigree to have transmitted the disorder in a clinically recognisable form. This has been confirmed since molecular studies became possible [76], and agrees with the excess of male transmission noted in adults by the early studies. At first sight these parental effects in myotonic dystrophy seem contradictory, and to disagree with the findings in other trinucleotide disorders; in HD, for example, the severe juvenile form is mainly paternally transmitted [77], in contrast to the maternally transmitted myotonic dystrophy. On closer examination, however, the paradox resolves itself, for the repeat length in juvenile HD (rarely over 100) [78] is comparable to that seen in the mildest cases of myotonic dystrophy [79], and completely different from that of congenital myotonic dystrophy. Thus, for an equivalent size range, the parent of origin effect, based on a greater tendency to expansion during male meioses, is comparable across the different dominantly inherited trinucleotide repeat diseases.

C. Other Genetic Aspects

Considering briefly the remaining points listed in Table 9-5 the *segregation ratio,* in terms of proportion of offspring affected, deserves a mention in view of suggestions from the study of different repeat lengths that there may be differential rates of transmission, possibly due to meiotic drive, and that these may also be sex-related.

The results of different studies remain inconsistent [80–82], but in terms of transmission of the disorder, in distinction to transmission of alleles in the normal range, there is no evidence for any distortion of the segregation ratio. A variety of early studies all found, once propositi had been deducted, values of 33–50% for the proportion of offspring affected, the variation probably reflecting the extent of search for minimal features and the age of offspring (see Harper [8], pp. 295–299).

Penetrance of the clinical phenotype in myotonic dystrophy now has to be reevaluated in the light of anticipation and the difference between generations; older studies which pooled the information between generations are of little meaning. Of greatest relevance, and clearest, is the situation for siblings where congenital myotonic dystrophy has occurred, where molecular analysis has confirmed the earlier findings of most individuals inheriting the mutation manifesting the disease during childhood [83]. The well-documented reduction of repeat length [84, 85], usually in male transmissions and occasionally into the asymptomatic range, probably underlies the few cases of true "skipped generations."

Genetic fitness is another aspect where the merging of data from different generations has largely invalidated earlier studies. Although all studies showed a significant reduction, it is clear that this ranges from being close to genetic lethality in the severe congenital cases to normal or close to normal in minimally affected gene carriers [86]. Since these minimal cases are commonly only ascertained through their affected offspring, they may give a spurious appearance of greater than normal fertility.

Homozygosity for the myotonic dystrophy mutation is exceptionally rare, but of great theoretical importance. A very few families have now been documented that make it clear that homozygosity does not significantly influence clinical manifestation, at least for small expansions [86]. This situation has also been found in Huntington's disease [87] and supports the view that the clinical pathology in both disorders does not result from simple loss of function.

D. Clinical Effects of Somatic Instability

Myotonic dystrophy, along with Fragile X syndrome, shows the most striking expansions in its specific trinucleotide repeat sequence, extending to several thousand repeats in some congenital cases and contrasting with the modest degree of expansion shown by the group of CAG repeat disorders, characterised by a translated glutamine repeat sequence appearing in the relevant proteins [75, 88]. This extreme instability is reflected in somatic instability also, shown by differences between tissues [89] and over a period of time [90]. Does this somatic instability have clinical consequences, such as predisposition to malignancy?

Such limited evidence as exists suggests probably that it does not. There is no evidence of any general increase of malignancy in myotonic dystrophy, except where there is a specific reason (e.g., pituitary tumours associated with endocrine disturbance). The unusual benign subcutaneous tumour calcifying epithelioma does show a specific association [91] and deserves investigation. There is no evidence for increase in atheroma [35] or other degenerative changes that might be an expected consequence of somatic genetic instability, and which are characteristic of inherited DNA repair defects.

E. Lessons from Other Clinical Phenotypes

The discovery of a specific trinucleotide repeat expansion as the underlying basis of myotonic dystrophy led rapidly to the analysis of numbers of patients worldwide. Not only did the same mechanism underlie almost all of the many thousands of patients studied, but a possible common origin of most of them was suggested by a shared wider haplotype [92, 93]. Most of these studies, however, uncovered a small number of individuals who appeared not to show any repeat sequence expansion; these are of particular interest in relation to the pathogenesis of the disease [94, 95]. While most of even these proved to be the result of technical errors or sample confusion, the small number of families with an apparent phenotype of myotonic dystrophy but no repeat expansion deserves close analysis.

Most of these families have proved on careful reassessment to represent a distinct disorder, now known as proximal myotonic myopathy (PROMM) [96, 97]. While the combination of myotonia, progressive muscle weakness, and cataract is indeed similar to myotonic dystrophy, as is the autosomal dominant inheritance and some of the changes in muscle pathology, there are clear distinguishing features that leave no doubt, now that larger numbers have been studied, that it is an entirely distinct disorder.

These include a mainly proximal distribution of muscle weakness very different to that usually seen in myotonic dystrophy, a mild and slowly progressive course (though cataract may occur relatively early), inconspicuous myotonia, and absence of other systemic features apart from cataract. There is no evidence of anticipation, childhood onset is not reported, and the genetic locus is clearly distinct from that of myotonic dystrophy, though it has not yet been mapped. Identification of the specific gene and its product will be of considerable relevance for myotonic dystrophy, since the pathological processes in the two disorders are likely to be related.

Although other families diagnosed clinically as myotonic dystrophy with no expansion have been reported, it is not clear that PROMM has been adequately excluded and at present no such family exists in which the phenotype has been generally accepted as identical to myotonic dystrophy, nor has any myotonic dystrophy family been found to show a molecular basis different to the normal trinucleotide repeat expansion. In this respect the situation is similar to that for Huntington's disease [98], but different from fragile X [99], where a small number of different defects have been reported in the gene, even though the great majority, like myotonic dystrophy, result from a trinucleotide repeat expansion. This supports the view that the myotonic dystrophy phenotype is not the result of a simple loss of function process, and it is also relevant that the very few cases of the disorder in sub-Saharan Africa, while suggestive of an independent origin, are still associated with the same basic mechanism of trinucleotide repeat expansion.

IV. CONCLUSION

Many of the puzzling clinical and genetic features of myotonic dystrophy can now be explained by the genetic instability consequent on trinucleotide repeat expansion. The great variability of age at onset, the anticipation, and the parent of origin effects can now be seen as direct results of this instability, while the extreme nature of these features in myotonic dystrophy reflects the greater degree of expansion and instability seen in the myotonic dystrophy repeat in comparison with other dominant inherited trinucleotide repeat disorders. With hindsight, the unusual clinical features and apparently anomalous inheritance pattern can be seen as directly pointing to some unusual molecular basis. This provides a valuable lesson to research workers on the importance of using clinical as well as laboratory information in uncovering the underlying basis of a disease, and of clinical and laboratory-based scientists working closely together to allow a synthesis of all available information.

References

1. Harley, H. G., Brook, J. D., Rundle, S. A., Harper, P. S., Crow, S., Reardon, W., Buckler, A. J., Housman, D. E., and Shaw, D. J. (1992). Expansion of an unstable DNA region and phenotypic variation in myotonic dystrophy. *Nature* **355,** 545.
2. Buxton, J., Shelbourne, P., Davies, J., Jones, C., van Tongeren, T., Aslandis, C., de Jong, P., Jansen, G., Anvret, M., Williamson, R., and Johnson, K. (1992). Detection of an unstable fragment of DNA specific to individuals with myotonic dystrophy. *Nature* **335,** 547.
3. Aslanidis, C., Jensen, G., Amemiya, C., Shutler, G., Tsilfidis, C., Manadeven, Chen, C., Allaman, J., Wormskamp, N. G. M., Vooljs, M., Buxton, J., Johnson, K., Smeets, H. J. M., Lennon, G. G., Carrano, A. V., Korneluk, R. G., Wieringa, B., and de Jong, P. J. (1992). Cloning of the essential myotonic dystrophy region and mapping of the putative defect. *Nature* **255,** 548.
4. Brook, J. D., McCurrach, M. E., Harley, H. G., Buckler, A. J., Church, D., Aburatani, Hunter, K., Stanton, V. P., Thirion, J-P., Hudson, T., Sohn, R., Zemelman, B., Snell, R. G., Rundle, S. A., Crow, S., Davies, J., Shelbourne, P., Buxton, J., Jones, C., Juvonen, V., Johnson, K., Harper, P. S., Shaw, D. J., and Housman, D. E. (1992). Molecular basis of myotonic dystrophy: expansion of a trinucleotide (CTG) repeat at the 3' end of a transcript encoding a protein kinase family member. *Cell* **68,** 799.
5. Fu, Y-H., Pizzuti, A., Fenwick, R. G., King, J., Rajnarayan, S., Dunne, P. W., Dubel, J., Nosser, G. A., Ashizawa, T., Dejong, P., Wieringa, B., Korneluk, R., Perryman, M. B., Epstein, H. F.,

and Caskey, C. T. (1992). An unstable triplet repeat in a gene related to myotonic dystrophy. *Science* **255,** 1256.
6. Mahadevan, M., Tsilfidis, C., Sabourin, L., Shutler, G., Amemiya, C., Jansen, G., Neville, C., Narang, M., Barcelo, J., Ohoy, K., Leblond, S., Arlemacdonald, J., Dejong, P. S., Wieringa, B., and Korneluk, R. G. (1992). Myotonic dystrophy mutation: an unstable CTG repeat in the 3' untranslated region of a candidate gene. *Science* **255,** 1253–1255.
7. Harper, P. S. (1989). "Myotonic Dystrophy," 2nd ed. Saunders, Philadelphia.
8. Harper, P. S., and Rudel, R. (1994). Myotonic dystrophy. *In* "Myology" (A. G. Engel *et al.,* Eds.), pp. 1192–1219. McGraw Hill, New York.
9. Harper, P. S. (1994). The myotonic disorders. *In* "Disorders of Voluntary Muscle" (J. E. Wallton *et al.*, Eds.), pp. 595–618. Churchill Livingstone, Edinburgh.
10. Steinert, H. (1909). Myopathologische Beitrage 1. Uber das klinische und anatomische Bild des Muskelschwunds der Myotoniker. *Dtsch. Z. Nervenheilkd.* **37,** 58–104.
11. Batten, F. E., and Gibb, H. P. (1909). Myotonia atrophica. *Brain* **32,** 187–205.
12. Rohrer, K. (1916). Ueber Myotonia Atrophica (Dystrophia Myotonica). *Dtsch. Z. Nervenheilkd.* **55,** 242–304.
13. Heidenhain, H. (1918). Uber progressive Vererbund der Muskulatur bei Myotonia Atrophica. *Beitr. Pathol.* **64,** 198–225.
14. Kolb, L. C. (1938). Congenital myotonia in goats. *Bull. Johns Hopkins Hospital* **63,** 221–237.
15. Bell, J. (1948). Dystrophia myotonica and allied diseases. *In* "Treasury of Human Inheritance 4." Cambridge University Press, Cambridge.
16. Thomasen, E. (1948). Myotonia. Universitetsforlaget, Aarhus.
17. Denny-Brown, D., and Nevin, S. (1941). The phenomenon of myotonia. *Brain* **64,** 1–18.
18. Adrian, R. H., and Bryant, S. H. (1974). On the repetitive discharge in myotonic muscles fibres. *J. Physiol.* **240,** 505–515.
19. Barchi, R. L. (1975). Myotonia. an evaluation of the chloride hypothesis. *Arch. Neurol.* **32,** 175–180.
20. George, A. L., Crackower, M. A., Abdalla, J. A., Hudson, A. J., Ebers, G. C. (1992). Molecular basis of Thomsen's disease (autosomal dominant myotonia congenita). *Nature Genet.* **3,** 305.
21. Ptacek, L. J., George, A. L., Jr., Barachi, R. L., Griggs, R. C., Riggs, J. E., Robertson, M., and Leppert, M. F. (1992). Mutations in an S4 segment of the adult skeletal muscle sodium channel gene cause paramyotonia congenita. *Neuron* **8,** 891.
22. Lipicky, R. J., Bryant, S. H. (1973). A biophysical study of the human myotonias. *In* "New Developments in Electromyography and Clinical Neurophysiology" (J. E. Desmedt, Ed.), Vol. 1, pp. 451–463. Karger, Basel.
23. Bryant, S. H. (1973). The electrophysiology of myotonia, with a review of congenital myotonia of goat. *In* "New Developments in Electromyography and Clinical Neurophysiology" (J. E. Desmedt, Ed.), Vol. 1, pp. 420–450. Karger, Basel.
24. Rudel, R., Lehmann-Horn, F., and Ricker, K. (1994). The nondystrophic myotonias. *In* "Myology" (A. G. Engel, Ed.), pp. 1291–1302. McGraw-Hill, New York.
25. Sarnat, H. B., Silbert, S. W. (1975). Maturational arrest of fetal muscle in neonatal myotonic dystrophy. *Arch. Neurol.* **33,** 466.
26. Phillips, M. F., and Harper, P. S. (1997). Cardiac disease in myotonic dystrophy. *Cardiovasc. Res.* **33,** 13–22.
27. Melacini, P., Villanova, C., Menegazzo, E., Novelli, G., Danieli, G., Rizzoli, G., Fasoli, G., Angelini, C., Buja, G., Miorella, M., Dallapiccola, B., and Dallavolta, S. (1995). Correlation between cardiac involvement and trinucleotide repeat length in myotonic dystrophy. J. *Am. Coll. Cardiol.* **25,** 239–245.
28. Jaspert, A., Fahsold, R., Grehl, and Claus, D. (1995). Myotonic dystropy—correlation of clinical symptoms with the size of the CTG trinucleotide repeat. *J. Neurol.* **242,** 99–104.
29. Harley, H. G., Rundle, S. A., MacMillan, J., Myring, J., Brook, J. D., Crow, S., Reardon, W., Fenton, I., Shaw, D. J., and Harper, P. S. (1993). Size of the unstable CTG repeat sequence in relation to phenotype and parental transmission in myotonic dystrophy. *Am. J. Hum. Genet.* **52,** 1164.
30. Tsilfidis, C., MacKenzie, A. E., Mettler, G., Barcelo, J., and Korneluk, R. G. (1992). Correlation between CTG trinucleotide repeat length and frequency of severe congenital myotonic dystrophy. *Nature Genet.* **1,** 192.
31. Salvatori, S., Biral, D., Furlan, S., and Marin, O. (1994). Identification and localization of the myotonic dystrophy gene product in skeletal and cardiac muscles. *Biochem. Biophys. Res. Commun.* **203,** 1365–1370.
32. Nguyen, H. H., Wolfe, J. T., Holmes, D. R., and Edwards, W. D. (1988). Pathology of the cardiac conduction system in myotonic dystrophy: a study of 12 cases. *J. Am. Coll. Cardiol.* **11,** 662–571.
33. Phillips, M. F., and Harper, P. S. (1997). Cardiac disease in myotonic dystrophy. Cardiovasc. Res. **33,** 13–22.
34. Brink, P. A., Ferreirn, A., Moolman, J. C., Weymar, H. W., van der Merew, P. L., and Corfield, V. A. (1995). Gene for progressive familial heart block type I maps to chromosome 19q13. *Circulation* **91,** 1633–1640.
35. Orndahl, G., Thulesius, O., Enestrom, S., and Dehlin, O. (1964). The heart in myotonic disease. *Acta Med. Scand.* **176,** 479–491.
36. O'Brien, T., Harper, P. S., and Newcombe, R. G. (1983). Blood pressure and myotonic dystrophy. *Clin. Genet.* **23,** 366–369.
37. Eckardt, V. F., Nix, W., Kraus, W., and Bohl, J. (1986). Esophageal motor function in patients with muscular dystrophy. *Gastroenterology* **90,** 628–635.
38. Hamel, R. J., Devriede, G., Arhan, P., Tetreault, J. P., Lemieux, B., and Scott, H. (1984). Functional abnormalities of the anal sphincters in patients with myotonic dystrophy: *Gastroenterology* **86**(6), 1469–1474.
39. O'Brien, T., and Harper, P. S. (1984). Course, prognosis and complications of childhood-onset myotonic dystrophy. *Dev. Med. Child. Neurol.* **26,** 62–67.
40. Harper, P. S. (1975a). Congenital myotonic dystrophy in Britain. 1. Clinical aspects. *Arch. Dis. Child.* **50,** 505–513.
41. Coccagna, G., Mantouant, M., and Parch, C. (1975). Alveolar hypoventilation and hypersomnia in myotonic dystrophy. *J. Neurol. Neurosurg. Psych.* **38,** 977.
42. Bird, T. D., Follett, C., and Greig, E. (1983). Cognitive and personality function in myotonic dystrophy. 1. Cognitive function, 2. Personality profiles. *J. Neurol. Neurosurg. Psych.* **46,** 971.
43. Reardon, W., MacMillan, J. C., Myring, J., Harley, H. G., Rundle, S. A., Beck, L., Harper, P. S., and Shaw, D. J. (1993). Cataract and myotonic dystrophy: The role of molecular diagnosis. *Br. J. Ophthalmol.* **77,** 579.
44. Dunne, P. W., Ma, L., Casey, D. L., and Epstein, H. F. (1996). Myotonic protein kinase expression in human and bovine lenses. Biochem. Biophys. Res. Commun. **225**(1), 281–288.
45. Morrone, A., Pegoraro, E., Angelini, C., Zammarchi, E., Marconi, G., and Hoffman, E. P. (1997). Patient muscle shows decreased insulin receptor RNA and protein consistent with abnormal insulin resistance. *J. Clin. Invest.* **99**(7), 1691–1698.
46. Barbosa, J., Nuttall, F. Q., Kennedy, W., and Goetz, F. (1974). Plasma insulin in patients with myotonic dystrophy. *N. Engl. J. Med.* **277,** 837.
47. Vlachopapadopoulo, E., Zachwieja, J. J., Gertner, J. M., Manzione, D., Bier, D. M., Matthews, D. E., Slonim, A. E. (1995). Metabolic and clinical response to recombinant human insulin-

like growth factor I in myotonic dystrophy—a clinical research center study. *J. Clin. Endocrinol. Metabol.* **80**(12), 3715–3723.
48. Vanier, T. M. (1960). Dystrophia myotonica in childhood. *Br. Med. J.* **2,** 1284.
49. Dyken, P. R., and Harper, P. S. (1973). Congenital dystrophia myotonica. *Neurology* **23,** 465.
50. Dunn, L. J., and Dierker, L. J. (1973). Recurrent hydramnios in association with myotonia dystrophica. *Obstet. Gynecol.* **42,** 104–106.
51. Rutherford, M. A., Heckmatt, J. Z., and Dubowitz, V. (1989). Congenital myotonic dystrophy: respiratory function at birth determines survival. *Arch. Dis. Child.* **64,** 191.
52. Farkas, E., Tome, F. M. S., Fardeau, M., Arsenio-Nunes, M. L., Dreyfus, P., and Diebler, M. F. (1974). Histochemical and ultrastructural study of muscle biopsies in 3 cases of dystrophia myotonica in the newborn child. *J. Neurol. Sci.* **21,** 273.
53. Fu, Y-H., Friedman, D. L., Richards, S., Pearlman, J. A., Gibbs, R. A., Pizzuti, A., Ashizawa, T., Perryman, M. B., Scarlato, G., Fenwick, R. G., Caskey, C. T. (1993). Decreased expression of myotonin-protein kinase messenger RNA and protein in adult form of myotonic dystrophy. *Science* **260,** 235.
54. Sabourin, L. A., Mahedevan, M. S., Narang, M., Lee, D. S. C., Surh, L. C., and Korneluk, R. G. (1993). Effect of the myotonic dystrophy (DM) mutation on mRNA levels of the DM gene. *Nature Genet.* **4,** 233.
55. Harper, P. S., and Dyken, P. R. (1975). Early onset dystrophia myotonica: evidence supporting a maternal environmental factor. *Lancet* **2,** 53.
56. Brunner, H. G., Nillesen, M., van Oost, B. A., Jansen, G., Wieringa, B., Rogers, H-H., and Smeets, H. J. M. (1992). Presymptomatic diagnosis of myotonic dystrophy. *J. Med. Genet.* **29,** 780.
57. Myring, H., Meredith, A. L., Harley, H. G., Koch, G., Norbury, G., Harper, P. S., and Shaw, D. J. (1992). Specific molecular prenatal diagnosis for the CTG mutation in myotonic dystrophy. J. Med. Genet. **29,** 785.
58. Harper, P. S., Harley, H. G., Reardon, W., Shaw, D. J. (1992). Anticipation in myotonic dystrophy: new light on an old problem. *Am. J. Hum. Genet.* **51,** 10–16.
59. Asherson, P., Walsh, C., Williams, J., Sargeant, M., Taylor, C., Clemments, A., Gill, M., Owen, M., McGuffin, P. (1994). Imprinting and anticipation: are they relevant to genetic studies of schizophrenia? *Br. J. Psych.* **164,** 619–624.
60. McInnis, M. G. (1996). Anticipation: an old idea in new genes. *Am. J. Hum. Genet.* **59,** 973–979.
61. Fraser, F. C. (1997). Trinucleotide repeats not the only cause of anticipation. *Lancet* **3350,** 459–460.
62. Morel, B. A. (1857). "Traite des degenerescences." JB Bailliere, Paris
63. Mott, F. W. (1910). Hereditary aspects of nervous and mental diseases. *Br. Med. J.* **2,** 1013–1020.
64. Mott, F. W. (1911). A lecture on hereditary and insanity. *Lancet* **1,** 1251–1259.
65. Greenfield, J. (1911). Notes on a family with myotonica atrophica and early cataract with report of an additional case of myotonia atrophica. *Rev. Neurol. Psych.* **9,** 169–181.
66. Fleischer, B. (1918). Uber myotonische Dystrophie mit katarakt. *Graefes Arch. Klin. Exp. Ophthalmol.* **96,** 91–133.
67. Bell, J. (1947). Dystrophia myotonica and allied diseases. *In* "Treasury of Human Inheritance" (L. S. Penrose, Ed.), Vol. 4, part 5, pp. 343–410. Cambridge University Press, Cambridge.
68. Penrose, L. S. (1948). The problem of anticipation in pedigrees of dystrophia myotonica. *Ann. Eugenics* **14,** 125–132.
69. Höweler, C. J. (1986). A clinical and genetic study in myotonic dystrophy. PhD thesis, University of Rotterdam, Rotterdam.
70. Höweler, C. J., Busch, H. F. M., Geraedts, J. P. M., Niermeijer, M. F., and Staal, A. (1989). Anticipation in myotonic dystrophy: fact or fiction. *Brain* **112,** 779–797.
71. Klein, D. (1958). La dystrophie myotonique (Steinert) et la myotonie congenitale (Thomsen) en Suisse. *J. Genet. Hum.* **1L**(Suppl.), 1–328.
72. Harper, P. S. (1975b). Congenital myotonic dystrophy in Britain. 2. Genetic basis. *Arch. Dis. Child.* **50,** 514–521.
75. Harper, P. S. (1996). New genes for old diseases: the molecular basis of myotonic dystrophy and Huntington's Disease. *J.R. Coll. Phys. London* **30**(3), 221–231.
76. Lopez de Munain, A., Cobo, A. M., Poza, J. J., Navarrete, D., Martorell, L., Palau, F., Emparanza, J. I., and Baiget, M. (1995). Influence of the sex of the transmitting grandparent in congenital myotonic dystrophy. *J. Med. Genet.* **32**(9), 689–691.
77. Merrit, A. D., Conneally, P. M., Rahman, N. F., and Drew, A. L. (1969). Juvenile Huntington's chorea. *In* "Progress in Neurogenetics" (A. Barbeau and J. R. Brunette, Eds.), pp. 645–650. Excerpta Medica,Amsterdam.
80. Leeflang, E. P., McPeek, M. S., and Arnheim, N. (1996). Analysis of meiotic segregation, using single-sperm typing: meiotic drive at the myotonic dystrophy locus. *Am. J. Hum. Genet.* **59**(4), 896–904.
81. Chakraborty, R., Stivers, D. N., Deka, R., Yu, .L. N., Shriver, M. D., and Ferrell, R. E. (1996). Segregation distortion of the CTG repeats at the myotonic dystrophy locus. *Am. J. Hum. Genet.* **59**(1), 109–118.
82. Inglehearn, C. F., and Gregory, C. Y. (1997). Meiotic drive at the myotonic dystrophy and the cone-rod dystrophy loci on chromosome 19q13.3. *Am. J. Hum. Genet.* **60**(6), 1562–1563. [Letter]
84. Lavedan, C., Hofmann-Radvanyi, H., Shelbourne, P., Rabes, J. P., Duros, C., Savoy, D., Dehaupas, I., Luce, S., Johnson, K., and Junien, C. (1993). Myotonic dystrophy: size and sex dependent dynamics of CTG meiotic instability and somatic mosaicism. *Am. J. Hum. Genet.* **52,** 875.
85. Brunner, H. G., Jansen, G., Nillesen, W., Nelen, M. R., de Die, C. E. M., Howeler, C. J., van Oost, B. A., Wieringa, B., Ropers, H-H., and Smeets, H. J. M. (1993). Reverse mutation in myotonic dystrophy. *N. Engl. J. Med.* **238,** 476.
86. Martorell, L., Illa, I., Rosell, Benitez, J., Sedano, M. J., and Baiget, M. (1996). Homozygous myotonic dystrophy: clinical and molecular studies of three unrelated cases. *J. Med. Genet.* **33**(9), 783–785.
87. The Huntington's Disease Collaborative Research Group (1993). A novel gene containing a trinucleotide repeat that is expanded and unstable on Huntington's disease chromosomes. *Cell* **72,** 971.
88. Perutz, M. (1996). Glutamine repeats and inherited neurodegenerative diseases: molecular aspects. *Curr. Biol.* **6,** 848–858.

CHAPTER 10

Genetic Studies of the Myotonic Dystrophy CTG Repeat

JAMES D. WARING The Solange Gauthier Karsh Laboratory, Research Institute, Children's Hospital of Eastern Ontario, Ottawa, Ontario, Canada K1H 8L1

ROBERT G. KORNELUK The Solange Gauthier Karsh Laboratory, Research Institute, Children's Hospital of Eastern Ontario, Ottawa, Ontario, Canada K1H 8L1; and Department of Microbiology and Immunology, University of Ottawa, Ottawa, Ontario, Canada K1H 8M5

I. INTRODUCTION

Myotonic dystrophy (DM) is an important human heritable neuromuscular disease due to its high incidence (approximately 1/8000), and frequent mortality in affected infants. In certain regions the incidence can reach 1/450 due to genetic founder effects [1]. In 1992, after years of effort, a positional cloning approach culminated in the simultaneous discovery by three groups of an unstable region on chromosome 19 as the genetic basis for DM [2–4]. A polymorphic region at 19q13.3 was found to be larger in DM patients, and the size of this expansion was correlated with the severity of the disease. These reports were quickly followed by the identification of an unstable $(CTG)_n$ trinucleotide repeat as the basis for this expansion [5–7]. Another disease was thereby added to a list of trinucleotide repeat disorders that at the time included spinal and bulbar muscular atrophy (SBMA; Kennedy's Disease), and fragile X mental retardation (FRAXA). The list has since quickly expanded to include Huntington's Disease (HD), spinocerebellar ataxia type I (SCA1), FRAXE mental retardation, dentatorubral-pallidoluysian atrophy (DRLPA), Machado-Joseph's disease (MJD; SCA3), Friedreich's ataxia (FA), SCA2, expansion at the fragile site FRA11B in some cases of Jacobsen syndrome, and most recently, SCA6 and SCA7. DM shares many features with these other diseases, but is also unique in many respects which have made it an interesting genetic model for study. In particular, the range of repeat expansions and the degree of somatic heterogeneity exceed that of the other trinucleotide repeat diseases. This is reflected in the wide range of severity and age of onset, and the extreme variability of disease expression.

The DM CTG repeat was found within the 3′ untranslated region (UTR) of a gene predicted to encode a serine–threonine kinase (designated myotonic dystrophy protein kinase, or DMPK), based on sequence homology. The best matches were originally to vertebrate cAMP-dependent kinases. The majority of the trinucleotide repeat disorders are caused by polyglutamine (CAG coding) expansions in their respective gene products, whose functions are often unknown (exceptions are the androgen receptor for SBMA [8], and the CACNA1A calcium channel in cases of SCA6 [9]). However, disease is primarily believed to be caused by a toxic gain of function conferred upon these proteins, an interpretation supported by mouse models [10]. For FRAXA, expansion of CGG repeats into the disease range in the 5′ UTR of the FMR-1 gene is associated with abnormal hypermethylation at an adjacent CpG island, which is also highly methylated on the inactivated X-chromosome in normal females [11]. This causes disease by transcriptional silencing, and to some degree translational inhibition, in affected males and a proportion of carrier females. Expansion at the FRA11B site sometimes causes deletions which truncate the 5′ portion of the CBL2 proto-oncogene [12]. For Friedreich's ataxia, a CAA repeat within the first intron interferes with splicing of a gene believed to specify phosphatidyl-4-phosphate 5-kinase activity [13]. The mechanism of disease remains unknown for DM. It has been particularly difficult to reconcile the position of the repeat expansion within the 3′-UTR of the DMPK gene with the dominant mode of inheritance. DMPK expression cannot be reduced below 50% of the normal level in affected patients, unless a position variegation effect also operates *in trans* to reduce the level of the normal allele. The substrate(s) for the DMPK protein, and the pathway in which it functions, have also remained elusive.

In the preceding section, the clinical nature of myotonic dystrophy was detailed. Also, the pronounced genetic anticipation so characteristic of transmission within DM families has been presented. These topics will serve as an excellent introduction for our discussion. Here, we will attempt to highlight studies of the DM CTG repeat, both normal and mutant length, which have shed light on aspects of DM transmission and pathogenesis. Much has been learned through these studies regarding the evolution and maintenance of DM, its prevalence in different populations, and influences over dynamic repeat behavior during transmission. Characterization of mutant repeat lengths have led to an appreciation of the emergence and progression of somatic and germline heterogeneity, and an attempt has been made to apply these findings toward a greater understanding of anticipation and disease expression. Finally, we will discuss exciting new research currently emerging regarding the biology of CTG repeat expansion-containing transcripts. The intriguing possibility that the CTG repeat expansion affects the expression of genes which neighbor DMPK will be addressed in another section.

II. DISEASE EVOLUTION

Characterization of the genomic region about the DM locus prior and subsequent to the identification of the gene revealed a highly conserved structure on disease chromosomes. The identification of the DM repeat was greatly facilitated by linkage disequilibrium of various polymorphic markers about the DM locus in pooled families. Observations during the 1980s placed the DM gene on chromosome 19, and demonstrated association with various alleles of the, e.g., complement C3, APOE, APOC2, and muscle creatine kinase genes. Further characterization of closer polymorphic markers demonstrated strong associations with various alleles in discrete DM populations, indicative of genetic founder effects [14, 15]. Linkage within a large, heterogeneous population was first observed for D19S63 in European families, a locus tightly linked to the DM region [16, 17], suggesting a common origin for these families. This was supported and extended in further studies with various markers in different populations, which showed similar results [18–20]. Strikingly, an intragenic Alu repeat insertion/deletion polymorphism was found, and the insertion allele found to be in complete association with the DM expansion in Caucasian and Japanese populations [15, 17, 20–22]. The insertion allele is the ancestral one, based on both its structure and its presence in primates [22, 23]. The normal frequency of the insertion allele is 0.51 in people of European ancestry [24], 0.45 in Japanese [15], and 0.71 in South African Negroids [25]. Only one exception to this association has been characterized to date, namely a Nigerian male affected with DM [26]. Thus the disease appeared to be the result of one or a few ancestral mutations. Furthermore, as DM is a dominant disease with a high frequency and low reproductive fitness (estimated at 0.7) [27], these data implied that a pool must exist from which new cases of DM arise. This is especially true as genetic anticipation would act to hasten the loss of DM mutations by increasing the loss of fitness in successive generations (one example is given in [28]).

Haplotype studies supported the idea that the evolution of DM chromosomes was a unique or rare event. The DM CTG repeat copy number varies from approximately 5 to 37 in normal individuals, and from 50 to several thousand repeats in individuals with disease. The distribution of normal alleles worldwide is trimodal, with peaks occurring at $(CTG)_5$, $(CTG)_{11\text{-}17}$, and $(CTG)_{\geq 19}$ [29]. $(CTG)_5$ is the most common allele in European and African populations, while $(CTG)_{11\text{-}17}$ alleles predominate in Japanese, Native American, and Tibetan populations [29, 30]. Allele sizes of 19 or longer are rare accounting for approximately 10% of the total, with no one greater than 1%. Imbert *et al.* [24] examined a French population for the distribution of CTG alleles in relation to the Alu insertion/deletion and D19S112 polymorphisms. CTG alleles of 5, and 19 or greater (including DM chromosomes), were found exclusively associated with the Alu insertion, while $(CTG)_{11\text{-}13}$ alleles were observed only on Alu deletion chromosomes. Furthermore, D19S112 showed a similar, marked disequilibrium pattern for $(CTG)_{19\text{-}36}$ and for DM alleles, implying a common origin. Neville *et al.* [18] extended this study using a series of intragenic or closely linked markers in a broader population, and identified four major and five derivative haplotypes. DM was in complete association with the most common haplotype (haplotype A, including the Alu insertion allele), representing approximately 50% of the population studied. Using the more distal marker D19S63, a similar allelic association was again seen for haplotype A individuals with $(CTG)_{\geq 19}$ alleles, and DM patients. This same general pattern has been seen in other studies [19, 31]. It can therefore be concluded that all cases of DM in these populations arose from a single ancestral chromosome, and that the $(CTG)_{19}$ alleles were derived from $(CTG)_5$ by one or a few events. Also, $CTG_{19\text{-}36}$ alleles represent the pool from which novel DM mutations arise.

The incidence of DM is very low in ethnic Africans, Cantonese, Thais, and Oceanians [32]. Studies on a South African Negroid population have demonstrated a greater haplotype diversity than in Caucasoids, and the exclusive association of the Alu deletion with $(CTG)_{11\text{-}13}$ was not seen [25]. There are also many fewer longer CTG alleles, consistent with the very low frequency of DM [33]. It has therefore been theorized that two founder haplotypes (CTG_5/Alu+, $CTG_{11\text{-}13}$/Alu-), from a diverse pool, left Africa in the migrations which established the Eurasian populations, and that the event which founded the $(CTG)_{19}$ alleles on one of them occurred subsequently, but before the various affected races diverged. Studies on other non-Caucasian populations have further refined this scheme of evolution [29, 34]. $(CTG)_5$ alleles on an Alu deletion background have been found, suggesting that the deletion allele may have originated first on $(CTG)_5$, and not $(CTG)_{11\text{-}13}$ chromosomes. In addition, larger alleles (17 and 27) on Alu deletion chromosomes were found, implying that $(CTG)_5$ alleles gave rise to $(CTG)_{\geq 19}$ multiple times. Based on this information, a slower, stepwise evolution of repeats to the larger, at-risk alleles has been proposed to be ongoing [35].

Linkage disequilibrium has been detected between the mutation and closely linked markers for other trinucleotide repeat diseases such as HD, FRAXA, SCA1, and MJD. At the haplotype level, two major haplotypes account for about 48% of HD mutant chromosomes [36]. Shorter (7 repeat) alleles at a polymorphic CCG tract adjacent to the CAG repeats [37, 38], and an amino acid deletion polymorphism (Δ2634) are strongly associated with longer normal CAG repeats and HD chromosomes [39–41]. For FRAXA, a few haplotypes are again preferentially associated with upper normal range repeat alleles and mutant chromosomes [42, 43]. The majority of normal FRAXA CGG alleles are interrupted by AGG triplets, whereas longer perfect, or uninterrupted, repeats are those which evolve to mutant alleles [44], and show the highest disequilibrium with disease haplotypes. These two diseases therefore appear to be similar to DM in that cases worldwide have arisen from a restricted pool, but more frequently.

Comparison of normal allele distributions supports this conclusion. An instability threshold for trinucleotide repeats ranging from 35 to 50 repeats is apparent from transmission data, after which mutation rates increase dramatically (see below). Normal alleles for the CAG coding-region diseases are typically distributed above 20 repeats (although this is complicated by the presence of interruptions for some disease genes), above the lower limit for the (minor) DM population from which mutant alleles emerge. Normal FRAXA repeats are distributed with a major mode about 29, and an additional minor mode about 37 [45]. An overlap can be seen between the upper normal range and the lower end of the mutation range, with repeats as small as 43 observed to be unstable in some transmissions, and repeats of 50–54 transmitted in a stable fashion in several meioses [45–47]. Furthermore, rare, meiotically unstable HD CAG alleles with a size of approximately 30–38 repeats, termed intermediate alleles, have been identified on the core disease haplotype. These were the source of HD in families with a *de novo* mutation, and have subsequently been documented in a number of transmissions [48]. These are believed to represent an unstable transition state from long normal alleles with incomplete penetrance. In contrast, alleles in the presumed transition class (30–50 repeats) are rare for DM. Repeats of 35 and 42 have been reported in patients diagnosed with DM [49, 50]. More recently, a 42

CTG repeat was found to expand over two generations into a full mutation in a Japanese family [51]. Therefore, while the major DM repeat allele modes (CTG_5, CTG_{11-14}) are immune to expansion, a greater percentage of the normal populations are close to the instability thresholds for other diseases.

III. MUTATION OF THE DM CTG REPEAT

Measurements of mutation rates have demonstrated increasing mutability with size for simple dinucleotide repeats [52]. Sequence analysis of normal and mutant repeats has supported the concept that the loss of stabilizing interruptions, as for example FRAXA, SCA1, SCA2, is the key event preceding expansion into the disease range [44, 53–56]. A single case has been reported of an unusual imperfect DM repeat which appears to have evolved as a (CCGCTG) hexamer, and exhibits greater stability than a perfect 27-repeat allele [57]. Therefore, the length of the DM repeat alone may be the driving force behind expansion. The association of DM with one haplotype (with one exception) may be a consequence of the fact that it is the ancient one. DM could then conceivably emerge on any haplotype, given sufficient time. The case of DM found in a Nigerian family on an Alu deletion background appears to be such a novel mutation, and demonstrates that the European DM haplotype is not required for the emergence or maintenance of larger repeat sizes [26].

However , some observations have been interpreted in favor of *cis* elements predisposing haplotypes to expansion for other trinucleotide repeat diseases. A positive correlation between the degree of mosaicism of HD CAG repeats in sperm (as compared with blood) and the likelihood of expansion during transmission was seen [58]. Reductions were seen where mosaicism was most limited. A DM patient who transmitted an apparent reduction to two of his offspring had a much smaller and more limited distribution of allele sizes in sperm compared with blood, assessed by single-cell typing using small-pool PCR (SP-PCR) [59]. While greater postzygotic expansion in the parental blood versus sperm likely masked the fact that small expansions were actually transmitted in this case (see below), there may be a relationship between the degree of mosaicism which develops in sperm and the likelihood of transmitting a large expansion. Mosaicism in sperm of SBMA patients is lower than that for the other $(CAG)_n$ disorders, and anticipation is correspondingly rarer [60]. Analysis of a CGG/GGG polymorphism found at the 3′ end of the MJD repeat found that the haplotype with the configuration [expanded $(CAG)_n$-CGG]/[normal $(CAG)_n$-GGG] was associated with greater intergenerational instability, and greater repeat variance in sperm as determined by SP-PCR, than the other genotypes [61, 62]. An intergenic conversion mechanism was suggested to account for these results. Furthermore, expansion rates estimated by SP-PCR for normal alleles of similar sizes vary greatly. For example, in SBMA, CAG repeats in the range of 28–31 have an expansion rate of 0.43%, 27 CTG repeats for DM have a rate of 6%, and 30 CAG repeats in HD have a rate of 9% (see below). There may be more to learn regarding chromosomal elements which augment or suppress instability from normal alleles at these different loci.

Imbert *et al.* [24] proposed that the unique or rare events which converted the ancestral $(CTG)_5$ to $(CTG)_{19}$ initiated a gradual drift to larger sizes. Infrequent conversion of these larger at risk alleles into the 30–50 range would greatly increase the likelihood of further mutation to $(CTG)_{>50}$, which then commits a family to eventual disease manifestation. CTG alleles in the 30–50 range were estimated to occur with a frequency of $1\text{-}2 \times 10^{-3}$, while alleles >50 were estimated at about 10^{-4}. The mutation rate for long normal alleles was placed between that for (CA) repeat microsatellites, and an estimated rate for SBMA normal alleles (which have a very similar distribution to the $(CTG)_{19\text{-}36}$ mode [63]): between 10^{-3} and 10^{-4}. Assuming a state of equilibrium, the replacement of eliminated mutant chromosomes suggests an efficient conversion to disease alleles of $(CTG)_{30\text{-}50}$, with a rate ranging from 0.01 to 0.5.

Large population screens to measure these rates have not been performed for DM. However, estimates of the rates at which normal alleles mutate have been made for disease genes using single-sperm typing. At the SBMA locus, $(CAG)_{28\text{-}31}$ alleles had an ~2.5-fold greater mutation rate than average-sized $(CAG)_{20\text{-}22}$ alleles (3.19 versus 1.31% [64]). However, contractions greatly outnumbered expansions for the larger class (2.9 and 0.31%, respectively), and primarily accounted for the increased mutability over the average size class (0.9 and 0.43%, respectively). In contrast, sperm from a patient (47 repeats) exhibited 66% expansions and 15% contractions [65]. Thus, in agreement with the limited transmission data available, an change in repeat number from 20 to 47 increased the contraction rate by about 9-fold, but the rate of expansion increased by greater than 200-fold. Similarly, average-sized alleles at the HD locus (15–18) exhibited only contractions (0.6%), while larger alleles began to show progressively increasing expansion rates [66]. A $(CAG)_{30}$ allele, near the boundary of normal and mutant alleles, showed 9% expansion and 3% contraction rates; an intermediate allele (36 repeats) had 42% expansion and 11% contraction rates; disease alleles (36–51) had expansion rates increasing to 97%

and contraction rates decreasing to 2%. Preliminary data for normal DMPK CTG alleles from two donors exhibited the same general pattern [64]. Although these data are limited by their restriction to males, and will not account for any effects of repeat length on fertility or viability, they again suggest that repeat length is the major driving force for expansion. Furthermore, these results suggest a fundamental distinction between expansion and contraction, with expansion being relatively rare until a critical size is reached.

The consequences of repeat expansion to the instability threshold vary. Repeats in the range of 50–100 result in no or very minimal disease for FRAXA and DM, whereas repeats above $(CAG)_{40}$ are often fully symptomatic for diseases with coding region expansions, such as HD or SBMA. An asymptomatic FRAXA premutation state has been defined for carrier females and normal transmitting males, with CGG repeat sizes of about 50 to 200. For DM, alleles in the 50–80 CTG repeat range were designated "protomutations," as patients will sometimes present with mild symptoms, generally cataracts at an advanced age [67]. DM and FRAXA repeats can reach several thousand in length, while large CAG expansions (>100) present within coding regions are rare, suggesting that this would result in a dominant lethal inactivation of the respective polyglutamine-containing protein products.

Once repeats evolve into the pre- or protomutation range, for FRAXA or DM, respectively, a strong relationship between repeat size and intergenerational expansion of the repeat is apparent. CGG repeats ≥90 result in expansion 100% of the time [45, 68]. A very similar degree of instability, with commitment to expansion for $(CTG)_{>80}$, was observed for DM [49, 67, 69]. Intergenerational increases of DM protomutations are much more frequent than transmissions without a change, although stable transmission of protomutations is not as likely to be seen after ascertainment of a clinically affected kindred due to anticipation, and may therefore still contribute to maintenance of the disease in the population.

IV. SEGREGATION DISTORTION

The removal of expanded alleles from pedigrees by decreased fitness has prompted the proposal that preferential transmission of longer alleles would be required to explain the maintenance of the at-risk $(CTG)_{\geq 19}$ pool. Several studies have found evidence of segregation distortion in favor of transmission of the longer CTG allele, but the data upon which sex carries this effect are conflicting. Carey *et al.* [70] examined meioses from normal heterozygotes with one allele ≥19, and found that the larger allele was transmitted 56.4% of the time overall, with a significant distortion for paternal transmissions. Genarelli *et al.* [71] documented transmissions from affected individuals, and found that the offspring inherited the mutant chromosome 58.1% of the time. Again, this distortion was more evident for paternal transmissions, and was primarily accounted for by transmissions to sons. A case was made for an increase in the DM gene pool via father-to-son transmission of (proto)mutations. However, Hurst *et al.* [72] reanalyzed the results of these two studies, and concluded that the findings of preferential male-specific transmission of longer alleles was not substantiated by the data. These studies were followed by two in which significant female-specific segregation distortion was seen: Shaw *et al.* [73] examined transmissions from normal heterozygotes, and observed 57.7% inheritance of the longer allele from mothers, while Chakraborty *et al.* [35] likewise found that transmission of the longer allele from mothers occurred 56.5% of the time. The possibility that distortion could be operating through meiotic drive was carefully studied by Leeflang *et al.* [74] in sperm samples from normal heterozygotes, taking into account possible effects of contamination and preferential amplification of different allele sizes. They saw no evidence for distortion from an equal ratio of longer and shorter alleles. They also point out that while the overall findings for distortion in previous studies were significant, the sex-specific differences in transmissions were not. Also, the study by Monckton *et al.* [59] found that there was no preferential production of longer alleles in sperm. Male-specific segregation distortion has been noted in pedigrees for other trinucleotide repeat disorders [75]. More study will be required to clarify which sex, if either, is subject to distortion of DM CTG transmission, or to identify a postfertilization mechanism through which this may be operating, in support of the idea that a pool of longer normal CTG repeats is maintained by preferential transmission for DM.

V. TRINUCLEOTIDE REPEAT MOSAICISM

Southern blotting reveals that DM expansion-containing genomic fragments from peripheral blood tissue have a diffuse or smeared appearance. Heterogeneity consistent with somatic mosaicism had previously been observed for FRAXA, although the patterns were typically composed of more discrete bands, with premutations and full mutations coexisting. Heterogeneity within, and differences in average repeat lengths between, tissues in the same individual was subsequently confirmed for DM [76–80]. Genotype–phenotype correlations have actively been sought for prognosis and prenatal testing decisions, but the classification of DM into different categories of severity shows broad overlaps for expansion sizes in blood tissue. It was therefore hoped that better correlations between clinical outcome and the degree of expansion in affected tissues might be observed. As noted by Thornton *et al.* [78], "The idea that muscle disease relates more directly to the CTG repeat amplification in muscle than leukocytes has intrinsic appeal . . . ". Surprisingly, expansion status in muscle appears to have little prognostic value. Muscle expansions are typically much larger than those in blood, sometimes up to 13-fold [76, 78, 81, 82]. While a weak correlation between the size of the expansion in muscle and in lymphocytes has sometimes been seen [78, 81], no connection between the size of muscle expansions and disease severity or progression is apparent [76, 81, 82]. Larger expansions in various tissues may be related broadly to the tissues affected by DM [80], but are clearly not sufficient for disease progression. For example, all skeletal muscles expansions were found to be generally large, with severely affected distal muscles not different from unaffected proximal ones [76, 78]. Overall, cardiac muscle expansions are often found to be the largest for the tissues studied [e.g., 78, 79], but almost all other tissues appear to have larger expansions than in blood. There has been some suggestion of a tissue-specific input upon expansion, as the smallest repeat size was found in the cerebellum not only for DM, but also for the CAG expansions in HD [83], MJD, and DRLPA [80].

Elucidation of the developmental timing of repeat expansion would shed light upon the mechanism of instability. Early studies documented little heterogeneity before 20 weeks of development. No variations were seen in tissues of fetuses less than 13 weeks old [79, 84, 85], or in one congenital DM (cDM) neonate [76], although it may be that in some of these cases the expansions were very large, at the limit of resolution for the detection of size differences. Only small variations were seen in different tissues of a 20-week-old fetus and in cDM neonates in other studies [77, 79, 86, 87]. However, in more recent studies, convincing distinctions were seen in different tissues for a 16-week-old fetus (with expansion sizes between 4.8 and 7.8 kb.), but not for a 13-week-old fetus [88]. Likewise, mosaicism was almost always seen in ≥16-week-old fetuses, but not at 13 weeks or before [89]. It was speculated that instability occurs at the onset of the second trimester during rapid growth of the fetus, and is suppressed during differentiation of the fetus in the first trimester. In the former study, lines were established from both the 13- and 16-week-old fetuses, and clones derived from these lines segregated discrete alleles. Some clones had a repeat size slightly larger than the parental line. Moreover, these lines showed significant expansion over time in culture, even though no tissue differences were seen for the 13-week-old fetus. A sigmoidal pattern of size gain was seen, with an apparent lag phase and a plateau. FRAXA repeats exhibited strikingly different behavior in the same experiments. A mosaic pattern of discrete bands was seen in fetal tissues, but this pattern did not vary between tissues [90]. Fibroblast lines cloned from a fully affected FRAXA male fetus again segregated unique CGG alleles, but the lines were stable in culture. These studies indicate that both repeats likely expand very early after conception, but CGG repeats become fixed while CTG repeats continue to be unstable. The authors suggest that this is due to the extended absence of methylation present about DM expansions after early embryogenesis, while CpG dinucleotides become hypermethylated in the presence of a full-mutation FRAXA repeat. This would allow the mismatch repair system to distinguish the parental from daughter strand if slippage occurred during replication. Suppression of instability during the first trimester may be due to a more efficient repair system, or possibly transient methylation adjacent to the DM CTG repeat. A mouse hypervariable minisatellite with the sequence GGCA also exhibited instability restricted to a very early period postconception [91].

Consistent with this proposal, it is now apparent that heterogeneity continues to increase postnatally for DM, at least in blood tissue. Mutant alleles have been seen to increase in size in lymphocytes sampled over time, and the gain can be correlated with the initial size of the repeat [92, 93]. It has also been a common observation that genomic fragments from a younger subject are limited in their heterogeneity, even for large expansions. Within the same age category, the degree of heterogeneity can again be correlated with the size of the expansion [92]. Adult monozygotic twins can develop clearly distinct expansion sizes in their lymphocytes [94]. In apparent contrast, muscle expansions do not appear to enlarge over time in adults [78, 81], and appear to be the same size in different muscle types [78, 80]. However it has been noted that the difference in size between muscle and lymphocyte expansions is smaller for younger patients [82], and increases with age. This suggests that muscle expansions grow rapidly postnatally and reach a plateau in early adulthood, consistent with the sigmoidal size gain described above. SP-PCR has demonstrated that the smears seen on Southern blot analysis are, in fact, composed of multiple unresolved alleles for DM [59, 92]. It was further verified that blood alleles increase in range, mean, and modal allele size over time [92]. In a comparison of fresh and cultured EBV-transformed lymphocytes from adult patients, no [77] or small differences [76] have been seen, with the transformed cells having the larger repeat in most cases in the latter study. This might be due to clonal selection of the B lymphocytes during transformation. SP-PCR analysis of clonal lines derived from these cultures demonstrates the accumulation over time of small mutational changes (expansions and contractions with no apparent bias) and rarer larger changes (biased toward contraction) [95]. Longer cultures of such clones may reveal an appreciable expansion bias as observed in fetal cell lines by Wohrle *et al.* [88]. Alternatively, the expansion bias observed *in vivo* [59] may depend upon other factors than cell division/DNA replication.

Heterogeneity has also been observed in the sperm of DM patients, and full mutations can be demonstrated by Southern blot analysis. The term "gonosomal mosaics" was coined to indicate this combined somatic and germline heterogeneity [79]. One important question regarding the timing of expansion is whether genetic anticipation can be appreciated strictly in terms of instability occurring postzygotically. No rearrangement of flanking markers is seen upon transmission, so unequal meiotic recombination can be excluded as a mechanism for expansion. It was found that repeat distributions in sperm often had greater variability, and were usually larger on average, than in the blood of the same individual [79]. However, sperm distributions were often smaller on average than in the blood of an offspring (consistent with anticipation). PCR analysis of sperm revealed a normal range reversion which had been transmitted to an offspring in one case [96], but not in two others [97]. Based on this low degree of overlap, it was suggested that it was unlikely that a rare sperm with a large expansion (or contraction) was the source of repeat distributions detected in the blood of offspring [79, 98]. However, SP-PCR analysis of sperm revealed, in some samples, an increased mean size relative to blood and skewed distribution patterns, interpreted to be consistent with an expansion bias during spermatogenesis [59]. These data suggest that passage of an expanded repeat through the male germline is heavily biased toward further expansion. In addition, overlaps in the distribution of parental sperm alleles with blood alleles in some offspring were seen. Therefore, anticipation may be a result of combined size increases from both germline and postzygotic expansion. In contrast, only premutations can be detected in sperm from FRAXA males [99]. Current data suggest that full mutations occur in oocytes prezygotically, and regress in sperm during early development.

A model has been proposed by Zheng *et al.* [100] which accounts for many features of dominant, trinucleotide repeat disorders manifesting anticipation. Mitotic instability is viewed as primarily a stochastic process, and sufficient to account for variable phenotype and penetrance. Here, the percentage of any tissue which contains a mutation-sized repeat, and therefore the likelihood that symptoms are manifested, depends upon how early in that tissue's development an expansion exceeding the threshold required for symptoms occurs. The point in development at which the expansions reach a threshold may be crucial, although not apparent until years later. Somatic and germ line expansion can be viewed as elements of the same process (see also [98]). This model does not take into account restricted time frames for expansion, selection against large expansions during spermatogenesis, or other tissue-specific differences which affect expansion. However, a unified process of mitotic instability does account for other features of DM, namely (1) anticipation would be more pronounced through male meiosis because of the greater number of divisions, thereby increasing the chance of expansion, and (2) a female giving birth to a severely affected child has an increased chance of her next DM child being severely affected, because the first birth establishes that the percentage of her oocytes containing sufficiently large mutations must be a high number.

VI. EFFECT OF SEX ON TRANSMISSION

Numerous studies following the identification of the DM mutation verified that greatly enhanced instability of repeats above the protomutation threshold was the basis for anticipation observed in pedigrees, and that the degree of expansion was positively correlated with the severity of disease and negatively correlated with the age of onset [17, 49, 50, 69, 77, 101–104]. Examination of pedigrees also revealed sex-specific influences in transmission patterns. No sex-specificity is seen in the rate of expansion transmission for DM, while expansion from the premutation to the symptomatic full mutation for FRAXA occurs only during maternal transmissions, the basis for the Sherman paradox. However, it has been noted for some time that the transmitting grandparent of affected DM patients is more often male [105]. In other words, the first asymptomatic generation which transmits an enlarged allele to an affected offspring shows a preponderance of males. Close inspection of transmission data reveals that expansions of small repeats (less than 100) are more pronounced when paternally transmitted [49, 98, 106]. There is a corresponding paucity of smaller alleles in the protomutation range for females; in some studies repeats of less than 60 are very rare [49, 67, 69, 77]. However, as paternal alleles increase in size, the magnitude of expansion diminishes. A negative correlation between paternal repeat size and the magnitude of intergenerational increases was demonstrated in many studies, while no such correlation was seen for maternal transmissions [50, 77, 107–110). Moreover, apparent repeat contractions begin to appear in paternal transmissions from long alleles. In a few rare cases, reversions into the normal range were seen (see below). One key study summarized data from many centers, and found that 6.4% of parent–child transmissions involved contractions (assessed using leukocyte DNA), the majority of which were paternally transmitted [111]. This limited ability to expand results in a maximum repeat size of 2000–3000 from male transmissions, and contractions become predominant over 2000 repeats.

However, anticipation was still apparent in approximately half of these reported contractions [111], contrary to expectation. This was much more pronounced for maternal transmissions, and in two cases gave rise to cDM [109]. It was suggested that problems of ascertainment may give a false notion of the degree of anticipation occurring, but was unlikely to be entirely accountable. An unusually high number of siblings with contractions was also noted. Comparison of the heterogeneity in germline and somatic tissues by SP-PCR has helped explain these findings [59]. Leukocyte distributions typically have a sharp lower boundary, which the authors suggest represents the inherited allele size. Distributions were skewed toward larger alleles, suggesting a highly directional process. In contrast, sperm exhibited a more normal distribution in two of three cases, with broad tails at both ends. The average allele size was increased relative to blood, consistent with anticipation, but rare alleles within the high normal range could be detected. No such reversions into the normal range were noted for blood, suggesting that this process is specific to germ cells. Surprisingly, a third case exhibited a somewhat opposite pattern, in which the sperm distribution showed relatively little variation, had a sharp lower boundary, and was smaller than that seen in blood. These patterns suggest that expansion from a progenitor allele proceeds at different rates between tissues and individuals, and that allele distributions broaden out and become more normal as expansion proceeds. This latter case transmitted an apparent contraction to two of his children, but in fact comparison of the father's sperm distribution with that of his children's blood suggests a small expansion was transmitted. This was presumably masked by the fact that continued somatic expansion had occurred in blood over the lifetime of the father. Together, these results suggest that many contractions are apparent, and not real, consistent with anticipation occurring in approximately half of them. True contractions from females may be even rarer, based on the higher rate of anticipation in these cases.

Transmissions which result in cDM cases are almost exclusively maternal. Many observations suggest that the absence of constraint on maternal as compared with paternal transmissions explains this exclusivity. First, the average expansion from maternal versus paternal transmissions is generally greater [50, 69, 77]. This may be primarily accounted for by transmissions to cDM cases [50, 101]. Second, intergenerational increases from longer CTG alleles are more frequent during maternal transmissions [50, 69] and expansions which result in cDM cases are often more significant than expansions which result in non-cDM cases [50, 77, 112]. Third, mothers of cDM offspring have larger repeats on average than mothers of non-cDM offspring [69, 77, 101, 112]. Furthermore, maternal repeats greater than $(CTG)_{100}$ were at >90% risk for very large expansions (>400 repeats) in one study [50]. In summary, two major sex-related dynamics create a typical scenario in which DM first emerges in a family during a male transmission, but the final stages of anticipation which ultimately result in congenital myotonic dystrophy are a consequence of female transmissions.

As referred to above, the initial instability of protomutations in males can be predicted from a stochastic model of mitotic expansion. Spermatogenesis requires many more cell divisions (from 50 to several hundred) than oogenesis (approximately 30), presumably increasing the chance of expansion of small repeats through male transmission. While this effect would become more negligible as average repeat sizes get longer, and further expansion more inevitable, anticipation should be more pronounced through male transmissions overall, in apparent contrast to the maternal specificity for cDM. The apparent negative selection operating on sperm when repeat sizes get longer may explain this discrepancy. Repeat distributions in sperm often have a lower average size than in blood for patients with larger expansions, with a maximum observed size in sperm of ~1000 repeats [79]. Also, no alleles greater than approximately 1000 repeats were seen in sperm by SP-PCR [59]. The mechanistic basis for this restriction, and whether it is similar to the (more severe) restriction operating in FRAXA sperm, is not known. In one case, haplotype analysis demonstrated a gene conversion event which substituted the mutant allele with the normal allele [113]. This was in a rare instance in which reduction was into the normal range, which has been reported from paternal transmissions only [103, 113, 107]. In other cases, the process which leads to expansion, such as unequal sister chromatid exchange or replication slippage, may be reversed, or there may be direct deletion of repeats. The apparent instability threshold of 30–50 repeats correlates well with the energy required to form hairpin loop structures [114]. Jansen *et al.* [79] suggested that longer repeat lengths allow the formation of elaborate mispaired structures, which may act as pause sites during replication, and therefore need to be resolved before gametogenesis can proceed.

Therefore, sperm may be unable to support the large expansions necessary for cDM, as opposed to some unique mechanism(s) which allow for large expansions in female transmissions [77, 108]. However, other factors have been considered for the maternal transmission of cDM, such as the lowered fertility of affected males, and the paucity of small alleles in affected females. Also, exceptional cases of very large transmissions from males have been seen which did not result in cDM (for example, [115]). A greater risk that an affected child would be cDM was noted for mothers who has already borne a cDM-affected child [27]. A related familial risk for cDM has also been defined. In addition, a greater risk that an affected child would be cDM was seen for mothers who manifested multisystemic disease prior to or during pregnancy [116]. However, as referred to above, these outcomes may simply be related to the likelihood of longer repeats present in the oocytes of mothers defined as at risk. An intrauterine factor has been proposed to play a role in the generation of cDM, and some experimental evidence provided using newborn rats as a test system [117, 118]. However, the mechanism of such a factor would have to involve an interplay between the CTG repeat status in the mother and the infant, and complete tolerance of the factor by all unaffected (normal CTG repeat) infants. Rare cases of paternally transmitted cDM have now been seen, which also places this in doubt [119–121]. The possibility that genomic imprinting affects the way CTG repeats expand postfertilization or affects the developing infant has also been considered, but no sex differences in methylation patterns [110, 122] or in mRNA expression patterns [123] have been seen.

VII. MUTANT TRANSCRIPTS AND DISEASE

To date, no case of DM has been attributed to any mutational defect other than the CTG repeat expansion. Although a number of cases clinically consistent with DM which have normal CTG alleles have been noted [124] these are likely phenocopies, as the remainder of the gene was normal. For FRAXA, rare mutations which ablate FMR1 expression have been noted, consistent with the notion that disease is a consequence of down-regulation of transcription due to abnormal hypermethylation of a CpG island adjacent to the expanded CGG repeat. The dominant nature of DM must therefore be due to either haploinsufficiency of the gene product or a dominant gain of function conferred on mutant transcripts, or possibly both. In this section, we will discuss recent studies which have assessed mutant transcript levels, processing, and localization in a attempt to elucidate the disease mechanism.

Early studies which assessed DMPK transcript levels by RT-PCR found substantially decreased levels of mutant transcripts, supporting the haploinsufficiency model. Fu *et al.* [125] used cDNA primed with both oligo-dT and random primers, and found total DMPK transcript levels were lowered in adult patient samples, although a wide range was seen in control samples. Specific primers were used which flanked the CTG repeat region in order to distinguish the alleles, and the reduction was found to be due to the mutant allele, with the levels detected showing a negative correlation with the expansion size (also [126]). More dramatic results were found when somatic cell hybrids containing chromosome 19 from an adult DM patient (113 repeats) or a normal control (13 repeats) were analyzed [127]. This approach allowed the level of mutant transcripts to be determined directly, and in the absence of any effects *in trans* of one allele on the other, which has been proposed (see below). Regions upstream of the CTG expansion, specific to total or spliced DMPK transcripts, were amplified using the appropriate primers from either oligo-dT or random-primed cDNA. A reduction of unprocessed mutant transcript to approximately 18% of the control levels was seen, and no processed transcript could be detected. Furthermore, Northern blot analysis of RNA prepared from a 20-week-old fetus (1.8-kb expansion), or from two cDM infants (6.4- and 8.0-kb expansions, respectively), also failed to detect expanded transcripts [128]. Normal allele levels in the DM fetus were reduced to 22% of that of a normal fetus, instead of the expected 50%, implying that the mutant allele can interfere with the normal allele *in trans.* In agreement, RT-PCR analysis of the fetal sample using random-primed cDNA amplified in three different regions (upstream, downstream, and across the repeats) again showed reduction of total DMPK transcript to approximately 30–40% of normal.

In contrast to these results, Sabourin *et al.* [86] found that total transcript levels in various cDM tissues, as analyzed by competitive RT-PCR and slot-blot analysis, were increased over controls, estimated at 2- to 4-fold for heart and brain, 4- to 6-fold for liver and lung, and 14-fold in brain, although the appropriate age-matched controls were not available. Wild-type and mutant alleles were distinguished using an informative biallelic polymorphism [21], and transcripts from both alleles were found to be expressed at equal levels. Likewise, another study using an allele-specific RT-PCR system found no difference in mature transcript levels for the mutant and normal alleles in adult muscle samples, regardless of repeat size [129]. Also, large mutant transcripts were detected by Northern blot in cDM patient fibroblasts (10 kb in size [86]), or in fibroblasts from a DM patient after conversion to muscle by MyoD (6 kb [130]).

Subsequent to these observations, two detailed studies used allele-specific quantitative RT-PCR systems to measure transcripts from both the precursor and mature fractions from a variety of sources. Wang *et al.* [131] analyzed both total and poly$(A)^+$ RNA in muscle from adult patients using DMPK-specific RT primers. In total RNA from patients' muscle, a reduction to approximately 50% of control levels was seen for DMPK transcripts, with considerable overlap between the two distributions, similar to results in Fu *et al.* [125]. However, a similar decline was seen in myopathic control samples. In contrast, there was a striking paucity of DMPK message in the poly$(A)^+$ fraction, with mutant message decreased to approximately 25% of the normal allele levels. The absolute levels of transcript indicated that 86% of the mutant transcript and 70% of the normal transcript were lost in mature mRNA (as opposed to about 10% in unaffected controls).

Krahe *et al.* [132] used a system specific for either spliced or unspliced message (using either DMPK-specific primers or a combination of oligo-dT and random primers for cDNA synthesis) to analyze muscle samples, cell lines, and somatic cell hybrids derived from both adult and cDM patients. They observed no significant deviation from a ratio of 1 for unprocessed mutant and normal allele transcripts from any source. Again, however, a reduction in mature mutant transcripts relative to normal transcripts (mean ratio 0.49) was observed, which could be correlated with the expansion size. Unlike Wang *et al.* [131], however, absolute levels of mature mRNA did not decline, except for a homozygous patient and in somatic cell hybrids. Total locus activity may therefore only decline in the absence of the wild-type allele, consistent with the results of Carango *et al.* [127], implying an interaction between alleles. Therefore, while these two studies differed in their conclusions regarding total and mature DMPK transcript levels, the total weight of the results strongly suggests that CTG expansions do not affect transcription, but rather some aspect of transcript metabolism. Krahe *et al.* [132] were not able to see any difference in stability or splicing patterns between mutant and normal transcripts, suggesting instead that export may be involved.

The validation of an effect of CTG expansions on posttranscriptional processing and a partial explanation for the discrepancies in transcript measurements appear to have come from studies of mutant transcript localization. The distribution of DMPK transcripts in fibroblasts and muscle biopsies from adult patients was determined by *in situ* hybridization. Mutant transcripts were found aggregated in discrete foci in the nuclei of affected cells [133]. The nature of these foci is not currently understood. They may be indicative of an irreversible association with nuclear factors or structures, or a structural interference with export due to aberrant folding of the transcripts induced by the repeats. In a subsequent study, it was confirmed by Northern blot analysis that mutant transcripts are primarily restricted to the nucleus, but also that the efficient extraction of mutant message is critically dependent upon the method used [134]. Further localization demonstrated that mutant transcripts were tightly associated with the nuclear matrix and therefore resistant to extraction. By this analysis, total wild-type message levels in DM cell lines were not different from controls, but mutant message levels varied from 32 to 71% of controls.

These results likely explain some of the discrepancies regarding mutant message levels in previous studies. It is also possible that the direct measurement of mutant transcript levels by RT-PCR in some experiments which involved priming near, or PCR through the CTG repeat tract, were affected by inefficient polymerization on repetitive sequence templates, as cautioned by Roses [135]. However, it now seems that both a decline in total mature DMPK message and an aberrant function conferred upon mutant DMPK transcripts could contribute to the pathogenesis of DM. This would be a novel disease mechanism, and a critical first clue toward the complete understanding of DM pathogenesis. The retention of mutant transcripts in these foci can explain the paucity of mutant message in the mature cytoplasmic fraction, and mechanisms can be envisioned by which the mutant DMPK message might affect transcript export both *in cis* and *in trans* [131, 132]. Recently, a decrease in insulin receptor mRNA and protein was measured in DM patient muscle biopsies [136].

An attempt to colocalize these foci with known splicing factors was not successful. However, gel shift assays designed to investigate factors which can bind various triplet repeats *in vitro* led to the identification of CUG binding proteins involved in two RNA–protein complexes [137, 138]. These activities (CUG-BP1, CUG-BP2) were distributed between the nucleus and the cytoplasm, depending upon the cell type. The heterogeneous nuclear ribonucleoprotein (hnRNP) hNab50 was detected in both of these complexes, and appears to consist of two isoforms responsible for CUG-BP1 and CUG-BP2. The authors suggested that the CTG expansion might greatly increase the available binding sites for hNab50, a member of a class of proteins thought to be essential for pre-mRNA processing, effectively titrating it in the various affected tissues. Another study has identified two proteins of unknown function (approximately 25 and 35 kDa) which bind CUG sequences, and were also speculated to be involved in pre-mRNA processing [139].

VIII. CONCLUSIONS

Studies over the past 5 years have largely delineated the genetic determinants underlying DM transmission patterns, disease severity, and the evolution and maintenance of mutations. In many ways, the opportunity to compare and contrast DM with the other trinucleotide repeat disorders has greatly facilitated progress in this area. Also, the biology of very large expansions can be studied in the case of DM and FRAXA, as repeats are more easily tolerated when in the untranslated regions of their respective transcripts. This should facilitate the determination of the mechanism of repeat expansion. At the other end of the "expansion spectrum," CAG coding region repeats appear to reach a threshold instability at a slightly smaller size, and small changes in repeat number have dramatic consequences in terms of the health of affected individuals. This should eventually aid in determining the function of the respective protein products, as a gain of function is almost certainly involved in disease.

At this stage, however, the etiology of DM remains far from clear. Research has therefore now reached a critical stage in which researchers are grappling with the pathological consequences of CTG expansions. In particular, the fundamental questions of whether fluctuations in DMPK levels are critical to disease, or indeed whether DM is a single- or multigene disorder (topics which will be discussed in other sections), are currently being addressed. Continued analysis of the processing and localization of DMPK transcripts, analysis of the expression of neighboring genes, and the current availability of specific antisera should allow the resolution of these issues in the near future. Furthermore, the identification of the substrate(s) and metabolic role of DMPK would allow us to examine affected tissues for a disturbance in this function. It has often been speculated that a distinct or additional mechanism operates in congenital DM over adult-onset disease, due to its unique features. Perhaps an irregularity in transcript metabolism and insufficiency of DMPK both operate when DM repeats get very long. As will also be discussed in a subsequent section, animal models will hopefully provide a complete organism for testing hypotheses, but much refinement to current approaches appears to be necessary. As vigorous research utilizing all of these diverse approaches is currently underway, a unified model of the basis of myotonic dystrophy is feasible, ultimately allowing the design of therapeutic approaches. In addition, interesting information on such diverse topics as muscle development, transcript metabolism, and kinase function will undoubtedly be gained during this endeavor.

References

1. Bouchard, G., Roy, R., Declos, M., Mathieu, J., and Kouladjian, K. (1989). Origin and diffusion of the myotonic dystrophy gene in the Saguenay region. *Can. J. Neurol. Sci.* **116,** 119–122.
2. Buxton, J., Shelbourne, P., Davies, J., Jones, C., van Tongeren, T., Aslanidis, C., de Jong, P., Jansen, G., Anvret, M., Riley, B., Williamson, R., and Johnson, K. (1992). Detection of an unstable fragment of DNA specific to individuals with myotonic dystrophy. *Nature* (*London*) **355,** 547–548.
3. Harley, H. G., Brook, J. D., Rundle, S. A., Crow, S., Reardon, W., Buckler, A. J., Harper, P. S., Housman, D. E., and Shaw, D. J. (1992). Expansion of an unstable DNA region and phenotypic variation in myotonic dystrophy. *Nature* (*London*) **355,** 545–546.
4. Aslanidis, C., Jansen, G., Amemiya, C., Shutler, G., Mahadevan, M., Tsilfidis, C., Chen, C., Alleman, J., Wormskamp, N. G. M., Vooijs, M., Buxton, J., Johnson, K., Smeets, H. J. M., Lennon, G. G., Carrano, A. V., Korneluk, R. G., Wierenga, B., and de Jong, P. J. (1992). Cloning of the essential myotonic dystrophy region and mapping of the putative defect. *Nature* (*London*) **355,** 548–551.
5. Brook, J. D., McCurrach, M. E., Harley, H. G., Buckler, A. J., Church, D., Aburatani, H., Hunter, K., Stanton, V. P., Thirion, J.-P., Hudson, T., Sohn, R., Zemelman, B., Snell, R. G., Rundle, S. A., Crow, S., Davies, J., Shelbourne, P., Buxton, J., Jones, C., Juvonen, V., Johnson, K., Harper, P. S., Shaw, D. J., and Housman, D. E. (1992). Molecular basis of myotonic dystrophy: expansion of a trinucleotide (CTG) repeat at the 3′ end of a transcript encoding a protein kinase family member. *Cell* **68,** 799–808.
6. Mahadevan, M., Tsilfidis, C., Sabourin, L., Shutler, G., Amemiya, C., Jansen, G., Neville, C., Narang, M., Barcelö, J., O'Hoy, K., Leblond, S., Earle-MacDonald, J., de Jong, P. J., Wieringa, B., and Korneluk, R. G. (1992). Myotonic dystrophy mutation: an unstable CTG repeat in the 3′ untranslated region of the gene. *Science* **255,** 1253–1255.
7. Fu, Y.-H., Pizzuti, A., Fenwick, R. G., King, J., Jr., Rajnarayan, S., Dunne, P. W., Dubel, J., Nasser, G. A., Ashizawa, T., de Jong, P. J., Wieringa, B., Korneluk, R., Perryman, M. B., Epstein, H. F., and Caskey, C. T. (1992). An unstable triplet repeat in a gene related to myotonic dystrophy. *Science* **255,** 1256–1258.
8. La Spada, A. R., Wilson, E. M., Lubahn, D. B., Harding, A. E., and Fischbeck, K. H. (1991). Androgen receptor gene mutations in X-linked spinal and bulbar muscular atrophy. *Nature* (*London*) **352,** 77–79.
9. Zhuchenko, O., Bailey, J., Bonnen, P., Ashizawa, T., Stockton, D. W., Amos, C., Dobyns, W. B., Subramony, S. H., Zoghbi, H. Y., and Lee, C. C. (1997). Autosomal dominant cerebellar ataxia (SCA6) associated with small polyglutamine expansions in the alpha 1A-voltage-dependent calcium channel. *Nat. Genet.* **15,** 62–69.
10. Burright, E. N., Orr, H. T., and Clark, H. B. (1997). Mouse models of human CAG repeat disorders. *Brain Pathol.* **7,** 965–977.

11. Sutherland, G. R., and Richards, R. I. (1995). Simple tandem DNA repeats and human genetic disease. *Proc. Natl. Acad. Sci. USA* **92,** 3636–3641.
12. Jones, C., Penny, L., Mattina, T., Yu, S., Baker, E., Voullaire, L., Langdon, W. Y., Sutherland, G. R., Richards, R. I., and Tunnacliffe, A. (1995). Association of a chromosome deletion syndrome with a fragile site within the protooncogene CBL2. *Nature (London)* **376,** 145–149.
13. Carvajal, J. J., Pook, M. A., dos Santos, M., Doudney, K., Hillerman, R., Minogue, S., Williamson, R., Hsuan, J. J., and Chamberlain, S. (1996). The Friedreich's ataxia gene encodes a novel phosphatidylinositol-4-phosphate 5-kinase. *Nat. Genet.* **14,** 157–162.
14. Cobo, A., Grinberg, D., Balcells, S., Vilageliu, L., Duarte-Gonzalez, R., and Baiget, M. (1992). Linkage disequilibrium detected between myotonic dystrophy and the anonymous marker D19S63 in the Spanish population. *Hum. Genet.* **89,** 287–291.
15. Yamagata, H., Miki, T., Ogihara, T., Nakagawa, M., Higuchi, I., Osame, M., Shelbourne, P., Davies, J., and Johnson, K. (1992). Expansion of unstable DNA region in Japanese myotonic dystrophy patients. *Lancet* **339,** 692.
16. Harley, H. G., Brook, J. D., Floyd, J., Rundle, S. A., Crow, S., Walsh, K. V., Thibault, M.-C., Harper, P. S., and Shaw, D. J. (1991). Detection of linkage disequilibrium between the myotonic dystrophy locus and a new polymorphic DNA marker. *Am. J. Hum. Genet.* **49,** 68–75.
17. Harley, H. G., Brook, J. D., Rundle, S. A., Crow, S., Reardon, W., Buckler, A. J., Harper, P. S., Housman, D. E., and Shaw, D. J. (1992). Expansion of an unstable DNA region and phenotypic variation in myotonic dystrophy. *Nature (London)* **355,** 545–546.
18. Neville, C. E., Mahadevan, M. S., Barceló, J. M., and Korneluk, R. G. (1994). High resolution genetic analysis suggests one ancestral predisposing haplotype for the origin of the myotonic dystrophy mutation. *Hum. Mol. Genet.* **3,** 45–51.
19. Lavedan, C., Hoffman-Radvanyi, H., Boileau, C., Bonaïti-Pellié, C., Savoy, D., Shelbourne, P., Duros, C., Rabes, J.-P., Dehaupas, I., Luce, S., Johnson, K., and Junien, C. (1994). French myotonic dystrophy families show expansion of a CTG repeat in complete linkage disequilibrium with an intragenic 1 kb expansion. *J. Med. Genet.* **31,** 33–36.
20. Goldman, A., Krause, A., Ramsay, M., and Jenkins, T. (1996). Founder effect and the prevalence of myotonic dystrophy in South Africans: molecular studies. *Am. J. Hum. Genet.* **59,** 445–452.
21. Mahadevan, M. S., Amemiya, C., Jansen, G., Sabourin, L., Baird, S., Neville, C. E., Wormskamp, N., Segers, B., Batzer, M., Lamerdin, J., de Jong, P., Wieringa, B., and Korneluk, R. G. (1993). Structure and genomic sequence of the myotonic dystrophy (DM kinase) gene. *Hum. Mol. Genet.* **2,** 299–304.
22. Mahadevan, M. S., Foitzik, M. A., Surh, L. C., and Korneluk, R. G. (1993). Characterization and polymerase chain reaction (PCR) detection of an *Alu* polymorphism in total linkage disequilibrium with myotonic dystrophy. *Genomics* **15,** 446–448.
23. Rubinsztein, D. C., Leggo, J., Amos, W., Barton, D. E., and Ferguson-Smith, M. A. (1994). Myotonic dystrophy CTG repeats and the associated insertion/deletion polymorphism in human and primate populations. *Hum. Mol. Genet.* **3,** 2031–2035.
24. Imbert, G., Kretz, C., Johnson, K., and Mandel, J.-L. (1993). Origin of the expansion mutation in myotonic dystrophy. *Nat. Genet.* **4,** 72–76.
25. Goldman, A., Ramsay, M., and Jenkins, T. (1995). New founder haplotypes at the myotonic dystrophy locus in Southern Africa. *Am. J. Hum. Genet.* **56,** 1373–1378.
26. Krahe, R., Eckhart, M., Ogunniyi, A. O., Osuntokun, B. O., Siciliano, M. J., and Ashizawa, T. (1995). De novo myotonic dystrophy mutation in a Nigerian kindred. *Am. J. Hum. Genet.* **56,** 1067–1074.
27. Harper, P. S. (1989). "Myotonic Dystrophy," 2nd ed. Saunders, London.
28. de Die-Smulders, C. E. M., Höweler, C. J., Mirandolle, J. F., Brunner, H. G., Hovers, V., Brüggenwirth, H., Smeets, H. J. M., and Geraedts, J. P. M. (1994). Anticipation resulting in the elimination of the myotonic dystrophy gene: a follow up study of one extended family. *J. Med. Genet.* **31,** 595–601.
29. Zerylnick, C., Torroni, A., Sherman, S. L., and Warren, S. T. (1995). Normal variation at the myotonic dystrophy locus in global human populations. *Am. J. Hum. Genet.* **56,** 123–130.
30. Davies, J., Yamagata, H., Shelbourne, P., Buxton, J., Ogihara, T., Nokelainen, P., Nakagawa, M., Williamson, R., Johnson, K., and Miki, T. (1992). Comparison of the myotonic dystrophy associated CTG repeat in European and Japanese populations. *J. Med. Genet.* **29,** 766–769.
31. Yamagata, H., Miki, T., Nakagawa, M., Johnson, K., Deka, R., and Ogihara, T. (1996). Association of CTG repeats and the 1-kb Alu insertion/deletion polymorphism at the myotonic protein kinase gene in the Japanese population suggests a common Eurasian origin of the myotonic dystrophy mutation. *Hum. Genet.* **97,** 145–147.
32. Ashizawa, T., and Epstein, H. F. (1991). Ethnic distribution of myotonic dystrophy gene. *Lancet* **338,** 642–643.
33. Goldman, A., Ramsay, M., and Jenkins, T. (1994). Absence of myotonic dystrophy in southern African negroids is associated with a significantly lower number of CTG trinucleotide repeats. *J. Med. Genet.* **31,** 37–40.
34. Deka, R., Majumder, P. P., Shriver, M. D., Stivers, D. N., Zhong, Y. X., Yu, L. M., Barrantes, R., Yin, S. J., Miki, T., Hundrieser, J., Bunker, C. H., McGarvey, S. T., Sakallah, S., Ferrell, R. E., and Chakraborty, R. (1996). Distribution and evolution of the CTG repeats at the myotonin protein kinase gene in human populations. *Genome Res.* **6,** 142–154.
35. Chakraborty, R., Stivers, D. N., Deka, R., Yu, L. M., Shriver, M. D., and Ferrell, R. E. (1996). Segregation distortion of the CTG repeats at the myotonic dystrophy locus. *Am. J. Hum. Genet.* **59,** 109–118.
36. MacDonald, M. E., Novelletto, A., Lin, C., Tagle, D., Barnes, G., Bates, G., Taylor, S., Allitto, B., Altherr, M., Myers, R., Lehrach, H., Collins, F. S., Wasmuth, J. J., Frontali, M., and Gusella, J. F. (1992). The Huntington's disease candidate region exhibits many different haplotypes. *Nat. Genet.* **1,** 99–103.
37. Rubinsztein, D. C., Leggo, J., Barton, D. E., and Ferguson-Smith, M. A. (1993). Site of (CCG) expansion in huntington gene. *Nat. Genet.* **5,** 214–215.
38. Andrew, S. E., Goldberg, Y. P., Theilmann, J., Ziesler, J., and Hayden, M. R. (1994). A CCG repeat polymorphism adjacent to the CAG repeat in the Huntington disease gene: implications for diagnostic accuracy and predictive testing. *Hum. Mol. Genet.* **3,** 65–67.
39. Squitieri, F., Andrew, S. E., Goldberg, Y. P., Kremer, B., Spence, N., Zeisler, J., Nichol, K., Theilmann, J., Greenberg, J., Goto, J., Kanazawa, I., Vesa, J., Peltonen, L., Almqvist, E., Anvret, M., Telenius, H., Lin, B., Napolitano, G., Morgan, K., and Hayden, M. R. (1994). DNA haplotype analysis of Huntington disease reveals clues to the origins and mechanisms of CAG expansion and reasons for geographic variations of prevalence. *Hum. Mol. Genet.* **3,** 2103–2114.
40. Rubinsztein, D. C., Leggo, J., Goodburn, S., Barton, D. E., and Ferguson-Smith, M. A. (1995). Haplotype analysis of the Δ2642

and $(CAG)_n$ polymorphisms in the Huntington's disease (HD) gene provides an explanation for the apparent 'founder' HD haplotype. *Hum. Mol. Genet.* **4,** 203–206.

41. Almqvist, E., Spence, N., Nichol, K., Andrew, S. E., Vesa, J., Peltonen, L., Anvret, M., Goto, J., Kanazawa, I., Goldberg, Y. P., and Hayden, M. R. (1995). Ancestral differences in the distribution of the δ2642 glutamic acid polymorphism is associated with varying CAG repeat lengths on normal chromosomes: insights in to the genetic evolution of Huntington disease. *Hum. Mol. Genet.* **4,** 207–214.
42. Richards, R. I., Holman, K., Friend, K., Kremer, E., Hillen, D., Staples, A., Brown, W. T., Goonewardena, P., Tarleton, J., Schwartz, C., and Sutherland, G. R. (1992). Evidence of founder chromosomes in fragile X syndrome. *Nat. Genet.* **1,** 257–260.
43. Oudet, C., Mornet, E., Serre, J. L., Thomas, F., Lentes-Zengerling, S., Kretz, C., Deluchat, C., Tejada, I., Boué, J., Boué, A., and Mandel, J. L. (1993). Linkage disequilibrium between the fragile X mutation and two closely linked CA repeats suggests that fragile X chromosomes are derived from a small number of founder chromosomes. *Am. J. Hum. Genet.* **52,** 297–304.
44. Kunst, C. B., and Warren, S. T. (1994). Cryptic and polar variation of the fragile X repeat could result in predisposing normal alleles. *Cell* **77,** 853–861.
45. Fu, Y.-H., Kuhl, D., Pizzuti, A., Pierretti, M., Sutcliffe, J. S., Richards, S., Verkerk, A. J. MA. JM. H., Holden, J. J. AJ. J., Fenwick, R. G., Jr., Warren, S. T., Oostra, B. A., Nelson, D. L., and Caskey, C. T. (1991). Variation of the CGG repeat at the fragile X site results in genetic instability: resolution of the sherman paradox. *Cell* **67,** 1047–1058.
46. Snow, K., Doud, L. K., Hagerman, R., Pergolizzi, R. G., Erster, S. H., and Thibodeau, S. N. (1993). Analysis of a CGG sequence at the FMR-1 locus in Fragile X families and in the general population. *Am. J. Hum. Genet.* **53,** 1217–1228.
47. Reiss, A. L., Kazazian, H. H., Krebs, C. M., McAughan, A., Boehm, C. D., Abrams, M. T., and Nelson, D. L. (1994). Frequency and stability of the fragile X premutation. *Hum. Mol. Genet.* **3,** 393–398.
48. Goldberg, Y. P., Kremer, B., Andrew, S. E., Theilmann, J., Graham, R. K., Squitieri, F., Telenius, H., Adam, S., Sajoo, A., Starr, E., Heiberg, A., Wolff, G., and Hayden, M. R. (1993). Molecular analysis of new mutations for Huntington's disease: intermediate alleles and sex of origin effects. *Nat. Genet.* **5,** 174–179.
49. Brunner, H. G., Brüggewirth, H. T., Nillesen, W., Jansen, G., Hamel, B. C. J., Hoppe, R. L. E., de Die, C. E. M., Höweler, C. J., van Oost, B. A., Wieringa, B., Ropers, H. H., and Smeets, H. J. M. (1993). Influence of sex of the transmitting parent as well as of parental allele size on the CTG expansion in myotonic dystrophy (DM). *Am. J. Hum. Genet.* **53,** 1016–1023.
50. Redman, J. B., Fenwick, R. G., Jr., Fu, Y.-H., Pizzuti, A., and Caskey, C. T. (1993). Relationship between parental trinucleotide GCT repeat length and severity of myotonic dystrophy in offspring. *JAMA* **269,** 1960–1965.
51. Yamagata, H., Miki, T., Sakoda, S.-I., Yamanaka, N., Davies, J., Shelbourne, P., Kubota, Y., Takenaga, S., Nakagawa, M., Ogihara, T., and Johnson, K. (1994). Detection of a premutation in Japanese myotonic dystrophy. *Hum. Mol. Genet.* **3,** 819–820.
52. Weber, J. L. (1990). Informativeness of human $(dC\text{-}dA)_n$-$(dG\text{-}dT)_n$ polymorphisms. *Genomics* **7,** 524–530.
53. Chung, M.-Y., Ranum, L. P. W., Duvick, L. A., Servadio, A., Zoghbi, H. Y., and Orr, H. T. (1993). Evidence for a mechanism predisposing to intergenerational CAG repeat instability in spinocerebellar ataxia type I. *Nat. Genet.* **5,** 254–258.
54. Imbert, G., Saudou, F., Yvert, G., Devys, D., Trottier, Y., Garnier, J.-M., Weber, C., Mandel, J.-L., Cancel, G., Abbas, N., Durr, A., Didierjean, O., Stevanin, G., Agid, Y., and Brice, A. (1996). Cloning of the gene for spinocerebellar ataxia 2 reveals a locus with high sensitivity to expanded CAG/glutamine repeats. *Nat. Genet.* **14,** 285–291.
55. Pulst, S.-M., Nechiporuk, A., Nechiporuk, T., Gispert, S., Chen, X.-N., Lopes-Cendes, I., Pearlman, S., Starkman, S., Orozco-Diaz, G., Lunkes, A., DeJong, P., Rouleau, G. A., Auburger, G., Korenberg, J. R., Figueroa, C., and Sahba, S. (1996). Moderate expansion of a normally biallelic trinucleotide repeat in spinocerebellar ataxia type 2. *Nat. Genet.* **14,** 268–276.
56. Sanpei, K., Takano, H., Igarashi, S., Sato, T., Oyake, M., Sasaki, H., Wakisaka, A., Tashiro, K., Ishida, Y., Ikeuchi, T., Koide, R., Saito, M., Sato, A., Tanaka, T., Hanyu, S., Takiyama, Y., Nishizawa, M., Shimizu, N., Nomura, Y., Segawa, M., Iwabuchi, K., Eguchi, I., Takahashi, H., and Tsuji, S. (1996). Identification of the spinocerebellar ataxia type 2 gene using a direct identification of repeat expansion and cloning technique, DIRECT. *Nat. Genet.* **14,** 277–284.
57. Leeflang, E. P., and Arnheim, N. (1995). A novel repeat structure at the myotonic dystrophy locus in a 37 repeat allele with unexpected high stability. *Hum. Mol. Genet.* **4,** 135–136.
58. Telenius, H., Almqvist, E., Kremer, B., Spence, N., Squitieri, F., Nichol, K., Grandel, U., Starr, E., Benjamin, C., Castaldo, I., Calabrese, O., Anvret, M , Goldberg, Y. P., and Hayden, M. R. (1995). Somatic mosaicism in sperm is associated with intergenerational $(CAG)_n$ changes in Huntington disease. *Hum. Mol. Genet.* **4,** 189–195.
59. Monckton, D. G., Wong, L.-J. C., Ashizawa, T., and Caskey, C. T. (1995). Somatic mosaicism, germline expansions, germline reversions, and intergenerational reductions in myotonic dystrophy males: small pool PCR analyses. *Hum. Mol. Genet.* **4,** 1–8.
60. Watanabe, M., Abe, K., Aoki, M., Yasuo, K., Itoyama, Y., Shoji, M., Ikeda, Y., Iizuki, T., Ikeda, M., Shizuka, M., Mizushima, K., and Hirai, S. (1996). Mitotic and meiotic stability of the CAG repeat in the X-linked spinal and bulbar muscular atrophy gene. *Clin. Genet.* **50,** 133–137.
61. Igarishi, S., Takiyama, Y., Cancel, G., Rogaeva, E. A., Sasaki, H., Wakisaka, A., Zhou, Y.-X., Takano, H., Endo, K., Sanpei, K., Oyake, M., Tanaka, H., Stevanin, G., Abbas, N., Dürr, A., Rogaev, E. I., Sherrington, R., Tsuda, T., Ikeda, M., Cassa, E., Nishizawa, M., Benomar, A., Julien, J., Weissenbach, J., Wang, G.-X., Agid, Y., St. George-Hyslop, P. H., Brice, A., and Tsuji, S. (1996). Intergenerational instability of the CAG repeat of the gene for Machado-Joseph disease is affected by the genotype of the normal chromosome: implications for the molecular mechanisms of the instability of the CAG repeat. *Hum. Mol. Genet.* **5,** 923–932.
62. Takiyama, Y., Sakoe, K., Soutome, M., Namekawa, M., Ogawa, T., Nakano, I., Igarishi, S., Oyake, M., Tanaka, H., Tsuji, S., and Nishizawa, M. (1997). Single sperm analysis of the CAG repeats in the gene for Machado-Joseph disease (MJD1): evidence for non-Mendelian transmission of the MJD1 gene and for the effect of the intragenic CGG/GGG polymorphism on the intergenerational instability. *Hum. Mol. Genet.* **6,** 1063–1068.
63. Edwards, A., Hammond, H. A., Jin, L., Caskey, C. T., and Chakraborty, R. (1992). Genetic variation at five trimeric and tetrameric tandem repeat loci in four human population groups. *Genomics* **12,** 241–253.
64. Zhang, L., Leeflang, E. P., Yu, J., and Arnheim, N. (1994). Studying human mutations by sperm typing: instability of CAG trinucleotide repeats in the human androgen receptor gene. *Nat. Genet.* **7,** 531–535.
65. Zhang, L., Fischbeck, K. H., and Arnheim, N. (1995). CAG repeat length variation in sperm from a patient with Kennedy's disease. *Hum. Mol. Genet.* **4,** 303–305.

66. Leeflang, E. P., Zhang, L., Tavaré, S., Hubert, R., Srinidhi, J., MacDonald, M. E., Myers, R. H., de Young, M., Wexler, N. S., Gusella, J. F., and Arnheim, N. (1995). Single sperm analysis of the trinucleotide repeats in the Huntington's disease gene: quantification of the mutation frequency spectrum. *Hum. Mol. Genet.* **4,** 1519–1526.
67. Barceló, J. M., Mahadevan, M. S., Tsilfidis, C., MacKenzie, A. E., and Korneluk, R. G. (1993). Intergenerational stability of the myotonic dystrophy protomutation. *Hum. Mol. Genet.* **2,** 705–709.
68. Heitz, D., Devys, D., Imbert, G., Kretz, C., and Mandel, J.-L. (1992). Inheritance of the fragile X syndrome: size of the fragile X premutation is a major determinant of the transition to full mutation. *J. Med. Genet.* **29,** 794–801.
69. Harley, H. G., Rundle, S. A., MacMillan, J. C., Myring, J., Brook, J. D., Crow, S., Reardon, W., Fenton, I., Shaw, D. J., and Harper, P. S. (1993). Size of the unstable CTG repeat sequence in relation to phenotype and parental transmission in myotonic dystrophy. *Am. J. Hum. Genet.* **52,** 1164–1174.
70. Carey, N., Johnson, K., Nokelainen, P., Peltonen, L., Savontaus, M.-L., Juvonen, V., Anvret, M., Grandell, U., Chotai, K., Robertson, E., Middleton-Price, H., and Malcolm, S. (1994). Meiotic drive at the myotonic dystrophy locus? *Nat. Genet.* **6,** 117–118.
71. Gennarelli, M., Dallapiccola, B., Baiget, M., Martorell, L., and Novelli, G. (1994). Meiotic drive at the myotonic dystrophy locus. *J. Med. Genet.* **31,** 980.
72. Hurst, G. D. D., Hurst, L. D., and Barrett, J. A. (1995). Meiotic drive and myotonic dystrophy. *Nat. Genet.* **10,** 132–133.
73. Shaw, A. M., Bernetson, R. A., Phillips, M. F., Harper, P. S., and Harley, H. G. (1995). Evidence for meiotic drive at the myotonic dystrophy locus. *J. Med. Genet.* **32,** 145.
74. Leeflang, E. P., McPeek, M. S., and Arnheim, N. (1996). Analysis of meiotic segregation, using single-sperm typing: meiotic drive at the myotonic dystrophy locus. *Am. J. Hum. Genet.* **59,** 896–904.
75. Ikeuchi, T., Igarashi, S., Takiyama, Y., Onodera, O., Oyake, M., Takano, H., Koide, R., Tanaka, H., Tsuji, S. (1996). Non-Mendelian transmission in dentatorubral-pallidoluysian atrophy and Machado-Joseph disease: the mutant allele is preferentially transmitted in male meiosis. *Am. J. Hum. Genet.* **58,** 730–733.
76. Ashizawa, T., Dubel, J. R., and Harati, Y. (1993). Somatic instability of CTG repeat in myotonic dystrophy. *Neurology* **43,** 2674–2678.
77. Lavedan, C., Hoffman-Radvanyi, H., Shelbourne, P., Rabes, J.-P., Duros, C., Savoy, D., Dehaupas, I., Luce, S., Johnson, K., and Junien, C. (1993). Myotonic dystrophy: size- and sex-dependent dynamics of CTG meiotic instability, and somatic mosaicism. *Am. J. Hum. Genet.* **52,** 875–883.
78. Thornton, C. A., Johnson, K., and Moxley III, R. T. (1994). Myotonic dystrophy patients have larger CTG expansions in skeletal muscle than in leukocytes. *Ann. Neurol.* **35,** 104–107.
79. Jansen, G., Willems, P., Coerwinkel, M., Nillesen, W., Smeets, H., Vits, L., Höweler, C., Brunner, H., and Wieringa, B. (1994). Gonosomal mosaicism in myotonic dystrophy patients: involvement of mitotic events in $(CTG)_n$ repeat variation and selection against extreme expansion in sperm. *Am. J. Hum. Genet.* **54,** 575–585.
80. Kinoshita, M., Takahashi, R., Hasegawa, T., Komori, T., Nagasawa, R., Hirose, K., and Tanabe, H. (1996). $(CTG)_n$ expansions in various tissues from a myotonic dystrophy patient. *Muscle Nerve* **19,** 240–242.
81. Anvret, M., Åhlberg, G., Grandell, U., Hedberg, B., Johnson, K., and Edström, L. (1993). Larger expansion of the CTG repeat in muscle compared to lymphocytes from patients with myotonic dystrophy. *Hum. Mol. Genet.* **2,** 1397–1400.
82. Zatz, M., Passos-Bueno, M. R., Cerqueira, A., Marie, S. K., Vainzof, M., and Pavanello, R. C. M. (1995). Analysis of the CTG repeat in skeletal muscle of young and adult myotonic dystrophy patients: when does the expansion occur? *Hum. Mol. Genet.* **4,** 401–406.
83. Telenius, H., Kremer, B., Goldberg, Y. P., Thielmann, J., Andrew, S. E., Zeisler, J., Adam, S., Greenberg, C., Ives, E. J., Clarke, L. A., and Hayden, M. R. (1994). Somatic and gonadal mosaicism of the Huntington disease gene CAG repeat in brain and sperm. *Nat. Genet.* **6,** 409–414.
84. Hecht, B. K., Donnelly, A., Gedeon, A. K., Byard, R. W., Haan, E. A., and Mulley, J. C. (1993). Direct molecular diagnosis of myotonic dystrophy. *Clin. Genet.* **43,** 276–285.
85. Norbury, G., Kiernan, E., and Miciak, A. (1993). Analysis of the expansion associated with myotonic dystrophy in various tissues from fetuses predicted to be at high risk. *J. Med. Genet.* **30,** 341.
86. Sabourin, L. A., Mahadevan, M. S., Narang, M., Lee, D. S. CD. S., Surh, L. C., and Korneluk, R. G. (1993). Effect of the myotonic dystrophy (DM) mutation on mRNA levels of the DM gene. *Nat. Genet.* **4,** 233–238.
87. Tachi, N., Ohya, K., Chiba, S., Sato, T., and Kikuchi, T. (1995). Minimal somatic instability of CTG repeat in congenital myotonic dystrophy. *Pediatr. Neurol.* **12,** 81–83.
88. Wöhrle, D., Kennerknecht, I., Wolf, M., Enders, H., Schwemmle, S., and Steinbach, P. (1995). Heterogeneity of DM kinase repeat expansion in different fetal tissues and further expansion during cell proliferation *in vitro*: evidence for a causal involvement of methyl-directed DNA mismatch repair in triplet repeat stability. *Hum. Mol. Genet.* **4,** 1147–1153.
89. Martorell, L., Johnson, K., Boucher, C. A., and Baiget, M. (1997). Somatic instability of the myotonic dystrophy $(CTG)_n$ repeat during human fetal development. *Hum. Mol. Genet.* **6,** 877–880.
90. Wöhrle, D., Hennig, I., Vogel, W., and Steinbach, P. (1993). Mitotic stability of fragile X mutations in differentiated cells indicates early post-conceptional trinucleotide repeat expansion. *Nat. Genet.* **4,** 140–142.
91. Gibbs, M., Collick, A., Kelly, R. G., and Jeffreys, A. (1993). A tetranucleotide repeat mouse minisatellite displaying substantial somatic instability during early preimplantation development. *Genomics* **17,** 121–128.
92. Wong, L.-C. J., Ashizawa, T., Monckton, D. G., Caskey, C. T., and Richards, C. S. (1995). Somatic heterogeneity of the CTG repeat in myotonic dystrophy is age and size dependent. *Am. J. Hum. Genet.* **56,** 114–122.
93. Martorell, L., Martinez, J. M., Carey, N., Johnson, K., and Baiget, M. (1995). Comparison of CTG repeat length expansion and clinical progression of myotonic dystrophy over a five year period. *J. Med. Genet.* **32,** 593–596.
94. López de Munain, A., Cobo, A. M., Huguet, E., Marri Massó, J. F., Johnson, K., and Baiget, M. (1994). CTG trinucleotide repeat variability in identical twins with myotonic dystrophy. *Ann. Neurol.* **35,** 374–375.
95. Ashizawa, T., Monckton, D. G., Vaishnav, S, Patel, B. J., Voskova, A., and Caskey, C. T. (1996). Instability of the expanded $(CTG)_n$ repeats in the myotonin protein kinase gene in cultured lymphoblastoid cell lines from patients with myotonic dystrophy. *Genomics* **36,** 47–53.
96. Giordano, M., De Angelis, M. S., Mutani, R., and Richiardi, P. M. (1994). Origin of a regressed myotonic dystrophy allele. *J. Med. Genet.* **31,** 130–132.
97. Brunner, H. G., Jansen, G., Nillesen, W., Nelen, M. R., de Die, C. E. M., Höweler, C. J., van Oost, B. A., Wieringa, B., Ropers,

H.-H., Smeets, H. J. M. (1993). Reverse mutation in myotonic dystrophy. *N. Engl. J. Med.* **328,** 476–480.

98. Wieringa, B. (1994). Myotonic dystrophy reviewed: back to the future? *Hum. Mol. Genet.* **3,** 1–7.

99. Reyniers, E., Vits, L., De Boulle, K., Van Roy, B., Van Velzen, D., de Graaff, E., Verkerk, A. J. M. H., Jorens, H. Z. J., Darby, J. K., Oostra, B., and Willems, P. J. (1993). The full mutation in the FMR-1 gene in male fragile X patients is absent in their sperm. *Nat. Genet.* **4,** 143–146.

100. Zheng, C.-J., Byers, B., and Moolgavkar, S. H. (1993). Allelic instability in mitosis: a unified model for dominant disorders. *Proc. Natl. Acad. Sci. USA* **90,** 10178–10182.

101. Tsilfidis, C., MacKenzie, A. E., Mettler, G., Barceló, J., and Korneluk, R. G. (1992). Correlation between CTG trinucleotide repeat length and frequency of severe congenital myotonic dystrophy. *Nat. Genet.* **1,** 192–195.

102. Hunter, A., Tsilfidis, C., Mettler, G., Jacob, P., Mahadevan, M., Surh, L., and Korneluk, R. G. (1992). The correlation of age of onset with CTG trinucleotide repeat amplification in myotonic dystrophy. *J. Med. Genet.* **29,** 774–779.

103. Shelbourne, P., Winqvist, R., Kunert, E., Davies, J., Leisti, J., Thiele, H., Bachmann, H., Buxton, J., Williamson, B., and Johnson, K. (1992). Unstable DNA may be responsible for the incomplete penetrance of the myotonic dystrophy phenotype. *Hum. Mol. Genet.* **1,** 467–473.

104. Ashizawa, T., Dubel, J. R., Dunne, P. W., Dunne, C. J., Fu, Y.-H., Pizzuti, A., Caskey, C. T., Boerwinkle, E., Perryman, M. B., Epstein, H. F., and Hejtmancik, J. F. (1992). Anticipation in myotonic dystrophy. II. Complex relationships between clinical findings and structure of the GCT repeat. *Neurology* **42,** 1877–1883.

105. Bell, J. (1947). Dystrophia myotonica and allied diseases. *In* "Treasury of Human Inheritance," Vol. 4. Cambridge University Press, Cambridge

106. López de Munain, A., Cobo, A. M., Poza, J. J., Navarrete, D., Martorell, L., Palau, F., Emparanza, J. I., and Baiget, M. (1995). Influence of the sex of the transmitting grandparent in congenital myotonic dystrophy. *J. Med. Genet.* **32,** 689–691.

107. Lavedan, C., Hoffman-Radvanyi, H., Rabes, J. P., Roume, J., and Junien, C. (1993). Different sex-dependent constraints in CTG length variation as explanation for congenital myotonic dystrophy. *Lancet* **341,** 237.

108. Mulley, J. C., Staples, A., Donnelly, A., Gedeon, A. K., Hecht, E. K., Nicholson, G. A., Haan, E. A., and Sutherland, G. R. (1993). Explanation for exclusive maternal origin for congenital form of myotonic dystrophy. *Lancet* **341,** 236–237.

109. Cobo, A. M., Baiget, M., Löpez de Munain, A., Poza, J. J., Emparanza, J. I., and Johnson, K. (1993). Sex-related difference in intergenerational expansion of myotonic dystrophy gene. *Lancet* **341,** 1159–1160.

110. Ashizawa, T., Dunne, P. W., Ward, P. A., Seltzer, W. K., and Richards, C. S. (1994). Effects of the sex of myotonic dystrophy patients on the unstable triplet repeat in their affected offspring. *Neurology* **44,** 120–122.

111. Ashizawa, T., Anvret, M., Baiget, M., Barceló, J. M., Brunner, H., Cobo, A. M., Dallapiccola, B., Fenwick, R. G., Jr., Grandell, J., Harley, H., Junien, C., Koch, M. C., Korneluk, R. G., Lavedan, C., Miki, T., Mulley, J. C., Löpez de Munain, A., Novelli, G., Roses, A. D., Seltzer, W. K., Shaw, D. J., Smeets, H., Sutherland, G. R., Yamagata, H., and Harper, P. S. (1994). Characteristics of intergenerational contractions of the CTG repeat in myotonic dystrophy. *Am. J. Hum. Genet.* **54,** 414–423.

112. Cobo, A. M., Poza, J. J., Martorell, L., López de Munain, A., Emparanza, J. I., and Baiget, M. (1995). Contribution of molecular analyses to the estimation of the risk of congenital myotonic dystrophy. *J. Med. Genet.* **32,** 105–108.

113. O'Hoy, K. L., Tsilfidis, C., Mahadevan, M. S., Neville, C. E., Barceló, J., Hunter, A. G. WA. G., and Korneluk, R. G. (1993). Reduction in size of the myotonic dystrophy trinucleotide repeat mutation during transmission. *Science* **259,** 809–812.

114. Gacy, A. M., Goellner, G., Juranic, N., Macura, S., and McMurray, C. T. (1995). Trinucleotide repeats that expand in human disease form hairpin structures *in vitro*. *Cell* **81,** 533–540.

115. Passos-Bueno, M. R., Cerqueira, A., Mainzof, M., Marie, S. K., and Zatz, M. (1994). Myotonic dystrophy: genetic, clinical, and molecular analysis of patients from 41 Brazilian families. *J. Med. Genet.* **32,** 14–18.

116. Koch, M. C., Grimm, T., Harley, H. G., and Harper, P. S. (1991). Genetic risks for children of women with myotonic dystrophy. *Am. J. Hum. Genet.* **48,** 1084–1091.

117. Harper, P. S., and Dyken, P. R. (1972). Early onset dystrophia myotonica-evidence supporting a maternal environmental factor. *Lancet* **2,** 53–55.

118. Farkas-Bargeton, E., Barbet, J. P., Dancea, S., Wehrle, R., Checouri, A., and Dulac, O. (1988). Immaturity of muscle fibres in the congenital form of myotonic dystrophy: its consequences and its origin. *J. Neurol. Sci.* **83,** 145–159.

119. Nakagawa, M., Yamada, H., Higuchi, I., Yamanashi, Y., Miki, T., Johnson, K., and Osame, M. (1994). A case of paternally inherited congenital myotonic dystrophy. *J. Med. Genet.* **31,** 397–400.

120. Bergoffen, J., Kant, J., Sladky, J., McDonald-McGinn, D., Zackai, E. H., and Fischbeck, K. H. (1994). Paternal transmission of congenital myotonic dystrophy. *J. Med. Genet.* **31,** 518–520.

121. Ohya, K., Tachi, N., Chiba, S., Sato, T., Kon, S., Kikuchi, K., Imamura, S., Yamagata, H., and Miki, T. (1994). Congenital myotonic dystrophy transmitted from an asymptomatic father with a DM-specific gene. *Neurology* **44,** 1958–1960.

122. Shaw, D. J., Chaudhary, S., Rundle, S. A., Crow, S., Brook, J. D., Harper, P. S., and Harley, H. G. (1993). A study of DNA methylation in myotonic dystrophy. *J. Med. Genet.* **30,** 189–192.

123. Jansen, G., Bartolomei, M., Kalscheuer, V., Merkx, G., Wormskamp, N., Mariman, E., Smeets, D., Ropers, H.-H., and Wieringa, B. (1993). No imprinting involved in the expression of DM-kinase mRNAs in mouse and human tissues. *Hum. Mol. Genet.* **2,** 1221–1227.

124. Thornton, C. A., Griggs, R. C., and Moxley, R. T. (1994). Myotonic dystrophy with no trinucleotide repeat expansion. *Ann. Neurol.* **35,** 269–272.

125. Fu, Y.-H., Friedman, D. L., Richards, S., Pearlman, J. A., Gibbs, R. A., Pizzuti, A., Ashizawa, T., Perryman, M. B., Scarlato, G., Fenwick, R. G., Jr., and Caskey, C. T. (1993). Decreased expression on myotonin-protein kinase messenger RNA and protein in adult form of myotonic dystrophy. *Science* **260,** 235–238.

126. Novelli, G., Gennarelli, M., Zelano, G., Pizzuti, A., Fattorini, C., Caskey, C. T., and Dallapiccola, B. (1993). Failure in detecting mRNA transcripts from the mutated allele in myotonic dystrophy muscle. *Biochem. Mol. Biol. Int.* **29,** 291–297.

127. Carango, P., Noble, J. E., Marks, H. G., and Funanage, V. L. (1993). Absence of myotonic dystrophy protein kinase (DMPK) mRNA as a result of a triplet repeat expansion in myotonic dystrophy. *Genomics* **18,** 340–348.

128. Hoffmann-Radvanyi, H., Lavedan, C., Rabès, J.-P., Savoy, D., Duros, C., Johnson, K., and Junien, C. (1993). Myotonic dystrophy: absence of CTG enlarged transcript in congenital forms, and low expression of the normal allele. *Hum. Mol. Genet.* **2,** 1263–1266.

129. Bhagwati, S., Ghatpande, A., and Leung, B. (1996). Normal levels of DM RNA and myotonin protein kinase in skeletal muscle from adult myotonic dystrophy (DM) patients. *Biochim. Biophys. Acta* **1317,** 155–157.
130. Otten, A. D., and Tapscott, S. J. (1995). Triplet repeat expansion in myotonic dystrophy alters the adjacent chromatin structure. *Proc. Natl. Acad. Sci. USA* **92,** 5465–5469.
131. Wang, J., Pegoraro, E., Menegazzo, E., Gennarelli, M., Hoop, R. C., Angelini, C., and Hoffman, E. P. (1995). Myotonic dystrophy: evidence for a possible dominant-negative RNA mutation. *Hum. Mol. Genet.* **4,** 599–606.
132. Krahe, R., Ashizawa, T., Abbruzzese, C., Roeder, E., Carango, P., Giacanelli, M., Funanage, V. L., and Siciliano, M. J. (1995). Effect of myotonic dystrophy trinucleotide repeat expansion on DMPK transcription and processing. *Genomics* **28,** 1–14.
133. Taneja, K. L., McCurrach, M., Schalling, M., Housman, D., and Singer, R. H. (1995) Foci of trinucleotide repeat transcripts in nuclei of myotonic dystrophy cells and tissues. *J. Cell. Biol.* **128,** 995–1002.
134. Davis, B. M., McCurrach, M. E., Taneja, K. L., Singer, R. H., and Housman, D. E. (1997). Expansion of a CUG trinucleotide repeat in the 3′ untranslated region of myotonic dystrophy kinase transcripts results in nuclear retention of transcripts. *Proc. Natl. Acad. Sci. USA* **94,** 7388–7393.
135. Roses, A. D. (1994). Muscle biochemistry and a genetic study of myotonic dystrophy. *Science* **264,** 587.
136. Morrone, A., Pegararo, E., Angelini, C., Zammarchi, E., Marconi, G., and Hoffman, E. P. (1997). RNA metabolism in myotonic dystrophy. *J. Clin. Invest.* **99,** 1691–1698.
137. Timchenko, L. T., Timchenko, N. A., Caskey, C. T., and Roberts, R. (1996). Novel proteins with binding specificity for DNA CTG repeats and RNA CUG repeats: implications for myotonic dystrophy. *Hum. Mol. Genet.* **5,** 115–121.
138. Timchenko, L. T., Miller, J. W., Timchenko, N. A., DeVore, D. R., Datar, K. V., Lin, L., Roberts, R., Caskey, C. T., and Swanson, M. S. (1996). Identification of a $(CUG)_n$ triplet repeat RNA-binding protein and its expression in myotonic dystrophy. *Nucleic Acids Res.* **24,** 4407–4414.
139. Bhagwati, S., Ghatpande, A., and Leung, B. (1996). Identification of two nuclear proteins which bind to RNA CUG repeats: significance for myotonic dystrophy. *Biochem. Biophys. Res. Commun.* **228,** 55–62.

Chapter 11

Biochemical Studies of DM Protein Kinase (DMPK)

Henry F. Epstein and Shida Jin
Department of Neurology, Baylor College of Medicine, Houston, Texas 77030

I. INTRODUCTION

The identification of the myotonic dystrophy locus (DM) by multiple groups in 1992 created a great deal of excitement [9, 26]. DM was of special interest for several reasons. DM is the first neuromuscular disease recognized to be an inherited trait [13]. DM was the most common form of adult muscular dystrophy [30]. DM was the third human disorder to be associated with triplet repeat expansion mutations [37, 68]. Most importantly from the viewpoint of this chapter, DM is the first inherited human disorder to be associated with a gene encoding a protein kinase.

A. Characterization of DM Locus and Predicted Product

The protein product of the DM locus was predicted on the basis of common amino acid sequence to be a member of the serine/threonine protein kinase family. With the then-available databases, the SP6 kinase, the cyclic AMP kinases (PKAs), and the protein kinase C group (PKCs) were considered the closest relatives of the catalytic domain near the amino terminal [9, 26]. Further analysis (Fig. 11-1) revealed a "myosin-like" coiled-coil region and a potential membrane-spanning helical tail in the more carboxyl regions in the full-length protein product [67]. More current and detailed analyses will be presented later.

B. General Aspects of Protein Kinases

Protein kinases represent one of the largest families of proteins and the genes encoding them. Protein kinases are present in all cellular forms of life. The genome of a typical multicellular organism appears to maintain over 1000 genes encoding distinct protein kinases. A protein kinase is chemically an ATP:protein phosphotransferase and requires Mg^{2+} as a cofactor and a second substrate, the protein or peptide. This latter requirement distinguishes all kinases from ATPases which as ATP:hydrolases do not require a transferring group. In addition to the serine/threonine kinase class, there are tyrosine protein kinases, mixed tyrosine/serine protein kinases, and histidine protein kinases. Within the serine/threonine class, there are several well-established, large groups including the PKAs, the PKCs, the calcium/calmodulin kinases, and the casein kinases. Many other smaller groups have now been recognized, including DMPK and its closest relatives.

The specificity of protein kinases is quite variable. The first such enzyme to be characterized was phosphorylase b kinase. This kinase shows unusually restricted substrate specificity; α-glucan (glycogen) phosphorylase and troponin I from skeletal muscle are the only effective substrates [59]. Most other protein kinases appear to be multifunctional; that is, they exhibit a broad range of substrate specificity in cell-free, purified component reactions. In addition to interacting with protein substrates, most protein kinases interact with other proteins for their own regulation. Regulatory activities include activation, inhibition, subcellular localization, and the formation of oligomeric complexes that provide a scaffold for effective enzyme:substrate interaction. Proteins which regulate protein kinases include other protein kinases, protein phosphatases, heterotrimeric G-proteins, 21-kDa GTPases, and various kinds of anchoring and scaffolding proteins. The latter kinds of proteins may enhance the specificity of interacting kinases by restricting their accessibility to potential substrates.

C. Relevant Clinical Characteristics

In evaluating the possible roles of DMPK in DM, it may be useful to consider which features of the clinical picture might be related to the action of a protein kinase. The hallmarks of the DM phenotype are pleiotropy and variability of expression. The major sites of pathology, their most common clinical signs, and the biological process most probably underlying this pathology are skeletal muscle (myotonia, muscle fiber membrane excitability), the eye lens (cataracts, developmental loss of lens fiber cell membranes), heart (conduction system

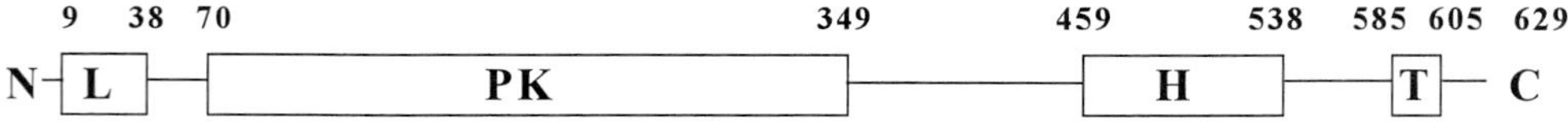

FIGURE 11-1 Schematic model of DMPK. N and C refer to the amino and carboxyl terminals. The recognized domains of DMPK are: L, leucine-rich repeat; PK, serine–threonine protein kinase catalytic domain; H, a-helical region which contains the coiled-coil, leucine zipper, and potential p21 GTPase binding regions; T, tail or ER membrane anchoring domain with its hydrophobic a-helix and flanking basic sequences. The numerals represent amino acid residues at the beginning or end of a domain. In the text, recombinant proteins are described by the domains that they contain. For example, LPKHT is full-length DMPK whereas PK is the catalytic domain alone.

blocks, fibroblast and adipocyte infiltration), and brain (mental retardation, multiple neuronal alterations). For each of these sites, the age of onset and severity of symptoms and signs can vary greatly. Virtually 100% of adults with DM show electrical myotonia and cataracts, but only 60 to 70% will show the cardiac abnormalities. Cardiac arrhythmias are the major cause of death in DM adults whereas respiratory distress secondary to muscle immaturity is the leading cause of death in neonatal DM.

A closer examination of the defects of skeletal muscle in DM reveals even more complexity. In addition to myotonia, skeletal muscle shows potential defects of differentiation and maturation in the immature fibers of the neonatally affected and the characteristic atrophy of Type I muscle fibers. Muscle weakness may arise from both the myotonia and the atrophy. About 50 to 60% of adult DM patients show insulin resistance at the level of their skeletal muscles although only a fraction develop frank diabetes [30].

How would the action of a protein kinase, or its absence, relate to this complex picture of pleiotropy and variability? It must be noted that mechanisms other than the actions of the kinase alone may be altered by the CTG expansion mutations in the sequence within the DMPK locus encoding the 3′ untranslated region and contribute significantly to the phenotype of DM. These possibilities will be discussed in other chapters. On the other hand, a potentially multifunctional serine/threonine protein kinase like DMPK which is expressed in many tissues and may have distinct forms associated with specific cell types and subcellular compartments could be related to much of the phenotypic spectrum in DM. Gain of function in DMPK may arise from altered synthesis of total DMPK, changes in the distinct isoforms arising from alternative splicing or initiation, and alterations in its regulation. These mechanisms will be discussed in more detail in the rest of this chapter.

II. MOLECULAR ANALYSIS OF DMPK STRUCTURAL DOMAINS

The close similarities in sequence between human DMPK and other kinase-encoding genes has caused some investigators to consider this group of putative enzymes "a myotonic dystrophy family of protein kinases" [75]. Conversely, no other identified putative protein kinase has all of the structural and functional properties of full-length human or murine DMPK. Each functionally or structurally defined domain in DMPK will be considered separately. In this section, we will discuss similarities and identities in sequence between DMPK and other protein kinases and their functional and structural significance.

A. Catalytic Domain

The region between residues 70 and 349 in human DMPK is closely related to many members of the family of serine/threonine protein kinases. Although originally thought to be most closely related to SP6 kinase, PKA, and PKC, DMPK now appears to represent a new group of enzymes within the serine/threonine family.

Fig 11-2 shows the relationships of sequence between several members of the "DMPK group" based on comparison of their catalytic protein kinase domains. All 11 of the previously described functional motifs of serine/threonine protein kinases are present [9]. The extents of identity and similarity of these sequences with respect to human DMPK are significantly greater than those to SP6, PKA, and PKC. The most important functional suggestion based on its similarities to known Rho A-binding kinases is that DMPK may also be activated by a member of the 21-kDa GTPase family. The DMPK group is very old phylogenetically; there are close relatives in fission yeast and *Neurospora* as well as the plant, nematode, and insect representatives described in Fig. 11-2 [34, 64, 74].

A tree of such relationships (Fig. 11-3) has been calculated on the basis of parsimony. The observed clustering is the result of phylogenetic as well as functional relationships such as between human and murine DMPK, or putatively functional as between the LET-502 protein of *C. elegans* and the rat ROK and human p160 [ROCK]. The human PK428 and *C. elegans* CEESP52F, although most closely related to human and murine DMPKs, are even more closely related to one another. This comparison suggests that the former two kinases might function more similarly to one another that to the DMPKs. Also, the catalytic domain of CEESP52F appears to be a significantly better homologue for DMPK in *C. elegans* than that of the putative RHO A kinase LET-502 (M. Shimizu, C. C. Bauer, and H. F. Epstein, unpublished results) [72].

B. Coiled-Coil, Zipper, and RHO A-Binding Domains

The carboxyl-terminal half of human DMPK, residues 352–629, is predicted to contain multiple stretches of α-helix by several algorithms (Fig. 11-4). Human and murine DMPK are 78.2% identical over the carboxyl half. The region between residues 459 and 538 is predicted by the stringent PAIRCOIL algorithm [5] to be capable of forming a coiled coil with a

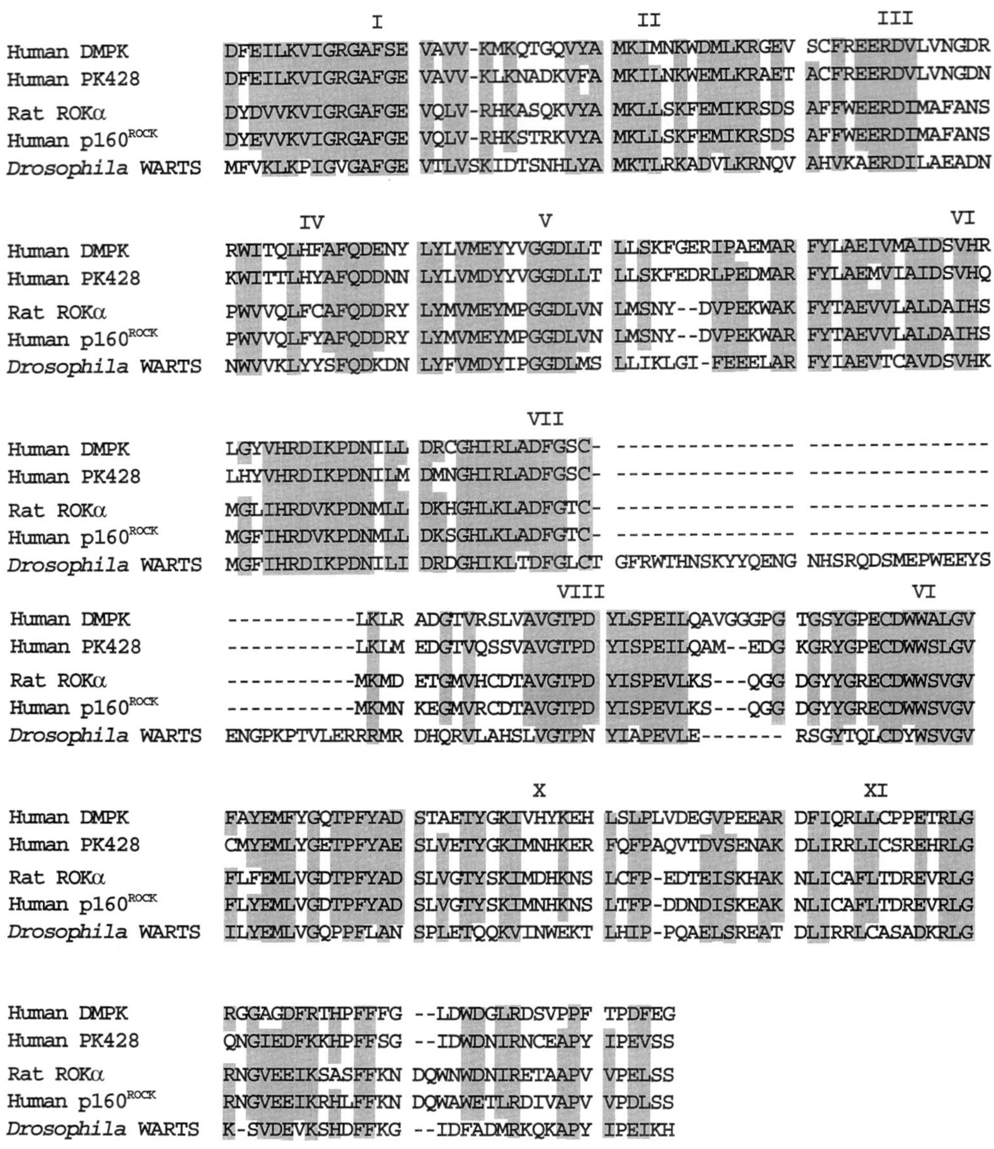

FIGURE 11-2 Comparison of the catalytic domain of DMPK to corresponding regions of related protein kinases. The alignment was performed according to PIMA [57, 58]. The roman numerals refer to the 11 subdomains of known serine–threonine protein kinases [29]. Identities or stringent similarities across 4 or 5 of the kinases are shaded. We set F = I = L = V; R = K, and D = E. The GenBank accession numbers are: Human DMPK, L19268; Human PK428, U59305; Rat ROKα, U38481; Human p160^{ROCK}, U43195; *Drosophila* WARTS, L39837.

suitably complementary α-helix and is 88.6% identical between human and mouse. Residues 503–530 in this coiled coil are compatible with a leucine zipper [73]. Such a zipper represents a shorter, less stable version of a true coiled coil and can be the basis of readily reversible associations between protein molecules [2].

This helical region also contains sequences with close similarities to functionally characterized domains which associate with RHO A in specific protein kinases and binding proteins [3, 38, 39, 70]. Potential RHO-interacting sequences include residues 415–514 and 482–538 which are 78.2% identical between human and murine DMPK (Fig. 11-5). Like the similarities of sequence in their catalytic domains, the RHO A-binding domains, putative or known, are a common feature of the DMPK group of protein kinases (Fig. 11-4). Whether RHO A or another 21-kDa GTPase protein is the physiologically significant regulator of DMPK has not been determined.

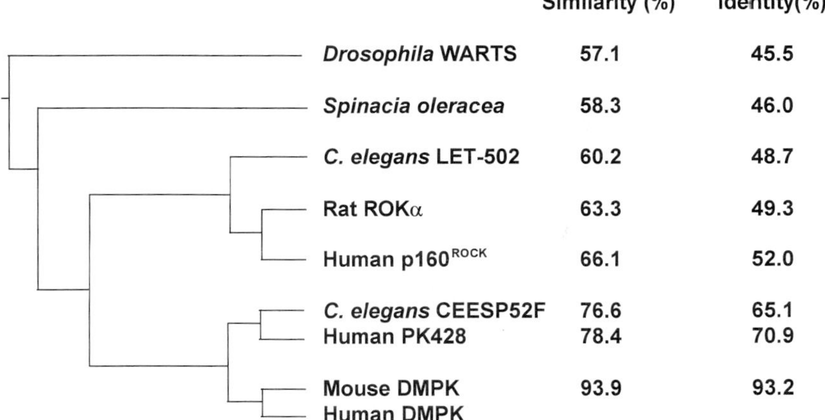

FIGURE 11-3 Tree showing sequence relationships of DMPK group of protein kinases. The amino acid sequences of the putative catalytic domains of these proteins were analyzed by PHYLIP [24], and the similarities and identities were calculated by the GCG program (Wisconsin Package Version 9.0, Genetics Computer Group, Madison, Wisconsin). The accession numbers for proteins (also see legend to Fig. 11-2) are: Mouse DMPK, X38014, X38015; *C. elegans,* CEESP52F; *C. elegans* LET-502, U85515; *Spinacia oleracea,* Z30330.

C. Amino-Terminal Leucine-Rich Repeat (LRR)

This structural motif has been identified in several proteins as a protein–protein interaction site [35]. The LRR structure has been determined by X-ray diffraction analysis of crystalline ribonuclease inhibitor. Amino acid residues 9–38 constitute a highly nonpolar region and contain an LRR (Fig. 11-6). The noncatalytic amino terminal region (residues 1–69) shows 92.8% identity between human and murine DMPKs which is not significantly different than the relationship between their catalytic domains. It is likely, therefore, that the LRR region is important for DMPK function. This region is unlikely to contain a well-defined leucine zipper or signal sequence, in contrast to some suggestions [67, 69].

D. Carboxyl-Terminal Membrane Anchor

In both human and murine DMPK, several algorithms predict a hydrophobic α-helix (Fig. 11-7). The amino acid sequence overlapping this helix in DMPK is very similar to transmembrane domains of human HMG CoA reductase and rat microsomal aldehyde dehydrogenase that anchor these enzymes to the cytoplasmic face of the endoplasmic reticulum [44]. Human and murine DMPKs are 62.5% identical for amino acid residues 585–605. Functional experiments to be discussed later suggest that full-length DMPK binds to endoplasmic reticulum [41, 69], and DMPK has been localized to the endoplasmic reticulum or membranes derived from it in a number of studies.

Just to the amino terminus of the anchoring domain, there is an as yet structurally and functionally uncharac-

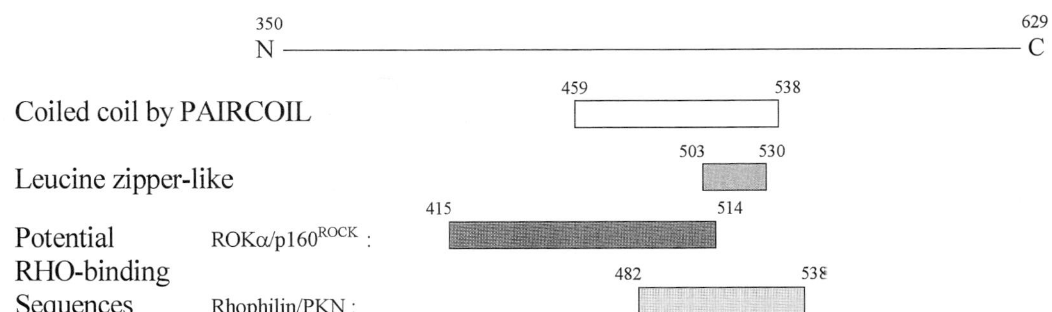

FIGURE 11-4 Analysis of the H region in DMPK. Different regions were analyzed: for coiled coil character by PAIRCOIL [5], for leucine zipper character [2], and for similarity to known RHO A binding sites as with the catalytic domains. The numerals depict the amino acid residues at the boundaries of the regions of interest.

Alignment of DMPK to RhoA-binding regions of Rhophilin and PKN (PRK)

```
Human DMPK        RQSLSREMEAIRTDNQNFASQLREAEAR..NRD...LEAHVRQLQERMELLQAEGATAVTGV
Mouse DMPK        RQSLSRELEAIRTANQNFSSQLQEAEVR..NRD...LEAHVRQLQERMEMLQAPGAAAITGV
Human PKN         RERLRREIR.KELKLKEGAENLRRATTD.LGRSLGPVELLLRGSSRRLDLLHQQLQELHAHV
Mouse Rhophilin   RARLHQQIS.KELRMRTGAENLYRATSNTWVRE..TVALELSYVNSNLQLLKEELAELSTSV
```

Alignment of DMPK to RhoA-binding regions of kinases belonging to DMPK family

```
Human DMPK    VPGPTPMEVEAEQLL EPHVQAPSLEPSVSP QDETAEVAVPAAVPA ..AEAEAEVTLRELQ
Mouse DMPK    VPDPTPMELEALQLP VSDLQGLDLQPPVSP PDQVAEEADLVAVPA PVAEAETTVTLQQLQ

ROKα          TSDVANLANEKEELN NKLKDTQEQLSKLKD EEISAA.AIKAQFEK .QLLTERTLKTQAV.
RHO-kinase    TSDVANLANEKEELN NKLKEAQEQLSRLKD EEISAA.AIKAQFEK .QLLTERTLKTQAV.
p160ROCK      TKDIEMLRKENEELN ERMRTAEEEYKLKKE EEIN...NLKAAFEK .NISTERTLKTQAV.

Human DMPK    EALEEEVLTRQSLS REMEAIRTDNQNFASQ LREAEARNRDLE
Mouse DMPK    EALEEEVLTRQSLS RELEAIRTANQNFSSQ LQEAEVRNRDLE

ROKα          NKLAEIMNRKEPVK RGSD..........TD VRRKEKENRKLH
RHO-kinase    NKLAEIMNRKEPVK RGND..........TD VRRKEKENRKLH
p160ROCK      NKLAEIMNRKD.FK IDRKKANT......QD LRKKEKENRKLQ
```

FIGURE 11-5 Comparison of the identities and stringent similarities of DMPK to the RHO A-binding domains of known interacting proteins. The analysis is the same as with Fig. 11-2. See identifying amino acid residues in Fig. 11-4.

terized region of 55 residues (536–590) in which there are only two substitutions, both highly conserved valine to isoleucine changes, between human and murine DMPK. This is the most highly conserved sequence in DMPK. Interestingly, the region which is encoded by exon 13 can be alternatively spliced to produce truncated DMPK in both humans and mice [33].

III. BIOCHEMICAL ANALYSIS OF DMPK FUNCTION

The key to most studies of biochemical function of DMPK has been the use of recombinant DNA technology to express enzyme protein in either bacterial or eukaryotic cell cultures. In the case of bacterial expression of DMPK, recombinant protein has been isolated and purified for use in direct biochemical experiments [17, 63]. In eukaryotic cells, expressed DMPK protein and its function have been studied primarily in whole cell and cell fractionation experiments [10, 42, 69]. In only one case, human heart DMPK, has the native enzyme been isolated, purified, and partially characterized [42]. A special application of recombinant expression has been the microinjection of DMPK mRNA into *Xenopus* oocytes for electrophysiological studies of potential interactions with membrane channels [47].

A. Recombinant Proteins

Different truncated and mutant versions of DMPK have also been created. For example, the protein kinase catalytic domain (PK), the catalytic and helical domains (PKH), and full-length including both the amino-terminal leucine-rich repeat and carboxyl-terminal tail domains (LPKHT) have all been constructed and expressed [10, 17, 42, 63, 69]. An important mutant is the lysine 100 substitution by alanine (K100A) which renders the enzyme inactive because the lysine is required for the binding of ATP to the catalytic center (S. Jin, A. Balasubramanyam, M. Sharma, U. P. Andley, and H. F. Epstein, manuscript in preparation; [69]).

More extensive purification and characterization has been performed with DMPK protein expressed from

```
Secondary structure prediction:
CF/GOR                α - helix         β - strand    turn              α - helix

DM kinase             M S A E V R L R R L Q Q L V L D P G F L G L E P L L D L L L G V H Q E L G A
                                        2     5   7     10  12        16        21    24
Conserved LRR motif                     L . . L . L . . * . L . . . L . . . . L . . L
```

FIGURE 11-6 Comparison of the L region in DMPK to the consensus LRR motif [35]. CF/GOR stands for Chou-Fasman/Garnier-Ogulthorpe-Robinson algorithms; and * means N or T at residue position 10.

Human DMPK	585	LSEALSLLLFAVVLSRAAALG	605
Mouse DMPK	587	LSEARCLLLFAAALAAAATLG	607
Human HMG CoA Reductase	121	LNEALPFFLLLIDLSRASALA	141
Rat msALDH	464	LQLLLLVCLVAVAAVIV	480

FIGURE 11-7 Comparison of T region of DMPK to known ER anchoring domains. Alignments, identities, and similarities are as with Figs. 11-2 and 11-5.

bacterial than eukaryotic cell cultures. The bacterial work has taken advantage of creating fusion proteins by recombinant techniques. The poly-His tag (Novagen) has been used with DMPK [17, 63]. The DMPK was affinity purified by liganding to nickel and competition with imidazole.

Other expression systems have also proved useful. DMPK has been expressed in the baculovirus/insect cell system [69], COS cells [10, 42], fibroblast lines [63], and in differentiating systems including lens (S. Jin, A. Balasubramanyam, M. Sharma, U. P. Andley, and H. F. Epstein, manuscript in preparation) and myogenic [10] cell lines. The purposes of these studies have ranged from determination of the appropriate gel mobility of DMPK to the effects of DMPK on the differentiation and proliferation of specific cell types.

The gel mobility of DMPK has proved to be a more complex issue than expected. Bacterially expressed proteins after SDS–PAGE and either direct Coomassie blue staining of the gel or antibody staining of a blot of the gel generally show a slower than predicted mobility (or higher than predicted M_r). For example, truncated forms (PKHs) that were predicted to be 55–56 kDa migrated as 65- to 66-kDa proteins [17, 27, 63]. On the other hand, full-length (LPKHT) or more markedly truncated forms (PK) migrated as expected at 72 and 45 kDa, respectively. DMPK expressed in various mammalian cells is more variable ranging from the proper M_r to significantly greater values. For full-length LPKHT constructs, 71- and 80-kDa proteins have been reported for expression in COS cells [42, 69]. For LPKH and LPKHT in baculovirus-infected SF21 insect cells, the proteins were reported as a doublet at 81, 82, and 85 kDa, respectively [69]. (In BC_3H1 cells with some myogenic potential, full-length LPKHT migrates close to expected at 69 kDa [10]).

The variability of recombinantly expressed protein in a spectrum of host cells raises important questions about the interpretations of molecular size in tissue samples. In fact, as we discuss below, such determinations have been controversial. Unfortunately, part of the problem is that a variety of different SDS–PAGE systems and distinct methods for determining M_rs have been used. Although not all of the variability can be explained in this manner, a good part of it may be.

A separate question is: what is the native molecular weight of DMPK? Sedimentation and gel filtration studies of bacterially expressed PKH indicate that the protein associates to form a 41S species with a molecular weight of 1.2×10^6. This size would suggest an oligomer of 21 DMPK monomers [17]. LPKHT and LPKH expressed in COS cells are described as high-molecular-weight and monomeric species, respectively, whereas both inserts expressed in insect cells through baculovirus infection produce high- (though distinct) molecular-weight forms [69]. The possible dependence of the intrinsic properties of DMPK monomer and its apparent self-association into oligomers upon specific intracellular interactions or modifications must be considered carefully.

B. Autophosphorylation

Recombinant DMPK expressed in either bacterial or eukaryotic cells shows significant autophosphorylation activity or transfer of phosphate from ATP to itself. Both bacterially expressed PK and PKH autophosphorylate [17]. In both cases, the phosphorylation reactions lead to a large apparent increase in M_r, consistent with an additional 10–20 kDa (Fig. 11-8). At least three phosphorylated species can be detected by SDS–PAGE in each case. The amino acids which become phosphory-

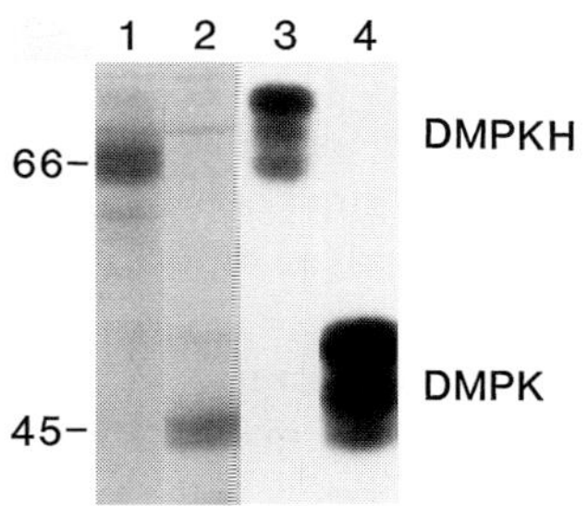

FIGURE 11-8 Autophosphorylation leads to changes in gel mobility of DMPK. The major protein bands of recombinant PKH and PK migrate with M_rs of 66,000 and 45,000, respectively. After autophosphorylation, several species appear over a 15,000 to 20,000 range of M_rs (1) PKH and (2) PK on 7.5% SDS–PAGE, Coomassie blue staining, (3) PKH and (4) PK after incubation with [γ-^{32}P]ATP and radioautography. Reprinted with permission from *Biochemistry* **33,** 10809–10814. Copyright 1994 American Chemical Society.

lated are serine and threonine. LPKH and LPKHT expressed in baculovirus-infected SF21 cells also autophosphorylate with the formation of multiple species that show significant shifts in gel mobility similar to those in Fig. 11-8 [69]. These results suggest that phosphorylation alone may shift the gel mobility of native DMPK and that multiple species of DMPK could coexist in physiological situations because of differential phosphorylation. The reported differences in the M_rs of recombinant proteins already described and those of protein bands from different tissues labeled with particular antibodies may be related at least partially to phosphorylation.

C. Transphosphorylation of Protein and Peptide Substrates

Recombinant DMPK expressed in either bacterial or eukaryotic cells shows transphosphorylation of protein and peptide substrates. The goals of these studies are to show that DMPK (1) is a true protein kinase, (2) is a serine–threonine kinase, (3) is a unique type of serine–threonine kinase, and (4) has physiologically significant substrates. The first three of these goals have been accomplished, and the last goal is the subject of ongoing studies.

The classical protein kinase substrate is the histone fraction or histone H1. DMPK constructs expressed in either recombinant bacteria or baculovirus-infected SF21 insect cells can transfer phosphate form ATP to histone [17, 69]. As with autophosphorylation, serine and threonine but not tyrosine become phosphorylated. Whether serine or threonine is the predominant target residue for transfer of phosphate appears to be variable [17, 63]. The inhibition of DMPK with the antikinase drugs H7 and H8 occurs only at significantly greater concentrations than those required for the inhibition of the known major classes of serine–threonine kinases including PKA and PKC [17]. These results indicate a functional difference in the catalytic action of DMPK from other groups of kinases which agrees with the structural distinctions based upon comparison of sequences.

The search for the physiological protein substrate or substrates is ongoing. It is important to recognize the possibility that in experiments with purified components there may not be any clearly predominant substrate. For example, RAF kinase, which in various cellular and genetic experiments clearly interacts with MEK kinase as an enzyme to substrate, only phosphorylates the latter 4- and 10-fold greater than the generic substrates for phosphorylation, myelin basic protein, and histone [25]. Recent experiments suggest that heterocomplexes of RAF kinase with the 14-3-3-1 protein may enhance its catalytic power [43].

The only published study concerning the phosphorylation of a potential physiological substrate by DMPK concerns the β-subunit of the voltage sensor channel, also known as the dihydropyridine receptor [63]. This protein is of great potential interest because of its roles in excitation–contraction coupling and as a calcium channel. A number of unpublished reports have indicated other membrane ion channels, DNA-binding transcription factors, and RNA-binding proteins as potential substrates of DMPK. Interestingly, no clear consensus phosphorylation site has yet been derived, although among the standard peptide substrates used for characterizing protein kinases, the sequence favored by PKAs shows the greatest activity with DMPK [63]. The phosphorylated PKA peptide is GRGLSLSR, and the phosphorylated β-subunit peptide is LRQSRLSSSK. From these results, the consensus sequence for DMPK substrate would be very broad. Indeed, conference reports on other potential substrates also suggest a broad substrate specificity. The possibility that there may be multiple physiologically significant substrates whose efficacy and specificity may be regulated by developmental, tissue, or transient cues must be considered seriously.

There is a single, unconfirmed report of tyrosine kinase activity associated with a skeletal muscle protein which cross-reacts with an anti-DM peptide antibody [21]. Recently, this group has suggested that DMPK is a mixed serine–threonine–tyrosine protein kinase (J. Puymirat, First International Myotonic Dystrophy Consortium Conference); however, this result has not been confirmed yet.

IV. CELLULAR INTERACTIONS OF DMPK

The potential roles of DMPK have been studied in three major areas of biological function: membrane excitation, membrane trafficking, and differentiation. These areas are clearly related to characteristic findings in DM such as myotonia, alterations in the architecture of muscle fibers and lens fiber cells, and the appearance of immature muscle fibers.

A. Membrane Excitation

The linking of myotonia to alterations in the excitability of the membrane of affected skeletal muscle fibers has been a central topic of research on the pathogenesis of DM [31]. Myotonia is the inability to relax a muscle after its voluntary contraction.

Experiments linking the expression of DMPK to alterations in the intrinsic properties of sodium channels have been performed ([47]; J. R. Miller, J. P. Mounsey, and J. R. Moorman, manuscript in preparation). As shown in Fig. 11-9, coexpression of DMPK results in a reduction in the amplitude of currents through normal skeletal muscle sodium channels expressed in *Xenopus* oocytes. A mutant sodium channel which lacks the known phosphorylation site in the inactivation gate is unaffected by DMPK coexpression. These experiments do not distinguish whether DMPK directly phosphorylates the sodium channel protein or works indirectly through another kinase. Experiments in progress indicate that the site of phosphorylation by DMPK in the sodium channel is AMKKLGSKKPQK (M. B. Perryman and J. R. Moorman, personal communication).

To further investigate these effects which may underlay the pathophysiological phenomenon of skeletal muscle myotonia in DM, Moorman and co-workers have investigated the molecular basis of the reduction in the measured amplitudes of sodium currents by studying the binding of radiolabeled saxitoxin (a specific, high-affinity inhibitor of muscle-type sodium channels) and recording the conductance of single channels by patch clamping RNA-injected oocytes. They have found that the number of saxitoxin-binding sites which represents the number of sodium channels was not changed; the probability of opening was reduced, but the conductance of single channels was unaffected as shown in Fig. 11-10 (J. R. Miller, J. P. Mounsey, and J. R. Moorman, manuscript in preparation). The authors suggest that the mechanism of this effect may involve reduction of the positive charge of the channel inactivation gate through phosphorylation.

B. Membrane Trafficking

The histopathological changes at the cellular and subcellular levels in muscle fibers [15] and lens fiber

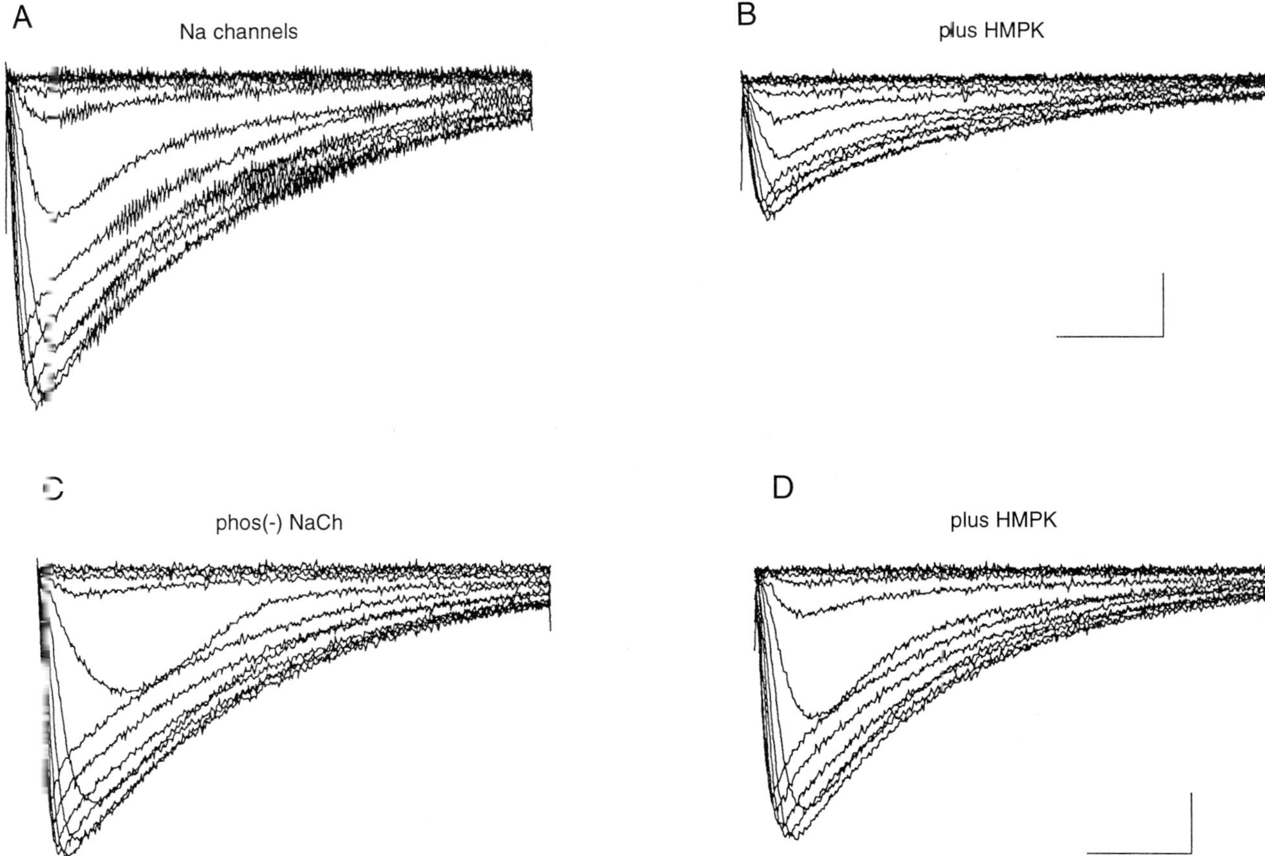

FIGURE 11-9 Modulation of skeletal muscle Na channels by DMPK. The currents are from *Xenopus* oocytes expressing rat skeletal muscle Na channels alone (A, C) or coexpressing human DMPK (HMPK) (B, D). The effect of DMPK coexpression on wild-type Na channels is a reduction in current amplitude. When a single phosphorylation site in the Na channel inactivation gate is disabled by a point mutation, coexpression of DMPK has no effect (C, D). Scale bars are 5 ms and 1 mA. Reproduced from *The Journal of Clinical Investigation* **95,** 2379–2384 (1995) by copyright permission of The American Society for Clinical Investigation.

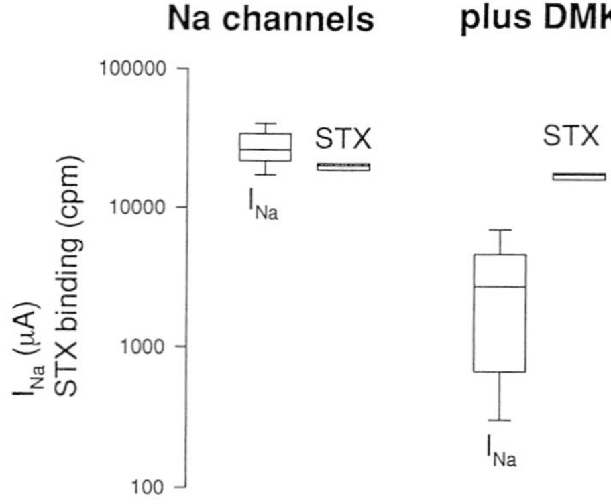

FIGURE 11-10 DMPK affects intrinsic properties of the Na channel. The effect is not mediated by a reduction in channel number, as the number of saxitoxin binding sites (STX) remains the same in oocytes expressing Na channels alone and in oocytes coexpressing sufficient DMPK to reduce current (I_{Na}) to 1/10th of the control amplitude (from J. R. Miller, J. P. Mounsey, and J. R. Moorman, manuscript in preparation).

cells [16] with DM and recent unpublished experiments on the effects of DMPK overexpression on cultured lens epithelial cells (S. Jin, A. Balasubramanyam, M. Sharma, U. P. Andley, and H. F. Epstein; manuscript in preparation) suggest that DMPK may contribute to the regulation of membrane transactions and trafficking.

In skeletal muscles which show severe myotonia and weakness in DM, there is a characteristic finding of "sarcoplasmic masses" [19]. These masses are usually near the plasma membrane of the muscle fiber and contain a variety of sarcoplasmic components including SR (sarcoplasmic reticulum), transverse tubular material, ER (endoplasmic reticulum), free ribosomes, and mitochondria. Myofibrillar structures including thick and thin filaments appear to be excluded. Antibody to recombinant DMPK labels these masses in muscle biopsies from DM patients [15].

In the lenses of DM patients, there are cataracts associated with characteristic polychromatic, iridescent opacities in the anterior and posterior subcapsular fibers which are detectable with the slit lamp biomicroscope [4]. Electron microscopic examination of lens fiber cells reveals multilamellar membrane inclusion bodies [14, 20] and possible semicrystalline aggregates of protein [40]. The exact origins of these membranes in terms of ER, Golgi apparatus (GA), or plasma membrane has not been determined.

Overexpression of DMPK in human and rabbit lens epithelial cell lines produces refractile bodies which dominate the cytoplasm and cell periphery in a manner reminiscent of the cataract opacities in DM (S. Jin, A. Balasubramanyam, M. Sharma, U. P. Andley, and H. F. Epstein, manuscript in preparation). Transfection with various truncated forms of DMPK and the catalytically inactive K100A mutant indicate that kinase activity is both necessary and sufficient to produce these bodies. These bodies contain protein disulfide isomerase, an ER resident protein, and carbohydrate markers of the GA. DMPK, normally a component of ER in these cells, is also in the bodies. Similar bodies are formed with colchicine treatment. These results suggest that trafficking of membranes and their cargo between the ER and GA is altered by DMPK overexpression. Recycling of ER vesicles from the GA is most likely the affected process.

C. Differentiation

The finding of hypoplastic, immature skeletal muscle fibers, and generalized hypotonia in infants severely affected with neonatal DM suggests that alteration of muscle differentiation and maturation is also a manifestation of the DM mutations [22]. When DMPK is overproduced in the BC_3H1 "myogenic" cell line which can express specific aspects and functions of differentiated skeletal muscle, differentiation is enhanced [10]. It should be noted that this is an effect of the kinase protein without coexpression of the mutation-containing 3′ region. On the other hand, "knockouts" of DMPK in mice do not produce any easily detectable loss of differentiation in skeletal muscle [32, 50]. A number of unpublished reports concerning DMPK expression in other cell lines confirms the observations with BC_3H1; however, overexpression of the complete gene or of the 3′ region in mice or cultured myogenic cells produced inhibition of muscle differentiation (R. Korneluk, First International Myotonic Dystrophy Consortium Conference). DMPK may indeed be able to activate regulatory elements of the myogenic program, but the relationship of DMPK protein, DMPK mRNA, and other interacting molecular species in skeletal muscle differentiation *in vivo* is presently unclear.

V. DMPK REDUNDANCY AND DEGENERACY

The deciphering of DMPK function and location in specific cell types may be complicated by a number of modifying phenomena. As already stated, DMPK is a member of a group of protein kinases, and it is likely that at least one of these enzymes is expressed in the same tissues as DMPK [75]. Both alternative splicing and initiation have been demonstrated [27, 33, 63], and there is evidence that distinct protein isoforms may be produced as a result [42, 69, 71]. Phosphorylation of DMPK leads to a change in its gel mobility, and multiple protein bands can be formed by this modification [17]. All of these phenomena may vary depending upon cell type, subcompartments, and physiological or pathological status. Most importantly, these potential modifiers of DMPK can act in combination. The result is a potential spectrum of DMPK forms which may have different functions and the presence of other kinases whose functions may overlap those of DMPK.

A. DMPK Group of Protein Kinases

It is very likely that the human genome can express several protein kinases with sufficient structural similarity and patterns of expression that their functions may partially or completely complement those of DMPK. This situation is called "functional redundancy." The extent of redundancy may vary between individual humans with distinct genetic backgrounds or between specific mouse strains and humans. Therefore, it is possible that some of the variability of expression in DM-affected persons may arise from different levels of redundant kinases. Also, it is possible that the apparently milder effects of DMPK knockouts in mice compared to clinical DM in humans may be due to species differences in the expression of potentially redundant kinases.

In humans, the sequences of the catalytic domains for DMPK and PK 428 are closely related with 78% similarity (see Fig. 11-3; [75]). Like DMPK, PK428 has a leucine-rich amino-terminal domain, and a hydrophobic region in its carboxyl-terminal region. Moreover, PK428 like DMPK is expressed in skeletal muscle, heart, and brain. The possible redundancy in function between the two enzymes has not been experimentally established.

Another potentially redundant kinase is human $p160^{ROCK}$. This enzyme is activated by the 21-kDa protein RHO. Its shows significant similarities to DMPK not only in its catalytic domain (66.1%) but also in its RHO binding region (Fig. 11-5). There are additional isoforms of this kinase as well which may also be redundant functionally. RHO A has been implicated in the formation of focal adhesion plaques which are important in many cell types and are necessary for muscle differentiation.

B. Alternative Initiation and Splicing

The existence of alternative forms of DMPK which may differ in their functional as well as physical properties is an important phenomenon. Both alternative initiation and splicing have been found with DM-encoded products [27, 33, 63, 69]. There is evidence that proteins corresponding to some of these alternatively expressed forms may be synthesized.

The principal evidence for alternative initiation of translation is that either native proteins or products of *in vitro* translation share the same carboxyl-terminal region antigens but exhibit two forms at 71/74 and 81/82 kDa [63, 69]. Using other antibodies, alternative forms at 55 and 72 kDa have been detected simultaneously in tissues and in cited (data not shown) *in vitro* translation experiments [63].

At the RNA level, cDNA cloning and analysis of the genomic DNA sequences for intron–exon boundaries [27, 33] have predicted the existence of multiple RNA forms. Four human spliceoforms have been detected by RT-PCR (V. Funanage, First International Myotonic Dystrophy Consortium Conference). Two of these forms would lead to DMPK without the carboxyl-terminal region (LPKH). LPKH would be consistent with the 64- to 67-kDa immunoreactive forms detected in various tissues [15, 16]; however, molecular characterization of the latter forms in cells or in cell-free experiments has not been performed.

C. Phosphorylation

There are no studies of differential phosphorylation of DMPK in whole cells. When human fibroblasts are grown in $[^{32}P]PO_4$, the immunoprecipitable 72-kDa but not the 55-kDa DMPK-related protein is phosphorylated [63]. The multimers seen in the autophosphorylation of purified recombinant DMPK are not detected; however, this discrepancy could be caused by different SDS–PAGE protocols. Whether the observed intracellular phosphorylation of DMPK is due to the action of other kinases or autophosphorylation is not known.

The observation made for recombinant DMPK [17, 69] and many other kinases and phosphoproteins that their apparent M_r is highly sensitive to their state of phosphorylation must always be considered as a potential source of degeneracy. Furthermore, DMPK phosphorylation may vary in either extent or target sites between cell types and developmental stages and be the basis of the variance in other properties of DMPK between these biological states.

D. Proteolytic Cleavage

Distinct forms of proteins may also be generated by proteolytic cleavage. For example, HMG CoA reductase is cleaved between its membrane anchoring domain and the rest of the enzyme. The catalytic and regulatory domains become solubilized [28]. A similar phenomenon may happen with DMPK (M. B. Perryman, personal communication). *In vitro* transcription and translation of full-length DMPK cDNA generates an 84-kDa form which is most likely full-length LPKHT. However, incubation of human cardiac extracts at 37°C leads to a decrease of the 84-kDa LPKHT and an increase in the 72-kDa form. The latter form is most likely LPKH and has lost its ER membrane anchoring tail.

E. Combination of Effects

The combination of multiple genes encoding DMPK-like kinases, alternative splicing and initiation of translation, and the degeneracy of phosphorylation can clearly act together to potentially create a spectrum of protein species. In evaluating either the biological effects of DMPK in an organism or cell line or the "correct M_r" of DMPK in a specific organ or cell population, these possible sources of redundancy and degeneracy must be considered.

VI. IMMUNOLOGICAL ANALYSIS OF DMPK

A major problem in DMPK research concerns the production and characterization of antibodies. At the present time, there is not a single antibody that has been demonstrated in published experiments to be monospecific to DMPK. A monospecific antibody would under defined conditions react only with DMPK protein in a given tissue. Some antibodies clearly react with DMPK but also with other proteins. Other antibodies produced to either recombinant DMPK or DMPK peptides may react with only one protein, but that protein has not been demonstrated to be a product of the DM locus.

For these reasons, basic chemical immunology and its application to evaluating specific antibody–antigen reactions will be reviewed briefly, and then the specific problems with putative anti-DMPK antibodies will be discussed.

A. Molecular Basis of Antibody–Antigen Reactions

The specificity and affinity of antibody–antigen reactions is based on the molecular interactions between the amino acid groups of the reactive center of the immunoglobulin and the chemical groups of the antigenic determinant (hapten, epitope).

In the case of polyclonal antibodies to a protein antigen, there are almost always multiple reactive immunoglobulins and multiple antigenic determinants. A commonly used example of such a reaction is the binding of "secondary" antibodies to "primary antibodies." Secondary antibodies are generally polyclonal and react with determinants on both the light and heavy chains of an immunoglobulin from a different species. Such an antibody will contain only antibodies specific to the primary immunoglobulin if it has been affinity purified. Affinity-purification is also advantageous because many antiserums will contain multiple specificities besides that to the desired immunogen and the antibody with desired specificity may represent only 1% or less of the total immunoglobulins. Affinity-purified antibodies will contain a high concentration of antibodies of desired specificity.

After affinity purification to the protein antigen, antibodies will still react with multiple determinants. There are several advantages of this property in a "secondary reagent." At least three determinants per antigen molecule are necessary for the formation of lattices in immunoprecipitation with bivalent antibodies [49]. The presence of multiple determinants and the bivalence of the immunoglobulin molecule also lead to enhancement of their intrinsic affinities for one another [12].

In monoclonal and many antipeptide antibodies, there is only one determinant or epitope per antigen polypeptide. Unless the antigen protein is oligomeric or polymeric, these antibodies will not precipitate antigen.

Monoclonal antibodies offer a model for the protein–protein interactions of antibody–antigen reactions generally. A relatively small number of amino acid residues (broadly 6–10) constitute most determinants or epitopes. Each of the chemical groups in the epitope may make a significant contribution to the energetics of the interaction. On the other hand, conservative substitutions for any one amino acid may lower the affinity but not prevent interaction. These interactions may or may not be sensitive to local conformation, the conformation of other regions, or the presence of modified residues at the site or in other regions.

B. Energetics of Antibody Specificity

The specificity of antibodies is in general a question of relative affinities. Very few antibodies to proteins will react with one and only one polypeptide sequence over a wide range of concentrations of antibody or antigen. However, within a specified range of concentrations, an antibody may bind exclusively to a single protein antigen, even in the presence of closely related isoforms [45]. For example, the differential affinity of an antibody for distinct protein isoforms may lead to differences in apparent specificity with respect to the isoforms at different ranges of antibody or antigen concentrations.

This phenomenon arises from the situation just discussed where only a small number of amino acids residues constitute an epitope, and there may be only one or two conservative substitutions in the epitope between the protein isoforms. A worst-case scenario is where the epitope is fully conserved between isoforms or merely related proteins. In this case, there may not be differential affinity. The best situation is where the gene encoding a specific isoform has recently acquired an insertion that encodes a uniquely antigenic determinant or a deletion which causes loss of an otherwise shared determinant.

In the case of polyclonal antibodies which are collections of monoclonal antibodies, each constituent antibody and its epitope may represent one of the situations just discussed. With polyclonal antibodies, the importance of rigorously determining the apparent affinity (average dissociation equilibrium constant) of the specific antibody–antigen reaction of interest and, thereby, the range of concentrations allowing this specificity must be emphasized even more than with monoclonal antibodies.

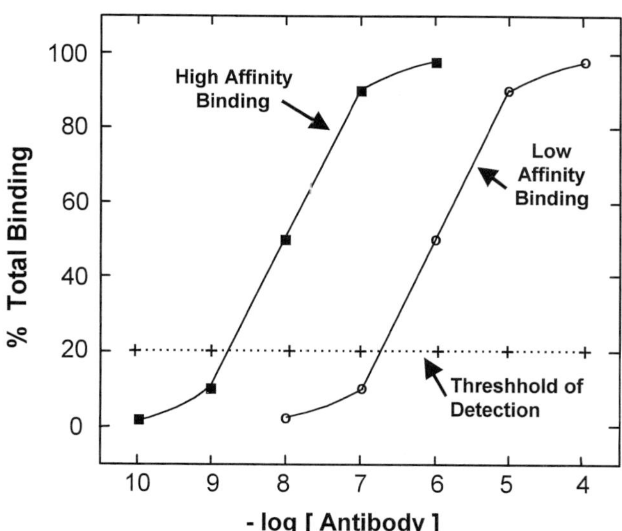

FIGURE 11-11 Hypothetical analysis of antibody specificity and detection. Binding of specific antigen to antibody at high affinity. (O) Binding of related antigen at low affinity. (+) Threshold of detection of antibody complexes. Note that different methods may have different sensitivities for the same antibody–antigen complexes. Therefore, at specific concentrations of antibody and antigen, their reaction may appear to be specific, but the reaction of other antigens may take place at higher concentrations of either reactant or may be detected by different techniques.

C. Detectability and Specificity

The ability to determine the specificity of antibody–antigen reactions depends upon the sensitivity of the method used to detect the complexes formed. Figure 11-11 shows a hypothetical case which relates the level of detection to the complexing of an antibody with its primary (high-affinity) and secondary (low-affinity) antigens. This comparison emphasizes that for most of the applications of anti-DMPK antibodies to the localization of DMPK , the sensitivity of detection is an important factor in determining apparent specificity. In considering this relationship, it must also be stressed that different detection systems for antibody–antigen complexes, for example, immunoblotting of SDS–PAGE and immunofluorescence microscopy of cells and tissues, may have different thresholds of detection with respect to antibody concentration [45].

D. Problems of Specificity with DMPK

The special problems of antibody specificity with DMPK arise partially from its structural similarities with other protein kinases including specific isoforms of DMPK as discussed with respect to potential redundancy and degeneracy and partially from the failure to recognize the potential intricacies of antibody–antigen interactions including problems of detection as well as reactivity.

It is very likely that protein isoforms of DMPK exist as the result of other genes encoding protein kinases with very similar structure to DMPK and as the result of alternative splicing and initiation. Depending upon the application, either of these mechanisms may lead to proteins which react with anti-DMPK antibodies. Differential phosphorylation of DMPK and its isoforms may also complicate analysis of immunological results by leading to altered M_rs or by altering the ability of DMPK species to react with anti-DMPK. These phenomena may also vary between specific tissues. For example, muscle may produce a pattern of DMPK isoforms distinct from that of brain. In addition, the extent and location of phosphorylation of DMPK catalyzed by other kinases and phosphatases may differ between developmental states of a particular tissue, as is plausible between myoblast and myotubes of differentiating skeletal muscle. In combination, the presence of multiple isoforms and multiple phosphoforms may lead to the reaction of very different DMPK species with different yet *bona fide* anti-DMPK antibodies. The effects of these phenomena upon the reactivity of anti-DMPK antibodies with DMPK species may or may not be resolved by the quantitative considerations already discussed.

The problem of cross-reaction of anti-DMPK antibodies with proteins that are more distantly related structurally or functionally than DMPK isoforms is one that should be resolvable by quantitative analysis. That is, the reactions of a given antibody with its immunogen or true DMPK protein can be compared in a quantitative fashion with the tissue protein in question. First, the establishment of a difference in apparent affinity between the two antigens indicates some chemical difference at the molecular level. The difference in affinity could then be the basis of establishing a range of concentrations in which only one of the two antigens interacts with antibodies or in which the extents of reaction of the antigens are very different.

E. Catalog of Anti-DMPK Antibodies

Each of the published anti-DMPK antibodies, its immunogen, its specificity, and how it was tested are listed. Only one of the antibodies has been shown not to react with an antigen in the DM knockout mice; however, this antibody reacts with other bands that do not disappear in the knockout as well. Lack of reaction in a knockout mouse is not an absolute criterion for specificity. A better genetically based criterion is a shift in reaction with a mutationally altered but present gene product [7, 45, 56, 66] because the absence of an immunoreactive could be a secondary phenomenon. Any or all of the antibodies below may be specific or nonspecific with respect to *bona fide* DM products. A difficult to achieve but highly stringent criterion for specificity is to demonstrate that the antigen(s) reacting with any of the antibodies listed is actually a DM product by peptide microsequencing. The molecular substructure for any of the immunoreactive protein bands has not been determined with respect to the predicted domains of DMPK.

1. Brewster *et al.* [8]. Rabbit antiserums were prepared against Peptides 1 and 2 conserved between human and murine catalytic domains: RLGYVHRDIKPDNIL and EHLSPLADTVVPEEAQWDL. Antipeptide 1 reacts with a 52-kDa protein band in brain, muscle, heart, and liver and with a 42-kDa band in brain. These bands do not react with reagents specific to PKA or Calmodulin/Calcium II kinase.

2. van der Ven *et al.* [67]. Antiserums were prepared in rabbits to three peptides, Pep 1, Pep 2, and Pep 3: RLGRDFEGATDTCNFD, GLRDSVPPFTPDFEGA, NGDRRWITQ. Anti-pep 1 and 2 react with an ~53-kDa protein band in heart and skeletal muscle. Anti-Pep 2 reacts with bacterially expressed PKH.

3. Fu *et al.* [27]. Also see Timchenko *et al.* [63], Bhaghwati *et al.* [6], and Tachi *et al.* [60]. Antiserums 254, 10033, and 8391 were prepared in rabbits to a peptide (possibly Pep 3 of van der Ven *et al.*), a PKH-like fusion protein, and the peptide PGTGSYGPECDW, respectively. React with 72- and 55-kDa bands in human fibroblasts. Antipeptide antisera react with their peptide immunogens and bacterially expressed LPKHT fusion protein.

4. Etongué-Mayer [21]. Antiserum to peptide: PSPRATDPPSHASRQY (residues 493–507) was prepared and affinity-purified to peptide. Antibodies react with peptide immunogen and with DMPK C-terminal fusion protein of Whiting *et al.* [71].

5. Koga *et al.* [36]. Antisera were prepared in rabbits to peptides DM1 and DM2 (DSTAETYGKIVHYKEH and EAEARNRDLEAHVRQ) and to DMPK fusion protein with bacterial maltose binding protein with insert from "middle" of exon 4 to "middle" of exon 14, ~PKH. Anti-DM2 reacts with 53- and 62-kDa proteins in muscle and heart, respectively, and with both in brain. All three antibodies react with bacterially expressed PKH-like DMPK.

6. Salvatori *et al.* [52]. Rabbit antiserums were prepared against PEP1: IREGAPLGVHLPFVGYSYSC. The antiserum reacted with single protein bands in the 50- to 54-kDa range from human , rabbit, and rat skeletal muscles.

7. Whiting *et al.* [71]. Also see Maeda *et al.* [42], Jansen *et al.* [32], and Waring *et al.* [69]. Antiserum was prepared in rabbits to the C-terminal 160 amino acids of DMPK. Reacts with a major 82-kDa protein band in human and rat skeletal muscles, a major 74-kDa band in human and rat heart and skeletal muscles, a minor 45-kDa band in human brain and human and rat heart and skeletal muscles, and a 66-kDa minor protein in human skeletal muscles. The antiserum reacts with bacterially expressed recombinant protein and wild-type mouse tissues (multiple bands). The major 74-kDa protein band is not detected in knockout mice, consistent with it being a product of the DM locus.

8. Saitoh *et al.* [51]. Antiserum in rabbits prepared against unspecified peptide derived from C-terminal region. Antibody reacts with 70-kDa recombinant protein in 70 kDa.

9. Dunne *et al.* [15]. Mouse monoclonal antibody against PKH. Reacts with 64-kDa protein band in human and rat skeletal muscles. Reacts with bacterially expressed recombinant PK (catalytic domain specific) and larger DMPK constructs.

10. Dunne *et al.* [16]. Antiserum prepared in rabbits to PKH. Reacts with 67-kDa protein band in human and rat eye lens. After adsorption to PK, reacts with

bacterially expressed PKH but not PK (helical domain specific).

VII. LOCALIZATION OF DMPK

The localization of DMPK, whether by cell fractionation or by microscopic methods, depends upon considerations of antibody specificity and detection. Judgment as to whether any of the localizations are rigorously correct must await highly stringent characterization of the antibodies under the conditions actually used. Since DMPK is a protein of relatively low abundance (<0.01% cell protein), it can generally be detected in homogenates or crude cell fractions only by highly sensitive immunoblotting techniques. The cytological methods which in all published papers have used light microscopy and in one case, electron microscopy, are based upon detection of antibody–antigen complexes. As listed above in the catalog of published anti-DMPK antibodies, the major organs in which DMPK has been detected immunologically are heart, skeletal muscle, brain, and eye lens. Although specialized differences exist at the tissue and cell levels between different organs, a commonality among all studies is localization at one or more membranes. In general, these membranes are derived from the endoplasmic reticulum.

A. Cell Fractionation

Several studies have investigated whether DMPK is concentrated in specific cell fractions which contain characteristic structures in a variety of cell types and tissues. In all studies, some form of DMPK cosedimented in the cytosol and with some membrane fraction. By reaction of the Whiting *et al.* [71] anti-C-terminal antibody with blots of human heart, most of the DMPK (80 and 71 kDa) was found in a low-speed supernatant (could contain high-molecular-weight particles) and a crude membrane pellet which contains, among other structures, intercalated discs [42]. Using the same antibody with blots of DMPK-transfected SF21 cells, LPKHT was found in the cytosol, light membrane, nuclear/ER, and purified nuclear fractions and LPKH was found in the cytosol and light membrane fractions [69]. By reaction of the 8391 anti-catalytic domain peptide in human fibroblasts, 55-kDa DMPK was found in nuclei and 72-kDa DMPK was found in the cytoplasm (unclear as to speed of centrifugation and therefore whether it contains other membrane fractions) [63]. By reaction of an anti-DMPK peptide antibody with a blot of DMPK-transfected L6 myogenic cells, DMPK was found in the light microsomal fraction which pellets at 100,000*g* and contains ER, trans-GA network, and the intermediate compartment [51].

B. Immunochemical Microscopic Localization

Nearly all of published studies have been at the light microscope level, with only one published electron microscope study. Immunochemical microscopic localization provides positive information on what can be detected; it cannot rigorously exclude what it cannot detect. One consideration is that all of the studies involve some extraction and fixation steps; therefore, localization will be detected only at precipitable structures and, in general, not to the true cytosol. Another consideration is that only antigens that are accessible and not buried by other molecular structures will be detected. Considering the caveats discussed, all of the microscopic studies confirmed localization of DMPK at some membrane or membranes in a variety of cell types. On the other hand, the exact membrane that is labeled within a cell type appears to depend upon the specific antibody used. For example, in the earliest studies with human and rat muscle, one group found with their anti-peptide antibodies that the plasma membrane and, specifically, the neuromuscular junctions (NMJs) are labeled [67], whereas another group which used an anti-peptide antibody reported that the SR-lateral cisternae are labeled [52]. Each of these conclusions has been confirmed by at least one other group using different antibodies. The NMJ localization has been confirmed with the specific anti-C-terminal antibody [71], and the SR-triad localization has been confirmed with a monospecific monoclonal antibody as shown in Fig. 11-12 [15]. It should be noted that distinct protein bands on immunoblots react with each of the antibodies used in these localizations.

The ER is a common site of localization, especially in cells that have not undergone terminal differentiation. ER or perinuclear localization has been detected in L6 myoblasts [51] and human and rat lens epithelial cells [16] as well as in unpublished reports in a variety of cell types including neurons. The SR and the specialized sarcolemmal sites are derived from the ER.

VIII. REGULATION OF DMPK

DMPK, like other protein kinases, not only serves to regulate its physiological substrates but also is a substrate or ligand for other regulatory proteins. Many pro-

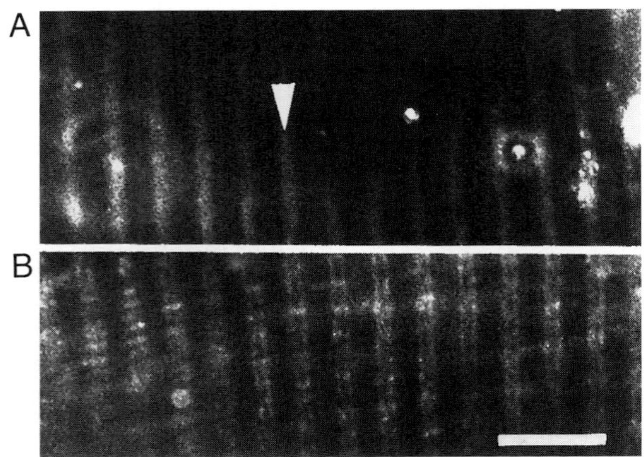

FIGURE 11-12 Localization of anti-DMPK antibody to SR in skeletal muscle. Monoclonal antibody to the PK domain labels the triad region (A) overlapping the lateral cisternae as labeled by rabbit antibody specific to the slow skeletal isoform of the SR Ca^{2+}-ATPase (B). Bar is 5 mm. Reproduced from *Cell Motil. Cytoskeleton* **33,** 52–63 (1996).

tein kinases are elements in cascades of protein kinases and phosphatases. Other protein kinases are regulated by p21 GTPase proteins and thereby linked to specific signal transduction pathways. As depicted schematically in Fig. 11-13, the participation of a protein kinase in such cascades and pathways may be enhanced by anchoring to specific subcellular sites and scaffolding with additional proteins to physically link elements so that they may functionally interact. Research in the regulation of DMPK is just beginning; therefore, it would appear timely to briefly discuss these different regulatory mechanisms in relation to DMPK.

A. P21 GTPase Proteins

The identities and similarities in predicted amino acid sequences between the helical region of DMPK and protein kinases and nonenzymatic proteins that bind to RHO A, a p21 GTPase, have been discussed already. Unpublished data suggest that RHO A may indeed interact with DMPK in cell-free and *in vitro* experiments. However, other p21 proteins may also interact with DMPK. Current research projects are focused on establishing the relative affinities of known p21 GTPases for DMPK, their effects on the phosphotransferase activities of DMPK, and on establishing their physiological relevance (W. Wang and H. F. Epstein, unpublished results).

B. Phosphorylation by Other Protein Kinases

The predicted amino acid sequence of DMPK contains consensus sites for phosphorylation by other protein kinases. Searches by the PROSITE algorithm reveal several potential phosphorylation sites. Three PKC sites are found at 225TVR227, 467TLR469, and

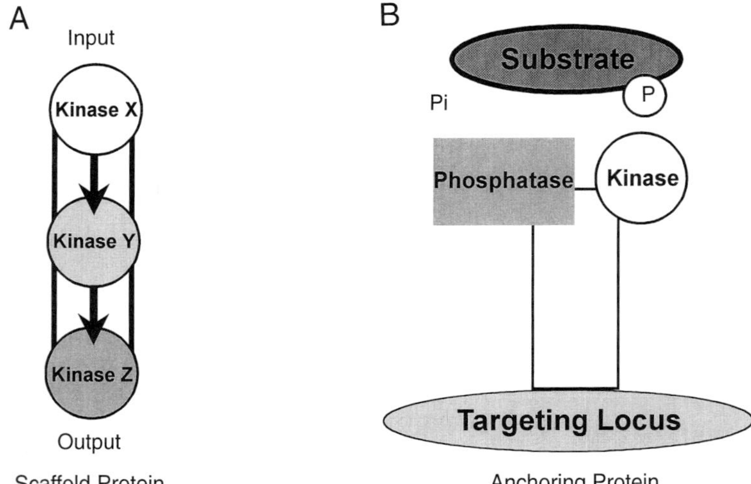

FIGURE 11-13 Schematic diagrams of scaffolding and anchoring of protein kinase complexes. (A) Scaffold protein links three protein kinases together in order to permit an effective cascade or pathway to function. (B) Anchoring protein links a protein kinase and a protein phosphatase with a targeting locus or specific subcellular site. This arrangement enhances the ability of the enzymes to bind their substrate protein and catalyze their specific reactions. Reproduced from *Cell* **85,** 9–12 (1996) by copyright permission of Cell Press.

540SPR542. Six Casein Kinase II sites are found at 282STAE285, 384TLSD387, 419TPME422, 443SPQD446, 467TLRE470, and 548SHLD551. These putative sites span the various structural and potentially functional domains of DMPK. Unpublished experiments suggest that RAF kinase may phosphorylate DMPK and bind to it in yeast two-hybrid assays (E. T. Walch, P. W. Dunne, H. F. Epstein; unpublished results). No systematic work has been published or presented in an unpublished manner on the biochemical or physiological significance of these sites or on the modification of DMPK by other protein kinases.

C. Dephosphorylation by Protein Phosphatases

Protein phosphatases play a significant role in the regulation of individual protein kinases, protein kinase cascades, and signal transduction pathways. In the case of multiple networks, it is the action of a specific protein phosphatase that provides the overriding regulatory switch [11]. Unpublished experiments suggest that both Type 1 and 2A protein phosphatases can remove phosphate from DMPK after autophosphorylation. Catalytic subunits of Type 2A phosphatase interact with PKH in yeast two-hybrid experiments (P. W. Dunne, L. Ma, H. F. Epstein, unpublished results). Further work is required to establish the biochemical and physiological significance of these interactions.

D. Anchoring to Specific Subcellular Sites

The T domain is a likely anchor of DMPK to membranes, specifically the ER, because of its close similarity to ER anchoring domains of human HMG-CoA reductase and rat microsomal aldehyde reductase (Fig. 11-7). The most direct experimental analysis has been by Korneluk and his colleagues in comparing the distribution of recombinant full-length LPKHT and truncated LPKH proteins upon their expression in SF21 insect cells [69]. LPKHT was present in membrane and nuclear fractions while LPKH was predominantly in the cytosol and light membrane fractions. Similar results have been obtained with similar constructs which were expressed in B3 human lens epithelial cells and localized by immunofluorescence using epitope tags (S. Jin, A. Balasubramanyam, M. Sharma, U. P. Andley, H. F. Epstein, manuscript in preparation). Perryman and his colleagues have reported that LPKH may be generated from LPKHT by proteolytic cleavage (M. B. Perryman, personal communication). In this case, proteolysis may serve as a regulatory mechanism for determining the fraction of DMPK that is bound to ER or free in the cytosol.

E. Scaffolding for Specific Protein–Protein Interactions

The DMPK-related complex of RAF kinase and RAS (ser/thr protein kinase and p21 GTPase) appears to interact with the 14-3-3 protein as a scaffold [46, 23]. In a biochemical experiment, ternary complexes of RAF kinase, histone H1, and DMPK showed enhanced phosphorylation of both DMPK and Histone H1 in comparison to either of the binary complexes (E. T. Walch, P. W. Dunne, and H. F. Epstein, unpublished results). Although speculative, the identification of a physiologically significant scaffold protein for the interactions of DMPK with its substrate protein(s) and the proteins regulating it would be an important finding.

IX. DOES DMPK PLAY A ROLE IN THE PATHOGENESIS OF DM?

The mechanism by which the CTG repeat expansion mutations produce the pathology of DM has not been elucidated. Indeed, much of the research has been controversial in that several distinct mechanisms have been proposed. However, it is possible that DM represents a new paradigm of inherited disease and of the molecular mechanisms underlying its pathogenesis. The responsible mutations may produce alterations in structure and function in several ways. Defects may occur in chromatin structure and function, in the transcription of DMPK RNA, in the expression of neighboring genes, especially the downstream gene encoding a homeodomain protein (DMAHP), in the binding of specific proteins to the expanded repeat at the level of either DNA or RNA, in the amount of DMPK synthesized, or of the proper mix of DMPK isoforms produced. It is important to emphasize that these proposed mechanisms are not mutually exclusive of one another. The last two hypotheses bear directly upon the theme of this chapter.

In this section, the role of protein kinases in human disease will be reviewed briefly. The complexities of the spectrum of DM phenotypes and of alterations in the amount, structure, or function of DMPK will be compared and analyzed. The question as to whether DM can exist without CTG repeat expansion mutations in the 3′ UTR of the DM locus and the potential significance of such a finding will be discussed.

A. The Role of Protein Kinases in Human Disease

There are multiple human diseases, both inherited and acquired, that are caused by alterations in the structure or function of protein kinases or their signaling pathways. The pathogenetic mechanisms and phenotypes of several of these diseases will be reviewed in terms of their relatedness to possible roles of DMPK in DM.

Coffin–Lowry Syndrome (CLS), like DM, is a highly pleiotropic and variably expressed disorder caused by any of several mutations in the DMPK-related ribosomal S6 kinase 2 (RSK-2) [65]. Although the disease locus is X-linked (Xp22.2), females can show clinical signs, suggesting a dominant mode of expression. Mental retardation, hypotonia, dysmorphisms of the face, limbs, neck, and spine, and small stature are observed. A serious characteristic finding is visceral neuropathy. Membrane inclusion bodies have been detected in patient fibroblasts by electron microscopy. RSK-2 is a serine/threonine protein kinase in the MAP signaling pathway that can be activated by protein hormones including nerve growth factor and epidermal growth factor. Responsible mutations include V75G, A227S, and R558stop (out of 720 codons) which destroy protein kinase activity. Overall CLS and RSK-2 share many features of DM and DMPK.

Ataxia Telangiectasia (A-T) is a rare, recessive autosomal disorder with a pleiotropic and variable phenotype. The disease locus, *ATM,* is on 11q22-23 and encodes a putative phosphoinositol-3′ kinase [53]. Patients with A-T may show characteristic ataxia (a neurological problem of balance), recurrent sinus and respiratory infections, and telangiectasia (dilated blood vessels in the sclerae of the eye and in the skin) [48]. Additional findings include cerebellar degeneration, insulin-resistant diabetes, enhanced sensitivity to X rays, and increased leukemias and lymphomas. If ATM does prove to be a true phosphoinositol-3′ kinase, it most likely plays a role in activating one of several protein kinase cascades.

The *RET* proto-oncogene encodes a receptor tyrosine kinase which is expressed in the neural crest and its derivatives [18]. Mutations in the gene produce a pleiotropic and variable phenotype which includes the multiple endocrine neoplasia type 2 syndromes (MEN 2), the related sporadic tumors thyroid carcinoma and pheochromocytoma, and both familial and sporadic Hirschsprung's Disease (congenital absence of enteric innervation in the colon or small intestine). The different phenotypes have been linked by findings in the patient with a C620R mutation of both severe Hirschsprung's disease (total colonic aganglionosis) and medullary thyroid carcinoma [55]. Another patient who was homozygous for the R313Q mutation due to consanguinity had the most severe Hirschsprung's phenotype with both small intestine and total ganglionic aganglionosis.

Alzheimer's Disease can be inherited as a familial disorder due to mutations in one of several loci, or, more commonly, it can be sporadic. Phosphorylation appears to be a critical part of the pathogenesis of this common neurodegenerative disorder although none of the known disease loci actually encode a protein kinase. Alzheimer's is primarily a disorder of the brain's cerebral cortex with associated deficits in memory and cognition but can ultimately be fatal. The brains of all patients show neuritic plaques containing β-amyloid in functionally relevant regions, and the brains of nearly all patients show neurofibrillary tangles. These histological structures contain paired helical filaments which are molecular assemblies composed of hyperphosphorylated *tau* protein subunits [54]. *Tau* protein is normally associated with axonal microtubules and is necessary for their stability. Experiments with embryonic rat hippocampal neurons (an affected population) show that the activity of *tau* protein kinase I is necessary for the induction of neurotoxicity by β-amyloid peptide [61].

This brief survey shows that classical genetic alterations in protein kinases or in other functions of signaling pathways can lead to disorders with highly variable, pleiotropic, and dominant phenotypes. This complexity of phenotype appears to be only partially the result of structurally distinct mutations. It is reasonable to consider, therefore, that the complex, dominant DM phenotype may be at least partially the result of alterations in the expression, function, or structure of DMPK. In addition, these alterations of DMPK may modify other mechanisms in the pathogenesis of DM.

B. Complex Disease and Complex Mechanisms

Although the physiological functions of DMPK and its roles in the pathogenesis of DM have not been established, it may still be useful to speculate upon them. DMPK is a serine/threonine protein kinase which is expressed in all of the organs affected in DM. Although still incomplete, the available evidence would suggest that DMPK is probably multifunctional, that is, it has more than one physiologically significant protein as substrate. Current evidence also suggests that DMPK is regulated, probably in several ways. It is plausible to assume that both the substrate targets of DMPK and its regulation may differ between specific types or developmental stages of cells. The pleiotropy of the DM phenotype are thus explicable in terms of DMPK. Since the protein products of multiple genes appear to interact with DMPK, it is also plausible that genetic or acquired variation in these other genes can affect DMPK activity. This mechanism can explain at least some of the individual variability in DM phenotype. These arguments are consistent with the findings described for the other kinase-related inherited disease: the CLS, A-T, and RET syndromes.

The remaining issues are: how do the CTG repeat expansion mutations in the 3′UTR of the DM locus produce alterations in DMPK, and why have no classical mutations of DMPK been found to produce DM? The latter question will be discussed in the next section. The question as to whether the repeat expansion mutations affect the synthesis of DMPK through either transcriptional or posttranscriptional mechanisms is still unresolved, and its resolution by further examination of DM patient material is unlikely. Experimental approaches to the problem are required, and their results may be useful as models for the explanation of DM in humans. A distinct possibility is that the known alternative splicing or initiation of DMPK or its RNA may be altered by the 3′UTR mutations. At least some of the alternative DMPK isoforms so produced are likely to perform different physiological functions, and therefore, changes in their relative amounts may lead to dominant effects.

C. DM without CTG Repeat Expansion?

If alterations in DMPK can explain even part of the pathogenesis of DM, why haven't more classical mutations which produce the DM phenotype been found? It is clear that similarly complex phenotypes can be produced by amino acid substitutions in RSK-2, another serine/threonine protein kinase. One answer is that the direct effects upon DMPK are only part of the pathogenetic mechanism in DM, and such pure kinase mutations would produce only a partial syndrome. The other answer is that classical mutations may occur less frequently than the repeat expansion mutations and that their detection is therefore more difficult.

The actual discovery of DM produced without expansion of the CTG repeat in the 3′UTR but with mutations elsewhere in the DM locus would resolve the issue. Whether or not DM has been shown to exist without the CTG repeat expansions is presently controversial (M. C. Koch, First International Myotonic Dystrophy Consortium Conference). She argues that such cases (about 2% of total DM) are really not DM, but related, more rare syndromes. On the other hand, experienced neuromuscular disease specialists have diagnosed patients without the expansion mutations as having DM [62, 1].

Do such patients have other mutations in the DM locus? Material from 14 such patients has been analyzed for other DM mutations without success (K. Johnson, R. Korneluk, and B.Wieringa, personal communications). However, the absence of proof is not proof of absence. Since such a finding is so critical to the DMPK hypothesis, our laboratory is reinvestigating the problem (M. Shimizu, T. Ashizawa, H. F. Epstein). If indeed some or all of the DM-like syndromes are associated with other loci, it will be equally important to uncover what they encode. The protein products of such genes could represent other elements in the protein kinase pathway containing DMPK.

References

1. Abruzzese, C., Krahe, R., Liguori, M., Tessarolo, D., Siciliano, M. J., Ashizawa, T., and Giacanelli, M. (1996). Myotonic dystrophy phenotype without expansion of $(CTG)_n$ repeat: an entity distinct from proximal myotonic myopathy (PROMM)? *J. Neurol.* **243,** 715–721.
2. Alber, T. (1992). Structure of the leucine zipper. *Curr. Opin. Genet. Dev.* **2,** 205–210.
3. Amano, M., Mukai, H., Ono, Y., Chihara, K., Matsui, T., Hamajima, Y., Okawa, K., Iwamatsu, A., and Kaibuchi, K. (1996). Identification of a putative target for RHO as the serine-threonine kinase protein kinase. *Science* **271,** 648–650.
4. Ashizawa, T., Hejtmancik, J. F., Liu, J., Perryman, M. B., Epstein, H. F., and Koch, D. D. (1992). Diagnostic value of ophthalmologic findings in myotonic dystrophy: comparison with risks calculated by haplotype analysis of closely linked restriction fragment length polymorphisms. *J. Med. Genet.* **42,** 55–60.
5. Berger, B., Wilson, D. B., Wolf, E., Toncher, T., Milla, M., and Kim, P. S. (1995). Predicting coiled coils by use of pairwise residue correlations. *Proc. Natl. Acad. Sci. USA* **92,** 8259–8263.
6. Bhagwati, S., Ghatpande, A., and Leung B. (1996). Normal levels of DM RNA and myotonin protein kinase in skeletal muscle from adult myotonic dystrophy (DM) patients. *Biochim. Biophys. Acta* **1317,** 155–157.
7. Bingham, P. M., Scott, M O., Wang, S., McPhail, M. J., Wilson, E. M., Garbern, J. Y., Merry, D. E., and Fischbeck, K. H. (1995). Stability of an expanded trinucleotide repeat in the androgen receptor gene in transgenic mice. *Nature Genet.* **9,** 191–196.
8. Brewster, B. S., Jeal, S., and Strong, P. N. (1993). Identification of a protein product of the myotonic dystrophy gene using peptide specific antibodies. *Biochem. Biophys. Res. Commun.* **194,** 1256–1260.
9. Brook, J. D., McCurrach, M. E., Harley, H. G., Buckler, A. J., Church, D., Aburatani, H., Hunter, K., Stanton, V. P., Thirion, J.-P., Hudson, T., Sohn, R., Zemelman, B., Snell, R. G., Rundle, S. A., Crow, S., Davies, J., Shelbourne, P., Buxton, J., Jones, C., Juovenen, V., Johnson, K., Harper, P. S., Shaw, D. J., and Housman, D. E. (1992). Molecular basis of myotonic dystrophy: expansion of a trinucleotide repeat (CTG) at the 3′ end of a transcript encoding a protein kinase family member. *Cell* **68,** 799–808.
10. Bush, E. W., Taft, C. S, Meixell, G. E., and Perryman, M. B. (1996). Overexpression of myotonic dystrophy kinase in BC_3H1

cells induces the skeletal muscle phenotype. *J. Biol. Chem.* **271,** 548–552.
11. Cohen, P. (1987). The structure and regulation of protein phosphatases. *Annu. Rev. Biochem.* **58,** 453–508.
12. Crothers, D. M., and Metzger, H. (1972). The influence of polyvalency on the binding properties of antibodies. *Immunochemistry* **9,** 341–357.
13. Curschmann, H. (1912). Uber familiare atrophische Myotonie. *Dtsch. 2. Nervenheilkd* **45,** 161–202.
14. Dark, A. J., and Streeten, B. W. (1977). Ultrastructural study of cataract in myotonica dystrophica. *Am. J. Ophthalmol.* **84,** 666–674.
15. Dunne, P. W., Ma, L., Casey, D. L., Harati, Y., and Epstein, H. F. (1996). Localization of myotonic dystrophy protein kinase in skeletal muscle and its alteration with disease. *Cell Motil. Cytoskel.* **33,** 52–63.
16. Dunne, P. W., Ma, L., Casey, D. L., and Epstein, H. F. (1996). Myotonic protein kinase expression in human and bovine lenses. *Biochem. Biophys. Res. Commun.* **225,** 281–288.
17. Dunne, P. W., Walch, E. T., and Epstein, H. F. (1994). Phosphorylation reactions of recombinant human myotonic dystrophy protein kinase and their inhibition. *Biochemistry* **33,** 10809–10814.
18. Eng, C., and Mulligan, L. M. (1997). Mutations of the *RET* protooncogene in the multiple endocrine neoplasia type 2 syndromes, related sporadic tumours, and Hirschsprung disease. *Hum. Mutat.* **9,** 97–109.
19. Engel, W. K. (1962). Chemocytology of striated annulets and sarcoplasmic masses in myotonic dystrophy. *J. Histochem. Cytochem.* **10,** 229–230.
20. Eshaghian, J., Murch, W. F., Goosens, W., and Rafferty, N. S. (1978). Ultrastructure of a young lens in Steinert myotonic dystrophy. *Invest. Ophthalmol. Vis. Sci.* **17,** 289–293.
21. Etongué-Mayer, P., Faure, R., Bouchard, J.-P., Thibault, M.-C., and Puymirat, J. (1994). The myotonin-protein kinase phosphorylates tyrosine residues in normal human skeletal muscle. *Biochem. Biophys. Res. Commun.* **199,** 89–92.
22. Farkas, E., Tomé, F. M. S., Fardeau, M., Arsenio-Nunes, M. L., Dreyfus, P., and Diebler, M. F. (1974). Histochemical and ultrastructural study of muscle biopsies in 3 cases of dystrophica myotonica in the newborn child. *J. Neurol. Sci.* **21,** 273–288.
23. Faux, M. C., and Scott, J. D. (1996). Molecular glue: kinase anchoring and scaffold proteins. *Cell* **85,** 9–12.
24. Felsenstein, J. (1989). PHYLIP: Phylogeny Inference Package (Version 3.2). *Cladistics* **5,** 164–166.
25. Force, T., Bonventre, J. V., Heidecker, G., Rapp, U., Avruch, J., and Kyriakis, J. M. (1994). Enzymatic characteristics of the c-Raf-1 protein kinase. *Proc. Natl. Acad. Sci. USA* **91,** 1270–1274.
26. Fu, Y.-H., Pizzuti, A., Fenwick, R. G., Jr., King, J., Rajnarayan, W., Dunne, P. W., Dubel, J., Nasser, G. A., Ashizawa, T., deJong, P., Wieringa, B., Korneluk, R., Perryman, M. B., Epstein, H. F., and Caskey, C. T. (1992). An unstable triplet repeat in a gene related to muscular dystrophy. *Science* **255,** 1256–1258.
27. Fu, Y-H., Friedman, D. L., Richards, S., Pearlman, J. L., Gibbs, R. A., Pizzuti, A., Ashizawa, T., Perryman, M. B., Scarlato, G., Fenwick, R. G., and Caskey, C. T. (1993). Decreased expression of myotonin-protein kinase mRNA and protein in adult form of myotonic dystrophy. *Science* **260,** 235–238.
28. Gil, G., Faust, J. R., Chin, D. J., Goldstein, J. L., and Brown, M. S. (1985). Membrane-bound domain of HMG CoA reductase is required for sterol-enhanced degradation of the enzyme. *Cell* **41,** 249–258.
29. Hanks, S. K., Quinn, A. M., and Hunter, T. (1988). The protein kinase: conserved features of primary structure and classification of family members. *Science* **241,** 42–52.
30. Harper, P. S. (1989). "Myotonic Dystrophy," 2nd ed. Saunders, London.
31. Hoffman, W. W., and De Nardo, G. L. (1968). Sodium flux in myotonic muscular dystrophy. *Am. J. Physiol.* **214,** 300–336.
32. Jansen, G., Groenen, P. J. T. A., Bächner, D., Jap, P. H. K., Coerwinkel, M., Oerlemans, F., van den Broek, W., Gohlsch, B., Pelte, D., Plump, J. J., Molenaar, P. C., Nederhoff, M. G. J., van Echteld, C. J. A., Dekker, M., Berns, A., Hameister, H., and Wieringa, B. (1996). Abnormal myotonic protein kinase levels produce only mild myopathy in mice. *Nature Genet.* **13,** 316–324.
33. Jansen, G., Mahaderan, M., Amemiya, C., Wormskamp, N., Segers, B., Hendriks, W., O'Hoy, K., Baird, S., Sabourin, L., Lennon, G., Jap, P. L., Iles, D., Coerwinkel, M., Hofker, M., Carrano, A. V., de Jong, P. J., Korneluk, R. G., and Wieringa, B. (1992). Characterization of the myotonic dystrophy region predicts multiple protein isoform-encoding mRNAs. *Nature Genet.* **1,** 261–266.
34. Justice, R. W., Zilian, O., Woods, D. F., Noll, M., and Bryant, P. J. (1995). The *Drosophila* tumor suppressor gene *warts* encodes a homolog of human myotonic dystrophy kinase and is required for the control of cell shape and proliferation. *Gene Dev.* **9,** 534–546.
35. Kobe, B., and Deisenhofer, J. (1994). The leucine-rich repeat: a versatile binding motif. *TIBS* **19,** 415–421.
36. Koga, R., Nakao, Y., Kurano, Y., Tsukahara, T., Nakamura, A., Ishiura, S., Nonaka, I., and Arahata, K. (1994). Decreased myotonin-protein kinase in the skeletal and cardiac muscles in myotonic dystrophy. *Biochem. Biophys. Res. Commun.* **202,** 577–585.
37. La Spada, A. R., Wilson, E. M., Lubahn, D. B., Harding, A. E., and Fischbeck, K. H. (1991). Androgen receptor gene mutations in X-linked spinal and bulbar muscular atrophy. *Nature* **352,** 77–79.
38. Leung, T., Munser, E., Tan, L., and Lim, L. (1995). A novel serine/threonine kinase binding the Ras-related RhoA GTPase which translocates the kinase to peripheral membranes. *J. Biol. Chem.* **270,** 29051–29054.
39. Leung, T., Chen, X.-Q., Manser, E., and Lim, L. (1996). The p160 RhoA-binding kinase ROKa is a member of a kinase family and is involved in the reorganization of the cytoskeleton. *Mol. Cell. Biol.* **16,** 5313–5327.
40. Lewis, R. A., Kretzer, F., Mehta, R., and Selezinka, W. (1982). Ultrastructure of a young lens in Steinert myotonic dystrophy. *In* "Current Concepts in Cataract Surgery." (J. M. Emery and A. C. Jacobson, Eds.), pp. 13–16. Appleton-Century-Crofts, New York.
41. Liscum, L., Finer-Moore, J., Stroud, R. M., Luskey, K. L., Brown, M. S., and Goldstein, J. L. (1985). Domain structure of 3-hydroxy-3-methyl-glutaryl coenzyme A reductase, a glycoprotein of the endoplasmic reticulum. *J. Biol. Chem.* **260,** 522–530.
42. Maeda, M., Taft, C. S., Bush, E. W., Holder, E., Bailey, W. M., Neville, H., Perryman, M. B., and Bies, R. D. (1995). Identification, tissue-specific expression, and subcellular localization of the 80- and 71-kDa forms of myotonic dystrophy kinase protein. *J. Biol. Chem.* **270,** 20246–20249.
43. Marshall, C. J. (1996). Raf gets it together. *Nature* **383,** 127–128.
44. Masaki, R., Yamamoto, A., and Tashiro, Y. (1994). Microsomal aldehyde dehydrogenase is localized to the endoplasmic reticulum via its carboxyl-terminal 35 amino acids. *J. Cell Biol.* **126,** 1407–1420.
45. Miller, D. M., III, Ortiz, I., Berliner, G. C., and Epstein, H. F. (1983). Differential localization of two myosins within nematode thick filaments. *Cell* **34,** 477–490.
46. Mochly-Rosen, D. (1995). Localization of protein kinases by anchoring proteins: a theme in signal transduction. *Science* **268,** 247–251.

47. Mounsey, P. S., Xu, P., John, J. E., Horne, L. T., Gilbert, J., Roses, A. D., and Moorman, J. R. (1995). Modulation of skeletal muscle sodium channels by human myotonin protein kinase. *J. Clin. Invest.* **95,** 2379–2384.

48. Nowak, R. (1995). Discovery of AT gene sparks biomedical research bonanza. *Science* **268,** 1700–1701.

49. Raffel, S. (1961). "Immunity," 2nd ed. Appleton-Century-Crofts, New York.

50. Reddy, S., Smith, D. B. J., Rich, M. M., Leferovich, J. M., Reilly, P., Davis, B. D., Than, K., Rayburn, H., Bronson, R., Cros, D., Balice-Gordon, R. J., and Housman, D. (1996). Mice lacking the myotonic dystrophy protein kinase develop a late onset progressive myopathy. *Nature Genet.* **13,** 325–335.

51. Saitoh, N., Sasagawa, N., Koike, H., Shimokawa, M., Surimachi, H., Ishiura, S., and Suzuki, K. (1996). Immunocytochemical localization of a full-length myotonin protein kinase in rat L6 myoblasts. *Neurosci. Lett.* **218,** 214–216.

52. Salvatori, S., Biral, D., Furlan, S., and Marin, O. (1994). Identification and localization of the myotonic dystrophy product in skeletal and cardiac muscles. *Biochem. Biophys. Res. Commun.* **203,** 1365–1370.

53. Savitsky, K., Bar-Shira, A., Gilad, S., Rotmon, G., Ziv, Y., Vanagaite, L, Tagle, D. A., Smith, S., Uziel, T., Sfez, S., Ashkenazi, M., Pecker, I., Frydman, M., Harnik, R., Patanjali, S. R., Simmons, A., Clines, G. A., Sartiel, A., Galti, R. A., Chessa, L., Sanal, O., Lavin, M. F., Jaspers, N. G. J., Taylor, A. M. R., Arlett, C. F., Miki, T., Weissman, S. M., Lovett, M., Collins, F. S., and Shiloh, Y. (1995). A single Ataxia Telangiectasia gene with a product similar to a PI-3 kinase. *Science* **268,** 1749–1753.

54. Selkoe, D. J. (1997). Alzheimer's disease: genotypes, phenotypes, and treatments. *Science* **275,** 630–631.

55. Seri, M., Yin, L., Barone, V., Bolino, A., Celli, I., Bocciardi, R., Pasini, B., Ceccherini, I., Lerone, M., Kristofferson, U., Larsson, L. T., Casasa, M., Cass, D. T., Abramowicz, M. J., Vanderwinden, J.-M., Kravcenkiene, I., Baric, I., Silengo, M., Martuccielo, G., and Romeo, G. (1997). Frequency of *RET* mutations in long- and short-segment Hirschsprung disease. *Hum. Mutat.* **9,** 243–249.

56. Servadio, A., Koshy, B., Armstrong, D., Antalfty, B., Orr, H. T., and Zoghbi, H. Y. (1995). Expression analysis of the ataxin-1 protein in tissues from normal and spinocerebellar ataxia type 1 individuals. *Nature Genet.* **10,** 94–98.

57. Smith, R. F., and Smith, T. F. (1990). Automatic generation of primary sequence patterns from sets of related protein sequences. *Proc. Natl. Acad. Sci. USA* **87,** 118–122.

58. Smith, R. F., and Smith, T. F. (1992). Pattern-induced multi-sequence alignment (PIMA) algorithm employing secondary structure-dependent gap penalties for comparative protein modeling. *Protein Eng.* **5,** 35–41.

59. Stull, J. T., Brostrom, C. O., and Krebs, E. G. (1972). Phosphorylation of the inhibitor component of troponin by phosphorylase kinase. *J. Biol. Chem.* **247,** 5272–5274.

60. Tachi, N., Kozuka N., Ohya, K., Chiba, S., and Kikuchi K. (1996). Immunocytochemical localization of myotonin protein kinase on muscle from patients with congenital myotonic dystrophy. *Histol. Histopathol.* **11,** 869–871.

61. Takashima, T., Noguchi, K., Sato, K., Hoshino, T., and Imahori, K. (1993). Tau protein kinase I is essential for amyloid b-protein-induced neurotoxicity. *Proc. Natl. Acad. Sci. USA* **90,** 7789–7793.

62. Thornton, C. A., Griggs, R. C. and Moxley, R. T. (1994). Myotonic dystrophy with no trinucleotide expansion. *Ann. Neurol.* **3,** 369–272.

63. Timchenko, L., Nastainczyk, W., Schneider, T., Patel, B., Hofmann, F., and Caskey, C. T. (1995). Full-length myotonin protein kinase (72 kDa) displays serine kinase activity. *Proc. Natl. Acad. Sci. USA* **92,** 5366–5370.

64. Toyn, J. H., and Johnston, L. H. (1993). Spolz is a limiting factor that interacts with the cell cycle protein kinases Dbf2 and Dbf20, which are involved in mitotic chromatid disjunction. *Genetics* **135,** 963–971.

65. Trivier, E., Cesare, D. D., Jacquot, S., Pannetier, S., Zackal, E., Young, I., Mandel, J.-L., Sassone-Corsi, P., and Hanauer, A. (1996). Mutations in the kinase Rsk-z associated with the Coffin-Lowry syndrome. *Nature* **384,** 567–570.

66. Trottier, Y., Devys, D., Imbert, G. Sandou, F., An, I., Lutz, Y., Weber, C., Agid, Y., Hirsch, E. C., and Mandel, J.-L. (1995). Cellular localization of the Huntington's disease protein and discrimination of the normal and mutated forms. *Nature Genet.* **10,** 104–110.

67. van der Ven, P. F. M., Jansen, G., van Kuppevelt, T. H. M. S. M., Perryman, M. B., Lupa, M., Dunne, P. W., ter Laak, H. J., Jap, P. H. K., Veerkamp, J. H., Epstein, H. F., and Wieringa, B. (1993). Myotonic dystrophy kinase is a component of neuromuscular junctions. *Hum. Mol. Genet.* **2,** 1889–1894.

68. Verkerk, A. J. M. H., Pierretti, M., Sutcliffe, J. S., Fu, Y.-H., Kuhl, D. P. A., Pizzuti, A., Reiner, O., Richards, S., Victoria, M. F., Zhang, F., Eussen, B. E., van Omenn, G. J. B., Kunst, C. B., Galjaard, H., Caskey, C. T., Nelson, D. L., Oostra, B. A., and Warren, S. T. (1991). Identification of a gene (*FMR-1*) containing a CGG repeat coincident with a breakpoint cluster region exhibiting length variation in fragile X syndrome. *Cell* **65,** 905–914.

69. Waring, J. D., Haq, R., Tamai, K., Sabourin, L. A., Ikeda, J.-E., and Korneluk, R. G. (1996). Investigation of myotonic dystrophy kinase isoform translocation and membrane association. *J. Biol. Chem.* **271,** 15187–15193.

70. Watanabe, G., Suito, Y., Madaule, P.., Ishizaki, T., Fujisawa, K., Morii, N., Mukai, H., Ono, Y., Kakizuka, A., and Narumiya, S. (1996). Protein kinase N (PKN) and PKN-related protein rhophilin as targets of small GTPase Rho. *Science* **271,** 645–648.

71. Whiting, E. J., Waring, J. D., Tamai, K., Somerville, M. J., Hincke, M., Staines, W. A., Ikeda, J.-E., and Korneluk, R. G. (1995). Characterization of myotonic dystrophy kinase (DMK) protein in human and rodent muscle and central nervous tissue. *Hum. Mol. Genet.* **4,** 1063–1072.

72. Wissmann, A., Ingles, J., McGhee, J. D., and Mains, P. E. (1997). *Caenorhabditis elegans* LET-502 is related to Rho-binding kinases and human myotonic dystrophy kinase and interacts genetically with a homolog of the regulatory subunit of smooth muscle myosin phosphatase to affect cell shape. *Gene Dev.* **11,** 404–422.

73. Woolfson, D. N., and Alber, T. (1995). Predicting oligomerization states of coiled coils. *Protein Sci.* **4,** 1596–1607.

74. Yarden, O., Plamann, M., Ebbole, D. J., and Yanofsky, C. (1992). *Cot-1,* a gene required for hyphal elongation in *Neurospora crassa,* encodes a protein kinase. *EMBO J.* **11,** 2159–2166.

75. Zhao, Y., Loyer, P., Li, H., Valentine, V., Kidd, V., and Kraft, A. S. (1997). Cloning and chromosomal localization of a novel member of the myotonic dystrbp3ophy family of protein kinases. *J. Biol. Chem.* **272,** 10013–10020.

Is Myotonic Dystrophy (DM) the Result of a Contiguous Gene Defect?

CATHERINE L. WINCHESTER AND KEITH J. JOHNSON
Division of Molecular Genetics, Institute of Biomedical and Life Sciences, University of Glasgow, Anderson College, Glasgow G11 6NU, Scotland

I. INTRODUCTION

Since the discovery of the dynamic $(CTG)_n$ repeat mutation in 1992, characteristics of its somatic and intergenerational instability and the correlation of genotype with disease phenotype have been determined. These insights have explained many of the genetic phenomena associated with DM and have allowed qualitatively predictive prenatal and presymptomatic diagnoses to be made. The basis for these is the fact that no mutation, other than an expanded $(CTG)_n$, has been found in DM patients. This is highly significant, given the prevalence of the disease and the number of patients for whom detailed molecular studies have been performed. However, identification of the causal factors in DM is complicated by the multisystemic nature of the disease and the variable expressivity of the symptoms. Transgenic mice have been generated by the overexpression or lack of expression of *DMPK,* to provide models with which to study DM disease mechanisms. However, the models show subsets of the disease symptoms and a complete DM phenotype has not been replicated. We propose that all these observations indicate that DM is unlikely to be caused by the dysfunction of the *DMPK* gene alone and that other genes contribute to the disease pathogenicity as a result of the $(CTG)_n$ repeat expansion.

The $(CTG)_n$ repeat is situated within a gene-rich region of chromosome 19q13. To date at least 129 genes have been mapped to this region, 25 of which are definitively mapped to 19q13.3, the chromosomal band in which *DMPK* is specifically located [1]. Two of these genes have been identified because of their close proximity to *DMPK* and therefore we hypothesise that they are the most likely additional candidates to be affected by the $(CTG)_n$ expansion mutation. *59* (*DMR-N9* in mice) flanks *DMPK* at its 5′ end and *DMAHP* (DM-associated homeodomain protein) is situated at the 3′ end (Fig. 12-1) [2–4]. Genes immediately flanking *59/DMR-N9* and *DMAHP* are yet to be identified, although exon trapping experiments have located exons from additional genes in the DM locus (Brook, Hamshere, and Alwazzan, personal communication). The identification of these genes has contributed additional components to the puzzle and a much more complex picture of the processes underlying DM is emerging. It is possible that amplification of the $(CTG)_n$ repeat causes subsequent effects at both the DNA level on expression of both *DMAHP* and *59/DMR-N9* and at the RNA level via *DMPK* mRNA. By potentially affecting at least two genes, the $(CTG)_n$ repeat may be acting simultaneously in some patients or tissues by at least two mechanisms. However, in order to understand how these genes contribute to the DM phenotype we need to identify all the genes in the locus, their functions, their interactions, and ultimately how they fit together to result in the complex range of disease symptoms. We have focused our attention on *DMAHP,* the gene immediately downstream of *DMPK* with the repeat in its promoter/enhancer, and therefore the primary candidate for disruption of function by the repeat-mediated changes in chromatin structure at the DM locus.

II. THE EFFECT OF THE $(CTG)_n$ REPEAT ON *DMAHP*

Human and mouse *DMAHP* genes were identified by sequence analysis of the extensive CpG island at the 3′ end of the *DMPK* gene [4]. Three exons were located that span a genomic region of approximately 5 kb, downstream of *DMPK* (Fig. 12-1). The promoter spans the intergenic region as well as the 3′ end of the *DMPK* gene, including the $(CTG)_n$ repeat. Therefore it is likely that the expansion of the $(CTG)_n$ repeat leads to the disruption of the promoter and consequently affects the expression of *DMAHP.*

CTG repeat tracts in promoters have previously been shown to repress transcription. For example, a $(CTG)_{25}$

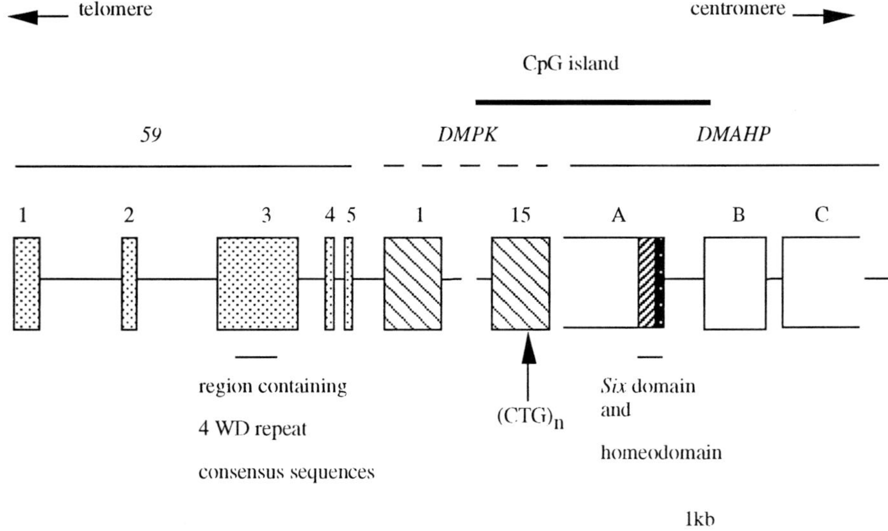

FIGURE 12-1 Genomic organisation of the human DMPK region of chromosome 19q13.3. A schematic diagram showing the proximity of human *59, DMPK,* and *DMAHP* genes. The genes are drawn telomere to centromere, the direction of transcription. The ~3.5-kb CpG island which extends from intron 14 of the *DMPK* gene to intron A of the *DMAHP* gene is represented by a thick black line. Exonic regions are represented by boxes and intronic and intergenic regions are represented by linking lines. Only exons 1 and 15 of the *DMPK* gene are drawn, linked by a dashed line. Exon 3 of *59* gene, the gene immediately flanking *DMPK* at the 5′ end, encodes amino acid sequences that are homologous to WD repeats of eukaryotic regulatory proteins. Exon A of *DMAHP,* the gene immediately flanking *DMPK* at the 3′ end, encodes a homeodomain that is homologous to the *Six* subfamily of homeodomain proteins. The promoter region of *DMAHP* is situated in the CpG island in the 3′ end of the *DMPK* gene.

tract in the promoter of mouse growth inhibitory factor/ metallothionein III gene repressed transcription of the gene *in vitro* [5]. The same repeat tract showed direction- and position-independent repression activity in the promoters of SV40, human metallothionein II_A gene, and glutathione *S*-transferase P gene [5]. The insertion of a $(CTG)_5$ tract in the promoter of the *dio1* gene of C3H mice resulted in repression of transcription and a consequent decrease in deiodinase activity that led to hyperthyroxinemia [6].

Repression of transcription has also been implicated by the discovery of the strong nucleosome assembly at expanded $(CTG)_n$ repeats [7–10]. Nucleosomes are known to repress transcription initiation over promoters which indicates that nucleosome formation at expanded $(CTG)_n$ repeats could lead to haploinsufficiency of *DMAHP*. Further evidence which implicates the involvement of DMAHP through the expansion of the $(CTG)_n$ repeat is the loss of DNase I hypersensitivity, which, along with nucleosome assembly, suggests that the local chromatin structure is altered. A number of alternative DNA conformations have been physically demonstrated for $(CTG)_n$ repeats [11–17] and predicted by computer modelling [12, 18]. Disorganisation of the chromatin structure around the $(CTG)_n$ repeat has implications for *DMAHP* as well as *DMPK* and other genes in the locus and could have a number of effects, including preventing the binding of regulatory or interacting proteins, as well as altering gene expression.

Finally, in a number of plant systems and *Drosophila*, the tandem expansions of transgenes cause euchromatin to form heterochromatin, resulting in the silencing of a gene or genes in the region [19, 20]. This is another possible mechanism by which the $(CTG)_n$ repeat expansion may have an effect at the transcriptional level on *DMAHP* and other genes at the DM locus.

The functions of normal-sized $(CTG)_n$ and $(CUG)_n$ repeats are currently being investigated and it is hoped that this will elucidate a possible role for the $(CTG)_n$ repeat in the pathogenicity of DM. Two single-stranded (ss) CTG-binding protein complexes have been identified in protein extracted from human brain [21]. Only one complex was specific for the $ss(CTG)_{10}$ oligonucleotide, whilst the other complex interacted with a $ss(CUG)_{10}$ oligonucleotide as well. A group of double-stranded (ds) $(CTG)_6$-, $ds(CTG)_{10}$-, and $ss(CTG)_8$-binding proteins have been identified in proteins extracted from HeLa cells [22]. Nuclear proteins of 21, 34, 37, 45, and 65 kDa were isolated from three $ds(CTG)_6$ complexes, whereas only one cytoplasmic $ss(CTG)_8$ complex was identified and it also interacted with a $(CUG)_8$ oligonucleotide. The interactions of the nuclear $ds(CTG)_6$ binding proteins were similar in lymphoblast cells from normal individuals and DM patients [23]. The effect on the binding of these proteins or other proteins to expanded CTG repeat tracts is currently being investigated. A nuclear protein extracted from HeLa cells has also been shown to specifically interact with a $ds(CGG)_{10}$ repeat of FRAXA [24].

It has recently been hypothesised that *DMPK* transcripts containing expanded $(CUG)_n$ repeats disrupt cellular RNA metabolism in a general manner. To date, the expression of the insulin receptor gene has been shown to be decreased in DM patients at the RNA and protein levels [25]. Although this has implications for an underlying pathological process in DM, the expression of other candidate genes, such as those corresponding to biochemical abnormalities associated with DM, will need to be explored.

A. Genomic Structure, Alternative Transcripts, and Putative Protein Products of *DMAHP*

The human and mouse *DMAHP* exonic sequences are highly homologous and share 86% identity at the amino acid level. The predicted amino acid sequence of the largest open reading frame shows significant homology to the *Six* homeodomain of a recently isolated subfamily of homeodomain proteins.

RT-PCR analysis of human and mouse *DMAHP* genes identified two alternative transcripts [4, 26]. One isoform contained exons A, B, and C and the other isoform had exon B spliced out (Fig. 12-2). The predicted proteins (A-B-C isoform of 82 kDa and A-C isoform of 36 kDa) contain the *Six* domain and the homeodomain, encoded by exon A, but differ at the C-terminus, as the splicing out of exon B causes the use of an alternative reading frame in exon C. The translated products of exon B and C do not show significant homology with any known proteins. However, there are proline-rich stretches which show low homology with the transactivation domains of some transcription factors.

Northern blotting and RT-PCR data have shown that the expression of *DMAHP* (Table 12-1) is similar to that of *DMPK*. Transcripts have been detected in a wide range of adult and embryonic tissues, including brain, heart, and skeletal muscle. Both isoforms appear to be similarly expressed in human tissues, but the exon A-B-C isoform is the major transcript in mice, with the exon A-C isoform only detected in mouse liver [26]. Transgenic analysis of the murine *dmahp* promoter activity showed that, despite variation in the mouse lines, expression was detected in some of the structures associated with the central nervous system, sensory organs, and the musculo-skeletal system [26].

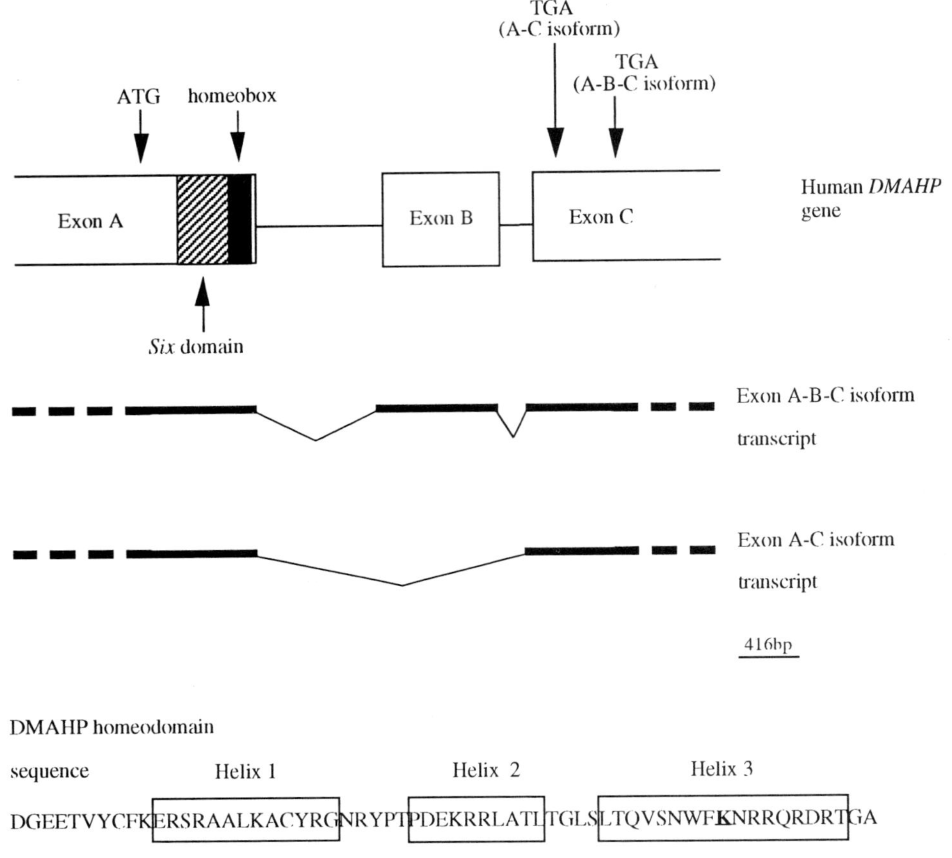

FIGURE 12-2 Structure of the human *DMAHP* gene, alternative transcripts and the homeodomain. A schematic diagram of the human *DMAHP* gene and the exon A-B-C and exon A-C alternative splice isoform transcripts. The dashed lines represent the unknown length of the 5′UTR or the 3′UTR and the bold lines represent transcribed regions. The sequence of the DMAHP homeodomain is shown with the amino acid residues constituting the structural helical domains boxed; the conserved lysine at position 9 of the recognition helix (helix 3) is in bold and underlined (K).

To date, nothing is known about the expression of *DMAHP* at the protein level or its putative function as a transcription factor. However, *in vitro* studies are in progress to identify the protein domains required for DNA binding and the specific DNA target sites (Sarah Harris, personal communication).

B. *Six* Subfamily of Homeodomain Encoding Genes

There are now at least six members of the *Six* subfamily of homeodomain protein encoding genes (Table 12-1), including a recently identified *C. elegans* gene (Accession No. z77670). The subfamily proteins are characterised by a distinctive homeodomain that has diverged considerably from previously described homeodomains, as no more than 30% identity is shared with other members of this gene family [27, 28]. A conserved lysine residue is found at position 9 of the *Six* homeodomain recognition helix [28] (Fig. 12-2). *Drososphila bicoid, goosecoid,* and *orthodenticle* genes and *C. elegans EgHbx 4* gene also contain a lysine at this position, which is thought to be crucial for DNA binding specificity. The region immediately flanking the 5′ end of the homeobox, encoding 110 amino acids, is also conserved in this subfamily of homeobox genes. The region has been called the *Six* domain and is thought also to be involved in DNA binding specificity. Other classes of homeodomain proteins, such as POU and PAX, also have additional DNA binding domains [29].

The *Six* subfamily genes are nonclustered homeodomain encoding genes, expressed in the anterior region of developing embryos. *Drosophila so* and murine *Six1, Six2,* and *Six3* genes have also been shown to be expressed anteriorally; *Six3* is thought to be the most anterior homeobox gene to date (Table 12-1). Another feature of the *Six* subfamily proteins is that apart from *Six1,* all the genes are expressed in the eye and are postulated to be involved in the determination and maintenance of eye formation [30]. Although *DMAHP* expression was not observed in the retina of promoter-

TABLE 12-1 Expression of the *Six* Subfamily of Homeobox Genes, in Order of Homology with Human and Mouse *DMAHP*

Gene	Expression data	Ref.
Human *DMAHP* (A-B-C isoform)	Transcripts detected in adult brain, fibroblast cell line, heart, and skeletal muscle, and DM patient skeletal muscle and fibroblast cell line	[4]
Human *DMAHP* (A-C isoform)	Transcripts detected in adult brain, fibroblast cell line, heart, and skeletal muscle, and DM patient skeletal muscle and fibroblast cell line	[4]
Mouse *DMAHP* (A-B-C isoform)	Transcripts detected in 12.5-day embryo head, limb, liver, kidney, and heart, adult brain, heart, kidney, liver, skeletal muscle, smooth muscle, and thymus and inner and outer nuclear layer, ganglion cell layer, and pigment epithelium of retina	[26]
Mouse *DMAHP* (A-C isoform)	Transcript detected in adult liver	[26]
Mouse *AREC3* (*Six4*)	Transcripts and protein detected in adult inner and outer nuclear layer, ganglion cell layer, and pigment epithelium of retina and in developing skeletal muscle	[30, 34]
Mouse *Six2*	Transcripts detected from embryonic d8.5 in head mesoderm and then in visceral smooth muscle, metanephros, genitalia, and a restricted cell population of the hind brain; expression in the developing limb tendons starts at day 11.5; transcripts also detected in adult inner and outer nuclear layer, ganglion cell layer, and pigment epithelium of retina	[28] [30]
Mouse *Six3*	Transcripts detected from embryonic day 6.5 in the anterior border and later in the anterior neural plate, tissues derived from it, and the ectoderm and in the developing eye; transcripts also detected in adult inner and outer nuclear layer, ganglion cell layer, and pigment epithelium of retina	[32] [30]
Human *SIX1*	Transcript detected in adult skeletal muscle	[33]
Mouse *Six1*	Transcripts detected from embryonic day 8.2–8.5 in the most anterior head mesoderm and then in mesodermally derived skeletal muscle, dorsal root ganglia, and pituitary gland; expression in the developing limb tendons starts at day 10.5	[28]
Drosophila sine oculis (*so*)	Transcripts and protein detected in all stages of the developing compound eye; RNA is transient and disappears by the ivagination of the optic lobe primordium	[27, 35]

transgenic embryonic mice, as reported for other members of the *Six* family genes, expression was observed in embryonic posterior lens capsules. This is of interest with regards to the DM phenotype, because of the formation of unique iridescent cataracts and the development of retinal symptoms. Recently it was shown that the transgenic expression of the murine *Six3* gene in fish embryos promoted ectopic lens formation in the area of the otic vesicle, which normally gives rise to the inner ear [31].

RT-PCR and Northern blotting analyses of the *Six* homeobox genes have identified, as for *DMAHP,* a number of alternatively spliced transcripts, all of which contained the homeobox [28, 30, 32–34]. Two of the mouse *Six3* splice isoforms that were analysed contained regions that used different reading frames [30].

The function of the protein products of the *Six* homeobox genes has been investigated in *Drosophila* mutants and by *in vitro* DNA binding analyses. A number of *Drosophila so* mutants have been identified and they all show defects in the visual system. The original so^1 mutant is eyeless but other *so* mutants cause abnormalities in the adult compound eye, the ocelli, the larval eye, and the optic lobes [27, 35]. DNA binding studies of recombinant so proteins have identified a 63-aa peptide that is sufficient for specific binding to a 6-bp target site, CGATAC [36]. The mutants, the predicted homeodomain, and the expression patterns suggest that the so protein plays an essential role in controlling the transcription of genes that are involved in the morphogenesis and pattern formation in the developing visual system.

AREC3 protein was originally identified by its binding to the Na^+, K^+ ATPase $\alpha 1$ subunit gene regulatory element (ARE) in myoblast cells [34]. The ARE is a positive regulatory region of the Na^+, K^+ ATPase $\alpha 1$ subunit gene, which is expressed in all tissues and is most important for cellular homeostasis. The use of recombinant GST- and GAL4- AREC3 proteins has enabled the DNA binding domain and the transactivation domains of AREC3 to be defined. The *Six* domain and the homeodomain are required for ARE-binding but the homeodomain showed a nonspecific binding activity or one of other sequence specificity [34]. The protein regions encoded by the remaining two exons of the *AREC3* gene had no effect on the activity or specificity of DNA binding, although the C terminus of the protein was identified as a transactivation domain. The DNA binding site was identified as GGNGNCNGGTTGC and therefore, despite so and AREC3 being members of a family of homeodomain proteins, the target binding sites are different. The identification of a DNA substrate and the expression patterns suggest that the AREC3 protein plays a role in controlling the transcription of genes involved in the developing visual system and ion channels.

Recombinant GST fusion proteins of the *Six* domain and the homeodomain of Six2, Six3, and Six5 (DMAHP) have also been used to determine DNA binding activity. Six2 and Six5 (DMAHP) bound specifically to the ARE consensus sequence of the Na^+, K^+ ATPase $\alpha 1$ subunit gene but Six3 interacted with an as yet unidentified DNA sequence [30]. The predicted homeodomain, expression patterns, and the preliminary evidence of DNA binding activity indicate that these proteins regulate the transcription of genes in specific pathways. It has been shown that *Six1* and *Six2* genes are expressed in a complementary manner during the development of limb tendons and *Six3* is expressed in the developing eye and anterior neural ridge.

C. Putative Roles for DMAHP in the Pathology of DM

DMAHP encodes a homeodomain protein and therefore its normal function is likely to be regulating gene expression during development and/or in adult homeostasis. A change in the expression of DMAHP as a result of the $(CTG)_n$ expansion mutation could therefore lead to the congenital abnormalities seen in CDM or cause errors in development that do not become apparent until adulthood. A characteristic feature of CDM is the presence of immature muscle fibres, indicating an arrest in development. Therefore one possibility is that dysfunction of *DMAHP* leads to the skeletal muscle problems incurred in CDM (Table 12-2). We can postulate that the massive $(CTG)_n$ repeat expansions seen in CDM during fetal development result in the acute congenital problems. The smaller but continually expanding repeat sizes of the classical adult-onset form of the disease develop because of less major disruption of muscle development, but may give rise to regionalised breakdown in muscle homeostasis where large repeats occur. *AREC3* and *Six1* have also been shown to be expressed in developing skeletal muscle, and the effect of the repeat may impinge on the normal function of these gene products if they are dependent on interactions with DMAHP to express their normal repertoire of transcriptional regulation.

Dysfunction of a homeobox gene, such as *DMAHP*, leading to disease is not an uncommon phenomenon; mutations in a number of homeobox genes have been shown to be responsible for human disorders, including congenital abnormalities (Table 12-3). For example, *PAX6* mutations have been shown to be involved in

TABLE 12-2 Candidate Contributors to the DM Phenotype (Excluding *DMPK*)

DM symptoms	Gene dysfunction predicted to be responsible	Ref.
Immature muscle fibres in CDM	*DMAHP*, expressed in skeletal muscle	[4, 26]
Skeletal muscle wasting in adult onset DM		
Cataracts	*DMAHP*, expressed in the posterior lens capsule	[26]
Myotonia	*DMAHP*, expressed in skeletal muscle and the protein binds to the Na^+, K^+ ATPase $\alpha 1$ subunit gene regulatory element	[4, 26, 30]
Facial ptosis	*DMAHP*, expressed in the ganglion of the trigeminal (V) and facial (VII) nerves	[26]
Cardiac conduction defects	*DMAHP*, expressed in the ganglion of the vagus (X) nerve	[26]
Respiratory problems	*DMAHP*, expressed in the ganglion of the vagus (X) nerve	[26]
Gut problems	*DMAHP*, expressed in the ganglion of the vagus (X) nerve	[26]
Testicular atrophy	*59/DMR-N9*, expressed in the testis; transcription termination altered when *DMPK* is knocked out	[2, 3, 43]
Cognitive problems	*59/DMR-N9*, expressed in the testis; transcription termination altered when *DMPK* is knocked out	[2, 3, 43]

congenital eye malformations. *DMAHP* homologues, *so*, *Six2*, *Six3*, and *AREC3*, are expressed in the developing eye. Although an eye phenotype is associated with DM, the cataracts develop late in the life of classical adult-onset patients and in presymptomatic carriers, which does not correlate with a congenital abnormality. However, it is possible that the pattern of dysfunction was laid down during development or occurred when regeneration of certain cell types was required later in life.

DMAHP shows the most homology to AREC3, which specifically binds to ARE. This is of interest with respect to DM because of the clinical symptom of myotonia. Na^+, K^+ ATPase is the enzyme responsible for maintaining the Na^+ and K^+ gradients across the cell membrane after the propagation of an action potential. Recombinant DMAHP protein has been shown to spe-

TABLE 12-3 Examples of Human Diseases Caused by Mutations in Homeobox Genes

Homeobox gene	Disease (mode of inheritance)	Phenotype	Ref.
HOXA13	Hand–foot–genital syndrome (autosomal dominant)	Reduced anterior and posterior digits; females have defects in Mullerian duct fusion and males have a displaced urethra	[45]
HOXA9	Myeloid leukaemia (translocational fusion)	Lymphocyte cancer	[46, 47]
HOXD13	Syndactyly (autosomal dominant)	Abnormality of hands and feet	[48]
PAX3	Waardenburg's syndrome (autosomal dominant)	Deafness and pigmentary disturbances, caused by defective function of the neural crest	[49]
PAX6	Aniridia, Peter's anomaly, and keratitis (autosomal dominant)	Congenital malformation of the eye, characterised by iris hypoplasia anterior segment malformations; eye disorder, characterised by corneal opacification and vascularization and by foveal hypoplasia	[50, 51]
PTC (patched)	Basal cell nevus syndrome (autosomal dominant)	Increased frequency of basal cell carcinoma, medulloblastoma, and developmental abnormalities; tendency towards polydactyly and syndactyly	[52, 53]
Sonic hedgehog	Holoprosencephaly (autosomal dominant)	Developmental defect of the forebrain and midface	[54, 55]

cifically bind to the ARE consensus sequence and therefore a change in DMAHP expression may be responsible for the repetitive discharges of action potential seen in DM. Interestingly, the promoter region of the *DMPK* gene contains a sequence very similar to that of the ARE binding site. A second sequence in the promoter of the *DMPK* gene is similar to that utilised by homeodomain proteins, such as bicoid, that have a conserved lysine residue at position 9 of the recognition helix. Such a residue is found in the recognition helix of the putative DMAHP amino acid sequence. It is possible that the altered DNA conformations seen in $(CTG)_n$ repeat sequences may affect the binding of DMAHP protein to the *DMPK* gene. Alternatively, these conformations may inhibit the expression of DMAHP so that there is less protein available to regulate *DMPK*. Maternally derived bicoid protein has recently been shown to regulate the expression of maternal *caudal* mRNA, as well as regulating the transcription of *Drosophila* embryonic genes [37, 38]. As DMAHP shares the conserved lysine residue in the recognition helix of bicoid, which is now thought to be critical for RNA binding as well as for DNA binding, DMAHP may also have RNA binding properties.

Mice transgenic for *dmahp* promoter constructs showed that *DMAHP* is expressed in a specific subset of cranial nerves and sensory tissues, as well as muscle, and it is possible that the expression pattern indicates the contribution of DMAHP to the DM phenotype. Expression was observed in the ganglia of cranial nerves V, VII, IX, and X, which supply sensory tissues of the mouth and ear, as well as the muscles for mastication and facial expression. Cranial nerve X, the vagus nerve, contains afferent fibres that are important in the reflex control of cardiovascular, respiratory, and alimentary functions. One can therefore speculate that dysfunction of *DMAHP* could lead to the facial ptosis or the cardiac, respiratory, or gut problems seen in DM.

Preliminary data have shown that the expression of DMAHP is down-regulated in DM patients (C. Thornton, personal communication). The synthesis of *DMAHP* RNA from chromosomes with expanded $(CTG)_n$ repeats was much less than the synthesis from normal chromosomes. The size of the repeat affected the level of transcription, as *DMAHP* transcription was lower from chromosomes with alleles with $(CTG)_{>100}$ than chromosomes with alleles with $(CTG)_{<100}$.

DMAHP may contribute to certain features of DM, and therefore point mutations in *DMAHP* may lead to a subset of symptoms. A change of expression of *DMAHP* in the cranial nerves could give myopathic faces but with no myotonia. There are also a number of disorders with symptoms similar to aspects of the DM phenotype, which map to the DM locus. For example, cardiac conduction defects are a problem in DM and a gene for progressive familial heart block type I maps to chromosome 19q13 [39]. Recently genes for nonsyndromic deafness (DFNA4) and Iraqi–Jewish atrophy plus have been shown to be linked to the *DMPK* CTG repeat [11, 40]. Mice transgenic for the *dmahp* promoter showed expression in the semicircular canals of the ear and in the ganglion of the vagus nerve which, amongst other organs, contributes sensory fibres to the wall of the acoustic canal and the tympanic membrane. It is possible that dysfunction of *DMAHP* alone may result in nonsyndromic deafness.

III. ADDITIONAL CANDIDATE GENES

59/DMR-N9 is currently being studied, as well as *DMPK* and *DMAHP,* as a contributor to the multisystemic DM phenotype because of its proximity to the $(CTG)_n$ repeat. As well as these three genes, one can also speculate on the possible involvement of other genes located in the vicinity of the $(CTG)_n$ repeat, whose expression may be altered by the gross disorganisation of the chromatin structure. It is possible that mutations in the single genes could lead to a subset of symptoms or disorders with overlapping phenotypes. The symptoms of nonsyndromic deafness, Iraqi–Jewish optic atrophy plus, and progressive familial heart block type I may be attributed to dysfunction of *DMAHP* (as described above), but it is possible they result from mutations in any of the genes in the DM locus. Other genes are implicated by the very small number of DM patients who lack $(CTG)_n$ repeat mutations [41, 42; A. Sermoni, personal communication]. The *DMPK* gene of a number of these patients has been fully sequenced and no point mutations have been identified. Therefore it is likely that their symptoms result from point mutations in other genes which mimic the disruption caused by the $(CTG)_n$ repeat expansion. However, to date no such point mutations have been identified in *DMAHP.*

A. *59/DMR-N9* as a Candidate Gene

59/DMR-N9 and *DMPK* are also closely linked in mouse and human sequences, with the distance from the polyA tail of *59/DMR-N9* to the 5′ end of the *DMPK* gene being approximately 500 bp in humans and 1.1 kb in mice (Fig. 12-2). The predicted amino acid sequences of the mouse and human genes share 92.7% identity and have homology with WD-repeat sequences [43]. WD-repeat sequences are found in a number of eukaryotic proteins that are involved in the control of cell division, transcription, pre-mRNA processing, cytoskeletal assembly, vesicle fusion, and signal transduction via G proteins [43]. This homology enables a number of functions to be predicted for the protein product of *59/DMR-N9.*

Northern blotting and *in situ* hybridisation analyses have identified *59* transcripts in baboon brain and heart and ubiquitous low levels of *DMR-N9* expression in all mouse embryonic tissues and in a number of adult tissues, with the highest levels in brain and testis (Table 12-4) [2, 3, 43].

To date there are no clinical data to suggest the involvement of *59/DMR-N9* in the pathology of DM, but the genes are highly expressed in testis and brain,

TABLE 12-4 Expression of *DMR-N9*

Developmental stage	Tissue transcripts detected in:	Ref.
Embryonic mouse	Placenta—foetal regions but not in the decidua 9.5 d.p.c.—all parts of the embryo 14.5 d.p.c.—high expression in neural tissues (telencephalon, retina, and spinal cord) 16.5 d.p.c.—high expression in telencephalon and mesencephalon, tongue, and rib muscles	[43]
Infant mouse	2 d.p.p—olfactory bulb, granular layer of the neopallic cortex	[43]
Adult mouse	Bladder, brain, heart, kidney, liver, lung, ovarium, spleen, seminal vesicle, skeletal muscle, testis, uterus	[2, 43]

tissues which are affected in DM [43]. However, a transgenic mouse with the 5′ end of the *DMPK* gene "knocked-out" affected the transcriptional termination, but not the levels of expression, of the *59/DMR-N9* gene, suggesting that *DMPK* and *59/DMR-N9* might be coregulated at the transcriptional level [44].

IV. MODEL

Whilst it is likely that a defect in *DMPK* expression/function at the RNA level can explain some of the observed changes in DM, it is becoming increasingly likely that the complex pattern of symptoms is the result of dysfunction of a number of genes. We propose a model to address some of the phenotypic anomalies that are unlikely to be explained by a defect in one gene and therefore indicate the involvement of additional genes such as *DMAHP* and *59/DMR-N9.*

Our model attempts to explain the three different disease classes, the variation in symptoms within these classes, and the multitude of organs affected. We hypothesise that the $(CTG)_n$ repeat expansion has a dominant negative effect at the *DMPK* mRNA level that is mediated by an alteration in the metabolism of *DMPK* transcripts and the interaction of $(CUG)_n$-binding proteins but also alters the local chromatin structure, which consequently affects the expression of *DMAHP* and possibly other genes in the locus. We propose that dysfunction of these genes is correlated with the size of the $(CTG)_n$ repeat. Therefore, small expansions, typical of minimally affected patients and asymptomatic carriers, result in minimal disruption of gene expression but a phenotype is seen in tissues such as lens that have a lower tolerance threshold. It is possible that only *DMPK* is disrupted by the smaller changes in $(CTG)_n$ repeat length, although minor changes in *DMAHP* expression in this tissue may be causative or contributory.

Larger expansions that are characteristic of adult onset disease could lead to greater changes in chromosomal topology and therefore have more dramatic effects on *DMPK* mRNA metabolism and protein binding and the dysfunction of additional genes. Alteration of *DMAHP* expression could result in the failure of muscle and cranial nerves to recruit sufficient progenitor cells during development *in utero* to sustain growth and development throughout adult life. The level of dysfunction does not reach a clinically significant threshold until adulthood and therefore symptoms are not seen until then. The ongoing somatic instability of the $(CTG)_n$ repeat throughout life may explain the variable expressivity as a result of different levels of disruption in tissues from the same patient. Alternatively or additionally they could cause the disruption of the function of other genes, which could lead to the development of further symptoms. Testicular atrophy and cognitive problems may be explained in this way by the additional dysfunction of *59* in the brain and testes.

Finally, the massive expansions seen in CDM patients could lead to gross chromosomal changes which affect a number of genes that have developmental importance. Indeed, clinically CDM is reminiscent of contiguous gene defects, with multisystemic involvement and mental retardation. The acute symptoms seen at birth are entirely different than those seen in the classical, adult-onset disease and are not simply an earlier, more severe form of the disease. Survival of the neonatal period often leads to a period of relevant improvement, before the onset of the typical adult symptoms of myotonia and muscle wasting.

V. CONCLUSION

With the increasing acceptance that despite its remarkable genetic homogeneity the DM phenotype is too complex to be explained by the dysfunction of a single gene, it is imperative that extremely rigorous, high-resolution gene expression studies are performed on tissues from patients, controls, and animal models. The spatial and temporal expression of the individual genes at the DM locus and their functions need to be determined to tease out the contributions of individual genes to the pathogenic processes involved in DM. If DMAHP functions as a DNA/RNA binding protein in the developing embryo, it is likely to be part of a cascade of regulatory proteins and its own transcription may be induced and controlled by the action of other homeodomain proteins. A detailed understanding of the expression of DMPK and DMAHP proteins and how the $(CTG)_n$ repeat mutation affects them is still required. Haploinsufficiency of these genes needs to be investigated in a spectrum of tissues from patients of a range of ages, $(CTG)_n$ repeat lengths, and phenotypes for the complete picture to be revealed. Until isoform and tissue-specific expression have been determined it will be impossible to understand the abnormalities that occur in the DM phenotype and whether DM is a single gene disorder. The effect of the $(CTG)_n$ repeat expansion on the normal function of the genes, the gene products, and their expression and cellular localisation need to be fully characterised, if possible at the single-cell level of resolution. Only when such detailed and exhaustive studies have been performed will we be able to say with any certainty which gene products are critical to the phenotype development, and begin to intervene clinically to allay the pathological processes.

References

1. Mohrenweiser, H., Olsen, A., Archibald, A., Beattie, C., Burmeister, M., Lamerdin, J., Lennon, G., Stewart, E., Stubbs, L., Weber, J. L., and Johnson, K. (1996). Report of the 3rd international workshop on human chromosome 19 mapping 1996. *Cytogenet. Cell Genet.* **74,** 162–181.
2. Jansen, G., Mahadevan, M., Amemiya, C., Wormskamp, N., Segers, B., Hendriks, W., O'hoy. K., Baird, S., Sabourin, L., Lennon, G., Jap, P. L., Iles, D., Coerwinkel, M., Hofker, M., Carrano, A. V., de Jong, P. J., Korneluk, R. G., and Wieringa, B. (1992). Characterization of the myotonic dystrophy region predicts multiple protein isoform- encoding mRNAs. *Nature Genet.* **1,** 261–266.
3. Shaw, D. J., McCurrach, M., Rundle, S. A., Harley, H. G., Crow, S. R., Sohn, R., Thirion, J. P., Hamshere, M. G., Buckler, A. J., Harper, P. S., Housman, D. E., and Brook, J. D. (1993). Genomic organization and transcriptional units at the myotonic dystrophy locus. *Genomics* **18,** 673–679.
4. Boucher, C. A., King, S. K., Carey, N., Krahe, R., Winchester, C. L., Rahman, S., Creavin, T., Meghji, P., Bailey, M. E. S., Chartier, F. L., Brown, S. D., Siciliano, M. J., and Johnson, K. J. (1995). A novel homeodomain encoding gene is associated with a large CpG island interrupted by the myotonic dystrophy unstable $(CTG)_{(n)}$ repeat. *Hum. Mol. Genet.* **4,** 1919–1925.
5. Imagawa, M., Ishikawa, Y., Shimano, H., Osada, S., and Nishihara, T. (1995). CTG triplet repeat in mouse growth inhibitory factor/metallothionein III gene promoter represses the transcriptional activity of the heterologous promoters. *J. Biol. Chem.* **270,** 20898–20900.
6. Maia, A. L., Berry, M. J., Sabbag, R., and Harney, J. W. (1995). Structural and functional differences in the *dio1* gene in mice with inherited type 1 deiodinase deficiency. *Mol. Endocrinol.* **9,** 969–980.
7. Wang, Y.-H., Amirhaeri, S., Kang, S., Wells, R. D., and Griffith, J. D. (1994). Preferential nucleosome assembly at DNA triplet repeats from the myotonic dystrophy gene. *Science* **265,** 669–671.
8. Otten, A. D., and Tapscott, S. J. (1995). Triplet repeat expansion in myotonic dystrophy alters the adjacent chromatin structure. *Proc. Natl. Acad. Sci.* **92,** 5465–5469.

9. Wang, Y.-H., and Griffith, J. (1995). Expanded CTG triplet blocks from the myotonic dystrophy gene create the strongest known natural nucleosome positioning elements. *Genomics* **25,** 570–573.
10. Godde, J. S., and Wolffe, A. P. (1996). Nucleosome assembly on CTG triplet repeats. *J. Biol. Chem.* **271,** 15222–15229.
11. Chen, X. A., Mariappan, S. V. S., Catasti, P., Ratliff, R., Moyzis, R. K., Laayoun, A., Smith, S. S., Bradbury, E. M., and Gupta, G. (1995). Hairpins are formed by the single DNA strands of the fragile X triplet repeats structure and biological implications. *Proc. Natl. Acad. Sci.* **92,** 5199–5203.
12. Gacy, A. M., Goellner, G., Juranic, N., Macura, S., and McMurray, C. T. (1995). Trinucleotide repeats that expand in human disease form hairpin structures in vitro. *Cell* **81,** 533–540.
13. Zhu, L. M., Chou, S. H., Xu, J. D., and Reid, B. R. (1995). Structure of a single cytidine hairpin loop formed by the DNA triplet GCA. *Nature Struct. Biol.* **2,** 1012–1017.
14. Mariappan, S. V. S., Garcia, A. E., and Gupta, G. (1996). Structure and dynamics of the DNA hairpins formed by tandemly repeated CTG triplets associated with myotonic dystrophy. *Nucleic Acids Res.* **24,** 775–783.
15. Pearson, C. E., and Sinden, R. R. (1996). Alternative structures in duplex DNA formed within the trinucleotide repeats of the myotonic dystrophy and fragile X loci. *Biochemistry* **35,** 5041–5053.
16. Petruska, J., Arnheim, N., and Goodman, M. F. (1996). Stability of intrastrand hairpin structures formed by the CAG/CTG class of DNA triplet repeats associated with neurological diseases. *Nucleic Acids Res.* **24,** 1992–1998.
17. Zheng, M. X., Huang, X. N., Smith, G. K., Yang, X. Y., and Gao, X. L. (1996). Genetically unstable CXG repeats are structurally dynamic and have a high propensity for folding. an NMR and UV spectroscopic study. *J. Mol. Biol.* **264,** 323–336.
18. Kuryavyi, V. V., and Jovin, T. M. (1995). Triad-DNA: a model for trinucleotide repeats. *Nature Genet.* **9,** 339–341.
19. Assaad, F. A., Tucker, K. L., and Signer, E. R. (1993). Epigenetic repeat-induced gene silencing (RIGS) in *Arabidopsis. Plant Mol. Biol.* **22,** 1067–1085.
20. Dorer, D. R., and Henikoff, S. (1994). Expansions of transgene repeats cause heterochromatin formation and gene silencing in Drosophila. *Cell* **77,** 993–1002.
21. Bhagwati, S., Ghatpande, A., and Leung, B. (1996). Identification of two nuclear proteins which bind to RNA CUG repeats: significance for myotonic dystrophy. *Biochem. Biophys. Res. Commun.* **228,** 55–62.
22. Timchenko, L. T., Timchenko, N. A., Caskey, C. T., and Roberts, R. (1996). Novel proteins with binding specificity for DNA CTG repeats and RNA CUG repeats: implications for myotonic dystrophy. *Hum. Mol. Genet.* **5,** 115–121.
23. Timchenko, L. T., Miller, J. W., Timchenko, N. A., Devore, D. R., Datar, K. V., Lin, L. J., Roberts, R., Caskey, C. T., and Swanson, M. S. (1996). Identification of a (CUG)(n) triplet repeat RNA binding protein and its expression in myotonic dystrophy. *Nucleic Acids Res.* **24,** 4407–4414.
24. Richards, R. I., Holman, K., Yu, S., and Sutherland, G. R. (1993). Fragile X syndrome unstable element, p(CCG)n, and other simple tandem repeat sequences are binding-sites for specific nuclear proteins. *Hum. Mol. Genet.* **2,** 1429–1435.
25. Morrone, A., Pegoraro, E., Angelini, C., Zammarchi, E., Marconi, G., and Hoffman E. P. (1997). RNA Metabolism in myotonic dystrophy. *Clin. Invest.* **99,** 1691–1698.
26. Heath, S. K., Carne, S., Hoyle, C., Johnson, K., and Wells, D. (1997). Characterisation of expression of *mDMAHP,* a homeodomain-encoding gene at the murine DM locus. *Hum. Mol. Genet.* **6,** 651–657.
27. Cheyette, B. N. R., Green, P. J., Martin, K., Garren, H., Hartenstein, V., and Zipursky, S. L. (1994). The Drosophila *sine oculis* locus encodes a homeodomain-containing protein required for the development of the entire visual system. *Neuron* **12,** 977–996.
28. Oliver, G., Wehr, R., Jenkins, N. A., Copeland, N. G., Cheyette, B. N. R., Hartenstein, V., Zipursky, S. L., and Gruss, P. (1995). Homeobox genes and connective tissue patterning. *Development* **121,** 693–705.
29. Gehring, W. J., Affolter, M., and Burglin, T. (1994). Homeodomain proteins. *Annu. Rev. Biochem.* **63,** 487–526.
30. Kawakami, K., Ohto, H., Takizawa, T., and Saito, T. (1996). Identification and expression of *six* family genes in mouse retina. *FEBS Lett.* **393,** 259–263.
31. Oliver, G., Loosli, F., Koster, R., Wittbrodt, J., and Gruss, P. (1996). Ectopic lens induction in fish in response to the murine homeobox gene *Six3. Mech.Dev.* **60,** 233–239.
32. Oliver, G., Mailhos, A., Wehr, R., Copeland, N. G., Jenkins, N. A., and Gruss, P. (1995). *Six3,* a murine homologue of the *sine oculis* gene, demarcates the most anterior border of the developing neural plate and is expressed during eye development. *Development* **121,** 4045–4055.
33. Boucher, C. A., Carey, N., Edwards, Y. H., Siciliano, M. J., and Johnson, K. J. (1996). Cloning of the human *SIX1* gene and its assignment to chromosome 14. *Genomics* **33,** 140–142.
34. Kawakami, K., Ohto, H., Ikeda, K., and Roeder, R. G. (1996). Structure, function and expression of a murine homeobox protein AREC3, a homologue of *Drosophila sine oculis* gene product, and implication in development. *Nucleic Acids Res.* **24,** 303–310.
35. Serikaku, M. A., and O'Tousa, J. E. (1994). *Sine oculis* is a homeobox gene required for Drosophila visual system development. *Genetics* **138,** 1137–1150.
36. Hazbun, T. R., Stahura, F. L., and Mossing, M. C. (1997). Site-specific recognition by an isolated DNA-binding domain of the sine oculis protein. *Biochemistry* **36,** 3680–3686.
37. Dubnau, J., and Struhl, G. (1996). RNA recognition and translational regulation by a homeodomain protein. *Nature* **379,** 694–699.
38. Rivera-Pomar, R., Niessing, D., Schmidt-Ott, U., Gehring, W. J., and Jackle, H. (1996). RNA binding and translational suppression by bicoid. *Nature* **379,** 746–749.
39. Brink, P. A., Ferreira, A., Moolman, J. C., Weymar, H. W., van der Merwe, P. L., and Corfield, V. A. (1995). Gene for progressive familial heart block type 1 maps to chromosome 19q13. *Circulation* **91,** 1633–1640.
40. Nystuen, A., Costeff, H., Elpeleg, O. N., Apter, N., Bonne-Tamir, B., Mohrenweiser, H., Haider, N., Stone, E. M., and Sheffield, V. C. (1997). Iraqi-Jewish kindreds with optic atrophy plus (3-methylglutaconic aciduria type 3) demonstrate linkage disequilibrium with the CTG repeat in the 3′ untranslated region of the myotonic dystrophy protein kinase gene. *Hum. Mol. Genet.* **6,** 563–569.
41. Meiner, A., Wolf, C., Carey, N., Okitsu, A., Johnson, K., Shelbourne, P., Kunath, B., Sauerman, W., Thiele, H., Kupferling, P., and Kuner, E. (1995). Direct molecular analysis of myotonic dystrophy in the German population: important considerations in genetic counselling. *J. Med. Genet.* **32,** 645–649.
42. Abbruzzese, C., Krahe, R., Liguori, M., Tessarolo, D., Siciliano, M. J., Ashizawa, T., and Giacanelli, M. (1996). Myotonic dystrophy phenotype without expansion of $(CTG)_n$ repeat: an entity distinct from proximal myotonic myopathy (PROMM)? *J. Neurol.* **243,** 715–721.
43. Jansen, G., Bachner, D., Coerwinkel, M., Wormskamp, N., Hameister, H., and Wieringa, B. (1995). Structural organization and developmental expression pattern of the mouse WD- repeat

gene DMR-N9 immediately upstream of the myotonic dystrophy locus. *Hum. Mol. Genet.* **4,** 843–852.

44. Jansen, G., Groenen, P., Bachner, D., Jap, P. H. K., Coerwinkel, M., Oerlemans, F., Vandenbroek, W., Gohlsch, B., Pette, D., Plomp, J. J., Molenaar, P. C., Nederhoff, M. G. J., Vanechteld, C. J. A., Dekker, M., Berns, A., Hameister, H., and Wieringa, B. (1996). Abnormal myotonic dystrophy protein kinase levels produce only mild myopathy in mice. *Nature Genet.* **13,** 316–324.
45. Mortlock, D. P., Post, L. C., and Innis, J. W. (1996). The molecular basis of hypodactyly (hd), a deletion in *HOXA13* leads to arrest of digital arch formation. *Nature Genet.* **13,** 28–289.
46. Borrow, J., Shearman, A. M., Stanton, V. P., Becher, R., Collins, T., Williams, A. J., Dube, I., Katz, F., Kwong, Y. L., Morris, C., Ohyashiki, K., Toyama, K., Rowley, J., and Housman, D. E. (1996). The t(7 11)(p15 p15) translocation in acute myeloid leukaemia fuses the genes for nucleoporin *NUP98* and class I homeoprotein *HOXA9. Nature Genet.* **12,** 159–167.
47. Nakamura, T., Largaespada, D. A., Lee, M. P., Johnson, L. A., Ohyashiki, K., Toyama, K., Chen, S. J., Willman, C. L., Chen, I. M., Feinberg, A. P., Jenkins, N. A., Copeland, N. G., and Shaughnessy, J. D. (1996). Fusion of the nucleoporin gene *NUP98* to *HOXA9* by the chromosome translocation t(7 11)(p15 p15) in human myeloid leukemia. *Nature Genet.* **12,** 154–158.
48. Muragaki, Y., Mundlos, S., Upton, J., and Olsen, B. R. (1996). Altered growth and branching patterns in synpolydactyly caused by mutations in *HOXD13. Science* **272,** 548–551.
49. Tassabehji, M., Read, A. P., Newton, V. E., Harris, R., Balling, R., Gruss, P., and Strachan, T. (1992). Waardenburg syndrome patients have mutations in the human homolog of the *PAX3* paired box gene. *Nature* **355,** 635–636.
50. Jordan, T., Hanson, I., Zaletayev, D., Hodgson, S., Prosser, J., Seawright, A., Hastie, N., and Vanheyningen, V. (1992). The human *PAX6* gene is mutated in two patients with aniridia. *Nature Genet.* **1,** 328–332.
51. Mirzayans, F., Pearce, W. G., Macdonald, I. M., and Walter, M. A. (1995). Mutation of the *PAX6* gene in patients with autosomal dominant keratitis. *Am. J. Hum. Genet.* **57,** 539–548.
52. Johnson, R. L., Rothman, A. L., Xie, J. W., Goodrich, L. V., Bare, J. W., Bonifas, J. M., Quinn, A. G., Myers, R. M., Cox, D. R., Epstein, E. H., and Scott, M. P. (1996). Human homolog of *patched,* a candidate gene for the basal cell nevus syndrome. *Science* **272,** 1668–1671.
53. Hahn, H., Wicking, C., Zaphiropoulos, P. G., Gailani, M. R., Shanley, S., Chidambaram, A., Vorechovsky, I., Holmberg, E., Unden, A. B., Gillies, S., Negus, K., Smyth, I., Pressman, C., Leffell, D. J., Gerrard, B., Goldstein, A. M., Dean, M., *et al.* (1996). Mutations of the human homolog of drosophila *patched* in the nevoid basal cell carcinoma syndrome. *Cell* **85,** 841–851.
54. Belloni, E., Muenke, M., Roessler, E., Traverso, G., Siegelbartelt, J., Frumkin, A., Mitchell, H. F., Doniskeller, H., Helms, C., Hing, A. V., Heng, H. H. Q., Koop, B., Martindale, D., Rommens, J. M., Tsui, L. C., and Scherer, S. W. (1996). Identification of *Sonic hedgehog* as a candidate gene responsible for holoprosencephaly. *Nature Genet.* **14** 353–356.
55. Roessler, E., Belloni, E., Gaudenz, K., Jay, P., Berta, P., Scherer, S. W., Tsui, L. C., and Muenke, M. (1996). Mutations in the human *sonic hedgehog* gene cause holoprosencephaly. *Nature Genet.* **14,** 357–360.

CHAPTER 13

Murine Models for Myotonic Dystrophy

DARREN G. MONCKTON Division of Molecular Genetics, Institute for Biomedical and Life Sciences, University of Glasgow, Anderson College, Glasgow G11 6NU, Scotland

TETSUO ASHIZAWA Department of Neurology, Baylor College of Medicine, Texas Medical Center, Houston, Texas 77030

MICHAEL J. SICILIANO Department of Molecular Genetics, University of Texas M.D. Anderson Cancer Center, Texas Medical Center, Houston, Texas 77030

I. INTRODUCTION

The 5 years since the predominant myotonic dystrophy (DM) mutation was identified as the expansion of a CTG repeat in the 3′ untranslated region (UTR) of the DM protein kinase gene (*DMPK*) [1–6] have been interesting times. The initial optimism that the identification of the genetic defect would quickly lead to an understanding of the disease process has been tempered by the increasing realization that DM is not a simple disorder. Indeed it would appear that each new piece of research raises more questions than it answers (see [7, 8]). Nonetheless, significant advances have been made slowly the apparently randomly assorted pieces of the puzzle are starting to fit together and a picture of understanding is beginning to emerge. Given the complexity of the disease, the difficulty in obtaining appropriate clinical material and the inadequacies of alternative model systems, murine models remain as an effective tool for attempting to dismantle the DM phenomenon. In this brief review we shall provide overviews of the current mouse models, interpret them in relationship to the theories underlying DM research, and speculate as to which additional models might best be developed to answer the many remaining uncertainties.

II. THE PROBLEM(S)

We shall first of all begin by attempting to distil what are the major questions that remain after the identification of the CTG repeat expansion as the major genetic defect in DM pedigrees, before we ask more detailed questions such as what, how, and why. We see the problem as dividing into two major questions: molecular genetics and molecular pathogenesis.

A. Molecular Genetics

The question of the molecular genetics of DM is an interesting one that has puzzled geneticists for around 80 years, since Fleischer gave the first account of genetic anticipation in DM pedigrees in 1918 [9]. For much of this time it was widely debated as to whether anticipation even existed, since no rational explanation for it could be made in the light of the knowledge of the day and earlier results were discounted as ascertainment bias [10]. Indeed, it was not until the in-depth study of Howeler *et al.*, in 1989 [11], just 3 years prior to the identification of the genetic defect, that anticipation was really accepted as a legitimate observation. Two years later, when a CGG expansion was identified as the underlying cause of fragile X syndrome, thus resolving the Sherman paradox [12], a potential molecular basis for the unusual genetics of DM became apparent. Thus when in 1992 the CTG repeat expansion was identified as the predominant genetic lesion in DM [1–6], a probable explanation for the unusual genetics became manifest. As originally discussed and subsequently confirmed, changes in allele length during intergenerational transmission underlie anticipation. A positive correlation of allele length with disease severity and an inverse correlation with age of onset (*e.g.,* [13–15]), coupled with an extreme bias toward expansion during intergenerational transmission, clearly underlies the anticipatory process (*e.g.*, [16–18]). A particularly high level of instability of small expansions in the male germline accounts for the excess of transmitting grandfathers in DM pedigrees (*e.g.*, [19–21]); the continued expansion of large expansions during female transmission accounts for the almost exclusively female origin for the transmission of congenital DM (*e.g.,* [22, 23]), while rare reversions in allele length can explain the incomplete penetrance of DM [21, 24–26]. Thus it would appear that the major "what?" of DM molecular genetics has been discovered. However, more fundamental questions such as "how?" and "why?," in addition to more detailed questions such as "where?" and "when?," remain largely unanswered. In particular, the molecular mechanisms underlying repeat expansion are not known, and the major genes modifying repeat stability between individuals and sexes are unidentified. Further, the exact timing of the intergenerational differences is not known. It is often assumed that germline differences arise during meiosis, but there is no direct evidence to confirm this.

In addition, fundamental questions regarding the population incidence of DM remain unresolved. Most important is how a disease for which alleles progress rapidly toward effective lethals over the course of three or four generations is maintained in the population. The very high levels of linkage disequilibrium observed in non-Negroid populations suggest that new mutations direct from all alleles are rare [27, 28]. More likely is the existence of a subset of expanded pre- or protomutation alleles that may be relatively stable [19, 29], although biased toward expansion into the full disease range at some point, or a pool of larger alleles within the normal size range that may behave similarly [27]. Nonetheless, both possibilities would still be expected to lead to a gradual loss of potential progenitor alleles from the gene pool. To counter such a loss it has been suggested that larger alleles may be preferentially transmitted through the germline [30]. Some evidence in support of segregation distortion has been offered, although this subject remains controversial and unresolved [21, 30–34].

Also, the relevance of somatic instability remains largely unexplored. It is now apparent that somatic instability is heavily biased toward further expansion and continuous throughout adult life [21, 35]. This process appears to proceed at a tissue-specific rate, with much larger expansions present in muscle relative to those observed in the blood DNA of the same individual [36, 37]. Although it would seem reasonable to assume that somatic expansion is associated with the progressive nature of the disease, such an association has not been formally established.

A further question that remains unanswered is the basis of the differences in repeat stability between the various loci associated with hypermutable triplet repeat expansions. Although most of these loci share the same basic repeat sequence, i.e., CAG/CTG, there are clearly significant differences in repeat stability. Most notable is that the DM repeat is far more unstable and gives rise to far larger expansion sizes than any of the other diseases, an observation that cannot be accounted for purely in terms of repeat number. For instance, Machado-Joseph disease (MJD) alleles of around 70 repeats are highly unstable with mutation rates around 90% per transmission, but with a restricted range of length changes usually in the order of less than 10 repeats [38–40]. In contrast, a similar sized DM allele might have a mutation rate approaching 100% per transmission [21] and an average length change of a gain of 180 repeats! Selection against germ cells carrying very large expansions does not appear to account for this observation since segregation distortion in favor of normal alleles in offspring from affected parents is not apparent in any of the diseases, and DM is the only disease in which gonadal atrophy occurs (see [41]), counter to what might be expected if large coding CAG expansions were particularly toxic to germ cells. Thus, it would appear that the observed differences in repeat stability between the different loci represent true differences in lability of the repeats, independent of repeat sequence and length. Therefore it remains that the DNA sequences flanking the repeats somehow modify their stability. The nature, location, and mode of action of such *cis*-acting modifiers are presently unknown, although evidence from the MJD locus [40, 42] and hypervariable minisatellite repeat loci [43, 44] suggest that *cis*-acting modifiers of repeat stability can lie very close to the repeat.

B. Molecular Pathogenesis

The DM phenotype is pleiotropic in nature, affecting a wide range of organ systems of variable expressivity both within and between families (see [41]). Although the major features such as myotonia and muscular atrophy are muscle-based, other symptoms include cataracts, hypersomnia, cardiac conduction defects, testicular atrophy, frontal balding in males, and reduced fertility in females (see [41]). Moreover, individuals with the most severe congenital form of the disease present with their own unique symptoms such as hypotonia and mental retardation before going on to develop the symptoms associated with the adult form of the disease (see [41]), indicating that very large expansions induce another set of symptoms, possibly mediated via an alternative pathway. Unfortunately, our understanding of the molecular pathogenesis underlying DM has not advanced with the same pace as the immediate insights gained into the molecular genetics of DM obtained since the identification of the CTG expansion within the 3′ UTR of the *DMPK* gene. Indeed, in terms of the molecular pathogenesis we still seem to be languishing at the primary "what?" question. Although the identification of the CTG expansion suggested immediate possibilities for possible pathogenic mechanisms, the proceeding years have not yielded confirmatory evidence in support of the simple mechanisms initially envisioned. Indeed, we have been forced to look for alternative explanations and come to accept that the molecular pathogenesis is unlikely to be simple, and probably involves multiple pathways. Since the CTG expansion lies in the 3′ UTR of the *DMPK* gene, a gain of function of a protein product mediated by the inclusion of additional coding repeats was immediately discounted. Thus, initial speculation centred on a direct effect of repeat expansion on *DMPK* expression levels, either over- [45] or underexpression [46]. Although this subject has been controversial too [47, 48], and no doubt complicated by the differences in specific technical procedures employed, the consensus of opinion suggests that the *DMPK* gene is expressed at approximately normal levels with a probable slight defect in message processing [49]. Thus, drastic changes in DMPK protein levels seem unlikely to account for the DM phenotype. Indeed, further evidence against a simple loss of function of *DMPK* is the absence of additional mutations in the *DMPK* gene associated with DM. If DM were a simple loss of function then additional mutations such as premature stops, frame shifts, and deletions in or around the *DMPK* gene might reasonably be expected. The absence of reports of such additional mutations argues in favor of a more complex pathogenic mechanism. In light of these results, attention has more recently diversified to consider alternative possible mechanisms. This work includes the identification of a second gene immediately downstream of the expansion, termed DM-associated homeodomain protein, *DMAHP* [50]. Although a direct role for *DMAHP* has yet to be established, preliminary data suggest that expression from the mutant chromosome may be downregulated in DM patient cells [51]. Thus, *DMAHP* remains an interesting candidate gene, particularly in light of recent expression data derived from a transgenic reporter experiment that indicates that *mDMAHP* is expressed in many of the tissues most dramatically affected in DM [52]. Other work has also centred on radically different pathogenic mechanisms that apportion a role to a gain of function of an RNA message from mutant *DMPK* genes that contain large expanded CUG repeat arrays. Evidence consistent with such a gain of function includes the identification of subnuclear foci of the *DMPK* mutant message [53], and patient-specific alterations in the levels of a characterised RNA CUG repeat binding protein [54] and accompanied global changes in mRNA levels from a variety of genes [55, 56].

III. WHY USE MOUSE MODELS?

Although significant advances have been made in the last few years in attempting to understand the DM phenomenon using analyses of patient-derived material, such an approach is inherently limited. These limitations include not only the difficulty in obtaining patient-derived material, but also the many other confounding factors associated with such samples, including differences in patient ages, allele sizes, genetic backgrounds, and environmental effects. Other model systems also have associated advantages and disadvantages. For instance, tissue culture of patient-derived cell lines provides a system in which somatic instability may be studied under controlled conditions [57, 58], but does not allow for analysis of germline events or embryonic development. Moreover, since no proven pathologically relevant measure of the DM phenotype has been developed at the cellular level it is very difficult to study pathogenic mechanisms *in vitro* in the absence of some understanding of what the pathologically important effects are. The generation of mouse models provides the unique opportunity of investigating aspects of the molecular genetics, such as embryonic development and male and female gametogenesis, and pathogenesis at the cellular, tissue, organ, and whole-animal level. Nonetheless, the power of mouse genetics still allows us to employ at least a partially reductionist approach and attempt to design different transgenics that may test various aspects of what in man, at least, is a very complex phenomenon. Additional obvious advantages include the ability to perform *in vivo* many experiments not feasible or ethically acceptable in humans and access to a much wider range of cells and tissues. Moreover, the mouse system in which crosses between other mouse strains can easily be set up provides an ideal opportunity for identifying important modifier genes in terms of both repeat stability and pathogenesis.

IV. MOUSE MODELS FOR DM GENETICS

Previous attempts to model unstable triplet repeats from some of the other triplet repeat diseases using simple transgenes in mice have failed to replicate the high levels of instability associated with expanded triplet repeats in humans [59–62]. The reasons underlying the lack of observed instability in these models remain unclear at the moment. Possible factors include the relatively small size of the repeat tracts inserted, a failure to transfer critical *cis*-acting sequence elements, and technical difficulties associated with detecting infrequent small length changes in multicopy transgene arrays. However, three recent reports have indicated that introduced transgenes (two from the DM locus [63, 64] and one from the Huntington's disease locus [65]) may replicate a number of the features associated with unstable DNA in humans, indicating that at least some aspects of the unstable DNA phenomenon may be amenable to modelling in mice. We shall briefly review the central findings of the two models based on transgenes derived from the DM locus; the HD-based model will be reviewed in Chapter 25.

One of the reasons suggesting that previous attempts to model unstable DNA in mice did not result in significant detectable instability was a failure to also transfer necessary critical *cis*-acting sequence elements in the relatively simple transgene constructs used. In an attempt to overcome this limitation Gourdon *et al.* [63] used a very large 45-kb genomic fragment from the DM region that includes the two genes immediately flanking the repeat, *DMPK* and *DMAHP,* as well as the *DMR-N9* gene [66] which lies just upstream of *DMPK* (Fig. 13-1A). The transgene was derived from a cosmid library of a mildly affected patient and included a DM CTG tract with 55 repeats. From this transgene seven independent transgenic lines were generated, containing from 1 to 4 copies of the transgene. Interestingly, all were shown to express all three human genes, although none have so far resulted in production of a phenotype in the mice. PCR analysis of the transmission of the repeat tracts from parent to offspring revealed that six out of seven of the lines were unstable with mutation rates up to 6.5% per repeat tract per transmission. Most of the repeat tract length changes were gains of a single repeat (~80%), but one gain each of 2 and 6 repeats was observed, in addition to a deletion of 1 repeat. Unfortunately the number of transmissions so far observed is too small to determine whether the apparent mutational differences between the lines are significant or whether there are differences between male and female transmission. Low-level small length change mosaicism was also observed in somatic tissues in some of these mice. The level of instability observed in these mice is about what might be expected for premutation [27] and small protomutation [19] sized DM alleles in humans, and similar to some endogenous mouse minisatellites [67–70].

Another of the factors suggested as limiting the instability observed in previous mouse models of triplet repeat stability is the relatively small size of repeat tracts used in the transgenes (up to 86 repeats). Thus, in our study we used transgenes derived from the DM human locus with a relatively large CTG tract of 162 repeats [64]. Such a repeat size would be expected to be extremely unstable in the germline of DM patients and show probable somatic instability as well. Since the rationale for the generation of our model was merely to model DNA repeat stability, our transgene, *Dmt*162, was very simple in structure, including only a relatively small amount of the human flanking DNA derived from the DM 3′ UTR and some 3′ flanking DNA, but none of the coding DNA from any of the nearby genes (Fig. 13-1A). We concentrated our analysis on four independent transgenic lines generated using the *Dmt*162 transgene: three single-copy and one bicopy. All four of these lines were shown to be hypermutable on transmission of the repeat from generation to generation, with mutation rates up to ~70% per transmission observed from hemizygous parents. However, striking differences between transgenic lines and sex of parent effects were apparent. Female transmissions were very similar in all the lines,

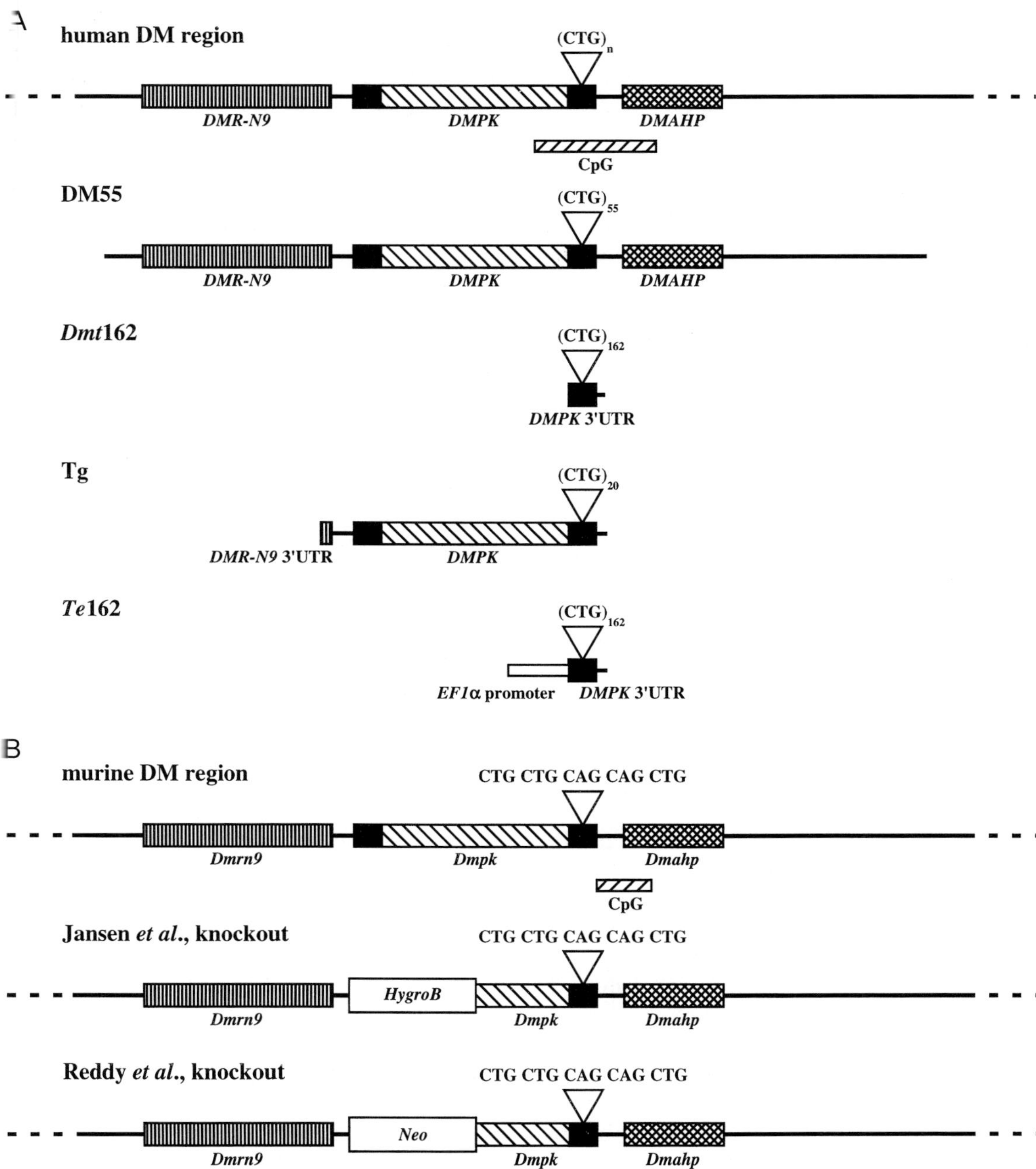

FIGURE 13-1 The myotonic dystrophy genomic region and mouse transgenes. (A) The human myotonic dystrophy genomic region and mouse transgenes. Shown is the human DM region including the upstream *DMR-N9* gene, *DMPK,* and the downstream *DMAHP* gene, the position of the CTG repeat and the CpG island. Also shown are the four transgene constructs used to generate transgenic mice: DM55 [63], *Dmt*162 [64], Tg [83], and *Te*162 [79]. (B) The mouse myotonic dystrophy genomic region and replacement alleles. Shown is the mouse DM region including the upstream *Dmrn9* gene, *Dmpk,* and the downstream *mDMAHP* gene, the position of the CTG repeat and the CpG island. Also shown are the two *Dmpk* replacement alleles used to generate *Dmpk* knockout mice by Jansen *et al.* [83] and Reddy *et al.* [84].

all occurring at about 60% per transmission and all biased toward, but not exclusively, small contractions of 1 to 7 repeats. In contrast, male transmission was characterized by a significant bias toward small expansions, mostly in the order of gains of 1 to 7 repeats, and extreme variation in overall rates from ~10 to ~70%.

Somatic instability was also observed in all lines and, interestingly, appeared to correlate with the levels of male germline instability. In addition to this general somatic instability, two hemizygous mice mosaic for two alleles, one parental and one mutant, were also detected (one a gain of 20 repeats, the other a contraction of 9 repeats). These most likely represent very early embryonic mutations that arise in the first one or two cell divisions postfertilization. Similar apparent early embryonic mutations have been shown to occur at two endogenous mouse minisatellites [69, 70] and for rearrangements of some tandemly arranged mouse transgenes [71, 72]. Similar very early embryonic mutations do not appear common in DM, although there is some evidence derived from twin studies that suggests that early embryonic mutations can occur [Ashizawa *et al.*, unpublished observations]. There is more evidence that instability of fragile X CGG repeat alleles may be limited to a similar very early window of embryonic development (see [73]). Alternatively, such events may not be early embryonic in origin, but represent instead the transmission of unrepaired progenitor/mutant heteroduplex DNA structures in the gametes.

One other interesting phenomenon was also observed during the analysis of the *Dmt*162 lines. Pedigree analysis revealed that in two of the lines the transgene was transmitted from hemizygous parents to offspring at greater than the expected 50%; with approximately two-thirds of offspring inheriting the transgene. In one line this segregation distortion was also associated with reduced litter sizes, ~50% of normal. Reduced litter sizes have been previously associated with a transgene integration-induced chromosomal translocation [74] and also with inferred segregation distortion [75]. Thus, it is possible that the segregation distortion in this line is a product of a similar chromosomal translocation. However, the other line is not associated with reduced litter sizes and the basis of the segregation distortion in that line has not yet been determined, but remains intriguing in light of the apparent segregation distortion at the DM [30, 32], MJD [42], and DRPLA [76] loci in man.

These results confirm that germline instability may be modelled in mice and that similar *cis*-acting, as demonstrated by interline differences, and *trans*-acting, as demonstrated by sex-specific differences, modifiers of repeat stability are operative. Nonetheless, the germline mutation rates observed for the *Dmt* transgenes, although very high, are not as high as might be expected for similar-sized alleles in humans, nor are the length changes observed as large. The reasons for this remain unclear at present, but could include transgene zygosity: all experiments so far have been performed with hemizygous mice within whom the transgene cannot pair during meiosis, which may be a critical process in germline instability. Alternatively, it may be associated with a failure to transfer critical *cis*-acting sequence elements. In this regard Gourdon *et al.*, report that they have recently constructed transgenic mice incorporating the same large cosmid-derived 45-kb fragment from the DM region but this time incorporating 320 CTG repeats, rather than the 55 repeats used previously [63]. It will be interesting to see whether the inclusion of such large tracts of the human flanking sequences will dramatically influence repeat stability of a large CTG tract. The nature of putative *cis*-acting elements remains unknown, but orientation, putatively with respect to replication origins, has been indicated as being important *in vitro* in *Escherichia coli* plasmids [77], yeast chromosomal insertions [78], and mammalian plasmid systems [Barber, Monckton, and Caskey, unpublished observations]. It seems to be a general assumption that repeat length change mutations arise during DNA replication, most likely via replication slippage. However, the highest levels of somatic instability in DM are seen in muscle, a tissue with a low rate of cell turnover and cell division, in contrast to the relatively low levels of instability observed in rapidly dividing systems such as white blood cells [21, 36, 37] and lymphoblastoid cell lines in tissue culture [58]. These data suggest that another factor other than cell division may be important in the mutation pathway. We have postulated that transcription through the region might be an important modulator of repeat stability, an hypothesis that would appear to fit with the essentially muscle specific expression pattern of the *DMPK* gene. To test this hypothesis we have also generated second generation transgenics in which the repeat tract is under the control of a ubiquitous promoter (see Section V.C for details) [79]. Preliminary results indicate that indeed transgenes in which the region is transcribed are more unstable than those in which it is not. Consistent with these observations, one of the mouse models for instability of the HD repeat also appears to correlate transcription levels with repeat stability [65].

V. MOUSE MODELS FOR DM PATHOGENESIS

As discussed earlier, exactly how repeat expansion gives rise to the DM phenotype is still unclear and a number of mechanisms remain plausible. Indeed, it is appearing increasingly likely that more than one mechanism may be involved. One of the advantages of using murine models is that we may test each of the possible mechanisms singly *in vivo*.

A. Mouse Models for Overexpression of Protein

It would seem reasonable to assume that the massive CTG expansion that occurs in DM patients may result in a change in the chromatin structure surrounding the expansion. Indeed, *in vitro* evidence indicating that CTG repeats are the strongest nucleosome positioning elements so far identified [80, 81] would suggest that this is likely, whilst the loss of a DNase1 hypersensitive site downstream of the repeat expansion in patients confirms this [82]. How such a structural change would affect transcriptional activity of the adjacent genes remains controversial (see details in Section II.B). Thus far, only one model overexpressing the human *DMPK* transcript in the mouse has been reported [83]. Jansen *et al.* used a 14-kb transgene (Tg, Fig. 13-1A) derived from the human DM locus that encompasses the whole of the *DMPK* genomic region, including a 20-repeat CTG tract and the last exon of the upstream *DMR-N9* gene. Using the Tg transgene, six lines of mice were generated incorporating from 2 to 25 copies of the transgene. Expression of human *DMPK* in these lines was shown to match the essentially muscle-specific pattern observed for endogenous murine *Dmpk*, but was only significantly increased over normal in lines with greater than 20 copies of the transgene and highest, ~5- to 10-fold normal, in line Tg26 which contains 25 copies. Outwardly these mice appeared essentially normal; no defects in skeletal muscle pathology or function were detected, nor indeed were histopathological changes detected in any of the other organ systems investigated, with the exception of the heart. Histopathological analysis of the heart in three independent Tg lines revealed a striking dysmorphology characterized by a bizarre whorling of the myocard, hypertrophy, patches of degeneration with vesicular nuclei and focal fibrosis. Nonetheless, cardiac function was apparently normal in these mice and how this relates, if at all, to the cardiac conduction defects and fatty fibrosis observed in DM patients remains unclear. In addition to the cardiac dysmorphology, the only other phenotype observed was an increased pre- and neonatal lethality. This effect was strongest, with up to 50% of Tg-positive offspring dying, when the transgene was transmitted by females who themselves became sick and occasionally died. The problem was less severe through male transmission, but still ~20% of Tg-positive offspring died, although the males remained fully fertile for over a year. The reasons for the neonatal mortality and maternal illness are not yet understood, nor indeed their relevance to the reduced fertility observed in DM females. However, the high neonatal mortality observed with congenital DM in humans is associated with respiratory distress [41] and would appear to be distinct from the pre- and neonatal lethality observed in these Tg mice.

B. Mouse Models for Loss of Gene Function

To test the alternative hypothesis that the DM phenotype is a product of a loss of function of *DMPK*, two groups have used homologous recombination in ES cells to generate null alleles of the murine *Dmpk* gene [83, 84]. Both groups knocked out the same region of the *Dmpk* gene, replacing exons one to seven, containing most of the kinase domain, with selectable markers (Fig. 13-1B). That true *Dmpk* null alleles had been generated was confirmed by northern and western analysis of *Dmpk* message and Dmpk protein in homozygous *Dmpk-/-* mice. Interestingly, Jansen *et al.* reported that insertion of the hygromycin cassette into the *Dmpk* gene also resulted in an alteration in message processing of the upstream gene, *Dmrn9*. The transcript size was increased from 2.5 to 4 kb, but no overall difference in protein levels was observed. An effect on the downstream *Dmahp* gene was not investigated, nor was any effect of the neomycin insertion used by Reddy *et al.* This simple observation highlights an important consideration in attempting to unravel the DM phenomenon namely, that the DM mutation lies in a very gene-rich region of the genome and it is likely that any major rearrangement in the region will affect more than one gene. Regardless, to study the phenotypic effects of loss of *Dmpk*, a range of body systems have been investigated in mice both heterozygous and completely deficient for *Dmpk*. Quite surprisingly, mice heterozygous for *Dmpk* deficiency and even those completely deficient are fully viable throughout development, displaying no overt symptoms whatsoever and being completely indistinguishable from wild-type mice. A careful examination of muscle function has, however, identified some abnormalities. Both groups have noticed an increase in the variability of fibre size in skeletal muscle, in particular an increase in the number of small fibres, although no obvious difference in the distribution of fibre types. Moreover, Reddy *et al.* correlated this finding with a progressive myopathy characterized by reduced muscle strength, increased regeneration (as measured by MyoD expression), degeneration, and fibrosis. In addition, they also identified patches of highly disorganized sarcomere in a subset of fibres. However, these observations were only really pronounced in completely *Dmpk*-deficient older mice (7 to 11 months old) and much less so in younger mice (3 to 4 months old) and heterozygotes. Moreover, none of the other features characteristic of DM in humans were observed including myotonia, heart conduction defects, cataracts, testicular atrophy in males, and fertility problems in females. These findings are intriguing and are certainly consistent with the hypothesis that loss of *DMPK* expression may be contributing to the DM phenotype in humans. Nonetheless, the phenotype is much less severe than might have been expected if this were really the only defect, especially given the dominant nature of the disease in humans. The absence of a severe muscle phenotype and many of the major features of DM does raise some questions. It is possible that these differences represent fundamental differences in human and mouse physiology, particularly concerning muscle function in two organisms that differ drastically in their muscle loads . A good example of a similar discrepancy is with the *mdx* mouse, which is the murine model for Duchenne's muscular dystrophy [85]. Despite a defect in the same gene, the phenotype in the *mdx* mouse is relatively mild and nowhere near as severe as the human disease. Thus it remains possible that these mice are a good model for DM, but with deficiencies reflecting basic physiological differences between men and mice. Alternatively, these results might suggest the DM phenotype is more complex and not simply a result of loss of function of *DMPK*. Moreover, we should not forget that *Dmpk* is, as defined by its expression pattern, essentially a muscle-specific gene that has been highly conserved between mouse and man, suggesting it serves an important role in muscle function. Thus, it would be surprising if complete loss of expression, or indeed overexpression, did not have some effect on muscle function. As such, the role of changes in *DMPK* expression levels and resultant DMPK protein levels in DM pathogenesis remain unclear.

C. Mouse Models for a Gain of Function of the mRNA

The CTG expansion in DM does not lie in the coding region of any of the genes, but is located in the 3′ UTR of the *DMPK* gene. The models reviewed above have concentrated on assessing the pathogenic consequences of changes in the levels of functional DMPK/Dmpk protein that the expansion may precipitate. An alternative possibility is that the *DMPK* RNA message derived from an expanded CTG repeat chromosome, and thus containing an expanded CUG repeat, may itself have a deleterious effect on some aspects of cellular/tissue function. To test this possibility directly we have generated transgenic mice that express a noncoding expanded repeat-containing minigene [79]. We have coupled the human EF1α promoter to the 3′ UTR of the human *DMPK* gene to produce the *Te* transgene (Fig. 13-1A). The *Te* transgene should drive ubiquitous expression in all mammalian cell types and produce a short noncoding mRNA message containing a large expanded CUG repeat array. From embryo injections using this transgene we derived eight founder mice, four female and four male. These founder mice were paired with wild-type mice in an attempt to establish permanent lines. All four females were fertile, although only one transmitted the transgene (line *Te*-G), suggesting the other three were mosaic for the integration event. It is not uncommon to generate mosaic founders during transgenesis; however, most still transmit through the germline. For all three of the mosaic transgenic founders not to transmit is thus unexpected and suggests that transgene positive cells may have been selected against during oogenesis. More substantively, however, females carrying the *Te*-G transgene are demonstrably of reduced fertility, any one female producing from only zero to four litters, compared to the six to eight that would be normally expected. Moreover, male progeny with the *Te*-G transgene have small testes, and are azoospermic and consequently infertile (transgene-negative litter mates are all of normal fertility). Of the four founder males, one produced pregnancies immediately and transmitted the transgene to half of the offspring, all of whom appear to be of normal fertility. The other three male founders did not produce pregnancies in multiple females. Histological examination of the testes from mice in four independent *Te* transgenic lines revealed similar abnormal testicular histology, including tubal shrinkage, Leydig cell hyperplasia, and hypogametogenesis (see Fig. 13-2), mirroring the testicular atrophy observed in human DM males. No other phenotype

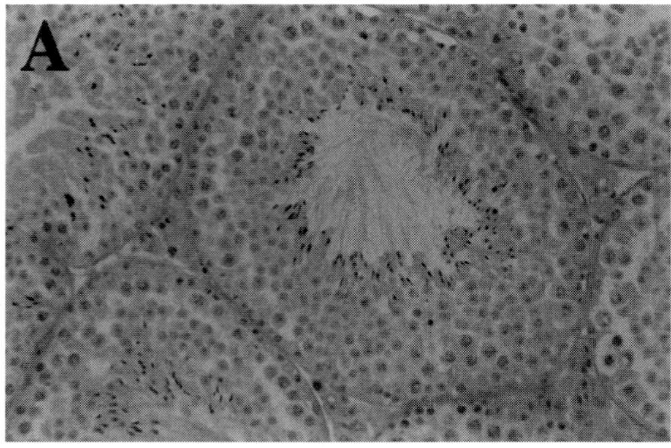

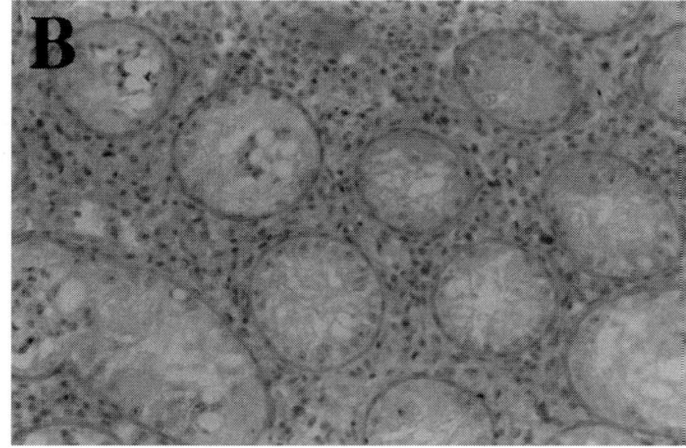

FIGURE 13-2 Testicular atrophy in CUG repeat expressing mice. (A) Normal adult mouse testes. Note the large tubules packed with spermatogenic cells and the small interstitial space. (B) Testes from a *Te* CUG repeat expressing adult mouse [79]. Note the reduced tubule size, loss of spermatogenic cells, and interstitial cell hyperplasia.

has been detected in these mice, including an absence of severe myopathy, myotonia, or cataracts. Thus, whether repeat expression is an important pathogenic process in other aspects of the DM phenotype remains undetermined. However, it should be considered that our transgene contains only ~160 CTG repeats. Such an allele would most likely be associated with classic adult onset DM in humans, with symptoms first observed in the third or fourth decade. Moreover, the repeat length observed in the muscle of DM patients is usually very much larger than that observed in other tissues. Since our mice were less than 1 year old when tested and we have no evidence for gross somatic expansion, an apparent lack of muscle specific symptoms may not be surprising. It is thus likely that even if an expressed expanded repeat is important in the DM muscle phenotype, a very large repeat may be required to produce a detectable effect in a short-lived species such as the mouse. Nonetheless, our results indicate that at least part of the DM phenotype results from a gain of function of an RNA message containing an expanded CUG array. Whether this is mediated via binding to the recently identified CUG-BP RNA binding protein [54] and precipitation into nuclear foci [53] has not yet been determined.

Table 13-1 Summary of the DM Phenotype in Humans and Those Features Observed in Current Mouse Models for DM

Phenotype	DM patients heterozygous for CTG expansion (Harper [41])	*Dmpk* knockout completely deficient (Reddy *et al.* [84])	*Dmpk* knockout completely deficient (Jansen *et al.* [83])	*DMPK* overexpressor hemizygous for transgene (Jansen *et al.* [83])	Expanded CUG RNA expressor hemizygous for transgene (Monckton *et al.* [79])
Survival	Decreased in adults	Normal	Normal	*Occasional death of females during pregnancy*	*Reduced life expectancy of adults?*
	Neonatal death due to respiratory distress in CDM	Absent	Absent	*Pre- and neonatal death, but not associated with respiratory distress*	Absent
Muscle strength	Distal weakness	*Reduced (in vitro)*	Normal	Normal	Normal
Muscle atrophy	Distal atrophy	Absent	Absent	Absent	Absent
Myotonia	Present	Absent	Absent	Absent	Absent
Muscle histology	Internal nuclei	*Some internal nuclei*	Normal	Normal	Normal
	Ring fibers	Absent	Absent	Absent	Absent
	Type I fiber atrophy	Normal	Normal	Normal	Normal
	Sarcoplasmic mass	Absent	Absent	Absent	Absent
	Variable fiber size	*Variable fiber size*	*Variable fiber size*	Normal	Normal
	Degeneration/ fibrosis/ regeneration	*Degeneration/ fibrosis/ regeneration*	*Degeneration/ fibrosis/ regeneration?*	Normal	Normal
Muscle ultrastructure	Abnormal sarcomere	*Abnormal sarcomere*	Not described	Not described	Not determined
	Mitochondrial abnormalities	Normal	Not described	Not described	Not determined
Electromyography	Myotonia	Normal	Normal	Normal	Normal
	Myopathic CMAP	Normal	Normal	Normal	Normal
Lens	Cortical opacities	Normal	Normal	Normal	Normal
Heart	Dilated cardiomyopathy	Not described	Normal	*Hypertrophic cardiomyopathy-like abnormality*	Normal
	Conduction defects	Not described	Not described	Normal ECG	Not determined
Testes and male fertility	Progressive testicular atrophy leading to infertility	Normal	Normal	Normal	*Gross testicular atrophy and complete sterility*
Female fertility	Reduced	Normal	Normal	Normal	*Reduced, progressive infertility*
Other systems	Pleiotropy	Not described	Not described	Normal (limited study)	Normal (limited study)

Note. Features in italics denote some similarity between the mouse model and the human condition.

VI. CONCLUDING REMARKS AND FUTURE DIRECTIONS

As discussed above a number of mouse models have been generated in various attempts to unravel some of the mysteries surrounding myotonic dystrophy. These studies have proven very enlightening, if not always in the directions that may have been initially envisioned. Models aimed at attempting to determine the molecular mechanisms underpinning anticipation and repeat stability have yielded encouraging results. Although the absolute levels of instability observed are not as high as observed in patients, instability does occur in the mouse models and seems to be under the influence of similar *cis*- and *trans*-acting genetic modifiers. As such the mouse system provides an excellent opportunity to dissect and identify genetic modifiers of instability. This approach will be especially aided by the recent increase in our understanding of the role of certain DNA repair genes in maintaining genomic instability in higher eukaryotes (see Chapters 36 and 49) and the rapidly increasing availability of mouse mutants for DNA repair genes (see Chapters 36 and 49). Progress in understanding how the repeat expansion actually gives rise to the DM phenotype has been equally fruitful. However, the results are not simple and no one mechanism would as yet appear to be responsible for the full DM phenotype (see Table 13-1). Rather, as with the patient-based analyses, the mouse models suggest that a number of mechanisms may play a role in various aspects of the DM phenotype, to a greater or lesser extent. Undoubtedly, however, the mouse models already generated and those that are being developed will play a major role in this continuing voyage of discovery. Models that are clearly required to shed further light on the situation and awaited with interest are knockouts of other candidate loss of function genes such as *Dmrn9* and *Dmahp*. Further analysis of the role of expressed CUG repeats is clearly warranted, most logically via the expression of a very large CUG tract in muscle, as is analysis of the role of the CUG-BP gene. Although this reductionist approach will be invaluable in the first instance to try and simplify what is clearly a complex situation, in the long term a mouse model for DM that covers all aspects of the phenomenon would be most useful. Thus, the generation of a mouse in which the cryptic CTG repeat in the homologous *Dmpk* gene (see Fig. 13-1B) is replaced with a large CTG tract would likely prove valuable. However, concerns have been raised that since the CpG island, which encompasses a large 3.6-kb region including the repeats in humans, is reduced in mice to a 1.8-kb region that does not include the repeat, effects on local chromatin structure and surrounding genes may not be the same [52]. In an attempt to obviate such a potential problem, Jansen *et al.* are attempting to replace the endogenous murine *Dmpk* 3′ UTR with a humanised expanded CTG repeat version [83]. The generation and continued analysis of mouse models for DM will allow further insights to the many unusual features associated with this debilitating disease and provide an excellent system in which therapeutic approaches may eventually be developed and tested.

References

1. Buxton, J., Shelbourne, P., Davies, J., Jones, C., Tongeren, T. V., Aslanidis, C., de Jong, P., Jansen, G., Anvret, M., Riley, B., Williamson, B., and Johnson, K. (1992). Detection of an unstable fragment of DNA specific to individuals with myotonic dystrophy. *Nature* **355,** 547–548.
2. Harley, H. G., Brook, J. D., Rundle, S. A., Crow, S., Reardon, W., Buckler, A. J., Harper, P. S., Housman, D., and Shaw, D. J. (1992). Expansion of an unstable DNA region and phenotypic variation in myotonic dystrophy. *Nature* **355,** 545–546.
3. Aslanidis, C., Jansen, G., Amemiya, C., Shutler, G., Mahadevan, M., Tsilfidis, C., Chen, C., Alleman, J., Wormskamp, N. G. M., Vooijs, M., Buxton, J., Johnson, K., Smeets, H. J. M., Lennon, G. G., Carrano, A. V., Korneluk, R. G., Wieringa, B., and de Jong, P. J. (1992). Cloning of the essential myotonic dystrophy region and mapping of the putative defect. *Nature* **355,** 548–550.
4. Fu, Y. H., Pizzuti, A., Fenwick, R. G., King, J., Rajnarayan, S., Dunne, P. W., Dubel, J., Nasser, G. A., Ashizawa, T., de Jong, P., Wieringa, B., Korneluk, R., Perryman, M. B., Epstein, H. F., and Caskey, C. T. (1992). An unstable triplet repeat in a gene related to myotonic muscular dystrophy. *Science* **255,** 1256–1258.
5. Brook, J. D., McCurrach, M. E., Harley, H. G., Buckler, A. J., Church, D., Aburatani, H., Hunter, K., Stanton, V. P., Thirion, J.-P., Hudson, T., Sohn, R., Zemelman, B., Snell, R. G., Rundle, S. A., Crow, S., Davies, J., Shelbourne, P., Buxton, J., Jones, C., Juvonen, V., Johnson, K., Harper, P. S., Shaw, D. J., and Housman, D. E. (1992). Molecular basis of myotonic dystrophy: expansion of a trinucleotide (CTG) repeat at the 3′ end of a transcript encoding a protein kinase family member. *Cell* **68,** 799–808.
6. Mahadevan, M., Tsilfidis, C., Sabourin, L., Shutler, G., Amemiya, C., Jansen, G., Neville, C., Narang, M., Barcelo, J., O'Hoy, K., Leblond, S., Earle-Macdonald, J., de Jong, P. J., Wieringa, B., and Korneluk, R. G. (1992). Myotonic dystrophy mutation: an unstable CTG repeat in the 3' untranslated region of the gene. *Science* **255,** 1253–1255.
7. Wieringa, B. (1994). Myotonic dystrophy reviewed: back to the future. *Hum. Mol. Genet.* **3,** 1–7.
8. Harris, S., Moncrieff, C., and Johnson, K. (1996). Myotonic dystrophy: will the real gene please step forward! *Hum. Mol. Genet.* **5,** 1417–1423.
9. Fleischer, B. (1918). Uber myotonische Dystrophie mit Katarakt. Albercht von Graefes. *Arch. Klin. Ophthalmol* **96,** 91–133.
10. Penrose, L. S. (1948). The problem of anticipation in pedigrees of dystrophia myotonica. *Ann. Eugenics* **14,** 125–132.
11. Howeler, C. J., Busch, H. F. M., Geraedts, J. P. M., Neirmeijer, M. F., and Staal, A. (1989). Anticipation in myotonic dystrophy: fact or fiction? *Brain* **112,** 779–797.
12. Fu, Y.-H., Kuhl, D. P. A., Pizzuti, A., Pieretti, M., Sutcliffe, J. S., Richards, S., Verkerk, A. J. M. H., Holden, J. J. A., Fenwick, R. G., Jr., Warren, S. T., Oostra, B. A., Nelson, D. L., and Caskey,

C. T. (1991). Variation of the CGG repeat at the fragile X site results in genetic instability: resolution of the Sherman paradox. *Cell* **67**, 1047–1058.

13. Novelli, G., Gennarelli, M., Menegazzo, E., Mostacciuolo, M. L., Pizzuti, A., Fattorini, C., Tessarolo, D., Tomelleri, G., Giacanelli, M., Danieli, G. A., Rizzuto, N., Caskey, C. T., Angelini, C., and Dallapiccola, B. (1993). (CTG)$_n$ triplet mutation and phenotype manifestations in myotonic dystrophy patients. *Biochem. Med. Metab. Biol.* **50,** 85–92.
14. Harley, H. G., Rundle, S. A., MacMillan, J. C., Myring, J., Brook, J. D., Crow, S., Reardon, W., Fenton, I., Shaw, D. J., and Harper, P. S. (1993). Size of the unstable CTG repeat sequence in relation to phenotype and parental transmission in myotonic dystrophy. *Am. J. Hum. Genet.* **52,** 1164–1174.
15. Hunter, A., Tsilfidis, C., Mettler, G., Jacob, P., Mahadevan, M., Surh, L., and Korneluk, R. (1992). The correlation of age of onset with CTG trinucleotide repeat amplification in myotonic dystrophy. *J. Med. Genet.* **29,** 774–779.
16. Ashizawa, T., Dubel, J. R., Dunne, P. W., Dunne, C. J., Fu, Y.-H., Pizzuti, A., Caskey, C. T., Boerwinkle, E., Perryman, M. B., Epstein, H. F., and Hejtmancik, J. F. (1992). Anticipation in myotonic dystrophy. II. Complex relationships between clinical findings and structure of the GCT repeat. *Neurology* **42,** 1877–1883.
17. Ashizawa, T., Dunne, P. W., Ward, P. A., Seltzer, W. K., and Richards, C. S. (1994). Effects of sex of myotonic dystrophy patients on the unstable triplet repeat in their affected offspring. *Neurology* **44,** 120–122.
18. Redman, J. B., Fenwick, R. G., Fu, Y.-H., Pizzuti, A., and Caskey, C. T. (1993). Relationship between parental trinucleotide GCT repeat length and severity of myotonic dystrophy in offspring. *J. Am. Med. Assoc.* **269,** 1960–965.
19. Barcelo, J. M., Mahadevan, M. S., Tsilfidis, C., MacKenzie, A. E. and Korneluk, R. G. (1993). Intergenerational stability of the myotonic dystrophy protomutation. *Hum. Mol. Genet.* **2,** 705–709.
20. Brunner, H. G., Bruggenwirth, H. T., Nillesen, W., Jansen, G., Hamel, C. J., Hoppe, R. L. ER. L., de Die, C. E. M., Howeler, C. J., van Oost, B. A., Wieringa, B., Ropers, H. H., and Smeets, H. J. M. (1993). Influence of sex of the transmitting parent as well as of parental allele size on the CTG expansion in myotonic dystrophy (DM). *Am. J. Hum. Genet.* **53,** 1016–1023.
21. Monckton, D. G., Wong, L.-J. C., Ashizawa, T., and Caskey, C. T. (1995). Somatic mosaicism, germline expansions, germline reversions and intergenerational reductions in myotonic dystrophy males: small pool PCR analyses. *Hum. Mol. Genet.* **4,** 1–8.
22. Tsilfidis, C., MacKenzie, A. E., Mettler, G., Barcelo, J., and Korneluk, R. G. (1992). Correlation between CTG trinucleotide repeat length and frequency of severe congenital myotonic dystrophy. *Nature Genet.* **1,** 192–195.
23. Lavedan, C., Hofmann-Radvanyi, H., Shelbourne, P., Rabes, J.-P., Duros, C., Savoy, D., Dehaupas, I., Luce, S., Johnson, K., and Junien, C. (1993). Myotonic dystrophy: size- and sex-dependent dynamics of CTG meiotic instability, and somatic mosaicism. *Am. J. Hum. Genet.* **52,** 875–883.
24. Shelbourne, P., Winquist, R., Kunert, E., Davies, J., Leisti, J., Thiele, H., Bachmann, H., Buxton, J., Williamson, B., and Johnson, K. (1992). Unstable DNA may be responsible for the incomplete penetrance of the myotonic dystrophy phenotype. *Hum. Mol. Genet.* **1,** 467–473.
25. Brunner, H. G., Jansen, G., Nillesen, W., Nelen, M. R., de Die, C. E. M., Howeler, C., van Oost, B. A., Wieringa, B., Ropers, H. H., and Smeets, H. J. M. (1993). Reverse mutation in myotonic dystrophy. *N. Eng. J. Med.* **328,** 476–480.
26. O'Hoy, K. L., Tsilfidis, C., Mahadevan, M. S., Neville, C. E., Barcelo, J., Hunter, A. G. W., and Korneluk, R. G. (1993). Reduction in size of the myotonic dystrophy trinucleotide repeat mutation during transmission. *Science* **259,** 809–812.
27. Imbert, G., Kretz, C., Johnson, K., and Mandel, J.-L. (1993). Origin of the expansion mutation in myotonic dystrophy. *Nature Genet.* **4,** 72–76.
28. Neville, C. E., Mahadevan, M. S., Barcelo, J. M., and Korneluk, R. G. (1994). High resolution genetic analysis suggests one ancestral predisposing haplotype for the origin of the myotonic dystrophy mutation. *Hum. Mol. Genet.* **3,** 45–51.
29. Yamagata, H., Miki, T., Sakoda, S.-I., Yamanaka, N., Ogihara, T., and Johnson, K. (1994). Detection of a premutation in Japanese myotonic dystrophy. *Hum. Mol. Genet.* **3,** 819–820.
30. Carey, N., Johnson, K., Nokelainen, P., Peltonen, P., Savontaus, M.-L., Juvonen, V., Chotai, K., Robertson, E., Middleton-Price, H., and Malcolm, S. (1994). Meiotic drive at the myotonic dystrophy locus? *Nature Genet.* **6,** 117–118.
31. Carey, N., Johnson, K., Nokelainen, P., Peltonen, L., Savontaus, M.-L., Juvonen, V., Anvret, M., Grandell, U., Chotai, K., Robertson, E., Middleton-Price, H. and Malcolm, S. (1995). Meiotic drive and myotonic dystrophy—Reply. *Nature Genet.* **10,** 133.
32. Chakraborty, R., Stivers, D. N., Deka, R., Yu, L. M., Shriver, M. D., and Ferrell, R. E. (1996). Segregation distortion of the CTG repeats at the myotonic dystrophy locus. *Am. J. Hum. Genet.* **59,** 109–118.
33. Hurst, G. D. D., Hurst, L. D., and Barrett, J. A. (1995). Meiotic drive and myotonic dystrophy. *Nature Genet.* **10,** 132–133.
34. Leeflang, E. P., McPeek, M. S., and Arnheim, N. (1996). Analysis of meiotic segregation, using single sperm typing: meiotic drive at the myotonic dystrophy locus. *Am. J. Hum. Genet.* **59,** 896–904.
35. Wong, L.-J. C., Ashizawa, T., Monckton, D. G., Caskey, C. T., and Richards, C. S. (1995). Somatic heterogeneity of the CTG repeat in myotonic dystrophy is age and size dependent. *Am. J. Hum. Genet.* **56,** 114–122.
36. Anvret, M., Ahlberg, G., Grandell, U., Hedberg, B., Johnson, K., and Edstrom, L. (1993). Larger expansions of the CTG repeat in muscle compared to lymphocytes from patients with myotonic dystrophy. *Hum. Mol. Genet.* **2,** 1397–1400.
37. Ashizawa, T., Dubel, J. R., and Harati, Y. (1993). Somatic instability of CTG repeat in myotonic dystrophy. *Neurology* **43,** 2674–2678.
38. Kawaguchi, Y., Okamoto, T., Taniwaki, M., Aizawa, M., Inoue, M., Katayama, S., Kawakami, H., Nakamura, S., Nishimura, M., Akiguchi, I., Kimura, J., Narumiya, S., and Kakizuka, A. (1994). CAG expansions in a novel gene for Machado-Joseph disease at chromosome 14q32.1. *Nature Genet.* **8,** 221–227.
39. Maruyama, H., Nakamura, S., Matsuyama, Z., Sakai, T., Dotu, M., Sobue, G., Seto, M., Tsujihata, M., Oh-i, T., Nishio, I., Sunohara, N., Takahashi, R., Hayashi, M., Nishino, I., Ohtake, T., Oda, T., Nishimura, M., Saida, T., Matsumoto, H., Baba, M., Kawaguchi, Y., Kakizuka, A., and Kawakami, H. (1995). Molecular features of the CAG repeats and clinical manifestation of Machado-Joseph disease. *Hum. Mol. Genet.* **4,** 807–812.
40. Igarashi, S., Takiyama, Y., Cancel, G., Rogaeva, E. A., Sasaki, H., Wakisaka, A., Zhou, Y.-X., Takano, H., Endo, K., Sanpei, K., Oyake, M., Tanaka, H., Stevanin, G., Abbas, N., Dürr, A., Rogaev, E. I., Sherrington, R., Tsuda, T., Ikeda, M., Cassa, E., Nishizawa, M., Benomar, A., Julien, J., Weissenbach, J., Wang, G.-X., Agid, Y., St. George-Hyslop, P. H., Brice, A., and Tsuji, S. (1996). Intergenerational instability of the CAG repeat of the gene for Machado-Joseph disease (MJD1) is affected by the genotype of the normal chromosome: implications for the molecular

mechanisms of the instability of the CAG repeat. *Hum. Mol. Genet.* **5,** 923–932.

41. Harper, P. S. (1989). "Myotonic Dystrophy," 2nd ed. W. B. Saunders, London.
42. Takiyama, Y., Sakoe, K., Soutome, M., Namekawa, M., Ogawa, T., Nakano, I., Igarashi, S., Oyake, M., Tanaka, H., Tsuji, S., and Nishizawa, M. (1997). Single sperm analysis of the CAG repeats in the gene for Machado-Joseph disease (MJD1): evidence for non-Mendelian transmission of the MJD1 gene and for the effect of the intragenic CGG/GGG polymorphism on the intergenerational instability. *Hum. Mol. Genet.* **6,** 1063–1068.
43. Monckton, D. G., Neumann, R., Guram, T., Fretwell, N., Tamaki, K., MacLeod, A., and Jeffreys, A. J. (1994). Minisatellite mutation rate variation associated with a flanking DNA sequence polymorphism. *Nature Genet.* **8,** 162–170.
44. Jeffreys, A. J., Tamaki, K., MacLeod, A., Monckton, D. G., Neil, D. L., and Armour, J. A. L. (1994). Complex gene conversion events in germline mutation at human minisatellites. *Nature Genet.* **6,** 136–145.
45. Sabourin, L. A., Mahadevan, M. S., Narang, M., Lee, D. S. C., Surh, L. C., and Korneluk, R. G. (1993). Effect of the myotonic dystrophy (DM) mutation on mRNA levels of the DM gene. *Nature Genet.* **4,** 233–238.
46. Fu, Y.-H., Friedman, D. L., Richards, S., Pearlman, J. A., Gibbs, R. A., Pizzuti, A., Ashizawa, T., Perryman, M. B., Scarlato, G., Fenwick, R. G., Jr., and Caskey, C. T. (1993). Decreased expression of myotonin-protein kinase messenger RNA and protein in adult form of myotonic dystrophy. *Science* **260,** 235–238.
47. Roses, A. D. (1994). Muscle biochemistry and a genetic study of myotonic dystrophy. *Science* **264,** 587.
48. Fu, Y.-H., Pearlman, J. A., Pizzuti, A., Fenwick, R. G., Jr., Friedman, D. L., Perryman, M. B., Richards, S., Gibbs, R. A., Ashizawa, T., Scarlato, G., and Caskey, C. T. (1994). Muscle biochemistry and a genetic study of myotonic dystrophy—Reply. *Science* **264,** 587–588.
49. Krahe, R., Ashizawa, T., Abbruzzese, C., Roeder, E., Carango, P., Giacanelli, M., Funanage, V. L., and Siciliano, M. J. (1995). Effect of myotonic dystrophy trinucleotide repeat expansion on *DMPK* transcription and processing. *Genomics* **28,** 1–14.
50. Boucher, C. A., King, S. K., Carey, N., Krahe, R., Winchester, C. L., Rahman, S., Creavin, T., Meghji, P., Bailey, M. E. S., Chartier, F. L., Brown, S. D., Siciliano, M. J., and Johnson, K. J. (1995). A novel homeodomain-encoding gene is associated with a large CpG island interrupted by the myotonic dystrophy unstable $(CTG)_n$ repeat. *Hum. Mol. Genet.* **4,** 1919–1925.
51. Thornton, C. A., Wymer, J. P., and Moxley, R. T. (1996). Expression of the *DMAHP* gene is surpressed in *cis* by the myotonic dystrophy (MTD) CTG repeat expansion. *Am. J. Hum. Genet.* **59S,** A161.
52. Heath, S. K., Carne, S., Hoyle, C., Johnson, K. J., and Wells, D. J. (1997). Characterisation of expression of *mDMAHP,* a homeodomain-encoding gene at the murine DM locus. *Hum. Mol. Genet.* **6,** 651–657.
53. Taneja, K. L., McCurrach, M., Schalling, M., Housman, D., and Singer, R. H. (1995). Foci of trinucleotide repeat transcripts in nuclei of myotonic dystrophy cells and tissues. *J. Cell Biol.* **128,** 995–1002.
54. Timchenko, L. T., Miller, J. W., Timchenko, N. A., Devore, D. R., Datar, K. V., Lin, L. J., Roberts, R., Caskey, C. T., and Swanson, M. S. (1996). Identification of a $(CUG)_{(n)}$ triplet repeat RNA-binding protein and its expression in myotonic-dystrophy. *Nucleic Acids. Res.* **24,** 4407–4414.
55. Wang, J., Pegoraro, E., Menegazzo, E., Gennarelli, M., Hoop, R. C., Angelini, C., and Hoffman, E. P. (1995). Myotonic dystrophy: evidence for a possible dominant-negative RNA mutation. *Hum. Mol. Genet.* **4,** 599–606.
56. Morrone, A., Pegoraro, E., Angelini, C., Zammarchi, E., Marconi, G., and Hoffman, E. P. (1997). RNA metabolism in myotonic dystrophy—Patient muscle shows decreased insulin receptor RNA and protein consistent with abnormal insulin resistance. *J. Clin. Invest.* **99,** 1691–1698.
57. Wöhrle, D., Kennerknecht, I., Wolf, M., Enders, H., Schwemmle, S., and Steinbach, P. (1995). Heterogeneity of DM kinase repeat expansion in different fetal tissues and further expansion during cell proliferation *in vitro*: evidence for a causal involvement of methyl-directed DNA mismatch repair in triplet repeat stability. *Hum. Mol. Genet.* **4,** 1147–1153.
58. Ashizawa, T., Monckton, D. G., Vaishnav, S., Patel, B. J., Voskova, A., and Caskey, C. T. (1996). Instability of the expanded (CTG)n repeats in the myotonin protein kinase gene in cultured lymphoblastoid cell lines from patients with myotonic dystrophy. *Genomics* **36,** 47–53.
59. Bingham, P. M., Scott, M. O., Wang, S., McPhaul, M. J., Wilson, E. M., Garben, J. Y., Merry, D. E., and Fischbeck, K. H. (1995). Stability of an expanded trinucleotide repeat in the androgen receptor gene in transgenic mice. *Nature Genet.* **9,** 191–196.
60. Burright, E. N., Clark, H. B., Servadio, A., Matilla, T., Feddersen, R. M., Yunis, W. S., Duvick, L. A., Zoghbi, H. Y., and Orr, H. T. (1995). *SCA1* transgenic mice: a model for neurodegeneration caused by an expanded CAG trinucleotide repeat. *Cell* **82,** 937–948.
61. Goldberg, Y. P., Kalchman, M. A., Metzler, M., Nasir, J., Zeisler, J., Graham, R., Koide, H. B., O'Kusky, J., Sharp, A. H., Ross, C. A., Jirik, F., and Hayden, M. R. (1996). Absence of disease phenotype and intergenerational stability of the CAG repeat in transgenic mice expressing the human Huntington disease transcript. *Hum. Mol. Genet.* **5,** 177–185.
62. Ikeda, H., Yamaguchi, M., Sugai, S., Aze, Y., Narumiya, S., and Kakizuka, A. (1996). Expanded polyglutamine in the Machado-Joseph disease protein induces cell-death *in vitro* and *in vivo.* *Nature Genet.* **13,** 196–202.
63. Gourdon, G., Radvanyi, F., Lia, A. S., Duros, C., Blanche, M., Abitbol, M., Junien, C., and Hofmann Radvanyi, H. (1997). Moderate intergenerational and somatic instability of a 55 CTG repeat in transgenic mice. *Nature Genet.* **15,** 190–192.
64. Monckton, D. G., Coolbaugh, M. I., Ashizawa, K., Siciliano, M. J., and Caskey, C. T. (1997). Hypermutable myotonic dystrophy CTG repeats in transgenic mice. *Nature Genet.* **15,** 193–96.
65. Mangiarini, L., Sathasivam, K., Mahal, A., Mott, R., Seller, M., and Bates, G. P. (1997). Instability of highly expanded CAG repeats in mice transgenic for the Huntington's disease mutation. *Nature Genet.* **15,** 197–200.
66. Jansen, G., Mahadevan, M., Amemiya, C., Wormskamp, N., Segers, B., Hendriks, W., O'Hoy, K., Baird, S., Sabourin, L., Lennon, G., Jap, P. L., Iles, D., Coerwinkel, M., Hofker, M., Carrano, A. V., de Jong, P. J., Korneluk, R. G., and Wieringa, B. (1992). Characterization of the myotonic dystrophy region predicts multiple protein isoform-encoding mRNAs. *Nature Genet.* **1,** 261–266.
67. Mitani, K., Takahashi, Y., and Kominami, R. (1990). A GGCAGG motif in minisatellites affecting their germline instability. *J. Biol. Chem.* **265,** 15203–15210.
68. Kelly, R. G., Bulfield, G., Collick, A., Gibbs, M., and Jeffreys, A. J. (1989). Characterization of a highly unstable mouse minisatellite locus: evidence for somatic mutation during early development. *Genomics* **5,** 844–856.
69. Kelly, R. G., Gibbs, M., Collick, A., and Jeffreys, A. J. (1991). Spontaneous mutation at the hypervariable mouse minisatellite locus Ms6-hm: flanking DNA sequence and analysis of germline

and early somatic mutation events. *Proc. R. Soc. Lon. B* **245,** 235–245.

70. Gibbs, M., Collick, A., Kelly, R. G., and Jeffreys, A. J. (1993). A tetranucleotide repeat mouse minisatellite displaying substantial somatic instability during early preimplantation development. *Genomics* **17,** 121–128.
71. Collick, A., Norris, M. L., Allen, M. J., Bois, P., Barton, S. C., Surani M. A., and Jeffreys, A. J. (1994). Variable germline and embryonic instability of the human minisatellite MS32 (D1S8) in transgenic mice. *EMBO J.* **13,** 5745–5753.
72. Collick, A., Drew, J., Penberth, J., Bois, P., Luckett, J., Scaerou, F., Jeffreys, A., and Reik, W. (1996). Instability of long inverted repeats within mouse transgenes. *EMBO J.* **15,** 1163–1171.
73. Wöhrle, D., Hennig, I., Vogel, W., and Steinbach, P. (1993). Mitotic stability of fragile X mutations in differentiated cells indicates early post-conceptional trinucleotide repeat expansion. *Nature Genet.* **4,** 140–142.
74. Mahon, K. A., Overbeek, P. A., and Westphal, H. (1988). Prenatal lethality in a transgenic mouse line is the result of a chromosomal translocation. *Proc. Natl. Acad. Sci. USA* **85,** 1165–1168.
75. Yu, Y E., Nemeth, M., Pathak, S., Meistrich, M. L., and Wong, P. K. (1998). Preliminary characterization of transgenic males harboring an unbalanced product of the reciprocal translocation T(3;10). *Mouse Genome.* [in press]
76. Ikeuchi, T., Igarashi, S., Takiyama, Y., Onodera, O., Oyake, M., Takano, H., Koide, R., Tanaka, H., and Tsuji, S. (1996). Non-Mendelian transmission in dentatorubral-pallidoluysian atrophy and Machado-Joseph disease: the mutant allele is preferentially transmitted in male meiosis. *Am. J. Hum. Genet.* **58,** 730–733.
77. Kang, S., Jaworski, A., Ohshima, K., and Wells, R. D. (1995). Expansion and deletion of CTG repeats from human disease genes are determined by the direction of replication in *E. coli. Nature Genet* **10,** 213–218.
78. Freudenreich, C. H., Stavenhagen, J. B., and Zakian, V. A. (1997). Stability of a CTG/CAG trinucleotide repeat in yeast is dependent on its orientation in the genome. *Mol. Cell. Biol.* **17,** 2090–2098.
79. Monckton, D. G., Ashizawa, T., Coolbaugh, M. I., Brook, G., Fortune, T., Gooch, C. L., Armstrong, D. L., Caskey, C. T., Meistrich, M., and Siciliano, M. J. (1997). Testicular atrophy and reduced female fertility in mice expressing a non-coding expanded CUG repeat containing myotonic dystrophy minigene. [Manuscript in preparation]
80. Wang, Y.-H., Amirhaeri, S., Kang, S., Wells, R. D., and Griffith, J. D. (1994). Preferential nucleosome assembly at DNA triplet repeats from the myotonic dystrophy gene. *Science* **265,** 669–671.
81. Wang, Y.-H., and Griffith, J. (1995). Expanded CTG triplet blocks from the myotonic dystrophy gene create the strongest known natural nucleosome positioning elements. *Genomics* **25,** 570–573.
82. Otten, A. D., and Tapscott, S. J. (1995). Triplet repeat expansion in myotonic dystrophy alters the adjacent chromatin structure. *Proc. Natl. Acad. Sci. USA* **92,** 5465–5469.
83. Jansen, G., Groenen, P. J. T. A., Bächner, D., Jap, P. H. K., Coerwinkel, M., Oerlemans, F., van den Broek, W., Gohlsch, B., Pette, D., Plomp, J. J., Molenaar, P. C., Nederhoff, M. G. J., van Echteld, C. J. A., Dekker, M., Berns, A., Hameister, H., and Wieringa, B. (1996). Abnormal myotonic dystrophy protein kinase levels produce only mild myopathy in mice. *Nature Genet.* **13,** 316–324.
84. Reddy, S., Smith, D. B. J., Rich, M. M., Leferovich, J. M., Reilly, P., Davis, B. M., Tran, K., Rayburn, H., Bronson, R., Cros, D., Balice-Gordon, R. J., and Housman, D. (1996). Mice lacking the myotonic dystrophy protein kinase develop a late onset progressive myopathy. *Nature Genet.* **13,** 325–335.
85. Bulfield, G., Siller, W. G., Wight, P. A. L., and Moore, K. J. (1984). X-chromosome-linked muscular-dystrophy (*mdx*) in the mouse. *Proc. Natl. Acad. Sci. USA* **81,** 1189–1192.

Part VI

Dentatorubral–Pallidoluysian Atrophy

CHAPTER 14

Clinical Aspects of DRPLA

IKUKO KONDO Department of Hygiene, Ehime University School of Medicine, Ehime 791-02, Japan

I. INTRODUCTION

Dentatorubral-pallidoluysian atrophy (DRPLA) (MIM. 125370) is an autosomal dominant disorder characterized by clinical features including a variable combination of dementia, euphoria, speech problems, ataxia, epilepsy, and involuntary movements presenting as chorea, tremor, and myoclonus. The prevalence of DRPLA is approximately six in one million in Japan, but this disease seems to be very rare except in Japan. Earlier onset is evident in successive generations (genetic anticipation).

Clinical features are classified into three forms depending on the age at onset. (1) For the juvenile form, myoclonus, epilepsy, and mental retardation are major symptoms; (2) for the early adult-onset form, cerebellar ataxia, choreoathetosis, epilepsy, and dementia are typical symptoms; (3) for the late-adult-onset form, dementia and ataxia are common. Clinical features in patients with DRPLA vary within the same family, but pathological changes in autopsied cases were quite similar. Major neuropathological changes consisting of combined degeneration of the dentatorubral and pallidoluysian systems. Using computerized tomography and magnetic resonance imaging, varying degrees of cerebellar and brain stem atrophy are visualized. T2-weighted MRI in patients with the late-adult-onset form shows symmetric high-signal lesions in the cerebral white matter, globus pallidus, midbrain, and pons. EEGs in patients with myoclonus epilepsy show diffuse slow spike and wave complexes.

Recently, an expansion of an unstable trinucleotide (CAG) repeat in the DRPLA gene on chromosome 12p13 was identified in patients with DRPLA. Normal alleles have 7 to 34 repeats, and DRPLA alleles have 51 to 88 repeats. There is a strong correlation between the ages of onset and the number of CAG repeat units. The paternally transmitted expanded alleles increase by more than 5 repeats, whereas maternal transmutations show either a decrease or an increase of fewer than 5 repeats. These molecular findings are useful in diagnosis of DRPLA and are important in understanding phenotypic variation and anticipation.

II. A HISTORICAL BACKGROUND OF DRPLA

In 1982, Naito *et al.* reported a new hereditary syndrome with the name "hereditary dentatorubral-pallidoluysian atrophy (DRPLA)" [1]. Three typical characteristics of this syndrome were: myoclonus epilepsy syndrome with or without cerebellar ataxia and/or choreoathetosis; dentatorubral-pallidoluysian atrophy; and autosomal dominant heredity [1]. Clinical features varied between patients, but major pathological changes included the degeneration of the dentatorubral and pallidoluysian systems [1]. This neuropathological combination has been reported sporadically in Caucasian patients [2–6]; and the term "dentatorubropallidoluysian atrophy" was first introduced by Smith in 1975 on the basis of neuropathologic findings in a middle-aged patient with ataxia and choreoathetic movements [2]. Since the two families with progressive myoclonus epilepsy were first described in 1972 by Naito *et al.* in Japan, who published the first English language account of the condition in 1982, similar clinicopathological studies in over 40 families had been reported in Japan, and the mode of inheritance seemed to be autosomal dominant [7–22]. Major pathological findings in these cases were dentatorubral and pallidoluysian atrophy [7–22]. At present, dentatorubral-pallidoluysian atrophy (DRPLA) (MIM. 125370, myoclonus epilepsy with choreoathetosis; Naito-Oyanagi disease; NOD) is established in clinical and molecular aspects [23].

DRPLA has a variable clinical presentation, but the pathology is routinely confined to cerebellofugal and pallidofugal systems. Diagnosis has been very difficult without pathological findings. Therefore, many clinical and pathological studies have been described in Japan, but there is only one report of a genetic linkage study in the literature [24].

In 1994, Koide *et al.* [25] and Nagafuchi *et al.* [26] identified independently a gene responsible for DRPLA (CTG-B37). The gene has been mapped on the short arm of chromosome 12 [26–28].The CTG-B37 is a partial cDNA clone carrying polymorphic trinucleotide (CAG) repeats isolated by Li *et al.* in 1993 [29].In the coding region of this gene [30], CAG trinucleotide repeats were expanded and unstable in families with DRPLA [25, 26]. The hereditary DRPLA occurs almost solely in Japanese, but some families with DRPLA have been reported in Europe [31, 32] and the United States [33, 34] after the gene was isolated [25, 26, 30]. Since the gene responsible for the fragile X syndrome, an X linked mental retardation, was identified in 1991 [35], the genes carrying unstable trinucleotide repeats have been isolated in several neuropsychiatric disorders including spinocerebellar atrophy [36] and Huntington's disease [37]. DRPLA is the seventh of this gene group.

III. GENETICS OF DRPLA

Family pedigrees of patients with DRPLA reveal that the mode of inheritance of this clinical condition is autosomal dominant as shown in Fig. 14-1. In three families, affected siblings from healthy parents have been reported (Fig. 14-2) [15, 17], and the penetrance is estimated to be approximately 90% in 40 families [1, 7–22]. Homozygotes for the disease gene have also been identified [38], and severity and onset are influenced by copy number of disease alleles [38], indicating that a double dose of expanded DRPLA allele enhances clinical severity. In addition, two homozygotes for intermediate CAG repeats (40 and 41 repeats) in the DRPLA gene were reported [39]. Their parents were consanguineous, but their mother (57 years of age) and the father (64 years of age) were neurologically normal [39]. Two siblings had not typical clinical symptoms of DRPLA, but spastic paraplegia was diagnosed when in their 30s [39].

Clinical features in patients with DRPLA vary even in the same family members. Therefore, diagnosis of DRPLA has been very difficult without information of family history and examination of family members. Anticipation refers to the progressively earlier onset and increase in disease severity in successive generations. In DRPLA, the strong anticipation has been confirmed in paternally transmitted cases [25, 26, 40, 41] similar to Huntington's disease [37]. In the first generation of a DRPLA family, ages of onset are generally over 50 years of age, and patients have cerebellar ataxia with or without dementia. In the subsequent generation, clinical symptoms start in affected siblings before the age of 40 years and symptoms are variable and more severe than those in the first generation [1, 7–22, 25, 26, 40, 41]. In the third generation, ages at onset range from 1 to 20 years and patients are diagnosed as having progressive myoclonus epilepsy (PME) [1, 42]. Anticipation is stronger with paternal transmission than with maternal transmission [40, 41]. The difference in age at onset between affected father and children was approximately 35 years, whereas the difference between affected mothers and children was about 15 years [40]. Enhancement of anticipation in paternal transmission is significant compared with maternal transmission.

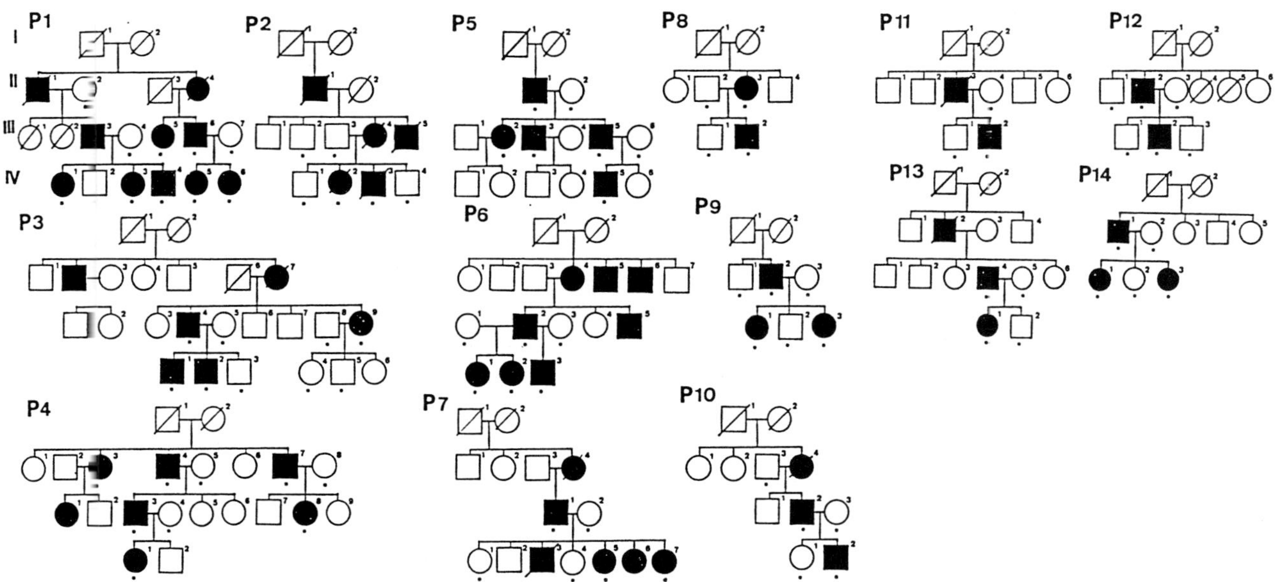

FIGURE 14-1 Family pedigrees of DRPLA. Closed squares and circles, affected; open squares and circles, normal; with slash, decided; with dot, examined.

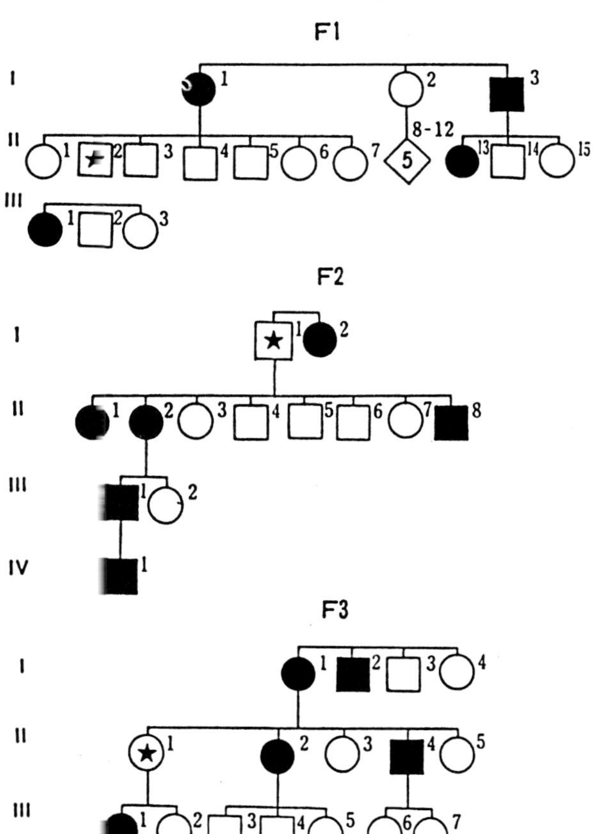

FIGURE 14 2 Three families with normal individuals (obligate carrier) with affected offspring. Closed squares and circles, affected; open squares and circles, normal; *obligate carrier.

The sex ratio was equal in affected individuals based on a study of clinical conditions, but the numbers of affected siblings exceed those of unaffected siblings, particularly in successive generations based on molecular analyses [43]. Since the incidence of miscarriage is not different in siblings with expanded DRPLA alleles from those in normal siblings, more prominent meiotic instability of the length in CAG trinucleotide repeats might be responsible for the difference of segregation of the expanded alleles [43].

IV. CLINICAL NEUROLOGY OF DRPLA

Clinical diagnosis of DRPLA depends on clinical signs and symptoms given in the history, those present in the patients, and those from the family history. Major clinical symptoms are mental retardation, epilepsy, myoclonus, ataxia, choreoathetosis, and dementia. Variable combinations of these symptoms have been observed in patients. Generally, patients with the early onset are more rapidly progressive, whereas patients with late-onset exhibit only mild ataxia and dementia.

A. Age at Onset

Precise onset of the disease is obscure in most patients. The distribution of ages at onset range from 1 to

75 years of age in the literature [1, 7–22, 25, 26, 40, 41]. The mean age at onset is 27 and 33 years in 169 cases in Niigata prefecture, the area first to report DRPLA families [15] and in the general population in Japan [44], respectively. A distribution of age at onset in DRPLA was characteristic (Fig. 14-3). There are three prominent peaks in age at onset; one is between 10 and 20 years, and the others are in the late 30s and 40s. Patients with age at onset under 20 years are usually defined as juvenile DRPLA, while those with age at onset over 50 years are considered to have the late-adult form of DRPLA. The cumulative frequency and curve in DRPLA show that 50% of patients have ages at onset under 25 years, and that 20% have ages at onset 45 (Fig. 14-4). Only 1.4% of patients are diagnosed at over 70 years of age (Fig. 14-4).

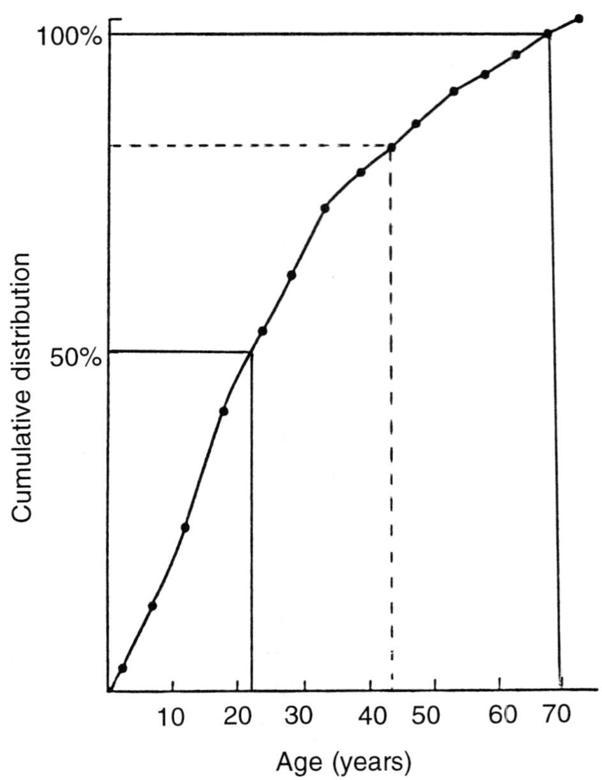

FIGURE 14-4 Cumulative frequency and curve of age at onset inpatients with DRPLA. Fifty percent of patients have symptoms before the age of 25 and only 1.4% of patients start clinical symptoms after 70 years of age.

B. Age at Death and Survival

DRPLA is a progressive disease and the duration of the disease is one of the important practical factors for patients and their relatives. Survival has been estimated in studies of age at onset and death already quoted in 90 cases. Age at death ranged from 5 to 79 years, and mean age at death was 42 [1, 7–22]. Average suffering duration was approximately 12 years. Causes of death were commonly severe generalized physiological disability and/or status epilepsy. Patients were completely dependent for all aspects of care at the last stage of the disease [1, 7–22].

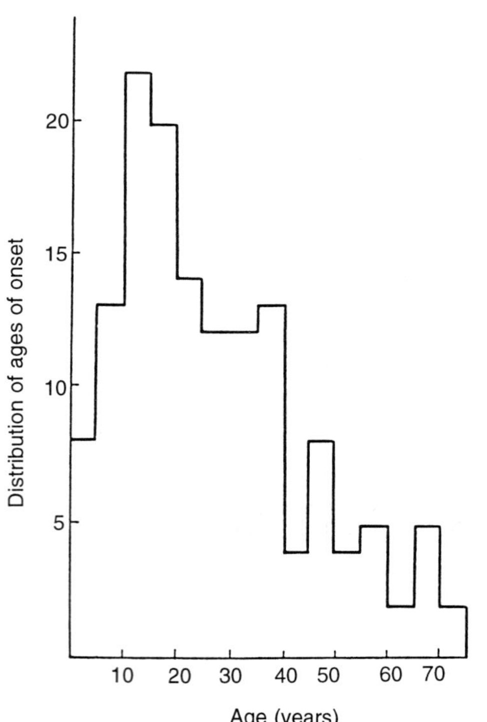

FIGURE 14-3 Distribution of ages at onset in patients with DRPLA.

C. Clinical Features in Relation to Age at Onset

In 1982, Naito and Oyanagi pointed out that the clinical features are quite varied in patients and in the generations [1]. Based on the clinical features and age of onset, DRPLA has been classified into three groups, summarized in Table 14-1. After the identification of the gene for DRPLA, relations between these variable clinical symptoms and the trinucleotide (CAG) repeat length in the gene have become evidently. The number of repeats is highly polymorphic with a distinct, bimodal distribution of normal and expanded alleles. In normal individuals' alleles, DNA sizes vary between 7 and 33 copies, whereas patients with DRPLA have expansions in the range of 53 to 88 repeats [25, 26].

TABLE 14-1 Clinical Symptoms in Three Forms of DRPLA

Symptoms	Juvenile form	Early-adult form	Late-adult form
Age of onset	Under 20s	Late 20s to 50s	Over 50s
Epilepsy	+++	+	–
Myoclonus	+++	+	–
Developmental delay	+++	–	–
Dementia	+++	+++	++
Ataxia	+	+++	+++
Choreoathetosis	+	+++	+++

D. Symptoms

The main clinical features of DRPLA are epilepsy, myoclonus, ataxia, choreoathetosis, and dementia. Combinations of these symptoms are different in patients with different age at onset. DRPLA can be classified into three forms based on clinical symptoms and age at onset.

1. Juvenile Form of DRPLA

Clinical signs and symptoms in patients with juvenile DRPLA are completely compatible with progressive myoclonus epilepsy (PME) syndrome. Major symptoms are mental retardation, seizures, myoclonus, and ataxia. Photoconvulsive responses are also observed [1, 42]. In most cases, mental retardation is noted before 5 years of age, when mild gait unsteadiness is first recognized. These symptoms are worsened gradually and relentlessly, until the patient finally becomes a vegetable after 10 years.

Clinical cause of a patient with typical symptoms of juvenile DRPLA is shown in Fig. 14-5. Unstable gait was noted at 4 years of age and convulsions started at the age of 6. At the age of 8, mental retardation, speech disturbance, mild ataxia, and epilepsy became remarkable and progressive myoclonus developed. Subsequently, tonic, clonic, or myoclonic seizures with photosensitivity frequently occurred despite treatment with anticonvulsant drugs and gait disturbance progressed gradually.

Pyramidal signs including pathological deep tendon reflexes became evident. At age 10, she became bedridden and exhibited vegetable state. At age 13, she died suddenly at home. Postmortem examination was not performed, but expanded CAG trinucleotide repeats in DRPLA gene was identified in her DNA from peripheral blood. Her cranial CT scans at age of 8 showed markedly thickened skull and moderate brain atrophy with dilated ventricles (Fig. 14-6). MRI findings at ages of 10 and 11 showed moderate to severe cerebellar atrophy with cortical atrophy and pons (Figs. 14-7, 14-8).

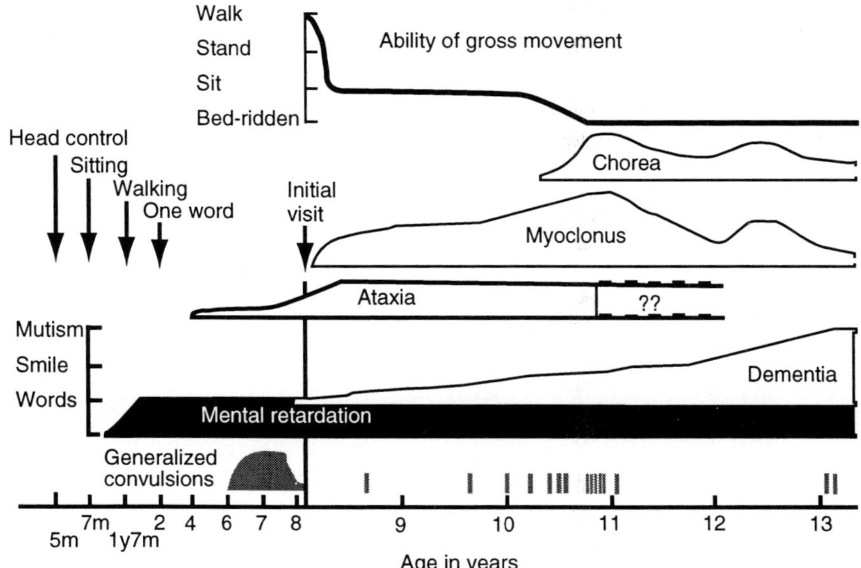

FIGURE 14-5 Clinical natural history of a patient with juvenile DRPLA. Mental development delay starts at 1 year and ataxia and general convulsions begin at 4 and 6 years, respectively. From 8 years, myoclonus and other symptoms gradually progress; patient is vegetative at the final stage of the disease.

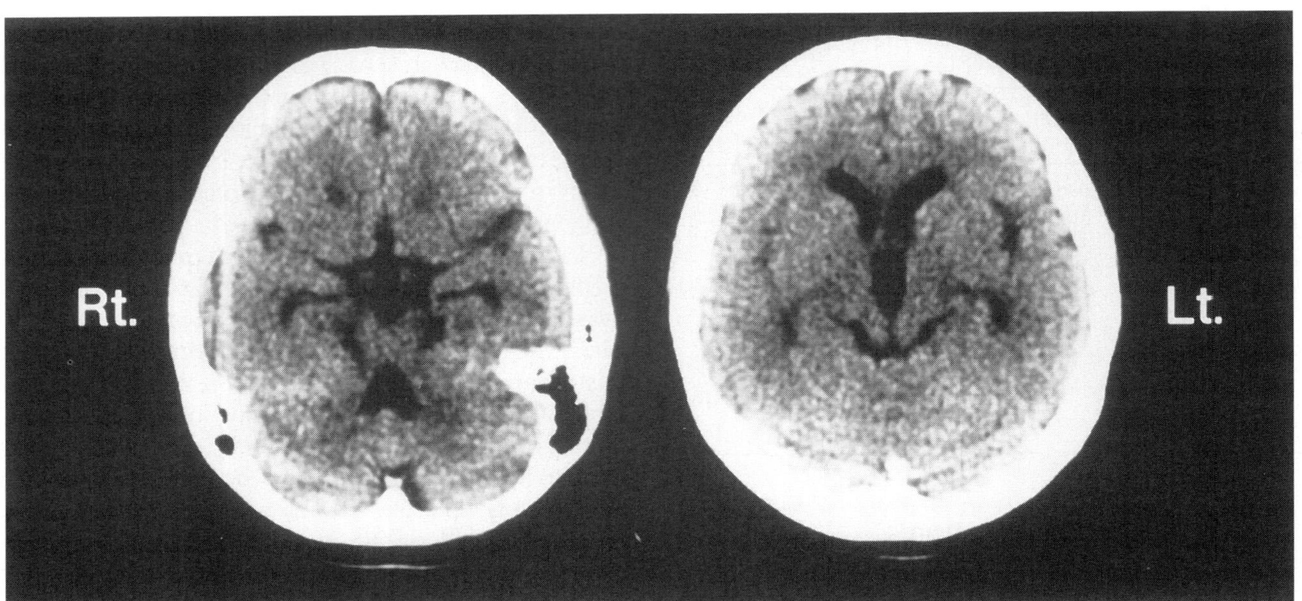

FIGURE 14-6 Cranial CT in a patient with DRPLA at 8 years old. Skull is diffusely thickened and cerebellar atrophy with dilated fourth ventricle is remarkable.

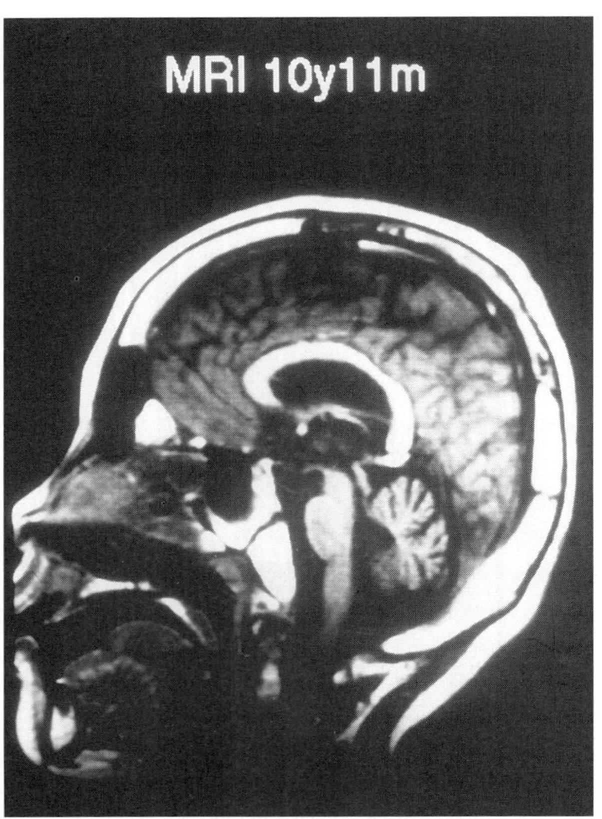

FIGURE 14-7 Saggital MRI in a patient with DRPLA at 10 years. Severe brain atrophy, especially of cerebrum and pons, is noted.

2. Early-Adult Onset Form of DRPLA

Patients with age at onset during their 20s to 50s usually have ataxia and choreoathetosis. Stuttered speech, ataxia, and finger tremor appear at age 30 and seizures occur later. Gait disturbance, adiadocokinesis, pyramidal signs, myoclonus, and dementia appear and are gradually progressive. Patients have EEG anomalies and generalized convulsions. Personal characteristics change gradually and patients sometimes become aggressive and euphoric. Clinical symptoms are widely varied in this form, but patients need total care at the final stage of the disease. Clinical natural histories of this form are sometimes similar to those in patients with Huntington's disease, and patients are often diagnosed as having Huntington's disease, especially in Caucasians [32–34]. This form was previously named a pseudo-Huntington form.

3. Late-Adult Onset Form of DRPLA

Ataxic gait and speech disturbance appear after 50 years of age and dementia is usually mild. Clinical features in patients with late-adult-onset form are characterized by ataxia, dysarthria, myoclonus, and choreoathetosis. Seizures are not usually associated.

E. Laboratory Investigation

1. Biochemical Laboratory Findings

Laboratory studies are usually normal, including studies on serum proteins, blood gases, lactate, pyr-

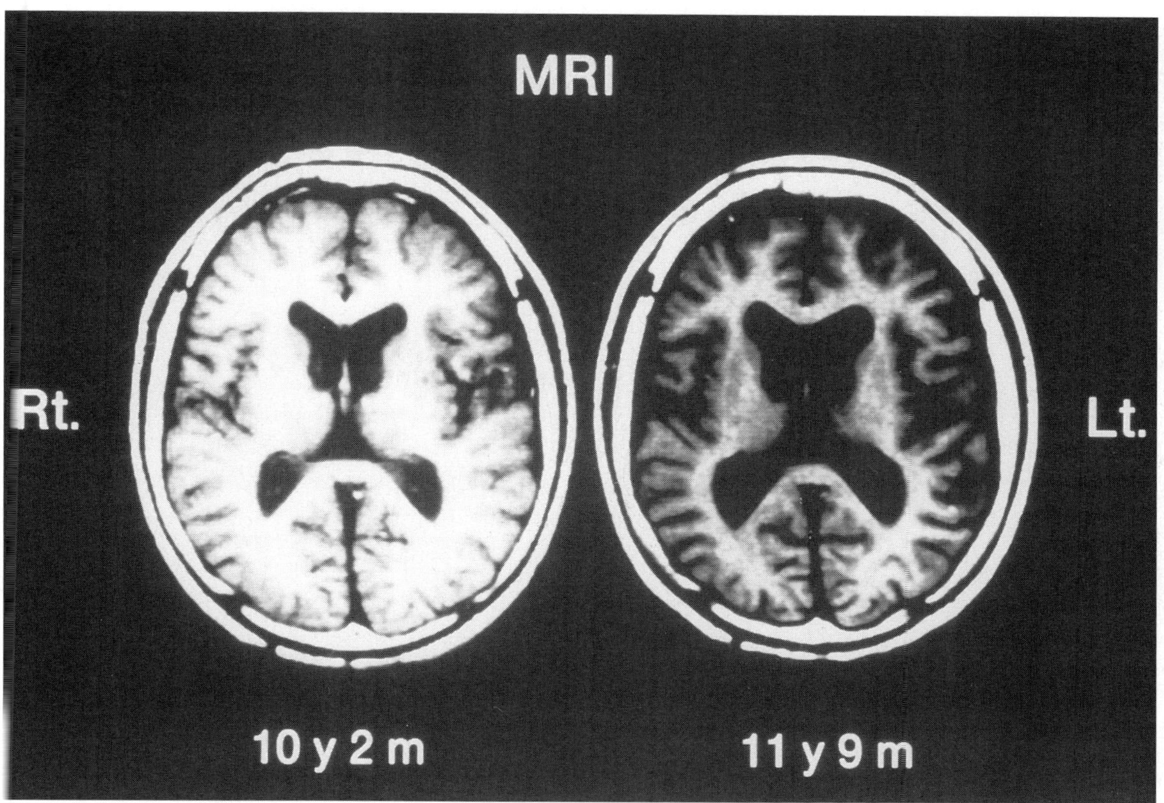

FIGURE 14-8 Axial T2-weighted MRI in a patient with DRPLA at 10 and 11 years old. Brain atrophy with dilated ventricles is progressively noted.

uvate, and ammonia. No abnormalities are seen in a screening test for leukocyte lysosomal enzymatic activities. Quantitative amino acid analysis of serum and urine is also normal. No abnormalities in cerebrospinal fluid (CSF) is detected. There is no biochemical diagnostic marker for diagnosis of DRPLA.

2. NEURORADIOLOGICAL FINDINGS

a. Computed Tomography (CT). Three common anomalies are present in cranial CT of the patients with DRPLA: (1) cerebellar atrophy associated with dilatation of the fourth ventricle, (2) atrophy of brain stem and cerebellum, and (3) atrophy of cerebrum with dilated lateral ventricles [45–47]. CT of a patient with juvenile form shows remarkable cerebellar atrophy with dilation of the fourth ventricle and atrophy of brain stem. Diffusely thickened skull gradually becomes evident and appears as cranioostosis [46]. Diffuse hyposignal lesions in the cerebral white matter are specific in patients with early adult-onset form.

b. Magnetic Resonate Imaging (MRI). MRI findings demonstrate severe cerebral and cerebellar atrophy accompanied by dilatation of the fourth ventricle in patients with the juvenile form (Fig. 14-8) [49]. Brain stem is severely atrophic. Characteristic MRI findings in late-adult-onset patients are observed in particular on T2-weighted images. The symmetric high-signal lesions in the cerebral white matter, globus pallidus, thalamus, midbrain, and pons in senile patients with DRPLA and these lesions cannot be detected on CT scans [45–48]. Younger patients with DRPLA did not show these signals [49].

c. Proton Magnetic Response Spectroscopy (^{1}H-MRS). ^{1}H-MRS has been studied in three patients with the juvenile form and one early adult-onset patient [51]. ^{1}H-MRS of parietal and basal ganglia regions clearly show three signals due to *N*-acetylaspartate (NAA), choline (Cho), and creatine (Cr) and no signal corresponding to lactate or any other metabolite. However, NAA/Cho and NAA/Cr obtained from both regions in patients with the juvenile form were markedly reduced compared not only with those from healthy controls, but also with a patient with the adult-onset form [52]. It has been suggested that the degree of regional neuron loss of the basal ganglia region closely correlated with the severity of the clinical features by

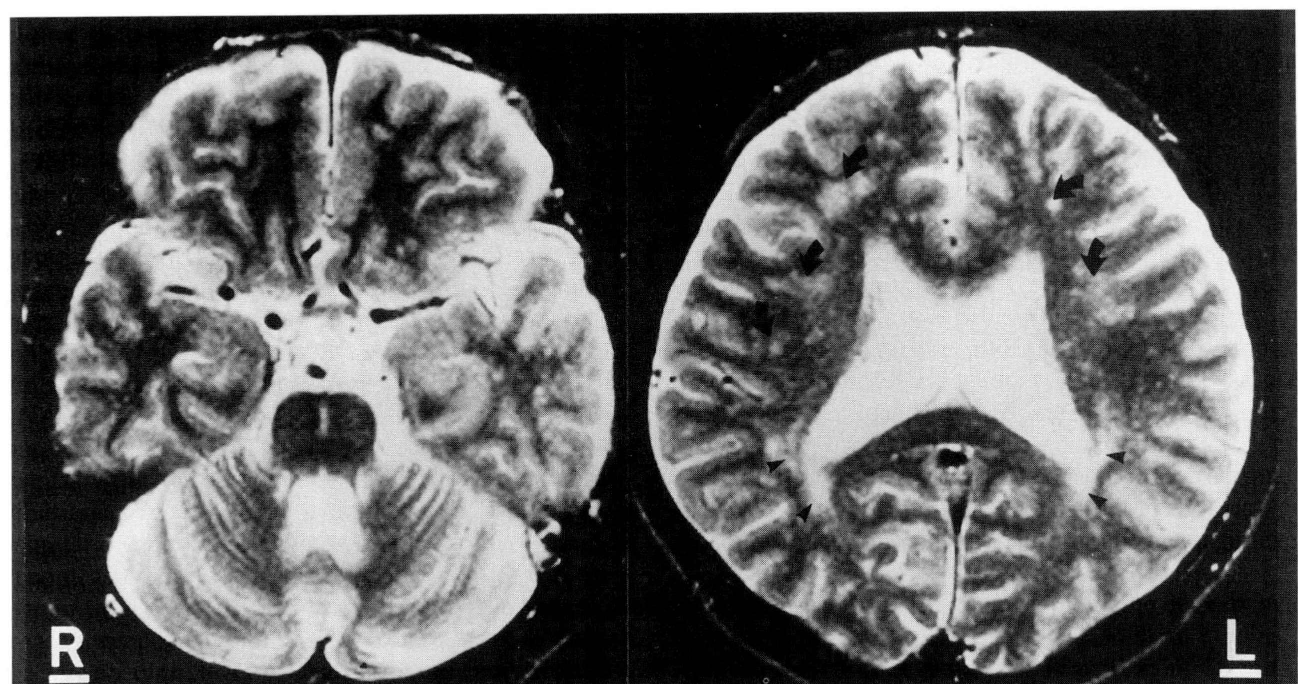

FIGURE 14-9 Axial T2-weighted spin-echo MRI sequence of a patient with juvenile DRPLA. Left, transaxial image through the level of superior pons shows mild cerebral and brain stem atrophy, as well as severe cerebellar atrophy. Right, transaxial image through the level of the body of the lateral ventricle shows some patchy hyperintensity (arrows) and periventricular hyperintensity (arrowheads). (Courtesy of Dr. Masahito Miyazaki, Tokushina University.)

clinicopathological studies. ^{1}H-MRS is a valuable tool to clarify the pathophysiological changes in patients with DRPLA.

d. Multimodal Evoked Potentials (MEPs). Brain stem auditory evoked potentials show reduced or absent brain stem components as well as delayed latencies in juvenile DRPLA (Figs. 14-10, 14-11) [53]. Short latency somatosensory evoked potentials (S-SEPs) also have prolonged central conduction time and reduced amplitude of cortical components. However, extremely enlarged flash visual evoked potentials with shortened latency even in the absence of giant SEPs are detected in young patients, but not in older patients [52].

3. ELECTROENCEPHALOGRAM (EEG)

EEG findings are quite different in juvenile patients than in patients with adult-onset forms. Over 90% of

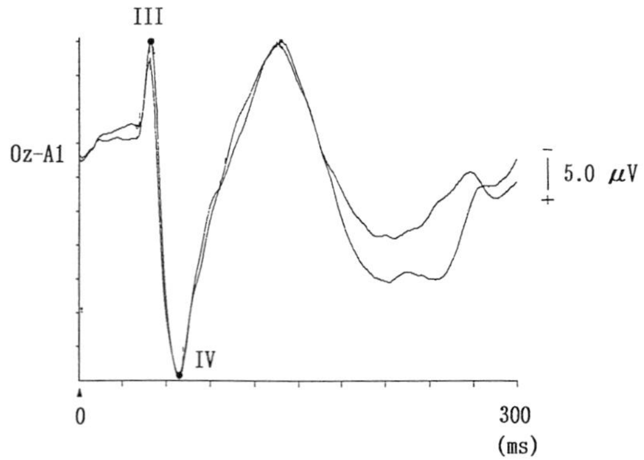

FIGURE 14-10 Evoked potentials in a patient with juvenile DRPLA. Note that the largest positive peak, designated "IV," is extremely enlarged and shortened. (Courtesy of Dr. Masahito Miyazaki, Tokushima University.)

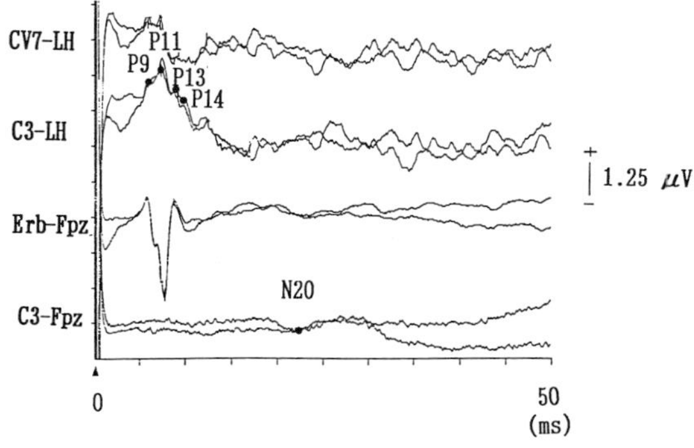

FIGURE 14-11 S-SEPs in a patient with juvenile DRPLA, waveforms elicited by right median nerve stimulation. Early components such as P9, P11, P13, and P14 are spared, but N20 are significantly delayed and reduced. (Courtesy of Dr. Masahito Miyazaki, Tokushima University.)

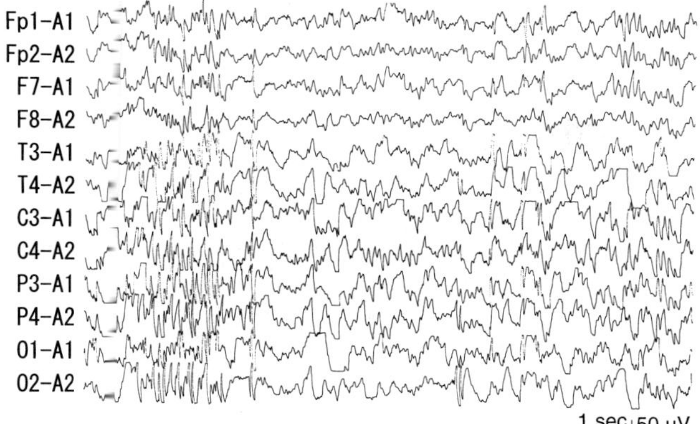

Figure 14-12 Interictal EEG of a patient with juvenile DRPLA during sleep at age 5 years, with frequent multifocal spike and wave complexes predominant in the right parietooccipital region, which tended to spread diffusely. (Courtesy of Dr. Haruo Hattori, Kyoto University.)

juvenile patients have seizures and EEG abnormalities consisting of random high-voltage slow waves,and paroxysmal atypical spike and wave discharges. Intraictal EEG during sleep shows diffuse multifocal spikes and spike and wave complexes predominant in the centritemporal regions (Figs. 14-12, 14-13 [49]. Photostimulation (18 f/s) initiates repetitive spikes and clinical seizures begin with (Fig. 14-14). These abnormal EEG patterns and spike and wave complexes disappear gradually with age (Fig. 14-15), and seizures become infrequent with neurological deterioration. However, photostimulation induces seizures and EEG findings do not change with age (Fig. 14-15).

In general, EEGs in patients with late-adult-onset form are normal and patients do not have convulsive seizures. However, 70% of patients with early adult-onset form sometimes have seizures and EEGs disclose irregular basic activities and spike waves. Aged patients are not usually photosensitive.

4. Pathological Findings

Since Naito *et al.* first reported this disorder in 1972 in Japan, neuropathological findings have been extensively reported [1, 7–22]. The brain including the brain stem and spinal cord is small. The mean brain weight is ap-

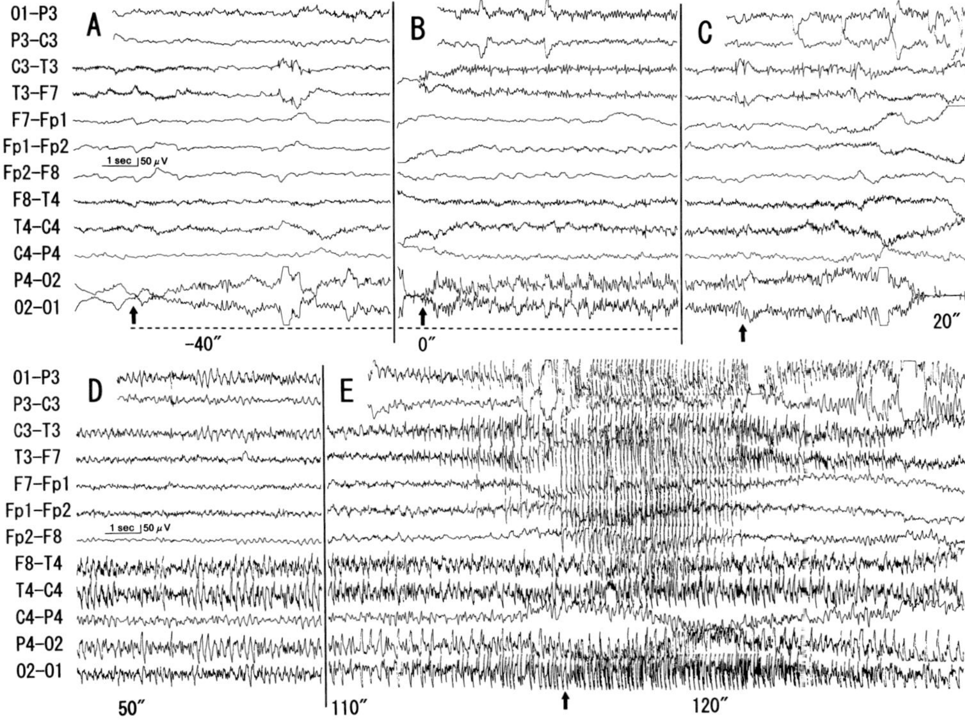

Figure 14-13 Ictal EEG of a patient with with juvenile DRPLA at age 13 years after photostimulation. (A) Photostimulation (broken line) at 8 Hz initiated repetitive spikes from the right occipital region (arrow). (B) Clinical seizure began with flexion of the left arm (arrow). (C) The head and eyes turned to the left 14 s later (arrow). (D) Slow tonic flexion of the left arm with conjugated deviation of the head and eyes to the left continued. (E) Generalized tonic–clonic convulsion began with diffuse paroxyama at 116 s (arrow). (Courtesy of Dr. Haruo Hattori, Kyoto University.)

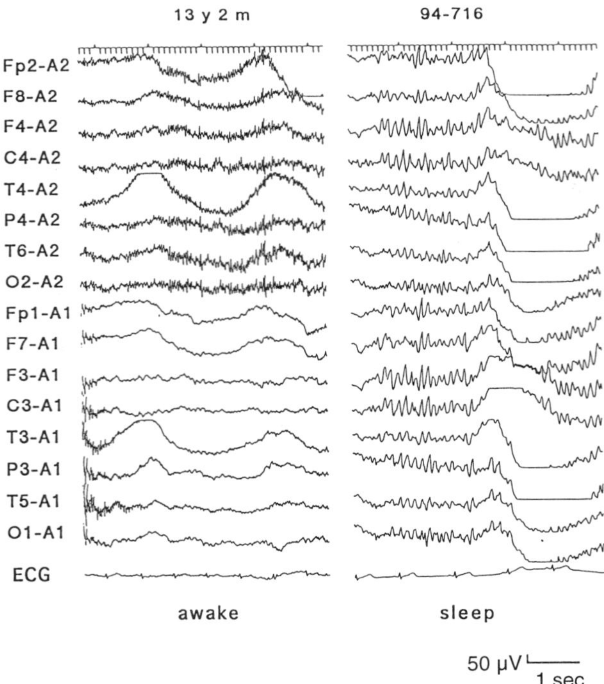

FIGURE 14-14 Interictal EEG of a patient with juvenile DRPLA at age 13 years. Continuous background activity during wake and sleep. No abnormal EEG pattern. (Courtesy of Dr. Tatsuya Ogiwara, Matsuyama Red Cross Hospital.)

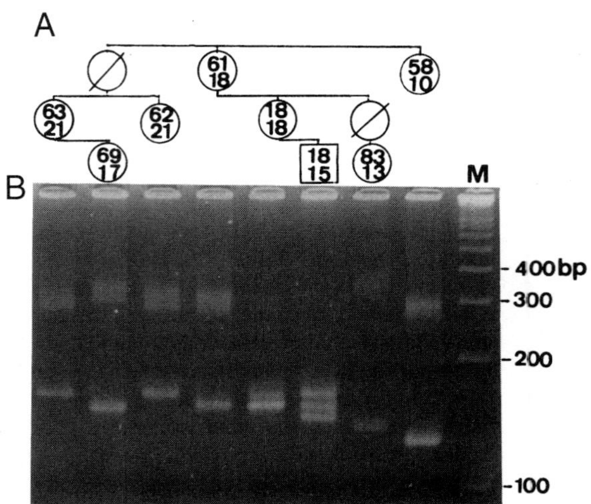

FIGURE 14-16 PCR products from genomic DNAs in a family with DRPLA. (A) A family tree of patients with DRPLA. CAG repeat number shown in each individual. (B) DNA pattern of PCR products amplified with primers introduced by Li *et al.* [29] in 2.5% agarose. M, 100-base-pair DNA size marker. Individuals with PCR products longer than 200 bp are affected and normal individuals have shorter DNA bands.

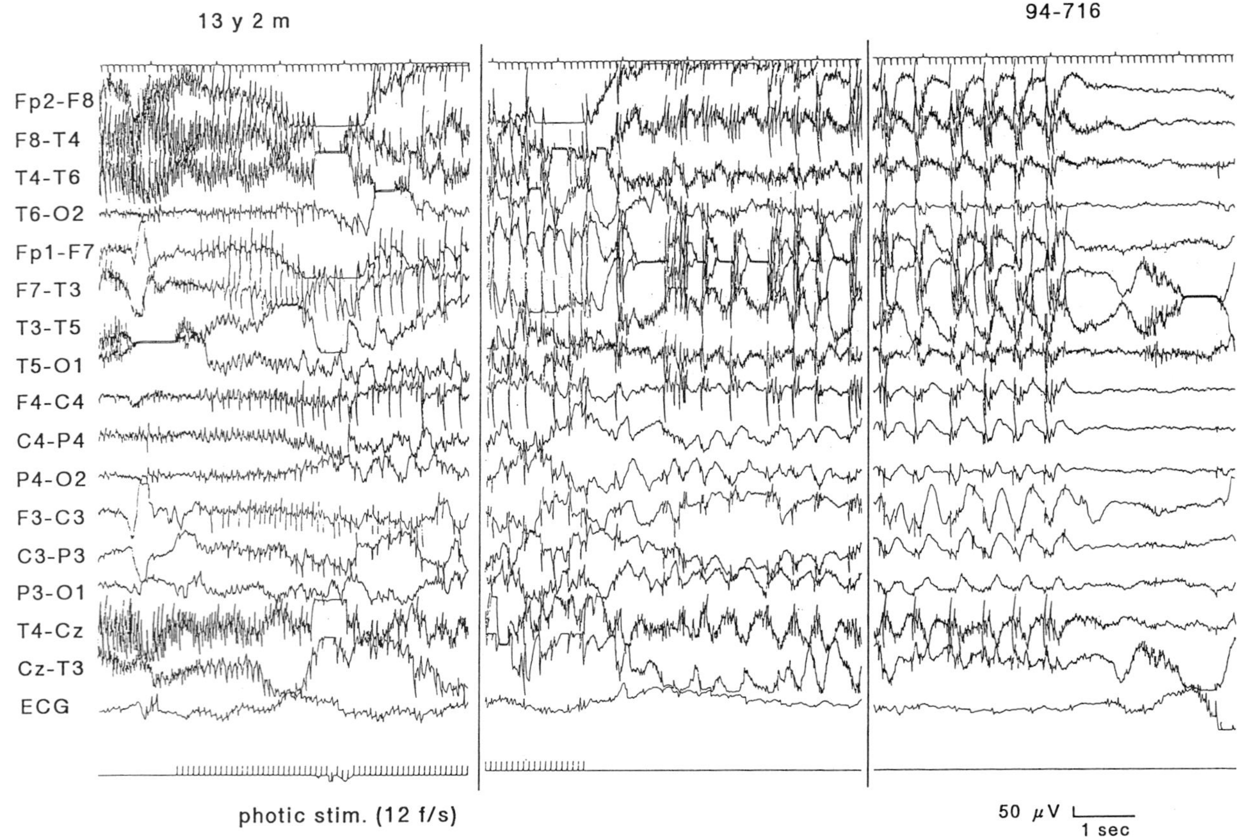

FIGURE 14-15 Ictal EEG after photostimulation in patient with juvenile DRPLA at age 13 years. Interictal EEG is shown in Fig. 14-14. EEG patterns and clinical seizures are similar to those in the patient shown in Fig. 14-13. (Courtesy of Dr. Tatsuya Ogiwara, Matsuyama Red Cross Hospital.)

proximately 1000 g in adult patients. The most striking findings in autopsy cases are (1) severe atrophy of the dentate nuclei with grumose degeneration, degeneration of the super cerebellar peduncles, and mild gliosis in the red nuclei, (2) severe atrophy of globus pallidus externa and subthalamic nuclei, (3) sparing of the striatum, and (4) atrophy of the brain stem predominantly at the tegmentum [16].Though the cerebral cortex and the thalamus are spared, the white matter of the brain shows slight to severe myelin pallor without gliosis in patients with adult-onset forms.

5. MOLECULAR ANALYSIS

After isolation of the gene responsible for DRPLA, the disease can be diagnosed based on molecular analysis. An expansion of trinucleotide (CAG) repeat number is easily defined using polymerase chain reaction (PCR) products from genomic DNAs. Normal alleles have 7 to 34 CAG repeats in Japanese population [25, 26, 41], and DNA sizes of PCR products are shorter than 200 bp using the appropriate primer-amplified CAG repeat region (Fig. 14-16) [29]. On the other hand, patients with DRPLA have longer DNA size products and repeat numbers range between 51 and 88 (Figs. 14-16, 14-17).

Several familial cases of DRPLA have been reported in Europe [31, 32] and the United States [33, 34], but epidemiologic studies have demonstrated that DRPLA is vary rare except in Japan. DRPLA mutation has been screened in 117 patients with cerebellar ataxia, from 94 families and 23 isolated cases by Dubourg *et al.*[54], but none of the patients carried this mutation in the French population [53]. Then, a founder effect in Japan was studied using two intragenic bimodal polymorphic markers in the DRPLA gene region [55]. One polymorphic system, named System A, is located in intron 1 and can be detected by PCR–SSCP (single-stranded conformation polymorphism); the other, named System B and located in intron 3, can be detected by PCR products digested with restriction enzyme *Bgl*II [54] (Fig. 14-18). Almost all patients with DRPLA have one haplotype with A1 and B1 alleles; frequencies of haplotype with A1 and B1 alleles are quite different in Japanese compared with those in Caucasians and Africans [55]. In Japanese, 50% have the haplotype with A1 and B1 alleles, but only 8% and 4% have this haplotype in Caucasians and Africans, respectively [55]. In our study, CAG repeat number in DRPLA is tightly associated with the B polymorphic alleles shown in Table 14-2 and alleles with over 18 repeats in the DRPLA region are associated with B1 haplotype. Expanded CAG repeat alleles (Table 14-3) have generally derived from alleles with B1 polymorphic marker, but the expanded DRPLA gene in one family was associated with B2 allele, suggesting that DRPLA is more unstable in Japanese than in other populations (Fig. 14-19).

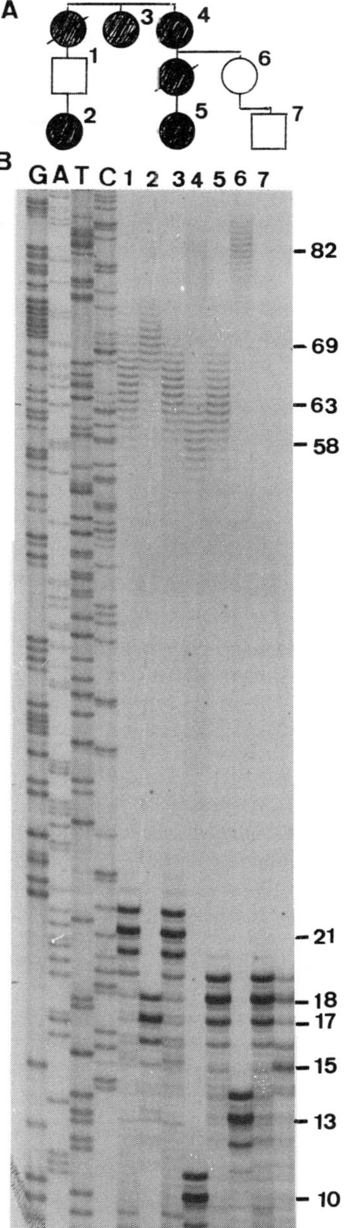

FIGURE 14-17 Family pedigree and autoradiograph of variable-length PCR products from family members with DRPLA. (A) A DRPLA family in three generations. (B) The PCR products, amplified with primers carrying flanking region of CAG repeats, were analyzed in a 5% denaturing polyacrylamide gel. A, C, T, and G are sequencing ladders obtained by using M13 DNA with a 41 RV primer. CAG repeat numbers are indicated on the right. CAG repeat numbers in affected individuals are 58, 63, 69, and 83, whereas CAG repeat numbers in normal alleles are less than 21.

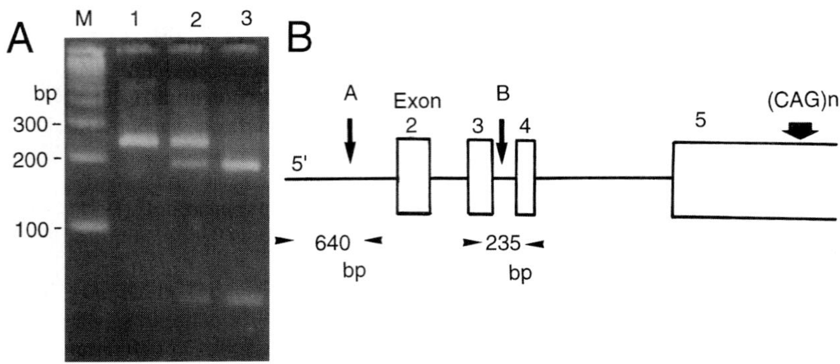

FIGURE 14-18 DNA pattern of PCR–*Eco*81 polymorphism (A) and scheme of DRPLA gene with three polymorphic regions (B). Three genotypes are shown in lines 1, 2, and 3: 1, homozygote for B1 allele of the B system; 2, heterozygote for B1 and B2; and 3, homozygote for B2 allele. M, DNA size marker. (B) A, A polymorphic system detected by PCR–SSCP; B, B polymorphic system detected by PCR–*Eco*81I; $(CAG)_n$, CAG trinucleotide repeat polymorphic site.

V. DIFFERENTIAL DIAGNOSIS

Clinical symptoms in DRPLA are commonly observed in many other neurodegenerative disorders. Patients with the juvenile form have very similar clinical symptoms to those of PME. Patients with the adult-onset form have symptoms of Huntington's disease (HD), spinocerebellar atrophy 1 (SCA1), Machado–Joseph disease (MJD), and benign adult familial myoclonus epilepsy (BAFME). At present, DRPLA can be simply diagnosed by means of PCR using genomic DNAs from peripheral blood. However, before DNA analysis is undertaken, clinical findings are very useful for differential diagnosis. Major clinical features of these diseases are summarized in Table 14-4.

TABLE 14-2 Distribution of DRPLA–B Polymorphism Genotypes in a Japanese Population

Genotypes	Observed No.	Expected No.	$\chi 2$
B1–B1	9	11.2	0.43
B1–B2	36	31.6	0.61
B2–B2	20	22.2	0.22
Total	65	65.0	1.26
Allele frequency	B1 = 0.415	B2 = 0.585	
Japanese[a]	B1 = 0.43	B2 = 0.56	
Caucasian[a]	B1 = 0.08	B2 = 0.92	
African[a]	B1 = 0.04	B2 = 0.96	

[a]Yanagisawa *et al.* [53].

A. Progressive Myoclonus Epilepsy (PME)

PME refers to a heterogeneous group of severe inherited epilepsies characterized by myoclonic seizures, generalized epilepsy, and progressive neurological deterioration including dementia and ataxia. In this group, there are four clinically recognized diseases: the Unverricht–Lundborg type, myoclonus epilepsy associated with ragged-red fibers (MERRF), Lafora disease, and sialidosis (cherry-red spot-myoclonus syndrome). Since modes of inheritance of these diseases are autosomal dominant or mitochondrially inherited, family history in patients is very useful to differentiate the diseases. In addition, a gene for PME of the Unverricht–Lundborg type is isolated and DNA mutations can be detected using PCR [55].

B. Huntington's Disease

Patients with the early adult-onset form are sometimes diagnosed as having Huntington's disease. In Cau-

TABLE 14-3 Haplotypes of $(CAG)_n$ and the B Polymorphism of DRPLA Gene in a Japanese Population

	DRPLA–B2	DRPLA–B1	Total
$(CAG)_{8-12}$	26	0	26
$(CAG)_{13-16}$	18	3	21
$(CAG)_{17}$	4	9	13
$(CAG)_{n>18}$	0	14	14
Total	48	26	74

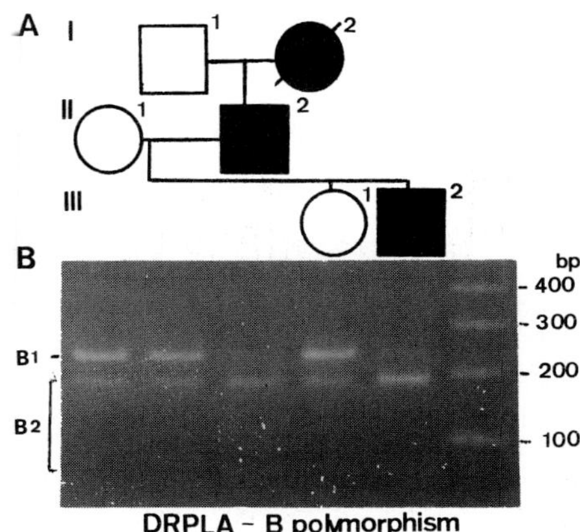

FIGURE 14-19 DNA pattern in a family with DRPLA. (A) Family tree of patient with DRPLA, (B) DNA pattern detected by PCR–RLFP. PCR products were digested with *Eco*81I and detected in a 2.5% agarose gel stained with ethidium bromide.

casians HD is the most known neurodegenerative genetic disease with a mode of autosomal dominant inheritance. Therefore, patients with DRPLA have been diagnosed as HD [31–34]. However, CT and MRI findings are different in HD and DRPLA, and family history is also helpful for differentiation of these diseases. DNA analysis is a fundamental diagnostic tool for HD [37].

TABLE 14-4 Major Clinical Features of Diseases in Differential Diagnosis of DRPLA

Syndrome	DRPLA	PME	HD	MJD	SCA1	BEFME
Age at onset of symptoms (years)	0–70	>20	10–50	20–40	40	20<
Mode of inheritance	AD	AR	AD	AD	AD	AD
Genetic anticipation	+	−	+	+	+	−
Mental retardation (dementia)	+	+	−	−	−	−
Ataxia	+	+	+	+	+	+
Epilepsy	+	−	+	−	−	+
Involuntary	myo cho athe	myo	cho	tre	tre	myo

Note. DRPLA, dentatorubral-pallidoluysian atrophy; PME, progressive myoclonus epilepsy; HD, Huntington's disease; MJD, Machado–Joseph disease; SCA1, spinocerebellar atrophy 1; BAFME, benign adult familial myoclonus epilepsy; AD, autosomal dominant; myo, myoclonus; cho, chorea; athe, athetosis; tre, tremor.

C. Machado–Joseph Disease (MJD)

Patients with the adult-onset form of DRPLA are diagnosed as having MJD. MJD is an autosomal dominant disease characterized by gait ataxia, limitation of eye movements, widespread fasciculations of muscles, and loss of reflexes in the lower limbs, followed by nystagmus, mild cerebellar tremor, and extensor plantae responses [55]. The disorder first manifests as ataxic gait after age 40. Postmortem examinations show loss of neurons and gliosis in the substantia nigra, and nuclei pons as well as the nuclei of the vertebral and cranial nerves, columns of Clarke, and anterior horns. A gene for MJD has been mapped to 14q24.3-q32 and the isolated gene responsible for MJD has trinucleotide (CAG) repeat in a coding region [56].

D. Spinocerebellar Atrophy 1 (SCA1)

SCA1 is an autosomal dominant neurodegenerative disorder. Gait ataxia and involuntary choreiform movements usually begin in the third or fourth decade of life. Patients have lower bulbar palsies, hyperreflexia, scanning and explosive speech, incoordination, and slow motor-nerve conduction. The basic defect consists of expansion of a trinucleotide CAG repeat mapped to chromosome 6p[36].

E. Benign Adult Familial Myoclonus Epilepsy (BAFME)

BAFME is an autosomal dominant disorder reported particularly in Japan [57]; clinical symptoms are mild myoclonus and abnormal EEGs. Epileptic seizures can be easily controlled by anticonvulsant drugs. Age at onset of the disease ranges from 30 to 50 years. Symptoms gradually progress, but neuropathological examinations have not shown any histological changes in autopsy cases. Clinical symptoms the in late-adult-onset form of DRPLA sometimes mimic symptoms of BAFME [58].

VI. CONCLUSION

Before a gene for DRPLA was isolated in 1994, diagnosis of DRPLA was clinically difficult and patients with DRPLA had been reported mostly in Japan. However, since a gene for DRPLA was identified, over 100 papers have been published in 2 years, and expanded CAG repeats in the DRPLA gene have been detected in Caucasians and Africans. At present, clinical features are slightly different in different ethnic groups. Therefore, further clinical findings and genetic information must be collected together to reveal fundamental abnormalities in DRPLA.

Acknowledgments

I thank my colleagues, in particular Drs. Tatsuya Ogiwara, Matuyama Red Cross Hospital, Masahito Miyazaki, Tokushima University School Of Medicine, Haruo Hattori, Kyoto University Graduate School of Medicine, and Eiichiro Uyama, Kumamoto University School of Medicine, who referred patients for molecular diagnosis of DRPLA and provided figures of EEG, CT, MRI, and SPECs for this chapter.

References

1. Naito, H., and Oyanagi, S. (1982). Familial myoclonus epilepsy and choreoathetosis: hereditary dentatorubral-pallidoluysian atrophy. *Neurology* **32,** 798–807.
2. Smith, J. K. (1975). Dentatorubralpallidoluysian atrophy. *In* "Handbook of Clinical Neurology" (F. J. Vinken, G. W. Bruyn, and J. M. B. V. De Hong, Eds.), Vol. 21, pp.519–534. North-Holland, Amsterdam.
3. Simith, J. K., Gonda, V. E., and Malamud, N. (1958). Unusual form of cerebellar ataxia: combined dentato-rubral and pallidoluysian degeneration. *Neurology* **8,** 205–209.
4. Titica, J., and van Bogaert, L. (1946). Heredo-degenerative hemiballismus: a contribution to the question of primary atrophy of the courus Luysii. *Brain* **69,** 251–263.
5. Verhaart, W. J. C. (1958). Degeneration of the brain stem reticular formation, other parts of the brain stem and the cerebellum: an example of heterogeneous systemic degeneration of the central nervous system. *J. Neuropathol. Exp. Neurol.* **17,** 382–391.
6. Neuman, M. A. (1959). Combined degeneration of globus pallidus and dentate nucleus and their projections. *Neurology* **9,** 403–408.
7. Oyanagi, S., Tanaka, M., Naito, H., Shirakawa, K., Saito, K., Nakamuram, N., and Ohama, E. (1976). A neurophalogical study of 8 autopsy cases of degenerative types of myoclobus epilepsy—with special reference to latent combination of degeneration of the pallido-luysian system. *Adv. Neurol. Sci.(Jpn)* **20,** 410–424.
9. Oyamagi, S., and Naito, H. (1977). A clinical and neuropathological study on four autopsy cases of degenerative type of myoclonus epilepsy with Mendelian dominant heredity. *Psych. Neurol. Jpn.* **52,** 113–129.
10. Naito, H., Tanaka, M., Hirose, Y., and Oyanagi, S. (1977). Clinicopathological study on two autopsied cases of degenerative type of myoclonus epilepsy with choreo-atethoid movement; Proposal of hereditary dentate and pallidal system atrophy. *Psych. Neurol. Jpn.* **79,** 193–204.
11. Takahata, N., Ito, K., Yoshimura, Y., Nishihori, K., and Suzuki, H. (1978). Familial chorea and myoclonus epilepsy. *Neurology* **28,** 913–919.
12. Hirayama, K., Iizuka, R., Maeda, K., and Watanabe, T. (1982). Clinicopathological study of dentatorubro-pallidoluysian atrophy. Part I: Its clinical form and analysis of symptomatology. *Adv. Neurol. Sci.(Jpn)* **25,** 225–236.
13. Maehara, K., Iizuka, R., and Hirayama, K. (1982). Clinicopathogical study for dentatorubral-pallidoluysian atrophy-Neuropathological study. *Adv. Neurol. Sci. (Jpn)* **26,** 1173–1189.
14. Iizuka, R., Hirayama, K., and Maehara, K. (1984). Dentato-rubro-pallidoluysian atrophy: a clinico-pathological study. *J. Neurol. Neurosurg. Psych.* **47,** 1288–1298.
15. Naito, H., Shirakawa, K., and Kondo, K. (1984). Myoclonus in dentatorubraropallidoluysian atrophy. *Adv. Neurol. Sci. (Jpn)* **28,** 733–742.
16. Iwabuchi, K., Amano, N., Yokoi, S., Nakano, T., and Yagishita, S. (1985). Two familial cases of dentato-rubro-pallidoluysian atrophy with pseudo-Huntington's chorea. *Clin. Neurol.* **25,** 1952–1969.
17. Nakano, T., Iwabuchi, K., Yagishita, S., Amano, N., Akagi, M., and Yamamoto, Y. (1985). An autopsy case of dentatorubopallidoluysian atrophy(DRPLA) clinically diagnosed as Huntington's chorea. *No to Shinkei (Jpn)* **37,** 767–774.
18. Maehara, K., and Iizuka, R. (1986). Clinico-pathological study of dentatro-rubro-pallido-luysian atrophy-Based on 38 reported cases-. *Clin.Neurol.(Jpn)* **26,** 405–411.
19. Morioka, E., Nakatsu, T., Kuroda, S., Yamamoto, N., Hosokawa, K., and Otsuki, S. (1987). An autopsy case of dentatorubro-pallidoluysian atrophy showing marked atrophy of the brain stem. *No to Shinkei (Jpn)* **39,** 769–773.
20. Takahashi, H., Ohama, E., Naito, H., Takeda, S., Nakashima, S., Makifuchi, T., and Ikuta, F. (1988). Hereditary dentato-rubral-pallidoluysian atrophy: clinical and pathological variations in a family. *Neurology* **38,** 1065–1070.
21. Takeda, S., Takahashi, H., and Ikuta, F. (1992). Dentatrorubropallidoluysian atrophy(DRPLA): comparative pathological study on clinical groups classified into juvenile, early adult and late adult types. *No to Shinkei (Jpn)* **44,** 111–116.
22. Yuasa, T. (1993). Hereditary dentatorubro-pallidoluysian atrophy(DRPLA): clinical studies on 45 cases. *Nippon-Rinsho (Jpn)* **51,** 3016–3023.
23. McKusick, V. A. (1994). *125379 Dentatorubral-pallidoluysian atrophy [DRPLA;myoclonus epilepsy with choreoathetosis; Naito-Oyanagi disease; NOD]. *In* "Mendelian Inheritance in Man, a Catalog of Human Genes and Genetic Diseases" (V. A. McKusick, Ed.), 11th ed., Vol.1, p.411. The Johns Hopkins University Press, Baltimore/London.
24. Kondo, I., Ohta, H., Yazaki, M., Ikeda, J. E., Gussela, J. F., and Kanazawa, I. (1990). Exclusion mapping of the hereditary dentatoruropallidoluysian atrophy gene from the Huntington's disease locus. *J. Med. Genet.* **27,** 105–108.
25. Koide, R., Ikeuchi, T., Onodera, O., Tanaka, H., Igarashi, S., Endo, K., Takahashi, H., Kondo, R., Ishikawa, A., Hayashi, T., Saito, M., Tomoda, A., Miike, T., Naito, H., Ikuta, F., and Tsuji, S. (1994). Unstable expansion of CAG repeat in hereditary dentatorubral-pallidoluysian atrophy(DRPLA). *Nature Genet.* **6,** 9–13.
26. Nagafushi, S., Yanagisawa, H., Sato, K., Shirayama, T., Ohsaki, E., Bundo, M., Takeda, T., Tadokoro, K., Kondo, I., Murayama, N., Tanaka, Y., Kikushima, H., Umino, K., Kurosawa, H., Fukukawa, T., Nihei, K., Inoue, T., Sano, A., Komure, O., Takahashi, M., Yoshizawa, T., Kanazawa, I., and Yamada, M. (1994). Denattorubral pallidoluysian atrophy expansion of an unstable CAG trinucleotide on chromosome 12p. *Nature Genet.* **6,** 14–18.
27. Takano, T., Yamanouchi, Y., Nagafuchi, S., and Yamada, M. (1996). Assignment of the dentatorubral and pallidoluysian atrophy(DRPLA) to 12p13.31 by in situ hybridization. *Genomics* **32,** 171–172.
28. Kuwano, A., Morimoto, Y., Nagai, T., Fukushima, Y., Ohashi, H., Hasegawa, T. and Kondo, I. (1996). Precise chromosoma locations of the genes for dentatodubral-pallidoluysian atrophy (DRPLA), von Willebrabd factor (F8vF), and parathyroid hormone-like hormone (PTHLT) in human chromosome 12p by deletion mapping. *Hum.Genet.* **97,** 95–98.

29. Li, S-H., McInnis, M. G., Margolis, R. L., Antonarakis, S. E., and Ross, C. A. (1993). Novel triplet repeat containing genes in human brain: cloning, expression and length polymorphism. *Genomics* **16,** 572–579.
30. Nagafuchi, S., Yanagisawa, H., Ohsaki, E., Shiroyama, T., Tadokoro, K., Inoue, T., and Yamada, M. (1994). Structure amd expression of the gene responsible for the triplet repeat disorder, dentatorubral and pallidoluysian atrophy (DRPLA). *Nature Genet.* **8,** 177–182.
31. Warner, T. T., Williams, L. D., Walker, R. W. H., Flinter, F., Robb, S. A., Bundey, S. E., Honavar, M., and Harding, A. E. (1995). A clinical and molecular genetic study of dentatorubropallidoluysian atrophy in four European families. *Ann. Neurol.* **37,** 452–459.
32. Neilsen, J. E., Sorensen, S. A., Hasholt, L., and Norremolle, A. (1996). Dentatorubral-pallidoluysian atrophy. clinical features of a five-generation Danish family. *Mov. Disord.* **11,** 533–541.
33. Potter, N. T., Meyer, M. A., Zimmerrman, A. W., Eisenstadt, M. L., and Andereson, I. J. (1995). Molecular and clinical findings in a family with denatorubral-pallidoluysian atrophy. *Ann. Neurol.* **37,** 272–277.
34. Potter, N. T. (1996). The relationship between (CAG)n repeat and age of onset in a family with dentatorubral-pallidoluysian atrophy (DRPLA): diagnostic implications of confirmatory and predictive testing. *J. Med. Genet.* **33,** 168–170.
35. Oberle, I., Rousseau, F., Heitz, D., Kretz, C., Devys, D., Hanauer, A., Boue, J., Bertheas, M. F., and Mandel, J. L. (1991). Instability of a 550-base pair DNA segment and abnormal methylation in fragile X syndrome. *Science* **252,** 1097–1102.
36. Orr H. T., Chung, M., Bnfi, S., Kwiatkowski, T. J., Jr., Servadio, A., Beaudet, A. L., NcCall, A. E., Duvick, L. A., Ranum, L. P. W., and Zoghbi, H. Y. (1993). Expansion of an unstable trinucleotide CAG repeat in spinocerebellar ataxia type 1. *Nature Genet.* **4,** 11–116.
37. The Huntington's Disease Collaborative Research Group (1993). A novel gene containing a trinucleotide repeat that is expanded and unstable on Huntington's disease chromosomes. *Cell* **72,** 971–983.
38. Sato, K., Kashihara, K., Okada, S., Ikeuchi, T., Tsuji, S., Shomoti, T., Morimoto, K., and Hayabara, T. (1995). Dose homozygosity advance the onset of dentatorubral-pallidoluysian atrophy? *Neurology* **45,** 1934–1936.
39. Kurohara, K., Kuroda, Y., Maruyama, H., Kawakami, H., Yukitake, M., Matsui, M., and Nakamura, S. (1997). Homozygosity for an allele carrying intermediate CAG repeats in the dentatorubral-pallidoluysian atrophy(DRPLA) gene results in spastic paraplegia. *Neurology* **48,** 1087–1090.
40. Sano, A., Yamauchi, N., Kakimoto, Y., Komure, O., Kawai, J., Hazama, F., Kuzume, K., Sano, N., and Kondo, I. (1994). Anticipation in hereditary dentatorubral-pallidoluysian atrophy. *Hum. Genet.* **93,** 699–702.
41. Komure, O., Sano, A., Nishino, N., Yamauchi, N., Ueno, S., Kondoh, K., Sano, N., Takahashi, M., Murayama, N., Kondo, I., Nagafuchi, S., Yamada, M., and Kanazawa, I. (1995). DNA analysis in hereditary dentatorubral-pallidoluysia atrophy: correlation between CAG repeat length and phenotypic variation and the the molecular bases of anticipation. *Neurology* **45,** 143–149.
42. Tomoda, A., Ikezawa, M., Ohtani, Y., Miike, T., and Kumamokto, T. (1992). Progressive myoclonus epilepsy:dentato-rubro-pallidoluysian atrophy(DRPLA) in childhood. *Brain Dev.* **13,** 266–269.
43. Ikeuchi, T., Igarashi, S., Takayama, Y., Onodera, O., Oyake, M., Tanaka, H., Koid, R., Tanaka, H., and Tsuji, S. (1996). Non-Mendelian transmission in dentatorubral-pallidoluysian atrophy and Machado-Joseph disease: the mutant allele is preferentially transmitted in male meiosis. *Am. J. Hum.Genet.* **58,** 730–733.
44. Hirayama, K., Takayanagi, T., Nakamura, R., Yanagisawa, N., Hattori, T., Kita, K., Yanagimoto, S., Uhita, M., Nagaika, M., and Satomori, Y. (1994). Spinocerebellar degenerations in Japan: a nationwide epidemiological and clinical study. *Acta Neurol. Scand. Suppl.* **153,** 1–22.
45. Ihara, Y., Namba, R., Nobukuni, K., Kawai, K., and Kuroda, S. (1991). Relation between cerebral white matter damage on CT and MRI and clinical data in DRPLA (pseudo-Huntington's form). *Clin. Neurol.(Jpn).* **31,** 815–820.
46. Uyama, E., Kondo, I., Uchino, M., Fukushima, T., Murayama, N., Kuwano, A., Inokuchi, N., Ohtani, Y., and Ando, M. (1995). Dentatorubral-pallidoluysian atrophy (DRPLA): clinical, genetic, and neuroradiologic studies in a family. *J. Neurol. Sci.* **130,** 146–153.
47. Miyazaki, M., Kato, T., Hashimoto, T., Harada, M., Kondo, I., and Kuroda, Y. (1996). MR of childhood-onset dentatorubral-pallidoluysian atrophy. *Am. J. Neuroradiol.* **16,** 1834–1836.
48. Imamura, A,, Ito, R., Fukutomi, O., Shimozawa, N., Nishimura, M., Suzuki, Y., Kondo, N.. Yamada, M., and Orii, T. (1994). High-intensity proton and T2-weighted MRI signals in the globus pallidus in juvenile-type of dentatorubral and a pllidoluysian atrophy. *Neuropediatrics* **25,** 234–237.
49. Miyazaki, M., Hashimoto, T., Yoneda, Y., Tayana, M., Harada, M., Miyoshi, H., Kuwano, N., Murayama, N., Kondo, I., and Kuroda, Y. (1996). Proton magnetic resonance spectroscopy on childhood-onset dentatorubral-pallidoluysian atrophy (DRPLA). *Brain Dev.* **18,** 142–146.
50. Hosokawa, S., Ichiya, Y., Kuwabara, Y., Ayabe, Z., Mitsuo, K., Goto, I., and Kato, M. (1987). Positron emission tomography in cases of chorea with different underlying diseases. *J. Neurol. Neurol. Psych..* **50,** 1284–1287.
51. Miyazaki, M., Hashimoto, T., Nakagawa, R., Yoneda, Y., Tayama, M., Kawano, N., Murayama, N., Kondo, I., and Kuroda, Y. (1996). Characteristic evoked potentials in childhood-onset dentatorubrl-pallidoluysian atrophy. *Brain Dev.* **18,**389–393.
52. Hattori, H., Higuchim, Y., Okuno, T., Asato, R., Fukumoto, M., and Kondo, I. (1997). Early-childhood progressive myoclonus epilepsy presenting as partial seizures in dentaorubral-pallidoluysian atrophy. *Epilepsia* **38,** 271–274.
53. Doboury, O., Durr, A., Chneiweiss, H., Stevanin, G., Cancel, G., Penet, C., Agid, Y., and Bride, A. (1995). Dose the ataxo-choreic form of DRPLA exist in Europe? Search of mutation in 120 families. *Rev. Neurol. Paris* **151,** 657–660.
54. Yanagisawa, H., Fujita, K., Nagafuchi, S., Nakahori, Y., Nakagome, Y., Akane, A., Senno, A., Komure, O., Kondo, I., Jin, D. K., Sorensen, S. A., Potter, N. T., Young, S. R., Nakamura, K., Nukina, N., Nagano, Y., Tadokoro, K., Okuyama, T., Miyashita, T., Inoue, T., Kanazawa, I., and Yamada, M. (1995). A unique origin and multistep process for the generation of expanded DRPLA triple repeats. *Hum. Mol. Genet.* **5,** 373–379.
55. Pennacchio, L. A., Lehesjoki, A-E., Stone, N. E., Willour, V. L., Virtaneva, K., Miao, J., Amato, E., Ramirez, L., Faham, M., Koskiniemi, M., Warrington, J. A., Norio, R., dela Chapelle, A., Cox, D. R., and Myers, R. M. (1996). Mutations in the gene encoding cystatin B in progressive myoclonus epilepsy (PME1). *Science* **271,** 1731–1734.
56. Kawaguchoi, Y., Okamoto, T., Taniwaki, T., Izawa, M., Inoue, M., Katayama, S., Kawakami, H., Nakamura, S., Nishimura, M., Akiguchi, I., Kimura, T., Narumiya, S., and Kakizaki, A. (1994). CAG expansions in a novel gene for Machado-Joseph disease at chromosome 14q32. *Nature Genet.* **8,** 221–228.
57. Yasuda, T. (1991). Benign adult familial myoclonus epilepsy (BAFME). *Kawasaki Med.J.* **17,** 1–13.
58. Kuwano, A., Takakubo, F., Morimoto, Y., Uyama, E., Uchino, M., Ando, M., Yasuda, T., Terao, A., Hayama, T., Koboyashi, R., and Kondo, I. (1996). Benign adult familial myoclonus epilepsy (BAFME): an autosomal dominant form not linked to the dentatorubral pallidoluysian atrophy (DRPLA) gene. *J. Med. Genet.* **33,**80–81.

CHAPTER 15

Molecular Genetics of Dentatorubral-Pallidoluysian Atrophy (DRPLA)

SHOJI TSUJI Department of Neurology, Brain Research Institute, Niigata University, Niigata 951, Japan

I. INTRODUCTION

Dentatorubral-pallidoluysian atrophy (DRPLA) is a rare autosomal dominant neurodegenerative disorder clinically characterized by various combinations of myoclonus, epilepsy, cerebellar ataxia, choreoathetosis, dementia, and psychiatric symptoms (MIM 125370) [46]. Although the term DRPLA was originally used by Smith *et al.* to describe a neuropathological condition associated with the most severe neuronal loss in the dentatorubral and pallidoluysian systems of the central nervous system when it was found in a sporadic case without familial occurrence [62, 63], the hereditary form of DRPLA was first described in 1972 by Naito and his colleagues [45]. They described two Japanese pedigrees showing an autosomal dominant mode of inheritance and a considerably broad spectrum of clinical features appearing even within the same family. In these pedigrees, patients with an onset before the age of 20 frequently exhibited myoclonus, seizure, and cerebellar ataxia, while adult-onset patients exhibited cerebellar ataxia, choreoathetosis, and dementia. Since Naito's report was published in 1972, reports on Japanese pedigrees showing similar clinical presentations have accumulated [1, 16, 18, 19, 27, 28, 47, 51, 65, 69]. Since a broad spectrum of clinical features can be observed even within the same pedigree, it soon became clear that these diverse clinical phenotypes constituted a single disease entity.

In 1982, Naito and Oyanagi compiled the findings from their clinical and neuropathological studies, and proposed the name "hereditary dentatorubral-pallidoluysian atrophy" for the conditions characterized by the following neuropathological, clinical, and genetic features [46]: (1) The dentatorubral and pallidoluysian systems are the major sites of neurodegeneration. (2) A variety of clinical syndromes, including myoclonus epilepsy syndrome, cerebellar ataxia, and choreoathetosis or chorea are detectable. (3) The mode of inheritance is autosomal dominant.

Since the description of the hereditary form of DRPLA by Naito and Oyanagi [45, 46], DRPLA has been found to occur almost exclusively in Japanese. The mode of inheritance of DRPLA is autosomal dominant with a high penetrance rate. The prevalence rate of DRPLA in Japanese has been estimated to be 0.2–0.7 in 100,000, which is comparable to that of Huntington's disease in Japan [26]. Although DRPLA has been reported to occur predominantly in Japanese individuals, several cases with quite similar clinical features have been described in other ethnic groups [10, 13, 72]. Since the discovery of the gene for DRPLA [33, 43], expansion of the CAG repeat of the DRPLA gene has been demonstrated by molecular analysis in European and North American families [4, 9, 48, 55, 75, 76]. Thus DRPLA seems not to be as geographically exclusive as previously thought.

The gene for DRPLA was discovered by two independent Japanese groups in 1994, and unstable expansions of CAG repeats in the protein-coding region were found to be the causative mutations for DRPLA [33, 43]. To date, seven diseases including bulbar and spinal muscular atrophy (SBMA) [36], Huntington's disease [71], SCA1 [50], DRPLA [33, 43], Machado–Joseph disease [32], SCA2 [25, 56, 58], and SCA6 [80] have been found to be caused by expanded CAG repeats, and it is expected that similar molecular mechanisms are also involved in many other hereditary neurodegenerative diseases in which the causative genes are as yet undetermined.

In this chapter, I will focus on the molecular genetic aspects of DRPLA with particular emphasis on the non-Mendelian aspect. I will also discuss the molecular mechanisms of neurodegeneration caused by expansions of CAG repeats.

II. IDENTIFICATION OF THE GENE FOR DRPLA

DRPLA is characterized by prominent anticipation with a strong parental bias on the degree of anticipation [22–24, 33, 43]. Paternal transmission results in more prominent anticipation (26–29 years/generation) than does maternal transmission (14–15 years/generation) (Table 15-1). Given the strong parental bias on the degree of anticipation observed in Huntington's disease (HD) [71] and spinocerebellar ataxia type 1 (SCA1) [50], we speculated that DRPLA must be a disease caused by unstable expansion of CAG repeats of an unidentified gene. To identify the gene for DRPLA, we pursued two strategies: to isolate cDNA clones carrying CAG repeats from human cerebellum cDNA libraries, and to search for genes which had already been reported to carry CAG repeats. When we began our effort to identify the gene for DRPLA, a very interesting paper was published, in which the isolation of various cDNA clones carrying CTG (the complementary strand of CTG is CAG) or CGG repeats from human brain cDNA libraries was described [38]. Among the cDNA clones described in this report, we were particularly interested in the CTG-B37 gene, which exhibited some of the same features as those described for the genes for HD and SCA1: (1) The CAG repeat is highly polymorphic among normal individuals. (2) The CAG repeat is located near the 5′ end of the coding sequence. (3) The CAG repeat is predicted to code for a polyglutamine tract on the basis of analysis of the open reading frame.

When we had analyzed the CAG repeat of the CTG-B37 gene as the possible causative gene for various dominant ataxias including DRPLA, we discovered that the CAG repeat was expanded in all the patients with a clinical or neuropathological diagnosis of DRPLA. The CAG repeat in normal individuals ranged in size from 6 to 35 repeat units, while the CAG repeat ranged from 54 to 79 repeat units in patients with DRPLA [22–24, 33] (Fig. 15-1).

III. STRUCTURE OF THE DRPLA GENE

The detailed structures of the full-length cDNA clones for the human DRPLA gene have been determined [43, 44, 49]. On the basis of nucleotide sequence

TABLE 15-1 Genetic Anticipation in DRPLA [22–24, 33]

	Paternal	Maternal	
Our cohort	-25.6 ± 2.4^a ($n = 27$)	-14.0 ± 4.0 ($n = 9$)	$P < 0.05$
Literature	-28.8 ± 1.9 ($n = 27$)	-14.8 ± 4.0 ($n = 11$)	$P < 0.05$

[a]Mean ± SEM.

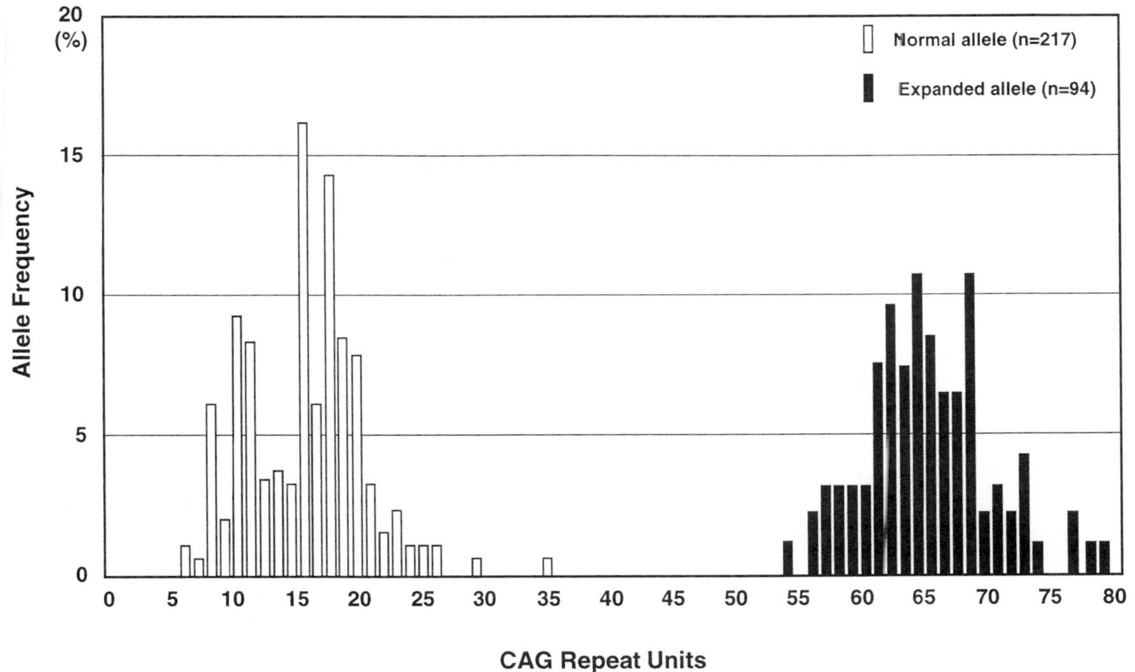

FIGURE 15-1 Size distributions of CAG repeats of the DRPLA gene in normal controls and DRPLA patients [22–24, 33].

analysis, the DRPLA cDNA is predicted to code for 1185 amino acids. The CAG repeat, which is expanded in DRPLA, is located 1462 bp downstream from the putative methionine initiation codon and is predicted to code for a polyglutamine tract. Interestingly, polyserine and polyproline tracts exist near the polyglutamine tract. In contrast to the polyglutamine tract, these polyserine and polyproline repeats are not highly polymorphic [49]. So far, no functional domains suggesting the physiological functions of the DRPLA gene have been identified, and the functions of the DRPLA gene remain to be elucidated.

The gene for DRPLA has recently been mapped to 12p13.31 by *in situ* hybridization [67]. The human DRPLA gene spans approximately 20 kbp and consists of 10 exons, and the CAG repeat is located in exon 5 [44].

Northern blot analysis revealed that a 4.7-kb transcript is widely expressed in various tissues including heart, lung, kidney, placenta, skeletal muscle, and brain without predilection in regions exhibiting neurodegeneration [43, 44, 49]. Reverse transcription-polymerase chain reaction (RT-PCR) analysis of mRNA extracted from various regions of autopsied brains of patients with DRPLA demonstrated that DRPLA mRNA from a mutant DRPLA gene with expanded CAG repeats is expressed at levels comparable to the wild-type DRPLA gene, suggesting that expansions of CAG repeats do not alter the transcriptional efficiency of the DRPLA gene [49]. The expression levels of the mutant DRPLA proteins were also analyzed by Western blot analysis, which indicates that mutant DRPLA proteins are expressed at levels similar to those of wild-type DRPLA proteins [79]. These studies strongly indicate that expansion of the CAG repeat does not affect the transcription or translation of the mutant DRPLA gene. Therefore, it seems likely that mutant DRPLA proteins with expanded polyglutamine tracts are toxic to neuronal cells, suggesting a "gain of toxic functions."

IV. EXPANSION OF CAG REPEATS OF THE DRPLA GENE IN PATIENTS WITH DRPLA—GENOTYPE AND PHENOTYPE CORRELATION

The discovery of the gene for DRPLA has made it possible to analyze the clinical diversities based on the size of expanded CAG repeats. As shown in Fig. 15-2, there is a tight inverse correlation between the age at DRPLA onset and the size of the expanded CAG repeat. To clarify the relationship between clinical features of DRPLA and the age at onset, we analyzed how the clinical features of DRPLA (i.e., ataxia, dementia or mental retardation, myoclonus, epilepsy, choreoathetosis, and psychiatric changes including character changes, delusions, or hallucinations) vary depending on the age at onset [22]. We found that ataxia and dementia are cardinal features irrespective of age at onset. Patients with onset before age 20 frequently ex-

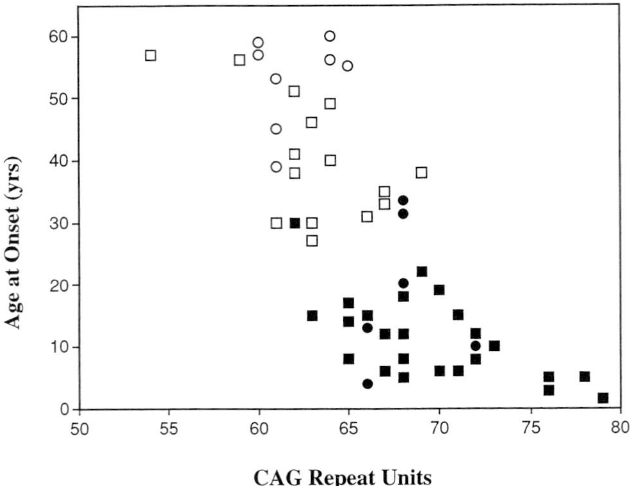

FIGURE 15-2 Correlation of the CAG repeat size with age at onset, parental origins, and clinical phenotype. (■)PME with paternal transmission, (●) non-PME with maternal transmission, (■) PME with paternal transmission, and (○) non-PME with maternal transmission [22–24, 33].

hibit myoclonus and epilepsy in addition to ataxia and dementia. The combination of these clinical features corresponds to the clinical phenotype of progressive myoclonus epilepsy (PME). On the other hand, patients with onset after age 20 frequently exhibit choreoathetosis and psychiatric changes in addition to ataxia and dementia. Since the age at onset is inversely correlated with the CAG repeat size, the above implies that various clinical features are strongly correlated with the CAG repeat size. Similar correlation of clinical features with the CAG repeat size has been demonstrated for other diseases caused by expansion of CAG repeats.

V. NON-MENDELIAN ASPECTS OF TRIPLET REPEAT DISEASES

The results of recent studies indicate that DRPLA as well as other diseases caused by expansion of unstable trinucleotide repeats exhibit many unusual phenomena which are difficult to interpret on the basis of traditional Mendelian inheritance. These include variable age at onset, genetic anticipation, appearance of apparently "sporadic cases", gene dosage effect and meiotic drive.

A. Variable Age at Onset

Figure 15-2 shows the correlation between the size of the expanded CAG repeat of the DRPLA gene and the age at onset of DRPLA [22, 33]. As has been demonstrated for other diseases caused by expansion of CAG repeats, a strong inverse correlation between the size of expanded CAG repeats and the age at onset was demonstrated for DRPLA. The inverse correlation implies that the size of the expanded CAG repeat is strongly related to the pathophysiologic processes. It should be noted that the size of the CAG repeat and the age at onset are more strongly correlated in the case of greatly expanded CAG repeats than for mildly expanded CAG repeats. This suggests that prediction of age at onset based on CAG repeat size is difficult especially for mild CAG repeat expansion and late onset. This should be carefully considered in genetic counseling.

B. Genetic Anticipation

As expected, we found that there is a considerable parental bias in the intergenerational change in the expanded CAG repeat size. Much larger intergenerational increase was observed for paternal transmission (5.8 ± 0.9 repeat units/generation, $n = 16$) compared to that of maternal transmission (1.3 ± 1.6 repeat units/generation, $n = 4$) [22–24, 33]. This phenomenon has also been described for HD [2, 11, 64, 71] and SCA1 [8]. As shown in Table 15-2, quite similar parental biases are present in DRPLA, HD, and SCA1. Among these three diseases, DRPLA is the one with the largest intergenerational increase in the size of CAG repeats, and with the most prominent anticipation. This strongly indicates that similar mechanisms must underlie the intergenerational expansion of the CAG repeats during male gametogenesis in diseases caused by CAG repeat expansions. In fact, it has been demonstrated that DNA from the sperm of patients with HD shows considerable variation in the size of expanded CAG repeats compared to DNA from somatic cells [70].

C. "Sporadic" Cases

The diagnosis of "sporadic" cases of triplet repeat diseases, i.e., cases of patients with no family history of similar diseases, has been extremely difficult due to a

TABLE 15-2 Intergenerational Changes in CAG Repeat Size [22–24, 33]

	Paternal transmission	Maternal transmission
DRPLA	5.8 ± 0.9 (n = 16)	1.3 ± 1.6 (n = 4)
MJD	3.2 ± 0.8 (n = 11)	1.2 ± 0.4 (n = 19)
HD	3.2 ± 0.6 (n = 29)	1.2 ± 0.3 (n = 42)
SCA1	3.2 ± 1.8 (n = 28)	−0.4 ± 0.4 (n = 16)

high penetrance in these diseases that makes diagnosis difficult unless the disease is proven to be transmitted as an autosomal dominant trait. We have so far encountered six sporadic cases of DRPLA, but in only one could we analyze the genomic DNA of both parents [22, 60]. The patient developed generalized epilepsy at the age of 26 and began exhibiting ataxia and choreoathetosis at age 37. She was found to carry an expanded CAG repeat of 62 repeat units in the DRPLA gene. Although neither her 66-year-old father nor her 63-year-old mother had any neurological abnormalities or epilepsy, her father was found to carry a mildly expanded CAG repeat of 57 repeat units in the DRPLA gene. With this intergenerational increase of 5 repeat units during paternal transmission, the patient was the first in her family to cross the phenotypic threshold.

This demonstrates that sporadic cases can occur even in the absence of a family history of the disease. In the above family, the intergenerational increase in the number of the CAG repeat units in paternal transmission resulted in the appearance of the sporadic case. In the case of HD, expansion from an "intermediate range" allele to a fully expanded allele has been described. Although for DRPLA no such cases have been described, many more sporadic cases should be analyzed to clarify the molecular mechanisms underlying the occurrence of sporadic cases.

D. Gene Dosage Effect

Recently we encountered a patient homozygous for the DRPLA mutation [22, 59]. The patient's parents are first cousins. The patient carries an expanded CAG repeat of 57 repeat units in the DRPLA gene as a homozygous state. Although neither his 73-year-old father nor his 71-year-old mother showed signs of DRPLA, both had mildly expanded CAG repeats of 57 repeat units as a heterozygous state. Individuals carrying a CAG repeat with 57 repeat units as a heterozygous state are predicted to develop symptoms at the age of around 50 or later [22–24]. The age of the patient at onset, however, was 17, and the patient exhibited a typical PME phenotype. Thus, the age of the patient at onset was much younger than that of typical patients heterozygous for the expanded allele with a CAG repeat of the same size. The clinical features in this case are much more severe than expected for those with heterozygous mutation and similar sizes of CAG repeats. Recently, Kurohara *et al.* described a family with the clinical features of autosomal recessive spastic paraplegia [35]. The affected individuals in this family showed spastic paraplegia with mild truncal ataxia and dysarthria but no dementia, epilepsy, myoclonus, or other involuntary movements. They were the products of a consanguineous marriage but the parents were neurologically normal. The analysis of CAG repeats of the DRPLA gene revealed that the patients were homozygous for an allele carrying an intermediate size of CAG repeats (41 or 40 repeats) in the DRPLA gene. This report raises the possibility that homozygosity for an intermediate allele of the DRPLA gene may exhibit a unique clinical phenotype of spastic paraparesis in addition to ataxia. These findings also suggest that there is a gene dosage effect in DRPLA. Recent studies suggest that the gene products with expanded polyglutamine tracts are toxic to neuronal cells [20]. Thus, neurotoxicity may be enhanced in homozygous patients.

A quite similar gene dosage effect was reported for MJD [37, 68]. In the case of HD, however, it was demonstrated that the age at onset and the disease severity do not differ between patients with a homozygous mutation and those with a heterozygous mutation of the HD gene [34, 77]. It is unclear whether or not the gene dosage effect invariably occurs in diseases caused by CAG repeat expansion. Analysis of a large number of cases of CAG repeat expansion diseases in which patients have homozygous mutations will be required before we can conclude that a gene dosage effect commonly occurs in diseases caused by CAG repeat expansions.

E. Meiotic Drive

From the results of segregation analysis of DRPLA pedigrees, we found that the segregation rate of transmission of a mutant DRPLA gene and that of a wild-type DRPLA gene were not equal in paternal transmission of DRPLA. As shown in Table 15-3, expanded alleles of the DRPLA gene are transmitted significantly more often than normal alleles in paternal transmission of DRPLA [21]. Such segregation distortion was not observed in cases of maternal transmission. A similar phenomenon was observed for MJD. This gender-specific segregation distortion is called "meiotic drive" [40, 57, 61].

Meiotic drive has been found to occur in several human diseases including retinoblastoma [42], split foot/hand disease [29], and retinal cone dystrophy [12]. In the case of triplet repeat diseases, meiotic drive also occurs in myotonic dystrophy [7, 17], though the validity of the statistical analysis method used to determine this has been disputed. Since genetic anticipation and meiotic instability of CAG repeats are frequently associated with paternal meiosis, similar molecular mechanisms may underlie meiotic drive, which is also specific to male meiosis.

TABLE 15-3 Segregation Distortion in DRPLA [21]

Disorder and parent of origin	No. affected (%) {male:female}	No. unaffected (%) {male:female}	Total {male:female}	$\chi 2$	Probability
DRPLA					
Father	78 (63) {37:41}	47 (37) {27:20}	125 {64:61}	7.69	$P < 0.01$
Mother	36 (42) {22:14}	50 (58) {24:26}	86 {46:40}	2.28	$0.1 < P < 0.2$
Total	114	97	211	1.37	$0.2 < P < 0.3$

VI. POPULATION GENETICS

As mentioned above, DRPLA occurs almost exclusively in Japanese. At the moment, for other ethic backgrounds, only a limited number of families have been identified by molecular analysis of the CAG repeat of the DRPLA gene [4, 9, 48, 55, 75, 76]. To investigate the molecular basis of the ethnic predilection, we analyzed the size distributions of the CAG repeats of the DRPLA gene in Japanese, African-American, and Caucasian subjects. We found that 7.4% of the Japanese alleles (n = 407 chromosomes) had more than 19 repeat units, whereas none of the Caucasian (n = 100 chromosomes) and only 1% of the African-American alleles (n = 103 chromosomes) had repeats of this size. These results strongly suggest that the larger alleles in the Japanese population are the source of the expanded CAG repeats, and the difference in the frequencies of the normal alleles with larger sizes may account for the ethnic predilection to DRPLA [5].

Yanagisawa *et al.* [78] performed detailed hapolotype analysis of the DRPLA gene using flanking markers and intragenic polymorphisms which they identified. They found that a common haplotype is observed frequently in Japanese DRPLA patients. Interestingly, intermediate alleles (19–35 repeats) are also frequently associated with the same haplotype [78]. These results strongly support the hypothesis that the mutant DRPLA gene arose from the intermediate alleles. The intermediate alleles may contain *cis*-elements which confer instability to the CAG repeat of the DRPLA gene.

VII. INSTABILITY OF CAG REPEATS—SOMATIC MOSAICISM

Somatic mosaicism has been detected in various regions of autopsied brains of patients with CAG repeat expansion diseases. The presence of smaller CAG repeats in the cerebellum than in other regions of the central nervous system was first detected in two cases of HD with unusually young ages at onset [70]. Since the cerebellum is the least involved brain region in HD, it was suggested that the presence of a smaller CAG repeat in the cerebellum than in other regions of the central nervous system may inhibit the pathological process in the cerebellum [70]. Subsequent studies on DRPLA [66, 73], however, revealed that the CAG repeats of the DRPLA gene are also smaller in the cerebellum than in other regions of the brain.

We dissected the cerebellum and cerebrum of autopsied DRPLA brains into the cortex and white matter, and extracted genomic DNA from these tissues. By the analysis of the size distributions of expanded CAG repeats of the DRPLA gene, we found that the expanded CAG repeats of the DRPLA gene are consistently smaller in the cerebellar and cerebral cortices than those in the cerebellar and cerebral white matters [66]. These results suggest that expanded CAG repeats of the DRPLA gene are smaller in neuronal cells than in glial cells. Hashida *et al.* also analyzed several brain regions from nine cases of DRPLA, three cases of MJD, and one case of SCA1 [15]. They also found that the expanded alleles were smaller in the cerebellar cortex than in other brain regions. The discrepancy in the expanded CAG repeat length between cerebellar cortex and other tissues was most prominent in DRPLA, and especially in cases of adult-onset DRPLA. Detailed analysis of microdissected cerebellar tissues demonstrated that the lower level of CAG repeat expansion in DRPLA cerebella was representative of CAG repeat expansion in the granule cells. The microdissected samples of the granular layer of the hippocampal formation, which is also densely packed with neuronal cells, revealed that the degree of CAG repeat expansion in this layer was similar to that in the cerebellum.

Taken together, these observations that expanded CAG repeats are smaller in the cerebellum, hippocampal formation, and cerebral cortex relative to other regions of the central nervous system, seem to reflect dense populations of neuronal cells in these regions. This may be related to the fact that neuronal cells are postmitotic cells and do not undergo cell division after maturation. The role of somatic mosaicism in the pathogenesis of DRPLA, however, remains to be elucidated.

VIII. MECHANISMS OF NEURODEGENERATION CAUSED BY CAG REPEAT EXPANSION

Although the entire structure of human DRPLA cDNA has been elucidated [43, 44, 49], the physiological functions of the gene product (DRPLA protein, DRPLAP) as well as the mechanisms of neurodegeneration caused by CAG repeat expansion remain unknown. To date, a number of hypotheses and theories regarding the molecular mechanisms of neurodegeneration caused by expansion of CAG repeats have been proposed, including the position effect hypothesis, the polar zipper theory, a hypothesis proposing the expanded polyglutamine tract as a substrate for transglutaminase, and a hypothesis of neurotoxicity caused by the expanded polyglutamine tracts [20, 30, 52–54].

To identify proteins which interact with the gene products and polyglutamine tracts has taken intensive effort using a yeast two-hybrid system. Among these proteins are huntingtin-associated protein 1 (HAP1) [39], huntingtin interacting protein I (HIP-I) [31, 74], glyceraldehyde dehydrogenase (GAPDH) [6], and calmodulin [3].

HAP1 was first isolated as the protein which may interact with huntingtin, the gene product of the Huntington's disease gene. Another protein, huntingtin interacting protein I (HIP-I), was recently found using a yeast two-hybrid system [31, 74]. It was shown that Hip-1 binds specifically to the N-terminus of human huntingtin, in both the two-hybrid screen and *in vitro* binding experiments. Interestingly, the huntingtin–HIP-1 interaction is restricted to the brain and is inversely correlated to the polyglutamine length in huntingtin [31]. HIP-I shares sequence homology and biochemical characteristics with sla2p, a protein essential for function of the cytoskeleton in *Saccharomyces cerevisiae.* Thus the HIP-I and huntingtin may play a functional role in the cell filament networks.

Burke *et al.* proposed that stronger binding of mutant DRPLAP, carrying an expanded polyglutamine tract to GAPDH, compared to wild-type DRPLAP may result in a decrease in GAPDH activity and energy failure due to the decreased activity in the glycolytic pathway in cells expressing mutant DRPLAP [6].

Although these results have revealed proteins which potentially interact with the gene products and polyglutamine tracts, these proteins are generally ubiquitously expressed in the central nervous system. Therefore the broad distribution of the proteins cannot explain the selective distribution patterns of neurodegeneration, which are unique to each neurodegenerative disease.

Recently, Ikeda *et al.* reported that COS cells transfected with truncated MJD cDNA with an expanded CAG repeat undergo apoptotic cell death [20]. They demonstrated that the expanded polyglutamine tract, but not a normal polyglutamine tract, is markedly toxic to COS cells. They also demonstrated that the expression of neither full-length wild-type nor mutant MJD cDNA does not exhibit the same apoptotic cell death. Mangiarini *et al.* recently demonstrated that transgenic mice carrying only an exon 1 with expanded CAG repeats exhibit a progressive neurological phenotype [41]. These results indicate that expression of truncated proteins which contain mostly expanded polyglutamine tracts seem to be highly toxic to cells. The results further raise the intriguing possibility that processed forms of proteins, which mostly contain expanded polyglutamine tracts, are toxic to cells and that a difference in processing patterns of the gene products among various regions in the central nervous system may be the reason for the different distributions of affected regions.

Ikeda *et al.* further demonstrated that expression of expanded polyglutamine tracts is toxic to Purkinje cells in the cerebellum of the transgenic mice [20]. They used an L1 promoter which allows exclusive expression of the transgene in Purkinje cells during the postnatal period. The results further confirm the toxicity of expanded polyglutamine tracts *in vivo.*

The results mentioned above, however, were obtained by high levels of forced expression of expanded polyglutamine tracts, and it remains unclear whether similar mechanisms of toxicity occur in the brains of patients with CAG repeat expansion diseases. Particularly, the molecular basis for selective distribution patterns of neuronal loss which are unique to each CAG repeat disease remains to be elucidated.

Recently, it has been suggested that huntingtin, the product of the gene for HD, may be the substrate for apopain [14]. Apopain is the human counterpart of the nematode cysteine protease death-gene product, CED-3 [14], and was proposed to be involved in the proteolytic processing of huntingtin. Goldberg *et al.* reported that the rate of cleavage of huntingtin increases with the length of the polyglutamine tract. It is intriguing to hypothesize that the truncated protein products containing the polyglutamine tract might be toxic to neuronal cells. Taken together, the above findings suggest that processing of gene products with an expanded polyglutamine tract and the toxicity of the processed products of the proteins may play key roles in the pathogenesis of neurodegeneration.

IX. CONCLUSION

As discussed in this chapter, there are a number of challenging issues regarding the molecular mechanisms of neurodegeneration caused by expansion of triplet repeats. Since the molecular structures of the causative genes for triplet repeat diseases have been fully elucidated, triplet repeat diseases no doubt offer the best model systems to investigate the pathophysiology of neurodegenerative diseases. It is strongly expected that creation of animal models of human diseases and development of cell culture systems for investigation of neuronal toxicity caused by triplet repeat expansion will be a crucial step for the development of therapeutic measures to treat triplet repeat diseases.

Acknowledgments

This study was supported in part by the Research for the Future Program from the Japan Society for the Promotion of Science (JSPS-RFTF96L00103), a Grant-in-Aid for Scientific Research on Priority Areas (Human Genome Program) from The Ministry of Education, Science, and Culture, Japan, a grant from the Research Committee for Ataxic Diseases, the Ministry of Health and Welfare, Japan, a grant for Surveys and Research on Specific Diseases, the Ministry of Health and Welfare, Japan, and special coordination funds from the Japanese Science and Technology Agency.

References

1. Akashi, T., Ando, S., Inose, T., Kamimura, H., and Mizushima, S. (1987). Dentato-rubro-pallido-luysian atrophy: A clinicopathological study. *Rinsho Seishin Igaku* **29,** 523–531. [in Japanese]
2. Andrew, S. E., Goldberg, Y. P., Kremer, B., Telenius, H., Theilmann, J., Adam, S., Starr, E., Squitieri, F., Lin, B., and Kalchman, M. A. (1993). The relationship between trinucleotide (CAG) repeat length and clinical features of Huntington's disease. *Nature Genet.* **4,** 398–403.
3. Bao, J., Sharp, A. H., Wagster, M. V., Becher, M., Schilling, G., Ross, C. A., Dawson, V. L., and Dawson, T. M. (1996). Expansion of polyglutamine repeat in huntingtin leads to abnormal protein interactions involving calmodulin. *Proc. Natl. Acad. Sci. USA* **93,** 5037–5042.
4. Burke, J. R., Wingfield, M. S., Lewis, K. E., Roses, A. D., Lee, J. E., Hulette, C., Pericak-Vance, M. A., and Vance, J. M. (1994). The Haw River syndrome: Dentatorubropallidoluysian atrophy (DRPLA) in an African-American family. *Nature Genet.* **7,** 521–524.
5. Burke, J. R., Ikeuchi, T., Koide, R., Tsuji, S., Yamada, M., Pericak-Vance, M. A., and Vance, J. M. (1994). Dentatorubral-pallidoluysian atrophy and Haw River syndrome. *Lancet* **344,** 1711–1712.
6. Burke, J. R., Enghild, J. J., Martin, M. E., Jou, Y. S., Myers, R. M., Roses, A. D., Vance, J. M., and Strittmatter, W. J. (1996). Huntingtin and DRPLA proteins selectively interact with the enzyme GAPDH. *Nature Med.* **2,** 347–350.
7. Carey, N., Johnson, K., Nokelainen, P., *et al.* (1995). Meiotic drive and myotonic dystrophy. (Reply). *Nature Genet.* **10,** 133.
8. Chung, M. Y., Ranum, L. P., Duvick, L. A., Servadio, A., Zoghbi, H. Y., and Orr, H. T. (1993). Evidence for a mechanism predisposing to intergenerational CAG repeat instability in spinocerebellar ataxia type I. *Nature Genet.* **5,** 254–258.
9. Connarty, M., Dennis, N. R., Patch, C., Macpherson, J. N., and Harvey, J. F. (1996). Molecular investigation of patients with Huntington's disease in Wessex reveals a family with dentatorubral and pallidoluysian atrophy. *Hum. Genet.* **97,** 76–78.
10. De Barsy, T. H., Myle, G., Troch, C., Matthys, R., and Martin, J. J. (1968). La Dyssynergie cerebelleuse myoclonique (R. Hunt): Affection autonome ou rariante du type degeneratif de l'epilepsie-myoclonie progressive (Unvericht-Lundborg) Appoche anatomoalinique. *J. Neurol. Sci.* **8,** 111–127.
11. Duyao, M., Ambrose, C., Myers, R., Novelletto, A., Persichetti, F., Frontali, M., Folstein, S., Ross, C., Franz, M., and Abbott, M. (1993). Trinucleotide repeat length instability and age of onset in Huntington's disease. *Nature Genet.* **4,** 387–392.
12. Evans, K., Fryer, A., Inglehearn, C., Duvall-Young, J., Whittaker, J. L., Gregory, C. Y., Butler, R., Ebenezer, N., Hunt, D. M., and Bhattacharya, S. (1994). Genetic linkage of cone-rod retinal dystrophy to chromosome 19q and evidence for segregation distortion. *Nature Genet.* **6,** 210–213.
13. Farmer, T. W., Wingfield, M. S., Lynch, S. A., Vogel, F. S., Hulette, C., and Katchinoff, B. (1989). Ataxia, chorea, seizures, and dementia. Pathologic features of a newly defined familial disorder. *Arch. Neurol.* **46,** 774–779.
14. Goldberg, Y. P., Nicholson, D. W., Rasper, D. M., Kalchman, M. A., Koide, H. B., Graham, R. K., Bromm, M., Kazemi-Esfarjani, P., Thornberry, N. A., Vaillancourt, J. P., and Hayden, M. R. (1996). Cleavage of huntingtin by apopain, a proapoptotic cysteine protease, is modulated by the polyglutamine tract. *Nature Genet.* **13,** 442–449.
15. Hashida, H., Goto, J., Kurisaki, H., Mizusawa, H., and Kanazawa, I. (1997). Brain regional differences in the expansion of a CAG repeat in the spinocerebellar ataxias - dentatorubral-pallidoluysian atrophy, Machado-Joseph disease, and spinocerebellar ataxia type 1. *Ann. Neurol.* **41,** 505–511.
16. Hirayama, K., Iizuka, R., Maehara, K., and Watanabe, T. (1981). Clinicopathological study of dentatorubropallidoluysian atrophy. Part I. Its clinical form and analysis of symptomatology. *Adv. Neurol.* **25,** 725–736. [in Japanese]
17. Hurst, G. D., Hurst, L. D., and Barrett, J. A. (1995). Meiotic drive and myotonic dystrophy. *Nature Genet.* **10,** 132–133.
18. Iizuka, R., Hirayama, K., and Maehara, K. A. (1984). Dentato-rubro-pallido-luysian atrophy: a clinico-pathological study. *J. Neurol. Neurosurg. Psych.* **47,** 1288–1298.
19. Iizuka, R., and Hirayama, K. (1986). Dentato-rubro-pallido-luysian atrophy. *In* "Handbook of Clinical Neurology" (P. J. Vinken, G. W. Bruyn, and H. L. Klawans, Eds.), Vol. 5, pp. 437–443. North-Holland, Amsterdam.
20. Ikeda, H., Yamaguchi, M., Sugai, S., Aze, Y., Narumiya, S., and Kakizuka, A. (1996). Expanded polyglutamine in the Machado-Joseph disease protein induces cell death in vitro and in vivo. *Nature Genet.* **13,** 196–202.
21. Ikeuchi, T., Igarashi, S., Takiyama, Y., Onodera, O., Oyake, M., Takano, H., Koide, R., Tanaka, H., and Tsuji, S. (1996). Non-mendelian transmission in dentatorubral-pallidoluysian atrophy and Machado-Joseph disease: the mutant allele is preferentially transmitted in male meiosis. *Am. J. Hum. Genet.* **58,** 730–733.
22. Ikeuchi, T., Koide, R., Tanaka, H., Onodera, O., Igarashi, S., Takahashi, H., Kondo, R., Ishikawa, A., Tomoda, A., Miike, T., Sato, K., Ihara, Y., Hayabara, T., Isa, F., Tanabe, H., Tokiguchi, S., Hayashi, M., Shimizu, N., Ikuta, F., Naito, H., and Tsuji, S. (1995). Dentatorubral-pallidoluysian atrophy: clinical features are closely related to unstable expansions of trinucleotide (CAG) repeat. *Ann. Neurol.* **37,** 769–775.
23. Ikeuchi, T., Koide, R., Onodera, O., Tanaka, H., Oyake, M., Takano, H., and Tsuji, S. (1995). Dentatorubral-pallidoluysian

atrophy (DRPLA). Molecular basis for wide clinical features of DRPLA. *Clin. Neurosci.* **3,** 23–27.

24. Ikeuchi, T., Onodera, O., Oyake, M., Koide, R., Tanaka, H., and Tsuji, S. (1995). Dentatorubral-pallidoluysian atrophy (DRPLA): close correlation of CAG repeat expansions with the wide spectrum of clinical presentations and prominent anticipation. *Semin. Cell. Biol.* **6,** 37–44.
25. Imbert, G., Saudou, F., Yvert, G., Devys, D., Trottier, Y., Garnier, J. M., Weber, C., Mandel, J. L., Cancel, G., Abbas, N., Durr, A., Didierjean, O., Stevanin, G., Agid, Y., and Brice, A. (1996). Cloning of the gene for spinocerebellar ataxia 2 reveals a locus with high sensitivity to expanded CAG/glutamine repeats. *Nature Genet.* **14,** 285–291.
26. Inazuki, G., Kumagai, K., and Naito, H. (1990). Dentatorubral-pallidoluysian atrophy (DRPLA): its distribution in Japan and prevalence rate in Niigata. *Seishin Igaku* **32,** 1135–1138.
27. Iwabuchi, K. (1987). Clinico-pathological studies on dentato-rubro-pallido-luysian atrophy (DRPLA). (In Japanese). *Yokohama Igaku* **38,** 91–301.
28. Iwabuchi, K., Amano, N., Yagishita, S., Sakai, H., and Yokoi, S. (1987). A clinicopathological study on familial cases of dentatorubro-pallidoluysian atrophy (DRPLA). *Clin. Neurol.* **27,** 1002–1012. [in Japanese]
29. Jarvik, G. P., Patton, M. A., Homfray, T., and Evans, J. P. (1994). Non-mendelian transmission in a human developmental disorder: split hand/split foot. *Am. J. Hum. Genet.* **55,** 710–713.
30. Kahlem, P., Terre, C., Green, H., and Djian, P. (1996). Peptides containing glutamine repeats as substrates for transglutaminase-catalyzed cross-linking—relevance to diseases of the nervous system. *Proc. Natl. Acad Sci. USA* **93,** 14580–14585.
31. Kalchman, M. A., Koide, H. B., Mccutcheon, K., Graham, R. K., Nichol, K., Nishiyama, K., Kazemiesfarjani, P., Lynn, F. C., Wellington, C., Metzler, M., Goldberg, Y. P., Kanazawa, I., Gietz, R. D., and Hayden, M. R. (1997). Hip1, a human homologue of s-cerevisiae sla2p, interacts with membrane-associated huntingtin in the brain. *Nature Genet.* **16,** 44–53.
32. Kawakami, H., Maruyama, H., Nakamura, S., Kawaguchi, Y., Kakizuka, A., Doyu, M., and Sobue, G. (1995). Unique features of the CAG repeats in Machado-Joseph disease. *Nature Genet.* **9,** 344–345.
33. Koide, R., Ikeuchi, T., Onodera, O., Tanaka, H., Igarashi, S., Endo, K., Takahashi, H., Kondo, R., Ishikawa, A., Hayashi, T., Saito, M., Tomoda, A., Miike, T., Naito, H., Ikuta, F., and Tsuji, S. (1994). Unstable expansion of CAG repeat in hereditary dentatorubral-pallidoluysian atrophy (DRPLA). *Nature Genet.* **6,** 9–13.
34. Kremer, B., Goldberg, P., Andrew, S. E., Theilmann, J., Telenius, H., Zeisler, J., Squitieri, F., Lin, B., Bassett, A., and Almqvist, E. (1994). A worldwide study of the Huntington's disease mutation. the sensitivity and specificity of measuring CAG repeats. *N. Engl. J. Med.* **330,** 1401–1406.
35. Kurohara, K., Kuroda, Y., Maruyama, H., Kawakami, H., Yukitake, M., Matsui, M., and Nakamura, S. (1997). Homozygosity for an allele carrying intermediate CAG repeats in the dentatorubral-pallidoluysian atrophy (DRPLA) gene results in spastic paraplegia. *Neurology* **48,** 1087–1090.
36. La Spada, A. R., Wilson, E. M., Lubahn, D. B., Harding, A. E., and Fischbeck, K. H. (1991). Androgen receptor gene mutations in X-linked spinal and bulbar muscular atrophy. *Nature* **352,** 77–79.
37. Lang, A. E., Rogaeva, E. A., Tsuda, T., Hutterer, J., and St George-Hyslop, P. (1994). Homozygous inheritance of the Machado-Joseph disease gene. *Ann. Neurol.* **36,** 443–447.
38. Li, S. H., McInnis, M. G., Margolis, R. L., Antonarakis, S. E., and Ross, C. A. (1993). Novel triplet repeat containing genes in human brain: cloning, expression, and length polymorphisms. *Genomics* **16,** 572–579.
39. Li, X. J., Sharp, A. H., Li, S. H., Dawson, T. M., Snyder, S. H., and Ross, C. A. (1996). Huntingtin-associated protein (HAP1): discrete neuronal localizations in the brain resemble those of neuronal nitric oxide synthase. *Proc. Natl. Acad. Sci. USA* **93,** 4839–4844.
40. Lyttle, T. W. (1993). Cheaters sometimes prosper: distortion of Mendelian segregation by meiotic drive. *Trends Genet.* **9,** 205–210.
41. Mangiarini, L., Sathasivam, K., Seller, M., Cozens, B., Harper, A., Hetherington, C., Lawton, M., Trottier, Y., Lehrach, H., Davies, S. W., and Bates, G. P. (1996). Exon 1 of the HD gene with an expanded CAG repeat is sufficient to cause a progressive neurological phenotype in transgenic mice. *Cell* **87,** 493–506.
42. Munier, F., Spence, M. A., Pescia, G., Balmer, A., Gailloud, C., Thonney, F., van Melle, G., and Rutz, H. P. (1992). Paternal selection favoring mutant alleles of the retinoblastoma susceptibility gene. *Hum. Genet.* **89,** 508–512.
43. Nagafuchi, S., Yanagisawa, H., Sato, K., Shirayama, T., Ohsaki, E., Bundo, M., Takeda, T., Tadokoro, K., Kondo, I., Murayama, N., Tnaka, Y., Kikushima, H., Umino, K., Kurosawa, H., Furukawa, T., Nihei, K., Inoue, T., Sano, A., Komure, O., Takahashi, M., Yoshizawa, T., Kanazawa, I., and Yamada, M. (1994). Expansion of an unstable CAG trinucleotide on chromosome 12p in dentatorubral and pallidoluysian atrophy. *Nature Genet.* **6,** 14–18.
44. Nagafuchi, S., Yanagisawa, H., Ohsaki, E., Shirayama, T., Tadokoro, K., Inoue, T., and Yamada, M. (1994). Structure and expression of the gene responsible for the triplet repeat disorder, dentatorubral and pallidoluysian atrophy (DRPLA). *Nature Genet* **8,** 177–182.
45. Naito, H., Izawa, K., Kurosaki, T., Kaji, S., and Sawa, M. (1972). Two families of progressive myoclonus epilepsy with Mendelian dominant heredity. *Psych. Neurol. Jpn* **74,** 871–897. [in Japanese]
46. Naito, N., and Oyanagi, S. (1982). Familial myoclonus epilepsy and choreoathetosis; hereditary dentatorubral-pallidoluysian atrophy. *Neurology* **32,** 789–817.
47. Naito, H., Ohama, E., Nagai, H., Wakabayashi, M., Morita, M., and Ikuta, F. (1987). A family of dentatorubropallidoluysian atrophy (DRPLA) including two cases with schizophrenic symptoms. *Psych. Neurol. Jpn* **89,** 144–158.
48. Norremolle, A., Nielsen, J. E., Sorensen, S. A., and Hasholt, L. (1995). Elongated CAG repeats of the B37 gene in a Danish family with dentato-rubro-pallido-luysian atrophy. *Hum. Genet.* **95,** 313–318.
49. Onodera, O., Oyake, M., Takano, H., Ikeuchi, T., Igarashi, S., and Tsuji, S. (1995). Molecular cloning of a full-length cDNA for dentatorubral-pallidoluysian atrophy and regional expressions of the expanded alleles in the CNS. *Am. J. Hum. Genet.* **57,** 1050–1060.
50. Orr, H. T., Chung, M. Y., Banfi, S., Kwiatkowski,, T. J., Jr., Servadio, A., Beaudet, A. L., McCall, A. E., Duvick, L. A., Ranum, L. P., and Zoghbi, H. Y. (1993). Expansion of an unstable trinucleotide CAG repeat in spinocerebellar ataxia type 1. *Nature Genet.* **4,** 221–226.
51. Oyanagi, S., and Naito, H. (1977). A clinico-neuropathological study on four autopsy cases of degenerative type of myoclonus epilepsy with Mendelian dominant heredity. (in Japanese). *Psych. Neurol. Jpn* **79,** 113–129.
52. Perutz, M. F., Johnson, T., Suzuki, M., and Finch, J. T. (1994). Glutamine repeats as polar zippers: their possible role in inherited neurodegenerative diseases. *Proc. Natl. Acad. Sci. USA* **91,** 5355–5358.
53. Perutz, M. F. (1995). Glutamine repeats as polar zippers: their role in inherited neurodegenerative disease. *Mol. Med.* **1,** 718–721.

54. Perutz, M. F. (1996). Glutamine repeats and inherited neurodegenerative diseases—molecular aspects. *Curr. Opin. Struct. Biol.* **6,** 848–858.
55. Potter, N. T. (1996). The relationship between (CAG)n repeat number and age of onset in a family with dentatorubral-pallidoluysian atrophy (DRPLA): diagnostic implications of confirmatory and predictive testing. *J. Med. Genet.* **33,** 168–170.
56. Pulst, S. M., Nechiporuk, A., Nechiporuk, T., Gispert, S., Chen, X. N., Lopes-Cendes, I., Pearlman, S., Starkman, S., Orozco-Diaz, G., Lunkes, A., DeJong, P., Rouleau, G. A., Auburger, G., Korenberg, J. R., Figueroa, C., and Sahba, S. (1996). Moderate expansion of a normally biallelic trinucleotide repeat in spinocerebellar ataxia type 2. *Nature Genet.* **14,** 269–276.
57. Raju, N. B., and Perkins, D. D. (1991). Expression of meiotic drive elements spore killer-2 and spore killer-3 in asci of neurospora tetrasperma. *Genetics* **129,** 25–37.
58. Sanpei, K., Takano, H., Igarashi, S., Sato, T., Oyake, M., Sasaki, H., Wakisaka, A., Tashiro, K., Ishida, Y., Ikeuchi, T., Koide, R., Saito, M., Sato, A., Tanaka, T., Hanyu, S., Takiyama, Y., Nishizawa, M., Shimizu, N., Nomura, Y., Segawa, M., Iwabuchi, K. and Eguchi, I., Tanaka, H., Takahashi, H., and Tsuji, S. (1996). Identification of the spinocerebellar ataxia type 2 gene using a direct identification of repeat expansion and cloning technique, DIRECT. *Nature Genet.* **14,** 277–284.
59. Sato, K., Kashihara, K., Okada, S., Ikeuchi, T., Tsuji, S., Shomori, T., Morimoto, K., and Hayabara, T. (1995). Does homozygosity advance the onset of dentatorubral-pallidoluysian atrophy? *Neurology* **45,** 1934–1936.
60. Shimizu, N., Yamami, T., Nakayama, M., Ikeuchi, T., Koide, R., and Tsuji, S. (1996). A sporadic case of dentatorubral pallidoluysian atrophy (DRPLA) with CAG repeat expansion but no clinical abnormalities in the father. *J. Neurol. Neurosurg. Psych.* **61,** 113–114.
61. Silver, L. M. (1993). The peculiar journey of a selfish chromosome: mouse t haplotypes and meiotic drive. *Trends Genet.* **9,** 250–254.
62. Smith, J. K., Gonda, V. E., and Malamud, N. (1958). Unusual form of cerebellar ataxia: combined dentato-rubral and pallido-Luysian degeneration. *Neurology* **8,** 205–209.
63. Smith, J. K. (1975). Dentatorubropallidoluysian atrophy. *In* "Handbook of Clinical Neurology" (P. J. Vinken and G. W. Bruyn, Eds.), Vol. 21, pp. 519–534. North-Holland, Amsterdam.
64. Snell, R. G., MacMillan, J. C., Cheadle, J. P., Fenton, I., Lazarou, L. P., Davies, P., MacDonald, M. E., Gusella, J. F., Harper, P. S., and Shaw, D. J. (1993). Relationship between trinucleotide repeat expansion and phenotypic variation in Huntington's disease. *Nature Genet.* **4,** 393–397.
65. Suzuki, S., Kamoshita, S., and Ninomura, S. (1985). Ramsay Hunt syndrome in dentatorubral-pallidoluysian atrophy. *Pediatr. Neurol.* **1,** 298–301.
66. Takano, H., Onodera, O., Takahashi, H., Igarashi, S., Yamada, M., Oyake, M., Ikeuci, T., Koide, R., Tanaka, H., Iwabuchi, K., and Tsuji, S. (1996). Somatic mosaicism of expanded CAG repeats in brains of patients with dentatorubral-pallidoluysian atrophy—cellular population-dependent dynamics of mitotic instability. *Am. J. Hum. Genet.* **58,** 1212–1222.
67. Takano, T., Yamanouchi, Y., Nagafuchi, S., and Yamada, M. (1996). Assignment of the dentatorubral and pallidoluysian atrophy (DRPLA) gene to 12p 13.31 by fluorescence in situ hybridization. *Genomics* **32,** 171–172.
68. Takiyama, Y., Igarashi, S., Rogaeva, E. A., Endo, K., Rogaev, E. I., Tanaka, H., Sherrington, R., Sanpei, K., Liang, Y., Saito, M., Tsuda, T., Takano, H., Ikeda, M., Lin, C., Chi, H., Kennedy, J. L., Lang, A. E., Wherrett, J. R., Segawa, M., Nomura, Y., Yuasa, T., Weissenbach, J., Yoshida, M., Nishizawa, M., Kidd, K. K., Tsuji, S., and St. George-Hyslop, P. H (1995). Evidence for inter-generational instability in the cag repeat in the MJD1 gene and for conserved haplotypes at flanking markers amongst japanese and caucasian subjects with Machado-Joseph disease. *Hum. Mol. Genet.* **4,** 1137–1146.
69. Tanaka, Y., Murobushi, K., Ando, S., Suwa, K., and Hayashi, M. (1977). Combined degeneration of the globus pallidus and the cerebellar nuclei and their efferent systems in two siblings of one family -Primary system degeneration of the globus pallidus and the cerebellar nuclei-. *Brain Nerve* **29,** 95–104. [in Japanese]
70. Telenius, H., Kremer, B., Goldberg, Y. P., Theilmann, J., Andrew, S. E., Zeisler, J., Adam, S., Greenberg, C., Ives, E. J., and Clarke, L. A. (1994). Somatic and gonadal mosaicism of the Huntington disease gene CAG repeat in brain and sperm. *Nature Genet.* **6,** 409–414.
71. The Huntington's Disease Collaborative Research Group. (1993). A novel gene containing a trinucleotide repeat that is expanded and unstable on Huntington's disease chromosomes. *Cell* **72,** 971–983.
72. Titica, J., and van Bogaert, L. (1946). Heredo-degenerative hemiballismus: a contribution to the question of primary atrophy of the corpus Luysii. *Brain* **69,** 251–263.
73. Ueno, S., Kondoh, K., Kotani, Y., Komure, O., Kuno, S., Kawai, J., Hazama, F., and Sano, A. (1995). Somatic mosaicism of CAG repeat in dentatorubral-pallidoluysian atrophy (DRPLA). *Hum. Mol. Genet.* **4,** 663–666.
74. Wanker, E. E., Rovira, C., Scherzinger, E., Hasenbank, R., Walter, S., Tait, D., Colicelli, J., and Lehrach, H. (1997). HIP-I: a huntingtin interacting protein isolated by the yeast two-hybrid system. *Hum. Mol. Genet.* **6,** 487–495.
75. Warner, T. T., Williams, L., and Harding, A. E. (1994). DRPLA in Europe. *Nature Genet.* **6,** 225.
76. Warner, T. T., Lennox, G. G., Janota, I. and Harding, A. E. (1994). Autosomal-dominant dentatorubropallidoluysian atrophy in the united kingdom. *Mov. Disord.* **9,** 289–296.
77. Wexler, N. S., Young, A. B., Tanzi, R. E., Travers, H., Starosta-Rubinstein, S., Penney, J. B., Snodgrass, S. R., Shoulson, I., Gomez, F., and Ramos Arroyo, M. A. (1987). Homozygotes for huntington's disease. *Nature* **326,** 194–197.
78. Yanagisawa, H., Fujii, K., Nagafuchi, S., Nakahori, Y., Nakagome, Y., Akane, A., Nakamura, M., Sano, A., Komure, O., Kondo, I., Jin, D. K., Sorensen, S. A., Potter, N. T., Young, S. R., Nakamura, K., Nukina, N., Nagao, Y., Tadokoro, K., Okuyama, T., Miyashita, T., and Inoue, T. (1996). A unique origin and multistep process for the generation of expanded DRPLA triplet repeats. *Hum. Mol. Genet.* **5,** 373–379.
79. Yazawa, I., Nukina, N., Hashida, H., Goto, J., Yamada, M., and Kanazawa, I. (1995). Abnormal gene product identified in hereditary dentatorubral-pallidoluysian atrophy (DRPLA) brain. *Nature Genet.* **10,** 99–103.
80. Zhuchenko, O., Bailey, J., Bonnen, P., Ashizawa, T., Stockton, D. W., Amos, C., Dobyns, W. B., Subramony, S. H., Zoghbi, H. Y., and Lee, C. C. (1997). Autosomal dominant cerebellar ataxia (SCA6) associated with small polyglutamine expansions in the a1a-voltage-dependent calcium channel. *Nature Genet.* **15,** 62–69.

CHAPTER 16

The Allelic Variant: Haw River Syndrome

JEFFERY M. VANCE Division of Neurology, Department of Medicine, Duke University Medical Center, Durham, North Carolina 27710

Since the 1840s an African-American family has grown and lived in a place of safety and isolation near the Haw River, in the central Piedmont of North Carolina (Fig. 16-1). The family suffered from a strange neurodegenerative disorder, leaving them with abnormal movements, psychoses, and dementia. Unknown and relatively neglected, in 1937 the attending physician of one family member wrote that "There is no name for this disease and this man should be sterilized." While social attitudes have changed in time, one thought did not: those groups caring for these patients believed it to be a unique disease, harbored by only one family, a true orphan disease. But recently, it has been shown that the disorder this family has transmitted for many generations, known as the Haw River Syndrome (HRS), is in fact the same molecular entity as dentatorubropallidoluysian atrophy (DRPLA), a disease found primarily in Japan. This finding is a testament of the power and sensitivity of modern genetics, perhaps leading to the eventual elimination of the "orphan" status of many similar isolated families.

The first formal description of the HRS was by Farmer and colleagues [1], who described the disease in five generations of the family (Fig. 16-2). This paper was a joint effort by the University of North Carolina, John Umstead State Hospital in Butner, North Carolina, and Duke University Medical Center. For years prior to this report, the groups had followed and treated single patients with the disorder. In each case, affected individuals lived by the Haw River, leading to the initial conclusion that the disease was due to toxins or other environmental causes. Given the small region in which the family resided, and its proximity to the Haw River, the river's name was given to the disorder. However, as the number of cases increased in the area, so did communication between centers caring for the patients. Soon it became clear that this neurodegenerative disorder was in fact hereditary, and all cases were members of the same large African-American family [2].

The symptoms of Haw River Syndrome are relatively uniform and include gait ataxia, dysarthria, chorea, intention tremor, dementia, and psychoses. The average age of onset is 15–30 years. Commonly, the initial presentation in the patients is of gait ataxia and dysarthria. Psychiatric manifestations of paranoia and delusions are also frequent. Chorea, tremor, seizures, and dementia follow, leading to death after a 15- to 25-year span of symptoms. Several patients were also noted to have a learning disability in school prior to developing definitive symptoms of the disorder, suggesting that in some individuals a mild mental retardation during childhood may be associated with the disease.

Significantly, Computed Assisted Tomography revealed no caudate atrophy in any HRS patients, failing to support the initial thoughts that HRS represented a variant of Huntington's disease. But an unusual finding was found in at least 10 patients. They all had calcification of the globus pallidus, a very rare phenomenon in the relatively young patients being studied [3] (Fig. 16-3). Subsequent imaging studies by Farmer and col-

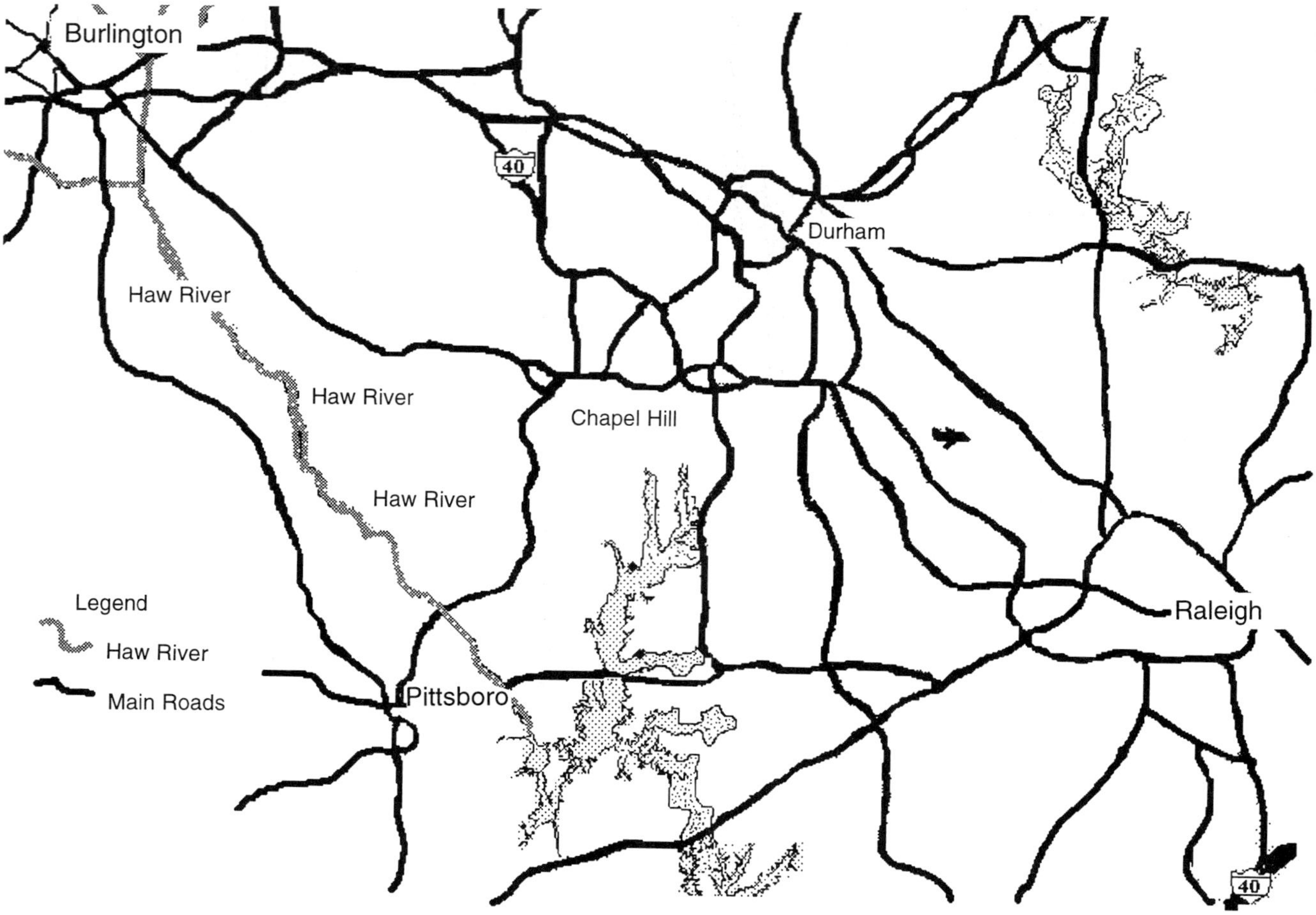

FIGURE 16-1 Map of the Haw River area in North Carolina.

leagues [3], using Magnetic Resonance Imaging (MRI), demonstrated extensive white matter demyelination in many of the patients as well (Fig. 16-4). Pathologic examination of two patients revealed very mild neuronal loss in the dentate nucleus, neuroaxonal dystrophy (NAD) of the nucleus gracilis, and demyelination of the centrum semiovale. Interestingly, the differential diagnosis by Farmer and colleagues included several entities: (1) Olivopontocerebellar atrophy (OPC) was considered due to the ataxia that occurred regularly in patients. However, these patients lacked the characteristic Purkinje cell abnormalities found in OPC and had seizures as a common manifestation of the illness. (2) Similarly, as mentioned previously, Huntington's disease was considered due to the presence of chorea and associated psychiatric features, but it was excluded due to the absence of caudate atrophy in any of the affected family members and, again, the aforementioned presence of seizures. (3) DRPLA was also considered, as many of the clinical manifestations overlapped this disorder as well. Nevertheless, the authors felt that the family's disorder was not DRPLA because of the rela-

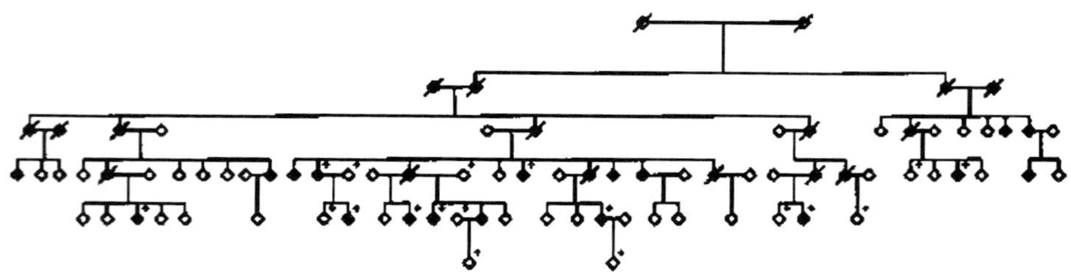

FIGURE 16-2 Pedigree of the Haw River family. Darker diamonds are affected individuals.

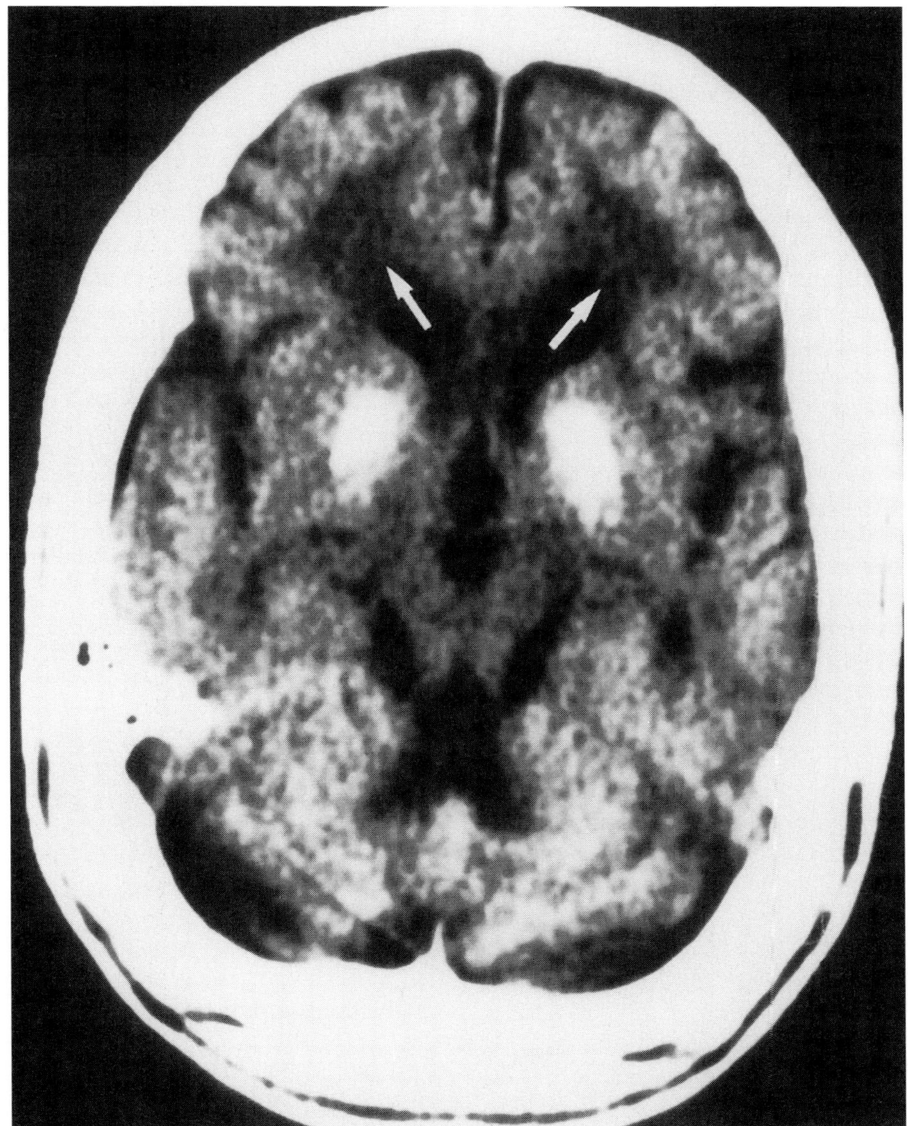

FIGURE 16-3 CT scan demonstrating calcification of the globus pallidus in a HRS patient. Arrows demonstrate lucency of white matter. Taken, with permission, from Farmer *et al., J. Neuroimaging* **1,** 123–128, 1991.

tively mild pathologic changes in the dentate nucleus and the significant incidence of calcification of the globus pallidus. In addition, the authors stated regarding the pathologic changes seen in an autopsied family member that "NAD and supratentorial white matter demyelination are not features of dentatorubropallidoluysian atrophy." Additionally, DRPLA was a disorder thought to occur primarily, if not exclusively, in Japan. Thus its presence in this African-American family was felt to be highly unlikely. This led authors to believe this was a unique disorder, perhaps limited to this single African-American family. It was subsequently given a Mendelian Inheritance in Man (MIM) number of 140340.

One of the authors of that paper [3] was Dr. Martha Wingfield, a staff neurologist at the University of North Carolina at Chapel Hill and John Umstead State Hospital, where many of the Haw River Patients were eventually sent, once they became too difficult for the family to manage. During her tenure at Umstead Hospital, Dr. Wingfield became devoted to the study of HRS. She often gave clothing and support to family members from her own personal funds. She became trusted and close to many members of this normally retiring family, enabling her to gather blood samples for research. The samples were taken to the Division of Neurology at Duke University Medical Center, in the hope that they would allow future study when more information became

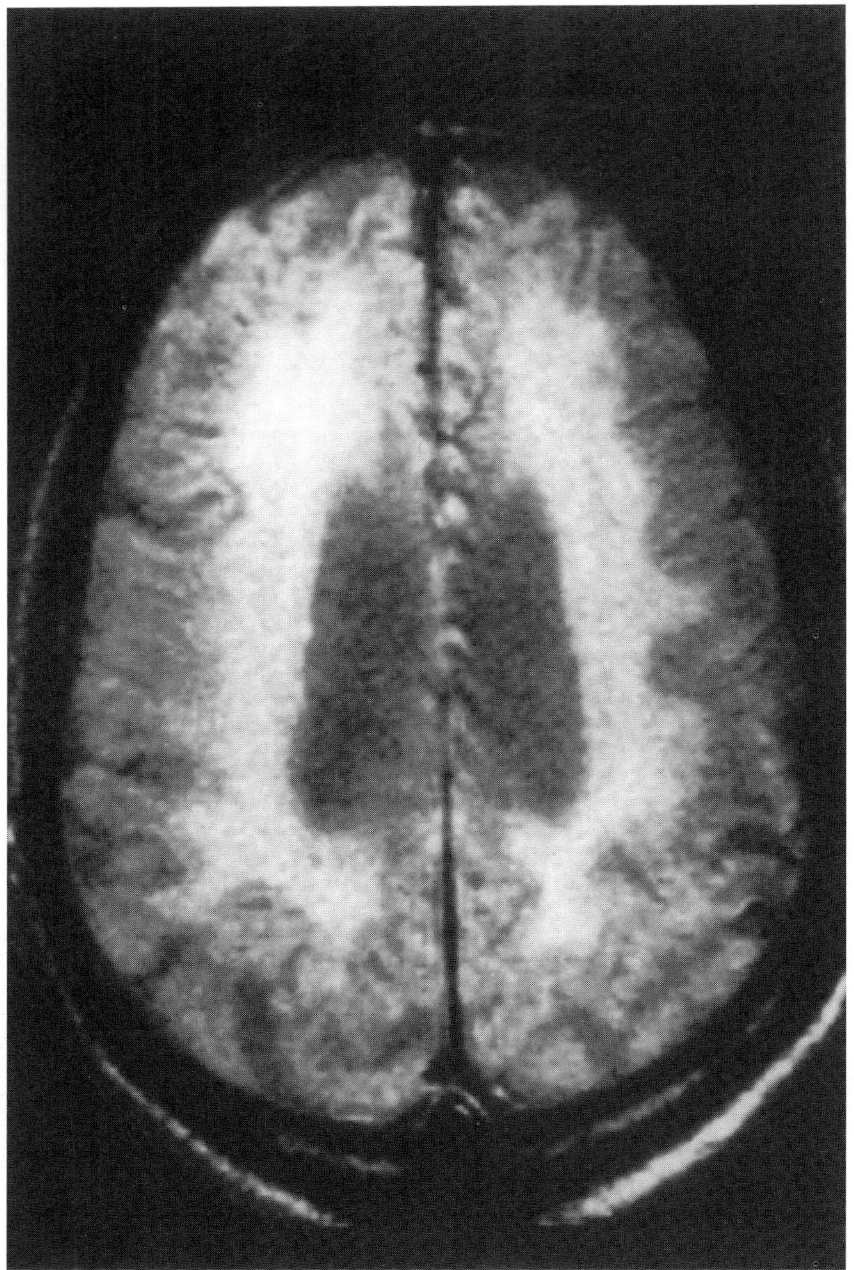

FIGURE 16-4 MRI of a Haw River syndrome patient demonstrating extensive white matter changes. Taken, with permission, from Farmer *et al., J. Neuroimaging* **1,** 123–128, 1991.

available to explore HRS. The eventual paper describing the defect in HRS is dedicated to this remarkable physician's quest to keep this family known to the research community.

In 1992, the defect for HD was reported, but this was quickly shown not to be the defect in HRS (personal communication). However, the reports by Koide and colleagues [4] and Nagafuchi and colleagues [5] in 1994 described the trinucleotide repeat defect in DRPLA in several Japanese patients. Despite the ethnic differences, the original differential diagnoses by Farmer and colleagues [1] led our laboratory to investigate the possibility that DRPLA and HRS shared the same defect [6]. Indeed, we found that the gene causing the defect in HRS did lie in the same chromosome 12 region as DRPLA. Further investigation demonstrated that the identical trinucleotide repeat expansion that caused DRPLA was also found in HRS, demonstrating that the two were indeed, at least molecularly, the same syndrome.

The size variation of the expanded alleles in the affected members of the HRS family was 63–68 repeats, similar to that of DRPLA. The number of individuals affected in the family is too small to determine if the age of onset is inversely correlated with the size of the repeat, as has been reported in all of the CAG repeats to date, including DRPLA.

The ethnic differences between the two entities are interesting. We [7] examined the normal CAG repeat frequency in normal populations of African-Americans, Caucasians, and Japanese. We found that while 7.4% of the Japanese alleles of 407 chromosomes fell in an intermediate range of 19–30 repeats, no Caucasian alleles greater than >18 were seen (100 chromosomes), and only 1/103 chromosomes in Africans fell within this region. This led us to postulate that the higher allele frequency in the 19- to 30-repeat range in the Japanese provided an intermediate, less stable pool of alleles that gave rise to the higher rate of DRPLA in this population. This is analogous to the racial differences reported in myotonic dystrophy in Africa [8]. Deka and colleagues [9] have supported this theory with studies of the allele expansion in DRPLA in a wider range of populations, including Chinese and Samoans. They also found that Japanese and Chinese had the highest level of repeats greater than 18 (22%) and Africans the lowest (1.1%).

Table 16-1 shows a comparison of the two disorders. Clinically, the two are certainly quite similar. The major clinical difference is that no patients with HRS have had myoclonic epilepsy (ME) or myoclonus, which is observed commonly in DRPLA. Pathologically, both HRS and DRPLA have neuronal loss in the dentate and subthalamic nuclei, but it is reported to be much milder in HRS (see below). However, Singer [10] argued against a separate disorder, believing the important point was that the dentate nucleus is constantly involved. He noted that Izuka and Hirayma [11] had previously reported Japanese patients with slight pathologic changes in this site, similar to those reported in HRS. Singer also questioned the significance of NAD, which he noted could be found in some normal individuals. He did agree that at that time the extensive white matter changes reported in HRS were unusual in DRPLA, but felt it "enlarged" the syndrome, which already had some demyelination reported in spinocerebellar and dorsal column tracts. In response, Farmer [12] stated that he believed NAD is unusual and is usually observed in older individuals then the HRS patients they studied. He maintained that the differences in both clinical presentation and pathology, including the calcification of the globus pallidus and white matter changes, warranted consideration of HRS as a separate entity.

Table 16-1 Clinical and Neuropathologic Comparison between HRS and DRPLA

Clinical	HRS	DRPLA
Generalized seizures	+++	++
Dementia	+++	++
Myoclonus +/− myoclonic epilepsy	−	++
Chorea	+++	+++
Neuropathological demyelination	++	+
Neuronal loss in globus pallidus	+	+++
Basal ganglia calcification	+++	+
Neuroaxonal dystrophy (NAD)	+++	+

So now that the molecular identity of HRS and DRPLA is known, the same question arises; is HRS a true allelic "variant" or was its unique status just the result of ascertainment bias? Like all questions benefiting from the wisdom of hindsight, it is, of course, a bit of an unfair question. Certainly, because of the rarity of reporting a single family, the initial autopsy findings reported in HRS were dependent on only two individuals. This always raises the possibility of chance observations that do not represent the larger group. However, recently, Suzuki [13] reviewed 11 non-Japanese cases to see if the neuropathology was indeed different from that described in Japan. Included in this review were two additional cases of HRS not previously described. She cites that they demonstrated "severe neuronal loss in the dentate nucleus, calcification of the globus pallidus and neuroaxonal dystrophy" that was indeed similar to previous cases. However, she cited that the relative "normality" of the subthalamic nucleus and red nucleus in all cases and the presence of calcification of the globus pallidus in such young individuals does suggest that HRS does appear "somewhat" different from most DRPLA cases, including the other non-Japanese cases examined. However, she points out that Takahata and colleagues reported [14] five familial cases in the Japanese population with similar mild changes in these areas in DRPLA. In addition, the white matter changes described in HRS have now been documented by several authors [15–17]. Thus, while unusual, most of the findings are not unique.

Similarly, the clinical differences observed, had they been in a Japanese family, may not have appeared as distinctive as they were in this African-American counterpoint. Indeed, Izuka and colleagues [18] have suggested that wide clinical variation may exist in the Japanese population as well, classifying DRPLA into three clinical forms: ataxic-choreoathetoid, pseudo-Huntington, and myoclonic-epilepsy. Thus, even within the Japanese population, there is significant variation in phenotype observed.

Therefore, HRS and DRPLA support the argument that we have too often in the past considered the action of one gene to be isolated, omnipotent in its effects upon phenotype. This is probably natural given the progression of human genetics from the easily identified, homogeneous enzymatic defects, to the more complex phenotypes of today. Surely, as each disorder's molecular mechanism is identified, we continually appreciate the complex interactions that occur producing a wide variation in phenotype. Indeed, phenotype similarity may be particularly difficult to appreciate, when the genetic admixture between two genetic groups, such as the Japanese and African-American populations, is relatively low, enhancing any genetic modifier effects.

Certainly phenotypic variability is commonly seen secondary to different mutation sites in the same gene. Potter and colleagues [19] have demonstrated that while several Caucasian pedigrees in the United States share a similar DRPLA haplotype with those of the Japanese, the HRS haplotype is different, suggesting that it is indeed of a different mutational origin. However, to date, no sequence differences have been seen between the HRS family and the DRPLA family in this gene, although the HRS gene has not been fully sequenced; so, haplotype differences remain a possible source of variation in this disorder. An additional line of evidence suggests that the interaction of the expanded DRPLA repeat with different genetic backgrounds may provide the richest milieu for phenotypic variation. The classification of Izuka and colleagues [18] would be compatible with a genetic modifier effect, as all of these families presumably carry the same A1-B1 DRPLA haplotype. Perhaps then, in the future, we should not be surprised to see additional orphans like HRS drawn back to their genomic family, outliers not because of their defect, but because of the different background in which they are expressed.

References

1. Farmer, T. W., Wingfield, M. S., Lynch, S. A., Vogel, F. S., Hulette, C., Katchinoff, B., and Jacobson, P. L. (1989). Ataxia, chorea, seizures, and dementia: pathologic features of a newly defined familial disorder. *Arch. Neurol.* **46,** 774–779.
2. Wolfe, S. (1994). The nursing challenge of Haw River. *RN* **57,** 43–46.
3. Farmer, T. W., Wingfield, M. S., Jacobson, P. L., Katchinoff, B. L., Lynch, S. A., and Curnes, J. T. (1991). Neuroimaging of a new familial disorder: ataxia, chorea, seizures and dementia. *J. Neuroimaging* **1,** 123–128.
4. Koide, R., Ikeuchi, T., Onodera, O., Tanaka, H., Igarashi, S., Endo, K., Takahashi, H., Kondo, R., Kshikawa, A., Hayashi, T., Saito, M., Tomoda, A., Miike, T., Naito, H., Ikuta, F., and Tsuji, S. (1994). Unstable expansion of CAG repeat in hereditary dentatorubral-pallidoluysian atrophy (DRPLA). *Nature Genet.* **6,** 9–13.
5. Nagafuchi, S., Yanagisawa, H., Sato, K., Shirayama, T., Ohsaki, E., Bundo, M., Takeda, T., Tadokoro, K., Kondo, I., Murayama, N., Tanaka, Y., Kikushima, H., Umino, K., Kurosawa, H., Furukawa, T., Nihei, K., Inoue, T., Sano, A., Komure, O., Takahashi, M., Toshizawa, T., Kanazawa, I., and Yamada, M. (1994). Dentatorubral and pallidoluysian atrophy expansion of an unstable CAG trinucleotide on chromosome 12p. *Nature Genet.* **6,** 14–18.
6. Burke, J. R., Wingfield, M. S., Lewis, K. E., Roses, A. D., Lee, J. E., Hulette, C., Pericak-Vance, M. A., and Vance, J. M. (1994). The Haw River Syndrome: dentatorubropallidoluysian atrophy (DRPLA) in an African-American family. *Nature Genet.* **7,** 521–524.
7. Burke, J. R., Ikeuchi, T., Koide, R., Tsuji, S., Yamada, M., Pericak-Vance, M. A., and Vance, J. M. (1994). Dentatorubral-pallidoluysian atrophy and Haw River syndrome. *Lancet* **344,** 1711–1712. [Letter]
8. Goldman, A., Ramsay, M., and Jenkins, T. (1995). New founder haplotypes at the myotonic dystrophy locus in southern Africa. *Am. J. Hum. Genet.* **56,** 1373–1378.
9. Deka, R., Miki, T., Shih-Jiun, Y., McGarvey, S. T., Shriver, M. D., Bunker, C. H., Raskin, S., Hundrieser, J., Ferrell, R. E., and Chakraborty, R. (1995). Normal CAG repeat variation at the DRLPA locus in world populations. *Am. J. Hum. Genet.* **57,** 508–511.
10. Singer, C. (1992). 'Haw River syndrome' or Dentato-Rubro-Pallido-Luysian Atrophy? *Arch. Neurol.* **49,** 13–14. [Letter]
11. Izuka, R., Hirayma, K. (1986). Dentato-Rubro-Pallidoluysian Atrophy. *In* and H. "Handbook of Clinical Neurology" (P. J. Vinkent, G. W. Bruyn, and H. L. Klawans, Eds.), pp. 437–443. Elsevier Science, Amsterdam.
12. Farmer, T. W. (1992). Reply: 'Haw River syndrome' or Dentato-Rubro-Pallido-Luysian Atrophy? *Arch. Neurol.* **49,** 13–14.
13. Suzuki, K. (1997). Neuropathy of non-Japanese DRPLA cases: review of reported cases in the literature. "DRPLA: From Clinical Neurology to Molecular Genetics" (S. Tsuji, A. Naitou, and S. Oyangi, Eds.), pp. 153–163. Igakusyoin, Tokyo. [in Japanese]
14. Takahata, N., Koichi, I., Yoshimura, Y., Nishihori, K., and Suzuki, H. (1979). Familial chorea and myoclonus epilepsy. *Neurology* **28,** 913–919.
15. Takahashi, H., Ohama, E., Naito, H., Takeda, S., Nakashima, S., Machifuchi, T., and Ikuta, F. (1988). Hereditary dentatoruburalpallidoluysian atrophy: clinical and pathologic variants in a family. *Neurology* **38,** 1065–1070.
16. Potter, N. T., Meyer, M. A., Zimmerman, A. W., Eisenstadt, M. I., and Anderson, I. J. (1995). Molecular and clinical findings in a family with dentatorubral-pallidoluysian. *Ann. Neurol.* **37,** 273–277.
17. Ihara, Y., Namba, R., Nobukuni, K., Kawai, K., and Kuroda, S. (1991). Relation between cerebral white matter damage on CT and MRI and clinical data in DRPLA (pseudo-Huntington form) [Japanese]. *Rinsho Shinkeigaku* **31,** 815–820. [in Japanese]
18. Iizuka, R., Hirayama, K., and Maehara, K. A. (1984). Dentatorubro-pallidoluysian atrophy, a clinico-pathological study. *J. Neurol. Neurosurg. Psych.* **47,** 1288–1298.
19. Potter, N. T., Yanagisawa, H., and Yamada, M. (1996). Different origins of expanded repeats for Haw River Syndrome and dentatorubro-pallidoluysian atrophy. *Lancet* **347,** 1271.

PART VII

Spinocerebellar Ataxia Type 1

CHAPTER 17

Clinical Aspects of Spinocerebellar Ataxia 1

S. H. SUBRAMONY AND P. J. S. VIG
Department of Neurology, University of Mississippi Medical Center, Jackson, Mississippi 39216

I. HISTORICAL INTRODUCTION

Marie [1] pointed out that not all inherited ataxias satisfied the clinical description of the familial ataxia originally described by Friedreich [2]. In particular, some of the inherited ataxias could be dominantly inherited and could have exaggerated deep tendon reflexes, as opposed to the areflexia seen with Friedreich's ataxia. In addition, such ataxias had onset later in life. The clinical and neuropathologic heterogeneity of inherited ataxias was further stressed by the work of researchers such as Holmes [3], Bell and Carmichael [4], and Greenfield [5]. Neuropathologic studies suggested that as opposed to Friedreich's ataxia, which was primarily a "spinal" ataxia, the dominant ataxias were either spinocerebellar or purely cerebellar in origin and often had "olivopontocerebellar atrophy" as their pathologic substrate. And yet, careful study of certain families with dominant ataxia, such as the Schut kindred, clearly documented the extreme variability in clinical phenotype that could be seen in such families [6]. This led to considerable difficulties on the part of clinicians to be able to distinguish Friedreich's ataxia from some of the dominant ataxias; many patients were diagnosed as having "variant Friedreich's ataxia." Other common diagnoses were olivopontocerebellar atrophy and Marie's ataxia.

In a letter to the New England Journal of Medicine in 1974, Yakura from Japan described a small family with dominant ataxia in which all affected members carried the same HLA markers as opposed to the unaffected members [7]. This was followed up by Drs. John Jackson and Robert Currier at the University of Mississippi Medical Center who established linkage to the HLA markers in a large family with dominant ataxia. The study suggested that the gene for this disease was located about 12 centimorgan from the HLA gene on chromosome 6p [8]. This was perhaps the first neurogenetic disease to be located to an autosome using classic linkage studies. This was followed by the confirmation of this locus using DNA markers by Zoghbi *et al.* and

Rich *et al.* and the eventual identification of the mutation as an expanded CAG repeat [9–11]. The original Mississippi family was shown to be linked to the same markers as well. SCA 1 is now known to be worldwide in occurrence.

II. PREVALENCE OF SCA 1 [12–21]

Table 17-1 summarizes the prevalence of SCA 1 in several recently published series. There is considerable variability in the occurrence of SCA 1 in different series; the figures range from 6 to 50% of cases of ADCA type I and 5 to 33% of all dominant ataxias. To some extent the prevalence depends on the type of families selected for the study. In addition, there appears to be some influence of ethnicity. Thus, the British and French as well as some of the Italian series have had a high proportion of SCA 1; the German series and the ethnically mixed U.S. series have had fewer SCA 1 patients; this also was true for the series from Montreal with a predominantly Portuguese patient collection. The major differences in these figures among geographically nearby countries is of interest. In Japan, Matsumura *et al.* [20] reported no SCA 1 among 47 families with dominant ataxia in the Kinki area; however, SCA 1 is known to be not uncommon in the Hokkaido and Tohoku areas of Japan.

III. CLINICAL FEATURES

A. Age of Onset, Duration of Illness, and Age at Death [6, 12–14, 16, 18, 19, 21–28]

Table 17-2 illustrates the range of the age of onset in SCA 1 from several large series of cases. Overall the

Table 17-1 Prevalence of SCA-1 in Different Series

Series	Ethnic background	% ADCA I	% All dominant
Ranum	U.S. (mixed)		14
Giunti	English, Italian	50	
Duborg	French	15	14
Ranum	U.S. (mixed)		14
Schols	German	11	
Silveira	Portuguese		5
Burk	German		17
Filla	Italian	6	
Matsumura	Japanese		0
Illarioshkin	East European	42	33

Table 17-2 Age of Onset in SCA-1

Series	CAG repeat	Mean A/O	A/O range	No. of patients	Type of series
Ranum	42–81		4–65	113	Multifamily
Giunti	46–66	34	17–53	103	Multifamily
Duborg	42–51	33 ± 9	21–52	42	Multifamily
Sasaki	39–63	36 ± 10	15–63	48	Multifamily
Schols			32–42	8	Multifamily
Filla		36 ± 9			Multifamily
Burk	47–52	34 ± 4		9	Multifamily
Illarioshkin	41–51	33 ± 7	22–45	11	Multifamily
Goldfarb	39–72	39	18–74		Pop. isolate

reported age of onset has ranged from 6 to 74 years. The mean age of onset has varied from 26 to 44.3 years. In many series, however, the mean age of onset is between 33 and 37 years. Many of the single family series also illustrate the range one can see within the same family. Ranum *et al.* suggested that there was some intrafamilial clustering in the age of onset with some families having mainly later onset, others having earlier onset, and still others a wide range in age of onset [12]. This wider variability is reflected to some extent in the larger range of CAG repeat numbers in these families. Poor penetrance has been suggested by the occurrence of asymptomatic older women who transmitted disease to offspring among the Siberian Iakut [27].

The duration of illness from onset to death has varied from 12 to 38 years (Table 17-3). There is considerable variation in the mean duration of illness. In some recent series this was reported to be over 20 years. In an early paper on the Schut family the mean duration of illness was less than 15 years. Two studies have noted that the mean duration of illness was somewhat longer in paternally inherited cases than in maternally inherited cases. The mean age at death in the Schut family was 36.8 years; however, in the more recent families reported by Zoghbi *et al.* [24] and Matilla *et al.* [28], the mean age of death was around 54 years.

Table 17-3 Duration of Illness in Years from Onset to Death in SCA-1

Series	Age at death	Mean duration	Range of duration
Nino	47 to 72		12 to 38
Zoghbi	12 to 70		5 to 10
Schut	37 (mean)	9 to 14	
Matilla	39 to 72	25 to 36	
Sasaki		21	13 to 33

B. Presenting Symptoms

Progressive difficulties with gait are the initial symptoms in the majority of patients. Patients often have simultaneous problems with slurred speech. The imbalance may initially become manifest in especially stressful situations such as going down steps or sudden turns. A sense of dizziness and difficulty with writing may be other presenting symptoms. Diplopia is not a common feature in SCA 1. Patients who usually perform highly skillful balancing acts such as skiing may notice difficulties even earlier with such activities.

C. Neurological Signs [6, 13, 14, 18, 19, 22, 24–30]

Table 17-4 lists the prevalence of a host of neurological signs seen in several series of patients with SCA 1. Overall, the phenotype of SCA 1 is fairly uniform and the evolution of clinical signs fairly predictable, and this is in agreement with the similarity of clinical findings noted in different series. There appears to be less intra- as well as interfamilial variability of phenotype in SCA 1 as compared to SCA 2 and SCA 3, diseases which have a similar age of onset and similar neurological signs. Some of the variability noted between different series as well as within families can be explained by a combined effect of the range of the expanded allele in the affected persons and the duration of the illness at which the examination was conducted. Some variability simply reflects the varying definitions of neurological signs that investigators have employed. The neurological abnormalities reflect the multisystem abnormalities that occur in this disease, predominantly affecting a variety of motor systems. The following simply lists the signs that may be observed in this disease; the subsequent section deals with the evolution of these signs.

1. Cerebellar Signs

Progressive ataxia occurs from the beginning and is universally present. It begins with gait ataxia manifested early on by difficulty with such tasks as standing on one leg and hopping, and later by inability to tandem walk. Gait ataxia then becomes frank with a broad-based gait and reeling and is accompanied by signs of limb ataxia with abnormalities of finger to nose and heel to shin maneuvers. This is accompanied by a progressive dysarthria that has the typical scanning quality of ataxic speech. Intention tremor can be seen but is usually not as severe as one can see with an isolated pancerebellar syndrome. Postural and action tremors are uncommon [22]. Progressive ataxia eventually results in loss of ambulation. By this time speech may become nearly unintelligible.

2. Oculomotor Difficulties

Many of the oculomotor signs seen in SCA 1 reflect cerebellar disease. Nystagmus is not universally reported and probably occurs in fewer than 50% of cases but tends to be more common among patients examined early in the disease [22]. Other oculomotor signs related to cerebellar disease include saccadic pursuit and square wave jerks. As the disease progresses there is slowing of saccadic velocity followed by limitation of gaze beginning with up gaze. With evolving saccadic abnormalities, nystagmus often disappears. Other oculomotor signs that can occur, usually late in the disease, include lid retraction and stare, large pupils [22], and blepharospasm. Impaired vestibulo-ocular reflex (VOR) and op-

TABLE 17-4 Prevalence of Neurological Signs in SCA-1

	Miss	Nino	Schut	Zoghbi	Spadaro	Duborg	Khati	Giunti	Burk	Sasaki	Filla
Ataxia	100	100	89	100	100	100	100	100	100	100	100
Gaze palsy	44		89	39	77	22	25	63	33	31	38
Nystagmus	61	45	58		23					37	0
Dysphagia	50				57	51	48	41	89	54	50
Inc. DTR	83	35	100	35	94	31	46	67	89	43	13
Ext. toes	83				0	40			33	83	38
Dec. DTR	28	12		12	20	5		15	11	20	6
Dec. sens	50	35	67	35	37	51	34	37	78	40	44
EP signs	0		Rare		11	20	15	14		3	
Inc. tone	33				14				78	23	75
Amyotrophy	17	15	78	15	51	5			11	20	6
Duration	97 ± 59					6 ± 6	87		103	98	9

tokinetic nystagmus (OKN) may also be seen. Diplopia is uncommon but has been reported [22].

3. Bulbar Difficulties

Dysarthria which has already been mentioned may be aggravated by additional difficulties with both upper and lower motor neuron problems related to bulbar muscles. Mild degrees of dysphagia characterized by coughing after eating or occasional overt choking may occur even early in the disease, but more severe dysphagia, leading to complications, typically occurs later in the disease. This is usually accompanied by atrophy and fasciculations over the tongue. There is often associated atrophy of facial and masticatory muscles and perioral fasciculations, though the latter tend to be more prominent and occur earlier in Machado-Joseph disease. Some impairment of cough can be seen even early in the disease but again major difficulties with cough and clearing secretions occur later.

4. Upper Motor Neuron Signs

Brisk deep tendon reflexes have been variably reported, but in our experience they are universal early in the disease. There may or may not be associated spasticity but most of the series in the literature report a variable proportion of patients as having increased tone. Extensor plantar responses become increasingly more common with more advanced disease.

5. Peripheral Nervous System Signs

A mild peripheral polyneuropathy develops usually midway into the illness. This is typified by distal symmetric sensory loss together with the disappearance of the ankle reflexes, but loss of ankle jerks is not universal. Generalized areflexia, such as can be seen in SCA 2, SCA3, and SCA 4, is uncommon in SCA 1. Further in the disease, some amyotrophy and muscle fasciculations can be seen. While some of the amyotrophy may be related to motor neuron or peripheral nerve pathology, other factors may include nutritional deprivation due to dysphagia as well as weight loss that may not be readily explained by known mechanisms.

6. Extrapyramidal Signs

Extrapyramidal signs tend to occur late in the disease in chair-bound patients and take the form of dystonic posturing and choreiform or athetoid movements. Parkinsonian features, pill-rolling tremor, and early occurrence of bradykinetic-rigid syndromes are not seen in SCA 1; such features suggest Machado-Joseph disease.

7. Other Signs

Clinically significant visual failure does not occur in SCA 1, and this sets it apart from SCA 7. Rare patients have had visual loss [31]. Many reports mention the occurrence of optic disc pallor and the occurrence of large pupils. A recent report has documented some visual loss, color vision defects, and reduced corneal endothelial cell density in patients with SCA 1 [32]. Some form of cognitive problems has been reported in anywhere from 5 to 73% of patients [14, 16, 21]. Kish *et al.* noted that despite advanced ataxia, most patients are able to take care of their personal hygiene and their immediate living surroundings [33]. The type of abnormalities noted on mental status examination included inappropriate laughing, easy tearfulness, reduced attention, impulsivity, and mild to moderate reduction of recent and remote memory. No language disturbance occurred. There were no apraxias or agnosias. Formal neuropsychological tests revealed significant declines compared to normals in a number of spheres involving verbal and nonverbal intelligence and frontal as well as memory systems, correlated with the severity of cerebellar ataxia. However, the cognitive decline did not have the same severity or profile that one encounters in Alzheimer's disease. Prominent cognitive problems were also noted among the childhood onset cases in the Texas kindred [24].

Some abnormalities of bladder function have been reported in less than 50% of cases [14, 18, 22]. This typically occurs in the later stages of the illness. Orthostatic hypotension occurred in 17% of patients in one series [22].

D. Evolution of the Neurological Signs

As one can see, a wide array of neurological abnormalities occurs in these patients. Typically, early disease is characterized by gait ataxia, dysarthria, brisk deep tendon reflexes, and mild limb ataxia. Nystagmus may be seen but saccade velocity is often normal. Ocular versions tend to be full. Later stages are characterized by increasing severity of limb and gait ataxia as well as dysarthria. Mild dysphagia develops as well as some reduction in the forcefulness of cough. Reflexes continue to be brisk in a generalized fashion. Saccades become slow and other abnormalities of oculomotor function such as pursuit abnormalities and hypo- and hypermetric saccades may be seen. An up-gaze palsy develops. Ambulation becomes increasingly impaired, leading to a chair-bound status usually 10 years after onset. Signs of a mild peripheral polyneuropathy are seen with loss of vibration sense distally in the legs and the later loss of ankle reflexes. Extensor plantar responses occur and some spasticity may be seen. Patients who are chair bound appear severely ataxic, often tending to slump in one direction, have severe dysarthria, assume dystonic postures with their arms and legs, and may have some general increase in tone. Some muscle wasting and fasciculations are seen and mild cognitive abnormalities are recognized. Bladder dysfunction becomes apparent at this stage and dysphagia becomes severe. Ocular motility becomes more and more impaired. There is facial and often tongue atrophy and cough is severely reduced, leading to frequent aspiration. This, together with a degree of respiratory muscle involvement, results in terminal events.

The rate of disability progression has been correlated with the number of CAG repeats in the expanded allele [34]. However, the occurrence of increasingly severe ataxia as well as many other associated clinical signs become more frequent with longer duration of disease. Some clinical signs such as facio-lingual atrophy, dysphagia, and skeletal muscle atrophy appear to have an independent correlation with the size of the repeat apart from the duration of the illness [22, 27]. Ophthalmoparesis may also be related to larger repeat size [27]. Patients with over 52 repeats become significantly disabled by 5 years after onset and this may be related to the motor neuron pathology that may be correlated with the large repeats. Homozygous individuals appear not to have any more severe disease than can be predicted by their larger alleles [27].

E. Differentiation from Other Dominant Ataxias

It is often impossible to clinically differentiate patients with SCA 1 from those with other types of dominant ataxias. This is especially true for the differential diagnosis between SCA 1, SCA 2, and SCA 3. SCA 5 and SCA 6 tend to have a slower progression rate, often compatible with a relatively normal life span, and the signs are often "purely cerebellar" with few added signs reflecting more widespread neuronal pathology. In general, the phenotype of SCA 1 is more homogeneous than that of SCA 2 and SCA 3 (Machado-Joseph disease, MJD). Patients with MJD often have prominent extrapyramidal signs early in the disease and on occasion exhibit little ataxia. Other signs that can occur with greater frequency or earlier in MJD include nystagmus, gaze palsy, and abnormalities of vestibulo-ocular reflex. Also, patients with MJD with a later onset may have ataxia and a severe polyneuropathy characterized by generalized areflexia and amyotrophy.

Similarly, patients with SCA 2 tend to develop early and severe abnormalities of saccade velocity as well as signs of polyneuropathy with loss of deep tendon reflexes. Such findings tend to be different from the typical findings in SCA 1.

IV. NEUROPHYSIOLOGICAL FINDINGS

Nerve conduction studies often confirm the presence of a polyneuropathy with reduced amplitudes of sensory responses and relatively preserved motor nerve conduction studies. These abnormalities tend to become more common in well-established disease. Some authors have noted motor nerve conduction abnormalities reflective of demyelination [16], but this has not been our experience. Central motor conduction studies are abnormal even early in the disease and correlate with signs of upper motor neuron disease [16]. Abnormalities of brain stem, visual, and somatosensory evoked potentials have been described as well [35]. The longer the pathway studied, the more likely that abnormalities would be found. This suggests that the neuronal pathology may be a length-dependent "dying-back" process. In this study the abnormalities of central motor conduction were more frequent in SCA 1 than in SCA 2; the accompanying sensory-motor neuropathy was more severe in SCA 2. Oculomotor recordings show slowing of saccades; however, this was not as uniform as in SCA 2 [18].

V. NEUROIMAGING FINDINGS

Imaging studies show pontocerebellar atrophy. We have noted some degree of superior cerebellar atrophy even very early in the disease. MRI morphometry shows significant atrophy of the cerebellar vermis and hemispheres as well as the pontine base, middle cerebellar peduncles, medulla, and cervical spinal cord. Statistical analysis showed that SCA 1 and SCA 3 patients could not be distinguished on the basis of morphometry but that SCA 2 patients had a smaller pontine base and middle cerebellar peduncles [18]. However, imaging studies in general do not allow us to identify the genotype among these disorders. Patients with SCA 6 tend to have isolated cerebellar atrophy. T2 weighted images may show some hyperintensity in the middle cerebellar peduncles in some patients [19]. PET studies have shown that the mean local cerebral metabolic rate for glucose (LCMRGLC) was significantly diminished in many areas including the cerebral cortex, caudate nucleus, putamen, thalamus, cerebellar hemispheres, and brain stem [36]. These findings did not distinguish SCA 1 patients from patients with multiple system atrophy. Studies with {11^c} flumazenil (FMZ) did not reveal any major changes in ligand distribution volume, suggesting relative preservation of benzodiazepine receptors.

VI. NEUROPATHOLOGY

On gross examination, the brain reveals olivopontocerebellar atrophy. Microscopic examination reveals severe loss of cerebellar Purkinje cells, olivary neurons, and pontine neurons that project to the cerebellum. The dorsal vermis of the cerebellum shows the most severe loss of Purkinje cells [37]. Many proximal axonal swellings (torpedoes) can be seen close to the Purkinje cell layer. However, the pathology is much more extensive than simple olivopontocerebellar atrophy. In the cerebellum, there is also loss in the granule cell layer and the dentate, cell loss being more prominent in the latter [37]. The flocculonodular lobe is partially spared [37]. The brain stem reveals loss of a variety of motor neurons such as the oculomotor, facial, and hypoglossal nuclei. There is mild loss of substantia nigra cells as well as cells in the caudate and putamen. Similarly, occasional cases show neuronal loss in the thalamus and subthalamic nucleus. Cell loss has also been described in the anterior horns, gracile nucleus, and cuneate nucleus. Sural nerve biopsies show loss of myelinated fibers. These cell losses are accompanied by appropriate loss of white matter tracts revealed by pallor on myelin stains. Thus, there is prominent loss of the middle cerebellar peduncles and pallor of the cerebellar white matter and of the posterior columns as well the spinocerebellar pathways. Interestingly, the upper motor neuron signs and the abnormal central conduction studies appear not to be correlated with any pathology in the pyramidal tract. In some of the reports, the degree of gliosis appears to be more severe than the degree of neuronal loss. Glial cytoplasmic inclusion bodies similar to those observed in multiple system atrophy were found in many areas in one patient reported by Gilman *et al.* [36]. These stained positively for tau and ubiquitin. Such findings

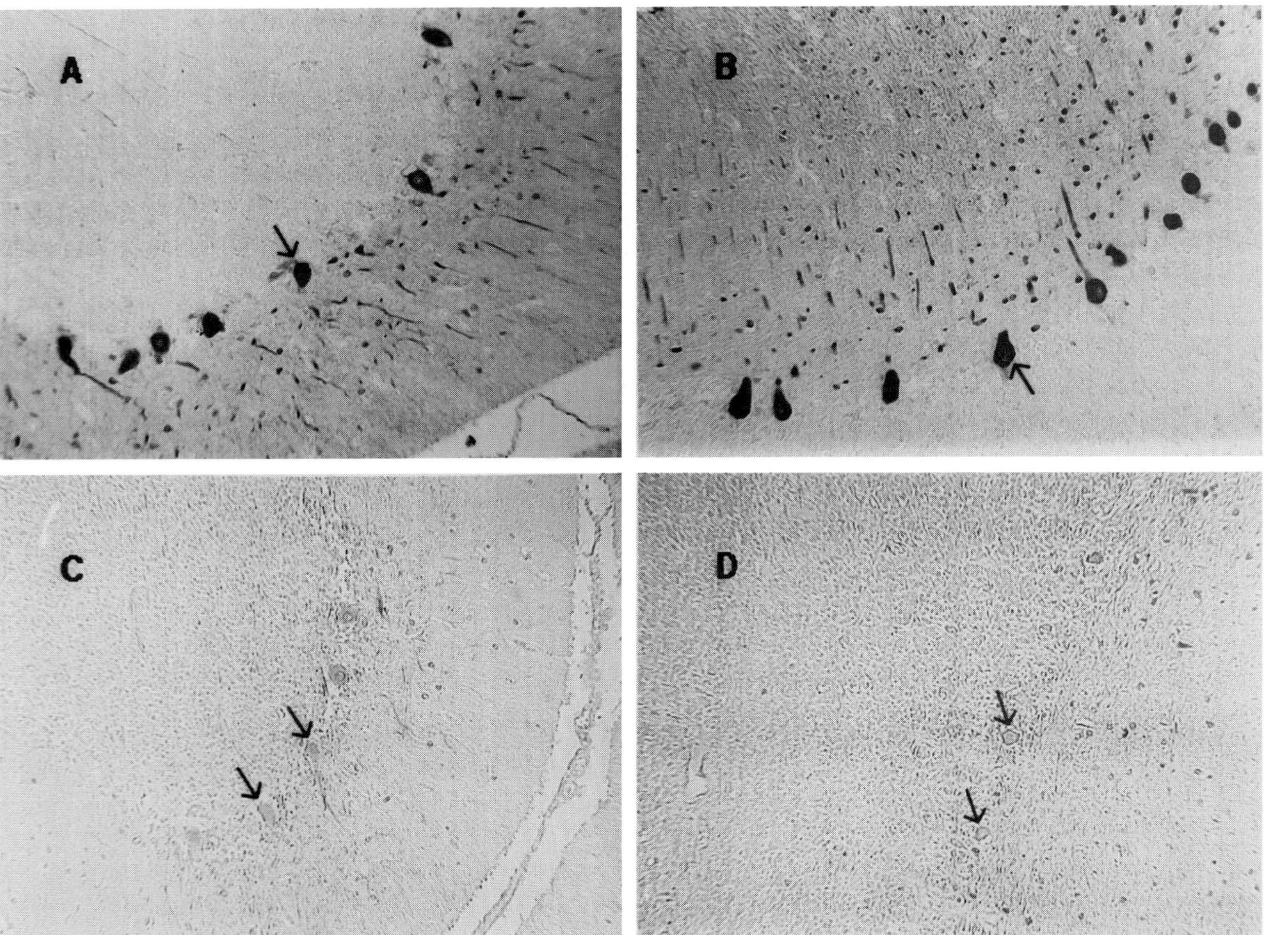

FIGURE 17-1 Parvalbumin immunocytochemistry of cerebellum. In control brains (A, B), the cerebellar Purkinje cells (arrows) exhibit a strong reaction. In cerebellum from SCA-1 patients (C, D), the surviving Purkinje cells (arrows) show little immunoreactivity to parvalbumin.

were not seen by Robitaille *et al.* in a more extensive collection of 11 brains [37]. A number of ultrastructural changes have been observed in degenerating Purkinje cells in cerebellar biopsy specimens from patients with SCA 1 [38]. These include increased electron density of the cells, the presence of myelin figures, and paucity of rough endoplasmic reticulum. There were also crystalloid inclusions and vermiform tubules with axons.

VII. NEUROCHEMICAL FINDINGS

A variety of neurochemical observations have been made in the autopsied brain tissue of patients with SCA 1. Such findings include the loss of aspartic acid in the cerebellar cortex, probably reflecting the loss of olivary neurons and the loss of GABA in the dentate, probably correlating with the loss of Purkinje cells. There were also abnormalities in the taurine content of the brain [39]. Interestingly, a severe loss of choline acetyltransferase has been reported in the cerebral cortex of patients with SCA 1; this enzyme is also diminished in the brains of patients with Alzheimer's disease [40]. Components of the α-ketoglutarate dehydrogenase complex which is involved in Kreb's cycle have been reported to be reduced in the cerebral and cerebellar cortex of SCA 1 patients [41]. This involved the E2 (dihydrolipoamide succinyl transferase) subunit of the enzyme but not the E1 or E3. Brain thiamin levels, both nonphosphorylated and phosphorylated, were not significantly reduced and the loss of E2 could not be explained by the loss of this cofactor [42]. These biochemical observations may simply reflect loss of neurons that contain these enzymes.

Immunocytochemistry for calcium binding proteins calbindin and parvalbumin has shown that the surviving Purkinje cells in the cerebella of patients with SCA 1 display reduced levels of parvalbumin immunoreactivity (Fig. 17-1). Calbindin immunoreactivity was preserved (Fig. 17-2) as was PV immunoreaction in other areas of the brain typically spared in SCA 1 [43]. These observations are of interest because abnormal function of

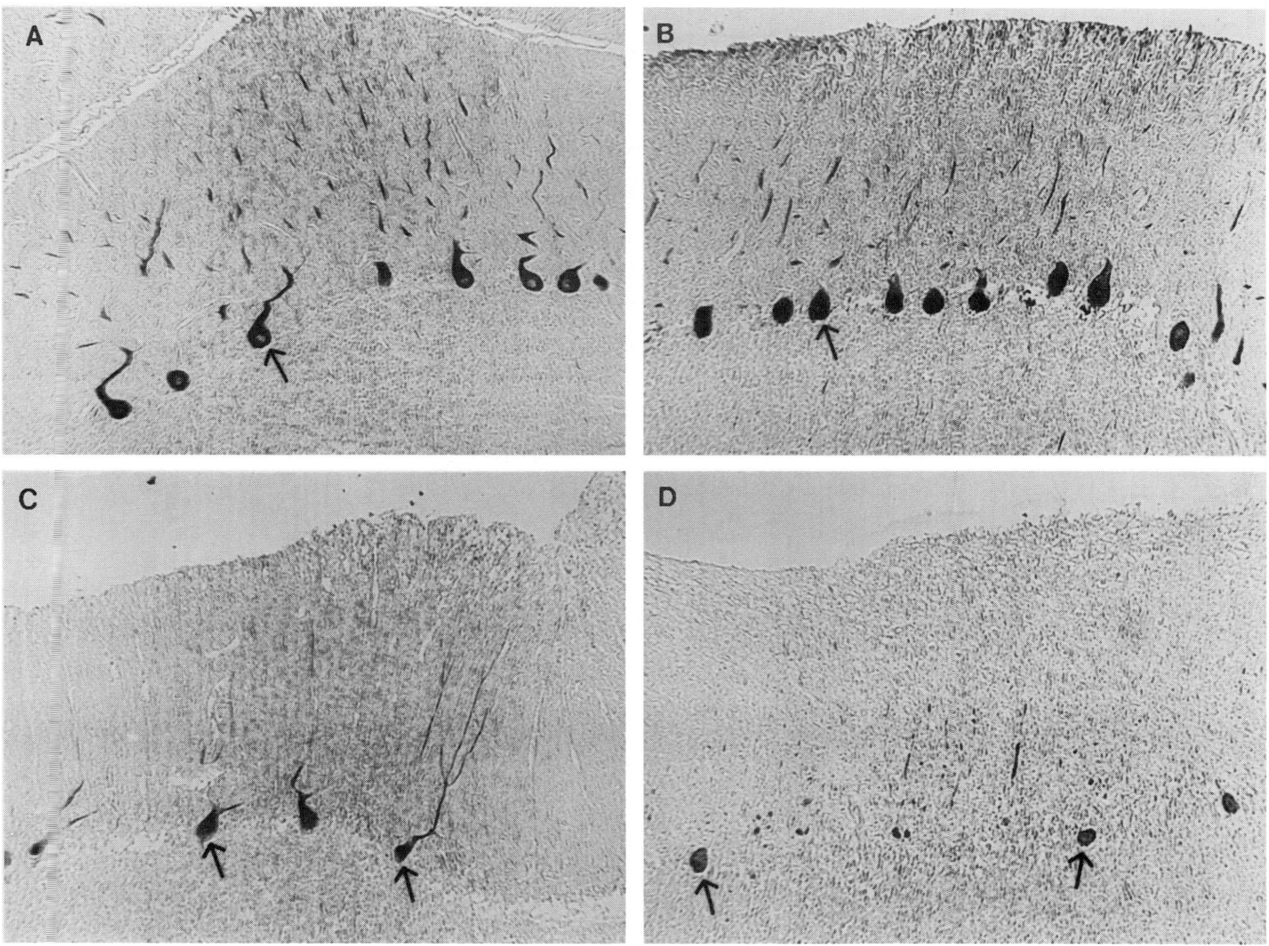

Figure 17-2 Calbindin immunocytochemistry of cerebellum. Purkinje cells (arrows) from control (A, B) and SCA-1 (C, D) brains. Both show similar immunoreactivity.

calcium binding proteins may result in susceptibility to neuronal death. We have recently been able to duplicate some of these findings in the Purkinje cells of SCA 1 transgenic mice [44].

References

1. Marie, P. (1893). Sur l'heredo-ataxie cerebelleuse. *Sem. Med. (Paris)* **13,** 444–447.
2. Friedreich, N. (1863). Uber degenerative Atrophic der spinalen Hinterstrange. *Virchows Arch. Pathol. Anat. Physiol.* **26,** 391–419, 433–459; **27,** 1–26.
3. Holmes, G. (1907). An attempt to classify cerebellar disease with a note on Marie's hereditary cerebellar ataxia. *Brain* **30,** 466–489.
4. Bell, J., and Carmichael, E. A. (1939). On hereditary ataxia and spastic paraplegia. *In* "The Treasury of Human Inheritance," Vol. IV, Part III, pp. 141–281. University Press, Cambridge.
5. Greenfield, J. G. (1954). "The Spino-Cerebellar Degenerations." Blackwell Scientific, Oxford.
6. Schut, J. W. (1950). Hereditary ataxia: clinical study through six generations. *Arch. Neurol. Psych.* **63,** 535–568.
7. Yakura, H., Wakisaka, A., Fujimoto, S., and Itakura, K. (1974). Hereditary ataxia and HL-A genotypes. *N. Engl. J. Med.* **291,** 154–155.
8. Jackson, J. D., Currier, R. D., Terasaki, P. I., and Morton, N. E. (1977). Spinocerebellar ataxia and HLA linkage. Risk prediction by HLA typing. *N. Engl. J. Med.* **296,** 1138–1141.
9. Zoghbi, H. Y., Sandkuyl, L. A., Ott, J., Daiger, S. P., Pollack, M., O'Brian, W. E., and Beaudet, A. L. (1989). Assignment of autosomal dominant spinocerebellar ataxia (SCA1) centrometric to the HLA region on the short arm of chromosome 6, using multilocus linkage analysis. *Am. J. Hum. Genet.* **44,** 255–263.
10. Rich, S. S., Wilkie, P., Schut, L., Vance, G., and Orr, H. T. (1987). Spinocerebellar ataxia: localization of an autosomal dominant locus between two markers on human chromosome 6. *Am. J. Hum. Genet.* **41,** 524–531.
11. Orr, H. T., Chung, M-y, Banfi, S., Kwiatkowski, T. J., Jr., Servadio, A., Beaudet, A. L., McCall, A. E., Duvick, L. A., Ranum, L. P. W., and Zoghbi, H. Y. (1993). Expansion of an unstable trinucleotide (CAG) repeat in spinocerebellar ataxia type I. *Nature Genet.* **4,** 221–226.
12. Ranum, L. P. W., Chung, M-y, Banfi, S., Bryer, A., Schut, L. J., Ramesar, R., Duvick, L. A., McCall, A. E., Subramony, S. H., Goldfarb, L., Gomez, C., Sandkuijl, L. A., Orr, H. T., and Zoghbi, H. Y. (1994). Molecular and clinical correlations in spinocerebellar ataxia type I (SCA-1): evidence for familial effects on age of onset. *Am. J. Hum. Genet.* **55,** 244–252.
13. Giunti, P., Sweeney, M. G., Spadaro, M., Jodice, C., Novelletto, A., Malaspina, P., Frontali, M., and Harding, A. E. (1994). The trinucleotide repeat expansion on chromosome 6p(SCA1) in autosomal dominant cerebellar ataxias. *Brain* **117,** 645–649.
14. Dubourg, O., Durr, A., Cancel, G., Stevanin, G., Chneiweiss, H., Agid, Y., and Brice, A. (1995). Analysis of the SCA-1 CAG repeat in a large number of families with dominant ataxia: clinical and molecular correlations. *Ann. Neurol.* **37,** 176–180.
15. Ranum, L. P. W., Lundgren, J. K., Schut, L. J., Ahrens, M. J., Perlman, S., Bird, T. D., Gomez, C., and Orr, H. T. (1995). Spinocerebellar ataxia type 1 and Machado-Joseph disease: incidence of CAG expansions among adult-onset ataxia patients form 311 families with dominant, recessive or sporadic ataxia. *Am. J. Hum. Genet.* **57,** 603–608.
16. Schols, L., Riess, O., Schols, S., Zeck, S., Amoiridis, G., Langkafel, M., Epplen, J. T., and Przuntek, H. (1995). Spinocerebellar ataxia type 1: clinical and neurophysiological characteristics in German kindreds. *Acta Neurol. Scand.* **92,** 478–485.
17. Silveira, I., Lopes-Cendes, I., Kish, S., Maciel, P., Gaspar, C., Coutinho, P., Botez, M. I., Teive, H., Arruda, W., Steiner, C. E., Pinto-Junior, W., Maciel, J. A., Jerin, S., Sack, G., Andermann, E., Sudarsky, L., Rosenberg, R., MacLeod, P., Chitayat, D., Babul, R., Sequeiros, J., and Rouleau, G. A. (1996). Frequency of spinocerebellar ataxia type 1, dentatorubropallidoluysian atrophy, and Machado-Joseph disease mutations in a large group of spinocerebellar ataxia patients. *Neurology* **46 (1),** 214–218.
18. Burk, K., Abele, M., Fetter, M., Dichgans, J., Skalej, M., Laccone, F., Didierjean, O., Brice, A., and Klockgether, T. (1996). Autosomal dominant cerebellar ataxia type 1: clinical features and MRI in families with SCA1, SCA2 and SCA3. *Brain* **119,** 1497–1505.
19. Filla, A., De Michele, G., Campanella, G., Perretti, A., Santoro, L., Serlenga, L., Ragno, M., Calabrese, O., Castaldo, I., De Joanna G., and Cocozza, S. (1996). Autosomal dominant cerebellar ataxia type I. Clinical and molecular study in 36 Italian families including a comparison between SCA1 and SCA2 phenotypes. *J. Neurol. Sci.* **142,** 140–147.
20. Matsumura, R., Takaganaagi, T., Murata, K., Futamura, N., and Fujimoto, Y. (1996). Autosomal dominant cerebellar ataxia in the Kinki area of Japan. *Jpn. J. Hum. Genet.* **41**(4), 399–406.
21. Illarioshkin, S. N., Slominsky, P. A., Ovchinnikov, I. V., Markova, E. D., Miklina, N. I., Klyushnikov, S. A., Shadrina, M., Vereshchagin, N. V., Limborskaya, S. A., and Ivanova-Smolenskaya, I. A. (1996). Spinocerebellar ataxia type 1 in Russia. *J. Neurol.* **243,** 506–510.
22. Sasaki, H., Fukazawa, T., Yanagihara, T., Hamada, T., Shima, K., Matsumoto, A., Hashimoto K., Ito, N., Wakisaka, A., and Tashiro, K. (1996). Clinical features and natural history of spinocerebellar ataxia type 1. *Acta Neurol. Scand.* **93,** 64–71.
23. Keats, B. J. B., Pollack, M. S., McCall, A., Wilensky, M. A.. Ward, L. J., Lu, M., and Zoghbi, H. Y. (1991). Tight linkage of the gene for spinocerebellar ataxia to D6S89 on the short arm of chromosome 6 in a kindred for which close linkage to both HLA and F13A1 is excluded. *Am. J. Hum. Genet.* **49,** 972–977.
24. Zoghbi, H. Y., Pollack, M. S., Lyons, L. A., Ferrell, R. E., Daiger, S. P., and Beaudet, A. L. (1988). Spinocerebellar ataxia: variable age of onset and linkage to human leukocyte antigen in a large kindred. *Ann. Neurol.* **23,** 580–584.
25. Nino, H. E., Noreen, H. J., Dubey, D. P., Resch, J. A., Namboodiri, K., Elston, R. C., and Yunis, E. J. (1980). A family with hereditary ataxia: HLA typing. *Neurology* **30,** 12–20.
26. Goldfarb, L. G., Chumakov, M. P., Petrov, P. A., Fedorova, N. I., and Gajdusek, D. C. (1989). Olivopontocerebellar atrophy in a large Iakut kinship in Eastern Siberia. *Neurology* **39,** 1527–1530.
27. Goldfarb, L. G., Vasconcelos, O., Platonov, F. A., Lunkes, A., Kipnis, V., Kononova, S., Chabrashvili, T., Vladimirtsev, V. A., Alexeev, V. P., and Gajduselk, D. C. (1996). Unstable triplet repeat and phenotypic variability of spinocerebellar ataxia type 1. *Ann. Neurol.* **39,** 500–506.
28. Matilla, T., Volpini, V., Genis, D., Rosell, J., Corral, J., Davalos, A., Molins, A., and Estivill, X. (1993). Presymptomatic analysis of spinocerebellar ataxia type 1 (SCA1) via the expansion of the SCA1 CAG repeat in a large pedigree displaying anticipation and parental male bias. *Human Mol. Genet.* **2**(12), 2123–2128.
29. Spadaro, M., Giunti, P., Lulli, P., Frontali, M., Jodice, C., Cappellacci, S., Morellini, M., Persichetti, F., Trabace, S., and Anastasi, R. (1992). HLA-linked spinocerebellar ataxia: a clinical and genetic study of large Italian kindreds. *Acta Neurol. Scand.* **85,** 257–265.

30. Khati, C., Stevanin, G., Durr, A., Chneiweiss, H., Belal, S., Seck, A., Cann, H., Brice, A., and Agid, Y. (1993). Genetic heterogeneity of autosomal dominant cerebellar ataxia type 1. Clinical and genetic analysis of 10 French families. *Neurology* **43**(6), 1131–1137.
31. Klostermann, W., Zuhlke, C., Heide, W., Kompf, D., and Wessel, K. (1997). Slow saccades and other eye movement disorders in spinocerebellar atrophy type 1. *J. Neurol.* **244**(2), 105–111.
32. Abe, T., Abe, K., Aoki, M., Itoyama, Y., and Tamai, M. (1997). Ocular changes in patients with spinocerebellar degeneration and repeated trinucleotide expansion of spinocerebellar ataxia type 1 gene. *Arch. Ophthalmol.* **115,** 231–236.
33. Kish, S. J., el-Awar, M., Schut, L., Leach, L., Oscar-Berman, M., and Freedman, M. (1988). Cognitive deficits in olivopontocerebellar atrophy: implications for the cholinergic hypothesis of Alzheimer's dementia. *Ann. Neurol.* **24,** 200–206.
34. Jodice, C., Malaspina P., Persichettie, F., Novelletto, A., Spadaro, M., Giunti, P., Morocutti, C., Terrenato, L., Harding, A. E., and Frontali, M. (1994). Effect of trinucleotide on repeat length and parental sex on phenotypic variation in spinocerebellar ataxia I. *Am. J. Hum. Genet.* **54,** 959–965.
35. Perretti, A., Santoro, L., Lanzillo, B., Filla, A., De Michele, G., Barbieri, F., Martino G., Ragno, M., Cocozza, S., and Caruso, G. (1996). Autosomal dominant cerebellar ataxia type I: multimodal electrophysiological study and comparison between SCA1 and SCA2 patients. *J. Neurol. Sci.* **142**(1–2), 45–53.
36. Gilman, S., Sima, A. A., Junck, L., Kluin, K. J., Koeppe, R. A., Lohman, M. E., and Little R. (1996). Spinocerebellar ataxia type 1 with multiple system degeneration and glial cytoplasmic inclusions. *Ann. Neurol.* **39,** 241–255.
37. Robitaille, Y., Schut, L., and Kish, S. J. (1995). Structural and immunocytochemical features of olivopontocerebellar atrophy caused by the spinocerebellar ataxia type 1 (SCA-1) mutation defines a unique phenotype. *Acta Neuropathol.* **90,** 572–581.
38. Landis, D. M. D., Resenberg, R. N., Landis, S. C., Schut, L., and Nylan, W. L. (1974). Olivopontocerebellar degeneration. Clinical and ultrastructural abnormalities. *Arch. Neurol.* **31,** 295–307.
39. Perry, T. L., Hansen, S., Currier, R. D., and Berry, K. (1978). Abnormalities in neurotransmitter amino acids in dominantly inherited cerebellar disorders. *Adv. Neurol.* **21,** 303–314.
40. Kish, S. J., Currier, R. D., Schut, L. J., Perry, T. L., and Morito, C. L. (1987). Brain choline acetyltransferase reduction in dominantly inherited olivopontocerebellar atrophy. *Ann. Neurol.* **22,** 272–275.
41. Mastrogiacomo, F., LaMarche, J., Dozic, S., Lindsay, G., Bettendorff, L., Robitaille, Y., Schut, L., and Kish, S. J. (1996). Immunoreactive levels of alpha-ketoglutarate dehydrogenase subunits in Friedreich's ataxia and spinocerebellar ataxia type 1. *Neurodegeneration* **5,** 27–33.
42. Bettendorff, L., Mastrogiacomo, F., LaMarche, J., Dozic, S., and Kish, S. J. (1996). Brain levels of thiamine and its phosphate esters in Friedreich's ataxia and spinocerebellar ataxia type 1. *Movement Disorders* **11**(4), 437–439.
43. Vig, P. J. S., Fratkin, J. D., Desaiah, D., Currier, R. D., and Subramony, S. H. (1996). Decreased parvalbumin immunoreactivity in surviving Purkinje cells of patient with spinocerebellar ataxia-1. *Neurology* **47,** 249–253.
44. Vig, P. J. S., Subramony, S. H., Burright, E. N., Fratkin, J. D., McDaniel, D. O., Desaiah, D., and Qin, Z. (1998). Reduced immunoreactivity to calcium-binding proteins in Purkinje cells precedes onset of ataxia in spinocerebellar ataxia-1 transgenic mice. *Neurology* **50,** 1–8.

CHAPTER 18

Clues about the Pathogenesis of SCA1 from Biochemical and Molecular Studies of Ataxin-1

BEENA T. KOSHY AND ANTONI MATILLA Departments of Pediatrics and Molecular and Human Genetics, Baylor College of Medicine, Houston, Texas 77030

HUDA Y. ZOGHBI Departments of Pediatrics and Molecular and Human Genetics, Baylor College of Medicine and Howard Hughes Medical Institute, Houston, Texas 77030

I. INTRODUCTION

Spinocerebellar ataxia type 1 (SCA1) is a dominantly inherited neurodegenerative disorder characterized by ataxia, dysarthria, progressive motor dysfunction, and dysphagia. In SCA1, there is selective degeneration of cerebellar Purkinje cells, neurons of cranial nerve nuclei, and inferior olive neurons [1]. The mutational basis of SCA1 is an expansion of a polymorphic CAG trinucleotide repeat which lies in the coding region of the gene [2]. Normal alleles contain 6–44 repeats whereas expanded alleles contain 40 or more repeats [3, 4]. Alleles with 21 or more repeats are interrupted by 1–3 CAT trinucleotide units encoding histidine. In contrast, SCA1 alleles contain a pure CAG tract [5, 6]. The CAG repeat is predicted to encode a glutamine tract in the protein ataxin-1. The length of the glutamine tract correlates inversely with the age of onset and directly with severity of the disease [7]. In this chapter, we summarize data on the characterization of the *SCA1* gene product, ataxin-1, and report the relevance of these data to understanding the molecular basis of the selective neurodegeneration in SCA1.

II. THE *SCA1* GENE PRODUCT

Ataxin-1 is a novel protein which does not share homologies with any other protein, nor does it contain any motifs that might provide insight about its function. Wild-type ataxin-1 is predicted to encode 792–829 amino acids depending on the number of glutamines [8]. The mouse homologue of the *SCA1* gene has been characterized and the murine protein is highly homologous to the human protein (89% identity) [8]. Of interest is the finding that the glutamine tract is virtually absent in the mouse and in its place there are two glutamines and three prolines [9].

A. Expression Patterns of Ataxin-1

Two antibodies were raised against different regions of ataxin-1, one to the C-terminal 269 amino acids [7] and the other to a peptide containing 34 residues (amino acids 164–197) near the N-terminus of the protein [10]. The C-terminal antibody, designated 11750V, detects a protein that has an electrophoretic mobility of 100 kDa on immunoblots containing protein extracts from lymphoblasts of unaffected individuals [7]. The wild-type human protein is predicted to be approximately 87 kDa; however, it clearly has an altered electrophoretic mobility which is most likely due to the polyglutamine tract. Murine ataxin-1 which lacks the glutamine tract migrates at a molecular weight corresponding to its predicted size. Similar findings were also observed for other polyglutamine-containing proteins implicated in neurodegeneration [11, 12]. Using the 11750V antiserum, lymphoblast extracts from SCA1 patients revealed the presence of two proteins, the 100-kDa wild-type form as well as a second protein of slower electrophoretic mobility corresponding to the mutant protein. The size of the protein translated from the mutant allele varied according to the size of the repeat expansion in the respective patients (Fig. 18-1). These data demonstrate that the expanded alleles are translated in lymphoblasts.

Immunoblot analysis using brain tissue extracts of SCA1 patients showed that both wild-type and mutant forms of ataxin-1 were present in the regions tested including the cerebellum (degenerates in SCA1) and

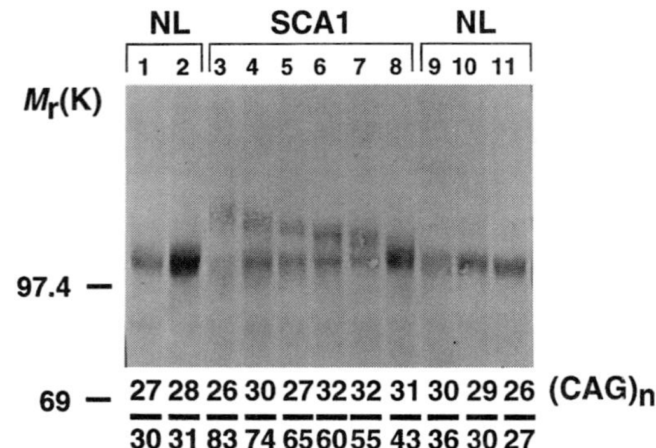

FIGURE 18-1 Expression analysis of ataxin-1 in lymphoblasts from normal (NL) and SCA1 individuals. Lanes 1, 2, and 9–11 contain protein extracts from normal individuals. Lanes 3–8 contain protein extracts from SCA1 patients. The size of the upper band corresponding to the mutant protein varies with the size of the CAG repeat in the SCA1 allele. The number of CAG repeats in both alleles is shown below each lane (reprinted with permission from Servadio *et al.* (1995), *Nature Genetics* **10,** 94–98).

the parietal cortex (spared in SCA1). The expression pattern of ataxin-1 was evaluated in peripheral tissues and the protein was detected in heart, skeletal muscle, liver, pancreas, and lung [7]. The level of the protein was much higher in the central nervous system tissue (~2- to 4-fold) compared to peripheral tissue. These findings contrast with the expression patterns of the *SCA1* transcript [8]. No significant differences were detected between the levels of *SCA1* mRNA in neuronal vs nonneuronal tissue. These data suggest that translational regulation may play an important role in the levels of ataxin-1 in various tissues.

B. Immunolocalization of Ataxin-1

Immunohistochemical analysis using the 11750V antiserum on brain tissue from a number of species including the rat, squirrel monkey, baboon, and humans revealed that ataxin-1 is localized to the nuclei of cortical neurons as well as the nuclei of neurons of the caudate, putamen, globus pallidus, pons, and the dentate nucleus of the cerebellum [7]. In cerebellar Purkinje cells, a distinct cytoplasmic staining was noted in addition to the strong nuclear staining. In nonneuronal tissues such as lymphoblasts, heart, skeletal muscle, and liver, the protein was localized to the cytoplasm. The immunolocalization patterns of ataxin-1 in the brain were analyzed in both normal individuals and SCA1 patients and no gross differences were detected [7].

C. Self-Association of Ataxin-1

Using the yeast two-hybrid system, Koshy *et al.* [13] demonstrated that ataxin-1 associates with itself and that this association can take place between either two wild-type molecules or two mutant ataxin-1 molecules (homodimerization), or between wild-type and mutant gene products (heterodimerization). Burright *et al.* [14] used a variety of deletion constructs in a yeast two-hybrid assay to identify the site of the association and to assess the effect of the length of the CAG repeat on the self-association. Burright and colleagues determined that the self-association of ataxin-1 is not dependent on the length of the polyglutamine tract and delineated the region responsible for self-association to amino acids 495–605 in a wild-type protein containing 30 repeats [14]. Sequence comparison of the human [2], rat [15], and mouse [9] ataxin-1 coding regions demonstrates that the self-association domain is highly conserved (>92% identity) and suggests that it may encode a functional domain of the protein.

III. STUDIES OF ATAXIN-1 IN GENETICALLY ENGINEERED MICE

A. *Sca1* Null Mice

To gain insight into the function of wild-type ataxin-1, Matilla *et al.* generated mice that lack the *Sca1* gene using standard gene targeting technology [16]. Mice nullizygous for the *Sca1* gene are viable and fertile and live a normal life span. Neither heterozygous nor homozygous mice develop ataxia as determined by cage behavior, foot print analysis, and bar crossing evaluation. These data argue that *SCA1* is not caused by a loss of function or haploinsufficiency of ataxin-1. The phenotype of *Sca1* null mice is characterized by a variety of neurobehavioral abnormalities which include decreased exploratory behavior and learning impairments.

B. Ataxin-1 Expression in *SCA1* Transgenic Mice

Transgenic animals expressing either the wild-type *SCA1* coding region with 30 CAG repeats (AO2) or the expanded protein with 82 repeats (BO5) under the control of the Purkinje specific promoter Pcp2 have been established (see Chapter 19). Mice carrying the mutant allele manifest motor learning deficiency by 6 weeks of age, and subsequently develop ataxia and Purkinje cell degeneration similar to the human disorder [17, 18]. Analysis of ataxin-1 expression in the transgenic mice was carried out using both immunohistochemistry and Western analysis. In immunohistochemical studies, using the anti-ataxin-1 antibody 11750V, strong staining of ataxin-1 is noted in the nuclei of Purkinje cells in mice carrying either the wild-type or the expanded allele. In contrast, immunoblot analysis using the same 11750V antibody failed to detect the protein encoded by the expanded allele [17], but detected human ataxin-1 derived from the transgene with 30 CAG repeats. In later experiments, homogenized cerebella from transgenic mice carrying the wild-type (AO2) or mutant alleles (BO5) on the *Sca1* null genetic background were used to prepare protein extracts either in a 2% SDS-based extraction buffer or in 7 M urea. A faint band corresponding to the mutant human ataxin-1 and a strong band corresponding to the wild-type human ataxin-1 were detected on immunoblots (Fig. 18-2a). The decrease in the intensity of the band from the mutant protein was remarkable given the levels of protein detected in BO5 mice bred over the *Sca1* null background using immunohistochemical analysis with the same antibody (Fig. 18-2b). In contrast, mutant ataxin-1 is easily detected by immunoblotting when using cerebellar extracts from patients. The main difference between the protein extracts from mice and patients is that the latter contains other unaffected neurons. These data suggest that mutant ataxin-1 has altered biochemical properties in the cell type where degeneration is prominent. Using either the N-terminal or carboxy-terminal antibodies in immunoblotting analysis, we did not detect degradation products that are specific to the mutant protein. Hence, one possible explanation for the discordance in the detection of ataxin-1 using Western versus immunohistochemical analyses is that the majority of the mutant protein aggregates and fails to enter the gels.

IV. SUBNUCLEAR LOCALIZATION OF ATAXIN-1

Skinner *et al.* characterized the subcellular localization of ataxin-1 in transgenic mice and found an interesting difference in the localization patterns of the wild-type and mutant protein in the AO2 and BO5 transgenic mice, respectively [10]. Human ataxin-1 derived from the transgene with either the 30 or 82 glutamines was localized to the nuclei but not the nucleoli of Purkinje cells. In mice carrying the wild-type gene, the protein stained diffusely throughout the nucleus as well as to multiple distinct subnuclear structures which were about 0.5 μm in size. In contrast, the mutant protein was localized to a single prominent structure (~2 μm), in addition to diffuse nuclear staining. Nuclei containing the large subnuclear structures often showed invaginations and ruffling of the nuclear membrane. The number of Purkinje cells containing the nuclear inclusion in the BO5 mice depended on the age of the animal. The fraction of cells containing the inclusion varied from 25% at 6 weeks of age to 90% at 12 weeks of age. Thus the presence of ataxin-1 in a distinct subnuclear structure preceded the onset of ataxia in the BO5 animals. Age of onset of ataxia in the BO5 animals as deduced by cage behavior is typically 12 weeks [17].

To determine if the nuclear distribution of ataxin-1 can be further characterized in a tissue culture system, Skinner *et al.* [10] examined COS-1 cells transfected with either the wild-type or mutant ataxin-1. Similar to the distribution pattern in the Purkinje cells, ataxin-1 localized to the nucleus in COS-1 cells but not to the nucleoli. Immunohistochemical analysis with the 11750V antibody revealed that 75% of cells transfected with the wild-type protein showed a diffuse staining of the nucleus in addition to numerous (typically more than 10) distinct subnuclear structures less than 1 μm in size, whereas 60% of cells transfected with the mutant protein had few (less than 5) large structures sometimes greater than 5 μm in size.

Although the role of different nuclear domains has not been well characterized, there is a large body of data, chiefly ultrastructural and immunohistochemical analyses that have shown that the mammalian cell nucleus is organized into functional domains [19, 20]. Skinner *et al.* [10] further refined the localization of ataxin-1 to various nuclear domains using antibodies to specific proteins that define various nuclear domains as molecular probes. Colocalization of ataxin-1 to various nuclear domains was assessed using several antibodies including the antibodies to the promyelocytic leukemia protein (PML) present in the promyelocytic oncogenic domains (POD) [21, 22], to SC-35, a component of the speckled domains [23], to p80 coilin, which characterizes the coiled bodies [24], and antibodies to the transcription factor BCL-6 [25], which distinguishes the transcription foci. No colocalization of either the wild-type or mutant protein was observed with antibodies to SC-35, p80 coilin, or BCL-6. But in COS-1 cells transfected with mutant ataxin-1 and doubly stained with anti-ataxin-1 and anti-PML antibodies to stain endogenous PML, an altered nuclear distribution of the PODs was noted. PODs were initially characterized using autoimmune antisera and are usually observed as small 0.5-μm structures with 10–20 foci per nucleus depending on the cell type. The POD is a macromolecular complex made up of different proteins. Three of the proteins have been identified, promyelocytic leukemia (PML), sp100, and NDP52. The PML protein identifies the POD and is associated with a translocation t(15;17) that results in acute promyelocytic leukemia (APL). In APL, the translocation results in a fusion of the PML protein to retinoic acid receptor (PML–RARα) with a disruption of the POD's nuclear staining pattern to a micropunctate pattern. Of interest is the fact that treatment with retinoic acid restores the PODs to their normal distribution and leads to a remission of APL. COS-1 cells transfected with the mutant ataxin-1 altered the distribution of the PODs. The typical 10–20 foci were replaced by a large staining focus that localized with the structures, formed by mutant ataxin-1. Occasionally, the wild-type protein was observed to colocalize with PML but it had no effect on the nuclear distribution of PODs [10]. The rare colocalization of wild-type ataxin-1 and PML occurs by chance given the abundance of both proteins in the nucleus.

Skinner and colleagues carried the localization studies a step further by examining if either the wild-type or mutant ataxin-1 was localized to the nuclear matrix since PML is known to associate with the matrix. The nuclear matrix (NM) was first described by Berezney and Coffey [26] and is operationally defined as the salt-resistant nonchromatin scaffolding of the nucleus. The NM holds most nuclear RNA and organizes the chromatin into loops and is believed to play an important role in organizing nucleic acid metabolism [27]. Nuclear matrices were prepared from ataxin-1-transfected COS-1 cells using *in situ* detergent extraction, DNase1 digestion, and serial salt extractions. The localization of the protein was examined with the 11750V antibody. Both the wild-type and mutant forms of ataxin-1 were noted to be attached to the nuclear matrix by immunohistochemistry. Skinner and collaborators [10] confirmed this by biochemical isolation of the nuclear matrix from both ataxin-1-transfected COS-1 cells and from the cerebella of *SCA1* transgenic animals. Western analysis of the isolated nuclear matrix revealed the presence of ataxin-1 [1].

Data from *SCA1* transgenic and *Sca1* null mice demonstrate that *SCA1* is caused by a gain of function mechanism [16, 17]. The new data on subcellular localization of ataxin-1 provide new insight and argue that in *SCA1* the mechanism of pathogenesis is possibly a dominant negative effect on another protein(s) that is perhaps retained in the large subnuclear structures that are formed by the mutant protein.

Strong evidence that subnuclear structures are linked to the pathogenesis in SCA1 comes from immunohistochemical analysis of brain tissue from SCA1 patients. A large single structure was seen in brain stem neurons of three SCA1 patients using either the N-terminal peptide antibodies or the 11750V antiserum (Fig. 18-3a). The subnuclear structures also stained strongly with anti-ubiquitin antibodies (Fig. 18-3b) [10]. At the present time, it is not clear whether the ubiquitin staining is evidence that ataxin-1 is targeted for degradation. Further, it is also unclear whether the presence of ubiquitin in the structures is only a signal for degradation or whether it has other unknown consequences independent of its role in the proteolytic pathway [28]. Recently,

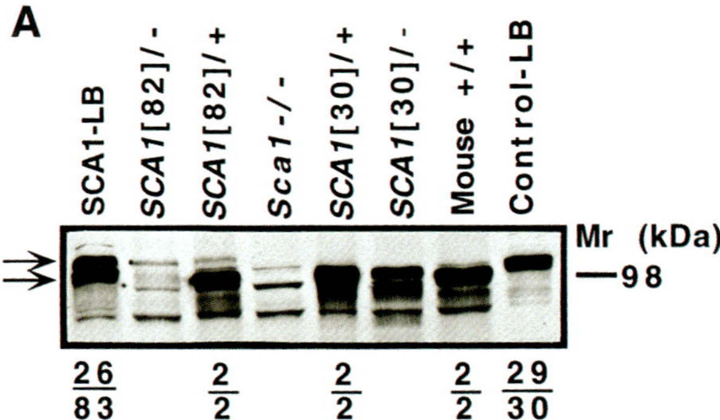

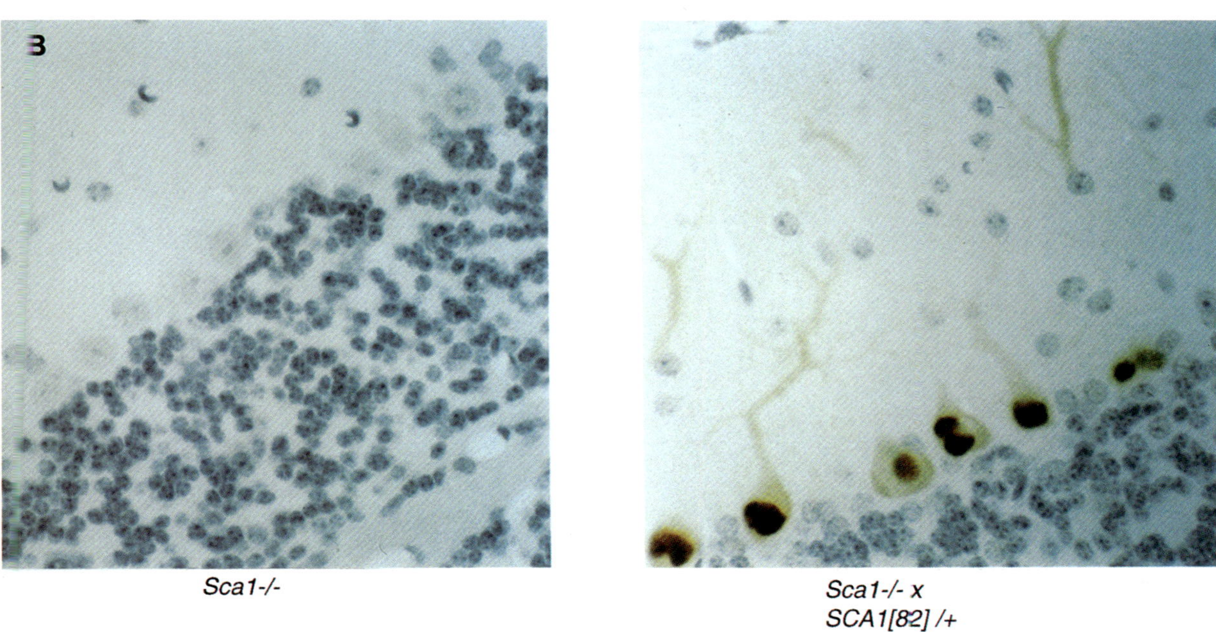

FIGURE 18-2 (a) Analysis of ataxin-1 expression in cerebella of *SCA1* transgenic mice. Depicted above each lane is the genotype of the mouse used to prepare cerebellar protein extracts. *SCA1*[82]/+ refers to the transgenic animal carrying the expanded allele with 82 CAG repeats in a wild-type background, *SCA1*[82]/- refers to the same transgenic animal but bred over the *Sca1* null background, *SCA1*[30]+ is transgenic for the *SCA1* coding region with 30 CAG repeats in a wild-type background, *SCA1*[30]/- is the same transgenic over the *Sca1* null background, -/- refers to *Sca1* null mice, +/+ refers to wild-type mouse, SCA1-LB refers to lymphoblast extract from an SCA1 patient and control-LB refers to lymphoblast extract from a normal individual. Beneath each lane is represented the CAG repeats in both alleles of the endogenous *SCA1* gene. (b) Immunolocalization of ataxin-1 in a paraffin-embedded section of the cerebellum from the *SCA1* transgenic mouse carrying the expanded allele with 82 CAG repeats over the *Sca1* null background. Intense staining is observed in the nucleus of the Purkinje cells with the anti-ataxin-1 antibody, 11750V. A much weaker staining is noted in the cytoplasm and the processes. No staining is seen in a corresponding section from an *Sca1* null mouse. The data in (a) and (b) demonstrate the disparity between ataxin-1 levels detected by Western analysis and immunohistochemistry in transgenic mice carrying the mutant *SCA1* allele. Original magnification 400×.

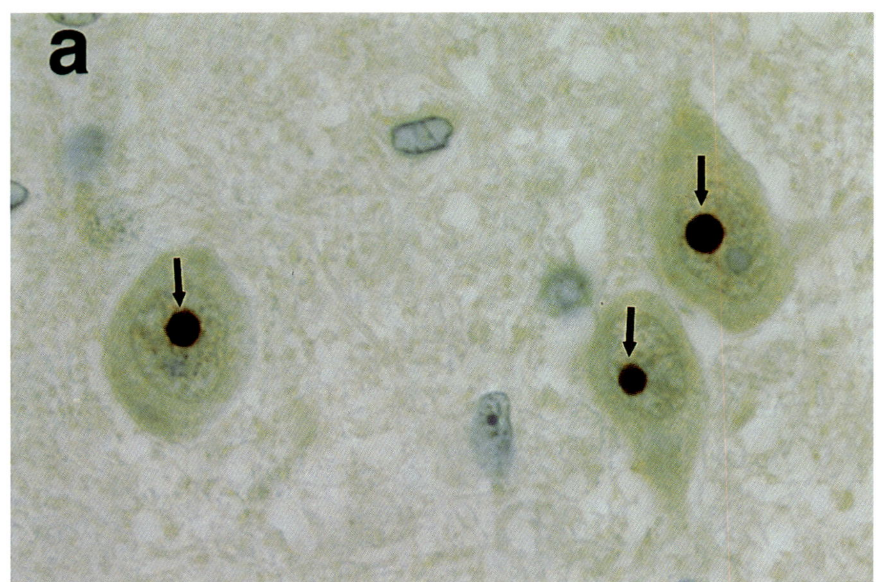

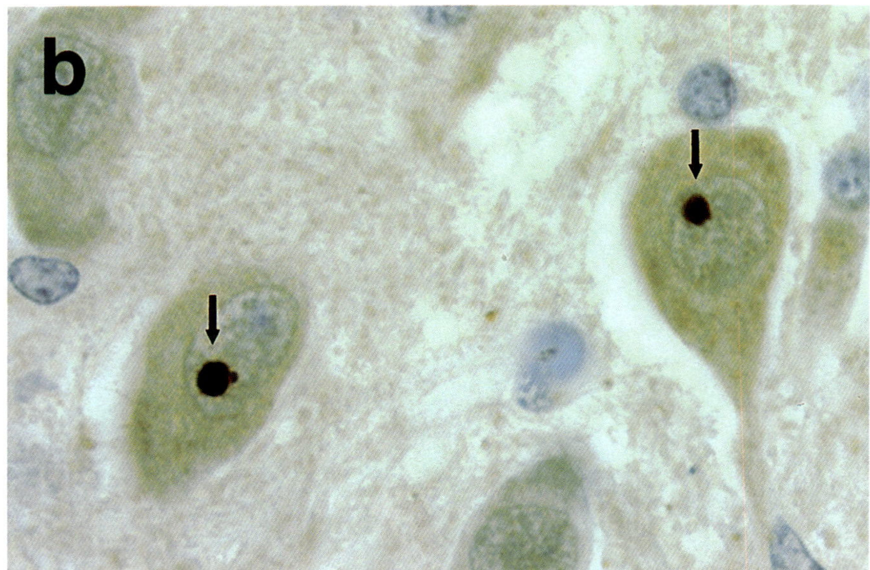

FIGURE 18-3 Immunohistochemical staining of pontine neurons from an SCA1 patient using anti-ataxin-1 (a) and anti-ubiquitin (b) antibodies. Intense staining for both ataxin-1 and ubiquitin is noted in a single large nuclear structure (arrows). Original magnification 2200×. Reprinted with permission from *Nature* **389**, 971–974. Copyright 1997 Macmillan Magazines Limited.

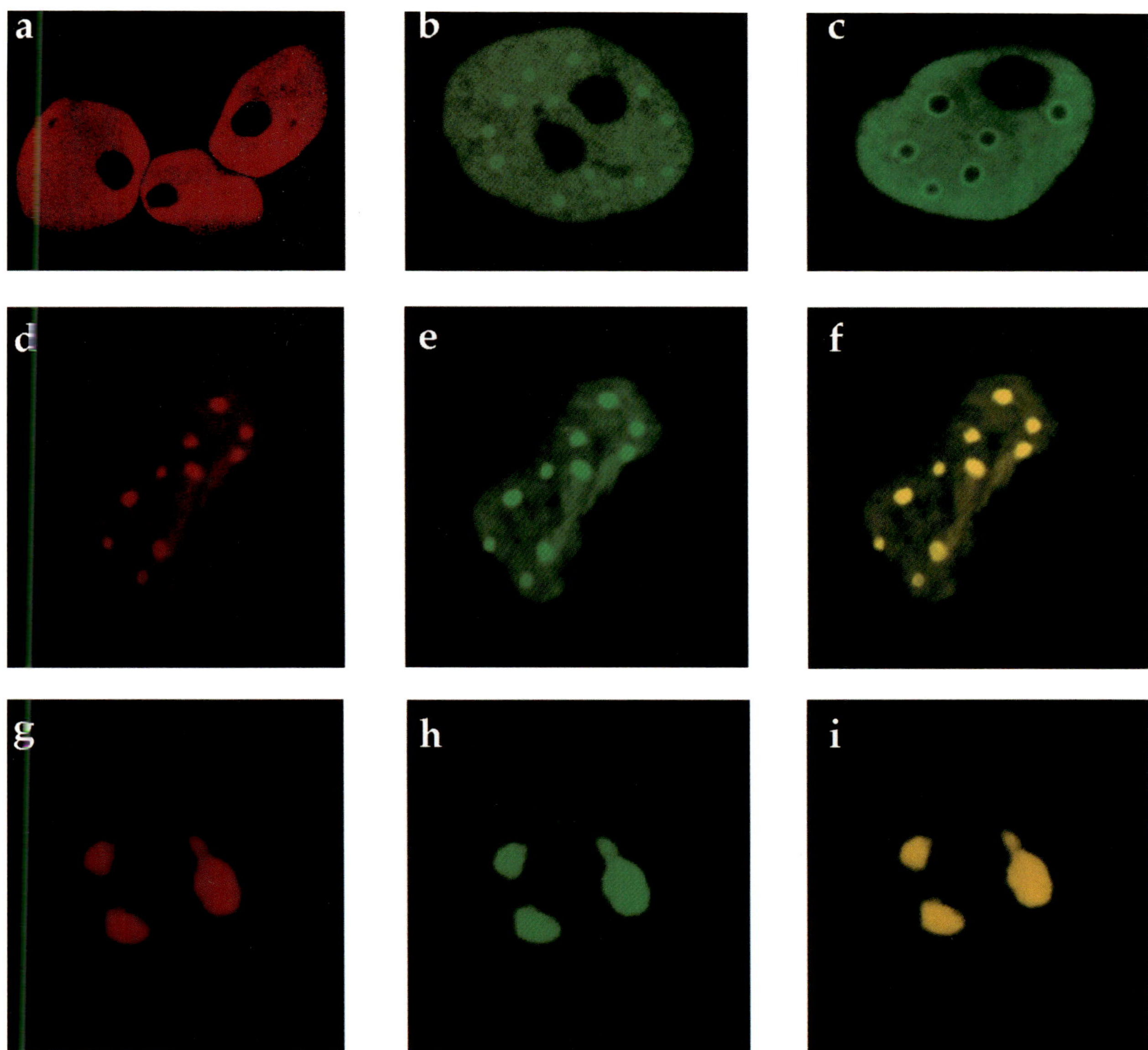

FIGURE 18-4 Co-localization of LANP and ataxin-1 in subnuclear structures by confocal laser immunofluorescence in COS7 cells. Anti-FLAG antibody (red) and ataxin-1 (green) were used to detect LANP and ataxin-1, respectively. The distribution of LANP when expressed alone in COS7 cells is homogenous in the nucleus (a). Ataxin-1 with either 30 glutamines (30Q) (b) or 82 glutamines (82Q) (c) shows both a diffuse nuclear staining pattern and discrete subnuclear structures in transiently transfected COS7 cells. Immunofluorescence in COS7 cells expressing both ataxin-1 containing either 30 or 82 glutamines and LANP demonstrates that LANP localizes to the subnuclear structures (d and g); (f) and (i) overlay demonstrating the co-localization of ataxin-1 and LANP. Original magnification 3500×. Reprinted with permission from *Nature* **389**, 974–978. Copyright 1997 Macmillan Magazines Limited.

ubiquitin-positive intranuclear inclusions have also been demonstrated in another polyglutamine disorder, Huntington disease [29]. In transgenic mice expressing exon 1 of the human HD gene, Davies *et al.* report the appearance of a neuronal intranuclear inclusion (NII) which shows huntingtin immunoreactivity at first followed by ubiquitin immunoreactivity. Preliminary studies characterizing the *SCA1* transgenic mice suggest that the intranuclear structures formed by ataxin-1 are initially reactive with only anti-ataxin antibodies followed by positive staining with anti-ubiquitin antibodies. Anti-ataxin-1 antibodies identify numerous nuclear structures as compared to the anti-ubiquitin antibodies which decorate a few structures in embedded sections (C. J. Cummings and H. Y. Zoghbi, unpublished data). Detailed analysis of the process of structure formation and ubiquitin reactivity in several mice will be done to confirm those results. Although the structures are related to disease, presently, we do not know if their presence is causative or secondary. Of interest is the finding that ubiquitin-positive neuronal inclusions occur in other neurological disorders including Alzheimer disease, Parkinson disease, and Pick disease [30]. The role of ubiquitin-positive deposits in all of these diseases is still unknown.

V. PROTEINS INTERACTING WITH ATAXIN-1

The data available to date support a dominant effect of mutant ataxin-1 but do not address the molecular basis of cell-specific degeneration in SCA1. One possible model would be that the expanded allele interacts with specific proteins leading to alteration in their function. To identify protein products that interact with wild-type ataxin-1, Koshy *et al.* [31] performed a two-hybrid screen with the wild-type *SCA1* gene containing 30 glutamines using a mouse embryonic library. Glyceraldehyde-3-phosphate dehydrogenase (GAPDH) was identified as an ataxin-1-interacting protein [13]. GAPDH is a widely expressed enzyme involved in glycolysis [31]. However, although GAPDH also interacted with the mutant protein, this interaction was not quantitatively different from the interaction with the wild-type protein as assessed by the β-galactosidase assay of yeast transformed with both ataxin-1 and GAPDH [13]. Since both GAPDH and ataxin-1 have a wide range of expression in all cell types, the biological significance of this interaction is unknown at this time. It is interesting that GAPDH also interacts with other glutamine-repeat-containing proteins, including the androgen receptor that is mutated in spinobulbar muscular atrophy [13], huntingtin, and the DRPLA protein [32]. At the present time, it is not clear how GAPDH fits into the pathway of neuronal degeneration caused by polyglutamine expansions. This is particularly puzzling because of the high abundance of GAPDH relative to the polyglutamine proteins mutated in neurodegenerative disease. A possible hypothesis to explain the role of GAPDH would be that a slow decline in the energy metabolism of a postmitotic neuronal cell can trigger the degenerative process.

More recently Matilla *et al.* identified another ataxin-1-interacting protein, the leucine-rich acidic nuclear protein (LANP) [33]. This protein was isolated in a yeast two-hybrid screen of a mouse brain library using the mutant ataxin-1 with 82 glutamines as bait. LANP is a member of a family of proteins which contain a leucine-rich repeat (LRR). Several independent investigators have isolated the same gene product as LANP from peripheral tissues with different names such as putative histocompatibility leukocyte antigens class II-associated protein I (PHAP-I) [34], nuclear phosphoprotein pp32 [34a], and mapmodulin [34b]. As the names suggest, the protein was also attributed different functions such as inhibition of protein phosphatase 2A [34c], inhibition of oncogene induced formation of transformed foci [34a], and more recently as a protein modulating the interaction between microtubules and microtubule associated proteins [34b].

The LRRs are a consensus sequence of typically 25 amino acids predominantly containing leucines or other aliphatic residues at positions 2, 5, 7, 12, 16, 21, and 24 with an asparagine, cysteine, or threonine at position 10 [35]. The LRR motifs are present tandemly within the protein with the number ranging from 1 to 30. These motifs are present in proteins with diverse functions and roughly half of them are involved in signal transduction pathways. Individual LRR units constitute β–α structural units with the parallel β-sheet being exposed for strong protein–protein interactions [35]. LANP has five LRRs at its N-terminus and a highly acidic cluster of amino acids at the C-terminus. Based on the crystal structure of porcine ribonuclease inhibitor protein which contains 15 LRRs, Matsuoka *et al.* [36] predict that LANP has a compactly folded head formed by the nonglobular LRRs and an extended tail. Thus LANP is most likely a molecule involved in protein–protein interactions with specific target molecules [36].

Both wild-type ataxin-1 with 30 glutamine residues and the mutant protein with the 82 glutamines bind LANP *in vivo* when cotransfected in yeast. However, the strength of the LANP-ataxin-1 interaction is significantly stronger with mutant ataxin-1 as compared to the wild-type protein. Comparison of β-galactosidase activity in yeast containing LANP or ataxin-1 with 2, 30, or 82 glutamines revealed that yeast transfected with mutant ataxin-1 and LANP had a 10-fold increase in β-galactosidase activity as compared to wild-type.

In situ hybridization studies by Matsuoka *et al.* [36] identified the cerebellar Purkinje cells and the granule neurons as regions with the highest levels of LANP mRNA expression. We assessed the levels of the LANP RNA in a variety of mouse tissues and found two isoforms of sizes 1.3 and 2.3 kb. Both isoforms showed the highest expression in the cerebellum, whereas in peripheral tissues, the 2.3 kb variant was detected at lower levels. Thus, the expression of LANP is restricted to a region of the brain that is affected in SCA1. In addition, the expression of both LANP and ataxin-1 in Purkinje cells peaks at postnatal day 14 after which the expression decreases to adult levels. Further, immunohistochemical analysis with the respective antisera revealed that both proteins, LANP and ataxin-1, are predominantly localized to Purkinje cells in the cerebellum with intense staining in the nuclei and weaker staining in the cytoplasm.

Recent data from Matilla *et al.* [33] demonstrate that LANP may be an excellent candidate to be a cell-specific target mediating pathogenesis in SCA1. Localization patterns of LANP were analyzed after transfection of COS7 cells with LANP alone or together with either wild-type or mutant ataxin-1. Immunohistochemical analysis of the distribution of LANP when transfected alone revealed a homogeneous nuclear staining pattern of the protein (Fig. 18-4a). When both ataxin-1 and LANP are coexpressed in COS7 cells, there is a dramatic alteration in the staining pattern, with LANP being colocalized with ataxin-1 in the subnuclear structures (Figs. 18-4d and 18-4g). This alteration in the staining pattern of LANP is seen in the presence of either wild-type or mutant ataxin-1 [33].

Matilla and collaborators [33] carried out nuclear matrix preparations from COS7 cells transfected with either ataxin-1 and LANP alone or cotransfected with the two proteins to examine the localization patterns of LANP in the presence and absence of ataxin-1. Using *in situ* nuclear matrix preparations of cells transfected with LANP alone, detergent extraction of the COS7 cells led to loss of LANP staining, indicating that by itself LANP does not localize to the nuclear matrix. In contrast, COS7 cells cotransfected with LANP and ataxin-1 showed that LANP was retained in the nuclear matrix along with ataxin-1 and localized to the subnuclear structures. This retention of LANP at the nuclear matrix was seen both with the wild-type and the mutant protein [33]. These data together with spatial and temporal expression patterns of LANP and ataxin-1 strongly support LANP as a cell-specific interactor that contributes to the progressive dysfunction of Purkinje cells. It is easy to envisage that in the presence of ataxin-1, LANP is localized to subnuclear structures which may keep it from exerting its normal function, thereby leading to a dominant negative action of ataxin-1 on LANP.

VI. CONCLUDING REMARKS

The exciting new biochemical and immunohistochemical data on ataxin-1 indicate that localization of the protein to subnuclear structures is likely to offer new clues to the understanding of the molecular basis of pathogenesis. It is noteworthy that although both wild-type and mutant ataxin-1 form nuclear structures, only the mutant protein forms large single structures in transgenic mice and patients with a concomitant change in the integrity of the nuclear membrane. Further, these changes in the nucleus precede the onset of ataxia in *SCA1* transgenic mice.

The formation of nuclear structures induced by a mutant polyglutamine protein has been demonstrated in Huntington disease and spinocerebellar ataxia type 3 (SCA3) [29, 37, 38]. This is particularly interesting since huntingtin and ataxin-3 are both normally cytoplasmic. In the disease state the protein undergoes a relocalization to the nucleus to a single nuclear structure. Immunohistochemistry on sections of SCA3 patients reveals nuclear structures only in neurons that undergo degeneration and not in neurons typically spared in the disease. Similarly in Huntington disease the nuclear inclusions are seen in affected brain regions. Furthermore, DiFiglia and colleagues demonstrated that an amino-terminal fragment of huntingtin (~40 kDa) appears to be the portion of the protein localizing to the nucleus [37]. The emerging theme in neurodegenerative disorders seems to be the presence of ubiquitin-positive nuclear structures. Future experiments need to address whether the structures trap proteins which are involved in vital cellular functions. Scherzinger *et al.*, using a GST–huntingtin protein containing exon 1 of the HD gene and variable number of CAG repeats, demonstrated that GST–huntingtin polyglutamine forms amyloid-like protein aggregates *in vitro* [39]. These data, taken together with the data from Huntington's disease and SCA1 transgenic mice, raise the possibility that polyglutamine-mediated neurodegenerative disorders may share pathogenetic mechanisms with other progressive neurologic diseases, such as Alzheimer's disease and Prion disorders. Expansions of the polyglutamine tracts may alter the conformation of the respective protein leading to structural changes and aggregation as observed for amyloid and Prion proteins.

What determines the cell specificity of the nuclear inclusions and degeneration in polyglutamine disorders is still unknown. For SCA1, LANP seems to be a plausible candidate to be a tissue-specific mediator of pathogenesis. Temporal and spatial expression as well as the finding that ataxin-1 alters the normal localization of LANP from a homogeneous nuclear distribution to subnuclear structures attached to the nuclear matrix make LANP an excellent candidate to be the tissue-specific interactor. Although the normal function of LANP is not known at the present time, the selective and temporal expression pattern in the cerebellum suggests a role in cerebellar morphogenesis. Mice overexpressing LANP in Purkinje cells as well as LANP null mice are being generated. Characterization of these mice will provide insight about the normal function of LANP. Furthermore, breeding these mice with the *SCA1* transgenic mice should help clinch the role of LANP in SCA1 pathology.

Acknowledgments

We are grateful to Dr. Keith Wilkinson (Emory University) for valuable suggestions and advice on the biochemical extraction of mutant ataxin-1. Work on SCA1 is supported by a grant from the NIH to H.Y.Z. (NS27699). A.M. was supported by a grant from the Spanish Ministerio de Educación Y Ciencia (PF 94 968798).

References

1. Koshy, B. T., and Zoghbi, H. Y. (1997). The CAG/polyglutamine tract diseases: gene products and molecular pathogenesis. *Brain Pathol.* **7,** 927–942.
2. Orr, H., Chung, M.-y., Banfi, S., Kwiatkowski, T. J., Jr., Servadio, A., Beaudet, A. L., McCall, A. E., Duvick, L. A., Ranum, L. P. W., and Zoghbi, H. Y. (1993). Expansion of an unstable trinucleotide (CAG) repeat in spinocerebellar ataxia type 1. *Nature Genet.* **4,** 221–226.
3. Matilla, T., Volpini, V., Genis, D., Rosell, J., Corral, J., Davalos, A., Molins, A., and Estivill, X. (1993). Presymptomatic analysis of spinocerebellar ataxia type 1 (SCA1) via the expansion of the *SCA1* CAG-repeat in a large pedigree displaying anticipation and parental male bias. *Hum. Mol. Genet.* **2,** 2123–2128.
4. Ranum, L. P. W., Chung, M.-Y., Banfi, S., Bryer, A., Schut, L. J., Ramesar, R., Duvick, L. A., McCall, A. E., Subramony, S. H., Goldfarb, L., Gomez, C., Sandkuijl, L. A., Orr, H. T., and Zoghbi, H. Y. (1994). Molecular and clinical correlations in spinocerebellar ataxia type 1 (SCA1): evidence for familial effects on the age of onset. *Am. J. Hum. Genet.* **55,** 244–252.
5. Chung, M.-Y., Ranum, L. P. W., Duvick, L., Servadio, A., Zoghbi, H. Y., and Orr, H. T. (1993). Analysis of the CAG repeat expansion in spinocerebellar ataxia type I: evidence for a possible mechanism predisposing to instability. *Nature Genet.* **5,** 254–258.
6. Chong, S. S., McCall, A. E., Cota, J., Subramony, S. H., Orr, H. T., and Zoghbi, H. Y. (1995). Gametic and somatic tissue-specific heterogeneity of the expanded *SCA1* CAG repeat in spinocerebellar ataxia type 1. *Nature Genet.* **10,** 344–350.
7. Servadio, A., Koshy, B., Armstrong, D., Antalfy, B., Orr, H. T., and Zoghbi, H. Y. (1995). Expression analysis of the ataxin-1 protein in tissues from normal and spinocerebellar ataxia type 1 individuals. *Nature Genet.* **10,** 94–98.
8. Banfi, S., Servadio, A., Chung, M.-y., Kwiatkowski, T. J., Jr., McCall, A. E., Duvick, L. A., Shen, Y., Roth, E. J., Orr, H. T., and Zoghbi, H. Y. (1994). Identification and characterization of the gene causing type 1 spinocerebellar ataxia. *Nature Genet.* **7,** 513–519.
9. Banfi, S., Servadio, A., Chung, M.-y., Capozzoli, F., Duvick, L. A., Elde, R., Zoghbi, H. Y., and Orr, H. T. (1996). Cloning and developmental expression analysis of the murine homolog of the spinocerebellar ataxia type 1 gene (*Sca1*). *Hum. Mol. Genet.,* **5,** 33–40.
10. Skinner, P. J., Koshy, B., Cummings, C., Klement, I. A., Helin, K., Servadio, A., Zoghbi, H. Y., and Orr, H. T. (1997). SCA1 pathogenesis involves alterations in nuclear matrix associated structures. *Nature* **389,** 971–974.
11. Aronin, N., Chase, K., Young, C., Sapp, E., Schwarz, C., Matta, N., Kornreich, R., Landwehrmeyer, B., Bird, E., Beal, M. F., Vonsattel, J.-P., Smith, T., Carraway, R., Boyce, F. M., Young, A. B., Penney, J. B.. and DiFiglia, M. (1995). CAG expansion affects the expression of mutant huntingtin in the Huntington's disease brain. *Neuron* **15,** 1193–1201.
12. Persichetti, F., Ambrose, C. M., Ge, P., McNeil, S. M., Srinidhi, J., Anderson, M. A., Jenkins, B., Barnes, G. T., Duyao, M. P., and Kananley, L. (1995). Normal and expanded Huntington's disease gene alleles produce distinguishable proteins due to translation across the CAG repeat. *Mol. Med.* **1,** 374–383.
13. Koshy, B., Matilla, T., Burright, E. N., Merry, D. E., Fischbeck, K. H., Orr, H. T. and Zoghbi, H. Y. (1996). Spinocerebellar ataxia type-1 and spinobulbar muscular atrophy gene products interact with glyceraldehyde-3-phosphate dehydrogenase. *Hum. Mol. Genet.* **5,** 1311–1318.
14. Burright, E. N., Davidson, J. D., Duvick, L. A., Koshy, B., Zoghbi, H. Y., and Orr, H. T. (1997). Identification of a self-association region within the *SCA1* gene product, ataxin-1. *Hum. Mol. Genet.* **6,** 513–518.
15. Gossen, M., Scmitt, I., Obst, K., Wahle, P., Epplen, J. T., and Riess, O. (1996). cDNA cloning and expression of rscal, the rat counterpart of the human spinocerebellar ataxia type 1 gene. *Hum. Mol. Genet.* **5**, 381–389.
16. Matilla, T., Roberson, E. D., Banfi, S., Morales, J., Armstrong, D. L., Burright, E. N., Orr, H. T., Sweatt, J. D., Zoghbi, H. Y., and Matzuk, M. M. (1997). Mice lacking ataxin-1 do not have ataxia but display neurobehavioral abnormalities and decreased PPF. [Submitted for publication]
17. Burright, E. N., Clark, H. B., Servadio, A., Matilla, T., Feddersen, R. M., Yunis, W. S., Duvick, L. A., Zoghbi, H. Y., and Orr, H. T. (1995). SCA1 transgenic mice: a model for neurodegeneration caused by an expanded CAG trinucleotide repeat. *Cell* **82,** 937–948.
18. Clark, H. B., Burright, E. N., Yunis, W. S., Larson, S., Wilcox, C., Hartman, B., Zoghbi, H. Y., and Orr, H. T. (1997). Cerebellar expression of a mutant allele of SCA1 in transgenic mice is associated with a motor learning deficiency and a subsequent progressive neurologic dysfunction with histological abnormalities. *J. Neurosci.* **17,** 7385–7395.
19. Spector, D. L. (1993). Macromolecular domains within the cell nucleus. *Annu. Rev. Cell Biol.* 265–315.
20. Penman, S. (1995). Rethinking cell structure. *Proc. Natl. Acad. Sci. USA* **92,** 5251–5257.
21. Weis, K., Rambaud, S., Lavau, C., Jansen, J., Carvalho, T., Carmo-Fonseca, M., Lamond, A., and Dejean, A. (1994). Retinoic acid

regulates aberrant nuclear localization of PML-RARa in acute promyelocytic leukemia cells. *Cell* **76,** 345–356.
22. Dyck, J. A., Maul, G. G., Miller, W. H., Chen, J. D., Kakizuka, A., and Evans, R. M. (1994). A novel macromolecular structure is a target of the promyelocyte-retinoic acid receptor oncoprotein. *Cell* **76,** 333–343.
23. Fu, X. D., and Maniatis, T. (1990). Factor required for mammalian spliceosome assembly is localized to discrete regions in the nucleus. *Nature* **343,** 437–441.
24. Raska, I., Andrade, L. E., Ochs, R. L., Chan, E. K., Chang, C. M., Roos, G. and Tan, E. M. (1991). Immunological and ultrastructural studies of the nuclear coiled body with autoimmune antibodies. *Exp. Cell Res.* **195,** 27–37.
25. Dhordain, P., Albagli, O., Ansieau, S., Koken, M. H., Deweindt, C., Quief, S., Lantoine, D., Leutz, A., Kerckaert, J., and Leprince, D. (1995). The BTB/POZ domain targets the LAZ3/BCL6 oncoprotein to nuclear dots and mediates homomerisation in vivo. *Oncogene* **11,** 2689–97.
26. Berezney, R., and Coffey, D. S. (1974). Identification of a nuclear protein matrix. *Biochem. Biophys. Res. Commun.* **60,** 1410–1417.
27. Berezney, R., Mortillaro, M. J., Ma, H., Wei, X., and Samarabandu, J. (1995). The nuclear matrix: a structural milieu for genomic function. *Int. Rev. Cytol.* **162A,** 1–65.
28. Ciechanover, A. (1994). The ubiquitin-proteasome proteolytic pathway. *Cell* **79,** 13–21.
29. Davies, S. W., Turmaine, M., Cozens, B. A., DiFiglia, M., Sharp, A. H., Ross, C. A., Scherzinger, E., Wanker, E. E., Mangiarini, L., and Bates, G. P. (1997). Formation of neuronal intranuclear inclusions underlies the neurological dysfunction in mice transgenic for the HD mutation. *Cell* **90,** 537–548.
30. Manetto, V., Perry, G., T., M., Mulvihill, P., Fried, V. A., Smith, H. T., Gambetti, P., and Autilio-Gambetti, L. (1988). Ubiquitin is associated with abnormal cytoplasmic filaments characteristic of neurodegenerative diseases. *Proc. Natl. Acad. Sci. USA* **85,** 4501–4505.
31. Harris, I., and Perham, R. N. (1963). Studies on glyceraldehyde 3-phosphate dehydrogenases. *Biochem. J.* **89,** 60.
32. Burke, J. R., Enghild, J. J., Martin, M. E., Jou, Y.-S., Myers, R. M., Roses, A. D., Vance, J. M., and Strittmatter, W. J. (1996). Huntingtin and DRPLA proteins selectively interact with the enzyme GAPDH. *Nature Med.* **2,** 347–350.
33. Matilla, T., Koshy, B., Cummings, C. J., Isobe, T., Orr, H. T., and Zoghbi, H. Y. (1997). The cerebellar leucine rich acidic nuclear protein interacts with ataxin-1. *Nature* **389,** 974–978.
34. Vaesen, M., Barnikol-Watanabe, S., Gotz, H., Awni, L. A., Cole, T., Zimmermann, B., Kratzin, H. D., and Hilschmann, N. (1994). Purification and characterization of two putative HLA class II associated proteins: PHAPI and PHAPII. *Biol. Chem. Hoppe Seyler* **375,** 113–126.
34a. Chen, T. H., Brody, J. R., Romantsev, F. E., Yu, J. G., Kayler, A. E., Vaneiff, E., Kuhajda, F. P., and Pasternack, G. R. (1996). Structure of pp32, an acidic nuclear protein which inhibits oncogene-induced formation of transformed foci. *Mol. Biol. Cell.* **7,** 2045–2056.
34b. Ulitzur, N., Rancaño, C., and Pfeffer, S. R. (1997). Biochemical characterization of mapmodulin, a protein that binds microtubule-associated proteins. *J. Biol. Chem.* **272,** 30577–30582.
34c. Li, M., Makkinje, A., and Damuni, Z. (1996). Molecular identification of I1PP2A, a novel potent heat-stable inhibitor protein of protein phosphatase 2A. *Biochem.* **35,** 6998–7002.
35. Kobe, B., and Deisenhofer, J. (1994). The leucine-rich repeat: a versatile binding motif. *TIBS* **19,** 415–421.
36. Matsuoka, K., Taoka, M., Satozawa, N., Nakayama, H., Ichimura, T., Takahashi, N., Yamakuni, T., Song, S.-Y., and Isobe, T. (1994). A nuclear factor containing the leucine-rich repeats expresses in murine cerebellar neurons. *Proc. Natl. Acad. Sci. USA* **91,** 9670–9674.
37. DiFiglia, M., Sapp, E., Chase, K. O., Davies, S. W., Bates, G. P., Vonsattel, J. P., and Aronin, N. (1997). Aggregation of huntingtin in neuronal intranuclear inclusions and dystrophic neurites in brain. *Science* **277,** 1990–1993.
38. Paulson, H. L., Perez, M. K., Trottier, Y., Trojanowsk, J. Q., Subramony, S. H., Das, S. S., Vig, P., Mandel, J.-L., Fischbeck, K. H., and Pittman, R. N. (1997). Intranuclear inclusions of expanded polyglutamine protein in spinocerebellar ataxia Type 3. *Neuron* **19,** 333–334.
39. Scherzinger, E., Lurz, R., Turmaine, M., Mangiarini, L., Hollenbach, B., Hasenbank, R., Bates, G. P., Davies, S. W., and Wanker, E. E. (1997). Huntingtin-encoded polyglutamine expansions form amyloid-like protein aggregates in vitro and in vivo. *Cell* **90,** 549–558.

CHAPTER 19

Murine Model of SCA1: Repeat Instability and Neurobiology

MICHAEL D. KAYTOR Institute of Human Genetics,
Department of Laboratory Medicine and Pathology,
University of Minnesota, Minneapolis, Minnesota 55455

HARRY T. ORR Institute of Human Genetics,
Department of Laboratory Medicine and Pathology
and Department of Biochemistry, University of Minnesota,
Minneapolis, Minnesota 55455

I. INTRODUCTION

The expansion of a CAG trinucleotide repeat within the coding region of genes is becoming an ever more prominent cause of neuromuscular disorders. To date seven neurodegenerative disorders are caused by this type of dynamic mutation. Among these are spinocerebellar ataxia type 1 (SCA1) [1], spinocerebellar ataxia type 2 (SCA2) [2–4], Machado-Joseph disease/spinocerebellar ataxia type 3 (MJD/SCA3) [5], spinocerebellar ataxia type 6 (SCA6) [6], Huntington's disease (HD) [7], spinal and bulbar muscular atrophy (SBMA) [8], and dentatorubral-pallidoluysian atrophy (DRPLA) [9, 10] / Haw River syndrome [11]. With the exception of SBMA, which is due to a CAG repeat expansion in the androgen receptor gene, these genes exhibit no homologies to any previously identified genes and the functions of their products remain unknown. Expansion of the CAG repeat into the affected range for these disorders is suggested to result in a gain of a deleterious function by the mutant protein. All of the disorders have an autosomal dominant pattern of inheritance (except

SBMA, which is X-linked), exhibit genetic anticipation, show an inverse correlation between age of onset and repeat length, and share a similar size range of repeat tracts for both normal and expanded alleles. Thus, a further speculation is that common mechanisms of pathogenesis and CAG repeat tract instability are shared among these disorders.

SCA1 is an autosomal dominant neurodegenerative disorder characterized pathologically by progressive degeneration of the cerebellum, brain stem, and the spinocerebellar tracts [12]. SCA1 patients present with progressive gait and limb ataxia, dysarthia, dysmetria, nystagmus, and variable degrees of muscle wasting. SCA1 is caused by the expansion of a CAG trinucleotide repeat in a novel gene located on chromosome 6p23 [1, 13]. The *SCA1* CAG repeat is highly polymorphic in the general population [1, 14–16]. Normal *SCA1* alleles have CAG tract sizes ranging from 6 to 44 repeats. Expanded *SCA1* alleles have repeat sizes ranging from 40 to 81 triplets.

The *SCA1* gene is widely expressed [1, 13], yet the disease shows selective and cell-specific pathological alterations [17]. In fact, most of the genes with disease-associated triplet repeat expansion are widely expressed, but each disease has a specific pattern of neuronal loss [18]. Therefore, explaining how selective populations of neurons are affected in each of these disorders remains a main focus of research.

II. *SCA1* CAG REPEAT INSTABILITY

A. Intergenerational Instability

A genetic hallmark of SCA1 is anticipation, a worsening of clinical symptoms and earlier age of onset in successive generations [19, 20]. The molecular basis of anticipation is the instability of the CAG repeat when it is transmitted from parent to offspring. Furthermore, there is an indirect correlation between the length of the CAG tract and the age of onset: the longer the CAG tract, the earlier the onset of symptoms [1, 16]. The *SCA1* CAG repeat is unstable in both maternal and paternal transmissions, with the largest expansions occurring almost exclusively upon paternal transmission [14]. This paternal bias for expansion is also evident in HD, MJD, and DRPLA [21–23]. In SCA1 and HD, this paternal bias for expansion is exemplified by juvenile-onset cases [20, 24, 25], which are all due to the paternal transmission of an affected allele. Although both repeat tract expansions and contractions have been observed in maternal transmission of mutant *SCA1* alleles, maternal transmissions exhibit a slight trend toward repeat tract contraction [14]. Accordingly, the trend toward contractions in maternal transmissions and expansion bias in paternal transmissions suggests that the mechanisms governing repeat instability differ in male and female transmissions. Moreover, the mechanisms underlying repeat tract expansions and those that affect tract contractions might also be different. Alternatively, the mechanisms that govern small expansions and contractions may be similar, but a unique mechanism may be responsible for the larger paternally derived expansions.

Further evidence suggesting that expanded CAG alleles are handled differently in male and female gametogenesis is the presence of transmission distortion (meiotic drive) of a mutant allele in certain CAG repeat expansion disorders. In both SCA3 and DRPLA, the mutant allele is preferentially transmitted during male meiosis [26]. An explanation for this non-Mendelian segregation is that the mutant allele has a selective advantage during male gametogenesis. While the molecular basis for such a hypothesis is difficult to understand, the bias for large expansions of mutant alleles and the more frequent transmission of expanded alleles in male meiosis suggest a common mechanism of segregation distortion and meiotic repeat instability in male meiosis. Perhaps the segregation distortion and the male meiotic instability are associated with the fidelity of the DNA replication machinery in replicating the CAG repeat tract. In contrast, a second study found that the mutant *SCA3* allele is preferentially transmitted in female meiosis, suggesting that the mechanisms that govern meiotic drive are complex [27]. Furthermore, this study failed to find evidence for transmission distortion in SCA1. Obviously, more comprehensive studies are required to definitively address this issue.

The difference in repeat instability observed in male and female transmissions is likely to be the increased number of cell divisions in spermatogenesis versus oogenesis. The increased number of cell divisions in male gamete maturation results in an increased number of meiotic replicative cycles [28], which could increase the likelihood of repeat instability. This is consistent with the accumulating evidence suggesting that trinucleotide repeat instability is linked to DNA replication [29–32]. Hence, the cumulative effects from additional cell cycles of replication-induced instability likely contribute to the larger paternally derived expansions.

B. Somatic Repeat Instability

In SCA1, the CAG repeat tract exhibits limited heterogeneity in somatic tissues, with variations in tract length on the order of only one or two triplets [33]. This is similar to what is observed for an expanded *HD* repeat [34]. In addition, limited somatic mosaicism has also been found in the brains of SCA1, HD, DRPLA, and MJD patients [35–37]. Thus, somatic variation in size of the repeat tract is substantially less than the alterations observed in intergenerational transmissions. This suggests that the mechanism(s) that govern somatic and intergenerational repeat instability are different. In addition, different DNA repair or recombination mechanisms may be active in meiosis, which may result in a wider spectrum of repeat changes. Interestingly, there is no evidence of somatic mosaicism in SBMA [38, 39]. One possible explanation for this is the relatively small intergenerational changes observed for mutant *SBMA* alleles. Therefore, a relationship may exist between meiotic and mitotic repeat instability such that substantial intergenerational repeat expansion may initiate somatic repeat instability.

In DRPLA patients with increasing age of death there is a broadening of the allele sizes in the cerebral cortex [35]. This observation may result from an increased number of cell divisions with age. For example, glial cells continue to proliferate and DNA replication within these cells may be responsible for the age-related variation in *DRPLA* CAG size. Interestingly, the expanded repeat in the cerebellar cortex of DRPLA patients did not show a correlation with age at death. This argues that a simple replication (cell division)-based mechanism does not provide the complete explanation. Age-related DNA repair mechanisms that are cell type- or brain region-specific may also have a role in somatic variation in repeat length.

C. Molecular Determinants of CAG Repeat Instability

At the *SCA1* locus there are two distinct classes of alleles: normal, stable alleles and expanded, unstable mutant alleles. The distribution of the number of CAG triplet repeat units on normal and affected chromosomes is generally nonoverlapping [1, 14–16]. The CAG tract size on normal *SCA1* chromosomes ranges from 6 to 44 triplets. Analysis of expanded *SCA1* chromosomes yielded a range from 40 to 81 triplets, which generally do not overlap the tract size on normal chromosomes. The one reported exception is a normal *SCA1* allele that has an expanded allele size, 44 repeats, but a novel repeat tract configuration [40]. This configuration consists of multiple CAT interruptions within the repeat tract, which presumably aid in maintaining the stability of the CAG tract [14]. These two separate classes suggest that once an allele has expanded into the mutant range it is not likely to contract back into the normal range.

The size of the trinucleotide repeat tract is clearly a critical determinant of repeat tract instability. An analysis of intergenerational instability of normal *SCA1* chromosomes found no variations in repeat number in over 1000 meioses [15, 16]. This is not the case for expanded alleles in which up to 65% of transmissions were unstable [14]. A similar study found that up to 70% of mutant *HD* transmissions resulted in alterations of CAG repeat number [21]. Interestingly, in HD there is a relationship between the length of the CAG repeat tract and its instability [21, 41]. The longer the repeat tract, the higher the frequency of unstable transmission. This supports the presence of a threshold level for repeat tract length beyond which instabilities occur.

In addition to repeat tract length, sequence configuration of the repeat tract plays an important role in stability versus instability. The vast majority of normal *SCA1* alleles contain one or more CAT interruptions within the CAG repeat tract. In one study, 123 out of 126 normal chromosomes had at least one CAT interruption [14]. The three normal alleles in this study that did not have a CAT interruption had repeat tract lengths of 21 (two instances) and 19 CAG triplets. In contrast, of 30 mutant chromosomes analyzed, all had an uninterrupted CAG repeat tract. These observations suggest that the loss of the CAT interruption in the *SCA1* repeat tract is an early step in the molecular mechanism of CAG repeat instability. This is not unlike other CAG repeat expansion disorders in which non-CAG triplets within the repeat tract are associated with tract stability. In *SCA2,* the loss of CAA interruptions are associated with repeat instability [2–4]. Furthermore, in *HD* the loss of a penultimate CAA repeat unit within the repeat tract is correlated with increased repeat instability [42].

Previous reports have suggested that the ability of triplet repeat tracts to form DNA secondary structures, i.e., hairpins, is associated with their propensity to destabilize [31, 43–45]. The formation of DNA secondary structures within GC-rich tandemly repeated DNA, when either strand of the DNA duplex is single-stranded, has been shown to occur for all triplet repeat sequences known to expand in human disease [46–48]. The formation of such structures during DNA replication is proposed to be a prerequisite for repeat tract expansions or contractions. Short repeat tracts or tracts with interruptions would not form hairpin structures and hence exhibit repeat tract stability. Loss of the repeat tract interruption in *SCA1* and *SCA2* may permit the repeat tract to form a secondary structure that is more susceptible to DNA polymerase slippage-mediated repeat tract instability. The mechanism that governs the loss of the interruption, which ultimately leads to allele expansion, is unclear. Certainly the length of uninterrupted repeat tract influences tract stability.

Most normal repeat tracts for the CAG repeat expansion disorders have repeat tract lengths of 25 or less, while expanded alleles are considerably longer [49]. Expanded alleles with longer repeat tracts are suggested to form stable hairpin structures. Therefore, longer repeat tracts may form structures of higher threshold stability and hence be more susceptible to repeat instability.

Accumulating evidence suggests that the chromosomal context of the CAG repeat tract may play a role in its stability. HD haplotype analysis indicates that mutant alleles arise more frequently on specific haplotypes [50]. MJD patients also share haplotypes at several markers surrounding the *MJD* gene which are infrequent in the normal population [51]. In addition, DNA polymorphisms in the *MJD* and *DRPLA* genes are associated with increased levels of intergenerational instability [52, 53]. A polymorphism within intronic sequence was associated with expanded *DRPLA* alleles. These findings suggest that DNA elements located in *cis* to the repeat tracts may affect their stability.

An additional factor that may have a role in repeat tract instability is an origin of DNA replication. A study of CAG repeat instability in yeast found that tract instability depended on the relative position of a DNA origin of replication [54]. When CTG, the reverse complement of CAG, was in the lagging-strand template the repeat tract was unstable. Most of the CAG instabilities were reductions (contractions) in repeat number. Conversely, in the opposite orientation the repeat was stable. In another study of CAG repeat instability in yeast a similar orientation dependence was observed using a different repeat stability assay [55]. Taken together these data strongly suggest a role for DNA replication in repeat tract alterations and support the proposed model of triplet repeat instability involving DNA polymerase slippage during replication.

III. CAG REPEAT INSTABILITY IN *SCA1* cDNA TRANSGENIC MICE

Recent studies on *HD* and myotonic dystrophy (*DM*) transgenic mice have found that triplet repeat tracts are unstable upon intergenerational transmission [56–58]. To construct transgenes these studies used various regions of the corresponding genomic loci. Two of these studies used independent genomic fragments containing the *DM* CTG repeat. DM is a trinucleotide repeat expansion disorder in which the CTG repeat is located in the 3′ untranslated region of the *DMPK* gene. In the third study, a 1.9-kb genomic fragment containing the 5′ end of the *HD* gene, including the CAG repeat, was used to generate transgenic mice [59]. These studies found both maternal and paternal intergenerational repeat instability. Both expansions and contractions, generally ranging from one to four repeat units, were observed. Interestingly, the large paternal expansions commonly observed in the human disorders were not present in the unstable transmissions detected in the transgenic mice. This suggests that the transgenic mice still do not possess all of the necessary elements required for the induction of the full spectrum of repeat tract alterations observed in human meiotic transmissions.

To uncover the molecular basis of *SCA1* triplet repeat instability, we analyzed both male and female transmissions of a normal or an expanded CAG repeat in transgenic mice. *SCA1* cDNA transgenic animals containing either an expanded allele (82 or 55 CAGs) or a normal interrupted allele with 30 triplet repeat units were generated (Fig. 19-1). Transgene expression was directed to the cerebellar Purkinje cells with the murine *Pcp-2* promoter [60]. In all of the transgenic lines used, transgene expression was detected in the cerebellum by Northern analysis. The single copy transgenic lines D02 (uninterrupted repeat tract, 82 repeats), E04 (uninterrupted repeat tract, 55 repeats), and C01 (interrupted repeat tract, 30 repeats) were used in the characterization of repeat instability.

In the *SCA1* cDNA transgenic mice, no variation in repeat size was observed between somatic tissues. To investigate intergenerational CAG repeat instability in these *SCA1* transgenic mice, both paternal and maternal transmissions were analyzed. Intergenerational repeat instability was found in *SCA1* cDNA transgenic animals only when the transgene was maternally transmitted [61] (Table 19-1). Furthermore, the instability in line D02 increased in magnitude as the transgenic mother grew older (Fig. 19-2). Interestingly, all of the instabilities were reductions (contractions) in CAG repeat number. These contractions ranged in size from one to nine CAG repeat units. This supports the suggestion that the mechanisms governing mitotic and meiotic repeat tract instability are distinct.

It is possible that the maternal repeat tract instability is due to the integration of the transgene into an inherently unstable region of the genome. However, this seems unlikely since we observed repeat instability in three independent transgenic lines. In addition to repeat instability in D02 animals, repeat instability was also observed in transgenic line B02 (Fig. 19-1). This line has an expanded repeat tract with 82 repeats but has 10 copies of the transgene and thus was not extensively characterized. A few unstable maternal transmissions were also observed in transgenic line E04 (uninterrupted repeat tract, 55 repeats) (Table 19-1). Repeat instability in this line was far less prominent than that observed in D02 animals, suggesting a relationship between repeat tract length and instability in the trans-

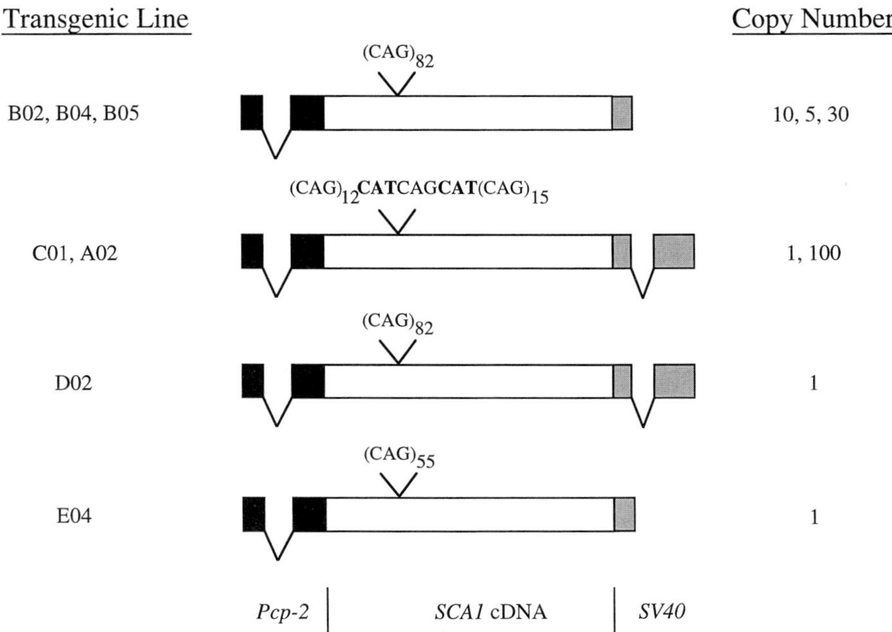

FIGURE 19-1 Schematic representation of *SCA1* cDNA transgenes. The transgenes are composed of the Purkinje cell specific promoter *Pcp-2* (solid area), approximately 3.2 kb *SCA1* cDNA (open area), and the SV40 polyadenylation signal (stippled area). Intronic sequences are depicted by a thin line in the *Pcp-2* promoter and SV40 sequences. The CAG repeat number and configuration of each repeat tract are indicated above each transgene. Also shown are the transgene copy numbers in each of the transgenic lines.

genic mice. Mice carrying a normal interrupted allele with 30 repeats (transgenic line C01) were also analyzed. No repeat instabilities were detected in either maternal or paternal transmissions of the C01 transgene. In summary, repeat tract instability was observed in all transgenic lines analyzed containing an uninterrupted, expanded, CAG repeat tract, while no repeat tract alterations were observed in the line carrying an interrupted, unexpanded, allele. Thus, both repeat tract length and configuration seem to be contributing factors to CAG repeat instability in transgenic mice.

To examine the temporal relationship between repeat instability and female gametogenesis, CAG tract instability was analyzed in unfertilized oocytes from young and aged transgenic mice [61]. This analysis suggested that instability occurs during oogenesis after DNA replication but prior to fertilization (Fig. 19-3). We did not observe CAG repeat tract instability in oocytes isolated from a young transgenic mother, but did observe CAG repeat tract instability in oocytes isolated from older transgenic mothers. This supports the hypothesis that as the oocytes are arrested in the diplotene stage of meiosis I they accumulate DNA damage over time and the repair of this damage leads to CAG instability in the form of repeat tract contractions. Thus, the repair of damage resulting from prolonged meiotic

TABLE 19-1 Maternal CAG Repeat Instability in *SCA1* cDNA Transgenic Mice

Line	Repeat tract	Parent	Transmissions: Total	Unstable	Frequency	Distribution of contractions: -1	-2	-3	-4	-5	-6	-7	-8	-9
D02	$(CAG)_{82}$	Male	86	3	0.03	3	—	—	—	—	—	—	—	—
		Female	213	143	0.67	54	44	20	11	6	4	1	1	2
E04	$(CAG)_{55}$	Male	46	1[a]	0.02	—	—	—	—	—	—	—	—	—
		Female	100	8	0.08	8	—	—	—	—	—	—	—	—
C01	$(CAG)_{12}$CATCAGCAT$(CAG)_{15}$	Male	58	0	0.0	—	—	—	—	—	—	—	—	—
		Female	60	0	0.0	—	—	—	—	—	—	—	—	—

[a]This unstable transmission was an expansion of 1 CAG unit.

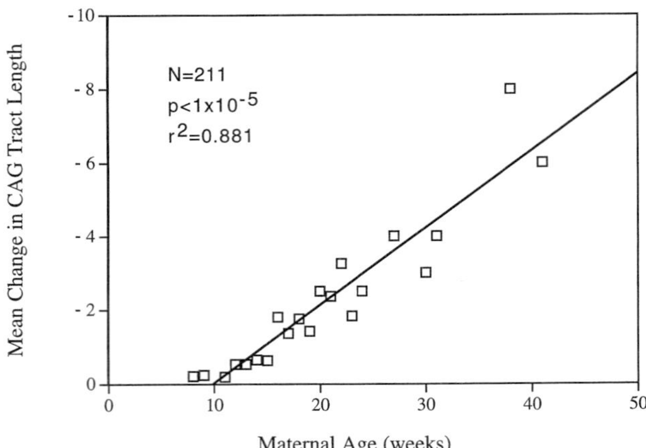

FIGURE 19-2 Comparison of the size of CAG contractions by maternal age. The age of D02 transgenic mothers at the time of birth of their litter was compared to the average CAG repeat tract change for that age. The analysis of 211 D02 transgenic offspring are shown. Also indicated are the *p* value and the correlation coefficient. A significant relationship between maternal age and repeat instability exists such that as transgenic mothers age, the magnitude of CAG repeat instability increases.

arrest would lead to larger contractions. Most postulated mechanisms of repeat instability involve DNA replication or recombination. It appears that in this case, CAG instability occurs independently of DNA replication because all DNA replication in murine oocytes has occurred by postnatal day 5 [62].

A. Mechanism of Repeat Instability in *SCA1* Transgenic Mice

A possible explanation for the failure to observe repeat instability in *SCA1* transgenic males may be that the transgene is differentially expressed in the gonads. However, this was not the case as transgene expression was not detected in either the ovaries or the testes of *SCA1* transgenic animals. Therefore, in the *SCA1* transgenic mice the mechanism of repeat instability is transcription independent. In contrast, in the *HD* transgenic mice the only transgenic line that exhibited repeat tract stability was the line that failed to express the transgene [56].

It is clear that the maternal repeat instability in the *SCA1* transgenic mice occurs independently of DNA replication since in oocytes the instability occurs well after meiotic DNA replication is completed [61]. Interestingly, this instability does depend on the age of the transmitting mother. This suggests an age-related, transcription- and replication-independent mechanism of repeat tract instability that involves multiple factors and pathways. The influential factors and pathways revealed in our study include the sex and age of the transmitting parent and the differences between female mitotic and meiotic division.

A recombination-based mechanism of repeat instability would be expected to produce both repeat tract expansions and contractions [29, 31]. In addition, homologous recombination occurs by the pachytene stage of the first meiotic division, which takes place prior to meiotic arrest [63, 64]. Thus, it seems unlikely that a recombination-based model accounts for the maternal repeat instability in the *SCA1* mice because only contractions in repeat number were found and there was a lack of repeat instability in oocytes isolated from young animals. Keeping in mind that oocytes are meiotically arrested by postnatal day 5, we propose a single-strand annealing mechanism to explain maternal repeat tract instability in the *SCA1* mice (Fig. 19-4). Single-strand

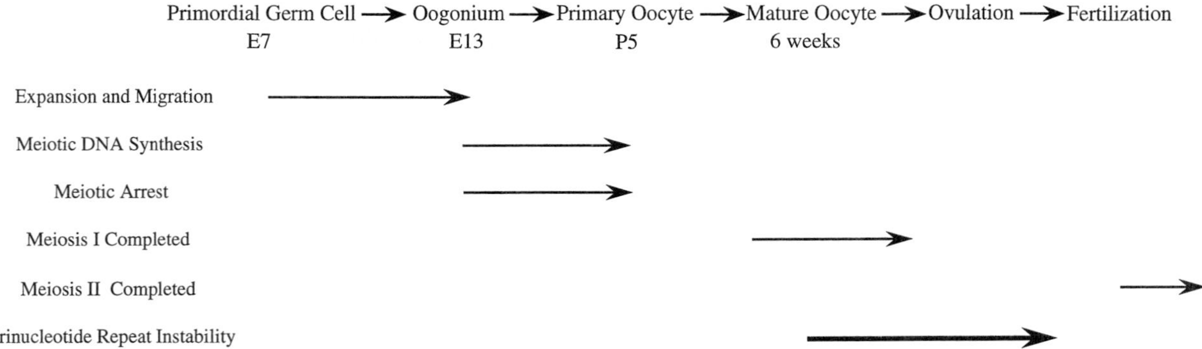

FIGURE 19-3 Oocyte development and CAG repeat tract instability. A population of primordial germ cells at embryonic day 7 (E7) become incorporated into the base of the allantois [62]. These primordial germ cells continue to divide and migrate, colonizing each gonad primordium by embryonic day 13 (E13). These female germ cells (oogonia) cease proliferation and enter meiosis, arresting at the diplotene stage of meiotic prophase I by postnatal day 5 (P5). Meiosis I is completed just prior to ovulation. Meiosis II is not completed until after fertilization. The failure to detect repeat instability in unfertilized oocytes from a 7-week-old female suggests CAG repeat instability is occurring after meiotic DNA replication but prior to fertilization.

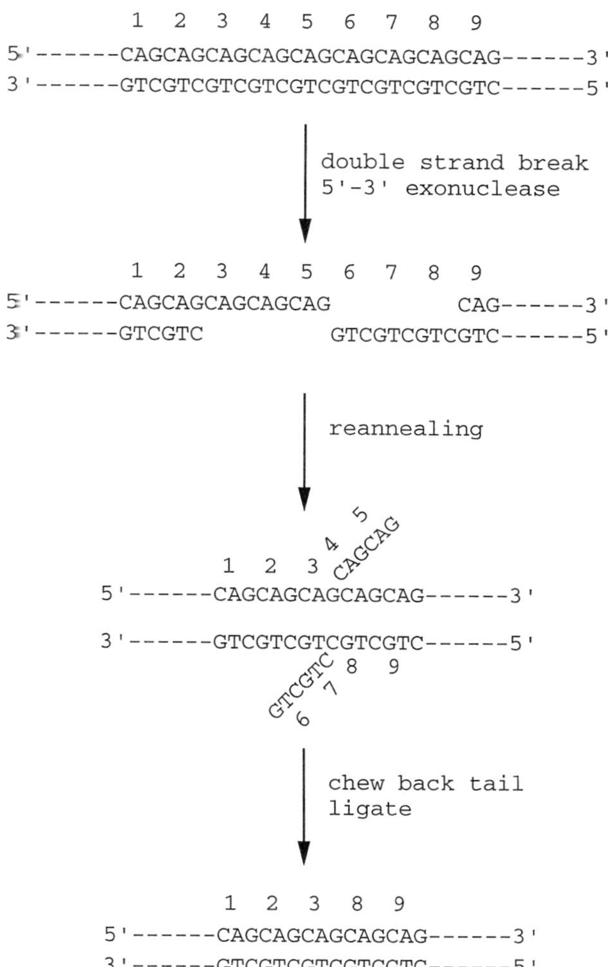

FIGURE 19-4 Model of CAG instability in arrested oocytes. A CAG repeat tract with nine units is shown. The complementary DNA strand (bottom strand) containing CTG is also shown. DNA damage produces a transient double strand break, which is processed by a 5′-3′ exonuclease. The single-strand overhangs promote reannealing of this intermediate DNA molecule. The remaining single-stranded tails are cleaved and the DNA molecule is ligated together. The net result of this modified single-strand annealing process is the loss of a number of CAG units. The length of the CAG reduction depends on the extent of recession of the double-strand break and the degree of reannealing.

annealing, sometimes called end-joining, is a mechanism of recombinational repair that is consistent with our observations [65–67]. In single-strand annealing, damaged DNA is excised and the DNA ends, facilitated by homologous pairing, are joined back together. This type of repair occurs most frequently in DNA that consists of multiple repeat units in a direct repeat orientation. Therefore, the net result of this process is the overall loss of some interstitial repeating units. The type of DNA damage responsible for initiating single-strand annealing is unclear. We postulate that in the *SCA1* transgenic mice oxidative DNA damage, which is known to increase with age [68], is responsible for initiating the destabilization process. If accurate, this model suggests that the CAG tract itself may be a hot-spot for the accumulation of DNA damage.

DNA mismatch repair (MMR) pathways may also play a role in the stabilization of triplet repeat tracts [69]. In fact, in yeast it has been found that the mismatch repair system prevents small changes in the repeat number, but does not have an effect on the larger repeat tract alterations [70]. Thus, it will be important to examine the role of MMR in DNA repair during oogenesis. This could be readily accomplished by crossing the *SCA1* mice with MMR knock-out mutant mice.

B. What's Missing for Repeat Instability in Mice?

To date, none of the studies on CAG transgenic mice have found large paternally derived expansions that are commonly observed in humans [56–58]. The failure to observe these large expansions may be explained by fundamental differences between mice and men. Perhaps cellular factors that influence repeat instability, particularly paternal expansions, produce their detrimental effects over time and the life span of a mouse is not sufficiently long for the induction of large expansions. Moreover, mice may lack certain cellular factors that are necessary for the induction of large paternally derived expansions. To detect the full spectrum of repeat instability in transgenic mice additional "human" stability factors may have to be incorporated into the transgenic models used to study triplet repeat instability.

The production of YAC transgenic mice, using a YAC encompassing the entire *SCA1* genomic locus, is one approach to address the importance of *cis*-acting DNA elements. In support of this strategy, *SBMA* YAC transgenic mice carrying an expanded allele of 45 repeats exhibit both repeat tract expansions and contractions upon intergenerational transmission [71]. Thus, we may shortly know if mice can be used to study the genetics of CAG intergenerational instability.

IV. MOLECULAR PATHOLOGY OF *SCA1* TRANSGENIC MICE

Typically, SCA1 is an adult-onset disorder that progresses over 2 to 3 decades before death usually resulting from bulbar dysfunction [12]. In SCA1 Purkinje cells of the cerebellar cortex, deep cerebellar nuclei, pontine nuclei, inferior olives, and some motor and cranial nerve nuclei are affected. The degree of neuronal loss is quite variable from patient to patient, with not all patients exhibiting neuronal loss in each of these regions. However, a consistent neuropathological finding among SCA1 patients is the loss of cerebellar Purkinje cells and loss of neurons in the inferior olivary nuclei [72].

The *SCA1* gene encodes a novel protein, ataxin-1 [1]. Both transcription and translation of a mutant *SCA1* allele have been shown in SCA1 patients [1, 73]. With the exception of cerebellar Purkinje cells, ataxin-1 is located in the nucleus of neuronal cells. In Purkinje cells it has both nuclear and cytoplasmic localization. In nonneuronal cells it is predominantly cytoplasmic. The murine *Sca1* gene, like the human gene, is widely expressed [74].

To elucidate mechanism(s) of pathogenesis caused by expanded polyglutamine tracts, *SCA1* transgenic mice have been created [75]. Transgene expression was directed to cerebellar Purkinje cells, a primary site of pathology in the human disorder [76] (via the murine *Pcp-2* promoter). Transgenic lines carrying either an expanded human *SCA1* allele (82 CAGs) or a normal unexpanded human allele (30 CAGs) were extensively characterized [75]. All of the B0 transgenic lines that expressed an expanded *SCA1* allele developed adult-onset ataxia (Fig. 19-1, Table 19-2). The single line that failed to develop ataxia also failed to express the transgene, demonstrating that expression is required for pathogenesis.

Both the level of transgene expression and the temporal expression pattern of the transgene seemed to determine the age of onset of ataxia in the B0 transgenic lines (Table 19-2). Of the transgenic lines carrying an expanded CAG repeat, B05 expressed the transgene most highly and had the earliest onset of ataxia. However, transgenic line B04 had the next youngest age of onset but expressed relatively low levels of transgene. In B04, *SCA1* transgene expression was detected as early as postnatal day 2, compared to postnatal day 10 in the other B0 lines. Thus, the developmental expression pattern of the transgene appears to be an important factor for determining the age of onset. Interestingly, in the mouse there is a transient burst of *Sca1* expression that is evident around postnatal day 14 (P14) [74]. This is the time at which the murine cerebellum is becoming physiologically functional. This suggests *Sca1* has a role in the latter stages of cerebellar development. The earlier age of onset in B04 mice, despite relatively low-level adult expression of the transgene, indicates that during development Purkinje cells may vary in terms of their susceptibility to the detrimental effects of expanded polyglutamine tracts.

TABLE 19-2 Ataxia and Transgene Expression in *Pcp-2/SCA1* Lines

Transgenic line[a]	Onset of ataxia	Expression	
		Adult[b]	Juvenile
A02	No ataxia (>2 years)	50×	ND
C01	No ataxia (>18 months)	10×	ND
B02	12–14 months	50×	P10
B04	16 weeks	10×	<P2
B05	12 weeks	100×	P10
D02	No ataxia (>1 year)	3×	ND
E04	No ataxia (>8 months)	100×	ND

Note. ND, not determined.
[a]*SCA1* transgenes are described in the legend to Fig. 1.
[b]Represents the fold increase above endogenous murine *Sca1* expression.

A. Behavioral Characterization of *SCA1* Transgenic Mice

Further behavioral and pathological characterization of these *SCA1* transgenic mice has provided insights into the disease pathogenesis and the molecular etiology of adult-onset ataxia. Two of the original transgenic lines created have been characterized further. Transgenic line B05 contains an expanded *SCA1* allele with 82 CAGs, while line A02 contains a normal interrupted *SCA1* allele with 30 repeats (Fig. 19-1). A02 animals displayed no neurologic abnormalities and were indistinguishable from age-matched littermates [75]. In contrast, B05 animals first became visibly ataxic as assessed by home cage behavior at 12 weeks of age. Initial phenotypic alterations included swaying of the head while walking and incoordination when attempting to stand on hind legs. As B05 transgenic animals got older they developed progressive loss of cerebellar function.

To more precisely assess the neurological effects of expanded *SCA1* alleles and normal alleles, *SCA1* transgenic mice were subjected to a battery of tests of motor skill and behavior [77]. Behavioral abnormalities were first evident at 5 weeks of age and increased with age. At 5 weeks, B05 animals exhibited motor deficiencies on a rotating rod apparatus. These animals performed as well as the controls on the rotating rod during the four trials of the first day. However, on the following 3 days of the test the performance of B05 animals was significantly lower than that of nontransgenic animals. This deficiency occurs at a time in which B05 animals are not visibly ataxic by home cage behavior.

To further assess neurologic affects of expanded polyglutamine tracts, B05 transgenic mice were analyzed for measurement of gait parameters, coordination on a bar cross apparatus, and behavior in an open field arena [77]. The appearance of gait abnormalities in B05 animals at 12 weeks of age corresponds to the first signs of ataxia by home cage behavior. Gait abnormalities increased with age, such that by 1 year of age B05 animals were unable to traverse a straight line, had a wider base with shorter steps, and walked with a shuffle gait instead of the normal alternating gait. To assess the fine motor coordination and balance capabilities of *SCA1* transgenic mice, animals were examined using a bar cross apparatus [78]. This test was carried out at 5 weeks of age when B05 animals first show deficiencies on the rotating rod apparatus. On the bar cross, 5-week-old B05 animals generally performed as well as nontransgenic littermates. Interestingly, B05 animals exhibited increased levels of spontaneous motor activity when compared to nontransgenic littermates. In addition, B05 animals exhibited increased levels of spontaneous motor activity in an open field test. Thus, deficiency on the rotating rod of B05 animals is not due to loss of balance or coordination, but rather suggest a motor learning deficit.

B. Histopathological Characterization of B05 *SCA1* Transgenic Mice

Immunohistochemical and morphometric analysis revealed that pathological changes accumulate over time in the *SCA1* transgenic mice (Fig. 19-5) [77, 79]. The cerebellar cortex of B05 animals developed normally. The first histologic changes detectable were the appearance of cytoplasmic vacuoles within Purkinje cell bodies around postnatal day 25 (P25). Electron microscopic analysis of these vacuoles indicated they were aqueous compartments formed by distentions of the endoplasmic reticulum. At the onset of motor learning impairment (5–6 weeks of age for B05 animals) occasional instances of loss of Purkinje cell proximal dendrites were observed (Fig. 19-5A).

By 12–15 weeks of age histologic alterations became more evident (Fig. 19-5B). There were substantial alterations in the complexity of the dendritic arborizations and shrinkage of the molecular layer. Interestingly, heterotopic Purkinje cells within the molecular layer were first apparent at this time. Although there were substantial alterations to the Purkinje cells, there was no detectable loss of Purkinje cells at this age. By 28 weeks of age B05 animals exhibited more dramatic pathological alterations (Fig. 19-5C). Many Purkinje cells had perikarya within the molecular layer. Purkinje cells with more than one primary dendrite were evident and dendritic spines were virtually nonexistent.

In B05 transgenic mice significant Purkinje cell loss was detected at 24 weeks of age, but not at 12 weeks of age [77]. Given that B05 mice are severely neurologically impaired by 15 weeks, it is apparent that Purkinje

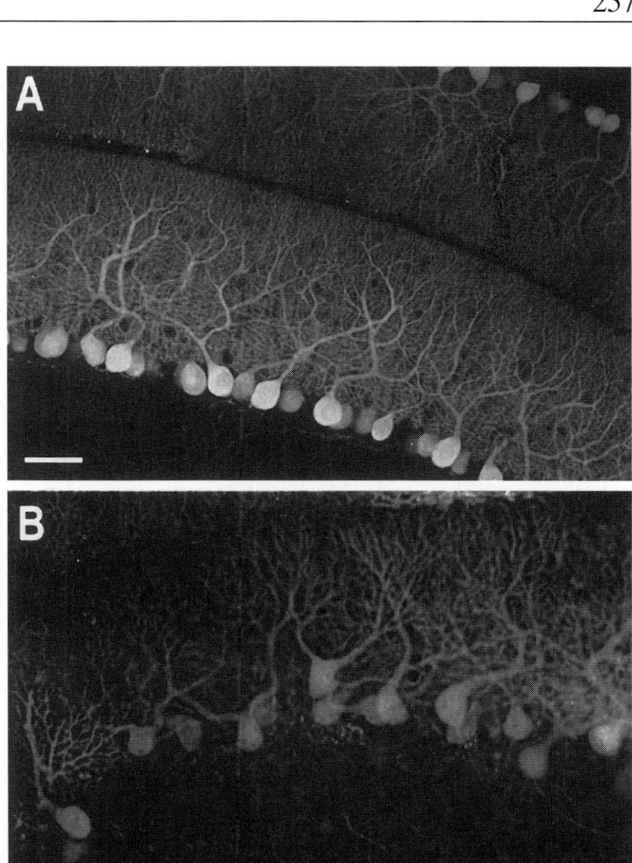

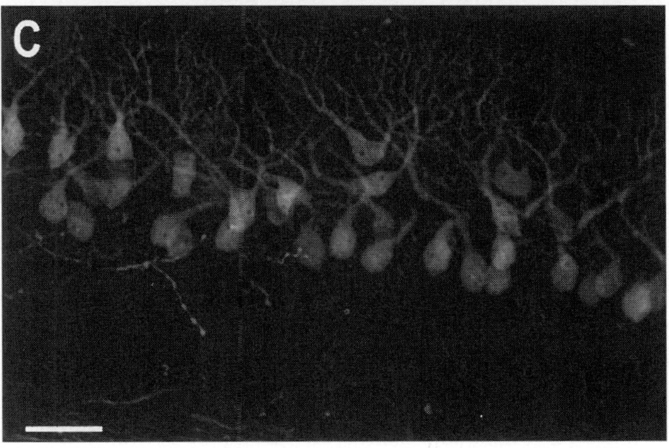

FIGURE 19-5 Calbindin immunohistochemistry of the cerebellum of *SCA1* (B05/+) transgenic mice at different stages of development showing progressive Purkinje cell changes. (A) As early as 6 weeks of age there is evidence of loss of proximal dendritic processes in some Purkinje cells. Distal arborization remains more complex (scale bar = 35 μm). (B) At 15 weeks there is significant alteration of dendritic morphology of Purkinje cells including heterotopic cell somata and loss of dendritic branches and spines (scale bar = 35 μm). (C) At 28 weeks, the loss of dendritic complexity is even more pronounced and there are more heterotopic Purkinje cells (scale bar = 35 μm). Reproduced with permission from *Brain Pathol.* **7,** 965–977 (1997).

cells become dysfunctional well before significant cell loss. Thus, the altered neurological phenotype is not simply the result of Purkinje cell loss, but rather is caused by Purkinje cell dysfunction which significantly precedes cell loss.

The appearance in B05 mice of Purkinje cells with their perikarya heterotypically located in the molecular layer is intriguing. Histologic examination of young animals revealed no ectopic Purkinje cells in the molecular layer (Fig. 19-5A). Thus, the mislocation of Purkinje cells is not due to abnormal Purkinje cell migration during cerebellar development. More likely, the malpositioned Purkinje cells are a result of disease progression. The early loss of synaptic input in the proximal parts of the dendritic tree, close to the perikaryon, may make it difficult for the cell to generate an action potential. A possible compensatory mechanism to maintain electrical activity may be the retraction of the nonfunctional dendritic trunk. This could be accomplished by the perikarya moving into the molecular layer, with concomitant elongation of the axon. This would allow the Purkinje cell to maintain contact with the parallel fibers located in the distal region of the molecular layer.

Importantly, Purkinje cell dendritic simplification and loss of spines have been observed in human SCA1 necropsy material [76, 80]. In addition, postmortem analysis of the cerebellar architecture has found the presence of ectopically located Purkinje cells in the molecular layer (H. B. Clark, personal communication). This provides pathological evidence that the *SCA1* transgenic mice are a useful model system for studying the molecular pathogenesis of SCA1.

A pathologic hallmark of SCA1 in humans is the occurrence of frequent proximal axonal dilations which have been called torpedo bodies [81]. These bodies have been suggested to result from abnormal axonal transport. Interestingly, in the B05 transgenic mice torpedo bodies are relatively rare. In contrast, there are more torpedo bodies in the cerebellar Purkinje cells of transgenic animals that overexpress a normal human *SCA1* allele. This suggests that the presence of the torpedo bodies in the B05 *SCA1* transgenic mice is not related to pathologic alterations that result in adult-onset ataxia.

Analysis of the *SCA1* transgenic mice has demonstrated that expression of ataxin-1 containing an expanded repeat tract in a cell type that is affected in the human disease leads to the induction of ataxia. Moreover, cell dysfunction and the appearance of phenotypic abnormalities precede significant Purkinje cell death in these transgenic mice. This strongly suggests that ataxia is not directly caused by the loss of Purkinje cells, but by Purkinje cell dysfunction at an earlier stage of the disease. Moreover, the cell loss that occurs later is likely a consequence of the earlier cell dysfunction.

V. CONCLUDING REMARKS

Study of the *SCA1* transgenic mice clearly demonstrates that directed expression of an expanded polyglutamine tract to cerebellar Purkinje cells is capable of inducing a neurologic phenotype and Purkinje cell pathology in transgenic mice. However, murine Purkinje cells appear to possess a reduced susceptibility to the effects of expanded polyglutamine tracts as compared to human neurons. To date, phenotypic and pathogenic alterations have been detected in *SCA1* transgenic mice that substantially overexpress mutant ataxin-1 relative to the endogenous murine protein. Transgenic mice that express a mutant protein with either 55 or 82 repeats at a level equal to murine ataxin-1 do not develop an altered phenotype or altered Purkinje cell morphology. This may be due to reduced sensitivity to the detrimental effects of expanded polyglutamine tracts or to the shortened life span of mice as compared to humans. This latter possibility is particularly relevant if there are cumulative effects of the exposure of neurons to expanded polyglutamine tracts.

In any case, *SCA1* transgenic studies indicate that Purkinje cell dysfunction occurs prior to cell death in the transgenic mice overexpressing an expanded allele of *SCA1*. Thus, the eventual Purkinje cell death is most likely an end-stage pathological aspect of the progressive cell dysfunction in the *SCA1* transgenic mice. These findings could have significant implications for the design of therapeutic treatments for SCA1 patients. If accurate, this suggests that preventing Purkinje cell dysfunction and not cell death would be a more effective strategy.

Given that an intriguing aspect of SCA1 genetics is anticipation, it is disappointing that the *SCA1* transgenic mice have yet to replicate the full spectrum of intergenerational repeat instability seen in humans. One possible explanation for this is that the cDNA constructs used to produce the transgenes lack essential *cis*-linked elements required for the induction of repeat instability. In support of this, three recent reports have found trinucleotide repeat instability in transgenic mice when regions of the corresponding genomic locus were used to construct the transgenes [56–58]. It is hoped that the generation and characterization of YAC transgenic mice will facilitate the elucidation of the role(s) that associated DNA elements have in triplet repeat instability. This should be accomplished through the generation of *SCA1* transgenic mice carrying an extended segment of the *SCA1* genomic locus. Until YAC transgenic mice are characterized, the analysis of repeat instability in the available lines should continue to provide clues into the mechanisms that govern repeat tract instability. Alternatively, mice may be less susceptible to the molecular aspects of repeat tract instability. The fact that murine homologs of trinucleotide repeat genes generally have repeat tracts with fewer triplets than found even on wild-type human alleles [74, 82, 83] is consistent with repeats in mice being more stable. In this regard mice may only be useful for studying components (i.e., maternal contractions) of the repeat tract instability normally observed in humans.

Regardless, the *SCA1* transgenic mice carrying an expanded allele did reveal an intriguing component of repeat instability in mammals. These mice demonstrated that during oogenesis, expanded *SCA1* CAG repeats become more unstable with advanced maternal age. Since this instability occurs well after meiotic DNA replication, it is likely to be due to the effects of postreplicative DNA repair mechanisms. Thus, while this maternal age-dependent instability may have little relevance to the instability of CAG repeats seen in individuals carrying an expanded *SCA1* allele, this maternal repeat instability in the *SCA1* mice could provide a useful means by which to examine DNA damage and repair during mammalian oogenesis.

References

1. Orr, H. T., Chung, M. Y., Banfi, S., Kwiatkowski, T. J., Jr., Servadio, A., Beaudet, A. L., McCall, A. E., Duvick, L. A., Ranum, L. P., and Zoghbi, H. Y. (1993). Expansion of an unstable trinucleotide CAG repeat in spinocerebellar ataxia type 1. *Nature Genet.* **4,** 221–226.
2. Sanpei, K., Takano, H., Igarashi, S., Sato, T., Oyake, M., Sasaki, H., Wakisaka, A., Tashiro, T., Ishida, Y., Ikeuchi, T., Koide, R., Saito, M., Sato, A., Tanaka, T., Hanyu, S., Takiyama, Y., Nishizawa, M., Shimizu, N., Nomura, Y., Segawa, M., Iwabuchi, K., Eguchi, I., Tanaka, H., Takahashi, H., and Tsuji, S. (1996). Identification of the spinocerebellar ataxia type 2 gene using a direct identification of repeat expansion and cloning technique, DIRECT. *Nature Genet.* **14,** 277–284.
3. Pulst, S.-M., Nechiporuk, A., Nechiporuk, T., Gispert, S., Chen, X.-N., Lopes-Cendes, I., Pearlman, S., Starkman, S., Orozco-Diaz, G., Lunkes, A., DeJong, P., Rouleau, G. A., Auburger, G., Korenberg, J. R., Figueroa, C., and Sahba, S. (1996). Moderate expansion of a normally biallelic trinucleotide repeat in spinocerebellar ataxia type 2. *Nature Genet.* **14,** 269–276.
4. Imbert, G., Saudou, F., Yvert, G., Devys, D., Trottier, Y., Garnier, J.-M., Weber, C., Mandel, J.-L. G., Cancel, G., Abbas, N., Durr, A., Didierjean, O., Stevanin, G., Agid, Y., and Brice, A. G. (1996). Cloning of the gene for spinocerebellar ataxia 2 reveals a locus with high sensitivity to expanded CAG/glutamine repeats. *Nature Genet.* **14,** 285–291.
5. Kawaguchi, Y., Okamoto, T., Yaniwaki, M., Aizawa, M., Inoue, M., Katayama, S., Kawakami, H., Nakamura, S., Nishimura, M., Akiguchi, I., Kimura, J., Narumiya, S., and Kakizuka, A. (1994). CAG expansion in a novel gene from Machado-Joseph disease at chromosome 14q32.1. *Nature Genet.* **8,** 221–227.
6. Zhuchenko, O., Bailey, J., Bonnen, P., Ashizawa, T., Stockton, D. W., Amos, C., Dobyns, W. B., Subramony, S. H., Zoghbi, H. Y., and Lee, C. C. (1997). Autosomal dominant cerebellar ataxia (SCA6) associated with small polyglutamine expansions in the α-1A-voltage-dependent calcium channel. *Nature Genet.* **15,** 62–69.
7. The Huntington's disease collaborative research group. (1993). A novel gene containing a trinucleotide repeat that is expanded and unstable on Huntington's disease chromosomes. *Cell* **72,** 971–983.
8. La Spada, A. R., Wilson, E. M., Lubahn, D. B., Harding, A. E., and Fischbeck, K. H. (1991). Androgen receptor gene mutations in X-linked spinal and bulbar muscular atrophy. *Nature* **352,** 77–79.
9. Koide, R., Ikeuchi, T., Onodera, O., Tanaka, H., Igarashi, S., Endo, K., Takahasi, H., Kondo, R., Ishikawa, A., Hayashi, T., Saito, M., Tomoda, A., Miike, T., Naito, H., Ikuta, F., and Tsuji, S. (1994). Unstable expansion of CAG repeat in hereditary dentatorubral-pallidoluysian atrophy (DRPLA). *Nature Genet.* **6,** 9–13.
10. Nagafuchi, S., Yanagisawa, H., Sato, K., Shirayama, T., Ohaski, E., Bundo, M., Takedo, T., Tadokoro, K., Kondo, I., Muruyama, N., Tanaka, Y., Kikushima, H., Umino, K., Kurosawa, H., Furukawa, T., Nihei, K., Inoue, T., Sano, A., Komure, O., Takahashi, M., Yoshizawa, T., Kanazawa, I., and Yamada, M. (1994). Dentatorubral and pallidoluysian atrophy: expansion of an unstable CAG trinucleotide on chromosome 12p. *Nature Genet.* **6,** 14–18.
11. Burke, J. R., Wingfield, M. S., Lewis, K. E., Roses, A. D., Lee, J. E., Hulette, C., Pericak-Vance, M. A., and Vance, J. M. (1994). The Haw River syndrome: dentatorubropallidoluysian atrophy (DRPLA) in an African-American family. *Nature Genet.* **7,** 521–524.
12. Zoghbi, H. Y., and Orr, H. T. (1995). Spinocerebellar ataxia type 1. *Semin. Cell Biol.* **6,** 29–35.
13. Banfi, S., Servadio, A., Chung, M. Y., Kwiatkowski, T. J., Jr., McCall, A. E., Duvick, L. A., Shen, Y., Roth, E. J., Orr, H. T., and Zoghbi, H. Y. (1994). Identification and characterization of the gene causing type 1 spinocerebellar ataxia. *Nature Genet.* **7,** 513–520.
14. Chung, M. Y., Ranum, L. P., Duvick, L. A., Servadio, A., Zoghbi, H. Y., and Orr, H. T. (1993). Evidence for a mechanism predisposing to intergenerational CAG repeat instability in spinocerebellar ataxia type I. *Nature Genet.* **5,** 254–258.
15. Jodice, C., Malasppina, P., Persichetti, F., Noveletto, A., Spadaro, M., Guinti, P., Morocutti, C., Terrenato, L., Harding, A. E., and Frontali, M. (1994). Effect of trinucleotide repeat length and parental sex on phenotypic variation in spinocerebellar ataxia 1. *Am. J. Hum. Genet.* **54,** 959–965.
16. Ranum, L. P. W., Chung, M.-y., Banfi, S., Bryer, A., Schut, L. J., Ramesar, R., Duvick, L. A., McCall, A., Subramony, S. H., Goldfarb, L., Gomez, C., Sandkuijl, L., Orr, H. T., and Zoghbi, H. Y. (1994). Molecular and clinical correlations in spinocerebellar ataxia type 1: evidence for familial effects on the age at onset. *Am. J. Hum. Genet.* **55,** 244–252.
17. Zoghbi, H. Y., and Ballabio, A. (1995). Spinocerebellar ataxia type 1. *In* "The Metabolic and Molecular Basis of Inherited Disease" (C. R. Scriver, A. L. Beaudet, W. S. Sly, and D. Valle, Eds.), pp. 4559–4568. McGraw Hill, New York.
18. Koshy, B. T., and Zoghbi, H. Y. (1997). The CAG/polyglutamine tract diseases: gene products and molecular pathogenesis. *Brain Pathol.* **7,** 927–942.

19. Schut, J. W. (1950). Hereditary ataxia: Clinical study through six generations. *Arch. Neurol. Psych.* **63,** 535–568.
20. Zoghbi, H. Y., Pollack, M. S., Lyons, L. A., Ferell, R. E., Daiger, S. P., and Beaudet, A. L. (1988). Spinocerebellar ataxia: variable age of onset and linkage to human leukocyte antigen in a large kindred. *Ann. Neurol.* **23,** 580–584.
21. Kremer, B., Almqvist, E., Theilmann, J., Spence, N., Telenius, H., Goldberg, Y. P., and Hayden, M. R. (1995). Sex-dependent mechanisms for expansions and contractions of the CAG repeat on affected Huntington disease chromosomes. *Am. J. Hum. Genet.* **57,** 343–350.
22. Maruyama, H., Nakamura, S., Matsuyama, Z., Sakai, T., Doyu, M., Sobue, G., Seto, M., Tsujihata, M., Oh-i, T., Nishio, T., Sunohara, N., Takahashi, R., Hayashi, M., Nishino, I., Ohtake, T., Oda, T., Nishimura, M., Saida, T., Matsumoto, H., Baba, M., Kawaguchi, Y., Kakizuka, A., and Kawakami, H. (1995). Molecular features of the CAG repeats and clinical manifestation of Machado-Joseph disease. *Hum. Mol. Genet.* **4,** 807–812.
23. Sano, A., Yamauchi, N., Kakimoto, Y., Komure, O., Kawai, J., Kuzume, K., Sano, N., and Kondo, I. (1994). Anticipation in hereditary dentatorubral-pallidoluysian atrophy. *Hum. Genet.* **93,** 699–702.
24. Haines, J. L., Schut, L. J., Weitkamp, L. R., Thayer, M., and Anderson, V. E. (1984). Spinocerebellar ataxia in a large kindred: age at onset, reproduction, and genetic linkage studies. *Neurology* **34,** 1542–1548.
25. Telenius, H., Kremer, H. P., Theilmann, J., Andrew, S. E., Almqvist, E., Anvret, M., Greenberg, C., Greenberg, J., Lucotte, G., Squitieri, F., Starr, E., Goldberg, Y. P., and Hayden, M. R. (1993). Molecular analysis of juvenile Huntington disease: the major influence on (CAG)n repeat length is the sex of the affected parent. *Hum. Mol. Genet.* **2,** 1535–1540.
26. Ikeuchi, T., Igarashi, S., Takiyama, Y., Onodera, O., Oyake, M., Takano, H., Koide, R., Tanaka, H., and Tsuji, S. (1996). Non-mendelian transmission in dentatorubral-pallidoluysian atrophy and Machado-Joseph disease: the mutant allele is preferentially transmitted in male meiosis. *Am. J. Hum. Genet.* **58,** 730–733.
27. Riess, O., Epplen, J. T., Amoiridis, G., Przuntek, H., and Schols, L. (1997). Transmission distortion of the mutant alleles in spinocerebellar ataxia. *Hum. Genet.* **99,** 282–284.
28. Drost, J. B., and Lee, W. R. (1995). Biological basis of germline mutation: Comparisons of spontaneous germline mutation rates among drosophila, mouse, and human. *Environ. Mol. Mutagen.* **25,** 48–64.
29. Kunkel, T. A. (1993). Slippery DNA and diseases. *Nature* **365,** 207–208.
30. Richards, R. I., and Sutherland, G. R. (1994). Simple repeat DNA is not replicated simply. *Nature Genet.* **6,** 114–116.
31. McMurray, C. T. (1995). Mechanisms of DNA expansion. *Chromosoma* **104,** 2–13.
32. Wells, R. D. (1996). Molecular basis of genetic instability of triplet repeats. *J. Biol. Chem.* **271,** 2875–2878.
33. Chong, S. S., McCall, A. E., Cota, J., Subramony, S. H., Orr, H. T., Hughes, M. R., and Zoghbi, H. Y. (1995). Gametic and somatic tissue-specific heterogeneity of the expanded SCA1 CAG repeat in spinocerebellar ataxia type 1. *Nature Genet.* **10,** 344–350.
34. Telenius, H., Kremer, B., Goldberg, Y. P., Theilmann, J., Andrew, S. E., Zeisler, J., Adam, S., Greenberg, C., Ives, E. J., Clarke, L. A., and Hayden, M. R. (1994). Somatic and gonadal mosaicism of the Huntington disease gene CAG repeat in brain and sperm. *Nature Genet.* **6,** 409–414. [published erratum appears in *Nature Genet.* (1994) **7,** 113].
35. Takano, H., Onodera, O., Takahashi, H., Igarashi, S., Yamada, M., Oyake, M., Ikeuchi, T., Koide, R., Tanaka, H., Iwabuchi, K., and Tsuji, S. (1996). Somatic mosaicism of expanded CAG repeats in brains of patients with dentatorubral-pallidoluysian atrophy: cellular population-dependent dynamics of mitotic instability. *Am. J. Hum. Genet.* **58,** 1212–1222.
36. Ueno, S., Kondoh, K., Kotani, Y., Komure, O., Kuno, S., Kawai, J., Hazama, F., and Sano, A. (1995). Somatic mosaicism of CAG repeat in dentatorubral-pallidoluysian atrophy (DRPLA). *Hum. Mol. Genet.* **4,** 663–666.
37. Lopes, C. I., Macial, P., Kish, S., Gaspar, C., Robitaille, Y., Clark, H. B., Koeppen, A. H., Nance, M., Schut, L., Silveira, I., Coutinho, P., Sequeiros, J., and Rouleau, G. A. (1996). Somatic mosaicism in the central nervous system in spinocerebellar ataxia type 1 and Machado-Joseph disease. *Ann. Neurol.* **40,** 199–206.
38. Spiegel, R., La Spada, A. R., Kress, W., Fischbeck, K. H., and Schmid, W. (1996). Somatic stability of the expanded CAG trinucleotide repeat in X-linked spinal and bulbar muscular atrophy. *Hum. Mutat.* **8,** 32–37.
39. Watanabe, M., Abe, K., Aoki, M., Yasuo, K., Itoyama, Y., Shoji, M., Ikeda, Y., Iizuka, T., Ikeda, M., Shizuka, M., Mizushima, K., and Hirai, S. (1996). Mitotic and meiotic stability of the CAG repeat in the X-linked spinal and bulbar muscular atrophy gene. *Clin. Genet.* **50,** 133–137.
40. Quan, F., Janas, J., and Popovich, B. W. (1995). A novel CAG repeat configuration in the *SCA1* gene: implications for the molecular diagnostics of spinocerebellar ataxia type 1. *Hum. Mol. Genet.* **4,** 2411–2413.
41. Ranen, N. G., Stine, O. C., Abbott, M. H., Sherr, M., Codori, A.-M., Franz, M. L., Chao, N. I., Chung, A. S., Pleasant, N., Callahan, C., Kasch, L. M., Ghaffari, M., Chase, G. A., Kazazian, H. H., Brandt, J., Folstein, S. E., and Ross, C. A. (1995). Anticipation and instability of IT-15 $(CAG)_N$ repeats in parent-offspring pairs with Huntington disease. *Am. J. Hum. Genet.* **57,** 593–602.
42. Goldberg, Y. P., McMurray, C. T., Zeisler, J., Almqvist, E., Sillence, D., Richards, F., Gacy, A. M., Buchanan, J., Telenius, H., and Hayden, M. R. (1995). Increased instability of intermediate alleles in families with sporadic Huntington disease compared to similar sized intermediate alleles in the general population. *Hum. Mol. Genet.* **4,** 1911–1918.
43. Kang, S., Jaworski, A., Ohshima, K., and Wells, R. D. (1995). Expansion and deletion of CTG repeats from human disease genes are determined by the direction of replication in E. coli. *Nature Genet.* **10,** 213–218.
44. Smith, G. K., Jie, J., Fox, G. E., and Gao, X. (1995). DNA CTG triplet repeats involved in dynamic mutations of neurologically related gene sequences form stable duplexes. *Nucleic Acids Res.* **23,** 4303–4311.
45. Pearson, C. E., and Sinden, R. R. (1996). Alternative structures in duplex DNA formed within the trinucleotide repeats of the myotonic dystrophy and fragile X loci. *Biochemistry* **35,** 5041–5053.
46. Gacy, A. M., Goellner, G., Juranic, N., Macura, S., and McMurray, C. T. (1995). Trinucleotide repeats that expand in human disease form hairpin structures in vitro. *Cell* **81,** 533–540.
47. Darlow, J. M., and Leach, D. R. (1995). The effects of trinucleotide repeats found in human inherited disorders on palindrome inviability in Escherichia coli suggest hairpin folding preferences in vivo. *Genetics* **141,** 825–832.
48. Petruska, J., Arnheim, N., and Goodman, M. F. (1996). Stability of intrastrand hairpin structures formed by the CAG/CTG class of DNA triplet repeats associated with neurological diseases. *Nucleic Acids Res.* **24,** 1992–1998.

49. Paulson, H. L., and Fischbeck, K. H. (1996). Trinucleotide repeats in neurogenetic disorders. *Annu. Rev. Neurosci.* **19,** 79–107.
50. Squitieri, F., Andrew, S. E., Goldberg, Y. P., Kremer, B., Spence, N., Zeisler, J., Nichol, K., Theilmann, J., Greenberg, J., Goto, J., Kanazawa, I., Vesa, J., Peltonen, L., Almqvist, E., Anvret, M., Telenius, H., Lin, B., Napolitano, G., Morgan, K., and Hayden, M. R. (1994). DNA haplotype analysis of Huntington disease reveals clues to the origins and mechanisms of CAG expansion and reasons for geographic variations of prevalence. *Hum. Mol. Genet.* **3,** 2103–2114.
51. Takiyama, Y., Igarashi, S., Rogaeva, E. A., Endo, K., Rogaev, E. I., Tanaka, H., Sherrington, R., Sanpei, K., Liang, Y., Saito, M., Tsuda, T., Takano, H., Ikeda, M., Lin, C., Chi, H., Kennedy, J. L., Lang, A. E., Wherrett, J. R., Segawa, M., Nomura, Y., Yuasa, T., Weissenbach, J., Yoshida, M., Nishizawa, M., Kidd, K. K., Tsuji, S., and George-Hyslop, P. H. S. (1995). Evidence for intergenerational instability in the CAG repeat in the *MJD1* gene and for conserved haplotypes at flanking markers amongst Japanese and Caucasian subjects with Machado-Joseph disease. *Hum. Mol. Genet.* **4,** 1137–1146.
52. Igarashi, S., Takiyama, Y., Cancel, G., Rogaeva, E. A., Sasaki, H., Wakisaka, A., Zhou, Y.-X., Takano, H., Endo, K., Sanpei, K., Oyake, M., Tanaka, H., Stevanin, G., Abbas, N., Durr, A., Rogaev, E. I., Sherrington, R., Tsuda, T., Ikeda, M., Cassa, E., Nishizawa, M., Benomar, A., Julien, J., Weissenbach, J., Wang, G.-X., Agid, Y., George-Hyslop, P. H. S., Brice, A., and Tsuji, S. (1996). Intergenerational instability of the CAG repeat of the gene for Machado-Joseph disease (*MJD1*) is affected by the genotype of the normal chromosome: implications for the molecular mechanisms of the instability of the CAG repeat. *Hum. Mol. Genet.* **5,** 923–932.
53. Yanagisawa, H., Fujii, K., Nagafuchi, S., Nakahori, Y., Nakagome, Y., Akane, A., Nakamura, M., Sano, A., Komure, O., Kondo, I., Jin, D. K., Sorensen, S. A., Potter, N. T., Young, S. R., Nakamura, K., Nukina, N., Nagao, Y., Tadokoro, K., Okuyama, T., Miyashita, T., Inoue, T., Kanazawa, I., and Yamada, M. (1996). A unique origin and multistep process for the generation of expanded DRPLA triplet repeats. *Hum. Mol. Genet.* **5,** 373–379.
54. Maurer, D. J., O'Callaghan, B. L., and Livingston, D. M. (1996). Orientation dependence of trinucleotide CAG repeat instability in Saccharomyces cerevisiae. *Mol. Cell. Biol.* **16,** 6617–6622.
55. Freudenreich, C. H., Stavenhagen, J. B., and Zakian, V. A. (1997). Stability of a CTG/CAG trinucleotide repeat in yeast is dependent on its orientation in the genome. *Mol. Cell. Biol.* **17,** 2090–2098.
56. Mangiarini, L., Sathasivam, K., Mahal, A., Mott, R., Seller, M., and Bates, G. P. (1997). Instability of highly expanded CAG repeats in mice transgenic for the Huntington's disease mutation. *Nature Genet.* **15,** 197–200.
57. Gourdon, G., Radvanyi, F., Lia, A.-S., Duros, C., Blanche, M., Abitbol, M., Junien, C., and Hofmann-Radvanyi, H. (1997). Moderate intergenerational and somatic instability of a 55-CTG repeat in transgenic mice. *Nature Genet.* **15,** 190–192.
58. Monckton, D. G., Coolbaugh, M. I., Ashizawa, K. T., Siciliano, M. J., and Caskey, C. T. (1997). Hypermutable myotonic dystrophy CTG repeats in transgenic mice. *Nature Genet.* **15,** 193–196.
59. Mangiarini, L., Sathasivam, K., Seller, M., Cozens, B., Harper, A., Hetherington, C., Lawton, M., Trattier, Y., Lehrach, H., Davies, S. W., and Bates, G. P. (1996). Exon 1 of the *HD* gene with an expanded CAG repeat is sufficient to cause a progressive neurological phenotype in transgenic mice. *Cell* **87,** 493–506.
60. Vandaele, S., Nordquist, D. T., Feddersen, R. M., Tretjakoff, I., Peterson, A. C., and Orr, H. T. (1991). Purkinje cell protein-2 regulatory regions and transgene expression in cerebellar compartments. *Genes Dev.* **5,** 1136–1148.
61. Kaytor, M. D., Burright, E. N., Duvick, L. A., Zoghbi, H. Y., and Orr, H. T. (1997). Increased trinucleotide repeat instability with advanced maternal age. *Hum. Mol. Genet.* **6,** 2135–2139.
62. Hogan, B., Beddington, R., Costantini, F., and Lacy, E. (1994). "Manipulating the Mouse Embryo: A Laboratory Manual." Cold Spring Harbor Laboratory, Cold Spring Harbor, NY.
63. von Wettstein, D., Rasmussen, S. W., and Holm, P. B. (1984). The synaptonemal complex in genetic segregation. *Annu. Rev. Genet.* **18,** 331–413.
64. Carpenter, A. T. C. (1987). Gene conversion, recombination nodules, and the initiation of meiotic synapsis. *Bioessays* **6,** 232–236.
65. Ozenberger, B. A., and Roeder, G. S. (1991). A unique pathway of double-strand break repair operates in tandomly repeated genes. *Mol. Cell. Biol.* **11,** 1222–1231.
66. Segal, D. J., and Carroll, D. (1995). Endonuclease-induced, targeted homolgous extrachromosomal recombination in Xenopus oocytes. *Proc. Natl. Acad. Sci. USA* **92,** 806–810.
67. Nicolas, A. L., Munz, P. L., and Young, C. S. (1995). A modified single-strand annealing model best explains the joining of DNA double-strand breaks mammalian cells and cell extracts. *Nucleic Acids Res.* **23,** 1036–1043.
68. Kaneko, T., Tahara, S., and Matsuo, M. (1996). Non-linear accumulation of 8-hydroxy-2′-deoxyguanosine, a marker of oxidized DNA damage, during aging. *Mutat. Res.* **316,** 277–285.
69. Pearson, C. E., Ewel, A., Acharya, S., Fishel, R. A., and Sinden, R. R. (1997). Human MSH2 binds to trinucleotide repeat DNA structures associated with neurodegenerative diseases. *Hum. Mol. Genet.* **6,** 1117–1123.
70. Schweitzer, J. K., and Livingston, D. M. (1997). Destabilization of CAG trinucleotide repeat tracts by mismatch repair mutations in yeast. *Hum. Mol. Genet.* **6,** 349–355.
71. La Spada, A. R. (1997). Trinucleotide repeat instability: Genetic features and molecular mechanisms. *Brain Pathol.* **7,** 943–963.
72. Gilman, S., Sima, A. F., Junck, L., Kluin, K. J., Koeppe, R. A., Lohman, B. A., and Little, R. (1996). Spinocerebellar ataxia type 1 with multiple system degeneration and glial cytoplasmic inclusions. *Ann. Neurol.* **39,** 241–255.
73. Servadio, A., Koshy, B., Armstrong, D., Antalffy, B., Orr, H. T., and Zoghbi, H. Y. (1995). Expression analysis of the ataxin-1 protein in tissues from normal and spinocerebellar ataxia type 1 individuals. *Nature Genet.* **10,** 94–98.
74. Banfi, S., Servadio, A., Chung, M.-Y., Capozzoli, F., Duvick, L. A., Elde, R., Zoghbi, H. Y., and Orr, H. T. (1996). Cloning and developmental expression analysis of the murine homolog of the spinocerebellar ataxia type 1 gene (*Sca1*). *Hum. Mol. Genet.* **5,** 33–40.
75. Burright, E. N., Clark, H. B., Servadio, A., Matilla, T., Feddersen, R. M., Yunis, W. S., Duvick, L. A., Zoghbi, H. Y., and Orr, H. T. (1995). SCA1 transgenic mice: a model for neurodegeneration caused by an expanded CAG trinucleotide repeat. *Cell* **82,** 937–948.
76. Koeppen, A. H. (1991). The Purkinje cell and its afferents in human hereditary ataxia. *J. Neuropathol. Exp. Neurol.* **50,** 505–514.
77. Clark, H. B., Burright, E. N., Yunis, W. S., Larson, S., Wilcox, C., Hartman, B., Matilla, A., Zoghbi, H. Y., and Orr, H. T. (1997). Purkinje cell expression of a mutant allele of *SCA1* in transgenic mice leads to disparate effects on motor behaviors followed by

a progressive cerebellar dysfunction and histological alterations. *J. Neurosci.* **17,** 7385–7395.

78. Gerlai, R., Friend, W., Becker, L., O'Hanon, D., Marks, A., and Roder, J. (1993). Female transgenic mice carrying multiple copies of the human gene for S100B are hyperactive. *Behav. Brain Res.* **55,** 51–59.
79. Burright, E. N., Orr, H. T., and Clark, H. B. (1997). Mouse models of human CAG repeat disorders. *Brain Pathol.* **7,** 965–977.
80. Ferrer, I., Genis, D., Davalus, A., Bernado, L., Sant, F., and Serano, T. (1994). The Purkinje cell in olivopontocerebellar atrophy. A golgi and immunohistochemical study. *Neuropathol. Appl. Neurobiol.* **20,** 38–46.
81. Koeppen, A. H., and Barron, K. D. (1984). The neuropathology of olivopontoceebellar atrophy. *In* "The Olivopontocerebellar Atrophies" (R. C. Duvoisin and A. Plaitakis, Eds.), pp. 13–38. Raven Press, New York.
82. Faber, P. W., King, A., van Rooij, H. C. J., Brinkmann, A. O., de Both, N. J., and Trapman, J. (1991). The mouse androgen receptor. *Biochem. J.* **278,** 269–278.
83. Barnes, G. T., Duyao, M. P., Ambrose, C. M., McNeil, S., Persichetti, F., Srinidhi, J., Gusella, J. F., and MacDonald, M. E. (1994). Mouse Huntington's disease gene homolog (Hdh). *Som. Cell Mol. Genet.* **20,** 87–97.

PART VIII

Other Dominant Spinocerebellar Ataxias

CHAPTER 20

Spinocerebellar Ataxia Type 2

STEFAN-M. PULST Division of Neurology, Los Angeles, California 90024

I. INTRODUCTION

What is now known as spinocerebellar ataxia type 2 (SCA2) was first described clinically as a distinct ethnic entity in a large, homogeneous population of patients with dominantly inherited cerebellar ataxia from the Holguin province of Cuba [1]. Its prevalence in Holguin was 41 per 100,000, much higher than that in the Western part of Cuba or in other parts of the world [2] It was speculated that the high incidence might be related to a founder effect or the interaction of a disease gene with an environmental toxin. Eighty-one patients from 20 different pedigrees were clinically examined. The majority were of Caucasian or Spanish ancestry. In addition to ataxic gait and other cerebellar findings, many patients had slow saccadic eye movements which in some had progressed to ophthalmoparesis. Tendon reflexes were brisk during the first years of life, but absent several years later. Seven autopsied cases showed a marked reduction of Purkinje cells.

An analysis of a larger sample of these patients confirmed the earlier analysis [3]. Age of onset showed a wide range from 2 to 65 years of age. Although there appeared to be a significant phenotypic overlap with other SCAs and great infrafamilial variability, the authors noted slow saccadic eye movements in many affected individuals. Muscle cramps unrelated to exercise were also common, and preceded other clinical manifestations in some individuals. Genotypic differentiation from the then only known ataxia locus, the SCA1 locus, on chromosome 6p did not occur until 1990 [4].

II. GENETIC LINKAGE ANALYSIS

In 1993, the SCA2 locus was mapped to CHR 12 in two ethnic populations. Using a genome-wide screen, Gispert *et al.* [5] identified a 20-cM interval on CHR 12q24.1 that contained the SCA2 locus. Pulst *et al.* [6] confirmed this location in a second pedigree of southern

Italian descent and demonstrated that SCA2 showed marked anticipation of disease onset (Fig. 20-1). Of 15 parent–child pairs, 14 showed earlier disease onset by at least 1 year. Using closely linked genetic markers it could be demonstrated that anticipation was not due to biased ascertainment based on later onset in asymptomatic gene carriers [6]. This observation strongly suggested that SCA2 was caused by an unstable DNA repeat.

Additional pedigrees from Italy, Tunisia, Austria and French-Canada, Martinique, and France [7–11] were also shown to have linkage to the new SCA2 locus. Several crucial recombination events in the Cuban pedigree finally limited SCA2 location to a 1-cM interval between markers D12S1328 and D12S1329 [12].

III. POSITIONAL CLONING OF THE SCA2 GENE

The SCA2 gene was recently identified independently by groups using three different approaches, one of which employed a positional cloning strategy. Pulst *et al.* [13] constructed a physical map of the critical region using P1 artificial chromosomes (PACs) and bacterial artificial chromosomes (BACs) and then identified CAG repeat-containing sequences.

Due to the lack of a YAC contig spanning the SCA2 candidate region a contig of PACs and BACs had to be generated. After several walking steps using STSs derived from PAC- and BAC-end sequences, overlap of the contigs was established by shared STS content and further confirmed by shared restriction fragments through hybridization of selected clones to Southern blots of *NotI*/*XbaI* digests of clones.

The dense localization of STSs allowed the precise positioning of YACs in the region. Y884_h_11 was the only YAC clone that was positive for both *D12S1332* and *D12S1333*. However, this clone contained an interstitial deletion that encompassed approximately 200 kb. A small portion of this deletion, which was subsequently shown to contain parts of the SCA2 gene, was not covered by any of the other YAC clones, but was contained in several PAC clones [14].

Anticipation observed in several chromosome 12-linked pedigrees as well as identification of a polyglutamine protein in Western blots of SCA2 patients [15] suggested that SCA2 was caused by CAG expansion. Clones containing CAG repeats were identified by hybridizing *XbaI*/*NotI* digests of a minimal tiling path of

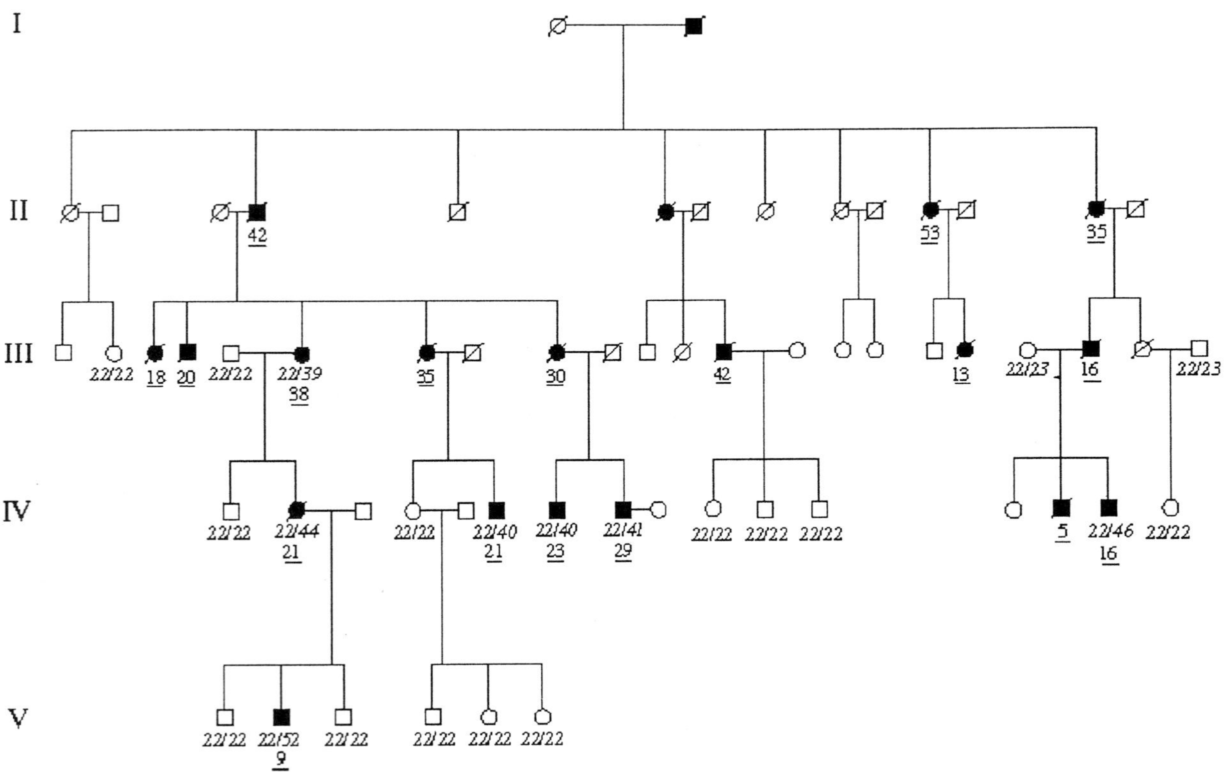

FIGURE 20-1 Anticipation of age of onset and meiotic SCA2 CAG repeat length instability in the FS pedigree (reprinted with permission from Adams *et al.* [23], *Neurology* **49**, 1163–1166, 1997). Age of onset is underlined.

clones with a $(CAG)_{10}$ oligonucleotide. Two CAG positive bands of distinct sizes were identified in the contig. Sequence analysis of one of these contained an extended CAG repeat that was twice interrupted. The repeat was embedded in a long open reading frame.

Amplification of the repeat region using flanking oligonucleotide primers showed two common alleles in normal individuals and the presence of a normal allele and an allele of increased size in all SCA2 patients from three independent pedigrees (Fig. 20-2). Sequence analysis of the normal alleles revealed that the two normal alleles had 22 and 23 repeats that were interrupted, whereas disease alleles were perfect repeats and ranged from 36 to 52 repeats [13].

IV. ALTERNATIVE WAYS OF GENE IDENTIFICATION

Simultaneous with positional cloning efforts, two groups employed novel ways to identify the SCA2 gene. Using a genomically based assay to detect expanded polyglutamines, designated DIRECT, Sanpei *et al.* [16] analyzed genomic DNAs from several patients with SCA2. The DNAs were digested with several restriction enzymes and Southern blots hybridized under high stringency conditions with a $(CAG)_{55}$ oligonucleotide. After determining the optimal hybridization conditions in a model system this approach was applied to DNAs from SCA2 patients. In patient DNAs digested with *Tsp*EI a 2.5-kb fragment was detected that was not present in unaffected family members or in controls. This fragment contained a CAG repeat with 42 repeat units and was mapped close to the SCA2 candidate region using a radiation hybrid panel. SCA2 patients from 10 independent pedigrees showed expansion of this repeat ranging from 35 to 59 units.

Using the 1C2 monoclonal antibody that recognizes long stretches of glutamines (Trottier *et al.*, [15]), Imbert *et al.* [17] identified clones in expression libraries generated from lymphoblastoid cDNAs from SCA2 and SCA7 patients. Of several positive clones, one clone contained 22 glutamines. Interestingly, this clone, which was later shown to encode the SCA2 cDNA, was identified in the library made from patients with SCA7 and contained the normal 22 glutamines. At the DNA level the repeat was not perfect and had two CAA interruptions. Primers flanking the repeat amplified expanded alleles in SCA2 patients ranging from 37 to 50 repeats.

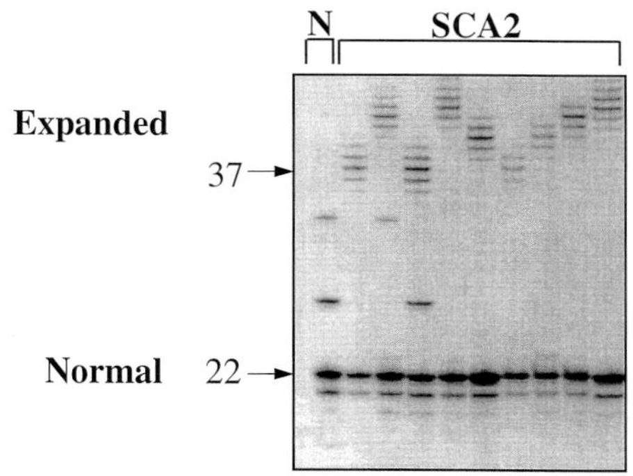

FIGURE 20-2 Expansion of the SCA2 repeat in chromosome 12-linked ataxia patients (SCA2). The DNA from a normal control (N) shows homozygosity for 22 repeats.

V. cDNA SEQUENCE

Using genomic clones as probes both Pulst *et al.* [13] and Sanpei *et al.* [16] identified several SCA2 cDNA clones from a fetal brain and a frontal cortex cDNA library, respectively. Both groups could not unequivocally identify the 5′ end of the SCA2 cDNA and used 5′ RACE and cloned RT-PCR products to identify additional cDNA sequence. Both groups predict the identical amino acid sequence with 1312 amino acids (22 glutamines [13]) and 1313 amino acids (23 glutamines [16]) with the CAG repeat coding for polyglutamine.

Sequence analysis of the 4-kb cDNA clone isolated by Imbert *et al.* [17] predicted a much shorter open reading frame. However, a second ORF in a different frame partially overlapped the first and predicted a larger protein if ribosomal frame shifting were to occur. Otherwise the cDNA sequence agreed with the sequence obtained by the other two groups. It is likely that the 4.0-kb cDNA clone represents a rare cDNA, since none of the other cDNA clones isolated by Pulst *et al.* [13] or Sanpei *et al.* [16] contained the second ORF. Furthermore, the protein recognized by the 1CE antibody [15] and by SCA2 specific antibodies (Pulst, unpublished) is more consistent with a protein of larger molecular weight.

The 5′ sequence of the SCA2 cDNA is extremely GC-rich and two potential ATG initiation codons can be identified. The most 5′ ATG is located 78 bp downstream of an in-frame stop codon. Usage of this translation initiation site predicts a protein of 140.1 kDa. The second ATG which has a better Kozak consensus sequence is located just 5′ to the CAG repeat and would result in a protein with relative MW of 125 kDa. Proteins observed by Western blot analysis and conservation of the 5′ ATG in the mouse (Nechiporuk *et al.*, unpublished) suggest that the 5′ ATG is the predominant site of translation initiation.

A. Sequence Homologies and Protein Domains

In analogy to the other SCA gene products, the SCA2 gene product has been designated ataxin-2. Homology

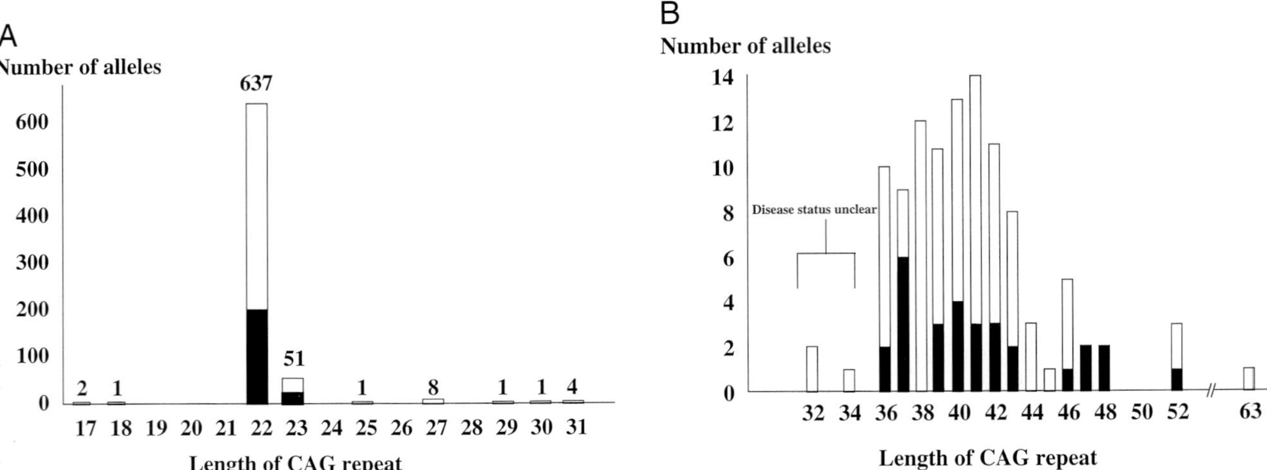

FIGURE 20-3 Distribution of SCA2 alleles on normal (A) and disease (B) chromosomes. Data compiled from Riess *et al.* [18] (open boxes) and Pulst *et al.* [13] (solid boxes). For the discussion of individuals with 32 and 34 repeats, see text.

searches using the cDNA and amino acid sequences have not identified homologies with proteins of known function. However, significant sequence homology was detected with a protein designated ataxin-2-related protein (A2RP) and the mouse SCA2 protein [13]. Despite the significant homologies, the polyglutamine tract in human ataxin-2 is not present in A2RP or in mouse ataxin-2, suggesting that it may not be important for ataxin-2 function. However, all acidic amino acids comprising a highly acid domain adjacent to the human polyglutamine domain are conserved in A2RP and mouse ataxin-2.

B. Expression Patterns

The SCA2 gene is widely expressed. On Northern blots, a 4.5-kb transcript is recognized in RNAs isolated from brain, heart, placenta, liver, skeletal muscle, and pancreas [13, 16, 17]. Little or no expression is seen in lung or kidney. The transcript is expressed throughout the brain. In RNAs isolated from SCA2 lymphoblastoid cell lines, expression of both the normal and expanded alleles is seen using reverse-transcribed PCR [13].

The SCA2 transcript in the mouse of identical size (Nechiporuk *et al.,* unpublished). Expression during mouse embryonic development with strong expression at days E11 and E12 suggests that ataxin-2 may have a role in normal embryogenesis.

C. Allele Ranges

The SCA2 CAG trinucleotide repeat is unusual in several aspects (Fig. 20-3). First, it is not highly polymorphic in normal individuals. Two alleles of 22 and 23 repeats account for >95% of alleles in most studies [13, 16, 18, 19]. Rare normal alleles ranging from 15 to 32 repeats have also been identified [16–18]. Second, normal alleles typically show one or two CAA interruptions. In contrast to the SCA1 gene, which contains CAT interruptions coding for histidine, the CAA interruptions do not interrupt the glutamine tract at the protein level. Thus, the biologic significance of intermediate alleles of 32 to 34 repeats may be difficult to determine (see below). Finally, the expansions on disease chromosomes are relatively small compared with SCA1 and SCA3. The most common disease alleles contain 37 to 39 repeats and are thus smaller than the longest normal alleles seen in the SCA3/MJD gene.

Disease alleles range from 36 to 59 repeats and carry perfect CAG repeats without interruption. It has not yet been resolved whether intermediate alleles exist similar to Huntington's disease. In a study of 241 apparently healthy octagenarians, alleles longer than the most common alleles with 22 and 23 repeats were extremely rare. Four alleles contained 27 repeats, and one allele each had 29 and 31 repeats [18]. When 842 patients with sporadic progressive ataxia were examined, 717 (82%) patients were homozygous for the 22-repeat allele. One patient had 30, four had 31, and two had 32 repeats [18]. One of the patients with 32 repeats was homozygous for an expansion in the Friedreich ataxia gene. An allele of 34 repeats was seen in the asymptomatic mother of a woman with SCA2. An allele with 32 repeats has also been seen in an asymptomatic 19-year-old whose symptomatic father carried an allele of 40 repeats [19]. The contracted allele had no CAA interruptions.

Several asymptomatic individuals in SCA2 pedigrees with 34 and 35 repeats on chromosomes carrying the disease haplotype have been identified [16, 17, 20]. However, these individuals are still too young to determine whether SCA2 might develop within the normal human life span.

SCA2 mutation can be found in patients without obvious family history of ataxia, although this has to be considered a rare event. Only 2 of 842 sporadic ataxia patients in the series of Riess *et al.* [18] had expansions of 41 and 49 repeats. In the series reported by Cancel *et al.* [19] 2 out of 90 patients with sporadic olivo-pontocerebellar atrophy had alleles with 37 and 39 repeats.

D. Anticipation and Meiotic Instability of the SCA2 Repeat

As in other diseases caused by unstable CAG DNA repeats, there is a clear inverse correlation between age of onset and repeat length. However, in contrast to other trinucleotide diseases, correlation is not linear and the best correlation is obtained with a negative exponential fit [13]. The widest range of age of onset is observed for fewer than 40 repeats. For example, the presence of 37 repeats was associated with ages of onset ranging from 20 to 60 years of age [13]. For larger repeat sizes, the variability is less and repeat sizes of >45 are almost always associated with disease onset under 20 years of age [13, 16–20]. Homozygosity for an expanded SCA2 allele does not appear to influence age of onset [16].

Initial observations in the Cuban pedigrees [3] and in the FS pedigree from southern Italy [6] did not point to consistent differences in the degree of anticipation depending on paternal or maternal inheritance. However, analysis of changes in CAG repeat sizes has indicated that large expansions are almost exclusively observed when the repeat is passed through the paternal germline [18, 20]. In contrast to these reports, Cancel *et al.* [19] did not find a paternal bias.

E. Frequency and Phenotype

The identification of the SCA2 CAG repeat has now widened the scope of phenotypic analysis to smaller pedigrees and has provided an estimate of SCA2 frequency in different sets of ethnic and geographic populations. SCA2 is a frequent cause of autosomal dominant ataxias. Geschwind *et al.* [20] found that, in an ethnically varied population in the UCLA ataxia clinic, SCA2 accounted for 13% of the autosomal dominant cerebellar ataxias (ADCAs) compared with 6% for SCA1 and 23% for SCA3 (Table 20-1). No common haplotype was found on disease chromosomes. In a large series from several ataxia clinics in Germany, SCA2 represented 14% of ADCA pedigrees [18]. A similar percentage (15%) was reported by Cancel *et al.* [19] in a set of 184 families from an ethnically and geographically diverse population.

Table 20-1 SCA2 Phenotype Compared with SCA1 and SCA3[a]

	SCA1	SCA2[a]	SCA2[b]	SCA3
Cerebellar dysfunction	100	100	100	100
Reduced saccadic velocity	50	71	92	10
Myoclonus	0	40	0	4
Dystonia or chorea	20	0	38	8
Pyramidal involvement	70	29	31	70
Peripheral neuropathy	100	94	44	80
Intellectual impairment	20	31	37	5

[a]Percentage of patients with a specific sign are indicated. Percentages for SCA1, SCA2[a], and SCA3 were modified from Riess *et al.* [18], and those for SCA2[b] were from Geschwind *et al.* [20].

In the original Cuban study, age of onset was from 2 to 65 years with a mean in the third to fourth decade. Earliest symptoms were gait ataxia often accompanied by leg cramps [3]. Greater than 50% of patients developed a kinetic or postural tremor, decreased muscle tone and tendon reflexes, and abnormal eye movements with slowed saccades progressing to supranuclear ophthalmoplegia. In families mapped to CHR12q24.1 by genetic linkage analysis, the SCA2 phenotype showed significant interfamilial variability both in the degree of anticipation and in the prominence of secondary symptoms at various stages of illness. Belal *et al.* [7] described a surprising 23% incidence of extrapyramidal signs in his family and Durr *et al.* [8] a 29% incidence of dementia.

Using direct analysis of the SCA2 repeat, Geschwind *et al.* [20] found almost universal presence of cerebellar ataxia and slow saccadic eye movements, but also a relatively high incidence of dystonia or chorea (38%) and dementia (37%). There was some indication of occurrence of specific phenotypes within families. Mild, primarily cerebellar symptoms appeared to segregate in families, whereas others had an early onset with dementia and chorea. One patient had been clinically diagnosed as having a typical MJD-like phenotype.

Similar findings were also reported by Cancel *et al.* [17] in a series of 111 patients from 32 families of diverse origins. Slow eye movements were seen in 56%, fasciculations in 25%, and dystonia in 9%. The authors also examined which findings were correlated with disease duration and which were correlated with increasing CAG repeat length. The size of the repeat was significantly larger in patients with dystonia, myoclonus, and myokymia, whereas both CAG length and duration influenced the frequency of decreased reflexes and vibration sense in the lower extremities, amyotrophy, fasciculations, and slow eye movements.

Schols *et al.* [17] found SCA2 expansion in 6 of 64 ADCA families of German ancestry. Clinical features were highly variable within and between families. Although no specific single feature was sufficient to distinguish SCA2 from other SCAs, slowed saccades, postural and action tremors, myoclonus, and hyporeflexia were more common than in SCA1 and SCA3. Similar to the study reported by Geschwind *et al.* [20] this study was remarkable in that all patients were examined by the same clinician, thus reducing interobserver variability.

Buerk [22] examined several SCA2 patients defined by linkage analysis and compared them with SCA1 and SCA3 patients. SCA2 patients had significantly slower saccadic speed (138°/s) than patients with SCA1 (244°/s) or SCA3 (347°/s). All eight SCA2 patients had saccadic velocities 2 standard deviations below the mean of a control group. MRI scans showed that the middle cerebellar peduncles and the pontine base were significantly smaller in SCA2 compared with SCA1 and SCA3 patients.

F. Neuropathology

Seven postmortem examinations have been reported in the Holguin population of Cuba [1]. There was a marked reduction in the number of cerebellar Purkinje cells. In silver preparations, Purkinje cell dendrites had poor arborization and torpedo-like formation of their axons as they passed through the granular layer. Parallel fibers were scanty. Granule cells were decreased in number, whereas Golgi and basket cells as well as neurons in the dentate and other cerebellar nuclei were well preserved. In the brainstem, there was marked neuronal loss in the inferior olive and pontocerebellar nuclei. Six of seven brains also had marked loss in the substantia nigra. In five spinal cords that were available for analysis marked demyelination was present in the posterior columns and to a lesser degree in the spinocerebellar tracts. Motor neurons and neurons in Clarke's column were reduced in size and number. Anterior and posterior roots were partially demyelinated, especially in lumbar and sacral segments.

Durr [8] reported autopsy findings in two patients from the Martinican families. In addition to the findings reported by Orozco *et al.* [1], they also noted severe gyral atrophy most prominent in the fronto-temporal lobes. The cerebral cortex was thinned, but without neuronal rarefaction. The cerebral white matter was atrophic and gliotic. Degeneration in the nigro-luysopallidal system again mainly involved the substantia nigra. One brain showed patchy loss in parts of the third nerve nuclei. Adams *et al.* [23] reported similar findings in one member of the FS pedigree (see Fig. 20-1). Nerve biopsy has shown moderate loss of large myelinated fibers [9].

VI. OUTLOOK

Although the clinical spectrum of SCA2 has been well defined, little is known about the function of the normal or mutated gene products. Preliminary Western blot analysis has indicated widespread expression of ataxin-2 (Huynh *et al.*, unpublished). Using fluorescently tagged ataxin-2 with 22 glutamines, a cytoplasmic localization is seen in COS cells (Scoles *et al.*, unpublished). It is not yet known whether an abnormal subcellular localization is detected in SCA2 postmortem brains or in cells transfected with expanded SCA2 alleles.

References

1. Orozco, G., Estrada R., Perry, T. L., Arana, J., Fernandez, R., Gonzalez-Quevedo, A., Galarraga, J., and Hansen, S. (1989). Dominantly inherited olivopontocerebellar atrophy from eastern Cuba: clinical, neuropathological, and biochemical findings. *J. Neurol. Sci.* **93,** 37–50.
2. Gudmundsson, K. (1969). The prevalence and occurrence of some rare neurological diseases in Iceland. *Acta Neurol. Scand.* **45,** 114–118.
3. Orozco, G., Fleites, A., Cordoves Sagaz, R., and Auburger, G. (1990). Autosomal dominant cerebellar ataxia: clinical analysis of 263 patients from a homogeneous population in Holguin, Cuba. *Neurology* **40,** 1369–1375.
4. Auburger, G., Orozco, G., Capote, R. F., Sanchez, S. G., Perez, M. P., del Cueto, M. E., Meneses, M. G., Farrall, M., Williamson, R., Chamberlain, S., and Baute, L. H. (1990). Autosomal dominant ataxia: genetic evidence for locus heterogeneity from a Cuban founder-effect population. *Am. J. Hum. Genet.* **46,** 1163–1177.
5. Gispert, S., Twells, R., Orozco, G., Brice, A., Weber, J., Herdero, L., Scheufler, K., Riley, B., Allotey, R. I., Nothers, C., Hillerman, R., Lunkes, A., Khati, C., Stevanin, G., Hernandez, A., Magariuno, C., Klockgether, T., Durr, A., Chneiweiss, H., Enczmann, J., Farrall, M., Beckmann, J., Mullan, M., Wernet, P., Agid, Y., Freund, H. J., Williamson, R., Auburger, G., and Chamberlain, S. (1994). Chromosomal assignment of the second (Cuban) locus for autosomal dominant cerebellar ataxia (SCA2) to chromosome 12q23-24.1 *Nature Genet.* **4,** 295–299.
6. Pulst, S.-M., Nechiporuk, A., and Starkman, S. (1993). Anticipation in spinocerebellar ataxia type 2. *Nature Genet.* **5,** 8–10.
7. Belal, S., Cancel, G., Stevanin, G., Hentati, F., Khati, C., Ben Hamida, C., Auburger, G., Agid, Y., Ben Hamida, M., and Brice, A. (1994). Clinical and genetic analysis of a Tunesian family with autosomal dominant cerebellar ataxia type 1 linked to the SCA2 locus. *Neurology* **44,** 1423–1426.
8. Durr, A., Smadja, D., Cancel, G., Lezin, A., Stevanin, G., Mikol, J., Bellance, R., Buisson, G. G., Chneiweiss, H., Dellanave, J., Agid, Y., Brice, A., and Vernant, J. C. (1995). Autosomal dominant cerebellar ataxia type 1 in Martinique (French West Indies):

clinical and neuropathological analysis of 53 patients from three unrelated SCA2 families. *Brain* **118,** 1573–1581.
9. Filla, A., DeMichele, G., Banfi, S., Santoro, L., Perretti, A., Cavalcanti, F., Pianese, L., Castaldo, I., Barrieri, F., Campanella, G., and Cocozza, S. (1995). Has spinocerebellar ataxia type 2 a distinct phenotype? Genetic and clinical study of an Italian family. *Neurology* **45,** 793–796.
10. Lopes-Cendes, I., Andermann, E., and Attig, E., Cendes, F., Wagner, M., Gerstenbrand F., Andermann, F., and Rouleau, G. A. (1994). Confirmation of the SCA-2 locus as an alternative locus for dominantly inherited spinocerebellar ataxia and refinement of the candidate region. *Am. J. Hum. Genet.* **54,** 774–781.
11. Nechiporuk, A., Lopes-Cendes, I., Nechiporuk, T., Starkman, S., Andermann, E., Rouleau, G. A., Weissenbach, J. S., Kort, E., and Pulst, S.-M. (1996). Genetic mapping of spinocerebellar ataxia type 2 gene on human chromosome 12. *Neurology* **46,** 1731–1735.
12. Gispert, S., Lunkes, A., Santos, N., Orozco, G., Ha-Hao, D., Ratzlaff, T., Agiar, J., Torrens, I., Heredero, L., Brice, A., Cancel, G., Stevanin, G., Vernant, J.-C., Dürr, A., Lepage-Lezin, A., Belal, S., Ben-Hamida, M., Pulst, S.-M., Rouleau, G., Weissenbach, J., LePaslier, D., Kucherlapati, R., Montgomery, K., Fukui, K., and Auburger, G. (1995). Localization of the candidate gene d-amino acid oxidase outside the refined 1-cM region of spinocerebellar ataxia 2. *Am. J. Hum. Genet.* **57,** 972–975.
13. Pulst, S.-M., Nechiporuk, A., Nechiporuk, T., Gispert, S., Chen, X., Lopes-Cendes, I., Perlman, S., Starkman, S., Orozco, G., Lunkes, A., DeJong, P., Rouleau, G., Auburger, G., Korenberg, J., Figueroa, C., and Sahba, S. (1996). Moderate expansion of a normally biallelic trinucleotide repeat in spinocerebellar ataxia type 2. *Nature Genet.* **14,** 269–276.
14. Nechiporuk, T., Nechiporuk, A., Sahba, S., Figueroa, K., Shibata H., Chen, Xiao-Ning, Korenberg, J. R., and Pulst, S. (1997). A high-resolution PAC and BAC map of the SCA2 region. *Genomics* **44,** 321–329.
15. Trottier, Y., Lutz, Y., Stevanin, G., Imbert, G., Devys, D., Cancel, G., Saudou, F., Weber, C., David, G., Laszlo, T., Agid, Y., Brice, A., and Mandel, J. L. (1995). Polyglutamine expansion as a pathological epitope in Huntington's disease and four dominant cerebellar ataxias. *Nature* **378,** 403–406.
16. Sanpei, K., Takano, H., Igarashi, S., Sato, T., Oyake, M., Sasaki, H., Wakisaka, A., Tashiro, K., Ishida, Y., Ikeuchi, T., Koide, R., Saito, M., Sato, A., Tanaka, T., Hanyu, S., Takiyama, Y., Nishizawa, M., Shimizu, N., Nomura, Y., Sagawa, M., Iwabuchi, K., Eguchi, L., Tanaka, H., Takahashi, H., and Tsuji, S. (1996). Identification of the spinocerebellar ataxia type 2 gene using a direct identification of repeat expansion and cloning technique, DIRECT. *Nature Genet.* **14,** 277–284.
17. Imbert, G., Saudou, F., Yvert, G., Devys, G., Trottier, Y., Garnier, J., Weber, C., Mandel, J. L., Cancel, G., Abbas, N., Durr, A., Didierjean, O., Stevanin, G., Agid, Y., and Brice, A. (1996). Cloning of the gene for spinocerebellar ataxia 2 reveals a locus with high sensitivity to expanded CAG glutamine repeats. *Nature Genet.* **14,** 285–291.
18. Riess, O., Laccone, F., Gispert, S., Schöls, L., Zühlke, C., Vieira-Saecker, A.-M., Herlt, S., Wessel, K., Epplen, J., Weber, B., Kreuz, F., Chahrokh-Zadeh, S., Meindl, A., Lunkes, A., Aguiar, J., Macek, Jr., M., Krebsova, A., Macek, Sr., M., Bürk, K., Tinschert, S., Schreyer, I., Pulst, S.-M., and Auburger, G. (1997). SCA2 trinucleotide expansion in German SCA patients. *Neurogenetics* **1,** 59–64.
19. Cancel, G., Dürr, A., Didierjean, O., Imbert, G., Bürk, K., Lezin, A., Belal, S., Benomar, A., Abada-Bendib, M., Vail, C., Guimarãe, J., Chneiweiss, H., Stevanin, G., Yvert, G., Abbas, N., Saudou, F., Lebre, A.-S., Yahyaoui, M., Hentati, F., Vernant, J.-C., Klockgether, T., Mandel, J.-L., Agid, Y., and Brice, A. (1997). Molecular and clinical correlations in spinocerebellar ataxia 2: a study of 32 families. *Hum. Mol. Genet.* **6,** 709–715.
20. Geschwind, D., Perlman, S., Figueroa, P., Treiman, L., and Pulst, S.-M. (1997). The prevalence and wide clinical spectrum of the spinocerebellar ataxia type 2 trinucleotide repeat in patients with autosomal dominant cerebellar ataxia. *Am. J. Hum. Genet.* **60,** 842–850.
21. Schols, S., Gispert, S., Vorgerd, M., Veira-Saecker, M., Blanke, P., Auburger, G., Amoinidis, G., Meves, S., Epplen, J., Przuntek, H., Pulst, S. M., and Riess, O. (1997). Spinocerebellar ataxia type 2: genotype and phenotype in German kindred. *Arch. Neurol.* **54,** 1073–1080.
22. Burk, K., Abele, M., Fetter, M., Dichgans, J., Skalej, M., Laccone, F., Didierjean, O., Brice, A., and Klockgether, T. (1996). Autosomal dominant cerebellar ataxia type 1: clinical features and magnetic resonance imaging in families with SCA1, SCA2 and SCA3. *Brain* **119,** 1497–1505.
23. Adams, C., and Pulst, S. M. (1997). Clinical and molecular analysis of a pedigree of southern Italian ancestry with spinocerebellar ataxia type 2. *Neurology* **49,** 1163–1166.

CHAPTER 21

Autosomal Dominant Cerebellar Ataxia with Macular Dystrophy (SCA7) Is Caused by a Highly Unstable CAG Repeat Expansion

GILLES DAVID AND NACER ABBAS — INSERM U289, Hôpital de la Salpêtrière, 75651 Paris Cedex 13, France

ALEXANDRA DÜRR — INSERM U289 and Fédération de Neurologie, Hôpital de la Salpêtrière, 75651 Paris Cedex 13, France

GIOVANNI STEVANIN AND GÉRALDINE CANCEL — INSERM U289, Hôpital de la Salpêtrière, 75651 Paris Cedex 13, France

GAËL YVERT, GEORGES IMBERT, FRÉDÉRIC SAUDOU, AND JEAN-LOUIS MANDEL — Institut de Génétique et de Biologie Moléculaire et Cellulaire (IGBMC), CNRS, INSERM, ULP, 67404 Illkirch Cedex, C. U. de Strasbourg, France

YVES AGID AND ALEXIS BRICE — INSERM U289 and Fédération de Neurologie, Hôpital de la Salpêtrière, 75651 Paris Cedex 13, France

I. INTRODUCTION

The association of cerebellar ataxia and retinal degeneration with autosomal dominant inheritance has been reported in a number of families with various denominations. Because of the strong concordance between cerebellar ataxia and retinal degeneration in these families, the phenotype was classified as a distinct form of autosomal dominant cerebellar ataxia, type III according to Konigsmark and Weiner [1] or type II according to Harding [2]. Several recent studies have characterized this phenotype in detail [3–5]. Marked anticipation with a parental sex bias suggested that the molecular basis of this disorder is an unstable mutation. Subsequently the ADCA II gene, SCA7, was mapped to chromosome 3p [6–8], and an abnormal protein of 130 kDa containing a polyglutamine expansion was detected in patients [9]. Finally a CAG repeat that is expanded and highly unstable in SCA7 patients was identified by positional cloning [10].

II. CLINICAL FEATURES

The age at onset varies widely within families with ADCA II. The mean age at onset is 29, but ranges from under 1 to over 60 years of age. Analysis of affected parent–child couples reveals a striking anticipation with a mean of 24 years. Mean anticipation is greater in paternal (28 years) than in maternal (18 years) transmissions. Anticipations of more than 50 years have been reported in some paternal transmissions, in which the child is affected many years before the father presents symptoms. As a result of the parental sex bias, most of the juvenile cases with onset below the age of 10 are paternally transmitted. Anticipation also results in a more rapid clinical course. Disease duration until death is reduced to a few years in early onset patients whereas disease durations up to 30 years have been reported in late-onset patients. Although anticipation is also observed in ADCA I families, which carry the SCA1, SCA2, or SCA3/MJD mutations, it never reaches the same extent. Incomplete penetrance is also a feature of ADCA II. Several obligate carriers have been identified that are still unaffected at 60 years of age or more, up to 82 [6, 11]. Although the penetrance is probably greater than 90%, incomplete penetrance complicates genetic counseling.

Although cerebellar ataxia and loss of vision can occur at the same time, in most patients cerebellar ataxia is the presenting symptom [3]. More than 30 years can elapse between the appearance of cerebellar symptoms and visual failure. In adults with onset over 30, ataxia is usually the presenting sign, whereas in patients with onset before 30, isolated decreased visual acuity, alone or associated with cerebellar ataxia, is the initial symptom. In both groups of patients, cerebellar ataxia is always associated with dysarthria. Various signs are associated with cerebellar ataxia in ADCA II (Table 21-1) [3–5, 10]. Most patients present with a pyramidal syndrome (increased reflexes and/or extensor plantar reflexes and/or lower limb spasticity), decreased vibration sense, dysphagia, and sphincter disturbances, as well as oculomotor abnormalities (supranuclear ophthalmoplegia and/or viscosity of eye movements). Extrapyramidal features (dystonia), myokymia, and mental impairment are infrequent, but hypoacousia, absent in other ADCAs, is present in one-fourth of the patients.

Loss of vision in ADCA II is progressive, leading to blindness with a bilateral and symmetrical pattern [4, 5]. Progressive macular degeneration first affects central

TABLE 21-1 Clinical Features of SCA7 Patients

Number of patients (families)	67 (19)
Mean age at onset (years)	29 (range 1–70)
Frequency of clinical signs:	
Cerebellar ataxia and dysarthria	+++
Decreased visual acuity	+++
Decreased saccade velocity	+++
Hyperreflexia in lower limbs	+++
Extensor plantar reflex	++
Lower limb spasticity	++
Decreased vibration sense	++
Sphincter disturbances	++
Decreased hearing	+
Lower limb amyotrophy	+
Extrapyramidal signs	+
Facial myokymia	+
Mental deterioration	+

Note. Frequency: +, 0–24%; ++, 25–74%; +++, 75–100%.

vision; peripheral vision is preserved at early stages. This explains why patients do not complain of visual symptoms before advanced visual failure, and why night vision is not impaired. Interestingly, dyschromatoptia in the blue–yellow axis is found years before visual failure becomes symptomatic. In contrast, fundoscopic abnormalities, consisting of a loss of the foveal reflex and progressive molting of pigment at the macula, are often delayed (Fig. 21-1). Secondary optic atrophy can often be detected in later stages. Electroretinogram shows abnormal scotopic responses, but photopic responses are preserved late. Visual evoked potentials are not discriminative for diagnosis purposes [4].

Brain imaging shows marked atrophy in the cerebellum, particularly in the vermis, and of the brain stem (Fig. 21-2), which may be associated with moderate atrophy of the cerebral cortex, which corresponds to postmortem anatomopathological observations of olivopontocerebellar atrophy [4, 5, 12]. In addition, the pregeniculate visual pathways are affected, probably as a consequence of retinal degeneration. Pathological examination of the retina shows early degeneration of the photoreceptors and the bipolar and the granular cells, particularly in the foveal and parafoveal regions. Later, the inner retinal layers are affected with patchy loss of epithelial pigment cells and penetration of pigmented cells in the retinal layers [12].

III. MAPPING OF THE SCA7 LOCUS TO CHROMOSOME 3p

After exclusion of loci involved in other ADCAs or in retinal degeneration [3–5, 13] a genome search was performed in families from various geographical origins. SCA7 was mapped to chromosome 3 by three independent groups [6–8]. According to these data, the candidate interval was located in a 12-cM region between markers D3S1312 and D3S1217. Analyses of additional families and the use of new microsatellite markers allowed further reduction of the candidate region [11, 14]. In parallel, a yeast artificial chromosome (YAC) contig

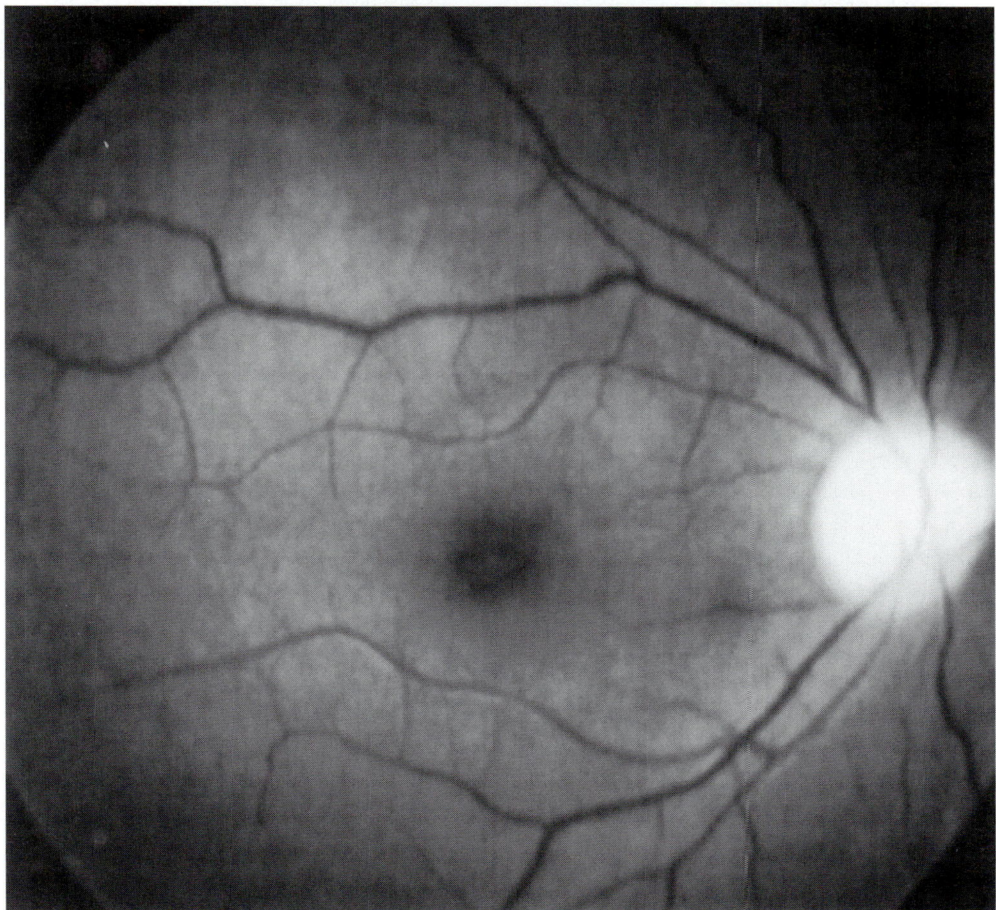

FIGURE 21-1 Fundal photograph of the right eye. Note the alteration of the foveal reflex and the increased pigmentation at the macula. The temporal part of the optic disc is pale. Photograph reproduced by the courtesy of Dr. M. H. Rigolet.

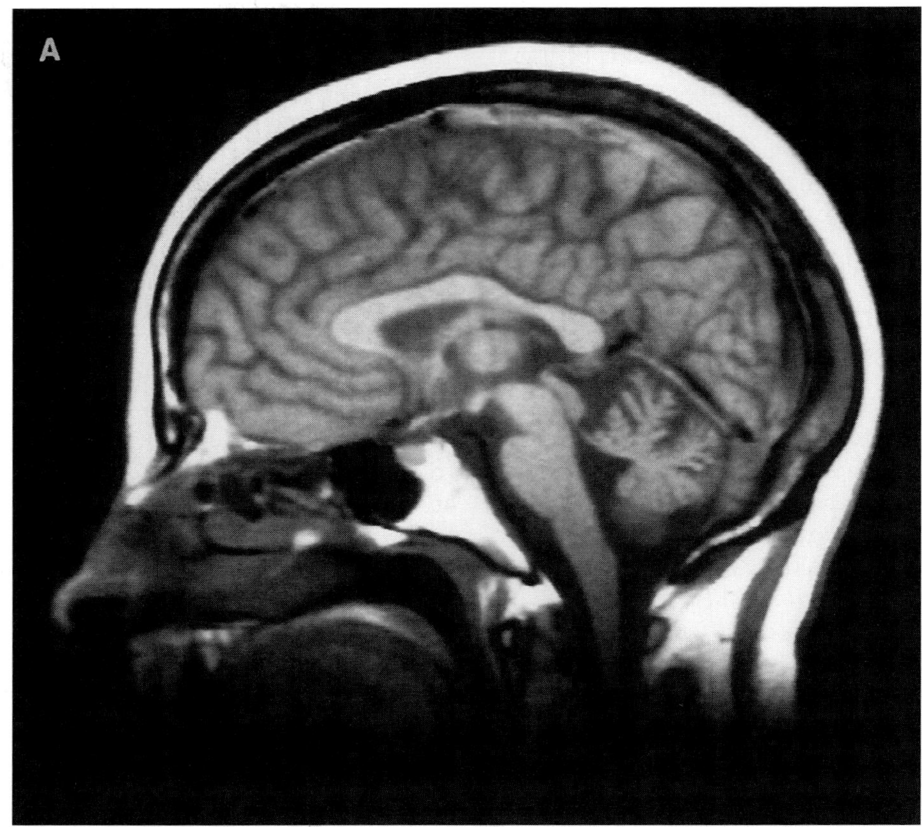
A

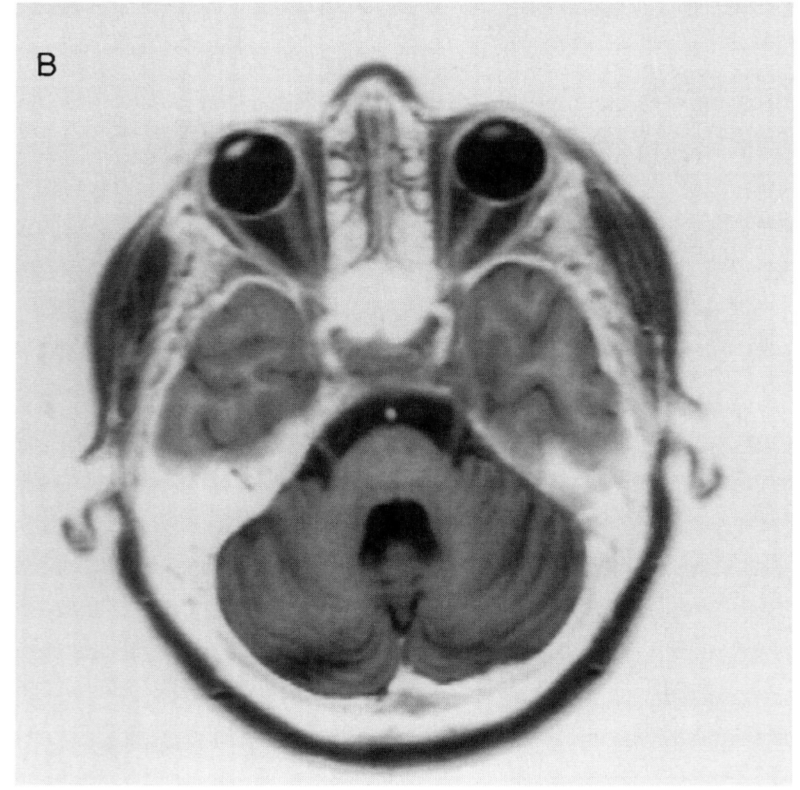
B

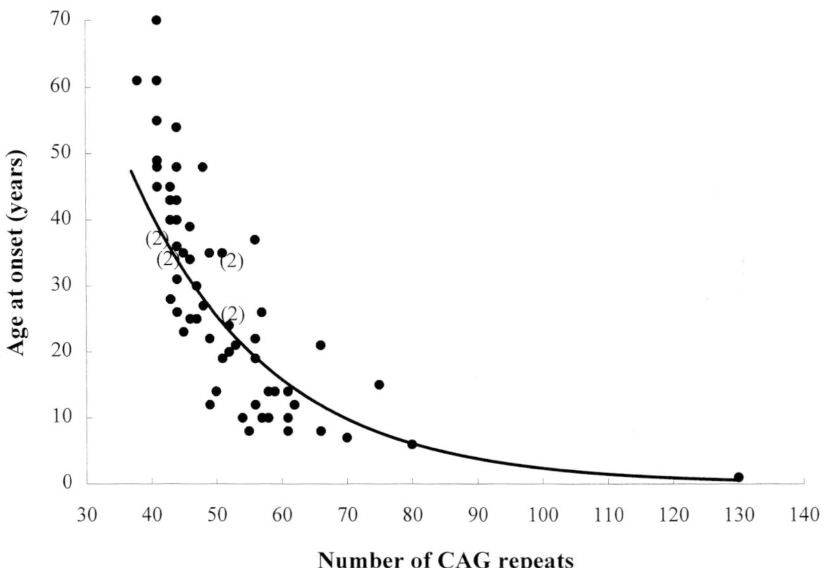

FIGURE 21-8 Negative correlation between the age at onset and the number of CAG repeats in expanded alleles from 61 SCA7 patients. Values in parentheses indicate the number of patients. The coefficient was calculated for an exponential regression ($r = -0.86$).

for helpful discussions. This research was supported by the Association Française contre les Myopathies (AFM), the Centre National de la Recherche Scientifique, the Institut National de la Santé et de la Recherche Médicale, the VERUM Foundation, the Groupement d'Etudes et de Recherches sur les Génomes (No. GREG9794), the CHRU of Strasbourg, the Association pour le Développement de la Recherche sur les Maladies Génétiques Neurologiques et Psychiatriques, the Association Française Retinitis Pigmentosa (AFRP), Biomed (No. CEE BMH4-CT960244), and Biomed Concerted Action (No. CEE BMH1-CT94-1243). G.D. was supported by the AFRP—Retina France and the Fédération des Aveugles et Handicapés Visuels de France. G.C. was supported by the AFM and G.I. by the Association Huntington France.

References

1. Konigsmark, B. W., and Weiner, L. P. (1970). The olivopontocerebellar atrophies: a review. *Medicine (Baltimore)* **49,** 227–241.
2. Harding, A. E. (1982). The clinical features and classification of the late onset autosomal dominant cerebellar ataxias. A study of 11 families, including descendants of the 'the Drew family of Walworth'. *Brain* **105,** 1–28.
3. Benomar, A., Le Guern, E., Dürr, A., Ouhabi, H., Stevanin, G., Yahyaoui, M., Chkili, T., Agid, Y., and Brice, A. (1994). Autosomal-dominant cerebellar ataxia with retinal degeneration (ADCA type II) is genetically different from ADCA type I. *Ann. Neurol.* **35,** 439–444.
4. Enevoldson, T. P., Sanders, M. D., and Harding, A. E. (1994). Autosomal dominant cerebellar ataxia with pigmentary macular dystrophy. A clinical and genetic study of eight families. *Brain* **117,** 445–460.
5. Gouw, L. G., Digre, K. B., Harris, C. P., Haines, J. H., and Ptacek, L. J. (1994). Autosomal dominant cerebellar ataxia with retinal degeneration: clinical, neuropathologic, and genetic analysis of a large kindred. *Neurology* **44,** 1441–1447.
6. Benomar, A., Krols, L., Stevanin, G., Cancel, G., Le Guern, E., David, G., Ouhabi, H., Martin, J. J., Dürr, A., Zaim, A., Ravise, N., Busque, C., Penet, C., Van Regemorter, N., Weissenbach, J., Yahyaoui, M., Chkili, T., Agid, Y., Van Broeckhoven, C., and Brice, A. (1995). The gene for autosomal dominant cerebellar ataxia with pigmentary macular dystrophy maps to chromosome 3p12-p21.1. *Nature Genet.* **10,** 84–88.
7. Gouw, L. G., Kaplan, C. D., Haines, J. H., Digre, K. B., Rutledge, S. L., Matilla, A., Leppert, M., Zoghbi, H. Y., and Ptacek, L. J. (1995). Retinal degeneration characterizes a spinocerebellar ataxia mapping to chromosome 3p. *Nature Genet.* **10,** 89–93.
8. Holmberg, M., Johansson, J., Forsgren, L., Heijbel, J., Sandgren, O., and Holmgren, G. (1995). Localization of autosomal dominant cerebellar ataxia associated with retinal degeneration and anticipation to chromosome 3p12-p21.1. *Hum. Mol. Genet.* **4,** 1441–1445.
9. Trottier, Y., Lutz, Y., Stevanin, G., Imbert, G., Devys, D., Cancel, G., Saudou, F., Weber, C., David, G., Laszlo, T., Agid, Y., Brice, A., and Mandel, J. (1995). Polyglutamine expansion as a pathological epitope in Huntington's disease and four dominant cerebellar ataxias. *Nature* **378,** 403–406.
10. David, G., Abbas, N., Stevanin, G., Durr, A., Yvert, G., Cancel, G., Weber, C., Imbert, G., Saudou, F., Antoniou, E., Drabkin, H., Gemmill, R., Giunti, P., Benomar, A., Wood, N., Ruberg, M., Agid, Y., Mandel, J., and Brice, A. (1997) Cloning of the SCA7 gene reveals a highly unstable CAG repeat expansion. *Nature Genet.* **17,** 65–70.
11. Krols, L., Martin, J. J., David, G., Van Regemorter, N., Benomar, A., Lofgren, A., Stevanin, G., Dürr, A., Brice, A., and Van Broeckhoven, C. (1997). Refinement of the locus for autosomal dominant cerebellar ataxia type II to chromosome 3p21.1-14.1. *Hum. Genet.* **99,** 225–232.
12. Martin, J. J., Van Regemorter, N., Krols, L., Brucher, J. M., de Barsy, T., Szliwowski, H., Evrard, P., Ceuterick, C., Tassignon, M. J., Smet-Dieleman, H., Hayez-Delatte, F., Willems, P. J., and Van Broeckhoven, C. (1994). On an autosomal dominant form

of retinal-cerebellar degeneration: an autopsy study of five patients in one family. *Acta Neuropathol.* (*Berlin*), **88,** 277–286.

13. Kumar, D., Blank, C. E., and Gelsthorpe, K. (1986). Hereditary cerebellar ataxia and genetic linkage with HLA. *Hum. Genet.* **72,** 327–332.
14. David, G., Giunti, P., Abbas, N., Coullin, P., Stevanin, G., Horta, W., Gemmill, R., Weissenbach, J., Wood, N., Cunha, S., Drabkin, H., Harding, A. E., Agid, Y., and Brice, A. (1996). The gene for autosomal dominant cerebellar ataxia type II is located in a 5-cM region in 3p12-p13: genetic and physical mapping of the SCA7 locus. *Am. J Hum. Genet.* **59,** 1328–1336.
15. Stevanin, G., Trottier, Y., Cancel, G., Dürr, A., David, G., Didierjean, O., Bürk, K., Imbert, G., Saudou, F., Abada-Bendib, M., Gourfinkel-An, I., Benomar, A., Abbas, N., Klockgether, T., Grid, D., Agid, Y., Mandel, J., and Brice, A. (1996). Screening for proteins with polyglutamine expansions in autosomal dominant cerebellar ataxias. *Hum. Mol. Genet.* **5,** 1887–1892.
16. Lindblad, K., Savontaus, M. L., Stevanin, G., Holmberg, M., Digre, K., Zander, C., Ehrsson, H., David, G., Benomar, A., Nikoskelainen, E., Trottier, Y., Holmgren, G., Ptacek, L. J., Anttinen, A., Brice, A., and Schalling, M. (1996). An expanded CAG repeat sequence in spinocerebellar ataxia type 7. *Genome Res.* **6,** 965–971.
17. Gerber, H. P., Seipel, K., Georgiev, O., Hofferer, M., Hug, M., Rusconi, S., and Schaffner, W. (1994). Transcriptional activation modulated by homopolymeric glutamine and proline stretches. *Science* **263,** 808–811.
18. La Spada, A. R., Wilson, E. M., Lubahn, D. B., Harding, A. E., and Fischbeck, K. H. (1991). Androgen receptor gene mutations in X-linked spinal and bulbar muscular atrophy. *Nature* **352,** 77–79.
19. Mandel, J. (1997). Breaking the rule of three. *Nature* **386,** 767–769.
20. Dürr, A., Stevanin, G., Cancel, G., Duyckaerts, C., Abbas, N., Didierjean, O., Chneiweiss, H., Benomar, A., Lyon-Caen, O., Julien, J., Serdaru, M., Penet, C., Agid, Y., and Brice, A. (1996). Spinocerebellar ataxia 3 and Machado-Joseph disease: clinical, molecular and neuropathological features. *Ann. Neurol.* **39,** 490–499.
21. Cancel, G., Dürr, A., Didierjean, O., Imbert, G., Bürk, K., Lezin, A., Belal, S., Benomar, A., Abada-Bendib, M., Vial, C., Guimaraes, J., Chneiweiss, H., Stevanin, G., Yvert, G., Abbas, N., Saudou, F., Lebre, A., Yahyaoui, M., Hentati, F., Vernant, J., Klockgether, T., Mandel, J., Agid, Y., and Brice, A. (1997). Molecular and clinical correlations in spinocerebellar ataxia 2: a study of 32 families. *Hum. Mol. Genet.* **6,** 709–715.
22. Dürr, A., and Brice, A. (1996). Genetics of movement disorders. *Curr. Opin. Neurol.* **9,** 290–297.
23. Sanpei, K., Takano, H., Igarashi, S., Sato, T., Oyake, M., Sasaki, H., Wakisaka, A., Tashiro, K., Ishida, Y., Ikeuchi, T., Koide, R., Saito, M., Sato, A., Tanaka, T., Hanyu, S., Takiyama, Y., Nishizawa, M., Shimizu, N., Nomura, Y., Sagawa, M., Iwabuchi, K., Eguchi, I., Tanaka, H., Takahashi, H., and Tsuji, S. (1996). Identification of the spinocerebellar ataxia type 2 gene using a direct identification of repeat expansion and cloning technique, DIRECT. *Nature Genet.* **14,** 277–284.
24. Duyao, M., Ambrose, C., Myers, R., Novelletto, A., Persichetti, F., Frontali, M., Folstein, S., Ross, C., Franz, M., Abbott, M., Gray, J., Conneally, P., Young, A., Penney, J., Hollingsworth, Z., Shoulson, I., Lazzarini, A., Falek, A., Koroshetz, W., Sax, D., Bird, E., Vonsattel, J., Bonilla, E., Alvir, J., Bickham Conde, J., Cha, J. H., Dure, L., Gomez, F., Ramos, M., Sanchez-Ramos, J., Snodgrass, S., de Young, M., Wexler, N., Moskowitz, C., Penchaszadeh, G., MacFarlane, H., Anderson, M., Jenkins, B., Srinidhi, J., Barnes, G., Gusella, J., and MacDonald, M. (1993). Trinucleotide repeat length instability and age of onset in Huntington's disease. *Nature Genet.* **4,** 387–392.
25. Kremer, B., Almqvist, E., Theilmann, J., Spence, N., Telenius, H., Goldberg, Y. P., and Hayden, M. R. (1995). Sex-dependent mechanisms for expansions and contractions of the CAG repeat on affected Huntington disease chromosomes. *Am. J. Hum. Genet.* **57,** 343–350.

CHAPTER 22

Genetic and Molecular Studies of Machado–Joseph Disease

MIKA NAKAMOTO, HANAKO IKEDA, AND AKIRA KAKIZUKA
The Fourth Department, Osaka Bioscience Institute, Osaka 565, Japan

I. INTRODUCTION

In 1972, Nakano *et al.* reported a large family of Portuguese–Azorean descent with a dominantly inherited ataxia, who migrated to Massachusetts in the late 19th and early 20th centuries. Most of the affected members started to manifest the illness as an ataxic gait in their 40s but the other neurological symptoms and signs varied considerably among the individuals. The affected region was speculated to be the cerebellum and brainstem. From the unique inheritance, symptoms, and pathologies, the disease was referred to as Machado disease from the family name [1]. In the same year, another family of Portuguese descent who manifested a dominantly inherited nigro-spino-dentatal degeneration with nuclear ophthalmoplegia was reported by Woods *et al.* [2]. In 1976, Rosenberg *et al.* reported a third large family, the Joseph family, with an autosomal dominant striato-nigral degeneration, who also migrated from the Azorean island of Flores to California [3]. All three familial neurodegenerations were thought to be different disorders at the time. In 1977, Romanul *et al.* reported a fourth family of Portuguese ancestry from an Azorean island with an autosomal-dominant ataxia [4]. Since all families were of Azorean origin and of autosomal inheritance, Romanul *et al.* suggested that the four families may suffer from the same disease; they explained the clinicopathological differences among and within the four families as phenotypic variations, and proposed to call the disease "Azorean disease" [4]. Later, several similar diseased states of non-Azorean origin were reported [5–11] and therefore the disorder has now been more preferentially called "Machado–Joseph disease" (MJD).

II. ISOLATION OF THE MJD RESPONSIBLE GENE, *MJD1*

A. Common Molecular Mechanisms for Inherited Neurodegenerative Disorders

As is the case with MJD, inherited neurodegenerative disorders manifest considerably diverse symptoms even within a single disease. From different point of views, however, the disorders share consistently observed common features. For examples, the inheritances are most often autosomal dominant. The pathology is neuronal cell loss and degenerations, although each disorder has its own susceptible regions in the central nervous system. The age of onset is generally after the 3rd or 4th decade. Furthermore, the symptoms often become worse and the ages of onset become earlier with each succeeding generation. This phenomenon is called "anticipation." These features have suggested that common molecular mechanisms underlie these inherited neurodegenerative disorders.

As an initial step in elucidating the molecular mechanism, extensive efforts have been made to map the disease locus for each disorder. In 1983, a clear linkage was identified between Huntington's disease (HD) and D4S10 on chromosome 4p16.3 [12]. Likewise, in 1991, a linkage was reported between spinocerebellar ataxia type 1 (SCA1) and D6S89 on chromosome 6p22-23 [13]. Following these discoveries, linkages of MJD to these loci were examined, but no correlations were observed [14–16]. These experiments denied the possibilities that MJD is an allelic disorder of HD or SCA1 and therefore supported the concept that MJD is a genetic entity distinct from HD or SCA1.

The first breakthrough was retrospectively performed in 1991 by La Spada *et al.* [17]. They reported that the *androgen receptor* (*AR*) gene is responsible for spinobulbar muscular atrophy (SBMA). SBMA is an X chromosome-linked recessive disorder manifesting as late-onset motor neuron degeneration as well as reduced male functions. With the latter phenotypes and the known evidence that the *AR* gene is on the X chromosome as clues, the *AR* genes from the affected family members were examined. In normal individuals, the *AR* genes contain approximately 20 CAG repeats in the coding region of the N-terminal portion, which encode glutamine tracts, whereas in patients with SBMA, the genes contain two to three times more CAG repeats. This discovery was not appreciated as being of general importance at the time; this CAG expansion was simply estimated to cause a kind of an AR functional loss, since the disorder was inherited in an X-linked-recessive manner and the several clinical phenotypes could be explained as a result of the reduced AR functions. In addition, anticipation was not clearly observed in SBMA. In 1993, the gene responsible for Huntington's disease (HD) was identified [18]. HD typically manifests all of the common features of the inherited neurodegenerative disorders. Every HD patient had another expanded CAG repeat in the newly isolated *HD* gene (*IT15*). The *HD* gene was a completely different gene from the *AR* gene except for the polyglutamine-coding CAG repeats, which are located in the first exon. Surprisingly, patients with longer CAG repeats manifested earlier disease onset and more severe clinical manifestations [18]. More surprisingly, the expanded CAG repeats seemed to be unstably transmitted to the next generations with a tendency for further elongation, which is now believed to be the genetic basis of "anticipation" [19]. Thus, other hereditary neurodegenerative disorders, especially with anticipation, such as dentatorubular pallidoluysian atrophy (DRPLA), MJD, and other autosomal-dominant spinocerebellar ataxias, were assumed to be caused by similar genetic mutations, probably CAG expansions. Indeed, in late 1993 and early 1994, novel CAG expansions were identified for SCA1 at chromosome 6p23 [20] and for DRPLA at chromosome 12p13 [21, 22].

B. Searching for Novel CAG-Containing Genes

On the assumption that other hereditary neurodegenerative disorders are also caused by yet unknown CAG expansions, we first started searching for novel CAG-containing genes. Using an oligonucleotide with 13 CTG repeats as a probe, we screened a human brain cDNA library and randomly isolated 30 clones containing CAG repeats [23]. One of them consists of 1776 base pairs with one long open reading frame (Fig. 22-1a) [23]. A CAG repeat is located at the relatively C-terminal portion of the open reading frame, and is predicted to be translated into a glutamine tract. There were 26 CAGs in this clone, including two variant sequences, CAA and AAG, at three positions. The amino acid sequence had no homology to any previously reported sequences except for the glutamine tract. Northern blot analysis revealed that transcripts were faintly detectable in all tissues examined with the exception

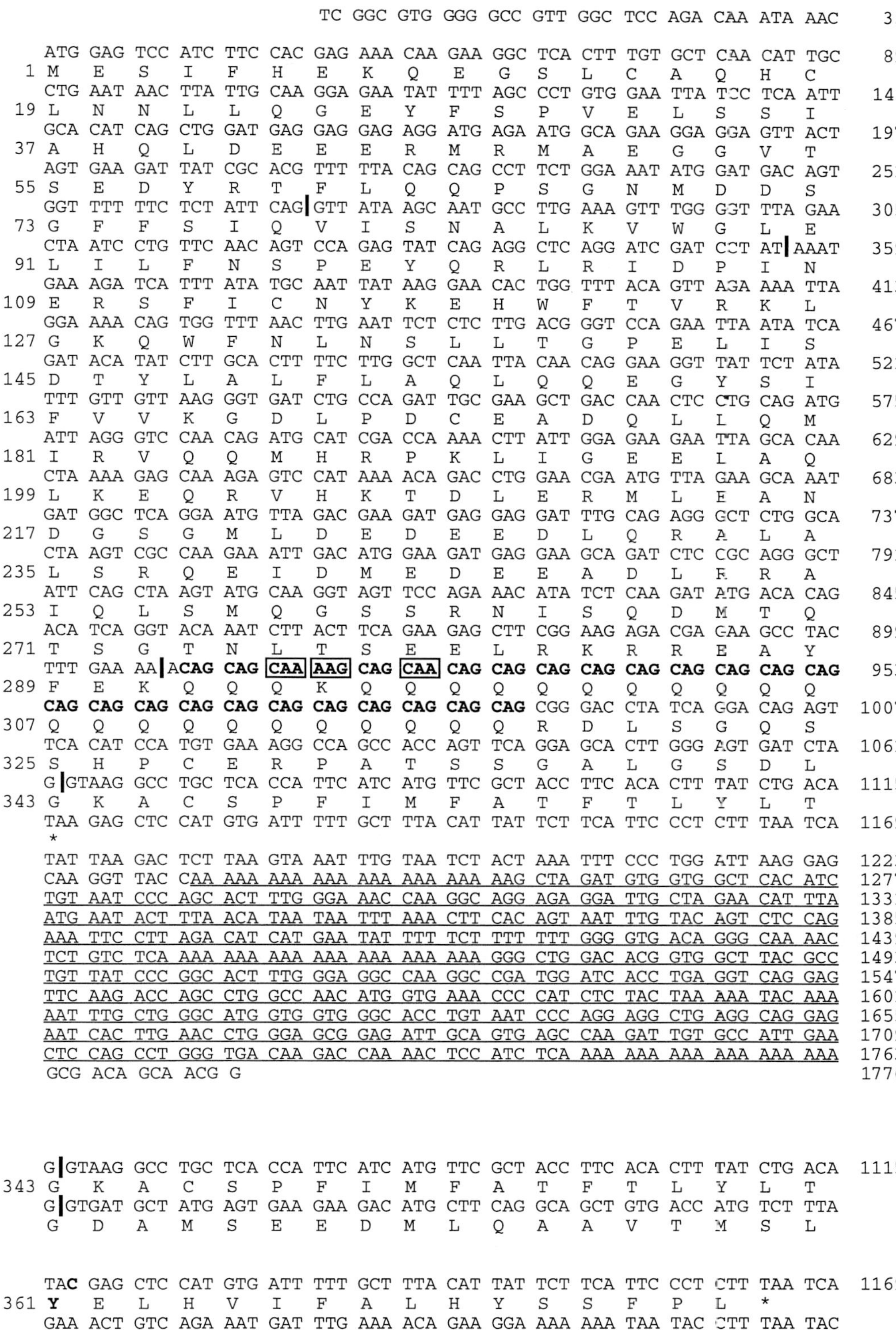

FIGURE 22-1 Nucleotide and amino acid sequences of *MJD1a* cDNA and its variant cDNAs isolated from a human brain cDNA library [23]. (a) The CAG repeat is shown in bold type. The variant triplets, CAA and AAG, are boxed. Human *Alu* repetitive sequences are underlined. The identified exon–intron boundaries are shown by bars. (b) Two additionally different C-terminal encoding portions were created by a polymorphic nucleotide change (A to C) or an alternative splicing. The A to C nucleotide change converts a stop codon to a tyrosine-coding codon and therefore adds 16 amino acids in the C-terminal to the MJD1a protein.

of the testis where a strong expression was evident as an approximately 2-kb mRNA. RT-PCR analysis of the human brain demonstrated expression of the two mRNAs with different lengths of the CAG repeats [23]. These results indicated that both alleles are expressed in the human brain, and the CAG repeat number of each allele is polymorphic. These two characteristics were also observed in the normal *HD, SCA1,* and *DRPLA* genes. We referred to the gene and cDNA as *MJD1* and *MJD1a,* respectively, in hopes that this gene is responsible for a neurodegenerative disorder such as MJD which was supposed to be a major hereditary neurodegeneration in Japan. Several other alternatively spliced MJD1 cDNAs were isolated. An alternative splicing occurred just downstream of the CAG repeat and created a different C-terminus of the MJD1a protein (Fig. 22-1b). In addition, the stop codon was found to be polymorphically changed to tyrosine-coding TAC in about 60% of the alleles in the Japanese (Fig. 22-1b) [24, unpublished observation].

For further investigation, we screened a human genomic library to isolate the corresponding genomic clones [23]. We obtained the MJD1 gene fragments together with three MJD1-related gene fragments. Other potential MJD1-related genes were named *MJD2, MJD3,* and *MJD4* [23]. By fluorescence *in situ* hybridization (FISH) analysis, these four clones were mapped at 14q32.1, 8q23, 14q21 and Xp22.1, respectively [23]. At just about the same time, the *MJD* locus was mapped to chromosome 14q24.3-32.1 by a linkage analysis on Japanese MJD families [25, 26]. Thus, the *MJD1* gene was a good candidate for the MJD responsible gene.

C. *MJD1* CAG Expansions in MJD Patients

Analysis of an *MJD1* gene fragment revealed intronic sequences just upstream of the CAG repeat (Fig. 22-2) [23]. We therefore made several pairs of oligonucleotides adjacent to the CAG repeats for the PCR analysis. Several PCR products were found to contain nucleotide substitutions in and adjacent to the CAG repeats (Fig. 22-2). As mentioned above, the CAG repeats had two variant triplets, CAA and AAG, at three positions (Figs. 22-1 and 22-2). A common substitution was observed on the CAA variant at the third position. This substitution, which converted the variant to a CAG triplet, was found in 7 out of 72 alleles from healthy volunteers [23]. All PCR products from patients and healthy volunteers had the two variants at precisely the first and the second positions. No other interruption was observed in either the normal or the expanded repeats. As mentioned above, MJD shows a wide variety of clinical phenotypes and is therefore very difficult to diagnose accurately in many cases. To avoid the analysis on misdiagnosed samples, we first analyzed the DNAs from an affected family, in which one member had pathologically confirmed MJD. Each genomic sample showed two PCR products representing both alleles of the *MJD1* gene, and all samples from three affected members demonstrated expanded PCR fragments as well as fragments with normal lengths (14 to 35 repeats) (Fig. 22-3). Expanded fragments were observed exclusively in affected individuals and were similar in size (69 to 72 repeats) in the three generations [23].

We next performed the analysis on patients clinically diagnosed with MJD but unconfirmed by pathological examination. Of nine MJD patients from seven families, eight showed expansions of between 68 to 79 repeats [23]. As expected, we found an inverse correlation between the age of onset of symptoms and repeat number: the patients with longer repeats demonstrated earlier onset of the disease [23]. We also counted CAG repeat numbers in 72 alleles from nonaffected individuals and estimated the normal range of repeats to be from 13 to 36 [23]. Subsequent analyses revealed that normal and

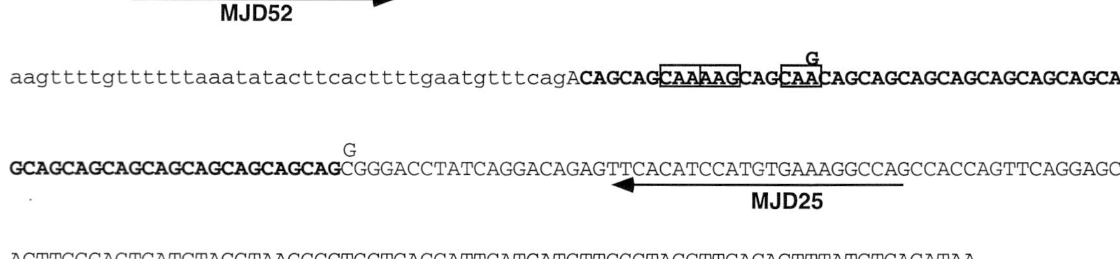

FIGURE 22-2 Genomic sequence of the *MJD1* gene surrounding the CAG repeat [23]. The CAG repeat is shown in bold type. The intron sequence is shown in lower case. The variant triplets, CAA and AAG, are boxed. Primer sequences used for PCR analyses are shown by arrows. Nucleotide substitutions found in at least two individuals are shown above the genomic sequence.

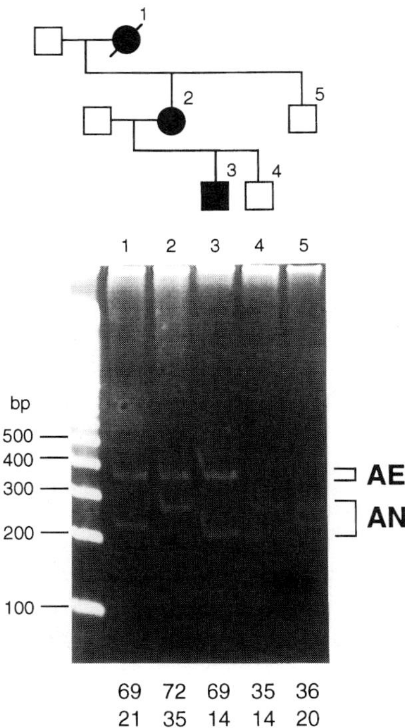

FIGURE 22-3 Detection of the CAG expansions in *MJD1* by genomic PCR analyses [23]. Samples from affected individuals with MJD and healthy individuals are shown by closed and open symbols, respectively. Circles and squares represent females and males, respectively. The slashed symbol represents a deceased patient. The estimated CAG numbers are presented below the lanes.

diseased CAG range from 12 to 41 [27–30] and from 62 to 84 [27, 28, 30] repeats, respectively.

III. MJD MOLECULAR GENETICS

The identification of the gene responsible for MJD has opened the way to several experiments for elucidating the genetic aspects of MJD.

A. Is the Founder Gene of MJD Single or Multiple?

One of the remarkable characteristics of the *MJD1* CAG repeats is the presence of a large gap between normal and affected ranges (see above). Maciel *et al.* concluded that the presence of the large gap is evidence of a single founder gene [27]. However, extensive haplotype analysis on diseased families from several different ethnic origins could not support the presence of the common founder chromosome [31–33], although another report supported its presence [34]. Therefore, it is an accepted idea now that the *MJD1* mutations have occurred independently in many parts of the world [28]. Consistently, MJD now has been proven to be the most frequently observed hereditary ataxia, with broad geographical distributions [28, 30, 35–39]. Giunti *et al.* reported that MJD accounts for 29% of patients with autosomal dominant cerebellar ataxia (ADCA) type I originating from the United Kingdom, India, Jamaica, Ghana, Brazil, and France [28]. Ranum *et al.* estimated the prevalence of MJD as 21% of 271 patients from 149 ADCA families by the genetic examination with worldwide ethnic backgrounds [36]. MJD has been thought to be prevalent in Japan. In an epidemiological study conducted in Japan before genetic evaluation was feasible, MJD was estimated to occupy 11% of all ADCA. However, the *MJD1* mutations were identified in 56% of 32 patients from 29 Japanese ADCA families [37]. Interestingly, 5 patients, who were clinically diagnosed as having the Menzel type of hereditary cerebellar ataxia, were confirmed to have the *MJD1* mutations [37].

B. Clinical Heterogeneities with the *MJD1* Mutations

Patients not only with the clinically Menzel type of hereditary cerebellar ataxia [37] but also with or without other clinical diagnoses have been shown to have the *MJD1* mutations occasionally. A typical example was reported from a German study. Basically no patients with autosomal dominant inheritance were diagnosed clinically as MJD in Germany, but the genetic examinations have revealed that the *MJD1* mutations are responsible for 19 of 38 families (50%) of German ADCA [35].

SCA3 was an another example; Stevanin *et al.* mapped the disease locus to chromosome 14q24.3-qter by linkage analysis in a non-SCA1/non-SCA2 family with spinocerebellar ataxias, and therefore named the disease SCA3 as a genetically distinctive category [40]. On the other hand, Twist *et al.* demonstrated the possible identities between MJD and SCA3 by showing that the *MJD* locus is located within 15 cM of the *SCA3* locus [33], although it is still controversial whether MJD and SCA3 are distinct clinical entities. By examining the *MJD1* gene, the *MJD1* CAG expansions were identified in members of a SCA3 family, demonstrating that SCA3 and MJD have genetically identical *MJD1* mutations [41].

An ataxo-choreic form of DRPLA might be another example. Cancel *et al.* mapped the clinically and neuropathologically diagnosed ataxo-choreic form of DRPLA to chromosome 14q24.3-qter, which was near the *MJD* locus [42]. The authentic *DRPLA* gene was proven to be located at chromosome 12p13 (see above). This form of DRPLA might be caused by the *MJD1* mutation.

Similarly, one family, which was diagnosed as DRPLA on the basis of neuropathologic findings, was proven to have the *MJD1* mutations [43].

Tuite *et al.* reported two Azorean patients manifesting L-dopa-responsive Parkinsonism with peripheral neuropathy, who were confirmed to have the expanded CAGs (61 and 71 repeats) in the *MJD1* gene [44]. This rare diseased state has been classified as MJD type IV (see below). In addition, it was proposed that the *MJD1* mutations could manifest spastic paraplegic symptoms [45].

C. Repeat Numbers and Clinical Manifestation

Such diverse phenotypes could not be explained by the CAG repeat length itself. Likewise, the contribution of the CAG repeat numbers to the ages of onset was estimated to be 46–48% [27, 46], though, of course, an inverse correlation between the CAG repeat numbers and the age of MJD onset is a well-accepted phenomenon [23, 28, 36]. The involvement of other genetic or environmental factors in determining the clinical presentation and evolution of the disease is easily imaginable. However, these factors are totally unknown. No significant correlation was found between the size of the CAG repeat on the normal allele and the age of onset [36].

MJD has been classified into four types (I to IV) from the clinical symptoms and ages of onset [47–52]: In type I, the predominant symptoms are pyramidal and extrapyramidal signs and the disease starts between the teens and the 30s; in type II, cerebellar and pyramidal signs become more dominant than extrapyramidal signs and the disease starts between the late teens and the 40s; in type III, cerebellar and peripheral nerve signs become characteristic and the onset occurs between the 40s and 60s; type IV is extremely rare, manifesting as Parkinsonism with peripheral neuropathy (see above). Since the major three types are partly based on the ages of onset, the average repeat was highest in type I (79.4 ± 1.0), followed by type II (74.6 ± 0.5) and type III (72.6 ± 1.1) [53]. Considering that the ages of onset and the degrees of symptoms are mostly correlated with the CAG repeat numbers, MJD as well as other CAG expanded diseases should be classified more appropriately according to repeat numbers. The repeat number-oriented classifications should be helpful in predicting the patients' disease processes and choosing the treatments.

D. Gene Dosage-Dependent Effects in MJD

Before the genetic confirmation was available, MJD patients suspected to have homozygous mutations were reported who developed earlier ages of onset and different clinical phenotypes than their affected parents [54]. Later, the enhancements of clinical symptoms and the earlier disease onset in the patients homozygous for the *MJD1* mutations were confirmed [55, 56], demonstrating that clinical symptoms of MJD are influenced in a mutated gene-dosage-dependent manner. Similar phenomena have been reported in DRPLA [57] and probably in SCA2 [58] and SCA6 [59]. In contrast, in HD such a gene dosage effect could not be observed [60]. Some unknown saturable mechanisms may exist in the disease mechanisms in HD. Controversially, the normal HD alleles were proposed to influence the age of onset in HD [61].

E. Gender Effects in MJD

Another controversial issue in this field is the effect of gender on onset. Kawakami *et al.* reported the enhanced earliness of the ages of onset of MJD in males compared to females in 14 affected Japanese sibling pairs [55]. A study on a large group of Japanese MJD patients confirmed the statistically significant gender effects [Kawakami *et al.*, personal communication]. However, analyses of large groups in Europe and North America did not reveal statistically significant differences of the ages of onset between males and females [62]. These discrepancies might depend on the ethnic origin of the patients or the assay conditions for the CAG repeat lengths: long CAG repeats migrate relatively slower than their actual sizes [29] and might therefore be underestimated at higher repeat numbers, where the differences of the ages of onset seemed less prominent [62].

The gender effects, if present, might explain a clinical feature of SBMA. SBMA is an exceptionally X-linked recessive disorder in inherited neurodegenerations with expanded CAGs in the responsible genes. By Lyon theory, genes on the X chromosome are randomly inactivated in the female, and theoretically 50% of the cells in the SBMA carrier female express the diseased *AR* gene with the expanded CAGs while the remaining 50% only express the normal *AR* gene. Such a chimerism might simply prevent the clinical manifestation of SBMA in the female, but it would not be difficult to imagine that the gender effect might further protect the females from the disease. In relation to this, estrogen has been recently reevaluated as a neuroprotective drug [63–68]. Therefore, the gender effect should be more extensively examined and clarified both in clinical cases and in animal models. Such examinations might be directly related to the development of the therapeutics for or the prevention of the diseases.

F. CAG Repeat Instabilities in MJD

Intergenerational instability is a hallmark of the expanded triplets of not only CAG but also CTG, CGG, and GAA, and is the only known molecular basis of anticipation. Importantly, no apparent instabilities have been observed in the normal ranges, generally below 40, of any of the triplets and longer expanded alleles would result in more unstable transmissions to the next generations. The CAG instability, especially for elongation, is so far more prominent in paternal transmission [19, 21 22, 61, 69–71]. In contrast, CTG, CGG, and GAA repeats seem more frequently elongated in maternal transmission [72–75]. Accordingly, in Japanese and Caucasian MJD, the diseased alleles were expanded with 3.2 ± 0.8 and 1.2 ± 0.4 repeats on the average in paternal and maternal transmissions, respectively [34]. The elongation of the CAG repeats in paternal transmission appeared to occur in spermatogenesis [34], and the expanded alleles seem to be more frequently transmitted to the next generations [76]. It is noteworthy that all of the intergenerationally unstable CAG repeats appear to be well-expressed in the testis: the CAG repeat instability might be coupled with the transcription. The effect of the normal *MJD* allele on the CAG instability has been proposed [77].

The numbers of the expanded CAG repeats differ in a tissue-specific manner in MJD as well [78, 79], which is called somatic mosaicism. Somatic mosaicism is commonly observed in other diseases with expanded CAGs, such as HD, DRPLA, and SCA1 [80–82]. Lopes-Cendes *et al.* demonstrated that the number of CAG repeats of *MJD1* was indistinguishable in various regions in the central nervous system (CNS) except for the cerebellum, where the repeat number is smallest [79]. So far, no clear pattern differences in the mosaicism have been observed among the diseases: the cerebellum always showed the smallest repeat number in MJD, HD, DRPLA, and SCA1 [80–82]. Therefore, the mosaicism may not be directly related to the specific neurological and pathological phenotypes for each disease.

IV. EXPERIMENTAL MODELS FOR MJD AND OTHER POLYGLUTAMINE DISEASES

A. A Cell Culture Model

To investigate the molecular events responsible for MJD, we expressed a range of different-sized MJD1a proteins in cultured COS cells (Fig. 22-4a) [83]. Expression of the retinoic acid receptor (RAR) was used as a control and coexpression of β-galactosidase was used to trace the transfected cells. No obvious differences in either morphology or numbers in X-gal positive cells (about 15%) were observed among the transfections with cDNAs for any of the full-length MJD1a proteins or a portion of the MJD1a protein including a normal-sized polyglutamine repeat ($Q_{35}C$). Likewise, no differences were observed with the RAR control. In contrast, very few X-gal positive cells were observed in cells transfected with a cDNA expressing an expanded polyglutamine repeat ($Q_{79}C$) (data not shown).

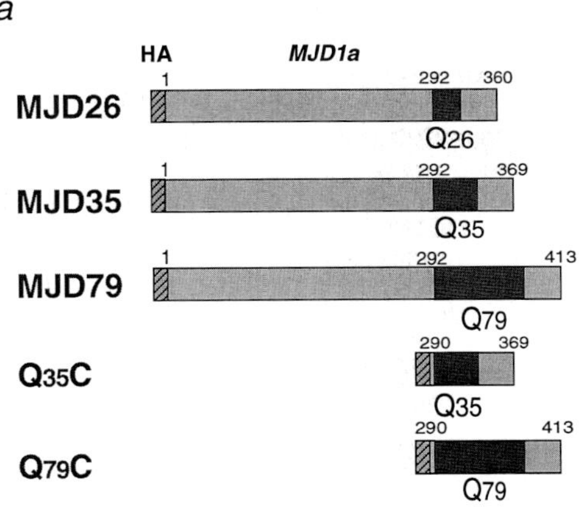

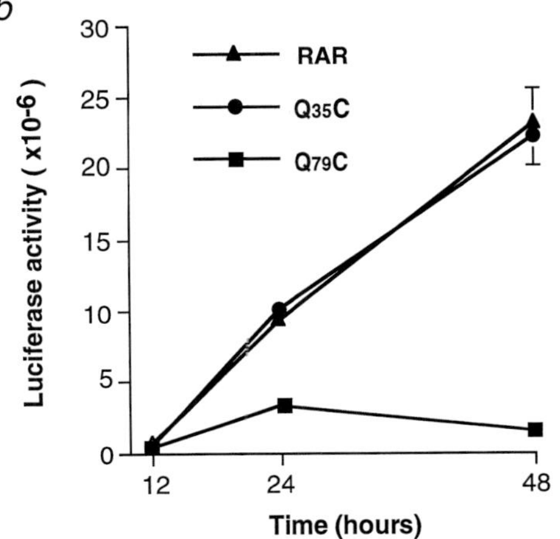

FIGURE 22-4 Expanded polyglutamines from MJD1a protein cause cell death [83]. (a) Schematic structures of the expressed proteins. The open reading frames are shown in boxes. The regions encoding the hemagglutinin epitope and polyglutamines are hatched and dotted, respectively. The numbers above each box represent the position of the amino acid counted from the original initiation methionine of the MJD1a protein. (b) A dramatic decrease was observed in the luciferase activities of the pCMXHA-$Q_{79}C$ transfected cells at 24 and 48 h, as compared with those of the pCMXHA-$Q_{35}C$ and -RAR transfections.

Immunocytochemical analysis of the transfected cells showed that the $Q_{35}C$, MJD26, MJD35, and MJD79 proteins were homogeneously stained in the cytoplasm, whereas the $Q_{79}C$ protein was stained in punctate patterns [83]. Importantly, some cells with the punctate staining appeared to be in the process of an apoptotic cell death, since the cells showed cytoplasmic fragmentation, condensed nuclei, and DNA fragmentation in the nuclei. Such apoptotic cells shrank and detached from the dishes. These apoptotic changes were exclusively observed in cells expressing the $Q_{79}C$ protein, but not in cells expressing other proteins. These observations indicated that the cells expressing $Q_{79}C$ proteins underwent apoptosis and the dead cells detached and therefore disappeared from the culture.

We undertook a more quantitative assay, in which we examined the relative cell numbers remaining in the culture after the transfection by monitoring luciferase activities (Fig. 22-4b) [83]. Consistent with the β-gal experiment, cells transfected with pCMXHA-$Q_{35}C$ and -RAR showed indistinguishable luciferase activities at 12, 24, and 48 h posttransfection. Until 12 h posttransfection, cells transfected with pCMXHA-$Q_{79}C$ demonstrated a luciferase activity comparable to those with pCMXHA-$Q_{35}C$ and -RAR. However, only a slight increase of luciferase activity was observed 24 h posttransfection and an apparent decrease of the activity 48 h posttransfection (Fig. 22-4b) [83]. These results demonstrated that the cells transfected with pCMXHA-$Q_{79}C$ were specifically lost from the culture during their period of expression, as a result of cell death. Subsequent experiments demonstrated that the expanded polyglutamine by itself is sufficient to induce cell death.

To examine whether the CAG expansion in the 3′ noncoding region could induce cell death, we inserted a cDNA fragment containing 79 repeats of CAG in the 3′ noncoding region of pCMXHA-$Q_{35}C$. The resultant plasmid was referred to as pCMXHA-$Q_{35}C$-$(CAG)_{79}$. Cells with pCMXHA-$Q_{35}C$-$(CAG)_{79}$ did not manifest any clear evidence supporting cell death [83]. These experiments provide evidence that the translation of the expanded polyglutamine is essential for cell death. These were the first clear demonstrations that the expanded CAG repeat must be translated into the polyglutamine to manifest a disease-related phenotype, an induction of apoptosis. We therefore proposed that the inherited neurodegenerative disorders with expanded CAG repeats in their responsible genes should be more appropriately be called "polyglutamine diseases" [83]. Interestingly, even in *Escherichia coli* the expression of the expanded polyglutamines displayed toxicity [84].

We next addressed the question of whether the observed cell death is dominant. COS cells transfected with equal amounts of pCMXHA-$Q_{79}C$ and pCMXHA-$Q_{35}C$, a situation which represents the state of heterozygotic patients with both normal and diseased *MJD1* alleles, exhibited an apparent but weaker cell death phenotype than cells with double the amount of pCMXHA-$Q_{79}C$ (Fig. 22-5a) [83]. These results clearly demonstrated that cell death induced by the expanded polyglutamine is dominant, as well as gene dosage-dependent. This observation is consistent with the clinical manifestations in MJD [55].

In the next experiments, the effect of the size of the expanded polyglutamine on cell death was examined. Expanded polyglutamine peptides with both 79 and 64 repeats induced cell death. The $Q_{79}C$ protein induced cell death more prominently than the $Q_{64}C$ protein (Fig. 22-5b) [83]. These results are consistent with one of the major clinical features of the polyglutamine diseases in which the longer expansions lead to more severe clinical manifestations resulting from greater neuronal loss in the diseased brain (see above).

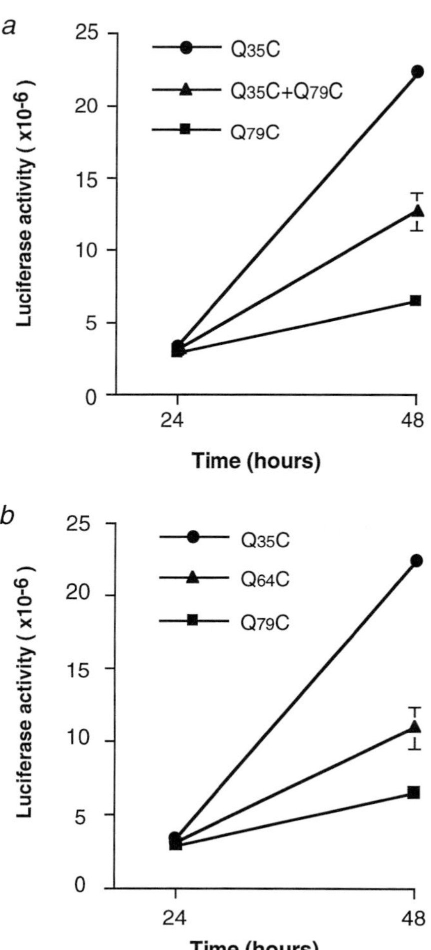

FIGURE 22-5 Characteristics of cell death induced by expanded polyglutamine [83]. (a) Dominant effect of the expanded polyglutamine. (b) Enhancement of cell death by the lengths of the expanded polyglutamine.

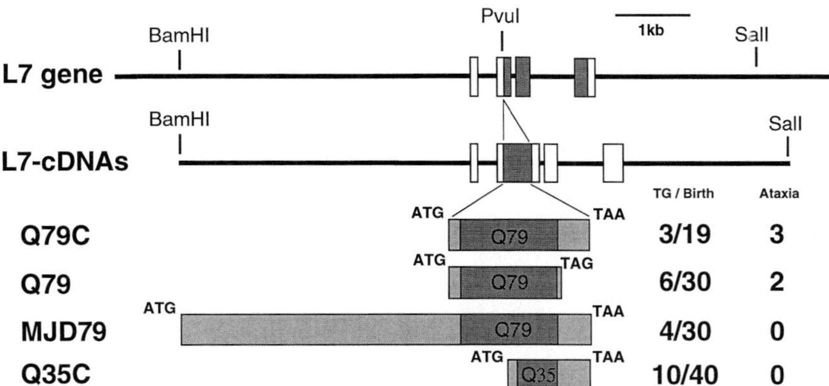

FIGURE 22-6 Transgene constructs for the expression in Purkinje cells *in vivo* [83]. Schematic structures of the *L7* gene and the transgenes. Exons are shown as boxes, and introns and flanking regions are shown as lines. Coding regions are shaded.

B. A Transgenic Mouse Model

To address whether the expanded polyglutamine can induce neuronal cell death *in vivo,* we took advantage of a previously developed system, in which exogenous DNA is expressed specifically in mouse Purkinje cells with the use of the *L7* gene and transgenic techniques [85, 86]. Figure 22-6 shows the series of constructs used in the transgenic experiments [83]. For these constructs the cDNA encoding the $Q_{79}C$, Q_{79}, $Q_{35}C$, or MJD79 protein was placed in the second exon just upstream of the coding region for the L7 protein. Among the 19, 30, 40, and 30 offspring, 3 animals were transgenic for the $Q_{79}C$, 6 for the Q_{79}, 10 for the $Q_{35}C$, and 4 for the MJD79 proteins, respectively. In addition, each $Q_{79}C$ mouse generated one transgenic F1 mouse. The size of the CAG repeats did not expand or shrink in the first and the second generations.

Evaluation of the transgenic mice indicated that all of the $Q_{79}C$ transgenic mice, their F1 offspring, and 2 out of 6 Q_{79} transgenic mice clearly demonstrated ataxic postures and gait disturbance [83]. This phenotype was characterized by the wide-based hindlimb stance (Fig. 22-7a) and frequent falling down when moving (Fig. 22-7b), and occurred as early as in the 4th week of age, apparently after full activation of the *L7* promoter. These transgenic mice were as active as their nontransgenic littermates, but they could not rear. In contrast, none of the mice transgenic for the $Q_{35}C$ and MJD79 proteins have demonstrated any phenotypes as of nearly 2 years of age. Strong ataxic phenotypes were more prominent with higher copy numbers of the $Q_{79}C$ transgene.

We next examined the brain sections of an 8-week-old ataxic $Q_{79}C$ mouse and a control littermate. By visual inspection, the affected cerebellum was very atrophic (Figs. 22-8a and 22-8b). Its volume was approximately one-eighth that of the nontransgenic littermate,

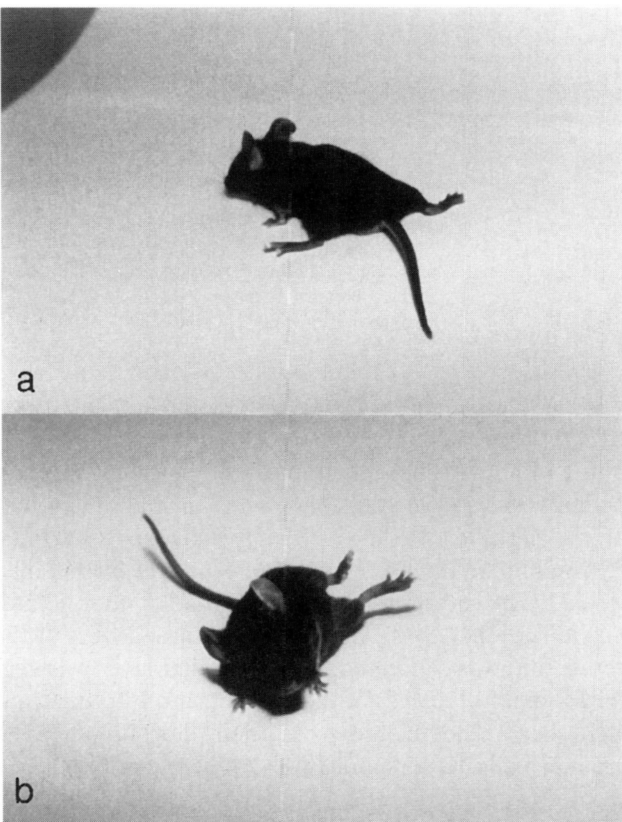

FIGURE 22-7 Ataxic phenotypes of polyglutamine transgenic mice [83]. Typical ataxic postures of a 6-week-old $Q_{79}C$ transgenic mouse are presented. The mouse shows wide-based hindlimb stance (a) and falling-down position on one side (b) which is frequently seen when moving.

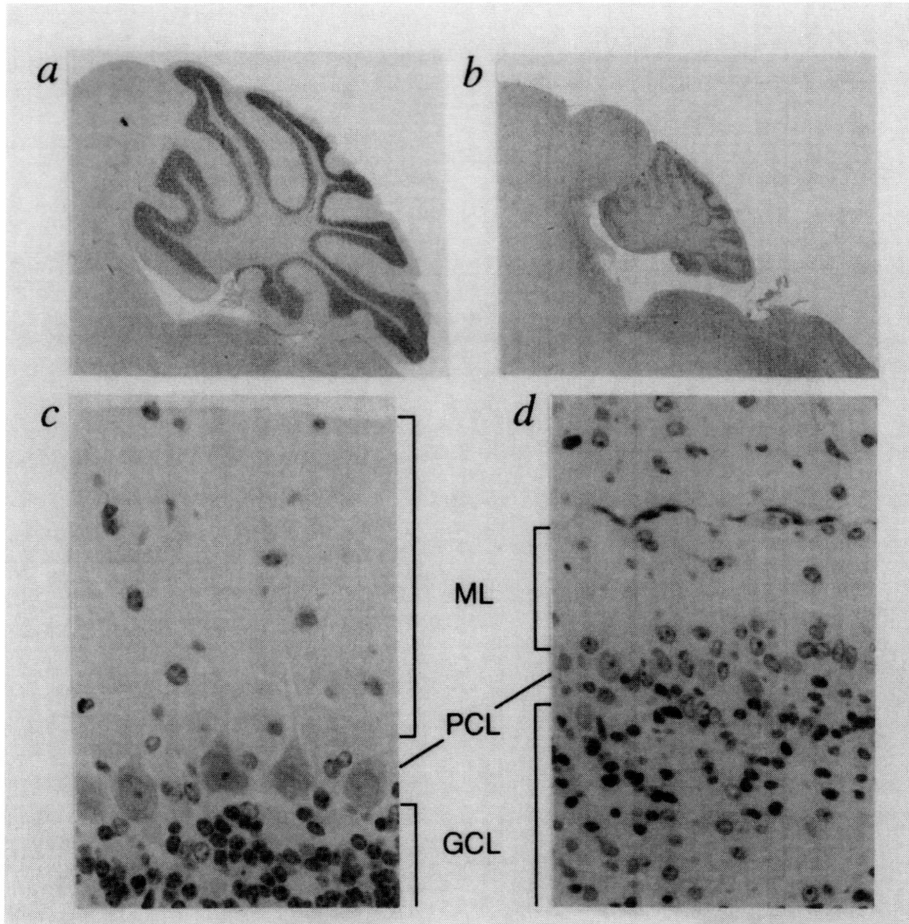

FIGURE 22-8 Histological analysis of the cerebellum of normal and transgenic mice [83]. Brain sections of an 8-week-old $Q_{79}C$ transgenic mouse (b, d) and an age-matched littermate are presented (a, c). Hematoxylin–eosin staining at low magnification (5×) (a, b) or at high magnification (100×) (c, d). ML, molecular layer; PCL, Purkinje cell layer; GCL, granule cell layer.

but it retained fundamental morphologies of the cerebellum: both the overall shape and the staining pattern of the granule cell layer were similar to the normal cerebellum, although the granule cell layer was stained weakly. The cerebrum appeared normal. Under greater magnification, it was evident that all three layers of the cerebellum were affected. The molecular layer was very thin compared to that of the control, and Purkinje cells were hardly detectable from their morphology. The granule cells were sparsely distributed and each cell seemed to have shrunk (Figs. 22-8c and 22-8d). Although we could detect the transgene expression in the affected cerebellum by RT-PCR, both the expressed protein and the Purkinje cells were barely detected by immunohistochemical staining with the anti-HA antibody or with an anti-calbindin antibody (see below), respectively.

The targeted Purkinje cells were examined in detail using the antibody against a Purkinje cell-specific marker, calbindin [87]. The normal Purkinje cells were clearly stained by the anti-calbindin antibody. However, consistent with hematoxylin–eosin staining, Purkinje cells in the affected mouse mostly disappeared and even the remaining Purkinje cells exhibited scarce dendrite formation, indicating that almost all Purkinje cells were affected. Given that the molecular and granule cell layers were not targeted, the observed morphologies in the two layers are likely to reflect the successive changes to the Purkinje cell loss. These morphological changes resemble those seen in a terminal clinical stage of cerebellar degeneration [88]. From all these data we concluded that the expanded polyglutamine is indeed causative for neuronal cell loss and degeneration *in vivo*.

V. PREDICTED MOLECULAR MECHANISMS FOR MJD AND OTHER POLYGLUTAMINE DISEASES

At present, seven inherited neurodegenerative disorders have been identified to be caused by the polyglutamine-coding CAG triplet expansions in the responsible genes [89]. The disorders include SBMA, HD, SCA1, DRPLA, MJD, SCA2, and SCA6 (Table 22-1). SBMA is an exceptionally X-linked recessive disorder (see above), and the others are of autosomal-dominant inheritance. All affected patients exhibit expanded CAGs of between 40 and 100 repeats, with the exception of SCA6 (Table 22-1). Polyglutamines are predicted to form β-pleated configurations [90]. Other β-pleated proteins such as βA4 and PrP^{Sc} have been demonstrated as causative agents in AD and prion disease, respectively, and both proteins also form proteinase-resistant aggregations [91–93]. These lines of evidence together with our experiments described above strongly indicate that the expanded polyglutamine itself forms a pathogenic β-pleated protein and causes the neurodegeneration through its aggregation. Given that polyglutamine-mediated aggregation is crucial in neurodegeneration, the parameters which determine the feasibility of the polyglutamines to aggregate would determine the age of onset and the clinical severity. These parameters would be easily postulated to be the concentration and the length of polyglutamines [89]. The differences of the ages of onset between SCA1 and MJD even with similar CAG repeats [94], for example, might simply reflect the difference in abundance of each polyglutamine. From this point of view, the uniqueness of SCA6 would be explained by the membrane localization of the *SCA6* gene product. The *SCA6* gene encodes α1A calcium channel [95]. Localization of a polyglutamine in such a restricted area as the cellular membrane easily raises its concentration and in turn increases their probability to come in contact with each other, which would allow relatively small expansions of 21 to 27 repeats of polyglutamines to easily aggregate and thereby cause pathogenesis. Of course, these are pure speculations. Nevertheless, this hypothesis could well explain many, if not all, aspects of the polyglutamine diseases [89].

Three animal models with diseased phenotypes have been established: SCA1, MJD, and HD models. To create SCA1 mice, a full-length SCA1 protein with 82 polyglutamines was targeted for expressed in mouse Purkinje cells, using the *L7/Pcp-2* gene promoter with the transgenic technique [87]. In SCA1 mice, ataxic phenotypes became evident 10 to 40 weeks after birth and only Purkinje cells were degenerated. The same *L7* gene promoter was used for the expression of the full-length MJD1 protein containing 79 polyglutamines, but no phenotypes were observed [83]. Purkinje cells are the major diseased region in SCA1, but are only mildly affected in MJD. Therefore, these two mouse models appeared to faithfully reproduce a phenotype of each disease. In other words, these experiments reproduced a clinical observation that Purkinje cells are sensitive to the diseased SCA1 protein but are refractory to the MJD1 protein even with expanded polyglutamines. However, the Purkinje cell-specific expression of 79 polyglutamines as a portion of MJD1 protein with short flanking amino acids did induce severe ataxic phenotypes as early as 4 weeks after birth, and all three cerebellar cell layers were severely affected, demonstrating that the precleaved protein is much more potent in inducing Purkinje cell death than the full-length protein [83]. From these results we proposed the potential region-specific cleavage of the proteins with the expanded polyglutamines as a mechanism which renders specific brain regions degenerated in each polyglutamine disease. Consistently, transgenic mice expressing exon 1 of *HD* gene with expanded CAG repeats mani-

Table 22-1 Inherited Neurodegenerative Disorders with Expanded Trinucleotide Repeats (from Ref. [89])

Diseases	Inheritances	Chromosomal positions	Genes	Triplets	Repeat numbers: Normal	Repeat numbers: Diseased
SBMA	XR	Xq13-q21	*Androgen receptor*	CAG	7–34	38–68
HD	AD	4p16.3	*HD (IT15)*	CAG	10–35	37–121
SCA1	AD	6p23	*SCA1*	CAG	6–39	43–82
DRPLA	AD	12p13	*DRPLA*	CAG	5–35	49–85
MJD(SCA3)	AD	14q32.1	*MJD1*	CAG	12–41	62–84
SCA2	AD	12q24.1	*SCA2*	CAG	14–31	35–59
SCA6	AD	19q13	*α1A calcium channel*	CAG	4–16	21–27
FA	AR	9q13	*X25*	GAA	6–42	200–1700

Note. XR, X-linked recessive; AD, autosomal dominant; AR, autosomal recessive

fested broad neurological phenotypes [96]. This exon 1 encodes 17 N-terminal amino acids, about 130 polyglutamines, and the following 52 additional amino acids. This protein was expressed under the control of the *HD* gene promoter. In accord with the broad expression of the *HD* gene, the whole brain was rather uniformly affected; the brain mass was on average reduced to about 80% of that of the age-matched littermates and the reduction seemed to be enhanced with aging. In addition, the volumes of other organs such as the liver, testis, ovary, and thymus were also slightly reduced [96]. These observations are consistent with the idea that expanded polyglutamines can induce apoptosis in the affected neuronal cells *in vivo*.

Further evidence was provided which supports the idea that the aggregation of polyglutamines is central in the pathogenesis. In the HD mice, age-dependent appearance of nuclear inclusion bodies has been elucidated by an antibody against the polyglutamine-flanking portion of the HD exon 1 protein, which appear just before the brain weight loss and the neurological signs [97]. The nuclear inclusions seem to consist of the HD exon 1 protein with 130 polyglutamines but not the normal mouse HD protein. Interestingly, an anti-ubiquitin antibody strongly recognizes the inclusions. Similar nuclear inclusions were reported in biopsy brain samples of the HD patient [98] and the MJD autopsy brain samples [99].

VI. CONCLUSIONS

The newly identified nuclear inclusions by the expanded polyglutamines further strengthen the crucial involvement of the expanded polyglutamine in the pathogenesis. The successes in reproducing similar, if not identical, diseased states in cultured cells and mice with the expanded CAG repeats have encouraged the researchers to undertake a challenge to discover common molecular mechanisms, which will lead to elucidation of the yet unanswered questions of these disorders and, more importantly, to finding novel treatments commonly effective in all polyglutamine diseases. Such a dream may come true in the very near future.

References

1. Nakano, K. K., Dawson, D. M., and Spence, A. (1972). Machado disease: a hereditary ataxia in Portuguese emigrants to Massachusetts. *Neurology* **22,** 49–55.
2. Woods, B. T., and Schaumburg, H. H. (1972). Nigro-spino-dentatal degeneration with nuclear ophthalmoplegia: a unique and partially treatable clinico-pathological entity. *J. Neurol. Sci.* **17,** 149–166.
3. Rosenberg, R. N., Nyhan, W. L., Bay, C., and Shore, P. (1976). Autosomal dominant striatonigral degeneration. *Neurology* **26,** 703–714.
4. Romanul, F. C. A., Fowler, H. L., Radvany, J., Feldman, R. G., and Feingold, M. (1977). Azorean disease of the nervous system. *N. Engl. J. Med.* **296,** 1505–1508.
5. Lima, L., and Coutinho, P. (1980). Clinical criteria for diagnosis of Machado-Joseph disease: report of a non-Azorean Portuguese family. *Neurology* **30,** 319–322.
6. Healton, E. B., Brust, J. C., Kerr, D. L., Resor, S., and Penn, A. (1980). Presumably Azorean disease in a presumably non-Portuguese family. *Neurology* **30,** 1084–1089.
7. Sakai, T., Ohta, M., and Ishino, H. (1983). Joseph disease in a non-Portuguese family. *Neurology* **33,** 74–80.
8. Suite, N. D., Sequeiros, J., and McKhann, G. M. (1986). Machado-Joseph disease in a Sicilian-American family. *J. Neurogenet.* **3,** 177–182.
9. Bharucha, N. E., Bharucha, E. P., and Bhabha, S. K. (1986). Machado-Joseph-Azorean disease in India. *Arch. Neurol.* **43,** 142–144.
10. Yuasa, T., Ohama, E., Harayama, H., Yamada, M., Kawase, Y., Wakabayashi, M., Atsumi, T., and Miyatake, T. (1986). Joseph's disease: clinical and pathological studies in a Japanese family. *Ann. Neurol.* **19,** 152–157.
11. Takiyama, Y., Ikemoto, S., Tanaka, Y., Mizuno, Y., Yoshida, M., and Yasuda, N. (1989). A large Japanese family with Machado-Joseph disease: clinical and genetic studies. *Acta Neurol. Scand.* **79,** 214–222.
12. Gusella, J. F., Wexler, N. S., Conneally, P. M., Naylor, S. L., Anderson, M. A., Tanzi, R. E., Watkins, P. C., Ottina, K., Wallace, M. R., and Martin, J. B. (1983). A polymorphic DNA marker genetically linked to Huntington's disease. *Nature* **306,** 234–238.
13. Zoghbi, H. Y., and Terrenato, L. (1991). The gene for autosomal dominant spinocerebellar ataxia (SCA1) maps telomeric to the HLA complex and is closely linked to the D6S89 locus in three large kindreds. *Am. J. Hum. Genet.* **49,** 23–30.
14. Forse, R. A., McLeod, P., Holden, J. J., and White, B. N. (1989). DNA marker studies show that Machado-Joseph disease is not an allele of the Huntington disease locus. *J Neurogenet.* **5,** 155–158.
15. Sasaki, H., Wakisaka, A., Tashiro, K., Hamada, T., and Katou, T. (1992). Linkage study of Machado-Joseph disease: genetic evidence for the locus different from SCA1. *Clin. Neurol.* **32,** 13–16.
16. Carson, W. J., Radvany, J., Farrer, L. A., Vincent, D., Rosenberg, R. N., MacLeod, P. M., and Rouleau, G. A. (1992). The Machado-Joseph disease locus is different from the spinocerebellar ataxia locus (SCA1). *Genomics* **13,** 852–855.
17. La Spada, A. R., Wilson, E. M., Lubahn, D. B., Harding, A. E., and Fischbeck, K. H. (1991). Androgen receptor gene mutations in X-linked spinal and bulbar muscular atrophy. *Nature* **352,** 77–79.
18. The Huntington's Disease Collaborative Research Group (1993). A novel gene containing a trinucleotide repeat that is expanded and unstable on Huntington's disease chromosomes. *Cell* **72,** 971–983.
19. Duyao, M., Ambrose, C., Myers, R., Novelletto, A., Persichetti, F., Frontali, M., Folstein, S., Ross, C., Franz, M., Abbott, M., Gray, J., Conneally, P., Young, A., Penney, J., Hollingsworth, Z., Shoulson, I., Lazzarini, A., Falek, A., Koroshetz, W., Sax, D., Bird, E., Vonsattel, J., Bonilla, E., Alvir, J., Conde, J. B., Cha, J.-H., Dure, L., Gomez, F., Ramos, M., Sanchez-Ramos, J., Snodgrass, S., de Young, M., Wexler, N., Moscowitz, C., Penchaszadeh, G., MacFarlane, H., Anderson, M., Jenkins, B., Srinidhi, J., Barnes, G., Gusella, J., and MacDonald, M. (1993). Trinucleotide repeat length instability and age of onset in Huntington's disease. *Nature Genet.* **4,** 387–392.

20. Banfi, S., Servadio, A., Chung, M. Y., Kwiatkowski, T. J., Jr., McCall, A. E., Duvick, L. A., Shen, Y., Roth, E. J., Orr, H. T., and Zoghbi, H. Y. (1993). Identification and characterization of the gene causing type 1 spinocerebellar ataxia. *Nature Genet.* **7,** 513–520.
21. Koide, R., Ikeuchi, T., Onodera, O., Tanaka, H., Igarashi, S., Endo, K., Takahashi, H., Kondo, R., Ishikawa, A., Hayashi, T., Saito, M., Tomoda, A., Miike, T., Naito, H., Ikuta, F., and Tsuji, S. (1994). Unstable expansion of CAG repeat in hereditary dentatorubral-pallidoluysian atrophy (DRPLA). *Nature Genet.* **6,** 9–13.
22. Nagafuchi, S., Yanagisawa, H., Sato, K., Shirayama, T., Ohsaki, E., Bundo, M., Takeda, T., Tadokoro, K., Kondo, I., Murayama, N., Tanaka, Y., Kikushima, H., Umino, K., Kurosawa, H., Furukawa, T., Nihei, K., Inoue, T., Sano, A., Komure, O., Takahashi, M., Yoshizawa, T., Kanazawa, I., and Yamada, M. (1994). Dentatorubral and pallidoluysian atrophy expansion of an unstable CAG trinucleotide on chromosome 12p. *Nature Genet.* **6,** 14–18.
23. Kawaguchi, Y., Okamoto, T., Taniwaki, M., Aizawa, M., Inoue, M., Katayama, S., Kawakami, H., Nakamura, S., Nishimura, M., Akiguchi, I., Kimura, J., Narumiya, S., and Kakizuka, A. (1994). CAG expansions in a novel gene for Machado-Joseph disease at chromosome 14q32.1. *Nature Genet.* **8,** 221–228.
24. Trottier, Y., Lutz, Y., Stevanin, G., Imbert, G., Devy, D., Cancel, G., Saudou, F., Weber, C., David, G., Tora, L., Agid, Y., Brice, A., and Mandel, J.-L. (1995). Polyglutamine expansion as a pathological epitope in Huntington's disease and four dominant cerebellar ataxias. *Nature* **378,** 403–406.
25. Takiyama, Y., Nishizawa, M., Tanaka, H., Kawashima, S., Sakamoto, H., Karube, Y., Shimazaki, H., Soutome, M., Endo, K., Ohta, S., Kagawa, Y., Kanazawa, I., Mizuno, Y., Yoshida, M., Yuasa, T., Horikawa, Y., Oyanagi, K., Nagai, H., Kondo, T., Inuzuka, T., Onodera, O., and Tsuji, S. (1993). The gene for Machado-Joseph disease maps to human chromosome 14q. *Nature Genet.* **4,** 300–304.
26. Takiyama, Y., Oyanagi, S., Kawashima, S., Sakamoto, H., Saito, K., Yoshida, M., Tsuji, S., Mizuno, Y., and Nishizawa, M. (1994). A clinical and pathologic study of a large Japanese family with Machado-Joseph disease tightly linked to the DNA markers on chromosome 14q. *Neurology* **44,** 1302–1308.
27. Maciel, P., Gaspar, C., DeStefano, A. L., Silveira, I., Coutinho, P., Radvany, J., Dawson, D. M., Sudarsky, L., Guimarãs, J., Loureiro, J. E. LJ. E., Nezarati, M. M., Corwin, L. I., Lopes-Cendes, I., Rooke, K., Rosenberg, R., MacLeod, P., Farrer, L. A., Sequeiros, J., and Rouleau, G. A. (1995). Correlation between CAG repeat length and clinical features in Machado-Joseph disease. *Am. J. Hum. Genet.* **57,** 54–61.
28. Giunti, P., Sweeney, M. G., and Harding, A. E. (1995). Detection of the Machado-Joseph disease/spinocerebellar ataxia three trinucleotide repeat expansion in families with autosomal dominant motor disorders, including the Drew family of Walworth. *Brain* **118,** 1077–1085.
29. Maruyama, H., Kawakami, H., and Nakamura, S. (1996). Reevaluation of the exact CAG repeat length in hereditary cerebellar ataxias using highly denaturing conditions and long PCR. *Hum. Genet.* **97,** 591–595.
30. Dürr, A., Stevanin, G., Cancel, G., Duyckaerts, C., Abbas, N., Didierjean, O., Chneiweiss, H., Benomar, A., Lyon-Caen, O., Julien, J., Serdaru, M., Penet, C., Agid, Y., and Brice, A. (1996). Spinocerebellar ataxia 3 and Machado-Joseph disease: clinical, molecular, and neuropathological features. *Ann. Neurol.* **39,** 490–499.
31. Sasaki, H., Wakisaka, A., Takada, A., Yoshiki, T., Ihara, T., Suzuki, Y., Hamada, T., Iwabuchi, K., Onari, K., Tada, J., Suzuki, T., and Tashiro, K. (1995). Mapping of the gene for Machado-Joseph disease within a 3.6-cM interval flanked by D14S291/D14S280 and D14S81, on the basis of studies of linkage and linkage disequilibrium in 24 Japanese families. *Am. J. Hum. Genet.* **56,** 231–242.
32. Stevanin, G., Cancel, G., Dürr, A., Chneiweiss, H., Dubourg, O., Weissenbach, J., Cann, H. M., Agid, Y., and Brice, A. (1995). The gene for spinal cerebellar ataxia 3 (SCA3) is located in a region of ~3cM on chromosome 14q24.3-q32.2. *Am. J. Hum. Genet.* **56,** 193–201.
33. Twist, E. C., Casaubon, L. K., Ruttledge, M. H., Rao, V. S., Macleod, P. M., Radvany, J., Zhao, Z., Rosenberg, R. N., Farrer, L. A., and Rouleau, G. A. (1995). Machado Joseph disease maps to the same region of chromosome 14 as the spinocerebellar ataxia type 3 locus. *J. Med. Genet.* **32,** 25–31.
34. Takiyama, Y., Igarashi, S., Rogaeva, E. A., Endo, K., Rogaev, E. I., Tanaka, H., Sherrington, R., Sanpei, K., Liang, Y., Saito, M., Tsuda, T., Takano, H., Ikeda, M., Lin, C., Chi, H., Kennedy, J. L., Lang, A. E., Wherrett, J. R., Segawa, M., Nomura, Y., Yuasa, T., Weissenbach, J., Yoshida, M., Nishizawa, M., Kidd, K. K., Tsuji, S., and George-Hyslop, P. H. S. (1995). Evidence for intergenerational instability in the CAG repeat in the MJD1 gene and for conserved haplotypes at flanking markers amongst Japanese and Caucasian subjects with Machado-Joseph disease. *Hum. Mol. Genet.* **4,** 1137–1146.
35. Schöls, L., Amoiridis, G., Langkafel, M., Büttner, T., Przuntek, H., Riess, O., Vieira-Saecker, A. M. M., and Epplen, J. T. (1994). Machado-Joseph disease mutations as the genetic basis of most spinocerebellar ataxias in Germany. *J. Neurol. Neurosurg. Psych.* **59,** 449–450.
36. Ranum, L. P. W., Lundgren, J. K., Schut, L. J., Ahrens, M. J., Perlman, S., Aita, J., Bird, T. D., Gomez, C., and Orr, H. T. (1995). Spinocerebellar ataxia type I and Machado-Joseph disease: Incidence of CAG expansions among adult-onset ataxia patients from 311 families with dominant, recessive, or sporadic ataxia. *Am. J. Hum. Genet.* **57,** 603–608.
37. Inoue, K., Hanihara, T., Yamada, Y., Kosaka, K., Katsuragi, T., and Iwabuchi, K. (1996). Clinical and genetic evaluation of Japanese autosomal dominant cerebellar ataxias; is Machado-Joseph disease common in the Japanese? *J. Neurol. Neurosurg. Psych.* **60,** 697–698.
38. Higgins, J. J., Nee, L. E., Vasconcelos, O., Ide, S. E., Lavedan, C., Goldfarb, L. G., and Polymeropoulos, M. H. (1996). Mutations in American families with spinocerebellar ataxia (SCA) type 3: SCA3 is allelic to Machado-Joseph disease. *Neurology* **46,** 208–213.
39. Burt, T., Currie, B., Kilburn, C., Lethlean, A. K., Dempsey, K., Blair, I., Cohen, A., and Nicholson, G. (1996). Machado-Joseph disease in east Arnhem Land, Australia: chromosome 14q32.1 expanded repeat confirmed in four families. *Neurology* **46,** 1118–1122.
40. Stevanin, G., Le Guern, E., Ravise, N., Chneiweiss, H., Durr, A., Cancel, G., Vignal, A., Boch, A. L., Ruberg, M., and Brice, A. (1994). A third locus for autosomal dominant cerebellar ataxia type 1 maps to chromosome 14q24.3-qter: evidence for the existence of a fourth locus. *Am. J. Hum. Genet.* **54,** 11–20.
41. Haberhausen, G., Damian, M. S., Leweke, F., and Müller, U. (1995). Spinocerebellar ataxia, type 3 (SCA3) is genetically identical to Machado-Joseph disease (MJD). *J. Neurol. Sci* **.132,** 71–75.
42. Cancel, G., Dürr, A., Stevanin, G., Chneiweiss, H., Duyckaerts, C., Serdaru, M., De Toffol, B., Agid, Y., and Brice, A. (1994). Is DRPLA also linked to 14q? *Nature Genet.* **6,** 8.
43. Sakai, T., Antoku, Y., Kawakami, H., Maruyama, H., Nakamura, S., and Tanaka, K. (1996). A family with Machado-Joseph disease,

previously diagnosed as dentatorubral-pallidoluysian atrophy. *Neurology* **46,** 1154–1156.

44. Tuite, P. J., Rogaeva, E. A., St. George-Hyslop, P. H., and Lang, A. E. (1995). Dopa-responsive Parkinsonism phenotype of Machado-Joseph disease: confirmation of 14q CAG expansion. *Ann. Neurol.* **38,** 684–687.
45. Sakai, T., and Kawakami, H. (1996). Machado-Joseph disease: a proposal of spastic paraplegic subtype. *Neurology* **46,** 846–847.
46. Lopes-Cendes, I., Silveira, I., Maciel, P., Gaspar, C., Radvany, J., Chitayat, D., Babul, R., Stewart, J., Dolliver, M., Robitaille, Y., Rouleau, G. A., and Sequeiros, J. (1996). Limits of clinical assessment in the accurate diagnosis of Machado-Joseph disease. *Arch. Neurol.* **53,** 1168–1174.
47. Coutinho, P., and Andrade, C. (1978). Autosomal dominant system degeneration in Portuguese families of the Azores Islands. *Neurology* **28,** 703–709.
48. Rosenberg, R. N. (1983). Dominant ataxias *In* "Genetics of Neurological and Psychiatric Disorders" (S. Kety, L. Rowland, R. Sidman, and S. Mattaysseed, Eds.), pp. 195–213 Raven Press, New York.
49. Barbeau, A., Roy, M., Cunha, L., de Vincente, A. N., Rosenberg, R. N., Nyhan, W. L., MacLeod, P. L., Chazot, G., Langston, L. B., Dawson, D. M., and Coutinho, P. (1984). The natural history of Machado-Joseph disease: an analysis of 138 personally examined cases. *Can. J. Neurol. Sci.* **11,** 510–525.
50. Rosenberg, R. N., and Grossman, A. (1989). Hereditary ataxia *In* "Neurologic Clinics" (W. G. Johnson, Ed.), Vol. 1, pp. 25–36. Saunders, Philadelphia.
51. Rosenberg, R. N. (1990). Autosomal dominant cerebellar phenotypes: the genotype will settle the issue. *Neurology* **40,** 1329–1331.
52. Conner, K. E., and Rosenberg, R. N. (1993). The hereditary ataxias *In* "The Molecular and Genetic Basis of Neurological Disease" (R. N. Rosenberg, S. Prusiner, S. DiMauro, R. L. Barchi, and L. M. Kunkel, Eds.), pp. 697–737. Butterworth-Heinemann, Boston.
53. Maruyama, H., Nakamura, S., Matsuyama, Z., Sakai, T., Doyu, M., Sobue, G., Seto, M., Tsujihata, M., Oh-i, T., Nishio, T., Sunohara, N., Takahashi, R., Hayashi, M., Nishino, I., Ohtake, T., Oda, T., Nishimura, M., Saida, T., Matsumoto, H., Baba, M., Kawaguchi, Y., Kakizuka, A., and Kawakami, H. (1995). Molecular features of the CAG repeats and clinical manifestation of Machado-Joseph disease. *Hum. Mol. Genet.* **4,** 807–812.
54. Lang, A. E., Rogaeva, E. A., Tsuda, T., Hutterer, J., and St. George-Hyslop, P. (1994). Homozygous inheritance of the Machado-Joseph disease gene. *Ann. Neurol.* **36,** 443–447.
55. Kawakami, H., Maruyama, H., Nakamura, S., Kawaguchi, Y., Kakizuka, A., Doyu, M., and Sobue, G. (1995). Unique features of the CAG repeats in Machado-Joseph disease. *Nature Genet.* **9,** 344–345.
56. Sobue, G., Doyu, M., Nakao, N., Shimada, N., Mitsuma, T., Maruyama, H., Kawakami, H., and Nakamura, S. (1996). Homozygosity for Machado-Joseph disease gene enhances phenotypic severity. *J. Neurol. Neurosurg. Psych.* **60,** 354–356.
57. Kurohara, K., Kuroda, Y., Maruyama, H., Kawakami, H., Yukitake, M., Matsui, M., and Nakamura, S. (1997). Homozygosity for an allele carrying intermediate CAG repeats in the dentatorubral-pallidoluysian atrophy (DRPLA) gene results in spastic paraplegia. *Neurology* **48,** 1087–1090.
58. Sanpei, K., Takano, H., Igarashi, S., Sato, T., Oyake, M., Sasaki, H., Wakisaka, A., Tashiro, K., Ishida, Y., Ikeuchi, T., Koide, R., Saito, M., Sato, A., Tanaka, T., Hanyu, S., Takiyama, Y., Nishizawa, M., Shimizu, N., Nomura, Y., Segawa, M., Iwabuchi, K., Eguchi, I., Tanaka, H., Takahashi, H., and Tsuji, S. (1996). Identification of the spinocerebellar ataxia type 2 gene using a direct identification of repeat expansion and cloning technique, DIRECT. *Nature Genet.* **14,** 277–284.
59. Matsuyama, Z., Kawakami, H., Maruyama, H., Izumi, Y., Komure, O., Udaka, F., Kameyama, M., Nishio, T., Kuroda, Y., Nishimura, M., and Nakamura, S. (1997). Molecular features of the CAG repeats of spinocerebellar ataxia 6 (SCA6). *Hum. Mol. Genet.* **6,** 1283–1287.
60. Wexler, N. S., Young, A. B., Tanzi, R. E., Travers, H., Starosta-Rubinstein, S., Penney, J. B., Snodgrass, S. R., Shoulson, I., Gomez, F., Arroyo, M. A. R., Penchaszadeh, G. K., Moreno, H., Gibbons, K., Faryniarz, A., Hobbs, W., Anderson, M. A., Bonilla, E., Conneally, P. M., and Gusella, J. F. (1987). Homozygotes for Huntington's disease. *Nature* **326,** 194–197.
61. Snell, R. G., MacMillan, J. C., Cheadle, J. P., Fenton, I., Lazarou, L. P., Davies, P., MacDonald, M. E., Gusella, J. F., Harper, P. S., and Shaw, D. J. (1993). Relationship between trinucleotide repeat expansion and phenotypic variation in Huntington's disease. *Nature Genet.* **4,** 393–403.
62. Dürr, A., Stevanin, G., Cancel, G., Abbas, N., Chneiweiss, H., Agid, Y., Feingold, J., and Brice, A. (1995). Gender equality in Machado-Joseph disease. *Nature Genet.* **11,** 118–119.
63. Behl, C., Widmann, M., Trapp, T., and Holsboer, F. (1995). 17-Beta estradiol protects neurons from oxidative stress-induced cell death in vitro. *Biochem. Biophys. Res. Commun.* **216,** 473–482.
64. Green, P. S., Gridley, K. E., and Simpkins, J. W. (1996). Estradiol protects against beta-amyloid (25–35)-induced toxicity in SK-N-SH human neuroblastoma cells. *Neurosci. Lett.* **218,** 165–168.
65. Behl, C., Skutella, T., Lezoualc'h, F., Post, A., Widmann, M., Newton, C. J., and Holsboer, F. (1997). Neuroprotection against oxidative stress by estrogens: structure-activity relationship. *Mol. Pharmacol.* **51,** 535–541.
66. Green, P. S., Bishop, J., and Simpkins, J. W. (1997). 17-Alpha-estradiol exerts neuroprotective effects on SK-N-SH cells. *J. Neurosci.* **17,** 511–515.
67. Sherwin, B. B. (1997). Estrogen effects on cognition in menopausal women. *Neurology* **48,** S21–S26.
68. Wickelgren, I. (1997). Estrogen stakes claim to cognition. *Science* **276,** 675–678.
69. Orr, H. T., Chung, M. Y., Banfi, S., Kwiatkowski, T. J., Jr., Servadio, A., Beauder, A. L., McCall, A. E., Duvick, L. A., Ranum, L. P. W., and Zoghbi, H. Y. (1993). Expansion of an unstable trinucleotide CAG repeat in spinocerebellar ataxia type I. *Nature Genet.* **4,** 221–226.
70. Jodice, C., Malaspina, P., Persichetti, F., Novelletto, A., Spadaro, M., Giunti, P., Morocutti, C., Terrenato, L, Narding, A. E., and Frontali, M. (1994). Effect of trinucleotide repeat length and parental sex on phenotypic variation in spinocerebellar ataxia 1. *Am. J. Hum. Genet.* **54,** 959–965.
71. Warner, T. T., Williams, L. D., Walker, R. W., Flinter, F., Robb, S. A., Bundey, S. E., Honavar, M., and Harding, A. E. (1995). A clinical and molecular genetic study of dentatorubropallidoluysian atrophy in four European families. *Ann. Neurol.* **37,** 452–459.
72. Tsilfidis, C., MacKenzie, A. E., Mettler, G., Barcelo, J., and Korneluk, R. G. (1992). Correlation between CTG trinucleotide repeat length and frequency of severe congenital myotonic dystrophy. *Nature Genet.* **1,** 192–195.
73. Harley, H. G., Rundle, S. A., MacMillan, J. C., Myring, J., Brook, J. D., Crow, S., Reardon, W., Fenton, I., Shaw, D. J., and Harper, P. S. (1993). Size of the unstable CTG repeat sequence in relation to phenotype and parental transmission in myotonic dystrophy. *Am. J. Hum. Genet.* **52,** 1164–1174.
74. Richards, R. I., and Sutherland, G. R. (1992). Dynamic mutations: a new class of mutations causing human disease. *Cell* **70,** 709–712.

75. Pianese, L., Cavalcanti, F., De Michele, G., Filla, A., Campanella, G., Calabrese, O., Castaldo, I., Monticelli, A., Cocozza, S. (1997). The effect of parental gender on the GAA dynamic mutation in the FRDA gene. *Am. J. Hum. Genet.* **60,** 460–463.
76. Ikeuchi, T., Igarashi, S., Takiyama, Y., Onodera, O., Oyake, M., Takano, H. Koide, R., Tanaka, H., and Tsuji, S. (1996). Non-Mendelian transmission in dentatorubral-pallidoluysian atrophy and Machado-Joseph disease: the mutant allele is preferentially transmitted in male meiosis. *Am. J. Hum. Genet.* **58,** 730–733.
77. Igarashi, S., Takiyama, Y., Cancel, G., Rogaeva, E. A., Sasaki, H., Wakisaka, A., Zhou, Y. X., Takano, H., Endo, K., Sanpei, K., Oyake, M., Tanaka, H., Stevanin, G., Abbas, N., Dürr, A., Rogaev, E. I., Sherrington, R., Tsuda, T., Ikeda, M., Cassa, E., Nishizawa, M., Benomar, A., Julien, J., Weissenbach, J., Wang, G. X., Agid, Y., St. George-Hyslop, P. H., Brice, A., and Tsuji, S. (1996). Intergenerational instability of the CAG repeat of the gene for Machado-Joseph disease (MJD1) is affected by the genotype of the normal chromosome: implications for the molecular mechanisms of the instability of the CAG repeat. *Hum. Mol. Genet.* **5,** 923–932.
78. Cancel, G., Abbas, N., Stevanin, G., Dürr, A., Chneiweiss, H., Néri, C., Duyckaerts, C., Penet, C., Cann, H. M., Agid, Y., and Brice, A. (1995). Marked phenotypic heterogeneity associated with expansion of a CAG repeat sequence at the spinocerebellar ataxia 3/Machado-Joseph disease locus. *Am. J. Hum. Genet.* **57,** 809–816.
79. Lopes-Cendes, I., Maciel, P., Kish, S., Gaspar, C., Robitaille, Y., Clark, H. B., Koeppen, A. H., Nance, A., Schut, L., Silveira, I., Coutinho, P., Sequeiros, J., and Roulear, G. A. (1996). Somatic mosaicism in the central nervous system in spinocerebellar ataxia type 1 and Machado-Joseph disease. *Ann. Neurol.* **40,** 199–206.
80. Telenius, H., Kremer, B., Goldberg, Y. P., Theilmann, J., Andrew, S. E., Zeisler, J., Adam, S., Greenberg, C., Ives, E. J., Clarke, L. A., and Hayden, M. R. (1994). Somatic and gonadal mosaicism of the Huntington disease gene CAG repeat in brain and sperm. *Nature Genet.* **6,** 409–414.
81. Ueno, S., Kondoh, K., Kotani, Y., Komure, O., Kuno, S., Kawai, J., Hazama, F., and Sano, A. (1995). Somatic mosaicism of CAG repeat in dentatorubral-pallidoluysian atrophy. *Hum. Mol. Genet.* **4,** 663–666.
82. Chong, S. S., McCall, A. E., Cota, J., Subramony, S. H., Orr, H. T., Hughes, M. R., and Zoghbi, H. Y. (1995). Gametic and somatic tissue-specific heterogeneity of the expanded SCA1 CAG repeat in spinocerebellar ataxia type 1. *Nature Genet.* **10,** 344–349.
83. Ikeda, H., Yamaguhci, M., Sugai, S., Aze, Y., Narumiya, S., and Kakizuka, A. (1996). Expanded polyglutamine in the Machado-Joseph disease protein induces cell death *in vitro* and *in vivo*. *Nature Genet.* **13,** 196–202.
84. Onodera, O., Roses, A. D., Tsuji, S., Vance, J. M., Strittmatter, W. J., and Burke, J. R. (1996). Toxicity of expanded polyglutamine-domain proteins in *Escherichia coli*. *FEBS Lett.* **399,** 135–139.
85. Oberdick, J., Levinthal, F., and Levinthal, C. (1988). A Purkinje cell differentiation marker shows a partial DNA sequence homology to the cellular sis/PDGF2 gene. *Neuron* **1,** 367–376.
86. Smeyne, R. J., Obendick, J., Schilling, K., Berrebi, A. S., Mugnaini, E., and Morgan J. I. (1991). Dynamic organization of developing Purkinje cells revealed by transgene expression. *Science* **254,** 719–721.
87. Burright, E. N., Clark, H. B , Servadio, A., Matilla, T., Feddersen, R. M., Yunis, W. S., Duvick, L. A., Zoghbi, H. Y., and Orr, H. T. (1995). SCA1 Transgenic mice: a model for neurodegeneration caused by an expanded CAG trinucleotide repeat. *Cell* **82,** 937–948.
88. Eadie, M. J. (1975). Cerebello-olivary atrophy *In* "Handbook of Clinical Neurology" (P. J. Vinken and G. W. Bruyn, Eds.), Vol. 20. North-Holland, Amsterdam.
89. Kakizuka, A. (1997). Degenerative ataxias: genetics, pathogenesis, and animal models. *Curr. Opin. Neurol.* **10,** 285–290.
90. Perutz, M. F., Johnson, T., Suzuki, M., and Finch, J. T. (1994). Glutamine repeats as polar zippers: their possible role in inherited neurodegenerative diseases. *Proc. Natl. Acad. Sci. USA* **91,** 5355–5358.
91. Hilbich, C., Kisters-Woike, B., Reed, J., Masters, C. L., and Beyreuther, K. (1991). Aggregation and secondary structure of synthetic amyloid beta A4 peptides of Alzheimer's disease. *J. Mol. Biol.* **218,** 149–163.
92. Jarrett, J. T., and Lansbury, P. T., Jr. (1993). Seeding "one-dimensional crystallization" of amyloid: a pathogenic mechanism in Alzheimer's disease and scrapie? *Cell* **73,** 1055–1058.
93. Pan, K. M., Baldwin, M., Nguyen, J., Gasset, M., Serban, A., Groth, D., Mehlhorn, I., Huang, Z., Fletterick, R. J., Cohen, F. E., and Prusiner, S. B. (1993). Conversion of α-helices into β-sheets features in the formation of the scrapie prion proteins. *Proc. Natl. Acad. Sci. USA* **90,** 10962–10966.
94. Zoghbi, H. Y. (1996). The expanding world of ataxins. *Nature Genet.* **14,** 237–238.
95. Zhuchenko O, Bailey, J., Bonnen, P., Ashizawa, T., Stockton, D. W., Amos, C., Dobyns, W. B., Subramony, S. H., Zoghbi, H. Y., and Lee, C. C. (1997). Autosomal dominant cerebellar ataxia (SCA6) associated with small polyglutamine expansions in the a1A-voltage-dependent calcium channel. *Nature Genet.* **15,** 62–69.
96. Mangiarini, L., Sathasivam, K., Seller, M., Cozens, B., Harper, A., Hetherington, C., Lawton, M., Trottier, Y., Lehrach, H., Davies, S. W., and Bates, G. P. (1996). Exon 1 of the HD gene with an expanded CAG repeat is sufficient to cause a progressive neurological phenotype in transgenic mice. *Cell* **87,** 493–506.
97. Davies, S. W., Turmaine, M., Cozens, B. A., DiFiglia, M., Sharp, A. H., Ross, C. A., Scherzinger, E., Wanker, E. E., Mangiarini, L., and Bates, G. P. (1997). Formation of neuronal intranuclear inclusions (NII) underlies the neurological dysfunction in mice transgenic for the HD mutation. *Cell* **90,** 537–548.
98. Roizin, L., Stellar, S., and Liu, J. C. (1979). Neuronal nuclear-cytoplasmic changes in Huntington's chorea: electron microscope investigations *In* "Advances in Neurology" (T. N. Chase, N. S. Wexler, and A. Barbeau, Eds.), Vol. 23, pp. 95–122. Raven Press, New York.
99. Paulson, H. L., Subramony, S. H., Das, S. S., Fratkin, J., McDaniel, O., and Pittman, R. N. (1997). Aberrant nuclear localization of the Machado-Joseph disease gene product in disease brain. *Neurology* **48,** A210.

Part IX

Huntington's Disease

CHAPTER 23

Huntington's Disease

RICHARD H. MYERS Department of Neurology, Boston University School of Medicine, Boston, Massachusetts 02114

KAREN S. MARANS Department of Neurology, Boston University School of Medicine, and VA Medical Center, Boston, Massachusetts 02130

MARCY E. MACDONALD Molecular Neurogenetics Unit, Massachusetts General Hospital, Charlestown, Massachusetts 02129

I. INTRODUCTION

In 1872, Dr. George Huntington published a seminal work titled *On chorea* [1]. This manuscript described the features of an illness observed among the members of three families residing in East Hampton, New York, on Long Island. Dr. Huntington's father and grandfather had been physicians practicing in the same community and thus he had the benefit of their observations of the ancestors of the contemporary members of the families.

Dr. Huntington called the disease "hereditary chorea" and his observation of the mode of inheritance is noteworthy because the basic tenants of genetics, as defined by Gregor Mendel, had not been recognized at the time. Dr. Huntington described the transmission of this disease in the following terms: "When either or both of the parents have shown manifestations of the disease . . . one or more of the offspring almost invariably suffer from the disease, if they live to adult age. But if by any chance these children go through life without it, the thread is broken and the grandchildren and great-grand children of the original shakers may rest assured that they are free from the disease." [1] This description provides a functional definition for the autosomal dominant mode of transmission with complete penetrance characteristic of Huntington's disease.

Dr. Huntington's treatise was of such clarity that it was reprinted in prominent neurology texts of the day and termed "Huntington's Chorea." More recently, the observation that some cases, and particularly those with a young onset, may present without chorea, has led to widespread adoption of the designation Huntington's Disease (HD) to refer to this affliction.

HD is a midlife-onset disease which strikes at a mean age of 40 years, but onset may vary from 4 to 80 years of age [2, 3]. It is a progressive neurodegenerative disease with primary neuropathological involvement in the basal ganglia. HD is invariably fatal, without periods of remission, and the course from onset to death averages 15 to 17 years [4]. The disease is characterized by involuntary choreiform movement [1], cognitive impairment [5], and personality disorder tending to depression [6], anger and temper outbursts, and personality change [3].

The *HD* gene was genetically linked to an anonymous marker located in 4p16.3 [7] and, following a 10-year molecular genetic search, was discovered to be an expanded and unstable CAG trinucleotide repeat [8]. The mutation extends a polymorphic stretch of CAG codons in the first exon of the *HD* gene, lengthening a segment of polyglutamine near the amino terminus of the *HD* protein which has been dubbed huntingtin. The mutation which leads to the expression of HD is an approximate doubling in the number of the triplet repeats from a normal number of about 18 to an expanded number of 40 or more.

II. EPIDEMIOLOGY OF HD

The prevalence of HD is estimated at 5 to 10 affected persons per 100,000 among individuals of European descent but is less common among other ethnic groups [2, 9]. Indeed, the prevalence of HD among the native African population is so low as to make an accurate estimate difficult, and its prevalence among Japanese and Asian populations is approximately one-tenth that observed in Caucasians [9]. Because, for each affected individual, there are an estimated two living gene carriers who are still too young to manifest symptoms, the prevalence of *HD* gene carriers in the general Caucasian population may be estimated at 15 to 30 per 100,000, and there are an equal number of at-risk siblings who are the fortunate ones who have not inherited the *HD* defect. Thus, approximately 1 in every 2500 births is an individual born at risk for the disease, making HD the most common of the CAG repeat expansion diseases. Genotype–phenotype correlations in HD, therefore, may permit the description of uncommon events such as new mutations or alleles with reduced penetrance which may be too rare to be observed in the other triplet repeat diseases.

III. ISOLATION OF THE *HD* MUTATION

The clinical and genetic characteristics of HD, including its distinct symptoms, midlife onset, unambiguous mode of inheritance, high penetrance, and prevalence in the general population, made it an ideal disease to approach using a nascent strategy based on genetic linkage. In 1983 the *HD* defect was mapped to the vicinity of an anonymous polymorphic DNA marker, *D4S10,* following the inheritance of restriction fragment length polymorphisms in two large HD kindred, of American and Venezuelan decent [7].

Genetic linkage to *D4S10* established that virtually all cases of HD were likely to arise from defects in the same gene [10]. These early linkage studies also demonstrated that *D4S10* was ~4 cM (~4% recombination) from *HD,* providing the basis for the molecular diagnosis of HD in asymptomatic at-risk individuals able to participate in a genetic linkage test. Furthermore, linkage to *D4S10* revealed the complete phenotypic dominance of HD. Affected homozygotes, possessing two copies of the *HD* defect, were discovered to be clinically indistinguishable from their heterozygous siblings with one copy of the mutation [11, 12].

The stage was set for the molecular genetic search that culminated in the cloning of the *HD* mutation in 1993 [13], a 10-year period that witnessed the development of numerous techniques and strategies that are now commonplace. A consortium of investigators employed a combination of genetic and physical mapping techniques to confine the mutation to a segment of ~2 million base pairs of 4p16.3, simultaneously generating genetic and physical maps of the region, developing overlapping clone sets, and refining the defect's location using recombination analysis in disease pedigrees and linkage disequilibrium between the disorder and genetic markers.

The location of the *HD* gene (Fig. 23-1) was pinpointed ultimately by an analysis of haplotypes formed by multiallele markers from across the region which revealed that although ~two-thirds of disease chromosomes had unrelated haplotypes and likely had independent origins, the remaining ~one-third shared a small (275 kb) region between *D4S95* and *D4S180.* This pro-

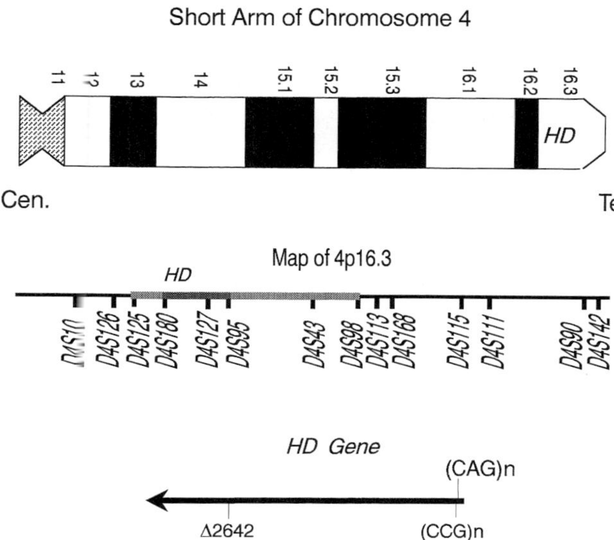

FIGURE 23-1 Location cloning of the *HD* mutation. The cytogenetic bands of the short arm of chromosome 4 (4p) are depicted with the location of the *HD* defect (*HD*) in 4p16.3 from genetic linkage and physical mapping studies indicated. The map of 4p16.3 provides an expanded view of this telomeric cytogenetic band (*thin black line*) to illustrate the location of the defect (*HD*) relative to the polymorphic DNA markers (*below line*) used to sequentially narrow the minimal *HD* genetic region (*thickened line*) by recombination analysis, linkage disequilibrium, and haplotype analysis. The location of the *HD* mutation (CAG_n) within the ~185 kb of 4p16.3 DNA spanned by the *HD* gene is shown relative to the 5′ to 3′ direction of transcription (*arrow*) and polyproline-encoding repeat (CCG_n) and Δ2642 polymorphisms.

vided strong evidence that this subset of disease chromosomes was likely to be descended from a common ancestral chromosome [14]. Together with Δ2642, a two-allele polymorphism in linkage disequilibrium with HD [15], a set of highly polymorphic DNA markers pinpointed a segment of ~150 kb on these major HD haplotype chromosomes as the most likely location of the *HD* gene. Analysis of candidate genes from this subregion led finally to the discovery of a polymorphic CAG repeat in the 5′ end of interesting transcript 15 (IT15) that was expanded and unstable on disease chromosomes [8]. The *HD* defect had been identified at long last and the molecular genetic search for its pathogenic mechanism began in earnest.

IV. THE *HD* TRINUCLEOTIDE REPEAT MUTATION

The pure stretch of CAG trinucleotide repeat that is expanded on disease chromosomes is located near the 5′ end of a novel 4p16.3 gene (Fig. 23-1), immediately adjacent to a broken array of CAG/CCG codons containing a mildly polymorphic (6 to 12 repeats), stably transmitted stretch of CCG triplets [8]. The results of initial genotype–phenotype correlation studies quickly revealed that most normal chromosomes and the majority (>90%) of disease chromosomes possess 7 CCG repeats [16, 17], whereas the adjacent array of pure CAG repeats was found to be highly polymorphic. Normal chromosomes possessed from 6 to 34 CAG repeats that are inherited in a Mendelian fashion and HD chromosomes from 39 to ~86 units that are inherited in a strikingly non-Mendelian manner. Moreover, rare alleles with 36–39 repeats were found in exceptional individuals, the unaffected elderly relatives of sporadic *de novo* cases of the disease [8]. These initial findings made it possible to offer a direct DNA test, the polymerase chain reaction (PCR) amplification determination of *HD* CAG repeat length, to many at-risk individuals in a variety of testing situations. The findings of subsequent genotype–phenotype studies performed in research and clinical diagnostic laboratories around the world have provided critical data for the generation of testable hypotheses for the mechanism of action of the *HD* mutation and have focused the discussion on the benefits and limitations inherent in the use of CAG repeat length as a clinical diagnostic tool.

Although it is highly polymorphic on normal and disease chromosomes, the CAG repeat lengths associated with the vast majority of chromosomes in either category possess similar repeat lengths, with the mean length on normal chromosomes at ~18 units and on disease causing chromosomes at ~48 repeats. Data derived from the bulk of chromosomes at around these lengths, therefore, provide the most certain genotype–phenotype findings, revealing trends that are associated with the majority of disease cases. These results reveal, for example, that the length of CAG repeat in the expanded range is strongly correlated with the age at onset of the disease, with onset age decreasing as the CAG repeat size increases. They also point out that any given expanded CAG repeat is associated with a broad range of onset ages, highlighting the importance of other factors in determining age at onset in any single individual. However, any and all repeat sizes have been observed on at least a few chromosomes and it is difficult, given the possibility of errors in genotyping and phenotyping data in large studies, to develop trends based on a few individual cases. This critical feature of the *HD* CAG repeat introduces large inherent uncertainties into the use of CAG repeat length as a diagnostic tool, especially for those rare individuals who possess allele lengths between the peaks of the distributions of alleles that are clearly not associated with disease and those that are invariably capable of provoking symptoms. Alleles in this range are now thought to exhibit reduced penetrance and the definitive answer to the question of whether any allele size will eventually provoke disease (given infinite life span) awaits a delineation of the undiscovered underlying pathogenic mechanism.

Currently four CAG repeat size intervals are recognized as associated with varying disease risk in HD (Fig. 23-2). These ranges have been defined by the U.S. HD Genetic Testing Group (USHDGTG) and are derived from information gleaned from more than 1000 HD tests and from published sources. Nevertheless, the disease-related risk pertaining to the repeats in the 29 to 39 range is often based upon fewer than 10 observations at each repeat size and thus these risk estimates can be expected to change with additional experience.

A. Normal: Repeat Sizes up to 28 Units

Persons with repeats in this range do not develop HD, nor has there been a confirmed instance of a child inheriting HD from a parent with a repeat in this range.

B. Nonpenetrant with Paternal Meiotic Instability: Repeats of 29 to 35 Units

Repeats in this range are rare and represent approximately 1% of expanded alleles seen in HD testing programs. There have been no confirmed reports of persons with repeats in this range expressing HD. There are, however, confirmed cases of paternally transmitted meiotic instability such that descendants of fathers with repeats in this range are known to have inherited an expanded allele in the clinical range. There are no confirmed instances of maternally transmitted meiotic instability in this range producing offspring with repeats in the clinical range.

C. Reduced Penetrance with Meiotic Instability: Repeats of 36 to 39 Units

Repeats in this range are rare and represent approximately 1 to 2% of expanded alleles seen in HD testing programs. Some persons with repeats in this range develop HD and others live into their late 90s without evidence of the disease [18–20]. There is evidence that penetrance increases with increasing allele size in this range. Penetrance has been roughly estimated at 25% for 36 repeats, 50% at 37 repeats, 75% at 38 repeats, and 90% at 39 repeats based upon USHDGTG data. These figures will undoubtedly change with additional information. There are several cases of paternally trans-

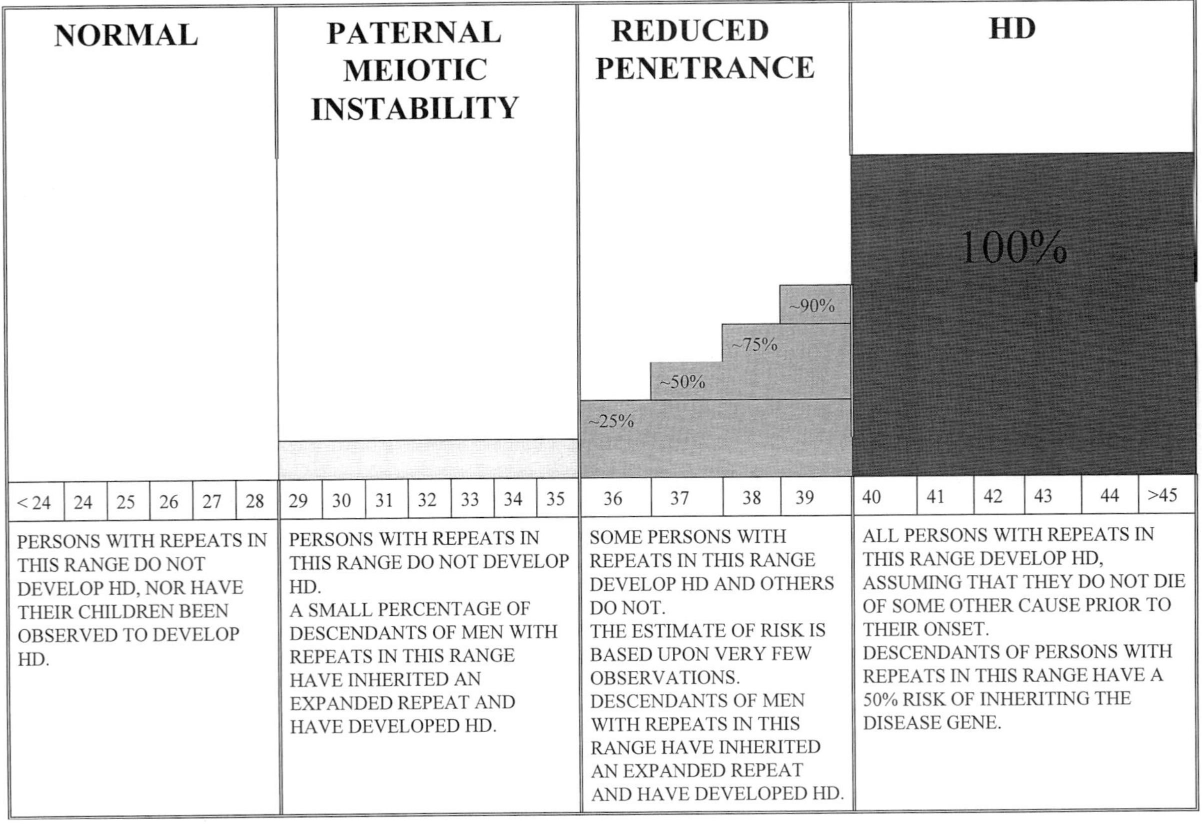

FIGURE 23-2 Repeat size and Huntington's disease risk.

mitted meiotic instability such that descendants of fathers with repeats in this range have inherited an expanded allele in the clinical range.

D. HD: Repeats of 40 Units or Larger

Currently, it is believed that all persons with repeats in the range of 40 or more will eventually develop HD. However, some individuals with repeats at the low end of this range are reported to exhibit initial symptoms at ages older than common life expectancy and, thus, there may be some reduced penetrance among carriers of 40 and 41 repeats [21]. The smallest alleles in disease patients are between 36 and 40 repeats while the longest allele observed in a normal individual is 39 units [22] (Table 23-1). There is an area of overlap in CAG repeat sizes, which represents the "reduced penetrance" range of 36 to 39 repeats. We do not view these as either "HD alleles" or "normal alleles." The term "intermediate repeat," coined in 1993 [18], has gained wide usage [23, 24], but this term has referred to the entire 29- to 39-repeat range and is, therefore, ambiguous. We propose that the terms "meiotic instability" range and "reduced penetrance" range be applied to the intervals of 29 to 35 and 36 to 39 repeats, respectively, to lessen the confusion.

V. CLINICAL CORRELATES

A. Chorea and Motor Impairment

The term chorea is of Greek origin and means dance. Chorea refers to a prominent feature of the gait disturbance characteristic of HD and the "dance-like" quality of movement. Choreic movements are involuntary, slow, and random and may involve any muscle group. Thus the involuntary aspect of the movement disorder in HD is that it involves random twitching of both upper and lower extremities and of both the left and right sides. These movements may begin as small fasciculations of the fingers, toes, or facial muscles early in the disease and evolve to large movements of the limbs and trunk as the illness progresses. Over the course of the disease, the movements may interfere with gait, speech, chewing, and swallowing, and other aspects of motor function associated with activities of daily living. However, the extent of the movement impairment may vary substantially for different persons and some may have little

TABLE 23-1 Smallest HD and Largest Normal Allele in Reported Series

Reference	Smallest HD allele	Largest normal allele	CCG adjusted	Country
Duyao *et al.* [25]	37 $N = 425$	34 $N = 545$	No	U.S.A.
Stine *et al.* [26]	36 $N = 114$	31 $N = 114$	No	U.S.A.
Andrew *et al.* [27]	38 $N = 360$	37 $N = 500$	No	Canada
Snell *et al.* [28]	36[a] $N = 421$	34 $N = 720$	No	United Kingdom
Barron *et al.* [29]	35 $N = 337$	33 $N = 795$	Yes	Scotland
Zühlke *et al.* [30]	40 $N = 352$	30 (35) $N = 200$	No	Germany
Zühlke *et al.* [31]	See above	33 $N = 513$	Yes	Germany
Craufurd and Dodge [32]	36[a] $N = 228$	34 $N = 401$	No	United Kingdom
De Rooji *et al.* [33]	37 $N = 180$	37 $N = 370$	No	Holland
Novelleto *et al.*, [34, 35]	37 $N = 195$	32 $N = 190$	Yes	Italy
Trottier *et al.* [36]	40 $N = 85$	37 $N = 145$	No	France
Kremer *et al.* [22][b]	36 $N = 995$	39 $N = 1595$	Yes	Multinational

[a]Alleles in the normal range attributed to HD phenocopies are not included.
[b]This sample may overlap with others reported in this table.

evidence of involuntary movement. In the late stages of HD, chorea may lessen and be replaced by increasing rigidity and dystonia.

In addition to the involuntary choreic movement impairment, there is a slowing of voluntary movement. This bradykinesia may be seen in tasks involving rapid alternating movements, eye movements, finger tap exercises, and tongue tap exercises. Some studies have proposed that the progression of impairment in voluntary movement may be more uniform across different stages of HD and may therefore represent a more consistently progressing feature for assessments of disease course than does severity of chorea.

The relation of motor impairment to CAG repeat size has not been widely studied. Andrew *et al.* [27] found no association between the repeat length and the type of clinical presentation (e.g., motor, mood, or cognitive disturbance as the initial feature).

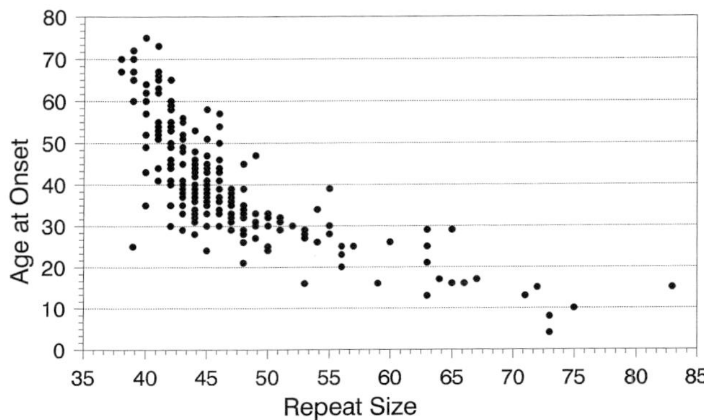

FIGURE 23-3 Relationship between the length of the expanded *HD* CAG repeat and age at neurologic onset of disease. The CAG repeat sizes for 220 persons HD diagnosed through the New England Huntington's Disease Research Center are presented in relationship to the age at onset of motor impairment. Repeat size is strongly related to age at onset. Onset age before age 20 is usually associated with a repeat size of more than 60 CAG units. Among persons with adult onset, the range in onset age for a given repeat is large and may vary by 35 years or more.

B. Age at Onset

The age at onset of Huntington's disease is highly variable. The earliest cases have been reported with onset ages of 2 to 3 years of age while the latest onsets approach 80 years or older. The average is at age 40 and is not related to the sex of the individual. Approximately 7% of all cases present before the age of 20 and this group of individuals has traditionally been termed "juvenile" onset HD [37].

The impact of the HD repeat upon phenotypic expression has been heavily investigated. The relationship between age at onset and repeat size is unequivocal and very strong but it is not linear; most investigators have noted that the best fit is a model in which the repeat predicts the $\log_{10}$ of onset age (see Refs. [25, 38]. For the 220 HD cases depicted in Fig. 23-3, sampled from the New England HD Center, the correlation between repeat size and log of onset age is $r = -0.84$ and accounts for about 70% of the variance in onset age. This high level of association between onset and repeat size is similar to that observed in published reports [21, 25, 38].

While the overall relationship of the repeat size to the onset age is strong, much of the effect is seen among the small fraction of individuals who have very large repeats and young ages at onset. Seven percent (n = 14) of this sample of 220 HD cases have an onset at age 20 or younger and nearly all ($n = 11$) of these individuals have repeats of 60 or more CAG units. When this small sample is removed from consideration, the correlation of repeat size to onset age reduces to $r = -0.68$, is linear, and explains about 45% of the variance in repeat size. Kremer *et al.* [39] noted that for late-onset cases, with onset at age 50 or older, the correlation of onset with repeat size is reduced ($r = -0.29$). The distribution of HD repeats is highly skewed, and approximately 50% of HD cases have repeats between 41 and 45 repeats. An additional 30% have repeats between 46 and 50 and 10% have repeats between 51 and 55 units. Approximately 90% of all HD cases have fewer than 55 repeats.

It has been noted that there is a ±18-year 95% confidence interval around the estimated onset age for a given repeat size [25]. Although one group [21] recently suggested that the confidence interval may be smaller than this estimate, all published reports confirm that the range in onset age for a given repeat size exceeds 30 years among persons with mid- and late-life disease. Thus, for more than 90% of all cases of HD the relationship of repeat size to onset age is not strong enough to be useful in predicting onset among persons who are not yet symptomatic.

C. Rate of Disease Progression

The course of the disease in HD is remarkably variable. There have been only a few studies of factors related to progression [40–43]. There is evidence that the age at onset is related to rate of progression [40] and it has been proposed that a single mechanism influences both the age of onset and the rate of disease progression [4]. Although not strong enough to predict onset presymptomatically, the HD repeat size appears to be the primary determinant of onset age and thus it is reasonable to expect that the repeat size is related to the rate of disease progression as mediated either by its effect upon onset age or by a direct effect upon disease progression. Another factor reported to be related to rate of progression is the sex of the affected parent, with offspring of affected fathers having a more rapid progression than offspring of affected mothers [40, 44]. In addition, lower body mass index early in the disease is related to more rapid disease progression [40]. While the sex of the affected parent and paternal transmission is now known to be related to the transmission of an expanded repeat, it is not known whether the repeat size influences weight loss or lower body mass index. No medication, dietary supplement, or other clinical intervention has been demonstrated to influence the rate of disease progression. A study of a set of monozygotic twins reared apart [45] suggests that progression is largely determined by genetic factors and the most likely determinant is the trinucleotide repeat size.

The rate of disease progression in HD has been little studied with respect to the repeat size and the results have not yielded a consensus. Although one early study [41] found no evidence that the rate of disease progression is modified by repeat size, recently Brandt *et al.* [42] and Illarioshkin *et al.* [43] have reported a significant association between repeat size and rate of disease progression. The participants in the Kieburtz *et al.* [41] study had been followed for a shorter period of time than had those in the Brandt *et al.* [42] study and this may account for the inability to detect an effect in the former analysis. These studies are important because, for example, trials for therapeutic interventions in HD need to consider the *HD* repeat in the randomization of study participants.

D. Neuropsychology

1. Cognitive Function in HD

While language functioning remains relatively intact, the most striking cognitive deficits seen in HD patients are in the areas of executive system functioning (e.g., strategies in problem solving and cognitive flexibility), short-term memory, and visuospatial functioning [5, 46–48]. The memory problem in HD patients is characterized by inconsistent retrieval of information [46, 49, 50]. In the early stages of the disease, recognition memory for verbal information is robust, while spontaneous recall of information may be more frequently impaired [5, 51].

Early in the disease process the cognitive deficits are relatively focal. A global, progressive subcortical dementia [52, 53] does not evolve until the disease advances significantly [48, 54–57]. Although this dementia is significant and debilitating in the later stages of the disease, it is qualitatively different from the cortical dementia seen in Alzheimer's disease [58–60].

2. Presymptomatic Cognitive Features

Previously, it was thought that individuals who were asymptomatic gene carriers for HD were indistinguishable both from those who were at risk but not gene carriers and from the general population. Although there remains some controversy in the literature there is evidence to suggest that subtle, but significant, changes may occur before what has traditionally been thought of as the onset of the clinical syndrome (i.e., the appearance of motor disturbance). Among persons undergoing genetic linkage studies, cognitive changes on tests of learning and memory, visuospatial skills, and frontal lobe functioning [61] were found presymptomatically. However, other studies [62–65] found no evidence of presymptomatic cognitive decline. Nevertheless, Blackmore *et al.* [64] found a significant group advantage among the presymptomatic gene-negative subjects. The gene-negative group average was higher than that of the presymptomatic gene-positive group on 30 of the 43 tasks administered.

Similar studies were conducted using direct genetic testing. Rosenberg *et al.* [66] found asymptomatic gene carriers to be inferior to the non-gene carriers on tests of attention, learning, and planning. Foroud *et al.* [67] not only found a significant difference in their subjects who were gene carriers, but found a significant correlation between the degree of cognitive deficit and the length of the carrier's trinucleotide (CAG) repeat.

A few of these studies had a very small number of subjects and may not have had sufficient power to detect differences between groups. More important, it may not be appropriate to treat all gene carriers as a homogenous group. Persons with longer CAG repeats may be more likely to be close to disease onset. Some of the ambiguity in this literature might be clarified by considering an estimate of proximity to onset based on the trinucleotide repeat length and the affected parents' age of onset. Controversy continues over whether the changes occur continuously and slowly over the lifetime of the individual or whether there is a sudden midlife change in functioning which coincides with or shortly precedes the onset of the motor symptoms [63].

3. CAG Relationship to Cognition

Andrew *et al.* [27] found no association between the CAG repeat length and the type of clinical presentation (e.g., motor, mood, or cognitive disturbance as the initial feature) and others have also failed to find a relationship between the repeat length and either the type of symptom onset or disease progression [68, 69] or the type of psychiatric involvement [70].

4. Personality Changes

Research findings on personality change in HD are ambiguous and contradictory. Perhaps some of the inconsistency arises out of the difficulty in establishing accurate psychiatric diagnostic criteria for patients with existing neurological diseases. Indeed, some HD patients who exhibit what are commonly regarded as the behavioral manifestations of the disease (e.g., impulsiveness, disinhibition, or suicidal behavior) have been inappropriately diagnosed as having a personality disorder [5]. In addition, over time both the diagnostic criteria for mental disorders and the methodology for investigating psychiatric symptoms have been modified. Chart reviews [71], self-report, and caretakers' reports [72] have all been used. However, as Burns *et al.* [72] pointed out, it may be problematic to rely on self-report measures when investigating syndromes in which dementia is a prominent feature. Finally, investigators have not typically defined their study samples according to the stage of the disease. Despite these problems, over the past few decades there have been valuable descriptions of the psychiatric and behavioral changes seen in HD patients.

A change in personality has been recognized as one of the cardinal features of subcortical dementia [52, 53]. Depression and apathy are commonly seen in both Parkinson's and Huntington's disease [73, 74]. Given that these affective changes can occur several years before the motor onset of HD [9, 73, 75, 76] they are considered to be organic in nature rather than an emotional response to a debilitating degenerative disease. In fact, some HD patients experience manic episodes in addition to depression [3, 71, 74]. Neither bipolar disorder nor mania is an expected functional reaction to living with HD.

Apathy, defined as "no emotion" [72], or "situational apathy" [77] was reported in 48 and 73% of HD subjects, respectively. Some investigators have reported that apathy is correlated with dementia [51, 74], while others [72, 78] have reported that there is no correlation. However, when Burns and colleagues [72] matched their HD and Alzheimer's disease samples on level of cognitive functioning, they found that apathy accompanies the dementia in HD to a greater extent than in Alzheimer's disease.

Irritability, often defined as being quick to anger, is another feature commonly reported in patients with HD [5, 9, 71, 72, 79]. Approximately 60% (64% [71], 58% [72]) of HD patients have been described as irritable. Burns *et al.* [72] found 59% of their sample to be aggressive. Folstein and Folstein [78] proposed that as a result of the disease process there may be dysfunction of the normal regulatory (dampening) mechanisms, which may result in an exacerbation of a premorbid aggressive personality trait.

There is a good deal of evidence for impulsive behavior [77, 80], antisocial behavior (ranging from criminal assaults and minor crimes to child abuse and neglect [9, 80–82]), and alcohol and substance abuse problems [5, 78, 81] associated with HD. The majority of alcohol use may be in the earlier stages (perhaps as a form of self-medication) and drops off as the disease progresses [5, 76]. However, King [83] did not find an increased level of alcoholism in HD.

In the early 1950s and 60s schizophrenia-like symptoms were reported as the predominant psychiatric feature of HD. However, Peyser and Folstein [84] attribute these reports to the psychiatric diagnostic criteria relied upon at the time and the lack of systematic empirical studies. Presently, affective disorders (including depression, mania, and bipolar disorder), not schizophrenia, are reported as the most common psychiatric syndrome in HD. Disorders related to schizophrenia have been reported to occur in between 5 and 10% of HD patients [51]. Hallucinations and paranoid delusions are reported by HD patients [5, 77, 78]. A high suicide rate among HD-affected persons has been noted [1, 6, 81, 85]. Schoenfeld *et al.* [6] found that the suicide rate in HD was 8 times the general population rate in Massachusetts. They noted that about half of the successful suicides were committed by individuals who had not yet been diagnosed, and concluded that suicide may occur more frequently in the earlier stages of the disease. White *et al.* [5] described patients who experienced command hallucinations regarding suicide, some who reported a compulsive urge to kill themselves, and some for whom suicidal ideation is secondary to their depression and demoralization related to the disease.

Folstein and her colleagues observed affective disorder more frequently in some HD families than in others [47, 86], and Webb and Trzepacz [76] reported that those who experienced psychiatric disturbances before chorea had a significantly older age of motor onset than those without this pattern of onset.

VI. NEUROPATHOLOGIC STUDIES

While the mutation may ultimately prove to affect cells and tissues in the periphery, the impact of the *HD* mutation is most obvious in the brain (Fig. 23-4). The neuropathology of HD has been heavily investigated with the earliest changes seen in the tail and in the paraventricular portions of the caudate nucleus, and in the dorsal part of the putamen [87]. In 1985, Vonsattel *et al.* [87], using postmortem brain specimens, developed a 5-point grading system (grade 0 to grade 4) for the pathological involvement based upon gross appearance of the basal ganglia. The system correlates well with microscopic neuropathologic changes in the disease. For example, they determined that in the late stages of HD (grade 4), 95% of the neurons in the caudate nucleus

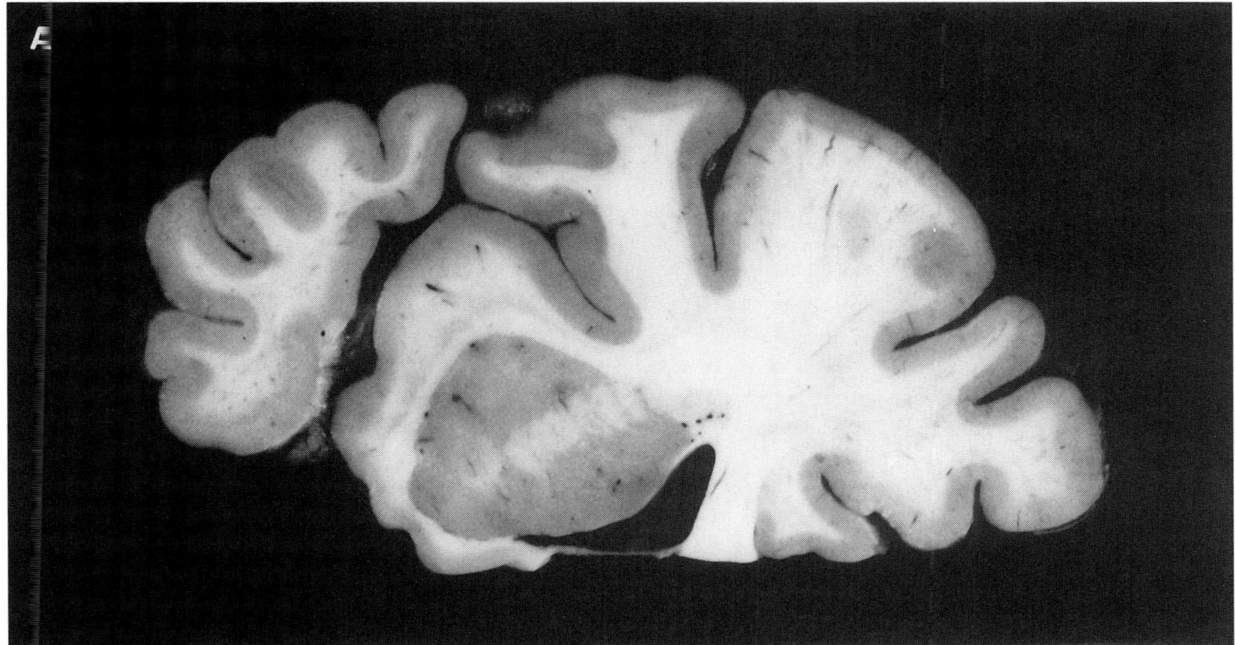

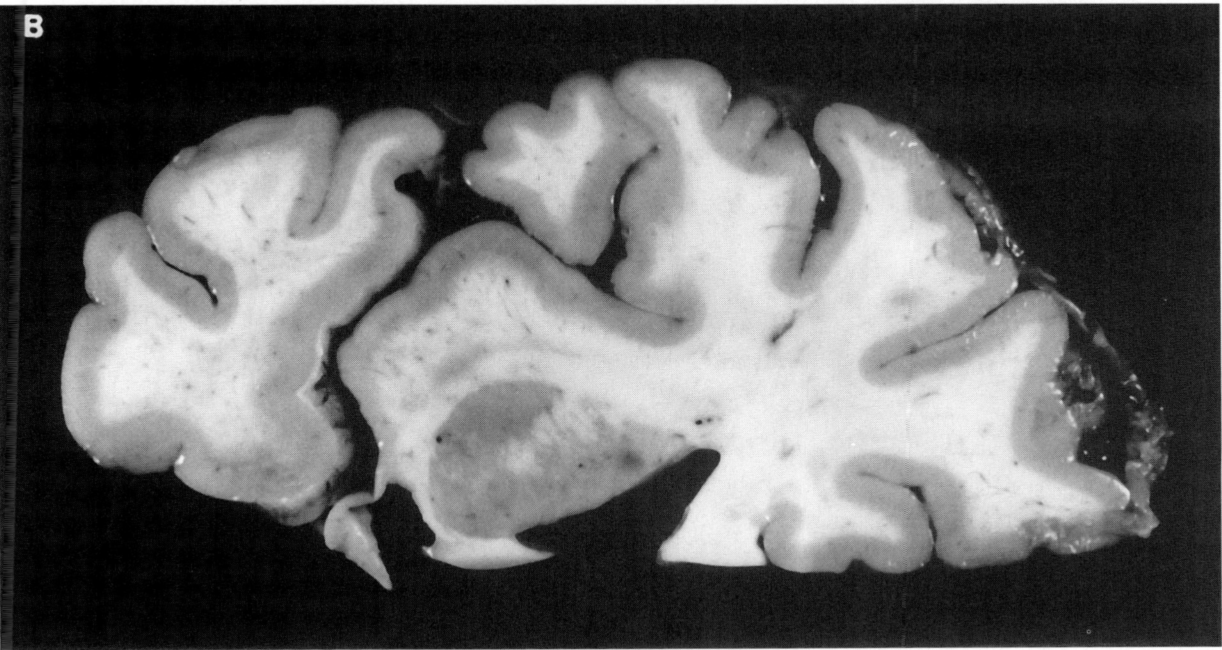

Figure 23-4 The *HD* mutation and neuropathology. Hemicoronal sections through the head of the caudate nucleus, nucleus accumbens, and putamen of normal (A) and neuropathologically graded HD postmortem (B–D) are shown. The progressive loss of neurons in Vonsattel grade 2 (B), grade 3 (C), and grade 4 (D) cases is evident in the basal ganglia from the degree of atrophy and abnormal dilation of the lateral ventricle. The thinning of the cerebral cortex which indicates loss of neurons in this region is more difficult to appreciate but is evident in the grade 3 and 4 cases.

are lost, whereas only one-third of the neurons were lost in the most mildly involved cases (grade 0 [80]).

De La Monte *et al.* [89] showed in morphometric studies that as the disease progresses, there are significant reductions in the cross-sectional area of the caudate, putamen, thalamus, and cerebral cortex. They also demonstrated that there is a severe loss of neurons in cerebral cortex and white matter seen across all stages of HD.

It is not known if the onset of neuropathological involvement coincides with the onset of overt symptoms in HD or whether the degeneration in HD is lifelong

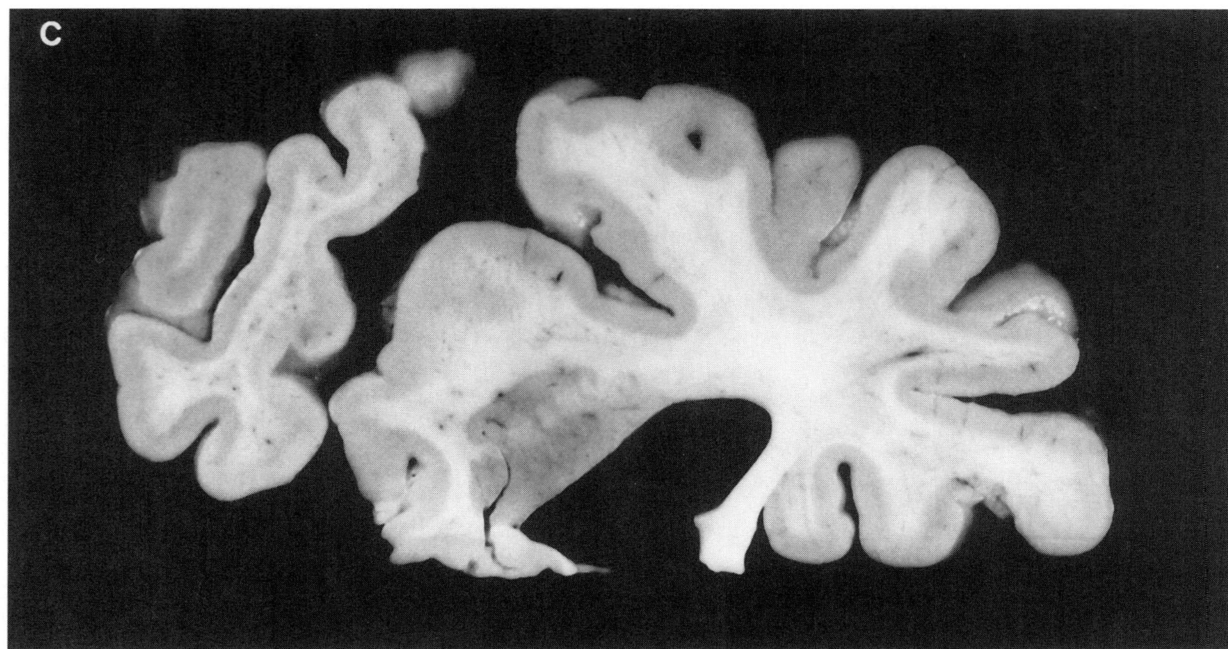

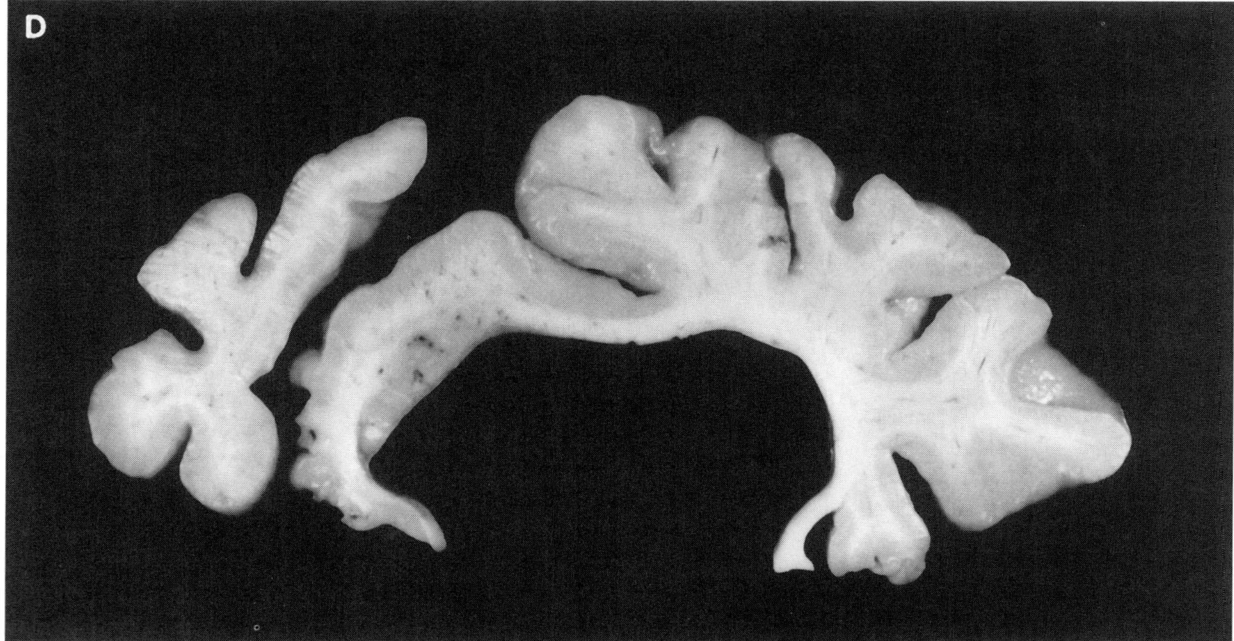

FIGURE 23-4 *(continued)*

and predates disease onset. The extent of pathological involvement in HD is heavily dependent upon the stage of the disease at the time of death of the individual [4, 87] such that persons with greater disease-related impairment have greater evidence of neuronal degeneration in affected brain regions.

Recent studies reveal that the extent of pathological involvement in HD is related to the HD repeat size [90, 91]. Quantitative measurement of neuronal cell loss (divided by time since birth) is strongly correlated with the HD repeat size ($r = 0.90$ [91]). This relationship accounts for 80% of the variance in neuronal cell loss. In addition, the studies of Penney *et al.* [91] and Furtado *et al.* [90] found that the disease expression is predicted for repeats of 35.5 units and above. This finding is consistent with the observation that 36 repeats is the smallest HD clinical allele. These studies support the view that the effects of the HD mutation progress from birth rather than from the time of disease manifestation [90, 91]. Thus there may be an underlying disease process which precedes the overt expression of the disease, and this process may be initiated from early in the development of the individual.

This observation is consistent with a model that postulates that HD is a consequence of the accumulation of a toxic effect of the *HD* gene over time [4]. The pace of the accumulation determines both the age at onset and the rate of disease progression. If the pace is rapid, the onset is early and the progression is rapid. If the pace is slow, the onset is late and the progression is slow [4, 88, 91]. However, this model does not easily account for the observation that HD homozygotes do not differ from their heterozygous siblings [11, 12]. It is difficult to see how accumulation of a toxic product can be made independent of the dosage of the mutant gene product. The underlying mechanism may yet hold a few surprises.

There are no published data addressing the impact of HD upon organs other than the brain. In an unpublished study of 122 HD cases coming to autopsy, organ weight was found to be significantly reduced for cases with increased severity of brain involvement (E. D. Bird, C. Hall and R. H. Myers, unpublished data). Weight loss was not uniform across all organs, and was most evident in the heart and liver and least evident in the kidney. The brain weight correlated poorly with measures of body mass index or relative body weight ($r = 0.17$), while the heart and kidney correlated more strongly ($r = 0.43$ and $r = 0.44$). The impact of HD on other organs is evident in the progression of the disease. However, it is less clear whether the impact is a direct consequence of effects of the HD gene upon these tissues or is merely a generalized effect of chronic disease.

A. Neuroimaging

There is a long history of investigating the relationship between the neuropathological effects of HD determined by neuroimaging methods and the clinical presentation of Huntington's disease. In the early 1980s, Sax *et al.* [92], using computed tomography (CT), derived index for the bicaudate ratio (BCR, distance between caudate nuclei divided by the distance between the frontal horns) to estimate caudate nucleus and basal ganglia volume loss. They demonstrated a significant association between the severity of chorea and increased BCR [92].

CT measures of the reductions in the caudate by using the BCR have been found to be positively correlated with deficits in eye movements, activities of daily living, Mini Mental State Examination [93], neuropsychological measures including Trials, Digit Symbol [93, 94], performance on short term memory tests, visuospatial tasks [92], and the Stroop [94].

Using magnetic resonance imaging (MRI), others were able to detect reductions in the thalamus and in cortical areas in addition to the well-documented caudate nucleus reductions. These changes have been noted in the medial temporal lobe [95, 96] and in the frontal cortex [97]. Interestingly, Harris *et al.* [98], using MRI, found that there was greater atrophy detected in the putamen than in the caudate nucleus in 15 patients with mild HD. In 1994, Aylward *et al.* [99] reported similar findings and suggested [100] that caudate atrophy may be a later pathologic hallmark in the disease whereas the putamen may atrophy in the early stages. Nevertheless, both earlier [87, 96] and more recent studies [97] have found that the caudate nucleus is more affected than the putamen in HD.

Positron emission computed tomography (PET) of ^{18}F-fluorodexyglucose (FDG) was used to detect decreases in glucose utilization in caudate and putamen in HD patients [101–103]. Decreases in glucose utilization have been found in HD patients who do not have demonstrable caudate atrophy [101, 102, 104, 105]. Martin *et al.* [106] also found significant decreases in cortical glucose metabolism. PET studies of the dopaminergic system have shown reductions in the D1 and D2 receptor densities by as much as 40% in the striatum of patients with HD [97, 103]. Ginovart *et al.* [97] also found reductions of the D1 receptors in the temporal cortex.

Jenkins *et al.* [107] hypothesized that the genetic defect in HD may be related to impaired energy metabolism. Using proton nuclear magnetic resonance (NMR) spectroscopy, they determined that lactate, a by-product of impaired oxidation of pyruvate, was elevated in HD patients. Elevated lactate concentrations were found in the basal ganglia, the occipital cortex [107], and the frontal cortex in HD patients [108] and in asymptomatic HD gene carriers.

VII. MODIFIERS OF AGE AT ONSET

In this section we will consider those factors other than the HD repeat size which may influence the age at onset of HD. Because onset age and rate of disease progression could be related, factors which influence onset might also be related to rate of progression. Alternatively, onset and progression might represent fundamentally distinct molecular and cellular processes that are influenced by different factors.

A. Sex of the Affected Parent

Age at onset in HD has long been recognized to be associated with the sex of the affected parent. Merritt *et al.* [37] first observed that the majority of cases with onset before the age of 21 inherited the *HD* gene from affected fathers. It was subsequently noted that the age at death was earlier for paternally transmitted HD cases [109] and that offspring of affected men of all ages at onset had significantly younger onsets than their fathers while offspring of affected women had an onset age similar to their mothers [110]. The meiotic instability of the *HD* repeat explains the observation of anticipation and the sex of parent effect in HD. Although meiotic instability is about the same during transmission from both sexes, there is a propensity toward repeat expansions, sometimes very large increases up to a doubling in size, in transmissions through the male germline.

There is some consensus that the sex of the affected parent may provide additional predictive power for onset age beyond that of the HD repeat, with the observation that offspring of affected fathers may have an onset that is younger than would be expected from the repeat size alone [38]. Nevertheless, the magnitude of this effect is small and the mechanism for this association, if real, is unknown.

B. The Normal Repeat

Early studies suggested that the normal chromosome, acting through either a "linked modifier" [111] or a "cis acting factor" [112, 113] may influence expression or onset of HD. However, modifiers other than the *HD* CAG repeat are unsubstantiated and all those reported appear to involve only a small percentage of the variance in onset age.

C. Haplotype of the HD Chromosome and the Δ2642 Codon

A commonly occurring HD haplotype was recognized for approximately one-third of all HD chromosomes [14]. Recently this haplotype has been recognized to be associated with two polymorphisms within the *HD* gene. The first is the number of proline-encoding CCG repeats immediately adjacent to the *HD* CAG repeat. Seven such repeats are seen on 93 to 95% of HD chromosomes [27, 16, 17] while this number is less common on normal chromosomes (51 to 67%). The second is the Δ2642 deletion codon [15] which is present on about one-third of HD chromosomes for affected individuals of European descent. The combination of 7 proline repeats plus the Δ2642 deletion codon is associated with larger *HD* alleles [27] and is found commonly among new mutation chromosomes. Persons carrying the Δ2642 deletion codon have been reported to have significantly larger average repeat sizes [113]. These polymorphisms may be in linkage disequilibrium with a mutation which influences repeat stability or disease expression.

VIII. INTERGENERATIONAL INSTABILITY

The sex of the affected parent has the strongest impact upon the instability of the trinucleotide. The HD allele is remarkably unstable in both maternal and paternal transmission. The estimated frequency of instability ranges from 70 to 80% [25, 31, 35, 36, 38] with a tendency for larger maternal [38] and paternal alleles [38, 114] to be more unstable than smaller alleles. Although maternally transmitted instability is virtually as common as paternally transmitted instability, the size and direction of the instability are very different depending upon the sex of the affected parent. For maternal transmissions, nearly equal numbers of expansions and contractions are seen and virtually all of the shifts range from 1 to 3 repeats in size. In contrast, paternal transmissions are more frequently expansions and in some cases the jumps in size can be dramatic (10 or more), even doubling the number of repeats. This striking phenomenon is evident in the wide range of expanded CAG repeat alleles represented in the sperm of many male HD patients [115]. While the basis of this instability is not known, the findings that the *HD* CAG repeat does not exhibit significant somatic instability and that monozygotic twins have the same *HD* CAG expansions argue that the critical event occurs in meiosis during the process of gametogenesis [114]. It is not clear whether the absence of large jumps in repeat size during transmission from females is due to the absence of a similar mechanism during oogenesis or whether such events are perhaps lethal and, therefore, do not result in viable progeny.

Thus, the findings that CAG repeat length is correlated with age at onset and that paternally transmitted repeats are more prone to larger increases in size than maternally transmitted *HD* repeats [25, 31, 35, 36, 38] provide an explanation for the pattern of anticipation (defined as an offspring with a substantially younger onset than the affected parent) in HD. Indeed, the longstanding observation that most juvenile onset HD is inherited through the male germline can be attributed to the dramatic increases in repeat size that occur during male gametogenesis.

IX. NEW MUTATIONS

Since the cloning of the HD gene in 1993, a number of mutation events leading to *de novo* disease expression have been described [18, 23, 33]. For 12 reported fresh mutation events (2 cases [18], 6 cases [23], 2 cases [33], 1 case [30], 1 case [117]) the sex of the parent transmitting the mutated allele could be identified. In each instance, a father with a repeat in the 29 to 39 range was identified as the transmitting parent. Confirmed *de novo* expression of HD has only been observed in the paternal transmission of an expanded allele.

Dürr *et al.* [117] report that 18 of 20 possible fresh mutations to HD had expanded repeats. However, in 16 of these 18, nonpaternity could not be ruled out. The two fathers in whom paternity was examined were determined to have repeats of 41 although they were reported as neurologically normal at the ages of 60 and 74. These two fathers appear to have alleles in the disease range but have delayed onsets. Among the remaining 16, 9 had a parent who died younger than age 60. In a similar report, Davis *et al.* [116] studied 44 potential *de novo* cases of HD but only 30 of these were confirmed to have expanded repeats. Paternity could be assessed in only 2, and for 1 of these nonpaternity was found. The frequency of nonpaternity may be high among suspected *de novo* cases. Indeed, most cases recently reported as potential fresh mutations to HD have not excluded (1) nonpaternity, (2) the possibility that a living unaffected parent has an as yet unexpressed clinical allele, (3) the early death of one or both parents, or (4) the possibility that a living unexamined parent may be symptomatic [23, 24]. It is likely that the majority of cases recently reported as potential new mutations have a hidden ancestry of the disease and the frequency of mutation to HD is very low.

Myers *et al.* [18] noted that the majority of *de novo* cases of HD occurred on the most common HD chromosomal haplotype [14]. This haplotype is now known to contain the Δ2642 codon deletion and a CCG repeat of 7. Almqvist *et al.* [113] have confirmed that the HD mutations are related to a chromosome with 7 CCGs. The following factors are related to fresh mutation in HD: (1) paternal transmission, (2) the repeat size is unusually large on the precursor chromosome, exceeding 28 units, (3) the most common HD haplotype including a CCG repeat of 7 and the Δ2642 codon deletion.

X. REDUCED PENETRANCE

The large alleles observed in relatives of individuals with fresh mutations to HD have been termed "intermediate" alleles [18, 23]. Intermediate alleles were originally defined as (1) larger than commonly observed in the general population, (2) not associated with disease expression, and (3) smaller than commonly seen in the HD range [18]. These alleles have been found to give rise to CAG repeat lengths in the disease causing range only in paternal transmission.

Paternally transmitted alleles in the range of 37 to 39 repeats were examined by McNeil *et al.* [20] and 40% (4 of 10) were found to be unstable; 3 expanded by 1 repeat and 1 contracted by 1 repeat. Goldberg *et al.* [24] examined the frequency of repeat instability in 18 paternal transmissions of alleles ranging from 29 to 35 repeats. The results of their study are similar to those of McNeil *et al.* [20]. They found that 44% were unstable (8 of 18), 7 of these were expansions, and most varied by 1 repeat (only 1 varied by as much as 3 repeats). These studies suggest that paternally transmitted alleles between 29 and 39 repeats are more unstable than those less than 29 but less unstable than those greater than 39.

Many observations of repeats in the 29- to 39-unit range have been made in the collateral branches of families with *de novo* mutation to HD. However, the frequency of disease expression among paternal transmissions of repeats of 29 to 39 units in the collateral lineages of fresh mutations cannot be determined with accuracy in this small sample. Although 40% of paternal transmissions of alleles from 29 to 39 repeats demonstrate meiotic instability in transmissions, the overwhelming majority of the descendants appear to survive to old age without evidence of HD [20].

XI. GENETIC TESTING

Genetic testing for Huntington's disease falls into three categories: (A) Presymptomatic testing of persons at risk for HD, (B) confirmation testing of persons with symptoms of HD, (C) prenatal testing to determine the risk for a fetus.

A. Presymptomatic Testing of Persons at Risk

When an individual has a family history of HD or a suspected family history of HD and a neurological evaluation cannot establish a diagnosis of HD (even in the presence of "soft signs"), the test is considered presymptomatic. The main consideration in presymptomatic testing is that the individual is making an informed choice concerning the options surrounding HD testing. Information is commonly delivered in two counseling sessions. Because of the emotional impact of testing, the individual is strongly encouraged to bring a friend or spouse to the counseling session.

Studies of the long-term emotional impact of testing [118] strongly support the use of testing protocols such as those recommended by the World Federation of Neurology [119–121].

B. Confirmation of Diagnosis

1. Following Diagnosis of HD

Misdiagnosis of HD is rarely made when an individual presents with a family history and characteristic symptoms, and when tests to rule out other illnesses are performed. Nevertheless, occasionally familial forms of Alzheimer's disease or of ataxia are misdiagnosed as HD and therefore the genetic test is appropriate for persons already carrying a diagnosis of HD, particularly if no relative has had an autopsy confirming HD. However, persons with a family history of HD for whom a definite diagnosis of HD cannot be made, with or without the presence of "soft signs," do not fall into this category.

2. Confirmation of Diagnosis in the Absence of Family History

Not infrequently, persons evaluated in neurologic movement disorder clinics present with symptoms which include HD in the differential diagnosis in spite of the absence of family history of the disease. Because the presence of new mutations to HD has now been established, genetic testing in these cases is appropriate.

In the above two situations, testing can be performed by simply submitting a blood sample to a certified DNA diagnostic laboratory.

C. Prenatal Testing

Prenatal testing follows a similar course to that defined for presymptomatic testing but there are several complicating factors which must be considered. Most pregnant individuals will not have undergone presymptomatic testing. Therefore, they will be inquiring about simultaneous presymptomatic and prenatal testing. In those instances where both the parent and the fetus are found to be HD gene carriers, the emotional impact is very profound. The experience of grieving, not only for the loss of the anticipated healthy baby but also for one's prospect for a healthy life, can be overwhelming. Unfortunately, there is little time to prepare couples for the potential adversities they may face. Consequently, prenatal testing may pose increased risk to the well-being of the individual and to the married couple.

A second concern is raised when persons are uncertain about their views on termination of a pregnancy. When a couple chooses not to terminate a carrier fetus, there may be significant implications for the child later in life with possible insurance or career discrimination. Couples who are uncertain about terminating may want to consider whether prenatal testing is useful.

XII. THE *HD* GENE

The ability to provide a molecular test for the presence of the *HD* mutation to people who are at risk for developing HD is a mixed blessing. There is currently no effective intervention for the disorder and treatment for affected patients is palliative. Sadly, for those asymptomatic individuals undergoing presymptomatic testing, who receive news that at some moment in the future the symptoms of HD will almost certainly begin, there is no intervention that can slow onset of the disorder. The molecular genetic approach which seeks to identify the underlying genetic mutation and, through knowledge of its effects on cellular processes, offers the hope of development of a rational clinical intervention. In this strategy it is first necessary to determine what the impact of the *HD* CAG expansion mutation is on the *HD* gene and its products.

A. Gene, mRNA, and Protein Products

The *HD* gene spans ~185 kb of DNA in 67 exons (Fig. 23-1) and is expressed as two major mRNA transcripts (13.5 and 10.5 kb) that encode the same ~350-kDa protein which has been dubbed huntingtin [8, 15]. The *HD* CAG repeat is located near the extreme 5′ end of the gene in exon 1, 17 codons from the ATG start of translation and, together with two penultimate CAACAG codons, it encodes a highly polymorphic segment of glutamines (typically 8 to 36 residues). Immediately adjacent, a degenerate stretch of CAG and CCG codons, including the slightly polymorphic CCG repeat [15], encodes a broken array of ~40 residues that are predominantly proline. Further 3′, in exon 58, the Δ2642 polymorphism that forms part of the major HD haplotype produces (on 5% of normal and ~40% of disease chromosomes) versions of huntingtin with 5 instead of 6 glutamate residues [14, 15].

The products of the *HD* gene are expressed in a wide variety of cells and tissues throughout development and in the adult. Both *HD* mRNA species are found in low to moderate abundance in all fetal and adult cells and tissues that have been examined, including the neuronal targets of the *HD* mutation [122–126]. The *HD* protein, with its highly polymorphic amino-terminal polyglutamine segment (due to translation of the CAG repeat) [127–130], does not show similarity to previously reported sequences except in the low-complexity amino-terminal polyglutamine–polyproline segment and a motif of unknown function ("HEAT" repeat) found in a number of unrelated proteins [131].

Huntingtin's wide pattern of expression in the cytoplasm of a variety of neuronal and nonneuronal cells and tissues does not readily account for the neuronal specificity of the *HD* defect [129, 130, 132–134]. Nevertheless, in the basal ganglia, the heterogeneous distribution of huntingtin in populations of medium- and large-sized neurons suggests that the cells that are the targets of the mutation express reasonably high levels of huntingtin [135, 136]. The selective vulnerability of striatal and cortical neurons to the *HD* mutation, however, is not simply explained by levels of huntingtin, as neuronal cells in other regions of the CNS also express high levels of huntingtin but are not affected in the disease. In the neuron, huntingtin is found in the cytoplasm and throughout the cell body, and in axons, dendrites, and perikarya. Its association with microtubules [132] and vesicles [134] has suggested a role in intracellular trafficking or neurotransmission, and recent experimental data implicate the protein in retrograde and anterograde transport [134].

The *HD* gene is conserved in evolution and homologues from mouse, rat, and pufferfish encode highly related ~350-kDa proteins [137–140]. Interestingly, the amino-terminal polyglutamine–proline-rich stretch which contains the site of the CAG repeat expansion in man is the least conserved segment with humans, having 2–36 glutamines, while the rat homolog has 8, mouse 7, and the pufferfish gene 4 consecutive glutamines. In the latter species, this region is not polymorphic and is encoded by a broken array of CAG and CAA codons, in contrast to the situation in man where the human gene, with its relatively long pure run of CAGs, is highly polymorphic. These findings suggest that although it is critically involved in causing HD in man, the exact length and composition of huntingtin's amino-terminal polyglutamine (and polyproline) segment is not crucial to huntingtin's inherent biochemical and physiological activities.

B. Huntingtin's Normal Function

Huntingtin's inherent biochemical and physiological functions are not known, but at the level of the whole organism they are vital, playing critical roles in normal embryonic development [141–143] and neurogenesis [144]. This may not be surprising given huntingtin's sequence conservation and ubiquitous expression in most (if not all) cell types of the adult and developing organism [140–142, 145]. Heterozygous loss of huntingtin does not cause any overt abnormalities in man or mouse [147, 148], but its complete absence in mice homozygous for targeted *Hdh*-inactivating mutations leads to death of these embryos *in utero* at ~embryonic day 7.5, just prior to the elaboration of the nervous system [141–143]. Studies of mice with targeted *Hdh* mutations that reduce the levels of huntingtin to somewhere between 0 and 50% reveal that huntingtin also plays a critical role later in embryogenesis during brain development [144]. Reduced levels of huntingtin lead to perinatal lethality and characteristic abnormalities in the head region, including gross abnormalities in the fore- and midbrain, while severely reduced levels produce a more dramatic version of this phenotype featuring abnormalities in skin, placement of the ear, and overt excencephaly. Thus, huntingtin is required at several stages of development and its critical function in neurogenesis raises the possibility that it may also be vital for cells in the mature adult CNS.

XIII. PATHOGENIC MECHANISM

The HD mutation could act in a variety of ways to produce HD's peculiar neuropathology, but hypothesized mechanisms will need to conform to the strong predictions that emerge from the results of the genotype–phenotype studies. Critical constraints are imposed by the findings that HD homozygotes are indistinguishable from typical HD heterozygotes [11, 12] and that the onset of disease and degree of neuropathology are correlated with the length of the expanded CAG repeat [25–25]. Thus, *HD* mutation must trigger neuropathology in a process that is saturated in a single dose's worth of mutant product but also one that is at the same time sensitive to CAG repeat length. The nature of the *HD* CAG expansion mutation does not readily explain the pathogenesis of the disorder and studies of its impact on the *HD* gene and its mRNA and protein products do not rigorously exclude any of several possible mechanisms. Nevertheless, accumulated evidence most strongly supports the notion that the mutation, which is translated into an elongated segment of glutamines near huntingtin's amino terminus, acts via a genetic gain of function mechanism conferring a novel deleterious property on the mutant huntingtin protein.

A. Mutant *HD* Gene Products

The HD mutation elongates the CAG repeat tract at the 5′ end of the *HD* mRNA but does not dramatically alter the gene's ubiquitous pattern of expression; both normal and mutant transcripts are expressed in the cells and tissues of heterozygous HD patients [15, 126]. These results argue that the CAG expansion does not produce a loss of huntingtin function at the level of the HD gene or its mRNA products, providing support for the idea that the defect acts at the level of an altered protein product.

The expanded CAG repeat is translated into an elongated polyglutamine segment near huntingtin's amino terminus and, in cells and tissues of HD patients, this feature of the mutant protein allows it to be distinguished from its normal counterpart. The lengthened glutamine stretch confers altered migration on SDS–PAGE [129, 130, 133, 134, 146, 147] and increased reactivity to monoclonal reagents directed at lengthy glutamine arrays [148]. However, it does not dramatically alter the expression pattern of the abnormal protein which closely mirrors that of its normal counterpart. In cells which are amenable to analysis, the alteration does not affect the protein's stability or cytoplasmic intracellular location in lymphoblastoid cells of HD heterozygotes or homozygotes [129, 146]. Studies of mutant huntingtin's pattern of expression and intracellular location in human postmortem brain are critical [132, 133, 136, 146, 147, 149] but have been hindered by the lack of an immunocytochemical reagent specific for the mutant protein. This has precluded the ability to distinguish mutant and normal versions of the protein in material from HD heterozygotes, but several studies have clearly demonstrated that the bulk of huntingtin immunoreactivity is not grossly altered in HD postmortem brain.

However, evidence is accumulating that at least a portion of the mutant protein is mislocalized in some neurons in HD postmortem brain. Antisera directed at the amino terminus of the protein have now revealed huntingtin immunoreactivity in dystrophic neurites and abnormal clusters (inclusions) in the cytoplasm and nucleus of neurons in the deep layers of the cerebral cortex and basal ganglia of HD postmortem brain [149]. At least a proportion of these inclusions are also anti-ubiquitin immunoreactive and appear to be the same as or similar to those described in an earlier study of HD postmortem brain [150]. It is not yet evident how these abnormal bodies form, nor is it clear whether the inclusions and dystrophic neurites are upstream or downstream of the critical molecular event that is triggered by the *HD* defect. Discovery of whether these abnormal huntingtin immunopositive structures are a cause or a result of HD pathology will certainly provide new insights into the pathogenic mechanism. Regardless of the answer to this question, however, these findings provide a new and extremely useful marker of neuronal cell pathology in HD.

B. Genetic Mechanisms

There are several possible genetic mechanisms by which mutant huntingtin might cause HD pathology, but it must do so in a manner that is saturated by a single dose's worth of abnormal protein in a process that is also influenced by the length of its amino-terminal polyglutamine segment. At the level of the mutant protein the mutation could act by influencing huntingtin's inherent activity in a loss of function scenario or dominant negative gain of function model or, alternatively, it could cause a genetic gain of function. The accumulated evidence, obtained mainly from studies which test the effects of manipulating the mouse HD gene homologue, strongly support the latter mechanism.

Simple loss of huntingtin activity is not likely to be involved because heterozygous *HD* inactivation by translocation in man [15] or by targeted mutations in the mouse gene [141, 142] does not lead to any abnormalities. Complete dominant negative inhibition of huntingtin's inherent activity is also unlikely because HD homozygotes are indistinguishable from their heterozygous siblings [11, 12] while homozygous inactivation of the mouse gene causes early embryonic lethality [141–143]. Similarly, scenarios involving a graded loss of huntingtin activity to somewhere between 0 and 50% have been eliminated with the observation that mice with targeted *Hdh* mutations which cause reduced levels of huntingtin exhibit gross defects in neurogenesis that are not observed in HD [144]. Direct evidence that the elongated polyglutamine does not reduce or eliminate huntingtin activity also derives from targeted mutations in the mouse gene which introduce an expanded CAG repeat. Mice with the Hdh^{Q50} mutation that extends the polyglutamine tract of the murine huntingtin protein from 7 to 50 residues do not display developmental abnormalities, even when expressing only a single dose's worth of the mutant huntingtin protein. The finding that a single Hdh^{Q50} allele is sufficient to rescue the deficits associated with complete or partially reduced levels of huntingtin provides a direct demonstration that the *HD* mutation does not comparably impair huntingtin's developmental activities [150]. Thus, the *HD* defect in man does not mimic the phenotypic consequences of complete or partial loss of huntingtin in the mouse, supporting the notion that it causes neuronal cell loss in the disorder by gain of function mechanism. The mutation could operate by increasing the protein's inherent activity or it may endow the mutant protein with an attribute that need not be related to huntingtin's physiological function.

Evidence from genotype–phenotype correlations in man suggests that a similar mechanism may also cause six other dominantly inherited neurodegenerative disorders due to expanded CAG repeats that lengthen a stretch of glutamine residues in unrelated proteins [151]. Attempts to model HD and the other CAG repeat disorders in the mouse by expressing mutant proteins from ectopic transgenes have yielded results that are difficult to interpret without a detailed knowledge of the pathogenic pathway in the human disease. In all cases, the transgenics misexpress the mutant human protein altering the spatial, temporal, or protein context of the mutation [152–155]. Nevertheless, dedicated expression of mutant ataxin-1 in Purkinje cells does kill these cerebellar neurons with consequent ataxia, demonstrating that the abnormal human protein can cause overt neuronal cell loss but not testing the cellular specificity of this effect [154]. Interestingly, similar dedicated expression of MJD1a protein did not kill Purkinje cells, although they did rapidly succumb to a truncated version of the mutant protein [153]. Transgenic mice expressing an amino-terminal segment of the *HD* protein with ~150 glutamines display weight loss, seizures, and premature death [155] but not the graded loss of neurons that is the hallmark of HD neuropathology. Importantly, these mice do display huntingtin immunopositive intranuclear inclusions similar to those observed in HD postmortem brain, raising the possibility that the pathogenic mechanism may involve the abnormal accumulation of an amino terminal fragment of the mutant huntingtin protein [156].

It has been difficult to obtain evidence of a selective accumulation of stable truncated huntingtin products in the cells and tissues of HD patients [146], but a stable, apparently truncated, huntingtin fragment has now been observed in nuclear extracts of an HD postmortem brain that exhibits huntingtin inclusions [149]. This exciting finding has served to spur efforts to follow the life cycle of mutant huntingtin in HD postmortem tissue and in transgenic mouse models of HD. Moreover, similar observations of abnormal inclusions in the postmortem brains of victims of other dominantly inherited neurodegenerative disorders make this scenario even more compelling.

XIV. PROSPECTS

Discovery of the *HD* mutation and knowledge gained from studies correlating CAG repeat length with the clinical and neuropathological features of the disorder have simplified the molecular diagnosis of HD, thereby improving clinical management of the disease. The development of rational therapeutic interventions that are based on discovery of the *HD* mutation, however, await elucidation of its underlying pathogenic mechanism. It is likely that the *HD* mutation and the polyglutamine encoding expanded CAG repeat defects that cause several other dominantly inherited neurodegenerative disorders share features of a similar polyglutamine-sensitive pathogenic process [151]. Ultimately, unraveling the molecular and cellular events in any one of these disorders may provide clues to the means by which the distinct protein context of an elongated polyglutamine array causes the specific pattern of neuronal cell loss that is seen in each of these diseases. In each case the mutant protein is likely to act via a gain of function mechanism, triggering the disease process in a manner that does not depend on its inherent function, although in each disorder the protein's activity may contribute to aspects of the disease phenotype. Results of genotype–phenotype correlations in these disorders reveal that, for a given expanded CAG repeat length, each mutant protein has a different capacity to trigger the pathogenic mechanism, arguing strongly that the protein context of the polyglutamine tract is a critical factor in the pathogenic mechanism. It is likely that the precise effect of the elongated polyglutamine in any given protein context will also depend on the concentration, localization, and normal function of the mutant protein. The neuronal cell specificity in each case may involve pathways that are common to many cell types or, alternatively, may be the result of physiological processes peculiar to neurons.

Of the many scenarios proposed, one of the most intriguing is the idea that a cleavage product of the mutant protein can accumulate to cause abnormal protein depositions in the nucleus. However, key features of this model must yet be examined experimentally to determine whether the cytoplasmic and intranuclear inclusions in postmortem brain material are a result of proteolytic cleavage events, whether the normal as well as the mutant versions of the proteins are involved, and, critically, whether the abnormal protein clusters are causative of the pathogenic process or are a downstream product.

Genotype–phenotype associations in HD continue to provide key features of the relationship of the *HD* mutation to the disease. These findings are changing the ability of the health care community to provide for members of HD families and are guiding efforts to produce critical cellular and animal models of the disease. As the most common of the polyglutamine neurodegenerative diseases, HD has also influenced the development of strategies aimed at understanding the other expanded CAG repeat disorders.

In the end, the ability to compare and contrast a number of different polyglutamine disorders promises to hasten the discovery of key steps in the pathway from CAG repeat expansion to specific neuronal cell pathology in HD. It is hoped that this knowledge will spur the development of rational therapeutic interventions for this debilitating disorder.

Acknowledgments

The authors are indebted to Dr. Jean-Paul Vonsattel and Mr. Lawrence Cherkas for providing Fig. 23-4. The authors' HD research is supported by NIH Grants NS16367 and NS32765, the Huntington's Disease Society of America, the Hereditary Disease Foundation, the Foundation for the Care and Cure of Huntington's Disease. and a grant from the Massachusetts Huntington's Disease Society of America to support K.S.M.

References

1. Huntington, G. (1872). On chorea. *Med. Surg. Rep.* **26,** 320–321. [Reprinted in *Adv. Neurol.* **1,** 33–35, 1972]
2. Harper, P. S. (1992).The epidemiology of Huntington's disease. *Hum. Genet.* **89,** 365–376.
3. Folstein, S. E. (1989). "Huntington's Disease: A Disorder of Families." The Johns Hopkins University Press, Baltimore.
4. Myers, R. H., Vonsattel, J. P., Stevens, T. J., Cupples, L. A., Richardson, E. P., Martin, J. B., and Bird, E. D. (1988). Clinical and neuropathological assessment of severity in Huntington's disease. *Neurology* **38,** 341–347.
5. White, R. F., Vasterling, J. J., Koroshetz, W., and Myers, R. (1992). Neuropsychology of Huntington's disease. *In* "Clinical Syndromes in Adult Neuropsychology: The Practitioners Handbook" (R. F. White, Ed.), pp. 213–251. Elsevier, New York.
6. Schoenfeld, M., Myers, R. H., Cupples, L. A., Berkman, B., Sax, D. S., and Clark, E. (1984). Increased rate of suicide among patients with Huntington's disease. *J. Neurol. Neurosurg. Psych.* **47,** 1283–1287.
7. Gusella, J. F., Wexler, N. S., Conneally P. M., Naylor, S. L., Anderson, M. A., Tanzi, R. E., Watkins, P. C., Ottina, K., Wallace, M. R., Sakaguchi, A. Y., Young, A. B., Shoulson, I., Bonilla, E., and Martin, J. B. (1983). A polymorphic DNA marker genetically linked to Huntington's disease. *Nature* **306,** 234–238.
8. The Huntington's Disease Research Collaborative Group. (1993). A novel gene containing a trinucleotide repeat that is expanded and unstable on Huntington's disease chromosomes. *Cell* **72,** 971–983.
9. Hayden, M. R. (1981). "Huntington's Chorea." Springer-Verlag, New York.
10. Conneally, P. M., Haines, J. L., Tanzi, R. E., Wexler, N. S., Penchaszadeh, G. K., Harper, P. S., Folstein, S. E., Cassiman, J. J., Myers, R. H., Young, A. B., Hayden, M. R., Falek, A., Tolosa, E. S., Crespi, S., Di Maio, L., Holmgren, G., Anvret, M., Kanazawa, I., and Gusella, J. F. (1989). Huntington disease: no evidence for locus heterogeneity. *Genomics* **5,** 304–308.
11. Wexler, N. S., Young, A. B., Tanzi, R. E., Travers, H., Starosta-Rubenstein, S., Penney, J. B., Snodgrass, S. R., Shoulson, I., Gomez, F., Ramos-Arroyo, M. A., Penchaszadeh, G., Moreno, R., Gibbons, K., Faryniarz, A., Hobbs, W., Anderson, M. A., Bonilla, E., Conneally, P. M., and Gusella, J. F. (1987). Homozygotes for Huntington's disease. *Nature* **326,** 194–197.
12. Myers, R. H., Leavitt, J., Farrer, L. A., Jagadeesh, J., McFarlane, H., Mark, R. J., and Gusella, J. F. (1989). Homozygote for Huntington's disease. *Am. J. Hum. Genet.* **45,** 615–618.
13. Gusella, J. F., and MacDonald, M. E. (1995). Huntington's Disease. *Seminars Cell Biol.* **6,** 21–28.
14. MacDonald, M. E., Novelletto, A., Lin, C., Tagle, D., Barnes, G., Bates, G., Taylor, S., Allitto, B., Altherr, M., Myers, R., Lehrach, H., Collins, F. S., Wasmuth, J. J., Frontali, M., and Gusella, J. F. (1992). The Huntington's disease candidate region exhibits many different haplotypes. *Nature Genet.* **1,** 99–103.
15. Ambrose, C. M., Duyao, M. P., Barnes, G., Bates, G. P., Lin, C. S., Srinidhi, J., Baxendale, S., Hummerich, H., Lehrach, H., Altherr, M., Wasmuth, J. J., Buckler, A., Church, D., Housman, D., Berks, M., Micklem, G., Durbin, R., Dodge, A., Read, A., Gusella, J. F., and MacDonald, M. E. (1994). Structure and expression of the Huntington's disease gene: evidence against simple inactivation due to an expanded CAG repeat. *Somat. Cell Mol. Genet.* **20,** 27–38.
16. Rubinsztein D. C., Barton D. E., Davison B. C. C., and Ferguson-Smith M. A. (1993). Analysis of the huntington gene reveals a trinucleotide-length polymorphism in the region of the gene that contains two CCG-rich stretches and a correlation between decreased age of onset of Huntington's disease and CAG repeat number. *Hum. Mol. Genet.* **2,** 1713–1715.
17. Andrew S. E., Goldberg Y. P., Theilmann J., Zeisler J., and Hayden M. R. (1994). A CCG repeat polymorphism adjacent to the CAG repeat in the Huntington disease gene: implications for diagnostic accuracy and predictive power. *Hum. Mol. Genet.* **3,** 65–67.
18. Myers, R. H., MacDonald, M. E., Koroshetz, W. J., Duyao, M. P., Ambrose, C. M., Taylor, S. A. M., Barnes, G., Srinidhi, J., Lin, S. S., Whaley, W. L., Lazzarini, A. M., Schwarz, M., Wolff, G., Bird, E. D., Vonsattel, J-P. G., and Gusella, J. F. (1993). *De novo* expansion of a $(CAG)_n$ repeat in sporadic Huntington's Disease. *Nature Genet.* **5,** 168–173.
19. Rubinsztein, D. C., Leggo, J., Coles, R., Almqvist, E., Biancalana, V., Cassiman, J. J., Chotai, K., Connarty, M., Craufurd, D., Curtis, A., Curtis, D., Davidson, M. J., Differ, A. M., Dode, C., Dodge, A., Frontali, M., Ranen, N. G., Stine, O. C., Sherr, M., Abbott, M. H., Franz, M. L., Graham, C. A., Harper, P. S., Hedreen, J. C., Jackson, A., Kaplan, J. C., Losekoot, M., MacMillan, J. C., Morrison, P., Trottier, Y., Novelletto, A., Simpson, S. A., Theilmann, J., Whittaker, J. L., Folstein, S. E., Ross, C. A., and Hayden, M. R. (1996). Phenotypic characterization of individuals with 30–40 CAG repeats in the Huntington disease (HD) gene reveals HD cases with 36 repeats and apparently normal elderly individuals with 36–39 repeats. *Am. J. Hum. Genet.* **59,** 16–22.
20. McNeil, S. M., Novelletto, A., Srinidhi, J., Barnes, G., Kornbluth, I., Altherr, M. R., Wasmuth, J. J., Gusella, J. F., MacDonald, M. E., and Myers, R. H. (1997). Reduced penetrance of Huntington's disease mutation. *Hum. Mol. Genet.* **6,** 775–779.
21. Brinkman, R. R., Mezei, M. M., Theilmann, J., Almqvist, E., and Hayden, M. R. (1997). The likelihood of being affected with Huntington disease by a particular age, for a specific CAG size. *Am. J. Hum. Genet.* **60,** 1202–1210.
22. Kremer, B., Goldberg, Y. P., Andrew, S. E., Theilmann, J., Telenius, H., Zeisler, J., Squitieri, F., Lin, B., Bassett, A., Almqvist, E., Bird, T. D., and Hayden, M. R. (1994). A worldwide study of the Huntington's disease mutation: the sensitivity and specificity of measuring CAG repeats. *N. Engl. J. Med.* **330,** 1401–1406.
23. Goldberg, Y. P., Kremer, B., Andrew, S. E., Theilmann, J., Graham, R. K., Squitieri, F., Telenius, H., Adam, S., Sajoo, A., Starr, E., Heiberg, A., Wolff, G., and Hayden, M. R. (1993). Molecular analysis of new mutations for Huntington's disease,

intermediate alleles and sex of origin effects *Nature Genet.* **5,** 174–179.

24. Goldberg, Y. P., McMurray, C. T., Zeisler, J., Almqvist, E., Sillence, D., Richards, F., Gacy, A. M., Buchanan, J., Telenius, H., and Hayden, M. R. (1995). Increased instability of intermediate alleles in families with sporadic Huntington disease compared to similar sized intermediate alleles in the general population. *Hum. Mol. Genet.* **4,** 1911–1918.
25. Duyao, M. P., Ambrose, C. M., Myers, R. H., Novelletto, A., Persichetti, F., Frontali, M., Folstein, S. E., Ross, C., Franz, M. L., Abbott, M., Gray, J., Conneally, P. M., Young, A., Penney, J., Hollingsworth, Z., Shoulson, I., Lazzarini, A. M., Falek, A., Koroshetz, W., Sax, D. S., Bird, E. D., Vonsattel, J. P., Bonilla, E., Alvir, J., Bickham Conde, J., Cha, J. H., Dure, L., Gomez, F., Ramos, M., Sanchez-Ramos, J., Snodgrass, S. R., de Young, M., Wexler, N. S., Barnes, G., Srinidhi, J., MacDonald, M. E., and Gusella, J. F. (1993) Trinucleotide repeat length: instability and age of onset in Huntington's disease. *Nature Genet.* **4,** 387–392.
26. Stine, C. O., Pleasant, N., Franz, M. L., Abbott, M. H., Folstein, S. E., and Ross, C. A. (1993). Correlation between the onset age of Huntington's disease and length of the trinucleotide repeat in IT-15. *Hum. Mol. Genet.* **2,** 1547–1549.
27. Andrew, S., Goldberg, P., Kremer, B., Telenius, H., Theilmann, J., Adam, S. H., Starr, E., Squitieri, F., Lin, B., Kalchman, M., Graham, R. K., and Hayden, M. R. (1993). The relationship between trinucleotide (CAG) repeat length and clinical features of Huntington's disease. *Nature Genet.* **4,** 398–403.
28. Snell, R. G., MacMillan, J. C., Cheadle, J. P., Fenton, I., Lazarou, L. P., Davies, P., MacDonald, M. E., Gusella, J. F., Harper, P. S., and Shaw, D. S. (1993). Relationship between trinucleotide repeat expansion and phenotypic variation in Huntington's disease. *Nature Genet.* **4,** 393–397.
29. Barron, L. H., Warner, J. P., Porteus, M., Holloway, S., Simpson, S., Davidson, R., and Brock, D. J. HD. J. (1993). A study of the Huntington's disease associated trinucleotide repeat in the Scottish population. *J. Med. Genet.* **30,** 1003–1007.
30. Zühlke, C., Riess, O., Schröder, K., Siedlaczck, I., Epplen, J. T., Engel, W., and Thies, U. (1993). Expansion of the $(CAG)_n$ repeat causing Huntington's disease in 352 patients of German origin. *Hum. Mol. Genet.* **2,** 1467–1469.
31. Zühlke, C., Olaf, R., Bockel, B., Lange, H., and Thies, U. (1993). Mitotic stability and meiotic variability of the $(CAG)_n$ repeat in the Huntington disease gene. *Hum. Mol. Genet.* **2,** 2063–2067.
32. Craufurd, D., and Dodge, A. (1993). Mutation size and age at onset in Huntington's disease. *J. Med. Genet.* **30,** 1008–1011.
33. De Rooji, K. E., De Koning, Gans, P. A. M., Skraastad, M. I., Belfroid, R. D. M., Vegter-Van Der Vlis, M., Roos, R. A. C., Bakker, E., Van Ommen, G-J. B., Dunnen, J. T. D., and Losekoot, M. (1993). Dynamic mutation in Dutch Huntington's disease patients: increased paternal repeat instability extending to within the normal size range. *J. Med. Genet.* **30,** 996–1002.
34. Novelletto, A, Persichetti, F., Sabbadini, G., Mandich, P., Bellone, E., Ajmar, F., Pergola, M., Del Senno, L., MacDonald, M. E., Gusella, J. F., and Frontali, M. (1994). Analysis of the trinucleotide repeat expansion in Italian families affected with Huntington disease *Hum. Mol. Genet.* **3,** 93–98.
35. Novelletto, A., Persichetti, F., Sabbadini, G., Mandich, P., Bellone, E., Ajmar, F., Squitieri, F., Campanella, G., Bozza, A., MacDonald, M. E., Gusella, J. F., and Frontali, M. (1994). Polymorphism analysis of the huntingtin gene in Italian families affected with Huntington disease. *Hum. Mol. Genet.* **3,** 1129–1132.
36. Trottier, Y., Biancalana, V., and Mandel, J-L. (1993). Instability of CAG repeats in Huntington's disease: relation to parental transmission and age of onset. *J. Med. Genet.* **31,** 377–382.
37. Merritt, A. D., Conneally, P. M., Rahman, N. F., and Drew, A. L. (1969). Juvenile Huntington's chorea, *In* "Progress in Neurogenetics" (A. Barbeau and T. R. Brunette, Eds.), pp. 645–650. Excerpta Medica, Amsterdam.
38. Ranen, N. G., Stine, C. O., Abbott, M. H., Sherr, M., Codori, A. M., Franz, M. L., Chao, N. I., Chung, A. S., Pleasant, N., Callahan, C., Kasch, L. M., Ghaffari, M., Chase, G. A., Kazazian, H. H., Brandt, J., Folstein, S. E., and Ross, C. A. (1995). Anticipation and instability of IT-15 (CAG)n repeats in parent-offspring pairs with Huntington's disease. *Am. J. Hum. Genet.* **57,** 593–602.
39. Kremer, B., Squitieri, F., Telenius, H., Andrew, S. E., Theilman, J., Spence, N., Goldberg, Y. P., and Hayden, M. R. (1993). Molecular analysis of late onset Huntington's disease. *J. Med. Genet.* **30,** 991–995.
40. Myers, R. H., Sax, D. S., Koroshetz, W. J., Mastromauro, C. A., Cupples, L. A., Kiely, D. K., Pettengill, F. K., and Bird, E. D. (1991). Factors associated with slow progression in Huntington's disease. *Arch. Neurol.* **48,** 800–804.
41. Kieburtz, K., MacDonald, M., Shih, C., Feigin, A., Steinberg, K., Bordwell, K., Zimmerman, C., Srinidhi, J., Sotack, J., Gusella, J., and Shoulson, I. (1994). Trinucleotide repeat length and progression of illness in Huntington's disease. *J. Med. Genet.* **31,** 872–874.
42. Brandt, J., Bylsma, F. W., Gross, R., Stine, O. C., Ranen, N., and Ross, C. A. (1996). Trinucleotide repeat length and clinical progression in Huntington's disease. *Neurology* **46,** 527–531.
43. Illarioshkin, S. N., Igarashi, S., Onodera, O., Markova, E. D., Nikolskaya, N. N., Tanaka, H., Chabrashwili, T. Z., Insarova, N. G., Endo, K., Ivanova-Smolenskaya, I. A., and Tsuji, S. (1994). Trinucleotide repeat length and rate of progression of Huntington's disease. *Ann. Neurol.* **36,** 630–635.
44. Shoulson, I., Odoroff, C., Oakes, D., Behr, J., Goldblatt, D., Caine, E., Kennedy, J., Miller, C., Bamford, K., and Rubin, A. (1989). A controlled clinical trial of baclofen as protective therapy in early Huntington's disease. *Ann. Neurol.* **25,** 252–259.
45. Sudarsky, L., Myers, R. H., and Walshe, T. M. (1983). Huntington's disease in monozygotic twins reared apart. *J. Med. Genet.* **20,** 408–411.
46. Butters, N., Wolfe, J., Martone, M., Granholm, E., and Cermak, L. S. (1985). Memory disorders associated with Huntington's disease: verbal recall, verbal recognition and procedural memory. *Neuropsychologia* **23,** 729–743.
47. Folstein, S. E. (1991). The psychopathology of Huntington's disease. *In* "Genes, Brain, and Behavior," pp. 181–191. Raven Press, New York.
48. Josiassen, R. C., Curry, L., Roemer, R. A., and DeBease, C. (1982). Patterns of intellectual deficit in Huntington's disease. *J. Clin. Neuropsychol.* **4,** 173–183.
49. Caine, E. D., Ebert, M. H., and Weingartner, H. (1977). An outline for the analysis of dementia: the memory disorder of Huntington's disease. *Neurology* **27,** 1087–1092.
50. Moss, M. B., Albert, M. S., Butters, N., and Payne, M. (1986). Differential patterns of memory loss among patients with Alzheimer's disease, Huntington's disease and Alcoholic Korsakoff's syndrome. *Arch. Neurol.* **43,** 239–246.
51. Shoulson, I. (1990). Huntington's disease: cognitive and psychiatric features. *Neuropsych. Neuropsychol. Behav. Neurol.* **3,** 15–22.
52. Albert, M. L., Feldman, R. G., and Willis, A. L. (1974). The 'subcortical dementia' of progressive supranuclear palsy. *J. Neurol. Neurosurg. Psych.* **37,** 121–130.
53. McHugh, P., and Folstein, M. (1975). Psychiatric syndromes of Huntington's chorea: a clinical and phenomenological study. In: "Psychiatric Aspects of Neurological Disease" (D. Benson and D. Blumer, Eds.). Grune and Stratton, New York.

54. Butters, N., Sax, D., Montgomery, K., and Tarlow, S. (1978). Comparison of the neuropsychological deficits associated with early and advanced Huntington's disease. *Arch. Neurol.* **35,** 585–589.
55. Bamford, K. A., and Caine, E. D. (1986). The neuropsychology of Huntington's disease: problems of clinical-pathological correlation in a progressive brain illness. *In* "Advances in Clinical Neuropsychology" (G. Goldstein and R. E. Tarter, Eds.). Plenum Press, New York.
56. Josiassen, R. C., Curry, L. M., and Mancall, E. (1983). Development of neuropsychological deficits in Huntington's disease. *Arch. Neurol.* **40,** 791–796.
57. Moses, J. A., Golden, C. J., Berger, P. A., and Wisniewski, A. M. (1981). Neuropsychological deficits in early, middle, and late stage Huntington's disease as measured by the Luria-Nebraska neuropsychological battery. *Int. J. Neurosci.* **14,** 95–100.
58. Paulsen, J. S., Butters, N., Sadek, J. R., Johnson, S. A., Salmon, D. P., Swerdlow, N. R., and Swenson, M. R. (1995). Distinct cognitive profiles of cortical and subcortical dementia in advanced illness. *Neurology* **45,** 951–956.
59. Lange, K. W., Sahakian, B. J., Quinn, N. P., Marsden, C. D., and Robbins, T. W. (1995). Comparison of executive and visuospatial memory function in Huntington's disease, and dementia of Alzheimer type matched for degree of dementia. *J. Neurol. Neurosurg. Psych.* **58,** 598–606.
60. Pillon, B., Deweer, B., Agid, Y., and Dubois, B. (1993). Explicit memory in Alzheimer's, Huntington's, and Parkinson's diseases. *Arch. Neurol.* **50,** 374–379.
61. Jason, G. W., Pajurkova, E. M., Suchowershy, O., Hewitt, J., Hilbert, C., Reed, J., and Hayden, M. R. (1988). Presymptomatic neuropsychological impairment in Huntington's disease. *Arch. Neurol.* **45,** 769–773.
62. Strauss, M. E., and Brandt, J. (1990). Are there neruopsychological manifestations of the gene for Huntington's disease in asymptomatic, at risk individuals? *Arch. Neurol.* **47,** 905–908.
63. Giordani, B., Berent, S., Boivin, M. J., Penney, J. B., Lehtinen, Sh., Markel, D., Hollingworth, Z., Butterbaugh, G., Hichwa, R. D., Gusella, J. F., and Young, A. B. (1995). Longitudinal neuropsychological and genetic linkage analysis of persons at risk for Huntington's disease. *Arch. Neurol.* **52,** 59–64.
64. Blackmore, L., Simpson, S. A., and Crawford, J. R. (1995). Cognitive performance in UK sample of presymptomatic people carrying the gene for Huntington's disease. *J. Med. Genet.* **32,** 358–362.
65. Rothlind, J., Brandt, J., Zee, D., Codori, A. M., and Folstein, S. (1993). Unimpaired verbal memory and oculomotor control in aymptomatic adults with the genetic marker for Huntington's disease. *Arch. Neurol.* **50,** 799–802.
66. Rosenberg, N. K., Sorensen, S. A., and Christensen, A-L. (1995). Neuropsychological characteristics of Huntington's disease carriers: a double blind study. *J. Med. Genet.* **32,** 600–604.
67. Foroud, T., Siemers, E., Kleindorfer, D., Bill D. J., Hodes, M. E., Norton, J. A., Conneally, P. M., and Christian, J. C. (1995). Cognitive scores in carriers of Huntington's disease gene compared to noncarriers. *Ann. Neurol.* **37,** 657–664.
68. Claes, S., Van Zand, K., Legius, E., Dom, R., Malfroid, M., Baro, F., Godderis, J., and Cassiman, J. (1995). Correlations between triplet repeat expansion and clinical features in Huntington's disease. *Arch. Neurol.* **113,** 749–753.
69. Zappacosta, B., Monza, D., Meoni, C., Ausoni, L., Soliveri, P., Gellera, C., Alberti, R., Mantero, M., Penati, G., Caraceni, T., and Girotti, F. (1996).Psychiatric symptoms do not correlate with cognitive decline, motor symptoms, or CAG repeat length in Huntington's disease. *Arch. Neurol.* **53,** 493–497.
70. Weigell-Weber, M., Schmid, W., and Spiegel, R. (1996). Psychiatric symptoms and CAG expansion in Huntington's disease. *Am. J. Med. Genet.* **67,** 53–57.
71. Pflanz, S., Besson, J. A. O., Ebmeier, K. P., and Simpson, S. (1991). The clinical manifestation of mental disorder in Huntington's disease: a retrospective case record study of disease progression. *Acta Psychi. Scand.* **83,** 53–60.
72. Burns, A., Folstein, S., Brandt, J., and Folstein, M. (1990). Clinical assessment of irritability, aggression, and apathy in Huntington and Alzheimer disease. *J. Nerv. Ment. Dis.* **178,** 20–26.
73. Mayeux, R., Stern, Y., Rosen, J., and Leventhal, J. (1981). Depression, intellectual impairment and Parkinson disease. *Neurology* **31,** 659–662.
74. McHugh, P. R. (1989). The neuropsychiatry of basal ganglia disorders: a triadic syndrome and its explanation. *Neuropsych. Neuropsychol. Behav. Neurol.* **2,** 239–247.
75. Folstein, S. E., Abbott, M. H., Chase, G. A., Jensen, B. A., and Folstein, M. F. (1983). The association of affective disorder with Huntington's disease in a case series and in families. *Psychol. Med.* **13,** 537–542.
76. Webb, M., and Trzepacz, P. T. (1987). Huntington's disease: correlations of mental status with chorea. *Biol. Psych.* **22,** 751–761.
77. Caine, E. D., and Shoulson, I. (1983). Psychiatric syndromes in Huntington's disease. *Am. J. Psych.* **140,** 728–733.
78. Folstein, S. E., and Folstein, M. F. (1983). Psychiatric features of Huntington's disease: recent approaches and findings. *Psych. Dev.* **2,** 193–205.
79. Folstein, S. E., Franz, M. L., Jensen, B. A., Chase, G. A., and Folstein, M. F. (1983). Conduct disorder and affective disorder among the offspring of patients with Huntington's disease. *Psychol. Med.* **13,** 45–52.
80. Caine, E. D., Hunt, R. D., Weingartner, H., and Ebert, M. H. (1978). Huntington's dementia: clinical and neuropsychological features. *Arch. Gen. Psych.* **35,** 377–384.
81. Dewhurst, K., Oliver, J. E., and McKnight, A. L. (1970). Sociopsychiatric consequences of Huntington's disease. *Br. J. Psych.* **116,** 255–258.
82. Oliver, J. E., and Dewhurst, K. E. (1969). Six generations of ill-used children in a Huntington's pedigree. *Postgrad. Med. J.* **45,** 757–760.
83. King, M. (1985). Alcohol abuse in Huntington's disease. *Psychol. Med.* **15,** 815–819.
84. Peyser, C. E., and Folstein, S. E. (1990). Huntington's disease as a model for mood disorders: clues from neuropathology and neurochemistry. *Mol. Chem.* **12,** 99–119.
85. Reed, E., Chandler, J. H., Hughes, E. M., and Davidson, R. T. (1958). Huntington's chorea in Michigan, demography and genetics. *Am. J. Hum. Genet.* **10,** 210–225.
86. Folstein, S. E., Chase, G. A., Wahl, W. E., McDonnell, A. M., and Folstein, M. F. (1987). Huntington disease in Maryland: clinical aspects of racial variation. *Am. J. Hum. Genet.* **41,** 168–179.
87. Vonsattel, J-P., Myers, R. H., Stevens, T., Ferrante, R. J., Bird, E. D., and Richardson, E. P. (1985). Neuropathological classification of Huntington's disease. *J. Neruopathol. and Exp. Neurol.* **44,** 559–577.
88. Myers, R. H., Vonsattel, J. P., Paskevich, P. A., Kiely, D. K., Stevens, T. J., Cupples, L. A., Richardson, E. P., and Bird, E. D. (1991). Decreased neuronal and increased oligodendroglial densities in Huntington's disease caudate nucleus. *J. Neuropathol. Exp. Neurol.* **50,** 729–742.
89. De La Monte, S. M., Vonsattel, J-P., and Richardson, E. P. (1988). Morphometric demonstration of atrophic changes in the

cerebral cortex, white matter, and neostriatum in Huntington's disease. *J. Neuropathol. Exp. Neurol.* **47,** 516–525.

90. Furtado, S., Suchowersky, O., Rewcastle, B., Graham, L., Klimek, M. L., and Garber, A. (1996). Relationship between trinucleotide repeats and neuropathological changes in Huntington's disease. *Ann. Neurol.* **39,** 132–136.
91. Penney, J. B., Vonsattel, J. P., MacDonald, M. E., Gusella, J. F., and Myers, R. H. (1997). CAG repeat number governs the development rate of pathology in Huntington's disease. *Ann. Neurol.* **41,** 689–692.
92. Sax, D. S., O'Donnell, B., Butters, N., Menzer, L., Montgomery, K., and Kayne, H. (1983). Computed tomographic, neurologic and neuropsychological correlates of Huntington's disease. *Int. J. Neurosci.* **18,** 21–36.
93. Starkstein, S., Brandt, J., Folstein, S., Strauss, M., Berthier, M. L., Pearlson, G. D., Wong, D., McDonnell, A., and Folstein, M. (1988). Neuropsychological and neuroradiological correlates in Huntington's disease. *J. Neurol. Neurosurg. Psych.* **51,** 1259–1263.
94. Bamford, K. A., Caine, E. D., Kido, D. K., Cox, C., and Shoulson, I. (1995). A prospective evaluation of cognitive decline in early Huntington's disease: functional and radiographic correlates. *Neurology* **45,** 1867–1873.
95. Simmons, J. T., Pastakia, B., Chase, T. N., and Shults, C. W. (1986). Magnetic resonance imaging in Huntington's disease. *Am. J. Neuroradiol.* **7,** 25–28.
96. Jernigan, T. L., Salmon, D. P., Butters, N., and Hesselink, J. R. (1991). Cerebral structure on MRI, Part II: Specific changes in Alzheimer's and Huntington's diseases. *Biol. Psych.* **29,** 68–81.
97. Ginovart, N., Lundin, A., Farde, L., Halldin, C., Backman, L., Swahn, C-G., Pauli, S., and Sedvall, G. (1997). PET study of the pre- and post-synaptic dopaminergic markers for the neurodegenerative process in Huntington's disease. *Brain* **120,** 505–514.
98. Harris, G. J., Pearlson, G. D., Peyser, C. E., Aylward, E. H., Roberts, J., Barta, R. J., Chase, G. A., and Folstein, S. E. (1992). Putamen volume reduction on magnetic resonance imaging exceeds caudate changes in mild Huntington's disease. *Ann. Neurol.* **31,** 69–75.
99. Aylward, E. H., Brandt, J., Codori, A. M., Mangus, R. S., Barta, P. E., and Harris, G. J. (1994). Reduced basal ganglia volume associated with the gene for Huntington's disease in asymptomatic at-risk persons. *Neurology* **44,** 823–828.
100. Aylward, E. H., Codori, A. M., Barta, P. E., Pearlson, G. D., Harris, G. J., and Brandt, J. (1996). Basal ganglia volume and proximity to onset in presymptomatic Huntington disease. *Arch. Neurol.* **53,** 1293–1296.
101. Kuhl, D. E., Phelps. M. E., Markham, C. H., Metter, E. J., Riege, W. H., and Winter, J. (1982). Cerebral metabolism and atrophy in Huntington's disease determined by 18FDG and computed tomographic scan. *Ann. Neurol.* **12,** 425–434.
102. Hayden, M. R., Martin, W. R. W., Stoessl, A. J., Clark, C., Hollenberg, S., Adam, M. J., Amman, W., Harrop, R., Rogers, J., Ruth, T., Sayre, C., and Pate, B. D. (1986). Positron emission tomography in the early diagnosis of Huntington's disease. *Neurology* **36,** 888–894.
103. Antonini, A., Leenders, K. L., Spiegel, R., Meier, D., Vontobel, P., Weigell-Weber, M., Sanchez, Pernaute, R., de Yebenez, J. G., Boesiger, P., Weindl, A., and Maguire, R. P. (1996). Striatal glucose metabloism and dopamine D2 receptor binding in asymptomatic gene carriers and patients with Huntington's disease. *Brain* **119,** 2085–2095.
104. Young, A. B., Penney, J. B., Starosta-Rubinstein, S., Markel, D. S., Berent, S., Giordani, B., Ehrenkaufer, R., Jewett, D., and Hichwa, R. (1986). PET Scan investigations of Huntington's disease: cerebral metabolic correlates of neurological features and functional decline. *Ann. Neurol.* **20,** 296–303.
105. Sax, D. S., Powsner, R., Kim, A., Tilak, S., Bhatia, R., Cupples, L. A., and Myers, R. (1996). Evidence of cortical metabolic dysfunction in early Huntington's disease by single-photon-emission computed tomography. *Mov. Disorders* **6,** 671–677.
106. Martin, W. R. W., Clar, C., Ammann, W., Stoessl, A. J., Shtybel, W., and Hayden, M. R. (1992). Cortical glucose metabolism in Huntington's disease. *Neurology* **42,** 223–229.
107. Jenkins, B. G., Koroshetz, W. J., Beal, F., and Rosen, B. (1993). Evidence for impairment of energy metabolism in vivo in Huntington's disease using localized 1H NMR spectroscopy. *Neurology* **43,** 2689–2695.
108. Harms, L., Meierkord, H., Timm, G., Pfeiffer, L., and Ludolph, A. C. (1997). Decreased N-acetyl-aspartate/choline ratio and increased lactate in the forntal lobe of patients with Huntington's disease: a proton magnetic resonance spectroscopy study. *J. Neurol. Neurosurg. Psych.* **62,** 27–30.
109. Bird, E. D., Caro, A. J., and Pilling, J. B. (1974). A sex related factor in the inheritance of Huntington's chorea. *Ann. Hum. Genet.* **37,** 255–260.
110. Myers, R. H., Cupples, L. A., Schoenfeld, M., D'Agostino, R. B., Terrin, N. C., Goldmarkher, N., and Wolf, P. A. (1985). Maternal factors in onset of Huntington's disease. *Am. J. Hum. Genet.* **37,** 511–523.
111. Farrer, L. A. Cupples, L. A., Wiater, P., Conneally, P. M., Gusella, J. F., and Myers, R. H. (1993). The normal HD allele, or a closely linked gene, influences age at onset of Huntington's disease. *Am. J. Hum. Genet.* **53,** 125–130.
112. Telenius, H., Almqvist, E., Kremer, B., Spence, N., Squitieri, F., Nichol, K., Grandell, U., Starr, E., Benjamin, C., Castaldo, I., Calabres, E. O., Anvret, M., Goldberg, Y. P., and Hayden, M. R. (1995). Somatic mosaicism in sperm is associated with intergenerational $(CAG)_n$ changes in Huntington's disease. *Hum. Mol. Genet.* **4,** 189–195.
113. Almqvist, E., Spence, N., Nichol, K., Andrew, S. E., Vesa, J., Peltonen, L., Anvret, M., Goto, J., Kanazawa, I., Goldberg, Y. P., and Hayden, M. R. (1995). Ancestral differences in the distribution of the 2642 glutamic acid polymorphism is associated with varying CAG repeat lengths on normal chromosomes: insights into the genetic evolution of Huntington's disease. *Hum. Mol. Genet.* **4,** 207–214.
114. Leeflang, E. P., Zhang, L., Tavare, S., Hubert, R., Srinidhi, J., MacDonald, M. E., Myers, R. H., Young, M. D., Wexler, N. S., Gusella, J. F., and Arnheim, N. (1995). Single sperm analysis of the trinucleotide repeats in the Huntington's disease gene: quantification of the mutation frequency spectrum. *Hum. Mol. Genet.* **4,** 1519–1526.
115. MacDonald, M. E., Barnes, G., Srinidhi, J., Duyao, M. P., Ambrose, C. M., Myers, R. H., Gray, J., Conneally, P. M., Young, A., Penney, J., Shoulson, I., Hollingsworth, Z., Koroshetz, W., Bird, E., Vonsattel, J. P., Bonilla, E., Moskowitz, C., Penchaszadeh, G., Brzustowicz, L., Alvir, J., Bickham Conde, J., Cha, J-H., Dure, L., Gomez, F., Ramos-Arroyo, M., Sanchez-Ramos, J., Snodgrass, S. R., de Young, M., Wexler, N. S., MacFarlane, H., Anderson, M. A., Jenkins, B., and Gusella, J. F. (1993). Gametic but not somatic instability of CAG repeat length in Huntington's disease. *J. Med. Genet* **30,** 982–986.
116. Davis, M. B., Bateman, D., Quinn, N. P., Marsden, C. D., and Harding A. E. (1994). Mutation analysis in patients with possible but apparently sporadic Huntington's disease. *Lancet* **344,** 714–717.
117. Dürr, A., Dodé, C., Hahn, V., Pêcheux, C., Pillon, B., Feingold, J., Kaplan, J.-C., Agid, Y., and Brice A. (1995). Diagnosis of "sporadic" Huntington's disease. *J. Neurol. Sci.* **129,** 51–55.

118. Taylor, C. A., and Myers R. H. (1997). Long term impact of Huntington disease linkage testing. *Am. J. Med. Genet.* **70,** 365–370.
119. Guidelines for the molecular genetics predictive testing Huntington's disease. (1994). *Neurology* **44,** 1533–1536.
120. Ethical Issues Policy Statement on Huntington's Disease Molecular Genetics Predictive Test. World Federation of Neurology: Research Committee Group on Huntington's Disease. (1990). *J. Med. Genet.* **27,** 34–38.
121. Hersh, S. M., Jones, R., Koroshetz, W. J., and Quaid, K. (1990). The neurogenetics genie: testing for the Huntington's disease mutation. *Neurology* **44,** 1369–1373.
122. MacDonald, M. E., Duyao, M., Calzonetti, T., Auerbach, A., Ryan, A., Barnes, G., White, W., Auerbach, W., Vonsattel, J-P., Gusella, J. F., and Joyner, A. L. (1996). Targeted inactivation of the mouse Huntington disease homologue *Hdh. In* "Cold Spring Harbor Symposia on Quantitative Biology," Vol LXI, pp. 627–638. Cold Spring Harbor Laboratory, Cold Spring Harbor, NY.
123. Li, S. H., Schilling, G., Young, W. S., 3D, Li, X. J., Margolis, R. L., Stine, O. C., Wagster, M. V., Abbott, M. H., Franz, M. L., Ranen, N. G., Folstein, S. E., Hedreen, J. C., and Ross, C. A. (1993). Huntington's disease gene (IT15) is widely expressed in human and rat tissues. *Neuron* **11,** 985–993.
124. Strong, T. V., Tagle, D. A., Valdes, J. M., Elmer, L. W., Boehm, K., Swaroop, M., Kaatz, K. W., Collins, F. S., and Albin, R. L. (1993). Widespread expression of the human and rat Huntington's disease gene in brain and nonneural tissues. *Nature Genet.* **5,** 259–265.
125. Landwehrmeyer, G. B., McNeil, S. M., Dure, L. S., 4th, Ge, P., Aizawa, H., Huang, Q., Ambrose, C. M., Duyao, M. P., Bird, E. D., Bonilla, E., de Young, M., Avila-Gonzales, A. J., Wexler, N. S., DiFiglia, M., Gusella, J. F., MacDonald, M. E., Penney, J. B., Young, A. B., and Vonsattel, J-P. (1995). Huntington's disease gene: regional and cellular expression in brain of normal and affected individuals. *Ann. Neurol.* **37,** 218–230.
126. Stine, O. C., Li, S. H., Pleasant, N., Wagster, M. V., Hedreen, J. C., and Ross, C. A. (1995). Expression of the mutant allele of IT-15 (the HD gene) in striatum and cortex of Huntington's disease patients. *Hum. Mol. Genet.* **4,** 15–18.
127. Jou, Y. S., and Myers, R. M. (1995). Evidence from antibody studies that the CAG repeat in the Huntington disease gene is expressed in the protein. *Hum. Mol. Genet.* **4,** 465–469.
128. Ide, K., Nukina, N., Goto, J., and Kanazawa, I. (1995). Abnormal gene product identified in Huntington's disease lymphocytes and brain. *Biochem. Biophys. Res. Commun.* **209,** 1119–1125.
129. Persichetti, F., Ambrose, C. M., Ge, P., McNeil, S. M., Srinidhi, J., Anderson, M. A., Jenkins, B., Barnes, G. T., Duyao, M. P., Kanaley, L., Wexler, N. S., Myers, R. H., Bird, E. D., Vonsattel, J. P., MacDonald, M. E., and Gusella, J. F. (1995). Normal and expanded Huntington's disease alleles produce distinguishable proteins due to translation across the CAG repeat. *Mol. Med.* **1,** 374–383.
130. Trottier, Y., Devys, D., Imbert, G., Saudou, F., An, I., Lutz, Y., Weber, C., Agid, Y., Hirsch, E. C., and Mandel, J. L. (1995). Cellular localization of the Huntington's disease protein and discrimination of the normal and mutated form. *Nature Genet.* **10,** 104–110.
131. Andrade, M. A., and Bork, P. (1995). HEAT repeats in the Huntington's disease protein *Nature Genet.* **11,** 115–116.
132. Gutekunst, C-A., Levey, A. I., Heilman, C. J., Whaley, W. L., Yi, H., Nash, N. R., Rees, H. D., Madden, J. J., and Hersch, S. M. (1995). Identification and localization of huntingtin in brain and human lymphoblastoid cell lines with anti-fusion protein antibodies. *Proc. Natl. Acad. Sci. USA* **92,** 8710–8714.
133. Sharp, A. H., Loev, S. J., Schilling, G., Li, S. H., Li, X. J., Bao, J., Wagster, M. V., Kotzuk, J. A., Steiner, J. P., Lo, A., Hedreen, J., Sisodia, S., Snyder, S. H., Dawson, T. M., Ryugo, D. K., and Ross, C. A. (1995). Widespread expression of Huntington's disease gene (IT15) protein product. *Neuron* **14,** 1065–1074.
134. DiFiglia, M., Sapp, E., Chase, K., Schwarz, C., Meloni, A, Young, C., Martin, E., Vonsattel, J-P., Carraway. R., Reeves, S. A., Boyce, F. M., and Aronin, N. (1995). Huntingtin is a cytoplasmic protein associated with vesicles in human and rat brain neurons. *Neuron* **14,** 1075–1081.
135. Kosinski, C. M., Cha, J-H., Young, A. B., Persichetti F., MacDonald, M. E., Gusella, J. F., Penney, J. B., and Standaert, D. G. (1997). Huntingtin in the neostriatum: selective accumulation in vulnerable neurons. *Ann. Neurol.* **144,** 239–247.
136. Ferrante, R. J., Gutekunst, C. A., Persichetti, F., Kowall, N., Gusella, J. F., Beal, M. F., MacDonald, M. E., and Hersch, S. M. (1997). Heterogeneous topographic and cellular distribution of huntingtin expression in the normal human neostriatum. *J. Neurosci.* **17,** 3052–3063.
137. Barnes, G. T., Duyao, M. P., Ambrose, C. M., McNeil, S., Persichetti, F., Srinidhi, J., Gusella, J. F., and MacDonald, M. E. (1994). Mouse Huntington's disease gene homolog (*Hdh*). *Somat. Cell Mol. Genet.* **20,** 87–97.
138. Lin, B., Nasir, J., MacDonald, H., Hutchinson, G., Graham, R. K., Rommens, J. M., and Hayden, M. R. (1994). Sequence of the murine Huntington disease gene: evidence for conservation, alternate splicing and polymorphism in a triplet (CCG) repeat [corrected]. *Hum. Mol. Genet.* **3,** 85–92. [Published erratum appears in *Hum. Mol. Genet.* **3,** 530, 1994]
139. Schmitt, I., Baechner, D., Megow, D., Henklein, P., Boulter, J., Hameister, H., Epplen, J. T., and Riess, O. (1995). Expression of the Huntington disease gene in rodents: Cloning the rat homologue and evidence for down regulation in non-neuronal tissues during development. *Hum. Mol. Genet.* **4,** 1173–1182.
140. Baxendale, S., Abdulla, S., Elgar, G., Buck, D., Berks, M., Micklem, G., Durbin, R., Bates, G. P., Brenner, S., Beck, S., and Lehrach, H. (1995). Comparative sequence analysis of the human and pufferfish Huntington's disease genes. *Nature Genet.* **10,** 67–76.
141. Duyao, M. P., Auerbach, A. B., Ryan, A., Persichetti, F., Barnes, G. T., McNeil, S. M., Ge, P., Vonsattel, J-P., Gusella, J. F., Joyner, A. L., and MacDonald, M. E. (1995). Homozygous inactivation of the mouse *Hdh* gene does not produce a Huntington's disease-like phenotype. *Science* **269,** 407–410.
142. Zeitlin, S., Liu, J-P., Chapman, D. L., Papaioannou, V. E., and Efstratiadis, A. (1995). Increased apoptosis and early embryonic lethality in mice nullizygous for the Huntington's disease gene homologue. *Nature Genet.* **11,** 155–162.
143. Nasir, J., Floresco, J. B., O'Kusky, J. R., Diewert, V. M., Richman, J. M., Zeisler, J., Borowski, A., Marth, J. D., Phillips, A. G., and Hayden, M. R. (1995). Targeted disruption of the Huntington's disease gene results in embryonic lethality and behavioral and morphological changes in heterozygotes. *Cell* **81,** 811–823.
144. White, J. K., Auerbach, W., Duyao, M. P., Vonsattel, J-P., Gusella, J. F., Joyner, A. L., and MacDonald, M. E. (1997). Huntingtin function is required for neurogenesis and is not impaired by the Huntington's disease CAG expansion. *Nature Genet.* [in press]
145. Bhide, P., Day, M., Sapp, E., Schwarcz, C., Sheth, A., Kim, J. Young, A. B., Penney, J., Golden, J., Aronin, N., and DiFiglia,

M. (1996). Expression of normal and mutant huntingtin ni the developing brain. *J. Neurosci.* **16,** 5523–5535.

146. Persichetti, F., Carlee, L., Faber, P. W., Mcneil, S. M, Ambrose, C. M. Srinidhi, J., Anderson, M. A., Barnes, G. T., Gusella, J. F., and MacDonald, M. E. (1996). Differential expression of normal and mutant Huntington's disease gene alleles. *Neurobiol. Dis.* 3, 183–190.
147. Aronin, N., Chase, K., Young, C., Sapp, E., Schwarcz, C., Matta, N., Kornreich, R., Landewehrmeyer, B., Bird, E., Beal, M. F., Vonsattel, J-P., Smith, T., Carraway, R., Boyce, F. M., Young, A. B., Penney, J. B., and DiFiglia, M. (1995). CAG expansion affects the expression of mutant huntingtin in the Huntington's disease brain. *Neuron* **15,** 1193–1201.
148. Trottier, Y., Lutz, Y., Stevanin, G., Imbert, G., Devys, D., Cancel, G., Saudou, F., Weber, C., David, G., Tora, L., Agid, E. C., Brice, A., and Mandel, J. L. (1995). Polyglutamine expansion as a pathological epitope in Huntington's disease and four dominant cerebellar ataxias. *Nature* **378,** 403–406.
149. DiFiglia, M., Sapp, E., Chase K. O., Davies, S. W., Bates, G. P., Vonsattel, J-P., and Aronin, N. (1997). Aggregation of huntingtin in neuronal intranuclear inclusions and dystrophic neurites in brain. *Science* **277,** 1990–1993.
150. Roizin, L., Stellar, S., and Liu, J. C. (1979) Neuronal nuclear-cytoplasmic changes in Huntington's chorea: electron microscope investigations. *In*: "Advances in Neurology, Huntington's Disease" (T. N. Chase, N. S. Wexler, and A. Barbeau, Eds.), Vol. 23, pp. 95–122. Raven Press, New York.
151. Gusella, J. F., Persichetti, F., and MacDonald, M. E. (1997). The genetic defect causing Huntington's disease: repeated in other contexts? *Mol. Med.* **4,** 238–246.
152. Bingham, P. M., Scott M. O., Wang, S., McPhaul, M. J., Wilson, E. M., Garbern, J. Y., Merry. D. E., and Fischbeck, K. H. (1995). Stability of an expanded trinucleotide repeat in the androgen receptor gene in transgenic mice. *Nature Genet.* **9,** 191–196. [Published erratum appears in *Nature Genet.* **10,** 249, 1995]
153. Ikeda, H., Yamaguchi, M., Sugai, S., Aze, Y., Narumiya, S., and Kakizuka, A. (1996). Expanded polyglutamine in the Machado-Joseph disease protein induces cell death in vitro and in vivo. *Nature Genet.* **13,** 196–202.
154. Burright, E. N., Clark, H. B., Servadio, A., Matilla, T., Fedderson, R. M., Yunis, W. S., Duvick, L. A., Zoghbi, H. Y., and Orr, H. T. (1995). SCA1 transgenic mice: a model for neurodegeneration caused by an expanded CAG trinucleotide repeat. *Cell* **82,** 937–948.
155. Mangiarini, L. E., Sathasivam, K., Seller, M., Cozens, B., Harper, A., Heterhington, C., Lawton, M., Trottier, Y., Lehrach, H., Davies, S. W., and Bates, G. P. (1996). Exon 1 of the HD gene with an expanded CAG rpeat is sufficient to cause a progressive neurological phenotype in transgenic mice. *Cell* **87,** 493–506.
156. Davies, S. W., Turmaine, M., Cozens, B. A., DiFiglia, M., Sharp, A. H., Ross, C. A., Scherzinger, E., Wanker, E., Mangiarini, L., and Bates, G. P. (1997). Formation of neuronal intranuclear inclusions (NII) underlies the neurological dysfunction in mice transgenic for the HD mutation. *Cell* **90,** 537–548.

CHAPTER 24

Molecular Pathogenesis of Huntington's Disease: Biochemical Studies of Huntingtin

AHM MAHBUBUL HUQ, ABIGAIL HACKAM, RONA K. GRAHAM, CHERYL L. WELLINGTON, AND MICHAEL R. HAYDEN
Centre for Molecular Medicine and Therapeutics, Vancouver, British Columbia, Canada; and Department of Medical Genetics, University of British Columbia, Vancouver, British Columbia, Canada

I. INTRODUCTION

HD is a progressive autosomal dominant neurodegenerative disease which usually affects approximately 1 in 10,000 individuals. It is characterized by chorea, cognitive deficit, psychiatric disturbances, and progression to death 10 to 15 years after onset [1, 2]. Onset is usually in middle age but may vary from childhood to late adulthood. Juvenile-onset HD, a rapidly progressing variant, occurs in about 5% of the patients who present with rigidity, spasticity, and intellectual decline before age 20 [1, 2].

II. HD GENE AND GENE PRODUCT

The HD gene is located on chromosome 4p16.3, encompasses 67 exons, and spans over 200 kb [3]. It is ubiquitously expressed as two transcripts 10.3 and 13.6 kb in length that differ in the size of the 3′ UTR [4]. Structural analysis of the promoter region is consistent with it being a housekeeping gene [5].

The HD gene lacks homology to any previously characterized gene and encodes a protein of 3144 amino acids with a predicted molecular mass of 348 kDa. The polyglutamine tract starts at residue 18 and is followed by a stretch of 29 consecutive prolines. The region downstream of the polyglutamine tract contains a HEAT repeat [6], a motif that is present in other cytoplasmic proteins such as elongation factor 3, the regulatory A subunit of protein phosphatase 2A, and TORI, a protein essential for cell cycle progression. The HEAT repeat consists of 40 loosely conserved amino acids repeated multiple times in tandem and is proposed to be involved in protein–protein interactions. It is present in proteins with a known role in transport processes such as VP15 that is involved in vesicle-mediated protein transport, importins that are involved in nuclear protein import/export pathways, and PSE1 that is involved in protein secretion. Huntingtin also contains a basic peptide region PIRRKGKEK (amino acids 1182–1190), which when fused to bacterial β-galactosidase is sufficient to localize the protein to the nucleus in human 293 cells [7] (Fig. 24-1). This peptide contains an N-terminal proline residue followed by a hexapeptide motif that contains 4 basic amino acids, and thus fulfills the sequence criteria proposed for a nuclear localization signal. The physiological significance of these findings for huntingtin is not clear.

The rodent and the pufferfish homologues of the HD gene have been cloned [8–11]. The human and the mouse genes are 90% homologous in the predicted coding region with a high degree of sequence identity in the 5′ and 3′ UTRs. The pufferfish homologue is also highly conserved with 69% identity to the HD gene at the nucleotide level. All 67 exons are conserved. The first 17 amino acids show 100% conservation between Fugu, human, and mouse peptide sequences (Fig. 24-1). The murine CAG tract encodes only seven glutamines and is interrupted by a CAA. The glutamine repeat in the pufferfish is 4 amino acids long and is encoded by two CAG and two CAA. The high degree of conservation across species suggest that the normal function of huntingtin is essential (Fig. 24-1). This has also been unequivocally demonstrated by embryonic lethality in mice deficient in huntingtin at about embryonic day 8 [12–14].

The mutation underlying HD is an expansion of a CAG/polyglutamine tract in the first exon [15]. The CAG repeat length is highly polymorphic in the population and the repeat size in the normal and affected ranges from 10 to 35 (median 18) and 36 to 121 (median 40), respectively [16, 17]. Adult-onset patients usually have an expansion from 40 to 55, whereas juvenile-onset patients have expansions above 60. There is a well established inverse correlation between CAG repeat length and age of onset [18]. Because transgenic mice expressing full-length HD mRNA but no protein are phenotypically normal at 18 months [19], it appears that translation of the protein is a requirement for disease phenotype.

The CAG in the HD gene is translated into an uninterrupted stretch of glutamine residues which when expanded alters migration on polyacrylamide gels [20–24]. Conformation changes resulting from a long polyglutamine stretch have also been demonstrated by the monoclonal antibody, 1C2, that selectively recognizes proteins with expanded polyglutamine [25].

Homopolymeric glutamine repeats were first discovered in homeotic proteins and later in other transcription factors [26]. Polyglutamine repeats dominated when residues were more than 14 in length [26]. On screening the database, 33 of 40 top scoring glutamine repeat containing proteins were transcription factors [27]. Based on the presence or absence of polyglutamine in the same protein in different species, differences in the size of the repeat in different species, differences in identity of the amino acids in the repeat, and absence of effect by deletion studies, it has been suggested that the glutamine repeats in various proteins are not essential for their normal physiological function, even though a function may evolve later [26]. For example, the N-terminal region of mouse IL-2 in mouse contains 5 to 21 glutamine residues, depending on the mouse strain [28, 29]. Human IL-2 does not contain any polyglutamine at the same region. Huntingtin, TATA binding protein, and androgen and glucocorticoid receptors have different lengths of the polyglutamine repeat at the same site in human and mouse [8, 30–33]. The C-terminal region of Sry contains 13 stretches of polyglutamine,

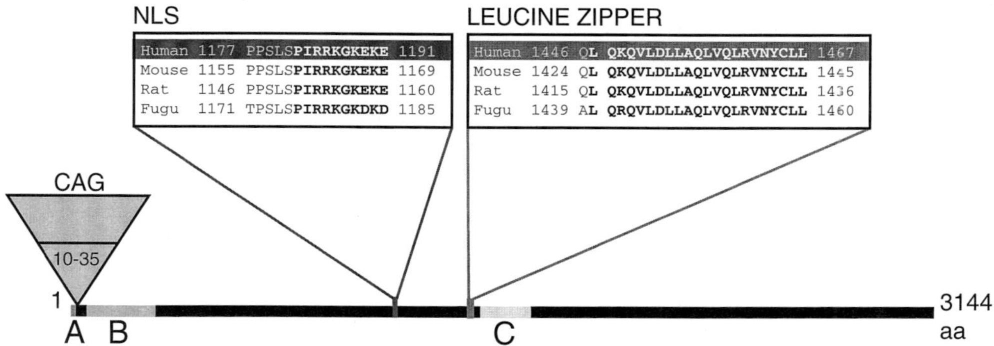

Percentage Identity and Homology to Human Huntingtin

	Overall		A:1-17 aa		B: 56-301 aa		C: 1462-1642 aa	
	Identity	Homo	Identity	Homo	Identity	Homo	Identity	Homo
Mouse	91	95	100	100	97	99	99	100
Rat	86	91	100	100	93	95	99	100
Fugu	80	90	100	100	82	92	95	98

Homo: Homology

FIGURE 24-1 Comparison of amino acids from different portions of human, mouse, rat, and pufferfish huntingtin showing the highly conserved regions of the protein. A putative leucine zipper domain and nuclear localization signal are also shown.

while in other species of old world mice and rat, it contains either no repeat or a repeat of other amino acids [34, 35]. Deletion of two of the three androgen polyglutamine tracts of human androgen receptor did not reduce its ability to activate the androgen receptor-sensitive gene [36, 37]. These observations suggest that polyglutamine tracts in various proteins are unlikely to act as functional motifs which would be disrupted by CAG expansion.

Several models to explain the pathogenesis of CAG expansion diseases have focused on the structural and biochemical properties of long glutamine tracts. Polyglutamine stretches are potential substrates for transglutaminases which may crosslink glutamines in the tract with lysine residues in other proteins, leading to the formation of protease-resistant γ-glutamyl ε-lysyl cross-links and isopeptides [38]. This hypothesis could account for the relatively late onset of HD by positing slow accumulation of abnormally transglutaminated protein until neurons suffer fatal injury. However, such cross-linked isopeptides have not been demonstrated in neurons that degenerate in HD.

Polyglutamine stretches tend to form polar zippers and aggregate together via hydrogen bonding [39, 40]. Experimental work with synthetic peptides and expression of glutamine tracts in bacteria indicates that such zippers do form *in vitro* [41]. Protein–protein interactions mediated by polyglutamine tracts may serve as a sink for transcription factors, many of which have polyglutamine tracts. Alternatively, polyglutamine aggregation might simply result in insoluble and toxic precipitates, analogous to the aggregates such as amyloid deposits implicated in the pathogenesis of other neurodegenerative diseases. Proteolytic cleavage of a GST–huntingtin fusion protein amyloid-like protein aggregates when the polyglutamine expansion is in the pathogenic range [42]. Similar aggregates have been found in the brains of mice transgenic for exon 1 of human HD gene carrying 115 to 156 CAG repeat expansions [42]. Huntingtin has been reported to be present in neuritic plaques, dystrophic neurites, and neurofibrillary tangles in Alzheimer's disease (AD) and in Pick bodies in Parkinson's disease (PD) [43]. In AD and PD the cytoplasmic inclusions seem to be made up largely of abnormal cytoskeletal proteins. It is interesting to note that one of the huntingtin-interacting proteins, HIP1, appears to be a cytoskeletal-associated protein. Another huntingtin-interacting protein, GAPDH, also interacts with amyloid precursor protein [44]. The presence of huntingtin in AD and PD pathological inclusions suggests that huntingtin is capable of being complexed in abnormal inclusions. One explanation for these observations is that huntingtin can form aggregates through polar zippers or other mechanisms and may be localized in inclusions in HD and neurodegenerative diseases.

III. PHENOCOPIES FOR HD

A very small minority of the patients have a clinical phenotype of HD without a CAG amplification in HD gene [45]. If these families mapped to the HD locus, it would indicate that another mutational mechanism besides CAG expansion can lead to the same disease phenotype. In at least three families, however, linkage studies excluded the HD locus as being responsible for this phenotype. These observations suggest that on rare occasions nonallelic heterogeneity may underlie the presentation of an HD-like phenotype. Delineation of these molecular mechanisms could yield additional insights into the pathogenesis of HD.

IV. MECHANISM OF DOMINANCE OF CAG EXPANSION IN HD

HD is inherited as an autosomal dominant trait. The mechanism of dominance in conditions associated with CAG expansion in triplet repeat diseases is not clear. Various mechanisms leading to loss or gain of function can be postulated as producing a dominant phenotype (Table 24-1). An example of loss of function leading to a dominant disorder in the context of a triplet repeat expansion in the coding region of the gene is seen in cleidocranial dysplasia, where polyalanine expansion leads to reduced normal function and clearly shows how triplet repeat expansion can disrupt normal function [46]. Another example is Spinobulbar muscular atrophy, where a feature of the phenotype is partial feminization including gynecomastia in males resulting from decreased function of the androgen receptor [37].

Several lines of evidence suggest that the expanded polyglutamine does not cause neuronal death by complete loss of normal function. First and most convincing is that individuals homozygous for CAG expansion in HD gene are liveborn [47, 48, 15]. However, mice with targeted disruption of both alleles are embryonic lethal [12–14]. Furthermore, comparisons of age of onset, clinical presentation, and course of the disorder do not show any significant difference between homozygotes and heterozygotes. The embryonic lethality in the HD knockout mice also indicates that the disease cannot be explained by a simple dominant negative effect whereby the mutant protein interacts with and inactivates the normal gene product, leading to a complete loss of function. These findings suggest a gain of function of the mutant protein through different potential mechanisms.

In contrast, mice heterozygous for the exon 5 knockout show neuronal loss in the subthalamic nucleus, increased motor activity, and subtle difference in learning ability [12]. These findings may indicate a role of huntingtin in the normal development of basal ganglia and also suggest that some of the effects of polyglutamine expansion may be caused by partial loss of function of the protein. Alternatively, these changes could be explained by a toxic gain of function of a putative truncated product in these mice.

TABLE 24-1 Mechanisms of Dominance

	Gain of function	
Loss of function	Normal function	Novel function
Reduced protein activity	Altered expression	Altered structural protein
Dominant negative effects	Increased activity	Toxic protein alterations

V. REGIONAL SELECTIVITY OF HD PATHOLOGY

The pathology in HD is restricted to the brain and involves neuronal loss and gliosis. There is no deposition of extracellular inclusion bodies, but recent studies indicate the presence of a nuclear inclusion body [49] in the HD brain. There are no specific peripheral changes. The caudate and putamen show the most dramatic changes, but there is overall atrophy of the brain which in advanced cases weighs 20 to 30% less than normal [50]. There is selective loss of neurons within the striatum, with near total degeneration of GABA-encephalin containing medium spiny neurons projecting to the globus pallidus and GABA-substance P containing spiny neurons projecting to the substantia nigra. The degeneration progresses in a posterior-anterior, dorso-ventral, and mediolateral direction [51] with relative preservation of interneurons [52]. There are both degenerative and regenerative changes in medium spiny neurons [53]. The prominence of neuronal loss in the striatum indicates that striatal projection neurons have some unique properties that make them particularly vulnerable to the toxic effects of mutant huntingtin. In the cerebral cortex, large neurons are mostly affected, especially in layer VI, with smaller amount of degeneration in layers III and V. The lateral tuberal nucleus of the hypothalamus is selectively atrophic [54] and there is moderate atrophy in the amygdala and some regions of the thalamus [55]. In some families there is prominent brain stem atrophy.

Several mechanisms have been postulated to explain the selective loss of neurons in HD despite the ubiquitous and widespread expression of huntingtin. First, most studies indicated that huntingtin is expressed in all neurons. However, some recent studies suggest that the selective vulnerability of spiny striatal neurons observed in HD is associated with a higher level of huntingtin expression, whereas the relative resistance of large to medium-sized aspiny neurons to degeneration is associated with lower levels of huntingtin expression (see Section VI.B.1).

Second, regional selectivity could also be due to restricted expression of a protein that has altered interaction with huntingtin with an expanded polyglutamine stretch. Several interacting proteins have been identified, but none is restricted in its distribution to sites of pathology and thus none so far adequately explain the regional selectivity (see Section VII).

Third, somatic instability of CAG repeat number has been described in brain of juvenile-onset HD patients, but there was no evidence for a causal relationship between neuron loss and somatic variability as heterogeneity in mutant protein expression occurred primarily dependent on glial contribution and mitotic activity [56]. In the cerebellum, where neurons are predominant, somatic heterogeneity is lowest [57].

Fourth, a polyglutamine containing truncated toxic product may result from tissue-specific cleavage. This is currently being studied using various nonneuronal and neuronal cell lines and tissues [174] (also see Sections VIII.B and VIII.C).

Finally, a normal decline in energy metabolism with increasing age may reduce the threshold of sensitivity to excitotoxic injury below a critical level in affected neurons of the HD brain. Excitotoxic mechanisms or mitochondrial dysfunction may act in concert with the mutant huntingtin since the striatum is particularly vulnerable to the local and systemic effects of inhibitors of mitochondrial function [58] and to toxic effects of excitatory amino acids [59] (also see Sections IX and X).

VI. TOPOGRAPHIC, CELLULAR, AND SUBCELLULAR DISTRIBUTION OF WILD-TYPE AND MUTANT HUNTINGTIN

A. RNA Studies

1. Regional and Cellular Distribution

Northern blot and *in situ* hybridization analyses indicates that the 10.3- and 13.6-kb transcripts are expressed in many tissues with higher expression of the longer transcript in brain [15, 4, 60, 61]. HD mRNA is expressed in both neural and nonneural tissues with high levels of expression in neurons, testes, ovaries, and lung [60–62]. Almost all neurons are labeled with no qualitative difference in mRNA expression in neurons of different brain regions. For example, HD mRNA is expressed in neocortex, striatum, and the deep cerebellar nuclei and cerebellar cortex [60]. The highest level of expression is apparent in regions with high neuronal density. HD mRNA levels are low in pancreas, smooth muscle, and liver [60]. During spermatogenesis, HD mRNA is expressed at variable levels, with immature spermatogonia expressing higher levels than maturing spermatids [60]. These studies suggest that selective neuronal loss in HD cannot be explained by differential expression of the HD gene in different tissues.

2. Developmental Expression

Northern and *in situ* hybridizations show that in rodents the HD gene is expressed in all developmental stages [10]. At day 14.5 p.c. the level of rat HD gene expression is similar in the brain and nonneuronal tissues. However, the expression of rat HD gene is markedly reduced in nonneuronal tissues in adult animals. Only in the testis RNA is the signal intensity of the 10.3-kb transcript as high as in the brain, whereas the 13.6-kb transcript was barely detectable.

B. Protein Studies

1. Regional and Cellular Distribution

Similar to RNA studies, most protein studies also indicate ubiquitous expression of huntingtin [20, 22]. In Western blots, an approximately 350-kDa protein is detected in rodent neural and nonneural tissues, in various human cell lines and adult human peripheral tissues and brain [20, 22] (Graham and Hayden, unpublished). Huntingtin expression is high in brain, moderate in testes, and low in other peripheral tissues (Graham and Hayden, unpublished) (Figs. 24-2 and 24-3). In brain, huntingtin is detected mainly in neurons [20, 22], and very few glia contain huntingtin. In testes, huntingtin is present in spermatogonia and spermatocytes, but is not present in mature spermatids (Graham and Hayden, unpublished).

Two recent studies show that huntingtin expression in the nervous system may be more heterogeneous than suggested by earlier studies [63, 64]. Both studies show that striatal interneurons that are spared in HD have low levels of expression of huntingtin. However, huntingtin immunoreactivity is primarily confined to neurons and neuropils in the matrix area in human [63], and predominantly in the striosomes (patch), with much less immunoreactivity in matrix in rat brain [64]. Neuropathological studies have suggested that striatal matrix is affected in HD to a greater extent than striosomes [51], but striosomes may be the earliest affected region [65]. Further investigation is needed to confirm whether huntingtin is expressed at higher levels in neurons that die selectively in HD.

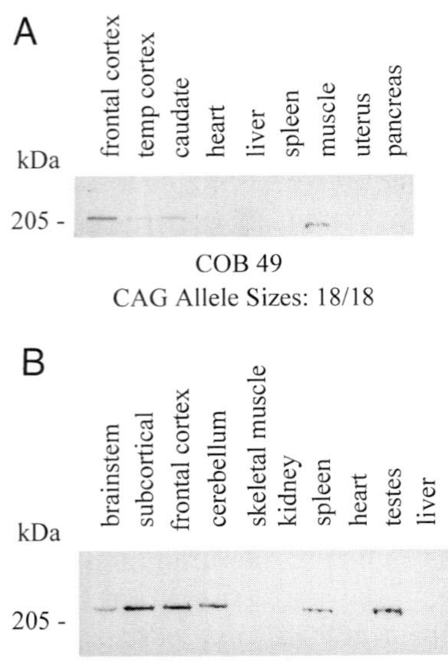

FIGURE 24-2 (A) Regional distribution of huntingtin in human control brain and peripheral tissues. Huntingtin expression was highest in brain and low in heart, liver, spleen, muscle, uterus, and pancreas. (B) Huntingtin expression in mouse brain and peripheral tissues. Expression was highest in brain and testes, moderate in spleen, but low in skeletal muscle, kidney, heart, and liver.

2. Developmental Expression

Selective neuronal loss in HD can be explained if the mutant huntingtin alters the normal development of striatum and cortex. An effect of the HD mutation on striatal maturation has been suspected because of the striking dorsoventral gradient of pathogenesis which suggests a greater vulnerability of dorsal projection neurons that mature later [51].

Huntingtin is detected at all stages of embryonic and postnatal brain development [66]. The protein expression in striatum increases significantly between postnatal day 7 and day 15, which marks a period of active neuronal differentiation and enhanced sensitivity to excitotoxic injury in the rodent striatum. Huntingtin immunoreactivity is present in neuronal perikarya in both developing and adult brain, but immunoreactive axons are seen only in adult brain. The emergence of huntingtin expression in axons of the mature neurons is relevant in that the projection neurons in both the cortex and the striatum are more affected than the interneurons in HD. Increased immunoreactivity in large neurons follows the caudal to rostral and dorsal to ventral gradient of neurogenesis. This suggests that an important factor regulating the increased expression of huntingtin may relate to the formation of target connections during and after the completion of neuronal migration. This study suggests a role of huntingtin in the normal maturation of neurons.

3. Mutant versus Wild-Type Huntingtin Expression

Mutant huntingtin can be distinguished from the wild-type protein on Western blot by virtue of its anomalous migration [20, 67–70] (Graham and Hayden, unpublished). However, similar regional distribution of both mutant and wild-type protein is evident (Figs. 24-2 and 24-3). While some have observed a mutant protein with reduced intensity, their is no evidence for reduced production or increased turnover of the mutant protein

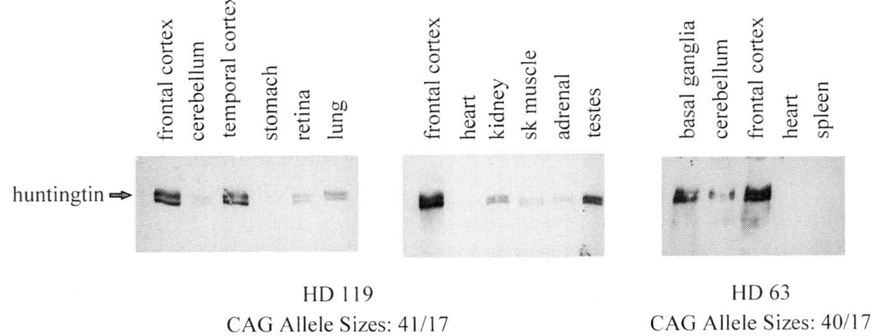

FIGURE 24-3 Huntingtin expression in HD brain and peripheral tissues. Immunoblots from the soluble fractions show the presence of both wild-type and mutant proteins. Frontal and temporal cortex, basal ganglia, and testes showed the highest level of expression, while low levels were detected in cerebellum, kidney, retina, and lung and minimal expression was seen in stomach, heart, skeletal muscle, spleen, and adrenal.

[67, 69] (Graham and Hayden, unpublished). The accessibility of different epitopes targeted by antibodies used in the experiments may be differently affected by the polyglutamine expansion.

4. SUBCELLULAR DISTRIBUTION

Determining the subcellular localization of huntingtin may be helpful in providing clues to its normal function. For example, the presence of huntingtin in the nucleus could indicate that it may be involved in nuclear processes such as transcription, replication, RNA splicing, mRNA transport, and nuclear organization.

Both nuclear and cytosolic localizations have been reported [71, 72, 7, 20–23] (Table 24-2). Using immunocytochemical and biochemical studies, huntingtin has been detected in both cytoplasm and nucleus in cultured human and mouse cells [71, 72, 7].

However, immunohistochemical studies in human, monkey, and rat show that huntingtin is present exclusively in the neuronal cytoplasm, enriched in nerve endings and associated with synaptic vesicles [20–23]. Biochemical data also suggest that wild-type and mutant huntingtin are present in cytosolic soluble fractions and microsomal membranes and associated with the cytoskeleton [21, 22, 73] (Graham and Hayden, unpublished) (Fig. 24-4). In these studies huntingtin was not observed in mitochondria or cell nuclei. In vesicle-enriched fractions, huntingtin immunoreactivity overlaps with the distribution of vesicle membrane proteins (SV2, transferrin receptor, and synaptophysin). These observations suggest a role of huntingtin in vesicle trafficking [21]. Several groups have also demonstrated an association of huntingtin with microtubules [22, 21]. Recent immunofluorescence studies suggest that huntingtin colocalizes with transferrin receptor and with coated vesicles in cytoplasm and is enriched in golgi apparatus and transgolgi network; the markedly punctate distribution of huntingtin in cortical and striatal neurons of some HD patients corresponds in part to multivesicular bodies which are part of an endosomal–lysosomal system [74–76].

These studies of subcellular distribution reveal an association of huntingtin with the cytoskeleton and cytosolic membrane and suggest a role in normal cytoskeletal function. A cytoskeletal association is also indicated by the discovery of huntingtin-interacting protein, HIP1, a protein with homology to yeast cytoskeletal associated protein *Sla2* [77]. Furthermore, huntingtin-associated protein, HAP1, is predominately a membrane-associated protein that shows a subcellular distribution

TABLE 24-2 Subcellular Localization of Huntingtin

Authors	Species	Position in huntingtin (aa) of epitopes recognized by antibodies	Subcellular location, immunocytochemistry	Subcellular location, Western
Hoogeveen *et al.* (1993)	Human	3114–3141	Cytoplasm, nucleus	
DiFiglia *et al.* (1995)	Human Rat	1–17 585–745 2911–3140	Cytoplasm, around membranes	Membranes, enriched in vesicle fractions
Sharp *et al.* (1995)	Rat	1–17 650–663	Cytoplasm	Cytosol, membranes, increased in vesicle
Gutekunst *et al.* (1995)	Rat, human, monkey	549–679	Cytoplasm, microtubule associated	
Trottier *et al.* (1995)	Human Rodent	181–810, 1247–1646 2146–2541, 2683–2979	Cytoplasm	
Bessert *et al.* (1995)	Human	1188–1204	Cytoplasm, nucleus	Cytosol, nucleus
Ide *et al.* (1995)	Human	1187–1207		Cytosol, membranes
Wood *et al.* (1996)	Human, rat, mouse	2110–2121		Cytosol in mouse, cytosol and membrane, in rat and human
De Rooij *et al.* (1996)	Human, mouse, cell lines	701–744 (2) 1929–2421 3114–3141 (2)	Cytoplasm, nucleus	Cytoplasm, nucleus
Davies *et al.* (1997)	Mouse	1–17	Cytoplasm, vesicular membranes, nucleus (exon 1 transgene)	

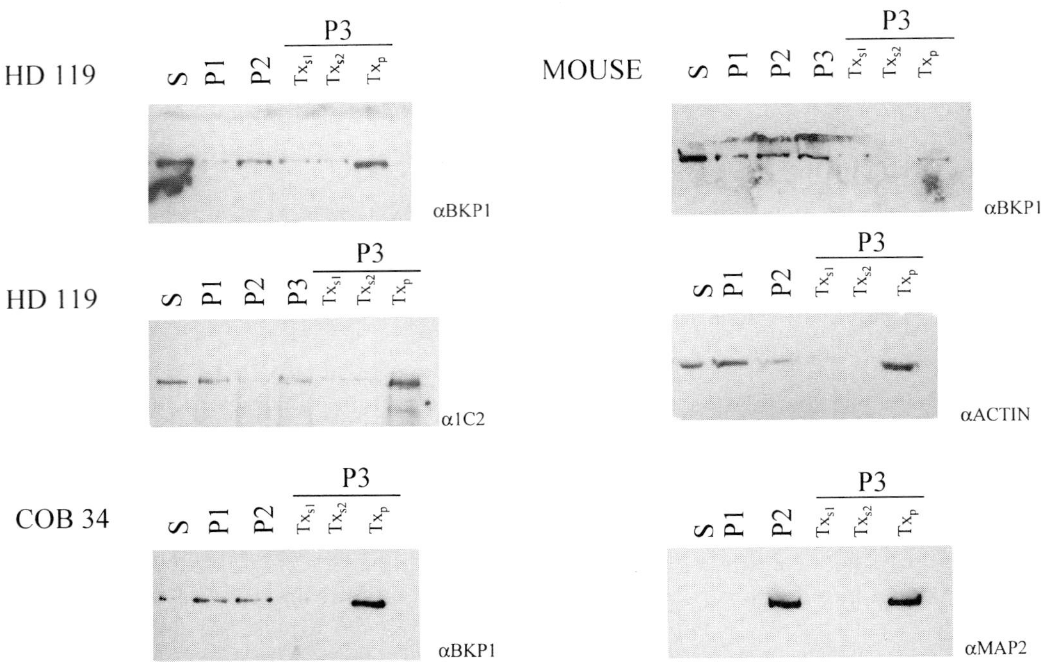

FIGURE 24-4 Western blot showing subcellular distribution of huntingtin in human control, adult-onset HD patient, and mouse cerebral cortex. Huntingtin immunoreactivity was predominant in the soluble and P3 membrane (high centrifugation pellet) in human but mostly in the soluble fractions in mice. No difference was observed between the distribution of the mutant and wild-type huntingtin. Cortex was separated into the cytosolic (S), nuclei and cell debris (P1), mitochondria and synaptosomes (P2), and microsomal and plasma membranes (P3) fractions; the cortical P3 membranes were extracted twice with Triton X-1000. Txs1 and Txs2 contain protein solubilized in the first and second extraction, respectively. Txp contained protein remaining on P3 membranes.

profile similar to synaptophysin, suggesting that HAP1 associates with the cytoskeleton or synaptic vesicles [78]. Huntingtin is also found in cytoplasmic inclusions in Alzheimer's and Parkinson's disease [43]. Those cytoplasmic inclusions, largely made up of cytoskeletal proteins, indicate that huntingtin may interact with cytoskeleton in forming inclusions.

5. PRESENCE OF MUTANT HUNTINGTIN IN INTRANUCLEAR INCLUSIONS

Mice transgenic for exon 1 of human HD gene carrying 115 to 156 CAG repeat expansions develop pronounced neuronal intranuclear inclusions [79] (see also Section XII.A). The inclusions contain an N-terminal fragment of huntingtin and ubiquitin and are seen prior to developing a neurological phenotype. In some homozygous mice, the inclusions contain fibrillar amyloid-like structures [42]. In control mice, huntingtin reactive products are not observed in the nucleus. Approximately 20% of the neurons but not glia contain inclusions. The nuclear inclusions are observed within neurons in the cerebral cortex, striatum, cerebellum, and the spinal cord, whereas other regions contain fewer neurons with nuclear inclusions. The mechanism of translocation of the N-terminal fragment to the nucleus or how it may cause neuronal dysfunction is not understood. Similar intranuclear accumulation of huntingtin in the cortex and striatum of postmortem HD brains has also been found [49].

A dense accumulation of huntingtin immunoreactive products is also found in multivesicular bodies in the cytoplasm of neurons more frequently in the exon 1 transgenic mice than control mice [79]. Perinuclear aggregates are also evident in HD brains and various cell lines expressing huntingtin [76]. When 293T cells are transfected with mutant HD cDNA and exposed to tamoxifen, they develop perinuclear aggregates dependent on the length of the polyglutamine repeat [175] (Fig. 24-5). Immunostaining with N- and C-terminal antibodies shows that these aggregates contain an N-terminal fragment of huntingtin. Also, the aggregates are observed less frequently in cells that are transfected with full-length huntingtin than cells that are transfected with an N-terminal fragment (see Section XI.D).

As mentioned earlier, huntingtin was also found in dystrophic neurites, neurofibrillary tangles, and plaques

through association with the cytoskeleton. What could these events be in the human brain? It is tempting to speculate that events at membrane-bound cytoplasmic organelles are prime targets and that HIP1 may function in the transport, assembly, or regulation of cytoplasmic organelles or in vesicle trafficking. Such functions may then be adversely impacted via aberrant interactions with huntingtin carrying expanded CAG repeats.

B. HAP1

Huntingtin-associated protein 1 (HAP1) is another protein identified by yeast two-hybrid screening whose interaction with huntingtin is modulated by CAG length [78]. In contrast to HIP1, the interaction between HAP1 and huntingtin is modulated in the opposite direction. HAP1 has increased affinity for huntingtin alleles with 44 glutamines, as opposed to an allele with 23 glutamines [78]. HAP1 does not interact with an atrophin-1 construct containing 21 CAG repeats, suggesting that huntingtin sequences surrounding the polyglutamine tract are necessary for the interaction with HAP1 [78].

In yeast, the first 230 amino acids of huntingtin are sufficient for an interaction with HAP1 [78]. *In vitro* binding experiments with a glutathione *S*-transferase (GST)–HAP1 fusion protein have demonstrated that both endogenous huntingtin and a transiently transfected huntingtin construct of 930 amino acids are specifically bound by GST–HAP1 [78]. Binding experiments using extracts from lymphoblastoid cell lines derived from HD patients confirm the correlation between repeat length and HAP1 interaction, as strongest binding is observed using extracts containing huntingtin alleles with 82 repeats. Huntingtin alleles with 44 or 22 or 19 repeats have progressively weaker interactions with HAP1 [78]. Coimmunoprecipitation experiments show that HAP1 can be precipitated with an antibody specific for HD either from whole rat brain or from HEK293 cells transiently transfected with HAP1 [78]. However, similar experiments with human brain tissue have so far been unsuccessful.

Northern blot analysis shows that expression of the 4.0-kb HAP1 mRNA is limited to the brain, with maximal expression in the rat striatum and olfactory bulb and weak expression in spinal cord, cerebellum, and cerebral cortex [78]. A partial human homologue of HAP1 is expressed maximally in caudate, subthalamic nucleus, cortex, and fetal brain. Interestingly, a variety of human HAP1 cDNA homologues were detected, suggesting the existence of multiple forms of human HAP1 or the presence of a family of HAP1-like transcripts. Western blot analysis of HAP1 from monkey and human brain regions confirms that HAP1 is selectively expressed with high levels observed in hippocampus, caudate, and cortex and lower levels in cerebellum [78]. Interestingly, although striatal neurons were found to contain low amounts of HAP1 mRNA, they react strongly with HAP1 antibodies in immunohistochemistry experiments [78]. Since the expression pattern of HAP1 does not correlate precisely with the regions selectively involved in HD, additional factors must once again be invoked to account for the selective neurodegeneration in HD.

Subcellular fractionation experiments showed that HAP1 is predominately a membrane-associated protein that shows a subcellular distribution profile similar to synaptophysin, suggesting that HAP1 associates with the cytoskeleton or synaptic vesicles [78]. Immunohistochemistry experiments show that HAP1 is exclusively cytoplasmic and associates with cytoplasmic granules with no apparent staining of the plasma membrane [78].

Although the primary sequence of HAP1 gives no clues as to its function, a combination of *in situ* and subcellular fractionation experiments have shown that HAP1 is expressed in several similar brain regions as nNOS or NADPH diaphorase, an enzyme involved in the synthesis of nitric oxide [78]. The notable exception to colocalized staining was observed in the cerebellum, with the granule cells staining intensely for nNOS and NADPH diaporase while HAP1 staining was much lighter [78].

The correlation between HAP1 and nNOS is intriguing, as nNOS containing neurons in striatum is selectively spared. It is unclear why nNOS containing neurons is spared in HD. There is evidence that neuronally derived NO may play a role in excitotoxic mechanisms. For example, mice with targeted disruptions in the gene for nNOS, or that are treated with NO inhibitors, are substantially more resistant to excitotoxic damage caused by intrastiatal injections of malonate or the systemic delivery of 3-nitroproprionic acid (3 NP) [88, 89]. However, both striatal and cortical nNOS containing neurons are resistant to several forms of induced apoptotic death [90] either in the presence or absence of NOS inhibitor nitroarginine. Thus, the basis of this resistance is likely separate from the production of nitric oxide or the detoxification of oxygen free radicals.

C. HIP2

A third huntingtin-interacting protein identified using a yeast two-hybrid screen is HIP2 [80]. Sequence analysis showed that HIP2 encodes the human ubiquitin conjugating enzyme, hE2-25K [91]. The identity of HIP2 as hE2-25K was confirmed by demonstrating that GST–HIP2 reacts strongly with affinity-purified antibodies raised against bovine E2-25K, while GST alone does not react [80]. E2 ubiquitinating enzymes have known roles in the turnover of abnormal proteins [91]. It is noteworthy that altered patterns of cellular ubiquitination have been observed in other neurodegenerative disorders including Alzheimer's disease, Parkinson's disease, and amyotrophic lateral sclerosis [92–94].

Similar to HIP1 and HAP1, binding of HIP2 to huntingtin requires sequences within the first 540 amino acids of huntingtin [80]. In contrast to HIP1 and HAP1, the interaction between HIP2 and huntingtin does not appear to be obviously influenced by CAG length when tested with alleles carrying 16 or 44 repeats [80]. It will be informative to test whether the binding of HIP2 is altered with polyglutamine lengths of 80 or 128 repeats.

The interaction between HIP2 and huntingtin was confirmed *in vivo* by demonstrating that endogenous huntingtin could be affinity-purified from cell lysates using GST–HIP2 [80]. Also, huntingtin is specifically detected in ubiquitin conjugates prepared from lysates of transformed lymphoblasts derived from an individual heterozygous for HD, showing that huntingtin is ubiquitinated *in vivo* [80].

E2-25K mRNA transcripts of 1.2 and 2.4 kb are expressed in all assessed human tissues and a 25-kDa protein detected by E2-25K-specific antibodies is ubiquitously expressed [80]. However, the 25-kDa protein is enriched in brain. Interestingly, two higher-molecular-weight immunoreactive species of 28 and 45 kDa are also detected in brain that may represent modified E2-25K proteins, additional members of the E2-25K family, or cross-reactive proteins. The expression profile of the 28-kDa band shows a striking parallel to the regional neuropathology in HD, being specifically enriched in striatum and cortex [80].

The correlation between selective expression of the presumed 28-kDa form of E2-25K and the specific neuronal loss in HD is especially intriguing. If this 28-kDa band turns out to be a form of E2 that is involved in the regulated catabolism of huntingtin, it is tempting to speculate that ubiquitination may play a role in the neuropathology of HD. For example, the number of ubiquitin-reactive neurites in HD brain is increased compared to controls [95]. This observation leads to the suggestion that ubiquitinated huntingtin may be degraded, giving rise to toxic protein fragments particularly when polyglutamine expansion is evident (see Section XII). Moreover, the presence of ubiquitin in the nuclear inclusions observed in the HD brain [49] suggests that the interaction with members of ubiquitin cascade may play an important role in the initiation and propagation of pathogenesis.

D. GAPDH

Glyceraldehyde-phosphate dehydrogenase (GAPDH) is a multifunctional enzyme with roles in glycolysis [96, 97] and in DNA repair and replication [98, 99]. GAPDH has been reported to interact with microtubules [100, 101], actin [102], RNA [103, 104], and amyloid precursor protein [44]. Using affinity chromatography methods, GAPDH has been shown to interact directly with the polyglutamine tracts of huntingtin and DRPLA [81]. Also, genetic methodology employing the yeast two-hybrid method showed interactions between GAPDH and the ataxin-1 and androgen receptor gene products [105].

The portion of GAPDH that interacts with ataxin-1 and androgen receptor contained amino acids 1–149 of the nicotinamide adenine dinucleotide (NAD) binding domain and only the first 21 amino acid residues of the catalytic domain, suggesting that the NAD binding domain is primarily involved in the interaction [105]. It is not known whether the interaction of ataxin-1 or androgen receptor with GAPDH disrupts the function of GAPDH, possibly by preventing the formation of active tetramers.

There is some discrepancy as to whether the interaction of GAPDH with the products of the trinucleotide repeat genes is modulated by CAG length. In yeast two-hybrid studies, CAG length does not affect binding [105]. However, in biochemical experiments using pure polyglutamine stretches immobilized on a solid support, GAPDH preferentially bound long lengths of polyglutamines [81]. Further experiments will be necessary to reach a consensus on whether polyglutamine length as well as flanking sequences influence the strength of interactions with GAPDH. However, it is particularly interesting that GAPDH appears to interact preferentially with smaller fragments of huntingtin as opposed to the full-length protein [81]. It will be interesting to determine whether amino-terminal fragments of huntingtin that may be produced by ubiquitin-mediated catabolism, or by caspases involved in apoptotic death (see Section VIII), have an increased interaction with GAPDH.

The interaction between GAPDH and the gene products from four different triplet expansion disorders suggests that that the fundamental biochemical defects in these diseases may be similar. The strong evidence for energy impairment in these neurodegenerative diseases (see below) makes the association with GAPDH particularly intriguing. Neurons are exquisitely sensitive to reductions in ATP generation and the direct interaction of these gene products with GAPDH may be the initial step in the impairment of glycolysis. Proton nuclear magnetic resonance studies show that the occipital cortex and basal ganglia of symptomatic HD patients have increased lactate concentrations [106], indicative of a block in respiration. HD and MJD patients also have increased lactate–pyruvate ratios in cerebrospinal fluid, further supporting impaired oxidative phosphorylation as playing a role in these disorders [106–108]. HD patients with elevated cortical lactic acid also had decreased phosphocreatinine (PCr)/inorganic phosphate ratios in the muscle at rest. These data imply abnormal energy metabolism in a site where huntingtin is expressed although no obvious pathology is found. However, the widespread distribution of GAPDH does not help explain the regional specificity of neuropathology in any of these diseases. Also, a recent study did not find any difference in GAPDH activity in HD brain compared to normal controls [109]. Thus the role of the huntingtin–GAPDH interaction in neurodegeneration in HD remains unclear.

E. EGF Receptor Signaling Complex (Ras-GAP, Grb2, Activated EGF Receptor)

Huntingtin possesses multiple proline-rich motifs in its sequence that resemble SH3 domain-binding motifs. Most SH3 domain-containing proteins are adapter proteins for tyrosine kinase receptor-mediated signaling. Liu *et al.* (1997) therefore examined the possible association of huntingtin with several SH3 domain-containing signaling molecules in the EGF receptor signaling complex by coimmunoprecipitation studies in human epithelial A431 and embryonic kidney 293 cells. They found that huntingtin associates with the EGF receptor signaling complex through binding to the SH3 domains of Grb2 and Ras–GAP *in vivo* and *in vitro.* Assembly of these huntingtin complexes is dependent on autophosphorylation and activation of the EGF receptor, suggesting that the formation of huntingtin–Grb2 and huntingtin–Ras–GAP complexes is part of a cellular signaling cascade mediated by the EGF receptor. Grb2 relays signals from activated EGF receptors to Ras/mitogen-activated protein kinase pathway by interacting with the nucleotide exchange factor sos [110, 111], whereas Ras–GAP down-regulates the level of Ras–GTP [112]. Since huntingtin interacts with both Grb2 and Ras–GAP *in vivo,* it may modulate the level of Ras–GTP and thereby the Ras-dependent signaling pathway. Elevation of GTP-bound Ras in cultured cells is correlated with increased apoptosis [113, 114]. Mice with null mutation of Ras–GAP show increased neuronal apoptosis in early embryonic development [115]. Further experiments are necessary to investigate whether mutant huntingtin containing expanded CAG would alter Ras-dependent signaling in neuronal cells. They also suggested that if this interaction is relevant, the high capacity for compensation within the tyrosine kinase receptor system may explain latency in the onset of HD.

F. Calmodulin

Potential protein–protein interactions of huntingtin from monkey brain using gel filtration and Western blot studies under nondenaturing conditions [83] revealed that huntingtin migrated with a large calmodulin–1000-kDa complex in the presence of calcium and partially dissociates in the absence of calcium. Huntingtin from HD patients' brains appeared to bind more avidly than huntingtin from control brain. Moreover, unlike wild-type huntingtin, the mutant huntingtin interacts indirectly with calmodulin in the absence of calcium. Calmodulin is an intermediate in many calcium-mediated signaling pathways. Many of the cellular processes such as vesicle transport or fusion are calcium dependent and alteration in the regulation of protein interactions in this process may have deleterious consequences.

VIII. APOPTOSIS IN HD

Although the precise biochemical mechanisms underlying cell loss in HD are not yet completely understood, the available evidence supports an apoptotic, rather than necrotic, mode of cell death. The lack of an inflammatory infiltration in HD brains argues strongly against necrotic cell death [2]. Animals heterozygous for a disruption in exon 5 of huntingtin have obviously apoptotic neurons. Huntingtin itself has been shown to be an apoptotic substrate [24]. HD brains show increased levels of DNA strand breaks typical of apoptotic cells as measured by TUNEL (terminal transferase-mediated deoxyuridine triphosphate nick end labeling) assays [116–118]. Finally, oxidative and excitotoxic stresses, both of which are known inducers of apoptotic death, have been implicated in HD [58]. However, because each of these observations is subject to methodological limitations, it should be noted that CAG expansion in huntingtin has not yet been shown to result unequivocally in apoptotic cell death. Such a demonstration will require the production of bona fide cell culture and animal models of HD. The major challenges in current research are to define the molecular pathways leading to cell loss in HD as a result of CAG expansion and to use this knowledge to devise novel strategies for therapeutic intervention.

A. The Role of Caspases in Apoptosis

Caspases (cysteinyl aspartic acid-specific proteases) are a family of proteases related to Ced3, the product of a gene required for programmed cell death in *C. elegans* [119–121]. Caspases are synthesized as inactive polypeptide zymogens that contain an amino-terminal prodomain followed by large and small catalytic subunits. A cascade of proteolysis, reminiscent of the coagulation cascade, releases the catalytic subunits that then associate into active heterotetramers consisting of two large and two small subunits.

To date, 11 mammalian caspases have been identified and are classified into families based on their sequence homology, function, and substrate specificity. Caspases-8 (FLICE) [122, 123], 10a (Mch4) [124], and 10b (FLICE-2) [125] are thought to act as apical enzymes that function to proteolytically activate downstream effector caspases. A common feature of this "activator" family of caspases is that their prodomains often contain motifs known as death effector domains (DEDs) that function in receptor-mediated apoptosis. Although similar DEDs have not been identified in the prodomains of caspase-2 (ICH-1) [126] or caspase-9 (ICE-LAP6, Mch6) [127, 128] it will be interesting to determine whether their extensive prodomains nevertheless contain regulatory information. Caspase-3 (apopain, Yama, CPP32) [129–131], caspase-7 (Mch3, ICE-LAP3, CMH-1) [132, 133], and caspase-6 (Mch2) [128] form the "effector" caspase family. These caspases have short prodomains and are thought to play direct roles in the dismantling or inactivating crucial cell constituents. A third caspase family is involved in cytokine maturation as well as apoptosis. This family includes caspase-1 (ICE) [134], caspase-4 (TX, ICH-2, ICE relII) [135, 136], and -5 (ICE relIII, TY) [137, 136].

Recently the optimal tetrapeptide substrates for each of the known caspases have been defined (N. A. Thornberry, personal communication) [138, 176]. These studies show that the optimal cleavage sites for the ICE-like caspases is (WL)EHD, which for the effector caspase family is DEXD and for the activator caspase family is (IVL)EXD.

Caspase-3 has been shown to be directly responsible for the proteolytic cleavage and inactivation of key homeostatic proteins during apoptosis [129–131]. These targets include poly(ADP-ribose) polymerase (PARP), an enzyme involved in genome surveillance and DNA repair in stressed cells [130, 131], and the 460-kDa catalytic subunit of DNA-dependent protein kinase (DNA-PK), an enzyme essential for DNA double-strand break repair [139]. Additional caspase-3 substrates include the U1-70K small ribonucleoprotein and sterol regulatory element binding proteins (SREBPs) [120, 140], actin [141], fodrin [142, 143, 120], and Gas2 [144].

The importance of caspase-3 as a crucial effector of neural apoptosis has been recently demonstrated by generating mice deficient in caspase-3 [145]. Mice homozygous for a targeted disruption in caspase-3 are smaller than their normal or heterozygous littermates, die either *in utero* or within 3 weeks of age, and exhibit grossly abnormal neural structures that result from a profound inability to lose neurons appropriately in development, a process that occurs by apoptosis [145]. The brains of these homozygous mice have such extensive ectopic cell masses that they often protrude through secondary defects in the skull. Although the brains of caspase-3-deficient mice have apparently normal composition and organization of neurons and glia, there is no evidence of apoptotic cells in their brains. These results show that elimination of caspase-3 activity blocks developmental neural apoptosis.

B. Cleavage of Huntingtin by Caspase-3

Huntingtin has recently been demonstrated to be a substrate for caspase-3. Goldberg *et al.* [24] have shown that incubation of *in vitro* translated huntingtin with apoptotic extract results in the formation of an 80-kDa cleavage product. In contrast, huntingtin remains intact when exposed to nonapoptotic extract or buffer-only conditions.

Inhibitor studies determined that the huntingtin cleavage activity in these apoptotic extracts was a member of the effector family of caspases. Proteolytic activities of the different caspase families can be distinguished by inhibitor profiles using selective tetrapeptive aldehydes designed to mimic the substrate P4-P1 amino acids at the site of cleavage. For example, Ac-DEVD-CHO specifically inhibits members of the effector family of caspases whereas Ac-YVAD-CHO is a potent inhibitor of the ICE-like caspases. Goldberg *et al.* [24] observed that cleavage of huntingtin is inhibited only with Ac-DEVD-CHO. In contrast, Ac-YVAD-CHO and CrmA, a baculovirus protein that specifically inhibits ICE-like caspases, are both ineffective at inhibiting huntingtin cleavage [24]. Together these results establish the effector family of caspases as being the enzymes responsible for huntingtin cleavage.

Purified recombinant caspase-3 was then shown to be an effective protease against *in vitro* translated huntingtin. Caspase-3 cleavage results in a fragment of approximately 80 kDa that is recognized by an antibody raised against the amino terminus of huntingtin. Kinetic analysis showed that huntingtin is a physiologically relevant substrate for caspase-3, as the efficiency of huntingtin cleavage is comparable to the efficiency of PARP cleavage over the same enzyme concentration range [24].

In transient transfection experiments, specific huntingtin cleavage products are observed to be the same size as the cleavage products generated *in vitro* by purified caspase-3 or in apoptotic extracts [24]. A COS cell line stably expressing a truncated huntingtin allele was also prepared. Treatment of this stable line with camptothecin, a topoisomerase I inhibitor known to induce apoptosis, results in production of an appropriately sized huntingtin cleavage product coincident with caspase-3 maturation and PARP cleavage [24]. These results show that huntingtin is cleaved in cells undergoing apoptosis. Further studies are necessary to determine whether cleavage of huntingtin occurs *in vivo*.

Although the specific caspase-3 cleavage site was not mapped in this study, there is a cluster of DXXD sites, the consensus cleavage site for caspase-3 [130] (Fig. 24-7), confined to a narrow region within the amino terminus. Cleavage at one of these sites would be expected to produce a protein with a molecular mass of between 50 and 60 kDa. However, as the predicted truncated fragment contains the polyglutamine stretch, it would be expected that this protein fragment would run at a higher molecular weight on SDS–PAGE. Clearly, the identification of the caspase-3 cleavage site by protein sequencing or by mutagenesis is an important goal for future experiments.

It is important to note, however, that these results do not distinguish whether caspase-3 cleavage of huntingtin is a cause or an effect of apoptosis. Although there is accumulating evidence that amino-terminal fragments of huntingtin may be toxic to cells, this has not yet been conclusively demonstrated. The *in vitro* and *in vivo* models currently being developed will provide the appropriate systems with which to test the hypothesis. If the toxic fragment hypothesis is validated, these models will also allow the biochemical mechanisms underlying the apoptotic cascade in HD to be understood.

C. Caspase Cleavage of Other Proteins Containing Polyglutamine Tracts

The demonstration that huntingtin is a substrate for caspase-3 directly connects the apoptotic machinery with a protein known to result in neurodegeneration. Seven other neurodegenerative diseases are also caused by expansion of a polyglutamine tract within their coding regions of predominately novel genes. Although the causative mutant proteins have little in common other than the CAG expansion, this group of diseases all result in a selective pattern of neuronal degeneration. These observations suggest that the biochemical mechanisms underlying the pathology of this group of diseases may have elements in common. For example, causative proteins of additional polyglutamine expansion diseases may also be substrates for caspase cleavage and result in the liberation of toxic protein fragments.

We have recently generated substrate specificity profiles for additional polyglutamine-containing proteins including atrophin-1 and androgen receptor. Like huntingtin, both proteins are cleaved in apoptotic extracts and cleavage is inhibited by Ac-DEVD-CHO but not

		510 (1846)		527 (1897)		
RAT	465	LQ-A**DSVD**LSGCDLT	------SAATD----	-G**DEED**ILSHSSSQF	S---------AVPSD	524
MOUSE	473	LQ-A**DSVD**LSGCDLT	------SAATD----	-G**DEED**ILSHSSSQF	S---------AVPPD	532
HUMAN	496	LQ-A**DSVD**LASCDLT	------SSATD----	-G**DEED**ILSHSSSQV	S---------AVPSD	555
FUGU	437	IQPG**DSVD**LSASSEQ	GGRGGGASASDTPES	PN**DEED**MLSRSSSCG	ANITPETVEDATPEN	496

		549 (1963)			585 (2071)	
RAT	525	PAM**DLND**--GTQA--	-SSPISDSSQTTTEG	PDSAVTPSDSSEIVL	**DGAD**SQYLGVQIGQP	569
MOUSE	533	PPM**DLND**--GTQP--	-SSPISDSSQTTTEG	PDSAVTPSDSSEIVL	**DGAD**SQYLGMQIGQP	577
HUMAN	556	PAM**DLND**--GTQA--	-SSPISDSSQTTTEG	PDSAVTPSDSSEIVL	**DGTD**NQYLGLQIGQP	600
FUGU	497	PAQEGRPVGGSGAYD	HSLPPSDSSQTTTEG	PDSAVTPSDVAELVL	DGSESQYSGMQIGTL	556

FIGURE 24-7 Amino acid and nucleotide positions of a cluster of potential caspase-3 cleavage sites in the N-terminal part of huntingtin from human, mouse, rat, and Fugu. The optimal caspase-3 cleavage tetrapeptide DEED at amino acid residue 527 is conserved in all four species.

by Ac-YVAD-CHO. In *in vitro* assays, all proteins are cleaved by purified caspase-3 albeit with different efficiencies. Of the three proteins tested so far, atrophin-1 is cleaved slightly more efficiently *in vitro* than huntingtin, whereas androgen receptor is a poor substrate relative to huntingtin (Fig. 24-8).

In addition, the caspase-substrate preference also varies in these *in vitro* experiments. Huntingtin is cleaved well by caspase-3, can be cleaved by high concentrations of caspase-1, and cannot be cleaved by either caspase-7 or caspase-10 (Fig. 24-8). It is interesting that the cleavage efficiency of huntingtin is so different with caspases-3 and -7, because their optimal tetrapeptide substrates are indistinguishable [138, 176]. These results suggest that the tertiary structure of the substrate protein plays an important role in enzyme recognition. In contrast, atrophin-1 is cleaved very well by both caspase-3 and -7 and moderately well by caspase-1 and -10. The androgen receptor is resistant to cleavage by all enzymes tested except for caspase-3, which is able to cleave small amounts of androgen receptor at high concentrations (Fig. 8).

It will be important to determine the cleavage sites for these proteins by directly sequencing the protein fragments produced by caspase cleavage. Comparisons of the structural features found within the proteolytic fragments produced can then be made with respect to their structures. Although many additional CAG-containing proteins remain to be tested, these preliminary results are suggestive of a general model in which caspase cleavage of proteins containing polyglutamine tracts may liberate toxic fragments. It will be important to test this model using appropriate *in vitro* and *in vivo* models of these diseases. It will also be interesting to compare the sensitivity to caspase cleavage with the regional distribution of the caspases within the brain in order to determine whether caspase-substrate specificity may play a role in the cell-type-specific neuronal loss.

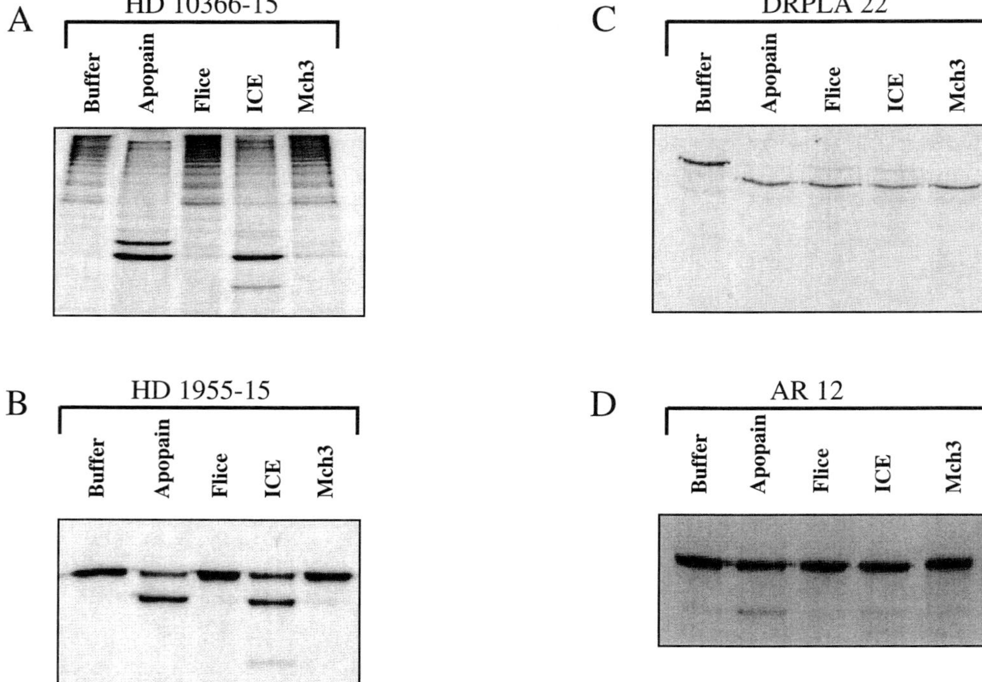

FIGURE 24-8 Proteins containing polyglutamine tracts are cleaved by a variety of caspases *in vitro.* cDNA constructs encoding proteins with polyglutamine tracts were transcribed and translated *in vitro* using Promega TNT reticulocyte lysate or wheat germ systems with $[^{35}S]$methionine. Radiolabeled reaction product (0.5 μl) was incubated with 10 nM purified caspase in cleavage buffer (50 mM Hepes/KOH, pH 7.0, 10% (w/v) sucrose, 2 mM EDTA, 0.1% (w/v) CHAPS, 5 mM EDTA at a final volume of 25 μl. Following incubation for 60 min at 37°C, the reaction was terminated with 7 μl of 5× SDS Laemmli loading buffer and denatured at 95°C for 3 min. Cleavage products were resolved on 7.5% SDS–polyacrylamide gels. Gels were fixed in 40% methanol/10% acetic acid and enhanced using Amplify (Amersham) and subjected to fluorography. (A) Full-length huntingtin containing 15 CAG repeats (10366-15); (B) truncated huntingtin with 15 CAG repeats (1955-15); (C) atrophin-1 with 22 CAG repeats (DRPLA-22); (D) androgen receptor with 12 CAG repeats (SBMA-12).

The results of these investigations will provide much-needed insights into the common and distinct pathogenic mechanisms resulting in these disorders.

IX. EXCITOTOXICITY IN HD

The excitotoxic hypothesis in HD proposes that endogenously produced excitatory amino acids (EAA), or closely related substances physiologically involved in neurotransmission, damage and kill neurons that are chronically exposed to their effects [52]. Neurons possessing receptors for EAA including NMDA, AMPA, and kainate receptors are affected. Since intrastriatal injection of KA could reproduce several aspects of neuropathology of HD, dysfunction of excitatory amino acids has been implicated in HD [146]. Administration of NMDA agonists such as quinolinic acid, l-homocysteic acid, and NMDA itself even more faithfully reproduce the structural and neurochemical selectivity observed in HD. Loss of NMDA receptor sites in a presymptomatic individual with HD has been reported [147]. Recently, using a polymorphic TAA repeat at the 3′ UTR of the GluR6 gene, Rubinstein *et al.* showed that genotypes at GluR6 kainate receptor locus are associated with variation in the age of onset of Huntington's disease [148]. Of the variance in the age of onset of HD that was not accounted for by the CAG repeats, 13% could be attributed to GluR6 genotype variation ($P = 0.008$). It is most likely that TAA repeats in 3′ UTR were acting as a neutral polymorphism in linkage disequilibrium with a functional variant in the GluR6 gene. However, linkage disequilibrium with a nearby gene cannot be formally ruled out.

However in HD, many apparently unaffected neurons in the frontal cortex, cerebellum, amygdala, and hippocampus contain EAA receptors. Thus other factors must be operative either to protect EAA receptor bearing neurons against the effect of mutant huntingtin or to render striatal or some other neuronal populations like the lateral tuberal nucleus particularly vulnerable. These factors could include alteration in the distribution of EAA receptors or alteration in the receptor function at prereceptor, receptor, or postreceptor levels.

X. MITOCHONDRIA, OXIDATIVE STRESS, AND ENERGY METABOLISM

Mitochondria and energy metabolism are known to play roles in HD and other neurodegenerative diseases including Parkinson's Disease, Alzheimer's Disease, cerebellar degeneration, and mitochondrial encephalopathies [58, 149, 59]. For example, defects in oxidative metabolism and mitochondrial electron transport have been demonstrated in HD patients [106, 107, 150, 151].

Results from animal studies also suggested a role of mitochondrial dysfunction in HD. Chronic administration of 3-NP, which is an irreversible inhibitor of succinic dehydrogenase, results in lesions in the caudate and putamen of primates that leads to spontaneous dystonia, dyskinesia, and cognitive deficits [152, 153]. 3-Nitropropionic acid produces lesions in rodents that simulate the neuropathologic features of HD. There is a progressive locomotor deterioration beginning with a hyperactive phase with no obvious striatal pathology that later resolves into a hypoactive phase with striatal pathology [154–158]. Similar to delayed neuronal loss in HD, the toxicity is age-dependent [159, 154, 156, 157]. In 3-NP-treated animals, neurons that stain positive for NADPH diaphorase are spared whereas spiny striatal neurons are depleted and show proliferative changes in dendrites similar to changes in HD [155–159]. The lesions are associated with increases in markers of free radical damage and are significantly attenuated in mice overexpressing SOD1. The NOS inhibitor 7-nitroinadazole blocks the lesions and is associated with a reduction of 3-nitrotyrosine. Taken together, these observations have been interpreted as evidence of mitochondrial dysfunction and oxidative stress producing selective lesions in striatum. However an alternative explanation that 3-NP causes apoptosis in striatal neurons via mechanisms other than excitotoxicity has also been proposed [90].

Mitochondrial dysfunction may be particularly relevant to neurodegenerative diseases because damaged mitochondria would be expected to have particularly deleterious effects in nonproliferating neural cells. PET studies indicate that basal ganglia and cerebral cortex have high levels of glucose utilization, while the cerebellum has lower metabolic rates. Mitochondrial DNA is extremely sensitive to mutagenic damage, not only because it is in close proximity to the production of reactive oxygen species, but also because it lacks protective histones and many repair pathways [160]. In quantitative studies of a commonly occurring 4977-bp deletion between ND5 and ATPase as a marker for mitochondrial DNA damage, the caudate, putamen, and substantia nigra had the highest levels of deletions (up to 10% in individuals more than 75 years of age), cerebral cortex, globus pallidus, and hippocampus had intermediate levels of deletions (2.5%), and the cerebellum and white matter had consistently low levels of deletions [161, 162]. It may be speculated that mitochondrial dysfunction secondary to DNA damage in particular cells may act in concert with the toxic effect of polyglutamine in causing neuronal death.

Impairments in mitochondrial energy metabolism may lead to a secondary excitotoxic state due to opening of NMDA receptor channels by interfering with ion pumps that act to maintain the potential across neural membranes [149]. Because much energy is required to fully repolarize synaptic membranes after a depolarizing stimulus, impairments in ATP production may lead to prolonged or subthreshold opening of voltage-gated-dependent Ca^{2+} channels or by decreasing the voltage-dependent protective Mg^{2+} block of NMDA channels [163]. As a result, cells with mitochondrial defects may have NMDA channels that can be activated by endogenous levels of glutamate, resulting in secondary excitotoxic death. Supporting this hypothesis is the observation that excitatory amino acids can block degeneration of cortical explants derived from 3-NP-treated animals [164].

XI. *IN VITRO* MODELS OF HD

A. Approaches to Establishing an *in Vitro* Model

Detailed understanding of the molecular basis of Huntington's disease and other polyglutamine disorders requires the development of appropriate model systems. As a complementary approach to transgenic mice expressing mutant huntingtin, various cell culture systems can be established as *in vitro* models of HD. Cells grown in culture have the advantage of being easily manipulated, allowing the control of multiple variables, and are far less laborious than generating transgenic mice. Transfected cells can be used to rapidly assess the influence of the disease protein, or an isolated expanded polyglutamine repeat, on various cell functions. Whereas transgenic mice facilitate the study of HD at the organ and organism levels, cell culture permits analysis of individual cells. Thus, one can follow the time course of pathology at the cellular level. Furthermore, the *in vitro* system can be used to model temporal steps in the disease pathway. For example, generation of a cleavage fragment may play a role in the pathogenesis of HD. Using cell culture, the direct effect of the cleavage fragment on cell viability can be analyzed. Finally, mechanisms involving protein misfolding, aggregation, aberrant associations, or novel activities can be readily assessed *in vitro.*

Although difficult to maintain, one of the best cell culture systems is primary neuronal cultures from brain regions known to be affected in HD. Alternative *in vitro* cellular models can be developed that express the gene of interest in a transient, stable, or inducible manner. Each expression system has particular advantages and disadvantages. Transient expression is very rapid, and although the expression in each cell in the transfection may vary, it works well for analyzing populations of cells rather than individual cells. This technique allows one to assess the effect of the introduced gene, with the only confounding variable being the transfection protocols, all of which are known to result in cell stress and a mixed population of transfected and untransfected cells.

Stable transfection has the advantage that it creates clonal cell populations, meaning that every cell expresses the gene of interest at a similar level. With careful screening, clonal cell lines can be obtained with various levels of expression based on different copy numbers of the integrated gene. This permits analysis of dosage-dependent effects of the introduced gene. Creation of a stable cell line is considerably more time consuming than the transient system, and the influence of the drug selection and plasmid integration cannot be predicted. There are also obvious drawbacks to selecting for constitutive expression of a potentially toxic gene, as this would result in a secondary selection for survival.

In contrast, inducible expression avoids the problem of toxic gene expression. Several inducible experimental systems have been described, but the method of choice remains the tetracycline-based system developed by Gossen and Bujard [165, 166]. The tetracycline-inducible system allows very tight and reversible control of gene expression. In contrast to the previously developed inducible systems that use metallothianine, estrogen receptor, or immediate early gene-inducible promoters, the bacterial-specific promoter upon which the tetracycline-inducible system is based is not present in the mammalian genome. This makes the tetracycline-inducible system highly selective for the gene of interest. Developing the inducible system is quite laborious as the current technology requires the creation of a double-stable cell line, first transfected with the tetracycline-regulated transactivator and then transfected with the tetracycline-responsive gene of interest. However, once established, the benefits of an inducible expression system may exceed those of the transient and stable expression systems.

B. Criteria for an *in Vitro* Model of HD

Due to the fact that cell culture systems strive to model HD in a homogeneous cell population, there are several criteria to consider in assessing whether an *in vitro* model system is appropriate. Any effects seen *in vitro* must be well-controlled to be able to attribute them to the introduction of the HD gene, rather than to the effects of transfection or selection. An effect of huntingtin expression observed *in vitro* that is proposed to be involved in *in vivo* pathology should exhibit CAG-length dependence. Since HD is a dominant disease there should be no difference when the wild-type allele is present or absent in the cell culture. To mimic the long-time course of HD, the *in vitro* model could exhibit a low-level build-up of a toxic effect up to a critical level, or involve a stochastic signal, such as damage leading to apoptosis. Since neuronal death is the end result of HD, it is important to assess whether the model system also exhibits cell death. Finally, any effect of huntingtin expression observed in nonneuronal cells should be reproducible in a striatal cell culture.

C. Other *in Vitro* Models

In vitro models have been used successfully to gain insight into the disease mechanism behind MJD. In one example, cell transfection was shown to be an appropriate model because a comparable phenotype was evident in transgenic mice using similar constructs. Ikeda *et al.* (1996) transfected a truncated ataxin-3 gene containing the normal (Q35) and mutant (Q79) number of glutamines into COS-7 monkey kidney cells [167]. By comparing the activities of a cotransfected luciferase gene, it was shown that cells that received the Q79 construct were lost from the culture due to cell death. The toxicity of the polyglutamine expansion was only evident in the truncated gene, and not in the context of the full-length ataxin-3 protein. Analysis of the expressed protein indicated that the mutant ataxin-3 formed insoluble complexes that appeared as punctate granules in the cytoplasm of transfected cells. Furthermore, cells that expressed the Q79 construct died apoptotically, with membrane blebbing and DNA fragmentation. Additional experiments demonstrated that the cell death phenotype was dominant and dose dependent, and the cell death effect increased with increasing size of the polyglutamine tract. The observation that a fragment causes pathological effects suggests that the full-length protein may mask the polyglutamine through protein conformation effects or by association with other proteins. Transgenic mice were also created that expressed the ataxin-3 fragment specifically in Purkinje cells. The cerebellum of the transgenic mice with the truncated Q79 construct exhibited severe cell atrophy, and the mice were ataxic. The pathological change observed in the transgenic mice containing an ataxin-3 fragment with polyglutamine expansion was consistent with the toxicity of the polyglutamine leading to cell death *in vitro*. Similar to the COS-7 cell transfections, mice transgenic for the full-length construct containing 79 glutamines did not exhibit neurodegeneration. Therefore, the cell culture system appropriately modeled an *in vivo* situation.

D. Progress toward an *in Vitro* Model of HD

As described earlier, the cleavage of huntingtin by the pro-apoptotic protease caspase-3, and DNA fragmentation studies in HD patient brains, suggested that inappropriate apoptosis plays a crucial role in HD. To investigate the involvement of HD and apoptosis, an *in vitro* model system was used in which apoptosis was induced. In this model, inducing apoptosis allows the downstream effects to be assessed in the absence of apoptotic insults that may be specific to neurons, such as excitotoxicity. HEK 293T cells (human embryonic kidney) were transiently transfected with full-length huntingtin cDNA and an N-terminal truncated huntingtin (1955 bp), both containing 0, 15, or 128 repeats. Apoptosis was induced by addition of tamoxifen and viability of the transfected cells was assessed. Mock and vector alone transfection served as controls (6% cell death ± 0% and 39% ± 13% using trypan blue exclusion, respectively) (mean ± SEM). Cells transfected with the truncated constructs exhibited higher proportions of death compared with cells that received full-length huntingtin containing the same polyglutamine length (Fig. 24-9). Transfection of the N-terminal construct with 128 repeats resulted in 88 ± 11% cell death, compared with 69 ± 7% cell death from transfection of the full-length protein with similar sized polyglutamine repeats ($P < 0.001$). Similar findings were seen when comparing truncated and full-length constructs with polyglutamine tracts of 0 and 15, respectively. In both truncated and full-length constructs there was a mild trend to increasing toxicity with longer polyglutamine tracts. For the truncated constructs containing 0, 15, and 128 polyglutamines, there was 76, 81, and 88% death, respectively. These results indicate that huntingtin sensitizes cells to the effects of an apoptosis-inducing agent. Cells expressing the N-terminal fragment appear to be more susceptible to apoptosis, possibly reflecting the time needed to produce a toxic fragment by cleavage of the full-length protein. Western blotting on cell lysates indicated that full-length huntingtin was cleaved to a size consistent with use of the caspase-3 cleavage sites. These data suggest that the production or expression of an N-terminal fragment increases susceptibility to apoptosis and accelerates cell death. Thus, the susceptibility to apoptosis of affected neurons in HD could be influenced

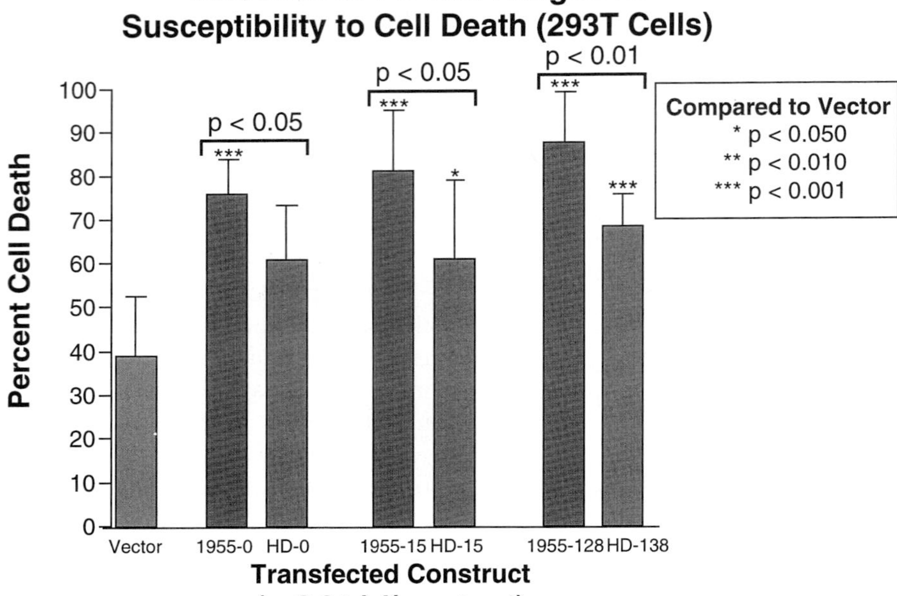

FIGURE 24-9 An N-terminal fragment of huntingtin increases the susceptibility of 293T cells to apoptosis. An human embryonic kidney cell line (293T) was transiently transfected with full-length huntingtin cDNA and an N-terminal truncated huntingtin (1955 bp), both containing 0, 15, or 128 repeats. Apoptosis was induced by addition of tamoxifen (35 μM) added 36–48 h posttransfection and viability of the transfected cells was assessed by trypan blue exclusion. Vector alone transfection served as control 39 ± 13% (mean ± SD). Cells transfected with the truncated constructs exhibited higher proportions of death compared with cells that received full-length huntingtin containing the same polyglutamine length. Transfection of the N-terminal construct with 128 repeats resulted in 88 ± 11% cell death, compared with 69 ± 7% cell death from transfection of the full-length protein with similar sized polyglutamine repeats ($P < 0.001$). In addition, in both truncated and full-length constructs there was a trend toward increasing toxicity with longer polyglutamine tracts. For the truncated constructs containing 0, 15, and 128 polyglutamines, there was 76, 81, and 88% death, respectively.

by huntingtin and may be the basis of the selective vulnerability.

Previous studies have proposed direct toxicity of the polyglutamine repeats [167]. However, huntingtin constructs that do not contain the polyglutamine tract also produce a toxic effect, although milder than constructs containing 15 or 128 glutamines. This indicates that the degree of cell death cannot be attributed entirely to the polyglutamine length. Instead, the susceptibility to cell death may also be conferred by other regions in the protein and, by implication, the proteins interacting in this region. This *in vitro* model can be used to test this hypothesis and to assess the effect of huntingtin-interacting proteins on cell susceptibility to apoptosis.

The N-terminal toxic fragment hypothesis requires that the fragment has novel properties that lead to cellular toxicity. Cleavage by caspase-3 may therefore result in a new interaction of activities of the huntingtin fragments. The mechanism could be analogous to cleavage of SREBP by caspase-3 that results in translocation of the N-terminal fragment from the cytoplasm to the nucleus, where it activates transcription [168]. As described earlier, there is evidence that the N-terminus of huntingtin can translocate to the nucleus, appearing as a densely stained circular inclusion [79]. Similarly, transfection of the N-terminal huntingtin fragment containing 128 repeats into 2-2 cells was observed to result in perinuclear aggregations (D. Martindale, Hayden, and Tufaro, personal communication), which may represent huntingtin trapped at the nuclear pores. Aggregates were observed less frequently in transfections with the full-length huntingtin containing 128 repeats and were rarely observed with transfection with huntingtin containing a normal number of repeats. Perinuclear aggregates were also present in 293T cells transfected with the N-terminal construct containing 128 repeats that had been exposed to tamoxifen (Hackam and Hayden, unpublished observations). Additional experiments need to be performed to determine the relationship between the formation of aggregates and cell death.

XII. THE TOXIC FRAGMENT HYPOTHESIS

The toxic fragment hypothesis postulates that huntingtin is cleaved within cells by a caspase or other protease into a short toxic protein fragment containing the polyglutamine tract. There is a growing body of evidence that pure polyglutamine tracts or protein fragments containing polyglutamine tracts may be toxic to cells through mechanisms that are not yet understood. The evidence supporting toxicity of proteins or protein fragments containing long polyglutamine tracts is summarized below.

A. Mice Transgenic for Huntingtin Exon 1

Mice transgenic for the human huntingtin promoter and first exon with a CAG repeat size of approximately 130 has recently been generated [169, 79]. Expression of this transgene produces a very truncated huntingtin fragment ending just downstream of the polyglutamine repeats. RT-PCR studies show that the transgene, when expressed, is observed in all tissues at levels ranging from 31 to 77% of the endogenous level. Western blot analysis confirms protein expression in brain and heart.

Five independent lines generated from a germline chimeric founder each display a complex progressive neurological phenotype that includes motor aspects such as clutching of the feet when picked up by the tail, tremor, handling-induced epileptic seizures, and involuntary stroking of the nose and face and of kicking or scratching with the hind limbs. Shaking of the trunk and a consistent loss of balance when turning, sitting on hind limbs, or reaching around to groom their backs is also observed [169]. Other features of the phenotype include weight loss despite food intake, characteristic vocalizations, urinary incontinence, and atrophied or vestigial reproductive systems that result in female sterility and reduced male fertility [169]. Progression of the phenotype correlates with gene dosage [169].

The overall body weight of the transgenic animals is approximately 30–40% lower than control littermates and the brains of the transgenic animals are approximately 19% smaller than control littermates. Subsequently, loss of brain weight was found to precede loss of body weight during disease progression [79].

Although the mice display a neurological phenotype, there is no obvious morphological evidence of neurodegeneration in the brains of these animals [169]. Rather, a recent detailed neuropathological investigation has shown that the neurons of affected animals are characterized by a remarkable nuclear inclusion with accompanying invaginations of the nuclear membrane and an increase in the nuclear pore density. These inclusions are present only in mice that go on to develop a neurological phenotype and herald the onset of neurological symptoms. The inclusions are never observed in control animals.

The inclusions are restricted to neurons and appear as single, circular regions within the nucleus that are easily distinguished from the nucleolus, sex chromatin, or accessory body of Cajal. The inclusions are slightly larger than the nucleolus. In some homozygous mice, the inclusions contain fibrillar amyloid-like structures [42]. Interestingly, striatal neurons of affected animals are also found to have extensive invaginations of the nuclear membrane and an apparent increase in the nuclear pore density.

The inclusions are immunoreactive against huntingtin, but only with an amino-terminal antibody which strongly suggests that formation of the inclusions is preceded by cleavage of huntingtin. It will be important to determine whether this is indeed the case, by determining the molecular weight of the huntingtin fragment present in the inclusions. It will also be interesting to test whether nuclear localization of the amino-terminal fragment is essential for onset of symptoms. It will be important to know how the fragment enters the nucleus and what it does once it gets there, and to determine whether similar changes occur outside the CNS.

Interestingly, the nuclear inclusions are also immunoreactive against ubiquitin. However, immunoreactivity against ubiquitin appears later than that against huntingtin, suggesting that ubiquitin involvement is a consequence of inclusion formation rather than a cause.

Although several types of transcription factor have polyglutamine tracts within their coding sequences, the nuclear localization of a huntingtin fragment opens up the possibility that transcription of a particular gene set may be altered as part of the pathogenic mechanism as has been suggested earlier by Goldberg *et al.* [170]. Davies *et al.* report that the nuclear inclusions observed in the transgenic mice are not immunoreactive against FosB or NGFI-A, two constitutively expressed neuronal transcription factors. Although these preliminary results suggest that the nuclear inclusions exclude transcription factors, it will nevertheless be important to test for altered gene expression upon the formation of the inclusion body.

Although their significance was not understood at the time, similar inclusions have been detected in neurons of biopsy material from HD patients and more recently from postmortem specimens [49]. These findings underscore the importance of the nuclear inclusions in the neurons of mice transgenic for huntingtin exon 1.

B. Targeted Disruption of the Murine HD Gene

Toxicity of an N-terminal huntingtin fragment containing the polyglutamine expansion was also raised by one of the three groups that generated targeted disruption of the murine HD gene, *Hdh* [12]. Although each group showed that homozygous disruption of huntingtin results in embryonic lethality [12–14], one group reported phenotypic differences in heterozygous animals. Animals heterozygous for a disruption in the huntingtin promoter show no behavioral or neuropathological defects [14]. In contrast, animals heterozygous for a targeted disruption in exon 5 of *Hdh* exhibit impaired performance on neurobehavioral tests [12]. Detailed morphometic analyses demonstrated that these animals have significant and specific neuronal loss in the subthalamic nucleus and globus pallidus [12]. Interestingly, electron micrographs of brain tissue obtained from these animals show neurons with the unmistakable hallmarks of apoptosis not seen in controls. The absence of any phenotype in heterozygous animals with promoter mutations and no possible protein product compared to animals with a targeted mutation in exon 5 and a phenotype suggests that this postulated truncated fragment could be toxic to specific cells.

Although the exon 4 knockout would also be expected to produce a truncated huntingtin fragment, these mice were reported to have no significant neuropathological changes [13]. These mice were not subjected to the detailed morphometric and behavioral analyses as were the exon 5 knockout mice, and it is possible that similar subtle deficits may not have been observed. Moreover, because the exon 4 and exon 5 knockout mice were generated in different genetic backgrounds [13, 12], it is also possible that additional differences between the two strains may account for the phenotypic differences between these animals. Alternatively, there may be true differences between these animals reflecting differences in the effect of truncated fragments of different sizes.

C. Polyglutamine Toxicity in MJD

As described earlier, the experiments performed by Ikeda *et al.* demonstrated that long poly-Q-containing constructs derived from MJD cDNA resulted in cell death in transfected Cos-7 cells [167]. Comparisons of the protein profiles by Western analysis show that lysates of cells transfected with MJD79 have high-molecular-weight smears resistant to boiling in SDS, suggesting that this protein becomes covalently modified and forms insoluble precipitates in dying cells [167]. Immunocytochemistry experiments show that while cells transfected with MJD alleles with shorter CAG lengths have uniform cytoplasmic staining, cells transfected with MJD79 show punctate staining patterns that appear to correlate morphologically with dying cells. The apoptotic effect is dominant, modulated by gene dosage, and requires translation into polyglutamine, as insertion of the CAG tract into the 3′ UTR of the expression plasmid abrogated the apoptotic potential [167].

When assayed *in vivo* under the control of the Purkinje cell-specific L7 promoter, transgenic mice containing a truncated MJD allele with 79 CAG repeats display a clear gait disturbance and ataxic wide-based hindlimb posture [167]. In contrast, neither the transgenic mice with a full-length MJD with 79 CAG repeats or a truncated MJD allele with 35 CAG repeats exhibited an ataxic phenotype. The affected transgenic mice cannot rear and frequently fell down when moving. Analysis of the brain shows severe atrophy of the cerebellum compared to normal littermates with all three layers being affected. In particular, the targeted Purkinje cells have far fewer dendrites and are much reduced in number. The granule cells appear shrunken and sparsely distributed, and the molecular layer is very thin [167]. Because only the Purkinje cells were targeted in this study, the changes in the other two cerebellar layers are thought to be secondary effects resulting from extensive Purkinje cell loss.

It is interesting that the full-length MJD protein with 79 repeats did not result in an ataxic phenotype in mice when targeted to Purkinje cells, whereas extensively truncated MJD alleles containing polyglutamine tracts with 79 repeats clearly promoted cell death in cultured cells and caused ataxia and cerebellar degeneration when specifically expressed in Purkinje cells *in vivo*. Why are the truncated proteins more effective in causing cell death? If toxicity depends on cleavage of the protein and subsequent exposure of the polyglutamine tract, one explanation could be that Purkinje cells lack factors that lead to or regulate the generation of cleaved, toxic fragments of the full-length MJD protein. Another possibility could be that the amino-terminal MJD sequences may limit exposure of the polyglutamine tract to other proteins in the cellular milieu of Purkinje cells and mitigate their toxic effects. In either case, these results support the hypothesis that protein fragments containing expanded polyglutamine tracts result in apoptotic cell death.

D. Polyglutamine Toxicity in *Escherichia coli*

Interestingly, GST–polyglutamine fusion proteins have been found to be toxic in *E. coli.* Onodera *et al.* (136) generated a series of GST fusion proteins containing varying lengths of polyglutamine tracts derived from DRPLA [171]. While expression of GST fusion proteins containing 10–35 glutamines was well tolerated by *E. coli,* expression of GST fusion proteins with 59 glutamines markedly inhibited their rate of growth [171]. Colony-forming assays demonstrated that the reduction in growth rate is caused by an inability to replicate, which is synonymous with cell death in prokaryotes. The toxicity is specific to polyglutamine, since frame-shifted versions of the GST fusion proteins that encode polyalanine tracts do not inhibit bacterial growth. Additionally, toxicity is not due to glutamine depletion since glutamine supplementation failed to rescue the cells.

While postmitotic neurons and rapidly dividing *E. coli* cells are clearly very different, the observation that polyglutamine tracts are toxic in simpler systems opens many possibilities for further studies. Model systems such as bacteria or yeast will likely be very useful in characterizing candidate pathways of polyglutamine-mediated toxicity and for testing the extent to which glutamine repeats act as polar zippers that modify protein–protein interactions [39].

E. Model for the Pathogenesis of HD

A crucial question is whether caspase-3 cleavage of huntingtin is a cause or an effect of neurodegeneration in HD. How can cleavage of huntingtin in cells already committed to apoptosis be involved in the initiation of the neuropathology? One model explaining how huntingtin may be both a caspase substrate and an initiator of apoptosis postulates that huntingtin is involved in a positive feedback loop that results in apoptosis (Fig. 24-6). For example, this model suggests that there is a basal level of caspase activity that may be related to the degree of stress in the cell, perhaps through an age-dependent decline in mitochondrial function. Although this basal caspase activity would be unable to trigger full-blown apoptosis, it may be sufficient to cleave small amounts of substrate. In particular, huntingtin with expanded CAG tracts may be especially vulnerable to cleavage by basal levels of activated caspases. If the cleavage products with expanded polyglutamine tracts are potentially toxic, as has been suggested for mutant huntingtin, even basal levels of cleavage may initiate a positive feedback loop by adding additional stress to the cell, resulting in additional caspase activation, additional cleavage, and eventual apoptosis.

Although this model is not yet fully supported by experimental data, there are several supporting pieces of evidence. For example, huntingtin is a known caspase substrate and alleles with increasing CAG lengths may be more susceptible to cleavage by basal levels of activated caspase-3 [24]. Thus, mutant huntingtin alleles may more easily generate toxic levels of the amino-terminal fragment over the lifetime of the organism. Additionally, the interaction between huntingtin and the ubiquitin-conjugating enzyme suggests that normal turnover of huntingtin via the ubiquitin pathway may be an alternative source of the toxic amino-terminal fragment, thus providing a second route of entry into the positive feedback loop. Finally, the demonstrated phenotypes in mice that express huntingtin fragments lend considerable weight to the toxic fragment hypothesis.

Although the presence of an expanded polyglutamine repeat is the single common factor in the toxic fragment hypothesis (see above), the amino acid sequences surrounding the polyglutamine tract in huntingtin clearly play a role in the toxic effect. Because these sequences contain the binding sites for the huntingtin-interacting proteins described to date, it will be important to determine the effect that interacting proteins have both on the generation of the toxic fragment and on the manifestation of its toxic effect.

Toxicity of the amino-terminal fragment of huntingtin may depend on its subcellular localization. For example, the fragment may result in toxicity only upon import into the nucleus and formation of a nuclear inclusion body. Understanding the process by which this inclusion body forms and the cellular consequences of its formation will clearly be an important area of investigation for the near future. Alternatively, toxicity of the amino-terminal fragment of huntingtin may be evident only with polyglutamine expansion. It is also possible that normal turnover of huntingtin by proteases, such as those involved in the ubiqutin pathway, may generate huntingtin fragments with toxic potential.

It will also be essential to define the age-related factors that contribute to pathogenesis. For example, age-dependent mitochondrial dysfunction is known to add stress to the cell [159, 154, 156, 172]. This age-related oxidative stress may contribute to initiation or propagation of the feedback loop by decreasing the threshold at which apoptosis is initiated.

Finally, it will also be important to determine the extent to which this model extends to the other diseases caused by polyglutamine expansion. Certainly, there is preliminary evidence that other proteins containing polyglutamine repeats are substrates for caspases. It will be interesting to determine whether, similar to huntingtin, there is an altered subcellular distribution for the fragments generated by caspase cleavage. As well, for each of these other polyglutamine expansion proteins, future research will no doubt elucidate the roles that their interacting proteins play in regional specificity of pathogenesis and in the pathogenic mechanism.

Testing the many facets of this hypothesis will require the development of sensitive *in vitro* and *in vivo* models of HD and other diseases associated with polyglutamine expansion. *In vitro* models will allow elucidation of the pathways leading to the generation of toxic fragments and of the biochemical mechanisms of toxicity. This information can then be used in the design of appropriate modeling of disease mechanism and disease treatment *in vivo*.

F. The Route to Therapy

If the hypothesis that huntingtin alleles with expanded polyglutamine tracts may contribute to the initiation of apoptotic death in selected neurons is validated, several novel therapeutic approaches can be envisioned. For example, agents designed to block caspase activation may be developed and tested for their efficacy in preventing neuronal cell loss in HD patients [173]. One potential caveat to this line of potential treatments may be that although the neurons may not die, they may remain under considerable stress and as such their functioning may not be completely normal. A second route to novel therapies may make use of knowledge gained from the study of huntingtin and its interacting proteins by counteracting the biochemical effects of CAG expansion. Because such agents will act well upstream of the apoptotic cascade, they may be more likely to result in normal neuronal function than agents designed to block the end stages of apoptosis.

XIII. CONCLUSION

There is still much to learn about how huntingtin with expanded polyglutamine tracts results in loss of particular neurons in the brain. One major area of investigation will focus on common pathways leading to cell death for diseases associated with polyglutamine expansion. We need to understand the biochemical ways in which long polyglutamine tracts lead to cell death by developing good *in vitro* and *in vivo* systems with which to probe the fundamental molecular consequences of polyglutamine expansion. Superimposed on these mechanistic questions is the need to understand the distinct patterns of regional specificity of neurodegeneration for each disease. Experiments to address these questions will rely heavily on technologies using *in vitro* systems and transgenic and knockout animals for each member of this fascinating group of diseases.

References

1. Hayden, M. (1981). "Huntington's Chorea." Springer-Verlag, Berlin.
2. Harper, P. (1991). "Huntington's Disease." W. B. Saunders, London.
3. Ambrose, C. M., Duyao, M. P., Barnes, G., Bates, G. P., Lin, C. S., Srinidhi, J., Baxendale, S., Hummerich, H., Lehrach, H., Altherr, M., Wasmuth, J., Buckler, A., Church, D., Housman, D., Berks, M., Micklem, G., Durbin, R., Dodge, A., Read, A., Gusella, J., and MacDonald, M. E. (1994). Structure and expression of the Huntington's disease gene: evidence against simple inactivation due to an expanded CAG repeat. *Somat. Cell Mol. Genet.* **20**(1), 27–38.
4. Lin, B., Rommens, J. M., Graham, R. K., Kalchman, M., MacDonald, H., Nasir, J., Delaney, A., Goldberg, Y. P., and Hayden, M. R. (1993). Differential 3′ polyadenylation of the Huntington disease gene results in two mRNA species with variable tissue expression. *Hum. Mol. Genet.* **2**(10), 1541–1545.
5. Lin, B., Nasir, J., Kalchman, M. A., McDonald, H., Zeisler, J., Goldberg, Y. P., and Hayden, M. R. (1995). Structural analysis of the 5′ region of mouse and human Huntington disease genes reveals conservation of putative promoter region and di- and trinucleotide polymorphisms. *Genomics* **25**(3), 707–715.
6. Andrade, M. A., and Bork, P. (1995). HEAT repeats in the Huntington's disease protein. *Nature Genet.* **11**(2), 115–116.
7. Bessert, D. A., Gutridge, K. L., Dunbar, J. C., and Carlock, L. R. (1995). The identification of a functional nuclear localization signal in the Huntington disease protein. *Mol. Brain Res.* **33**(1), 165–173.
8. Lin, B., Nasir, J., MacDonald, H., Hutchinson, G., Graham, R. K., Rommens, J. M., and Hayden, M. R. (1994). Sequence of the murine Huntington disease gene: evidence for conservation, alternate splicing and polymorphism in a triplet (CCG) repeat. *Hum. Mol. Genet.* **3**(1), 85–92.
9. Barnes, G. T., Duyao, M. P., Ambrose, C. M., McNeil, S., Persichetti, F., Srinidhi, J., Gusella, J. F., and MacDonald, M. E. (1994). Mouse Huntington's disease gene homolog (Hdh). *Somat. Cell Mol. Genet.* **20**(2), 87–97.
10. Schmitt, I., Bachner, D., Megow, D., Henklein, P., Hameister, H., Epplen, J. T., and Riess, O. (1995). Expression of the Huntington disease gene in rodents: cloning the rat homologue and evidence for downregulation in non-neuronal tissues during development. *Hum. Mol. Genet.* **4**(7), 1173–1182.
11. Baxendale, S., Abdulla, S., Elgar, G., Buck, D., Berks, M., Micklem, G., Durbin, R., Bates, G., Brenner, S., Beck, S., and Leharch, H. (1995). Comparative sequence analysis of the human and pufferfish Huntington's disease genes. *Nature Genet.* **10**(1), 67–76.
12. Nasir, J., Floresco, S. B., O'Kusky, J. R., Diewert, V. M., Richman, J. M., Zeisler, J., Borowski, A., Marth, J. D., Phillips, A. G., and Hayden, M. R. (1995). Targeted disruption of the Huntington's disease gene results in embryonic lethality and behavioral and morphological changes in heterozygotes. *Cell* **81**(5), 811–823.

13. Duyao, M. P., Auerbach, A. B., Ryan, A., Persichetti, F., Barnes, G. T., McNeil, S. M., Ge, P., Vonsattel, J. P., Gusella, J. F., Joyner, A. L., and MacDonald, M. E. (1995). Inactivation of the mouse Huntington's disease gene homolog Hdh. *Science* **269**(5222), 407–410.
14. Zeitlin, S., Liu, J. P., Chapman, D. L., Papaioannou, V. E., and Efstratiadis, A. (1995). Increased apoptosis and early embryonic lethality in mice nullizygous for the Huntington's disease gene homologue. *Nature Genet.* **11**(2), 155–163.
15. HDCRG. (1993). A novel gene containing a trinucleotide repeat that is expanded and unstable on Huntington's disease chromosomes. The Huntington's Disease Collaborative Research Group. *Cell* **72**(6), 971–983.
16. Kremer, B., Goldberg, P., Andrew, S. E., Theilmann, J., Telenius, H., Zeisler, J., Squitieri, F., Lin, B., Bassett, A., Almqvist, E., Bird, T. D., and Hayden, M. R. (1994). A worldwide study of the Huntington's disease mutation. The sensitivity and specificity of measuring CAG repeats. *N. Engl. J. Med.* **330**(20), 1401–1406.
17. Andrew, S. E., Goldberg, Y. P., Kremer, B., Telenius, H., Theilmann, J., Adam, S., Starr, E., Squitieri, F., Lin, B., Kalchman, M. A., Graham, R. K., and Hayden, M. R. (1993). The relationship between trinucleotide (CAG) repeat length and clinical features of Huntington's disease. *Nature Genet.* **4**(4), 398–403.
18. Brinkman, R. R., Mezei, M. M., Theilmann, J., Almqvist, E., and Hayden, M. R. (1997). The likelihood of being affected with Huntington disease by a particular age, for a specific CAG size. *Am. J. Hum. Genet.* **60**(5), 1202–1210.
19. Goldberg, Y. P., Kalchman, M. A., Metzler, M., Nasir, J., Zeisler, J., Graham, R., Koide, H. B., O'Kusky, J., Sharp, A. H., Ross, C. A., Jirik, F., and Hayden, M. R. (1996). Absence of disease phenotype and intergenerational stability of the CAG repeat in transgenic mice expressing the human Huntington disease transcript. *Hum. Mol. Genet.* **5**(2), 177–185.
20. Trottier, Y., Devys, D., Imbert, G., Saudou, F., An, I., Lutz, Y., Weber, C., Agid, Y., Hirsch, E. C., and Mandel, J. L. (1995). Cellular localization of the Huntington's disease protein and discrimination of the normal and mutated form. *Nature Genet.* **10** 1), 104–110.
21. DiFiglia, M., Sapp, E., Chase, K., Schwarz, C., Meloni, A., Young, C. Martin, E., Vonsattel, J. P., Carraway, R., Reeves, S. A., Boyce, F. M., and Aronin, N. (1995). Huntingtin is a cytoplasmic protein associated with vesicles in human and rat brain neurons. *Neuron* **14**(5), 1075–1081.
22. Gutekunst, C. A., Levey, A. I., Heilman, C. J., Whaley, W. L., Yi, H., Nash, N. R., Rees, H. D., Madden, J. J., and Hersch, S. M. (1995). Identification and localization of huntingtin in brain and human lymphoblastoid cell lines with anti-fusion protein antibodies. *Proc. Natl. Acad. Sci. USA* **92**(19), 8710–8714.
23. Sharp, A. H., Loev, S. J., Schilling, G., Li, S. H., Li, X. J., Bao, J., Wagster, M. V., Kotzuk, J. A., Steiner, J. P., Lo, A., Hedreen, J., Sisodia, S., Snyder, S. H., Dawson, T. M., Ryugo, D. K., and Ross, C. A. (1995). Widespread expression of Huntington's disease gene (IT15) protein product. *Neuron* **14**(5), 1065–1074.
24. Goldberg, Y. P., Nicholson, D. W., Rasper, D. M., Kalchman, M. A., Koide, H. B., Graham, R. K., Bromm, M., Kazemi-Esfarjani, P., Thornberry, N. A., Vaillancourt, J. P., and Hayden, M. R. (1996). Cleavage of huntingtin by apopain, a proapoptotic cysteine protease, is modulated by the polyglutamine tract. *Nature Genet.* **13**(4), 442–449.
25. Trottier, Y., Lutz, Y., Stevanin, G., Imbert, G., Devys, D., Cancel, G., Saudou, F., Weber, C., David, G., Tora, L., Agid, Y., Brice, A., and Mandel, J.-L. (1995). Polyglutamine expansion as a pathological epitope in Huntington's disease and four dominant cerebellar ataxias. *Nature* **378**(6555), 403–406.
26. Green, H., and Wang, N. (1994). Codon reiteration and the evolution of proteins. *Proc. Natl. Acad. Sci. USA* **91**(10), 4298–4302.
27. Seipel, K., Georgiev, O., Gerber, H. P., and Schaffner, W. (1994). Basal components of the transcription apparatus (RNA polymerase II, TATA-binding protein) contain activation domains: is the repetitive C-terminal domain (CTD) of RNA polymerase II a "portable enhancer domain"? *Mol. Reprod. Dev.* **39**(2), 215–225.
28. Yokota, T., Arai, N., Lee, F., Rennick, D., Mosmann, T., and Arai, K. (1985). Use of a cDNA expression vector for isolation of mouse interleukin 2 cDNA clones: expression of T-cell growth-factor activity after transfection of monkey cells. *Proc. Natl. Acad. Sci. USA* **82**(1), 68–72.
29. Taniguchi, T., Matsui, H., Fujita, T., Takaoka, C., Kashima, N., Yoshimoto, R., and Hamuro, J. (1983). Structure and expression of a cloned cDNA for human interleukin-2. *Nature* **302**(5906), 305–310.
30. Hoffman, A., Sinn, E., Yamamoto, T., Wang, J., Roy, A., Horikoshi, M., and Roeder, R. G. (1990). Highly conserved core domain and unique N terminus with presumptive regulatory motifs in a human TATA factor (TFIID). *Nature* **346**(6282), 387–390.
31. Tamura, T., Sumita, K., Fujino, I., Aoyama, A., Horikoshi, M., Hoffmann, A., Roeder, R. G., Muramatsu, M., and Mikoshiba, K. (1991). Striking homology of the 'variable' N-terminal as well as the 'conserved core' domains of the mouse and human TATA-factors (TFIID). *Nucleic Acids Res.* **19**(14), 3861–3865.
32. Chang, C., and Kokontis, J. (1988). Identification of a new member of the steroid receptor super-family by cloning and sequence analysis. *Biochem. Biophys. Res. Commun.* **155**(2), 971–977.
33. Lubahn, D. B., Joseph, D. R., Sar, M., Tan, J., Higgs, H. N., Larson, R. E., French, F. S., and Wilson, E. M. (1988). The human androgen receptor: complementary deoxyribonucleic acid cloning, sequence analysis and gene expression in prostate. *Mol. Endocrinol.* **2**(12), 1265–1275.
34. Whitfield, L. S., Lovell-Badge, R., and Goodfellow, P. N. (1993). Rapid sequence evolution of the mammalian sex-determining gene SRY. *Nature* **364**(6439), 713–715.
35. Tucker, P. K., and Lundrigan, B. L. (1993). Rapid evolution of the sex determining locus in Old World mice and rats. *Nature* **364**(6439), 715–717.
36. Simental, J. A., Sar, M., Lane, M. V., French, F. S., and Wilson, E. M. (1991). Transcriptional activation and nuclear targeting signals of the human androgen receptor. *J. Biol. Chem.* **266**(1), 510–518.
37. Mhatre, A. N., Trifiro, M. A., Kaufman, M., Kazemi-Esfarjani, P., Figlewicz, D., Rouleau, G., and Pinsky, L. (1993). Reduced transcriptional regulatory competence of the androgen receptor in X-linked spinal and bulbar muscular atrophy. *Nature Genet.* **5**(2), 184–188.
38. Green, H. (1993). Human genetic diseases due to codon reiteration: relationship to an evolutionary mechanism. *Cell* **74**(6), 955–956.
39. Perutz, M. F., Johnson, T., Suzuki, M., and Finch, J. T. (1994). Glutamine repeats as polar zippers: their possible role in inherited neurodegenerative diseases. *Proc. Natl. Acad. Sci. USA* **91**(12), 5355–5358.
40. Perutz, M. (1994). Polar zippers: their role in human disease. *Protein Sci.* **3**(10), 1629–1637.
41. Stott, K., Blackburn, J. M., Butler, P. J., and Perutz, M. (1995). Incorporation of glutamine repeats makes protein oligomerize: implications for neurodegenerative diseases. *Proc. Natl. Acad. Sci. USA* **92**(14), 6509–6513.

42. Scherzinger, E., Lurz, R., Turmaine, M., Mangiarini, L., Hollenbach, B., Hasenbank, R., Bates, G. P., Davies, S. W., Lehrach, H., and Wanker, E. E. (1997). Huntingtin-encoded polyglutamine expansion form amyloid-like protein aggregates in vitro and in vivo. *Cell* **90,** 549–558.
43. Jones, A. L., Wood, J. D., and Harper, P. S. (1997). Huntington-disease - advances in molecular and cell biology. *J. Inherited Metabol. Dis.* **20**(2), 125–138.
44. Schulze, H., Schuler, A., Stuber, D., Dobeli, H., Langen, H., and Huber, G. (1993). Rat brain glyceraldehyde-3-phosphate dehydrogenase interacts with the recombinant cytoplasmic domain of Alzheimer's beta-amyloid precursor protein. *J.Neurochem.* **60**(5), 1915–1922.
45. Andrew, S. E., Goldberg, Y. P., Kremer, B., Squitieri, F., Theilmann, J., Zeisler, J., Telenius, H., Adam, S., Almquist, E., Anvret, M., Lucotte, G., Stoessl, A. J., Campanella, G., and Hayden, M. R. (1994). Huntington disease without CAG expansion: phenocopies or errors in assignment? *Am. J. Hum. Genet.* **54**(5), 852–863.
46. Mundlos, S., Otto, F., Mundlos, C., Mulliken, J. B., Alysworth, A. S., Albright, S., Lindhout, D., Cole, W. G., Henn, W., Knoll, J. H. M., Owen, M. J., Mertelsmann, R., Zabel, B. U., and Olsen, B. R. (1997). Mutations involving the transcription factor CBFA1 cause cleidocranial dysplasia. *Cell* **89,** 773–779.
47. Wexler, N. S., Young, A. B., Tanzi, R. E., Travers, H., Starosta-Rubinstein, S., Penney, J. B., Snodgrass, S. R., Shoulson, I., Gomez, F., Ramos Arroyo, M. A. R., Penchaszadeh, G. K., Moreno, H., Gibbons, K., Faryniarz, A., Hobbs, W., Anderson, M. A., Bonilla, E., Conneally, P. M., and Gusella, J. F. (1987). Homozygotes for Huntington's disease. *Nature* **326**(6109), 194–197.
48. Myers, R. H., Leavitt, J., Farrer, L. A., Jagadeesh, J., McFarlane, H., Mastromauro, C. A., Mark, R. J., and Gusella, J. F. (1989). Homozygote for Huntington disease. *Am. J. Hum. Genet.* **45**(4), 615–618.
49. DiFiglia, M., Sapp, E., Chase, K. O., Davies, S. W., Bates, G. P., Vonsattel, J. P., and Aronin, N. (1997). Aggregation of huntingtin in neuronal intranuclear inclusions and dystrophic neurites in brain. *Science.* [in press]
50. Hayden, M., Kremer B. (1995) *In* "The Metabolic and Molecular Bases of Inherited Disease" (C. R. Scriver, A. L. Beaudet, W. S. Sly, and D. Valle, Eds.), Vol. III, 27th ed., pp. 4483–4510. McGraw-Hill, New York.
51. Vonsattel, J. P., Myers, R. H., Stevens, T. J., Ferrante, R. J., Bird, E. D., and Richardson, E., Jr. (1985). Neuropathological classification of Huntington's disease. *J. Neuropathol. Exp. Neurol.* **44**(6), 559–577.
52. DiFiglia, M. (1990). Excitotoxic injury of the neostriatum: a model for Huntington's disease. *Trends Neurosci.* **13**(7), 286–289.
53. Graveland, G. A., Williams, R. S., and DiFiglia, M. (1985). Evidence for degenerative and regenerative changes in neostriatal spiny neurons in Huntington's disease. *Science* **227**(4688), 770–773.
54. Kremer, H. P., Roos, R. A., Dingjan, G., Marani, E., and Bots, G. T. (1990). Atrophy of the hypothalamic lateral tuberal nucleus in Huntington's disease. *J. Neuropathol. Exp. Neurol.* **49**(4), 371–382.
55. Zech, M., Roberts, G. W., Bogerts, B., Crow, T. J., and Polak, J. M. (1986). Neuropeptides in the amygdala of controls, schizophrenics and patients suffering from Huntington's chorea: an immunohistochemical study. *Acta Neuropathol.* **71**(3–4), 259–266.
56. Takano, H., Onodera, O., Takahashi, H., Igarashi, S., Yamada, M., Oyake, M., Ikeuci, T., Koide, R., Tanaka, H., Iwabuchi, K., and Tsuji, S. (1996). Somatic mosaicism of expanded CAG repeats in brains of patients with dentatorubral-pallidoluysian atrophy - cellular population-dependent dynamics of mitotic instability. *Am. J. Hum. Genet.* **58**(6), 1212–1222.
57. Telenius, H., Kremer, B., Goldberg, Y. P., Theilmann, J., Andrew, S. E., Zeisler, J., Adam, S., Greenberg, C., Ives, E. J., Clarke, L. A., and Hayden, M. R. (1994). Somatic and gonadal mosaicism of the Huntington disease gene CAG repeat in brain and sperm. *Nature Genet.* **6**(4), 409–414.
58. Beal, M. F. (1995). Aging, energy, and oxidative stress in neurodegenerative diseases. *Ann. Neurol.* **38**(3), 357–366.
59. Coyle, J. T., and Puttfarcken, P. (1993). Oxidative stress, glutamate, and neurodegenerative disorders. *Science* **262**(5134), 689–695.
60. Strong, T. V., Tagle, D. A., Valdes, J. M., Elmer, L. W., Boehm, K., Swaroop, M., Kaatz, K. W., Collins, F. S., and Albin, R. L. (1993). Widespread expression of the human and rat Huntington's disease gene in brain and nonneural tissues. *Nature Genet.* **5**(3), 259–265.
61. Li, S. H., Schilling, G., Young, W. D., Li, X. J., Margolis, R. L., Stine, O. C., Wagster, M. V., Abbott, M. H., Franz, M. L., Ranen, N. G., Folstein, S. E., Hedreen, J. C., and Ross, C. A. (1993). Huntington's disease gene (IT15) is widely expressed in human and rat tissues. *Neuron* **11**(5), 985–993.
62. Landwehrmeyer, G. B., McNeil, S. M., Dure, L. t., Ge, P., Aizawa, H., Huang, Q., Ambrose, C. M., Duyao, M. P., Bird, E. D., Bonilla, E., de Young, M., Avila-Gonzales, A. J., Wexler, N. S., MacDonald, M. D., Penney, J. B., Young, A. B., and Vonsattel, J.-P. (1995). Huntington's disease gene: regional and cellular expression in brain of normal and affected individuals. *Ann. Neurol.* **37**(2), 218–230.
63. Ferrante, R. J., Gutekunst, C. A., Persichetti, F., McNeil, S. M., Kowall, N. W., Gusella, J. F., MacDonald, M. E., Beal, M. F., and Hersch, S. M. (1997). Heterogeneous topographic and cellular distribution of huntingtin expression in the normal human neostriatum. *J. Neurosci.* **17**(9), 3052–3063.
64. Kosinski, C. M., Young, A. B., Persichetti, F., MacDonald, M., Gusella, J. F., Penny, J. B., Jr., and Standaert, D. G. (1997). Huntingtin immunoreactivity in the rat neostriatum: differential accumulation in projection and interneurons. *Exp. Neurol.* **144,** 239–247.
65. Hedreen, J. C., and Folstein, S. E. (1995). Early loss of neostriatal striosome neurons in Huntington's disease. *J. Neuropathol. Exp. Neurol.* **54**(1), 105–120.
66. Bhide, P. G., Day, M., Sapp, E., Schwarz, C., Sheth, A., Kim, J., Young, A. B., Penney, J., Golden, J., Aronin, N., and DiFiglia, M. (1996). Expression of normal and mutant huntingtin in the developing brain. *J. Neurosci.* **16**(17), 5523–5535.
67. Aronin, N., Chase, K., Young, C., Sapp, E., Schwarz, C., Matta, N., Kornreich, R., Landwehrmeyer, B., Bird, E., Beal, M. F., Vonsattel, J.-P., Smith, T., Carraway, R., Boyee, F. M., Young, A. B., Penney, J. B., and DiFiglia, M. (1995). CAG expansion affects the expression of mutant Huntingtin in the Huntington's disease brain. *Neuron* **15**(5), 1193–1201.
68. Persichetti, F., Ambrose, C. M., Ge, P., McNeil, S. M., Srinidhi, J., Anderson, M. A., Jenkins, B., Barnes, G. T., Duyao, M. P., Kanaley, L., Wexler, N. S., Myers, R. H., Bird, E. D., Vonsattel, J.-P., MacDonald, M. E., and Gusella, J. F. (1995). Normal and expanded Huntington's disease gene alleles produce distinguishable proteins due to translation across the CAG repeat. *Mol. Med.* **1**(4), 374–383.
69. Persichetti, F., Carlee, L., Faber, P. W., Mcneil, S. M., Ambrose, C. M., Srinidhi, J., Anderson, M., Barnes, G. T., Gusella, J. F., and Macdonald, M. E. (1996). Differential expression of normal

and mutant huntingtons disease gene alleles. *Neurobiol. Dis.* **3**(3), 183–190.
70. Ide, K., Nukina, N., Masuda, N., Goto, J., and Kanazawa, I. (1995). Abnormal gene product identified in Huntington's disease lymphocytes and brain. *Biochem. Biophys. Res. Commun.* **209**(3), 1119–1125.
71. Hoogeveen, A. T., Willemsen, R., Meyer, N., de Rooij, K. E., Roos, R. A., van Ommen, G. J., and Galjaard, H. (1993). Characterization and localization of the Huntington disease gene product. *Hum. Mol. Genet.* **2**(12), 2069–2073.
72. De Rooij, K. E., Dorsman, J. C., Smoor, M. A., Den Dunnen, J. T. and Van Ommen, G. J. (1996). Subcellular localization of the Huntington's disease gene product in cell lines by immunofluorescence and biochemical subcellular fractionation. *Hum. Mol. Genet.* **5**(8), 1093–1099.
73. Wood, J. D., Macmillan, J. C., Harper, P. S., Lowenstein, P. R., and Jones, A. L. (1996). Partial characterisation of murine huntingtin and apparent variations in the subcellular localisation of huntingtin in human, mouse and rat brain. *Hum. Mol. Genet.* **5**(4), 481–487.
74. Sapp, E., Schwarz, C., Chase, K., Bhide, P., Young, A. B., Penny, J., Vonsattel, L. P., Aronin, N., DiFiglia, M. (1996). Altered neuronal expression and intracellular trafficking of huntingtin in the huntington's disease brain. *Soc. Neurosci. Abst.* **22,** 226.
75. Velier, J., Schwarz, C., Young, C., Fallon, J., Hyman, B., Martin, E. ..., Hughes, S., Vallee, R., Aronin, N., DiFiglia, M. (1996). Wildtype and mutant huntingtin localize to the golgi complex and to the vesicles in the peripheral cytoplasm in fibroblasts of control and HD patients. *Soc. Neurosci. Abst.* **22,** 226.
76. Sapp, E., Schwarz, C., Chase, K., Bhide, P. G., Young, A. B., Penney, J., Vonsattel, J. P., Aronin, N., and DiFiglia, M. (1997). Huntingtin localization in brains of normal and Huntington's disease patients. *Ann. Neurol.* **43,** 604–612.
77. Kalchman, M. A., Koide, H. B., McCutcheon, K., Graham, R. K., Nichol, K., Nishiyama, K., Kazemi-Esfarjani, P., Lynn, F. C., Wellington, C., Metzler, M., Goldberg, Y. P., Kanazawa, I., Gietz, R. D., and Hayden, M. R. (1997). HIP1, a human homologue of S. cerevisiae Sla2p, interacts with membrane-associated huntingtin in the brain. *Nature Genet.* **16**(1), 44–53.
78. Li X. J., Li, S. H., Sharp, A. H., Nucifora, F., Jr., Schilling, G., Lanahan, A., Worley, P., Snyder, S. H., and Ross, C. A. (1995). A huntingtin-associated protein enriched in brain with implications for pathology. *Nature* **378**(6555), 398–402.
79. Davies, S., Turmaine, M., Cozens, B. A., DiFiglia, M., Sharp, A. H., Ross, C. A., Scherzinger, E., Wanker, E. E., Mangiarini, L., and Bates, G. P. (1997). Formation of neuronal intranuclear inclusions (NII) underlies the neurological dysfunction in mice transgenic for the HD mutation. *Cell* **90,** 537–548.
80. Kalchman, M. A., Graham, R. K., Xia, G., Koide, H. B., Hodgson, J. G., Graham, K. C., Goldberg, Y. P., Gietz, R. D., Pickart, C. M., and Hayden, M. R. (1996). Huntingtin is ubiquitinated and interacts with a specific ubiquitin-conjugating enzyme. *J. Biol. Chem.* **271**(32), 19385–19394.
81. Burke, J. R., Enghild, J. J., Martin, M. E., Jou, Y. S., Myers, R. M., Roses, A. D., Vance, J. M., and Strittmatter, W. J. (1996). Huntingtin and DRPLA proteins selectively interact with the enzyme GAPDH. *Nature Med.* **2**(3), 347–350.
82. Wanker, E. E., Rovira, C., Scherzinger, E., Hasenbank, R., Walter, S., Tait, D., Colicelli, J., and Lehrach, H. (1997). HIP-I: a huntingtin interacting protein isolated by the yeast two-hybrid system. *Hum. Mol. Genet.* **6**(3), 487–495.
83. Bao, J., Sharp, A. H., Wagster, M. V., Becher, M., Schilling, G., Ross, C. A., Dawson, V. L., and Dawson, T. M. (1996). Expansion of polyglutamine repeat in huntingtin leads to abnormal protein interactions involving calmodulin. *Proc. Natl. Acad. Sci. USA* **93**(10), 5037–5042.
84. Liu, Y. F., Deth, R. C., and Devys, D. (1997). SH3 domain-dependent association of huntingtin with epidermal growth factor receptor signaling complexes. *J. Biol. Chem.* **272**(13), 8121–8124.
85. Holtzman, D. A., Yang, S., and Drubin, D. G. (1993). Synthetic-lethal interactions identify two novel genes, SLA1 and SLA2, that control membrane cytoskeleton assembly in Saccharomyces cerevisiae. *J. Cell Biol.* **122**(3), 635–644.
86. Munn, A. L., Stevenson, B. J., Geli, M. I., and Riezman, H. (1995). end5, end6, and end7: mutations that cause actin delocalization and block the internalization step of endocytosis in Saccharomyces cerevisiae. *Mol. Biol. Cell* **6**(12), 1721–1742.
87. Na, S., Hincapie, M., McCusker, J. H., and Haber, J. E. (1995). MOP2 (SLA2) affects the abundance of the plasma membrane H(+)-ATPase of Saccharomyces cerevisiae. *J. Biol. Chem.* **270**(12), 6815–6823.
88. Dawson, V. L., Kizushi, V. M., Huang, P. L., Snyder, S. H., and Dawson, T. M. (1996). Resistance to neurotoxicity in cortical cultures from neuronal nitric oxide synthase-deficient mice. *J. Neurosci.* **16**(8), 2479–2487.
89. Schulz, J. B., Matthews, R. T., Jenkins, B. G., Ferrante, R. J., Siwek, D., Henshaw, D. R., Cipolloni, P. B., Mecocci, P., Kowall, N. W., Rosen, B. R., and Beal, M. F. (1995). Blockade of neuronal nitric oxide synthase protects against excitotoxicity in vivo. *J. Neurosci.* **15**(12), 8419–8429.
90. Behrens, M. I., Koh, J. Y., Muller, M. C., and Choi, D. W. (1996). Nadph diaphorase-containing striatal or cortical neurons are resistant to apoptosis. *Neurobiol. Dis.* **3**(1), 72–75.
91. Chen, Z. J., Niles, E. G., and Pickart, C. M. (1991). Isolation of a cDNA encoding a mammalian multiubiquitinating enzyme (E225K) and overexpression of the functional enzyme in Escherichia coli. *J. Biol. Chem.* **266**(24), 15698–15704.
92. Schiffer, D., Attanasio, A., Chio, A., Migheli, A., and Pezzulo, T. (1994). Ubiquitinated dystrophic neurites suggest corticospinal derangement in patients with amyotrophic lateral sclerosis. *Neurosci. Lett.* **180**(1), 21–24.
93. Sugiyama, H., Hainfellner, J. A., Yoshimura, M., and Budka, H. (1994). Neocortical changes in Parkinson's disease, revisited. *Clin. Neuropathol.* **13**(2), 55–59.
94. Taddei, N., Liguri, G., Sorbi, S., Amaducci, L., Camici, G., Nassi, P., Cecchi, C., and Ramponi, G. (1993). Cerebral soluble ubiquitin is increased in patients with Alzheimer's disease. *Neurosci. Lett.* **151**(2), 158–161.
95. Cammarata, S., Caponnetto, C., and Tabaton, M. (1993). Ubiquitin-reactive neurites in cerebral cortex of subjects with Huntington's chorea: a pathological correlate of dementia? *Neurosci. Lett.* **156**(1–2), 96–98.
96. Fox, J., and Dandliker, W. B. (1956). A study of some physiological properties of glyceraldehyde-3-phosphate dehydrogenase. *J. Biol. Chem.* **218,** 53–57.
97. Harris, J., and Perham, R. N. (1963). Studies on glyceraldehyde-3-phosphate dehydrogenases. *Biochem. J.* **89,** 60–64.
98. Meyer-Siegler, K., Rahman-Mansur, N., Wurzer, J. C., and Sirover, M. A. (1992). Proliferative dependent regulation of the glyceraldehyde-3-phosphate dehydrogenase/uracil DNA glycosylase gene in human cells. *Carcinogenesis* **13**(11), 2127–2132.
99. Ronai, Z. (1993). Glycolytic enzymes as DNA binding proteins. *Int. J. Biochem.* **25,** 1073–1076.
100. Huitorel, P. D. (1985). Bundling of microtubules by glyceraldehyde-3-phosphate dehydrogenase and its modulation by ATP. *Eur. J. Biochem.* **150,** 265–269.

101. Somers, M., Engelborghs, Y., and Baert, J. (1990). Analysis of the binding of glyceraldehyde-3-phosphate dehydrogenase to microtubules, the mechanism of bundle formation and the linkage effect. *Eur. J. Biochem.* **193**(2), 437–444.
102. Mejean, C., Pons, F., Benyamin, Y., and Roustan, C. (1989). Antigenic probes locate binding sites for the glycolytic enzymes glyceraldehyde-3-phosphate dehydrogenase, aldolase and phosphofructokinase on the actin monomer in microfilaments. *Biochem. J.* **264,** 671–677.
103. Nagy, E., and Rigby, W. F. (1995). Glyceraldehyde-3-phosphate dehydrogenase selectively binds AU-rich RNA in the NAD(+)-binding region (Rossmann fold). *J. Biol. Chem.* **270**(6), 2755–2763.
104. Singh, R., and Green, M. R. (1993). Sequence-specific binding of transfer RNA by glyceraldehyde-3-phosphate dehydrogenase. *Science* **259**(5093), 365–368.
105. Koshy, B., Matilla, T., Burright, E. N., Merry, D. E., Fischbeck, K. H., Orr, H. T., and Zoghbi, H. Y. (1996). Spinocerebellar ataxia type-1 and spinobulbar muscular atrophy gene products interact with glyceraldehyde-3-phosphate dehydrogenase. *Hum. Mol. Genet.* **5**(9), 1311–1318.
106. Koroshetz, W. J., Jenkins, B. G., Rosen, B. R., and Beal, M. F. (1997). Energy metabolism defects in Huntington's disease and effects of coenzyme Q10. *Ann. Neurol.* **41**(2), 160–165.
107. Gu, M., Gash, M. T., Mann, V. M., Javoy-Agid, F., Cooper, J. M., and Schapira, A. H. (1996). Mitochondrial defect in Huntington's disease caudate nucleus. *Ann. Neurol.* **39**(3), 385–389.
108. Matsuishi, T., Sakai, T., Nagamitsu, S., Shoji, H., Ueda, N., Kaneko, S., Kano, T., Iwashita, H., and Kato, H. (1996). Decreased cerebrospinal fluid levels of substance p in Machado-Joseph disease. *J. Neurol. Sci.* **142**(1–2), 107–110.
109. Browne, S. E., Bowling, A. C., Macgarvey, U., Baik, M. J., Berger, S. C., Muqit, M., Bird, E. D., and Beal, M. F. (1997). Oxidative damage and metabolic dysfunction in huntingtons-disease - selective vulnerability of the basal ganglia. *Ann. Neurol.* **41**(5), 646–653.
110. Lowenstein, E. J., Daly, R. J., Batzer, A. G., Li, W., Margolis, B., Lammers, R., Ullrich, A., Skolnik, E. Y., Bar-Sagi, D., and Schlessinger, J. (1992). The SH2 and SH3 domain-containing protein GRB2 links receptor tyrosine kinases to ras signaling. *Cell* **70**(3), 431–442.
111. Egan, S. E., Giddings, B. W., Brooks, M. W., Buday, L., Sizeland, A. M., and Weinberg, R. A. (1993). Association of Sos Ras exchange protein with Grb2 is implicated in tyrosine kinase signal transduction and transformation. *Nature* **363**(6424), 45–51.
112. Ellis, C., Moran, M., McCormick, F., and Pawson, T. (1990). Phosphorylation of GAP and GAP-associated proteins by transforming and mitogenic tyrosine kinases. *Nature* **343**(6256), 377–381.
113. Chen, C. Y., and Faller, D. V. (1996). Phosphorylation of Bcl-2 protein and association with p21Ras in Ras-induced apoptosis. *J. Biol. Chem.* **271**(5), 2376–2379.
114. Wang, H. G., Millan, J. A., Cox, A. D., Der, C. J., Rapp, U. R., Beck, T., Zha, H., and Reed, J. C. (1995). R-Ras promotes apoptosis caused by growth factor deprivation via a Bcl-2 suppressible mechanism. *J. Cell Biol.* **129**(4), 1103–1114.
115. Henkemeyer, M., Rossi, D. J., Holmyard, D. P., Puri, M. C., Mbamalu, G., Harpal, K., Shih, T. S., Jacks, T., and Pawson, T. (1995). Vascular system defects and neuronal apoptosis in mice lacking ras GTPase-activating protein. *Nature* **377**(6551), 695–701.
116. Dragunow, M., Faull, R. L., Lawlor, P., Beilharz, E. J., Singleton, K., Walker, E. B., and Mee, E. (1995). In situ evidence for DNA fragmentation in Huntington's disease striatum and Alzheimer's disease temporal lobes. *Neuroreport* **6**(7), 1053–1057.
117. Portera-Cailliau, C., Hedreen, J. C., Price, D. L., and Koliatsos, V. E. (1995). Evidence for apoptotic cell death in Huntington disease and excitotoxic animal models. *J. Neurosci.* **15**(5 Pt 2), 3775–3787.
118. Thomas, L. B., Gates, D. J., Richfield, E. K., O'Brien, T. F., Schweitzer, J. B., and Steindler, D. A. (1995). DNA end labeling (TUNEL) in Huntington's disease and other neuropathological conditions. *Exp. Neurol.* **133**(2), 265–272.
119. Ellis, H. M., and Horvitz, H. R. (1986). Genetic control of programmed cell death in the nematode C. elegans. *Cell* **44**(6), 817–829.
120. Martin, S., and Green, D. R. (1995). Protease activation during apoptosis: death by a thousands cuts? *Cell* **82,** 349–352.
121. Yuan, J.-Y., Shaham, S., Ledoux, S., Ellis, H. J. M., and Horvitz, H. R. (1993). The C. elegans cell death gene ced-3 encodes a protein similar to mammalian interleukin-1 beta converting enzyme. *Cell* **75,** 641–652.
122. Boldin, M., Goncharov, M., Golstev, Y. V., and Wallach, D. (1996). Involvement of MACH, a novel MOR/FADD-interacting protease, in Fas/Apo-1 and NF receptor-induced cell death. *Cell* **85,** 803–815.
123. Muzio, M., Chinnaiyan, A. M., Kischkel, F. C., O'Rouke, K., Shevchenko, N. J., Scaffdi, C., Bretz, J. D., Zhang, M., and Gentz, R. (1996). Flice, a novel FADD-homologous ICE/CED-like protease, is recruited to the CD95 (Fas/Apo-1) death-inducing signal complex. *Cell* **85,** 817–827.
124. Fernandes-Alnemri, T., Armstrong, R. C., Krebs, J., Srinivasula, S. M., Wang, L., Bullrich, F., Fritz, L. C., Trapani, J. A., Tomaselli, K. J., Litwack, G., and Alnemri, E. S. (1996). In vitro activation of CPP32 and Mch3 by Mch4, a novel human apoptotic cysteine protease containing two FADD-like domains. *Proc. Natl. Acad. Sci. USA* **93**(15), 7464–7469.
125. Vincenz, C., and Dixit, V. M. (1997). Fas-associated death domain protein interleukin-1beta-converting enzyme 2 (FLICE2), an ICE/Ced-3 homologue, is proximally involved in CD95- and p55-mediated death signaling. *J. Biol. Chem.* **272**(10), 6578–6583.
126. Wang, L., Miura, M., Bergeron, L., Zhu, H., and Yuan, J. (1994). Ich-1, an ICE/Ced-3 related gene encodes both positive and negative regulators of programmed cell death. *Cell* **78,** 739–750.
127. Duan, H., Orth, K., Chinnaiyan, A. M., Poirier, G. G., Froelich, C. J., He, W. W., and Dixit, V. M. (1996). ICE-LAP6, a novel member of the ICE/Ced-3 gene family, is activated by the cytotoxic T cell protease granzyme B. *J. Biol. Chem.* **271**(28), 16720–16724.
128. Srinivasula, S., Fernandes-Alnemri, T., Zangrill, J., Robertson, N., Armstrong, R. C., Wang, L., Trapani, J. A., Tomaselli, K. J., Litwack, G., and Alnemri, E. S. (1996). The Ced-3/Interleukin 1 beta converting enzyme like homolog Mch6 and the lamin cleaving enzyme Mch2 alpha substrates for the apoptotic mediator CPP32. *J. Biol. Chem.* **271**(43), 27099–27106.
129. Fernandes-Alnemri, T., Litwack, G., and Alnemri, E. S. (1994). CPP32, a novel human apoptotic protein with homology to Caenorhabditis elegans cell death protein Ced-3 and mammalian interleukin-1 beta-converting enzyme. *J. Biol. Chem.* **269**(49), 30761–30764.
130. Nicholson, D. W., Ali, A., Thornberry, N. A., Vaillancourt, J. P., Ding, C. K., Gallant, M., Gareau, Y., Griffin, P. R., Labelle, M., Lazebnik, Y. A., Munday, N. A., Raju, S. M., Smulson, M. E., Yamin, T.-T., Yu, V. L., and Miller, D. K. (1995). Identification and inhibition of the ICE/CED-3 protease necessary for mammalian apoptosis. *Nature* **376**(6535), 37–43.

131. Tewari, M., Quan, L. T., O'Rourke, K., Desnoyers, S., Zeng, Z., Beidler, D. R., Poirier, G. G., Salvesen, G. S., and Dixit, V. M. (1995). Yama/CPP32 beta, a mammalian homolog of CED-3, is a CrmA-inhibitable protease that cleaves the death substrate poly(ADP-ribose) polymerase. *Cell* **81**(5), 801–809.
132. Duan, H., Chinnaiyan, A. M., Hudson, P. L., Wing, J. P., He, W. W., and Dixit, V. M. (1996). ICE-LAP3, a novel mammalian homologue of the Caenorhabditis elegans cell death protein Ced-3 is activated during Fas- and tumor necrosis factor-induced apoptosis. *J. Biol. Chem.* **271**(3), 1621–1625.
133. Fernandes-Alnemri, T., Takahshi, A., Armstrong, R., Krebs, J., Fritz, L., Tomaseli, K. J., Wang, L., Yu, Z., Croce, C. M., Salveson, G., Earnshaw, W. C., Litwack, G., and Alnemri, E. S. (1995). Mch3, a novel human apoptotic cysteine protease highly related to CPP32. *Cancer Res.* **55**(24), 6045–6052.
134. Cerretti, D., Kozolsky, C. J., Mosley, B., Nelson, N., Ness, K. V., Greenstreet, T. A., March, C. J., Kronheim, S. R., Duck, T., Cannizzaro, L. A., Heubner, K., and Black, R. A. (1992). Molecular cloning of the interleukin-1 beta converting enzyme. *Science* **256,** 97–100.
135. Faucheu, C., Diu, A., Chan, A. W., Blanchet, A. M., Miossec, C., Herve, F., Collard-Dutilleul, V., Gu, Y., Aldape, R. A., Lippke, J. A., Rocher, C., Su, M.-S., Livingston, D. J., Hercend, T., and Lalanne, J.-L. (1995). A novel human protease similar to the interleukin-1 beta converting enzyme induces apoptosis in transfected cells. *EMBO J.* **14**(9), 1914–1922.
136. Munday, N., Vaillancourt, J. P., Ali, A., Casano, F. J., Miller, D. K., Molineaux, S. M., Yamin, T-T., Yu, V. L., and Nicholson, D. W. (1995). Molecular cloning and pro-apoptotic activity of ICErelII and ICErelIII, members of the ICE/CED-3 family of cysteine proteases. *J. Biol. Chem.* **270**(26), 15870–15876.
137. Faucheu, C., Blanchet, A. M., Collard-Dutilleul, V., Lalanne, J. L., and Diu-Hercend, A. (1996). Identification of a cysteine protease closely related to interleukin-1 beta-converting enzyme. *Eur. J. Biochem.* **236**(1), 207–213.
138. Talanian, R. V., Quinlan, C., Trautz, S., Hackett, M. C., Mankovich, J. A., Banach, D., Ghayur, T., Brady, K. D., and Wong, W. W. (1997). Substrate specificities of caspase family proteases. *J. Biol. Chem.* **272,** 9677–9682.
139. Casciola-Rosen, L., Nicholson, D. W., Chong, T., Rowan, K. R., Thornberry, N. A., Miller, D. K., and Rosen, A. (1996). Apopain/CPP32 cleaves proteins that are essential for cellular repair: a fundamental principle of apoptotic death. *J. Exp. Med.* **183**(5), 1957–1964.
140. Wang, X., Sato, R., Brown, M. S., Hua, X., and Goldstein, J. L. (1994). SREBP-1, a membrane-bound transcription factor released by sterol-regulated proteolysis. *Cell* **77**(1), 53–62.
141. Song, Q., Wei, Lees-Miller, S., Alnemri, E., Watters, D., Lavin, and M. F. (1997). Resistance of actin to cleavage during apoptosis. *Proc. Natl. Acad. Sci. USA* **94**(1), 157–162.
142. Cryns, V., Bergeron, L., Zhu, H., Li, H., and Yuan, J. (1996). Specific cleavage of alpha-fodrin during Fas- and tumor necrosis factor-induced apoptosis is mediated by an interleukin 1- beta converting enzyme/CED-3 protease distinct from the poly(ADP-ribose)polymerase protease. *J. Biol. Chem.* **271**(49), 31277–31282.
143. Greidinger, E., Miller, D. K., Yamin, T. T., Casciola-Rosen, L., and Rosen, A. (1996). Sequential activation of three distinct ICE-like activities in Fas-ligated Jurkat cells. *FEBS Lett.* **390**(3), 299–303.
144. Brancolini, C., Benedetti, M., and Schneider, C. (1995). Microfilament reorganization during apoptosis: the role of Gas2, a possible substrate for ICE-like proteases. *EMBO J.* **14,** 5179–5190.
145. Kuida, K., Zheng, T. S., Na, S., Kuan, C., Yang, D., Karasuyama, H., Rakic, P., and Flavell, R. A. (1996). Decreased apoptosis in the brain and premature lethality in CPP32-deficient mice. *Nature* **384**(6607), 368–372.
146. Coyle, J. T., Ferkany, J. W., and Zaczek, R. (1983). Kainic acid: insights from a neurotoxin into the pathophysiology of Huntington's disease. *Neurobehav. Toxicol. Teratol.* **5**(6), 617–624.
147. Albin, R. L., Young, A. B., Penney, J. B., Handelin, B., Balfour, R., Anderson, K. D., Markel, D. S., Tourtellotte, W. W., and Reiner, A. (1990). Abnormalities of striatal projection neurons and N-methyl-D-aspartate receptors in presymptomatic Huntington's disease. *N. Engl. J. Med.* **322**(18), 1293–1298.
148. Rubinsztein, D. C., Leggo, J., Chiano, M., Dodge, A., Norbury, G., Rosser, E., and Craufurd, D. (1997). Genotypes at the GluR6 kainate receptor locus are associated with variation in the age of onset of Huntington disease. *Proc. Natl. Acad. Sci. USA* **94**(8), 3872–3876.
149. Beal, M. F., Hyman, B. T., and Koroshetz, W. (1993). Do defects in mitochondrial energy metabolism underlie the pathology of neurodegenerative diseases? *Trends Neurosci.* **16**(4), 125–131.
150. Jenkins, B. G., Koroshetz, W. J., Beal, M. F., and Rosen, B. R. (1993). Evidence for impairment of energy metabolism in vivo in Huntington's disease using localized 1H NMR spectroscopy. *Neurology* **43**(12), 2689–2695.
151. Parker, W. D., Boyson, S. J., Luder, A. S., and Parks, J. K. (1990). Evidence for a defect in NADH: ubiquinone oxidoreductase (complex I) in Huntington's disease. *Neurology* **40**(8), 1231–1234.
152. Brouillet, E., Hantraye, P., Ferrante, R. J., Dolan, R., Leroy-Willig, A., Kowall, N. W., and Beal, M. F. (1995). Chronic mitochondrial energy impairment produces selective striatal degeneration and abnormal choreiform movements in primates. *Proc. Natl. Acad. Sci. USA* **92**(15), 7105–7109.
153. Erecinska, M., and Nelson D. (1994). Effects of 3-nitropropionic acid on synaptosomal energy and transmitter metabolism; relevance to neurodenerative brain diseases. *J. Neurochem.* **63**(3), 1033–1041.
154. Borlongan, C. V., Koutouzis, T. K., Freeman, T. B., Cahill, D. W., and Sanberg, P. R. (1995). Behavioral pathology induced by repeated systemic injections of 3-nitropropionic acid mimics the motoric symptoms of Huntington's disease. *Brain Res.* **697**(1–2), 254–257.
155. Borlongan, C. V., Koutouzis, T. K., Randall, T. S., Freeman, T. B., Cahill, D. W., and Sanberg, P. R. (1995). Systemic 3-nitropropionic acid: behavioral deficits and striatal damage in adult rats. *Brain Res. Bull.* **36**(6), 549–556.
156. Brouillet, E., Jenkins, B. G., Hyman, B. T., Ferrante, R. J., Kowall, N. W., Srivastava, R., Roy, D. S., Rosen, B. R., and Beal, M. F. (1993). Age-dependent vulnerability of the striatum to the mitochondrial toxin 3-nitropropionic acid. *J. Neurochem.* **60**(1), 356–359.
157. Koutouzis, T. K., Borlongan, C. V., Freeman, T. B., Cahill, D. W., and Sanberg, P. R. (1994). Intrastriatal 3-nitropropionic acid: a behavioral assessment. *Neuroreport* **5**(17), 2241–2245.
158. Palfi, S., Ferrante, R. J., Brouillet, E., Beal, M. F., Dolan, R., Guyot, M. C., Peschanski, M., and Hantraye, P. (1996). Chronic 3-nitropropionic acid treatment in baboons replicates the cognitive and motor deficits of Huntington's disease. *J. Neurosci.* **16**(9), 3019–3025.
159. Beal, M. F., Brouillet, E., Jenkins, B. G., Ferrante, R. J., Kowall, N. W., Miller, J. M., Storey, E., Srivastava, R., Rosen, B. R., and Hyman, B. T. (1993). Neurochemical and histologic characterization of striatal excitotoxic lesions produced by the mitochondrial toxin 3-nitropropionic acid. *J. Neurosci.* **13**(10), 4181–4192.

160. Wallace, D. C. (1992). Mitochondrial genetics: a paradigm for aging and degenerative diseases? *Science* **256**(5057), 628–632.
161. Corral-Debrinski, M., Horton, T., Lott, M. T., Shoffner, J. M., Beal, M. F., and Wallace, D. C. (1992). Mitochondrial DNA deletions in human brain: regional variability and increase with advanced age. *Nature Genet.* **2**(4), 324–329.
162. Soong, N.-W., Hinton, D. R., Cortpassi, G., and Arnheim, N. (1992). Mosaicism for a specific mitochondrial DNA mutation in adult human brain. *Nature Genet.* **2,** 318–323.
163. Erecinska, M., and Dagani F. (1990). Relationships between the neuronal sodium/potassium pump and energy metabolism. Effects of K, Na, and adenosine triphosphate in isolated brain synaptosomes. *J. Neurochem.* **63**(3), 1033–1041.
164. Ludolph, A., Seelig, M. O., Ludoph, A., Novitt, P., Allen, C. N., Spencer, P. S., and Sabri, M. I. (1992). 3-Nitropropionic acid decreases cellular energy levels and causes neuronal degradation in cortical explants. *Neurodegeneration* **1,** 21–28.
165. Gossen, M., and Bujard, H. (1992). Tight control of gene expression in mammalian cells by tetracycline-responsive promoters. *Proc. Natl. Acad. Sci. USA* **89**(12), 5547–5551.
166. Gossen, M., Freundlieb, S., Bender, G., Muller, G., Hillen, W., and Bujard, H. (1995). Transcriptional activation by tetracyclines in mammalian cells. *Science* **268**(5218), 1766–1769.
167. Ikeda, H., Yamaguchi, M., Sugai, S., Aze, Y., Narumiya, S., and Kakizuka, A. (1996). Expanded polyglutamine in the Machado-Joseph disease protein induces cell death in vitro and in vivo. *Nature Genet.* **13**(2), 196–202.
168. Wang, X., Zelenski, N. G., Yang, J., Sakai, J., Brown, M. S., and Goldstein, J. L. (1996). Cleavage of sterol regulatory element binding proteins (SREBPs) by CPP32 during apoptosis. *EMBO J.* **15**(5), 1012–1020
169. Mangiarini, L., Sathasivam, K., Seller, M., Cozens, B., Harper, A., Hetherington, C., Lawton, M., Trottier, Y., Lehrach, H., Davies, S. W., and Bates, G. P. (1996). Exon 1 of the HD gene with an expanded CAG repeat is sufficient to cause a progressive neurological phenotype in transgenic mice. *Cell* **87**(3), 493–506.
170. Goldberg, Y. P., Telenius, H., and Hayden, M. R. (1994). The molecular genetics of Huntington's disease. *Curr. Opinion Neurol.* **7**(4), 325–332.
171. Onodera, O., Roses, A. D., Tsuji, S., Vance, J. M., Strittmatter, W. J., and Burke, J. R. (1996). Toxicity of expanded polyglutamine-domain proteins in escherichia coli. *FEBS Lett.* **399**(1–2), 135–139.
172. Koutouzis, T. K., Borlongan, C. V., Scorcia, T., Creese, I., Cahill, D. W., Freeman, T. B., and Sanberg, P. R. (1994). Systemic 3-nitropropionic acid: long-term effects on locomotor behavior. *Brain Res.* **646**(2), 242–246.
173. Nicholson, D. W. (1996). ICE/CED3-like proteases as therapeutic targets for the control of inappropriate apoptosis. *Nature Biotechnol.* **14,** 297–301.
174. Wellington, C. L., Ellerby, L. M., Hackam, A. S., Margolis, R. L., Trifiro, M. A., Singaraja, R., McCutcheon, K., Salvesen, G. S., Propp, S. S., Bromm, M., Rowland, K. J., Zhang, T., Rasper, D., Roy, S., Thornberry, N., Pinsky, L., Kakizuka, A., Ross, C. A., Nicholson, D. W., Bredesen, D. E., and Hayden, M. R. (1998). Caspase cleavage of gene products associated with triplet expansion disorders generates truncated fragments containing the polyglutamine tract. *J. Biol. Chem.* [In press].
175. Martindale, D., Hackam, A., Wieczorek, A., Ellerby, L., Wellington, C., McCutcheon, K., Singaraja, R., Kazemi-Esfarjani, P., Devon, R., Kim, S. U., Bredesen, D. E., Tufaro, F., and Hayden, M. R. (1998). Length of polyglutamine tract influences localization and frequency of intracellular aggregates. *Nature Genet.* **18,** 150–154.
176. Thornberry, N. A., Rano, T. A., Peterson, E. P., Rasper, D. M., Timkey, T., Garcia-Calvo, M., Houtzager, V. M., Nodstorm, P. A., Roy, S., Vaillancourt, J. P., Chapman, K. T., and Nicholson, D. W. (1997). A combinatorial approach defines specificities of members of the caspase family and granzyme B. functional relationships established for key mediators of apoptosis. *J. Biol. Chem.* **272,** 17907–17911.

CHAPTER 25

Murine Models of Huntington's Disease

GILLIAN P. BATES Medical and Molecular Genetics, United Medical and Dental Schools, Guy's Hospital, London SE1 9RT, United Kingdom

ERICH E. WANKER Max-Planck-Institüt für Moleculare Genetik, Berlin (Dahlem), Germany

STEPHEN W. DAVIES Department of Anatomy and Developmental Biology, University College London, London WC1E 6BT, United Kingdom

I. INTRODUCTION

Huntington's disease (HD) belongs to the class of neurodegenerative disorders caused by a CAG/polyglutamine (polygln) expansion [1]. It is an autosomal dominant disorder with an onset of symptoms generally in midlife although this can range from early childhood to greater than 70 years. Anticipation is predominantly associated with male transmission, with the result that 70% of juvenile cases inherit the disease from their father. The symptoms have an emotional, motor, and cognitive component. A detailed description of all aspects of HD can be found in Harper (1991) [2] and Myers (Chapter 23, this volume). Chorea is a characteristic feature of the motor disorder and other motor abnormalities include dystonia, bradykinesia, oculomotor dysfunction, fine motor incoordination, dysarthria, and dysphagia. The adult form of the disease frequently progresses to an akinetic state. Cerebellar dysfunction, upper motor neuron abnormalities, tremor, epilepsy, and myoclonus are rare except in the juvenile form which commonly presents with a "Parkinson-like rigidity." The emotional disorder is complex, frequently showing depression and irritability and the cognitive component comprises a subcortical dementia.

HD is associated with severe atrophy to the caudate nucleus which is often reduced to a rim of tissue, and there are subtle changes in the cerebral cortex [3]. Vonsattel and colleagues classified the extent of the striatal atrophy into 5 grades: grade 0 in which no gross or generalised microscopic abnormalities consistent with HD are seen, progressing to grade 4 in which the most extreme atrophy is observed [3]. All grades show a 30% reduction in HD brain weight associated with 20–30% areal reductions in cerebral cortex, white matter, hippocampus, amygdala, and thalamus [4]. It has been proposed that shrinkage of these structures occurs early in the disease process and is not progressive. Gliosis was not apparent and the neuronal density appeared to be normal [4]. In contrast, the caudate, putamen, and globus pallidus progressively degenerate with prolonged survival. This localised progressive atrophy is associated with reactive astrocytosis [3, 5]. It has been suggested that although grade 0 brains have no generalised striatal pathology, scattered islands of reactive astrocytes corresponding to the striosomal compartments of the striatum are present [6]. This early pathology has been reported in affected individuals with a 3- to 5-year history of chorea, although grade 0 pathology has been observed after up to 13 years of chorea [3, 6, 7].

The HD mutation is an expanded CAG repeat [1] with normal and mutant allele size ranges of CAG_{6-39} and CAG_{35-180} repeats, respectively [8–11]. A negative correlation between age of onset and repeat length is most pronounced for juvenile HD, for which the longest repeats have been observed. The majority of adult-onset cases are associated with repeats in the 40–55 range, and expansions of over 70 invariably cause the juvenile form of the disease [12–14]. Repeat instability is observed in the majority of transmissions. However, large expansions are almost exclusively seen upon paternal transmission, a consequence of the greater repeat instability observed in sperm and accounting for the sex bias observed in the anticipation [12, 15]. Single sperm typing indicated a mutation frequency of greater than 90% in disease alleles greater than 38 repeats with a propensity to expansions [16]. Somatic repeat stability has also been identified although this is relatively modest. Comparison of repeat size in a large panel of HD tissues showed instability to be most pronounced in regions of the CNS [17] with a tendency to expansion except in the case of the cerebellum, in which apparent contractions were consistently observed [17, 18]. Of non-CNS tissues, liver, muscle, lung, and testis were noted to show an intermediate level of instability.

The HD gene is 170 kb in size and contains 67 exons and the mutation is located in exon 1 [19]. This entire genomic region has been sequenced [20]. Two ubiquitously expressed transcripts have been identified, of 10,366 bp (IT15) and 13,711 bp, arising from differential polyadenylation [21]. The huntingtin protein products arising from expanded alleles have been identified in protein extracts from HD patients [22–25] indicating that the mutation does not block transcription or translation. Despite selective neuronal vulnerability, the HD transcript is widely expressed in brain and peripheral tissues [26, 27]. Within the brain there is a widespread predominantly neuronal distribution [26–28] that is present by 20 weeks gestation [29]. Immunohistochemistry, electron microscopy, and subcellular fractionations have shown that huntingtin is primarily a cytosolic protein, a fraction of which is associated with vesicles and/or microtubules, suggesting that it plays a functional role in cytoskeletal anchoring or vesicle transport [30–33]. Within the striatum, the pattern of selective neuronal vulnerability is mirrored by huntingtin expression levels, the projection neurons staining more intensely on immunocytochemistry than the interneurons [34, 35].

CAG/polygln repeat expansion is the mutation causing at least six other inherited neurodegenerative disorders, namely spinal and bulbar muscular atrophy (SBMA) [36], dentatorubral-pallidoluysian atrophy (DRPLA) [37, 38], and spinocerebellar ataxia (SCA) types 1 [39], 2 [40–42], 3 [43], and 6 [44]. They are autosomal dominant (with the exception of X-linked SBMA) and the normal and expanded CAG/polygln repeat ranges are largely comparable (SCA6 presents the exception). The proteins harbouring the polygln stretches are mostly novel and otherwise unrelated. They can tolerate a large variation in the size of the polygln tracts in the normal range, but upon a certain size these tracts become pathogenic. In all cases the proteins are widely or ubiquitously expressed, but despite extensively overlapping expression patterns, the neuronal cell death is relatively specific and can differ markedly [45], although in the juvenile forms of these diseases, the neuropathology becomes more widespread and less distinct.

One approach to uncover both the normal and toxic functions of these polygln-containing proteins is to identify and characterise proteins with which they interact. A number of such huntingtin interactors have now been isolated which include HAP 1 [46], HIP-1 [47, 48], a specific ubiquitin-conjugating enzyme (hE2-25K) [49], and GAPDH [50]. GAPDH has also been reported to interact with atrophin 1 [50], ataxin 1, and the androgen receptor [51]. In all cases, whether these proteins play a role in the pathogenic pathways has yet to be determined. Huntingtin has been shown to be specifically cleaved by apopain, a cysteine protease with a key role in the proteolytic events leading to apoptosis [52]. It has been suggested that apopain cleavage results in an N-terminal huntingtin fragment containing the polygln expansion and that this truncated protein is toxic to certain cells. This model is supported in part by the observation that the C-terminus of the ataxin 3 protein, containing a polygln expansion, is more toxic to COS cells in transient transfection assays than the entire ataxin 3 protein carrying an identical expansion [53].

The HD mutation is thought to act by a gain of function which is supported by the identification of an antibody that specifically detects pathogenic polygln expansions [30]. This may be indicative of a conformational change occurring upon a certain size threshold, possibly facilitating novel interactions. It has been proposed that polyglns can form β-pleated sheets termed polar zippers which would potentially enable polygln tracts to interact [54, 55]. A polygln expansion may modulate interactions with other polygln-containing proteins. Alternatively, it could result in the polymerisation and the eventual precipitation of huntingtin, as supported by the recent observation that N-terminal fragments of huntingtin containing pathogenic expansions form amyloid-like aggregates *in vitro* [56].

II. MURINE MODELS

The isolation of the HD gene made it possible to apply transgenic technology to generate, for the first time, a representative model of this disease. A combination of approaches to mouse modelling has provided insights into the mechanism by which the polygln expansion acts. The first mouse models of polygln disease in which a progressive neurological phenotype is apparent have now been described. From these first reports, it is clear that it will be possible to model both the molecular events causing the disease and the CAG repeat instability in the mouse. The current models are already helping to uncover the molecular basis of polygln disease and are likely to aid in the identification of factors important in governing the disease-specific patterns of neuropathology.

A. Conservation of the HD Gene

The conservation of the HD gene has been studied in human and mouse [57, 58], rat [59], and the pufferfish (*Fugu rubripes*) [20]. Comparative sequence analysis has been conducted across cDNAs in the case of human, mouse, and rat and across the entire gene in the case of human and *Fugu*. The human and mouse cDNAs were found to be 91% identical at the protein level and 86% identical at the DNA level [57, 58]. This degree of similarity was too high to identify conserved domains within the protein that may have a functional importance. Therefore, the comparative sequence analysis was extended to the more distantly related *Fugu*. The *Fugu* HD gene (*FrHD*) was found to be 23 kb in length which is 7.4-fold smaller than the human gene of 170 kb and is in keeping with the relative sizes of the human and *Fugu* genomes [20]. The *Fugu* coding sequence is 69% identical to the human sequence at the nucleotide level and all 67 exons and exon/intron boundaries are conserved.

The polygln repeat is conserved across all species and ranges from 4 glns in *Fugu* to 7 in mouse and 8 in rat, increasing to between 8 and 41 in humans [9, 10]. The polygln repeat is encoded by $(CAG)_2(CAA)_2$ in *Fugu*, by $(CAG)_2CAA(CAG)_4$ in mouse, by $(CAG)_2CAA(CAG)_5$ in rat, and by $(CAG)_nCAACAG$ in humans. Whilst this conservation suggests a functional importance for the polygln repeat, it must be able to tolerate wide variations in the number of gln residues. The proline-rich region immediately downstream from the polyglns is largely conserved between mouse, rat, and human but absent in *Fugu*.

B. Knockouts of the Mouse *Hhd* Gene

In recent years the ability to knock out the function of a mouse gene by homologous recombination in ES cells has become standard technology [60]. This approach is frequently used to explore the normal function of a gene and also in disease modelling. A heterozygous knockout may be expected to model a disease that occurs through haplo-insufficiency and a homozygous null should produce the phenotype caused by the loss of function of a gene arising from either a homozygous recessive or a dominant negative mutation.

Three research groups independently generated knockouts of the mouse *Hdh* gene [61–63]. In each case, different regions of *Hdh* were replaced with a neo selectable marker. Nasir *et al.* [61] deleted approximately half of exon 5 plus intron 4 sequences (Hdh^{ex5}), Duyao *et al.* [62] removed exon 4 and exon 5 (Hdh^{ex4+5}), and Zeitlin *et al.* [63] targeted the promoter and exon 1 (Hdh^{prex1}). In all cases, intercrosses between heterozygous knockouts demonstrated that the absence of a functional copy of *Hdh* leads to embryonic lethality. Embryos do not progress beyond the e7.5 egg cylinder stage (Fig. 25-1). Resorptions of Hdh^{prex1} and Hdh^{ex4+5} occurred between e8.5 and e10.5 and Hdh^{ex5} at e8.5 (and possibly e7.5). Abnormalities were observed in both the embryonic and extraembryonic tissues. Zeitlin *et al.* [63] showed by vital staining and TUNEL (TdT-mediated dUTP-biotin nick end labeling) that there was an increase in the number of cells undergoing apoptosis in the mutant embryos. The authors propose that lack of huntingtin results in excessive programmed cell death that eliminates a subset of critically located cells. This has a detrimental effect on patterning and leads to sub-

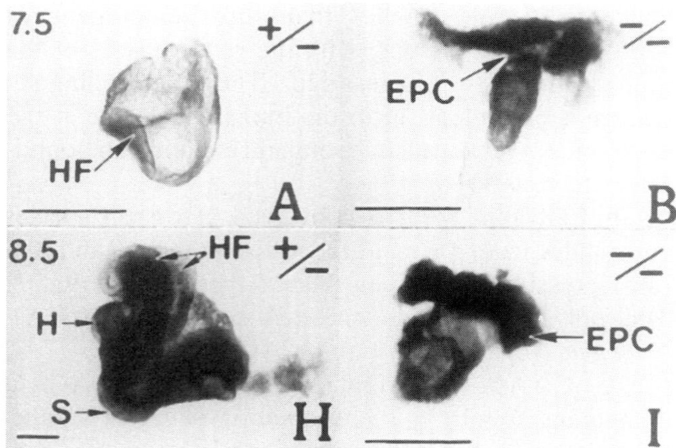

FIGURE 25-1 Phenotype of heterozygous (±) and nullizygous (−/−) *Hdh* knockout embryos. (A) Heterozygous embryo dissected out of the decidua. Headfolds have begun to form at the anterior end of the embryo. (B) Homozygous littermate of the embryo in (A). The embryo is small and underdeveloped. (H) Heterozygous embryo (E8.5) dissected out of decidua. Prominent headfolds, heart, and somites are visible. (I) Homozygous littermate to embryo in (H). Abbreviations: EPC, ectoplacental cone; HF, headfolds; H, heart; S, somites. The scale bar is 500 mm. Reproduced from *Cell* **81,** 811–823, 1995, with the permission of Cell Press.

sequent disorganisation on formation of the mesoderm layer. They conclude that the function of huntingtin may be to counterbalance an apoptotic pathway.

In two of these studies, heterozygous mice expressing only one copy of *Hdh* were phenotypically normal [62, 63]. In contrast, Nasir *et al.* [61] reported that heterozygotes showed increased motor activity and cognitive deficits with a significant neuronal loss in the subthalamic nucleus. To explain the discrepancy between these results it has been suggested that the targeted allele may produce a truncated protein which could conceivably cause a dominant effect in the heterozygous mice. Nasir *et al.* [61] detected an mRNA transcribed from the targeted allele in which exon 5 has been skipped, generating a frame shift with a stop codon immediately downstream of the targeting event. This would be predicted to produce a truncated protein of approximately 20 kDa and the authors report that such a band was present on Western blots from heterozygous mouse brains.

In combination, these reports indicate that the HD mutation does not act through a dominant negative loss of function or through haplo-insufficiency.

C. Approaches toward the Generation of a Murine Model of HD

CAG/polygln expansions most probably act through a dominant gain of function mechanism. Therefore, it would seem likely that a mouse model of HD could be generated by creating mice that are transgenic for the HD mutation irrespective of the presence of two normal copies of the mouse *Hdh* gene. Constructs used for transgenesis commonly include human or mouse cDNAs under the control of appropriate promoters or the entire human gene together with its endogenous control elements. Genomic clones are frequently more successful in generating transgenic models than cDNAs as they are more likely to direct an expression profile that mimics the endogenous gene. The large size of the HD gene (170 kb) necessitates that genomic constructs are prepared and manipulated in the form of yeast artificial chromosomes (YACs). Using YAC technology, Hodgson *et al.* [64] have successfully generated mice that are transgenic for the normal human HD gene. They have crossed the YAC HD transgene onto an *Hdh* nullizygous background and shown that the human YAC can rescue the embryonic lethal phenotype. This indicates that the transgene is expressed appropriately and predicts that the introduction of a mutant version of the human YAC would generate a model of HD. An alternative approach is offered by "knock-ins" involving the replacement of the seven polygln residues in the mouse *Hdh* gene with an expanded version of a size that would be pathogenic in humans.

The puffer fish (*F. rubripes*) has a compact genome of 400 Mbp which is approximately 7.5-fold smaller than the human genome [65]. It contains a similar number of genes but is deficient in intergenic, intronic, and dispersed repetitive sequences. The *Fugu* genome is becoming established as the model vertebrate genome for the identification and characterisation of novel human genes and conserved regulatory sequences. It has also been proposed that *Fugu* genes may provide natural minigenes for the production of transgenic mice. Should this be possible, pronuclear injection of *Fugu* genes as cosmids may provide an alternative to using YACs containing human or mouse genes. The manipulation and introduction of the clones would be correspondingly simpler whilst retaining some of the advantages of working with a genomic clone.

The *Fugu* homologue of the HD gene (*FrHD*) has been used to test this possibility [65a]. The human and *Fugu* HD genes cover 170 and 23 kb, respectively, and have previously been sequenced in their entirety [20]. Despite the absence of conserved promoter sequences, the *Fugu* promoter was found to be functional in mouse cells. However, a detailed RT-PCR analysis across the entire 10-kb transcript showed the presence of several aberrant splice forms which would be incompatible with the production of the *Fugu* huntingtin protein. The *Fugu* HD gene was found to be incorrectly processed in mouse cells both *in vitro* and *in vivo* which sheds doubt on the usefulness of *Fugu* genes for transgenesis [65a].

The first murine model of HD to be reported has used standard transgenic approaches to introduce exon 1 of the HD gene carrying highly expanded CAG repeats into the mouse germline [66]. This model is described in detail in Section III.

III. MICE TRANSGENIC FOR THE HD MUTATION DEVELOP A DOMINANT NEUROLOGICAL PHENOTYPE

A. Microinjection Fragments and Transgenic Lines

The microinjection construct comprised a genomic fragment containing the 5′ end of the human HD gene (Fig. 25-2) including approximately 1 kb of promoter sequences, exon 1 carrying highly expanded CAG repeats, and 262 bp of intron 1 [66]. A single male founder was recovered who carried five separate transgene integration events and from whom five unique transgenic lines were established as summarised in Table 25-1. The transgene repeat expansions are considerably larger than those associated with the juvenile form of HD. The founder was backcrossed to CBA × C57BL/6 F1 hybrids and the genotypes of 321 offspring indicated that he was a germline chimera. The integration sites segregated independently but were only recovered in certain combinations, demonstrating that one set of germ cells contained the R6/1, R6/2, and R6/5 transgenes and another contained the R6/0 and R6/T transgenes. Two further lines, HDex6 and HDex27, were established from independent founders using the equivalent genomic fragment with $(CAG)_{18}$ repeat tracts.

Table 25-1 Summary of the Genomic Organisation of the Transgene Integration Site for Each of the Five Transgenic Lines Established from the R6 Founder [66]

Transgenic line	Transgene copy number	CAG repeat (N)[a]
R6/0	Single integrant	142
R6/1	Single integrant	113
R6/2	One intact copy	144
R6/5	Four intact copies	128–156
R6/T	One very truncated copy	—

[a] Repeat sizes are those most likely to have been present in the founder. However, as a consequence of germ line instability [67] there is a range of repeat sizes associated with lines R6/1, R6/2, and R6/5.

B. Expression Analysis of the Transgenes

The observation of a phenotype in some of these lines (see below) prompted us to look for expression of the transgene. The R6 transgenic lines had initially been established as a repeat stability study and the microinjection fragment had not been designed as an expression construct. The transgene mRNA was shown to be ubiquitously expressed in a panel of 18 tissues and brain regions in all lines except R6/0 (line R6/T carried such a highly truncated transgene that it was not included in the expression analysis). Northern blots of total brain RNA using an intron 1 probe showed that the intronic sequences were present in the transgene message. This result was extremely informative as it indicated that the human exon 1 sequence had not spliced to any mouse exons conveniently situated in the 3′ flanking mouse DNA. The presence of human intron 1 sequences in the transgene message therefore confirmed the presence of a stop codon at the beginning of intron 1 and predicted that the transgene protein would contain the first 69 amino acids of the huntingtin protein in addition to the polygln tract encoded

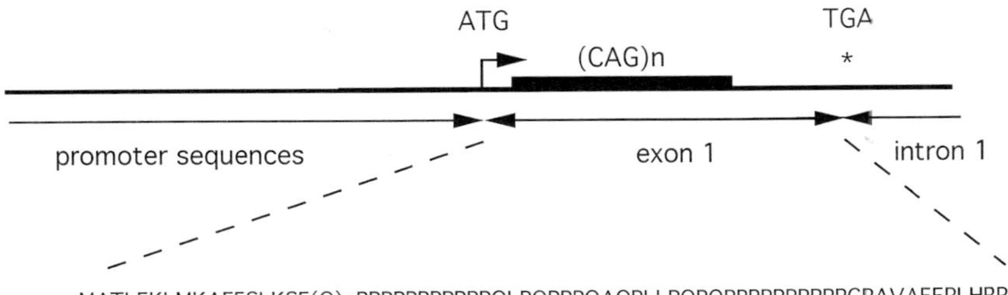

Figure 25-2 Schematic representation of the microinjection fragment used to generate the R6 transgenic lines. The fragment is approximately 2 kb in size and contains promoter sequences, exon 1, and 262 bp of intron 1. ATG marks the initiation of translation and TGA, the stop codon immediately at the beginning of intron 1. The amino acid sequence of the protein arising from translation of this fragment is shown.

TABLE 25-2 Correlation between Repeat Size, Expression of the Transgene, and the Phenotype [66, 72]

Transgenic line	CAG repeat size	RNA	Protein	Phenotype
R6/1	115	+	+	+
R6/2	145	+	+	+
R6/5	128–156	+	+	+
R6/0	142	–	–	–
HDex6	18	+	+	–
HDex27	18	+	+	–

by the CAG repeat (Fig. 25-2). The transgene protein was detected by immunoblot in all tissues and brain sections tested for all lines except R6/0. The expression analysis of the R6 and HDex lines is summarised in Table 25-2.

C. Stability of the CAG Repeat in the Transgenic Mice

The R6 lines show both intergenerational and somatic repeat instability [66, 67]. The repeats are clearly unstable on transmission in lines R6/1, R6/2, and R6/5; however, this is less clear for line R6/0 as the changes observed in this line could be accounted for by errors in sizing. R6/5 was the only line in which an extensive comparison of instability on both male and female transmission was conducted, and the repeats had a tendency to increase on male transmission and decrease on female transmission [67]. This trend was supported by the intergenerational instability observed in the other lines (R6/2 females are sterile and therefore only male transmission could be studied in this line). This pattern of instability more closely correlates to that seen in SCA1 [68] than in HD in which there is not such an overt trend toward contraction on maternal transmission [12, 14, 17]. The CAG expansions introduced into these mice are considerably larger than are normally seen in HD patients. The change in size of the repeat on transmission in the mice is smaller than would be expected from comparison with size changes associated with highly expanded CAG repeats seen in humans [12, 14, 17]. The discrepancy in the degree of instability between humans and mice may reflect the difference in their life span, a model supported by the observation that the size of the intergenerational expansion increased with the age of the transmitting male in line R6/2.

Somatic instability was detected in lines R6/1, R6/2, and R6/5 but not in line R6/0 (Fig. 25-3). In all three lines, onset of instability was at approximately 6 weeks and the CAG repeat range increased with the age of the mouse. This argues against a pathogenic role for repeat instability as the age of onset of symptoms in these lines differs markedly. The pattern of instability was more widespread in some lines than others although on the whole it was first present and most prominent in brain regions. Peripheral tissues that consistently showed instability included liver and kidney. Overall the somatic instability was comparable to that described in individuals carrying CAG expansions [17, 18, 68–71]. The major difference between line R6/0, in which instability was not apparent, and the other lines was the absence of transgene expression. This is probably due to gene silencing by a position effect as the R6/0 transgene has clearly integrated into a region of unusual genomic structure [66].

The absence of instability in line R6/0 as compared to the other R6 lines could have mechanistic implications. It cannot be caused by differences in *trans*-acting factors which are likely to be invariant between lines. Line R6/0 differs from the other three only at the site of integration and it cannot be ruled out that important *cis*-acting sequences are absent from this site. However, given the comparable levels of instability seen in three of the R6 lines it seems likely that such sequences, if they exist, are present on the transgene itself. It is known that the R6/0 transgene has integrated into a genomic region that is probably acting to silence the expression of the transgene which raises the possibility that the instability is linked to transgene expression. This may be more a consequence of the integration site being in a region of open chromatin leading to DNA damage rather than through a mechanism directly linked to transcription.

D. The R6 Lines Develop a Progressive Neurological Phenotype

Lines R6/1, R6/2, and R6/5 have been found to develop a progressive neurological phenotype. A pheno-

FIGURE 25-3 Illustration of the CAG repeat somatic instability seen in the R6 lines. The repeats were amplified by PCR using a fluorescent primer and sized on an ABI373 sequencer using the Genescan and Genotyper software packages [67]. In each case the Genescan trace arising from a range of tissues at the age at which the mouse was culled is compared to the trace obtained from tail DNA taken at 3 weeks (top row). The size of the major peaks in the tail traces are R6/1, 115; R6/2, 145; R6/0, 142; R6/5, range from 123 to 156. The R6/5 line contains four copies of the CAG repeat and the difference in the tail trace between the two R6/5 mice arises from germline instability. It is clear that even after 38 weeks there is no evidence for somatic instability in line R6/0. Reproduced from *Hum. Mol. Genet.* **6**, 1663–1637, 1997, by permission of Oxford University Press.

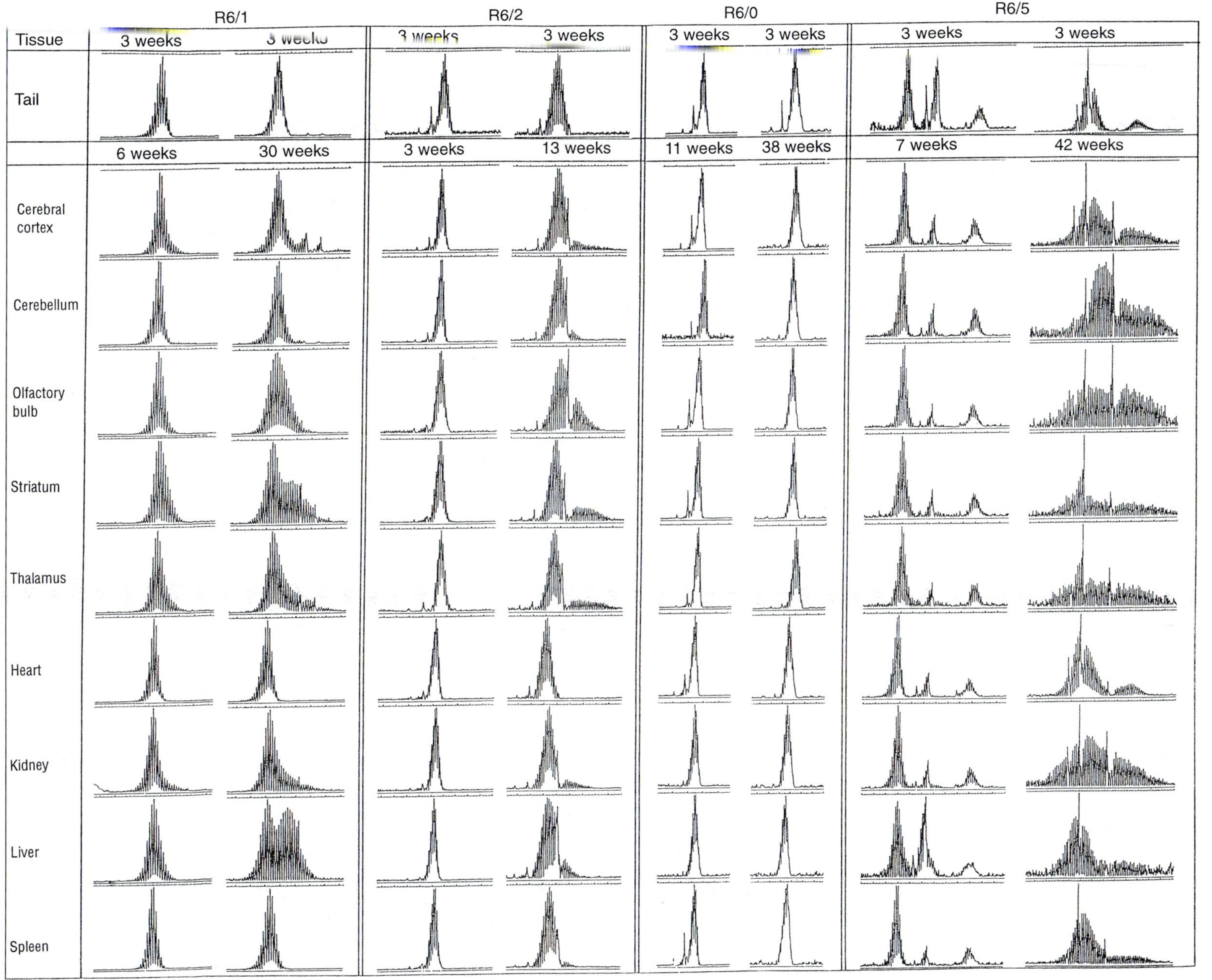

R6/1
R6/2
R6/0
R6/5
Tissue
3 weeks
3 weeks
3 weeks
3 weeks
3 weeks
3 weeks
3 weeks
3 weeks
Tail
6 weeks
30 weeks
3 weeks
13 weeks
11 weeks
38 weeks
7 weeks
42 weeks
Cerebral cortex
Cerebellum
Olfactory bulb
Striatum
Thalamus
Heart
Kidney
Liver
Spleen

type has not been observed in line R6/0 in which the transgene is not expressed, nor in the HDex lines transgenic for the same construct carrying 18 repeats (Table 25-2).

Line R6/2 has been characterised most extensively. The mice carry repeat expansions ranging from 141 to 157 repeats, the variability having arisen from germline instability [67]. At weaning the R6/2 transgenes are indistinguishable from their normal littermates and they develop a progressive, complex neurological phenotype with onset age of approximately 2 months. The progression of the disease is rapid and the mice deteriorate over the following month. The movement disorder includes an irregular gait, resting tremor, rapid, abrupt, irregularly timed shuddering movements, stereotypic grooming movements, and in some cases, epileptic seizures. Coincident with the onset of the movement disorder, body weight plateaus and begins to fall such that by 12 weeks, the weight of the transgenes is approximately 60% that of controls. In males, this occurs from 8 weeks (Fig. 25-4) and in females it begins a week or two later. Females are sterile and loss of fertility in males is accompanied by severe atrophy of the testes occurring from 8 weeks. The transgenic mice appear to urinate more frequently as indicated by an increase in the wetting of the bedding at one end of the cage.

The R6 genotypes and the associated onset age of the phenotype are summarised in Table 25-3. The phenotype is sensitive to genotype dosage in that as the number of transgenes per mouse increases, the age of onset decreases and the disease progression becomes more rapid. The most severely affected genotype recovered in the F1 generation was that hemizygous for each of the R6/1, R6/2, and R6/5 integration events. These mice were affected prior to weaning at 3 weeks of age and did not live beyond 7 weeks. They were recovered at a lower frequency than expected in the F1 generation and it is possible that a number were lost as neonatal deaths. Consequently, overexpression of this transgene could be lethal. A phenotype was not observed in R6/5 hemizygotes at well over 1 year of age and R6/5 homozygotes have an age of onset of approximately 9 months.

TABLE 25-3 Age at Onset of the Phenotype as Associated with a Range of R6 Genotypes [66]

Genotype	Age at onset
R6/1 + R6/2 + R6/5	<3 weeks
R6/1 + R6/2	3–4 weeks
R6/2 + R6/5	6–7 weeks
R6/2	~2 months
R6/1 + R6/5	3–4 months
R6/1	4–5 months
R6/5 homozygotes	9 months

E. Neuronal Intranuclear Inclusions (NII) Underlie the Neurological Dysfunction

Initial neuropathological analysis included comparison of Nissl-stained serial 40-mm sections of R6/2 and littermate control brains taken at 12 weeks in either the coronal or horizontal planes throughout the entire brain and spinal cord [66]. The morphology of the central nervous system appeared normal with no focal areas of malformation or neurodegeneration. There appeared to

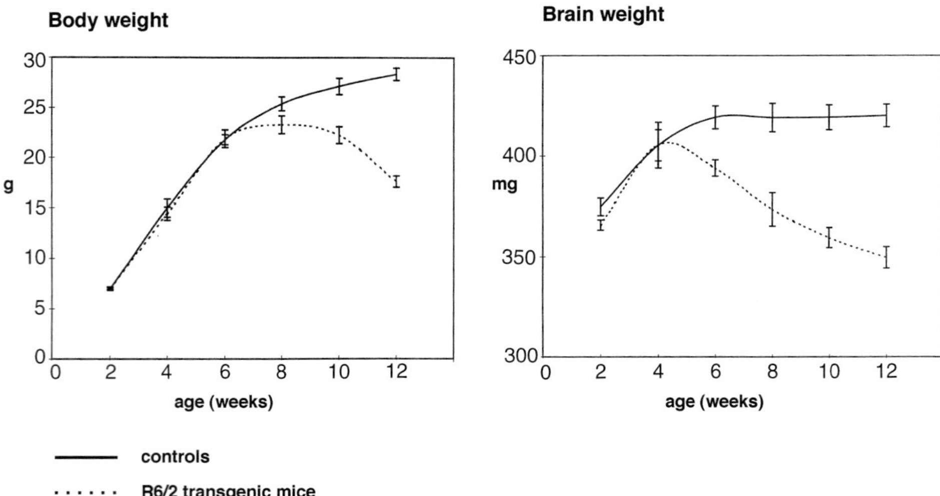

FIGURE 25-4 Comparison of the changes in body weight and brain weight in a series of R6/2 transgenic males. Reproduced from *Cell* **90,** 537–548, 1997, with the permission of Cell Press.

be no evidence of neuronal cell loss, oligodendrocyte loss, reactive gliosis, or inflammatory change. A normal distribution of astrocytes and ramified microglia cells was observed in the absence of any indication of increased reactivity of astrocyte staining or the presence of rounded microglia or infiltrating macrophages.

At 12 weeks the only difference between R6/2 and control brains was that the brains from the transgenic animals were approximately 20% smaller than controls. This areal reduction was uniform throughout all CNS structures and yet the neuronal density appeared normal, consistent with findings in HD brains [4]. A longitudinal study of the comparative weights of R6/2 and control brains showed a progressive reduction in the transgene brain weight from a time between 4 and 6 weeks to approximately 20% of controls at 12 weeks [72]. The loss in brain weight clearly precedes the loss in body weight (Fig. 25-4).

The first evidence of a specific neuropathological change in the transgene brains came from immunocytochemistry with antibodies raised against the N-terminus of huntingtin [72]. In adult control mouse brain these antibodies labeled the entire grey matter with neuronal labeling localised to the cytoplasm of the cell bodies, dendrites, and axons and traversing the neuropil. This is consistent with the diffuse cytoplasmic staining previously reported for huntingtin in rat, primate, and human brain [23, 31–33]. Immunoelectron microscopy in the normal adult mouse showed huntingtin to be scattered through the cytoplasm, associated with vesicle membranes with small patches of labeling in dendrites, unmyelinated axon fibres, axon terminals, and synaptic contacts. This pattern of staining is again comparable to that reported in the rat, monkey, and normal human brain [32, 33]. In normal mice huntingtin was never found in the nucleus.

In symptomatic mice from lines R6/1, R6/2, and R6/5, a densely stained solitary circular inclusion was seen in certain neuronal nuclei which we termed a neuronal intranuclear inclusion (NII). This was never present in brains from normal mice, lines HDex6 and HDex27 carrying 18 CAG repeats, line R6/0 in which the transgene is not expressed, or line R6/5 hemizygotes which do not develop a phenotype. Line R6/2 has again undergone the most extensive analysis. The largest inclusions were found in the cerebral cortex, striatum, cerebellar Purkinje cells, and spinal cord with many fewer in the hippocampus, thalamus, globus pallidus, and substantia nigra. Inclusions were never seen in an astrocyte, oligodendrocyte, or microglial cell. Within the striatum, NII are not seen in the various classes of interneurons and may be limited to the projection cells. Immuno EM localised the NII to a single defined region of the nucleus that is distinct from the nucleolus, the accessory body of Cajal and Barr bodies in female mice. Currently, the NII have only been found to be immunoreactive for N-terminal huntingtin and ubiquitin antibodies, each of which detect an even staining. Antibodies to more C-terminal huntingtin epitopes do not colocalise with the inclusions, suggesting that the endogenous mouse protein is not present in the NII. The presence of NII can be detected prior to any other neuropathological or phenotypic changes, being present within the cortex at 3.5 weeks and striatum at 4.5 weeks. They become immunoreactive with ubiquitin antibodies at 5–6 weeks and there is a progressive increase in the size and staining density for both huntingtin and ubiquitin. The inclusions can be seen at the ultrastructural level at 8 weeks and associated nuclear changes (as described below) are apparent by 10–12 weeks.

Ultrastructure analysis of NII from R6/2 brains [72] showed them to have a circular pale structure with a fine granular and occasional filamentous morphology that is devoid of a membrane (Fig. 25-5). However, the morphology of NII from a 17-month-old R6/5 homozygous mouse brain showed more evidence of high-molecular-weight fibrous structures with randomly oriented filaments, 5–10 nm in diameter and often measuring up to 250 nm in length [56]. Morphometric analysis indicated that the NII occupies approximately 1% of the volume of the nucleus as compared to approximately 0.33% occupied by the nucleolus. Prominent nuclear invaginations, normally seen in only 2–10% striatal neurons, were present in almost every neuron containing an NII and there was also an apparent increase in the clustering and number of nuclear pores. There was no significant change in the size of the nucleolus or the nucleus in cells containing an NII. The formation of nuclear membrane invaginations and an increase in the nuclear pore density is a response to axonal injury, and successful regeneration is associated with an increase in the nucleolar size. The cell may be attempting the first stages of an ultimately unsuccessful regeneration process. All three of these ultrastructural nuclear changes have previously been reported in EM studies from HD patients. The most prominent ultrastructural change reported in biopsies of the cerebral cortex and caudate nucleus of 18 patients with HD [73] is the presence of a nuclear inclusion strikingly similar to those identified in the transgenic mice (Fig. 25-5). A marked increase in nuclear membrane indentations [74] and in the density of nuclear pores [75] has also been described. More recently, NII have been identified by immunocytochemistry with N-terminal huntingtin and ubiquitin antibodies and by EM in postmortem HD brains [75a, 75b]. The NII found in patients were not identified with more C-terminal huntingtin antibodies.

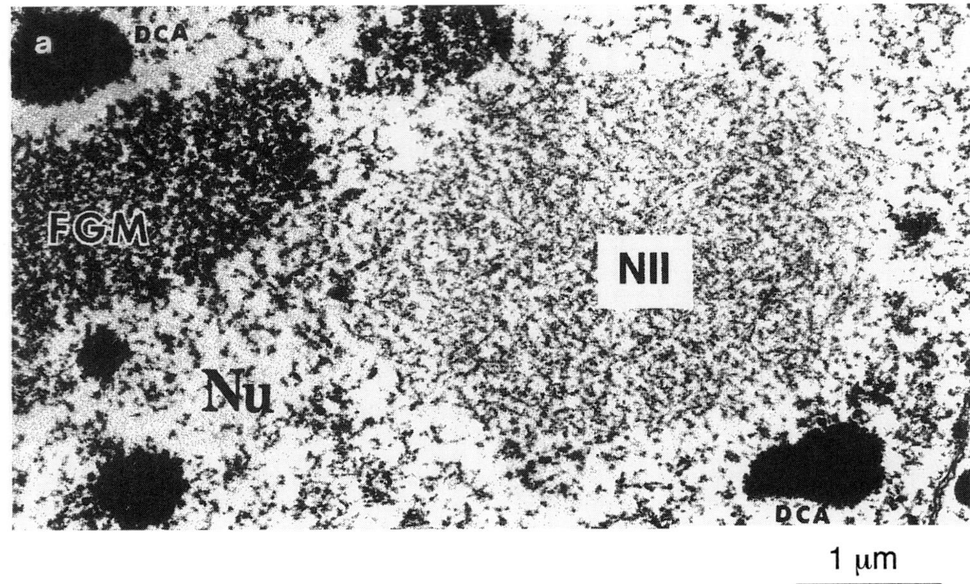

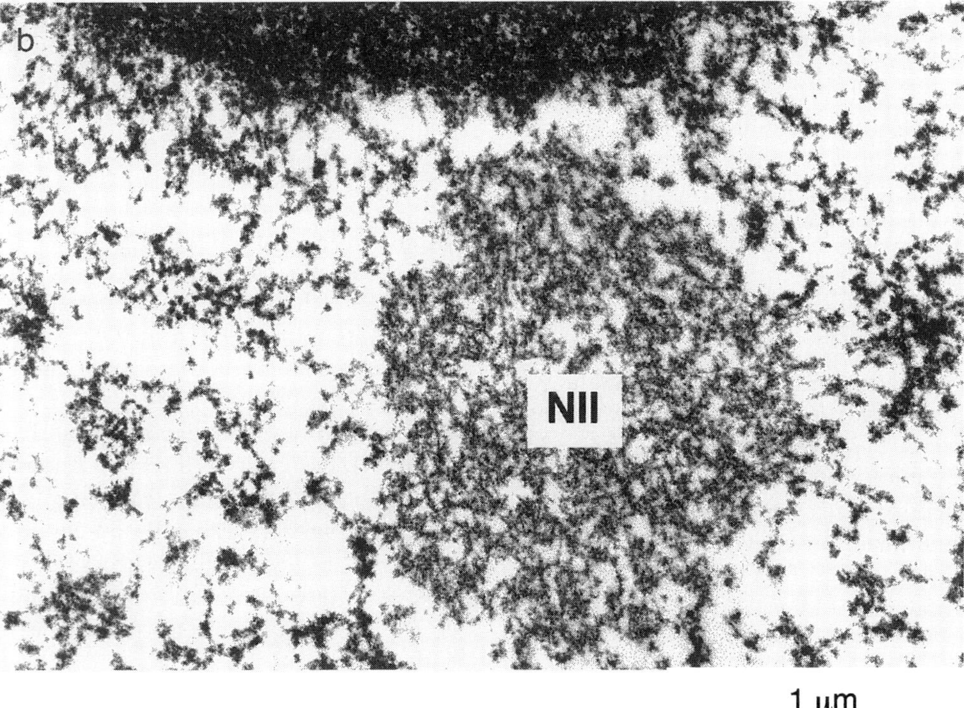

FIGURE 25-5 Electron micrographs showing the morphology of NII. (a) EM micrograph of caudate biopsy material from a 44-year-old HD patient who had been symptomatic for 12 years. The NII was described as a metamorphosis of filaments and fine granules. FGM, fibrillogranular mass; DCA, dense chromatin aggregates; Nu, nucleoplasm. (b) EM micrograph of a striatal neuron from an R6/5 homozygous transgenic mouse of 15 months of age. The NII has a strikingly similar morphology to that in (a). Note also the clumped appearance of the surrounding chromatin. Reproduced from "Advances in Neurology," Vol. 23, pp. 95–122, 1979, with the permission of Lippincott–Raven Publishers.

F. The Transgene Protein Can Form Amyloid-Like Protein Aggregates *in Vitro*

The NII have thus far only been found to be immunoreactive for N-terminal and ubiquitin antibodies, and recent *in vitro* binding assays indicate that the formation of aggregates may not require further interactions [56]. GST fusion proteins corresponding to exon 1 of the HD gene (the transgene protein) were found to form aggregates *in vitro* when containing highly expanded repeats of 83 and 122 glns (GST–HD83 and GST–HD122), whereas the protein with 51 glutamines (GST–HD51) was soluble under these conditions. Removal of the GST tag by factor Xa or trypsin proteolytic cleavage renders the protein more insoluble, allowing the cleaved protein to form aggregates when carrying the pathogenic expansion of 51 glns but not when carrying normal expansions (20 and 30 glns). Electron microscopic analysis of protein fractions obtained by proteolytic cleavage of GST–HD51 showed numerous clusters of high-molecular-weight fibrils and ribbon-like structures, ultrastructurally indistinguishable from purified amyloids (Fig. 25-5). The fibrils were not uniform in length. Most were 1–5 mm long and clustered into bundles, but single shorter fibrils (100–200 nm) were also detected. Fibrils were generally 10 nm in diameter. Frequently, clots were seen on one or both ends of the fibrils corresponding to partially cleaved GST (Fig. 25-6d). In strong contrast to GST–HD51, the GST–HD20 and HD30 proteins did not show any tendency to form ordered high-molecular-weight structures, either with or without protease treatment. The fibrils were stained with Congo red and when examined under polarised light, green colour and birefringence was detected which is indicative of β-amyloids [76, 77]. It has been previously demonstrated by X-ray diffraction studies that synthetic peptides containing polyglns form β-sheets strongly held together by hydrogen bonds [54]. Perutz [55] proposed that elongated polyglns may form stable hairpins when the number of glns exceeds 41. He suggested that at a certain critical length the loss of entropy during the formation of the hairpin may become negligible and the system would be strongly stabilised by a gain of entropy due to the liberated water molecules [55].

The isolation of nuclear fractions from the brain and kidney of 12-week R6/2 transgenic mice demonstrated the presence of structures similar to the *in vitro* aggregates in the brains of symptomatic transgenic mice [56]. Western blotting with N-terminal huntingtin and ubiquitin antibodies identified the presence of a high-molecular-weight band on a Western blot in the nuclear fractions similar to that obtained by proteolytic cleavage of protein GST–HD51. The high-molecular-weight immunoreactive band was not seen in the cytoplasmic fraction and not at all in fractions from control mice. In transgenic mice, approximately 10- to 20-fold more high-molecular-weight protein was present in brain than in kidney. It is possible that the transgene NII are composed exclusively of transgene protein aggregates held together by β-sheet structures through polygln interactions. Either the structure of an expanded polygln or its molecular interactions present a target for ubiquitination that does not result in protein degradation. Ubiquitination can occur in the nucleus as ubiquitin-activating enzyme E1 is localised to the nucleus in a cell-cycle-dependent manner [78, 79]. Ubiquitin molecules would be covalently bound to lysine residues at the N or C terminus of the transgene protein and would not necessarily interfere with the formation or stability of the polygln β-sheets.

IV. MECHANISTIC IMPLICATIONS FOR THE MOLECULAR BASIS OF HUNTINGTON'S DISEASE

The transgene protein is translocated to the nucleus, ubiquitinated, and starts to form aggregates, but the order in which these events occur is unknown. The ability to detect NII by immunocytochemistry with huntingtin before they can be detected with ubiquitin may suggest that the aggregates form prior to ubiquitination. However, the homogeneous nature of the ubiquitin staining tends to argue against this. The transgene protein does not contain a nuclear localisation signal and the mechanism by which it is transported to the nucleus is not known. It is possible that it is translocated by an as yet unidentified interacting protein or, alternatively, the possibility that a polygln expansion can itself act as a nuclear localisation signal has not been ruled out. It is not known whether the transgene protein aggregates initiate in the cytoplasm, or only form once the protein has entered the nucleus. The progressive increase in size of the NII suggests that the transgene protein continuously enters the nucleus and that at any one time there will be within the nucleus a fraction of transgene protein that has not yet aggregated with an inclusion.

The observation of NII in certain key neurons within the transgenic mouse brains prior to the observation of any other neuropathological or phenotypic change places the nucleus as the primary site of pathogenicity and suggests that the formation of NII represents the primary molecular event underlying the neurological phenotype. It is clear that the presence of an NII can bring about pronounced neurological dysfunction in the absence of cell death. There is evidence to suggest that the presence of the transgene protein in the nucleus,

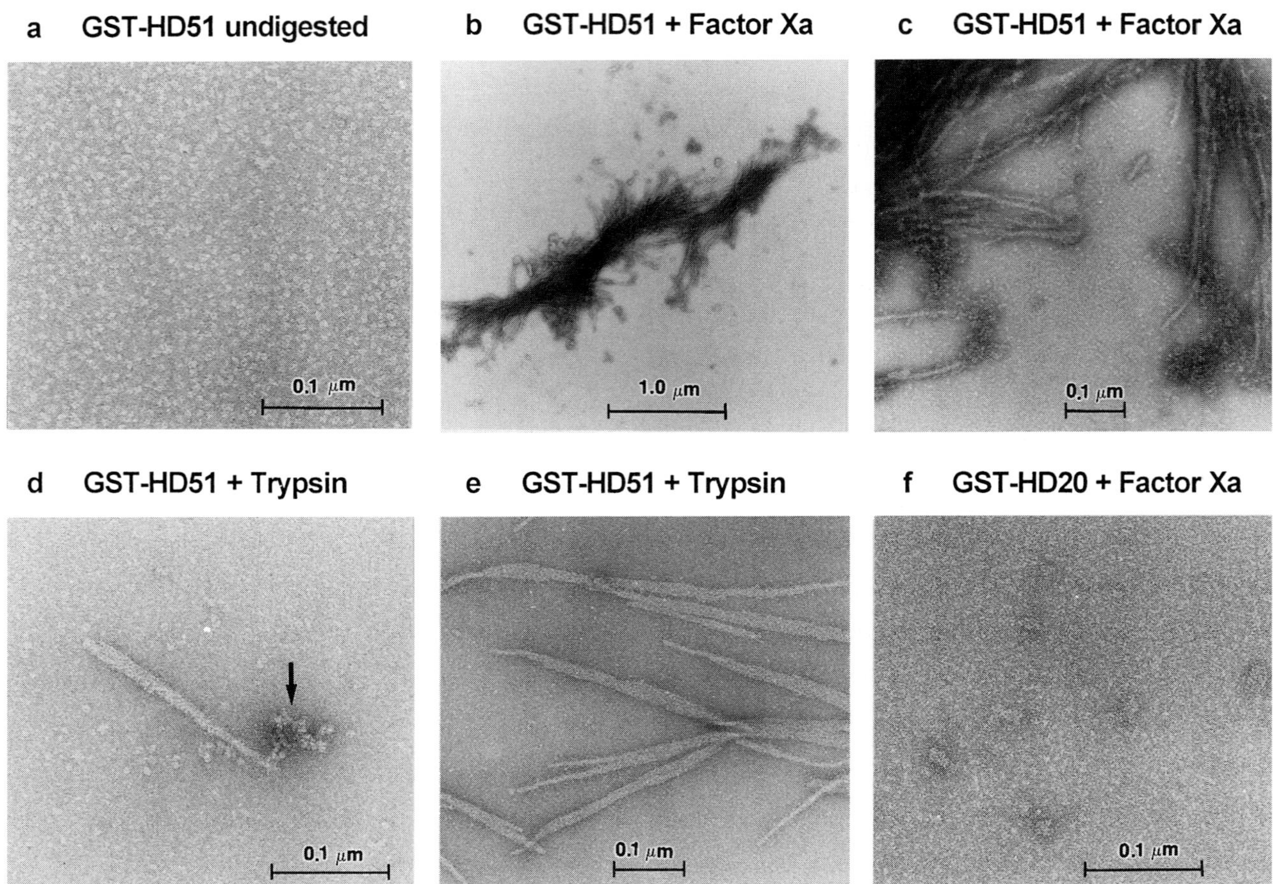

FIGURE 25-6 Electron micrographs of native GST–HD proteins and their factor Xa and trypsin cleavage products. GST fusion proteins were affinity-purified, protease-treated, negatively stained with uranyl acetate, and viewed by electron microscopy. The undigested proteins appear as homogeneous round particles (a). Removal of the GST tag with factor Xa results in the formation of amyloid-like fibrils and intermediate structures (b + c). After partial cleavage (3 h) of GST–HD51 with trypsin, the fibrils are associated with terminal clots (d) whereas complete digestion (16 h) produces only fibrils (e). Removal of the GST tag from GST–HD20 shows evidence of some clumping (f). Reproduced from *Cell* **90,** 549–558, 1997, with the permission of Cell Press.

either before or after complexing with an NII, has a selective effect on the transcription levels of specific neurotransmitter receptors [80]. It is possible that the expanded polygln interacts with other transcription factors; indeed the TATA binding protein contains a polygln tract of a size corresponding to the lower end of the pathogenic range of several of the polyglutamine diseases [30]. It cannot be discounted that the presence of an NII specifically perturbs transcription levels. Immunocytochemistry shows that the constitutive transcription factors Fos B and NGFIA do not immunoreact with NII [72], suggesting either that the NII is tightly bound to DNA/chromatin, preventing access to these transcription factors, or that it excludes DNA/chromatin from the region of the nucleus in which it is located.

In HD patients, the NII can be detected with N-terminal antibodies but not with those raised against more C-terminal portions of huntingtin [75a, 75b]. This may suggest that in HD patients, huntingtin becomes cleaved and it is only the N-terminus carrying the polygln expansion that enters the nucleus. Demonstration that an N-terminal 80-kDa huntingtin fragment is cleaved from the full-length protein by apopain, a cysteine protease [52], indicates that the N-terminus of huntingtin is primarily accessible for proteases and distinct proteolytic cleavage products can be formed. The identification of NII that are immunoreactive for antibodies to ataxin 3 and ubiquitin in specific neuronal nuclei of SCA3 [81] postmortem brains and SCA1 [82] suggests that the formation of NII may represent the pathogenic basis of all polygln neurodegenerative disease. In addition, the demonstration that the C-terminus of ataxin 3 is more toxic to COS cells in transient transfections than the full-length protein carrying the same polygln expansion [53] adds further weight to a toxic fragment hypothesis and suggests that proteolytic cleavage may also be important in this disease.

V. CONCLUSION

The recent identification of NII in HD patients by immunocytochemistry indicates that the R6 lines represent an extremely useful model for the elucidation of the molecular basis of HD. Although the R6 lines contain only exon 1 of the HD gene and express a very truncated form of huntingtin, the selective neuronal vulnerability observed in HD is at least to some extent mirrored by the presence of NII in the transgenic lines. Therefore, the R6 lines are also likely to be informative with respect to the factors that specify the selective and differing patterns of neuropathology observed in these disorders. In addition, the early onset and rapid progression of the phenotype in line R6/2 make it an especially suitable resource for testing the efficacy of potential therapeutic agents.

Acknowledgments

We thank Marian DiFiglia, Christopher Ross, Henry Paulson, and Harry Orr for helpful discussions and Adrienne Knight for proofreading the manuscript. The authors are supported by funds from the MRC, Wellcome Trust, DFG, Hereditary Disease Foundation, and the Special Trustees of Guy's Hospital.

References

1. HDCRG (1993). A novel gene containing a trinucleotide repeat that is unstable on Huntington's disease chromosomes. *Cell* **72,** 971–983.
2. Harper, P. S. (1991). "Huntington's Disease," 22nd ed. W. B. Saunders, London.
3. Vonsattel, J.-P., Myers, R. H., Stevens, T. J., Ferrante, R. J., Bird, E. D., and Richardson, E. P. (1985). Neuropathological classification of Huntington's disease. *J. Neuropathol. Exp. Neurol.* **44,** 559–577.
4. de la Monte, S. M., Vonsattel, J.-P., and Richardson, E. P. (1988). Morphometric demonstration of atrophic changes in the cerebral cortex, white matter and neostriatum in Huntington's disease. *J. Neuropathol. Exp. Neurol.* **47,** 516–525.
5. Myers, R. H., Vonsattel, J. P., Paskevich, P. A., Kiely, D. K., Stevens, T. J., Cupples, L. A., Richardson, E. P., and Bird, E. D. (1991). Decreased neuronal and increased oligodendroglial densities in Huntington's disease caudate nucleus. *J. Neuropathol. Exp. Neurol.* **50,** 729–742.
6. Hedreen, J. C., and Folstein, S. E. (1995). Early loss of early neostriatal neurons in Huntington's disease. *J. Neuropathol. Exp. Neurol.* **54,** 105–120.
7. Myers, R. H., Vonsattel, J. P., Stevens, T. J., Cupples, L. A., Richardson, E. P., Martin, J. B., and Bird, E. D. (1988). Clinical and neuropathological assessment of severity in Huntington's disease. *Neurology* **38,** 341–347.
8. Stine, O. C., Pleasant, N., Franz, M. L., Abbott, M. H., Folstein, S. E., and Ross, C. A. (1993). Correlation between the onset age of Huntington's disease and length of the trinucleotide repeat in IT-15. *Hum. Mol. Genet.* **2,** 1547–1549.
9. Rubinsztein, D. C., Leggo, J., Coles, R., Almqvist, E., Biancalana, V., Cassiman, J.-J., Chotai, K., Connarty, M., Crauford, D., Curtis, A., Curtis, D., Davidson, M. J., Differ, A.-M., Dode, C., Dodge, A., Frontali, M., Ranen, N. G., Stine, O. C., Sherr, M., Abbott, M. H., Franz, M. L., Graham, C. A., Harper, P. S., Hedreen, J. C., Jackson, A., Kaplan, J.-C., Losekoot, M., MacMillan, J. C., Morrison, P., Trottier, Y., Novelletto, A., Simpson, S. A., Theilmann, J., Whittaker, J. L., Folstein, S. E., Ross, C. A., and Hayden, M. R. (1996). Phenotypic characterisation of individuals with 30–40 CAG repeats in the Huntington's disease (HD) gene reveals HD cases with 36 repeats and apparently normal elderly individuals with 36–39 repeats. *Am. J. Hum. Genet.* **59,** 16–22.
10. Barron, L. H., Warner, J. P., Porteous, M., Holloway, S., Simpson, S., Davidson, R., and Brock, D. J. H. (1993). A study of the Huntington's disease associated trinucleotide repeat in the Scottish population. *J. Med. Genet.* **30,** 1003–1007.
11. Sathasivam, K., Amaechi, I., Mangiarini, L., and Bates, G. P. (1997). Identification of an HD patient with a $(CAG)_{180}$ repeat expansion and the propagation of highly expanded CAG repeats in lambda phage. *Hum. Genet.* **99,** 692–695.
12. Duyao, M., Ambrose, C., Myers, R., Novelletto, A., Persichetti, F., Frontali, M., Folstein, S., Ross, C., Franz, M., Abbott, M., Gray, J., Conneally, M. P., Young, A., Penney, J., Hollingsworth, Z., Shoulson, I., Lazzarini, A., Falek, A., Koroshetz, W., Sax, D., Bird, E., Vonsattel, J., Bonilla, E., Alvir, J., Conde, J. B., Cha, J.-H., Dure, L., Gomez, F., Ramos, M., Sanchez-Ramos, J., Snodgrass, S., de Young, M., Wexler, N., Moscowitz, C., Penchaszadeh, G., MacFarlane, H., Anderson, M., Jenkins, B., Srinidhi, J., Barnes, G., Gusella, J., and MacDonald, M. (1993). Trinucleotide repeat length instability and age of onset in Huntington's disease. *Nature Genet.* **4,** 387–392.
13. Telenius, H., Kremer, H. P. H., Theilmann, J., Andrew, S. E., Almquist, E., Anvret, M., Greenberg, C., Greenberg, J., Lucotte, G., Squitieri, F., Starr, E., Goldberg, Y. P., and Hayden, M. R. (1993). Molecular analysis of juvenile Huntington disease: the major influence on $(CAG)_n$ repeat length is the sex of the affected parent. *Hum. Mol. Genet.* **2,** 1535–1540.
14. Ranen, N. G., Stine, O. C., Abbott, M. H., Sherr, M., Codori, A.-M., Franz, M. L., Chao, N. I., Chung, A. S., Pleasant, N., Callahan, C., Kasch, L. M., Ghaffari, M., Chase, G. A., Kazazian, H. H., Brandt, J., Folstein, S. E., and Ross, C. A. (1995). Anticipation and instability of IT15 $(CAG)_n$ repeats in parent-offspring pairs with Huntington's disease. *Am. J. Hum. Genet.* **57,** 593–602.
15. Telenius, H., Almquist, E., Kremer, B., Spence, N., Squitieri, F., Nichol, K., Grandell, U., Starr, E., Benjamin, C., Castaldo, I., Calabrese, O., Anvret, M., Goldberg, Y. P., and Hayden, M. R. (1995). Somatic mosaicism in sperm is associated with intergenerational $(CAG)_n$ changes in Huntington's disease. *Hum. Mol. Genet.* **4,** 189–195.
16. Leeflang, E. P., Zhang, L., Tavare, S., Hubert, R., Srinidhi, J., MacDonald, M. E., Myers, R. H., de Young, M., Wexler, N. S., Gusella, J. F., and Arnheim, N. (1995). Single sperm analysis of the trinucleotide repeats in the Huntington's disease gene: quantification of the mutation frequency spectrum. *Hum. Mol. Genet.* **4,** 1519–1526.
17. Telenius, H., Kremer, B., Goldberg, Y. P., Theilmann, J., Andrew, S. E., Zeisler, J., Adam, S., Greenberg, C., Ives, E. J., Clarke, L. A., and Hayden, M. R. (1994). Somatic and gonadal mosaicism of the Huntington disease gene CAG repeat in brain and sperm. *Nature Genet.* **6,** 409–413.
18. Aronin, N., Chase, K., Young, C., Sapp, E., Schwarz, C., Matta, N., Kornreich, R., Landwehrmeyer, B., Bird, E., Beal, M. F., Vonsattel, J.-P., Smith, T., Carraway, R., Boyce, F. M., Young, A. B., Penney, J. B., and DiFiglia, M. (1995). CAG expansion

affects the expression of mutant huntingtin in the Huntington's disease brain. *Neuron* **15,** 1193–1201.

19. Ambrose, C. M., Duyao, M. P., Barnes, G., Bates, G. P., Lin, C. S., Srinidhi, J., Baxendale, S., Hummerich, H., Lehrach, H., Altherr, M., Wasmuth, J., Buckler, A., Church, D., Housman, D., Berks, M., Micklem, G., Durbin, R., Dodge, A., Read, A., Gusella, J., and MacDonald, M. E. (1993). Structure and expression of the Huntington's disease gene: evidence against simple inactivation due to an expanded CAG repeat. *Somat. Cell Mol. Genet.* **20,** 27–38.
20. Baxendale, S., Abdulla, S., Elgar, G., Buck, D., Berks, M., Micklem, G., Durbin, R., Bates, G., Brenner, S., Beck, S., and Lehrach, H. (1995). Comparative sequence analysis of the human and pufferfish Huntington's disease genes. *Nature Genet.* **10,** 67–75.
21. Lin, B., Rommens, J. M., Graham, R. K., Kalchman, M., MacDonald, H., Nasir, J., Delaney, A., Goldberg, Y. P., and Hayden, M. R. (1993). Differential 3′ polyadenylation of the Huntington disease gene results in two mRNA species with variable tissue expression. *Hum. Mol. Genet.* **2,** 1541–1545.
22. Jou, Y.-S., and Myers, R. M. (1995). Evidence from antibody studies that the CAG repeat in the Huntington disease gene is expressed in the protein. *Hum. Mol. Genet.* **4,** 465–469.
23. Trottier, Y., Devys, D., Imbert, G., Sandou, F., An, I., Lutz, Y., Weber, C., Agid, Y., Hirsch, E. C., and Mandel, J.-L. (1995). Cellular localisation of the Huntington's disease protein and discrimination of the normal and mutated forms. *Nature Genet.* **10,** 104–110.
24. Ide, K., Nobuyuki, N., Masuda, N., Goto, J., and Kanazawa, I. (1995). Abnormal gene product identified in Huntington's disease lymphocytes and brain. *Biochem. Biophys. Res. Commun.* **209,** 1119–1125.
25. Schilling, G., Sharp, A. H., Loev, S. J., Wagster, M. V., Li, S.-H., Stine, O. C., and Ross, C. A. (1995). Expression of the Huntington's disease (IT15) protein product in HD patients. *Hum. Mol. Genet.* **4,** 1365–1371.
26. Li, S. H., Schilling, G., Young, W. S., Li, X. J., Margolis, R. L., Stine, O. C., Wagster, M. V., Abbott, M. H., Franz, M. L., Ranen, N. G., Folstein, S. E., Hedreen, J. C. and Ross, C. A. (1993). Huntington's disease gene (it-15) is widely expressed in human and rat tissues. *Neuron* **11,** 985–993.
27. Strong, T. V., Tagle, D. A., Valdes, J. M., Elmer, L. W., Boehm, K., Swaroop, M., Kaatz, K. W., Collins, F. S., and Albin, R. L. (1993). Widespread expression of the human and rat Huntington's disease gene in brain and nonneuronal tissues. *Nature Genet.* **5,** 259–263.
28. Landwehrmeyer, G. B., McNeil, S. M., Dure, L. S., Ge, P., Aizawa, H., Huang, Q., Ambrose, C. M., Duyao, M. P., Bird, E. D., Bonilla, E., de Young, M., Avila-Gonzales, A. J., Wexler, N. S., DiFiglia, M., Gusella, J. F., MacDonald, M. E., Penney, J. B., Young, A. B., and Vonsattel, J.-P. (1995). Huntington's disease gene: regional and cellular expression in brain of normal and affected individuals. *Ann. Neurol.* **37,** 218–230.
29. Dure, L. S., Landwehrmeyer, G. B., J. Golden, McNeil, S., Ge, P., Aizawa, H., Huang, Q., Ambrose, C. M., Duyao, M. P., Bird, E. D., DiFiglia, M., Gusella, J. F., MacDonald, M. E., Penney, J. B., Young, A. B., and Vonstattel, J.-P. (1994). IT15 gene expression in fetal human brain. *Brain Res.* **659,** 33–41.
30. Trottier, Y., Lutz, Y., Stevanin, G., Imbert, G., Devys, D., Cancel, G., Sandou, F., Weber, C., David, G., Tora, L., Agid, Y., Brice, A., and Mandel, J.-L. (1995). Polyglutamine expansion as a pathological epitope in Huntington's disease and four dominant cerebellar ataxias. *Nature* **378,** 403–406.
31. Sharp, A. H., Loev, S. J., Schilling, G., Li, S.-H., Li, X.-J., Bao, J., Wagster, M. V., Kotzuk, J. A., Steiner, J. P., Lo, A., Hedreen, J., Sisodia, S., Snyder, S. H., Dawson, T. M., Ryugo, D. K., and Ross, C. A. (1995). Widespread expression of Huntington's disease gene (IT15) protein product. *Neuron* **14,** 1065–1074.
32. DiFiglia, M., Sapp, E., Chase, K., Schwarz, C., Meloni, A., Young, C., Martin, E., Vonstattel, J.-P., Carraway, R., Reeves, S. A., Boyce, F. M., and Aronin, N. (1995). Huntingtin is a cytoplasmic protein associated with vesicles in human and rat brain neurons. *Neuron* **14,** 1075–1081.
33. Gutekunst, C.-A., Levey, A. I., Heilman, C. J., Whaley, W. L., Yi, H., Nash, N. R., Rees, H. D., Madden, J. J., and Hersch, S. M. (1995). Identification and localisation of huntingtin in brain and human lymphoblastoid cell lines with anti-fusion protein antibodies. *Proc. Natl. Acad. Sci. USA* **92,** 8710–8714.
34. Kosinski, C. M., Cha, J. H., Young, A. B., Persichetti, F., MacDonald, M., Gusella, J. F., Penney, J. B., and Standaert, D. G. (1997). Huntingtin immunoreactivity in the rat neostriatum: differential accumulation in projection and interneurons. *Exp. Neurol.* **144,** 239–247.
35. Ferrante, R. J., Gutekunst, C.-A., Persichetti, F., McNeil, S. M., Kowall, N. W., Gusella, J. F., MacDonald, M. E., Beal, M. F., and Hersch, S. M. (1997). Heterogeneous topographic and cellular distribution of huntingtin expression in the normal human neostriatum. *J. Neurosci.* **17,** 3052–3063.
36. La Spada, A. R., Wilson, E. M., Lubahn, D. B., Harding, A. E., and Fischbeck, K. H. (1991). Androgen receptor gene mutations in X-linked spinal and bulbar muscular atrophy. *Nature* **352,** 77–79.
37. Koide, R., Ikeuchi, T., Onodera, O., Tanaka, H., Igarashi, S., Endo, K., Takahashi, H., Kondo, R., Ishikawa, A., Hayashi, T., Saito, M., Tomoda, A., Miike, T., Naito, H., Ikuta, F., and Tsuji, S. (1994). Unstable expansion of CAG repeat in hereditary dentatorubral-pallidoluysian atrophy (DRPLA). *Nature Genet.* **6,** 9–13.
38. Nagafuchi, S., Yanagisawa, H., Sato, K., Shirayama, T., Ohsaki, E., Bundo, M., Takeda, T., Tadokoro, K., Kondo, I., Murayama, N., Tanaka, Y., Kikushima, H., Umino, K., Kurosawa, H., Furukawa, T., Nihei, K., Inoue, T., Sano, A., Komure, O., Takahashi, M., Yoshizawa, T., Kanazawa, I., and Yamada, M. (1994). Dentatorubral and pallidoluysian atrophy expansion of an unstable CAG trinucleotide on chromosome 12p. *Nature Genet.* **6,** 14–18.
39. Orr, H. T., Chung, M., Banfi, S., Kwiatkowski, T. J., Jr., Servadio, A., Beaudet, A. L., McCall, A. E., Duvick, L. A., Ranum, L. P. W., and Zoghbi, H. Y. (1993). Expansion of an unstable trinucleotide CAG repeat in spinocerebellar ataxia type 1. *Nature Genet.* **4,** 221–226.
40. Imbert, G., Sandou, F., Yvert, G., Devys, D., Trottier, Y., Garnier, J.-M., Weber, C., Mandel, J.-L., Cancel, G., Abbas, N., Durr, A., Didierjean, O., Stevanin, G., Agid, Y., and Brice, A. (1996). Cloning of the gene for spinocerebellar ataxia 2 reveals a locus with high sensitivity to expanded CAG/glutamine repeats. *Nature Genet.* **14,** 285–291.
41. Pulst, S.-M., Nechiporuk, A., Nechiporuk, T., Gispert, S., Chen, X.-N., Lopes-Cendes, I., Pearlman, S., Starkman, S., Orozco-Diaz, G., Lunkes, A., DeJong, P., Rouleau, G. A., Auburger, G., Korenberg, J. R., Figueroa, C., and Sahba, S. (1996). Moderate expansion of a normally biallelic trinucleotide repeat in spinocerebellar ataxia type 2. *Nature Genet.* **14,** 269–276.
42. Sampei, K., Takano, H., Igarashi, S., Sato, T., Oyake, M., Sasaki, H., Wakisaka, A., Tashiro, K., Ishida, Y., Ikeuchi, T., Koide, R., Saito, M., Sato, A., Tanaka, T., Hanyu, S., Takiyama, Y., Nishizawa, M., Shimizu, N., Nomura, Y., Segawa, M., Iwabuchi, K., Eguchi, I., Tanaka, H., Takahashi, H., and Tsuji, S. (1996). Identification of the spinocerebellar ataxia type 2 gene using a direct identification of repeat expansion and cloning technique, DIRECT. *Nature Genet.* **14,** 277–284.

43. Kawaguchi, Y., Okamoto, T., Taniwaki, M., Aizawa, M., Inoue, M., Katayama, S., Kawakami, H., Nakamura, S., Nishimura, M., Akiguchi, I., Kimura, J., Narumiya, S., and Kakizuka, A. (1994). CAG expansions in a novel gene for Machado-Joseph disease at chromosome 14q32.1. *Nature Genet.* **8,** 221–228.

44. Zhuchenko, O., Bailey, J., Bonnen, P., Ashizawa, T., Stockton, D., Amos, C., Dobyns, W. B., Subramony, S. H., Zoghbi, H. Y., and Li, C. C. (1997). Autosomal dominant cerebellar ataxia (SCA6) associated with small polyglutamine expansions in the alpha 1A-voltage dependent calcium channel. *Nature Genet.* **15,** 62–69.

45. Ross, C. A. (1995). When more is less: pathogenesis of glutamine repeat neurodegenerative diseases. *Neuron* **15,** 493–496.

46. Li, X.-J., Li, S.-H., Sharp, A. H., Nucifora, F. C., Schilling, G., Lanahan, A., Worley, P., Snyder, S. H., and Ross, C. A. (1995). A huntingtin-associated protein enriched in brain with implications for pathology. *Nature* **378,** 398–402.

47. Wanker, E. E., Rovira, C., Scherzinger, E., Hasenbank, R., Walter, S., Tait, D., Colicelli, J., and Lehrach, H. (1997). HIP-1: a huntingtin interacting protein isolated by the yeast two-hybrid system. *Hum. Mol. Genet.* **6,** 487–495.

48. Kalchman, M. A., Koide, H. B., McCutcheon, K., Graham, R. K., Nichol, K., Nishiyama, K., Kazemi-Esfariani, P., Lynn, F. C., Wellington, C., Metzler, M., Goldberg, Y. P., Kanazawa, I., Gietz, R. D., and Hayden, M. R. (1997). *HIP1,* a human homologue of *S. cerevisiae* Sla2p, interacts with mambrane-associated huntingtin in the brain. *Nature Genet.* **16,** 44–53.

49. Kalchman, M. A., Graham, R. K., Xia, G., Koide, H. B., Hodgson, J. G., Graham, K. C., Goldberg, Y. P., Gietz, R. D., Pickart, C. M., and Hayden, M. R. (1996). Huntingtin is ubiquinated and interacts with a specific ubiquitin-conjugating enzyme. *J. Biol. Chem.* **271,** 19385–19394.

50. Burke, J. R., Enghild, J. J., Martin, M. E., Jou, Y.-S., Myers, R. M., Roses, A. D., Vance, J. M., and Strittmatter, W. J. (1996). Huntingtin and DRPLA proteins selectively interact with the enzyme GAPDH. *Nature Med.* **2,** 347–350.

51. Banfi, S., Servadio, A., Chung, M., Kwiatkowski, T., McCall, A. E., Duvick, L., Shen, Y., Roth, E. J., Orr, H. T., and Zoghbi, H. Y. (1994). Identification and characterisation of the gene causing type 1 spinocerebellar ataxia. *Nature Genet.* **7,** 513–520.

52. Goldberg, Y. P., Nicholson, D. W., Rasper, D. M., Kalchman, M. A., Koide, H. B., Graham, R. K., Bromm, M., Kazemi-Esfarjani, P., Thornberry, N. A., Vaillancourt, J. P., and Hayden, M. R. (1996). Cleavage of huntingtin by apopain, a proapoptotic cysteine protease, is modulated by the polyglutamine tract. *Nature Genet.* **13,** 442–449.

53. Ikeda, H., Yamaguchi, M., Sugai, S., Aze, Y., Narumiya, S., and Kakizuka, A. (1996). Expanded polyglutamine in the Machado-Joseph disease protein induces cell death *in vitro* and *in vivo.* *Nature Genet.* **13,** 196–202.

54. Perutz, M. F., Johnson, T., Suzuki, M., and Finch, J. T. (1994). Glutamine repeats as polar zippers: their possible role in inherited neurodegenerative diseases. *Proc. Natl. Acad. Sci. USA* **91,** 5355–5358.

55. Perutz, M. F. (1996). Glutamine repeats and inherited neurodegenerative diseases: molecular aspects. *Curr. Opin. Struct. Biol.* **6,** 848–858.

56. Scherzinger, E., Lurz, R., Turmaine, M., Mangiarini, L., Hollenbach, B., Hasenbank, R., Bates, G. P., Davies, S. W., Lehrach, H., and Wanker, E. E. (1997). Huntingtin encoded polyglutamine expansions form amyloid-like protein aggregates *in vitro* and *in vivo.* *Cell* **90,** 549–558.

57. Barnes, G. T., Duyao, M. P., Ambrose, C. M., McNeil, S., Persichetti, F., Srinidhi, J., Gusella, J. F., and MacDonald, M. E. (1994). Mouse Huntington's disease gene homolog (Hdh). *Somat. Cell Mol. Genet.* **20,** 87–97.

58. Lin, B., Nasir, J., MacDonald, H., Hutchinson, G., Graham, R. K., Rommens, J. M., and Hayden, M. R. (1994). Sequence of the murine Huntington's disease gene: evidence for conservation, and polymorphism in a triplet (CCG) repeat alternate splicing. *Hum. Mol. Genet.* **3,** 85–92.

59. Schmitt, I., Bachner, D., Megow, D., Henklein, P., Hameister, H., Epplen, J. T., and Riess, O. (1995). Expression of the Huntington disease gene in rodents: cloning of the rat homologue and evidence for downregulation in non-neuronal tissues during development. *Hum. Mol. Genet.* **4,** 1173–1182.

60. Joyner, A. L. (1993) Gene targeting. A practical approach. *In* "The Practical Approach Series" (D. Rickwood and B. D. Hames, Eds.). Oxford University Press, New York.

61. Nasir, J., Floresco, S. B., O'Kusky, J. R., Diewert, V. M., Richman, J. M., Zeisler, J., Borowski, A., Marth, J. D., Phillips, A. G., and Hayden, M. R. (1995). Targeted disruption of the Huntington's disease gene results in embryonic lethality and behavioral and morphological changes in heterozygotes. *Cell* **81,** 811–823.

62. Duyao, M. P., Auerbach, A. A., Ryan, A., Persichetti, F., Barnes, G. T., McNeil, S. M., Ge, P., Vonstattel, J.-P., Gusella, J. F., Joyner, A. L., and MacDonald, M. E. (1995). Inactivation of the mouse Huntington's disease gene homolog *Hdh. Science* **269,** 407–410.

63. Zeitlin, S., Liu, J.-P., Chapman, D. L., Papaioannou, V. E., and Estradiatis, A. (1995). Increased apoptosis and early embryonic lethality in mice nullizygous for the Huntington's disease gene homologue. *Nature Genet.* **11,** 155–163.

64. Hodgson, J. G., Smith, D. J., McCutcheon, K., Koide, H. B., Nishiyama, K., Dinulos, M. B., Stevens, M. E., Bissada, N., Nasir, J., Kanazawa, I., Disteche, C. M., Rubin, E. M., and Hayden, M. R. (1996). Human huntingtin derived from YAC transgenes compensates for loss of murine huntingtin by rescue of the embryonic lethal phenotype. *Hum. Mol. Genet.* **5,** 1875–1885.

65. Brenner, S., Elgar, G., Sandford, R., Macrae, A., Venkatesh, B., and Aparicio, S. (1993). Characterisation of the puffer fish (Fugu) genome as a compact model vertebrate genome. *Nature* **366,** 265–268.

65a. Sathasivam, K., Baxendale, S., Mangiarini, L., Bertaux, F., Kanazawa, I., Lehrach, H., and Bates, G. P. (1997). Aberrant processing of the *Fugu* HD (*FrHD*) mRNA in mouse cells and in transgenic mice. *Hum. Mol. Genet.* **6,** 2141–2149.

66. Mangiarini, L., Sathasivam, K., Seller, M., Cozens, B., Harper, A., Hetherington, C., Lawton, M., Trottier, Y., Lehrach, H., Davies, S. W., and Bates, G. P. (1996). Exon 1 of the Huntington's disease gene containing a highly expanded CAG repeat is sufficient to cause a progressive neurological phenotype in transgenic mice. *Cell* **87,** 493–506.

67. Mangiarini, L., Sathasivam, K., Mahal, A., Mott, R., Seller, M., and Bates, G. P. (1997). Instability of highly expanded CAG repeats in transgenic mice is related to expression of the transgene. *Nature Genet.* **15,** 197–200.

68. Chong, S. S., McCall, A. E., Cota, J., Subramony, S. H., Orr, H. T., Hughes, M. R., and Zoghbi, H. Y. (1995). Gametic and somatic tissue-specific heterogeneity of the expanded SCA1 CAG repeat in spinocerebellar ataxia type 1. *Nature Genet.* **10,** 344–350.

69. Ueno, S., Kondoh, K., Kotani, Y., Komure, O., Kuno, S., Kawai, J., Hazama, F., and Sano, A. (1995). Somatic mosaicism of CAG repeat in dentatorubral-pallidoluysian atrophy (DRPLA). *Hum. Mol. Genet.* **4,** 663–666.

70. Takano, H., Onodera, O., Takahashi, H., Igarashi, S., Yamada, M., Oyake, M., Ikeuchi, T., Koide, R., Tanaka, H., Iwabuchi, K., and Tsuji, S. (1996). Somatic mosaicism of expanded CAG repeats

in brains of patients with dentatorubral-pallidoluysian atrophy: cellular population-dependent dynamics of mitotic instability. *Am. J. Hum. Genet.* **58,** 1212–1222.

71. Tanaka, F., Sobue, G., Doyu, M., Ito, Y., Yamamoto, M., Shimada, N., Yamamoto, K., Riku, S., Hshizume, Y., and Mitsuma, T. (1996). Differential pattern of tissue-specific somatic mosaicism of expanded CAG trinucleotide repeat in dentatorubral-pallidoluysian atrophy, Machado-Joseph disease, and X-linked recessive spinal and bulbar muscular atrophy. *J. Neurol. Sci.* **135,** 43–50.
72. Davies, S. W., Turmaine, M., Cozens, B. A., DiFiglia, M., Sharp, A. H., Ross, C. A., Scherzinger, E., Wanker, E. E., Mangiarini, L., and Bates, G. P. (1997). Formation of neuronal intranuclear inclusions (NII) underlies the neurological dysfunction in mice transgenic for the HD mutation. *Cell* **90,** 537–548.
73. Roizin, L., Stellar, S., and Liu, J. C. (1979). Neuronal nuclear-cytoplasmic changes in Huntington's chorea: electron microscope investigations. *In* "Advances in Neurology" (T. N. Chase, N. S. Wexler, and A. Barbeau, Eds.), Vol 23, pp. 95–122. Raven Press, New York.
74. Roos, R. A. C., and Bots, G. T. A. M. (1983). Nuclear membrane indentations in Huntington's chorea. *J. Neurol. Sci.* **61,** 37–47.
75. Tellez-Nagel, I., Johnson, B., and Terry, R. D. (1974). Studies on brain biopsies of patients with Huntington's chorea. *J. Neurocytol.* **3,** 308–332.

75a. DiFiglia, M., Sapp, E., Chase, K. O., Davies, S. W., Bates, G. P., Vonsattel, J.-P., and Aronin, N. (1997). Aggregation of huntingtin in neuronal intranuclear inclusions and dystrophic neurites in brain. *Science* **277,** 1990–1993.

75b. Becher, M. W., Kotzuk, J. A., Sharp, A. H., Davies, S. W., Bates, G. P., Price, D. L., and Ross, C. A. (1988). Intranuclear neuronal inclusions in Huntington's disease and dentatorubral pallidoluysian atrophy: correlation between the density of inclusions and IT15 CAG repeat length. *Neurobiol. Dis.* [In press]

76. Glenner, G. G. (1980). Amyloid deposits and amyloidosis. *N. Engl. J. Med.* **302,** 1283–1292, 1333–1343.
77. Caputo, C. B., Fraser, P. E., Sobel, I. E., and Krischner, D. A. (1992). Amyloid-like properties of a synthetic peptide corresponding to the carboxy terminus of b-amyloid protein precursor. *Arch. Biochem. Biophys.* **292,** 199–205.
78. Grenfell, S. J., Trausch-Azar, J. S., Handley-Gearhart, P. M., Ciechanover, A., and Schwartz, A. L. (1994). Nuclear localisation of the ubiquitin-activating enzyme, E1, is cell-cycle-dependent. *Biochem J.* **300,** 701–708.
79. Stephen, A. G., Trausch-Azar, J. S., Ciechanover, A., and Schwartz, A. L. (1996). The ubiquitin-activating enzyme E1 is phosphorylated and localised to the nucleus in a cell cycle-dependent manner. *J. Biol. Chem.* **271,** 15608–15614.
80. Cha, J.-H., Kosinski, C., Kerner, J. A., Alsdorf, S. A., Mangiarini, L., Davies, S. W., Penney, J. B., Bates, G. P., and Young, A. (1998). Altered brain neurotransmitter receptors in transgenic mice expressing a portion of an abnormal Huntington's disease gene. *Proc. Natl. Acad. Sci. USA.* [In press]
81. Paulson, H. L., Perez, M. K., Trottier, Y., Trojanowski, J. Q., Subramony, S. H., Das, S. S., Vig, P., Mandel, J.-L., Fischbeck, K. H., and Pittman, R. N. (1997). Intranuclear inclusions of expanded polyglutamine protein in spinocerebellar ataxia type 3. *Neuron* **19,** 1–20.
82. Skinner, P. J., Koshy, B. T., Cummings, C. J., Klement, I. A., Helin, K., Servadio, A., Zoghbi, H. Y., and Orr, H. T. (1997). Ataxin-1 with an expanded glutamine tract alters nuclear matrix-associated structures. *Nature* **389,** 971–974.

PART X

Friedreich's Ataxia

CHAPTER 26

Friedreich's Ataxia

MASSIMO PANDOLFO Département de Médecine, Université de Montréal; Department of Neurology and Neurosurgery, McGill University; and Centre Hospitalier Universitaire de Montréal, Centre de Recherche Louis-Charles Simard, Montréal, Québec H2L 4M1, Canada

MICHEL KOENIG Faculté de Médecine, Université Louis Pasteur (ULP), Hôpitaux Universitaires de Strasbourg; and Institut de Génétique et de Biologie Moléculaire et Cellulaire, INSERM/CNRS/ULP, 67404 Illkirch, France

I. INTRODUCTION AND HISTORY

In 1863, Nicholaus Friedreich, Professor of Medicine in Heidelberg, described a "degenerative atrophy of the posterior columns of the spinal cord" leading to progressive ataxia, sensory loss, and muscle weakness, often associated with scoliosis, foot deformity, and cardiopathy [1, 2]. The disease might afflict several individuals in a sibship, but parents were never affected. Some critics, particularly Charcot, suspected that Friedreich's patients had tabes, a form of neurosyphilis. However, after Friedreich published additional cases in 1876 and 1877 [3, 4], it became generally accepted that he had described a new disease entity. Friedreich was able to pinpoint all the essential clinical and pathological features of the disease. He just missed the loss of deep tendon reflexes, and that was because Erb described these reflexes only in 1885. The new disease was given the name Friedreich's ataxia (currently abbreviated as FRDA) in 1882 by Brousse [5]. In 1890, Ladame [6] already reported more than 100 cases. These pioneering studies were done in the context of the late 19th and early 20th century flourishing of neuropathology and clinical neurology that gave shape to the modern nosological structure of these disciplines. Inevitably, progress occurred through debates and disputes. It is not surprising that the subsequent identification of clinically similar diseases and the presence of cases that could not be easily classified somewhat blurred the definition of FRDA for many years [7]. Only in the late 1970s a renewed interest in the disease prompted a reevaluation of the literature and the analysis of a large series of patients to establish clear diagnostic criteria. The landmark studies were done first by the Québec Collaborative Group [8], then by Harding [9]. Recessive inheritance was firmly established as an essential feature of FRDA [8–11]. The Québec Collaborative Group, after evaluating 50 patients of French Canadian ancestry, subdivided them into four groups: typical FRDA (33 patients), incomplete FRDA (3 patients, all from the same sibship), atypical FRDA (6 patients), and non-FRDA (8 patients). The clinical features of the typical FRDA group were identified and proposed as diagnostic criteria. Harding [9] felt, however, that such criteria, although appropriate for advanced cases, were too strict to allow diagnosis of early cases. She analyzed 115 patients from 90 families, some at an early stage of the disease, and proposed the diagnostic criteria shown in Table 26-1, which include certain signs and symptoms that may not be present at the onset, but have to manifest as the diseases evolves. These studies, as well as more recent ones [12, 13], identified a degree of variability in the clinical features of FRDA, including age at onset, rate of progression, severity, and extent of disease involvement. As remarked by Harding [9], such variability, sometimes occurring even within the same sibship [14, 15], is greater than in most other recessive neurological diseases. Patients may be confined to a wheelchair in their early teens, or may still be ambulant in their late thirties. Cardiac complications may be minimal or absent, or so severe as to cause premature death. Only some patients develop skeletal abnormalities, optic atrophy, diabetes mellitus, and sensorineural deafness. Atypical cases, with an overall FRDA-like phenotype but missing one or more essential features of typical FRDA, are sometimes observed in sibships along with typical cases [15], indicating that they represent extreme examples of the clinical spectrum of FRDA. But other atypical cases cluster in families, clouding classification. Examples include: Acadian FRDA, observed in a specific population of French origin living in North America, which has a milder course than classical FRDA and is rarely accompanied by a cardiomyopathy [16, 17]; late-onset Friedreich's ataxia (LOFA), a disease with all FRDA features but onset after 25 years of age [18, 19]; and Friedreich's ataxia with retained reflexes (FARR), a variant in which tendon reflexes in the lower limbs are preserved [20]. After the identification of the FRDA gene and of its most common mutation, the unstable hyperexpansion of a GAA triplet repeat polymorphism [21], genotype–phenotype correlations became possible, clarifying these issues and bringing important consequences on the diagnostic criteria for FRDA.

TABLE 26-1 Diagnostic Criteria for FRDA According to Harding [9]

Autosomal recessive inheritance

Onset before age 25

Within 5 years from onset
- Limb and truncal ataxia
- Absent tendon reflexes in the legs
- Extensor plantar responses
- Motor NCV > 40 m/sec in upper limbs with small or absent SAPs

After 5 years since onset
- As above plus dysarthria

Additional criteria, not essential for diagnosis, present in >2/3 of cases
- Scoliosis
- Pyramidal weakness of the legs
- Absent reflexes in upper limbs
- Distal loss of joint position and vibration sense in lower limbs
- Abnormal EKG

Other features, present in <50% of cases
- Nystagmus
- Optic atrophy
- Deafness
- Distal weakness and wasting
- Pes cavus
- Diabetes

II. CLINICAL AND PATHOLOGICAL ASPECTS OF FRIEDREICH'S ATAXIA

A. Epidemiology

FRDA is the most common of the hereditary ataxias in the Caucasian population. Except in areas where other ataxic disorders are exceptionally prevalent because of a founder effect, it generally accounts for half of the overall heredodegenerative ataxia cases, and for three-quarters of those with onset before age 25 [22]. The disease seems to have a fairly similar prevalence of around 2×10^{-5} in almost all studied Caucasian populations [10, 22–25], with local clusters due to a founder effect, as those observed in Rimouski, Québec [26], and in Kathikas-Arodhes, Cyprus [27]. FRDA is almost inexistant in Japan (S. Tsuji, M. Watanabe, and N. Tachi, personal communications).

B. Pathology

1. Central Nervous System

FRDA causes a characteristic pattern of central nervous system (CNS) pathology [28, 29]. Friedreich [1–4, 30] identified the degeneration of the posterior columns of the spinal cord as the hallmark of the disease. The posterior columns contain the central branches of the axons of large dorsal root ganglia (DRG) sensory neurons. These axons extend without interruption to the brainstem, forming the gracile (Goll) and cuneate (Burdach) tracts, the former originating at the lumbosacral level, the latter from cervicothoracic segments. The appearance of the posterior columns is shrunken, grayish, and translucent. Demyelination, loss of fibers, and fibrillary gliosis are more severe in the Goll than in the Burdach tract, i.e., the fibers originating more caudally are more severely affected. Atrophy is also observed in the spinocerebellar tracts, the dorsal being more affected than the ventral. Clarke's column, where the spinocerebellar tracts originate, shows severe loss of neurons. Therefore, the sensory systems providing information to the brain and cerebellum about the position and speed of body segments, particularly the lower limbs, are severely compromised in FRDA. Motor neurons in the ventral horns are well preserved, but the long crossed and uncrossed corticospinal motor tracts are atrophied. The pattern of atrophy of the long tracts of fibers suggests a "dying back" process [31], particularly clear in the case of corticospinal fibers. These are severely atrophic at the lumbar level, much less so in the cervical cord and in the brainstem, and of normal appearance in the cerebral peduncula. Such process implies that the primary biochemical defect in FRDA acts first or more severely on axons than on cell bodies, at least in some types of neurons, leading to a more severe involvement of the long sensory and motor pathways.

In the brainstem, what appears to be trans-synaptic degeneration, with intense gliosis, can be observed in the gracile and cuneate nuclei, where the dorsal column tracts terminate. The medial lemnisci, which continue the central sensory pathway after these nuclei, often show shrinking and loss of myelin, particularly in their ventral portion, deriving from the gracile nuclei. In most cases, the sensory pathways originating from the cranial nerves also show myelin pallor and loss of fibers, including the entering roots of the V, IX, and X nerves, the descending trigeminal tracts, and the solitary tracts. The auditory and vestibular systems are sometimes affected, with demyelination and loss of entering fibers from the VIII nerve, and atrophy of the lateral vestibular nuclei, of the cochlear nuclei, of the superior olives, and of the inferior colliculi. The accessory cuneate nuclei, corresponding to Clarke's column in the spinal cord, are markedly atrophic. Cranial motor nuclei and the medial longitudinal bundles appear to be spared.

Cerebellar atrophy is not a characteristic of FRDA. The cerebellar cortex shows only mild loss of Purkinje cells and occasional axonal torpedoes, particularly in the superior lamellae, late in the disease course. Quantitative analysis of synaptic terminals, however, suggests a loss of contacts over Purkinje cells bodies and proximal dendrites, probably reflecting the loss of cerebellar afferents [32]. Conversely, the deep cerebellar nuclei, where cerebellar efferents originate, are severely affected with marked neuronal loss and gliosis in the dentate nucleus. As a consequence, the superior cerebellar pedunculi appear markedly atrophic, and the red nuclei, where many fibers from the dentate terminate, may show some gliosis. The basal ganglia may show in some cases a moderate cell loss in the external pallidus and subthalamic nuclei. The thalamus, striatum, and substantia nigra do not appear to be directly involved by the disease. The cerebral cortex appears normal, with the exception of a loss of large pyramidal cells in the primary motor areas. Optic nerves and tracts show a variable degree of involvement. When present, atrophy of the visual system may extend to the lateral geniculate bodies [33].

In summary, the neuropathology of FRDA is characterized by the atrophy of the central sensory pathways carrying to the brain and cerebellum crucial information for the correct execution of movement and for equilibrium, of the cerebellar efferent pathway, and of the distal portion of the corticospinal motor tracts. As described below, the consequent malfunction of each of these systems contributes to the characteristic clinical picture of FRDA.

Finally, since many patients with FRDA die as a consequence of heart disease, it is not uncommon to observe widespread hypoxic changes and focal infarcts in the CNS.

2. Peripheral Nervous System

The presence of sensory axonal polyneuropathy is a hallmark of FRDA [8, 9, 31, 34–37]. The motor component of peripheral nerves is well preserved. In the DRG, loss of large primary sensory neurons is an invariable and early finding. Proliferation of capsule cells, forming clumps called "Residualknötchen," and an irregular appearance of intragangliar fibers are also observed. Atrophy of the central branches of the axons of the affected primary sensory cells causes thinning of the dorsal roots, particularly at the lumbosacral level, and atrophy of the peripheral branches causes loss of large myelinated fibers from peripheral nerves. The fine, unmyelinated fibers are well preserved, and interstitial connective tissue is increased.

3. Heart

The heart is clinically or subclinically affected in the vast majority of FRDA patients [8, 9, 12, 13, 28, 38]. Enlargement of the heart is the typical finding, with thickening of ventricular walls and interventricular septum [38, 39]. After a long disease course, however, the gross appearance becomes more that of a dilatative cardiomyopathy [40]. Pericardial adhesions are often found. Microscopically, hypertrophic cardiomyocytes are intermingled with fibers undergoing atrophy or granular degeneration and with normal appearing ones. Connective tissue is increased, with diffuse and focal inflammatory cell infiltration. Intracellular iron deposits in cardiomyocytes [41, 42] are an invariable, specific finding in FRDA cardiomyopathy, and may reflect the basic biochemical defect [43].

4. Other Organs

About 10% of FRDA patients have diabetes mellitus and show a loss if islet cells without the signs of autoimmune aggression found in type I diabetes [44].

Skeletal abnormalities are very common in FRDA. Scoliosis is observed in more than half of the patients, mostly as a double thoracolumbar curve [45, 46]. As a rule, a kyphotic curve is associated. The kyphoscoliosis of FRDA resembles the idiopathic form of scoliosis [47]. Although it is often suggested that it derives from muscular imbalance due to the neurologic disease, this has never been convincingly demonstrated. Pes cavus, pes equinovarus, and clawing of the toes are also very common and thought to be secondary to nerve degeneration [8, 9].

Amyotrophy of small hand muscles and of distal leg and foot muscles is a common feature of the disease [9].

C. Clinical Aspects

1. Onset

Typical FRDA is said to have its onset around puberty, but wide variations are observed [8, 9, 12, 13, 48, 49]. Some patients show symptoms well before puberty, even at age 2–3 [50, 51], and a few others have a very late onset in adult life. In the "typical FRDA" patients described by the Québec Collaborative Group [8] onset was before age 20. In her diagnostic criteria, Harding [9] extended this limit to age 25. The status of later onset cases (late-onset Friedreich's ataxia, LOFA) remained uncertain until the molecular defect was identified, when it became clear that LOFA patient carry mutations in the same gene as typical FRDA cases. Age of onset is more variable among families than within families [9, 14], but large variations within a sibship are occasionally observed. These used to be explained by the effect of modifier genes or environmental factors, and are now at least in part explained by the dynamic nature of the mutation [21].

Gait instability (65%) or generalized clumsiness (25%) are the usual initial symptoms. Occasionally, non-neurological manifestations, such as scoliosis (5%) or cardiomyopathy (5%), precede the onset of ataxia [8, 9, 12, 13].

2. Neurological Signs and Symptoms

The cardinal neurological feature of FRDA is a progressive, unremitting ataxia (Fig. 26-1). As already noticed by Friedreich, it is a mixed cerebellar–sensory ataxia, whose pathological basis is the degeneration of both sensory pathways and cerebellar afferent and efferent pathways. Most commonly, it begins with clumsiness in gait and frequent falls (truncal ataxia). Limb incoordination, dysmetria, and intention tremor then follow. Progression is steady, until patients lose the ability to perform fine motor activities, to walk, stand, and eventually sit without support. Speech is affected within 5 years after onset [8, 9, 12, 13], and becomes more and more indistinct as disease progresses. Dysarthria in FRDA patients consists of slow, jerky speech with sudden utterances [52, 53].

Muscle tone is most often decreased, sometimes normal, and rarely increased. Muscular weakness is common and progressive, particularly at the lower limbs, usually affecting the proximal muscles first [54]. It is thought to derive, at least in part, from the degeneration of the corticospinal tracts. However, when patients become wheelchair-bound, they still have significant strength in their lower limbs, on average 70% of normal, clearly indicating that ataxia and not weakness is the primary cause for loss of ambulation in FRDA [54]. Atrophy of distal lower limb muscles and of small hand muscles can often be observed, even early in the course of the disease [9]. In the later stages, when patients are wheelchair-bound, disuse atrophy occurs.

Sensory loss is another cardinal manifestation of FRDA. It is the consequence of the degeneration of large sensory fibers in peripheral nerves and in the central sensory pathways. Loss of position and vibration sense is commonly found at onset and invariably after 2 years. Vibration sense loss may sometimes precede the appearance of ataxia. Perception of light touch, pain, and temperature are initially normal, but tend to decrease in most patients with advanced disease.

Another consequence of the sensory axonal neuropathy, the loss of tendon reflexes at least in the lower limbs, has been considered an obligatory feature that

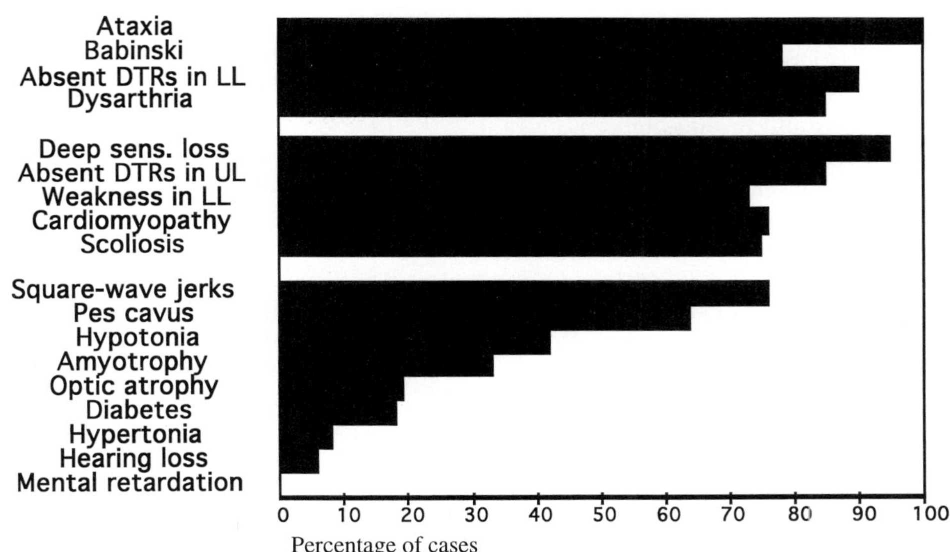

FIGURE 26-1 Clinical findings in an unselected series of 27 consecutive patients with a positive FRDA molecular test. DTR, deep tendon reflexes; LL, lower limbs.

needs to be present at onset to establish the diagnosis [8, 9]. Tendon reflexes are lost in all four limbs by almost all patients within a few years. Extensor plantar responses, reflecting the pyramidal tract involvement, were also considered an obligatory sign. However, a minority of patients with a proven molecular diagnosis of FRDA have elicitable tendon reflexes for several years after diagnosis (Friedreich's ataxia with retained reflexes, FARR); a few even have exaggerated reflexes with spasticity. Other patients, also with a proven molecular diagnosis, have normal plantar responses. Probably, the relative weight of sensory neuropathy, causing areflexia and hypotonia, and of pyramidal tract degeneration, causing extensor plantar responses and spasticity, can vary from patient to patient. In most cases, a typically mixed picture results, but sometimes one component may obscure the other. The current molecular studies are trying to shed light on this variability, as described in the following sections.

The most common abnormality of ocular movements in FRDA is fixation instability with square-wave jerks [55]. In addition, patients may show various combinations of cerebellar, vestibular, and brainstem oculomotor signs. Ophthalmoparesis is not observed and nystagmus is uncommon.

About 30% of the patients develop optic atrophy, with or without visual impairment, particularly in the later stages of disease [8, 9, 13, 56–58]. Sensorineural hearing loss, also more common with advanced disease, affects about 20% of the patients [59, 60]. Optic atrophy and sensorineural hearing loss tend to be associated in the same patients, and with diabetes, more often than expected by chance alone [9, 61], probably as a sign of more severe, widespread disease. They can progress to the point of rendering the patient functionally blind and deaf.

3. Heart Disease

Cardiomyopathy may significantly contribute to disability in FRDA and even cause premature death as a consequence of arrhythmias and heart failure [62]. In a significant portion of FRDA patients, however, heart disease remains asymptomatic [8, 9, 12, 13, 63–68]. Cardiac involvement is more often observed in patients with earlier age of onset [69]. When symptoms occur, the most common complaints are shortness of breath (40% of patients) and palpitations (11%) [9]. Typical electrocardiographic (EKG) abnormalities include widespread T-wave inversions and signs of ventricular hypertrophy. Conduction disturbances and supraventricular ectopic beats (in 10% of patients) are occasional findings [12, 13, 67, 68]. Atrial fibrillation is a negative prognostic sign [36], as it precedes fatal cardiac complications by 6 months or less in almost a quarter of the cases in which these occur. EKG changes vary in time, with occasional normal recordings: this has led to underestimation of their frequency. If repeated recordings are obtained, however, EKG seems to be the most sensitive test for FRDA cardiomyopathy. Echocardiography and Doppler-echocardiography can demonstrate concentric hypertrophy of the ventricles (62%) or asymmetric septal hypertrophy (29%), along with diastolic function abnormalities [9, 70]. As the disease progresses, the transition from a hypertrophic to a dilatative type of cardiomyopathy has been documented [40].

4. Diabetes Mellitus

About 10% of FRDA patients have diabetes mellitus, and an additional 20% have carbohydrate intolerance. A detailed study of glucose and insulin metabolism in FRDA patients revealed a deficiency in arginine-stimulated insulin secretion in all cases, including normotolerant individuals [71], suggesting that beta cells are invariably affected by the primary genetic defect of FRDA. In addition, peripheral insulin resistance may also contribute to the development of diabetes in FRDA [72]. Diabetes usually appears after the neurological symptoms. It may be revealed by the appearance of characteristic symptoms, like polyuria and polydipsia, or by glucose testing. Although some cases may be initially controlled by oral hypoglycemic drugs, insulin dependence is eventually the rule. Diabetes may aggravate the neurological picture by adding diabetic neuropathic complications, as distal sensory loss. It is therefore important to regularly check FRDA patients for the development of diabetes, as it may increase the burden of disease, and even promote potentially fatal complications. Family studies suggested that having a sib with diabetes increases the risk for an individual with FRDA to develop diabetes up to 45% [14]. These subjects should be tested for glucose tolerance every 6 months.

5. Other Manifestations

Most patients have skeletal abnormalities even in the early phase of the disease. Kyphoscoliosis affects 85–100% of patients and may be severe in 10%, particularly when onset is before puberty. It is usually slowly progressive, bearing little relation with the degree of muscle weakness. It may cause pain and cardiorespiratory problems. Pes cavus and pes equinovarus, although found in more than 50% patients, are not typical of FRDA, being even more common in other neuropathic diseases as Charcot-Marie-Tooth disease (CMT). Autonomic disturbances, most commonly cold and cyanotic legs and feet, appear with increasing frequency as the disease advances [73]. Parasympathetic abnormalities, including decreased heart rate variability parameters, have been reported [74]. Urgency of micturition is rare.

6. Laboratory Investigations

While current attention is focused on iron metabolism in FRDA [43], no data are yet available indicating if and how any related abnormality can be revealed by laboratory tests in the living patient. To date, most laboratory investigations are carried out for differential diagnostic purposes. Lipoproteins, vitamin E, lactate, pyruvate, urinary organic acids, serum very long chain fatty acids (VLCFA), serum phytanic acid, and leukocyte and/or fibroblast lysosomal enzymes are all normal in FRDA. These tests allow exclusion of the diagnoses of a- or ipo-β-lipoproteinemie (Bassen-Kornzweig and Tangier diseases), isolated vitamin E deficiency, mitochondrial disorders, adrenoleukodystrophy or adrenomyeloneuropathy, Refsum disease, and a number of lysosomal storage diseases, all of which may pose a differential diagnosis problem, particularly at the early stage.

7. Neuroimaging

The current structural and functional CNS imaging technologies can not only reveal neuropathological details with extraordinary definition, but also detect metabolic dysfunctions associated with the disease process.

CNS structural imaging in FRDA patients, by magnetic resonance imaging (MRI) or CT scanning, closely reflects the findings of postmortem neuropathological studies. The spinal cord is the most severely affected structure, appearing much more compromised than any other part of the CNS, including the cerebellum. This pattern of involvement contrasts with other inherited degenerative ataxias [75], in particular with the group of early onset ataxias with retained reflexes (EOCA), which have a similar age of onset and are also recessively inherited. In EOCA, contrary to FRDA, the cerebellum is much more atrophic than the cervical spinal cord. Thinning of the cervical spinal cord can be detected on sagittal and axial images in almost all FRDA patients [75–79]. MR signal abnormalities in the posterior and lateral columns can also be observed, consistent with the degenerative process affecting these structures [79]. As stated, intracranial structures, including brainstem, cerebellum, and cerebrum, are less evidently affected, but they are not completely spared. Accurate assessment of regional atrophy on CT scans indicates that cerebral hemispheres, brainstem, and cerebellum are all smaller in FRDA patients compared to healthy controls, and the extent of this diffuse atrophy correlates with clinical severity [78]. In addition, mild vermian and lobar cerebellar atrophy clearly occurs in more severe and more advanced cases [77, 80]. Blood flow in the cerebellum, as assessed by TC-HMPAO SPECT, appears to be markedly decreased, more than expected for the degree of atrophy [80]. Interestingly, PET scans reveal an increased glucose metabolism in the brain of FRDA patients who are still ambulatory. This finding appears to be specific for FRDA [81]. As disease progresses and the patients lose their ability to walk, glucose consumption decreases and eventually becomes subnormal [78, 81]. Although not yet fully explained, this abnormality may relate to mitochondrial dysfunction, as suggested by the most recent molecular studies [43].

8. Neurophysiological Investigations

Electromyographic and electroneurographic studies in FRDA reveal the underlying dying-back axonal sensory neuropathy. Sensory action potentials (SAPs) in peripheral nerves are severely reduced or absent, even early in the course of the disease [82–84]. Motor and sensory nerve conduction velocities (NCVs) are within or just below the normal range, a feature that helps distinguish an early case of FRDA from a case of demyelinating hereditary sensorimotor neuropathy, as CMT. Dispersion and delay of somatosensory evoked potentials (SEPs), observed in all patients, indicate degeneration of both peripheral and central sensory fibers [13]. Brainstem auditory evoked potentials (BAEPs) also progressively deteriorate in all patients, beginning from the most rostral component, wave V [13, 85]. Visual evoked potentials (VEPs) are commonly (50–90%) reduced in amplitude, but P 100 latency is not increased [13, 56–58, 85]. Analysis of motor evoked potentials by magnetic stimulation reveals slowing of central motor conduction in all cases [86].

9. Variant Phenotypes

Reports have occasionally appeared of patients not fulfilling one or two diagnostic criteria for FRDA, but whose clinical picture was in other respects compatible with the diagnosis, and for whom no satisfactory alternative diagnosis could be found. Affected sibs of these patients sometimes had typical FRDA, and sometimes shared the same "variant" phenotype. The Québec Collaborative Group described "incomplete FRDA" and "atypical FRDA" patient groups, but could not clarify the relation between their disease and typical FRDA [8]. Only after the FRDA gene was mapped to chromosome 9, locus homogeneity with typical FRDA was demonstrated by linkage analysis in clinical variants as LOFA [18, 19], FARR [20], and Acadian FRDA [87]. Even more variability has been recognized after direct genetic testing became possible, including very atypical phenotypes as pure sensory ataxia [88].

LOFA patients are defined as having onset after age 25. They have an overall milder, slowly evolving disease. In a recent clinical study, we found only 2 LOFA patients out of 6 (17%) that were confined to a wheelchair after a disease duration of more than 12 years, compared to 86% of the typical FRDA patients ($P = 0.01$ by Fisher exact test) [61]. Compared to typical FRDA patients, LOFA patients also have less skeletal abnormalities. The frequency of cardiomyopathy in LOFA was found to be similar to typical FRDA in some studies [19, 61], but significantly lower in others [69].

FARR patients have retained deep tendon reflexes in the lower limbs after the onset of neurological symptoms. It is possible that this situation may be transitory, as all FARR patients examined in a recent study were in an early stage of the disease [61]. In addition, they tend to be mild cases with slightly later onset and slower course than most typical FRDA patients [20, 61].

A FRDA variant has been described among the descendants of French colonizers who settled in Eastern Canada during the 17th century. These people, known as Acadians, left their colony after the British conquest in the mid 1700s. Many moved to Louisiana, where they became known as "Cajuns" (a corruption of Acadians). Some returned to the Canadian Atlantic provinces. The disease is thought to have been introduced into this population by very few individuals, a so called "founder effect." Acadian FRDA is milder than typical FRDA, it evolves more slowly, and it affects the heart less commonly and less severely [16]. We recently confirmed these differences by comparing 44 Acadian patients to 100 typical FRDA patients, as we found a slightly later age of onset, a slower disease progression, and a much lower prevalence of cardiomyopathy among the Acadians. All other clinical manifestations did not significantly differ in frequency [61]. Locus homogeneity between Acadian and typical FRDA was demonstrated by linkage analysis [87]. After the FRDA gene was cloned, molecular testing of Acadian patients revealed the same dynamic mutation (GAA triplet repeat expansion) found in the vast majority of typical patients [61]. The specificity of the Acadian phenotype is still unexplained at the molecular level. Because, as a consequence of the founder effect, a major marker haplotype is present on most Acadian FRDA chromosomes [89], the effect of some *cis*-acting sequence variant is suspected, but such variant, if really present, has not yet been identified. A subtype of Acadian FRDA has recently been described in four patients from two New Brunswick families. They had a progressive ataxia evolving less aggressively than typical FRDA, retained reflexes and spasticity, and no cardiomyopathy. This rare subtype of Acadian FRDA has been called spastic ataxia of Acadian type (SPA-Acadian). Molecular testing revealed GAA triplet repeat expansions in the FRDA gene. The FRDA chromosomes from these two families carried a unique marker haplotype, again suggesting the effect of a *cis*-acting sequence variant. In a third family, patients were heterozygous for the SPA-Acadian haplotype and the common Acadian FRDA haplotype. They had absent reflexes and no spasticity, closely resembling Acadian FRDA patients in all respects [17].

D. Prognosis and Treatment

FRDA is a progressive disease that inevitably leads to increasing disability. Patients lose their ability to walk on average 15 years after onset, but variability is very large [9, 90]. Early onset and left ventricular hypertrophy appear to be predictors of a faster rate of progression of the disease [61, 90]. The burdens of neurological impairment, cardiomyopathy, and occasionally diabetes cause a shortened life expectancy [90]. Older studies found that most patients died in their 30s, but survival may be significantly prolonged with treatment of cardiac symptoms, particularly arrhythmias, by antidiabetic treatment, and by preventing and controlling complications resulting from prolonged disability. Carefully assisted patients may live several more decades. No cure is available, and there are currently no treatments that affect the degenerative process.

Even symptomatic pharmacological treatment has not been effective. Apart from anecdotal evidence, at best marginal and transitory improvement has been reported in some, but not all, controlled trials. Attempts were made to use drugs that affect neurotransmitters acting in the cerebellar circuitry and whose levels are modified, usually reduced, by the degenerative process. Tested drugs include cholinergic agonists, such as physostigmine [91] and choline, neuropeptides such as TRH [92, 93], serotoninergic agonists such as 5-OH tryptophan [94, 95] and buspirone, and the dopaminergic drug amantadine [96–98].

Physical therapy can help patients in dealing with their neurological deficit. Rehabilitation programs should include exercises aimed at maximizing the residual capacity of motor control. Orthopedic interventions are sometimes necessary, including surgical correction of severe scoliosis and, in patients who can still walk, of foot deformity.

Important advances in understanding the pathogenesis of FRDA came after the gene was cloned, with the initial characterization of its protein product, frataxin. The possible role of frataxin in controlling mitochondrial iron homeostasis [43] suggests ways of pharmacological intervention that may be worth considering, particularly if more basic research data confirm and extend the initial reports. Although not a cure, such interventions may act on the primary degenerative process and possibly slow down the disease course. Gene therapy is also being investigated, but is a long-term option.

III. ISOLATION AND ANALYSIS OF THE FRIEDREICH'S ATAXIA GENE

A. Mapping and Cloning of the FRDA Gene

During the 1970s and early 80s, many attempts were made to identify the primary biochemical defect in FRDA [99–101]. Several abnormalities were proposed as the primary defect, including lipoamide dehydrogenase deficiency [102–105], pyruvate carboxylase deficiency [106], mitochondrial malic enzyme deficiency [107–111], increased urinary taurine excretion [112, 113], and abnormal lipids [114]. None of these observations has been confirmed.

A positional cloning approach was then undertaken to identify the defective gene in FRDA [115–118]. The primary mapping of the locus was accomplished in 1988 by S. Chamberlain and her group at St. Mary's Hospital in London [119]. They examined 20 families with at least three affected sibs according to Harding's criteria, and demonstrated linkage without recombination to the anonymous marker D9S15, on chromosome 9. Soon thereafter, a study by J.L. Mandel's group in Strasbourg found another anonymous marker, D9S5, to be also at no recombination from FRDA [120]. Linkage was soon confirmed in families of various ethnic background [87, 121–123]. Physical assignment of D9S5 and D9S15 to 9q13-q21.1, just below the pericentromeric heterochromatin, was obtained by *in situ* hybridization [124–127]. Physical linkage between D9S15 and D9S5 was found by PFGE, and YAC and cosmid clones spanning more than 1 Mb around these markers were isolated [128–130]. Expansion of the analyzed family set and use of additional markers [131–133] allowed the determination of the gene order as cen-FRDA-D9S5-D9S15-tel [134, 135]. Analysis of linkage disequilibrium detected allelic association between FRDA and D9S5, in agreement with the recombination data [136]. Locus order was also confirmed by homozygosity analysis in inbred families and by extended haplotype analysis in a group with founder effect, the Louisiana Acadians ("Cajuns") [89]. Efforts then concentrated to extend the cloned region centromeric to D9S5, isolate new polymorphic markers to refine the genetic analysis [137–139], and eventually bracket the disease locus between flanking markers. A critical interval of 450 kb was initially defined by analysis of recombination events in Tunisian, Yugoslavian, and Spanish families [139, 140]. The progressive isolation of additional polymorphic markers then allowed the reduction of the critical interval to 360 kb [141] and finally to 240 kb by homozygosity analysis of the same families [142].

B. Isolation of Candidate Genes

Figure 26-2 shows the transcriptional map of the 9q13 region where the FRDA gene is localized. Six genes, distributed over 800 kb of genomic DNA, were identified during the FRDA positional cloning project. From telomere to centromere, these genes are: X11, X123, X104/CSFA1/ZO-2, X25 (FRDA), PRKACG, and STM7. X11 [143, 144] encodes a transmembrane protein of unknown function expressed in the CNS; X123 [141] is a small gene not resembling any known sequence; X104/CSFA1/ZO-2 [141, 145] encodes a 160-kDa pro-

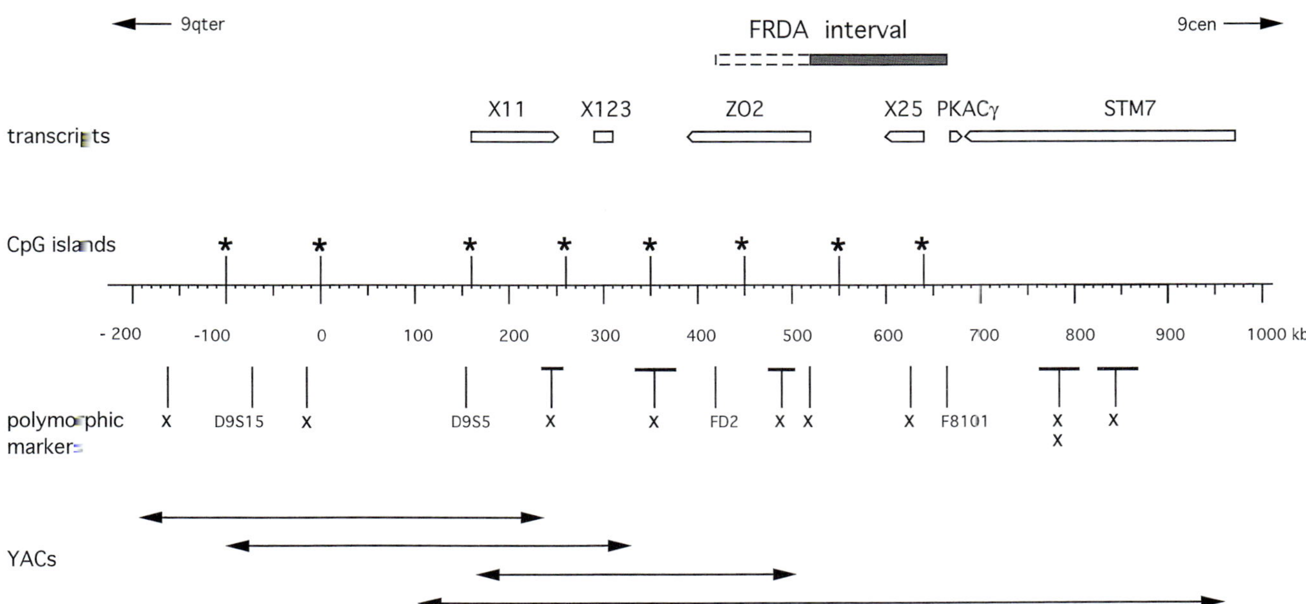

FIGURE 26-2 The Friedreich ataxia region on chromosome 9q13. Some polymorphic markers that were critical to map the FRDA gene are indicated. Other markers are indicated by Xs The candidate region was eventually narrowed down to 240 kb between FD2 and F8101. Absence of mutations in the ZO2 gene of patients further reduced the critical interval to 150 kb (shaded bar). The position of CpG islands is indicated by asterisks. A transcriptional map is shown above the physical map. Arrowheads indicate the direction of transcription, when known. Modified from Koenig *et al.* (1996) *Med. Sci.* **12,** 431–435, with permission.

tein that localizes in the tight junctions of epithelial cells [146] and shows similarities to another tight junction protein, ZO-1, to erythrocyte p55, to the *Drosophila dlg* gene and to a synapse-associated protein from rat brain PSD-95/SAP90; X25 [21] is the FRDA gene, described below; PRKACG [142] is a small intronless gene encoding the testis-specific gamma catalytic subunit of cAMP-dependent protein kinase; STM7 [147] encodes a phosphatidyl inositol 4-phosphate 5-kinase (PI4P5K), an enzyme involved in the phosphoinositide signaling pathway [148]. A combination of molecular biology techniques, including exon amplification, computer prediction from random sequences, and direct cDNA selection, was utilized to identify these genes.

C. Structure of the FRDA Gene

Initially called X25, the FRDA gene was found starting from a putative exon independently identified by exon amplification and by computer analysis of random cosmid sequences [21]. It is composed of seven exons spread over 85 kb of genomic DNA (Fig. 26-3). The first five exons, numbered 1 to 5a, are localized within a 40-kb interval, then a large intron of 30 kb precedes the sixth exon, exon 5b, and the seventh exon, exon 6, is further 15 kb downstream. Transcription goes in the centromere to telomere direction. The 5′ end of the gene, including its first exon, is associated with an unmethylated CpG island. CpG islands are DNA regions rich in CG dinucleotides, which mark DNA segments containing genes. Methylation of cytosines in CpG islands is an important way of regulating, usually repressing, gene expression. Six of the seven exons of the FRDA gene contain protein coding sequence. Exons 1 to 4 are invariably joined together during the processing of the FRDA transcript. Variability is then introduced in the 3′ end of the mature mRNA by the alternative usage of exon 5a or 5b as fifth exon, leading to the synthesis of different protein isoforms differing at their carboxy terminus. Based on Northern blot analysis, exon 5a-containing transcripts are much more common in all examined tissues; they encode for a novel protein baptized frataxin [21] (see Section V). Further variability is introduced by the possible addition of exon 6 to exon 5b-containing transcripts, which occurs when a donor splice site consensus sequence in exon 5b is used instead of a polyadenylation site a few base pairs downstream. However, no changes in the protein sequence are introduced by the usage of exon 6, which is entirely noncoding.

Some authors proposed that frataxin is encoded by a larger transcriptional unit than originally thought, and that its gene is in fact part of the neighboring STM7 gene, which is transcribed in the same direction [148]. The claim was based on fragments containing sequences

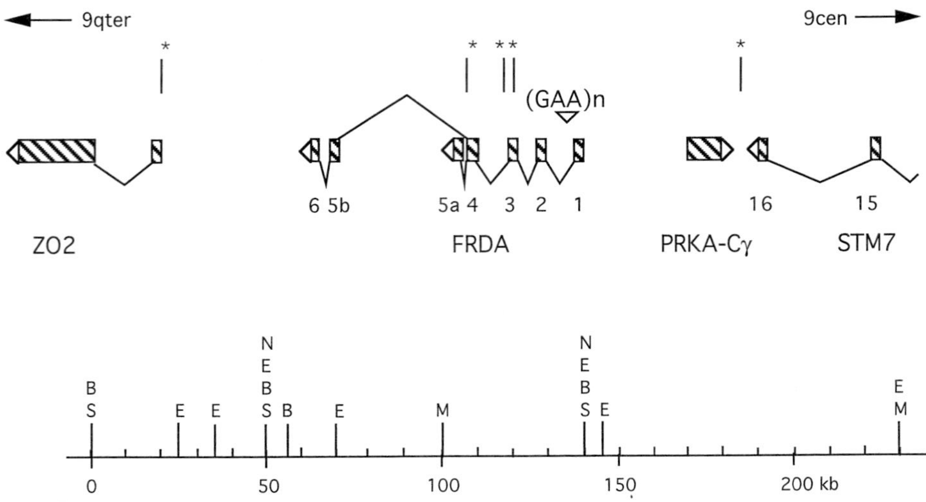

FIGURE 26-3 Structure of the FRDA gene, and schematic representation of the splicing of its exons. Exons are shown as hatched boxes. The trinucleotide repeat [(GAA)n] that underwent expansion in FRDA patients is located 1.4 kb downstream from exon 1 (see the following section). Part of the flanking genes is shown. The corresponding physical map is shown below. N, E, B, and S: *Not*I, *Eag*I, *Bss*HII, and *Sac*II restriction sites, respectively. Cluster of these rare cutter sites indicates the presence of a CpG island. Polymorphic markers used for linkage disequilibrium analysis (see Table 26-3) are indicated by asterisks. Modified from M. Koenig and J.-L. Mandel (1997) *Curr. Opinion Neurobiol.* **7**, 689–694, with permission.

from both frataxin and STM7 genes obtained after forced exon connection by reverse-transcription PCR (RT-PCR) between the two genes. A number of discordant observations, including the lack of any other evidence for the presence of such fragments, as could be obtained by RNase protection analysis or by the isolation of a cDNA, and the size and intracellular localization of the protein recognized by anti-frataxin antibodies (see Section V.B), argue against this hypothesis [149, 150].

D. Expression of the FRDA Gene

The FRDA gene shows tissue-specific expression [21]. Among human adult tissues, highest RNA levels are found in heart, intermediate levels in liver, skeletal muscle, and pancreas, and minimal in other tissues, including whole brain. A 1.3-kb major transcript can be identified, in agreement with the predicted size of an exon 5a-containing mRNA. Fainter bands of 1.05, 2.0, 2.8, and 7.3 kb are also detected in heart. The 1.05- and 2.0-kb bands correspond to exon 5b and exon 5b+6 containing isoforms, respectively. The larger bands are most likely to derive from a transcribed pseudogene (M. Pandolfo, unpublished results). Within the human adult central nervous system (CNS), the 1.3-kb transcript is expressed in the spinal cord, at lower levels in the cerebellum, and very little in the cerebral cortex (human dorsal root ganglia were not tested). Overall, expression of X25 appears to be high in the primary sites of degeneration in FRDA, both within and outside the CNS.

The developmental expression of the gene has been investigated in embryo and adult mouse by RNA *in situ* hybridization [151, 152]. Expression is negligible until embryonic day 10 (E10), and then progressively increases, becoming very clear at E14. Frataxin RNA is found within the CNS, both in proliferating cells of the ependimal layer (periventricular zone) and in more mature cells in the developing forebrain. In the adult, brain expression is restricted to the ependimal layer, choroid plexus, and the granular layer of the cerebellum. The spinal cord shows a characteristic pattern of expression, highest in the thoracolumbar region, starting around E12.5. Surprisingly with respect to the pathology, expression is higher in the anterior horns than in the posterior horns. The frataxin gene is prominently transcribed in large neuronal cells in the dorsal root ganglia (DRG) by E12.5 up to adult life.

In the mouse developing embryo, the frataxin gene is also prominently expressed in extraneural tissues such as heart and in tissues that are apparently not affected in FRDA patients, such as liver, muscle, thymus, skin, developing teeth, and brown fat [21, 151, 152] (the latter tissue, present in newborns, is particularly rich in mitochondria). Frataxin is highly expressed in mouse adult and fetal kidney but little if any in human adult kidney, which is not a site of pathology in FRDA [21, 151]. Apart from differences between mouse and man, the discrepancy between frataxin pattern of expression which is broader than the FRDA sites of pathology might be accounted in part by the nondividing nature of the affected cell types (neurons, cardiomyocytes, beta cells of the pancreas), meaning that they cannot be replaced when then die, unlike liver, skeletal muscle, and cells expressing frataxin during fetal development. Further knowledge on the molecular mechanisms underlying degeneration in FRDA is needed to answer these issues.

IV. GENE MUTATIONS IN FRDA

A. Point Mutations

Point mutations in the frataxin gene are a rare cause of FRDA [21]. Only about 2% of the FRDA chromosomes carry sequence changes resulting in the premature truncation of frataxin, or in an amino acid change of likely functional significance. However, the identification of these point mutations has been essential in establishing that X25 is the FRDA gene. Five point mutations have been fully characterized so far [21, 149, 153] (Table 26-2). In a French family with two affected sibs, a T → G transversion in exon 3 changed a leucine codon (TTA) into a stop codon (TGA) (L106X). In a Spanish family with one affected member, an A → G transition disrupted the acceptor splice site preceding exon 4 (385-2A → G), changing the invariant AG into a GG. This mutation is predicted to result in skipping of exon 4 with frame-shift of the coding sequence over exon 5. In five patients from three Southern Italian families, an amino acid change, isoleucine to phenylalanine (I154F), was found in an invariant position within the portion of the highly conserved domain encoded in exon 4. A point mutation in exon 1, involving the ATG (methionine) codon used as translation start site (ATG → ATT), was found in one family. No specific phenotypic feature characterizes these point mutation cases [21, 149]. A fifth mutation, causing an amino acid change, glycine to valine (G130V), was found in three sibs affected with atypical Friedreich ataxia [153]. Patients presented with peripheral neuropathy and sensory ataxia (posterior column ataxia) but no or little cerebellar ataxia and no cardiomyopathy. In all these cases, affected individuals are heterozygous for their point mutation, with a normal frataxin coding sequence on the other homologue of chromosome 9.

TABLE 26-2 Point Mutations Found in Friedreich's Ataxia

Nucleotide change	Predicted effect on coding sequence	Geographic origin	Ref.
3G — T	M1I (incorrect initiation)	Germany	[149]
316T → G	L106X (protein truncation)	France	[21]
385-2A → G	splice (exon 4 skipping)	Spain	[21]
389G → T	G130V	U.S.A.	[153]
460A → T	I154F	Italy	[21]

B. GAA Trinucleotide Repeat Expansion

FRDA is a member of the group of inherited diseases caused by large triplet repeat expansions. In the vast majority of FRDA chromosomes (98%), an abnormal GAA repeat expansion occurs within the first intron of the gene encoding frataxin [21]. This intron is 12 kb in size, and the triplet repeat is localized about 1.4 kb after exon 1, in the middle of a repetitive sequence of the Alu-Sx family (Fig. 26-4). The GAA repeat apparently derived from a poly-A expansion of the canonical A_5TACA_6 sequence linking the two halves of the Alu sequence. Normal chromosomes bear less than 40–42 triplets, while FRDA chromosomes have from 80 to >1200. Other diseases in the triplet repeat-associated group are X-linked or dominantly inherited, and include expansions of otherwise stable CGG, CAG, or CTG repeat polymorphisms, located within the 5′ untranslated region, coding sequence, and 3′ untranslated region of their respective disease-related genes. The triplet expansion in FRDA has three novel, and so far unique, features: it involves a GAA repeat sequence, is located in an intron, and is associated with an autosomal recessive disease. It results in a prolonged tract of purines on one strand and pyrimidines on its complement. Progressive myoclonus epilepsy (EPM1) is another autosomal recessive disease due to a large expansion, though in this case a dodecanucleotide (CCCCGCCCCGCG) repeat sequence is involved [154–156].

C. Detection and Diagnostic Value of Expanded GAA Repeats

Two techniques may be used to detect the expanded GAA repeats associated with FRDA: Southern blot (SB) analysis and polymerase chain reaction (PCR). SB is optimally sensitive to demonstrate heterozygotes as well as homozygotes for expanded GAA repeats. By digesting genomic DNA with the restriction enzyme *Bsi*HKAI, a normal fragment of about 2400 bp containing the GAA repeat can be detected by a frataxin exon 1 probe [157]. Expanded repeats generate larger bands. The main disadvantages of the SB technique are the

CTTAGAAAATGGATTTCCTGGCAGGACGCGGTGGCTCATGCCCATAATCTCAGCACTTTGGGAGGCCTAG

GAAGGTGGATCACCTGAGGTCCGGAGTTCAAGACTAACCTGGCCAACATGGTGAAACCCAGTATCTACTA

AAAAATACAAAAAAAAAAAAAAAAAA**GAAGAAGAAGAAGAAGAAGAAGAAGAA**AATAAAGAAAAGT

TAGCCGGGCGTGGTGTCGCGCGCCTGTAATCCCAGCTACTCCAGAGGCTGCGGCAGGAGAATCGCTTGAG

CCCGGGAGGCAGAGGTTGCATTAAGCCAAGATCGCCCAATGCACTCCGGCCTGGGCGACAGAGCAAGACT

CCGTCTCAAAAAATAATAATAATAAATAAAAATAAAAAATAAAATGGATTTCCCAGCATCTCTGGAAAAA

FIGURE 26-4 Genomic sequence flanking the GAA triplet repeat {151}. The most frequent allele (9 GAA) is shown. Alu sequence is underlined. The GAA repeat is preceded by a $(A)_{18}$ polyA tract and followed by the AATAAAGAAAAG sequence, which are not part of the Alu consensus sequence. The duplicated sequences flanking the Alu repeat are boxed. Modified from [21] with permission.

larger amount of DNA required for the analysis and the longer procedure compared to PCR. Several primer pairs have been developed to amplify the GAA repeat [21, 158]. In all cases, the amplification protocol requires special precautions to enhance the efficiency and specificity of the reaction, particularly for heterozygote detection. However, despite some technical difficulties, PCR remains the method of choice to test for expanded GAA repeats. PCR is effective in accurately determining the size of expansions in homo- and heterozygotes, requires little DNA, and is rapidly carried out.

Detection of the expansion mutation provides thus a most useful diagnostic test. The length of expansion has, however, limited value for individual prognosis (see Section IV.G). Finding of a heterozygous mutation in a patient should lead to the search for a point mutation in the other allele, but it should be kept in mind that the frequency of expansion heterozygotes in the normal Caucasian population is about 1 in 90 [159, 160]. Based on the frequency of point mutations in FRDA chromosomes in Caucasians, in the absence of stratification only 4 in 10,000 FRDA patients would be expected to carry a point mutation on both alleles, assuming that homozygosity for a mutation inactivating totally the frataxin gene is not lethal. Absence of the expansion therefore almost excludes the FRDA diagnosis, at least in outbred Caucasian populations. It is possible that in some other populations, point mutations may be more prevalent. It should be noted that the expansion mutation has not yet been detected in Japanese patients with recessive ataxia (S. Tsuji, M. Watanabe, and N. Tachi, personal communications). Rare nonallelic heterogeneity may be present [161].

D. Instability of the Expanded GAA Triplet Repeat

When examining a family for the FRDA expanded repeat, it is common to see it changing in size when transmitted from parent to child [21, 61, 157, 158]. Instability during parent–offspring transmission can also be indirectly demonstrated by the detection of two distinct alleles in affected children of consanguineous parents, who are expected to be homozygous-by-descent at the FRDA locus [21]. Expansions and contractions of expanded GAA repeats can both be observed. In triplet repeat diseases with dominant inheritance, the dynamic nature of the mutation is reflected in anticipation, where increase in expansion sizes in successive generations correlates with earlier onset, and usually more rapid progression of disease. Recessive inheritance precludes the occurrence of anticipation, but variation in the size of FRDA expanded alleles does underlay phenotypic variation, regardless of anticipation, as detailed below (Section IV.G). Differences in stability between paternal and maternal transmission of the expanded alleles cannot be studied in patients, because establishing the parental origin of each expanded allele would require complex manipulations, as the isolation of each chromosome 9 homologue in a separate somatic cell hybrid. Such differences are best studied in carrier children, where parental origin of expanded allele can easily be determined by linkage analysis. Recent results suggest that fully expanded alleles most often contract during paternal transmission, but are equally likely to further expand or contract during maternal transmission [162, 163], a result also supported by sperm analysis [162]. In this regard, FRDA resembles the other diseases associated with very large expansions in noncoding regions, such as fragile X and myotonic dystrophy, while smaller expansions of CAG repeats in coding regions, found in dominant ataxias or Huntington disease, tend to undergo size increases during paternal transmission.

Mitotic instability, leading to somatic mosaicism for expansion sizes, adds to meiotic instability in some diseases associated with large triplet repeat expansions, such as myotonic dystrophy [164]. The phenomenon can be observed in FRDA as well [61]. Analysis of GAA expansions reveals ample variations in different cell types or tissues from the same patient. Furthermore, heterogeneity among cells occurs at a variable degree in different tissues. For instance, cultured fibroblasts and cerebellar cortex show very little heterogeneity in expansion sizes among cells, lymphocytes are more heterogeneous, and most brain regions show a quite complex pattern of allele sizes, indicating extensive cellular heterogeneity [165]. While some of these differences could be accounted for by a major period of instability during the first weeks of embryonic development, one is also prompted to conclude that the GAA expanded repeats are inherently more stable in some cell types [165]. In general, it is clear that determining the size of a patient's expansions in peripheral blood lymphocytes, from which DNA is usually obtained, only provides a single sample of the overall repeat size distribution occurring within that patient, and therefore only an approximate estimate of expansion sizes in affected tissues.

E. Population Dynamics of the Expanded GAA Repeat

The estimated frequency of GAA expansion carriers is about 1:90 in the Caucasian population [10, 23, 24, 26, 27, 159, 160], making it the most common triplet repeat expansion identified to date. FRDA heterozygotes do not appear to have any specific health problem that could cause reduced fitness [9, 166]. This makes the natural history of the mutation at the population level strikingly different from any other known disease due to trinucleotide expansions. In fragile X and myotonic dystrophy, where expansions of comparable size occur in noncoding sequences, carriers have severe early onset disease and a strong reproductive disadvantage. Large expansions in these diseases are newly formed from unstable alleles of intermediate sizes, resulting in the phenomenon of anticipation. In FRDA, large expanded alleles are transmitted by asymptomatic carriers, and new expansion events in heterozygotes would have no consequence at the phenotypic level. Only the much rarer homozygotes, who have FRDA, are less likely to reproduce. GAA expansions are therefore maintained in the population with minimal negative selective pressure.

How did the FRDA-associated expansions originate? Analysis of the polymorphism of the normal GAA repeat, and linkage disequilibrium analysis, indicate that expansions evolved from short repeats through multiple steps. The normal GAA repeat in the Caucasian population shows two main classes of alleles [159, 160, 167] (Fig. 26-5). About 80% of alleles have 7 to 12 GAAs (the most frequent allele carry 9 repeats). A second heterogeneous mode of 16 to 34 GAAs accounts for 17% of alleles. The study of 5 close flanking or intragenic polymorphic DNA markers, in association with the expansion alleles and the 2 modes of normal alleles, revealed a very striking linkage disequilibrium [159], much stronger than in previous studies with fewer and/or more distant markers [131, 133, 136, 168]. Three haplotypes, which can be derived from each other by a single event (mutation or recombination affecting one of the polymorphic markers), were found associated with 85% of the expanded alleles (>100 GAAs) and 75% of the large normal alleles (16 to 34 GAAs) while they account for only 4% of the small normal alleles (7 to 12 GAAs) (Table 26-3). Examination of minor haplotypes suggests in fact that 95% of large normal alleles (in Caucasians) originated from a single ancient founder event affecting GAA length, and that these large normal alleles, after a number of small increases in size because of slippage events, eventually reached the threshold for instability.

This represents a mechanism very similar to that proposed for myotonic dystrophy [169], where almost all large normal alleles (19–30 CTGs) in the Caucasian population stem from a single founder event and serve as a reservoir for further expansion (Fig. 26-6). However, the myotonic dystrophy expansions are selected against while FRDA expansions carried by heterozygotes are probably not counterselected, leaving open the possibility that FRDA mutations have derived from large normal alleles either by recurrent events (as in myotonic dystrophy) or by a few secondary founder events.

The first hypothesis seems correct, since intermediate unstable alleles (premutations) have been found in a few cases of extreme instability (changes from 34 to 65 pure GAA to expansions of 300 to 650 repeats, in a single generation) [159, 160, 167]. It is interesting that a length of 34 repeats is close to the instability threshold for other triplet repeats as well, such as CGG and CAG repeats. This is probably the threshold above which these sequences can form secondary structures promoting strand displacement during DNA replication, leading to reiterative synthesis resulting in large size increases. In FRDA, these dramatic single-step size changes might represent exceptional cases selected by the occurrence of an affected child, since not all parental transmissions of a pure GAA premutation or of a small expansion showed dramatic instability [159, 167]. Also, several very large normal alleles (from 33 to 55 triplets) that contain interrupting $(GAGGAA)_{5\text{-}10}$ repeats were found [159, 167], which may confer stability, as in other trinucleotide repeat diseases, such as FRAXA, SCA1, and SCA2 [26, 166, 170, 171]. In conclusion, the FRDA-

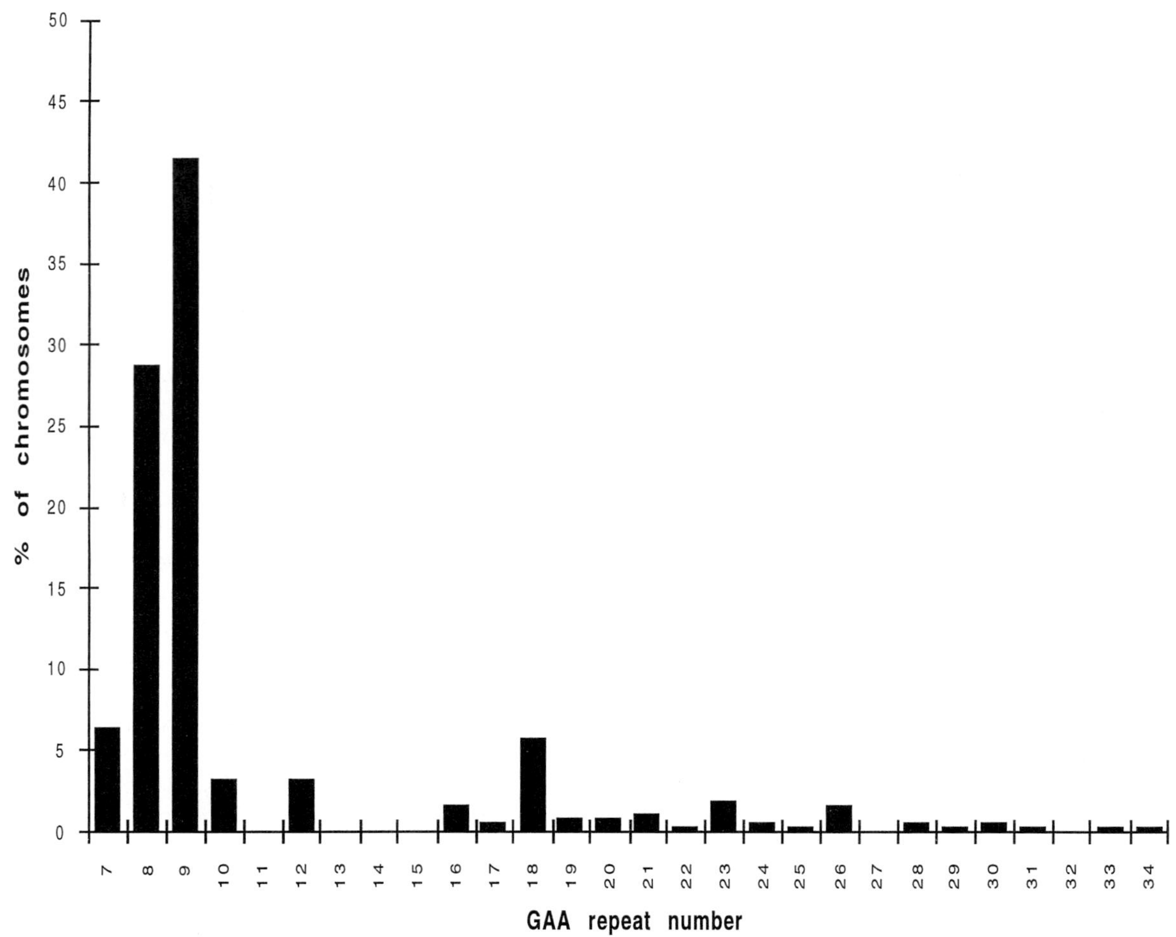

FIGURE 26-5 The distribution of the GAA repeat sizes observed in control Caucasian chromosomes is bimodal. Most (83%) alleles contain around 9 repeats (7–12), and 17% are large normal alleles of 16 repeats or more. Modified from [159] with permission.

associated GAA repeat seems to undergo similar mechanisms of stepwise expansion (Fig. 26-3), and to be stabilized by interruptions, such as several other disease-associated trinucleotide repeats. This suggests that some general properties are intrinsic to this entire category of simple sequence repeats, despite sequence motif differences.

F. Molecular Mechanism of the GAA Expansion

1. Effect of the Expanded GAA Repeat on Frataxin Gene Expression

The expanded GAA repeat has been shown to exert its disease-causing effect by suppressing FRDA gene expression [21, 149]. This loss-of-function pathogenetic mechanism is in accordance with the recessive nature of the disease. FRDA can therefore be defined as a deficiency of frataxin, the protein encoded by the mutated gene.

Evidence of frataxin deficiency has been obtained both at the RNA and at the protein level. Individuals with FRDA show a severe reduction in the level of mature frataxin mRNA, when tested using a reproducible, quantitative assay such as ribonuclease (RNase) protection [149]. Heterozygous carriers show intermediate levels between affected individuals and healthy controls [149, 153]. All portions of frataxin mRNA are reduced in abundance at the same extent, and no evidence of partially processed transcripts is found. These data suggest inhibition of transcription as the most likely mechanism leading to frataxin deficiency, but other possibilities, such as abnormal excision of the expansion containing intron, cannot be excluded at this time. Reduction in abundance of the transcripts is not linked to abnormal methylation of the CpG island containing exon 1, as indicated by analysis with methylation-sensitive restriction enzymes, contrary to the case of Fragile X syndrome [21]. Western blot analysis of CNS, lymphoblast, and skeletal muscle samples from FRDA

TABLE 26-3 Frequencies (in %) of Haplotypes Associated with Small Normal (SN), Large Normal (LN), and Expanded (E) GAA Repeat Alleles (from Ref. [159])

	GAA alleles		
Haplotype[a]	SN	LN	E
AT2CC	0.7	45.6	50.9
AT3CC	0.0	8.8	20.8
AT2CT	2.9	21.1	14.2
CT1CC	10.8	0.0	2.8
xTxCx	31.7	93.0	96.2
xCxCx	17.3	3.6	3.8
xCxTx	50.0	3.6	0.0
xTxTx	0.7	0.0	0.0
n^b	139	57	106

[a]Haplotypes are constructed from five differents markers with alleles A or C, C or T, 1 to 6, and C or T, respectively. Position of the markers along the genomic map is shown in Fig. 26-3. Full haplotypes are only shown for some of them. Partial haplotypes, where *x* means any allele, are shown in the remaining cases.

[b]*n* indicates the number of independently analyzed chromosomes.

patients confirmed a severe frataxin deficiency [150]. The 18-kDa band recognized by anti-frataxin antibodies appears to be very faint, but still detectable, in patient samples, in agreement with the finding of reduced but not absent frataxin mRNA.

Preliminary data indicate that the residual amount of frataxin mRNA and protein is inversely proportional to expansion sizes. From what is still a limited number of observations, it appears that a significant amount (20–30% of normal) of mRNA and protein is produced by expanded alleles containing less than 300 triplets [150]. Current data do not yet allow a precise quantification of this correlation, or identification of a repeat size above which the suppressive effect on frataxin expression becomes maximal. However, the existence of a graded effect of expansions on the residual frataxin level is already clear, providing a biological basis for any correlation between expansion sizes and phenotypic features.

2. PROPERTIES OF THE EXPANDED GAA REPEAT

Some properties of a long GAA repeat may offer some clues about how it may affect gene expression. The repeat is a polypurine·polypyrimidine (R·Y) sequence, i.e., a DNA segment with all purines (R) in one strand, and all pyrimidines (Y) in the other strand. Such sequences are known to adopt non-B DNA structures, particularly intramolecular triple helices (Fig. 26-7) [172]. This was shown to occur for GAA repeats containing 38 or more triplets in supercoiled plasmids, both at pH 4.5 and at pH 8.1 [173]. Under physiological conditions, the most likely conformation is an R-R·Y intramolecular triplex structure, where the purine-rich strand of the Watson-Crick R·Y DNA duplex dissociates and winds back down the major groove of the DNA helix pairing in an antiparallel orientation with the central purine-rich strand, via reverse Hoogsteen hydrogen

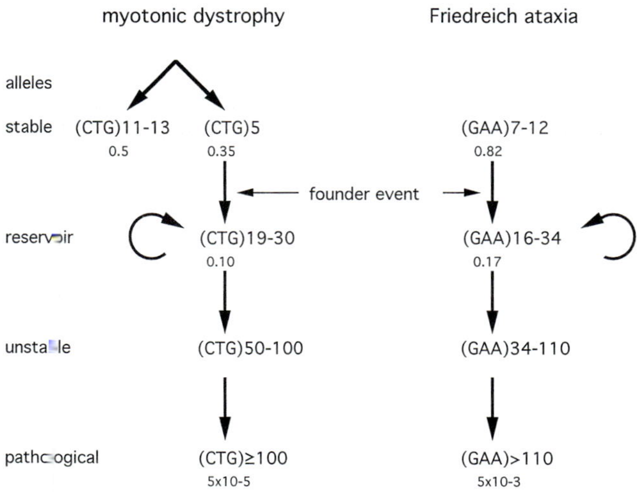

FIGURE 26-6 Compared evolution of the Friedreich ataxia GAA repeat {246} and the myotonic dystrophy CTG repeat {159}. The frequency of each class of allele is given below the length range. Polymerase slippage (circular arrow) is assumed to generate variability within the "reservoir" class. Data from [159] and [169].

A. DNA Triplex

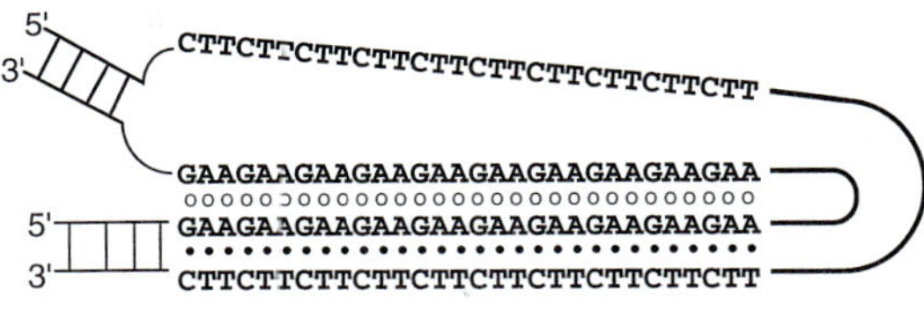

B. DNA•RNA Triplex

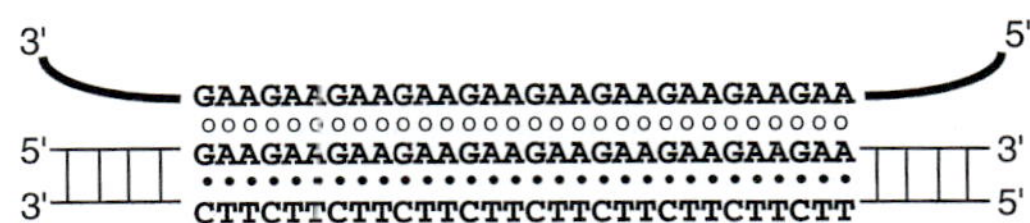

FIGURE 26-7 Model of an intramolecular R-R · Y triple helix formed by a GAA trinucleotide repeat (courtesy of Dr. K. Ohshima) [175].

bonds (-). Triplex structures have been shown to effectively inhibit transcription [174–176] making this mechanism a particularly attractive possibility. The spontaneous formation of an intramolecular triplex by an R·Y sequence, however, requires DNA supercoiling [172]. A wave of local negative supercoiling is known to be generated during transcriptional elongation *in vivo,* behind the polymerase [177, 178]. This could trigger the formation of an intramolecular triplex [179], potentially blocking further transcription. A 500-bp sequence, containing three interspersed R·Y tracts, immediately upstream of the rat *GAP-43* gene, was shown to behave as a transcriptional diode, selectively suppressing transcription of purine-rich RNA, possibly because the triple helix formed after transcription of this strand is more stable at physiological pH than the one predicted to form following transcription of the pyrimidine-rich strand [174]. Preliminary data (K. Ohshima *et al.*, manuscript in preparation) indicate that if GAA·TTC repeats of different lengths (from 20 to 270 triplets) and orientations are inserted into the intron of a reporter gene and transiently transfected into COS-7 cells, a length-dependent reduction in the level of the reporter gene mRNA occurs when rGAA is transcribed, resembling the situation in FRDA. Only a much slighter reduction occurs when a rUUC tract of similar length is transcribed. Based on these results, and on the above-described previous studies, our working hypothesis is that the GAA triplet expansion in FRDA is likely to adopt a transcriptionally mediated non-B DNA structure under conditions consistent with those in the nucleus. Although the evidence at present is not definitive, this would be the first example of a directional blockade to transcriptional elongation associated with a human genetic disease and would define a novel mutational mechanism.

G. Phenotype–Genotype Correlations

Before the discovery of the FRDA gene and of a GAA expansion as its most common mutation, mutation heterogeneity was thought to explain the clinical variability of FRDA. The effect of modifier genes and/or environmental factors was considered a plausible way to explain variability within families. The dynamic nature of the GAA expansion now offers an unexpected way to explain some phenotypic heterogeneity. As for all other triplet repeat disorders, regardless of their pathogenetic mechanisms, the size of the expanded repeat is also inversely related to age of onset and disease severity in FRDA [61, 157, 158, 163, 180]. As FRDA patients have two expanded alleles, correlation was expected to be more complex that in dominant or X-linked triplet repeat diseases, where patients have only one expanded allele. Earlier age of onset, earlier age when confined in wheelchair, more rapid rate of disease progression, and presence of nonobligatory disease manifestations such scoliosis [157], cardiomyopathy [157, 181], and diabetes [158], indicative of more widespread degeneration, all showed the best correlation with the size of the smaller repeat (GAA-1). Figure 26-8 is a scattergram showing the relation between age at onset and GAA-1 in a sample of 140 FRDA patients [157]. Detailed echocardiographic studies revealed significant correlations between ventricular septal thickness, posterior wall thickness, and left ventricular mass index with GAA-1 length [181]. As detailed in the preceding section, the suppressive effect of the GAA expansion on frataxin expression is proportional to the length of the repeat, at least up to 300–400 triplets, providing a biological basis for the observed genotype–phenotype correlation. The contribution of the size of the larger allele (GAA-2) to the phenotype is more difficult to evaluate, as GAA-2 is not independent from GAA-1 in many respects (size limit imposed by GAA-1, consanguinity and founder events, etc.). In all studies, correlation coefficients between disease severity parameters and expansion sizes in peripheral blood lymphocytes were not very high ($r = -0.69$ to -0.75), indicating the existence of additional sources of phenotypic variation. Somatic mosaicism for expansion sizes may be one of these factors. As analysis of lymphocytes only provides one sample of the repeat size distribution occurring within a patient, correlations with the phenotype can only be approximate. A remarkable finding in genotype–phenotype studies has been the realization of the wide clinical vari-

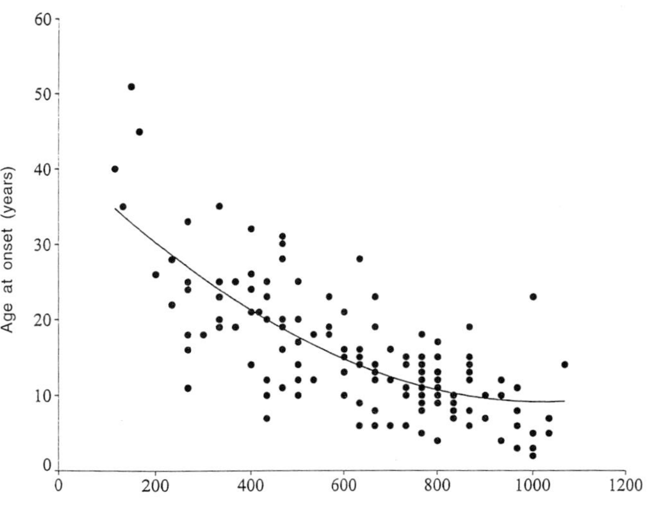

FIGURE 26-8 Scattergram of age of onset vs GAA-1 size in 140 FRDA patients (from Ref. [157]). The patient with 1000 GAA repeats and onset at 23 years of age had mosaicism for a third expanded fragment of 670 repeats, which might in part explain his late age at onset of disease.

ability in individuals who carry a GAA expansion on both copies of their frataxin genes on chromosome 9. These include several subjects who show an atypical phenotype not fulfilling the FRDA diagnostic criteria, particularly because of late onset (LOFA), retained reflexes (FARR), flexor plantar responses, lack of cardiomyopathy, unusually slow evolution with no dysarthria many years after onset, or a combination thereof. This became particularly clear when unselected series of patients started to be tested for diagnostic purposes. One study also clearly revealed the existence of clinical variability in FRDA which cannot be accounted for by GAA repeat length variations [61]. This was indicated by the analysis of Acadian FRDA, including SPA-Acadian, and of FARR cases. All patients in these groups were found to carry two expanded GAA repeats. However, despite their phenotypic peculiarities, the distribution of repeat sizes in the Acadian and FARR groups did not significantly differ from the typical FRDA group. The founder effect in Acadians suggests that the distinctiveness of the Acadian phenotype is linked to the FRDA gene itself, rather than to modifying genes which happen to be prevalent in that population. FARR has not been associated to any specific marker haplotype, but patients tend to cluster in families, and the related form, SPA-Acadian, is associated with a specific haplotype [17], suggesting that these variants also are linked to some variation within the FRDA gene. However, this has yet to be identified, as no variation in the frataxin coding sequence, as well as no change in the intron sequences directly involved in the splicing process, could be identified in Acadian, SPA-Acadian, or FARR patients [61]. Differences might be present in the as yet undefined frataxin promoter/regulatory regions, or within the expanded repeat sequence, and determine higher residual frataxin levels in tissues such as heart or peripheral nerves.

In conclusion, clinical variability in FRDA appears to be in part the consequence of the dynamic nature of the causative mutation, and in part to be related to yet unidentified polymorphisms within or near the FRDA gene. The concurrence of environmental factors and of modifier genes is of course still likely to add to the variability deriving from the disease gene itself.

V. BIOCHEMICAL STUDIES OF FRATAXIN

A. Primary Structure and Phylogeny of Frataxin

Frataxin is the protein encoded by the FRDA gene. The major mRNA isoform (exons 1–5a) encodes a 210-amino-acid protein and the putative exon 1–5b isoform found in heart and muscle would encode a 171-amino-acid protein [21]. Frataxin amino acid sequence does not resemble that of any protein of known function. However, similar proteins are made by as different species as the mouse [151], the roundworm *C. elegans,* and baker's yeast [21] (Fig. 26-9). The region of sequence conservation corresponds to the domain encoded by exons 3, 4, and 5a of human frataxin, but a particular segment at the boundary between exons 4 and 5a maintains an almost identical amino acid sequence in all investigated species. Interestingly, sequence comparisons showed the presence of more distant homologues in γ purple (Gram-negative), but not in Gram-positive bacteria [182] (Fig. 26-1). According to a prevailing theory, mitochondria derive from an ancestral intracellular symbiont related to γ purple bacteria. The frataxin gene might therefore derive from the ancestral mitochondrial genome and have moved to the nuclear genome during evolution, as was the case for most originally mitochondrial genes [182]. This would imply that frataxin is a mitochondrial protein. Mitochondrial proteins have sequence features that are recognized by a mitochondrial import system. Computer analysis of frataxin amino acid sequence predicted a mitochondrial targeting signal at the N-terminus of the yeast and mouse homologues [151] (a domain that is absent in the bacterial homologues), adding independent support for a mitochondrial localization. The computer prediction was ambiguous for the corresponding human sequence.

B. Subcellular Localization of Frataxin

The subcellular localization of human and yeast frataxin was analyzed by epitope tagging experiments, where frataxin is produced in transfected cells from expression vectors that add a peptidic tag at the C terminus of frataxin, in order to allow recognition of the N-terminal targeting signal. Mitochondrial localization was demonstrated by colocalization with well established mitochondrial markers, such as the subunit II of the cytochrome *c* oxidase, porin, MitoTracker, and CMXRos [43, 151, 183, 184]. Tagging of N-terminal fragments of human frataxin allowed the mitochondrial targeting sequence to be narrowed down to the first 20 amino acids, in agreement with computer predictions for the yeast and mouse proteins [150]. The confirmation that mitochondrial localization is not an artifact of the tagging experiments was obtained by analysis of endogenous frataxin with specific monoclonal antibodies. Western blot analysis with antibodies directed against frataxin C-terminus showed a single band of about 18 kDa in brain, cerebellum, spinal cord, skeletal muscle, and lymphoblasts, and an additional weaker and smaller

FIGURE 26-9 Alignment of frataxin from various species and with bacterial homologues. Pluses (+) indicate identical amino acids, and dots indicate related amino acids. Note the highly conserved domain from amino acids 141 to 167 of human frataxin. Point mutations found in patients (see Table 26-2) are indicated on top of the sequences. I, M1I mutant; X, L106X; -2, 385-2A → G; V, G130V; F, I154F. Boundaries of segments encoded by different exons in man are shown as vertical lines.

band in heart [150]. This is less than the predicted size of the exons 1–5a encoded frataxin, based on the amino acid sequence (23.5 kDa). N-terminal antibodies failed to detect the endogenous frataxin, indicating a post-translational processing involving the removal of the N-terminal signal peptide that explains the reduced size of mature frataxin. The C-terminal antibody was used to locate endogenous frataxin within the mitochondria both by immunofluorescence and after subcellular fractionation by differential centrifugation. Immunoelectron microscopy results indicate that frataxin, which has no hydrophobic transmembrane segment, is nevertheless associated with mitochondrial membranes and crests (Fig. 26-10) [150].

C. Mutants of the Yeast Frataxin Homologue (YFH1)

Disruption of a yeast gene by homologous recombination is a powerful tool to analyze its function that may be conserved throughout evolution. Three independent groups observed that deletion of the yeast frataxin gene, *YFH1* (Yeast Frataxin Homolog 1) causes a respiratory-deficient phenotype. Disruptants were obtained by the insertion of the *HIS3* auxotrophic marker into the open reading frame of *YFH1* (YDL120w). The *Δyfh1* (*yfh1::HIS3*) strain was unable to grow on rich medium containing glycerol/ethanol (YPGE) as the carbon source, showed limited growth on medium containing a fermentable carbon source (YPD), and could not grow on any media at 37°C, suggesting that *Δyfh1* was unable to carry out oxidative phosphorylation. Accordingly, *Δyfh1* showed a severe reduction in oxygen consumption, even in cells grown in rich medium. Loss of respiratory competence is caused by accumulation of mitochondria deficient rho$^-$ clones [43, 151, 183, 185], as demonstrated by analysis of mitochondrial DNA, by using a rho$^-$ tester strain, and by transforming *Δyfh1* wild-type diploids with a *YFH1* plasmid. When diploids were sporulated, the phenotype of the haploid *Δyfh1*

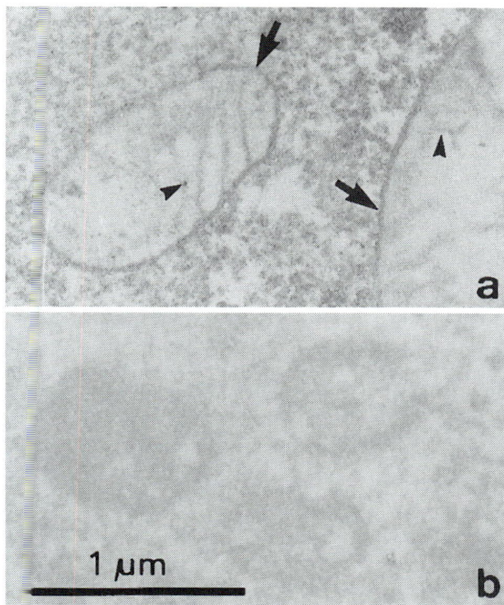

FIGURE 26-10 Immunoelectron microscopy reveals that frataxin is associated with mitochondrial membranes in transfected HeLa cells (courtesy of Drs. V. Campuzano and C. Hindelang). (a) Preembedding mAb 1G2 peroxidase immunodetection labels the mitochondrial external membranes (arrows) and crests (arrowheads). (b) Negative control without primary antibody.

progeny was fully corrected by the plasmid. After plasmid removal with fluoro-orotic acid, the respiratory-deficient phenotype appeared and could not be corrected by retransforming with the same *YFH1* plasmid, indicating that permanent, heritable mitochondrial damage had occurred. The yeast frataxin gene was independently isolated as a multicopy suppressor able to rescue a yeast mutant strain unable to grow on iron-limited medium [43]. In addition, the *Δyfh1* mutant was hypersensitive to oxidative stress, as demonstrated by enhanced sensitivity to H_2O_2, iron, and copper compared to wild-type cells [43, 185]. This led to investigation of iron metabolism in *Δyfh1* yeast. Measurement of iron content revealed a doubling of total cell iron content in *Δyfh1* and a 10-fold higher mitochondrial iron concentration than wild-type cells [43, 185]. No differences were observed in the mitochondrial content of copper or calcium. Such an increase in mitochondrial iron, compared to a moderate increase in total cell iron, indicates a decrease in cytosolic iron content. This results in a marked induction (10- to 50-fold by Northern blot analysis) of the high-affinity iron transport system, which consists of a ferroxidase (Fet3p) and permease (Ftr1p) normally not expressed in cells that are iron replete. Notably, the AFT1-1^{up} strain, which has constitutive expression of the high-affinity iron transport system due to a mutation in the transcriptional regulator *AFT1* and shows intracellular iron levels comparable to *Δyfh1*, does not show an increase in mitochondrial iron nor a respiratory growth defect [43]. This indicates that the accumulation of mitochondrial iron is specifically associated with the deletion of *YFH1,* and is not simply a consequence of increased cellular iron uptake. Accumulation of mitochondrial iron in *Δyfh1* renders this strain hypersensitive to oxidative stress, most likely as a consequence of the Fenton reaction (Fe^{2+}-catalyzed production of hydroxyl radical). Yeast cells can survive without mitochondria when grown in the appropriate media, and spontaneously generate rho$^-$ mutants. Therefore, it is possible that rho$^-$ mutants of *Δyfh1* yeast cells gain a selective advantage because they are protected against oxidative damage by lack of respiration, so they eventually overtake the culture. Further study of the biochemical defect in the yeast mutant should reveal the function of the frataxin homolog in the homeostasis and intracellular trafficking of iron.

D. Biochemical Defect in FRDA Patients

If the function of human frataxin is similar to that of the yeast protein, mitochondrial iron accumulation and oxidative damage should be occurring in FRDA patients' cells. However, even if this is the case, cells from FRDA patients clearly do not have the same phenotype as *Δyfh1* yeast cells. This may be explained by several fundamental differences between human and yeast cells. First, human cells in the body cannot survive without mitochondria and do not spontaneously form rho$^-$ mutants. Indeed, patients with FRDA do not have loss or deletions of mitochondrial DNA (A. Rötig and E. Shubridge, personal communications). In addition, human patients do not have a complete deficiency of frataxin, while *Δyfh1* yeast cells do. Therefore, in the human disease the phenotype is less dramatic and is probably due to a slow, progressive accumulation of peroxidation products of cellular proteins, lipids, and nucleic acids, which will eventually cause mitochondrial dysfunction and cellular atrophy. Affected tissues, such as heart and CNS, are rich in mitochondria. The dying-back neuropathy of FRDA could be explained by peroxidative damage and energy deprivation in long axons, mitochondria-rich structures that have a great dependence on sufficient energy to maintain their integrity. A possible role of apoptotic cell death also remains to be investigated. What is the evidence that iron accumulation and oxidative damage occur in FRDA? So far, iron deposits have been observed in heart myocytes of FRDA patients [42], but observations are lacking in the CNS. Several mitochondrial abnormalities were proposed as the primary defect in FRDA, including lipoamide dehydrogenase deficiency [102], pyruvate carboxylase deficiency [106], mitochondrial malic enzyme deficiency [107], and respiratory chain deficiency [186]. None of these observations has been confirmed or was consistently found in all patients. These studies were limited to tissues (liver, fibroblasts, muscle) that do not show signs of pathology, despite their high mitochondrial and frataxin content. Very recently, Rötig *et al.* demonstrated that a reproducible respiratory chain deficiency is limited to affected tissues such as heart [187]. More precisely, they identified selective deficiencies of the respiratory complexes I, II, and III and of both mitochondrial and cytosolic aconitase activities. The common link between all these enzymes and complexes is that they contain iron–sulfur (Fe–S) clusters in their active sites [187]. Their inactivation is direct proof of oxidative stress in FRDA-affected tissues, since Fe–S proteins are remarkably sensitive to free radicals [188]. In addition, cytosolic aconitase is involved in iron homeostasis and is converted to iron-responsive element binding protein (IRE-BP) when cytosolic iron decreases [189]. The loss of cytosolic aconitase activity observed in FRDA might therefore reflect a decrease of cytosolic iron content (a situation similar to the one found in the yeast *Δyfh1* mutant, see Section V.C) and subsequent switch of aconitase into IRE-BP. Further biochemical analysis of appropriate target tissues may be difficult in man, and a mouse model of FA (by knockout of the frataxin gene) should be extremely valuable to understand frataxin function, and the pathological consequences of its deficiency.

E. Perspectives

The mitochondrial localization of frataxin and mitochondrial iron accumulation in the yeast mutant and possibly also in affected tissues of patients brings FRDA pathology under new light. The mitochondrial localization of frataxin presumably explains the shared clinical features, including ataxia, peripheral neuropathy, cardiomyopathy, and diabetes, with some diseases of the mitochondrial genome, such as MERRF (myoclonic epilepsy and ragged red fibers) [190], NARP (neuropathy, ataxia, and retinitis pigmentosa) [191], and Leigh syndrome [192]. Mitochondrial iron accumulation and Fe–S protein inactivation suggest that increased production of free radicals and increased oxidative stress is involved in FRDA pathology. Several other neurodegenerative disorders were found to be due to reduced protection against oxidative stress, such as a form of familial lateral amyotrophic sclerosis, a motor-neuron degenerative disease caused by mutations in the superoxide dismutase gene (SOD1) [193], and vitamin E deficiencies caused by mutations in the α-tocopherol transfer protein gene [194] or in the microsomal triglyceride transfer protein gene [195]. Defect of the α-tocopherol transfer protein results in severe isolated vitamin E deficiency, in the absence of any other vitamin or metabolic abnormality, and results in a clinical presentation (ataxia with isolated vitamin E deficiency, AVED) very similar to Friedreich ataxia [196–198]. Vitamin E (or its major active form α-tocopherol) is a major liposoluble antioxidant molecule, protecting biological membranes against lipid peroxidation [199]. It is striking to notice that two inherited entities, one resulting from reduced protection against free radicals (AVED) and the other resulting presumably from increased free radical toxicity (Friedreich ataxia), present with similar neurodegenerative specificities (however, unlike FRDA, AVED usually has no heart involvement). Friedreich's ataxia may therefore serve as a paradigm for the growing list of neurodegenerative diseases caused by free radical toxicity. The next challenge is to understand frataxin function and how its loss results in iron accumulation in mitochondria of some, but probably not all, tissues. Several groups are looking for proteins that interact with frataxin. Mouse knockout models of frataxin deficiency are being developed. The prospect of understanding the pathogenesis of FRDA now appears within reach, and promises to uncover previously unsuspected regulatory processes involved in the fine tuning of intracellular iron homeostasis. For the patients and their families, the possibility of developing a rational treatment for the disease may turn out to be a legitimate hope, and no longer just an unrealistic dream.

Acknowledgments

Work in the authors' laboratories was supported by grants from the National Institute of Neurological Diseases and Stroke (NINDS), the Muscular Dystrophy Association (MDA), U.S.A. (to M.P.), and from the Centre National de la Recherche Scientifique, the Institut National de la Santé et de la Recherche Médicale, the Association Française contre les Myopathies, and the Ministère de la Recherche et de la Technologie (to M.K.). We thank L. Montermini, V. Campuzano, M. Cossée, S. Jiralerspong, L. Cova, K. Ohshima, H. Koutnikova, J.-L. Mandel, A. Vescovi, S. Cocozza, A. Richter, P. I. Patel, S. I. Bidichandani, E. Andermann, S. Stifani, P. Duquette, J. Kaplan, F. Foury, A. Brice, A. Dürr, A. Rötig, and P. Rustin, for sharing unpublished information and for useful suggestions and comments.

References

1. Friedreich, N. (1863). Über degenerative Atrophie der spinalen Hinterstränge. *Virchows Arch. Pathol. Anat.* **27,** 1–26.
2. Friedreich, N. (1863). Über degenerative Atrophie der spinalen Hinterstränge. *Virchows Arch. Pathol. Anat.* **26,** 433–459.
3. Friedreich, N. (1876). Über ataxie mit besonderer berücksichtigung der hereditären formen. *Virchows Arch. Pathol. Anat.* **68,** 145–245.

4. Friedreich, N. (1877). Über ataxie mit besonderer berücksichtigung der hereditären formen. *Virchows Arch. Pathol. Anat.* **70,** 140–142.
5. Brousse, M. (1882). De l'ataxie héréditaire. *Thèse de Montpellier*
6. Ladame, P. (1890). Friedreich's disease. *Brain* **13,** 467–537.
7. Bell, J., and Carmichael, E. A. (1939). On hereditary ataxia and spastic paraplegia. *Treas. Hum. Inherit.* **4,** 141–281.
8. Geoffroy, G., Barbeau, A., Breton, G., Lemieux, B., Aube, M., Leger, C., and Bouchard, J. P. (1976). Clinical description and roentgenologic evaluation of patients with Friedreich ataxia. *Can. J. Neurol. Sci.* **3,** 279–286.
9. Harding, A. E. (1981). Friedreich's ataxia: a clinical and genetic study of 90 families with an analysis of early diagnosis criteria and intrafamilial clustering of clinical features. *Brain* **104,** 589–620.
10. Skre, H. (1975). Friedreich's ataxia in western Norway. *Clin. Genet.* **7,** 287–298.
11. Harding, A. E., and Zilkha, K. J. (1981). 'Pseudo-dominant' inheritance in Friedreich's ataxia. *J. Med. Genet.* **18,** 285–287.
12. Filla, A., De Michele, G., Caruso, G., Marconi, R., and Campanella, G. (1990). Genetic data and natural history of Friedreich's disease: a study of 80 Italian patients. *J. Neurol.* **237,** 345–351.
13. Muller-Felber, W., Rossmanith, T., Spes, C., Chamberlain, S., Pongratz, D., and Deufel, T. (1993). The clinical spectrum of Friedreich's ataxia in German families showing linkage to the FRDA locus on chromosome 9. *Clin. Invest.* **71,** 109–114.
14. Winter, R. M., Harding, A. E., Baraitser, M., and Bravery, M. B. (1981). Intrafamilial correlation in Friedreich's ataxia. *Clin. Genet.* **20,** 419–427.
15. Filla, A., De Michele, G., Cavalcanti, F., Santorelli, F., Santoro, L., and Campanella, G. (1991). Intrafamilial phenotype variation in Friedreich's disease: possible exceptions to diagnostic criteria. *J. Neurol.* **238,** 147–150.
16. Barbeau, A., Roy, M., Sadibelouiz, M., and Wilensky, M. A. (1984). Recessive ataxia in Acadians and "Cajuns". *Can. J. Neurol. Sci.* **11,** 526–533.
17. Richter, A., Poirier, J., Mercier, J., Julien, D., Morgan, K., Roy, M. Gosselin, F., Bouchard, J. P., and Melancon, S. B. (1996). Friedreich ataxia in Acadian families from eastern Canada: clinical diversity with conserved haplotypes. *Am. J. Med. Genet.* **64,** 594–601.
18. Klockgether, T., Chamberlain, S., Wullner, U., Fetter, M., Dittmann, H., Petersen, D., and Dichgans, J. (1993). Late-onset Friedreich's ataxia. Molecular genetics, clinical neurophysiology, and magnetic resonance imaging. *Arch. Neurol.* **50,** 803–806.
19. De Michele, G., Filla, A., Cavalcanti, F., Di Maio, L., Pianese, L., Castaldo, I., Calabrese, O., Monticelli, A., Varrone, S., Campanella, G., and Coccoza, S. (1994). Late onset Friedreich's disease: clinical features and mapping of mutation to the FRDA locus. *J. Neurol. Neurosurg. Psych.* **57,** 977–979.
20. Palau, F., De Michele, G., Vilchez, J. J., Pandolfo, M., Monros, E., Cocozza, S., Smeyers, P., Lopez-Arlandis, J., Campanella, G., Di Donato, S., and Filla, A. (1997). Early-onset ataxia with cardiomyopathy and retained tendon reflexes maps to Friedreich's ataxia locus on chromosome 9q. *Ann. Neurol.* **37,** 359–362.
21. Campuzano, V., Montermini, L., Moltó, M. D., Pianese, L., Cossée, M., Cavalcanti, F., Monros, E., Rodius, F., Duclos, F., Monticelli, A., Zara, F., Cañizares, J., Koutnikova, H., Bidichandani, S., Gellera, C., Brice, A., Trouillas, P., De Michele, G., Filla, A., de Frutos, R., Palau, F., Patel, P. I., Di Donato, S., Mandel, J.-L., Cocozza, S., Koenig, M., and Pandolfo, M. (1996). Friedreich ataxia: autosomal recessive disease caused by an intronic GAA triplet repeat expansion. *Science* **271,** 1423–1427.
22. Harding, A. E. (1983). Classification of the hereditary ataxias and paraplegias. *Lancet* **1,** 1151–1155.
23. Romeo, G., Menozzi, P., Ferlini, A., Fadda, S., Di Donato, S., Uziel, G., and Lucci, B. (1983). Incidence of Friedreich ataxia in Italy estimated from consanguinous marriages. *Am. J. Hum. Genet.* **35,** 523–529.
24. Leone, M., Brignolio, F., Rosso, M. G., Curtoni, E. S., Moroni, A., Tribolo, A., and Schiffer, D. (1990). Friedreich's ataxia: a descriptive epidemiological study in an Italian population. *Clin. Genet.* **38,** 161–169.
25. Lopez-Arlandis, J. M., Vilchez, J. J., Palau, F., and Sevilla, T. (1995). Friedreich's ataxia: an epidemiological study in Valencia, Spain, based on consanguinity analysis. *Neuroepidemiology.* **14,** 14–19.
26. Bouchard, J. P., Barbeau, A., Bouchard, R., Paquet, M., and Bouchard, R. W. (1979). A cluster of Friedreich's ataxia in Rimouski, Quebec. *Can. J. Neurol. Sci.* **6,** 205–208.
27. Dean, G., Chamberlain, S., and Middleton, L. (1988). Friedreich's ataxia in Kathikas-Arodhes, Cyprus. *Lancet* **1,** 587.
28. Hewer, R. L. (1968). Study of fatal cases of Friedreich's ataxia. *Br. Med. J.* **3,** 649–652.
29. Lamarche, J. B., Lemieux, B., and Lieu, H. B. (1984). The neuropathology of "typical" Friedreich's ataxia in Quebec. *Can. J. Neurol. Sci.* **11,** 592–600.
30. Friedreich, N. (1863). Uber degenerative Atrophie der spinalen Hinterstränge. *Virchows Arch. Pathol. Anat.* **26,** 391–419.
31. Said, G., Marion, M. H., Selva, J., and Jamet, C. (1986). Hypotrophic and dying-back nerve fibers in Friedreich's ataxia. *Neurology* **36,** 1292–1299.
32. Koeppen, A. H. (1991). The Purkinje cell and its afferents in human hereditary ataxia. *J. Neuropathol. Exp. Neurol.* **50,** 505–514.
33. Carroll, W. M., Kriss, A., Baraitser, M., Barrett, G., and Halliday, A. M. (1980). The incidence and nature of visual pathway involvement in Friedreich's ataxia. A clinical and visual evoked potential study of 22 patients. *Brain* **103,** 413–434.
34. Hughes, J. T., Brownell, B., and Hewer, R. L. (1968). The peripheral sensory pathway in Friedreich's ataxia. An examination by light and electron microscopy of the posterior nerve roots, posterior root ganglia, and peripheral sensory nerves in cases of Friedreich's ataxia. *Brain* **91,** 803–818.
35. Ouvrier, R. A., McLeod, J. G., and Conchin, T. E. (1982). Friedreich's ataxia. Early detection and progression of peripheral nerve abnormalities. *J. Neurol. Sci.* **55,** 137–145.
36. Harding, A. E. (1984). "The Hereditary Ataxias and Related Disorders." Churchill Livingstone, London.
37. Jitpimolmard, S., Small, J., King, R. H., Geddes, J., Misra, P., McLaughlin, J., Muddle, J. R., Cole, M., Harding, A. E., and Thomas, P. K. (1993). The sensory neuropathy of Friedreich's ataxia: an autopsy study of a case with prolonged survival. *Acta Neuropathol. (Berlin)* **86,** 29–35.
38. Gottdiener, J. S., Hawley, R. J., Maron, B. J., Bertorini, T. F., and Engle, W. K. (1982). Characteristics of the cardiac hypertrophy in Friedreich's ataxia. *Am. Heart J.* **103,** 525–531.
39. Pasternac, A., Krol, R., Petitclerc, R., Harvey, C., Andermann, E., and Barbeau, A. (1980). Hypertrophic cardiomyopathy in Friedreich's ataxia: symmetric or asymmetric? *Can. J. Neurol. Sci.* **7,** 379–382.
40. Casazza, F., and Morpurgo, M. (1996). The varying evolution of Friedreich's ataxia cardiomyopathy. *Am. J. Cardiol.* **77,** 895–898.
41. Lamarche, J. B., Cote, M., and Lemieux, B. (1980). The cardiomyopathy of Friedreich's ataxia morphological observations in 3 cases. *Can. J. Neurol. Sci.* **7,** 389–396.
42. Lamarche, J. B., Shapcott, D., Côté, M., and Lemieux, B. (1993). Cardiac iron deposits in Friedreich's ataxia. *In* "Handbook of

Cerebellar Diseases" (R. Lechtenberg, Ed.), pp. 453–458. Marcel Dekker, New York.
43. Babcock, M., de Silva, D., Oaks, R., Davis-Kaplan, S., Jiralerspong, S., Montermini, L., Pandolfo, M., and Kaplan, J. (1997). Regulation of mitochondrial iron accumulation by Yfh1, a putative homolog of frataxin. *Science* **276,** 1709–1712.
44. Schoenle, E. J., Boltshauser, E. J., Baekkeskov, S., Landin Olsson, M., Torresani, T., and von Felten, A. (1989). Preclinical and manifest diabetes mellitus in young patients with Friedreich's ataxia: no evidence of immune process behind the islet cell destruction. *Diabetologia* **32,** 378–381.
45. Allard, P., Dansereau, J., Thiry, P. S., Geoffroy, G., Raso, J. V., and Duhaime, M. (1982). Scoliosis in Friedreich's ataxia. *Can. J. Neurol. Sci.* **9,** 105–111.
46. Labelle, H., Tohme, S., Duhaime, M., and Allard, P. (1986). Natural history of scoliosis in Friedreich's ataxia. *J. Bone Joint Surg. [Am].* **68,** 564–572.
47. Aronsson, D. D., Stokes, I. A., Ronchetti, P. J., and Labelle, H. B. (1994). Comparison of curve shape between children with cerebral palsy, Friedreich's ataxia, and adolescent idiopathic scoliosis. *Dev. Med. Child. Neurol.* **36,** 412–418.
48. Campanella, G., Filla, A., De Falco, F., Mansi, D., Durivage, A., and Barbeau, A. (1980). Friedreich's ataxia in the south of Italy: a clinical and biochemical survey of 23 patients. *Can. J. Neurol. Sci.* **7,** 351–357.
49. D'Angelo, A., Di Donato, S., Negri, G., Beulche, F., Uziel, G., and Boeri, R. (1980). Friedreich's ataxia in northern Italy: I. Clinical, neurophysiological and in vivo biochemical studies. *Can. J. Neurol. Sci.* **7,** 359–365.
50. Ulku, A., Arac, N., and Ozeren, A. (1988). Friedreich's ataxia: a clinical review of 20 childhood cases. *Acta Neurol. Scand.* **77,** 493–497.
51. De Michele, G., Di Maio, L., Filla, A., Majello, M., Cocozza, S., Cavalcanti, F., Mirante, E., and Campanella, G. (1996). Childhood onset of Friedreich ataxia: a clinical and genetic study of 36 cases. *Neuropediatrics* **27,** 3–7.
52. Gentil, M. (1990). Dysarthria in Friedreich disease. *Brain Lang.* **38,** 438–448.
53. Cisneros, E., and Braun, C. M. (1995). Vocal and respiratory diadochokinesia in Friedreich's ataxia. Neuropathological correlations. *Rev. Neurol. (Paris)* **151,** 113–123.
54. Beauchamp, M., Labelle, H., Duhaime, M., and Joncas, J. (1995). Natural history of muscle weakness in Friedreich's Ataxia and its relation to loss of ambulation. *Clin. Orthop.* 270–275.
55. Spieker, S., Schulz, J. B., Petersen, D., Fetter, M., Klockgether, T., and Dichgans, J. (1995). Fixation instability and oculomotor abnormalities in Friedreich's ataxia. *J. Neurol.* **242,** 517–521.
56. Kirkham, T. H., and Coupland, S. G. (1981). An electroretinal and visual evoked potential study in Friedreich's ataxia. *Can. J. Neurol. Sci.* **8,** 289–294.
57. Livingstone, I. R., Mastaglia, F. L., Edis, R., and Howe, J. W. (1981). Visual involvement in Friedreich's ataxia and hereditary spastic ataxia. A clinical and visual evoked response study. *Arch. Neurol.* **38,** 75–79.
58. Rabiah, P. K., Bateman, J. B., Demer, J. L., and Perlman, S. (1997). Ophthalmologic findings in patients with ataxia. *Am. J. Ophthalmol.* **123,** 108–117.
59. Ell, J., Prasher, D., and Rudge, P. (1984). Neuro-otological abnormalities in Friedreich's ataxia. *J. Neurol. Neurosurg. Psych.* **47,** 26–32.
60. Cassandro, E., Mosca, F., Sequino, L., De Falco, F. A., and Campanella, G. (1986). Otoneurological findings in Friedreich's ataxia and other inherited neuropathies. *Audiology* **25,** 84–91.
61. Montermini, L., Richter, A., Morgan, K., Justice, C. M., Julien, D., Castelloti, B., Mercier, J., Poirier, J., Capazzoli, F., Bouchard, J. P., Lemieux, B., Mathieu, J., Vanasse, M., Seni, M. H., Graham, G., Andermann, F., Andermann, E., Melançon, S., Keats, B. J. B., Di Donato, S., and Pandolfo, M. (1997). Phenotypic variability in Friedreich ataxia: role of the associated GAA triplet repeat expansion. *Ann. Neurol.* **41,** 675–682.
62. Leone, M., Rocca, W. A., Rosso, M. G., Mantel, N., Schoenberg, B. S., and Schiffer, D. (1988). Friedreich's disease: survival analysis in an Italian population. *Neurology* **38,** 1433–1438.
63. Hartman, J. M., and Booth, R. W. (1960). Friedreich's ataxia: a neurocardiac disease. *Am. Heart J.* **60,** 716–720.
64. Boyer, S. H., Chisholm, A. W., and McKusick, V. A. (1962). Cardiac aspects of Friedreich's ataxia. *Circulation* **25,** 493–505.
65. Harding, A. E., and Hewer, R. L. (1983). The heart disease of Friedreich's ataxia. A clinical and electrocardiographic changes in 30 cases. *Q. J. Med.* **52,** 489–502.
66. Pentland, B., and Fox, K. A. (1983). The heart in Friedreich's ataxia. *J. Neurol. Neurosurg. Psych.* **46,** 1138–1142.
67. Child, J. S., Perloff, J. K., Bach, P. M., Wolfe, A. D., Perlman, S., and Kark, R. A. (1986). Cardiac involvement in Friedreich's ataxia: a clinical study of 75 patients. *J. Am. Coll. Cardiol.* **7,** 1370–1378.
68. Alboliras, E. T., Shub, C., Gomez, M. R., Edwards, W. D., Hagler, D. J., Reeder, G. S., Seward, J. B., and Tajik, A. J. (1986). Spectrum of cardiac involvement in Friedreich's ataxia: clinical, electrocardiographic and echocardiographic observations. *Am. J. Cardiol.* **58,** 518–524.
69. Maione, S., Giunta, A., Filla, A., De Michele, G., Spinelli, L., Liucci, G. A., Campanella, G., and Condorelli, M. (1997). May age onset be relevant in the occurrence of left ventricular hypertrophy in Friedreich's ataxia? *Clin. Cardiol.* **20,** 141–145.
70. Morvan, D., Komajda, M., Doan, L. D., Brice, A., Isnard, R., Seck, A., Lechat, P., Agid, Y., and Grosgogeat, Y. (1992). Cardiomyopathy in Friedreich's ataxia: a Doppler-echocardiographic study. *Eur. Heart J.* **13,** 1393–1398.
71. Finocchiaro, G., Baio, G., Micossi, P., Pozza, G., and Di Donato, S. (1988). Glucose metabolism alterations in Friedreich's ataxia. *Neurology* **38,** 1292–1296.
72. Fantus, I. G., Seni, M. H., and Andermann, E. (1993). Evidence for abnormal regulation of insulin receptors in Friedreich's ataxia. *J. Clin. Endocrinol. Metab.* **76,** 60–63.
73. Margalith, D., Dunn, H. G., Carter, J. E., and Wright, J. M. (1984). Friedreich's ataxia with dysautonomia and labile hypertension. *Can. J. Neurol. Sci.* **11,** 73–77.
74. Pousset, F., Kalotka, H., Durr, A., Isnard, R., Lechat, P., Le Heuzey, J. Y., Thomas, D., and Komajda, M. (1996). Parasympathetic activity in Friedrich's ataxia. *Am. J. Cardiol.* **78,** 847–850.
75. Riva, A., and Bradac, G. B. (1995). Primary cerebellar and spinocerebellar ataxia an MRI study on 63 cases. *J. Neuroradiol.* **22,** 71–76.
76. Wessel, K., Schroth, G., Diener, H. C., Muller-Forell, W., and Dichgans, J. (1989). Significance of MRI-confirmed atrophy of the cranial spinal cord in Friedreich's ataxia. *Eur. Arch. Psych. Neurol. Sci.* **238,** 225–230.
77. Wullner, U., Klockgether, T., Petersen, D., Naegele, T., and Dichgans, J. (1993). Magnetic resonance imaging in hereditary and idiopathic ataxia. *Neurology* **43,** 318–325.
78. Junck, L., Gilman, S., Gebarski, S. S., Koeppe, R. A., Kluin, K. J., and Markel, D. S. (1994). Structural and functional brain imaging in Friedreich's ataxia. *Arch. Neurol.* **51,** 349–355.
79. Mascalchi, M., Salvi, F., Piacentini, S., and Bartolozzi, C. (1994). Friedreich's ataxia: MR findings involving the cervical portion of the spinal cord. *Am. J. Roentgenol.* **163,** 187–191.

80. Giroux, M., Septien, L., Pelletier, J. L., Dueret, N., and Dumas, R. (1994). Decrease in cerebellar blood flow in patients with Friedreich's ataxia: a TC-HMPAO SPECT study of three cases. *Neurol. Res.* **16,** 342–344.
81. Gilman, S., Junck, L., Markel, D. S., Koeppe, R. A., and Kluin, K. J. (1990). Cerebral glucose hypermetabolism in Friedreich's ataxia detected with positron emission tomography. *Ann. Neurol.* **28,** 750–757.
82. McLeod, J. G. (1971). An electrophysiological and pathological study of peripheral nerves in Friedreich's ataxia. *J. Neurol. Sci.* **12,** 333–349.
83. Peyronnard, J. M., Bouchard, J. P., and Lapointe, M. (1976). Nerve conduction studies and electromyography in Friedreich's ataxia. *Can. J. Neurol. Sci.* **3,** 313–317.
84. Ackroyd, R. S., Finnegan, J. A., and Green, S. H. (1984). Friedreich's ataxia. A clinical review with neurophysiological and echocardiographic findings. *Arch. Dis. Child.* **59,** 217–221.
85. Vanasse, M., Garcia-Larrea, L., Neuschwander, P., Trouillas, P., and Mauguiere, F. (1988). Evoked potential studies in Friedreich's ataxia and progressive early onset cerebellar ataxia. *Can. J. Neurol. Sci.* **15,** 292–298.
86. Mondelli, M., Rossi, A., Scarpini, C., and Guazzi, G. C. (1995). Motor evoked potentials by magnetic stimulation in hereditary and sporadic ataxia. *Electromyogr. Clin. Neurophysiol.* **35,** 415–424.
87. Chamberlain, S., Shaw, J., Wallis, J., Rowland, A., Chow, L., Farrall, M., Keats, B., Richter, A., Roy, M., Melançon, S., Deufel, T., Berciano, J., and Williamson, R. (1989). Genetic homogeneity at the Friedreich ataxia locus on chromosome 9 . *Am. J. Hum. Genet.* **44,** 518–521.
88. Berciano, J., Combarros, O., De Castro, M., and Palau, F. (1997). Intronic GAA triplet repeat expansion in Friedreich's ataxia presenting with pure sensory ataxia. *J. Neurol.* **244,** 390–391.
89. Sirugo, G., Keats, B., Fujita, R., Duclos, F., Purohit, K., Koenig, M., and Mandel, J. L. (1992). Friedreich ataxia locus in Louisiana Acadians: Demonstration of a founder effect by analysis of microsatellite-generated extended haplotypes. *Am. J. Hum. Genet.* **50,** 559–566.
90. De Michele, G., Perrone, F., Filla, A., Mirante, E., Giordano, M. De Placido, S., and Campanella, G. (1996). Age of onset, sex, and cardiomyopathy as predictors of disability and survival in Friedreich's disease: a retrospective study on 119 patients. *Neurology* **47,** 1260–1264.
91. Kark, R. A., Budelli, M. M., and Wachsner, R. (1981). Double-blind, triple-crossover trial of low doses of oral physostigmine in inherited ataxias. *Neurology* **31,** 288–292.
92. Le Witt, P. A., and Ehrenkranz, J. R. (1982). TRH and spinocerebellar degeneration. *Lancet* **2,** 981. [Letter]
93. Filla, A., De Michele, G., Di Martino, L., Mengano, A., Iorio, L. Maggio, M. A., and Campanella, G. (1989). Chronic experimentation with TRH administered intramuscularly in spinocerebellar degeneration. Double-blind cross-over study in 30 subjects. *Rev. Neurol.* **59,** 83–88.
94. Wessel, K., Hermsdorfer, J., Deger, K., Herzog, T., Huss, G. P., Kompf, D., Mai, N., Schimrigk, K., Wittkamper, A., and Ziegler, W. (1995). Double-blind crossover study with levorotatory form of hydroxytryptophan in patients with degenerative cerebellar diseases. *Arch. Neurol.* **52,** 451–455.
95. Trouillas, P., Serratrice, G., Laplane, D., Rascol, A., Augustin, E., Barroche, G., Clanet, M., Degos, C. F., Desnuelle, C., Dumas, E., Michel, R., Viallet, F., Warter J.-M., and Adeline, P. (1995). Levorotatory form of 5-hydroxytryptophan in Friedreich's ataxia. Results of a double-blind drug-placebo cooperative study. *Arch. Neurol.* **52,** 456–460.
96. Peterson, P. L., Saad, J., and Nigro, M. A. (1988). The treatment of Friedreich's ataxia with amantadine hydrochloride. *Neurology.* **38,** 1478–1480.
97. Filla, A., De Michele, G., Orefice, G., Santorelli, F., Trombetta, L., Banfi, S., Squitieri, F., Napolitano, G., Puma, D., and Campanella, G. (1993). A double-blind cross-over trial of amantadine hydrochloride in Friedreich's ataxia. *Can. J. Neurol. Sci.* **20,** 52–55.
98. Botez, M. I., Botez-Marquard, T., Elie, R., Pedraza, O. L., Goyette, K., and Lalonde, R. (1996). Amantadine hydrochloride treatment in heredodegenerative ataxias: a double blind study. *J. Neurol. Neurosurg. Psych.* **61,** 259–264.
99. Barbeau, A. (1976). Friedreich's ataxia 1976: an overview. *Can. J. Neurol. Sci.* **3,** 389–397.
100. Barbeau, A. (1978). Friedreich's ataxia 1978: an overview. *Can. J. Neurol. Sci.* **5,** 161–165.
101. Barbeau, A. (1982). Friedreich's disease 1982: etiologic hypotheses a personal analysis. *Can. J. Neurol. Sci.* **9,** 243–263.
102. Blass, J. P., Kark, R. A. P., and Menon, N. K. (1976). Low activities of the pyruvate and oxoglutarate dehydrogenase complexes in five patients with Friedreich's ataxia. *N. Engl. J. Med.* **295,** 62–67.
103. Kark, R. A., and Rodriguez-Budelli, M. (1979). Pyruvate dehydrogenase deficiency in spinocerebellar degenerations. *Neurology* **29,** 126–131.
104. Kark, R. A., Budelli, M. M., Becker, D. M., Weiner, L. P., and Forsythe, A. B. (1981). Lipoamide dehydrogenase: rapid heat inactivation in platelets of patients with recessively inherited ataxia. *Neurology* **31,** 199–202.
105. Robinson, B. H., Sherwood, W. G., Kahler, S., O'Flynn, M. E., and Nadler, H. (1981). Lipoamide dehydrogenase deficiency. *N. Engl. J. Med.* **304,** 53–54.
106. Dijkstra, U. J., Willems, J. L., Joosten, E. M., and Gabreels, F. J. (1983). Friedreich ataxia and low pyruvate carboxylase activity in liver and fibroblasts. *Ann. Neurol.* **13,** 325–327.
107. Stumpf, D. A., Parks, J. K., Eguren, L. A., and Haas, R. (1982). Friedreich ataxia: III. Mitochondrial malic enzyme deficiency. *Neurology* **32,** 221–227.
108. Stumpf, D. A., Parks, J. K., and Parker, W. D. (1983). Friedreich's disease: IV. Reduced mitochondrial malic enzyme activity in heterozygotes. *Neurology* **33,** 780–783.
109. Chamberlain, S., and Lewis, P. D. (1983). Normal mitochondrial malic enzyme levels in Friedreich's ataxia fibroblasts. *J. Neurol. Neurosurg. Psych.* **46,** 1050–1051.
110. Gray, R. G., and Kumar, D. (1985). Mitochondrial malic enzyme in Friedreich's ataxia: failure to demonstrate reduced activity in cultured fibroblasts. *J. Neurol. Neurosurg. Psych.* **48,** 70–74.
111. Fernandez, R. J., Civantos, F., Tress, E., Maltese, W. A., and De Vivo, D. C. (1986). Normal fibroblast mitochondrial malic enzyme activity in Friedreich's ataxia. *Neurology* **36,** 869–872.
112. Lemieux, B., Barbeau, A., Beroniade, V., Shapcott, D., Breton, G., Geoffroy, G., and Melancon, S. (1976). Aminoacids metabolism in Friedreich's ataxia. *Can. J. Neurol. Sci.* **3,** 373–378.
113. Lemieux, B., Giguère, R., Barbeau, A., Melancon, S., and Shapcott, D. (1978). Taurine in cerebrospinal fluid in Friedreich's ataxia. *Can. J. Neurol. Sci.* **5,** 125–129.
114. Walker, J. L., Chamberlain, S., and Robinson, N. (1980). Lipids and lipoproteins in Friedreich's ataxia. *J. Neurol. Neurosurg. Psych.* **43,** 111–117.
115. Chamberlain, S., Walker, J. L., Sachs, J. A., Wolf, E., and Festenstein, H. (1979). Non-association of Friedreich's ataxia and HLA based on five families. *Can. J. Neurol. Sci.* **6,** 451–452.
116. Koeppen, A. H., Goedde, H. W., Hirth, L., Benkmann, H. G., and Hiller, C. (1980). Genetic linkage in hereditary ataxia. *Lancet* **1,** 92–93.

117. Chamberlain, S., Worrall, C. S., South, S., Shaw, J., Farrall, M., and Williamson, R. (1987). Exclusion of the Friedreich ataxia gene from chromosome 19. *Hum. Genet.* **76,** 186–190.

118. Keats, B. J., Ward, L. J., Lu, M., Krieger, S., Wilensky, M. A., Forster-Gibson, C. J., Roy, M., Monte, M., Barbeau, A., Simpson, N. E., Eiberg, H., Tippett, P., Williamson, R., and Chamberlain, S. (1987). Linkage studies of Friedreich ataxia by means of blood-group and protein markers. *Am. J. Hum. Genet.* **41,** 627–634.

119. Chamberlain, S., Shaw, J., Rowland, A., Wallis, J., South, S., Nakamura, Y., von Gabain, A., Farrall, M., and Williamson, R. (1988). Mapping of mutation causing Friedreich' ataxia to human chromosome 9. *Nature* **334,** 248–250.

120. Fujita, R., Agid, Y., Trouillas, P., Seck, A., Tommasi-Davenas, C., Driesel, A. J., Olek, K., Grzeschik, K. H., Nakamura, Y., Mandel, J.-L., and Hanauer, A. (1989). Confirmation of linkage of Friedreich ataxia to chromosome 9 and identification of a new closely linked marker. *Genomics* **4,** 110–111.

121. Richter, A., Melancon, S., Farrall, M., and Chamberlain, S. (1989). Friedreich's ataxia: confirmation of gene localization to chromosome 9 in the Quebec French Canadian population. *Cytogenet. Cell Genet.* **51,** 1066.

122. Ross, D. A., McLeod, J. G., and Nicholson, G. A. (1989). Linkage studies with Friedreich's ataxia in Australian pedigrees. *Cytogenet. Cell Genet.* **51,** 1069.

123. Chamberlain, S., Shaw, J., Wallis, J., Wilkes, D., Farrall, M., and Williamson, R. (1989). Linkage analysis of DNA markers around the Friedreich's ataxia locus on chromosome 9. *Cytogenet. Cell Genet.* **51,** 975.

124. Hanauer, A., Chery, M., Fujita, R., Driesel, A. J., Gilgenkrantz, S., and Mandel, J. L. (1990). The Friedreich ataxia gene is assigned to chromosome 9q13-q21 by mapping of tightly linked markers and shows linkage disequilibrium with D9S15. *Am. J. Hum. Genet.* **46,** 133–137.

125. Raimondi, E., Antonelli, A., Driesel, A. J., and Pandolfo, M. (1990). Regional localization by in situ hybridization of a human chromosome 9 marker tightly linked to the Friedreich's ataxia locus. *Hum. Genet.* **85,** 125–126.

126. Shaw, J., Lichter, P., Driesel, A. J., Williamson, R., and Chamberlain, S. (1990). Regional localisation of the Friedreich ataxia locus to human chromosome 9q13-q21.1. *Cytogenet. Cell Genet.* **53,** 221–224.

127. Raimondi, E., Bernasconi, P., Moralli, D., Fujita, R., Uziel, G., Di Donato, S., De Carli, L., and Pandolfo, M. (1991). Localization of DNA probes tightly linked to the Friedreich' s ataxia locus by in situ hybridization in a case of pericentric inversion of chromosome 9. *Hum. Genet.* **86,** 525–528.

128. Wilkes, D., Shaw, J., Anand, R., Riley, J., Winter, P., Wallis, J., Driesel, A. G., Williamson, R., and Chamberlain, S. (1991). Identification of CpG islands in a physical map encompassing the Friedreich's ataxia locus. *Genomics* **9,** 90–95.

129. Fujita, R., Hanauer, A., Vincent, A., Mandel, J. L., and Koenig, M. (1991). Physical mapping of two loci (D9S5 and D9S15) tightly linked to Friedreich ataxia locus (FRDA) and identification of nearby CpG islands by pulse-field gel electrophoresis. *Genomics* **10,** 915–920.

130. Fujita, R., Sirugo, G., Duclos, F., Abderrahim, H., Le Paslier, D., Cohen, D., Brownstein, B. H., Schlessinger, D., Mandel, J. L., and Koenig, M. (1992). A 530 kb YAC contig tightly linked to the Friedreich ataxia locus contains five CpG clusters and a new highly polymorphic microsatellite. *Hum. Genet.* **89,** 531–538.

131. Fujita, R., Hanauer, A., Sirugo, G., Heilig, R., and Mandel, J. L. (1990). Additional polymorphisms at marker loci D9S5 and D9S15 generate extended haplotypes in linkage disequilibrium with Friedreich ataxia. *Proc. Natl. Acad. Sci. USA* **87,** 1796–1800.

132. Wallis, J., Williamson, R., and Chamberlain, S. (1990). Identification of a hypervariable microsatellite polymorphism within D9S15 tightly linked to Friedrich's ataxia. *Hum. Genet.* **85,** 98–100.

133. Pandolfo, M., Sirugo, G., Antonelli, A., Weitnauer, L., Ferreti, L., Leone, M., Dones, I., Cerino., A, Fujita, R., Hanauer, A., Mandel, J.-L., and Di Donato., S (1990). Friedreich ataxia in Italian Families: genetic homogeneity and linkage disequilibrium with the marker Loci D9S5 and D9S15. *Am. J. Hum. Genet.* **47,** 228–235.

134. Belal, S., Kyproula, P., Sirugo, G., Ben Hamida, C., Panos, I., Hentati, F., Beckmann, J., Koenig, M., Mandel, J. L., Ben, H. M., and Middleton, L. T. (1992). Study of large inbred Friedreich ataxia families reveals a recombination between D9S15 and the disease locus. *Am. J. Hum. Genet.* **51,** 1372–1376.

135. Chamberlain, S., Farrall, M., Shaw, J., Wilkes, D., Carvajal, J., Hillermann, R., Doudney, K., Harding, A. E., Williamson, R., Sirugo, G., Fujita, R., Koenig, M., Mandel, J.-L., Palau, F., Monros, E., Vilchez, J., Prieto, F., Richter, A., Vanasse, M., Melancon, S., Coccoza, S., Redolfi, E., Cavalcanti, F., Pianese, L., Filla, A., Di Donato, S., and Pandolfo, M. (1993). Genetic recombination events which position the Friedreich Ataxia locus proximal to the D9S15/D9S5 linkage group on chromosome 9q. *Am. J. Hum. Genet.* **52,** 99–109.

136. Sirugo, G., Cocoza, S., Mandel, J. L., Brice, A., Cavalcanti, F., De Michele, G., Dones, I., Filla, A., Koenig, M., Lorenzetti, D., Monticelli, A., Pianese, L., Redolfi, E., Rousseau, F., Di Donato, S., and Pandolfo, M. (1993). Linkage disequilibrium analysis of Friedreich's Ataxia in 140 Caucasian families: positioning of the disease locus and evaluation of allelic heterogeneity. *Eur. J. Hum. Genet.* **1,** 133–143.

137. Pandolfo, M., Munaro, M., Cocozza, S., Redolfi, E. M., Pianese, L., Cavalcanti, F., Monticelli, A., and Di Donato, S. (1993). A dinucleotide repeat polymorphism (D9S202) in the Friedreich's ataxia region on chromosome 9q13-q21.1. *Hum. Mol. Genet.* **2,** 822.

138. Pianese, L., Cocozza, S., Campanella, G., Castaldo, I., Cavalcanti, F., De Michele, G., Filla, A., Monticelli, A., Munaro, M., Redolfi, E., Varrone, S., and Pandolfo, M. (1994). Linkage disequilibrium between FD1-D9S202 haplotypes and the Friedreich's ataxia locus in a central-southern Italian population. *J. Med. Genet.* **31,** 133–135.

139. Rodius, F., Duclos, F., Wrogemann, K., Le Paslier, D., Ougen, P., Billault, A., Belal, S., Musenger, C., Brice, A., Dürr, A., Mignard, C., Sirugo, G., Weissenbach, J., Cohen, D., Hentati, F., Ben Hamida, M., Mandel, J. L., and Koenig, M. (1994). Recombinations in individuals homozygous by descent localize the Friedreich Ataxia locus in a cloned 450-kb interval. *Am. J. Hum. Genet.* **54,** 1050–1059.

140. Monros, E., Smeyers, P., Rodius, F., Cañizares, J., Moltó, M. D., Vilchez, J., Pandolfo, M., Lopez-Arlandis, J., de Frutos, R., Prieto, F., Koenig, M., and Palau, F. (1994). Refined mapping of Friedreich ataxia locus by identification of recombinant events in patients homozygous by descent. *Eur. J. Hum. Genet.* **2,** 291–299.

141. Duclos, F., Rodius, F., Wrogemann, K., Mandel, J.-L., and Koenig, M. (1994). The Friedreich ataxia region: characterization of two novel genes and reduction of the critical region to 300 kb. *Hum. Mol. Genet.* **3,** 909–914.

142. Montermini, L., Rodius, F., Pianese, L., Moltó, M. D., Cossée, M., Campuzano, V., Cavalcanti, F., Monticelli, A., Palau, F., Gyapay, G., Wenhert, M., Zara, F., Patel, P. I., Cocozza, S., Koenig, M., and Pandolfo, M. (1995). The Friedreich ataxia criti-

cal region spans a 150-kb interval on chromosome 9q13. *Am. J. Hum. Genet.* **57,** 1061–1067.

143. Duclos, F., Boschert, U., Sirugo, G., Mandel, J.-L., Hen, R., and Koenig, M. (1993). Gene in the region of the Friedreich ataxia locus encodes a putative transmembrane protein expressed in the nervous system. *Proc. Natl. Acad. Sci. USA* **90,** 109–113.
144. Duclos, F., and Koenig, M. (1995). Comparison of primary structure of a neuron-specific protein, X11, between human and mouse. *Mamm. Genome* **6,** 57–58.
145. Pandolfo, M., Pizzuti, A., Redolfi, E., Munaro, M., Di Donato, S., Cavalcanti, F., Filla, A., Monticelli, A., Pianese, L., and Cocozza, S. (1994). Isolation of a new gene in the Friedreich ataxia candidate region on human chromosome 9 by cDNA direct selection. *Biochem. Med. Metab. Biol.* **52,** 115–119.
146. Jesaitis, L. A., and Goodenough, D. A. (1994). Molecular characterization and tissue distribution of ZO-2, a tight junction protein homologous to ZO-1 and the Drosophila discs-large tumor suppressor protein. *J. Cell Biol.* **124,** 949–961.
147. Carvajal, J., Pook, M. A., Doudney, K., Hillermann, R., Wilkes, D., Al-Mahdawi, S., Williamson, R., and Chamberlain, S. (1995). Friedreich ataxia: a defect in signal transduction? *Hum. Mol. Genet.* **4,** 1411–1419.
148. Carvajal, J., Pook, M. A., dos Santos, M., Doudney, K., Hillermann, R., Minogue, S., Williamson, R., Hsuan, J. J., and Chamberlain, S. (1996). The Friedreich's ataxia gene encodes a novel phosphatidylinositol -4- phosphate 5-kinase. *Nature Genet.* **14,** 157–162.
149. Cossée, M., Campuzano, V., Koutnikova, H., Fischbeck, K. H., Mandel, J.-L., Koenig, M., Bidichandani, S., Patel, P. I., Moltó, M. D., Cañizares, J., de Frutos, R., Pianese, L., Cavalcanti, F., Monticelli, A., Cocozza, S., Montermini, L., and Pandolfo, M. (1997). Frataxin fracas. *Nature Genet.* **15,** 337–338.
150. Campuzano, V., Montermini, L., Lutz, Y., Cova, L., Hindelang, C., Jiralerspong, S., Trottier, Y., Kish, S. J., Faucheux, B., Trouillas, P., Authier, F. J., Dürr, A., Mandel, J.-L., Vescovi, A. L., Pandolfo, M., and Koenig, M. (1997). Frataxin is reduced in Friedreich ataxia patients and is associated with mitochondrial membranes. *Hum. Mol. Genet.* **6,** 1771–1780.
151. Koutnikova, H., Campuzano, V., Foury, F., Dollé, P., Cazzalini, O., and Koenig, M. (1997). Studies of human, mouse and yeast homologues indicate a mitochondrial function for frataxin. *Nature Genet.* **16,** 345–351.
152. Jiralerspong, S., Liu, Y., Montermini, L., Stifani, S., and Pandolfo, M. (1997). Frataxin shows developmentally regulated tissue-specific expression in the mouse embryo. *Neurobiol. Dis.* **4,** 103–113.
153. Bidichandani, S., Ashizawa, T., and Patel, P. I. (1997). Atypical Friedreich ataxia caused by compound heterozygosity for a novel missense mutation and the GAA triplet-repeat expansion. *Am. J. Hum. Genet.* **60,** 1251–1256.
154. Lafrenière, R. G., Rochefort, D. L., Chrétien, N., Rommens, J. M., Cochius, J. I., Kälviäinen, R., Nousiainen, U., Patry, G., Farrell, K., Söderfeldt, B., Federico, A., Hale, B. R., Cossio, O. H., Sorensen, T., Pouliot, M. A., Kmiec, T., Uldall, P., Janszky, J., Pranzatelli, M. R., Andermann, F., Andermann, E., and Rouleau, G. A. (1997). Unstable insertion in the 5′ flanking region of the cystatin B gene is the most common mutation in progressive myoclonus epilepsy type 1, EPM1. *Nature Genet.* **15,** 298–302.
155. Virtaneva, K., D'Amato, E., Miao, J., Koskiniemi, M., Norio, R., Avanzini, G., Franceschetti, S., Michelucci, R., Tassinari, C. A., Omer, S., Pennacchio, L. A., Myers, R. M., Dieguez-Lucena, J. L., Krahe, R., de la Chapelle, A., and Lehesjoki, A. E. (1997). Unstable minisatellite expansion causing recessively inherited myoclonus epilepsy, EPM1. *Nature Genet.* **15,** 393–396.
156. Lalioti, M. D., Scott, H. S., Buresi, C., Rossier, C., Bottani, A., Morris, M. A., Malafosse, A., and Antonarakis, S. E. (1997). Dodecamer repeat expansion in cystatin B gene in progressive myoclonus epilepsy. *Nature* **386,** 847–851.
157. Dürr, A., Cossée, M., Agid, Y., Campuzano, V., Mignard, C., Penet, C., Mandel, J.-L., Brice, A., and Koenig, M. (1996). Clinical and genetic abnormalities in patients with Friedreich's ataxia. *N. Engl. J. Med.* **335,** 1169–1175.
158. Filla, A., De Michele, G., Cavalcanti, F., Pianese, L., Monticelli, A., Campanella, G., and Cocozza, S. (1996). The relationship between trinucleotide (GAA) repeat length and clinical features in Friedreich ataxia. *Am. J. Hum. Genet.* **59,** 554–560.
159. Cossée, M., Schmitt, M., Campuzano, V., Reutenauer, L., Moutou, C., Mandel, J.-L., and Koenig, M. (1997). Evolution of the Friedreich's ataxia trinucleotide repeat expansion: Founder effect and premutations. *Proc. Natl. Acad. Sci. USA* **94,** 7452–7457.
160. Epplen, C., Epplen, J. T., Frank, G., Miterski, B., Santos, E. J. M., and Schöls, L. (1997). Differential stability of the (GAA)n tract in the Friedreich ataxia gene. *Hum. Genet.* **99,** 834–836.
161. Kostrzewa, M., Klockgether, T., Damian, M. S., and Müller, U. (1997). Locus heterogeneity in Friedreich ataxia. *Neurogenetics* **1,** 43–47.
162. Pianese, L., Cavalcanti, F., De Michele, G., Filla, A., Campanella, G., Calabrese, O., Castaldo, I., Monticelli, A., and Cocozza, S. (1997). The effect of parental gender on the GAA dynamic mutation in the FRDA gene. *Am. J. Hum. Genet.* **60,** 463–466.
163. Monros, E., Moltó, M. D., Martinez, F., Cañizares, J., Blanca, J., Vilchez, J. J., Prieto, F., de Frutos, R., and Palau, F. (1997). Phenotype correlation and intergenerational dynamics of the Friedreich ataxia GAA trinucleotide repeat. *Am. J. Hum. Genet.* **61,** 101–110.
164. Anvret, M., Ahlberg, G., Grandell, U., Hedberg, B., Johnson, K., and Edstrom, L. (1993). Larger expansions of the CTG repeat in muscle compared to lymphocytes from patients with myotonic dystrophy. *Hum. Mol. Genet.* **2,** 1397–1400.
165. Montermini, L., Kish, S. J., Jiralerspong, S., Lamarche, J. B., and Pandolfo, M. (1997). Somatic mosaicism for the Friedreich's ataxia GAA triplet repeat expansions in the central nervous system. *Neurol.* **49,** 606–610.
166. Harding, A. E. (1994). "The Hereditary Ataxias and Related Disorders." Churchill Livingstone, Edinburgh.
167. Montermini, L., Andermann, E., Richter, A., Pandolfo, M., Cavalcanti, F., Pianese, L., Iodice, L., Farina, G., Monticelli, A., Turano, M., Filla, A., De Michele, G., and Cocozza, S. (1997). The Friedreich ataxia GAA triplet repeat: premutation and normal alleles. *Hum. Mol. Genet.* **6,** 1261–1266.
168. Monros, E., Cañizares, J., Moltó, M. D., Rodius, F., Montermini, L., Cossée, M., Martinez, F., Smeyers, P., Prieto, F., de Frutos, R., Koenig, M., Pandolfo, M., Bertranpetit, J., and Palau, F. (1996). Evidence for a common origin of most Friedreich ataxia chromosomes in the Spanish population. *Eur. J. Hum. Gen.* **4,** 191–198.
169. Imbert, G., Kretz, C., Johnson, K., and Mandel, J.-L. (1993). Origin of the expansion mutation in myotonic dystrophy. *Nature Genet.* **4,** 72–76.
170. Kunst, C. B., and Warren, S. T. (1994). Cryptic and polar variation of the fragile X repeat could result in predisposing normal alleles. *Cell* **77,** 853–861.
171. Imbert, G., Saudou, F., Yvert, G., Devys, D., Trottier, Y., Garnier, J. M., Weber, C., Mandel, J. L., Cancel, G., Abbas, N., Durr, A., Didierjean, O., Stevanin, G., Agid, Y., and Brice, A.

(1996). Cloning of the gene for spinocerebellar ataxia 2 reveals a locus with high sensitivity to expanded CAG/glutamine repeats. *Nature Genet.* **14,** 285–291.

172. Wells, R. D., Collier, D. A., Hanvey, J. C., Shimizu, M., and Wohlrab, F. (1988). The chemistry and biology of unusual DNA structures adopted by oligopurine.oligopyrimidine sequences. *FASEB. J.* **2,** 2939–2949.
173. Ohshima, K., Kang, S., Larson, J. E., and Wells, R. D. (1996). Cloning, characterization, and properties of seven triplet repeat DNA sequences. *J. Biol. Chem.* **271,** 16773–16783.
174. Grabczyk, E., and Fishman, M. C. (1995). A long purine-pyrimidine homopolymer acts as a transcriptional diode. *J. Biol. Chem.* **270,** 1791–1797.
175. Cooney, M., Czernuszewicz, G., Postel, E. H., Flint, S. J., and Hogan, M. E. (1988). Site-specific oligonucleotide binding represses transcription of the human c-myc gene in vitro. *Science* **241,** 456–459.
176. Duval-Valentin, G., Thuong, N. T., and Helene, C. (1992). Specific inhibition of transcription by triple helix-forming oligonucleotides. *Proc. Natl. Acad. Sci. USA* **89,** 504–508.
177. Liu, L. F., and Wang, J. C. (1987). Supercoiling of the DNA template during transcription. *Proc. Natl. Acad. Sci. USA* **84,** 7024–7027.
178. Wu, H. Y., Shyy, S. H., Wang, J. C., and Liu, L. F. (1988). Transcription generates positively and negatively supercoiled domains in the template. *Cell* **53,** 433–440.
179. Htun, H., and Dahlberg, J. E. (1989). Topology and formation of triple-stranded H-DNA. *Science* **243,** 1571–1576.
180. Lamont, P. J., Davis, M. B., and Wood, N. W. (1997). Identification and sizing of the GAA trinucleotide repeat expansion of Friedreich's ataxia in 56 patients. Clinical and genetic correlates. *Brain* **120,** 673–680.
181. Isnard, R., Kalotka, H., Dürr, A., Cossée, M., Schmitt, M., Pousset, F., Thomas, D., Brice, A., Koenig, M., and Komajda, M. (1997). Correlation between left ventricular hypertrophy and GAA trinucleotide repeat length in Friedreich's ataxia. *Circulation* **95,** 2247–2249.
182. Gibson, T. J., Koonin, E. V., Musco, G., Pastore, A., and Bork, P. (1996). Friedreich's ataxia protein: bacterial homologs point to mitochondrial dysfunction. *Trends Neurosci.* **19,** 465–468.
183. Wilson, R. B., and Roof, D. M. (1997). Respiratory deficiency due to loss of mitochondrial DNA in yeast lacking the frataxin homologue. *Nature Genet.* **16,** 352–357.
184. Priller, J., Scherzer, C. R., Faber, P. W., MacDonald, M. E., and Young, A. B. (1997). Frataxin gene of Friedreich's ataxia is targeted to mitochondria. *Ann. Neurol.* **42,** 265–269.
185. Foury, F., and Cazzalini, O. (1997). Deletion of the yeast homologue of the human gene associated with Friedeich's ataxia elicits iron accumulation in mitochondria. *FEBS Lett.* **411,** 373–377.
186. Schöls, L., Reichmann, H., Amoiridis, G., Seibel, P., Wagener, S., Seufert, S., and Przuntek, H. (1996). Mitochondrial disorders in degenerative ataxias. *Eur. J. Neurol.* **3,** 55–60.
187. Rötig, A., deLonlay, P., Chretien, D., Foury, F., Koenig, M., Sidi, D., Munnich, A., and Rustin, P. (1997). Frataxin gene expansion causes aconitase and mitochondrial iron-sulfur protein deficiency in Friedreich ataxia. *Nature Genet.* **17,** 215–217.
188. Fridovitch, I. (1995). Superoxide radical and superoxide dismutases. *Annu. Rev. Biochem.* **64,** 97–112.
189. Kaptain, S., Downey, W. E., Tang, C., Philpott, C., Haile, D., Orloff, D. G., Harford, J. B., Rouault, T. A., and Klausner, R. D. (1991). A regulated RNA binding protein also possesses aconitase activity. *Proc. Natl. Acad. Sci. USA* **88,** 10109–10113.
190. Fukuhara, N., Tokiguchi, S., Shirakawa, K., and Tsubaki, T. (1980). Myoclonus epilepsy associated with ragged-red fibres (mitochondrial abnormalities): disease entity or a syndrome? Light-and electron-microscopic studies of two cases and review of literature. *J. Neurol. Sci.* **47,** 117–133.
191. Holt, I. J., Harding, A. E., Petty, R. K. H., and Morgan-Hughes, A. (1990). A new mitochondrial disease associated with mitochondrial DNA heteroplasmy. *Am. J. Hum. Genet.* **46,** 428–433.
192. Pastores, G. M., Santorelli, F. M., Shanske, S., Gelb, B. D., Fyfe, B., Wolfe, D., and Willner, J. P. (1994). Leigh syndrome and hypertrophic cardiomyopathy in an infant with a mitochondrial DNA point mutation (T8993G). *Am. J. Med. Genet.* **50,** 265–271.
193. Rosen, D. R., Siddique, T., Patterson, D., Figlewicz, D. A., Sapp, P., Hentati, A., Donaldson, D., Goto, J., O'Regan, J. P., Deng, H. X., Rahmani, Z., Krizus, A., McKenna-Yasek, D., Cayabyab, A., Gaston, S. M., Berger, R., Tanzi, R. E., Halperin, J. J., Herzfeldt, B., Van den Bergh, R., Hung, W. Y., Bird, T., Deng, G., Mulder, D. W., Smyth, C., Laing, N. G., Soriano, E., Pericak-Vance, M. A., Haines, J., Rouleau, G. A., Gusella, J. S., Horvitz, H. R., and Brown, R. H., Jr. (1993). Mutation in Cu/Zn superoxide dismutase gene are associated with familial amyotrophic lateral sclerosis. *Nature* **362,** 59–62.
194. Ouahchi, K., Arita, M., Kayden, H. J., Hentati, F., Ben Hamida, M., Sokol, R., Arai, H., Inoue, K., Mandel, J.-L., and Koenig, M. (1995). Ataxia with isolated vitamin E deficiency is caused by mutations in the a-tocopherol transfer protein. *Nature Genet.* **9,** 141–145.
195. Sharp, D., Blinderman, L., Combs, K. A., Kienzle, B., Ricci, B., Wager-Smith, K., Gil, C. M., Turck, C. W., Bouma, M.-E., Rader, D. J., Aggerbeck, L. P., Gregg, R. E., Gordon, D. A., and Wetterau, J. R. (1993). Cloning and gene defects in microsomal triglycerides transfer protein associated with abetalipoproteinaemia. *Nature* **365,** 65–69.
196. Stumpf, D. A., Sokol, R., Bettis, D., Neville, H., Ringel, S., Angelini, C., and Bell, R. (1987). Friedreich's disease: V. Variant form with vitamin E deficiency and normal fat absorption. *Neurology* **37,** 68–74.
197. Ben Hamida, M., Belal, S., Sirugo, G., Ben Hamida, C., Panayides, K., Ioannou, P., Beckmann, J., Mandel, J.-L., Hentati, F., Koenig, M., and Middleton, L. (1993). Friedreich's ataxia phenotype not linked to chromosome 9 and associated with selective autosomal recessive vitamin E deficiency in two inbred Tunisian families. *Neurology* **43,** 2179–2183.
198. Ben Hamida, C., Doerflinger, N., Belal, S., Linder, C., Reutenauer, L., Dib, C., Gyapay, G., Vignal, A., Le Paslier, D., Cohen, D., Pandolfo, M., Mokini, V., Novelli, G., Hentati, F., Ben Hamida, M., Mandel, J.-L., and Koenig, M. (1993). Localization of Friedreich ataxia phenotype with selective vitamin E deficiency to chromosome 8q by homozygosity mapping. *Nature Genet.* **5,** 195–200.
199. Di Mascio, P., Murphy, M. E., and Sies, H. (1991). Antioxidant defense systems: the role of carotenoids, tocopherols, and thiols. *Am. J. Clin. Nutr.* **53,** 194S–200S.

PART XI

Candidate Disorders Revealing Anticipation

CHAPTER 27

Anticipation and Psychiatric Disorders

MELVIN G. MCINNIS AND RUSSELL L. MARGOLIS
Johns Hopkins University School of Medicine, Baltimore, Maryland 21287

I. INTRODUCTION

With the discovery, beginning in 1991, that a number of diseases are caused by trinucleotide repeat expansion mutations, considerable interest has focused on genetic anticipation, a term implying increasing severity or decreasing age of onset of a disease in successive generations within a pedigree. Anticipation is a hallmark (though not a universal feature) for this group of disorders. Now that at least one molecular mechanism associated with this clinical phenomenon has been established, the hunt for diseases with anticipation has been carried to all areas of medicine [1]. The possibility that anticipation may be present in psychiatric diseases has attracted particular attention. On the one hand, as noted in Section II, the historical roots for the concept of anticipation derived from the study of psychiatric disorders. On the other hand, despite strong evidence supporting the genetic basis of these disorders, the causative or susceptibility factors remain unknown, and any clue is of great value.

We will focus our discussion on bipolar affective disorder, schizophrenia, and autism. These three disorders have coherent core phenotypes about which a general consensus exists and appear to be at least partially genetic in etiology. Classic bipolar disorder consists of cyclical swings of mood from depression (low mood, anhedonia, sleep and appetite disturbance, psychomotor abnormalities, fatigue, diminished self-attitude, and cognitive slowing) and mania (elation or irritability, grandiosity, impulsivity, increased energy and activity, and rapidity of speech and cognition). Schizophrenia is characterized by gross disturbances of perception (hallucinations) and cognition (delusions and thought disorder), as well as the "negative" symptoms of amotivation and paucity of thought. The critical features of autism are impairment in social interaction and communication accompanied by repetitive and stereotypical behavior beginning by age 3.

The evidence for a genetic basis of these diseases stems from adoption, twin, and family studies. Adoption studies in bipolar disorder [2, 3] and schizophrenia [4] have established that the rate of illness is significantly higher among biological relatives of affected adopted probands than in their adoptive relatives. In bipolar disorder, for instance, the concordance rate for illness between monozygotic (MZ) twins is approximately 70 and 20% in dizygotic (DZ) twins [5]; family studies of bipolar disorder estimate the overall risk of affective disorder to a first degree relative of a proband with bipolar disorder to be in the range of 15% [6]. Folstein and Rutter [7], in their twin study of autism, found a 36% strict pairwise concordance rate for MZ twins compared to 0% in DZ twins. Considering a broader definition of the phenotype to include cognitive abnormalities brought the concordance rate to 82% for MZ twins and 10% for DZ twins. Siblings of autistic probands have also demonstrated an increased rate of cognitive abnormalities [8]. Twin and family studies also clearly support the presence of genetic factors in schizophrenia [9, 10], as well as in the anxiety disorders (panic disorder, phobias, and obsessive–compulsive disorder) [11, 12]. The search for underlying genes in the psychiatric disorders has employed the traditional methods of genetic linkage analysis. Despite numerous reports of linkage to bipolar disorder [13–15] and schizophrenia [16–26] no predisposing genes have yet been characterized.

There are several reasons why the identification of genes predisposing to psychiatric illness has proven to be difficult and why the study of anticipation in psychiatric disorders is important. Many psychiatric geneticists believe that diseases such as autism, bipolar disorder, and schizophrenia may be caused by many genes of small effect. The mode of inheritance of these disorders is unknown [27, 28]. Even the phenotype of these disorders remains less than perfectly clear, despite intense efforts to operationalize the definitions [29, 30]. No method for external validation of what remains a clinical judgment has yet been discovered. The phenotypic variability across or even within families is an ever present challenge; fundamentally, the limits and boundaries of the phenotypes are not known. For instance, in the study of bipolar disorder it is debated whether recurrent major depression represents a genetic form of the illness or not [31]. It is likely that many instances of recurrent major depression will represent a genetic variation of bipolar disorder, perhaps a variant with an earlier onset. Other cases will be genetically distinct from bipolar disorder, perhaps representing an "environmental" form of the illness or a completely different genotype. Thus we are left with a "Catch 22" situation: defining a valid phenotype depends on knowing the genotype, but establishing the genotype is impossible without knowing the phenotype. The study of inheritance patterns that have led to the identification of genes in other disorders offers a new approach to the study of psychiatric genetics, and a potential way around the current dilemma.

II. ANTICIPATION AND PSYCHIATRIC DISORDER: A HISTORICAL PERSPECTIVE

The term "genetic anticipation" evolved from the study of physical and mental degeneration that began with Morel [32] in the mid-19th century, who noted that many forms of ill health appeared to become worse in succeeding generations, including psychiatric diseases. Morel considered the environmental influences that might facilitate degeneration, such as poor nutrition and other external forces beyond human control. The "Law of Anticipation" was first introduced as a concept in the context of psychiatric disease in the beginning of this century by Sir Frederick Mott [33, 34], who studied 420 parent–offspring pairs in the asylums of London. The diagnostic methods were limited to the general accounts of the keepers of the asylums. Vague, nondescript terms were used in the diagnosis such as "insanity," "chronic alcoholism," and "imbecility." The social and political climate at the time led to a concern over the natural and moral decay of nations [35], a logical consequence of genetic anticipation that led to the acceptance of Mott's work as dogma. Galton, for instance [36], advocated the establishment of a "genetic card" to be carried by all members of society for comparison with prospective mates to ensure that the stock would improve in quality rather than continuing the downward spiral of decay. Mott [34] describes families wherein the severity of illness prevented the succeeding generation from procreating; he viewed this as an advantage. National pride creeps in to Mott's summary discussion wherein he concludes:

> . . . as far as this old English stock is concerned, it has as much sap and vitality and power as it had two centuries ago; and that with due pruning of rotten branches , and due hoeing up of weeds, which will grow about the roots, the like products will be yielded again. The weeds to which I refer are mainly three: the first of them is dishonesty, the second is sentimentality, and the third is luxury.

Despite these shortcomings, Mott's work [34] is among the earliest of the genetic efforts in psychiatric disease, and his conclusion that hereditary predisposition is an important factor in the etiology of psychiatric disease has stood the test of time.

After Mott, the concept of anticipation was largely wreaked on the pseudo-science of Nazi eugenics, in which the concept of anticipation was invoked as a justification or efforts to improve racial purity. Aside from the postwar moral aversion to anticipation, the scientific basis of the concept was brought into question. Penrose [37] forcefully argued that the observations of Fleischer [38] and Bell [39], consistent with anticipation in myotonic dystrophy, were likely to be due to biases of ascertainment and that anticipation in myotonic dystrophy had no biological basis. The concept did not regain a respectable position in the field of genetics until the report of Howeler *et al.* [40] demonstrated the presence of anticipation in myotonic dystrophy in a systematically ascertained patient sample. The final step in the revival of anticipation, however, only arose with the establishment of the biological mechanism underlying at least some instances of anticipation. The result has been an enthusiastic search for the presence of anticipation in a plethora of human diseases [1].

III. BIPOLAR DISORDER AND ANTICIPATION

The first study to test the hypothesis that anticipation might be present in bipolar disorder examined 34 bipolar pedigrees ascertained as part of a genetic linkage project [41]. Age of onset and disease severity (measured by episodes of illness per year) were compared between generations under four sampling schemes: (1) random pairs, (2) all possible pairs, (3) random transmitting pairs (i.e., parent–offspring), and (4) all transmitting pairs. Both measures showed a significantly earlier age of onset and increased disease frequency in the younger generation. Life table analyses showed a significant decrease in survival to first mania or depression from the first to the second generation. The data were also evaluated to take into account various biases, including those pointed out by Penrose many years earlier, that might lead, falsely, to the conclusion that anticipation is present. The additional analyses included removing the subjects with substance abuse in the younger generation, since early substance abuse can advance the onset of psychiatric disorder. This change had no effect on the results, and neither did other analyses, which took into account decreased fertility of affected individuals, deaths of affected individuals prior to study, and unreliable reporting of onset age. The cohort effect [42] was studied with bivariate analysis and the results remained significant. While all of the biases of ascertainment initially described by Penrose [37] could not be definitely excluded, the data were consistent with the presence of anticipation in bipolar disorder.

The presence of anticipation in affective disorder has been supported in two studies from Sweden. Nylander *et al.* [43] studied 14 Swedish bipolar families and found a significant difference in age of onset and disease severity between generations, consistent with anticipation. There was no difference in age of onset between offspring of affected fathers and mothers. The same group [44] later studied unipolar depression in 31 parent–offspring pairs and found that the younger generation had on average a 15-year younger age of onset compared to the older and 1.5 times more episodes of illness. The authors addressed a particularly pernicious bias of ascertainment by excluding those pairs in which the parent had an age of onset greater than the age of the offspring at the time of ascertainment, addressing the concern of Penrose that such pairs are by virtue of the ascertainment process more likely to fall into the study. The results remained significant in favor of anticipation.

Recently, a Romanian group [45] has examined 115 bipolar probands, 455 first degree relatives, and 313 second degree relatives and included only BP I subjects in the anticipation analysis. The probands were randomly selected to avoid the ascertainment bias. The authors used a sampling scheme similar to that of McInnis *et al.* [41]. The age at onset was found to be 6–10 years younger in probands with affected parents or uncles/aunts. Two-thirds of these families showed anticipation under both a broad and narrow definition of affection status in the parents' generation. The age at investigation was younger in probands showing anticipation. Anticipation was found only in probands inheriting the disorder from the paternal side. The authors noted that there is an inevitable association between young current age and young age at onset, which could result in spurious anticipation effects, an argument implied by Penrose [37]. They point out that this bias of ascertainment cannot account for the observed anticipation. Their finding that anticipation is present only in paternal transmission suggests that imprinting may also be present in bipolar disorder.

IV. SCHIZOPHRENIA AND ANTICIPATION

There have been several studies that have addressed the presence of anticipation in schizophrenia. It has been observed that the affected offspring of parents with schizophrenia have an earlier age of onset [28, 46]. The reason for this is unknown but with the identification of a molecular mechanism for anticipation the obvious hypothesis of unstable DNA at least lends a testable theory to the enigma. The studies of anticipation in schizophrenia have at least confirmed the original age of onset findings of a younger age of onset in the offspring of affected parents. As in bipolar disorder the transmission is generally non-Mendelian and no genes have been identified as yet that predispose to schizophrenia.

While previous investigators made observations consistent with anticipation, the first systematic study of anticipation and schizophrenia was done by Penrose [47]. He collected data on all first admissions to psychiatric hospitals in Ontario, Canada, between 1926 and 1943. He organized the data into 13 subgroups of diagnoses and analyzed the data within the diagnosis groups as well as comparing the age of first hospitalization between generations. The not unreasonable assumption was that age of first hospitalization is a measure of age of onset, since the first psychotic episode usually results in hospitalization. Overall, Penrose found following mean age of first hospitalization: fathers, 53.96 years; mothers, 47.18 years; sons, 33.06 years; and daughters, 35.43 years. Penrose dismisses the concept of anticipation by noting:

> This finding, which in one form or another, is very characteristic of mental hospital data, has in the past been attributed to a tendency for progressive degeneration or anticipation of diseases in succeeding generations. Such an explanation, which is not in accordance with the concepts of modern genetics, is unnecessary, because more likely explanations are close at hand.

The explanations to which Penrose refers are of course the biases of ascertainment (Table 27-1) which are described in his critique of the notion of anticipation in myotonic dystrophy. In addition he notes further potential biases when addressing psychiatric illness and includes the fertility bias—early onset mental illness tends to prevent parenthood.

The study by Penrose was revisited by Bassett and Husted [48], who reanalyzed the data in a manner that directly tested the hypothesis of anticipation but also endeavored to evaluate the biases of ascertainment and fertility. The authors found clear evidence of anticipation in the Penrose sample; 88% of 137 pairs showed an intergenerational age of onset difference with a median of 15 years earlier in the younger generation. Including affective disorders in the diagnoses expanded the sample to 331 pairs and again 88% of pairs showed evidence of anticipation. It is noteworthy that 84% of the affective disorder was in the older generation, suggesting that the phenotype progressed from affective disorder to schizophrenia. This is consistent with the controversial theory, expounded by Crow [49], that affective disorder and schizophrenia are not two separate disorders but are on opposite ends of a continuum.

Bassett and Husted [48] also evaluated 111 aunt/uncle–niece/nephew pairs from the Penrose data to examine the magnitude of the fertility bias. They argue that this removes the bias that results from the selection of parents who are by definition reproductively fit. Nieces and nephews had an age of onset similar to that of the offspring, but the aunts/uncles had an earlier age of onset compared to the parents. This suggested an effect of ascertainment and fertility biases, but not enough to account for the anticipation, since 75% of avuncular pairs showed anticipation with a median intergenerational age of onset difference of 8 years. The authors conclude that various biases influence the data set but that anticipation is nonetheless present.

Other work is consistent with the findings of Bassett and Husted [48]. Asherson *et al.* [50] examined the age of onset in 29 families, finding a mean difference of 11.1 years between the older and the younger generation. The affected phenotype included schizophrenia, schizoaffective disorder, and unspecified functional psychosis. The number of intergenerational pairs is not specified nor is it clear if the proband was included in the analyses, an important point since inclusion of the usually severely ill proband tends to enhance the finding of anticipation. There was no evidence of genetic imprinting, as offspring of affected mothers had a similar age of onset as offspring of affected fathers. Chotai *et al.* [51] studied 19 parent–offspring pairs obtained from 14 two-generation families with schizophrenia. The mean age of onset for the parent generation was 37.3 years and that for the offspring generation was 20.8 years. The mean difference of 16.5 years supports the presence of anticipation in schizophrenia. Thibaut *et al.* [52] studied 26 of 48 schizophrenic families ascertained for a genetic linkage study. The sampling scheme described by McInnis *et al.* [41] was used. In the most rigorous sampling scheme, random intergenerational pairs, there was a median difference in age of onset of 11 years, and there was a significant difference in age of onset in all sampling schemes. A significantly shorter interval before onset of illness in younger generations was also apparent by

TABLE 27-1 Biases of Ascertainment

Biases (according to Penrose [47])
Selection of affected parents in whom the onset of the disease is late.
Selection of affected offspring in whom the onset of the disease is early.
Selection of cases with simultaneous onset in parents and offspring.
Weakness of real correlation of ages of onset in parents and offspring.
Additional biases
Substance abuse in younger children.
Reduced fertility in older generation.
Social influences in older generation.
Secular trends, e.g., Cohort Effect [42].
Decreased memory of illness episodes, especially in older generation.
Increased awareness of illness in physicians and family members.

survival analysis. Omitting those offspring in the analysis whose age of onset was younger than the mean age at birth of the first child in transmitting parents did not affect the findings.

Bassett and Honer [53] studied first admission to psychiatric hospital with the diagnosis of schizophrenia. This study examined eight extended nonconsanguiniuos families ascertained for a genetic linkage study. The results were that significantly more individuals were hospitalized with psychosis, the illness appeared to become more severe, and the age of first hospitalization was younger in succeeding generations, consistent with genetic anticipation in all families. However, while living subjects were interviewed directly, this study relied on family history information and records for much of the data from older generations.

In a systematically gathered sample (avoiding some of the biases inherent in samples used for linkage studies), Gorwood *et al.* [54] examined age of onset in 97 schizophrenic subjects belonging to 24 pedigrees with affected members in at least two generations. The ascertainment was limited to a 1-year period within a limited geographical area of Reunion Island (Indian Ocean). A method of calculating expected age at onset was used that took into consideration the age of interview of the younger subject and the distribution of the age of onset of the parents. The younger generation of patients had a mean corrected age of onset of 21.80 years that was earlier than the predicted age of onset (24.95 years). Both expected and predicted age of onset were significantly different from the age of onset of the parental generation (32.21 years). Excluding the proband and families with bilineal matings and factoring in reduced fertility did not alter the results. No cohort effect, a secular trend toward younger age of onset in succeedingly younger cohorts, was detected.

Johnson *et al.* [55] reported evidence of anticipation in schizophrenic families identified through the NIMH Genetics Initiative. Families were included that had at least two affected members in successive generations and were not bilineal. Affectation diagnoses included schizophrenia, schizo-affective disorder–depressed, and psychosis not otherwise specified. Three indices of age of onset were used and included age of first psychotic symptoms, age of first psychiatric treatment, and age of first hospitalization. Disease severity was measured by several different indices consisting of the number of hospitalizations and global ratings of positive and negative symptom severity as measured on standardized rating scales. The four sampling schemes suggested by McInnis *et al.* [41] were tested. Anticipation was demonstrated for age of onset, regardless of the index or sampling scheme used. Anticipation was not detected for disease severity. Analyses that took into account drug use and diminished fecundity did not affect the results.

Studies of schizophrenia, with the caveat that complete elimination of ascertainment bias may be impossible, consistently demonstrate a younger age of onset in succeeding generations. In addition, it is clear that a fertility bias exists, as affected individuals are less likely to procreate if they are affected at an early age. No evidence of the cohort effect has been found and there appears to be no imprinting effect from the affected parent.

V. AUTISM AND ANTICIPATION

There have been no formal studies of anticipation in autism. However, several lines of evidence suggest that autism may also exhibit this phenomenon. Psychopathology has been described in the parents of autistic children and has included an overall increased rate of nonpsychotic mental disorder [56] and depression [57]. Family studies of autistic probands and their families provide strong evidence of cognitive abnormalities in the parents of autistic children. Piven *et al.* [58] found increased incidence of anxiety disorders, including generalized anxiety and social phobia among parents of autistic children, and depression was three times more frequent than in a control group of parents of Down syndrome children, a finding that has been replicated [59]. More recently, Piven *et al.* [8] found higher rates of social deficits in the parents of autistic children, as measured by limited friendships, little spontaneous show of affection, limited "to and fro" conversation, odd stereotypical behavior, and rigidity and perfectionism. The authors do not address the issue of anticipation, but it is clear that the boundaries of the phenotype have yet to be delineated and the phenomena observed in the parents may represent a milder form of the illness.

Rett syndrome is a neurodevelopmental disorder which is characterized by a period of normal development followed by a developmental stagnation and regression [60]. The disorder is usually sporadic, but rare pedigrees with apparent nonpenetrance in obligate carriers have been described. These families arguably show evidence of anticipation and were tested for the presence of using the repeat expansion detection method [61]. There was no evidence of expansion in 6 familial or 26 sporadic cases of Rett syndrome [62].

VI. TRINUCLEOTIDE REPEATS AND ANTICIPATION

A. Evidence from Known Trinucleotide Repeat Expansion Disorders

How might the phenomenon of anticipation in psychiatric disorders be explained? The obvious possibility is that at least one of the mutations involved in the etiology of each of these disorders is an unstably transmitted trinucleotide repeat expansion. The most important line of reasoning supporting this notion is that anticipation is present in almost all of the diseases caused by trinucleotide repeat expansions. However, there is other support as well. First, it appears that the brain is vulnerable to repeat expansion. As noted elsewhere in this volume, eight diseases characterized by neuronal degeneration are caused by expansion of a CAG repeat; the repeat expansion in each disease occurs in a different gene. Strikingly, these genes are expressed throughout the body, yet the pathology caused by the expansion mutation is limited to certain brain regions [63]. It therefore appears that the brain is uniquely vulnerable to CAG repeat expansions. In addition, CAG and CGG repeats are frequently found in genes encoding proteins that regulate neurodevelopment and neuroplasticity. Expansion mutations in such genes might have very specific effects on brain development or function. For instance, the protein encoded by the fragile X (A subtype) gene binds to RNA, and may serve to regulate protein formation during brain development [64]. Second, psychiatric symptoms are almost always part of the syndrome of diseases caused by repeat expansion. Either mental retardation or dementia is a feature of all but one of the repeat expansion diseases. In Huntington's disease, the rate of affective disorder is quite high, obsessive–compulsive disorder is not unusual, and schizophrenia has been observed [65]. In dentatorubropallidoluysian atrophy (DRPLA), up to 74% of patients may have psychiatric syndromes, including mania, depression, and schizophrenia [66]. In the fragile X syndrome, affective, schizotypal, and autistic syndromes are common [67].

Third, repeat expansion is consistent with patterns of inheritance other than anticipation observed in psychiatric disorder. Bipolar disorder, schizophrenia, and autism each cluster in families, but the pattern of inheritance is obscure. Twin studies in each disorder show significant, but not 100%, concordance. Presumably through somatic mosaicism, twins discordant, at least in symptom severity, have been reported in trinucleotide repeat expansion disorders. For instance, Kruyer [68] reported a pair of monozygotic male twins clinically discordant for fragile X, explained by a difference in repeat length. Reiss *et al.* [69] reported a pair of female monozygotic twins with fragile X syndrome in which one twin had an IQ of 105 and the other an IQ of 47. Similarly, there has been at least one report of monozygotic twins with myotonic dystrophy who had different length CTG repeat expansion [70].

Trinucleotide repeat expansion mutation can give rise to sporadic cases and incomplete penetrance, both patterns commonly seen in psychiatric disorders. In a number of the expansion mutation diseases, individuals exist who carry an allele of intermediate length, not long enough to cause disease in them, but capable of expanding during meiosis to cause disease in their offspring. Therefore, disease may arise seemingly *de novo* in a branch of a pedigree [71, 72]. The opposite phenomenon of repeat contraction also occasionally occurs [73]; cases have been described in which individuals affected with fragile X and myotonic dystrophy have transmitted normal length repeats to their offspring, so that a phenotype is no longer expressed in a branch of a pedigree. In fragile X, normal males can transmit the mutation to their daughters, but their sisters and the daughters of their sisters are almost never affected, while 35% of the sisters of affected males are symptomatic. Fully expanded mutations of the CGG repeat in the CBL2 protooncogene sometimes cause no phenotype, while in other individuals the expansion leads to a deletion of the terminal portion of chromosome 11 with consequent mental retardation and other complications [74, 75].

Symptom severity and the clinical phenotype varies dramatically in at least 10 of the known expansion mutation disorders, a function of gender (in fragile X), repeat length, somatic variations in repeat length, and other unidentified environmental or genetic factors. Of particular relevance, some patients with fragile X, Huntington's disease, and DRPLA experience one of the psychiatric disorders noted above, but others do not. Variability of a similar extent can be seen in psychiatric disorders. The classic example is affective disorder, in which symptoms can vary from occasional mild, nonincapacitating episodes of depression to near constant mania complete with hallucinations, delusions, and out-of-control behavior [76].

B. Evidence from Repeat Expansion Detection

The presence of trinucleotide repeat expansion in DNA from patients with psychiatric disorders has been investigated in several ways. Repeat expansion detection (RED) has provided the most intriguing evidence. This technique, first described by Schalling and colleagues [61] and discussed in further detail in a separate chapter, essentially consists of repeated annealing and denaturation of a radiolabeled (CAG)n oligonucleotide to genomic DNA. At each annealing step, a ligase joins oligonucleotides that have hybridized to adjacent locations on a CAG/CTG repeat. After multiple cycles, the length of the ligated oligonucleotide reflects the length of the longest repeat in the genomic DNA. The technique does not localize the expansion, nor can it be used for all types of repeat. Nonetheless, results from several studies using RED have suggested that CAG/CTG repeat expansions may be associated with bipolar disorder and schizophrenia. Linblad *et al.* [77] compared familial and sporadic cases of bipolar disorder against age- and sex-matched controls from the same populations (Belgian and Swedish). Mean repeat length was longer in the bipolar subjects, and 15% of the subjects (versus 4% of controls) had repeat lengths greater then 210 base pairs. O'Donovan [78] made similar findings, and in addition found no association between age of onset and repeat length but a larger repeat length in females compared to males, leading to the suggestion that an expanded repeat might exist on the X chromosome. Similarly, Morris [79] found longer repeats in female subjects with schizophrenia. In an analysis of 152 schizophrenic and 143 bipolar patients and matched controls, including 62 probands (41 bipolar and 21 schizophrenic) with DNA from affected parents, O'Donovan [80] continued to demonstrate a shift toward longer RED products in both schizophrenia and bipolar disorder. However, no size difference was apparent in parent–offspring pairs, the shift in the patient groups was equal for males and females, and there was no correlation between age of onset and repeat length.

Other studies using RED have not found expansions in schizophrenia or bipolar disorder. Vincent *et al.* [81] found no differences in repeat length between bipolar subjects and controls and only marginal differences between schizophrenic subjects and controls. No differences at all were detected when a longer probe was used, and no correlation was found between age of onset and repeat size. Petronis *et al.* [82] also found no evidence of expansion by RED in Eastern Canadian schizophrenia pedigrees. As discussed elsewhere, RED results must be interpreted with caution. The sensitivity of the technique to pick up small expansions may be limited, and it cannot be used for CCG/CGG repeats. Also, interpretation of RED results is complicated, like all association experiments, by population differences in repeat lengths.

C. Test of Expansions in Specific Repeats

As an alternative to RED, detection of trinucleotide repeat expansion in psychiatric disorder has also been attempted by examining the length of specific repeats. At least five approaches have been used. Some studies have simply looked at a number of repeats, usually those that are known to be polymorphic, in a number of subjects with schizophrenia or bipolar disorder [83–85]. A second group of studies has looked at specific CAG/CTG repeats only in subjects in whom long repeats were suspected based on RED results [86]. Neither approach has yet yielded a clear candidate mutation.

In a third approach, specific repeats in which expansions cause one of the known neuropsychiatric diseases have been tested in various other psychiatric disorders. Since early DRPLA may resemble schizophrenia, a number of studies have tested for atrophin-1 expansions in schizophrenia [87–90]; no expansions have thus far been identified. Based on evidence that the fragile X phenotype often includes autistic-like features, at least one group has examined multiplex autistic families for expansions at the FMR1, FRAXE, and FRAXF sites [91]. In 19 pedigrees with two affected brothers, no repeat expansions were found. Ashworth *et al.* [92] studied 23 schizophrenic families for the instability at the FMR1 locus and in one family a fragile X premutation was found in one individual with schizophrenia and developmental delay was a mosaic for the full and premutation. This raises the possibility that FMR-1 mutations can modify the clinical phenotype of schizophrenia. Fourth, repeat-containing genes of particular biological interest have also been tested. Variations in the CAG repeat in the gene encoding the TATA binding protein [93] were not associated with disease. An allele of GluR6 (a glutamate receptor subunit) in a subject with schizophrenia had one more AAT repeat than the largest control allele, but the extra repeat did not appear to segregate with the disorder in other family members [84]. A 5′ untranslated CAG repeat in a human gene homologous to a *C. elegans* cell-fate determining gene was expanded in one individual with a movement disorder and bipolar disorder type II; linkage of the expansion to either phenotype remains unclear [94]. Schizophrenia has been tentatively linked to the spinocerebellar ataxia type 1 (SCA1, ataxin-1) locus [95, 96], but this linkage remains less than definitive [90].

Fifth, in what may emerge as a harbinger of future endeavors, loci to which psychiatric disorders have been tentatively linked have been examined for expansion mutations. Speight *et al.* [97] found no expansions in eight CAG/CTG loci on chromosome 4 in bipolar or schizophrenic patients. O'Hara *et al.* [98] analyzed three CAG trinucleotide repeat loci (D6S1014, D6S1015, and D6S1058) in 30 families with schizophrenia. Linkage to chromosome 6 has been reported in schizophrenia [99]. No unusually long alleles that would suggest abnormal expansion were observed, nor were there size differences between the offspring and parental generations. Based on linkage of bipolar disorder to chromosome 18 [13, 15, 100], Breschel *et al.* [101] conducted a search for CAG repeats on chromosome 18, resulting in the identification of a highly polymorphic repeat (CTG-18.1) on an intron of SEF2-1, a helix-loop-helix DNA binding protein involved in transcriptional regulation.

The repeat is unstably expanded in 3% of the population, but does not segregate with bipolar disorder.

VII. ALTERNATIVE MOLECULAR EXPLANATIONS FOR ANTICIPATION

The various approaches for detecting trinucleotide repeat expansion mutations in bipolar affective disorder, schizophrenia, and autism have yet to yield definitive answers. Given that anticipation is a real phenomenon, might there be other genetic explanations for it and the other non-Mendelian features of psychiatric disorders? Linkage studies in these disorders have rapidly increased in power and sophistication, and have failed to yield consistent well-demarcated regions of interest. The opinion among most psychiatric geneticists is that bipolar disorder, schizophrenia, and autism are oligogenic in etiology [102]; if a single gene of major effect were present the current studies would have identified it. One might imagine a scenario in which multiple genes aggregate in a successive generation and lead to a more severe phenotype. Other molecular explanations for anticipation might include accumulation of abnormal mitochondria or changes in the length of a microdeletion or insertion. Whether any of these phenomena is relevant to psychiatric disorders is purely speculative.

VIII. CONCLUSION

The identification of the underlying molecular mechanism, trinucleotide repeat expansion, for anticipation in a number of diseases with neuropsychiatric features has spurred a renewed interest in the study of anticipation in the classical psychiatric disorders. The preponderance of evidence suggests, even after considering biases of ascertainment and environmental effects, that anticipation is present in both schizophrenia and bipolar disorder. The intergenerational age of onset difference in these disorders appears to in the range of 8 to 12 years. Anticipation may be influenced by imprinting, with paternal transmission resulting in a more pronounced effect.

These findings have both clinical and research implications. From the standpoint of genetic counseling, the findings are not so robust as to provide clear-cut or absolute guidelines. However, it is warranted to raise the possibility that offspring may have an earlier onset or more severe disorder than their parents. Since diagnosis and treatment of these disorders at an early stage are of utmost importance, families at least can be on the lookout for early symptoms. The more immediately relevant implication of these findings is to the direction of psychiatric genetic research.

So far, the genetic etiologies of schizophrenia, bipolar disorder, autism, and other psychiatric conditions have been remarkably elusive. Certainly the finding of anticipation suggests that trinucleotide repeats expansions, or perhaps other unstable segments of DNA, may be involved. However, other mechanisms, such as the accumulation of multiple independent mutations or the influence of viruses, may provide an alternative explanation for anticipation. Regardless of the molecular cause of anticipation, it seems likely that it will provide a clue for generating candidate genes to complement the current large-scale, and as yet unsuccessful, efforts of traditional genetic linkage studies.

References

1. McInnis, M. G. (1996). Anticipation—an old idea in new genes. *Am. J. Hum. Genet.* **59,** 973–979.
2. Mendlewicz, J., and Rainer, J. D. (1977). Adoption study supporting genetic transmission in manic-depressive illness. *Nature* **268,** 327–329.
3. Von Knorring, A. L., Cloninger, C. R., Bohman, M., and Sogvardsson, S. (1983). An adoption study of depressive disorder and substance abuse. *Arch. Gen. Psych.* **40,** 943–950.
4. Heston, L. L. (1966). Psychiatric disorders in foster home reared children of schizophrenic mothers. *Br. J. Psych.* **112,** 819–825.
5. Bertelsen, A., Harvald, B., and Hauge, M. (1977). A Danish twin study of manic depressive disorders. *Br. J. Psych.* **130,** 330–351.
6. Gershon, E. S., Hamovit, J., Guroff, J. J., Dibble, E., Leckman, J. F., Sceery, W., Targum, S. D., Nurnberger, J. I., Jr., Goldin, L. R., and Bunney, W. E., Jr. (1982). A family study of schizoaffective, bipolar I, bipolar II, unipolar, and normal control probands. *Arch. Gen. Psych.* **39,** 1157–1167.
7. Folstein, S., and Rutter, M. (1977). Infantile autism: a genetic study of 21 twin pairs. *J. Child Psychol. Psych.* **18,** 297–321.
8. Piven, J., Palmer, P., Jacobi, D., Childress, D., and Arndt, S. (1997). Broader autism phenotype: evidence from a family history study of multiple-incidence autism families. *Am. J. Psych.* **154,** 185–190.
9. Gottesman, I. I., and Shields, J. (1976). A critical review of recent adoption, twin, and family studies of schizophrenia: behavioral genetics perspectives. *Schizophr. Bull.* **2,** 360–401.
10. Goldsmith, H. H., Gottesman, I. I., and Lemery, K. S. (1997). Epigenetic approaches to developmental psychopathology. *Dev. Psychopathol.* **9,** 365–387.
11. Kendler, K. S., Silberg, J. L., Neale, M. C., Kessler, R. C., Heath, A. C., and Eaves, L. J. (1991). The family history method: whose psychiatric history is measured? *Am. J. Psych.* **148,** 1501–1504.
12. Kendler, K. S., Neale, M. C., Kessler, R. C., Heath, A. C., and Eaves, L. J. (1992). Generalized anxiety disorder in women. A population-based twin study *Arch. Gen. Psych.* **49,** 267–272.
13. Stine, O. C., Xu, J., Koskela, R., McMahon, F. J., Gschwend, M., Friddle, C., Clark, C. D., McInnis, M. G., Simpson, S. G., Breschel, T. S., Vishio, E., Riskin, K., Feilotter, H., Chen, E., Chen, S., Folstein, S. E., Meyers, D. A., Botstein, D., Marr, T. G., and DePaulo, J. R. (1995). Evidence for linkage of bipolar disorder to chromsome 18 with a parent-of-origin effect. *Am. J. Hum. Genet.* **57,** 1384–1394.

14. Freimer, N. B., Reus, V. I., Escamilla, M. A., McInnes, L. A., Spesny, M., Leon, P., Service, S. K., Smith, L. B., Silva, S., Rojas, E., Gallegos, A., Meza, L., Fournier, E., Baharloo, S., Blankenship, K., Tyler, D. J., Batki, S., Vinogradov, S., Weissenbach, J., Barondes, S. H., and Sandkuijl, L. A. (1996). Genetic mapping using haplotype, association and linkage methods suggests a locus for severe bipolar disorder (BPI) at 18q22-q23. *Nature Genet.* **12,** 436–441.
15. DeBruyn, A., Souery, D., Mendelbaum, K., Mendlewicz, J., and Van Broeckhoven, C. (1996). Linkage analysis of families with bipolar illness and chromosome 18 markers. *Biol. Psych.* **39,** 679–688.
16. Schizophrenia Linkage Collaborative Group for Chromosomes 3, 6 and 8. (1996). Additional support for schizophrenia linkage on chromosomes 6 and 8: a multicenter study. *Am. J. Med. Genet.* **67,** 580–594.
17. Coon, H., Holik, J., Hoff, M., Reimherr, F., Wender, P., Myles-Worsley, M., Waldo, M., Freedman, R., and Byerley, W. (1994). Analysis of chromosome 22 markers in nine schizophrenia pedigrees. *Am. J. Med. Genet.* **54,** 72–79.
18. Coon, H., Jensen, S., Holik, J., Hoff, M., Myles-Worsley, M., Reimherr, F., Wender, P., Waldo, M., Freedman, R., and Leppert, M. (1994). Genomic scan for genes predisposing to schizophrenia. *Am. J. Med. Genet.* **54,** 59–71.
19. Gill, M., Vallada, H., Collier, D., Sham, P., Holmans, P., Murray, R., McGuffin, Nanko, S., Owen, M., Antonarakis, S., Housman, D., Kazazian, H., Nestadt, G., Pulver, A. E., Straub, R. E., MacLean, C. J., Walsh, D., Kendler, K. S., DeLisi, L., Polymeropoulos, M., Coon, H., Byerley, W., Lofthouse, R., Gershon, E., and Read, C. M. (1996). A combined analysis of D22S278 marker alleles in affected sib-pairs: support for a susceptibility locus for schizophrenia at chromosome 22q12. Schizophrenia Collaborative Linkage Group (Chromosome 22). *Am. J. Med. Genet.* **67,** 40–45.
20. Vallada, H. P., Gill, M., Sham, P., Lim, L. C., Nanko, S., Asherson, P., Murray, R. M., McGuffin, P., Owen, M., and Collier, D. (1995). Linkage studies on chromosome 22 in familial schizophrenia. *Am. J. Med. Genet.* **60,** 139–146.
21. Polymeropoulos, M. H., Coon, H., Byerley, W., Gershon, E. S., Goldin, L., Crow, T. J., Rubenstein, J., Hoff, M., Holik, J., and Smith, A. M. (1994). Search for a schizophrenia susceptibility locus on human chromosome 22. *Am. J. Med. Genet.* **54,** 93–99.
22. Lasseter, V. K., Pulver, A. E., Wolyniec, P. S., Nestadt, G., Meyers, D., Karayiorgou, Housman, D., Antonarakis, S., Kazazian, H., and Kasch, L. (1995). Follow-up report of potential linkage for schizophrenia on chromosome 22q: Part 3 *Am. J. Med. Genet.* **60,** 172–173.
23. Pulver, A. E., Karayiorgou, M., Wolyniec, P. S., Lasseter, V. K., Kasch, L., Nestadt, Antonarakis, S., Housman, D., Kazazian, H. H., and Meyers, D. (1994). Sequential strategy to identify a susceptibility gene for schizophrenia: report of potential linkage on chromosome 22q12-q13.1: Part 1. *Am. J. Med. Genet.* **54,** 36–43.
24. Moises, H. W., Yang, L., Li, T., Havsteen, B., Fimmers, R., Baur, M. P., Liu, X., and Gottesman, I. I. (1995). Potential linkage disequilibrium between schizophrenia and locus D22S278 on the long arm of chromosome 22. *Am. J. Med. Genet.* **60,** 465–467.
25. Schwab, S. G., Eckstein, G. N., Hallmayer, J., Lerer, B., Albus, M., Borrmann, M., Lichtermann, D., Ertl, M. A., Maier, W., and Wildenauer, D. B. (1997). Evidence suggestive of a locus on chromosome 5q31 contributing to susceptibility for schizophrenia in German and Israeli families by multipoint affected sib-pair linkage analysis. *Mol. Psych.* **2,** 156–160.
26. Straub, R. E., MacLean, C. J., O'Neill, F. A., Walsh, D., and Kendler, K. S. (1997). Support for a possible schizophrenia vulnerability locus in region 5q22-31 in Irish families. *Mol. Psych.* **2,** 148–155.
27. MacKinnon, D. F., Jamison, K. R., and DePaulo, J. R. (1997). Genetics of manic Depressive Illness. *Annu. Rev. Neurosci.* **20,** 355–373.
28. Gottesman, I. I. (1991). "Schizophrenia Genesis. The Origins of Madness." W. H. Freeman, New York.
29. Spitzer, R. L., and Endicott, J. (1979). "Schedule for Affective Disorders and Schizophrenia—Life-time Version (Sads-L)." New York State Psychiatric Institute, New York.
30. Nurnberger, J. I., Blehar, M. C., Kaufmann, C. A., York-Cooler, C., Simpson, S. G., Harkavy-Friedman, J., Severe, J. B., Malaspina, D., Reich, T., and & collaborators from the NIMH Genetics Initiative (1994). The Diagnostic Interview for Genetic Studies: rationale, unique features, and training *Arch. Gen. Psych.* **51,** 849–859.
31. Winokur, G., Coryell, W., Keller, M., Endicott, J., and Leon, A. (1995). A family study of manic-depressive (bipolar I) disease. Is it a distinct illness separable from primary unipolar depression? *Arch. Gen. Psych.* **52,** 367–373.
32. Morel, B. A. (1857). "Traite Des Degenerescences." J. B. Bailliere, Paris.
33. Mott, F. W. (1910). Hereditary aspects of nervous and mental diseases. *Br. Med. J.* **2,** 1013–1020.
34. Mott, F. W. (1911). A lecture on heredity and insanity. *Lancet* **1,** 1251–1259.
35. Pick, D. (1989). "Faces of Degeneration." Cambridge University Press, Cambridge.
36. Galton, F. (1909). "Essays in Eugenics." The Eugenics Education Society., London.
37. Penrose, L. S. (1948). The problem of anticipation in pedigrees of dystrophia myotonica. *Ann. Eugenics* **14,** 125–132.
38. Fleischer, B. (1918). Uber myotonische Dystrophie mit Katarakt: Eine heriditare, familiare Degeneration. *Archiv. fur Ophthalmologie.* **96,** 91–133.
39. Bell, J. (1947). Dystrophia myotonica and allied diseases. Pearson, K. *In* "The Treasury of Human Inheritance" (K. Pearson, Ed.), Vol. 5, p. 343. DuLaw, London.
40. Howeler, C. J., Busch, H. F. M., Geraedts, J. P. M., Niermeijer, M. F., and Staal, A. (1989). Anticipation in myotonic dystrophy: Fact or fiction? *Brain* **112,** 779–797.
41. McInnis, M. G., McMahon, F. J., Chase, G., Simpson, S. G., Ross, C. A., and DePaulo, J. R., Jr. (1993). Anticipation in bipolar affective disorder. *Am. J. Hum. Genet.* **53,** 385–390.
42. Gershon, E. S., Hamovit, J. H., Guroff, J. J., and Nurnberger, J. I. (1987). Birth-cohort changes in manic and depressive disorders in relatives of bipolar and schizoaffective patients. *Arch. Gen. Psych.* **44,** 314–319.
43. Nylander, P. O., Engstrom, C., Chotai, J., Wahlstrom, J., and Adolfsson, R. (1994). Anticipation in Swedish families with bipolar affective disorder *J.Med.Genet.* **31,** 686–689.
44. Engstrom, C., Thornlund, A. S., Johansson, E. L., Langstrom, H., Chotai, J., Adolfsson, R., and Nylander, P. O. (1995). Anticipation in uinpolar affective disorder. *J. Affect. Dis.* **35,** 31–40.
45. Grigoroiu-Serbanescu, M., Wickramaratne, P. J., Hodge, S. E., Milea, S., and Mihailescu, R. (1997). Genetic anticipation and imprinting in bipolar illness. *Br. J. Psych.* **170,** 162–166.
46. Kay, D. W. K. (1963). Late paraphrenia and its bearing on the aetiology of schziphrenia. *Acta Psych. Scand.* **39,** 159–169.
47. Penrose, L. S. (1991). Survey of cases of familial mental illness. *Eur. Arch. Psych. Neurol. Sci.* **240,** 315–324.

48. Bassett, A. S., and Husted, J. (1997). Anticipation or ascertainment bias in schizophrenia - Penrose's familial mental illness sample. *Am. J. Hum. Genet.* **60,** 630–637.
49. Crow, T. J. (1990). The continuum of psychosis and its genetic origins. The sixty-fifth Maudsley lecture. *Br. J. Psych.* **156,** 788–797.
50. Asherson, P., Walsh, C., Williams, J., Sargeant, M., Taylor, C., Clements, A., Gill, M., Owen, M., and McGuffin, P. (1994). Imprinting and anticipation. Are they relevant to genetic studies of schizophrenia? *Br. J. Psych.* **164,** 619–624.
51. Chotai, J., Engstrom, C., Ekholm, B., son Berg, M.-L. J., Adolfsson, R., and Nylander, P. O. (1995). Anticipation in Swedish families with schizophrenia. *Psychi. Genet.* **5,** 181–186.
52. Thibaut, F., Martinez, M., Petit, M., Jay, M., and Campion, D. (1995). Further evidence for anticipation in schizophrenia. *Psych. Res.* **59,** 25–33.
53. Bassett, A. S., and Honer, W. G. (1994). Evidence for anticipation in Schizophrenia. *Am. J. Hum. Genet.* **54,** 864–870.
54. Gorwood, P., Leboyer, M., Falissard, B., Jay, M., Rouillon, F., and Feingold, J. (1996). Anticipation in schizophrenia: new light on a controversial problem *Am. J. Psych.* **153,** 1173–1177.
55. Johnson, J. E., Cleary, J., Ahsan, H., Friedman, J. H., Malaspina, D., Cloninger, CR, Faraone, S. V., Tsuang, M. T., Kaufmann, C. A. (1997). Anticipation in schizophrenia: biology or bias. *Am. J. Med. Genet.* **74,** 275–280.
56. Lotter, V. (1966). Epidemiology of autistic conditions in young children: I prevalence. *Soc. Psych.* **1,** 124–137.
57. DeLong, G. R., and Dwyer, J. T. (1988). Correlation of family history with specific autistic subgroups: Asperger's syndrome and bipolar affective disease. *J. Autism Dev. Disord.* **18,** 593–600.
58. Piven, J., Chase, G. A., Landa, R., Wzorek, M., Gayle, J., Cloud, D., and Folstein, S. (1991). Psychiatric disorders in the parents of autistic individuals. *J. Am. Acad. Child Psych.* **30,** 471–478.
59. Smalley, S. L., McCracken, J., and Tanguay, P. (1995). Autism, affective disorders, and social phobia. *Am. J. Med. Genet.* **60,** 19–26.
60. Hagberg, B., Aicardi, J., Dias, K., and Ramos, O. (1983). A progressive syndrome of autism, dementia, ataxia, and loss of purposeful hand use in girls: Rett's syndrome: report of 35 cases. *Ann. Neurol.* **14,** 471–479.
61. Schalling, M., Hudson, T. J., Buetow, K., and Housman, D. E. (1993). Direct detection of novel expanded trinucleotide repeats in the human genome *Nature Genet.* **4,** 135–139.
62. Hofferbert, S., Schanen, N. C., Budden, S. S., and Francke, U. (1997). Is Rett-syndrome caused by a triplet repeat expansion? *Neuropediatrics* **28,** 179–183.
63. Ross, C. A. (1995). When more is less: pathogenesis of glutamine repeat neurodegenerative diseases *Neuron* **15,** 493–496.
64. Siomi, H., Siomi, M. C., Nussbaum, R. L., and Dreyfuss, G. (1993). The protein product of the fragile X gene, FMR1, has characteristics of an RNA-binding protein. *Cell* **74,** 291–298.
65. Shiwach, R. (1994). Psychopathology in Huntington's disease patients. *Acta Psych. Scand.* **90,** 241–246.
66. Ikeuchi, T., Onodera, O., Oyake, M., Koide, R., Tanaka, H., and Tsuji, S. (1995). Dentatorubral-pallidoluysian atrophy (DRPLA): close correlation of CAG repeat expansions with the wide spectrum of clinical presentations and prominent anticipation. *Semin. Cell Biol.* **6,** 37–44.
67. Thompson, N. M., Gulley, M. L., Rogeness, G. A., Clayton, R. J., Johnson, C., Hazelton, Cho, C. G., and Zellmer, V. T. (1994). Neurobehavioral characteristics of CGG amplification status in fragile X females. *Am. J. Med. Genet.* **54,** 378–383.
68. Kruyer, H., Mila, M., Glover, G., Carbonell, P., Ballesta, F., and Estivill, X. (1994). Fragile X syndrome and the (CGG)n mutation: two families with discordant MZ twins. *Am. J. Hum. Genet.* **54,** 437–442.
69. Reiss, A. L., Abrams, M. T., Greenlaw, R., Freund, L., and Denckla, M. B. (1995). Neurodevelopmental effects of the FMR-1 full mutation in humans. *Nature Med.* **1,** 159–167.
70. Lopez, de Munain, Cobo, A. M., Huguet, E., Marti, Masso, J. F., Johnson, K., and Baiget, M. (1994). CTG trinucleotide repeat variability in identical twins with myotonic dystrophy. *Ann. Neurol.* **35,** 374–375.
71. McNeil, S. M., Novelletto, A., Srinidhi, J., Barnes, G., Kornbluth, I., Altherr, M. R., Wasmuth, J. J., Gusella, J. F., MacDonald, M. E., and Myers, R. H. (1997). Reduced penetrance of the Huntington's disease mutation. *Hum. Mol. Genet.* **6,** 775–779.
72. Alford, R. L., Ashizawa, T., Jankovic, J., Caskey, C. T., and Richards, C. S. (1996). Molecular detection of new mutations, resolution of ambiguous results and complex genetic counseling issues in Huntington disease. *Am. J. Med. Genet.* **66,** 281–286.
73. Leeflang, E. P., Zhang, L., Tavare, S., Hubert, R., Srinidhi, J., MacDonald, M. E., Myers, R. H., de Young, M., Wexler, N. S., and Gusella, J. F. (1995). Single sperm analysis of the trinucleotide repeats in the Huntington's disease gene: quantification of the mutation frequency spectrum. *Hum. Mol. Genet.* **4,** 1519–1526.
74. Jones, C., Slijepcevic, P., Marsh, S., Baker, E., Langdon, W. Y., Richards, R. I., and Tunnacliffe, A. (1994). Physical linkage of the fragile site FRA11B and a Jacobsen syndrome chromosome deletion breakpoint in 11q23.3 *Hum. Mol. Genet.* **3,** 2123–2130.
75. Jones, C., Penny, L., Mattina, T., Yu, S., Baker, E., Voullaire, L., Langdon, W. Y., Sutherland, G. R., Richards, R. I., and Tunnacliffe, A. (1995). Association of a chromosome deletion syndrome with a fragile site within the proto-oncogene CBL2 *Nature* **376,** 145–149.
76. Goodwin, F. K., and Jamison, K. R. (1990). "Manic Depressive Illness." Oxford University Press, New York.
77. Lindblad, K., Nylander, P. O., DeBruyn, A., Souery, D., Zander, C., Engstrom, C., Holmgren, G., Hudson, T., Chotai, J., Mendlewicz, J., Van Broeckhoven, C., Schalling, M., and Adolfsson, R. (1995). Detection of expanded CAG repeats in Bipolar Affective Disorder using the repeat exapnsion detection (RED) method. *Neurobiol. Dis.* **2,** 55–62.
78. O'Donovan, M. C., Guy, C., Craddock, N., Murphy, K. C., Cardno, A. G., Jones, L. A., Owen, M. J., and McGuffin, P. (1995). Expanded CAG repeats in schizophrenia and bipolar disorder. *Nature Genet.* **10,** 380–381.
79. Morris, A. G., Gaitonde, E., Mckenna, P. J., Mollon, J. D., and Hunt, D. M. (1995). CAG repeat expansions and schizophrenia— association with disease in females and early age-at-onset. *Hum. Mol. Genet.* **4,** 1957–1961.
80. O'Donovan, M. C., Guy, C., Craddock, N., Bowen, T., McKeon, P., Macedo, A., Maier, Wildenauer, D., Aschauer, H. N., Sorbi, S., Feldman, E., Mynett-Johnson, L., Claffey, E., Nacmias, B., Valente, J., Dourado, A., Grassi, E., Lenzinger, E., Heiden, A. M., Moorhead, S., Harrison, D., Williams, J., McGuffin, P., and Owen, M. J. (1996). Confirmation of association between expanded CAG/CTG repeats and both schizophrenia and bipolar disorder. *Psychol. Med.* **26,** 1145–1153.
81. Vincent, J. B., Klempan, T., Parikh, S. S., Sasaki, T., Meltzer, H. Y., Sirugo, G., Cola, P., Petronis, A., and Kennedy, J. L. (1996). Frequency analysis of large CAG/CTG trinucleotide repeats in schizophrenia and bipolar affective disorder. *Mol. Psych.* **1,** 141–148.
82. Petronis, A., Bassett, A. S., Honer, W. G., Vincent, J. B., Tatuch, Y., Sasaki, T., Ying, D. J., Klempan, T. A., and Kennedy, J. L. (1996). Search for unstable DNA in schizophrenia families with

evidence for genetic anticipation *Am. J. Hum. Genet.* **59,** 905–911.

83. Sasaki, T., Billett, E., Petronis, A., Ying, D., Parsons, T., Macciardi, F. M., Meltzer, H. Y., Lieberman, J., Joffe, R. T., Ross, C. A., McInnis, M. G., Li, S. H., and Kennedy, J. L. (1996). Psychosis and genes with trinucleotide repeat polymorphism *Hum. Genet.* **97,** 244–246.
84. Jain, S., Leggo, J., DeLisi, L. E., Crow, T. J., Margolis, R. L., Li, S. H., Goodburn, S., Walsh, C., Paykel, E. S., Ferguson-Smith, M. A., Ross, C. A., and Rubinsztein, D. C. (1996). Analysis of thirteen trinucleotide repeat loci as candidate genes for schizophrenia and bipolar affective disorder *Am. J. Med. Genet.* **67,** 139–146.
85. Petronis, A., Heng, H. H., Tatuch, Y., Shi, X. M., Klempan, T. A., Tsui, L. C., Ashizawa, T, Surh, L. C., Holden, J. J., and Kennedy, J. L. (1996). Direct detection of expanded trinucleotide repeats using PCR and DNA hybridization techniques. *Am. J. Med. Genet.* **67,** 85–91.
86. Bowen, T., Guy, C., Speight, G., Jones, L., Cardno, A., Murphy, K., McGuffin, P., Owen, M. J., and O'Donovan, M. C. (1996). Expansion of 50 CAG/CTG repeats excluded in schizophrenia by application of a highly efficient approach using repeat expansion detection and a PCR screening set *Am. J. Hum. Genet.* **59,** 912–917.
87. Lesch, K. P., Stober, G., Balling, U., Franzek, E., Li, S. H., Ross, C. A., Newman, M., Beckmann, H., and Riederer, P. (1994). Triplet repeats in clinical subtypes of schizophrenia: variation at the DRPLA (B 37 CAG repeat) locus is not associated with periodic catatonia. *J. Neural. Transm.* **98,** 153–157.
88. Guedj, F., Cao, Q., Cravchik, A., Ram, A., Badner, J., Gershon, E. S., and Gejman, P. V. (1996). Analysis of DRPLA trinucleotide repeats in schizophrenia *Psych. Genet.* **6,** 33–34.
89. Brando, L. J., Yolken, R., Herman, M. M., Kleinman, J. E., Ross, C. A., and Torrey, E. F. (1996). Analysis of the DRPLA triplet repeat in brain tissue and leukocytes from schizophrenics *Psych. Genet.* **6,** 1–5.
90. Morris-Rosendahl, D. J., Burgert, E., Uyanik, G., Mayerova, A., Duval, F., Macher, J. P., and Crocq, M. A. (1997). Analysis of the CAG repeats in the SCA1 and B37 genes in schizophrenic and bipolar I disorder patients: tentative association between B37 and schizophrenia. *Am. J. Med. Genet.* **74,** 324–330.
91. Holden, J. J., Wing, M., Chalifoux, M., Julien-Inalsingh, C., Schutz, C., Robinson, P., Szatmari, P., and White, B. N. (1996). Lack of expansion of triplet repeats in the FMR1, FRAXE, and FRAXF loci in male multiplex families with autism and pervasive developmental disorders. *Am. J. Med. Genet.* **64,** 399–403.
92. Ashworth, A., Abusaad, I., Walsh, C., Nanko, S., Murray, R. M., Asherson, P., McGuffin, P., Gill, M., Owen, M. J., and Collier, D. A. (1996). Linkage analysis of the fragile X gene FMR-1 and schizophrenia: no evidence for linkage but report of a family with schizophrenia and an unstable triplet repeat. *Psych. Genet.* **6,** 81–86.
93. Rubinsztein, D. C., Leggo, J., Crow, T. J., DeLisi, L. E., Walsh, C., Jain, S., and Paykel, E. S. (1996). Analysis of polyglutamine-coding repeats in the TATA-binding protein in different human populations and in patients with schizophrenia and bipolar affective disorder. *Am. J. Med. Genet.* **67,** 495–498.
94. Margolis, R. L., Stine, O. C., McInnis, M. G., Ranen, N. G., Rubinsztein, D. C., Leggo, J., Jones Brando, L. V., Kidwai, A. S., Loev, S. J., Breschel, T. S., Callahan, C., Simpson, S. G., DePaulo, J. R., McMahon, F. J., Jain, S., Paykel, E. S., Walsh, C., DeLisi, L. E., Crow, T. J., Torrey, E. F., Ashworth, R. G., Macke, J. P., Nathans, J., and Ross, C. A. (1996). A human homologue of the *C. elegans* cell fate-determining gene *mab-21* that contains a highly polymorphic (CAG)n trinucleotide repeat: cDNA cloning, expression, and mapping to chromosome 13q13. *Hum. Mol. Genet.* **21,** 607–616.
95. Wang, S., Detera-Wadleigh, S. D., Coon, H., Sun, C. E., Goldin, L. R., Duffy, D. L., Byerley, W. F., Gershon, E. S., and Diehl, S. R. (1996). Evidence of linkage disequilibrium between schizophrenia and the SCa1 CAG repeat on chromosome 6p23. *Am. J. Hum. Genet.* **59,** 731–736.
96. Pujana, M. A., Martorell, L., Volpini, V., Valero, J., Labad, A., Vilella, E., and Estivill, X. (1997). Analysis of amino-acid and nucloetide variants in the spinocerebellar ataxia type 1 (SCA1) gene in schizophrenia patients. *Hum. Genet.* **99,** 772–775.
97. Speight, G., Guy, C., Bowen, T., Asherson, P., McGuffin, P., Craddock, N., Owen, M. J., and O'Donovan, M. C. (1997). Exclusion of CAG/CTG trinucleotide repeat loci which map to chromosome 4 in bipolar disorder and schizophrenia. *Am. J. Med. Genet.* **74,** 204–206.
98. Ohara, K., Tani, K., Tsukamoto, T., Ino, A., Nagai, M., Suzuki, Y., Xu, H. D., Xu, DS, Wang, Z. C. (1997). Three CAG trinucleotide repeats on chromosome 6 (D6S1014, D6S1015, and D6S1058) are not expanded in 30 families with schizophrenia. *Neuropsychopharmacology* **17,** 279–283.
99. Turecki, G., Rouleau, G. A., Joober, R., Mari, J., and Morgan, K. (1997). Schizophrenia and chromosome 6p. *Am. J. Med. Genet.* **74,** 195–198.
100. Berrettini, W. H., Ferraro, T. N., Goldin, L. R., Weeks, D. E., Detera-Wadleigh, S., Nurnberger, J. I., Jr., and Gershon, E. S. (1994). Chromosome 18 DNA markers and manic-depressive illness: evidence for a susceptibility gene *Proc. Natl. Acad. Sci. USA* **91,** 5918–5921.
101. Breschel, T. S., McInnis, M. G., Margolis, R. L., Sirugo, G., Corneliussen, B., Simpson, S. G., McMahon, F. J., MacKinnon, D. F., Xu, J. F., Pleasant, N., Huo, Y., Ashworth, R. G., Grundstrom, T., Kidd, K. K., DePaulo, J. R., Jr., and Ross, C. A. (1997). A novel, heritable, expanding CTG repeat in an intron of the SEF2-1 gene on chromosome 18q21.1. *Hum. Mol. Genet.* **11,** 1155–1163.
102. McGuffin, P., and Murray, R. M. (1991). "The New Genetics of Mental Illness." Butterworth-Heinemann, Oxford.

CHAPTER 28

Genetic Anticipation in Neurological and Other Disorders

ANDREW D. PATERSON, DAVID M. J. NAIMARK, JOHN B. VINCENT, JAMES L. KENNEDY, AND ARTURAS PETRONIS
Neurogenetics Section, Clarke Institute of Psychiatry, Toronto, Ontario M5T 1R8, Canada

I. INTRODUCTION TO GENETIC ANTICIPATION

Genetic anticipation is characterized by an earlier age at diagnosis (AAD), and/or an increase in the severity of a genetic disease in subsequent generations. An increase in the proportion of affected individuals in subsequent generations, or "increasing penetrance," is also occasionally described as a feature of genetic anticipation. Of these aspects, the dynamics of age at diagnosis between generations have most commonly been studied. The phenomenon of genetic anticipation was first described last century, when it was termed "degeneration" [1]. A historical review of anticipation has been provided elsewhere [2]. There has been ongoing debate as to whether anticipation is a real phenomenon or an artifact resulting from ascertainment biases [3, 4]. The turning point for genetic anticipation was provided by two studies, each of which highlighted important features of anticipation. The first, on Huntington's disease (HD) [4], demonstrated a differential degree of anticipation depending on the sex of the disease-transmitting parent. This effect of the sex of the transmitting parent, termed genomic imprinting, was suggested as a criteria to distinguish true from artifactual anticipation [4]. The second study, which renewed interested in anticipation, used a prospective ascertainment design for myotonic dystrophy (DM) [5], and thus accounted for some of the ascertainment biases suggested previously [3].

Over the last few years there has been a resurgence of interest in anticipation following the discovery of trinucleotide repeat expansions (TRE) as the molecular mutation underlying anticipation in a number of Mendelian disorders (see Chapters 3, 7–10, 14–17, 20, 21, 23). We will not discuss further anticipation in these unstable DNA diseases where the genes have been cloned, but will rather concentrate on descriptions of anticipation in other, mainly complex, disorders. For a few of these disorders, traditional mutations have been described in some families, but in the majority the genes involved are as yet unidentified. Our focus highlights the potential heuristic value and complexities of the study of genetic anticipation.

A. The Heuristic Value of Genetic Anticipation for the Identification of Disease Genes

Among the fundamental difficulties in identifying genes which confer modest risk for complex disorders are the reduced power of both parametric and nonparametric linkage analyses in the face of genetic heterogeneity [6, 7]. This is confounded by the potential complexities of epistatic gene–gene and gene–environment interactions. A more powerful approach recently proposed involves disequilibrium mapping across the genome, but this approach requires both large sample sizes and thousands of intragenic markers which have yet to be identified [6]. Faced with these problems, some researchers have focused their attention on the possibility of using the search for genetic anticipation in an attempt to accelerate the identification of genes. The current paradigm, based on the TRE diseases identified to date (see above), states that evidence for anticipation in a disorder implies the presence of a TRE. A number of molecular genetic techniques have been developed or adapted to detect a TRE without any preliminary data regarding the location of the disease gene: cDNA screening [8], repeat expansion detection (RED) technique [9], DIRECT [10], Southern blot hybridization-based approach [11], and fluorescent *in situ* hybridization [12]. Of note is the DIRECT approach which was employed to clone *SCA2* [13].

Some authors have been skeptical about the heuristic value of genetic anticipation. In a study of anticipation in DM it was suggested that evidence for anticipation can only be definitively assessed when the genes involved have been found [14]. This argument can be extended to other diseases because of the potential biases discussed below (Section III.A).

B. The Unstable DNA Hypothesis of Complex Diseases

The unstable DNA hypothesis of complex diseases has been expounded most clearly for major psychosis [15], but can be applied to a wide variety of complex diseases where there is evidence for anticipation in an attempt to identify genes which confer relative risk in at least a proportion of cases. Briefly, the unstable DNA hypothesis states that the presence of anticipation in a genetic disorder is due to TRE as the underlying molecular cause. The advantage of the unstable DNA hypothesis is its theoretical ability to account for a number of non-Mendelian features seen in complex disorders. Such non-Mendelian features include incomplete penetrance that occurs when individuals possessing the disease haplotype are unaffected. This may result from unstable DNA being below a certain threshold of expansion for expression of the phenotype. Furthermore, the phenomenon of variable expressivity can also be explained by unstable DNA where differing degrees of expansion above certain thresholds may be associated with alternative phenotypes. The discordance of monozygotic twins, a feature of many complex disorders, may be explained by the differential expansion of unstable DNA in the postzygotic period. Somatic mosaicism of a TRE has already been described for some unstable DNA diseases (fragile X [16], DM [17], and HD [18]) and may also add to incomplete penetrance, expressivity, and the discordance of MZ twins. Another attractive feature of the unstable DNA hypothesis of complex diseases is its potential to account for sporadic cases, which account for the majority of cases in many complex disorders and which cannot be adequately explained by traditional genetics. It could be hypothesized that an expansion of unstable DNA occurs in the germ cells of one parent and results in disease in the proband. The expansion may revert back to normal range in the probands' germ cells resulting in absence of disease in the proband's offspring.

II. ANTICIPATION: REVIEW OF PUBLISHED STUDIES

This section provides a summary of studies that have been published featuring genetic anticipation in a wide range of diseases. Published articles were obtained from a Medline search performed using the keyword "genetic anticipation" covering the period from 1966 to June 1997, Online Mendelian Inheritance in Man [19], and from references cited in publications relating to genetic anticipation. Ascertainment criteria, analysis, and interpretation are covered, and are presented in Table 28-1 for a number of neurological disorders, and in Table 28-2 for other disorders. For four neurological disorders (spinocerebellar ataxia 7, spastic paraplegia, facioscapulohumeral dystrophy, and Parkinson's disease) a more detailed review of the studies performed to date are included, and where available the results of linkage studies and other DNA studies including RED analysis are also incorporated. We have tried to provide a brief summary, although with a wide range of diseases demonstrating evidence for anticipation it is quite possible that we have overlooked published work. Our primary intention is to give an overview of the range of diseases and studies which highlight relevant ascertainment and statistical issues. Online Mendelian Inheritance in Man [19] is recommended for additional up-to-date clinical and molecular details about these disorders.

A. Neurological Disorders with Evidence for Anticipation

1. Spinocerebellar Ataxia 7

Eight families with autosomal dominant cerebellar ataxia associated with visual failure secondary to a pigmentary macular dystrophy, also termed spinocerebellar ataxia 7 (SCA7), have been described [20]. Presentation in two-thirds of patients was due to ataxia and the remainder presented with either visual failure or both. Anticipation was apparent in the offspring of affected fathers, and transmission of the disease to severe, infantile-onset cases was always from an affected father; this is similar to the situation found in HD [4]. Four other families with SCA7 demonstrated anticipation [21]. Individuals from the two youngest generations were excluded from the analysis in an attempt to avoid ascertainment bias. These families showed linkage to chromosome 3p12-p21.1. These findings were consistent with several other studies which found linkage to 3p14-21.1, and also demonstrated anticipation [22–24]. A further two families with SCA7 have been described, with anticipation described in the largest family regarding both age at diagnosis and severity as determined by disease duration until death [25]. Also, this study [25] replicated the previous finding of genomic imprinting [20], with paternal transmissions associated with a greater degree of anticipation compared to maternal ones. As predicted from the anticipation studies, application of the RED technique to DNA from individuals with SCA7 from eight families demonstrated a significant increase in the size of the RED (CAG)n repeat size in affected individuals compared to unaffected spouses [26]. Based on the correlation between RED product and repeat size for MJD/SCA3, individuals affected with SCA7 are predicted to have about 64 CAG repeats [26].

2. Autosomal Dominant Spastic Paraplegia

Bruyn *et al.* described a large Dutch spastic paraplegia (SPG) family demonstrating possible anticipation in successive generations [27]. Another study of 23 families with SPG found that the mean age at diagnosis in offspring was earlier than parents [28]. No evidence for genomic imprinting in these families was detected. Both SPG3, linked to chromosome 14q [29, 30], and SPG4, linked to chromosome 2p, provide support for anticipation [31, 32]. A recent report described an interesting SPG pedigree where anticipation has been suggested, and in which four of the six SPG patients have comorbid epilepsy [33]. A correlation between the age at onset in parents and the degree of anticipation has led some workers to reject anticipation in SPG [28, 32]. They attributed the observed anticipation to ascertainment biases where the children of young parents have not yet reached the age of risk [28, 32]. This contention has been used previously [3], but it has been convincingly argued that the presence of such a correlation does not exclude anticipation and is in fact a separate phenomenon [34]. No molecular studies investigating unstable DNA have been reported as yet for SPG.

3. Facioscapulohumeral Muscular Dystrophy

A study of 19 multiplex families with facioscapulohumeral muscular dystrophy (FSHD) provided 28 parent–child pairs which showed a significant earlier age at diagnosis using a paired *t* test in the younger generation [35]. Analysis was also performed after excluding the proband from each family in an attempt to reduce ascertainment bias and the anticipation findings remained significant. The authors discussed the uncertainty of the age at onset in the earlier generations, but noted that generally the parents were less severely affected than their children. No difference in the severity according to the gender of the transmitting parent was found. Another study found anticipation in the familial cases as determined by AAD [36]. A further attempt to assess anticipation in FSHD used a measure of disease severity and studied 93 patients from 62 kindreds [37]. Of the patients evaluated, 27 inherited the disease maternally as opposed to 17 paternally; however, the mean rate of disease progression did not differ between these groups. Comparison of severity was performed using serial assessments of muscle strength using quantitative isometric myometry (QMT) scores in 23 parent–child pairs, all from families who have shown evidence of linkage to chromosome 4. Using QMT scores normalized for age, gender, and height, and adjusted for disease duration, the offspring were more severely affected in 15 of the pairs, the parent was more severely affected in 7 pairs, and 1 pair was equally affected; this was interpreted as supporting anticipation in FSHD.

Molecular studies of patients with FSHD appear quite complicated and no firm conclusions can be drawn about the exact molecular genetic defect at this stage. However, it has been clearly shown that an *Eco*RI fragment of probe D4F104S1 is shorter than 28 kb in individuals with FSHD [36]. This is due to deletions of 3.3-kb tandemly repeated units (D4Z4); such deletions are associated with FSHD. A report of 30 Japanese FSHD patients from 11 families found that all but 1 FSHD patient had a small (<28 kb) *Eco*RI fragment, and in the FSHD families the small fragment cosegregated with the disease. All of the 8 apparently sporadic cases had small *Eco*RI fragments; however, 6% of controls also had small *Eco*RI fragments. A correlation was found between the size of the *Eco*RI fragment and age at diagnosis when familial and sporadic cases were combined. However, the clinical severity of the sporadic cases was similar despite the fact that they possessed a range of fragment sizes. It was noted that individuals with new mutations and earliest onset had the shortest fragments. Since each family had a specific small *Eco*RI fragment associated with the disease, the anticipation observed within a family could not be attributed to the size of the fragment alone. No intergenerational instability of the fragment has been identified in affected families.

Another study also found a correlation in a combined sample of sporadic and familial cases between smaller *Eco*RI fragments and earlier age at onset ($r = 0.56$, $p < 0.001$) [38]. This observation has been used to suggest that this fragment may account for a part of the variance in age at onset in FSHD. The authors utilized the age at loss of ambulation as the age-at-onset time-point and noted a possible trend to younger age at onset with successive generations, but describe that they are not able to exclude the possibility that this results from differential recall. It is interesting to remark that between affected members in the same family, the size of the deletion remains constant, yet anticipation is apparent.

More recently, a novel candidate gene, *FRG1,* that maps 100 kb centromeric to the repeated units has been cloned and is being subjected to detailed analysis in FSHD [39]. Regarding the repeat region at the telomere of 4q, it is interesting to note that an exchange of the 3.3-kb repeat units between chromosomes 4q35 and 10q26 in at least 20% of the normal population was detected [40]. It is suggested that such a high frequency of exchanges means that the 3.3-kb repeats do not contain part of the FSHD1 gene [40].

4. Parkinson's Disease

Because of the existence of similarities between familial Parkinson's disease (PD) and TRE disorders regarding selective neurodegeneration, a number of groups have looked at PD families for anticipation. In one study, out of 137 patients with PD, 21 probands had an affected parent, aunt, or uncle, and the age at onset of affected family members differed significantly between generations using a i test [41]. A large Italian kindred with 60 affected individuals found 46 parent–child pairs, and although there was an initial suggestion of anticipation, this fell below conventional significance when they analyzed the 25 pairs for whom the authors were certain about the age at onset [42]. Eleven families from these previous studies [41, 42] which demonstrated evidence for anticipation were genotyped using RED [43]. No difference in the (CAG)n RED product size was found between both affected individuals from these families, and also in unrelated PD cases compared to controls. The authors state that they are able to exclude large CAG repeats in the majority of PD in families, but suggest the possibility of other types of repeats being involved. Alternatively they suggest that the observed anticipation may result from other mechanisms. Another study ascertained 100 consecutive PD cases of whom 9 had at least one living affected relative, and observed anticipation of onset age in new generations in some families [44]. Three other groups have reported families which are consistent with anticipation [45–47]. A separate study of 13 PD pedigrees looked at age at diagnosis in 33 pairs including second-degree relatives (aunt/uncle, niece/nephew) [48]. Using this approach they found support for anticipation but the proband was sometimes included twice in different pairs. When the pairs involving the proband were removed, a nonsignificant trend toward anticipation was found. It was suggested that the results may be biased by the presence of young members who had not yet become affected. After reviewing other published PD pedigrees these workers concluded that population-based studies are required for the clear demonstration of anticipation. Linkage to chromosome 4q of some families with autosomal dominant PD has been described. In these families, substitution mutations in the α-synuclein gene which cosegregate with disease have been identified recently [49].

5. Other Neurological Disorders

Table 28-1 summarizes studies of anticipation performed on other neurological disorders.

B. Anticipation in Nonneurological Diseases

For nonneurological disorders, an overview of published studies relating to anticipation is provided in Table 28-2.

Table 28-1 Other Neurological Diseases Reported to Demonstrate Anticipation

Disease and ref.	Ascertainment[a]	Sample and size[b]	Statistical analysis[c]	Anticipation (years)[d]	Imprinting[e]
Autosomal dominant progressive external ophthalmoplegia with hypogonadism; Melberg *et al.* [50]	Case report	1 Family	Observation	+	ND
Autosomal dominant pure cerebellar ataxia; Ishikawa *et al.* [51]	Families	8 Families	Paired test	3	ND
Cataracts with motor neuronopathy, short stature, and skeletal abnormalities; Slavotinek *et al.* [52]	Case report	1 Family	Observation	+	ND
Cerebellar ataxia and bipolar affective disorder; Piqueras *et al.* [53]	Case report	1 Family	Observation	+	ND
Familial periodic cerebellar ataxia without myokymia; Teh *et al.* [54]	Case report	2 Families	Observation	+	ND
Glaucoma; Dhir *et al.* [55]	Case report	1 Family	Observation	+	ND
Idiopathic torsion dystonia					
Labuda *et al.* [56]	Clinic	27 P–C p	Paired test	+	Maternal
Cheng *et al.* [57]	Clinic	49 Families	Paired test	+	ND
Meniere's disease; Morrison [58]	Families	41	Observation	+	ND
Oculopharyngeal muscular dystrophy; Blumen *et al.* [59]	Clinic	29 Families	Observation	+	ND
Olivopontocerebellar atrophy; Dai [60]	Clinic	36 Families	Observation	+	ND
Otosclerosis; Morrison [61]	Clinic	45 P–C p	Compared means	6	ND
Restless legs syndrome; Trenkwalder *et al.* [62]	Case report	1 Family	Paired test	+	ND
Rolandic epilepsy and speech dyspraxia; Scheffer *et al.* [63]	Case report	1 Family	Observation	+	ND
Scapuloperoneal spinal muscular atrophy; Isozumi *et al.* [64]	Case report	1 Family	Observation	+	ND
Spinocerebellar ataxia 5; Ranum *et al.* [65]	Case report	1 Family	Observation	+	ND
Subarachnoid hemorrhage; Bromberg *et al.* [66]	Familial cases and literature review	23	Compared means	+	ND
Von Hippel-Landau syndrome; Weslowski *et al.* [67]	Case report	1 Family	Observation	+	ND

[a]Studies were categorized based on stated or inferred ascertainment: case report, families, clinic (from a clinical sample), literature review.
[b]If the main ascertainment was family-based, then the number of families is given; otherwise, the number of parent–child pairs (P–C p) is provided.
[c]"Observation" is used if no statistical tests were performed. Otherwise, tests are classified into "compared means" (comparison of mean age at diagnosis between generations) or "paired test" (the use of a pairwise statistical test).
[d]The number of years of anticipation between generations is given; otherwise, (+) denotes the presence of anticipation, (−) the absence of anticipation.
[e]Maternal or paternal signifies that such transmissions are associated with a greater degree of anticipation; ND, not determined.

III. DISCUSSION

There are several major issues concerning both ascertainment and statistical problems related to the analysis of anticipation. This has led some authors to conclude that whether anticipation is genuine for a specific disease is a matter of opinion [112].

A. Ascertainment Issues

Important potential biases in the analysis of data for anticipation may be introduced by ascertainment; this has been discussed elsewhere in detail [3, 4]. We comment here on a number of additional biases. The majority of studies investigating anticipation were opportunis-

Table 28-2 Nonneurological Diseases with Evidence Suggesting Anticipation

Disease	Ascertainment[a]	Sample type and size[b]	Statistical analysis[c]	Anticipation (years)[d]	Imprinting[e]
Autosomal dominant cyclic hematopoiesis					
Palmer *et al.* [68]	Families	9 Families	Observation of severity	+	ND
Autosomal dominant polycystic kidney disease					
Fick *et al.* [69]	Families	86 Families	Families with degree of anticipation >10 years	49%	Maternal
Gerbeth *et al.*, 1995 [70]	Families	74 P–C p	Paired test	ns + in 15 families	ND
Simon *et al.*, 1996 [71]		39 Families			ND
Blau syndrome					
Raphael *et al.* [72]	Case report	1 Family	Observation	+	ND
Breast cancer					
Jacobsen [73]	Registry	21 M–D p	Compare means	11	ND
Smithers [74]	Clinic	25 M–D p	Observation	?	ND
Morse [75]	Clinic	13 M–D p	Compare means	10.5	ND
Haagensen [76]	Clinic	18 M–D p	Compare means	4	ND
Bucalossi and Veronesi [77]	Registry	58 M–D p	Compare means		ND
Paterson *et al.* [78]	Literature review	218 M–D p	Paired test	6–9	ND
Vehmanen *et al.* [79]	BRCA2 families		—	–	ND
Thorlacius *et al.* [80]	BRCA2 families	12 M–D p	Paired test	10 (n.s.)	ND
Crohn's disease					
Polito *et al.* [81]	Clinic, registry families	27 P–C p	Paired test and regression	10	Paternal
		32 P–C p		15	
Satsangi *et al.* [82]	Families	77 P–C p	Compare means	16	ND
Familial adenomatous polyposis					
Veale [83]	Families	10–16 P–C p	Compare means and correlation	–	ND
Presciuttini *et al.* [84]	Registry	6 Families	Regression	15	ND
Familial amyloidotic polyneuropathy type I					
Drugge *et al.* [85]	Families	35 Families	Observation	+	Maternal
Tashima *et al.* [86]	Families	20 P–C p	Observation	+	–
Familial primary pulmonary hypertension					
Loyd *et al.* [87]	Families	24 Families	Paired test	10	Maternal
Familial total anomalous pulmonary venous return					
Bleyl *et al.* [88]	Case report	1 Family	Observation of penetrance	+	ND
Hereditary nonpolyposis colorectal cancer					
Vasen *et al.* [89]	Families	41 Families	Paired test	13	ND
Rodriguez-Bigas *et al.* [90]	Registry	40 Families	Observation	5	ND
Tsai *et al.* [91]	Registry	326 Families	Survival analysis	ns	ND
Holt-Oram syndrome					
Newbury-Ecob *et al.* [92]	Cases	28 Families	Paired test	+	Maternal
Hypertrophic cardiomyopathy					
Gregor and Cerny [93]	Families	105 Families	Observation	+	ND

Leukemia					
Horwitz *et al.* [94]	Case + literature review	9 Families	Paired test	28	ND
Melanoma					
Martijn *et al.* [95]	Clinic records	3 P–C p	Observation	+	ND
Goldstein *et al.* [96]	Families	23 Families		5–17	ND
Goldstein *et al.* [97]	Families	23 Families	No. and thickness of tumor	–	ND
Multiple endocrine neoplasia Type I					
Giraud *et al.* [98]	Case report	1 Family	Observation	+	ND
Multiple hamartoma (Cowden syndrome)					
Hanssen *et al.* [99]	Report and review	1 Family	Observation	+	Maternal
Neuroblastoma					
Plon [100]	Case report	1 Family	Increasing penetrance	+	ND
Ovarian cancer					
Piver *et al.* [101]	Registry	279 M–D Pairs	Paired test	8.6	ND
Goldberg [102]	Registry	131 Families	Survival analysis	+	ND
Paraganglioma					
Baysal *et al.* [103]	Families		Paired test	+	+
Primary biliary cirrhosis					
Brind *et al.* [104]	Clinic	8 M–D p	Paired test	16	ND
Psoriasis					
Theeuwes and Morhenn [105]	Literature review		Observation	+	Paternal
Rheumatoid arthritis					
Deighton *et al.* [106]	Clinic and families	153 P–C p 15 Families	Paired test	+	Maternal
McDermott *et al.* [107]	Registry families	63 P–C p	Paired test	+	–
Testicular cancer					
Raghavan *et al.* [108]	Case report	1 F–S p	Observation	13	ND
Deickmann *et al.* [109]	Literature review	17 F–S p	Compare means	15	ND
Heimdal *et al.* [110]	Registry	7 F–S p	Compare means	16	ND
Waardenburg syndrome type 1					
Pierpont *et al.* [111]	Case	1 Family	Observation of penetrance	+	ND

[a]Studies were categorized based on stated or infered ascertainment: case report, families, clinic (from a clinical sample), literature review.

[b]If the main ascertainment was family-based, then the number of families is given; otherwise, the number of parent–child or mother–daughter pairs (P–C p; M–D p) is provided.

[c]"Observation" is used if no statistical tests were performed. Otherwise, tests are classified into: "compare means" (comparison of mean age at diagnosis between generations) or "paired test" (the use of a pairwise statistical test); other tests are specified.

[d]The number of years of anticipation between generations is given; otherwise (+) denotes the presence of anticipation, (−) the absence of anticipation; ND, not determined; ns, not significant.

[e]Maternal or paternal signifies that such transmissions are associated with a greater degree of anticipation; (−) no evidence for imprinting; ND, not determined.

tic, in that they used families collected for other reasons: either referral to specialist centers or, in particular, recruitment for genetic linkage studies. There may be preferential ascertainment of families for linkage studies where there are many living cases over several generations, and it has been suggested that this may produce pedigree structures that mimic genetic anticipation [113]. A representative sample of the population of all familial cases is important in anticipation studies (see below).

An additional complexity in the study of common disorders is the occurrence of sporadic (nongenetic) cases in relatives from families which also have genetic disease (also called phenocopies). It is a well recognized phenomenon for many diseases that sporadic cases have a later mean age at diagnosis than familial cases, although the reason for this general observation is not clear. The presence of phenocopies in families with genetic disease could potentially produce apparent anticipation because the parental generation would be more likely to be affected with sporadic disease due to their more advanced age. Moreover, the occurrence of two sporadic cases in one family may simulate a genetic situation. Once germline mutations responsible for familial cases are identified, then testing of all affected pedigree members to identify such phenocopies can exclude such a bias.

Although for the majority of diseases the severity of symptoms increases as a disease progresses, for some diseases this may not be the case. One example is autosomal dominant cyclic hematopoiesis, for which symptoms have been shown to improve generally with age [114]. Thus the generational effect observed [68] may partially be explained by an age-dependent reduction in severity.

Changes in risk factors for a disease over time, termed secular trends or cohort effects, may mimic anticipation. An ongoing population-based registry is the only rigorous method for the investigation of cohort effects. If there is no cohort effect, then any observed anticipation may well be real. However, if there is evidence for a cohort effect, it becomes rather complicated to separate out the contributions of cohort effect and real anticipation.

For cancers in particular, screening may result in earlier diagnosis in subsequent generations. Additional vigilance may occur if there is a family history of disease, and both of these may increase the chance that a diagnosis will be made earlier in additional family members when they become affected; this may mimic anticipation. One approach to investigate these potential confounding factors would be to examine differences in the stage of disease at diagnosis between generations.

It has now become clear that in order to overcome these biases, population-based random samples of families are necessary for anticipation studies [115, 116]. It has been suggested that the ideal design for the study of anticipation requires ascertainment of probands independent of family history and to then follow them prospectively for many years until both parents and children are through the age of risk [115]. Unfortunately, relatively few epidemiological studies contain information relating to the age at diagnosis of affected relatives, and therefore most of these studies cannot be used for population-based studies of anticipation. This drawback also applies to a number of proband-based disease registries. Additional concerns arise here with a substantial proportion of data missing from such studies. For rare disorders, a population approach would be extremely expensive and ascertain only a few pedigrees. Thus ascertainment from specialist referral is most feasible for rare disorders, although it should be kept in mind that this may introduce bias with particularly severe cases being preferentially ascertained. As a compromise to the proposed ideal design, it has been suggested that other types of data should be used for anticipation, specifically either studies of severity or studies involving whole families where unaffected siblings of both the parents' and children's generation are included [115]. These authors suggest that it is possible to overcome the bias due to differential timing of the diagnostic assessment by employing survival analysis in this situation (Section III.B.3). An alternative statistically appropriate approach requires ascertainment of families at only a single time-point; analysis is performed which takes into account the differential timing of diagnosis between generations (see Section III.B.4) [116].

B. Statistical Considerations

1. SAMPLING APPROACHES

Population-based ascertainment is likely to produce a predominance of families with single parent–child pairs, and occasionally multiplex families. A concern with such ascertainment relates to the appropriate analysis for multiplex families regarding the sampling of numerous parent–child pairs from a large single family. In such a situation, problems may arise when the same individual is used as a parent in multiple transmissions to affected children. This results in a number of nonindependent pairs. Any error in the age at onset in the parent may excessively weight the results. A strategy of random sampling from the pedigree has been suggested to deal with this [117]. Another possible approach is to average the age of diagnosis in siblings of the pedigree to produce a "parent-averaged offspring" pair. A further problem relates to whether it is acceptable to include an individual as both a parent and a child in such analysis. Some workers who have used ascertainment of a type other than population-based have excluded the proband from the analysis in an attempt to reduce ascertainment biases. Although this may help reduce ascertainment biases, it does not remove them.

2. Pairwise Statistical Tests

In addition to ascertainment and sampling, an important issue regards appropriate statistical tests. As can be seen from Tables 28-1 and 28-2, many of the studies on anticipation have used some form of pairwise statistical tests to compare the age at diagnosis in parents and children. However, it has recently been shown that using such an approach may lead to erroneous results [115]. The essence of the problem relates to the difference in ages between generations, which means that a proportion of children have not reached the age at diagnosis for the disease at the time of ascertainment. This leads to a so-called right-truncation of distribution of the children's age of onset. In brief, the simulation studies generated parent–child pairs, their ages of onset, and an age at interview for each parent–child pair [115]. The occurrence of children unaffected at the time of simulated interview resulted in rejection of that pair, and further pairs were simulated until a total of 100 pairs were obtained in which both parent and child were affected. This study demonstrated that when parent–child pairs are interviewed at a relatively early age (i.e., before the majority of children have reached the mean AAD) this results in an elevation of the type I error, and in some situations this can be as high as 100% [115].

3. Survival Analysis

Survival analysis has been proposed as an approach for investigation of anticipation, using the age of diagnosis as the endpoint [115]. Unaffected individuals are also included and it is suggested that this overcomes the bias due to differential timing of the diagnostic assessment [115]. Survival analysis has been used by some groups but information from unaffected individuals has generally not been included [91, 94]. Others argue that survival analysis is not appropriate for complex diseases, such as schizophrenia, where the majority of offspring will not become affected, but are treated as "censored" in survival analysis [118]. Nevertheless, it is clear that a differential degree of censoring between generations does not affect Kaplan-Meier analysis of survival curves. However, the magnitude of the difference between the survival curves of different generations is not necessarily directly related to the degree of anticipation [119]. Other factors which may influence the results of survival analysis include the true mean AAD for the parental generation, the variance of the AAD distribution, and the mean age at ascertainment [119].

4. New Statistical Approaches

A new test has recently been developed by Huang and Vieland [116] which takes into account the right truncation of the children's generation age at onset distribution and thus reduces the elevated type I error resulting from the use of traditional pairwise statistics [115]. The method computes the "conditional likelihood" of the data which takes the right truncation into consideration, uses an iterative method to find maximum likelihood estimates for means and variances of the AAD for parents and children, and then uses those estimates to test for anticipation. The method focuses on the AAD and age at interview (AAI) for each member of a parent–child pair (i.e., four observed age variables per pair) where the AAD for the parent and child of a given pair are x_1 and x_2 and the ages of interview are c_1 and c_2, respectively. The key to the conditional likelihood calculation involves finding the joint probability of the observed age values for each pair, $P(x_1, x_2, c_1, c_2)$ which is

$$P(X_1 \le x_1, X_2 \le x_2, C_1 \le c_1, C_2 \le c_2 \mid X_1 \le C_1, X_2 \le C_2),$$

where X_1, X_2, C_1, and C_2 represent all possible values of the AAD and AAI for parents and children, respectively, which are less than the observed values, and where $P(\mathrm{A} \mid \mathrm{B})$ is the probability of "A" given that "B" has occurred. In order to compute $P(x_1, x_2, c_1, c_2)$, it is necessary to assume that the AAD and AAI are independent and also to specify the underlying joint AAD distribution function, $f(x_1, x_2)$, and joint AAI distribution function $g(c_1, c_2)$. A reasonable assumption for the form of f is the bivariate normal (see below). However, the AAI distribution function, $g(c_1, c_2)$, is not generally known and may lead to errors if misspecified. An alternative is to compute the conditional probability, $P_c\,(x_1, x_2 \mid c_1, c_2)$, which is

$$P(x_1, x_2 \mid x_1 \le c_1, x_2 \le c_2, c_1, c_2).$$

The latter probability depends only on $f(x_1, x_2)$ and $F(c_1, c_2)$ where $F(c_1, c_2)$ is the integral of $f(c_1, c_2)$.

A univariate normal distribution is defined by two parameters, μ and σ, the mean and standard deviation. The equation for the ubiquitous "bell curve" is the probability density function for a univariate normal which can be labeled $f(x)$. If the values of $f(x)$ are summed together for increasing values of x (i.e., the bell curve is integrated), a sigmoid cumulative density function is produced which is denoted by $F(x)$. Similarly, a bivariate normal distribution (i.e., the joint distribution of two normal variables) is defined by five parameters, μ_1, μ_2, σ_1, σ_2, and ρ, the two means, two standard deviations, and correlation coefficient. It is assumed that $f(x_1, x_2)$ and $F(c_1, c_2)$ conform to a bivariate normal probability density and cumulative density function respectively. Given values of μ_1, μ_2, σ_1, σ_2, ρ, $f(x_1, x_2)$, and $F(c_1, c_2)$ and therefore $P_c(x_1, x_2 \mid c_1, c_2)$ may be calculated. Since $P_c(x_1, x_2 \mid c_1, c_2)$ is the probability of a single parent–child pair, the likelihood of all of the observed parent–child pairs is the product of each of the $P_c(x_1, x_2 \mid c_1, c_2)$ values. An iterative method is used to find estimates of μ_1, μ_2, σ_1, σ_2, and ρ which maximize the likelihood of the observed data. These maximum likelihood estimates can then be used to test for anticipation. For example, if the maximum likelihood estimate of (μ_1-μ_2) is significantly greater than zero, then the data support the presence of anticipation. Thus, this new method defines a conditional likelihood which takes into account the right truncation of the data, uses iterative likelihood methods to find estimates for means and variances of the AAD for parents and children, and then uses those adjusted estimates to test for anticipation.

Thus far, this approach has been tested on bipolar affective disorder families [116] ascertained for linkage analysis, which have previously been reported to demonstrate evidence for anticipation using simple paired statistics [117]. The results of this new test show that the evidence for anticipation is reduced from $p = 0.0001$ using simple paired statistics to a p value of 0.014 using the new test [116]. Simulations have shown that this new test has an empirical type I error rate of 5% when performed at a nominal value of $\alpha = 5\%$, even when samples as small as 25 pairs are used [116]. This test, if validated in other studies, may be particularly advantageous since it does not require prospective data.

A similar approach to the above test [116] has been suggested by Gorwood *et al.* [118, 120] but there is a potential bias with this formulation. For each child, an expected age of onset, Eaao, is computed which is equal to the average AAD of all of the parents whose observed AAD is less than the child's age at interview. The child's Eaao is subtracted from his or her observed AAD. This difference forms the basis for a paired t test: under the null hypothesis that if the parental and offspring generations have the same AAD distribution, the Eaao should be the same as the observed AAD. However, the latter assumption is only true if the parents and children are sampled from the whole population of those at risk. If the parent–child pairs are sampled because both members are affected, then the observed AAD distribution in the parental generation will be right truncated with respect to the population distribution. This will result in Eaao values which are biased "to the left" (i.e., toward younger values). Thus, for any given observed AAD for a child, the difference between observed AAD and the Eaao will be systematically greater than zero. This will tend to result in an elevated type I error.

C. Genomic Imprinting in Diseases Demonstrating Genetic Anticipation

For monogenic disorders which demonstrate anticipation the presence of genomic imprinting may add biological support to the observed anticipation. For example, in HD the degree of anticipation in paternal transmissions is on average 6–7 years, as opposed to 1 year in maternal transmissions [4]. It has been argued that ascertainment biases should influence maternal and paternal transmissions to the same degree and that an appreciable difference in the degree of anticipation depending on the sex of the disease-transmitting parent favors the idea of true anticipation [4]. However, for complex diseases it has been argued that the absence of genomic imprinting does not allow rejection of evidence for anticipation [112]. Various susceptibility loci may be differentially imprinted and mask parental effects. In this situation, it may only be possible to determine the presence of genomic imprinting in a proportion of families when genetic linkage or a disease gene is identified for such a fraction. This allows for the division of a genetically heterogeneous sample into homogenous groups on the basis of molecular data. Another complexity which may result in an apparent parent-of-origin effect is sex differences in the AAD distribution. Differences in the AAD distribution for males and females may produce apparent genomic imprinting when the data are separated into mother–offspring and father–offspring pairs. Division of the data depending on the sex of the disease-transmitting parent as well as offspring is necessary in such a situation. Despite the potential heuristic advantage of studying genomic imprinting in diseases with evidence for anticipation, parent-of-origin effects have generally not been investigated (Tables 28-1 and 28-2).

D. Results of Molecular Studies

The diversity of results from molecular studies searching for TREs is interesting. In spinocerebellar ataxia 7 it appears likely that the mutation is a TRE, whereas traditional mutations have been identified recently in some families with Parkinson's disease. However, in diseases such as facioscapulohumeral muscular dystrophy the interpretation of the anticipation and molecular results is unclear. In many disorders reported to demonstrate anticipation, such as spastic paraplegia, the presence of TRE remains to be examined. Repeat expansion detection (RED) [9] has been used as screening approach in an attempt to identify TRE mutations in a few disorders where there is evidence for anticipation including familial periodic cerebellar ataxia without myokymia [54], and familial total anomalous pulmonary venous return [88]. These studies have reported an absence of cosegregation of a large CAG RED product with disease. An exception to this is a study performed in testicular cancer [121]. DNA from testicular tumor cell lines was analyzed using the RED technique, and evidence for larger CAG tracts was observed. Furthermore, individuals from five testicular cancer families were screened using RED and in all families there was an increase in the average RED product [121]. In one particular family, CAG tracts estimated at greater than 204 bp were identified in three of five affected members. However, one affected individual from another arm of this pedigree did not demonstrate a similar RED product

Of interest is the recent finding by our group that the majority of individuals with (CAG)n RED products greater than or equal to 270 bp have large repeat tracts at a locus termed 7,6A, on chromosome 18 [122]. Large repeats at this locus initially appear to have no pathogenic implications [122]. It may be that this locus is responsible for many of the large RED products that have been observed in some studies.

It is of note that the correlation between the age at diagnosis and (CAG)n in Huntington's disease can account for only approximately half the observed variance in the age at diagnosis of the disorder [123]. This raises the question of whether other as yet unidentified genetic or environmental factors are involved in influencing the age at diagnosis in this disorder. A similar situation may apply to many other disorders for which evidence for anticipation has been described.

E. Repetitive DNA Sequences in Recessive Neurological Diseases

Recently the discovery of unstable DNA in two recessive neurological disorders has widened the possibility of types of diseases that may be due to unstable DNA. Recessive diseases are generally not suitable for studies of anticipation since there are very few families with affected individuals from different generations. Therefore the search for unstable DNA in such diseases initially did not appear to be relevant. However, the unexpected revelation of a GAA repeat expansion in the autosomal recessive disorder, Friedreich's ataxia [124], has shown that TRE mutations are not limited to dominant phenotypes but in some cases may result in recessive conditions. Furthermore, the recent discovery of a metastable minisatellite repeat in autosomal recessive progressive myoclonus epilepsy of Unverricht-Lundborg type (EPM1) has highlighted that other repetitive DNA sequences apart from TRE may be pathogenic. An expansion of a dodecamer (12-mer) repeat in the upstream region of the *CSTB* gene is responsible for the majority of cases of EPM1 [125–127]. A correlation of age at diagnosis with repeat sizes has not been investigated yet. Two CEPH pedigrees revealed preferential expansion of alleles from the paternal line [127]. This finding has led to the suggestion that other repeat expansions could be responsible in recessive diseases, particularly where conventional mutations have been identified only in a small proportion of cases [128]. It is interesting to note that the expanded repeat motif in Friedreich's ataxia is not GC-rich as is the case for other TRE diseases, and that the repeat motif in EPM1 is both GC-rich and more complex than a simple TRE.

IV. SUMMARY

We emphasize two main points. First, cancers, as well as neurological, developmental, and autoimmune diseases, are part of the wide spectrum of diseases which have been studied regarding the phenomena of genetic anticipation. In general, the conclusions that have been reached are that there is some evidence for anticipation in many of these diseases. A population-based approach has not been used for ascertainment in the vast majority of these, and thus the conclusions reached have to be viewed with caution. Regarding statistical approaches, it is clear that simple pairwise analysis in most cases is not appropriate because this tends to elevate the type I error [115]. Therefore, it has not been possible to demonstrate or refute anticipation unequivocally in any of these disorders, and this may arise from the number of potential ascertainment biases which might be present.

Our second point is that genetic anticipation may be a widespread phenomenon reaching far beyond unstable DNA diseases. Genetic or pathophysiological dissimilarities between diseases presented here and unstable DNA diseases do not exclude genetic anticipation a priori. Furthermore, the presence of stable mutations in diseases such as breast cancer, familial polyposis coli, Parkinson's disease, multiple endocrine neoplasia type 1, and autosomal dominant polycystic kidney disease does not necessarily allow for rejection of the possibility of anticipation. The mechanisms of genetic anticipation may not be limited to TRE and it is possible that there are other molecular causes of anticipation, although this is currently unclear [78]. Five years ago, in their editorial dedicated to anticipation in unstable DNA diseases, Sutherland and Richards wrote: "Clinical observations are important. The fact that they may be ideologically unpopular or that there is no known mechanism to explain them should not be used as reason to deny their validity" [129]. A similar idea has been expressed recently in the third edition of Vogel and Motulsky's "Human Genetics: Problems and Approaches" [130]. They state that anticipation in Huntington's disease and myotonic dystrophy previously appeared incompatible with Mendelism, and that anticipation was explained away using sophisticated, but unfounded statistical arguments. With hindsight, they suggest that this is an important philosophical lesson for human genetics. Such a lesson may have relevance to other non-Mendelian phenomena that we do not currently understand.

Acknowledgments

Thanks to the numerous colleagues with whom we have had interesting discussions relating to genetic anticipation. We thank Dr. Glen Sunohara for his helpful comments, and Richard Van Holst for proofreading. Also, thanks are due to Drs. J. Goldberg, B. Teh, Y. Y. Tsai, and J. Huang for kindly providing us with copies of manuscripts prior to publication. In particular we thank Dr. Jian Huang and Dr. Li Hsu for their productive discussions. We also thank the librarians at the Farrar Library, Clarke Institute of Psychiatry, for their assistance. A. D. P. and D. M. J. N. are recipients of fellowships from the Medical Research Council of Canada. J. B. V. is a recipient of a fellowship from the Medical Research Council/Schizophrenia Society of Canada. A. P. is supported by a grant from OMHF/OFOS and is an OMHF New Investigator. J. L. K. is a recipient of a NARSAD Independent Investigator award.

References

1. Morel (1857). "Traite des degenerescences de l'espece humaine." Bailliere, Paris.
2. McInnis, M. G. (1996). Anticipation: an old idea in new genes. *Am. J. Hum. Genet.* **59,** 973–979.
3. Penrose, L. S. (1948). The problem of anticipation in pedigrees of dystrophia myotonica. *Ann. Eugenics* **14,** 125–132.
4. Ridley, R. M., Frith, C. D., Crow, T. J., and Conneally, P. M. (1988). Anticipation in Huntington's disease is inherited through the male line but may originate in the female. *J. Med. Genet.* **25,** 589–595.
5. Höweler, C. J., Busch, H. F. M., Geraedts, J. P. M., Niermeijir, M. F., and Staal, A. (1989). Anticipation in myotonic dystrophy: fact or fiction? *Brain* **112,** 779–797.
6. Risch, N., and Merikangas, K. (1996). The future of genetic studies of complex human diseases. *Science* **273,** 1516–1517.
7. Risch, N., and Merikangas, K. (1997). Genetic analysis of complex diseases. *Science* **275,** 1329–1330.
8. Li, S. H., McInnis, M. G., Margolis, R. L., Antonarakis, S. E., and Ross, C. A. (1993). Novel triplet repeat containing genes in human brain: cloning, expression, and length polymorphisms. *Genomics* **16,** 572–579.
9. Schalling, M., Hudson, T. J., Buetow, K. H., and Housman, D. E. (1993). Direct detection of novel expanded trinucleotide repeats in the human genome. *Nature Genet.* **4,** 135–139.
10. Sanpei, K., Igarashi, S., Eguchi, I., Takiyama, Y., Tanaka, H., and Tsuji, S. (1995). Direct detection of expanded (CAG/CTG) repeats in the myotonin-protein kinase genes of myotonic dystrophy patients using a high-stringency hybridization method. *Biochem. Biophys. Res. Commun.* **212,** 341–346.
11. Petronis, A., Heng, H. H. Q., Tatuch, Y., Shi, X-M., Tsui, L-C., Ashizawa, T., Surh, L. C., Holden, J. J. A., and Kennedy, J. L. (1996). Direct detection of expanded trinucleotide repeats using DNA hybridization techniques. *Am. J. Med. Genet.* **67,** 85–91.
12. Haaf, T., Sirugo, G., Kidd, K. K., and Ward, D. C. (1996). Chromosomal localization of long trinucleotide repeats in the human genome by fluorescence in situ hybridization. *Nature Genet.* **12,** 183–185.
13. Sanpei, K., Takano, H., Igarashi, S., Sato, T., Oyake, M., Sasaki, H., Wakisaka, A., Tashiro, K., Ishida, Y., Ikeuchi, T., Koide, R., Saito, M., Sato, A., Tanaka, T., Hanyu, S., Takiyama, Y., Nishizawa, M., Shimizu, N., Nomura, Y., Segawa, M., Iwabuchi, K., Eguchi, I., Tanaka, H., Takahashi, H., and Tsuji, S. (1996). Identification of the spinocerebellar ataxia type 2 gene using a direct identification of repeat expansion and cloning technique, DIRECT. *Nature Genet.* **14,** 277–284.
14. Ashizawa, T., Dunne, C. J., Dubel, J. R., Perryman, M. B., Epstein, H. F., Boerwinkle, E., and Hejtmancik, J. F. (1992). Anticipation in myotonic dystrophy: I. Statistical verification based on clinical and hapolotype findings. *Neurology* **42,** 1871–1877.
15. Petronis, A., and Kennedy, J. L. (1995). Unstable genes-unstable mind? *Am. J. Psych.* **152,** 164–172.
16. Devys, D., Biancalana, V., Rousseau, F., Boue, J., Mandel, J. L., and Oberle, I. (1992). Analysis of full fragile X mutations in fetal tissues and monzygotic twins indicated that abnormal methylation and somatic heterogeneity are established early in development. *Am. J. Med. Genet.* **43,** 208–216.
17. Thornton, C. A., Johnson, K., and Moxley 3rd, R. T. (1994). Myotonic dystrophy patients have larger CTG expansions in skeletal muscle than in leukocytes. *Ann. Neurol.* **35,** 104–107.
18. Telenius, H., Kremer, B., Goldberg, Y. P., Theilmann, J., Andrew, S. E., Zeisler, J., Adam, S., Greenberg, C., Ives, E. J., Clarke, L. A., and Hayden, M. R. (1994). Somatic and gonadal mosaicism of the Huntington disease gene CAG repeat in brain and sperm. *Nature Genet.* **6,** 409–414.
19. Online Mendelian Inheritance in Man (OMIM). Center for Medical Genetics, Johns Hopkins University (Baltimore, MD) and National Center for Biotechnology Information, National Library of Medicine (Bethesda, MD), 1996. World Wide Web URL: http://www3.ncbi.nlm.nih.gov/omim/
20. Enevoldson, T. P., Sanders, M. D., and Harding, A. E. (1994). Autosomal-dominant cerebellar-ataxia with pigmentary macular dystrophy a clinical and genetic-study of 8 families. *Brain* **117,** 445–460.
21. Benomar, A., Krols, L., Stevanin, G., Cancel, G., LeGuern, E., David, G., Ouhabi, H., Martin, J. J., Durr, A., Zaim, A., Ravise, N., Busque, C., Penet, C., Van Regemorter, N., Weissenbach, J., Yahyaoui, M., Chkili, T., Agid, Y., Van Broeckhoven, C., and Brice. A, (1995). The gene for autosomal dominant cerebellar ataxia with pigmentary macular dystrophy maps to chromosome 3p12-p21.1. *Nature Genet.* **10,** 84–88.
22. Gouw, L. G., Kaplan, C. D., Haines, J. H., Digre, K. B., Rutledge, S. L., Matilla, A., Leppert, M., Zoghbi, H. Y., and Ptacek, L. J. (1995). Retinal degeneration characterizes a spinocerebellar ataxia mapping to chromosome 3p. *Nature Genet.* **10,** 89–93.
23. Holmberg, M., Johansson, J., Forsgren, L., Heijbel, J., Sandgren, O., and Holmgren, G. (1995). Localization of autosomal dominant cerebellar ataxia associated with retinal degeneration and

anticipation to chromosome 3p12-p21.1. *Hum. Mol. Genet.* **4,** 1441–1445.

24. Jobsis, G. J., Weber, J. W., Barth, P. G., Keizers, H., Baas, F., van Schooneveld, M. J., van Hilten, J. J., Troost, D., Geesink, H. H., and Bolhuis, P. A. (1997). Autosomal dominant cerebellar ataxia with retinal degeneration (ADCA II): clinical and neuropathological findings in two pedigrees and genetic linkage to 3p12-p21.1. *J. Neurol. Neurosurg. Psych.* **62,** 367–371.
25. David, G., Giunti, P., Abbas, N., Coullin, P., Stevanin, G., Horta, W., Gemmill, R., Weissenbach, J., Wood, N., Cunha, S., Drabkin, H., Harding, A. E., Agid, Y., and Brice, A. (1996). The gene for autosomal dominant cerebellar ataxia type II is located in a 5-cM region in 3p12-p13: genetic and physical mapping of the SCA7 locus. *Am. J. Hum. Genet.* **59,** 1328–1336.
26. Lindblad, K., Savontaus, M. L., Stevanin, G., Holmberg, M., Digre, K., Zander, C., Ehrsson, H., David, G., Benomar, A., Nikoskelainen, E., Trottier, Y., Holmgren, G., Ptacek, L. J., Anttinen, A., Brice, A., and Schalling, M. (1996). An expanded CAG repeat sequence in spinocerebellar ataxia type 7. *Genome Res.* **6,** 965–971.
27. Bruyn, R. P., van Deutekom, J., Frants, R. R., and Padberg, G. W. (1993). Hereditary spastic paraparesis. Clinical and genetic data from a large Dutch family. *Clin. Neurol. Neurosurg.* **95,** 125–129.
28. Dürr, A., Brice, A., Serdaru, M., Rancurel, G., Derouesne, C., Lyon-Caen, O., Agid, Y., and Fontaine, B. (1994). The phenotype of pure autosomal-dominant spastic paraplegia. *Neurology* **44,** 1274–1277.
29. Gispert, S., Santos, N., Damen, R., Voit, T., Schulz, J., Klockgether, T., Orozco, G., Kreuz, F., Weissenbach, J., and Auburger, G. (1995). Autosomal dominant familial spastic paraplegia: reduction of the FSP1 candidate region on chromosome 14q to 7 cM and locus heterogeneity. *Am. J. Hum. Genet.* **56,** 183–187.
30. Raskind, W. H., Pericak-Vance, M. A., Lennon, F., Wolff, J., Lipe, H. P., and Bird, T. D. (1997). Familial spastic paraparesis: Evaluation of locus heterogeneity, anticipation, and haplotype mapping of the SPG4 locus on the short arm of chromosome 2. *Am. J. Med. Genet.* **74,** 26–36.
31. Burger, J., Metzke, H., Paternotte, C., Schilling, F., Hazan, J., and Reis, A. (1996). Autosomal dominant spastic paraplegia with anticipation maps to a 4-cM interval on chromosome 2p21-p24 in a large German family. *Hum. Genet.* **98,** 371–375.
32. Dürr, A., Davoine, C. S., Paternotte, C., von Fellenberg, J., Cogilnicean, S., Coutinho, P., Lamy, C., Bourgeois, S., Prud-homme, J. F., Penet, C., Mas, J. L., Burgunder, J. M., Hazan, J., Weissenbach, J., Brice, A., and Fontaine, B. (1996). Phenotype of autosomal dominant spastic paraplegia linked to chromosome 2. *Brain* **119,** 1487–1496.
33. Webb, S., Flanagan, N., Callaghan, N., and Hutchinson, M. (1997). A family with hereditary spastic paraparesis and epilepsy. *Epilepsia* **38,** 495–499.
34. Hodge, S. E., and Wickrameratne, P. (1995). Statistical pitfalls in detecting age-of-onset anticipation: the role of correlation in studying anticipation and detecting ascertainment bias. *Psych. Genet.* **5,** 43–47.
35. Zatz, M., Marie, S. K., Passos-Bueno, M. R., Vainzof, M., Campiotto, S., Cerqueira, A., Wijmenga, C., Padberg, G., and Frants, R. (1995). High proportion of new mutations and possible anticipation in Brazilian facioscapulohumeral muscular dystrophy. *Am. J. Hum. Genet.* **56,** 99–105.
36. Goto, K., Lee, J. H., Matsuda, C., Hirabayashi, K., Kojo, T., Nakamura, A., Mitsunaga, Y., Furukawa, T., Sahashi, K., and Arahata, K. (1995). DNA rearrangements in Japanese facioscapulohumeral muscular-dystrophy patients—clinical correlations. *Neuromusc. Disorders* **5,** 201–208.
37. Tawil, R., Forrester, J., Griggs, R. C., Mendell, J., Kissel, J., McDermott, M., King, W., Weiffenbach, B., Figlewicz, D., and The FSH-DY Group (1996). Evidence for anticipation and association of deletion size with severity in facioscapulohumeral muscular dystrophy. *Ann. Neurol.* **39,** 744–748.
38. Lunt, P. W., Jardine, P. E., Koch, M. C., Maynard, J., Osborn, M., Williams, M., Harper, P. S., and Upadhyaya, M. (1995). Correlation between fragment size at D4F104S1 and age at onset or at wheelchair use, with a possible generational effect, accounts for much phenotypic variation in 4q35-facioscapulohumeral muscular dystrophy (FSHD). *Hum. Mol. Genet.* **4,** 951–958.
39. van Deutekom, J. C. T., Lemmers, R. J. L. F., Grewal, P. K., van Geel, M., Romberg, S., Dauwerse, H. G., Wright, T. J., Padberg, G. W., Hofker, M. H., Hewitt, J. E., and Frants, R. R. (1996). Identification of the first gene (*FRG1*) from the FSHD region on human chromosome 4q35. *Hum. Mol. Genet.* **5,** 581–590.
40. van Deutekom, J. C. T., Bakker, E., Lemmers, R. J. L. F., van der Wielen, M. J. R., Bik, E., Hofker, M. H., Padberg, G. W., and Frants, R. R. (1996). Evidence for subtelomeric exchange of 3.3 kb tandemly repeated units between chromosomes 4q35 and 10q26: implications for genetic counselling and etiology of FSHD1. *Hum. Mol. Genet.* **5,** 1997–2003.
41. Payami, H., Bernard, S., Larsen, K., Kaye, J., and Nutt, J. (1995). Genetic anticipation in Parkinson's disease. *Neurology* **45,** 135–138.
42. Golbe, L. I., DiIorio, G., Sanges, G., Lazzarini, A. M., LaSala, S., Bonavita, V., and Duvoisin, R. C. (1996) Clinical genetic analysis of Parkinson's disease in the Contursi kindred. *Ann. Neurol.* **40,** 767–775.
43. Carero-Valenzuela, R., Lindblad, K., Payami, H., Johnson, W., Schalling, M., Stenroos, E. S., Shattuc, S., Nutt, J., Brice, A., and Litt, M. (1995). No evidence for association of familial Parkinson's disease with CAG repeat expansion. *Neurology* **45,** 1760–1763.
44. Bonifati, V., Fabrizio, E., Vanacore, N., Demari, M., and Meco, G. (1995). Familial Parkinson's-disease—a clinical genetic-analysis. *Can. J. Neurol. Sci.* **22,** 272–279.
45. Plante-Bordeneuve, V., Taussig, D., Thomas, F., Ziegler, M., and Said, G. (1995). A clinical and genetic study of familial cases of Parkinson's disease. *J. Neurol. Sci.* **133,** 164–172.
46. Markopoulou, K., Wszolek, Z. K., and Pfeiffer, R. F. (1995). A Greek-American kindred with autosomal dominant, levodopa-responsive Parkinsonism and anticipation. *Ann. Neurol.* **38,** 373–378.
47. Morrisson, P. J., Godwin-Austen, R. B., and Raeburn, J. A. (1996). Familial autosomal dominant dopa responsive Parkinson's disease in three living generations showing extreme anticipation and childhood onset. *J. Med. Genet.* **33,** 504–506.
48. Maraganore, D. M., Schaid, D. J., Rocca, W. A., and Harding, A. E. (1996). Anticipation in familial Parkinson's disease: a reanalysis of 13 United Kingdom kindreds. *Neurology* **47,** 1512–1517.
49. Polymeropoulos, M. P., Lavedan, C., Leroy, E., Ide, S. E., Dehejia, A., Dutra, A., Pike, B., Root, H., Rubenstein, J., Boyer, R., Stenroos, E. S., Chandrasekharappa, S., Athanassiadou, A., Papapetropoulos, T., Johnson, W. G., Lazzarini, A. M., Duvoisin, R. C., Di Iorio, G., Globe, L. I., and Nussbaum, R. L. (1997). Mutation in the α-synclein gene indetified in families with Parkinson's disease. *Science* **276,** 2045–2047.
50. Melberg, A., Arnell, H., Dahl, N., Stalberg, E., Raininko, R., Oldfors, A., Bakall, B., Lundberg, P. O., and Holme, E. (1996).

Anticipation of autosomal dominant progressive external ophthalmoplegia with hypogonadism. *Muscle Nerve* **19,** 1561–1569.

51. Ishikawa, K., Mizusawa, H., Saito, M., Tanaka, H., Nakajima, N., Kondo, N., Kanazawa, I., Shoji, S., and Tsuji, S. (1996). Autosomal dominant pure cerebellar ataxia. A clinical and genetic analysis of eight Japanese families. *Brain* **119,** 1173–1182.
52. Slavotinek, A. M., Pike, M., Mills, K., and Hurst, J. A. (1996). Cataracts, motor system disorder, short stature, learning difficulties, and skeletal abnormalities: a new syndrome? *Am. J. Med. Genet.* **62,** 42–47.
53. Piqueras, J. F., Santos, J., Visedo, G., Perez de Castro, I., Puertollano, R., Montejo, J., Ramo Tello, C., and Valle, J. (1995). Familial cosegregation of manic-depressive illness and a form of hereditary cerebellar ataxia. *Am. J. Med. Genet.-Neuropsych. Genet.* **60,** 206–209.
54. Teh, B. T., Silburn, P., Lindblad, K., Betz, R., Boyle, R., Schalling, M., and Larsson, C. (1995). Familial periodic cerebellar ataxia without myokymia maps to a 19-cM region on 19p13. *Am. J. Hum. Genet.* **56,** 1443–1449.
55. Dhir, S. P., Mohan, K., Prakash, S., and Jain, I. S. (1983). Anticipation in hereditary open angle glaucoma. *Ophthalm. Paediatr. Genet.* **2,** 109–111.
56. Labuda, M. C., Fletcher, N. A., Korczyn, A. D., Inzelberg, R., Harding, A. E., and Pauls, D. L. (1993). Genomic imprinting and anticipation in idiopathic torsion dystonia. *Neurology* **43,** 2040–2043.
57. Cheng, J. T., Liu, A., Wasmuth, J., Liu, B. P., and Truong, D. (1996). Clinical evidence of genetic anticipation in adult-onset idiopathic dystonia. *Neurology* **47,** 215–219.
58. Morrison, A. W. (1995). Anticipation in Meniere's disease. *J. Laryngol. Otol.* **109,** 499–502.
59. Blumen, S. C., Nisipeanu, P., Sadeh, M., Asherov, A., Tome, F. M., and Korczyn, A. D. (1993). Clinical features of oculopharyngeal muscular dystrophy among Bukhara Jews. *Neuromusc. Disorders* **3,** 575–577.
60. Dai, Z. H. (1991). Olivopontocerebellar atrophy: clinical analysis of 100 cases Chinese. *Chin. J. Neurol. Psych.* **24,** 111–113.
61. Morrison, A. W. (1967). Genetic factors in otosclerosis. *Ann. R. Coll. Surgeons Eng.* **41,** 202–237.
62. Trenkwalder, C., Seidel, V. C., Gasser, T., and Oertel, W. H. (1996). Clinical symptoms and possible anticipation in a large kindred of familial restless legs syndrome. *Movement Disorders* **11,** 389–394.
63. Scheffer, I. E., Jones, L., Pozzebon, M., Howell, R. A., Saling, M. M., and Berkovic, S. F. (1995). Autosomal dominant rolandic epilepsy and speech dyspraxia: a new syndrome with anticipation. *Ann. Neurol.* **38,** 633–642.
64. Isozumi, K., DeLong, R., Kaplan, J., Deng, H. X., Iqbal, Z., Hung, W. Y., Wilhelmsen, K. C., Hentati, A., Pericak-Vance, M. A., and Siddique, T. (1996). Linkage of scapuloperoneal spinal muscular atrophy to chromosome 12q24.1-q24.31. *Hum. Mol. Genet.* **5,** 1377–1382.
65. Ranum, L. P., Schut, L. J., Lundgren, J. K., Orr, H. T., and Livingston, D. M. (1994). Spinocerebellar ataxia type 5 in a family descended from the grandparents of President Lincoln maps to chromosome 11. *Nature Genet.* **8,** 280–284.
66. Bromberg, J. E., Rinkel, G. J., Algra, A., van Duyn, C. M., Greebe, P., Ramos, L. M., and van Gijn, J. (1995). Familial subarachnoid hemorrhage: distinctive features and patterns of inheritance. *Ann. Neurol.* **38,** 929–934.
67. Wesolowski, D. P., Ellwood, R. A., Schwab, R. E., and Farah, J. (1981). Hippel-Lindau syndrome in identical twins. *Br. J. Radiol.* **54,** 982–986.
68. Palmer, S. E., Stephens, K., and Dale, D. C. (1996). Genetic anticipation in autosomal dominant cyclical hematopoesis. *Am. J. Hum. Genet.* **59S,** A37.
69. Fick, G. M., Johnson, A. M., and Gabow, P. A. (1994). Is there evidence for anticipation in autosomal-dominant polycystic kidney disease? *Kidney Int.* **45,** 1153–1162.
70. Gerberth, S., Ritz, E., Zeier, M., and Stier, E. (1995). Anticipation of age at renal death in autosomal dominant polycystic kidney disease (ADPKD)? *Nephrol. Dialysis Transplant.* **10,** 1603–1606.
71. Simon, P., Le Goff, J. Y., Ang, K. S., Charassse, C., Lecacheux, P., and Cam, G. (1996). Epidemiologic data, clinical-features and prognosis of adult polycystic kidney-disease in a French area. *Nephrologie* **17,** 123–130.
72. Raphael, S. A., Blau, E. B., Zhang, W. H., and Hsu, S. H. (1993). Analysis of a large kindred with Blau syndrome for HLA, autoimmunity, and sarcoidosis. *Am. J. Dis. Children* **147,** 842–848.
73. Jacobsen, O. (1946). "Heredity in Breast Cancer." Nyt Nordisk Forlag, Arnold Busck, Copenhagen.
74. Smithers, D. W. (1948). Family histories of 459 patients with cancer of the breast. *Br. J. Cancer* **2,** 163–167.
75. Morse, P. D. (1951). The hereditary aspect of breast cancer in mother and daughter. *Cancer* **4,** 745–748.
76. Haagensen (1956). "Diseases of the Breast," p. 335. W.B. Saunders, p. 335.
77. Bucalossi, P., and Veronesi, U. (1957). Some observations on cancer of the breast in mothers and daughters. *Br. J. Cancer* **11,** 337–347.
78. Paterson, A. D., Kennedy, J. L., and Petronis, A. (1996). Evidence for genetic anticipation in non-Mendelian diseases. *Am. J. Hum. Genet.* **59,** 264–268.
79. Vehmanen, P., Friedman, L. S., Eerola, H., Sarantaus, L., Pyrhonen, S., Ponder, B. A. J., Muhonen, T., and Nevenlinna, H. (1997). A low proportion of BRCA2 mutations in Finnish breast cancer families. *Am. J. Hum. Genet.* **60,** 1050–1058.
80. Thorlacius, S., Sigurdsson, S., Bjarnadottir, H., Olafsdottir, G., Jonasson, J. G., Tryggvadottir, L., Tulinius, H., and Eyfjord, J. E. (1997). Study of a single BRCA2 mutation with high carrier frequency in a small population. *Am. J. Hum. Genet.* **60,** 1079–1084.
81. Polito, J. M., Rees, R. C., Childs, B., Mendeloff, A. I., Harris, M. L., and Bayless, T. M. (1996). Preliminary evidence for genetic anticipation in Crohn's disease. *Lancet* **347,** 798–800.
82. Satsangi, J., Grootscholten, C., Holt, H., and Jewell, D. P. (1996). Clinical patterns of familial inflammatory bowel disease. *Gut* **38,** 738–741.
83. Veale, A. M. O. (1965). "Intestinal Polyposis." Eugenics Laboratory Memoirs XL, Cambridge University Press, Cambridge.
84. Presciuttini, S., Veresco, L., Sala, P., Gismondi, V., Rossetti, C., Bafico, A., Ferrara, G. B., and Bertario, L. (1994). Age of onset in familial adenomatous polyposis: heterogeneity with families and among APC mutations. *Ann. Hum. Genet.* **58,** 331–342.
85. Drugge, U., Andersson, R., Chizari, F., Danielsson, M., Holmgren, G., Sandgren, O., and Sousa, A. (1993). Familial amyloidotic polyneuropathy in Sweden: a pedigree analysis. *J. Med. Genet.* **30,** 388–392.
86. Tashima, K., Ando, Y., Tanaka, Y., Uchino, M., and Ando, M. (1995). Change in the age of onset in patients with familial amyloidotic polyneuropathy type I. *Int. Med.* **34,** 748–750.
87. Loyd, J. E., Butler, M. G., Foroud, T. M., Conneally, P. M., Phillips 3rd, J. A., and Newman, J. H. (1995). Genetic anticipation and abnormal gender ratio at birth in familial primary pulmonary hypertension. *Am. J. Respir. Crit. Care Med.* **152,** 93–97.

88. Bleyl, S., Nelson, L., Odelberg, S. J., Ruttenberg, H. D., Otterud, B., Leppert, M., and Ward, K. (1995). A gene for familial total anomalous pulmonary venous return maps to chromosome 4p13-q12. *Am. J. Hum. Genet.* **56,** 408–415.
89. Vasen, H. F., Taal, B. G., Griffieon, G., Nagengast, F. M., Cats, A., Menko, F. H., Oskam, W., Kleibeuker, J. H., Offerhaus, G. J. A., and Meera Khan, P. (1994). Clinical heterogeneity of familial colorectal cancer and its influence on screening protocols. *Gut* **35,** 1262–1266.
90. Rodriguez-Bigas, M. A., Lee, P. H., O'Malley, L., Weber, T. K., Suh, O., Anderson, G. R., and Petrelli, N. J. (1996). Establishment of a hereditary non-polyposis colorectal cancer registry. *Dis. Colon Rectum* **39,** 649–653.
91. Tsai, Y. Y., Petersen, G. M., Booker, S. V., Bacon, J. A., Hamilton, S. R., and Giardiello, F. M. (1997). Evidence against genetic anticipation in familial colorectal cancer. *Genet. Epid.* **14,** 435–446.
92. Newbury-Ecob, R. A., Leanage, R., Raeburn, J. A., and Young, I. D (1996). Holt-Oram syndrome: a clinical genetic study. *J. Med. Genet.* **33,** 300–307.
93. Gregor, P., and Cerny, M. (1992). A genetic study in hypertrophic cardiomyopathies (Czech population). *Cor et Vasa* **34,** 218–226.
94. Horwitz, M., Goode, E. L., and Jarvik, G. P. (1996). Anticipation in familial leukemia. *Am. J. Hum. Genet.* **59,** 990–998.
95. Martijn, H., Oldhoff, J., Oosterhuis, J. W., Schraffordt Koops, H., and Vermey, A. (1981). Familial malignant melanomas Dutch. *Nederlands Tijdschrift voor Geneeskunde* **125,** 1194–1198.
96. Goldstein, A. M., Fraser, M. C., Clark, W. H., Jr., and Tucker, M. A. (1994). Age at diagnosis and transmission of invasive melanoma in 23 families with cutaneous malignant melanoma/dysplastic nevi. *J. Natl. Cancer Inst.* **86,** 1385–1390.
97. Goldstein, A. M., Clark, W. H., Jr., Fraser, M. C., and Tucker, M. A. (1996). Apparent anticipation in familial melanoma. *Melanoma Res.* **6,** 441–446.
98. Giraud, S., Choplin, H., Teh, B. T., Lespinasse, J., Jouvet, A., Labat-Moleur, F., Lenoir, G., Hamon, B., Hamon, P., and Calender, A. (1997). A large multiple endocrine neoplasia type 1 family with clinical expression suggestive of anticipation. *J. Clin. Endocrinol. Metab.* **82,** 3487–3492.
99. Hanssen, A. M., Werquin, H., Suys, E., and Fryns, J. P. (1993). Cowden syndrome: report of a large family with macrocephaly and increased severity of signs in subsequent generations. *Clin. Genet.* **44,** 281–286.
100. Plon, S. E. (1997). Anticipation in pediatric malignancies. *Am. J. Hum. Genet.* **60,** 1256–1257.
101. Piver, M. S., Goldberg, J. M., Tsukada, Y., Mettlin, C. J., Jishi, M. F., and Natarajan, N. (1995). Characteristics of familial ovarian cancer: a report of the first 1,000 families in the Gilda Radner Ovarian Cancer Registry. *Eur. J. Gynaecol. Oncol.* **18,** 169–176.
102. Goldberg, J. M., Piver, M. S., Jishi, M. F., and Blumenson, L. (1997). Age at onset of ovarian cancer in women with a strong family history of ovarian cancer. *Gynaecol. Oncol.* **66,** 3–9.
103. Baysal, B. E., Farr, J. E., Rubinstein, W. S., Galus, R. A., Johnson, K. A., Aston, C. E., Myers, E. N., Johnson, J. T., Carrau, R., Kirkpatrick, S. J., Myssiorek, D., Singh, D., Saha, S., Gollin, S. M., Evans, G. A., James, M. R., and Richard 3rd, C. W. (1997). Fine mapping of an imprinted gene for familial nonchromaffin paragangliomas, on chromosome 11q23. *Am. J. Hum. Genet.* **60,** 121–132.
104. Brind, A. M., Bray, G. P., Portmann, B. C., and Williams, R. (1995). Prevalence and pattern of familial disease in primary biliary cirrhosis. *Gut* **36,** 615–617.
105. Theeuwes, M., and Morhenn, V. (1995). Allelic instability in the mitosis model and the inheritance of psoriasis. *J. Am. Acad. Dermatol.* **32,** 44–52.
106. Deighton, C., Heslop, P., McDonagh, J., Walker, D., and Thomson, G. (1994). Does genetic anticipation occur in familial rheumatoid arthritis? *Ann. Rheum. Dis.* **53,** 833–835.
107. McDermott, E., Khan, M. A., and Deighton, C. (1996). Further evidence for genetic anticipation in familial rheumatoid arthritis. *Ann. Rheum. Dis.* **55,** 475–477.
108. Raghavan, D., Jelihovsky, T., and Fox, R. M. (1980). Father-son testicular malignancy. Does genetic anticipation occur? *Cancer* **45,** 1005–1009.
109. Deickmann, K. P., Becker, T., Jonas, D., and Bauer, H. W. (1987). Inheritance and testicular cancer. *Oncology* **44,** 367–377.
110. Heimdal, K., Olsson, H., Tretli, S., Flodgren, P., Borresen, A. L., and Fossa, S. D. (1996). Familial testicular cancer in Norway and southern Sweden. *Br. J. Cancer* **73,** 964–969.
111. Pierpont, J. W., St. Jacques, D., Seaver, L. H., and Erickson, R. P. (1995). A family with unusual Waardenburg syndrome type I (WSI), cleft lip (palate), and Hirschsprung disease is not linked to PAX 3. *Clin. Genet.* **47,** 139–143.
112. Petronis, A., Sherrington, R. P., Paterson, A. D., and Kennedy, J. L. (1995). Genetic anticipation in schizophrenia: pro and con. *Clin. Neurosci.* **3,** 76–80.
113. King, M. C. (1997). Leaving Kansas . . . finding genes in 1997. *Nature Genet.* **15,** 8–10.
114. Palmer, S. E., Stephens, K., and Dale, D. C. (1996). Genetics, phenotype and natural history of autosomal dominant cyclical hematopoesis. *Am. J. Med. Genet.* **66,** 413–422.
115. Heiman, G. A., Hodge, S. E., Wickrameratne, P., and Hsu, H. (1996). Age-at-interview bias in anticipation studies: computer simulations and an example with panic disorder. *Psych. Genet.* **6,** 61–66.
116. Huang, J., and Vieland, V. (1997). A new statistical test for age-of-onset anticipation: application to bipolar disorder. *Genet. Epid.* **14,** 1091–1096.
117. McInnis, M. G., McMahon, F. J., Chase, G. A., Simpson, S. G., Ross, C. A., and DePaulo, J. R. (1993). Anticipation in bipolar affective disorder. *Am. J. Hum. Genet.* **53,** 385–390.
118. Gorwood, P., Leboyer, M., Falissard, B., Jay, M., Rouillon, F., and Feingold, J. (1996). Anticipation in schizophrenia: new light on a controversial problem. *Am. J. Psych.* **153,** 1173–1177.
119. Paterson, A. D., Naimark, D. M. J., Petronis, A., and Kennedy, J. L. (1997). Survival analysis does not reliably detect genetic anticipation. *Am. J. Hum. Genet.* **61S,** A208.
120. Gorwood, P. (1997). Anticipation in schizophrenia: new light on a controversial problem. *Am. J. Psych.* **154,** 590. [Corrections]
121. King, B. L., Peng, H. Q., Goss, P., Huan, S., Bronson, D., Kacinski, B. M., and Hogg, D. (1997). Repeat expansion detection analysis of (CAG)n tracts in tumor cell lines, testicular tumors, and testicular cancer families. *Cancer Res.* **57,** 209–214.
122. Breschel, T. S., McInnis, M. G., Margolis, R. L., Sirugo, G., Coreliussen, B., Simpson, S. G., McMahon, F. J., MacKinnon, D. F., Xu, J. F., Pleasant, N., Huo, Y., Ashworth, R. G., Grundstrom, C., Grundstrom, T., Kidd, K. K., DePaulo, J. R., and Ross, C. A. (1997). A novel, heritable, expanding CTG repeat in an intron of the SEF2–1 gene on chromosome 18q21.1. *Hum. Mol. Genet.* **6,** 1855–1863.
123. Albin, R. L., and Tagle, D. A. (1995). Genetics and molecular biology of Huntington's disease. *Trends Neurosci.* **18,** 11–14.
124. Campuzano. V., Montermini, L., Molto, M. D., Pianese, L., Cossee, M., Cavalcanti, F., Monros, E., Rodius, F., Duclos, F., Monticelli, A., Zara, F., Canizares, J., Koutnikova, H., Bidichandani, S. I., Gellera, C., Brice, A., Trouillas, P., De Michele, G., Filla, A., De Frutos, R., Palau, F., Patel, P. I., Di Donato, S., Mandel, J. L., Cocozza, S., Koenig, M., and Pandolfo, M. (1996).

Friedreich's ataxia: autosomal recessive disease caused by an intronic GAA triplet repeat expansion. *Science* **271,** 1423–1427.

125. Lafreniére, R. G., Rochefort, D. L., Chretien, N., Rommens, J. M., Cochius, J. I., Kalviainen, R., Nousiainen, U., Patry, G., Farrell, K., Soderfeldt, B., Federico, A., Hale, B. R., Cossio, O. H., Sorensen, T., Pouliot, M. A., Kmiec, T., Uldall, P., Janszky, J., Pranzatelli, M. R., Andermann, F., Andermann, E., and Rouleau, G. A. (1997). Unstable insertion in the 5′ flanking region of the cystatin B gene is the most common mutation in progressive myoclonus epilepsy type 1, EPM1. *Nature Genet.* **15,** 298–302.

126. Virtaneva, K., D'Amato, E., Miao, I. M., Koskiniemi, M., Norio, R., Avanzini, G., Franceschetti, S., Michelucci, R., Tassinari, C. A., Omer, S., Pennacchio, L. A., Myers, R. M., Dieguez-Lucena, J. L., Krahe, R., de la Chapelle, A., and Lehesjoki, A. E. (1997). Unstable minisatellite expansion causing recessively inherited myoclonus epilepsy, EPM1. *Nature Genet.* **15,** 393–396.

127. Lalioti, M. D., Scott, H. S., Buresi, C., Rossier, C., Bottani, A., Morris, M. A., Malafosse, A., and Antonarakis, S. E. (1997). Dodecamer repeat expansion in cystatin B gene in progressive myoclonus epilepsy. *Nature* **386,** 847–851.

128. Mandel, J. L. (1997). Human genetics - Breaking the rule of three. *Nature* **386,** 767–769.

129. Sutherland, G. R., and Richards, R. I. (1992). Anticipation legitimized: unstable DNA to the rescue. *Am. J. Hum. Genet.* **51,** 7–9.

130. Vogel, F., and Motulsky, A. G. (1997). "Human Genetics: Problems and Approaches," 3rd edition, pp. 143–144. Springer, Berlin.

PART XII

Approaches to Detect Unstable Trinucleotide Repeat Loci

CHAPTER 29

Detection of Unstable Trinucleotide Repeat Loci: Genome and cDNA Screening

RUSSELL L. MARGOLIS Laboratories of Molecular Neurobiology and Genetic Neurobiology, Department of Psychiatry, Johns Hopkins University School of Medicine, Baltimore, Maryland 21287

CHRISTOPHER A. ROSS Laboratories of Molecular Neurobiology and Genetic Neurobiology, Department of Psychiatry, Department of Neuroscience, and Program in Cellular and Molecular Medicine, Johns Hopkins University School of Medicine, Baltimore, Maryland 21287

I. INTRODUCTION

Since 1991, trinucleotide repeat expansion has been demonstrated to cause 12 diseases, and there is evidence that this list will grow longer [1–3]. It is useful to classify these disorders according to the location of the repeat in the relevant gene. The expanding repeats in Huntington's disease [4], spinal and bulbar muscular atrophy [5], dentato-rubral and pallido-luysian atrophy [6, 7], and spinocerebellar ataxia types 1 [8], 2 [9–11], 3 (Machado–Joseph disease, MJD) [12], and 6 [13] consist of CAG triplets, which are transcribed and translated into glutamine. Four or five other diseases are caused by expansions of repeats that are transcribed but not translated: expansions of CCG or GGC repeats in 5′ untranslated regions cause the A [14–16], E [17], and perhaps F [18] forms of the fragile X syndrome and some cases of Jacobsen's syndrome [19], and a 3′

untranslated CTG repeat expands to cause myotonic dystrophy [20–22]. Finally, one disease, Friedreich's ataxia, is caused by expansion of an untranscribed GAA repeat located in an intron [23].

The search for additional trinucleotide repeats that may cause disease has led to a number of novel methods of expansion detection discussed elsewhere in this volume, including RED (repeat expansion detection) [24], DIRECT (direct identification of repeat expansion and cloning technique) [11], and detection of proteins with expanded glutamines repeats [2]. However, a widely used method for detecting candidate repeats remains screening of DNA clones with primers or probes specific for a given repeat, with the goal of first finding a repeat, than determining through mapping or direct testing for expansion if it is relevant to a given disease. The discovery that repeats associated with disease are found in translated, transcribed but not translated, and untranscribed portions of genes has led to separate efforts to find repeats in cDNA and in genomic DNA.

Based on the characteristics of the known expansion mutation diseases, a number of groups have pursued the hypothesis that other disorders stem from expansion of unstable DNA [1, 2]. This in turn has led to a search for previously unidentified genes containing trinucleotide repeats. The general approach has been to test these newly found genes for expansion mutations in three overlapping categories of disease: (1) hereditary neurodegenerative disorders, (2) disorders with suspected anticipation, and (3) disorders mapped to a location near a suspected trinucleotide repeat.

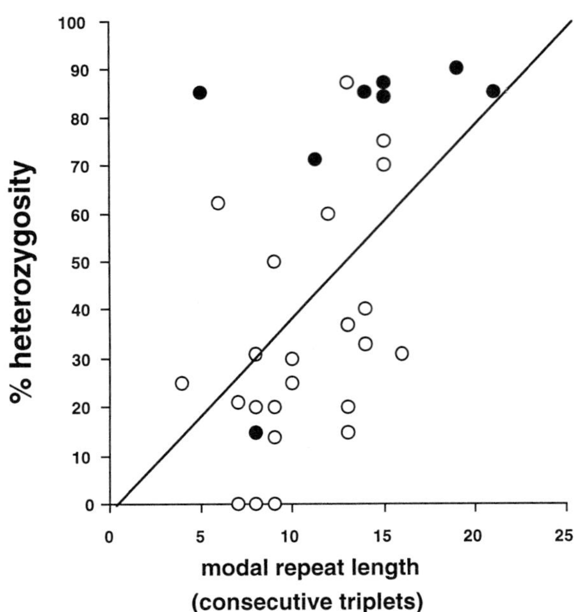

FIGURE 29-1 Heterozygosity vs modal repeat length for CAG/CTG repeats. Solid circles indicate repeats known to undergo expansion mutation and cause disease. Unfilled circles indicate other CAG repeats [26, 32] in which modal repeat length and heterozygosity have been established. The two variables are marginally correlated (r^2 = 0.28). There is a clear, but not invariant, tendency for expansion mutation to occur in highly polymorphic repeats. Reproduced with permission [32], copyright Springer-Verlag.

II. cDNA CLONING STRATEGIES

A. Methodological Considerations

Screening cDNA libraries with oligonucleotide probes specific for a given type of repeat has been used extensively to identify genes containing trinucleotide repeats [25–32, 13]. Much of this effort has focused on CAG repeats, but genes with CCG, AAT, and CCA repeats have also been found using this method [25, 26, 33, 34]. Most of the repeats that expand to cause disease are highly polymorphic in the normal population, and the length of an uninterrupted repeat is correlated with the extent of repeat length polymorphism (Fig. 29-1), so obtaining long uninterrupted repeats is the usual object of these studies. Typically, probes have consisted of radiolabeled oligonucleotides containing between 7 and 20 triplets of the repeat (or the reverse complement of the repeat) of interest. Probes of 10–15 triplets, combined with very high stringency hybridization (e.g., 50% formamide at 60°C for CAG repeats) and wash conditions (e.g., 70°C in 0.2× SSC for CAG repeats), seem optimal [32]. An end-labeled (CIG)8 probe has been successfully used to screen a cDNA library [35], and libraries have also been successfully screened using biotinylated probes subsequently captured by magnetic beads [30].

A number of considerations guide the choice of cDNA library (for a more detailed discussion of this issue, see, among others [36–40]). First, library complexity is always a concern, and is becoming more so as increasing numbers of cDNAs are entered into GenBank. Complexity is partly a function of the quality of any given library, but is also dependent on the tissue studied. A wider variety of genes are expressed in some tissues, such as brain, than in other tissues. A second consideration is the location of the pathology of the disease of interest. Those studying neurodegenerative disorders have tended to search for repeats in brain cDNA libraries. On the other hand, the search for a possible role of repeat expansion in diabetes mellitus has led to screens of human skeletal muscle [41] and pancreatic islet cDNA libraries [42]. Third, some genes will only be expressed at a certain stage of development, so that screening cDNA from tissues at various developmental stages may be necessary.

Fourth, human cDNA has the advantage that repeats in humans tend to be longer than in other species, and hence may be easier to detect [43–46]. Also, it is possible that some genes of potential interest may be found only in humans. On the other hand, construction of cDNA libraries from fresh rodent or nonhuman primate tissue avoids the problem of RNA degradation that is always a threat when human tissues are collected. Fifth, libraries with longer inserts lead to more informative cDNA clones, but potentially increase the amount of sequencing necessary to identify a repeat. To minimize this difficulty, a PCR primer that consists of the repeat of interest with a degenerate 3′ nucleotide can be used to obtain sequence flanking the repeat. A primer designed from this flanking sequence can then be used to sequence through the repeat and into the flanking region on the far side of the repeat, thereby establishing the length of the repeat and providing sufficient sequence to design PCR primers spanning the repeat [47].

B. Results of Screening cDNA Libraries

The strategy of generating candidate repeat expansion genes by screening cDNA libraries for genes with repeats has so far led to the discovery of the etiology of three diseases, each caused by a CAG expansion mutation. The gene for DRPLA, initially known as CTG-B37 and now termed atrophin-1, was the first triplet repeat disease gene identified by this method. Atrophin-1 was initially found in a series of CAG and CCG clones specifically identified as candidate genes for neurological and psychiatric disorders [26, 48]. Several features of atrophin-1, described in the initial report, made it an attractive candidate for a neurodegenerative disorder: it contained a CAG repeat encoding glutamine, the repeat length was highly variable in the normal population, it was expressed in the human brain, and it was located on chromosome 12. The CAG repeat in atrophin-1 led Koide *et al.* [6] to test for atrophin-1 expansions in a variety of hereditary ataxias, leading to their discovery that atrophin-1 expansions are associated with DRPLA. Nagafuchi *et al.* used the same information and their own data linking DRPLA to chromosome 12 to simultaneously demonstrate that an expansion of atrophin-1 results in DRPLA [7].

The gene for Machado–Joseph disease (SCA3) was found by the same strategy [12]. The similarity of the inheritance (autosomal dominant with anticipation, linkage to chromosome 14q) and the phenotype (progressive neurodegeneration) to other CAG expansion diseases identified up to that time, particularly SCA1, DRPLA, and HD, was apparent and led to a search for a CAG expansion as the causative mutation. Screening a cDNA library for CAG repeats yielded a clone (MJD1) on chromosome 14 that contained a polymorphic CAG repeat ultimately proven to be expanded in MJD.

SCA6 was also identified by this method [13], except that in this case the disease had not previously been distinguished as a separate entity from other hereditary ataxias. Again, a number of gene fragments containing CAG repeats were found by screening a cDNA library. Repeat length was tested in a pool of individuals with autosomal dominant cerebellar ataxias of unknown etiology, leading to the finding that the repeat in a gene later determined to be the α_{1A}-voltage-dependent calcium channel was expanded in eight pedigrees.

A number of interesting genes not yet related to a disease have been identified through these candidate gene searches. A search for cDNAs containing CCA repeats turned up the same voltage-dependent calcium channel that was later found to also contain a CAG repeat that expands to cause SCA6 [33]. Genes of interest identified during searches for CAG repeats include a human member of the *numb* family and a cDNA that encodes 40 consecutive glutamines, a length sufficient to cause disease if it were present in other proteins [32]. A cDNA containing a CAG repeat, originally found in a random screen of a retinal library, turned out to be a portion of a gene on chromosome 13q13 encoding a protein highly homologous to the *C. elegans* cell-fate-determining protein *mab21*. The repeat is in the 5′ untranslated region and is highly polymorphic. An unstably transmitted expansion of the repeat has been found in two pedigrees without a clear relationship to a phenotype [49, 50].

C. Alternative Strategies

As an alternative to screening cDNA libraries with repeat-specific probes, Carney *et al.* [51] developed a PCR method for detecting transcribed repeats, termed random rapid amplification of cDNA ends (RRACE). This modification of "traditional" RACE begins with reverse transcription of total RNA using a primer with a specific 5′ sequence and a random 3′ sequence. This product is then amplified using (CAG)8 and a sequence that will hybridize to the 5′ sequence introduced during reverse transcription as the two PCR primers. Products are then purified and subcloned, resulting in a cloned insert with a repeat at one end.

III. CLONING OF TRINUCLEOTIDE REPEATS FROM GENOMIC DNA

A. Rationale for Genomic Cloning

The cDNA cloning strategy is based on the assumption that repeats expanding to cause disease will be found in gene transcripts. The discovery that an intronic GAA expansion results in Friedreich's ataxia [23] upset that notion. The report that a C-G-rich repeat of 12 bases in the region 5′ to exon 1 of the cystatin B gene causes progressive myoclonus epilepsy (EPM1) [52, 53] confirmed that all repeat expansions are not in transcribed regions, and also demonstrated that diseases may arise from expansions of microsatellites or minisatellites other than trinucleotide repeats. Another problem is that most cDNA libraries tend to underrepresent repeats. GC-rich repeats are more difficult for polymerases and reverse transcriptase to replicate, with the result that cDNAs may end in the middle of such regions. Similarly, the 5′ end of transcripts, often the site of CCG repeats, will inevitably be underrepresented in most cDNA libraries, partly from the inherent skew toward incompletely transcribed RNA during RNA harvesting and partly because the synthesis of cDNA from RNA in many libraries is primed from the 3′ polyA mRNA tail.

B. Screening the Entire Genome

These considerations have led some investigators to search for repeats in genomic DNA libraries, including libraries of the complete genome, libraries limited to a single chromosome, and collections of clones spanning specific regions of interest. The most systematic search for genomic trinucleotide repeats yet undertaken used oligonucleotide probes to screen P1 clones, marker-enriched small insert clones, and cosmid clones from the entire genome. A total of 338 sequence tagged sites (STSs) containing CAG/CTG repeats were identified, 299 of which were assigned to chromosomes [54]; 45 of 141 repeats with at least 7 consecutive triplets were polymorphic based on testing in four unrelated individuals. A goal of this project (and a side benefit of other efforts to clone repeats from cDNA or genomic DNA) was to generate markers for use in genomic mapping. Trinucleotide repeats within a gene may be particularly helpful in linkage analysis of disease. For instance, Behcet disease has been closely linked to MICA (major histocompatibility complex class I chain-related gene family, type A) based on a polymorphic trinucleotide repeat within the transmembrane portion of the open reading frame [55].

Other methods have also been used to detect repeats within genomic DNA. Boffa *et al.* [56] designed a system to detect specific DNA sequences, including repeats, within transcriptionally active genes. They extracted nuclei from cells in culture and isolated transcriptionally active chromatin fragments from the nuclei. These fragments were then probed with a peptide nucleic acid (PNA), a nucleic acid homologue in which the phosphate-sugar backbone of the nucleic acid is replaced by a peptide backbone. The resulting molecule can still hybridize to DNA, but the absence of negatively charged phosphate groups enables it to penetrate DNA duplexes and hybridize to its complementary strand. In this case, a PNA containing a repeating CTG unit was used to identify regions of transcribed DNA with CAG repeats, which could then be isolated and sequenced. A genomic differential display method has also been devised that may assist in detecting trinucleotide repeats [57]. In essence, PCR is performed using restriction fragments of genomic DNA that have been preselected for the presence of a chosen sequence, such as a repeat. Repeats have also been detected using fluorescent *in situ* hybridization (FISH) [58]. The utility of these new methods in detecting repeats not found by other methods remains unknown.

C. Screening Specific Chromosomes

The strategy of screening genomic DNA derived from a single chromosome for repeats has been used by Breschel *et al.* [59] as part of an effort to identify candidate genes for bipolar affective disorder. A chromosome 18 specific library was probed with a CAG and identified a clone (termed CTG 18.1) with a CTG repeat. The repeat was mapped to 18q21, within the SEF1 gene. The repeat is highly polymorphic. In occasional families the repeat expands to approximately 87–250 triplets, and in a few cases to thousands of triplets. These expansions have not been linked to a phenotype, but efforts are underway to assess the effect of the expansion on gene expression and function.

D. Screening Contigs Spanning Regions Linked to Disease

Searching for repeats in genomic clones has proven of considerable value in speeding the positional cloning of genes linked to expansion mutation diseases. This strategy was first used by one of the three groups [21] that identified the CTG expansion underlying myotonic dystrophy. GC-rich oligonucleotide 21-mer probes were used to screen a cosmid library derived from YAC clones spanning the region of interest on chromosome 19q13. The (GCT)7 probe hybridized to two overlapping cosmids, and subcloning and sequencing revealed the CTG repeat expanded in affected individuals. Similarly, in an effort to identify SCA2, Pulst *et al.* [10] created a contig of P1-artificial chromosomes (PACs) and bacterial artificial chromosomes (BACs) spanning the region on 12q24.1 to which SCA2 had been linked. After restriction digest, the clones in this contig were screened for the presence of various trinucleotide repeats. Two CAG-positive bands were detected, one of which, after subcloning and sequencing, proved to contain a long (though interrupted) CAG repeat within an open reading frame. This repeat was subsequently shown to be expanded in individuals with SCA2. At present, almost all positional cloning efforts for neurodegenerative disorders or diseases with anticipation take advantage of this kind of strategy.

IV. REPEAT CHARACTERIZATION AND HETEROZYGOSITY ANALYSIS

Which newly cloned repeats are most likely to undergo expansion mutation? Repeat length is an obvious factor, since 10 of the 12 repeats that expand to cause disease have a modal length of greater than 10 consecutive triplets. Analysis of FMR1 [60] (the gene in which a CGG expansion results in the A from of the fragile X syndrome), SCA1 [61], and SCA2 [9–11] suggest that perfect repeats are more likely to undergo expansion than repeats containing interruptions. In addition, almost all repeats that expand to cause disease are highly polymorphic in the normal population, with heterozygosity scores in the 70–90% range. However, since the CAG repeat of the SCA2 gene [9–11] and the CCG repeat of the CBL2 gene [19] (in which an expansion is associated with deletion of chromosome 11q and the Jacobsen syndrome phenotype) are minimally polymorphic, it is impossible to completely exclude a repeat as an expansion candidate simply on the basis that it is not very polymorphic. Perhaps the most important clue that a repeat is associated with a disease is its localization by genetic linkage or physical mapping strategies to a chromosomal region to which a disease has also been linked.

Cloning and characterization of repeats has also facilitated analysis of repeat length polymorphism. In particular, the heterozygosity at a given repeat length is greater for AAT/TTA repeats than for CCA/GGT repeats, while CCA/GGT repeats are no less polymorphic than CAG/CTG repeats (Fig. 29-2, Table 29-1). Given that a particular CAG/CTG repeat, and perhaps CCG/CGG repeat, appears more likely to expand if it is highly polymorphic, the high degree of polymorphism in AAT/TTA and CCA/GGT repeats suggest the possibility that some of these may also undergo expansion mutation.

V. THE FUTURE OF REPEAT CLONING

One of the consistent findings of the various efforts to clone genes with trinucleotide repeats is that many of these genes remain undetected, or at least are not present in GenBank. Each new cloning effort invariably finds new genes of interest. In part this reflects the 3′ bias in many of the expressed sequence tags (ESTs) collected as part of the human genome project. Now that intronic repeats are known to cause disease, genomic cloning efforts to identify repeats will take equal prominence to cDNA cloning. The possibility that repeats with combinations of other nucleotides may expand to cause disease cannot as of yet be excluded, and it is now clear that minisatellite repeats can expand to cause disease. The strategy of finding previously undiscovered genes that contain repeats and testing for expansions in these repeats remains a powerful method for detecting disease genes.

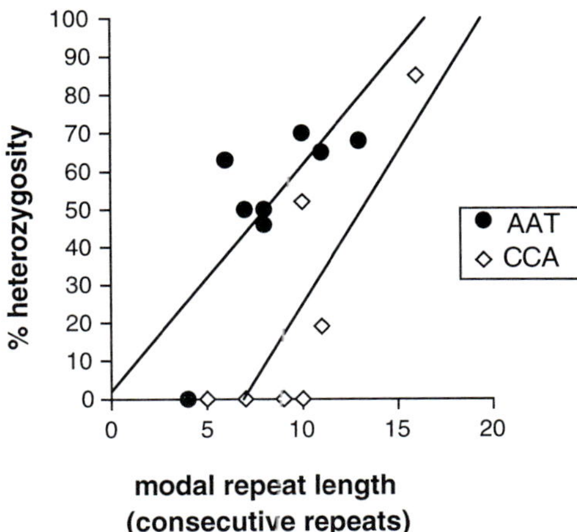

FIGURE 29-2 Solid circles indicate AAT/ATT repeats, and open diamonds indicate CCA/TGG repeats. Repeat length is moderately correlated with heterozygosity (AAT/ATT: r^2 = 0.32; CCA/TGG: r^2 = 0.46). For a given repeat length, AAT/ATT repeats tend to be more polymorphic than CCA/TGG repeats.

TABLE 29-1 Relationship between Repeat Length and Polymorphism

Repeat type	% Heterozygosity	Modal repeat length	Ratio of heterozy/length
AAT/ATT	51.5 ± 22.7	8.4 ± 2.9	6.0 ± 2.9**
CCA/TGG	22.3 ± 33.7	9.7 ± 3.4	1.75 ± 2.5
CAG/CTG	30.1 ± 27.4	10.1 ± 3.4	2.87 ± 2.6

Note. The ratio of heterozygosity to modal repeat length was calculated for each type of repeat to provide an index of length polymorphism per triplet. AAT/ATT repeats are more variable at a given length than CCA/TGG or CAG/CTG repeats. All values are ± standard deviation. One-way manova: F = 5.67, p = 0.007. Tukey HSD post hoc test for samples of unequal size: AAT/ATT vs CCA/TGG, p = .013; AAT/ATT vs CAG/CTG, p = 0.061; CCA/TGG vs CAG/CTG, p = 0.71.

Acknowledgments

This work was supported in part by NIH MH02175-10A1, NIH NS34172, NIH MH50763, and a NARSAD Established Investigator Award to C.A.R.

References

1. Ross, C. A., McInnis, M. G., Margolis, R. L., and Li, S.-H. (1993). Genes with triplet repeats: candidate mediators of neuropsychiatric disorders. *Trend Neurosci.* **16,** 254–260.
2. Trottier, Y., Lutz, Y., Stevanin, G., Imbert, G., Devys, D., Cancel, G., Saudou, F., Weber, C., David, G., and Tora, L., Agid, Y., Brice, A., and Mandel J.-L. (1995). Polyglutamine expansion as a pathological epitope in Huntington's disease and four dominant cerebellar ataxias. *Nature* **378,** 403–406.
3. McInnis, M. G. (1996). Anticipation: an old idea in new genes. *Am. J. Hum. Genet.* **59,** 973–979.
4. Huntington's Disease Collaborative Research Group. (1993). A novel gene containing a trinucleotide repeat that is expanded and unstable on Huntington's disease chromosomes. *Cell* **72,** 971–983.
5. La Spada, A. R., Wilson, E. M., Lubahn, D. B., Harding, A. E., and Fischbeck, K. H. (1991). Androgen receptor gene mutations in X-linked spinal and bulbar muscular atrophy. *Nature* **352,** 77–79.
6. Koide, R., Ikeuchi, T., Onodera, O., Tanaka, H., Igarashi, S., Endo, K., Takahashi, H., Kondo, R., Ishikawa, A., Hayashi, T., Saito, M., Tomoda, A., Miike, T., Naito, H., Ikuta, F., and Tsuji, S. (1994). Unstable expansion of CAG repeat in hereditary dentatorubral-pallidoluysian atrophy (DRPLA). *Nature Genet.* **6,** 9–12.
7. Nagafuchi, S., Yanagisawa, H., Sato, K., Shirayama, T., Ohsaki, E., Bundo, M., Takeda, T., Tadokoro, K., Kondo, I., Murayama, N., Tanaka, Y., Kikushima, H., Umino, K., Kurosawa, H., Furukawa, T., Nihei, K., Inoue, T., Sano, A., Osamu, K., Takahashi, M., Yoshizawa, T., Kanazawa, I., and Yamada, M. (1994). Expansion of an unstable CAG trinucleotide on chromosome 12p in dentatorubral and pallidoluysian atrophy. *Nature Genet.* **6,** 14–17.
8. Orr, H. T., Chung, M.-Y., Banfi, S., Kwiatkowski, T. J., Servadio, A., Beaudet, A. L., McCall, A. E., Duvick, L. A., Ranum, L. P. W., and Zoghbi, H. Y. (1993). Expansion of an unstable trinucleotide repeat in spinocerebellar ataxia type 1. *Nature Genet.* **4,** 221–226.
9. Imbert, G., Saudou, F., Yvert, G., Devys, D., Trottier, Y., Garnier, J.-M., Weber, C., Mandel, J.-L., Cancel, G., Abbas, N., Durr, A., Didierjean, O., Stevanin, G., Agid, Y., and Brice, A. (1996). Cloning of the gene for spinocerebellar ataxia 2 reveals a locus with high sensitivity to expanded CAG/glutamine repeats. *Nature Genet.* **14,** 285–291.
10. Pulst, S.-M., Nechiporuk, A., Nechiporuk, T., Gispert, S., Chen, X.-N., Lpes-Cendes, I., Pearlman, S., Starkman, S., Orozco-Diaz, G., Lunkes, A., DeJong, P., Rouleau, G. A., Auburger, G., Korenberg, J. R., Figueroa, C., and Sahba, S. (1996). Moderate expansion of a normally biallelic trinucleotide repeat in spinocerebellar ataxia type 2. *Nature Genet.* **14,** 269–276.
11. Sanpei, K., Takano, H., Igarashi, S., Sato, T., Oyake, M., Sasaki, H., Wakisaka, A., Tashiro, K., Ishida, Y., Ikeuchi, T., Koide, R., Saito, M., Sato, A., Tanaka, T., Hanyu, S., Takiyama, Y., Nishizawa, M., Shimizu, N., Nomura, Y., Segawa, M., Iwabuchi, K., Eguchi, I., Tanaka, H., Takahashi, H., and Tsuji, S. (1996). Identification of the spinocerebellar ataxia type 2 gene using a direct identification of repeat expansion and cloning technique, DIRECT. *Nature Genet.* **14,** 277–284.
12. Kawaguchi, Y., Okamoto, T., Taniwaki, T., Aizawa, M., Inoue, M., Katayama, S., Kawakami, H., Nakamura, S., Nishimura, M., Akiguchi, I., Kimura, J., Narumiya, S., and Kakizuka, A. (1994). CAG expansions in a novel gene for Machado-Joseph disease at chromosome 14q32.1. *Nature Genet.* **8,** 221–228.
13. Zhuchenko, O., Bailey, J., Bonnen, P., Ashizawa, T., Stockton, D. W., Amos, C., Dobyns, W. B., Subramony, S. H., Zoghbi, H. Y., and Lee, C. C. (1997). Autosomal dominant ataxia (SCA6) associated with small polyglutamine expansions in the a1A-voltage-dependent calcium channel. *Nature Genet.* **15,** 62–69.
14. Fu, Y.-H., Kuhl, D. P. A., Pizzuti, A., Pieretti, M., Sutcliffe, J. S., Richards, S., Verkerk, A. J. M. H., Holden, J. J. A., Fenmwick, R. G., Warren, S. T., Oostra, B. A., Nelson, D. L., and Caskey, C. T. (1991). Variation of the CGG repeat at the Fragile X site results in genetic instability: resolution of the Sherman paradox. *Cell* **67,** 1047–1058.
15. Verkerk, A. J. M. H., Pieretti, M., Sutcliffe, J. S., Fu, Y.-H., Kuhl, D. P. A., Pizzuti, A., Reiner, O., Richards, S., Victoria, M. F., Zhang, F., Eussen, B. E., van Ommen, G.-J. B., Blonden, L. A. J., Riggins, G. J., Chastain, J. L., Kunst, C. B., Galjaard, H., Caskey, C. T., Nelson, D. L., Oostra, B. A., and Warren, S. T. (1991). Identification of a gene (FMR-1) containing CGG repeat coincident with a breakpoint cluster region exhibiting length variation in Fragile X syndrome. *Cell* **65,** 905–914.
16. Yu, S., Pritchard, M., Kremer, E., Lynch, M., Nancarrow, J., Baker, E., Holman, K., Mulley, J. C., Warren, S. T., Schlessinger, D., Sutherland, G. R., and Richards, R. I. (1992). Fragile X genotype characterized by an unstable region of DNA. *Science* **252,** 1179–1981.
17. Knight, S. J. L., Flannery, A. V., Hirst, M. C., Campbell, L., Christodoulou, Z., Phelps, S. R., Pointon, J., Middletonprice, H. R., Barnicoat, A., Pembrey, M. E., Holland, J., Oostra, B. A., Bobrow, M., and Davies, K. E. (1993). Trinucleotide repeat amplification and hypermethylation of a CpG island in FRAXE mental retardation. *Cell* **74,** 127–134.
18. Ritchie, R. J., Knight, S. J. L., Hirst, M. C., Grewal, P. K., Bobrow, M., Cross, G. S., and Davies, K. E. (1994). The cloning of FRAXF: trinucleotide repeat expansion and methylation at a third fragile site in distal Xqter. *Hum. Mol. Genet.* **3,** 2115–2121.
19. ones, C., Penny, L., Mattina, T., Yu, S., Baker, E., Voullaire, L., Langdon, W. Y., Sutherland, G. R., Richards, R. I., and Tunnacliffe, A. (1995). Association of a chromosome deletion syndrome with a fragile site with the proto-oncogene CBL2. *Nature* **376,** 145–149.
20. Brook, J. D., McCurrach, M. E., Harley, H. G., Buckler, A. J., Church, D., Aburatani, H., Hunter, K., Davies, J., Shelbourne, P., Buxton, J., Jones, C., Juvonen, V., Johnson, K., Harper, P. S., Shaw, D. J., and Housman, D. E. (1992). Molecular basis of myotonic dystrophy: expansion of a trinucleotide (CTG) repeat at the 3′ end of a transcript encoding a protein kinase family member. *Cell* **68,** 799–808.
21. Fu, Y.-H., Pizzuti, A., Fenwick, R. G., King, J., Rajnarayan, S., Dunne, P. W., Dubel, J., Nasser, G. A., Ashizawa, T., De Jong, P., Wieringa, B., Korneluk, R., Perryman, M. B., Epstein, H. F., and Caskey, C. T. (1992). An unstable triplet repeat in a gene related to myotonic muscular dystrophy. *Science* **255,** 1256–1258.
22. Mahadevan, M., Tsilfidis, C., Sabourin, L., Shutler, G., Amemiya, C., Jansen, G., Nelville, C., Narang, M., Barcelo, J., O'Hoy, K., Leblond, S., Earle-MacDonald, J., DeJong, P. J., Wieringa, B., and Korneluk, R. G. (1992). Myotonic dystrophy mutation: an unstable CTG repeat in the 3′ untranslated region of the gene. *Science* **255,** 1253–1255.
23. Campuzano, V., Montermini, L., Molto, M. D., Pianese, L., Cossee, M., Cavalcanti, F., Monros, E., Rodius, F., Duclos, F., Monticelli, A., Zara, F., Canizares, J., Koutnikova, H., Bidichandani, S. I., Gellera, C., Brice, A., Trouillas, P., Demichele, G.,

Filla, A., Defrutos, R., Palau, F., Patel, P. I., Didonato, S., Mandel, J. L., Cocozza, S., Koenig, M., and Pandolfo, M. (1996). Friedreich's ataxia: autosomal recessive disease caused by an intronic GAA triplet repeat expansion. *Science* **271,** 1423–1427.

24. Schalling, M., Hudson, T. J., Buetow, K. H., and Housman, D. E. (1993). Direct detection of novel expanded trinucleotide repeats in the human genome. *Nature Genet.*. **4,** 135–139.
25. Riggins, G. J., Lokey, L. K., Chastain, J. L., Leiner, H. A., Sherman, S. L., Wilkinson, K. D., and Warren, S. T. (1992). Human genes containing polymorphic trinucleotide repeats. *Nature Genet.* **2,** 186–191.
26. Li, S.-H., McInnis, M. G., Margolis, R. L., Antonarakis, S. E., and Ross, C. A. (1993). Novel triplet repeat containing genes in human brain: cloning, expression, and length polymorphisms. *Genomics* **16,** 572–579.
27. Jiang, J.-X., Deprez, R. H. L., Zwarthoff, E. C., and Riegman, P. H. J. (1995). Characterization of four novel CAG repeat-containing clones. *Genomics* **30,** 91–93.
28. Tsuji, S., Igarashi, S., Takiyama, Y., Onodera, O., and Tanaka, H. (1995). Isolation and characterization of human cDNA clones containing polymorphic CAG trinucleotide repeats from human cerebellum, fetal brain, striatum, and nigra cDNA libraries. *Am. J. Hum. Genet.* **57,** A152.
29. Neri, C., Albanese, V., Lebre, A.-S., Holbert, S., Saada, C., Bougueleret, L., Meier-Ewert, S., Le Gall, I., Millasseau, P., Bui, H., Giudicelli, C., Massart, C., Guillou, S., Gervy, P., Poullier, E., Rigault, P., Weissenbach, J., Lennon, G., Chumakov, I., Dauseet, J., Lehrach, H., Cohen, D., and Cann, H. M. (1996). Survey of CAG/CTG repeats in human cDNAs representing new genes: candidates for inherited neurological disorders. *Hum. Mol. Genet.* **5,** 1001–1009.
30. Reddy, P. H., Stockburger, E., Wilderson, J., Ellison, J., Gillevet, P., and Tagle, D. A. (1996). Oligo capture of CAG repeat containing cDNAS from adult human brain. *Am. J. Hum. Genet.* **59,** A281.
31. Bulle, F., Chiannilkulchai, N., Pawlak, A., Weissenbach, J., Gyapay, G., and Guellaen, G. (1997). Identification and chromosomal localization of human genes containing CAG/CTG repeats expressed in testis and brain. *Genome Res.* **7,** 705–715.
32. Margolis, R. L., Abraham, M. R., Gatchell, S. B., Li, S.-H., Kidwai, A. S., Breschel, T. S., Stine, O. C., Callahan, C., McInnis, M. G., and Ross, C. A. (1997). cDNAs with long CAG trinucleotide repeats from human brain. *Hum. Genet.* **100,** 114–122.
33. Margolis, R. L., Breschel, T. S., Li, S.-H., Kidwai, A. S., Antonarakis, S. E., McInnis, M. G., and Ross, C. A. (1995). Identification and characterization of cDNA clones containing CCA trinucleotide repeats derived from human brain. *Somat. Cell Mol. Genet.* **21,** 279–284.
34. Margolis, R. L., Breschel, T. S., Li, S.-H., Kidwai, A. S., McInnis, M. G., and Ross, C. A. (1995). Polymorphic (AAT)n trinucleotide repeats derived from a human brain cDNA library. *Hum. Genet.* 495–496.
35. Phillips, K. L., Gartrell, D. M., Roses, A. D., and Lee, J. E. (1993). A triplet repeat polymorphism in a gene expressed in human hypothalmus. *Hum. Mol. Genet.* **2,** 1332.
36. Sambrook, J., Fritsch, E. F., and Maniatis, T. (1989). "Molecular Cloning: A Laboratory Manual." Cold Spring Harbor Laboratory, Cold Spring Harbor, NY.
37. Sainz, J. (1993). Human cDNA libraries and brain banks. *J. Neural. Trans.* **39**(Suppl.), 135–141.
38. McCarrey, J. R., and Williams, S. A. (1994). Construction of cDNA libraries from limiting amounts of material. *Curr. Opin. Biotech.* **5,** 34–39.
39. Maser, R. L., and Calvet, J. P. (1995). Analysis of differential gene expression in the kidney by differential cDNA screening, subtractive cloning, and mRNA differential display. *Sem. Nephrol.* **15,** 29–42.
40. Ausubel, F. M., Brent, R., Kingston, R. E., Seidman, J. G., Smith, J. A., and Struhl, K. (1997). "Current Protocols in Molecular Biology." Wiley, New York.
41. Yamagata, K., Takeda, J., Menzel, S., Chen, X., Eng, S., Lim, L. R., Concannon, P., Hanis, C. L., Spielman, R. S., Cox, N. J., and Bell, G. I. (1996). Searching for NIDDM susceptibility genes—studies of genes with triplet repeats expressed in skeletal muscle. *Diabet.* **39,** 725–730.
42. Aoki, M., Koryani, L., Riggs, A. C., Wasson, J., Chiu, K. C., Vaxillaire, M., Froguel, P., Gough, S., Liu, L., Donis-Keller, H., and Permutt, M. A. (1996). Identification of trinucleotide repeat-containing genes in human pancreatic islets. *Diabet.* **45,** 157–164.
43. Rubinsztein, D. C., Amos, W., Leggo, J., Goodburn, S., Jain, S., Li, S.-H., Margolis, R. L., Ross, C. A., and Ferguson-Smith, M. A. (1995). Microsatellite evolution--evidence for directionality and variation between species. *Nature Genet.* **10,** 337–343.
44. Loev, S. J., Margolis, R. L., Young, W. S., Schillimg, G., Li, S.-H., Ashworth, R. G., and Ross, C. A. (1995). Sequence and expression of the rat atrophin-1 (DRPLA) gene. *Neurobiol. Dis.* **2,** 129–138.
45. Gossen, M., Schmitt, I., Obst, K., Wahle, P., Epplen, J. T., and Riess, O. (1996). cDNA cloning and expression of RSCA1, the rat counterpart of the human spinocerebellar ataxia type 1 gene. *Hum. Mol. Genet.* **5,** 381–389.
46. Banfi, S., Servadio, A., Chung, M.-y., Capozolli, F., Duvick, L. A., Elde, R., Zoghbi, H. Y., and Orr, H. T. (1996). Cloning and developmental expression analysis of the murine homolog of the spinocerebellar ataxia type 1 gene (Sca1). *Hum. Mol. Genet.* **5,** 33–40.
47. Margolis, R. L., Li, S.-H., and Ross, C. A. (1993). A rapid method for sequencing trinucleotide repeats. *Nucleic Acids Res.* **21,** 4983–4984.
48. Ross, C. A., McInnis, M. G., Margolis, R. L., and Li, S.-H. (1993). Genes with triplet repeats: candidate mediators of neuropsychiatric disorders. *Trend Neurosci.* **16,** 254–260.
49. Potter, N. T. (1997). Meiotic instability associated with the CAGR1 trinucleotide repeat at 13q13. *J. Med. Genet.* **34,** 411–413.
50. Rosenblatt, A., Ranen, N. G., Rubinsztein, D. C., Stine, O. C., Margolis, R. L., Wagster, M. V., Becher, M. W., Rosser, A. E., Leggo, J., Hodges, J. R., French-Constant, C. K., Sherr, M., Franz, M. L., Abott, M. H., and Ross, C. A. Characteristics of patients with clinical features similar to Huntington's disease, but negative for expanded CAG repeats in huntingtin. [Submitted for publication]
51. Carney, J. P., McKnight, C., VanEpps, S., and Kelley, M. R. (1995). Random rapid amplification of cDNA ends (RRACE) allows for cloning of multiple novel human cDNA fragments containing (CAG)n repeats. *Gene* **155,** 289–292.
52. Lafreniere, G. R., Rochefort, D. L., Chretine, N., Rommens, J. M., Cochius, J. I., Kalviainen, R., Nousiainen, U., Patry, G., Farrell, K., Soderfeldt, B., Federico, A., Hale, B. R., Cossio, O. H., Sorensen, T., Pouliot, M. A., Kmiec, T., Uldall, P., Janszky, J., Pranzatelli, M. R., Andermann, F., Andermann, E., and Rouleau, G. A. (1997). Unstable insertion in the 5′ flanking region of the cystatin B gene is the most common mutation in progressive myoclonus epilepsy type 1, EPM1. *Nature Genet.* **15,** 298–302.
53. Lalioti, M. D., Scott, H. S., Buresi, C., Rossier, C., Bottani, A., Morris, M. A., Malafosse, A., and Antonarakis, S. E. (1997). Dodecamer repeat expansion in cystatin B gene in progressive myoclonus epilepsy. *Nature* **386,** 847–851.

54. Gastier, J. M., Pulido, J. C., Sunden, S., Brody, T., Buetow, K. H., Murray, J. C., Weber, J. L., Hudson, T. J., Sheffield, V., and Duyk, G. M. (1995). Survey of trinucleotide repeats in the human genome--assessment of their utility as genetic markers. *Hum. Mol. Genet.* **4,** 1829–1836.

55. Mizuki, N., Ota, M., Kimura, M., Ohno, S., Ando, H., Katsuyama, Y., Yamazaki, M., Watanabe, K., Goto, K., Nakamura, S., Bahram, S., and Inoko, H. (1997). Triplet repeat polymorphism in the transmembran region of the MICA gene: a strong association of six GCT repetitions with Behcet disease. *Proc. Natl. Acad. Sci. USA* **94,** 1298–1303.

56. Boffa, L. C., Carpaneto, E. M., and Allfrey, V. G. (1995). Isolation of active genes containing CAG repeats by DNA strand invasion by a peptide nucleic acid. *Proc. Natl. Acad. Sci. USA* **92,** 1901–1905.

57. Broude, N. E., Chandra, A., and Smith, C. L. (1997). Differential display of genome subsets containing specific interspersed repeats. *Proc. Natl. Acad. Sci. USA* **94,** 4548–4553.

58. Haaf, T., Sirugo, G., Kidd, K. K., and Ward, D. C. (1996). Chromosomal localization of long trinucleotide repeats in the human genome by fluorescence in situ hybridization. *Nature Genet.* **12,** 183–185.

59. Breschel, T. S., McInnis, M. G., Margolis, R. L., Sirugo, G., Corneliussen, B., Simpson, S. G., McMahon, F. J., MacKinnon, D. F., Xu, J. F., Plesant, N., Huo, Y., Ashworth, R. G., Grundstrom, C., Grundstrom, T., Kidd, K. K., DePaulo, J. R., and Ross, C. A. A novel, heritable, expanding CTG repeat in an intron of SEF2-1 localizes to human chromosome 18q21.1. [Submitted for publication]

60. Hirst, M., Grewal, P., and Davies, K. E. (1994). Precursor arrays for triplet repeat expansion at the fragile X locus. *Hum. Mol. Genet.* **3,** 1553–1560.

61. Chung, M., Ranum, L. P. W., Duvick, L. A., Servadio, A., Zoghbi, H. Y., and Orr, H. T. (1993). Evidence for a mechanism predisposing to intergenerational CAG repeat instability in spinocerebellar ataxia type 1. *Nature Genet.* **5,** 254–258.

CHAPTER 30

RED Technology

MARTIN SCHALLING, KERSTIN LINDBLAD, QIU-PING YUAN, CATHERINE E. BURGESS, AND CECILIA ZANDER
Neurogenetics Unit, Department of Molecular Medicine, Karolinska Hospital, Stockholm, Sweden

TOM HUDSON
Genome Center, Whitehead Institute, Massachusetts Institute of Technology, Cambridge, Massachusetts 02139

I. INTRODUCTION

An expansion can occur in any of more than a thousand repeats present in the human genome [1]. In the disease state, such expansions often contain CAG/CTG or CGG/CCG repeat sequences. However, expansions resulting in disease may occur in motifs other than CAG/CTG and CGG/CCG [2] as exemplified by the interruption of gene processing by a GAA repeat expansion in Friedreich's ataxia [3]. In principle, any simple repeat sequence within or near a gene could disrupt the function of that gene if marked changes in the repeat size occurred. In many diseases associated with repeat expansions, an unusual pattern of inheritance is displayed. Typically, the phenotype changes from one generation to the next with an earlier age of onset and/or more severe symptoms, a phenomenon termed anticipation (Chapters 27 and 28). Anticipation has also been observed in a number of disorders not yet identified at the molecular level.

In such disorders, repeat expansions could be involved in the pathological process. It is possible to search for such expansions using the repeat expansion detection (RED) method [4].

The RED method has been used to detect several of the expanded repeats that cause disease [4–6]. RED [4] can detect expanded trinucleotide repeats anywhere in the genome as well as in cloned material without information regarding genomic location of the repeat. The RED method, consequently, makes it possible to identify potentially pathological repeats in unmapped disorders. Since the method uses the DNA expansion itself as template, it is not necessary for the sequence to be expressed either as RNA or as protein. Association studies using the RED method have shown an increased presence of CAG/CTG repeat expansions in bipolar affective disorder as well as in schizophrenia [7–9] (see below).

In the RED method, genomic DNA serves as a template for repeat-specific oligonucleotides. A thermostable ligase is used to ligate two or more oligonucleotides that have annealed at adjacent bases of a repeat sequence in genomic DNA. A pool of multimers is generated through multiple rounds of cycling (Fig. 30-1). This is a linear amplification process requiring several hundred cycles of ligation/denaturation. Following gel electrophoresis, the products are visualized by blotting onto a membrane and hybridization with a ^{32}P-labeled repeat probe complementary to the multimer (Fig. 30-2). The maximum product size observed corresponds to the size of the longest repeat sequence present in the genome tested. The basic principle of the RED method is illustrated in Fig. 30-1.

II. MATERIALS AND METHODS

A. DNA Template

A standard phenol/chloroform extraction method or QIAamp Blood Kit (Qiagen Inc., Chatsworth, CA) is used for DNA preparation. Avoid high salt concentrations as this may inhibit the RED reaction.

B. Oligonucleotides

The RED procedure has been optimized for use with a $(CTG)_{10}$ oligonucleotide [10]. The oligonucleotide length may need to be increased if the GC content of the repeat motif to be analyzed is low. Oligonucleotides of different sizes may be used when screening for several motifs simultaneously [2]. It is important that oligonucleotides are phosphorylated and purified for the RED reaction to work. Any residue of shorter molecules from the synthesis will reduce the signal markedly.

C. Enzymes, Isotopes, Buffers, and Solutions

We use Ampligase (Epicentre Technologies, Madison, WI) with the supplied buffer for thermal cycling and terminal deoxynucleotidyl transferase (Amersham, Little Chalfont, UK) with the supplied buffer for end-labeling of hybridization probes. Any detection system can be used, but we prefer an isotope containing low levels of DTT such as [^{32}P]dATP (NEG 012Z, NEN DuPont Medical, Wilmington, DE) (6000 Ci/mmol).

In addition, ATP (10 mM), 1× TBE, TE^{-4}, SSC, SDS, gel loading dye: 100% formamide + 0.1% xylene cyanol + 0.1% bromophenol blue, 6% denaturing polyacrylamide/6 M urea gel, Sequagel-6 (National Diagnostics, Atlanta, GA), and Amersham Rapid Hybe (Amersham) (or any conventional hybridization solution) are needed for experiments.

D. Apparatus and Supplies

We have been using the GeneAmp PCR System 9600 (Perkin Elmer Cetus, Norwalk, CT), PTC-200 (MJ Research, Watertown, MA), or Rapidcycler (Idaho Technology, Idaho Falls, ID). We have observed less efficient amplification using oil-covered reactions. In addition, Whatman 3-mm filter paper, Hybond N+ membrane (Amersham) or any similar membrane, DuPont Reflection, NEF 495 X-ray film, and intensifying screens or a similar product is needed.

III. RED PROTOCOL

The principle of RED is presented in Fig. 30-1.

A. Reaction Mixture

Reaction mixture should contain genomic DNA template, 1 μg /reaction (0.5–5 μg); TE-4, if not included in genomic DNA, 1 μl/reaction; 10× Ampligase buffer, 0.5–1 μl/reaction; oligonucleotide (5′-phosphorylated), 50 ng/reaction; Ampligase, 15 U/reaction; H_2O, 10 μl final volume.

B. Amplification Conditions

RED may be performed in any top-heated automatic thermocycler. The following conditions are optimized

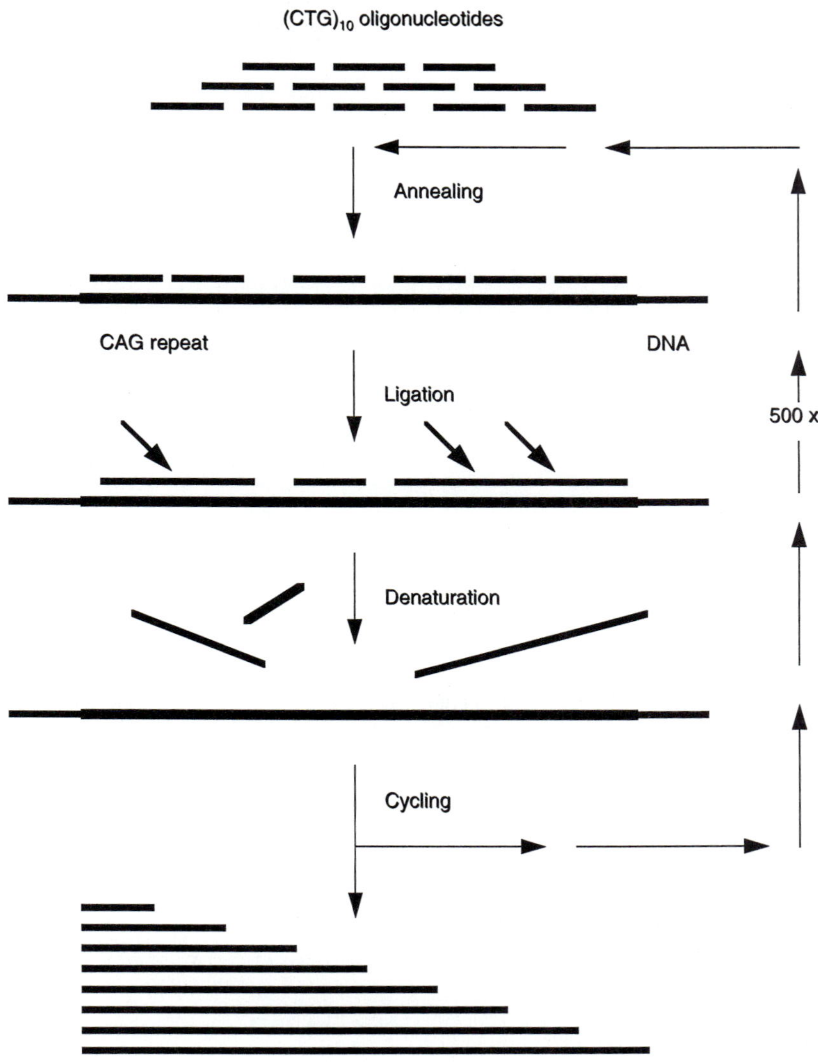

FIGURE 30-1 Linear amplification of expanded repeat sequences in genomic DNA using the RED method. DNA is heat denatured and an oligonucleotide complementary to the repeat motif of interest is allowed to anneal to the template. Ligation occurs when two oligonucleotides anneal at adjacent bases. The reaction is repeated several hundred times on a thermocycler yielding a pool of multimers, where the longest product corresponds to the longest repeat in the template.

for the Perkin Elmer 9600 and a $(CTG)_{10}$ oligonucleotide [10]: primary denaturing at 95°C for 5 min, followed by 500 cycles of annealing/ligation (80°C/20 s) and denaturation (94°C/10 s). Faster cycling is possible if a capillary thermal cycler is used, for example the Rapidcycler (Idaho Technology), although addition of 10 mg/ml BSA is required. We have generated strong and reproducible signal with 40 min cycling time (200 cycles at 0 s denaturation, 5 s annealing/ligation) on the Rapidcycler. It is important to be aware that changing the reaction conditions might reduce the specificity of the method. As in any other amplification-based method, artifacts may occur when using a lower annealing/ligation temperature, a higher DNA concentration, or a lower buffer concentration.

IV. ELECTROPHORESIS AND HYBRIDIZATION

A. Electrophoresis

Use a 6% polyacrylamide/6 M urea gel with a wide tooth comb. Preelectrophorese at 90 W for 20 min. Heat denature RED products in 0.5× gel loading dye for 5 min and load sample on gel. Electrophorese until xylene cyanol (xc) has migrated 12 cm into gel. Separate plates and discard gel below 16 cm to avoid probe hybridization to excess oligonucleotides.

B. Blotting

Place a wet sheet of Hybond N+ membrane on the gel, and overlay with three dry 3MM sheets cut to fit, the top glass plate, and a weight for 2 h. Thereafter, immobilize the DNA on the membrane by cross-linking. Alternatively, 2 h of electroblotting at 2 A in 1× TBE will give a high degree of transfer provided that gel and membrane are in close contact.

C. Labeling of Probe

We prefer labeling at the 3′ end, effectively permitting addition of multiple [^{32}P]dATPs to each molecule, yielding a high specific activity. Mix 8.7 μl H_2O, 5 μl 5× Co buffer, and 2.5 μl oligonucleotide (50 ng/μl) on ice. Add 7 μl [^{32}P]dATP and 1.8 μl TdT enzyme [11]. Incubate for 1 h at 37°C. Add 500 μl 0.1 M Tris, pH 8.0, to stop the reaction. Probe should be labeled to a specific activity of 2–9 × 10^9 cpm/μg.

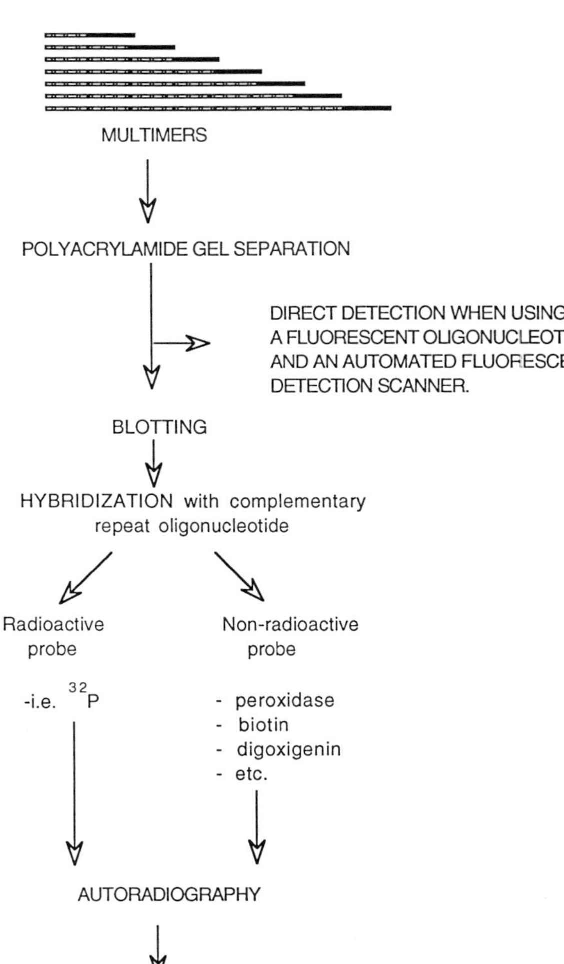

FIGURE 30-2 Detection of multimers. Amplified samples are size separated on a polyacrylamide gel and transferred to a membrane before hybridization with a complementary probe. Detection may also be performed directly using a fluorescent oligonucleotide and an automated fluorescent detection scanner.

D. Hybridization

Prehybridize membrane in Amersham Rapid Hybe solution for 20 min; add labeled probe and hybridize for 1 h (60°C for a $(CAG)_{10}$). Alternatively, conventional hybridization mixtures can be used for an overnight hybridization. Wash the membrane for 10 min at room temperature and 30 min at 60°C in 1× SSC, 0.1% SDS with at least one change of wash solution.

V. DETECTION

Expose membranes overnight (or up to 7 days) to X-ray film at −70°C using intensifying screens. A phosphorimager may be used to obtain results within a few hours. Alternatively, laser-based fluorescent detection may be used during electrophoresis on a Pharmacia ALF (Pharmacia Biotech, Uppsala, Sweden) or similar instrument. Nonradioactive detection systems such as the digoxigenin system may also be used, as illustrated in Fig. 30-2.

VI. RESULTS

A. General Considerations

Products are detected as a ladder of bands (see Fig. 30-2). The band with the highest molecular weight represents the largest repeat expansion in that particular genome. A "base line" ligation product formed by di- or trimers of the oligonucleotide used should be seen in all lanes as the human genome contains many short repeat sequences. Absence of such a product should be regarded as a reaction failure in need of troubleshooting. A sample with a known repeat expansion should be included as a positive control to deduce RED product sizes in the samples analyzed.

CAG/CTG is the most common trinucleotide sequence motif found in the genome. It is also the motif that has been most frequently associated with disease. Approximately 29% of the normal population show CAG/CTG expansions at or longer than 180 bp with no apparent phenotype [7]. Two loci, ERDA1 on chromosome 17q21.3 [11a] and CTG18.1 on chromosome 18 [11b], produce 95% of these expansions and can thus be analyzed by PCR. The combination of RED with PCR of these two loci significantly enhances the ability to detect pathogenic expansions. However, some individuals will show RED products unrelated to a disease phenotype even in families segregating a pathological repeat. It is often possible to separate several repeats in a given family by segregation analysis. In addition, the relative intensity of each band in the DNA ladder can be analyzed to distinguish the superimposition of two repeat alleles within one lane [5]. This is possible only if two repeat expansions segregating in one family are of different sizes.

Recent studies have shown that the distribution of CAG/CTG expansions differs from one ethnic group to another [11a, 12]. For example, the Northern European population retains the 120-bp expansion as its most common size (70%) whereas the Chinese population is equally distributed between 120 and 270 bp. This makes it particularly important to ethnically match control groups in association studies.

B. Results from RED Analysis of Spinocerebellar Ataxia Type 7

SCA7 belongs to a group of more than 10 different autosomal dominant cerebellar ataxias that are characterized by degeneration of pathways involving the cerebellum and other brain regions. The clinical phenotypes include ataxia, dysarthria, dysmetria, intention tremor, and, in some forms, psychiatric disturbances including psychotic episodes and depression. The fact that the clinical symptomatology overlaps greatly within this group makes genetic diagnostics particularly useful in confirmation of diagnosis and assessment of prognosis. Anticipation is a hallmark for most of these disorders, and in several a CAG repeat expansion has been identified as the cause of the disease (Chapters 17, 20–22).

SCA7 has a somewhat particular phenotype including macular degeneration leading to blindness [13]. SCA7 maps to chromosome 3p14-21.1 and is characterized by genetic anticipation including both a younger age of onset and a more severe phenotype in successive generations [14–16]. Based on the fact that other dominant ataxias displaying anticipation have been associated with CAG trinucleotide repeat expansions, we undertook a RED analysis of 31 affected and 36 unaffected individuals, 14 of which were spouses [6]. In all families genetic linkage had been confirmed to chromosome 3p. We screened all individuals for CAG repeat expansions using a $(CTG)_{10}$ oligonucleotide. We observed that all affected individuals displayed RED products corresponding to more than 50 copies of the repeat. The repeats segregated with the phenotype within the families, and a χ^2 analysis revealed an association between the repeat expansion and the clinical phenotype ($p < 0.000001$). We observed several cases with a larger RED product in the affected offspring correlating with a more severe clinical presentation. We could therefore determine that the SCA7 phenotype resulted from a CAG trinucleotide repeat expansion. The average size of the repeat was estimated to be 64 copies based on a previous study of the correlation between PCR-based and RED-based size analysis [5]. Interestingly, a number of unaffected offspring in SCA7 families carried expanded CAG repeats as well as the affected haplotype, giving a strong indication that these individuals will be at risk of developing the disease in the future. While this work was in progress an anti-polyglutamine antibody was developed and used to detect a protein product in lymphocytes from several disease states including SCA7 [17], giving added support to our analysis. These results have recently been confirmed through the cloning of the gene harboring the expanded repeat in SCA7 [18].

C. Results from RED Analysis of Bipolar Disorder

There are a number of studies indicating that bipolar affective disorder (BPAD) has quite a marked genetic component [19–21]. The disease affects about 1% of the population and has been linked to multiple loci in the genome. Most of the linkage studies have identified different loci, and while some consensus is now emerging on a few loci, there is still a lot of heterogeneity suggested by the present data. The inheritance patterns seem rather complex [22–24] and there is definite evidence for anticipation in BPAD [25–27]. Thus, there is a significant decrease in survival to first mania or depression from the first to the second generation and a significantly increased disease severity. Based on these observations, we undertook a large association study using the RED technique to study probands of BPAD families as well as sporadic cases from two European regions, the north of Sweden and Belgium [7]. A major finding in this study was a significant ($p = 0.0006$) shift toward larger CAG/CTG allele sizes in both the Swedish and the Belgian BPAD patient population ($n = 123$) as compared to controls ($n = 274$). When all BPAD patients were analyzed together, a markedly increased frequency of CAG repeat expansions (47%) compared to controls (29%) was seen. In this study, we took great care to include ethnically, sex-, and age-matched controls as there have been indications both of sex-based differences in the frequency of repeat expansions [9] and ethnically based differences in repeat expansion distribution [11a, 14]. We also included both an aged control group without any sign or history of psychiatric illness and healthy age- and sex-matched controls. The distribution of repeat expansions was identical in the different control populations and they were therefore pooled into one group for the study. This study [7] provided the first evidence for a CAG/CTG trinucleotide repeat expansion in a major psychiatric disorder and was soon replicated by others [8]. Expanded alleles at two loci, ERDA1 [11a] and CTG18.1 [11b], correlate with the majority of large RED products. Because the CAG/CTG expansions appear to be involved in the clinical expression of BPAD and could be the molecular basis for explaining the phenomenon of anticipation observed, we expanded our analysis to 14 multigenerational families from the north of Sweden (Lindblad *et al.,* submitted). We observed repeat expansions in 12 of the 14 families, and in these 12 families, 81% of the affected offspring had expansions larger than 120 bp. Long CAG/CTG RED products were observed primarily in individuals with an early age of onset. Most of the RED products ≥270 nt correlated with expansions at the CTG18.1 locus. We are thus able to conclude that, at least in a population from the north of Sweden CAG/CTG repeat expansions are overrepresented in families with affective disorder. There also seems to be some genetic heterogeneity as some families segregating affective disorder lack RED products above 120 bases in size. This could naturally be accounted for by genetic heterogeneity alone or by the fact that a shorter repeat is segregating in some of the families in a range difficult to detect with the RED method.

D. Results from Analysis of Childhood-Onset Schizophrenia

There has been some uncertainty with regard to a role for expanded repeat sequences in schizophrenia. To maximize detectability, we choose to work with a highly select group of childhood-onset cases, as well as age-, sex-, and ethnically matched controls [28]. The hypothesis was that early onset cases would represent particularly long, and therefore more easily detectable, repeat expansions. We found an increased incidence of expansions in the disease group, and also observed a shift toward larger-sized products among the childhood cases as compared to previously published results from adult samples [29]. Many of the large RED products correlate with expansions at either the ERDA1 [11a] or CTG18.1 [11b] loci.

VII. TROUBLESHOOTING GUIDE

A. Empty Tubes Following RED Cycling

Select PCR tubes carefully. Due to the long cycling procedure, evaporation can be a problem. We have found that Continental LP tubes work well.

B. Light Areas on Autorad

Poor contact or too much buffer during blotting may obscure part of the ladder present in a given lane.

C. Dark Autorad without Visible Lanes

The probe has randomly hybridized to the whole filter. This can be caused by a spill of RED reaction sample with a large excess of oligonucleotide into the upper buffer tank during gel loading. Flush tips in a separate container and be extremely careful when loading the wells, more so than when loading sequencing reactions or microsatellites.

D. Dark Sample Lanes, Lack of Products

This indicates reaction failure. This could be due to the quality of the DNA. Check the level of degradation and the actual concentration of high-molecular-weight DNA on an agarose gel. High salt concentration may reduce the efficiency of the Ampligase. Try lowering the Ampligase buffer concentration to 0.5×. A low pH also lowers the reaction yield and may cause reaction failure. Include 1 μl buffered TE^{-4}, pH 8.0, in the RED reaction if the DNA is in an unbuffered solution. Weak bands could also be due to poor handling of the reagents. We have noticed that the Ampligase buffer is sensitive to repeated thawing.

VIII. RECENT DEVELOPMENTS

A. Locus-Specific RED

When the sequence flanking a repeat is known, locus-specific RED can be used. In locus-specific RED, a longer oligonucleotide complementary to the region adjacent to the repeat is included in the RED reaction. This will generate both locus-specific and nonspecific RED products. When detected on film, this will appear as a ladder of double bands. The doublet with the highest molecular weight corresponds to the length of the repeat in the chosen locus. Any additional singlet bands indicate expansions at a separate, undefined locus. This form of detection makes possible the determination of repeat size at a specified disease locus for diagnostic purposes. In the future, this process could be performed more rapidly using laser-based semiautomatic detection systems.

B. The RED Method as a Cloning Tool

Given that a number of disorders are characterized by anticipation, it is likely that many arise from trinucleotide expansions. Several studies in our and other laboratories show that the RED technique is a powerful tool for detection of novel repeat expansions associated with disease. Once the disease-associated expansion has been found, the focus of gene cloning efforts can be limited to trinucleotide repeat-containing genes. Two strategies for the cloning of expanded repeats are presented in chapters 31 and 32. We and others have developed methods to clone directly these repeats using RED as a selection tool [11a, 29, 30]. We have taken the approach of DNA fractionation to reduce the complexity of genomic DNA and subsequent screening with the RED technique to identify repeat positive fractions. This pool can then be fractionated again and screened for repeat positive fractions. Assuming that there are 1000 CAG repeats above 7 repeats (permitting stable hybridization) in length within the genome, reduction by a factor of 20–30 suffices to resolve individual bands on a gel or, alternatively, spots in a two-dimensional gel-based separation system. These bands or spots are then identified and cloned into a suitable vector system, sequenced and subsequently PCR-tested on patient material. This strategy is now being applied to a number of different disorders.

Acknowledgments

This study was funded by the EC Biomed 2, Swedish Medical Research Council, Scottish Rite, Svenska Läkarsällskapet, Thuring stiftelse, Söderström-Königska stiftelse, and funds from the Karolinska Institute and Karolinska Hospital. M.S. is the recipient of the Ireland Award, a NARSAD Established Investigator Award.

References

1. Riggins, G. J., Lokey, L. K., Chastain, J. L., Leiner, H. A., Sherman, S. L., Wilkinson, K. D., and Warren, S. T. (1992). Human genes containing polymorphic trinucleotide repeats. *Nature Genet.* **2,** 186–191.

2. Lindblad, K., Zander, C., Schalling, M., and Hudson, T. (1994). Growing triplet repeats. *Nature Genet.* **7,** 124.

3. Campuzano, V., Montermini, L., Molto, M. D., Pianese, L., Cossée, M., Cavalcanti, F., Monros, E., Rodius, F., Duclos, F., Monticelli, A., Zara, F., Cañizares, J., Koutnikova, H., Bidichandani, S. I., Gellera, C., Brice, A., Trouillas, P., De Michele, G., Filla, A., De Frutos, R., Palau, F., Patel, P. I., Di Donato, S., Mandel, J.-L., Cocozza, S., Koenig, M., and Pandolfo, M. (1996). Friedreich's ataxia: autosomal recessive disease caused by an intronic GAA triplet repeat expansion. *Science* **271,** 1423–1427.

4. Schalling, M., Hudson, T. J., Buetow, K. H., and Housman, D. E. (1993). Direct detection of novel expanded trinucleotide repeats in the human genome. *Nature Genet.* **7,** 135–139.

5. Lindblad, K., Lunkes, A., Maciel, P., Stevanin, G., Zander, C., Klockgether, T., Ratzlaff, T., Brice, A., Rouleau, G. A., Hudson, T., Auburger, G. and Schalling, M. (1996). Mutation detection in Machado-Joseph disease using Repeat Expansion Detection. *Mol. Med.* **2,** 77–85.

6. Lindblad, K., Savontaus, M.-L., Stevanin, G., Holmberg, M., Digre, K., Zander, C., Ehrsson, H., David, G., Benomar, A., Nikoskelainen, E., Trottier, Y., Holmgren, G., Ptacek, L. J., Anttinen, A., Brice, A., and Schalling, M. (1996). An expanded CAG repeat sequence in spinocerebellar ataxia type 7. *Genome Res.* **6,** 965–971.

7. Lindblad, K., Nylander, P.-O., De Bruyn, A., Sourey, D., Zander, C., Engström, C., Holmgren, G., Hudson, T., Chotai, J., Mendlewicz, J., Van Broeckhoven, C., Schalling, M. and Adolfsson, R. (1995). Detection of expanded CAG repeats in bipolar affective disorder using the repeat expansion detection (RED) method. *Neurobiol. Dis.* **2,** 55–62.

8. O'Donovan, M. C., Guy, C., Craddock, N., Murphy, K. C., Cardino, A. G., Jones, L. A., Owen, M. J., and McGuffin, P. (1995). Expanded CAG repeats in schizophrenia and bipolar disorder. *Nature Genet.* **10,** 380–381.

9. Morris, A. G., Gaitonde, E., McKenna, P. J., Mollon, J. D., and Hunt, D. M. (1995). CAG repeat expansions and schizophrenia: association with disease in females and with early age-of-onset. *Hum. Mol. Genet.* **4,** 1957–1961.

10. Zander, C., Thelaus, J., Lindblad, K., Karlsson, M., Schalling, M. Multivariate analysis of factors influencing Repeat Expansion Detection. *Genome Research* [In press]

11. Schalling, M., Friberg, K., Seroogy, K., Riederer,P., Bird, E., Schiffmann, S. N., Mailleux, P., Vanderhaeghen, J.-J., Kuga, S., Goldstein, M., Luppi, P. H., Jouvet, M., and Hökfelt, T. (1990). Analysis of expression of cholecystokinin in dopamine cells in the ventral mesencephalon of several species and in humans with schizophrenia. *Proc. Natl. Acad. Sci. USA* **87,** 8427–8431.

11a. Nakamoto, M., Takebayashi, H., Kawaguchi, Y., Narumiya, S., Taniwaki, M., Nakamura, Y., Ishikawa, Y., Akiguchi, I., Kimura, J., and Kakizuka, A. (1997). A CAG/CTG expansion in the normal population. *Nature Genet.* **17,** 385–386.

11b. Breschel, T. S., McInnis, M. G., Margolis, R. L., Sirugo, G.,

Corneliussen, B., Simpson, S. G., McMahon, F. J., MacKinnon, D. F., Xu, J. F., Pleasant, N., Huo, Y., Ashworth, R. G., Grundstrom, T., Kidd, K. K., DePaulo, J. R., and Ross, C. A. (1997). A novel, heritable, expanding CTG repeat in an intron of the SEF2-1 gene on chromosome 18q21.1. *Hum. Mol. Genet.* **6,** 1855–1863.

12. Sirugo, G., Deinard, A., Kidd, J., and Kidd, K. (1997). Survey of maximum CTG/CAG repeat lengths in humans and nonhuman primates: total genome scan in populations using the repeat expansion detection method. *Hum. Mol. Genet.* **6,** 403–408.
13. Anttinen, A., Nikoskelainen, E., Marttila, R. J., Grenman, R., Falck, B., Aarnisalo, E., and Kalimo, H. (1986). Familial olivopontocerebellar atrophy with macular degeneration: a separate entity among the olivopontocerebellar atrophies. *Acta Neurol. Scand.* **73,** 180–190.
14. Benomar, A., Krols, L., Stevanin, G., Cancel, G., Davdi, G., Ouhabi, H., Maratin, J.-J., Dürr, A., Zaim, A., Ravisé, N., Busque, C., Penet, C., Van Regemorteer, N., Weissenbach, J., Yahyaoui, M., Chkili, T., Agid, Y., Van Brockhoven, C., and Brice, A. (1995). The gene for autosomal dominant cerebellar ataxia with pigmentary macular dystrophy maps to chromosome 3p12-p21.1. *Nature Genet.* **10,** 84–88.
15. Gouw, L. G., Kaplan, C. D., Haines, J. H., Digre, K. B., Rutledge, S. L., Matilla, A., Leppert, M., Zoghbi, H. Y., and Ptacek, L. J. (1995). Retinal degeneration characterizes a spinocerebellar ataxia mapping to chromosome 3p. *Nature Genet.* **10,** 89–93.
16. Holmberg, M., Johansson, J., Forsgren, L., Heijbel, J., Sandgren, O., and Holmgren, G. (1995). Localization of autosomal dominant cerebellar ataxia associated with retinal degeneration and anticipation to chromosome 3p12-p21.1. *Hum. Mol. Genet.* **4,** 1441–1445.
17. Trottier, Y., Lutz, Y., Stevanin, G., Imbert, G., Devys, D., Cancel, G., Saudou, F., Weber, C., David, G., Tora, L., Agid, Y., Brice, A., and Mandel, J.-L. (1995). Polyglutamine expansion as a pathological epitope in Huntington's disease and four dominant cerebellar ataxias. *Nature* **378,** 403–406.
18. David, G., Abbas, N., Stevanin, G., Dürr, A., Yvert, G., Cancel, G., Weber, C., Imbert, G., Saudou, F., Antoniou, E., Drabkin, H., Gemmill, R., Giunti, P., Benomar, A., Wood, N., Ruberg, M., Agid, Y., Mandel, J.-L., and Brice, A. (1997). Cloning of the SCA7 gene reveals a highly unstable CAG repeat expansion. *Nature Genet.* **17,** 65–70.
19. Mendlewicz, J. and Rainer, J. D. (1977). Adoption study supporting genetic transmission in manic-depressive illness. *Nature* **268,** 327–329.
20. Bertelsen, A., Harvald, B., and Hauge, M. A. (1977). Danish twin study of manic depressive disorders. *Br. J. Psych.* **130,** 330–351.
21. Rice, J., Reich, T., Andreason, N. C., Endicott, J., Van Eerdewegh, M., Fishman, R., Hirschfeld, R. M. A., and Klerman, G. L. (1987). The familial transmission of bipolar illness. *Arch. Gen. Psych.* **44,** 441–447.
22. Goldin, L. R., Cox, N. J., Pauls, D. L., Gershon, E. S., and Kidd, K. K. (1984). The detection of major loci by segregation and linkage analysis: a simulation study. *Genet. Epidemiol.* **1,** 285–296.
23. Orr, H. T., Chung, M., Banfi, S., Kwiatkowski, T. J., Jr., Servadio, A., Beaudet, A. L., McCall, A. E., Duvick, L. A., Ranum, L. P. W., and Zoghbi, H. Y. (1993). Expansion of an unstable trinucleotide (CAG) repeat in spinocerebellar ataxia type 1. *Nature Genet* **4,** 221–226.
24. Risch, N. (1990). Genetic linkage and complex diseases, with special reference to psychiatric disorders. *Genet. Epidemiol.* **7,** 3–16.
25. McInnis, M. G., Mcmahon, F. J., Chase, G. A., Simpson, S. G., Ross, C. A., and DePaulo, J. P., Jr. (1993). Anticipation in bipolar affective disorder. *Am. J. Hum. Genet.* **53,** 385–390.
26. McInnis, M. G. (1996). Anticipation: an old idea in new genes. *Am. J. Hum. Genet.* **59,** 973–979.
27. Nylander, P.-O., Engström, C., Chotai, J., Wahlström, J., and Adolfsson, R. (1994). Anticipation in Swedish families with bipolar affective disorder. *J. Med. Genet.* **31,** 686–689.
28. Burgess, C. E., Lindblad, K., Krain, A., Sidransky, E., Yuan, Q.-P., Long, R. T., Breschel, T., McInnis, M., Ross, C. A., Lee, P., Ginns, E., Lenane, M., Kumra, S., Jacobsen, L., Rapoport, J. L., and Schalling, M. (1998). Large CAG/CTG repeats are associated with childhood onset schizophrenia. *Mol. Psychiatr.* [In press]
29. Lindblad, K., Burgess, C. E., Yuan, Q.-P., Lakkis, L., Hudson, T. J., Warren, S. T., and Schalling, M. (1997). A strategy for isolation of sequence flanking trinucleotide repeat expansion loci. *Am. J. Hum. Genet.* **61,** A313 (1832).
30. Koob, M. D., Benzoq, K. A., Bird, T. D., Day, J. W., Moseley, M. L., Ranum, L. P. W. (1998). RAPID cloning of expanded trinucleotide repeat sequences from genomic DNA. *Nature Genet.* **18,** 72–75.

CHAPTER 31

Selective Recognition of Proteins with Pathological Polyglutamine Tracts by a Monoclonal Antibody

YVON TROTTIER Institut de Génétique et de Biologie Moléculaire et Cellulaire (IGBMC), CNRS/INSERM/ULP, Illkirch-Strasbourg, France

GABRIELLE ZEDER-LUTZ Institut de Biologie Moléculaire et Cellulaire (IBMC), CNRS/ULP-9021, Strasbourg, France

JEAN-LOUIS MANDEL Institut de Génétique et de Biologie Moléculaire et Cellulaire (IGBMC), CNRS/INSERM/ULP, Illkirch-Strasbourg, France

I. INTRODUCTION

Polyglutamine expansion (encoded by a CAG repeat) in specific proteins is one of the most intriguing pathological mechanisms causing adult-onset neurodegenerative disorders. While the normal gene products tolerate, without any detectable adverse effect, a rather wide variation in size of a polyglutamine tract (polygln) (typically between 10 and 35 glns), beyond a threshold (35–40 glns in five of the eight known diseases) the proteins acquire pathogenic properties. To date in all the eight diseases known to share this mechanism, there is a strong inverse correlation between the length of the polygln tract and the age of onset of clinical symptoms: each added gln residue beyond the threshold results, on average, in a 1.5 to 2 years earlier onset. At the DNA level, the expanded CAG repeats are found to be unstable upon transmission from one generation to the next (except for the shorter pathologic repeats in SCA6 [52]), with a clear tendency to expansion. These two features account for the anticipation phenomenon (increased severity in successive generations) that is observed at various degrees in polygln expansion diseases.

Polygln expansion was first identified in the androgen receptor leading to the X-linked spino-bulbar muscular atrophy (SBMA or Kennedy's disease), a different phenotype than those caused by loss of function mutations in this gene [2, 24]. Polygln expansions have since then been found in a growing class of autosomal dominant degenerative disorders: Huntington's disease (HD), five spinocerebellar ataxias (SCA1–3, SCA6, and SCA7), and dentatorubral-pallidoluysian atrophy (DRPLA) [3, 19, 22, 23, 33, 36, 40, 44, 49, 52]. The functions of the proteins involved in these disorders are unknown, except for the transcription factor activity of the androgen receptor (SBMA) and for the SCA6 gene that encodes a brain-specific subunit of a calcium channel [2, 52]. In the other diseases, affected proteins share no similarity either with proteins of known function or among themselves, giving no clue about putative functional domains. Moreover the various proteins are found in either cytoplasm or nucleus or in both compartments, suggesting that they do not share a common function in the cell.

Polygln expansion diseases are all characterized by neuronal cell death, but in different brain regions, leading to different clinical manifestations [42]. The selective neuronal death shows no obvious correlation with the pattern of expression of the corresponding proteins, since other cells expressing the pathological proteins do not appear affected. The mechanism of cell death due to polygln expansion has not yet been elucidated, but is likely to be common to all diseases (with the possible exception of SCA6 [52]) and is thought to confer a novel property (gain of toxic function) to the affected protein (see [37, 42] for reviews).

We will present here evidence that polygln expansion generates a specific alteration in the conformation of the target proteins, creating or stabilizing an epitope detected by a monoclonal antibody (1C2). The polygln length-dependent properties of the epitope parallel the effect on disease severity. This conformation change may have various consequences like favoring self-interaction or interaction with new proteins, modifying maturation, phosphorylation, turnover or processing of the mutated protein. Evidence for an intranuclear accumulation of polygln expansion containing polypeptides in target neurons has been recently reported in an HD transgenic mice model and in the brain of patients with SCA3 [4, 38]. These findings support a self-aggregation model [45]. We will also discuss applications of the 1C2 antibody to the screening for other putative polyglutamine expansion diseases, and to the further understanding of the pathological mechanism.

II. SPECIFIC RECOGNITION OF POLYGLUTAMINE EXPANSIONS BY A MONOCLONAL ANTIBODY

Polygln stretches are present in many eukaryotic proteins, particularly in several transcription factors [7] and in proteins of unknown function identified by screens for CAG repeats in cDNA libraries [21, 26, 34]. In humans, the TATA-binding protein (TBP), a general transcription factor, also has a polymorphic polygln stretch with the most common allelic form of 38 glns, and rare variants having 42 glns [11, 20]. This represents the longest known polygln stretch in a human protein that is not associated to a disease phenotype (very rare, apparently normal, alleles have been also reported at the SCA3 locus, with 43 and 44 glns) [18].

A monoclonal antibody (1C2) raised against the TBP protein was developed by Dr. L. Tora and the epitope mapping analysis indicated that it recognizes its polygln-containing region [25]. We were interested to know whether this antibody could recognize other proteins containing an expanded polygln stretch, like huntingtin. Using specific anti-huntingtin antibodies for Western blotting (WB) analysis, we and others showed that the migration of huntingtin is strictly dependent on the size of the polygln, allowing separation of a mutated huntingtin from a normal one [5, 46, 50]. The 1C2 antibody showed the unexpected property of recognizing selectively the mutated huntingtins, while the normal ones gave very faint or no signal [51]. Moreover the ability of 1C2 to detect mutated huntingtin was clearly dependent on the polygln size; in a semiquantitative assay, proteins containing a long stretch (60–85 glns) gave a much higher signal intensity (about 20- to 40-fold) than those with 39 and 40 glns (representing the smaller pathological alleles) (Fig. 31-1). Under certain conditions of detection (very long exposure of the blot), huntingtins having polygln in the upper normal range (≥28) were faintly detected, but not those with smaller polygln tracts. Thus, the efficiency of detection of pathological alleles appeared to parallel the severity of the disease.

The recognition properties of 1C2 were confirmed for other glutamine-repeat diseases, as the antibody selectively detected the mutated ataxin-1 and ataxin-3, respectively, in lymphoblastoid cell lines (LCL) from SCA1 and SCA3 patients. Even for TBP, the allelic forms with 38–42 glns are recognized more strongly than those with 32–33 glns, while those with ≤30 glns

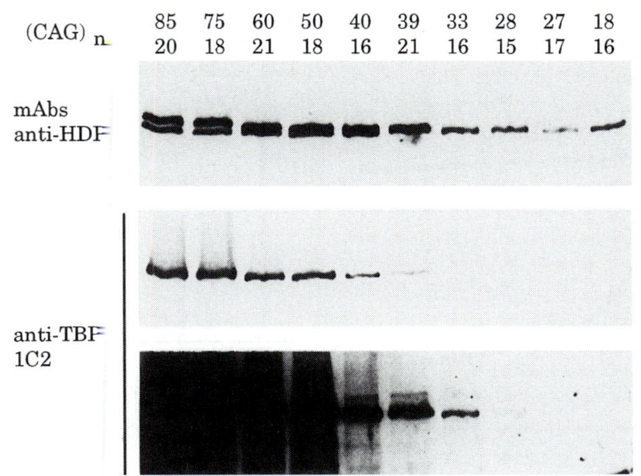

FIGURE 31-1 Preferential detection by 1C2 of huntingtin with expanded polyglutamine. Lymphoblastoid cell extracts from HD and normal individuals are ordered by decreasing length of polyglutamine with respect to the larger allele. Using specific anti-huntingtin monoclonal antibodies, allelic huntingtins are detected with equivalent signal intensities. With 1C2 antibody, the intensity of signal of the mutated huntingtin increases with the size of the expansion, from 39 to 85 glns. A gross overexposure of the blots shows a signal for huntingtin with a large normal alleles of 33 glns, and a very faint one for 28 glns (bottom). (Reprinted with permission from *Nature* ***378***, 403–406, copyright 1995 Macmillan Magazines Limited).

were not detected under the experimental conditions used [51].

The 1C2 antibody was used to test the hypothesis of polygln expansion in patients with other neurological diseases, with well-documented or suggested anticipation, including familial spastic paraplegia (FSP) and spinocerebellar ataxia [51]. On WB analysis of LCL from patients, 1C2 detected a 130-kDa protein specific to patients with a form of spinocerebellar ataxia associated with retinal degeneration (locus SCA7 that had been mapped on chromosome 3 [1, 13, 17]), and a 150-kDa protein specific to patients with chromosome 12-linked spinocerebellar ataxia patients [9] (SCA2, clinically indistinguishable from the SCA1 and SCA3 forms), but no novel protein was detected in FSP LCLs. Moreover, in two SCA2 families, the signal intensity of the 150-kDa protein increased in the next generation, and appeared to be inversely correlated with patient's age at disease onset, suggesting that the polygln stretch contained in the 150-kDa protein increases in size through parental transmission and accounts for the anticipation phenomenon observed in these families.

The 1C2 antibody was then used to screen two cDNA expression libraries from SCA2 and SCA7 LCLs, respectively. No clones corresponding to the expected mutated alleles were recovered. The absence of such clones might be due to a toxic effect of expanded polygln in *Escherichia coli,* as reported by Onodera *et al.* [35], but this possibility has not been experimentally tested in the phage system used. We, however, observed that 1C2 was able to detect smaller gln repeats (above 11 glns) under the bacteriophage plaque screening conditions, which are less stringent than those for Western blot (no SDS and reducing agent). This allowed us to identify five new polygln coding cDNAs, including one corresponding to a normal allele of the SCA2 gene [19].

III. ANALYSIS OF THE INTERACTION BETWEEN 1C2 AND POLYGLUTAMINES

The recognition properties of the 1C2 antibody could be explained in at least two ways. First, 1C2 may have a higher affinity for polyglutamine expansions than for normal stretches. Since the same "primary sequence" epitope, the repeated glns, is present in both normal and expanded stretches, a higher affinity could be accounted for by the presence of a specific conformation epitope (different than that of the normal stretch) that would appear in polyglns of >35 glns, and that might be stabilized or more predominant in longer stretches. Second, expansion of a polyglutamine epitope may facilitate a bivalent binding of the antibody, and thus stabilize the interaction. The two hypotheses are, however, not mutually exclusive.

To test these hypotheses, a kinetic analysis of the interaction between the F(ab) fragment of 1C2 antibody and polygln tracts was performed under native conditions using a surface plasmon resonance biosensor (BIAcore, Pharmacia) [32]. The F(ab) fragment showed a much stronger affinity for an expansion of 73 glns than for a normal polygln stretch (15 glns). Notably, the complex formed between the F(ab) (that has a single antigen binding site) and the polygln expansion is remarkably stable, and it dissociates 100 times more slowly than that formed with a normal polygln stretch (Fig. 31-2). Similar kinetics of interaction were obtained on BIAcore using the whole 1C2 antibody, indicating that a bivalent interaction is not a major factor in the detection process (G.Z.-L., unpublished results). These data indicate that the expanded polyglutamine adopts a "new" conformation, which is better recognized by 1C2.

In several experimental protocols where interaction was tested using near to native conditions (BIAcore, ELISA, or bacteriophage screening), the 1C2 antibody was clearly able to recognize polygln stretches of 11–20 glns. Even in the initial epitope mapping experiment, where a very high density of polypeptides was used, a polypeptide containing Q6 gave a positive signal. This contrasts with the apparent threshold of detection (~33 glns) found under the denaturing conditions of

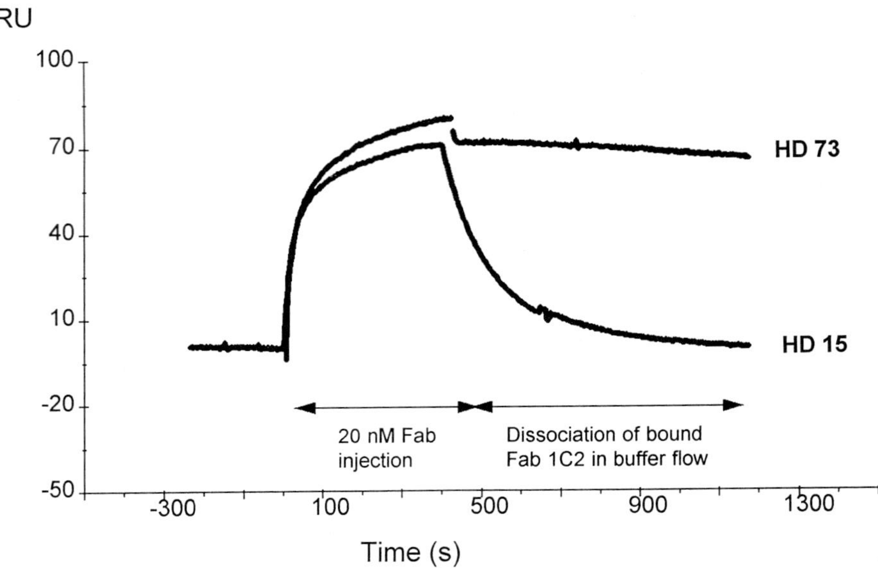

FIGURE 31-2 Sensorgrams of the interaction of a F(ab) fragment of 1C2 with histidine-tagged truncated huntingtins having 15 or 73 glns. The curves show the response from the surface plasmon resonance detector (expressed in Resonance Units, RU), and illustrate the striking difference in the dissociation of the F(ab)/antigen complex between a normal stretch (HD15) and a pathological one (HD73). Similar data were obtained when using the whole monoclonal antibody instead of the F(ab) fragment. Around 50 pg of each target protein was immobilized on a nickel–NTA surface. HD15 and HD73 correspond to the amino terminus of huntingtin with 15 glns (545-aa fragment) and 73 glns (603-aa fragment), respectively, and an in-frame histidine tag sequence at the amino terminus.

WB, where the target protein is unfolded with SDS, a reducing agent, and by heating. It is possible that the epitope recognized by 1C2 is unstable under denaturing conditions until the polygln stretch reaches a critical length, or alternatively, that denaturing conditions favor the formation or stabilization of the epitope.

IV. POTENTIAL USES OF THE 1C2 ANTIBODY

Given its properties, the 1C2 antibody may be a useful tool for identifying new proteins/genes involved in other polyglutamine-expansion neurodegenerative disorders, with proven or suggested anticipation, as already shown for SCA2 and SCA7 [51]. This might be the case for other spinocerebellar ataxias (SCA5 localized on 11cen [41], and SCA4 on 16q [6]), autosomal dominant FSP linked to chromosome 14q and 2p [8, 14–16], certain families with bipolar affective illness or schizophrenia, and rare forms of Parkinson's disease [43]. It should be emphasized that the presence of anticipation in psychiatric disorders is rather controversial, as it might well be caused by ascertainment bias [30, 31]. Also, given the known high genetic heterogeneity of these diseases, a putative expansion mutation is very unlikely to constitute a major genetic cause. It could, however, either constitute a high relative risk in a small proportion of patients, or a low relative risk in a larger set of patients.

We found no evidence for expanded polyglutamine detected by 1C2 in LCLs from patients with the FSP form linked to chromosome 2 [14] and with an unlocalized autosomal dominant FSP form [51]. Several groups tried without success to identify new types of polygln expansion in patients with sporadic or unassigned autosomal dominant cerebellar ataxia [28, 47]. One should note, however, that while a positive result proves the mechanism, false-negative findings may be obtained if the pathological protein is poorly expressed or absent in the cells analyzed, or if the polyglutamine is too small to be detected with high sensitivity. For instance, patients with SCA6 mutations would be missed, as the target protein is brain specific, and also probably because of the relatively small size of pathologic expansions (21 to 30 glns) [52].

Typically on our WB analysis, 1C2 revealed TBP (at about 49 kDa) and an unknown ~230-kDa nuclear protein in all LCLs analyzed. Pathologic polygln-containing proteins like huntingtin or ataxin-3 were detected with equal or much higher signal than that of TBP. However ataxin-1 with less than 50 glns was not detected, most probably due to its low expression in LCLs [51]. Under some conditions (for instance, high concentration of antibody, very long exposure of the blot, and/or avoiding stringent washes), several additional nuclear proteins of different sizes are faintly detected in normal controls, and some of them even show a variability of detection, which could perhaps be explained by polymorphic polyglutamine repeats (Y.T., unpublished observations). These proteins may mask a specific signal or cause problems of interpretation unless a cytoplasmic cell fraction is analyzed. In any case, in the search of a putative polygln expansion protein, analysis of early onset cases is warranted, given the strong polyglutamine length-dependent properties of the antibody.

The antibody is also a good reagent to clone candidate genes encoding polygln, as proved in our phage screening, where we obtained five novel genes [19]. The use of denaturing conditions might help to increase the specificity for pathological alleles, when screening phage libraries with 1C2. An appropriate positive control would be required, however, to check whether the phage system used allows expression of a protein with a long polygln stretch, and indeed yields a much stronger signal by immunoscreening than for a normal allele.

The 1C2 antibody can be used to selectively follow the localization and the fate of the abnormal proteins in brain samples of patients deceased from a polygln expansion disease, and in cellular and animal models for polygln expansion toxicity. In immunohistochemical applications, it may be necessary to use very high dilutions of the mAb, to avoid (or decrease) detection of TBP or other endogenous nuclear proteins that react with 1C2 [12]. Indeed, the antibody was helpful for the characterization of transgenic mice expressing a mutated huntingtin fragment (with about 150 glns) that show a phenotype comparable to the clinical symptoms found in HD [29]. Davies *et al.* [4] recently observed in these mice abnormal intranuclear inclusions, containing the short mutated huntingtin fragment, that are specifically present in neurons (especially in brain regions corresponding to those affected in HD patients), and that appear prior to the onset of neurological manifestations. Similar intranuclear inclusions have been detected in brain from SCA3 patients, using both anti-ataxin 3 antibodies and 1C2 [38]. These abnormal structures are also labeled by anti-ubiquitin antibodies [4, 38]. Expanded polyglutamines can aggregate *in vitro,* forming amyloid-like fibrils, a phenomenon that could account for the observations of nuclear inclusions [45]. In a study of brains from HD patients, it was also reported that mutated huntingtin (as detected by 1C2) showed a more variable staining intensity in neurons than huntingtin detected by other antibodies in control brains. [12]. This suggests that differential accumulation of mutated huntingtin (or at least of fragments containing the 1C2 epitope) could account for the fact that even in a given region, all neurons do not die at the same time. All together, these recent observations indicate that accumulation/aggregation of the pathological, possibly truncated, protein constitutes an important step in the pathogenesis of this class of diseases.

V. CONCLUSION

Our results with 1C2 indicate that, past the pathological threshold, a conformational change occurs within the polygln stretch. The length-dependent recognition of the antibody, which mimics the clinical severity of the disease, suggests that the occurrence or the stability of this conformation increases with the size of the expansion, and that this pathological epitope is directly involved in the disease mechanism. It could for instance facilitate aggregation and formation of intranuclear inclusions. The presence of this novel conformational epitope might also alter the conformation of adjacent domains of the target proteins, and this may have functional consequences like favoring interaction with other proteins or modifying maturation, phosphorylation, or processing of the mutated protein. For instance, the presence of this pathological epitope might account for the polygln length-dependent interaction with the HAP1 protein [27], or alter the cleavage by apopain (caspase 3) as reported for huntingtin [10]. Such functional alterations could play a role in the differential clinical phenotypes of the various diseases.

Perutz and colleagues [39, 48] showed that even short polyglns can form stable oligomers involving β-pleated sheets (a sort of "polar zipper"). It remains to be demonstrated whether such model structure may account for the narrow and similar threshold observed in five of the eight polygln expansion diseases, and for the dramatic effect of increased polygln length. X-ray crystallographic analysis of very long polyglutamine stretches is thus warranted. Toward this goal, we are presently attempting to cocrystallize a polygln containing protein fragment with the 1C2 F(ab). The abnormal polygln epitope may constitute a good target to screen for therapeutic agents that could interfere with the toxic effect of polygln expansion proteins. Indeed it would be very interesting to see whether a single-chain Fv antibody derived from 1C2 can interfere, when targeted to neurons, with the pathological manifestations that occur in available mouse models for HD, SCA1, or SCA3. This could open the way to new therapeutic approaches in polygln diseases.

Acknowledgments

We thank Dr. L. Tora for generously providing the 1C2 antibody, Y. Lutz for assistance in antibody production and in immunological tests, and C. Weber for technical assistance. We thank Drs. A. Brice, D. Devys, G. Imbert, A. Lunkes, G. Stevanin, and G. Yvert for useful discussions and for collaboration on various aspects of the work reported here, and Dr. M. H. V. Van Regenmortel for support and advice on the BIAcore experiments. This work was supported by grants from CNRS, INSERM, EEC (BMH4-CT96-0244), GREG, and CHRU of Strasbourg. Y.T. is supported by a fellowship from the Hereditary Disease Foundation (U.S.A.).

References

1. Benomar, A., Krols, L., Stevanin, G., Cancel, G., Leguern, E., David, G., Ouhabi, H., Martin, J. J., Durr, A., Zaim, A., Ravise, N., Busque, C., Penet, C., Vanregemorter, N., Weissenbach, J., Yahyaoui, M., Chkili, T., Agid, Y., Vanbroeckhoven, C., and Brice, A. (1995). The gene for autosomal dominant cerebellar ataxia with pigmentary macular dystrophy maps to chromosome 3p12-p21.1. *Nature Genet.* **10,** 84–88.
2. Brooks, B. P., and Fischbeck, K. H. (1995). Spinal and bulbar muscular atrophy: a trinucleotide-repeat expansion neurodegenerative disease. *Trends n Neurosci.* **18,** 459–461.
3. David, G., Abbas, N., Stevanin, G., Durr, A., Yvert, G., Cancel, G., Weber, C., Imbert, G., Saudou, F., Antoniou, E., Drabkin, H., Gemmill, R., Giunti, P., Benomar, A., Wood, N., Ruberg, M., Agid, Y., Mandel, J.-L., and Brice, A. (1997). Cloning for the spinocerebellar ataxia 7 gene (SCA7): highly unstable CAG repeat expansion in ADCA type II. *Nature Genet.* **17,** 65–70.
4. Davies, S. W., Turmaine, M., Cozens, B. A., Difiglia, M., Sharp, A. H., Ross, C. A., Scherzinger, E., Wanker, E. E., Mangiarini, L., and Bates, G. P. (1997). Formation of neuronal intranuclear inclusions underlies the neurological dysfunction in mice transgenic for the HD mutation. *Cell* **90,** 537–548.
5. Difiglia, M., Sapp, E., Chase, K., Schwarz, C., Meloni, A., Young, C., Martin, E., Vonsattel, J.-P., Carraway, R., Reeves, S. A., Boyce, F. M., and Aronin, N. (1995). Huntingtin is a cytoplasmic protein associated with vesicules in human and rat brain neurons. *Neuron* **14,** 1075–1081.
6. Flanigan, K., Gardner, K., Alderson, K., Galster, B., Otterud, B., Leppert, M. F., Kaplan, C., and Ptacek, L. J. (1996). Autosomal dominant spinocerebellar ataxia with sensory axonal neuropathy (SCA4): clinical description and genetic localization to chromosome 16q22.1. *Am. J. Hum. Genet.* **59,** 392–399.
7. Gerber, H. P., Seipel, K., Georgiev, O., Hofferer, M., Hug, M., Rusconi, S., and Schaffner, W. (1994). Transcriptional activation modulated by homopolymeric glutamine and proline stretches. *Science* **263,** 808–811.
8. Gispert, S., Santos, N., Damen, R., Voit, T., Schulz, J., Klockgether, T., Orozco, G., Kreuz, F., Weissenbach, J., and Auburger, G. (1995). Autosomal dominant familial spastic paraplegia: reduction of the FSPI candidate region on chromosome 14q to 7 cM and locus heterogeneity. *Am. J. Hum. Genet.* **56,** 183–187.
9. Gispert, S., Twells, R., Orozco, G., Brice, A., Weber, J., Heredero, L., Scheufler, K., Riley, B., Allotey, R., Nothers, C., Hillermann, R., Lunkes, A., Khati, C., Stevanin, G., Hernandez, A., Magarino, C., Klockgether, T., Durr, A., Chneiweiss, H., Enczmann, J., Farrall, M., Beckmann, J., Mullan, M., Wernet, P., Agid, Y., Freund, H.-J., Williamson, R., Auburger, G., and Chamberlain, S. (1993). Chromosomal assignment of the second locus for autosomal dominant cerebellar ataxia (SCA2) to chromosome 12q23-24.1. *Nature Genet.* **4,** 295–299.
10. Goldberg, Y. P., Nicholson, D. W., Rasper, D. M., Kalchman, M. A., Koide, H. B., Graham, R. K., Bromm, M., Kazemi-Esfarjani, P., Thornberry, N. A., Vaillancourt, J. P., and Hayden, M. R. (1996). Cleavage of huntingtin by apopain, a proapoptotic cysteine protease, is modulated by the polyglutamine tract. *Nature Genet.* **13,** 442–449.
11. Gostout, B., Liu, Q., and Sommer, S. S. (1993). "Cryptic" repeating triplets of purines and pyrimidines (cRRY(i)) are frequent and polymorphic: analysis of coding cRRY(i) in the proopiomelanocortin (POMC) and TATA-binding protein (TBP) genes. *Am. J. Hum. Genet.* **52,** 1182–1190.
12. Gourfinkel-An, I., Cancel, G., Trottier, Y., Devys, D., Tora, L., Lutz, Y., Imbert, G., Saudou, F., Stevanin, G., Agid, Y., Brice, A., Mandel, J.-L., and Hirsch, E. C. (1997). Differential distribution of the normal and mutated form of huntingtin in the human brain. *Ann. Neurol.* **42,** 712–719.
13. Gouw, L. G., Kaplan, C. D., Haines, J. H., Digre, K. B., Rutledge, S. L., Matilla, A., Leppert, M., Zoghbi, H. Y., and Ptacek, L. J. (1995). Retinal degeneration characterizes a spinocerebellar ataxia mapping to chromosome 3p. *Nature Genet.* **10,** 89–93.
14. Hazan, J., Fontaine, B., Bruyn, R. P. M., Lamy, C., van Deutekom, J. C. TJ. C., Rime, C.-S., Dürr, A., Melki, J., Lyon-Caen, O., Agid, Y., Munnich, A., Padberg, G. W., de Recondo, J., Frants, R. R., Brice, A., and Weissenbach, J. (1994). Linkage of a new locus for autosomal dominant familial spastic paraplegia to chromosome 2p. *Hum. Mol. Genet.* **3,** 1569–1573.
15. Hazan, J., Lamy, C., Melki, J., Munnich, A., de Recondo, J., and Weissenbach, J. (1993). Autosomal dominant familial spastic paraplegia is genetically heterogeneous and one locus maps to chromosome 14q. *Nature Genet.* **5,** 163–167.
16. Hentati, A., Pericak-Vance, M. A., Lennon, F., Wasserman, B., Hentati, F., Juneja, T., Angrist, M. H., Hung, W.-Y., Boustany, R.-M., Bohlega, S., Iqbal, Z., Huether, C. H., Ben Hamida, M., and Siddique, T. (1994). Linkage of a locus for autosomal dominant familial spastic paraplegia to chromosome 2p markers. *Hum. Mol. Genet.* **3,** 1867–1871.
17. Holmberg, M., Johansson, J., Forsgren, L., Heijbel, J., Sandgren, O., and Holmgren, G. (1995). Localization of autosomal dominant cerebellar ataxia associated with retinal degeneration and anticipation to chromosome 3p12-p21.1. *Hum. Mol. Genet.* **4,** 1441–1445.
18. Hsieh, M., Tsai, H. F., Lu, T. M., Yang, C. Y., Wu, H. M., and Li, S. Y. (1997). Studies of the CAG repeat in the Machado-Joseph disease gene in Taiwan. *Hum. Genet.* **100,** 155–162.
19. Imbert, G., Saudou, F., Yvert, G., Devys, D., Trottier, Y., Garnier, J. M., Weber, C., Mandel, J. L., Cancel, G., Abbas, N., Dürr, A., Didierjean, O., Stevanin, G., Agid, Y., and Brice, A. (1996). Cloning of the gene for spinocerebellar ataxia 2 reveals a locus with high sensitivity to expanded CAG/glutamine repeats. *Nature Genet.* **13,** 285–291.
20. Imbert, G., Trottier, Y., Beckman, J., and Mandel, J.-L. (1994). The gene for the TATA-binding protein (TBP) that contains a highly polymorphic protein coding CAG repeat maps to 6q27. *Genomics* **21,** 667–668.
21. Jiang, J.-X., Lekanne Deprez, R., Zwarthoff, E., and Riegman, P. (1995). Characterization of Four Novel CAG Repeat-Containing cDNAs. *Genomics* **30,** 91–93.
22. Kawaguchi, Y., Okamoto, T., Taniwaki, M., Aizawa, M., Inoue, M., Katayama, S., Kawakami, H., Nakamura, S., Nishimura, M., Akiguchi, I., Kimura, J., Narumiya, S., and Kazizuka, A. (1995). CAG expansions in a novel gene for Machado-Joseph disease at chromosome 14q32.1. *Nature Genet.* **8,** 221–228.
23. Koide, R., Ikeuchi, T., Onodera, O., Tanaka, H., Igarashi, S., Endo, K., Takahashi, H., Kondo, R., Ishikawa, A., Hayashi, T., Saito, M., Tomoda, A., Miike, T., Naito, T., Ikuta, F., and Tsuji, S. (1994). Unstable expansion of CAG repeat in hereditary

dentatorubral-pallidoluysian atrophy (DRPLA). *Nature Genet.* **6,** 9–13.

24. La Spada, A. R., Wilson, E. M., Lubahn, D. B., Harding, A. E., and Fischbeck, K. H. (1991). Androgen receptor gene mutations in X-linked spinal and bulbar muscular atrophy. *Nature* **352,** 77–79.
25. Lescure, A., Lutz, Y., Eberhard, D., Jacq, X., Krol, A., Grummt, I., Davidson, I., Chambon, P., and Tora, L. (1995). The N-terminal domain of the human TATA-binding protein plays a role in transcription from TATA-containing RNA polymerase II and III promoters. *EMBO J.* **13,** 1166–1175.
26. Li, S. H., McInnis, M. G., Margolis, R. L., Antonarakis, S. E., and Ross, C. A. (1993). Novel triplet repeat containing genes in human brain: cloning, expression, and length polymorphisms. *Genomics* **16,** 572–579.
27. Li, X. J., Li, S. H., Sharp, A. H., Nucifora, F. C., Jr., Schilling, G., Lanahan, A., Worley, P., Snyder, S. H., and Ross, C. A. (1995). A huntingtin-associated protein enriched in brain with implications for pathology. *Nature* **378,** 398–402.
28. Lopes-Cendes, I., Gaspar, C., Trottier, Y., Mandel, J.-L., and Rouleau, G. A. (1996). Searching for proteins containing polyglutamine expansions in a large group of spinocerebelar ataxia patients. *Am. Soc. Hum. Genet.* A1559 [Abstract]
29. Mangiarini, L., Sathasivam, K., Seller, M., Cozens, B., Harper, A., Hetherington, C., Lawton, M., Trottier, Y., Lehrach, H., Davies, S. W. and Bates, G. P. (1996). Exon 1 of the HD gene with an expanded CAG repeat is sufficient to cause a progressive neurological phenotype in transgenic mice. *Cell* **87,** 493–506.
30. McInnis, M. G., McMahon, F. J., Chase, G. A., Simpson, S. G., Ross, C. A., and De Paulo, J. R., Jr. (1993). Anticipation in bipolar affective disorder. *Am. J. Hum. Genet.* **53,** 385–390.
31. McInnis, M. G. (1996). Anticipation: an old idea in new genes. *Am. J. Hum. Genet.* **59,** 973–979.
32. Myszka, D. G. (1997). Kinetic analysis of macromolecular interactions using surface plasmon resonance biosensors. *Curr. Opin. Biotechnol.* **8,** 50–57.
33. Nagafuchi, S., Yanagisawa, H., Sato, K., Shirayama, T., Ohsaki, E., Bundo, M., Takeda, T., Tadokoro, K., Kondo, I., Murayama, N., Tanaka, Y., Kikushima, H., Umino, K., Kurosawa, H., Furukawa, T., Nihei, K., Inoue, T., Sano, A., Komure, O., Takahashi, M., Yoshizawa, T., Kanazawa, I., and Yamada, M. (1994). Dentatorubral and pallidoluysian atrophy expansion of an unstable CAG trinucleotide on chromosome 12p. *Nature Genet.* **6,** 14–18.
34. Neri, C., Albanese, V., Lebre, AS., Holbert, S., Saada, C., Bougueleret, L., Meier-Ewert, S., LeGall, I., Millasseau, P., Bui, H., Giudicelli, C., Massart, C., and Guillou, S. (1996). Survey of CAG/CTG repeats in human cDNAs representing new genes: candidates for inherited neurological disorders. *Hum. Mol. Genet.* **5,** 1001–1009.
35. Onodera, O., Roses, A. D., Tsuji, S., Vance, J. M., Strittmatter, W. J., and Burke, J. R. (1996). Toxicity of expanded polyglutamine-domain proteins in Escherichia coli. *FEBS Lett.* **399,** 135–139.
36. Orr, H. T., Chung, M.y., Banfi, S., Kwiatkowski, T. J., Jr., Servadio, A., Beaudet, A. L., McCall, A. E., Duvick, L. A., Ranum, L. P., and Zoghbi, H. Y. (1993). Expansion of an unstable trinucleotide CAG repeat in spinocerebellar ataxia type 1. *Nature Genet.* **4,** 221–226.
37. Paulson, H. L., and Fischbeck, K. H. (1996). Trinucleotide repeats in neurogenetic disorders. *Annu. Rev. Neurosci.* **19,** 79–107.
38. Paulson, H. L., Perez, M. K., Trottier, Y., Trojanowski, J. Q., Subramony, S. H., Das, S. S., Vig, P., Mandel, J. L., Fischbeck, K. H., and Pittman, R. N. (1997). Intranuclear inclusions of expanded polyglutamine protein in spinocerebellar ataxia type 3. *Neuron* **19,** 333–344.
39. Perutz, M. F., Johnson, T., Suzuki, M., and Finch, J. T. (1994). Glutamine repeats as polar zippers: their possible role in inherited neurodegenerative diseases. *Proc. Natl. Acad. Sci. USA* **91,** 5355–5358.
40. Pulst, S. M., Nechiporuk, A., Nechiporuk, T., Gispert, S., Chen, X. N., Lopes-Cendes, I., Pearlman, S., Starkman, S., Orozco-Diaz, G., Lunkes, A., DeJong, P., Rouleau, G. A., Auburger, G., Korenberg, J. R., Figueroa, C., and Sahba, S. (1996). Moderate expansion of a normally biallelic trinucleotide repeat in spinocerebellar ataxia type 2. *Nature Genet.* **13,** 269–276.
41. Ranum, L. P. W., Schut, L. J., Lundgren, J. K., Orr, H. T., and Livingston, D. M. (1994). Spinocerebellar ataxia type 5 in a family descended from the grandparents of president Linclon maps to chromosome 11. *Nature Genet.* **8,** 280–284.
42. Ross, C. (1995). When more is less: pathogenesis of glutamine repeat neurodegenerative diseases. *Neuron* **15,** 493–496.
43. Ross, C. A., McInnis, M. G., Margolis, R. L., and Li, S.-H. (1993). Genes with triplets: candidate mediators of neuropsychiatric disorders. *Trends Neurosci.* **16,** 254–260.
44. Sanpei, K., Takano, H., Igarashi, S., Sato, T., Oyake, M., Sasaki, H., Wakisaka, A., Tashiro, K., Ishida, Y., Ikeuchi, T., Koide, R., Saito, M., Sato, A., Tanaka, T., Hanyu, S., Takiyama, Y., Nishizawa, M., Shimizu, N., Nomura, Y., Segawa, M., Iwabuchi, K., Eguchi, I., Tanaka, H., Takahashi, H., and Tsuji, S. (1996). Identification of the spinocerebellar ataxia type 2 gene using a direct identification of repeat expansion and cloning technique, DIRECT. *Nature Genet.* **13,** 277–284.
45. Scherzinger, E., Lurz, R., Turmaine, M., Mangiarini, L., Hollenbach, B., Hasenbank, R., Bates, G. P., Davies, S. W., Lehrach, H., and Wanker, E. E. (1997). Huntingtin-encoded polyglutamine expansions form amyloid-like protein aggregates in vitro and in vivo. *Cell* **90,** 549–558.
46. Sharp, A. H., Loev, S. J., Schilling, G., Li, S. H., Li, X. J., Bao, J., Wagster, M. V., Kotzuk, J. A., Steiner, J. P., Lo, A., Hedreen, J., Sisodia, S., Snyder, S. H., Dawson, T. M., Ryugo, D. K., and Ross, C. A. (1995). Widespread expression of Huntington's disease gene (IT15) protein product. *Neuron* **14,** 1065–1074.
47. Stevanin, G., Trottier, Y., Cancel, G., Dürr, A., David, G., Didierjean, O., Bürk, K., Imbert, G., Saudou, F., Abada, M., Gourfinkel-An, I., Benomar, A., Abbas, N., Klockgether, T., Grid, D., Agid, Y., Mandel, J. L., and Brice, A. (1996). Screening for proteins with polyglutamine expansions in autosomal dominant cerebellar ataxias. *Hum. Mol. Genet.* **5,** 1887–1892.
48. Stott, K., Blackburn, J. M., Butler, P. J. G., and Perutz, M. (1995). Incorporation of glutamine repeats makes protein oligomerize: implication for neurodegenerative diseases. *Proc. Natl. Acad. Sci. USA* **92,** 6509–6513.
49. The Huntington's Disease Collaborative Research Group (1993). A novel gene containing a trinucleotide repeat that is expanded and unstable on Huntington's disease chromosomes. *Cell* **72,** 971–983.
50. Trottier, Y., Devys, D., Imbert, G., Saudou, F., An, I., Lutz, Y., Weber, C., Agid, Y., Hirsch, E. C., and Mandel, J. L. (1995). Cellular localization of the Huntington's disease protein and discrimination of the normal and mutated form. *Nature Genet.* **10,** 104–110.
51. Trottier, Y., Lutz, Y., Stevanin, G., Imbert, G., Devys, D., Cancel, G., Saudou, F., Weber, C., David, G., Tora, L., Agid, Y., Brice, A., and Mandel, J.-L. (1995). Polyglutamine expansion as a pathological epitope in Huntington's disease and four dominant cerebellar ataxias. *Nature* **378,** 403–406.
52. Zhuchenko, O., Bailey, J., Bonnen, P., Ashizawa, T., Stockton, D. W., Amos, C., Dobyns, W. B., Subramony, S. H., Zoghbi, H. Y., and Lee, C. C. (1997). Autosomal dominant cerebellar ataxia (SCA6) associated with small polyglutamine expansions in the alpha 1A-voltage-dependent calcium channel. *Nature Genet.* **15,** 62–69.

CHAPTER 32

DIRECT Technologies[1]

K. SANPEI AND S. TSUJI
Department of Neurology, Brain Research Institute, Niigata University, Niigata 951, Japan

I. INTRODUCTION

The expansion of unstable CAG trinucleotide repeats in causative genes has been identified as a common pathogenic mechanism in a growing number of hereditary neurodegenerative diseases, including spinal and bulbar muscular atrophy (SBMA) [1], Huntington's disease (HD) [2], spinocerebellar ataxia type 1 (SCA1) [3], dentatorubral-pallidoluysian atrophy (DRPLA) [4, 5], Machado–Joseph disease (MJD) [6], spinocerebellar ataxia type 2 (SCA2) [7–9], spinocerebellar ataxia type 6 (SCA6) [10], and spinocerebellar ataxia type 7 (SCA7) [11]. From the viewpoint of clinical genetics, these diseases are characterized by anticipation, i.e., accelerating age at onset and increasing disease severity in successive generations, and the broad spectra of the clinical presentations. It has been discovered that anticipation is a result of intergenerational increase in the size of expanded CAG repeats. These observations suggest that many hereditary neurodegenerative diseases characterized by anticipation and the broad spectrum of the clinical presentations are likely to be caused by the unstable expansion of CAG repeats.

The CAG repeats in the genes of normal individuals are, in general, highly polymorphic, not exceeding 35 repeats in size, whereas the pathologically expanded CAG repeats range in size from 37 to as many as 100 repeats. SCA6 is the only exception, in which the expanded the CAG repeats range from 21 to 26. These results imply that a method which allows the selective and sensitive detection of CAG repeats consisting of more than 35 repeats would facilitate the search for the causative genes for a number of neurodegenerative diseases.

A technique called RED (repeat expansion detection) was developed to detect expanded trinucleotide repeats in the genomic DNA. The RED technique employs a ligation-chain reaction to detect the presence of expanded trinucleotide repeats. When the genomic DNA contains expanded trinucleotide repeats, the ligation-chain reaction generates concatemers of the oli-

[1]Detailed protocols and reagents described in this chapter are available upon request to the authors.

gonucleotide consisting of trinucleotide repeats such as $(CTG)_{17}$. Although the RED technique is highly sensitive to detect the presence of expanded trinucleotide repeats [12], the major and frustrating drawback of the RED technique is that it cannot be applied to the cloning of the genomic fragments that contain the expanded trinucleotide repeats.

With this background, we have devised a novel and robust technique, the direct identification of repeat expansion and cloning technique (DIRECT), which allows selective detection of the expanded CAG repeats by genomic Southern blot and the subsequent cloning of the genomic segments containing the expanded CAG repeats. We applied DIRECT to clone the gene responsible for spinocerebellar ataxia type 2 (SCA2). Anticipation had already been documented in SCA2 [13], suggesting that the causative mutation involves the expansion of CAG repeats.

II. PRINCIPLE OF DIRECT IDENTIFICATION OF REPEAT EXPANSION AND CLONING TECHNIQUE (DIRECT)

As shown in Fig. 32-1, if we use a DNA probe containing a long trinucleotide repeat, $(CAG)_{55}$ for example, such a probe would hybridize to an expanded CAG repeat (>35 repeats) but not to CAG repeats with the size within normal ranges (<35 repeats) under stringent hybridization conditions. This is because the T_m (melting temperature) of the hybrids between the probe and the genomic DNA segment would be different depending on the number of bases engaged in the hybrid formation (Fig. 32-1). Once a genomic fragment containing an expanded trinucleotide repeat is detected using Southern blotting hybridization analysis employing this probe, under stringent hybridization conditions, the genomic segment containing the expanded trinucleotide repeat can easily be cloned from the agarose gel using standard procedures. In developing DIRECT, therefore, the following two factors were crucial. First, the hybridization conditions should be sufficiently stringent such that only pathologically expanded CAG repeats (>35 repeats), but not CAG repeats of normal alleles, are detected. Second, the probe should be sufficiently sensitive to allow the detection of a single-copy genomic segment containing pathologically expanded CAG repeats by genomic Southern blotting hybridization analysis.

III. METHODS

A. Preparation of Blots

1. Digest high-molecular-weight genomic DNA (10–15 μg) with various restriction enzymes (100 U)

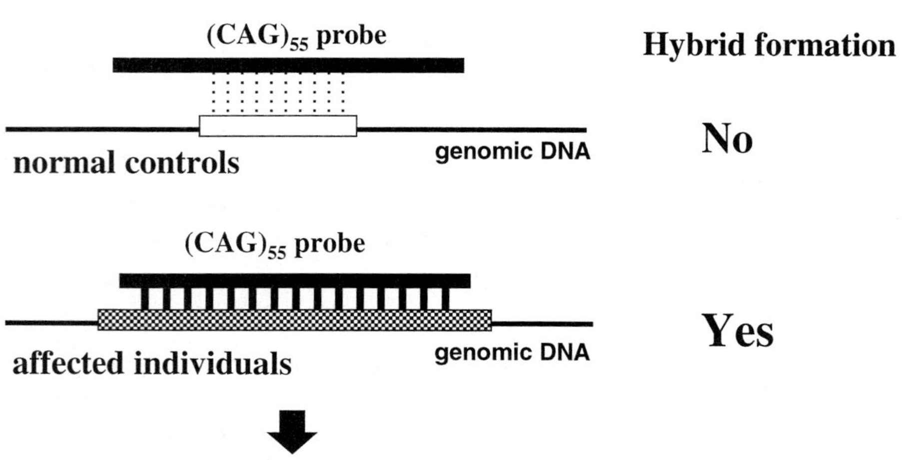

FIGURE 32-1 Schematic illustration of the principle of selective hybridization. The melting temperature of hybrids (T_m) between the $(CAG)_{55}$ probe and the genomic DNA segment containing a CAG repeat varies depending on the number of bases engaged in the hybrid formation. Therefore, under stringent hybridization conditions, it is possible to determine hybridization conditions at which only the hybrid between the $(CAG)_{55}$ probe and the genomic DNA segment containing expanded CAG repeats (>35 repeats) is formed, while the hybrid between the $(CAG)_{55}$ probe and the genomic DNA segment containing CAG repeats of normal lengths (<35 repeats) is not formed.

in 200 μl of appropriate buffers. Aliquots of the solution should be run through agarose gels to make sure that the digestion is complete, as well as to estimate the concentration of genomic DNA in each solution.

2. Ethanol precipitate the digested genomic DNA and redissolve in 30 μl of TE. Load the solution onto agarose gels and run through 0.8% agarose gels overnight. For better resolution, electrophoresis at 2 V/cm gel length is recommended. It is important that each lane contains equal amounts of genomic DNA for the comparison of signal intensities between lanes.

3. Transfer to a nitrocellulose membrane (BA-S85 Ref. No. 439196, Schleicher & Schuell) by a standard procedure of Southern blotting.

4. Bake the nitrocellulose membrane at 80°C in a vacuum oven for 2 h.

B. Preparation of $(CAG)_{55}$ Probe with a High Specific Radioactivity

The genomic segment that contained 55 CAG repeats was first cloned into a plasmid vector from a DRPLA patient. Using primers (primer 1: 5′-CAC CAC CAG CAA CAG CAA CA-3′, and primer 2: 5′-biotin-GGC CCA GAG TTT CCG TGA TG- 3′) designed based on the sequences flanking the CAG repeat of the DRPLA gene, PCR was performed in the presence of [α-^{32}P]dATP (see Fig. 32-2). The method allowed the incorporation of 59 molecules of [α-^{32}P]dATP per molecule of the DNA probes, exclusively into the sense strand. In preliminary experiments, we found that the presence of flanking sequences in addition to the CAG repeat (55 repeats) does not substantially affect the hybridization conditions (data not shown).

1. Vacuum dry 12.5 μl of 9.25 MBq of [α-^{32}P]dATP (222 Tbq/mmol) in a 0.5-ml Eppendorf tube.

Solution for PCR

Template (p-2093[a] (0.73 ng/μl))	0.4 μl
Primer 1 (20 μM)	0.4 μl
2 (20 μM)	0.4 μl
dCTP (4 mM)	0.4 μl
dGTP (4 mM)	0.4 μl
TTP (4 mM)	0.4 μl
4.9 M N,N,N-trimethylglycine	8.0 μl
10× reaction buffer (100 mM Tris-HCl (pH 8.3), 500 mM KCl, 15 mM $MgCl_2$)	1.6 μl
Distilled water	3.6 μl
Total volume	15.6 μl
Taq DNA polymerase (5.0 U/μl)	0.4 μl

[a]p-2093: plasmid containing a genomic fragment of the CAG repeat (55 repeats) and the flanking sequences of DRPLA gene [14].

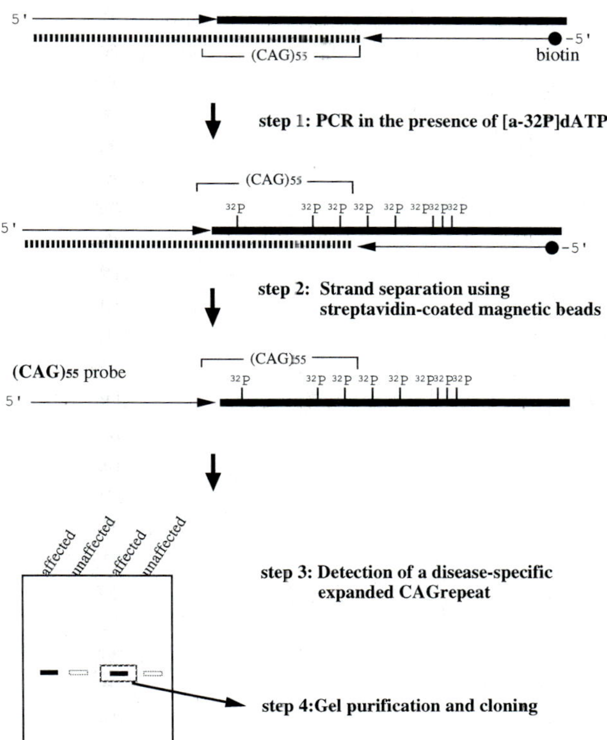

FIGURE 32-2 Schematic illustration of the direct identification of repeat expansion and cloning technique (DIRECT). A single-stranded $(CAG)_{55}$ probe with a high specific radioactivity was prepared by internal labeling with [α-^{32}P]dATP in the polymerase chain reaction (PCR) (step 1) followed by strand separation using streptavidin-coated magnetic beads (step 2). Disease-specific expanded CAG repeats were selectively detected by Southern blotting hybridization analysis using the $(CAG)_{55}$ probe under stringent hybridization conditions (step 3). The genomic DNA segment containing the expanded CAG repeat was purified from the gel and cloned into a λZAPII vector (Stratagene) (step 4).

2. Add the above solution to the tube containing the dried [α-^{32}P]dATP, and then add Taq DNA polymerase.

1. PCR CYCLE

After an initial 2-min denaturation at 94°C, PCR was performed for 30 cycles, each cycle consisting of denaturation at 94°C for 1 min, annealing at 54°C for 1 min, and extension at 72°C for 3 min, followed by a final extension at 72°C for 10 min.

2. STRAND SEPARATION USING MAGNETIC BEADS

To further improve the sensitivity of the probe, we considered that single-stranded DNA probes would be much better than double-stranded DNA probes. To enable the use of double-stranded DNA as the probe, the double-stranded DNA probe was denatured into single-stranded DNA probes before hybridization. The presence of the complementary strand in the hybridization solution, however, consumes the probe molecules and the effective probe concentration would be much lowered. To overcome this problem, we prepared single-stranded DNA probe using strand separation. One of the primers used in the PCR had been biotinilated. Due to the base composition of the template DNA, [α-^{32}P]dATP was incorporated exclusively into the unbiotinilated strand. The biotinilated strand was then easily separated from the complementary strand with the use of streptavidin-coated magnetic beads. The length of the CAG repeats of the probe (55 repeats) was arbitrarily selected with the consideration that a relatively long CAG repeat probe would be appropriate to achieve a clear "cut-off" for the detection of expanded CAG repeats.

1. Resuspend the Dynabeads M-280 Streptavidin (Dynal A. S, Oslo, Norway) by gently shaking the vial to obtain a homogeneous suspension.
2. Add 100 μl of Dynabeads M-280 Streptavidin to a 1.5-ml Eppendorf tube and place the tube in the Dynal MPC (Magnetic Particle Concentrator) for at least 30 s.
3. Remove the supernatant by aspiration with a pipette while the tube remains in the Dynal MPC.
4. Take the tube from the Dynal MPC. Gently resuspend the Dynabeads M-280 Streptavidin in 100 μl of PBS (pH 7.4), containing 0.1% BSA (bovine serum albumin) along the internal surface of the tube.
5. Place the tube in the Dynal MPC for at least 30 seconds and repeat step 4.
6. Remove the tube from the Dynal MPC and add 100 μl of PBS (pH 7.4), containing 0.1% BSA.
7. Add 20 μl of the washed Dynabeads M-280 Streptavidin into a new 1.5-ml Eppendorf TUBE and place the tube in the Dynal MPC. Remove the supernatant with a pipette while keeping the tube in the Dynal MPC.
8. Resuspend the Dynabeads in 20 μl of 1× binding and washing buffer (B&W buffer) and mix gently. Remove the supernatant with a pipette while keeping the tube in the Dynal MPC. 2× concentrated B&W buffer: 10 mM Tris-HCl (pH 7.5), 1 mM EDTA, 2.0 M NaCl.
9. Resuspend the Dynabeads in 40 μl 2× B&W buffer.
10. Add 40 μl of the prewashed beads to 16 μl of solution containing the PCR products.
11. Incubate the PCR products with the Dynabeads for 15 min keeping the beads suspended by gentle rotation of the tube. Place the tube containing the immobilized product in a Dynal MPC and remove the supernatant with a pipette.
12. Wash the Dynabeads with 40 μl of 1× B&W buffer.
13. Place the tube in a Dynal MPC and remove the supernatant. Resuspend the Dynabeads in 16 μl of a freshly prepared 0.1 M NaOH solution.
14. Incubate the solution for 10 min at room temperature.
15. Using the Dynal MPC, collect the Dynabeads on the side of the tube and transfer the supernatant solution containing the labeled CAG strand to a new tube (tube A).
16. Wash the Dynabeads with 34 μl 0.1 M NaOH.
17. Take the supernatant solution and transfer to the tube A to combine the solution containing the $(CAG)_{55}$ probe.

C. Prehybridization

Prehybridization solution contains 2.75× SSPE (1× SSPE = 150 mM NaCl, 10 mM NaH_2PO_4, 1 mM EDTA), 50% formamide, 5× Denhardt's solution, and 100 ng/ml sheared salmon sperm DNA.

Incubate the nitrocellulose membranes in an appropriate amount of prehybridization solution in a heat-sealable hybridization bag for 2 h at 62°C.

D. Hybridization

Hybridization solution contains 2.75× SSPE (1× SSPE = 150 mM NaCl, 10 mM NaH_2PO_4, 1 mM EDTA), 50% formamide, 5× Denhardt's solution, 100 ng/ml sheared salmon sperm DNA, and $(CAG)_{55}$ probe (6×10^6 cpm/ml) (the $(CAG)_{55}$ probe should be prepared freshly for each experiment).

Replace the prehybridization solution with the hybridization solution containing the $(CAG)_{55}$ probe and incubate for 18 h at 62°C with gentle shaking.

E. Washes

Wash the membranes twice in 1× SSC (150 mM NaCl, 15 mM sodium citrate) containing 0.5% SDS with gentle shaking for each 1 h at room temperature, and finally in SSC–0.5% SDS for 30 min at 65°C.

F. Exposure to Films

The filters were autoradiographed to Kodak Bio Max MS films for 16 h at −70°C using an MS intensifying screen. An exposure time of 16 h is generally sufficient to obtain strong signals. Note that the stringency of the hybridization conditions (the temperature is particularly important) should be monitored using a membrane containing cloned DNA (50 pg) carrying various lengths of CAG repeats (9, 23, 43, and 51 repeat units) in each experiment. In general the cutoff points can be adjusted by changing the temperature slightly for hybridization.

Strong signals should be obtained for the 50 pg of cloned DNA with 2-h exposure to Kodak Bio Max MS films (Fig. 32-7e). If the signals were weak under these conditions, it was difficult to obtain good signals for genomic DNA.

G. Cloning of the Genomic DNA Segment Containing Pathologically Expanded CAG Repeats

Since it was previously reported that the CAG repeats in plasmid vectors can be unstable during propagation through *Escherichia coli* [15], we selected a l-phage vector with the consideration that the CAG repeats might be more stable in λ-phage vectors than in plasmid vectors (Fig. 32-3).

1. Digest 270 μg of genomic DNA from an affected individual in 3600 μl with 1800 U of an appropriate enzyme.
2. Ethanol precipitate and resuspend the DNA in 360 μl of TE.
3. Make an agarose gel (the porosity of agarose gel varies with the size of target DNA) using a two-well comb (Bio-Rad 170-4345) and apply 180 μl into each well.
4. Electrophoresis should be performed at approximately 2 V/cm gel length for a minimum of 10 h. The optimum time required for good resolution should be determined for each experiment.
5. Determine the expected position for the genomic fragment of interest based on the mobility of size markers, and then cut out the gel segments. (Make an effort to avoid damage to the DNA by minimizing the duration of UV exposure to the gel.)
6. Place gel pieces in a dialysis bag (Spectra/Por 6 Moleculaporous Dialysis Membrane, MWCO: 15000, Spectrum, Houston, TX) and add 20 ml of 1× TAE buffer.
7. Electroelute the DNA fragments at 40 V for 2 h.
8. Reverse the polarity and electrophorese for an additional minute.

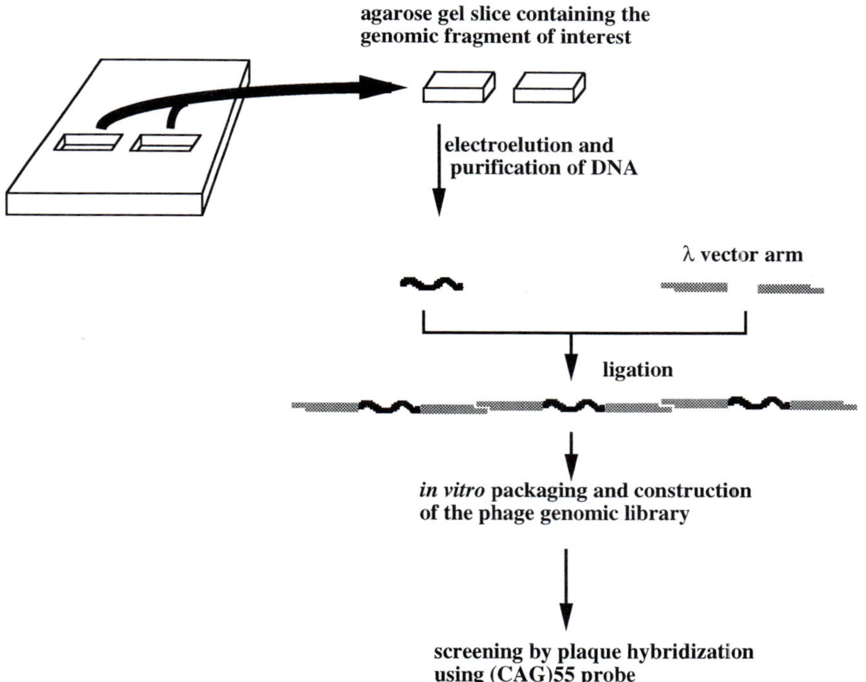

FIGURE 32-3 Schematic illustration of the construction of a λ-phage genomic DNA library. Restriction fragments containing expanded CAG repeats were purified through agarose gel electrophoresis and ligated to λ-phage arms which were digested by appropriate enzyme(s). The resultant phage library was screened using the $(CAG)_{55}$ probe under the stringent hybridization conditions.

9. Carefully take the solution out of the bag and transfer to a 50-ml conical tube. Wash the bag with an additional 10 ml of TE and combine with solution in the 50 ml conical tube.

10. Spin the tube to remove gel pieces. Transfer the supernatant to a new tube.

11. Concentrate the solution to approximately 1 ml using Centricon-100 (Amicon, Inc., Beverly, MA).

12. Extract with an equal volume of phenol-chloroform solution.

13. Extract with an equal volume of chloroform.

14. Add 2 μg of λZAPII vector arm digested with an appropriate enzyme and treat with alkaline phosphatase.

15. Ligation and *in vitro* packaging are performed by standard procedures.

16. Screen the phage genomic library by plaque hybridization using the $(CAG)_{55}$ probe under hybridization conditions identical to those used for the genomic Southern blot hybridization (described above).

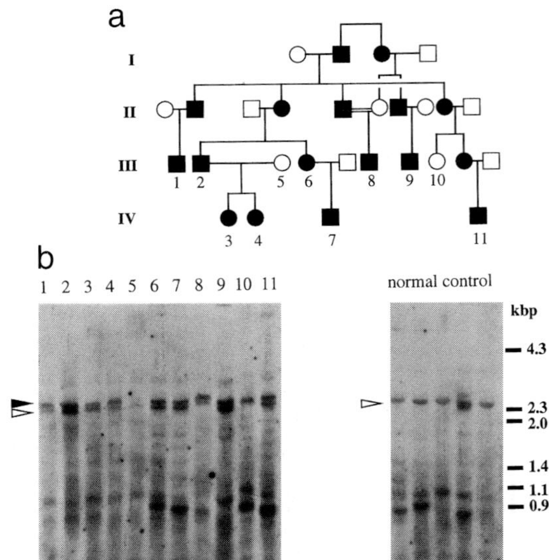

FIGURE 32-4 Detection of pathologically expanded CAG repeats in the SCA2 gene of patients with SCA2. In the pedigree chart (a), affected males and females are represented by filled squares and filled circles, respectively. Unaffected males and females are represented by open squares and open circles, respectively. Genomic DNA from members of a four-generation SCA2 family were analyzed using DIRECT. Only the lanes containing genomic DNA isolated from the affected individuals showed a 2.5-kbp *Tsp*EI fragment (black arrowhead) in addition to the 2.4-kbp fragment (white arrowhead) that was observed for all of the individuals including normal controls (b).

IV. DISCUSSION

The major advantage of DIRECT is that it allows the cloning of expanded trinucleotide repeats, which overcomes the drawback of RED [12]. We have successfully applied DIRECT to cloning of the causative gene for SCA2 [8]. As shown in Fig. 32-4, an extra *Tsp*EI fragment was observed in all the affected individuals of a single pedigree, while such a band was not observed in unaffected family members or normal controls. We have tested various restriction enzymes (*Tsp*EI, *Rsa*I, *Sal*I–*Eco*RI, *Sac*I–*Eco*RI, *Pvu*II, *Hin*dIII, *Eco*RI, *Xba*I, *Eco*RI–*Hin*dIII, *Eco*RI–*Eco*52I, *Eco*RI–*Xho*I, *Eco*RI–*Spe*I, *Eco*RI–*Xba*I, and *Pst*I–*Pvu*II) using genomic DNA of affected individuals of the family, and detected extra bands specific to affected individuals in genomic DNA digested with *Tsp*EI, *Rsa*I, *Sac*I–*Eco*RI, and *Sal*I–*Eco*RI (Fig. 32-5). However, we could not detect such extra bands specific to the affected individuals in genomic DNA digested with *Pvu*II, *Hin*dIII, *Eco*RI, *Xba*I, *Eco*RI–*Hin*dIII, *Eco*RI–*Eco*52I, *Eco*RI–*Xho*I, *Eco*RI–*Spe*I, *Eco*RI–*Xba*I, or *Pst*I–*Pvu*II. These results indicate that the following three points are crucial for the successful application of DIRECT. (1) Choice of pedigrees, (2) choice of restriction enzymes, and (3) quality control of hybridization (see Fig. 32-7).

A. Choice of Pedigrees

DIRECT allows the detection of pathologically expanded trinucleotide repeats in the genomic DNA of affected individuals. Therefore, we will not require the chromosomal localization of the disease genes in advance, which indicates that we will not require linkage analysis. In other words, DIRECT can be applied to small pedigrees where linkage analysis is not feasible. In principle, expanded trinucleotide repeats can be identified even in one affected individual. As shown in Fig. 32-6, however, DIRECT visualizes multiple bands even in unaffected individuals. This means that even normal individuals may contain CAG repeats larger than 35 repeats. In fact, we have recently identified such a CAG repeat ranging from 10 to 90 repeats in normal individuals (unpublished observation). These bands detected in normal individuals are occasionally polymorphic in size of restriction fragments, which may account for the fact that expanded trinucleotide repeats have occasionally been detected in normal individuals using the RED technique [12]. Based on our experience, we recommend the analysis of multiple affected and unaffected members to confirm that the band detected by DIRECT exhibits perfect cosegregation with the disease.

B. Choice of Restriction Enzymes

There are no *a priori* suggestions for the choice of restriction enzymes. For cloning purposes, it is desirable

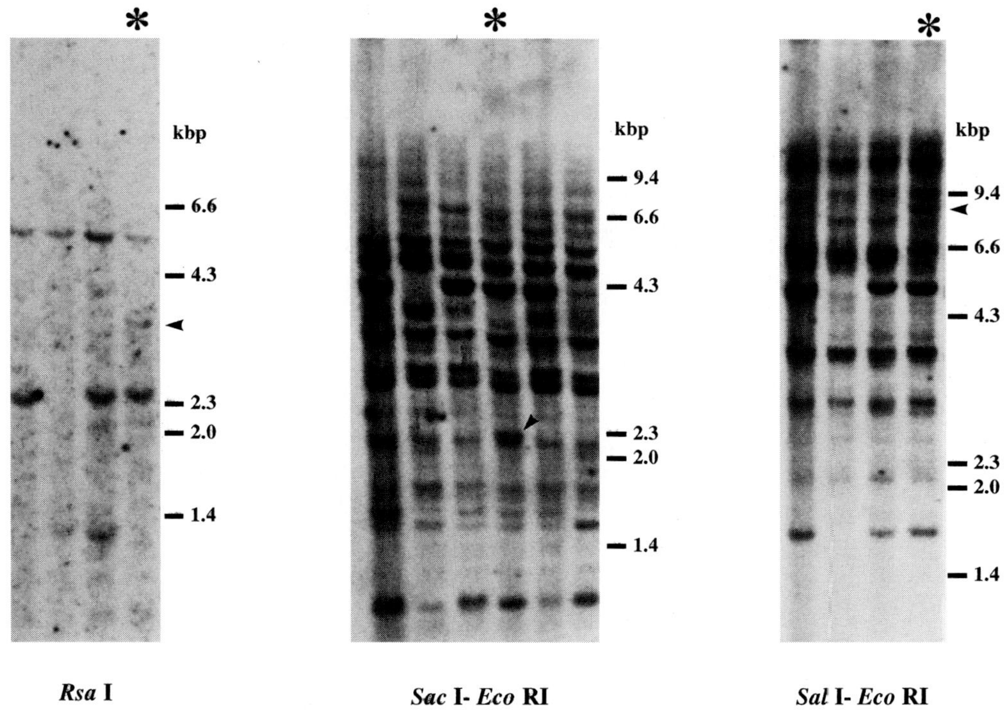

FIGURE 32-5 Detection of extra bands in the genomic DNA of a patient with SCA2. The genomic DNA from a patient with SCA2 (asterisk) as well as controls was digested with *Rsa*I, *Sac*I–*Eco*RI, and *Sal*I–*Eco*RI and blotted to a nitrocellulose membrane, followed by stringent hybridization to the $(CAG)_{55}$ probe. An extra band (arrowhead) is present exclusively in the SCA2 patient.

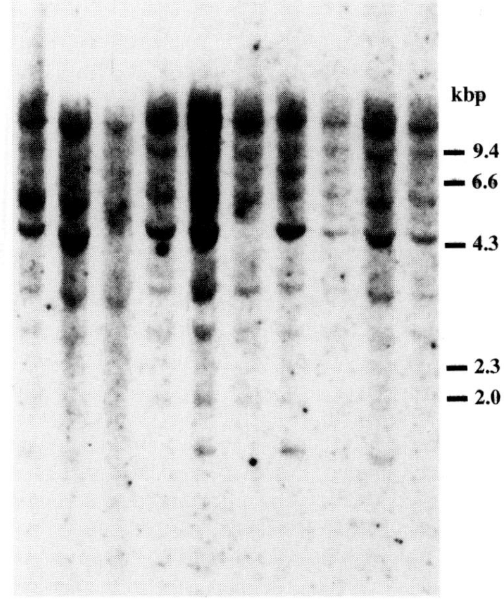

FIGURE 32-6 Multiple bands were detected by the $(CAG)_{55}$ probe in genomic DNAs of normal controls. Genomic DNA was digested with *Eco*RI and blotted to a nitrocellulose membrane followed by stringent hybridization to the $(CAG)_{55}$ probe. Multiple bands were detected even in normal individuals.

that the size of the restriction fragments containing the expanded trinucleotide repeat does not exceed several kbp. It is also desired that the restriction fragments have cohesive ends which can easily be cloned into cloning vectors. Therefore, we strongly recommend trying as many restriction enzymes as possible to find the enzymes which generate restriction fragments of appropriate sizes. Restriction enzymes which have tetrametric recognition sites, such as *Sau*3AI and *Tsp*EI, are good candidates, since these enzymes generate smaller fragments than those enzymes possessing hexametric recognition sites and only a few bands are detected by the $(CAG)_{55}$ probe in normal individuals.

C. Quality Control of Hybridization

To apply DIRECT successfully, quality control of hybridization is also crucial. Particularly, the specific radioactivity must be high enough to detect a single copy genomic DNA fragment, and the cutoff point should be strictly controlled by adjusting the hybridization conditions. For this purpose, we recommend the performance of a controlled hybridization using a membrane containing cloned DNA (50 pg) carrying various lengths of CAG repeats (9, 23, 43, and 51 repeats). (We used a

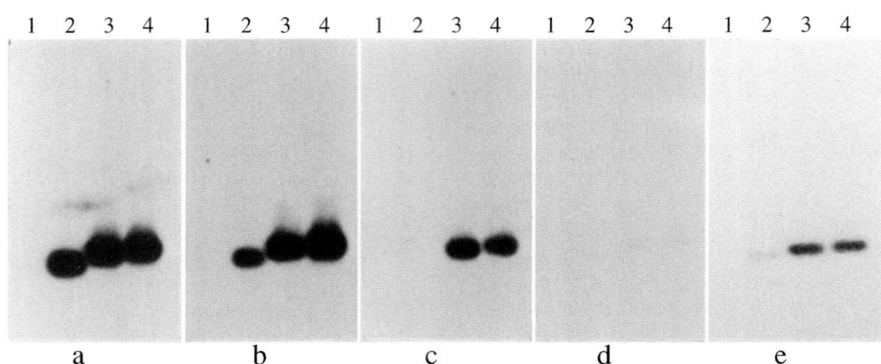

FIGURE 32-7 Quality control of hybridization conditions. Ten nanograms of PCR products derived from four androgen receptor genes containing $(CAG)_n$ (n = 9, 23, 43, and 51) was run through 1% agarose gel and blotted to a nitrocellulose membrane. Hybridizations under various conditions were performed to determine the optimum conditions for employing a probe that was generated using terminal deoxynucleotidyl transferase [14]. (a) hybridization in 3× SSPE and 50% formamide at 52°C; (b) in 3× SSPE and 60% formamide at 52°C; (c) in 3× SSPE and 50% formamide at 62°C; (d) in 3× SSPE and 60% formamide at 62°C; (e) cloned DNA of androgen receptor gene (50 pg) carrying various lengths of CAG repeats (9, 23, 43, and 51 repeats) were run through 1% agarose gel and blotted to a nitrocellulose membrane followed by hybridization to the $(CAG)_{55}$ probe in 2.75× SSPE and 50% formamide at 62°C.

cloned 1.3-kbp androgen receptor gene fragment containing the CAG repeat). If the specific radioactivity of the probe is sufficiently high, strong signals are obtained with 2 h exposure to Kodak Bio Max MS films. For adjusting the cutoff points, the hybridization temperature should be adjusted so that strong signals are obtained for 43 or 51 repeats but not for 9 or 23 repeats. Alternatively the concentration of the ionic strength (the concentration of SSPE) can be adjusted. The discovery of the mutation of SCA6 demonstrated that expanded CAG repeats range from 21 to 26. This observation raises the possibility that all the CAG repeat diseases do not necessarily show expanded CAG repeats larger than 35. Therefore, in applying DIRECT, multiple cutoff points may require testing, since some diseases may show cutoff points of around 20 as was demonstrated in SCA6.

V. CONCLUSION

We have demonstrated that DIRECT is a robust method for cloning genes for triplet repeat diseases. Since DIRECT does not require prior knowledge of the location of the causative genes, DIRECT should be applied to the identification of causative genes of many hereditary neurodegenerative diseases which show genetic anticipation or a broad spectra of clinical presentations [16–18].

References

1. La Spada, A. R., Wilson, E. M., Lubahn, D. D., Harding, A. E., and Fischbeck, K. H. (1991). Androgen receptor gene mutations in X-linked spinal and bulbar muscular atrophy. *Nature* **352,** 77–79.
2. The Huntington's Disease Collaborative Research Group. (1993). A novel gene containing a trinucleotide repeat that is expanded and unstable on Huntington's disease chromosomes. *Cell* **72,** 971–983.
3. Orr, H. T., Chung, M. Y., Banfi, S., Kwiatokiwski, T. J., Jr., Servadio, A., Beaudet, A. L., McCall, A. E., Duvick, L. A., Ranum, L. P. W., and Zoghbi, H. Y. (1993). Expansion of an unstable trinucleotide CAG repeat in spinocerebellar ataxia type 1. *Nature Genet.* **4,** 221–226.
4. Koide, R. Ikeuchi, T., Onodera, O., Tanaka, H., Igarashi, S., Endo, K., Takahashi, H., Kondo, R., Ishikawa, A., Hayashi, T., Saito, M., Tomoda, A., Miike, T., Naito, H., Ikita, F., and Tsuji, S. (1994). Unstable expansion of CAG repeat in hereditary dentatorubral-pallidoluysian atrophy (DRPLA) *Nature Genet.* **6,** 9–13.
5. Nagafuchi, S., Yanagisawa, H., Sato, K., Shirayama, T., Ohsaki, E., Bundo, M., Takeda, T., Tadokoro, K., Kondo, I., Murayama, N., Tanaka, Y., Kikushima, H., Umino, K., Kurosawa, H., Furukawa, T., Nihei, K., Inoue, T., Sano, A., Komure, O., Takahishi, M., Yoshizawa, T., Kanazawa, I., and Yamada, M. (1994). Dentatorubral and pallidoluysian atrophy expansion of an unstable CAG trinucleotide on chromosome 12p. *Nature Genet.* **6,** 14–18.
6. Kawaguchi, Y., Okamoto, T., Taniwaki, M., Aizawa, M., Inoue, M., Katayama, S., Kawakami, H., Nakamura, S., Nishimura, M., Akiguchi, Y., Kimura, J., Narumiya, S., and Kakizuka, A. (1994). CAG expansion in a novel gene for Machado-Joseph disease at chromosome 14q32.1. *Nature Genet.* **8,** 221–228.
7. Pulst, S. M., Nechiporuk, T., Gispert, S., Chen, X. N., Lopes-Cendes, I., Pearlman, S., Starkman, S., Oronzco, D. G., Lunkes, A., Dejong, P., Rouleau, G. A., Auburger, G., Korenberg, Jr., Figueroa, C., and Sahba, S. (1996). Moderate expansion of a normally biallelic trinucleotide repeat in spinocerebellar ataxia type 2. *Nature Genet.* **14,** 269–276.
8. Sanpei, K., Takano, H., Igarashi, S., Sato, T., Oyake, M., Sasaki, H., Wakisaka, A., Tashiro, K., Ishida, Y., Ikeuchi, T., Koide, R., Saito, M., Sato, A., Tanaka, T., Hanyu, S., Takiyama, Y., Nishizawa, M., Shimizu, N., Nomura, Y., Segawa, M., Iwabuchi, K.,

Eguchi, I., Tanaka, H., Takahashi, H., and Tsuji, S. (1996). Identification of the spinocerebellar ataxia type 2 gene using a direct identification of repeat expansion and cloning technique, DIRECT. *Nature Genet.* **14,** 277–284.

9. Imbert, G., Saudou, F., Yvert, G., Devys, D., Trottier, Y., Garnier, J. M., Weber, C., Mandel, J. L., Cancel, G., Abbas, N., Durr, A., Didierjean, O., Stevanin, G., Agid, Y., and Brice, A. (1996). Cloning of the gene for spinocerebellar ataxia 2 reveals a locus with high sensitivity to expanded CAG/glutamine repeats. *Nature Genet.* **14,** 285–291.
10. Zhuchenko, O., Bailey, J., Bonnen, P., Ashizawa, T., Stockton, D. W., Amos, C., Dobyns, W. B., Subramony, S. H., Zoghbi, H. Y., and Lee, C. C. (1997). Autosomal dominant cerebellar ataxia (SCA6) associated with small polyglutamine expansion in the a1A votage-dependent calcium channel. *Nature Genet.* **15,** 62–69.
11. David, G., Abbas, N., Stebanin, N., Durr, A., Yvert, G., Cancel, Weber, C., Invert, G., Saudou, F., Antoniou, E., Drabkin, H., Gemmill, R., Giunti, P., Bevomar, A., Wood, N., Ruberg, M., Agid, Y., Mandel, J. L., and Brice, A. (1997). Cloning of the SCA7 gene reveals a highly unstable CAG repeat expansion. *Nature Genet.* **17,** 65–70.
12. Schalling, M., Hudson, T. J., Buetow, K. H., and Housman, D. E. (1993). Direct detection of novel expanded trinucleotide repeats in the human genome. *Nature Genet.* **4,** 135–139.
13. Gispert, S., Twells, R., Oronzco, G., Brice, A., Weber, J., Heredero, L., Scheufler, K., Riley, B., Allotey, R., Nothers, C., Hillermann, Lunkes, A., Khati, C., Stevanin, G., Hernandez, A., Magarino, C., Klockgether, T., Durr, A., Chneiweiss, H., Enczmann, J., Farrall, M., Beckmann, J., Mullan, M., Wernet, P., Agid, Y., Freund, H. J., Williamson, R., Auburger, G., and Chamberlain, S. (1993). Chromosomal assignment of the second locus for autosomal dominant cerebellar ataxia (SCA2) to chromosome 12q23–24.1. *Nature Genet.* **4,** 295–299.
14. Sanpei, K., Igarashi, S., Eguchi, I., Takiyama, Y., Tanaka, H., and Tsuji, S. (1995). Direct detection of expanded (CAG/CTG) repeats in the myotonin-protein kinase genes of myotonic dystrophy patients using a high-stringency hybridization method. *BBRC* **212,** 341–346.
15. Kang, S., Jaworski, A., Ohshima, K., and Wells, R. D. (1995). Expansion and deletion of CTG repeats from human disease gene are determined by the direction of replication in *E. coli. Nature Genet.* **10,** 213–218.
16. Ranum, L. P. W., Schut, L. J., Lundgren, J. K., Orr, H. T., and Livingston, D. M. (1994). Spinocerebellar ataxia type 5 in a family descended from the grandparents of President Lincoln maps to chromosome 11. *Nature Genet.* **8,** 280–284.
17. Basset, A. S., and Honer, W. G. (1994). Evidence for anticipation in schizophrenia. *Am. J. Hum. Genet.* **54,** 864–870.
18. McInnis, M. G. McMahon, F. J., Chase, G. A., Simpson, S. G., Ross, C. A., and DePaulo, Jr. (1993). Anticipation in bipolar affective disorder. *Am. J. Hum. Genet.* **53,** 385–390.

Part XIII

Systems for the Study of Genetic Instabilities

CHAPTER 33

Systems for the Study of Genetic Instabilities: *Escherichia coli*

ALBINO BACOLLA, RICHARD P. BOWATER,[1] AND ROBERT D. WELLS

Department of Biochemistry and Biophysics, Institute of Biosciences and Technology, Texas A&M University, Texas Medical Center, Houston, Texas 77030

I. GENETIC INSTABILITIES OF REPETITIVE DNA AND HUMAN DISEASES

Simple repetitive DNA sequences are dispersed throughout natural genomes and have an intrinsic genetic instability which results in frequent length changes [1, 2]. The rate of genetic change that occurs in these sequences is related to their number of direct repeats and, therefore, a mutated product has a different potential for mutation compared to its predecessor; this phenomenon has been termed dynamic mutation [3, 4].

The mechanisms for evolution of repetitive sequences have been much debated [5, 6] and their genetic instability has been proposed to occur via a variety of pathways which are known to modify the genetic material. Slippage of the DNA strands at the replication fork is the favored mechanism to produce changes within simple repeats [6–8], but direct evidence that this occurs inside cells has not been obtained. Other possible mechanisms include unequal exchange during recombination between homologous DNA sequences [9] or gene con-

[1] Present address: Imperial Cancer Research Fund, Clare Hall Laboratories, South Mimms, Hertfordshire EN6 3LD, United Kingdom.

version [10]. It is clear that the evolution of repeating sequences is complex, and it is likely that interactions occur between these different pathways.

Genetic instabilities within repetitive DNA sequences have been linked to a variety of human diseases [4, 11]. Initial associations between simple repeats and human diseases were shown for triplet repeats that were (G + C)-rich. Recently, however, genetic instabilities of different types of repeats (nontriplet motifs) have been correlated with some human disorders [12, 13]. Here, we limit our discussion to microsatellite sequences, which are tandem (direct) repeats with a high degree of repetition and 1–5 bp in their unit structure.

The links to human diseases have produced a surge in studies into the molecular mechanisms causing genetic instabilities of simple repeating sequences. In this review we focus on investigations into the genetic stability of simple repeats in prokaryotes, highlighting the similarities and differences compared to other organisms. Finally, the advantages and disadvantages of using this system will be discussed.

A. Microsatellite Instability and Cancer

All organisms have biochemical systems that provide a high fidelity of genome replication and thus prevent the origin of mutated phenotypes [14]. Mechanisms are in operation to give a high fidelity during DNA synthesis, and also to correct mistakes that are made during replication. Deficiencies within these systems have the potential to produce many errors within genome sequences and this situation has been termed the "mutator phenotype" since it would produce an organism prone to mutations. The progression of many human tumors through multiple stages [15] suggests that their development requires a number of genetic alterations. The existence of a mutator phenotype would increase the potential for tumors to arise and, therefore, may be a primary event for the occurrence of some tumors [16].

Length changes are frequently observed within simple repeating sequences in all genomes. However, elevated frequencies of length changes occur under some conditions of reduced replication fidelity. A fundamental system involved in maintaining genomic integrity is that first identified in *Escherichia coli* as methyl-directed mismatch repair [17]. A number of experiments have shown a similar mismatch repair system in eukaryotes, with a high degree of conservation throughout all organisms [14, 18–20]. Upon inactivation of this system of DNA repair, increased heterogeneities have been observed at some unit lengths of simple repetitive DNA (e.g., mono- and dinucleotides) in bacterial systems [21, 22] and in yeast [23, 24]. Since there is an increased rate of mutation throughout the whole genome under these conditions, these observations suggest that the level of heterogeneity at simple repeating sequences is a good indicator of the overall level of DNA stability in cells.

Experimental evidence that deficient DNA repair could cause human tumors was obtained when cell lines from inherited and sporadic human cancers were shown to undergo an increased frequency of length changes in specific dinucleotide repeats [25–27]. It was later shown that this increased genetic instability was due to defects in mismatch repair proteins [28–30]. Numerous studies have confirmed that defective mismatch repair is responsible for the elevated microsatellite instability and increased mutation rate in some cancers (for recent reviews see [31, 32]). It is particularly notable that mice containing homozygous null mutations in some mismatch repair genes are viable, but develop tumors and exhibit elevated microsatellite instability [33–36]. However, while these transgenic mice studies strengthen the link between microsatellite instability and susceptibility to cancers, they also show that high rates of mutagenesis in multiple tissues are compatible with normal development.

The associations of defective mismatch repair and elevated microsatellite instability are particularly strong for hereditary nonpolyposis cancer, one of the most common inherited disorders known [37, 38]. Elevated microsatellite instability has also been observed in other genes relating to cancer, including BAX, a gene that promotes apoptosis [39]. Presently, it is not clear if an increased mutation rate at microsatellites can be used as an absolute marker for defective mismatch repair. Although some cancers with elevated microsatellite instability do not carry a defect in the known mismatch repair genes, only a subset of genes involved in human mismatch repair have been identified [19, 32]. Also, the effects of other DNA repair systems on microsatellite instability in humans is unknown, and it is conceivable that these will play a role in tumor development.

Genetic instabilities within mono- and dinucleotide repeats increase for longer runs of consecutive repeats and, therefore, are decreased by interruptions to the repeat sequence [7, 14]. These observations are consistent with the hypothesis that slipped-strand mispairing during DNA synthesis generates misaligned intermediates. These parameters are intrinsic to the DNA repeat, but it is also known that flanking sequences can influence the genetic stability of simple repeat sequences [14]. As discussed below in relation to triplet repeat sequences (TRS), these observations suggest that many factors, including DNA repair, replication, and transcription, affect the genetic stability of microsatellite sequences.

[10] proposed gene conversion events to explain germline mutations at human minisatellites; however, these two mechanisms are not mutually exclusive.

Recent investigations [82] revealed the relationship between cell growth and deletions of CTG · CAG triplet repeats in plasmids. Long CTG · CAG repeats in plasmids can influence cell growth which effects the observed expansions and deletions. At extended growth periods, the observed frequencies of deletion were dramatically increased if the cells passed through stationary phase before subculturing. High frequencies of deletions were observed because of a growth advantage of cells containing plasmids with deleted triplet repeats. These [82] are the first observations to show a direct influence between a plasmid-based DNA sequence or structure and factors controlling bacterial growth. Other related studies [83] showed that transcription promotes deletions of long CTG · CAG triplet repeats from human neuromuscular disease genes. This elevated genetic instability was detected because active transcription into the triplet repeat influenced the growth transitions of the host cell, allowing advantageous growth for cells harboring plasmids with deleted repeat sequences. The variety of deleted products observed in separate cultures suggested that transcription altered the metabolism of the DNA in a manner that produced random length changes in the repeat sequence. For cultures containing plasmids without active transcription into the triplet repeat, or those maintained in exponential growth, deletions occurred within the repeat at a lower frequency (5- to 20-fold). In these incubations, the extent of deletions was proportional to the number of cell divisions and many repeat lengths were observed within each culture, suggesting that the decrease in average repeat length at long incubation times was due to multiple small deletions. These observations showed that deletions within long CTG · CAG repeats contained on plasmids in *E. coli* occur via more than one pathway, and their level of genetic instability is altered by the enzymatic processes occurring upon the DNA. Also, this work suggests a role for the involvement of transcription in DNA slippage which gives rise to the genetic instabilities.

Furthermore, single-stranded DNA-binding protein enhances the stability of CTG · CAG triplet repeats in *E. coli* [84]. Studies were conducted with mutants that lack SSB protein in order to evaluate the possible involvement of this protein which is an important component in DNA replication, repair, and recombination. SSB can prevent the formation of DNA secondary structures. Replication can pause at sites of potential DNA secondary structure and pause sites are associated with template misalignment mutagenesis. The potential for slippage and for DNA secondary structure formation within triplet repeat sequences may contribute to the instabilities of these sequences observed in individuals with triplet repeat diseases (described above). With a biochemical assay for stability, Rosche *et al.* [84] showed that the absence of single-stranded DNA-binding protein leads to an increase in the frequency of large deletions within the TRS.

B. CTG · CAG Is Preferentially Expanded

Ohshima *et al.* [85] discovered that the CTG · CAG triplet repeat is the dominant genetic expansion product. This extraordinary discovery was made possible by the successful cloning and characterization of all 10 TRS [86, 87]. The relative capacity of the 10 TRS to be expanded in *E. coli* [85] was explored with a competition study. Surprisingly, the CTG · CAG triplet repeat was expanded at least nine times more frequently than any of the other nine triplets [85]. Low levels of expansion were found also for GTG · CAC, GTC · GAC, CGG · CCG, and GAA · TTC. Thus, the structure of the CTG · CAG repeat and/or its utilization by the DNA synthetic systems *in vivo* must be quite different from that of the other triplets. The surprising discovery that CTG · CAG triplet repeats are the dominant expansion products, as found (reviewed in [70]) in clinical samples from human hereditary diseases, suggests the importance of DNA structural properties [70, 88, 89]. Other investigations have revealed that duplex CTG · CAG and CGG · CCG repeats have unorthodox properties including nucleosome assembly [90, 91] and their capacity to cause DNA polymerases to pause within the repeat sequences [92, 93], as well as conformational features as revealed by helical repeat and polyacrylamide gel migrations [94, 95]. Further elucidation of the CTG · CAG repeat structural features along with the genetic factors responsible for expansion may explain why most triplet repeat hereditary disease genes contain CTG · CAG repeats (reviewed in [70]). Although other triplet repeats are found in the human genome [96], the lengths are shorter (generally <15 repeats) than found for these disease genes.

CTG · CAG and CGG · CCG have topological properties of writhe and flexibility that have not been described for other DNAs (see Chapter 38). Also, the former TRS is unusual in its high binding affinity for nucleosomes and is stabilized by mismatch repair-deficient cells. The later two properties are not shared with CGG · CCG. Future work will be required to evaluate the role of these behaviors in the preferential expansion of CTG · CAG.

C. Site of Expansion

Kang *et al.* [97] described an investigation aimed at identifying the region of CTG · CAG triplet repeats that are preferentially expanded. Analysis of expanded regions using the interrupting CTA triplet sequence as a location marker within the CTG · CAG tract revealed that the expansion of large CTG · CAG repeats is one event rather than an accumulation of multiple small expansions and that the expansions occur more frequently in the region distal from the replication origin. Also, it was shown that a loss of interruptions increases the expansion frequency. Thus, the instability of large triplet repeats in hereditary diseases occurs by a mechanism different from the instability in microsatellite sequences caused by defects in mismatch repair systems for certain sporadic cancers and hereditary nonpolyposis colorectal cancers.

D. Mismatch Repair

Mismatch repair-deficient *E. coli* [20] were studied in order to further elucidate the factors involved in genetic instabilities as well as DNA structural issues *in vivo* [98]. Long CTG · CAG repeats are stabilized in ColE1-derived plasmids in *E. coli* containing mutations in the methyl-directed mismatch repair genes (*mutS, mutL,* or *mutH*). When plasmids containing $(\text{CTG} \cdot \text{CAG})_{180}$ were grown for about 100 generations in *mutS, mutL,* or *mutH* strains, 60 to 85% of the plasmids contained a full-length repeat, whereas in the parent strain only about 20% of the plasmids contained the full-length repeat. The deletions occur only in the $(\text{CTG} \cdot \text{CAG})_{180}$ insert, not in DNA flanking the repeat. While many products of the deletions are heterogeneous in length, preferential deletion products of about 140, 100, 60, and 20 repeats were observed. The *E. coli* mismatch repair proteins apparently recognize three-base loops formed during replication and then generate long single-stranded gaps where stable hairpin structures may form which can be bypassed by DNA polymerase during the resynthesis of duplex DNA (Fig. 33-2). Direct experiments will be required to test the veracity of this model. Similar studies were conducted with plasmids containing CGG · CCG repeats; no stabilization of these triplets was found in the mismatch repair mutants. The reason for this was unclear but may be due to the rate of formation and the stability of hairpins in CTG · CAG and CGG · CCG repeats as designated by question marks in Fig. 33-2. Since procaryotic and human mismatch repair proteins are similar [20], and since several carcinoma cell lines which are defective in mismatch repair show instability of simple DNA microsatellites [20, 98], these mechanistic investigations in a bacterial cell may provide insights into the molecular basis for some human genetic diseases.

E. Fragile X CGG · CCG

A series of inserts containing 6 to 240 copies of CGG · CCG were stably cloned in plasmids [99]. Several factors influence the stability (deletions and expansions) of the inserts: repeat length, the presence of interruptions, the orientation of the insert relative to the unidirectional replication origin, *E. coli* host strains, the location of the insert, and the copy number of the vector. The instability varies strongly with the length of the insert; longer tracts of CGG · CCG repeats show a greater degree of instability compared to shorter inserts. Furthermore, the effect of length on DNA polymerase pausing was also observed during synthesis of the repeat *in vitro* when the Klenow fragment of DNA polymerase I was used; lengths of greater than 61 repeats showed stronger pausing sites, occurring at repeat number 30 (away from the CGG · CCG start site), when CCG was the template strand. This phenomenon was also observed with CTG · CAG triplet repeats [92]. These results suggest that, at a critical length, the CGG sequence adopts a non-B conformation(s) [95, 100] which blocks DNA polymerase progression, leading to the idling and subsequent slippage to give expanded products and hence provide the molecular basis for this non-Mendelian genetic process.

The canonical human FMR-1 repeat carries 30 CGG · CCG triplets interrupted by two AGG triplets at the 10th and 20th repeat. Fragile X carriers carry longer repeats (50 to 200) that contain long stretches of uninterrupted CGG · CCG triplets which predispose this sequence to hyperexpansion in successive generations. Affected individuals have longer methylated repeats (230 to 2000) [101]. Our results indicate that the presence of interruptions greatly enhances the stability of the CGG · CCG tract in *E. coli.* Other studies on the alleles derived from human patients show the presence of stable and unstable CGG · CCG triplets of similar size, suggesting that a feature other than length, but intrinsic to the repeat, was responsible for stability. This supported the observations of Eichler *et al.* [102] who found that lengths of >33 uninterrupted CGG · CCG triplets showed marked instability, regardless of total repeat length, suggesting that the loss of the AGG interruptions is an important mutational event in the generation of alleles predisposed to the fragile X syndrome.

As mentioned above, another important factor dictating stability is the orientation of the CGG · CCG-containing insert. Our results indicate that the triplet repeat was stably maintained in vectors if the CGG strand was in the leading template strand (orientation I) with respect to the origin of replication. However, if CCG fell in the leading template strand (orientation II), the insert was highly destabilized (depending on the

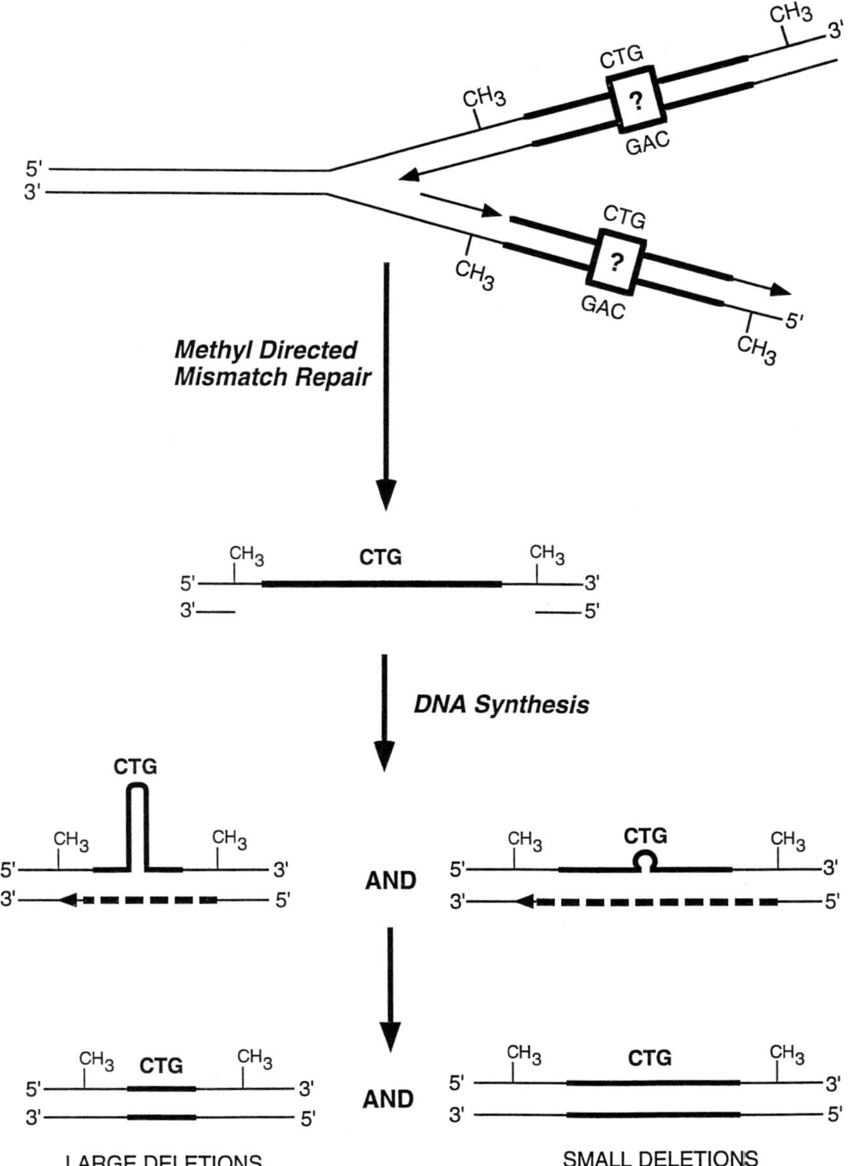

FIGURE 33-2 Models for the involvement of *E. coli* MMR proteins in the enhancement of destabilization of $(CTG)_n$ triplet repeats *in vivo*. Reprinted with permission from Jaworski *et al., Proc. Natl. Acad. Sci. USA* **92,** 11019–11023 [98]. Copyright 1995, National Academy of Sciences, U.S.A.

lengths undergoing deletions and expansions. As in the case of CTG · CAG-repeating sequences, the frequency of expansion and deletion of the CGG · CCG triplet repeats is influenced by the direction of replication [71] which involves an asymmetric DNA polymerase complex that simultaneously replicates the leading as well as the lagging strand [79]. Replication-dependent deletion between direct repeats occurs preferentially in the lagging strand due to the unequal probability of forming hairpins [103]. Therefore, the deletion of the insert (in orientation II) can be explained by the propensity of the CGG template strand to form a stable hairpin [73–78] which is bypassed by the replication machinery during resynthesis of the DNA. On the other hand, expansions within the tract are likely due to strand realignment through slippage of the complementary strands during pausing (described above) to generate a folded and elongated nascent DNA on the leading strand [71]. Deletions were the most abundant species detected, but expansions were also visible when pRW3024 [(i.e., $(CGG \cdot CCG)_{24}$ in orientation II)] was propagated in *E. coli* DH5α; the bands differed from each other by one repeating CGG · CCG unit, suggesting the involvement of slipped structures during replication. This method allowed the cloning of the expanded and deleted products (6 to 49 repeats) in orientation I and their propagation in *E. coli* SURE to give a stable DNA preparation.

F. DNA Polymerase Pausing and Hairpin Formation

The pausing of DNA synthesis *in vitro* and hairpin formation at specific loci in double-stranded CTG · CAG and CGG · CCG triplet repeats [92, 93] is important with respect to our understanding of the types of DNA structures formed by these sequences, the capacity of DNA metabolizing enzymes to utilize these sequences *in vitro,* and the formation of hairpins as intermediates in genetic instabilities. This subject is discussed in the chapter in this book by Ohshima and Wells (Chapter 45).

G. Molecular Similarities between Humans and *Escherichia coli*

The studies described above on a genetically and biochemically tractable system for elucidating the molecular mechanisms responsible for expansion, and thus anticipation, represent a significant advance. Several remarkable molecular similarities exist including the following [104]:

- Genetic instability (expansions and deletions) of triplet repeat sequences (CTG · CAG, CGG · CCG, or AAG · CTT).
- Longer repeats are more unstable than shorter sequences.
- CTG · CAG is preferentially expanded in *E. coli*; this repeat sequence was found in the majority of the triplet repeat diseases.
- Repeat sequence imperfections (polymorphisms) stabilize long tracts of TRS.
- Similar types of imperfections (polymorphisms) are found (e.g., the polypurine · polypyrimidine motif in the Friedreich's ataxia AAG · CTT repeat sequence is maintained).
- The lengths of the smallest deletion products in *E. coli* (10–20 triplet repeats) approximate the lengths found in normal humans.
- DNA polymerases from humans and *E. coli* pause in long CTG · CAG, CGG · CCG, and AAG · CTT sequences, thus rendering them susceptible to mutations.

Hence, certain features of the molecular processes related to the involvement of TRS in human hereditary diseases may be elucidated effectively in simple cellular systems. Obviously, a number of other developmental and neurobiological questions can only be solved in higher eucaryotic cells. Thus, some features of the concept of "unstable genes—unstable mind" [105] may be tractable in genetically defined systems in mice, microbes, and molecules.

H. Summary of Factors That Influence Genetic Instability

As elaborated above, several factors are known that influence the stability of long TRS in plasmids in *E. coli.* These factors include the following:

- Genetic make-up of host cells. The absence of recA is important for the cloning of long TRS and, as discussed above, the presence of SSB protein is significant [84].
- Growth conditions. Several factors related to growth conditions, including media and not permitting the cells to go through stationary phase, are also critical [82, 83].
- Generations of cells. Since the deletion and expansion behavior appears to be due to DNA replication, the number of generations of cells [71, 97] is important.
- Transcription. Active transcription through the TRS enhances deletions [83].
- *Mut S, H,* and *L.* The absence of these mismatch repair functions enhances the stability since it is likely that short loops which are formed by DNA slippage are recognized and cleaved with deletions as the products [98].
- Nucleotide excision repair. It is possible that nucleotide excision repair [106] may also be involved in the observed instabilities for CTG · CAG (P. Parniewski and R. D. Wells, unpublished).
- Vector and cloning location. The type of vector and the location of cloning of a TRS in the vector is important and may relate to copy number. The instabilities may be due, in part, to the proximity to DNA polymerase I–III switch sites [98].
- Sequence of insert. Certain types of sequences are more unstable than others. For example, the fragile-X CGG · CCG sequence appears to be quite unstable whereas several other sequences are somewhat more stable in the form of long tracts [86, 87, 99].
- Length of insert. The length of the insert is critical; shorter lengths (30–50 TRS) are rather more stable, but longer sequences (>200 TRS) are quite unstable [71, 86, 87, 97–99].

- Interruptions in insert. The presence of interruptions, or polymorphisms as recognized in human genetics, is stabilizing to long TRS. This behavior has been observed with virtually all of the 10 TRS studied to date [86, 87, 99].
- Orientation of insert. The orientation of the insert is important for favoring deletions or expansions as described above [71, 86, 87, 99].

III. NONREPLICATION-BASED INSTABILITIES DUE TO SLIPPAGE

A. Replication-Independent Instabilities and Slippage

As described above, *E. coli* has proven to be a valuable system for the study of basic mechanisms underlying triplet repeat instabilities. The sizes of these replication-dependent expansions and contractions are substantial, ranging from 20 to >100 repeats. In addition to these large expansions and deletions, small shifts of ±1 to ~±7 repeat units were observed in the position of (CTA · TAG) marker interruptions when $(CTG \cdot CAG)_n$ tracts were sequenced following subculturing in *E. coli*. Thus, these occasional sequence polymorphisms, which were the result of cloning of human sequences into plasmid DNA, provided a valuable means for following small slipped-register expansions and deletions (SSED). Several observations suggest that SSED was the consequence of small slippages occurring along the duplex DNA rather than at the replication fork, and, therefore, that this instability involved mechanisms different than those associated with errors of replication. For consideration of the nonreplication-based slippage (Fig. 33-3A), we assume that the complementary DNA strands are not nicked and, hence, are not free to rotate around each other, as in closed, circular, or genomic DNA. However, since nicks exist at the replication fork in Fig. 33-3B, free rotation is possible in this case.

First, SSED was observed in TRS tracts that had undergone large expansions or deletions [97, 107]. Second, contrary to the stabilization observed for the $(CTG \cdot CAG)_{175}$ insert in the MMR deficient cells, SSED occurred more frequently in MMR^- *E. coli* mutants [107]. Third, SSED took place on TRS tracts whose total number of (CTG · CAG) repeats was identical before and after subcultivation. Fourth, SSED and large deletions were also observed when CTG · CAG was cloned in human lymphoblastoid cell lines [108], indicating that these two types of instabilities occur independently from the host genetic background. Thus, assuming that both large and small instabilities are due to slippage, we note the differences between the types of hairpin loops that form on an unnicked double-stranded DNA template (replication-independent), versus those that originate at a replication fork (replication-dependent) (Fig. 33-3).

Considering a comparison of nonreplication-based slippage versus replication-based slippage (Fig. 33-3), first, slippage on unnicked DNA produces hairpin loops on both strands (strands 1 and 2 in Fig. 33-3A) but only one strand (strands 3 and 6, but not their complementary strands 5 and 4 in Fig. 33-3B) may fold at a replication fork. Second, both stable and unstable structures can coexist on unnicked duplex DNA, whereas the replication fork may be biased toward the formation of stable hairpins. In addition, hairpins of up to ~120 base pairs (~40 repeats) are proposed to form at a replication fork [71, 97], whereas much smaller structures (≤7 repeats) are suspected to loop-out on unnicked, duplex DNA. Third, the resolution of hairpins back to a regular B-helix can occur through three mechanisms: strand denaturation and renaturation, nucleolytic attack, and second round of replication (which acts as a specialized form of strand denaturation and renaturation). Of these, strand denaturation/renaturation and nucleolytic attack yield identical results for replication-dependent and replication-independent slippage. However, resolution through the second round of replication leads to unmodified DNA lengths in the case of hairpins formed on unnicked duplex DNA, whereas it introduces variations in DNA lengths when hairpins occurred at the previous replication fork (Fig. 33-3C).

B. Mechanisms for SSED

What mechanisms are responsible for SSED? From the preceding description it is clear that since neither denaturation/renaturation nor the subsequent round of replication alter repeat length for the replication-independent slippage (Fig. 33-3A), SSED should include endonucleolytic steps on hairpin structures formed on unnicked DNA.

This point is illustrated in Fig. 33-4, where two pathways are presented that yield a small, slipped-register expansion (left side) or deletion (right side). The top molecule represents a $(CTG \cdot CAG)_{14}$ tract that contains the naturally occurring CTA · TAG interruption at position 7. This tract is similar to, but shorter than, the $(CTG \cdot CAG)_n$ fragments on which SSED was observed [97, 107]. Following strand-dissociation, slippage produces two hairpins, one composed of CTGCTG (CTG loop), the other of CAGCAG (CAG loop). Misalignment also splits the polymorphic CTA · TAG, so that now CTA pairs with CAG and TAG opposes CTG, which gives rise to two mismatched base pairs, A · C

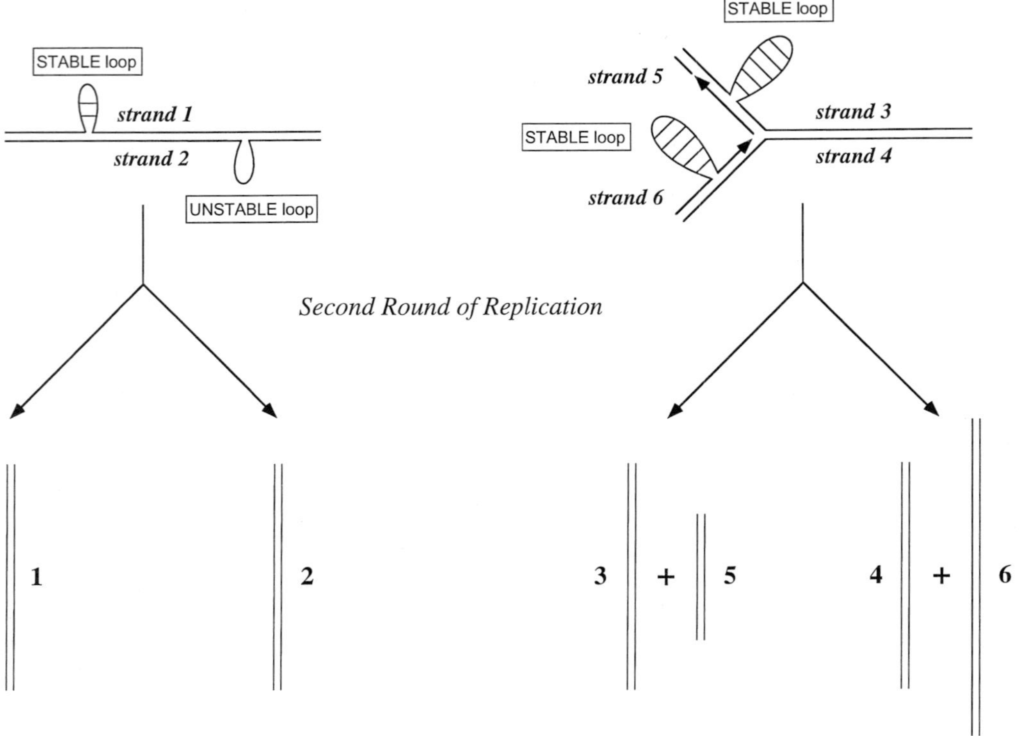

FIGURE 33-3 Comparison of slippage on unnicked double-stranded DNA and at the replication fork. (A) A segment of double-stranded DNA is shown whose right and left sides extend to form an unnicked, continuous chain, such as in a circular molecule or a long chromosome. *Strand 1* and *strand 2* designate top and bottom strands, respectively, of N number of nucleotides. The two loops represent hairpins formed as a result of strand slippage by units of 3 or multiples of 3 (Fig. 33-4). Horizontal bars in the top loop signify that this hairpin contains hydrogen bonded bases, and therefore has an ordered, duplex-like structure. On the contrary, the loop on the bottom strand designates a set of nucleotides that are unable to self hybridize, and therefore do not form a stable structure. Following replication, the length of the duplicated *strand 1* and *strand 2* is still N. We presume that the loops do not pose an impediment for passing DNA polymerases. (B) *Strand 3* and *strand 4* are as *strand 1* and *strand 2* in (A), respectively, but at the time of replication. *Strand 5* is the lagging and *strand 6* the leading strand. Arrows indicate the free 3′ ends of the growing chains. Only stable loops form, since these are able to compete with the normal, linear duplex. Therefore, only *strand 3* (top strand) and *strand 6* (copy of bottom strand) loop-out. At the end of the replication process, *strand 5* is $N - X$ nucleotides long, X being the number of bases in the opposing hairpin. Similarly, *strand 6* is X nucleotides longer than *strand 4* due to the extra bases in the hairpin. After a second round of replication, the copied *strand 3* and *strand 4* are N base pairs long, whereas *strand 5* is $N - X$, and *strand 6* is $N + X$ base pairs in length. For simplicity, no slippage is assumed to occur during the second round of replication. (C) Tabulation of the lengths of the DNA segments after the second round of replication.

and G · T, respectively. Thus, resolution of this structure into a regular B-helix requires the repair of hairpin loops as well as of mismatches. The left pathway assumes that an incision is inflicted opposite to each loop (arrows) which, followed by DNA synthesis and ligation to fill in the gaps, leads to expansion. Repair of the mismatched base pairs may be accomplished on either strand, but only the use of the bottom strand as a template results in a shift to the right of the original interruption (position 9 vs 7).

The pathway on the right envisions cleavage of the hairpin loops at their base, followed by ligation; this generates a two-triplet repeat deletion. Again, only repair of the mismatched bases by using the bottom strand

C

Lengths of DNA Segments after Second Round of Replication

Strand	*Length*
1	N
2	N
3	N
4	N
5	$N - X$
6	$N + X$

FIGURE 33-3 (*continued*)

as a template results in a concomitant shift to the left of the CTA · TAG polymorphism (position 5 vs 7). Similar results may also be obtained by a combination of these two pathways—for example an incision opposite to the CAG loop plus an excision of the CTG loop—if a second round of replication is included in the overall reaction. Recombination could also give similar products.

C. Nuclease Activities at Slipped Structures

Which enzymes are postulated to recognize and cleave hairpin loops? As stated above, disruption of the MMR complex in *E. coli* increases SSED, indicating that this repair system is capable of processing non-B DNA structures [107]. *Escherichia coli* strains deficient in nucleotide excision repair (NER) proteins (UvrB) [106] also showed a higher incidence of SSED on $(CTG \cdot CAG)_n$ [107] (P. Parniewski and R. D. Wells, unpublished work), proving that at least two systems, which are involved in the maintenance of genome integrity, act to resolve and correct errors associated with slippage at $(CTG \cdot CAG)_n$ tracts. Since the MMR and NER systems may interact [109, 110], the DNA repair pathways may be similar. A third class of enzymes that was shown to cleave both hairpins and mismatches, albeit of different sequence compositions, was the topoisomerases [111–113]. This nicking activity was followed by other nuclease and polymerase reactions, as well as by illegitimate recombination events [114, 115], suggesting that, unlike MMR and NER, topoisomerase-mediated endonucleolytic attack is associated with increased genomic instability [116], and thus with SSED.

Slippage as well as some topoisomerase activities are stimulated by negative supercoiling [117, 118]. In addition, $(CTG \cdot CAG)_n$, as well as $(CGG \cdot CCG)_n$ sequences, are more flexible than random DNA [95] and, as a consequence, are predicted to accumulate abnormally high levels of superhelical densities [100]. Altogether, these data lead us to speculate that one of the mechanisms through which SSED may occur is as follows. Transient surges of negative supercoiling, generated during translocation of helix-tracking enzymes such as the DNA and RNA polymerase complexes, are particularly large at $(CTG \cdot CAG)_n$, and this causes the repeat sequence to undergo slippage and form hairpins. Topoisomerases, which are then attracted to the sites, bind the non-B DNA structures and introduce phosphodiester breaks, which are then processed and repaired by other enzymes (i.e., polymerases). During this repair process, SSED is introduced, due to the asymmetry in DNA lengths (Fig. 33-4). Obviously, much further work will be required to elucidate the genetic and biochemical steps in this process.

IV. PROSPECTS FOR THE FUTURE

Enormous progress has been made in our understanding of the molecular mechanisms of genetic instabilities in the model system *E. coli.* A wide range of factors related to DNA replication play an integral role in deletions and expansions. Whereas it is realized that developmental problems or neurobiological phenomena will not be revealed in this simple system, the power of combining genetics and biochemistry in *E. coli* makes it an attractive model for studying molecular processes. However, much remains to be learned; we are at the threshold of investigating the molecular processes in this simple system.

At this point, we cannot even be certain that all of the fundamental mechanisms that generate instability have been identified. Moreover, the mechanism of slippage as the proposed cause of polymorphisms has not been directly proven [7, 70, 119]. In this regard, data should be obtained on the kinetic equilibria of duplex melting as a function of the free energies of supercoiling for the triplet repeats.

The future is very bright indeed for unraveling these important questions. Since the molecular mechanisms of the fundamental processes of replication, repair, etc., are the same in *E. coli,* yeast, mice, and humans, it is likely that this work will be of lasting importance as we work toward developing therapeutic strategies for alleviating the suffering of these debilitating diseases.

Acknowledgments

This work was supported by National Institutes of Health Grant GM52982 and a grant from the Robert A. Welch Foundation. A portion of this review has been published elsewhere [120] in a somewhat different format.

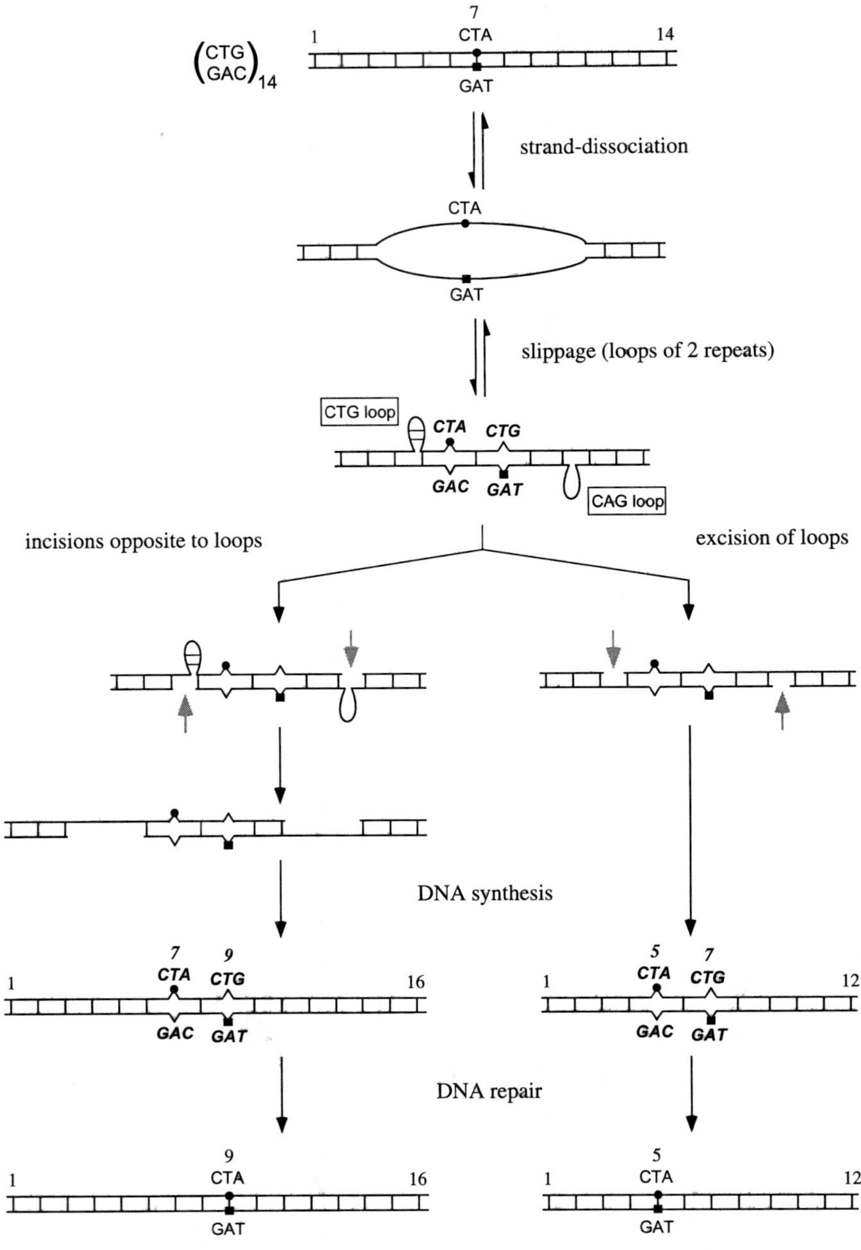

FIGURE 33-4 Mechanisms for small, slipped register-mediated expansions and deletions (SSED). A duplex DNA is shown that contains 14 units of CTG · CAG repeats (vertical bars). The sequence is interrupted at the 7th position by CTA · TAG to simulate the polymorphism of the human tracts cloned in *E. coli.* Misalignment occurs after strand-separation and subsequent reannealing, splitting the base pairing between CTA and TAG. As a result, CTA at position 7 (top strand) pairs with CAG at position 5 (bottom strand), and TAG at position 7 (bottom strand) pairs with CTG at position 9 (top strand). Thus, slippage creates two mismatched base pairs, A · C and G · T, and two looped-out structures, one composed of CTGCTG on the top strand, the other being CAGCAG on the bottom strand. These hairpins are repaired through two mechanisms. (Left) Incisions are inflicted opposite to the loops (arrows) and the resulting gaps are filled in and ligated. This creates a small slipped-register expansion. The bottom strand is then used as a template to correct the A · C and G · T mismatches, and thus the CTA · TAG shifts to the right by two repeats. (Right) The hairpins are excised at their base (arrows), and the nicks are sealed, leading to a small, slipped-register deletion. A · C and G · T mismatches are repaired by using the bottom strand as a template, resulting in a shift to the left of the CTA · TAG interruption. If the top strand is used to correct the mismatches, expansion (left side) and deletion (right side) will be maintained, but the position of the CTA · TAG interruption will be unchanged (position 7).

References

1. Charlesworth, B., Sniegowski, P., and Stephan, W. (1994). The evolutionary dynamics of repetitive DNA in eucaryotes. *Nature* **371,** 215–220.
2. Tautz, D., and Schlötterer, C. (1994). Simple sequences. *Curr. Opin. Genet. Dev.* **4,** 832–837.
3. Richards, R. I., and Sutherland, G. R. (1992). Dynamic mutations: a new class of mutations causing human disease. *Cell* **70,** 709–712.
4. Sutherland, G. R., and Richards, R. I. (1995). Simple tandem DNA repeats and human genetic disease. *Proc. Natl. Acad. Sci. USA* **92,** 3636–3641.
5. Dover, G. (1995). Slippery DNA runs on and on and on. *Nature Genet.* **10,** 254.
6. Hancock, J. M. (1996). Simple sequences and the expanding genome. *Bioessays* **18,** 421–425.
7. Levinson, G., and Gutman, G. A. (1987). Slipped-strand mispairing: a major mechanism for DNA sequence evolution. *Mol. Biol. Evol.* **4,** 203–221.
8. Lustig, A. J., and Petes, T. D. (1993). Genetic control of simple sequence stability in yeast. *In* "Genome Rearrangement and Stability" (K. E. Davies and S. T. Warren, Eds.), pp. 79–106. Cold Spring Harbor Laboratory, Cold Spring Harbor, NY.
9. Smith, G. P. (1973). Unequal crossover and the evolution of multigene families. *Cold Spring Harbor Symp. Quant. Biol.* **38,** 507–513.
10. Jeffreys, A. J., Tamaki, K., MacLeod, A., Monckton, D. G., Neil, D. L., and Armour, J. A. (1994). Complex gene conversion events in germline mutation at human minisatellites. *Nature Genet.* **6,** 136–145.
11. Krontiris, T. G. (1995). Minisatellites and human disease. *Science* **269**, 1682–1683.
12. Buard, J., and Jeffreys, A. J. (1997). Big, bad minisatellites. *Nature Genet.* **15,** 327–328.
13. Mandel, J.-L. (1997). Breaking the rule of three. *Nature* **386,** 767–769.
14. Umar, A., and Kunkel, T. A. (1996). DNA-replication fidelity, mismatch repair and genome instability in cancer cells. *Eur. J. Biochem.* **238,** 297–307.
15. Vogelstein, B., and Kinzler, K. W. (1993). The multistep nature of cancer. *Trends Genet.* **9,** 138–141.
16. Loeb, L. A. (1994). Microsatellite instability: marker of a mutator phenotype in cancer. *Cancer Res.* **54,** 5059–5063.
17. Modrich, P. (1991). Mechanisms and biological effects of mismatch repair. *Annu. Rev. Genet.* **25,** 229–253.
18. Fishel, R., and Kolodner, R. D. (1995). Identification of mismatch repair genes and their role in the development of cancer. *Curr. Opin. Genet. Dev.* **5,** 382–395.
19. Kolodner, R. D. (1995). Mismatch repair: mechanisms and relationship to cancer susceptibility. *Trends Biochem. Sci.* **20,** 397–401.
20. Modrich, P., and Lahue, R. (1996). Mismatch repair in replication fidelity, genetic recombination and cancer biology. *Annu. Rev. Biochem.* **65,** 101–133.
21. Freund, A. M., Bichara, M., and Fuchs, R. P. (1989). Z-DNA-forming sequences are spontaneous deletion hot spots. *Proc. Natl. Acad. Sci. USA* **86,** 7465–7469.
22. Levinson, G., and Gutman, G. A. (1987). High frequencies of short frameshifts in poly-CA/TG tandem repeats borne by bacteriophage M13 in *Escherichia coli* K-12. *Nucleic Acids Res.* **15,** 5323–38.
23. Strand, M., Earley, M. C., Crouse, G. F., and Petes, T. D. (1995). Mutations in the msh3 gene preferentially lead to deletions within tracts of simple repetitive DNA in *Saccharomyces cerevisiae. Proc. Natl. Acad. Sci. USA* **92,** 10418–10421.
24. Strand, M., Prolla, T. A., Liskay, R. M., and Petes, T. D. (1993). Destabilisation of tracts of simple repetitive DNA in yeast by mutations affecting DNA mismatch repair. *Nature* **365,** 274–276.
25. Ionov, Y., Peinado, M., Malkhosyan, S., Shibata, D., and Perucho, M. (1993). Ubiquitous somatic mutations in simple repeated sequences reveal a new mechanism for colonic carcinogenesis. *Nature* **363,** 558–561.
26. Thibodeau, S. N., Bren, G., and Schaid, D. (1993). Microsatellite instability in cancer of the proximal colon. *Science* **260,** 816–819.
27. Aaltonen, L. A., Peltomäki, P., Leach, F. S., Sistonen, P., Pylkkänen, L., Mecklin, J. P., Järvinen, H., Powell, S. M., Jen, J., Hamilton, S. R., Petersen, G. M., Kinzler, K. W., Vogelstein, B., and de la Chapelle, A. (1993). Clues to the pathogenesis of familial colorectal cancer. *Science* **260,** 812–816.
28. Parsons, R., Li, G.-M., Longley, M. J., Fang, W.-H., Papadopoulos, N., Jen, J., de la Chapelle, A., Kinzler, K. W., Vogelstein, B., and Modrich, P. (1993). Hypermutability and mismatch repair deficiency in RER+ tumor cells. *Cell* **75,** 1227–1236.
29. Leach, F. S., Nicolaides, N. C., Papadopoulos, N., Liu, B., Jen, J., Parsons, R., Peltomäki, P., Sistonen, P., Aaltonen, L. A., Nyström-Lahti, M., Guan, X.-Y., Zhang, J., Meltzer, P. S., Yu, J.-W., Kao, F.-T., Chen, D. J., Cerosaletti, K. M., Fournier, R. E. K., Todd, S., Lewis, T., Leach, R. J., Naylor, S. L., Weissenbach, J., Mecklin, J.-P., Jarvinen, H., Petersen, G. M., Hamilton, S. R., Green, J., Jass, J., Watson, P., Lynch, H. T., Trent, J. M., de la Chapelle, A., Kinzler, K. W., and Vogelstein, B. (1993). Mutations of a mutS homolog in hereditary nonpolyposis colorectal cancer. *Cell* **75,** 1215–1225.
30. Fishel, R., Lescoe, M. K., Rao, M. R. SM. R., Copeland, N. G., Jenkins, N. A., Garber, J., Kane, M., and Kolodner, R. (1993). The human mutator gene homolog MSH2 and its association with hereditary nonpolyposis colon cancer. *Cell* **75,** 1027–1038.
31. Kinzler, K. W., and Vogelstein, B. (1996). Lessons from hereditary colorectal cancer. *Cell* **87,** 159–170.
32. Eshelmann, J. R., and Markowitz, S. D. (1996). Mismatch repair defects in human carcinogenesis. *Hum. Mol. Genet.* **5,** 1489–1494.
33. Baker, S. M., Bronner, C. E., Zhang, L., Plug, A. W., Robatzek, M., Warren, G., Elliott, E. A., Yu, J., Ashley, T., Arnheim, N., Flavell, R. A., and Liskay, R. M. (1995). Male mice defective in the DNA mismatch repair gene PMS2 exhibit abnormal chromosome synapsis in meiosis. *Cell* **82,** 309–319.
34. de Wind, N., Dekker, M., Berns, A., Radman, M., and te Riele, H. (1995). Inactivation of the mouse Msh2 gene results in mismatch repair deficiency, methylation tolerance, hyperrecombination, and predisposition to cancer. *Cell* **82,** 321–330.
35. Reitmair, A. H., Schmits, R., Ewel, A., Bapat, B., Redston, M., Mitri, A., Waterhouse, P., Mittrucker, H. W., Wakeham, A., Liu, B., Thomason, A., Griesser, H., Gallinger, S., Ballhausen, W. G., Fishel, R., and Mak, T. W. (1995). Msh2 deficient mice are viable and susceptible to lymphoid tumours. *Nature Genet.* **11,** 64–70.
36. Narayanan, L., Fritzell, J. A., Baker, S. M., Liskay, R. M., and Glazer, P. M. (1997). Elevated levels of mutation in multiple tissues of mice deficient in the DNA mismatch repair gene Pms2. *Proc. Natl. Acad. Sci. USA* **94,** 3122–3127.
37. de la Chapelle, A., and Peltomäki, P. (1995). Genetics of hereditary colon cancer. *Annu. Rev. Genet.* **29,** 329–348.
38. Marra, G., and Boland, C. R. (1995). Hereditary nonpolyposis colorectal cancer: the syndrome, the genes, and historical perspectives. *J. Natl. Cancer Inst.* **87,** 1114–1125.
39. Rampino, N., Yamamoto, H., Ionov, Y., Li, Y., Sawai, H., Reed, J. C., and Perucho, M. (1997). Somatic frameshift mutations in

the BAX gene in colon cancers of the microsatellite mutator phenotype. *Science* **275,** 967–969.

40. Willems, P. J. (1994). Dynamic mutations hit double figures. *Nature Genet.* **8,** 213–215.
41. Ashley, C. T., and Warren, S. T. (1995). Trinucleotide repeat expansion and human disease. *Annu. Rev. Genet.* **29,** 703–728.
42. Warren, S. T. (1996). The expanding world of trinucleotide repeats. *Science* **271,** 1374–1375.
43. Paulson, H. L., and Fischbeck, K. H. (1996). Trinucleotide repeats in neurogenetic disorders. *Annu. Rev. Neurosci.* **19,** 79–107.
44. Timchenko, L. T., and Caskey, C. T. (1996). Trinucleotide repeat disorders in humans: discussions of mechanisms and medical issues. *FASEB J.* **10,** 1589–1597.
45. Reddy, P. S., and Housman, D. E. (1997). The complex pathology of trinucleotide repeats. *Curr. Opin. Cell Biol.* **9,** 364–372.
46. Li, X.-J., Li, S.-H., Sharp, A. H., Nucifora, F. C., Jr., Schilling, G., Lanahan, A., Worley, P., Snyder, S. H., and Ross, C. A. (1995). A huntingtin-associated protein enriched in brain with implications for pathology. *Nature* **378,** 389–402.
47. Burke, J. R., Enghild, J. J., Martin, M. E., Jou, Y. S., Myers, R. M., Roses, A. D., Vance, J. M., and Strittmatter, W. J. (1996). Huntingtin and DRPLA proteins selectively interact with the enzyme GAPDH. *Nature Med.* **2,** 347–350.
48. Kalchman, M. A., Koide, H. B., McCutcheon, K., Graham, R. K., Nichol, K., Nishiyama, K., Kazemi-Esfarjani, P., Lynn, F. C., Wellington, C., Metzler, M., Goldberg, Y. P., Kanazawa, I., Gietz, R. D., and Hayden, M. R. (1997). HIP1, a human homologue of *S. cerevisiae* Sla2p, interacts with membrane-associated huntingtin in the brain. *Nature Genet.* **16,** 44–53.
49. Kunst, C. B., and Warren, S. T. (1994). Cryptic and polar variation of the fragile X repeat could result in predisposing normal alleles. *Cell* **77,** 853–861.
50. Eichler, E. E., Holden, J. J., Popovich, B. W., Reiss, A. L., Snow, K., Thibodeau, S. N., Richards, C. S., Ward, P. A., and Nelson, D. L. (1994). Length of uninterrupted CGG repeats determines instability in the FMR1 gene. *Nature Genet.* **8,** 88–94.
51. Chung, M.-Y., Ranum, L. P. WL. P., Duvick, L. A., Servadio, A., Zoghbi, H. Y., and Orr, H. T. (1993). Evidence for a mechanism predisposing to intergenerational CAG repeat instability in spinocerebellar ataxia type I. *Nature Genet.* **5,** 254–258.
52. Zoghbi, H. Y. (1996). The expanding world of ataxins. *Nature Genet.* **14,** 237–238.
53. Telenius, H., Kremer, B., Goldberg, Y. P., Theilmann, J., Andrew, S. E., Zeisler, J., Adam, S., Greenberg, C., Ives, E. J., Clarke, L. A., and Hayden, M. R. (1994). Somatic and gonadal mosaicism of the Huntington disease gene CAG repeat in brain and sperm. *Nature Genet.* **6,** 409–414.
54. Duyao, M., Ambrose, C., Myers, R., Novelletto, A., Persichetti, F., Frontali, M., Folstein, S., Ross, C., Franz, M., Abbott, M., Gray, J., Conneally, P., Young, A., Penney, J., Hollingsworth, Z., Shoulson, I., Lazzarini, A., Falek, A., Koroshetz, W., Sax, D., Bird, E., Vonsattel, J., Bonilla, E., Alvir, J., Conde, J. B., Cha, J.-H., Dure, L., Gomez, F., Ramos, M., Sanchez-Ramos, J., Snodgrass, S., de Young, M., Wexler, N., Moscowitz, C., Penchaszadeh, G., MacFarlane, H., Anderson, M., Jenkins, B., Srinidhi, J., Barnes, G., Gusella, J., and MacDonald, M. (1993). Trinucleotide repeat length instability and age of onset in Huntington's disease. *Nature Genet.* **4,** 387–392.
55. Monckton, D. G., Wong, L. J., Ashizawa, T., and Caskey, C. T. (1995). Somatic mosaicism, germline expansions, germline reversions and intergenerational reductions in myotonic dystrophy males: small pool PCR analyses. *Hum. Mol. Genet.* **4,** 1–8.
56. Dutch-Belgian Fragile X Consortium (1994). FMR1 knockout mice: a model to study Fragile X mental retardation. *Cell* **78,** 23–33.
57. Jansen, G., Groenen, P. J. T. A., Bachner, D., Jap, P. H. K., Coerwinkel, M., Oerlemans, F., van den Broek, W., Gohlsch, B., Pette, D., Plomp, J. J., Molenaar, P. C., Nederhoff, M. G. J., van Echteld, C. J. A., Dekker, M., Berns, A., Hameister, H., and Wieringa, B. (1996). Abnormal myotonic dystrophy protein kinase levels produce only mild myopathy in mice. *Nature Genet.* **13,** 316–324.
58. Reddy, S., Smith, D. B., Rich, M. M., Leferovich, J. M., Reilly, P., Davis, B. M., Tran, K., Rayburn, H., Bronson, R., Cros, D., Balice-Gordon, R. J., and Housman, D. (1996). Mice lacking the myotonic dystrophy protein kinase develop a late onset progressive myopathy. *Nature Genet.* **13,** 325–335.
59. Duyao, M. P., Auerbach, A. B., Ryan, A., Persichetti, F., Barnes, G. T., McNeil, S. M., Ge, P., Vonsattel, J.-P., Gusella, J. F., Joyner, A. L., and MacDonald, M. E. (1995). Inactivation of the mouse Huntington's disease gene homologue Hdh. *Science* **269,** 407–410.
60. Zeitlin, S., Liu, J. P., Chapman, D. L., Papaioannou, V. E., and Efstratiadis, A. (1995). Increased apoptosis and early embryonic lethality in mice nullizygous for the Huntington's disease gene homologue. *Nature Genet.* **11,** 155–163.
61. Nasir, J., Floresco, S. B., O'Kusky, J. R., Diewert, V. M., Richman, J. M., Zeisler, J., Borowski, A., Marth, J. D., Phillips, A. G., and Hayden, M. R. (1995). Targeted disruption of the Huntington's disease gene results in embryonic lethality and behavioral and morphological changes in heterozygotes. *Cell* **81,** 811–823.
62. Goldberg, Y. P., Kalchman, M. A., Metzler, M., Nasir, J., Zeisler, J., Graham, R., Koide, H. B., O'Kusky, J., Sharp, A. H., Ross, C. A., Jirik, F., and Hayden, M. R. (1996). Absence of disease phenotype and intergenerational stability of the CAG repeat in transgenic mice expressing the human Huntington disease transcript. *Hum. Mol. Genet.* **5,** 177–185.
63. Burright, E. N., Clark, H. B., Servadio, A., Matilla, T., Feddersen, R. M., Yunis, W. S., Duvick, L. A., Zoghbi, H. Y., and Orr, H. T. (1995). SCA1 transgenic mice: a model for neurodegeneration caused by an expanded CAG trinucleotide repeat. *Cell* **82,** 937–948.
64. Ikeda, H., Yamaguchi, M., Sugai, S., Aze, Y., Narumiya, S., and Kakizuka, A. (1996). Expanded polyglutamine in the Machado-Joseph disease protein induces cell death in vitro and in vivo. *Nature Genet.* **13,** 196–202.
65. Bingham, P. M., Scott, M. O., Wang, S., McPhaul, M. J., Wilson, E. M., Garbern, J. Y., Merry, D. E., and Fischbeck, K. H. (1995). Stability of an expanded trinucleotide repeat in the androgen receptor gene in transgenic mice. *Nature Genet.* **9,** 191–196.
66. Mangiarini, L., Sathasivam, K., Seller, M., Cozens, B., Harper, A., Hetheringthon, C., Lawton, M., Trottier, Y., Lehrach, H., Davies, S. W., and Bates, G. P. (1996). Exon 1 of the HD gene with an expanded CAG repeat is sufficient to cause a progressive neurological phenotype in transgenic mice. *Cell* **87,** 493–506.
67. Monckton, D. G., Coolbaugh, M. I., Ashizawa, K. T., Siciliano, M. J., and Caskey, C. T. (1997). Hypermutable myotonic dystrophy CTG repeat mouse transgenes. *Nature Genet.* **15,** 193–196.
68. Gourdon, G., Radvanyi, F., Lia, A.-S., Duros, C., Blanche, M., Abitbol, M., Junien, C., and Hofmann-Radvanyi, H. (1997). Moderate intergenerational and somatic instability of a 55-CTG repeat in transgenic mice. *Nature Genet.* **15,** 190–192.
69. Mangiarini, L., Sathasivam, K., Mahal, A., Mott, R., Seller, M., and Bates, G. P. (1997). Instability of highly expanded CAG repeats in mice transgenic for the Huntington's disease mutation. *Nature Genet.* **15,** 197–200.
70. Wells, R. D. (1996). Molecular basis of genetic instability of triplet repeats. *J. Biol. Chem.* **271,** 2875–2878.

71. Kang, S., Jaworski, A., Ohshima, K., and Wells, R. D. (1995). Expansion and deletion of CTG triplet repeats from human disease genes are determined by the direction of replication. *Nature Genet.* **10,** 213–218.
72. Jaworski, A., Higgins, N. P., Wells, R. D., and Zacharias, W. (1991). Topoisomerase mutants and physiological conditions control supercoiling and Z-DNA formation *in vivo. J. Biol. Chem.* **266,** 2576–2581.
73. Mitas M., Yu, A., Dill, J., and Haworth, I. S. (1995). The trinucleotide sequence $d(CGG)_{15}$ forms a heat-stable hairpin containing $G^{syn} \cdot G^{anti}$ base pairs. *Biochemistry* **34,** 12803–12811.
74. Chen X., Mariappan, S. V. S., Catasti, P., Ratliff, R., Moyzis, K., Laayoun, A., Smith, S. S., Bradbury, E. M., and Gupta, G. (1995). Hairpins are formed by the single DNA strands of the fragile X triplet repeats: structure and biological implications. *Proc. Natl. Acad. Sci. USA* **92,** 5199–5203.
75. Gacy A. M., Goellner, G., Juranic, N., Macura, S., and McMurray, C. T. (1995). Trinucleotide repeats that expand in human disease form hairpin structures *in vitro. Cell* **81,** 533–540.
76. Mitchell, J. E., Newbury, S. F., and McClellan, J. A. (1995). Compact structures of $d(CNG)_n$ oligonucleotides in solution and their possible relevance to fragile X and related human genetic diseases. *Nucleic Acids Res.* **23,** 1876–1881.
77. Fry, M., and Loeb, L. A. (1994). The fragile X syndrome $d(CGG)_n$ nucleotide repeats form a stable tetrahelical structure. *Proc. Natl. Acad. Sci. USA* **91,** 4950–4954.
78. Smith, G. K., Jie, J., Fox, G. E., and Gao, X. (1995). DNA CTG triplet repeats involved in dynamic mutations of neurologically related gene sequences form stable duplexes. *Nucleic Acids Res.* **23,** 4303–4311.
79. Wells, R. D., and Sinden, R. R. (1993). Defined ordered sequence DNA, DNA structure, and DNA-directed mutation. *In* "Genome Analysis," Vol. 7, "Genome Rearrangement and Stability" (K. Davies and S. Warren, Eds.), pp 107–138. Cold Spring Harbor Laboratory, Cold Spring Harbor, NY.
80. Freudenreich, C. H., Stavenhagen, J. B., and Zakian, V. A. (1997). Stability of a CTG/CAG trinucleotide repeat in yeast is dependent on its orientation in the genome. *Mol. Cell. Biol.* **17,** 2090–2098.
81. Maurer, D. J., O'Callaghan, B. L., and Livingston, D. M. (1996). Orientation dependence of trinucleotide CAG repeat instability in *Saccharomyces cerevisiae. Mol. Cell. Biol.* **16,** 6617–6622.
82. Bowater, R. P., Rosche, W. A., Jaworski, A., Sinden, R. R., and Wells, R. D. (1996). Relationship between *Escherichia coli* growth and deletions of CTG · CAG triplet repeats in plasmids. *J. Mol. Biol.* **264,** 82–96.
83. Bowater, R. P., Jaworski, A., Larson, J. E., Parniewski, P., and Wells, R. D. (1997). Transcription increases the deletion frequency of CTG · CAG triplet repeat sequences from human neuromuscular disease genes in *E. coli. Nucleic Acids Res.* **25,** 2861–2868.
84. Rosche, W. A., Jaworski, A., Kang, S., Kramer, S. F., Larson, J. E., Giedroc, D. P., Wells, R. D., and Sinden, R. R. (1996). Single strand DNA binding protein enhances the stability of CTG triplet repeats in *Escherichia coli. J. Bacteriol.* **178,** 5042–5044.
85. Ohshima, K., Kang, S., and Wells, R. D. (1996). CTG triplet repeats from human hereditary disease are dominant genetic expansion products in *E. coli. J. Biol. Chem.* **271,** 1853–1856.
86. Ohshima, K., Kang, S., Larson, J. E., and Wells, R. D. (1996). Cloning, characterization, and properties of seven triplet repeat DNA sequences. *J. Biol. Chem.* **271,** 16773–16783.
87. Ohshima, K., Kang, S., Larson, J. E., and Wells, R. D. (1996). TAA · TTA triplet repeats in plasmids form a non-hydrogen bonded structure. *J. Biol. Chem.* **271,** 16784–16791.
88. Sinden, R. R. (1994). "DNA Structure and Function." Academic Press, San Diego.
89. Wells, R. D. (1988). Unusual DNA structures. *J. Biol. Chem.* **263,** 1095–1098.
90. Wang, Y.-H., Amirhaeri, S., Kang, S., Wells, R. D., and Griffith, J. (1994). DNA triplet repeats from the myotonic dystrophy gene are preferential nucleosome assembly sites *in vitro. Science* **265,** 669–671.
91. Wang, Y.-H., Gellibolian, R., Shimizu, M., Wells, R. D., and Griffith, J. (1996). Long repeating CCG triplet repeat blocks exclude nucleosomes: a possible mechanism for the nature of fragile sites in chromosomes. *J. Mol. Biol.* **263,** 511–516.
92. Kang, S., Ohshima, K., Shimizu, M., Amirhaeri, S., and Wells, R. D. (1995). Pausing of DNA synthesis *in vitro* at specific loci in CTG and CGG triplet repeats from human hereditary diseases. *J. Biol. Chem.* **270,** 27014–27021.
93. Ohshima, K., and Wells, R. D. (1997). Hairpin formation during DNA synthesis primer realignment *in vitro* in triplet repeat sequences from human hereditary disease genes. *J. Biol. Chem.* **272,** 16798–16806.
94. Chastain, P. D., Eichler, E. E., Kang, S., Nelson, D. L., Levene, S. D., and Sinden, R. R. (1995). Anomalously rapid electrophoretic mobility of DNA containing triplet repeats associated with human disease genes. *Biochemistry* **34,** 16125–16131.
95. Bacolla, A., Gellibolian, R., Shimizu, M., Amirhaeri, S., Kang, S., Ohshima, K., Larson, J. E., Harvey, S. C., Stollar, B. D., and Wells, R. D. (1997). Flexible DNA: genetically unstable CTG · CAG and CGG · CCG from human hereditary neuromuscular disease genes. *J. Biol. Chem.* **272,** 16783–16792.
96. Gastier, J. M., Pulido, J. C., Sunden, S., Brody, T., Buetow, K. H., Murray, J. C., Weber, J. L., Hudson, T. J., Sheffield, V. C., and Duyk, G. M. (1995). Survey of trinucleotide repeats in the human genome: assessment of their utility as genetic markers. *Hum. Mol. Genet.* **4,** 1829–1836.
97. Kang, S., Ohshima, K., Jaworski, A., and Wells, R. D. (1996). CTG triplet repeats from the myotonic dystrophy gene are expanded in *E. coli* distal to the replication origin as a single large event. *J. Mol. Biol.* **258,** 543–547.
98. Jaworski, A., Rosche, W. A., Gellibolian, R., Kang, S., Shimizu, M., Sinden, R. R., and Wells, R. D. (1995). Mismatch repair in *Escherichia coli* enhances instability *in vivo* of $(CTG)_n$ triplet repeats from human hereditary diseases. *Proc. Natl. Acad. Sci. USA* **92,** 11019–11023.
99. Shimizu, M., Gellibolian, R., Oostra, B. A., and Wells, R. D. (1996). Cloning, characterization, and properties of plasmids containing CGG triplet repeats from the fragile X gene. *J. Mol. Biol.* **258,** 614–626.
100. Gellibolian, R., Bacolla, A., and Wells, R. D. (1997). Triplet repeat instability and DNA topology: an expansion model based on statistical mechanics. *J. Biol. Chem.* **272,** 16793–16797.
101. Warren, S. T., and Nelson, D. L. (1994). Advances in molecular analysis of fragile X syndrome. *J. Am. Med. Assoc.* **271,** 536–542.
102. Eichler, E. E., Hammond, H. A., Macpherson, J. N., Ward, P. A., and Nelson, D. L. (1995). Population survey of the human FMR1 CGG repeat substructure suggests biased polarity of the loss of AGG interruption. *Hum. Mol. Genet.* **4,** 2199–2208.
103. Trinh, T. Q., and Sinden, R. R. (1991). Preferential DNA secondary structure mutagenesis in the lagging strand of replication in *E. coli. Nature* **352,** 544–547.
104. Wells, R. D. (1997). Triplet repeat diseases studied in man, microbes, and molecules. *Am. J. Psych.* **154,** 887.
105. Petronis, A., and Kennedy, J. L. (1995). Unstable genes-unstable mind? *Am. J. Psych.* **152,** 164–172.

106. Sancar, A. (1996). DNA excision repair. *Annu. Rev. Biochem.* **65,** 43–81.
107. Wells, R. D., Pluciennik, A., Parniewski, P., Bacolla, A., Gellibolian, R., and Jaworski, A. (1997). [Submitted]
108. Ashizawa, T., Monckton, D. G., Vaishnav., S., Patel, B. J., Voskova, A., and Caskey, C. T. (1996). Instability of the expanded $(CTG)_n$ repeats in the myotonin protein kinase gene in cultured lymphoblastoid cell lines from patients with myotonic dystrophy. *Genomics* **36,** 47–53.
109. Mellon, I., Rajpal, D. K., Koi, M., Boland, C. R., and Champe, G. N. (1996). Transcription-coupled repair deficiency and mutations in human mismatch repair genes. *Science* **272,** 557–560.
110. Kirkpatrick, D. T., and Petes, T. D. (1997). Repair of DNA loops involves DNA-mismatch and nucleotide-excision repair proteins. *Nature* **387,** 929–931.
111. Froelich-Ammon, S. J., Gale, K. C., and Osheroff, N. (1994). Site-specific cleavage of a DNA hairpin by topoisomerase II. DNA secondary structure as a determinant of enzyme recognition/cleavage. *J. Biol. Chem.* **269,** 7719–7725.
112. Yeh, Y.-C., Liu, H.-F., Ellis, C. A., and Lu, A.-L. (1994). Mammalian topoisomerase I has base mismatch nicking activity. *J. Biol. Chem.* **269,** 15498–15504.
113. Kaiser, V. L., and Ripley, L. S. (1995). DNA nick processing by exonuclease and polymerase activities of bacteriophage T4 DNA polymerase accounts for acridine-induced mutation specificities in T4. *Proc. Natl. Acad. Sci. USA* **92,** 2234–2238.
114. Bae, Y.-S., Kawasaki, I., Ikeda, H., and Liu, L. F. (1988). Illegitimate recombination mediated by calf thymus DNA topoisomerase II *in vitro. Proc. Natl. Acad. Sci. USA* **85,** 2076–2080.
115. Shuman, S. (1992). DNA strand transfer reactions catalyzed by vaccinia topoisomerase I. *J. Biol. Chem.* **267,** 8620–8627.
116. Wang, J. C. (1996). DNA topoisomerases. *Annu. Rev. Biochem.* **65,** 635–692.
117. Koo, H.-S., Wu, H.-Y., and Liu, L. F. (1990). Effects of transcription and translation on gyrase-mediated DNA cleavage in *Escherichia coli. J. Biol. Chem.* **265,** 12300–12305.
118. Madden, K. R., Steward, L., and Champoux, J. J. (1995). Preferential binding of human topoisomerase I to superhelical DNA. *EMBO J.* **14,** 5399–5409.
119. Monckton, D. G., and Caskey, C. T. (1995). Unstable triplet repeat diseases. *Circulation* **91,** 513–520.
120. Wells, R. D., Bacolla, A., and Bowater, R. P. (1998). Instabilities of triplet repeats: factors and mechanisms. *In* "Trinucleotide Diseases and Instability" (Ben A. Oostra, Ed.). Springer-Verlag, New York. [In press]
121. Samadashwily, G. M., Raca, G., and Mirkin, S. M. (1997). Trinucleotide repeats affect DNA replication *in vivo. Nature Genet.* **17,** 298–304.

CHAPTER 34

Genetic Instabilities in Yeast

SUE JINKS-ROBERTSON, CHRISTOPHER GREENE, AND WENLIANG CHEN

Department of Biology and Graduate Program in Genetics and Molecular Biology, Emory University, Atlanta, Georgia 30322

I. INTRODUCTION

The role of genetic instability in both inherited and sporadic human diseases has been acknowledged for years, but only recently has the inherent instability of the eukaryotic genome been fully appreciated. Two discoveries in particular have been instrumental in bringing aspects of genetic instability to the forefront of biomedical research. First, it was discovered that the expansion of naturally occurring trinucleotide repeats is the genetic basis for a number of inherited neurological diseases such as fragile X syndrome, myotonic dystrophy, and Huntington's disease (reviewed in [1]). Second, it was demonstrated that the microsatellite instability characteristic of tumor cells from patients with hereditary nonpolyposis colon cancer (HNPCC) is due to a defect in the postreplicative mismatch repair system (reviewed in [2]). Although genetic instabilities are clearly important in human disease, studying the molecular mechanisms of these instabilities is very difficult in mammalian cells. Many investigators thus have turned to model genetic systems, one of the most useful of which has been the yeast *Saccharomyces cerevisiae.*

DNA sequences prone to genetic instability have been termed At Risk Motifs or ARMs [3]. Genetic assays available in yeast are particularly well-suited to the characterization of ARMs and to the identification of gene products that act in *trans* to stabilize ARMs. This chapter will focus on the genetic tools available for studying genetic instability in *S. cerevisiae,* as well as the insights gained from relevant studies. Two types of genetic instabilities will be considered: replication-associated instabilities and recombination-associated instabilities. Instabilities that involve the process of DNA replication can lead to various categories of mutations, but most often are manifest as either the expansion/

contraction of simple tandem repeats or the deletion/duplication of sequences between nontandem repeats. Replication-associated genetic instabilities in yeast are likely to be directly relevant to issues of trinucleotide repeat stability in human cells. Recombination-associated genetic instabilities result either from homologous recombination or from nonhomologous (illegitimate) end-joining events. Both of these processes are important pathways for repairing broken chromosomes and can generate chromosomal rearrangements such as deletions, duplications, inversions, and translocations. Although the role, if any, of recombination in trinucleotide repeat instability is unclear at present, recombination is important for shaping overall genome structure and likely contributes to the genetic instability characteristic of tumor cells.

II. GENETIC INSTABILITIES ASSOCIATED WITH DNA REPLICATION

The fidelity of DNA replication is determined at three steps: (1) the rate with which DNA polymerase incorporates incorrect nucleotides, (2) the efficiency with which the polymerase-associated exonucleolytic proofreading activity removes incorrect nucleotides, and (3) the efficiency with which the postreplicative mismatch repair (MMR) machinery removes errors that escape proofreading [4]. Although base substitution events are an important class of replication error, the focus here will be specifically on those errors that involve the addition or deletion of nucleotides. Addition/deletion events often cause frameshift mutations, which are specifically those sequence changes that do not occur in multiples of three nucleotides and hence change the reading frame of the gene in question. Frameshift events can be uniquely detected with a selective system that employs the reversion of an auxotrophic frameshift allele. Prokaryotic studies indicate that frameshift mutations generally originate from slipped mispairing between repeated sequences during DNA replication and hence are templated by the DNA sequence itself (for a review see [5]). As first noted by Streisinger *et al.* [6] for mononucleotide runs, the transient dissociation of the nascent DNA strand from the template strand during DNA replication, followed by misalignment of the complementary sequences, will lead to a deletion event if the extrahelical base(s) is on the template strand and an insertion event if the extrahelical base(s) is on the nascent strand (see Fig. 34-1). Addition/deletion mutations do not necessarily involve tandem repeats, but may also occur through slippage and mispairing between distant direct repeats. Finally, quasipalindromic sequences or imperfect direct repeats may direct complex mutations in which base substitutions accompany a frameshift mutation. In addition to slippage and mispairing during the replication of simple repeats, genetic instability also could result from unequal recombination within a repeat array [7]. Either a gene conversion or a crossover event between misaligned homologous chromosomes or sister chromatids can result in a change in the number of repeats.

Frameshift intermediates are removed either by exonucleolytic proofreading, which removes terminal nucleotides that are incorrectly base paired with the template, or by the postreplicative MMR system. In eukaryotes there are three DNA polymerases that are important for DNA synthesis during genome duplication: DNA polymerases α, δ, and ε which are encoded by the yeast *POL1, POL3,* and *POL2* genes, respectively [8]. Although the precise roles of DNA polymerases δ and ε in leading versus lagging strand replication are not clear, each has an associated 3′ to 5′ exonuclease activity that effects proofreading and is important for mutation avoidance. In contrast, DNA polymerase α does not have an associated exonuclease activity. It is, however, associated in a complex with primase and thus is assumed to be responsible for the synthesis of RNA primers on both the leading and lagging strands.

Replication errors that escape polymerase proofreading are detected and repaired by postreplicative mismatch repair. The best understood MMR system is the methyl-directed mismatch repair system of *Escherichia coli* (for a review see [9]), and most of what is known about MMR in eukaryotes is based on this system. In *E. coli* a MutS homodimer binds directly to mismatched bases, MutH binds to hemimethylated *dam* sites, and a MutL homodimer promotes interaction between the MutS/DNA complex and MutH, thereby activating a latent MutH endonuclease activity that cleaves the newly replicated, unmethylated strand, thus marking it for removal. In eukaryotes there are multiple homologs of the bacterial MutS and MutL proteins (no MutH homologs have been identified) and, as noted previously, defects in these proteins are a causative factor in HNPCC. In *S. cerevisiae,* homologs of both MutS (Msh1p-6p) and MutL (Pms1p and Mlh1p-3p) have been identified and characterized (for reviews see [10, 11]). Msh2p, Msh3p, Msh6p, Pms1p, and Mlh1p are important for correcting mismatches arising during nuclear DNA replication and recombination, while Msh1p corrects mismatches that arise in mitochondrial DNA. In contrast, Msh4p and Msh5p are meiotic-specific proteins that have no apparent effect on mismatch repair, but rather affect crossing-over. The roles of Mlh2p and Mlh3p have not been established. Mlh1p and Pms1p can form a heterodimer and mutations in either gene

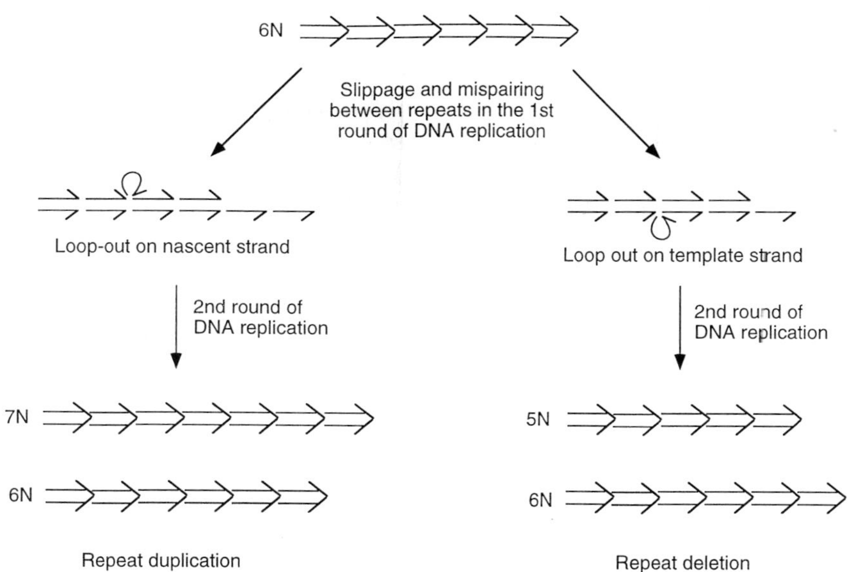

FIGURE 34-1 Replication slippage in tandem repeats generates insertions and deletions of the repeat unit.

completely eliminate replication-associated MMR. Although the Mlh1p/Pms1p heterodimer is presumed to be the active form for editing replication intermediates, recent work indicates that Mlh1p, but not Pms1p, may be important in promoting meiotic crossing-over [12]. Yeast strains deleted individually for *MSH3* or *MSH6* have a weak mutator phenotype, whereas simultaneous deletion of *MSH3* and *MSH6* yields a strong mutator phenotype indistinguishable from that of a *msh2Δ* strain [13–15]. These *in vivo* data are the basis of a model in which Msh2p forms a heterodimer with either Msh3p or Msh6p; Msh2p/Msh3p appears to recognize only insertion/deletion mismatches, while Msh2p/Msh6p exhibits a preference for base substitution mismatches.

A. Deletions/Insertions within Tandem Repeats

Tandem repeats of the same sequence are referred to as either microsatellites or minisatellites, depending on the size of the repeating unit. The inability of the yeast MMR machinery to remove replication slippage intermediates in repeat units of 16 bp or more has led to the suggestion that repeat lengths of less than approximately 15 bp should be classified as microsatellites and those longer than 15 bp as minisatellites [15]. Two major systems have been developed specifically to study the stability of simple tandem repeats in yeast [16]. In the first system a synthetic fragment containing a defined simple repeat (e.g., a poly(GT) tract of 14 repeat units) is cloned into a polylinker region located near the border between the yeast *LEU2* and bacterial *lacZ* portions of a fusion protein gene, thereby disrupting the reading frame of the gene. A change in the number of simple repeats can restore the correct reading frame, resulting in the production of a full-length *LEU2–lacZ* fusion protein. Such events can be identified as blue yeast colonies on X-gal medium. In the second system, repeats are cloned in-frame into the polylinker region of a *LEU2–URA3* fusion protein gene. Changes in the number of simple repeats can disrupt the reading frame, resulting in Ura$^-$ segregants, which can be identified on medium containing 5-fluoroorotic acid, a compound that is toxic to Ura$^+$ cells [17]. The *LEU2–URA3* system is of more general usefulness than the *LEU2–lacZ* system because it potentially can detect a wider range of slippage events (there are more ways to disrupt a reading frame than to restore one) and because it employs a strong selection rather than a colony color screen. Following phenotypic identification of altered tract lengths, the actual change in the number of repeats can be determined either by DNA sequencing or by PCR.

1. MONONUCLEOTIDE REPEATS

The stability of an 18-bp poly(G) tract within the *LEU2–URA3* fusion construct has been examined in both wild-type and MMR-defective strains [15]. In a wild-type strain the rate of tract alteration, as measured by the rate of Ura$^-$ segregants, is approximately 10^{-5}. The rate of tract alteration increases 6300-, 130-, 30-, and 4600-fold in *msh2, msh3, msh6,* and *msh3 msh6* mutants, respectively. These data, in agreement with other studies, indicate that 1-bp loops are recognized by both the Msh2p–Msh3p and the Msh2p–Msh6p complexes, and that the complexes likely compete for the repair of 1-bp loops. Whereas +1 events are much more common than −1 events in the wild-type background, there is a strong bias for −1 events in the *msh3* and *msh6* mutants. The spectrum of events in a *msh2* background could not be determined due to the extremely high mutation rate. The −1 frameshift bias in the MMR-defective strains suggests either that the MMR system removes −1 frameshift intermediates more efficiently than it removes +1 frameshift intermediates, or that there is a another system that removes +1 frameshift intermediates more efficiently than −1 frameshift intermediates. Streisinger and Owen [18], in fact, suggested that proofreading might preferentially remove extrahelical bases on the nascent strand (the +1 frameshift intermediates), but this has not been examined experimentally.

Tran *et al.* [19] have systematically increased the length of a poly(A) tract inserted at a fixed position within the *LYS2* coding sequence. The poly(A) tracts ranged in size from 4 to 14 bp and created either +1 or −1 frameshift alleles. The rates of Lys^+ reversion events occurring within the tracts were determined in wild-type, *msh2* mutant, and proofreading-defective yeast strains. A comparison of tract instability in *msh2* strains relative to that in the wild-type strains indicates that the ratio increases as the tract length increases. A similar comparison of tract instability as a function of length in the proofreading-defective strains relative to the wild-type strain reveals a relatively constant ratio. These data indicate that with short tracts, both MMR and proofreading contribute to the removal of frameshift intermediates, whereas with long tracts, most of the stability results from removal of frameshift intermediates by the MMR system rather than by exonucleolytic proofreading. These *in vivo* results are entirely consistent with *in vitro* results demonstrating that the efficiency of proofreading decreases as the length of a mononucleotide run increases [20].

The inverse relationship between proofreading efficiency and tract length has been attributed to the potential distance between the extrahelical base and the polymerase active center [20]. The longer a mononucleotide run, the further away from the active center the extrahelical base potentially can migrate, and hence the lower the probability that it will be detected and exonucleolytically removed. Although the instability of a given monotonic run clearly increases as the length of the run increases [19], an examination of random frameshift mutations occurring within a defined 150-bp interval of the *LYS2* gene indicates that factors other than run length also are important [21]. These other factors have not been systematically examined but might be related to the composition of the run (AT versus GC), the sequence of bases on the leading versus the lagging strand during DNA replication (i.e., whether the A's on the leading versus the lagging strand), and the surrounding sequence context.

The extreme instability of monotonic runs in MMR-defective yeast suggests that genes containing such runs should be particularly susceptible to mutation in MMR-defective tumor cells. An examination of mutations in tumor suppressor genes in HNPCC cells indicates that monotonic runs are indeed very strong mutational hotspots [22–25].

2. Dinucleotide Repeats

The most common microsatellite in eukaryotic genomes is poly(GT) [26], and all of the yeast work done thus far has utilized this particular repeat. Both the *LEU2–lacZ* and *LEU2–URA3* systems have been used to demonstrate that poly(GT) tracts containing approximately 15 repeats are highly unstable in wild-type yeast cells [16, 27, 28]. The rate of tract alteration is about 10^{-5} per cell division when the microsatellite is on a plasmid; the stability improves slightly when the tract is integrated into a chromosome. Approximately 90% of the tract alterations are additions or deletions of single repeats, and there is a strong bias for +2 additions over −2 deletions. At least in the case of poly(GT) tracts, two lines of evidence indicate that the instability is due to replication slippage rather than unequal recombination. First, although meiotic levels of recombination are generally several orders of magnitude higher than mitotic levels [29], the instability of poly(GT) does not increase during meiosis [27]. Second, the *RAD52* gene is required for all recombination in yeast [29], and yet the stability of poly(GT) is not altered in *rad52* mutant strains [16]. Although instability of the poly(GT) tract appears to result from replication slippage, it is interesting to note that the tract becomes 5- to 10-fold more unstable if it is highly transcribed [30]. Inversion of the poly(GT) tract within the fusion protein does not affect its stability [16], which indicates that replication slippage for this particular dinucleotide repeat is not a leading versus lagging strand phenomenon.

Given that the instability of the poly(GT) microsatellite is attributed to replication slippage, the relative contributions of exonucleolytic proofreading and MMR to microsatellite stability have been examined. A poly(GT) tract is destabilized 100- to 300-fold by mutations in *MSH2,* the major MutS homolog, and by mutations in either *PMS1* or *MLH1,* the major MutL homologs [27, 28]. Mutations in *MSH3* or *MSH6,* the other two MutS homologs involved in MMR, destabilize the poly(GT) repeat only about 30- and 5-fold, respectively, while mutations in both genes destabilize the tract several hundred-fold [14, 15, 28]. In the MMR-defective strains, with the exception of the *msh6* mutant, all tract alterations are insertions or deletions of a single repeat unit. In addition, the bias for +2 insertions seen in wild-type strains is lost. In the *pms1* and *mlh1* strains, +2 and −2 events occur at about equivalent frequencies, whereas −2 events predominate in *msh2* and *msh3* mutants. The biases in wild-type and MMR-defective strains presumably reflect both the relative occurrence and the repair efficiencies of loop-outs on the nascent versus template strands, which lead to insertions and deletions, respectively. In contrast to the strong stabilizing effect of MMR proteins, exonucleolytic proofreading plays a relatively minor role in the stability of a poly(GT) tract. A mutation that eliminates the exonuclease activity of pol δ destabilizes the tract only 5- to 10-fold, while a comparable mutation in pol ε has no effect on tract stability [27]. The relatively minor role of proofreading in the maintenance of microsatellite stability has been attributed to the potential migration of the extrahelical loop away from the polymerization center in long tandem repeats (see Section I.A.1 above).

As documented for mononucleotide runs, the stability of a poly(GT) tract is dependent upon its length. The stabilities of poly(GT) tracts containing 15, 33, 51, 99, or 105 bp of repeated sequence have been examined in the context of the *LEU2–URA3* fusion construct [31]. In both wild-type and *msh2* strains, the rate of tract instability increases as the tract length increases; the relationship is greater than linear, but is not exponential. Although long tracts are much more unstable than short tracts, the instability of all tracts is elevated to roughly the same extent in a *msh2* background relative to a wild-type background. The similar *msh2*:wild-type ratio for all tract lengths argues that the number of repeats affects polymerase slippage rather than the efficiency of mismatch repair. In addition to the dramatic increase in tract instability as a function of the number of GT repeats, the spectrum of tract alterations also changes as the tract length increases. For the 15 and 33-bp poly(GT) tracts in wild-type strains, greater than 90% of the events are additions or deletions of a single repeat unit, and there is a two-fold bias of insertions relative to deletions. With the longer tracts, a larger proportion of the frameshift events involves more than two repeat units. For the 99- and 105-bp tracts, all of the events that involve a single repeat are +2 events. In all of the *msh2* strains, most frameshift events involve a single repeat unit, and events involving more than two repeat units are very rare.

Studies of the trinucleotide repeat human diseases indicate that the tract purity greatly impacts microsatellite stability [32, 33]. In yeast, the impact of tract purity on microsatellite stability has been examined by inserting a single AT or CT dinucleotide into a 51-bp poly(GT) tract [34]. Each of the variant repeats stabilizes the poly(GT) tract approximately fivefold, and in each case, the stabilization is not dependent on a functional MMR system.

In addition to the proofreading and MMR proteins, the yeast homolog of the mammalian FEN-1 (flap endonuclease and five prime exonuclease) protein also has a strong stabilizing effect on poly(GT) sequences [35]. FEN-1 has both endonuclease and exonuclease activities; it functions in the removal of the last ribonucleotide at the 5′ ends of Okazaki fragments as well as displaced 5′ ends during lagging strand synthesis [36]. The yeast FEN-1 homolog is encoded by the *RAD27/RTH1* gene, and mutations in this gene elevate poly(GT) instability several hundred-fold, which is comparable to the elevation observed in *msh2* mutants. In contrast to the bias for −2 events seen in MMR-defective strains, there is a strong bias for +2 events in *rad27* mutants. Although it was originally suggested that FEN-1 functions in mismatch repair [35], subsequent analyses have indicated that the protein functions in a distinct mutation avoidance pathway [37], which will be described in Section I.C below.

3. Trinucleotide Repeats

The expansion and contraction of a CAG repeat, which is one of the repeats important in the human trinucleotide expansion diseases [1], have been modeled in yeast. Because the *LEU2–lacZ* and *LEU2–URA3* fusion protein systems developed to examine microsatellite instability require a shift in reading frame to detect tract alterations, these systems obviously are not useful for examining trinucleotide repeat stability. CAG repeats are extremely unstable in yeast, however, which has allowed alterations in tract length to be detected by simply examining tract sizes in sibling cells derived from a single colony [38, 39]. In this technique a yeast colony is dispersed to single cells, the cells are plated and allowed to form colonies, and the DNA from the colonies is analyzed for changes in tract length by either PCR or Southern blotting. Rates of tract alteration as high as 3×10^{-2} changes per cell per generation have been reported [38]. Although this method can detect all alterations within the trinucleotide tract being examined, it is a very tedious technique. As an alternative to the physical assay, a system employing a colony color screen has recently been reported [40]. This system takes advantage of the stringent spacing required for the *Schizosaccharomyces pombe adh1* promoter to direct proper transcription initiation of an *S. cerevisiae* gene [41]. In this system a poly(CAG) tract containing 50 repeats was inserted between the *adh1* promoter TATA element and the transcriptional start site for the *S. cerevis-*

iae ADE8 gene. In an *ade2* strain background, production of Ade8p results in a red colony color, whereas cells lacking Ade8p form white colonies. The presence of the full-length CAG tract prevents expression of Ade8p (colonies are white) and contraction of the 50-repeat tract to between 8 and 38 total repeats restores Ade8p expression (colonies are red). This assay is quantitative and is very easy to perform, but can only detect large deletions; expansions or changes of just a few repeat units cannot be detected. Although the utility of the *adh1–ADE8* system for assaying trinucleotide repeat stability is limited, it should nevertheless prove very useful for identifying mutants with altered tract stabilities. Potential mutants can be identified by screening mutagenized colonies for changes in the frequency of red sector production.

As described above for the mononucleotide and dinucleotide repeats, the instability of a poly(CAG) repeat is correlated with the number of repeats, with longer tracts being more unstable than shorter tracts [38]. In the nonselective, physical system, which can detect expansions as well as contractions, deletion events dominate the spectrum of length changes and most deletions removed multiple repeat units [38, 39]. Mutations that eliminate mismatch repair destabilize trinucleotide repeats, although the magnitude of the destabilization is much less than that seen with the mono- and dinucleotide repeats [42]. In contrast to the large changes in CAG tract size seen in wild-type strains, small changes of only 1–2 repeats predominate in MMR-defective cells. As with the mono and dinucleotide repeats, there is a bias for deletions over insertions in MMR-defective cells. It should be noted that the *adh1–ADE8* system, which detects only large deletions, has not detected an involvement of the MMR system in poly(CAG) stability [40]. This seemingly anomalous result can be attributed to the failure of MMR proteins to recognize and correct loops of greater than approximately 15 nucleotides (see Section I.A.4 below). As with the dinucleotide poly(GT) repeats, mutations in the *RAD52* gene do not appear to alter trinucleotide repeat stability [40], indicating that the instability does not involve recombination.

As has been observed in *E. coli* [43], the stability of a CAG trinucleotide repeat in yeast is dependent upon its orientation relative to the direction of DNA replication [38, 39]. When positioned next to a defined origin of replication on a chromosome, poly(CAG) is much more unstable when CAG is on the leading strand template (and CTG on the lagging strand template) than when CAG is on the lagging strand template (and CTG on the leading strand template). The orientation dependence of poly(CAG) stability has been attributed to the ability of CTG repeats to form more stable intrastrand hairpin structures than CAG repeats *in vitro* [44]. It is noteworthy that all of the trinucleotide repeats implicated in human disease (i.e., CAG, CGG, and GAA) can form hairpin structures [45]. During DNA replication, the lagging strand template has considerable single-stranded character, which could allow the formation of an extruded CTG hairpin. Replication across the base of the hairpin would result in deletions on the nascent CAG strand (Fig. 34-2A). This mechanism of deletion formation, although somewhat different from the simple replication slippage proposed for other microsatellites, is quite similar to inverted repeat-promoted deletions between short nontandem direct repeats (see Section I.B below). It should be noted that one yeast study failed to detect an orientation dependence for poly(CAG) stability [40]. In this study, however, the direction of the replication fork movement through the CAG repeat was not determined and it is possible that the direction of fork movement was random.

The CAG tract alterations observed in yeast have been either changes of a small number of repeat units or large deletions. The small changes likely result from simple replication slippage. If one assumes a minimal length of sequence is necessary for stable intrastrand secondary structure, then a hairpin-associated mechanism of deletion formation can explain the predominance of large deletions within trinucleotide repeats in wild-type yeast strains. The small hairpins required for the generation of intermediate size deletions may not be thermodynamically stable, which could account for the paucity of these deletions relative to very large deletions. A similar argument has been proposed to explain the requirement for a critical number of trinucleotide repeats before expansion occurs in humans [46].

It has been suggested that one effect of CAG expansion may be the induction of a heterochromatic state which would silence adjacent genes [47]. In support of this possibility, long CAG repeats have been shown to be potent nucleosome assembly sites *in vitro* [48]. In order to examine possible trinucleotide-induced silencing of adjacent sequences in yeast, a 130-repeat poly(CAG) tract was inserted both upstream and downstream of the yeast *URA3* gene [39]. Although the *URA3* gene has been used extensively to study silencing in yeast [49], no poly(CAG)-induced silencing of *URA3* could be detected.

Loci of poly(CCG) repeats in the human genome are fragile sites that exhibit increased chromosome breakage under cell culture conditions that slow DNA replication [1]. In yeast, placement of a poly(CTG) tract between direct repeats increases the frequency of recombination between the repeats in a tract length-dependent manner, suggesting that CTG tracts may be sites of double-strand breaks [49a]. Consistent with this interpretation, the frequency of recombination between direct repeats increases when DNA replication is slowed

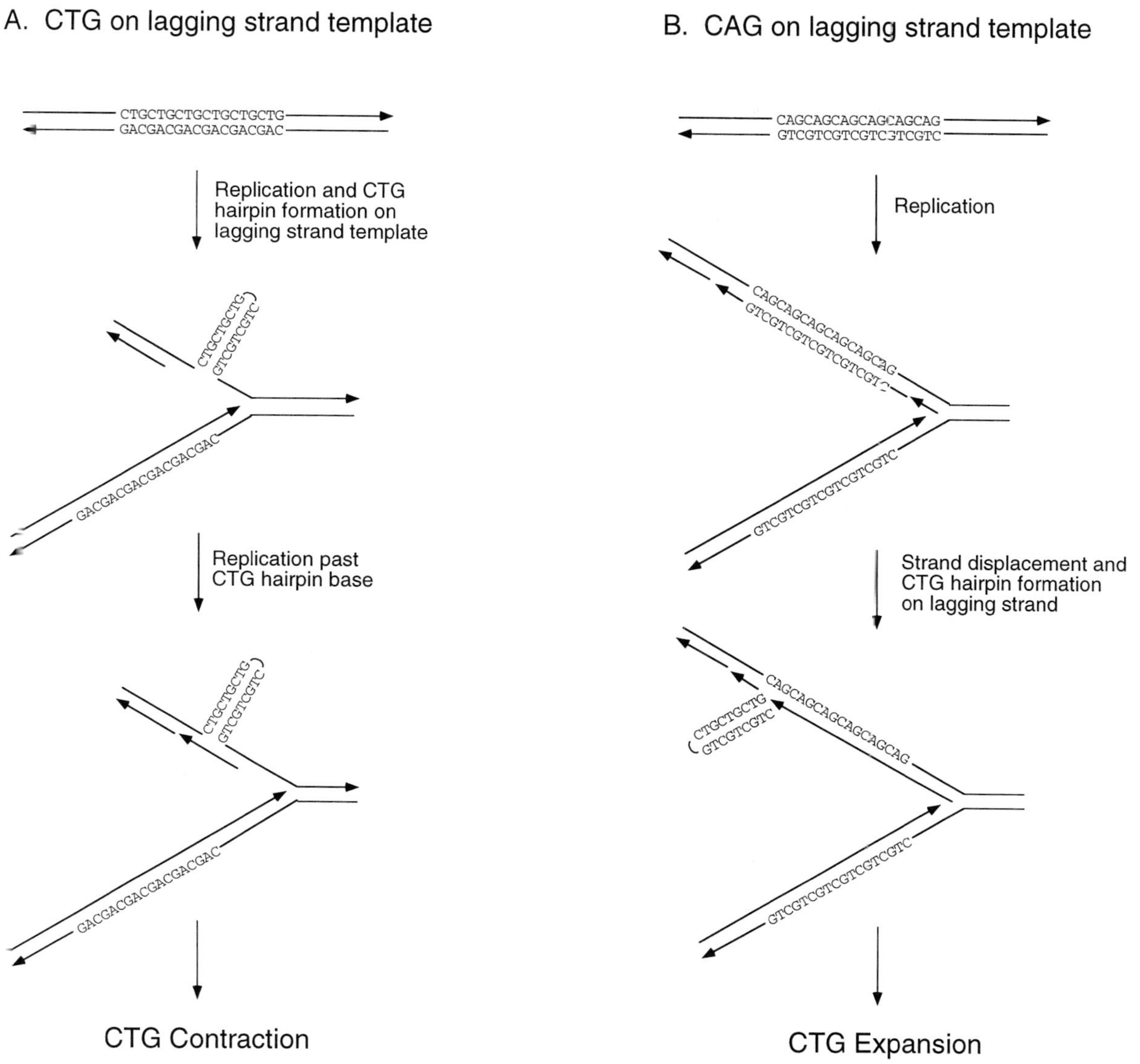

FIGURE 34-2 Models for the contraction and amplification of the CTG trinucleotide repeat.

with hydroxyurea, and double-strand breaks can be detected physically on the relevant yeast chromosome. Finally, mutational elimination of the yeast FEN-1 homolog increases the instability of a poly(CTG) tract (see Figure 34-2 and Section II.C for a discussion of FEN-1 and trinucleotide repeat stability).

4. HIGHER ORDER DIRECT REPEATS

The effect of repeat unit size on the stability of micro- and minisatellites has been examined systematically both in wild-type strains and in strains defective in various MutS homologs [15]. In these experiments the *LEU2–URA3* fusion construct was used to measure the stabilities of repeat units of 1, 2, 4, 5, 8, 10, 11, 13, 16, or 20 bp. Because the size of the extrahelical loop in frameshift intermediates should be the size of the repeat unit (or multiples of the repeat unit), a comparison of tract stabilities in wild-type versus MMR-defective mutants allows one to deduce the recognition specificities of specific MutS homologs. Based on such analyses, Msh2p and Msh3p are involved in the repair of loops up to 13 nt but do not repair loops 16 nt or longer; Msh6p only repairs loops of 1 or 2 nt. As noted previously, it has been proposed that this MMR-specific cutoff in loop recognition be used to define the lengths of microsatellites versus minisatellites [15].

B. Deletions with Endpoints in Nontandem Direct Repeats

Deletions between nontandem, short direct repeats (<100 bp) is most likely due to replication slippage and is stimulated by the presence of intervening inverted repeats; the inverted repeats either can be tandem (palindromes) or can be separated by several kilobases. Direct repeats that are longer than approximately 100 bp promote deletion events primarily via homologous recombination mechanisms, and will be discussed in Section III.A in relation to recombination-associated genetic instabilities.

Tandem inverted repeats are very unstable in bacterial cells and stimulate deletion events involving nearby short direct repeats [5, 50]. Palindromes can potentially form intrastrand hairpins, and such hairpins are assumed to promote replication slippage between short direct repeats either by bringing repeats physically closer together or by perturbing the process of replication. Two studies have examined the stability of palindromes in yeast. In one study, a palindrome having the potential to form a 47-bp hairpin was inserted out-of-frame into a *LEU2–lacZ* fusion construct, and colonies producing a full-length fusion protein were identified [51]. In the other study, sequences capable of forming either 30- or 80-bp hairpins were inserted between the transcription and translation start sites of either the yeast *URA3* or *HIS4* gene. The presence of the palindrome apparently prevents translation of the mRNA, allowing deletion events involving the palindromes to be identified by selecting for prototrophs [52]. In both studies the palindromes were very unstable, with deletions that restored function of the reporter genes appearing at rates of 10^{-4}–10^{-5} per cell division. Most of the deletions examined had endpoints in 4- to 9-bp direct repeats that either flanked or were near the base of the presumptive palindrome-promoted hairpin. The rate of deletion formation was not altered in *rad52* mutants [51, 52] nor during meiosis [52], but was elevated in some strains harboring conditional mutations in DNA polymerase α [52]. This behavior supports the notion that deletion formation is due to hairpin-promoted polymerase slippage between short direct repeats, and argues against a direct role for recombination in deletion events involving palindromes and short direct repeats. The rate of deletion formation was shown to be affected by the length of the palindrome (longer palindromes promote higher rates of deletion formation than do shorter palindromes) as well as the chromosomal location of the palindrome [52]. This latter observation may be related to the direction of replication fork movement through the palindrome (see below). Finally, separating the inverted repeats by 25–60 bp decreased the rate of deletion formation approximately 10-fold [52].

The association of deletion formation with long, nontandem inverted repeats in yeast was first demonstrated in relation to prototrophic "reversion" of Tn5-induced *lys2* mutations [53, 54]. Tn5 has 1.5-kb inverted repeats flanking 6 kb of unique sequence, and duplicates a 9-bp target sequence upon integration. *lys2::Tn5* revertants result from either precise or imprecise excision of the transposon, and the deletion endpoints in all cases are within short direct repeats near the outside ends of the inverted repeats [54, 55]. Mutations that increase the frequency of Tn5 excision have been identified (originally referred to as *tex* mutations for Tn5 excision [56]); one of the corresponding mutant alleles (*tex1-1*) is a conditional allele of the DNA polymerase δ gene and is commonly referred to as *pol3-t*. Reversion of *lys2::Tn5* is elevated 100-fold in *pol3-t* mutants at semipermissive temperatures, an observation that directly implicates DNA replication in Tn5 excision events. Mutations in *RAD50* or *RAD52* block the stimulation of Tn5 excision in *pol3-t* mutants, suggesting that the encoded recombination-associated proteins may be needed to form/stabilize a hairpin intermediate, or to anneal/transfer the slipped 3′ end in this system. In *E. coli,* deletion events involving hairpin intermediates occur more often during lagging strand than during leading strand synthesis [57], and a similar argument has been made for Tn5 excision events in yeast [54]. Because the lagging strand template has more single-stranded character than the leading strand template, it is assumed to be more likely than the leading strand template to experience hairpin formation and promote deletions. In addition to promoting intrachromosomal deletion events, Tn5 also stimulates interchromosomal recombination in yeast [55]. It has been suggested that interchromosomal recombination and replication slippage are simply alternative outcomes for a replication fork stalled by a hairpin structure in the template strand [55].

Although initial studies of Tn5 excision from the *lys2::Tn5* allele required restoration of Lys2p function, this functional constraint has been eliminated by inserting the *URA3* gene into the central portion of Tn5 and selecting for Ura$^-$ segregants on 5-FOA medium [55]. Removal of the Lys2p functional constraint has allowed detection of many more imprecise excision events; deletion breakpoints cluster near the ends of the Tn5 insert and all are within short direct repeats. Many of the Ura$^-$Lys$^-$ segregants retain portions of the Tn5 and such "insertion" isolates have been useful for isolating subsequent deletions that restore Lys2p function [58]. Two nonpalindromic insertions (InsD and InsE of 31 and 61 bp, respectively) have been particularly useful in characterizing deletions that do not involve inverted repeats or simple tandem repeats. As before, deletion events have endpoints in small direct repeats that either are identical or contain a single mismatch. The spectrum of deletion events involving these small repeats is altered when the direction of replication fork movement through the repeats is reversed, suggesting leading versus lagging strand deletion specificities [58]. In addition, experiments with InsD and InsE *lys2* alleles have indicated that the yeast MMR machinery can efficiently recognize loops of 7 nt but not loops of 31 nt [59], a result consistent with tandem repeat studies [15]. Experiments in *pol3-t* mutants suggest a role for *RAD52* and *RAD50* in deletion formation when the direct repeats are separated by nonpalindromic sequences [58]. The involvement of *RAD52/RAD50* in deletion formation between short direct repeats in some systems [54, 58] but not others [51, 52] may be related to the distance between the repeats and/or the ability of the intervening sequence to form hairpins.

C. Generation of Tandem Duplications

The yeast *RAD27/RTH1* gene encodes the homolog of the mammalian FEN-1 protein, which is involved in removal of the 5′ ends of Okazaki fragments during DNA replication and may also process branched DNA repair intermediates [36]. Although the high level of dinucleotide repeat instability in *rad27* mutants led to the suggestion that FEN-1 is part of the mismatch repair machinery in yeast [35], subsequent analyses of mutation spectra at the *LYS2* and *CAN1* loci revealed the accumulation of novel insertion mutations in *rad27* mutants [37]. The *rad27* spectra are dominated by 5- to 18-bp duplications flanked by 3- to 12-bp direct repeats, a type of mutation that is extremely rare in wild-type cells. In addition, mutations in *RAD27* cause a hyperrecombination phenotype and exhibit synthetic lethality when combined with a mutation in either *RAD51* or *RAD52*, indicating that *rad27* mutants accumulate lesions that are normally processed by the double-strand break (DSB) repair pathway. Kolodner and colleagues have suggested that the failure to process the 5′ ends of Okazaki fragments leads to displacement synthesis, which produces a persistent 5′ flap. The persistent 5′ flap could slip out-of-register and mispair with the template strand, which would give rise to the observed duplications at the next round of DNA replication. Alternatively the presence of the 5′ flap might facilitate breakage of the intact template strand, which would be single-stranded opposite the flap junction, thus generating a broken chromosome. The synthetic lethality between *rad27* and *rad51/rad52* suggests that such DSB formation is frequent. The 5′ flap, because of its displacement by DNA synthesis, constitutes a duplication of sequences on the other side of the DSB. Subsequent out-of-register pairing between short repeats at the ends of the broken molecules ("mutagenic" single-strand annealing) will duplicate the sequence between the short repeats. The proposed mechanisms not only explain the duplications that arise in *rad27* mutants, but also can account for the predominance of insertion mutations within dinucleotide repeats in *rad27* mutants. Finally the mechanisms proposed to operate in *rad27* mutants may explain the origin of spontaneous mutations such as the highly revertible *cyc1-96* allele, which is a 19-bp duplication flanked by 5-bp direct repeats [60].

Gordenin *et al.* [3] have proposed a model for trinucleotide repeat expansion that incorporates the novel mutagenic properties of yeast strains defective for FEN-1, as well as the ability of trinucleotide repeats to form stable hairpins (see Fig. 34-2B). The model proposes that the presence of a trinucleotide repeat at the 5′ end of an Okazaki fragment can give rise to a flap that folds into a hairpin structure. Because FEN-1 diffuses onto single-stranded ends only and will not load onto duplex ends [36], it is hypothesized that the formation of a hairpin may completely block flap removal by FEN-1. Subsequent processing of the hairpin flap by the slippage-mispairing or the DSB repair mechanisms proposed above for *rad27* mutants would result specifically in expansion of the trinucleotide repeat. Based on hairpin thermal stabilities, one would predict a more stable hairpin flap formation if the poly(CTG) tract is on the lagging strand, which would place poly(CAG) on the lagging strand template. It should be noted that this is the reverse of the orientation-specificity seen with poly(CAG) contractions in yeast (see above), which places poly(CTG) on the lagging strand template. The hairpin flap model assumes that strand displacement synthesis occurs to produce a 5′ flap and that the flap includes the trinucleotide repeat; the inclusion of a trinucleotide repeat in the flap may be related to its distance from an origin of replication. The rarity of trinucleotide expansions in yeast could be related to a general lack of 5′ flap formation, or simply could reflect inappropriate tract placement relative to a replication origin. It has recently been demonstrated that the instability of a poly(CTG) tract (both expansions and contractions) is increased in a *rad27* mutant, implicating a role for unprocessed Okazaki fragments in trinucleotide repeat diseases [49a].

Gene amplification of single-copy sequences in higher eukaryotes often involves the formation of large palindromes [61]. The formation of a palindromic duplication has been described recently in yeast and has been shown to involve short inverted repeats [62]. Specifically, palindromic molecules can be generated from non-palindromic plasmids by engineering a double-strand break adjacent to short inverted repeats. The inverted repeats presumably promote hairpin formation at the break site, and replication through the hairpin duplicates the entire molecule in inverted orientation. Palindrome formation is dependent on the *RAD52* gene, suggesting that at least part of the yeast homologous recombination machinery is required to effect a critical step in palindrome formation.

III. GENETIC INSTABILITIES ASSOCIATED WITH HOMOLOGOUS RECOMBINATION

Homologous recombination is an important mechanism for repairing DNA damage in mitotically dividing cells, and is essential in most organisms for ensuring proper chromosome segregation during meiosis. Yeast and other fungi have historically served as model organisms for studying homologous recombination because of the availability of tetrad analysis, which allows genetic and molecular characterization of all products derived from a single meiotic division. Tetrad analysis in fungi led to the discovery of the non-Mendelian segregations termed gene conversion and postmeiotic segregation (PMS), and the realization that such aberrant segregations are associated with the process of crossing-over. In gene conversion, one allele is converted to another allele, which gives rise to a 6:2 or 2:6 segregation pattern instead of the normal 4:4 meiotic segregation pattern of alleles. PMS is signaled by a 5:3 or 3:5 meiotic segregation pattern, and reflects the persistence of heteroduplex DNA in which the single strands composing a duplex are genetically distinct. In current recombination models (see [63]), an exchange of single strands between different chromosomes gives rise to heteroduplex DNA, which, if left unrepaired, is manifested genetically as a PMS event. The repair of mismatches within heteroduplex is accomplished by the same mismatch repair (MMR) machinery that removes replication errors, and such repair can result in a gene conversion event. The resolution of a heteroduplex recombination intermediate requires strand cleavage, which gives rise to crossing-over between markers that flank the region of heteroduplex formation about 50% of the time.

The yeast *S. cerevisiae* has a very active meiotic recombination system, with 1 cM corresponding to approximately 3 kb of DNA (reviewed in [29]). Although meiotic rates of recombination are generally several orders of magnitude greater than mitotic levels, DNA transformed into yeast integrates efficiently into genomic DNA by homologous recombination greater than 90% of the time. Early yeast gene targeting experiments led to the discovery that broken molecules are highly recombinagenic, and this observation was crucial to the proposal of the DSB repair model of recombination [64]. DSBs are thought to initiate most recombination in yeast, and presumably are important for initiating recombination in other eukaryotes as well. Whereas spontaneous mitotic DSBs likely derive from random DNA damage, meiotic DSBs are produced enzymatically [65] at nonrandom sites [66]. Consistent with the notion that the number of initiating lesions determines the amount of recombination, mitotic recombination can be induced to very high levels by enzymatic introduction of a DSB into one of the recombining partners [67]. This is generally done using the endogenous HO endonuclease, whose normal function is to initiate mating type switching in yeast.

Homologous recombination usually involves like sequences at identical positions on homologous chromosomes (allelic recombination) but may also involve dispersed repeated sequences (ectopic recombination). In contrast to allelic recombination, which generally functions to promote genome stability, ectopic recombination is potentially a potent source of genetic instability. As with allelic recombination, the rate of ectopic recombination in yeast is higher in meiosis than in mitosis and there is a strong association between gene conversion and crossing-over [68, 69]. Ectopic recombination can involve direct repeats, inverted repeats, or repeats positioned on nonhomologous chromosomes and crossing-over between repeats can produce deletions/duplications, inversions, or translocations, respectively (see Fig. 34-3). Gene conversion events between dispersed repeats do not give rise to genome rearrangements, but are important for homogenizing repeats as well as for creating new alleles.

The detection of ectopic recombination events generally requires a selective system, which precludes the easy use of many naturally occurring repeated sequences in yeast. Most studies, therefore, make use of artificial duplications, which are constructed by inserting defined sequences into preselected positions within the yeast genome, a process that is greatly facilitated by the efficiency of yeast transformation and targeting. A typical duplication consists of auxotrophic heteroalleles that can recombine to produce a selectable, prototrophic gene. Ectopic recombination can be examined in either haploid or diploid yeast strains, and both mitotic and meiotic events can be analyzed. In the discussion that follows the focus will be on the genetic systems used to study ectopic recombination in yeast and the types of ectopic interactions that have been documented. In addition, physical aspects of the repeated sequences that can influence the rate of interaction will be addressed. Recombination events involving extrachromosomal substrates (i.e., plasmids) will not be considered since these events are not directly relevant to issues of genome stability.

A. Recombination between Direct Repeats

Recombination events involving direct repeats can lead to deletion of the region between the repeats as well

A. Direct repeats

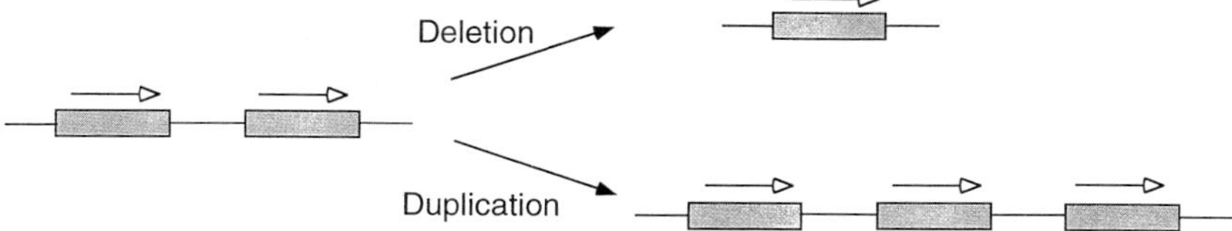

B. Inverted repeats

C. Repeats on nonhomologous chromosomes

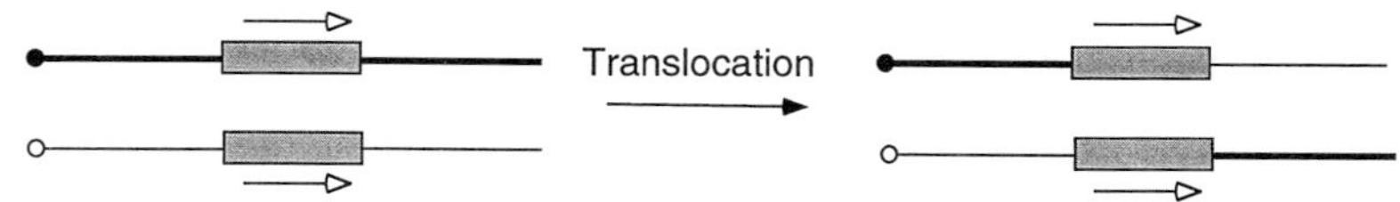

FIGURE 34-3 Genetic rearrangements resulting from homologous recombination between repeated sequences.

as to duplication of this region (Fig. 34-3A). Whether a direct repeat-associated rearrangement is the result of replication slippage versus recombination is likely related to the size of the repeats as well as to the distance between the repeats. Replication slippage requires dissociation of the nascent and template strands and subsequent reassociation in an out-of-register mode, so one might expect the probability of slippage to decrease as the repeat size increases and as the distance between repeats increases. Although recombination rates likewise appear to decrease as the distance between repeats increases [70], there is a positive correlation between repeat size and recombination rate [71]. The most direct way to distinguish between recombination versus replication slippage in yeast is that the former is generally dependent on the *RAD52* gene, whereas the latter is not (exceptions are the *RAD52*-dependence of some slippage events in *pol3-t* mutants (see Section II.B above) and the *RAD52*-independence of single strand annealing events [72]). A comprehensive review of the genetic control of direct repeat (and inverted repeat) recombination can be found in Klein [72] and will not be discussed here.

Direct repeats in yeast are generally derived by homologous integration of a nonreplicative, circular plasmid into genomic DNA (for examples, see [73, 74]). As shown in Fig. 34-4A, if the plasmid contains a *his3* recombination substrate with two mutations (*his3-x* and *his3-y*) and a selectable marker (*URA3*), recombination with a wild-type, chromosomal *HIS3* gene yields Ura^+ transformants in which a duplication of different, mutant *his3* genes flanks the remainder of the plasmid. Recombination between the *his3* heteroalleles to produce a wild-type *HIS3* gene can be identified by plating cells in the absence of histidine. In general, the frequency of spontaneous mitotic recombination between direct repeats is in the 10^{-4} range.

As illustrated in Fig. 34-4B, there are many types of recombination that can occur between direct repeats, and a given event can involve either repeats on the same chromatid (intrachromatid event) or repeats on sister chromatids. Two general classes of His^+ recombinants can be identified. His^+ recombinants in the first class result from a simple gene conversion event and hence retain the plasmid sequences between the flanking *his3* direct repeats. Gene convertants thus are Ura^+ as well as His^+. The second class of His^+ recombinants results from crossing-over and is Ura^-; the plasmid sequences between the original repeats are lost and only a single copy of *his3* sequences remains. Although unequal sister chromatid exchange produces Ura^- recombinants that are identical to those arising via intrachromatid crossing-over (a "pop-out" event), a His^+Ura^+ recombinant containing a triplication of *his3* sequences (and a direct duplication of the entire plasmid) is uniquely diagnostic of sister chromatid exchange. If both a posi-

A

Constructing direct repeats

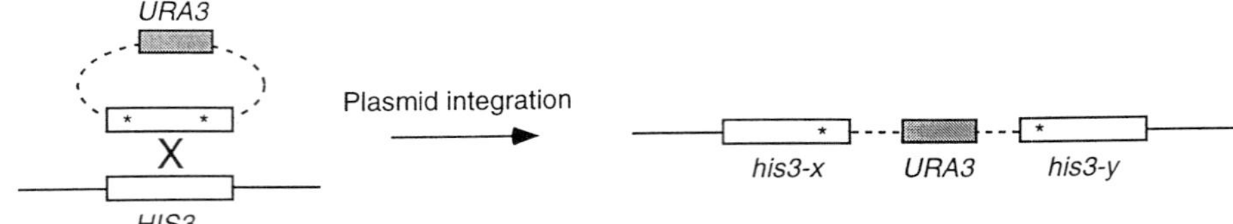

B

Direct repeat interactions

1. Gene conversion

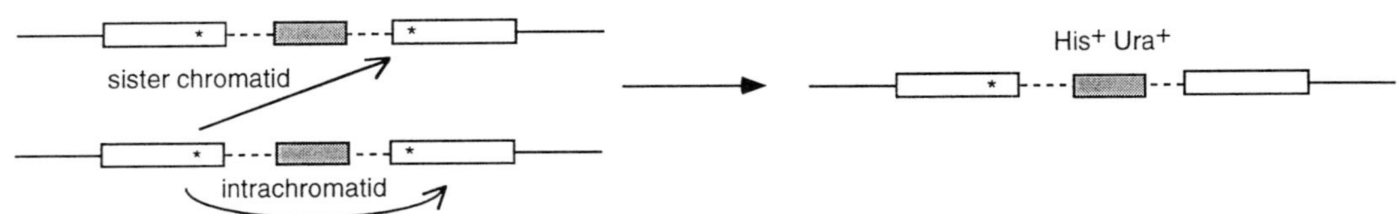

2. Crossing-over

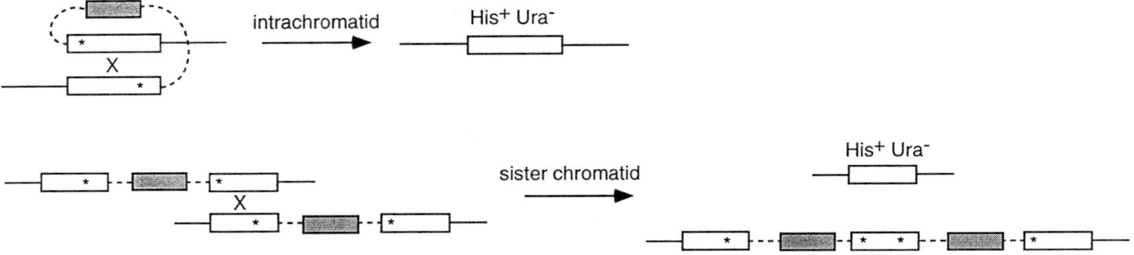

3. Single strand annealing (SSA)

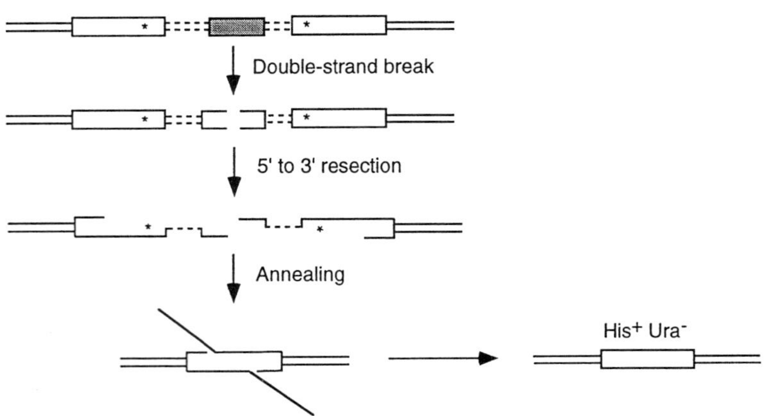

FIGURE 34-4 Construction of and interactions between direct repeats.

tive and negative selection exist for the selectable marker used to detect plasmid integration, then pop-outs/crossovers can be selected directly, without the requirement for a His$^+$ phenotype. One such widely used selectable marker is the *URA3* gene, which can be selected for on uracil-deficient medium and selected against on medium containing 5-fluoroorotic acid (5-FOA [17]).

If one is only interested in detecting the crossover type of recombinant, the system shown in Fig. 34-4A can be modified so that the copy of *his3* upstream of the plasmid sequences is a 3′ truncation and the copy of *his3* downstream of the plasmid sequences is a 5′ truncation [75]. Either intrachromatid or sister chromatid crossing-over deletes the region between the repeats and reconstitutes a full-length *HIS3* gene. If one is specifically interested in sister chromatid interactions, a selective system has been developed that allows detection of only sister chromatid events [76]. Truncations also are used in this system, but in this case the 5′ truncation is located upstream of the plasmid sequences and the 3′ truncation downstream of the plasmid sequences. Intrachromatid exchanges produce a *his3* gene truncated at both ends whereas sister chromatid exchanges produce a full-length, selectable *HIS3* gene.

The relative frequency of intrachromatid versus sister chromatid interactions is likely relevant to issues of genome stability. Recombination events that occur before genome duplication must be intrachromatid, but those occurring after genome duplication can be either intrachromatid or sister chromatid. Experiments in yeast indicate that sister chromatid interactions are indeed favored over intrachromatid interactions [77].

In addition to standard gene conversion and crossing-over, there exists a third, mechanistically distinct homologous recombination mechanism that always results in deletion of the sequences between the interacting direct repeats. This third mechanism is termed single strand annealing (SSA) and is illustrated in Fig. 34-4B. In SSA, the ends of a broken molecule are exonucleolytically processed to long 3′ overhangs. If direct repeats flank the site of the DSB, then complementary single strands will be exposed and annealing of the single strands can occur. Subsequent processing to remove nonhomologous, unpaired ends restores an intact chromosome. SSA was first proposed in mammalian cells [78], but was subsequently documented in yeast by physically following the fate of a defined HO-induced DSB [79]. In such experiments, HO expression is induced in a population of cells, which results in the synchronous production of a defined DSB in the majority of the cells. Repair of the DSB can be followed by either Southern blot analysis or PCR [67]. Such analyses allow the production of single-stranded ends as well as the production of gene conversion versus SSA-type recombinants to be monitored. Analyses of gene conversion versus SSA in various yeast mutants have revealed different genetic requirements for the two types of recombination [80]. Although it is not possible to directly demonstrate the occurrence of SSA in the repair of spontaneous DNA lesions, the genetic requirements of direct repeat recombination indicate that SSA is an important repair mechanism [72].

Recombination rates between direct (or inverted) repeats positioned several kilobases apart on the same chromosome are at least 100-fold higher than recombination rates between the same repeats when they are positioned on different (either homologous or nonhomologous) chromosomes [71]. This intrachromosomal preference may reflect the efficiency of a localized mitotic homology search that involves the same or a sister chromatid versus a genome-wide homology search. Since physical linkage between repeats appears to greatly facilitate recombination, an interesting issue concerns the relationship between recombination rate and the distance between interacting repeats. Although this has not been systematically examined for the gene conversion/crossover recombination pathway, available data suggest an inverse relationship between repeat distance and recombination rate [70]. In the case of SSA, sequences close to the site of a DSB are used preferentially for repair relative to more distal sites [81].

The direct repeat interactions observed between artificial duplications in yeast are likely to be relevant to mutational processes in higher eukaryotes. Although gene conversion is difficult to directly demonstrate in mammalian cells, the acquisition by one gene of multiple sequence changes characteristic of another gene is considered to be formal proof of a conversion event. Such sequence transfers between human globin genes have been documented [82], and it has been suggested that gene conversion might be responsible for the somatic hypermutation of immunoglobulin genes [83]. Steroid 21-hydroxylase deficiency can result from interactions between the gene for this enzyme and a nonfunctional pseudogene; both gene conversion and crossing-over have been implicated as sources of mutation [84]. Duplications resulting from crossing-over between repetitive Alu elements generate mutations in the Duchenne muscular dystrophy gene [85] and in the low-density lipoprotein gene [86], loss of which results in familial hypercholesterolemia. Finally, the reciprocal products of unequal crossing-over between direct repeats appear to be responsible for two distinctly different neurological diseases. Charcot-Marie Tooth disease type 1A results from a duplication of the region between the repeats, whereas hereditary neuropathy with liability to pressure palsies results from a deletion of the region between the repeats [86a].

B. Recombination between Inverted Repeats

Crossing-over between inverted repeats results in the inversion of the region between the repeats (Fig. 34-3B), which can disrupt or perturb expression of an overlapping or nearby gene. Artificial inverted repeats can be constructed in yeast using gene targeting techniques similar to those used to construct direct repeats. A plasmid containing inverted repeats can be integrated into a yeast chromosome, which results in the inverted repeats being flanked by integration-generated direct repeats (e.g., see [87]). Alternatively, standard one-step transplacement procedures can be used to generate inverted repeats that are not flanked by direct repeats [71]. As with most direct repeat systems, the inverted repeats usually are composed of auxotrophic heteroalleles of a selectable marker, and recombination events are detected by selecting for prototrophic segregants. Prototrophic recombinants can result from either gene conversion or crossing-over, and the two classes of recombinants can be distinguished by PCR or Southern blot analysis (e.g., see [71, 87]). Inversion events also can be specifically selected by placing a regulatory element for an adjacent gene within the invertible segment, so that expression of the adjacent gene is controlled by the orientation of the invertible segment [88]. Finally, an inversion-specific assay has been developed in which the inversion event reconstitutes a full-length selectable marker [89].

Inverted repeats potentially participate in fewer types of recombination events than do direct repeats, which generally makes the analysis of inverted repeat recombination more straightforward than direct repeat recombination. In those cases where the same repeats have been positioned in both direct and inverted orientation, however, the overall level of recombination is similar for the two orientations [71]. As with direct repeats, both intrachromatid and sister chromatid gene conversion between inverted repeats can give rise to prototrophs. In contrast to direct repeats, however, only intrachromatid crossing-over between inverted repeats produces viable recombinants; a sister chromatid exchange between inverted repeats produces a dicentric chromosome and an acentric fragment, and so constitutes a lethal event. Although inversion of the region between inverted repeats is generally assumed to be diagnostic of an intrachromatid crossover, it should be noted that inversion also can result from gene conversion between sister chromatids [90]. In addition to gene conversion and crossing-over, single strand annealing between direct repeats occurs in response to a double-strand break; similar events involving inverted repeats cannot repair a broken chromosome. Although repeat-associated inversions as causative factors in human disease appear to be rare, disruption of the factor VII gene via an inversion is responsible for some cases of Hemophilia A [91].

C. Recombination between Repeats on Nonhomologous Chromosomes

Crossing-over between repeated sequences on nonhomologous chromosomes (heterochromosomal repeats) gives rise to reciprocal translocations if the repeats are in the same orientation relative to the centromeres on their respective chromosomes (Fig. 34-3C). If the repeats are not in the same orientation with respect to their centromeres, then crossing-over generates an acentric fragment and a dicentric chromosome. As with artificial direct and inverted repeats, heterochromosomal repeats generally are constructed by targeting auxotrophic heteroalleles to defined chromosomal locations, and recombinants are detected by selecting for prototrophic segregants on appropriate medium. Although gene conversion can be distinguished from crossing-over by genetic criteria (see [68]), physical techniques are more often used. Southern blot analysis can be used to determine the linkage relationships of restriction sites that flank the duplicated sequences [68, 69] or chromosome rearrangements can be detected by CHEF analysis [92]. Alternatively, more specialized systems have been developed that allow one to distinguish gene conversion from crossing-over simply by examining the phenotype of a recombinant [71, 93].

Heterochromosomal recombination is a surprisingly efficient process in yeast, and as in other types of recombination, there is a strong association between gene conversion and crossing-over. In mitosis, heterochromosomal recombination between 5- to 10-kb repeats embedded in nonhomologous chromosomes occurs at essentially the same rate as allelic recombination events involving completely homologous chromosomes [70]. Although heterochromosomal recombination is induced to high levels during meiosis, as is allelic recombination, the rate of heterochromosomal recombination is consistently lower than the rate of meiotic allelic recombination [92]. Very interestingly, the rate of heterochromosomal recombination in meiosis is influenced by the position of the repeats along the arms of their respective chromosomes [92]. In most organisms, meiotic allelic crossovers are essential for ensuring the proper disjunction of homologous chromosomes; crossovers physically link homologs so that they become properly attached to opposite poles of the meiotic spindle. In contrast to the genome stabilizing effect of allelic crossovers, heterochromosomal crossing-over in meiosis is associated with chromosome nondisjunction [94].

D. Mechanisms for Limiting Recombination between Dispersed Repeats

Studies using artificial duplications in yeast indicate that allelic interactions are generally favored over ectopic interactions in meiosis. In mitosis, sequences that reside on nonhomologous chromosomes can recombine as well as allelic sequences, and repeats that reside on the same chromosome usually recombine better than allelic sequences. Given the apparent efficiency of recombination events involving artificial repeats and the abundance of naturally occurring repeats in the genomes of higher eukaryotes, a key issue concerns how eukaryotic genome rearrangements are maintained at an acceptably low level. One possibility is that naturally occurring repeats are simply "cold" for recombination. The observation that Ty elements in yeast undergo much less meiotic recombination than similarly sized artificial repeats is consistent with this possibility [95]. In addition to being relatively poor recombination substrates, two additional features of repeated elements may effectively limit interactions between them: (1) the length of homology shared by the interacting sequences (substrate size) and (2) the degree of DNA sequence identity between the interacting sequences (substrate homology). Sequences that are similar but not identical are referred to as being homeologous as opposed to homologous. Experiments in yeast indicate that both substrate length and substrate homology are important for limiting ectopic interactions. The limitation on recombination exerted by substrate size is assumed to be primarily an intrinsic property of the recombination machinery. The limitation on recombination exerted by substrate homology derives not only from properties of the recombination machinery, but also from action of the MMR system.

The effect of substrate size on homologous recombination can be examined by systematically varying the degree of overlap between two interacting sequences. Experiments done in bacterial cells indicate that there is a linear relationship between substrate size and the rate of recombination; this relationship is the basis of the concept of a minimal efficient processing segment (MEPS) [96]. The MEPS corresponds to the minimal length of perfect homology needed to initiate efficient recombination; below the MEPS, recombination can still occur, but it becomes very inefficient. Because any pair of substrates can be considered as an overlapping series of MEPS, the number of MEPS increases as a linear function of substrate length. In *E. coli,* the MEPS has been estimated to be approximately 50 bp [96].

In yeast, the relationship between substrate size and recombination rate has been examined using two types of assays. First, the length of one substrate can be held constant and 5′ and/or 3′ sequences can be deleted from the other substrate. Second, the rate of interchromosomal crossing-over within a variably sized interval can be obtained by assaying for altered linkage relationships of flanking genetic markers. Whereas the first system can be used to measure both gene conversion and crossover rates, the latter system is specific for crossovers. For repeats greater than 1 kb in size, the frequency of mitotic interchromosomal recombination is positively correlated with repeat size, whereas the frequency of intrachromosomal events plateaus at a constant level [97]. For repeats less than 1 kb in size, there is a roughly linear relationship between substrate size and the overall mitotic recombination rate, although crossovers constitute a smaller proportion of total events as the substrate size decreases [71]. This relationship holds for direct repeats, inverted repeats, and repeats on nonhomologous chromosomes and has been used to estimate a MEPS of approximately 200 bp in yeast. With plasmid substrates of sub-MEPS size, there appears to be an exponential relationship between mitotic recombination rate and substrate size [98]. In meiosis, the MEPS in yeast has been estimated to be 150–200 bp [99].

The effect of substrate homology on chromosomal recombination rates can be examined using either naturally occurring repeated sequences [93, 100] or artificially constructed repeats [89, 101, 102]. An advantage of using artificial versus endogenous repeats is that standard molecular techniques (e.g., site-directed mutagenesis or mutagenic PCR) can be used to vary the degree of sequence identity between artificially constructed repeats. In addition to using repeats of defined identity, the effect of sequence divergence on meiotic recombination has been examined either by importing whole chromosomes from other yeast species into *S. cerevisiae,* or by forming hybrid strains between different yeast species [103–105]. In considering meiotic studies, it is important to note that crossovers are essential for homolog disjunction; inhibition of meiotic crossing-over translates into chromosome nondisjunction and spore inviability.

The concept of mismatch correction by specialized proteins was critical to the development of current recombination models, which posit that resolution of heteroduplex intermediates leads to crossing-over. It is interesting to note that the first yeast MMR gene (*PMS1*) was in fact identified by the high level of postmeiotic segregation in mutants, rather than as a result of its associated mitotic mutator phenotype [106]. In addition to mismatch correction, much recent attention has focused on the role of the yeast MMR machinery in regulating recombination between nonidentical sequences. The involvement of MMR proteins in regulating homeologous recombination was discovered in crosses between *E. coli* and *Salmonella typhimurium,* whose genomes are diverged by approximately 20% at the nucleotide level. Disabling the MMR system in the recipient strain increases homeologous recombination several thousand-fold, indicating that MMR proteins exert a potent antirecombination function [107]. The MMR system presumably acts by recognizing mismatches in heteroduplex recombination intermediates, but precisely how this recognition impacts the process of recombination is not understood. It has been suggested either that the processing of mismatches by MMR proteins destroys the recombination intermediate [108], or that MMR proteins impede the progress of recombination [109, 110]. A recent bacterial study indicates that MMR proteins do not destroy preformed heteroduplex molecules containing multiple mismatches [111], which indicates that the antirecombination activity likely derives from a nondestructive, MMR-imposed recombination block. MMR proteins could impede recombination by blocking an early strand assimilation step between nonidentical substrates, by blocking subsequent branch migration, or by affecting how the intermediate is resolved. The antirecombination activity of MMR proteins is likely to be biologically relevant and might effectively limit ectopic recombination. MMR-defective cells thus would be expected to have higher levels of genetic instability resulting from genome rearrangements as well as higher mutation rates resulting from replication errors. In addition, the antirecombination activity of the MMR machinery may aid in speciation processes by lowering the reproductive fitness of hybrid organisms.

Numerous yeast studies have examined the impact of substrate identity on mitotic chromosomal recombination rates, and all have uniformly found that sequence divergence inhibits recombination [89, 93, 101, 102, 112, 113]. Although sequence divergence clearly impedes recombination, the involvement of the MMR machinery in regulating homeologous recombination has not been absolute. At least part of the discrepancy likely results from the many different types of substrate systems used to assay homeologous recombination. A plasmid–chromosome assay system, for example, detected no involvement of the MMR machinery in regulating recombination [102], whereas two different chromosomal systems clearly demonstrated an antirecombination role for MMR proteins [89, 101]. An additional possibility that must be considered when attempting to compare results derived using different substrate systems is that MMR proteins might impact gene conversion versus crossing-over quite differently. One system that failed to detect an antirecombination role for MMR proteins was only able to monitor gene conversion events [102], whereas the strongest antirecombination effects have been observed with an intron-based system, which only detects crossovers [89]. Of particular note is the fact that a single mismatch within a 350-bp region of perfect homology impacts recombination in a MMR-dependent manner in the intron-based system [114]. Experiments with this system also have demonstrated that the apparent antirecombination effect of MMR system is related to the degree of sequence identity between the substrates [89, 114]. At low levels of sequence divergence (<5%), the MMR machinery alone inhibits recombination whereas at higher levels of sequence divergence (>5%) recombination is inefficient even in the absence of MMR activity. For those systems in which the MMR machinery fails to impact homeologous recombination, it has been suggested that short heteroduplexes between highly diverged sequences may be stabilized by replication rather than by branch migration. In contrast to branch migration through homeologous sequences, replication would yield perfectly matched duplexes, which would exclude extensive involvement of the MMR machinery [102].

In addition to substrate considerations, yeast and other eukaryotes possess numerous MutS and MutL homologs (see above), only some of which may be involved in regulating mitotic recombination. Early mitotic studies utilized *pms1* mutants [100, 115], whereas recent studies have tended to concentrate on the antirecombination role of the MutS homologs [89, 101]. Msh2p has the strongest antimutator activity of the MutS homologs, and it also appears to have the strongest antirecombination activity [89, 101]. In contrast to the equivalent roles of Msh2p and Pms1p in terms of antimutator activity, Msh2p has a much stronger antirecombination activity than does Pms1p [89]. It thus appears that although the recognition/processing of mismatched replication intermediates requires concerted action of both MutS and MutL homologs, MutS homologs are able to act alone to block recombination between diverged sequences.

As in mitosis, the MMR machinery regulates meiotic interactions between diverged sequences. In contrast to the use of short repeats in mitotic studies, however, meiotic studies have generally made use of entire diverged chromosomes [104, 116] or genomes [105]. Sequence divergence between whole chromosomes is a potent barrier to meiotic crossing-over in yeast [105, 116]; as expected, the reduction in crossing-over translates into high levels of chromosome nondisjunction and spore inviability [105]. Mutational elimination of MMR proteins improves both meiotic crossing-over and spore viability [104, 105], which indicates an antirecombination role for the MMR system. In contrast to mitotic experiments, meiotic tetrad dissection is capable of detecting very subtle differences in patterns of spore viability and recombination. Such analyses have suggested distinct antirecombination roles for individual MMR proteins [104] that have not been detected using available mitotic methods.

One issue that remains to be addressed concerns the relative sequence homology requirements of meiotic versus mitotic recombination. On the one hand, one might expect there to be more stringent requirements for sequence identity in meiosis since genome rearrangements directly impact gamete viability. On the other hand, successful meiosis demands very efficient crossing-over, which may mean more relaxed homology requirements. It will be important to compare directly the relative antirecombination activity of the yeast MMR machinery in mitosis versus meiosis, and this will necessitate using the same recombination substrates for parallel mitotic and meiotic analyses.

IV. GENETIC INSTABILITIES ASSOCIATED WITH ILLEGITIMATE RECOMBINATION

In contrast to the extensive DNA sequence identity required to facilitate homologous recombination, illegitimate recombination can be considered to be a nonhomologous process that involves little, if any, identity at the nucleotide level. Both homologous and illegitimate recombination, however, constitute important mechanisms for repairing eukaryotic DNA damage that results in double-strand breaks. Double-strand break repair by homologous recombination is a very high-fidelity repair process, whereas illegitimate recombination is inherently mutagenic, with short sequences often being added or deleted at the recombination breakpoints. The relative repair roles of homologous versus illegitimate recombination can be inferred from the efficiency with which exogenously introduced DNA integrates into a homologous genomic target sequence. In yeast cells, gene targeting occurs with greater than 90% efficiency and hence has become an invaluable tool for genome manipulation [117]. In stark contrast to the situation in yeast, gene targeting in mammalian cells occurs with an efficiency of less than 1% [118]. The inefficiency of gene targeting in mammalian cells is likely related to the remarkable capacity for joining DNA ends, which is perhaps best exemplified by the obligatory end-joining that occurs during the somatic rearrangement [V(D)J recombination] of immunoglobulin loci [119]. A mechanistic similarity between illegitimate recombination and V(D)J recombination is suggested by the observation that mutations in DNA-dependent protein kinase (DNA-PK) reduce both DNA repair capacity, which presumably occurs by predominantly illegitimate recombination, and the efficiency of V(D)J recombination [120, 121]. Programmed immunoglobulin rearrangements can result in aberrant end-joining of immunoglobulin to nonimmunoglobulin sequences, and such events are associated with human leukemias [122].

Although illegitimate recombination in yeast is rare relative to homologous recombination, several systems have been developed that allow the specific detection and analysis of illegitimate events. These systems generally remove the opportunity for homologous recombination either by eliminating homologous sequences or by mutationally eliminating Rad52p, a protein which is essential for homologous recombination. The first such assay developed was a transformation-based assay in which yeast cells containing a complete deletion of the *URA3* locus were transformed with a DNA fragment containing this locus [123]. Stable Ura$^+$ transformants should be obtained only if the *URA3*-containing fragment integrates into the yeast genome; in the absence of genomic homology, integration can only occur via illegitimate recombination. Experiments with this system led to the discovery of a particular type of illegitimate recombination termed restriction enzyme-mediated (REM) recombination [123]. In REM, the restriction enzyme used to generate the transforming DNA fragment apparently enters yeast cells and makes double-strand breaks at genomic recognition sites. The overhangs produced by the restriction enzyme *in vivo* provide sufficient microhomology to allow transient base pairing with the ends of the transforming DNA and subsequent integration of the fragment into the genomic restriction site. Even in the absence of a restriction enzymes, however, illegitimate integration of the transforming DNA fragment into the yeast genome often involves microhomologies [123–125], which is very reminiscent of end-joining events in mammalian cells [118]. REM and other illegitimate recombination events are not dependent on Rad1p, Rad51p, Rad52p, or Rad57p, but are dependent on Rad50p [125]; a similar dependence of illegitimate events on Rad50p also has been found using other assay systems (see below). In addition to the illegitimate events involving microhomology between the genomic target and the ends of the transforming fragment, a number of events involve no obvious microhomology, but rather occur next to preferred topoisomerase I cleavage sites [124, 126]. Aberrant topoisomerase cleavage/rejoining reactions also have been implicated as a source of illegitimate recombination events in mammalian cells [127]. Finally, transforming fragments can be ligated *in vivo* at a low frequency to mitochondrial DNA sequences containing a fortuitous origin of replication, thus allowing the *URA3* gene to be maintained extrachromosomally [124].

Illegitimate recombination between yeast chromosomal sequences has been examined following the creation of a chromosomal double-strand break in *rad52* strains. The double-strand break either can be made at a specific site using the HO endonuclease, or can be made at random sites by mechanical breakage of conditionally dicentric chromosomes [128]. Because continued cell division depends on repair of the break, one need only select for colony formation in order to detect illegitimate recombination events. Such chromosomal repair events occur with an efficiency of 1% or less and have been termed nonhomologous end joining (NHEJ). Two classes of HO-initiated NHEJ events have been detected, both of which involve 1- to 6-bp microhomologies at the joints [128, 129]. In the first class the HO generated ends are degraded, yielding small deletion events. In the second class, the 3′ overhangs generated by HO cleavage are annealed out of alignment, which generates 2- to 3-bp insertions at the cut site. NHEJ events do not require Rad1p, Rad2p, Rad51p, Rad52p, Rad54p, or Rad57p, but are dependent on Rad50p, Xrs2p, and Mre11p [129]. The insertion events are preferentially lost in *rad50, xrs2,* or *mre11* mutants, suggesting that the corresponding proteins may be involved in stabilizing the mitotically generated broken ends and preventing degradation. It should be noted that these same proteins are necessary for the formation of Spo11-produced meiotic double-strand breaks [65]. In addition to the interesting genetic requirements of NHEJ, the efficiency of the events exhibits cell cycle dependency [129]. HO-generated double-strand breaks not only can be repaired by NHEJ, but also can be repaired by the insertion of foreign DNA at the breakpoint. The foreign DNA observed thus far has been derived from the endogenous Ty retrotransposon and presumably represents the capture of cDNA fragments by broken chromosomes [130, 131]. Such capture may be relevant to the broad genomic dispersal of retroelements in higher eukaryotes.

In addition to the assays for analyzing illegitimate recombination events involving chromosomal sequences, illegitimate events also have been studied using replicating plasmids. In the first type of plasmid-based system, the deletion of plasmid-encoded genetic markers conferring drug sensitivity is identified on appropriate selective medium [132]. Such illegitimate deletions have similar genetic requirements as the chromosomal events described above, with the exception that they exhibit Rad52p-dependence. In the second type of plasmid-based system, a plasmid containing a selectable marker is linearized prior to transformation into yeast; successful replication of the plasmid in yeast requires recircularization *in vivo.* Recircularization occurs predominantly via direct ligation if enzyme-generated cohesive ends are present. If cohesive ends are not present, then NHEJ occurs with deletion endpoints in microhomologies of 2–16 bp [133, 134].

The involvement of DNA-PK in end-joining events in mammalian cells has prompted an examination of a similar involvement in illegitimate recombination events in yeast. DNA-PK is composed of Ku, a heterodimer of 70- and 80-kDa proteins that binds to DNA ends, and a catalytic subunit that has kinase activity. Yeast has homologs of the Ku proteins, but has no obvious homolog of the catalytic subunit of DNA-PK. Disruption of the yeast gene encoding either Ku70 or Ku80 sensitizes haploid cells to DNA damaging agents in some assays [135–137] but not in others [133, 134]. In spite of this apparent discrepancy, disruption of either Ku gene uniformly enhances the sensitivity of *rad52* strains to agents that cause double-strand breaks [133–137]. This latter observation indicates that Ku is involved in an alternative, Rad52p-independent DNA repair pathway. Whereas Rad52p is essential for DNA repair via homologous recombination, transformation experiments with linearized plasmids indicate that Ku is important for direct end-joining events that do not involve degradation [133, 134]. In addition, genetic studies examining recircularization of linearized plasmids have demonstrated that Ku and Rad50p are in the same epistasis group [136]. In addition to their roles in end joining events, the yeast Ku proteins also appear to stabilize the ends of yeast chromosomes; disruption of the gene for either the Ku70 or Ku80 homolog results in shortened telomeres [134, 138].

The general picture that emerges from the illegitimate recombination studies is that there are two end-joining pathways in yeast, both of which involve microhomologies in the range of 1–6 bp. One pathway is dependent on Ku, Rad50p, Xrs2p, and Mre11p and involves the direct ligation of ends with single stranded overhangs. Ligation of restriction enzyme-generated ends appears to be conservative whereas misalignment of HO-generated ends can generate small insertions. In either case, Ku, Rad50p, Xrs2p, and Mre11p presumably protect the broken ends from nucleolytic processing. In the second end-joining pathway, nucleolytic processing results in the NHEJ type of events with deletion endpoints in microhomologies. Interestingly, the presence of Ku decreases the efficiency of joining blunt-ended restriction fragments [133, 134]. This inhibition presumably reflects the protection of the blunt ends by Ku; some nucleolytic degradation is necessary to generate the single strand overhangs that appear to be necessary for end joining.

Two hybrid experiments have uncovered an unexpected interaction between Ku70 and Sir4p [139], a protein involved in silencing at telomeres and at the *HM* mating type loci in yeast. Both illegitimate recombination and NHEJ require Sir4p (as well as Sir2p and Sir3p, but not Sir1p), and epistasis analysis indicates that Ku70 and Sir4p act in the same pathway [139]. It has been speculated that Ku70 binds to the ends of DSBs and then recruits the Sir proteins, which might put the DNA ends into an inactive chromatin structure that promotes end-joining. Completion of the end-joining reaction obviously requires the action of a ligase, and recent reports indicate that the yeast homolog of mammalian ligase IV is the enzyme that catalyzes ligation of the broken ends [140, 141]. Future studies in yeast will no doubt reveal additional molecular and genetic details of illegitimate recombination and such studies should be directly applicable to issues of illegitimate recombination and gene targeting in higher eukaryotes.

V. SUMMARY

Genetic instabilities are the basis of many human diseases, and systems developed in yeast have been important for elucidating the underlying causes of these instabilities. Much of the utility of yeast derives from the ease with which the genome can be manipulated and the availability of mutations in genes whose products are involved in the processes of DNA replication, repair, and recombination. Studies in yeast, for example, have revealed the types of sequences that are frequently involved in replication slippage events as well as the repair functions that normally operate to remove mutagenic slippage intermediates. Replication slippage results in duplications and deletions within simple repeat arrays, as well as duplications and deletions of unique regions that are bordered by short direct repeats. Yeast replication studies are directly relevant to aspects of mutagenesis in higher cells and likely will be relevant to the amplification of simple repeats that occurs in human trinucleotide repeat diseases. In addition to instabilities associated with DNA replication, recombination-associated instabilities also have been examined in yeast. Recombination-associated genome rearrangements arise either via homologous recombination between repeated sequences or via nonhomologous recombination, which usually involves end-joining of broken DNA molecules. Recombination not only is a source of medically relevant and evolutionarily important genome rearrangements, but also is critical to issues of efficient gene targeting in mammalian cells. Studies done in yeast clearly have contributed greatly to our current understanding of genetic instabilities, and it is anticipated that this model organism will continue to provide an important system for examining the basic mechanisms modeling eukaryotic genome structure.

Acknowledgments

We thank Tom Petes and Dmitri Gordenin for helpful comments on the manuscript. This work is supported by grants from the National Science Foundation and the National Institutes of Health to S.J.R.

References

1. Ashley, C. T., and Warren, S. T. (1995). Trinucleotide repeat expansion and human disease. *Annu. Rev. Genet.* **29,** 703–278.
2. Umar, A., and Kunkel, T. A. (1996). DNA-replication fidelity, mismatch repair and genome instability in cancer cells. *Eur. J. Biochem.* **238,** 297–307.
3. Gordenin, D. A., Kunkel, T. A., and Resnick, M. A. (1997). Repeat expansion—all in a flap? *Nature Genet.* **16,** 116–118.
4. Schaaper, R. M. (1993). Base selection, proofreading, and mismatch repair during DNA replication in *Escherichia coli. J. Biol. Chem.* **268,** 23762–23765.
5. Ripley, L. S. (1990). Frameshift mutation: Determinants of specificity. *Annu. Rev. Genet.* **24,** 189–213.
6. Streisinger, G., Okada, Y., Emrich, J., Newton, J., Tsugita, A., Terzaghi, E., and Inouye, M. (1966). Frameshift mutations and the genetic code. *Cold Spring Harbor Symp. Quant. Biol.* **31,** 77–84.
7. Smith, G. P. (1973). Unequal crossover and the evolution of multigene families. *Cold Spring Harbor Symp. Quant. Biol.* **38,** 507–513.
8. Sugino, A. (1995). Yeast DNA polymerases and their role at the replication fork. *Trends Biochem. Sci.* **20,** 319–323.
9. Modrich, P., and Lahue, R. (1996). Mismatch repair in replication fidelity, genetic recombination and cancer biology. *Annu. Rev. Biochem.* **65,** 101–133.
10. Kolodner, R. (1996). Biochemistry and genetics of eukaryotic mismatch repair. *Genes Dev.* **10,** 1433–1442.
11. Crouse, G. F. (1998). Mismatch repair systems in *Saccharomyces cerevisiae. In* "DNA Damage and Repair," Vol. 1, "DNA Repair in Prokaryotes and Lower Eukaryotes" (J. A. Nickoloff and M. F. Hoekstra, Eds.), pp. 411–448. Humana Press, Totowa, NJ.
12. Hunter, N., and Borts, R. H. (1997). Mlh1 is unique among mismatch repair proteins in its ability to promote crossing-over during meiosis. *Genes Dev.* **11,** 1573–1582.
13. Marsischky, G. T., Filosi, N., Kane, M. F., and Kolodner, R. (1996). Redundancy of *Saccharomyces cerevisiae MSH3* and *MSH6* in *MSH2*-dependent mismatch repair. *Genes Dev.* **10,** 407–420.
14. Johnson, R. E., Kovvali, G. K., Prakash, L., and Prakash, S. (1996). Requirement of the yeast *MSH3* and *MSH6* genes for *MSH2*-dependent genomic stability. *J. Biol. Chem.* **271,** 7285–7288.
15. Sia, E. A., Kokoska, R. J., Dominska, M., Greenwell, P., and Petes, T. D. (1997). Microsatellite instability in yeast: dependence on repeat unit size and DNA mismatch repair genes. *Mol. Cell. Biol.* **17,** 2851–2858.

16. Henderson, S. T., and Petes, T. D. (1992). Instability of simple sequence DNA in *Saccharomyces cerevisiae. Mol. Cell. Biol.* **12,** 2749–2757.
17. Boeke, J. D., Trueheart, J., Natsoulis, G., and Fink, G. R. (1987). 5-Fluoroorotic acid as a selective agent in yeast molecular genetics. *Methods Enzymol.* **154,** 164–175.
18. Streisinger, G., and Owen, J. (1985). Mechanisms of spontaneous and induced frameshift mutation in bacteriophage T4. *Genetics* **109,** 633–659.
19. Tran, H. T., Keen, J. D., Kricker, M., Resnick, M. A., and Gordenin, D. A. (1997). Hypermutability of homonucleotide runs in mismatch repair and DNA polymerase proofreading yeast mutants. *Mol. Cell. Biol.* **17,** 2859–2865.
20. Kroutil, L. C., Register, K., Bebenek, K., and Kunkel, T. A. (1996). Exonucleolytic proofreading during replication of repetitive DNA. *Biochemistry* **35,** 1046–1053.
21. Greene, C. N., and Jinks-Robertson, S. (1997). Frameshift intermediates in homopolymer runs are removed efficiently by yeast mismatch repair proteins. *Mol. Cell. Biol.* **17,** 2844–2850.
22. Huang, J., Papadopoulos, N., McKinley, A. J., Farrington, S. M., Curtis, L. J., Wyllie, A. H., Zheng, S., Willson, J. K. V., Markowitz, S. D., Morin, P., Kinzler, K. W., Vogelstein, B., and Dunlop, M. G. (1996). *APC* mutations in colorectal tumors with mismatch repair deficiency. *Proc. Natl. Acad. Sci. USA* **93,** 9049–9054.
23. Markowitz, S., Wang, J., Myeroff, L., Parsons, R., Sun, L., Lutterbaugh, J., Fan, R. S., Zborowska, E., Kinzler, K. W., Vogelstein, B., Brattain, M., and Willson, J. K. V. (1995). Inactivation of the type II TGF-β receptor in colon cancer cells with microsatellite instability. *Science* **268,** 1336–1338.
24. Rampino, N., Yamamoto, H., Ionov, Y., Li, Y., Sawai, H., Reed, J. C., and Perucho, M. (1997). Somatic frameshift mutations in the *BAX* gene in colon cancers of the microsatellite mutator phenotype. *Science* **275,** 967–969.
25. Souza, R. F., Appel, R., Yin, J., Wang, S., Smolinski, K. N., Abraham, J. N., Zou, T. T., Shi, J. Q., Lei, J., Cottrell, J., Cymes, K., Biden, K., Simms, L., Leggett, B., Lynch, P. M., Fraizer, M., Powell, S. M., Harpaz, N., Sugimura, H., Young, J., and Meltzer, S. J. (1996). Microsatellite instability in the insulin-like growth factor II receptor gene in gastrointestinal tumors. *Nature Genet.* **14,** 255–257.
26. Hamada, H., Petrino, M. G., and Kakunaga, T. (1982). A novel repeated element with Z-DNA-forming potential is widely found in evolutionarily diverse eukaryotic genomes. *Proc. Natl. Acad. Sci. USA* **79,** 6465–6469.
27. Strand, M., Prolla, T. A., Liskay, R. M., and Petes, T. D. (1993). Destabilization of tracts of simple repetitive DNA in yeast by mutations affecting DNA mismatch repair. *Nature* **365,** 274–276.
28. Strand, M., Earley, M. C., Crouse, G. F., and Petes, T. D. (1995). Mutations in the *MSH3* gene preferentially lead to deletions within tracts of simple repetitive DNA in *Saccharomyces cerevisiae. Proc. Natl. Acad. Sci. USA* **92,** 10418–10421.
29. Petes, T. D., Malone, R. E., and Symington, L. S. (1991). Recombination in yeast. *In* "The Molecular Biology of the Yeast *Saccharomyces*: Genome Dynamics, Protein Synthesis and Energetics" (J. R. Broach, J. R. Pringle, and E. W. Jones, Eds.), pp. 407–521. Cold Spring Harbor Laboratory, Cold Spring Harbor, NY.
30. Wierdl, M., Greene, C. N., Datta, A., Jinks-Robertson, S., and Petes, T. D. (1996). Destabilization of simple repetitive DNA sequences by transcription in yeast. *Genetics* **143,** 713–721.
31. Wierdl, M., Dominska, M., and Petes, T. D. (1997). Microsatellite instability in yeast: dependence on the length of the microsatellite. *Genetics* **146,** 769–779.
32. Kunst, C. B., and Warren, S. T. (1994). Cryptic and polar variation of the fragile X repeat could result in predisposing normal alleles. *Cell* **77,** 853–861.
33. Chong, S. S., McCall, A. E., Cota, J., Subramony, S. H., Orr, H. T., Hughes, M. R., and Zoghbi, H. Y. (1995). Gametic and somatic tissue-specific heterogeneity of the expanded *SCA1* CAG repeat in spinocerebellar ataxia type 1. *Nature Genet.* **10,** 344–350.
34. Petes, T. D., Greenwell, P. W., and Dominska, M. (1997). Stabilization of microsatellite sequences by variant repeats in the yeast *Saccharomyces cerevisiae. Genetics* **146,** 491–498.
35. Johnson, R. E., Kovvali, G. K., Prakash, L., and Prakash, S. (1995). Requirement of the yeast *RTH1* 5′ to 3′ exonuclease for the stability of simple repetitive DNA. *Science* **269,** 238–240.
36. Lieber, M. R. (1997). The FEN-1 family of structure-specific nucleases in eukaryotic DNA replication, recombination and repair. *BioEssays* **19,** 233–240.
37. Tishkoff, D. X., Filosi, N., Gaida, G. M., and Kolodner, R. D. (1997). A novel mutation avoidance mechanism dependent on *S. cerevisiae RAD27* is dinstinct from DNA mismatch repair. *Cell* **88,** 253–263.
38. Maurer, D. J., O'Callaghan, B. L., and Livingston, D. M. (1996). Orientation dependence of trinucleotide CAG repeat instability in *Saccharomyces cerevisiae. Mol. Cell. Biol.* **16,** 6617–6622.
39. Freudenreich, C. H., Stavenhagen, J. B., and Zakian, V. A. (1997). Stability of a CTG/CAG trinucleotide repeat in yeast is dependent on its orientation in the genome. *Mol. Cell. Biol.* **17,** 2090–2098.
40. Miret, J. J., Pessoa-Brandao, L., and Lahue, R. S. (1997). Instability of CAG and CTG trinucleotide repeats in *Saccharomyces cerevisiae. Mol. Cell. Biol.* **17,** 3382–3387.
41. Furter-Graves, E. M., and Hall, B. D. (1990). DNA sequence elements required for transcription initiation of the *Schizosaccharomyces pombe ADH* gene in *Saccharomyces cerevisiae. Mol. Gen. Genet.* **223,** 407–416.
42. Schweitzer, J. K., and Livingston, D. M. (1997). Destabilization of CAG trinucleotide repeat tracts by mismatch repair mutations in yeast. *Hum. Mol. Genet.* **6,** 349–355.
43. Kang, S., Jaworski, A., Ohshima, K., and Wells, R. D. (1995). Expansion and deletion of CTG repeats from human disease genes are determined by the direction of replication in *E. coli. Nature Genet.* **10,** 213–218.
44. Petruska, J., Arnheim, N., and Goodman, M. F. (1996). Stability of intrastrand hairpin structures formed by the CAG/CTG class of DNA triplet repeats associated with neurological diseases. *Nucleic Acids Res.* **24,** 1992–1998.
45. Mitas, M. (1997). Trinucleotide repeats associated with human disease. *Nucleic Acids Res.* **25,** 2245–2253.
46. McMurray, C. T. (1995). Mechanisms of DNA expansion. *Chromosoma* **104,** 2–13.
47. Otten, A. D., and Tapscott, S. J. (1995). Triplet repeat expansion in myotonic dystrophy alters the adjacent chromatin structure. *Proc. Natl. Acad. Sci. USA* **92,** 5465–5469.
48. Wang, Y.-H., and Griffith, J. (1995). Expanded CTG triplet blocks from the myotonic dystrophy gene create the strongest known natural nucleosome positioning elements. *Genomics* **25,** 570–573.
49. Aparicio, O. M., Billington, B. L., and Gottschling, D. E. (1991). Modifiers of position effect are shared between telomeric and silent mating-type loci in *S. cerevisiae. Cell* **66,** 1279–1287.

49a. Freundenreich, C. H., Kantrow, S. M., and Zakian, V. A. (1998). Expansion and length-dependent fragility of CTG repeats in yeast. *Science* **279,** 853–856.

50. Weston-Hafer, K., and Berg, D. E. (1991). Limits to the role of palindromy in deletion formation. *J. Bacteriol.* **173,** 315–318.
51. Henderson, S. T., and Petes, T. D. (1993). Instability of a plasmid-borne inverted repeat in *Saccharomyces cerevisiae. Genetics* **134,** 57–62.
52. Ruskin, B., and Fink, G. R. (1993). Mutations in *POL1* increase the mitotic instability of tandem inverted repeats in *Saccharomyces cerevisiae. Genetics* **134,** 43–56.
53. Gordenin, D. A., Trofimova, M. V., Shaburova, O. N., Pavlov, Y. I., Chekuolene, Y. V., Proscyavichus, Y. Y., Sasnauskas, K. V., and Janulaitis, A. A. (1988). Precise excision of bacterial transposon Tn5 in yeast. *Mol. Gen. Genet.* **213,** 388–393.
54. Gordenin, D. A., Malkova, A. L., Peterzen, A., Kulikov, V. N., Pavlov, Y. I., Perkins, E., and Resnick, M. A. (1992). Transposon Tn5 excision in yeast: influence of DNA polymerases α, δ, and ε and repair genes. *Proc. Natl. Acad. Sci. USA* **89,** 3785–3789.
55. Gordenin, D. A., Lobachev, K. S., Degtyareva, N. P., Malkova, A. L., Perkins, E., and Resnick, M. A. (1993). Inverted DNA repeats: a source of eukaryotic genomic instability. *Mol. Cell. Biol.* **13,** 5315–5322.
56. Gordenin, D. A., Proscyavichus, Y. Y., Malkova, A. L., Trofimova, M. V., and Peterzen, A. (1991). Yeast mutants with increased bacterial transposon Tn5 excision. *Yeast* **7,** 37–50.
57. Trinh, T. Q., and Sinden, R. R. (1991). Preferential DNA secondary structure mutagenesis in the lagging strand of replication in *E. coli. Nature* **352,** 544–545.
58. Tran, H. T., Degtyareva, N. P., Koloteva, N. N., Sugino, A., Masumoto, H., Gordenin, D. A., and Resnick, M. A. (1995). Replication slippage between distant short repeats in *Saccharomyces cerevisiae* depends on the direction of replication and the *RAD50* and *RAD52* genes. *Mol. Cell. Biol.* **15,** 5607–5617.
59. Tran, H. T., Gordenin, D. A., and Resnick, M. A. (1996). The prevention of repeat-associated deletions in *Saccharomyces cerevisiae* by mismatch repair depends on size and origin of deletions. *Genetics* **143,** 1579–1587.
60. Das, G., Consaul, S., and Sherman, F. (1988). A highly revertible *cyc1* mutant of yeast contains a small tandem duplication. *Genetics* **120,** 57–62.
61. Fried, M., Feo, S., and Heard, E. (1991). The role of inverted duplications in the generation of gene amplification in mammalian cells. *Biochem. Biophys. Acta* **1090,** 143–155.
62. Butler, D. K., Yasuda, L. E., and Yao, M.-C. (1996). Induction of large DNA palindrome formation in yeast: implications for gene amplification and genome stability in eukaryotes. *Cell* **87,** 1115–1122.
63. Stahl, F. W. (1994). The Holliday junction on its thirtieth anniversary. *Genetics* **138,** 241–246.
64. Szostak, J. W., Orr-Weaver, T. L., Rothstein, R. J., and Stahl, F. W. (1983). The double-strand-break repair model for recombination. *Cell* **33,** 25–35.
65. Keeney, S., Giroux, C. N., and Kleckner, N. (1997). Meiosis-specific DNA double-strand breaks are catalyzed by Spo11, a member of a widely conserved protein family. *Cell* **88,** 375–384.
66. Baudat, F., and Nicolas, A. (1997). Clustering of meiotic double-strand breaks on yeast chromosome III. *Proc. Natl. Acad. Sci. USA* **94,** 5213–5218.
67. Haber, J. E. (1995). *In vivo* biochemistry: physical monitoring of recombination induced by site-specific endonucleases. *BioEssays* **17,** 609–620.
68. Jinks-Robertson, S., and Petes, T. D. (1986). Chromosomal translocations generated by high-frequency meiotic recombination between repeated yeast genes. *Genetics* **114,** 731–752.
69. Lichten, M., Borts, R. H., and Haber, J. E. (1987). Meiotic gene conversion and crossing over between dispersed homologous sequences occurs frequently in *Saccharomyces cerevisiae. Genetics* **115,** 233–246.
70. Lichten, M., and Haber, J. E. (1989). Position effects in ectopic and allelic mitotic recombination in *Saccharomyces cerevisiae. Genetics* **123,** 261–268.
71. Jinks-Robertson, S., Michelitch, M., and Ramcharan, S. (1993). Substrate length requirements for efficient mitotic recombination in *Saccharomyces cerevisiae. Mol. Cell. Biol.* **13,** 3937–3950.
72. Klein, H. L. (1995). Genetic control of intrachromosomal recombination. *BioEssays* **17,** 147–159.
73. Aguilera, A., and Klein, H. L. (1988). Genetic control of intrachromosomal recombination in *Saccharomyces cerevisiae.* I. Isolation and genetic characterization of hyper-recombination mutations. *Genetics* **119,** 779–790.
74. Jackson, J. A., and Fink, G. R. (1981). Gene conversion between duplicated genetic elements in yeast. *Nature* **292,** 306–311.
75. Schiestl, R. H., and Prakash, S. (1988). *RAD1,* an excision repair gene of *Saccharomyces cerevisiae,* is also involved in recombination. *Mol. Cell. Biol.* **8,** 3619–3626.
76. Fasullo, M., and Davis, R. W. (1987). Recombinational substrates designed to study recombination between unique and repetitive sequences *in vivo. Proc. Natl. Acad. Sci. USA* **84,** 6215–6219.
77. Kadyk, L. C., and Hartwell, L. H. (1992). Sister chromatids are preferred over homologs as substrates for recombinational repair in *Saccharomyces cerevisiae. Genetics* **132,** 387–402.
78. Lin, F.-L., Sperle, K., and Sternberg, N. (1984). Model for homologous recombination during transfer of DNA into mouse L cells: role for DNA ends in the recombination process. *Mol. Cell. Biol.* **4,** 1020–1034.
79. Fishman-Lobell, J., Rudin, N., and Haber, J. E. (1992). Two alternative pathways of double-strand break repair that are kinetically separable and independently modulated. *Mol. Cell. Biol.* **1292,** 1303.
80. Ivanov, E. L., Sugawara, N., Fishman-Lobell, J., and Haber, J. E. (1996). Genetic requirements for the single-strand annealing pathway of double-strand break repair in *Saccharomyces cerevisiae. Genetics* **142,** 693–704.
81. Sugawara, N., and Haber, J. E. (1992). Characterization of double-strand break-induced recombination: homology requirements and single-stranded DNA formation. *Mol. Cell. Biol.* **12,** 563–575.
82. Hill, A. V. S., Nicholls, R. D., Thein, S. L., and Higgs, D. R. (1985). Recombination within the human embryonic zeta-globin locus: a common zetz-zeta chromosome produced by gene conversion of the pseudo-zeta gene. *Cell* **42,** 809–819.
83. Maizels, N. (1989). Might gene conversion be the mechanism of somatic hypermutation of mammalian immunoglobulin genes? *Trends Genet.* **5,** 4–8.
84. Tusie-Luna, M. T., Ramirex-Jimenez, S., Ordonez-Sanchez, M. L., Cabella-Villegas, J., Altairano-Bustamante, N., Calzada-Leon, R., Robles-Valdes, C., Mendoza-Morfin, F., Mendez, J. P., and Teran-Garcia, M. (1996). Low frequency of deletion alleles in patients with steroid 21-hydroxylase deficiency in a Mexican population. *Hum. Genet.* **98,** 376–379.
85. Hu, X., Ray, P. N., and Worton, R. G. (1991). Mechanisms of tandem duplication in the Duchenne muscular dystrophy gene include both homologous and nonhomologous intrachromosomal recombination. *EMBO J.* **9,** 2471–2477.
86. Lehrman, M. A., Goldstein, J. L., Russell, D. W., and Brown, M. S. (1987). Duplication of seven exons in LDL receptor gene caused by Alu-Alu recombination in a subject with familial hypercholesterolemia. *Cell* **48,** 827–835.
86a. Reiter, L. T., Murakami, T., Koeuth, T., Pentao, L., Muzny, D. M., Gibbs, R. A. and Lupski, J. R. (1996). A recombination

hotspot responsible for two inherited peripheral neuropathies is located near a mariner transposon-like element. *Nature Genet.* **12,** 288–297.

87. Rattray, A. J., and Symington, L. S. (1994). Use of a chromosomal inverted repeat to demonstrate that the *RAD51* and *RAD52* genes of *Saccharomyces cerevisiae* have different roles in mitotic recombination. *Genetics* **138,** 587–595.
88. Willis, K. K., and Klein, H. L. (1987). Intrachromosomal recombination in *Saccharomyces cerevisiae*: reciprocal exchange in an inverted repeat and associated gene conversion. *Genetics* **117,** 633–643.
89. Datta, A., Adjiri, A., New, L., Crouse, G. F., and Jinks-Robertson, S. (1996). Mitotic crossovers between diverged sequences are regulated by mismatch repair proteins in *Saccharomyces cerevisiae. Mol. Cell. Biol.* **16,** 1085–1093.
90. Rothstein, R., Helms, C., and Rosenberg, N. (1987). Concerted deletions and inversions are caused by mitotic recombination between delta sequences in *Saccharomyces cerevisiae. Mol. Cell. Biol.* **7,** 1198–1207.
91. Lakich, D., Kazazian, H. H., Antonarakis, S. E., and Gitschier, J. (1993). Inversions disrupting the factor VIII gene are a common cause of severe haemophilia A. *Nature Genet.* **5,** 236–241.
92. Goldman, A. S. H., and Lichten, M. (1996). The efficiency of meiotic recombination between dispersed sequences in *Saccharomyces cerevisiae* depends upon their chromosomal location. *Genetics* **144,** 43–55.
93. Harris, S., Rudnicki, K. S., and Haber, J. E. (1993). Gene conversions and crossing over during homologous and homeologous ectopic recombination in *Saccharomyces cerevisiae. Genetics* **135,** 5–16.
94. Jinks-Robertson, S., Sayeed, S., and Murphy, T. (1997). Meiotic crossing-over between nonhomologous chromosomes affects chromosome segregation in yeast. *Genetics* **146,** 69–78.
95. Kupiec, M., and Petes, T. D. (1988). Meiotic recombination between repeated transposable elements in *Saccharomyces cerevisiae. Mol. Cell. Biol.* **8,** 2942–2954.
96. Shen, P., and Huang, H. V. (1986). Homologous recombination in *Escherichia coli*: dependence on substrate length and homology. *Genetics* **112,** 441–457.
97. Yuan, L.-W., and Keil, R. L. (1990). Distance-independence of mitotic intrachromosomal recombination in *Saccharomyces cerevisiae. Genetics* **124,** 263–273.
98. Ahn, B.-Y., Dornfeld, K. J., Fagrelius, T. J., and Livingston, D. M. (1988). Effect of limited homology on gene conversion in a *Saccharomyces cerevisiae* plasmid recombination system. *Mol. Cell. Biol.* **8,** 2442–2448.
99. Hayden, M. S., and Byers, B. (1992). Minimal extent of homology required for completion of meiotic recombination in *Saccharomyces cerevisiae. Dev. Genet.* **13,** 498–514.
100. Bailis, A. M., and Rothstein, R. (1990). A defect in mismatch repair in *Saccharomyces cerevisiae* stimulates ectopic recombination between homeologous genes by an excision repair dependent process. *Genetics* **126,** 535–547.
101. Selva, E. M., New, L., Crouse, G. F., and Lahue, R. S. (1995). Mismatch correction acts as a barrier to homeologous recombination in *Saccharomyces cerevisiae. Genetics* **139,** 1175–1188.
102. Porter, G., Westmoreland, J., Priebe, S., and Resnick, M. A. (1996). Homologous and homeologous intermolecular gene conversion are not differentially affected by mutations in the DNA damage or mismatch repair genes *RAD1, RAD50, RAD51, RAD52, RAD54, PMS1* and *MSH2. Genetics* **143,** 755–767.
103. Resnick, M. A., Zgaga, Z., Hieter, P., Westmoreland, J., Fogel, S., and Nilsson-Tillgren, T. (1992). Recombinational repair of diverged DNAs: a study of homeologous chromosomes and mammalian YACs in yeast. *Mol. Gen. Genet.* **234,** 65–73.
104. Chambers, S., Hunter, N., Louis, E. J., and Borts, R. H. (1996). The mismatch repair system reduces meiotic homeologous recombination and stimulates recombination-dependent chromosome loss. *Mol. Cell. Biol.* **16,** 6110–6120.
105. Hunter, N., Chamberg, S. R., Louis, E. J., and Borts, R. H. (1996). The mismatch repair system contributes to meiotic sterility in an interspecific yeast hybrid. *EMBO J.* **15,** 1726–1733.
106. Williamson, M. S., Game, J. C., and Fogel, S. (1985). Meiotic gene conversion mutants in *Saccharomyces cerevisiae.* I. Isolation and characterization of *pms1-1* and *pms1-2. Genetics* **110,** 609–646.
107. Matic, I., Rayssiguier, C., and Radman, M. (1995). Interspecies gene exchange in bacteria: the role of SOS and mismatch repair systems in evolution of species. *Cell* **80,** 507–515.
108. Radman, M. (1988). Mismatch repair and genetic recombination. *In* "Genetic Recombination" (R. Kucherlapati and G. R. Smith, Eds.), pp. 169–192. American Society for Microbiology, Washington, D. C.
109. Alani, E., Reenan, R. A. G., and Kolodner, R. D. (1994). Interaction between mismatch repair and genetic recombination in *Saccharomyces cerevisiae. Genetics* **137,** 19–39.
110. Worth, L., Jr., Clark, S., Radman, M., and Modrich, P. (1994). Mismatch repair proteins MutS and MutL inhibit RecA-catalyzed strand transfer between diverged DNAs. *Proc. Natl. Acad. Sci. USA* **91,** 3238–3241.
111. Westmoreland, J., Porter, G., Radman, M., and Resnick, M. A. (1997). Highly mismatched molecules resembling recombination intermediates efficiently transform mismatch repair proficient *Escherichia coli. Genetics* **145,** 29–38.
112. Negritto, M. T., Wu, X., Kuo, T., Chu, S., and Bailis, A. M. (1997). Influence of DNA sequence identity on efficiency of targeted gene replacement. *Mol. Cell. Biol.* **17,** 278–286.
113. Resnick, M. A., Skaanild, M., and Nilsson-Tillgren, T. (1989). Lack of DNA homology in a pair of divergent chromosomes greatly sensitizes them to loss by DNA damage. *Proc. Natl. Acad. Sci. USA* **86,** 2276–2280.
114. Datta, A., Hendrix, M., Lipsitch, M., and Jinks-Robertson, S. (1997). Dual roles for DNA sequence identity and the mismatch repair system in the regulation of mitotic crossing-over in yeast. *Proc. Natl. Acad. Sci. USA* **94,** 9757–9762.
115. Priebe, S. D., Westmoreland, J., Nilsson-Tillgren, T., and Resnick, M. A. (1994). Induction of recombination between homologous and diverged DNAs by double-strand gaps and breaks and role of mismatch repair. *Mol. Cell. Biol.* **14,** 4802–4814.
116. Nilsson-Tillgren, T., Gjermansen, C., Holmberg, S., Petersen, M. C. L., and Kielland-Brandt, M. C. (1986). Analysis of chromosome V and the *ILV1* gene from *Saccharomyces carlsbergensis. Carlsberg Res. Commun.* **51,** 309–326.
117. Rothstein, R. J. (1983). One-step gene disruption in yeast. *Methods Enzymol.* **101,** 202–211.
118. Roth, D., and Wilson, J. (1988). Illegitimate recombination in mammalian cells. *In* "Genetic Recombination" (R. Kucherlapati and G. R. Smith, Eds.), pp. 621–653. American Society for Microbiology, Washington, DC.
119. Bogue, M., and Roth, D. B. (1996). Mechanism of V(D)J recombination. *Curr. Opin. Immunol.* **8,** 175–180.
120. Blunt, T., Finnie, N. J., Tacciolo, G. E., Smith, G. C. M., Demengeot, J., Gottlieb, T. M., Mizuta, R., Varghese, A. J., Alt, F. W., Jeggo, P. A., and Jackson, S. P. (1995). Defective DNA-dependent protein kinase activity is linked to V(D)J recombina-

tion and DNA repair defects associated with the murine *scid* mutation. *Cell* **80,** 813–823.

121. Liang, F., Romanienko, P. J., Weaver, D. T., Jeggo, P. A., and Jasin, M. (1996). Chromosomal double-strand break repair in Ku80-deficient cells. *Proc. Natl. Acad. Sci. USA* **93,** 8929–8933.
122. Yabumoto, K., Adasada, T., Muramatsu, M., Kadowaki, N., Hayashi, T., Fukuhara, S., and Okuma, M. (1996). Rearrangement of the 5′ cluster region of the BCL2 gene in lymphoid neoplasm: a summary of nine cases. *Leukemia* **10,** 970–977.
123. Schiestl, R. H., and Petes, T. D. (1991). Integration of DNA fragments by illegitimate recombination in *Saccharomyces cerevisiae. Proc. Natl. Acad. Sci. USA* **88,** 7585–7589.
124. Schiestl, R. H., Dominska, M., and Petes, T. D. (1993). Transformation of *Saccharomyces cerevisiae* with nonhomologous DNA: illegitimate integration of transforming DNA into yeast chromosomes and in vivo ligation of transformating DNA to mitochondrial DNA sequences. *Mol. Cell. Biol.* **13,** 2697–2705.
125. Schiestl, R. H., Zhu, J., and Petes, T. D. (1994). Effect of mutations in genes affecting homologous recombination on restriction enzyme-mediated and illegitimate recombination in *Saccharomyces cerevisiae. Mol. Cell. Biol.* **14,** 4493–4500.
126. Zhu, J., and Schiestl, R. H. (1996). Topoisomerase I involvement in illegitimate recombination in *Saccharomyces cerevisiae. Mol. Cell. Biol.* **16,** 1805–1812.
127. Champoux, J. J., and Bullock, P. A. (1988). Possible role for the eucaryotic type I topoisomerase in illegitimate recombination. *In* "Genetic Recombination" (R. Kucherlapati and G. R. Smith, Eds.), pp. 655–666. American Society for Microbiology, Washington, DC.
128. Kramer, K. M., Brock, J. A., Bloom, K., Moore, J. K., and Haber, J. E. (1994). Two different types of double-strand breaks in *Saccharomyces cerevisiae* are repaired by similar *RAD52*-independent, nonhomologous recombination events. *Mol. Cell. Biol.* **14,** 1293–1301.
129. Moore, J. K., and Haber, J. E. (1996). Cell cycle and genetic requirements of two pathways of nonhomologous end-joining repair of double-strand breaks in *Saccharomyces cerevisiae. Mol. Cell. Biol.* **16,** 2164–2173.
130. Teng, S.-C., Kim, B., and Gabriel, A. (1996). Retrotransposon reverse-transcriptase-mediated repair of chromosomal breaks. *Nature* **383,** 641–644.
131. Moore, J. K., and Haber, J. E. (1996). Capture of retrotransposon DNA at the sites of chromosomal double-strand breaks. *Nature* **383,** 644–646.
132. Tsukamoto, Y., Kato, J., and Ikeda, H. (1996). Effects of mutations of *RAD50, RAD51, RAD52,* and related genes on illegitimate recombination in *Saccharomyces cerevisiae. Genetics* **142,** 383–391.
133. Boulton, S. J., and Jackson, S. P. (1996). *Saccharomyces cerevisiae* Ku70 potentiates illegitimate DNA double-strand break repair and serves as a barrier to error-prone DNA repair pathways. *EMBO J.* **15,** 5093–5103.
134. Boulton, S. J., and Jackson, S. P. (1996). Identification of a *Saccharomyces cerevisiae* Ku80 homologue: roles in DNA double strand break rejoining and in telomeric maintenance. *Nucleic Acids Res.* **24,** 4639–4648.
135. Siede, W., Friedl, A. A., Dianova, I., Eckardt-Schupp, F., and Friedberg, E. C. (1996). The *Saccharomyces cerevisiae* Ku autoantigen homologue affects radiosensitivity only in the absence of homologous recombination. *Genetics* **142,** 91–102.
136. Milne, G. T., Jin, S., Shannon, K. B., and Weaver, D. T. (1996). Mutations in two Ku homologs define a DNA end-joining repair pathway in *Saccharomyces cerevisiae. Mol. Cell. Biol.* **4189,** 4198.
137. Mages, G. J., Feldmann, H. M., and Winnaker, E-L. (1996). Involvement of the *Saccharomyces cerevisiae HDF1* gene in DNA double-strand break repair and recombination. *J. Biol. Chem.* **271,** 7910–7915.
138. Porter, S. E., Greenwell, P. W., Ritchie, K. B., and Petes, T. D. (1996). The DNA-binding protein Hdf1p (a putative Ku homologue) is required for maintaining normal telomere length in *Saccharomyces cerevisiae. Nucleic Acids Res.* **24,** 582–585.
139. Tsukamoto, Y., Kato, J-I., and Ikeda, H. (1997). Silencing factors participate in DNA repair and recombination in *Saccharomyces cerevisiae. Nature* **388,** 900–903.
140. Teo, S-H., and Jackson, S. P. (1997). Identification of *Saccharomyces cerevisiae* DNA ligase IV: involvement in DNA double-strand break repair. *EMBO J.* **16,** 4688–4795.
141. Wilson, T. E., Grawunder, U., and Lieber, M. R. (1997). Yeast DNA ligase IV mediates nonhomologous DNA end-joining. *Nature* **388,** 495–498.

CHAPTER 35

Systems for the Study of Triplet Repeat Instability: Cultured Mammalian Cells

PETER STEINBACH, DORIS WÖHRLE, DIETER GLÄSER, AND WALTHER VOGEL
Department of Medical Genetics, University of Ulm, 89070 Ulm, Germany

I. INTRODUCTION

This chapter focuses on ongoing rather than completed studies of triplet repeat instability using cultured cells of humans and other mammalian species as a model system. Experimental studies have concentrated on repeat expansions in fragile X mental retardation and myotonic dystrophy.

The identification of unstable DNA sequences as the underlying cause of genetic disease is a fascinating new development in human genetics raising a number of important questions addressing the understanding of both the mechanisms and the effects of this new type of mutation [1–5]. Which factors contribute to disease pathogenesis by triggering or modifying the unstable behaviors of these repeat sequences? When and where may expansion of smaller repeats to disease-causing alleles occur? What are the direct consequences of such expansions setting off a molecular pathway leading to disease? Shedding light on the unstable behavior of expanded triplet repeats also promises to provide us with valuable insights into complex basic processes of molecular genetics and cell biology, such as DNA replication and repair. Unstable triplet repeats may become a paradigm of studies of genomic stability and instability in early development.

Trinucleotide repeat expansion mutations have been identified in a number of human genetic diseases, including spinal and bulbar muscular atrophy (SBMA) [6, 7], fragile X syndrome [8–12], myotonic dystrophy (DM) [13–17], Huntington's disease (HD) [18], spinocerebellar ataxia type 1 (SCA1) [19, 20], spinocerebellar ataxia type 2 (SCA2) [21, 22], Machado–Joseph disease or spinocerebellar ataxia type 3 (SCA3) [23], hereditary dentatorubral-pallidoluysian atrophy (DRPLA) [24, 25] and Friedreich's ataxia (FRDA) [26]. Mutational expansions of triplet repeats were also detected in the gene FMR2 predisposing to fragile X E mental retardation [27–31] and in the CBL2 proto-oncogene predisposing to Jacobson syndrome associated with fragile site 11B [32]. Other fragile sites, including fragile X F [33, 34] and fragile site 16A [35], have also been identified to correspond to expanded trinucleotide repeats, but these sites are not associated with genetic disease.

A. Physical Features of Instability

Despite the fact that the affected genes and resulting disease pathology due to expansion differ, repeat instability in disease shows many common features. These concern both the transmissions through the germline to successive generations and the patterns of expansions found in different somatic tissues.

The disease alleles are generated upon transmission from parent to offspring with a probability occurring only when there is a critical repeat length in a parent, i.e., a risk allele or a "premutation" [14, 18, 36–40]. Above that threshold, the repeat length can change bidirectionally by generally more than 10 copies with the rate of expansion far exceeding the rate of contraction. Expansion to a disease allele may also depend on the sex of the transmitting parent. In the "polyglutamine diseases" SBMA, HD, SCA1, SCA3, and DRPLA, the expanded CAG repeats show more frequent and stronger instability upon transmission from a male [7, 20, 24, 25, 41–46]. In myotonic dystrophy, intergenerational variation is more important through female germ line whereas a tendency for repeat contraction was observed almost exclusively upon transmission from males with expansions larger than 500 CTG triplets [4, 47]. In fragile X syndrome, full expansions of the CGG repeat in patients are exclusively found on X chromosomes which were received from a heterozygous mother [10, 48, 49]. With increasing length of the repeating unit there is genetic anticipation, an increase in the probability and severity of the disease in successive generations. Smaller changes of repeat size usually by 1–3 copies are associated with relatively equal rate of insertion and deletion. This feature of instability is associated with normal variation of repeat sizes and probably also with expansion of repeats of intermediate sizes to a risk allele [10, 38, 43, 50].

In somatic tissues, repeat instability is generally characterized by size variation among different cells. Somatic mosaicism has been reported in cases of largely expanded $(CCG)_n$, $(CGG)_n$, $(CTG)_n$, and $(GAA)_n$ repeats [27, 32, 34, 35, 51–53], but does also exist in patients with expanded CAG repeats where the repeat copy numbers are much lower [26, 36, 42, 54, 55]. Marked heterogeneity of expansion among different tissues like blood, brain, skin, or muscle is another physical feature of repeat instability [42, 55–58]. These changes can only result from mitotic activity. Mosaicism of triplet repeat sizes was also found in sperm of carrier fathers where it could have resulted from mitotic instability at germ cell proliferation and/or from unequal exchanges of triplet units at meiosis [54, 59–65]. There is evidence that somatic heterogeneity of repeat expansion is age dependent as larger sizes and broader ranges of expansions are found in patients with higher ages [20, 42, 55, 58, 66–68].

B. Proposed Mechanisms of Repeat Expansion

As disease resulting from unstable expansion of a triplet repeat is observed upon transmission to a child, focus was primarily placed on mechanisms acting at meiosis. Now there is ample evidence that the expanded disease alleles are often, if not exclusively, generated at mitosis. They may, in some cases, only be generated at particular stages of development of both germ cells and somatic tissues. A proposed meiotic mechanism of repeat expansion was unequal crossing over or unequal sister chromatid exchange in which recombination results in an unequal distribution of repeat units [69, 70]. The predictions of this model were, however, not upheld by genetic data [50, 71–75], indicating that the effects of such a mechanism, if any, are strongly overridden by another mechanism. Polymerase or DNA slippage is a widely recognized mechanism giving rise to repeat instability at DNA replication [76–81]. But slippage, in such a simple form, is not sufficient to create the large size gains observed in triplet repeat expansion diseases and also cannot explain why expansion rate is much larger than contraction rate. As previously discussed [82], large expansion requires slippage to a matching repeat separated by more than a few repeating units and quickly becomes energetically unfavorable unless the energy difference between the duplex and the slipped stages is minimized. The latter can be achieved by formation of alternative DNA structures. Probably involved in the process of large-scale expansion are hairpin structures or unimolecular foldbacks containing a number of hydrogen bonds between mismatched base pairs [83–89]. Hairpin structures may be formed in any biological process requiring unwinding and reannealing of single-stranded triplet repeat tracts, particularly at DNA synthesis.

Hairpin-mediated slippage (Fig. 35-1) has been proposed to occur when the polymerase and its attached nascent (primer) strand dissociates from the replication complex and when the slippage event is stabilized by hairpin formation. As the lifetime of the slipped stage would also be increased by hairpin formation, the likelihood is increased that polymerase adds a number of additional repeat units to the nascent strand (Fig. 35-1a). One may, however, doubt whether polymerase will ever dissociate from its contacts to other proteins at the fork or whether the entire fork complex will slip. It is, therefore, difficult to imagine how slippage of the nascent strand can occur.

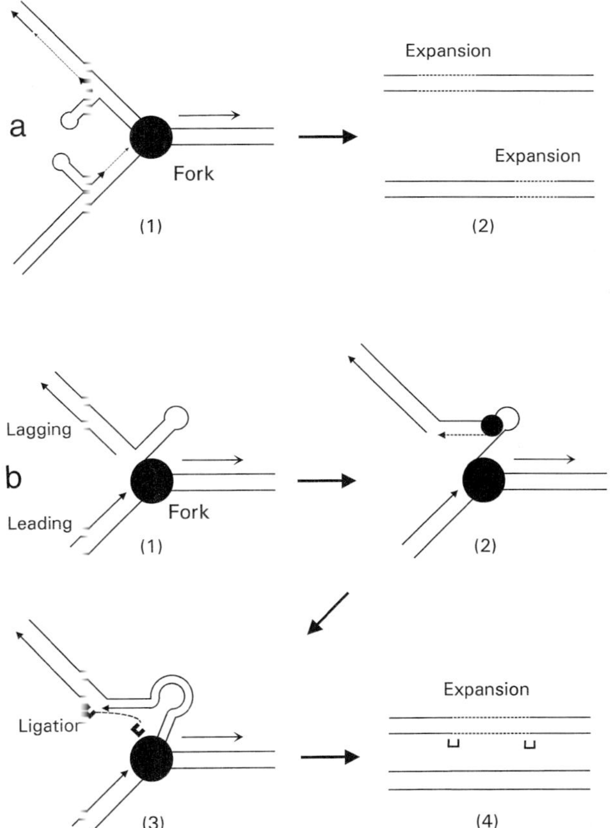

FIGURE 35-1 Hairpin-mediated DNA slippage and triplet repeat expansion. (a) DNA slippage during replication by hairpin formation of a nascent strand. The polymerase and its attached strand dissociate from the replication complex and the slippage event is stabilized by hairpin formation. Polymerase then adds a number of additional repeat copies to the nascent strand. An expanded novel allele could, in theory, result at the next round of DNA replication but is much more likely to occur by template correction at postreplicative DNA mismatch repair. (b) Lagging-strand model (adapted from McMurray [82]). Replication of the lagging strand is proposed to be initially blocked by double-stranded hairpin structures which are formed within the template by the expanded repeat and stably persist until the structure is relieved by additional DNA synthesis initiated at a single-stranded loop in the hairpin structure. This process is followed by integration of the extra DNA by ligation between the previous and the next Okazaki fragment, resulting in an increase of repeat length by template correction.

An alternative hypothesis of hairpin-mediated DNA slippage (Fig. 35-1b), has been provided with a "lagging strand model" [82]. During replication on the lagging strand, single-stranded DNA occurs naturally since the lagging strand exists as a single-stranded molecule for a significant time period. Immediately following the generation of the single-stranded DNA, hairpin formation of the repetitive DNA could occur, yielding a double-stranded molecule that is not recognized by single-stranded binding proteins (SSBP) normally preventing structure formation at DNA replication. Such hairpins will, therefore, lead to a block for replication on the lagging strand at the base of the slipped strand DNA structure. Another function of SSBP is polymerase targeting on the lagging strand by making protein–protein contact with polymerase. As SSBP can bind within the single-stranded loop of a hairpin, replication may be initiated within the loop region generating a short, loop-initiated fragment which subsequently becomes ligated to the preceding Okazaki fragment. Breaking of hydrogen bonds by loop-initiated replication leads to resolution of the hairpin structure and relief of the initial block of replication. Then normal synthesis of the succeeding Okazaki fragment can proceed from the fork. When fork-initiated polymerase displaces the extra copy of repetitive DNA, corresponding to some proportion of the hairpin, it becomes integrated between the two adjacent Okazaki fragments and stabilizes by hydrogen bonding to the template.

Subsequent to hairpin-mediated DNA slippage, an expanded novel allele could, in theory, not result until the next cell cycle. This would require persistence of the hairpin structure through the completion of the S, G2, and M phases until the next round of replication. Mutational expansion is, however, more likely to occur by template correction at postreplicative DNA mismatch repair (Section III.A).

Triplet repeat expansion was also suggested to result from reiterative DNA synthesis induced by alternative structures such like hairpins. These could form ahead of the replication complex, acting as barriers to the progression of the replication fork and leading to delayed replication. Such a mechanism could give rise to very large expansions in a single round of replication [77]. This scheme is, however, not consistent with the observation that expansion by insertion only involves a pure tract of the repetitive sequence. Expansion does not include either adjacent single-copy sequences or even those parts of some repeats containing one or two punctuations of a different triplet, e.g., the 5′-region of the CGG repeat of fragile X syndrome which usually contains AGG interruptions [90, 91].

Another mechanism that could give rise to unidirectional changes of repeat size on a large scale is gene conversion. Such events are more likely to occur in the presence of an expanded repeat and may be facilitated by hairpin or other alternative structures in a single-stranded DNA molecule allowing for loop–loop interactions. Rare observations of regression to normal size alleles in myotonic dystrophy and fragile X syndrome have been attributed to discontinuous gene conversion [92, 93]. In Machado–Joseph disease (SCA3) there is indirect evidence that interaction between homologous alleles at a polymorphic CGG/GGG triplet is involved in the intergenerational instability of the CAG repeat [45]. There is also a possibility that the formation of cruciform structures, triple-stranded structures, or loops in the double-stranded DNA helices of expanded repeats facilitates interaction among and unequal pairing of sister chromatids resulting in complex recombination during the S, G2, or M phase of the cell cycle. This mechanism might result either in further expansion or in contraction by deletion of repeat units. Such a mechanism has been described in the "recombination gap repair model" proposed by Jansen *et al.* [94].

Further mechanisms of repeat expansion have been proposed to explain the features of repeat instability observed when *Escherichia coli* and yeast cells were used as model systems [95]. It remains to be shown whether similar mechanisms are effective in mammalian cells and do contribute to mutational expansion of triplet repeats in human families.

II. STUDYING MITOTIC INSTABILITY OF TRIPLET REPEATS IN CULTURED MAMMALIAN CELLS

The nature of individual mutational events responsible for the instability of triplet repeats in dividing cells, and the factors contributing to stability or instability, can be studied most readily in cultured cells. Different approaches were used to address different study questions. Although the behavior of expanded repeats may be determined by a number of factors, actual research has concentrated on the effects of size, structure, and methylation. There may also be gene-dependent effects of expansions on cell proliferation leading to changes of mutation patterns in the cell population. Most of these aspects can already be elucidated using cultured human fibroblasts. This system, whose simplicity is one of its advantages, serves as an experimental model of the behavior of diploid cells at proliferation and has been used to study expanded alleles of triplet repeats when going through many rounds of DNA replication (Section II.A). From heterogeneous primary cultures, fibroblasts with individual expansion have been separated by dilution cloning demonstrating the clonal maintenance of some triplet repeats and the clonal instability of others (Section II.B). Using immortalized mammalian cell lines, such as somatic cell hybrids and lymphoblastoid cell lines, the advantages of indefinite proliferation can be used to study the behavior of individual alleles at long-term proliferation, on different genetic backgrounds, and in rapidly dividing cells (Section II.C). All these culture systems were originally established from heterogeneous material or were derived from heterogeneous primary cultures. Therefore, selection advantages or disadvantages of particular cell types, particular genotypes, or particular clones in the investigated cell population have to be taken into consideration.

A. Heterogeneous Populations of Human Fibroblasts

Cultures of fibroblasts or fibroblastoid cell types, defined as fibroblast cultures for the purposes of this chapter, were established from freshly dissected biopsies of various embryonic or adult tissues and were maintained in standard culture medium supplemented with fetal calf serum and antibiotics. In this simple system, only those cells of a given tissue will be retained that are able to proliferate under the particular *in vitro* conditions. Fibroblasts are able to divide *in vitro* every 20–24 h. The number of population doublings is limited to about 50. Cultured fibroblasts have been used to study the behaviors of largely expanded triplet repeats in the FMR1 gene of fragile X syndrome and in the DMPK gene of myotonic dystrophy.

When analyzed on Southern blots of DNA isolated from primary cultures or early subcultures (passages), these expansions most frequently show up as more or less clear bands of different sizes, or as diffuse heterogeneous smears usually including a "midpoint" of maximum signal density. Expansions seen as single and relatively sharp bands do also exist, for example in young patients with congenital myotonic dystrophy, but they are exceptions rather than the rule. Cultures showing heterogeneous patterns of repeat expansions have served as a model system to address questions on the behavior of such patterns during proliferation of diploid somatic and germ cells. This approach gave ample evidence that the patterns of expansion do change upon cell proliferation, due to repeat expansion and in some cases probably also owing to selection biases, and that methylation of expanded triplet sequences is a major determinant of repeat stability.

1. INSTABILITY OF UNMETHYLATED EXPANSIONS

Determination of the methylation status of expanded triplet repeats is difficult and would, in most instances, require application of sophisticated methods such as chemical sequencing [96, 97]. Therefore, data on the methylation patterns of normal and expanded triplet repeats are still limited [35]. It seems, however, reasonable to assume that methylation of trinucleotide repeats is restricted to large expansions of sequences containing CpG dinucleotides. This is (1) because most of the methylation in the human genome is found on such dinucleotides and (2) because expanded $(GGC)_n \cdot (GCC)_n$ repeats form alternative DNA structures including C · C mismatches at the CpG sites which are exceptional substrates stimulating the enzymatic activity of the human methyltransferase to become much more efficiently methylated than normal Watson-Crick paired CpGs [84, 98]. Such sequences could be exceptional centers of *de novo* methylation from which methylation then spreads on adjacent CpG dinucleotides [27, 32, 34, 35, 99]. A similar structural basis allowing for *de novo* CpNpG methylation seems to be absent from expanded $(AGC)_n \cdot (GCA)_n$ repeats.

a. Instability of $(CTG)_n$ Expansions in Myotonic Dystrophy Fibroblast cultures were established from tissue specimen of a 16-week-old fetus who was prenatally diagnosed to have a maternally inherited CTG repeat expansion of about 1800 triplet units. On Southern blot analysis of primary cultures, this large expansion was seen as a clear single band, indicating sufficient homogeneity of expansion among these cells. This was a prerequisite to allow for studying its behavior on continuing cell proliferation [57]. After 15 doublings of the population, corresponding to an average of 15 rounds of DNA replication per cell, an increase in the length of the expanded repeat became evident on Southern analysis (Fig. 35-2). At 40 doublings, the CTG repeats of the mutated alleles had gained about 470 triplets. Between 24 and 33 doublings, the rate of repeat expansion was highest and was about 18 triplets per cell cycle. There was no evidence that this *in vitro* repeat expansion, which was highly synchronous in the cell population, was accompanied by contraction of repeats in a significant proportion of cells. In another fibroblast culture, established from tissue of a fetus with a smaller expansion of only 200 CTG triplets, a very similar behavior was observed with a gain of about 50 triplets at 19.5 doublings *in vitro* and a maximum expansion rate of 8 repeat units per cycle.

This unstable behavior of expanded CTG repeat tracts in the DMPK gene of myotonic dystrophy was directly observed and measured in an *in vitro* system of cells which, by Northern analysis, showed no detectable transcription of the gene involved. Confounding of the results by effects of the mutated gene is, therefore, probably excluded. Repeat expansion in this system is concluded to be related to mechanisms acting most probably at DNA replication and possibly also at post-replicative DNA repair. These mechanisms should also be able to add relatively high numbers of copies to the repeat in a single round of replication, without simultaneous production of a significant number of deleted alleles.

There is ample evidence that the unstable expansive behavior observed in this culture system does also occur *in vivo* in a very similar manner. A marked heterogeneity of expansions among different tissues in adults and fetuses with expansions in the DMPK gene has been described [47, 57, 94, 100, 101]. Adult patients were reported to have, compared to blood, larger expansions in muscle where the gene is predominantly expressed [56, 102]. There is also evidence that expansion of CTG repeats does proceed *in vivo* in patients and is associated with increasing heterogeneity of repeat size among cells [66, 67, 103], giving rise to very broad heterogeneous smears of expansions on Southern blot analysis of some older patients [104].

b. Instability of Unmethylated $(CGG)_n$ Expansions in Fragile X Syndrome In fragile X mental retardation, most affected patients have large expansions of the $(CGG)_n$ repeat situated in the 5′ untranslated region of the FMR1 gene [37, 105]. These expansions are designated "full mutations" and include more than 220 repeat units. Full mutation usually coincides with hypermethylation involving the expanded repeat [96, 99], the preceding promoter region [97], and adjacent single copy sequences downstream of the repeat [8, 99]. The disease alleles result from expansion of a "premutation." This is an expanded allele of usually 60 to 220 copies that is not methylated and is found in phenotypically normal carriers including normal transmitting males (see Chapter 3 in this book).

Transmitting males have been identified with normal or borderline intelligence and with molecular genetic evidence of full mutations [106–109]. In some of these "high functioning" males, full mutation was apparently not associated with methylation. In one such male, Taylor *et al.* [110] reported a high degree of heterogeneity of mutation sizes among different tissues, suggesting that these unmethylated full mutations may be somati-

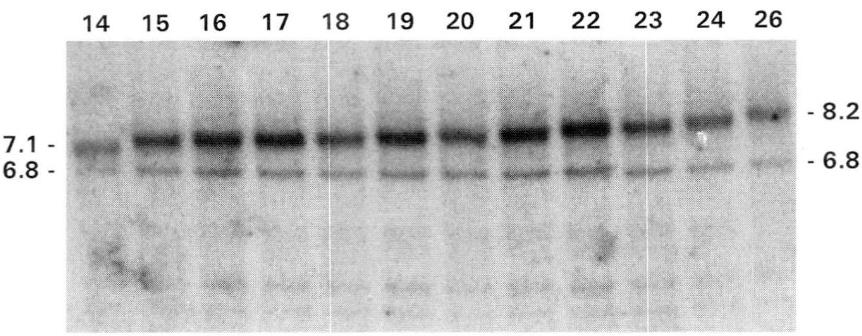

FIGURE 35-2 Further expansion of CTG repeats during proliferation of fibroblasts from a DM fetus. Passage numbers are given above each lane. In this experiment, each passage corresponded to 1.5 population doublings. DNA was digested with *Pst*I and hybridized to probe pM10M6 [14]. Reprinted from Wöhrle *et al., Hum. Mol. Genet.* **4,** 1147–1153, 1995, by permission of Oxford University Press.

cally unstable. We obtained a skin biopsy of another high functioning male with unmethylated full mutations and established a fibroblast culture to investigate the behavior of these expansions *in vitro.* Restriction analysis with methylation-sensitive enzymes, *Eag*I, *Fnu*4HII, and *Hpa*II, gave no evidence of methylation in a significant proportion of cells. Upstream promoter sequences and sequences on the 3′ side of the repeat were, without detectable exceptions, not resistant to cleavage with these enzymes. Immunochemical analysis of FMR1 protein (FMRP) revealed a mosaic pattern of gene expression in the cultured fibroblasts, including cells with normal expression, cells with reduced expression, and cells staining negative for FMRP [104]. Southern blot analysis of DNA from early subcultures revealed a heterogeneous smear of expansions ranging continuously from 48 to 1600 repeat units but containing some segments with higher signal densities. This pattern was unusual for fragile X but resembled that found in older patients with myotonic dystrophy and, therefore, suggested that these expansions would not be stably maintained at cell proliferation [104].

The mutation patterns were analyzed on *Hin*dIII blots of DNA isolated from successive subcultures (Fig. 35-3). Despite the difficulty of visualizing the unusually faint patterns, long exposures of the signals yielded sufficiently clear results. The behaviors of these CGG repeat expansions at cell proliferation *in vivo* were strikingly different from the findings obtained on a similar *in vitro* experiment using fibroblasts of an affected full mutation male (Fig. 35-3a). With increasing number of population doublings, the signals of larger expansions were lost from the mutation pattern while smaller ones were retained. The larger the expansions, the earlier they became undetectable at continuing cell proliferation.

The observed changes in the mutation pattern (Figs. 35-3b, 35-3c) do correspond exactly to the expectation, if the expansions are unstable and if each individual allele is expanded during cell proliferation. This instability would lead to a shift toward larger sizes and to an increase of size heterogeneity, making the pattern more

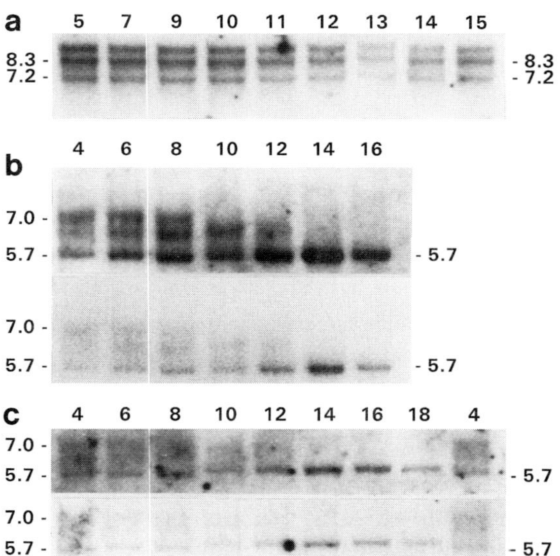

FIGURE 35-3 Mitotic behavior of methylated and unmethylated full mutations of fragile X syndrome during proliferation of human fibroblasts. Southern analysis of *Hin*dIII-digested DNA samples, hybridized to probe Ox1.9 [111]. Fragment sizes are given in kb. Passage numbers are given above the lanes. (a) Stability of methylated full mutations of a fragile X positive fetus during cell proliferation. A passage corresponded to about 1.5 doublings of the cell population. This figure is shown for the purpose of comparison. Reprinted from Wöhrle *et al., Hum. Mol. Genet.* **4,** 1147–1153, 1995, by permission of Oxford University Press. (b, c) Instability of unmethylated full mutations in the fibroblasts of a high-functioning fragile X male [104]. In this experiment, the passage numbers corresponded to the numbers of population doublings. The results of two different radioautographic exposures are shown. Note the disappearance of larger expansions. In (c) a comparison of the 18th and the 4th passage gave evidence for an *in vitro* expansion of the shortest expansion during cell proliferation.

and more diffuse with the larger alleles being more affected than the smaller ones. Such a shift of repeat size associated with an increase of size heterogeneity was, indeed, observed in another experiment, when, after somatic cell hybridization, single expansions were analyzed after their separation into different clones (Section II.C).

As the largest expansions were the first signals that became undetectable during cell proliferation, the changed patterns are also suggestive of a selection disadvantage against cells with larger alleles. Feng *et al.* [112] have recently shown that the transcripts of unmethylated alleles with larger repeats are insufficiently translated into FMRP [113]. This protein is upregulated at cell proliferation [114–116] and probably has a function as a translation factor [117]. Therefore, reducing of FMRP production with increasing size of unmethylated CGG repeats may well lead to a decrease of the potency of cells to proliferate.

In this context, cultured human cells carrying unmethylated full mutations of fragile X syndrome may provide valuable information on disease pathogenesis which may be related to a function of the mutated gene in a given population of cells. For example, the unknown "contraction" mechanism may correspond to a selection mechanism acting during proliferation of spermatogonia in the absence of a second normal active FMR1 allele. Full mutations have been detected in the germ cells of full mutation male fetuses but are found to be "replaced" by unmethylated premutations in adult life [60, 118, 119]. A selection advantage of spermatogonia with unmethylated premutations has been proposed [60, 120, 121]. The number of cell divisions for male gamete production ranges between 50 and several hundred during the effective fertile life span [122]. It is unknown whether full mutations are unstable during germ cell proliferation generating a small proportion of cells with shorter or contracted repeats. If so, such cells will probably accumulate in the population due to a selection advantage provided by a more efficient translation of FMRP. There is evidence for a slow progressive selection for leukocytes with a normal FMR1 allele on the active X chromosome [123].

In summary, investigation of expanded triplet repeats in normal fibroblast cultures containing heterogeneous cell populations revealed that these mutations are unstable during cell proliferation. Large gains of repeat sizes were observed but there was no evidence of any obvious size reduction. If repeat contractions occur, they may be either rare events or rather small mutations remaining undetected unless there is a selection pressure leading to accumulation of contracted alleles in the cell population. These conclusions seem to apply to expansions which are probably never methylated or which remained unmethylated as an exception. Methylated expansions, surprisingly, were found to be extremely stable during cell proliferation in the same *in vitro* system.

2. Mitotic Stability of Methylated $(CGG)_n$ Expansions in Fragile X Syndrome

In contrast to expansions of the CTG repeat in myotonic dystrophy and to large CGG repeat expansions in high functioning fragile X males, the methylated full mutations of mentally retarded fragile X males usually show up as patterns of multiple bands of expanded FMR1 alleles on Southern analysis. Nevertheless, large expansions of fragile X full mutations were initially believed to be unstable at both meiosis and mitosis. No mechanism of repeat expansion at meiosis has, however, been proposed whose predictions could fit to the genetic data that fragile X mutations as well as other expanded triplet repeat sequences are frequently found embedded in a conserved haplotype background of a founder chromosome [72, 73]. Also, the concept of mitotic instability had to be modified in that these expanded repeats obviously are not unstable at any mitosis throughout the patients' life. This modification became necessary when different groups reported their findings on comparative analysis of full mutation patterns in different tissues of fragile X fetuses and in affected monozygotic twins [51, 52]. Identical patterns of mutations were found in different fetal tissues as well as in white blood cells of adult monozygotic twins who shared circulation at embryogenesis. These findings suggested that the patterns of mutations were generated after formation of the zygote at an early stage of embryonic development and subsequently acquired substantial mitotic stability. In all these cases analysis of restriction sites of methylation-sensitive enzymes suggested methylation and inactivation of the gene promoter [97] which is mostly, if not always, associated with complete methylation of the expanded CGG repeat [96, 99].

The mitotic behavior of apparently methylated full mutations was studied in a fibroblast culture established from a tissue specimen of a fragile X fetus who showed homogeneity of expansion patterns among all tissues examined [53]. The patterns were stably maintained during cell proliferation since no changes were detectable when the cells were grown to 36 population doublings (Fig. 35-3a). The high degree of stability of these full mutations was in contrast to the results obtained on similar experiments with other expanded, but probably unmethylated, repeats. This led to the conclusion that the high degree of mitotic stability would be due to complete methylation of the expanded CGG repeats. Focus was then placed on the contribution of methyl-directed DNA mismatch repair to the stability and instability of expanded triplet repeats [57].

Another important finding of this *in vitro* experiment with apparently methylated full fragile X mutations has now been recognized. There was no loss of any bands from the mutation pattern during cell proliferation indicating absence of any significant selection against cells with particular repeat sizes. This is again different from the findings obtained on full mutations of high functioning fragile X males (Fig. 35-3b) and further suggests that this particular behavior could, at least in the situation of hemizygosity, depend on inactivity of methylated expanded alleles in the absence of cells with an active gene.

B. Separation of Single Expanded Alleles by Dilution Cloning of Fibroblasts

To investigate whether or not the size of particular repeats is maintained in a clonal fashion and to identify factors contributing to stability or instability, homogeneous cell populations were grown from cells with one particular expanded repeat. Such clones were obtained after dilution plating of primary fibroblast cultures previously established from tissue biopsies of fragile X and myotonic dystrophy individuals. In this culture system, the genetic background of separated alleles is not changed, as is the case in clones of somatic cell hybrids (Section II.C.1). There is, however, the disadvantage that the life span remaining after isolation of a fibroblast clone is sometimes too short to allow for a number of different experiments, and that the homogeneity of the isolated cell population cannot be increased or made sure by subcloning. The latter could be done, e.g., with lymphoblastoid cell lines (Section II.C.2). Cloning by dilution plating usually requires more than 20 rounds of DNA replication until DNA amounts sufficient for Southern analysis are obtained.

Using the dilutation plating approach on cultured fibroblasts, we were able to present the first direct evidence that the full mutation patterns of fragile X syndrome, previously assumed to be due to continual mitotic instability in dividing cells, actually reflect a mosaic of cells carrying different alleles with each individual repeat size being stably maintained in progeny cells establishing a clone [53]. It is now recognized that the conclusions drawn from these results may only apply to methylated expansions, since the genomic *Eco*RI fragments, carrying the stable expansions, were resistant to cleavage with methylation-sensitive enzymes *Eag*I and *Hpa*II [53]. The same dilution cloning approach was used in DM, with the primary intention to confirm the fragile X results on another type of expanded triplet repeat. But, surprisingly, the sizes of CTG repeat expansions were not maintained clonally [57].

Fibroblasts were first cloned from an adult DM patient. The primary culture was heterogeneous, showing a group of expanded alleles normally distributed around a mean expansion of about 6.3 kb. In blood leukocytes, the expansion was about four times smaller. Isolated clones showed different expansions whose sizes were equal to or larger than the main expansion seen in the uncloned heterogeneous culture. This finding proved that the heterogeneous smears of CTG repeat expansions resulted from somatic mosaicism of cells with different repeat sizes, as do the mutation patterns found in other triplet repeat diseases. However, the bands of expansions that were separated into individual clones had a blurred appearance which was suggestive of continuing mitotic instability. In another experiment, clones were isolated from a fibroblast culture of a DM fetus who presented a large homogeneous expansion of about 7 kb. This cloning resulted in the isolation of expansions (but not of contractions) which were significantly larger than 7 kb and did, therefore, confirm previous findings that CTG repeats of myotonic dystrophy show unstable expansion at continuing cell proliferation.

In summary, dilution cloning of fibroblasts allowed for the demonstration that (1) somatic mosaicism generated by mitotic instability of triplet repeats is the origin of both continuous size distributions and discontinuous patterns of expansions and (2) the sizes of some expanded triplet repeats are stably maintained in a clonal fashion whereas the sizes of other repeats are not, and (3) gave some evidence that instability at cell proliferation and mitosis predominantly results in repeat expansion. The latter is, however, only seen if such clones could be maintained through sufficiently high numbers of cell divisions. In contrast to fibroblasts, clones of permanent cells seem to be a better system to address this particular question of expansion versus contraction rates, and to evaluate the particular patterns of repeat size distribution resulting from mitotic instability of an isolated single expansion.

C. Permanent Cell Lines

Two different types of cell lines have been used to study the somatic instability of single alleles with an expanded triplet repeat in permanent cells. These were somatic cell hybrids, constructed after fusion of human fibroblasts with spontaneously immortalized rodent cells (Section II.C.1), and lymphoblastoid cell lines obtained by Epstein–Barr virus (EBV) transformation of patients' B-cells (Section II.C.2). Although these systems would, due to infinite growth of cells, also be suited to investigate mutation rates of smaller triplet repeats [54], experimental data have only been available from studies of larger expansions of fragile X syndrome (unpublished results) and of myotonic dystrophy [124].

1. SOMATIC CELL HYBRIDS

Clones of somatic cell hybrids, representing the product of a single fusion event, are obtained in selection medium by limiting dilution of the cell concentration to permit attachment of single cells to the culture dish and growing into colonies which are then individually subcultured. By this procedure, the behavior of FMR1 alleles with particular repeat sizes was followed at infinite proliferation. In contrast to fibroblast clones, the somatic cell hybrid system naturally includes a normal active FMR1 homologue on the rodent X chromosome, which is expressed at a high level in the permanent host cells (Fig. 35-4a). This gene will compensate for any selection disadvantages possibly caused by a decrease of human FMRP expression upon *in vitro* expansion of active alleles (see Section II.A.1). The somatic cell hybrid system will, at least in theory, permit detection of each individual mutation of unstable fragile X triplet repeats, and would allow for the characterization of mutation patterns resulting from accumulation of unstable products.

In our own laboratory hybrid analysis has been carried out on affected males with discontinuous patterns of (methylated) full expansions, on transmitting males and females with (unmethylated) premutations, and on a high functioning fragile X male with a continuous pattern of apparently unmethylated full mutations. The expansions of the latter proband were previously examined in cultured fibroblasts (Section II.A.1). In each experiment, human fibroblasts were fused with hypoxanthine-phosphoribosyltransferase (HPRT)-deficient permanent mouse cells (RAG or A9 cells) by means of polyethylene glycol (PEG) treatment and HPRT-positive clones were selected in hypoxanthine aminopterin thymidine (HAT) medium. Independent of the original distribution of repeat sizes, Southern analysis of clones (10^{-6}– 10^{-8} cells) revealed that single expansions of discrete sizes were isolated.

Large premutations, when isolated into different hybrid clones, showed up as single bands. Different clones obtained from the same fusion experiment did not con-

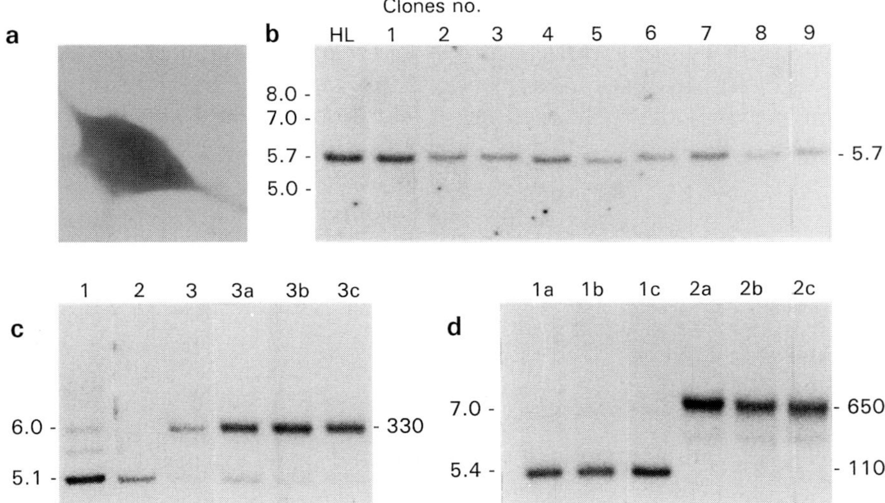

FIGURE 35-4 Somatic cell hybrid analysis of fragile X pre- and full mutations at Southern analysis. DNA was digested with *Hind*III and hybridized to probe Ox1.9 [111]. (a) The permanent murine host cells (RAG line) express the protein FMRP in the cytoplasm, as shown by immunocytochemistry according to Willemsen *et al.* [125]. (b) Hybrid clones obtained after fusion of fibroblasts from a transmitting male carrying a premutation including ca. 220 CGG triplets. (HL) Uncloned cells from the hybrid line. In the clones, the sizes of isolated premutation fragments vary between 5.6 and 5.9 kb. (c) Hybrid analysis of full mutations isolated from fibroblasts of a heterozygous female. The lanes are as follows. (1) Mutation pattern in the donor's fibroblasts, (2) hybrid clone with isolated normal allele. (3, 3a) Clone with an isolated full mutation fragment containing 330 CGG triplets. (3b, c) Clone 3 after proliferation through approximately 15 (3b) and 21 additional rounds of replication (3c). (d) Lanes 1a–1c: Hybrid analysis of a smaller premutation (about 110 CGGs) from a normal transmitting male (1a) showing no significant instability during 15 (1b) and 30 additional cell cycles (1c). (2a–2c) Hybrid analysis of a full mutation (650 CGGs) from an affected fragile X male (2a). There was no change of repeat size at further cell proliferation during approximately 15 (2b) and 30 further cell divisions (2c).

tain identical repeat sizes, indicating somatic instability of (unmethylated) premutations in the donors' fibroblast cultures (Fig. 35-4b). This finding was confirmed by PCR analysis (data not shown). Hybrid analysis of full mutations from fragile X males and females showed single expansions on homogeneous bands (Figs. 35-4c, 35-4d). These full-mutation fragments did not change in the hybrid clones during further cell proliferation.

From the continuous pattern of full mutations, presented by the high functioning fragile X male, discrete length alleles were isolated into hybrid clones. They were, however, not homogeneous as they showed up as a diffuse band of expansions normally distributed around a major allele (Fig. 35-5). When these clones were allowed to further proliferate, the diffuse bands became faint smears, indicating a simultaneous increase of both average repeat size and size heterogeneity. Cloning by somatic cell hybridization obviously resulted in isolation of large expansions that are unstable at cell proliferation with the majority of accumulating unstable products resulting from expansion. As the CGG repeats of these unstable alleles are most probably unmethylated, these observations support the hypothesis that stability of expanded CGG repeats is mediated by DNA methylation.

In summary, the somatic cell hybrid approach, applied to repeat expansions of fragile X syndrome, led to reproduction of the findings previously obtained on

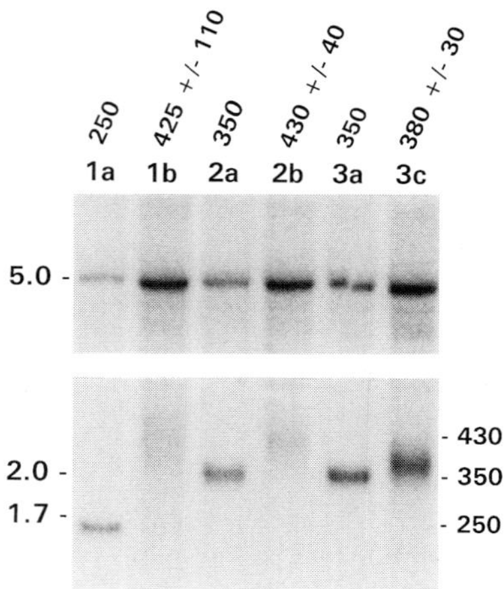

FIGURE 35-5 Instability of unmethylated fragile X full mutations upon somatic cell hybrid analysis. The donor of the mutated alleles was the high functioning fragile X male whose fibroblasts were examined previously (Figs. 35-3b, 35-3c). Three expansions, one with 250 CGGs and two with 350 CGGs each, were isolated into different hybrid clones (1a, 2a, 3a). In these alleles, the sequence of the *Eag*I site in the gene promoter was unmethylated (not shown). In all three clones the isolated repeats experienced further expansion during 30 further doublings of the cells (1b, 2b, 3b). The mean sizes and the distribution of expanded fragments are given above these lanes. DNA was digested with *Pst*I and hybridized to probe Ox0.55 [111]. Fragment sizes (decimals) are given in kb.

repeat analysis using cultured fibroblasts and to some additional notions. In particular, these experiments gave further evidence that methylated expansions of CGG repeats are mitotically stable while unmethylated repeats, including pre- and full mutations, are unstable with the unmethylated alleles giving rise to expansion rather than contraction. The somatic cell hybrid system allowed for a direct observation of fragile X repeat expansions for the first time. This model system is now being used to study the somatic instability in more detail.

2. Lymphoblastoid Cell Lines

Epstein-Barr virus transformation of B-lymphocytes from patients affected by Machado–Joseph disease (SCA3) gave direct evidence for somatic heterogeneity of CAG repeat expansion in this particular tissue [54]. To establish another model system, in which somatic instability of expanded CTG repeats could be studied including direct determination of mutation rates of individual alleles isolated into cell lines of single-cell origins, lymphoblastoid cell lines were established from two patients with myotonic dystrophy [124]. On DNA analysis of peripheral blood leukocytes, both DM patients presented average expansion sizes of about 900 CTG triplets. In the transformed B-cells, however, the average repeat size was significantly larger (about 1270 units) in one DM patient but not in the other. This initial shift was probably due to selection of B-lymphocytes from the total leukocyte population involved in EBV transformation. In contrast to our previous demonstration of continual further expansion of CTG repeats with 1900 and 200 triplets (Section II.A.1), the repeat sizes did not significantly change in the lymphoblastoid cell lines during 29 further passages.

Somewhat contradicting results were, however, obtained when lymphoblastoid cell lines of single-cell origin were established by cloning which involved multiple steps of limiting dilution. These cells are anchorage-independent, growing in fluid suspension culture, and do not form colonies attached as a monolayer to the culture surface. To eliminate the probability of multiple input cells at the time of the primary cloning, additional rounds of cloning were performed to finally obtain populations with each individual cell being a descendent of a single progenitor with one particular repeat allele. This cloning procedure demonstrated a significant size heterogeneity of expansions in the original cell lines. However, despite this heterogeneity, most of the clones apparently derived from progenitor cells with only about 230 triplets. This size was much smaller than the sizes previously measured in the majority of the primary cells. After approximately 20 rounds of DNA replication two types of mutations were detected in the descendents of the isolated progenitor cells. First, there were frequent mutations yielding small size changes of the expanded allele resulting in a normal distribution around the progenitor allele (mean ± 3 standard deviations). Such a distribution could result from simple polymerase slippage to a matching repeat separated by not more than a few repeating units [76, 78–80]. The second type of mutations was characterized by infrequent large repeat size changes showing a bias toward contractions (as well as some apparent PCR artifacts).

The pattern of somatic instability of expanded repeats in myotonic dystrophy, which has been shown previously to be characterized by an ongoing increase of repeat size both *in vivo* [66, 103] and *in vitro* during proliferation of fibroblasts [57], was not reflected in the lymphoblastoid cell lines. No explanation has been found for this discrepancy. Similar biases toward contraction have also been observed upon analysis of CTG/CAG repeat tracts on plasmids in *E. coli* [126] and on chromosomes in yeast strains with mutations in the mismatch repair system [127]. Therefore, host cell genetic factors may be responsible and the findings on lymphoblastoid cell lines might be complicated by unknown confounding influences of this particular *in vitro* environment or of the EBV transformation. The latter has also been known to occasionally result in aberrant methylation patterns.

In summary, in contrast to other experimental systems of cultured mammalian cells that were used to study unstable triplet repeat expansion, lymphoblastoid cell lines seem to be complicated by confounding influences of unknown origin. An excess of repeat expansion over contraction is a feature common to all unstable triplet repeats that give rise to human disease. This feature was demonstrated in cultures of other mammalian cells but was not obvious during proliferation of lymphoblastoid cell lines with largely expanded CTG repeats of myotonic dystrophy. Experiments with other repeat expansions should demonstrate whether this discrepancy was due to the particular model system or could reflect a property of the particular repeat.

III. CONCLUSIONS

In an increasing number of human genetic disorders, unstable transmission of triplet repeat sequences is identified as a sufficient cause to set off the pathway of pathogenetic events leading to disease manifestation. Some of these unstable sequences have been studied in a number of *in vitro* systems, including *E. coli* [126–129] and yeast cells [127, 130–132], which significantly contributed to elucidation of basic molecular mechanisms of sequence expansion and repeat instability. To understand the disease pathology, however, much more information is required, particularly on factors contributing to mutational expansion, which occurs during germ cell proliferation and/or early postzygotic development, or throughout embryonic, fetal, and adult life and in any tissue. Other factors to be identified and characterized are responsible for different behaviors of expanded repeats at different stages of human development. These and other questions have been addressed in ongoing studies using cultured cells of humans and other mammalian species as a model system.

Repeat expansion has been known to be a consequence of the repeat itself, i.e., its size, type, substructure, and its ability to form alternative DNA structures [82–88]. Using cultured mammalian cells as a model system, two additional factors have been identified which probably contribute significantly to the behavior of expanded repeats in proliferating populations of human cells. These factors are the methylation pattern of the expanded sequence (Section III.A) and the function of the mutated human gene at cell proliferation (Section III.B).

A. Methylation and Mechanisms

De novo methylation of triplet repeat sequences by human methyltransferase occurs on largely expanded CGG repeats of full mutation fragile X males [96, 99], and probably on expanded $(GCC)_n \cdot (GGC)_n$ sequences of other genes or fragile sites as well [35). There is evidence that this modification is induced by structure formation of expanded CCG strands [83] and is, as far as we know, absent from expanded repeats with other triplet sequences. Extensive DNA methylation may subsequently occur, by spreading from such a center of *de novo* methylation, in the flanking single copy sequences of largely expanded repeats [27, 34, 35, 133]. CpG-rich sequences adjacent to a triplet repeat may, alternatively, be a part of a DNA structure formed by expanded triplet repeats [82, 134] and may, by this mechanism, itself become a substrate of *de novo* methylation [135].

In fragile X syndrome, methylation of the CGG repeat correlates with methylation of the adjacent gene promoter and with gene silencing [96, 97, 99]. High functioning fragile X males are exceptions in that most if not all of the mutated alleles in their adult cells are unmethylated in spite of full expansion. These nonretarded fragile X males could have received a relatively large, unstable premutation which was, however, not a substrate of a developmental specific *de novo* methylation [104].

As shown by genomic sequencing [96, 97], methylation associated with full fragile X mutation involves each individual CpG dinucleotide of the expanded CGG repeat and almost all such dinucleotides in the adjacent promoter region. Such a degree of hypermethylation will not become significantly reduced either by experimental demethylation using azacytidine or by random errors of maintenance methylase that may occur at cell proliferation *in vitro.* Analysis of the mitotic behavior of expanded repeats from affected full mutation fragile X males during *in vitro* proliferation of both fibroblasts (Sections II.A.2, II.B) and somatic cell hybrids (Section II.C.1) clearly showed that the repeat size in an individual cell is stably maintained at mitosis. This behavior was very much unlike that of other repeats, which are likely to be unmethylated, i.e., expanded CTG repeats of DM patients and full mutations of a high functioning fragile X male with CGG repeats embedded in the unmethylated genomic environment of active alleles (Sections II.A.1, II.C.1). Hypermethylation was, therefore, proposed to cause stability of largely expanded and previously unstable triplet repeats [57, 136]. This conclusion is biologically plausible.

A basic biological process giving rise to stability of repetitive DNA sequences is methyl-directed DNA mismatch repair (see Chapter 36). After semiconservative replication of DNA, the template strand is used to correct mismatches resulting from replication errors such as triplet repeat expansions occurring in the nascent strand. Identification of the correct sequence by repair protein relies on methyl signals that are in a transient state of hemimethylation only present on the template strand. Different mechanisms of triplet repeat expansion, which is likely to occur predominantly at DNA replication, have been proposed (Section I.B). All these would, however, only result in mutational expansion of an allele if the template is corrected to the nascent strand.

Some expanded CGG triplet repeats are hypermethylated. This is usually the case in affected fragile X patients with full mutations. Due to the presence of multiple methyl signals, strand-specific mismatch repair will be very effective. This probably explains the high degree of mitotic stability of these expansions. Other expanded triplet repeats (fragile X premutations, expansions of triplet repeats not containing CpG dinucleotides) are likely to be unmethylated. If unmethylated repeats are sufficiently large, no methyl signal will be found in appropriate distance to a heteroduplex. Then template correction will frequently occur due to misdirection of DNA mismatch repair and will result in somatic instability of the expanded unmethylated sequence.

Further evidence for a causal relation between repeat stability and DNA methylation has been obtained in a recent study of methylation mosaics in fragile X syndrome [104]. The characteristics of unstable expansion, i.e., fuzzy bands or smears of signals on Southern analysis (Fig. 35-5), were strongly associated with a coincidence of a repeat size of more than 130 triplet units and absence of methylation from the *Eag*I site. The involvement of postreplicative DNA mismatch repair in the stabilization of repeat sequences is also evidenced by previous findings that instability of tandem repeats including repetitive trinucleotide sequences occurs in a variety of human tumors and is in many cell lines associated with altered DNA mismatch repair proteins [137–144]. This led to the conclusion that most of the mutated alleles occurring at cell proliferation resulted from mistakes made during DNA replication. A possible mechanism widely recognized as a source of errors at replication of tandem repeats is DNA slippage [76]. Considering this mechanism, different models have been proposed by which large-scale expansions of triplet repeat tracts to disease causing mutations could occur. These models were developed to explain the physical features of repeat expansion which are common among different diseases [82, 94]. However, common features do not necessarily imply that there is only one mechanism underlying repeat expansion in a given situation. We would suggest that unstable expansion will never have such a single causation but would be due to different mechanisms that are usually effective in a simultaneous manner contributing to combined effects. This view keeps a number of models in discussion although they are by themselves not sufficient to explain all the features of triplet repeat instability (Fig. 35-6).

"Simple" (i.e., non-hairpin-mediated) polymerase slippage could always be effective in giving rise to frequent but small changes of repeat sizes in either direction. If the progenitor repeat is not methylated and is not stabilized by effective methyl-directed DNA mismatch repair, simple slippage will result in mutated alleles in a normal distribution around a progenitor allele (Fig. 35-6a). Such distributions of unstable alleles were observed after cloning of cells from patients with CTG repeat expansions of myotonic dystrophy and from a high-functioning fragile X male with apparently unmethylated full mutations of the CGG repeat. Simple polymerase slippage can probably not account for large changes in repeat copy numbers and also cannot explain the general finding that expansion far exceeds the rate of contraction. These two features of repeat, which were clearly reflected at continuing proliferation of cultured human fibroblasts and somatic cell hybrids, are explained by an elegant model proposed by McMurray [82]. This is repeat expansion mediated by hairpin formation at the lagging strand (Fig. 35-1b).

DNA replication of the lagging strand is proposed to be initially blocked by double-stranded hairpin structures [145] which are formed within the template by the

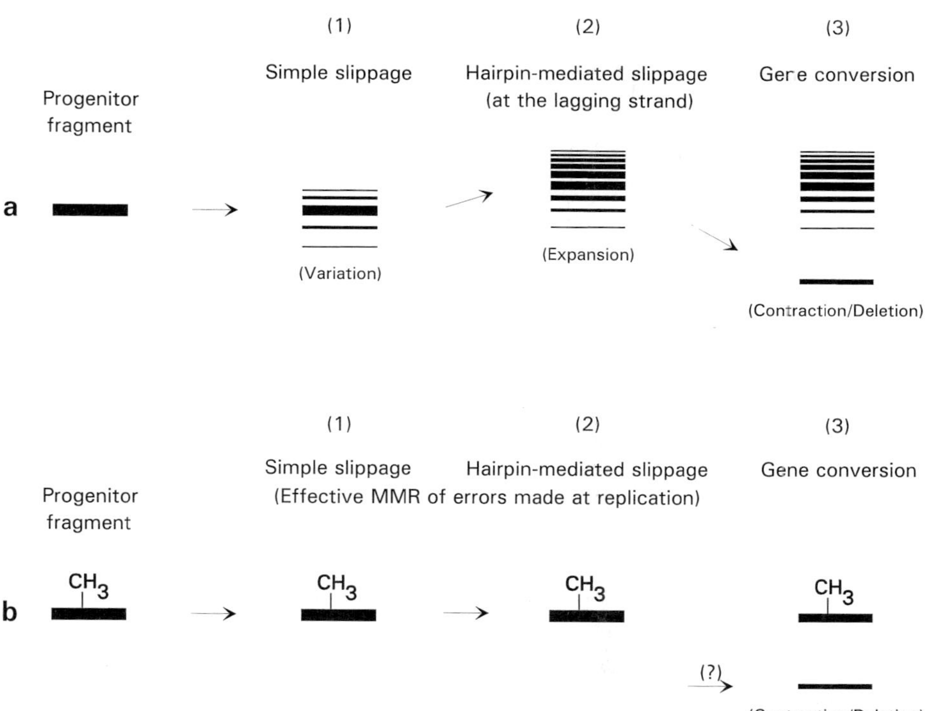

FIGURE 35-6 Patterns of repeat instability showing up on Southern analysis of cultured mammalian cells are proposed to result from different mechanisms contributing to combined effects. (a) Unmethylated progenitor alleles are postulated to always be a substrate of simple DNA slippage leading to fuzzy bands of expansions due to variation of repeat size in either direction (1). Larger alleles are more prone to variation. Above a threshold of size, hairpin-mediated slippage (probably occurring at the lagging strand) is proposed to give rise to repeat expansion at larger scales (2). Simple slippage leading to size variation of each individual allele is, however, still effective. This would result in expansion combined with a skewed distribution of alleles, and, finally, in a continuous smear of expanded fragments. Large contractions of previously expanded repeats probably also occur at cell proliferation by gene conversion or other complex mechanism leading to deletion of repeat copies and/or of sequences around the repeat (3). (b) If the expanded repeat includes numerous methyl-signals (CH_3), DNA slippage (1, 2) is postulated to be inefficient as these signals would direct the postreplicative methyl-directed DNA mismatch repair (MMR) to prevent repeat expansion which would otherwise occur by template correction. Large deletions (3) are expected to occur as these may not be prevented by MMR [127]. (?) In the rare cases observed previously, the methylation status of the progenitor repeats prior to deletion is not known.

expanded repeat and stably persist until the structure is relieved by additional DNA synthesis initiated at a single-stranded loop in the hairpin structure. This process is followed by integration of the extra DNA between the previous and the next Okazaki fragment resulting in an increase of repeat length by template correction. The length increase in copy numbers would be relatively large, being some fraction of the hairpin unit. It could, therefore, reach about half of the threshold number of triplets required to form a hairpin of sufficient stability in a single round of replication. This number depends on the sequence of the repeated triplet and was estimated to be between 32 and 48 for $(CGG)_n$, $(CTG)_n$, and $(CAG)_n$ [85]. Hairpin-mediated repeat expansion at DNA replication is consistent with genetic data in that punctuations within pure tracts of a repeat lead to stabilization of the repetitive sequence. Such punctuations are AGG interruptions in the CGG repeat of fragile X [90], CAT interruptions in the CAG repeats of SCA1 [20], CAA interruptions in the CAG repeat of SCA2 [21, 22], and CTG and CTT interruptions in the CCG repeat of FRA16A [35]. Such interruptions would reduce the stability of hairpin structures [85]. McMurray's model also predicts that expansion would be much more frequent than contraction by deletion of repeat units. The latter mutation could only occur if the block of replication by hairpin formation is not effective in that the template sequence involved in structure formation is skipped by polymerase. There is, however, some evidence that deletion of expanded repeats by skipping of a template structure could exceptionally occur.

About 1% of fragile X patients are mosaic for a full mutation and a *de novo* deletion of the CGG repeat and flanking sequences [146–150]. In some cases the *Eag*I site of the gene promoter was not involved in the deletion and was found to be unmethylated, in contrast to the full-mutation alleles of the same probands. To explain this difference, De Graaff *et al.* [147] discussed as one possibility that regression of a methylated full-mutation allele could be followed by "passive" demethylation. As, however, methylated full mutations have been shown to be mitotically stable, it is more likely that an unmethylated and unstable allele was deleted early in development and was no longer able to formation of structures inducing *de novo* methylation [83]. Sequence analysis of the deleted alleles revealed heterogeneity of deletion junctions. Data suggest, however, that prior to excision, structure formation of a single DNA strand by base-pairing between distant but complementary sequence elements, outside or inside the repeat, could have occurred [147]. The use of flanking repetitive sequences to form a hairpin structure including an unstable triplet repeat has been shown in Huntington's disease where a CGG repeat region immediately adjacent to the unstable CAG repeat forms the base of the hairpin structure but is itself not implicated in repeat expansion [82, 134].

Large contractions of previously expanded triplet repeats, not associated with either deletion of flanking sequences or somatic mosaicism, have also been reported. In myotonic dystrophy, transmission of the affected chromosome 19 from a parent to a child is relatively frequently accompanied by a decrease in length of the triplet repeat [4, 94]. Most of these descendents have the father as the transmitting parent. In the absence of evidence of either somatic or germline mosaicism, affected fathers carrying an expanded CTG repeat passed their mutated chromosomes to their progeny but with a novel, normal-sized allele differing in length from either of the fathers' alleles [62, 92, 151]. One of these cases [92] has been intensively analyzed for the segregation of alleles at numerous polymorphic marker loci in and around the DMPK gene. Evidence has been obtained that complex gene conversion is the mechanism by which the novel allele was generated. As two different stretches that had been derived from the father's normal chromosome were present within the novel allele, the likeliest explanation of this event is discontinuous gene conversion. A similar finding has been reported from a fragile X family [93]. A daughter of a female carrier inherited the fragile X premutation chromosome. The CGG repeat sequence, however, and an intragenic polymorphic marker showed the normal maternal alleles while two other intragenic markers as well as more distant markers showed the haplotype associated with the maternal mutation. The extended haplotype of this chromosome most likely resulted from a complex recombination event in the mother, with two stretches of the premutated chromosome being simultaneously transferred and replaced by sequences of the normal chromosome.

Gene conversion has been proposed to occur more frequently in the presence of a previously expanded repeat [82, 94]. Alternative structures formed by expanded triplet repeats could mediate interaction between homologous chromosomes at the S, G2, or M phase of the cell cycle and thereby facilitate complex recombination events within repetitive regions. Gene conversion of very large expanded repeats could, in some situations, result in an increased rate of mutational deletion. The behavior of unstable repeat expansions at continual mitosis in cultured cells did, however, suggest that shorter repeats may accumulate in a proliferating cell population if the mutated gene has a function at cell proliferation.

B. Selection Biases at Cell Proliferation

Another factor probably contributing to unstable behavior (changes of mutation size) of expanded repeats at proliferation of germ cells and other cells, is the function of the gene itself, which is interfered with by repeat expansion and may not be compensated for by a normal homologue or products of other genes with similar functions. In myotonic dystrophy the largest expansions are found in muscle [56, 67, 100, 102]. Somatic mosaicism in brain of HD patients has been proposed to result during progressive neuronal degeneration and gliosis [68]. The function of an expanded gene in germ cells may influence stability and parental sex biases of disease transmission.

Regardless of repeat expansion, the FMR1 gene of fragile X syndrome is transcribed if the gene promoter is not methylated [97, 152]. Translation of the gene product (FMRP) is, however, markedly reduced in the presence of larger repeats, probably because of an impediment to the linear migration of the 40 S ribosomal subunit along the 5′ untranslated mRNA sequence by structure formation in the expanded repeat [112]. According to recent evidence, FMRP plays a role in cell proliferation [114–116] and, most likely, functions as an RNA transporter involved in translational regulation [117].

Changes upon continual *in vitro* proliferation were observed in a heterogeneous population of fibroblasts showing a smear of apparently unmethylated and mitotically unstable expansions in the FMR1 gene associated with significant expression of FMRP in the majority of these cells. While cells with smaller repeat sizes were retained, cells with larger expansions were apparently lost very quickly from the population with the signals of the largest expansions being the first to disappear from the mutation pattern (Fig. 35-3b). This unstable behavior of the cell population may reflect a selection disadvantage of cells with larger expansions. In view of the particular functions of FMRP at cell proliferation and protein translation, this disadvantage could be a direct consequence of reduced FMRP production but would more probably result from a decrease in the overall rate of translation. Such a decrease could result if the expanded and structurally altered mRNA molecules are competitively bound to the ribosomes but cannot be translated and have a dominant negative effect on cell function.

Factors giving rise to selection biases in the presence of unstable triplet repeats at cell proliferation *in vivo* might well be responsible for hitherto unexplained processes of "contractions" of previously expanded sequences or of "replacements" of larger expansions by smaller ones. For instance, it would be worthwhile to investigate whether the sex difference of repeat expansion to full fragile X mutation could be related to a selective loss of unmethylated full mutations after prolonged proliferation of spermatogonia in the absence of a normal active homologue.

Acknowledgments

The authors are supported by grants from the Deutsche Forschungsgemeinschaft (DFG) and from the University Hospital of Ulm, Germany.

References

1. Richards, R. I., and Sutherland, G. R. (1992). Dynamic mutations: a new class of mutations causing human disease. *Cell* **70**, 709–712.
2. Mandel, J. L. (1994). Trinucleotide disease on the rise. *Nature Genet.* **7,** 453–455.
3. Willems, P. J. (1994). Dynamic mutations hit double figures. *Nature Genet.* **8,** 213–215.
4. Wieringa, B. (1994). Commentary: myotonic dystrophy reviewed: back to the future? *Hum. Mol. Genet.* **3,** 1–7.
5. Zoghbi, H. Y. (1996). The expanding world of ataxins. *Nature Genet.* **14,** 237–238.
6. La Spada, A. R., Wilson, E. M., Lubahn, D. B., Harding, A. E., and Fischbeck, K. H. (1991). Androgen receptor gene mutations in X-linked spinal and bulbar muscular atrophy. *Nature* **352,** 77–79.
7. La Spada, A. R., Roling, D. B., Harding, A. E., Warner, C. L., Spiegel, R., Hausmanowa-Petrusewicz, I., Yee, W. C., and Fischbeck, K. H. (1992). Meiotic stability and genotype-phenotype correlation of the trinucleotide repeat in X-linked spinal and bulbar muscular atrophy. *Nature Genet.* **2,** 301–304.
8. Oberlé, I., Rousseau, F., Heitz, D., Kretz, C., Devys, D., Hanauer, A., Boué, J., Bertheas, M., and Mandel, J. L. (1991). Amazing instability of a 550 bp DNA segment and abnormal methylation in fragile X syndrome. *Science* **252,** 1097–1102.
9. Verkerk, A. J. M. H., Pieretti, M., Sutcliffe, J. S., Fu, Y. H., Kuhl, D. P. A., Pizzuti, A., Reiner, O., Richards, S., Victoria, M. F., Zhang, F., Eussen, B. W., Van Ommen, G. J. B., Blonden, L. A. J., Riggins, G. J., Chastain, J. L., Kunst, C. G., Galjaard, H., Caskey, C. T., Nelson, D. L., Oostra, B. A., and Warren, S. T. (1991). Identification of a gene (FMR-1) containing a CGG repeat coincident with a breakpoint cluster region exhibiting length variation in fragile X syndrome. *Cell* **65,** 905–914.
10. Fu, Y. H., Kuhl, D. P. A., Pizzuti, A., Pieretti, M., Sutcliffe, J. S., Richards, S., Verkerk, A. J. M. H., Holden, J. J. A., Fenwick, R. G. Jr., Warren, S. T., Oostra, B. A., Nelson, D. L., and Caskey, C. T. (1991). Variation of the CGG repeat at the fragile X site results in genetic instability: resolution of the Sherman paradox. *Cell* **67,** 1047–1058.
11. Kremer, E. J., Yu, S., Pritchard, M., Nagajara, R., Heitz, D., Lynch, M., Baker, E., Hyland, V. J., Little, R. D., Wada, M., Toniolo, D., Vincent, A., Rousseau, F., Schlessinger, D., Sutherland, G. R., and Richards, R. J. (1991). Isolation of a human DNA sequence which spans the fragile X. *Am. J. Hum. Genet.* **49,** 656–661.
12. Yu, S., Pritchard, M., Kremer, E., Lynch, M., Nancarrow, J., Baker, E., Holman, K., Mulley, J. C., Warren, S. T., Schlessinger, D., Sutherland, G. R., and Richards, R. I. (1991). Fragile X genotype characterized by an unstable region of DNA. *Science* **252,** 1179–1181.
13. Aslanidis, C., Jansen, G., Amemiya, C., Shutler, G., Mahadevan, M., Tsilfidis, C., Chen, C., Alleman, J., Wormskamp, N. G. M., Vooijs, M., Buxton, J., Johnson, K., Smeets, H. J. M., Lennon, G. G., Carrano, A. V., Korneluk, R. G., Wieringa, B., and De Jong, P. J. (1992). Cloning of the essential myotonic dystrophy region and mapping of the putative defect. *Nature* **355,** 548–551.
14. Brook, J. D., McCurrach, M. E., Harley, H. G., Buckler, A. J., Church, D., Aburatani, H., Hunter, K., Stanton, V. P., Thirion, J. P., Hodson, T., Sohn, R., Lemelman, B., Snell, R. G., Rundle, S. A., Crow, S., Davies, J., Shelbourne, P., Buxton, J., Jones, C., Juvonen, V., Johnson, K., Harper, P. S., Shaw, D. J., and Housman, D. E. (1992). Molecular basis of myotonic dystrophy: expansion of a trinucleotide (CTG) repeat at the 3′ end of a transcript encoding a protein kinase family member. *Cell* **68,** 799–808.
15. Buxton, J., Shelbourne, P., Davies, J., Jones, C., Van Tongeren, T., Aslanidis, C., De Jong, P., Jansen, G., Anvret, M., Riley, B., Williamson, R., and Johnson, K. (1992). Detection of an unstable fragment of DNA specific to individuals with myotonic dystrophy. *Nature* **355,** 547–548.
16. Fu, Y. H., Pizzutti, A., Fenwick, R. G., Jr., King, J., Rajnarayan, S., Dunne, P. W., Dubel, J., Nasser, G. A., Ashizawa, T., De Jong, P., Wieringa, B., Korneluk, R., Perryman, M. B., Epstein, H. F., and Caskey, C. T. (1992). An unstable triplet repeat in a gene related to the myotonic muscular dystrophy. *Science* **255,** 1256–1258.
17. Harley, H. G., Brook, J. D., Rundle, S. A., Crow, S., Reardon, W., Buckler, A. J., Harper, P. S., Housman, D. E., and Shaw, D. J. (1992). Expansion of an unstable DNA region and phenotypic variation in myotonic dystrophy. *Nature* **355,** 545–546.
18. The Huntington's Disease Collaborative Research Group (1993). A novel gene containing a trinucleotide repeat that is expanded and unstable on Huntington's disease chromosomes. *Cell* **72,** 971–983.
19. Orr, H. T., Chung, M., Banfi, S., Kwiatkowski, T. J., Jr., Servadio, A., Beaudet, A. L., McCall, A. E., Duvick, L. A., Ranum,

L. P. W., and Zoghbi, H. Y. (1993). Expansion of an unstable trinucleotide CAG repeat in spinocerebellar ataxia type I. *Nature Genet.* **4,** 221–226.

20. Chung, M., Ranum, L. P. W., Duvick, L. A., Servadio, A., Zoghbi, H. Y., and Orr, H. T. (1993). Evidence for a mechanism predisposing intergenerational CAG repeat instability in spinocerebellar ataxia type I. *Nature Genet.* **5,** 254–258.
21. Imbert, G., Saudou, F., Yvert, G., Devys, D., Trottier, Y., Garnier, J. M., Weber, C., Mandel, J. L., Cancel, G., Abbas, N., Dürr, A., Didierjean, O., Stevanin, G., Agid, Y., and Brice, A. (1996). Cloning of the gene for spinocerebellar ataxia 2 reveals a locus with high sensitivity to expanded CAG/glutamine repeats. *Nature Genet.* **14,** 285–291.
22. Pulst, S. M., Nechiporuk, A., Nechiporuk, T., Gispert, S., Chen, X. N., Lopes-Cendes, I., Pearlman, S., Starkman, S., Orozco-Diaz, G., Lunkes, A., Dejong, P., Rouleau, G. A., Auburger, G., Korenberg, J. R., Figueroa, C., and Sahba, S. (1996). Moderate expansion of a normally biallelic trinucleotaide repeat in spinocerebellar ataxia type 2. *Nature Genet.* **14,** 269–276.
23. Kawaguchi, Y., Okamoto, T., Taniwaki, M., Aizawa, M., Inoue, M., Katayama, S., Kawakami, H., Nakamura, S., Mishimura, M., Akiguchi, I., Kimura, J., Marumiya, S., and Kakizuka, A. (1994). CAG expansions in a novel gene for Machado-Joseph disease at chromosome 14q32.1. *Nature Genet.* **8,** 221–227.
24. Nagafuchi, S., Yanagisawa, H., Sato, K., Shirayama, T., Ohsaki, E., Bundo, M., Takeda, T., Tadokoro, K., Kondo, I., Murayama, N., Tanaka, Y., Kikushima, H., Umino, K., Kurosawa, H., Furukawa, T., Nihei, K., Inoue, T., Sano, A., Komure, O., Takahashi, M., Yoshizawa, T., Kanawawa, I., and Yamada, M. (1994). Dentatorubral and pallidoluysian atrophy expansion of an unstable CAG trinucleotide on chromosome 12p. *Nature Genet.* **6,** 14–18.
25. Koide, R., Ikeuchi, T., Onodera, O., Tanaka, H., Igarashi, S., Endo, K., Takahashi, H., Kondo, R., Ishikawa, A., Hayashi, T., Saito, M., Tomodea, A., Miike, T., Naito, H., Ikuta, F., and Tsuji, S. (1994). Unstable expansion of CAG repeat in hereditary dentatorubral-pallidoluysian atrophy (DRPLA). *Nature Genet.* **6,** 9–13.
26. Campuzano, V., Montermini, L., Moltò, M. D., Pianese, L., Cossée, M., Cavalcanti, F., Monros, E., Rodius, F., Duclos, F., Monticelli, A., Zara, F., Cañizares, J., Koutnikova, H., Bidichandani, S. I., Gellera, C., Brice, A., Trouillas, P., De Michele, G., Filla, A., De Frutos, R., Palau, F., Patel, P. I., Di Donato, S., Mandel, J. L., Cocozza, S., Koenig, M., and Pandolfo, M. (1996). Friedreich's ataxia: autosomal recessive disease caused by an intronic GAA triplet repeat expansion. *Science* **271,** 1423–1427.
27. Knight, S. J. L., Flannery, A. V., Hirst, M. C., Campbell, L., Christodoulou, Z., Phelps, S. R., Pointon, J., Middleton-Price, H. R., Barnicoat, A., Pembrey, M. E., Holland, J., Oostra, B. A., Bobrow, M., and Davies, K. E. (1993). Trinucleotide repeat amplification and hypermethylation of a CpG island in FRAXE mental retardation. *Cell* **74,** 127–134.
28. Knight, S. J. L., Voelckel, M. A., Hirst, M. C., Flannery, A. V., Moncla, A., and Davies, K. E. (1994). Triplet repeat expansion at the FRAXE locus and X-linked mild mental handicap. *Am. J. Hum. Genet.* **55,** 81–86.
29. Chakrabarti, L., Knight, S. J. L., Flannery, A. V., and Davies, K. E. (1996). A candidate gene for mild mental handicap at the FRAXE fragile site. *Hum. Mol. Genet.* **5,** 275–282.
30. Gécz, J., Gedeon, A. K., Sutherland, G. R., and Mulley, J. C. (1996). Identification of the gene FMR2, associated with FRAXE mental retardation. *Nature Genet.* **13,** 105–108.
31. Gu, Y., Shen, Y., Gibbs, R. A., and Nelson, D. L. (1996). Identification of FMR2, a novel gene associated with the FRAXE CCG repeat and CpG island. *Nature Genet.* **13,** 109–113.
32. Jones, C., Penny, L., Mattina, T., Yu, S., Baker, E., Voullaire, L., Landon, W. Y., Sutherland, G. R., Richards, R. I., and Tunnacliffe, A. (1995). Association of a chromosome deletion syndrome with a fragile site within the proto-oncogene CBL2. *Nature* **376,** 145–149.
33. Hirst, M. C., Barnicoat, A., Flynn, G., Wang, Q., Daker, M., Buckle, V. J., Davies, K. E., and Bobrow, M. (1993). The identification of a third fragile site, FRAXF, in Xq27-28 distal to both FRAXA and FRAXE. *Hum. Mol. Genet.* **2,** 197–200.
34. Parrish, J. E., Oostra, B. A., Verkerk, A. J. M. H., Richards, C. S., Reynolds, J., Spikes, A. S., Shaffer, L. G., and Nelson, D. L. (1994). Isolation of a GCC repeat showing expansion in FRAXF, a fragile site distal to FRAXA and FRAXE. *Nature Genet.* **8,** 229–235.
35. Nancarrow, J. K., Kremer, E., Holman, K., Eyre, H., Dogget, N. A., Le Paslier, D., Callen, D. F., Sutherland, G. R., and Richards, R. I. (1994). Implications of FRA16A structure for the mechanism of chromosomal fragile site genesis. *Science* **264,** 1938–1941.
36. Biancalana, V., Serville, F., Pommier, J., Julien, J., Hanauer, A., and Mandel, J. L. (1992). Moderate instability of the trinucleotide repeat in spino bulbar muscular atrophy. *Hum. Mol. Genet.* **1,** 255–258.
37. Rousseau, F., Heitz, D., Biancalana, V., Blumenfeld, S., Kretz, C., Boué, J., Tommerup, N., Van der Hagen, C., DeLozier-Blanchet, C., Croquette, M. F., Gilgenkrantz, S., Jalbert, P., Voelckel, M. A., Oberlé, I., and Mandel, J. L. (1991). Direct diagnosis by DNA analysis of the fragile X syndrome of mental retardation. *N. Engl. J. Med.* **325,** 1673–1681.
38. Reiss, A. L., Kazazian, H. H., Krebs, C. M., McAughan, A., Boehm, C. D., Abrams, M. T., and Nelson, D. L. (1994). Frequency and stability of the fragile X premutation. *Hum. Mol. Genet.* **3,** 393–398.
39. Goldberg, Y. P., Kremer, B., Andrew, S. E., Theilmann, J., Graham, R. K., Squitieri, F., Telenius, H., Adam, S., Sajoo, A., Starr, E., Heiberg, A., Wolff, G., and Hayden, M. R. (1993). Molecular analysis of new mutations for Huntington's disease: intermediate alleles and sex of origin effects. *Nature Genet.* **5,** 174–179.
40. Epplen, C., Epplen, J. T., Frank, G., Miterski, B., Santos, E. J. M., and Schöls, L. (1997). Differential stability of the $(GAA)_n$ tract in the Friedreich ataxia (STM7) gene. *Hum. Genet.* [in press]
41. Duyao, M., Ambrose, C., Myers, R., Novelletto, A., Persichetti, F., Frontale, M., Folstein, S., Ross, C., Franz, M., Abbott, M., Fray, J., Conneally, P., Young, A., Penney, J., Hollingsworth, Z., Shoulson, I., Lazzarini, A., Falek, A., Koroshetz, W., Sax, D., Bird, E., Vonsattel, J., Bonilla, E., Alvir., J., Bickham, C., Cha, J. H., Dure, L., Gomez, F., Ramos, M., Sanchez-Ramos, J., Snodgrass, S., De Young, M., Wexler, N., Moscowitz, C., Penchaszadeh, G., MacFarlane, H., Anderson, M., Jenkins, B., Srinidhi, J., Barnes, G., Gusella, J., and MacDonald, M. (1993). Trinucleotide repeat length instability and age of onset in Huntington's disease. *Nature Genet.* **4,** 387–392.
42. Telenius, H., Kremer, B., Goldberg, Y. P., Theilmann, J., Andrew, S. E., Zesler, J., Adam, S., Greenberg, C., Ives, E. J., Clarke, L. A., and Hayden, M. R. (1994). Somatic and gonadal mosaicism of the Huntington disease gene CAG repeat in brain and sperm. *Nature Genet.* **6,** 409–414.
43. Goldberg, Y. P., McMurray, C. T., Zeisler, J., Almqvist, E., Sillence, D., Richards, F., Gacy, A. M., Buchanan, J., Telenius, H., and Hayden, M. (1995). Increased instability of intermediate alleles in families with sporadic Huntington's disease compared

to similar sized intermediate alleles in the general population. *Hum. Mol. Genet.* **4,** 1911–1918.

44. Kremer, B. , Theilmann, J., Almqvist, E., Spence, N., Telenius, H., Goldberg, Y. P., and Hayden, M. R. (1995). Sex dependent mechanisms for expansions and contractions of the CAG repeat on affected Huntington disease chromosomes. *Am. J. Hum. Genet.* **57,** 343–350.
45. Igarashi, S., Takiyama, Y., Cancel, G., Rogaeva, E. A., Sasaki, H., Wakisaka, A., Zhou, Y. X., Takano, H., Endo, K., Sanpei, K., Oyake, M., Tanaka, H., Stevanin, G., Abbas, N., Dürr, A., Rogaev, E. I., Sherrington, R., Tsuda, T., Ikeda, M., Cassa, E., Nishizawa, M., Benomar, A., Julien, J., Weissenbach, J., Wang, G. X., Agid, Y., St. George-Hyslop, P., Brice, A., and Tsuij, S. (1995). Intergenerational instability of the CAG repeat of the gene for Machado-Joseph disease (MJD1) is affected by the genotype of the normal chromosome: implications for the molecular mechanisms of the instability of the CAG repeat. *Hum. Mol. Genet.* **5,** 923–932.
46. Ikeuchi, T., Koide, R., Tanaka, H., Onodera, O., Igarashi, S., Takanashi, H., Kondo, R., Ishikawa, A., Tomoda, A., Miike, T., Sato, K., Ihara, Y., Haybara, T., Isa, F., Tanabe, H., Tokiguchi, S., Hayashi, M., Shimizu, N., Ikuta, F., Naito, H., and Tsuij, S. (1995). Dentatorubral-pallidoluysian atrophy (DRPLA): clinical features are closely related to unstable expansions of trinucleotide (CAG) repeat. *Ann. Neurol.* **37,** 769–775.
47. Lavedan, C., Hofmann-Radvanyi, H., Shelbourne, P., Rabes, J. P., Duros, C., Savoy, D., Dehaupas, I., Luce, S., Johnson, K. and Junien, C. (1993). Myotonic dystrophy: size- and sex-dependent dynamics of CTG meiotic instability, and somatic mosaicism. *Am. J. Hum. Genet.* **52,** 875–883.
48. Sherman, S. L., Morton, N. E., Jacobs, P. A., and Turner, G. (1984). The marker (X) syndrome: a cytogenetic and genetic analysis. *Ann. Hum. Genet.* **48,** 21–37.
49. Sherman, S. L., Jacobs, P. A., Morton, N. E., Froster-Iskenius, U., Howard-Peebles, P. N., Nielsen, K. B., Partington, M. W., Sutherland, G. R., Turner, G., and Watson, M. (1985). Further segregation analysis of the fragile (X) syndrome with special reference to transmitting males. *Hum. Genet.* **69,** 289–299.
50. Imbert, G., Kretz, C., Johnson, K., and Mandel, J. L. (1993). Origin of the expansion mutation in myotonic dystrophy. *Nature Genet.* **4,** 72–76.
51. Devys, D., Biancalana, V., Rousseau, F., Baul, J., Mandel, J. L., and Oberlé, I. (1992). Analysis of full fragile X mutations in fetal tissues and monozygotic twins indicate that abnormal methylation and somatic heterogeneity are established early in development. *Am. J. Med. Genet.* **43,** 208–216.
52. Wöhrle, D., Hirst, M. C., Kennerknecht, I., Davies, K. E., and Steinbach, P. (1992). Genotype mosaicism in fragile X fetal tissues. *Hum. Genet.* **89,** 114–116.
53. Wöhrle, D., Hennig, I., Vogel, W., and Steinbach, P. (1993). Mitotic stability of fragile X mutations in differentiated cells indicates early post-conceptional trinucleotide repeat expansion. *Nature Genet.* **4,** 140–142.
54. Cancel, G., Abbas, N., Stevanin, G., Dürr, A., Chneiweiss, H., Néri, C., Duyckaerts, C., Penet, C., Cann, H. M., Agid, Y., and Brice, A. (1995). Marked phenotypic heterogeneity associated with expansion of a CAG repeat sequence at the spinocerebellar ataxia 3/Machado-Joseph disease locus. *Am. J. Hum. Genet.* **57,** 809–816.
55. Ueno, S., Kondoh, K., Kotani, Y., Komure, O., Kuno, S., Kawai, J., Hazama, F., and Sano, A. (1995). Somatic mosaicism of CAG repeat in dentatorubral-pallidoluysian atrophy (DRPLA). *Hum. Mol. Genet.* **4,** 663–666.
56. Anvret, M., Ahlberg, G., Grandell, U., Hedberg, B., Johnson, K., and Edström, L. (1993). Larger expansions of the CTG repeat in muscle compared to lymphocytes from patients with myotonic dystrophy. *Hum. Mol. Genet.* **2,** 1397–1400.
57. Wöhrle, D., Kennerknecht, I., Wolf, M., Enders, H., Schwemmle, S., and Steinbach, P. (1995). Heterogeneity of DM kinase repeat expansion in different fetal tissues and further expansion during cell proliferation in vitro: evidence for a causal involvement of methyl-directed DNA mismatch repair in triplet repeat stability. *Hum. Mol. Genet.* **4,** 1147–1153.
58. Takano, H., Onodera, O., Takahashi, H., Igarashi, S., Yamada, M., Oyake, M., Ikeuchi, T., Koide, R., Tanaka, H., Iwabuchi, K., and Tsuij, S. (1996). Somatic mosaicism of expanded CAG repeats in brains of patients with dentatorubral-pallidoluysian atrophy: cellular population-dependent dynamics of mitotic instability. *Am. J. Hum. Genet.* **58,** 1212–1222.
59. Zhang, L., Leeflang, E. P., Yu, J., and Arnheim, N. (1994). Studying human mutations by sperm typing: instability of CAG trinucleotide repeats in the human androgen receptor gene. *Nature Genet.* **7,** 531–535.
60. Reyniers, E., Vits, L., De Boulle, K., Van Roy, B., Van Velzen, D., De Graaff, E., Verkerk, A. J. M. H., Jorens, H. Z. J., Darby, J. K., Oostra, B., and Willems, P. J. (1993). The full mutation in the FMR-1 gene of male fragile X patients is absent in their sperm. *Nature Genet.* **4,** 143–146.
61. Mornet, E., Chateau, C., Hirst, M. C., Thepot, F., Taillandier, A., Cibois, O., and Serre, J. L. (1996). Analysis of germline variations at the FMR1 CGG repeat shows variation in the normal-premutation borderline range. *Hum. Mol. Genet.* **5,** 821–825.
62. Brunner, H. G., Jansen, G., Nillesen, W., Nelen, M. R., De Die, C. E. M., Höweler, C. J., Van Oost, B. A., Wieringa, B., Ropers, H. H., and Smeets, H. J. M. (1993). Brief report: reverse mutation in myotonic dystrophy. *N. Engl. J. Med.* **328,** 476–480.
63. Giordano, M., De Angelis, M. S., Mutani, R., and Richiardi, P. M. (1994). Origin of a regressed myotonic dystrophy allele. *J. Med. Genet.* **31,** 130–132.
64. Telenius, H., Almqvist, E., Kremer, B., Spence, N., Squitieri, F., Nichol, K., Grandell, U., Starr, E., Benjamin, C., Castaldo, I., Calabrese, O., Anvret, M., Goldberg, Y. P., and Hayden, M. R. (1995). Somatic mosaicism in sperm associated with intergenerational $(CAG)_n$ changes in Huntington disease. *Hum. Mol. Genet.* **4,** 189–195.
65. Leeflang, E. P., Zhang, L., Taveré, S., Hubert, R., Srinidhi, J., MacDonald, M. E., Myers, R. H., De Young, M., Wexler, N. S., Gusella, J. F., and Arnheim, N. (1995). Single sperm analysis of the trinucleotide repeats in the Huntington's disease gene: quantification of the mutation frequency spectrum. *Hum. Mol. Genet.* **4,** 1519–1526.
66. Wong, L. J. C., Ashizawa, T., Monckton, D. G., Caskey, C. T., and Richards, C. S. (1995). Somatic heterogeneity of the CTG repeat in myotonic dystrophy is age and size dependent. *Am. J. Hum. Genet.* **56,** 114–122.
67. Zatz, M., Passos-Bueno, M. R., Cerqueira, A., Marie, S. K., Vainzof, M., and Pavanello, R. C. M. (1995). Analysis of the CTG repeat in skeletal muscle of young and adult myotonic dystrophy patients: when does the expansion occur. *Hum. Mol. Genet.* **4,** 401–406.
68. Benitez, J., Robledo, M., Ramos, C., Ayuso, C., Astarloa, R., Yébenes, J. G., and Brambati, B. (1995). Somatic stability in chorionic villi samples and other Huntington fetal tissues. *Hum. Genet.* **96,** 229–232.
69. Smith, G. P. (1976). Evolution of repeated DNA sequences by unequal crossover. *Science* **191,** 528–535

70. Nussbaum, R. L., Airhart, S. D., and Ledbetter, D. H. (1986). Recombination and amplification of pyrimidine-rich sequences may be responsible for initiation and progression of the Xq27 fragile site: an hypothesis. *Am. J. Med. Genet.* **23,** 715–722.
71. Tanaka, F., Doyu, M., Ito, Y., Matsumoto, M., Mitsuma, T., Abe, K., Aoki, M., Itoyama, Y., Fischbeck, K. H., and Sobue, G. (1996). Founder effect in spinal and bulbar muscular atrophy (SBMA). *Hum. Mol. Genet.* **5,** 1253–1257.
72. Richards, R. I., Holman, K., Friend, K., Kremer, E., Hillen, D., Staples, A., Brown, W. T., Goonewardena, P., Tarleton, J., Schwartz, C., and Sutherland, G. R. (1992). Evidence of founder chromosomes in fragile X syndrome. *Nature Genet.* **1,** 257–260.
73. Oudet, C., Von Koskull, H., Nordström, A. M., Peippo, M., and Mandel, J. L. (1993). Striking founder effect for the fragile X syndrome in Finland. *Eur. J. Hum. Genet.* **1,** 181–189.
74. Rubinsztein, D. C., Leggo, J., Goodburn, S., Barton, D. E., and Ferguson-Smith, M. A. (1995). Haplotype analysis of the Δ 2642 and $(CAG)_n$ polymorphisms in the Huntington's disease (HD) gene provides an explanation for an apparent "founder" HD haplotype. *Hum. Mol. Genet.* **4,** 203–206.
75. Yanagisawa, H., Fujii, K., Nagafuchi, S., Nakahori, Y., Nakagome, Y., Akane, A., Nakamura, M., Sano, A., Komure, O., Kondo, I., Jin, D. K., Sorensen, S. A., Potter, N. T., Young, S. R., Nakamura, K., Nukina, N., Nagao, Y., Tadokoro, K., Okuyama, T., Miyashita, T., Inoue, T., Kanazawa, I., and Yamada, M. (1996). A unique origin and multistep process for the generation of expanded DRPLA triplet repeats. *Hum. Mol. Genet.* **5,** 373–379.
76. Levinson, G., and Gutman, G. A. (1987). Slipped-strand mispairing: a major mechanism for DNA sequence evolution. *Mol. Biol. Evol.* **4,** 203–221.
77. Sinden, R. R., and Wells, R. D. (1992). DNA structure, mutations, and human genetic disease. *Curr. Opin. Biotechnol.* **3,** 612–622.
78. Kunkel, T. A. (1993). Slippery DNA and disease. *Nature* **365,** 207–208.
79. Modrich, P. (1991). Mechanisms and biological effects of mismatch repair. *Annu. Rev. Genet.* **25,** 229–253.
80. Modrich, P. (1994). Mismatch repair, genetic instability, and cancer. *Science* **266,** 1959–1960.
81. Wells, R. D. (1996). Molecular basis of genetic instability of triplet repeats. *J. Biol. Chem.* **271,** 2875–2878.
82. McMurray, C. T. (1995). Mechanisms of DNA expansion. *Chromosoma* **104,** 2–13.
83. Smith, S. S., Laayoun, A., Lingeman, R. G., Baker, D. J., and Riley, J. (1994). Hypermethylation of telemere-like foldbacks at codon 12 of the human c-Ha-ras gene and the trinucleotide repeat of the FMR-1 gene of fragile X. *J. Mol. Biol.* **243,** 143–151.
84. Chen, X., Mariappan, S. V. S., Catasti, P., Ratcliff, R., Moyzis, R. K., Laayoun, A., Smith, S. S., Bradbury, E. M., and Gupta, G. (1995). Hairpins are formed by the single DNA strands of the fragile X triplet repeats: structure and biological implications. *Proc. Natl. Acad. Sci. USA* **92,** 5199–5203.
85. Gacy, A. M., Goellner, G., Juranic, N., Macura, S., and McMurray, C. T. (1995). Trinucleotide repeats that expand in human disease form hairpin structures in vitro. *Cell* **81,** 533–540.
86. Mitas, M., Yu, A., Dill, J., and Haworth, I. S. (1995). The trinucleotide repeat sequence $d(CGG)_{15}$ forms a heat-stable hairpin containing $G^{syn} \cdot G^{anti}$ base pairs. *Biochemistry* **34,** 12803–12811.
87. Smith, G. K., Jie, J., Fox, E. G., and Gao, X. (1995). DNA CTG triplet repeats involved in dynamic mutations of neurologically related gene sequences form stable duplexes. *Nucleic Acids Res.* **23,** 4303–4311.
88. Pearson, C. E., and Sinden, R. R. (1996). Alternative structures in duplex DNA formed within the trinucleotide repeats of the myotonic dystrophy and fragile X loci. *Biochemistry* **35,** 5041–5053.
89. Petruska, J., Arnheim, N., and Goodman, M. F. (1996). Stability of intrastrand hairpin structures formed by the CAG/CTG class of DNA triplet repeats associated with neurological diseases. *Nucleic Acids Res.* **24,** 1992–1998.
90. Snow, K., Tester, D. J., Kruckeberg, K. E., Schaid, D. J., and Thibodeau, S. N. (1994). Sequence analysis of the fragile X trinucleotide repeat: implications for the origin of the fragile X mutation. *Hum. Mol. Genet.* **3,** 1543–1551.
91. Hirst, M. C., Grewal, P. K., and Davies, K. E. (1994). Precursor arrays for triplet repeat expansion at the fragile X locus. *Hum Mol Genet* **3,** 1553–1560.
92. O'Hoy, K. L., Tsilfidis, C., Mahadevan, M. S., Neville, C. E., Barcelo, J., Hunter, A. G. W., and Korneluk, R. G. (1993). Reduction in size of the myotonic dystrophy trinucleotide repeat mutation during transmission. *Science* **259,** 809–811.
93. Van den Ouweland, A. M. W., Deelen, W. H., Kunst, C. B., Giovanucci Uzielli, M. L., Nelson, D. L., Warren, S. T., Oostra, B. A., and Halley, D. J. J. (1994). Loss of mutation at the FMR1 locus through multiple exchanges between maternal X chromosomes. *Hum. Mol. Genet.* **3,** 1823–1827.
94. Jansen, G., Willems, P., Coerwinkel, M., Nillesen, W., Smeets, H., Vits, L., Höweler, C., Brunner, H., and Wieringa, B. (1994). Gonosomal mosaicism in myotonic dystrophy patients: involvement of mitotic events in $(CTG)_n$ repeat variation and selection against extreme expansion in sperm. *Am. J. Hum. Genet.* **54,** 575–585.
95. Gordenin, D. A., Kunkel, T. A., and Resnick, M. A. (1997). Repeat expansion—all in a flap? *Nature Genet.* **16,** 116–118.
96. Hornstra, I. K., Nelson, D. L., Warren, S. T., and Yang, T. P. (1993). High resolution methylation analysis of the FMR1 gene trinucleotide repeat region in fragile X syndrome. *Hum. Mol. Genet.* **2,** 1659–1265.
97. Schwemmle, S., De Graaff, E., Deissler, H., Gläser, D., Wöhrle, D., Kennerknecht, I., Just, W., Oostra, B. A., Doerfler, W., Vogel, W., and Steinbach, P. (1997). Characterization of FMR1 promoter elements by in vivo footprinting analysis. *Am. J. Hum. Genet.* **60,** 1354–1362.
98. Mariappan, S. V. S., Catasti, P., Chen, X., Ratcliff, R., Moyzis, R. K., Bradbury, E. M., and Gupta, G. (1996). Solution structures of the individual single strands of the fragile X DNA triplets $(GCC)_n \cdot (GGC)_n$. *Nucleic Acids Res.* **24,** 784–792.
99. Hansen, R. S., Gartler, S. M., Scott, C. R., Chen, S. H., and Laird, C. D. (1992). Methylation analysis of CGG sites in the CpG island of the human FMR1 gene. *Hum. Mol. Genet.* **1,** 571–587.
100. Thornton, C. A., Johnson, K., and Moxley, R. T. (1994). Myotonic dystrophy patients have larger CTG expansions in skeletal muscle than in leukocytes. *Ann. Neurol.* **35,** 104–107.
101. Martorell, L., Johnson, K., Boucher, C. A., and Baiget, M. (1997). *Hum. Mol. Genet.* **6,** 877–880.
102. Peterlin, B., Logar, N., and Zidar, J. (1996). CTG repeat analysis in lymphocytes, muscles and fibroblasts in patients with myotonic dystrophy. *Pflügers Arch. Eur. J. Physiol.* **432** [Suppl.], R199–R200.
103. Monckton, D. G., Wong, L. J., Ashizawa. T., and Caskey, C. T. (1995). Somatic mosaicism, germline expansions, germline reversions and intergenerational reductions in myotonic dystrophy males: small pool PCR analyses. *Hum. Mol. Genet.* **4,** 1–8.
104. Wöhrle, D., Salat, U., Gläser, D., Mücke, J., Meisel-Stosiek, M., Schindler, D., Vogel, W., and Steinbach, P. (1997). Unusual patterns of mutations in high functioning fragile X males and

other cases may result from instability of expanded CGG repeats in the absence of methylation. *J. Med. Genet.* **35,** 103–111.

105. Steinbach, P., Wöhrle, D., Tariverdian, G., Kennerknecht, I., Barbi, G., Edlinger, H., Enders, H., Go;u;tz-Sothmann, M., Heilbronner, H., Hosenfeld, D., Kircheisen, R., Majewski, F., Meinecke, P., Passarge, E., Schmidt, A., Seidel, H., Wolff, G., and Zankl, M. (1993). Molecular analysis of mutations in the gene FMR-1 segregating in fragile X families. *Hum. Genet.* **92,** 491–498.
106. Loesch, D. Z., Huggins, R., Hay, D. A., Gedeon, A. K., Mulley, J. C., and Sutherland, G. R. (1993). Genotype-phenotype relationships in fragile X syndrome: a family study. *Am. J. Hum. Genet.* **53,** 1064–1073.
107. Hagerman, R. J., Hull, E. C., Safanda, J. F., Carpenter, I., Staley, L. W., O'Connor, R. A., Seydel, C., Mazzocco, M. M. M., Snow, K., Thibobeau, S. N., Kuhl, D., Nelson, D. L., Caskey, C. T., and Taylor, A. K. (1994). High functioning fragile X males: demonstration of an unmethylated fully expanded FMR-1 mutation associated with protein expression. *Am. J. Med. Genet.* **51,** 298–308.
108. Wang, Z., Taylor, A. K., and Bridge, J. A. (1996). FMR1 fully expanded mutation with minimal methylation in a high functioning fragile X male. *J. Med. Genet.* **33,** 376–378.
109. Lachiewicz, A. M., Spiridigliozzi, G. A., McConkie-Rosell, A., Burgess, D., Feng, Y., Warren, S. T. and Tarleton, J. (1996). A fragile X male with a broad smear on Southern blot analysis representing 100–500 CGG repeats and no methylation at the EagI site of the FMR-1 gene. *Am. J. Med. Genet.* **64,** 278–282.
110. Taylor, A. K., Tassone, F. , Dyer, P. N., and Hagerman R. J. (1996). FMR1 mutation analysis in multiple tissues: dramatic differences between tissues of a high functioning but not of a typical fragile X male. Fifth International Fragile X Conference, Portland, OR.
111. Nakahori, Y., Knight, S. J., Holland, J., Schwantz, C., Roche, A., Tarleton, J., Wong, S., Flint, T. J., Froster-Iskenius, U., Bentley, D., Davies, K. E., and Hirst, M. C. (1991). Molecular heterogeneity of the fragile X syndrome. *Nucleic Acids Res.* **19,** 4355–4359.
112. Feng, Y., Zhang, F., Lokey, L. K., Chastain, J. L., Lakkis, L., Eberhart, D., and Warren, S. T. (1995). Translational suppression by trinucleotide repeat expansion at FMR1. *Science* **268,** 731–734.
113. Verheij, C., Bakker, C. E., De Graaff, E., Keulemans, J., Willemsen, R., Verkerk, A. J. M. H., Galjaard, H., Reuser, A. J. J., Hoogeveen, A. T., and Oostra, B. A. (1993). Characterization and localization of the FMR1 gene product associated with fragile X syndrome. *Nature* **363,** 722–724.
114. Devys, D., Lutz, Y., Rouyer, N., Bellocq, J. P., and Mandel, J. L. (1993). The FMR1 protein is cytoplasmic, most abundant in neurons and appears normal in carriers of fragile X premutation. *Nature Genet.* **4,** 335–340.
115. Khandjian, E. W., Fortin, A., Thibodeau, A., Tremblay, S., Côté, F., Devys, D., Mandel, J. L., and Rousseau, F. (1995). A heterogeneous set of FMR1 proteins is widely distributed in mouse tissues and is modulated in cell culture. *Hum. Mol. Genet.* **4,** 783–789.
116. Khandjian, E. W., Corbin, F., Woerly, S., and Rousseau, F. (1996). The fragile X mental retardation protein is associated with ribosomes. *Nature Genet.* **12,** 91–93.
117. Eberhart, D. E., Malter, H. E., Feng, Y., and Warren, S. T. (1996). The fragile X mental retardation protein is a ribonucleoprotein containing both nuclear localisation and nuclear export signals. *Hum. Mol. Genet.* **8,** 1089–1091.
118. Rousseau, F., Heitz, D., Tarleton, J., Macpherson, J., Malmgren, H., Dahl, N., Barnicoat, A., Mathew, C., Mornet, E., Tejada, I., Maddalena, A., Spiegel, R., Schinzel, A., Marcos, J., Schorderet, D. F., Schaap, T., Maccioni, L., Russo, S., Jacobs, P. A., Schwartz, C., and Mandel, J. L. (1994). A multicenter study on genotype-phenotype correlations in the fragile X syndrome, using direct diagnosis with probe StB12.3: the first 2,253 cases. *Am. J. Hum. Genet.* **55,** 225–237.
119. Malter, H. E., Iber, J. C., Willemsen, R., De Graaff, E., Tarleton, J. C., Leisti, J., Warren, S. T., and Oostra, B. A. (1997). Characterization of the full fragile X syndrome mutation in fetal gametes. *Nature Genet.* **15,** 165–169.
120. Bächner, D., Steinbach, P., Wöhrle, D., Just, W., Vogel, W., Hameister, H., Manca, A , and Poustka, A (1993). Enhanced Fmr-1 expression in testis. *Nature Genet.* **4,** 115–116.
121. Bächner, D., Manca, A., Steinbach, P., Wöhrle, D., Just, W., Vogel, W., Hameister, H., and Poustka, A. (1993). Enhanced expression of the murine FMR1 gene during germ cell proliferation suggests a special function in both the male and the female gonad. *Hum. Mol. Genet.* **2,** 2043–2050.
122. Edwards, J. H. (1989). Familiarity, recessivity and germline mosaicism. *Ann. Hum. Genet.* **53,** 33–47.
123. Rousseau, F., Heitz, D., Oberlé, I., and Mandel, J. L. (1991). Selection in blood cells from female carriers of the fragile X syndrome: inverse correlation between age and proportion of active X chromsomes carrying the full mutation. *J. Med. Genet.* **28,** 830–836.
124. Ashizawa, T., Monckton, D. G., Vaishnav, S., Patel, B. J., Voskova, A., and Caskey, C. T. (1996). Instability of expanded $(CTG)_n$ repeats in the mytonin protein kinase gene in cultured lymphoblastoid cell lines from patients with myotonic dystrophy. *Genomics* **36,** 47–53.
125. Willemsen, R., Mohkamsing, S., De Vries, B., Devys, D., van den Ouweland, A., Mandel, J. L., Galjaard, H., and Oostra, B. (1996). Rapid antibody test for fragile X syndrome. *Lancet* **345,** 1147–1148.
126. Kang, S., Jaworski, A., Ohshima, K., and Wells, R. D. (1995). Expansion and deletion of CTG repeats from human disease genes aredetermined by the direction of replication in E. coli. *Nature Genet.* **10,** 213–218.
127. Schweitzer, J. K., and Livingston, D. M. (1997). Destabilization of CAG trinucleotide repeat tracts by mismatch repair mutations in yeast. *Hum. Mol. Genet.* **6,** 349–355.
128. Ohshima, K., Kang, S., and Wells, R. D. (1996). CTG triplet repeats from human hereditary diseases are dominant genetic expansion products in Escherichia coli. *J. Biol. Chem.* **271,** 1853–1856.
129. Kang, S., Ohshima, K., Jaworski, A., and Wells, R. D. (1996). CTG triplet repeats from the myotonic dystrophy gene are expanded in Escherichia coli distal to the replication origin as a single large event. *J. Mol. Biol.* **258,** 542–547.
130. Strand, M., Prolla, T. A., Liskay, R. M., and Petes, T. D. (1993). Destabilization of a simple repetitive DNA in yeast by mutation affecting mismatched repair. *Nature* **365,** 274–276.
131. Maurer, D. J., O'Callaghan, B. L., and Livingston, D. M. (1996). Orientation dependence of trinucleotide CAG repeat instability in yeast. *Mol. Cell. Biol.* **16,** 6617–6622.
132. Tran, H. T., Gordenin, D. A., and Resnick, M. A. (1996). The prevention of repeat-associated deletions in Saccharomyces cerevisiae by mismatch repair depends on the size and origin of deletions. *Genetics* **143,** 1579–1587.
133. Steinbach, P., Gläser, D., Vogel, W., Wolf, M., and Schwemmle, S. (1997). The DMPK gene of severely affected myotonic dystrophy patients is hypermethylated proximal to the largely expanded CTG repeat. *Am. J. Hum. Genet.* **62.** [In press]

134. Rubinsztein, D. C., Barton, D. E., Davison, B. C. C., and Ferguson-Smith, M. A. (1993). Analysis of the Huntington gene reveals a trinucleotide-length polymorphism in the region of the gene that contains two CCG-rich stretches and a correlation between decreased age of onset of Huntington's disease and CAG repeat number. *Hum. Mol. Genet.* **2,** 1713–1715.
135. Schwemmle, S. (1998). A model of the causal relationship between expansion of a triplet repeat and methylation of an adjacent CpG island. *Am. J. Med. Genet.* [In press]
136. Wöhrle, D., Schwemmle, S., and Steinbach, P. (1996). DNA methylation and triplet repeat stability. *Am. J. Med. Genet.* **64,** 1–8.
137. Aaltonen, L. A., Peltomäki, P., Leach, F. S., Sistonen, P., Pylkkänen, L., Mecklin, J. P., Järvinen, H., Powell, S. M., Jen, J., Hamilton, S. R., Petersen, G. M., Kinzler, K. W., Vogelstein, B., and De la Chapelle, A. (1993). Clues to the pathogenesis of familial colorectal cancer. *Science* **260,** 812–816.
138. Thibodeau, S. N., Bren, G., and Shaid, D. (1993). Microsatellite instability in cancer of the proximal colon. *Science* **260,** 816–819.
139. Wooster, R., Cleton-Jansen, A. M., Collins N., Mangion, J., Cornelis, R. S., Cooper, C. S., Gusterson, B. A., Ponder, B. A. J., Von Deimling, A., Wiestler, O. D., Cornelisse, C. J., Devilee, P., and Stratton, M. R. (1994). Instability of short tandem repeats (microsatellites) in human cancers. *Nature Genet.* **6,** 152–156.
140. Liu, B., Nicolaides, N. C., Markowitz, S., Willson, J. K. V., Parsons, R. E., Jen, J., Papadopoulos, N., Peltomäki, P., De la Chapelle, A., Hamilton, S. R., Kinzler, K. W., and Vogelstein, B. (1995). Mismatch repair gene defects in sporadic colorectal cancers with microsatellite instability. *Nature Genet.* **9,** 48–55.
141. Speicher, M. R. (1996). Microsatellite instability in human cancer. *Oncology Res.* **7,** 267–275.
142. Eshleman, J. R., and Markowitz, S. D. (1996). Mismatch repair defects in human carcinogenesis. *Hum. Mol. Genet.* **5,** 1489–1494.
143. Kolodner, R. (1996). Biochemistry and genetics of eukaryotic mismatch repair. *Genes Dev.* **10,** 1433–1442.
144. Fulchignioni-Lataud, M. C., Olchwang, S., and Serre, J. L. (1997). The fragile X CGG repeat shows a marked level of instability in hereditary non-polyposis colorectal cancer patients. *Eur. J. Hum. Genet.* **5,** 89–93.
145. Usdin, K., and Woodford, K. J. W. (1995). CGG repeats associated with DNA instability and chromosome fragility form structures that block DNA synthesis in vitro. *Nucleic Acids Res.* **23,** 4202–4209.
146. De Vries, B. B. A., Wiegers, A. M., De Graaff, E., Verkerk, A. J. M. H., Van Hemel, J. O., Halley, D. J. J., Fryns, J. P., Curfs, L. M. G., Niermeijer, M. F., and Oostra, B. A. (1993). Mental status and fragile X expression in relation to FMR-1 gene mutation. *Eur. J. Hum. Genet.* **1,** 72–79.
147. De Graaff, E., Rouillard, P., Willems, P. J., Smits, A. P. T., Rousseau, F., and Oostra, B. A. (1995). Hotspot for deletions in the CGG repeat region of FMR1 in fragile X patients. *Hum. Mol. Genet.* **4,** 45–49.
148. Milà, M., Castellvi-Bel, S., Sánchez, A., Lázaro, C., Villa, M., and Estivill, X. (1996). Mosaicism for the fragile X syndrome full mutation and deletions within the CGG repeat of the FMR1 gene. *J. Med. Genet.* **33,** 338–340.
149. Schmucker, B., Ballhausen, G., and Pfeiffer, R. A. (1996). Mosaicism of a microdeletion of 486 bp involving the CGG repeat of the FMR1 gene due to misalignment of GTT tandem repeats at chi-like elements flanking both breakpoints and a full mutation. *Hum. Genet.* **98,** 409–414.
150. Quan, F., Grompe, M., Jakobs, P., and Popovitch, B. W. (1995). Spontaneous deletion in the FMR1 gene in a patient with fragile X syndrome and cherubism. *Hum. Mol. Genet.* **4,** 1681–1684.
151. Shelbourne, P., Winqvist, R., Kunert, E., Davies, J., Leisti, J., Thiele, H., Bachmann, H., Buxton, J., Williams, B., and Johnson, K. (1992). Unstable DNA may be responsible for the incomplete penetrance of the myotonic dystrophy phenotype. *Hum. Mol. Genet.* **1,** 467–473.
152. Sutcliffe, J. S., Nelson, D. L., Zhang, F., Pieretti, M., Caskey, C. T., Saxe, D., and Warren, S. T. (1992). DNA methylation represses FMR-1 transcription in fragile X syndrome. *Hum. Mol. Genet.* **1,** 397–400.

CHAPTER 36

Genetic Instability in Transgenic Mice Deficient in DNA Mismatch Repair

PETER M. GLAZER[1] Departments of Therapeutic Radiology and Genetics, Yale University School of Medicine, New Haven, Connecticut 06520

SUSAN E. ANDREW AND FRANK R. JIRIK Biomedical Research Center and Department of Medicine, University of British Columbia, Vancouver, British Columbia, Canada V6T 1Z3

LATHA NARAYANAN Departments of Therapeutic Radiology and Genetics, Yale University School of Medicine, New Haven, Connecticut 06520

I. DNA MISMATCH REPAIR

DNA mismatches, including base mispairs and loop-outs of various sizes, can come about by several mechanisms, including replication errors, repair errors, or as intermediates in recombinational processes [1, 2]. Repair of such errors is critical to the maintenance of genomic integrity. The *Escherichia coli* MutHLS system serves as the model for our understanding of the post-replicative MMR pathway [1, 2] . The primary function of the MutHLS system in *E. coli* and the related systems in yeast and mammalian cells is thought to be in the repair of errors arising in DNA replication, hence the term postreplicative MMR. In *E. coli,* as elucidated by

[1]To whom correspondence should be addressed at Department of Therapeutic Radiology, Yale University School of Medicine, P.O. Box 208040, New Haven, CT 06520-8040. Fax: (203) 737-2630. E-mail: peter_glazer@qm.yale.edu.

Modrich and Lahue [2] and others, recognition of mismatches by MutS is followed by MutL binding and recruitment of MutH, which mediates a strand-specific cleavage. Exonuclease activity creates a gap, which is filled in by repair synthesis, eliminating the mismatch. A key to this pathway is a mechanism for strand discrimination. In *E. coli,* the transient undermethylation of the daughter strand distinguishes it from the parental one (2). In mammalian cells, strand breaks are thought to be the signal (3), although further support for this hypothesis is needed. Other systems of MMR also exist, including so-called short patch repair pathways. In mammalian cells, for example, there is a glycosylase specific for G:T or G:U mismatches [4, 5]. Comprehensive reviews of MMR have recently been published [1, 2].

II. DNA MISMATCH REPAIR AND CANCER

The human syndrome of hereditary nonpolyposis colorectal cancer, HNPCC, has been linked to germline mutations in the human homologs of *E. coli* and yeast DNA MMR genes [1, 2, 6–10]. This syndrome is associated with a high risk of early-onset colon cancer, as well as cancers of the endometrium and other sites [2]. Recent work has revealed that mammalian cells contain multiple homologs of the *E. coli* MutS and MutL proteins, including MSH2, MSH3/DUG1/MRP1, and GTBP/MSH6 [1, 2, 7, 11, 12] and MLH1, PMS2, and PMS1 [1, 2, 8–10], respectively. Mutations in the *MSH2, MLH1, PMS1,* and *PMS2* genes have been detected in HNPCC patients [1, 2, 7–10]. However, *MSH2* and *MLH1* mutations appear to be much more frequently associated with the disease than are mutations in the other genes [1], an observation that has yet to be explained.

Biochemical studies using mammalian or yeast cell extracts or partially purified proteins have shown that the MutS homologs can associate in heterodimers in at least two combinations (MSH2 with MSH3 and MSH2 with GTBP/MSH6, yielding MutSα and MutSβ complexes, respectively), each with different activities and mismatch substrate preferences (MSH2/MSH6 for single base pair mismatches and MSH2/MSH3 for single base mispairs and insertion/deletion mismatches), suggesting both redundancy and divergence of function [1, 13, 14]. Studies with the MutL homologs suggest that MLH1 and PMS2 form a functional heterodimer [15–17]; other possible MutL complexes have yet to be characterized.

In addition, evidence is emerging that the MMR gene homologs may participate in various other cellular functions, such as transcription-coupled repair [18], recombination [19], and even cell cycle regulation [20]. Consequently, a deficiency in one of these homologs may disrupt genome stability in multiple ways. Although MMR deficiency has been associated with diminished transcription-coupled repair, the mechanism for this is unknown, and the relationship of MMR with other repair processes beyond this has not been fully explored. Studies of *in vitro* repair of DNA substrates containing a compound lesion consisting of a mispair and a bulky adduct detected no influence of MMR on the level of nucleotide excision repair (NER) activity [21]. Whether or not MMR proteins can influence other repair pathways, a separate issue is whether MMR may play a role in correcting repair-related misincorporation errors. Repair processes are known to involve polymerases with measurable error rates [22], and it is expected that repair errors therefore occur. The potential role of MMR in correcting such errors remains to be determined.

III. TRANSGENIC MOUSE KNOCK-OUT MODELS OF MMR DEFICIENCY

A. Distinct Patterns of Carcinogenesis

Knock-out mouse models carrying targeted disruptions of selected mouse MMR gene homologs have been developed, including *MSH2* [23–26], *MLH1* [27], *PMS1* (S. M. Baker, T. Prolla, and R. M. Liskay, unpublished), and *PMS2* [28]. The *MSH2, MLH1,* and *PMS2* nullizygous animals are prone to early-onset carcinogenesis. However, the distribution of tumors in each of the knock-out models is distinct. The *MSH2* nullizygotes are prone to lymphomas, sarcomas, and carcinomas of the small intestine and skin [23–26]. The *MLH1* knock-out mice develop lymphomas, sarcomas, and small intestine carcinomas. Only lymphomas and sarcomas have been seen in the *PMS2* nullizygotes [28], in spite of close surveillance, and the *PMS1*-deficient animals have not shown any increase in cancer incidence up to age 1.5 years (Liskay, personal communication; data not yet published).

None of the mice are prone to carcinomas of the colon, as are HNPCC patients. However, there is some overlap with the subsets of HNPCC designated the Lynch II and the Muir-Torre syndromes, in which patients are also susceptible to carcinomas of the small intestine and skin (among other sites), respectively [2]. It has been suggested that the differences in tumor patterns between the knock-out mice and HNPCC patients may reflect, in part, differences in the sequences of critical target genes that may be more or less susceptible to mutation in the two species [2]. For example, defects in the human type II TGF-β receptor are common in human HNPCC-related tumors, with the responsible mutation found to occur in an (A)10 repeat [29]. The mouse receptor gene lacks the (A)10 hot spot. Other explanations, such as differences in lifestyle and genotoxic exposures, are certainly plausible.

Most affected patients are heterozygous for mutations at one of the MMR gene loci, and carcinogenesis is thought to require loss of heterozygosity (LOH) in a somatic cell lineage. This requirement for an LOH step may also account for the above difference between mice and humans. However, one set of HNPCC patients was found to be deficient in MMR in their nonmalignant tissue, due to heterozygosity for a presumably dominant negative allele of *PMS2* [30]. These patients therefore resemble the nullizygous mice in having MMR deficiency without the need for LOH. Nonetheless, they showed the expected human, not mouse, pattern of cancer predisposition.

Other differences in the respective mouse knock-out phenotypes also exist. *MSH2* knock-outs are fertile [23–26], but *PMS2* nullizygous males are sterile [28], as are *MLH1* nullizygous males and females [27]. Involvement of PMS2 and MLH1 in meiotic DNA metabolism has been suggested [27, 28].

B. Analyses of Genetic Instability Due to MMR Deficiency

1. The Need for *in Vivo* Forward Mutation Assays

The *MSH2, MLH1,* and *PMS2* knock-out mice have all shown aspects of MMR deficiency in either *in vitro* assays of cell-free extracts or PCR-based assays for instability of repeated sequences [23–28]. However, these assays measure only selected aspects of genetic stability and repair; the phenotypic differences with respect to tumorigenesis (and fertility) suggest that other important, but possibly subtle, differences exist in the roles of the respective factors in the maintenance of genomic stability. For example, cells deficient in MSH6 (GTBP) are less prone to instability of simple sequence repeats than are MSH2-deficient cells [12]. However, these assays report only expansion or contraction of simple repeated sequences, representing just one specific subset of all possible genetic changes. Forward mutation assays, in contrast, are capable of detecting a wide range of mutations.

Such assays have been carried out using the selectable *HPRT* and *APRT* loci in colon cancer-derived cell lines, some of which are deficient in MMR. These studies have revealed some significant differences among the mutation patterns in different lines [31–35]. However, such tumor cell lines are likely to contain multiple genetic abnormalities besides MMR gene defects, and some of these might affect DNA metabolism, complicating interpretation of the experiments. For example, one colon cancer cell line often used in such analyses, DLD-1, is known to express a mutant form of polymerase δ [34]. In addition, such experiments are not able to explore factors that may be important in a whole animal, such as tissue differences.

Hence, the full impact of MMR gene inactivation on genomic integrity *in vivo* remains to be elucidated. For this reason, our groups have applied a highly sensitive lambda shuttle vector-based assay for *in vivo* mutagenesis to examine genetic instability in mice with targeted disruptions of the *MSH2* and *PMS2* genes.

2. Transgenic Mice for Analysis of Mutagenesis *in Vivo*

The analysis of mutations *in vivo* (i.e., in whole animals) has been facilitated by the construction of transgenic mice carrying lambda shuttle vector DNA in their genome [36–39]. The use of lambda phage as a shuttle vector for mutation detection was first developed in mouse fibroblasts [40] and was subsequently applied to transgenic mice [36–39]. The lambda DNA is introduced into the mouse cell or mouse oocyte DNA by standard transfection or microinjection techniques [39], and rescue of the lambda vector DNA from within the mouse DNA is accomplished by incubation of the mouse DNA in lambda *in vitro* packaging extracts (Fig. 36-1). These extracts can identify, cut out, and package the lambda DNA from within the mouse DNA into viable phage particles for growth and analysis in bacteria [40]. In this way, mutations in the reporter gene that occurred in the mice can be detected using *E. coli* and phage genetic techniques [40]. Transgenic mice have been used to study *in vivo* mutagenesis by a variety of genotoxic agents [36–38, 41–43], and they have been proposed as test systems for toxicologic analysis [44].

3. Use of the *supF* Gene as a Mutation Reporter Gene

In addition to the commercially available mice carrying either the *lacI* or *lacZ* reporter transgenes [44], we established a transgenic model carrying the *supF* amber suppressor tyrosine tRNA gene as a reporter [39, 40, 45–51]. There are several advantages of the *supF* system: (1) there is extensive experience with this gene in mammalian cell shuttle vector-based mutagenesis experiments; (2) the tRNA gene offers a target sensitive to mutations at most base pairs; (3) the coding sequence is only 85 base pairs, and so it is amenable to sequence analysis; and (4) sequences within the *supF* gene can be

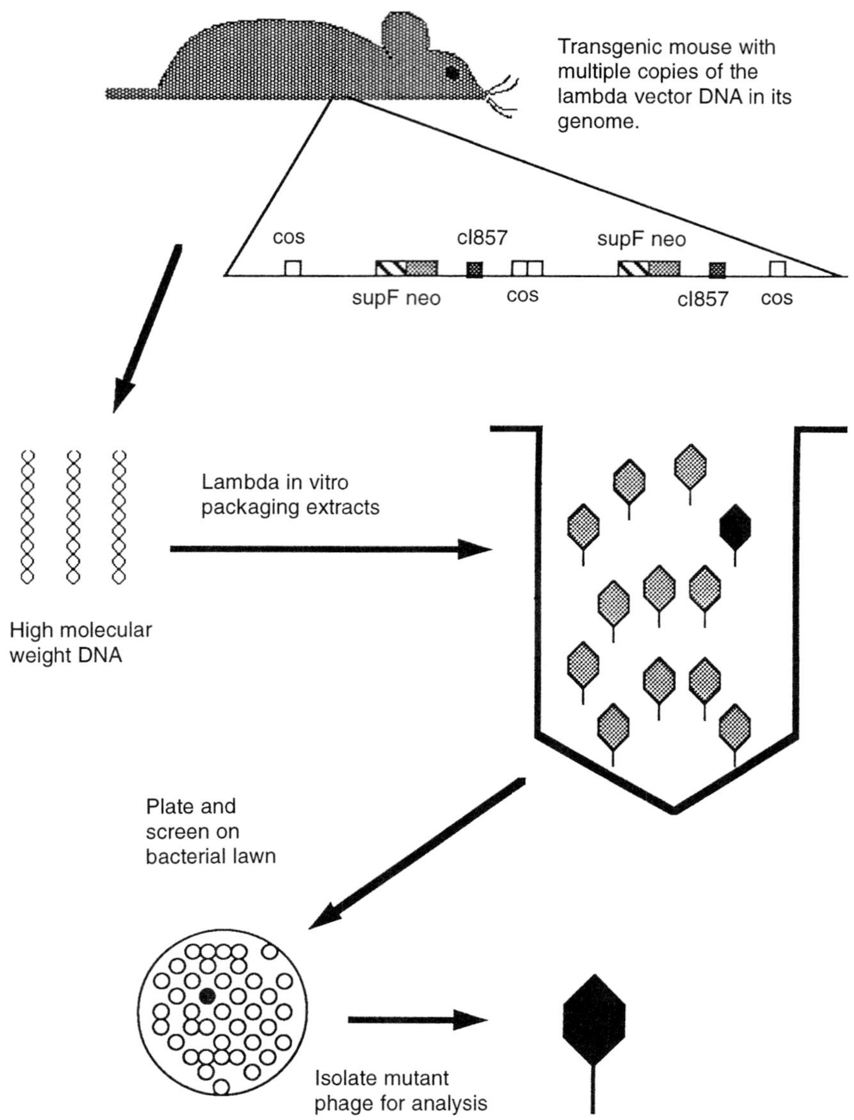

FIGURE 36-1 Strategy for rescue of lambda shuttle vector DNA from the transgenic mice. High-molecular-weight DNA prepared from the mice is incubated in lambda *in vitro* packaging extracts which can identify, cut out, and package the lambda DNA from within the mouse DNA into viable lambda particles for growth and analysis in *E. coli*.

modified by site-directed techniques in order to create novel reporter genes with altered sensitivities to spontaneous and induced mutagenic processes [52, 53].

In addition, evidence exists demonstrating the heterogeneity of mutagenesis and repair in mammalian cells, and sites throughout the genome have been found which show variability in susceptibility to mutagenesis [54]. Also, sequence motifs obviously vary among genes, rendering them differentially sensitive to the various pathways of mutagenesis. For example, the *lacI* gene contains a high number of CpG sites, at which methylation occurs in mammalian cells. Such methylation promotes deamination of cytosine, leading C to T transitions. This feature of *lacI* may account for some of the initial results found with the *MSH2* knock out mice, which were analyzed using the *lacI* reporter (see below). Hence, excessive reliance on one or two transgenic models may yield misleading results biased by the peculiarities of the particular transgene loci. In the long run, analyses of genetic instability in repair-deficient mice may require examination of mutation patterns in a variety of *in vivo* mouse models.

4. LIMITATIONS OF EXISTING TRANSGENIC MODELS

While lambda-containing transgenic mice have proven useful as tools to study *in vivo* mutagenesis, several problems have been identified with this approach.

The reporter transgenes are all bacterial or phage genes (*lacI, lacZ, supF, gpt, cI,* and *cII,* among others) and are not expressed in mouse cells. Hence, the effects of transcription on repair and mutagenesis cannot be assessed. This is particularly important in light of the differential repair of transcribed genes due to the pathway of transcription-coupled repair [55]. There is also emerging evidence that high levels of transcription may promote mutagenesis [56]. Further, MMR factors have been found to play a role in transcription-coupled repair [18], and so our analysis of the consequences of mismatch-repair deficiency *in vivo* may be influenced by the lack of transcription.

In the existing models, when the identical mutation is recovered more than once from a specific tissue in a mouse, it is not possible to determine whether the identical mutations are siblings, arising from the same mutational event and amplified by replication, or are truly independent, produced by separate events. As a result of this uncertainty, potentially significant mutation data are commonly excluded from analysis. Consequently, mutation hot spots are often underestimated, and, furthermore, the detection of hot spots becomes an artificial function of the number of animals and samples the investigator is willing to evaluate.

5. Construction of Transgenic Mice Carrying Lambda Phage Shuttle Vector DNA

Several lines of *supF* transgenic mice have been constructed for the purpose of studying *in vivo* mutagenesis. These lines each varied in several respects: vector copy number, locus of integration, phage rescue efficiency, and background level of spontaneous mutations. Also, the most recently derived line, 3340, carries the new *supFG1* gene instead of the standard *supF* gene. This gene was engineered to contain extended poly G:C base pair regions (see Fig. 36-3), rendering it particularly sensitive to replication slippage errors. The 3340 line has proven quite useful because it has a low background of spontaneous mutations and a reasonable efficiency of vector rescue. Another line, 1139, containing the standard *supF* gene, has also been used for *in vivo* mutation studies. It is interesting because it has a high frequency of spontaneous deletion mutagenesis at the transgene locus [49]. It has also turned out to be valuable as a mutation reporter model, particularly in comparison to the 3340 mice. Most of our work to probe the effects of *PMS2* deficiency has been performed with the 3340 mice, although the 1139 animals have also been used [51].

6. Generation of *PMS2*-Deficient, *supF* Hybrid Transgenic Mice

To determine the effect of *PMS2* inactivation on genomic integrity *in vivo,* we constructed hybrid transgenic mice carrying targeted disruptions at the *PMS2* loci along with the chromosomally integrated λsupF shuttle vector. To carry out a more comprehensive analysis, both the 1139 mice (carrying the standard *supF* gene [49]) and the 3340 mice (with the *supFG1* gene modified to contain extended simple sequence repeats (Fig. 36-2) [53]) were used. Male *supF*-positive mice were bred with *PMS2* heterozygous females in order to construct hybrid transgenic mice. The F1 mice were crossed to produce F2 mice that were wild type, heterozygous, or nullizygous at the *PMS2* loci and that also carried the *supF* transgene.

7. Analysis of Spontaneous Mutagenesis in *PMS2*-Deficient Mice: Elevated Levels of Mutation in Multiple Tissues

Three independent sets of 3340/*PMS2* male litter mates were identified for analysis at 12 weeks of age. The mice were maintained under standard animal husbandry conditions in the same cage. There was no known exposure to any mutagenic or genotoxic agent. DNA was prepared from several different tissues of the mice and used for vector rescue (via *in vitro* packaging) and reporter gene analysis (Table 36-1).

The frequency of mutations in the wild-type mice was in the range of 1 to 2×10^{-5}, consistent with previous observations of baseline mutation frequencies in such transgenic animal systems (Table 36-1; [36, 38, 49]). However, the nullizygote samples consistently showed 100-fold or more elevations in mutation frequencies in all tissues and mice tested ([51]; Table 36-1).

Mice heterozygous for the *PMS2* disruption did not show increased genetic instability (Table 36-1), at least in this initial analysis, indicating that a single functional allele may provide near normal repair activity and that *PMS2*-related genetic instability may therefore be recessive. However, we cannot rule out the possibility that the heterozygotes do have a low level of genetic instability, since slight increases in mutation frequency (such as twofold or less) are difficult to reliably detect in this assay, even with the analysis of large numbers of animals and phage. In light of the apparent participation of mismatch repair factors in aspects of DNA repair other than mismatch correction [18, 57, 58], differences between the heterozygotes and the wild-type may eventually be revealed by studies of induced mutagenesis.

The heterozygous mice mimic the situation in most patients with the HNPCC syndrome, who are also heterozygous at one of the mismatch repair loci in their somatic cells [1, 6]. Their tumors, however, are characterized by loss or inactivation of the second allele, leading to genomic instability. Such loss of heterozygosity events should also be possible in mice, thus predisposing the affected cell lineages to *supF* mutations. However, the low mutation frequencies in the heterozygotes sug-

A

3340/*PMS2* nullizygotes

Mouse F2-19 (skin, liver, and colon)

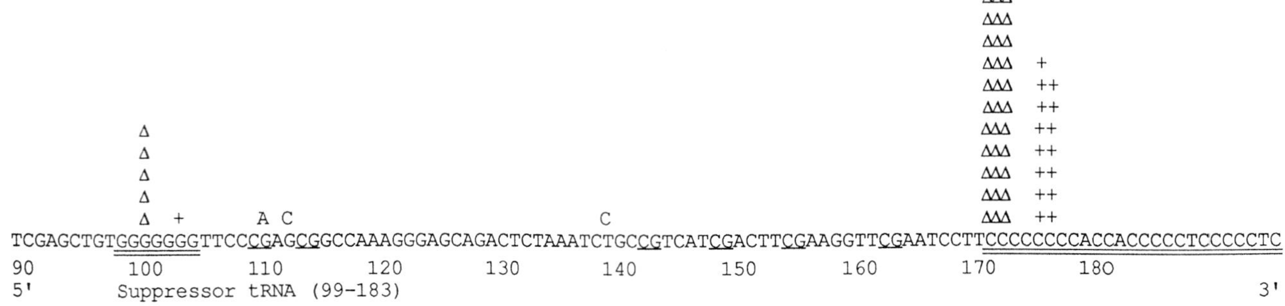

Mouse F2-29 (skin, liver, colon)

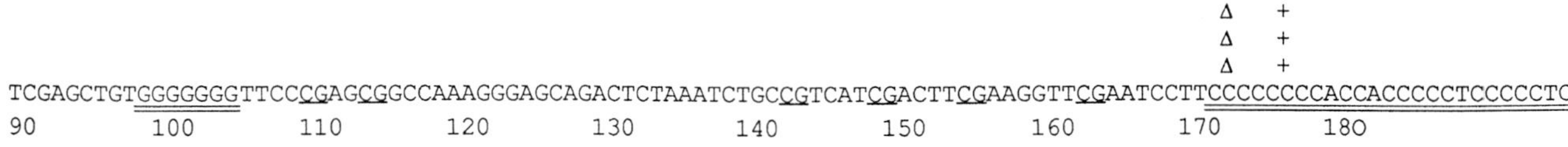

Mouse F3-11 (skin)

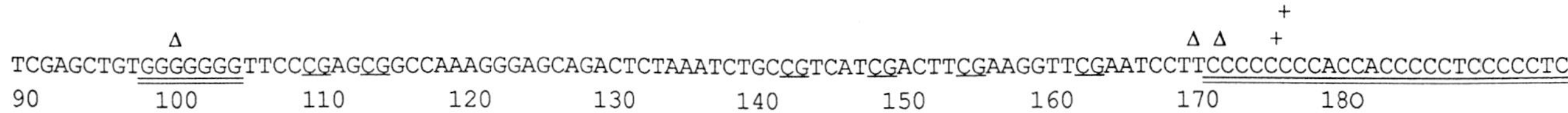

B

1139/*PMS2* nullizygotes

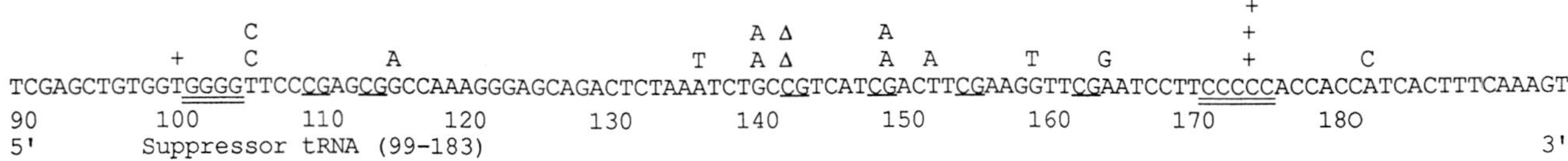

1139 wild type

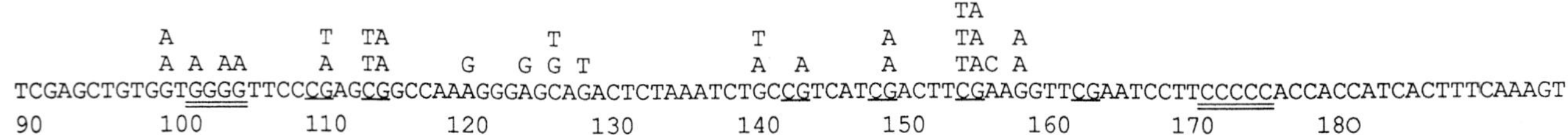

FIGURE 36-2 Sequences of reporter gene mutations in *PMS2* nullizygous mice. (A) Mutations in the *supFG1* gene in 3340/*PMS2* nullizygotes. Mutations from three different mice from independent litters are shown, with the tissues of origin indicated. Base substitutions are listed above the original sequence, and single base pair deletions or insertions are indicated by the symbols (Δ) or (+), respectively, above the corresponding site. Mononucleotide repeat sequences are highlighted by double underlining. CpG sequences, potential sites of cytosine methylation, are noted by a single underline. (B) Mutations in the *supF* gene in 1139/*PMS2* nullizygotes and in wild-type 1139 mice, as indicated.

gest that such events are rare. Further studies on this point are needed, however.

In the 1139/*PMS2* hybrids, the nullizygotes showed only a 5-fold overall increase in mutation frequency (Table 36-2). However, the 1139 mice were previously found to have an unusually high frequency of spontaneous deletions due to locus-specific effects at the transgene integration site [49]. When only *supF* point muta-

TABLE 36-1 Spontaneous Mutagenesis of *supFG1* Transgene in *PMS2*-Deficient 3340 Mice

Mouse/tissue	Mutation frequency ($\times 10^{-5}$)	Mutants/total
Wild type		
Mouse F2-17		
Skin	≤2	0/72,900
Liver	4	6/170,260
Mouse F3-9		
Skin	≤2	0/72,240
Colon	2	1/45,620
Total	2	7/361,110
Heterozygous		
Mouse F2-18		
Skin	5	5/92,730
Liver	4	5/131,500
Spleen	≤2	0/75,630
Colon	≤2	0/78,620
Mouse F2-4		
Skin	1	1/87,690
Mouse F3-13		
Skin	2	1/51,600
Total	2	12/517,760
Nullizygous		
Mouse F2-19		
Skin	237	302/127,700
Liver	132	55/41,700
Spleen	226	33/14,620
Colon	286	201/70,360
Brain	270	53/19,610
Lung	206	35/16,980
Mouse F2-29		
Skin	177	27/15,270
Spleen	182	27/14,850
Colon	187	37/19,780
Mouse F3-11		
Skin	173	58/33,590
Liver	118	54/45,650
Total	210	882/420,110

tions were considered, the nullizygotes again showed a substantial elevation in mutation frequency of approximately 25-fold relative to the wild-type mice [51].

8. MUTATION PATTERNS: INFLUENCE OF SEQUENCE CONTEXT AND MONONUCLEOTIDE REPEATS

The mutations arising in the nullizygous mice were analyzed at the sequence level (Fig. 36-2). Almost all of the mutations in the 3340/*PMS2* nullizygotes (96%) were single base pair deletions or insertions within G:C base pair repeats at positions 99–105 and 172–179. The eight G:C base pair run at 172–179 is a particular hot spot. (This stretch constitutes a subset of a longer G:C-rich region which extends beyond the end of the tRNA gene at position 183.) This pattern was seen in multiple tissues in three different nullizygous mice from three different litters.

In the 1139 mice, single base pair insertions and deletions were less prominent (33% of the total) but were still more frequent than in the wild-type mice (0 out of 30). The difference in mutation pattern between the 3340 and the 1139 mice can be attributed in part to the differences in the reporter genes. *SupFG1* in 3340 mice has 2 runs of seven and eight G:C base pairs, respectively. The longest repeat sequence in the *supF* gene in the 1139 mice, however, is a five G:C base pair stretch at 171–176. We propose that this shorter region is less susceptible to slippage replication errors [59] and is therefore less dependent on *PMS2* function for stability.

By comparison between the results in the *supF* and *supFG1* reporter genes, we can conclude that seven or more mononucleotides are particularly unstable *in vivo*. This is consistent with the frequent finding of frameshift mutations within a 10 base pair poly A:T run in the TGF-β receptor gene in a series of colon cancer cell lines [29, 60]. Such long mononucleotide runs are also particularly common within promoters and introns. For example, the *c-myc* promoter contains several extended

TABLE 36-2 Spontaneous Mutagenesis of the *supF* Transgene in *PMS2*-Deficient 1139 Mice

Mouse/tissue	Mutation frequency ($\times 10^{-5}$)	Mutants/total	% Point mutations	% Deletions	Frequency of point mutations ($\times 10^{-5}$)
Wild type					
Skin	39	199/511,950			
Liver	29	47/199,010			
Spleen	35	82/232,580			
Total	35	328,943,540	6	94	2
Nullizygous					
Skin	117	163/138,910			
Liver	139	165/118,860			
Spleen	154	151/98,360			
Total	135	479/356,120	42	58	57

runs of G:C base pairs [61], and the binding site for the Sp1 transcription factor is highly G:C rich [62]. We would predict that these regulatory regions would be especially unstable in the setting of *PMS2* deficiency, leading to accumulating abnormalities in gene expression and thereby contributing to aberrant cellular function and eventually carcinogenesis.

The base substitutions in the nullizygotes were predominately transitions. Since there were so few base substitutions in the 3340/*PMS2* nullizygotes (only 3 out of 74), the main comparison to be made is between the 1139/*PMS2* nullizygotes and the wild-type mice (Table 36-3). Overall, the spectra of base substitutions are similar except that the proportion of C-to-T transitions at CpG sites in the nullizygotes is only 20%, compared to 47% in the wild type. Since CpG sequences are sites of cytosine methylation in mammalian cells, transitions at these positions are thought to arise from the propensity of 5-methylcytosine to deaminate to thymidine. Although such mutations do occur in the *PMS2* nullizygotes, they are underrepresented relative to those in the wild-type animals, suggesting that *PMS2* is not essential for repair of these lesions. Rather, other MutL homologs may have a role in the repair of these mispairs. Also, it is known that there is an alternate repair pathway involving a glycosylase specific for the G:U and G:T mismatches that are produced by deamination [5, 63]. The paucity of methylcytosine-related transitions in the nullizygotes is not due to undermethylation of the transgene sequences. The *supF* transgenes in both the 3340 and 1139 mice are each heavily methylated, as determined by comparison of susceptibility of the mouse genomic DNA to *Hpa*II and *Msp*I restriction.

TABLE 36-3 Mutations in *PMS2* Nullizygous Mice

Mutation	3340/PMS2 nullizygotes	1139/PMS2 nullizygotes	1139 Wild-type mice
C:G → T:A	1	5	23
T:A → C:G	1	3	2
C:G → A:T	0	1	3
C:G → G:C	1	0	1
T:A → A:T	0	2	0
T:A → G:C	0	1	1
C:G → T:A ! CpG	1	2	14
+1 Insertion	24	4	0
−1 Deletion	47	2	0
Total	74	18	30

C. Genetic Instability in Transgenic Mice Deficient in *Msh2*

1. A Novel *lacI* Mutation-Detection System

A transgenic line (BC-1) carrying a concatamerized lambda-phage shuttle vector including the *lacI* gene was established for *in vivo* mutation analyses [64]. Similar to the commercially available (Stratagene) Big Blue *lacI* transgenic, the BC-1 construct contains *lacI* as the mutational target gene. In contrast to existing *lacI* strains, the BC transgene *lacI* target was situated within a fragment containing a rearranged murine immunoglobulin heavy chain gene locus, which in turn was inserted into lambda-phage arms to permit retrieval of the transgene. The development of an additional *lacI* transgenic strain permits an appraisal of the potential modifying effects that factors such as the character of sequences flanking the *lacI* target or the chromosomal integration site of the target might have on mutation frequency.

The BC-1 transgenic mouse carries a similar number of integrated transgenes as Stratagene's Big Blue transgenic mouse and packages with approximately the same efficiency [37, 64]. Thus, BC-1 is well suited for collecting the numbers of plaque-forming units required for mutation frequency determinations.

2. Generation of MSH2-Deficient, *lacI* Hybrid Transgenic Mice

To investigate the consequences of MSH2 deficiency on the tissue-specific spontaneous mutation frequency and spectrum, BC-1 was crossed with $MSH2^{-/-}$ mice that were generated by gene targeting technology [23].

3. Analysis of Spontaneous Mutagenesis in MSH2-Deficient Mice: Elevated Mutation Frequency in Multiple Tissues

Three BC-1/$MSH2^{+/-}$ heterozygotes, three BC-1/$MSH2^{-/-}$ homozygotes, and three BC-1 controls ($MSH2^{+/+}$) were analyzed at 3 weeks of age and mutation frequencies were determined for three different tissues in each animal: small intestine and thymus, the source of most of the tumors in the $MSH2^{-/-}$ mice (23, 25), and brain, due to its relatively low mitotic activity at this age and also selected as a tissue not prone to tumor development in these mice.

LacI mutation frequencies from the controls were in the range of $3–5 \times 10^{-5}$, consistent with previous spontaneous mutation frequencies in BC-1 mice and other transgenic reporter systems [38, 39, 64]. Mutation frequencies for *MSH2* heterozygotes were similar to those of controls, suggesting that heterozygosity for a *MSH2* mutation, like heterozygosity for *PMS2* [51], fails to significantly handicap MMR under normal conditions (Table 36-4). The mutation frequencies in homozygotes, however, were elevated in all three tissues evaluated (Table 36-4). Mutation frequencies in brain were increased (4.8-fold) as compared to controls, although not to the extent seen in small intestine and thymus, where the increases were 11.0- and 15.2-fold, respectively. If the higher mutation frequencies seen in small intestine and thymus were reflective of the proliferating cell populations in these tissues, then the small intestine mutation frequencies observed, for example, likely represent an underestimate of gut epithelial cell frequencies, as whole small intestine (including smooth muscle) was included in the analysis. The variability in mutation frequency between tissues (Table 36-4) may have been reflective of the mitotic history of each of the tissues (thymus < small intestine < brain). However, testing of additional animals will be necessary to establish the validity of this observation.

The elevated mutation frequency in $MSH2^{-/-}$ mice (4.8- to 15.2-fold) is consistent with the increase in mutation frequency observed in *mutS Escherichia coli* [65, 66] and MSH2-deficient *Saccharomyces cerevisiae* [67, 68]. High *hprt* gene mutation frequencies were also observed in murine MSH2-deficient embryonic stem cells [26], and in human tumor cell lines with MMR deficiencies [31, 32, 58, 69].

4. Spectrum of *lacI* Mutations in MSH2-Deficient Mice Compared to Controls

To determine how the spectrum of mutations was altered due to MMR deficiency, *lacI* mutants from

TABLE 36-4 Spontaneous Mutation Frequencies in the *lacI* Transgene in *Msh2* Mice

Genotype	Tissue	Animal	Total pfu	Number of mutants	Mutation frequency $\times 10^{-5}$
Wild type BC-1 control	Small intestine	1	286,860	11	3.8
		2	261,540	4	1.5
		3	265,320	11	4.1
		Mean			3.1 ± 1.4
	Thymus	1	292,890	13	4.4
		2	264,180	8	3.0
		3	285,680	5	1.8
		Mean			3.1 ± 1.3
	Brain	1	213,940	10	4.7
		2	259,400	9	3.5
		3	236,340	15	6.3
		Mean			4.8 ± 1.4
Heterozygote Msh2 +/−	Small intestine	1	253,560	7	2.8
		2	299,340	7	2.3
		Mean			2.6 ± 0.4
	Thymus	1	259,080	5	1.9
		2	257,900	17	6.6
		Mean			4.3 ± 3.3
	Brain	1	216,240	14	6.5
		2	222,740	13	5.8
		3	223,060	7	3.1
		Mean			5.1 ± 1.7
Nullizygote Msh2 −/−	Small intestine	1	282,640	93	32.9
		2	269,980	99	36.7
		3	246,440	80	32.5
		Mean			34.0 ± 2.3
	Thymus	1	300,460	89	29.6
		2	295,020	166	56.3
		3	281,640	155	55.0
		Mean			47.0 ± 15.0
	Brain	1	210,380	56	26.6
		2	226,680	39	17.2
		3	216,880	55	25.4
		Mean			23.1 ± 5.1

Note: Taken, with permission, from Andrew *et al., Oncogene* **15,** 123–129, 1997.

MSH2$^{-/-}$ and control animals were sequenced. Of the spontaneous control mutations, transitions were most frequently observed (63.0%), followed by transversions (26.1%), deletions (6.5%), and frameshifts (4.3%) (Fig. 36-3), consistent with previous observations in non-DNA repair-deficient hosts [64]. Analysis of the total number of *MSH2*$^{-/-}$ mutations revealed that transitions again represented approximately the same fraction of total mutations (64.2%), while transversions had decreased (13.2%), and frameshifts had increased (18.9%) (Fig. 36-3). The total number of transitions, transversions, frameshifts, and deletions sequenced from *MSH2*$^{-/-}$ animals (all tissues combined) is significantly different compared to the proportion of transitions, transversions, frameshifts, and deletions sequenced from control animals. The percentage of G:C-to-A:T transitions at CpG sites was similar in control and *MSH2*$^{-/-}$ mice (39.1 and 32.1% of mutations, respectively) (Fig. 36-3).

The mutational spectrum of the *lacI* genes recovered from the mice differed from the bacterial *lacI* spectrum [70], indicating that the mutations described had been acquired within the murine host. Sectored plaques, arising from bacterial processing of DNA damage acquired within the murine host, were not included in the results. Importantly, control and *MSH2*$^{-/-}$ DNA yielded equal frequencies of sectored plaques, demonstrating that the elevated mutation frequency in *MSH2*$^{/-}$ animals did not simply stem from unrepaired DNA damage within the rescued phage genomes.

The *lacI* mutation spectrum in *MSH2*$^{/-}$ mice was similar to that of *mutS E. coli,* characterized by a predominance of base-pair substitutions over frameshift mutations [65, 71, 72]. In contrast to one study where only 37% of mutations were C-to-T transitions [65], we saw such changes at a frequency of 45.3%, with 70.8% occurring at CpG sites. This was in keeping with CpG dinucleotides being an important source of spontaneous mutations in mammals, due to G:T mismatch formation following deamination of 5-methyl cytosine residues [73].

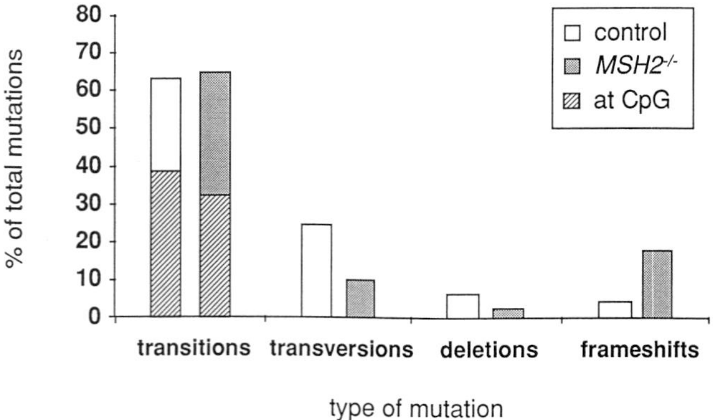

FIGURE 36-3 *LacI* gene mutation spectrum from the three tissues obtained from control and *MSH2*$^{-/-}$ homozygote mice, corrected for possible clonality. Reprinted, with permission, from Andrew *et al., Oncogene* **15,** 123–129, 1997.

Interestingly, frameshift mutations were not observed at all mononucleotide repeats in *MSH2*$^{-/-}$ animals. For example, the run of 5 adenine residues at position 135–139, but not that at 1006–1010, was frequently mutated. This might be attributable to the finding that mutations in the amino-terminal DNA binding domain are most likely to inactivate the *lacI* repressor [74]. In keeping with this observation, the majority of mutants recovered were localized to this region of the *lacI* gene.

The hot spot for frameshift mutations seen at *lacI* position 135–139 in *MSH2*$^{-/-}$ mice was also the most common frameshift site in *mutS*-deficient bacteria [65]. This finding may be indicative of the similarities in both DNA polymerase fidelity and MMR across widely divergent species when presented with the same DNA target sequence. It was also consistent with this site being a mutational hot spot. While position 180 was a common site for transition mutations in the eukaryotic hosts, no changes were observed at this position in MutS-deficient bacteria [65]. Alternatively, mutations at position 93, a minor hot spot in bacteria, also occurred in the *MSH2*$^{-/-}$ mice. It is interesting to speculate that transition mutations at position 180 in the eukaryotic host likely resulted from unrepaired G:T mismatches generated by 5-methyl cytosine deamination, while mutations at position 93 in both eukaryotes and prokaryotes arose from DNA polymerase misincorporations.

5. IMPLICATIONS OF MSH2 DEFICIENCY ON MUTATION

The finding of unique mutation sites in the *lacI* genes of controls as compared to the *MSH2*$^{-/-}$ mice was consistent with the actions of at least two independent processes in determining the mutational spectrum, the first being the frequency of sequence-dependent DNA polymerase errors and the second reflecting the capacity of specific mismatches, within a given sequence context, to serve as substrates for MSH2-initiated recognition and repair.

The predominance of point mutations within *lacI* genes obtained from *MSH2*-deficient mice suggested that single-base mispairs (especially transitions), and not frameshifts, were the most common DNA polymerase errors in such genes and that such errors are normally efficiently repaired. This was in keeping with evidence from bacteria indicating a preference of the proofreading activity of *pol III* for transversions over transitions [75]. The occurrence of slippage of one nucleotide (−1 more frequently than +1) at homonucleotide stretches was the second most common class of mutations within the *lacI* gene of MMR-deficient animals. Interestingly, all but one of the frameshifts occurred at a repeat of 3 or more mononucleotides, with 5/11 (48%) occurring at a 5 adenine repeat.

The finding of frameshifts at a 3 mononucleotide repeat suggests that many genomic sites, in addition to those containing highly repetitive dinucleotide tracts, are at risk of frameshift mutations in MSH2 deficiency. This observation, along with the finding of an elevated frequency of transitions, indicated that all genes are potential targets for mutation in MSH2 deficiency. Interestingly, the majority of somatic mutations in the p53 gene of colorectal cancers are base substitutions [76], with over 50% of these being transitions at CpG dinucleotides [77]. Thus, in addition to mutations in genes containing mononucleotide repeats, such as that encoding TGF-β receptor [29], MMR deficiency potentially increases the frequency of damage to key growth control genes typically inactivated by point mutations. Indeed, in HNPCC tumors, mutations within APC and p53 genes are 40% frameshifts and 60% point mutations; 70% of APC mutations in another study are frameshifts, with the remainder being point mutations [78].

6. Comparison with PMS2-Deficient *supF* Mice

The mutation frequency of *supF/PMS2*$^{-/-}$ mice was approximately 25- to 100-fold higher than normal, substantially higher than the *lacI/MSH2*$^{-/-}$ mutation frequency [51]. The mutants recovered from the *PMS2*$^{-/-}$ also differed from that of the *MSH2*$^{-/-}$ mice, revealing a higher proportion of frameshift mutations and fewer base substitutions, especially transitions at CpG sites, in keeping with the repetitive sequences within the *supF* target genes employed. As *MSH2* and *PMS2* operate within the same MMR pathway, the observed differences highlight the critical importance of the target gene in determining both mutation frequency and spectrum. Alternatively, the differences could, in part, be due to MSH2-specific functions not involving MutL homologs. Complementary crosses of mice deficient in specific MMR components and transgenic reporter gene strains will be necessary to clarify these issues further.

IV. CONCLUSIONS

Using a shuttle vector-based forward mutation assay, high somatic mutation frequencies in *PMS2* and *MSH2* nullizygous mice of up to 100-fold above the wild-type background were detected. The nullizygous mice showed increased mutations in all tissues tested, including skin, liver, spleen, colon, brain, and lung. However, the animals are not predisposed to cancer in all tissues. The PMS2 mice develop only lymphomas and sarcomas [28]. The *MSH2* mice display a wider tissue distribution of neoplasms, including lymphomas, sarcomas, and carcinomas of the skin and small intestine [23–26], yet, the MSH2 mice still show high levels of mutagenesis in many tissues not prone to cancer. This disparity shows that mutagenesis is just one element in carcinogenesis, highlighting the complexity of cancer etiology. Additional factors could include tissue damage and aberrant mitogenic stimuli. However, the particular factors that influence the tissue distribution of neoplasms in the mice remain to be determined.

In HNPCC patients, the range of tumors is also limited, but it is different from that seen in the mice, with epithelial cancers of the colon most common. One explanation for this difference could be that most of the HNPCC patients are genotypically heterozygotes, and inactivation of the second mismatch repair gene allele is a critical step in the progression to neoplasm [1, 6, 79]. Tissue-specific factors that vary between species, such as differential exposure to diet-derived genotoxic agents in the colon, may play an important part in promoting the loss of heterozygosity.

Several unusual patients, however, have been found to lack mismatch repair activity in their nonneoplastic tissues as well as in their cancers [30, 80]. Two of these were determined to be heterozygous at the *PMS2* locus, but in each case the mutant allele was found to code for a truncated protein with apparent dominant negative activity. In terms of *PMS2*-related mismatch repair function, the phenotype of these patients is therefore similar to that of the *PMS2*-nullizygous mice, yet they developed colorectal carcinomas, not lymphomas or sarcomas. This difference suggests that there may be additional species-specific factors influencing tumor distribution. Determining whether these factors are genetic in nature (such as differences in the controls on growth and differentiation) or reflect environmental and life-style differences will require further study.

References

1. Kolodner, R. (1996). Biochemistry and genetics of eukaryotic mismatch repair. *Genes Dev.* **10,** 1433–1442.
2. Modrich, P., and Lahue, R. (1996). Mismatch repair in replication fidelity, genetic recombination, and cancer biology. *Annu. Rev. Biochem.* **65,** 101–133.
3. Hare, J. T., and Taylor, J. H. (1985). One role for DNA methylation in vertebrate cells is strand discrimination in mismatch repair. *Proc. Natl. Acad. Sci. USA* **82,** 7350–7354.
4. Wiebauer, K., and Jiricny, J. (1989). *In vitro* correction of G:T mispairs to G:C pairs in nuclear extracts from human cells. *Nature* **339,** 234–236.

5. Wiebauer, K., and Jiricny, J. (1990). Mismatch-specific thymine DNA glycosylase and DNA polymerase B mediate the correction of G:T mispairs in nuclear extracts from human cells. *Proc. Natl. Acad. Sci. USA* **87,** 5842–5845.
6. de la Chapelle, A., and Peltomaki, P. (1995). Genetics of hereditary colon cancer. *Annu. Rev. Genet.* **29,** 329–348.
7. Leach, F. S., Nicolaides, N. C., Papadopoulos, N., Liu, B., Jen, J., Parsons, R., Peltomaki, P., Sistonen, P., Aaltonen, L. A., Nystrom-Lahti, M., Guan, X., Zhang, J., Metzler, P., Yu, J., Kao, F., Chen, D., Cerosaletti, K., Fournier, R., Todd, S., Lewis, T., Leach, R., Naylor, S., Weissbach, J., Mecklin, J., Arvinen, H. J., Petersen, G., Hamilton, S., Green, J., Jass, J., Wattson, P., Lynch, H., Trent, J., de la Chapelle, A., Kinzler, K., and Vogelstein, B. (1993). Mutations of a mutS homolog in hereditary nonpolyposis colorectal cancer. *Cell* **75,** 1215–1225.
8. Bronner, C. E., Baker, S. M., Morrison, P. T., Warren, G., Smith, L. G., Lescoe, M. K., Kane, M., Earabino, C., Lipford, J., Lindblom, A., Tannergard, P., Bollag, R., Godwin, A., Ward, D., Nordenskjold, M., Fishel, R., Kolodner, R., and Liskay, R. M. (1994). Mutation in the DNA mismatch repair gene homologue hMLH1 is associated with hereditary non-polyposis colon cancer. *Nature* **368,** 258–261.
9. Nicolaides, N. C., Papadopoulos, N., Liu, B., Wei, Y. F., Carter, K. C., Ruben, S. M., Rosen, C. A., Haseltine, W. A., Fleischmann, R. D., Fraser, C. M., Adams, M., Venter, J., Dunlop, M., Hamilton, S., and Petersen, G. (1994). Mutations of two PMS homologues in hereditary nonpolyposis colon cancer. *Nature* **371,** 75–80.
10. Papadopoulos, N., Nicolaides, N. C., Wei, Y. F., Ruben, S. M., Carter, K. C., Rosen, C. A., Haseltine, W. A., Fleischmann, R. D., Fraser, C. M., Adams, M. D., Venter, J., Hamilton, S., Petersen, G., Watson, P., Lynch, H., Peltomaki, P., Mecklin, J., de la Chapelle, A., Kinzler, K., and Vogelstein, B. (1994). Mutation of a mutL homolog in hereditary colon cancer. *Science* **263,** 1625–1629.
11. Fishel, R., Lescoe, M. K., Rao, M. R., Copeland, N. G., Jenkins, N. A., Garber, J., Kane, M., and Kolodner, R. (1993). The human mutator gene homolog MSH2 and its association with hereditary nonpolyposis colon cancer *Cell* **75,** 1027–1038. [Published erratum appears in *Cell* **77**(1), 167, 1994]
12. Papadopoulos, N., Nicolaides, N. C., Liu, B., Parsons, R., Lengauer, C., Palombo, F., D'Arrigo, A., Markowitz, S., Willson, J. K., Kinzler, K. W., Jiricny, J., and Vogelstein, B. (1995). Mutations of GTBP in genetically unstable cells. *Science* **268,** 1915–1917.
13. Palombo, F., Gallinari, P., Iaccarino, I., Lettieri, T., Hughes, M., D'Arrigo, A., Truong, O., Hsuan, J. J., and Jiricny, J. (1995). GTBP, a 160-kilodalton protein essential for mismatch-binding activity in human cells. *Science* **268,** 1912–1914.
14. Drummond, J. T., Li, G. M., Longley, M. J., and Modrich, P. (1995). Isolation of an hMSH2-p160 heterodimer that restores DNA mismatch repair to tumor cells. *Science* **268,** 1909–1912.
15. Prolla, T. A., Pang, Q., Alani, E., Kolodner, R. D., and Liskay, R. M. (1994). MLH1, PMS1, and MSH2 interactions during the initiation of DNA mismatch repair in yeast. *Science* **265,** 1091–1093.
16. Prolla, T. A., Christie, D. M., and Liskay, R. M. (1994). Dual requirement in yeast DNA mismatch repair for MLH1 and PMS1, two homologs of the bacterial mutL gene. *Mol. Cell. Biol.* **14,** 407–415.
17. Li, G. M., and Modrich, P. (1995). Restoration of mismatch repair to nuclear extracts of H6 colorectal tumor cells by a heterodimer of human MutL homologs. *Proc. Natl. Acad. Sci. USA* **92,** 1950–1954.
18. Mellon, I., Rajpal, D. K., Koi, M., Boland, C. R., and Champe, G. N. (1996). Transcription-coupled repair deficiency and mutations in human mismatch repair genes. *Science* **272,** 557–560.
19. Worth, L., Jr., Clark, S., Radman, M., and Modrich, P. (1994). Mismatch repair proteins MutS and MutL inhibit RecA-catalyzed strand transfer between diverged DNAs. *Proc. Natl. Acad. Sci. USA* **91,** 3238–41.
20. Hawn, M. T., Umar, A., Carethers, J. M., Marra, G., Kunkel, T. A., Boland, C. R., and Koi, M. (1995). Evidence for a connection between the mismatch repair system and the G2 cell cycle checkpoint. *Cancer Res.* **55,** 3721–3725.
21. Mu, D., Tursun, M., Duckett, D. R., Drummond, J. T., Modrich, P., and Sancar, A. (1997). Recognition and repair of compound DNA lesions (base damage and mismatch) by human mismatch repair and excision repair systems. *Mol. Cell. Biol.* **17,** 760–769.
22. Kunkel, T. A., Bebenek, K., Roberts, J. D., Fitzgerald, M. P., and Thomas, D. C.. (1989). Analysis of fidelity mechanisms with eukaryotic DNA replication and repair proteins. *Genome* **31,** 100–103.
23. Reitmair, A. H., Schmits, R., Ewel, A., Bapat, B., Redston, M., Mitri, A., Waterhouse, P., Mittrucker, H. W., Wakeham, A., Liu, B., Thomson, A., Greisser, H., Gallinger, S., Ballhausen, W., Fishel, R., and Mak, T. W. (1995). MSH2 deficient mice are viable and susceptible to lymphoid tumours. *Nature Genet.* **11,** 64–70.
24. Reitmair, A. H., Redston, M., Cai, J. C., Chuang, T. C., Bjerknes, M., Cheng, H., Hay, K., Gallinger, S., Bapat, B., and Mak, T. W. (1996). Spontaneous intestinal carcinomas and skin neoplasms in Msh2-deficient mice. *Cancer Res.* **56,** 3842–3849.
25. Reitmair, A. H., Cai, J. C., Bjerknes, M., Redston, M., Cheng, H., Pind, M. T., Hay, K., Mitri, A., Bapat, B. V., Mak, T. W., and Gallinger, S.. (1996). MSH2 deficiency contributes to accelerated APC-mediated intestinal tumorigenesis. *Cancer Res.* **56,** 2922–2926.
26. de Wind, N., Dekker, M., Berns, A., Radman, M., and te Riele, H. (1995). Inactivation of the mouse Msh2 gene results in mismatch repair deficiency, methylation tolerance, hyperrecombination, and predisposition to cancer. *Cell* **82,** 321–330.
27. Baker, S. M., Plug, A. W., Prolla, T. A., Bronner, C. E., Harris, A. C., Yao, X., Christie, D. M., Monell, C., Arnheim, N., Bradley, A., Ashley, T., and Liskay, R. M. (1996). Involvement of mouse Mlh1 in DNA mismatch repair and meiotic crossing over. *Nature Genet.* **13,** 336–342.
28. Baker, S. M., Bronner, C. E., Zhang, L., Plug, A. W., Robatzek, M., Warren, G., Elliott, E. A., Yu, J., Ashley, T., Arnheim, N., Flavell, R. A., and Liskay, R. M. (1995). Male mice defective in the DNA mismatch repair gene PMS2 exhibit abnormal chromosome synapsis in meiosis. *Cell* **82,** 309–319.
29. Markowitz, S., Wang, J., Myeroff, L., Parsons, R., Sun, L. Z., Lutterbaugh, J., Fan, R. S., Zborowska, E., Kinzler, K. W., Vogelstein, B., Brattain, M., and Willson, J. K. V. (1995). Inactivation of the type II TGF-b receptor in colon cancer cells with microsatellite instability. *Science* **268,** 1336–1338.
30. Parsons, R., Li, G. M., Longley, M., Modrich, P., Liu, B., Berk, T., Hamilton, S. R., Kinzler, K. W., and Vogelstein, B. (1995). Mismatch repair deficiency in phenotypically normal human cells. *Science* **268,** 738–740.
31. Bhattacharyya, N. P., Skandalis, A., Ganesh, A., Groden, J., and Meuth, M. (1994). Mutator phenotypes in human colorectal carcinoma cell lines. *Proc. Natl. Acad. Sci. USA* **91,** 6319–6323.
32. Eshleman, J. R., Lang, E. Z., Bowerfind, G. K., Parsons, R., Vogelstein, B., Willson, J. K. V., Veigl, M. L., Sedwick, W. D., and Markowitz, S. D. (1995). Increased mutation rate at the *hprt* locus accompanies microsatellite instability in colon cancer. *Oncogene* **10,** 33–37.
33. Eshleman, J. R., Markowitz, S. D., Donover, P. S., Lang, E. Z., Lutterbaugh, J. D., Li, G. M., Longley, M., Modrich, P., Veigl, M. L., and Sedwick, W. D. (1996). Diverse hypermutability of

multiple expressed sequence motifs present in a cancer with microsatellite instability. *Oncogene* **12,**1425–1432.

34. Bhattacharyya, N. P., Ganesh, A., Phear, G., Richards, B., Skandalis, A., and Meuth, M. (1995). Molecular analysis of mutations in mutator colorectal carcinoma cell lines. *Hum. Mol. Genet.* **4,**2057–2064.
35. Phear G., Bhattacharyya. N. P., and Meuth. M. (1996). Loss of heterozygosity and base substitution at the *APRT* locus in mismatch-repair-proficient and -deficient colorectal carcinoma cell lines. *Mol. Cell. Biol.* **16,**6516–6523.
36. Gossen, J. A., de Leeuw, W. J., Tan, C. H., Zwarthoff, E. C., Berends, F., Lohman, P. H., Knook, D. L., and Vijg, J. (1989). Efficient rescue of integrated shuttle vectors from transgenic mice: a model for studying mutations in vivo. *Proc. Natl. Acad. Sci. USA* **86,**7971–7975.
37. Kohler, S. W., Provost, G. S., Fieck, A., Kretz, P. L., Bullock, W. C., Putman, D. L., Sorge, J. A., and Short, J. M. (1991). Analysis of spontaneous and induced mutations in transgenic mice using a lambda ZAP/lacI shuttle vector. *Environ. Mol. Mutagen.* **18,**316–321.
38. Kohler, S. W., Provost, G. S., Fieck, A., Kretz, P. L., Bullock, W. C., Sorge, J. A., Putman, D. L., and Short, J. M.. (1991). Spectra of spontaneous and mutagen-induced mutations in the lacI gene in transgenic mice. *Proc. Natl. Acad. Sci. USA* **88,** 7958–7962.
39. Summers, W. C., Glazer, P. M., and Malkevich, D.. (1989). Lambda phage shuttle vectors for analysis of mutations in mammalian cells in culture and in transgenic mice. *Mutat. Res.* **220,** 263–268.
40. Glazer, P. M., Sarkar, S. N., and Summers, W. C. (1986). Detection and analysis of UV-induced mutations in mammalian cell DNA using a lambda phage shuttle vector. *Proc. Natl. Acad. Sci. USA* **83,** 1041–1044.
41. Gossen, J. A., de Leeuw, W. J., Verwest, A., Lohman, P. H., and Vijg, J. (1991). High somatic mutation frequencies in a LacZ transgene integrated on the mouse X-chromosome. *Mutat. Res.* **250,** 423–429.
42. Winegar, R. A., Lutze. L. H., Hamer. J. D., O'Loughlin. K. G., and Mirsalis. J. C. (1994). Radiation-induced point mutations, deletions and micronuclei in lacI transgenic mice. *Mutat. Res.* **307,**479–487.
43. Hoorn, A. J., Custer, L. L., Myhr, B. C., Brusick, D., Gossen, J., and Vijg, J. (1993). Detection of chemical mutagens using Muta Mouse: a transgenic mouse model. *Mutagenesis* **8,** 7–10.
44. Gorelick, N. J., Tindall, K. R., and Glickman, B. W. (1996). Introduction: state of the art in transgenic animals in mutation research. *Environ. Mol. Mutagen.* **28,** 295–298.
45. Gunther, E. J., Murray, N. E., and Glazer, P. M. (1993). High efficiency, restriction-deficient in vitro packaging extracts for bacteriophage lambda DNA using a new E. coli lysogen. *Nucleic Acids Res.* **21,** 3903–3904.
46. Gunther, E. J., Yeasky, T. M., Gasparro, F. P., and Glazer, P. M. (1995). Mutagenesis by 8-methoxypsoralen and 5-methylangelicin photoadducts in mouse fibroblasts: mutations at cross-linkable sites induced by monoadducts as well as cross-links. *Cancer Res.* **55,** 1283–1288.
47. Yuan, J., Yeasky, T. M., Rhee, M. C., and Glazer, P. M. (1995). Frequent T:A→G:C transversions in X-irradiated mouse cells. *Carcinogenesis* **16,** 83–8.
48. Yuan, J., Yeasky, T. M., Havrem P. A., and Glazerm P. M. (1995). Induction of p53 in mouse cells decreases mutagenesis by UV radiation. *Carcinogenesis,* 2295–2300.
49. Leach, E. G., Gunther, E. J., Yeasky, T. M., Gibson, L. H., Yang-Feng, T. L., and Glazer, P. M. (1996). Frequent spontaneous deletions at a shuttle vector locus in transgenic mice. *Mutagenesis* **11,** 49–56.
50. Reynolds, T. Y., Rockwell, S., and Glazer, P. M. (1996). Genetic instability induced by the tumor microenvironment. *Cancer Res.* **56,** 5754–5757.
51. Narayanan, L., Fritzell, J. A., Baker, S. M., Liskay, R. M., and Glazer, P. M. (1997). Elevated levels of mutation in multiple tissues of mice deficient in the DNA mismatch repair gene, *Pms2. Proc. Natl. Acad. Sci. USA* **94,** 3122–3127.
52. Parris, C. N., Levy, D. D., Jessee, J., and Seidman, M. M. (1994). Proximal and distal effects of sequence context on ultraviolet mutational hotspots in a shuttle vector replicated in xeroderma cells. *J. Mol. Biol.* **236,** 491–502.
53. Wang, G., Levy, D. D., Seidman, M. M., and Glazer, P. M. (1995). Targeted mutagenesis in mammalian cells mediated by intracellular triple helix formation. *Mol. Cell. Biol.* **15,** 1759–1768.
54. Leach, E. G., Narayanan, L., Havre, P. A., Gunther, E. J., Yeasky, T. M., and Glazer, P. M. (1996). Tissue specificity of spontaneous point mutations in l supF transgenic mice. *Environ. Mol. Mutagen.* **28,** 459–464.
55. Hanawalt, P. C. (1994). Transcription-coupled repair and human disease. *Science* **266,** 1957–8.
56. Datta, A., and Jinks-Robertson, S. (1995). Association of increased spontaneous mutation rates with high levels of transcription in yeast. *Science* **268,** 1616–9.
57. Branch, P., Aquilina, G., Bignami, M., and Karran, P. (1993). Defective mismatch binding and a mutator phenotype in cells tolerant to DNA damage. *Nature* **362,** 652–654.
58. Kat, A., Thilly, W. G., Fang, W. H., Longley, M. J., Li, G. M., and Modrich, P. (1993). An alkylation-tolerant, mutator human cell line is deficient in strand-specific mismatch repair. *Proc. Natl. Acad. Sci. USA* **90,** 6424–6428.
59. Streisinger, G., Okada, Y., Emrich, J., Newton, J., Tsugita, A., Terzah, E., and Inoye, M. (1966). Frameshift mutations and the genetic code. *Cold Spring Harbor Symp. Quant. Biol.* **31,** 77–84.
60. Parsons, R., Myeroff, L. L., Liu, B., Willson, J. K., Markowitz, S. D., Kinzler, K. W., and Vogelstein, B. (1995). Microsatellite instability and mutations of the transforming growth factor beta type II receptor gene in colorectal cancer. *Cancer Res* **55,** 5548–50.
61. Postel, E. H., Flint, S. J., Kessler, D. J., and Hogan, M. E. (1991). Evidence that a triplex-forming oligodeoxyribonucleotide binds to the c-myc promoter in HeLa cells, thereby reducing c-myc mRNA levels. *Proc. Natl. Acad. Sci. USA* **88,** 8227–31.
62. Briggs, M. R., Kadonaga, J. T., Bell, S. P., and Tjian, R. (1986). Purification and biochemical characterization of the promoter-specific transcription factor, Sp1. *Science* **234,** 47–52.
63. Neddermann, P., and Jiricny, J. (1994). Efficient removal of uracil from G:U mispairs by the mismatch-specific thymine DNA glycosylase from HeLa cells. *Proc. Natl. Acad. Sci. USA* **91,** 1642–1646.
64. Andrew, S. E., Pownall, S., Fox, J., Hsiao, L., Hambleton, J., Penney, J. E., Kohler, S. W., and Jirik, F. R. (1996). A novel lacI transgenic mutation-detection system and its application to establish baseline mutation frequencies in the scid mouse. *Mutat. Res.* **357,** 57–66.
65. Schaaper, R. M., and Dunn, R. L. (1987). Spectra of spontaneous mutations in *Escherichia coli* strains defective in mismatch correction: the nature of *in vitro* DNA replication errors. *Proc. Natl. Acad. Sci. USA* **84,** 6220–6224.
66. Dohet, C., Wagner, R., and Radman, M. (1985). Repair of defined single basepair mismatches in *Escherichia coli. Proc. Natl. Acad. Sci. USA* **82,** 503–505.
67. Reenan, R. A., and Kolodner, R. D. (1992). Isolation and characterization of two Saccharomyces cerevisiae genes encoding homo-

logs of the bacterial HexA and MutS mismatch repair proteins. *Genetics* **132,** 963–973.

68. Strand, M., Prolla, T. A., Liskay, R. M., and Petes, T. D. (1993). Destabilization of tracts of simple repetitive DNA in yeast by mutations affecting DNA mismatch repair. *Nature* **365,** 274–276.
69. Branch, P., Hampson, R., and Karran, P. (1995). DNA mismatch binding defects, DNA damage tolerance, and mutator phenotypes in human colorectal carcinoma cell lines. *Cancer Res.* **55,** 2304–2309.
70. Schaaper, R. M., Danforth, B. N., and Glickman, B. W. (1986). Mechanisms of spontaneous mutagenesis: an analysis of the spectrum of spontaneous mutation in the Escherichia coli lacI gene. *J. Mol. Biol.* **189,** 273–284.
71. Cox, E. C. (1976). Bacterial mutator genes and the control of spontaneous mutation. *Annu. Rev. Genet.* **10,** 135–156.
72. Leong, P. M., Hsia, H. C., and Miller, J. H.. (1986). Analysis of spontaneous base substitutions generated in mismatch-repair-deficient strains of Escherichia coli. *J. Bacteriol.* **168,** 412–416.
73. Cooper, D. N., and Youssoufian, H. (1988). The CpG dinucleotide and human genetic disease. *Hum. Genet.* **78,** 151–155.
74. Gu, M., Ahmed, A., Wei, C., Gorelick, N., and Glickman, B. W. (1994). Development of a lambda-based complementation assay for the preliminary localization of lacI mutants from the Big Blue mouse: implications for a DNA-sequencing strategy. *Mutat. Res.* **307,** 533–540.
75. Sloane, D. L., Goodman, M. F., and Echols, H. (1988). The fidelity of base selection by the polymerase subunit of DNA polymerase III holoenzyme. *Nucleic Acids Res.* **16,** 6465–6475.
76. Beroud, C., Verdier, F., and Soussi, T. (1996). p53 gene mutation: software and database. *Nucleic Acids Res.* **24,** 147–150.
77. Fearon, E. R., and Jones, P. A. (1992). Progressing toward a molecular description of colorectal cancer development. *FASEB J.* **6,** 2783–2790.
78. Huang, J., Papadopoulos, N., McKinley, A. J., Farrington, S. M., Curtis, L. J., Wyllie, A. H., Zheng, S., Willson, J. K. Markowitz, S. D., Morin, P., Kinzler, K. W., Vogelstein, B., and Dunlop, M. G. (1996). APC mutations in colorectal tumors with mismatch repair deficiency. *Proc. Natl. Acad. Sci. USA* **93,** 9049–9054.
79. Modrich, P. (1994). Mismatch repair, genetic stability, and cancer. *Science* **266,** 1959–1960.
80. Hamilton, S. R., Liu, B., Parsons, R. E., Papadopoulos, N., Jen, J., Powell, S. M., Krush, A. J., Wood, P. A., Taqi, F., Booker, S. V., Petersen, G. M., Offerhaus, G. J. A., Tersmette, A. C., Giardiello, F. M., Voogelstein, B., and Kinzler, K. W. (1995). The molecular basis of Turcot's syndrome. *N. Engl. J. Med.* **332,** 839–847.

CHAPTER 37

Human Germline Mutation Analysis by Single Genome PCR: Application to Dynamic Mutations

ESTHER P. LEEFLANG Molecular Biology Program, University of Southern California, Los Angeles, California 90089

SIMON TAVARÉ Molecular Biology Program, University of Southern California, Los Angeles, California 90089; and Department of Mathematics, University of Southern California, Los Angeles, California 90089

PAUL MARJORAM Department of Mathematics, University of Southern California, Los Angeles, California 90089

RAJI GREWAL, CAROLYN O. S. NEAL, AND NORMAN ARNHEIM[1] Molecular Biology Program, University of Southern California, Los Angeles, California 90089

[1]To whom correspondence should be addressed.

I. THE DYNAMIC MUTATION PROCESS

Trinucleotide repeat disease alleles can undergo "dynamic" mutations in which repeat number can change when a disease gene is transmitted from an affected parent to an offspring (reviewed in [1]). The molecular basis of dynamic mutation is of great fundamental interest and stands in contrast to the static nature of classical nucleotide substitution mutations when transmitted through families. The discovery of trinucleotide repeat diseases has stimulated interest in studying length mutations in microsatellite repeats with the aim of understanding fundamental aspects of the expansion and contraction processes.

In the dynamic mutation process, repeat instability can be influenced by the number of repeats, the sex of the transmitting parent, and the "purity" of the repeat tract [1]. A role for DNA sequences linked and unlinked to the repeat region has been proposed [2–4]. Additional factors such as age and even environmental influences may also be involved.

Studies on model systems involving trinucleotide repeats cloned in *Escherichia coli,* yeast, and mice [5–10] have not yet been able to reproduce all the mutation properties typical of the transmission of many large human disease alleles. Thus, understanding the contribution of the different factors involved in germline instability of trinucleotide repeats in humans must be grounded in a description of the human mutation process.

II. ADVANTAGES OF USING SINGLE SPERM GENOMES FOR ANALYSIS OF GERMLINE MICROSATELLITE MUTATIONS: COMPARISON TO FAMILY STUDIES

The study of microsatellite mutations in humans is traditionally based on family studies and is subject to several limitations. Since only a small number of offspring are available from any one family, the number of available germline transmissions is relatively small and is insufficient to estimate accurately the mutation frequency in the parents. To obtain meaningful estimates, data must be pooled from many different families. Another potential problem is that family studies based on analysis of DNA derived from lymphoblastoid cell lines have dinucleotide repeat mutation frequencies higher than in untransformed lymphocytes [11].

The finding that PCR can be used to determine the genotype at a locus in a single human diploid or sperm cell using PCR [12] or a diluted molecule from sperm [13] led to alternative approaches to studying human germline recombination and mutation. Single genome analysis of gametes is a more precise way of measuring mutation frequencies. The large sample sizes afforded by single genome analysis make it possible to measure accurately the mutation frequency and size distribution of mutant sperm (mutation spectrum) at trinucleotide repeat loci in a single individual. Since the meiotic products themselves are studied directly, mutations occurring during the culture of lymphoblastoid lines and the effects of biological selection following gametogenesis or fertilization are eliminated. The ability to study mutations in a single individual makes it easier to design experiments to determine the role that different factors play in contributing to microsatellite repeat instability.

III. METHODS OF SINGLE GENOME STUDY

A germline mutation is defined in reference to the size of the allele inherited at the time of fertilization. In many of the trinucleotide repeat diseases this can easily be determined by analysis of the allele size in somatic tissues of that individual. In some of the diseases, however (e.g., Frax A and DM), this is not so easily accomplished since there can be extensive allele size heterogeneity in somatic tissues [1]. Single genome analysis of trinucleotide repeat germline mutations has been carried out in a number of ways (see below). Which method to use will depend on many factors including any problems associated with amplifying the particular locus, the mutation frequency, and the size distribution of the mutations.

A. Single Sperm Analysis

One method involves isolation of single sperm cells (using flow cytometry, micromanipulation, or dilution) followed by lysis, PCR, and allele length analysis using electrophoresis. Single sperm typing has been carried out at the HD, SBMA, SCA-1, DM, SCA-3, and Frax A loci including normal, premutation, and disease alleles [14–20].

B. Small Pool PCR

An alternative approach is to use small pool PCR (SP-PCR) [21]. Dilutions of DNA purified from tens of millions of sperm are made such that each sample contains 5–10 genomes. In such a diluted sample, PCR product from a mutant genome will have an altered electrophoretic mobility compared to PCR product from the unmutated genomes. This method can more easily detect mutations involving the addition or deletion of many trinucleotide repeats, but one or two repeat changes in large alleles are more difficult to verify due to overlapping of the PCR "stutter" bands. SP-PCR has been applied to the germline analysis of Frax A premutation alleles [22] and both premutations and disease causing DM alleles [23].

C. Single Molecule Dilution

More extensive DNA dilutions, such that each sample carries on average a single or a fraction of a genome, can also be made. This approach has been used successfully to study dinucleotide repeat mutations in human tumor cells and tissues of mice lacking DNA mismatch repair [24–27] as well as trinucleotide repeat mutations in somatic tissues of patients with Huntington's disease [17]. If the mutation frequency is high, the sperm DNA should be diluted to well below single copy level to avoid samples containing two mutations. Compared to single sperm typing, a larger number of PCR assays are required since many aliquots will contain no template molecules. An advantage is that it requires no special technologies for sperm isolation.

D. Total Sperm PCR

Crude analysis of mutation spectra can be made in single individuals by carrying out PCR on total DNA samples purified from semen [28, 29]. Based on the extent of the smear of PCR products after electrophoresis, the degree of triplet repeat mutation can be evaluated on a gross level, with maximum, minimum, and modal values of allele size deduced from densitometric analysis. One advantage of this method is that the general level of stability at a locus can be estimated in a very large number of individuals with relatively little effort. Problems with this approach include the lack of detail in the mutation spectra which are critical for some analyses (see below). For example, faint signals derived from a small subset of the sperm population may be obscured by the inherent PCR stutter from more abundant allele sizes. Perhaps more serious is the possibility that data from amplification of total sperm DNA will be strongly biased toward smaller alleles due to their competitive advantage during PCR.

IV. *IN VITRO* PCR ARTIFACTS ARE MINIMAL

Single genome analysis requires many PCR cycles for detection of PCR product. Because of the possibility that PCR artifacts might result in *in vitro* generated mutations, specific control experiments have been designed to assess the likelihood of such artifacts.

A. Contamination

One possible artifact is that any observed mutation might result from contamination by DNA template derived from another individual or previously amplified PCR product. Experiments that yield positive results from no DNA template control samples should be viewed with caution. Contamination of the sample DNA with the genomic DNA from another individual can be examined in the case of single sperm typing by simultaneously amplifying tightly linked informative markers [17, 30].

B. PCR Stutter

Another possible artifact is that mutations may result from events that take place during PCR. Amplification of most di- and trinucleotide repeats using total genomic DNA as template typically results in the appearance of minor PCR products with altered repeat lengths. This could be the consequence of somatic heterogeneity in the tissue from which the DNA was isolated or DNA replication slippage events during amplification. That PCR alone is capable of producing "stutter" is documented by the stutter patterns detected when using a single DNA molecule as template [17, 19, 20, 31]. Only PCR artifacts that occur in the first few cycles have a chance of producing detectable amounts of product which could be counted as a true *in vivo* mutation. Such false mutations are more likely to be observed as contractions than expansions since a smaller template generated by a PCR slippage error could increase in relative frequency due to a selective advantage during amplification.

C. Amplification Bias

If the PCR assay detects unmutated allele sizes more readily than mutated allele sizes a bias in the mutation frequency and mutation spectrum will result. This could occur if many mutants were expanded beyond a size possible for efficient amplification. A specific criterion can be used to test for such a bias. According to Mendel's laws, approximately 50% of single sperm genomes from an individual heterozygous for a normal allele and a disease allele would be expected to contain the normal allele. The disease allele and all mutations derived from it should make up the remaining 50%. The detection of a significantly large excess of the normal allele could suggest that not all mutations are detectable. In the case of HD [17] there does not appear to be any bias. This might not be the case for very large Frax A or DM alleles, which are more difficult to amplify. An alternative explanation for a deviation from Mendel's laws in sperm typing data is segregation distortion [18, 30]. Specific methods for detecting segregation distortion in sperm typing data have been developed [30].

D. Experimental Approaches to PCR Artifact Detection

An experimental strategy has been used [19, 32] to determine whether a single sperm identified as a mutant was misclassified due to an artifact during amplification or resulted from a true germline mutation event (Fig. 37-1). Single sperm are amplified for 6 PCR cycles. Half of the PCR product is removed from each sample and saved. The reaction is continued without any interruption with the remaining material and the final products analyzed to detect samples with a mutation. If a sperm with a mutation is identified, the corresponding PCR product that had been saved after 6 cycles can be analyzed. If the original sperm had a mutation, the molecules in the saved portion (half the duplexes produced after 6 cycles, or approximately 32 molecules maximally) should have the mutated allele. However, if the sample originally contained a normal molecule and a mutation occurred in the first PCR cycle (or later), a mixture of normal and mutant molecules would be present. Thus, to distinguish between *in vitro* and *in vivo* mutations, PCR is carried out on single molecule dilution aliquots from the saved sample that had undergone the first 6 PCR cycles. If normal molecules are present in these aliquots, the original sperm must have been normal. We have never found evidence for PCR artifacts contributing to misidentification of a sperm as being normal or mutant.

Another way of assessing the possible role of PCR artifacts in single genome analysis is to compare the size distribution of sperm mutations with the size distribution of mutations in somatic DNA from the same individual as determined by single molecule dilution experiments [17]. For example, analysis of HD sperm [17] from one donor with 51 CAG repeats in his somatic DNA showed a mutation frequency of 99% with a mean change of +21 repeats. Single molecule dilution analysis on somatic lymphoblastoid DNA from this donor was also carried out. Among the 30 disease-length molecules examined, seven events with a change of −1 repeats and three events with a change of +1 repeats were detected for a mean change of −0.13 repeat. Whether the variation in allele size in somatic DNA is due to PCR artifact, some variation in allele size among individual lymphoblast cells (possibly occurring during cell culture) or due to the inherent error in the exact measurement of molecular weight of the PCR product cannot be determined. Similar studies were carried out on sperm and somatic DNA from a donor with 36 repeats [17] which also showed that somatic variation was restricted compared to that of the germline. We conclude that the observed variation in single sperm allele sizes is due to germline mutation events.

V. SINGLE SPERM STUDIES ON SBMA

SBMA is a rare disease caused by CAG repeat expansions in the coding region of the X-linked human androgen receptor (AR) gene [33]. Clinically, SBMA is characterized by adult-onset proximal muscular atrophy with bulbar involvement and slow progression [34]. Affected individuals have more than 40 repeats.

A. Mutability of Non-Disease-Causing Alleles

The first studies to analyze trinucleotide repeat germline mutations using single genome analysis [19] focused

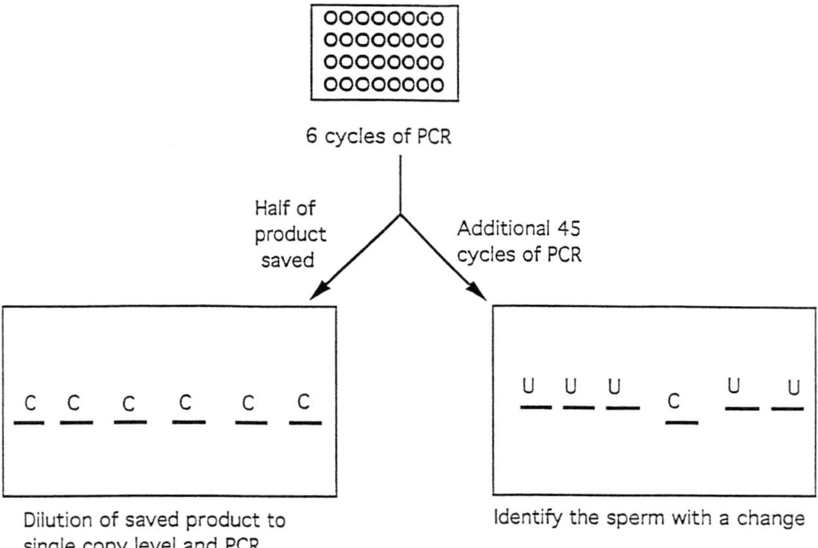

FIGURE 37-1 Outline of the control experiment showing that mutations detected by sperm typing do not result from PCR artifacts. Six cycles of PCR are carried out on individual sperm in a microtiter plate. Half the volume from each well of the microtiter plate is taken and saved. PCR on the remaining sample in each well is continued. Electrophoresis of PCR product (lower right) will reveal whether any sperm is unmutated (U) or underwent a mutation (C = contraction). The saved aliquot from the well exhibiting the contraction is diluted to single copy level and amplified. Following electrophoresis (lower left) only mutated PCR products are expected if the original sperm itself had been mutated *in vivo*. Reprinted by permission of Nature Publishing.

on how disease alleles might arise from normal alleles at the AR locus. The mutation frequency of normal alleles with the average repeat number in the population (20–22 repeats) was compared to the mutation frequency of alleles at the highest end of the normal range (28–31 repeats) but below the repeat number that causes disease.

Of 685 informative sperm from three individuals with alleles in the average size range (20–22 repeats), 1.33% were found to be mutated. Among 1253 informative sperm from four normal individuals with alleles in the 28–31 repeat range, the mutation frequency was 3.21%, a significantly larger proportion.

Most mutations were contractions. Contractions among the 28- to 31-repeat alleles occurred at a frequency of 2.9% compared to a frequency of 0.9% for the alleles of average size. This difference is also statistically significant. On the other hand, the expansion frequencies of the two allele size classes were much smaller and not significantly different from each other (0.31 and 0.43%, respectively). About half of the contractions and expansions observed involved one or two repeat changes.

It is interesting to note that the great excess of contractions over expansions (9:1) typical of the large (28–31 repeats) normal alleles at the AR locus would make it difficult for a normal allele to reach the repeat number required to become a disease allele. This may explain the lack of success in detecting alleles in the 33–39 repeat range in studies of normal populations. This would also explain the low prevalence of SBMA and why, compared to Huntington's disease, no new mutations have been documented for SBMA [35]. Additionally, the sexual dysfunction associated with SBMA may also limit the number of 33–39 repeat alleles introduced into the population by contractions of disease-causing alleles in affected individuals (see Section V.B).

B. Instability of Disease-Causing Alleles

Studies on 1538 single sperm from two SBMA patients sperm ([20] and Grewal *et al.*, unpublished studies) revealed mutation frequencies of 80% (49 repeat individual) and 81% (47 repeat individual). The mutation spectrum for one individual [20] is shown in Fig. 37-2. The expansion mutation frequencies (66 versus 55%) and mean change in allele size for expansions (+2.7 versus +2.1 CAG repeats) were not significantly different between the two patients.

These sperm typing results are consistent with the limited data on transmissions of SBMA alleles pre-

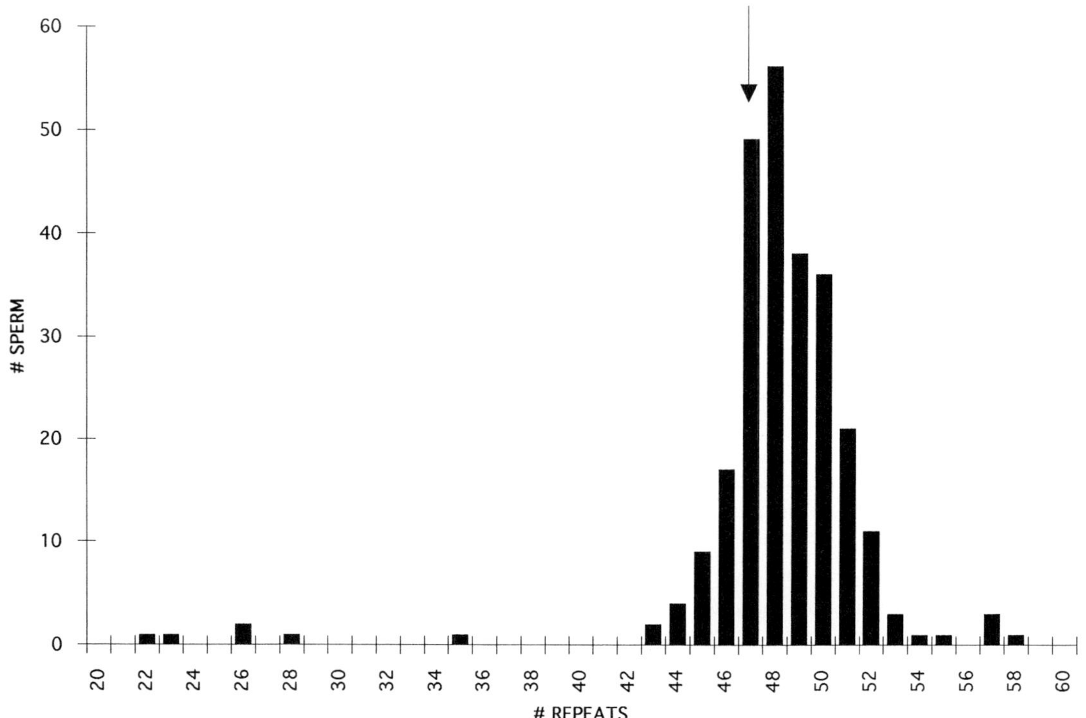

FIGURE 37-2 Distribution of allele size changes in single sperm from a patient with SBMA. The length of the donors SBMA allele in somatic cells was 47 repeats. Arrow denotes the length of the allele in somatic cells. Reprinted by permission of Nature Publishing.

viously reported in pedigrees [34, 36, 37]. In these families, only 17 transmissions via affected fathers are recorded. Eighteen percent showed no change, 12% had one repeat contractions, and 70% had expansions with an average gain of three repeats. Although the family data are generally consistent with the sperm typing data, a more detailed comparison can only be made if larger numbers of family studies with paternal transmissions become available. This will be difficult given the low incidence of the disease.

C. Comparison of the AR Locus Mutation Properties between Normal and Disease Alleles

The trinucleotide repeat mutation frequencies at the AR locus of normal individuals can be compared to those of the two SBMA patients. As the CAG repeat number increases from normal to the disease range, the average frequency of contraction increases about 20-fold. The average expansion frequency increases disproportionately by about 165-fold (from 0.37 to 61%). Since the contraction frequency does not decrease proportionally with the increase in the expansion frequency, the two mutational events show an independence that may be due to distinct molecular mechanisms [19]. For example, an additional mutation process may engage only when allele size exceeds a threshold (disease-causing alleles) and lead primarily, unlike the situation for normal alleles where contractions predominate, to expansions.

VI. SINGLE SPERM STUDIES ON DM

A study on three sperm donors with large normal alleles at the DM locus was carried out for comparison with the AR locus [19]. The donors were heterozygous for one allele at the high end of the normal range and the 5-repeat allele predominant in the general population. The single sperm were simultaneously amplified for a tightly linked polymorphic marker to distinguish which of the two DM alleles underwent a mutation event. Based on 788 observations on all three individuals, no mutations were confirmed in the 5-repeat allele class. The mutation frequency estimate of a 20-repeat DM allele (based on 249 observations) revealed contraction and expansion frequencies (0.8 and 0.4%, respectively), quite similar to the results for the 20- to 22-repeat allele size class at the AR locus. Analysis of a 27-repeat DM allele revealed a mutation behavior quite different from normal alleles of a similar size at the AR locus. Based on 277 observations, the mutation frequency of the 27 repeat allele was estimated to be 6%. Even more surprising was the finding that expansions were 5.5 times more frequent than contractions. Finally, a 37-repeat allele that was expected to exhibit an even higher mutation frequency showed only 1% expansions. Unlike the 27-repeat allele, this larger allele underwent more contractions (2.6%) than expansions.

The observation that the individual with 27 repeats at the DM locus has an expansion frequency greater than the donor with 37 repeats was unexpected. We subsequently determined the DNA sequence of the 37-repeat allele. Unlike all other alleles reported at the DM locus we found that this allele had the structure $(CTG)_4(CCGCTG)_{16}CTG$ [31]. This structure could explain why the 37-repeat allele has a lower mutation frequency than the 27-repeat allele. Interrupted repeat tracts, as seen in normal alleles of SCA1 [38], SCA2 [39–41], and Frax A [14, 42, 43], play an important role in reducing instability.

VII. SINGLE SPERM STUDIES ON HD

Huntington's disease (HD) is a dominant hereditary neurodegenerative disorder, usually with an onset in middle age, and associated with progressive disordered movements, decline in cognitive function, and emotional disturbance. A large, worldwide study [44] showed that the repeat number at the HD locus in normal individuals ranges from 10 to 29 repeats. New disease alleles presumably arise as a result of a gradual increase of normal alleles into a repeat number range (29–36 repeats) referred to herein as intermediate size alleles or reduced penetrance alleles [45–48]. These alleles can, in a single generation, increase in size so as to cause clinical symptoms. Disease alleles can be very much more unstable, especially in paternal transmissions.

A. Instability of Normal HD Alleles

Data on 475 sperm from five normal HD alleles [17] in the 15–18 repeat range estimate a mutation frequency of 0.6% resulting from one- or two-repeat contractions, similar to a frequency estimate based on pedigree analysis (0.2%) [49]. Both estimates are uncertain considering the small number of mutations observed. Note also that sperm typing estimates measure the mutation frequency of specifically chosen allele sizes. Measuring the mutation frequency for a particular allele size based on pooling data from many different families segregating for the same size allele will almost always lack sufficient numbers for an accurate estimate. If the data for different allele sizes are pooled, the estimate of the mutation frequency is not allele size specific but represents an average mutation frequency based on the average of the allele sizes represented in the families.

B. Instability of Intermediate Size HD Alleles

Alleles with as few as 29 repeats have been found in fathers of offspring clinically affected with HD from families with no previous history of the disease. It is clear, however, that the definition of a new mutation depends upon whether the father does or does not exhibit a clinical phenotype (for a discussion see [16, 45]).

The instability of an intermediate allele carried by a father can be examined by sperm typing. Among 163 sperm analyzed from an unaffected individual with a 36-repeat allele and in his fifth decade [17], 16% of the gametes derived from the 36-repeat allele repeat (Fig. 37-3A) had expanded into the allele size range expected to give a clinical phenotype (defined for this purpose as a ≥38-repeat allele). The chance of having a clinically affected child is, for this man, 8%. Recent sperm typing studies [16] on an individual heterozygous for a 35-repeat allele indicate a comparable risk of 2.3%. Studies on two other individuals with 29 and 30 repeats [50] showed the risk of having a clinically affected child as 0.1 and 0.3%, respectively.

The risk of a male carrier of an intermediate allele having a child with HD may vary among individuals as a consequence of *cis*-acting DNA sequences, related to their haplotype [51], that affect mutation frequency (see [16]). Regardless, the risk to any particular father carrying an intermediate allele of having an affected child can be determined directly using single molecule analysis.

C. Mutation Frequency of Disease-Causing Alleles

Studies on three sperm donors heterozygous for alleles with repeat numbers >39 have been published [17]. Among these individuals, 316 normal alleles were detected, while 287 HD alleles were observed; this is not statistically different from the expected 1:1 segregation. Thus, both large and small alleles can be amplified with almost equal efficiency.

Among the 287 sperm carrying disease causing alleles from these three individuals, 96% differed in size from the donors' somatic HD allele; 93% were expansions and 3% were contractions. Two-thirds of all the samples were studied to distinguish between alter-

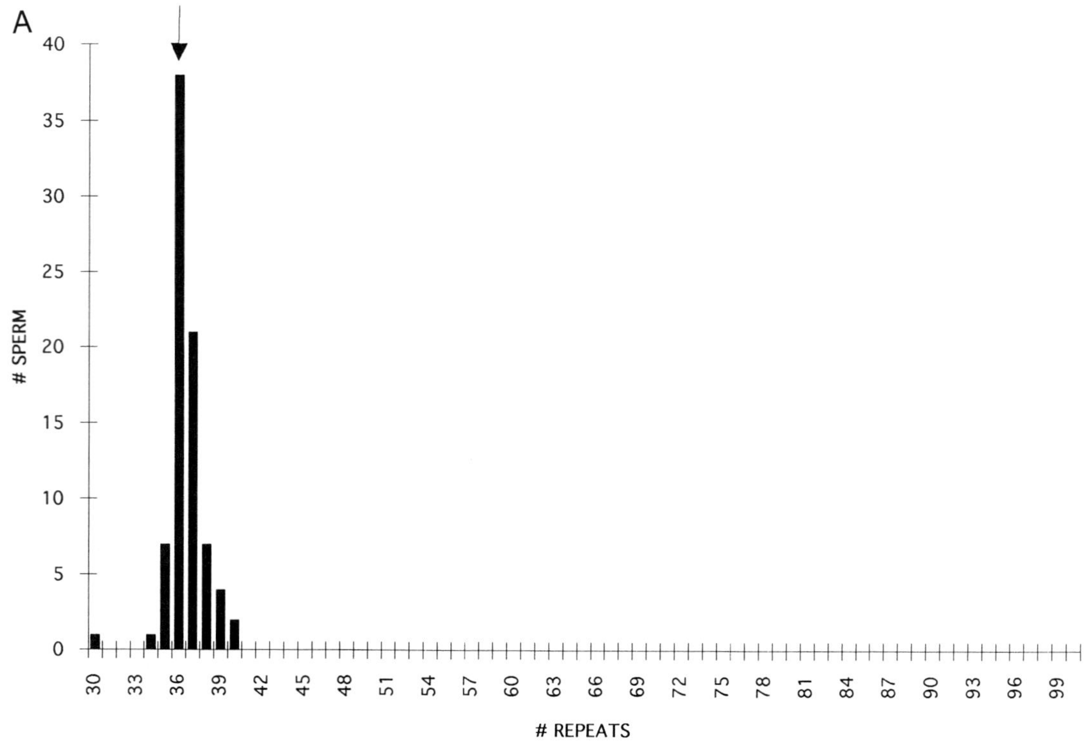

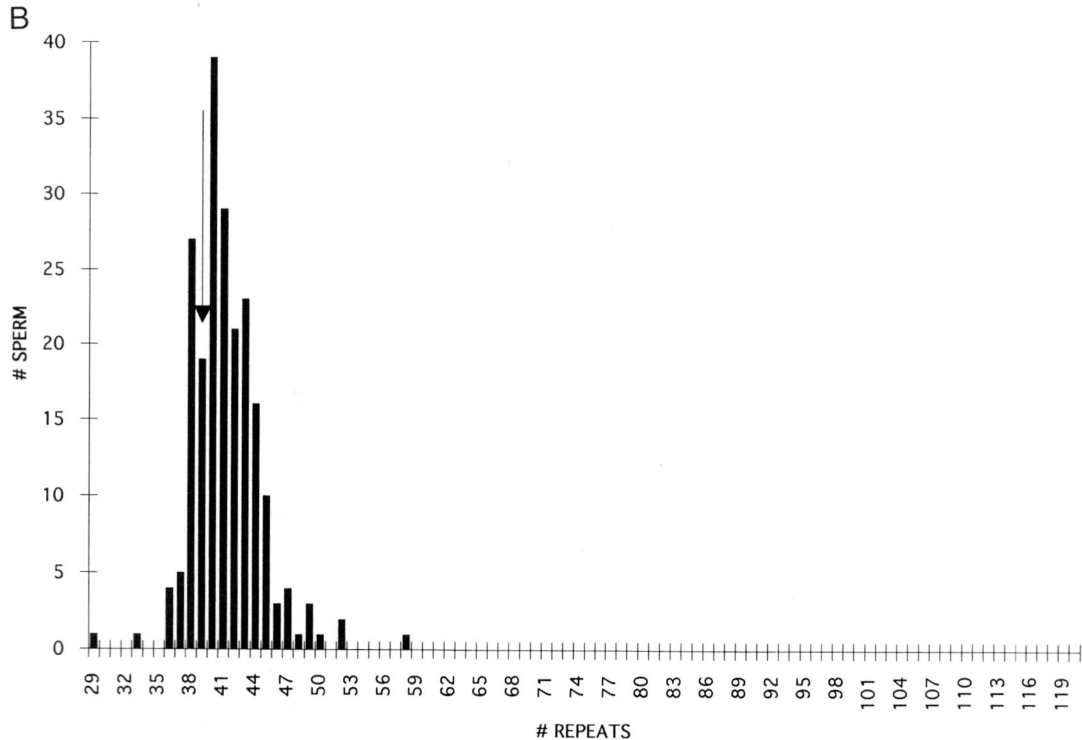

FIGURE 37-3 Mutation spectra of sperm donated by four individuals heterozygous for a 36 (A), 39 (B), and 49 (C) and 51 (D) repeat HD allele. A sample from the individual with 39 repeats but taken 2 years later is also shown (E). Arrows denote the length of the donor's HD allele in somatic cells determined by analysis of lymphoblast DNA. Figures A–D reproduced from *Hum. Mol. Genet.* **4,** 1519–1526, 1995, by permission of Oxford University Press.

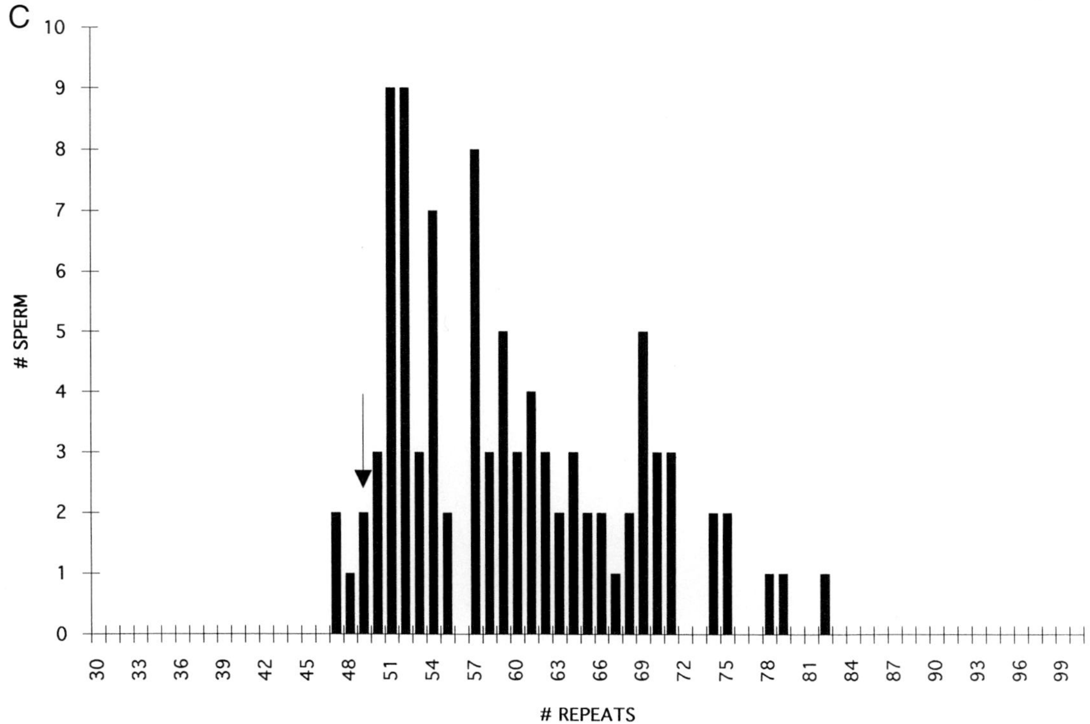

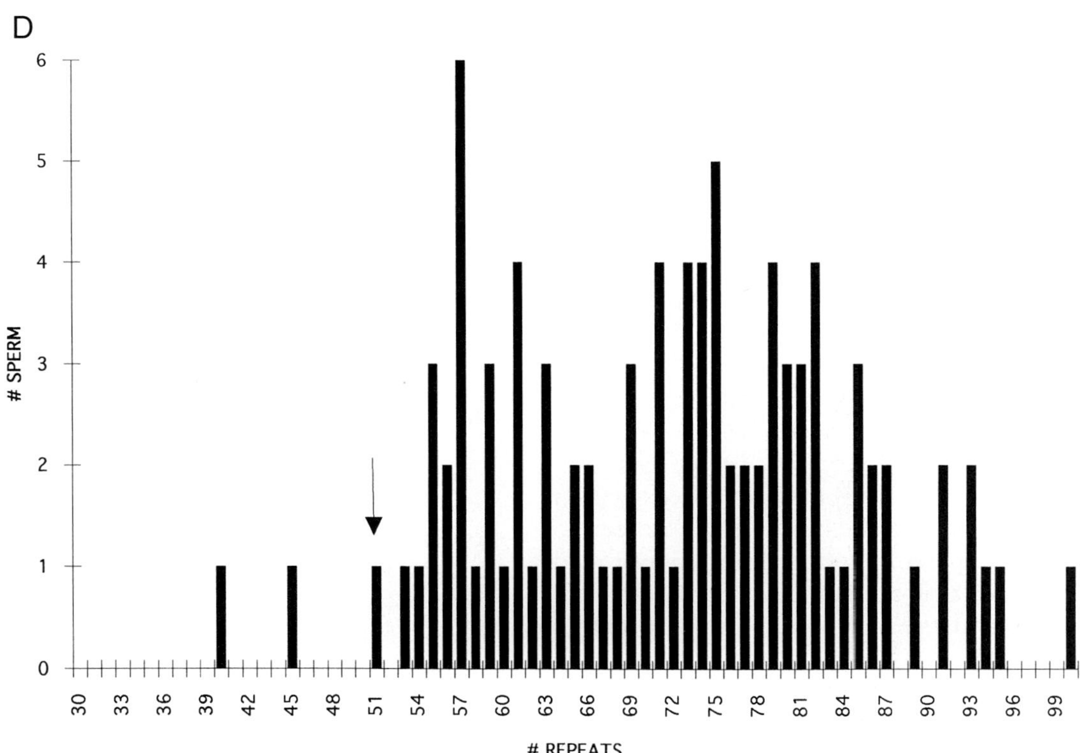

FIGURE 37-3 *(continued)*

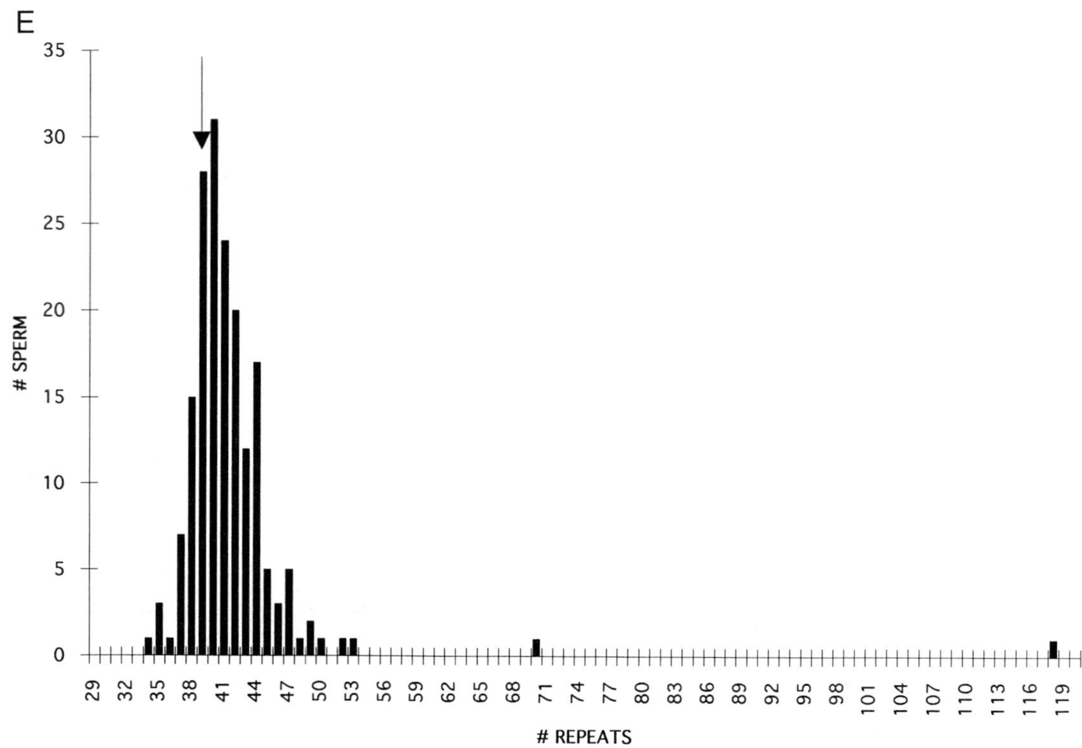

FIGURE 37-3 *(continued)*

ations in the number of CAG repeats and in the less polymorphic and immediately adjacent CCG repeats [52, 53]. No mutations in CCG repeat number were detected.

D. The Size Distribution of Mutations Derived from Disease-Causing Alleles

The CAG repeat number distribution in single sperm donors with 39, 49, and 51 repeat tracts are shown in Figs. 37-3B–37-3D, respectively [17]. There is a significant change in the size distribution of the expansion mutations with increasing allele size. It can be seen to progress from an apparently normal distribution around the somatic DNA size in the case of the 36-repeat allele (Fig. 37-3A), to a markedly more uniform distribution of expanded alleles up to twice the size of the somatic DNA in the case of the larger alleles. Contraction sizes were generally limited to 6 repeats or less.

E. Mutation Spectra of Two Samples Taken from the Same Individual

We were also able to estimate the constancy of HD allele size distributions on a time scale of 2 years by analyzing sperm samples taken at age 63 (Fig. 37-3B) and age 65 (Fig. 37-3E) [54] from the donor with 39 repeats. The mutation spectra show virtually identical patterns except for two sperm in the older sample that had allele sizes considerably larger than those seen in the younger sample.

F. Effects on Mutation of *Cis* Polymorphisms and Parental Origin

Recent analysis of instability in MJD families reported that heterozygosity for a single nucleotide polymorphism adjacent to the repeated region influences the instability of the disease-causing allele [2]. A tightly linked polymorphism exists adjacent to the HD CAG tract [52, 53]. Based on data from 26 HD sperm donors [54] we divided our samples into those that were homozygous for the $(CCG)_7$ allele (7 donors) and those that were heterozygous for $(CCG)_7/(CCG)_{10}$ alleles (19 donors). We found no significant difference between the two groups with respect to mean change in repeat number following mutation (unpublished). Presumably neither the CCG polymorphism, nor any other polymorphisms in linkage disequilibrium with it, contribute significantly to the instability properties of the CAG repeat tracts.

An analysis of the mean change in repeat number based on grouping the HD alleles according to whether they were paternally (7 donors) or maternally (18 donors) inherited was also made. The results (unpublished) reveal that there was no major effect of parental imprinting on instability.

G. Comparison of HD Sperm Typing Data to Family Transmissions

The published single sperm typing data on the HD alleles [17] can be compared with HD paternal transmissions studied in families. In five reports [46, 47, 55, 56] exact repeat number changes can be determined in 141 paternal transmissions. Of these, 72% showed a contraction or expansion mutation. The single sperm data [17] gave a 96% average mutation frequency. Expansions account for 97% of the sperm mutations and 91% (range: 83–100%) of the mutations detected in paternal family transmissions. An average of 5.6 repeats was added for each expansion observed in the pedigrees (range: 3.0 to 9.0) while 12.1 repeats was the average size of an expansion detected in sperm. The differences between the sperm and family data could result from the small number of HD sperm donors studied and the fact that two of the three had an HD allele size significantly larger than the average paternal HD allele size in the families. There may also be differences among the populations studied. Some of the differences between the sperm and family data could also reflect selection against sperm with very large alleles in terms of survival, the ability to carry out fertilization, or postzygotic selection.

H. Comparison of SBMA and HD Disease Allele Instability

Single sperm data from the two SBMA individuals with 47 and 49 CAG repeats showed an 81% average mutation frequency (61% expansions and 20% contractions). The change in the number of repeats averaged 2.4. The largest expansion seen was +11. The sperm data on the HD allele closest in size (49 CAG repeats) showed a 95% expansion and a 3% contraction frequency with an average of 10.8 repeats added. The largest expansion was +33 repeats. Even donors with a smaller number of HD repeats can have a higher mutation frequency, a greater average change in repeat number per mutation, and a broader distribution in mutant allele size than the 47- and 49-repeat SBMA alleles.

The difference in the instability at these two loci is just one more demonstration that repeat number is not the only important factor that contributes to the measured triplet expansion frequency (see [1]). The differences in mutation behavior between the loci remain to be explained.

VIII. ANALYSIS OF SPERM TYPING DATA

A. Elucidating the Contribution of Different Variables to the Trinucleotide Repeat Mutation Frequency

The ability to estimate accurately the mutation frequency for a given trinucleotide repeat allele requires a large sample and can usually be achieved only by single genome methods such as sperm typing. Up to the present, data on only a small number of sperm donors have been published for any one disease. However, if a large sample of individuals could be studied, then the factors that influence the trinucleotide repeat mutation frequency can be examined using routine epidemiological designs. Standard statistical methods could be used to study the relative effects on germline mutation frequency of repeat number, *cis* DNA sequences, unlinked or linked polymorphisms, parental origin of the examined allele, or indeed any other variable. For example, a study of men from the large Venezuelan HD cohort [54] would effectively eliminate any influence of *cis* DNA sequences as a variable since all of the HD alleles derive from the same founder chromosome [57].

B. Understanding the Trinucleotide Repeat Mutation Process

The sperm typing data may also be used to gain insight into the molecular details of the trinucleotide repeat mutation process. One approach is to create a realistic probabilistic model of the mutation process based on current biochemical information [17]. The success of any model can be judged on the basis of how well it describes the actual sperm typing data. Such models can assist in formulating hypotheses to be tested by biochemical, biophysical, and genetic studies. One important question is how the number of repeats influences the trinucleotide repeat mutation rate. For example, mutations in large disease alleles show greater changes in repeat number than smaller disease alleles. A simple explanation is that larger alleles have more triplets and, given a constant mutation rate per triplet during DNA replication, the likelihood of mutation is greater. A number of factors that must be considered when trying to estimate the trinucleotide repeat mutation rate follow.

C. Definition of the Mutation Rate

Estimates of the nucleotide substitution mutation rate for cells in culture using standard methods are defined in terms of mutations per cell division. In humans, mutation rates are often defined in terms of mutations per generation. Unlike "classical" mutations, not all sperm carrying a trinucleotide repeat mutation have necessarily undergone the same number of mutation events. Thus, a sperm which has gained 10 repeats compared to somatic DNA could have undergone the expansion due to a single mutation event during meiosis or have experienced a mixture of multiple, but smaller, expansion and contraction events occurring during mitosis over many germline cell divisions. This uncertainty complicates any estimate of a mutation rate. Indeed, the definition of mutation rate in this case is ambiguous. It could refer to the probability of a new allele being transmitted per human generation, the probability that a new allele will be generated per cell division, or the probability that as each triplet is being replicated a change in repeat number will occur. These ambiguities require new analytical methods to be devised.

D. Molecular Mechanism of Mutation

Both contraction and expansion events in a repeated sequence can result from unequal reciprocal recombination between homologs or sister chromatids. In either case, equal numbers of contractions and expansions should be observed, an expectation that is not fulfilled by the trinucleotide repeat data. Nonreciprocal recombination mechanisms leading to changes in repeat number [58] cannot be excluded, although data from yeast suggest that any role for recombination is unlikely (see [59]). Mutations due to replication slippage are a more likely possibility [60] and recent experimental evidence on microsatellite repeat mutations in yeast and studies on human colon cancer also support this type of mechanism [6, 7, 61–65].

E. Details of the Molecular Mechanism of a DNA Replication Slippage Event

The proximal cause of a slippage event, the number of repeats added or deleted per slippage event, and the number of slippage events that take place as the polymerase traverses the repeated region is unknown. In the case of trinucleotide repeat sequences, mutations may be promoted by the ability of the repeated DNA to form secondary structures (reviewed in [5, 66]). Slippage could result from formation of an intrastrand secondary structure on the replication template. If the replication complex were to "skip over" the folded region, the number of repeats would be reduced. Alternatively, elongation of the nascent strand might be impeded by such a structure, leading to backwards slippage of the nascent strand, additional triplet incorporation, and an expansion mutation.

Slippage mutations could occur when either the leading or lagging strand serves as DNA replication template. Arguments supporting the importance of slippage on the lagging strand include the observation that significant instability at trinucleotide repeat containing loci is first manifested when the length of the repeated region approaches the size of an average mammalian Okazaki fragment [67]. Data on the strand preference of some mutation processes in *E. coli* also support a lagging strand model (reviewed in [43]).

Once DNA replication is completed, mutations can be subject to DNA repair. Yeast and human tumor cells (reviewed in [65]) as well as mice [24–26, 68] that are deficient in DNA mismatch repair have increased dinucleotide repeat mutation frequencies. How DNA repair specifically influences trinucleotide repeat mutations in mammals is not known.

F. Meiotic or Mitotic Origin of Mutations

Trinucleotide repeat mutations could occur during meiosis or throughout the mitotic divisions of the germline. If mutations occur mitotically, the number of cell divisions over which a mutation event may have occurred needs to be estimated. One approach is to consider the age of the donor at the time the sperm sample was collected. Before spermatogenesis begins at puberty (assumed to begin at age 13), the spermatogonial stem cells have undergone an estimated 34 divisions since formation of the zygote. After puberty the stem cells divide approximately 23 times a year [69].

G. Mathematical Modeling

Probabilistic models of molecular mutation mechanisms may be used in conjunction with observed mutation spectra to study the details of the mutation process. Any such model includes a number of unknown parameters, for example those defining the number of triplets added or lost during an expansion or contraction event, and the rate of contraction or expansion events during replication. Comparison of the observed mutation spectra with the spectra predicted by the model can be used to estimate the parameter values that best match the model to the data. Many different numerical methods can be used to accomplish this (for an example that uses maximum likelihood estimation, see [17]). For a given mutation model, even the "best-fitting" one may not provide a good description of the underlying data. To assess this issue, it is usual to compare the best-fitting spectrum with the observed spectrum, both qualitatively and quantitatively, and look for gross inconsistencies. Such inconsistencies indicate the inappropriateness of one or more assumptions in the model, or omission of a key biological feature. Simulation of synthetic data from the fitted model often provides valuable insight into issues involving adequacy of fit, especially when dealing with discrete events such as repeat number, where particular allele sizes may be represented rather infrequently in the data.

It is important to emphasize that in modeling of this sort there is always a trade-off between simplification and the (putative) biological realities. It is clearly not useful to fit a model with 50 parameters to (say) 20 data points. Thus parsimonious description of the biological issues is required. It is also worth emphasizing that recent developments in computing power and in computational statistics have to some extent freed us from many of the drawbacks of models chosen for analytical or mathematical convenience.

Once a variety of different models of mutation have been fitted to the data, it is often possible to distinguish among them statistically. For example, in another report [17] we fitted two models that assumed mutations occur at the last mitotic division before meiosis. The models differed with respect to the magnitude of change in repeat number that accompanied a single mutation event. By comparing the variance in repeat number predicted under the model with the variance observed in the data we found that small incremental changes are more consistent with the data than very large changes.

In recent unpublished work [54], we have been investigating models that address the issue of whether a single mutation process, occurring during the mitotic divisions of spermatogenesis, can explain mutation spectra taken from a large sample of HD patients. Such a model has to follow the cell division history of a repeat region as it evolves from fertilization through the spermatogonial stem cell cycle to meiosis. The details of the mutation mechanism itself also have to be specified; we assumed a slippage mechanism that might be different on the two template strands. We described the probabilistic mechanism of repeat change during replication of a triplet in terms of three parameters: one gives the chance of a ±1-repeat slippage event, another the chance of a large expansion, and a third that describes the features of such an expansion. Coupled with the estimates discussed above for the number of mitotic divisions in the history of a typical sperm, we can use the statistical methods outlined previously to estimate the three parameters and assess the adequacy of the fit.

We compared this model to one in which all mutations arise at a single mitotic division prior to meiosis. Do the data allow us to distinguish between these two models? From a statistical perspective, we found that both models provide an adequate fit to the data; statistical considerations alone could not separate the models. However, biological considerations support a model with a multigenerational mutation process rather than the single generation process. This example emphasizes the time-honored observation that statistical adequacy of a model does not imply biological adequacy.

Once a model has been found that adequately summarizes a complex data set such as the HD mutation spectra, it could be used for prediction. For example, the model that allows mutations to arise during any mitotic division could be used to estimate the mutation spectra of a given individual at different ages. The same sort of consideration can be used, in theory at least, to provide an experimental approach to testing such a model: take sperm samples from a given individual over a number of years, and compare the observed spectra with those predicted by the model. As yet, such data are not available.

Finally, we note that a model might also suggest other biological features that warrant further investigation. For example, our recent HD work points to the possible role of two different DNA repair mechanisms that correct "small" and "large" slippage loops.

References

1. Ashley, C. T., and Warren, S. T. (1995). Trinucleotide repeat expansion and human disease. *Annu. Rev. Genet.* **29,** 703–728.
2. Igarashi, S., Takiyama, Y., Cancel, G., Rogaeva, E. A., Sasaki, H., Wakisaka, A., Zhou, Y. X., Takano, H., Endo, K., *et al.* (1996). Intergenerational instability of the CAG repeat of the gene for Machado-Joseph Disease (MJD1) is affected by the genotype of the normal chromosome: implications for the molecular mechanisms of the instability of the CAG repeat. *Hum. Mol. Genet.* **5,** 923–932.
3. Nolin, S. L., Lewis, F. A., Ye, L. L., Houck, G. E., Glicksman, A. E., Limprasert, P., Li, S. Y., Zhong, N., Ashley, A. E., *et al.* (1996). Familial transmission of the FMR1 CGG repeat. *Am. J. Hum. Genet.* **59,** 1252–1261.
4. Murray, A., Macpherson, J. N., Pound, M. C., Sharrock, A., Youings, S. A., Dennis, N. R., McKechnie, N., Linehan, P., Morton, N. E., and Jacobs, P. A. (1997). The role of size, sequence and haplotype in the stability of FraxA and FraxE alleles during transmission. *Hum. Mol. Genet.* **6,** 173–184.
5. Wells, R. D. (1996). Molecular basis of genetic instability of triplet repeats. *J. Biol. Chem.* **271,** 2875–2878.
6. Schweitzer, J. K., and Livingston, D. M. (1997). Destabilization of CAG trinucleotide repeat tracts by mismatch repair mutations in yeast. *Hum. Mol. Genet.* **6,** 349–355.

7. Maurer, D. J., Ocallaghan, B. L., and Livingston, D. M. (1996). Orientation dependence of trinucleotide CAG repeat instability in Saccharomyces cerevisiae. *Mol. Cell. Biol.* **16,** 6617–6622.
8. Monckton, D. G., Coolbaugh, M. I., Ashizawa, K. T., Siciliano, M. J., and Caskey, C. T. (1997). Hypermutable myotonic dystrophy CTG repeats in transgenic mice. *Nature Genet.* **15,** 193–196.
9. Gourdon, G., Radvanyi, F., Lia, A. S., Duros, C., Blanche, M., Abitbol, M., Junien, C., and Hofmannradvanyi, H. (1997). Moderate intergenerational and somatic instability of a 55-CTG repeat in transgenic mice. *Nature Genet.* **15,** 190–192.
10. Mangiarini, L., Sathasivam, K., Mahal, A., Mott, R., Seller, M., and Bates, G. P. (1997). Instability of highly expanded CAG repeats in mice transgenic for the Huntington's disease mutation. *Nature Genet.* **15,** 197–200.
11. Weber, J., and Wong, C. (1993). Mutation of human short tandem repeats. *Hum. Mol. Genet.* **2,** 1123–1128.
12. Li, H., Gyllensten, U., Cui, X., Saiki, R., Erlich, H., and Arnheim, N. (1988). Amplification and analysis of DNA sequences in single human sperm and diploid cells. *Nature* **335,** 414–417.
13. Jeffreys, A. J., Wilson, V., Neumann, R., and Keyte, J. (1988). Amplification of human minisatellites by the polymerase chain reaction: towards DNA fingerprinting of single cells. *Nucleic Acids Res.* **16,** 10953–10971.
14. Kunst, C., Leeflang, E., Iber, J., Arnheim, N., and Warren, S. (1997). The effect of FMR1 CCG repeat interruptions on mutation frequency as measured by sperm typing. *J. Med. Genet.* **34,** 627–631.
15. Chong, S. S., McCall, A. E., Cota, J., Subramony, S. H., Orr, H. T., Hughes, M. R., and Zoghbi, H. Y. (1995). Gametic and somatic tissue-specific heterogeneity of the expanded SCA1 CAG repeat in spinocerebellar ataxia type-1. *Nature Genet.* **10,** 344–350.
16. Chong, S. S., Almqvist, E., Telenius, H., Latray, L., Nichol, K., Bourdelatparks, B., Goldberg, Y. P., Haddad, B. R., Richards, F., Sillence, D., Greenberg, C. R., Ives, E., Van den Engh, G., Hughes, M. R., and Hayden, M. R. (1997). Contribution of DNA Sequence and CAG Size to mutation frequencies of intermediate alleles for Huntington disease: evidence from single sperm analyses. *Hum. Mol. Genet.* **6,** 301–309.
17. Leeflang, E. P., Zhang, L., Tavare, S., Hubert, R., Srinidhi, J., Macdonald, M. E., Myers, R. H., Deyoung, M., Wexler, N. S., *et al.* (1995). Single sperm analysis of the trinucleotide repeats in the Huntington's disease gene: quantification of the mutation frequency spectrum. *Hum. Mol. Genet.* **4,** 1519–1526.
18. Takiyama, Y., Sakoe, K., Soutome, M., Namekawa, M., Ogawa, T., Nakano, I., Igarashi, S., Oyake, M., Tanaka, H., Tsuji, S., and Nishizawa, M. (1997). Single sperm analysis of the CAG repeats in the gene for Machado-Joseph disease (MJD1): evidence for non-Mendelian transmission of the MJD1 gene and for the effect of the intragenic cgg/ggg polymorphism on intergenerational instability. *Hum. Mol. Genet.* **6,** 1063–1068.
19. Zhang, L., Leeflang, E. P., Yu, J., and Arnheim, N. (1994). Studying human mutations by sperm typing: instability of CAG trinucleotide repeats in the human androgen receptor gene. *Nature Genet.* **7,** 531–535.
20. Zhang, L., Fischbeck, K. H., and Arnheim, N. (1995). CAG repeat length variation in sperm from a patient with Kennedy's disease. *Hum. Mol. Genet.* **4,** 303–305.
21. Monckton, D., Neumann, R., Guram, T., Fretwell, N., Tamaki, K., Macleod, A., and Jefferys, A. (1994). Minisatellite mutation-rate variation associated with a flanking DNA-sequence polymorphism. *Nature Genet.* **8,** 162–170.
22. Mornet, E., Chateau, C., Hirst, M. C., Thepot, F., Taillandier, A. C.-O., and Serre, J. L. (1996). Analysis of germline variation at the FMR1 CGG repeat shows variation in the normal-premutated borderline range. *Hum. Mol. Genet.* **5,** 821–825.
23. Monckton, D. G., Wong, L.-J. C. A. T., and Casky, C. T. (1995). Somatic mosaicism, germline expansions, germline reversions and intergenerational reductions in myotonic dystrophy males: small pool PCR analysis. *Hum. Mol. Genet.* **4,** 1–8.
24. Baker, S. M., Bronner, C. E., Zhang, L., Plug, A. W., Robatzek, M., Warren, G., Elliott, E. A., Yu, J., Ashley, T., Arnheim, N., *et al.* (1995). Male mice defective in the DNA mismatch repair gene PMS2 exhibit abnormal chromosome synapsis in meiosis. *Cell* **82,** 309–319.
25. Baker, S. M., Plug, A. W., Prolla, T. A., Bronner, C. E., Harris, A. C., Yao, X., Christie, D. M., Monell, C., Arnheim, N., Bradley, A., Ashley, T., and Liskay, R. M. (1996). Involvement of mouse Mlh1 in DNA mismatch repair and meiotic crossing over. *Nature Genet.* **13,** 336–342.
26. Edelmann, W., Cohen, P. E., Kane, M., Lau, K., Morrow, B., Bennett, S., Umar, A., Kunkel, T., Cattoretti, G., Chaganti, R., Pollard, J. W., Kolodner, R. D., and Kucherlapati, R. (1996). Meiotic pachytene arrest in MLH1-deficient mice. *Cell* **85,** 1125–1134.
27. Shibata, D., Navidi, W., Salovaara, R., Li, Z. H., and Aaltonen, L. A. (1996). Somatic microsatellite mutations as molecular tumor clocks. *Nature Med.* **2,** 676–681.
28. MacDonald, M. E., Barnes, G., Srinidhi, J., Duyao, M. P., Ambrose, C. M., Myers, R. H., Gray, J., Conneally, P. M., Young, A., Penney, J., Shoulson, I., Hollingsworth, Z., Koroshetz, W., Bird, E., Vonsattel, J. P., Bonilla, E., Moscowitz, C., Penchaszadeh, G., Brzustowicz, L., Alvir, J., Bickham Conde, J., Cha, J.-H., Dure, L., Gomez, F., Ramos-Arroyo, M., Sanchez-Ramos, J., Snodgrass, S. R., de Young, M., Wexler, N. S., MacFarlane, H., Anderson, M. A., Jenkins, B., and Gusella, J. F. (1993). Gametic but not somatic instability of CAG repeat length in Huntington's disease. *J. Med. Genet.* **30,** 982–986.
29. Telenius, H., Kremer, B., Goldberg, Y. P., Theilmann, J., Andrew, S. E., Zeisler, J., Adam, S., Greenberg, C., Ives, E. J., Clarke, L. A., *et al.* (1994). Somatic and gonadal mosaicism of the Huntington disease gene CAG repeat in brain and sperm *Nature Genet.* **6,** 409–414. [Published erratum appears in *Nature Genet.* **7**(1) 113, 1994]
30. Leeflang, E., McPeek, M., and Arnheim, N. (1996). Analysis of meiotic segregation using single-sperm typing: meiotic drive at the myotonic dystrophy locus. *Am. J. Hum. Genet.* **59,** 896–904.
31. Leeflang, E., and Arnheim, N. (1995). A novel repeat structure at the Myotonic Dystrophy locus in a 37 repeat allele with unexpectedly high stability. *Hum. Mol. Genet.* **4,** 135–136.
32. Cortopassi, G., and Arnheim, N. (1990). Detection of a specific mitochondrial DNA deletion in tissues of older humans. *Nucleic Acids Res.* **18,** 6927–6933.
33. La Spada, A., Wilson, E., Luban, D., Harding, A., and Fischbeck, K. (1991). Androgen receptor gene mutations in X-linked spinal and bulbar muscular atrophy. *Nature* **352,** 77–79.
34. Shimada, N., Sobue, G., Doyu, M., Yamamoto K, Yasuda, T., Mukai, E., Kachi, T., and Mitsuma, T. (1995). X-linked recessive bulbospinal neuronopathy: clinical phenotypes and CAG repeat size in androgen receptor gene. *Muscle Nerve* **18,** 1378–1384.
35. Tanaka, F., Doyu, M., Ito, Y., Matsumoto, M., Mitsuma, T., Abe, K., Aoki, M., Itoyama, Y., Fischbeck, K., and Sobue, G. (1996). Founder effect in spinal and bulbar muscular atropy (SBMA). *Hum. Mol. Genet.* **9,** 1253–1257.
36. La Spada, A., Roling, D., Harding, A., Warner, C., Spiegel, R., Hausmanowa-Petrusewicz, I., Yee, W.-C., and Fischbeck, K. (1992). Meiotic stability and genotype-phenotype correlation of the trinucleotide repeat in X-linked spinal and bulbar muscular atrophy. *Nature Genet.* **2,** 301–304.

37. Biancalana, V., Serville, F., Pommier, J., Julien, J., Hanauer, A., and Mandel, J. (1992). Moderate instability of the trinucleotide repeat in spinobulbar muscular atrophy. *Hum. Mol. Genet.* **4,** 255–258.
38. Chung, M. Y., Ranum, L. P. W., Duvick, L. A., Servadio, A., Zoghbi, H. Y., and Orr, H. T. (1993). Evidence for a mechanism predisposing to intergenerational CAG repeat instability in spinocerebellar ataxia type-I. *Nature Genet.* **5,** 254–258.
39. Sanpei, K., Takano, H., Igarashi, S., Sato, T., Oyake, M., Sasaki, H., Wakisaka, A., Tashiro, K., Ishida, Y., *et al.* (1996). Identification of the spinocerebellar ataxia type-2 gene using a direct identification of repeat expansion and cloning technique, DIRECT. *Nature Genet.* **14,** 277–284.
40. Pulst, S. M., Nechiporuk, A., Nechiporuk, T., Gispert, S., Chen, X. N., Lopescendes, I., Pearlman, S., Starkman, S., Orozcodiaz, G., *et al.* (1996). Moderate expansion of a normally biallelic trinucleotide repeat in spinocerebellar ataxia type-2. *Nature Genet.* **14,** 269–276.
41. Imbert, G., Saudou, F., Yvert, G., Devys, D., Trottier, Y., Garnier, J. M., Weber, C., Mandel, J. L., Cancel, G., *et al.* (1996). Cloning of the gene for spinocerebellar ataxia-2 reveals a locus with high-sensitivity to expanded CAG/glutamine repeats. *Nature Genet.* **14,** 285–291.
42. Eichler, E. E., Holden, J. J. A., Popovich, B. W., Reiss, A. L., Snow, K., Thibodeau, S. N., Richards, C. S., Ward, P. A., and Nelson, D. L. (1994). Length of uninterrupted CGG repeats determines instability in the FMR1 gene. *Nature Genet.* **8,** 88–94.
43. Kunst, C. B., and Warren, S. T. (1994). Cryptic and polar variation of the fragile-X repeat could result in predisposing normal alleles. *Cell* **77,** 853–861.
44. Kremer, B., Goldberg, P., Andrew, S. E., Theilmann, J., Telenius, H., Zeisler, J., Squitieri, F., Lin, B. Y., Bassett, A., *et al.* (1994). A worldwide study of the Huntington's disease mutation: the sensitivity and specificity of measuring Cag repeats. *N. Engl. J. Med.* **330,** 1401–1406.
45. McNeil, S. M., Novelletto, A., Srinidhi, J., Barnes, G., Kornbluth, I., Altherr, M. R., Wasmuth, J. J., Gusella, J. F., Macdonald, m. E., *et al.* (1997). Reduced penetrance of the Huntington's disease mutation. *Hum. Mol. Genet.* **6,** 775–779.
46. Zuhlke, C., Riess, O., Bockel, B., Lange, H., and Thies, U. (1993). Mitotic stability and meiotic variability of the (CAG)n repeat in the Huntington disease gene. *Hum. Mol. Genet.* **2,** 2063–2067.
47. Legius, E., Cuppens, H., Dierick, H., Van, Z.-K., Dom, R., Fryns, J. P., Evers, K.-G., Decruyenaere, M., Demyttenaere, K., Marynen, P., *et al.* (1994). Limited expansion of the (CAG)n repeat of the Huntington gene: a premutation (?). *Eur. J. Hum. Genet.* **2,** 44–50.
48. Goldberg, Y. P., Kremer, B., Andrew, S. E., Theilmann, J., Graham, R. K., Squitieri, F., Telenius, H., Adam, S., Sajoo, A., *et al.* (1993). Molecular analysis of new mutations for Huntington's disease—intermediate alleles and sex of origin effects. *Nature Genet.* **5,** 174–179.
49. Novelletto, A., Persichetti, F., Sabbadini, G., Mandich, P., Bellone, E., Ajmar, F., Pergola, M., Del, S.-L., MacDonald, M. E., Gusella, J. F., *et al.* (1994). Analysis of the trinucleotide repeat expansion in Italian families affected with Huntington disease. *Hum. Mol. Genet.* **3,** 93–98.
50. Leeflang, E. P., Fan, F., Losekoot, M., and Van Ommen, G. J. B. (1995). Single sperm analysis of the transition from high normal to diseased triplet repeat length in the Huntington's disease locus. *Am J. Hum. Genet.* **57,** 140.
51. MacDonald, M. E., Novelletto, A., Lin, C., Tagle, D., Barnes, G., Bates, G., Taylor, S., Allitto, B., and Altherr, M. (1992). The Huntington's disease candidate region exhibits many different haplotypes. *Nature Genet.* **1,** 99–103.
52. Rubinsztein, D. C., Leggo, J., Barton, D. E., and Ferguson-Smith, M. A. (1993). Site of (CCG) polymorphism in the HD gene. *Nature Genet.* **5,** 214–215.
53. Andrew, S. E., Goldberg, Y. P., Theilmann, J., Zeisler, J., and Hayden, M. R. (1994). A CCG repeat polymorphism adjacent to the CAG repeat in the Huntington disease gene, implications for diagnostic accuracy and predictive testing. *Hum. Mol. Genet.* **3,** 65–67.
54. Leeflang, E. P., Tavare, S., Marjoram, P., Neal, C. O. S., Srinidhi, J., MacFarlane, H., MacDonald, M. E., Gusella, J. F., de Young, M., Wexler, N. S., and Arnheim, N. (1997). Dynamic mutation rates at the Huntington's disease locus. [Unpublished data]
55. Duyao, M., Ambrose, C., Myers, R., Novelletto, A., Persichetti, F., Frontali, M., Folstein, S., Ross, C., Franz, M., Abbott, M., Gray, J., Conneally, P., Young, A., Penney, J., Hollingsworth, Z., Shoulson, I., Lazzarini, A., Falek, A., Koroshetz, W., Sax, D., Bird, E., Vonsattel, J., Bonilla, E., Alvir, J., Bickham Conde, J., Cha, J.-H., Dure, L., Gomez, F., Ramos, M., Sanchez-Ramos, J., Snodgrass, S., de Young, M., Wexler, N., Moscowitz, C., Penchaszadeh, G., MacFarlane, H., Anderson, M., Jenkins, B., Srinidhi, H., Barnes, G., Gusella, J., and MacDonald, M. (1993). Trinucleotide repeat length instability and age of onset in Huntington's disease. *Nature Genet.* **4,** 387–392.
56. Trottier, Y., Biancalana, V., and Mandel, J. L. (1994). Instability of CAG repeats in Huntington's Disease: relation to parental transmission and age of onset. *J. Med. Genet.* **31,** 377–382.
57. Wexler, N. S., Rose, E. A., and Housman, D. E. (1991). Molecular approaches to hereditary diseases of the nervous system: Huntington's disease as a paradigm. *Annu. Rev. Neurol.* **14,** 503–529.
58. Jansen, G., Willems, P., Coerwinkel, M., Nillesen, W., Smeets, H., Vits, L., Howeler, C., Brunner, H., and Wieringa, B. (1994). Gonosomal mosaicism in myotonic dystrophy patients: involvement of mitotic events in (CTG)n repeat variation and selection against extreme expansion in sperm. *Am. J. Hum. Genet.* **54,** 575–585.
59. Wierdl, M., Dominska, M., and Petes, T. D. (1997). Microsatellite instability in yeast: dependence on the length of the microsatellite. *Genetics* **146,** 769–779.
60. Streisinger, G., Okada, Y., Emrich, J., Newton, J., Tsugita, A., Terzaghi, E., and Inouye, M. (1966). Frameshift mutations and the genetic code. Cold Spring Harbor Symp. Quant. Biol. **31,** 77–84.
61. Aaltonen, L. A., Peltomaki, P., Leach, F. S., Sistonen, P., Pylkkanen, L., Mecklin, J. P., Jarvinen, H., Powell, S. M., Jen, J., Hamilton, S. R., *et al.* (1993). Clues to the pathogenesis of familial colorectal cancer. *Science* **260,** 812–816.
62. Peltomaki, P., Aaltonen, L. A., Sistonen, P., Pylkkanen, L., Mecklin, J. P., Jarvinen, H., Green, J. S., Jass, J. R., Weber, J. L., Leach, F. S., *et al.* (1993). Genetic mapping of a locus predisposing to human colorectal cancer. *Science* **260,** 810–812.
63. Ionov, Y., Peinado, M. A., Malkhosyan, S., Shibata, D., and Perucho, M. (1993). Ubiquitous somatic mutations in simple repeated sequences reveal a new mechanism for colonic carcinogenesis. *Nature* **363,** 558–561.
64. Strand, M., Prolla, T. A., Liskay, R. M., and Petes, T. D. (1993). Destabilization of tracts of simple repetitive DNA in yeast by mutations affecting DNA mismatch repair. *Nature* **365,** 274–276. [Published erratum appears in *Nature* **368**(6471), 569, 1994]
65. Modrich, P., and Lahue, R. (1996). Mismatch repair in replication fidelity, genetic recombination and cancer biology. *Annu. Rev. Biochem.* **65,** 101–133.

66. Mitas, M. (1997). Trinucleotide repeats associated with human disease. *Nucleic Acids Res.* **25,** 2245–2253.
67. Richards, R. I., and Sutherland, G. R. (1994). Simple repeat DNA is not replicated simply. *Nature Genet.* **6,** 114–116.
68. De Wind, N., Dekker, M., Berns, A., Radman, M., and te Riele, H. (1995). Inactivation of the mouse MSH2 gene results in mismatch repair deficiency, methylation tolerance, hyperrecombination, and predisposition to cancer. *Cell* **82,** 321–330.
69. Drost, J. B., and Lee, W. R. (1995). Biological basis of germline mutation: comparisons of spontaneous germline mutation rates among drosophila, mouse, and human. *Environ. Mol. Mutagen.* **25,** 48–64.

Part XIV

Biophysical and Structural Studies on Triplet Repeat Sequences

Biophysical and Structural Studies on Triplet Repeat Sequences: Duplex Triplet Repeat Structures

ROBERT GELLIBOLIAN[1] AND ALBINO BACOLLA

Institute of Biosciences and Technology, Center for Genome Research, Texas A&M University, Department of Biochemistry and Biophysics, Texas Medical Center, Houston, Texas 77030

[1]Present address: The Salk Institute for Biological Studies, Gene Expression Laboratory, 10010 N. Torrey Pines Rd., La Jolla, CA 9203?. Fax: (619) 457-2762.

I. NON-B-DNA CONFORMATIONS

A. Background

It has been recognized for many years that DNA adopts various conformations. For example, the solution of the first crystal structure of a base pair revealed that the hydrogen bonding between an adenine and a thymine [1] was not the one proposed by Watson and Crick [2]. Instead, it involved positions on the purine (such as N7) that were unfulfilled in the "canonical" double-helical, right-handed B-form of the DNA, and which are now recognized as Hoogsteen (and reverse-Hoogsteen) hydrogen bonds.

There are numerous ways in which adenine, thymine, guanine, and cytosine can associate. These include the interactions between a purine (R) and a pyrimidine (Y), as well as between homo- and heteropurines, and homo- and heteropyrimidines [3]. In addition, the chemical bond connecting the bases to their deoxyribose moiety (glycosidic bond) is a flexible hinge about which a base can pivot. The consequence of these various base–base associations, pivoting, and other degrees of freedom is that a family of distinct and complex macroscopic structures may form, with new members being discovered constantly. These non-B-DNA structures (Fig. 38-1 and Table 38-1) have been treated comprehensively elsewhere [4, 5] and are briefly reviewed below.

B. Left-Handed Z-DNA

Alternating purine–pyrimidines $(RY \cdot RY)_n$ such as $d(GC \cdot GC)_n$ and, to a lesser extent, $d(AC \cdot GT)_n$ forms both right-handed and left-handed (Z-form) double helices. The switch in handedness is caused by ~180° pivoting of alternating bases (i.e., the G residues in $d(GC \cdot GC)_n$) about their glycosidic bond accompanied by concomitant movements of the complementary base (i.e., the C residues) with the retention of Watson–Crick hydrogen bonds [6]. The changes in the macroscopic shape between B- and Z-DNA are dramatic. Z-DNA is a potent immunogen, and patients afflicted with systemic lupus erythematosus or rheumatoid arthritis also produce autoantibodies that recognize Z-DNA *in vitro* [7]. *In vivo,* the transition from B to Z is favored by negative supercoiling, and methylation of the cytosine residues [8, 9]. Transcription is associated with hypersupercoiling of the DNA template [10] and, as expected, favors the formation of left-handed Z-segments [11, 12].

Although a required physiological role for Z-DNA is yet to be demonstrated, the body of evidence about its formation *in vivo* [4, 12, 13] suggests that Z-DNA participates in the regulation of gene expression [14, 15] and, more generally, in those biological processes associated with transient surges of local negative supercoiling, such as replication, recombination, and mismatch repair.

C. Three-Stranded DNA (Triplex DNA, H-DNA)

Three-stranded nucleic acids were recognized very early in the history of DNA structure [5, 16, 17]. The addition of a third strand to a right-handed, duplex DNA requires that one strand of the B-DNA contains successive purine bases, and therefore triplex DNA is formed by $(R \cdot Y)_n$ sequences with preferably mirror repeat symmetry (Fig. 38-1 and Table 38-1). As elegantly shown by nuclear magnetic resonance [18], the third strand hydrogen bonds in the major groove with the purine residues *via* their unfulfilled groups, namely the Hoogsteen positions. Triplex DNA is also facilitated by negative supercoiling and methylation of cytosine residues, and is stabilized by pH and divalent metal ions [5]. A triplex variant has been described, nodule DNA, in which two intramolecular triplexes are connected by single-stranded loops [19, 20].

$(R \cdot Y)_n$ (and also $(RY \cdot RY)_n$) sequences stimulate homologous recombination [21]; they are found at hot spots for gene conversion events [22], and interfere physically with the progression of DNA-tracking enzymes such as the polymerases [23, 24]. $(R \cdot Y)_n$ tracts are overrepresented in eukaryotic genomes [25], where they often occur in the proximity of $(RY \cdot RY)_n$ motifs [26–29]. Thus, like Z-DNA, the physiological role of triplex DNA may be that of modulating supercoiling-assisted biological processes.

D. Four-Stranded DNA (Tetraplex DNA, G4-DNA)

Tetraplex DNA is a structure formed by the sequences specific for telomeres, the ends of linear chromosomes [30]. Telomeres are synthesized by the RNA-

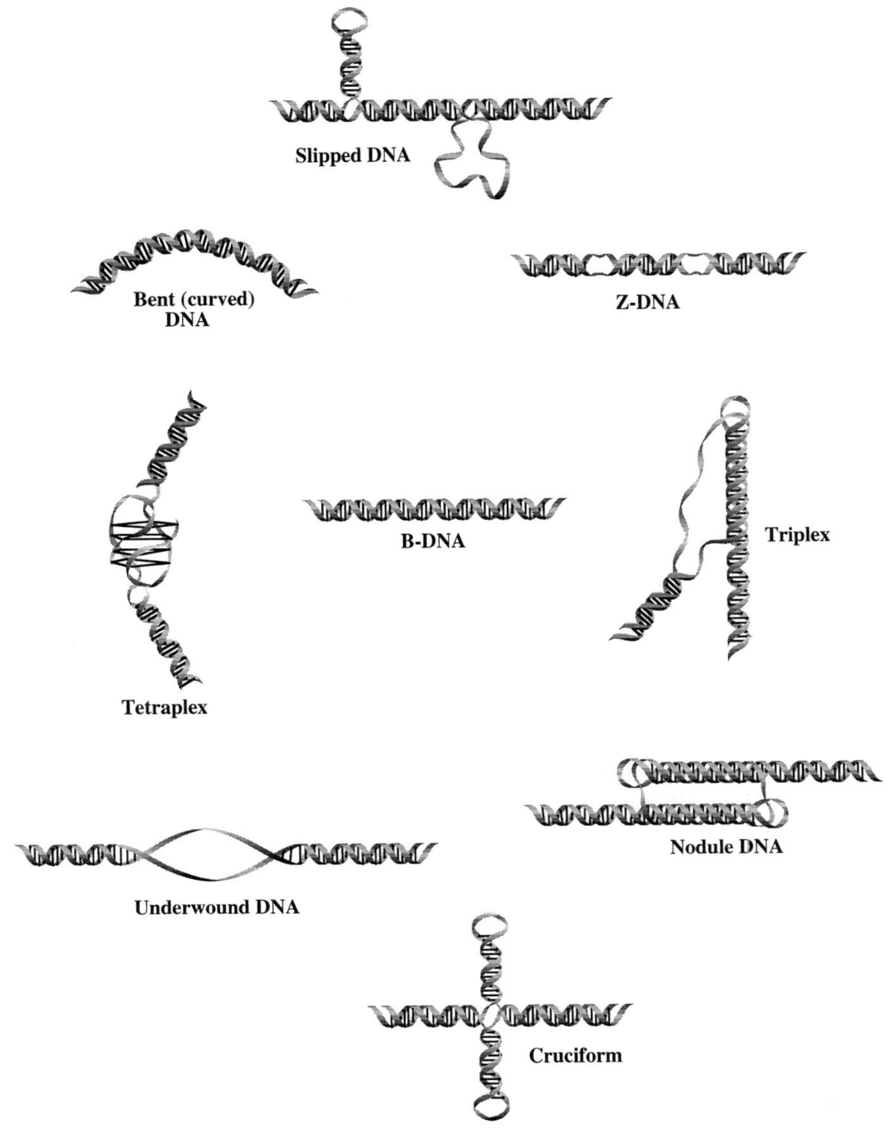

FIGURE 38-1 Non B-DNA structures.

dependent polymerase telomerase and serve two major roles: to protect the ends of linear chromosomes from nucleolytic degradation, and to prevent the loss of genetic information due to lagging-strand synthesis, as leading-strand replication reaches the ends of chromosomes. The 3′-termini of telomeres are single-stranded (~12–16 nt long) G-rich tandem motifs, such as (TTAGGG), (GGGT), and (GGGGTTTT). *In vitro,* oligonucleotides corresponding to these termini form tetra-stranded complexes, characterized by stacks of four guanines per plane interacting through Hoogsteen hydrogen bonds (31, 32).

Proteins have been isolated that bind G4-nucleic acids. Of these, the nuclease encoded by the yeast *KEM1/SEP1* gene displays a specific G4-DNA directed cleavage, 5′ to the tetraplex structure [33]. A *kem1/sep1* null mutation causes a shortening of the telomeres, and leads to premature senescence [34]. Thus, the physiological role of tetraplex DNA is believed to be that associated with the structure of telomeres and their function.

E. Cruciforms

A cruciform is the two-arm structure resulting from the Watson–Crick pairing of two halves of an inverted repeat. An inverted repeat is a set of bases followed on the same strand by the complementary sequence arranged in the opposite orientation (Fig. 38-1 and Table 38-1). The two halves of an inverted repeat have

TABLE 38-1 Non-B-DNA Structures and Sequence-Specific Duplex Conformational Features

Structure/duplex feature	Sequence features	Example	Stabilizing factors
Left-handed Z-DNA	$(RY \cdot RY)_n$	$(GC)_{40} \cdot (GC)_{40}$ or $(AC)_{40} \cdot (GT)_{40}$	Negative supercoiling, high ionic strength, Me^{2+}, methylation of C
Triplex DNA (H-DNA)	$(R \cdot Y)_n$ (mirror repeats)	$(AAG)_{18} \cdot (CTT)_{18}$	Negative supercoiling, Me^{2+}, methylation of C, pH 4.5–7.0
Nodule DNA (bitriplex)	$(R \cdot Y)_n$ (mirror repeats)	$(GA)_{37} \cdot (TC)_{37}$	Negative supercoiling, Me^{2+}, pH 7.5
Tetraplex DNA (G4-DNA)	G_{4-10} of telomere ends	$(G_4T_2)_2$ *or* $(G_4T_4)_2$	Na^+ and K^+
Cruciforms	Inverted repeats	<u>GTCCAGTA</u>TACTGGAC	Negative supercoiling
Slipped DNA	Direct repeats	$(GATGTC)_{80} \cdot (GACATC)_{80}$ *or* $(CTG)_{130} \cdot (CAG)_{130}$	Negative supercoiling, relatively low G-C content
Parallel DNA	Oligonucleotides with A-T	$5'$-$AT_2A_3T_4A_5T_6$-$3'$ $5'$-$TA_2T_3A_4T_5A_6$-$3'$	Reverse (*trans*)-Watson-Crick pairing
Bent (curved) DNA	Repeating phased A-tracts	$(A_{4-6}N_{6-4})_n \cdot (N_{6-4}T_{4-6})_n$	Supercoiling not required
Anisomorphic DNA	Multiple direct repeats of 12-bp G-rich motifs in HSV-1	$(R_{12} \cdot Y_{12})_{14-19}$	Negative supercoiling, pH 5.0, Mg^{2+} or Ca^{2+} plus >50 mM NaCl
Flexible and writhed DNA	CTG · CAG and CGG · CCG triplet repeats	$(CTG \cdot CAG)_n$ and $(CGG \cdot CCG)_n$	Intrinsic conformational property, supercoiling not required
Underwound DNA	A-T-rich regions	$(TTA \cdot TAA)_{18-90}$	Negative supercoiling, higher temperature, low salt, no Me^{2+}
Cubic DNA	Synthetic oligonucleotides		

Note. The structures and sequence-specific duplex characteristics listed in the first column are described in the text. The sequence features and some of the examples are also described. The underlined sequence denotes one of the two inverted repeats. Some of the stabilizing factors (i.e., high ionic strength, low pH) are not discussed in the text because they may play a less important role in the stabilization of the structures *in vivo*.

complementary bases, and therefore can hybridize to form an intramolecular B-type helix. The extrusion of a cruciform from a normal duplex DNA requires strand separation, a process that is greatly facilitated by negative supercoiling. Thus, the same biological processes that are associated with the formation of Z- and triplex DNA also induce cruciforms [4].

In addition to other roles, cruciform extrusion has been proposed to assist in establishing an "open DNA conformation" for the assembly of prereplicating complexes at replication origins [35].

F. Slipped DNA

Slipped DNA occurs when a set of nucleotides on one strand of the DNA can hybridize to more than one set on the opposing strand (reviewed in [4]). For example, CTGCTG and the complementary CAGCAG may hydrogen bond in three ways:

```
CTGCTG         CTGCTG         CTGCTG
GACGAC  GACGAC                  GACGAC
```

Thus, any direct repeat may form slipped structures. Slipped DNA is highly relevant to triplet repeat sequences, and therefore is treated separately in this volume [36].

G. Other Structures and Features of B-Type Double Helices

The structures described so far are profound alterations of the conventional B-type helix. Other conformational properties arise in the normal right-handed, antiparallel double-helix from specific sequence motifs that may elicit biological consequences. For example, bent DNA is a well-characterized feature of duplex DNA (Table 38-1, Fig. 38-1). Repeated runs of A (i.e., $A_{3\text{-}5}$) evenly phased with the helical repeat give a curved appearance to the overall DNA molecule [37–40]. Other motifs, such as GGGCCC · GGGCCC, also induce helix bending [41]. In some instances, as observed in the mitochondrial DNA from trypanosomatids, molecules of a few hundred base pairs spontaneously form minicircles, due to the phased A-tracts [39]. Bent DNA has been shown to be an intrinsic architectural element that participates in the regulation of a variety of processes such as transcription, replication, recombination, and nucleosome assembly [37, 42, 43]; the obligatory biological role for bent DNA may be stronger than for any of the other unusual DNA structures. The physical nature of other types of conformational deformations, such as those giving rise to anisomorphic DNA at the G-rich tandem repeats from herpes simplex virus type-1 [44], is yet to be determined.

Considering triplet repeats, $(CTG \cdot CAG)_n$ and $(CGG \cdot CCG)_n$ sequences are highly flexible and are able to efficiently absorb negative superhelical tension [45, 46]. On the contrary, $(TTA \cdot TAA)_n$ tracts are destabilized by negative superhelical stress, which causes the complementary strands to separate [47]. These issues are discussed further below.

Parallel, double-stranded DNA may be formed by oligonucleotides containing exclusively A and T residues. In this orientation, the T residues are hydrogen bonded to A through O2, rather than O4, in an arrangement that is known as *trans*-Watson–Crick base pairing, as opposed to the "canonical" *cis*-Watson–Crick base pair [48–50]. *In vivo,* no evidence has been obtained for the existence of parallel DNA; however, given their stability, parallel nucleic acids are possible in a nuclear environment.

Finally, the ability of double-stranded oligonucleotides with single-stranded ends to form cubic structures has been investigated *in vitro,* in an attempt to use DNA as a biological carrier for the delivery of drugs and other molecules [51].

In summary, it is clear from this overview that DNA structure responds exquisitely to both sequence-specific motifs and negative supercoiling. These two factors are analyzed in more detail below in association with the genetic instability of triplet repeats.

II. THE FLEXIBILITY AND WRITHE OF $(CTG \cdot CAG)_n$ AND $(CGG \cdot CCG)_n$

A. DNA as a Flexible Molecule

The structure of DNA is dynamic. In addition to the various conformations described above, the three-dimensional shape of the double helix changes constantly in solution, due to thermal energy. This motion, or flexibility, is manifested through (a) fluctuations in the helical twist, which change the length of the helical turns, and (b) transient curvatures along the duplex. Contrary to the static bends described above for the A-tracts and the $G_3C_3 \cdot G_3C_3$ motifs, these curvatures take place randomly along the chain; they are transient, occur in all directions, and span an ample range of angles.

The structure of DNA is also altered during its physiological functions. Thus, for example, the complementary helices are unwound and separated during transcription and replication, whereas the duplex follows a curved trajectory in nucleosomes. In many of these interactions, supercoiling is also introduced into the DNA, as pointed out earlier. Duplex DNA opposes such thermal fluctuations and structural alterations through its bending and torsional restoring forces (moduli) which, being intrinsic properties of the molecule, are constant parameters.

It is evident from this description that in order to elucidate the dynamics of protein–DNA interactions, as well as the structural consequences that originate from topological stress, knowledge is required about the restoring forces.

B. Restoring Forces of DNA

Several methods have been employed for measuring the bending and torsional moduli of duplex DNA. The earliest measurements were conducted in the late 1970s–early 1980s by methods such as the sedimentation velocity [52], fluorescence depolarization anisotropy [53–56], linear dichroism [57, 58], transient electric birefringence [59], electron paramagnetic resonance [60], laser light scattering [61], kinetics of ring closure [62], and equilibrium distribution of topological isomers [63, 64]. Extraction of the restoring forces from the data obtained by these methods entailed the use of the theories that underlie the techniques. Several of these theories were further developed in later years [65–68], and this had a twofold effect. On the one hand, reanalysis of some of the experimental data provided greater accuracy of the results and, on the other hand, it enabled the direct calculation of both constants from a single experiment, which was not possible before. In more recent years, additional determinations were conducted with the kinetics of ring closure [69–71], fluorescence depolarization [72, 73], and transient electric dichroism [74], whereas new techniques, such as cryoelectron microscopy [75] and Monte Carlo simulation [76–79], were also utilized.

The results from these analyses, for a variety of DNA sequences, are compiled in Table 38-2. These data were obtained from several laboratories over more than two decades on DNA molecules prepared and characterized by different procedures, and this has contributed to experimental variability. For example, in numerous cases the samples were a collection of DNA fragments of different lengths, which affects the results. Also, in studies where repetitive sequences were analyzed, the question remains whether the molecules were in their canonical duplex B-form or, rather, alternative conformations such as left handed Z-form and slipped structures may have also been present, which influences the results. Thus, rather than providing a source for rigorous com-

TABLE 38-2 Restoring Forces of DNA

DNA	β ($\times 10^{-19}$ erg · cm)	α ($\times 10^{-19}$ erg · cm)	P (Å)	Methods	References
Random DNA	2.1 ± 1.0 (23)	2.1 ± 0.4 (20)	530 ± 100 (20)	FDA, EM, MC, FD, LD, CK, LLS, SM, TD, EPR, SV, NMR, TEB	[52–79]
$(CTG \cdot CAG)_n$	2.3	1.1	278	CK	[45]
$(CGG \cdot CCG)_n$	2.4	1.3	315	CK	[45]
$(G \cdot C)_n$	1.2, 4.0	3.6–6.9	884–1700	FDA	[54, 55]
$(GC \cdot GC)_n$	1.4, 1.5	3.4	840	FDA, LLS	[54, 56, 61]
$(AC \cdot GT)_n$	0.1	2.4–3.4	595–850	FDA	[55]
$(A \cdot T)_n$	0.2, 1.3	1.7–2.0	425–510	FDA	[54, 55]
rA · rU	1.53	N.D.	N.D.	FDA	[54]
$(AT \cdot AT)_n$	0.9	0.8–1.0	200–250	FDA, TED	[54, 74]
dA · dU	0.8	N.D.	N.D.	FDA	[54]
70% (A + T)	N.D.	3.3	820	EM	[75]
dA · rU · rU	2.6	N.D.	N.D.	FDA	[54]
Bent DNA	N.D.	1.4	357	GE	[81]
Me-C duplex DNA	2.3	1.6	391	CK	[71]

Note. The torsional modulus β, the bending modulus α, and the persistence length P are listed. P is related to α by $\alpha = K_B TP$, where K_B is the Boltzmann constant and T is 293.15°K (20°C). The abbreviations used for the methods are as follows: FDA, fluorescence depolarization anisotropy (fluorescence anisotropy decay); EM, cryoelectron microscopy; MC, Monte Carlo simulation; FD, flow dichroism; SM, statistical mechanics; CK, circularization kinetics; TD, topoisomer distribution; EPR, electron paramagnetic resonance; SV, sedimentation velocity; NMR, nuclear magnetic relaxation; TEB, transient electric birefringence; GE, two-dimensional gel electrophoresis; LD, linear dichroism; LLS, laser light-scattering; TED, transient electric dichroism. For random DNA, the mean ± SD is reported, with the number of observations in parentheses. Whenever a range was given in the references, the mean value was used for the computations. Me-C, duplex DNA fragments containing 5′-methylated cytosine residues.

parisons, the compilation in Table 38-2 offers a general perspective of the work in the field, and enables interesting conclusions to be made while highlighting some of the difficulties. For DNA of general sequence (random), the torsional modulus β and the bending modulus α have equal values (2.1×10^{-19} erg · cm). However, since the S.D. associated with β is high (1.0 vs 0.4), which reflects the contributions from early measurements, β may be higher than reported here. On the other hand, the persistence length P (related to α) is close to 500 Å, which is a commonly accepted value.

Table 38-2 also lists the restoring forces for G/C-rich and A/T-rich ordered sequences, including the triplet repeats $(CTG \cdot CAG)_n$ and $(CGG \cdot CCG)_n$ [45]. Because only one or two determinations are available for these ordered sequences, the single values are given. For the TRS, alternating $(GC \cdot GC)_n$, $(rA \cdot rU)_n$, and methylated cytosine-duplex DNA, the torsional moduli are within the normal range. In fact, the results obtained by circularization kinetics on the TRS (2.3 and 2.4 $\times 10^{-19}$ erg · cm) [45] and methylated DNA (2.3 $\times 10^{-19}$ erg · cm) [71], and by fluorescence anisotropy on $(GC \cdot GC)_n$ (1.50 and 1.36 $\times 10^{-19}$ erg · cm) and $(rA \cdot rU)_n$ (1.53×10^{-19} erg · cm) [54, 56], were close to those measured by the respective techniques on random DNA (2.0 to 2.4 $\times 10^{-19}$ erg · cm with CK; 1.43 and 1.29 $\times 10^{-19}$ erg · cm with FDA, respectively). For $(A \cdot T)_n$, $(AT \cdot AT)_n$, $(dA \cdot dU)_n$, and $(AC \cdot GT)_n$, the torsional moduli were lower than normal. On the contrary, β was high in the triple-stranded $(dA \cdot rU \cdot rU)_n$ (2.65 vs 1.43 $\times 10^{-19}$ erg · cm) [54]. The data for $(G \cdot C)_n$ are discordant.

Overall, these results suggest that deoxyribo A/T-rich sequences are easy to undertwist (or overtwist) (see also [80]), whereas triple-stranded nucleic acids are torsionally rigid. Also, G/C-rich sequences appear quite normal, albeit different compositions may have different properties.

The persistence lengths span a wide range for the ordered sequences. For the TRS, P were lower (278 and 315 Å) than for random DNA (530 Å) [45], whereas P was reported to be higher (840 to 1700 Å) for the other G/C-rich sequences, $(G \cdot C)_n$ and $(GC \cdot GC)_n$ [55, 61]. P was not affected by methylation (389 vs 391 Å), indicating that this modification does not alter the bending flexibility of the DNA chain [71]. DNA containing phased A-tracts (bent) was characterized by a low persistence length (357 Å) [81]. However, since this is a reflection of the static curvature of the molecule, the derived α does not represent the restoring bending force for this DNA. For the A/T-rich sequences, low persistence lengths were reported for $(A \cdot T)_n$ (425 to 510 Å) [55] and $(AT \cdot AT)_n$ (200 to 250 Å) [74], whereas a high persistence length was measured on repetitive DNA containing 70% A + T (820 Å) [75].

Therefore, as pointed out earlier, this analysis shows that whereas the determinations of the restoring forces for random DNA have yielded reasonable consensus values, those on repetitive DNA are still too scarce and often imprecise [82]—possibly because of structural variations associated with the sequences—to enable thorough comparisons.

C. Flexibility of TRS

$(CTG \cdot CAG)_n$ and $(CGG \cdot CCG)_n$ were found to be particularly flexible in their bending motion by circularization kinetics [45]. These conclusions were also supported by data obtained from apparent helical repeat determinations, topoisomer formation [45], and permutation analyses [36]. The topological consequences that such sequences impart on the writhe (three-dimensional shape) of plasmid DNA are illustrated schematically in Fig. 38-2. Attempts to correlate DNA flexibility with the morphology of base pairs have recently been made, based on the fact that in each dinucleotide step, i.e., AG · CT, TA · TA, and CC · GG, 1 base pair can move relative to its neighbor. The extent and directions of such movements can be calculated from NMR and crystal structures of duplex oligonucleotides and, indeed, ranks of flexibility have been constructed [83–86]. For the TRS, $(CGG \cdot CCG)_n$ and $(CTG \cdot CAG)_n$ occupy the 3rd and 4th position, respectively, out of 12, whereas $(TTA \cdot TAA)_n$, being 12th, is the least flexible [45]. Therefore, the prediction that $(CGG \cdot CCG)_n$ and $(CTG \cdot CAG)_n$ are highly flexible agrees with our experimental data, whereas the prediction that $(TAA \cdot TTA)_n$ is rigid is in concert with the measurements conducted by cryoelectron microscopy [75] and the results from two-dimensional gel electrophoresis showing that this duplex is destabilized by negative supercoiling [47].

What are the biological implications of increased or decreased flexibility? Among others, two topological properties of the DNA may be considered: looping and supercoiling. When linear DNA fragments are circularized, such as during the kinetics of ring closure, the efficiency of the reaction depends on the length of the linear molecules as well as on their fractional helical turn. In fact, up to a certain length, longer molecules show higher efficiency of cyclization. However, within this pattern, linear fragments with an integral number of helical turns cyclize about 10 times more efficiently than comparable molecules with an additional half a helical turn (~5 base pairs). This is due to the fact that the latter molecules need to under- or overtwist by half a turn in order to align their ends. Figure 38-3A shows this behavior during the circularization of DNA fragments containing increasing numbers of CGG · CCG repeats [45].

The energetics associated with length and twist may also be observed *in vivo* whenever DNA looping mediates the interaction between distally bound proteins.

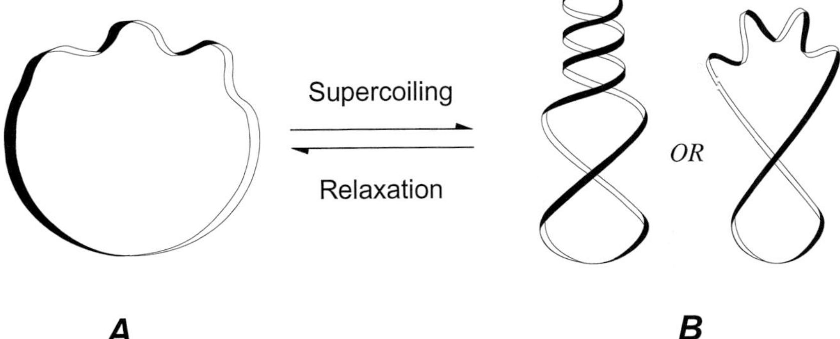

FIGURE 38-2 Cartoon of flexible and writhed DNA. A circular, double-stranded DNA molecule is shown by a closed ribbon. In this representation the winding of the DNA helix is ignored and the direction in space of the helix axis (which corresponds to the ribbon axis), or writhe, is emphasized. (A) Relaxed DNA. Duplex DNA changes shape continuously due to thermal energy; however, the flexible TRS sequences undergo more movement than the rest of the plasmid, and thus are associated with a greater writhe. This is illustrated by the oscillating segments on an otherwise flat, circular ribbon. (B) Supercoiled DNA. When supercoiling is introduced into a circular DNA molecule, tertiary turns are present. On the left drawing, tertiary turns are shown by the four plectonemic crossings of black segments over the white segments. The flexible TRS are identified as the three tighter crossings at the top. The drawing on the right shows TRS supercoiling as three solenoidal (toroidal) turns (top). Both the right-handed plectonemic and the left-handed solenoidal coils correspond to negative writhing numbers and, therefore, to negative supercoiling. However, it should be noted that, whereas plectonemic supercoiling is favored on DNA of random composition, the form adopted by TRS remains to be elucidated.

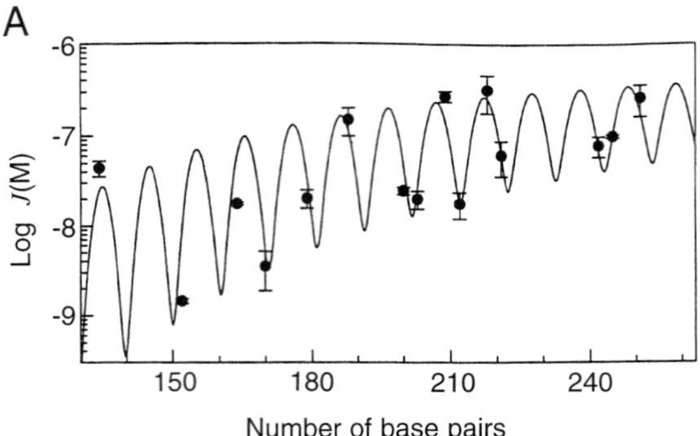

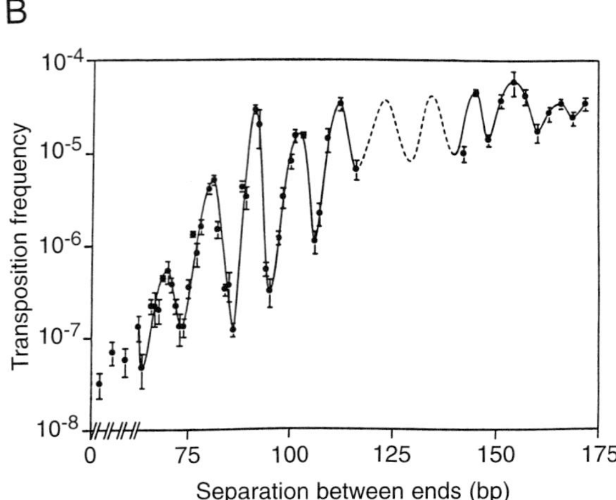

FIGURE 38-3 (A) Circularization kinetics on DNA fragments containing CGG · CCG repeats. Linear DNA fragments containing 34 to 73 CGG · CCG repeats and cohesive ends were reacted with T4 DNA ligase, and the probability of ring-closure (molar J-factor) was calculated. The logarithm of the molar J-factor (log J(M)) was then plotted as a function of DNA length. The experimental values were connected by a theoretical fit that enabled the determination of the bending and torsional moduli, as well as the helical repeat h^0. h^0 was 10.35 bp/turn [45]. The peaks in the oscillating curve represent the J-factors for molecules with an integer number of helical turns, whereas the minima indicate the J-factors of fragments with an additional half a helical turn. Reprinted with permission from Bacolla *et al., JBC,* **272,** 16783–16792 (1997). (B) Transposition frequency of IS*50*. IS*50* transposon was fused with plasmid DNA (pIS*50*.11-*n* and pFMA50-*n*) that contained inserts of various lengths (from 9 to 172 bp) between the outside and the inside ends of IS*50*. Transposition frequencies were determined by scoring for the appearance of chloramphenicol and nalixidic acid resistant (Cam^r and Nal^r) progeny, after donor cells containing the IS*50* transposon (Cam^r) were mated with Nal^r recipient cells [87]. Reprinted with permission from Goryshin *et al., Proc. Natl. Acad. Sci. USA* **91,** 10834–10838, 1994 [87]. Copyright 1994, National Academy of Sciences, U.S.A.

Figure 38-3B shows the effect of DNA length on the transposition frequency of the bacterial IS*50* transposable element [87]. In this system, two transposase complexes bind the ends of the transposable element and then dimerize to carry out a "cut and paste" reaction. The similarities in the length- and twist-dependence between cyclization and transposition are indeed striking. In fact, the periodicity of 10.4 bp/turn found in both types of experiments is in excellent agreement.

Thus, an increased bending flexibility should facilitate looping, and therefore accelerate the reactions by giving higher values on the *y* axes, whereas a greater twisting flexibility should alleviate the dependence on the fractional helical turn, and result in shallower oscillations. Common nuclear processes such as recombination and replication are assisted by DNA looping [88] and, therefore, the kinetics of these reactions are expected to be accelerated by long tracts of flexible DNA, such as the $(CTG \cdot CAG)_n$ and $(CGG \cdot CCG)_n$ sequences.

Wrapping of DNA around histone octamers should also be facilitated by a flexible sequence. Indeed, $(CTG \cdot CAG)_n$ was found to be the best DNA to assemble into nucleosomes [89–92]. However, since different results were obtained for $(CGG \cdot CCG)_n$ [93–95], it is apparent that several features associated with bending and/or sequence-dependent interactions with histone octamers all contribute to nucleosome assembly.

III. EXPANSION MODEL BASED ON STATISTICAL MECHANICS

A. Thermodynamics and Statistical Mechanics

The whole of classical thermodynamics is based on the empirical laws which led to the birth of concepts such as temperature, internal energy, and entropy as functions of state. At no point in the development of these laws, or their consequences, does any reference need to be made to the atomic theory that matter consists of ultimate particles. Furthermore, thermodynamics provides no insight into the possible origin of its laws or any means of making a direct calculation of the thermodynamic properties of a substance or molecule such as DNA, at equilibrium. As a result, statistical mechanics was developed to advance knowledge in both of these directions and make possible the derivation of new formulae to calculate the thermodynamic properties (i.e., free energy and entropy) of molecules.

Equilibrium is defined as the state of a system in which its various macroscopic properties (i.e., temperature, free energy, and entropy) do not change with time. However, as soon as the interpretation of these properties are extended at the molecular level, it must be realized that the equilibrium is of a dynamic rather than static character. This stems from thermal fluctuations within the molecules as well as the motions and collisions occurring between them. The essential problem, then, becomes that of obtaining information about the "most probable" state of a molecule by averaging over all of its states, at equilibrium, using the tools of statistical mechanics.

In this section, we show how the application of statistical mechanics to CTG · CAG and CGG · CCG sequences not only has helped us in finding a correlation between the presence of defined premutation lengths and the greater propensity for instability, but also provided for the formulation of a model for the mechanism of TRS expansion.

B. Statistical Mechanics of DNA Molecules

Inside the nucleus, the DNA molecule exists in a highly compacted form. In eukaryotic systems, the first level of compaction occurs in the nucleosome(s), where ~145 base pairs are wrapped around a histone core with a radius of 45 Å [96]. Moreover, the double-stranded DNA genomes of large bacteriophages are condensed inside the phage heads, possibly in the form of coaxial spools, the smallest of which has a radius of 60 Å [97]. This level of compaction suggests that the DNA is approximately circular and highly bent. Consequently, an understanding of the energetics of DNA packaging requires knowledge about the thermodynamics of bending and twisting of this molecule.

The number of conformational states of a DNA molecule increases in proportion to the length of DNA due to the ease in deformation of the helix axis (i.e., flexibility) under thermal fluctuations. But, aside from length, different sequence compositions along the helix will also affect the degree of flexibility. For example, a linear DNA molecule with complementary cohesive ends will invariably produce a population of topological isomers when circularizing into its covalently closed, circular form (e.g., cccDNA) under ligation conditions [98, 99]. The distribution of topoisomers in the population falls within a Gaussian envelope whose width is defined by the variance in linking number (*i.e.*, $<(\Delta Lk)^2>$, where Lk is the linking number and is defined as the number of times one strand crosses the other) with each isomer differing from its neighbor by one helical turn. This fluctuation necessitates the application of statistical mechanical calculations on the ensemble of DNA configurational states, leading to information regarding the most probable values for the thermodynamic properties. With this in mind, standard theoretical equations have been developed by Shimada and Yamakawa [65] in which the DNA is treated as helical worm-like chain (HWC). The model is composed of three parameters: (a) the bending modulus (α), (b) the twisting modulus (β), and (c) the Kuhn segment of the DNA (λ^{-1}). Knowledge about all three will allow one to calculate the degree of bending (variance in writhe, $<(Wr)^2>$) and twisting (variance in twist, $<(\Delta Tw)^2>$) fluctuations.

C. Statistical Mechanics of Triplet Repeat Sequences

In Section II, we discussed the flexible nature of CTG · CAG and CGG · CCG TRS. The results were based on experimental data obtained using sequences from 224 to 245 bp in length [45]. Despite the wealth of empirical data, this spectrum is clearly insufficient to see the effects of length from the premutation to full mutation range (~50 to >200 for *DM* and ~60 to >230 for *FRAXA*). However, cloning of long CTG · CAG and CGG · CCG sequences and subsequent analyses were impractical for the following reasons: (a) long stretches of repeats are very unstable [100], and (b) the method requires knowledge of the exact number of base pairs in the sequence. With the highly repetitive nature and length of the sequences in question, this was a problem. Hence, in order to see the effect(s) of length on the variances of writhe, twist, and linking number at lengths of up to 10 kbp, we extrapolated the theoretical equation(s) of Shimada and Yamakawa [65, 68] developed for the helical worm-like chain to encompass this length using the parameters (e.g., λ^{-1}; Kuhn segment, α; bending modulus, β; torsional modulus) extracted experimentally for short stretches of $(CTG \cdot CAG)_n$ and $(CGG \cdot CCG)_n$ repeats [45].

D. Variances of Writhe and Twist

Table 38-3 summarizes the results obtained by statistical mechanical calculations of the difference in the variance of writhe (a statistical measure of the fluctuations of the helix axis) normalized to chain length, L (i.e., $\Delta<(Wr)^2>/L$) between B-DNA and $(CTG \cdot CAG)_n$ as well as for B-DNA and $(CGG \cdot CCG)_n$. In essence, $\Delta<(Wr)^2>/L$ [e.g., $(<(Wr)^2>/L)_{TRS} - (<(Wr)^2>/L)_{B\text{-}DNA}$] computes the net differences in writhe when molecules of B-DNA and CTG · CAG of equal length are compared. This difference peaks at ~730 bp (~243 repeats) for $(CTG \cdot CAG)_n$ and ~780 bp (~260 repeats) for $(CGG \cdot CCG)_n$ with a greater difference seen for $(CTG \cdot CAG)_n$ [46]. Furthermore, since the variances in twist and writhe both play a significant role in the distribution of topoisomeric species, their relative contribution to supercoiling was also calculated. As seen in Table 38-3, for B-DNA, the contribution of writhe to Lk is 70%, whereas that of CGG · CCG and CTG · CAG is ~78 and 79%, respectively.

TABLE 38-3

	Length at which $\Delta<(Wr)^2>/L$ reaches a maximum[a] (bp)	$<(\Delta Tw)^2>$[b] (%)	$<(Wr)^2>$[c] (%)	ζ[d]
B-DNA	N/A	30	70	2.307
$(CTG \cdot CAG)_n$	730 ($n \cong 243$)	21	79	3.806
$(CGG \cdot CCG)_n$	780 ($n = 260$)	22	78	3.448

Note. Summary of calculations for the variances in writhe ($<Wr)^2>$), twist ($<(\Delta Tw)^2>$), and normalized writhe ($<(Wr)^2>/L$, where L is the contour length of the DNA and is equal to $n_{bp} \times$ 3.4 Å). $<(Wr)^2>$ represents the variations in the fluctuations of the helix axis (*i.e.*, bending) whereas $<(Wr)^2>/L$ is the normalized variation in writhe with respect to the length of the DNA (*i.e.*, variation of the helical axis per unit length of the DNA). $<(\Delta Tw)^2>$ represents the angular variations caused by twisting of the bases along the DNA's helical axis leading to under- or overwinding of the molecule. This table is a compilation of data from Bacolla *et al.* [45].

[a]The difference is between B-DNA and $(CTG \cdot CAG)_n$ and between B-DNA and $(CGG \cdot CCG)_n$.

[b]$<(\Delta Tw)^2> = (1 + \sigma)/2\pi^2$, where σ is Poisson's ratio ($\sigma = (\alpha/\beta) - 1$).

[c]$<(Wr)^2> = 0.095\ L^3[19.47 + L^2]^{-1} \exp[-6.8\ L^{-2.5}] + 0.00385\ L^2 \exp(-L) \times [1 + 1.092\ L + 1.076\ L^2 + 0.2788\ L^3]$, where $L = (n_{bp} \times 3.4$ Å).

[d]$\zeta = <(Wr)^2>/<(\Delta Tw)^2>$. The values reported represent the ratio at a DNA length of 10 kbp.

$<(\Delta Lk)^2>$ is related to the variance of writhe ($<(Wr)^2>$) and twist ($<(\Delta Tw)^2>$) [101] by:

$$<(\Delta Lk)^2> = <(Wr)^2> + <(\Delta Tw)^2> \quad (1)$$

The estimated torsional modulus, β, for the TRS segments [45] was found to be similar to that of B-DNA (i.e., 2.0–2.4 × 10^{-19} erg · cm) [62–64, 69], indicating that both $(CTG \cdot CAG)_n$ and $(CGG \cdot CCG)_n$ supercoil more efficiently than B-DNA due to an increased contribution from writhe.

E. Supercoiling Free Energy of CTG · CAG and CGG · CCG

The free energy of supercoiling, $\Delta G_{\Delta Lk}$, is related to ΔLk by $\Delta G_{\Delta Lk} = K(\Delta Lk)^2$ [101], where K, the apparent twisting coefficient, decreases with increasing DNA length [63, 64]. Since ΔLk and writhe are linearly related, the latter has a direct influence on the supercoiling free energy (i.e., $\Delta G_{\Delta Lk}$). Why is it important to calculate the free energy? The efficiency of many cellular processes such as transcription and replication are intimately related to the level of supercoiling (hence free energy) available within a particular DNA domain [10, 102, 103]. Also, the level of instability for a $(CTG \cdot CAG)_n$ or $(CGG \cdot CCG)_n$ tract depends on its length and orientation relative to the origin of replication [104, 105]. We propose that it is the accumulation of supercoiling within the TRS which leads to the instability.

TABLE 38-4 Summary of Calculations of the Supercoiling Free Energy ($\Delta G_{\Delta Lk}$) for B-DNA, $(CTG \cdot CAG)_n$, and $(CGG \cdot CCG)_n$ TRS

	Free energy at 10 kbp (kcal/mol)[a]	$\Delta\Delta G_{\Delta Lk}$[b] (kcal/mol)	Length of TRS having minimal $\Delta\Delta G_{\Delta Lk}$
B-DNA	606	N/A	N/A
$(CTG \cdot CAG)_n$	399	−723	500 bp (167 repeats)
$(CGG \cdot CCG)_n$	443	−558	540 bp (180 repeats)

Note. $\Delta G_{\Delta Lk}$ was calculated using the formula of Shimada and Yamakawa [68]: $\Delta G_{\Delta Lk} = \Gamma_0/[1 + 2\pi^2<(Wr)^2>/(1 + \sigma)\lambda L]$ where, $\Gamma_0 = \pi^2/\lambda\ell_{bp}(1 + \sigma)$, $<(Wr)^2>$ is the variance in writhe, $\ell_{bp} = 3.4$ Å, and λ^{-1} = Kuhn segment = $2P$ with P being the persistence length of the DNA. $\Delta\Delta G_{\Delta Lk}$ shows the difference in supercoiling free energy between the TRS and B-DNA at every point along the DNA, up to 10 kbp.

[a]The following formula was used to calculate $\Delta G_{\Delta Lk}$: $n_{bp}K/RT$ [101]. The calculations were performed with RT = 0.5825 at $T = 20°C$.

[b]$\Delta\Delta G_{\Delta Lk} = (\Delta G_{\Delta Lk})_{TRS} - (\Delta G_{\Delta Lk})_{B\text{-}DNA}$.

As summarized in Table 38-4, $\Delta G_{\Delta Lk}$ reaches a constant value of 606 kcal/mol at 10 kbp for B-DNA, 443 kcal/mol for $(CGG \cdot CCG)_n$, and 399 kcal/mol for $(CTG \cdot CAG)_n$. The values for the TRS are 27 and 34%, respectively, lower than random sequence DNA for chains longer than 3 kbp. Below 2 kbp, $\Delta G_{\Delta Lk}$ rises sharply and approaches 1980 kcal/mol at 0 length for all DNAs [45]. $\Delta\Delta G_{\Delta Lk}$, on the other hand, becomes progressively more negative at increasing DNA lengths, reaching a minimum of −723 kcal/mol at 500 bp (167 repeats) for $(CTG \cdot CAG)_n$ and −558 kcal/mol at 540 bp (180 repeats) for $(CGG \cdot CCG)_n$, suggesting that $(CTG \cdot CAG)_n$ and $(CGG \cdot CCG)_n$ require less energy to supercoil (writhe) than random DNA. In addition, the difference in free energy of supercoiling (i.e., $\Delta\Delta G_{\Delta Lk}$) between TRS and random DNA is not uniform with length, but reaches a maximum at about 500–540 bp. At this n_{bp}, the variance in linking number, obtained from the relation $<(\Delta Lk)^2> = RT/2K$, equals 0.100 for B-DNA, whereas it is 0.157 for $(CGG \cdot CCG)_n$ and 0.190 for $(CTG \cdot CAG)_n$, an increase of 57 and 90%, respectively. This analysis confirms that there is an optimal length at which $(CTG \cdot CAG)_n$ and $(CGG \cdot CCG)_n$ writhe more favorably than at shorter or longer lengths [45].

F. Supercoiling Free Energy and TRS Instability in Pre- to Full-Mutation Alleles

All of the calculations that involved the evaluation of $<(Wr)^2>$, namely $\Delta(<(Wr)^2>/L)$ and $\Delta G_{\Delta Lk}$, indicate that whereas the TRS have a greater ability to supercoil than random sequence DNA, tracts of $(CTG \cdot CAG)_n$ or $(CGG \cdot CCG)_n$ 500 to 540 bp long (167 to 183 repeats as estimated from $\Delta G_{\Delta Lk}$) have the greatest tendency to writhe when compared with the same lengths of random B-DNA. We refer to this range as the region of "hyperflexibility." The increased hyperflexibility of the TRS coincides with the length of repeats (180–200 units) that demarcates the premutation from the full mutation range in fragile X and myotonic dystrophy and also coincides with the repeat size that leads to far greater expansions (hundreds of repeats) in offspring [106–108]. In other words, the largest accumulation of supercoiling energy tends to occur at lengths of 167–183 repeats. The presence of this optimal length for the partitioning of writhe in the TRS is significant, considering its surprisingly close correspondence with the repeat size of 180–200 that distinguishes both the premutation from the full mutation length and the occurrence of small versus large expansions in the *FRAXA* and *DM* loci [106–108].

This correspondence of the region of hyperflexibility with the premutation to full mutation threshold makes it tempting to speculate that triplet repeat expansion is associated with the supercoiling of DNA. No doubt, the poorly understood molecular and cellular events involved with DNA slippage, genetic instabilities, anticipation, and alterations in gene expression that elicit changes in development which are recognized as disease syndromes are complex. We do not propose that DNA structure alone is responsible (transcription, replication, and single strand binding proteins may also contribute to instability), but the dynamic as well as static conformational features of the TRS play a key role. Below, we discuss how this hyperflexible region of the TRS is involved in priming the initial events leading to genetic instability within the locus and subsequent repeat expansion.

G. Influence of DNA Flexibility on Genetic Expansion of CTG · CAG and CGG · CCG

Cellular processes such as transcription and replication dramatically alter the local superhelical densities of DNA due to the unwinding of the helices [10, 102, 103]. Also, the extent of instability for a $(CTG \cdot CAG)_n$ or a $(CGG \cdot CCG)_n$ tract depends on its length and orientation relative to the origin of replication in *Escherichia coli* [100, 104]. We propose that high levels of supercoiling (writhe) accumulate within the TRS (Fig. 38-2) during processes such as transcription and replication which lead to instability. Some of these events are outlined in Fig. 38-4. Translocation of the DNA polymerase complex generates positive supercoils ahead of the replication fork, and possibly negative supercoils behind it, on the leading strand (step A) [10, 101]. Positive and negative supercoils partition preferentially within a tract of TRS (rather than within random B-DNA sequences) due to its higher flexibility (step B). This partitioning is influenced by the length of the TRS with segments of 400 to 600 bp accommodating the greatest levels of writhe as opposed to the same lengths of random DNA. The TRS-localized increases in positive superhelicity hinder their efficient removal by topoisomerases, thus decreasing the processivity of the polymerase complex. Pausing [109, 110] of the enzymatic complex would then allow reiterative DNA synthesis [111] to take place (step C), thus leading to expanded daughter strands. Alternatively, the polymerase complex may dissociate from the template (step D), and allow the diffusion of positive and negative supercoil domains. This diffusion is accompanied with the release of a newly synthesized strand(s) from its parent strand(s), which causes hairpins [110] to form in the daughter strand(s). This leads to expansion, and then synthesis resumes (E) and the process is repeated. Thus, the unorthodox capability of CTG · CAG and CCG · CGG to serve as a sink for supercoil density may play an integral role in the expansion process.

IV. GAA · TTC STRUCTURES RELATED TO FRIEDREICH'S ATAXIA

A. Friedreich's Ataxia

Friedreich's ataxia (FRDA) was first described in 1863 as a familial form of ataxia with onset under the age of 20. FRDA is the most common of the early onset ataxias and accounts for at least 50% of the cases of hereditary ataxia; the prevalence in Europe and the United States is approximately 1 per 10,000 (reviewed in [112–114]). Males and females are equally affected. The usual presenting features are ataxia of gait, but this is occasionally proceeded by scoliosis or cardiac symptoms [112–114]. Essential criteria proposed for the diagnosis of FRDA are progressive limb and gait ataxia developing before the age of 25, autosomal recessive inheritance, absent tendon reflexes in the legs, and electrophysiological evidence of an axonal sensory neuropathy. Dysarthria, areflexia, pyramidal weakness of the legs, extensor plantar responses, and distal loss of joint position and vibration sense are not found in all patients at the same time of presentation but are eventually universal. Heart muscle disease is found in at least 2/3 of the patients and diabetes mellitus occurs in 10% of the patients. Furthermore, 10–20% have impaired glucose tolerance. Chapter 26 in this volume reviews the disease and molecular biology of Friedreich's ataxia in detail.

The prognosis of FRDA is variable with more than 95% of the patients being wheelchair bound by the age of 45; on average, they lose the ability to walk 15 years after the onset of symptoms. Pathological changes involve mainly the spinal cord with degeneration of the posterior columns and spinocerebellar tracts. The treatment of FRDA has been disappointing [112–114].

Major discoveries in this autosomal recessive disease were reported in the past 7 years including the mapping of the mutated gene to chromosome 9q13-q21.1 in 1988–1990 and the recent realization that it is caused by an intronic GAA · TTC triplet repeat expansion in most cases [115]. The gene, X25, was identified in the critical region for the FRDA chromosomal locus which encodes a 210-amino-acid protein, frataxin, that has a number of homologs in species as distant as *C. elegans* and yeast. Five exons were spread over 40 kb and alternative splicing was observed [115]. Northern blot analysis from different human tissues revealed the highest level of expression of X25 in heart, intermediate levels in liver, skeletal muscle, and pancreas, and minimal levels in other tissues including whole brain. A 1.3-kb major transcript was identified. In screening 184 FRDA patients, three point mutations that introduced changes in the X25 gene product were identified. These conservative missense mutations were shown to be disease-causing. However, the GAA expansion from 100 to more than 900 GAA · TTC repeats was shown in 98% of the cases to be the responsible mutation. PCR analysis of an additional 98 normal controls did not show any expansion and revealed that the GAA · TTC repeat is polymorphic, varying from 7 to 22 units.

The molecular function of frataxin is unclear but a major discovery was just reported at the First Annual Montreal International Neurology Meeting (May 29–June 1, 1997) by Kaplan *et al.* [116]. These workers characterized a yeast gene with extensive homology to the human frataxin protein which encodes a mitochondrial protein involved in iron homeostasis. Frataxin also is a mitochondrial protein. Heart failure is the most common cause of death in FRDA and myocardial cells from FRDA patients exhibit subnormal mitochondrial enzyme activity and show stainable iron deposits [116].

FRDA patients appear to have either undetectable or extremely low X25 mRNA amounts when compared to carriers or unrelated controls [115]. Hence, Pandolfo *et al.* hypothesized that the cause is due to either an abnormal RNA processing or the interference of the expanded GAA repeat with the transcriptional machinery. Further analyses on a collaborative basis between the Pandolfo and Wells laboratories indicated the likelihood of the latter hypothesis [117]. The phenotypic variability in FRDA has recently been studied in a group of 100 patients with typical disease as well as three groups of patients with atypical clinical presentations [118]. Overall, the phenotypic spectrum of FRDA appeared to be wider than defined by the currently used diagnostic criteria and includes two types of variability [118, 119].

Very recently, Chamberlain *et al.* [120] reported that X25 comprises part of the STM7 gene, contributing to at least four splice variants, and the identification of new coding sequences. Functional analysis of the STM7 recombinant protein corresponding to the reported 2.7-kilobase transcript has demonstrated phosphatidylinositol-4-phosphate-5-kinase activity, supporting the idea that the disease is caused by a defect in the phosphoinositide pathway, possibly affecting the vesicular trafficking or synaptic transmission. However, the types of data used in this investigation and their analyses have been questioned [121]. Furthermore, the very recent demonstration of frataxin as an iron metabolism protein and the correlation with human systems [116] also raises questions concerning the Chamberlain work [120].

FRDA is the only triplet repeat disease recognized to date that is caused by an expansion of GAA · TTC. The fragile X syndromes are caused by an expansion of CGG · CCG, and myotonic dystrophy, Huntington's disease, Kennedy's disease, etc., are caused by expansions of CAG · CTG (reviewed in [106–108, 114]).

B. Model for Reduction of X25 RNA Abundance

Triplexes are three-stranded nucleic acid structures (usually DNA) formed at tracts of oligopurines and oligopyrimidines (reviewed in [4, 5, 17, 122] and described above) (Fig. 38-1 and Table 38-1). Since thorough investigations were conducted in the 1980s on triplexes, substantial information is available on the effect of sequence and the type of R · Y sequences required, the effects of pH and methylation of C residues, the types of bi-triplexes (nodule DNA) formed, the effect of interposing non-R · Y sequences, the influence of environmental factors on the stabilization of the four triplex isomers, the effect of stabilization by intercalating agents, and related factors [4, 5, 17, 122–134].

This laboratory has previously demonstrated that the GAA · TTC sequence in FRDA forms triplexes [126–134] in the lengths that were available to us at that time (8–58 repeats). Hence, the possible relationship of this unorthodox DNA conformation to the gene expression properties (reduced abundance of mRNA) of the X25 gene is enticing. The formation of a triplex is known to

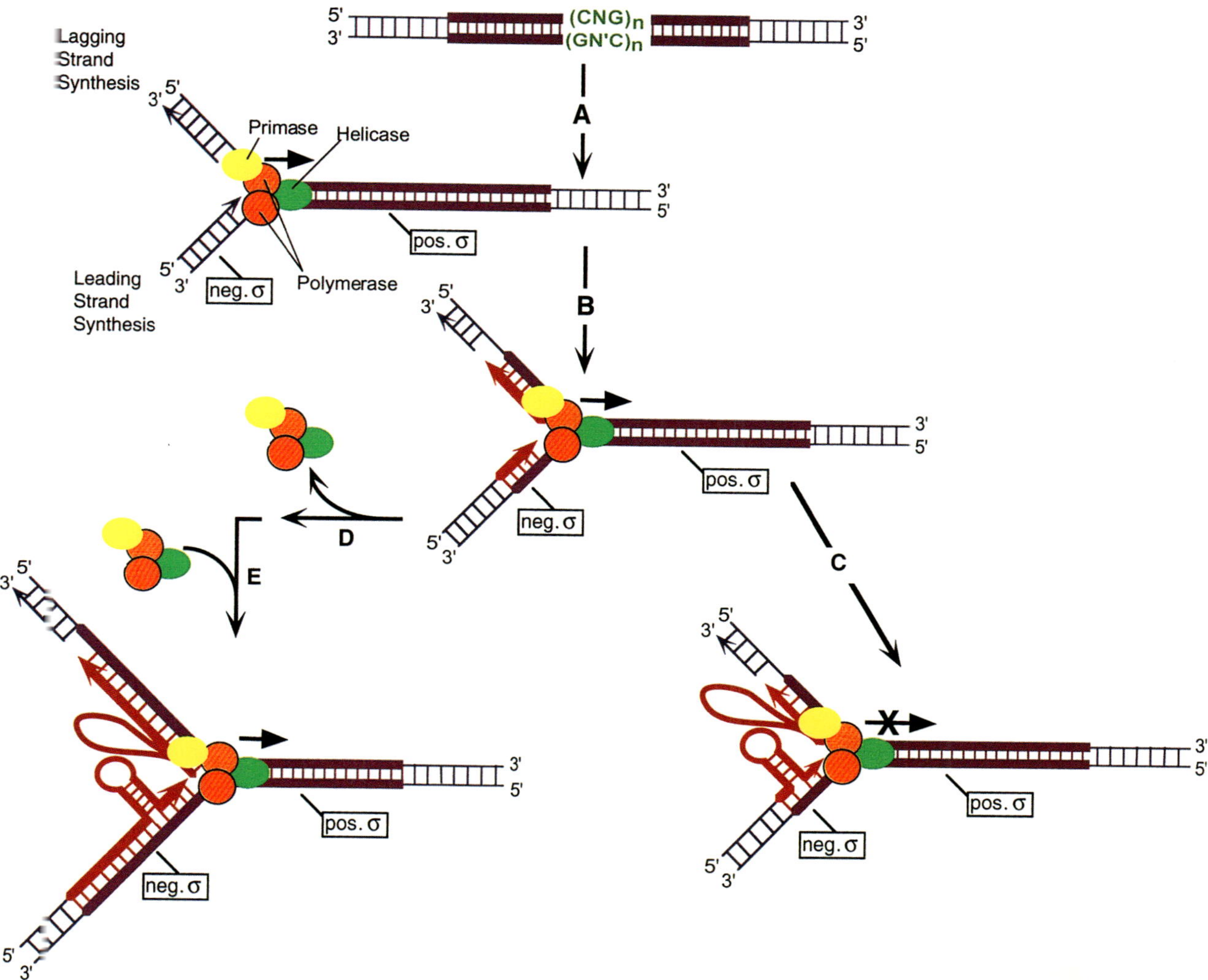

FIGURE 38-4 Model for the expansion of $(CTG{\cdot}CAG)_n$ and $(CGG{\cdot}CCG)_n$ TRS. Thick solid lines indicate TRS regions, whereas thinner lines correspond to the random sequence flanking DNA. N represents nucleotide T or G, and N′ represents nucleotide A or C. pos a and neg s refer to positive and negative supercoiling respectively. Reprinted with permission from reference [46].

inhibit transcription [5, 17, 122, 135, 136] and the presence of a stable RNA · DNA hybrid during transcription has been characterized (reviewed in [5]) in an immunoglobulin switch region. Postel *et al.* [135] demonstrated that a triplex-forming oligodeoxyribonucleotide binds to the *c*-myc promoter in HeLa cells and thereby reduces *c*-myc mRNA levels. Thus, triplex formation can occur between an exogenous oligonucleotide and duplex DNA in the nucleus of treated cells. Furthermore, the *in vitro* transcription of a poly dA · poly dT-containing sequence is inhibited by interaction between the template and its transcripts. This recent work is particularly relevant due to the demonstration that the poly dT-containing template forms an RNA · DNA triplex with its mRNA, blocking further transcription.

We hypothesized [134] (Fig. 38-5) that the mechanism of the reduction of abundance of mature X25 mRNA in individuals with FRDA is the formation of an intermolecular triplex between the GAA · TTC in the first X25 intron and the RNA segment with the rAAG tract removed by splicing. Prior work [17, 135, 136] showed that the presence of a triplex inhibits transcription. In the case of long rAAG tracts (100 or more repeats) from FRDA cases, the triplex may be sufficiently stable thermodynamically to cause the reduction in abundance of the FRDA mature mRNA, whereas for shorter rAAG stretches from normal individuals (6–20 repeats), the triplex may be unstable and will not cause an inhibition. This hypothesis is consistent with the clinical observations that patients with longer GAA · TTC repeats (350–600 repeats) are more severely afflicted than patients with shorter repeats (150–250 repeats). Possible pairing schemes (reverse Hoogsteen) are shown (Fig. 38-5) for the T · A · A and C · G · G triads. It should be emphasized that this model has not been subjected rigorously to experimental testing but is the topic of current work in this laboratory.

V. CLONING OF ALL 10 TRIPLET REPEAT SEQUENCES

The cloning of triplet repeat sequences of differing lengths, orientations, extents, and types of interruptions, etc., is important for a wide variety of further investigations (described in this book) both in human as well as other eukaryotic systems and in prokaryotic models. Hence, considerable effort has been devoted to these methodologies.

A. CTG · CAG

Since this sequence is the mutation responsible for the majority of the hereditary neurological diseases described herein, more effort has been expended on the cloning of this sequence than for other sequences. In general, the gene involved in the disease has been subcloned from patient materials and then subsequently recloned in a wide variety of vectors. In fact, this is the strategy originally employed by this laboratory [89, 104]. Also, since we have established an effective system for investigating expansions and deletions of CTG · CAG [104, 137–139], this has provided a wealth of plasmids with CTG · CAG tracts ranging from 15 to 400 repeats. These clones have been invaluable for a wide range of investigations (described herein).

In summary, virtually all CTG · CAG plasmids have been prepared by the expansion and deletion system in *E. coli* [89, 104, 137–139] and have been characterized by DNA sequencing on both strands.

B. CGG · CCG

The FMR-1 gene for this human fragile X syndrome contains a genetically unstable CGG · CCG region in the 5′ nontranslated region (see reviews in this book). In general, other laboratories have derived clones from the FMR-1 gene of patients. Work in this laboratory has taken two approaches. First, a family of plasmids containing pure CGG · CCG repeats where the repeats ranged from 6 to 32 was prepared by a combination of standard cloning methodologies with synthetic oligonucleotides along with the procedures [104, 137–139] for generating expansions and deletions *in vivo* in *E. coli.* It was not possible to obtain longer inserts by this method. However, the direct cloning of cDNA sequences from human patients has served as a source of longer TRS. The gift of a clone (RN2) from Dr. Ben A. Oostra was crucial for obtaining clones that contain long tracts of CGG · CCG. We isolated a fragment containing approximately 80 interrupted CGG · CCG repeats and made oligomers of this fragment which were subsequently cloned by standard procedures. Using these approaches, it was possible to obtain inserts that contained 80, 160, 240, and even longer tracts of interrupted CGG · CCG repeats. In our experience, the fragile X sequence is the most genetically unstable sequence compared to the other nine TRS. Again, DNA sequencing provided a rigorous characterization of the inserts [100].

C. AAT · ATT

AAT · ATT is interesting since a survey of triplet repeats in the human genome revealed that it is the

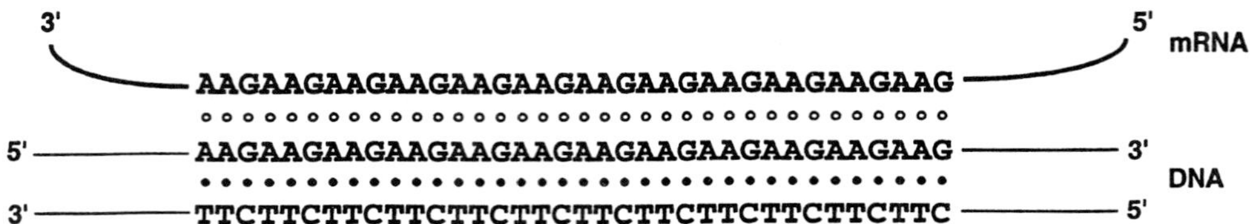

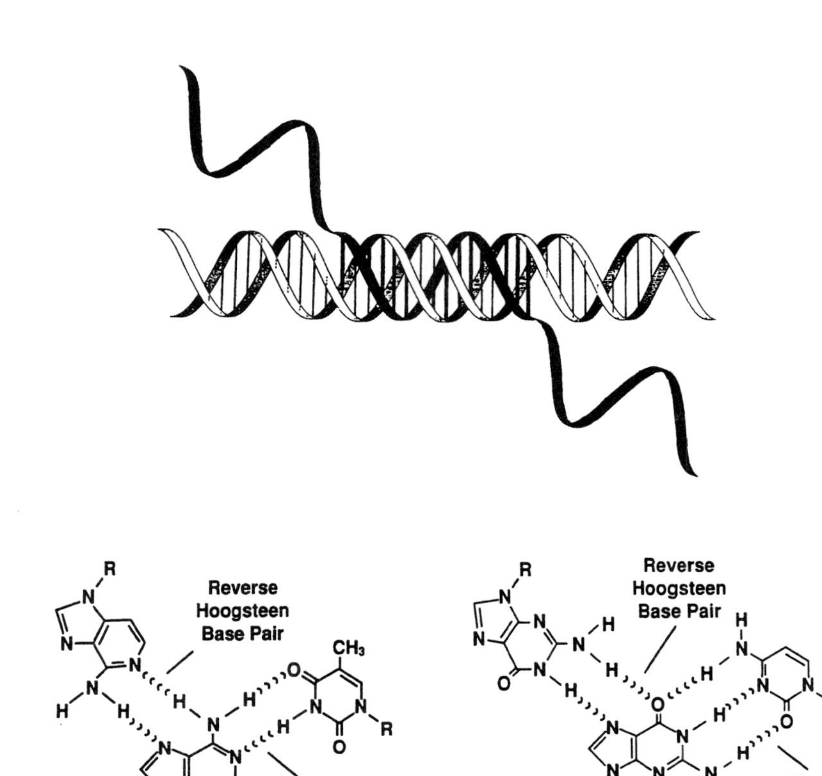

FIGURE 38-5 Model for reduction of mRNA abundance in Friedreich's ataxia syndrome. (top and center) Relationship between RNA and duplex chromosomal DNA; (bottom) possible base pairing relationships for A · A · T and G · G · C. Reproduced with permission from Ohshima *et al., J. Biol. Chem.* **271,** 16773–16783, 1996 [134].

most abundant (by at least fourfold) and is the most frequently polymorphic (reviewed in [47, 134]). Also, this microsatellite sequence is very abundant in introns but is uncommon in exons and has been implicated in the regulation of transcription. A 270-bp tract of this sequence is found upstream of a variant surface glycoprotein gene of *Trypanosoma brucei* (reviewed in [47]). We subcloned the TAA · TTA sequence from clones containing the *T. brucei* sequence and prepared a family of shorter length sequences by the deletion strategy reported previously [137–139]. Accordingly, inserts with 18, 30, 60, and 90 TRS were prepared and characterized by DNA sequencing.

D. The Other Seven TRS

This laboratory has cloned the other seven TRS, namely, GTA · TAC, GAT · ATC, GTT · AAC, CAC · GTG, AGG · CCT, TCG · CGA, and AAG · CTT [134]. Only the last sequence has been shown to be related to a neurological disease gene. Novel cloning strategies with chemically synthesized oligonucleotides were employed to clone these seven TRS. This approach in conjunction with *in vivo* expansion studies in *Escherichia coli* enabled the preparation of at least 81 plasmids containing the repeat sequences with lengths of approximately 16 up to 158 triplets in both orientations with varying extents of polymorphisms. The inserts were characterized by DNA sequencing as well as physical and chemical probe analyses.

VI. STRUCTURAL STUDIES ON THE OTHER EIGHT TRIPLET REPEATS

A. Overview

To date, more than 10 human hereditary diseases have been identified to be associated with the expansion of TRS [106–108]. Two of the TRS, namely CTG · CAG and CGG · CCG, have been cloned and characterized as described above. In this section, we will discuss some of the results obtained with the remaining 8 possible TRS (i.e., GTA · TAC, GAT · ATC, GTT · AAC, CAC · GTG, AGG · CCT, TCG · CGA, AAG · CTT, and TTA · TAA). Only 3 (i.e., CTG · CAG, CGG · CCG, and AAG · CTT) of the 10 TRS have been found to be associated with neurodegenerative diseases in humans, but studies on all 10 are important since more TRS may be identified in the future to be linked to critical disease genes. Also, the behaviors of these 7 TRS serve as controls for our understanding of the properties of the 3 disease-related sequences. Therefore, it is crucial to obtain information on the structural characteristics of all 10 duplex sequences.

B. Pausing of DNA Polymerase in the TRS

Unusual DNA conformation(s) have been known to inhibit DNA synthesis by DNA polymerases [140–145]. This enables the use of primer extension as a probe for the detection of the DNA secondary structure in question, in this case being that of the TRS. Studies conducted by Ohshima *et al.* [47, 134] on all 10 TRS show that DNA polymerases require double-stranded DNA, but not necessarily one which is supercoiled. Furthermore, pausing of the polymerases depended strongly on the temperature of preincubation, and length of the TRS (see Chapter 45 in this book for a detailed description). Table 38-5 summarizes some of the data for the effects of length and sequence composition on DNA polymerase pausing.

C. DNA Structural Analysis of the Eight TRS

Ohshima and co-workers used two powerful techniques to identify the type of DNA secondary conformation in the TRS, namely chemical probing and two-dimensional agarose gel electrophoresis [4, 5, 17, 122–134]. The structural characterization of (CTG · CAG)$_n$ and (CGG · CCG)$_n$ has already been discussed in the previous sections using circularization kinetics, helical repeat determinations, and PAGE [45, 46]. However, chemical probe analyses and two-dimensional agarose gel electrophoresis did not reveal the presence of any aberrant secondary structure(s) within these two TRS [45]. This was a highly unusual result and revealed the absence of unpaired loop regions (such as in cruciforms or triplexes) as well as the absence of perturbed bases (such as at B–Z junctions with left-handed Z-DNA). These probes have been widely used for characterizing other types of unusual DNA structures and it was surprising, indeed, when the rigorous utilization of these methodologies was unsuccessful with CTG · CAG and CGG · CCG. Thus, this gave rise to the need to utilize other techniques [45, 46]. Table 38-6 summarizes the data obtained for the eight TRS using chemical modification and two-dimensional gel electrophoresis.

Chemical modification of a supercoiled plasmid containing (TTA · TAA)$_{18}$ was performed using OsO_4, diethylpyrocarbonate (DEPC) and chloroacetaldehyde (CAA) to evaluate the presence of unpaired bases within the sequence [47]. The modified bases (OsO_4 modifies T whereas DEPC and CAA modify the A) were detected by the inhibition of the DNA polymerase progression at the corresponding modification site. Furthermore, the sequence showed strong reactivity on both strands throughout the entire repeating tract of TTA · TAA, indicating the possibility of unpaired regions in both the top and bottom strands [47], over the entire length of the repeat. Ohshima and colleagues also investigated the effect of length (from 18 to 90 repeats) and superhelical density (-s) on the extent as well as the pattern of modification of the bases. The results they obtained by quantitation of the amount of OsO_4 modification as a function of superhelical density clearly showed that for linear DNA containing TTA · TAA repeats and supercoiled plasmids with $s = -0.025$, virtually no modification was observed. However, the degree of OsO_4 reactivity increased sigmoidally in proportion to the level of torsional tension introduced in the circular molecule(s) [47]. Similarly, as presented in Table 38-6, AGG · CCT and AAG · CTT also showed reactivity [134] toward the 5′-half of the TRS at low pH, indicating the formation of intramolecular triplexes.

Two-dimensional agarose gel electrophoretic analyses showed that only three triplet repeats, namely TTA · TAA, AGG · CCT, and AAG · CTT, showed

TABLE 38-5 Summary of DNA Polymerase Pausing Patterns for All 10 TRS

TRS	n	Template: Top strand	Template: Bottom strand	Repeat at which pausing occurs: Top strand	Repeat at which pausing occurs: Bottom strand
CTG · CAG	26	–	–	–	–
	75	+	+	5′ & $(CTG)_{37}$	5′ & $(CAG)_{30}$
	130	++	++	5′ & $(CTG)_{37}$	5′ & $(CAG)_{30}$
CGG · CCG	26	–	–	–	–
	80	+	–	$(CGG)_{29–31}$	–
	160	++	–	$(CGG)_{29–31}$	–
TTA · TAA	30	–	–	–	–
	60	–	–	–	–
	90	–	–	–	–
GTA · TAC	61	–	–	–	–
	77	–	–	–	–
AGG · CCT	16	+	+	3′ Half	Throughout
	30	++	++	3′ Half	Throughout
GAT · ATC	62	–	–	–	–
	91	–	–	–	–
TCG · CGA	46	–	–	–	–
	98	++	–	$(TCG)_{29–33}$	–
GTT · AAC	42	–	–	–	–
	74	–	–	–	–
CAC · GTG	74	–	–	–	–
	115	–	–	–	–
AAG · CTT	16	+	+	Throughout	3′ Half

TABLE 38-6 Two-Dimensional Gel and Chemical Probing Analyses for TRS

Triplet repeat	n	2-D gel relaxation: pH 8.3	2-D gel relaxation: pH 4.5	Chemical probe modification: pH 8.3	Chemical probe modification: pH 4.5
$(TTA \cdot TAA)_n$	18	Yes	—	Yes	—
	30	Yes	—	Yes	—
	60	—	—	Yes	—
	90	—	—	Yes	—
$(GTA \cdot TAC)_n$	45	No	—	No	—
$(GAT \cdot ATC)_n$	44	No	—	No	—
$(GTT \cdot AAC)_n$	26	—	—	No	No
	42	No	—	—	—
$(CAC \cdot GTG)_n$	16	—	—	No	No
	46	No	No	—	—
$(AGG \cdot CCT)_n$	16	No	Yes	No	Yes
	30	No	Yes	—	—
	53	No	—	—	—
$(TCG \cdot CGA)_n$	16	—	—	No	No
	46	No	—	—	—
$(AAG \cdot CTT)_n$	16	No	Yes	No	Yes
	38	Yes	Yes	No	Yes
	58	Yes	Yes	No	Yes

supercoil induced transitions. The relaxation of TTA · TAA occurred throughout the entire region of the repeat suggesting the formation of an open non-H-bonded region. Figure 38-6 shows a model of the structure formed by the TTA · TAA repeat. Based on chemical modification and 2-D gel data (Table 38-6), Ohshima *et al.* [47] hypothesized that the opening of the helix created a non-H-bonded single-stranded DNA region throughout the length of the repeat. This is not influenced by the length of the TRS, but was dependent on superhelical density.

For AGG · CCT, the relaxation due to the triplex occurs at pH 4.5 but not 8.3, whereas for AAG · CTT both extremes allow for the formation of an intramolecular triplex [134]. The mechanism(s) leading to the expansion of the AAG · CTT repeats is not known, but is likely to be interesting since this repeat forms a triplex and is the mutation responsible for Friedreich's ataxia (see Section IV above).

VII. PROSPECTS FOR THE FUTURE

Our comprehension of the role of triplet repeat DNA structural issues related to hereditary neurological diseases is in its infancy. The discoveries described above give rise to a wealth of implications for future investigations. On occasion, investigators inquire about the relevance of DNA conformational problems to gene expression and disease etiology. From our common knowledge, we understand the relevance of the shape of a dolphin to its aquatic gymnastics as well as the shape of a bird to its flight characteristics. Accordingly, the shape of a region of DNA is intimately related to its behaviors and gene expression properties, although they may be less than completely clear at the present time. A good example is the intrinsically writhed and flexible conformation of CTG · CAG and its preferential expansion in a genetically defined system, relative to the other nine triplet repeat sequences [137]. Obviously,

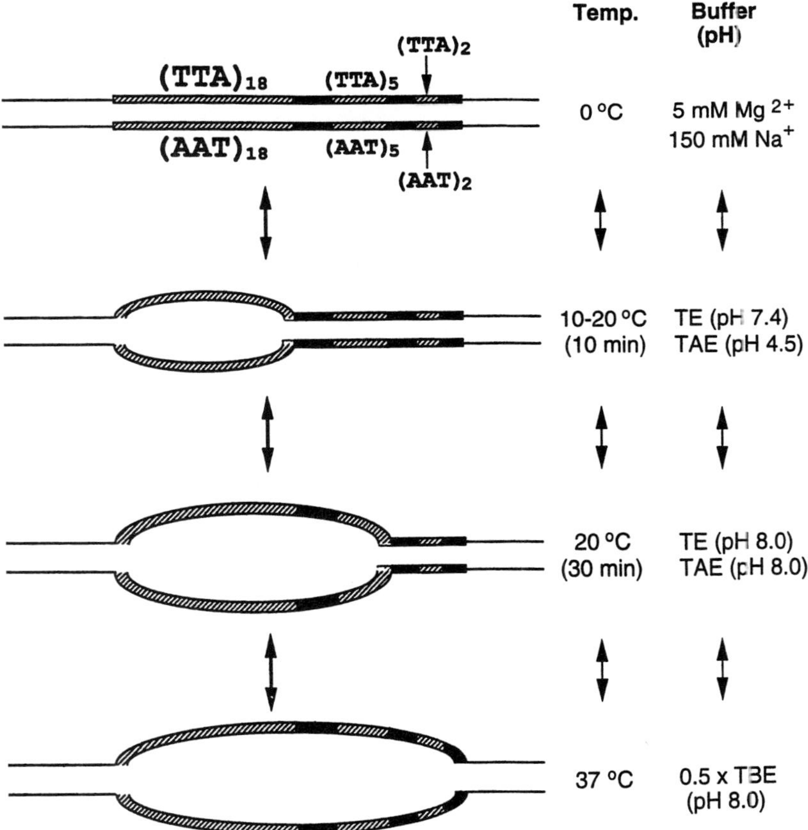

FIGURE 38-6 Model for the unpairing of the TAA · TTA repeats and its neighboring (A + T)-rich sequences. The (A + T)-rich region, including the $(TAA \cdot TTA)_5$ and $(TAA \cdot TTA)_2$ are melted at elevated temperatures or under differing buffer conditions. The strand containing TTA · TAA repeat sequences is cross-hatched. The (A + T)-rich regions are shaded. Reprinted with permission from Ohshima *et al.*, *J. Biol. Chem.* **271,** 16784–16791, 1996 [134].

the three-dimensional structure and properties of this sequence are unique relative to the other nine cases and/or the interaction of this conformation with DNA synthetic enzymes *in vivo* gives rise to this biological observation. Interestingly, a similar conclusion that the "repeat itself may be the important factor determining repeat instability" has been deduced by other workers in primate systems [146].

The neurological community was generally unfamiliar with DNA structural problems prior to approximately 1995; at this time considerable work was just being published regarding DNA hairpin structures (reviewed in [147]) and the involvement of DNA structure in genetic instabilities [104, 137]. Most workers in the field currently believe that DNA structural properties are important determinants related to gene expression and expansion–deletion but also readily acknowledge the involvement of a variety of replication, recombination, and accessory factors.

The discovery of flexible and writhed CTG · CAG and CGG · CGG [45, 46] gives rise to a number of issues.

· This new DNA structure contains no unpaired or perturbed bases as revealed by chemical and enzymatic probe analyses and has ~10.4 bp per turn. This structure is a right-handed B conformation but has the intrinsic property of being substantially more flexible than random DNA sequences and is writhed.

· The conformation has been proposed [45, 46] to serve as a sink for negative supercoil density. This concept of a "soft spot" in DNA has not yet been rigorously and directly proven.

· The relationship to DNA polymerase pausing is uncertain. Interestingly, Mirkin *et al.* [148] have shown pausing of DNA synthesis on long CTG · CAG and CGG · CCG sequences *in vivo* in *E. coli.*

· The role of flexible and writhed DNA in strand slippage has been proposed but remains to be proven regarding its role in genetic expansion.

· The role of this structural feature in transcription stalling [149] has yet to be revealed. A method must be developed for converting the flexible and writhed conformation to a normal right-handed B structure with ordinary properties in order to test these concepts.

· Are the other eight TRS sequences flexible and writhed? No studies have yet been conducted by circularization kinetics on fragments containing the other eight sequences nor have helical repeat determinations been conducted. Polyacrylamide gel electrophoretic determinations [134] suggest that it is unlikely that the other eight sequences share this property. Likewise, the structural calculations from crystallographic analyses are consistent with this notion.

· Our comprehension of the molecular or atomic features that cause the intrinsic flexibility and writhing have yet to be broached experimentally. Stacking interactions were not considered previously and only dinucleotide step analyses were performed [45].

· The relationship of DNA bending of these sequences and histone interactions is rather unclear. On the one hand, CTG · CAG sequences found near the 3′ end of the myotonic dystrophy (DM) locus have been found to form unusually stable nucleosomes [89, 90] at lengths of >75 copies of the TRS, a length compatible with the threshold for the disease. This suggested a model in which the generation of a hyperstable heterochromatin might profoundly alter the local chromatin structure, resulting in (a) DNA polymerase pausing, DNA template slippage, and subsequent expansion of the repeats, and (b) inhibition of transcription through the octamer-TRS core [89]. On the other hand, studies conducted on CGG · CCG repeats near the 5′ end of the fragile X gene (FMR-1) indicate that these sequences have a lesser propensity to associate with histones [93–95]. Given that both TRS are intrinsically flexible, there is no molecular explanation associated with this drastic differential binding affinity, indicating that a feature other than flexibility plays a role in wrapping the TRS around the histone core.

· Flexible and writhed DNA should be considered differently than triplexes or left-handed Z-DNA regarding *in vivo* functions since no transition from a right-handed B conformation exists. This new conformation is an intrinsic property of the DNA and is not a conformer in equilibrium with the normal DNA structure.

· The effect of methylation of CGG · CCG on the flexible and writhed conformation has not been investigated yet.

· The influence of sequence interruptions (both in terms of types and locations) have not been investigated nor has the role of ligand interactions such as intercalators.

Concerning the GAA · TTC sequence and structure related to Friedreich's ataxia, the following uncertainties are evident:

· What is the structure of long GAA · TTC sequences? Whereas this may be one single large triplex, it may also be a family of smaller triplexes or a completely new type of conformation.

· Will rAAG form a triplex with long GAA · TTC sequences and what is the effect of the lengths of these two tracts on this behavior? No direct investigations have yet been conducted.

· What is the effect of these structures on transcription?

· What is the mechanism of expansion of the FRDA sequence? Some investigators have inferred that short oligonucleotides of these sequences do not form hairpins. Thus, the mechanism remains to be identified.

· If a triplex is the dominant conformational feature, should we think about GGA · TCC sequences forming triplexes and having an involvement with other neurological diseases?

Considering the other seven TRS including AAT · ATT, no disease relationship has yet been identified. However, in the event that one of these sequences is found to be the mutation causing a hereditary neurological disease, this situation will change "overnight." As with the Friedreich's ataxia sequence, the interest level will skyrocket.

In summary, we are in the very early stages of our comprehension of the involvement of interesting DNA conformational features in the etiology of several devastating neurological diseases. Among other questions, investigators wonder why only these three TRS are associated with hereditary neurological disease genes. Likewise, why are these TRS expanded in the disease gene loci but not in other locations. Also, the sex of parent effects remain to be clarified. Furthermore, a recurring question is why hereditary neurological disease genes are involved with triplet repeats but not other types of hereditary diseases. The future will be intriguing indeed as answers to these questions unfold.

Acknowledgments

This work was supported by grants from the National Institutes of Health (GM52982) and the Robert A. Welch Foundation to Robert D. Wells. The authors express their appreciation to R.D.W. for his support and encouragement. In addition, they express appreciation to Dr. Stephen C. Harvey for advice regarding structural issues and to Dr. Naoaki Sakamoto for some of the artwork.

References

1. Hoogsteen, K. (1963). The crystal and molecular structure of a hydrogen-bonded complex between 1-methylthymine and 9-methyladenine. *Acta Crystallogr.* **16,** 907–916.
2. Watson, J. D., and Crick, F. H. CF. H. (1953). The structure of DNA. *Cold Spring Harbor Symp. Quant. Biol.* **XVIII,** 123–131.
3. Saenger, W. (1984). "Principles of Nucleic Acid Structure." Springer-Verlag, New York.
4. Sinden, R. R. (1994). "DNA Structure and Function." Academic Press, San Diego.
5. Soyfer, V. N., and Potaman, V. N. (1996). "Triple-Helical Nucleic Acids." Springer-Verlag, New York.
6. Wang, A. H.-J., Quigley, G. J., Kolpak, F. J., van Boom, J. H., van der Marel, G., and Rich, A. (1979). Molecular structure of a left-handed double-helical DNA fragment at atomic resolution. *Nature* **282,** 680–686.
7. Stollar, B. D. (1992). Immunochemical analyses of nucleic acids. *Prog. Nucleic Acid Res. Mol. Biol.* **42,** 39–77.
8. Rahmouni, A. R., and Wells, R. D. (1989). Stabilization of Z DNA *in vivo* by localized supercoiling. *Science* **246,** 358–363.
9. Zacharias, W., Jaworski, A., and Wells, R. D. (1990). Cytosine methylation enhances Z-DNA formation *in vivo. J. Bacteriol.* **172,** 3278–3283.
10. Wang, J. C. (1996). DNA topoisomerases. *Annu. Rev. Biochem.* **65,** 635–692.
11. Wittig, B., Dorbic, T., and Rich, A. (1991). Transcription is associated with Z-DNA formation in metabolically active permeabilized mammalian cell nuclei. *Proc. Natl. Acad. Sci. USA* **88,** 2259–2263.
12. Lukomski, S., and Wells, R. D. (1994). Left-handed Z-DNA and *in vivo* supercoil-density in the *Escherichia coli* chromosome. *Proc. Natl. Acad. Sci. USA* **91,** 9980–9984.
13. Jaworski, A., Hsieh, W.-T., Blaho, J. A., Larson, J. E., and Wells, R. D. (1987). Left-handed DNA *in vivo. Science* **238,** 773–777.
14. Herbert, A., and Rich, A. (1996). The biology of left-handed Z-DNA. *J. Biol. Chem.* **271,** 11595–11598.
15. Müller, V., Takeya, M., Brendel, S., Wittig, B., and Rich, A. (1996). Z-DNA-forming sites within the human β-globin gene cluster. *Proc. Natl. Acad. Sci. USA* **93,** 780–784.
16. Felsenfeld, G., Davies, D. R., and Rich, A. (1957). Formation of a three-stranded polynucleotide molecule. *J. Am. Chem. Soc.* **79,** 2023–2024.
17. Frank-Kamenetskii, M. D., and Mirkin, S. M. (1995). Triplex DNA structures. *Annu. Rev. Biochem.* **64,** 65–95.
18. Radhakrishnan I., and Patel, D. J. (1994). DNA triplexes: solution structures, energetics, interactions, and function. *Biochemistry* **33,** 11405–11416.
19. Kohwi-Shigematsu, T., and Kohwi, Y. (1991). Detection of triple-helix related structures adopted by poly(dG)-poly(dC) sequences in supercoiling plasmid DNA. *Nucleic Acids Res.* **19,** 4267–4271.
20. Panyutin, I. G., and Wells, R. D. (1992). Nodule DNA in the $(GA)_{37} \cdot (CT)_{37}$ insert in superhelical plasmids. *J. Biol. Chem.* **267,** 5495–5501.
21. Rooney, S. M., and Moore, P. D. (1995). Antiparallel, intramolecular triplex DNA stimulates homologous recombination in human cells. *Proc. Natl. Acad. Sci. USA* **92,** 2141–2144.
22. Fitch, D. H. A., Mainone, C., Goodman, M., and Slightom, J. L. (1990). Molecular history of gene conversions in the primate fetal γ-globin genes. Nucleotide sequences from the common gibbon, *Hylobates lar. J. Biol. Chem.* **265,** 781–793.
23. Baran, N., Lapidot, A., and Manor, H. (1991). Formation of DNA triplexes accounts for arrests of DNA synthesis at $d(TC)_n$ and $d(GA)_n$ tracts. *Proc. Natl. Acad. Sci. USA* **88,** 507–511.
24. Grabczyk, E., and Fishman, M. C. (1994). A long purine-pyrimidine homopolymer acts as a transcriptional diode. *J. Biol. Chem.* **270,** 1791–1797.
25. Behe, M. J. (1995). An overabundance of long oligopurine tracts occurs in the genome of simple and complex eukaryotes. *Nucleic Acids Res.* **23,** 689–695.
26. Margot, J. B., Demers, G. W., and Hardison, R. C. (1989). Complete nucleotide sequence of the rabbit β-like globin gene cluster. *J. Mol. Biol.* **205,** 15–40.

27. Shehee, W. R., Loeb, D. D., Adey, N. B., Burton, F. H., Casavant, N. C., Cole, P., Davies, C. J., McGraw, R. A., Schichman, S. A., Severynse, D. M., Voliva, C. F., Weyter, F. W., Wisely, G. B., Edgell, M. H., and Hutchison III, C. A. (1989). Nucleotide sequence of the BALB/c mouse β-globin complex. *J. Mol. Biol.* **205,** 41–62.
28. Weinreb, A., Collier, D. A., Birshtein, B., and Wells, R. D. (1990). Left-handed Z-DNA and intramolecular triplex formation at the site of an unequal sister chromatid exchange. *J. Biol. Chem.* **265,** 1352–1359.
29. Bianchi, A., Wells, R. D., Heintz, N. H., and Caddle, M. S. (1990). Sequences near the origin of replication of the DHFR locus of the chinese hamster ovary cells adopt left-handed Z-DNA and triplex structures. *J. Biol. Chem.* **265,** 21789–21796.
30. Zakian, V. A. (1989). Structure and function of telomeres. *Annu. Rev. Genet.* **23,** 579–604.
31. Kang, C., Zhang, X., Ratliff, R., Moyzis, R., and Rich, A. (1992). Crystal structure of four-stranded *Oxytricha* telomeric DNA. *Nature* **356,** 126–131.
32. Smith, F. W., and Feigon, J. (1992). Quadruplex structure of *Oxytricha* DNA oligonucleotides. *Nature* **356,** 164–168.
33. Liu, Z., and Gilbert, W. (1994). The yeast *KEM1* gene encodes a nuclease specific for G4 tetraplex DNA: implication of *in vivo* functions for this novel DNA structure. *Cell* **77,** 1083–1092.
34. Liu, Z., Lee, A., and Gilbert, W. (1995). Gene disruption of a G4-DNA-dependent nuclease in yeast leads to cellular senescence and telomere shortening. *Proc. Natl. Acad. Sci. USA* **92,** 6002–6006.
35. Pearson, C. E., Zorbas, H., Price, G. B., and Zannis-Hadjopoulos, M. (1996). Inverted repeats, stem-loops, and cruciforms: significance for initiation of DNA replication. *J. Cell. Biochem.* **63,** 1–22.
36. Sinden, R. R., and Pearson, C. E. (1997). Slipped strand DNA, dynamic mutations, and human disease. *In* "Genetic Instabilities and Hereditary Neurological Disorders" (R. D. Wells and S. T. Warren, Eds.) Academic Press, San Diego. [Chapter 39, this volume]
37. Hagermann, P. J. (1990). Sequence-directed curvature of DNA. *Annu. Rev. Biochem.* **59,** 755–781.
38. Olson, W. K., and Zhurkin, V. B. (1996). Twenty years of DNA bending. *In* "Biological Structure and Dynamics" (R. H. Sarma and M. H. Sarma, Eds.), pp. 341–370. Adenine Press, New York.
39. Griffith, J., Bleyman, M., Rauch, C. A., Kitchin, P. A., and Englund, P. T. (1986). Visualization of the bent helix in kinetoplast DNA by electron microscopy. *Cell* **46,** 717–724.
40. Koo, H.-S., Wu, H.-M., and Crothers, D. M. (1986). DNA bending at adenine · thymine tracts. *Nature* **320,** 501–506.
41. Brukner, I., Susic, S., Dlakic, M., Savic, A., and Pongor, S. (1994). Physiological concentration of magnesium ions induces a strong macroscopic curvature in GGGCCC-containing DNA. *J. Mol. Biol.* **236,** 26–32.
42. Ohyama, T. (1996). Bent DNA in the human adenovirus type 2 E1A enhancer is an architectural element for transcription stimulation. *J. Biol. Chem.* **271,** 27823–27828.
43. Wada-Kiyama, Y., and Kiyama, R. (1995). Conservation and periodicity of DNA bend sites in the human β-globin gene locus. *J. Biol. Chem.* **270,** 12439–12445.
44. Wohlrab, F., and Wells, R. D. (1989). Slight changes in conditions influence the family of non-B-DNA conformations of the herpes simplex virus type 1 DR2 repeats. *J. Biol. Chem.* **264,** 8207–8213.
45. Bacolla, A., Gellibolian, R., Shimizu, M., Amirhaeri, S., Kang, S., Ohshima, K., Larson, J. E., Harvey, S. C., Stollar, B. D., and Wells, R. D. (1997). Flexible DNA: genetically unstable CTG · CAG and CGG · CCG from human hereditary neuromuscular disease genes. *J. Biol. Chem.* **272,** 16783–16792.
46. Gellibolian, R., Bacolla, A., and Wells, R. D. (1997). Triplet repeat instability and DNA topology: an expansion model based on statistical mechanics. *J. Biol. Chem.* **272,** 16793–16797.
47. Ohshima, K., Kang, S., Larson, J. E., and Wells, R. D. (1996). TTA · TAA triplet repeats in plasmids form a non-H bonded structure. *J. Biol. Chem.* **271,** 16784–16791.
48. Jovin, T. M., Rippe, K., Ramsing, N. B., Klement, R., Elhorst, W., and Vojtisková, M. (1990). Parallel stranded DNA. *In* "Structure and Methods" (R. H. Sarma and M. H. Sarma, Eds.), Vol. 3, pp. 155–174. Adenine Press, New York.
49. van de Sande, J. H., Ramsing, N. B., Germann, M. W., Elhorst, W., Kalisch, B. W., Kitzing, E. V., Pon, R. T., Clegg, R. C., and Jovin, T. M. (1988). Parallel stranded DNA. *Science* **241,** 551–557.
50. Germann, M. W., Kalisch, B. W., Pon, R. T., and van de Sande, J. H. (1990). Length-dependent formation of parallel-stranded DNA in alternating AT segments. *Biochemistry* **29,** 9426–9432.
51. Chen, J. H., and Seeman, N. C. (1991). Synthesis from DNA of a molecule with the connectivity of a cube. *Nature* **350,** 631–633.
52. Kovacic, R. T., and van Holde, K. E. (1977). Sedimentation of homogeneous double-stranded DNA molecules. *Biochemistry* **16,** 1490–1498.
53. Thomas, J. C., and Schurr, J. M. (1983). Fluorescence depolarization and temperature dependence of the torsion elastic constant of linear f29 deoxyribonucleic acid. *Biochemistry* **22,** 6194–6198.
54. Millar, D. P., Robbins, R. J., and Zewail, A. H. (1982). Torsion and bending of nucleic acids studied by subnanosecond time-resolved fluorescence depolarization of intercalated dyes. *J. Chem. Phys.* **76,** 2080–2094.
55. Hogan, M., LeGrange, J., and Austin, B. (1983). Dependence of DNA helix flexibility on base composition. *Nature* **304,** 752–754.
56. Fujimoto, B. S., Shibata, J. H., Schurr, R. L., and Schurr, J. M. (1985) Torsional dynamics and rigidity of fractionated poly(dGdC). *Biopolymers* **24,** 1009–1022.
57. Rizzo, V., and Schellman, J. (1981). Flow dichroism of T7 DNA as a function of salt concentration. *Biopolymers* **20,** 2143–2163.
58. Diekmann, S., Hillen, W., Morgeneyer, B., Wells, R. D., and Pörschke, D. (1982). Orientation relaxation of DNA restriction fragments and the internal mobility of the double helix. *Biophys. Chem.* **15,** 263–270.
59. Hagerman, P. J. (1981). Investigation of the flexibility of DNA using transient electric birefringence. *Biopolymers* **20,** 1503–1535.
60. Hurley, I., Osei-Gyimah, P., Archer, S., Scholes, C. P., and Lerman, L. S. (1982). Torsional motion and elasticity of the deoxyribonucleic acid double helix and its nucleosomal complexes. *Biochemistry* **21,** 4999–5009.
61. Thomas, T. J., and Bloomfield, V. A. (1983). Chain flexibility and hydrodynamics of the B and Z forms of poly(dG-dC)·poly(dG-dC). *Nucleic Acids Res.* **11,** 1919–1930.
62. Shore, D., and Baldwin, R. L. (1983). Energetics of DNA twisting. I. Relation between twist and cyclization probability. *J. Mol. Biol.* **170,** 957–981.
63. Shore, D., and Baldwin, R. L. (1983). Energetics of DNA twisting. II. Topoisomer analysis. *J. Mol. Biol.* **170,** 983–1007.
64. Horowitz, D. S., and Wang, J. C. (1984). Torsional rigidity and length dependence of the free energy of DNA supercoiling. *J. Mol. Biol.* **173,** 75–91.
65. Shimada, J., and Yamakawa, H. (1984). Ring-closure probabilities for twisted wormlike chains. Application to DNA. *Macromolecules* **17,** 689–698.

66. Yamakawa, H., and Fujii, M. (1984). Dynamics of helical worm-like chains. V. Nuclear magnetic relaxation. *J. Chem. Phys.* **81,** 997–1014.
67. Yoshizaki, T., Fujii, M., and Yamakawa, H. (1985). Dynamics of helical worm-like chains. VI. Fluorescence depolarization. *J. Chem. Phys.* **82,** 1003–1013.
68. Shimada, J., and Yamakawa, H. (1985). Statistical mechanics of DNA topoisomers. The helical worm-like chain. *J. Mol. Biol.* **184,** 319–329.
69. Taylor, W. H., and Hagerman, P. J. (1990). Application of the method of phage T4 DNA ligase-catalyzed ring-closure to the study of DNA structure. II. NaCl-dependence of DNA flexibility and helical repeat. *J. Mol. Biol.* **212,** 363–376.
70. Kahn, J. D., Yun, E., and Crothers, D. M. (1994). Detection of localized DNA flexibility. *Nature* **368,** 163–166.
71. Hodges-Garcia, Y., and Hagerman, P. J. (1995). Investigation of the influence of cytosine methylation on DNA flexibility. *J. Biol. Chem.* **270,** 197–201.
72. Wu, P., Song, L., Clendenning, J. B., Fujimoto, B. S., Benight, A. S., and Schurr, J. M. (1988). Interaction of chloroquine with linear and supercoiled DNAs. Effect on the torsional dynamics, rigidity, and twist energy parameter. *Biochemistry* **27,** 8128–8144.
73. Selvin, P. R., Cook, D. N., Pon, N. G., Bauer, W. R., Klein, M. P., and Hearst, J. E. (1992). Torsional rigidity of positively and negatively supercoiled DNA. *Science* **255,** 82–85.
74. Chen, H. H., Rau, D. C., and Charney, E. (1985). The flexibility of alternating dA-dT sequences. *J. Biomol. Struct. Dyn.* **2,** 709–719.
75. Bednar, J., Furrer, P., Katritch, V., Stasiak, A. Z., Dubochet, J., and Stasiak, A. (1995) Determination of DNA persistence length by cryo-electron microscopy. Separation of the static and dynamic contributions to the apparent persistence length of DNA. *J. Mol. Biol.* **254,** 579–594.
76. Hagerman, P. J. (1985). Analysis of the ring-closure probabilities of isotropic wormlike chains: application to duplex DNA. *Biopolymers* **24,** 1881–1897.
77. Levene, S. D., and Crothers, D. M. (1986). Ring closure probabilities for DNA fragments by Monte Carlo simulation. *J. Mol. Biol.* **189,** 61–72.
78. Levene, S. D., and Crothers, D. M. (1986). Topological distributions and the torsional rigidity of DNA. A Monte Carlo study of DNA circles. *J. Mol. Biol.* **189,** 73–83.
79. Hagerman, P. J., and Ramadevi, V. A. (1990). Application of the method of phage T4 DNA ligase-catalyzed ring-closure to the study of DNA structure. I. Computational analysis. *J. Mol. Biol.* **212,** 351–362.
80. Dlakic, M., and Harrington, R. E. (1995). Bending and torsional flexibility of G/C-rich sequences as determined by cyclization assays. *J. Biol. Chem.* **270,** 29945–29952.
81. Ulanovsky, L., Bodner, M., Trifonov, E. N., and Choder, M. (1986) Curved DNA: design, synthesis, and circularization. *Proc. Natl. Acad. Sci. USA* **83,** 862–866.
82. Fujimoto, B. S., and Schurr, J. M. (1990). Dependence of the torsional rigidity of DNA on base composition. *Nature* **344,** 175–178.
83. Gorin, A. A., Zhurkin, V. B., and Olson, W. K. (1995). B-DNA twisting correlates with base-pair morphology. *J. Mol. Biol.* **247,** 34–48.
84. El Hassan, M. A., and Calladine, C. R. (1995). The assessment of the geometry of dinucleotide steps in double-helical DNA; a new local calculation scheme. *J. Mol. Biol.* **251,** 648–664.
85. Young, M. A., Ravishanker, G., Beveridge, D. L., and Berman, H. M. (1995). Analysis of local helix bending in crystal structures of DNA oligonucleotides and DNA protein complexes. *Biophys. J.* **68,** 2454–2468.
86. El Hassan, M. A., and Calladine, C. R. (1996). Propeller-twisting of base-pairs and the conformational mobility of dinucleotide steps in DNA. *J. Mol. Biol.* **259,** 95–103.
87. Goryshin, I. Y., Kil, Y. V., and Reznikoff, W. S. (1994). DNA length, bending, and twisting constraints on IS*50* transposition. *Proc. Natl. Acad. Sci. USA* **91,** 10834–10838.
88. Schleif, R. (1992). DNA looping. *Annu. Rev. Biochem.* **61,** 199–223.
89. Wang, Y.-H., Amirhaeri, S., Kang, S., Wells, R. D., and Griffith, J. D. (1994). Preferential nucleosome assembly at DNA triplet repeats from the myotonic dystrophy gene. *Science* **265,** 669–671.
90. Wang, Y.-H., and Griffith, J. (1995). Expanded CTG triplet blocks from the myotonic dystrophy gene create the strongest known natural nucleosome positioning elements. *Genomics* **25,** 570–573.
91. Otten, A. D., and Tapscott, S. J. (1995). Triplet repeat expansion in myotonic dystrophy alters the adjacent chromatin structure. *Proc. Natl. Acad. Sci. USA* **92,** 5465–5469.
92. Godde, J. S., and Wolffe, A. P. (1996). Nucleosome assembly on CTG triplet repeats. *J. Biol. Chem.* **271,** 15222–15229.
93. Wang, Y.-H., Gellibolian, R., Shimizu, M., Wells, R. D., and Griffith, J. (1996). Long CCG triplet repeat blocks exclude nucleosomes: a possible mechanism for the nature of fragile sites in chromosomes. *J. Mol. Biol.* **263,** 511–516.
94. Wang, Y.-H., and Griffith, J. (1996). Methylation of expanded CCG triplet repeat DNA from fragile X syndrome patients enhances nucleosome exclusion. *J. Biol. Chem.* **271,** 22937–22940.
95. Godde, J. S., Kass, S. U., Hirst, M. C., and Wolffe, A. P. (1996). Nucleosome assembly on methylated CGG triplet repeats in the fragile X mental retardation gene 1 promoter. *J. Biol. Chem.* **271,** 24325–24328.
96. Klug, A., Rhodes, D., Smith, J., Finch, J. T., and Thomas, J. O. (1980). A low resolution structure for the histone core of the nucleosome. *Nature* **287,** 509–516.
97. Earnshaw, W. C., and Harrison, S. C. (1977). DNA arrangement in isometric phage heads. *Nature* **268,** 598–602.
98. Depew, R. E., and Wang, J. C. (1975). Conformational fluctuations of DNA helix. *Proc. Natl. Acad. Sci. USA* **72,** 4275–4279.
99. Pulleyblank, D. E., Shure, M., Tang, D., Vinograd, J., and Vosberg, H. P. (1975). Action of nicking-closing enzyme on supercoiled and nonsupercoiled closed circular DNA: formation of a Boltzmann distribution of topological isomers. *Proc. Natl. Acad. Sci. USA* **72,** 4280–4284.
100. Shimizu, M., Gellibolian, R., Oostra, B. A., and Wells, R. D. (1996). Cloning, characterization and properties of plasmids containing CGG triplet repeats from the FMR-1 gene. *J. Mol. Biol.* **258,** 614–626.
101. Cozzarelli, N. R., and Wang, J. C. (1990). *In* "DNA Topology and Its Biological Effects." Cold Spring Harbor Laboratory, Cold Spring Harbor, NY.
102. Liu, L. F., and Wang, J. C. (1987). Supercoiling of the DNA template during transcription. *Proc. Natl. Acad. Sci. USA* **84,** 7024–7027.
103. Rahmouni, A. R., and Wells, R. D. (1992). Direct evidence for the effect of transcription on local DNA supercoiling *in vivo*. *J. Mol. Biol.* **223,** 131–144.
104. Kang, S., Jaworski, A., Ohshima, K., and Wells, R. D. (1995). Expansion and deletion of CTG repeats from human disease genes are determined by the direction of replication in *E. coli*. *Nature Genet.* **10,** 213–218.
105. Freudenreich, C. H., Stavenhagen, J. B., and Zakian, V. A. (1997). Stability of a CTG/CAG trinucleotide repeat in yeast is dependent on its orientation in the genome. *Mol. Cell. Biol.* **17,** 2090–2098.

106. Paulson, H. L., and Fischbeck, K. H. (1996). Trinucleotide repeats in neurogenetic disorders. *Annu. Rev. Neurosci.* **19,** 79–107.
107. Warren, S. T., and Ashley, C. T., Jr. (1995). Triplet repeat expansion mutations: the example of fragile X syndrome. *Annu. Rev. Neurosci.* **18,** 77–99.
108. Ashley, C. T., Jr., and Warren, S. T. (1995). Trinucleotide repeat expansion and human disease. *Annu. Rev. Genet.* **29,** 703–728.
109. Kang, S., Ohshima, K., Shimizu, M., Amirhaeri, S., and Wells, R. D. (1995). Pausing of DNA synthesis *in vitro* at specific loci in CTG and CGG triplet repeats from human hereditary disease genes. *J. Biol. Chem.* **270,** 27014–27021.
110. Ohshima, K., and Wells, R. D. (1997). Hairpin formation during DNA synthesis primer realignment *in vitro* in triplet repeat sequences from human hereditary disease genes. *J. Biol. Chem.* **272,** 16798–16806.
111. Burd, J. F., and Wells, R. D. (1970). Effect of incubation conditions on the nucleotide sequence of DNA products of unprimed DNA polymerase reactions. *J. Mol. Biol.* **53,** 435–459.
112. Harding, E. (1993). Clinical features and classification of inherited ataxias. *In* "Advances in Neurology" (A. E. Harding and T. Deufel, Eds.), Vol. 61, pp. 1–14. Raven Press, New York.
113. Greenfield, J. G. (1992). Disease of the basal ganglia, cerebellum and motor neurons, Friedreich's ataxia. *In* "Greenfield's Neuropathology" (J. H. Adams, J. A. N. Corseillis, and L. W. Duchen, Eds.), pp. 1015–1018, 4th edition. Wiley, New York.
114. Zoghbi, H. Y. (1996). Heritable Ataxias. *In* "Principles of Child Neurology," Chap. 69, pp. 1543–1561. McGraw Hill, New York.
115. Campuzano, V., Montermini, L., Molto, M. D., Pianese, L., Cossee, M., Cavalcanti, F., Monros, E., Rodius, F., Duclos, F., Monticelli, A., Zara, F., Canizares, J., Koutnikova, H., Bidichandani, S. I., Gellera, C., Brice, A., Trouillas, P., de Michele, G., Filla, A., De Frutos, R., Palau, F., Patel, P. I., Di Donato, S., Mandel, J.-L., Cocozza, S., Koenig, M., and Pandolfo, M. (1996). Friedreich's ataxia: autosomal recessive disease caused by an intronic GAA triplet repeat expansion. *Science* **271,** 1423–1427.
116. Babcock, M., De Silva, D., Oaks, R., Davis-Kaplan, S., Jiralerspong, S., Montermini, L., Pandolfo, M., and Kaplan, J. (1997). Regulation of mitochondrial iron accumulation by Yfh1p, a putative homolog of frataxin. *Science* **276,** 1709–1712.
117. Bidichandani, S. I., Montermini, L., Molto, M. D., Ohshima, K., Wells, R. D., Patel, P. I., and Pandolfo, M. (1998). Transcriptional suppression in Friedreich ataxia. [Manuscript in preparation]
118. Montermini, L., Richter, A., Morgan, K., Macies, C., Julien, D., Castellotti, B., Mercier, J., Poirier, J., Capozzoli, F., Bouchard, J.-P., Lemieux, B., Mathieu, J., Vanasse, M., Seni, M.-H., Graham, G., Andermann, F., Andermann, E., Melancon, S. B., Keats, B. J. B., Di Donato, S., and Pandolfo, M. (1997). Phenotypic variability in Friedreich ataxia: role of the associated GAA triplet repeat expansion. *Annals of Neur.* **41,** 675–682.
119. Filla, A., De Michele, G., Cavalcanti, F., Pianese, L., Monticelli, A., Campanella, G., and Cocozza, S. (1996). The relationship between trinucleotide (GAA) repeat length and clinical features in Friedreich ataxia. *Am. J. Hum. Genet.* **59,** 554–560.
120. Carvajal, J. J., Pook, M. A., Dos Santos, M., Doudney, K., Hillermann, R., Minogue, S., Williamson, R., Hsuan, J. J., and Chamberlain, S. (1996). The Friedreich's ataxia gene encodes a novel phosphatidylinositol-4-phosphate 5-kinase. *Nature Genet.* **14,** 157–162.
121. Cossee, M., Campuzano, V., Koutnikova, H., Fischbeck, K., Mandel, J.-L., Koenig, M., Bidichandani, S. I., Patel, R. I., Molte, M. D., Canizares, J., De Frutos, R., Pianese, L., Cavalcanti, F., Monticelli, A., Cocozza, S., Montermini, L., and Pandolfo, M. (1997). Frataxin fracas. *Nature Genet.* **15,** 337–338.
122. Wells, R. D., Collier, D. A., Hanvey, J. C., Shimizu, M., and Wohlrab, F. (1988). The chemistry and biology of unusual DNA structures adopted by oligopurine · oligopyrimidine sequences. *FASEB J.* **2,** 2939–2949.
123. Guieysse, A.-L., Praseuth, D., Grigoriev, M., Harel-Bellan, A., and Helene, C. (1996). Detection of covalent triplex with human cells. *Nucleic Acids Res.* **24,** 4210–4216.
124. Bacolla, A., Ulrich, M. J., Larson, J. E., Ley, T. J., and Wells, R. D. (1995). An intramolecular triplex in the human γ-globin 5′-flanking region is altered by point mutations associated with hereditary persistence of fetal hemoglobin. *J. Biol. Chem.* **270,** 24556–24563.
125. Xu, G., and Goodridge, A. G. (1996). Characterization of a polypyrimidine/polypurine tract in the promoter of the gene for chicken malic enzyme. *J. Biol. Chem.* **271,** 16008–16019.
126. Hanvey, J. C. , Klysik, J., and Wells, R. D. (1988). Influence of DNA sequence on the formation of non-B right-handed helices in oligopurine · oligopyrimidine inserts in plasmids. *J. Biol. Chem.* **263,** 7386–7396.
127. Hanvey, J. C., Shimizu, M., and Wells, R. D. (1988). Intramolecular DNA triplexes in supercoiled plasmids. *Proc. Natl. Acad. Sci. USA* **85,** 6292–6296.
128. Shimizu, M., Hanvey, J. C., and Wells, R. D. (1989). Intramolecular DNA triplexes in supercoiled plasmids: I. Effect of loop size on formation and stability. *J. Biol. Chem.* **264,** 5944–5949.
129. Hanvey, J. C., Shimizu, M., and Wells, R. D. (1989). Intramolecular DNA triplexes in supercoiled plasmids: II. Effect of base composition and non-central interruptions on formation and stability. *J. Biol. Chem.* **264,** 5950–5956.
130. Hanvey, J. C., Shimizu, M., and Wells, R. D. (1989). Site-specific inhibition of *Eco*RI restriction/modification enzymes via DNA triple helix. *Nucleic Acids Res.* **18,** 157–161.
131. Shimizu, M., Hanvey, J. C., and Wells, R. D. (1990). Multiple non-B-DNA conformations of polypurine · polypyrimidine sequences in plasmids. *Biochemistry* **29,** 4704–4713.
132. Kang, S., Wohlrab, F., and Wells, R. D. (1992). Metal ions cause the isomerization of certain intramolecular triplexes. *J. Biol. Chem.* **267,** 1259–1264.
133. Kang, S., Wohlrab, F., and Wells, R. D. (1992). GC rich flanking tracts decrease the kinetics of intramolecular DNA triplex formation. *J. Biol. Chem.* **267,** 19435–19442.
134. Ohshima, K., Kang, S., Larson, J. E., and Wells, R. D. (1996). Cloning, characterization, and properties of seven triplet repeat DNA sequences. *J. Biol. Chem.* **271,** 16773–16783.
135. Postel, E. H., Flint, S. J., Kessler, D. J., and Hogan, M. E. (1991). Evidence that a triplex-forming oligodeoxyribonucleotide binds to the c-myc promoter in HeLa cells, thereby reducing c-myc mRNA levels. *Proc. Natl. Acad. Sci. USA* **88,** 8227–8231.
136. Morgan, A. R., and Wells, R. D. (1968). Specificity of the three-stranded complex formation between double-stranded DNA and single-stranded RNA containing repeating nucleotide sequences. *J. Mol. Biol.* **37,** 63–80.
137. Ohshima, K., Kang, S., and Wells, R. D. (1996). CTG triplet repeats from human hereditary disease are dominant genetic expansion products in *E. coli*. *J. Biol. Chem.* **271,** 1853–1856.
138. Kang, S., Ohshima, K., Jaworski, A., and Wells, R. D. (1996). CTG triplet repeats from the myotonic dystrophy gene are expanded in *E. coli* distal to the replication origin as a single large event. *J. Mol. Biol.* **258,** 543–547.
139. Bowater, R. P., Rosche, W. A., Jaworski, A., Sinden, R. R., and Wells, R. D. (1996). Relationship between *Escherichia coli* growth and deletions of CTG · CAG triplet repeats in plasmids. *J. Mol. Biol.* **264,** 82–96.

140. Lapidot, A., Baran, N., and Manor, H. (1989). $(dT\text{-}dC)_n$ and $(dG\text{-}dA)_n$ tracts arrest single stranded DNA replication *in vitro*. *Nucleic Acids Res.* **17,** 883–900.
141. Krasilnikov, A. S., Panyutin, I. G., Samadashwily, G. M., Cox, R., Lazurkin, Y. S., and Mirkin, S. M. (1997). Mechanisms of triplex-caused polymerization arrest. *Nucleic Acids Res.* **25,** 1339–1346.
142. Dayn, A., Samadashwily, M., and Mirkin, S. M. (1992). Intramolecular DNA triplexes: unusual sequence requirements and influence on DNA polymerization. *Proc. Natl. Acad. Sci. USA* **89,** 11406–11410.
143. Weisman-Shomer, P., Dube, D. K., Perrino, F. W., Stokes, K., Loeb, L. A., and Fry, M. (1989). Sequence specificity of pausing by DNA polymerases. *Biochem. Biophys. Res. Commun.* **164,** 1149–1156.
144. Woodford, K. J., Howel, R. M., and Usdin, K. (1994). A novel K(+)-dependent DNA synthesis arrest site in a commonly occurring sequence motifs in eukaryotes. *J. Biol. Chem.* **269,** 27029–27035.
145. Usdin, K., and Woodford, K. J. (1995). CGG repeats associated with DNA instability and chromosome fragility form structures that block DNA synthesis *in vitro*. *Nucleic Acids Res.* **23,** 4202–4209.
146. Ashizawa, T., Monckton, D. G., Vaishnav, S., Patel, B. J., Voskova, A., and Caskey, C. T. (1996). Instability of the expanded (CTG)n repeats in the myotonin protein kinase gene in cultured lymphoblastoid cell lines from patients with myotonic dystrophy. *Genomics* **36,** 47–53.
147. Mitas, M. (1997). Trinucleotide repeats associated with human disease. *Nucleic Acids Res.* **25,** 2245–2253.
148. Mirkin, S. M. (1997). Personal communication.
149. Parsons, A., Isban, M., and Sinden, R. R. (1997). In vitro transcriptional pausing in triplet repeat-containing DNA from myotonic dystrophy and fragile X loci. [Submitted for publication]

CHAPTER 39

Slipped Strand DNA, Dynamic Mutations, and Human Disease

CHRISTOPHER E. PEARSON[1] and Richard R. Sinden[2]

Center for Genome Research, Institute of Biosciences and Technology, Department of Biochemistry and Biophysics, Texas A&M University, Texas Medical Center, Houston, Texas, 77030

[1]Present address: Department of Genetics, Hospital for Sick Children and Department of Molecular and Medical Genetics, University of Toronto, 555 University Ave., Toronto, Ontario, Canada M5G 1X8.

[2]To whom correspondence should be addressed. Fax: (713) 677-7689. E-mail: rsinden@ibt.tamu.edu.

I. INTRODUCTION

A. Triplet Repeats, Repeat Length, Repeat Purity, and Human Disease

Recently, the etiology of at least 12 human genetic diseases, including myotonic dystrophy (DM), fragile X syndrome (FRAXA, and FRAXE), Jacobsen syndrome (FRA11B), spinocerebellar ataxia types 1, 2, 3, 6, and 7 (SCA1, SCA2, SCA3, SCA6, and SCA7), spinal and bulbar muscular atrophy (SBMA), Huntington's disease (HD), dentatorubral-pallidoluysian atrophy (DRPLA, also known as Haw River syndrome), and Friedreich's ataxia (FRDA) have been traced to genetic variation in the lengths of $(CTG)_n{\cdot}(CAG)_n$, $(CGG)_n{\cdot}(CCG)_n$, or $(GAA)_n{\cdot}(TTC)_n$ triplet repeats in DNA (Table 39-1). (For review of the genetic and clinical aspects of these diseases see [1–5] and other reviews in this book). In addition to the disease-linked fragile sites FRAXA, FRAXE, and FRA11B (Jacobsen syndrome), expansions of CGG trinucleotide repeats have also been associated with the chromosomal fragile site FRAXF [6] and the autosomal fragile site FRA16A [7, 8]. The importance of the length of the repeat tract is evident for many of the diseases/instabilities: In normal individuals these loci contain a short length of triplet repeats (usually 5–37), which is polymorphic within the population. Increases in the lengths of the triplet repeats to 50–120 are associated with disease symptoms. Some of the unstable loci (FRAXA, FRAXE, DM, HD, FRDA) are associated with genetically unstable premutation/protomutation/intermediate alleles. In DM, FRAXA, and FRDA, expanded lengths of the triplet repeats to 50–120 are associated with genetically unstable protomutation/premutation alleles. Expansion to thousands of repeats can occur at these loci. There is a strong dependence of the probability of mutation upon the number of repeats which has lead to the appropriate term of "dynamic mutation" to describe this process [9].

The mechanism by which repeat expansion occurs is completely unknown. Many proposed models for instability involve alternative DNA structures within the triplet repeats resulting in aberrant DNA replication, recombination, or repair. Although slipped strand DNA structures could theoretically form within long runs of a triplet repeat (or any tandem repeat for that matter), the first conclusive biophysical evidence for such structures has only recently been published [10, 11]. In this review we have discussed the current state of knowledge about slipped strand structures, most of which has only begun to be revealed. The importance of a thorough knowledge of slipped strand structures may be crucial to understanding the mechanism of their genetic insta-

TABLE 39-1 Unstable Trinucleotide Repeats in Humans

Disorder/site	Repeat	Normal length	Premutation protomutation intermediate allele	Disease/expansion	Interruption	Interruption loss, influence on stability
FRAXA	$(CGG)_n$	6–52	59–230	230–2000 (pure)	AGG	+, threshold ≥34 pure
FRAXE	$(CCG)_n$	4–39	? (31–61)	200–900	—	NA
FRAXF	$(CGG)_n$	7–40	?	306–1008	$(GCC\ GTC)_{3\text{-}4}$	—
FRA16A	$(CCG)_n$	16–49	?	1000–1900	Complex*	+
FRA11B	$(CGG)_n$	11	80	100–1000	—	NA
SMBA	$(CAG)_n$	14–32	?	40–55	—	NA
DM	$(CTG)_n$	5–37	50–80	80–1000; congenital, 2000–3000	*	NA*
HD	$(CAG)_n$	10–34	36–39	40–121	*	NA*
SCA1	$(CAG)_n$	6–39	—	40–81 (pure)	CAT	+, threshold ≥40 pure
SCA2	$(CAG)_n$	14–31	—	34–59 (pure)	CAA	+
SCA3/MJD	$(CAG)_n$	13–44	?	60–84	*	NA*
SCA6	$(CAG)_n$	4–18	?	21–28	—	NA
SCA7	$(CAG)_n$	7–17	?	38–130	—	NA
DRPLA/HRS	$(CAG)_n$	7–25	?	49–75	*	NA
FRDA	$(GAA)_n$	6–29	? (>34–40)	200–900	(GAG GAA)	+, threshold >34–40

*Indicates that the reader should refer to the text (or literature) for a discussion of the interruptions and flanking sequences in these repeats. (?) A possible mutagenic intermediate length. Not all diseases are associated with a premutation/protomutation clinical or repeat length status. (—) None. (+) Positive influence of loss of interruptions on greater instability. NA, not applicable.

bility. As these structures are proposed key intermediates in many processes of repeat polymorphism and mutation, it is integral to understand their formation and structural characteristics.

B. A Historical View of Slipped Strand DNA and Reiterative Synthesis

Soon after the publication of the canonical double helical structure of deoxyribonucleic acid [12] the minds of some scientists were performing mental gymnastics with the possible conformational permutations of this structure. There are many possible alternative DNA structures, and these are dependent upon the primary sequence of the DNA. These structures include cruciforms [13], triple-stranded DNAs [14], and four-stranded (quadruplex) DNAs [15], as well as DNA structure containing slippage between two strands [16]. Direct repeated sequences and tracts of tandemly repeated sequences have the potential to form slipped strand structures through interstrand base pairing in an out-of-register fashion. In 1966 Streisinger suggested that such structures may form under physiological conditions to serve as intermediates in the processes of mutagenesis [17]. Some of the earliest DNA sequences mistakenly thought to adopt slipped strand DNAs were very likely triplex-forming DNA sequences [18–23].

The reiterative synthesis of simple repeated nucleic acids was observed first with bacterial preparations of RNA polymerase [24], then with DNA polymerase [25]. This reiterative synthesis is thought to be accomplished by a DNA replication-associated process known as replication slippage. Several labs have contributed to the understanding of this process [26–31]. Below we present a review of the structural dynamics of tandemly repeated DNAs, and the possible means by which length alterations may occur. A special emphasis is placed upon trinucleotide repeats and eukaryotic systems with a focus upon the instabilities observed in the above-mentioned severe human genetic disorders. We do not cover other well characterized alternative DNA structures such as cruciforms, Z-DNA, triplex DNA, quadruplex DNA, and parallel DNA; for these the reader is referred to some excellent reviews [32–34].

C. Interstrand Motions in Tandemly Repeated Sequences

1. Slipping, Sliding, Creeping, Shifting, Bubble and Branch Migration

From the initial experiments demonstrating both the *de novo* and the reiterative synthesis of simple repeating nucleotides [25] it seemed evident that the two strands, a short primer and short template strand, must be slipping with respect to each other in an out-of-register fashion. Strands that slipped in the 5' direction would result in 3′ recessed ends which are substrates for extension by DNA polymerases. Since all polymerases extend only in the 3′ to 5′ direction, strands that slip in the 3′ direction resulting in a 5′ recessed ends would not be active as templates for extension. The precise details of such slipping are not known. It is possible, although unlikely, that each of the base pair hydrogen bonds between the two strands is completely broken, the strands are shifted, then all hydrogen bonds are reformed (Fig. 39-1B). This mechanism, which we term sliding, is unlikely as it would require a great deal of energy, requiring not just the breakdown and reformation of bonds but also the actual translational movement of the two strands. This is in contrast with Holliday (4-way DNA) junction branch migration [35], or soliton wave/bubble migration [36], where there is juxtaposed simultaneous breaking and forming of base pairs (Figs. 39-1A and 39-1E).

Lehman and colleagues [37, 38] described the unusual conformational mobility of poly (dA)·(dT). Upon the addition of short oligomers of dT (230 nucleotides) to long polymers of dA (3000 nucleotides) the dT oligos could be ligated together by DNA ligase. The efficiency of the reaction suggested that the oligo dT rather than being fixed in a randomly positioned location upon the poly dA was able to move along the poly dA to reach a location where its 5′-phosphoryl end would be juxtaposed to the 3′-hydroxyl end of another oligo dT where ligase could efficiently form a phosphodiester bond between the two dT oligomers. From a variety of experiments the authors suggested that the likely mechanism of oligo dT motion upon the poly dA was probably through a "creep" [28] (or partial dissociation) type of motion (Fig. 39-1C) as opposed to an "exchange" or sliding type mechanism (Fig. 39-1B). Whereas sliding would require the complete dissociation and possible exchange of the oligo dT to another poly dA, creeping would only involve the partial denaturation, out-of-register renaturation, and eventual positional shift of the oligo dT along the poly dA.

If one assumes that no intrastrand interactions are occurring in the slipped-out repeat strands (such as may occur in a hairpin) (Fig. 39-2A) and that the ends of the slip-out are available for reassociation with the complementary repeat strand, the slip-out can effectively branch migrate along the complementary strand (Fig. 39-1D). This is very much the same as D-loop migration, Holliday junction branch migration, or bubble migration (Fig. 39-1, compare A, D, and E). Such loop migration would entail only the energy necessary for the breaking and reforming of the interstrand hydrogen bonds ahead

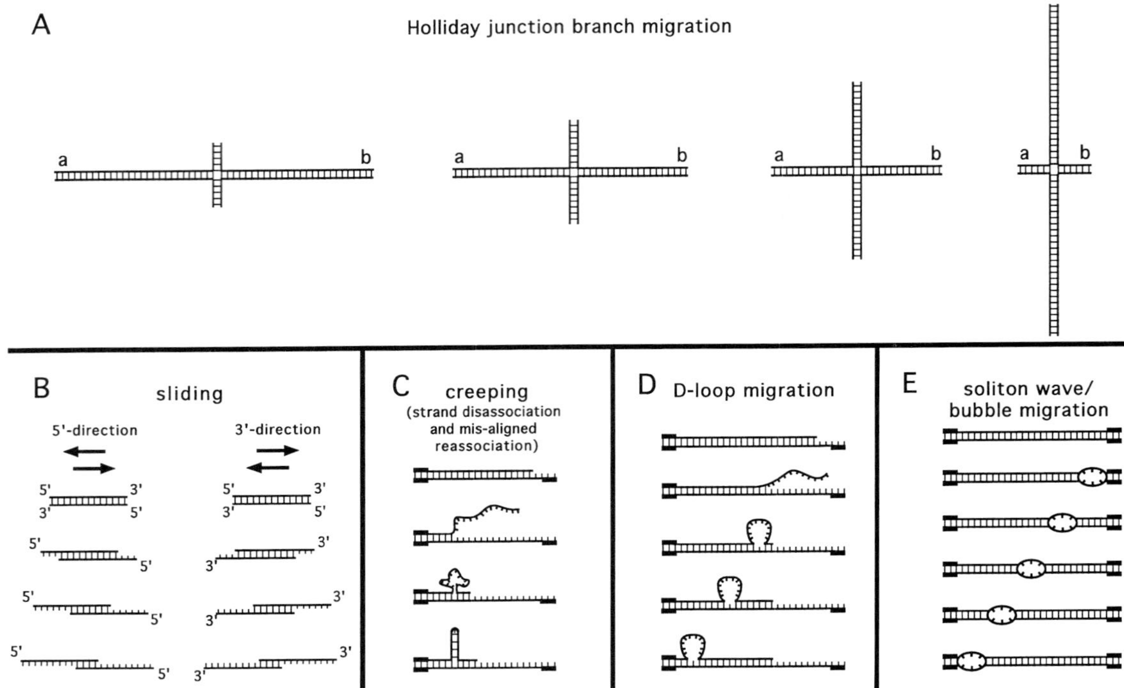

FIGURE 39-1 Models of interstrand motions and strand slippage. (A) Holliday junction branch migration. (B) Sliding. (C) Creeping (partial denaturation and out-of-register reassociation). (D) D-loop migration. (E) Soliton wave of a denaturation bubble. For discussion see text of Section I.C.

of and behind the loop-out, respectively. However, if there are intrastrand interactions occurring within the slipped-out repeats, then migration of the slip-out would be a high-energy-demanding process—requiring that the slip-out intrastrand base pairs as well as the base pairs at the slip-out junction be broken and reformed (Fig. 39-2A). Such interstrand sliding motions would move at single repeat unit increments. Sliding during DNA replication, especially in a nascent lagging strand, may result in some of the somatic and population polymorphism's of triplet (or other) repeat tracts which arise as ± 1–3 repeats. We do not believe that larger alterations (large deletions and expansions) arise from such a process, as it is too demanding both kinetically and biophysically. We presently favor the strand dissociation/reassociation (creep) model (Fig. 39-1C) for the formation of expansion and deletion intermediates. As discussed below, the creeping type mechanism of slippage (Figs. 39-1C, 39-2D) would allow for length alterations at both the 3′ and the 5′ ends of repeat tracts (Fig. 39-2E).

Although it is unlikely that sliding (Fig. 39-1B) occurs during *in vitro* reiterative synthesis, some mechanism must produce 3′ recessed ends that can be extended by DNA polymerase. The exact mechanism through which this occurs is not clear. Rokita and Fredes-Romero [39] observed an unusual susceptibility of $d(CG)_6$ to degradation. Product analysis indicated that $d(CG)_6$ was readily converted to $d(CG)_5$ and $d(CG)_4$. The degradation of $d(CG)_6$ proceeded at an intermediate and linear first-order rate, and could proceed to consume at least 75% of the starting material, thereby indicating that the whole population of DNA molecules were susceptible rather than a specific fraction. Simple end-fraying of these sequences was ruled out since all of the products of degradation were those that remained in-register (i.e., they were all shorter by increments of a CpG unit). Melting temperatures (T_m) are usually presented as a measure of the net stability of a particular duplex formation. The melting temperatures of each of the oligos $d(CG)_4$, $d(CG)_5$, and $d(CG)_6$ were 60, 71, and >80°C, respectively. However, their relative sensitivities to S1 digestion were apparently the same. In addition, a 14-base-pair oligo composed of nonrepetitive heterogeneous sequence with a T_m of 61°C, considerably less than that of the similar length $d(CG)_6$ oligo, was not sensitive to S1 digestion. Thus, the relative T_m values were poor predictors of susceptibility to structural alteration. It is interesting that each of the $n = 4, 5$, and 6 $d(CG)_n$ oligos displayed similar kinetics of digestion. This suggests that the mechanism of structural alteration is through intermediates that are similar for each and likely to be independent of length. One would expect if complete denaturation and renaturation were required for out-of register alignment to occur (such as in sliding; Fig. 39-1B) that the energy requirement and hence the kinetics would be different for the different length intermediates. But since the ki-

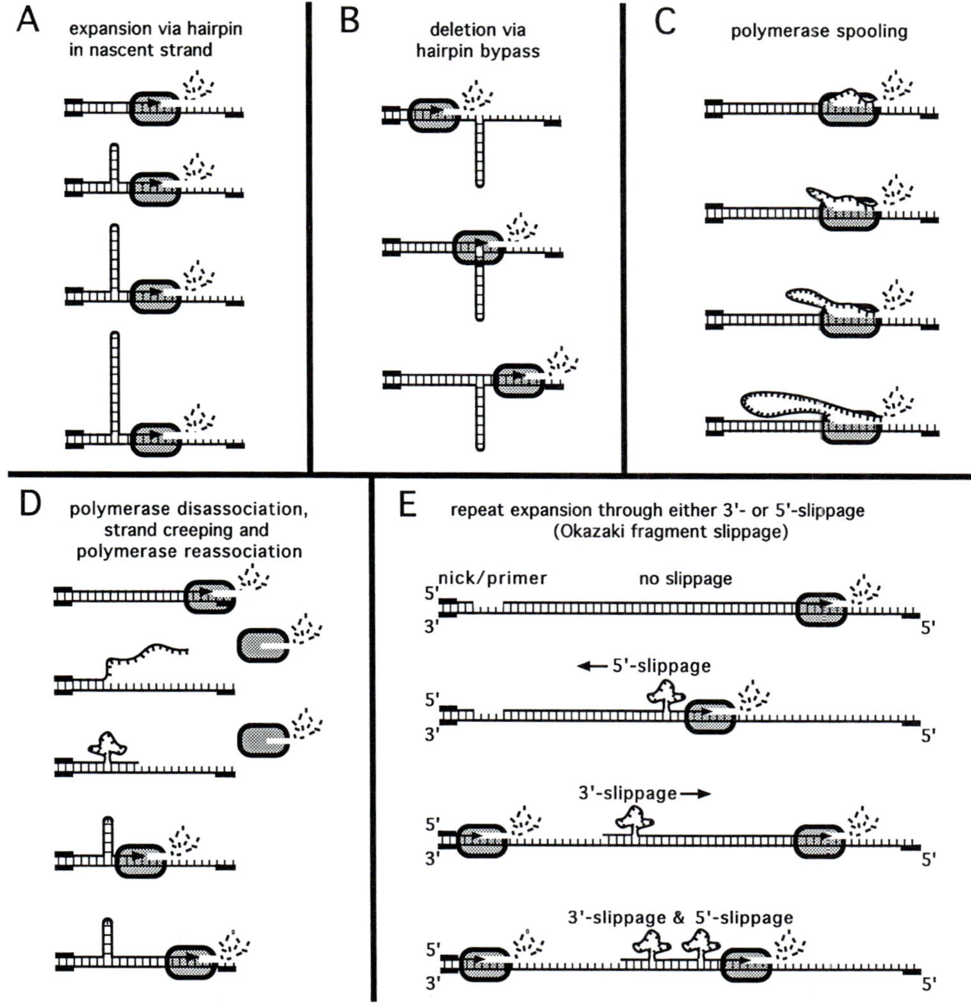

FIGURE 39-2 Models of strand slippage and expansion. (A) Expansion via hairpin formation in nascent strand. (B) Deletion via hairpin formation in template strand. (C) Polymerase spooling (polymerase-mediated strand slippage). (D) Polymerase dissociation followed by creeping type strand slippage and polymerase reassociation. (E) Expansion through slippage of the 3′ or 5′ ends in the 5′ and 3′ directions, respectively. For discussion see text of Sections I.C and I.E.

netics were similar this suggests that a process similar to creeping slippage (Fig. 39-1C) is responsible for the enhanced susceptibility of the $(CG)_n$ sequences. The formation and migration of a single unpaired CG unit could result in slipped oligos that would upon S1 digestion result in products having lengths of $n - 1$, $n - 2$, etc. It seems unlikely that intrastrand G–C interactions within a single slipped-out CG repeat would hinder its migration. Such migration of a denatured bubble has been observed in tracts [40, 41]. We note that S1 nuclease is unable to recognize single base bulges and the recognition of two or more extra bases can be affected by sequence and/or structural context.

2. Free-Ends of DNA and Slippage

The process of alternative triplet (or other repeat tract) structure formation may be greatly facilitated by the presence of a free end *within* the repeat tract. Free ends are known to enhance the transition of duplex-to-hairpin ([42–44] and references therein) and duplex-to-random coil [42, 43, 45, 46] as well as the reverse reactions. The theory of the excluded volume effect during DNA strand renaturation states that, "For a random coil polymer, the ends of the molecule are further from the center of mass than is the center of the molecule. If there is an excluded volume effect, the rate of renaturation will be faster at the end than at the center of a single-stranded DNA molecule" [47]. Thus, the excluded volume effect takes into account both the length of the molecule (or the denatured region) and DNA topology (the volume). Due to the excluded volume effect during DNA renaturation, free ends are known to reanneal more rapidly than sequences not proximal to an end [48]. In the absence of replication, free ends occur at both nicks and gaps in the DNA. In the special case of trinucleotide repeats where each of the individual strands (CAG, CTG, CGG, and CCG) can form stable intrastrand structures (hairpins), a free end within the repeat tract may permit considerable slippage. Free ends within the repeat tract would occur during the process of replication. Creeping type slippage (Figs. 39-1C, 39-2D) can occur either at the 3′ primer end or at the 5′ tail end (Fig. 39-2E). Polymerase extension of the slipped 3′ primer end or gap-filling of the slipped 5′ end would result in expansion products (see below discussion of Okazaki fragment slippage; Section I.E).

3. Repeat Expansion—Slippage Synthesis

Schlotterer and Tautz [30] reported that the rate of *in vitro* reiterative synthesis of several di- and trinucleotide repeats is dependent upon the sequence of the template, which suggests a sequence-specific slippage rate. The rate of synthesis of several di- and trinucleotide repeats is independent of the length of the fragments being synthesized; this suggests that the actual mechanism of slippage involves relatively short regions of the DNA, likely occurring proximal to the growing end (3′) of the nascent molecule. This mechanism is identical to creeping type slippage (Figs. 39-1C, 39-2D). This conclusion is similar to the S1 nuclease results of Rokita and Fredes-Romero [39], as described above (see Section I.C.1). Slippage probably involves the production of bulges or slip-outs of the nascent strand just behind (5′ of) the growing end (see Fig. 39-2D) rather than complete strand separation and out-of register realignment. This may be the mechanism that is responsible for small changes in repeat numbers that are typical of normal polymorphic variations, or in mismatch repair-deficient mutator cells. The large expansions observed in some of the triplet repeat diseases may be accounted for by multiple rounds of such incremental slippages (see below discussion of Okazaki fragment slippage; Section I.E).

4. Polymerases and Short Bubbles

The presently accepted model for the *in vitro* reiterative synthesis of polymeric DNAs from oligomeric repeats maintains that the reaction rate is determined by both the rate of slippage (DNA rearrangement) and the rate at which the DNA–polymerase complex dissociates and reassociates. However, it is evident that the specific properties of a particular polymerase can have a serious influence on the mechanism of reiterative synthesis [30, 49]. Schlotterer and Tautz assayed a range of DNA polymerases for their ability to synthesize a variety of di-, tri-, and tetranucleotide repeats [30]. Clearly different polymerases have different potentials to function in reiterative synthesis. In addition, the structure of the synthesized products differs between different polymerases: the intact *Escherichia coli* PolI or the Klenow fragment and the T4 polymerases result in distinct DNA product sizes, while the T7 or the Taq polymerases result in a smear of DNA products.

5. Polymerase-Mediated Strand Slippage

Using 2-aminopurine-labeled model DNAs and time-resolved fluorescence spectroscopy Hochstrasser *et al.* [50] have analyzed the extent of melting at the primer–template junction that occurs within the Klenow DNA–polymerase. This allowed the direct observation in solution of local melting of the double helix within the enzyme. In the absence of protein, the DNAs displayed several conformational states of the DNA termini, including the fully paired and stacked bases, as well as three unpaired states of the terminal base. The state with the shortest lifetime was that of the fully unstacked and unpaired base. The other unpaired states involved partial stacking interactions. Addition of Klenow polymerase to these model DNAs caused a significant degree of melting in the duplex termini, indicating that the polymerase induced end-fraying of the template–primer. Interestingly, the polymerase increased the lifetime of the unpaired and unstacked terminal base while decreasing the lifetime of the fully paired and stacked conformation. These results agree with the cocrystal structural results of the Klenow fragment and duplex DNA which indicate that there are at least four single-stranded nucleotides bound in an extended conformation within the enzyme ([50] and references therein). Whether the exact number of unpaired bases within the enzyme in solution is similar to that observed in the crystal is not known. These results indicated that unpairing of the nascent strand from the template does occur within the polymerase. Such unpairing while replicating the triplet repeats may result in slippage of the nascent strand and/or the enzyme. Both the polymerase and nascent strand, which is loosely bound to template strand, may slide back along the template strand, pause, then resynthesize the template DNA (Fig. 39-2C). In effect the unpaired portion of the DNA strands within the polymerase may serve as a lubricant for slippage allowing for polymerase spooling. Such an interaction may very well be specific to the polymerase as well as to certain types of tandem repeats (i.e., both the sequence and length of repeat unit). Such slippage may be preferentially dependent upon the sequence of the nascent versus the template strand, and depend on whether replication is accomplished by a leading or lagging strand polymerase. Several labs, in prokaryotic and eukaryotic organisms, have observed a bias for leading or lagging strand replication on the stability (deletions) of trinucleotide repeats [51–54].

Electron microscopic analysis of the products of reiterative synthesis revealed that the DNAs were multibranched containing numerous hairpins [55, 56]. These authors proposed several possibilities for the structure of the products, including snap-back synthesis. Hairpin formation behind a polymerase would also be consistent with their findings. Seemingly, the duplex DNAs that served as templates slipped in minor increments, synthesis occurred, and the process repeated itself. Such a mechanism would be in agreement with the model of "spooling-out slippage" (slippage occurring just behind the polymerase, as shown in Fig. 39-2C).

D. Primer–Template Misalignment: A Source of Spontaneous Mutation

Primer–template misalignment is generally acknowledged as a major source of spontaneous genetic mutations. Spontaneous mutations can occur through errors in DNA replication causing base substitutions, frameshifts, and concerted or multiple mutational events [57–59]. Deletions or duplications between direct repeats can occur by template misalignment of the 3′ end of the nascent DNA strand forward or backward, respectively [60, 61]. The ability of DNA to fold into specific DNA secondary structures can influence the spectrum and frequency of deletion and duplication mutations associated with direct repeats [62–64]. Many spontaneous mutations associated with human genetic disease likely involve primer–template misalignment [65].

For primer–template misalignment to occur presumably the polymerase must pause or dissociate from the DNA template. Pause sites are usually associated with hotspots for spontaneous template misalignment mutagenesis [66–68]. Moreover, replication typically pauses at sites of potential DNA secondary structure [69–73], structures that are inherently mutagenic [34]. Specifically, DNA secondary structures, such as cruciforms, are known to be genetically unstable in bacteria and possibly in human cells [74, 75], where they are likely deleted by a mechansm involving primer–template misalignment stabilized by a hairpin stem [74]. Primer–template slippage can occur with an elevated frequency when stabilized by DNA secondary structure (hairpin) formation, and this can be a strand-specific phenomena [63, 64, 76]. Rosche *et al.* [77] have recently shown that the correction of a quasipalindrome occurs predominantly by an intermolecular strand switch that is preferential for the leading strand. Intermolecular strand switching has been shown to occur *in vitro* [78], and with a high frequency *in vivo* at a long inverted repeat cloned 300 bp from the origin of DNA replication in a pUC-based plasmid [79]. Complex frameshift mutations, which are consistent with an intermolecular strand switch occurring during DNA replication, have been shown to occur after *in vitro* polymerization by *E. coli* DNA polymerase I [80]. Thus primer–template misalignment during DNA replication can be the molecular basis for many types of spontaneous mutations in DNA.

Genetic instability associated with triplet repeats is generally acknowledged to be associated with primer–template slippage, as this is the simplest model by which short deletions, expansions, or base substitutions might occur. The massive expansion of triplet repeats observed in the full mutation of fragile X syndrome, myotonic dystrophy, or Friedrich's ataxia is not easily explained by simple primer–template slippage (or by genetic recombination). The expansion of a repeat tract by a factor of 10 or more is a type of mutation event that has not, prior to its discovery in humans, been previously identified in bacteria or nonhuman eukaryotes. For expansion associated with simple primer template misalignment, multiple rounds of slippage would necessarily have to occur for a length increase of more than a factor of 2. Likewise, multiple rounds of recombination would also have to occur for massive expansion. Reiterative DNA synthesis, which does involve multiple rounds of primer template slippage, is one proven *in vitro* phenomenon by which short templates of DNA produce greatly expanded products [25, 55, 56]. We have previously proposed a model for reiterative DNA synthesis at trinucleotide repeats that could lead to expansion [81, 82]. The model proposed that the formation of some alternative DNA structure may be the initiating event that temporarily blocks the progression of a DNA replication fork, thus leading to reiterative DNA synthesis. Following reiterative DNA synthesis leading to the expansion of repeats, the block is alleviated or the 3′ end of the nascent primer eventually overcomes the block and continues synthesis (Fig. 39-3). Such an event may occur either at the leading or lagging strands. There may be no mismatch repair system in cells that recognizes the exceptionally large loop-out region, and following a second round of DNA synthesis, the expansion mutation becomes incorporated in the duplex genome.

E. Free Ends, Nicks, Gaps, and Okazaki Slippage

As mentioned above, free ends within repeat tracts may facilitate slippage (see Section I.C.2). Free-ends within the repeat tract would occur during the process of chromosome duplication. At the replication fork free ends occur specifically at the 3′ growing end of either the leading or lagging strands. Free ends are also present at the 5′ tail end of the lagging nascent strand where

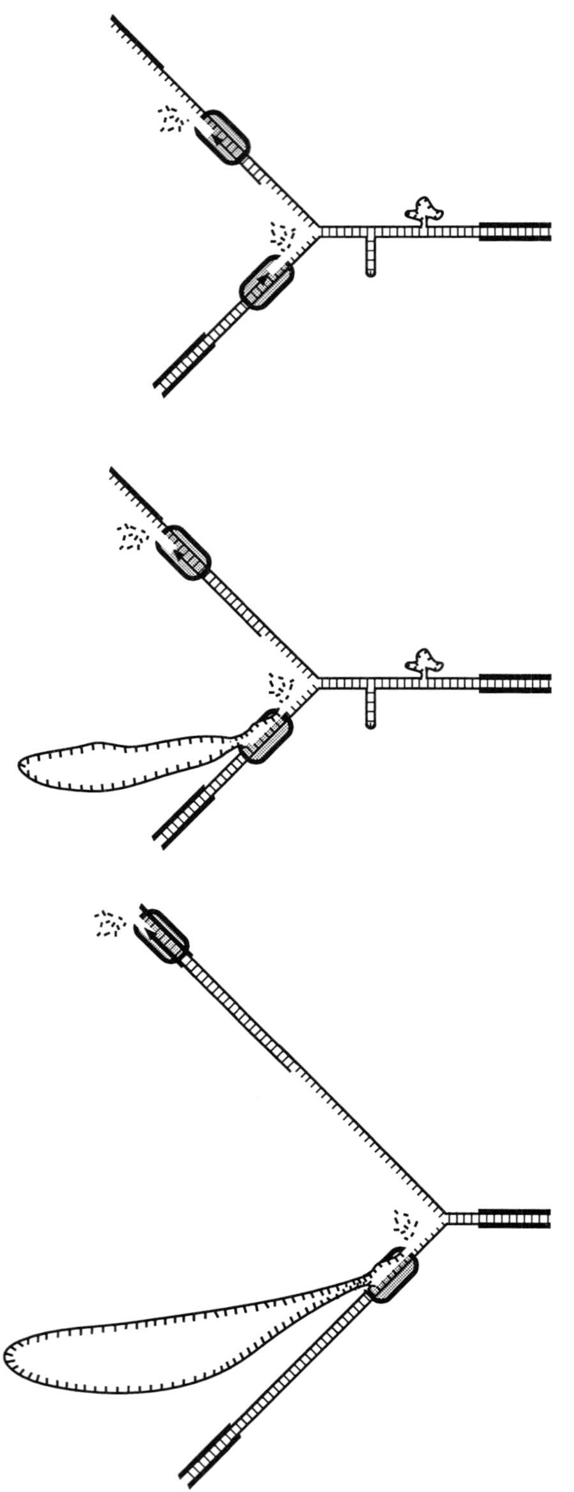

FIGURE 39-3 Mutation through primer–template misalignment. Some structure within the repeat tract (shown here as slipped strand DNA) may cause the arrest of replication fork progression. This arrest may induce the misalignment of the primer and allow for polymerase-spooling reiterative synthesis within the repeat tract. Eventually the blockage is removed or passed by DNA polymerase, possibly by misalignment across the blockage. Upon a subsequent round of replication, the expansion is incorporated into the genome. We have shown only expansion of the leading strand, although it may occur also at the lagging strand. For discussion see text of Section I.D.

RNA priming initiated the Okazaki fragment. Multiple rounds of RNA priming occur on the lagging strand of replication forks. The length of Okazaki fragments for mammalian cells has been reported to range from 25 to 300 nucleotides with a median of 105 nucleotides [83, 84], or may be formed by the ligation of shorter (40 nucleotides) precursor chains and can reproducibly be as short as 40 nucleotides [84, 85]. It may not be pure coincidence that the size of the proto-/premutation repeat tracts (50–60 repeat units) is close to that of an Okazaki fragment [86]. The longer the repeat tract, the more likely that the initiation of Okazaki fragment synthesis occurs within the repeat tract, and the higher the probability of formation of an alternative DNA structure. Slippage of the nascent leading strand resulting in expansion must occur at the 3′ growing end. However, creeping type slippage (Figs. 39-1C, 39-2D) of the nascent lagging strand (Okazaki fragment) can occur either at the 3′ growing end or at the 5′ tail end (Fig. 39-2E), in the 5′ and 3′ directions, respectively. The case of 3′ slippage of the Okazaki fragment would require that the addition of the expanded repeats be synthesized onto the slipped strand. In the case of slippage at the 5′ tail end of an Okazaki fragment the synthesis of the expanded repeats would merely be a form of gap-filling.

F. DNA Turnover Mutation

A possible form of mutation that may occur on repeated sequences is termed "DNA turnover mutation" [87]. This mechanism is independent of chromosome duplication and would result in unscheduled DNA synthesis, localized to the mutated region(s). DNA turnover mutagenesis may occur during certain developmental windows, in certain cell or tissue types, in resting cells such as oocytes, spermatocytes, nerve cells, or muscle cells. This form of mutation may be the result of recombinogenic or error-prone repair processes. A form of this mutation mechanism would be the error-prone repair of accumulated nicks in the DNA. The free-ends composing the nick may denature and incorrectly renature (Fig. 39-4A). Either the primer (3′) free-end may realign incorrectly (Fig. 39-4A) or the template strands (5′) may slip relative to each other (Fig. 39-4B). A given nick slippage of the 3′ strand or slippage of the 5′ end can result in expansion (Figs. 39-2E, 39-4A, 39-4B). Nicks may also

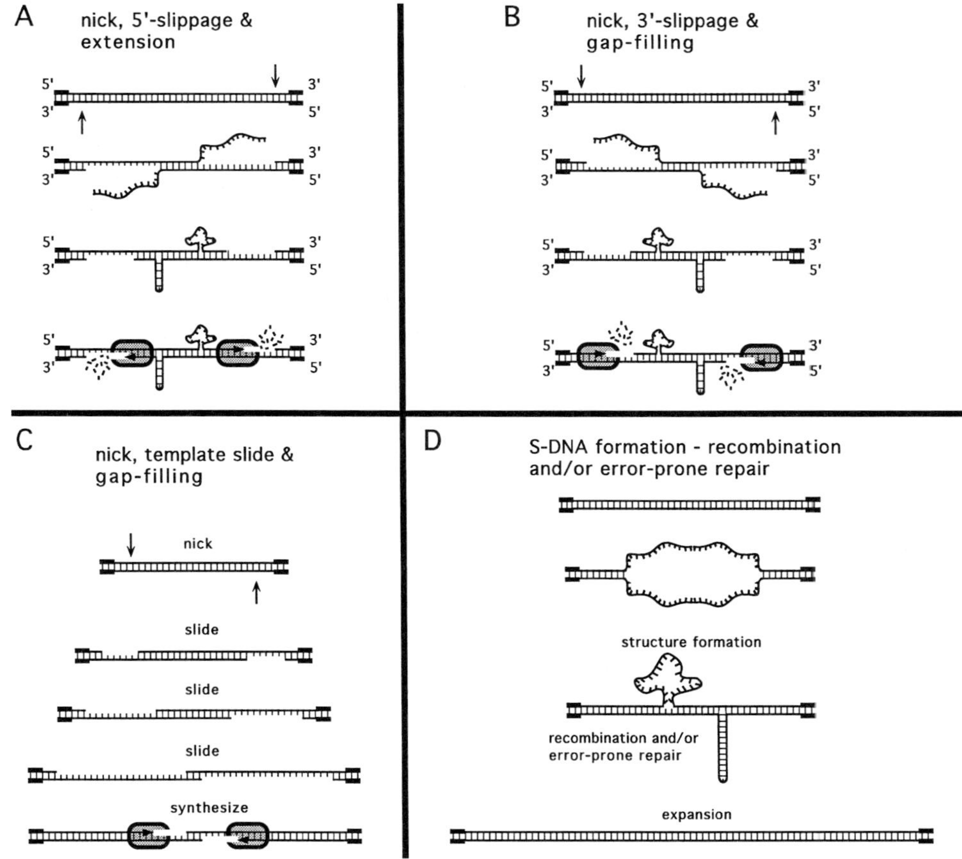

FIGURE 39-4 Models of DNA turnover mutagenesis. (A) Accumulated nicks within the repeat tract can allow slippage of the 3′ end, out-of-register realignment, followed by polymerase extension of the slipped strand. (B) Accumulated nicks within the repeat tract can allow slippage of the 5′ end, out-of-register realignment, followed by polymerase gap-filling. (C) Accumulated nicks within the repeat tract can allow strand sliding, followed by gap-filling. (D) Formation of S-DNA in the repeat tracts in the absence of genome duplication or nicking. These structures may induce recombination or error-prone repair mechanisms. Multiple rounds of this may result in repeat expansions. For discussion see text of Section I.E.

facilitate interstrand sliding (Fig. 39-4C; see Section I.E), which may also result in expansions. In either of these situations repair replication would fill in the gap resulting in expanded products. Alternatively, as discussed below (see Section II.B; see also [10]) slipped strand homoduplex DNAs (S-DNAs) may occur within the repeat tract following reannealing of the two denatured single strands, through soliton-like breathing of the duplex, or passage of a transcription complex (Fig. 39-4D). During certain developmental stages (possibly specific to diseased loci/chromosomes) the repeat sequences may be constantly undergoing the processes of structure formation and mismatch repair/recombination [10] (Fig. 39-4D). This would increase the possibility for replication errors during repair/recombination and may result in repeat length alterations. We proposed such a mechanism for triplet repeat instability [10], whereby multiple rounds of triplet repeat structure formation and error-prone repair could result in repeat expansion. This type of mutation may very well explain some of the observed tissue-specific repeat length differences at certain disease loci. It may also explain some of the sex differences and the time windows for expansion.

II. SUMMARY OF WORK FROM THE SINDEN LAB

As noted in the Introduction, it was soon after the discovery of the double helical structure of the DNA that Gary Felsenfeld proposed the possible existence of slipped strand DNA structures [16]. However, until only very recently [10, 11] no biophysical evidence had been presented for the existence of such structures. As these structures are key intermediates in many processes of genomic polymorphism and mutation it is integral to understand their formation and structural characteristics. Below we review our understanding of these novel structures. For all of our experiments, except where indicated, we have used genomic (or cDNA) clones of the human triplet repeat loci. The role of the nonrepetitive sequences flanking the triplet repeat tracts in either genetic instability or structure formation is yet to be determined.

A. Slippage Models

The rehybridization of an unwound tract of direct repeats can result in homoduplex Slipped stranded DNA (Figs. 39-5A–39-5D), which we have termed S-DNA [10]. During replication, slippage of the DNA polymerase and the nascent strand backward on the template strand, structures containing an excess of repeats on the nascent strand would form, resulting in expansion products (Fig. 39-6). Slippage of the DNA polymerase and the nascent strand forward on the template strand would result in the formation of structures containing an excess of repeats on the template strand, giving rise to deletion products. Both expansion and deletion processes will give rise to replicated duplex DNAs having a stretch of mismatched repeats (a heteroduplex region) (Fig. 39-6). We have termed this structure Slipped Intermediate heteroduplex DNA, or SI-DNA [11]. The expected products (in the absence of repair) of two rounds of replication are shown in Fig. 39-6. For simplicity we have depicted slippage of only the lagging strand, although it may also occur in the leading strand. The resultant alternative structures, S-DNA or SI-DNA, may interact with various cellular proteins and may be involved in the instability associated with triplet repeats.

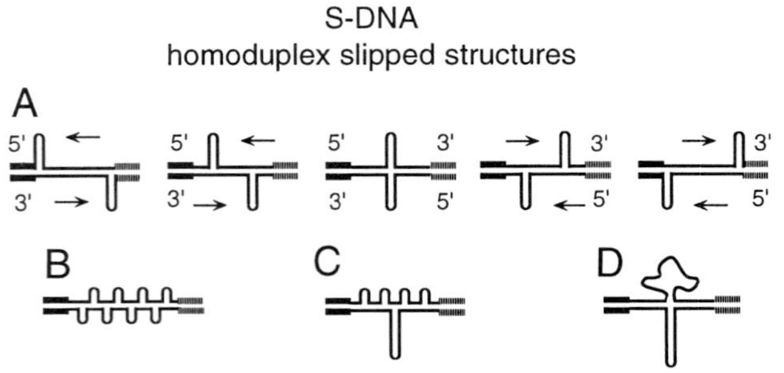

S-DNA
homoduplex slipped structures

FIGURE 39-5 Homoduplex slipped strand structures (S-DNA). (A) Various structural isomers of slipped strand DNAs are expected due to the multiple modes of out-of-register mispairings that are possible. (B) Many intrastrand DNA interactions can occur on each strand, or on only a specific strand (C). (D) Alternatively, one strand may adopt a more single-stranded like structure (random-coil) while the other forms a more stable intrastrand structure. For discussion see text of Sections II.A and II.B.

B. S-DNA: Homoduplex Slipped Strand DNA

In order to study slipped strand DNAs we devised a protocol to induce their formation, which we have termed "reduplexing" [10] (Fig. 39-7). Briefly, $(CTG)_n{\cdot}(CAG)_n$ repeat-containing plasmids (from the DM locus or the SCA1 locus) were linearized by restriction digestion and ^{32}P-end-labeled on either the 5′-end of the CAG strand or the 3′-end of the CTG strand (for the $(CGG)_n{\cdot}(CCG)_n$ containing FRAXA genomic clones, labeling was on the CGG or CCG strand) (Fig. 39-8). Following radiolabeling complementary DNA strands were separated by alkali denaturation and then reannealed by neutralization and incubation at 68°C. This reduplexing protocol has been described in detail [10]. Following reduplexing, the samples were digested with a second restriction enzyme to liberate the repeat-containing inserts. DNA samples were then electrophoretically separated on high-resolution polyacrylamide gels.

Slipped strand structures formed by out-of-register mispairing will contain two structural features. First, they should contain interstrand Watson–Crick base pairs formed between the complementary slipped repeat tracts. Second, slipped strand DNAs should contain slipped-out regions comprised of triplet repeats. Slipped-out regions formed from a long direct repeat, within which intrastrand base pairing is not possible, may form a random coil structure. In the case of CTG, CAG, CGG, and CCG repeats, the potential exists for intrastrand hairpin formation ([88–96]; reviewed in [97] and in this book), which will create a three-way junction. Another alternative entails the interaction of the two slipped-out regions containing sequences complementary to each other [98, 99]. All of these junctions will distort the axial trajectory of a DNA molecule sufficiently to permit their detection by electrophoretic techniques [100–102].

Comparison of the polyacrylamide gel electrophoresis patterns of reduplexed $(CTG)_n{\cdot}(CAG)_n$ repeats revealed that the reduplexed plasmid DNAs contained a set of anomalously slow migrating products not present in the linear nontreated plasmids [10, 11]. An autoradiographic exposure of the reduplexed DM $(CTG)_{30}{\cdot}(CAG)_{30}$-containing plasmid revealed that 39% of a 203-bp *Hin*dIII/*Eco*RI repeat-containing fragment migrated as a series of closely spaced distinct products between 232 and 410 bp, with a single region of greatest intensity migrating as 287 bp (Fig. 39-8B, compare lanes 1 and 2). For reduplexed DM $(CTG)_{50}{\cdot}(CAG)_{50}$-containing plasmid, 70% of the 263-bp repeat-containing fragment migrated slower

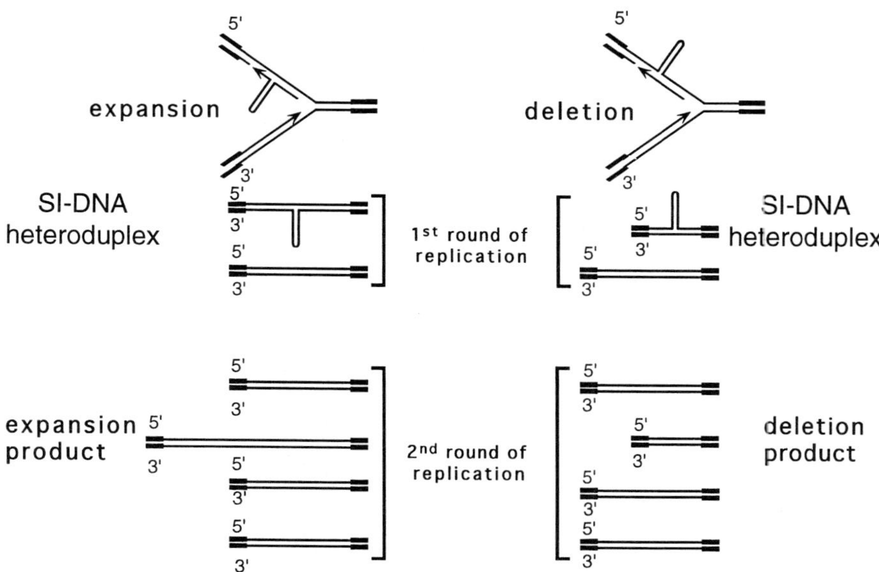

FIGURE 39-6 Heteroduplex slipped strand structures (SI-DNA). Replication-induced formation of heteroduplex slipped intermediates DNAs (SI-DNAs). Slippage and extension of the nascent strand will give rise to an expansion of repeats on that strand, resulting in an SI-DNA heteroduplex; replication past an intrastrand structure in the template strand will result in a contraction of repeats in that strand, resulting in an SI-DNA heteroduplex. Either through repair mechanisms or a subsequent round of replication, the expansions or deletions are incorporated into the genome. We have shown only slippage of the lagging strand, although it may also occur on the leading strand. For discussion see text of Sections II.D and II.E.

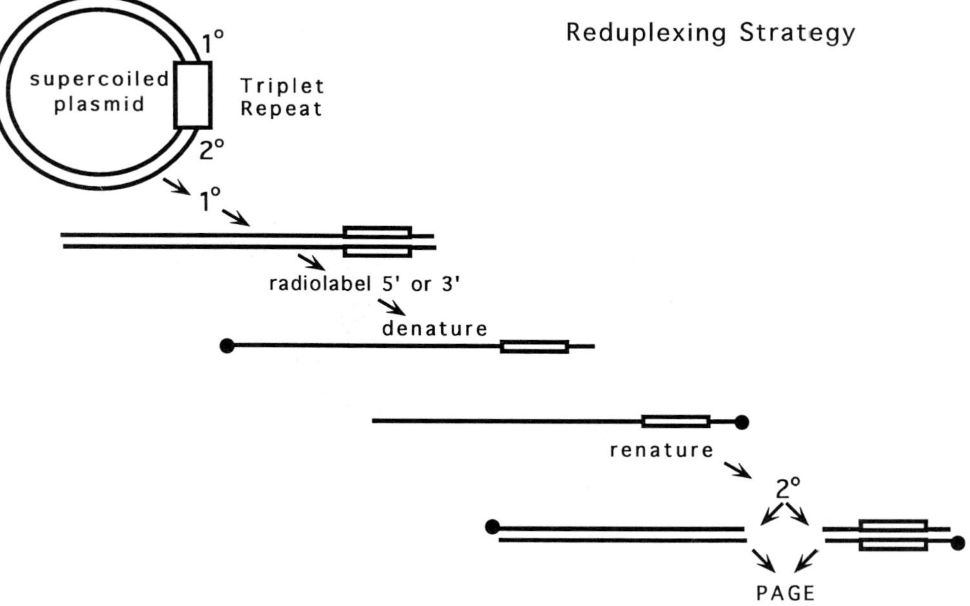

FIGURE 39-7 Reduplexing protocol. To induce the formation of either S-DNA or SI-DNA, repeat-containing plasmids were linearized by restriction digestion, radiolabeled on either the 3′ or 5′ ends, denatured by either alkali (pH 13) or temperature (>85°C), and then allowed to reanneal at 68°C for 2–3 h [10]. The repeat-containing fragment was liberated through a second restriction digestion, and products were analyzed on polyacrylamide gels. For the production of S-DNAs, a single plasmid with a given length of repeats were used. For the production of SI-DNA equimolar amounts of two plasmids differing only in the number of repeats were mixed and treated as above. For discussion see text of Sections II.B and II.D (see also [10, 11]).

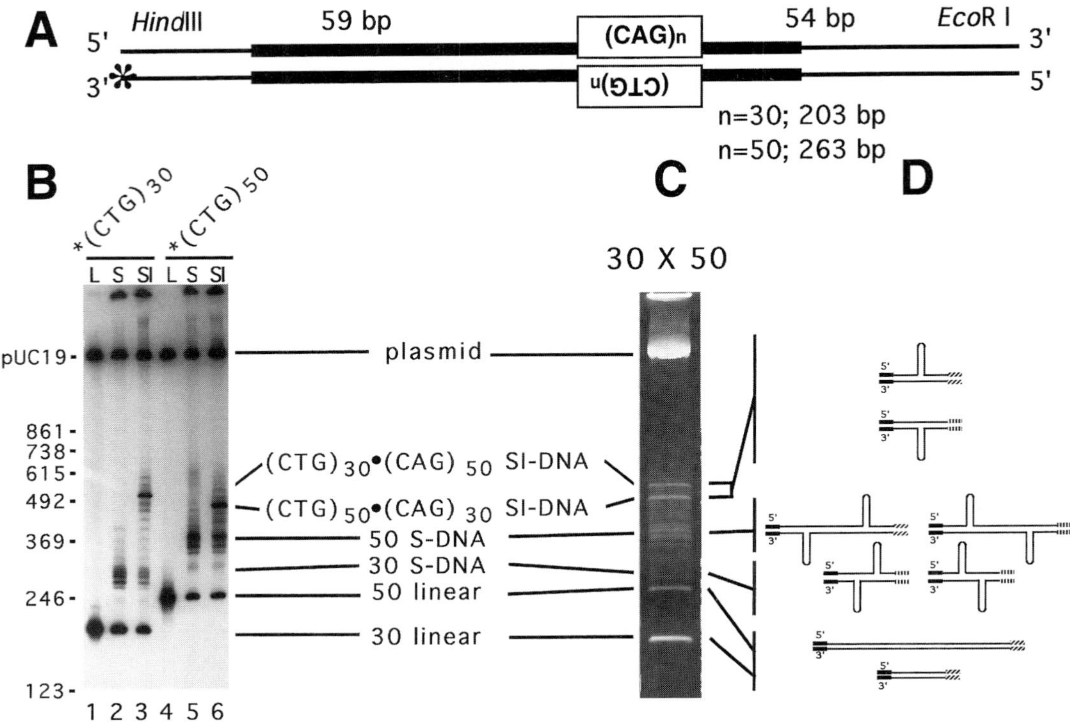

FIGURE 39-8 Reduplexing/heteroduplexing DM DNA fragments results in slowly migrating DNAs. (A) Map of the *Hind*III/*Eco*RI $(CTG)_n \cdot (CAG)_n$-containing fragments from DM genomic clones, where $n = 30$ and $n = 50$. Human nonrepetitive flanking sequences (positions 357 to 375 and 391 to 433) are in bold, while the thin lines represent plasmid vector sequences. Positions of the unique 3′ CTG radiolabels are indicated (*). Linear nontreated (LNT) DNAs of $n = 30$ and $n = 50$ plasmids (lanes 1 and 4, respectively), reduplexed DNAs (S) from $n = 30$ and $n = 50$ plasmids (lanes 2 and 5, respectively), heteroduplexed DNAs (SI) of $n = 30$ and $n = 50$ plasmids (lanes 3 and 6, respectively), and the **Hind*III/*Eco*RI pUC19 vector band are indicated. (B) Products were separated on a 4% polyacrylamide gel, dried and exposed for autoradiography, or stained with ethidium (C). (D) A schematic of the electrophoretically separated DNA structures. Positions of the 123-bp ladder are shown. For discussion see text of Sections II.B and II.D (see also [10, 11]).

than the linear form as a series of closely spaced distinct products between 295 and 696 bp, with a single region of greatest intensity migrating as 370 bp (Fig. 39-8B, compare lanes 4 and 5). A larger percentage of the DNA containing the expanded repeat length ($n = 50$) formed S-DNA structures than DNA containing repeat lengths found in normal individuals ($n = 30$) (Fig. 39-8B, compare lanes 2 and 5).

The series of anomalously slow migrating products suggested that reduplexing had induced the formation of multiple alternative, non-B-DNA structures in the repeat-containing restriction fragments. For each repeat-containing plasmid, the number of novel bands, their relative mobilities, and relative intensities were indistinguishable when either the CAG or the CTG strands were radiolabeled. This suggests that the novel products were composed of both CTG- and CAG-containing strands. The occurrence of the slow migrating DNAs was dependent upon reduplexing and independent of DNA concentration during renaturation. This result argues against the formation of complexes containing multiple restriction fragments. Reduplexing plasmids lacking repeat tracts did not result in any novel anomalous bands [103, 104]. Similarly, FRAXA genomic clones revealed anomalously slow migrating DNAs induced by reduplexing, and these products were composed of both the CGG- and the CCG-containing strands. The distribution of S-DNA isomers was smaller in the FRAXA (CGG)·(CCG)-containing than in the DM (CTG)·(CAG)-containing plasmids. Unlike the formation of many alternative DNA structures (e.g. cruciforms, Z-DNAs, and triplex DNAs), the formation of the reduplex-induced alternative structures within the trinucleotide repeats did not require DNA supercoiling.

1. Effect of Repeat Length on Structure Formation

Genetic analyses of DNA from individuals having DM or FRAXA diseases indicate that there is a close relationship between the number of repeats and both the age of onset and the severity of the disease symptoms (as mentioned in the Introduction). This length effect is also extended directly to the mutability of the repeats. We further investigated the effect of repeat length on the propensity to form S-DNA using a larger array of DM repeat sizes, ranging from 17 up to 255 repeats [10, 11, 104a]. We observed both an increase in the relative amount of novel products and an increase in the complexity of the pattern of novel bands for the reduplexed DM fragments as the number of (CTG)·(CAG) repeats increased. Following gel electrophoresis, S-DNA formation was measured as a densitiometric percentage of the total amount of repeat-containing fragment. The propensity to form structures jumped from 2% for normal 17 repeats to 70% as the repeat tract increased to the expanded length of 50 or more repeat units (Fig. 39-9). This jump in S-DNA formation occurred between the repeat lengths typical of normal individuals and intermediate allele lengths found in proto- or premutant DM individuals who display mild disease symptoms. These data indicated that with larger repeat tracts the percentage of the DNA fragment that actually forms the novel structures increased, indicating a direct relationship between the number of repeats and the propensity to form slipped structures. This association of repeat tract length on the propensity to form S-DNA structures correlates the association of repeat tract length with genetic disease and dynamic mutability. Thus, both slipped structure formation and mutability of trinucleotide repeats are dynamic.

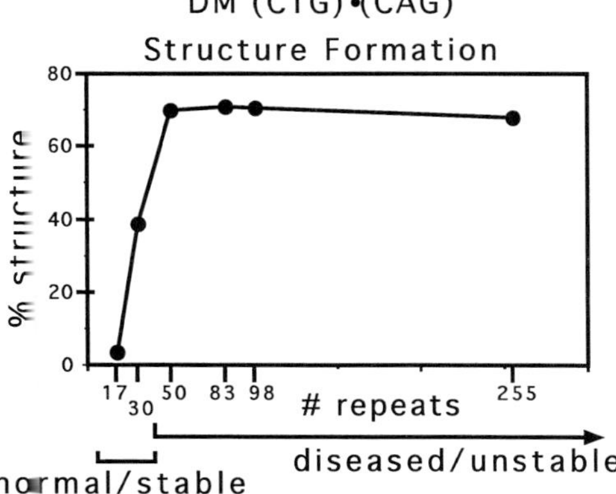

FIGURE 39-9 Propensity of S-DNA formation in DM (CTG)·(CAG) repeats as a function of repeat length. Following gel electrophoresis, the amount of S-DNA formed by each repeat length was measured. The percentage of the repeat-containing fragment is shown as a function of repeat length. For discussion see text of Sections I.B and III.I.1.

2. S-DNAs Are Slipped Homoduplexes

From the above analysis it seemed that the S-DNA structures we had identified were veritable homoduplex slipped strand structures. However, these repeat-containing plasmids are unstable during propagation in *E. coli* in that the number of repeats can vary among the products of a plasmid preparation ([10, 51, 104a, 105]; C. E. Pearson, unpublished results). Although the deletion products are a small proportion of the plasmid preparation, the length heterogeneity within a given plasmid preparation presents the possibility that some of the slowly migrating DNA products may be due to *heteroduplexes* of complementary strands having different numbers of repeats, rather than *homoduplexes.* We took advantage of the fact that in supercoiled plasmids the two DNA strands of a given molecule are catenated and even when denatured cannot separate from each other [10]. Reduplexing of the supercoiled triplet-containing plasmids was followed by double restriction digestion and analysis on polyacrylamide gels and revealed a pattern of slow migrating products indistinguishable from those obtained by the reduplexing of the linearized DNAs. Alkali denaturation of DNA does not affect the integrity of the phosphate backbone ([45, 46]; C. E. Pearson, unpublished results). Thus, the formation of the slow migrating products from supercoiled plasmids indicates that they are the result of alternative interactions between the two complementary strands of the same double-stranded molecule, rather than the result of interactions between complementary strands from different duplex DNA molecules. This result has been confirmed by reduplexing an isolated restriction fragment containing *only* 50 (CTG)·(CAG) repeats; a similar pattern of S-DNA products was observed [104a].

3. Alternative Structures Map within the Trinucleotide Repeats

Mapping of both the $(CTG)_n{\cdot}(CAG)_n$ and $(CGG)_n{\cdot}(CCG)_n$ S-DNA structures was accomplished by selective restriction digestion and gel analysis [10]. The source of the anomalous slow migration mapped to within a few nucleotides on either side of the triplet repeat tract. Furthermore, the enzymes used for mapping were unable to restrict single-stranded DNA; therefore, the complete digestion of the S-DNAs indicated that the nonrepetitive DNA flanking the repeat tracts was in a linear duplex conformation.

4. Alternative Structures Exhibit Remarkable Stability

That the homoduplexed slipped structures were observed at all was surprising to us, as we expected that the two strands might realign by creeping, i.e., migration of the slip-out regions, until two loops coincide and reform a completely base-paired linear duplex. To investigate the biophysical stability of the S-DNAs, we examined their electrophoretic migration following gel purification. Electrophoretic analysis of each of the isolated major S-DNAs formed by $(CTG)_{50}\cdot(CAG)_{50}$ or $(CTG)_{255}\cdot(CAG)_{255}$ revealed a pattern of bands identical to those which were excised from the gel [10], thus indicating that the S-DNA structures were stable through electroelution, multiple buffer changes, and ethanol precipitation, as well as phenol extraction and strong vortexing. Incubation of these same samples overnight at 37 or 55°C did not alter their eletrophoretic band patterns, indicating that the S-DNAs were biophysically stable up to 55°C. However, a 60-min incubation at 85°C did result in some interconversion of the isolated major S-DNA isomer to other isomers, as well as some linear duplex. The remarkable biophysical stability may result from intrastrand hairpins formed within the CTG, and possibly CAG, strands. This intrastrand interaction would prevent loop movement since base pairs would simultaneously have to be broken in the hairpin as well as the Watson–Crick paired interstrand duplex region.

5. In S-DNA the CAG Strand Is Preferentially Sensitive to Mung Bean Nuclease

Each of the slipped-out repeats and possibly the duplex–slip-out junction should have some single-stranded character (even if hairpins are formed) and thus should be sensitive to attack by single-strand nucleases. We used mung bean nuclease to test for the presence of single-stranded regions in the S-DNAs [10]. Single-stranded DNAs, unwound regions, cruciforms, and hairpin loops are specifically recognized by mung bean nuclease, at neutral pH. We analyzed the mung bean nuclease sensitivity of each of the complementary strands of $(CTG)_{50}\cdot(CAG)_{50}$ or $(CTG)_{255}\cdot(CAG)_{255}$ S-DNAs relative to their linear controls. The linear DNAs displayed no specific digestion. In both $(CTG)_{50}\cdot(CAG)_{50}$ and $(CTG)_{255}\cdot(CAG)_{255}$ S-DNAs, mung bean nuclease preferentially digested the CAG strands. The preferential nuclease digestion was specific for the S-DNA structure, and the digestion occurred within the repeat tracts. The preferential sensitivity of the CAG strand to mung bean nuclease reflects its having a greater single-strand character than the CTG strand, possibly by forming a less stable hairpin or loop (Fig. 39-5D). This is consistent with the demonstration that, although both the CTG and the CAG strands could form hairpins, CTG forms a more stable one ([88–96] reviewed in [97] and in this book). It is also possible that the structures formed by the CAG- and CTG-containing strands may be different. In a given alternative structure, the number of slipped-out regions with single-strandedness on the CAG strand may be greater than those on the CTG strand (Fig. 39-5C). The more numerous slipped-out regions with single-strandedness on the CAG strand would result in more mung bean nuclease sensitive sites.

Short CTG, CAG, CGG, and CCG single-stranded oligonucleotides can independently assume noncomplementary duplex hairpin or collapsed structures ([88–96] reviewed in [97] and in this book). Therefore, in duplex DNA the repeat tracts of each strand may form completely independent (intrastrand) hairpin structures, while the mixed-sequence DNA flanking the repeat tracts is linear duplex. However, other possibilities such as random coil, compact/collapsed, or intrastrand quadruplex conformations cannot be ruled out. Alternatively, the slipped regions of each strand may interact with each other, resulting in interstrand knotted structures [98, 99, 106, 107].

C. Repeat Interruptions, Genetic Instability, and S-DNA

As discussed in more detail below, in three of the triplet disease loci, SCA1, SCA2, and FRAXA, the repeat tracts are normally interrupted (see sections III.H and III.I). In normal individuals, the $(CAG)_n$ tract of SCA1 contains 1–3 CAT interruptions, with <17 pure CAGs (see below, Section III.J.3). The stability threshold length for SCA1 at which increased instability and disease transmission occurs is 40 pure (CAG) repeats [108, 109]. All SCA1 patients have uninterrupted expanded $(CAG)_n$ tracts. In normal individuals, the $(CGG)_n$ tract of FRAXA contains 1–3 AGG interruptions [110–116] (see below, Section III.J.1), and the stability threshold length at which increased instability and disease transmission occurs is 34 pure CGG repeats [110]. In both of these diseases, the presence of the interruptions confers increased genetic stability to the repeat tract, such that upon genetic transmission the interrupted tracts are less likely to expand. Loss of the interruptions, resulting in a longer length of the pure tract, correlates with genetic instability and disease, suggesting that loss of the interruptions is an important mutagenic step [107, 114]. The reasons for the stabilizing effect of the interruptions are not well understood. Interruptions may interfere with the formation of slipped stranded DNAs, and thereby diminish the possibility for mutations. To address this question, we investigated the effect of interruptions on the formation of S-DNA using SCA1 and FRAXA repeats, with lengths above and below the stability thresholds [117, 117a].

We analyzed the formation of S-DNA in various lengths of pure or interrupted SCA1 cDNA clones to understand the effect of repeat length and CAT interruptions on S-DNA formation [117, 117a]. The lengths analyzed spanned the stability threshold lenth of 40 repeats Similar to the DM DNAs, the reduplexed pure SCA1 $(CAG)_n\cdot(CTG)_n$ DNAs ($n = 30$–74) resulted in a series of slow-migrating S-DNA isomers, indicating a heterogeneous population of S-DNAs. Again, similar to the DM DNAs [10], the percentage of S-DNA formed increased with longer repeat lengths. The number of major S-DNA isomers formed for longer repeat tracts increased with increasing repeat length, indicating an increase in structural complexity with increased tract length. Comparison of the propensity of S-DNA formation by pure and interrupted SCA1 repeat tracts below the threshold length ($n = 30$ for both) revealed slightly less S-DNA for the interrupted repeat. In addition, the population of S-DNA isomers was different between pure and interrupted repeat tracts of similar size. Comparison of the propensity of S-DNA formation by pure and interrupted SCA1 repeat tracts above the threshold length ($n = 49$ and 44, respectively) revealed about twofold less S-DNA for the interrupted repeat with fewer S-DNA isomers than for pure repeats. The reduction in the number of S-DNA isomers formed in the interrupted tracts likely results from a reduction in the number and location of stable hairpins formed compared with hairpins formed in the pure tracts, which can occur throughout the repeat tract. As described in more detail in the Discussion (see Section III.I.2), this may be one mechanism by which the CAT interruptions provide genetic protection to the length stability of the repeat tracts.

We also analyzed S-DNA formation in pure or AGG-interrupted FRAXA genomic clones containing repeat lengths spanning the stability threshold length (of 34 repeats) [117, 117a]. The amount of S-DNA formed and the structural complexity increased with increasing length of pure $(CGG)_n$. The presence of AGG interruptions reduced S-DNA formation. The effect of multiple interruptions was more severe than the effect of a single interruption in two repeat tracts of similar length. One might expect that the propensity to form S-DNA may be determined by the length of the pure uninterrupted segment of a repeat tract. In such a case, $(CGG)_{17}$ and $(CGG)_9AGG(CGG)_{17}$ may have the same propensity to form S-DNA. However, $(CGG)_9AGG(CGG)_{27}$ had the same propensity for S-DNA formation as a $(CGG)_{17}$ pure tract, while $(CGG)_{21}$ had a greater propensity to form S-DNA than $(CGG)_9AGG(CGG)_{27}$. It seems that the presence of AGG interruptions within the $(CGG)_n$ tracts "poisons" the ability to form slipped structures [116a]. As described in more detail in the Discussion (see Section III.I.2), this may be one mechanism by which the AGG interruptions provide genetic protection to the length stability of the repeat tracts.

D. SI-DNA: Slipped Strand Intermediates (Heteroduplexes)

Slipped Intermediates (SI-DNA) are composed of two complementary strands containing different numbers of repeats [11]. Note, however, that SI-DNA heteroduplexes are not true sequence heterologies since both strands contain complementary repeating sequences. Thus, these structures more closely reflect slipped trinucleotide intermediates formed *in vivo* as intermediates to expansion or deletion.

To produce SI-DNA we used two genomic clones derived from the human DM locus that have identical nonrepetitive flanking sequences and 30 or 50 triplet repeats (Fig. 39-8A), corresponding to the normal and expanded alleles, respectively. For the production of SI-DNA, equimolar amounts of linear 30 and 50 triplet repeat-containing plasmids were mixed and heteroduplexed. Following renaturation, DNAs were digested to liberate a shorter DNA fragment (*Eco*RI/*Hin*dIII) containing the SI-DNA. Polyacrylamide electrophoretic analysis revealed that reduplexing resulted in a reproducible pattern of slow migrating products not present in the nontreated samples. The ethidium-stained pattern of the 30×50 heteroduplex mixture is shown (Fig. 39-8C). The ethidium-stained products of the $^*30 \times 50$ or $^*50 \times 30$ heteroduplexing reactions (where the * represents the labeled $(CTG)_n$-containing strand) revealed that both $n = 30$ and $n = 50$ linear and S-DNA products were formed. The two major slower-migrating products and the fainter closely spaced products, which only formed in the reactions that contained both the $n = 30$ and $n = 50$ parent DNAs, are the heteroduplex SI-DNAs (Fig. 39-8B, lanes 3 and 6; Fig. 39-8C, 39-8D). The ethidium-stained pattern was identical, regardless of whether the $n = 30$ or the $n = 50$ $(CTG)_n$-containing strand was radiolabeled. By selectively radiolabeling the $(CTG)_{30}$- or $(CTG)_{50}$-containing strands it is apparent that the uppermost major band represents the $^*(CTG)_{30}\cdot(CAG)_{50}$ SI-DNA heteroduplex (Fig. 39-8D), while the faster-migrating major band represents the $^*(CTG)_{50}\cdot(CAG)_{30}$ SI-DNA heteroduplex (Fig. 39-8D). As with the S-DNA structures formed from reannealing single parent DNAs, there appear to be several structural isomers, with the major product presumably representing the most energetically favorable conformation. The difference in the structures of the SI-DNAs containing slipped-out CTG or CAG strands is sufficient to confer different electrophoretic mobilities to the sister heteroduplexes, with limited overlap. We have previously demonstrated the electrophoretic resolution of two sister three-way junctions of identical sequence composition [103].

The distinct bands formed by each of the sister SI-DNAs strongly suggested that the excess 20 repeats on the $n = 50$ strand was contained on a single slip-out, and each of the various isomers represented a different point of nucleation/extrusion along the complementary $n = 30$ strand. However, further data are necessary to prove this.

As with both the linear nontreated repeat-containing DNAs and the major S-DNAs, each of the major (sister) SI-DNAs was biophysically stable following gel purification as they demonstrated minimal interconversion back to linear or other forms [10, 11]. However, unlike either the linear nontreated repeat-containing DNAs or the major S-DNAs, both the isolated SI-DNAs demonstrated limited (<0.5%) spontaneous self-association to very-slow-migrating products, whose intensity varied from one preparation to another. It is likely that these represent intermolecular multimeric associations similar to the higher ordered structures of $(CA)_n{\cdot}(TG)_n$ and $(GA)_n{\cdot}(TC)_n$ tracts observed by Belotserkovskii and Johnson [118, 119]. SI-DNAs can also be formed with SCA1 (CAG)·(CTG) clones, as well as with the FRAXA (CGG)·(CCG) clones [C. E. Pearson, unpublished].

E. Mismatch Repair and S-DNA or SI-DNA

As mentioned above, slippage of the DNA polymerase and the nascent strand backward or forward on the template strand would result in the formation of slipped intermediate heteroduplex structures (SI-DNAs) containing an excess of repeats on the nascent strand, resulting in expansion or deletion products, respectively (Fig. 39-6; see also Section II.A). If these structures are not recognized and repaired by cellular systems upon a subsequent round of replication they will result in length alterations in the repeat tract.

1. Human Mismatch Repair Proteins and Triplet Stability

hMSH2 is the human homologue of the *E. coli* MutS mismatch repair proteins [120]. In *E. coli,* MutS initiates mismatch repair through binding to DNA at the site of a mispair [121]. The human hMSH2 protein also binds specifically to duplex oligonucleotides containing single base pair mismatchs as well as short (5–16 nucleotide) insertion/deletion loops [122, 123]. It is likely that this protein plays a role similar to that of the bacterial MutS. Predisposition to Hereditary Non-Polyposis Colorectal Cancer (HNPCC) can be transmitted as mutations in human mismatch repair genes *hMSH2* or *hMLH1.* Human tumor cell lines that are deficient in *hMSH2* display a generalized increase in spontaneous mutation rates, instability in simple repeat (microsatellite) sequences, and yield extracts that are defective for single base pair and insertion/deletion loop mismatch repair *in vitro* ([124], reviewed in [125]).

2. Mismatch Repair-Deficient Cells and Trinucleotide Repeat Stability

For our initial experiments to address the possible role of human mismatch repair proteins on trinucleotide repeat instability we used various human cell lines. Both the DM and FRAXA repeat loci in hMSH2-deficient human cells (LoVo) or hMLH1-deficient cells (HTC116) did not display large repeat expansions and contained repeat lengths typical of normal individuals [126], indicating that an hMSH2 deficiency in LoVo cells does not result in large changes in triplet tract lengths. Our results have since been confirmed by another lab [127]. Other groups have shown that the (CTG)·(CAG) repeats of the myotonic dystrophy and androgen receptor genes do display length heterogeneities in several mismatch repair deficient human cell lines (including LoVo cells) and other malignant cells of the mutator phenotype [128–130]. In fact, it has been demonstrated that triplet repeats are unstable to a degree greater than that observed for mono- and dinucleotide repeat tracts [128–130]. A major finding was that the relative frequency of insertions versus deletions differed depending on the sequence of the repeat tract [128]. Together these points indicate that there is ample evidence for a physiological connection of mismatch repair to triplet repeat instabilities. Additional proteins or protein–protein complexes ([131, 132] reviewed in [125]) may participate in inhibition of large triplet expansions; these may affect the recognition and binding of different DNA substrates.

The observed lack of large trinucleotide expansion in mismatch repair-deficient human tumor cell lines may not directly address the possibility of tissue specificity, altered expression of *hMSH2* during development, the contribution(s) of transcription-coupled repair, or cells undergoing high rates of recombination as causes of trinucleotide expansion. Furthermore, while HNPCC patients do not appear to be predisposed to the diseases associated with expansion of trinucleotide repeat sequences, it should be noted that they are heterozygous carriers, while only homozygous or hemizygous mutation(s) of the mismatch repair genes result in a mismatch repair deficiency.

3. hMSH2 Structure-Specific Binding

To address a possible role of human mismatch repair in trinucleotide repeat expansion we investigated the binding of hMSH2 to the slipped strand structures expected to be intermediates in trinucleotide expansions as a model for the fundamental recognition process that is required in mismatch repair [11]. To be certain of the structural specificity of possible binding interactions we gel-purified each of the $n = 30$ and $n = 50$ major S-DNAs, and major sister SI-DNAs as well as their respective linear forms. Using a band-shift assay we were unable to detect hMSH2 binding to either of the linear DNAs. However, hMSH2 bound to each of the S-DNAs and to one of the SI-DNAs [11]. hMSH2 bound with different affinities for each of the triplet repeat DNAs: $(CTG)_{50}{\cdot}(CAG)_{50}$ S-DNA > $(CTG)_{30}{\cdot}(CAG)_{30}$ S-DNA ~ $(CTG)_{30}{\cdot}(CAG)_{50}$ SI-DNA > $(CTG)_{50}{\cdot}(CAG)_{30}$ SI-DNA ≫ $(CTG)_{50}{\cdot}(CAG)_{50}$ linear = $(CTG)_{30}{\cdot}(CAG)_{30}$ linear. There appears to be a direct relationship between hMSH2 binding and the number of trinucleotide repeats. This relation parallels the effect of repeat length on instability *in vivo*. The K_D of hMSH2 binding to these trinucleotide repeat structures was in the high nanomolar range (50–200 nM), which is similar to the mispair binding affinity exhibited by the *E. coli* MutS protein and previously reported hMSH2 mispair binding affinity [121–123]. Moreover, hMSH2 appears to bind the SI-DNAs in a structure- or sequence-specific manner in which there is preferential binding to slipped-out CAG repeats. Binding competition experiments confirm that the binding of MSH2 to S-DNA and SI-DNAs is structure-specific and is similar to the binding of hMSH2 to single base pair, insertion/deletion loops, and mismatched nucleotides.

Using synthetic $(CTG)_{15}$ and $(CAG)_{15}$ oligonucleotides as binding substrates, hMSH2 bound more efficiently to the $(CAG)_{15}$ oligonucleotide than to the $(CTG)_{15}$ oligonucleotide [11], thus confirming the preferential binding of the $*(CTG)_{30}{\cdot}(CAG)_{50}$ SI-DNA, which has an excess of 20 CAG repeats. Apparently there is an intrinsic difference in the recognition of CAG trinucleotide repeat sequences over CTG trinucleotide repeat sequences. These results also suggest that on either of the homoduplex S-DNAs it is the slipped-out CAG, and not the slipped-out CTG, strand that is being preferentially bound by hMSH2.

The structure(s) within the S-DNA and SI-DNA substrates that are recognized by hMSH2 are unknown. However these structures are likely to be somewhat different than the insertion/deletion loop substrates that were previously shown to be efficiently recognized by the hMSH2 protein [122, 123, 131, 132]. The slipped-out region(s) formed by SI-DNAs are significantly larger than that used in previous experiments. There are a number of possible DNA structures that might be recognized by MSH2 within these slipped strand substrates, including: (i) the junction of the interstrand region with the slipped-out strand, (ii) a CAG or CTG hairpin/loop, (iii) an A·A, T·T, C·C, C·T, A·C, G·G, A·G, or G·T mismatch(es) contained in perfectly or imperfectly annealed CTG or CAG hairpins, or (iv) the tip of an intrastrand hairpin. The binding to the $(CAG)_{15}$ and $(CTG)_{15}$ oligos can exclude the three-way junction as the only recognition site since these oligonucleotides cannot form such structures.

III. DISCUSSION

A. Sequence Specificity for S-DNA Formation

The results reviewed here provide some of the first biophysical evidence for slipped strand DNA structures. S-DNA forms during reannealing of DNA containing $(CTG)_n{\cdot}(CAG)_n$ or $(CGG)_n{\cdot}(CCG)_n$ repeat tracts. The propensity for S-DNA formation increases with increasing length, while sequence interruptions inhibit S-DNA formation. Reduplexing DNAs lacking repeat tracts or containing a variety of inverted repeats did not result in any novel anomalous bands [10, 103, 104]. While tracts of direct repeats are requisite for the formation of S-DNA, it is not known if direct, but not tandem, repeats will form S-DNA.

We note that comparisons of the propensity to form slipped strand structures and the structures formed can only be made between plasmids having identical flanking sequences and vector backbones. Each of the clones within a given set (DM, SCA1, or FRAXA) are identical but for their repeat tract length and pattern of interruptions. Significantly, each contains the regions of nonrepetitive flanking sequences found in their respective chromosomal sites. It is only possible to compare the propensity of structure formation and the electrophoretic pattern of products formed between plasmids of a given set (containing differing length or interruption pattern), because nonrepetitive sequences flanking the repeat can influence the percentage and pattern of S-DNA products formed [C. E. Pearson and R. R. Sinden, unpublished results].

Johnston and colleagues have shown that denaturation and renaturation of the dinucleotide repeats

$(GA)_{37}\cdot(TC)_{37}$ or $(CA)_{30}\cdot(TG)_{30}$ resulted in a minor but reproducible amount of novel products [118, 119]. These products all migrate in acrylamide as dimers, trimers, tetramers, and higher order aggregates. The mispaired aggregates of $(GA)_{37}\cdot(TC)_{37}$ or $(CA)_{30}\cdot(TG)_{30}$ are likely to be formed by mispairing of multiple (>2) strands [118, 119]. Reduplexing of DNAs containing either the dinucleotide repeats $(GA)_{37}\cdot(TC)_{37}$ or $(CA)_{30}\cdot(TG)_{30}$ did not result in the formation of S-DNA structures [C. E. Pearson, unpublished results]. The mispaired aggregates of dinucleotide repeats obtained by Belotserkovski and Johnston [118] differ significantly from the S-DNA structures formed from triplet repeats which migrate slower than the linear monomer, but faster than that expected of a dimer or higher ordered aggregate. This indicates that S-DNA is the result of bistranded intramolecular interactions.

B. Structural Variability of S-DNA and SI-DNA

A high degree of structural variability in slipped strand DNA (i.e., many individual isomers) is expected by the very nature of the repeating units, which could form many different out of register mispairings. This is in contrast to defined conformations such as cruciforms, Z-DNA, or intramolecular triplex structures in which the alternative helix structure is clearly defined by the primary DNA sequence [32–34]. In long triplet repeat tracts, the length of the slipped-out regions as well as their position can vary as illustrated in Figs. 39-5A–39-5D. Varying extents of slippage may occur in either the 3′ or 5′ direction, and the number of slipped-out regions per repeat-containing strand can also vary (Fig. 39-5A). For either reduplexed $(CTG)_n\cdot(CAG)_n$ or $(CGG)_n\cdot(CCG)_n$ S-DNAs a high degree of conformational polymorphism was evident. Due to both the length of the repeat tracts and the broad spectrum of different subspecies of structures being formed, the fine details of the structures cannot readily be determined. Recent results through electron microscopic analyses confirm both the monomeric nature and the structural heterogeneity of both S-DNAs and SI-DNAs [104a].

C. Structural Transitions with S-DNA and SI-DNA Isomers

We have previously shown that the $(CTG)_n\cdot(CAG)_n$ S-DNAs are remarkably stable under physiological conditions. As described above (Fig. 39-8; see Sections II.B.1 and II.B.4) a $(CTG)_n\cdot(CAG)_n$ sample containing only 50 repeats can form a variety of structural S-DNA isomers, and individual isomers are stable to purification protocols. This structural stability is also maintained over a surprisingly wide temperature range (up to 55°C). This indicates that there does not appear to be a rapid interconversion between the various structural isomers of S-DNA nor between those and the linear form. This differs greatly for the case of a bulged base in a repeat tract: Woodson and Crothers [41] analyzed a series of oligonucleotides having internal $(G)_n\cdot(C)_n$ tracts, specifically creating duplex hybrids where one strand contained four G residues and the other contained three C residues. The extra G residue, which could be positioned at any of multiple points along its complementary C tract, appeared to branch migrate throughout the whole of the tract. Similar results were found for an extra C residue in a $(G)_n\cdot(C)_n$ tract [40]. Thus, in the case of an extra (bulged) base in a mononucleotide repeat, there is rapid breaking and forming of hydrogen bonds to permit migration of the unpaired residue along its complement tract. In this situation there is only a single base, and hence no possibility for impeding intrastrand interactions. A single bulged CAG, CTG, CGG, or CCG repeat, although each is palindromic, may not be sufficiently long to have strong intrastrand (C–G) interactions to impede its migration. Single-stranded oligonucleotides composed of CTG, CAG, CGG, and CCG repeats can form hairpins *in vitro* with as few as 3–5 repeats [88, 90, 94, 96].

D. Models for S-DNA Formation

For S-DNA formation to occur, denaturation of the two strands must first occur (Fig. 39-10). The entropic contribution to slipped structure formation has been described [133]. Following denaturation, there are several events that might occur leading to duplex DNA formation or to S-DNA formation. First, the two DNA strands may simply undergo complete renaturation to a fully duplex linear molecule (Fig. 39-10). The first pathway that would lead to S-DNA formation (Pathway I) involves nucleation within the triplet repeat tract, but in an out-of-register fashion. As the "zippering" process of renaturation occurs outward from the site of nucleation, two equal-length loops of triplet DNA will form in the complementary strands at each end of the repeat tract (Fig. 39-10, Pathway I). The size of the loops will be equal to the distance from the site of nucleation to the site of proper base pairing. If denatured DNA containing CTG or CGG repeats were slowly cooled, nucleation would likely occur within the repeat tracts due to the high G+C content of CTG and CGG triplets, and S-DNA formation by Pathway I might be favored.

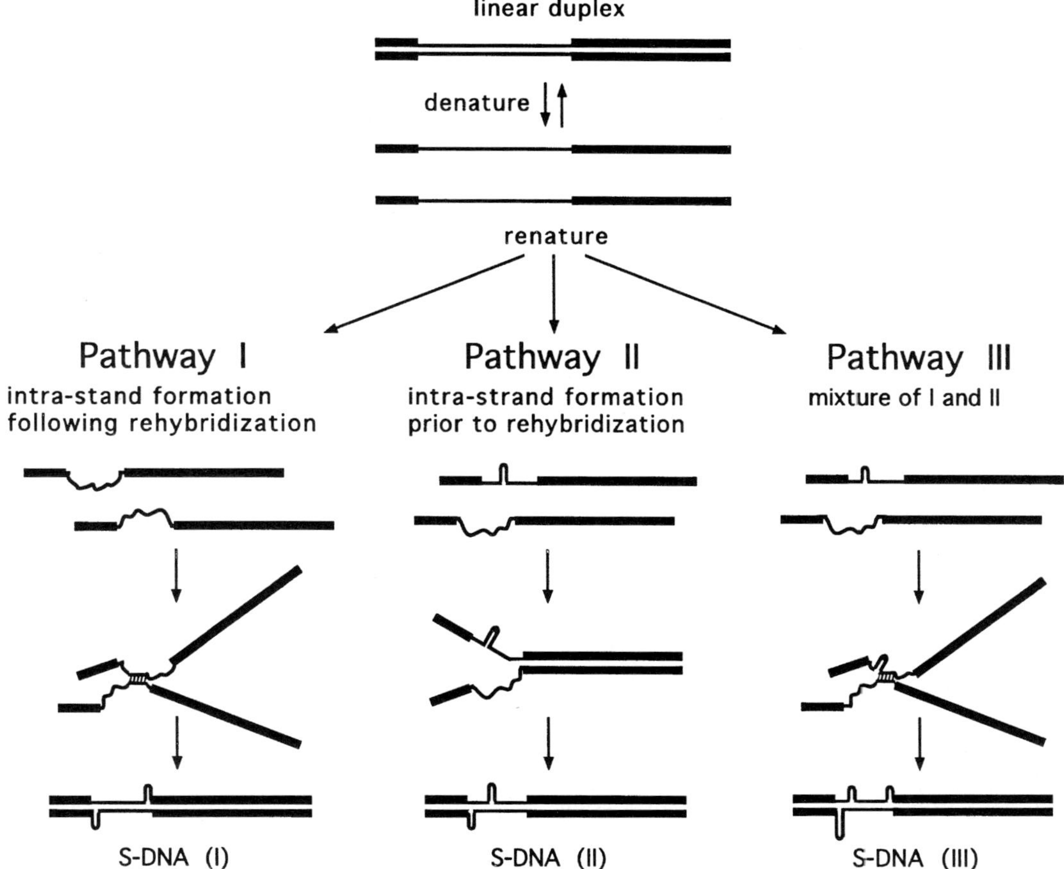

FIGURE 39-10 Pathways for S-DNA formation. Various pathways for the formation of S-DNA are possible. Pathway I involves the out-of-register nucleation within the repeat tract followed by strand zippering. Pathway II involves the formation of an intrastrand structure in the repeat tract, and then interstrand nucleation and strand zippering. Pathway III involves a mixture of Pathways I and II, where there is both intrastrand structure formation and out-of-register nucleation within the repeat tract. Each of the different pathways would form distinct S-DNA structures. For discussion see text of Section III.D.

Pathway I might be favored for (CTG)·(CAG) tracts under the renaturation condition we used previously (alkali denaturation followed by rapid renaturation and incubation at 68°C in relatively low ionic strength [10]). Since the T_m for a CTG hairpin at low ionic strength is below 68°C, S-DNA formation by Pathway I might be consistent with the heterogeneous population of S-DNA isomers observed for the (CTG)·(CAG) repeat-containing DNAs.

Pathway II involves formation of a hairpin in one strand prior to nucleation between the complementary strands. There is a high probability that the single-stranded S-DNA loops will form mispaired hairpins as characterized for single CTG-, CAG-, CGG-, and CCG-containing strands ([88–96], reviewed in [97] and in this book). This may especially be the case for $(CGG)_n$ triplet repeats, where the hairpins are particularly stable. If nucleation occurs outside the repeat tract in the flanking DNA, then zippering will occur across the hairpin forming a three-way junction. The second loop (equal in length to the first) will form at the end of the repeat tract on the opposite side from where nucleation occurred (Fig. 39-10, Pathway II). Because the complementary triplet sequences are likely to have different hairpin stabilities (the probability of hairpin formation being CTG > GAC ~ CAG > GTC and, reportedly, CGG > CCG [92, 93, 96], hairpins are more likely to form in one strand than the other. Given the 75°C T_m of the CGG hairpin [92], S-DNA formation within a (CGG)·(CCG) tract may well occur by Pathway II. The two putative S-DNA isomers found in the interrupted $(CGG)_{54}$ tract [10] may have arisen via formation by Pathway II. It is important to mention that S-DNA formation by Pathway I or II predict ultimately different S-DNA structures.

There is a third pathway by which S-DNA formation could occur. Pathway III involves a combination of Pathways I and II—hairpin formation and out-of-register misalignment. The formation of an initial short hairpin followed by out-of-register misalignment would drive the formation of another short hairpin in the same strand, as well as a longer hairpin in the opposite strand (Fig. 39-10, Pathway III). Pathway III is more complex, and some rare, unusual molecules observed in the EM may be the result of structure formation by this pathway [104a].

E. Biological Relevance of S-DNA Formation

The process where single-stranded DNAs reanneal to become double-stranded occurs in a variety of biological processes such as transcription, recombination, repair, and replication. As discussed above and below the formation of heteroduplex SI-DNAs can occur during DNA polymerization (see Sections II.D and III.F). However, other mechanisms for the formation of homoduplex S-DNAs are possible. The passage of an RNA polymerase may result in an altered structure of the DNA template. We note that all of the unstable triplet repeats are transcribed, although some are in the 5′-UTR, 3′-UTR, coding, or even intronic regions. The DNA template is in fact unwound to a certain degree by the bound RNA polymerase. Following the passage of the RNA polymerase reannealing of the two DNA strands may not occur perfectly and the triplet repeat DNAs may rehybridize with out-of-register base pairing, thus resulting in S-DNA.

S-DNA may also form in unwound regions in the DNA. Long stretches (>10 kb) of single-stranded DNA have been reported in eukaryotic organisms including human cells ([134] and references therein). Some of these unwound regions were associated with DNA replication [135], although long unwound regions not associated with replication have also been reported [134, 136]. Single-stranded regions have also been associated with recombination, early development [135], and meiosis [137, 138]. Alternatively, a relatively short unwound stretch of DNA, possibly originating from a DUE (DNA unwinding element) [139] may be transmitted along the two DNA strands in a soliton-like fashion [36] (Fig. 39-1E, Section I.C.1). When such a small unwound bubble reaches a trinucleotide repeat tract it may extend to a rather large unwound region, which could upon reassociation result in S-DNA.

F. Torsionally Independent Structure Formation and Stability

Primer–template misalignment during DNA replication is believed to be a major source of genome instability, particularly at repetitive DNA elements ([63, 64, 81], and see Section I.D). Neither S-DNA or SI-DNA triplet repeat structures described here require unrestrained superhelical tension for formation and either can form in linear (relaxed) DNA fragments. This characteristic is key to their participation in replication-dependent mutations/alterations. At the replication fork there may be no superhelical torsion between the nascent and template strands. Since S-DNA and SI-DNA can form in the absence of superhelical tension and exhibit remarkable thermostability, they may form at a replication fork either behind the replication complex between the nascent and template strands and/or ahead of the replication machinery between the two template strands.

G. Cytogenetic Chromosome Fragility: Repeat Sequence, Length, Methylation, and DNA Structure

In normal and premutant lengths of the FRAXA repeat the repeat and the associated CpG island are unmethylated. However, the repeat and CpG island are usually fully or partially methylated in the fully expanded disease alleles [140]. Oostra and colleagues [141] found two normal males who had fully expanded *FMR1* CGG repeats; the repeats were not methylated, the protein was expressed, and the fragile site was cytogenetically observable. This indicates that methylation is not necessary for fragile site expression: expansion is sufficient. It also indicates that methylation is required for transcriptional repression of the *FMR1* gene. Loss of transcription of the *FMR1* gene and lack of functional gene product are the cause of the FRAXA disease symptoms. Fragile site expression at FRAXA was observed by another group to be independent of CGG methylation status [142].

There are similarities between the DM and the folate-sensitive rare fragile sites FRAXA, FRAXE, FRAXF, FRA16A repeat expansion, such that they are composed of similar unstable trinucleotide repeats and in each case the repeats can expand to >1000 copies, and in some cases >2000. Despite these similarities, several attempts to cytogenetically observe chromosomal fragile site expression at the normal or fully expanded DM loci failed to detect fragility [143, 144]. The ability to cytogenetically express the folate-sensitive rare fragile sites (FRAXA, FRAXE, FRAXF, FRA16A) appears to be specific to CGG repeat expansions.

In the case of FRAXA it is clear that methylation of the repeat is a requisite for disease onset; this is due to the associated de-repression of *FMR1* gene expression. This suggests that the methylation is induced by the expansion of the trinucleotide repeat. The FRA16A expands to 3.0–5.7 kb (100–1900 repeats), and like FRAXA, FRAXE, and FRAXF it appears to also be methylated when expanded [7]. These authors argue that since normal chromosome 16s do not display methylation at either of the FRA16A loci, methylation of the FRA16A repeat is a result of expansion and not a requisite for expansion. This may suggest that methylation may not play a role in the mutation mechanism. Indeed, expansion of the FRA16A repeat also does not occur normally. An alternative proposal, not considered by these authors, is that the primary event may well have been aberrant methylation which subsequently led to genetic instability and expansion. However, since the finer details of methylation, demethylation, and remethylation occurring during meiosis and early embryogenesis (reviewed in [145, 146]) at individual unstable repeat tracts are not known, a complete understanding of the role that methylation may play in the mechanism of instability is limited.

DNA methylation is known to have serious effects upon the potential of certain sequences to form alternative (non-B-DNA) structures (reviewed in [147]). Methylation of certain sequences can induce or further enhance preexisting curvature. Base methylation facilitates the B-DNA to Z-DNA transition, base methylation increases the stability of intermolecular triplex-DNA, and methylation of bases at the centers of inverted repeats can either impair or enhance cruciform extrusion rates. It is surprising that methylation of the plasmid-borne FRAXA $(CGG)_n \cdot (CCG)_n$ did not observably alter its secondary structure [52]. These analyses were performed on supercoiled substrates and detection was through two-dimensional electrophoresis [52]. Behn-Krappa and Doerfler [31] reported the *in vitro* ability of various lengths of oligonucleotides to participate in reiterative synthesis. They found that methylation of the $(CCG)_n$ or $(CGG)_n$ templates decreased the amount and the full length of the synthesized products. This suggests that template methylation does not enhance reiterative synthesis. The human methyl transferase, like other methyl transferases, can use distorted DNA helices as substrates for cytosine methylation [148]. It has been suggested that the expanded repeats may more easily form alternative structures and that these structures may serve as substrates for methylation [88].

H. Repeat Interruptions and Genetic Instability

1. Triplet Repeats

In three of the triplet disease loci, SCA1, SCA2, and FRAXA, the repeat tracts are normally interrupted with CAT, CAA, and AGG, respectively. In normal individuals, the $(CGG)_n$ tract of FRAXA contains 1–3 AGG interruptions [110–116] (Section III.J.1). In normal individuals, the $(CAG)_n$ tract of SCA1 contains 1–3 CAT interruptions ([108, 109], reviewed in [2]) (Section III.J.3). In normal individuals, the $(CAG)_n$ tract of SCA2 contains 1–3 CAA interruptions ([149–151; reviewed in [5]) (Section III.J.4). In each of these diseases, the presence of the interruptions confers increased genetic stability to the repeat tract, such that upon genetic transmission the interrupted tracts are less likely to expand. Loss of the interruptions, resulting in a longer length of the pure tract, correlates with genetic instability and disease. In many diseases the instability and disease onset occurs within a window of between 35 and 40 pure repeats. For FRAXA, the stability threshold length at which increased instability and disease transmission occurs is 34 pure CGG repeats [110]. There is a "gray zone" length of the FRAXA $(CGG)_n$ repeat, where the total lengths range from 34 to 55 repeats (Section III.J.1). Individuals having these gray zone repeat lengths where the repeat tract is interrupted are normal and the tract is stably transmitted to offspring. However, individuals having these gray zone repeat lengths, where the pure CGG tract is >33 repeats, are unstably transmitted (Section III.J.1). Thus, the presence or absence of AGG interruptions can seriously influence the genetic stability of the $(CGG)_n$ repeat tract. For SCA1, there are several classes of the $(CAG)_n$ tract, while the majority (98%) of nonaffected individuals have 1–3 CAT interruptions and <17 pure CAG repeats (Section III.J.3). For SCA1, the stability threshold length at which increased instability and disease transmission occurs is 40 pure CAG repeats [108, 109]. All SCA1 patients have uninterrupted expanded $(CAG)_n$ tracts. Interruptions within the $(CAG)_n$ tract in the DM locus [152], the $(CGG)_n$ tract of the autosomal fragile site FRA16A [7, 8], and the $(GAA)_n$ tract in the Friedreich's ataxia loci [153, 154] may also confer increased genetic stability to these repeats. The reasons for the stabilizing effect of the interruptions are not well understood. Our results suggest that interruptions may interfere with the formation of slipped stranded DNAs, and thereby diminish the possibility for mutations associated with these DNA structures (see Figs. 39-11 and 39-12, Sections II.C and III.I.2) [117, 117a].

2. Other Interrupted Repeats

Interruptions within other non-disease-associated tandemly repeated tracts have been observed to have effects on their genetic stability. Below we present a few examples of these. The $(CA)_n \cdot (TG)_n$ repeats are particularly abundant within the human genome [155]. Since these sequences exhibit length polymorphisms

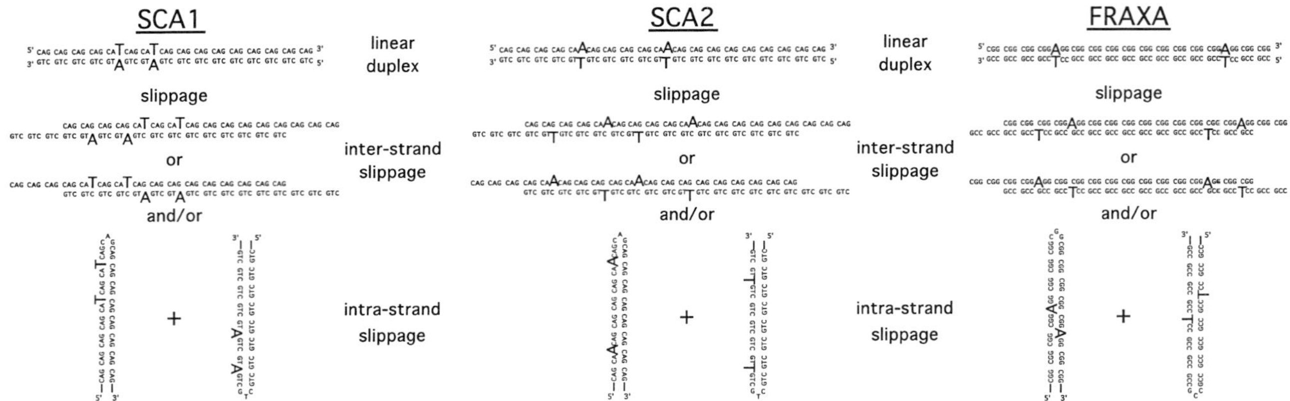

FIGURE 39-11 Models for the quantitative reduction of S-DNA in interrupted repeat tracts. The reduced amount of S-DNA formed by the interrupted repeat tracts may be due to the reduced biophysical stability of the mismatched interruptions contained within inter- or intrastrand slipped DNAs. For discussion see text of Sections II.C and III.I.2 (see also [117, 117a]).

they are commonly used as genomic markers as they are informative as to the number of repeat units. An analysis of 100 $(CA)_n \cdot (TG)_n$ repeats within the human genome led to three separate categories of the tracts: those that are pure and uninterrupted (most of the tracts fell into this category, 64%); interrupted repeats (25%), and compound repeat sequences with adjacent tandem simple repeat of a different sequence (11%) [155]. Per-

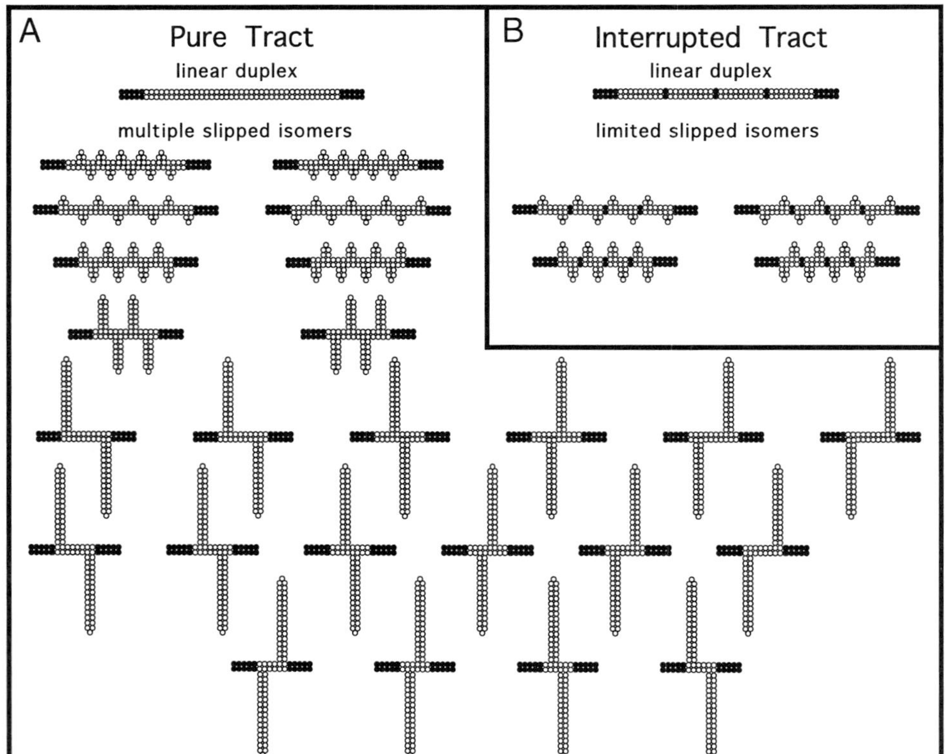

FIGURE 39-12 Model for the qualitative reduction of S-DNA isomers in interrupted repeat tracts. For both SCA1 and FRAXA the open circles represent (CAG)·(CTG) and (CGG)·(CCG) repeats, respectively, while the filled circles represent the (CAT)·(ATG) and (AGG)·(CCT) interruptions. (A) The pure tracts can have slipped-out nucleation points throughout the repeat tract. (B) The reduction in S-DNA isomers formed in the interrupted tracts may be due to a limitation on the positions of, and hence the number and length of, slipped-out repeats. For discussion see text of Sections II.C and III.I.2 (see also [117, 117a]).

fect repeats with less than 10 units displayed very little if any variation between individuals, while repeat tracts with 16–21 units displayed the most variability. Interrupted dinucleotide tracts were less variable even when relatively long. In fact the variability of a given interrupted tract could be predicted based upon the length of the longest run of uninterrupted repeats [155]. These results demonstrate that long lengths of pure repeats are unstable relative to shorter or interrupted lengths.

A recent scan of the first eight completely sequenced yeast *Saccharomyces cerevisiae* chromosomes identified an array of trinucleotide repeat tracts ranging in length from 4 to 130 repeats [156]. Interestingly the longest pure tract was only 19 repeats long while all greater than 19 were interrupted with nonrepeat units. Examination of different laboratory strains of yeast revealed that at any given repeat there was length polymorphism for only the pure repeat tracts which varied in repeat numbers between 6 and 18 repeats. The interrupted repeats did not display any length variations. The lack of length polymorphs in the interrupted tracts suggests that the interruptions in these repeats offer genetic stability to the tracts.

I. DNA Structure and the Mechanism of Genetic Instability

1. Association of Tract Length

Genetic analyses of DNA from individuals with myotonic dystrophy or fragile X indicate that there is a close relationship between the number of repeats and potential for mutation (reviewed in other chapters in this book). We observed an increase in the complexity of the pattern of alternative DNA structures for the reduplexed DM, SCA1, and FRAXA fragments as the number of repeats increased from lengths typical of normal individuals ($n = 17$) to full mutation lengths ($n = 255$) found in diseased individuals. The relationship of repeat tract length with the propensity to form S-DNA is not linear. The propensity of S-DNA formation increases dramatically between 30 and 50 repeats. In the case of DM $(CTG)_n \cdot (CAG)_n$ the propensity to form S-DNA increased from 2% for normal 17 repeats to 70% as the repeat tract increased to the diseased length of 50 or more repeat units. This increase in S-DNA formation occurred in DNAs having repeat lengths typical of intermediate allele length alleles found in proto- or premutant individuals. These lengths are more prone to genetic instability. A similar, but not identical, length dependency upon S-DNA formation was observed for SCA1 and FRAXA [117, 117a].

2. Association of Tract Purity

The tendency for triplet repeat tracts to expand increases above a certain threshold length; most triplet repeats exhibit instability in the range of 35–40 repeats. The stability threshold lengths for FRAXA, SCA1, and FRDA are >34, >40, and 34–40 repeats, respectively [108, 110, 153, 154] (Sections III.J.1, III.J.3, and III.J.8, respectively). Sequence interruptions within the tract can genetically stabilize the repeats. We have shown that repeat interruptions "poison" the ability of repeat tracts to form S-DNA (see Figs. 39-11 and 39-12, section II.C) [117, 117a]. Specifically, CAT or AGG interruptions in the $(CAG)_n$ and $(CGG)_n$ repeat tracts, respectively, have both quantitative and qualitative effects upon structure formation: reducing the amount of total structure formed and altering the relative amounts of specific slipped isomers. CAT or AGG interruptions in $(CAG)_n$ and $(CGG)_n$ repeat tracts, respectively, may offer genetic stability by several mechanisms: (1) the interruptions may maintain an in-frame register of Watson–Crick base pairing (i.e., anchoring) (Fig. 39-11) and thereby inhibit interstrand slippage between the two strands; and/or (2) the interruptions may inhibit intrastrand interactions (Fig. 39-11); and/or (3) the interruptions may reduce the opportunity for slippage (Fig. 39-12). In the first possibility, interstrand slippage would be disfavored as it would result in mismatched interruptions reducing the thermal stability between complementary strands (Fig. 39-11). In the second possibility, S-DNA involves both inter- and intrastrand base pairings; thermodynamic predictions suggest that hairpins formed with interrupted CAG, CTG, CGG, and CCG repeats would be destabilized [89] (Fig. 39-11). This destabilization may tend to reduce the formation and stability of S-DNA. Single-stranded oligonucleotides composed of CAG, CTG, CGG, or CCG repeats can form hairpins with as few as 3–5 repeats under certain conditions [88, 90, 94, 96] and longer repeat tracts may form multiple, short hairpins rather than fewer long hairpins [93]. These first two explanations may account for the quantitative reduction in the amount of S-DNA formed. The third explanation may account for the qualitative effect of interruptions upon S-DNA formation. Both the increased structural complexity with increasing length of pure repeats and the reduction in the number of S-DNA isomers with interrupted tracts suggest that fewer slip-out nucleation points result in fewer opportunities for slippage in the interrupted tracts (Fig. 39-12). Interruptions might limit hairpin formation to the short stretches of pure repeats (Fig. 39-12). In this instance the amount of repeat tract contained in a slip-out would be shorter than would be possible in pure tracts. Shorter amounts of repeat tract contained within slip-outs may permit only short genetic alterations in tract lengths. Together the quantitative and qualitative effects of interruption upon slipped structure formation provide strong evidence for a structural destabilizing effect of interruptions on both interstrand and intrastrand structure formation. At the replication fork, interruptions would inhibit slippage between nascent and template strands and/or hairpin formation within a single strand. The formation of stable alternative DNA structures during replication can result in mutagenesis (see Sections I.D and I.E) [63, 64] making replication slippage an attractive model for triplet instability. Our results suggest that a first line of defense afforded by the interruptions may be via inhibition of slippage.

Another possible stabilizing mechanism of the CAT and AGG interruptions is through a protein-mediated system—possibly mismatch repair [11]. The mismatches formed by interstrand slippage of DNA containing interruptions (Fig. 39-11) may be recognized, resulting in abortive replication and perfect strand realignment. Intrastrand hairpins containing mismatched interruptions would result in different substrates for repair compared to hairpins formed from pure tracts (Fig. 39-11). These mismatched interruptions may, in the sequence context of the hairpin, be very efficiently recognized and repaired. Mismatch repair proteins are involved in mismatch repair, transcription-coupled repair, and abortive recombination [121]. The role of mismatch repair in interruption-mediated repeat stability may be secondary to that of slippage inhibition as recent evidence in yeast indicates that interruptions stabilize dinucleotide repeats independent of mismatch repair [157].

The human mismatch repair protein, hMSH2, binds to both slipped homoduplexes (S-DNA) and to slipped intermediate heteroduplexes (SI-DNA) [11] formed with pure $(CTG)_n \cdot (CAG)_n$ tracts, and binds preferentially to slipped-out CAG relative to slipped-out CTG repeats [11]. Slipped intermediates having an excess of CTG repeats may escape detection more frequently than those with CAG repeats. This asymmetry may indicate that it is the CTG hairpin, rather than the CAG hairpin, which determines trinucleotide instability at neurodegenerative disease loci. These results appear to parallel the observation that certain palindromic sequences can escape repair in yeast cells [158]. An alternative possibility is that hMSH2 either alone or in a complex form may stabilize or protect slipped strand CAG loops from repair, thus allowing replication to finalize an expansion. It is also possible that hMSH2 binding to the slipped strand CAG loops exposes the slipped-out CTG strand to resolvases. Such asymmetric recognition may account for strand-specific differences in the replication fidelity [63, 64, 159, 160], inequality in mutation rates between complementary DNA strands [159, 160], and perhaps the polar mutability of trinucleotide repeat tracts observed *in vivo* [112]. Further studies on the possible role of mismatch repair and other protein-mediated DNA metabolizing systems are needed.

J. Disease Loci and Repeat Tract Sequence Variation

Below we have limited our discussion of the repeat sequences to only the loci that contain repeat interruptions or particular stability-influencing flanking sequences. Such sequence alterations may affect structure formation. For a more detailed description of each disease see other reviews in this book.

1. FRAXA

FRAXA is the most frequent cause of inherited mental retardation. The inheritance of FRAXA is consistent with an X-linked dominant disorder. The mutation responsible for the fragile site has been localized to the instability of a triplet repeat within the 5′-untranslated region of the *FMR1 gene* (Fig. 39-13). The distribution of the repeat lengths in the normal population ranges from 6 to 52. Sequence analysis of normal individuals revealed that the $(CGG)_n$ repeat is cryptic, in that it is normally interrupted at the 5′ end by AGG units, most often occurring at every 10th, 20th, and 30th repeat (Fig. 39-13). The most common allele among the nonaffected population contains 29–30 repeats having 2–3 AGG interruptions. Premutation lengths of 60–230 repeats in carrier females have a high probability of expansion upon transmission to offspring. These premutation lengths have also been detected in normal males but are stably transmitted, hence the term normal transmitting males. Full mutation, symptomatic individuals contain 230–1000 repeats. Only the fully expanded repeat tracts in diseased individuals have an associated methylation of the adjacent CpG island and trinucleotide repeat tract [140].

The absence of the interruptions resulted in a longer pure tract of $(CGG)_n$, occurring primarily at the 3′ end. Nelson and colleagues identified a length of 34 uninterrupted CGG repeats as the genetic instability threshold [110]. This is the length of pure CGG repeats at which there is an increased probability of tract length alterations upon genetic transmission. Seemingly the presence of AGG units provides genetic stability to the $(CGG)_n$ repeat and loss of an AGG may be an important mutational event in the generation of unstable alleles. The AGG interruptions may provide genetic protection by impairing slippage of the strands. Since not all alleles within the high-end normal range (46 repeats) are unsta-

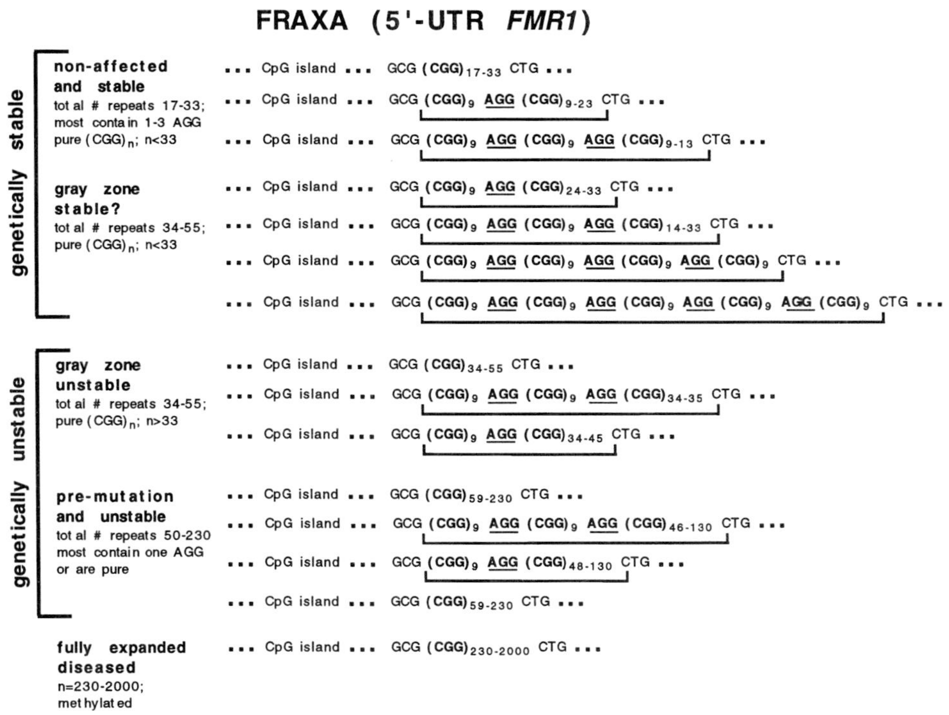

FRAXA (5'-UTR *FMR1*)

genetically stable	**non-affected and stable** total # repeats 17-33; most contain 1-3 AGG pure $(CGG)_n$; n<33	... CpG island ... GCG $(CGG)_{17\text{-}33}$ CTG ... ... CpG island ... GCG $(CGG)_9$ AGG $(CGG)_{9\text{-}23}$ CTG ... ... CpG island ... GCG $(CGG)_9$ AGG $(CGG)_9$ AGG $(CGG)_{9\text{-}13}$ CTG ...
	gray zone stable? total # repeats 34-55; pure $(CGG)_n$; n<33	... CpG island ... GCG $(CGG)_9$ AGG $(CGG)_{24\text{-}33}$ CTG ... ... CpG island ... GCG $(CGG)_9$ AGG $(CGG)_9$ AGG $(CGG)_{14\text{-}33}$ CTG ... ... CpG island ... GCG $(CGG)_9$ AGG $(CGG)_9$ AGG $(CGG)_9$ AGG $(CGG)_9$ CTG ... ... CpG island ... GCG $(CGG)_9$ AGG $(CGG)_9$ AGG $(CGG)_9$ AGG $(CGG)_9$ AGG $(CGG)_9$ CTG ...
genetically unstable	**gray zone unstable** total # repeats 34-55; pure $(CGG)_n$; n>33	... CpG island ... GCG $(CGG)_{34\text{-}55}$ CTG ... ... CpG island ... GCG $(CGG)_9$ AGG $(CGG)_9$ AGG $(CGG)_{34\text{-}35}$ CTG ... ... CpG island ... GCG $(CGG)_9$ AGG $(CGG)_{34\text{-}45}$ CTG ...
	pre-mutation and unstable total # repeats 50-230 most contain one AGG or are pure	... CpG island ... GCG $(CGG)_{59\text{-}230}$ CTG ... ... CpG island ... GCG $(CGG)_9$ AGG $(CGG)_9$ AGG $(CGG)_{46\text{-}130}$ CTG ... ... CpG island ... GCG $(CGG)_9$ AGG $(CGG)_{48\text{-}130}$ CTG ... ... CpG island ... GCG $(CGG)_{59\text{-}230}$ CTG ...
	fully expanded diseased n=230-2000; methylated	... CpG island ... GCG $(CGG)_{230\text{-}2000}$ CTG ...

FIGURE 39-13 FRAXA repeat tract interruptions and genetic stability. The unstable repeat tract lengths and configurations are indicated by brackets. The most common allele among nonaffected individuals has 29–30 repeats with 2–3 AGG interruptions. The $(CGG)_n$ tract in between AGG interruptions can contain 5–12 repeats, but usually contains 9 or 10 repeats (see [110–116]). The absence or loss of AGG interruptions (underlined) within the CGG tract is associated with expansion, disease, and greater instability. The stability threshold ($n > 33$ pure repeats) is the length of pure CGG repeats at which there is an increased probability for length alterations upon genetic transmission [110]. The indicated CpG island as well as the repeat tract are methylated on chromosomes containing the expansion. For discussion see text of Sections III.J.1, III.H.1, II.C, and III.I.2.

bly transmitted, this indicates that although the length of the repeat tract is important, it does not appear to be the only factor responsible for genetic instability [161]. The protective mechanism of the AGG interruptions may very well be due to their ability to impair slippage between complementary strands (see Figs. 39-11 and 39-12, Sections II.C and III.I.2) [117, 117a].

2. FRA16A

FRA16A has also been localized to an unstable $(CCG)_n$-like repeat [7, 8]. The FRA16A expands to 3.0–5.7 kb (1000–1900 repeats), and like FRAXA, FRAXE and FRAXF it appears to also be methylated when expanded [7]. The FRA16A repeat is a complex polymorphism, with four distinct simple tandem repeat regions (Fig. 39-14). Variations in each region were observed. However, the region displaying the greatest length variability within cytogenetically normal chromosomes was a $(CCG)_n$ tract which when centrally interrupted by a CTG unit displayed limited length variability [8]. The presence of the CTG interruption in the $(CCG)_n$ tract may be analogous to the situation of interruptions within the FRAXA, SCA1, and SCA2 repeats. The reduced length variations of interrupted repeat tracts may be correlated with a reduced ability to structurally form slipped structures (see Figs. 39-11 and 39-12, Sections II.C and III.I.2) [117, 117a].

3. SCA1

Spinocerebellar ataxia 1 is autosomal dominant, is also associated with the expansion of a CAG repeat in the coding region of the *ataxin-1* gene, and encodes a polyglutamine tract (reviewed in [2], and in this book). The SCA1 CAG tract is normally interrupted by 1–3 nonrepeat CAT units (Fig. 39-15). The CAT codons code for histidine. Thus, in normal individuals the polyglutamine tract is interrupted by 1–3 histidine residues. The CAG repeat in nonaffected individuals contains from 6 to 39 repeats while the diseased individuals have 40 to 81 repeats. The genetic stability threshold length is 40 repeats [108]. All expanded diseased alleles are pure, suggesting that the loss of the CAT interruptions leads to instability andexpansion. Clearly, the lengths of the high-end normal interrupted alleles are closely

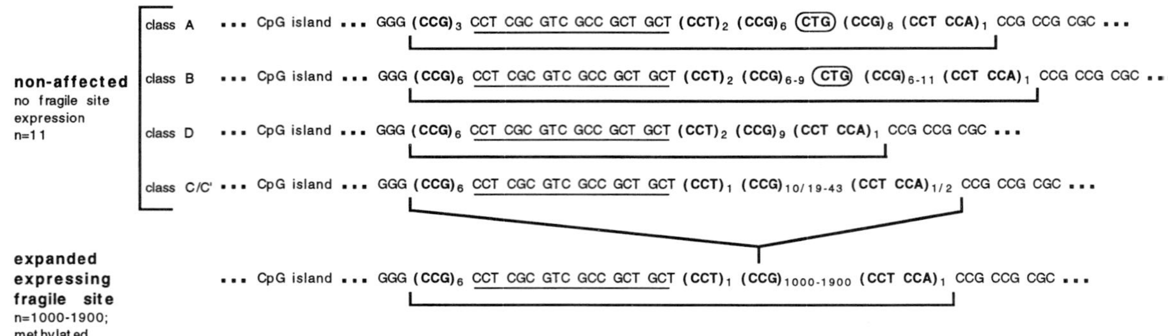

FIGURE 39-14 FRA16A repeat tract interruptions and genetic stability. The unstable repeat tract lengths and configurations are indicated by brackets. The most common allele among nonaffected individuals has 11 repeats. The absence or loss of the CTG (circled) interruption within the CGG tract is associated with greater instability and fragile site expression. The indicated CpG island as well as the repeat tract is methylated on chromosomes containing the expansion. For discussion see text of Sections III.H.2 and III.G (see also [7, 8]).

juxtaposed to the low-end expanded pure diseased alleles. No overlap between repeat lengths of normal and diseased individuals exists as is the case for HD and FRAXA. However, Popovich and colleagues [162] observed an unusually long SCA1 repeat tract that was surprisingly transmitted stably. This repeat tract had 44 repeats, which is above the disease length of 40 repeats. Upon sequence analysis, this clone is multiply interrupted with CAT units (see Fig. 39-15, class 4). Apparently the presence of the interruptions seems to permit the stable genetic transmission of the long tract.

The interrupted SCA1 repeats of nonaffected individuals display strikingly high intergenerational stability, such that in over 1000 meioses not one alteration was observed [2]. Chong *et al.* [109] performed single sperm PCR as well as low-copy genome analyses to

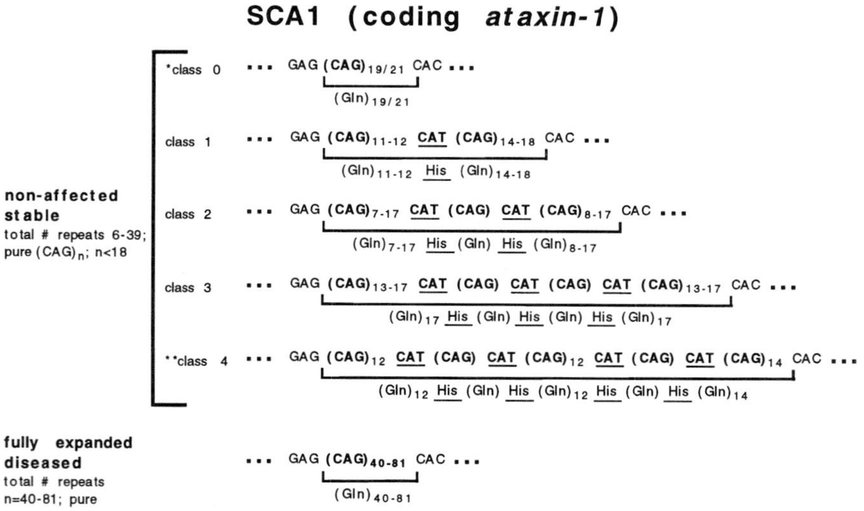

SCA1 (coding *ataxin-1*)

non-affected stable
total # repeats 6-39; pure (CAG)n; n<18

*class 0 ... GAG (CAG)19/21 CAC ...
(Gln)19/21

class 1 ... GAG (CAG)11-12 CAT (CAG)14-18 CAC ...
(Gln)11-12 His (Gln)14-18

class 2 ... GAG (CAG)7-17 CAT (CAG) CAT (CAG)8-17 CAC ...
(Gln)7-17 His (Gln) His (Gln)8-17

class 3 ... GAG (CAG)13-17 CAT (CAG) CAT (CAG) CAT (CAG)13-17 CAC ...
(Gln)17 His (Gln) His (Gln) His (Gln)17

**class 4 ... GAG (CAG)12 CAT (CAG) CAT (CAG)12 CAT (CAG) CAT (CAG)14 CAC ...
(Gln)12 His (Gln) His (Gln)12 His (Gln) His (Gln)14

fully expanded diseased
total # repeats n=40-81; pure

... GAG (CAG)40-81 CAC ...
(Gln)40-81

FIGURE 39-15 SCA1 repeat tract interruptions and genetic stability. The unstable repeat tract lengths and configurations are indicated by brackets. The amino acids coded by the repeat tract are also shown. The most common allele among nonaffected individuals is the class 2 configuration. *Only a few cases of class 0 have been reported [108]. **Only a single case of the class 4 (stably transmitted above the threshold length allele) has been reported [162]. The absence or loss of all the CAT interruptions (underlined) within the CAG tract is associated with expansion, disease, and greater instability. The stability threshold ($n >$ 39) is the length at which there is both disease and an increased probability for length alterations upon genetic transmission. For discussion see text of Sections III.J.1, III.H.3, II.C, and III.I.2 (see also [2, 108, 109, 162]).

investigate the length heterogeneity of SCA1 repeat lengths in different tissues. The nonaffected interrupted alleles did not display length variation, while the expanded pure alleles did. Length heterogeneity was observed in a variety of cells including the sperm, various parts of the brain, and lymphocytes. There was no apparent correlation of tissues displaying the largest repeats and those associated with the neuropathology. However, using DNAs from individuals with closely matched lengths of SCA1 repeats, a high-end normal interrupted allele of 39 repeats, and a low-end expanded pure allele of 40 repeats they demonstrated that expanded allele displayed length heterogeneity while the nonaffected allele was homogenous. Although the length difference between these two alleles was only a single repeat unit, the presence of the CAT interruptions seemed to provide genetic stability to the 39-repeat tract. The protective mechanism of the CAT interruptions may very well be due to their ability to impair slippage between complementary strands (see Figs. 39-11 and 39-12, Sections II.C and III.I.2) [117, 117a].

4. SCA2

Spinocerebellar ataxia 2 is autosomal dominant and is also associated with the expansion of a CAG repeat, a tract in the *ataxin-2* gene which encodes a polyglutamine tract ([149–151], reviewed in [5]). Similar to the SCA1 repeat, the SCA2 CAG tract is normally interrupted by nonrepeat units (Fig. 39-16), in this case CAA units. The CAA codons, like CAG codons, code for glutamine; hence, at the level of the protein, the presence or absence of interruptions in the SCA2 repeat does not affect the purity of the polyglutamine tract. The CAG repeat in nonaffected individuals contains from 14 to 31 repeats while the diseased individuals have 34 to 59 repeats [163]. The level of the expanded alleles is surprisingly low. No overlap between repeat lengths of normal and diseased individuals exists.

Interestingly, individuals with CAG lengths between the high end of the nonaffected range ($n = 24$) and the low end of expanded disease lengths ($n = 35$) repeats are rare. This may indicate that there is a premutant or intermediate length suggestive of a step-wise mechanism of expansion [164]. In nonaffected individuals, the SCA2 repeat displays remarkable stability upon transmission and very little mosaicism. This may be due to the presence of multiple interruptions interspersed throughout the repeat tract. The protective mechanism of the CAA interruptions may very well be due to their ability to impair slippage between complementary strands (see Figs. 39-11 and 39-12, Sections II.C and III.I.2) [117, 117a].

5. MJD/SCA3

Machado–Joseph disease (MJD) [165], identical to SCA3, is also caused by the expansion of a CAG tract which is found in the coding region of the *MJD1* (*ataxin-3*) gene (Fig. 39-17). Nonaffected individuals have from

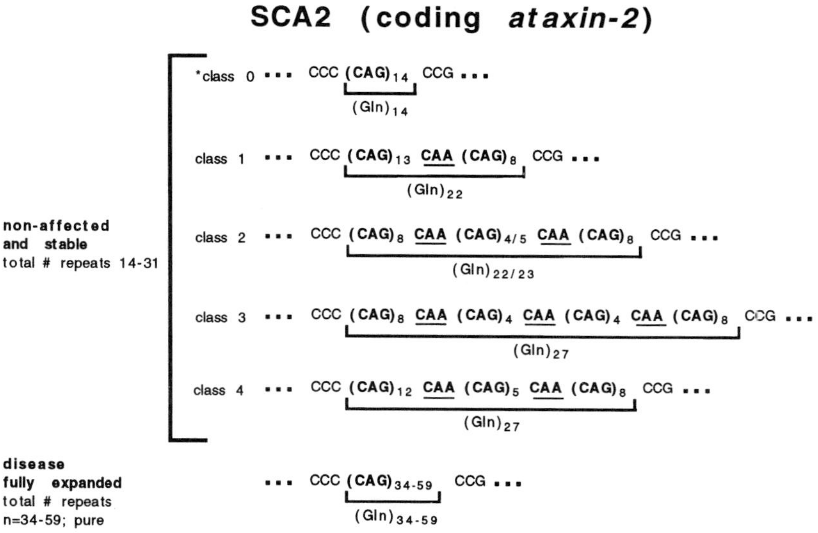

SCA2 (coding *ataxin-2*)

non-affected and stable, total # repeats 14-31:

*class 0 ... CCC $(CAG)_{14}$ CCG ... — $(Gln)_{14}$

class 1 ... CCC $(CAG)_{13}$ CAA $(CAG)_{8}$ CCG ... — $(Gln)_{22}$

class 2 ... CCC $(CAG)_{8}$ CAA $(CAG)_{4/5}$ CAA $(CAG)_{8}$ CCG ... — $(Gln)_{22/23}$

class 3 ... CCC $(CAG)_{8}$ CAA $(CAG)_{4}$ CAA $(CAG)_{4}$ CAA $(CAG)_{8}$ CCG ... — $(Gln)_{27}$

class 4 ... CCC $(CAG)_{12}$ CAA $(CAG)_{5}$ CAA $(CAG)_{8}$ CCG ... — $(Gln)_{27}$

disease, fully expanded, total # repeats n=34-59; pure:

... CCC $(CAG)_{34\text{-}59}$ CCG ... — $(Gln)_{34\text{-}59}$

FIGURE 39-16 SCA2 repeat tract interruptions and genetic stability. The unstable repeat tract lengths and configurations are indicated by brackets. The amino acids coded by the repeat tract are also shown. The most common allele among nonaffected individuals is the class 2 configuration. *Only a single case of the class 0 (short pure tract allele) has been reported [162]. The absence or loss of all the CAA interruptions (underlined) within the CAG tract is associated with expansion, disease, and greater instability. For discussion see text of Sections III.J.1, III.H.4, II.C, and III.I.2 (see also [5, 149–151, 163, and 164]).

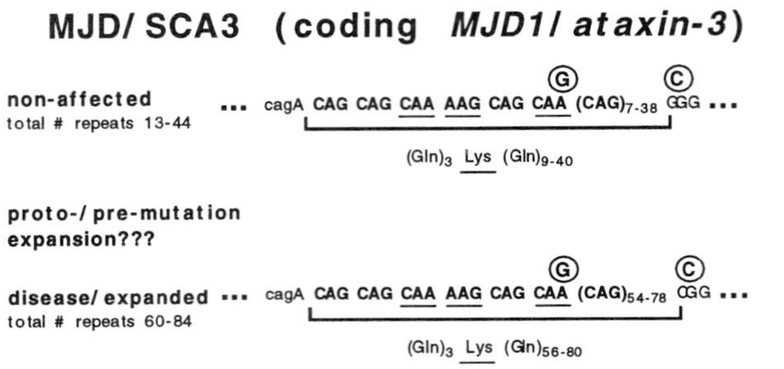

MJD/ SCA3 (coding *MJD1/ ataxin-3*)

non-affected ... cagA CAG CAG CAA AAG CAG CAA $(CAG)_{7\text{-}38}$ GGG ...
total # repeats 13-44
$(Gln)_3$ Lys $(Gln)_{9\text{-}40}$

proto-/ pre-mutation expansion???

disease/ expanded ... cagA CAG CAG CAA AAG CAG CAA $(CAG)_{54\text{-}78}$ CGG ...
total # repeats 60-84
$(Gln)_3$ Lys $(Gln)_{56\text{-}80}$

FIGURE 39-17 MJD/SCA3 repeat tract interruptions and genetic stability. The unstable repeat tract lengths and configurations are indicated by brackets. The amino acids coded by the repeat tract are also shown. The repeat tract interruptions (underlined) and the less frequently found variant bases (circled) are indicated. The gap in repeat tract lengths between the nonaffected and expanded disease lengths suggests that there may be a pre- or protomutation intermediate length. For discussion see text of Section III.H.5 (see also [165–168]).

13 to 44 repeats, while diseased individuals have from 60 to 84 repeats. The MJD repeat tract is not pure as it contains two variant repeat sequences, CAA and AAG, which are found at the third and fourth positions, respectively, and another CAA unit at the sixth position (Fig. 39-17). The third variant repeat, the CAA at the sixth position, was found in some nonaffected individuals (<6.8%) to be a CAG unit [165–167]. The presence of these interruptions results in a glutamine tract interrupted by a lysine residue (AAG), while the CAA units, like CAG, also code for glutamine (Fig. 39-17). The CAA and AAG interruptions are present, and at the same positions, in both nonaffected and expanded diseased alleles; thus it is the 3′ end of the CAG tract which undergoes expansion. The interruptions within the 5′ end of the MJD CAG tract do not offer genetic stability to the repeat as do the CAT, CAA, and AGG interruptions in the SCA1, SCA2, and FRAXA repeats, respectively. This may not be too surprising as they are present at the very beginning of the tract.

The first nonrepeat base at the 3′ end of the CAG tract can be a G or a C (Fig. 39-17). Usually this terminal base is a G in humans and other species [166–168]. This terminal base is a C in 91% of expanded alleles [167] and in 55% of chromosomes of nonaffected individuals with 27–40 repeats [166]. Most MJD repeat tracts with <26 repeats were followed by a G. Clearly, the presence of a C rather than a G is associated with the numbers of CAG [166]. Igarishi *et al.* [167] found that of the two MJD alleles present in the genome of both nonaffected and diseased individuals the larger (or expanded) repeat tract of the two alleles was consistently associated with a terminal C residue. In addition, they found that diseased individuals having the expanded tract in *cis* with a terminal C and the normal repeat in *cis* with a G resulted in significantly greater intergenerational instabilities than diseased individual with the expanded tract in *cis* with a terminal C and the normal repeat in *cis* with a C or individuals with the expanded tract in *cis* with a terminal G and the normal repeat in *cis* with a G. This result strongly implies that there is an interallelic interaction between the expanded MJD chromosomes and the normal chromosomes which results in intergenerational instability. This suggests that gene conversion may be the mechanism responsible for intergenerational instability at the MJD loci. Gene conversion events have also been suggested to result in contraction of expanded triplet repeats at the DM [169] and FRAXA loci [170]. Alternatively, it may be that the terminal C residue is the result of the expansion event rather than a predisposed association. That is, the terminal base may very well have been either a G or a C residue, but following a mutagenic expansion event it may be that it tends to be misincorporated as a C residue.

6. DM

Myotonic dystrophy is autosomal dominant and is associated with the expansion of a CTG repeat in the 3′-untranslated region of the *DMPK* gene. DM displays the signs typical of genetic anticipation: both an earlier age of onset and an increase in severity of clinical symptoms with successive generations. Accompanied by the genetic anticipation is an increase in the number of the CTG repeats (Fig. 39-18). The CTG repeat in nonaffected individuals contains from 5 to 35 repeats. In adult-onset (classical) patients the repeat has expanded to 80–1000 repeats. Congenital DM, where individuals who survive birth develop disease symptoms by age 10, demonstrates repeat lengths of >2000. There are very few individuals having high-end normal lengths (n = 35–80) of DM CTG repeats. Analysis of a large series

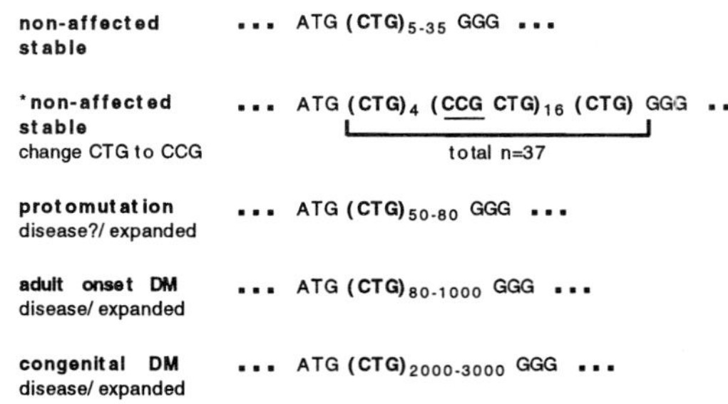

FIGURE 39-18 DM repeat tract interruptions and genetic stability. The unstable repeat tract lengths and configurations are indicated by brackets. *Only a single case of the long, stably transmitted, and multiply interrupted allele has been reported [152]. For discussion see text of Section III.H.6 (see also [152, 171, 172]).

of DM parent–child pairs revealed the protomutation length of the DM allele to be 50–80 repeats [171, 172]. This length was found to be transmitted with either no change or mild expansions. Premutation lengths found in FRAXA have not been associated with disease symptoms. However, in the case of DM mild disease symptoms have been associated with some, but not all, individuals having protomutation repeat lengths of n = 50–80 [171, 172]. It is not known if the asymptomatic individuals with the protomutation DM lengths will eventually develop disease symptoms. Hence the term protomutation was chosen for these DM allele lengths [171].

While searching for high-end normal lengths of the DM repeat a stably transmitted allele of 37 repeats was identified [152] (see Fig. 39-18). Through single sperm typing analysis, this individuals allele was found to exhibit a higher expansion mutation frequency than an individual with 20 repeats but lower than that of an individual with 27 repeats. This unexpected finding prompted these authors to clone and sequence the repeat of the individual with 37 repeats. Sequence analysis revealed a compound array repeat, where seemingly a CTG had been changed to a CCG and the resultant hexamer (CCG·CTG)$_n$ formed a new repeating unit (Fig. 39-18). The hexamer is a degenerate version of the CTG repeat and can be considered as interruptions within the CTG repeat. Thus, the interruptions in this individuals DM repeat permitted the stable transmission of the unusually long repeat length. Although this repeat configuration has been identified in only a single individual for DM locus, the presence of interruptions is analogous to the situation of interruptions within the FRAXA SCA1, and SCA2 repeats. The increased genetic stability of this interrupted allele may very well be due to impaired slippage between complementary strands (see Figs. 39-11 and 39-12, Sections II.C and III.I.2) [116, 116a].

7. HD

Huntingtons disease is autosomal dominant, is associated with the expansion of a CAG repeat in the coding region of the *IT15* (*huntingtin*) gene, and encodes a polyglutamine tract (Fig. 39-19). The CAG repeat in nonaffected individuals contains from 10 to 34 repeats while the diseased individuals have 40 to 121 repeats.

Downstream of the unstable CAG repeat, separated by 12 nucleotides, is a (CCG)$_n$ tract, encoding a polyproline tract in the *huntingtin* protein (Fig. 39-19). Although originally thought to be invariant [173], this (CCG)$_n$ also displays length polymorphisms ranging from 7 to 12 repeats [174–176]. The majority of individuals with (CAG)$_n$ lengths between 30 and 39, and HD patients

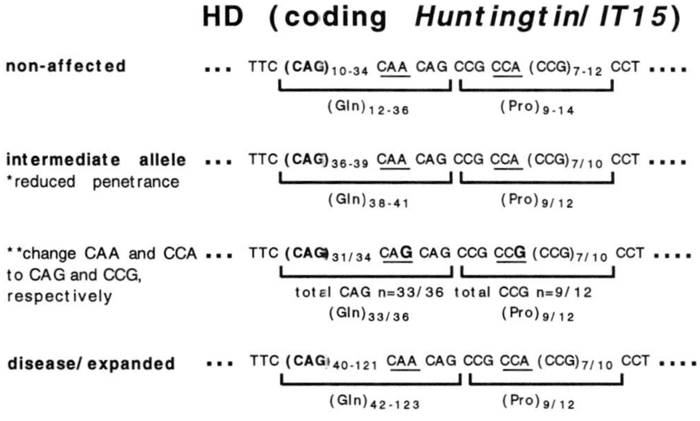

HD (coding *Huntingtin/ IT15*)

non-affected ... TTC (CAG)$_{10\text{-}34}$ CAA CAG CCG CCA (CCG)$_{7\text{-}12}$ CCT ....
(Gln)$_{12\text{-}36}$ (Pro)$_{9\text{-}14}$

intermediate allele ... TTC (CAG)$_{36\text{-}39}$ CAA CAG CCG CCA (CCG)$_{7/10}$ CCT ....
*reduced penetrance
(Gln)$_{38\text{-}41}$ (Pro)$_{9/12}$

**change CAA and CCA to CAG and CCG, respectively ... TTC (CAG)$_{31/34}$ CAG CAG CCG CCG (CCG)$_{7/10}$ CCT ....
total CAG n=33/36 total CCG n=9/12
(Gln)$_{33/36}$ (Pro)$_{9/12}$

disease/ expanded ... TTC (CAG)$_{40\text{-}121}$ CAA CAG CCG CCA (CCG)$_{7/10}$ CCT ....
(Gln)$_{42\text{-}123}$ (Pro)$_{9/12}$

FIGURE 39-19 HD repeat tract interruptions and genetic stability. The unstable repeat tract lengths and configurations are indicated by brackets. The amino acids coded by the repeat tract are also shown. The repeat "interruptions" (underlined) between the disease-associated CAG tract and the distal CCG tract are indicated. *A single family having alleles that had no "interruptions" between the two repeat tracts has been reported [177, 178]. For discussion see text of Section III.H.7 (see also [173–182]).

with expanded (CAG)$_n$ tracts, had (CCG)$_n$ tracts containing 7 or, less frequently, 5, 9, or 10 repeats [176–178]. Although the HD CCG repeat is polymorphic and is within the coding region of a neurologically important protein, disease-associated expansions of this tract have not been observed. In FRAXA, FRAXE, FRAXF, FRA16B, and FRA11A it is the expansion of a polymorphic (CGG)·(CCG) repeat. It may be that the short length of the (CCG)$_n$ tract in HD is below that which will result in hyperexpansion.

The sequence of the region between the (CAG)$_n$ and the (CCG)$_n$ tracts suggests a degeneracy of each of the tracts. Changing the CAA, which immediately follows the (CAG)$_n$, to a CAG and the CCA, which is immediately before the (CCG)$_n$, to a CCG would result in increased longer uninterrupted CAG and CCG tracts (Fig. 39-19). These base changes would not affect the sequence of the *huntingtin* protein; however, they may alter the stability of each of the DNA repeats. In fact, a family in which both these sequence changes occurred has been reported [177]. The CAA and CCA units between the (CAG)$_n$ and (CCG)$_n$ tracts may be considered as "interruptions." In this family the sequence flanking the CAG repeat appears to have an effect upon its genetic stability [178]. However, it is not clear if the increased genetic instability conferred by the loss of both "interruptions" was due to (1) the purity of the (CAG)$_n$ tract, (2) the purity of the (CCG)$_n$ tract, or (3) the juxtaposition of pure (CAG)$_n$ and (CCG)$_n$ tracts. A follow-up report on this family demonstrated through single sperm analyses two intermediate allele (IA) individuals with similar lengths of (CAG)$_n$ having either the wild-type interrepeat sequence or the mutant pure juxtaposed repeats [178]. Indeed, the mutant pure repeats displayed more contractions or expansions in sperm relative to the wild type, indicating a greater instability of the juxtaposed pure repeats. It should be noted that this mechanism of "interruption" loss does not appear to be a common process [177, 178]. Analyses of other IA individuals did not reveal any changes in either the CAA or CCA interrepeat units [180]; hence at this point this mechanism of interruption loss cannot be generalized to the HD repeat tract.

Individuals with CAG lengths between the high-end of the nonaffected range and the low-end of the diseased range (n = 35–39) repeats are rare. Several reports of sporadic/*de novo* expansions of the HD CAG repeat have been reported ([177–182], and references therein). Individuals having these lengths were termed intermediate alleles (IAs) [177, 178]. HD IAs may represent premutations since they are larger than those commonly observed in the general population ($n > 29$), and are shorter than low-end lengths found in HD patients ($n < 40$). However, in the case of HD IAs (n = 36–39) there is an apparent phenotypic gray zone clearly indicating variable penetrance at lengths in this range [177–179]. Reflecting recent results, McNeil *et al.* [180] have suggested a more suitable term for these intermediate alleles as "reduced penetrance." The term protomutation, used for rare individuals having DM $(CAG)_n$ repeat lengths between 50 and 100 repeats [171], may fall under the same category as HD IAs.

8. FRDA

In addition to (CTG)·(CAG) and (CGG)·(CCG) repeats a third unstable triplet repeat, (GAA)·(TTC), was found to be associated with a human disease [183]. Expansion of this triplet in the *X25* (*frataxin*) gene is associated with Friedreich's ataxia. This is the only unstable triplet found so far to exist within the intronic region of the disease gene. In addition, the disease is autosomal recessive, thus requiring instability (or knock-out) at both alleles. The repeat tract in the majority of normal individuals has 8–9 repeats, with a few having between 16 and 34 repeats [153, 154, 183] (Fig. 39-20), while diseased individuals have between 112 and 1700 repeats. There are very few FRDA alleles with repeats between the high-end normal lengths and the low-end disease lengths. However, recent evidence indicates that there may be a premutation/protomutation length for FRDA [153, 154]. Some individuals having 42 to 60 repeats which were stably transmitted were

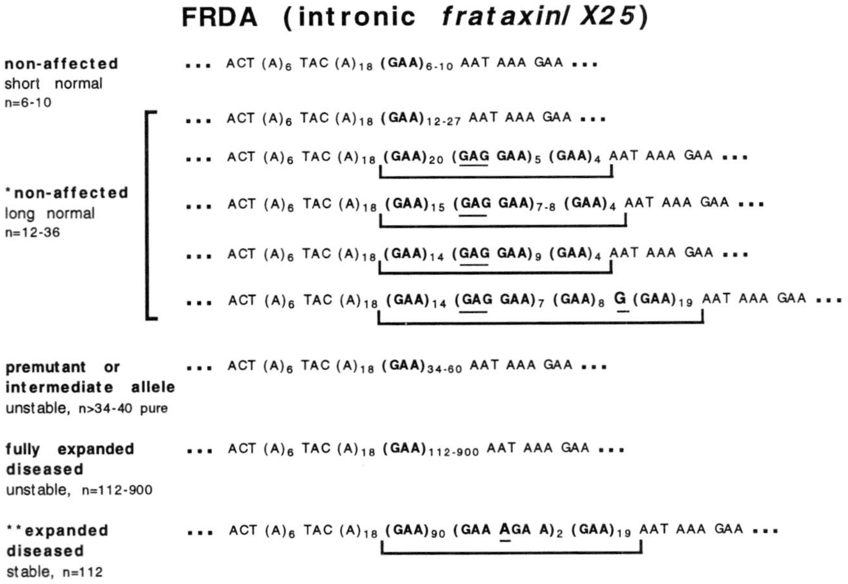

FRDA (intronic *frataxin/ X25*)

non-affected short normal n=6-10	... ACT $(A)_6$ TAC $(A)_{18}$ $(GAA)_{6\text{-}10}$ AAT AAA GAA ...
***non-affected** long normal n=12-36	... ACT $(A)_6$ TAC $(A)_{18}$ $(GAA)_{12\text{-}27}$ AAT AAA GAA ...
	... ACT $(A)_6$ TAC $(A)_{18}$ $(GAA)_{20}$ $(\underline{GAG}\ GAA)_5$ $(GAA)_4$ AAT AAA GAA ...
	... ACT $(A)_6$ TAC $(A)_{18}$ $(GAA)_{15}$ $(\underline{GAG}\ GAA)_{7\text{-}8}$ $(GAA)_4$ AAT AAA GAA ...
	... ACT $(A)_6$ TAC $(A)_{18}$ $(GAA)_{14}$ $(\underline{GAG}\ GAA)_9$ $(GAA)_4$ AAT AAA GAA ...
	... ACT $(A)_6$ TAC $(A)_{18}$ $(GAA)_{14}$ $(\underline{GAG}\ GAA)_7$ $(GAA)_8$ $\underline{G}$ $(GAA)_{19}$ AAT AAA GAA ...
premutant or intermediate allele unstable, n>34-40 pure	... ACT $(A)_6$ TAC $(A)_{18}$ $(GAA)_{34\text{-}60}$ AAT AAA GAA ...
fully expanded diseased unstable, n=112-900	... ACT $(A)_6$ TAC $(A)_{18}$ $(GAA)_{112\text{-}900}$ AAT AAA GAA ...
****expanded diseased** stable, n=112	... ACT $(A)_6$ TAC $(A)_{18}$ $(GAA)_{90}$ $(GAA\ \underline{A}GA\ A)_2$ $(GAA)_{19}$ AAT AAA GAA ...

FIGURE 39-20 FRDA repeat tract interruptions and genetic stability. The unstable repeat tract lengths and configurations are indicated by brackets. The disease-associated GAA tract is preceded by $(A)_6TAC(A)_{16}$, which likely represents a degenerate version of the $(A)_5TAC(A)_5$ canonical *Alu* sequence [154]. The majority of nonaffected individuals have short lengths of the $(GAA)_n$ tract (n = 6–10). The in-frame and out-of-frame variant repeats (underlined) are indicated. *The last repeat configuration in this series is a single case of a stable 55-repeat allele identified in a healthy carrier [153]. **A single case of a stable n = 112 expanded repeat allele was identified in an FRDA patient [153]. For discussion see text of Section III.H.8 (see also [153, 154, 183]).

found to have repeats containing GAG interruptions (Fig. 39-20). These interruptions may inhibit slippage in a similar fashion as do AGG or CAT interruptions within the FRAXA and SCA1 repeats (see Figs. 39-11 and 39-12, Sections II.C and III.I.2) [117, 117a]. In addition, Cossee *et al.* [153] identified a mild pathological expansion ($n = 112$ repeats) which was stable through four transmissions; sequence analysis of this expanded allele showed that it too was interrupted (Fig. 39-20). Clearly, interruptions within GAA trinucleotide repeats can provide genetic stability to the tracts.

IV. SUMMARY: STRUCTURAL CHARACTERISTICS OF S-DNA AND SI-DNA

We have detected homoduplex (S-DNA) and heteroduplex (SI-DNA) slipped strand structures formed within the (CTG)·(CAG)- and the (CGG)·(CCG)-containing fragments of the DM, SCA1, and FRAXA loci [10, 11, 104a, 117, 117a]. These are the first reports of an alternative DNA structure formed in disease-relevant lengths of complementary trinucleotide repeats in the absence of supercoiling. The novel structures have the following characteristics, each of which are consistent with slipped strand structures:

1. Both homoduplex slipped stranded S-DNA and slipped intermediate heteroduplex SI-DNA are composed of complementary strands containing triplet repeats.
2. S-DNA structures are formed from complementary strands having equal numbers of repeats.
3. SI-DNA structures are formed between strands having different numbers of repeats.
4. In both S-DNA and SI-DNA the structural anomaly occurs within the repeat tract.
5. In both S-DNA and SI-DNA the structural anomaly is subtended by linear duplex nonrepetitive DNAs.
6. The formation of S-DNA or SI-DNA does not require superhelical tension, and consequently they can form at a replication forks and nicked or gapped DNAs.
7. The slipped structures form very easily and are remarkably stable under physiological conditions, and consequently may form *in vivo.*
8. In S-DNA the CAG strand is more susceptible to mung bean nuclease digestion than its CTG complement, indicating a greater single-stranded character of the CAG strand.
9. Electron microscopic analysis of S-DNA and SI-DNA confirms both the monomeric nature and the structural heterogeneity, such that inter- and intrastrand interactions can occur throughout the repeat tract.
10. S-DNA is truly slipped out-of-register mispairing between the two complementary strands.
11. As repeat length increases there is both an increase in the propensity to form the S-DNA, and an increase in its conformational complexity (more isomers).
12. Nonrepeat interruptions in either $(CAG)_n$ or $(CGG)_n$ tracts can reduce the propensity for structure formation.
13. Nonrepeat interruptions reduce the number and type of slipped isomers formed.
14. The human mismatch repair protein, hMSH2, preferentially binds to the slipped-out CAG strand, relative to the CTG strand, in both homoduplex S-DNA and slipped intermediate heteroduplex SI-DNA.

The (CTG)·(CAG) and the (CGG)·(CCG) S-DNAs were similar to the extent that each is contained within the repeat tracts, flanked by linear duplex nonrepeating sequences, and they can be composed of complementary strands of equal lengths.

V. CONCLUSIONS

The effect of both the length and the purity of the repeat tract on the propensity of S-DNA formation correlates with their effect on genetic stability in human diseases. Our results provide strong evidence for the participation of an alternative DNA structure in the mechanism of trinucleotide repeat instability. Specifically implicating S-DNA and SI-DNA as these structures. In addition, our results suggest that the first line of defense afforded by the non-repeat interruptions seems to be via inhibition of slippage. These characteristics are consistent with the possibility that in the human chromosome these alternative structures may form without requiring different numbers of repeats on the two strands at a particular locus.

References

1. Ashley, C. T., Jr., and Warren, S. T. (1995). Trinucleotide repeat expansion and human disease. *Annu. Rev. Genet.* **29,** 703–728.
2. Zoghbi, H. Y., and Orr, H. T. (1995). Spinocerebellar ataxia type 1. *Cell Biol.* **6,** 29–35.
3. Paulson, H. L., and Fischbeck, K. H. (1996). Trinucleotide repeats in neurogenetic disorders. *Annu. Rev. Neurosci.* **19,** 79–107.

4. Guesella, J. F., and MacDonald, M. E. (1996). Trinucleotide instability: a repeating theme in human inherited disorders. *Annu. Rev. Med.* **47,** 201–209.
5. Zoghbi, H. Y. (1996). The expanding world of ataxins. *Nature Genet.* **14,** 237–238.
6. Parrish, J. E., Oostra, B. A., Verkerk, A. J. M. H., Richards, C. S., Reynolds, J., Spikes, A. S., Shaffer, L. G., and Nelson, D. L. (1994). Isolation of GCC repeat showing expansion in FRAXF, and fragile site distal to FRAXA and FRAXE. *Nature Genet.* **8,** 229–235.
7. Nancarrow, J. K., Kremer, E., Holman, K., Eyre, H., Doggett, N. A., LePasiler, D., Callen, D. F., Sutherland, G. R., and Richards, R. I. (1994). Implications of FRA16A for the mechanism of chromosomal fragile site genesis. *Science* **264,** 1938–1941.
8. Nancarrow, J. K., Holman, K., Mangelsdorf, M., Hori, T., Denton, M., Sutherland, G. R., and Richards, R. I. (1995). Molecular basis of $(CCG)_n$ repeat instability at the FRA16A fragile site locus. *Hum. Mol. Genet.* **4,** 367–372.
9. Richards, R. I., and Sutherland, G. R., (1992). Dynamic mutations: a new class of mutations causing human disease. *Cell* **70,** 709–712.
10. Pearson, C. E., and Sinden, R. R. (1996). Alternative structures in duplex DNA formed within the trinucleotide repeats of the myotonic dystrophy and fragile X loci. *Biochemistry* **35,** 5041–5053.
11. Pearson, C. E., Ewel, A., Acharya, S., Fishel, R. A., and Sinden, R. R. (1997). Human MSH2 binds to trinucleotide repeat DNA structures associated with neurodegenerative diseases. *Hum. Mol. Genet.* **6,** 1117–1123.
12. Watson, J. D., and Crick, F. (1953). Molecular structure of nucleic acids: a structure for deoxyribose nucleic acid. *Nature* **171,** 737–738.
13. Platt, J. R. (1955). Possible separation of intertwined nucleic acid chains by transfer-twist. *Proc. Natl. Acad. Sci. USA* **41,** 181–183.
14. Felsenfeld, G., Davies, D. R., and Rich, A. (1957). Formation of a three stranded polynucleotide molecule. *J. Am. Chem. Soc.* **79,** 2023–2024.
15. Gellert, M., Lipsett, M. N., and Davies, D. R. (1962). Helix formation by guanylic acid. *Proc. Natl. Acad. Sci. USA* **48,** 2013–2018.
16. Felsenfeld, G. (1958). Theoretical studies on the interaction of synthetic polyribonucleotides. *Biochim. Biophys. Acta* **29,** 133–144.
17. Streisinger, G., Okada, Y., Emrich, J., Newton, J., Tsugita, A., Terzaghi, E., and Inouye, M. (1966). Frameshift mutations and the genetic code. *Cold Spring Harb. Symp. Quant. Biol.* **31,** 77–84.
18. Hentschel, C. C. (1982). Homocopolymer sequences in the spacer of a sea urchin histone gene repeat are sensitive to S1 nuclease. *Nature* **295,** 714–716.
19. Mace, H. A. F., Pelham, H. R. B. and Travers, A. A. (1983). Association of an S1 nuclease-sensitive structure with short direct repeats 5′ of Drosophila heat shock genes. *Nature* **304,** 555–557.
20. Htun, H., Lund, E., and Dahlberg, J. E. (1984). Human U1 RNA genes contain an unusual nuclease S1 cleavage site within the conserved 3′ flanking region. *Proc. Natl. Acad. Sci. USA* **81,** 7288–7292.
21. McKeon, C. Schmidt, A., and de Crombrugghe B. (1984). A sequence conserved in both the chicken and mouse $\alpha 2$(I). collagen promoter contains sites sensitive to S1 nuclease. *J. Biol. Chem.* **259,** 6636–6640.
22. Shen, C.-K. J. (1984). Superhelicity induces hypersensitivity of a human polypyrimidine polypurine DNA sequence in the human $\alpha 2$–$\alpha 1$ globin intergenic region to S1 nuclease digestion—high resolution mapping of the clustered cleavage sites. *Nucleic Acids Res.* **11,** 7899–7910.
23. Yu, Y.-T., and Manley, J. L. (1986). Structure and function of the S1 nuclease-sensitive site in the adenovirus late promoter. *Cell* **45,** 743–751.
24. Chamberlin, M., and Berg, P. (1962). Deoxyribonucleic acid-directed synthesis of ribonucleic acid by an enzyme from *Escherichia coli. Proc. Natl. Acad. Sci USA* **48,** 81–94.
25. Kornberg, A. Bertsch, L. L., Jackson, J. F., and Khorana, H. G. (1964). Enzymatic synthesis of deoxyribonucleic acid. XVI. Oligonucleotides as templates and the mechanism of their replication. *Proc. Natl. Acad. Sci USA* **51,** 315–323
26. Wells, R. D., Jacob, T. M., Narang, S. A., and Khorana, H. G. (1967). Studies on polynucleotides. LXIX. Synthetic deoxyribonucleotides as templates for the DNA polymerase of *Escherichia coli*: DNA-like polymers containing repeating trinucleotides. *J. Mol. Biol.* **27,** 237–263.
27. Wells, R. D., Buchi, H., Kossel, H., Ohtsuka, E., and Khorana, H. G. (1967). Studies on polynucleotides LXX. Synthetic deoxyribonucleotides as templates for the DNA polymerase of *Escherichia coli*: DNA-like polymers containing repeating tetranucleotides. *J. Mol. Biol.* **27,** 265–272.
28. Baldwin, R. L. (1968). Kinetics of helix formation and slippage of the dAT copolymer, *In* "Symposium on Molecular Associations in Biology" (B. Pullman, Ed.), pp. 145–162. Academic Press, New York.
29. Ratliff, R. L., Hoard, D. E., Hayes, N., Smith, D. A., and Gray, D. M. (1976). Preparation and properties of the repeating sequence polymers $d(A–I–C)_n$-$d(I–C–T)_n$ and $d(A–C–G)_n$-$d(G–C–T)_n$. *Biochemistry* **15,** 168–176.
30. Schlotterer, C., and Tautz, D. (1992). Slippage synthesis of simple sequence DNA. *Nucleic Acids Res.* **20,** 211–215.
31. Behn-Krappa, A., and Doerfler, W. (1994). Enzymatic amplification of synthetic oligoribonucleotides: Implications for triplet repeat expansions in the human genome. *Hum. Mutat.* **3,** 19–24.
32. Pearson, C. E., Zorbas, H., Price, G. B., and Zannis-Hadjopoulos, M. (1996). Inverted repeats, stem-loops, and cruciforms— significance for initiation of DNA replication. *J. Cell Biochem.* **63,** 1–22.
33. Soyfer, V. N., and Potaman, V. N. (1995). "Triple-Helical Nucleic Acids." Springer, New York.
34. Sinden, R. R. (1994). "DNA Structure and Function." Academic Press, San Diego.
35. Panyutin, I. G., Biswas, I., and Hsieh, P. (1995). A pivotal role for the structure of the Holliday junction in DNA branch migration. *EMBO J.* **14,** 1819–1826.
36. Englander, S. W., Kallenbach, N. R., Heeger, A. J., Krumhansl, J. A., and Litwin, S. (1980). Nature of the open state in long polynucleotide double helices: possibility of soliton excitations. *Proc. Natl. Acad. Sci. USA* **77,** 7222–7226.
37. Olivera, B. M., and Lehman, I. R. (1968). Enzymic joining of polynucleotides. III. The polydeoxyadenylate polydeoxythymidylate homopolymer pair. *J. Mol. Biol.* **36,** 261–274.
38. Olivera, B. M., Scheffler, I. E., and Lehman, I. R. (1968). Enzymic joining of polynucleotides. IV. Formation of a circular deoxyadenylate deoxythymidylate copolymer. *J. Mol. Biol.* **36,** 275–285.
39. Rokita, S. E., and Romero-Fredes, L. (1989). Facile interconversion of duplex structures formed by copolymers of d(CG). *Biochemistry* **28,** 9674–9679.
40. Woodson, S. A., and Crothers, D. M. (1987). Proton nuclear resonance studies on bulge-containing DNA oligonucleotides from a mutational hot-spot sequence. *Biochemistry* **26,** 904–912.
41. Woodson, S. A., and Crothers, D. M. (1988). Preferential location of bulged guanosine internal to a G·C tract by 1H NMR. *Biochemistry* **27,** 436–445.

42. Scheffler, I. E., Elson, E. L., and Baldwin, R. L. (1970). Helix formation by d(TA) oligomers. I. Hairpin and straight-chain. *J. Mol. Biol.* **36,** 292–304.
43. Scheffler, I. E., Elson, E. L., and Baldwin, R. L. (1970). Helix formation by d(TA) oligomers. II. Analysis of the helix–coil transitions of linear and circular oligomers. *J. Mol. Biol.* **48,** 145–171.
44. Xodo, L. E., Manzini, G., Quadrifoliglio, F., Yathindra, N., van der Marel, G. A., and van Boom, J. H. (1989). A facile duplex–hairpin interconversion through a cruciform intermediate in a linear DNA fragment. *J. Mol. Biol.* **205,** 777–781.
45. Pouwels, P. H., Knijnenburg, C. M., van Rotterdam, J., and Cohen, J. A. (1968). Structure of the replicative form of bacteriophage phiX174. VI. Studies on alkali-denatured double-stranded phiX DNA. *J. Mol. Biol.* **32,** 169–182.
46. Rush, M. G., and Warner, R. C. (1970). Alkali denaturation of covalently closed circular duplex deoxyribonucleic acid. *J. Biol. Chem.* **245,** 2704–2708.
47. Wetmur, J. G. (1971). Excluded volume effects on the rate of renaturation of DNA. *Biopolymers* **10,** 601–613.
48. Kinberg-Calhoun, J., and Wetmur, J. G. (1981). Circular, but not circularly permuted, deoxyribonucleic acid reacts slower than linear deoxyribonucleic acid with complementary linear deoxyribonucleic acid. *Biochemistry* **20,** 2645–2650.
49. Ji, J., Clegg, N. J., Peterson, K. R., Jackson, A. L., Laird, C. D., and Loeb, L. A. (1996). *In vitro* expansion of GGG:GCC repeats: identification of the preferred strand of expansion. *Nucleic Acids Res.* **24,** 2835–2840.
50. Hochstrasser. R. A., Carver, T. E., Sowers, L. C., and Millar, D. P. (1994). Melting of a DNA helix terminus within the active site of a DNA polymerase. *Biochemistry* **33,** 11971–11979.
51. Kang, S., Jaworski, A., Ohshima, K., and Wells, R. D. (1995). Expansion and deletion of CTG repeats from human disease genes are determined by the direction of replication in *E. coli. Nature Genet.* **10,** 213–218.
52. Shimizu, M., Gellibolian, R., Oostra, B. A., and Wells, R. D. (1996). Cloning, characterization and properties of plasmids containing CGG triplet repeats from the FMR-1 gene. *J. Mol. Biol.* **258,** 614–626.
53. Maurer, D. J., O'Callaghan, B. L., and Livingston, D. M. (1996). Orientation dependence of trinucleotide CAG repeat instability in *Saccharomyces cerevisiae. Mol. Cell. Biol.* **16,** 6617–6622.
54. Freudenreich, C. H., Stavenhagen, J. B., and Zakian, V. A. (1997). Stability of a CTG/CAG trinucleotide repeat in yeast is dependent on its orientation in the genome. *Mol. Cell. Biol.* **17,** 2090–2098.
55. Shildkraut, C. L., Richardson, C. C., and Kornberg, A. (1964). Enzymic synthesis of deoxyribonucleic acid. XVII. Some unusual properties of the product primed by native DNA templates. *J. Mol. Biol.* **9,** 24–45.
56. Inman, R. B., Shildkraut, C. L., and Kornberg, A., (1965). Enzymic synthesis of deoxyribonucleic acid. XX. Electron microscopy of products primed by native templates. *J. Mol. Biol.* **11,** 285–292.
57. Ripley, L. S. (1991). Concerted mutagenesis: Its potential impact on interpretation of evolutionary relationships. *In* "Molecular Evolution of the Major Histocompatibility Complex" (J. Klein and D. Klein, Eds.), pp. 63–94. Springer-Verlag, Berlin/Heidelberg.
58. Kunkel, T. A. (1990). Misalignment-mediated DNA synthesis errors. *Biochemistry* **29,** 8003–8011.
59. Levinson, G., and Gutman, G. A. (1987). Slipped-strand mispairing: a major mechanism for DNA sequence evolution. *Mol. Biol. Evol.* **4,** 203–221.
60. Ripley, L. S. (1990). Frameshift mutation: determinants of specificity. *Annu. Rev. Genet.* **24,** 189–213.
61. Drake, J. W., Glickman, B. W., and Ripley, L. S. (1983). Updating the theory of mutation. *Am. Sci.* **71,** 621–630.
62. Glickman, B. W., and Ripley, L. S. (1984). Structural intermediates of deletion mutagenesis: a role for palindromic DNA. *Proc. Natl. Acad. Sci. USA* **81,** 512–516.
63. Trinh, T. Q., and Sinden, R. R. (1991). Preferential DNA secondary structure mutagenesis in the lagging strand of replication in *E. coli. Nature* **352,** 544–547.
64. Trinh, T. Q., and Sinden, R. R. (1993). The influence of primary and secondary structure DNA structure in deletion and duplication between direct repeats in *Escherichia coli. Genetics* **134,** 409–422
65. Cooper, D. N., and Krawczak, M. (1993). "Human Gene Mutation." Bios Scientific, Oxford, UK.
66. Bebenek, K., Abbotts, J., Roberts, J. D., Wilson, S. H., and Kunkel, T. A. (1989). Specificity and mechanism of error-prone replication by human immunodeficiency virus-1 reverse transcriptase. *J. Biol. Chem.* **264,** 16948–16956.
67. Papanicolaou, C., and Ripley, L. S. (1991). An *in vitro* approach to identifying specificity determinants of mutagenesis mediated by DNA misalignments. *J. Mol. Biol.* **221,** 805–821.
68. Bebenek, K., Abbotts, J., Wilson, S. H., and Kunkel, T. A. (1993). Error-prone polymerization by HIV-1 reverse transcriptase. *J. Biol. Chem.* **268,** 10324–10334.
69. Kaguni, L. S., and Clayton, D. A. (1982). Template-directed pausing in *in vitro* DNA synthesis by DNA polymerase a from *Drosophila melanogaster* embryos. *Proc. Natl. Acad. Sci. USA* **79,** 983–987.
70. Weaver, D. T., and DePamphilis, M. L. (1984). Role of palindromic and nonpalindromic sequences in arresting DNA synthesis *in vitro* and *in vivo. J. Mol. Biol.* **180,** 961–986.
71. Brinton, B. T., Caddle, M. S., and Heintz, N. H. (1991). Position and orientation-dependent effects of a eukaryotic Z-triplex DNA motif on episomal DNA replication in COS-7 cells. *J. Biol. Chem.* **266,** 5153–5161.
72. Samadashwily, G. M., Dayn, A., and Mirkin, S. (1993). Suicidal nucleotide sequences for DNA polymerization. *EMBO J.* **12,** 4975–4983.
73. Baran, N., Lapidot, A., and Manor, H. (1991). Formation of DNA triplexes accounts for arrests of DNA synthesis at $d(TC)_n$ and $d(GA)_n$ tracts. *Proc. Natl. Acad. Sci. USA* **88,** 507–511.
74. Sinden, R. R., Zheng, G., Brankamp, R. G., and Allen, K. N. (1991). On the deletion of inverted repeated DNA in *Escherichia coli*: effects of length, thermal stability, and cruciform formation *in vivo. Genetics* **129,** 991–1005.
75. Kramer, P. R., Stringer, J. R., and Sinden, R. R. (1996). Stability of an inverted repeat in a human fibrosarcoma cell. *Nucleic Acids Res.* **24,** 4234–4241.
76. Rosche, W. A., Trinh, T. Q., and Sinden, R. R. (1995). Differential DNA secondary structuremediated deletion mutation in the leading and lagging strands. *J. Bacteriol.* **177,** 4385–4391.
77. Rosche, W. A., Trinh, T. Q., and Sinden, R. R. (1997). Leading strand specific spontaneous mutation corrects a quasi–palindrome by an intermolecular strand switch mechanism. *J. Mol. Biol.* **269,** 176–187.
78. Lechner, R. L., Engler, M. J., and Richardson, C. C. (1983). Characterization of strand displacement synthesis catalyzed by Bacteriophage T7 DNA polymerase. *J. Biol. Chem.* **258,** 11174–11184.
79. Ohshima, A., Inouye, S., and Inouye, M. (1992). *In vivo* duplication of genetic elements by the formation stem-loop DNA with-

out an RNA intermediate. *Proc. Natl. Acad. Sci. USA* **89,** 1016–1020.

80. Papanicolaou, C., and Ripley, L. S. (1989). Polymerase-specific differences in the DNA intermediates of frameshift mutagenesis. *In vitro* synthesis errors of *Escherichia coli* DNA polymerase I and its large fragment derivative. *J. Mol. Biol.* **207,** 335–353.
81. Sinden, R. R., and Wells, R. D. (1992). DNA structure, mutations, and human genetic disease. *Curr. Opin. Biotechnol.* **3,** 612–622.
82. Wells, R. R., and Sinden, R. R. (1993). Defined ordered sequence DNA, DNA structure, and DNA-directed mutation. In "Genome Analysis, Vol 7, Genome Rearrangement and Stability" (K. Davies and S. T. Warren, Eds.), pp. 107–138. Cold Spring Harbor Laboratory, Cold Spring Harbor, NY.
83. Burhans, W. C., Vassilev, L. T., Caddle, M. S., Heintz, N. H., and DePamphilis, M. L. (1990). Identification of an origin of bidirectional DNA replication in mammalian chromosomes. *Cell* **62,** 955–965.
84. Burhans, W. C., Vassilev, L. T., Wu, J., Sogo, J. M., Nallaseth, F. S., and DePamphilis, M. L. (1991). Emetine allows identification of origins of mammalian DNA replication by imbalanced DNA synthesis, not through conservation nucleosome segregation. *EMBO J.* **10,** 4351–4360.
85. Nethanel, T., Reisfeld, S., Dinter-Gottlieb, G., and Kaufmann, G. (1988). An Okazaki piece of SV40 may be synthesized by ligation of shorter precursor chains. *J. Virol.* **62,** 2867–2873
86. Richards, R. I., and Sutherland, G. R. (1994). Simple repeat DNA is not replicated simply. *Nature Genet.* **6,** 114–116.
87. Bridges, B. A. (1997). DNA turnover and mutation in resting cells. *BioEssays* **19,** 347–352.
88. Chen, X., Mariappan, S. V. S., Catasti, P., Ratliff, R., Moyzis, R. K., Laayoun, A., Smith, S. S., Bradbury, E. M., and Gupta, G. (1995). Hairpins are formed by the single strands of the fragile X triplet repeats: structure and biological implications. *Proc. Natl. Acad. Sci. USA* **92,** 5199–5203.
89. Gacy, A. M., Goellner, G., Juranic, N., Macura, S., and McMurray, C. T. (1995). Trinucleotide repeats that expand in human disease form hairpin structures *in vitro. Cell* **81,** 533–540.
90. Gao, X., Huang, X., Smith, G. K., Zheng, M., and Liu, H. (1995). A new antiparallel duplex motif of DNA CCG repeats that is stabilized by extrahelical bases symetrically localized in the minor groove. *J. Am. Chem. Soc.* **117,** 8883–8884.
91. Mitas, M., Yu, A., Dill, J., and Haworth, I. S. (1995). The trinucleotide repeat sequence $d(CGG)_{15}$ forms a heat-stable hairpin containing Gsyn.Ganti base pairs. *Biochemistry* **34,** 12803–12811.
92. Mitas, M., Yu, A., Dill, J., Kamp, T. J., Chambers, E. J., and Haworth, I. S. (1995). Hairpin properties of single-stranded DNA containing a GC-rich triplet repeat: $(CTG)_{15}$. *Nucleic Acids Res.* **23,** 1050–1059.
93. Petruska, J., Arnheim, N., and Goodman, M. F. (1996). Stability of intrastrand hairpin structures formed by the CAG/CTG class of DNA triplet repeats associated with neurological diseases. *Nucleic Acids Res.* **24,** 1992–1998.
94. Smith, G. K., Jie, J., Fox, G. E., and Gao, X. (1995). DNA CTG triplet repeats involved in dynamic mutations of neurobiologically related gene sequences form stable duplexes. *Nucleic. Acids Res.* **23,** 4303–4311.
95. Yu, A., Dill, J., Wirth, S. S., Huang, G., Lee, V. H., Haworth, I. S., and Mitas, M. (1995). The trinucleotide repeat sequence $d(GTC)_{15}$ adopts a hairpin conformation. *Nucleic Acids Res.* **23,** 2706–2714.
96. Zheng, M. X., Huang, X. N., Smith, G. K., Yang, X. Y., and Gao, X. L. (1996). Genetically unstable CXG repeats are structurally dynamic and have a high propensity for folding—an NMR and UV spectroscopic study. *J. Mol. Biol.* **264,** 323–336.
97. Pearson, C. E., and Sinden, R. R. (1998). Trinucleotide repeat DNA structures: dynamic mutations from dynamic DNA. *Curr. Opin. Struct. Biol.* **8**(3). [In press].
98. Coggins, L. W., and O'Prey, M. (1989). DNA tertiary structures formed *in vitro* by misaligned hybridization of multiple tandem repeat sequences. *Nucleic Acids Res.* **17,** 7417–7426.
99. Coggins, L. W., O'Prey, M., and Akhter, S. (1992). Intrahelical pseudoknots and interhelical associations mediated by mispaired human minisatllite DNA sequences *in vitro. Gene* **121,** 279–285.
100. Hsieh, C.-H., and Griffith, J. D. (1989). Deletions of bases in one strand of duplex DNA, in contrast to single-base mismatches, produc highly kinked molecules: Possible relevance to the folding of single-stranded nucleic acids. *Proc. Natl. Acad. Sci. USA* **86,** 4833–4837.
101. Rice, J. A., and Crothers, D. M. (1989). DNA bending by the bulge defect. *Biochemistry* **28,** 4512–4516.
102. Lilley, D. M. J. (1995). Kinking of DNA and RNA by base bulges. *Proc. Natl. Acad. Sci. USA* **92,** 7140–7142.
103. Pearson, C. E., Ruiz, M. T., Price, G. B., and Zannis-Hadjopoulos, M. (1994). Cruciform DNA binding protein in HeLa cell extracts. *Biochemistry* **33,** 14185–14196.
104. Pearson, C. E., Zannis-Hadjopoulos, M., Price, G. B., and Zorbas, H. (1994). A novel type of interaction between a cruciform DNA structure and a cruciform binding protein from HeLa cells. *EMBO J.* **14,** 1571–1580.
104a. Pearson, C. E., Wang, Y.-H., Griffith, J. D., and Sinden, R. R. (1998). Structural analysis of slipped-strand DNA (S-DNA) formed in $(CTG)_n \cdot (CAG)_n$ repeats from the myotonic dystrophy locus. *Nucleic Acids Res.* **26,** 816–823.
105. Bowater, R. P., Rosche, W. A., Jaworski, A., Sinden, R. R., and Wells, R. D. (1996). Relationship between *Escherichia coli* growth and deletions of CTG·CAG triplet repeats in plasmids. *J. Mol. Biol.* **264,** 82–96.
106. Studier, F. W. (1969). Effects of the conformation of single-stranded DNA on renaturation and aggregation. *J. Mol. Biol.* **41,** 199–209.
107. Broker, T. R. Soll, L., and Chow, L. T. (1977). Underwound loops in self-renatured DNA can be diagnostic of inverted duplications and translocated sequences. *J. Mol. Biol.* **113,** 579–589
108. Chung, M.-Y., Ranum, L. P. W., Duvick, L. A., Servadio, A., Zoghbi, H. Y., and Orr, H. T. (1993). Evidence for a mechanism predisposing to intergenerational CAG repeat instability in spinocerebellar ataxia type I. *Nature Genet.* **5,** 254–258.
109. Chong, S. S., McCall, A. E., Cota, J., Subramony, S. H., Orr, H. T., Hughes, M. R., and Zoghbi, H. Y. (1995). Gametic and somatic tissue-specific heterogeneity of the expanded SCA1 CAG repeat in spinocerebellar ataxia type 1. *Nature Genet.* **10,** 344–350.
110. Eichler, E. E., Holden, J. J. A., Popovich, B. W., Reiss, A. L., Snow, K., Thibodeau, S. N., Richards, C. S., Ward, P. A., and Nelson, D. L. (1994). Length of uninterrupted CGG repeats determines instability in the FMR1 gene. *Nature Genet.* **8,** 88–94.
111. Hirst, M. C., Grewal, P. K., and Davies, K. E. (1994). Precursor arrays for triplet repeat expansion at the fragile X locus. *Hum. Mol. Genet.* **3,** 1553–1560.
112. Kunst, C. B., and Warren, S. T. (1994). Cryptic and polar variation of the fragile X repeat could result in predisposing normal alleles. *Cell* **77,** 853–861.
113. Snow, K., Tester, D. J., Kruckeberg, K. E., Schaid, D. J., and Thibodeau, S. N. (1994). Sequence analysis of the fragile X trinucleotide repeat: implications for the origin of the fragile X mutation. *Hum. Mol. Genet.* **9,** 1543–1551.

114. Zhong, N., Yang, W., Dobkin, C., and Brown, W. T. (1996). Fragile X gene instability: anchoring AGGs and linked microsatellites. *Am. J. Med. Genet.* **59,** 351–361.
115. Eichler, E. E., Hammond, H. A., Macpherson, J. N., Ward, P. A., and Nelson, D. L. (1995). Population survey of the human FMR1 CGG repeat substructure suggests biased polarity for the loss of AGG interruptions. *Hum. Mol. Genet.* **4,** 2199–2208.
116. Eichler, E. E., Macpherson, J. N., Murray, A., Jacobs, P. A., Chakravarti, A., and Nelson, D. L. (1996). Haplotype and interspersion analysis of the FMR1 CGG repeat identifies two different mutational pathways for the origin of the fragile X syndrome. *Hum. Mol. Genet.* **5,** 319–330.
117. Pearson, C. E. Eichler, E. E., Lorenzetti, D., Acharya, S., Kramer, P. R., Kramer, S. F., Nelson, D. L., Zoghbi, H. Y., and Sinden, R. R. (1996). Slipped strand structures in neurodegenerative disease-associated trinucleotide repeats: a role for human mismatch repair and cryptic interruptions. *Am. J. Hum. Genet.,* **59**(4, Suppl.), 244.
117a. Pearson, C. E., Eichler, E. E., Lorenzetti, D., Kramer, S. F., Zoghbi, H. Y., Nelson, D. L., and Sinden, R. R. (1998). Interruptions in the triplet repeats of SCA1 and FRAXA reduce the propensity and complexity of slipped strand DNA (S-DNA) formation. *Biochem.* **37,** 2701–2708.
118. Belotserkovskii, B. P., and Johnston, B. H. (1996). Polypropylene tube surfaces may induce denaturation and multimerization of DNA. *Science* **271,** 222–223.
119. Belotserkovskii, B. P., and Johnston, B. H. (1996). Surface-dependent spontaneous denaturation and multimerization of double stranded DNA. *In* "Biological Structure and Dynamics, Proceedings of the Ninth Conversation." R. H. Sarma and M. H. Sarma, Eds.). pp. 157–164.
120. Fishel, R., Lescoe, M. K., Rao, M. R. S., Copeland, N. G., Jenkins, N. A., Garber, J., Kane, M., and Kolodner, R. (1993). The human mutator gene homolog MSH2 and its association with hereditary nonpolyposis colon cancer. *Cell* **75,** 1027–1038.
121. Modrich, P., and Lahue, R. (1996). Mismatch repair in replication fidelity, genetic recombination, and cancer biology. *Annu. Rev. Biochem.* **65,** 101–133.
122. Fishel, R., Ewel, A., and Lescoe, M. K. (1994). Purified human MSH2 protein binds to DNA containing mismatched nucleotides. *Cancer Res.* **54,** 5539–5542.
123. Fishel, R., Ewel, A., Lee, S., Lescoe, M. K., and Griffith, J. (1994). Binding of mismatched microsatellite DNA sequences by the human MSH2 protein. *Science* **266,** 1403–1405.
124. Umar, A., Boyer, J. C., and Kunkel, T. A. (1994). DNA loop repair by human cell extracts. *Science* **266,** 814–816.
125. Umar, A., and Kunkel, T. A. (1996). DNA-replication fidelity, mismatch repair and genome instability in cancer cells. *Eur. J. Biochem.* **238,** 297–307.
126. Kramer, P. R., Pearson, C. E., and Sinden, R. R. (1996). Stability of triplet repeats of myotonic dystrophy and fragile X loci in human mutator mismatch repair cell lines. *Hum. Genet.* **98,** 151–157.
127. Goellner, G. M., Tester, D., Thibodeau, S., Almqvist, E., Goldberg, Y. P., Hayden, M. R., and McMurray, C. T. (1997). Different mechanisms underlie DNA instability in Huntington disease and colorectal cancer. *Am. J. Hum. Genet.,* **60,** 879–890.
128. Shibata, D., Peinado, M. A., Ionov, Y., Malkhosyan, S., and Perucho, M. (1994). Genomic instability in repeated sequences is an early somatic event in colorectal tumorigenesis that persists after transformation. *Nature Genet.* **6,** 273–281.
129. Wooster, R., Cleton-Jansen, A. M., Collins, N., Mangion, J., Cornelis, R. S., Cooper, C. S., Gusterson, B. A., Ponder, B. A. J., vonDeimling, A., Wiestler, O. D., Cornelisse, C. J., Devilee, P., and Stratton, M. R. (1994). Instability of short tandem repeats (microsatellites) in human cancers. *Nature Genet.* **6,** 152–156.
130. Huddart, R. A., Wooster, R., Horwich, A., and Cooper, C. S. (1995). Microsatellite instability in human testicular germ cell tumours. *Br. J. Cancer* **72,** 642–645.
131. Drummond, J. T., Li, G.-M., Longley, M. J., and Modrich, P. (1995). Isolation of an hMSH2–p160 heterodimer that restores DNA mismatch repair to tumor cells. *Science* **268,** 1909–1912.
132. Acharya, S., Wilson, T., Gradia, S., Kane, M. F., Guerrette, S., Marsischky, G. T., Kolodner, R., and Fishel, R. (1996). hMSH2 forms specific mispair-binding complexes with hMSH3 and hMSH6. *Proc. Natl. Acad. Sci. USA* **93,** 13629–13634.
133. Harvey, S. C. (1997). Slipped structures in DNA triplet repeat sequences: entropic contribution to genetic instabilities. *Biochemistry* **36,** 3047–3049.
134. Bjursell, G., Gussander, E., and Lindahl, T. (1979). Long regions of single-stranded DNA in human cells. *Nature* **280,** 420–423.
135. Micheli, G., Baldari, C. T., Carri, M. T., Di Cello, G., and Buongiorno-Nardelli, M. (1982). An electron microscope study of chromosomal DNA replication in different eukaryotic systems. *Exp. Cell. Res.* **137,** 127–140.
136. Henson, P. (1978). The presence of single-stranded regions in mammalian DNA. *J. Mol. Biol.* **119,** 487–506.
137. Klein, H. L., and Byers, B. (1978). Stable denaturation of chromosomal DNA from *Saccharomyces cerevisiae* during meiosis. *J. Bacteriol.* **134,** 629–635.
138. Conrad, M. N., and Newlon, C. S. (1983). Stably denatured regions in chromosomal DNA from *cdc2 Saccharomyces cerevisiae* cell cycle mutant. *Mol. Cell. Biol.* **3,** 1665–1669.
139. Umek, R. M., and Kowalski, D. (1988). The ease of DNA unwinding as a determinant of initiation at yeast replication origins. *Cell* **52,** 559–567.
140. Hornstra, I. K. Nelson, D. L., Warren, S. T., and Yang, T. P. (1993). High resolution methylation analysis of the FMR1 gene trinucleotide repeat region in fragile X syndrome. *Hum. Mol. Genet.* **2,** 1659–1665.
141. Smeets, H. J. M., Smits, A. P. T., Verheij, C. E., Theelen, J. P. G., Willemsen, R., van de Burgt, I., Hoogeveen, A. T., Oosterwijk, J. C., and Oostra, B. A. (1995). Normal phenotype in two brothers with a full FMR1 mutation. *Hum. Mol. Genet.* **4,** 2103–2108.
142. Mingroni-Netto, R. C., Fernandes, J. G., and Vianna-Morgante, A. M. (1994). Relationship of expansion of CGG repeats and X-inactivation with expression of Fra(X)(q27.3) in heterozygotes. *Am. J. Med. Genet.* **51,** 443–446.
143. Jalal, S. M., Lindor, N. M., Michels, V. V., Buckley, D. D., Hoppe, D. A., Sarkar, G., and Dewald, G. W. (1993). Absence of chromosome fragility at 19q13.3 in patients with myotonic dystrophy. *Am. J. Med. Genet.* **46,** 441–443.
144. Wenger, S. L., Giangreco, C. A., Tarleton, J., and Wessel, H. B. (1996). Inability to induce fragile sites at CTG repeats in congenital myotonic dystrophy. *Am. J. Med. Genet.* **66,** 60–63.
145. Kafri, T., Gao, X., and Razin, A. (1993). Mechanistic aspects of genome-wide demethylation in the preimplantation mouse embryo. *Proc. Natl. Acad. Sci. USA* **90,** 10558–10562.
146. Brandeis, M., Ariel, M., and Cedar, H. (1993). Dynamics of DNA methylation during development. *Bioessays* **15,** 709–713.
147. Zacharias, W. (1993). Methylation of cytosine influences DNA structure. *EXS.* **64,** 27–38.
148. Laayoun, A., and Smith, S. S. (1995). Methylation of slipped duplexes, snapbacks and cruciforms by human DNA(cytosine-5) methyltransferase. *Nucleic Acids Res.* **23,** 1584–1589.
149. Imbert, G., Saudou, F., Yvert, G., Devys, D., Trottier, Y., Garnier, J.-M., Wever, C., Mandel, J.-L., Cancel, G., Abbas, N.,

Durr, A., Didierjean, O., Stevanin, G., Agid, Y., and Brice, A. (1996). Cloning of the gene for spinocerebellar ataxia 2 reveals a locus with high sensitivty to expanded CAG/glutamine repeats. *Nature Genet.* **14,** 285–291.

150. Pulst, S.-M., Nechiporuk, A., Nechiporuk, T., Gispert, S., Chen, X.-N., Lopes-Cendes, I., Pearlman, S., Starkman, S., Orozco-Diaz, G., Lunkes, A., DeJong, P., Rouleau, G. A., Auburger, G., Korenberg, J. R., Figueroa, C., and Sahba, S. (1996). Moderate expansion of a normally biallelic trinucleotide repeat in spinocerebellar ataxia type 2. *Nature Genet.* **14,** 269–276.

151. Sanpei, K., Takano, H., Igarashi, S., Sato, T., Oyake, M., Sasaki, H., Wakisaka, A., Tashiro, K., Ishida, Y., Ikeuchi, T, Koide, R., Saito, M., Sato, A., Tanaka, T., Hanyu, S., Takiyama, Y., Nishizawa, M., Shimizu, N., Nomura, Y., Segawa, M., Iwabuchi, K., Eguchi, I., Tanaka, H., Takahashi, H., and Tsuji, S. (1996). Identification of the spinocerebellar ataxia type 2 gene using a direct identification of repeat expansion and cloning technique, DIRECT. *Nature Genet.* **14,** 277–284.

152. Leeflang, E. P., and Arnheim, N. (1995). A novel repeat structure at the myotonic locus in a 37 repeat allele with unexpectedly high stability. *Hum. Mol. Genet.* **4,** 135–136.

153. Cossee, M., Schmitt, M., Campuzano, V., Reutenauer, L., Moutou, C., Mandel, J-L. and Koenig, M. (1997). Evolution of the Friedreich's ataxia trinucleotide repeat expansion: Founder effect and premutations. *Proc. Natl. Acad. Sci.* **94,** 7452–7457.

154. Montermini, L., Anderman, E., Richter, A., Pandolfo, M., Cavalcanti, F., Pianese, L., Iodice, L., Farina, G., Monticelli, A., Turano, M., Filla, A., De Michele, G., and Cocozza, S. (1997). The Friedreich ataxia GAA triplet repeat: premutation and normal alleles. *Hum. Mol. Genet.* **6,** 1261–1266.

155. Weber, J. L. (1990). Informativeness of human (dC–dA)·(dG–dT)·polymorphisms. *Genomics* **7,** 524–530.

156. Richard, G. F., and Dujon, B. (1996). Distribution and variability of trinucleotide repeats in the genome of the yeast *Saccharomyces cerevisiae. Gene* **174,** 165–174

157. Petes, T. D., Greenwell, P. W., and Dominska, M. (1997). Stabilization of microsatellite sequences by variant repeats in the yeast *Saccharomyces cerevisiae. Genetics* **146,** 491–498

158. Nag, D. K., White, M. A., and Petes, T. D. (1989). Palindromic sequences in heteroduplex DNA inhibit mismatch repair in yeast. *Nature* **340,** 318–320.

159. Wu, C.-I., and Maeda, N. (1987). Inequality in mutation rates of the two strands of DNA. *Nature* **327,** 169–170.

160. Bohr, V. A. (1991). Gene specific DNA repair. *Carcinogenesis* **12,** 1983–1992.

161. Reiss, A. L., Kazazian, H. H., Jr., Krebs, C. M., McAughan, A., Boehm, C. D., Abrams, M. T., and Nelson, D. L. (1994). Frequency and stability of the fragile X premutation. *Hum. Mol. Genet.* **3,** 393–398.

162. Quan, F., Janas, J., and Popovich, B. W. (1996). A novel repeat configuration in the SCA1 gene—implications for the molecular diagnostics of spinocerebellar ataxia type 1. *Hum. Mol. Genet.* **4,** 2411–2413.

163. Cancel, G., Durr, A., Didierjean, O., Imbert, G., Burk, K., Lezin, A., Belal, S., Benomar, A., Abada-Bendib, M., Vial, C., Guimaraes, J., Chneiweiss, H., Stevanin, G., Yvert, G., Abbas, N., Saudou, F., Lebre, A.-S., Yahyaoui, M., Hentati, F., Vernant, J.-C., Klockgether, T., Mandel, J.-L., Agid, Y., and Brice, A. (1997). Molecular and clinical correlations in spinocerebellar ataxia 2: a study of 32 families. *Hum. Mol. Genet.* **6,** 709–715.

164. Geschwind, D. H., Perlman, S., Figueroa, C. P., Treiman, L. J., and Pulst, S. N. (1997). The prevalence and wide clinical spectrum of the spinocerebellar ataxia type 2 trinucleotide repeat in patients with autosomal dominant cerebellar ataxia. *Am. J. Hum. Genet.* **60,** 842–850.

165. Kawaguchi, Y., Okamoto, T., Taniwaki, M., Aizawa, M., Inoue, M., Katayama, S., Kawakami, H., Nakamura, S., Nishimura, M., Akigushi, I., Kimura, J., Narumiya, S., and Kakizuka, A. (1994). CAG expansions in a novel gene for Machado-Joseph disease at chromosome 14q32.1. *Nature Genet.* **8,** 221–228.

166. Limprasert, P., Nouri, N., Heyman, R. A., Nopparatana, C., Kamonslip, M., Deninger, P. L., and Keats, J. B. (1996). Analysis of CAG repeat of the Machado-Joseph gene in human, chimpanzee and monkey populations: a variant nucleotide is associated with the number of CAG repeats. *Hum. Mol. Genet.* **5,** 207–213.

167. Igarashi, S., Takiyama, Y., Cancel, G., Rogaeva, E. A., Sasaki, H., Wakisaka, A., Zhou, Y.-X., Takano, H., Endo, K., Sanpei, K., Oyake, M., Tanaka, H., Stevanin, G., Abbas, N., Durr, A., Rogaev, E. I., Sherrington, R., Tsuda, T., Ikeda, M., Cassa, E., Nishizawa, M., Benomar, A., Julien, J., Weissenbach, J., Wang, G.-X., Agid, Y., St. George-Hyslop, P. H., Brice, A., and Tsuji, S. (1996). Intergenerational instability of the CAG repeat of the gene for Machado Joseph disease (*MJD1*) is affected by the genotype of the normal chromosome: implications for the molecular mechanisms of the instability of the CAG repeat. *Hum. Mol. Genet.* **5,** 923–932.

168. Matsumura, R., Takayanagi, T., Murata, K., Futamura, N., Hirano, M., and Ueno, S. (1996). Relationship of $(CAG)_n$ configuration to repeat instability of the Machado-Joseph disease gene. *Hum. Genet.* **98,** 643–645.

169. O'Hoy, K. L., Tsilfidis, C., Mahadevan, M. S., Neville, C. E., Barcelo, J., Hunter, A. G. W. and Korneluk, R. G. (1993). Reduction in size of the myotonic dystrophy trinucleotide repeat mutation during transmission. *Science* **259,** 809–812.

170. Ouweland, A. M. W., Deelen, W. H., Kunst, C. B., Uzielli, M.-L. G., Nelson, D. L., Warren, S. T., Oostra, B. A. and Halley, D. J. J. (1994). Loss of mutation at the FMR1 locus through multiple exchanges between maternal X chromosomes. *Hum. Mol. Genet.* **3,** 1823–1827.

171. Barcelo, J. M., Mahadevan, M. S., Tsilfidis, C., MacKenzie, A. E., and Korneluk, R. G. (1993). Intergenerational stability of the myotonic dystrophy protomutation. *Hum. Mol. Genet.* **2,** 705–709.

172. Yamagata, H., Miki, T., Sakoda, S.-I., Yamanaka, N., Davies, J., Shelbourne, P., Kubota, R., Takenaga, S., Nakagawa, M., Ogihara, T., and Johnson, K. (1994). Detection of a premutation in Japanese myotonic dystrophy. *Hum. Mol. Genet.* **3,** 819–820.

173. Huntingtons Collaborative (1993). A novel gene containing a trinucleotide repeat that is expanded and unstable on Huntington's disease chromosomes. *Cell* **72,** 971–983.

174. Rubinsztein, D. C., Barton, D. E., Davison, B. C. C. and Ferguson-Smith, M. A. (1993). Analysis of the *huntingtin* gene reveals a trinucleotide-length polymorphism in the region of the gene that contains two CCG-rich stretches and a correlation between decreased age of onset of Huntington's disease and CAG repeat number. *Hum. Mol. Genet.* **2,** 1713–1715.

175. Rubinsztein, D. C., Leggo, J., Barton, D. E., Ferguson-Smith, M. A. (1993). Site of (CCG) polymorphism in the HD gene. *Nature Genet.* **5,** 214–215.

176. Andrew, S. E., Goldberg, Y. P., Theilmann, J., Zeisler, J., and Hayden, M. R. (1994). A CCG repeat polymorphism adjacent to the CAG repeat in the Huntington disease gene: implications for diagnostic accuracy and predictive testing. *Hum. Mol. Genet.* **3,** 65–67.

177. Goldberg, Y. P., McMurray, C. T., Zeisler, J., Almqvist, E., Sillence, D., Richards, F., Gacy, A. M., Buchanan, J., Telenius, H., and Hayden, M. R. (1995). Increased instability of intermedi-

ate alleles in families with sporadic Huntington disease compared to similar sized intermediate alleles in the general population. *Hum. Mol. Genet.* **4,** 1911–1918.

178. Chong, S. S., Almquist, E., Telenius, H., LaTray, L., Nichol, K., Bourdelat-Parks, B., Goldberg, Y. P., Haddad, B. R., Richards, F., Sillence, D., Greenberg, C. R., Ives, E., Van den Engh, G., Hughes, M. R., and Hayden, M. R. (1997). Contribution of DNA sequence and CAG size to mutation frequencies of intermediate alleles for Huntington disease: evidence from single sperm analyses. *Hum. Mol. Genet.* **6,** 301–309.

179. Rubinsztein, D. C., Leggo, J., Coles, R., Almqvist, E., Biancalana, V., Cassiman, J.-J., Chotai, K., Connarty, M., Craufurd, D., Curtis, A., Curtis, D., Davidson, M. J., Differ, A.-M., Dode, C., Dodge, A., Frontali, M., Ranen, N. G., Stine, O. C., Sherr, M., Abbott, M. H., Franz, M. L., Graham, C. A., Harper, P. S., Hedreen, J. C., Jackson, A., Kaplan, J.-C., Losekoot, M., MacMillan, J. C., Morrison, P., Trottier, Y., Novelletto, A., Simpson, S. A., Theilmann, J., Whittaker, J. L., Folstein, S. E., Ross, C. A., and Hayden, M. R. (1996). Phenotypic characterization of individuals with 30–40 CAG repeats in the Huntington disease (HD) gene reveals HD cases with 36 repeats and apparently normal elderly individuals with 36–39 repeats. *Am. J. Hum. Genet.* **59,** 16–22.

180. McNeil, S. M., Novelletto, A., Srinidhi, J., Barnes, G., Kornbluth, I., Altherr, M. R., Wasmuth, J. J., Gusella, J. F., MacDonald, M. E., and Myers, R. H. (1997). Reduced penetrance of the Huntington's disease mutation. *Hum. Mol. Genet.* **6,** 775–779.

181. Myers, R. H., MacDonald, M. E., Koroshetz, W. J., Duyao, M. P., Ambrose, C. M., Taylor, S. A. M., Barnes, G., Srinidhi, J., Whaley, W. L., Lazzarini, A. M., Schwarz, M., Wolff, G., Bird, E. D., Vonsattel, J.-P. G., and Gusella, J. F. (1993). *De novo* expansion of a $(CAG)_n$ repeat in sporadic Huntington's disease. *Nature Genet.* **5,** 168–173.

182. Goldberg, Y. P., Kremer, B., Andrew, S. E., Theilmann, J., Graham, R. K., Squitieri, F., Telenius, H., Adam, S., Sajoo, A., Starr, E., Heilberg, A., Wolff, G., and Hayden, M. R. (1993). Molecular analysis of new mutations for Huntington's disease: intermediate alleles and sex of origin effects. *Nature Genet.* **5,** 174–179.

183. Campuzano, V., Montermini, L., Molto, M. D., Pianese, L., Cossee, M., Cavalcanti, F., Monros, E., Rodius, F., Duclos, F., Monticelli, A., Zara, F., Canizares, J., Koutnikova, H., Bidichandani, S. I., Gellera, C., Brice, A., Trouillas, P., De Michele, G., Filla, A., De Frutos, R., Palau, F. , Patel, P. I., Di Donato, S., Mandel, J.-L., Cocozza, S., Koenig, M., and Pandolfo, M. (1996). Friedreich's ataxia: autosomal recessive disease caused by an intronic GAA triplet repeat expansion. *Science* **271,** 1423–1427.

CHAPTER 40

Structure and Dynamics of Single-Stranded Nucleic Acids Containing Trinucleotide Repeats

XIAOLIAN GAO Department of Chemistry, University of Houston, Houston, Texas; and Department of Biochemistry and Biophysical Sciences, University of Houston, Houston, Texas

XUENING HUANG Department of Chemistry, University of Houston, Houston, Texas

G. KENNETH SMITH Department of Biochemistry and Biophysical Sciences, University of Houston, Houston, Texas

MINXUE ZHENG Department of Chemistry, University of Houston, Houston, Texas

I. INTRODUCTION

The compilation of this book concerning triplet repeat (TR)- [1] related hereditary neurodegenerative diseases testifies to the rapid growth of a new research area, in which the combined efforts of various fields of biological studies are directed to the understanding of the mechanisms and the molecular bases of these diseases. Although some of the TR-related neurological diseases are well-known [2], such as fragile-X syndrome and Huntington's disease, the other lesser known diseases, such as Friedreich's ataxia, appear to stem from similar etiologies by presenting an expanded number of particular TRs. The facts that there are only certain types of triplet repeats, i.e., CXG (X = A, C, G, T) and GAA/TTC, found within the disease-related genes and that the increasing number of repeats correlates with the onset and severity of the diseases are rather intriguing. This defined genetic pattern links DNA sequences and their chain length with the causes of neurological disorders and has led many to ponder the role of DNA structure and dynamics in the process of TR expansion [3–5]. For instance, it is conceivable that certain DNA motifs have the potential to form local structures and, thus, cause strand slippage in cellular replication or recombination processes, leading to chain length expansion or deletion. These localized structures may block replication enzymes [6] or present unique features for protein recruitment [7–13], leading to abnormal gene function [14, 15]. The focus of this article is NMR and UV characterization of DNA CXG and GXC (X = A and T) TR oligonucleotides in solution. These studies provide molecular details concerning strand association and base pairing of the studied sequences and form a basis for future studies of the properties and behavior of long stretches of TR sequences.

The rapid accumulation of the three-dimensional resolution structures of high-order DNA oligonucleotides in solution and in crystalline states [16–18] has impacted our views of DNA architecture. A variety of sequence alignments including antiparallel, parallel, diagonal antiparallel, or diagonal parallel alignments, in combination with *syn*- or anti-glycosidic base orientation, have been identified in various multistranded helices [16, 17]. Structural studies have been extended to detail the arrangement of non-Watson–Crick (W–C) base pairs (mismatches) in various sequence contexts. It is well established that mismatched base pairs can form in either parallel or antiparallel fashion stabilized by H bonds [19]. Unusual mismatch base pairs and their conformations, such as antiparallel tandem sheared G◇A or purine◇purine mismatches [20, 21], an antiparallel A◇A mismatch in tandem with a G◇A mismatch [22], and parallel, continuous C^+◇C base pairs [23], have recently been characterized. These results demonstrated an increasingly expanding array of structural topologies that are adopted by cellular DNA and improved our information base for elucidation of the structure–function relationship of DNA molecules.

An area which has received a resurgence of attention is the folding topology and structure of nucleic acid repeat sequences. These are sequences containing tracks of 1–6 nucleotides repeated in tandem interspersed throughout human genomic DNA [24, 25]. These sequences exhibit chain length variations (most frequently expansions) in successive generations, a form of gene mutation that has been linked to neurological disorders [26], tumor formation and growth (telomeric sequences) [27], and prostate cancer (shorter CAG repeats) [28]. Research concerning the structure and function of microsatellite repeat sequences has revealed novel, complex folding motifs adopted by N_1G_n or N_1C_n (N = A or T; l, n = 2–4) in chromosomal telomeric sequences of several eukaryotic species [17]. T_1C_n sequences are present as interdigitated tetraplexes formed by antiparallel association of two parallel duplexes. T_nG_2, T_nG_3, and T_nG_4 repeats tend to associate into four stranded structures, which differ greatly in strand orientation and conformational details [17]. The most striking structure of a N_1G_n sequence, d(T_2G_4)$_4$, is an intramolecular, quadruple-turn quadruplex (versus a commonly seen triple-turn quadruplex) [29]. The structures of these G- or C-rich sequences highlights the unique molecular surface of telomeric DNA and its possible interactions with specific recognition proteins. These telomeric DNA–protein interactions play a critical role in controlling cell aging and tumor growth processes [30]. Although G-rich like telomeric DNA, the pentanucleotide repeat centromeric sequences, d(TGGAA)$_n$ or (AATGG)$_n$, (n = 2–6) have been found to form antiparallel duplexes rather than quadruplexes [31, 32]. The most interesting feature of these duplexes is the formation of tandem mismatches within the GGA motif. This structure is stabilized by sheared G◇A base pairs and also by either a G◇G mismatch [31] or interstrand stacking of unpaired G residues [32]. Another example of repeat DNA is dinucleotide repeats, such as $d(CT)_n$, which have been observed in antiparallel duplexes in a pH range of 2.9–7.5 [33], and $d(GA)_n$ sequences, which adopt a homopairing scheme forming G◇G and A◇A pairs when forced into a parallel orientation or when the repeat number exceeds 10 [34, 35]. Only in the last few years has the trinucleotide repeats or TR sequences attracted intensive attention (*vide infra*).

II. AN OVERALL REVIEW

The progress of biophysical studies of DNA TR sequences has been greatly accelerated since the discovery of the genetic link between these DNA sequences and a number of neurological disorders. Although only limited high-resolution structural information of TR sequences exists, it has been recognized that certain repeat sequences may form stable structures which, while lacking perfect W–C complementarity, may be essential to cellular function [3, 36]. Indeed, our initial examination of the CXG and the GXC TR oligonucleotides revealed many distinct NMR spectral features, indicating the presence of versatile structures containing normal W–C base pairs as well as various unusual forms of base pairing and alignment. In this section, we will first summarize the results of TR sequence studies reported by several research groups. The discussion will be followed by the results obtained in this laboratory [37–39]. The content of this section is restricted to sequences consisting of 2–30 repeats, with an emphasis on repeat number, n, <15. Structural and conformational studies of longer TR sequences have been thoroughly discussed by Wells and Sinden in this book. Interested readers are also referred to Chapter 41 for additional NMR studies of TR sequences.

A. The Various Factors That Influence the Experimental Results

Studies using a variety of methods for *in vitro* characterization of TR oligonucleotides have generated controversial results [4, 36]. This is not surprising considering a plethora of factors which may influence conformations and structures of TR sequences. These factors include: (a) Sequence design. For instance, $d(CCG)_n$ may be permuted into $d(CGC)_n$ or $d(GCC)_n$ and these changes, at the oligomer level, lead to different structures and/or strand alignments. Furthermore, nucleotides interrupting TRs may cause significant alteration in the nature of the interactions between residues and strands, resulting in formation of structures which are different from those formed by noninterrupted TR sequences. (b) Experimental models. The interpretations of spectroscopic data, such as UV and CD data, are model-dependent. However, it is often that such information is not available and thus, assumptions are used. Nucleic acid structures are highly sensitive to environmental factors, such as pH, temperature, sample concentration, ion species, and ion concentration. Incorrect assumptions may be misleading. A typical example is that a system present in a conformational equilibrium often produces ambiguous physiochemical parameters. (c) Sample purity. Long, homologous sequences and G-rich sequences are notoriously difficult to be purified. Sample heterogeneity is a common problem and tends to generate incomprehensible results. (d) Sample preparation. Slow kinetics of conformation changes has been observed in some sequences, such as $d(CGG)_n$. Therefore, experimental observations are affected by the "history" of the samples studied, such as freezing, boiling, or preconditioning samples. If not carefully controlled, the results of the experiments may not accurately reflect the equilibrium properties of the molecule studied. (e) Experimental conditions. NMR experiments require samples which are 50–100 times more concentrated than those used in UV measurements, which, in turn, use more concentrated samples than gel electrophoresis. Different experimental methods use different media, such as gel matrix for electrophoresis, which can greatly alter sample conditions, thus affecting experimental results. Despite these complications, the intense effort in the studies of TR sequences has led to significant advancement in our understanding of the basic behaviors of these molecules. Several structural models have been proposed to explain the experimental results.

B. The Hairpin Models

$d(CXG)_n$ and $d(GXC)_n$ (n = 10, 15, 25, or 30) sequences containing a short stretch of complementary or partially self-complementary oligonucleotides on either end of each sequence have been extensively studied in the presence of Tris-HCl (50 mM), $MgCl_2$ (2–10 mM), NaCl (~50 mM), spermine (0.1 mM), and EDTA (0.1 mM) at pH 7.0–8.5 [36]. The general strategy of these studies is to probe the location of flexible residues within a long sequence using chemical and enzymatic reagents that are sensitive to single-stranded residues in combination with electrophoretic assays. The chain folding and base pair formation can then be derived from electrophoretic gel mobility patterns. Typically, flexible, non-base-paired A, C, G, and T residues (i.e., those in loops or in extrahelical forms) are identified by their reaction with diethyl pyrocarbonate, hydroxyl amine, dimethyl sulfate, and $KMnO_4$, respectively. The reactions are followed by hydrolysis to produce chain-cleaved fragments. P1 nuclease is also used to cleave single-stranded regions, which are detected by characteristic electrophoretic gel patterns. These studies have detected unstacked residues, which are supposedly W–C base paired, in the central region of the TR sequence studied, whereas other residues are protected. These patterns are consistent with intramolecular folding into hairpins, which contain loop residues and a base-paired duplex stem. The noncomplementary residues, A, C, G, or T, in the stem region of the hairpin sequences manifest different conformational features. The T residues in CTG and GTC repeats and the A residues in GAC repeats are not exposed to chemical modifications and thus, these residues are considered in a stacked conformation. The A residues in CAG repeats and G residues in CGG repeats are partially exposed to chemical modifications and, thus, these residues are relatively more flexible and are not well-stacked in the helix.

The relative stabilities of these hairpin structures were compared by electrophoretic mobility melting profiles and the melting temperatures, T_m, follow a descending order of $d(CGG)_n$ (75°C) > $d(GAC)_n$ (49°C) ~ $d(CTG)_n$ (48°C) > $d(GTC)_n$ (38°C) ~ $d(CAG)_n$ (38°C) > $d(CCG)_n$ (30°C) (n = 15, metal ion concentration = 2 mM, pH 8.5). The longer sequences showed a similar trend: $d(CGG)_n$ (75°C) > $d(GAC)_n$ (54°C) ~ $d(CTG)_n$ (52°C) > $d(CAG)_n$ (50°C) (n = 25, 0.1 M NaCl, 0.1 mM EDTA, pH 7.0) [40]. UV studies conducted in solutions containing 0.17 M sodium phosphates, 1 mM EDTA, pH 7.0, however, revealed a somewhat different order: $d(CTG)_n$ (66°C) > $d(GAC)_n$ (62°C) > $d(CAG)_n$ (60°C) > $d(GTC)_n$ (53°C) (n = 10) [41]. The T_m's for $d(CAG)_n \cdot d(CTG)_n$ and $d(GAC)_n \cdot d(GTC)_n$ are 83°C. Although consistent in that $d(CTG)_n$ duplexes are more stable than $d(GTC)_n$ duplexes, there are discrepancies in the relative stabilities of $d(GAC)_n$ versus $d(CTG)_n$ duplexes and $d(CAG)_n$ versus $d(GTC)_n$ duplexes. These differences, however, may be explained by the pH dependence of the structures of the GAC repeats and the difficulties in T_m derivation of the CAG repeats due to their broad temperature transitions (*vide infra*) [38, 39].

Hairpin structure formation by the $d(CXG)_n$ oligonucleotides has also been advocated in several other publications [40–42]. $d(CTG)_{25}$ and $d(CAG)_{25}$ reportedly form unimolecularly folded hairpins [40], based on their concentration-independent NMR spectra and the observation of stable W–C C · G base imino proton (NH) and broad loop NH resonances. NOEs between G and T NHs were detected in the $d(CTG)_{25}$ hairpin and, thus, the T residues were considered to be stacked in the helix as T◇T mismatches. Further evidence of hairpin formation was demonstrated by complex formation of antibiotic actinomycin D with a homoduplex $d[G(CTG)_nC]$ or $d[G(CAG)_nC]$ [43]. The n = 2 homoduplexes bind two molecules of the antibiotics and the complexes thus formed were unable to form complementary heteroduplexes upon heat-induced reannealing. These complexes have been proposed as potential models for the stabilization of cruciform formation.

C. Noncanonical DNA Structures Proposed for TR Sequences

1. Parallel Homoduplexes [18]

NMR studies have recently demonstrated that a heptamer strand d(CGACGAC) containing the CGA or $n = 2$ GAC triplet repeats [44] and several DNA oligonucleotides containing a 5′-CGA trinucleotide segment [34, 39, 45, 46] form parallel homoduplexes under acidic pH. It was suggested that the CGA is a motif capable of promoting parallel duplex formation by initial protonation of the 5′-terminal C [45]. The apparent pK_a for C protonation in the parallel d(CGACGAC) duplex is as high as 6.8. The structures of these duplexes have been determined in high resolution to reveal salient features of the parallel strand recognition in DNA. As part of our effort to elucidate the properties of the CXG and GXC triplet repeats, we have systematically examined $d(GAC)_n$ (n = 2–4) and $d(GAC)_3$-xx-$d(GAC)_3$. The structure of the parallel $d(GAC)_3$ duplex has been determined using NMR restrained distance geometry (DG) and molecular dynamics simulation (MDS) methods. The structure of the corresponding antiparallel duplex has also been modeled. These results and the comparison of the parallel $d(GAC)_n$ with the antiparallel $d(GAC)_n$ homoduplex will be discussed in a later section.

2. Quadruplexes

In addition to the duplex forms discussed above, quadruplexes formed by $d(CGG)_n$ and 5Me-$d(CGG)_n$ were proposed to be responsible for the chain length expansion of the CGG TR sequences [47–49]. Electrophoretic gel mobility studies of the native sequences demonstrated that in nondenaturing polyacrylamide gel these sequences migrated, in a monomeric form, either slower [$d(CGG)_7$, 5Me-$d(CGG)_{5\text{-}7}$] [47] or faster [$d(CGG)_{20}$] [48] than random marker sequences. These structures are stabilized by K^+ ion and G N7, and C N3 and N4 are protected from chemical modification, which are indicative of H-bond formation at these sites. These features are typically observed in quadruplexes and thus were used to support the conclusion that $d(CGG)_n$ may form quadruplexes. An NMR structural study [49] of d($\underline{\underline{G}}CG\underline{G}TTT\underline{\underline{G}}CG\underline{G}$), which contains two interrupted

CGG trinucleotides, in the presence of 0.15 *M* NaCl revealed the formation of a quadruplex consisting of dimerization of intramolecular hairpins. In the hairpin the underlined G residues align to form alternating *syn*- (double underline) and *anti*-G◇G mismatches and the central CG residues pair to form W–C base pairs. Dimerization of the hairpins results in formation of two stable CGCG quartets, H-bonded in the major groove, flanked by G_4 quartets. This set of unique quadruple base alignments involving CGG trinucleotides suggests the possibility that longer CGG TR sequences may form stable quadruplexes. However, their existence in $n > 1$ CGG repeat oligonucleotides remains to be proved.

3. Compact and Triad-DNA Model

The hairpin and quadruplex models have been challenged by alternative structural models. An unspecified compact structure, which is neither a quadruplet nor a hairpin was presented based on the studies of $d(CXG)_n$ ($n = 5$) and related oligonucleotides using chemical modification reactions, electrophoresis, and UV spectroscopy [50]. These studies argued that gel mobility retardation does not necessarily prove quartet formation, since it is known that intramolecular RNA quadruplex exhibits accelerated mobility. Additionally, chemical protection may be due to a variety of structural reasons and thus is not conclusive of a particular structure formation. A rather striking structural model, triad-DNA, was proposed based upon computational modeling results [51]. This structure features a double helix formed by stacked base triples, each of which consists of two bases from one strand and one base from the other strand (Fig. 40-1A). Thus, each trinucleotide segment in the duplex forms two stacked base triples, resulting in a zigzag backbone conformation. The formation of triad base triples as proposed may be limited to CGG and CAG repeats, which would form base triples similar to antiparallel purine-purine-pyrimidine base triples found in many triplexes [16]. The corresponding triad base triples produced by two CCG or two CTG strands would not be stable.

III. STRATEGIES AND EXPERIMENTS USED IN CHARACTERIZATION OF TR OLIGONUCLEOTIDES

A. Strategies Applied in the Studies of Triplet Repeat Sequences

A fundamental step toward the understanding of the molecular mechanisms of triplet expansion is to gain

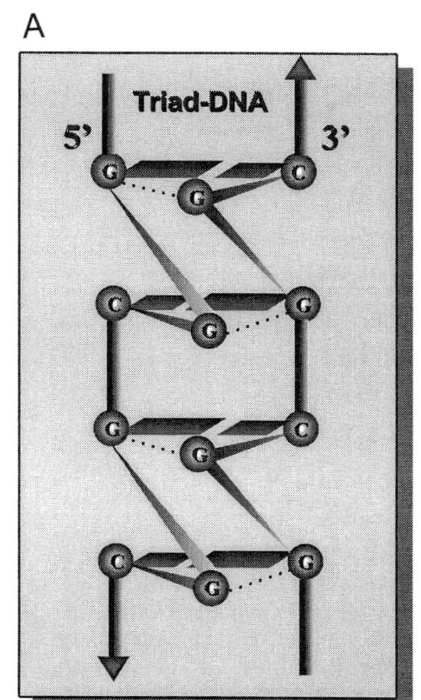

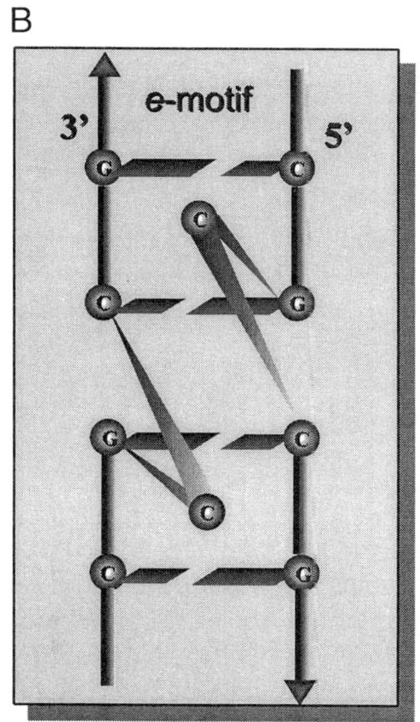

FIGURE 40-1 Schematic drawings of unusual DNA structures. (A) The triad-DNA structure proposed by Kuryavyi and Jovin [51]. Dashed lines indicate interstrand H-bonding similar to that observed in a DNA triplex [16]. (B) The *e*-motif DNA duplex formed by $d(CCG)_2$ [37]. A major difference between (A) and (B) is that extrahelical residues are in the major groove and form H bonds in (A) and are in the minor groove and make hydrophobic contacts in (B).

insights into the structural preferences and solution properties of TR sequences. Toward this end we have investigated a series of d(CXG)$_n$ (n = 2–20), d(GXC)$_n$ (n = 2–6), and their permuted or analogous sequences have been examined in this laboratory (Table 40-1) [37–39]. In the following, we discuss the rationale and general methods used in these studies and summarize the major results. The studies of these seemingly simple single-stranded TR sequences required a combined use of NMR, UV, and other spectroscopic or biophysical methods. These sequences are non-self-complementary and, thus, possible structures formed are not restricted to canonical helices. Furthermore, since TR sequences are highly repetitive, the spectral resonances are narrowly distributed, and thus spectral resolution is rather limited. To circumvent these problems, our studies of the CXG and GXC TR sequences were designed to begin with a minimal number of repeats (n = 2–4). The studies of these truncated versions of the longer TR sequences establish spectroscopic benchmarks, which then permit spectral analyses of the structural and dynamic properties of longer TR sequences. Specifically, high-resolution NMR methods have been applied to determine the strand alignment (parallel or antiparallel), base orientation (*syn* or *anti*) and pairing (H-bonding format), relative stability, and dynamics of TR oligonucleotides. These results were analyzed in combination with those obtained from UV and electrophoresis experiments to determine the strand stoichiometry and relative stability of these sequences. In situations in which even short TR oligonucleotides did not give sufficient spectral resolution, information has been derived from analog substituted sequences, such as inosine (I) substitution as a G analog, the backbone linker sequences, and deuterium-substituted sequences. These analogous sequences greatly simplified NMR spectra, providing an effective means for detailed characterization of TR sequences.

TABLE 40-1 List of the Sequences Studied[a]

Sequence	n
CCG	
d(CCG)$_n$	2–10, 15
r(CCG)$_n$	2
d(CGC)$_n$	2
3′-d(GCC)$_n$5′-xx-5′d(CCG)$_n$-3′	2
d(CGCCGAC)$_n$	1
d(GACGCCG)$_n$	1
^{15}N-labeled d(CCG)$_n$[b]	5
d(CGCCG)$_n$	1
d(GCC)$_n$	2
d(CCG)$_n$-xx-d(CCG)$_n$[1]	2
3′-(CGCCG)$_n$-5′-xx-5′-(GCCGC)$_n$-3′	1
d(CGCCGACGAC)$_n$	1
d(GACGACGCCG)$_n$	1
1,4-Me- or 2,5-Me-d(CCG)$_n$	2
CGG	
d(CGG)$_n$	2–10, 16
d(CIGCGGCGG)$_n$	1
d(CGGCIGCGG)$_n$	1
Deteurated d(CGG)$_n$[c]	5
d(CGG)$_n$T	4, 5, 7, 8–12, 14, 20
d(CGICGGCGG)$_n$	1
d(CGGCGICGG)$_n$	1
GAC	
d(GAC)$_n$	2, 3, 4
d(GAC)$_n$-xx-d(GAC)$_n$	3
d(GAT)$_n$	3
r(GAC)$_n$	3
r(GAU)$_n$	3
GTC	
d(GTC)$_n$	2, 3, 4
CTG	
d(CTG)$_n$	2–5, 8
CAG	
d(CAG)$_n$	2–5, 8
Others	
d(GAA)$_n$	2
d(AAG)$_n$	2

[a]List of the sequences studied in this laboratory. x, Triethyleneglycol phosphate linker; d, DNA; r, RNA; Me, methyl.

[b]The NH_2 group on the 7th C is ^{15}N-labeled; sample prepared by L. Sowers (National Medical Center).

[c]The H8 in all G residues is deteurated except that of G8, G9, and G15.

B. Nuclear Magnetic Resonance (NMR) Experiments

1. 1D EXCHANGEABLE ^{1}H SPECTRA

Most NMR experiments were performed in aqueous buffer solutions containing 0.1 *M* NaCl, 10 m*M* sodium phosphate, and 0.1 m*M* Na^+-EDTA, adjusted to the desired pH. NMR spectra were recorded in either 90% H_2O–10% D_2O (referred as H_2O spectra) or D_2O. H_2O spectra detect NH resonances from G, T, and protonated A and C residues and amino proton (NH_2) resonances from A, C, and G residues (Fig. 40-2). The chemical shifts of these proton resonances are indicative of

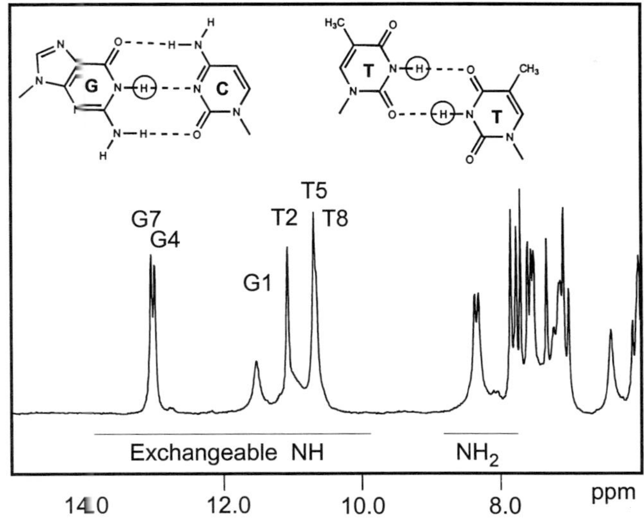

FIGURE 40-2 Representative 1D ^1H spectrum of the $d(GTC)_3$ duplex. C · G and T◇T base pair drawings are included on top of the figure and circled protons are G and T NHs. Spectrum was recorded in 90% H_2O–10% D_2O for observation of exchangeable protons [i.e., NH resonances of G and T residues and NH_2 resonances of A, C, and G residues] as illustrated in the figure.

the type of the H-bonding (Table 40-2). An example of the one-dimensional (1D) H_2O spectrum of the $d(GTC)_3$ duplex is displayed in Fig. 40-2. In this spectrum, the resonances appearing at 13 ppm are indicative of G NH in C · G W–C base pairs, similar to what is observed in a canonical DNA duplex. The T NHs involved in A · T W–C base pairs usually resonate at a slightly downfield position (13.0–14.5 ppm) with respect to G NHs. ^1H resonances observed in 11–12 ppm region are often due to non-W–C base pairs, such as G-quartets [46]. In the spectrum shown in Fig. 40-2 several resonances were observed in the 10.5–11 ppm region. These resonances are typical of NH resonances from non-base-paired G and T residues and were assigned to T residues in T◇T mismatch. H-bond formation involving NH_2 resonances can be detected by their well-resolved chemical shifts (Fig. 40-2). Protonation of base moieties induce downfield shift of base proton resonances. Protonation is also likely to be associated with distinct resonances at 15–16 ppm. A summary of the empirical correlations between type of H bonds and chemical shifts of exchangeable proton resonances is given in Table 40-2.

H_2O spectra also provide valuable information about structural dynamics and stability. A crude assessment of relative structural stabilities can be obtained from pH and temperature profiles of NH resonances. At elevated pH or temperatures, the exchange rates of NHs with H_2O accelerate, leading to signal line broadening. Although line broadening is subject to effective catalysis by acids or bases [52], by comparison, under comparable conditions significant line broadening is indicative of destabilization of H-bonded base pairs and the structures stabilized by H bonds. A plot of linewidth versus temperature generates a transition curve, from which T_m of a particular structure can be derived. These correlations were illustrated in our studies of the CTG TR sequences [38].

2. 2D EXCHANGEABLE ^1H SPECTRA

NOESY spectra recorded in H_2O contain correlations linking NH and NH_2 to each other and to other base and sugar protons. W–C base pairs (Fig. 40-3A) are associated with characteristic NOE patterns which

TABLE 40-2 NH and NH_2 Chemical Shifts and the Correlation to the Type of H Bonds

Type of H-bonding association[a]	Resonance	ppm[b]	Resonance	ppm[b]
W-C C · G base pair	G NH	12–13.5	C NH_2	~6.8, ~8.2
W-C A · T base pair	T NH	13–14.5		
G quartet	G NH	11–12	G NH_2	~7.5, ~9.0
G opposite to G, not paired	G NH	10.5–11		
T◇T mismatch	T NH	10.5–11.5		
A◇A mismatch, parallel			A NH_2	~6.8, ~9.0
C◇C mismatch, parallel	C^+ NH	15–16	C^+ NH_2	~8.2, ~9.3
G◇G mismatch, parallel	G NH	~10.5		
C◇I base pair	I NH	~15	C NH_2	~7.0, ~8.4
I◇G mismatch	G NH	~10.7	I NH	~12.0

[a]The list includes only the types of H-bonding discussed in this chapter. All base pairs are antiparallel unless otherwise noted.
[b]Typical range of chemical shifts.

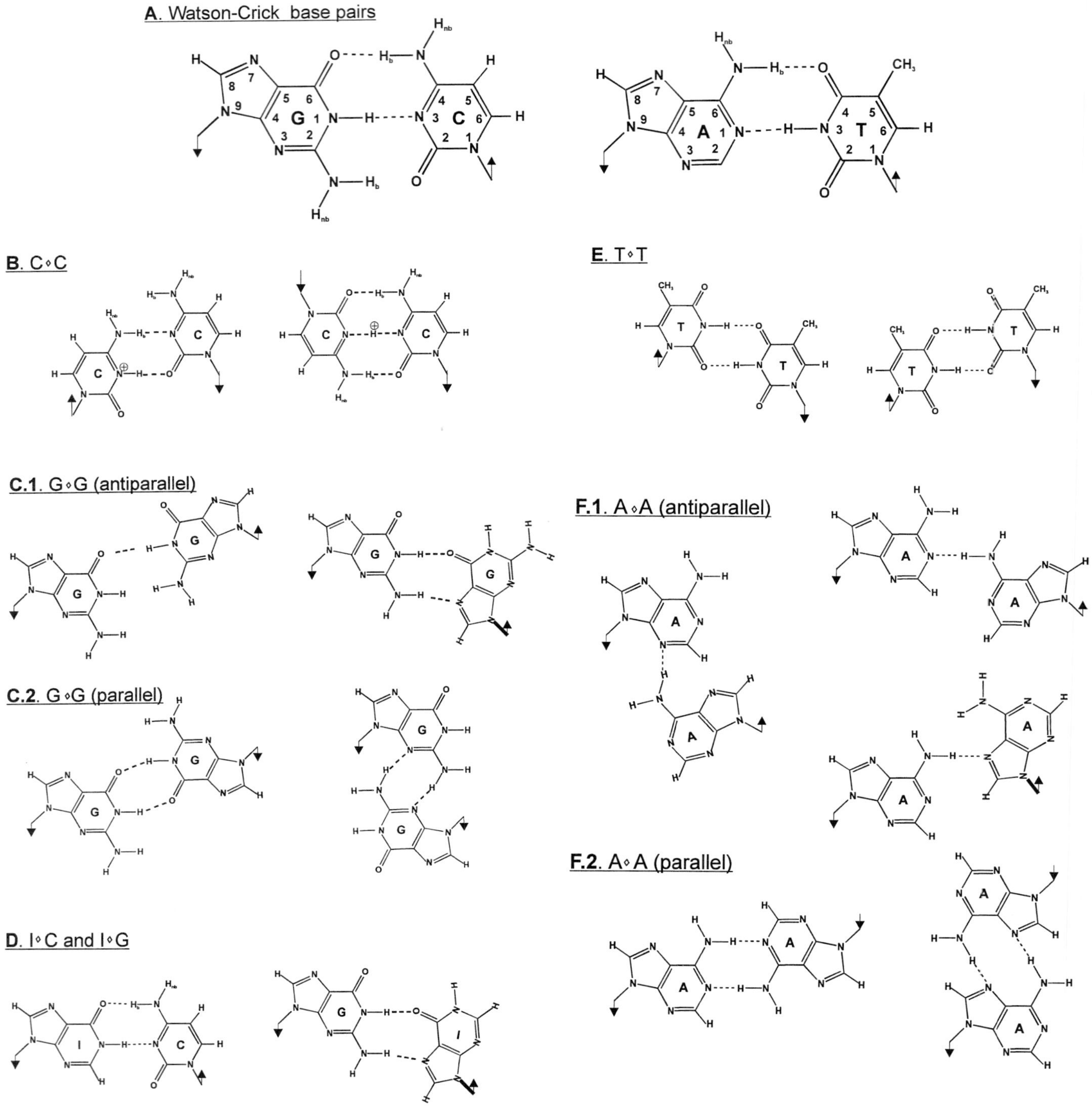

A. Watson-Crick base pairs

B. C∘C

E. T∘T

C.1. G∘G (antiparallel)

F.1. A∘A (antiparallel)

C.2. G∘G (parallel)

F.2. A∘A (parallel)

D. I∘C and I∘G

FIGURE 40-3 Drawings of base pairs discussed in the text. Dashed lines indicate H bonds; arrows indicate the 5′-3′ strand orientation; heavier drawing of the glycosidic bonds indicate a *syn*-conformation. (A) W–C base pairs, with atomic numbering system. (B) C◇C mismatches. (C) G◇G mismatches. (D) I◇C and I◇G mismatches. (E) T◇T mismatches. (F) A◇A mismatches.

link NH in one spectral dimension to NH_2 (C · G pair) or to H2 (A · T pair) in the second dimension. C NH_2 (NH_2 of other bases are usually too mobile to be observed) is usually marked by strong NOEs from its H-bonded (H_b) and non-H (H_{nb}) resonances. C NH_2 is also connected to base H5 (Fig. 40-4) of the same residue and

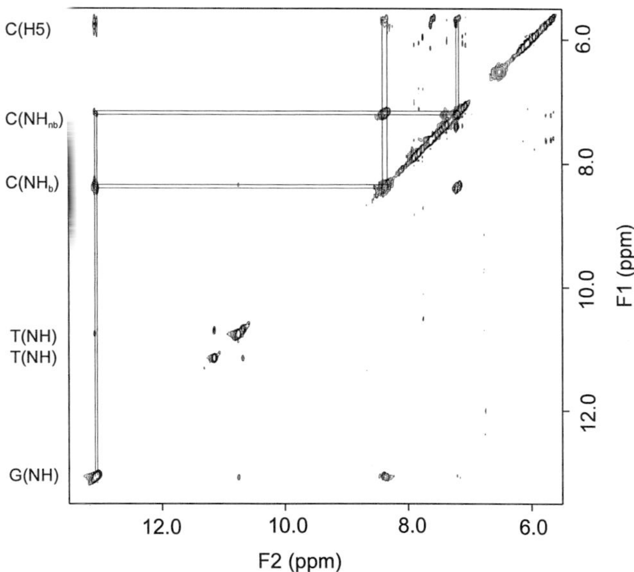

FIGURE 40-4 Representative expanded 2D NOESY of the d(GTC)$_3$ duplex. Spectrum was recorded in 90% H_2O–10% D_2O for observation of exchangeable protons. Chemical shift assignments of the cross peaks are given at left of the vertical axis. The correlations between NH (F2 axis) and NH_2 (F1 axis) resonances and between C NH_2 (F2 axis) and its own H5 (F1 axis) resonances are displayed. These NOE connectivities are associated with W–C base pairs.

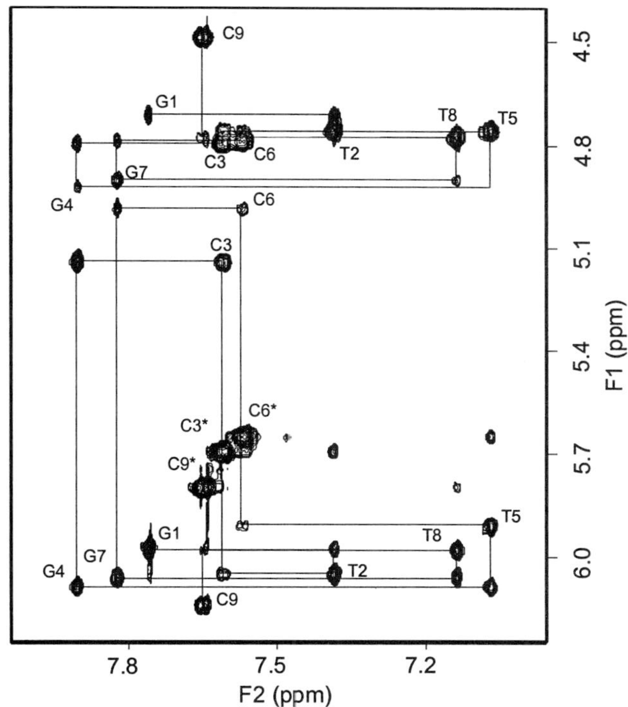

FIGURE 40-5 Representative 2D NOESY of the d(GTC)$_3$ duplex, recorded in D_2O for observation of nonexchangeable protons. The spectral region contains cross peaks correlating base H8, H2, or H6 (F2 axis) to H1′ and H3′ (F1 axis) through intra- (labeled with residue numbers) and interresidue contacts. This spectral region and that containing cross peaks correlating base H8 or H6 to H2′ and H2′ provide key information on sequential connectivities, which are used to define helical conformations and structures.

thus provides reliable residue-specific assignments for C · G base pairs. For multistranded structures, such as a triplex and a quadruplex, different NOE patterns, such as those connecting NH and/or NH_2 with interstrand sugar or base proton resonances, can be detected [16, 53]. These unusual spectral features are diagnostic of nonduplex structures.

3. 2D NONEXCHANGEABLE ^{1}H SPECTRA: NOESY

A set of spectra recorded in D_2O detect interactions between nonexchangeable protons. An example NOESY spectrum of the d(GTC)$_3$ duplex is displayed in Fig. 40-5. In this expanded spectral region, which covers base and sugar proton resonances in the F2 and F1 dimensions, respectively, the correlations between base and sugar protons of the same and sequential residues can be identified. These NOE connectivities are consistent with the formation of a right-handed helix containing stacked base moieties. The absence of these peaks, aberrant changes in relative 2D cross peak intensities, and the presence of new NOE cross peaks correlate to formation of unusual structures.

4. 2D NONEXCHANGEABLE ^{1}H SPECTRA: TOCSY, DQF-COSY, AND COSY-35

These spectra register covalent bond connectivities between protons, mostly between those of sugar moieties (Fig. 40-6). The assignments of COSY (all types listed above) cross peaks are used to assist NOESY analysis and to derive scalar coupling constants, which are the basis for elucidation of sugar and backbone conformations. As illustrated in Fig. 40-6, the coupling cross peak patterns are representative of different sugar puckers. Quantitative and qualitative analyses of COSY cross peaks have been described in numerous reviews and articles [54, 55].

5. ^{31}P SPECTRA

^{31}P spectra were used extensively in our investigation of single-stranded CXG and GXC TR sequences. Information obtained from ^{31}P NMR data includes ^{31}P chemical shifts and intensities and patterns of 2D ^{1}H-^{31}P coupling cross peaks. A 1D ^{31}P spectrum (Fig. 40-7A) provides a quick, reliable assessment of the global structures of oligonucleotides since ^{31}P chemical shifts are a sensitive indicator of unusual backbone conformations. ^{31}P resonances of a canonical oligonucleotide duplex appear in a narrow 0.7 ppm region, whereas increase in chemical shift dispersion reflects

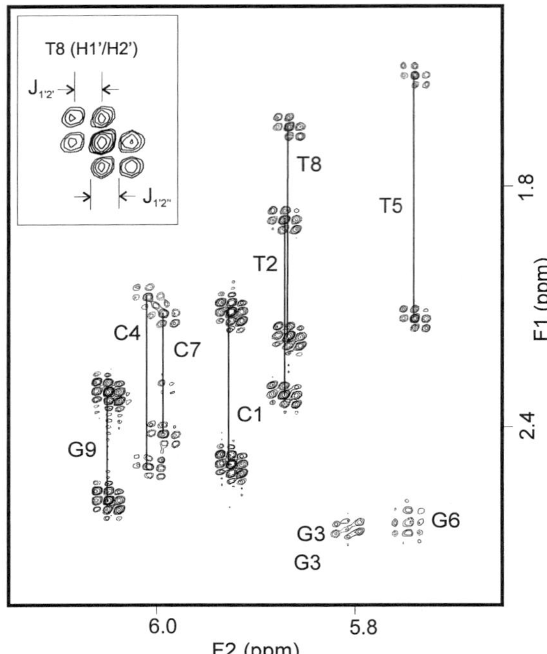

FIGURE 40-6 Representative 2D COSY-35 of the d(CTG)$_3$ duplex. The expanded plot displays scalar coupling (J) correlations between H1′ (F2 axis) and H2′/H2′ (F2 axis). The multiplicity of each cross peak permits calculation of coupling constants. J values of H1′-H2′ and H1′-H2′ ($J_{1'2'}$ and $J_{1'2'}$) can be quantitatively derived as shown in the inserted plot.

distortions in backbone geometry [56, 57]. The assignments of the shifted resonances help to characterize the helices and define the α and ζ backbone angles (Fig. 40-7A). Complementary to 1D ^{31}P spectral analysis, detailed analyses of well-resolved cross peaks in the ^{1}H-^{31}P COSY spectra (Fig. 40-7B) provide specific information about backbone torsion angles. ^{31}P of the backbone phosphodiester is coupled to the H3′ of the preceding and H5″/H5′ of the subsequent sugar residues (chain direction defined from 5′ to 3′). ^{31}P is sometimes found to couple to the H4′ of the subsequent residue, if backbone angles P-O5′-C5′-C4′-H4′ follow a W geometry. The intensities and patterns of these coupling cross peaks are determined by ^{1}H-^{31}P (active coupling) and ^{1}H-^{1}H (passive coupling) coupling constants [54, 58]. This information, in combination with empirical backbone angle relationships [59] and NOE data, defines correlations between 1D and 2D ^{31}P spectral data and local as well as global backbone conformations of oligonucleotides.

C. Temperature-Dependent UV Measurements

UV thermal melting was applied to compare the relative stabilities of the CXG and GXC TR sequences, to examine the stability effects of pH and cations, and to investigate the chain-length-dependent structural transitions. Because UV analysis is model dependent [60, 61], the assistance of high-resolution structural information obtained from NMR analysis is critical for correct interpretation of certain unusual UV results. For instance, the relative stability of d(CAG)$_n$ with respect to other TR sequences has been ranked differently (see discussion in Section II.B). NMR studies indicate that the d(CAG)$_n$ sequences contain highly flexible A residues under a range of pH conditions (pH 5.4–7.9) and their global structures are polymorphic. Therefore, the melting profiles of d(CAG)$_n$ largely deviate from that of a two state transition, and thus the derived T_m are subject to large variations, leading to controversial results. In contrast, the melting curve of single-stranded d(CTG)$_8$, displayed in Fig. 40-8, is cooperative. The UV melting curves recorded for the TR sequences were compared; the broad, undefined transition curves identify the sequences which do not follow a simple two-conformational-state temperature transition. The thermal stability of these sequences can only be compared on a qualitative basis. Cooperative transition curves are analyzed according to the known relationships [60, 62, 63] to give T_m and free energy of structural formation at 37°C ($\Delta G_{37°C}$). Comparisons of the T_m's determined by UV and NMR methods indicate that these values are comparable, although NMR samples are 50–100 times more concentrated than UV samples.

D. Circular Dichroism (CD) Spectroscopic Experiments and Electrophoresis

Some of the CXG and GXC TR oligonucleotides have been examined by electrophoresis and CD spectroscopy. Electrophoretic mobilities were compared to assist the determination of the stoichiometry of strand associations (nondenaturing gel conditions) and inspect sample purity (denaturing gel conditions) [39].

IV. d(CCG)$_n$ AND THE FORMATION OF THE NOVEL *e*-MOTIF DUPLEXES [37]

A. NMR Evidence for an Unusual Conformation Adopted by d(CCG)$_2$

CCG repeats may possibly align in three orientations (Fig. 40-9), each of which would be stabilized by similar numbers of W–C base pairs and antiparallel C$^+$◇C, or by stable parallel C$^+$◇C and G◇G base pairs (Figs. 40-

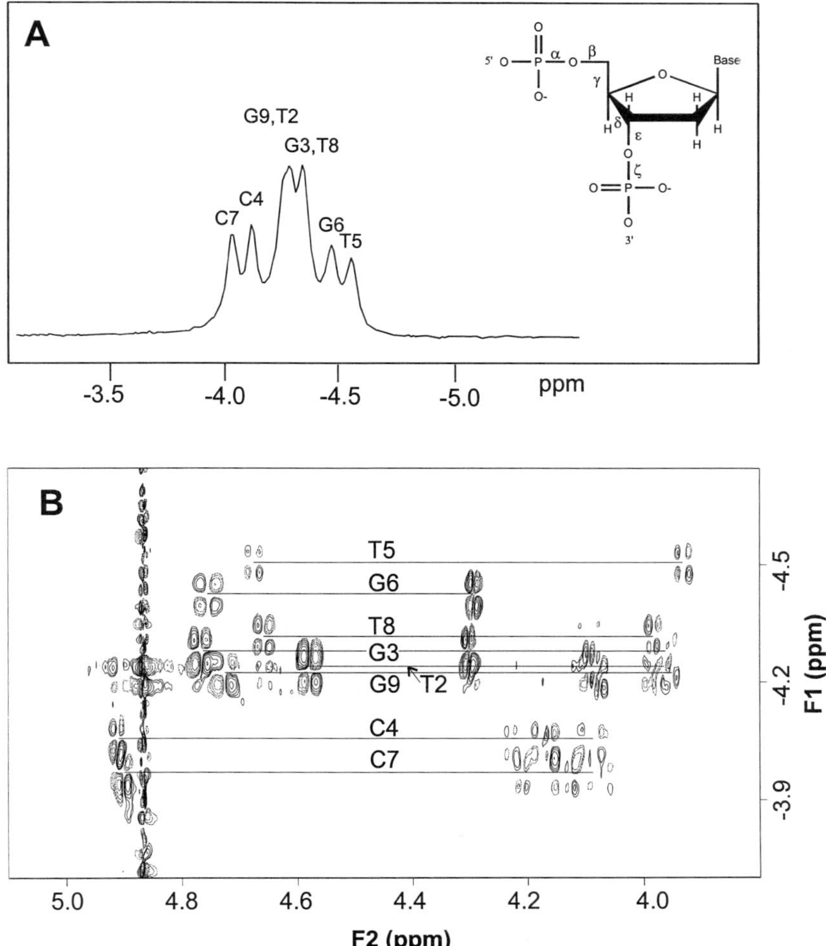

FIGURE 40-7 Representative ^{31}P spectra of the d(CTG)$_3$ duplex. (A) 1D ^{31}P spectrum and sugar backbone drawing illustrating backbone angle definitions recorded at 10°C, pH 6.3. Resonances are labeled according to 5′-P units (i.e., T2 = C1pT2). (B) 2D ^{1}H-^{31}P COSY spectrum. The solid lines in the expanded 2D spectrum correlate ^{31}P in the F1 dimension with H3′ of preceding residue and H5′, H5′, and H4′ of subsequent residue in the F2 dimension. The ^{31}P chemical shift and pattern of each cross peak are characteristic of backbone conformations.

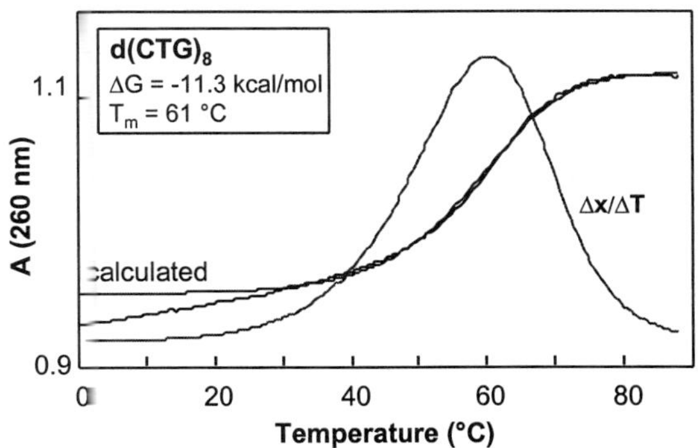

FIGURE 40-8 Representative UV melting profile of d(CTG)$_8$. The absorption was monitored at 260 nm to give hyperchromicity, T_m, and $\Delta G_{37°C}$ for conformational transitions of DNA sequences. Experimental and calculated (labeled) curves are overlaid. Derivative of the melting curve (Dx/DT) is in arbitrary units and scaled to plot size.

3B and 40-3C). d(CCG)$_n$ (n = 2–15) and several permuted and analogous sequences (Table 40-1) have been examined. The ^{1}H spectra of d(CCG)$_2$ displayed distinct spectral patterns (Fig. 40-10) with exchangeable W–C NH resonances at the 13 ppm region and C NH_2 resonances at ~9 ppm, which were downfield-shifted compared to the average 8.5 ppm for W–C NH_2 (Fig. 40-10A). The line width of the NH resonances as a function of temperature gave an NMR T_m of 17°C (Table 40-3). The most striking spectral feature was in the ^{31}P spectrum, which showed well-resolved resonances in a wide 2.2 ppm region compared to the 0.7 ppm chemical shift dispersion normally observed in canonical DNA duplexes (Fig. 40-10B). The perturbed d(CCG)$_2$ proton resonances relax unusually slowly as measured by an average T_1 (nonselective longitude relaxation time constant) = 2.9 s as opposed to 2.0–2.2 s for normal DNA

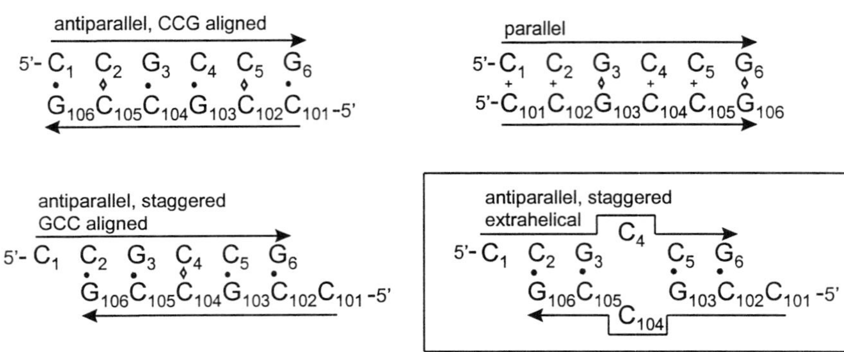

FIGURE 40-9 Possible alignments of the CCG TR sequences. The alignment adopted by the d(CCG)$_2$ duplex, as determined by NMR, is boxed. Mismatched bases are indicated by a diamond symbol and protonation is indicated by a plus sign.

duplexes of a similar size. These results clearly indicated an unusual conformation adopted by the (CCG)$_2$ sequence. The complete assignments of ^{1}H and ^{31}P resonances of d(CCG)$_2$ and 2D NMR spectral analysis have been reported [37]. d(CCG)$_2$ displayed NMR signals corresponding to a single hexamer strand, indicating formation of a symmetrical structure; 20% polyacrylamide nondenaturing gel electrophoresis showed that d(CCG)$_2$ and d(CCG)$_3$ migrate slightly faster than the corresponding complementary hexamer and nonamer duplexes at 4°C. This result, coupled with the absence of interstrand NOEs as typically observed in quadruplex structures, excludes the possible formation of higher order structures. The NH resonances at 13.04 and 13.48 ppm, assigned to G3 and G6 NH resonances, are consistent with formation of four W–C G · C base pairs (Fig. 40-10D). NOESY cross peaks show sequential HN-HN and intra-base-pair HN-NH2 interactions between G and C residues. These observations, especially the presence of the HN-HN NOE between G3 and G106 residues, established the staggered alignment of the d(CCG)$_2$ duplex in 5′-GCC · GCC-3′ units with the noncomplementary C residues sandwiched between a 5′-G and a 3′-C (Figs. 40-9 and 40-10D).

Sequential connectivities of nonexchangeable proton resonances of d(CCG)$_2$ can be traced throughout the sequence. Detailed analysis of the NMR spectra of d(CCG)$_2$ exhibited unconventional long-range NOE connectivities, as illustrated in Figs. 40-10C and 40-10D. The strand orientation and the differentiation of intrastrand from interstrand NOEs were verified by analyzing NMR spectra of a linker sequence, 5′-d(CCG)$_2$-xx-d(CCG)$_2$ (x = triethyleneglycol). This sequence exhibited spectral patterns similar to those of d(CCG)$_2$, although the twofold symmetry in the spectra was disturbed. These data confirm that the C4 residues in the symmetrical strands of the d(CCG)$_2$ and 5′-d(CCG)$_2$-xx-d(CCG)$_2$ duplexes do not stack-in to form a C$^+$◇C mismatch but project into the minor groove pointing their base moieties in the 5′-direction (Fig. 40-11).

B. Structure of the *e*-Motif d(CCG)$_2$ Duplex

The most salient feature of the d(CCG)$_2$ duplex is the simultaneous bisection of the minor groove by the two extrahelical C residues, one facing the other from opposite strands. The structure of d(CCG)$_2$ was elucidated using DG and NMR restrained MDS methods [64] based on a large number of NOEs and dihedral angle restraints. The core of the d(CCG)$_2$ duplex is composed of a W–C base-paired, right-handed complementary tetramer (rather than a hexamer) helix, CG-CG · CG-CG, while the two extrahelical C residues in the minor groove are symmetrically interposed between the CG dinucleotide units (Fig. 40-11). The base moieties of the projected C residues are tilted in a 5′-direction with their hydrophilic edges exposed to the solvent. This unusual alignment of the CCG repeats is accompanied by significant changes in backbone and helical conformations. The helix is untwisted and features wider minor and major grooves compared to canonical DNA helices. The distorted helix of d(CCG)$_2$ causes the distinctions in groove depth and width between the major and minor grooves, apparent in canonical A- and B-form DNA duplexes, to be insignificant [65].

The conformation of the noncomplementary C residues in the *e*-motif d(CCG)$_2$ duplex may have implications to 5-methyl (5-Me) transferring to C residues in the CCG TR sequences. Recently, the crystal structures of two DNA–enzyme complexes, involving *Hha*I and *Hae*III DNA C-5-methyltransferase and that of the DNA

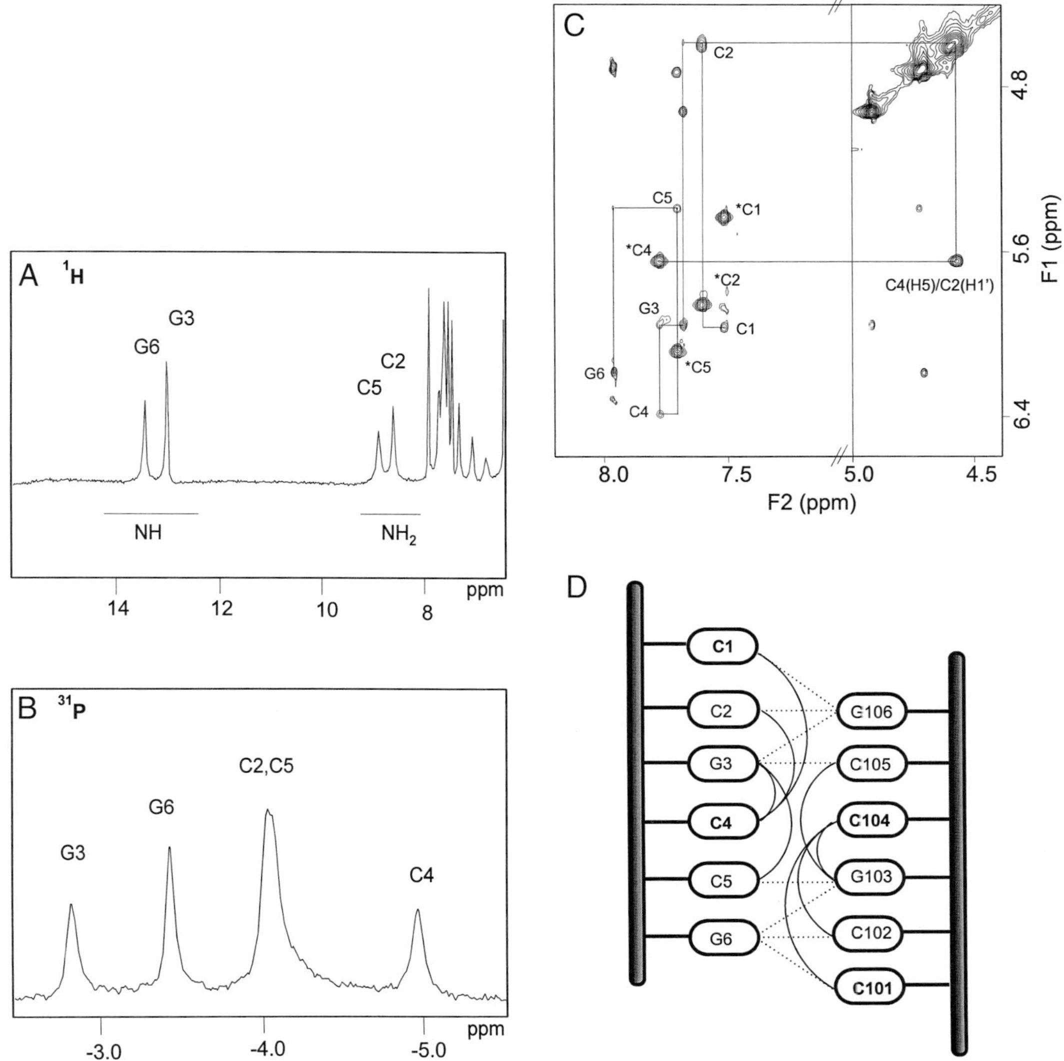

FIGURE 40-10 ^{1}H and ^{31}P 1D NMR spectra of the d(CCG)$_2$ duplex. (A) 1D ^{1}H spectrum showing W–C base-paired NH and NH$_2$ resonances. (B) ^{31}P spectrum displaying large chemical shift dispersion. ^{31}P numbering follows 5′-P convention (i.e., C2 is C1-p-C2). (C) NOESY spectrum (120 ms mixing time) displaying base to sugar sequential connectivities. An important long range NOE linking proton resonances of C5 and C2 residues is shown in the figure. (D) Schematic drawings of long-range, interresidue NOEs detected in d(CCG)$_2$. Dashed lines indicate H bonds. Non-W–C base-paired C residues are in bold.

repair enzyme endonuclease III, have revealed a new aspect of protein–DNA recognition (reviewed in [66]). The formation of these complexes involves the displacement of H_2O in protein active sites with an extrahelical base (an extrahelical C in methyltransferases or an extrahelical T in endonuclease III). The extrahelical prepositioning of the C residue (underlined) in CCG repeats may be a prelude to favorable enzymatic reactions.

The structure of the *e*-motif reveals a novel helical scaffold in which a new strand of helix is generated in the minor groove by extrahelical residues in the duplex (Fig. 40-1B). The nature of the interactions between the third strand and the duplex is hydrophobic and bases are tilted with respect to the W–C base pairs (Fig. 40-11B). This result validates the basic concept of triad-DNA [51] (Fig. 40-1A) in that each strand of a duplex may contribute one extrahelical residue per trinucleotide to result in shortened helix (Fig. 40-1). However, there are major differences in structural details between the proposed triad-DNA model and the *e*-motif duplex. The former is based on base triple formation occurring in the major groove of the duplex, whereas the latter is

TABLE 40-3 NMR Spectral Information of NH Resonances of the n = 2 and 3 TR Sequences Studied[a]

Sequence	δ (ppm)[b]	$T_{m,nmr}$(°C)[c]	Experimental conditions and spectral features[d]
$(CCG)_2$	13.0, 13.5	17	*e* Motif with GCC type of alignment; ^{31}P spans ~2.0 ppm.
$(CCG)_3$	9.0, 9.8, 11.8, 13–13.5, 15.2	23, 29	*e* Motif is present along with other forms; *e* motif prevails at pH 7.8.
$(CGG)_2$	10.5, 11.2, 13.0–13.2	17, 27	Spectral patterns are similar to those of $(CGG)_3$.
$(CGG)_3$	10.5, 11.1, 12.9–13.2	27, 37	From pH 6.4 to 8.6, the spectral region of nonexchangeable protons is similar. At low temperature, ^{31}P spans 1.5 ppm.
	11.0, 13.0–13.2	35	Observed at pH 8.6.
	~11.0 (very broad)	⩾33	Observed at pH 4.9.
$(CAG)_2$	~11, ~12.6	0	Antiparallel Watson–Crick base paired duplex.
$(CAG)_3$	10.9, 12.5–7	15	Antiparallel Watson–Crick base paired duplex.
$(CTG)_2$	10.8–11.1, 13.1	>0	Antiparallel duplex.
$(CTG)_3$	10.8–11.1, 13.1	27	Antiparallel duplex.
$(GAC)_2$	9.1–9.2, 10.5	13–27	Parallel duplex, pH 6.3–5.0.
$(GAC)_3$	9.0–9.3, 10.3–10.4	28–50	Parallel duplex, pH 6.3–4.0.
$(GAC)_3$	12.5	>12	Antiparallel duplex, pH 8.3.
$(GTC)_2$	10.8 (broad)	<5	Antiparallel duplex.
$(GTC)_3$	10.8, 12.5	20	Antiparallel duplex.

[a]Unless specially indicated, NMR sample concentrations were 0.6–3.0 m*M* dissolved in solutions containing 0.1 *M* NaCl, 10 m*M* sodium phosphate, and 0.1 m*M* EDTA at pH 6.3.

[b]Most spectra were collected at low temperatures where stable NHs were observed. Chemical shift values reported are downfield from 9 ppm.

[c]Derived from NMR melting curves which record linewidth as a function of temperature. T_m (estimated error limit of ±1°C) is the temperature where the first derivative of the curve approaches maximum. Multiple transitions are indicated by more than one T_m.

[d]Major structural features under specified experimental conditions.

featured by hydrophobic interactions between extrahelical and helical residues in the minor groove.

C. The Polymorphic Conformations and Dynamic Properties of Longer $d(CCG)_n$

The analyses of the NMR spectra of $d(CCG)_n$ (n = 3–5, 10, 15) identified the presence of unusual duplex motifs in long repeat sequences and coexistence of the *e*-motif with other structural forms (Huang and Gao, unpublished results). These studies were to examine the basic properties of CCG repeats as a function of chain length, which may provide hints to the molecular mechanisms of TR expansions. Experiments, conducted at various pH and temperatures, clearly detected chain-length-dependent variations in spectral appearance. The n = 3–5 sequences existed in multiple forms, including the *e*-motif and a form containing protonated C residues (detected a signal at >15 ppm). As chain length was increased, NMR signals became uniformly broader and the resonance signals unique to the *e*-motif gradually diminish; the W–C base-paired NH resonances, however, remained as the major species. Because of extensive spectral overlaps, high-resolution analysis of these homologous sequences is not feasible at this time. This difficulty is being addressed by selective ^2H, ^{13}C, and/or ^{15}N stable isotope labeling, so that the base pairing, specific interactions and, thus, the sequence folding, can be directly discerned from longer CCG TR oligonucleotides.

D. Permuted and Methylated CCG Repeats

The various CCG related sequences (Table 40-1) were examined primarily by 1D NMR to demonstrate sequence effects and to explore other possible stable structures (Huang and Gao, unpublished results). Indeed, these studies revealed distinctly different spectra for sequences which differ from each other by a single base. For example, $d(GCC)_2$ and $d(CGC)_2$, which are permuted sequences of $d(CCG)_2$, exhibited spectra which do not resemble those of the *e*-motif. d(CGCCG) exhibited yet another type of spectral pattern (well-resolved spectra containing W–C NH resonances and a sharp peak at 15.5 ppm), suggesting the coexistence of C^+ residues and W–C base pairs. These examples

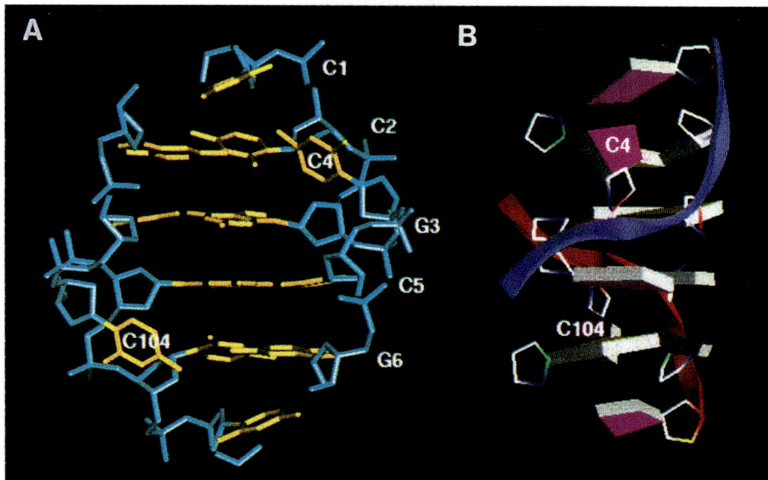

FIGURE 40-11 Structure of the *e*-motif d$(CCG)_2$ duplex. (A) A representative structure derived from distance geometry and restrained molecular dynamics simulations. The stick drawing is a view into the minor groove demonstrating the two extrahelical C residues located therein. Backbones are shown in light blue and base moieties are shown in yellow. The duplex contains a twofold symmetry and thus, only one strand is numbered. Extrahelical C residues are indicated for both strands. (B) The block-ribbon drawing (generated using GRASP) is a side view demonstrating helical stacking between G·C base pairs and the terminal C bases, the orientation of the extrahelical C in the minor groove, and the right-handed, significantly distorted helical backbone.

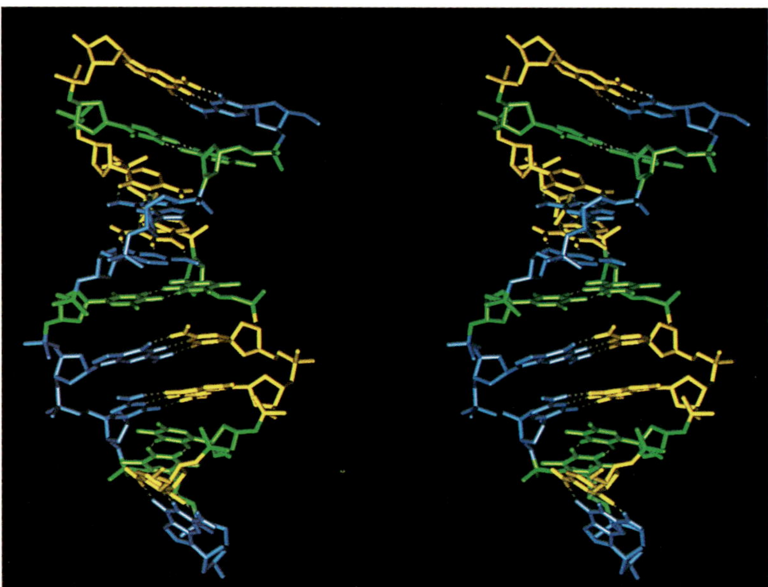

FIGURE 40-15 Structure of the d$(CTG)_3$ duplex elucidated using NMR restrained molecular dynamics simulations. The two strands are colored in yellow and blue, respectively. T T mismatched pairs are shown in green. Dashed lines indicate H bonds between two strands.

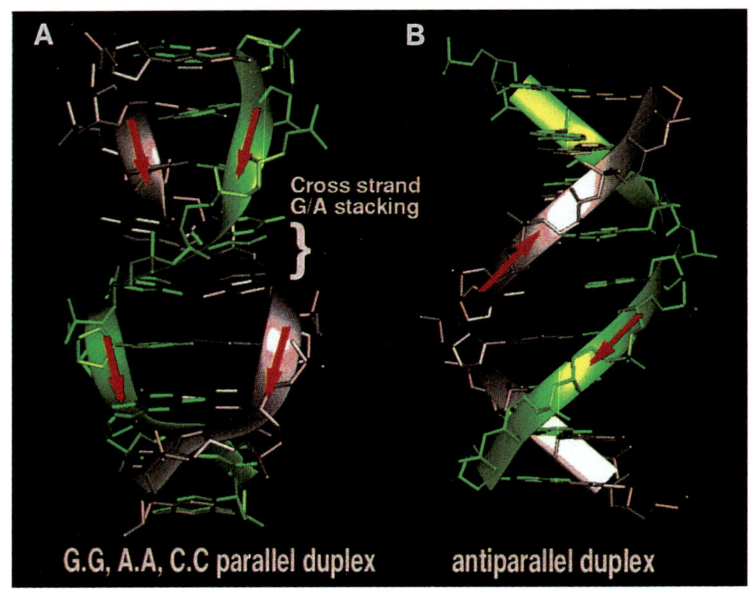

FIGURE 40-18 Structures of the $d(GAC)_3$ duplexes. Strand orientations $5' \rightarrow 3'$ are labeled using arrows. (A) NMR structure of the parallel $d(GAC)_3$ duplex. (B) Model structure of the antiparallel $d(GAC)_3$ duplex. Comparison of A and B demonstrates the two dramatically different structures adopted by $d(GAC)_3$ at neutral and high pH, respectively.

illustrate that when CCGCCG is permuted or truncated by only one residue, the *e*-motif form is no longer the most stable form. The conformational preferences of different CCG repeat oligomers reveal the delicate balance of sequence requirements for stable structure formation.

The CCG repeat oligonucleotides ($n = 2$) containing methylated C residues (5Me-C, Table 40-1) have been examined by 1D and 2D NMR spectral analyses (Huang and Gao, unpublished results). These methylated oligonucleotides were to model natural CCG repeats which are sensitive regions for *in vivo* 5-methylation. The NMR analyses of 5Me-C-d(CCG)$_2$ (methylation at 1 and 4, or at 2 and 5 positions) indicate that these analogous sequences adopt structures similar to the *e*-motif duplex and that methylation induces conformational flexibility of the resultant duplexes. Further studies are required to understand the structural correlation of sequence-dependent hypermethylation of the CCG TR sequences in the genes undergoing triplet expansion.

V. d(CGG)$_n$ AND THE FORMATION OF ANTIPARALLEL DUPLEXES CONTAINING ONE FLEXIBLE G RESIDUE IN EACH TRIPLET [39]

G-rich sequences tend to associate into multistranded complexes and for this reason, it has been proposed that the stable structure formation in the CGG repeats involves quadruplexes [47–49]. Our NMR studies of d(CGG)$_n$ (n = 3–5, 7–9, 12, 16, 20) demonstrated that NMR spectra of these CGG TR sequences are similar and that there were major spectral changes at different temperatures [67]. At low temperatures spectral features which are consistent with quadruplexes were detected. These features disappeared at elevated temperatures [at ~27°C for d(CGG)$_3$] and the spectral patterns observed at near physiological conditions are consistent with antiparallel duplexes as the dominant form (Table 40-3). The basic spectral patterns of the CGG repeat sequences appeared insensitive to changes in Na^+ concentration (0.1–1 *M*) and pH (neutral to basic). At acidic pH (pH 5), d(CGG)$_3$ exhibited additional broad peaks at ~11 and ~9 ppm while the signals observed at basic pH were still visible. The broad peaks were not immediately reversible even at elevated temperatures (>30°C). This observation is consistent with what was described as acidic condition promoted quadruplex aggregation of d(CGG)$_4$ in the presence of K^+ ion [68].

d(CGG)$_3$ has been studied in great detail. The antiparallel duplex formation of the CGG repeat strands, one of the several possible strand associations (Fig. 40-12), was determined from the presence of W–C base-paired G NH and C NH_2 resonances from NMR spectral analyses. Our data show that d(CGG)$_3$ forms a dynamic duplex and do not support a major presence of high-order structures of d(CGG)$_n$ under biologically relevant conditions. Additionally, the gel electrophoretic mobility of d(CGG)$_n$ (n = 3, 4) under nondenaturing conditions is faster than that of the corresponding complementary duplex and is comparable to other TR oligonucleotide duplexes of the same length. The elucidation of the alignment of CGG repeats was assisted by I substituted d(CGG)$_3$ analogous sequences (Table 40-1). The H-bonded NH resonance of I in an I · C pair (Fig. 40-3D) normally appears at ~15 ppm [69] and shows an NOE to the H2 resonance of the same base pair. This was observed in the I2- or I5-substituted

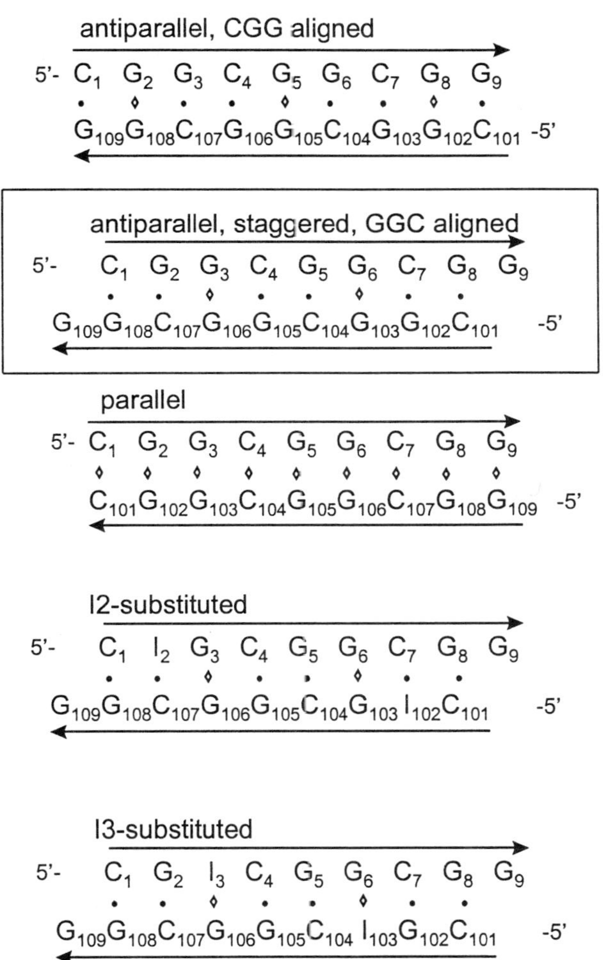

FIGURE 40-12 Possible sequence alignments of d(CGG)$_n$ and inosine substituted(CGG)$_n$. Sequence alignment observed in d(CGG)$_3$ is boxed. Mismatched bases are indicated by a diamond symbol. The actual sequence alignment observed in the d(CGG)$_3$ duplex is boxed.

$d(CGG)_3$ (Fig. 40-12). In contrast, NH resonances of I3 and I6 in CGG repeats were found at ~12 ppm, a typical position for a non-hydrogen-bonded I NH. Furthermore, $d(CGG)_3$ and I-substituted $d(CGG)_3$ sequences exhibited similar spectral patterns, indicating that I-substitutions do not cause major structural rearrangement. Based upon these observations, we conclude that the first G in the CGG triplet is base paired with a C in the opposite strand, while the second G is not base paired and is mobile. These results establish that the preferred alignment for the CGG repeats is GGC with the central G unpaired and sandwiched in between a 5′-G and a 3′-C (Fig. 40-12). In both CCG and CGG sequences, the unpaired C or G residues are preferably situated between 5′-G and 3′-C residues rather than between 5′-C and 3′-G residues.

Although the dynamic motions of the $d(CGG)_3$ duplex prohibit its high-resolution structure elucidation, the major features of the duplex can be distinguished. The $d(CGG)_3$ helix is not perturbed compared to canonical complementary duplexes as judged from ^{31}P chemical shifts. The two intervening W–C C · G base pairs in the helix are well stacked. Mismatched G residues (underlined) in CG<u>G</u> repeats [G3 and G6 in $d(CGG)_3$] are flexible since they do not form H-bonded base pairs. These noncomplementary G residues [G3 and G6 in $d(CGG)_3$] are confined neither in a *syn*-conformation nor in a mismatched base-paired form, most likely undergoing dynamic exchange among glycosidic conformational isomers. In comparison the I3 residue in the CGICGGCGG duplex adopts a *syn*-conformation to form a I(*syn*) · G(anti) mismatched pair (Fig. 40-3D). Structural perturbations induced by formation of the I(*syn*) · G(anti) base pair appear to be local when the imino proton and nonexchangeable proton spectra of the d(CGG)3 duplex and the CGIC-GGCGG duplex are compared. Presently, the studies of the CGG repeats have been extended to longer sequences. G H8 deuterated $d(CGG)_5$ has been synthesized and analyzed to provide information on sequence folding and base pairing (Table 40-1; Huang and Gao, unpublished results).

VI. d(CTG)$_n$ AND d(GTC)$_n$: THE FORMATION OF DUPLEXES CONTAINING T◇T MISMATCHES [38]

A. Characterization of the d(CTG)$_n$ Duplex

$d(CTG)_{2\text{-}4,6}$ sequences and their complementary duplexes have been examined by NMR, UV, and electrophoresis. Among these sequences, the formation of the $d(CTG)_3$ duplex (Fig. 40-13) is characterized by the presence of W–C base paired G NH at 13.2 ppm and mismatched T NH at ~11 ppm. The majority of the ^1H and ^{31}P resonances were assigned based on well-defined 1D and 2D NMR spectra. These results reveal a twofold symmetry in the duplex and essentially one set of resonances for each T residue in $d(CTG)_3$, indicating fast exchange between the two T◇T mismatch pair forms (Fig. 40-3E). The antiparallel duplex structure of $d(CTG)_3$ is stable and invariant at pH 4.5–8.8 and in 0.1–4.0 *M* NaCl.

An interesting spectral feature of the $d(CTG)_3$ homoduplex is the stable NH resonance of the central T◇T mismatch, which exhibited intensity and linewidth comparable to those of W–C base-paired G NHs even at pH 8.5 (Fig. 40-14). The exchange rates of T NH resonances for T◇T in the $d(CTG)_3$ duplex and for A · T in the d(CAG) · $d(CTG)_3$ duplex, measured

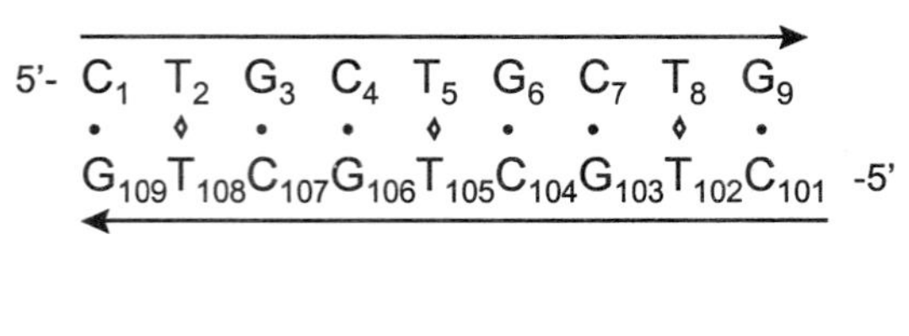

FIGURE 40-13 Sequence alignment of $d(CTG)_3$ and $d(GTC)_3$. Mismatched bases are indicated by a diamond symbol.

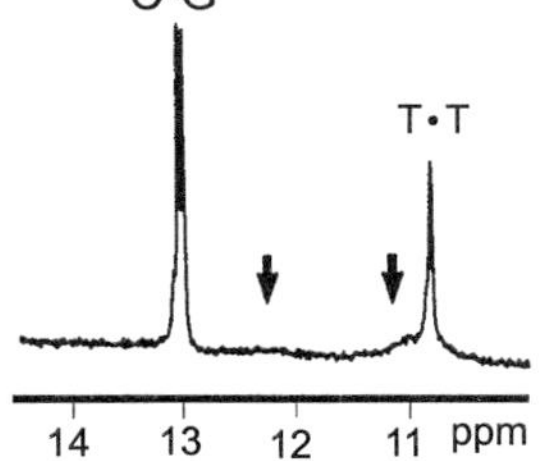

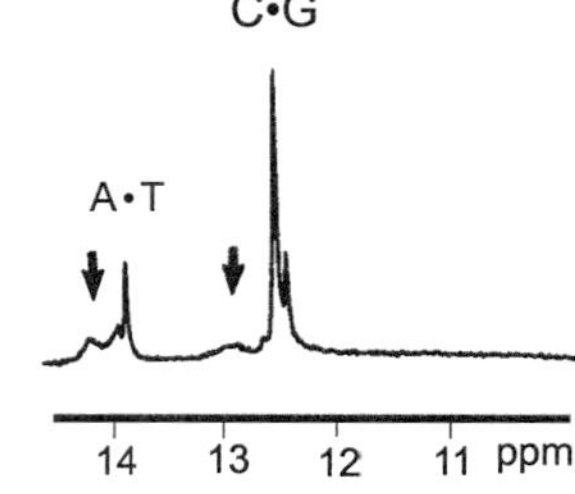

FIGURE 40-14 1D H_2O spectra of (A) the homoduplex $d(CTG)_3$ and (B) the complementary duplex $d(CTG)_3$ · $d(CAG)_3$, recorded at pH 8.5 demonstrating stable NH resonances of the T◇T mismatch. Arrows indicate the absence of terminal NH resonances, while NH of the T◇T mismatch in the $d(CTG)_3$ duplex exhibited a sharp signal. The NH linewidth of T◇T and A · T is comparable.

by NMR in the absence of added catalyst and at neutral pH, are comparable (Smith and Gao, unpublished results). To understand how multiple T◇T mismatches can be accommodated in repeat sequences, 2D NMR spectra were analyzed in detail. These results indicate that d(CTG)$_3$ forms a generally B-type duplex with T bases stacked in the helix, similar to a previously reported duplex containing a single CTG [70]. Non-B-form characteristics, however, were detected from the relative intensities of sequential NOEs at the CT and TG steps. The presence of tandem T◇T mismatches does not cause major perturbation in sugar phosphate backbone conformations, which are in the C2′-endo and canonical backbone torsion angle family.

B High-Resolution Structure of the d(CTG)$_3$ Duplex

The structure of the d(CTG)$_3$ duplex was elucidated from a large set of restraints using NMR restrained MDS, relaxation matrix refinement, and MDS under hydration conditions. A stereo view of the d(CTG)$_3$ duplex is shown in Fig. 40-15. This duplex features an alternating pattern characteristic of the repeat sequence. Although no H-bonding restraints were applied for T◇T mismatches during the calculation, the two T bases in the opposite strands are H-bonded through NHs and carbonyl oxygen atoms (Fig. 40-3E). The mismatched T residues of opposite strands are not equivalent, because the base moiety of one T residue is pushed toward the major groove. This distorted orientation is supported by the NOEs between the adjacent T NH and the C NH_2 resonances. The formation of the T◇T mismatch is accompanied by a reduction in the cross strand C1′-C1′ separation from 10.7 Å (for an A · T base pair in canonical DNA duplexes) to 8.6 ± 0.3 Å. This narrower groove width in d(CTG)$_3$ suggests that bridging water may not be involved in the T◇T mismatch, as observed in a crystalline pyrimidine·pyrimidine, C◇U, mismatch [71].

C. Comparison of the d(GTC)$_n$ Duplex with the d(CTG)$_n$ Duplex [39]

d(GTC)$_n$ (n = 2–4) sequences have been studied by NMR (Figs. 40-2, 40-4, and 40-5) using methods similar to what were used for studying d(CTG)$_n$ sequences [72]. Unlike d(CTG)$_2$, d(GTC)$_2$ does not form a duplex even at low temperatures. Stable duplex formation was detected with d(GTC)$_3$ which displayed NH of W–C C · G base pairs and T◇T mismatches (Figs. 40-2, 40-3E, and 40-13). Sequential NOE analysis (Figs. 40-4 and 40-5) demonstrates that base pairs and mismatches are stacked in the helix and T◇T mismatches are in fast exchange between the two H-bonding forms. The NH of the T residues are well-resolved and melted cooperatively as do the G NH resonances. Although these spectral features of the d(GTC)$_3$ duplex are reminiscent of those of the d(CTG)$_3$ duplex, major structural distortions in d(GTC)$_3$ are evident in backbone torsion angles. The ^{31}P resonances of this duplex disperse over a 1.6 ppm region compared to 0.5 ppm for the d(CTG)$_3$ duplex. The ^{31}P resonances in CpG are significantly downfield-shifted and those in GpT are slightly upfield-shifted. These shifted resonances signify that the backbone conformation of the d(GTC)$_3$ duplex deviates from that of the canonical DNA duplex and from that of the d(CTG)$_3$ duplex, despite both having the same sequence composition and containing T◇T mismatches. These results reveal that the accommodation of T◇T mismatches in between G and C residues or in between G · C and C · G base pairs in GTC requires additional energy for structural readjustment compared to CTG repeats. This may account for the lower stability of the d(GTC)$_n$ duplexes compared to the d(CTG)$_n$ complexes, illustrating the critical role of nearest neighbor interactions in stabilizing DNA structure.

VII. d(CAG)$_n$ AND THE FORMATION OF DYNAMIC DUPLEXES [39]

d(CAG)$_2$ does not form stable structures even at 0°C, but stable NH resonances from W–C C · G base pairs were detected for d(CAG)$_3$ (Table 40-3), indicating formation of an antiparallel duplex (Fig. 40-16A). The

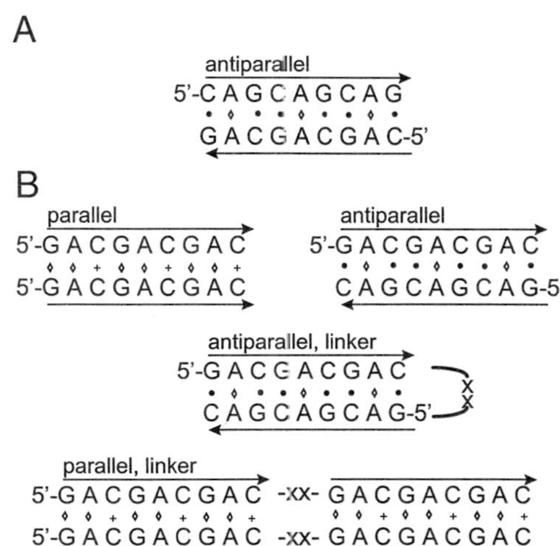

FIGURE 40-16 Possible sequence alignments of (A) d(CAG)$_n$ and (B) d(GAC)$_n$. Mismatched bases are indicated by a diamond symbol. Protonated, mismatched C bases are indicated by a plus symbol.

majority of the ^{1}H resonances of d(CAG)$_3$ have been assigned and the spectral patterns are consistent with the formation of a right-handed antiparallel duplex with little perturbation in backbone conformations (^{31}P chemical shifts are within 0.8 ppm). The dynamic features of the duplex were readily identified from the broader resonances compared to those of other TR repeats and from the diminished intensities of sequential NOEs. Despite the presence of stable W–C C · G base pairs, the A residues in CAG are not base paired and conformationally unstable. pH (7.9 to 5.4) has no effect on the overall appearance of NMR spectra of d(CAG)$_3$. However, the chemical shifts of A H2 resonances are sensitive to pH (downfield-shifted by ~0.1 ppm), indicating protonation occurring on the A base moieties. Although the dynamic nature of the CAG repeats renders the high-resolution structure determination unreliable, the observed solution behavior of this TR sequence should have important biological implications.

VIII. d(GAC)$_n$ AND THE STRUCTURAL TRANSITION FROM A PARALLEL TO AN ANTIPARALLEL DUPLEX [72]

A. pH Dependence of the Conformations of d(GAC)$_n$

Wang and his co-workers demonstrated that 5′-CGA drives formation of parallel duplexes under low and near neutral pH conditions [44]. d(GAC)$_n$ (n = 2–4) and d(GAC)$_3$-xx-d(GAC)$_3$ have been examined in this laboratory to determine whether repetitive GAC trinucleotides will form parallel structures and, if so, the chain-length-dependence of such structures. At pH 6.8 or 7.8 d(GAC)$_3$ exhibited well-resolved, distinctly different NMR spectra [72], which have been completely assigned. These analyses indicate that d(GAC)$_3$ forms a parallel duplex with an apparent pK_a of 7.2 for protonation of mismatched C, which is about 2.9 units higher than that of a monomeric C (Fig. 40-3B). The parallel duplex contains parallel G◇G, A◇A, and C+◇C pairs (Fig. 40-3), which are characterized by several spectral features: the absence of W–C G NH resonance at 12–13 ppm, the presence of a very broad C+ proton resonance at 15.5 ppm, G NH resonances at 10.3 ppm, and C+ NH$_2$ at ~9 ppm. The backbone of the parallel d(GAC)$_3$, which displayed a 2 ppm ^{31}P chemical shift dispersion, is significantly different from canonical DNA duplexes. A large number of NOEs important for defining the parallel base pairing and base alignment have been identified. The longer d(GAC)$_3$-xx-d(GAC)$_3$

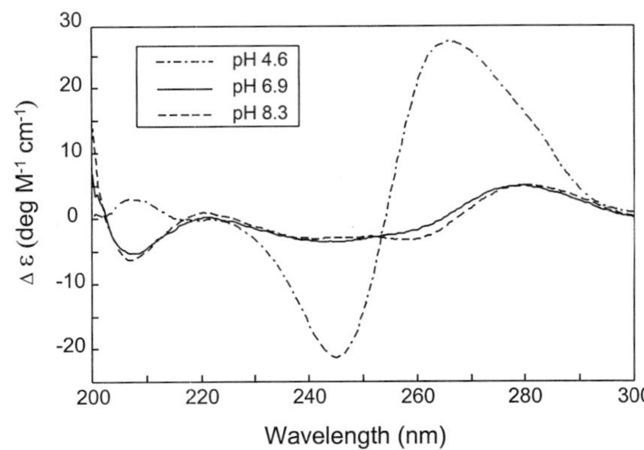

FIGURE 40-17 CD spectra (22°C) of d(GAC)$_3$-xx-d(GAC)$_3$ as a function of pH.

also readily undergo parallel ↔ antiparallel structural transition as identified by NMR and CD [73] (Figs. 40-16B and 40-17). However, the apparent pK_a of the longer sequence is decreased to 5.6, suggesting that longer GAC repeats tend to fold back to form antiparallel duplexes rather than a long stretch of parallel duplexes.

B. Structural Features of the Antiparallel d(GAC)$_n$ Duplex and Comparison with CXG Repeats

The high-pH d(GAC)$_n$ repeats are antiparallel duplexes accompanied by a minor populated form, which is not the parallel duplex and has yet to be identified [72]. The analysis of the NMR spectra of d(GAC)$_3$ has been completed. The absence of *syn*-conformations for A residues and the NOE connectivities between mismatched A bases in the duplex unambiguously identify wobble mismatched A◇A pairs, which involve H-bond formation *via* the W–C edge of A (Fig. 40-3E). The ^{31}P spectrum of the antiparallel d(GAC)$_3$ duplex indicates that the backbone is less distorted compared to the parallel duplex. However, the ^{31}P resonances of the GpA step are downfield-shifted, possibly due to some mild adjustments of the backbone α and ζ angles in order to accommodate A◇A mismatched pairs.

GAC TR sequences are related to CAG (identical composition, residues are connected in a reverse order) and GGC (A → G conversion, both are purine bases) TR sequences. The stable mismatched A◇A pair in the d(GAC)$_3$ duplex contradicts the unstable, mobile conformations adopted by the A or G residues in the disease-related d(CAG)$_3$ or d(CGG)$_3$ duplex. The GAC and CAG sequences also display different pH-dependent profiles, suggesting that protonation of C residues is not the only factor stabilizing the parallel duplex. The molecular basis for the observed remarkably different behaviors in these related TR sequences remains to be investigated.

C. Structure of the Parallel d(GAC)$_n$ Duplex

The structure of the parallel d(GAC)$_3$ duplex has been solved to high-resolution using NMR restrained computational methods, such as DG and MDS (Fig. 40-18) [72]. The duplex exhibits a twofold symmetry around the helical axis and contains two symmetrical grooves. The most intriguing feature of the duplex is the cross-strand stacking between the parallelly oriented G and A residues, attributable to the symmetrical scissoring transformation (positive ~3-Å displacement of the G base and negative ~−3-Å displacement of the A base in relation to the helical axis) between the G and A bases of the same strand and interlocking between the two strands at the GA step. The structure of the repeat d(GAC)$_3$ duplex is consistent with that reported for d(TCGA)$_2$ [46] and d(CGACGAC)$_2$ [44]. This result, in combination with NMR analysis of longer d(GAC) repeats, indicates that GAC TRs are a remarkably stable parallel duplex motif.

IX. RELATIVE STABILITIES OF d(CXG)$_n$ AND d(GXC)$_n$ OLIGONUCLEOTIDES [39]

The thermal stability of the CXG and GXC sequences was examined using UV temperature-dependent measurements. These experiments were performed under well-controlled conditions using information derived from NMR to avoid conditions that would lead to multiple conformations. The UV melting profiles of d(CXG)$_n$ and d(GXC) (n = 4) are displayed in Fig. 40-19. Inspection of these profiles suggests that simple comparisons of the T_m values of these sequences may lead to incomplete or premature conclusions. The UV melting profiles of the n = 4 TR sequences suggest that their melting behavior is strongly sequence-dependent. The observed melting profiles fall into two groups: (1) the four CXG sequences exhibit two or multiple conformational melting transitions with the transitions of d(CAG)$_4$ being particularly broad (Fig. 40-19); (2) the two GXC sequences melt cooperatively with a single transition temperature (Fig. 40-19) with improved cooperativity at increasing salt concentrations (0.1–1 *M* NaCl) (Zheng and Gao, unpublished data). When the T_m's of the second transition are compared, that of the d(CGG)$_4$ duplex is higher than those of d(CTG)$_4$ and d(CCG)$_4$, which in turn are higher than those of d(GAC)$_4$ and d(GTC)$_4$ [note pH conditions used for d(CCG)$_4$ and d(GAC)$_4$ (Fig. 40-19)]. The T_m of d(CAG)$_4$ is highly uncertain, since the melting process is dominated by the random motions of noncomplementary A residues. This order of decreasing T_m corresponds to that determined by melting analyses of n = 15 or longer repeats using electrophoresis in that d(CGG)$_n$ forms the most stable structures and d(CTG)$_n$ is more stable than d(GTC)$_n$. Besides the chain length differences, the discrepancy in the relative stabilities of CCG and GAC may be attributed to pH used in the reported experiments, since the conformations of both sequences are sensitive to pH changes. The conformation of the CCG repeats appears to be better defined at pH ~8.0 as detected by NMR, while the antiparallel form of the GAC repeats requires higher pH to compete with the parallel form.

The free energy values ($\Delta G_{37°C}$) derived from UV curve fitting for the CXG and GXC sequences follow a linear relationship to T_m. The comparison of these results with those of the corresponding complementary duplexes reveals an interesting trend: the differences in T_m or ΔG between the corresponding homo- and heteroduplexes are much smaller for the CXG TR sequences than for the GAC/GTC sequences. For instance, ΔT_m and $\Delta\Delta G$ between the homo-d(CTG)$_4$ duplex and the d(CAG)$_4$ · d(CTG)$_4$ are 5°C and 3.6 kcal/mol, respectively, while ΔT_m and $\Delta\Delta G$ between d(GTC)$_4$ and d(GAC)$_4$ · d(GTC)$_4$ are 46°C and 8.4 kcal/mol, respectively. The temperature and free energy differences indicate that the probability of homo-d(CTG)$_4$ duplex versus d(CAG)$_4$ · d(CTG)$_4$ formation is ~1/15, whereas homo-d(GTC)$_4$ duplex versus d(GAC)$_4$ · d(GTC)$_4$ is ~1/600. These results suggest that formation of structured domains by single-stranded TR sequences within W–C base-paired duplex helical regions in general would not be a favored process. Nonetheless, this process is possible if the intermediate structures can be stabilized through intermolecular interactions, such as specific recognition by proteins. The formation of homo-stranded structures is much more likely to occur with the CXG sequences than with the GXC sequences.

X. CHAIN LENGTH DEPENDENCE OF THE STABILITIES OF d(CXG)$_n$: IMPLICATIONS TO FOLDING

The CXG TR sequences exhibit characteristically different chain-length-dependent melting profiles. UV melting results of the CTG and the CCG repeats (n

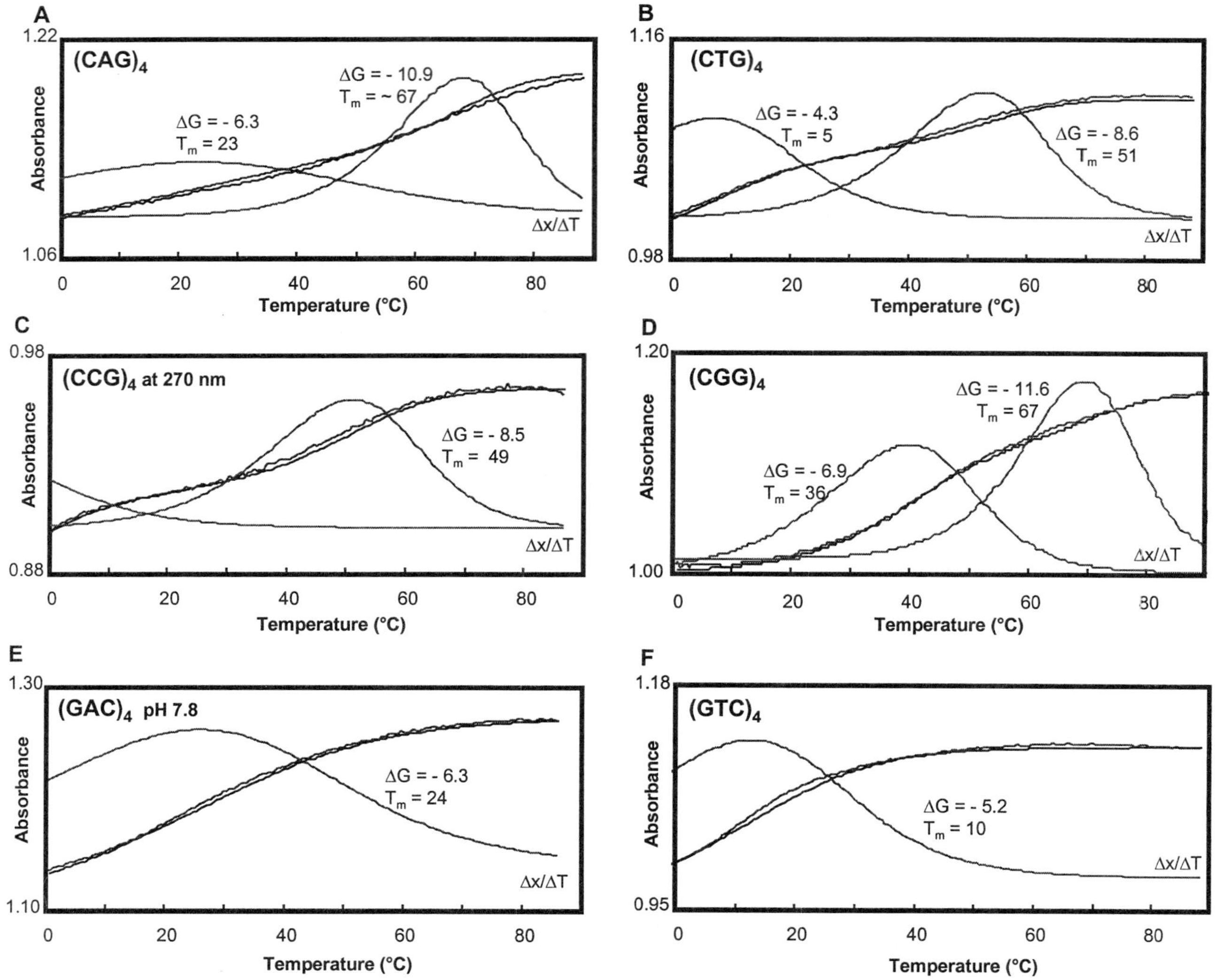

FIGURE 40-19 UV melting profiles of (A) d(CAG)$_n$, (B) d(CTG)$_n$, (C) d(CCG)$_n$, (D) d(CGG)$_n$, (E) d(GAC)$_n$, and (F) d(GTC)$_n$ (n = 4) sequences. T_m (°C) and $\Delta G_{37°C}$ (kcal/mol), derived from curve fitting (calculated profiles and derivative curves are shown in overlay plots), are indicated in the figure. All CXG repeats (A–D) display biphasic temperature transitions and have a second T_m much higher than T_m of d(GAC)$_n$, and d(GTC)$_n$ (E and F). Derivative of the melting curve (Dx/DT) is in arbitrary units and scaled to plot size.

= 3–8) are summarized in Fig. 40-20. The melting of d(CTG)$_n$ is cooperative when n = 2, 3, and 8 (Figs. 40-8) and T_m increases with chain length from ~5 to 61°C. However, T_m of d(CCG)$_n$ (n = 2–10, 15) levels off at a chain length of n = 4 and remained at 45–50°C for sequences as long as n = 15. The chain-length-dependent melting of the CGG repeats (n = 2–10, 12, 14, 16, and 20) has been carefully characterized. This series of UV spectra exhibited yet another type of pattern [67]. These results demonstrate the importance of the interactions between immediate neighboring residues [60, 61] and the concerted effects of the sequence and chain length on relative stabilities and melting behaviors of TR sequence. The structural

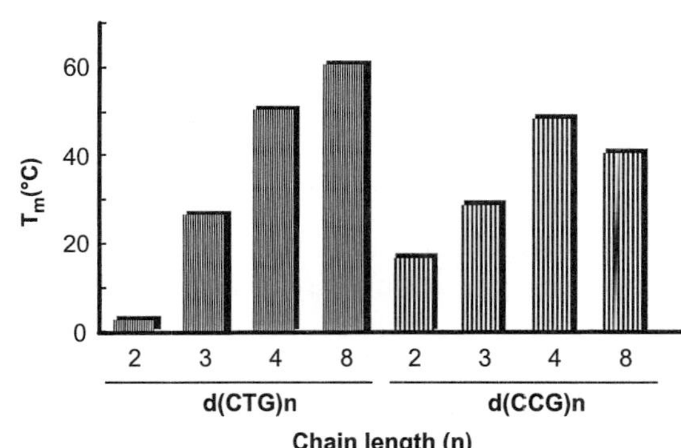

FIGURE 40-20 Chain-length-dependent T_m of the CTG and the CCG duplexes derived from NMR and UV melting curves.

basis of the observed chain length dependence of the various TR sequences is presently under investigation.

Another UV spectral feature is the multiconformational versus cooperative melting profiles observed for the $n = 4$ CXG and GXC sequences, respectively. The first transitions in the biphasic UV melting curves of $d(CCG)_4$ and $d(CTG)_4$ (Fig. 40-19) are concentration independent. These results, in combination with NMR analyses, suggest that at lower temperatures the $n = 4$ CXG TR sequences may exist in a hairpin ↔ duplex equilibrium, while intramolecular folding of GAC and GTC repeats may occur at a longer length. This assessment is consistent with previously discussed lower stability of the homoduplex structures of GAC and GTC TR sequences than those of the CXG sequences. Therefore, the CXG sequences not only are more dynamic, but also have a higher propensity to form folded structures (note that folded structures may or may not be hairpins for longer repeats).

XI. EFFECT OF METAL IONS

Our studies, thus far, have not found major conformational changes for the CXG sequences upon altering ion species or ion concentrations. The Li^+, Na^+, or K^+ forms of $d(CGG)_n$ (n = 3 and 8) were generated and these samples were dissolved in solutions containing 0.1–1 M Li^+, Na^+ or K^+ ions; 1D NMR spectra of these sequences, recorded at 0 and 25°C over a long period of time, the basic features of NMR spectra, i.e., chemical shifts and temperature transitions, were observed to be similar to those previously described for the $d(CGG)_3$ sequence. Our preliminary studies on the effects of Mg^{2+} using a $d(CGG)_3/K^+$ sample showed no change in NMR spectral appearance. These results suggest that in the presence of various concentrations of K^+, which often promotes formation of quadruplexes by specific coordination within G quartets [74], CGG TR sequences remain in an antiparallel duplex form. NMR studies of salt concentration dependence of the CTG repeats demonstrate that the $d(CTG)_3$ duplex is not disrupted by high salt concentrations (4 M NaCl) [38].

In addition to the NMR experiments, UV melting was monitored as a function of salt concentration for several TR sequences, which include $d(CGG)_8$ (0.1 and 1 M NaCl or KCl) and $d(GAC)_4$ (0.1 and 1 M NaCl). These experiments demonstrate a positive effect on T_m at increased salt concentrations, as normally observed in DNA complementary duplexes.

XII. SUMMARY AND FUTURE PROSPECTS

NMR and a variety of biophysical experiments have provided compelling evidence for the presence of novel, stable structures and yet dynamic solution properties exhibited by the seemingly simple CXG TR sequences. Both CXG and GXC TR sequences form duplexes under near physiological conditions. These duplexes are stabilized by C · G W–C base pairs that are embedded in between noncomplementary bases present in every third position along the helix. The closely related CXG sequences all contain linear GC steps and differ only in the residues interrupting the repeat dinucleotides. Yet, the four CXG sequences assume vastly different structures and solution properties. GAC and CAG are related by reversal in sequence connectivity. However, the structures and solution properties of the two sequences are fundamentally different in several aspects. Among these are their pH sensitivity and structural dynamics.

CCG and the methylated CCG repeats may be present in an *e*-motif form, which is a novel duplex that differs from any known DNA structure in its stable, symmetrical disposition of the extrahelical C residues in the minor groove. There is a surprising similarity between the *e*-motif $d(CCG)_2$ duplex and the substrate DNA duplex for *Hae*III methyltransferase in their extrahelical alignment of a C residue and the uninterrupted stack of G · C base pairs at sites immediately adjacent to the extrahelical C. It is known that the rate-determining step in C methylation is enzyme binding and, thus, the presence of an extrahelical C population should lower the energy required for enzyme binding. The extrahelical C residues impart high flexibility in CCG duplexes of long chain length, making C residues more accessible by methyltransferases. Therefore, the NMR structure of the CCG repeats may provide a clue for the reported hypermethylation associated with the CCG repeat expansion.

The dynamic features of the disease-related CXG duplexes are remarkable in that the A residues in $d(CAG)_n$, C in $d(CCG)_n$, and G in $d(CGG)_n$ do not form mismatch pairs in stable stacked conformation. Rather, these residues resonate among a range of conformations. The four CXG repeats differ in their characteristic dynamic alignments of mismatched bases, imparting distinct global structures and local details. This conformational flexibility manifested by the CXG repeats may play a critical role in their recognition by specific enzymes which may be a key step in triplet expansion mechanisms.

Our studies have demonstrated distinctions in structures and biophysical properties between the GXC TR sequences and the human disease-related CXG TR sequences. The thermal stability studies of the CXG and the GXC TR oligonucleotides indicate that conformational motions of the CXG TR sequences do not cause a large reduction in energy. The energy barrier for strand separation in complementary duplexes to form homo-stranded intermediate structures is expected to be much lower for the CXG than for the GXC sequences. The genetically unstable CXG repeats form conformationally flexible duplexes, which have a high propensity to undergo linear ↔ folded structural transitions. In contrast, the mismatched bases in the GAC and GTC duplexes adopt stable, base-paired conformations.

The studies of TR sequences have led to many questions for future consideration. The unusual chain-length-dependent UV melting profiles for CCG and CGG repeats reflect noncanonical patterns of chain folding, which cannot be explained by known DNA structural models, such as linear hairpin folding models. This assessment corresponds with the analysis of longer repeats. Based on the electrophoresis results of reannealed CTG · CAG TR repeats, Sinden and Pearson have proposed the presence of S-DNA, a novel, complex, slipped DNA structure that consists of multiple folded motifs [75]. The base pair details (specificity and alignment) of short oligonucleotides have not been directly proven in long TR sequences. The basic properties of interspersed TR sequences, such as $d(CGG)_9AGG(CGG)_9$ found in normal genes but not in mutated genes [76], as compared to their homologous counterparts are largely unknown. The effect of important divalent ions, such as Mg^{2+}, remains to be investigated [77]. Structure investigations should be extended to the newly discovered GAA/TTC repeats [78, 79], to TR RNA transcripts, and to possible protein–TR DNA and protein–TR RNA complexes [12–14]. Most importantly, to understand the origin of TR dynamic expansion, the biophysical studies should closely follow *in vitro* and *in vivo* biological studies of intricate TR expansion processes.

Acknowledgments

The 600 MHz NMR spectrometer at the University of Houston is funded by the W. M. Keck Foundation. Acknowledgment is made to NIH (R01GM54652, R29GM59957) and the Robert A. Welch Foundation (E-1270) for financial support, and the W. M. Keck Center for Computational Biology for computer resource support. Dr. J. Jie has made important contributions to the structure elucidation of the $d(CTG)_3$ duplex. We thank J. Look (Rice University) and X. Yang for their help in UV data collection and processing, and Dr. W. Bloch (Perkin Elmer Applied Biosystems) for communicating UV results prior to publication and for providing some of the $d(CGG)_nT$ repeats. X.G. and G.K.S. are grateful for the support from Professor G. E. Fox. G.K.S. was a NASA Fellow student.

References

1. Kunst, C. B., Zerylnick, C., Karickhoff, L-, Eichler, E., Bullard, J., Chalifoux, M., Holden, J. J. A., Torroni, A., Nelson, D. L., and Warren, S. T. (1996). FMR1 in global populations. *Am. J. Hum. Genet.* **58,** 513–522.
2. Sinden, R. R., and Wells, R. D. (1992). DNA structure, mutations, and human genetic disease. *Curr. Opin. Biotech.* **3,** 612–622. [References cited therein].
3. Wells, R. D. (1996). Molecular basis of genetic instability of triplet repeats. *J. Biol. Chem.* **271,** 2875–2878.
4. Harvey, S. (1997). Slipped structures in DNA triplet repeat sequences: entropic contributions to genetic instabilities. *Biochemistry* **36,** 3047–3049.
5. Kang, S., Ohshima, K., Shimizu, M., and Wells, R. D. (1995). Pausing of DNA synthesis *in vitro* at specific loci in CTG and CGG triplet repeats from human hereditary disease genes. *J. Biol. Chem.* **270,** 27014–27021.
6. Pearson, C. E., Ewel, A., Acharya, S., Fishel, R. A., and Sinden, R. R. (1997). Human MSH2 binds to trinucleotide repeat DNA structures associated with neurodegenerative diseases. *Hum. Mol. Genet.* [in press].
7. Deissler, H., Behn-Krappa, A., and Doerfler, W. (1996). Purification of nuclear proteins from human Hela cells that bind specifically to the unstable tandem repeat $(CGG)_n$ in the human FMR1 gene. *J. Biol. Chem.* **271,** 4237–4334.
8. Timchenko, L. T., Timchenko, N. A., Caskey, C. T., and Roberts, R. (1996). Novel proteins with binding specificity for DNA CTG repeats and RNA CUG repeats: implications for myotonic dystrophy. *Hum. Mol. Genet.* **5,** 115–121.
9. Timchenko, L. T., Miller, J. W., Timchenko, N. A., DeVore, D. R., Datar, K. V., Lin, L., Roberts, R., Caskey, T., and Swanson, M. S. (1996). Identification of a $(CUG)_n$ triplet repeat RNA binding protein and its expression in myotonic dystrophy. *Nucleic Acids Res.* **24,** 4407–4414.
10. Yano-Yanagisawa, H., Li, Y., Wang, H., and Kohwi, Y. (1995). Single-stranded DNA binding proteins isolated from mouse brain recognize specific trinucleotide repeat sequences in vitro. *Nucleic Acids Res.* **23,** 2654–2660.
11. Zhao, Y., Cheng, W., Gibb, C. L. D., Gupta, G., and Kallenbach, N. R. (1996). HMG box proteins interact with multiple tandemly repeated $(GCC)_n$ · (GGC)m DNA sequences. *J. Biomol. Struct. Dynam.* **14,** 235–238.
12. Yeakley, J. M., Morfin, J. P., Rosenfeld, M. G., and Fu, X. D. (1996). A complex of nuclear proteins mediates SR protein binding to a purine-rich splicing enhancer. *Proc. Natl. Acad. Sci. USA* **93,** 7582–7587.
13. Wang, Y. H., Amirhaeri, S., Kang, S., Wells, R. D., and Griffith, J. D. (1994). Preferential nucleosome assembly at DNA triplet repeats from the myotonic dystrophy gene. *Science* **265,** 669–670.
14. Boyer, J. C., Umar, A., Risinger, J. I., Lipford, J. R., Kane, M., Yin, S., Barrett, J. C., Kolodner, R. D., and Kunkel, T. A. (1995). Microsatellite instability, mismatch repair deficiency, and genetic defects in human cancer lines. *Cancer Res.* **55,** 6063–6010.
15. Radhakrishnan, I., and Patel, D. J. (1994). DNA triplexes, hydration sites, energetics, interactions and function. *Biochemistry* **33,** 11405–11416.
16. Rhode, D., and Giraldo, R. (1995). Telomere structure and function. *Curr. Opin. Struct. Biol.* **5,** 311–322.
17. Germann, M. W., Zhou, N, van de Dande, J. H., and Vogel, H. J. (1995). Parallel-stranded duplex DNA: an NMR perspective. *Methods Enzymol.* **261,** 207–225.
18. Saegner, W. (1984). "Principles of Nucleic Acid Structure," p. 120. Springer-Verlag, New York.
19. Li, Y., Zon, G., and Wilson, W. D. (1991). NMR and molecular modeling evidence for a GA mismatch base pair in a purine-rich DNA duplex. *Proc. Natl. Acad. Sci. USA* **88,** 26–30.

20. Chou, S. H., and Reid, B. R. (1997). Sheared purine × purine pairing in biology. *J. Mol. Biol.* **267,** 1055–1067.
21. Maskos, K., Gunn, B. M., LeBlanc, B. A., and Morden, K. M. (1993). NMR study of G · A and A · A pairing in d(GCGAATAAGCG)$_2$. *Biochemistry* **32,** 3585–3595.
22. Gehring, K., Leroy, J.-L., and Gueron, M. (1993). A telomeric DNA structure with protonated cytosine · cytosine base pairs. *Nature* **263,** 561–565.
23. Charlesworth, B., Sniegowski, P., and Stephan, W. (1994). The evolutionary dynamics of repetitive DNA in eukaryotes. *Nature* **371,** 215–220.
24. Nowak, R. (1994). Mining treasures from 'junk DNA.' *Science* **263,** 608–610.
25. Richards, R., and Sutherland, G. R. (1992). Heritable unstable DNA sequences. *Nature Genet.* **1,** 7–9.
26. Blackburn, E. (1990). Structure and function of telomeres. *Nature* **350,** 569–573.
27. Giovannucci, E., Stampfer, M. J., Krithivas, K., Brown, M., Brufsky, A., Talcott, J., Hennekens, C. H., and Kantoff, P. W. (1997). The CAG repeat within the androgen receptor gene and its relationship to prostate cancer. *Proc. Natl. Acad. Sci. USA* **94,** 3320–3323.
28. Wang, Y., and Patel, D. J. (1994). Solution structure of the Tetrahymena telomeric repeat d(T_2G_4)$_4$ G-tetraplex. *Structure* **2,** 1141–1156.
29. Harley, C. B. (1995). Telomeres and aging: fact, fancy, and the future. *J. NIH Res.* **7,** 64–68.
30. Cattasi, P., Gupta, G., Garcia, A. E., Ratliff, R., Hong, L., Yau, P., Moyzis, R. K., and Bradbury, E. M. (1994). Unusual structures of the tandem repetitive DNA sequences located at human centromeres. *Biochemistry* **33,** 3819–3830.
31. Chou, S.-H., Zhu, L., and Reid, B. R. (1994). The unusual structure of the human centromere (GGA)$_2$ motif. Unpaired guanosine residues stacked between sheared G · A pairs. *J. Mol. Biol.* **244,** 259–268.
32. Jaishree, T. N., and Wang, A. H.-J. (1994). Conformations of the alternating (C-T)$_n$ sequence under neutral and low pH. *FEBS Lett.* **337,** 139–144.
33. Robinson, H., van Boom, J. H., and Wang, A. H.-J. (1994). 5′-CGA motif induces other sequences to form homobase-paired parallel-stranded DNA duplex: the structure of (G-A)$_n$ derived from four DNA oligomers containing (G-A)$_3$ sequence. *J. Am. Chem. Soc.* **116,** 1565–1566.
34. Dolinnaya, N. G., Ulku, A., and Fresco, J. R. (1997). Parallel-stranded liner homoduplexes of d(A^+-G)$_n$ > 10 and d(A-G)$_n$ > 10 manifesting the contrasting ionic strength sensitivities of poly(A^+-A^+) and DNA. *Nucleic Acids Res.* **25,** 1100–1107.
35. Mitas, M. (1997). Trinucleotide repeats associated with human diseases. *Nucleic Acids Res.* **25,** 2245–2254.
36. Gao, X., Huang, X., Smith, G. K., Zheng, M., and Liu, H. (1995). A new antiparallel duplex motif of DNA CCG repeats that is stabilized by extrahelical bases symmetrically located in the minor groove. *J. Am. Chem. Soc.* **117,** 8883–8884.
37. Smith, G. K., Jie, J., Fox, G. E., and Gao, X. (1995). DNA CTG triplet repeats involved in dynamic mutations of neurologically related gene sequences from stable duplexes. *Nucleic Acids Res.* **23,** 4303–4311.
38. Zheng, M., Huang, X., Smith, G. K., Yang, X., and Gao, X. (1996). High folding propensity and duplex formation of single stranded DNA triplet repeats related to hereditary neurodegenerative disease. *J. Mol. Biol.* **264,** 323–336.
39. Gacy, A. M., Goeliner, G., Juranic, N., Macura, S., and McMurray, C. T. (1995). Trinucleotide repeats that expand in human disease form hairpin structures in vitro. *Cell* **81,** 533–540.
40. Pertruska, J., Arnheim, N., and Goodman, M. F. (1996). Stability of intrastrand hairpin structures formed by the CAG/CTG class of DNA triplet repeats associated with neurological diseases. *Nucleic Acids Res.* **24,** 1992–1999.
41. Nadel, Y., Weisman-Shomer, P., and Fry, M. (1995). The fragile X syndrome single strand d(CGG)$_n$ nucleotide repeats readily fold back to form unimolecular hairpin structures. *J. Biol. Chem.* **270,** 28970–28977.
42. Lian, C., Robinson, H., and Wang, A. H.-J. (1996). Structure of actinomycin D bound with (GAAGCTTC)$_2$ and (GATGCTTC)$_2$ and its binding to the (CAG)$_n$: (CTG)$_n$ triplet sequence as determined by NMR analysis. *J. Am. Chem. Soc.* **118,** 8791–8801.
43. Robinson, H., and Wang, A. H.-J. (1993). 5′-CGA sequence is a strong motif for homo base-paired parallel-stranded DNA duplex as revealed by NMR analysis. *Proc. Natl. Acad. Sci. USA* **90,** 5224–5228.
44. Robinson, H., van der Marel, G. A., van Boom, J. H., and Wang, A. H.-J. (1992). Unusual DNA conformation at low pH revealed by NMR: parallel-stranded DNA duplex with homo base pairs. *Biochemistry* **31,** 10510–10517.
45. Wang, Y., and Patel, D. J. (1994). Solution structure of the d(TCGA) duplex at acidic pH. A parallel-stranded helix containing C^+ · C, G · G and A · A pairs. *J. Mol. Biol.* **242,** 508–526.
46. Fry, M., and Loeb, L. (1994). The fragile X syndrome d(CGG)$_n$ nucleotide repeats form a stable tetrahelical structure. *Proc. Natl. Acad. Sci. USA* **91,** 4950–4954.
47. Usdin, K., and Woodford, K. J. (1995). CGG repeats associated with DNA instability and chromosome fragility form structures that block DNA synthesis in vitro. *Nucleic Acids Res.* **23,** 4203–4209.
48. Kettani, A., Kumar, R. A., and Patel, D. J. (1995). Solution structure of a DNA quadruplex containing the fragile X syndrome triplet repeat. *J. Mol. Biol.* **254,** 638–656.
49. Michell, J. E., Newbury, S. F., and MaClellan, J. A. (1995). Compact structure of d(CNG)$_n$ oligonucleotides in solution and their possible relevance to fragile X and related human genetic diseases. *Nucleic Acids Res.* **23,** 1876–1881.
50. Kuryavyi, V., and Jovin, T. M. (1995). Triad-DNA: a model for trinucleotide repeats. *Nature Genet.* **9,** 339–341.
51. Gueron, M., Kochoyan, M., and Leroy, J.-L. (1987). A single mode of DNA base-pair opening drives imino proton exchange. *Nature* **328,** 89–92.
52. Feigon, J., Koshlap, K. M., and Smith, F. W. (1995). ^{1}H NMR spectroscopy of DNA triplexes and quadruplexes. *Methods Enzymol.* **261,** 225–255.
53. Majumdar, A., and Hosur, R. V. (1992). Simulation of 2D NMR spectra for determination of solution conformations of nucleic acids. *Prog. NMR Spect.* **24,** 109–158.
54. van de Ven, F. J., and Hilbers, C. W. (1988). Nucleic acid and nuclear magnetic resonance. *Eur. J. Biochem.* **178,** 1–38.
55. Gorenstein, D. G. (1984). Phosphorus-31 chemical shifts and spin-spin coupling constants: Principles and empirical observations. *In* "Phosphorus-31 NMR. Principles and Application" (D. G. Gorenstein, Ed.), pp. 7–56. Academic Press, New York.
56. Gorenstein, D. G. (1992). ^{31}P NMR of DNA. *Methods Enzymol.* **211,** 254–285.
58. van Wijk, J., Huckriede, B. D., Ippel, J. H., and Altona, C. (1992). Furanose sugar conformations in DNA from NMR coupling constants. *Methods Enzymol.* **211,** 286–306.
58. Saenger, W. (1984). "Principles of Nucleic Acid Structure," pp. 51–104. Springer-Verlag, New York.
59. Breslauer, K. J. (1994). Extracting thermodynamic data from equilibrium melting curves for oligonucleotide order-disorder transi-

tions in protocols for oligonucleotide conjugates. *Methods Mol. Biol.* **26,** 347–372.
60. Turner, D. H. (1996). Thermodynamics of base pairing. *Curr. Opinion Struct. Biol.* **6,** 299–304. [References cited]
61. Petersheim, M., and Turner, D. H. (1983). Base stacking and base pairing contributions to helix stability. *Biochemistry* **22,** 256–263.
62. UV analysis mainly used a curve fitting program developed in this laboratory. Thermodynamic parameters of cooperative transitions derived from this program are nearly identical to those determined using the MeltWin program (Ref. 61). The major advantage of using the Excel software is that fitting results are graphically visualized and results for the non-ideal curves appear to be more reasonable.
63. Brunger, A. T. (1993). "X-PLOR Version 3.1: A System for X-ray Crystallography and NMR." Yale University Press, New Haven, CT.
64. Huang, X., and Gao, X. (1997). Manuscript in preparation.
65. Verdine, G. L., and Bruner, S. D. (1997). How do DNA repair proteins locate damaged bases in the genome. *Chem. Biol.* **4,** 329–334.
66. Huang, X., and Gao, X. (1997). Structure and dynamics of DNA CCG and CGG trinucleotide repeats. [Submitted for publication]
67. Chen, F. M. (1995). Acid-facilitated supramolecular assembly of G-quadruplexes in $d(CGG)_4$. *J. Biol. Chem.* **270,** 23090–23096.
68. Uesugi, S., Oda, Y., Ikehara, M., Kawase, Y., and Ohtsuka, E. (1987). Identification of I:A mismatch base-pairing structure in DNA. *J. Biol. Chem.* **262,** 6965–6968.
69. Arnold, F. H., Wolk, S., Cruz, W. P., and Tinoco, I., Jr. (1987). Structure, dynamics, and thermodynamics of mismatched DNA oligonucleotide duplexes $d(CCCAGGG)_2$ and $d(CCCTGGG)_2$. *Biochemistry* **26,** 4068–4075.
70. Cruse, W. B. T., Saludjian, P., Biala, E., Strazewski, P., Prange, T., and Kennard, O. (1994). Structure of a mispaired RNA double helix at 1.6 Å resolution and implications for the prediction of RNA secondary structure. *Proc. Natl. Acad. Sci. USA* **91.** 4160–4164.
71. Zheng, M. (1997). Studies of the structure and stability of DNA trinucleotide repeat sequence by nuclear magnetic resonance spectroscopy. Thesis, University of Houston.
72. Johnson, Jr., W. C. (1996). Determination of the conformation of nucleic acids by electronic CD. *In* "Circular Dichroism and the Conformational Analysis of Biomolecules" (G. D Fasman, Ed.), pp. 433–468. Plenum Press, New York.
73. Saenger, W. (1984). "Principles of Nucleic Acid Structure," pp. 315–320. Springer-Verlag, New York.
74. Pearson, C. E., and Sinden, R. R. (1996). Alternative structures in duplex DNA formed within the trinucleotide repeats of the myotonic dystrophy and fragile X loci. *Biochemistry* **35,** 5041–5053.
75. Eichler, E. E., Holden, J. J., Popovich, B. W., Reiss, A. L., Snow, K., Thibodeau, S. N., Richards, C. S., Ward, P. A., and Nelson, D. L. (1994). Length of uninterrupted CGG repeats determine instability in the FMR1 gene. *Nature Genet.* **8,** 88–94.
76. Unpublished spectrophotometric melting curves (W. Bloch, Perkin Elmer Applied Biosystems) reveal that $d(CGG)_n$ shows two stable secondary structures, one of which requires the specific presence of both Mg^{2+} and K^+ and has a T_m above 100°C; the other shows no Mg^{2+} binding. The much weaker $d(CCG)_n$ secondary structure is strongly stabilized by Mg^{2+}. Mg^{2+} has almost no effect on $d(CTG)_n$ and $d(CAG)_n$ melting.
77. Our preliminary NMR examination of GAAGAA and AAGAAG hexamers in solutions containing 0.1 *M* NaCl at neutral pH indicates that unlike the CXG repeats, these hexamers are not structured; no H-bonded NH resonances were detected in the 10–16 ppm spectral region. It is possible that this TR sequence requires a longer minimal chain length to form structures or the formed structure is less stable than those of the CXG TR sequences.
78. Described as unpublished results in [Mitas, M. (1997). Trinucleotide repeats associated with human diseases. *Nucleic Acids. Res.* **25,** 2245–2254.] that $d(GAA)_{15}$ forms a hairpin.

CHAPTER 41

Structural Studies on the Unstable Triplet Repeats

S. V. SANTHANA MARIAPPAN, XIAN CHEN, AND PAOLO CATASTI
Life Sciences Division, LS-8, MS M880, Los Alamos National Laboratory, Los Alamos, New Mexico 87545

E. MORTON BRADBURY
Life Sciences Division, LS-DO, MS M880, Los Alamos National Laboratory, Los Alamos, New Mexico 87545; and Department of Biological Chemistry, School of Medicine, University of California at Davis, Davis, California 95616

GOUTAM GUPTA
Theoretical Biology and Biophysics, T-10, MS K710, Los Alamos National Laboratory, Los Alamos, New Mexico 87545

I. INTRODUCTION

The expansions of triplet DNA repeats define a new type of mutation in the human genome [1, 2]. The genetic instability due to triplet expansion is associated with many genetically inherited neurological disorders [3–6]. Figures 41-1 and 41-2 show different triplet repeats in different genetic disorders and their associated genes. These triplet repeats are located most frequently inside the noncoding regions (upstream, downstream, intron) of genes and less frequently inside the coding regions. All of the unstable triplet repeats identified so far are GC-rich of the form (CXG)/(CX′G), where X and X′ are complementary to each other: for example, GCC/GCC in the fragile X syndrome (FraX) (Fig. 41-1A, [3]), CTG in myotonic dystrophy (DM) (Fig. 41-1B, [4]), and CAG in Huntington's disease (HD) (Fig. 41-1C, [5]) belong to this category. The only exception is the GAA/TTC repeat associated in Friedreich's ataxia (FRDA) (Fig. 41-2, [6]).

Expansions of DNA triplet repeats (or any repeat) may involve one of the three mechanisms [7–9]: (i) unequal crossover, i.e., crossover between tracts misaligned by an integral number of repeats; (ii) slippage during DNA replication, i.e., during replication, the primer and template strand transiently dissociate and the slippage of the strands can then result in either expansion or deletion; and (iii) misalignment followed by excision repair, i.e., a mutagenic alternative DNA secondary structure may be formed during or after replication, which is excised and repaired leading to either deletion or expansion. Although these mechanisms have the potential to explain genetic instabilities associated with disease-related triplets, the exact mechanism has yet to be proven. However, it is generally accepted that the key step in triplet expansion is the formation of non-Watson–Crick DNA structures during replication or crossover recombination [10, 11]. Hence, the identification and characterization of these unusual DNA structures formed by triplet repeats are of crucial importance in understanding the mechanism of expansion.

Unusual DNA structures include hairpins, cruciforms and junctions, and intramolecular triplexes and tetraplexes. Once they are identified and completely characterized, it is necessary to determine whether these unusual structures are preferentially stabilized in longer repeats. Finally, it has to be determined by *in vitro* and *in vivo* assays whether these structures can, indeed, cause DNA slippage structures during replication.

In this chapter, we provide experimental evidence in support of the hypothesis that unusual DNA structures are responsible for the expansions of the disease-related triplets and their associated genetic instabilities. For this, we have performed the following experiments: (i) we have characterized the unusual DNA structures by nondenaturing gel electrophoresis of short and long

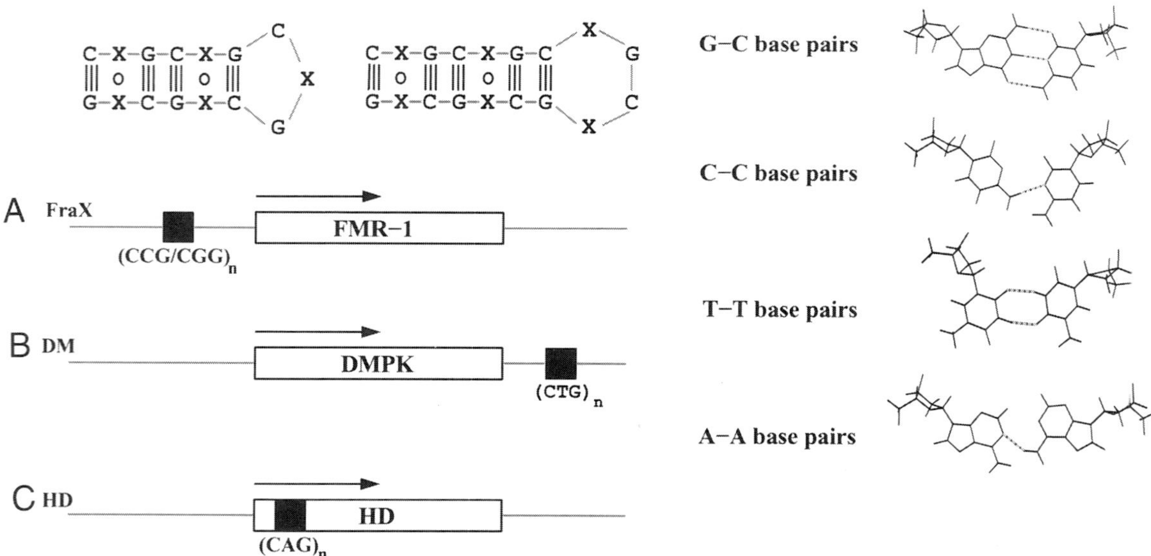

FIGURE 41-1 GC-rich triplet repeats and their locations with respect to their associated genes: (A) CCG upstream of the FMR1 gene, (B) CTG downstream of the DMPK gene, and (C) CAG inside the exon of the HD gene. Hairpin structures formed by the $(CXG)_n$ triplet repeats have three-nucleotide loops for odd repeat numbers and four-nucleotide loops for even repeat numbers. Note that mismatches have orientations similar to the flanking G-C pairs and therefore can be easily embedded into the structure.

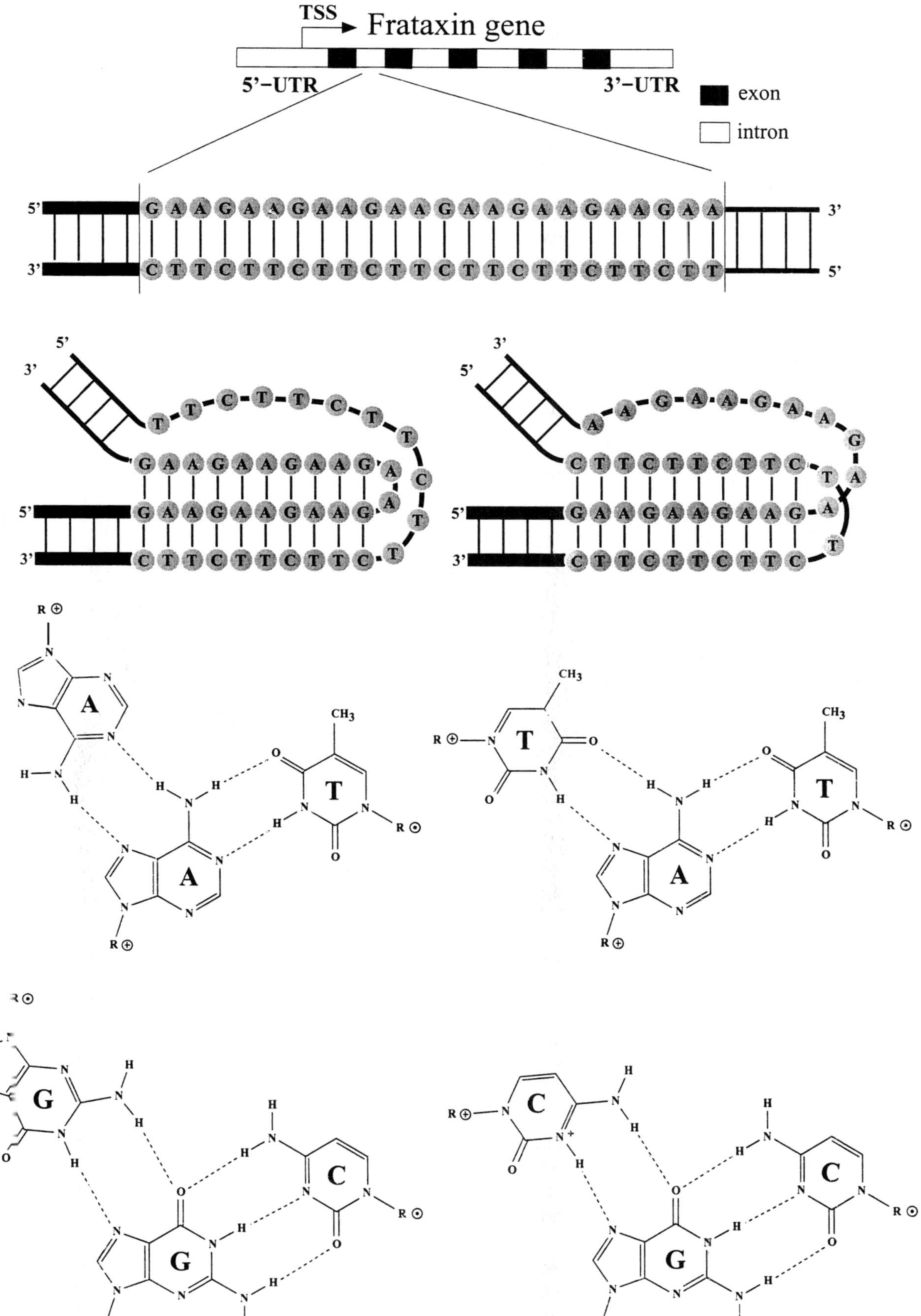

FIGURE 41-2 Possible triplexes formed by the GAA/TTC repeats inside the first intron of the frataxin gene. The folding of the TTC strand leads to a triplex with $C^{+} \cdot G \cdot C$ and $T \cdot A \cdot T$ triads and, therefore, this structure is more stable at acidic pH. The folding of the GAA strand leads to a triplex with $G \cdot G \cdot C$ and $A \cdot A \cdot T$ triads.

triplet repeats, (ii) we have determined the high-resolution structures of short triplet repeats by homonuclear (^1H-^1H) and hetero (^{15}N-^1H) NMR spectroscopy, and (iii) we have detected these unusual DNA structures by an *in vitro* replication assay using various DNA polymerases and their accessory proteins. Two classes of unusual DNA structures are discussed, i.e., the (CXG) triplet repeats that tend to form hairpin structures (Figs. 41-1A–41-1C) and the GAA/TTC triplet repeats that tend to form triplexes (Fig. 41-2). For clarity, the structural studies are discussed separately for each triplet repeat.

Many structural investigations on triplet repeats have been carried out by us and other laboratories [12–48]. The work by other laboratories essentially falls into five distinct categories: (i) detailed thermodynamic analyses that correlate the stabilities of various (CXG) hairpins with their repeat lengths [34], (ii) a combination of low-resolution NMR and thermodynamic and genetic analyses that provides an elegant explanation of how the genetic instability in the triplet repeats results from unusual DNA structures and not from a deficiency in the mismatch repair system as found for the dinucleotide instability in colon cancers [11, 24–27], (iii) gel mobility studies on long tracts of triplet repeats that show the presence of multiple slipped structures [18, 37], (iv) reconstitution experiments that show differential abilities of different triplet tracts to form nucleosomes [e.g., $(CTG)_n$ forms better positioned and more stable nucleosomes than $(GCC)_n$] [29, 30, 40, 41, 106], and (v) *in vitro* and *in vivo* replication and mismatch repair assays with long tracts of triplet repeats that demonstrate the presence of slippage structures [28, 31, 39, 42–45, 47, 48]. These data are extremely important since they provide clear indication that the disease-related triplet repeats can form stable unusual DNA structures which may also be present during their replication. As explained below, our efforts [13, 32, 33, 46, 105] complement the literature by providing high-resolution NMR structural details of the hairpins or triplexes formed by the triplet repeats (see Figs. 41-1 and 41-2). Also our replication assay clearly distinguishes the hairpin-induced slippage structures from the triplex-induced slippage structures.

Homonuclear (^1H-^1H) and (^{15}N-^1H) heteronuclear NMR spectroscopy give the following structural details of the hairpins and triplexes: (i) exact base pairing schemes, (ii) precise chainfolding, and (iii) interactions involving nucleotides in the loop and in the stem. These structural details enable us to explain how single interruptions in the GCC (or CAG) repeat or in the GAA/TTC repeat confer genetic stability by lowering the stability of the hairpin or the triplex, respectively. In addition, the local structure of the CpG sites in the stem of the $(GCC)_n$ hairpin helps explain why this hairpin is a better substrate for methylation by the human methyltransferase than either the Watson–Crick duplex, $(GCC)_n \cdot (GGC)_n$, or the $(GGC)_n$ hairpin.

For our *in vitro* replication assay we have selected repeat lengths, $n < 40$, such that only one copy or a few copies of the alternative structures are formed in the template. Thus, the nature of the replication product in the *in vitro* assay truly reflects the nature of the alternative structure. For example, the formation of a hairpin in the template causes a replication bypass and a reduction in the length of the replication product that corresponds to length of the hairpin. On the other hand, if a triplex is formed in the template replication arrest occurs in the middle of the repeat and the point of arrest indicates the length of the triplex. Note that the CXG repeats tend to form hairpin-induced slippage structures whereas the GAA/TTC repeats tend to form triplex-induced slippage structures. However, for $n \gg 40$, the presence of multiple copies of the alternative structures may lead to a higher order template structure resulting in replication arrests irrespective of whether the individual units are hairpins or triplexes. Apart from distinguishing a hairpin from a triplex, we can also determine the stability of a hairpin or a triplex formed by triplet repeats in the template as a function of its length. Note that in our replication assay the presence of the growing complementary strand, DNA polymerase, and structure-destabilizing proteins such as single-strand binding proteins and other ATP-dependent accessory proteins tend to destabilize a hairpin or a triplex. Although not proven directly, it is reasonable to predict that the same slippage structures will be present either in the template or in the growing chain during replication of these triplet repeats *in vivo*. Hairpin- or triplex-induced DNA slippage structures in the template should lead to deletion whereas the slippage in the growing chain should cause expansion.

II. FRAGILE X TRIPLET REPEAT, $(GCC)_n/(GGC)_n$: STRUCTURAL BASIS FOR EXPANSION AND CpG METHYLATION

The fragile X syndrome is the most common X-linked mental disorder, accounting for 50% of all reported cases [49]. The fragile X syndrome was originally identified by the presence of a microscopic gap or constriction, termed a fragile site, in the long arm of the X chromosome at Xq27.3 in affected individuals by culturing these cells under conditions of folate deficiency [50]. Recently, the gene associated with fragile X syndrome has been isolated and is called FMR1 (fragile X mental retardation-1). The FMR1 gene shows three important features in individuals affected with fragile X syndrome [49–53 : (i) the 5′ untranslated region of the gene contains the triplet repeats of (GGC/GCC) which are massively expanded, (ii) the CpG islands inside the triplet repeat are hypermethylated, and (iii) the expression of the FMR1 gene is either considerably reduced or completely suppressed. The expansion of GGC/GCC triplet repeat and the associated hypermethylation are probably the cause of the suppression of the FMR1 gene and the fragile sites in the X chromosomes. The number of GGC/GCC repeats in normal phenotypes varies between 6 and 53 with 29 occurring most frequently. Premutation alleles have between 54 and 200 repeats, whereas full mutation alleles have more than 200 repeats (Fig. 41-1A). The risk of expansion to the full mutation is dependent on the size of the premutation allele. If the repeat number is small (50–70 copies) then the risk is low, and if the number of copies is high (>90) the risk is close to 100%. The risk of expansion depends also on the purity of the repeat; a single base interruption [e.g., $(GCC)_9 \cdot TCC \cdot (GCC)_9$] in the original repeat sequence reduces the risk [54].

In this section, we first show that the individual single strands of the fragile X repeat, i.e., $(GCC)_n$ and $(GGC)_n$, can form hairpin structures. We then describe the results of an *in vitro* replication assay that demonstrates the presence of hairpin-induced slippage structures. We also show by a methylation assay why the $(GCC)_n$ hairpin-induced slippage structure is an excellent substrate for the human methyltransferase, the enzyme that methylates the Cs at the CpG sites. Finally, we propose a structure-based mechanism of how expansion and hypermethylation can cause suppression of the FMR1 gene and the onset and progression of the fragile X syndrome.

A. Structural Characterization of $(GCC)_n$ by Gel Electrophoresis

Theoretically, at neutral pH, the two individual strands of the fragile X repeat can form either a mismatched homoduplex or a monomeric hairpin. The homoduplex and the stem of the hairpin of the $(GCC)_n$ strands involve Watson–Crick G · C pairs and mismatched C · C pairs. Note that the hairpin of $(GCC)_n$ should have half the length but approximately the same cross-section as the homoduplex (i.e., $[(GCC)_n]_2$) or the Watson–Crick duplex (i.e., $(GCC)_n \cdot (GGC)_n$]. Therefore, the duplex is expected to show about half the gel mobility of the corresponding hairpin. The electrophoretic mobilities of $(GCC)_n$ in a nondenaturing (15%) polyacrylamide gel reveal the presence of only hairpins for repeat lengths, $n > 5$ at both 5 and 200 mM NaCl concentration [32].

B. Structural Characterization of $(GCC)_n$ by NMR

The imino proton spectra of $(GCC)_5$ and $(GCC)_6$ at 5°C and at pH 6.3 show the presence of G-imino protons within 13.4–13.1 ppm that correspond to Watson–Crick G · C pairs as well as a broad envelop around 11.0 ppm that corresponds to loop G-imino protons. The temperature-dependent imino proton profile of $(GCC)_{5,6}$ reveals that the loop G-imino signals disappear above 5°C. Deconvolution of the areas under the imino signals indicates the presence of four G · C pairs in $(GCC)_5$ and five G · C pairs in $(GCC)_6$ which are consistent with either a blunt hairpin or a slipped hairpin. For example, a blunt hairpin of $(GCC)_5$ should have the G1 · C15 pair whereas the slipped hairpin of $(GCC)_5$ should have the unpaired C15. See Fig. 41-3A for descriptions of slipped and blunt hairpins of $(GCC)_5$. In order to distinguish between the slipped and blunt hairpin, imino proton spectra have been recorded at 5°C for the analogs, $(GCC)_4GC$ (Fig. 41-3B) and $G(GCC)_5$ (Fig. 41-3C). If $(GCC)_5$ formed a blunt hairpin, the removal of G15 should show the loss of G1 · C15 pair in the imino spectrum of $(GCC)_4GC$ whereas if $(GCC)_5$ formed a slipped hairpin, the removal of G15 should leave the imino spectrum of $(GCC)_4GC$ unaltered which is exactly what we have observed [32]. Again the addition of a 5′ G (i.e., G0 in $G(GCC)_5$ of Fig. 41-3C) should lead to an increase in the total number of G · C pairs for a slipped hairpin whereas the same modification for a blunt hairpin should have no change in the imino spectrum. In fact, $G(GCC)_5$ shows an increase in the number of imino protons corresponding to G · C pairs [32]. Hence, the imino spectrum of $G(GCC)_5$ is also consistent with a slipped hairpin structure of $(GCC)_5$. Note that the same base pairing pattern is preserved in the stems of slipped hairpins formed by $(GCC)_5$ and $(GCC)_6$ (Fig. 41-3D). However, the number of nucleotides in the loop is different in the two cases: as shown in Fig. 41-3, four nucleotides are present in the loop of the slipped $(GCC)_5$ hairpin while three nucleotides are present in the loop of the slipped $(GCC)_6$.

More direct evidence for a blunt or a slipped $(GCC)_n$ hairpin is obtained by monitoring the pairing of the C at the CpG step in this triplet repeat, i.e., in a blunt hairpin this C should be G · C-paired whereas in a slipped hairpin it should be C · C-paired. We have identified the pairing of the C at CpG site of the $(GCC)_n$ hairpin by performing ^{15}N-^{1}H HSQC (heteronuclear sin-

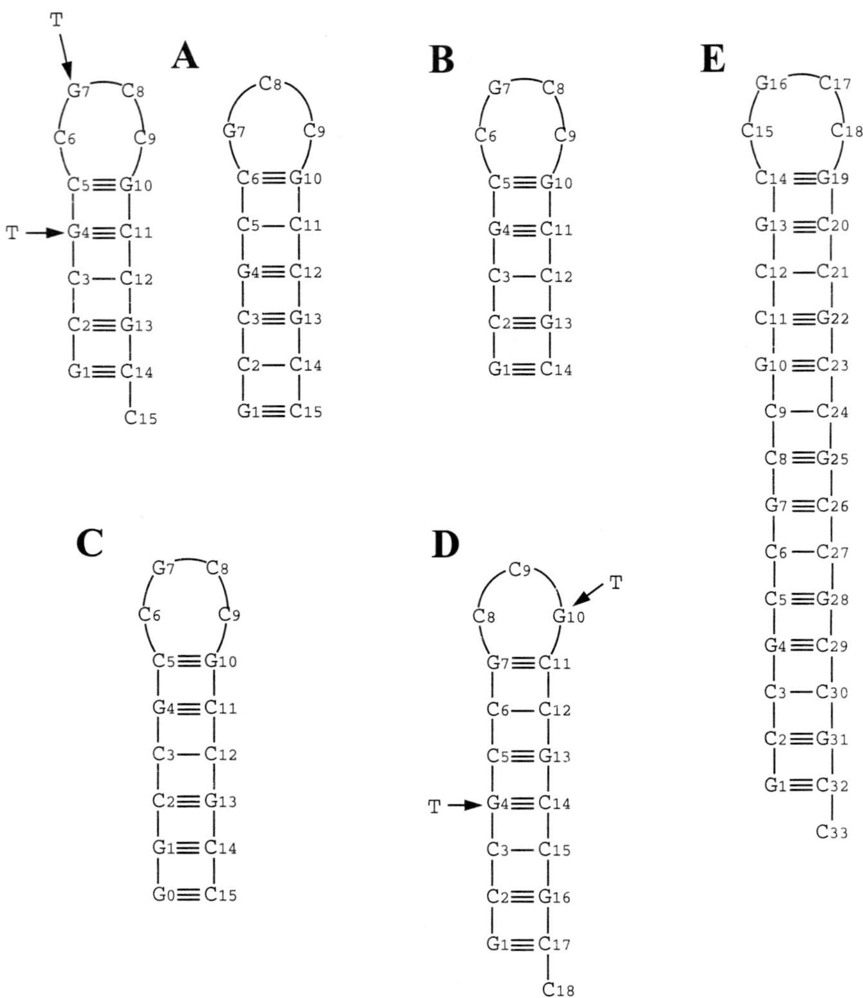

FIGURE 41-3 The hairpin structures of (A) $(GCC)_5$, (B) $(GCC)_4GC$, (C) $G(GCC)_5$, (D) $(GCC)_6$, and (E) $(GCC)_{11}$. The blunt and the slipped hairpins are shown for $(GCC)_5$. NMR data are only consistent with the slipped hairpin structures of $(GCC)_{5,6,11}$. Loop-G signals around 11.0 ppm are observed in all three systems. With increasing temperature loop-G resonance is the first to disappear, e.g., above 5°C in $(GCC)_5$. Subsequently the imino resonances from the terminal G · C pairs in the stem disappear at 35°C. The effect of single site G→T substitutions are also studied. The H-bonding and open-closure of the Cs in the C · G and C · C pairs in stem and the Cs in the loop are studied by specific ^{15}N4-labeling of the Cs.

gle quantum coherence) spectroscopy on three oligomers [105]. Two of them are $(GCC)_5$ sequences and both are ^{15}N4 (amino)-labeled at single sites (one at C11 and the other at C3). The third is a 7-base-pair-long duplex, (C1<u>G2C3C4G5C6</u>G7)$_2$ with ^{15}N4-labeling at C3 and C4. In the duplex, the central 5 (underlined) base pairs mimic the building block of the stem of a $(GCC)_n$ hairpin. Also, in the duplex C3 is G · C-paired whereas C4 is C · C-paired and this allows unambiguous identifications of the ^{15}N4/^{1}H signals of Cs in the G · C and C · C pairs. The ^{15}N-^{1}H HSQC spectrum of $(GCC)_5$ with ^{15}N-labeling at C11 shows a pair of crosspeaks as expected from a C in a G · C pair. This is only consistent with a slipped $(GCC)_5$ hairpin (and not with a blunt hairpin). Again, the ^{15}N-^{1}H HSQC spectrum of $(GCC)_5$ with ^{15}N-labeling at C3 shows a single crosspeak as expected from a C in a C · C pair which is only consistent with a slipped $(GCC)_5$ hairpin.

We have also studied the interaction and exchange properties of the C · C pair in a slipped $(GCC)_5$ hairpin [105]. For this we have incorporated ^{15}N4-labels at C2 and C11 (both involved in G · C pairs), at C3 and C12 (both involved in C · C pairs), and at C6 (in the loop). The loop amino signal of C is upfield-shifted with respect to the amino signal of C from the C · C pair. The ^{15}N-^{1}H-^{1}H HMQC-NOESY experiments reveal NOEs between the amino protons of C in the C · C pair and the imino protons from the neighboring G · C pairs in the stem. This proves that the C · C pair is internally stacked in the stem of the $(GCC)_5$ hairpin. The pH- and temperature-dependent ^{15}N-^{1}H HSQC experiments reveal that the amino protons of the C · C pair exchanges more rapidly than those of the G · C pair but slower than those that belong to the C in the loop. The ^{15}N-^{1}H HSQC spectrum of $(GCC)_{11}$ in which two consecutive Cs in the stem are $^{15}N4$-labeled also confirms that the Cs at the CpG steps of the stem are C · C-paired whereas the Cs at the GpC steps are G · C-paired.

Detailed analyses of the NOESY at 25, 50, 75, 100, 125, 200, and 500 ms of mixing and the DQF-COSY data of the slipped $(GCC)_{5,6}$ hairpins reveal that all the nucleotides adopt (*C2′-endo, anti*) conformation [32]. The presence of continuous sequential interactions involving both exchangeable and nonexchangeable protons reconfirms that the C · C pairs in these hairpins are internally stacked. The Cs in the C · C pairs are not protonated since in $(GCC)_{5,6,11}$ we have observed no imino signal from protonated Cs within the pH range 6–7 [32]. The C · C pairs probably involves a single H bond between amino (N4) donor and imino or carbonyl (N3 or O2) acceptor. This leads to two possibilities in which either of the two Cs can act as a proton donor or an acceptor. As previously shown, the C · C pairs in these hairpins are more susceptible to open-closure than the G · C pairs. In addition, weaker intra- and inter-nucleotide NOESY cross-peaks at the C · C pairs of the $(GCC)_5$ and $(GCC)_6$ hairpins indicate the presence of local flexibility. In 400-ps unrestrained molecular dynamics, the C3 · C12 pair in the $(GCC)_5$ hairpin can undergo local periodic sliding motions between the two degenerate H-bonding states without violating local or distant NOE constraints. Such a sliding motion makes Cs in the C · C pairs intrinsically more flexible than Cs in the G · C pairs. As discussed later, the flexibility of the C · C pair at the CpG step of the $(GCC)_n$ hairpins imparts an exceptional substrate specificity for the human methyltransferase. Figures 41-4A and 41-4B show the ensemble-averaged slipped hairpin structures of $(GCC)_5$ and $(GCC)_6$ as derived from the NMR data.

Gao and co-workers [20] and Mitas and co-workers [55] proposed an "E-motif" for $(CCG)_n$, in which the Cs at the CpG sites are C · G-paired. Gao and co-workers have based their prediction on the slipped duplex structure of $(CCG)_2$ whereas Mitas and co-workers based their prediction on gel electrophoresis, P1 digestion, and chemical modification studies, $5'a(CCG)_{15}a'3'$, where a and a′ are complementary to each other. We are concerned that pronounced end-effects significantly distort the structures in the short $(CCG)_n$ (n = 2 or 3) duplexes studied by Gao and co-workers whereas the overinterpretation of the experimental data of the $(CCG)_{15}$ sequence with sticky (a and a') flanks mars the structural conclusions derived by Mitas and co-workers.

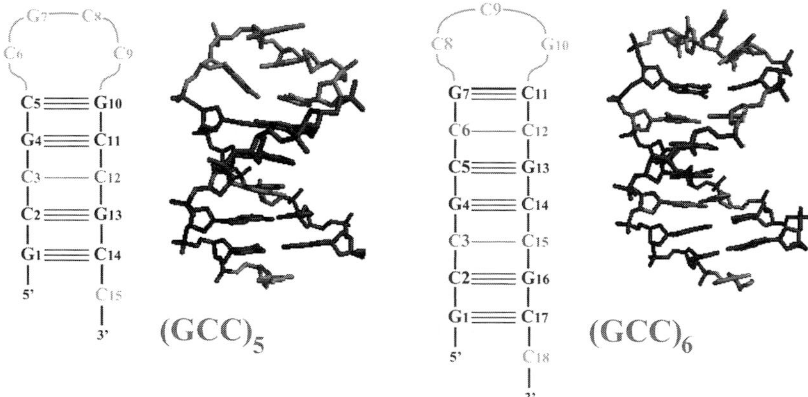

FIGURE 41-4 Schematic and three-dimensional structures of the C-rich strands of the fragile X triplet repeats: $(GCC)_5$ and $(GCC)_6$. The structures are derived using 1D/2D proton NMR spectroscopy and averaged over 100 sampled structures. Both $(GCC)_5$ and $(GCC)_6$ form slipped hairpins with a 3′ overhanging C. Both structures fold so as to maximize G · C Watson–Crick base pairs. Single hydrogen bonded C · C mismatches are present at the CpG steps in the stem. All nucleotides are in (*C2′-endo, anti*) conformations. $(GCC)_5$ has four nucleotides in the loop, whereas $(GCC)_6$ has only three. Analyses of the 2D NMR COSY and NOESY data lead to about 200 NOEs for each structure; 100 structures compatible with the distance constraints are extracted from the 400-ps restrained MD trajectory and energy minimized. All the structures belong to the same cluster with an average Mean Square Deviations (MSD) of 0.6 $Å^2$ for $(GCC)_5$ and 0.7 $Å^2$ for $(GCC)_6$.

Therefore, we have studied $(GCC)_{5,6,7,\&11}$ which all form stable hairpins under physiological salt concentrations. In all these hairpins, the Cs at the CpG sites of the stem are C · C-paired and this rules out the possibility of an "E-motif" for $(CCG)_n$.

C. Structural Characterization of $(GGC)_n$ by Gel Electrophoresis

We have carried out gel mobility studies of $(GGC)_n$ for $n = 5$, 6, 7, and 11. For short repeat lengths, i.e., $(GGC)_{5,6,7}$, two populations have been observed: a homoduplex (the major population) and a hairpin (the minor population). Higher DNA and salt concentrations favor the duplex population. However, for longer repeat lengths, i.e., $(GGC)_{>11}$, the hairpin is the predominant population at all DNA and salt concentrations. Therefore, the gel data [32] unambiguously demonstrate that the $(GGC)_n$ strands of the fragile X repeat are capable of forming hairpin structures when the repeat number is large (i.e., $n > 11$).

D. Structural Characterization of $(GGC)_n$ by NMR

Although gel electrophoresis indicates the presence of both hairpin and duplex structures for $(GGC)_n$, only duplex structures are predominantly present for $n =$ 4–11 under NMR solution conditions (DNA concentrations being two orders of magnitude higher in NMR experiments). We have determined high-resolution structures of $(GGC)_{4,5,6}$ by NMR [32] since the duplex structure adequately models the stem of the hairpin formed by longer $(GGC)_n$ sequences. Figure 41-5 schematically describes the $(GGC)_4$ duplex and its analogs that we have studied to determine the pairing scheme in the duplex.

A detailed analysis of NMR data reveals that $(GGC)_4$, $(GGC)_5$, and $(GGC)_6$ all form duplexes with a 6-base-pair-long structural repeat,

$$G1^{anti}\text{-}G2^{anti}\text{-}C3^{anti}\text{-}G4^{anti}\text{-}G5^{syn}\text{-}C6^{anti}\text{-}G7^{anti}$$

$$C1^{anti}\text{-}G2^{syn}\text{-}G3^{anti}\text{-}C4^{anti}\text{-}G5^{anti}\text{-}G6^{anti}\text{-}C7^{anti}$$

Two symmetric O6—H-N1 H bonds are present in the $G^{anti} \cdot G^{syn}$ pairing [32, 56]. Figure 41-6A shows the ensemble-averaged structure of the $[(GGC)_4]_2$ duplex that is consistent with the NMR data. We used a molecular modeling approach to construct the hairpin structures of the G-rich strands. The stem of the hairpin is constructed on the basis of the NMR data of the duplex and then the two arms of the stem are connected by an

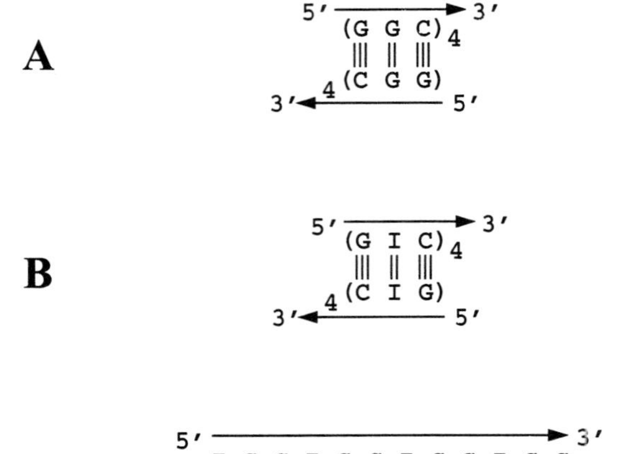

FIGURE 41-5 The H-bonding schemes in the (A) $(GGC)_4$, (B) $(GIC)_4$, and (C) $(IGC)_4$ duplexes. NMR studies on $(GGC)_4$ and its analogs with G → I substitutions help us to prove that the G · G base pairing in $(GGC)_4$ is through the imino protons.

energetically stable loop segment. Figure 41-7B shows the proposed energy-minimized hairpin model of $(GGC)_9$ in which the stem structure is consistent with the NMR data of $[(GGC)_4]_2$ duplex.

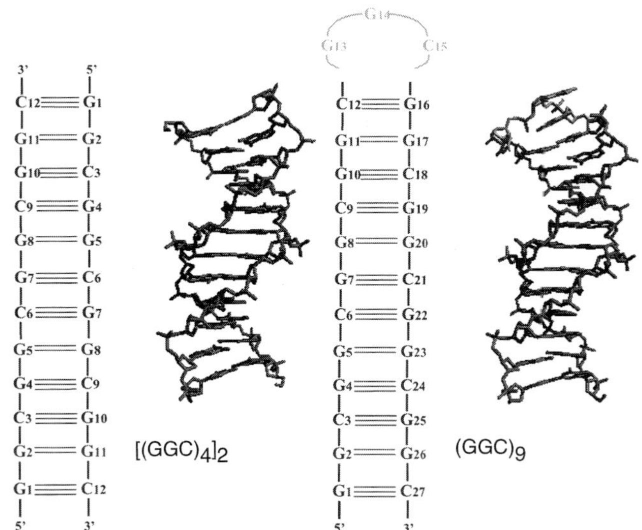

FIGURE 41-6 (left) Three-dimensional structure of $[(GGC)_4]_2$ averaged over 100 structures derived using 1D/2D NMR spectroscopy and restrained molecular dynamics simulations; 200 distance constraints for the restrained MD simulations are estimated from the mixing-time-dependent NOESY data using full-relaxation matrix analysis. (B, right) Three-dimensional average structure of $(GGC)_9$ hairpin from the 100 hairpin structures determined by restrained MD simulations. The distance constraints for the stem used in the molecular dynamics simulations were the same as those of $[(GGC)_4]_2$ duplex and no constraints are imposed on the loop.

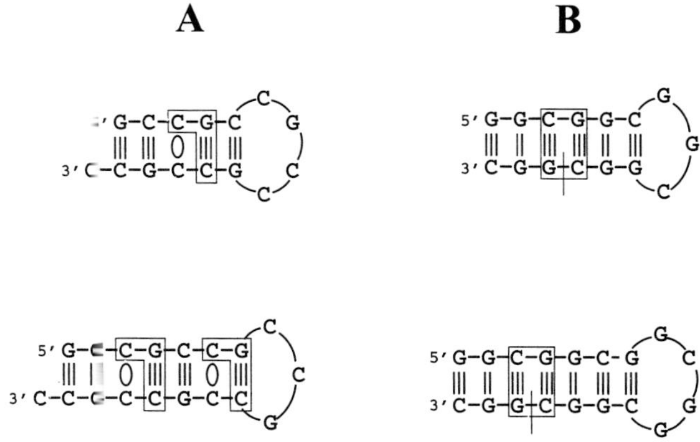

FIGURE 41-7 The nature of CpG sites in (A) $(GCC)_{5,6}$ and (B) $(GGC)_{5,6}$ hairpins. Note that the Cs at the CpG sites in the stem are C · C-paired in the case of $(GCC)_n$ hairpins, whereas in the $(GGC)_n$ hairpins the same Cs are G · C-paired.

Mitas and co-workers [21] have also suggested a $G^{syn} \cdot G^{anti}$ pairing for the 5′a$(CGG)_{15}$a3′ hairpin. However, Gao and co-workers [19] have concluded that there is no G G pairing in the short $(CGG)_{2,3}$ duplexes since they could not observe any imino signal due to a G · G pair. Probably the short length and less than optimal DNA and salt concentrations hinder the formation of a uniformly paired duplex structure.

E. Structural Differences in the Hairpins Formed by the $(GCC)_n$ and $(GGC)_n$ Strands

NMR and gel electrophoresis data show that the individual $(GCC)_n$ and $(GGC)_n$ strands of the fragile X repeat can form hairpin structures under physiological conditions [13, 32]. The $(GCC)_n$ strand can form a hairpin even for short repeats ($n > 5$) whereas the $(GGC)_n$ strand requires longer repeats ($n > 11$). Also as shown in Fig. 41-7, the CpG sites are different in these two hairpins. In the $(GCC)_n$ hairpins the Cs at the CpG sites in the stem are C · C-paired whereas in the $(GGC)_n$ hairpins the same Cs are G · C-paired. This difference in the local CpG structures in these two hairpins affects their substrate efficiencies for the human methyltransferase because in the catalytic process, the most efficient configuration of the target CpG has been postulated to be the one that involves the C in a C · C pair and the G in a G · C pair [57–59]. As shown in Fig. 41-7, the $(GCC)_n$ hairpins have exactly the same CpG configuration that is preferred by the human methyltransferase whereas the $(GGC)_n$ hairpins have both C and G of the CpG site in G · C pairs. We have carried out a methylation assay [14] with the human methyltransferase on $(GCC)_n$ and $(GGC)_n$ hairpins and the corresponding Watson–Crick duplex, $(GCC)_n \cdot (GGC)_n$ for n = 5, 6, 7, 10, 11, 15, 18, and 21. Our results show that for a given repeat length the substrate efficiency of the $(GCC)_n$ hairpin is about 5 times higher than the Watson–Crick duplex, $(GCC)_n \cdot (GGC)_n$. The substrate efficiency of the $(GGC)_n$ hairpin is even lower than the Watson–Crick duplex, $(GCC)_n \cdot (GGC)_n$. It is to be noted that the catalytic domains of methyltransferases are conserved through evolution from bacteria to humans [60]. Also, it has been shown by X-ray crystallography that in the activated (substrate–bacterial methyltransferase) complex, the C of CpG is in a "flipped out" conformation [61, 62]. Since they are C · C-paired, the Cs of CpG in the $(GCC)_n$ hairpin will flip out more easily than the same Cs that are G · C-paired in either the Watson–Crick duplex, $(GCC)_n \cdot (GGC)_n$, or the $(GGC)_n$ hairpin which would account for the higher substrate efficiency of the $(GCC)_n$ hairpin than either the Watson–Crick duplex, $(GCC)_n \cdot (GGC)_n$, or the $(GGC)_n$ hairpin.

F. Evidence for Hairpin-Induced Slippage Structures of the Fragile X Repeat: An *in Vitro* Replication Assay

We have performed *in vitro* replication of M13 single-stranded DNA templates by *Taq* polymerase assays for the intrinsic preference of hairpin formation by the $(GCC)_n$ or $(GGC)_n$ strands in presence of its complementary Watson–Crick partner. Figure 41-8 outlines the experimental design (for details, see [63]). Triplet repeats, $(GCC)_n$ or $(GGC)_n$ [n = 8 or 21], are inserted into the single-stranded M13 phage vectors [M13mp18 or M13mp19]. The replication (or primer extension in our case) is carried out by a 17-nucleotide-long primer that attaches to the template 40 nucleotides away from the insert. The replication is stopped by using dideoxy terminators. The replication product is sequenced on an Applied Biosystem Auto-Sequencer. If the insert in the DNA template forms a hairpin, the chain elongation during replication should either continue past the base of the hairpin or stop at the beginning of the hairpin; either result, as shown in Fig. 41-8, will result in a shorter replication product depending upon the size of the hairpin in the insert sequence. If the replication arrests at the insert then the replication product will contain only the initial flanking sequence and part of the insert. If the replication skips the hairpin then the replication product will contain both flanking sequence and a shortened insert sequence. In the absence of a hairpin in the insert, the replication product should correspond to the entire length of the DNA template including the $(GCC)_n$ or $(GGC)_n$ insert and the flanking sequences.

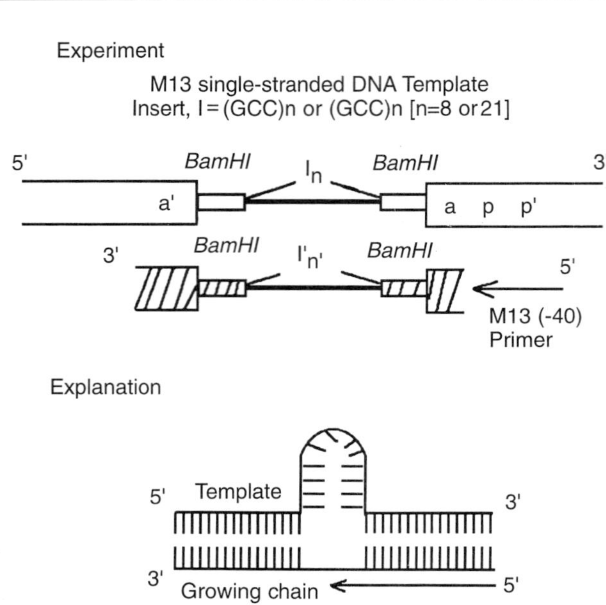

Figure 41-8 An *in vitro* replication assay using M13 single-stranded DNA with $(CCG)_n$ or $(CGG)_n$ inserts: the experimental protocol and the explanations of the replication bypass due to hairpin formation in the insert. In the discussion of these results, $(GCC)_n$ or $(CCG)_n$ [$(CGG)_n$ or $(GGC)_n$] are used interchangeably. For the $(GCC)_{21}$, a finite portion of the insert escapes replication at 45, 60, and 72°C since the skipped portion forms a hairpin. The length of the hairpin decreases upon increasing the reaction temperature. When $(GCC)_8$ or $(GGC)_{8,21}$ are used as inserts, there is no hairpin formation at 45, 60, and 72°C.

In all cases, the majority of the replication products belong to one size category. A complete replication product [i.e., $(GGC)_8$ and the two flanking sequences] is obtained for the $(GCC)_8$ insert in the template for a reaction temperature of 60°C which melts the $(GCC)_8$ hairpin. For $(GCC)_{21}$, a finite length of the $(GCC)_{21}$ insert is always bypassed within a temperature range of 45–72°C and the length of the bypass increases with decreasing temperature. However, at a reaction temperature of 85°C, the entire length of the $(GCC)_{21}$ insert is replicated. This observation is explained by the formation of a hairpin structure in the $(GCC)_{21}$ insert in the template. Note that the ends of the hairpin stem can still be replicated because they fray into unpaired single strands. The extent of end-fraying increases with increasing temperature and as a result, the length of the central $(GCC)_{21}$ insert bypassed by *Taq* polymerase decreases with increasing temperature. In this assay, unless the replication traverses the full length of the insert, $(GCC)_n$, complete Watson–Crick complementarity between the insert and the replicated DNA is not achieved. During replication, this strand asymmetry (i.e., lack of perfect complementarity) facilitates hairpin-folding of the $(GCC)_n$ insert. This hairpin may be bypassed if the rate of replication is much faster than the rate of decay of the $(GCC)_n$ hairpin. Therefore, within the range of reaction temperature, 45–72°C, if a residual $(GCC)_n$ hairpin is long enough to be stable, then it may escape replication. This kind of bypassing of a hairpin during replication is analogous to the deletion of a hairpin formed by extrachromosomal palindromic sequences observed during the replication of yeast DNA.

A very different result is obtained when the complementary triplet repeats, $(GGC)_{8,21}$, are inserted into the M13 single-stranded DNA template. Replication bypasses are not observed for either $(GGC)_8$ or $(GGC)_{21}$ within 45–85°C. This is consistent with our earlier observation [13] that the $(GGC)_n$ strand has a lower propensity for hairpin formation than its complementary partner, $(GCC)_n$. Fry and Loeb [12] observed a slowly migrating species in the native gel of the individual $(GGC)_n$ strand. This species, which constituted less than 40% of the total population, was assumed to be a G-quartet structure although from our NMR and gel mobility data we have found no evidence of such a structure [13, 32]. In general, a G-quartet structure in the M13 template blocks replication; for example, the formation of multiply folded G-quartet structure by the insulin-linked polymorphic region (ILPR), $(ACAG_4TGTG_4)_n$, leads to replication arrest at the beginning of the insert and this replication arrest is not normally released even in the presence of replication accessory proteins (*Escherichia coli,* SSB/human RP-A, helicase, etc.—see [93]). However, in the case of a $(GGC)_{21}$ insert in the M13 template a complete replication product, including the insert and the flanking sequences, is observed. This rules out the formation of G-quartet structure during replication. It may be pointed out that hairpin or G-quartet structure of the $(GGC)_n$ strand may be detected in the replication assay for longer lengths (n) or in the presence of KCl.

The individual $(GCC)_n$ strand can form a hairpin structure even for $n = 5$. However, in the presence of its complementary strand, $(GCC)_n$ requires a sufficiently long n (21 in our case) for the formation of a hairpin. Similarly, the individual $(GGC)_n$ strand forms a hairpin structure for $n > 11$. However, in the presence of its complementary strand, the $(GCC)_n$ strand does not form a hairpin even for $n = 21$. Both the $(GCC)_n$ and $(GGC)_n$ strands require even longer n for hairpin formation when replication protein A (RP-A), helicase, etc., are also present during replication [64]. Replication accessory proteins tend to unwind self-assembled single-stranded structures in a partially sequence-specific manner [65]. Nonetheless, since single-stranded regions are created during replication there is always a finite probability of hairpin formation by the $(GCC)_n$ or the $(GGC)_n$ strand of the FraX repeats.

G. Extremely High Methylation Efficiencies of the Hairpin-Induced Slippage Structures: A Methylation Assay

Our *in vitro* replication assay shows that the slippage structures are essentially three-way junctions in which the $(GCC)_n$ hairpin has the potential to slip and slide on the two Watson–Crick duplex arms (Fig. 41-9). Such a process may be facilitated by substrate–enzyme interactions at 37°C. In these three-way junctions, the potential for multiple locations of the $(GCC)_n$ hairpin on the Watson–Crick duplex allows the G · C-paired CpG sites in the Watson–Crick duplex to be converted into the C · C-paired CpG sites in the stem of the hairpin. Therefore, in a mobile three-way junction a larger number of C · C-paired CpG sites will be recruited for methylation than for a fixed hairpin formed by an excess of the $(GCC)_n$ strand. To test this hypothesis [63], we have constructed completely mobile (Fig. 41-9A), partially mobile (Fig. 41-9B), and immobile three-way junctions (Fig. 41-9C). In the completely mobile three-way junctions (Fig. 41-9A), the (GGC) strands of shorter lengths are annealed with the (GCC) strands of longer lengths: for example, $(GGC)_{10}$ · $(GCC)_{15}$, $(GGC)_{10}$ · $(GCC)_{18}$, $(GGC)_{10}$ · $(GCC)_{21}$, and $(GGC)_{15}$ · $(GCC)_{21}$. In the partially mobile three-way junctions (Fig. 41-10B), the free ends of the Watson–Crick duplex are covalently closed by two T_4 loops; therefore, in these single-stranded

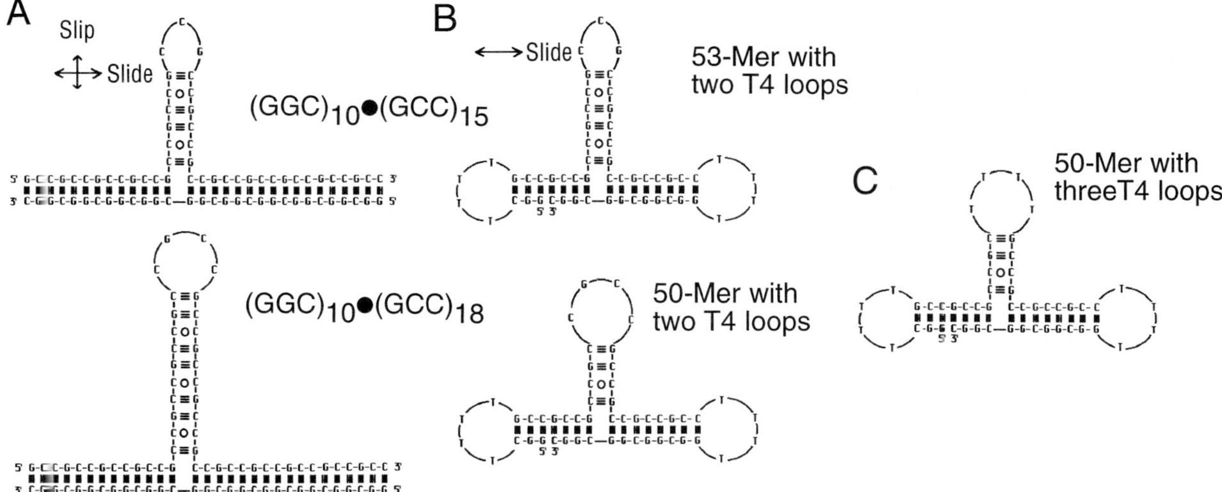

FIGURE 41-9 Three types of three-way junctions: (A) completely mobile, (B) partially mobile, and (C) immobile. (A) Completely mobile three-way junctions are formed by annealing GCC and GGC strands of unequal lengths (e.g., $(GGC)_{10}$ · $(GCC)_{15}$, $(GGC)_{10}$ · $(GCC)_{18}$, $(GGC)_{10}$ · $(GCC)_{21}$, and $(GGC)_{15}$ · $(GCC)_{21}$). Three-way junctions are created by the hairpin formation involving the excess of the longer GCC strands. The loop in the hairpin has either three or four nucleotides depending upon whether the longer GCC strand has odd (e.g., $(GGC)_{10}$ · $(GCC)_{15}$, upper panel) or even (e.g., $(GGC)_{10}$ · $(GCC)_{18}$, lower panel) number of repeats in excess. Note that the Cs at the CpG sites of the hairpin are C · C-paired (and not G · C-paired); this makes the hairpin a better substrate for methylation than the Watson–Crick duplex. Both slipping and sliding of the $(GCC)_n$ hairpin are possible in all of these three way junctions leading to the conversions of low-affinity Watson–Crick CpG sites of methylation into high-affinity hairpin CpG sites of methylation. In addition, after methylation when the hairpin slides it may return a methylated

*C**pG**
Gp**C**

to the Watson–Crick duplex which creates a hemimethylated (and high-affinity) CpG site. Sliding (and slipping) of the hairpin double methylation will cause a more efficient double methylation site,

*C**pG**
Gp**C***

*C is the 5 methyl cytosine. Note that the presence of the TCC interruption in the loop of the $(GCC)_n$ hairpin will restrict sliding while the TCC interruption in the stem of the $(GCC)_n$ hairpin will destabilize the three-way junction. (B) Two partially mobile three-way junctions: 53- and 50-mer. The ends of the Watson–Crick regions are connected by two T_4 loops which essentially prevent slipping of the hairpin. (C) An analog of the 50-mer in (B) in which the (CGCC) loop is replaced by a T_4 loop; this prevents slipping and sliding.

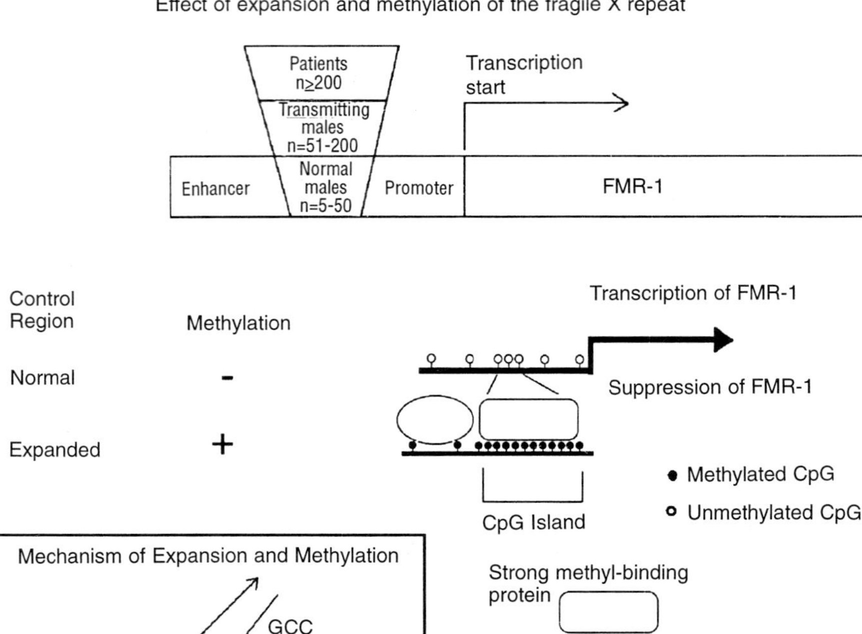

FIGURE 41-10 A schematic representation of the FMR1 gene. The location of the fragile X repeat, $(GCC)_n \cdot (GGC)_n$, is shown in the 5′ untranslated region although the exact location and the interrelationship of the promoter and the enhancer are not yet known. The ranges of repeat numbers, *n,* are shown for normal, transmitting, and affected males. Therefore, from normal to affected males, the control region undergoes a transition from low-density CpG to high-density CpG island due to the massive expansion and the subsequent methylation of the fragile X repeat. The high-density methylated CpG sites attract strong methyl binding proteins that cannot be dissociated by transcription activator proteins or RNA polymerase and this leads to the suppression of the FMR1 gene. (Inset) A sketch of how the presence of the transient and mobile three-way junction can cause both expansion and methylation. The intrinsic preference of hairpin formation by the GCC strand leads to a three-way junction with a Watson–Crick anchor. When the GGC strand acts as the template, such a three-way junction leads to expansion if the GCC hairpin escapes repair. As shown later, these three-way junctions are also excellent substrates for CpG methylation by the human methyltransferase.

three-way junctions the $(GCC)_n$ hairpin can only slide (and not slip). The partially mobile three-way junctions (Fig. 41-9B) are expected to be less efficient in recruiting new CpG sites in the hairpin conformation than the completely mobile three-way junctions (Fig. 41-9A). In the immobile three-way junctions (Fig. 41-9C) the loop segment of the $(GCC)_n$ hairpin is replaced by a T_4 loop in addition to closure of the free ends of the Watson–Crick duplex by two T_4 loops. The structures of the DNA substrates for methylation have been characterized by combining nondenaturing gel electrophoresis, digestion studies using single-strand-specific P1 nuclease, and NMR studies of the exchangeable imino protons [14]. The possibility of the three-way junctions for $(GGC)_{10} \cdot (GCC)_{15}$ has also been independently established by Kallenbach and coworkers [64] by nondenaturing gel electrophoresis and digestion by *Exo*VII (an enzyme that cleaves single-strand tails at the end of the duplexes). Due to the lack of a single-stranded tail (see Fig. 41-8A), the three-way junction of $(GGC)_{10} \cdot (GCC)_{15}$ was found to be resistant to digestion by *Exo*VII.

The rate of methylation by the human methyltransferase is obtained by measuring the tritium count on cytosines transferred from the tritiated cofactor, *Ado-Met* [57–59]. Table 41-1 lists the rates of methylation for different three-way junctions; two $(GCC)_n$ hairpins and a Watson–Crick duplex are included as controls.

TABLE 41-1 Rates of Methylation by the Human Methyltransferase

Substrates	Rate (fmol/min)	Number of CpG/substrate	Relative rate[a]
Three-way junctions			
Immobile (Fig. 41-9C)			
50-mer with three T_4 loops	4.0	9	1.0
Partially mobile (Fig. 41-9B)			
50-mer with two T_4 loops	76.6	11	19.2
53-mer with two T_4 loops	92.4	12	23.1
Completely mobile (Fig. 41-9A)			
$(GGC)_{10} \cdot (GCC)_{15}$	76.3	23	25.4
$(GGC)_{10} \cdot (GCC)_{18}$	192.5	26	48.1
$(GGC)_{10} \cdot (GCC)_{21}$	364.9	29	91.2
$(GGC)_{15} \cdot (GCC)_{21}$	298.5	34	74.6
Controls			
Watson–Crick duplex			
$(GGC)_{15} \cdot (GCC)_{15}$	29.6	28	7.4
Hairpins			
$(GCC)_{10}$	30.2	9	7.6
$(GCC)_{21}$	159.7	20	39.9

[a]Scaled with respect to the rate for the 50-mer with four T_4 loops.

The rates are scaled for the enzyme to DNA ratio. The effective rate of methylation is determined by the initial substrate–enzyme recognition, the kinetics of the transition to the activated state, and the subsequent release of the product. For the initial recognition, the methyltransferase requires the target CpG site and additional flanking base pairs [60]. The actual size and the sequence of the recognition element distinguish one methyltransferase from another although the catalytic mechanism involving the "flipped-out C" remains the same for all the enzymes [62]. Hence, once the Watson–Crick duplex or the hairpin is above a critical size and has the correct recognition element, the kinetics of transition to the activated state essentially determines the rate of methylation. In the completely mobile three-way junctions, the effective rate of methylation is governed by the following factors: (i) the number of the CpG sites in the Watson–Crick duplex, (ii) the number of CpG sites in the hairpin, and (iii) the rate of interconversion of the Watson–Crick CpG sites to the hairpin CpG sites due to the slipping and sliding of the $(GCC)_n$ hairpin. The third factor creates a greater number of high-affinity hairpin CpG methylation sites in the mobile three-way junctions than in a hairpin of fixed length. In addition, after methylation, if the $(GCC)_n$ hairpin slips or slides, it generates hemimethylated CpG sites in the flanking Watson–Crick duplexes which are again better substrates for methylation than the unmethylated CpG sites [58, 59]. Therefore, due to the presence of high-affinity hairpin CpG sites and hemimethylated Watson–Crick CpG sites, the completely mobile three-way junctions are expected to be much better methylation substrates than either the single $(GCC)_n$ hairpin or the Watson–Crick $(GCC)_n \cdot (GGC)_n$ duplex.

The importance of the mobility of the $(GCC)_n$ hairpin in the three-way junctions becomes evident from comparisons of the rates of methylation for the partially immobile three-way junctions with two T_4 loops (see Fig. 41-9B) with the rate of methylation for the completely immobile junction (see Fig. 41-9C). Although these two types of substrates have an almost equal number of CpG sites, the partially mobile three-way junction is 20 times more efficient than the immobile three-way junction (Table 41-1). This difference is attributed to the fact that the $(GCC)_n$ hairpin is able to slide in the partially mobile three-way junctions (see Fig. 41-9B) whereas it is completely locked in the immobile three-way junction (see Fig. 41-9C). Also, note that the three-way junctions with the highest probability of slipping and sliding, namely, $(GGC)_{10} \cdot (GCC)_{21}$ and $(GGC)_{15} \cdot (GCC)_{21}$, also have the highest rates of methylation.

The three-way junction, $(GGC)_{10} \cdot (GCC)_{21}$, is about 14 times more efficient as a substrate than the Watson–Crick $(GCC)_{15} \cdot (GGC)_{15}$ duplex although these two substrates have almost the same number of CpG sites. The observed difference in methylation in these two substrates results from the differences in their structure and dynamics. The rate of methylation for $(GGC)_{10} \cdot (GCC)_{21}$ is over twice that of the single $(GCC)_{21}$. This observation argues against the possibility of a reaction mechanism in which $(GGC)_{10} \cdot (GCC)_{21}$ dissociates into single $(GCC)_{21}$ and $(GGC)_{10}$ hairpins. If this were true, the rate of methylation for $(GGC)_{10} \cdot (GCC)_{21}$ should equal the rate of methylation for the single $(GCC)_{21}$ hairpin because, as previously mentioned, the $(GGC)_{10}$ hairpin does not get methylated by the human methyltransferase [13]. Therefore, the presence and mobility of the $(GCC)_n$ hairpin make the three-way junctions (see Fig. 41-9A) better substrates for methylation by the human methyltransferase than either the single $(GCC)_n$ hairpin or the Watson–Crick $(GCC)_n \cdot (GGC)_n$ duplex.

H. A Mechanism of Suppression of the FMR1 Gene in Fragile X Syndrome

Figure 41-10 describes a molecular mechanism in for the expansion and hypermethylation of the fragile X triplet repeats based upon our high-resolution NMR, *in vitro* replication, and methylation data. The intrinsic preference of hairpin formation by the GCC strand initiates mobile three-way junctions during replication that provide a molecular basis for the repeat expansion and hypermethylation of the CpG island inside the fragile X repeat. The resulting high density of methylated CpG islands provides binding sites for methyl-CpG-binding proteins [67–71] leading to the suppression of the FMR1 gene.

Since **TCC/GGA** interruptions confer stability to the FraX repeat [54], we have performed 1D NMR studies to examine the effect of the **T**CC interruptions inside the stem and the loop of the $(GCC)_n$ hairpins. The imino protons of the $(GCC)_n$ hairpins with and without **T**CC interruptions show different temperature-dependence profiles, thereby also suggesting (qualitatively) differences in their stabilities. The imino protons of the $(GCC)_5$ hairpin disappears at 40°C. The imino protons of the $(GCC)_5$ hairpin with a **T**CC interruption in the stem disappear at 20°C whereas the imino protons of the $(GCC)_5$ hairpin with a **T**CC interruption in the loop disappear at 35°C. It appears that the **T**CC interruption in the stem causing two consecutive mismatches significantly destabilizes the $(GCC)_n$ hairpin [63]. Hence, such an interruption should weaken the possibility of slippage during replication and allow stable transmission of the fragile X repeats over generations. On the other hand, the **T**CC interruption in the loop causes a marginal difference in the stability of the $(GCC)_n$ hairpin [63]. However, such an interruption involving a CTCC loop (like the T_4 loop) will restrict the mobility of the $(GCC)_n$ hairpin in the three-way junction (see Figs. 41-9B and 41-9C). Even if formed during replication, the immobile three-way junction with a **T**CC loop will be efficiently repaired. In addition, as shown in Table 41-1, an immobile three-way junction is a poor substrate for methylation. Therefore, our studies help us visualize how the **TCC/GGA** interruptions protect against the expansion and hypermethylation of the fragile X repeat.

III. MYOTONIC DYSTROPHY (DM) TRIPLET REPEATS, $(CTG)_n$: ROLE OF THE HAIRPIN STRUCTURES IN EXPANSION AND ABNORMAL EXPRESSION OF THE DMPK GENE

Myotonic dystrophy (DM) is an autosomal dominant disorder characterized primarily by myotonia and progressive weakness and is the most common adult-onset muscular dystrophy [1, 2, 4]. The rare congenital form of DM is associated with profound hypotonia and mental retardation. The gene that causes myotonic dystrophy (DMPK-myotonic dystrophy protein kinase) has recently been identified [72, 73]. The DMPK gene also contains regions of strong homology to cAMP-dependent protein kinase. The 3′ untranslated region of the gene contains CTG triplet repeats (Fig. 41-1B). The length of this triplet repeat sequence is highly polymorphic with the number of copies varying from 5 to 37 in normal individuals. The carriers and affected individuals have more than 39 copies and in some cases it expands beyond 2000 copies. The degree of expansion correlates with the severity of the disease and in a few cases reverse mutations have been observed resulting in the reduction in the number of CTG triplet sequence to the normal population range with the concomitant disappearance of the DM symptoms. This suggests that DM is primarily caused by expansions of the CTG repeat at the 3′UTR of the DMPK gene.

Structural studies by gel electrophoresis and high-resolution NMR are discussed in this section (for details, see [33]). It is shown that the $(CTG)_n$ repeats form stable hairpin structures under physiological salt concentrations even for short repeat lengths (i.e., $n = 5$ or 6). High-resolution NMR spectroscopy allows detailed analyses of the structures and dynamics of these DNA hairpins. Our preliminary data also suggest that similar hairpins are also possible at the RNA level. The presence of hairpins in the 3′UTR of the DMPK mRNA may be of biological significance for various reasons. For example, it has been shown in *E. coli* that putative hairpin sequences in the 3′UTR of a gene can mediate efficient termination of mRNA transcription [74]. Similarly, it has also been shown that hairpin forming sequences, which are evolutionarily conserved (from Xenopus to humans), are present in the 3′UTR of the histone genes and these sequences are involved in mRNA processing and/or in mRNA transport from the nucleus to the cytoplasm [75]. The transport of the DMPK mRNA is particularly relevant for DM since in normal phenotypes the mRNA is efficiently transported from the nucleus whereas in disease phenotypes the mRNA remains bound to the nuclear matrix proteins [76]. Therefore, it is important to understand the structural difference between the 3′UTR $(CUG)_n$ repeats in the mRNA of normal phenotypes and those (expanded) repeats in disease phenotypes.

A. Structural Characterization of $(CTG)_n$ by Gel Electrophoresis

Figure 41-11 shows the possible structural forms of $(CTG)_{5,6}$ and their analogs. Note that $(CTG)_{5,6}$ can adopt either a monomeric hairpin or a mismatched duplex. The nondenaturing gel electrophoretic mobility data of $(CTG)_{5,6}$ distinguish these two possibilities. $(CTG)_{5,6}$ migrate faster than the 10-base-pair duplex. This suggests the presence of unimolecular hairpins. The $(CTG)_5$ hairpin is expected to migrate like a 7/8-base-

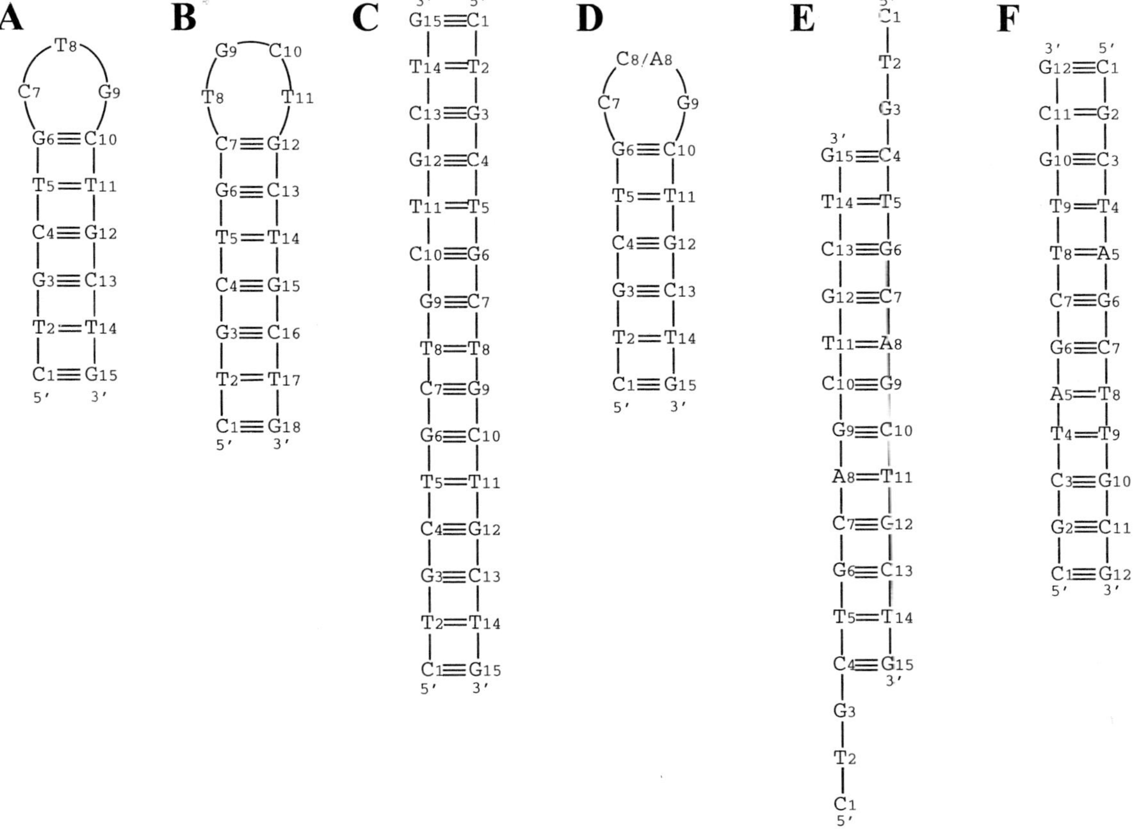

FIGURE 41-11 Secondary structures formed by various $(CTG)_n$ analogs. Hairpins: (A) $(CTG)_5$, (B) $(CTG)_6$, (D) $(CTG)_2CCG(CTG)_2$ and $(CTG)_2CAG(CTG)_2$, and (F) CGCTAGCTTGCG. Homoduplexes: (C) $(CTG)_5$, (E) $(CTG)_2CAG(CTG)_2$, and (F) CGCTAGCTTGCG. The oligomer in (F) has been shown to form two T · T pairs with two H bonds. A comparative NMR study on $(CTG)_{5,6}$ also shows the presence of T · T pairs with two H bonds. The $T_8 \rightarrow C_8$ substitution in $(CTG)_5$ does not alter the hairpin structure whereas the $T_8 \rightarrow A_8$ induces a hairpin-to-duplex equilibrium.

pair-long duplex whereas the $(CTG)_6$ hairpin should migrate like a 9/10-base-pair-long duplex. Also, similar gel patterns are observed under two different (i.e., 5 and 200 mM) NaCl concentrations. In addition, hairpins still remain the predominant conformation even when the DNA concentrations of $(CTG)_{5,6}$ are raised from 0.25 to 25 mM [33].

B. Structural Characterization of $(CTG)_n$ by NMR

Analyses of the NOESY and DQF-COSY data also reveal that all the constituent nucleotides in $(CTG)_{5,6}$ hairpins adopt (*C2′-endo, anti*) conformations with two H-bonded T · T pairs in the stem [33, 77]. Observation of a few (but key) interproton distance constraints defining intraloop and loop–stem interactions in the $(CTG)_5$ hairpin allows us to distinguish among four different (CTG) loop conformations: (i) three bases in the 3′ side of the stem, (ii) one base in the 5′ with two bases in the 3′ side of the stem, (iii) two bases in the 5′ with one base in the 3′ side, and (iv) three bases in the 5′ side of the stem; 200-ps-restrained MD simulations have been done separately using each model as a starting configuration. The structures derived from these four models show difference only in the single-stranded loop segments of the hairpins. In model (i), all the three bases in the loop are stacked with the 3′ side of the stem. In model (ii), T8 and G9 are stacked with each other on the 3′ side while C7 is stacked on the 5′ side of the stem. Model (iii) has G9 stacked with the 3′ side of the stem while C7 and T8 are stacked in the 5′ side of the stem. In model (iv), C7, T8, and G9 are all stacked with the 5′ side of the stem although G9 is partially flipped out of the stacked array. Model (i) shows better agreement with the distance constraints in the loop. Figure 41-12 shows the lowest energy structure of $(CTG)_5$ belonging to the family of models (i).

Figure 41-12 shows the lowest energy structure for $(CTG)_6$ that best satisfies the NMR constraints. In this structure, the four nucleotides in the (TGCT) loop are divided equally on each side of the stem (i.e., two nucleotides on each side). This results in a T · T pair in

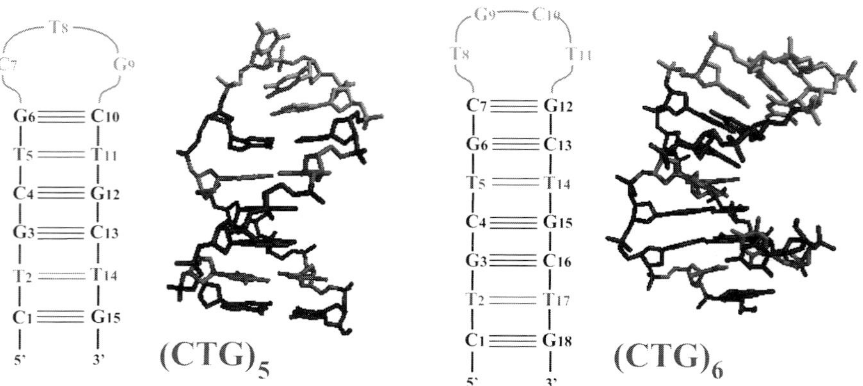

FIGURE 41-12 Three-dimensional average structures of $(CTG)_5$ and $(CTG)_6$. The structures are derived using 1D/2D proton NMR spectroscopy. Both $(CTG)_5$ and $(CTG)_6$ form blunt hairpins. Both structures fold so as to maximize G-C Watson–Crick base pairs. Two hydrogen bonded T · T mismatches are present in the stem. All nucleotides are in (*C2'-endo, anti*) conformations. $(CTG)_5$ has three nucleotides in the loop, whereas $(CTG)_6$ has four. Analyses of 2D NMR COSY and NOESY data lead to 200 NOEs for each structure. NOEs were converted into distances using full-relaxation matrix analysis; 100 structures compatible with the distance constraints were generated. All the structures belong to the same cluster with an average MSD lower than 1.0 Å^2 for each structure.

the loop although we do not have any experimental evidence in favor of such a pairing. It is possible that the T · T pair in the loop of $(CTG)_6$ opens and closes so fast on the NMR time scale that the imino signal is not observed.

For $(CTG)_n$, we have chosen $n = 5$ and 6 for our NMR studies since at these repeat lengths a hairpin is the only conformation under physiological salt concentrations [33]. Gao and co-workers [19] have studied the short $(CTG)_{2,3}$ duplexes, probably to more accurately determine the structures of the stem of a $(CTG)_n$ hairpin. However, they have not been able to determine whether the T · T pairs in the duplex are singly or doubly H-bonded. We have combined pH and temperature studies on various $(CTG)_n$ sequences to show that the T · T pairs are doubly H-bonded [33]. Mitas and co-workers [15] have performed gel electrophoresis, P1 digestion, and chemical modification studies on $5'a(CTG)_{15}a'3'$. It is not surprising that the modified $(CTG)_{15}$ sequence tends to form a hairpin structure since the flanking a and a' impose a constraint for hairpin folding. Although their system of choice has been biased, Mitas and co-workers have reached qualitatively the right conclusion regarding the $(CTG)_n$ sequences. However, they have managed to overinterpret their data to arrive at several wrong conclusions about the finer details of the $(CTG)_n$ hairpins. They have suggested that the T · T pairs are singly H-bonded. They have also proposed an inaccurate loop structure of the $(CTG)_n$ hairpins.

C. Site-Specific Dynamics of the $(CTG)_n$ Hairpins

We have calculated the order parameters [33, 78–80] for different interproton vectors in the $(CTG)_{5,6}$ hairpins by MD simulations hairpins with only hydrogen bonding constraints (and with no NOE constraints). The calculated values are compared with those estimated from the experimental cross-relaxation constants. Theoretical and experimental values of the order parameters, S^2, and the apparent correlation times, τ_{app}, are computed for interproton vectors with fixed distances such as H6-H5 in cytosines and H2'-H2" in sugars. NOE intensities for these two vectors only reflect the dynamics of the corresponding cyotosines and sugars in the $(CTG)_{5,6}$ hairpins. On a scale of 1 to 0, $S^2 = 1$ implies extreme rigidity and $S^2 = 0$ implies extreme flexibility. Similarly a small value of τ_{app} implies extreme flexibility whereas a high value of τ_{app} implies extreme rigidity. The cytosines in the loop, i.e., C7 in $(CTG)_5$ and C10 in $(CTG)_6$, are most flexible. In the $(CTG)_{5,6}$ hairpins, the Cs in the interior of the stems are least flexible. However, the Cs at the two termini of the stem are moderately flexible.

Analyses of S^2 and τ_{app} of various intrasugar H2'-H2" dipolar interactions indicate that in $(CTG)_5$, G3 and G12 (both in the stem) are the least flexible (Table 41-2). Theoretical calculations reveal that within the loop of the $(CTG)_5$ hairpin, C7 and T8 are less flexible than G9. Sugars corresponding to mismatches in the $(CTG)_5$ hairpin are also more flexible than those from

TABLE 41-2

Interaction	Base position	σ (s^{-1})[a]	τ_{app} (ns)
H5–H6	$(CTG)_5$		
	C7	0.48	2.1
	C1/C10	0.53	2.2
	C4/C13	0.61	2.5
	$(CTG)_6$		
	C1	0.66	2.7
	C10[b]		
H2′-H2″	$(CTG)_5$		
	T8	0.86	1.0
	T5/T11/T14	0.58	0.9
	$(CTG)_6$		
	G9/G18	0.34	0.8
	C1/C16	0.86	1.0
	C4/C7	0.80	1.0
	T2/T11	0.80	1.0
	C10/T8	0.98	1.1
	T5/T14/T17	0.46	0.9

Note. The methodology of the estimation of σ and τ_{app} from the NOESY data at 0–125 ms of mixing is described in Refs. [33] and [78–80].

[a]10% error in the estimated σs.

[b]NOE not observed up to 125 ms of mixing.

Watson-Crick pairs. Similar features are also observed for $(CTG)_6$.

D. Possible Role of the $(CUG)_n$ Hairpins in the Processing/Transport of the DMPK mRNA

The intrinsic propensity of hairpin formation by the $(CTG)_n$ sequence may also manifest itself at the level of mRNA. The formation of RNA hairpins by the $(CUG)_n$ sequences on the 3′ untranslated side of the DMPK gene may either halt the transcription machinery or provide a specific target for protein binding in the posttranscriptional mRNA processing and/or mRNA transport. It has been reported [81] that the levels of precursor mRNAs from the normal and DM alleles show no difference. However, the posttranscriptional processing of the normal and DM alleles are quite different in that the mRNA maturation is severely impaired when $(CUG)_n$ triplets are expanded in disease phenotypes. As stated earlier, it has also been demonstrated that the precursor mRNA remains bound to the nuclear matrix in DM phenotypes [76]. These data agree with our hypothesis that a few $(CUG)_n$ hairpins enable the formation of specific RNA–protein complexes required for efficient termination of transcription and for posttranscriptional mRNA processing or transport. This specificity is impaired when the $(CUG)_n$ triplets are expanded or the formation of multiple hairpins in expanded repeats allows several single-stranded loops in different hairpins to bind to the nuclear matrix proteins which are specific for single-stranded regions.

IV. HUNTINGTON'S DISEASE (HD) REPEAT, $(CAG)_n$: UNUSUAL HAIRPIN STRUCTURES AND THEIR ROLE IN EXPANSION

Huntington's disease (HD) is an autosomal dominant syndrome linked to chromosome 4p [82, 83]. Movement disorder, emotional disorder, and dementia are the common symptoms in HD. Recently, the gene responsible for HD has been identified from chromosome 4p, which is referred to as IT15 (Interesting Transcript 15). The HD gene contains a CAG triplet repeat sequence inside the first exon (Fig. 41-1C). Normal individuals have between 11 and 34 copies with a median of 19, whereas affected individuals have 37 and 86 copies with a median of 45. Patients with longer repeats have an earlier age of onset with a high correlation between the length of the repeat and the age of onset. The onset pattern of the disease suggests a possible role of expansion of the CAG repeat in Huntington's disease.

In this section, we describe the hairpin and homoduplex structures formed by the $(CAG)_n$ repeats (for details, see [46]). For shorter repeat numbers (i.e., $n = 5$ or 6) the $(CAG)_n$ repeats form homoduplex structures whereas for longer repeats (i.e., $n = 10$ or 11) they form doubly folded hairpins, i.e., hairpins with two single-stranded loops. Single H-bonded A · A pairs are present in the homoduplexes and in the hairpins. We have also performed an *in vitro* replication assay to show the presence of a hairpin for $(CAG)_{21}$ in the template.

A. Hairpin and Homoduplex Structures of $(CAG)_5$ and $(CAG)_6$

1. GEL ELECTROPHORESIS

Figure 41-13 shows the schematic representations of hairpin and mismatched duplex structures of $(CAG)_5$ and $(CAG)_6$. Previously, we have shown for $(GCC)_{5,6}$ and $(CTG)_{5,6}$ that although the hairpin folding is different for odd (i.e., $n = 5$) and even (i.e., $n = 6$) repeat numbers the base-pairing scheme of the stem remains the same. In Fig. 41-13, the $(CAG)_{5,6}$ hairpins are assumed to have similar folding patterns for odd and even repeat numbers. The gel mobility of $(CAG)_5$ and

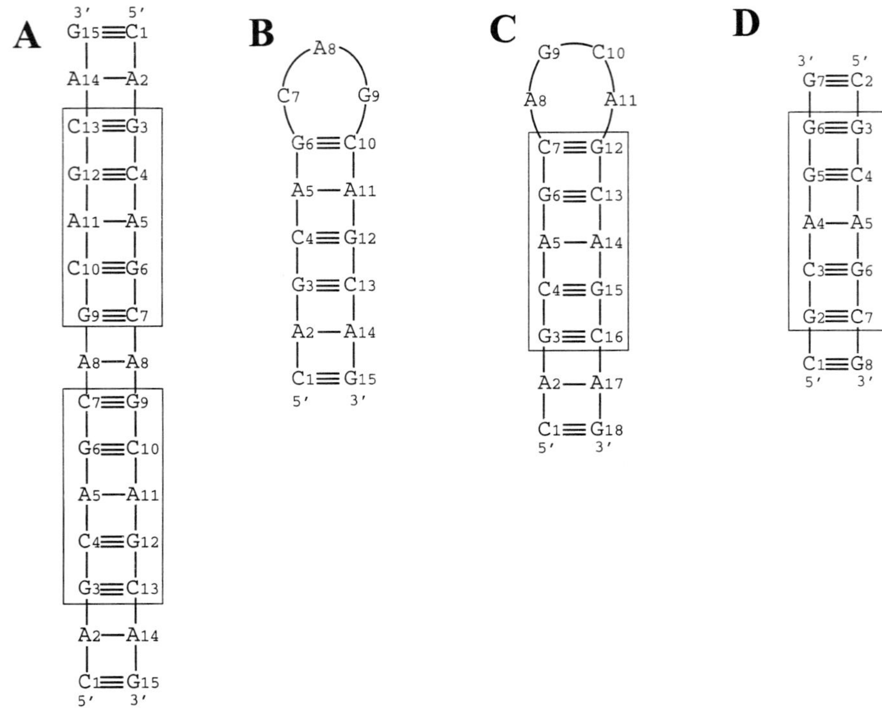

FIGURE 41-13 (A) $[(CAG)_5]_2$ duplex, (B) $(CAG)_5$ hairpin, (C) $(CAG)_6$ hairpin, and (D) 7 base pair duplex, $[(CGCAGCG)]_2$. The nucleotides are numbered from 5′ to 3′ direction. The 5-base-pair consensus structural motif is indicated by boxes in all the structures except for $(CAG)_5$ hairpin. Note that the hairpins with odd number of copies have three nucleotides, whereas hairpins with even number of copies have four nucleotides in the loop. The 7-base-pair duplex is chosen in order to characterize the detailed conformation of all the five nucleotides in the 5-base-pair structural motif. The terminal 5′ and 3′ G · C pairs are added to arrest the end-fraying of the 5-base-pair structural motif.

$(CAG)_6$ at neutral pH under different NaCl concentrations (20, 150, and 500 mM) show that $(CAG)_5$ forms both hairpin and homoduplex structures as major and minor species, respectively, whereas $(CAG)_6$ forms exclusively a hairpin structure. It appears that within the range of 20–500 mM NaCl concentrations, the $(CAG)_6$ hairpin is thermodynamically more favorable than the $(CAG)_5$ hairpin. The gel mobility data at three different DNA concentrations (5, 10, and 20 mM) for $(CAG)_5$ and $(CAG)_6$ also reveal that they exist predominantly in the hairpin form at 5 mM DNA and increasing the DNA concentration increases the relative population of the homoduplex.

2. HIGH-RESOLUTION STRUCTURE OF $[(CGCAGCG)]_2$ AND ITS RELEVANCE TO $[(CAG)_5]_2$

Within the range of DNA concentrations required for NMR experiments, we are unable to trap exclusively the hairpin conformation of $(CAG)_5$. Hence, our structural studies are restricted to the self-assembled $[(CAG)_5]_2$ duplex, which also adequately models the stem of the $(CAG)_5$ hairpin (Fig. 41-13). We have also carried out structural studies on the 7-base-pair duplex, $(CGCAGCG)_2$ which shares a common $[(GCAGC)]_2$ motif with the $[(CAG)_5]_2$ duplex or the stem of the hairpin. This enabled us to determine the exact stereochemistry of each nucleotide in the common motif and the exact nature of the A · A pairing.

Measurements of thermodynamic stabilities of various pur · pur/pyr · pyr/pyr · pur mispairs in a model duplex suggest that the A · A base pair is thermodynamically the least stable of all mispairs [86]. However, recent ^{1}H NMR and UV-melting studies indicate that A · A base pairs can be incorporated into a DNA duplex without globally distorting the structure and a without a drastic reduction in the stability [87]. Therefore, it is of interest to determine the nature of the pairing and stacking of the adenines in the $[(CAG)_5]_2$ duplex. For this, we have carried out heteronuclear (^{15}N-^{1}H) NMR experiments on the $[(CA^*G)_5]_2$ duplex where A*s represent ^{15}N6-labeled adenines. These experiments on measurements on the $[(CA^*G)_5]_2$ duplex and the

NOESY experiments on the $[(CGCAGCG)]_2$ duplex reveal single H-bonded A · A base pairing in both the duplexes.

A detailed analyses of NOESY data at 25, 50, 75, 100, 125, 200, and 500 ms of mixing show that the A · A mismatch in the middle of the $(CGCAGCG)]_2$ duplex disrupts neither the local B-DNA geometry nor the overall structure. Figure 41-14A shows the average of the 100 energy-minimized structures $[(CGCAGCG)]_2$. Gervais *et al.* [87] also made similar observations for an 11-base-pair duplex related to the K-ras gene. Note also that the distance between the hydrogen-bonded amino proton of one adenine and the H2 of the other is close in space (~2.7 Å), consistent with the observation of an NOE for $N6H_2$-H2 dipolar interaction.

Because of resonance overlap and low signal-to-noise-ratio, we have not been able to derive many independent distance constraints to determine the hairpin and duplex structures of $(CAG)_5$. Hence they have been modeled by using the same distance constraints determined for $[(CGCAGCG)]_2$. The duplex and the stem of the hairpin are divided into overlapping structural blocks on the basis of the 5-base-pair $[(GCAGC)]_2$ motif as observed in the middle of the 7-base-pair duplex. The intra- and sequential distance constraints of the individual blocks and the sequential distance constraints

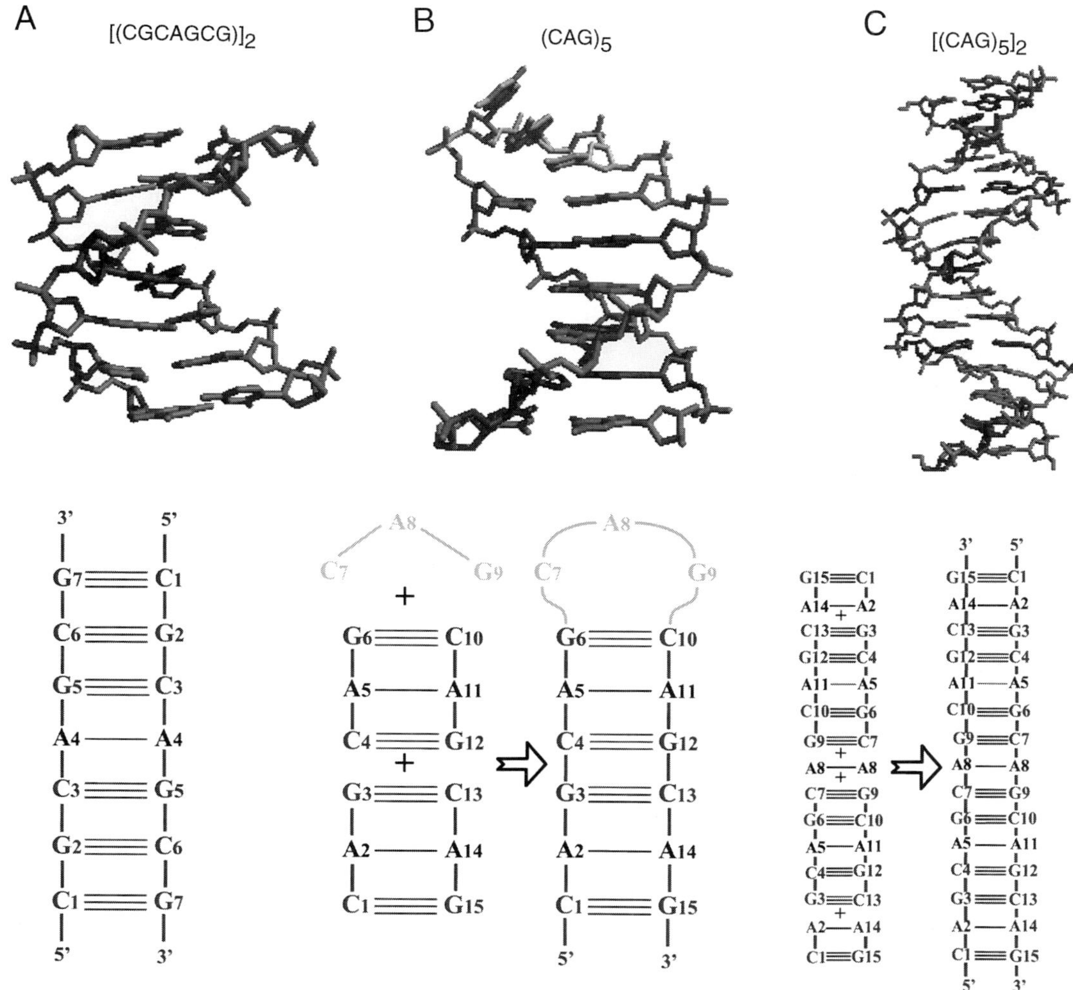

FIGURE 41-14 Energy minimized structures of (left) $(CGCAGCG)_2$, (middle) $[(CAG)_5]_2$, and (right) $(CAG)_5$ hairpin. The initial structures are constructed for duplex and hairpin in the same way as for $(CGCAGCG)_2$. Because of severe resonance overlap, we are unable to derive distance constraints either for the hairpin or for the duplex of $(CAG)_5$. The stem of the hairpin and the duplex are divided into overlapping motifs of $(GCAGC)_2$ duplex. Distance constraints for the 5-base-pair-long motif are from the NMR data of the duplex, $(CGCAGCG)_2$. The overlapping structural blocks defined by the 5-base-pair motif are shown below the structure of duplex and hairpin. No constraints are added for the interproton distances involving the loop nucleotides.

of the connecting steps of the individual blocks are the same as those of $[(CGCAGCG)]_2$. The distance constraints, assigned in this way together with the hydrogen bonding constraints, have been employed for the constrained minimization using AMBER (version 4.0); 20,000 conjugate gradient steps were used for minimization. Figures 41-14B and 41-14C show the duplex and hairpin structures of $(CAG)_5$ modeled by restrained minimization. In modeling the hairpin structures, no distance constraints have been imposed on the loop nucleotides. Similar structural features of the 5-base-pair structural motif present in the $[(CGCAGCG)]_2$ prevail in the duplex and the stem of the hairpin.

Based upon their replication and mismatch repair assays, Wells and co-workers [28] have suggested that the (CAG) repeats form less stable hairpins than the (CTG) repeats. It has been argued that the As in the $(CAG)_n$ hairpins are extrahelical, which according to Wells and co-workers is in agreement with the NMR data of Gao and co-workers [22]. However, the NMR data of Gao and co-workers show continuous sequential NOE connectivity in the $(CAG)_3$ duplex. This rules out the possibility of extrahelical As in the duplex. Moreover, Gao and co-workers [22] did not identify the H-bonding nature of the two As facing each other in the duplex. Our NMR studies on the $(CAG)_5$ duplex and the model (CG-CAGCG) duplex unequivocally show that the As in these duplexes are intrahelical and singly H-bonded.

B. Effect of Repeat Length on $(CAG)_n$ Structures

Nondenaturing gel electrophoresis, digestion by the single-strand-specific P1 nuclease, and 1D NMR studies have been carried out on $(CAG)_{10}$ and $(CAG)_{11}$ in order to determine the effects of repeat length, *n,* on the structures formed by the $(CAG)_n$ repeats. Under all solution conditions, $(CAG)_{10}$ and $(CAG)_{11}$ remain exclusively in the hairpin conformation. Figures 41-15A–41-15C show three possible structures of $(CAG)_{10}$ that would be consistent with the electrophoretic mobility data. Note that a $(CAG)_{10}$ hairpin with one single-stranded loop (Fig. 41-15A) has 9 G · C pairs (the terminal one is susceptible to end-fraying), whereas a $(CAG)_{10}$ hairpin with two single-stranded loops (Fig. 41-15C) has 8 G · C pairs (none is susceptible to end-fraying). We have carried out P1 digestions to probe the single-stranded regions in the $(CAG)_{10}$ and $(CAG)_{11}$ hairpins which reveal the presence of hairpin folding of $(CAG)_{10}$ and $(CAG)_{11}$ because substantial portions of these sequences remain undigested at a low enzyme to DNA ratio. However, what is more interesting is the observation of three protected fragments, i.e., digests of 22–24, 14–16, and 7–8 nucleotides for $(CAG)_{10}$ and digests of 23–25, 15–17, and 8–9 nucleotides for $(CAG)_{11}$. As shown in Fig. 41-15C, of the three fragments the longest one (over 22 nucleotides) is not expected if $(CAG)_{10}$ and $(CAG)_{11}$ form hairpins with one single-stranded loop. However, if $(CAG)_{10}$ and $(CAG)_{11}$ form hairpins with two single-stranded loops, fragments (>22 nucleotides) are expected after P1 digestion. Important evidence for the formation of double hairpin structures comes from the ^{1}H NMR spectra of $(CAG)_{10}$ and $(CAG)_{11}$ at different temperatures and pHs. In $(CAG)_{10}$, the presence of four G-imino signals within 12.0–13.0 ppm with intensity distribution (1 : 1 : 2 : 4) is consistent with 8 G · C base pairs. The G-imino resonances at 10.75 and 10.95 ppm disappear above 10°C and they are highly sensitive to pH. These two Gs may, therefore, belong to two different loops. Thus, the imino proton profile of $(CAG)_{10}$ is consistent with a hairpin containing two single stranded loops (Fig. 41-15C). The presence of two imino proton resonances for the guanines in the loop is also consistent with a $(CAG)_{10}$ hairpin with a single loop of 6 nucleotides (Fig. 41-15B). However, for such a hairpin only 7 (and not 8) G · C base pairs are expected.

Although, the 5′ and 3′ ends come close to each other in doubly looped $(CAG)_{10,11}$ hairpins, it does not mean that hairpin formation would have to be punctuated beyond $n = 10/11$ since several doubly looped hairpins of length 10 or 11 may be linked by double-stranded spacers as shown in Fig. 41-15D. Organization of several such 10- or 11-repeat-long hairpins may occur for longer repeat lengths. This would suggest that the stability of longer $(CAG)_n$ repeats largely depends upon the stability of individual doubly looped $(CAG)_{10,11}$ hairpins. Interestingly, Petruska *et al.* [34] made a similar observation in that the stability of $(CAG)_{10}$ is quite similar to that of $(CAG)_{30}$. Organization of several doubly looped $(CAG)_{10,11}$ hairpins may also describe the slippage structure during replication of long $(CAG)_n$ repeats (see Fig. 41-15D).

C. $(CAG)_n$ Hairpins Cause Slippage during Replication: An *in Vitro* Assay

As previously described in Fig. 41-8, we have used *in vitro* replication of M13 single-stranded DNA templates by *Taq* polymerase to assay for the intrinsic preference of hairpin formation by the $(CAG)_n$ strands in presence of its complementary Watson–Crick partner. Triplet repeats, $(CAG)_n$ or $(CTG)_n$ (n = 8 or 21), are inserted into the single-stranded M13 phage vectors (M13mp18 or M13mp19).

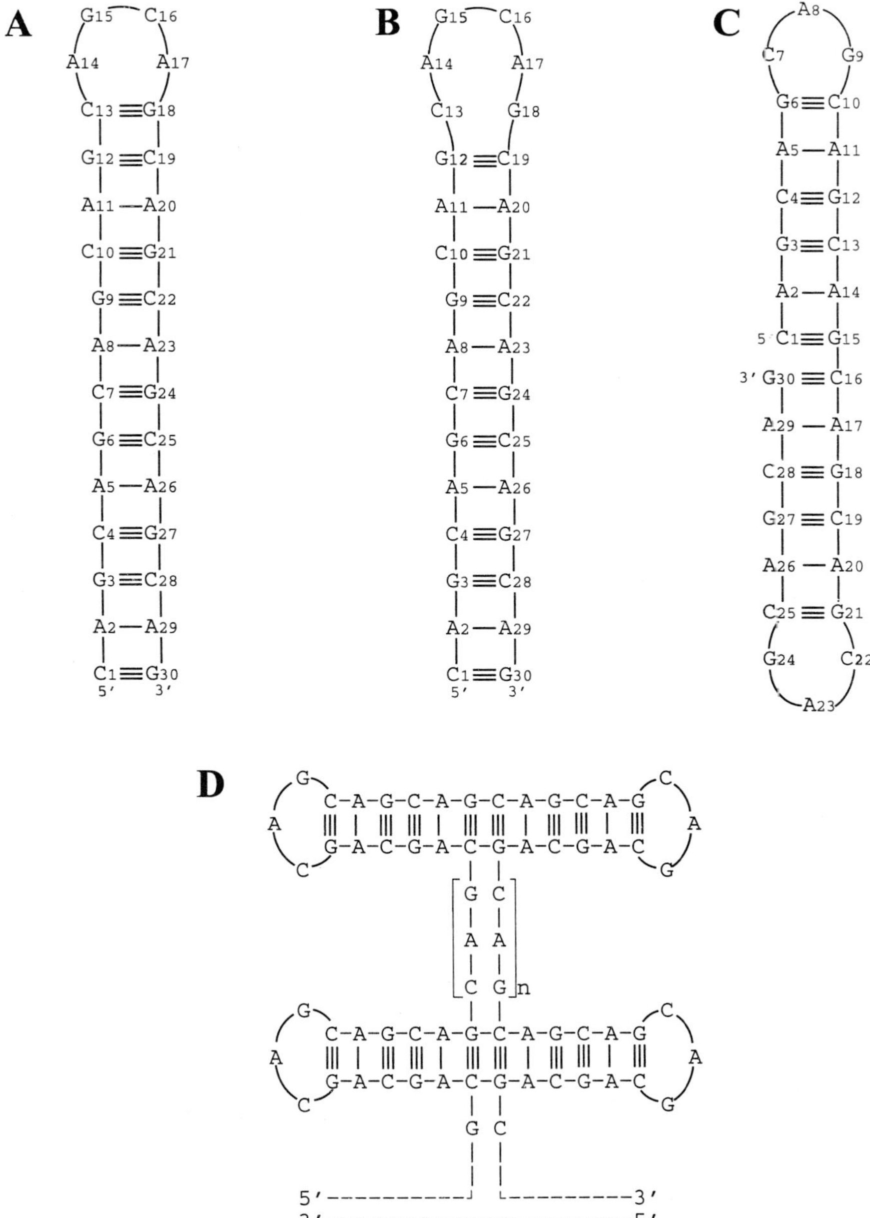

FIGURE 41-15 Three possible hairpin structures of $(CAG)_{10}$: (A) a singly folded hairpin with four nucleotides in the loop, (B) a singly folded hairpin with six nucleotides in the loop, and (C) a doubly folded hairpin with three nucleotides in each loop. The last loop folding is supported by gel electrophoresis, P1 digestion, and NMR data. Exchange properties of the loop imino signals are characterized by monitoring them at different pH and temperatures [46, 84–85]. $(CAG)_{11}$ also forms a doubly folded hairpin. (D) Possible arrangement of the doubly folded $(CAG)_n$ hairpins in a slippage structure.

In this replication assay, the majority of the replication product sequences belong to one size category. The full length of the insert and the flanking sequences are replicated for $(CAG)_8$ at a reaction temperature of 60°C, which should completely melt the $(CAG)_8$ hairpin. For $(CAG)_{21}$, 17 out of 21 repeats in the insert and the flanking sequences before and after two *Bam*H1 restriction sites are replicated, i.e., the remaining 4 repeats in the insert are bypassed. This observation is explained by the formation of a hairpin structure by the $(CAG)_{21}$

insert in the template. The extent of end-fraying increases with increasing temperature and as a result, the length of the central $(GCC)_{21}$ insert that is bypassed by *Taq* polymerase decreases with increasing temperature. In our replication assay, a repeat length (n) of 21 is required to demonstrate the formation of a hairpin although the individual $(CAG)_n$ strands showed exclusive presence of hairpins for $n = 10–11$. Therefore, the presence of the complementary strand and the polymerase pushes the critical threshold of n required for hairpin formation by the $(CAG)_n$ strand to a higher value. This threshold value of n may be higher when in addition to the polymerase all the replication accessory proteins (i.e., helicase, single-strand binding proteins) are also present during chain elongation.

V. FRIEDREICH'S ATAXIA (FRDA), TRIPLET REPEATS, $(GAA)_n/(TTC)_n$: CHARACTERIZATIONS OF THE TRIPLEX STRUCTURES BY NMR SPECTROSCOPY AND *IN VITRO* REPLICATION

Friedreich's ataxia (FRDA) is an autosomal recessive degenerative disease. The disease susceptible gene, $\chi 25$, have recently been identified in the FRDA locus on 9q13-q21.1 [6, 88, 89]. $\chi 25$ encodes for a 210-amino-acid-long protein called frataxin. FRDA is associated with the expansion of GAA/TTC triplet in the first intron of the $\chi 25$ gene (Fig. 41-2). The number of GAA/TTC repeats in normal chromosomes varies between 7 and 22, whereas FRDA alleles ranges between 201 and 1186. Usually there is no detectable mRNA expression of the frataxin gene in FRDA alleles. Point mutations that either hinder intron–exon splicing or abruptly terminate the mRNA synthesis of $\chi 25$ also lead to FRDA. This suggests that FRDA is a single-gene disorder.

Unlike the GC-rich $(CXG)_n$ triplet repeats which are capable of forming hairpin structures with X · X mispairs in the stem, the individual strands of $(GAA)_n/(TTC)_n$ do not form hairpin structures. Instead, they tend to form triplex structures as shown Fig. 41-2. Note that the folding of the longer TTC strand in this triplex leads to the formation of $C^+ \cdot G \cdot C$ and $T \cdot A \cdot T$ triads (Fig. 41-2). Here we discuss the determination of the high-resolution structure of such a triplex by homonuclear and heteronuclear NMR experiments. We also demonstrate by an *in vitro* replication assay the presence of similar triplexes with pyr · pur · pyr triads when long $(TTC)_n$ repeats (i.e., $n = 28–36$) are present in the template. In addition, by similar assays we show the presence of triplexes with pur · pur · pyr triads (see Fig. 41-2) when long $(GAA)_n$ repeats (i.e., $n = 28–36$) are present in the template although these triplexes are less stable than those with pur · pur · pyr triads.

Previously, Wells and co-workers performed systematic studies (reviewed in [90]) on various triplex-forming sequences including $(GAA)_n/(TTC)_n$. The triplex-forming ability of the $(GAA)_n/(TTC)_n$ repeat was studied by monitoring the induction of superhelicity in circular plasmids containing this repeat. It was shown that the $(GAA)_n/(TTC)_n$ repeat (for $n > 40$) formed a triplex with pyr · pur · pyr triads even at neutral pH. Here we show that a triplex of $(GAA)_n/(TTC)_n$ containing as few as only six triads (two $C^+ \cdot G \cdot C$ and four $T \cdot A \cdot T$) is also stable at neutral pH.

A. Structural Characterization of a Triplex with pyr · pur · pyr Triads by NMR Spectroscopy

The possibility of a triplex structure for the $(GAA)_n/(TTC)_n$ repeat becomes obvious because a finite population of a triplex is observed even in the 1D NMR spectra of the exchangeable imino (NH) and amino (NH_2) protons of [1 : 1] $(GAA)_3 \cdot (TTC)_3$ at pH 7.0 and 15°C. The presence of the NH_2 signals from the protonated $C^+ \cdot G$ Hoogsteen pairs indicates the presence of $C^+ \cdot G \cdot C$ triads as expected in a triplex. In addition, the NH region is split into two groups, one belonging to the Watson–Crick duplex (the major population) and the other belonging to the triplex (the minor population). With (1 : 2) $(GAA)_3 \cdot (TTC)_3$ stoichiometry, the spectrum of the NH and NH_2 protons shows the exclusive presence of a triplex conformation at neutral pH. A triplex conformation is also exclusively present at neutral pH when $(GAA)_3$ is mixed with $(TTC)_7$ in equimolar amounts. A triplex structure for [1 : 1] $(GAA)_3 \cdot (TTC)_7$ implies that the $(TTC)_7$ strand probably folds with a T_2 loop. However, the spectral qualities in (1 : 2) $(GAA)_3 \cdot (TTC)_3$ and [1 : 1] $(GAA)_3 \cdot (TTC)_7$ do not allow the determination of high-resolution triplex structures by 2D NMR. We have, therefore, chosen an intramolecularly folded triplex, $(GAAGAA)T_4(TTCTTC)$-$T_4(CTTCTT)$, that models the folding of the $(TTC)_n$ strand with a T_4 loop (see Fig. 41-16 and [91]). In an actual system (see Fig. 41-2), either a T_2 or a T_2CT_2 loop is present. The spectral quality is dramatically improved in this unimolecular triplex.

A set of 295 interproton distance constraint has been derived by analyzing WATERGATE NOESY at 100 and 200 ms of mixing and NOESY in D_2O at 25, 50, 75, 100, 125, 200, and 400 ms of mixing. NOE data reveal that

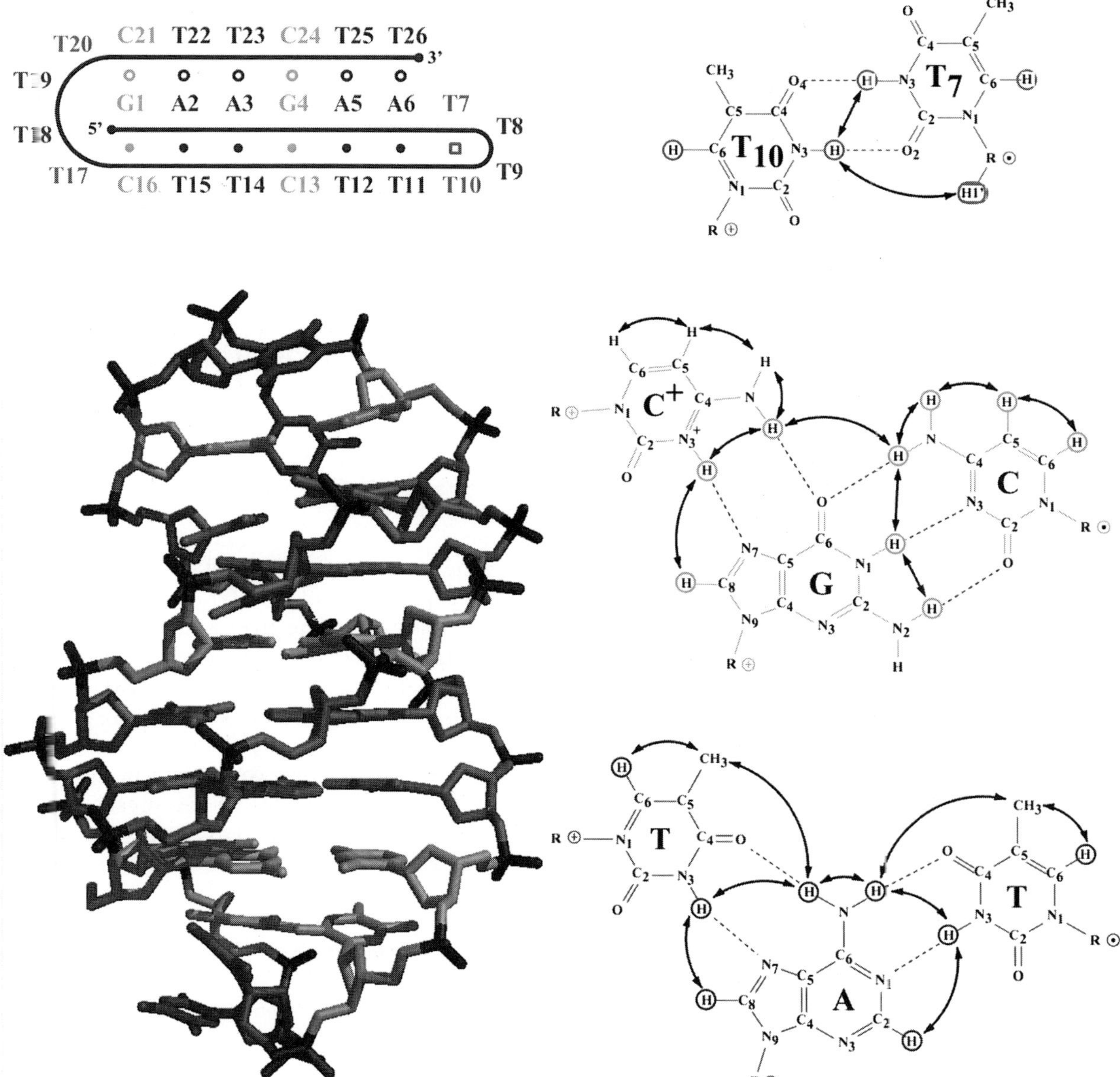

FIGURE 41-16 Schematic chainfolding, descriptions of H-bonding in the T · T pair and in the $C^+ \cdot G \cdot C$ and T · A · T triads, and a three-dimensional average structure of the triple helix forming sequence $(GAA)_2T_4(TTC)_2T_4(CTT)_2$ derived using 1D/2D NMR spectroscopy. The stem of the triplex contains four T · A · T triads, and two $C^+ \cdot G \cdot C$. The stem is connected by two T_4 loops, each of them containing one T · T base pair. All nucleotides are in (*C2′-endo, anti*) conformations. Analyses of the 2D NMR data of $(GAA)_2T_4(TTC)_2T_4(CTT)_2$ lead to about 300 NOEs which are converted into distances using full-relaxation matrix analysis; 20 structures compatible with the distance constraints are generated. All the structures were found to belong to the same cluster with an average MSD of 0.52 $Å^2$.

all the nucleotides are in (*C2′-endo, anti*) conformations. We have also determined [91] the specific interactions of the amino protons belonging to the 4 Cs in this triplex. For this, we have ^{15}N4-labeled the 4 Cs in (GAAGAA)T_4(TTCTTC)T_4(CTTCTT) and performed (^{15}N-^{1}H-^{1}H) HMQC-NOESY experiments. Figure 41-16 shows the stereo view of the superimposition of 20 minimized structures. Figure 41-16 also shows the nature of $T_7 \cdot T_{10}$ pair that prevents the end-fraying of the $T_{11} \cdot A_6 \cdot T_{26}$ triad of the triplex. The (TTCTTC) arm involved in Watson–Crick pairing and the (CTTCTT) arm involved in Hoogsteen pairing show several interarm NOEs with the (GAAGAA) arm that uniquely lock the structure of the stem. Therefore, the structure in Fig. 41-16 represents a quantitative model for the triplex with pyr · pur · pyr triads that can be formed by the GAA/TTC repeats.

B. Identification of Triplexes with pyr · pur · pyr and pur · pur · pyr Triads by an *in Vitro* Replication Assay

Because our NMR studies unequivocally demonstrate that GAA/TTC repeats can form triplex structures, it is obvious that the formation of such structures during replication would cause slippage and replication. We have demonstrated the presence of such structures during replication by an *in vitro* replication using the single-stranded M13 DNA template. This assay is different from the one described in Fig. 41-8 which is performed in presence of *Taq* polymerase at various temperatures (45–80°C) and in absence of any replication accessory proteins. As shown in Fig. 41-17, in this room temperature assay, three different proteins (or protein assemblies) are used: (i) DNA polymerase that extends the primer on the single-stranded template, (ii) the single-strand binding protein (SSBP, such as the *E. coli* SSB or the human RP-A) that binds the polymerase on the 3′ side of the primer and tends to destabilize any secondary structure in the template, and (iii) the accessory and ATP-dependent protein complex that binds the polymerase on the 5′ side of the primer. This multicomponent protein–DNA complex ensures efficient chain elongation and processivity in DNA replication. Abnormal replication in this assay indicates that the replication machinery is unable to perform template-directed synthesis due to the presence of secondary (unusual) structure in the template. In fact, the nature of the replication product reflects the nature of the unusual structure present in the template. For example, a replication bypass is expected if a DNA repeat in the template forms a hairpin (as shown previously in Fig. 41-8). On the other hand if a DNA repeat in the template forms a triplex, a replication arrest is expected in the middle of the repeat (see Fig. 41-17). Finally, if a DNA repeat in the template forms a G-quartet, a replication arrest is expected in the beginning of the repeat. As previously stated, this assay has been successfully used to demonstrate the presence of (i) simple hairpins in the fragile X triplet repeats, (ii) triplex in FRDA triplet repeats, and (iii) hairpin G-quartet and hairpin i-motif in the insulin minisatellite [92–94].

As shown in Fig. 41-17, the slippage of $(TCC)_n$ in the template may cause the formation of a triplex with pyr · pur · pyr triads whereas the slippage of $(GAA)_n$ in the template may lead to the formation of a triplex with pur · pur · pyr triads. Since the triplex with pyr · pur · pyr triads contains $C^+ \cdot G \cdot C$, an acidic pH is likely to enhance the stability of such a triplex. Also in this triplex, the second pyr strand is located in the major groove of the pur strand of the Watson–Crick duplex and one of the H bonds in the major groove involves N7 of purines (see Fig. 41-2). Therefore, when $(TTC)_n$ is present in the template the use of 7-deaza GTP (instead of dGTP) in the precursor pool should eliminate the possibility of a H bond involving N7 of G and, therefore, should drastically lower the stability of the triplex with pyr · pur · pyr triads.

In the triplex with pur · pur · pyr triads (see Fig. 41-2), the second pur strand is located in the major groove of the first pur strand of the Watson–Crick duplex. Note that the formation of this triplex requires no protonation. However, here also N7 in the first pur strand is involved in H-bonding. Therefore, the presence of $(7\text{-deaza-GAA})_n$ in the template instead of $(GAA)_n$ should eliminate one of the two H-bonding potentials of Gs in the major groove and should, therefore, drastically lower the stability of such a triplex.

The replication of $(TTC)_8$ in the M13 single-stranded DNA template gives a full replication product containing $(GAA)_8$ when the primer extension is carried out in the presence of the T_7 DNA polymerase within the pH range 5.5–8.0. Similar results are obtained with the T_4 DNA polymerase and the *E. coli* Klenow fragment. However, the replication pattern of $(TTC)_{36}$ in the template in the presence of the T_4 DNA polymerase, the T_7 DNA polymerase, or the *E. coli* Klenow fragment at pH 5.5–7.0 shows a strong replication arrest in the middle of the $(TCC)_{36}$ tract in the template (i.e., at $n = 18$). Similarly, when $(TCC)_{28}$ is present in the template, a strong replication arrest is observed at $n = 14$. Therefore, it appears that in our assay at least 28 repeats of (TTC) are necessary to cause the formation of a triplex during replication. The addition of either the *E. coli* SSB/human RP-A or the T_4 SSB (i.e., the product of gene 32) and the T_4 accessory proteins (i.e., the products of genes 44/62 and gene 45) to the replication assay releases the arrest for $(TTC)_{28,36}$ in the template. The replication arrest in the $(TTC)_{36}$ template is also released when 7-deaza G is used in the precursor pool. This is consistent with a triplex with a H bond through N7 of G in the major groove (see Fig. 41-2).

A replication arrest occurs in the middle of the $(GAA)_{28,36}$ tract (i.e., at $n = 14$ or 17) only for the T_4 polymerase. This is consistent with the formation of a triplex with pur · pur · pyr triads as shown in Fig. 41-2; such a triplex is detected in our replication assay only for $(GAA)_n$ templates with repeat number 28 or larger. Like the triplex with pyr · pur · pyr triads, this triplex is also unwound when either the *E. coli* SSB/human RP-A or the T_4 SSB plus other T_4 accessory proteins is

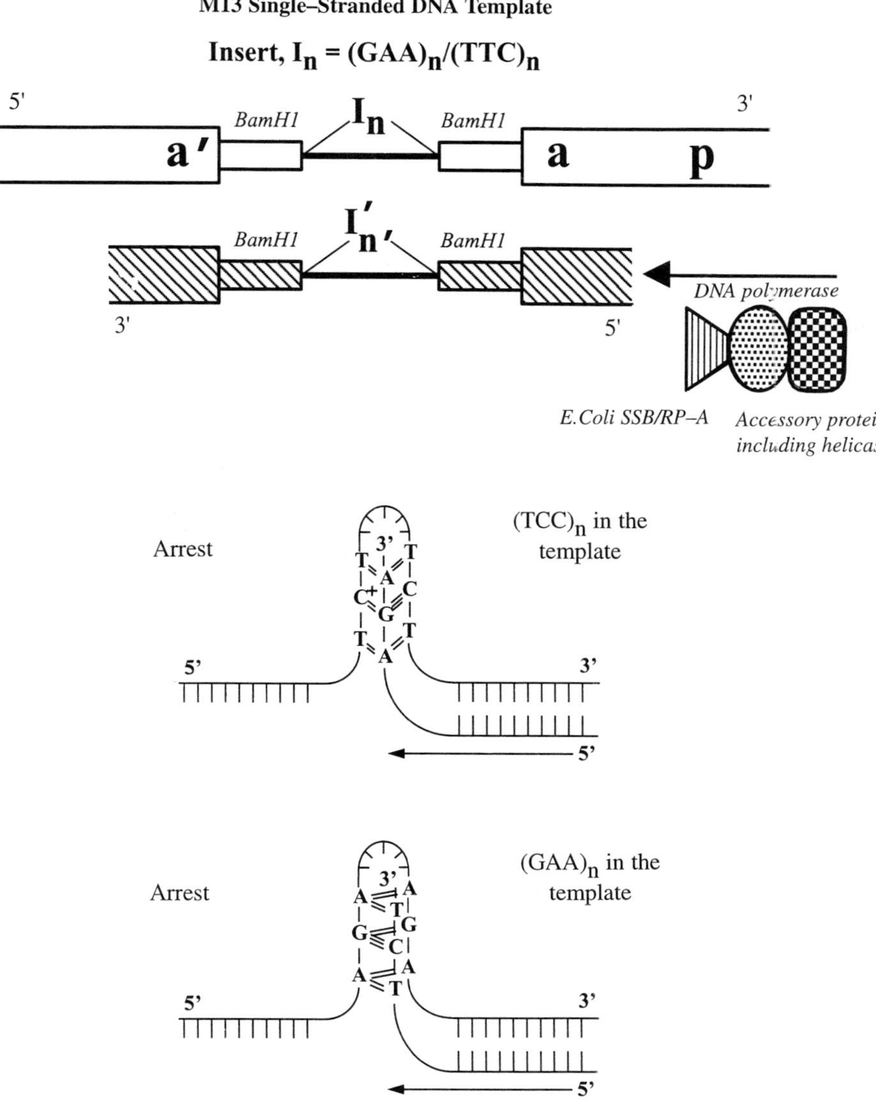

FIGURE 41-17 An *in vitro* replication assay with $(GAA)_n$ or $(TTC)_n$ inserts in the single stranded M13 DNA templates. The folding of $(GAA)_n$ or $(TTC)_n$ creates a triplex with pur · pur · pyr or pyr · pur · pyr triads.

added during replication. The replication of (7-deaza -GAA)$_{36}$ in the template shows no replication arrest. This indicates that in this triplex the Gs of the second pur strand is interacting with the Gs of the first pur strand through H bonds involving N7 (see Fig. 41-2). However, this triple helix is less stable than the triplex with pyr · pur · pyr triads since in the presence of the T_7 DNA polymerase no replication arrest is observed when $(GAA)_{28,36}$ are present in the template whereas as previously described strong replication arrests are observed for the same enzyme when $(TTC)_{28,36}$ are present in the template.

VI. CONCLUDING REMARKS

A. On the Mutagenic Unusual DNA Structures

Even before the discovery of the triplet related diseases, Sinden and Wells [95–97] proposed that unusual DNA structures may cause slippage during replication, resulting in deletions or expansions in the genomic DNA. After the discovery of the triplet-related diseases and the accumulation of a vast body of sequence, biochemical, and genetic data, the possibility of unusual

DNA structures as "dynamic mutagens" could be put to rigorous tests. In fact, Wells and co-workers [97] have attempted to explain these data associated with various triplets on the basis of their ability to form hairpin, triplex, or noodle DNA structures. In her review McMurray [11] provides key experimental support for the hypothesis that the molecular mechanism for (CXG) triplet-related neurological disorders is different from that for colon cancers [98–101]. In the triplet-related diseases, large expansions in the triplet repeats are caused by the formation of unusual DNA structures, whereas in colon cancers small deletions or expansions in the dinucleotide repeats are caused by the defects in the mismatch repair system. The experimental support comes from a combination of thermodynamic and genetic analyses [11]. It appears that the stabilities of the $(CXG)_n$ hairpins are length dependent. The stability vs length correlation agrees well with the disease susceptibility vs length correlation. This means that stable unusual DNA structures may form during replication or recombination and that the longer the triplet repeat, the higher the probability of the formation of unusual DNA structures. This structural feature of the triplet repeats is quite distinct from that of the dinucleotide repeats in colon cancer. In other words, slippage in the dinucleotide repeats may involve a small number of repeats. Perhaps such a slippage is efficiently corrected by the mismatch repair system in normal cells. However, in colon cancer cells a genetic defect in the mismatch repair system causes small deletions or expansions. Slippage of the triplet repeats, on the other hand, involves a large number of repeats which may not be corrected by the mismatch repair system. Indeed, it has been shown that the genetic instability in the triplet repeat can be blocked by the mismatch repair system only when the repeat length is small [102–104]. The repair system becomes ineffectual when the repeat number of the triplets exceeds a certain critical threshold.

1. On the $(CXG)_n$ Hairpins

In these hairpins, the X · X mismatches are periodically present in the stem of the hairpins. The pattern of base pairing remains the same for odd and even numbers of repeats. This is brought about by the three-nucleotide loop in the $(CXG)_n$ hairpin with an odd number of repeats and the four-nucleotide loop in the $(CXG)_n$ hairpins with an even number of repeats. Generally, the longer the repeat, the higher the probability of hairpin formation. However, different (CXG) repeats have different length requirements for hairpin formation. For example, the individual strands of $(CCG)_n$ or $(CTG)_n$ can form hairpins starting at $n = 5$ or 6, whereas the individual strands of $(CGG)_n$ or $(CAG)_n$ form hairpins starting only at $n = 10$ or 11. It appears that when X is A or G, a homoduplex conformation of $(CXG)_n$ is preferred to a hairpin conformation for shorter ($n < 10$) repeat lengths. At larger repeat lengths, the equilibrium shifts toward the hairpin. *In vitro* replication assay also demonstrates that $(CCG)_n$ forms a hairpin at $n = 21$ whereas $(CGG)_n$ does not form a hairpin at $n = 21$. Note that in the presence of the complementary strand (i.e., in our replication assay), a larger repeat number is required for hairpin formation than for the individual strand, i.e., $n = 5$ is required for the individual $(CCG)_n$ strand whereas $n = 21$ is required for $(CCG)_n$ in the replication assay. Although in the replication assay we detect the presence of an unusual DNA structure in the template, the same structure can also be present in the growing chain during replication. For example, if in the presence of a hairpin the $(CCG)_n$ template causes a deletion then the formation of the same hairpin in the growing $(CCG)_n$ chain should cause expansion. This is exactly what Laird and co-workers [39] have observed in their replication assay with various DNA polymerases. In addition they have observed that the $(CGG)_n$ strand is less prone to expansion than the $(CCG)_n$ strand which is again consistent with our finding that the $(CCG)_n$ strand more readily forms a hairpin structure than the $(CGG)_n$ strand.

2. On the Triplexes Formed by the Friedreich's Ataxia Triplet Repeats

Our studies on the GAA/TTC repeats show the possibility of two types of triplexes: one with pyr · pur · pyr triads and the other with pur · pur · pyr triads. The triplex with pyr · pur · pyr triads is more stable than the one with pur · pur · pyr triads. Also in the triplex with pyr · pur · pyr triads the second pyr strand runs parallel in the major groove of the Watson–Crick duplex whereas in the triplex with pur · pur · pyr triads the second pur strand runs anti-parallel to the Watson–Crick duplex. In accordance with the data of McMurray and co-workers [26], we also find the possibility of a stable triplex with pur · pur · pyr triads when the second pur strand runs parallel to the Watson–Crick duplex. We are currently in the process of acquiring high-resolution NMR data of an intramolecular triplex with pur · pur · pyr triads. The formation of such a DNA–RNA triplex inside the intron with the GAA/TTC repeats may abruptly halt transcription. In addition, the intramolecular DNA triplex (GAAGAA)T_4(TTCTTC-)T_4(CTTCTT) is a good model to study the effect of sequence change in the triplex with pyr · pur · pyr triads. It is shown that the analog (GAAG**G**A)T_4(TT**C**CTC)T-$_4$(CTTC**C**T) which models the GAA to GGA substitution, lowers the stability of the triplex at neutral pH. Note that the same substitutions in the middle for a GAA repeat confer stability to the repeat.

Acknowledgments

This work was supported by the Human Genome Project of the Office of Health and Environmental Research (OHER) of the Department of Energy and OHER support to E. M. B. We thank Ms. Sue Thompson for synthesis and purification of various DNA oligomers used in this work. We thank Dr. Clifford Unkefer for maintenance of the 500 MHz Bruker DRX instrument.

References

1. Caskey, C. T., Pizzuti, A., Fu, Y-H., Fenwick, R. G., and Nelson, D. L. (1992). Triplet repeat mutations in human disease. *Science* **256,** 784–789.
2. Ross, C. A., McInnis, M. G., Margolis, R. L., and Li, S-H. (1993). Genes with triplet repeats: candidate mediators of neuropsychiatric disorders. *TINS* **16,** 254.
3. Bell, M. V., Hirst, M. C., Nakahori, Y., MacKinnon, R. N., Roche, A., Flint, T. J, Jacobs, P. A., Tommerup, N., Tranebjaerg L., Froster-Iskenius, U., Kerr, B., Turner, G., Lindenbaum, R. H., Winter, R., Pembrey, M., Thibodeau, S., and Davies, K. E. (1991). Physical mapping across the Fragile X: hypermethylation and clinical expression of the Fragile X syndrome. *Cell* **64,** 861–866.
4. Mahadevan, M., Tsilfidis, C., Sabourin, L., Shutler, G., Amemiya, C., Jansen, G., Neville, C., Marang, M., Barcelo, J., O'Hoy, K., Leblond, S., Earle-Macdonald, J., De Jong, P. J., Wieringa, B., and Korneluk, R. G. (1992). Myotonic dystrophy mutation: an unstable CTG repeat in the 3′ untranslated region of the gene. *Science* **255,** 1253–1255.
5. Duyao, M., Ambrose, A., Myers, R., Novelleto, A., Persichetti, F., Frontali, M., Folstein, S., Ross, C., Franz, M., Abbott, M., Cray, J., Conneally, P., Young, A., Penney, J., Hollingsworth, Z., Shoulson, I., Lazzarini, A., Falek, A., Koroshetz, W., Sax, D., Bird, E., Vonsattel, J., Bonilla, Alvir, J., Conde, J. B., Cha, J-H., Dure, L., Gomez, F., Ramos, M., Sanchez-Ramos, J., Snodgrass, S., De Young, M., Wexler, N., Moscowitz, C., Penchaszadeh, G., MacFarlane, H., Anderson, M., Jenkins, B., Srinidhi, J., Barnes, G., Gusella, J., and MacDonald, M. (1993). Trinucleotide repeat length instability and age of onset in Huntington's disease. *Nature Genet.* **4,** 387–397.
6. Campuzano, V., Montermini, L, Molto, M. D., Pianese, L., Cossee, M., Cavalcanti, F., Monros, E., Rodius, F., Duclos, F., Monticelli, A., Zara, F., Canizres, J., Koutnikova, H., Bidichandani, S. I., Gellera, C., Brice, A., Trouillas, P., De Michele, G., Filla, A., De Fruots, R., Palau, F., Patel, P. I., Di Donato, S., Mandel, J-L., Cocozza, S., Koenig, M., and Pandolfo, M. (1996). Friedreich's ataxia: autosomal recessive disease caused by an intronic GAA triplet repeat expansion. *Science* **271,** 1423–1427.
7. Richards, R. I., and Sutherland, G. R. (1994). Simple repeat DNA is not replicated simply. *Nature Genet.* **6,** 114–116.
8. Dover G. (1995). Slippery DNA runs on and on and on . . . *Nature Genet.* **10,** 254–256.
9. Richards, R. I., and Sutherland, G. R. (1992). Dynamic mutations causing human disease. *Cell* **70,** 709–712.
10. Wells, R. D. (1996). Molecular basis of genetic instability of triplet repeats. *J. Biol. Chem.* **271,** 2875–2878.
11. McMurray, C. T. (1995). Mechanisms of DNA expansion. *Chromosoma* **4,** 2–13.
12. Fry, M., and Loeb, L. A. (1994). The Fragile X syndrome $d(CGG)_n$ nucleotide repeats form a stable tetrahelical structure. *Proc. Natl. Acad. Sci. USA* **91,** 4950–4954.
13. Chen, X., Mariappan, S. V. S., Catasti, P., Ratliff, R., Moyzis, R. K., Laayoun, A., Smith, S. S., Bradbury, E. M., and Gupta, G. (1995). Hairpins are formed by the single DNA strands of the fragile X triplet repeats: structure and biological implications. *Proc. Natl. Acad. Sci. USA* **92,** 5199–5203.
14. Mariappan, S. V. S., Chen, X., Castasti, P., Ratliff, R., Moyzis, R. K., Laayoun, A., Smith, S. S., Bradbury, E. M., and Gupta, G. (1996). Hairpin and junction structures of fragile X triplets. *In* "Proceedings of the 9th Conversation in Biomolecular Stereodynamics, June 1995, Albany, NY" (R. H. Sarama and M. H. Sarma, Eds.), pp. 105–119. Adenine Press.
15. Mitas, M., Yu, A., Dill, J., Kamp, T. J., Chambers, E. J., and Haworth, I. S. (1995). Hairpin properties of single-stranded DNA containing a GC-rich triplet repeat: $(CTG)_{15}$. *Nucleic Acids Res.* **23,** 1050–1059.
16. Yu, A., Dill, J., Wirth, S. S., Huang, G., Lee, V. H., Howorth, I. S., and Mitas, M. (1995). The trinucleotide repeat sequence $d(GTC)_{15}$ adopts a hairpin conformation. *Nucleic Acids Res.* **23,** 2706–2714.
17. Joworski, A., Rosche, W. A., Gellibolian, R., Kangk, S., Shimizu, M., Bowater, R. P., Sinden, R. R., and Wells, R. D. (1995). Mismatch repair in Escherichia coli enhances instability of $(CTG)_n$ triplet repeats from human hereditary diseases. *Proc. Natl. Acad. Sci. USA* **92,** 111019–11023.
18. Chastain, P. D., II, Eichler, E. E., Kang, S., Nelson, D. L., Levene, S. D., and Sinden, R. R. (1995). Anomalous rapid electrophoeretic mobility of DNA containing triplet repeats associated with human disease genes. *Biochemistry* **34,** 16125–16131.
19. Smith, G. K., Jie, J., Fox, G. E., and Gao, X. (1995). DNA CTG triplet repeats involved in dynamic mutations of neurologically related gene sequences form stable duplexes. *Nucleic Acids Res.,* **23,** 4303–4311.
20. Gao, X., Huang, X., Smith, G. K., Zheng, M., and Liu, H. (1995). New antiparallel duplex motif of DNA CCG repeats that is stabilized by extrahelical bases symmetrically located in the minor groove. *J. Am. Chem. Soc.* **95,** 1517–8883.
21. Mitas, M., Yu, A., Dill, J., and Haworth, I. S. (1995). The trinucleotide repeat sequence $d(CGG)_{15}$ forms a heat-stable hairpin containing $G^{syn}G^{anti}$ base pairs. *Biochemistry* **34,** 12803–12811.
22. Aheng, M., Huang, X., Smith, G. K., Yang, X., and Gao, X. (1996). Genetically unstable CXG repeats are structurally dynamic and have a high propensity for folding. An NMR and UV spectroscopic study. *J. Mol. Biol.* **264,** 323–336.
23. Kettani, A., Kumar, R. A., and Patel, D. J. (1995). Solution structure of a DNA quadruplex containing the Fragile X syndrome triplet repeat. *J. Mol. Biol.* **95,** 2822–2836.
24. Gacy, A. M., Goellner, G., Juranic, N., Macura, S., and McMurray, C. T. (1995). Trinucleotide repeats that expand in human disease form hairpin structures in vitro. *Cell* **8,** 533–540.
25. Goldberg, Y. P., McMurray, C. T., Zeisler, J., Almqvist, E., Sillence, D., Richards, F., Gacy, A. M., Buchanan, J., Telenius, H., and Hayden, M. R. (1995). Increased instability of intermediate alleles in families with sporadic Huntington disease compared to similar sized intermediate alleles in the general population. *Hum. Mol. Genet.* **4,** 1911–1918.
26. McMurray, C. T., Gacy, A. M., Goellner, G., Spiro, C., Dyer, R., Mikesell, M., Yao, J., Johnson, A. J., Juranic, N., Macura, S., Richter, A., and Melancon, S. B. (1997). DNA structures associated with class I expansion of GAA in Friedreich's ataxia. *In* "Proceedings of the 10th Conversation in Biomolecular Stereodynamics, June 1997, Albany, NY" (R. H. Sarama and M. H. Sarma, Eds.). Adenine Press. [In press]

27. Goellner, G. M., Tester, D., Thibodeau, S., Almqvist, E., Goldberg, Y. P., Hayden, M. R., and McMurray, C. T. (1997). Different mechanisms underlie DNA instability in Huntington's disease and colorectal cancer. *Am. J. Hum. Genet.* **60,** 879–890
28. Kang, S., Jaworski, A., Ohshima, K., and Wells, R. D. (1995). Expansion and deletion of CTG repeats from human disease genes are determined by the direction of replication in E. coli. *Nature Genet.* **10,** 213–218.
29. Otten, A. D., and Tapscott, S. J. (1995). Triplet repeat expansion in myotonic dystrophy alters the adjacent chromatin structure. *Proc. Natl. Acad. Sci. USA* **92,** 5465–5469.
30. Wang, Y.-H., and Griffith, J. (1994). Expanded CTG triplet blocks from the myotonic dystrophy gene create the strongest known natural nucleosome positioning elements. *Genomics* **25,** 570–573.
31. Kang, S., Ohshima, K., Shimize, M., Amirhaeri, S., and Wells, R. D. (1995). Pausing of DNA synthesis in vitro at specific loci in CTG and CCG triplet repeats form human hereditary disease genes. *J. Biol. Chem.* **270,** 27104–27021.
32. Mariappan, S. V. S., Catasti, P., Chen, X., Ratliff, R., Moyzis, R. K., Bradbury, E. M., and Gupta, G. (1996). Solution structures of the individual single strands of the fragile X DNA triplets $(GCC)_n$- $(GGC)_n$. *Nucleic Acids Res.* **24,** 784–792.
33. Mariappan, S. V. S., Garcia, A. E., and Gupta, G. (1996). Structure and dynamics of the DNA hairpins formed by tandemly repeated CTG triplets associated with myotonic dystrophy. *Nucleic Acids Res.* **24,** 775–783.
34. Petruska, J., Arnheim, N., and Goodman, M. F. (1996). Stability of intrastrand hairpin structures formed by the CAG/CTG class of DNA triplet repeats associated with neurological diseases. *Nucleic Acids Res.* **24,** 1992–1998.
35. Nadel, Y., Weisman-Shomer, P., and Fry, M. (1995). The fragile X syndrome single strand $d(CCG)_n$ nucleotide repeats readily fold back to form unimolecular hairpin structures. *J. Biol. Chem.* **270,** 28970–28977.
36. Yu, A., Dill, J., and Mitas, M. (1995). The purine-rich trinucleotide repeat sequences $d(CAG)_{15}$ and $d(GAC)_{15}$ form hairpins. *Nucleic Acids Res.* **23,** 4055–4057.
37. Pearson, C. E., and Sinden, R. R. (1996). Alternative structures in duplex DNA formed within the trinucleotide repeats of the myotonic dystrophy and Fragile X loci. *Biochemistry* **35,** 5041–5053.
38. Lian, C., Robinson, H., and Wang, A. H. J. (1996). Structure of ACTINOMYCIN D binding with (GAAGCTTC)2 and (GATGCTTC)2 and its binding to $(CAG)_n \cdot$ (CTG)n triplet sequence determined by NMR analysis. *J. Am. Chem. Soc.*
39. Ji, J., Clegg, N. J., Peterson, K. R., Jackson, A. L., Laird, C. D., and Loeb, L. A. (1996). In vitro expansion of GGC : GCC repeats: identification of the preferred strand of expansion. *Nucleic Acids Res.* **24,** 2835–2840.
40. Wang, Y.-H., Gellibolian, R., Shimizu, M., Wells, R. D., and Griffith, J. (1996). Long CCG triplet repeat blocks exclude nucleosomes: a possible mechanism for the nature of fragile sites in chromosomes. *J. Mol. Biol.* **263,** 511–516.
41A. Wang, Y.-H., and Griffith, J. (1996). Methylation of expanded CCG triplet repeat DNA from Fragile X syndrome patients enhances nucleosome exclusion. *J. Biol. Chem.* **271,** 22937–22940.97.
41B. Wang, Y.-H., and Griffith J. (1995). Expanded CTG triplet blocks from the myotonic dystrophy gene create the strongest known natural nucleosome positioning elements. *Genomics* **25,** 570–573.
42. Schweitzer, J. K., and Livingston, D. M. (1997). Destabilization of CAG trinucleotide repeat tracts by mismatch repair mutations in yeast. *Hum. Mol. Genet.* **6,** 349–355.
43. Kang, S., Ohshima, K., Jaworski, A., and Wells, R. D. (1996). CTG triplet repeats from the myotonic dystrophy gene are expanded in Escherichia coli distal to the replication origin as a single large event. *J. Mol. Biol.* **258,** 543–547.
44. Rosche, W. A., Jaworski, A., Kang, S., Kramer, S. F., Larson, J. E., Geidroc, D. P., Wells, R. D., and Sinden, R. R. (1996). Single-stranded DNA-binding protein enhances the stability of CTG triplet repeats in Escherichia coli. *J. Bacteriol.* **178,** 5042–5044.
45. Ohshima, K., Kang, S., and Wells, R. D. (1996). CTG triplet repeats from human hereditary diseases are dominant genetic expansion products in Escherichia coli. *J. Biol. Chem.* **271,** 1853–1856.
46. Mariappan, S. V. S., Silks, L. A., III, Chen, X., Springer, P. A., Wu, R., Moyzis, R. K., Bradbury, E. M., Garcia, A. E., and Gupta, G. (1997). Solution structure of the Huntington's disease DNA triplets, $(CAG)_n$. *J. Biol. Struct. Dynamics.*
47. Brahmachari, S. K., Meera, G., Sarkar, P. S., Balugurumoorthy, B., Tripathi, J., Raghavan, S., Shaligram, U., and Pataskar, S. S. (1995). Simple repetitive sequences in the genome: structural and functional significance. *Electrophoresis* **16,** 1705–1714.
48. Brahmachari, S. K., Sarkar, P. S., Shaligram, U., Matsuddi, M., Raghavan, S., Bhandari, Pataskar, S. S., Narayan, M., and Quasar, S. P. (1997). Genome instability: structural basis of triplet repeat expansion and genetic disorder. *In* "Proceedings of the 10th Conversation in Biomolecular Stereodynamics, June 1997, Albany, NY" (R. H. Sarama and M. H. Sarma, Eds.). Adenine Press. [In press]
49. Pieretti, M., Zhang, F., Fu, Y.-H., Warren, S. T., Oostra, B. A., Caskey, C. T., and Nelson, D. L. (1991). Absence of expression of the FMR-1 gene in fragile X syndrome. *Cell* **66,** 817–822.
50. Laird, C., Jaffe, E., Karpen, G., Lamb, M., and Nelson, R. Fragile sites in human chromosomes as regions of late-replicating DNA. *TIG* **3,** 274–280.
51. Laird, C. D. (1987). Proposed mechanism of inheritance and expression of the human fragile-X syndrome of mental retardation. *Genetics* **117,** 587–599.
52. Oberle, I. Rousseau, F., Heitz, D., Kretz, C., Devys, D., Hanauer., A., Boue, J., Bertheas, M. F. and Mandel, J. L. (1991). Instability of a 550-base pair DNA segement and abnormal methylation in fragile X syndrome. *Science* **252,** 1097–1102.
53. Gecz, J. Gedeon, A. K., Sutherland, G. R., and Mulley, J. C. (1996). Identification of the gene FMR2, associated with FRAXE mental retardation. *Nature Genet.* **13,** 105–110.
54. Eichler, E. E., Holden, J. J. A., Popovich, B. W., Reiss, A. L., Snow, K. Thibodeau, S. N., Richards, C. S., Ward, P. A., and Nelson, D. L. (1994). Length of uninterrupted CGG repeats determines instability in the FMR1 gene. *Nature Genet.* **8,** 88–92.
55. Cognet, J. A. H., Gabarro-Arpa, J., Bret, M. L., van der Marel, G. A., van Boom, J. H., and Fazakerley, G. V. (1991). Solution conformation of an oligonucleotide containing G · G mismatches determined nuclear magnetic resonance and molecular mechanics. *Nucleic Acids Res.* **19,** 6771–6779.
56. Yu, A., Barron, M. D., Romero, R. M., Christy, M., Gold, B., Dai, J., Gray, D. M., Haworth, I. S., and Mitas, M. (1997). At physiological pH, d(CCG)15 forms a hairpin containing protonated cytosines and a distorted helix. *Biochemistry* **36,** 3687–3699.
57. Smith, S. S., Kaplan, B. E., Sowers, L. C., and Newman, E. M. (1992). Mechanism of human methyl-directed DNA methyltransferase and the fidelity of cytosine methylation. *Proc. Natl. Acad. Sci. USA* **89,** 4744–4748.
58. Baker, D. J., Kan, J. L. C., and Smith, S. S. (1988). Recognition of structural perturbations in DNA by human DNA (cytosine-5) methyltransferase. *Gene* **74,** 207–210.

59. Baker, D. J., Laayoun, A., and Smith, S. S. (1993). Transition state analogs as affiity labels for human DNA methyltransferases. *Biochem. Biophys. Res. Commun.* **196,** 864–871.
60. Posfai, J., Bhagwat, A. S., Posfai, G., and Roberts, R. J. (1989). Predictive motifs derived from cytosine methyltransferases. *Nucleic Acids Res.* **17,** 2421–2435.
61. Klimasauskas, S., Kumar, S., Roberts, R. J., and Cheng, X. (1994). Hhal methyltransferase flips its target base out of the DNA helix. *Cell* **76,** 357–369.
62A. Roberts, R. J. (1995). On base flipping. *Cell* **82,** 639–645.
62B. Chen, X., Mariappan, S. V. S., R., Moyzis, R. K., Bradbury, E. M., and Gupta, G. (1997). Hairpin induced slippage and hypermethylation of the fragile X DNA triplets. *J. Biomol. Struct. Dynamics.* [in press]
63A. Depamphilis, M. L., and Wasserman, P. M. (1980). Replication of eukaryotic chromosomes: a close-up view of the replication fork. *Annu. Rev. Biochem.* **49,** 627–666.
63B. Stillman, B. (1994). Smart machines at the DNA replication fork. *Cell* **78,** 725–728.
64. Zhao, J., Cheng, W., Gibb, C. L. D., Gupta, G., and Kallenbach, N. R (1996). HMG box proteins interact with multiple tandemly repeated $(GCC)_n$ · (GGC)n DNA sequences. *J. Biomol. Struct. Dynamics* **14,** 235–238.
65. Antequera, F., and Bird, A. (1993). Number of Cp'G island and genes in human and mouse. *Proc. Natl. Acad. Sci. USA* **90,** 11995–11999.
66. Bird, A. P. (1986). CpG-rich island and the function of DNA methylation. *Nature* **321,** 209–213.
67. Boyes, J., and Bird, A. (1992). Repression of genes by DNA methylation depends on CpG density and promoter strength: evidence for involvement of a methyl-CpG binding protein. *EMBO J.* **11,** 327–333.
68. Meehan, R. R., Lewis, J. D., McKay, S. Kleiner, E. L., and Bird, A. (1989). Identification of a mammalian protein that binds specifically to DNA containing methylated CpGs. *Cell* **58,** 499–507.
69. Lewis J. D., Meehan, R. R., Henzel, W. J., Maurer-Fogy, I., Jappesen, P., Klein, F., and Bird, A. (1992). Purification, sequence, and cellular localization of a novel chromosomal protein that binds to methylated DNA. *Cell* **69,** 905–914.
70. Lie, L. F., and Wang, J. C. (1987). Supercoiling of the DNA template during transcription. *Proc. Natl. Acad. Sci. USA* **84,** 7024–7027.
71. Mahadevan, M., Tsilfidis, C., Sabourin, L., Shutler, G., Amemiya, C., Jansen, G., Neville, C., Narang, M., Barcelo, J., O'Hoy, K., Leblond, S., Earle-Macdonald, J., De Jong, P. J., Wieringa, B., and Korneluk, R. G. (1992). Myotonic dystrophy mutation: an unstable CTG repeat in the 3′ untranslated region of the gene. *Science* **255,** 1253–1255.
72. Fu, Y. H., Pizzuti, A., Fenwick, R. G., Jr., King, J., Rajnarayan, S., Dunne, P. W., Dubel, J., Nasser, G. A., Ashizawa, T., De Jong, P., Wieringa, B., Korneluk, R., Perryman, M. B. , Epstein, H. F., and Caskey, C. T. (1992). An unstable triplet repeat in a gene related to myotonic muscular dystrophy. *Science* **255,** 1256–1258.
73. Brook, J. D., McCurrach, M. E., Harley, H. G., Buckler, A. J., Church, D., Aburatani, H., Hunter, K., Stanton, V. P., Thirion, J., Hudson, T., Sohn, R., Zemelman, B., Snell, R. G., Rundle, S. A., Crow, S., Davies, J., Shelbourne, P., Buxton, J., Jones, C., Juvonen, V., Johnson, K., Harper, P. S., Shaw, D. J., and Housman, D. E. (1992). Molecular basis of myotonic dystrophy: expansion of a trinucleotide (CTG) repeat at the 3′ end of a transcript encoding a protein kinase family member. *Cell* **68,** 799–808.
74. Briat, J-F., Bollag, G., Kearney, C. A., Molineuz, I., and Chamberlin, M. J. (1987). Tau factor from Escherichia coli mediates accurate and efficient termination of transcription at the bacteriophage T3 early termination site in vitro. *J. Mol. Biol.* **198,** 43–49.
75. Eckner, R., and Birnstiel, M. (1992). Evolutionary conserved multiprotein complexes interact with the 3' untranslated region of histone transcripts. *Nucleic Acids Res.* **20,** 1023–1030.
76. Taneja, K. L., McCurrach, M., Schalling, M., Housman, D., and Singer, R. H. (1995). Foci of trinucleotide repeat transcripts in nuclei of myotonic dystrophy cells and nuclei. *J. Cell Biol.* **128,** 995–1002.
77. Kouchakdjian, M., Li, B. F. L., Swan, P. F., and Patel, D. J. (1988). Pyrimidine · pyrimidine base-pair mismatches in DNA: a nuclear magnetic resonance study of T · T pairing at neutral pH and C · C pairing at acidic pH in dodecnucleotide DNA duplexes. *J. Mol. Biol.* **202,** 139–155.
78. Lane, A. N., and Forster, M. J. (1989). Determination of internal dynamics of deoxyriboses in the DNA hexamer d(CGTACG)2 by 1H NMR. *Eur. Biophys. J.* **17,** 221–232.
79. Lipari, G., and Szabo, A. (1982). Model-free approach to the interpretation of nuclear magnetic resonance relaxation in macromolecules. 1. Theory and range of validity. *J. Am. Chem. Soc.* **104,** 4546–4558.
80. Lipari, G., and Szabo, A. (1982). Model-free approach to the interpretation of nuclear magnetic resonance relaxation in macromolecules. 2. Analysis of experimental results. *J. Am. Chem. Soc.* **104,** 4559–4570.
81. Krahe, R., Ashizawa, T., Abbruzzese, C., Roeder, E., Garango, P., Giacaelli, M., Funanage, V., and Siciliano, M. J. (1995). Effect of myotonic dystrophy trinucleotide repeat expansion on DMPK transcription and processing. *Genomics* **28,** 1–14.
82. The Huntington's Disease Collaborative Research Group (1993). A novel gene containing a trinucleotide repeat that is expanded and unstable on Huntington's disease chromosomes. *Cell* **72,** 971–983.
83. Leeflang, E. P., Zhang, L., Tavare, S., Hubert, R., Srinidhi, J., MacDonald, M. E., Myers, R., de Young, M., Wexler, N. S., Gusella, J. F., and Arnheim, N. (1995). Single sperm analysis of the trinucleotide repeats in the Huntington's disease gene: quantification of the mutation frequency spectrum. *Hum. Mol. Genet.* **4,** 1519–1526.
84. van de Ven, F. J. M., and Hilbers, C. W. (1988). Nucleic acids and nuclear magnetic resonance. *Eur. J. Biochem.* **270,** 1–38.
85. Orbons, L. P. M., van der Marel, G. A., van Boom, J. H., and Altona, C. (1987). An NMR study of the polymorphous behavior of the mismatched octamer d(m^5C-G-m^5C-G-T-G-m^5C-G) in solution: the B, Z, and hairpin forms. *J. Biomol. Structure Dynamics* **4,** 939–963.
86. Ikuta, S., Takagi, K., Wallace, R. B., and Itakura, K. (1987). Dissociation kinetics of 19 base-paired oligonucleotide-DNA duplexes containing different single mismatched base pairs. *Nucleic Acids Res.* **15,** 797–811.
87. Gervais, V., Cognet, J. A. H., Le Bret, M., Sowers, L. C., and Fazakerley, G. V. (1995). Solution structure of two mismatches A · A and T · T in the K-ras gene context by nuclear magnetic resonance and molecular dynamics. *Eur. J. Biochem.* **228,** 279–290.
88. Cossee, M., Schimtt, M., Campuzano, V., Reutenauer, L., Moutou, C., Mandel, J. L., and Koenig, M. (1997). Evolution of the Friedreich's ataxia trinucleotide repeat expansion. *Proc. Natl. Acad. Sci. USA* **94,** 7452–7457.
89. Bidichandani, S. I., Ashizawa, T., and Patel, P. (1997). Atypical Friedreich's ataxia caused by compound heterozygosity for a

novel missense mutation and the GAA-triplet repeat expansion. *Am. J. Hum. Genet.* **60,** 1251–1256.

90A. Filla, A., de Michele, G., Cavalcanti, F., Pianese, L., Monticelli, A., Campanella G., and Cocozza, S. (1996). The relationship between trinucleotide (GAA)repeat length and clinical features in Friedreich ataxia. *Am. J. Hum. Genet.* **59,** 554–560.

90B. Wells, R. D., Collier, D. A., Hanvey, J. C., Shimizu, M., and Wohlrab, F. (1988). The chemistry and biology of unusual DNA structures adopted by oligopurine · oligopyrimidine sequences. *FASEB J.* **2,** 2939–2949.

91. Mariappan, S. V. S., Catasti, P., Silks, L. A., Bradbury, E. M., and Gupta, G. (1997). High resolution structure of triplex formed by the GAA/TTC triplets associated with the Friedreich's Ataxia. [Submitted for publication]

92. Baran, N., Lapidot, A., and Manor, H. (1991). Formation of DNA triplexes accounts for arrests of DNA synthesis at d(TC) and d(GA) tracts. *Proc. Natl. Acad. Sci.USA* **88,** 507–511.

93. Catasti, P., Chen, X., Moyzis, R. K., Bradbury, E. M., and Gupta, G. (1997). Structure-function correlations of the insulin-linked polymorphic region. *J. Mol. Biol.* **263,** 534–545.

94. Catasti, P., Chen, X., Deaven, L. L., Moyzis, R. K., Bradbury, E. M., and Gupta, G. (1997). Cytosine-rich strands of the insulin minisatellite adopt hairpins with intercalated cyotosine+ · cytosine pairs. *J. Mol. Biol.* **272,** 369–382.

95. Wells, R. D. (1988). Unusual DNA structures. *J. Biol. Chem.* **263,** 1095–1098.

96. Sinden, R., R., and Wells, R. D. (1992). DNA structure, mutations, and human genetic disease. *Biotech.* **3,** 612–622.

97. Trinh, T. Q., and Sinden, R. R. (1991). Preferential DNA secondary structure mutagenesis in the lagging strand of replication in E. coli. *Nature* **352,** 544–547.

98. Thibodeau, S. N., Bren, G., and Schaid, D. (1993). Microsatellite instability in cancer of the proximal colon. *Science* **260,** 816–819.

99. Fishel, R., Lescoe, M. K., Rao, M. R. S., Copeland, N. G., Jenkins, N. A., Garber, J., Kane, M., and Kolodner, R. (1993). The human mutator gene homolog MSH2 and its association with hereditary nonpolyposis colon cancer. *Cell* **75,** 1027–1038.

100. Bonner, C. E., Baker, S. M., Morrison, P. T., Waren, G., Smith, L. G., Lescoe, M. K., Kane, M., Erabino, C., Lipford, J., Lindblom, A., Tannergard, Pl., Bollag, R. J., Godwin, A. R., Ward, D. C., Nordenskjeid, M., Fishel, R., Kolodner, R., and Liskay, R. M. (1994). Mutation in the DNA mismatch repair gene homologue hMLH1 is assosicated with hereditary non-polyposis colon cancer. *Nature* **368,** 258–261.

101. Papadopoulos, N, Nicolaides, N. C., Wei, Y-F., Ruben, S. M., Carter, K. C., Rosen, C. A., Haseltine, W. A., Fleischmann, R. D., Fraser, C. M., Adams, M. D., Venter, J. C., Hamilton, S. R., Petersen, G. M., Watson, P., Lynch, G. T., Peltomaki, P., Mecklin, J-P., de la Chapelle, A., Kinzler, K. W, and Vogelstein, B. (1994). Mutation of a *mutL* homolog in hereditary colon cancer. *Science* **263,** 1625–1629.

102. Kramer, P. R., Pearson, C. E., and Sinden, R. R. (1996). Stability of triplet repeats of myotonic dystrophy and fragile X loci in human mutator repair cell lines. *Hum. Genet.* **98,** 151–157.

103. Fishel, R., Ewel, A., Lee, S., Lescoe, M. K., and Griffith, J. (1994). Binding of mismatched microsatellite DNA sequences by the human MSH2 protein. *Science* **266,** 1403–1405.

104. Schweitzer, J. K., and Livingston, D. M. (1997). Destabilization of CAG trinucleotide repeat tracts by mismatch repair mutations in yeast. *Hum. Mol. Genet.* **6,** 349–355.

105. Mariappan, S. V. S., Silks, L. A., Bradbury, E. M., and Gupta, G. (1997). Hairpins of fragile X GCC DNA triplet form single hydrogen-bonded C · C mispairs: selective ^{15}N4-labeled cytosines and isotope-edited nuclear magnetic resonance spectroscopy. [Submitted for publication]

106. Godde, J. S., and Wolffe, A. P. (1996). Nucleosome assembly on CTG triplet repeats. *J. Biol. Chem.* **271,** 15222–15229.

Chapter 42

Nucleosome Analyses and Diseases of Chromatin Structure

JACK D. GRIFFITH AND YUH-HWA WANG

Lineberger Comprehensive Cancer Center, University of North Carolina at Chapel Hill, Chapel Hill, North Carolina 27599

I. INTRODUCTION

The eukaryotic chromosome is formed by a progressive condensation of the DNA, beginning with its assembly into a string of beads or nucleosomes [1]. Each nucleosome consists of 146 base pairs of DNA wrapped about a histone protein octamer and their presence near genetic control elements can strongly influence gene expression (reviewed in [2,3]). This string of beads is further condensed by the binding of histone H1 into chromatin fibers ~30 nm in diameter that contain 7 nucleosomes (nearly 1500 base pairs) for each coil of the fiber. The 30-nm fibers are bound to the nuclear matrix possibly once every 40~100 kb creating distinct chromatin domains. Further compaction of the 30-nm fiber gives rise to the condensation levels observed by light microscopy in metaphase chromosomes.

While a large body of evidence exists about how eukaryotic DNA is transcribed and replicated *in vitro* in the absence of the histones, many aspects about how

Note. For nomenclature of triplet repeat sequences, CCG · CGG designates a duplex sequence of repeating CCG which also can be written CGC or GCC; CGG, the complementary strand, may also be written as GGC or GCG. The orientation is 5′ to 3′ for both designations of the antiparallel strands. In some cases for simplicity, the duplex insert is referred to by the first three letters (i.e., CCG for CCG · CGG).

the replication and transcription machinery deals with DNA highly condensed into chromatin remain unclear. Recent studies [4] have revealed that when a specific gene is assembled into chromatin, the level of basal transcription may be depressed while the level of sequence-specific transcription may remain high or become elevated, thus having the effect of greatly increasing the overall transcriptional response when the DNA is chromatinized. While these effects likely occur at the level of the placement of individual nucleosomes along promoter elements, the way in which longer range chromatin condensation may influence gene expression *in vivo* remains poorly understood.

It has been shown that each nucleosome within a nucleosomal array is able to interact with adjacent nucleosomes (reviewed in [5]). These researchers [6, 7] provided evidence that saturated arrays (DNA occupied by histone octamers packed tightly together) can form a more stable folded structure than subsaturated templates. Bradbury and colleagues [8] demonstrated that nucleosome arrays inhibit both initiation and elongation of transcripts by bacteriophage T7 RNA polymerase and that the extent of inhibition is directly proportional to the number of nucleosome cores assembled on the DNA template. Since the effect of arrays of nucleosomes is to greatly amplify the effects of chromatin structure on replication or transcription, the net result of expansion of a DNA sequence that generated unusually stable nucleosomes could be dramatic. In this case, expansion of such a DNA element from a size encompassing one nucleosome to a size that would generate an array of five or six hyperstable nucleosomes could produce a chromatin element with very strong effects *in vivo*.

The ability of a variety of DNA sequences to form nucleosomes has been examined. Repeated tracts of 4 to 6 adenines in phase with the helix produce sequence-directed bends, and bent DNA preferentially assembles into nucleosomes [9, 10], presumably due to the greater ease of wrapping of the DNA about the histone core. The sequence element $(A/T)_3NN(G/C)_3NN$ creates an anisotropically flexible wedge which also facilitates wrapping DNA about the histone octamer [11]. The 5S RNA gene from several species contains such flexible wedges and exhibits very strong sequence-directed nucleosome positioning [12–15]. Indeed Shrader and Crothers [11] were able to create synthetic nucleosome positioning elements in which nucleosome formation was energetically favored over the 5S DNA by spacing up to 10 wedges in phase with the helix.

Sequences also exist that inhibit nucleosome formation and *in vivo* such sequences, were they to occur in promoter or enhancer regions, could play an extremely important role in maintaining access to the DNA. DNA–RNA hybrids [16], left handed Z-DNA [17], and DNA containing poly(dA : dT) runs have all been shown to exclude nucleosome formation *in vitro* [18–20]. Recently Iyer and Struhl [21] found that poly(dA:dT) tracts in the yeast his3 promoter stimulate Gcn4-activated transcription *in vivo* due to nucleosome exclusion and increased accessibility of Gcn4-binding sites

II. NUCLEOSOME POSITIONING AND CTG REPEAT DISEASES

Expansion of repeating tracts of CTG nucleotide triplets have been implicated in both the loss-of-function and gain-of-function triplet diseases; these findings are described in detail in other chapters in this volume. In the gain-of-function neurological diseases the CTG expansions code for additional polyglutamines (CAG codes for glutamine) that are incorporated into a protein whose altered function appears to elicit neurological dysfunction in certain cells. Here the sizes of the expanded tracts are relatively small (36 to 121) as contrasted to the much larger expanded tracts observed in the loss of function diseases, such as DM. In DM the triplet which expands is also CTG and expanded tracts as large as 5000 repeats have been observed. It is generally assumed that while the fundamental molecular mechanisms that lead to triplet expansion are the same for either class of disease, the pathological effects of these expanded tracts are very different for DM as contrasted to diseases involving proteins with large polyglutamine domains.

The unusual properties of DM may point to the way(s) that the expanded triplet tract exerts its deleterious effect in the cell. Features that must be taken into account include the failure to find any cases of DM that result from point mutations or deletions within the DMPK gene in the way that such mutations in the FMR-1 gene have been seen to give rise to a fragile X syndrome. Further, general clinical observations consistently suggest that as the size of the CTG triplet block in the downstream region of the DM protein kinase gene increases, the complexion of the disease appears to change significantly. This may imply that as the triplet block further expands, more than one gene may become compromised. Current thinking has involved models of altered chromatin structure (see below), RNA molecules that may form aberrant structures or bind new cellular proteins, or RNA that may not be processed in the normal route. Below, work is summarized showing that expanded CTG triplet blocks can greatly alter chromatin structure.

A. Nucleosome Positioning by Long CTG Repeat Blocks

Knowledge that the threshold for the triplet diseases frequently approximates the amount of DNA in a nucleosome, and the general understanding that any short repeating sequence may amplify sequence signals for nucleosome positioning or exclusion, led Wang *et al.* [22] to examine the ability of long repeating tracts of $(CTG)_n$ to assemble into chromatin. This work was carried out using a simple, but well-established, method of chromatin assembly involving mixing DNA with purified histone octamers (H2a, H2b, H3, H4) in 2 M NaCl followed by a progressive lowering of the salt concentration to 0.6 M or less by step-wise dilution or dialysis [10, 11, 22–25]. In the approach of by Wang *et al.* [22] a minimal amount of histone octamers was used such that each DNA would, on the average, contain only a single assembled nucleosome. This simplified measurement of the nucleosome position since the foreshortening due to multiple nucleosomes would have greatly complicated analysis. In this study direct electron microscopy (EM) was utilized to measure the position of several hundred individual nucleosomes along single DNAs and the data were summed to create a nucleosome positioning map. To distinguish one end of the linear DNA from the other, one DNA end was labeled with a biotin–streptavadin tag.

For such analysis it was important to also consider the effects of magnesium and supercoiling of the template DNA. Numerous DNA structural transitions are facilitated by negative supercoiling or influenced by heating in the presence of magnesium. A supertwisted plasmid, pSH2 isolated from *Escherichia coli,* containing 130 CTG repeats derived from a myotonic dystrophy patient was treated with magnesium and heat and then reconstituted into chromatin. The DNA was then linearized to place the CTG triplets in pSH2 between 32 and 43 map units along the DNA (Fig. 42-1). The resulting nucleosome map (Fig. 42-2A) revealed that 48% of all the nucleosomes were present in the region between 30 and 45 map units. In contrast, the map for the Bluescript vector lacking the CTG insert (Fig. 42-2B) showed a uniform distribution of nucleosomes over the length of the DNA with only 10% of the nucleosomes present in the region between 30 and 45 map units. The use of linear or supercoiled DNA that had not been treated with heat or magnesium still resulted in 29% of all nucleosomes localized to the triplet repeat. Further analysis using a set of six pUC19-based plasmids (pretreated with heat and magnesium) containing 26 to 250 contiguous CTG triplet repeats generated in the laboratory of R. D. Wells showed that as the size of the triplet block increased in size, the efficiency of nucleosome formation at the repeats also increased up to a repeat size of 180. These data are summarized in a normalized form in Fig. 42-2C.

These data provided the first direct evidence that repeating CTG triplets can serve as a strong nucleosome positioning element. Indeed, this increase in strength of nucleosome formation occurred over the range where the triplet blocks change from normal, to the gray zone, to disease-causing. The EM observations therefore suggest that expanded blocks may create unusually stable nucleosomes that could alter local chromatin structure. Microscopy, however, was unable to provide precise measurements of the strength of the CTG repeat blocks as nucleosome positioning elements, data that would be essential to models based on these observations.

A method for measuring the strength of nucleosome positioning elements has been developed by Shrader and Crothers [11], termed competitive nucleosome reconstitution. A schematic illustration of this method is shown in Fig. 42-3. In such assays, a small amount of the specific DNA to be tested (0.1 μg) is labeled with ^{32}P and then mixed with a 200-fold excess of unlabeled calf thymus DNA (20 μg) and ~5 μg calf thymus histone octamers [10] in a solution containing 2 M NaCl. The salt is slowly lowered to 0.1 M NaCl and the mixture electrophoresed on a 5% polyacrylamide gel to separate free DNA from the nucleosome-assembled DNA. The amount of DNA in each band is quantified. Usually three or four separate but identical experiments are carried out for each set of DNAs. In such studies, a sequence found in the *Xenopus borealis* somatic 5S RNA gene is frequently used as a standard against which other sequences are measured. This 5S element was previously the strongest known natural nucleosome positioning element. Using this method, Wang and Griffith [26] compared DNAs containing 75 and 130 CTG repeats to the *X. borealis* somatic 5S RNA gene. In such studies it is important that the DNAs being compared are of the same size. Thus two pairs of DNAs, ~285 and ~450 bp in size, were compared.

Results of such an analysis for the sets of DNAs of ~285 and ~450 bp are shown in Fig. 42-4. The nucleosome-assembled DNA appears as a retarded band. One major retarded band was observed in both 5S RNA gene samples while two retarded bands were found in the ~285-bp $(CTG)_{75}$-containing DNA and possibly three closely spaced retarded bands were present in the ~450-bp $(CTG)_{130}$-containing DNA. The multiple retarded bands represent DNAs associated with 2 or 3 nucleosomes. The results revealed that the ratio of the fraction of the nucleosome-assembled DNA to free

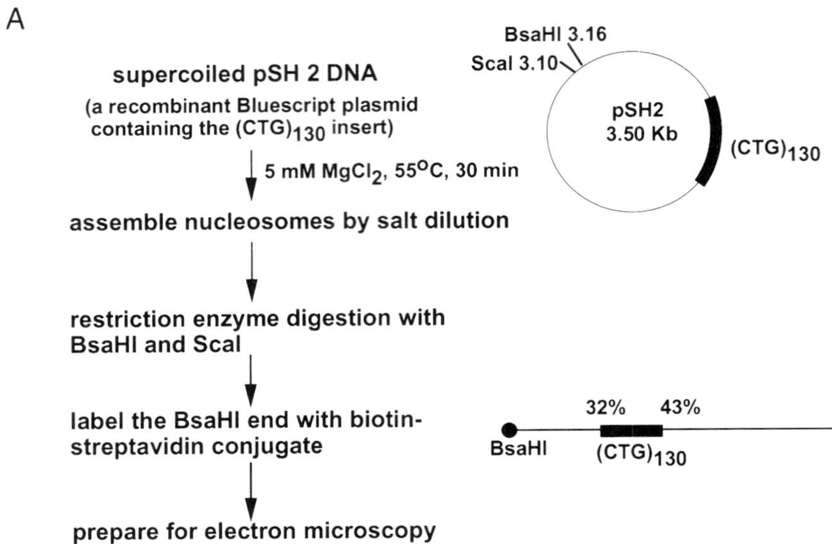

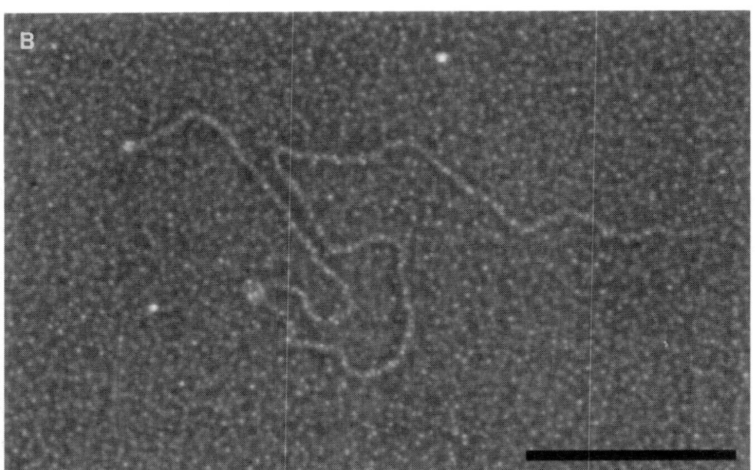

FIGURE 42-1 (A) Scheme for nucleosome assembly on supertwisted plasmid DNA and end labeling with streptavadin–biotin complexes. (B) Visualization of nucleosomes assembled on streptavadin-labeled pSH2 DNA containing a $(CTG)_{130}$ repeat derived from myotonic dystrophy patients. The DNA (curved filament) contains streptavadin protein bound to one end and one nucleosome assembled near the center of the DNA. In the absence of added histone, no nucleosome-like objects are observed. Reconstitution of the DNA with histones is described in this chapter. Preparation for EM including rotary shadow cast with tungsten is described in [22]. Bar, 130 nm.

DNA is 5.6 ± 0.4-fold higher in the DNA containing the $(CTG)_{75}$ block as compared to the 5S RNA gene. The ~450-bp DNA containing the $(CTG)_{130}$ block is 8.7 ± 0.7-fold stronger than the 5S RNA gene (472 bp) in nucleosome assembly.

These assays allow one to calculate the free energy difference for nucleosome formation. The difference in free energy between the $(CTG)_{75}$ DNA and the 5S RNA gene was found to be 1015 + 43 cal/mol and for the $(CTG)_{130}$ DNA, the difference was 1276 ± 45 cal/mol. In Fig. 42-5, these results are combined with data obtained by Shrader and Crothers [11] and further work described below. To match the different studies, the free energy for the vector DNA for each set of experiments was taken as zero.

These studies demonstrated that when CTG repeat blocks grow to sizes of $n = 75$ or greater they become the strongest known natural nucleosome positioning elements, and (at $n = 130$) are nearly 10 times stronger than the 5S element. The difference in binding free energy between the strongest and average DNA elements is seen to span a range of 100-fold or greater. However, the total

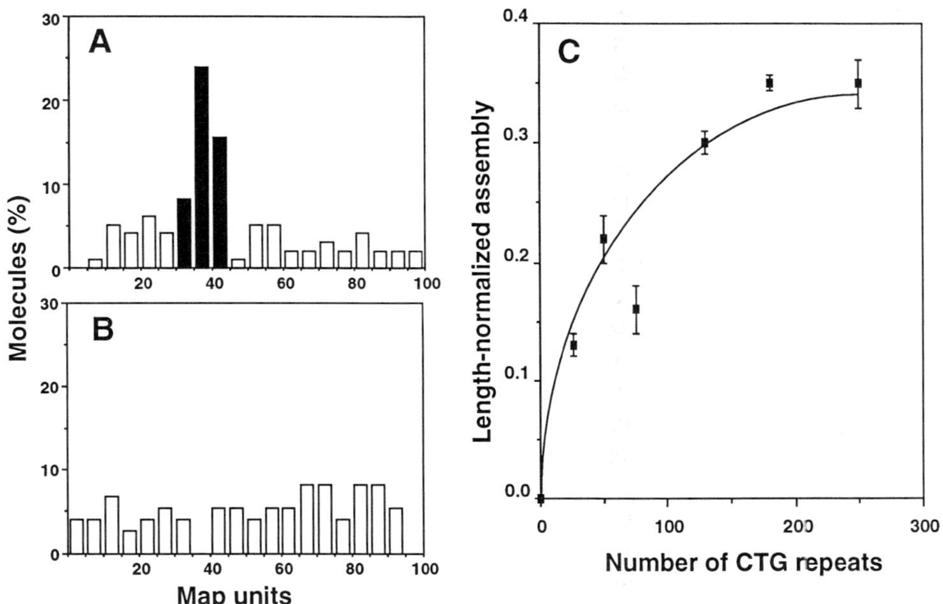

FIGURE 42-2 Distribution of nucleosomes reconstituted onto: (A) closed circular pSH2 DNA [which contains the $(CTG)_{130}$ repeat], (B) closed circular Bluescript vector. Reconstitution and preparation for EM were performed as described [22]. At least 100 DNA molecules containing single nucleosomes were photographed and the position of each nucleosome from the streptavidin-labeled end measured, and a histogram (length broken into 20 segments with each percentage of length being 1 map unit) showing the location of the nucleosomes along the DNA was generated. The black bars indicate the position of the CTG repeat sequences along the DNA. (C) Strength of sequence-directed nucleosome positioning correlates with the size of the repeated CTG block. Data from Fig. 3 of Wang *et al.* [22] were normalized as described.

energy difference measured in high salt is relatively low, 3 kcal/mol [22]. Thus whether or not such differences measured in high salt would be physiologically relevant remained in question. In the cell, at physiological ionic strength and in the presence of histone H1 and nucleosome assembly factors these differences could be great enough to dictate nucleosome positioning. The strongest evidence that these free energy differences do represent physiologically important differences was to come from

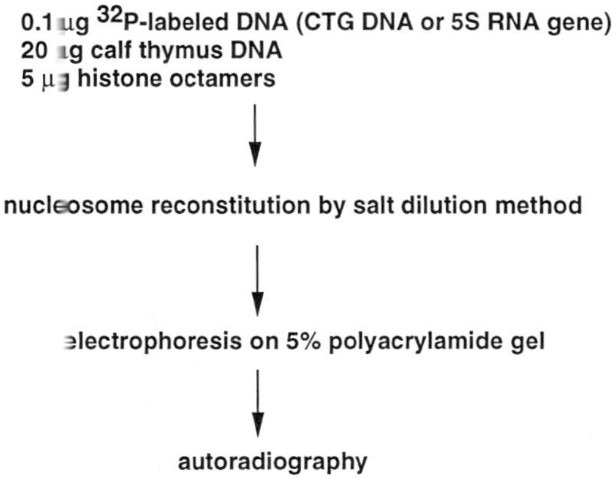

0.1 μg ^{32}P-labeled DNA (CTG DNA or 5S RNA gene)
20 μg calf thymus DNA
5 μg histone octamers

↓

nucleosome reconstitution by salt dilution method

↓

electrophoresis on 5% polyacrylamide gel

↓

autoradiography

FIGURE 42-3 Scheme for competitive nucleosome reconstitution.

model studies of the CCG triplet repeat from fragile X expansions as described below.

Godde and Wolffe [27] also examined nucleosome assembly on $(CTG)_n$ repeats by competitive nucleosome reconstitution and mapped translational and rotational positions by nuclease and hydroxyl radical cleavage. They confirmed the observation of Wang and Griffith [26] that CTG repeats create sites for preferential nucleosome assembly. However, they did not observe the increase in efficiency of nucleosome assembly with increasing size of the repeat from 10 to 55 and 62 copies, which may be too short to display the effect. Probing of rotational position demonstrated a clear 10- to 11-bp repeating pattern over the DNA containing six copies of CTG repeat and the pattern remains similar in $(CTG)_{55}$ DNA. Moreover, the $(CTG)_6$-containing DNA favors a single translational position with the CTG sequence bordering the nucleosomal dyad whereas the $(CTG)_{55}$ DNA showed several nucleosome positions.

B. Biological Implications of Long Arrays of Hyperstable Chromatin

The *in vitro* nucleosome reconstitution studies described above clearly show that repeating CTG triplets

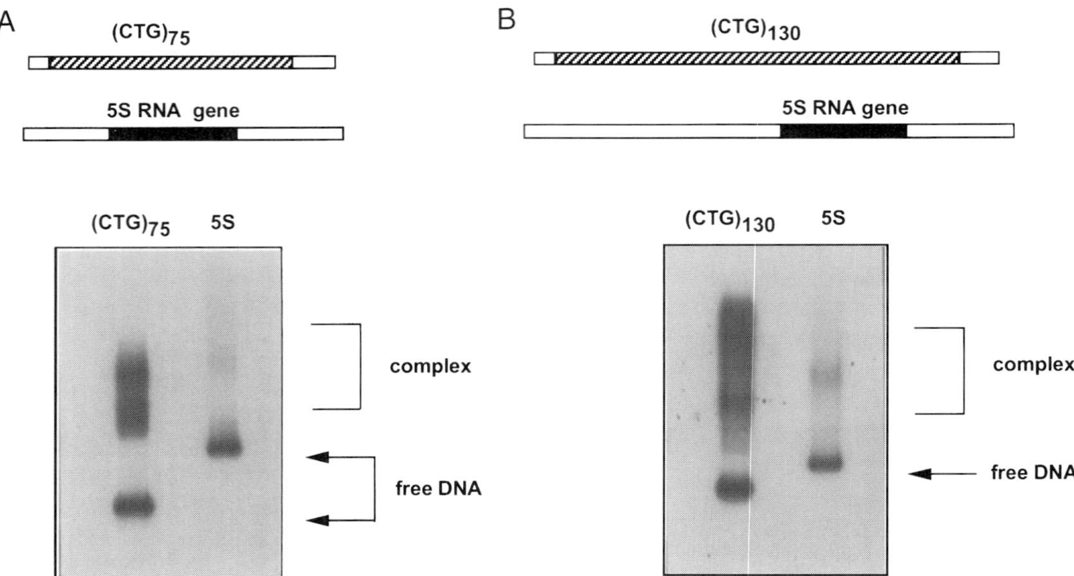

FIGURE 42-4 Competitive nucleosome reconstitutions with CTG triplet containing DNA. (A) The ~285-bp $(CTG)_{75}$-containing fragment or the 303-bp 5S RNA gene, and (B) the ~450-bp $(CTG)_{130}$ DNA or the 472-bp 5S RNA gene labeled with ^{32}P (100 ng) were mixed with 20 μg of unlabeled calf thymus DNA and ~5 μg calf thymus histone octamers in a solution containing 2 M NaCl, 100 μg/ml of BSA, and 0.1% Nonidet P-40. The total DNA added is about fourfold molar excess over the histone octamers. The salt was slowly lowered and the mixture then electrophoresed on a 5% polyacrylamide gel. Samples were visualized by autoradiography.

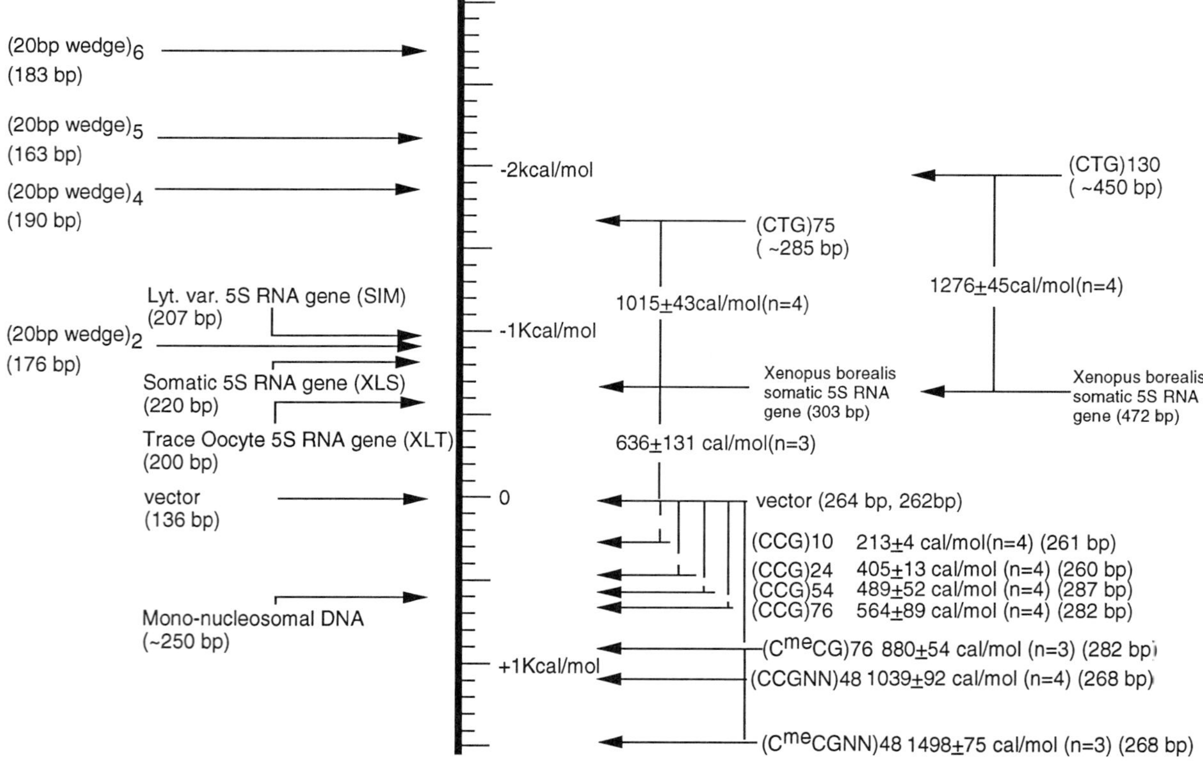

FIGURE 42-5 Comparison of free energies derived from several nucleosome positioning sequences. Data which are at the left of the free energy scale bar were adapted from Shrader and Crothers [11]. The sequence of the 20-bp anisotropically flexible wedge is TCGGGTTTAGAGCCTGTAAC. Results from studies of Wang and Griffith [26, 46, 59] are listed at the right side of the bar with the free energies of the vector DNA (pUC19 fragments) defined as 0 cal/mol.

can generate very unusual chromatin structure. When the repeat tracts grow into the size range in which DM disease symptoms appear, these repeat tracts can generate chromatin segments with stabilities that are possibly 10 times greater than the previously strongest known natural nucleosome positioning element, the 5S DNA. Furthermore, Otten and Tapscott [28] have carried out *in vivo* mapping of nuclease hypersensitive sites with DNA samples of fibroblasts and myoblasts from three unrelated myotonic dystrophy patients in which the DM alleles are heterozygote and the expansion is ~6 kb. Compared with the observations from an unaffected individual and the wild-type allele, the expanded allele showed loss of the DNaseI hypersensitive site located just 3′ of the CTG repeat and resistance to restriction endonuclease cleavage adjacent to the DNase I hypersensitive site. Recently, this hypersensitive site has been mapped to an enhancer region which regulates the expression of the DM locus-associated homeodomain protein (DMAHP) gene [29]. In cells of DM patients with a loss of the hypersensitive site, the amount of DMAHP transcript was reduced compared to the controls and the amount of transcript from the expanded allele was also greatly lowered compared to that from the wild-type allele. These results elegantly demonstrate that the expanded CTG repeat produces an altered chromatin structure possibly resulting from the strong nucleosome positioning by the CTG repeat. The altered chromatin structure might suppress the transcription of the DMAHP gene which leads to the pathogenesis of DM.

Although the precise physical reason why such repeating triplets generate such highly stable chromatin remains unclear, a number of possible biological implications of the appearance of such abnormally stable chromatin segments are worth mentioning.

The appearance of a segment of hyperstable chromatin could in itself promote further expansion of the triplet tract. During the replication of eukaryotic DNA, the replication machinery must remove the DNA template from the histone octamer prior to, or possibly in concert with, the helicase action that separates the two DNA strands. Were the replication machinery to be severely hindered in this action by the presentation of an array of hyperstable nucleosomes, then the likelihood of polymerase slippage as the replication fork transited the triplet tract would be expected to be much greater.

Just as arrays of hyperstable nucleosomes might impede the replication machinery, they might also slow or stop transcription complexes. When RNA polymerase stalls on DNA the nascent RNA chain is frequently cleaved, inactivating the RNA. Whether or not expanded CTG tracts in DM protein kinase gene result in lowered mRNA levels is currently in question and is discussed in detail in other chapters in this volume. It is possible that such an effect may only be evident when the tracts become longer than those investigated in some of the studies reported to date.

Finally, a global effect of arrays of hyperstable nucleosomes may be to generate domains of inactive chromatin. Such elements could inactivate genes either upstream or downstream of the domain and possibly over distances of many kilobases. Domains generated by an array of hyperstable nucleosomes would be similar to chromatin barriers that have been shown to silence genes moved into their proximity. Given the recent findings that the CTG tract in the DMPK gene is also in the upstream region of the DMAHP gene [29, 30], this presents one of the most attractive models of how this expanded CTG tract could exert its biological effect. It could be imagined that as the CTG tract grows, the size and thus strength of the array of hyperstable nucleosomes increases greatly; with each additional nucleosome added the effect of this very unusual hyperstable chromatin boundary could grow from just interference with transcription of the DMAHP gene to interference also with the DMPK gene and finally to interfering with all three genes including the upstream 59 protein gene. The clinical observations that the presentation of DM differs depending on the size of the CTG tract may reflect a progressive involvement of additional genes as the tract increases in size.

III. NUCLEOSOME POSITIONING AND EXPANDED CCG REPEATS

Fragile sites are chromosomal loci that stain poorly, contain gaps, and are frequent sites of DNA strand breakage. Thus they represent regions of unstable chromatin structure. More than 100 separate fragile sites have been identified in the human genome, and classified as common or rare; further, they are divided according to the agents used to identify them [31]. Fragile sites are highly conserved and appear to have played key roles in the step-wise evolution of primate chromosomes. Recent work suggests that they are major sites of drug-induced chromosomal amplification [31]. A number of fragile sites have been cloned, the majority being members of the class termed rare, folate-sensitive sites which are seen by light microscopy when cells are exposed to a folate starvation regime. Sequence analysis of 5 rare, folate-sensitive sites (FRAXA, FRAXE, FRAXF, FRA16A, and FRA11B) revealed long blocks of repeating CCG triplets together with the frequent methylation of nearby CpG islands [33–39]. Another fragile site, FRA16B, was recently sequenced and found to involve a highly A/T-rich element demonstrating that not all fragile sites involve repeating CCG triplets.

Expansion of the triplet block in the 5′ untranslated region of the FMR-1 gene [40, 41] at FRAXA from 20 to 50 repeats in most individuals to >200 CCG repeats has been linked to the FraX disease as described in detail in this volume. The expanded triplet block is the site of preferential breakage at FRAXA [33, 34]. The FRAXE site on the X chromosome [36] is correlated with a rare form of mental retardation, and the FRA11B site is associated with Jacobsen's syndrome. Here, a portion of the long arm of chromosome 11 is lost, implying a link between this fragile site and chromosome breakage [39].

While the properties of the fragile X syndrome are described in detail elsewhere in this volume, several points merit emphasis. First, unlike DM, which may involve a complex inaction of several genes, FraX appears to result from the inactivation of just the FMR-1 gene. This is illustrated by the rare cases of FraX shown to be due to point mutations or deletions in the FMR-1 gene as contrasted to the more common triplet expansion. Second, there is a strong correlation between the size of the triplet repeat and the severity of the disease and also the purity of the repeat tract. Thus moderate-sized pure CCG repeat tracts are more deleterious than longer repeat tracts containing interruptions. Finally there is also a strong correlation between the severity of the disease symptoms and the level of methylation of the region. Thus any model for the molecular basis of the FraX will have to account for these observations.

A. Discovery of Nucleosome Exclusion by Repeating CCG Tracts

The discovery of hyperstable nucleosome formation by long tracts of CTG triplet repeats spurred a parallel analysis of DNAs containing long CCG triplet repeat tracts. This work was further impelled by the knowledge that the fragile sites containing CCG triplet blocks exhibit properties of unstable chromatin suggesting that significant effects on chromatin structure due to the presence of these repeats might be found. Indeed it had been speculated earlier [42] that fragile sites may result from the inability of DNA to fold compactly during metaphase.

In the work of Wang *et al.* [43] a series of plasmids generated in the laboratory of R. D. Wells containing different sized tracts of repeating CCG triplets were employed. The plasmid pRW3376 contains 76 tandem CCG repeats [44]; this DNA was reconstituted with purified histone octamers, linearized, and one DNA end-labeled with streptavadin to place the $(CCG)_{76}$ repeat block between 18 and 26% from the marked end in a manner similar to that illustrated in Fig. 42-1A. A nucleosome map was then prepared. Inspection of the data showed that only 2.6% of the DNAs had a nucleosome located between 15 and 20 map units from the marked end and no molecules had nucleosomes between 20 and 25 map units. This was in contrast to other segments of 5 map units length which showed from 5 to 10% of the DNA with nucleosomes assembled. The parent vector showed no regions of nucleosome exclusion. These direct EM observations provided the first evidence that repeating CCG triplets could lead to nucleosome exclusion as contrasted to nucleosome positioning observed with repeating CTG triplets. Control experiments showed that legitimate nucleosomes were being formed on normal sequence DNA in the same reactions.

Competitive nucleosome reconstitution was then used in this study to measure the energetics of nucleosome formation for DNA fragments containing CCG repeats of increasing size (Fig. 42-6A). The ratio of nucleosome assembled DNA to free DNA for a pUC19 fragment and similar length fragments containing 10, 24, 54, and 76 CCG repeats was determined by these workers. It was found that as the length of the repeat block increased, the efficiency of nucleosome assembly decreased. While a 261-bp DNA containing 10 tandem CCG repeats showed a 1.4-fold lower efficiency (30% decrease) in nucleosome assembly relative to the pUC19 fragment, the fragment containing 76 tandem CCG repeats was 2.6 ± 0.4-fold less effective (62% decrease), corresponding to a 564 ± 89 cal/mol difference in free energy (Fig. 42-5). Compared to the free energy value for a similar sized DNA containing 75 CTG repeats (see above), this amounts to a 40-fold difference between the two repeating triplet DNAs. These observations clearly paralleled the *in vivo* observation of fragile sites being regions of unstable chromatin.

While the physical basis for repeating CTG triplets forming hyperstable nucleosomes remains unclear, experiments with model variants of the CCG motif provided clues into why this expanded triplet may exclude nucleosomes and, further, revealed the presence of "microfragile" sites in eukaryotic genomes.

B. Model Studies and the Discovery of Microfragile Sites Involving the Repeat $[(C/G)_3NN]_n$

Consideration of the properties of the sequence motif in the 5S RNA gene, $(G/C)_3NN(A/T)_3NN$, which forms very strong nucleosomes [11, 45], led Wang and Griffith [46] to suggest that DNA containing long repeats in the form of $(G/C)_3NN(G/C)_3NN$ might exclude nucleosomes. This latter motif differs from the 5S RNA gene element in that each time minor groove compression is

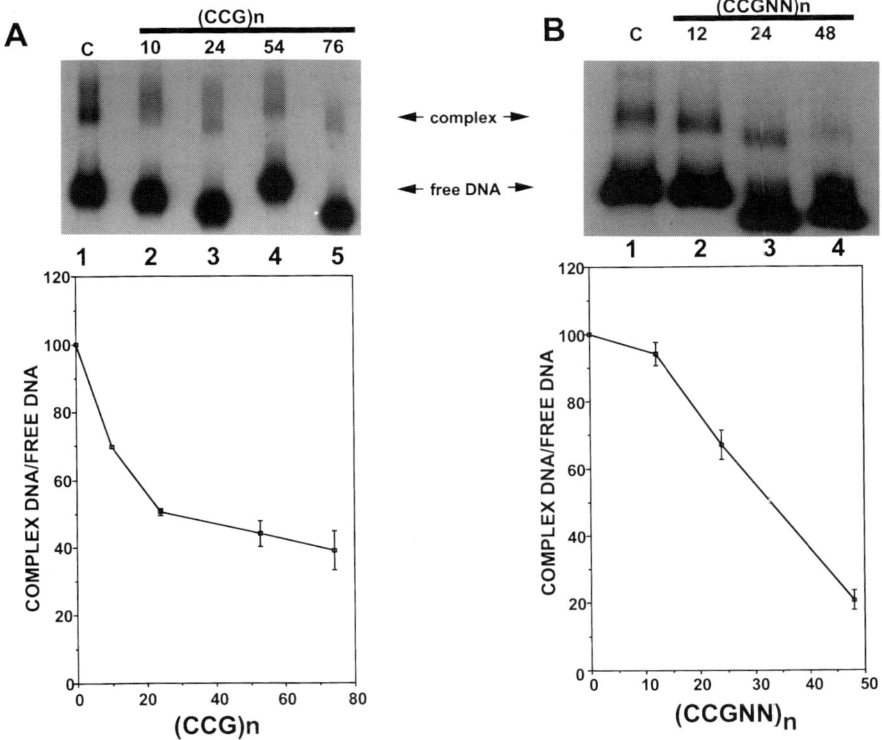

FIGURE 42-6 Competitive nucleosome reconstitution with CCG- and CCGNN-containing DNAs. (A, upper panel) Autoradiogram of a competitive nucleosome reconstitution experiment. Lane 1, the 262-bp pUC19 fragment; lanes 2–5, DNA fragments containing 10, 24, 54, and 76 CCG repeats, respectively. Note: The $(CCG)_{54}$ and $(CCG)_{76}$ repeat blocks contain AGG and CAG interruptions [44]. (A, lower panel) Dependence of nucleosome assembly on the length of the repeat block. (B, upper panel) Autoradiogram of a competitive nucleosome reconstitution experiment. Lane 1, the 262-bp pUC19 fragment; lanes 2–4, DNA fragments containing 12, 24, and 48 CCGNN repeats, respectively. (B, lower panel) Dependence of nucleosome assembly on the length of the repeat block. DNA preparation, reconstitution of the fragments with histones, and gel electrophoresis were as described [43, 46]. Each DNA was reconstituted in three separate but identical experiments, and the fraction of DNA in the nucleosome-assembled and nucleosome-free DNA bands was measured by a PhosphorImager.

required, the DNA presents the histone octamer with a wedge that favors bending into the major groove. Shrader and Crothers [11] had noted that a triplet of As or Ts preferentially bends into the minor grove of DNA while a triplet of Gs or Cs bends preferentially into the major grove. Thus DNA containing a nucleosome-sized tract of repeating CCGNN pentanucleotides, which is a member of the $(G/C)_3NN(G/C)_3NN$ motif family, should resist nucleosome formation. Pure repeating CCG triplet tracts are also a member of this general family and if nucleosome exclusion were observed for DNA containing repeating CCGNN pentanucleotides, this would provide an explanation for nucleosome exclusion by repeating CCG triplets.

To test the hypothesis that long tracts of $(G/C)_3NN$ repeats will exclude nucleosomes, three pGEM3zf(+)-based recombinant plasmids, $p(CCGNN)_{12}$, $p(CCGNN)_{24}$, and $p(CCGNN)_{48}$, containing, respectively, 12, 24, and 48 tandem copies of the CCGNN repeat with the Ns varied and rich in As and Ts, were generated. EM was then utilized together with *in vitro* nucleosome reconstitution onto supertwisted $p(CCGNN)_{48}$ plasmid DNA to examine the propensity of the inserts to assemble into nucleosomes as described above. Subsequent to assembly the DNA was linearized placing the CCGNN block in $p(CCGNN)_{48}$ between 27 and 34% from the tagged end. Analysis revealed strong nucleosome exclusion over the $(CCGNN)_{48}$ insert with only 2% of all nucleosomes mapped located between 25 and 35 map units, whereas in other segments of this length ~10% or more of the DNA contained a nucleosome. In contrast the parent pGEM3zf(+) vector showed no regions of strong nucleosome exclusion.

Competitive nucleosome reconstitution assays were then employed as described [26] to quantify the degree of nucleosome exclusion over the insert in these plasmids. DNAs ~260 bp in length containing $(CCGNN)_n$ blocks of 12, 24, or 48 repeats were generated. As the length of the CCGNN repeat sequence increased from 12 to 48 repeats within a ~260-bp segment, the ability of the DNA to exclude nucleosomes increased in proportion to the length of the repeat block (Fig. 42-6B). Analysis showed that the 268-bp DNA containing the $(CCGNN)_{48}$ repeat is 4.9 ± 0.6-fold less efficient in nucleosome assembly than the same sized pUC19 fragment. Indeed, a 261-bp DNA containing 12 tandem CCGNN repeats assembled nucleosomes 94% as efficiently as a 262-bp pUC19 fragment. When these results were combined with results from earlier studies, it was estimated that the $(CCGNN)_{48}$ repeat block is ~78-fold less efficient than a similar length block of repeating CTG triplets. The difference in free energy between the $(CCGNN)_{48}$ fragment and the pUC19 DNA is 937 ± 79 cal/mol and ~2600 cal/mol relative to the CTG_{75} repeating element (Fig. 42-5).

Based on the findings described above, a computer search was carried out using the GenBank database to examine the prevalence of the general $[(G/C)_3NN]$ motif [46]. Searches against 240 bp of a $[(G/C)_3NN]_{48}$ continuously repeating sequence revealed many matches, and 75 examples showed >85% sequence match over 200-bp of this motif (Table 42-1). Many of these were present in or near the control regions for eukaryotic genes. Of the 75 with the greatest number of sequence matches, 31 genes were noted to contain the $(G/C)_3NN$ motif in the 5′ region upstream of the coding sequences. Such regions would be loci where nucleosome exclusion would be expected to provide favored access to sequence-specific proteins engaged in the regulation of gene expression. It was also observed that at least 20 of these 31 genes lack a TATA box in the promoter region and are "TATA-less" genes. In two of these genes, the 5′ regions have been mapped for nuclease hypersensitive sites. The −530 to −300 nt upstream control region of the human dihydrofolate reductase gene contains the $[G/C)_3NN]_{48}$ motif (with 87% sequence match) and this overlaps with a hypersensitive region mapped *in vivo* [47]. In the gene for the human ETS-2 nuclear phosphoprotein, there is a sequence upstream of the gene between nts −195 and +45 containing the $[G/C)_3NN]_{48}$ motif (with 86% sequence match). The promoter region of this gene maps between nts −159 and +141. Here studies of nucleosome positioning in the promoter region using S1 nuclease revealed a hypersensitive region from nts −150 to −50 [48]. This 100-bp hypersensitive region exactly comaps with the region containing the $[G/C)_3NN]_{48}$ motif.

The discovery of long DNA sequences in the human genome with strong matches to the $[(G/C)_3NN]_n$ motif and the finding that, in two cases, these sites correspond to regions shown by others to be spared of nucleosomes have several important implications with regard to the topic of this chapter. First, they provide the most compelling evidence to date that the differences in free energy measured *in vitro* in high salt (which are in the range of a few kilocalories) are biologically significant. Indeed the two most common human triplet diseases, DM and FraX, involve triplet expansions which bracket the range from strongest to weakest nucleosome assembly elements. The second and broader implication of this work is that "microfragile" sites composed of long repeats of $(C/G)_3NN$ pentanucleotides present in the 5′ control regions of genes lacking TATA boxes may be a common occurrence in eukaryotic genomes. Here these elements would serve to loosen the chromatin structure, providing access to the DNA for the transcriptional machinery.

In FraX, AGG interruptions have been shown to stabilize the CGG repeats against expansion [49–54]. Comparison of the data presented the papers of Wang and Griffith [46] reveals a greater resistance to nucleosome assembly for the $(CCGNN)_{48}$ DNA as contrasted to the similar sized $(CCG)_{76}$ block. This suggests that the interruptions in the latter DNA may significantly impair the ability of this overall element to exclude nucleosomes. Thus the well-characterized effects of these interruptions (see other chapters in this volume) could be to moderate the ability of these long repeat blocks to alter the local chromatin structure.

C. Methylation of CCG Enhances Nucleosome Exclusion

In eukaryotic cells, methylation of CpG dinucleotides by DNA methyltransferase has been found to directly inhibit gene expression (reviewed in [55]). This enzyme places a methyl group on the cytosine residues of 5′CpG3′ dinucleotides and has been shown to play an important role in embryonic development [56]. The inhibition of transcription by methylation may be mediated by the binding of a methyl-CpG binding protein to DNA sequences containing methylated CpG dinucleotides [57, 58]. In the context of FraX, methylation of the expanded triplet repeat block has been found to result in much more pronounced disease symptoms relative to expansions of the same size that are not heavily methylated (see other chapters in this volume). These observations prompted studies to determine how methylation of the CCG tracts would influence their already poor ability to assemble into chromatin.

In Wang and Griffith [59] SssI methylase was used to place a methyl group on the C^5 position of cytosines within the 5′CpG3′ dinucleotides. Two sets of experiments were carried out in this study. In one, a 262-bp fragment from pUC19 was compared to a 282-bp DNA

TABLE 42-1 Examples of Genes Containing Regions Showing ≥85% Sequence Matches to the $[(G/C)_3NN)]_{48}$ Sequence over 240 bp Taken from a GenBank Search[a]

Location	Gene or encoded protein	Sequence matches[b]
5′ end (upstream of coding sequences)	Human glutamate dehydrogenase	222/235
	Human arginosuccinate lyase	213/230
	Human insulin receptor	214/237
	Wheat alpha amylase	208/233
	Chicken hsp 90	205/231
	Human dihydrofolate reductase	205/228
	Pig nuclear factor 1	201/225
	Rat insulin-like growth factor binding protein	202/232
	Rabbit metallothionine	195/229
	Mouse neu proto-oncogene	222/236
	Human heamatopoietic cell specific protein	215/235
	Human fibronectin	205/233
	Rat phosphorylase kinase catalytic subunit	199/230
	Human vitronectin protein	196/229
	Mouse S16 ribosomal protein	202/232
	Rat neu oncogene	214/231
	Human gastrin releasing peptide	202/228
	Human ETS 2 oncogene	204/231
	Rat nucleolin	199/230
Coding region	Chicken c-fos proto-oncogene	208/231
	Chicken tropoelastin	206/224
	Human collagen-like protein	201/217
	Herpes simplex 1 UL 18	204/230
	EBV nuclear antigen 3C	204/234
	Human thymidine kinase	199/233
	Chicken ubiquitin	202/231
	Human translocation fusion protein (E2A-Prl)	202/232
	Human p120	199/233
3′ end (downstream of coding sequences)	Human cytochrome p450	207/232
	Mouse histone H2A.X	209/235
	Human lamin B2	206/232
	Chicken p53 oncoprotein	193/215
	Human alpha 1 collagen type 1	204/232

[a]At least 75 genes in the GenBank were found to have ≥85% sequence matches to $[(G/C)_3NN]_{48}$.

[b]The first number indicates the number of bases showing sequence matches with $[(G/C)_3NN]_{48}$ and the second number refers to the length of this sequence.

containing 76 tandem CCG repeats. In the second, the same pUC19 fragment was compared to a 268-bp fragment containing 48 contiguous CCGNN repeats as described above. The number of possible methylation sites for the pUC19 fragment, the CCG-containing DNA, and the CCGNN-containing DNA is 40, 164, and 102, respectively. Here the level of methylation was estimated from the cleavage pattern by the methylation-sensitive restriction endonuclease *Hha*I. Competitive nucleosome reconstitution was then used to measure the energetics of nucleosome formation over the CCG and CCGNN repeats as compared to a pUC 19 DNA of the same size. This analysis (Fig. 42-7A) showed that DNA methylation had no measurable influence on nucleosome formation for the pUC 19 fragment, an observation noted in several previous studies [60–62]. In contrast, the $(CCG)_{76}$ DNA methylated at 85% of the available sites was 2.0 ± 0.2-fold less effective in nucleosome assembly compared to the same $(CCG)_{76}$ DNA fragment that was unmethylated, and 4.4 ± 0.4-fold less effective relative to the pUC19 fragment. The difference in free energy between the methylated (at 85%) and the unmethylated $(CCG)_{76}$ DNA is 405 ± 44 cal/mol. With respect to the pUC19 DNA, the methylated (at 85%) $(CCG)_{76}$ DNA is 880 ± 54 cal/mol less favorable (Fig. 42-5). Similar results were obtained when the pUC19 fragment was compared to the 268-bp fragment containing 48 tandem CCGNN repeats. The $(CCGNN)_{48}$ DNA fragment methylated at 62% of the available sites was 2.1 ± 0.3- and 12.6 ± 1.6-fold less effective in nucleosome formation compared to the unmethylated $(CCGNN)_{48}$ DNA and the pUC 19 frag-

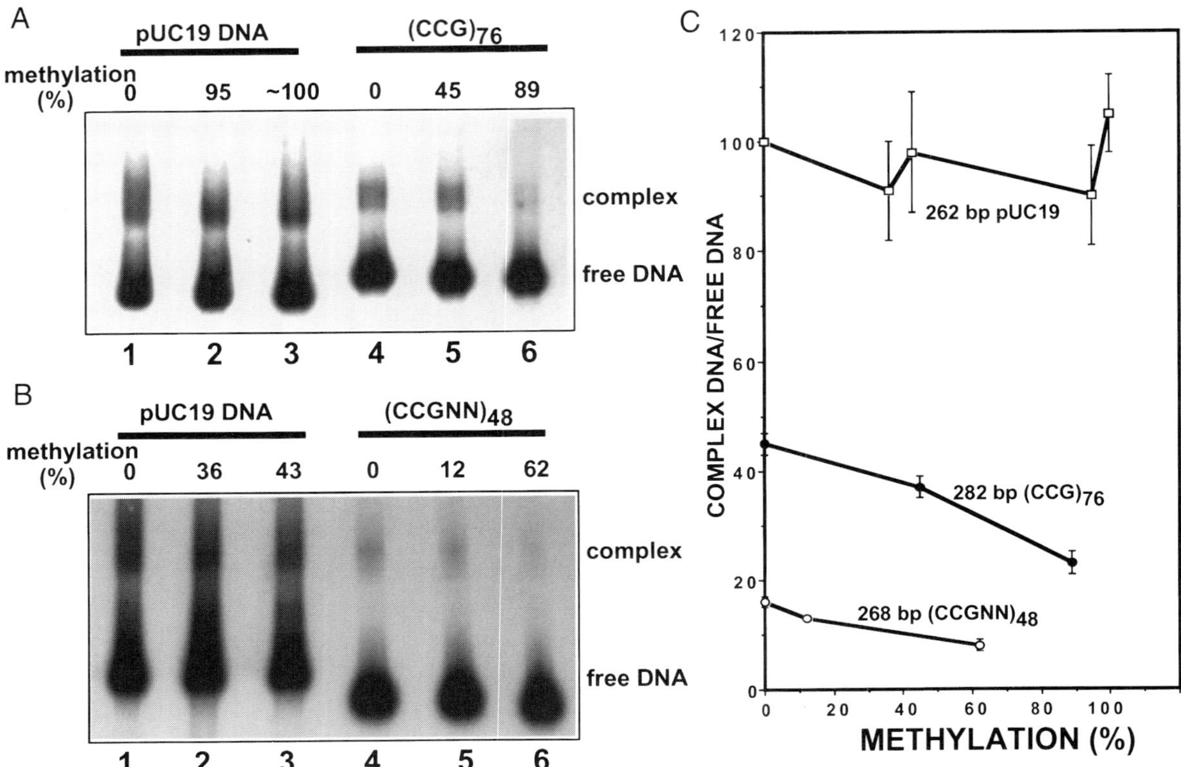

FIGURE 42-7 Competitive nucleosome reconstitution with methylated DNAs. Competitive nucleosome reconstitution was carried out comparing (A) a 262-bp pUC19 fragment to a 282-bp DNA containing 76 tandem CCG repeats at three different levels of methylation. Lanes 1–3, the pUC19 fragments methylated at 0, 95, and ~100% of the available sites, respectively; lanes 4–6, the $(CCG)_{76}$ DNA fragment methylated at 0, 45, and 89% of the available sites, respectively. (B) Comparison of the 262-bp pUC19 fragment to a 268-bp DNA containing 48 tandem CCGNN repeats at three different levels of methylation. Lanes 1–3, the pUC19 fragments methylated at 0, 36, and 43% of the available sites, respectively; lanes 4–6, the $(CCGNN)_{48}$ DNA methylated at 0, 12, and 62% of the available sites, respectively. (C) Dependence of nucleosome assembly on methylation of CpG dinucleotides. Data from the competitive nucleosome reconstitution experiments are shown. Each DNA was reconstituted in three separate but identical experiments, and the fraction of DNA in the nucleosome-assembled and nucleosome-free DNA bands was measured by PhosphorImager scanning.

ment, respectively (Figs. 42-5, 42-7B). These efficiencies correspond to 425 ± 93 and 1498 ± 75 cal/mol differences in free energy, respectively. The greatest difference was observed with a fragment containing the $(CCGNN)_{48}$ repeat block. When this DNA was highly methylated, it was nearly 13-fold less effective in nucleosome assembly as compared to a mixed sequence DNA of the same size.

In Wang and Griffith [59] a general model was presented for the role of CCG repeats and methylation in the generation and expression of fragile sites containing these triplet repeats. It was proposed that long repeating CCG triplet blocks inhibit chromatinization of DNA which would increase the accessibility of these regions to DNA methyltransferase. Next, methylation of the repeats would further repress chromatinization and make the DNA more available for binding by methyl CpG binding proteins which have been shown to repress transcription. The net effect of these changes would be the expression of the fragile site and the genetic repression of nearby genes such as the FMR-1 gene.

Wolffe and colleagues [63] also examined the ability of CCG repeats to assemble into nucleosomes and found that DNA containing a short CCG repeat (13 copies), upon methylation, showed preferential nucleosome assembly with affinity similar to the $(CTG)_{10}$-containing DNA. However, in a longer CCG DNA (74 copies), methylation represses the ability to assemble nucleosomes to a level similar to mixed sequence DNA. However, the mechanism for the different effects of methylation and repeat number on nucleosome assembly is not clear.

IV. SUMMARY AND CONCLUSIONS

The studies summarized in this chapter raise the possibility of there being a novel class of diseases that might be termed "diseases of chromatin structure." The hypothesis is that chromatin structure, in particular the formation of nucleosomes along DNA, has been selected through evolution such that nearly all DNA sequences have an affinity for chromatinization that resides within a rather narrow window of free energy values. This may be very important to the control of gene expression and the packaging of DNA into chromosomes. If, however, certain sequences were to arise and expand in size such that these sequences exhibited either exceptionally high nucleosome stability or exceptionally low nucleosome stability, then pathological effects might result. If this hypothesis is borne out by further experimentation, the two most common expanded-triplet related human diseases, DM and FraX, will have provided examples in which nucleosome stability is either pathologically high (DM) or low (FraX). Whether or not there are diseases of chromatin structure, studies of the effect of these simple repeats in forming chromatin and, in the future, the ability of such chromatinized templates to be replicated or transcribed will contribute greatly to our knowledge of the role of chromatin structure in fundamental biological processes.

Acknowledgment

This work was supported by a grant from the NIH (GM 31819).

References

1. Griffith, J. D. (1975). DNA structure: evidence from electron microscopy. *Science* **187,** 1202–1203.
2. Thoma, F. (1992). Nucleosome positioning. *Biochim. Biophys. Acta* **1130,** 1–19.
3. Wolffe A. P. (1994). Transcription: in tune with the histones. *Cell* **77,** 13–16.
4. Sheridan, P., Sheline, C., Cannon, K., Voz, M., Pazin, M., Kadonga, J., and Jones, K. (1995). Activation of the HIV-1 enhancer by the LEF-1 HMG protein on nucleosome assembled DNA in vitro. *Genes Dev.* **9,** 2090–2104.
5. Fletcher, T. M., and Hansen, J. C. (1996). The nucleosomal array: structure/function relationships. *Crit. Rev. Eukaryotic Gene Exp.* **6,** 149–188.
6. Garcia Ramirez, M., Dong, F., and Ausio, J. (1992). Role of histone "tails" in the folding of oligonucleosomes depleted of histone H1. *J. Biol. Chem.* **267,** 19587–19595.
7. Fletcher, T. M., Serwer, P., and Hansen, J. C. (1994). Quantitative analysis of macromolecular conformational changes using agarose gel electrophoresis: application to chromatin folding. *Biochemistry* **33,** 0859–10863.
8. O'Neil, T. E., Roberge, M., and Bradbury, E. M. (1992). Nucleosome arrays inhibit both initiation and elongation of transcripts by bacteriophage T7 RNA polymerase. *J. Mol. Biol.* **223,** 67–78.
9. Trifonov, E. N. (1985). Curved DNA. *CRC Crit. Rev. Biochem.* **19,** 89–106.
10. Hsieh, C.-H., and Griffith, J. D. (1988). The terminus of SV40 replication and transcription contains a sharp sequence-directed curve. *Cell* **52,** 535–544.
11. Shrader, T. E., and Crothers, D. M. (1989). Artifical nucleosome positioning sequences. *Proc. Natl. Acad. Sci. USA* **86,** 7418–7422.
12. Simpson, R. T., and Stafford, D. W. (1983). Structural features of a phased nucleosome core particle. *Proc. Natl. Acad. Sci. USA* **80,** 51–55.
13. Rhodes, D. (1979). Nucleosome cores reconstituted from poly (dA-dT) and the octamer of histones. *Nucleic Acids Res.* **6,** 1805–1816.
14. Ramsay, N. (1986). Deletion analysis of a DNA sequence that positions itself precisely on the nucleosome core. *J. Mol. Biol.,* **189,** 179–188.
15. Gottesfeld, J. M. (1987). DNA sequence-directed nucleosome reconstitution on 5S RNA genes of *Xenopus lavevis. Mol. Cell. Biol.* **7,** 1612–1622.
16. Dunn, K., and Griffith, J. D. (1979). The presence of RNA in a double helix inhibits its interaction with histone protein. *Nucleic Acids Res.* **8,** 555–566.
17. Nickol, J., Behe, M., and Felsenfeld, G. (1982). Effect of the B-Z transition in poly(dG-m^{5}dC) · poly(dG-m^{5}dC) on nucleosome formation. *Proc. Natl. Acad. Sci. USA* **79,** 1771–1775.
18. Simpson, R. T., and Kunzler, P. (1979). Chromatin and core particles formed from the inner histones and synthetic polydeoxyribonucleotides of defined sequence. *Nucleic Acids Res.* **6,** 1387–1415.
19. Kunkel, G. R., and Martinson, H. G. (1981). Nucleosomes will not form on double-stranded RNA or over poly(dA):poly(dT) tracts in recombinant DNA. *Nucleic Acids Res.* **9,** 6869–6888.
20. Prunell, A. (1982). Nucleosome reconstitution on plasmid-inserted poly(dA):poly(dT). *EMBO J.* **1,** 173–179.
21. Iyer, V., and Struhl, K. (1995). Poly(dA:dT), a ubiquitous promoter element that stimulates transcription via its intrinsic DNA structure. *EMBO J.* **14,** 2570–2579.
22. Wang, Y.-H., Amirhaeri, S., Kang, S., Wells, R. D., and Griffith, J. D. (1994). Preferential nucleosome assembly at DNA triplet repeats from the myotonic dystrophy gene. *Science* **265,** 669–671.
23. Hayes, J. J., Tullius, T. D., and Wolffe, A. P. (1990). The structure of DNA in a nucleosome. *Proc. Natl. Acad. Sci. USA* **87,** 7405–7409.
24. Hayes, J. J., Clark, D. J., and Wolffe, A. P. (1991). Histone contributions to the structure of DNA in the nucleosome. *Proc. Natl. Acad. Sci. USA* **88,** 6829–6833.
25. Hayes, J. J., Bashkin, J., Tullius, T. D., and Wolffe, A. P. (1991). The histone core exerts a dominant constraint on the structure of DNA in a nucleosome. *Biochemistry* **30,** 8434–8440.
26. Wang, Y.-H., and Griffith, J. D. (1995). Expanded CTG triplet blocks from the myotonic dystrophy gene create the strongest known natural nucleosome positioning elements. *Genomics* **25,** 570–573.
27. Godde, J. S., and Wolffe, A. P. (1996). Nucleosome assembly on CTG triplet repeats. *J. Biol. Chem.* **271,** 15222–15229.
28. Otten, A. D., and Tapscott, S. J. (1995). Triplet repeat expansion in myotonic dystrophy alters the adjacent chromatin structure. *Proc. Natl. Acad. Sci. USA* **92,** 5465–5469.
29. Klesert, T. R., Otten, A. D., Bird, T. D., and Tapscott, S. J. (1997). Trinucleotide repeat expansion at the myotonic dystrophy locus reduces expression of DMAHP. *Nature Genet.* **16,** 402–406.
30. Harris, S., Moncrieff, C., and Johnson, K. (1996). Myotonic dystrophy: will the real gene please step forward! *Hum. Mol. Genet.* **5,** 1417–1423.
31. Sutherland, G. R. (1991). Chromosomal fragile sites. *GATA* **8,** 161–166.

32. Coquelle, A., Pipiras, E., Toledo, F., Buttin, G., and Debatisse, M. (1997). Expression of fragile sites triggers intrachromosomal mammalian gene amplification and sets boundaries to early amplicons. *Cell* **89,** 215–225.
33. Kremer, E. J., Pritchard, M., Lynch, M., Yu, S., Holman, K., Baker, E., Warren, S. T., Schlessinger, D., Sutherland, G. R., and Richards, R. I. (1991). Mapping of DNA instability at the fragile X to a trinucleotide repeat sequence. *Science* **252,** 1711–1718.
34. Verkerk, A. J. M. H., Pieretti, M., Sutcliffe, J. S., Fu, Y.-H., Kuhl, D. P. A., Pizzuti, A., Reiner, O., Richards, S., Victoria, M. F., Zhang, F., Eussen, B. E., van Ommen, G.-J. B., Blonden, L. A. J., Riggins, G. J., Chastain, J. L., Kunst, C. B., Galaard, H., Caskey, C. T., Nelson, D. L., Oostra, B. A., and Warren, S. T. (1991). Identification of a gene (FMR-1) containing a CGG repeat coincident with a breakpoint cluster region exhibiting length variation in fragile X syndrome. *Cell* **65,** 905–914.
35. Hornstra, I K., Nelson, D. L., Warren, S. T., and Yang, T. P. (1993). High resolution methylation analysis of the FMR-1 gene trinucleotide repeat region in fragile X syndrome. *Hum. Mol. Genet.* **2,** 1659–1665.
36. Knight, S., Flannery, A. V., Hirst, M. C., Campbell, L., Christodoulou, Z., Phelps, S. R., Pointon, J., Middleton-Price, H. R., Barnicoat, A., Pembrey, M. E., Holland, J., Oostra, B. A., Bobrow, M., and Davies, K. E. (1993). Trinucleotide repeat amplification and hypermethylation of a CpG island in FraXE mental retardation. *Cell* **74,** 127–134.
37. Parrish, J. E., Oostra, B. A., Verkerk, A. J. M. H., Richards, C. S., Reynolds, J., Spikes, A. S., Shaffer, L. G., and Nelson, D. L. (1994). Isolation of a GCC repeat showing expansion in FRAXAF, a fragile site distal to FRAXA and FRAXE. *Nature Genet.* **8,** 229–235.
38. Nancarrow, J. K., Kremer, E., Holman, K., Eyre, H., Doggett, N. A., Paslier, D. L., Callen, D. F., Sutherland, G. R., and Richards, R. I. (1994). Implication of FRA16A structure for the mechanism of chromosomal fragile site genesis. *Science* **264,** 1938–1941.
39. Jones, C., Penny, L., Mattina, T., Yu, S., Baker, E., Voullaire, L., Langdon, W. Y., Sutherland G. R., Richards, R. I., and Tunnacliffe, A. (1995). Association of a chromosome deletion syndrome with a fragile site within the proto-oncogene CBL2. *Nature* **376,** 145–149.
40. Nelson, D. L. (1995). The fragile X syndromes. *Semin. Cell Biol.* **6,** 5–11.
41. Sutherland, G. R., and Richards, R. I. (1995). The molecular basis of fragile site in human chromosomes. *Curr. Opin. Genet. Dev.* **5,** 323–327.
42. Chaudhuri, J. P. (1972). On the origin and nature of achromatic lesions. *Chromosome Today* **3,** 147–151.
43. Wang, Y.-H., Gellibolian, R., Shimizu, M., Wells, R. D., and Griffith, J. (1996). Long CCG triplet repeat blocks exclude nucleosomes: a possible mechanism for the nature of fragile sites in chromosomes. *J. Mol. Biol.* **263,** 511–516.
44. Shimizu, M., Gellibolian, R. Oostra, B. A., and Wells, R. D. (1996). Cloning, characterization, and properties of plasmids containing CGG triplet repeats from the FMR-1 gene. *J. Mol. Biol.* **258,** 614–626.
45. Shrader, T. E., and Crothers, D. M. (1990). Effects of DNA sequence and histone-histone interactions on nucleosome placement. *J. Mol. Biol.* **216,** 69–84.
46. Wang, Y.-H., and Griffith, J. (1996). The $[(G/C)_3NN]_n$ motif: a common DNA repeat that excludes nucleosomes. *Proc. Natl. Acad. Sci. USA* **93,** 8863–8867.
47. Shimada, T., Inokuchi, K., and Nienhuis, A. W. (1986). Chromatin structure of the human dihydrofolate reductase gene promoter. *J. Biol. Chem.* **261,** 1445–145.
48. Mavrothalassitis, G. J., Watson, D. K., and Papas, T. S. (1990) The human ETS-2 gene promoter: molecular dissection and nuclease hypersensitivity. *Oncogene* **5,** 1337–1342.
49. Eichler, E. E., Holden, J. J., Popovich, B. W., Reiss, A. L., Snow, K., Thibodeau, S. N., Richards, C. S., Ward, P. A., and Nelson, D. L. (1994). Length of uninterrupted CCG repeats determines instability in the FMR-1 gene. *Nature Genet.* **8,** 8–94.
50. Hirst, M. C. Grewal, P. K., and Davies, K. E. (1994). Precursor arrays for triplet repeat expansion at the fragile X locus. *Hum. Mol. Genet.* **3,** 1553–1560.
51. Kunst, C., and Warren, S. (1994). Cryptic and polar variation of the Fragile X repeat could result in predisposing normal alleles. *Cell* **77,** 853–861.
52. Reiss, A. L., Kazazian, H. H., Jr., Krebs, C. M., McAughan, A., Boehm, C. D., Abrams, M. T., and Nelson, D. L. (1994). Frequency and stability of the fragile X syndrome. *Hum. Mol. Genet.* **3,** 393–398.
53. Snow, K., Tester, D. J., Kruckeberg, K. E. Schaid, D. J., and Thibodeau, S. N. (1994). Sequence analysis of the fragile X trinucleotide repeat: implications for the origin of the fragile X mutation. *Hum. Mol. Genet.* **3,** 1543–1551.
54. Eichler, E. E., Hammond, H. A., MacPherson, J. N., Ward, P. A., and Nelson, D. L. (1995). Population survey of the human FMR1 CGG repeats structure suggests biased polarity for the loss of AGG interruptions. *Hum. Mol. Genet.* **4,** 2199–2208.
55. Tate, P. H., and Bird, A. P. (1993). Effects of DNA methylation on DNA-binding proteins and gene expression. *Curr. Opin. Genet. Dev.* **3,** 226–231.
56. Li, E., Bestor, T. H., and Jaenisch, R. (1992). Targeted mutation of the DNA methyltransferase gene results in embryonic lethality. *Cell* **69,** 915–926.
57. Boyes, J., and Bird, A. (1991). DNA methylation inhibits transcription indirectly via a methyl-CpG binding protein.*Cell* **64,** 1123–1134.
58. Boyes, J., and Bird, A. (1992). Repression of genes by DNA methylation depends on CpG density and promoter strength: evidence for involvement of a methyl-CpG binding protein. *EMBO J.* **11,** 327–333.
59. Wang, Y.-H., and Griffith, J. (1996). Methylation of expanded CCG triplet repeat DNA from fragile X syndrome patients enhances nucleosome exclusion. *J. Biol. Chem.* **271,** 22937–22940.
60. Felsenfeld, G., Nickol, J., Behe, M., McGhee, J. D., and Jackson, D. (1982). Methylation and chromatin structure. *Cold Spring Harbor Symp. Quant. Biol.* **47,** 577–584.
61. Drew, H. R., and McCall, M. J. (1987). Structural analysis of a reconstituted DNA containing three histone octamers and histone H5. *J. Mol. Biol.* **197,** 485–511.
62. Englander, E. W., Wolffe, A. P., and Howard, B. H. (1993). Nucleosome interactions with a human Alu element. Transcriptional repression and effects of template methylation. *J. Biol. Chem.* **268,** 19565–19573.
63. Godde, J. S., Kass, S. U., Hirst, M. C., and Wolffe, A. P. (1996). Nucleosome assembly on methylated CGG triplet repeats in the fragile X mental retardation gene 1 promoter. *J. Biol. Chem.* **271,** 24325–24328.

Part XV

Replication–Repair of TRS

CHAPTER 43

Studies of DNA Polymerases in Replication-Based Repeat Expansion

SAMUEL H. WILSON, RAKESH K. SINGHAL,[1] AND BARBARA Z. ZMUDZKA[2]
Department of Health and Human Sciences, National Institutes of Health, National Institute of Environmental Health Sciences, Research Triangle Park, North Carolina 27709

I. INTRODUCTION

The mechanism of short direct repeat expansion and heterogeneity in the human population is unknown, both at the molecular level and at the level of mitotic vs. meiotic events. However, we know that genetic disorders associated with triplet repeat expansion are transmitted through germline cells, and that direct repeat iteration number can change from generation to generation. It will be useful, therefore, to consider probable meiotic events, such as gap-filling DNA synthesis during recombination, as well as such mitotic events as leading and lagging strand DNA replication, and DNA repair.

The human health significance of triplet repeat sequences in genomic DNA has been realized at a time when we are beginning to appreciate the importance of simple repeat sequences to the accuracy of DNA replication. This underscores the importance of detailed study of replication mechanisms of short direct repeat sequences. Such sequences occur abundantly in human genomic DNA, and are represented by di-, tri-, and tetranucleotide repeats, among others. Any understand-

[1]Present address: Perinatology Center, Department of Pediatrics, The New York Hospital–Cornell Medical Center, New York, NY 10021.

[2]Present address: Center for Devices and Radiological Health, FDA, HFZ-114, Rockville, MD 20857.

ing of the instability in these sequences from generation to generation, and their heterogeneity throughout the human population, is only beginning to emerge, but the need to learn more about the fundamental replication properties of these sequences is clear.

During DNA replication non-semiconservative replicative events sometimes occur [1–3]. Many of the well-studied examples involve the process of nascent primer misalignment or relocation, relative to the template immediately ahead of the growing primer strand. Primer relocation (PRL) has been implicated in single-base mismatch mutagenesis [6, 7], in deletions and insertions, and in frameshift mutagenesis [8–14]. Indeed, PRL mechanisms are increasingly recognized as an important mechanism leading to genome heterogeneity and instability in both prokaryotes and eukaryotes [6, 9, 12]. This mechanism has been shown to account for frameshifts leading to deletions and insertions in the genomic DNA of bacteriophage T4 and to expansions of repetitive sequence elements in the genome of bacteriophage T7 [8–13]. In both cases, the corresponding purified DNA polymerase produces *in vitro* replication products that are identical to the genomes of bacteriophage isolated after *in vivo* replication [9, 11]. Similarly, nucleotide mismatch mutations are introduced by DNA polymerases *in vitro* by a mechanism of primer realignment named Streisinger slippage (for reviews see [1, 3, 6]). Early studies with purified *Escherichia coli* DNA polymerase I and *Micrococcus luteus* DNA polymerase had established that polymerization can occur by a primer slippage mechanism where the replication product is produced in a quantity far greater than that of the template DNA added to the reaction mixture (reviewed in [2]). Similarly, "net" DNA synthesis by DNA polymerase b resulting in the production of DNA in amounts considerably exceeding the amount of single-stranded DNA templates has been reported [15, 16]. These studies introduced the concept that under certain reaction conditions, a DNA polymerase that is traditionally considered "template-directed" can synthesize DNA in a non-semiconservative fashion. Taken together, these studies indicate that base-pairing at the 3′ end of the primer during DNA replication can be a dynamic process with major biological consequences: primer alignment across the base of a hairpin structure in the template can lead to deletion, and template switching at inverted repeat sequences in a replication fork can lead to insertions. Localized breathing of the 3′ end of the growing primer strand can lead to a misaligned, yet base-pair-annealed, primer 3′ end, thus facilitating frameshifts and base substitutions. Similarly, primer breathing and loop-out realignment across a simple direct repeat sequence, such as $(CTG)_n$, could lead to triplet repeat expansion.

Replication across template regions containing runs of the same nucleotide, or short direct repeat sequences such as $(CTG)_n$, $(CGG)_n$, $(CAG)_n$, or $(CGGT)_n$ provides an ideal system to study PRL mutagenesis. Mechanistic features required for primer realignment mutagenesis are not known. Yet, base-pair complementarity and annealing at the primer 3′ end is a strong preference for DNA polymerases and it is, therefore, reasonable to presume that sequence complementarity is a prerequisite for primer relocation mutagenesis. This idea provides a straightforward working framework for study of primer realignment mutagenesis involving simple direct repeat sequences.

DNA polymerases appear to be intimately involved in the primer relocation mutagenesis mechanism. Their strong preference for base-pair complementary annealing at the 3′ end of the primer provides sequence specificity to the PRL mechanism. Second, if a polymerase is unable to extend a growing primer strand, because of a block in the template or because of intrinsic sequence/conformation properties of the template · primer, the primer will have a greater opportunity for relocation, assuming that relocation depends on the amount of time given for the growing primer strand to search for a realignment intermediate. Support for this idea has come from studies on mutagenesis by the HIV-1 reverse transcriptase [17–19].

Our results with the HIV-1 reverse transcriptase, published in 1989, were the first to indicate that termination sites for processive polymerization along a template are hot spots for mutagenesis [17]. Since reinitiation by DNA polymerase on the terminated primer and continued synthesis beyond the termination site were required for product formation in our system, we proposed that PRL may occur during the reinitiation event [17–19]. This could happen by a polymerase-mediated event, either by allowing thermal breathing of the primer 3′ end and then replicating off of the realigned primer, or by actively destabilizing the annealing of the primer 3′ end. We found that highly processive replication across a frameshift hot spot prevents mutagenesis [18, 19]. Therefore, replication events leading to distributive synthesis, versus processive synthesis, are candidates for producing PRL mutagenesis in human cellular DNA. These distributive synthesis events may include: lagging strand synthesis during replication by DNA polymerase α, short patch repair synthesis (gap-filling) by DNA polymerase β, and leading-strand synthesis during replication in cases where strong termination sites exist in the leading-strand template, i.e., damaged residues, hairpins [20], terminator sequences [18], or unusual DNA conformations [21]. This latter possibility was recently described for so-called "arrest sites" in a DNA template-containing dinucleotide repeats of $(TC)_n$ and $(GA)_n$ during *in vitro* replication by Klenow polymerase or the Taq DNA polymerase. It was proposed that these dinucleotide repeat sequences are associated with triple helix formation and that such unusual DNA structures terminate processive replication and provide for pause sites along the template [21].

Studies devoted to the detailed replication properties of templates containing short direct repeat sequences are enlightening. Such sequences are abundant in natural DNA and are known to be sites of genome instability [24–28] Repeat sequences are often associated with non-B conformations in DNA, including such relatively aberrant structures as cruciforms and hairpins in single-stranded DNA molecules [22, 29, 30]. Short repeat motifs are found as clustered DNA elements and as sequences widely dispersed in the human genome [31]. In addition, they are found in protein coding regions, and also in regions that have no apparent protein-coding function. Amplification of certain triplet repeat sequences has been associated with inherited diseases, such as myotonic muscular dystrophy, fragile X syndrome, and Kennedy's disease [22, 31–37], among others.

Molecular mechanisms during DNA synthesis are strong candidates for causative events in the origin of triplet repeat expansion. Unfortunately, mechanisms accounting for triplet expansion are not yet understood, even at the level of purified human DNA polymerases. Experiments with these enzymes establish a framework for understanding how triplet repeat expansion may occur. The influence of template DNA structure on template utilization by purified DNA polymerases is not well understood in terms of DNA sequence and DNA conformation properties. Yet, it is clear that DNA polymerases are exquisitely sensitive to alterations in template and primer sequence. It is unlikely that these enzymes are immune to alterations in global β-DNA conformation that are potentially caused by a series of direct triplet repeats.

II. RESULTS

We have conducted studies of *in vitro* replication of templates containing short direct repeats. These results and others from our laboratory, both published and unpublished, indicate that primer relocation mutagenesis occurs during *in vitro* DNA replication, and will be an interesting type of mutagenesis to study in the case of triple direct repeat-containing templates.

A. Strand Displacement and Template Switching by DNA Polymerase β

We initially evaluated the question of β-polymerase utilization of plasmid DNA containing a single nick.[3] The aim was to determine if β-pol can conduct strand-displacement synthesis on such a substrate and, secondly, to determine the size of product molecules produced. We chose to use the plasmid pML2 that was nicked at a single site with restriction enzyme *Eco*RI in the presence of ethidium bromide. This substrate material was purified after the nicking reaction and then incubated with increasing quantities of purified human DNA polymerase β and the other components for DNA synthesis, including magnesium as a divalent cation. We were surprised to observe that virtually no DNA synthesis occurred at β-polymerase molar concentrations less than the concentration of the pML2 DNA substrate. However, at concentrations of β-pol of approximately 200-fold over the concentration of substrate nicked DNA, we observed abundant DNA synthesis. Product characterization revealed that the enzyme conducted strand displacement for a short region into the nick (about 20 nucleotides) and then conducted template switching. Synthesis then proceeded to the end of the displaced strand, whereupon a build-up of product molecules occurred. These product molecules are explained by a panhandle-type structure as illustrated for T7 DNA polymerase products by Lechner *et al.* [9].

The analysis of this structure involved restriction enzyme cutting of the product using *Eco*RI. The newly synthesized DNA in these structures was found to be consistent with "snap-back" synthesis as already discussed in detail by Richardson and associates for T7 DNA polymerase replication products [9] and by Ohshima and Wells for triplet repeat containing sequences [40]. Direct Maxam–Gilbert sequencing of this material confirmed that it resulted from template switching and synthesis to the end of the template strand, yet there was no sequence of complementary DNA for primer annealing on the displaced strand longer than three nucleotide residues.

Longer periods of synthesis and/or addition of larger quantities of enzyme to the reaction mixture facilitated formation of novel types of product molecules. These product molecules were extremely long as revealed on denaturing gels. However, the molecules could be cut with *Eco*RI, leaving a homogeneous product representing the extended primer from the *Eco*RI site to the end of the displaced strand.[4] We conclude from these results that pol-β conducted snap-back synthesis both at the end of the displaced strand and immediately 5′ of the *Eco*RI site. The overall product molecule was an accordion-like expansion of sequences between the *Eco*RI site and the end of the displaced strand. This expansion represented a sequence not present in the original template DNA, but nevertheless related to the

[3] B. Z. Zmudzka and S. H. Wilson, unpublished observations.

[4] Ibid.

sequence of the original template DNA. These experiments serve as a point of reference for potential *in vitro* reaction products formed by DNA polymerase β, since the activity of β-polymerase may be triggered by DNA damage and subsequent excision-repair pathways. *In vitro* reactions modeling these events will be interesting to consider as a function of the presence of triplet direct repeat sequences.

B. DNA Polymerase Processivity

An area of study that may be highly relevant to the problem of repeat expansion is the effect of template sequence on the processivity of DNA polymerases. When DNA polymerizing enzymes conduct one cycle of synthesis on natural DNA templates-binding, processive DNA synthesis, and dissociation—the template position for termination of DNA synthesis is nonrandom: Termination occurs preferentially at certain nucleotide positions referred to as termination sites and termination probabilities at different template positions can vary by orders of magnitude [23, 38]. Although the mechanism of termination is not well understood, the process can be studied experimentally because product molecules accumulate in a reaction mixture at chain-lengths corresponding to termination sites, and these accumulated molecules can be easily quantified after sequencing gel electrophoresis [23]. The specific nucleotide position where termination preferentially occurs (i.e., a termination site) is presumably governed by a combination of factors, such as enzyme/template · primer contacts, effects of the incoming dNTP on stabilizing the pretransition state complex, and stabilization of the pyrophosphate leaving group. All of these factors could influence polymerase transit time across a given nucleotide position, and it is possible that the longer the polymerase remains at a specific template position the greater the likelihood that the enzyme will dissociate from the template at that position [39].

Eukaryotic DNA polymerases *in vitro* exhibit an inverse correlation between processivity (i.e., lack of termination) and frameshift mutation frequency in an M13mp2 mutagenesis system: the most processive mammalian enzyme, DNA polymerase γ, shows the lowest frameshift frequency; the least processive enzyme, DNA polymerase β, shows the highest propensity for frameshifts; and DNA polymerase α, a moderately processive enzyme, shows an intermediate frameshift frequency [14, 41]. The HIV-1 RT, a processive DNA polymerase, shows a strong tendency to terminate DNA synthesis at certain nucleotide positions in the M13mp2 template system [17]. Further, the mutation spectrum of the HIV-1 RT on the M13mp2 system shows frameshifts within "runs" of three or more of the same nucleotide, and we found that each of these runs has at least one termination site or position with relatively high probability of terminating DNA synthesis. These results indicated that template runs containing a termination site are hot spots for frameshift mutations by the HIV-1 RT [17].

In our study to further examine effects of template sequence on termination, we engineered single-base changes in the M13mp2 template in a region generally free of template secondary structure. We find that single-base changes in the template can influence termination, supporting the idea that HIV-1 RT can recognize template sequence. The sequence of the template · primer stem is important for termination, whereas the sequence of the single-stranded template ahead of the primer is not. For example, a single A to G change in an A-rich sequence strongly suppresses termination. This effect corresponds to 12- and 20-fold reductions in termination probability, respectively, at the -2 and -6 positions, relative to the base opposite the 3′ end of the primer as position 0. C is the strongest antiterminator residue, followed by G. A is a terminator residue and exerts effect when it is in the -1, -2, -3, -5, and -6 positions of the template · primer stem. The single-base C to A change in a T-rich region of the template produced a 5- to 10-fold increase in termination probability at these positions. When the single-base change was in the 0, -4, or -7 positions, there was no effect on termination. The results indicate that a DNA polymerase can behave as a sequence-specific double-stranded DNA binding protein, in that it can distinguish base residues in the template · primer stem [18].

C. Replication of Direct Repeat-Containing Templates *in Vitro*

To evaluate the capacity of DNA polymerases to replicate single-stranded templates containing triplet repeat sequences, we created a family of synthetic oligonucleotide templates and primers. These templates contained nine iterations of the direct repeats CTG, GTC, or GGC downstream from a primer binding site. The primers were labeled in the 5′ end in order to follow DNA polymerase reaction products.

After *in vitro* DNA replication reactions were conducted with purified DNA polymerase β or *E. coli* DNA polymerase I large fragment, reaction products were analyzed by sequencing gel electrophoresis and autoradiography.[5] Replication products by β-pol failed to reach the end of the template and accumulated at a

[5]R. Singhal and S. H. Wilson, unpublished observations.

length corresponding to extension to the center of the direct repeat sequences. Yet, the quantity of enzyme added to the incubation mixture and the time of incubation were appropriate for synthesis to the end of the template in the case of heteropolymeric DNA templates. Therefore, replication through the GGC sequence was difficult for β-pol. A small amount of replication product longer than template length also was observed in the reaction mixture. Similarly, with polymerase I products accumulated throughout the range of repeat sequences in the template, indicating that the enzyme had difficulty in replicating through the GGC tandem repeat sequence. In addition, products piled up at the end of the template as expected, and there was accumulation of product molecules somewhat longer than template length. Similar analyses of replication by β-polymerase across the repeat sequences CTG and GTC was conducted. In this case, replication of the GTC sequence proceeded in an unencumbered fashion, and products accumulated corresponding to the full-length template. No products longer than template length were observed. By contrast, in the case of replication of the CTG repeat sequence, molecules accumulated roughly in the center of the repeat element. These results indicate that structural features of the CTG direct repeat prevent complete replication by β-polymerase, suggesting the presence of unusual DNA structure in the reaction mixture. β-Polymerase was essentially unable to extend off of the product molecules accumulating. This is evident because β-pol functions as a distributive or nonprocessive enzyme; therefore, all of the nascent product molecules initiated in the reaction had been elongated to the pause site length and had accumulated at that point. The fact that such strong polymerase pausing exists with the CTG replication products suggests that DNA structural features of this sequence contribute to primer realignment mutagenesis.

III. CONCLUDING REMARKS

Much progress has been made during the past several years in understanding the impact of repeat sequences in DNA templates. The results reported here and elsewhere [40] indicate that replication-based expansion and contraction can be mediated by information imbedded in the triplet repeat-bearing sequences of templates.

References

1. Streisinger, G., Okada, Y., Emrich, J., Newton, J., Tsugita, A. Terzaghi, E., and Inouye, I. (1966). Frameshift mutation and the genetic code. *Cold Spring Harbor Symp. Quant. Biol.* **31,** 77–86.
2. Burd, J. F., and Wells, R. D. (1970). Effect of incubation conditions on the nucleotide sequence of DNA products of unprimed DNA polymerase reactions. *J. Mol. Biol.* **53,** 435–459.
3. Ripley, L. S. (1982). Model for the participation of quasi palindromic DNA sequences in frameshift mutation. *Proc. Natl. Acad. Sci. USA* **79,** 4128–4132.
4. Levinson, G., and Gutman, G. A. (1987). Slipped-strand mispairing: a major mechanism for DNA sequence evolution. *Mol. Biol. Evol.* **4,** 203–221.
5. Schlötterer, C., and Tautz, D. (1992). Slippage synthesis of simple sequence DNA. *Nucleic Acids Res.* **20,** 211–215.
6. Kunkel, T. A. (1992). DNA replication fidelity. *J. Biol. Chem.* **267,** 18251–18254.
7. Kunkel, T. A. (1990). Misalignment-mediated DNA synthesis errors. *J. Am. Chem. Soc.* **29,** 8003–8011.
8. Engler, M. J., and Richardson, C. C. (1983). Bacteriophage T7 DNA replication: synthesis of lagging strands in a reconstituted system using purified proteins. *J. Biol. Chem.* **258,** 11197–11205.
9. Lechner, R. L., Engler, M. J., and Richardson, C. C. (1970). Characterization of strand displacement synthesis catalyzed-by bacteriophage T7 DNA polymerase. *J. Mol. Biol.* **53,** 435–459.
10. Lechner, R. L., and Richardson, C. C. (1970). Effect of incubation conditions on the nucleotide sequence of DNA products of unprimed DNA polymerase reactions. *J. Mol. Biol.* **53,** 435–459.
11. Papanicolaou, C., and Ripley, L. S. (1991). An *in vitro* approach to identifying specificity of mutagenesis mediated by DNA misalignments. *J. Mol. Biol.* **221,** 805–821.
12. Ripley, L. S. (1990). Frameshift mutation: determinants of specificity. *Annu. Rev. Genet.* **24,** 189–213.
13. Glickman, B. W., and Ripley, L. S. (1984). Structural intermediates of deletion mutagenesis: a role for palindromic DNA. *Proc. Natl. Acad. Sci. USA* **81,** 512–516.
14. Kunkel, T. A. (1985). The mutational specificity of DNA polymerase-β during *in vitro* DNA synthesis. production of frameshift, base substitution and deletion mutations. *J. Biol. Chem.* **260,** 5887–5796.
15. Siedlecki, J. A., Nowak, R., Soltyk, A., and Zmudzka, B. (1981). Net DNA synthesis catalysed by calf thymus DNA polymerase b. *Acta Biochim. Polonica* **28,** 157–173.
16. Nowak, R., Kulik, J., and Siedlecki, J. (1987). The ability of DNA polymerase b to synthesize DNA beyond the gap with displacement of the non-replicated strand. *Acta Biochim.ica Polonica* **34,** 205–215.
17. Bebenek, K., Abbotts, J., Roberts, J., Wilson, S. H., and Kunkel, T. A. (1989). Specificity and mechanism of error-prone replication by HIV reverse transcriptase. *J. Biol. Chem.* **264,** 16948–16956.
18. Abbotts, J., Bebenek, K., Kunkel, T. A., and Wilson, S. H. Mechanism of HIV-1 reverse transcriptase: termination of processive synthesis on a natural DNA template is influenced by the sequence of the template · primer stem. *J. Biol. Chem.* [In press]
19. Bebenek, K., Abbotts, J., Wilson, S. H., and Kunkel, T. A. Error-prone polymerization by HIV-1 reverse transcriptase: contribution of template · primer misalignment, miscoding and termination probability tomutational hot spots. *J. Biol. Chem.* [In press]
20. Weaver, D. T., and Depamphilis, M. L. (1984). Role of palindromic and non-palindromic sequences in arresting DNA synthesis *in vitro* and *in vivo. J. Mol Biol.* **180,** 961–986.
21. Baran, N., Lapidot, A., and Manor, H. (1991). Formation of DNA triplexes accounts for arrests of DNA synthesis at $d(TC)_n$ and $d(GA)_n$ tracts. *Proc. Natl. Acad. Sci. USA* **88,** 507–511.
22. Sinden, R. R., and Wells, R. D. (1992). DNA structure, mutations, and human genetic disease. *Curr. Opin. Biotechnol.* **3,** 612–622.
23. Detera, S. D., Becerra, S. P., Swack, J., and Wilson, S. H. (1981). Studies on the mechanism of DNA polymerase a: nascent chain

elongation, steady state kinetics and the initiation phase of DNA synthesis. *J. Biol. Chem.* **256,** 6933–6943.

24. Selker, E. U. (1990). Premeiotic instability of repeated sequences in *Neurospora crassa. Annu. Rev. Genet.* **24,** 579–613.
25. Krawczak, M., and Cooper, D. N. (1991). Gene deletions causing human genetic disease: mechanisms of mutagenesis and the role of the local DNA sequence environment. *Hum. Genet.* **86,** 425–441.
26. Edwards, A., Hammond, H. A., Jin, L., Caskey, C. T., and Chakraborty, R. (1992). Genetic variation at five trimeric and tetrameric tandem repeat loci in four human population groups. *Genomics* **12,** 241–253.
27. Trinh, T. Q., and Sinden, R. R. The influence of primary and secondary DNA structure in deletion and duplication between direct repeats in *Escherichia coli. Genetics.* [In PRESS]
28. Klysik, J., Stirdivant, S. M., and Wells, R. D. (1982). Left handed DNA: cloning, characterization, and instability of inserts containing different lengths of (dC-dG) in *Escherichia coli. J. Biol. Chem.* **257,** 10152–10158.
29. Wells, R. D. (1988). Unusual DNA strucures. *J. Biol. Chem.* **263,** 1095–1098.
30. Wohlrab, F., and Wells, R. D. (1989). Slight changes in conditions influence the family of non-B-DNA conformations of the herpes simplex virus type 1 DR2 repeats. *J. Biol. Chem.* **264,** 8207–8213.
31. Caskey, C. T., Pizzuti, A., Fu, Y-H., Fenwick, R. G., Jr., and Nelson, D. L. (1992). Triplet repeat mutations in human disease. *Science* **256,** 784–789.
32. Harley, H. G., Brook, J. D., Rundle, S. A., Crow, S., Reardon, W., Buckler, A. J., Harper, P. S., Housman, D. E., and Shaw, D. J. (1992). Expansion of an unstable DNA region and phenotypic variation in myotonic dystrophy. *Nature* **355,** 545–546.
33. Aslanidis, C., Jansen, G., Amemiya, C., Shutler, G., Mahadevan, M., Tsilfidis, C., Chen, C., Alleman, J., Wormskamp, N. G. M., Vooijs, M., Buxton, J., Johnson, K., Smeets, H. J. M., Lennon, G. G., Carrano, A. V., Korneluk, R. G., Wieringa, B., and Jong, P. J. (1992). Cloning of the essential myotonic dystrophy region and mapping of the putative defect. *Nature* **355,** 548–551.
34. Fu-Y-H., Kuhl, D. P. A., Pizzuti, A., Pieretti, M. Sutcliffe, J. S., Richards, S., Verkerk, A. J. M. H., Holden, J. J. A., Fenwick, R. G., Jr., Warren, S. T., Obstra, B. A., Nelson, D. L., and Caskey, C. T. (1991). Variation of the CGG repeat at the fragile X site results in genetic instability: resolution of the Sherman paradox. *Cell* **67,** 1047–1058.
35. Kremer, E. J., Pritchard, M., Lynch, M., Yu, S., Holman, K., Baker E., Warren, S. T., Schlessinger, D., Sutherland, G. R., and Richards, R. I. (1991). Mapping of DNA instability at the fragile X to a trinucleotide repeat sequence $p(CCG)_n$. *Science* **252,** 1711–1714.
36. Yu, S., Pritchard, M., Kremer, E., Lynch, M., Nancarrow, J., Baker, E., Holman, K., Mulley, J. C., Warren, S. T., Schlessinger, D., Sutherland, G. R., and Richards, R. I. (1991). Fragile X genotype characterized by an unstable region of DNA. *Science* **252,** 1179–1181.
37. La Spada, A. R., Wilson, E. M., Lubahn, D. B., Harding, A. E., and Fischbeck, K. H. (1991). Androgen receptor gene mutations in X-linked spinal and bulbar muscular atrophy. *Nature* **352,** 77–79.
38. Detera, S. D., and Wilson, S. H. (1982). Studies on the mechanism of *Escherichia coli* DNA polymerase I large fragment: chain termination and modulation by poly-nucleotides. *J. Biol. Chem.* **257,** 9770–9780.
39. Abbotts, J., SenGupta, D. N., Zon, G., and Wilson, S. H. (1988). Studies on the mechanism of *Escherichia coli* DNA polymerase I large fragment: effect of template sequence and substrate variation on termination of synthesis. *J. Biol. Chem.* **263,** 15094–15103.
40. Ohshima, K., and Wells, R. D. (1987). Hairpin formation during DNA synthesis primer realignment *in vitro* in triplet repeat sequences from human hereditary disease genes. *J. Biol. Chem.* **272,** 16978–16806.
41. Kunkel, T. A. (1985). The mutational specificity of DNA polymerases-α and -γ during *in vitro* DNA synthesis. *J. Biol. Chem.* **260,** 12866–12874.

DNA Replication Errors Involving Strand Misalignments

LISA C. KROUTIL[1] AND THOMAS A. KUNKEL[2]

Laboratory of Molecular Genetics, National Institute of Environmental Health Sciences, Research Triangle Park, North Carolina 27709

I. INTRODUCTION

The low spontaneous mutation rate in normal human cells [1] results partly from the high fidelity of replication of the six billion nucleotides of the human genome. This extraordinary accuracy results from three major steps to prevent copying errors. Discrimination at the replication fork partly reflects the inherently high nucleotide selectivity of DNA polymerases. Occasional polymerase errors may be excised prior to further chain elongation by $3' \rightarrow 5'$ exonucleolytic activity. Errors that escape proofreading may later be corrected by postreplication mismatch repair, which selectively corrects errors in the newly synthesized strand.

The amount of discrimination against point mutations from these steps has been estimated in *Escherichia coli* by measuring mutation rates in strains selectively defective in one or more of these steps [2]. The spontaneous mutation rate of the *lacI* gene is $\sim 10^{-10}$ mutations per base pair replicated per generation. A mutant lacking postreplication mismatch repair has a much higher rate, such that mismatch repair reduces base substitution rates by 20- to 400-fold, depending on the type of substitution. The mutation rate of approximately 10^{-7} in the

[1]Present address: Department of Chemistry, Beloit College, Beloit, WI 53511.

[2]To whom correspondence should be addressed. (919) 541-2644. E-mail: kunkel@niehs.nih.gov.

strain lacking mismatch repair can be considered (with some simplifying assumptions, see [2]) to be the fidelity of chromosomal replication. Analysis of a double mutant lacking mismatch repair and defective in the 3′ → 5′ exonuclease activity of the ε subunit of the replicative DNA polymerase III holoenzyme suggests that proofreading contributes between 40- and 200-fold to replication fidelity, with the balance (factors of 2×10^5 to 2×10^6) representing the base selectivity of the replication machinery. These values represent an excellent starting point for considering the relative contributions of these processes to replication fidelity in eukaryotic cells.

Since eukaryotic DNA mismatch repair has been the subject of several recent reviews (e.g., [3–5] and also see Chapter 36), we will not consider mismatch repair further here. Instead, we review what is known about how the first two of these processes functions in eukaryotes. We focus primarily on studies of model DNA replication *in vitro,* referring the reader to other chapters in this book that pertain more directly to disease implications and observations *in vivo.* We begin with a brief discussion of principles for base substitution fidelity ascertained from studies with DNA polymerases. We then review pathways for replication errors involving DNA strand misalignments, since these intermediates may yield the addition/deletion mutations found in diseases. This includes a discussion of the frameshift fidelity of DNA polymerases, including those for which structural information is available. We also discuss what is known about the fidelity of replication *in vitro* by the human replication apparatus, and end with a recently proposed hypothesis to explain expansions associated with triplet repeat diseases.

II. DISCRIMINATION AGAINST BASE SUBSTITUTION ERRORS

Of the three processes mentioned above, the single greatest contribution to replication fidelity for base substitution errors comes from the high selectivity of DNA polymerases. As described with model prokaryotic and viral DNA polymerases, this selectivity occurs during at least three points in a DNA polymerization reaction cycle (also see reviews by Carroll and Benkovic [6], Echols and Goodman [7], Kunkel [8], and Johnson [9]). Deoxyribonucleoside triphosphates bind to the complex formed between the polymerase and the template–primer. Some discrimination results from the fact that incorrect dNTPs bind much less avidly than do correct dNTPs. Selectivity can also result from what is inferred to be a conformational change in the ternary complex to position the dNTP for subsequent phosphodiester bond formation. Here the polymerase may lock onto the template–primer · dNTP complex much more rapidly for correct base pairs that can adopt Watson–Crick geometry than for incorrect base pairs that cannot, ultimately leading to a much faster rate of incorporation of correct nucleotides. Finally, discrimination against incorrect incorporation can directly be due to a strong reduction in the rate of phosphodiester bond formation itself. For the model prokaryotic and viral polymerases examined to date, the relative importance of these three steps varies over a considerable range and depends on the polymerase studied, the composition of the mispair, and the surrounding sequence context. Although the base substitution fidelity of eukaryotic DNA polymerases has been measured (e.g., see [10]), the contributions of the individual steps in the reaction cycle to the fidelity of these DNA polymerases remain to be established.

Many DNA polymerases contain intrinsic 3′ → 5′ exonuclease activities capable of removing nucleotides from a primer terminus. With the Klenow polymerase, a slow step has been detected after chemistry and prior to pyrophosphate release [6], providing an opportunity for removal of misinserted nucleotides. Once a nucleotide is incorporated and pyrophosphate has been released, the complex can enter the next cycle of polymerization. However, the rate of correct incorporation onto a terminal mispair is much slower than is the rate of correct incorporation onto a correctly paired terminus. This provides an opportunity for proofreading of mispairs. As discussed below, this is a critical junction in the reaction cycle for determining frameshift fidelity during copying of repetitive DNA.

Prokaryotic polymerases containing associated 3′ → 5′ exonuclease activities have average base substitution error rates of about 10^{-6} to 10^{-7}. The contribution of proofreading to these rates, as estimated from several experimental approaches, is on average about 100-fold, but may vary over a wide range, from only a few-fold to almost 1000-fold. This reflects enzyme- and sequence-specific influences, especially the different rate constants for extension from the 12 possible mispairs. Such proofreading differences are expected based on the idea originally proposed by Brutlag and Kornberg [11] that a terminus containing a terminal mispair has a higher probability of being single-stranded ("frayed") than does a correctly paired terminus. A frayed end will preferentially bind to the exonuclease active site, which prefers single-stranded DNA. Similarly, a matched, double-stranded terminus will preferentially bind to the polymerase active site, which prefers double-stranded DNA (for review, see Joyce and Steitz [12]). Because the stability of the duplex region of the template–primer will depend on its DNA sequence, proofreading efficiency is expected to differ in different sequence contexts having differing stabilities. Moreover, the degree of fraying needed to allow single-stranded DNA to bind to the exonuclease active site may vary, depending on the distance between the polymerase and exonuclease active sites. This distance could be greater for some enzymes than others, leading to enzyme-mediated differences in proofreading efficiency. This may be highly relevant to proofreading of misaligned, repetitive-sequence substrates.

III. MODELS FOR REPLICATION ERRORS INVOLVING TEMPLATE–PRIMER MISALIGNMENTS

The idea of strand slippage at a DNA end was put forth in the classical paper by Streisinger *et al.* [13] to explain the observation that frameshift mutations in bacteriophage T4 frequently appear in regions of repeating bases. Numerous subsequent studies in T4 and other systems (e.g., reviewed by Ripley [14]) confirm that repeat sequences are hot spots for addition and deletion mutations. These, and even some substitution errors, may be initiated by template–primer misalignments during replication. During replication of repetitive sequences, strand slippage may result in misaligned intermediates stabilized by correct base pairs (Fig. 44-1A). Subsequent polymerization leads to deletion if the unpaired nucleotide(s) is in the template strand (Fig. 44-1A) or to addition if the unpaired nucleotide(s) is in the primer strand (not shown). This model has received much attention recently due to the connection between repetitive sequence stability and diseases, including cancer (e.g., see [5]) and hereditary degenerative diseases (several chapters in this book). However, obtaining clear insights into misalignment-mediated replication errors can be problematic. Unlike base substitutions, where the nucleotide substituted is precisely known, one does not know exactly which nucleotide(s) is added or deleted when frameshifts occur in repeated sequences. Moreover, while direct misinsertion yields a base substitution intermediate involving a covalent bond, misaligned intermediates depend on disruption and reformation of hydrogen bonds, and the multiple possible intermediates can potentially interconvert. Thus, tests of models for misalignment-mediated replication errors are not unequivocal. It is certainly true that less is currently known concerning discrimination against frameshifts than against substitutions during individual steps in the polymerization reaction.

Nevertheless, there is good evidence that replication errors do in fact occur via misaligned intermediates. For example, the Streisinger slippage model predicts that the error rate should increase as the length of the run increases, because the potential number of correct base pairs that could stabilize the misaligned intermediates increases, as does the number of potential misaligned intermediates that can form (Fig. 44-1A). Furthermore, the longer the run, the greater the distance between the extra nucleotide and the 3′-OH primer terminus, potentially reducing interference by the extra base during phosphodiester bond formation within the enzyme active site. As predicted, frameshift error rates during replication, expressed per nucleotide polymerized to correct for differences in the number of nucleotides in runs of different lengths, do increase as the length of a homopolymeric run increases [15–19]. Moreover, error rates for one-base deletions in homopolymeric runs by pol β [20] and HIV-1 RT (Fig. 44-2) decrease when the template sequence is altered to either shorten or eliminate a repetitive sequence. This effect, observed in a simple replication reaction, is reminiscent of the stability imposed on triplet repeat sequences by interruptions with nonrepetitive nucleotides. For example, interrupted alleles are less polymorphic for repeat length variations [21] and unstable alleles show loss of imperfections within repetitive sequences [22–24].

Following slippage, correct incorporation of a nucleotide with subsequent realignment before continued incorporation can generate a terminal mispair (Fig. 44-1B). This can yield a base substitution, but one initiated by slippage rather than misinsertion. This model, reviewed in more detail in [10], has been termed dislocation mutagenesis, by analogy to a dislocated shoulder joint that pops out of alignment but ultimately resumes a normal position. This process is not necessarily limited to substitutions occurring at the ends of runs. Interruptions within runs might also be prone to base substitutions that eliminate the imperfection. As illustrated in Fig. 44-3, synthesis could proceed to some point within a repeat sequence prior to the imperfection. Were slippage to occur followed by incorporation of a limited number of nucleotides, a realignment could generate a mispair at the site of the imperfection, ultimately yielding a substitution that eliminates the imperfection in the repeat sequence. This could be particularly important for the formation of trinucleotide repeat alleles prone to the large expansions associated with disease.

Another model for deletion/addition replication errors is direct misincorporation followed by template–primer rearrangement to provide a correct terminal base pair for continued polymerization (Fig. 44-1C). The resulting misalignment ultimately leads to a frameshift error, but it is initiated by misinsertion rather than strand slippage. This model was suggested by observations (reviewed in [15]) indicating that "difficult-to-

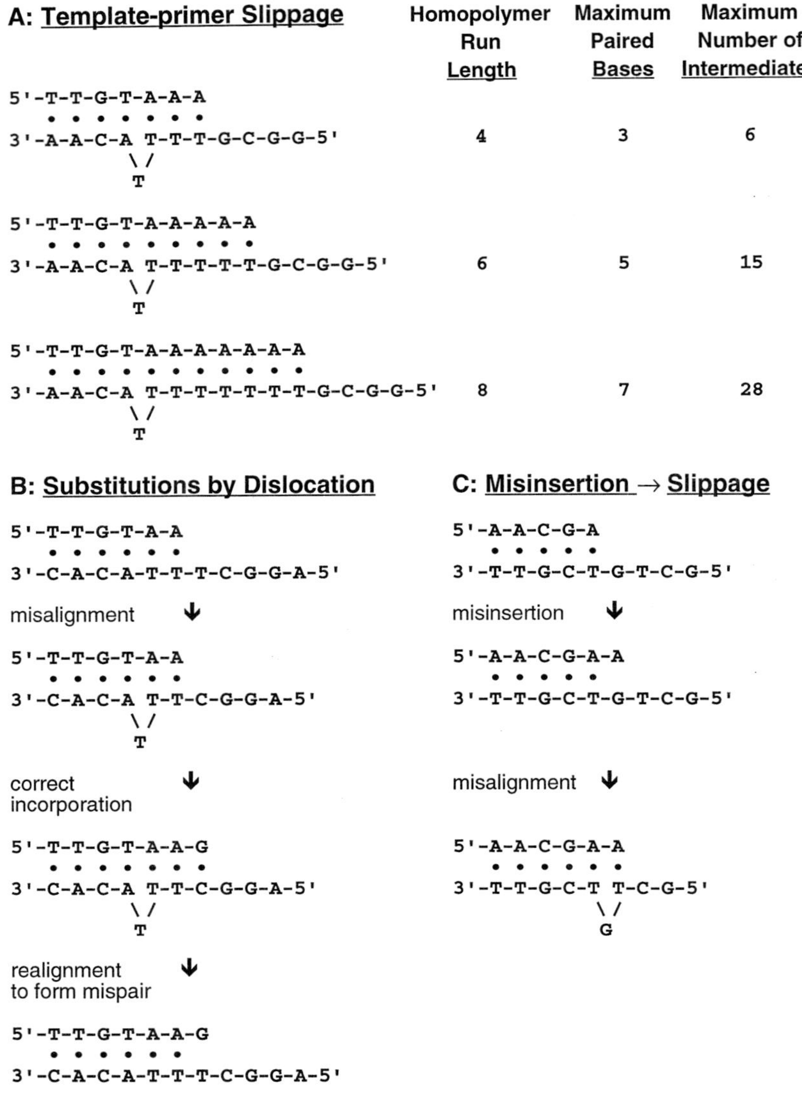

FIGURE 44-1 Pathways for errors involving misaligned template–primers. See text for description. Reproduced from [10], with permission.

extend" mispairs may realign such that synthesis proceeds from a substrate containing an extra nucleotide in the template strand but a correct base pair at the terminus. The model is supported by studies of replication of undamaged DNA by several DNA polymerases ([25] and references therein). The concept of "difficult-to-extend" termini also led to the suggestion [26] that incorporation opposite damaged templates might yield frameshifts by this mechanism. Studies involving replication of DNA containing several different lesions [27–30] reveal that DNA damage does indeed yield frameshift mutations by incorporation followed by primer relocation. Hypothetically, this mechanism is possible at any template position and is not limited to the production of single-base frameshifts. Thus, frameshift errors may be initiated by this process at trinucleotide repeat sequences, possibly including DNA damage-induced primer relocation.

IV. RATES OF SINGLE-BASE DELETIONS AND ADDITIONS BY DNA POLYMERASES

Quantitative addition/deletion error rates for DNA polymerases when copying many of the microsatellite sequences known to be unstable in cancer and the hereditary degenerative diseases (e.g., di- and trinucleotide repeats) have not yet been reported. However, the frameshift fidelity of DNA synthesis *in vitro* catalyzed by numerous DNA polymerases has been examined using the *lacZ* α-complementation gene in M13mp2

A. Original Template

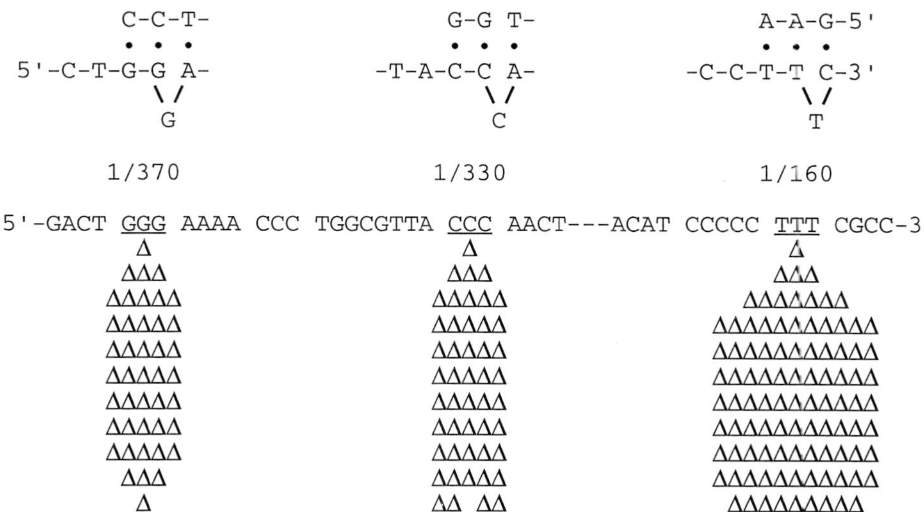

B. Modified Template

C-C-T-
5'-C-T-G-G A-
G

G-T-
-T-A-C-T A-
C

A-G-5'
-C-C-A-T C-3'
T

1/190 ≤1/12,000 ≤1/12,000

5'-GACT GGG AAAA CCC AGGCATTA CTC AACT---ACAT CCCCC ATT CGCC-3'

FIGURE 44-2 Effect of interrupting repetitive sequences on error rates at HIV-1 reverse transcriptase one-nucleotide deletion hot spots. Shown in part A are three sequences in M13mp2 template DNA that are hot spots for single-base deletions when copied by HIV-1 RT. Above each three-base run (underlined) are diagrams of potential misaligned frameshift intermediates and the observed deletion error rates, expressed as errors per detectable nucleotide polymerized. Part B shows the results on frameshift error rates following modification to eliminate one run (CCC → CTC) and to shorten another (TTT → ATT), while the third (GGG) remained unchanged. Each open triangle represents an individual sequenced M13mp2 mutant recovered from the HIV-1 RT copying reactions with either of the two templates. The results clearly show that shortening or eliminating the run strongly increased frameshift fidelity. Adapted from [34], with permission.

DNA as a template. This 250-base sequence contains many noniterated nucleotides as well as several homopolymeric runs of two to five nucleotides (see [31] for target description). It is therefore an appropriate target to investigate frameshift fidelity during polymerization and to probe the misalignment models discussed above.

The average frameshift error rates during DNA synthesis *in vitro* catalyzed by the five template-dependent eukaryotic DNA polymerases are shown in Table 44-1. These are rates for single-base frameshift errors, and are expressed per detectable nucleotide polymerized. DNA polymerase α is one of at least three DNA poly-

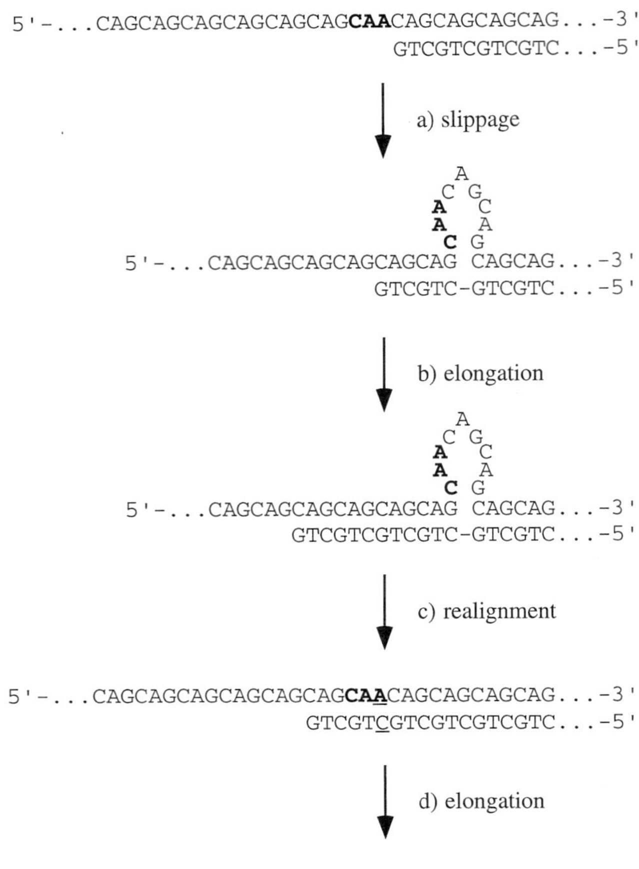

FIGURE 44-3 Model for slippage-initiated substitution mutagenesis to eliminate an interruption within a repetitive sequence. See text for description. Following step (d), replication of the bottom strand would yield progeny with an uninterrupted triplet repeat sequence.

TABLE 44-1 Eukaryotic DNA Polymerase Error Rates for Single Base Frameshifts

DNA polymerase	Proofreading	Error rate ($\times 10^{-6}$)[a]	Ref.
Pol α	No	50	[20,32,49]
Pol β	No	900	[20]
Pol δ	Yes	18[b]	[49]
Pol ε	Yes	5[c]	[49]
Pol γ	Yes	2.4	[20]

[a] In several instances measurements have been made with enzyme preparations from multiple sources. The values given are averages.
[b] Reactions contained PCNA to stimulate synthesis.
[c] This value was obtained in a reaction containing dGMP to inhibit proofreading; thus, it represents a minimal estimate to the accuracy of this enzyme.

merases required for replication of the nuclear genome in eukaryotic cells, and functions to initiate replication at origins and Okazaki fragments on the lagging strand. As isolated from several sources, pol α has an average one-base frameshift error rate of 50×10^{-6}. Similar values have been obtained when the fidelity of the yeast p180 catalytic subunit alone is compared to that of the four-subunit DNA polymerase α–RNA primase complex [32]. Thus, highly purified DNA polymerase α is not particularly accurate relative to the high fidelity required to replicate eukaryotic genomes. This may partly result from the fact that most purified preparations of pol α lack an intrinsic and/or associated proofreading exonuclease activity that could enhance fidelity (but see references in [10]). However, the observed error rate may more than suffice *in vivo* if pol α is only responsible for synthesis of a small number of nucleotides from an RNA primer. Mistakes made here could also be removed during the RNA primer excision–replacement synthesis reaction (see section below on excision of flaps). Moreover, pol α fidelity estimates are thus far limited to synthesis initiated from exogenously supplied DNA primers. It is possible that the fidelity of RNA-primed DNA synthesis could be higher (or lower) than present data suggest.

DNA polymerase β, the smallest of the eukaryotic DNA polymerases, is readily purified as a single subunit enzyme lacking proofreading activity. Among its several possible functions *in vivo,* one is gap-filling synthesis during base excision repair of lesions in DNA. As purified from several sources, pol β has the lowest frameshift fidelity among the eukaryotic DNA polymerases (Table 44-1). Low-fidelity synthesis by pol β is consistent with a modest catalytic role *in vivo,* i.e., filling gaps of one or a few nucleotides during base excision repair. Alternatively, pol β may have higher accuracy than current estimates suggest. Thus far, pol β fidelity has been measured using template–primers containing long single-stranded template regions, where synthesis is distributive rather than processive. Recently, pol β has been shown to catalyze processive synthesis on a template adjacent to a 5′-phosphoryl end (see Chapter 43). As this type of synthesis may more closely resemble that occurring during base excision repair, it will be interesting to determine the frameshift fidelity of pol β using substrates containing single-stranded gaps of one or a few nucleotides. It is also possible that pol β fidelity will be influenced by the additional proteins required for base excision repair. X-ray crystallographic structural information is now available for pol β bound to a variety of substrates, providing an excellent opportunity to probe models for production of frameshift errors.

DNA polymerase δ plays a central role in eukaryotic replication, perhaps performing the bulk of chain elongation during replication of the nuclear genome. Evidence also suggests that it is required for both nucleotide excision repair and postreplication repair of mismatched and misaligned substrates. Its average frameshift error rate in the presence of PCNA (which is required to obtain gap-filling synthesis) is 18×10^{-6} (Table 44-1). The fact that pol δ is more accurate than pol α or pol β is consistent with the proofreading of frameshift intermediates by its associated $3' \rightarrow 5'$ exonuclease activity, which has properties expected of an editing exonuclease (reviewed in [33]). The third essential DNA polymerase for replication of the nuclear genome, pol ε, also has an associated exonuclease activity and high frameshift fidelity (Table 44-1), as does DNA polymerase γ, the mitochondrial replicative polymerase (Table 44-1).

Even when copying the same sequence, exonuclease deficient DNA polymerases can have different frameshift error rates (e.g., compare pol α to pol β in Table 44-1). In fact, polymerase error rates for the same mistake can vary as much as 1000-fold (see [15]), illustrating the important role of polymerase-specific protein · substrate interactions in determining polymerase selectivity against frameshift errors. Even for the same DNA polymerase, frameshift fidelity is also highly sequence-dependent. As mentioned above, the frameshift error rates of several exonuclease-deficient enzymes are higher at reiterated sequences than at noniterated nucleotides and increase with increasing homopolymeric run length, consistent with the strand slippage model discussed above. The nucleotide composition of the run itself also affects the frameshift error rate. Thus, some exonuclease-deficient DNA polymerases are less accurate when copying template pyrimidine runs than template purine runs (e.g., see Table 1 in [15]). This may reflect weaker stacking between adjacent pyrimidines than adjacent purines, leading to more frequent formation of misaligned intermediates. Frameshift fidelity can vary over a wide range even for homopolymeric runs of the same length and nucleotide composition, depending on the surrounding sequence (e.g., see Table VII in [34]). Explanations for such effects await a more detailed understanding of precisely how DNA polymerases interact with template–primers.

One rule that emerges from studies of frameshifts within runs is that one-base deletion error rates are generally higher than are one-base addition error rates. As noted by Streisinger and Owen [35], this could be due to the equilibrium between aligned versus misaligned configurations. Hypothetically, the intermediate for a one-base addition error, containing an extra nucleotide in the primer, has one less base pair relative to the intermediate for a one-base deletion error, which contains an extra nucleotide in the template strand (e.g., see Fig. 2 in [15]). Even for intermediates stabilized by an equal number of correct base pairs, the formation of a one-base addition intermediate requires disruption of one more base pair than does formation of a one-base deletion intermediate. Alternatively or in addition, one-base deletion intermediates may not be constrained by the DNA polymerase to the same extent as one-base addition intermediates. It remains to be determined whether the polymerase error rate bias toward deletions rather than additions holds for frameshift mutations in other repetitive sequences, such as dinucleotide and trinucleotide repeats.

V. COMPLEX FRAMESHIFT ERRORS BY DNA POLYMERASES

DNA polymerases also commit errors involving the loss or gain of more than a single nucleotide. Many of these can be explained by slippage between directly repeated DNA sequences separated by a variable number of intervening nucleotides. The model is illustrated in Fig. 44-4 (right) using as an example a 317-base deletion generated by pol β [36]. It involves synthesis of nine nucleotides starting from the provided 3′-OH primer and proceeding through a C-C-C-G-C template repeat, and then disruption of these five G · C primer-terminal base pairs. This is followed by rearrangement of the DNA with reformation of five hydrogen-bonded G · C base pairs involving the newly made DNA and the second copy of the direct repeat, 317 bases downstream. This intermediate may be further stabilized by three G · C base pairs that can potentially form a stem within the resulting loop of the heteroduplex intermediate. This process is similar to forming a frameshift in a homopolymeric run, but involves more nucleotides, greater distances, and a larger misaligned heteroduplex. Continued synthesis from the misaligned intermediate and subsequent expression of the nascent strand yields the deletion.

In that same study, pol β also generated a complex mutant in which 123 bases were deleted and three base changes were present at one end of the deletion. This complex deletion can be explained by a transient misalignment model involving four blocks of DNA sequences spread over the entire 390-base single-stranded gap (Fig. 44-4, left). The model is formally equivalent to that for base substitutions by dislocation in that a misalignment occurs, followed by a limited amount of correct incorporation from the misaligned intermediate. In this case, the correct incorporation involves strand displacement synthesis. This is particularly interesting given the more recent evidence suggesting the involvement of a displaced strand in expansion mutagenesis (see the last section of this chapter). This limited dis-

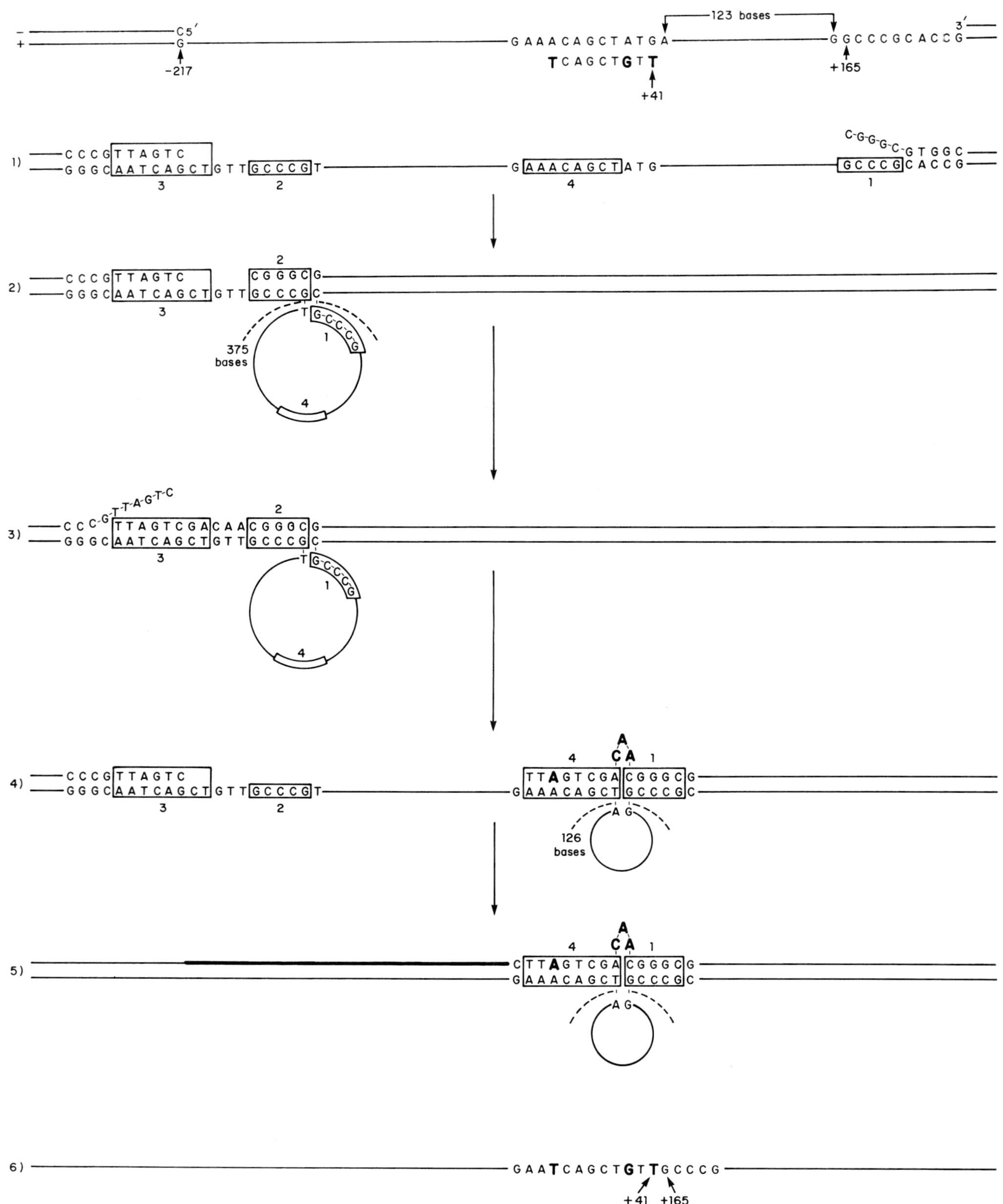

FIGURE 44-4 Models for simple and complex deletion mutations by a DNA polymerase. See text for description. Modified from [36], with permission.

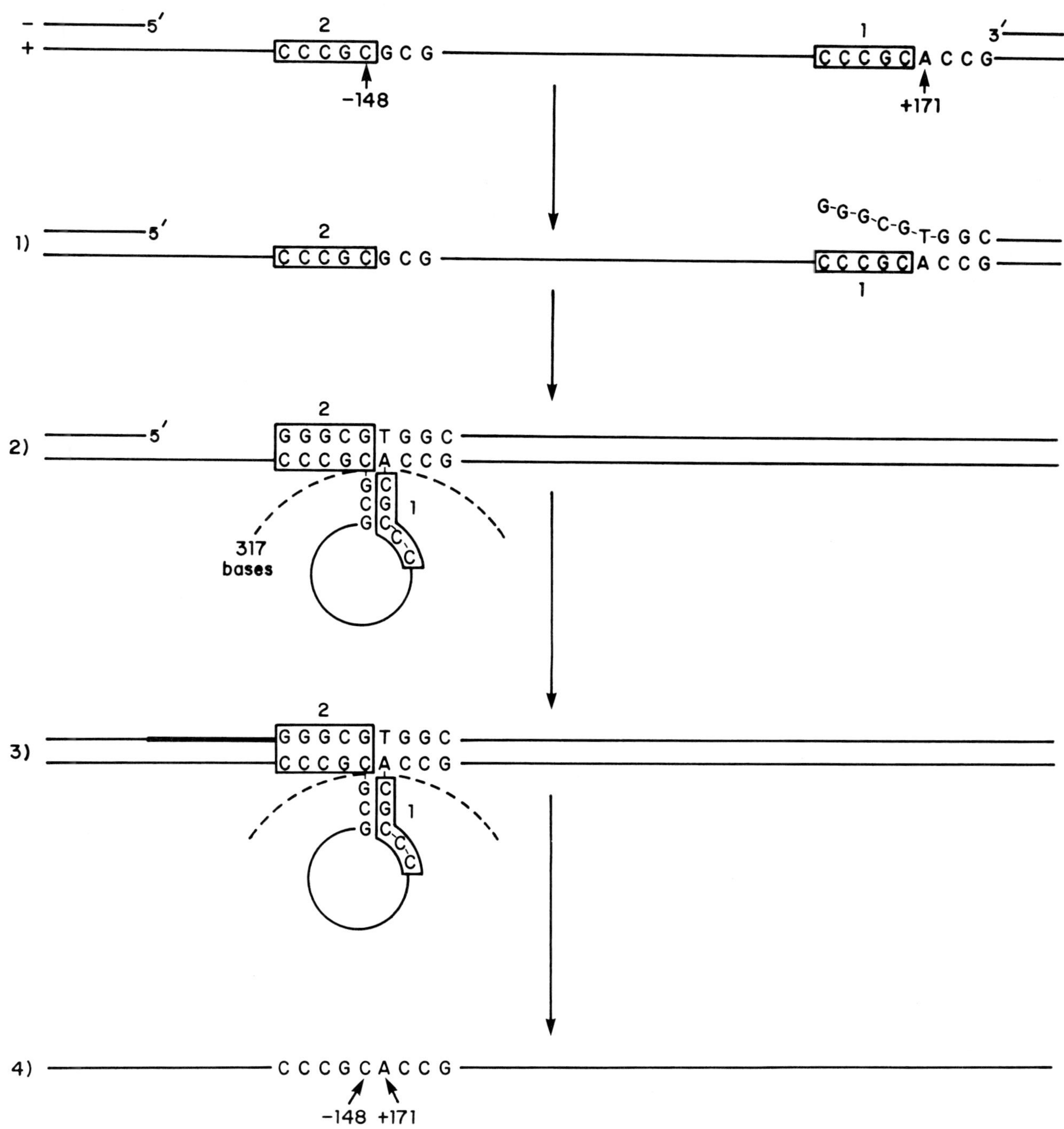

FIGURE 44-4 *(continued)*

placement synthesis is followed by realignment, then continued synthesis from the second heteroduplex intermediate containing an internal A · A mispair and 3 extra nucleotides in the primer strand. Subsequent replication of the newly synthesized strand yields the observed complex frameshift mutation. The result is that a block of 11 nucleotides has been moved from one location to a new position 253 nucleotides distant.

Models to explain the origins of simple and complex frameshift mutations recovered from reactions catalyzed by other polymerases have also invoked transient misalignment, but to different positions. In those studies, aberrant synthesis was suggested to involve strand-switching (e.g., see Fig. 4 in [37]) or primer loop-back (e.g., see Fig. 4 in [32], Fig. 3 in [37], and Fig. 3 in [38]). Consistent with previous models for generating complex frameshifts recovered *in vivo* (for a review see [14]), the ends of some of these complex errors produced *in vitro* suggest the involvement of unusual DNA structures that cause pausing during replication (see Chapter 33 for further details).

Just as for one-base frameshifts, the frequency of polymerization-dependent errors involving more than one nucleotide depends on several variables. For example, duplication errors are much less frequent than are deletion errors, perhaps partly reflecting the larger number of base pairs which must be disrupted to generate a duplication intermediate. This fact is highly relevant in considering the origin of large expansions in triplet repeat diseases (see Section X below). While extensive deletion error specificity data are not available for most polymerases, in a collection of 51 deletion mutants generated by yeast DNA polymerase I that could be explained by a direct repeat model, an increase in error rate was observed as the length of the direct repeat increased from three to five [32]. Beyond this length the error rate did not increase. This relationship may reflect the frequency of initial formation of the misalignment and/or the requirement for a stable intermediate that is acceptable to the polymerase for continued synthesis. Finally, the error rate may depend not only on the length of the direct repeat, but also on its position and nucleotide composition and on the polymerase. Unusual specificities are readily apparent from the limited data available. For example, the 317-base deletion shown in Fig. 44-4 was generated at 100-fold higher frequency by DNA polymerase β [39] than by exonuclease-deficient Klenow fragment polymerase [40]. Among the deletion mutants generated by yeast DNA polymerase I, the 317-nucleotide deletion was recovered 20 times, while deletions between a variety of other direct repeats of identical length and similar nucleotide composition were not seen once.

VI. FRAMESHIFT FIDELITY, PROCESSIVITY, AND POLYMERASE ACCESSORY PROTEINS

One property that correlates with frameshift fidelity in homopolymeric runs is processivity, the number of nucleotides incorporated per DNA polymerase association–dissociation with the template–primer. Pol α is both more accurate and more processive than is pol β (Table 44-1 and [41]), suggesting that higher processivity correlates with higher frameshift fidelity This correlation has been examined in detail with other DNA polymerases. For example, proofreading-deficient HIV-1 reverse transcriptase is highly inaccurate for one-base frameshifts within some but not all template runs, and hot spots for frameshift errors are those template positions where the probability of termination of processive synthesis is high [42]. When changes were introduced into the sequences flanking these hot spots, increases or decreases in frameshift error rates were observed that correlated with concomitant increases or decreases in termination of processive synthesis within the run [34]. Also, DNA synthesis by exonuclease-deficient T7 DNA polymerase is highly inaccurate for addition errors in homopolymeric runs when the polymerase operates without thioredoxin [16], an accessory protein that confers high processivity to the polymerase catalytic subunit. These data thus reveal a clear pattern wherein low processivity correlates with low frameshift fidelity, and are consistent with the idea that the formation and/or utilization of misaligned template–primers is increased during the dissociation–reinitiation phase of a polymerization reaction. Since most replication complexes contain accessory proteins that enhance DNA polymerase processivity, the fact that the absence of an accessory protein known to confer high processivity to a DNA polymerase affects the fidelity of replication of repetitive DNA suggests that perturbations in processivity might promote repetitive sequence instability *in vivo.*

VII. STRUCTURE–FUNCTION ANALYSIS OF DNA POLYMERASE FRAMESHIFT FIDELITY

Large DNA polymerase-specific and sequence-specific influences on frameshift fidelity may eventually be understood by investigating the interactions between these enzymes and the template–primer and incoming dNTP. A wealth of new information on the structures of DNA polymerases as determined by X-ray crystallography reveals that they share a common general structure in which three subdomains are organized in the shape of a right hand (for a review see [12, 43] and references therein). While a complete review of polymerase structures is beyond the scope of this chapter, we now briefly describe studies of HIV-1 reverse transcriptase (RT) and Klenow fragment DNA polymerase that illustrate how structural information can be used to probe frameshift fidelity.

HIV-1 RT is a heterodimer of a p66 subunit having both DNA polymerase and RNase H activity and a p51 subunit derived by C-terminal proteolysis of p66 and lacking RNase H activity. Both subunits have the fingers, palm, and thumb subdomains common to all polymerases described to date [44]. In p66, these subdomains form a cleft for binding the template–primer, with the palm containing polymerase catalytic residues at the bottom of the cleft and the fingers and thumb forming the sides. A fourth subdomain connects the polymerase to the RNase H subdomain in p66, but occludes the cleft in p51, thus preventing its contribution to polymerization activity in the heterodimer. The structure of an RT-template–primer complex [45] shows that the DNA is bound to the p66 subunit between the polymerase and RNase H active sites. The DNA is A-form-like near the polymerase active site and is B-form closer to the RNase H active site, with a 40–45° bend at the junction. This bend is in the region where α helix H of the thumb interacts with the duplex DNA.

Molecular dynamics modeling based on structural information suggests five amino acid residues of potential importance for binding to the duplex template-primer (Fig. 1 in [46]). Four (Q258, G262, W266, and Q269) are in α helix H and one (I94) is in β-sheet 5b of the palm. These residues appear to form a track which interacts in the minor groove, over a distance from the second through the sixth base-pair from the 3′-OH primer terminus, just where extra nucleotides might reside in misaligned frameshift intermediates (Fig. 44-1). When these five residues were each changed to alanine, the resulting mutant RTs had reduced template–primer binding affinity, reduced processivity, and reduced frameshift fidelity for one-base frameshifts in homopolymeric runs [46, 47]. The properties and the structural information suggest that HIV-1 RT, and possibly other RTs that have conserved all five amino acids at these positions [46], contains a discrete structural element, which we have designated a minor groove binding track (MGBT). We suggest that this MGBT interacts with the minor groove of the duplex template–primer stem adjacent to the polymerase active site and is important for controlling nucleic acid binding and the rate of replication slippage errors.

The possible importance to frameshift fidelity of polymerase interactions with the minor groove of the duplex template–primer is also suggested by studies of Klenow fragment DNA polymerase. This enzyme contains a flexible 50-amino-acid subdomain at the tip of the thumb which includes two alpha helices suggested to interact with the duplex template–primer [48]. A Klenow polymerase mutant containing a 24-amino-acid deletion that removes a portion of the tip of the thumb has relatively normal dNTP binding and catalytic rate, yet its DNA binding affinity, ability to conduct processive synthesis, and frameshift fidelity are all strongly reduced [19]. The addition error rate increases as the length of the reiterated sequence increases, indicative of errors initiated by template–primer strand slippage. Thus, these observations suggest a role for the tip of the thumb of Klenow polymerase in determining DNA binding, processivity, and frameshift fidelity, perhaps by tracking the minor groove of the duplex DNA. Similar studies are possible with other DNA polymerases, including those involved in replication and repair in eukaryotic cells.

VIII. PROOFREADING OF FRAMESHIFT INTERMEDIATES

The fact that DNA polymerases α and β, which lack associated exonuclease activity, are less accurate than are pol δ, ε, and γ, which have intrinsic $3' \rightarrow 5'$ exonucleases (Table 44-1), suggests that frameshift errors are proofread. The frameshift fidelity of pol ε is reduced [49] in reactions containing a high concentration of dNTPs (favoring polymerization over exonuclease) and dGMP (an inhibitor of proofreading exonucleases), further indicating that frameshift errors by this enzyme are proofread. Comparison of the frameshift fidelity of wild-type versus exonuclease-deficient derivatives of several polymerases [16, 18, 40, 50] provides more direct evidence that frameshift intermediates at non-run sequences, homopolymeric runs of up to five base pairs, and for deletions of large number of nucleotides between direct repeats are all subject to exonucleolytic proofreading.

As mentioned above, proofreading enhances base substitution fidelity up to several hundred-fold. The efficiency of proofreading primarily depends on kinetic partitioning of the terminal base pair between the polymerase and exonuclease active sites (reviewed in [9]). For a correctly paired template–primer terminus, polymerization is greatly favored over exonucleolytic cleavage (Fig. 44-5, top). In contrast, polymerization from a terminal mismatch is much slower (Fig. 44-5, line 2), providing additional time for exonucleolytic proofreading. However, the situation for frameshift errors arising in repetitive sequences is different. The observation that error rates increase with increasing run length implies that the maximum number of correct base pairs in the repeated sequence stabilizes the intermediate and that the unpaired nucleotide is the greatest distance possible from the polymerase active site, presenting the least possible kinetic barrier to further elongation. This logic suggests that an unpaired nucleotide in a short repetitive sequence may be somewhat similar to a terminal mismatch in terms of fraying and kinetic partitioning (Fig. 44-5, line 3), and thus may be edited efficiently. However, for misalignment in a longer run, the primer terminus may fray less and partition more in favor of the polymerase rather than the exonuclease (Fig. 44-5, line 4), resulting in decreased proofreading efficiency.

This hypothesis was tested by comparing the frameshift error rates of proofreading-proficient T7, T4, and *Pyrococcus furiosis* DNA polymerases to their exo-

Correct Incorporation

AG
k_{exo}
$0.2\ s^{-1}$
T
AG
k_{pol}
$250\ s^{-1}$
TC
AG

Incorrect Incorporation

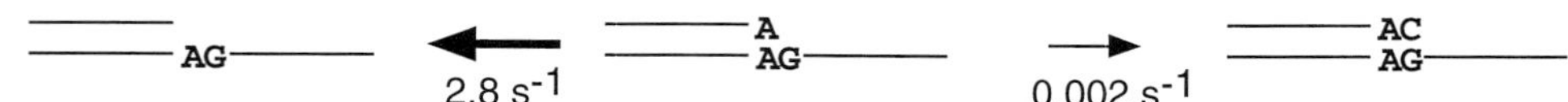

Frameshift in a Short Run

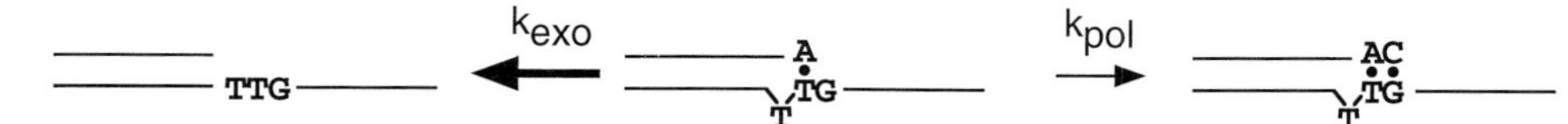

Frameshift in a Long Run

FIGURE 44-5 Kinetic partitioning of misinsertion and misalignment errors. The rate constants shown are for T7 DNA polymerase as described in [65]. Bold arrows indicate a fast rate of polymerization or hydrolysis relative to the opposite reaction. Reproduced with permission from [18]. Copyright 1996, American Chemical Society.

nuclease-deficient derivatives, for +1 and −1 base errors in homopolymeric repeat sequences of three to eight base pairs [18]. All three exonuclease-deficient polymerases produced frameshift errors at rates that increased as a function of run length, consistent with the involvement of misaligned intermediates. While their wild-type counterparts were all more accurate, indicating that frameshift intermediates are corrected by proofreading, the contribution of the exonuclease to fidelity decreased substantially as the length of the homopolymeric run increased. For example, proofreading enhances the frameshift fidelity of T7 DNA polymerase in a run of three A · T base pairs by 160-fold, similar to its contribution to base substitution fidelity (Fig. 44-6). However, in a run of eight consecutive A · T base pairs, the exonuclease only enhances frameshift fidelity by sixfold (Fig. 44-6). A similar pattern was observed with T4 and *Pfu* DNA polymerases.

These data reveal that two of the three general processes that determine replication fidelity, DNA polymerase selectivity and exonucleolytic proofreading efficiency, are less efficient during replication of repetitive sequences and diminish as a function of an increasing number of repeat units (Fig. 44-6). This implies that the fidelity of replication of such sequences will depend heavily on postreplication repair processes to reduce rates of addition and deletion mutations *in vivo*. It further implies that inactivation of such repair will elevate mutation rates in repetitive sequences to a greater extent than in nonrepetitive sequences. Several recent observations *in vivo* are consistent with these ideas. The instability of microsatellite sequences observed in certain tumor cells strongly correlates with mutations in known mismatch repair genes and with loss of insertion/deletion mismatch repair activity in extracts of such cells (for a review see [3–5]). Mutations that inactivate the proofreading exonuclease of yeast DNA polymerases δ and ε only slightly decrease the stability of GT repeats *in vivo,* while mutations that inactivate postreplication mismatch repair decrease dinucleotide repeat stability by several hundred-fold compared to the wild-type yeast strain [51]. Yeast mutation rates in homopolymeric runs of A · T base pairs are much more strongly elevated as a consequence of a mutation in the mismatch repair gene *MSH2* than as a result of a mutation in the intrinsic exonuclease of pol ε [52]. In that study, the effect on mutation rate increases with increasing run length, demonstrating, for example, that *MSH2*-dependent mismatch repair contributes 10,000-fold to the stability of a run of 14 A · T base pairs. This is a much larger

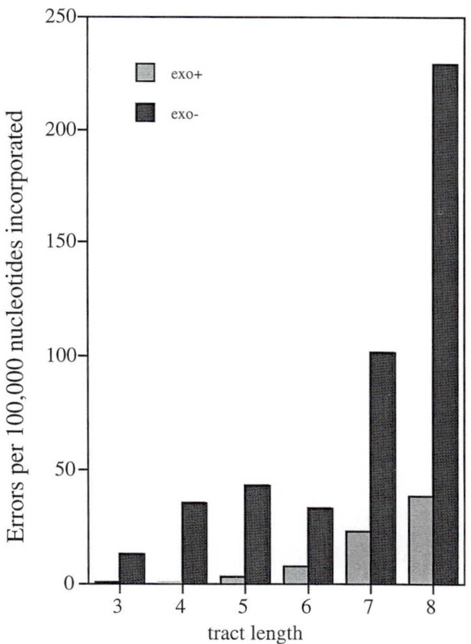

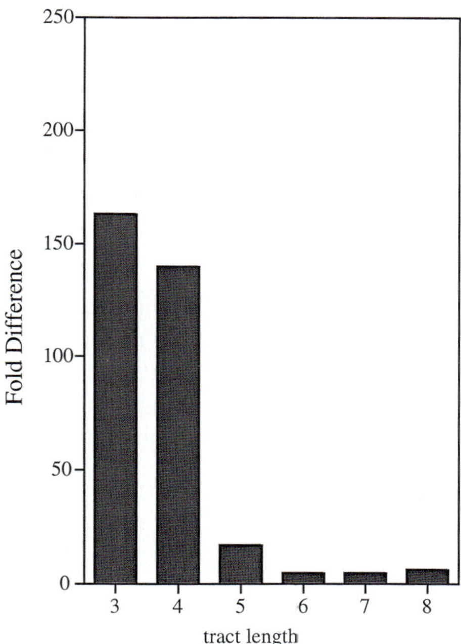

FIGURE 44-6 T7 DNA polymerase error rates for one nucleotide deletions as a function of T · A base pair run length. See text for description. Left panel, single-base deletion error rates for wild-type and exonuclease-deficient T7 DNA polymerases; right panel, ratio of mutant to wild-type polymerase error rates.

effect than observed for mutation rates in nonrepetitive sequences. A similar trend has been observed [53] for replication of mono- and dinucleotide repeat sequences in strains of *E. coli* deficient in mismatch repair (*mutS*) or in mismatch repair plus proofreading by the ε subunit of the replicative DNA polymerase III (*mutSdnaQ*).

IX. THE FIDELITY OF THE HUMAN REPLICATION COMPLEX

DNA replication requires the concerted action of many proteins in addition to DNA polymerases. Thus, understanding the origins of frameshift errors during replication requires studies of replication fidelity with this multiprotein replication machine. This has been possible using systems that replicate double-stranded DNA *in vitro*. One such system for studying human genomic replication uses the SV40 origin of replication. Circular, double-stranded DNA substrates containing the SV40 origin can be fully replicated by the proteins present in primate cells, where only the addition of SV40 T antigen is needed to initiate replication at the origin. For studies *in vitro*, replication can be performed in extracts of cells grown in culture or by reconstitution of a replication reaction with purified proteins.

SV40 origin-containing DNA is replicated in human HeLa and simian CV-1 cell extracts *in vitro* with high fidelity. Error rates vary from $\leq 6.2 \times 10^{-6}$ to $\leq 0.1 \times 10^{-6}$, depending on the substitution or frameshift error being monitored (reviewed in [10]). These are "less than or equal to" values because no replication errors are detected with undamaged DNA when replication reactions are performed with equimolar dNTP concentrations. Investigating this high replication fidelity has required manipulation of reaction components in order to obtain replication errors. Approaches that have been used include adding an excess of one dNTP to force specific types of errors, adding a dNMP to inhibit proofreading, or damaging the DNA substrate to obtain mutagenic translesion synthesis. In several cases, measurements have been made with two substrates, with the origin asymmetrically placed on opposite sides of the reporter gene. In this situation, one can estimate fidelity for the same error in the same sequence context, but replicated either by the leading or lagging strand replication machinery.

Several questions regarding frameshift fidelity have been addressed in this model replication system. To determine if proofreading of frameshift intermediates occurs during replication, reactions were performed using conditions to limit proofreading activity. The results [54] suggest that proofreading contributes substantially to replication fidelity for one-nucleotide deletion errors. Two *in vitro* studies have addressed the fidelity of leading versus lagging strand replication. One [54] found less than a twofold difference in error rate for deletion of 1 T · A base pair from a run of 5 T · A base pairs, but a significant difference in the pattern of 1-base deletions over a 41-base-pair target. Results of a second study [55] suggested a difference in the fidelity of replication of a run of 5 C · G base pairs. Thus, the frameshift error rate and error specificity may be different for the leading and lagging strand replication machinery, at least for some types of frameshift errors. Replication of the two strands is highly asymmetric, providing unequal opportunities to make mistakes. Replication of the two strands may be performed by different DNA polymerases or perhaps the same polymerase but with a different complement of accessory proteins. This could yield differences in misalignment rates and/or ability to extend rather than proofread misaligned template-primers. Differential proofreading provides one mechanism to explain differences in leading- and lagging-strand replication fidelity. Since the assignment of the leading- and lagging-strand DNA polymerases during eukaryotic replication is not yet definitive, proofreading on the two strands could be carried out by any of several exonucleases. Replication on the leading strand is highly processive as compared to discontinuous synthesis of Okazaki fragments on the lagging strand, which involves more than one DNA polymerase and/or one or more switches between enzymes as well as the synthesis and eventual replacement of RNA primers. The possible influence of mismatch repair on replication fidelity in extracts should also be considered (for further discussion see [3, 10]).

A large number of genes in eukaryotic cells either control or catalyze the repair of a wide variety of physical and chemical insults, some resulting from normal cellular processes (e.g., deamination, depurination, oxidative stress, alkylation) and others from exposure to the external environment. When these repair systems fail, lesions may persist in DNA or in dNTP precursor pools, leading to lesion-induced replication infidelity. As an example, the ability of the replication complex in a HeLa cell extract to bypass site-specific *N*-2-acetylaminofluorene (AAF) adducts has been examined, with a focus on lesion-dependent frameshift errors [56, 57]. The major effect of the AAF adduct was inhibition of replication, with termination occurring immediately before incorporation opposite the adduct. Among the replicated products was a higher proportion of those representing replication of the undamaged strand, consistent with the possibility that the first fork to encounter the lesion became uncoupled, i.e., replication of the damaged strand ceased while replication of the undamaged strand continued. Product analysis suggested that translesion bypass had occurred and that two-base deletion errors had been generated in a dinucleotide repeat sequence, most likely by a mechanism involving correct incorporation opposite the lesion followed by slippage. The results clearly demonstrated that the mutagenic potential of the AAF lesion depended both on the local DNA sequence environment and on whether it was replicated by the leading or lagging strand replication apparatus. In a second study of this type [58], replication of DNA containing psoralen-induced monoadducts was also found to generate deletions, in this case deletions of a single A · T base pair, most frequently in homopolymeric runs. The results suggested that the lagging strand replication apparatus was less accurate than the leading strand replication apparatus for these adduct-dependent deletions. These studies are consistent with the possibility that some of the instability of repetitive sequences in human diseases could reflect mutagenic translesion DNA replication.

X. A MODEL FOR EXPANSION OF TRIPLET REPEATS

Among the types of errors recovered from the model DNA synthesis reactions described above, large deletions are common while large duplications (expansions) are not observed. This bias can be reconciled by the fact that a larger number of base pairs must be disrupted to form an expansion intermediate than is required for a large deletion intermediate, and suggests that large expansions of triplet repeat sequences may not result from slippage of the 3′-OH primer terminus at the growing fork. How then can a replication-based model account for expansions? Some replication models have invoked realignment of DNA in an Okazaki fragment during synthesis of the lagging strand, mediated by repeat sequences that form hairpins or other non-B-DNA secondary structures (for a review see Chapter 33). To yield expansions, such structures should exist in the newly synthesized strand, as opposed to deletions predicted from structures in the template strand. This suggests that the nascent strand should be single stranded at some point, and that the size of the region dissociated from the template may be at least the size of the expanded region. This requires disruption of a large number of base pairs during replication. How might this occur?

An interesting possibility was suggested by the observation that duplications arise in a yeast strain containing a null mutation in the *RAD27* gene [59]. This gene, and its mammalian homologue *FEN1,* encodes an endonuclease that removes a 5′ flap generated by displacement synthesis when DNA polymerase encounters the 5′ end of a downstream Okazaki fragment (Fig. 44-7a, [60, 61]). The duplication rate in this strain is increased over that in a wild-type strain by at least 1000-fold [59], suggesting that single-stranded flaps are generated frequently dur-

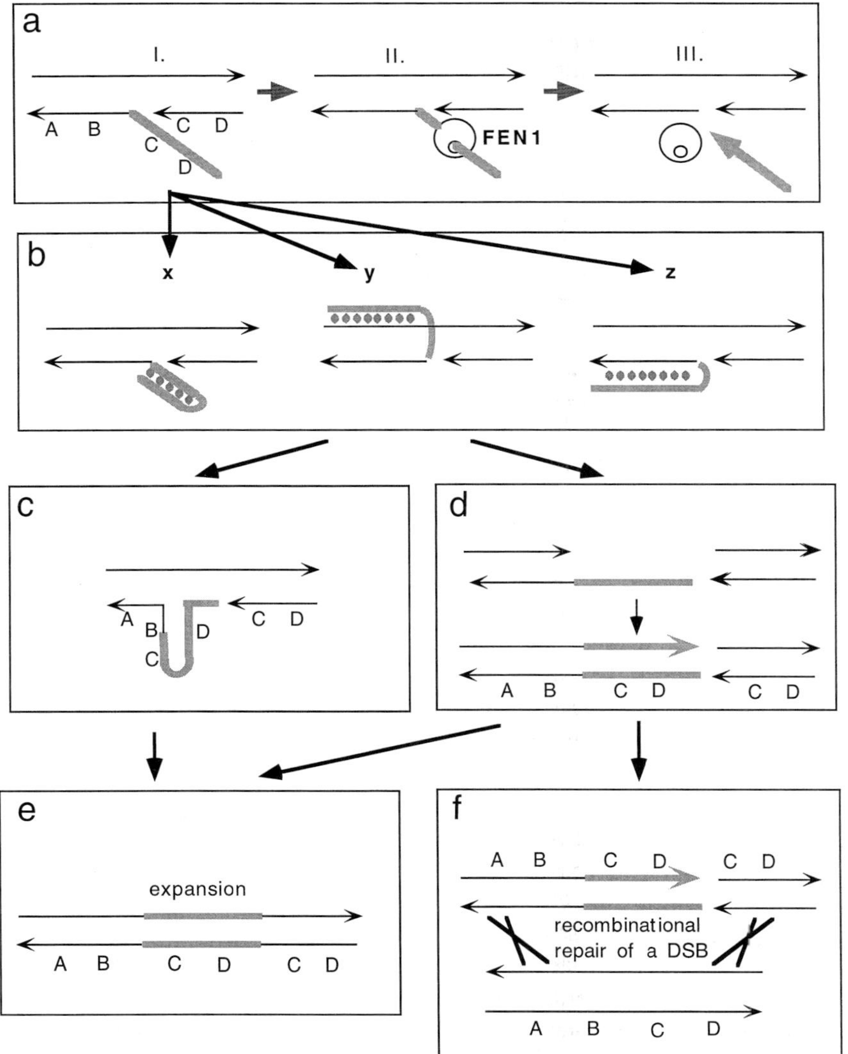

FIGURE 44-7 Expansion via a FEN1-resistant flap in an Okazaki fragment. Reproduced from [63], with permission. (a) Removal of the displaced 5′-flap of an Okazaki fragment by the structure specific endonuclease FEN1. Arrowheads on DNA strands correspond to 3′ ends. A, B, C, and D define different DNA segments. (I) Synthesis of the upstream Okazaki fragment results in the displacement of the 5′-end of the downstream Okazaki fragment to generate a flap (thick gray line). Note that DNA segments C and D were copied twice by DNA polymerase. (II) The FEN1 endonuclease is loaded at the 5′-end of the flap and slides down to the base of the flap. (III) The flap is removed by the FEN1 endonuclease. (b) A single-stranded flap could give rise to a non-B DNA structure depending on its sequence. Noncanonical base pairing is indicated by gray dots. (x) Hairpins formed by a single-strand flap. (y) An intramolecular triplex formed by parallel Hoogsteen base pairing between a G-rich strand of a duplex and a C-rich single flap. (z) An intramolecular triplex formed by antiparallel Hoogsteen base pairing between a G-rich strand of a duplex and a G-rich flap. (c) Realignment of a displaced flap on the template. (d) Conversion of a stalled replication fork into a double-strand break [59]. Recessed end(s) at a break could be converted into blunt end(s) by DNA synthesis. (e) Intramolecular repair via end joining of blunt or partially recessed ends, leading to duplications (expansions). End joining is a major mode of DSB repair in mammalian cells, but is much less frequent in yeast and in *E. coli*. (f) Intermolecular repair via recombination. Precise recombination repair with a sister DNA molecule (i.e., the sister chromatid synthesized on the leading template) will not lead to genetic change. Recombination involving a repeat in the same chromosome, an allele in a homologous chromosome, or a repeat elsewhere in the genome (ectopic recombination) could lead to deletions, duplications, loss of heterozygosity, or interchromosomal rearrangements.

ing normal Okazaki fragment processing. The *rad27* mutant also had an increased rate of recombination and could only survive if able to repair breaks in double-stranded DNA.

In a discussion of mutator specificity and disease [62], it was suggested that the duplication mechanism proposed by Tishkoff *et al.* [59] might account for repeat sequence expansion. Studies of FEN1 substrate specificity *in vitro* indicate that the entire flap must be single-stranded for FEN1 to act as an endonuclease. For example, if the flap is partially double-stranded or has protein bound, flap removal is blocked. This led to a model [63] wherein sequence-specific repeat expansions may result from formation of FEN1-resistant secondary structures when expansion-prone sequences are located in the flap (Fig. 44-7b). DNA sequences observed to expand are known to form structures *in vitro* that are potentially resistant to FEN1 (discussed in the legend to Fig. 44-7b and reviewed in Chapter 33).

Unexcised flaps could be processed into expansions in several ways. Realignment could result in a loop (Fig. 44-7c) similar to the loop proposed in other realignment models (Chapter 33). Expansion could occur via a double-strand break followed by mutagenic repair (Figs. 44-7d, 44-7f). Based on the high rates of recombination in a yeast rad27 (FEN1) mutant and the requirement of a functional double-strand break repair pathway for survival of this mutant, Tishkoff *et al.* [59] proposed that an unexcised flap causes replication to be slowed or stalled, leading to a double-strand break. If the break occurs inside a repeated region and is processed to a gap that extends beyond the repeats, recombination repair would not produce mutations (Fig. 44-7f). If the break is internal to the repeats (not shown), homologous recombination could lead to deletions or expansions. In contrast, repair by end-joining (Fig. 44-7e) would lead only to expansions, which could explain the observed expansion bias. Possibly, the lower rate of end-joining versus homologous recombination in yeast and bacteria compared to mammalian cells could account for the relatively low rates of triplet repeat expansion in these model microorganisms (see Refs. in [63]). Differences in the rates of end-joining versus homologous recombination between tissues, stages of development, or sexes might explain germline and sex specificity for some expansions (e.g., see [64]). Several factors might affect expansion rates of FEN1-resistant sequences. There may be FEN1-independent pathways of flap removal, since there are eukaryotic exonucleases with specificity similar to FEN-1 [61]. Some of these participate in nucleotide excision repair and do not require a free 5′-end in single stranded DNA. Since they could account for infrequent processing of a FEN1-resistant flap, mutations in such genes could lead to increased expansion rates. It is also possible that bound proteins could stabilize flap structures, thereby reducing even further the likelihood of FEN1 action at a flap.

References

1. Loeb, L. A. (1991). Mutator phenotype may be required for multistage carcinogenesis. *Cancer Res.* **51,** 3075–3079.
2. Schaaper, R. M. (1993). Base selection, proofreading, and mismatch repair during DNA replication in *Escherichia coli. J. Biol. Chem.* **268,** 23762–23765.
3. Umar, A., and Kunkel, T. A. (1996). DNA replication fidelity, mismatch repair and genome instability in cancer cells. *Eur. J. Biochem.* **238,** 297–307.
4. Modrich, P., and Lahue, R. (1996). Mismatch repair in replication fidelity, genetic recombination and cancer biology. *Annu. Rev. Biochem.* **65,** 101–133.
5. Kolodner, R. (1996). Biochemistry and genetics of eukaryotic mismatch repair. *Genes Dev.* ***10***, 1433–1442.
6. Carroll, S. S., and Benkovic, S. J. (1990). Mechanistic aspects of DNA polymerases: *Escherichia coli* DNA polymerase I (Klenow fragment) as a paradigm. *Chem. Rev.* **90,** 1291–1307.
7. Echols, H., and Goodman, M. F. (1991). Fidelity mechanisms in DNA replication. *Annu. Rev. Biochem.* **60,** 477–511.
8. Kunkel, T. A. (1992). DNA replication fidelity. *J. Biol. Chem.* **267,** 18251–18254.
9. Johnson, K. A. (1993). Conformational coupling in DNA polymerase fidelity. *Annu. Rev. Biochem.* **62,** 685–713.
10. Roberts, J. D., and Kunkel, T. A. (1996). Fidelity of DNA replication. *In* "DNA Replication in Eukaryotic Cells" (M. L. DePamphilis, Ed.), pp. 217–247. Cold Spring Harbor Laboratory Press, Cold Spring Harbor, NY.
11. Brutlag, D., and Kornberg, A. (1972). Enzymatic synthesis of deoxyribonucleic acid. XXXVI. A proofreading function for the 3′ → 5′ exonuclease activity in deoxyribonucleic acid polymerases. *J. Biol. Chem.* **247,** 241–248.
12. Joyce, C. M., and Steitz, T. A. (1994). Function and structure relationships in DNA polymerases. *Annu. Rev. Biochem.* **63,** 777–822.
13. Streisinger, G., Okada, Y., Emrich, J., Newton, J., Tsugita, A., Terzaghi, E., and Inouye, M. (1966). Frameshift mutations and the genetic code. *Cold Spring Harbor Symp. Quant. Biol.* **31,** 77–84.
14. Ripley, L. S. (1990). Frameshift mutation: determinants of specificity. *Annu. Rev. Genet.* **24,** 189–213.
15. Kunkel, T. A. (1990). Misalignment-mediated DNA synthesis errors. *Biochemistry* **29,** 8003–8011.
16. Kunkel, T. A., Patel, S. S., and Johnson, K. A. (1994). Error-prone replication of repeated DNA sequences by T7 DNA polymerase in the absence of its processivity subunit. *Proc. Natl. Acad. Sci. USA* **91,** 6830–6834.
17. Bebenek, K., Beard, W. A., Casas-Finet, J. R., Kim, H.-R., Darden, T. A., Wilson, S. H., and Kunkel, T. A. (1995). Reduced frameshift fidelity and processivity of HIV-1 reverse transcriptase mutants containing alanine substitutions in helix H of the thumb subdomain. *J. Biol. Chem.* **270,** 19516–19523.
18. Kroutil, L. C., Register, K., Bebenek, K., and Kunkel, T. A. (1996). Exonucleolytic proofreading during replication of repetitive DNA. *Biochemistry* **35,** 1046–1053.
19. Minnick, D. T., Astatke, M., Joyce, C., and Kunkel, T. A. (1996). A thumb subdomain mutant of the large fragment of *Escherichia coli* DNA polymerase I with reduced DNA binding affinity, processivity, and frameshift fidelity. *J. Biol. Chem.* **271,** 24954–24961.

20. Kunkel, T. A. (1986). Frameshift mutagenesis by eukaryotic DNA polymerases *in vitro*. *J. Biol. Chem.* **261,** 13581–13587.
21. Nancarrow, J. K., Holman, K., Mangelsdorf, M., Hori, T., Denton, M., Sutherland, G. R., and Richards, R. I. (1995). Molecular basis of p(CCG)n repeat instability at the FRA16A fragile site locus. *Hum. Mol. Genet.* **4,** 367–372.
22. Eichler, E. E., Holden, J. J. A., Popovich, B. W., Reiss, A. L., Snow, K., Thibodeau, S. N., Richards, C. S., Ward, P. A., and Nelson, D. L. (1994). Length of uninterrupted CGG repeats determine instability in FMR1 gene. *Nature Genet.* **8,** 88–94.
23. Kunst, C. B., and Warren, S. T. (1994). Cryptic and polar variation of the fragile X repeat could result in predisposing normal alleles. *Cell* **77,** 853–861.
24. Snow, K., Tester, D. J., Kruckeberg, K. E., Schaid, D. J., and Thibodeau, S. N. (1994). Sequence analysis of the fragile X trinucleotide repeat: implications for the origin of the fragile X mutation. *Hum. Mol. Genet.* **3,** 1543–1551.
25. Bebenek, K., Roberts, J. D., and Kunkel, T. A. (1992). The effects of dNTP pool imbalances on frameshift fidelity during DNA replication. *J. Biol. Chem.* **267,** 3589–3596.
26. Kunkel, T. A., and Soni, A. (1988). Exonuclease proofreading enhances the fidelity of DNA synthesis by chick embryo DNA polymerase-γ. *J. Biol. Chem.* **263,** 4450–4459.
27. Wang, C.-I., and Taylor, J.-S. (1992). *In vitro* evidence that UV-induced frameshift and substitution mutations at T tracts are the result of misalignment-mediated replication past a specific thymine dimer. *Biochemistry* **31,** 3671–3681.
28. Shibutani, S., and Grollman, A. P. (1993). On the mechanism of frameshift (deletion) mutagenesis *in vitro*. *J. Biol. Chem.* **268,** 11703–11710.
29. Lindsley, J. E., and Fuchs, R. P. (1994). Use of single-turnover kinetics to study bulky adduct bypass by T7 DNA polymerase. *Biochemistry* **33,** 764–772.
30. Napolitano, R. L., Lambert, I. B., and Fuchs, R. P. (1994). DNA sequence determinants of carcinogen-induced frameshift mutagenesis. *Biochemistry* **33,** 1311–1315.
31. Bebenek, K., and Kunkel, T. A. (1995). Analyzing the fidelity of DNA polymerases. *Methods Enzymol.* **262,** 217–232.
32. Kunkel, T. A., Hamatake, R. K., Motto-Fox, J., Fitzgerald, M. P., and Sugino, A. (1989). Fidelity of DNA polymerase I and the DNA polymerase I-DNA primase complex from *Saccharomyces cerevisiae*. *Mol. Cell. Biol.* **9,** 4447–4458.
33. Bambara, R. A., and Jessee, C. B. (1991). Properties of DNA polymerases δ and ε, and their roles in eukaryotic DNA replication. *Biochim. Biophys. Acta* **1088,** 11–24.
34. Bebenek, K., Abbotts, J., Wilson, S. H., and Kunkel, T. A. (1993). Error-prone polymerization by HIV-1 reverse transcriptase. Contribution of template-primer misalignment, miscoding, and termination probability to mutational hot spots. *J. Biol. Chem.* **268,** 10324–10334.
35. Streisinger, G., and Owen, J. E. (1985). Mechanism of spontaneous and induced frameshift mutation in bacteriophage T4. *Genetics* **109,** 633–659.
36. Kunkel, T. A., and Soni, A. (1988). Mutagenesis by transient misalignment. *J. Biol. Chem.* **263,** 14784–14789.
37. Papanicolaou, C., and Ripley, L. S. (1989). Polymerase-specific differences in the DNA intermediates of frameshift mutagenesis. *In vitro* synthesis errors of *Escherichia coli* DNA polymerase I and its large fragment derivative. *J. Mol. Biol.* **207,** 335–353.
38. Schaaper, R. M., Koffel-Schwartz, N., and Fuchs, R. P. P. (1990). N-Acetoxy-N-acetyl-2-aminofluorene-induced mutagenesis in the lacI gene of *Escherichia coli*. *Carcinogenesis* **11,** 1087–1095.
39. Kunkel, T. A. (1985). The mutational specificity of DNA polymerase-β during *in vitro* DNA synthesis: production of frameshift, base substitution, and deletion mutations. *J. Biol. Chem.* **260,** 5787–5796.
40. Bebenek, K., Joyce, C. M., Fitzgerald, M. P., and Kunkel, T. A. (1990). The fidelity of DNA synthesis catalyzed by derivatives of *Escherichia coli* DNA polymerase I. *J. Biol. Chem.* **265,** 13878–13887.
41. Kunkel, T. A. (1985). The mutational specificity of DNA polymerases-α and -γ during *in vitro* DNA synthesis. *J. Biol. Chem.* **260,** 12866–12874.
42. Bebenek, K., Abbotts, J., Roberts, J. D., Wilson, S. H., and Kunkel, T. A. (1989). Specificity and mechanism of error-prone replication by human immunodeficiency virus-1 reverse transcriptase. *J. Biol. Chem.* **264,** 16948–16956.
43. Wang, J., Sattar, A. K. M. A., Wang, C. C., Karam, J. D., Konigsberg, W. H., and Steitz, T. A. (1997). Crystal structure of a pol a family replication DNA polymerase from bacteriophage RB69. *Cell* **89,** 1087–1099.
44. Kohlstadt, L. A., Wang, J., Friedman, J. M., Rice, P. A., and Steitz, T. A. (1992). Crystal structure at 3.5 Å resolution of HIV-1 reverse transcriptase complexed with an inhibitor. *Science* **256,** 1783–1790.
45. Jacobo-Molina, A., Ding, J., Nanni, R. G., Clark, A. D., Jr., Lu, X., Tantillo, C., Williams, R. L., Kramer, G., Ferris, A. L., Clark, P., Hizi, A,. Hughes, S. H., and Arnold, E. (1993). Crystal structure of human immunodeficiency virus type I reverse transcriptase complexed with double-stranded DNA at 3.0 Å resolution shows bent DNA. *Proc. Natl. Acad. Sci. USA* **90,** 6320–6324.
46. Bebenek, K., Beard, W. A., Darden, T. A., Li, L., Prasad, R., Luxon, B. A., Gorenstein, D. G., Wilson, S. H., and Kunkel, T. A. (1997). A minor groove binding track in reverse transcriptase. *Nature Struct. Biol.* **4,** 194–197.
47. Beard, W. A., Stahl, S. J., Kim, H.-R., Bebenek, K., Kumar, A., Strub, M.-P., Becerra, S. P., Kunkel, T. A., and Wilson, S. H. (1994) Structure/function studies of HIV-1 reverse transcriptase: alanine scanning mutagenesis of an α-helix in the thumb subdomain. *J. Biol. Chem.* **269,** 28091–28097.
48. Beese, L. S., Derbyshire, V., and Steitz, T. A. (1993). Structure of DNA polymerase I Klenow fragment bound to duplex DNA. *Science* **260,** 352–355.
49. Thomas., D. C., Roberts, J. D., Sabatino, R. D., Myers, T. W., Downey, K. M., So, A. G., Bambara, R. A., and Kunkel, T. A. (1991). Fidelity of mammalian DNA replication and replicative DNA polymerases. *Biochemistry* **30,** 11751–11759.
50. Cai, H., Yu, H., McEntee, K., Kunkel, T. A., and Goodman, M. F. (1995). Purification and properties of wild type and exonuclease-deficient DNA polymerase II from *Escherichia coli*. *J. Biol. Chem.* **270,** 15327–15335.
51. Strand, M., Prolla, T. A., Liskay, R. M., and Petes, T. D. (1993). Destabilization of tracts of simple repetitive DNA in yeast by mutations affecting DNA mismatch repair. *Nature* **365,** 274–276.
52. Tran, H. T., Keen, J. D., Kricker, M., Resnick, M. A., and Gordenin, D. A. (1997). Hypermutability of homonucleotide runs in mismatch repair and DNA polymerase proofreading yeast mutants. *Mol. Cell. Biol.* **17,** 2859–2865.
53. Strauss, B. S., Sagher, D., and Acharya, S. (1997). Role of proofreading and mismatch repair in maintaining the stability of nucleotide repeats in DNA. *Nucleic Acids Res.* **25,** 806–813.
54. Roberts, J. D., Nguyen, D., and Kunkel, T. A. (1993). Frameshift fidelity during replication of double-stranded DNA in HeLa cell extracts. *Biochemistry* **32,** 4083–4089.
55. Izuta, S., Roberts, J. D., and Kunkel, T. A. (1995). Replication error rates for T · dGTP, G · dGTP and A · dGTP mispairs: evidence for differential proofreading by leading and lagging strand DNA replication complexes. *J. Biol. Chem.* **270,** 2595–2600.

56. Thomas, D. C., Veaute, X., Kunkel, T. A., and Fuchs, R. P. P. (1994). Mutagenic replication in human cell extracts of DNA containing site-specific N-2-acetylaminofluorene adducts. *Proc. Natl. Acad. Sci. USA* **91,** 7752–7756.
57. Thomas, D. C., Veaute, X., Fuchs, R. P. P., and Kunkel, T. A. (1995). Frequency and fidelity of translesion synthesis of site-specific N-2-acetylaminofluorene adducts during DNA replication in a human cell extract. *J. Biol. Chem.* **270,** 21226–21233.
58. Thomas, D. C., Svoboda, D. L., Vos, J.-M. H., and Kunkel, T. A. (1996). Mutagenic translesion bypass of psoralen monoadducts during DNA replication in a HeLa cell extract. *Mol. Cell. Biol.* **16,** 2537–2544.
59. Tishkoff, D. X., Filosi, N., Gaida, G. M., and Kolodner, R. D. (1997). A novel mutation avoidance mechanism dependent on *S. cerevisiae* RAD27 is distinct from DNA mismatch repair. *Cell* **88,** 253–263.
60. Bambara, R. A., Murante, R. S., and Henricksen, L. A. (1997). Enzymes and reactions at the eukaryotic DNA replication fork. *J. Biol. Chem.* **272,** 4647–4650.
61. Lieber, M. R. (1997). The FEN-1 family of structure-specific nucleases in eukaryotic DNA replication, recombination and repair. *Bioessays* **19,** 233–240.
62. Kunkel, T. A., Resnick, M., and Gordenin, D. (1997). Mutator specificity and disease: jumping over the FENce. *Cell* **88,** 155–158.
63. Gordenin, D., Kunkel, T. A., and Resnick, M. (1997). Repeat expansion—all in a flap? *Nature Genet.* **16,** 116–118.
64. Ashley, C. T., Jr., and Warren, S. T. (1995). Trinucleotide repeat expansion and human disease. *Annu. Rev. Genet.* **29,** 703–728.
65. Wong, I., Patel, S. S., and Johnson, K. A. (1991). An induced-fit kinetic mechanism for DNA replication fidelity: direct measurement by single turnover kinetics. *Biochemistry* **30,** 526–537.

CHAPTER 45

In Vitro DNA Synthesis of Triplet Repeat Sequences

KEIICHI OHSHIMA[1] and Robert D. Wells
Institute of Biosciences and Technology, Texas A&M University, Department of Biochemistry and Biophysics, Texas Medical Center, Houston, Texas 77030

With Appendix by

HIROSHI HIASA AND KENNETH J. MARIANS
Molecular Biology Program, Memorial Sloan-Kettering Cancer Center, New York, New York 10021

Appendix: Replication Fork Progression through CTG · CAG Triplet Repeats *in Vitro*

---

[1]Present address: Research Center Louis-Charles Simard, CHUM, Pavillon Notre-Dame, 1560 Sherbrooke St. East, Montreal, Quebec, Canada H2L 4M1. Fax: (514) 896-4762.

I. SYNTHESIS OF TRIPLET REPEAT SEQUENCES *IN VITRO*

The non-Mendelian expansion of triplet repeat sequences (TRS) including CTG · CAG, CGG · CCG, and GAA · TTC is responsible for the etiology for several neurological diseases including Huntington's disease, the fragile X syndrome, and Friedreich's ataxia [1–6]. The molecular mechanisms for the expansion of the TRS are unclear but are likely due, at least in part, to replication-based processes [7–12]. It has been suggested that the slippage of the complementary strands of the TRS by units of three during DNA replication may be responsible for expansions as well as deletions [3, 4, 6, 9, 13, 14].

Abnormal DNA replication may be due to DNA sequence features as well as protein–DNA and protein–protein interactions. DNA secondary structures give rise to unusual replication including the stalling of polymerases which may result in DNA misalignment and lead to mutations (deletions and insertions). Also, template sequence-related stalling rather than a dependence on secondary structures (hairpins) was found in homopolymeric nucleotide runs [15–18]. The stalling of DNA polymerases results in termination of DNA synthesis to reduce the abundance of normal replicated products. In fact, defective helicase activities were suggested to be the cause of Bloom's and Warner's syndromes [19, 20].

Herein, we describe the *in vitro* and *in vivo* replication and transcription properties of TRS that give rise to stalling (pausing), hairpin formation, and expansions. Second, we review the DNA replication behaviors *in vitro* which are caused principally by other specific DNA sequences that form unusual DNA structures (cruciforms (hairpins), triplexes, and tetraplexes) which give rise to stall sites. Third, we present an overview of the biological literature on slippage-mediated misalignment, DNA polymerase stalling, and strand switching.

A. DNA Polymerase Stalling

The stalling by DNA polymerases in long TRS was discovered by accident [21] during the primer extension analyses for determining the location of chemical probe reactivities [22]. These studies were underway to evaluate the structural properties of long tracts of CTG · CAG and CGG · CCG. Serendipitously, we discovered [21] that DNA polymerases *in vitro* aberrantly stall at the junctions of the vectors with the TRS and within the repeat tracts. Kang *et al.* [21] investigated the DNA syntheses of CTG · CAG triplets ranging from 17 to 180 and CGG · CCG repeats from 9 to 160 repeats in length *in vitro*. Primer extensions using the Klenow fragment of *Escherichia coli* DNA polymerase I, the modified T7 DNA polymerase (Sequenase), or the human DNA polymerase beta stalled strongly at polymerase-specific loci in the CTG · CAG repeats. In general, the strongest stall sites were located near the intersection of the repeat tract with the vector and approximately 90–135 bp within the tract. The stallings were abolished by heating at 70°C; hence, we concluded that an unusual DNA structure stabilized by hydrogen bonding was responsible for the stalling. As the length of the triplet repeats in duplex DNA, but not in single-stranded DNA, was increased, the magnitude of stalls increased. The location of the stall sites was determined by the distance between the site of primer hybridization and the beginning of the triplet repeats. These measurements gave an early clue to the possible formation of hairpins [9].

CGG · CCG triplet repeats also showed similar, but not identical, patterns of stalling. Specifically, the stall sites could not be abolished by temperature, even as high as 90°C. Thus, we concluded that the structure that blocked the polymerase movement must be more thermally stable for this sequence than the conformation in CTG · CAG. These results indicate that appropriate lengths of the triplets exist in an unusual conformation, the flexible and writhed structure [22, 23], that blocks DNA polymerase progression. The resultant idling polymerase may catalyze slippages to give expanded sequences and hence provide the molecular basis for the non-Mendelian expansion of these repeat sequences. As documented elsewhere in this review, examples are numerous of unusual structures (such as triplexes) which inhibit DNA synthesis.

Further investigations were conducted on seven additional TRS [7] and on TTA · TAA repeats [8]. No stalling was observed within TTA · TAA [8] which is consistent with the fact that it does not adopt a flexible and writhed conformation and with the thermolabile properties of this insert.

In general, stalling was observed on TCG · CGA [7], the sequence isomer of CTG · CAG, which has the same property for the stalling of DNA polymerases as CTG · CAG and CGG · CCG. Thus, out of all 10 triplet repeats, these 3 repeats may form the flexible and writhed conformation [22] which is apparently responsible for the stalling. Stalling was also observed for AGG · CCT and AAG · CTT, presumably due to the adoption of intramolecular triplexes with G*G · C and A*A · T base triads since the primer extension analyses were performed at pH 7.5 in the presence of magnesium ion and these results are diagnostic for triplexes. Biophysical studies demonstrated the triplex-forming capability of these two inserts [7]. Furthermore, no stallings at all were found for the other four triplet repeats which are the following: GTA · TAC, GAT · ATC, GTT · AAC, and CAC · GTG. Hence, the conformational properties of the DNA sequences are intimately linked to their *in vitro* DNA replication behaviors.

B. Hairpin Formation

Additional studies on the products formed by the stalled DNA polymerases revealed the formation of small hairpin structures [9]. As described above, strong DNA synthesis stall sites were found at specific locations within the long tracts (>~70 repeats) of CTG · CAG and CGG · CCG as well as TCG · CGA by primer extensions *in vitro* using DNA polymerases (the Klenow fragment of *E. coli* DNA polymerase I, the modified T7 DNA polymerase (Sequenase), and human DNA polymerase beta). We isolated and analyzed the products of stalled synthesis found at approximately 30–45 triplets from the beginning of the TRS [9]. DNA sequence analyses revealed that the stalled products contained short tracts of homogeneous TRS (6–12 repeats) in the middle of the sequence corresponding to the flanking region of the primer–template sequence. The sequence at the 3′ side terminated at the end of the primer, indicating that the primer molecule had served as a template. In addition, chemical probe and polyacrylamide gel electrophoretic analyses revealed that stalled products existed in hairpin structures. Ohshima and Wells postulated that these products of the DNA polymerases were caused by the existence of an unusual DNA conformation, flexible and writhed DNA [22], within the TRS, during the *in vitro* DNA synthesis, enhancing the slippages and the hairpin formations in the TRS due to primer realignment. The consequence of these steps is DNA synthesis to the end of the primer and termination. Obviously, primer realignment, including hairpin formation, may play an important intermediate role in the replication of TRS *in vivo* to elicit genetic expansions.

A model for the stalling of DNA polymerases during *in vitro* DNA synthesis is shown in Fig. 45-1. This model proposes that the strong stallings were caused by termination of DNA synthesis due to the hairpin formation of the TRS followed by primer realignment. First, the DNA polymerase might encounter a flexible and writhed TRS [22, 23], which impedes the progression of DNA synthesis. Stallings in the proximal regions of the TRS (~12 triplet units) could be intermediate stalled products during the DNA synthesis [9]; however, since they have not been sequenced, their identity is uncertain. Second, the idling of the impeded DNA polymerase might enable DNA slippage in the nascent strand. Third, the hairpin which was formed might allow for primer realignment, creating a functional primer 3′-end. Fourth, as the template is now switched, the DNA chain would be elongated using the nascent strand as a template. Finally, DNA synthesis was terminated at the end of the primer-nascent strand molecule. Hence, the stalling of DNA polymerase in the proximal region (~12 triplet units) caused by an unusual TRS conformation [22, 23] is critical for template switching.

The discovery of hairpins formed *in vitro* is important for confirming, at least partially, our *in vivo* model of genetic instabilities [10–12, 24, 25] (described in detail in Chapter 33). Other investigations with relatively short oligonucleotides postulated models for hairpin structures for certain TRS from NMR, thermodynamic, and other types of measurements (see Chapters 40 and 41).

C. Relationship to Stalling Results in Other Systems

The Appendix to this chapter describes investigations on CTG · CAG triplet repeat sequences of various lengths (55–255 repeats) in the *ori*C-directed system. Supercoiled plasmids containing the intact triplet repeats were purified and used as templates in the *in vitro* *ori*C DNA replication system using a plasmid vector pHH12. The triplet repeats were inserted in the orientation such that the CTGs were in the leading strand template. After analysis on native and alkaline agarose gel electrophoresis, the triplet repeats had no effect on monomer accumulation via either the Topo III- or Topo IV-catalyzed decatenation pathway during replication, nor was the size distribution of either the leading or lagging strand products affected. These results suggested that the presence of the TRS elicited neither strong stalling of the replication forks nor premature termination of DNA replication in any gross sense during *ori*C DNA replication. These workers explained the difference between their data in the *ori*C system and the Wells lab results with purified enzymes by contrasting a DNA polymerase elongating on a single-stranded template vs a replication fork moving through duplex DNA [see Appendix].

Alternatively, Mirkin and co-workers [26] have conducted investigations on $(CTG \cdot CAG)_{70}$ as well as a family of different lengths of CGG · CCG TRS *in vivo* in *E. coli* using the Brewer–Fangman methodologies for determining synthesis arrest. These workers discovered a stall site at the TRS for CTG · CAG and for longer lengths of the fragile X sequence but not for shorter lengths. The data were interpreted as supporting our stalling results [9, 21] and are in concert with pub-

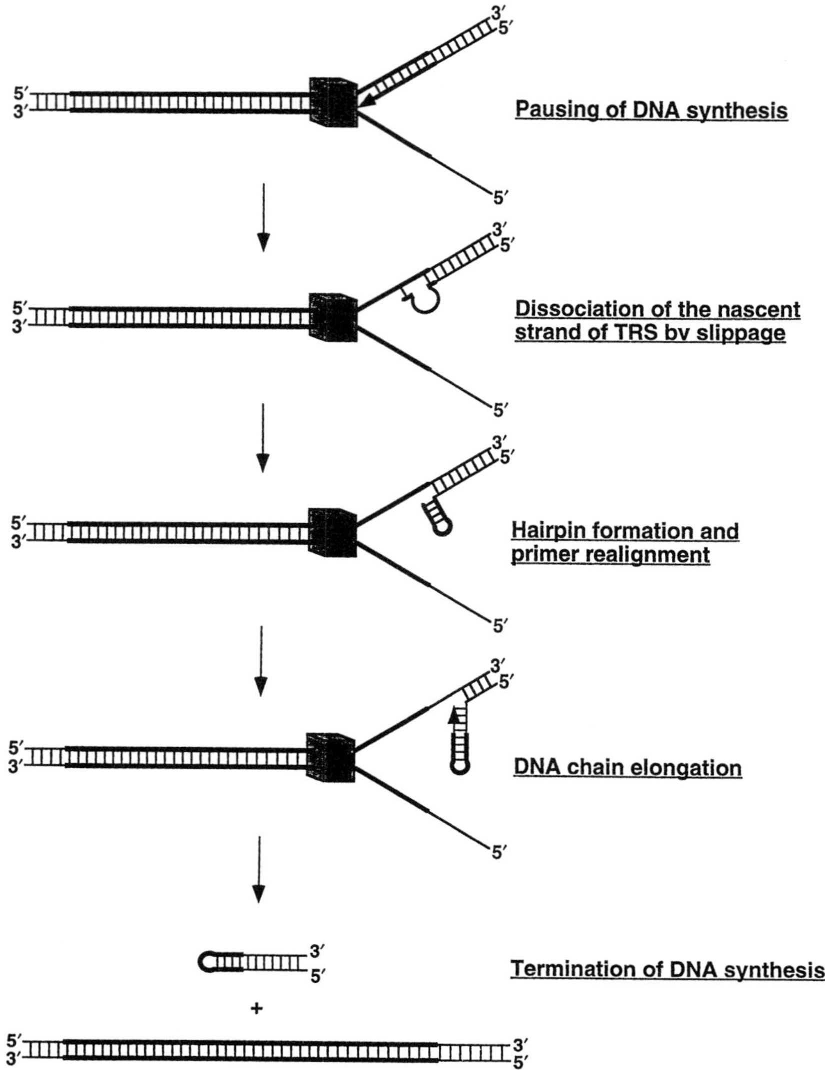

FIGURE 45-1 Model for the stalling of DNA polymerases during *in vitro* DNA synthesis. Reprinted with permission from Ref. [9], K. Ohshima and R. D. Wells, *J. Biol. Chem.* **272,** 16798–16806, 1997.

lished models for replication–expansion of triplet repeat sequences [10–12, 24, 25]. It is possible that the apparent differences between these three types of results may relate to the types of origins which are employed and the DNA polymerases which are carrying out the synthetic processes. In the case of the *in vitro* stalling data [9, 21] the template–primers are prepared *in vitro* and their properties may be subject to their method of preparation. Obviously, these investigations were conducted with highly purified DNA polymerases and, hence, a number of auxiliary protein factors, such as helicases, single-stranded DNA binding proteins, and topoisomerases, are absent. In the *ori*C system, a complex of proteins which is believed to be involved in DNA fork replication is present and this system clearly mimics more closely the *in vivo* circumstance on a duplex template. Alternatively, the *E. coli in vivo* system [26] is biologically intact. However, since these plasmids are replicated under the unidirectional control of the ColE1 origin, replication is initiated by leading strand synthesis by DNA polymerase I of at least 200–400 bp of DNA from a long RNA primer. During the initial phase of replication by polymerase I, the lagging template strand forms a single-stranded D loop. After dissociation of polymerase I, subsequent elongation is carried out by DNA polymerase III. In the absence of chloramphenicol or mutations in polymerase III, strand switching, presumably by polymerase I, can occur *in vivo* up to 600–900 bp from the origin of replication [27]. The plasmids used by Mirkin *et al.* [26] contain the triplet repeats cloned 1.5 kbp from the origin of replication. Hence, these workers believe that DNA polymerase III replicates the TRS in their system. Clearly, further studies will be required to decipher these differences between the three systems. In any case, it is obvious that the replication behaviors of these TRS are not orthodox and simple [4].

In vitro DNA synthesis studies have also been conducted on relatively shorter synthetic oligonucleotides containing TRS. Usdin and Woodford [31] observed the potassium-dependent DNA polymerase blockage of *in vitro* DNA synthesis within a $(CGG)_{20}$ tract. The strong arrests were seen at the cytosines of the template in the proximal half of the repeat units. The blockage was strand-dependent as well as length-dependent; no inhibition of DNA synthesis was found when the opposite strand $(CCG)_{20}$ was the template and when a template contained fewer than 13 consecutive CGG repeats. These arrests were not polymerase-dependent, since they were seen with different DNA polymerases including the modified T7 DNA polymerases (Sequenase), the Klenow fragment of *E. coli* DNA polymerase I, Taq polymerase, avian myeloblastosis virus (AMV) reverse transcriptase, and the T4 DNA polymerase. These workers proposed that the inhibition of DNA polymerization might be due to the formation of intrastrand tetraplex structures formed by hydrogen bonding between guanines (Fig. 45-2) on the CGG template. The reasons are as follows: (a) Replacement of 7-deaza-dGTP for dGTP showed no K^+-dependent blockage, indicating the N7 position of guanines was required for formation of the block of DNA synthesis; (b) in the chemical modifications with dimethylsulfate (DMS) and bromoacetaldehyde (BAA) for an oligonucleotide containing $(CCG)_{20}$, the guanines within the repeat tract were protected from DMS modification in the presence of K^+, indicating that the N7 position of each guanine was involved in a K^+-dependent hydrogen bonded structure. For BAA modification, except for the center of the cytosines (position C11), the cytosines were protected, indicating pairing between cytosines, and (c) fast gel mobility of an oligonucleotide containing $(CGG)_{20}$ was found in a nondenaturing polyacrylamide gel electrophoresis experiment and the methylation of guanines by DMS eliminated the anomalous mobility, suggesting the formation of a structure involving hydrogen bonding at the N7 positions of guanines. They also speculated that a K^+-independent structure represents a folded intermediate in the pathway of formation of the K^+-dependent tetraplex. This structure is likely to be a hairpin since all guanines within a $(CGG)_{20}$ tract were modified by DMS and the cytosines were protected from BAA modification except for the center (Fig. 45-2). As described above, we also observed a similar type of

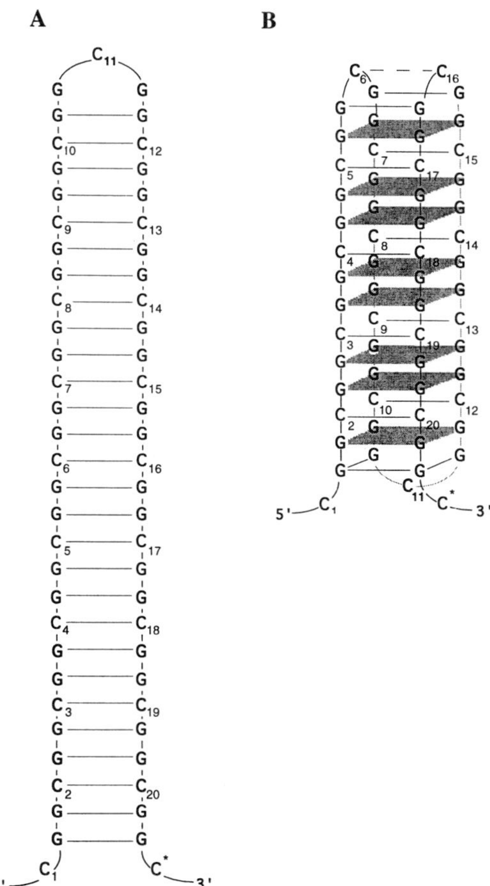

FIGURE 45-2 Models for the K^+-independent (A) and K^+-dependent (B) structures formed by a $(CGG)_{20}$ tract. Reprinted from Ref. [31], K. Usdin and K. J. Woodford, *Nucleic Acids Res.* **23,** 4202–4209, 1995, by permission of Oxford University Press.

blockage at the beginning of much longer tracts of CGG · CCG repeats without K^+, using the Klenow fragment of *E. coli* DNA polymerases I, Sequenase, and the human polymerase beta [21]. The long repeat sequences (>80 repeats) [9, 21] used in our work form a flexible and writhed conformation [22, 23]; this structure is quite different from the one proposed by Usdin and Woodford [31] for short oligomers.

D. Effect of Secondary Structure-Destabilizing Agents on Stalling

Mytelka and Chamberlain [28] found a stalling of DNA polymerase *in vitro* mostly at the central consensus Py-G-C sequence in some G-C-rich regions which were not necessarily adjacent to hairpins in the DNA templates. The stalling was eliminated by the presence of betaine in the reaction mixture. Betaine is a zwitterionic osmoprotectant found in many halophilic organisms that has been found to alter the DNA stability of GC-rich regions; in the presence of this molecule, the GC-rich regions melt at temperatures similar to AT-rich regions [28, 29]. Thus, these workers suggested that the polymerase stalling was due to the GC-rich sequence per se, which slows down the DNA polymerization. Also, the addition of 2 *M* betaine enabled the amplification of 75 repeats of CTG · CAG from the myotonic dystrophy gene, which was amplified only with difficulty under normal PCR conditions [28].

We performed primer extension analyses to investigate if betaine would reduce or eliminate the polymerase stallings found in the distal region of the TRS described above [9]. The primer–template complexes for $(CGG \cdot CCG)_{160}$ were incubated in the presence of 2 *M* betaine before the DNA elongation reaction by the Klenow fragment of *E. coli* DNA polymerase I. The stalling was not significantly influenced. Higher concentrations of betaine (3.3 and 5.2 *M*) were not informative due to the general inhibition of DNA polymerization. Hence, betaine did not alter the secondary structure of the DNA [22, 23]. This result is in agreement with the observation that the stallings on CGG · CCG were heat-resistant up to 90°C [21].

Alternatively, different results were found with the *E. coli* single-stranded DNA binding protein (SSB) which stabilizes plasmids containing long tracts of TRS in *E. coli* [30]. SSB is known to prevent the formation of secondary structures such as hairpins and triplexes [30]; these data are described in Section II. In contrast to the betaine data, the presence of SSB decreased the stalling [9], suggesting that SSB may stabilize the TRS by preventing the formation of an unusual DNA secondary structure [22, 23]. These observations using SSB suggest that the stalling was probably due to formation of an unusual DNA secondary structure rather than the effect of the high GC-rich sequences per se.

In other experiments with TRS oligonucleotides [31], the potassium-dependent blockage of DNA synthesis for a $(CGG)_{20}$ tract was not influenced by the presence of the gene 32 product.

E. Effect of Interruptions on Stalling

A consideration of the effects of interruptions (polymorphisms) in tracts of CTG · CAG, CGG · CCG, and GAA · TTC is important since they have been statistically correlated with the stability of the triplets [32–34]; the loss of interruptions increases the genetic instability. Interruptions of AGG in CGG · CCG repeats were found in the normal FMR1 gene whereas the lack of interruptions is known to be predisposed to expansions [33]. We have evaluated *in vivo* the effect of interruptions on the frequency of expansion of TRS in *E. coli* [10]; a comparison of an uninterrupted $(CTG \cdot CAG)_{130}$ with a sequence of identical length that contains one CTA polymorphism, derived from the myotonic dystrophy gene, showed that the uninterrupted sequence was expanded five times more frequently than the interrupted sequence. Thus, interruptions decrease the frequency of expansion in this biological model system. This may be due to the inhibition of complementary strand DNA slippages in the Okazaki fragment and the resultant inability to form an unusual DNA structure [11, 31, 35, 36], rather than destabilization of an unusual DNA structure which impedes DNA polymerization [22, 23, 31]. Therefore, the stalling of DNA polymerase might not be influenced by the interruptions per se, but may be a result of DNA slippage. The reader is referred to Chapter 33 for a description of the mechanism for frameshift mutations and DNA slippages in triplet repeat expansions.

In other investigations *in vitro* with relatively short oligonucleotides, Usdin and Woodford [31] observed that the blockage of DNA synthesis in the presence of potassium was not influenced by the presence of an AGG interruption in a $(CGG)_{20}$ tract, suggesting the contribution of cytosines is small for the formation of the proposed tetraplex structure [31]. However, they predicted that *in vivo* interruptions might be more important for the destabilization of DNA structures.

F. Expansion and DNA Synthesis *in Vitro*

The *in vitro* synthesis of long-chain DNA-like polymers of repeating nucleotide sequences was accomplished using the complementary chemically synthesized oligonucleotides (~8–12 nucleotides in length) along with all four deoxyribonucleoside triphosphates and an appropriate DNA polymerase. Duplex DNAs of repeating homo-, di-, tri-, and tetranucleotide sequences were prepared by these methods ([37–40] and reviewed in [41, 42]) (Fig. 45-3). In fact, these duplex DNAs were of paramount importance in solving the genetic code (reviewed in [41, 42]). The molecular mechanism for "expanding" the short oligonucleotides into long-chain duplex DNAs of identical nucleotide sequences was never fully unraveled. However, it was always believed [37–40] that the slippage of the complementary repeating sequence primers along with reiterative synthesis, slippage, synthesis, etc., accounted for the DNA chain elongation (Fig. 45-4). It was furthermore realized in the 1960s that reiterative RNA synthesis could also occur, in addition to DNA synthesis by strand slippage [41, 42].

Schlotterer and Tautz [43, 44] recently reinvestigated the rate of *in vitro* DNA synthesis for all 10 TRS using double-stranded DNA templates containing short tracts of TRS (~15 repeats) and a DNA polymerase (either

$$d(TG)_6 + d(AC)_6 + \left\{ \begin{array}{c} dTTP \\ dATP \\ dCTP \\ dGTP \end{array} \right\} \rightarrow \text{Poly d-TG:CA}$$

$$d(TTC)_4 + d(AAG)_3 + \left\{ \begin{array}{c} dTTP \\ dATP \\ dCTP \\ dGTP \end{array} \right\} \rightarrow \text{Poly d-TTC:GAA}$$

$$d(TATC)_3 + d(TAGA)_2 + \left\{ \begin{array}{c} dTTP \\ dATP \\ dCTP \\ dGTP \end{array} \right\} \rightarrow \text{Poly d-TATC:GATA}$$

FIGURE 45-3 Types of reactions catalyzed by DNA polymerase I to generate long-chain DNA-like polymers of defined repeating nucleotide sequences. Reprinted with permission from Ref. [114], Khorana et al., *Cold Spring Harbor Symp. Quant. Biol.* **31**, 39–49, 1966.

the Klenow fragment of *E. coli* DNA polymerase I or the T4 or T7 DNA polymerases). These workers showed that the AAT · ATT repeats elongated the fastest and gave the longest products whereas GGC · GCC was the slowest and gave the shortest products. The elongation efficiency, which probably reflects the rate of complementary strand slippage, was suggested to be dependent on the AT content of the sequences involved.

Although GGC · GCC was the repeat sequence to be elongated the slowest [43], Ji *et al.* [45] observed highly expanded products from a double-stranded DNA containing (GGC · GCC)$_5$ using several different DNA polymerases including the Taq DNA polymerase, the Klenow fragment of *E. coli* DNA polymerase I, Sequenase, DNA polymerase α and β, and the HIV reverse transcriptase; the Taq polymerase gave the longest expansion products. These workers suggested that the capacity for TRS expansion was a property of the individual DNA polymerases [45] in addition to the sequence effects. Also, these different observations with CGG · CCG repeats [43, 45] may have been due to the fact that Ji *et al.* [45] used ~50 times higher DNA polymerase concentrations than used by Schlotterer and Tautz [43].

Interestingly, using PCR conditions with (CGG)$_{17}$, (CCG)$_{17}$, (CTG)$_{17}$, (CAG)$_{17}$, and (TAA)$_{17}$, Behn-Krappa and Doerfler [46] showed that the single-stranded template containing (TAA)$_{17}$ gave the lowest elongation rate for *in vitro* DNA synthesis, a result opposite to the trend previously reported by Schlotterer and Tautz [43]. Obviously, the difference in reaction conditions (*in vitro* DNA replication vs PCR conditions) may have been responsible for this difference.

The *in vitro* DNA synthesis of TRS by Schlotterer and Tautz [43] was likely to be influenced by the differential abilities for complementary strand slippage for each of the 10 TRS. The *in vitro* studies [43, 45, 46] were performed using short DNA fragments containing only TRS without any flanking sequences at the ends. It is likely that this system would give more slippage of complementary strands (Fig. 45-4A) than those containing nonrepetitive sequences at either or both ends which lock the TRS into register (Fig. 45-4B). In the *in vivo* replication circumstance, we would expect that only the newly synthesized DNA fragment on the lagging strand template (Okazaki fragment) [4] would slip, but the leading strand template–nascent DNA strands would not have the same propensity for slippage since they are locked into register (Fig. 45-4B). Accordingly, we have proposed that additional factors such as hairpins must be considered for the expansion *in vivo*; these hairpins would be a result of the slippage process [7–12].

As described elsewhere in this book (see Chapter 33), this laboratory has established an *in vivo* expansion–deletion system in *E. coli*. Studies with all 10 TRS revealed that the CTG · CAG repeats were the dominant expansion products in *E. coli* [24]. CGG · CCG and GTC · GAC were the second most efficiently expanded. The molecular basis of expansions vs deletions of CTG · CAG and CGG · CCG repeats was explained on the basis of the preferential stabilization of loop structures during replication [6, 10–12] (Fig. 45-4B). Thus, long GC-rich TRS including CGG, CCG, CTG, CAG, GTC, and GAC may be prone to form stable hairpin structures [35, 36]; these hairpins might be induced by DNA slippage in the newly synthesized strands which would then be involved in the *in vivo* expansion process [9, 11]. The topic of hairpin formation in short oligonucleotides containing these sequences is described elsewhere in this book (Chapters 40 and 41). Thus, one might deduce that TRS with a lower ability to form hairpins [35, 36] would not give TRS expansions *in vivo* [7, 24]. Interestingly, McMurray *et al.* [47] have stated that the strands in GAA · TTC do not form hairpins easily. However, this sequence is expanded *in vivo* [24]; furthermore, a large number of plasmids have been constructed containing long tracts of GAA · TTC (up to 400 repeats) by this expansion method [115]. Thus, elucidation of the mechanism of expansion of the GAA · TTC repeat will be fascinating. This sequence is important since its expansion to 100–1200 repeats is the mutation responsible for Friedreich's ataxia (FRDA) ([1, 6], and Chapter 26).

G. Expansion and DNA Polymerase Stalling *in Vivo*

What is the relationship between the stalling of DNA polymerases *in vitro* and the expansions of TRS *in vivo*? Whereas a direct answer to this question is not available,

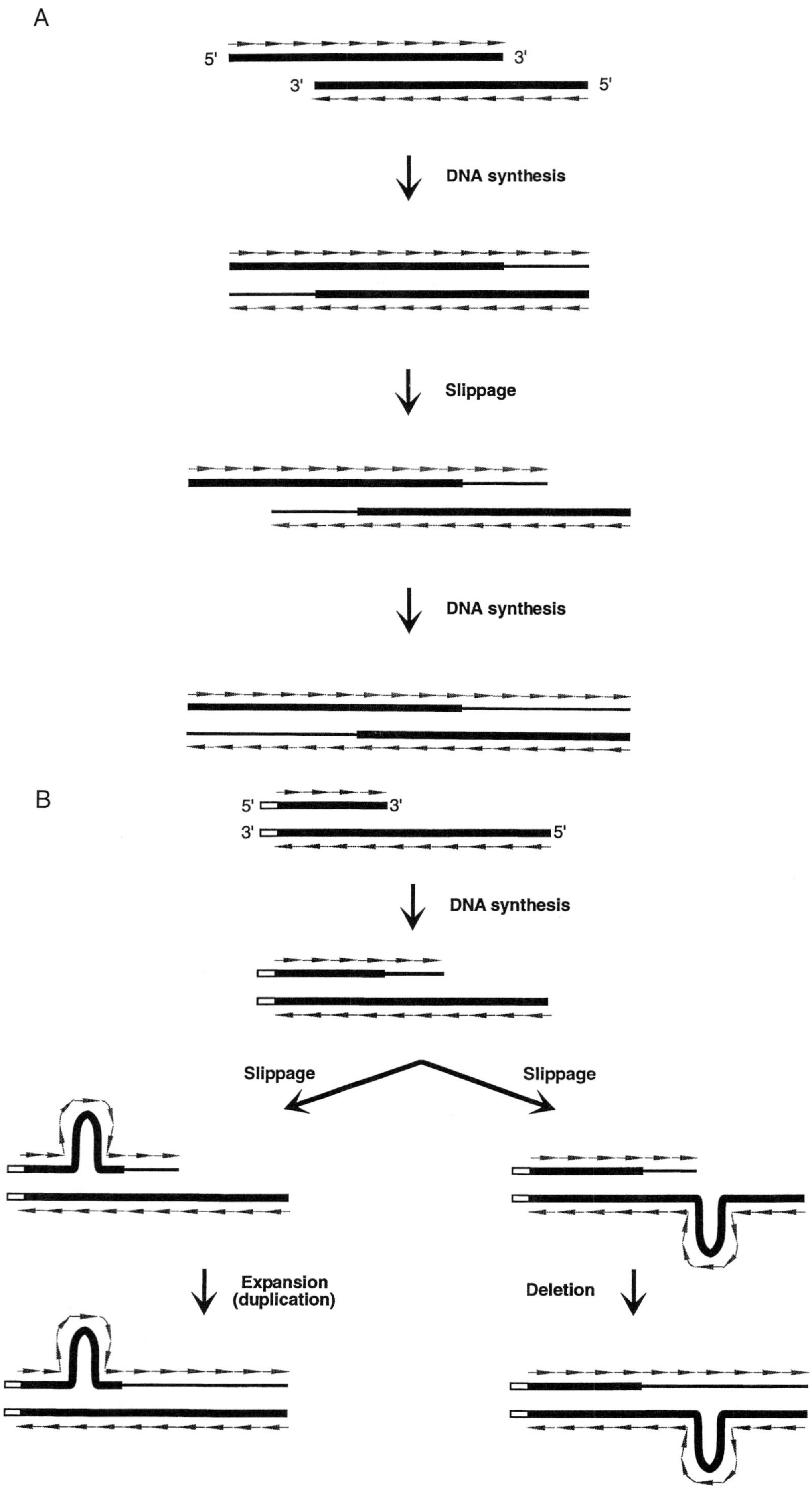

A
5'
3'
3'
5'
DNA synthesis
Slippage
DNA synthesis
B
5'
3'
3'
5'
DNA synthesis
Slippage
Slippage
Expansion
(duplication)
Deletion

it may be interesting to note that the GC-rich sequences including CTG · CAG, CGG · CCG, and GTC · GAC are expanded *in vivo* and do show the length-dependent stalling of DNA polymerases *in vitro* [7–9, 21] which was due to hairpin formation caused by primer realignment [9]. Furthermore, stall sites have been observed for CTG · CAG and CGG · CCG TRS *in vivo* in *E. coli* [26]. Other work has clearly indicated that stalled sites for DNA polymerases are hot spots for mutations caused by DNA misalignment [48–50] and also DNA double-strand break points [51] which would lead to germline mutations including duplications and deletions in minisatellites by recombination [14, 52, 53]. Thus, it is possible that the stalling of DNA polymerases *in vivo* might give rise to genetic expansion through replication and/or recombination. The potential involvement of other factors including nucleosome binding [54–58], methyl-directed mismatch repair [25, 59–62], TRS binding proteins [63–66], and methylation of CGG · CCG [46, 67–69] remains to be fully elucidated.

H. Effect of Proteins and Actinomycin D

The involvement of sequence-specific ligands and replication–termination proteins which bind to specific sequences may impede DNA replication [70–73]. The sequence-specific binding of actinomycin D to single-stranded DNA molecules inhibits DNA synthesis *in vitro* [70]. Also, the progression of DNA replication forks was inhibited by the binding of sequence-specific proteins to a DNA consensus sequence [71–73] at the replication terminus sites on replicons. Hence, a substantial amount of literature exists on the activity of specific proteins which impede DNA synthesis. Unfortunately, we are at the early stages in our comprehension of the role of these or other factors in the formation of termination sites.

I. Transcription Stalling

Parsons *et al.* [74] have studied a synchronized *in vitro* transcription system to monitor the inherent ability of RNA polymerase II to transcribe genomic DNA from normal $(CTG)_{17}$, carrier $(CTG)_{50}$, and affected $(CTG)_{255}$ individuals, and, as a control, DNA corresponding to the normal $(CGG)_{54}$ fragile X repeat tract. Core RNA polymerase II efficiently transcribed all repeat units irrespective of repeat length or orientation. However, approximately 50% of polymerases transiently adopted an elongation stall at the entrance to the CTG repeat. The dwell half-life of these complexes was 10 ± 1 s. Furthermore, the elongation rate within the CTG repeat was inherently slow. The average transcription rates within the CTG, CCG, CGG, and CAG tracts were 170, 250, 300, and 410 nt/min, respectively. These differences correlated with changes in the sequence-specific transient pausing pattern within the CNG repeat tracts; in general, individual nucleotide incorporation rates were slower after incorporation of a pyrimidine. Finally, and unexpectedly, approximately 3% of the run-off transcription products generated on linear $(CTG)_{17}$ and $(CAG)_{17}$ templates were discrete, approximately 15 nt longer than expected RNAs whose synthesis was inhibited by addition of the transcriptional elongation factor SII. Hence, it is apparent that the capacity of a template sequence and conformation to influence transcript elongation is important, as described above for DNA polymerases.

Alternatively, no transcription stalling was observed *in vivo* or in *in vitro* in T7 RNA polymerase systems [75]. $(CTG \cdot CAG)_n$ (where $n = 17$, 30, and 130) were cloned into pET28a(+) which has three major advantages for transcription studies. First, the transcriptional unit is in the opposite orientation from the other genes on this plasmid which results in a very low read-through transcription. Second, the very strong T7 terminator is present at the end of the region for T7 transcription. Third, RNA polymerase can be induced in the appropriate strains to transcribe the T7 region under induction by IPTG and can be superregulated by the level of lysozyme produced by the strain. Northern blot analyses were conducted in this *in vivo* system which showed no accumulation of incomplete or aberrant synthesis of the T7 region for all lengths of repeats analyzed as well as the control vector [75]. Additional experiments were conducted *in vitro* with T7 RNA polymerase. The plas-

FIGURE 45-4 Slippage of the complementary template and primer strands in DNA synthesis. (A) *In vitro* DNA chain elongation with synthetic oligonucleotide template–primers to synthesize long-chain duplex sequences with simple repeating homo-, di-, tri-, and tetranucleotide motifs (Fig. 45-3). The ends as well as the centers of the complementary strands are free to slip, probably by an inch worm mechanism with the formation of hairpin loops on both strands, due to their repetitive nature [37–44]. (B) *In vitro* DNA chain elongation for a simple repeat sequence within a recombinant plasmid. Since the end (the plasmid vector sequence represented as an unfilled tract on the left) is not free to slip, expansions or deletions may ensue in the repeating sequence insert, depending on the propensity for the newly synthesized chain or the template, respectively, to adopt a hairpin structure. Gray arrows represent the repeating sequence motifs. For a description of similar considerations with respect to *in vivo* expansions and deletions, see Chapter 33.

mids which were employed transcribed the CAG repeat-containing strand; this is the same orientation that gave the strongest pausing in the work with RNA polymerase II [74]. Reactions at 37, 25, or 15°C and at a wide range of time points showed no evidence for CUG-induced stalling or premature termination *in vitro*. The only major band was the full-length product. Hence, these results are substantially different from the data obtained with RNA polymerase II and, hence, suggest that the properties of the individual polymerases may be important, as described above for the expansion capacities of the different DNA polymerases. Further studies will be required to evaluate the reasons for this behavior.

Interestingly, cDNA clones of the rat polymeric immunoglobin receptor gene contain the sequence $(GGA)_{24\text{-}26}(GAA)_{9\text{-}37}$; studies on the RNA products reveal the presence of a product formed by slippage [76]. Leffert *et al.* discovered that the RNA formed in this system was longer by several repeat units than the cDNA template. The only apparent mechanism for this behavior is the slippage of the transcript–RNA polymerase complex relative to the template DNA. Similar results have been described in the literature for other repeating sequences, even in the mid-1960s [reviewed in 76]. Thus, the RNA polymerase II slippage described above is not unexpected.

Considering *in vivo* studies, Bowater *et al.* [77] reported that the induction of transcription into long CTG · CAG repeats contained on plasmids in *E. coli* increases the frequency of deletions within the repeat sequences. This elevated genetic instability was detected because active transcription into the triplet repeat influenced the growth transitions of their host cells, allowing advantageous growth for cells harboring plasmids with deleted repeat sequences. The variety of deletion products observed in separate cultures suggests that transcription altered the metabolism of the DNA in a manner that produced random length changes in the repeat sequence. For cultures containing plasmids without active transcription into the triplet repeat, or those maintained in exponential growth, deletions occurred within the repeat at a lower frequency (5- to 20-fold lower). In these incubations the extent of deletions was proportional to the number of cell divisions and many repeat lengths were observed within each culture, suggesting that the decrease in average repeat length at long incubation times was due to multiple small deletions. These observations show that deletions within long CTG · CAG repeats contained on plasmids in *E. coli* occur via more than one pathway, and their level of genetic instability is altered by the enzymatic processes occurring upon the DNA. Thus, although the mechanism of the transcription effect is uncertain, there can be no doubt of the profound influence of its stimulation of the deletion frequency of long TRS.

II. STRUCTURE- AND SEQUENCE-DEPENDENT STALLING

This section will review prior investigations documenting the role of unusual DNA conformations (hairpins, triplexes, and tetraplexes) on stalling of DNA synthesis. Hence, these studies represent the background upon which the TRS stalling investigations were conducted.

A. Hairpin Structures (Cruciforms)

Substantial prior work has documented the involvement of hairpin-loop structures as well as triplexes and tetraplexes in the stalling or pausing of DNA replication. Hairpin structures are formed at inverted repeat sequences (reviewed in Chapter 38). These inverted repeat sequences are sometimes referred to as palindromes; however, this nomenclature is not correct since, strictly speaking, a palindrome is a sequence which can be written backwards to give the same word (i.e., rotor or hannah). Cruciforms (hairpin structures) consist of stem and loop regions. It was realized approximately 17 years ago that DNA polymerization was impeded *in vitro* by the formation of a hairpin loop in the DNA template [78–80] in several different systems, both single-stranded and double-stranded templates. In the presence of a helix destabilizing protein (i.e., gene 32 protein), the stalling was decreased to give an increased rate of chain growth for the replicating molecules [78]. Also, an increase in the reaction temperature enhanced the replication rate [78]. These results suggested that the secondary structure of the DNA template was the cause of the kinetic arrest sites. Although DNA polymerases arrest at hairpin-forming sequences *in vitro*, DNA polymerases do not always stall *in vivo* at the sites predicted for formation of hairpin structures [81].

B. Triplexes

Triplexes are three-stranded nucleic acid structures that are either of the intramolecular type (folded back conformations in supercoiled DNAs) or of the intermolecular type (formed by the interaction of an oligonucleotide with a duplex DNA tract [reviewed in this book by Gellibolian and Bacolla]. It may be noteworthy to distinguish between the triplex DNA conformations and repeating triplet repeat sequences, such as CTG · CAG. The presence of triplexes inhibits DNA synthesis due to the blockage of the progression of the enzyme by the third strand in the major groove which is paired to the Watson–Crick duplex by Hoogsteen base pairs. Several excellent papers have documented the triplex inhibition of DNA synthesis; the polymerization is influenced by pH, ionic strength, and temperature [82] as expected from the known stabilities of triplexes.

Prior investigations [83] demonstrated that homopurine · homopyrimidine tracts [$(TC)_{27}$ and $(GA)_{27}$] in a single-stranded DNA molecule inhibited DNA synthesis *in vitro* in the center of this sequence. The stalling was eliminated by replacement of dATP or dGTP by 7-deaza dATP or 7-deaza dGTP as the nucleotide substrate, respectively, and addition of single-stranded DNA-binding proteins [83]. These results indicated that the newly synthesized DNA strand was involved in triplex formation, probably due to folding back of the template strand when the newly synthesized chain reached the center of the sequence. In other investigations [84, 85], SV40 viral DNA replication *in vivo,* with the same GA · TC sequences cloned into the viral genome, also showed the stalling of DNA replication.

For double-stranded DNA templates, in contrast to single-stranded templates, triplex structures are formed in supercoiled DNA which are either the H-y3 or the H-y5 isomer (see Chapter 38 for a description of these structures). DNA polymerases stalled at the border of the triplex structures and the arrest sites differed between these triplex isomers [86]. A different approach was performed using a nicked double-stranded DNA template containing a triplex-forming sequence in which the polymerase-driven triplex formation was observed; when the purine-rich strand was displaced, the polymerization was inhibited in the middle of the sequence. Alternatively, no termination occurred when the pyrimidine-rich strand was displaced [87].

It may be noted that the antisense strategy of using an oligonucleotide to inhibit replication [88] or transcription [89] by formation of an intermolecular triplex is enjoying widespread utility.

The origin of replication of the dihydrofolate reductase locus of CHO cells contains a tract of repeating sequences where a Z-DNA helix neighbors a triplex conformation with no intervening base pairs. The sequence is $(GC)_5(AC)_{18}(AG)_{21}$. Interestingly, DNA polymerization *in vitro* was more severely impeded in the triplex to Z-DNA direction than in the Z-DNA to triplex direction [90]. This result is not unexpected since triplexes have a pronounced directionality whereas this behavior is absent with the Z-DNA conformation.

C. Tetraplexes

The structure of tetraplexes was reviewed elsewhere in this book (see Chapter 38). Guanine-rich (G-rich) sequences form a four-stranded DNA tetraplex structure in the presence of certain alkali metal cations such as Na and Rb and the structure is stabilized also in the presence of potassium [91] (Fig. 45-4). Usdin and Woodford performed primer extension analyses using several DNA polymerases with single-stranded templates containing G-rich sequences such as $G_{16}CG(GGT)_2GG$ in the promoter region of the chicken β-globin gene [92, 93], and with $(TGG)_{20}$ [94], $(CAGGG)_8$ [94], and $(CGG)_{20}$ [31]. They found potassium-dependent arrests of DNA polymerase within the G-rich sequences, suggesting that the replication block was due to formation of an intramolecular tetraplex in the template, which required non-Watson–Crick base interactions between guanines and were stabilized by the presence of potassium. The potassium-dependent stability might be explained by the size-selective binding in the central cavity of the G quartets [91]. Other studies on tetraplexes were described in Section I.

In summary, considering the substantial body of stalling–pausing data with cruciforms (hairpins), triplexes, and quadraplexes (tetraplexes), it is not surprising that stalling and hairpin formation were found with some of the TRS (CTG · CAG, CGG · CCG, and GAA · TTC). The results with the FRDA sequence (GAA · TTC) are easy to explain based on its known behavior of adopting a triplex. However, CTG · CAG and CGG · CCG have the intrinsic structure of a flexible and writhed right-handed B-helix. This conformation has not been recognized previously [22, 23]. Hence, further work will be required to elucidate the role of the properties of this structure regarding replication and transcription.

III. STRAND SLIPPAGE AND DNA POLYMERASE STALLING

This section will review some of the biological literature on slippage-mediated misalignment, DNA polymerase stalling, and strand switching.

A. Slippage-Mediated DNA Misalignment

Frameshift mutations by strand slippage were initially proposed by Streisinger *et al.* in 1966 [95]. DNA mutations including deletions and the addition of nucleotides can be initiated by strand slippage followed by template–primer realignment [96–98]; slippage in the template strand leads to a deletion whereas slippage in the newly synthesized strand leads to an expansion (reviewed in detail in Chapter 33). DNA strand slippage is likely to occur in reiterative sequences including mono-, di-, tri-, and tetranucleotide repeat tracts [96–99]. In fact, oligonucleotides containing these repeat sequences used as templates produced long molecules by *in vitro* DNA synthesis (Figs. 45-3 and 45-4) [37–46].

Direct repeat sequences also induced slippage-mediate DNA misalignment [100–103].

Slippage-mediated nucleotide deletions and insertions have been suggested to cause mutations in human chromosomes [103–106]. The types of repetitive sequences include dinucleotide repeats, which are linked to hereditary cancers (see Chapter 49) and TRS expansion diseases [1–6, 14], which can lead to the instability of a chromosome.

B. Misalignment and DNA Polymerase Stalling

A DNA polymerase stalled site is a hot spot for DNA mutagenesis [48–50, 100]. As described above, DNA polymerase stalling is observed when a DNA polymerase encounters a barrier which is principally due to the formation of any of several DNA secondary structures. The arrest of DNA polymerase could lead to primer–template realignment, especially in repetitive sequences which result in deletions and insertions.

Direct repeats have been characterized with a single-stranded DNA template containing inverted repeats between direct repeats; primer extension analyses produced deletions of one unit of the direct repeats as well as the inverted repeat [100]. This observation suggested that the DNA polymerase might stall at the hairpin structure formed by the inverted repeats and the resultant primer–template realignment after slippage might give rise to annealing of the newly synthesized strand to the homologous second direct repeat [100].

C. Strand Switching

Strand switching from one template to the other was observed with RNA templates, as catalyzed by reverse transcriptases which had been stalled [107, 108]. Defined ordered DNA sequences containing inverted repeats [103, 109–111] and direct repeats [112, 113] can lead to duplications and deletions by primer–template switching during DNA replication. A plasmid containing a 34-bp inverted repeat sequence inserted in a pUC19 derivative produced a fragment containing symmetrical flanking sequences besides the inverted repeats during DNA replication in *E. coli* [110]. DNA sequences and restriction analyses revealed that this fragment was generated by the formation of a hairpin structure at the inverted repeat followed by primer–template switching in the newly synthesized strand. The resynthesis was terminated at the starting site of the replication or the termination site of the lagging strand synthesis [110].

IV. PROSPECTS FOR THE FUTURE

Our comprehension of the DNA replication behavior *in vitro* and *in vivo* of TRS is in the very early stages. To date, survey-type experiments have been conducted to lay out the basic behaviors. Interestingly, the replication of simple TRS is not simple, as predicted in 1994 [4].

An effective way to demonstrate the state of our knowledge may be to identify some of the unknowns. These include the following:

· What is the behavior of different DNA polymerases (DNA polymerase I or III) on these sequences?

· What is the relationship, if any, between the flexible and writhed structure and stalling along with hairpin formation; alternatively, are other unusual DNA structures involved?

· Is DNA replication the only mechanism for expansion? Might recombination and/or frameshift mutations be involved?

· Is DNA polymerase stalling–hairpin formation a consequential molecular component in expansion in human cells *in vivo*?

· Is DNA slippage the mechanism for the expansion process?

· Does DNA structure cause the stalling and hairpin formation due to preexisting conformations, are these structures formed during renaturation of the template–primers, or are they formed during the DNA synthetic process?

· What are the molecular reasons for the apparent differences between the Marians data in the *ori*C system as compared to the Wells lab *in vitro* data with purified enzymes and the Mirkin *et al.* data in living *E. coli* cells? Furthermore, what is the relationship of these data with human DNA synthesis processes?

· What is the involvement of proteins that specifically bind to flexible and writhed TRS which participate in replication? What factors promote or inhibit hairpin formation?

· What is the molecular basis for the involvement of transcription in expansion *in vivo*?

Obviously, it is clear from this partial list that we are in the very early stages of our understanding of the molecular processes responsible for genetic instability.

Acknowledgments

This work was supported by a grant from the N.I.H. (GM52982) and the Robert A. Welch Foundation. We thank E. A. Lawson for critically reading the manuscript and R. R. Iyer for assistance in the preparation of a figure.

References

1. Paulson, H. L., and Fischbeck, K. H. (1996). Trinucleotide repeat in neurogenetic disorders. *Annu. Rev. Neurosci.* **19,** 79–107.
2. Mandel, J.-L. (1997). Breaking the rule of three. *Nature* **386,** 767–769.
3. Kunkel, T. A. (1993). Slippery DNA and diseases. *Nature* **365,** 207–208.
4. Richards, R. I., and Sutherland, G. R. (1994). Simple repeat DNA is not replicated simply. *Nature Genet.* **6,** 114–116.
5. Dover, G. (1995). Slippery DNA runs on and on and on. *Nature Genet.* **10,** 254–256.
6. Wells, R. D. (1996). Molecular basis of genetic instability of triplet repeats. *J. Biol. Chem.* **271,** 2875–2878.
7. Ohshima, K., Kang, S., Larson, J. E., and Wells, R. D. (1996). Cloning characterization, and properties of seven triplet repeat DNA sequences. *J. Biol. Chem.* **271,** 16773–16783.
8. Ohshima, K., Kang, S., Larson, J. E., and Wells, R. D. (1996). TTA · TAA triplet repeats in plasmids form a non-H bonded structure. *J. Biol. Chem.* **271,** 16784–16791.
9. Ohshima, K., and Wells, R. D. (1997). Hairpin formation during DNA synthesis primer realignment *in vitro* in triplet repeat sequences from human hereditary disease genes. *J. Biol. Chem.* **272,** 16798–16806.
10. Kang, S., Ohshima, K., Jaworski, A., and Wells, R. D. (1996). CTG triplet repeats from the myotonic dystrophy gene are expanded in *Escherichia coli* distal to the replication origin as a single large event. *J. Mol. Biol.* **258,** 543–547.
11. Kang, S., Jaworski, A., Ohshima, K., and Wells, R. D. (1995). Expansion and deletion of CTG repeats from human disease genes are determined by the direction of replication in *E. coli. Nature Genet.* **10,** 213–210.
12. Shimizu, M., Gellibolian, R., Oostra, B. A., and Wells, R. D. (1996). Cloning, characterization and properties of plasmids containing CGG triplet repeats from the FMR-1 gene. *J. Mol. Biol.* **258,** 614–626.
13. Wells, R. D., and Sinden, R. R. (1993). Defined ordered sequence DNA, DNA structure, and DNA-directed mutation. *In* "Genome Analysis," Vol. 7, "Genome Rearrangement and Stability" (K. Davies and S. Warren, Eds.), pp 107–138. Cold Spring Harbor Laboratory, Cold Spring Harbor, NY.
14. Gordenin, D. A., Kunkel, T. A., and Resnick, M. A. (1997). Repeat expansion-all in a flap? *Nature Genet.* **16,** 116–118.
15. Weaver, D. T., and DePamphilis, M. L. (1982). Specific sequences in native DNA that arrest synthesis by DNA polymerase α. *J. Biol. Chem.* **257,** 2075–2086.
16. Klarmann, G. J., Schauber, C. A., and Preston, B. D. (1993). Template-directed pausing of DNA synthesis by HIV-1 reverse transcriptase during polymerization of HIV-1 sequences *in vitro. J. Biol. Chem.* **268,** 9793–9802.
17. Abbotts, J., Bebenek, K., Kunkel, T. A., and Wilson, S. H. (1993). Mechanism of HIV-reverse transcriptase. *J. Biol. Chem.* **268,** 10312–10323.
18. Weinman-Shomer, P., Dube, D. K., Perrino, F. W., Stokes, K., Loeb, L. A., and Fry, M. (1989). Sequence specificity of pausing by DNA polymerases. *Biochem. Biophys. Res. Commun.* **164,** 1149–1156.
19. Ellis N. A., Groden, J., Ye, T.-Z., Straughen, J., Lennon, D. J., Ciocci, S., Proytcheva, M., and German, J. (1995). The Bloom's syndrome gene product is homologous to RecQ helicases. *Cell* **83,** 655–666.
20. Yu, C.-E., Oshima, J., Fu, Y.-H., Wijsman, E. M., Hisama, F., Alisch, R., Matthews, S., Nakura, J., Miki, T. Ouais, S., Martin, G. M., Mulligan, J., and Schellenberg, G. D. (1996). Positional cloning of the Warner's syndrome gene. *Science* **272,** 258–262.
21. Kang, S., Ohshima, K., Shimizu, M., Amirhaeri, S., and Wells, R. D. (1995). Pausing of DNA synthesis *in vitro* at specific loci in CTG and CGG triplet repeats from human hereditary disease genes. *J. Biol. Chem.* **270,** 27014–27021.
22. Bacolla, A., Gellibolian, R., Shimizu, M., Amirhaeri, S., Kang, S., Ohshima, K., Larson, J. E., Harvey, S. C., Stollar, D. B., and Wells, R. D. (1997). Flexible DNA: Genetically unstable CTG · CAG and CGG · CCG from human hereditary neuromuscular disease genes. *J. Biol. Chem.* **272,** 16783–16792.
23. Gellibolian, R., Bacolla, A., and Wells, R. D. (1997). Triplet repeat instability and DNA topology: an expansion model based on statistical mechanics. *J. Biol. Chem.* **272,** 16793–16797.
24. Ohshima, K., Kang, S., and Wells, R. D. (1996). CTG triplet repeats from human hereditary diseases are dominant genetic expansion products in *Escherichia coli. J. Biol. Chem.* **271,** 1853–1856.
25. Jaworski, A., Rosche, W. A., Gellibolian, R., Kang, S., Shimizu, M., Bowater, R. P., Sinden, R. R., and Wells, R. D. (1995). Mismatch repair in *Escherichia coli* enhances instability of $(CTG)_n$ triplet repeats from human hereditary diseases. *Proc. Natl. Acad. Sci. USA* **92,** 11019–11023.
26. Samadashwily, G. M., Raca, G., and Mirkin, S. (1997). Trinucleotide repeats affect DNA replication *in vivo. Nature Genet.* **17,** 298–304.
27. Backman, K., Betlach, M., Boyer, H. W., and Yanofsky, S. (1978). Genetic and physical studies on the replication of ColE1-type plasmids. *Cold Spring Harbor Symp. Quant. Biol.* **43,** 69–76.
28. Mytelka, D. S., and Chamberlain, M. J. (1996). Analysis and suppression of DNA polymerase pauses associated with a trinucleotide consensus. *Nucleic Acids Res.* **24,** 2774–2781.
29. Rees, W. A., Yager, T. D., Korte, J., and von Hippel, P. H. (1993). Betaine can eliminate the base pair composition dependence of DNA melting. *Biochemistry* **32,** 137–144.
30. Rosche, W. A., Jaworski, A., Kang, S., Kramer, S. F., Larson, J. E., Geidroc, D. P., Wells, R. D., and Sinden, R. R. (1996). Single-stranded DNA-binding protein enhances the stability of CTG triplet repeats in *Escherichia coli. J. Bacteriol.* **178,** 5042–5044.
31. Usdin, K., and Woodford, K. J. (1995). CGG repeats associated with DNA instability and chromosome fragility form structures that block DNA synthesis *in vitro. Nucleic Acids Res.* **23,** 4202–4209.
32. Chung, M.-Y., Ranum, L. P. W., Duvick, L. A., Servadio, A., Zoghbi, H. Y., and Orr, H. T. (1993). Evidence for a mechanism predisposing to intergenerational CAG repeat instability in spinocerebellar ataxia type I. *Nature Genet.* **5,** 254–258.
33. Eichler, E. E., Holden, J. J. A., Popovich, B. W., Reiss, A. L., Snow, K., Thibodeau, S. N., Richards, C. S., Ward, P. A., and Nelson, D. L. (1994). Length of uninterrupted CGG repeats determines instability in the *FMR1* gene. *Nature Genet.* **8,** 88–94.
34. Montermini, L., Andermann, E., Labuda, M., Richter, A., Pandolfo, M., Cavalcanti, F., Pianese, L., Iodice, L., Farina, G., Monticelli, A., Turano, M., Filla, A., De Michele, G., and Cocozza, S. (1997). The Friedreich ataxia GAA triplet repeat: premutation and normal alleles. *Hum. Mol. Genet.* **8,** 1261–1266.
35. Mitas, M. (1997). Trinucleotide repeats associated with human disease. *Nucleic Acids Res.* **25,** 2245–2253.
36. McMurray, C. T. (1995). Mechanisms of DNA expansion. *Chromosoma* **104,** 2–13.
37. Kornberg, A., Bertsch, L. L., Jackson, J. F., and Khorana, H. G. (1964). Enzymatic synthesis of deoxyribonucleic acid, XVI. Oligonucleotides as templates and the mechanism of their replication. *Proc. Natl. Acad. Sci. USA* **51,** 315–323.

38. Wells, R. D., Ohtsuka, E., and Khorana, H. G. (1965). Studies on polynucleotides. L. Synthetic deoxyribopolynucleotides as templates for the DNA polymerase of *Escherichia coli*: a new double-stranded DNA-like polymer containing repeating dinucleotide sequences. *J. Mol. Biol.* **14,** 221–240.
39. Wells, R. D., Jacob, T. M., Narang, S. A., and Khorana, H. G. (1967). Studies on polynucleotides, LXIX. Synthetic deoxyribopolynucleotides as templates for the DNA polymerase of *Escherichia coli*: DNA-like polymers containing repeating trinucleotide sequences. *J. Mol. Biol.* **27,** 237–263.
40. Morgan, A. R., Coulter, M. B., Flintoff, W. F., and Paetkau, V. H. (1974). Enzymatic synthesis of deoxyribonucleic acids with repeating sequence. A new repeating trinucleotide deoxyribonucleic acid, $d(T\text{-}C\text{-}C)_n \cdot d(G\text{-}G\text{-}A)_n$. *Biochemistry* **13,** 1596–1603.
41. Wells, R. D., and Wartell, R. M. (1974). The influence of nucleotide sequence on DNA properties. Biochemistry Series One. *Biochem. Nucleic Acids* **6,** 41–64.
42. Wells, R. D., Blakesley, R. W., Burd, J. F., Chan, H. W., Dodgson, J. B., Hardies, S. C., Horn, G. T., Jensen, K. F., Larson, J., Nes, I. F., Selsing, E., and Wartell, R. M. (1977). The role of DNA structure in genetic regulation. *Crit, Rev. Biochem.* **4,** 305–340.
43. Schlötterer, C., and Tautz, D. (1992). Slippage synthesis of simple sequence DNA. *Nucleic Acids Res.* **20,** 211–215.
44. Tautz, D., and Schlötterer, C. (1994). Simple sequences. *Curr. Opin. Genet. Dev.* **4,** 832–837.
45. Ji, J., Clegg, N. J., Peterson, K. R., Jackson, A. L., Laird, C. D., and Loeb, L. A. (1996). *In vitro* expansion of GGC:GCC repeat: identification of the preferred strand expansion. *Nucleic Acids Res.* **24,** 2835–2840.
46. Behn-Krappa, A., and Doerfler, W. (1994). Enzymatic amplification of synthetic oligodeoxyribonucleotides: implications for triplet repeat expansions in the human genome. *Hum. Mutat.* **3,** 19–24.
47. Gacy, A. M., Goellner, G. M., Spiro, C., Dyer, R., Mikesell, M., Yao, J. Z., Johnson, A. J., Juranic, N., Macura, S., Richter, A., Melancon, S. B., and McMurray, C. T. (April 1–6, 1997). DNA structures associated with class I expansion of GAA in Friedreich's ataxia. Poster number CS3–103, presented at Santa Fe meeting on "Unstable Triplets, Microsatellites, and Human Disease" (J. Griffith, R. D. Wells, and D. L. Nelson, organizers).
48. Papanicolaou, C., and Ripley, L. S. (1991). An *in vitro* approach to identifying specificity determinants of mutagenesis mediated by DNA misalignments. *J. Mol. Biol.* **221,** 805–821.
49. Bierne, H., Ehrlich, S. D., and Michel, B. (1991). The replication termination signal *terB* of the *Escherichia coli* chromosome is a deletion hot spot. *EMBO J.* **10,** 2699–2705.
50. Bebenek, K., Abbotts, J., Roberts, J. D., Wilson, S. H., and Kunkel, T. A. (1989). Specificity and mechanism of error-prone replication by human immunodeficiency virus-1 reverse transcriptase. *J. Biol. Chem.* **264,** 16948–16956.
51. Michel, B., Ehrlich, S. D., and Uzest, M. (1997). DNA double-strand breaks caused by replication arrest. *EMBO J.* **16,** 430–438.
52. Burad, J., and Jeffreys, A. J. (1997). Big, bad minisatellites. *Nature Genet.* **15,** 327–328.
53. Warren, S. T. (1997). Polyalanine expansion in synpolydactyly might result from unequal crossing-over of HOXD13. *Science* **275,** 408–409.
54. Wang, Y.-H., Amirhaeri, S., Kang, S., Wells, R. D., and Griffith, J. D. (1994). Preferential nucleosome assembly at DNA triplet repeats from the myotonic dystrophy gene. *Science* **265,** 669–671.
55. Wang, Y.-H., and Griffith, J. (1995). Expanded CTG triplet blocks from the myotonic dystrophy gene create the strongest known natural nucleosome positioning elements. *Genomics* **25,** 570–573.
56. Wang, Y.-H., Gellibolian, R., Shimizu, M., Wells, R. D., and Griffith, J. (1996). Long CCG triplet repeat blocks exclude nucleosomes: a possible mechanism for the nature of fragile sites in chromosomes. *J. Mol. Biol.* **263,** 511–516.
57. Wang, Y.-H., and Griffith, J. (1996). Methylation of expanded CCG triplet repeat DNA from fragile X syndrome patients enhances nucleosome exclusion. *J. Biol. Chem.* **271,** 22937–22940.
58. Wang, Y.-H., and Griffith, J. D. (1996). The $[(G/C)_3NN]_n$ motif: A common DNA repeat that excludes nucleosomes. *Proc. Natl. Acad. Sci. USA* **93,** 8863–8867.
59. Strand, M., Prolla, T. A., Liskay, R. M., and Petes, T. D. (1993). Destabilization of tracts of simple repetitive DNA in yeast by mutations affecting DNA mismatch repair. *Nature* **365,** 274–276.
60. Miret, J. J., Pessoa-Bandao, L., and Lahue, R. S. (1997). Instability of CAG and CTG trinucleotide repeats in *Saccahromyces cerevisiae. Mol. Cell. Biol.* **17,** 3382–3387.
61. Schweitzer, J. K., and Livingston, D. M. (1997). Destabilization of CAG trinucleotide repeat tracts by mismatch repair mutations in yeast. *Hum. Mol. Genet.* **6,** 349–355.
62. Pearson, C. E., Ewel, A., Acharya, S., Fishel, R. A., and Sinden, R. R. (1997). Human MSH2 binds to trinucleotide repeat DNA structures associated with neurodegenerative diseases. *Hum. Mol. Genet.* **6,** 1117–1123.
63. Richards, R. I., Holman, K., Yu, S., and Sutherland, G. R. (1993). Fragile X syndrome unstable element, $p(CGG)_n$, and other simple tandem repeat sequences are binding sites for specific nuclear proteins. *Hum. Mol. Genet.* **2,** 1429–1435.
64. Yano-Yanagisawa, H., Li, Y., Wang, H., and Kohwi, Y. (1995). Single-stranded DNA binding proteins isolated from mouse brain recognize specific trinucleotide repeat sequences *in vitro*. *Nucleic Acids Res.* **23,** 2654–2660.
65. Timchenko, L. T., Timchenko, N. A., Caskey, C. T., and Roberts, R. (1996). Novel proteins with binding specificity for DNA CTG repeats and RNA CUG repeats: implications for myotonic dystrophy. *Hum. Mol. Genet.* **5,** 115–121.
66. Deissler, H., Behn-Krappa, A., and Doerfler, W. (1996). Purification of nuclear proteins from human HeLa cells that bind specifically to the unstable tandem repeat $(CGG)_n$ in the human FMR1 gene. *J. Biol. Chem.* **271,** 4327–4334.
67. Hansen, R. S., Canfield, T. K., Lamb, M. M., Gartler, S. M., and Laird, C. D. (1993). Association of fragile X syndrome with delayed replication of the *FMR1* gene. *Cell* **73,** 1403–1409.
68. Hansen, R. S., Canfield, T. K., Fjeld, A. D., Mumm, S., Laird, C. D., and Gartler, S. M. (1997). A variable domain of delayed replication in *FRAXA* fragile X chromosomes: X inactivation-like spread of late replication. *Proc. Natl. Acad. Sci. USA* **94,** 4587–4592.
69. Fry, M., and Loeb, L. A. (1994). The fragile X syndrome $d(CGG)_n$ nucleotide repeats form a stable tetrahelical structure. *Proc. Natl. Acad. Sci. USA* **91,** 4950–4954
70. Rill, R. L., and Hecker, K. H. (1996). Sequence-specific actinomycin D binding to single-stranded DNA inhibits HIV reverse transcriptase and other polymerases. *Biochemistry* **35,** 3525–3533.
71. Hill, T. A., Tecklenburg, M. L., Pelletier, A. J., and Kuempel, P. L. (1989). *tus,* the trans-acting gene required for termination of DNA replication in *Escherichia coli,* encodes a DNA-binding protein. *Proc. Natl. Acad. Sci. USA* **86,** 1593–1597.
72. Sista, P. R., Mukherjee, S., Patel, P., Khatri, G. S., and Bastia, D. (1989). A host-encoded DNA-binding protein promotes termination of plasmid replication at a sequence-specific replication terminus. *Proc. Natl. Acad. Sci. USA* **86,** 3026–3030.
73. Hidaka, M., Kobayashi, T., Takenaka, S., Takeya, H., and Horiuchi, T. (1989). Purification of a DNA replication terminus (*ter*)

site-binding protein in *Escherichia coli* and identification of the structural gene. *J. Biol. Chem.* **264,** 21031–21037.

74. Parsons, M. A., Izban, M. G., and Sinden, R. R. (1997). Transcription by RNA polymerase II through triplet repeat-containing DNA from the human myotonic dystrophy and fragile X loci. *J. Biol. Chem.*, submitted.

75. Lawson, E. A., and Wells, R. D. (1997). Unpublished work.

76. Koch, K. S., Gleiberman, A. S., Aoki, T., Leffert, H. L., Feren, A., Jones, A. L., and Fodor, E. J. (1995). Discordant expression and variable numbers of neighboring GGA- and GAA-rich triplet repeats in the 3′ untranslated regions of two groups of messenger RNAs encoded by the rat polymeric immunoglobulin receptor gene. *Nucleic Acids Res.* **23,** 1098–1112.

77. Bowater, R. P., Jaworski, A., Larson, J. E., Parniewski, P., and Wells, R. D. (1997). Transcription increases the deletion frequency of CTG · CAG triplet repeat sequences from human neuromuscular disease genes in *E. coli. Nucleic Acids Res.* **25,** 2861–2868.

78. Huang, C.-C., and Hearst, J. E. (1980). Pauses at positions of secondary structure during *in vitro* replication of single-stranded fd bacteriophage DNA by T4 DNA polymerase. *Anal. Chem.* **103,** 127–139.

79. Huang, C.-C., and Hearst, J. E. (1981). Fine mapping of secondary structures of fd phage DNA in the region of the replication origin. *Nucleic Acids Res.* **9,** 5587–5599.

80. Bedinger, P., Munn, M., and Alberts, B. M. (1989). Sequence-specific pausing during *in vitro* DNA replication on double-stranded DNA templates. *J. Biol. Chem.* **264,** 16880–16886.

81. Weaver, D. T., and DePamphilis, M. L. (1984). The role of palindromic and non-palindromic sequences in arresting DNA synthesis *in vitro* and *in vivo. J. Mol. Biol.* **180,** 961–986.

82. Krasilnikov, A. S., Panyutin, I. G., Samadashwily, G. M., Cox, R., Lazurkin, Y. S., and Mirkin, S. M. (1997). Mechanisms of triplex-caused polymerization arrest. *Nucleic Acids Res.* **25,** 1339–1346.

83. Baran, N., Lapidot, A., and Manor, H. (1991). Formation of DNA triplexes accounts for arrests of DNA synthesis at $d(TC)_n$ and $d(GA)_n$ tracts. *Proc. Natl. Acad. Sci. USA* **88,** 507–511.

84. Rao, B. S., Manor, H., and Martin, R. G. (1988). Pausing in simian virus 40 DNA replication by a sequence containing $(dG\text{-}dA)_{27} \cdot (dT\text{-}dC)_{27}$. *Nucleic Acids Res.* **16,** 8077–8094.

85. Rao, B. S. (1994). Pausing of simian virus 40 DNA replication fork movement in vivo by $(dG\text{-}dA)_n \cdot (dT\text{-}dC)_n$ tracts. *Gene* **140,** 233–237.

86. Dayn, A., Samadashwily, G. M., and Mirkin, S. M. (1992). Intramolecular DNA triplexes: unusual sequence requirements and influence on DNA polymerization. *Proc. Natl. Acad. Sci. USA* **89,** 11406–11410.

87. Samadashwily, G. M., Dayn, A., and Mirkin, S. M. (1993). Suicidal nucleotide sequences for DNA polymerization. *EMBO J.* **12,** 4975–4983.

88. Giovannangeli, C., Thuong, N., and Hélène, C. (1993). Oligonucleotide clamps arrest DNA synthesis on a single-stranded DNA target. *Proc. Natl. Acad. Sci. USA* **90,** 10013–10017.

89. Hélène, C., and Toulmé, J.-J. (1990). Specific regulation of gene expression by antisense, sense and antigene nucleic acids. *Biochim. Biophys. Acta* **1049,** 99–125.

90. Brinton, B. T., Caddle, M. S., and Heintz, N. H. (1991). Position and orientation-dependent effects of a eukaryotic Z-triplex DNA motif on episomal DNA replication in COS-7 cells. *J. Biol. Chem.* **266,** 5153–5161.

91. Sen, D., and Gilbert, W. (1990). A sodium-potassium switch in the formation of four-stranded G4-DNA. *Nature* **344,** 410–414.

92. Woodford, K. J., Howell, R. M., and Usdin, K. (1994). A novel K^+-dependent DNA synthesis arrest site in a commonly occurring sequence motif in eukaryotes. *J. Biol. Chem.* **269,** 27029–27035.

93. Howell, R. M., Woodford, K. J., Weitzmann, M. N., and Usdin, K. (1996). The chicken β-globin gene promoter forms a novel "cinched" tetrahelical structure. *J. Biol. Chem.* **271,** 5208–5214.

94. Weitzmann, M. N., Woodford, K. J., and Usdin, K. (1997). DNA secondary structures and the evolution of hypervariable tandem arrays. *J. Biol. Chem.* **272,** 9517–9523.

95. Streisinger, G., Okada, Y., Emrich, J., Newton, J., Tsugita, A., Terzaghi, E., and Inoue, M. (1966). Frameshift mutations and the genetic code. *Cold Spring Harbor Symp. Quant. Biol.* **31,** 77–84.

96. Kunkel, T. A. (1990). Misalignment-mediated DNA synthesis errors. *Biochemistry* **29,** 8003–8011.

97. Kunkel, T. A. (1992). DNA replication fidelity. *J. Biol. Chem.* **267,** 18251–18254.

98. Ripley, L. S. (1990). Frameshift mutation: determinants of specificity. *Annu. Rev. Genet.* **24,** 189–213.

99. Farabaugh, P. J. (1978). Genetic studies of the *lac* repressor VII. On the molecular nature of spontaneous hotspots in the *lacI* gene of *Escherichia coli. J. Mol. Biol.* **126,** 847–863.

100. Canceill, D., and Ehrlich, S. D. (1996). Copy-choice recombination mediated by DNA polymerase III holoenzyme from *Escherichia coli. Proc. Natl. Acad. Sci. USA* **93,** 6647–6652.

101. d'Alencon, E., Petranovic, M., Michel, B., Noirot, P., Aucouturier, A., Uzest, M., and Ehrlich, S. D. (1994). Copy-choice illegitimate DNA recombination revisited. *EMBO J.* **13,** 2725–2734.

102. Albertini, A. M., Hofer, M., Calos, M. P., and Miller, J. H. (1982). On the formation of spontaneous deletions: the importance of short sequence homologies in the generation of large deletions. *Cell* **29,** 319–328.

103. Trinh, T. Q., and Sinden, R. R. (1993). The influence of primary and secondary DNA structure in deletion and duplication between direct repeats in *Escherichia coli. Genetics* **134,** 409–422.

104. Oron-Karni, V., Filon, D., Rund, D., and Oppenheim, A. (1997). A novel mechanism generating short deletion/insertions following slippage is suggested by a mutation in the human a_2-globin gene. *Hum. Mol. Genet.* **6,** 881–885.

105. Casimir, C. M., Bu-Ghanim, H. N., Rodaway, A. R. FA. R., Bentley, D. L., Rowe, P., and Segal, A. W. (1991). Autosomal recessive chronic granulomatous disease caused by deletion at a dinucleotide repeat. *Proc. Natl. Acad. Sci. USA* **88,** 2753–2757.

106. Fitches, A. C., May, S. J., and Olds, R. J. (1996). A novel antithrombin gene mutation: slippage and mispairing as a mechanism of genetic disease. *Pathology* **28,** 339–342.

107. Buiser, R. G., DeStefano, J. J., Mallaber, L. M., Fay, P. J., and Bambara, R. A. (1991). Requirements for the catalysis of strand transfer synthesis by retroviral DNA polymerases. *J. Biol. Chem.* **266,** 13103–13109.

108. DeStefano, J. J., Mallaber, L. M., Rodriguez-Rodriguez, L., Fay, P. J., and Bambara, R. A. (1992). Requirements for strand transfer between internal regions of heteropolymer template by human immunodeficiency virus reverse transcriptase. *J. Virol.* **66,** 6370–6378.

109. Lechner, R. L., Engler, M. J., and Richardson, C. C. (1983). Characterization of strand displacement synthesis catalyzed by bacteriophage T7 DNA polymerase. *J. Biol. Chem.* **258,** 11174–11184.

110. Ohshima, A., Inouye, S., and Inouye, M. (1992). *In vivo* duplication of genetic elements by the formation of stem-loop DNA without an RNA intermediate. *Proc. Natl. Acad. Sci. USA* **89,** 1016–1020.

111. Bi, X., and Liu, L. F. (1996). DNA rearrangement mediated by inverted repeats. *Proc. Natl. Acad. Sci. USA* **93,** 819–823.
112. Bi, X., and Liu, L. F. (1996). A replicational model for DNA recombination between direct repeats. *J. Mol. Biol.* **256,** 849–858.
113. Gordenin, D. A., Malkova, A. L., Peterzen, A., Kulikov, V. N., Pavlov, Y. I., Perkins, E., and Resnick, M. A. (1992). Transposon Tn5 excision in yeast: influence of DNA polymerases α, δ, and ε, and repair genes. *Proc. Natl. Acad. Sci. USA* **89,** 3785–3789.
114. Khorana, H. G., Buchi, H., Ghosh, H., Gupta, N., Jacob, T. M., Kossel, H., Morgan, R., Narang, S. A., Ohtsuka, E., and Wells, R. D. (1966). Polynucleotide synthesis and the genetic code. *Cold Spring Harbor Symp. Quant. Biol.* **31,** 39–49 (1966).
115. Ohshima, K., Montermini, L., Wells, R. D., and Pandolfo, M. (1998). Inhibitory effects of expanded GAA · TTC triplet repeats from intron I of the friedreich ataxia gene on transcription and replication *in vivo*. *J. Biol. Chem.* [In press]

Appendix: Replication Fork Progression Through CTG · CAG Triplet Repeats *in Vitro*

I. INTRODUCTION

Recent studies on human hereditary neuromuscular and neurodegenerative diseases have revealed that the expansion of a simple nucleotide triplet repeat is associated with these diseases (reviewed in [1–5]). Although intense investigation has been focused on the mechanism by which expansion occurs, it is not yet understood.

Kang *et al.* [6] demonstrated that both expansion and deletion of triplet repeats could be observed during plasmid replication in *E. coli.* Based on these studies it was suggested that polymerase slippage during DNA replication could be the mechanism underlying expansion and deletion of triplet repeats.

In an effort to duplicate these observations *in vitro,* we have studied the effect of the presence of CTG · CAG triplet repeats during DNA replication using a DNA replication system reconstituted with *E. coli* replication proteins.

II. *oriC*-DIRECTED REPLICATION OF TEMPLATES CARRYING CTG · CAG TRIPLET REPEATS *IN VITRO*

A series of minichromosomes was constructed by replacing the regions containing the pBR322 replication origin and the ampicillin-resistant gene of pBROTB535 type II [7], which carries *oriC,* with pUC19 (pHH12) and its derivatives containing CTG · CAG triplet repeats (50–255 repeats) (Fig. 45-6A). Plasmid DNAs were prepared from *E. coli* K38*tus:kan*^{R} cells grown continuously without passage through stationary phase. During the cloning and plasmid preparation processes, we observed deletion of triplet repeats. As expected, longer triplet repeats were less stable (data not shown).

Supercoiled (form I) DNA molecules containing intact CTG · CAG triplet repeats (50–255 repeats) were purified and used as DNA templates in the *in vitro oriC* DNA replication system. Experiments were performed using a vector plasmid, pHH12, and its derivatives containing 50, 100, 175, and 255 CTG · CAG repeats. Triplet repeats were inserted in the orientation such that the CTGs were in the leading-strand template. We were unable to prepare plasmids DNAs where the CTGs were in the opposite orientation in large scale and with sufficient purity for use as templates because of the extreme instability of the sequences when the CTGs were present in the lagging-strand template. Only the results using 255 triplet repeats are shown here. Results obtained with shorter tracks of repeats were identical.

The DNA products from *oriC* replication reactions using either pHH12 or pHH12-CTG_{255} as templates were fractionated by native (Fig. 45-5A) and alkaline (Fig. 45-5B) agarose gel electrophoresis. Triplet repeats had no effect on monomer accumulation *via* either the Topo III- or Topo IV-catalyzed decatenation pathway during DNA replication (Fig. 45-5A), nor was the size distribution of either the leading- or lagging-strand products affected (Fig. 45-5B). These results suggested that the presence of the triplet repeats in the template elicited neither strong pausing of the replication forks nor premature termination of DNA replication in any gross sense during *oriC* DNA replication.

To directly examine the effect of triplet repeats on replication fork progression, we analyzed the size of the leading strands generated in the presence of Tus, the replication fork arrest protein. The minichromosomes used in these studies consist of pUC19 DNA carrying *oriC* and two *TerB* sequences that bind Tus, the replication fork arrest protein [8, 9]. The *TerB* sequences are separated by 1 kb and are oriented to exclude the passage of replication forks between them (Fig. 45-6A). Thus, in the presence of Tus, bidirectional replication from *oriC* will generate two distinctly sized leading-strands, one 2.5 kb in length and the other 2.8 kb in

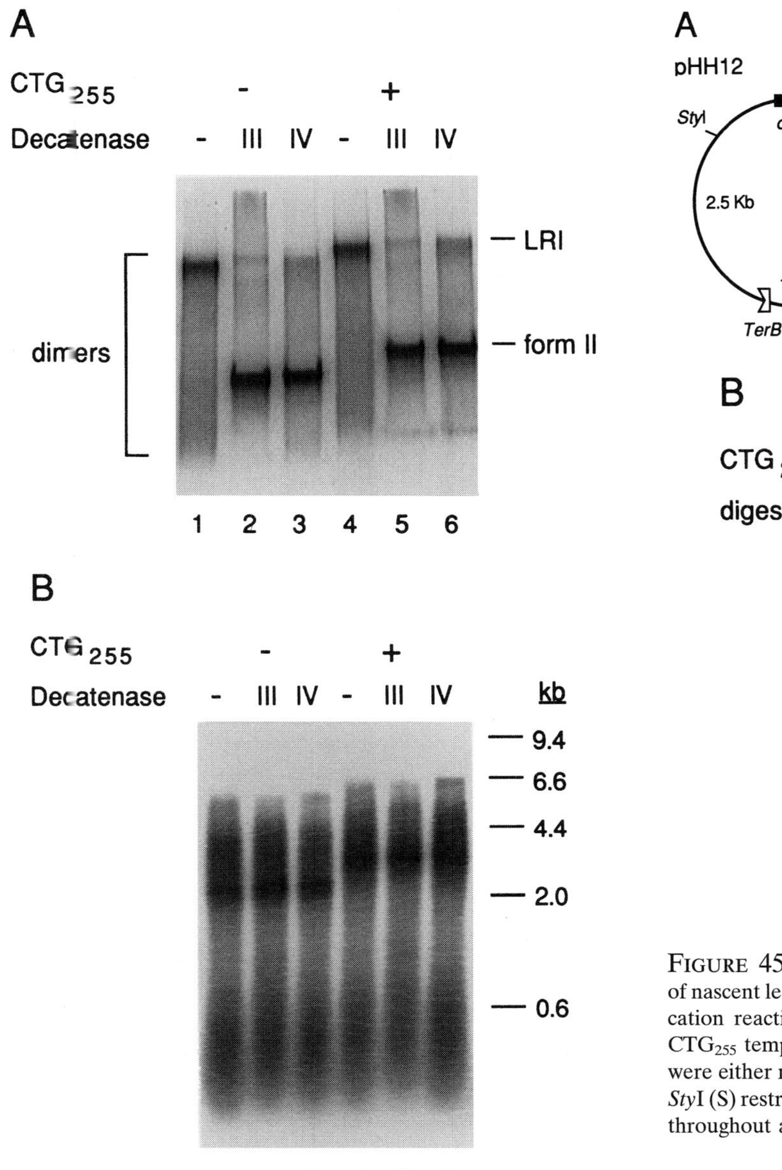

FIGURE 45-5 (A) Replication products of CTG_{255}-containing templates as analyzed by native agarose gel electrophoresis. Replication reactions contained DNA gyrase, either the pHH12 or pHH12-CTG_{255} template as indicated, and no additional topoisomerase, topoisomerase III, or topoisomerase IV. LRI, late replicative intermediate. Dimers, multiply linked form II:form II DNA dimers. (B) Replication products of CTG_{255}-containing templates as analyzed by alkaline agarose gel electrophoresis. The same reaction mixtures described for (A) were analyzed on denaturing gels.

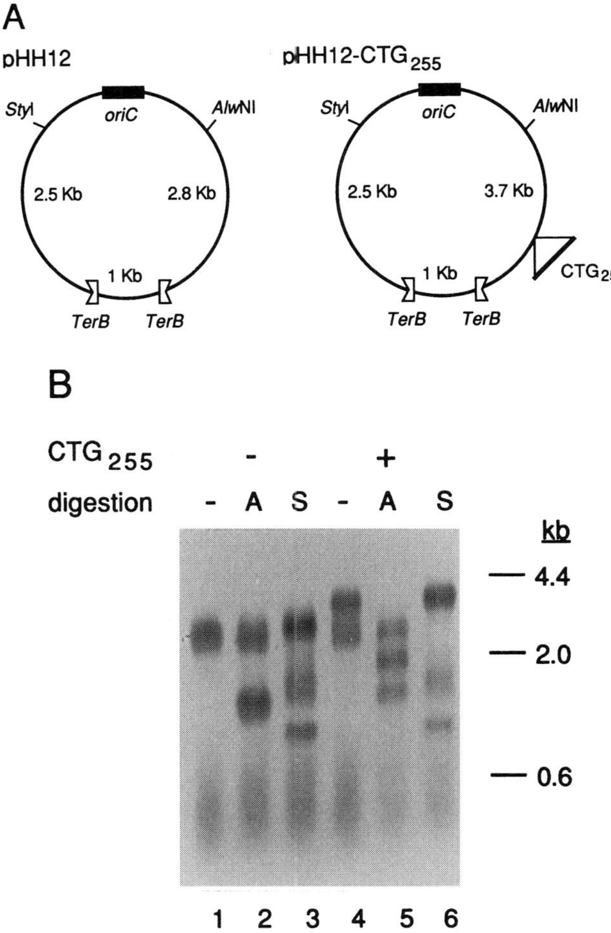

FIGURE 45-6 (A) Map of plasmid templates. (B) Identification of nascent leading strands by restriction endonuclease analysis. Replication reactions contained Tus and either the pHH12 or pHH12-CTG_{255} template as indicated. After replication, the DNA products were either not treated (−) or digested with either the *Alw*NI (A) or *Sty*I (S) restriction endonuclease and then analyzed by electrophoresis throughout an alkaline agarose gel.

length [7, 10]. The insertion of CTG_{255} changes the size of the full-length, clockwise-moving leading strand from 2.8 to 3.7 kb in length (Fig. 45-6A). The CTG tract is 0.5 kb from the *TerB* sequence. Thus, any pausing or termination of replication induced by the triplet repeats would manifest itself by the presence of leading strands that were roughly 3.2 kb in length.

The DNA products from *oriC* replication reactions in the presence of Tus using either the pHH12 template or the pHH12-CTG_{255} template were fractionated by alkaline agarose gel electrophoresis (Fig. 45-6B). The DNA products with the pHH12 template migrated as two populations, one as an apparently unique, but broad, band at about 2.5–3 kb and one distribution centered at 0.5 kb (Fig. 45-6B, lane 1). The DNA products with the pHH12-CTG_{255} template migrated as three populations, two distinct bands at 2.5 and 3.7 kb and one distribution centered at around 0.5 kb (Fig. 45-6B, lane 4).

The distinct bands in each case could be shown to represent the leading strands by restriction enzyme digestions. As schematized in Fig. 45-6A, digestion of the DNA products with *Alw*NI should cleave the clockwise-moving leading strand. On the other hand, digestion of the DNA products with *Sty*I should cleave the counter-clockwise-moving leading strand. The predicted cleavages were observed (Fig. 45-6B). In the case of the pHH12 template, *Alw*NI digested the 2.8-kb band and left the 2.5-kb band intact (Fig. 45-6B, lane 2), whereas *Sty*I digested the 2.5-kb band and left the 2.8-kb band intact (Fig. 45-6B, lane 3). These results confirmed that DNA replication proceeded bidirectionally on this plasmid DNA.

A similar confirmation was obtained with the DNA products from the pHH12-CTG_{255} template. Here, *Alw*NI digested the 3.7-kb band (Fig. 45-6B, lane 5), whereas *Sty*I digested the 2.5-kb band (Fig. 45-6B, lane 6). No distinct clockwise-moving leading strands were observed in the 3.2-kb size range, indicating that if there were either replication fork pausing or arrest at the triplet repeats, it happened at a frequency that was too low to detect by this analysis.

III. ROLLING-CIRCLE DNA REPLICATION WITH TEMPLATES CARRYING CTG · CAG TRIPLET REPEATS

During standard *oriC* DNA replication, the active DNA template replicates only once. Deletion of triplet repeats in *E. coli* depended on the generation time. Thus, it is possible that instability of triplet repeats would only be observed as a result of successive rounds of DNA replication.

Our previous studies have demonstrated that the primase concentration in the *oriC* replication reaction modulates the mode of DNA replication [7]. High concentrations of primase are required for the proper assembly of the replication forks and bidirectional DNA replication. At low concentrations of primase, leading- and lagging-strand synthesis are uncoupled and DNA replication proceeds unidirectionally. Under these conditions, undirectional DNA replication can convert to rolling-circle DNA replication, which generates multigenome-length DNA products. We have used this rolling-circle DNA replication system to examine the stability of the triplet repeats after multiple rounds of DNA replication.

Kinetic analysis of *oriC* replication at a low concentration of primase (Fig. 45-7) showed that the presence of 255 CTG · CAG triplet repeats on the template had no obvious effect on the incorporation of labeled precursor into acid-insoluble product during rolling-circle replication. The rates of DNA synthesis using either the pHH12 template or the pHH12-CTG_{255} template were

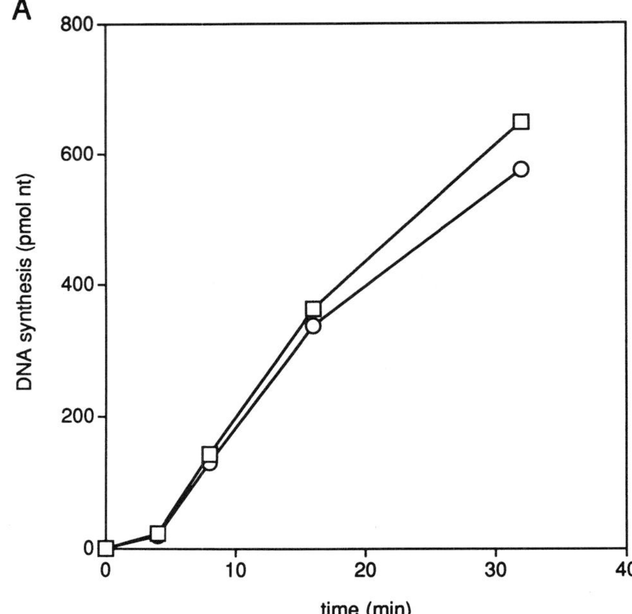

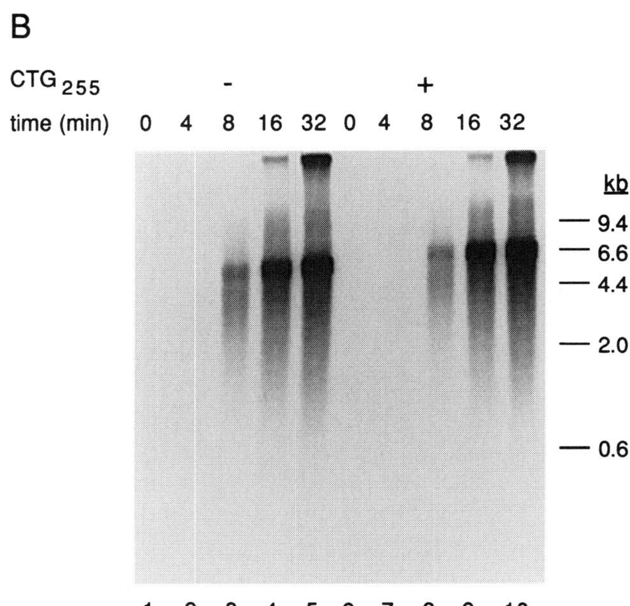

FIGURE 45-7 (A) Kinetics of DNA synthesis during rolling circle replication. □, pHH12; ○, pHH12-CTG_{255}. (B) Rolling circle replication reaction products as analyzed by alkaline agarose gel electrophoresis. Replication reactions contained 1/10 the usual concentration of primase and either the pHH12 or pHH12-CTG_{255} template as indicated.

identical (Fig. 45-7A). Product analysis on alkaline gels (Fig. 45-7B) did not reveal any differences between the DNA products made from either the pHH12 template or the pHH12-CTG_{255} template.

We anticipated that replication products with discrete sizes would accumulate if the replication fork paused at the CTG · CAG repeats during rolling-circle DNA replication. However, this was not the case. As shown in Fig. 45-7B, only longer, multigenome-length replication products accumulated. These results suggested that replication forks could negotiate through a long CTG · CAG repeat multiple times without pausing during rolling-circle DNA replication.

IV. THE DIFFERENCES BETWEEN DNA POLYMERASES AND REPLICATION FORKS

The data presented here indicate that, in large part, replication forks can progress through long tracts of triplet repeats without error. On the other hand, Wells and his co-workers [11, 12] have clearly shown that these triplet repeats can be roadblocks for DNA polymerases and halt DNA strand synthesis. These results appear to be a paradox. However, if one considers the differences between a DNA polymerase elongating on a single-stranded template and a replication fork moving through duplex DNA, the results can be explained.

Rearrangements of triplet repeat sequences to produce deletions as a result of nascent DNA synthesis likely require a reorganization of the secondary structure of the template sequence to induce a polymerase pause. At the replication fork, the only place where extensive single-stranded template exists is on the lagging-strand side. On the leading strand, the polymerase interacts with the replication fork helicase *via* a protein–protein interaction [13]; thus it is very likely that the unwound leading-strand template is passed directly from the helicase to the polymerase. On the lagging strand, there is always an amount of single-stranded template available that is equivalent to the size of the Okazaki fragment that is undergoing synthesis.

Thus, a DNA polymerase elongation assay resembles the situation on the lagging strand at the replication fork. The polymerase is advancing along a template that is already single-stranded. This would allow the triplet repeats to assume an alternative secondary structure that would be capable of impeding the polymerase in the absence of a helicase to unwind the DNA. This is also, presumably, why CTG repeats in the lagging-strand template are very unstable, generating deletions as predicted by the slipped-mispairing model [14, 15] and why we could not obtain templates with the CTGs in this orientation. Expansions of the triplet repeat sequences at the replication fork would require a rearrangement of the repeat sequences in the nascent leading-strand. Because these sequences are essentially completely double-stranded when synthesized, their rearrangement would be considerably less frequent than rearrangements in the lagging-strand template.

Results presented here demonstrate that replication forks are typically capable of progressing through regions of the template containing CTG · CAG triplet repeats and synthesizing DNA strands unimpeded. It is possible, however, that a small population of replication forks might not replicate these regions accurately. These rare events would probably not be detectable in the *in vitro* DNA replication assay, although they may be sufficient to generate deletions and expansions of triplet repeats *in vivo*. Alternatively, there may be cellular factors that bind to the triplet repeats and either induce isomerization of their secondary structure or cause replication fork pausing, thereby exacerbating the instability of the sequences.

References

1. Davies, K., and Warren, S. (Eds.) (1993). "Genome Analysis," Vol. 7. Cold Spring Harbor Laboratory, Cold Spring Harbor, NY.
2. Bates, G., and Lehrach, H. (1994). *BioEssays* **6,** 277–284.
3. Sutherland, G. R., and Richards, R. I. (1995) *Proc. Natl. Acad. Sci. USA* **92,** 36736–36741.
4. Panzer, S., Kuhl, D. P. A., and Caskey, C. T. (1995). *Stem Cells* **13,** 3636–3641.
5. Krahe, R., and Ashizawa, T. (1995). *In* "Hypervariable Genetic Markers" (J. Wertherall and D. Growth, Eds.), pp. 29–60. CRC Press, Boca Raton, FL.
6. Kang, S., Jaworski, A., Ohshima, K., and Wells, R. D. (1995). *Nature Genet.* **10,** 213–218.
7. Hiasa, H., and Marians, K. J. (1994). *J. Biol. Chem.* **269,** 6058–6063.
8. Pelletier, A. J., Hill, T. M., and Kuempel, P. L. (1989). *J. Bacteriol.* **171,** 1739–1741.
9. Hidaka, M., Akiyama, M., and Horiuchi, T. (1988). *Cell* **55,** 467–475.
10. Hiasa, H., and Marians, K. J. (1994). *J. Biol. Chem.* **269,** 26959–26968.
11. Kang, S., Ohshima, K., Shimizu, M., Amirhaeri, S., and Wells, R. D. (1995). *J. Biol. Chem.* **270,** 27014–27021.
12. Ohshima, K., and Wells, R. D. (1997). *J. Biol. Chem.* **272,** 16798–16806.
13. Kim., S., Dallmann, H. G., McHenry, C. S., and Marians, K. J. (1996). *Cell* **84,** 643–650.
14. Wells, R. D., and Sinden, R. R. (1993). *In* "Genome Analysis," Vol. 7 (K. R. Davis and S. T. Warren, Eds.), pp. 107–138. Cold Spring Harbor Laboratory, Cold Spring Harbor, NY.
15. Richards, R. I., and Sutherland, G. R. (1994). *Nature Genet.* **6,** 114–116.

PART XVI

Oligoglutamine Effects

Amino Acid Repeats in Proteins and the Neurological Diseases Produced by Polyglutamine

HOWARD GREEN Department of Cell Biology, Harvard Medical School, Boston, Massachusetts 02115
PHILIPPE DJIAN CNRS, Centre de Recherche sur l'Endocrinologie Moléculaire et le Développement, 92190 Meudon, France

I. CODON REPEATS IN EUKARYOTES

The study of homogeneous amino acid repeats in proteins has emphasized the role of polyglutamine-containing proteins in producing human diseases of the central nervous system, but amino acid repeats are found in (presumably normal) proteins of all eukaryotic organisms. The particularly long stretches of polyglutamine in disease-producing proteins of the human must therefore be considered as an aberration of a process of repeat generation that is widespread in animals, plants, and microorganisms. Repeated codon duplications extend the primary sequence of the encoded protein; subsequent divergence of the nucleotide sequence, a process for which there is clear evidence, can then adapt the protein to a function. While an expanded polyCAG is the direct cause of the human diseases, the evolutionary significance of polyCAG extends to the new sequences generated from it by nucleotide substitutions. It is necessary to consider:

1. that some codons are more commonly repeated than others;
2. the repeat lengths that are encountered;
3. how originally homogeneous repeats are modified;
4. that some regions within a segment of codon repeats are subject to further expansion, whereas other regions are quite stable.

Triplet repeats also occur outside coding regions and some of these repeats produce human disease; we will confine ourselves here to repeats that are translated into polyamino acid sequences in proteins.

II. POLYGLUTAMINE, THE MOST COMMON AMINO ACID REPEAT: THE ENCODING polyCAG AND ITS MODIFICATION

While absent from prokaryotic proteins, homogeneous amino acid repeats of substantial length are not rare in eukaryotic proteins. The nature of the repeated amino acids and the responsible codons was the subject of earlier studies [1–5].

We have searched the approximately 90,000 sequences stored in the PIR-Protein database for proteins containing uninterrupted amino acid repeats of more than 15 residues, excluding the polyglutamine-containing proteins associated with seven known human diseases of the nervous system. The FIND patterns program within the GCG program (Wisconsin package) was used to obtain the repeats, which were sorted into seven categories of length (Table 46-1). Glutamine is overall the most common reiterant, accounting for 40% of the total. Nearly half the amino acids give rise to no repeats of over 15 residues. Repeats of hydrophobic amino acids are uncommon but proline and alanine are found in numerous repeats of over 15 residues. Glutamine, serine, asparagine, proline, and aspartic acid occur in repeats of more than 30 residues, glutamine account-

TABLE 46-1 Uninterrupted Amino Acid Repeats in Proteins

	Number of repeats of each residue							
Amino acid residue	16–20	21–25	26–30	31–35	36–40	41–45	46–50	Total
Q	32	11	4	2	2	0	0	51
S	8	5	1	1	0	1	0	16
N	6	6	3	1	0	0	0	16
G	10	0	0	0	0	0	0	10
P	4	2	2	0	0	0	1	9
E	4	3	1	0	0	0	0	8
D	3	3	2	0	0	1	0	9
T	5	1	0	0	0	0	0	6
A	4	1	0	0	0	0	0	5
R	0	0	1	0	0	0	0	1
L	1	0	0	0	0	0	0	1
V,F,Y,M,C,H,I,K,W	0	0	0	0	0	0	0	0
Total	77	32	14	4	2	2	1	132

Note. Duplicate entries and orthologous proteins in different organisms have been eliminated. The few proteins containing multiple segments of uninterrupted repeats have been classified according to the longest segment.

ing for four of the nine examples. It would appear highly significant that with the exception of a rare polymorphic form of TFIID (TBP), which contains 42 repeats [6], there is no protein of animals, plants, or microorganisms that contains over 40 glutamine repeats, a value commonly exceeded in the abnormal proteins associated with human central nervous system disease. The few homopolymers of over 40 repeats were polyserine, polyaspartic acid, and polyproline (one example of each). These polyamino acids have not been associated with central nervous system diseases, but expansion of polyalanine in the HOXD13 gene from the normal value of 15 (barely excluded from the range of Table 46-1) to 22 or more is the cause of synpolydactyly [7].

A summary of the codon sequences corresponding to the glutamine repeats of Table 46-1 is given in Table 46-2. Four of these are in human genes. In most cases, there has been evolution of the nucleotide sequence encoding the glutamine repeats by silent (synonymous) substitution. Only two polyglutamine sequences are encoded exclusively by CAG's (the glucocorticoid receptor and interleukin 2), and only two exclusively by CAA's (protein kinase (EC 2.7.1.37) 2 and membrane protein YOR267C); the vast majority contain mixtures of the two codons, so that the maximal length of uninterrupted triplets is much shorter than that of the encoded polyglutamine. These substitutions would tend to prevent the hairpin formation/slipped-strand mispairing that is thought to lead to the polyCAG expansions in the neurological diseases (see below). If the original reiterant was of the present-day preponderant codon, then 30 were originally CAG reiterants, 19 were CAA reiterants, and 2 are indeterminant. As a potential cause of human disease, the protein TBP seems a better candidate than the others, but its maximum length of polyCAG is only 21 triplets (see below).

The presence of numerous synonymous codons is not always the result of independent substitutions, but can occur by duplications of a segment already containing a synonymous substitution; for example, in regulatory protein nit4, 24 consecutive glutamines of a total of 27 are encoded by $(CAG\ CAA)_{12}$.

It has been noted that the rate of nucleotide substitution in the polyglutamine-encoded repeat region of the *Drosophila* mastermind gene (*mam*) is much higher than elsewhere in the gene [8]. In the *mam* gene of *Drosophila virilis* [9], a sequence of 57 triplets encodes only glutamines and histidines (Fig. 46-1). Of these codons, 31 are CAG, 13 are CAA, 9 are CAT, and 4 are CAC. They are interspersed with no evident periodicity that might suggest that they were the result of secondary duplications. All deviations from polyCAG can be explained by 26 independent nucleotide substitutions occurring only in third-base positions of the original polyCAG. In the homologous gene of *Drosophila melanogaster,* over half of the same sequence is present and again histidines are the only substitutes for glutamines; as they are not located at the same positions as in *D. virilis,* they are the result of independent replacements. It is unlikely that an evolutionary process in which all substitutions are in the third-base position and histidine is the only amino acid replacement could be the result of entirely random mutations. The beginnings of the same process can be seen in the genes for SCA1, SCA2, and DRPLA.

The Sry gene of *Mus musculus* and related species contains several series of CAG repeats, of which the first two are of 11 and 5 codons (Fig. 46-2). In the isologous gene of *Mastomys hildebrantii,* a related rodent, the nucleotide sequence is very similar, but the insertion of a C four codons upstream of the first CAG repeat sequence has shifted the frame, so that the CAG codons corresponding to the first two segments of repeats of *M. musculus* become GCA's, encoding alanine. Downstream, a single nucleotide is deleted from the *Mastomys* sequence and the initial reading frame is restored. In this way, a single insertional event has changed the encoded repeat sequence. Although glutamine-rich regions of proteins have known functions [10], there are very few examples of a function for uninterrupted polyglutamine [11] (but see Gerber *et al.* [12]). The example of *Mastomys* implies that none of three polyglutamine sequences in the Sry protein of *M. musculus* has any function, since in *M. hildebrantii,* the sequence of the first two is changed to polyalanine and the third is completely absent.

The CYC8 gene of yeast [13], identical to *SSN6* (Table 46-2), possesses a segment whose sequence is $(CAN\ GCN)_{32}\ (CAA)_6\ CAG\ (CAA)_{20}\ (CAG)_3\ CAA$, where N is usually A. It seems likely that a GCA codon was generated from CAG CAA by insertion/deletion and a resulting frame shift. The codon pair CAA GCA was then extensively duplicated and both triplets were further substituted. The order of the frame shift and the duplication is opposite to that of the Sry gene.

III. TWO GENES WHOSE CODING REGIONS WERE DERIVED FROM polyCAG

A. Involucrin

1. ORIGIN OF THE CODING REGION

The coding region of the involucrin gene is confined to a single exon and its most abundant codon is CAG, which accounts for 23% of total codons in the human

TABLE 46-2 Triplets Encoding Glutamine Repeats

	Length PolyQ	No. of CAG's	Maximum length polyCAG	No. of CAA's	Maximum length polyCAA
Odd paired -D.mel. A49839	16	9	3	7	2
Probable RNA/SSDNA bindg.HMD1 -S.cer. S60122	16	4	1	12	8
Gap protein hunchback -D.mel. S05548	16	8	4	8	3
*Neurogenic locus mam -D.mel. A33106	16	13	8	3	2
Kinase SLT2-(EC2.7.1.-) -S.cer. S43737	16	11	3	5	1
NAB3 protein-S.cer. S48529	16	4	1	12	8
Protein tyrosine-phosphatase -D.mel. A41622	16	13	7	3	2
Protein kinase (EC2.7.1.37)2 -D.dis. JQ1150	16	0	0	16	16
Intrastrand crosslink recognition protein -S.cer. S37849	16	6	5	10	4
Protein kinase CDC28 -S.cer. OKBYS1	16	7	2	9	4
*Regulatory protein zeste -D.vir. A42020	17	11	5	6	3
Transcription factor UGA35 -S.cer. S48485	17	9	3	8	3
Homeotic abdominal A -D.mel. A35915	17	12	4	5	1
Regulatory protein elf-1 -D.mel. S06206	17	13	6	4	1
Female sterile homeotic protein -D.mel. A43742	17	10	7	7	4
Hypothetical YHR161C -S.cer. S466771	17	6	1	11	2
BMH2 protein -S.cer. S51250	17	4	1	13	6
Notch protein -D.mel. A24420	17	13	4	4	1
Alpha-beta gliadin precursor-Wheat EEWTA	18	9	9	9	9
Kinase related protein, sevenless -D.mel. A35774	18	11	4	7	4
Ovarian protein -D.mel. S24577	18	8	3	10	3
Involucrin -M.mus. A49377	19	18	18	1	1
Glucocorticoid receptor -R.rat QRRTG	19	19	19	0	0
Alpha-fetoprotein enhancer binding -H.sap. A41948	19	6	3	13	10
Hypothetical protein YIL105C -S.cer. S48467	19	9	5	10	7
Homeotic protein Abdominal-A homolog -S.gre. S13803	20	16	9	4	2
Hypothetical protein YDR228C -S.cer. S59435	20	12	6	8	4
Chromogranin A precursor R. nor. P10354	20	16	16	4	4
*Homeotic protein prospero -D.mel S24548	20	14	3	6	2
Probable membrane protein YOR267C -S.cer. S67164	20	0	0	20	20
CCAAT/enhancer binding protein, C/EBP -D.mel A43481	20	14	4	6	2
Hypothetical protein 1-A.gam.-S27770	20	17	14	3	3
Interleukin 2-M.mus I68871	21	21	21	0	0
Transcription factor Oct-3 -H.sap. S29334	21	16	6	5	1
Hypothetical protein (AAC3) -D.dis. S05357	21	2	1	19	8
Transcription factor btd -D.sp. S39356	22	13	4	9	3
ANP1 protein -S.cer. S50508	22	8	4	14	5
Merozoite surface antigen 1 precursor -P.viv. A39401	23	2	1	21	15
Hypothetical protein YDR145w -S.cer. S57972	23	11	5	12	6
Transcription activator GAL11 -S.cer. S66736	23	10	2	13	3
HBRM protein -H.sap. S39580	23	19	13	4	2
RAE-28 -M.mus. I53172	24	18	8	6	2
Bib protein -D.mel. S09699	25	17	5	8	3
Regulatory protein nit-4 -N.cra. A41696	27	14	2	13	1
Polyomavirus enhancer binding protein 2 -M.mus. A48233	28	21	9	7	2

(continues)

TABLE 46-2 (*continued*)

	Length PolyQ	No. of CAG's	Maximum length polyCAG	No. of CAA's	Maximum length polyCAA
Hypothetical protein 2 -c.cap. S31574	30	17	6	13	4
Hypothetical protein YMR164C-S. cer. S54522	30	12	2	18	6
*SSN6 protein -S.cer. S25365	31	4	3	27	20
*Mopa box -M.mus. A26892	33	22	6	11	3
Regulatory protein SNF5 -S.cer. RGBYS5	37	18	8	19	6
Transcription initiation factor IID(TBP) -H.sap. TWHU2D	38	32	18	6	3

Note. Asterisks indicate presence of more than a single segment of repeats; only the longest segment is listed.

	H	**Q**	**Q**	**Q**	**Q**	**Q**	**H**	**Q**
D.vir	CAT	CAG	CAG	CA**A**	CAG	CAG	CA**C**	CAG
D.mel	GAT	CA**T**	CA**C**	CAG	CA**A**	CAG	CA**A**	CAG
	D	**H**	**H**	**Q**	**Q**	**Q**	**Q**	**Q**
	Q	**H**	**Q**	**Q**	**Q**	**Q**	**H**	**Q**
D.vir	CA**A**	CA**T**	CAG	CAG	CAG	CAG	CA**T**	CAG
D.mel	CA**C**	CAG	CA**C**	CAG	CAG	CAG	CAG	CAG
	H	**Q**	**H**	**Q**	**Q**	**Q**	**Q**	**Q**
	Q	**Q**	**Q**	**H**	**Q**	**Q**	**H**	**Q**
D.vir	CAG	CAG	CA**A**	CA**T**	CAG	CA**A**	CA**T**	CAG
D.mel	CA**A**	CA**A**	–	–	–	–	–	–
	Q	**Q**						
	Q	**H**	**Q**	**Q**	**Q**	**Q**	**Q**	**H**
D.vir	CAG	CA**T**	CAG	CAG	CA**A**	CAG	CAG	CA**T**
D.mel	–	–	–	–	–	–	–	–
	Q	**Q**	**Q**	**Q**	**H**	**Q**	**Q**	**Q**
D.vir	CAG	CAG	CAG	CAG	CA**T**	CAG	CAG	CAG
D.mel	–	–	–	–	–	–	–	–
	Q	**H**	**Q**	**Q**	**Q**	**Q**	**Q**	**Q**
D.vir	CA**A**	CA**C**	CAG	CA**A**	CAG	CA**A**	CA**A**	CA**A**
D.mel	–	–	–	–	–	–	–	–
	H	**H**	**H**	**Q**	**Q**	**Q**	**Q**	**Q**
D.vir	CAT	CA**C**	CA**C**	CAG	CAG	CAG	CA**A**	CA**A**
D.mel	–	–	–	–	–	–	–	–
	Q	**G**	**G**	**G**				
D.vir	CA**A**	GGC	GGC	GGT				
D.mel	–	GGC	GGC	GGA				
		G	**G**	**G**				

FIGURE 46-1 Interruption of the polyCAG region of the *mastermind* gene of *Drosophila virilis* and *melanogaster* by third-base substitutions. Sequence of *D. virilis* begins with codon 364 of the complete sequence (Accession No. M92914). Nucleotide substitutions (bold) do not correspond in the two species and must be independent. All amino acid replacements are histidines. Dashes indicate codons missing from *D. melanogaster*.

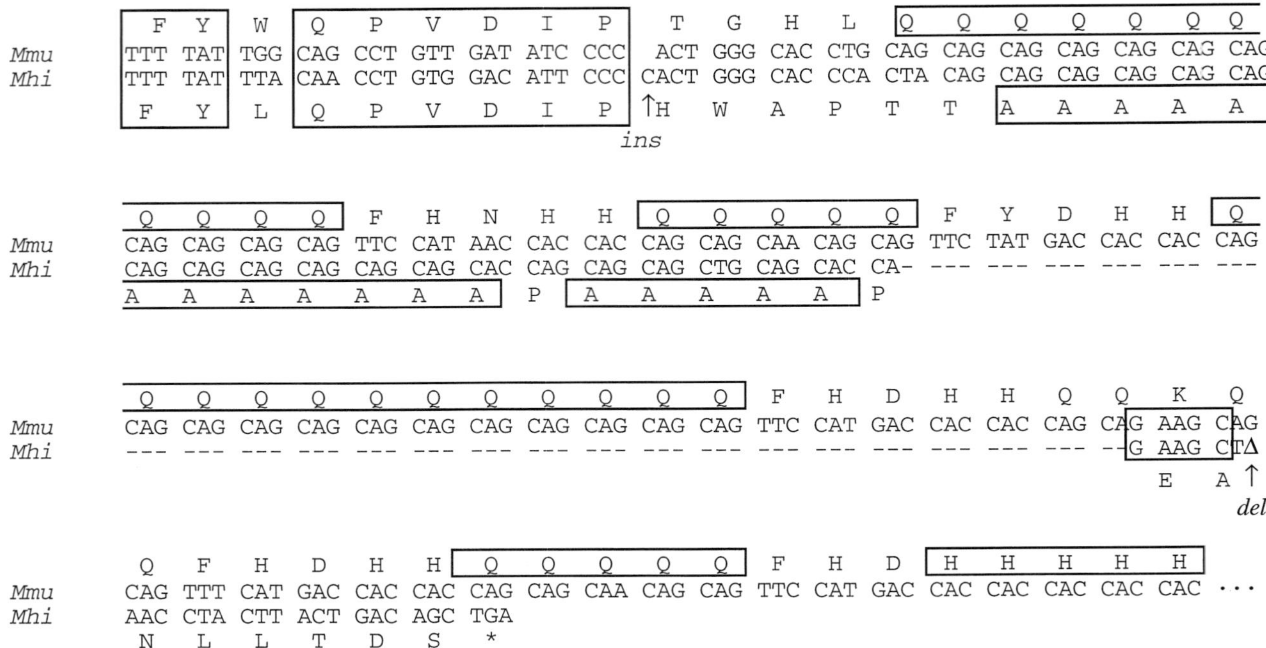

FIGURE 46-2 The 3′ end of the Sry gene of *Mus musculus* and *Mastomys hildebrantii*. Both genes encode several groups of amino acid repeats. In *Mastomys,* insertion of a single C changes the frame and the encoded repeated amino acid from glutamine to alanine. The original frame is restored by deletion of a C downstream of the alanine repeats (from [125]).

gene [14, 15]. If codons that could have been derived from CAG by a single nucleotide substitution (GAG, CTG, AAG, CAC, CAA, CAT, and CCG) are added, the total accounts for 65% of the coding region. Since glutamine residues are essential to the function of involucrin as a substrate for transglutaminase-catalyzed cross-linking [16, 17], the coding region of the gene at an early stage of its evolution probably consisted of polyCAG, which was later modified by nucleotide substitution. This process results in a low T content of the sense strand of the coding region (10% in the human gene). Because of the nucleotide substitutions, there are usually no long sequences of uninterrupted polyCAG. In the human gene, for example, the longest such sequence is $(CAG)_5$. However, the mouse gene has a sequence of $(CAG)_{18}$, of which 14 are of recent origin, as they are not present in the rat [18]. The duplication of CAG codons in the involucrin gene has occurred in both ancient and recent evolution of the gene.

2. CONTINUING ADDITION OF 10-CODON REPEATS IN THE HIGHER PRIMATES

The two features that make the involucrin gene relevant to the diseases of polyglutamine are the particular significance of glutamine, as discussed, and the prominence of repeat addition. Although all mammalian involucrin genes contain a segment of short CAG-rich tandem repeats in the coding region, the common lineage of the higher primates departed from the pattern found in other mammals [19, 20] by generating a new segment of repeats, each consisting of 10 codons, of which 3 were CAG's (Fig. 46-3). The process of repeat addition to this region continued throughout the subsequent divergence of the higher primates. Because of the presence of marker nucleotides in the repeats, it has been possible to trace this process in detail and to show that repeat addition has been vectorial. The early region, which is the 3′ end of the present-day segment of repeats, was generated in the common lineage of all higher primates (Fig. 46-3) prior to the divergence of the new-world and old-world forms [21], which occurred about 40 million years ago. After this divergence, further repeats were added to both old- and new world sublineages (middle region), the hot spot for repeat addition moving in the 5′ direction. When the old-world monkeys diverged from the apes, about 20 million years ago, each sublineage continued to add repeats independently. Then, after final divergence of present-day species, most, including the human, continued to add repeats (the late region). It is clear that the process of building a new segment of repeats in the involucrin gene is coextensive with the history of the higher primates, as it began in the common lineage nearly 65 million years ago and has continued in most sublineages up to the present time [15].

The human late region, not shared by other primates, contains predominantly two types of repeat, B and B^s, which are distinguished by two marker nucleotides. B

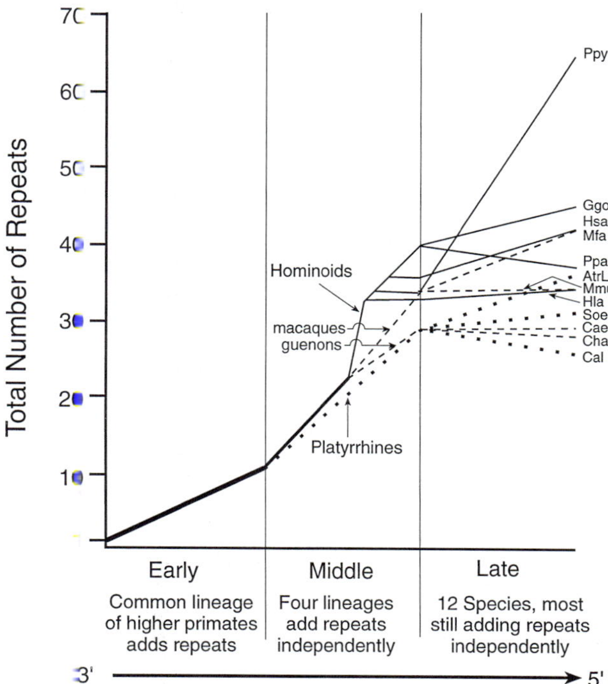

FIGURE 46-3 Vectorial repeat addition in the involucrin gene of higher primates. The early region, consisting of a total of 10 ten-codon repeats of consensus sequence AAG CAC CTG GAG CAG CAG GAG GGG CAG CTG (A repeat) or GAG CTC CCA GAG CAG CAG GAG GGG CAG CTG (B repeat) was completed in the common lineage of the higher primates (thick line). Subsequent repeats were added mainly at a location close to the 5′ end of the segment of previously added repeats. As a result, the segment of repeats grew from the 3′ to the 5′ end. Line of intermediate thickness indicates common lineage of Old World (catarrhine) primates (from [21]). *Ppy, Pongo pygmaeus; GgoL, Gorilla gorilla,* large allele; *Hsa, Homo sapiens; Mfa, Macaca fascicularis; Ppa, Pan paniscus; AtrL, Aotus trivirgatus,* large allele; *Mmu, Macaca mulatta; Hla, Hylobates lar; Soe, Saguinus oedipus; Cae, Cercopithecus aethiops; Cha, Cercopithecus hamlyni; Cal, Cebus albifrons.*

repeats lie immediately 5′ of the B^s repeats (Fig. 46-4A). Polymorphism of repeat number can result from duplication of either repeat type [22, 23]. Caucasian, African, and one class of East Asian alleles have different numbers of B repeats, B^s repeats, or both. Although the mutational hot spot for repeat addition lies within this region in all humans, there are racial differences in the frequency with which the two repeat types are duplicated. African alleles nearly always have only two B^s repeats, but frequently have seven or more B repeats (Fig. 46-4B). In contrast, Caucasian alleles frequently have three B^s repeats, but rarely more than six B repeats. East Asian alleles differ from both in having a high frequency of six B repeats combined with three B^s repeats. The location of the hot spot is therefore influenced by the genetic background in which the duplications occur.

In summary, after beginning as polyCAG the involucrin gene evolved by nucleotide substitution and in the higher primates by site-specific serial addition of CAG-rich repeats. In genes which have suitable marker nucleotides dividing regions of homogeneous triplet repeats, it is also common to find that there is a hot spot for repeat addition (see below).

B. GRP-1

A cDNA encoding a 172-amino-acid nuclear protein (GRP-1) has recently been sequenced [24]. Fifty percent of this protein consists of glutamine residues, but these are clustered in the central part of the molecule. The regions encoding the N-terminal 31 amino acids and the C-terminal 21 amino acids do not contain glutamine codons, but in the central region, CAG's account for 78 of the 120 codons (65%). Another feature of this region is the regular substitution of CAG by CAC. There are 22 repeats of CAG CAG CAC and 6 additional repeats of the same sequence containing a single additional nucleotide substitution (CAA CAG CAC or GAG CAG CAC). These repeats were generated by a mutation CAG → CAC, followed by successive duplications of the nine-nucleotide segment containing 2 glutamine codons and 1 histidine codon, followed by a few further substitutions. The total amount of the present-day sequence originating from polyCAG may be calculated by adding these 31 histidine codons and the 6 other substituted codons mentioned to the CAG's, giving a total of 115 or 96% of the codons of the central region. This entire region, accounting for 70% of the total protein, was generated from polyCAG by repeated duplications and substitutions. The examples of involucrin and GRP-1 show how polyCAG has been used as a first stage in the evolutionary expansion of coding region.

IV. HUMAN GENES WITH CAG REPEATS WHOSE EXPANSION PRODUCES DISEASE OF THE CENTRAL NERVOUS SYSTEM: ADDITION OF REPEATS AT SPECIFIC LOCATIONS AS A TREND IN PRIMATE EVOLUTION

A. The Region Containing polyCAG and the T Content of Sequence Downstream of the polyCAG

The polyglutamine region and encoding nucleotide sequence of seven disease-producing proteins is given in Figure 46-5. In the normal alleles of huntingtin, the androgen receptor, SCA3, SCA6, and DRPLA, the number of uninterrupted CAG repeats varies from 10 or fewer to

A 5′ --- B ▼B ▼B ▼B ▼B B ↓B^S B^S --- 3′

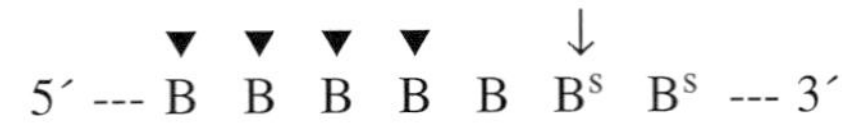

B

	$2B^S$		$3B^S$		All Other Repeat Patterns
	6B (% of Total)	7B (% of Total)	5B (% of Total)	6B (% of Total)	(% of Total)
Caucasian	45	3.3	**38**	2.5	11.5
African	47	**23**	2.1	6.4	21.3
East Asian	0	0	0	**87.0**	13.0

FIGURE 46-4 Recent additions of repeats to the late region of the involucrin gene in different human races. (A) Most of the human late region is composed of B repeats and variant B^S repeats, which differ from B repeats by two marker nucleotides in the third codon (TCT). There is polymorphism of number of both of these types. Arrow indicates the B^S repeat that has been duplicated in some alleles to produce a total of three B^S repeats, and arrowheads indicate one or more B repeats that have been duplicated to produce a total of as many as nine. (B) Incidence of most distinctive repeat patterns in different human lineages. The most lineage-specific repeat pattern (shown in bold) is $5B3B^S$ for Caucasians, $6B2B^S$ for Africans, and $6B3B^S$ for one lineage of East Asians. For further details, see [23].

a maximum of about 36. In the normal alleles of SCA1, the sequence of polyCAG is separated into two regions by other codons CAT or CAT CAG CAT (as shown in Fig. 46-5), or CAT $(CAG\ CAT)_2$ [25]. In SCA2, the polyCAG is separated into three regions by two intervening CAA codons [26–28]. The presence of histidine codons flanking the polyCAG region of the DRPLA gene, and to a lesser extent of the SCA1 gene, recalls the conspicuous histidine codons interrupting the polyCAG's of the *Drosophila mam* gene (Fig. 46-1) and the GRP-1 gene.

The sequences flanking the polyCAG in the various genes bear no resemblance to each other. Such regions may have begun as polyCAG, but have been modified by a high rate of nucleotide substitution [29, 30]. If a sequence of polyCAG undergoes nucleotide substitution, the resulting codons should retain a low frequency of T's,

Huntingtin	GCC	TTC	GAG	TCC	CTC	AAG	TCC	TTC	CAG	CAA	CAG	CCG	CCA	CCG	CCT	CCT	CAG	CTT	CCT
(120)	**A**	**F**	**E**	**S**	**L**	**K**	**S**	**F**	**Q_{11-34}**	**Q**	**Q**	**P**	**P**	**P_7**	**P**	**P**	**Q**	**L**	**P**
Androgen Receptor	GCA	CCT	CCC	GGT	GCC	AGT	TTG	CTG	CAG	CAA	GAG	ACT	AGC	CCC	AGG	CAG	GGT	GAG	GAT
(32,121)	**S**	**P**	**P**	**G**	**A**	**S**	**L**	**L_3**	**Q_{18-31}**	**Q**	**E**	**T**	**S**	**P**	**R**	**Q_6**	**G**	**E**	**D**
SCA1	CAC	AAG	GCT	GAG	CAG	CAT	CAG	CAT	CAG	CAC	CTC	AGC	AGG	GCT	CCG	GGG	CTC	ATC	ACC
(122)	**H**	**K**	**A**	**E**	**Q_{7-17}**	**H**	**Q**	**H**	**Q_{8-18}**	**H**	**L**	**S**	**R**	**A**	**P**	**G**	**L**	**I**	**T**
SCA2	ACC	ATG	TCG	CTG	AAG	CCC	CAG	CAA	CAG	CAA	CAG	CCG	CCC	GCG	GCT	GCC	AAT	GTC	CGC
(26-28)	**T**	**M**	**S**	**L**	**K**	**P**	**Q_8**	**Q**	**Q_{4-13}**	**Q**	**Q_{5-8}**	**P_2**	**P**	**A**	**A**	**A**	**N**	**V**	**R**
SCA3	TTT	GAA	AAA	CAG	CAA	AAG	CAG	CAA	CAG	CGG	GAC	CTA	TCA	GGA	CAG	AGT	TCA	CAT	CCA
(123)	**Y**	**E**	**K**	**Q_2**	**Q**	**K**	**Q**	**Q**	**Q_{13-36}**	**R**	**D**	**L**	**S**	**G**	**Q**	**S**	**S**	**H**	**P**
SCA6	CGT	AAG	GCC	GGC	TCG	GGG	CCC	CCG	CAG	GCG	GTG	GCC	AGG	CCG	GGC	CGG	GCG	GCC	ACC
(58)	**R**	**K**	**A**	**G_2**	**S**	**G**	**P**	**P**	**Q_{4-16}**	**A**	**V**	**A**	**R**	**P**	**G**	**R**	**A**	**A**	**T**
DRPLA	CAT	CAC	CAT	CAC	CAG	CAA	CAG	CAA	CAG	CAT	CAC	GGA	AAC	TCT	GGG	CCC	CCT	GGA	GCA
(124)	**H**	**H**	**H**	**H_2**	**Q**	**Q**	**Q**	**Q**	**Q_{7-23}**	**H**	**H**	**G**	**N**	**S**	**G**	**P**	**P_3**	**G**	**A**

FIGURE 46-5 The polyglutamine regions of seven proteins producing disease of the human nervous system. These regions are divided by mutations in SCA1, SCA2, and SCA3. The proteins corresponding to these three genes are often designated as ataxins and the DRPLA protein as atrophin.

TABLE 46-3 Low T Content of Sequence Downstream of polyCAG

	No. of nucleotides	No. of T's	T content of region (%)	T content of entire coding region (%)
huntingtin	120	8	6.7	23.2
AR	231	27	11.7	19.3
SCA1	72	7	9.7	15.7
SCA2	78	9	11.5	20.3
SCA3	60	9	15.0	23.9
SCA6	192	12	6.2	18.5
DRPLA	78	12	15.4	17.3

as in the entire coding region of the involucrin gene. There is such evidence for short regions of at least six of the seven genes of central nervous system diseases (Table 46-3), in which the T content of sequences of between 60 and 231 nucleotides located immediately downstream of the polyCAG is low compared with the T content of the entire coding region of the gene. The deficit in T is substantial for all the genes except DRPLA. The divergence of these codons from CAG's should tend to prevent their further duplication by slipped-strand mispairing. New duplications will therefore be confined to the region of homogeneous triplet repeats.

B. The Trend toward CAG Addition in the Huntingtin Gene of Higher Primates

The 5′ end of the first exon of the huntingtin gene of four higher primates, the mouse, and the puffer fish are shown in Fig. 46-6. There are two segments rich in codons for a single amino acid. Segment I consists of glutamine codons, whose expansion in the human gives rise to Huntington disease. At this site, the gibbon and the great apes show a tendency toward increase in number of CAG repeats (7–9), but normal humans have a modal value of 17, and therefore have surpassed the other species in the trend to CAG addition. These repeats in normal alleles are of much more recent origin than the CAG repeats in the genes listed in Table 46-2.

C. The Trend toward CAG Addition in the Androgen Receptor Gene of Higher Primates

At codon 58 of the human androgen receptor gene (Fig. 46-7), there begins a sequence of repeated CAG's (segment I). Four nonprimate mammals have 2–11 repeats. At the same position, the gibbon has only 4 repeats but the gorilla and the chimpanzee have 13–17 and 13–16, respectively. As in the case of the huntingtin gene, the trend is most developed in the human, which has a modal number of 21 [31, 32]. The CAG additions are highly localized, since in segment II, located slightly downstream, the human has no more CAG repeats than the other species, and in segment III, beginning at codon 194, the human has fewer CAG repeats than in the other species. Segment III has diverged considerably from polyCAG; for example, in the mouse gene, numerous mutations in the original sequence of 20 polyCAG's to CAA's and CAC's have reduced the maximum length of poly CAG to 3. Segment IV of the human gene contains a series of 17–20 consecutive glycine repeats encoded by GGC. In the only other species available

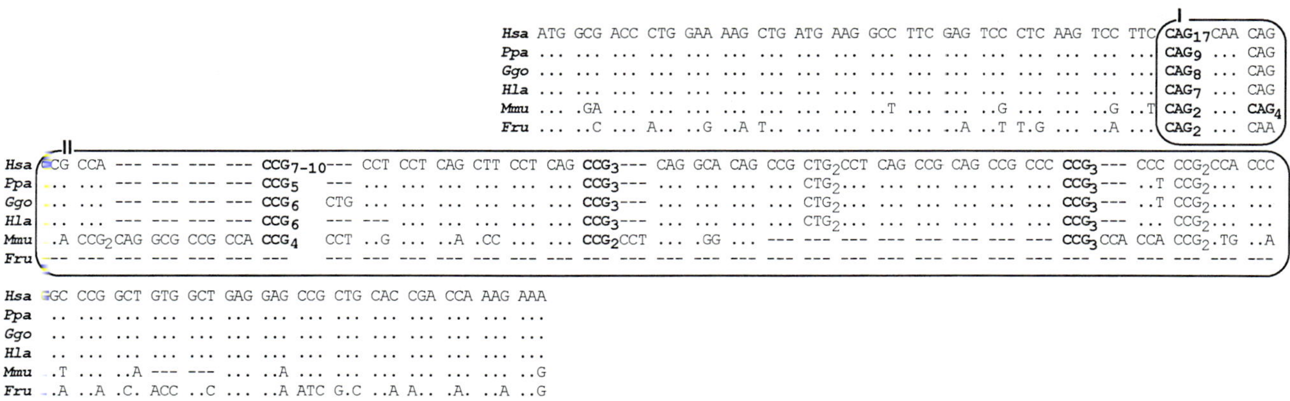

FIGURE 46-6 Sites of codon repeats in huntingtin genes of different species. The 5′ end of the first exon of six species is shown: *Hsa, Homo sapiens* [120]; *Ppa, Pan paniscus,* the chimpanzee; *Ggo, Gorilla gorilla; Hla, Hylobates lar,* the gibbon [30]; *Mmu, Mus musculus,* the house mouse [126]; and *Fru, Fugo rubripes,* the puffer fish [127]. Segment I encodes exclusively glutamine and segment II encodes largely proline. Within segment II there are three groups of short repeats. Dots indicate identity of nucleotides to those of the human sequence. CAG repeat number in segment I of the human is the modal value from [29]. (---) Absence of corresponding codons.

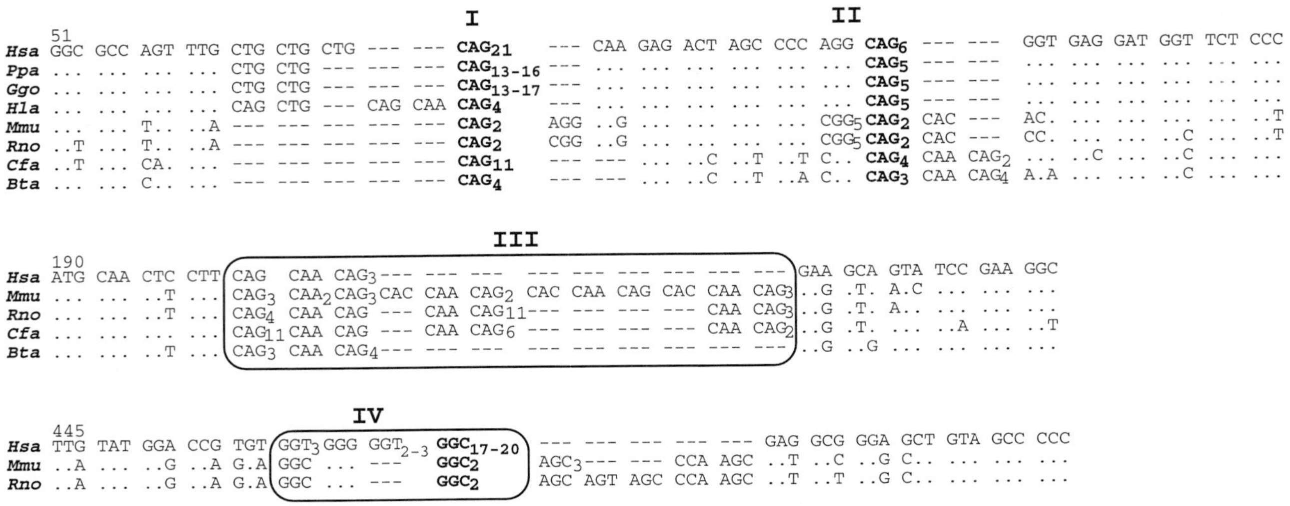

```
                                            I                                                  II
     51
Hsa GGC GCC AGT TTG CTG CTG CTG --- --- CAG21      --- CAA GAG ACT AGC CCC AGG CAG6    --- ---  GGT GAG GAT GGT TCT CCC
Ppa ... ... ... ... CTG CTG --- --- --- CAG13-16   --- ... ... ... ... ... ... CAG5    --- ---  ... ... ... ... ... ...
Ggo ... ... ... ... CTG CTG --- --- --- CAG13-17   --- ... ... ... ... ... ... CAG5    --- ---  ... ... ... ... ... ...
Hla ... ... ... ... CAG CTG --- CAG CAA CAG4       --- ... ... ... ... ... ... CAG5    --- ---  ... ... ... ... ... ...
Mmu ... ... T.. ..A --- --- --- --- --- CAG2       AGG ..G ... ... ... ... CGG5 CAG2   CAC ---  AC. ... ... ... ... ..T
Rno ..T ... T.. ..A --- --- --- --- --- CAG2       CGG ..G ... ... ... ... CGG5 CAG2   CAC ---  CC. ... ... ..C ... ..T
Cfa ..T ... CA. ... --- --- --- --- --- CAG11      --- --- ... ..C ..T ..T C.. CAG4    CAA CAG2 ... ..C ... ..C ... ...
Bta ... ... C.. ... --- --- --- --- --- CAG4       --- --- ... ..C ..T ..A C.. CAG3    CAA CAG4 A.A ... ... ..C ... ...

                                               III
     190
Hsa ATG CAA CTC CTT (CAG   CAA  CAG3 --- --- ---   --- --- --- --- --- ---  ) GAA GCA GTA TCC GAA GGC
Mmu ... ... ..T ... |CAG3  CAA2 CAG3 CAC CAA CAG2  CAC CAA CAG CAC CAA CAG3 | ..G .T. A.C ... ... ...
Rno ... ... ..T ... |CAG4  CAA  CAG  --- CAA CAG11 --- --- --- --- CAA CAG3 | ..G .T. A.. ... ... ...
Cfa ... ... ... ... |CAG11 CAA  CAG  --- CAA CAG6  --- --- --- --- CAA CAG2 | ..G .T. ... ..A ... ..T
Bta ... ... ..T ... (CAG3  CAA  CAG4 --- --- ---   --- --- --- --- --- ---  ) ..G ..G ... ... ... ...

                              IV
     445
Hsa TTG TAT GGA CCG TGT (GGT3 GGG GGT2-3 GGC17-20 ) --- --- --- --- --- GAG GCG GGA GCT GTA GCC CCC
Mmu ..A ... ..G ..A G.A |GGC  ... ---    GGC2     | AGC3 --- --- CCA AGC ..T ..C ..G C.. ... ... ...
Rno ..A ... ..G ..A G.A (GGC  ... ---    GGC2     ) AGC AGT AGC CCA AGC ..T ..T ..G C.. ... ... ...
```

FIGURE 46-7 Sites of codon repeats in the androgen receptor gene of different species. Abbreviations for species as in Figs. 46-3 and 46-6. *Rno, Rattus norvegicus* [121]; *Cfa, Canis familiaris* [128]; *Bta, Bovis taurus* (Z75313). Other sources of sequence: *Hsa* [121, 129], apes [30], *Ggo* (L49357 and L49348), *Ppa* (L49351 and L49352), *Mmu* (M37890).

for comparison, the rat and mouse, there are only two consecutive GGC codons at this position.

V. ADDITIONAL EXAMPLES OF REPEAT ADDITIONS AT SPECIFIC LOCATIONS

A. The Polyproline-Encoding Segment of the Huntingtin Gene

In contrast to segment I, containing the polyCAG subject to expansion (Fig. 46-6), segment II is rich in proline codons which, in the human, account for at least 74% of the total. The most common proline codon is CCG. As the puffer fish completely lacks segment II and the mouse lacks only a small part of it, most of this segment was probably generated in the common mammalian lineage or earlier and the rest completed in the common primate lineage. Segment II probably began by a mutation CAG → CCG in the adjacent segment I, and evolved by repeated duplications and further mutations.

Segment II has three groups of CCG repeats. The second and third groups of CCG repeats are of the same length in the five mammalian species. The first group of CCG repeats, located closest to the polyCAG repeats, is the most interesting. It is of nearly constant length in the mouse and the apes (four to six triplets), but is a little longer in most humans (seven triplets). It is remarkable that further repeat additions have occurred only in human alleles that do not have an expansion of the adjacent polyCAG sequence of segment I. The results of two studies, "The Edinburgh Group" [33] and "The Vancouver Group" [34], are summarized in Table 46-4. Thirty-one percent of normal human alleles has 10 consecutive CCG's. In contrast, the Huntington disease alleles contain either no example of 10 CCG's [33], or at most a few [34], the overall average frequency being 3.3%. Whatever may be the cause of the difference between the two studies, it is clear that the incidence of repeat addition at this location in both studies is far

TABLE 46-4 Number of CCG Repeats in huntingtin Genes of Normal and Huntington Disease Families

No. of CCG repeats	Normal alleles				Huntington's disease alleles			
	Edin.	Vanc.	Total	%	Edin.	Vanc.	Total	%
5	1	0	1	0.13	0	0	0	0
7	344	137	481	62.2	130	105	235	96.3
8	16	0	16	2.1	1	0	1	0.4
9	31	5	36	4.7	0	0	0	0
10	176	61	237	30.7	0	8	8	3.3
11	0	1	1	0.13	0	0	0	0
12	0	1	1	0.13	0	0	0	0
			773	100			244	100

lower in Huntington disease alleles than in normal alleles. In Huntington disease alleles, the mutational hot spot for repeat addition is confined to the CAG repeats of segment I, whereas in normal alleles there is a hot spot within the first block of CCG repeats of the adjacent segment II. In a population of normal alleles, lower numbers of CCG repeats were also correlated with higher numbers of CAG repeats [29]. Lineage-specific repeat addition in the huntingtin gene appears to resemble race-specific repeat addition in the involucrin gene.

B. The Genes for SCA1 and SCA2

The normal polyCAG sequence of SCA1 is divided into two regions by an intervening sequence consisting of CAT or CAT CAG CAT or CAT (CAG CAT)$_2$ [25]. The second and third of these sequences were evidently generated from the first by addition of one and two units, respectively, of CAG CAT. The introduction of histidine codons, so evident in the *Drosophila mam* gene, has been important in the stabilization of the human SCA1 gene, since disease-producing expansion of the CAG repeats occurs only when these histidine codons are absent [25]. Nevertheless, on each side of the intervening sequence containing CAT, the polyCAG sequence is subject to modest repeat addition, the upstream polyCAG containing from 7 to 17 repeats and the downstream from 8 to 18. While this difference between the two ranges has been found significant [25], the hot spot for repeat addition extends over both regions.

It has been mentioned that in SCA2, the polyCAG sequence is divided into three regions by two intervening CAA codons [26–28]. The upstream polyCAG uniformly contains 8 triplets, but the other regions are polymorphic, the middle containing 4–13 and the downstream containing 5–8. In this gene, the hot spot for repeat addition is located outside the region of the upstream repeats; it is most active in the middle region, but there is a lower frequency of addition in the downstream region. The small number of silent substitutions likely prevents disease-producing expansion but, like the T substitutions in SCA1, does not prevent small-scale repeat addition.

C. The DRPLA Gene

The polyQ regions encoded by the DRPLA gene of the mouse and the human are shown in Fig. 46-8. Whereas the human has an uninterrupted sequence of poly Q (normally about 15), the mouse poly Q sequence has been divided into two regions by the substitution of a proline codon for a glutamine codon, as the result of two substitutions CAG → CCA. It has been shown [35] that in region 1 of different species of *Mus,* the upstream glutamines are highly variable in number, different species possessing from three to eight consecutive glutamine residues, but region 2 is of constant length. Clearly, the hot spot for repeat addition in the mouse is sharply defined and is located 5′ of the proline codon interrupting the two segments of polyCAG. Although the region of homogenous CAG$_{15}$ of the human is subject to expansion, it is impossible to identify the location of the added repeats. As in the case of the *Drosophila mam* gene, histidine codons are prominent in the regions bordering the polyCAG region of DRPLA in both human and mouse.

D. The Gene for the TATA-Binding Protein (TBP)

Although human TBP has not been associated with disease, it contains a sequence of 25–42 glutamine residues near the N-terminus [6]. A pure polyCAG is present in two regions, II and IV, separated and flanked by CAA codons (Table 46-5). The most common alleles (type A) contain a constant number of nine repeats at region II and a variable number at region IV. Although these regions are separated by only three codons, the repeat-adding process has evidently been confined to region IV, where the modal number of repeats is slightly different in Caucasians, Asians, and Blacks. In a less common allele (type B), the repeat number varies less in region IV, but also varies in region II. Type B alleles are much more common in Caucasians than in Asians and Blacks. The similarity to involucrin is striking in the rather precisely located hot spot for repeat addition in type A alleles, in the additional site of repeat addition

```
                                          Region 1      Region 2
Mouse   S T A H P A A P T - H H H H   Qn        P    Q3    H H H G N S G

Human   S T A H P P V S T H H H H H ←——— Qn ———→  - H H G N S G
```

FIGURE 46-8 Comparison of the polyglutamine regions in the DRPLA protein of mouse and human. Addition of glutamine residues in different mouse species occurs 5′ of the intervening proline.

TABLE 46-5 Triplets Encoding Glutamine Repeats in TBP

	I CAG_3 CAA_3	II CAG_n	III CAA CAG CAA	IV CAG_n	V CAA CAG
A (12 alleles)	1	9	1	9–20	1
B (8 alleles)	1	7–11	1	16–21	1

Note. The codons of the glutamine repeats have been grouped in five regions [6]. Regions II and IV are polymorphic. Entries give the number of repeats ($_n$).

in type B alleles, and in the different locations of repeat addition in different human races.

E. The HOXD13 Gene

Elongation of a polyalanine region in the HOXD13 protein of the human results in synpolydactyly [7]. The normal allele contains a series of 15 consecutive alanine codons (most commonly GCG), located not far from the 5′ end of the coding region (Fig. 46-9). Three independent disease-producing insertions were identified, each adding 7–10 alanine codons to the previously existing segment of repeats. Because of the presence of two marker nucleotides, it is possible to identify the nature of the mutational event that led to the increase in alanine codons. It seems clear that each of the three mutations occurred as a single duplication of 7, 8, or 10 codons, centered at almost the same location, slightly upstream of the center of the wild-type sequence [7]. The added repeat unit is about the same size as the repeats added in the involucrin gene.

F. The CBFA1 Gene

The polyoma virus enhancer binding protein or PEBP2 [36] contains, near the N-terminal end, a sequence of glutamine repeats followed by a sequence of alanine repeats. The same protein, independently discovered and called CBFA1, is essential for bone formation [37, 38]. A number of different mutations in the gene have been found to cause the human disease cleidocranial dysplasia [39]. In one of these mutations, a duplication within the region encoding the 17 consecutive alanine codons added 10 additional alanine codons. In another mutation, an insertion within the region encoding 23 polyglutamine codons led to a frame shift, changing the last three CAG codons to AGC (alanine), and most of the wild-type alanine repeats to glycine repeats. The contribution of this alteration in repeat sequence to the phenotype of cleidocranial dysplasia is not clear because the frame shift also resulted in termination further downstream.

Different mutations in this gene, which are all dominant, whether they affect the alanine repeats or not, result in a phenotype of skeletal abnormalities resembling that of mice bearing a single ablated gene [37]. In contrast to the diseases due to expanded polyglutamine, this seems to point to a loss of function due to haploinsufficiency as the cause of the skeletal abnormalities resulting from mutations of CBFA1.

VI. FEATURES OF REPEAT EXPANSION NOT EXPLAINED BY HAIRPIN FORMATION/SLIPPED-STRAND MISPAIRING

The most common explanation for expansion of homogeneous repeats is hairpin formation/slipped-strand

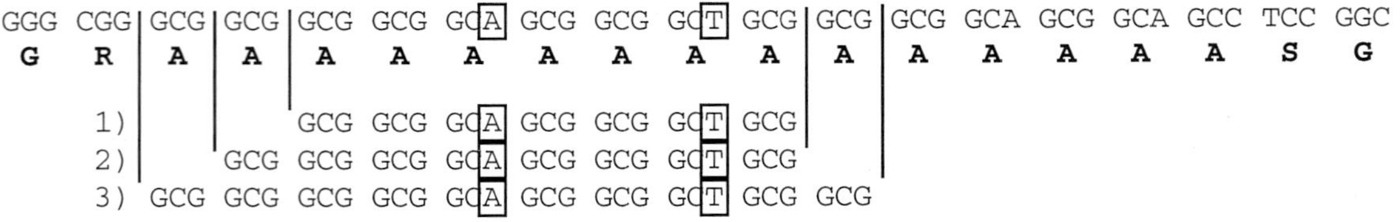

FIGURE 46-9 Duplications in the polyalanine region of HOXD13. Upper line shows nucleotide and encoded amino acid sequence of the part of the human HOXD13 gene encoding the wild-type polyA region. The predominant alanine codon is GCG but two-third base substitutions, A and T (boxed), serve as markers. The three different mutations associated with synpolydactyly [7] are the result of a single duplication of (1) 7 codons, (2) 8 codons, and (3) 10 codons, centered very close to the GCA codon.

mispairing [40–42]. However, this process alone does not explain the following features of repeat expansion:

1. The precise location of newly added repeats, commonly, as in the involucrin gene, near the 5′ end of the existing segment of repeats;
2. In alleles causing most of the diseases listed in Fig. 46-5, repeat length is more stable in the germ line of the female than in that of the male, since expansions transmitted from the male are larger and more frequent (for references, see [43]); only SCA2 shows repeat instability whether transmitted from male or female line [26].
3. Repeat expansion in a given sequence in the human and stability of the sequence in the mouse. An SCA1 allele, which had expanded from 53 to 82 repeats in a single paternal transmission, was stable when introduced as a transgene in the mouse [44]. Similarly, an androgen receptor gene unstable in the human was stable as a transgene [45].
4. Coding regions in the genes of the seven neurological diseases contain expanded polyCAG in the sense strand and polyCTG in the antisense strand. No example is known of the reverse. Polyleucine is not found in repeats of over 20 residues in any protein (Table 46-1). In *Escherichia coli,* polyCTG:GAC is more readily expanded than any other triplet repeats [46], and this is thought to require that the polyCTG form hairpins in the newly synthesized lagging strand during replication [42]. Unless this strand was always the anti-sense strand, the near absence of all but short polyCTG from the sense strand of eukaryotic coding regions is not explained by hairpin formation/slipped strand mispairing. PolyCAG forms hairpins with almost the same stability as polyCTG [47].

These features of expansion indicate that, while formation of hairpins [48] and slipped-strand mispairing are important aspects of the process of expansion, its more specific features remain unexplained.

VII. POLYGLUTAMINE AND NEURONAL DESTRUCTION

A. Expanded Polyglutamine Produces Dominant Gain of Function by Neuronal Proteins

In all seven diseases associated with proteins containing an expanded polyglutamine sequence, the allele bearing the expanded sequence is dominant over the normal allele. Dominant inheritance can be explained by gain of function or by a dominant negative mutation. These can be distinguished if inactivating mutations do not produce the same phenotype as an expansion. For example, the absence of a single huntingtin allele does not produce Huntington disease [49]. A transgene encoding only a small fragment of huntingtin [50] was able to produce typical central nervous system disease in the mouse. It seems unlikely that a fragment containing only 17 amino acids, apart from the polyglutamine, could have retained enough of the function of a protein of over 3000 amino acids to permit the fragment to act as a dominant negative.

Although the function of huntingtin remains unknown, the protein is essential in early embryogenesis, as disruption of the two alleles in mice causes death of the embryos at gastrulation [51, 52]. Since homozygotes for Huntington disease alleles have no developmental defects and have no more serious disease than heterozygotes [53], the disease must be the consequence of a pure gain of function that does not greatly impair the normal function of the protein.

When the androgen receptor gene is mutated [54] or deleted [55], the resulting androgen insensitivity causes testicular feminization in males, but not central nervous system disease, whereas expansion of the polyCAG sequence causes SBMA, together with a reduced androgen sensitivity that may be explained by reduced competence of the receptor as a transcription factor [56, 57]. It is clear that the polyCAG expansion causes both a gain of function and a partial loss of function, but the neurological disease is the consequence of the gain of function.

SCA6 is caused by a small polyglutamine expansion within the gene encoding the α_{1A} calcium channel subunit [58]. The polyglutamine is located close to the C terminus, within the cytoplasmic part of the protein. Other dominant mutations, such as chain terminations that delete much of the coding region, cause the disease episodic ataxia type 2 [59]. This disease has similarities to SCA6, but the ataxia is transitory and responsive to drugs. Polyglutamine expansion in the cytoplasmic region therefore causes more serious disease than the deletions that include this region and much of the membrane-spanning region. It is likely that the polyglutamine expansion affects the Purkinje cells by a gain of function, whereas the milder effects of the other mutations are due to a loss of function. The cytoplasmic region, including the polyglutamine, is altogether lacking in some presumably normal alternatively spliced isoforms.

In summary, the polyglutamine expansions of seven unrelated proteins, all resulting in disease of the central nervous system, exert their effects by a gain of function. If polyglutamine of excessive length is lethal to neurons, the normal function of the protein containing it may be of little relevance to the disease, as is evident for SBMA and Huntington disease. Variation between diseases in the minimal size of a cell-lethal polyglutamine may depend on how exposed the polyglutamine is at the surface of the protein and where the protein is located in the cell.

B. Evidence for Cell Lethality as the Mechanism of Polyglutamine-Induced Disease

Whether in the human disease or in its simulation in mice bearing a transgene encoding an expanded polyCAG, the presence of the protein containing expanded polyglutamine is thought to be cell-lethal. This would explain why in the polyglutamine-induced diseases there is extensive neuronal loss. In the case of Huntington disease, apoptosis rather than necrosis has been proposed as the mechanism of lethality [60]. What requires explanation in all of these diseases is the slowness of the lethality, which generally requires many years. It is necessary to think of a process of cell damage that is continuous, but that can be sufficiently compensated by repair processes to permit viability for a long period of time. One such form of damage could be the formation of protein aggregates, not fully compensated by proteolysis.

With the exception of the α_{1A} calcium channel protein, each of the disease-producing proteins containing an expanded polyglutamine is present in a variety of tissues, but the toxicity of the protein is evident only in neurons of the central nervous system. In mice transgenic for a fragment of the Huntington disease gene containing little more than the promoter region and the polyCAG sequence, expression of the transgene was ubiquitous and yet affected only the central nervous system [50]; however, it is not clearly demonstrated that other cell types suffer no lethal consequences at all because we do not know the tolerance of tissues for cell lethality, particularly when the lost cells can be replaced by proliferation. For example, the absence of SBMA in females bearing an androgen receptor with expanded polyglutamine presumably means that their loss of approximately half as many neurons as in males produces no evident disease.

The growth of COS cells, an epithelial cell type, was found to be inhibited by expression of a cDNA encoding a fragment of the SCA3 protein containing an expanded polyglutamine sequence (79 triplets), whereas the same fragment containing a short polyglutamine sequence (35 triplets) had no appreciable effect [61]. It was concluded on the basis of indirect evidence that the reduced growth rate was due to cell killing. It has even been reported that expression in *E. coli* of a GST fusion protein containing a long stretch of polyglutamine inhibits cell growth [62], but there was no direct evidence of cell lethality.

C. Affected Parts of the Brain in Relation to Extent of Cell Lethality

The most severely affected areas of the brain vary in each of the diseases: the caudate nucleus and putamen in Huntington disease [63, 64], lower motor neurons and primary sensory ganglion cells in SBMA [65], the cerebellar Purkinje cells and related neurons in SCA1 [44], the dentatorubral and pallidoluysian systems in DRPLA, etc. The neuropathology of Huntington disease has been more extensively studied than that of the other diseases and it has been established that although the caudate and putamen are the most affected areas, neuronal loss is widespread in the nervous system (for references see [66]). Even regions usually thought to be little affected, such as the cerebellum [67], the hypothalamus, and the spinal cord (for references see [68]), show extensive neuronal loss. For instance, the number of neurons in the lateral tuberal nucleus of the hypothalamus in randomly chosen Huntington disease patients was as low as 5% of its usual value [69]. Evidently there has to be severe neuronal loss before clinical symptoms become apparent. As a result, the regions of the brain giving rise to motor dysfunction must depend not only on the degree of cell lethality, but also on how much neuronal loss can be tolerated.

D. Protein Aggregation as the Cause of Cell Lethality

1. Evidence for Molecular Aggregation of Proteins Containing Expanded Polyglutamine

Although huntingtin is a very large protein (345 kDa) and an expanded polyglutamine sequence increases its molecular weight by at most a few kilodaltons, the normal and mutated forms found in some patients can be resolved by electrophoresis in polyacrylamide gels. Extracts of huntingtin prepared from affected areas of the brain showed diminution in the abundance of the mutant protein and its replacement by a smear extending above its expected position [70, 71]. This suggests that aggregates formed preferentially from the protein bearing the expanded polyglutamine. The aggregates were stable to heating in the presence of ionic detergent and reducing agent. Although the aggregates formed from huntingtin extended over a broad range of size, there was no evidence for the presence of discrete bands of polymer that could be formed by huntingtin alone (the smallest possible being a dimer of 700 kDa); the aggregates formed by mutant huntingtin are presumably the result of interaction with other proteins.

Aggregation of the expanded huntingtin was linked to the toxic effect of the protein, since it occurred in the most affected regions of the brain such as caudate/putamen and cortex, but was not found in less affected regions such as hippocampus and cerebellum [70]. The decrease in amount of the expanded protein and the smearing that replaced it were more prominent in proteins with long expansions than in those with short expansions.

Similar results have been obtained with SCA3 [61]. In transfected COS cells making a fragment of ataxin 3 with 35 consecutive glutamines, the fragment did not aggregate but formed a discrete band of expected size after electrophoresis in polyacrylamide gels; in contrast, a fragment containing 79 consecutive glutamine residues formed a smear of high molecular weight.

2. Subcellular Localization of These Proteins

huntingtin is found in the cytosol [71, 72]. It has also been described as associated with microtubules [72], with synaptophysin, a protein of synaptic vesicles [71], and with the vesicles themselves [73]. However, when a truncated huntingtin bearing an expanded polyglutamine was expressed from a transgene, the product localized in a nuclear inclusion, whereas the endogenous mouse protein did not [74]. The authors suggest that the nucleus is the primary site of pathogenesis. The DRPLA protein, atrophin [75], and ataxins 2 and 3 [76] are predominantly cytosolic, but nuclear localization of ataxin 3 in SCA3 has been reported [77]. The androgen receptor is both nuclear and cytoplasmic [78]. Ataxin-1 is nuclear in all neuronal cells, but in Purkinje cells, the most susceptible neurons in SCA1, it is found in the cytoplasm as well, both normally and in SCA1 [79]. The α_1 Ca^{2+} channel associated with SCA6 is an integral membrane protein.

E. Mechanisms of Aggregation of Proteins Containing Expanded Polyglutamine

1. Specific Aggregation

A number of proteins, some of which are brain-specific, have been found to interact with huntingtin. It has been suggested that the binding or lack of binding of a specific protein by huntingtin bearing expanded polyglutamine has a role in the pathogenesis of the disease. Binding proteins including calmodulin [80] ubiquitin-conjugating enzyme [81], glyceraldehyde phosphate dehydrogenase [82, 83], HIP1 (huntingtin-interacting protein) [84, 85], and HAP1 (huntingtin-associated protein) [86, 87]. Protein aggregates containing calmodulin were reported of size greater than 1000 kDa [80], but apparently no insoluble aggregates were found. Another pair of proteins, TRIP-1 and TRIP-2, are specific to the brain and bind to polyCAG and some other triplet repeats in single-stranded DNA [88].

These proteins were all identified by affinity methods. HAP1 and glyceraldehyde phosphate dehydrogenase bound more effectively to huntingtin that contained expanded polyglutamine than to huntingtin that contained a normal-length polyglutamine, but in the case of HIP1, the binding to normal huntingtin was more effective. It was suggested that specific binding of HAP1 to huntingtin might explain neuronal specificity of the disease, since HAP1 is confined to the brain, whereas huntingtin is widely distributed [86]. However, HAP1 is most abundant in the olfactory bulb, the supraoptic nucleus and the pedunculopontine nucleus, regions less affected in Huntington disease [64]. Since both calmodulin and glyceraldehyde phosphate dehydrogenase are present in numerous tissues, their interaction with huntingtin or atrophin cannot explain selective loss of neurons. In view of the disease-producing effects of what are essentially polyCAG transgenes [50, 61] any accessory protein would have to interact directly with expanded polyglutamine and should lack specificity for any of the seven diseases.

2. Polar Zipper Formation

Polyglutamine sequences of more than a few residues in length are highly insoluble, because they form hydrogen-bonded polar zippers between their main chain and side chain amides [89, 90]. Incorporated into a small protein, Q_{10} resulted in the formation of dimers and trimers [91]. Cell lethality might be explained by the formation of aggregates resulting from polar zipper formation, but such aggregates should be composed of multimers of the protein itself. Apart from SCA6, the disease-producing proteins appear to be soluble even when they contain a polyQ of greater than 40 residues. Presumably the large molecular size and tertiary structure of these proteins would tend to oppose the formation of polar zippers. Although polyasparagine also ought to be able to form polar zippers, none appear to have been described.

3. Transglutaminase-Mediated Cross-Linking

Transglutaminase catalyzes the formation of aggregates by introducing intermolecular isopeptide cross-links between glutamine and lysine residues of polypeptides. Peptide-bound asparagine is not a substrate for transglutaminase [92, 93].

The relation between cell death and transglutaminase has been the subject of numerous papers. Increases in the amount of transglutaminase have been associated with the action of various agents that are lethal to cells [94–97]. In some cases, but not in others, there is an increase in the amount of transglutaminase mRNA [98, 99]. Proof is lacking that the cell death is the result of enzymatic catalysis of isopeptide bond formation.

Cellular transglutaminase activity is latent, since the Ca^{2+} concentration of cells (10^{-6} *M*) is too low to activate the enzyme. In cell death, whether programmed or due to toxic agents, the Ca^{2+} concentration of the cytosol may be expected to rise, owing to its liberation from the mitochondria or endoplasmic reticulum, or to loss of the plasma membrane barrier to the much higher extracellular Ca^{2+} concentration. For this reason, ε-(γ-glutamyl) lysine cross-links may form as part of the lethal process, but not be the cause of cell death. Approximately 5×10^{-4} of the hepatocytes of normal rat liver possess envelope-like structures cross-linked by glutamyl-lysine bonds [100]. Similar structures and increased transglutaminase activity are found in fibrosarcoma cells with reduced capacity to metastasize [101] and in forms of programmed cell death.

Transglutaminase may also be activated when the cell receives some form of injury which is not lethal but which results in a rise in cytosolic Ca^{2+}. The proteins of hepatocytes suffering chemical injury have an increased number of glutamyl-lysine cross-links [94], and free ε-(γ-glutamyl) lysine has been isolated from cultures of hepatocytes [102], indicating proteolysis of cross-linked proteins. In ethanol toxicity, the cytoplasm of hepatocytes acquires Mallory bodies, consisting of keratins and other proteins [103] rendered insoluble by glutamyl-lysine cross-links [104]. Presumably the hepatocytes possessing Mallory bodies can live for a long period but may eventually succumb either to other effects of chemical injury or to the effects of the cross-linking.

In the case of the central nervous system, transient rises in Ca^{2+} are physiological. Mathematical modelling has led to the conclusion that compartmentalization of Ca^{2+} entry is most probably a general principle for all calcium-dependent cellular functions [105]. Aggregation of channels permits local rises of Ca^{2+} to as much as two orders of magnitude above the micromolar concentration in the resting cytosol. This is a concentration only one order of magnitude below the optimal concentration for the activity of transglutaminase. A more recent calculation has led to the conclusion that the Ca^{2+} excess in the cytoplasmic zone up to 10 nm from the pore of an open Ca^{2+} channel may be up to three orders of magnitude higher than the normal cytosolic Ca^{2+} [106], or at the optimal level for transglutaminase.

Transglutaminase is found in all parts of the brain [107]. The enzyme is cytosolic, but has been reported to be enriched in synaptosomes [107, 108] and a role has been proposed for transglutaminase in synaptic neurotransmitter release [108]. A brain transglutaminase purified to homogeneity has been thought to differ in some particulars from ordinary tissue transglutaminase [109]. Transglutaminase is able to cross-link various brain proteins, including neurofilament proteins [110].

Introduction of human tissue transglutaminase into mouse cells by transfection produced cytological changes and a modest increase in glutamyl-lysine cross-links in cell protein [111]. When the transglutaminase activity of a line of human neuroblastoma cells was increased about 50-fold by transfection of a transglutaminase gene, the growth of the cell line was impaired and the cells showed cytologic abnormalities [112]. High levels of transglutaminase activity in lymphoid cells of HIV-infected patients have been associated with increased cell death and an increase in the free glutamyl-lysine concentration in the blood [113].

Cross-links of ε-(γ-glutamyl) lysine can be isolated from normal human brain (experiments of J. Wilson, D. Selkoe, and L. Lorand, cited in Lorand (1996) [114]). It has also been shown that high-frequency electrical stimulation of hippocampal slices resulted in the appearance of glutamyl-lysine cross-links in hippocampal proteins [115]. In regions close to the site of stimulation, the quantity of ε-(γ-glutamyl) lysine found was over 200 times the detection level, whereas none could be detected in the absence of electrical stimulation. The stimulation must have produced an influx of Ca^{2+} to a concentration able to activate the endogenous transglutaminase.

In SCA6, the polyglutamine expansion in the cytoplasmic region of the channel protein may be quite small compared with what is necessary in order for other polyglutamine-containing proteins to produce disease; this might be explained by the fact that the polyglutamine in SCA6 is located at the cytoplasmic side of the Ca^{2+} channels, where Ca^{2+} reaches the highest concentration.

As a model for the polyglutamine-induced diseases, it was shown that peptides containing polyglutamine of 5–18 residues are excellent substrates of transglutaminase [116]. Lengthening the polyglutamine increased not only the reactivity of the peptide, but also the reactivity of each glutamine residue. The nature of the amino acids flanking the polyglutamine appeared to be irrelevant, as long as they were charged sufficiently to keep the polyglutamine substrate in solution. The peptide $R_5Q_{18}R_5$ formed insoluble, cross-linked aggregates with a polyvalent lysine donor, such as polylysine, or with brain proteins. While externally added transglutaminase increased the cross-linking, brain extract itself possessed sufficient enzyme to form the aggregates. Glutamyl-lysine cross-links could be isolated from the aggregates after proteolytic digestion. Q_{10} and Q_{62} fused to the glutathione *S*-transferase protein (GST) are also excellent substrates of transglutaminase [117]. These findings strengthened the hypothesis [11] that the dominant gain of function resulting from expanded polyglutamine in proteins is greatly improved quality as a substrate of transglutaminase.

Aggregates stabilized by cross-linking would be subject to degradation by proteolytic enzymes. This is consistent with the recent finding that in transgenic mice the nuclear inclusion resulting from expanded polyglutamine soon accumulates ubiquitin [74]. Residual ε-(γ-glutamyl) lysine is resistant to proteases, but it can be cleaved by γ-glutamylamine cyclotransferase, an enzyme present in brain [118]. In order to become a substrate of the cyclotransferase, the ε-(γ-glutamyl) lysine must be proteolytically liberated from the cross-linked polypeptide chain, and as highly cross-linked aggregates are resistant to proteolysis [119], the necessary free ε-(γ-glutamyl) lysine may not be available for the cyclotransferase reaction. A small amount of cross-linked protein resulting from a short polyglutamine sequence may be effectively dealt with by proteolytic breakdown, leaving the ε-(γ-glutamyl) lysine to be excreted, but when the polyglutamine is expanded and provides a better transglutaminase substrate, the increased amount of cross-linked protein may exceed cellular capacity for proteolysis and therefore lead to lethality. The affected neurons cannot be replaced, as there are no precursors of this terminally differentiated cell type.

Acknowledgments

The authors acknowledge the valuable assistance of Mrs. Shahla Movahedi. The research of the authors is aided by grants from the National Cancer Institute (H.G.), and from the Centre National de la Recherche Scientifique and the Association Française contre les Myopathies, France (P.D.).

Note added in proof. The specific loss, in the affected part of the brain, of monomeric-expanded huntingtin and its appearance in aggregates has been duplicated by the action of transglutaminase *in vitro* (P. Kahlem, H. Green, and P. Djian, *Molecular Cell,* March 1998).

References

1. Hancock, J. M. (1993). Evolution of sequence repetition and gene duplications in the TATA-binding protein TBP (TFIID). *Nucleic Acids Res.* **21,** 2823–2830.
2. Green, H., and Wang, N. (1994). Codon reiteration and the evolution of proteins. *Proc. Natl. Acad. Sci. USA* **91,** 4298–4302.
3. Stallings, R. L. (1994). Distribution of trinucleotide microsatellites in different categories of mammalian genomic sequence: implications for human genetic diseases. *Genomics* **21,** 116–121.
4. Karlin, S., and Burge, C. (1996). Trinucleotide repeats and long homopeptides in genes and proteins associated with nervous system disease and development. *Proc. Natl. Acad. Sci. USA* **93,** 1560–1565.
5. Richard, G.-F., and Dujon, B. (1996). Distribution and variability of trinucleotide repeats in the genome of the yeast *Saccharomyces cerevisiae. Gene* **174,** 165–174.
6. Gostout, B., Liu, Q., and Sommer, S. S. (1993). "Cryptic" repeating triplets of purines and pyrimidines (cRRY(i)) are frequent and polymorphic: analysis of coding cRRY(i) in the proopiomelanocortin (POMC) and TATA-binding protein (TBP) genes. *Am. J. Hum. Genet.* **52,** 1182–1190.
7. Muragaki, Y., Mundlos, S., Upton, J., and Olsen, B. R. (1996). Altered growth and branching patterns in synpolydactyly caused by mutations in HOXD13. *Science* **272,** 548–551.
8. Newfeld, S. J., Schmid, A. T., and Yedvobnick, B. (1993). Homopolymer length variation in the *Drosophila* gene *mastermind. J. Mol. Evol.* **37,** 483–495.
9. Newfeld, S. J., Smoller, D. A., and Yedvobnick, B. (1991). Interspecific comparison of the unusually repetitive *Drosophila* locus *mastermind. J. Mol. Evol.* **32,** 415–420.
10. Hoey, T., Weinzierl, R. O. J., Gill, G., Chen, J.-L., Dynlacht, B. D., and Tjian, R. (1993). Molecular cloning and functional analysis of Drosophila TAF110 reveal properties expected of coactivators. *Cell* **72,** 247–260.
11. Green, H. (1993). Human genetic diseases due to codon reiteration: relationship to an evolutionary mechanism. *Cell* **74,** 955–956.
12. Gerber, H.-P., Seipel, K., Georgiev, O., Hofferer, M., Hug, M., Rusconi, S. and Schaffner, W. (1994). Transcriptional activation modulated by homopolymeric glutamine and proline stretches. *Science* **263,** 808–811.
13. Trumbly, R. J. (1988). Cloning and characterization of the *CYC8* gene mediating glucose repression in yeast. *Gene* **73,** 97–111.
14. Eckert, R. L., and Green, H. (1986). Structure and evolution of the human involucrin gene. *Cell* **46,** 583–589.
15. Green, H., and Djian, P. (1992). Consecutive actions of different gene-altering mechanisms in the evolution of involucrin. *Mol. Biol. Evol.* **9,** 977–1017.
16. Rice, R. H., and Green, H. (1979). Presence in human epidermal cells of a soluble protein precursor of the cross-linked envelope: activation of the cross-linking by calcium ions. *Cell* **18,** 681–694.
17. Simon, M., and Green, H. (1988). The glutamine residues reactive in transglutaminase-catalyzed cross-linking of involucrin. *J. Biol. Chem.* **263,** 18093–18098.
18. Djian, P., Phillips, M., Easley, K., Huang, E., Simon, M., Rice, R. H., and Green, H. (1993). The involucrin genes of the mouse and the rat: study of their shared repeats. *Mol. Biol. Evol.* **10,** 1136–1149.
19. Tseng, H., and Green, H. (1988). Remodeling of the involucrin gene during primate evolution. *Cell* **54,** 491–496.
20. Tseng, H., and Green, H. (1990). The involucrin genes of pig and dog: comparison of their segments of repeats with those of prosimians and higher primates. *Mol. Biol. Evol.* **7,** 293–302.
21. Djian, P., and Green, H. (1992). The involucrin gene of old world monkeys and other higher primates: synapomorphies and parallelisms resulting from the same gene-altering mechanism. *Mol. Biol. Evol.* **9,** 417–432.
22. Simon, M., Phillips, M., and Green, H. (1991). Polymorphism due to variable number of repeats in the human involucrin gene. *Genomics* **9,** 576–580.
23. Djian, P., Delhomme, B., and Green, H. (1995). Origin of the polymorphism of the involucrin gene in Asians. *Am. J. Hum. Genet.* **56,** 1367–1372.
24. Cox, G. W., Taylor, L. S., Willis, J. D., Melillo, G., White III, R. L., Anderson, S. K., and Lin, J.-J. (1996). Molecular cloning and characterization of a novel mouse macrophage gene that

encodes a nuclear protein comprising polyglutamine repeats and interspersing histidines. *J. Biol. Chem.* **271,** 25515–25523.

25. Chung, M.-y., Ranum, L. P. W., Duvick, L. A., Servadio, A., Zoghbi, H. Y., and Orr, H. T. (1993). Evidence for a mechanism predisposing to intergenerational CAG repeat instability in spinocerebellar ataxia type 1. *Nature Genet.* **5,** 254–258.
26. Imbert, G., Saudou, F., Yvert, G., Devys, D., Trottier, Y., Garnier, J.-M., Weber, C., Mandel, J.-L., Cancel, G., Abbas, N., Durr, A., Didierjean, O., Stevanin, G., Agid, Y., and Brice, A. (1996). Cloning of the gene for spinocerebellar ataxia 2 reveals a locus with high sensitivity to expanded CAG/glutamine repeats. *Nature Genet.* **14,** 285–291.
27. Sanpei, K., Takano, H., Igarashi, S., Sato, T., Oyake, M., Sasaki, H., Wakisaka, A., Tashiro, K., Ishida, Y., Ikeuchi, T., Koide, R., Saito, M., Sato, A., Tanaka, T., Hanyu, S., Takiyama, Y., Nishizawa, M., Shimizu, N., Nomura, Y., Segawa, M., Iwabuchi, K., Eguchi, I., Tanaka, H., Takahashi, H., and Tsuji, S. (1996). Identification of the spinocerebellar ataxia type 2 gene using a direct identification of repeat expansion and cloning technique, DIRECT. *Nature Genet.* **14,** 277–284.
28. Pulst, S.-M., Nechiporuk, A., Nechiporuk, T., Gispert, S., Chen, X.-N., Lopes-Cendes, I., Pearlman, S., Starkman, S., Orozco-Diaz, G., Lunkes, A., DeJong, P., Rouleau, G. A., Auburger, G., Korenberg, J. R., Figueroa, C., and Sahba, S. (1996). Moderate expansion of a normally biallelic trinucleotide repeat in spinocerebellar ataxia type 2. *Nature Genet.* **14,** 269–276.
29. Rubinsztein, D. C., Amos, W., Leggo, J., Goodburn, S., Ramesar, R. S., Old, J., Bontrop, R., McMahon, R., Barton, D. E., and Ferguson-Smith, M. A. (1994). Mutational bias provides a model for the evolution of Huntington's disease and predicts a general increase in disease prevalence. *Nature Genet.* **7,** 525–530.
30. Djian, P., Hancock, J. M., and Chana, H. S. (1996). Codon repeats in genes associated with human diseases: fewer repeats in the genes of non-human primates and nucleotide substitutions concentrated at the sites of reiteration. *Proc. Natl. Acad. Sci. USA* **93,** 417–421.
31. Edwards, A., Hammond, H. A., Jin, L., Caskey, C. T., and Chakraborty, R. (1992). Genetic variation at five trimeric and tetrameric tandem repeat loci in four human population groups. *Genomics* **12,** 241–253.
32. La Spada, A. R., Wilson, E. M., Lubahn, D. B., Harding, A. E., and Fischbeck, K. H. (1991). Androgen receptor gene mutations in X-linked spinal and bulbar muscular atrophy. *Nature* **352,** 77–79.
33. Barron, L. H., Rae, A., Holloway, S., Brock, D. J. H., and Warner, J. P. (1994). A single allele from the polymorphic CCG rich sequence immediately 3′ to the unstable CAG trinucleotide in the IT15 cDNA shows almost complete disequilibrium with Huntington's disease chromosomes in the Scottish population. *Hum. Mol. Genet.* **3,** 173–175.
34. Andrew, S. E., Goldberg, Y. P., Theilmann, J., Zeisler, J., and Hayden, M. R. (1994). A CCG repeat polymorphism adjacent to the CAG repeat in the Huntington disease gene: implications for diagnostic accuracy and predictive testing. *Hum. Mol. Genet.* **3,** 65–67.
35. Oyake, M., Onodera, O., Shiroishi, T., Takano, H., Takahashi, Y., Kominami, R., Moriwaki, K., Ikeuchi, T., Igarashi, S., Tanaka, H., and Tsuji, S. (1997). Molecular cloning of murine homologue dentatorubral-pallidoluysian atrophy (DRPLA) cDNA: strong conservation of a polymorphic CAG repeat in the murine gene. *Genomics* **40,** 205–207.
36. Ogawa, E., Maruyama, M., Kagoshima, H., Inuzuka, M., Lu, J., Satake, M., Shigesada, K., and Ito, Y. (1993). PEBP2/PEA2 represents a family of transcription factors homologous to the products of the *Drosophila* runt gene and the human *AML1* gene. *Proc. Natl. Acad. Sci. USA* **90,** 6859–6863.
37. Otto, F., Thornell, A. P., Crompton, T., Denzel, A., Gilmour, K. C., Rosewell, I. R., Stamp, G. W. H., Beddington, R. S. P., Mundlos, S., Olsen, B. R., Selby, P. B., and Owen, M. J. (1997). *Cbfa1*, a candidate gene for Cleidocranial Dysplasia syndrome, is essential for osteoblast differentiation and bone development. *Cell* **89,** 765–771.
38. Komori, T., Yagi, H., Nomura, S., Yamaguchi, A., Sasaki, K., Deguchi, K., Shimizu, Y., Bronson, R. T., Gao, Y.-H., Inada, M., Sato, M., Okamoto, R., Kitamura, Y., Yoshiki, S., and Kishimoto, T. (1997). Targeted disruption of *Cbfa1* results in a complete lack of bone formation owing to maturational arrest of osteoblasts. *Cell* **89,** 755–764.
39. Mundlos, S., Otto, F., Mundlos, C., Mulliken, J. B., Aylsworth, A. S., Albright, S., Lindhout, D., Cole, W. G., Henn, W., Knoll, J. H. M., Owen, M. J., Mertelsmann, R., Zabel, B. U., and Olsen, B. R. (1997). Mutations involving the transcription factor CBFA1 cause cleidocranial dysplasia. *Cell* **89,** 773–779.
40. Levinson, G., and Gutman, G. A. (1987). Slipped-strand mispairing: a major mechanism for DNA sequence evolution. *Mol. Biol. Evol.* **4,** 203–221.
41. Schlotterer, C., and Tautz, D. (1992). Slippage synthesis of simple sequence DNA. *Nucleic Acids Res.* **20,** 211–215.
42. Wells, R. D. (1996). Molecular basis of genetic instability of triplet repeats. *J. Biol. Chem.* **271,** 2875–2878.
43. Paulson, H. L., and Fischbeck, K. H. (1996). Trinucleotide repeats in neurogenetic disorders. *Annu. Rev. Neurosci.* **19,** 79–107.
44. Burright, E. N., Clark, H. B., Servadio, A., Matilla, T., Federsen, R. M., Yunis, W. S., Duvick, L. A., Zoghbi, H. Y., and Orr, H. T. (1995). *SCA1* transgenic mice: a model for neurodegeneration caused by an expanded CAG trinucleotide repeat. *Cell* **82,** 937–948.
45. Bingham, P. M., Scott, M. O., Wang, S., McPhaul, M. J., Wilson, E. M., Garbern, J. Y., Merry, D. E., and Fischbeck, K. H. (1995). Stability of an expanded trinucleotide repeat in the androgen receptor gene in transgenic mice. *Nature Genet.* **9,** 191–196.
46. Ohshima, K., Kang, S., and Wells, R. D. (1996). CTG triplet repeats from human hereditary diseases are dominant genetic expansion products in *Escherichia coli*. *J. Biol. Chem.* **271,** 1853–1856.
47. Mitas, M. (1997). Trinucleotide repeats associated with human disease. *Nucleic Acids Res.* **25,** 2245–2253.
48. Gacy, A. M., Goellner, G., Juranic, N., Macura, S., and McMurray, C. T. (1995). Trinucleotide repeats that expand in human disease form hairpin structures in vitro. *Cell* **81,** 533–540.
49. Ambrose, C. M., Duyao, M. P., Barnes, G., Bates, G. P., Lin, C. S., Srinidhi, J., Baxendale, S., Hummerich, H., Lehrach, H., Altherr, M., Wasmuth, J., Buckler, A., Church, D., Housman, D., Berks, M., Micklem, G., Durbin, R., Dodge, A., Read, A., Gusella, J., and MacDonald, M. E. (1994). Structure and expression of the Huntington's disease gene: evidence against simple inactivation due to an expanded CAG repeat. *Somat. Cell Mol. Genet.* **20,** 27–38.
50. Mangiarini, L., Sathasivam, K., Seller, M., Cozens, B., Harper, A., Hetherington, C., Lawton, M., Trottier, Y., Lehrach, H., Davies, S. W., and Bates, G. P. (1996). Exon 1 of the *HD* gene with an expanded CAG repeat is sufficient to cause a progressive neurological phenotype in transgenic mice. *Cell* **87,** 493–506.
51. Nasir, J., Floresco, S. B., O'Kusky, J. R., Diewert, V. M., Richman, J. M., Zeisler, J., Borowski, A., Marth, J. D., Phillips, A. G., and Hayden, M. R. (1995). Targeted disruption of the Huntington's Disease gene results in embryonic lethality and behavioral and morphological changes in heterozygotes. *Cell* **81,** 811–823.

52. Duyao, M. P., Auerbach, A. B., Ryan, A., Persichetti, F., Barnes, G. T., McNeil, S. M., Ge, P., Vonsattel, J.-P., Gusella, J. F., Joyner, A. L., and MacDonald, M. E. (1995). Inactivation of the mouse Huntington's disease gene homolog Hdh. *Science* **269,** 407–410.
53. Wexler, N. S., Young, A. B., Tanzi, R. E., Travers, H., Starosta-Rubinstein, S., Penney, J. B., Snodgrass, S. R., Shoulson, I., Gomez, F., Ramos Arroyo, M. A., Penchaszadeh, G. K., Moreno, H., Gibbons, K., Faryniarz, A., Hobbs, W., Anderson, M. A., Bonilla, E., Conneally, P. M., and Gusella, J. F. (1987). Homozygotes for Huntington's disease. *Nature* **326,** 194–197.
54. McPhaul, M. J., Marcelli, M., Tilley, W. D., Griffin, J. E., and Wilson, J. D. (1991). Androgen resistance caused by mutations in the androgen receptor gene. *FASEB J.* **5,** 2910–2915.
55. Quigley, C. A., Friedman, K. J., Johnson, A., Lafreniere, R. G., Silverman, L. M., Lubahn, D. B., Brown, T. R., Wilson, E. M., Willard, H. F., and French, F. S. (1992). Complete deletion of the androgen receptor gene: definition of the null phenotype of the androgen insensitivity syndrome and determination of carrier status. *J. Clin. Endocrinol. Metab.* **74,** 927–933.
56. Mhatre, A. N., Trifiro, M. A., Kaufman, M., Kazemi-Esfarjani, P., Figlewicz, D., Rouleau, G., and Pinsky, L. (1993). Reduced transcriptional regulatory competence of the androgen receptor in X-linked spinal and bulbar muscular atrophy. *Nature Genet.* **5,** 184–188.
57. Kazemi-Esfarjani, P., Trifiro, M. A., and Pinsky, L. (1995). Evidence for a repressive function of the long polyglutamine tract in the human androgen receptor: possible pathogenetic relevance for the $(CAG)_n$-expanded neuronopathies. *Hum. Mol. Genet.* **4,** 523–527.
58. Zhuchenko, O., Bailey, J., Bonnen, P., Ashizawa, T., Stockton, D. W., Amos, C., Dobyns, W. B., Subramony, S. H., Zoghbi, H. Y., and Lee, C. C. (1997). Autosomal dominant cerebellar ataxia (SCA6) associated with small polyglutamine expansions in the α_{1A}-voltage-dependent calcium channel. *Nature Genet.* **15,** 62–69.
59. Ophoff, R. A., Terwindt, G. M., Vergouwe, M. N., van Eijk, R., Oefner, P. J., Hoffman, S. M. G., Lamerdin, J. E., Mohrenweiser, H. W., Bulman, D. E., Ferrari, M., Haan, J., Lindhout, D., van Ommen, G.-J. B., Hofker, M. H., Ferrari, M. D., and Frants, R. R. (1996). Familial hemiplegic migraine and episodic ataxia type-2 are caused by mutations in the Ca^{2+} channel gene CACNL1A4. *Cell* **87,** 543–552.
60. Portera-Cailliau, C., Hedreen, J. C., Price, D. L., and Koliatsos, V. E. (1995). Evidence for apoptotic cell death in Huntington disease and excitotoxic animal models. *J. Neurosci.* **15,** 3775–3787.
61. Ikeda, H., Yamaguchi, M., Sugai, S., Aze, Y., Narumiya, S., and Kakizuka, A. (1996). Expanded polyglutamine in the Machado-Joseph disease protein induces cell death *in vitro* and *in vivo*. *Nature Genet.* **13,** 196–202.
62. Onodera, O., Roses, A. D., Tsuji, S., Vance, J. M., Strittmatter, W. J., and Burke, J. R. (1996). Toxicity of expanded polyglutamine-domain proteins in *Escherichia coli*. *FEBS Lett.* **399,** 135–139.
63. Tellez-Nagel, I., Johnson, A. B., and Terry, R. D. (1974). Studies on brain biopsies of patients with Huntington's chorea. *J. Neuropathol. Exp. Neurol.* **33,** 308–332.
64. Ross, C. A. (1995). When more is less: pathogenesis of glutamine repeat neurodegenerative diseases. *Neuron* **15,** 493–496.
65. Sobue, G., Hashizume, Y., Mukai, E., Hirayama, M., Mitsuma, T., and Takahashi, A. (1989). X-linked recessive bulbospinal neuronopathy. *Brain* **112,** 209–232.
66. Sharp, A. H., and Ross, C. A. (1996). Neurobiology of Huntington's disease. *Neurobiol. Dis.* **3,** 3–15.
67. Jeste, D. V., Barban, L., and Parisi, J. (1984). Reduced Purkinje cell density in Huntington's disease. *Exp. Neurol.* **85,** 78–86.
68. Vonsattel, J.-P., Myers, R. H., Stevens, T. J., Ferrante, R. J., Bird, E. D., and Richardson, E. P., Jr. (1985). Neuropathological classification of Huntington's disease. *J. Neuropathol. Exp. Neurol.* **44,** 559–577.
69. Kremer, H. P. H., Roos, R. A. C., Dingjan, G., Marani, E., and Bots, G. T. A. M. (1990). Atrophy of the hypothalamic lateral tuberal nucleus in Huntington's disease. *J. Neuropathol. Exp. Neurol.* **49,** 371–382.
70. Schilling, G., Sharp, A. H., Loev, S. J., Wagster, M. V., Li, S.-H., Stine, O. C., and Ross, C. A. (1995). Expression of the Huntington's disease (IT15) protein product in HD patients. *Hum. Mol. Genet.* **4,** 1365–1371.
71. Trottier, Y., Devys, D., Imbert, G., Saudou, F., An, I., Lutz, Y., Weber, C., Agid, Y., Hirsch, E. C., and Mandel, J.-L. (1995). Cellular localization of the Huntington's disease protein and discrimination of the normal and mutated form. *Nature Genet.* **10,** 104–110.
72. Gutekunst, C.-A., Levey, A. I., Heilman, C. J., Whaley, W. L., Yi, H., Nash, N. R., Rees, H. D., Madden, J. J., and Hersch, S. M. (1995). Identification and localization of huntingtin in brain and human lymphoblastoid cell lines with anti-fusion protein antibodies. *Proc. Natl. Acad. Sci. USA* **92,** 8710–8714.
73. DiFiglia, M., Sapp, E., Chase, K., Schwarz, C., Meloni, A., Young, C., Martin, E., Vonsattel, J.-P., Carraway, R., Reeves, S. A., Boyce, F. M., and Aronin, N. (1995). huntingtin is a cytoplasmic protein associated with vesicles in human and rat brain neurons. *Neuron* **14,** 1075–1081.
74. Davies, S. W., Turmaine, M., Cozens, B. A., DiFiglia, M., Sharp, A. H., Ross, C. A., Scherzinger, E., Wanker, E. E., Mangiarini, L., and Bates, G. P. (1997). Formation of neuronal intranuclear inclusions underlies the neurological dysfunction in mice transgenic for the HD mutation. *Cell* **90,** 537–548.
75. Yazawa, I., Nukina, N., Hashida, H., Goto, J., Yamada, M., and Kanazawa, I. (1995). Abnormal gene product identified in hereditary dentatorubral-pallidoluysian atrophy (DRPLA) brain. *Nature Genet.* **10,** 99–103.
76. Trottier, Y., Lutz, Y., Stevanin, G., Imbert, G., Devys, D., Cancel, G., Saudou, F., Weber, C., David, G., Tora, L., Agid, Y., Brice, A., and Mandel, J.-L. (1995). Polyglutamine expansion as a pathological epitope in Huntington's disease and four dominant cerebellar ataxias. *Nature* **378,** 403–406.
77. Paulson, H. L., Das, S. S., Crino, P. B., Perez, M. K., Patel, S. C., Gotsdiner, D., Fischbeck, K. H., and Pittman, R. N. (1997). Machado-Joseph disease gene product is a cytoplasmic protein widely expressed in brain. *Ann. Neurol.* **41,** 453–462.
78. Zhou, Z.-X., Wong, C.-I., Sar, M., and Wilson, E. M. (1994). The androgen receptor: an overview. *Recent Prog. Horm. Res.* **49,** 249–274.
79. Servadio, A., Koshy, B., Armstrong, D., Antalffy, B., Orr, H. T., and Zoghbi, H. Y. (1995). Expression analysis of the ataxin-1 protein in tissues from normal and spinocerebellar ataxia type 1 individuals. *Nature Genet.* **10,** 94–98.
80. Bao, J., Sharp, A. H., Wagster, M. V., Becher, M., Schilling, G., Ross, C. A., Dawson, V. L., and Dawson, T. M. (1996). Expansion of polyglutamine repeat in huntingtin leads to abnormal protein interactions involving calmodulin. *Proc. Natl. Acad. Sci. USA* **93,** 5037–5042.
81. Kalchman, M. A., Graham, R. K., Xia, G., Koide, H. B., Hodgson, J. G., Graham, K. C., Goldberg, Y. P., Gietz, R. D., Pickart, C. M., and Hayden, M. R. (1996). huntingtin is ubiquitinated and interacts with a specific ubiquitin-conjugating enzyme. *J. Biol. Chem.* **271,** 19385–19394.

82. Burke, J. R., Enghild, J. J., Martin, M. E., Jou, Y.-S., Myers, R. M., Roses, A. D., Vance, J. M., and Strittmatter, W. J. (1996). huntingtin and DRPLA proteins selectively interact with the enzyme GAPDH. *Nature Med.* **2,** 347–350.
83. Ishitani, R., Sunaga, K., Hirano, A., Saunders, P., Katsube, N., and Chuang, D.-M. (1996). Evidence that glyceraldehyde-3-phosphate dehydrogenase is involved in age-induced apoptosis in mature cerebellar neurons in culture. *J. Neurochem.* **66,** 928–935.
84. Wanker, E. E., Rovira, C., Scherzinger, E., Hasenbank, R., Walter, S., Tait, D., Colicelli, J., and Lehrach, H. (1997). HIP-1: a huntingtin interaction protein isolated by the yeast two-hybrid system. *Hum. Mol. Genet.* **6,** 487–495.
85. Kalchman, M. A., Koide, H. B., McCutcheon, K., Graham, R. K., Nichol, K., Nishiyama, K., Kazemi-Esfarjani, P., Lynn, F. C., Wellington, C., Metzler, M.,Goldberg, Y. P., Kanazawa, I., Gietz, R. D., and Hayden, M. R. (1997). *HIP1*, a human homologue of *S. cerevisiae Sla2p*, interacts with membrane-associated huntingtin in the brain. *Nature Genet.* **16,** 44–53.
86. Li, X.-J., Li, S.-H., Sharp, A. H., Nucifora, F. C., Jr., Schilling, G., Lanahan, A., Worley, P., Snyder, S. H., and Ross, C. A. (1995). A huntingtin-associated protein enriched in brain with implications for pathology. *Nature* **378,** 398–402.
87. Li, X.-J., Sharp, A. H., Li, S.-H., Dawson, T. M., Snyder, S. H., and Ross, C. A. (1996). huntingtin-associated protein (HAP1): discrete neuronal localizations in the brain resemble those of neuronal nitric oxide synthase. *Proc. Natl. Acad. Sci. USA* **93,** 4839–4844.
88. Yano-Yanagisawa, H., Li, Y., Wang, H., and Kohwi, Y. (1995). Single-stranded DNA binding proteins isolated from mouse brain recognize specific trinucleotide repeat sequences *in vitro.* *Nucleic Acids Res.* **23,** 2654–2660.
89. Perutz, M. F., Johnson, T., Suzuki, M., and Finch, J. T. (1994). Glutamine repeats as polar zippers: their possible role in inherited neurodegenerative diseases. *Proc. Natl. Acad. Sci. USA* **91,** 5355–5358.
90. Perutz, M. F. (1996). Glutamine repeats and inherited neurodegenerative diseases: molecular aspects. *Curr. Opin. Struct. Biol.* **6,** 848–858.
91. Stott, K., Blackburn, J. M., Butler, P. J. G., and Perutz, M. (1995). Incorporation of glutamine repeats makes protein oligomerize: implications for neurodegenerative diseases. *Proc. Natl. Acad. Sci. USA* **92,** 6509–6513.
92. Folk, J. E., and Cole, P. W. (1965). Structural requirements of specific substrates for guinea pig liver transglutaminase. *J. Biol. Chem.* **240,** 2951–2960.
93. Chung, S. I., Shrager, R. I., and Folk, J. E. (1970). Mechanism of action of guinea pig liver transglutaminase. *J. Biol. Chem.* **246,** 6424–6435.
94. Fesus, L., Thomazy, V., and Falus, A. (1987). Induction and activation of tissue transglutaminase during programmed cell death. *FEBS Lett.* **224,** 104–108.
95. Fukuda, K., Kojiro, M., and Chiu, J.-F. (1993). Induction of apoptosis by transforming growth factor-β1 in the rat hepatoma cell line McA-RH7777: a possible association with tissue transglutaminase expression. *Hepatology* **18,** 945–953.
96. Boehm, M. F., Zhang, L., Zhi, L., McClurg, M. R., Berger, E., Wagoner, M., Mais, D. E., Suto, C. M., Davies, P. J. A., Heyman, R. A., and Nadzan, A. M. (1995). Design and synthesis of potent retinoid X receptor selective ligands that induce apoptosis in leukemia cells. *J. Med. Chem.* **38,** 3146–3155.
97. Lu, X., Xie, W., Reed, D., Bradshaw, W. S., and Simmons, D. L. (1995). Nonsteroidal antiinflammatory drugs cause apoptosis and induce cyclooxygenases in chicken embryo fibroblasts. *Proc. Natl. Acad. Sci. USA* **92,** 7961–7965.
98. Furuya, Y., and Isaacs, J. T. (1993). Differential gene regulation during programmed death (Apoptosis) *versus* proliferation of prostatic glandular cells induced by androgen manipulation. *Endocrinology* **133,** 2660–2666.
99. Sabourin, L. A., and Hawley, R. G. (1990). Suppression of programmed death and G_1 arrest in B-cell hybridomas by interleukin-6 is not accompanied by altered expression of immediate early response genes. *J. Cell. Physiol.* **145,** 564–574.
100. Fesus, L., Thomazy, V., Autuori, F., Ceru, M. P., Tarcsa, E., and Piacentini, M. (1989). Apoptotic hepatocytes become insoluble in detergents and chaotropic agents as a result of transglutaminase action. *FEBS Lett.* **245,** 150–154.
101. Knight, C. R. L., Rees, R. C., and Griffin, M. (1991). Apoptosis: a potential role for cytosolic transglutaminase and its importance in tumour progression. *Biochim. Biophys. Acta* **1096,** 312–318.
102. Fesus, L., Tarcsa, E., Kedei, N., Autuori, F., and Piacentini, M. (1991). Degradation of cells dying by apoptosis leads to accumulation of $\varepsilon(\gamma$-glutamyl) lysine isodipeptide in culture fluid and blood. *FEBS Lett.* **284,** 109–112.
103. Franke, W. W., Denk, H., Schmid, E., Osborn, M., and Weber, K. (1979). Ultrastructural, biochemical, and immunologic characterization of Mallory bodies in livers of Griseofulvin-treated mice. *Lab. Invest.* **40,** 207–220.
104. Zatloukal, K., Fesus, L., Denk, H., Tarcsa, E., Spurej, G., and Bock, G. (1992). High amount of ε-(γ-glutamyl) lysine cross-links in Mallory bodies. *Lab. Invest.* **66,** 774–777.
105. Simon, S. M., and Llinas, R. R. (1985). Compartmentalization of the submembrane calcium activity during calcium influx and its significance in transmitter release. *Biophys. J.* **48,** 485–498.
106. Stern, M. D. (1992). Buffering of calcium in the vicinity of a channel pore. *Cell Calcium* **13,** 183–192.
107. Gilad, G. M., and Varon, L. E. (1985). Transglutaminase activity in rat brain: characterization, distribution, and changes with age. *J. Neurochem.* **45,** 1522–1526.
108. Pastuszko, A., Wilson, D. F., and Erecinska, M. (1986). A role for transglutaminase in neurotransmitter release by rat brain synaptosomes. *J. Neurochem.* **46,** 499–508.
109. Ohashi, H., Itoh, Y., Birckbichler, P. J., and Takeuchi, Y. (1995). Purification and characterization of rat brain transglutaminase. *J. Biochem.* **118,** 1271–1278.
110. Selkoe, D. J., Abraham, C., and Ihara, Y. (1982). Brain transglutaminase: *in vitro* crosslinking of human neurofilament proteins into insoluble polymers. *Proc. Natl. Acad. Sci. USA* **79,** 6070–6074.
111. Gentile, V., Thomazy, V., Piacentini, M., Fesus, L., and Davies, P. J. A. (1992). Expression of tissue transglutaminase in Balb-C 3T3 fibroblasts: effects on cellular morphology and adhesion. *J. Cell Biol.* **119,** 463–474.
112. Melino, G., Annicchiarico-Petruzzelli, M., Piredda, L., Candi, E., Gentile, V., Davies, P. J. A., and Piacentini, M. (1994). Tissue transglutaminase and apoptosis: sense and antisense transfection studies with human neuroblastoma cells. *Mol. Cell. Biol.* **14,** 6584–6596.
113. Amendola, A., Gougeon, M.-L., Poccia, F., Bondurand, A., Fesus, L., and Piacentini, M. (1996). Induction of "tissue" transglutaminase in HIV pathogenesis: evidence for high rate of apoptosis of $CD4^+$ T lymphocytes and accessory cells in lymphoid tissues. *Proc. Natl. Acad. Sci. USA* **93,** 11057–11062.
114. Lorand, L. (1996). Neurodegenerative diseases and transglutaminase. *Proc. Natl. Acad. Sci. USA* **93,** 14310–14313.
115. Friedrich, P., Fesus, L., Tarcsa, E., and Czeh, G. (1991). Protein cross-linking by transglutaminase induced in long-term potentiation in the CA1 region of hippocampal slices. *Neuroscience* **43,** 331–334.

116. Kahlem, P., Terre, C., Green, H., and Djian, P. (1996). Peptides containing glutamine repeats as substrates for transglutaminase-catalyzed cross-linking: relevance to diseases of the nervous system. *Proc. Natl. Acad. Sci. USA* **93,** 14580–14585.
117. Cooper, A. J. L., Sheu, K.-F. R., Burke, J. R., Onodera, O., Strittmatter, W. J., Roses, A. D., and Blass, J. P. (1997). Polyglutamine domains are substrates of tissue transglutaminase: does transglutaminase play a role in expanded CAG/poly-Q neurodegenerative diseases? *J. Neurochem.* **69,** 431–434.
118. Fink, M. L., Chung, S. I., and Folk, J. E. (1980). γ-Glutamylamine cyclotransferase: specificity toward ε-(L-γ-glutamyl)-L-lysine and related compounds. *Proc. Natl. Acad. Sci. USA* **77,** 4564–4568.
119. Folk, J. E., and Finlayson, J. S. (1977). The ε-(γ-glutamyl) lysine crosslink and the catalytic role of transglutaminases. *Adv. Protein Chem.* **31,** 1–133.
120. Huntington's Disease Collaborative Research Group (1993). A novel gene containing a trinucleotide repeat that is expanded and unstable on Huntington's disease chromosomes. *Cell* **72,** 971–983.
121. Chang, C., Kokontis, J., and Liao, S. (1988). Structural analysis of complementary DNA and amino acid sequences of human and rat androgen receptors. *Proc. Natl. Acad. Sci. USA* **85,** 7211–7215.
122. Orr, H. T., Chung, M.-y., Banfi, S., Kwiatkowski, T. J., Jr., Servadio, A., Beaudet, A. L., McCall, A. E., Duvick, L. A., Ranum, L. P. W., and Zoghbi, H. Y. (1993). Expansion of an unstable trinucleotide CAG repeat in spinocerebellar ataxia type 1. *Nature Genet.* **4,** 221–226.
123. Kawaguchi, Y., Okamoto, T., Taniwaki, M., Aizawa, M., Inoue, M., Katayama, S., Kawakami, H., Nakamura, S., Nishimura, M., Akiguchi, I., Kimura, J., Narumiya, S., and Kakizuka, A. (1994). CAG expansions in a novel gene for Machado-Joseph disease at chromosome 14q32.1. *Nature Genet.* **8,** 221–228.
124. Koide, R., Ikeuchi, T., Onodera, O., Tanaka, H., Igarashi, S., Endo, K., Takahashi, H., Kondo, R., Ishikawa, A., Hayashi, T., Saito, M., Tomoda, A., Miike, T., Naito, H., Ikuta, F., and Tsuji, S. (1994). Unstable expansion of CAG repeat in hereditary dentatorubral-pallidoluysian atrophy (DRPLA). *Nature Genet.* **6,** 9–13.
125. Tucker, P. K., and Lundrigan, B. L. (1993). Rapid evolution of the sex determining locus in Old World mice and rats. *Nature* **364,** 715–717.
126. Lin, B., Nasir, J., MacDonald, H., Hutchinson, G., Graham, R. K., Rommens, J. M., and Hayden, M. R. (1994). Sequence of the murine Huntington disease gene: evidence for conservation, and polymorphism in a triplet (CCG) repeat alternate splicing. *Hum. Mol. Genet.* **3,** 85–92.
127. Baxendale, S., Abdulla, S., Elgar, G., Buck, D., Berks, M., Micklem, G., Durbin, R., Bates, G., Brenner, S., Beck, S., and Lehrach, H. (1995). Comparative sequence analysis of the human and pufferfish Huntington's disease genes. *Nature Genet.* **10,** 67–76.
128. Shibuya, H., Nonneman, D. J., Huang, T. H.-M., Ganjam, V. K., Mann, F. A., and Johnson, G. S. (1993). Two polymorphic microsattelites in a coding segment of the canine androgen receptor gene. *Anim. Genet.* **24,** 345–348.
129. Tilley, W. D., Marcelli, M., Wilson, J. D., and McPhaul, M. J. (1989). Characterization and expression of a cDNA encoding the human androgen receptor. *Proc. Natl. Acad. Sci. USA* **86,** 327–331.

CHAPTER 47

Pathogenesis of Polyglutamine Neurodegenerative Diseases: Toward a Unifying Mechanism

CHRISTOPHER A. ROSS Department of Psychiatry and Neuroscience and Program in Cellular and Molecular Medicine, School of Medicine, The Johns Hopkins University, Baltimore, Maryland 21205

RUSSELL L. MARGOLIS Department of Psychiatry, School of Medicine, The Johns Hopkins University, Baltimore, Maryland 21205

MARK W. BECHER Department of Pathology, School of Medicine, The Johns Hopkins University, Baltimore, Maryland 21205

JONATHAN D. WOOD, SIMONE ENGELENDER, AND ALAN H. SHARP Department of Psychiatry, School of Medicine, The Johns Hopkins University, Baltimore, Maryland 21205

I. INTRODUCTION: GAIN OF GLUTAMINE, GAIN OF FUNCTION

Eight neurodegenerative diseases are currently known to be caused by expanding polyglutamine repeats (Table 47-1). Evidence from a number of different sources indicates that the major pathogenic mechanisms of all these diseases involve a gain of an adverse property by polyglutamine. The evidence for this contention will be reviewed, focusing mainly on Huntington's disease (HD), the disorder which has been most studied, but also drawing on evidence from the other disorders [1–14]. The data will be organized according to the methods used to obtain them—a scheme which loses the historical flow, but permits direct comparison of data from the different diseases.

II. CLINICAL GENETICS

All of the glutamine repeat diseases whose genes are located on autosomes show dominant transmission. Dominant transmission is often associated with a gain-of-function pathogenesis, while recessive transmission is often associated with loss of function pathogenesis. HD long served as a textbook example of autosomal dominant inheritance. While the discovery of the triplet repeat expansion has modified our understanding of some aspects of its genetics, HD still appears to show classic genetic dominance. It is similar in its clinical presentation whether the patient has one or two copies of the mutated gene [15, 16]. The age of onset for homozygotes for the HD mutation generally tends to be within the range of age of onset for cases with only one copy of comparable repeat length (though perhaps tending toward earlier onset). By contrast, in the other glutamine repeat diseases patients with two copies of the expanded repeat often have an age of onset substantially earlier than would be predicted by the length of the longer repeat alone [17].

No cases of HD (or, with the possible exception of SCA6, any of the other disorders in Table 47-1) have been identified with deletions or point mutations in any of the causative genes. This is in contrast to Fragile X syndrome and Freidreich's ataxia, which involve loss of gene function of FMR-1 and Frataxin, respectively. Patients with deletions and point mutations of these genes have disease phenotypes similar to those of patients with the triplet repeat expansions [18–22]. One individual has disruption of one allele of the HD gene, due to a balanced translocation between chromosome 4 and chromosome 12 with a breakpoint in the HD gene at exon 40–41. In mid-adulthood, this individual does not have an HD-like phenotype, even though expression of huntingtin is only 50% of normal [23, 24].

The situation for SBMA is more complicated since it is on the X chromosome. While it shows an X-linked recessive pattern of inheritance at the clinical level, it is possible that at the cellular level it may be considered to be dominant. Because females have one X chromosome randomly inactivated in every cell, 50% of neurons would not have expression from the expanded allele. Fifty percent of neurons without the mutated protein

TABLE 47-1 Glutamine Repeat Diseases

Disease	Gene	Regions most affected
Huntington's disease (HD)	Huntingtin or IT15	Striatum and cortex
Dentatorubral and pallidoluysian atrophy (DRPLA)	Atrophin-1	Basal ganglia and cerebellum
Spinocerebellar ataxia-1 (SCA1)	Ataxin-1	Cerebellum and basal ganglia
SCA2	Ataxin-2	Cerebellum and brainstem
SCA-3 (Machado–Joseph disease)	Ataxin-3	Cerebellum and basal ganglia
SCA-6[a]	P/Q Calcium channel	Cerebellar Purkinje cells
SCA-7	Ataxin-7	Cerebellum and retina
Spinal and bulbar muscular atrophy (Kennedy's disease or SBMA)	Androgen receptor	Spinal motor neurons

[a]May involve a different mechanism (see text).

may be sufficient to maintain essentially normal function. Thus, the inheritance could still be compatible with a gain-of-function pathogenesis at the molecular level. Consistent with this idea is the observation that deletions or other mutations which inactivate the androgen receptor do not cause SBMA but rather cause a different syndrome of androgen insensitivity [18, 25].

For all of the disorders (except SCA6) there is a consistent range of CAG repeats which is never associated with pathology and a range of repeats which is essentially always associated with the development of symptoms within a normal lifetime. SCA6 may involve a different pathogenesis (see Section IV below). For all the other disorders, 33 or fewer CAG repeats do not result in disease. The thresholds for the different disorders are somewhat different and for some of the disorders there is a range of incomplete penetrance, but in general more than about 35 to 40 CAG repeats can cause disease. For all of the disorders, there is an association between longer CAG repeats and earlier age of onset, with greater than 60 or 70 repeats usually associated with onset before adulthood. Any explanation of the pathogenesis of the disorders will necessarily need to explain these relationships.

III. PATHOLOGY

The pathology of all the glutamine repeat disorders involves selective neuronal vulnerability. Unlike the other triplet repeat disorders, such as fragile X syndrome or myotonic dystrophy, no clear pathological changes have been described outside of the nervous system. All of the glutamine repeat disorders show neuronal loss and gliosis without inflammation or prominent deposition of extracellular material. The areas affected show considerable overlap, but also differences [10, 26, 27]. Areas affected in different patterns in the different disorders include deep layers of the cerebral cortex, basal ganglia, brainstem nuclei, cerebellar dentate nucleus, Purkinje cells of the cerebellum, and spinal and bulbar motor neurons. By contrast in none of the disorders is there prominent pathology in the hippocampus or the basal forebrain, areas severely affected in Alzheimer's disease.

Huntington's disease is characterized by selective atrophy of the caudate and putamen or corpus striatum of the basal ganglia. In this region, medium spiny projection neurons are severely lost while several populations of large interneurons are relatively spared [28–31]. Similarly in the cerebral cortex there is selective neuronal vulnerability with layers III, V, and VI (especially the pyramidal neurons) affected and layers II and IV relatively spared [32, 33]. Other areas of the brain are more mildly or inconsistently affected.

For HD, enough cases have been studied to establish a clear relationship between repeat length and the neuropathology. Cases with longer repeat lengths have more severe pathology in the basal ganglia [34–36]. In addition, cases with longer repeats tend to have more pathology in regions affected in the other glutamine repeat diseases. Nevertheless, there still appears to be selective vulnerability and differential sensitivity. In both HD and DPRLA, juvenile cases have prominent losses of Purkinje cells, which are normally only mildly affected in adult cases in both diseases. However, the striatum is relatively unaffected even in juvenile DPRLA, while the dentate nucleus of the cerebellum is relatively unaffected even in juvenile HD [10].

While the atrophy and neuronal cell loss of HD has been extensively studied, there has been less attention focused on the morphology of the surviving neurons. One might expect that these neurons would be shrunken and atrophic. However, this is not necessarily the case. Neurons studied with the Golgi impregnation method show evidence of "regenerative" or "plastic" changes, including increased numbers of dendrites, increased numbers of long recurved dendrites, and greater density and size of spines on dendrites [37, 38]. Thus, a complete understanding of the pathogenesis of HD will need to explain these regenerative changes as well as neuronal cell death.

It is also unclear to what extent symptoms are caused by neuronal death or whether they relate to functional changes prior to death. In most cases symptomatic HD is reflected in neuronal loss at autopsy. However, occasional cases arise of individuals with clinically diagnosed HD, but no discernible neuronal loss. These have been termed "grade zero" in the severity scale of Vonsattel *et al.* [36]. There is no appreciable gliosis or neuronal loss—therefore, it is possible that symptoms in these patients arise from functional changes rather than actual neuronal loss.

IV. NEUROTOXICITY

Even before the discovery of the CAG repeat expansions causing these diseases, animal models had been created using neurotoxins. NMDA receptor agonists such as quinolinic acid mimic HD pathology in the striatum with loss of medium spiny projection neurons and sparing of cholinergic and NADPH diaphorase neurons [29, 39]. Metabotropic glutamate receptors may also be involved [40].

Metabolically compromised neurons may be more sensitive to excitotoxicity than normal neurons, a phenomenon described as "indirect excitotoxicity" [41, 42]. Peripheral injections into rodents or primates of several mitochondrial toxins including 3-nitroproprionic acid (3NPA) reproduce aspects of the pattern of striatal pathology seen in HD [42, 43]. Several other metabolic toxins have similar effects, including iodoacetate, an inhibitor of the enzyme glyceraldehyde 3-phosphate dehydrogenase (GAPDH) [44]. Other metabolic poisons cause preferential toxicity in different regions of the brain, often those affected in glutamine repeat disease. For instance, cyanide, a complex four mitochondrial inhibitor, causes toxicity in several subcortical regions [45] affected in spinocerebellar ataxias.

These neurotoxological experiments have suggested that several pathways may be involved in cell death. For instance, both excitotoxicity and metabolic poisoning may result in free radical damage [42, 46–48]. Intracellular calcium, released by glutamate receptors, calcium channels, or IP3 receptors, can mediate cell death.

This latter mechanism may be directly relevant to SCA6. In this disease, the glutamine repeat is in a splice isoform of the neuronal P-type calcium channel [49–51]. This channel is greatly enriched in cerebellar Purkinje cells. The disease may result from alteration in channel function, causing abnormal regulation of calcium influx, leading to toxicity.

Neurotoxic stimuli may give rise to either necrosis or apoptosis. Some stimuli can cause either form of cell death, depending on dose and tissue course. Apoptosis may be the relevant mechanism for these neurodegenerative diseases. Apoptosis is a controlled event regulated by cellular machinery, in contrast to necrosis, defined as cell swelling and lysis, which occurs when toxins overcome cellular homeostatic mechanisms. Apoptosis plays an essential role in normal development, serving to prune and shape a variety of structures including neurons in the central nervous system [52]. The process is triggered by aspartate proteases including the proteases termed caspases [53, 54]. The enzyme GAPDH and other metabolic enzymes may also be involved in triggering apoptosis of some cells [55].

Studies of HD brains are consistent with some of these neurotoxicologic mechanisms. NMR spectroscopy of HD patients provides evidence compatible with metabolic compromise of neurons in HD [56, 57]. Postmortem studies have demonstrated marked biochemical defects of mitochondrial complex II and complex III activity, and moderate defects of complex IV activity in mitochondria isolated from the brain tissue of individuals with HD [58, 59]. Evidence of free radical activation is also found in HD postmortem tissue [59].

V. TARGETED GENE DELETIONS

The creation of mice with targeted deletions of some of the CAG repeat genes is beginning to define the effects of loss-of-function of these genes on brain development and function. Mice with targeted deletions of both copies of the HD gene die in early embryonic development, shortly before implantation [60–62]. In one of these studies, it was found that the embryos had increased numbers of apoptotic cell profiles suggesting the possibility that a normal function of huntingtin might be to inhibit apoptosis [62]. Animals with heterozygous deletions of the HD gene have been reported either to be normal or to have subtle brain abnormalities. In the study of Nasir *et al.* [60], mice heterozygous for a deletion of exon five of the HD gene had fewer neurons in the subthalamic nucleus and the globus pallidus, which led the authors to suggest that the HD gene may have a role in inhibiting cell death of these structures during development.

These studies suggest that HD is not the result of a simple loss of function. However, they cannot rule out a dominant-negative mechanism leading to a gene dosage of between 0 and 50%. More recent studies have addressed this issue. In the process of making a "knock-in" model of HD, the MacDonald group [63] has created mice with one allele of the HD gene deleted and one allele with an insertion of a neo cassette. The neo cassette interferes with transcription of the HD gene and leads to levels of transcription between 0 and 50% of normal. Development of these animals proceeds well beyond the implantation stage; however, the embryos are abnormal. There are a number of defects described, but most striking are exencephaly and abnormal brain development, including ectopic neurons in the lateral ventricles. This phenotype is similar to that of the targeted disruption of caspase three [64], again raising the question of a role for huntingtin in regulation of neuronal cell death. In addition, mice with one allele deleted but one allele with a Q50 repeat have normal development, indicating that the repeat expansion does not cause a loss of whatever function is necessary for normal development. Taken together, these experiments do not support a primary role for loss of function in the pathogenesis of HD. However, they do not exclude the possibility that loss of function may contribute in some way to the phenotype.

Mice with targeted deletion of the SCA1 gene have also been created [65]. Heterozygous animals are normal; homozygous animals have behavioral and learning deficits, but no evidence for the progressive neuronal degeneration seen in SCA1. Thus, this model argues strongly against loss of function being critical for the pathogenesis of SCA1.

Mice with targeted deletions of the androgen receptor have androgen insensitivity syndrome, but no evidence for neuronal degeneration as in SBMA. Knockouts of the other CAG repeat genes have not yet been created.

VI. BIOCHEMISTRY AND CELL BIOLOGY OF THE GENE PRODUCTS

A. mRNA

The mRNA for the HD gene is widely expressed in both the brain and peripheral tissues [66–68]. Like the HD message, the DPRLA mRNA is widely expressed [69–72] with no enrichment in areas affected in the disease. Similarly, messages for the SCA1, SCA2, and SCA3 genes are also widely expressed with no clear enrichment in affected tissues. The mRNA for SCA1 may show some enrichment in Purkinje cells, the cells which are most selectively affected [73, 74]. However, this study was conducted in rodents, and the distribution of expression in humans has not been studied in detail yet.

B. Protein Products

The huntingtin protein, like the message, is widely expressed with both a soluble and particulate localization with cells [75–81]. These studies used antibodies directed at several portions of the protein including the predicted N-terminus, demonstrating that the CAG repeat is translated into polyglutamine. Immunocytochemical and subcellular fractionation studies indicate that huntingtin is present throughout neurons, in perikarya, dendrites, and nerve terminals, with a generally cytoplasmic localization. It was not detected to any discernible degree in the nucleus (except in one study [82, 83]). There is association with cytoskeletal elements and vesicle populations [81, 84] within cells, with some apparent selectivity to endosomal compartments. Huntingtin is detected at all stages of embryonic and postnatal brain development [85]. More recent studies focusing on the striatum have suggested that huntingtin is enriched in the medium spiny neurons, the neurons which are selectively vulnerable in HD, and not expressed, or expressed at lower levels, in the large interneurons [86]. This may contribute to the selective vulnerability of the medium spiny neurons. The protein appears to be enriched in membranes of the Golgi complex [87]. In fibroblasts, huntingtin is associated with tubulovesicular membranes in the perinuclear region and with vesicles widely scattered throughout the cytoplasm.

Atrophin-1, the DRPLA protein product, is also widely expressed. In has a generally cytoplasmic distribution in neurons [88–90].

Ataxin-1, the SCA1 gene product, is widely expressed [91]. In peripheral tissues, it is mainly cytoplasmic. However, in neurons, especially Purkinje cells—the cells most affected in SCA1—it has a predominantly nuclear localization.

Ataxin-3 is widely expressed and is generally cytoplasmic, though with some differences in expression in different neurons [92]. For instance, in the striatum, it is enriched in large neurons.

The androgen receptor of SBMA is also relatively widely expressed, though not as ubiquitiously as some of the other messages. In addition, there may be some selectivity of expression in spinal motor neurons. This is especially seen when using androgen receptor binding techniques [93].

C. Functional Assays

The androgen receptor is the only CAG repeat disease gene product with a known function. The effect of altering the glutamine repeat length on different measures of androgen receptor activity can be assayed. However, the data appear to be somewhat contradictory, with some reports finding normal and other reports finding altered transactivational capacity [94–96].

VII. INTERACTING PROTEINS

A. Huntingtin Interactors

Intensive effort has been devoted to finding huntingtin's interaction partners, to give clues to its normal function and, hopefully, to the pathogenesis of the disease. A number of interactors have been identified. They are beginning to shed light on huntingtin's normal functions; however, their role in the pathogenesis of the disease is still uncertain.

One huntingtin interactor was termed HAP1 for huntingtin-associated protein 1 [97]. The strength of the interaction appeared to be modestly increased with increasing glutamine repeat length, and, unlike huntingtin, HAP1 is enriched in the brain, consistent with a possible role in pathogenesis of the disorder. However, its expression pattern in the brain is not restricted to regions affected in HD [97, 98]. HAP1, like huntingtin, has no significant homology to other known proteins.

Therefore, more recent studies have attempted to find interacting proteins for HAP1. Several interesting proteins have recently been identified. The middle third of HAP1 interacts with the p150Glued component of dynactin [99], as well as several other cytoskeletal-related proteins. The p150Glued protein is believed to have a role in the regulation of the function of dynein, a major retrograde motor protein in neurons [100, 101]. Both huntingtin and HAP1 undergo rapid axonal transport [102].

The N-terminus of HAP1 interacts with a protein called Duo, which is highly homologous to a protein called Trio [103, 104]. Duo was independently identified as an interactor with a transmembrane protein in vesicles in the secretory pathway, and termed P-CIP-10 or kalirin [24]. Duo contains a guanine nucleotide exchange factor domain capable of activating the Ras-like GTPase Rac1, which in turn can regulate neuronal cytoskeletal functions. These interactions are consistent with a postulated normal function for huntingtin in vesicle transport and other cytoskeletal interactions. However, the significance of these interactions for pathogenesis of HD is still speculative (see Section XII below).

Huntingtin also interacts with a protein called HIP1 in a glutamine repeat length-dependent fashion [105, 106]. HIP1 is homologous to the yeast protein Sla2p [107]. Mutations in this protein interfere with endocytosis. HIP1 is also homologous to a portion of talin, which is involved in membrane cytoskeletal interactions. Biochemical evidence supports the association between HIP1 and the cytoskeleton. There is a striking decrease in the strength of the interaction between HIP1 and huntingtin when the glutamine repeat in huntingtin is expanded. These interactions support the idea that huntingtin is involved with vesicle transport and the cytoskeleton, especially in relation to endocytosis. The expansion of the glutamine repeat may alter these processes.

Huntingtin, as well as the other glutamine repeat proteins, can interact with the metabolic enzyme GAPDH [108]. However, it is unclear if this interaction is dependent on the length of the glutamine repeat [109]. Nevertheless, since GAPDH has been implicated in neuronal cell death, this interaction is of potential interest.

Huntingtin also interacts with a ubiquitin conjugating enzyme (termed HIP2 or E2-25K) [110]. The interaction with HIP2 does not appear to be dependent on the length of the glutamine repeat. Huntingtin can be ubiquitinated, though it is not clear if it is via this enzyme.

B. Atrophin-1 Interactors

Atrophin-1 interacts with several WW-domain containing proteins [111]. The atrophin-1 sequence has PY motifs which are likely to mediate these interactions. Two of the atrophin-1 interactors contain PDZ domains. These domains have been found in proteins which interact with membrane receptors and target them to specific postsynaptic sites [112, 113]. Three of the atrophin-1 interactors contain HECT domains, which are found in E3 ubiquitin ligase enzymes, including Nedd-4 [114]. A yeast mutant of one of these enzymes (Rsp5) is deficient in endocytosis of a membrane protein [115]. The phenotype of this mutant is very similar to that of the Sla2p mutant of the HIP1 interactor of huntingtin. Thus, huntingtin and atrophin-1 may have some functional similarities.

C. Ataxin-1 Interactors

Recent data have uncovered an extremely interesting interactor for the ataxin-1 protein. This is a protein called leucine-rich acidic nuclear protein (LANP) [116–118]. The interaction is dramatically stronger when the ataxin-1 protein has an expanded glutamine repeat. The localization of LANP in the brain fits well with the regions most severely affected in SCA1. LANP is most highly expressed in nuclei of cerebellar Purkinje cells, the cells most affected in SCA1. It is also present in deep cerebellar nuclei, brainstem, and thalamus, which are also affected in SCA1. As described below, the cellular localization of this protein makes it a very good candidate for involvement in the pathogenesis of SCA1.

D. Other Interactions

The molecular interactions of the androgen receptor are well described for its ligand-binding domain and its DNA-binding domain. However, the function of the N-terminus, which contains the glutamine repeat, is still uncertain. This region of the glucocorticoid receptor is believed to be involved in interactions with cell type-specific transcription factors [119] and presumably the androgen receptor has a similar molecular organization. However, interacting proteins have not yet been identified. Protein interacting partners of ataxin-2, ataxin-3, and ataxin-7 have also not yet been identified.

VIII. POLYGLUTAMINE

A. Possible Functions of Polyglutamine

Many proteins contain stretches of polyglutamine. However, the normal function of these glutamine repeats in proteins is unknown. CAG repeats were originally identified in drosophila and termed opa-repeats, and were later found also in mouse [120, 121]. Proteins containing glutamine repeats often appear to have a role in the regulation of development and neurogenesis [122, 123]. Glutamine repeats are more common than repeats of other single amino acids [124]. However, the lengths of the repeats are poorly conserved in homologous genes from different species. For instance, the mouse HD gene encodes only seven repeats and the puffer fish homologue has only four repeats [125–127]. Similarly, the glutamine repeats in rodent atrophin-1 are shorter than those in humans. The rat atrophin-1 glutamine repeat is only five consecutive glutamines long, followed by four glutamine–proline pairs, and the mouse repeat is also interrupted by proline [70, 128].

Glutamine-rich regions have been described in the factor interaction domains of transcription factors, though it is not clear whether glutamine repeats function similarly to glutamine-rich regions. However, a number of proteins with glutamine repeats are transcription factors. Altering the size of polyglutamine repeats appears to alter transactivation [129]. The glutamine repeats probably do not bind to DNA, but may be involved in protein–protein interactions.

B. Polar Zipper Hypothesis

One hypothesis for the role of glutamine repeats in human disease comes from molecular modeling studies suggesting that glutamine repeats can interact via a so-called "polar zipper." Perutz proposed that two antiparallel beta strands of polyglutamine can be linked together by hydrogen bonds between their main chain and side chain amides [130–132]. These interactions might cause the proteins to aggregate, and then to precipitate within cells. Circular dichroism, electron microscopic, and X-ray defraction studies of synthetic peptides and an engineered protein incorporating glutamine repeats were consistent with the formation of beta strands and possibly beta sheets by these glutamine repeats [130–132]. Recent evidence from *in vitro* animal model and postmortem studies are consistent with this possibility (see below).

C. Transglutaminase Hypothesis

Another hypothesis for how glutamine repeats could form an insoluble precipitate within cells is the transglutaminase hypothesis [50]. This involves covalent modification of glutamines via an isopeptide linkage to lysine. However, whether this occurs with polyglutamine stretches *in vivo* is unknown.

IX. *IN VITRO* DATA

There is still no direct structural evidence for polar zipper formation by polyglutamine. However, recent *in vitro* studies indicate that a short truncation (containing exon 1) of the huntingtin protein containing the expanded glutamine repeat can aggregate *in vitro* to form amyloid-like fibrils [133]. This region was chosen because of the mouse model results described below [134, 135]. In the *in vitro* study, a filter assay was devised to demonstrate that this truncated polyglutamine-containing fragment would aggregate if the polyglutamine were in the range which causes disease, but not if the polyglutamine repeat were in the normal range. This aggregated material had an amyloid-like fibrillar appearance using negatively stained electron microscopy. Furthermore, the protein aggregates showed green birefringence when stained with Congo red and examined using polarized light microscopy. Thus, this study provides striking evidence for aggregation of this polyglutamine truncated fragment and supports the polar zipper hypothesis.

The size of the huntingtin truncation was important in these experiments. The protein was generated as a glutathione *S*-transferase (GST) fusion. If the truncated huntingtin were not cleaved from the GST, it would only aggregate when the polyglutamine was very long, i.e., greater than 80 repeats. When it was cleaved from GST, however, it would aggregate the more usual range of expanded lengths, for instance 51. These data are consistent with the possibility that full-length huntingtin might be less toxic than a proteolytic fragment of huntingtin.

A. Proteolytic Cleavage

Huntingtin can be cleaved *in vitro* by the apoptosis-related protease caspase-3 (also termed apopain or CPP32) [136]. There are consensus cleavage sites for this enzyme at around position 513 and 53 and huntingtin produced by *in vitro* transcription and translation can be cleaved by purified caspase-3 *in vitro*. Strikingly, cleavage is accelerated by an expanded polyglutamine repeat, consistent with relevance for the disease.

B. Relationship to Age of Onset Curves

These data suggest the possibility that polyglutamine-containing proteins may cause disease by aggregation, possibly due to a polar zipper-mediated mechanism. This could be a pathogenic mechanism common to all the polyglutamine diseases. Differences among the diseases would be explained by the different protein contexts of the polyglutamine in each disease. The GST fusion protein experiments [133] give results compatible with this idea, in that the GST fusion protein appears more soluble than the truncated huntingtin by itself. Furthermore, the data on the relationship between ages of onset and repeat lengths are consistent with the idea that there are different "length–response" curves in the different diseases [9, 17]. SCA2 and SCA7 have a very steep relationship between glutamine repeat length and age of onset, with a relatively low threshold of 35–36, and juvenile onset for repeat lengths in the 60s. By contrast, SCA3 may have a higher threshold, with all cases observed, so far, having repeat lengths of >60, and juvenile onset only for repeats in the 70s–80s. SCA1 and HD appear to be intermediate.

X. CELL MODELS

The suggestion that aggregation may be a common mechanism for toxicity of polyglutamine containing proteins has provided an important biochemical insight. However, in order to understand the pathogenesis of the diseases, it is necessary to see if this can take place in cells, and if so, what effect it might have. Polyglutamine has been reported to be toxic to bacteria [72], but whether the mechanism is relevant to mammalian cells is uncertain.

One mammalian cell model involved transient expression in COS-7 cells of truncated portions of the ataxin-3 protein [137]. The truncated constructs coded for a glutamine repeat (either normal or expanded) with only a small segment of flanking sequence. Cells transfected with a full-length ataxin-3 construct survived and replicated normally regardless of the length of the glutamine. By contrast, cells transfected with the truncated construct grew normally only if the repeat length was normal. The truncated protein with a long repeat resulted in lack of growth and, using several assays, apoptotic cell death. Cells with the expanded truncated construct could be labeled for transfected protein in a punctate distribution within the cytoplasm, suggestive of aggregation, while cells transfected with normal length repeat or the full-length protein had label with a cytoplasmic diffuse distribution. Furthermore, on Western blots, protein expressed from the truncated expanded construct appeared to have an abnormally high molecular weight, suggesting that it may have been present in an aggregated form.

Recent experiments with both ataxin-3 and ataxin-1 indicate that aggregates may form in the nucleus. Transient transfection of HEK-293 cells with a truncated portion of ataxin-3 with an expanded CAG repeat resulted in large cytoplasmic structures resembling vacuoles which concentrated a lysosomal marker [92]. In addition, there were smalller structures both within and around the nucleus. These structures could be labeled with the 1C2 monoclonal antibody, demonstrating the presence of the expanded glutamine repeats. By contrast, full-length ataxin-3 with an expanded repeat had a diffuse cytoplasmic localization, as did full-length or truncated ataxin-3 with the normal length repeat [92]. In these experiments, the truncated construct with the expanded repeat could induce cell death. Western blot data indicated that this construct resulted in the formation of aggregates, as represented by material at the top of the resolving gel and at the top of the stacking gel, presumably material which was insoluble and could not enter the gel. In these experiments, when the truncated and full-length ataxin-3 constructs were cotransfected, the truncated construct could recruit full-length protein into the aggregates, as demonstrated by both immunofluorescence and Western blot analysis.

The most detailed biochemical information regarding the nuclear inclusions comes from studies of ataxin-1 [116, 138]. Ataxin-1 has a nuclear localization signal, and its distribution in cells can include the nucleus, as well as the cytoplasm. In neurons, including Purkinje cells, its normal localization is predominantly nuclear. COS cells were transfected with full-length ataxin with normal or expanded repeats. The ataxin-1 with normal repeats had a diffuse localization within the nucleus with some small punctate dots, and very occasional large punctate dots. By contrast, the expanded glutamine repeat ataxin-1 was predominantly localized within large punctate dots within the nucleus. Biochemical analysis indicated that these were associated with the nuclear matrix. Cells transfected with the expanded ataxin-1 construct showed altered subnuclear localization of the PML protein.

The yeast two-hybrid experiments described above (Section VII) showed that ataxin-1 can associate with the protein LANP in a manner dramatically dependent on the length of the glutamine repeat. Furthermore, cell transfection experiments indicate that ataxin-1 strikingly colocalizes with LANP in transfected COS cells, further suggesting that LANP may be involved in the pathogenesis of the disorder. LANP defines zones called nuclear bodies. By contrast, mutant ataxin-1 did not colocalize with markers of other nuclear subdomains including the SC35 protein, coilin, or BCL6. Coexpression of LANP with mutated ataxin-1 resulted in localization to salt-resistant nuclear structures associated with the nuclear matrix [116].

XI. TRANSGENIC ANIMAL MODELS

Transgenic animal models have provided some of the most striking evidence for the gain-of-function mechanism and the nuclear inclusions.

A pioneering animal model of HD was constructed using exon one of huntingtin with a very long expanded repeat [139]. These animals developed behavioral syndromes strikingly similar to HD, including incoordination, abnormal involuntary movements, seizures, and weight loss. However, unlike HD patients, they have not shown neuronal cell loss, though there is gliosis in the dorsal striatum, where it is often seen in the earliest stages of HD (Davies, personal communication). These data are consistent with the idea that neuronal dysfunction, rather than neuronal death, may be responsible for many of the symptoms of the polyglutamine diseases.

Study of these mice with antibodies to the N-terminus of huntingtin revealed that they develop intranuclear inclusions containing the truncated huntingtin transgene product, but not the endogenous huntingtin protein [135]. A careful study of the time course of the changes in these animals indicated that the intranuclear inclusions are present just before the animals have behavioral symptoms or brain or body weight loss. The intranuclear inclusions are clearly distinct from the nucleolus. They are not separated by a membrane from the rest of the nucleus, as shown using careful EM techniques. Thus, these observations do not prove causality, but they are certainly consistent with the possibility that the intranuclear inclusions underlie the neurologic dysfunction in these mice.

A mouse model of SCA1 was created using full-length ataxin-1 with expanded polyglutamine repeats, under the control of a Purkinje cell-specific promotor [140] These animals developed ataxia, Purkinje cell abnormalities, and Purkinje cell death in a manner very reminiscent of SCA1. Analysis of cell number indicated that there was no significant loss of Purkinje cells at the time when the animals had prominent behavioral symptoms [65]. Thus dysfunction rather than death of neurons must be important for the development of symptoms.

More recent studies of these ataxin-1 transgenic mice have shown that they develop nuclear inclusions prior to the development of symptoms [138]. The inclusions could be detected with antibodies to ataxin-1 or antibodies to ubiquitin. Ataxin-1 is normally present in small punctate structures in the nucleus; in the animals with the expanded repeat, ataxin-1 is present in large discrete structures resembling inclusions. The percentage of cells with the large inclusions increased from 25 at 6 weeks to 90 at 12 weeks—a time course consistent with a pathogenic role for these inclusions.

A similar transgenic model was also created using this promotor and the ataxin-3 protein. These animals developed Purkinje cell death, but only when the transgene contained the truncated ataxin-3 protein, not the full-length ataxin-3 protein. The cellular localization of the ataxin-3 protein was not reported [137]. Since Purkinje cells are not a prime target of the pathology in MJD, it is conceivable that they do not possess the machinery for truncation of the ataxin-3 protein, while other cells which are normally sensitive may possess this machinery. However, this is speculation at the moment.

XII. PATIENT MATERIALS

Ultimately all data in cell models and animal models must be compared to study of patients. The recent discovery of intranuclear inclusions in patient material validates the observations in cell culture and transgenic animal models.

The most data are available for HD. Study of lymphoblasts derived from patients had shown that the huntingtin protein with the expanded glutamine repeat was stable, with a half-life similar to that of the normal protein, with no obvious difference in its subcellular localization [141]. Studies of postmortem brains with antibodies not directed against the N-terminus had indicated that there were subtle differences in the distribution of the huntingtin protein in patients with HD, compared to brains from control individuals, with increased labeling in a lysosomal-like component [142]. Studies of brains from patients showed that the protein with the expanded glutamine repeat can be detected on Western blots as a band migrating more slowly than the normal allele [78, 79, 143, 144].

In contrast, studies of huntingtin using antibodies directed at the N-terminus have revealed the intranuclear inclusions [145, 146]. The intranuclear inclusions are present in neurons but not in glia. They are most abundant in the cortex and the caudate, the areas most affected in HD, though they are also seen at lower frequency in other regions such as the red nucelus, amygdala, hippocampus, and dentate nucleus of the cerebellum. Within the caudate, they are seen in medium-sized neurons, most of which are likely to be medium spiny neurons (the neurons which are affected), but not large interneurons (the neurons which are relatively spared). Within the cortex, they are enriched in layers III, V, and VI, the areas which are affected, and less frequent in layers II and IV, the areas which are relatively unaffected. Furthermore, the density of inclusions is significantly correlated with the length of the CAG repeat in IT15 [145, 146]. They were not present in the brain from one presymptomatic individual with the HD mutation [145].

The inclusions can be labeled with antibodies to the N-terminus of huntingtin, but not by at least two antibodies directed at internal epitopes. The inclusions can also be labeled with antibodies to ubiquitin, suggesting that the huntingtin protein is misfolded or otherwise abnormal and has been targeted for degradation but cannot be removed by proteolysis.

Western blot analysis of HD brain revealed the presence of an N-terminal fragment of huntingtin of around 40 kDa [145]. This was seen only in HD brain and not in control brain and was enriched in a nuclear fraction.

Ultrastructural analysis of the inclusions indicated that they were composed of a mixture of granules, straight and tortuous filaments, and masses of parallel and randomly oriented fibrils [145], but not enclosed by an intracellular membrane. This is very similar to the appearance of the inclusions seen in the transgenic HD model [135]. Similar structures had been reported in previous biopsy studies of HD cortex before the availability of antibodies to huntingtin [147].

In addition, huntingtin was present in aggregates in dystrophic neurites in HD brains [148]. These were present predominantly in cortical layers V and VI and appeared to be contained within neurofilament-labeled axonal processes. Previous studies have indicated that dystrophic neurites form as a result of dysfunction of retrograde axonal transport.

Strikingly similar intranuclear inclusions have recently been reported in SCA3 [148]. The inclusions were seen best in the ventral pons, a major target of pathology in SCA3. They were also seen less frequently in other regions known to be affected in SCA3, including the substantia nigra, globus pallidus, dorsal medulla, inferior olivary complex, and dentate nucleus. They were not found in regions typically spared in the disease such as the cerebral cortex, hippocampus, striatum, and Purkinje cell layer of the cerebellum. In most neurons, there was a single inclusion but in some neurons there were two or three in the same nucleus. The inclusions could be labeled either with a monoclonal antibody specific for ataxin-3 (1H9) or with the monoclonal antibody 1C2, which specifically recongizes expanded glutamine repeats. In addition, as in HD, they could be labeled with antibodies to ubiquitin. However, they were unlabeled with a variety of antibodies to cytoskeletal proteins such as actin, tublin, microtubule-associated protein, and neurofilaments.

Inclusions have also been seen in SCA1 [138]. In this case, inclusions could be labeled with antibodies to either the N- or the C-terminus of ataxin-1, indicating that full-length protein was present, though not excluding the possibility of truncation of ataxin-1 as well. As in SCA3, the inclusions were best seen in the pons; less material was available for detailed analysis of the anatomic distribution in this disorder.

Analysis of DRPLA material indicated the presence of very similar nuclear inclusions in neurons of the cerebellar dentate nucleus, as well as in the caudate and cerebral cortex [146].

XIII. A UNIFYING MODEL

The recent data suggest a unifying model for the pathogenesis of glutamine repeat diseases. Figure 47-1 indicates the application of the model to HD, taking into account data from the other diseases [149]. Several of the steps are speculative and different forms of evidence are available from biochemistry, cell models, animal models, and patient material. The model helps integrate the *in vitro* and the cellular data; however, many questions remain.

One key question is what determines the cell type specificity of pathology in the different diseases. For SCA1, part of the answer may involve the interaction between ataxin-1 and the LANP protein. This interaction is dramatically increased when ataxin-1 has an expanded glutamine repeat. LANP is highly expressed in Purkinje cells, the cells most severely affected in the disease, and less highly expressed in other cells affected in the disease; there is little expression in unaffected areas. Whether other (possibly homologous) proteins help explain the cell type specificity for the other diseases is still unclear. The pattern of expression of HAP1 does not appear to explain the cell death in HD. Furthermore, there does not appear as dramatic a differential binding to the mutated protein as for ataxin-1. Thus, it is quite possible that other associated proteins are yet to be identified for HD and the other glutamine repeat diseases.

The intracellular location of the inclusions is also striking. In patient material and the transgenic animals, the inclusions are present in neuronal nuclei. For ataxin-1, this location is not surprising since the protein is normally present in the nucleus. However, this is not the case for HD or SCA3. The mechanism by which the protein enters the nucleus is uncertain. It is also uncertain whether this mechanism has any specificity for the cells affected in the disorder. In addition, the role of dystrophic neurites, or other abnormal accumulations of polyglutamine, is still unknown.

The role of proteolytic cleavage is unknown. Proteolytic cleavage may be critical for HD, since the inclusions

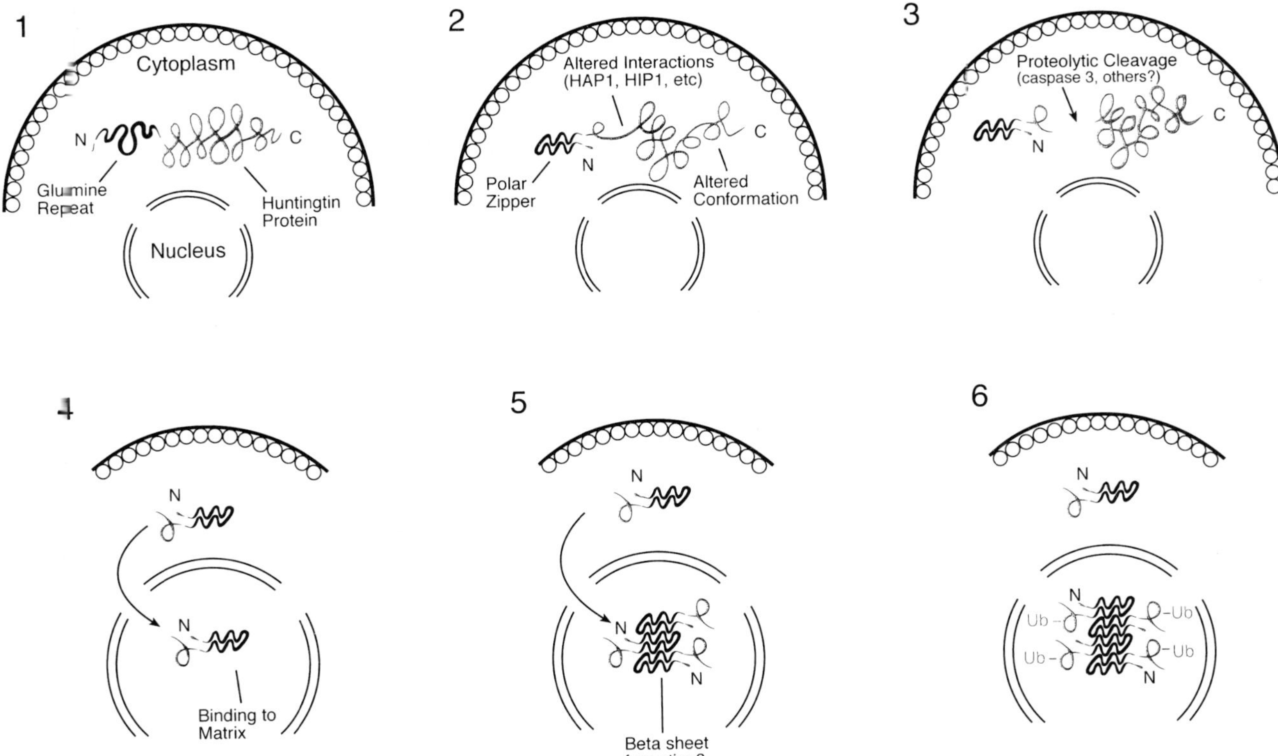

FIGURE 47-1 Tentative model for pathogenesis of polyglutamine diseases, focusing on HD. The first step in the pathogenesis is when the glutamine repeat undergoes a conformational change, presumably due to hairpin formation via a polar zipper. This cannot happen for the normal protein (1), but only for the protein with an expanded glutamine repeat (2). Conformational changes within the protein alter interactions with other proteins such as HAP1 and HIP1, possibly resulting in pathological events within the cytoplasm. The protein with the expanded glutamine repeat undergoes proteolytic cleavage, via caspase-3 or other enzymes (3). The N-terminal fragment is translocated to the nucleus (4) where it may bind to the nuclear matrix (for ataxin-1, binding to LANP takes place in the nuclear matrix) forming a nidus for additional molecules to associate via a beta pleated sheet (5). As additional huntingtin fragments aggregate, the proteins are modified by ubiquitinization (6), but cannot be removed, forming the intranuclear neuronal inclusions. Reprinted from Ross, C. A., *Neuron* **19,** 1147–1150, 1997, with permission. Copyright Springer-Verlag.

contain only the N-terminal antigen and not other internal antigens. (It is still possible that the full-length protein is present, though the other antigens are masked, but this appears less likely.) Since huntingtin is so large, it may be that truncation facilitates the nuclear translocation. Furthermore, it is uncertain to what extent caspase-3, which can cleavage huntingtin *in vitro,* is the enzyme which cleaves it *in vivo*. Perhaps some combination of caspses, ubiquitin-dependent proteases, or other proteases, are involved.

It is unclear whether events in the cytoplasm also play a role in pathogenesis. The altered interaction between huntingtin and other proteins such as HAP1 and HIP1 might have effects which contribute to the cellular phenotype of HD. For instance, abnormal interactions among huntingtin, HAP1, and dynactin might lead to dysfunction in retrograde transport, and the formation of dystrophic neurites.

Also puzzling is the question of the role of the nuclear inclusions in causing neuronal dysfunction and death. In the animal models of both HD and SCA1, the neuronal inclusions clearly precede the development of symptoms. However, this does not prove that they are causal. It is not yet clear how the intranuclear inclusions might alter cell functioning. In HD patients, the neuronal inclusions are present in cells in the affected regions and not in unaffected cells. Furthermore, they are more dense in brains from patients with long glutamine repeats than in brains from patients with less dramatically expanded glutamine repeats. Thus, they certainly appear to be striking markers of pathology in HD. However, again, this does not prove a causal role. One way to test both the role of proteolysis and possibly the role of formation of the intranuclear inclusions might be to mutate sites of proteolytic cleavage in cDNAs prior to injection for transgenic animal models or cell models. Other cDNA mutagenesis studies might examine the role of nuclear localization signals. The role of mitochondrial dysfunction is also unclear: is this caused by the nuclear inclusions (perhaps via alteration in transcription of nuclear encoded mitochondrial genes), or via some other mechanism?

Thus, the striking advances this past year have raised a whole new set of questions. Further elaborations of the cell and animal models will be important to answer these questions.

References

1. Miyazaki, M., Hashimoto, T., Yoneda, Y., Tayama, M., Harada, M., Miyoshi, H., Kawano, N., Murayama, N., Kondo, I., and Kuroda, Y. (1996). Proton magnetic resonance spectroscopy in childhood-onset dentatorubral-pallidoluysian atrophy (DRPLA). *Brain Dev.* **18,** 142–146.
2. LaSpada, A. R., and Clark, A. W. (1997). Inherited neurodegenerative disorders caused by CAG/polyglutamine tract expansions: symposium introduction. *Brain Pathol.* **7,** 877–880.
3. Gusella, J. F., and MacDonald, M. E. (1996). Trinucleotide instability: a repeating theme in human inherited disorders. *Annu. Rev. Med.* **47,** 201–209.
4. Housman, D. (1995). Gain of glutamines, gain of function? *Nature Genet.* **10,** 3–4.
5. Bates, G. (1996). Expanded glutamines and neurodegeneration—a gain of insight. *BioEssays* **18,** 175–178.
6. Warren, S. T., and Nelson, D. L. (1993). Trinucleotide repeat expansions in neurologic disease. *Curr. Opin. Neurobiol.* **3,** 752–759.
7. Paulson, H. L., and Fishbeck, K. H. (1996). Trinucleotide repeats in neurogenetic disorders. *Annu. Rev. Med.* **19,** 79–107.
8. Nance, M. A. (1997). Clinical aspects of CAG repeat diseases. *Brain Pathol.* **7,** 881–900.
9. Timchenko, L. K., and Caskey, C. T. (1996). Trinucleotide repeat disorders in humans: discussions of mechanisms and medical issues. *FASEB J.* **10,** 1589–1597.
10. Ross, C. A., Becher, M. W., Colomer, V., Engelender, S., Wood, J. D., and Sharp, A. H. (1997). Huntington's disease and dentatorubral-pallidoluysian atrophy: proteins, pathogenesis, and pathology. *Brain Pathol.* **7,** 1003–1016.
11. Sharp, A. H., and Ross, C. A. (1996). Neurobiology of Huntington's disease. *Neurobiol. Dis.* **3,** 3–15.
12. Tsuji, S. (1997). Molecular genetics of triplet repeats: unstable expansion of triplet repeats as a new mechanism for neurodegenerative diseases. *Int. Med.* **36,** 3–8.
13. Cha, J-H., and Dure IV, L. S. (1994). Trinucleotide repeats in neurologic diseases: an hypothesis concerning the pathogenesis of Huntington's disease, Kennedy's disease, and spinocerebellar ataxia type 1. *Life Sci.* **54,** 1459–1464.
14. Ross, C. A., Margolis, R. L., Rosenblatt, A., Ranen, N. G., Becher, M. W., and Aylward, E. (1997). Reviews in molecular medicine: Huntington disease and the related disorder, dentatorubral-pallidoluysian atrophy (DRPLA). *Medicine* **76,** 305–338.
15. Wexler, N. S., Young, A. B., Tanzi, R. E., Travers, H., Starosta-Rubinstein, S., Penney, J. B., Snodgrass, S. R., Shoulson, I., Gomez, F., Ramos Arroyo, M. A., Penchaszadeh, G. K., Moreno, H., Gibbons, K., Faryniarz, A., Hobbs, W., Anderson, M. A., Bonilla, E., Conneally, P. M., and Gusella, J. F. (1987). Homozygotes for Huntington's disease. *Nature* **326,** 194–197.
16. Myers, R. H., Leavitt, J., Farrer, L. A., Jagadeesh, J., McFarlane, H., Mastomauro, C. A., Mark, R. J., and Gusella, J. F. (1989). Homozygote for Huntington disease. *Am. J. Hum. Genet.* **45,** 615–618.
17. Gusella, J. F., Persichetti, F., and MacDonald, M. E. (1997). The genetic defect causing Huntington's disease: repeated in other contexts? *Mol. Med.* **3,** 238–246.
18. LaSpada, A. R., Paulson, H. L., and Fishbeck, K. H. (1994). Trinucleotide repeat expansion in neurologic disease. *Ann. Neurol.* **36,** 814–822.
19. Feng, Y., Zhang, F., Lokey, L. K., Chastain, J. L., Lakkis, L., Eberhart, D., and Warren, S. T. (1995). Translational suppression by trinucleotide repeat expansion at FMR1. *Science* **268,** 731–734.
20. Hirst, M., Grewal, P., Flannery, A., Slatter, R., Maher, E., Barton, D., Fryns, J-P., and Davies, K. (1995). Two new cases of FMR1 deletion associated with mental impairment. *Am. J. Hum. Genet.* **56,** 67–74.
21. Montermini, L., Richter, A., Morgan, K., Justice, C. M., Julien, D., Castellotti, B., Mercier, J., Poirier, J., Capozzoli, F., Bouchard, J-P., Lemieux, B., Mathieu, J., Vanasse, M., Seni, M-H., Graham, G., Andermann, F., Andermann, E., Melacon, S. B., Keats, B. J. B., Di Donato, S., and Pandolfo, M. (1997). Phenotypic variability in Friedreich ataxia: role of the associated GAA triplet repeat expansion. *Ann. Neurol.* **41,** 675–682.
22. Campuzano, V., Montermini, L., Molto, M. D., Pianese, L., Cossee, M., Cavalcanti, F., Monros, E., Rodius, F., Duclos, F., Monticelli, A., Zara, F., Canizares, J., Koutnikova, H., Bidichandani, S., Gellera, C., Brice, A., Trouillas, P., DeMichele, G., Filla, A., DeFrutos, R., Palau, F., Patel, P. I., Pragna, I., DiDonato, S., Mandel, J-L., Cocozza, S., Koenig, M., and Pandolfo, M. (1996). Friedreich's ataxia: autosomal recessive disease caused by an intronic GAA triplet repeat expansion. *Science* **271,** 1423–1427.
23. Ambrose, C. M., Duyao, M. P., Barnes, G., Bates, G. P., Lin, C. S., Srinidhi, J., Baxendale, A., Church, D., Houseman, D., Berks, M., Micklem, G., Durbin, R., Dodge, A., Read, A., Gusella, J., and MacDonald, M. E. (1994). Structure and expression of the Huntington's disease gene: evidence against simple inactivation due to an expanded CAG repeat. *Somat. Cell Mol. Genet.* **20,** 27–28.
24. Alam, M. R., Johnson, R. C., Darlington, D. N., Hand, T. A., Mains, R. E., and Eipper, B. A. (1997). Kalirin, a cytosolic protein with spectrin-like and GDP/GTP exchange factor-like domains that interacts with peptidylglycine alpha-amidating monooxygenase, an integral membrane peptide-processing enzyme. *J. Biol. Chem.* **272,** 12667–12675.
25. Quigley, C. A., Friedman, K. J., Johnson, A., LaFreniere, R. G., Silverman, L. M., Lubahn, D. B., Brown, T. R., Wilson, E. M., Willard, H. F., and French, F. S. (1992). Complete deletion of the androgen receptor gene: definition of the null phenotype of the androgen insensitivity syndrome and determination of carrier status. *J. Clin. Endocrinol. Metab.* **74,** 927–933.
26. Ross, C. A. (1995). When more is less: pathogenesis of glutamine repeat neurodegenerative diseases. *Neuron* **15,** 493–496.
27. Robitaille, Y., Lopes-Cendes, I., Becher, M., Rouleau, G., and Clark, A. W. (1997). The neuropathology of CAG repeat diseases: review and update of genetic and molecular features. *Brain Pathol.* **7,** 901–926.
28. Kowall, N. W., Ferrante, R. J., and Martin, J. B. (1987). Patterns of cell loss in Huntington's disease. *Trends Neurosci.* **10,** 24–29.
29. Ferrante, R. J., Kowall, N. W., Beal, M. F., Richardson, Jr., E. P., Bird, E. D., and Martin, J. B. (1985). Selective sparing of a class of striatal neurons in Huntington's disease. *Science* **230,** 561–563.
30. Ferrante, R. J., Beal, M. F., Kowall, N. W., Richardson, E. P., and Martin, J. B. (1987). Sparing of acetylcholinesterase-containing striatal neurons in Huntington's disease. *Brain Res.* **411,** 162–166.

31. Albin, R. L., Reiner, A., Anderson, K. D., Dure IV, L. S., Handelin, B., Balfour, R., Whetsell, W. O., Penney, J. B., and Young, A. B. (1992). Preferential loss of striato external pallidal projection neurons in presymptomatic Huntington's disease. *Ann. Neurol.* **31,** 425–430.
32. Sotrel, A., Paskevich, P. A., Kiely, D. K., Bird, E. D., Williams, R. S., and Myers, R. H. (1991). Morphometric analysis of the prefrontal cortex in Huntington's disease. *Neurology* **41,** 1117–1123.
33. Hedreen, J. C., Peyser, C. E., Folstein, S. E., and Ross, C. A. (1991). Neuronal loss in layers V and VI of cerebral cortex in Huntington's disease. *Neurosci. Lett.* **133,** 257–261.
34. Persichetti, F., Srinidhi, J., Kanaley, L., Ge, P., Myers, R. H., D'Arrigo, K., Barnes, G. T., MacDonald, M. E., Vonsattel, J-P., Gusella, J. F., and Bird, E. D. (1996). Huntington disease CAG trinucleotide repeats in pathologically confirmed post-mortem brains. *Neurobiol. Dis.* **1,** 159–166.
35. Furtado, S., Suchowersky, O., Rewcastle, B., Graham, L., Klimek, M. L., and Garber, A. (1996). Relationship between trinucleotide repeats and neuropathological changes in Huntington disease. *Ann. Neurol.* **39,** 132–136.
36. Vonsattel, J. P., Myers, R. H., Stevens, T. J., Ferrante, R. J., Bird, E. D., and Richardson, E. P. (1985). Neuropathological classification of Huntington's disease. *J. Neuropathol. Exp. Neurol.* **44,** 559–577.
37. Graveland, G. A., Williams, R. S., and DiFiglia, M. (1985). A Golgi study of the human neostriatum: neurons and afferent fibers. *J. Comp. Neurol.* **234,** 317–333.
38. Sotrel, A., Williams, R. S., Kaufmann, W. E., and Myers, R. H. (1993). Evidence for neuronal degeneration and dendritic plasticity in cortical pyramidal neurons of Huntington's disease: a quantitative Golgi study. *Neurology* **43,** 2088–2986.
39. Beal, M. F., Kowall, N. W., Ellison, D. Z. W., Mazurek, M. F., Swartz, K. J., and Martin, J. B. (1986). Replication of the neurochemical characteristics of Huntington's disease by quinolinic acid. *Nature* **321,** 168–171.
40. Beal, M. F., Finn, S. F., and Brouillet, E. (1993). Evidence for the involvement of metabotropic glutamate receptors in striatal excitotoxin lesions in vivo. *Neurodegeneration* **2,** 81–91.
41. Albin, R. L., and Greenamyre, J. T. (1992). Alternative excitotoxic hypotheses. *Neurology* **42,** 733–738.
42. Beal, M. F. (1992). Does impairment of energy metabolism result in excitotoxic neuronal death in neurodegenerative illnesses? *Ann. Neurol.* **31,** 119–130.
43. Brouillet, E., Hantraye, P., Ferrante, R. J., Dolan, R., Leroy-Willig, A., Kowall, N. W., and Beal, M. F. (1993). Chronic mitochondrial energy impairment produces selective striatal degeneration and abnormal choreiform movement in primates. *Proc. Natl. Acad. Sci. USA* **60,** 356–359.
44. Matthews, R. T., Ferrante, R. J., Jenkins, B. G., Browne, S. E., Goetz, K., Berger, S., Chen, Y-C., and Beal, M. F. (1997). Iodacetate produces striatal excitotoxic lesions. *J. Neurochem.* **69,** 285–289.
46. Dawson, T. M., Dawson, V. L., and Snyder, S. H. (1992). A novel neuronal messenger molecule in the brain: the free radical, nitric oxide. *Ann. Neurol.* **32,** 297–311.
47. Lafon-Cazal, M., Pietri, S., Culcasi, M., and Bockaert, J. (1993). NMDA-dependent superoxide production and neurotoxicity. *Nature* **364,** 535–537.
48. Schulz, J. B., Henshaw, D. R., Siwek, D., Jenkins, B. G., Ferrante, R. J., Cipolloni, P. B., Kowall, N. W., Rosen, B. R., and Beal, M. F. (1995). Involvement of free radicals in excitotoxicity in vivo. *J. Neurochem.* **64,** 2239–2247.
49. Zhuchenko, O., Bailey, J., Bonnen, P., Ashizawa, T., Stockton, D. W., Amos, C., Dobyns, W. B., Subramony, S. H., and Zoghbi, H. Y. (1997). Autosomal dominant cerebellar ataxia (SCA6) associated with small polyglutamine expansions in the alpha1A-voltage-dependent calcium channel. *Nature Genet.* **15,** 62–69.
50. Green, H. (1993). Human genetic diseases due to codon reiteration: relationship to an evolutionary mechanism. *Cell* **74,** 955–956.
51. Llinas, R., Sugimori, M., Hillman, D. E., and Cherksey, B. (1992). Distribution and functional significance of the P-type, voltage-dependent Ca2+ channels in the mammalian central nervous system. *Trends Neurosci.* **15,** 351–355.
52. Jacobson, M. D., Weil, M., and Raff, M. C. (1997). Programmed cell death in animal development. *Neuropathol. Appl. Neurobiol.* **88,** 347–354.
53. Nagata, S. (1997). Apoptosis by death factor. *Cell* **88,** 355–365.
54. Tewari, M., Quan, L. T., O'Rourke, K., Desnoyers, S., Zeng, Z., Beidler, D. R., Poirer, G. G., Salvesen, G. S., and Dixit, V. M. (1995). Yama/CPP32 beta, a mammalian homolog of CED-3, is a CrmA-inhibitable protease that cleaves the death substrate poly (ADP-ribose) polymerase. *Cell* **81,** 801–809.
55. Ishitani, R., Sunaga, K., Hirano, A., Saunders, P., Katsube, N., and Chuang, D-M. (1996). Evidence that glyceraldehyde-3-phosphate dehydrogenase is involved in age-induced apoptosis in mature cerebellar neurons in culture. *J. Neurochem.* **66,** 928–935.
56. Jenkins, B., Koroshetz, W., Beal, M. F., and Rosen, B. (1993). Evidence for an energy metabolism defect in Huntington's disease using localized proton spectroscopy. *Neurology* **43,** 2689–2695.
57. Jenkins, B. G., Brouillet, E., Chen, Y-C., Storey, E., Schulz, J. B., Kirschner, P., Beal, M. F., and Rosen, B. R. (1996). Non-invasive neurochemical analysis of focal excitotoxic lesions in models of neurodegenerative illness using spectroscopic imaging. *J. Cereb. Blood Flow Metab.* **16,** 450–461.
58. Gu, M., Gash, M. T., Mann, V. M., Javoy-Agid, F., Cooper, J. M., and Schapira, A. H. V. (1996). Mitochondrial defect in Huntington's disease caudate nucleus. *Ann. Neurol.* **39,** 385–389.
59. Browne, S. E., Bowling, A. C., MacGarvey, U., Baik, M. J., Berger, S. C., Muqit, M. M. K., Bird, E. D., and Beal, M. F. (1997). Oxidative damage and metabolic dysfunction in Huntington's disease: selective vulnerability of the basal ganglia. *Ann. Neurol.* **41,** 646–653.
60. Nasir, J., Floresco, S. B., Okusky, J. R., Diewert, V. M., Richman, J. M., Zeisler, J., Borowski, A., Math, J. D., Phillips, A. G., and Hayden, M. R. (1995). Targeted disruption of the Huntington disease gene results in embryonic lethality and behavioral and morphological changes in heterozygotes. *Cell* **81,** 811–823.
61. Duyao, M. P., Auerbach, A. B., Ryan, A., Persichetti, F., Barnes, G. T., McNeil, S. M., Ge, P., Vonsattel, J. P., Gusella, J. F., Joyner, A. L., and MacDonald, M. E. (1995). Inactivation of the mouse Huntington's disease gene homolog Hdh. *Science* **269,** 407–410.
62. Zeitlin, S., Liu, J-P., Chapman, D. L., Papaioannou, V. E., and Efstratiadis, A. (1995). Increased apoptosis and early embryonic lethality in mice nullizygous for the Huntington's disease gene homologue. *Nature Genet.* **11,** 155–163.
63. White, J. K., Auerbach, W., Duyao, M. P., Vonsattel, J-P., Gusella, J. F., Joyner, A. L., and MacDonald, M. E. (1997). Huntingtin function is not impaired by the Huntington's disease CAG expansion mutation. *Nature Genet.* **17,** 404–410.
64. Kuida, K., Zheng, T. S., Na, S., Kuan, C-Y., Yang, D., Karasuyama, H., Rakic, P., and Flavell, R. A. (1996). Decreased apoptosis in the brain and premature lethality in CPP32-deficient mice. *Nature* **384,** 368–372.

65. Clark, H. B., Burright, E. N., Yunis, W. S., Larson, S., Wilcox, C., Hartman, B., Matilla, A., Zoghbi, H. Y., and Orr, H. T. (1997). Purkinje cell expression of a mutant allele of SCA1 in transgenic mice leads to disparate effects on motor behaviors, followed by a progressive cerebellar dysfunction and histological alterations. *J. Neurosci.* **17,** 7385–7395.
66. Li, S-H., Schilling, G., Young III, W. S., Margolis, R. L., Stine, O. C., Wagster, M. V., Abbott, M. H., Franz, M. L., Ranen, N. G., Folstein, S. E., Hedreen, J. C., and Ross, C. A. (1993). Huntington's disease is widely expressed in human and rat tissues. *Neuron* **11,** 985–993.
67. Strong, T. V., Tagle, D. A., Valdes, J. M., Elmer, L. W., Boehm, K., Swaroop, M., Kaatz, K. W., Collins, F. S., and Albin, R. L. (1993). Widespread expression of the human and rat Huntington disease gene in brain and nonneuronal tissues. *Nature Genet.* **5,** 259–265.
68. Landwehrmeyer, G. B., McNeil, S. M., Dure, L. S., Ge, P., Aizawa, H., Huang, Q., Ambrose, C. M., Duyao, M. P., Bird, E. D., Bonilla, E., deYoung, M., Avila-Gonzales, A. J., Wexler, N. S., DiFiglia, M., Gusella, J. F., MacDonald, M. D., Penney, J. B., Young, A. B., and Vonsattel, J-P. (1995). Huntington's disease gene: regional and cellular expression in brain of normal and affected individuals. *Ann. Neurol.* **37,** 218–230.
69. Margolis, R. L., Li, S-H., Young, W. S., Wagster, M. V., Stine, O. C., Kidwai, A. S., Ashworth, R. G., and Ross, C. A. (1996). DRPLA gene (Atrophin-1) sequence and mRNA expression in human brain. *Mol. Brain Res.* **36,** 219–226.
70. Loev, S. J., Margolis, R. L., Young, W. S., Li, S-H., Schilling, G., Ashworth, R. G., and Ross, C. A. (1995). Rat DRPLA disease gene homologue: presence of alternating proline and glutamine residues adjacent to the glutamine repeat. *Neurobiol. Dis.* **2,** 129–138.
71. Nagafuchi, S., Yanagisawa, H., Ohsaki, E., Shirayama, T., Tadokoro, K., Inoue, T., and Yamada, M. (1994). Structure and expression of the gene responsible for the triplet repeat disorder, dentatorubral and pallidoluysian atrophy (DRPLA). *Nature Genet.* **8,** 177–182.
72. Onodera, O., Roses, A. D., Tsuji, S., Vance, J. M., Strittmatter, W. J., and Burke, J. R. (1996). Toxicity of expanded polyglutamine-domain proteins in Escherichia coli. *FEBS Lett.* **135,** 139.
73. Banfi, S., Servadio, A., Chung, M-Y., Capozzoli, F., Duvick, L. A., Elde, R., Zoghbi, H. Y., and Orr, H. T. (1996). Cloning and developmental expression analysis of the murine homolog of the spinocerebellar ataxia type 1 gene (Sca1). *Hum. Mol. Genet.* **5,** 35–40.
74. Gossen, M., Schmitt, I., Obst, K., Wahle, P., Epplen, J. T., and Riess, O. (1996). cDNA cloning and expression of RSCA1, the rat counterpart of the human spinocerebellar ataxia type 1 gene. *Hum. Mol. Genet.* **5,** 381–389.
75. DiFiglia, M., Sapp, E., Chase, K., Schwarz, C., Meloni, A., Young, C., Martin, E., Vonsattel, J-P., Carraway, R., Reeves, S. A., Boyce, F. M., and Aronin, N. (1995). Huntingtin is a cytoplasmic protein associated with vesicles in human and rat brain neurons. *Neuron* **14,** 1075–1081.
77. Jou, Y-S., and Myers, R. M. (1995). Evidence from antibody studies that the CAG repeat in the Huntington's disease gene is expressed in the protein. *Hum. Mol. Genet.* **4,** 465–469.
78. Sharp, A. H., Loev, S. J., Schilling, G., Li, S-H., Li, X-J., Bao, J., Wagster, M. V., Kotzuk, J. A., Steiner, J. P., Lo, A., Hedreen, J., Sisodia, S., Snyder, S. H., Dawson, T. M. Ryugo, D. K., and Ross, C. A. (1995). Widespread expression of the Huntington's disease gene (IT15) protein product. *Neuron* **14,** 1065–1074.
79. Trottier, Y., Lutz, Y., Stevanin, G., Imbert, G., Devys, D., Cancel, G., Saudou, F., Weber, C., David, G., Tora, L., Agid, Y., Brice, A., and Mandel, J-L. (1995). Polyglutamine expansion as a pathological epitope in Huntington's disease and four dominant cerebellar ataxias. *Nature* **378,** 403–406.
80. Wood, J. D., MacMillan, J. C., Harper, P. S., Lowenstein, P. R., and Jones, A. L. (1996). Partial characteristics of murine huntingtin and apparent variations in the subcellular localisation of huntingtin in human, mouse and rat brain. *Hum. Mol. Genet.* **5,** 481–487.
81. Tukamoto, T., Nukina, N., Ide, K., and Kanazawa, I. (1997). Huntington's disease gene product, huntingtin, associates with microtubules in vitro. *Mol. Brain Res.* [In press]
82. Hoogeveen, A. T., Willemsen, R., Meyer, N., DeRooij, K. E., Roos, R. A. C., van Ommen, G.-J. B., and Galjaard, H. (1993). Characterization and localization of the Huntington disease gene product. *Hum. Mol. Genet.* **2,** 2069–2073.
83. DeRooij, K. E., Dorsman, J. C., Smoor, M. A., Den Dunnen, J. T., and Van Ommen, G.-J. B. (1996). Subcellular localization of the Huntington's disease gene product in cell lines by the immunofluorescence and biochemical subcellular fractionation. *Hum. Mol. Genet.* **5,** 1093–1099.
84. Gutekunst, C. A., Levey, A. I., Heilman, C. J., Whaley, W. L., Yi, H., Nash, N. R., Rees, H. D., Madden, J. J., and Hersch, S. M. (1995). Identification and localization of huntingtin in brain and human lymphoblastoid cell lines with anti-fusion protein antibodies. *Proc. Natl. Acad. Sci. USA* **92,** 8710–8714.
85. Bhide, P. G., Day, M., Sapp, E., Schwarz, C., Sheth, A., Kim, J., Young, A. B., Penney, J., Golden, J., Aronin, N., and DiFiglia, M. (1996). Expression of normal and mutant huntingtin in the developing brain. *J. Neurosci.* **16,** 5523–5535.
86. Ferrante, R. J., Gutekunst, C-A., Persichetti, F., McNeil, S. M., Kowall, N. W., Gusella, J. F., MacDonald, M. E., Beal, M. F., and Hersh, S. M. (1997). Heterogeneous topographic and cellular distribution of huntingtin expression in the normal human neostriatum. *J. Neurosci.* **17,** 3052–3063.
87. Velier, J., Schwarz, C., Young, C., Fallon, J., Hyman, B., Martin, E. J., Hughes, S., Vallee, R., Aronin, N., and DiFiglia, M. (1996). Wild-type and mutant huntingtin localize to the golgi complex and to vesicles in the peripheral cytoplasm in fibroblasts of control and HD patients. *Soc. Neurosci. Abstr.* **22,** 226. [Abstract]
88. Knight, S. P., Richardson, M. M., Osmand, A. P., Stakkestad, A., and Potter, N. T. (1997). Expression and distribution of the dentatorubral-pallidoluysian atrophy gene product (atrophin-1/DRPLA) in neuronal and non-neuronal tissues. *J. Neurol. Sci.* **146,** 12–26.
89. Nishiyama, K., Nakamura, K., Murayama, S., Yamada, M., and Kanazawa, I. (1997). Regional and cellular expression of the dentatorubral-pallidoluysian atrophy gene in brains of normal and affected individuals. *Ann. Neurol.* **41,** 599–605.
90. Yazawa, I., Nukina, N., Hashida, H., Goto, J., Yamada, M., and Kanazawa, I. (1995). Abnormal gene product identified in hereditary dentatorubral-pallidoluysian atrophy (DRPLA) brain. *Nature Genet.* **10,** 99–103.
91. Servadio, A., Koshy, B., Armstrong, D., Antalffy, B., Orr, H. T., and Zoghbi, H. Y. (1995). Expression analysis of the ataxin-1 protein in tissues from normal and spinocerebellar ataxia type 1 individuals. *Nature Genet.* **10,** 94–98.
92. Paulson, H. L., Das, S. S., Crino, P. B., Perez, M. K., Patel, S. C., Gotsdiner, D., Fischbeck, K. H., and Pittman, R. N. (1997). Machado-Joseph disease gene product is a cytoplasmic protein widely expressed in brain. *Ann. Neurol.* **41,** 453–462.
93. Sar, M., and Stumpf, W. E. (1977). Androgen concentration in motor neurons of cranial nerve and spinal cord. *Science* **197,** 77–79.

94. Brooks, B. P., Paulson, H. L., Merry, D. E., Salazar-Grueso, E. F., Brinkmann, A. O., Wilson, E. M., and Fischbeck, K. H. (1997). Characterization of an expanded glutamine repeat androgen receptor in a neuronal cell culture system. *Neurobiol. Dis.* **4,** 313–323.

95. Chamberlain, N. L., Driver, E. D., and Miesfeld, R. L. (1994). The length and location of CAG trinucleotide repeats in the androgen receptor N-terminal domain affect transactivation function. *Nucleic Acids Res.* **22,** 3181–3186.

96. Kazemi-Esfarjani, P., Trifiro, M. A., and Pinsky, L. (1995). Evidence for a repressive function of the long polyglutamine tract in the human androgen receptor: possible pathogenetic relevance for the (CAG)n-expanded neuropathies. *Hum. Mol. Genet.* **4,** 523–527.

97. Li, X-J., Li, S-H., Sharp, A. H., Nucifora, F. C., Jr., Schilling, G., Lanahan, A., Worely, P., Snyder, S. H., and Ross, C. A. (1995). A huntingtin associated protein enriched in brain with implications for pathology. *Nature* **378,** 398–402.

98. Li, X-J., Sharp, A. H., Li, S-H., Dawson, T. M., Snyder, S. H., and Ross, C. A. (1996). Huntingtin associated protein (HAP1): discrete neuronal localizations in brain resemble nitric oxide synthase. *Proc. Natl. Acad. Sci. USA* **93,** 4839–4844.

99. Engelender, S., Sharp, A. H., Colomer, V., Tokito, M. K., Lanahan, A., Worley, P., Holzbaur, E. L. F., and Ross, C. A. (1997). Huntingtin associated protein (HAP1) interacts with dynactin p150Glued and other cytoskeletal related proteins. *Hum. Mol. Genet.* **6,** 2205–2212.

100. Sweeney, H. L., and Holzbaur, E. L. F. (1996). Mutational analysis of motor proteins. *Annu. Rev. Physiol.* **58,** 751–792.

101. Waterman-Storer, C. M., Kuznetsov, S., Karki, S., Tabb, J. S., Weiss, D. G., Langford, G. M., and Holzbaur, E. L. F. (1997). The interaction between cytoplasmic dynein and dynactin is required for fast axonal transport. *Proc. Natl. Acad. Sci. USA.* [In press]

102. Block-galarza, J., Chase, K. O., Sapp, E., Vaughn, K. T., Valee, R. B., DiFiglia, M., and Aronin, N. (1997). Fast transport and retrograde movement of huntingtin and HAP1 in axons. *Neuroreport* **8,** 2247–2251.

103. Debant, A., Serra-Pages, C., Seipel, K., O'Brien, S., Tang, M., Park, S-H., and Streuli, M. (1996). The multidomain protein Trio binds to LAR transmembrane tyrosine phosphatase, contains a protein kinase domain, and has separate rac-specific and rho-specific guanine nucleotide exchange factor domains. *Proc. Natl. Acad. Sci. USA* **93,** 5466–5471.

104. Colomer, V., Engelender, S., Sharp, A. H., Duan, K., Cooper, J. K. Lanahan, A., Lyford, G., Worley, P., and Ross, C. A. (1997). Huntingtin-associated protein 1 (HIP1) binds to a trio-like polypeptide, with a rac1 guanine nucleotide exchange factor domain. *Hum. Mol. Genet.* **6,** 1519–1525.

105. Kalchman, M. A., Koide, H. B., McCutcheon, K., Graham, R. K., Nichol, K., Nishiyama, K., Kazemi-Esfarjani, P., Lynn, F. C., Wellington, C., Metzler, M., Goldberg, Y. P., Kanazawa, I., Gietz, R. D., and Hayden, M. R. (1997). HIP1, a human homolog of S. cerevisiae Sla2P, interacts with membrane-associated huntingtin in the brain. *Nature Genet.* **16,** 44–53.

106. Wanker, E. E., Rovira, C., Scherzinger, E., Hasenbank, R., Walter, S., Tait, D., Colicelli, J., and Lehrach, H. (1997). HIP-1: a huntingtin interacting protein isolated by the yeast two-hybrid system. *Hum. Mol. Genet.* **8,** 487–495.

107. Holtzman, D. A., Yang, S., and Drubin, D. G. (1993). Synthetic-lethal interactions identify two novel genes, SLA1 and SLA2, that control membrane cytoskeleton assembly in Saccharomyces cerevisiae. *J. Cell. Biol.* **122,** 635–644.

108. Burke, J. R., Enghild, J. J., Martin, M. E., Joy, Y. S., Myers, R. M., Roses, A. D., Vance, J. M., and Strittmatter, W. J. (1996). Huntingtin and DRPLA proteins selectively interact with the enzyme GAPDH, *Nature Med.* **347,** 350.

109. Koshy, B., Matilla, T., Burright, E. N., Merry, D. E., Fischbeck, K. H., Orr, H. T., and Zoghbi, H. Y. (1996). Spinocerebellar ataxia type-1 and spinobulbar muscular atrophy gene products interact with glyceraldehyde-3-phosphate dehydrogenase. *Hum. Mol. Genet.* **5,** 1311–1318.

110. Kalchman, M. A., Graham, R. K., Xia, G., Koide, H. B., Hodgson, J. G., Graham, K. C., Goldberg, Y. P., Gietz, R. D., Pickart, C. M., and Hayden, M. R. (1996). Huntingtin is ubiquitinated and interacts with a specific ubiquitin conjugating enzyme. *J. Biol. Chem.* **271,** 19385–19394.

111. Wood, J. D., Yuan, J., Margolis, R. L., Colomer, V., Duan, K., Kushi, J., Kaminsky, Z., Kleiderlein, Jr., J. J. Sharp, A. H., and Ross, C. A. (1997). Atrophin-1, the DRPLA gene product, interacts with two families of WW domain-containing proteins. *Am. J. Hum. Genet.* **61,** A324.

112. Dong, H., O'Brien, R. J., Fung, E. T., Lanahan, A. A., Worley, P. F., and Huganir, R. L. (1997). GRIP: a synaptic PDZ domain-containing protein that interacts with AMPA receptors. *Nature* **386,** 279–282.

113. Huibregtse, J. M., Scheffner, M., Beaudenon, S., and Howley, P. M. (1995). A family of proteins structurally and functionally related to the E6-AP ubiquitin-protein ligase. *Proc. Natl. Acad. Sci. USA* **92,** 2563–2567.

114. Kumar, S., Tomooka, Y., and Noda, M. (1992). Identification of a set of genes with developmentally down-regulated expression in the mouse brain. *Biochem Biophys. Res. Commun.* **185,** 1155–1161.

115. Hein, C., Springae, J-Y., Volland, C., Haguenauer-Tspas, R., and Andre, B. (1995). NPI1, an essential yeast gene involved in induced degradation of Gap1 and Fur4 permeases, encodes the Rsp5 ubiquitin-protein ligase. *Mol. Med.* **18,** 77–87.

116. Matilla, A., Koshy, B., Cummings, C. J., Isobe, T., Orr, H. T., and Zoghbi, H. Y. (1997). The cerebellar leucine rich acidic nuclear protein (LANP) interacts with ataxin-1. *Nature* **389,** 974–978.

117. Matsuoka, K., Taoka, M., Satozawa, N., Nakayama, H., Ichimura, T., Takahashi, N., Yamakuni, T., Song, S. Y., and Isobe, T. (1994). A nuclear factor containing the leucine-rich repeats expressed in murine cerebellar neurons. *Proc. Natl. Acad. Sci. USA* **91,** 9670–9674.

118. Kobe, B., and Deisenhofer, J. (1995). Proteins with leucine-rich repeats. *Curr. Opin. Struct. Biol.* **5,** 409–416.

119. Adler, A. J., Danielsen, M., and Robins, D. M. (1992). Androgen-specific gene activation via a consensus glutocorticoid response element is determined by interaction with nonreceptor factors. *Proc. Natl. Acad. Sci. USA* **89,** 11660–11663.

120. Wharton, K. A., Yedvobnick, B., Finnerty, V. G., and Artavanis-Tsakonas, S. (1985). Opa: a novel family of transcribed repeats shared by the notch locus and other developmentally regulated loci in D. melanogaster. *Cell* **40,** 55–62.

121. Duboule, D., Haenlin, M., Galliot, B., and Mohier, E. (1987). DNA sequences homologous to the Drosophila opa repeat are present in murine mRNAs that are differentially expressed in fetuses and adult tissues. *Mol. Cell. Biol.* **7,** 2003–2006.

122. Karlin, S., and Burge, C. (1996). Trinucleotide repeats and long homopeptides in genes and proteins associated with nervous system disease and development. *Proc. Natl. Acad. Sci. USA* **93,** 1560–1565.

123. Ross, C. A., McInnis, M. G., Margolis, R. L., and Li, S-H. (1993). Genes with triplet repeats: candidate mediators of neuropsychiatric disorders. *Trends Neurosci.* **16,** 254–260.

124. Green, H., and Wang, N. (1994). Codon reiteration and the evolution of proteins. *Proc. Natl. Acad. Sci. USA* **91,** 4298–4302.
125. Barnes, G. T., Duyao, M. P., Ambrose, C. M., McNeil, S., Persichetti, F., Srinidhi, J., Gusella, J. F., and MacDonald, M. E. (1994). Mouse Huntington's disease gene homolog (Hdh). *Somat. Cell Mol. Genet.* **20,** 87–97.
126. Baxendale, S., Abdulla, S., Elgar, G., Buck, D., Berks, M., Micklem, G., Durbin, R., Bates, G., Brenner, S., Beck, S., and Lehrach, H. (1995). Comparative sequence analysis of the human and pufferfish Huntington's disease genes. *Nature Genet.* **10,** 67–76.
127. Lin, B., Rommens, J. M., Graham, R. K., Kalchman, M., MacDonald, M., Nasir, J., Delaney, A., Goldberg, Y. P., and Hayden, M. R. (1993). Differential 3′ polyadenylation of the Huntington's disease gene results in two mRNA species with variable tissue expression. *Hum. Mol. Genet.* **2,** 1541–1545.
128. Oyake, M., Onodera, O., Toshihiko, S., Takano, H., Takahashi, Y., Kominami, R., Moriwaki, K., Ikeuchi, T., Igarashi, S., Tanaka, H., and Tsuji, S. (1997). Molecular cloning of murine homologue dentatorubral-pallidolysian atrophy (DRPLA) cDNA: strong conservation of a polymorphic CAG repeat in the murine gene. *Genomics* **40,** 205–207.
129. Gerber, H-P., Seipel, K., Georgiev, O., Hofferer, M., Hug, M., Rusconi, S., and Schaffern, W. (1994). Transcriptional activation modulated by homopolymeric glutamine and proline stretches. *Science* **263,** 808–811.
130. Perutz, M. F. (1996). Glutamine repeats and inherited neurodegenerative diseases: molecular aspects. *Curr. Opin. Struct. Biol.* **6,** 848–858.
131. Perutz, M. (1994). *In "Protein Science."* Cambridge University Press, London.
132. Stott, K., Blackburn, J. M., Butler, P. J. G., and Perutz, M. (1995). Incorporation of glutamine repeats makes protein oligomerize: implications for neurodegenerative diseases. *Proc. Natl. Acad. Sci. USA* **92,** 6509–6513.
133. Scherzinger, E., Lurz, R., Turmaine, M., Mangiarini, L., Hollenbach, B., Hasenbank, R., Bates, G. P., Davies, S. W., Lerach, H., and Wanker, E. E. (1997). Huntingtin-encoded polyglutamine expansions form amyloid-like protein aggregates in vitro and in vivo. *Cell* **90,** 549–558.
135. Davies, S. W., Turmaine, M., Cozens, B. A., DiFiglia, M., Sharp, A. H., Ross, C. A., Scherzinger, E., Wanker, E. E., Mangiarini, L., and Bates, G. P. (1997). Formation of neuronal intranuclear inclusions (NII) underlies the neurological dysfunction in mice transgenic for the HD mutation. *Cell* **90,** 537–548.
136. Goldberg, Y. P., Kalchman, M. A., Zeisler, J., Graham, R., Koide, H. B., Ross, C. A., Sharp, A. H., O'kusky, J., Jirik, F., and Hayden, M. R. (1996). Transgenic mice containing the human Huntington disease gene: absence of disease phenotype in intergenerational stability of the CAG repeat. *Hum. Mol. Genet.* **5,** 177–185.
137. Ikeda, H., Yamaguchi, M., Sugai, S., Aze, Y., Narumiya, S., and Kakizuka, A. (1996). Expanded polyglutamine in the Machado-Joseph disease. *Nature Genet.* **13,** 196–202.
138. Skinner, P. J., Koshy, B., Cummings, C. J., Klement, I. A., Helin, K., Servadio, A., Zoghbi, H. Y., and Orr, H. T. (1997). SCA1 pathogenesis involves alterations in nuclear matrix associated structures. *Nature* **389,** 971–974.
139. Mangiarini, L., Sathasivam, S., Seller, M., Cozens, B., Harper, A., Hetherington, C., Lawton, M., Trottier, Y., Lehrach, H., Davies, S. W., and Bates, G. P. (1996). Exon 1 of the HD gene with an expanded CAG repeat is sufficient to cause a progressive neurological phenotype in transgenic mice. *Cell* **87,** 493–506.
140. Burright, E. N., Clark, H. B., Servadio, A., Matilla, T., Feddersen, R. M., Yunis, W. S., Duvick, L. A., Zoghbi, H. Y., and Orr, H. T. (1995). SCA1 transgenic mice: a model for neurodegeneration caused by an expanded CAG trinucleotide repeat. *Cell* **82,** 937–948.
141. Persichetti, F., Ambrose, C. M., Ge, P., McNeil, S. M., Srinidhi, J., Anderson, M. A., Jenkins, B., Barnes, G. T., Duyao, M. P., Kanaley, L., Wexler, N. S., Myers, R. H., Bird, E. D., Vonsattel, J-P., MacDonald, M. E., and Gusella, J. F. (1995). Normal and expanded Huntington disease gene alleles produce distinguishable proteins due to translation across the CAG repeat. *Mol. Med.* **1,** 374–383.
142. Sapp, E., Schwarz, C., Chase, K., Bhide, P. G., Young, A. B., Penney, J., Vonsattel, J. P., Aronin, N., and DiFiglia, M. (1997). Huntingtin localization in brains of normal and Huntington's disease patients. *Ann. Neurol.* **42,** 604–612.
143. Schilling, G., Sharp, A. H., Loev, S. J., Wagster, M. V., Li, S-H., Stine, O. C., and Ross, C. A. (1995). Expression of the Huntington's disease (IT15) protein product in HD patients. *Hum. Mol. Genet.* **4,** 1365–1371.
144. Aronin, N., Chase, K., Young, C., Sapp, E., Schwartz, C., Matta, N., Kornreich, R., Landwehrmeyer, B., Bird, E., Beal, M. F., Vonsattel, J-P., Smith, T., Carraway, R., Boyce, F. M., Young, A. B., Penney, J. B., and DiFiglia, M. (1995). CAG Expansion affects the expression of mutant huntingtin in the Huntington's disease brain. *Neuron* **15,** 1193–1201.
145. DiFiglia, M., Sapp, E., Chase, K. O., Davies, S. W., Bates, G. P., Vonsattel, J. P., and Aronin, N. (1997). Aggregation of huntingtin in neuronal intranuclear inclusions and dystrophic neurites in brain. *Science* **277,** 1990–1993.
146. Becher, M. W., Kotzuk, J. A., Sharp, A. H., Davies, S. W., Bates, G. P., Price, D. L., and Ross, C. A. (1997). Intranuclear neuronal inclusions in Huntington's disease and dentatorubral and pallidoluysian atrophy: correlation between the density of inclusions and IT15 CAG triplet repeat length. *Neurobiol. Dis.* **4,** 387–395.
147. Roizin, L., Stellar, S., and Liu, J. C. (1979). *In* "Advances in Neurology" (T. N. Chase, N. S. Wexler, and A. Barbeau, Eds.), pp. 95–122. Raven Press, New York.
148. Paulson, H. L., Perez, M. K., Trottier, Y., Torjanowski, J. Q., Subramony, S. H., Das, S. S., Vig, P., Mandel, J-L., Fischbeck, K. H., and Pittman, R. N. (1997). Intranuclear inclusions of expanded polyglutamine protein in spinocerebellar ataxia type 3. *Neuron* **19,** 333–344.
149. Ross, C. A. (1997). Intranuclear neuronal inclusions: a common pathogenic mechanism for glutamine-repeat neurodegenerative disorders? *Neuron* **19,** 1147–1150.

Part XVII

Genome Instability Not Involving Triplet Repeats

CHAPTER 48

Expansion of a 12-mer Repeat in Progressive Myoclonus Epilepsy

STYLIANOS E. ANTONARAKIS Department of Genetics and Microbiology, University of Geneva Medical School, and Division of Medical Genetics, Cantonal Hospital of Geneva, Geneva, Switzerland

I. INTRODUCTION

The progressive myoclonus epilepsies are a group of rare hereditary disorders characterized by generalized epileptic seizures, myoclonus (brief contraction of a muscle or a group of muscles), and progressive neurological deterioration including ataxia and dementia [6]. Table 48-1 shows the genetic disorders that belong to this category [7, 8].

The Unverricht–Lundborg type of Progressive Myoclonus Epilepsy (EPM1; OMIM 354800) [38, 58], also known as Baltic myoclonus and Mediterranean myoclonus, is an autosomal recessive disorder characterized by severe stimulus-sensitive myoclonus, and generalized tonic–clonic seizures. The age of onset is between 6 and 18 years of age, peaking around age 11. The disorder has shown progressive course and characteristic electroencephalogram [21, 22]. The severity of the symptoms and the rate of progression vary between and within families [43]. Mild ataxia, intention tremor, and dysarthria eventually develop in the patients. The myoclonus is resistant to treatment, whereas the generalized seizures respond to antiepileptic therapy (valproate and clonazepam).

The syndrome was first described by Unverricht in 1891 in Estonia [58] and Lundborg in Sweden in 1903 [38]. The incidence of EPM1 in Finland is approximately 1 in 20,000 [43]. It is also found in the Western Mediterranean (Southern France, North Africa) [19, 25]. There is frequent consanguinity in the parents of affected individuals.

The EPM1 of Unverricht–Lundborg type is pathologically distinct from the "Lafora body" myoclonus epilepsy [10, 49] because of the absence of Lafora inclusion bodies (acid mucopolysaccharide inclusions of varying size in the neuronal cytoplasm and many other

TABLE 48-1 Progressive Myoclonus Epilepsies

Disease	Inheritance	Locus	OMIM	Gene	OMIM	Chromosome
1. Unverricht-Lundborg	AR	EPM1	254800	CSTB	601145	21q22.3
2. Lafora disease	AR	MELF	254780	?		6q23–q25 [51]
3. Myoclonus epilepsy with ragged-red fiber	Mitochondrial	MERRF	545500	MTTK	590060	Mitochondrial DNA
4. Neuronal ceroid lipofusionosis						
Infantile	AR	CLN1	256730	PPT	600722	1p32
Late infantile	AR	CLN2	204500	CLN2		12p15
Juvenile	AR	CLN3	204200	CLN3		16p12.1
Adult	AR and AD	CLN4	204300	?		?
Late infantile, variant	AR	CLN5	256731	?		13q21.1-932
5. Sialidoses						
Types I and II	AR		256550	NEU		6p21.3

cells). The most consistent finding is a marked loss of Purkinje cells in the cerebellum, neuronal loss in medial thalamus and spinal cord, and proliferation of Bergmann glia. There may also be a loss of cells in the granular cell layer [20–22, 42]. Membrane-bound vacuoles with clear contents in eccrine cells have recently been found [16].

II. LINKAGE MAPPING OF EPM1 TO 21q22.3

Lehesjoki *et al.* [34] used DNA from 12 Finnish families with EPM1 to map the phenotype by linkage analysis. Each of these families contained two or three affected children. The analysis was done before the

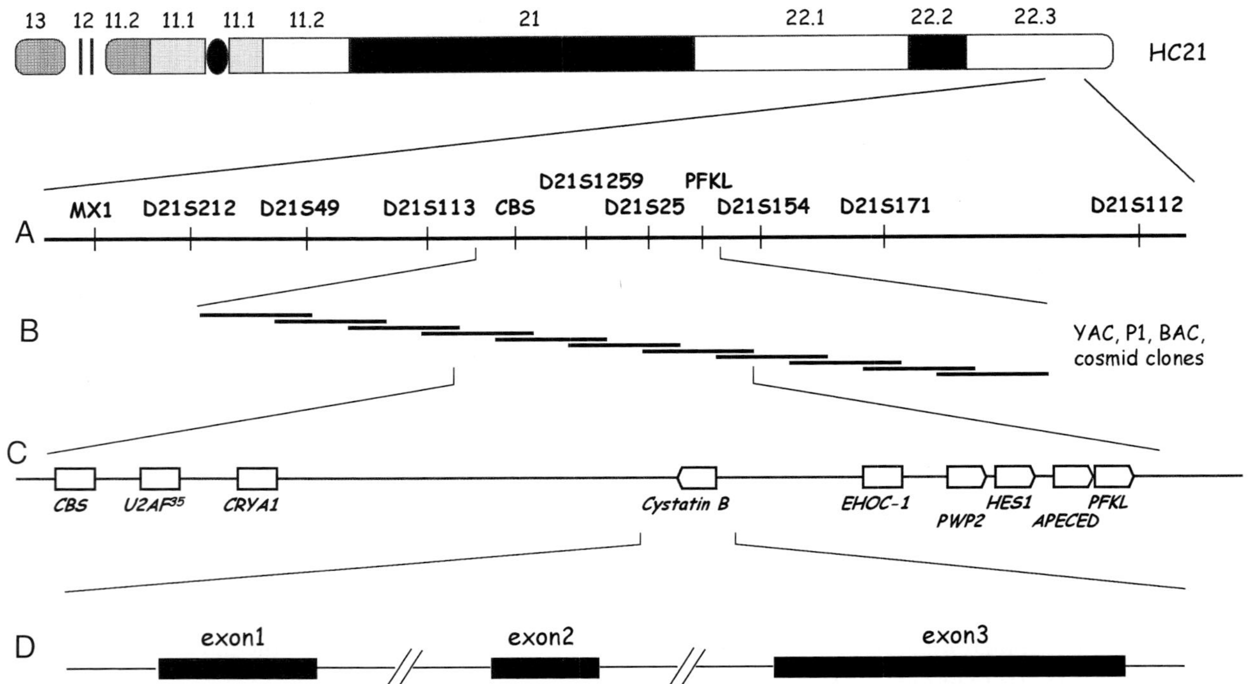

FIGURE 48-1 Schematic representation of the distal region 21q22.3 of chromosome 21 and the position of CSTB (cystatin B) gene. (A) Linkage map of the 21q22.3 region showing the location of polymorphic markers. (B) Physical map of the region. (C) Genes identified in the region. When the orientation of transcription is known, the gene symbols are shown with arrows. (D) Genomic organization of the CSTB gene.

availability of numerous dinucleotide repeats and therefore markers detectable by Southern blotting were used. After testing 64 markers and excluding approximately 25% of the genome, linkage was first detected with marker CRI-L4Z7 (D21S112) which was one of the most informative VNTR (variable number of tandem repeats) polymorphisms (Fig. 48-1A). D21S112 was mapped to chromosome 21 [18]. The maximum lod score obtained with D21S112 was 6.91 at a recombination fraction of 0.034. Further analysis using additional markers on chromosome 21q22.3 and multipoint linkage suggested that the most likely location of EPM1 in these Finnish families was an interval of 11.8 cM around markers D21S49, D21S154, and D21S112. Subsequently, additional studies using samples from non-Finnish families with EPM1 demonstrated that there is locus homogeneity in this disorder since all pedigrees showed linkage to markers on 21q22.3. Malafosse *et al.* [40] used 6 EPM1 families from the Mediterranean and found a maximum lod score of 6.12 at $\theta = 0.00$ with another chromosome 21q22.3 marker D21S113. Lehesjoki *et al.* [35–37] and Cochius *et al.* [15] further confirmed that families with EPM1 from the United States, Canada, Japan, Iran, and Turkey also map to 21q22.3. The marker that resulted in the maximum lod score 5.63 at recombination fraction of 0.00 in the latter study was a dinucleotide repeat polymorphism within the PFKL gene. A large family from Valais, Switzerland, showed a maximum lod score of 4.6 at $\theta = 0.00$ for markers in the PFKL region [28].

The availability of a dense linkage map [41], a preliminary physical map of chromosome 21 [14], and a considerable number of patients from the same ethnic group (Finnish) enabled a fine mapping of EPM1 by linkage disequilibrium. Analysis of 88 EPM1 and 62 normal Finnish chromosomes for several dinucleotide polymorphisms at 21q22.3 showed that there was a considerable linkage disequilibrium of certain polymorphic alleles with the elusive EPM1 gene. For example, 96% of EPM1-bearing chromosomes were marked with allele 4 of D21S1259, while only 40% of the normal chromosomes contained allele 4 for this marker. Linkage disequilibrium and polymorphism haplotype analysis indicated with strong statistical significance that the EPM1 locus mapped to a 175-kb interval between markers D21S2040 and D21S1259 [37, 59].

III. POSITIONAL CLONING OF THE EPM1 GENE

Since there was no biochemical clue as to the pathophysiology of the disease, positional cloning strategies were employed by several investigators to clone the gene, mutations of which would result in EPM1. A complete array of overlapping clones in bacteria (BACs and cosmids) (Fig. 48-1B) was developed by Stone *et al.* [52] and other laboratories [26] and several genes were identified by exon trapping, cDNA selection, and nucleotide sequence sampling [11–13, 63] (Figs. 48-1B, 1C). Mutation analysis using DNAs from affected individuals excluded several positional candidate genes as being responsible for EPM1 [29, 30, 50, 61, 62] (Fig. 48-1C). cDNA sequences from the cystatin B gene (CSTB; Genbank L03558, U46692, SwissProt P04080) which has been cloned in 1990 but not mapped to human chromosomes were isolated by Pennacchio *et al.* [44] and mapped in the EPM1 critical region (Fig. 48-1D). They found mutations in cystatin B in a few patients with EPM1 and thus provided the first convincing evidence that this gene is responsible for this disorder [44].

IV. POINT MUTATIONS IN CSTB GENE

The gene for cystatin B (CSTB) is small and contains only three exons (Figs. 48-1D and 48-2). Sequence analysis from DNAs of many patients with EPM1 revealed

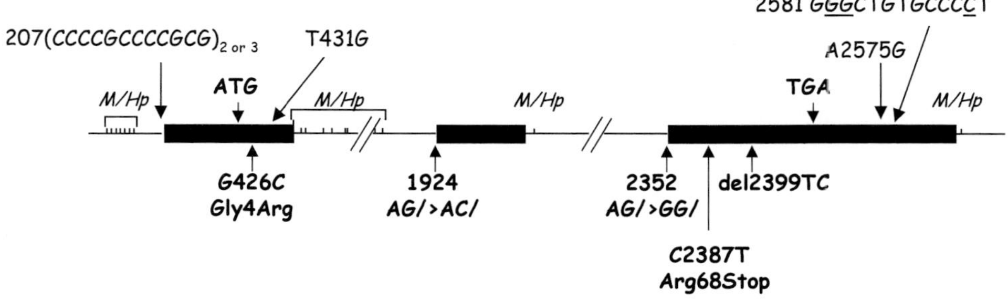

FIGURE 48-2 Schematic representation of the CSTB gene. Below the gene, the point mutations that cause EPM1 are shown. Above the gene, the polymorphic variants in normal individuals are shown. The *Msp*I (*M*) and *Hpa*II (*Hp*) restriction sites are also depicted.

TABLE 48-2 Point Mutations in CSTB in EPM1

Exon 1	GGG → CGG Gly4Arg	[31]
Intron 1 splice acceptor	ag/GTG → ac/GTG	[44]
Intron 2 splice acceptor	ag/GTG → gg/GTG	[31]
Exon 3	CGA → TGA Arg68Stop	[44]
Exon 3	del 2399TC	[31]

five point mutations that are likely to cause abnormal gene expression and protein function [31, 32, 44]. These mutations are shown in Fig. 48-2 and are listed in Table 48-2.

Surprisingly, these mutations account only for approximately 10% of the CSTB alleles examined in EPM1 patients. From a total of 236 EPM1 alleles studied, point mutations were found in 23 alleles [27, 31, 32, 44, 60]. It is unlikely that other nucleotide substitutions were missed because the gene is small and all the areas including introns have been examined by SSCP and direct sequencing. The mutation associated with the most common Finnish haplotype was not among these nucleotide substitutions and remained a mystery.

V. DNA POLYMORPHISMS IN THE CSTB GENE

In addition, several nucleotide polymorphisms have been identified associated with both normal and EPM1-bearing chromosomes. Most polymorphisms are shown in Fig. 48-2 and are the following [31]: (i) Nucleotide substitution 431T or G in the third nucleotide of a codon and does not result in an amino acid change, (ii) 2575 A or G in the 3′UTR, (iii) addition of 2Gs after nucleotide 2581, and (iv) insertion of C after nucleotide 2589 in the 3′UTR (the numbering is as per [44]).

The most significant polymorphism, however, is the presence of two or three copies of the dodecamer sequence (CCCCGCCCCGCG) in the 5′ flanking region of the gene [31]. Figure 48-3 shows this dodecamer polymorphism. Analysis of DNAs of normal Caucasian individuals revealed that the frequency of the three-copy allele is 66% while that of the two-copy allele was 33%. The frequency of the alleles in North Africans is 47 and 53% for two or three copies, respectively [32].

VI. CYSTATIN B

Type I cystatins (or stefins; reviewed by Turk and Bode [56]) are molecules of about 100 amino acids with neither disulfate bonds nor carbohydrate groups that are inhibitors of cysteine proteases. Cysteine proteases compose a group of proteolytic enzymes that cleave peptide bonds by use of a catalytic cysteine residue; those inhibited by cystatins are members of the papain superfamily [46]. The name cystatin was applied by Barrett [5] to a protein from chicken egg white that inhibits various cysteine proteases. The human cystatin B [47] is a number of the cystatin family (Fig. 48-4). The amino acid sequences of rat, mouse, and bovine cystatin B are known [45, 48, 57]. Although they are primarily

```
ATG ATG TGC GGG GCG CCC TCC GCC ACG CAG CCG GCC ACC GCC GAG ACC CAG CAC ATC GCC
 M   M   C   G   A   P   S   A   T   Q   P   A   T   A   E   T   Q   H   I   A

GAC CAG GTG AGG TCC CAG CTT GAA GAG AAA GAA AAC AAG AAG TTC CCT GTG TTT AAG GCC
 D   Q   V   R   S   Q   L   E   E   K   E   N   K   K   F   P   V   F   K   A

GTG TCA TTC AAG AGC CAG GTG GTC GCG GGG ACA AAC TAC TTC ATC AAG GTG CAC GTC GGC
 V   S   F   K   S   Q   V   V   A   G   T   N   Y   F   I   K   V   H   V   G

GAC GAG GAC TTC GTA CAC CTG CGA GTG TTC CAA TCT CTC CCT CAT GAA AAC AAG CCC TTG
 D   E   D   F   V   H   L   R   V   F   Q   S   L   P   H   E   N   K   P   L

ACC TTA TCT AAC TAC CAG ACC AAC AAA GCC AAG CAT GAT GAG CTG ACC TAT TTC TGA
 T   L   S   N   Y   Q   T   N   K   A   K   H   D   E   L   T   Y   F  Stop
```

FIGURE 48-4 Nucleotide sequence of the cDNA and predicted amino acid sequence of the human cystatin B.

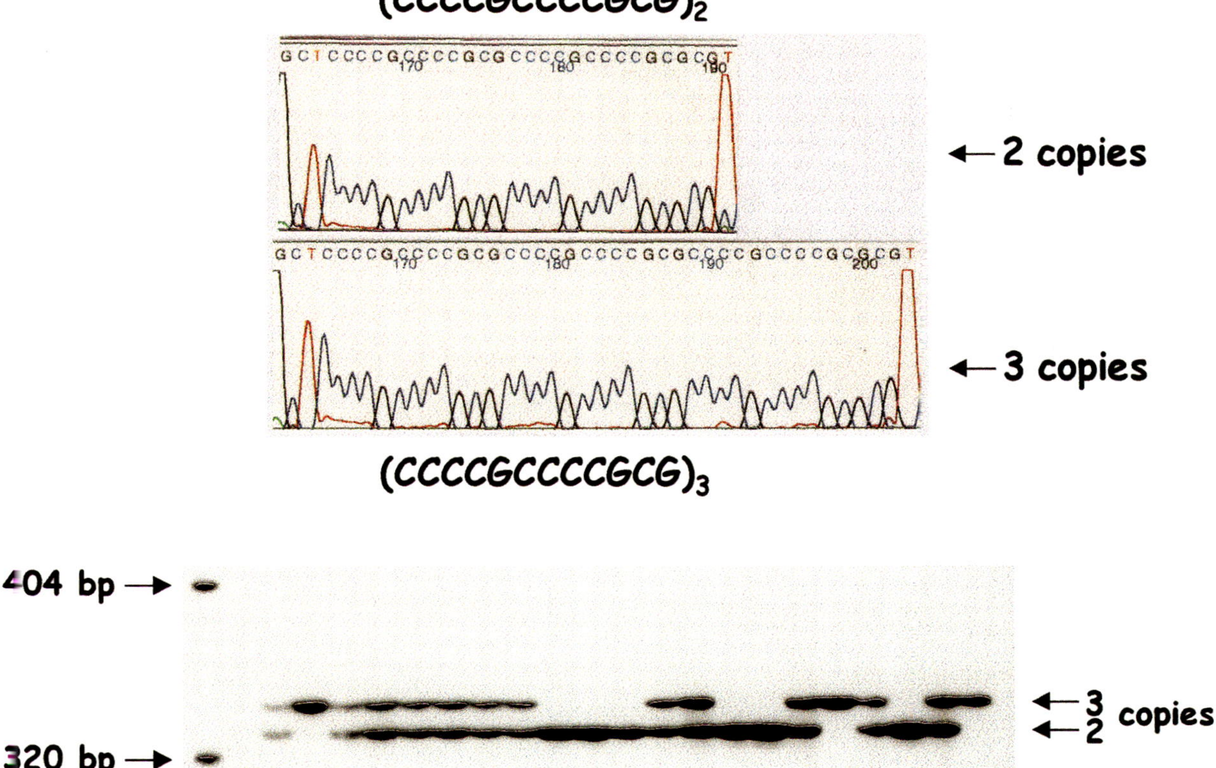

FIGURE 48-3 Nucleotide sequence of alleles with two or three copies of the 12-mer (CCCCGCCCCGCG) repeat in the 5′ flanking region of CSTB. The bottom shows the autoradiogram of the PCR-amplified region that contains the polymorphic repeat in different individuals. Homozygotes for two or three repeats and heterozygotes are depicted.

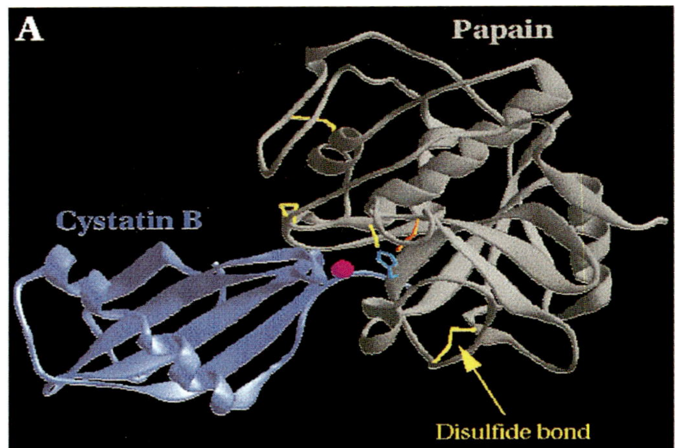

426 G>C
Gly4Arg

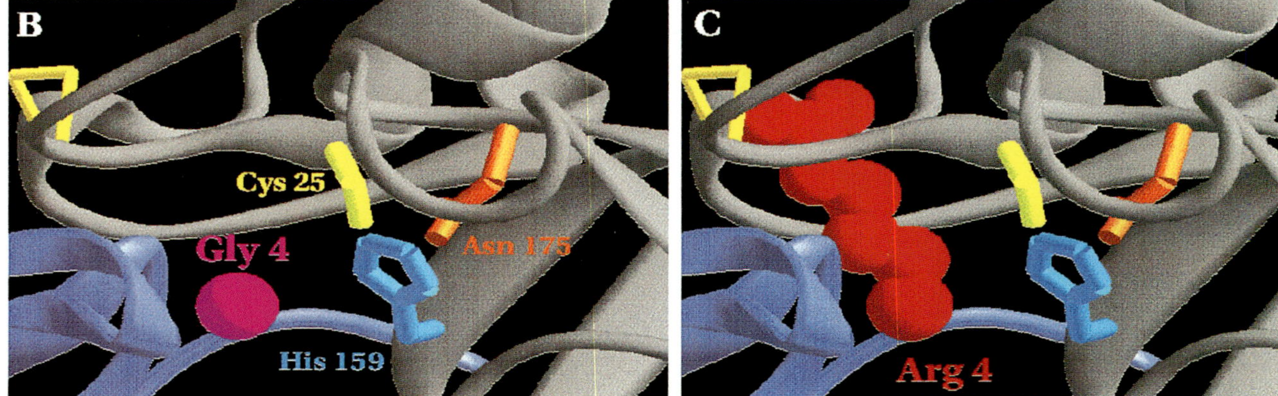

FIGURE 48-5 "Ribbon" representation of the cystatin B–papain complex. (A) Full enzyme–inhibitor complex. (B and C) Magnification of the active site in the wild-type and the Arg4 mutant cystatin B, respectively. The location of the Gly4 in wild-type cystatin B is shown as a sphere in (B) (position of the alpha carbon); the side chain residues are indicated in (C) as a group of spheres. Note that the size of the Arg4 side chain is in steric conflict with the binding site on papain. This is likely to result in reduced inhibitory activity of the Arg4 cystatin B mutant.

intracellular proteins, they have also been detected in extracellular fluids [1]. mRNA for cystatin B has been identified by Northern blot analysis in all tissues examined [44].

The crystal structures of chicken cystatin B and the complex cystatin B–papain have been determined [9, 53]. Chicken cystatin B consists of a five stranded antiparallel β-sheet (A to E) wrapped around a five-turn α-helix. The most amino-terminal part of cystatin B (Met1 to Pro6 equivalent in human), and the loops connecting the β-sheets B to C and D to E were shown to interact with the papain active site (PPB entry 1STF) [53]. It is of interest that the mutation Gly4Arg observed in a homozygous patient with EPM1 (see Table 48-2; [31] is located within 4 Å of the closest papain residue, Asp158, and also within <5 Å of Cys25 in the papain active site. The Gly4Arg substitution is therefore likely to cause major steric hindrance and strongly affect the papain-binding capacity of cystatin B (Fig. 48-5). Moreover, Gly4 is conserved in cystatin proteins from many species [46, 56].

Truncation experiments with chicken and human cystatin B proteins showed that N-terminal deletions which included Gly4 (or the equivalent Gly9 in chicken) result in a 5000-fold decrease in binding affinity to papain [2, 39, 55]. It is likely that the Gly4Arg substitution creates a cystatin B molecule that does not bind efficiently to its natural substrates and therefore does not have inhibitory activity. However, it is unclear whether the pathogenesis of EPM1 in the patient with this mutation is a result of reduced inhibitory activity or/and decreased stability of mutant cystatin B.

VII. DODECAMER REPEAT EXPANSION IN CSTB

The mystery of the most common mutation in the CSTB in EPM1 patients remained unsolved even after the complete nucleotide sequencing of the gene in a large number of patients. Southern blot analysis of DNAs from EPM1 patients with unknown CSTB mutations revealed that restriction fragments containing the entire CSTB gene were larger than normal [27, 32, 60]. The larger *Eco*RI fragments were of varying size between patients, segregated with the EPM1 phenotype in families, and were not present in many normal alleles examined. Moreover, all the EPM1 patients with a known point mutation in one allele were heterozygous for a normal and a larger restriction fragment and study of their families showed that the point mutation was present in the normal size fragment [27, 32, 60]. Further analyses localized the region to between nucleotides −400 and −80 upstream of the ATG translation initiation codon of CSTB as the area of the abnormally larger sequences that were associated with the phenotype of the majority of EPM1 patients. Genomic clones in bacterial plasmids containing the abnormal region were unstable, showing high frequency of insert loss. Sequencing of these cloned DNA fragments revealed not genomic rearrangement, translocation, or new sequences. Instead, under special conditions, the nucleotide sequence clearly showed a long stretch of uninterrupted perfect repetition of the dodecamer $(CCCCGCCCCGCG)_n$ sequence [32] (Fig. 48-6). In the initial sequencing experiments more than 50 repeats were observed in the clones studied. Subsequent analysis of DNAs from all available

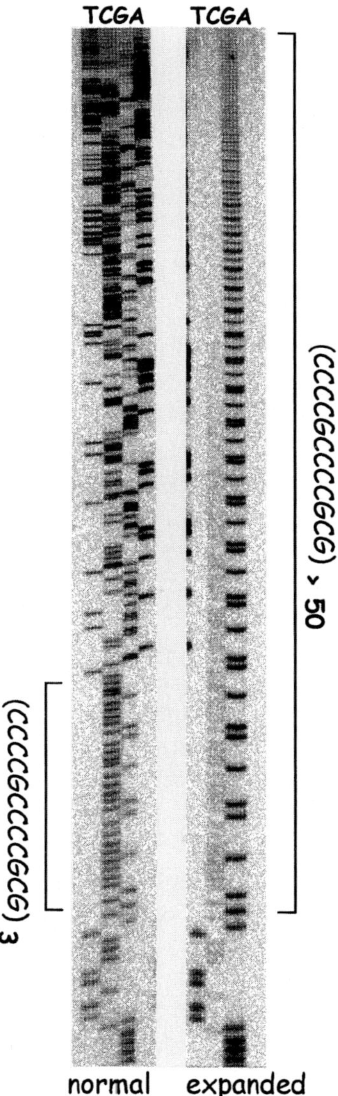

FIGURE 48-6 Nucleotide sequence analysis of a normal and a mutant clone that contain the region of the dodecamer repeat in the 5′ flanking region of the CSTB gene. The mutant clone contains more than 50 identical, uninterrupted copies of the 12-mer (CCCCGCCCC-GCG) repeat.

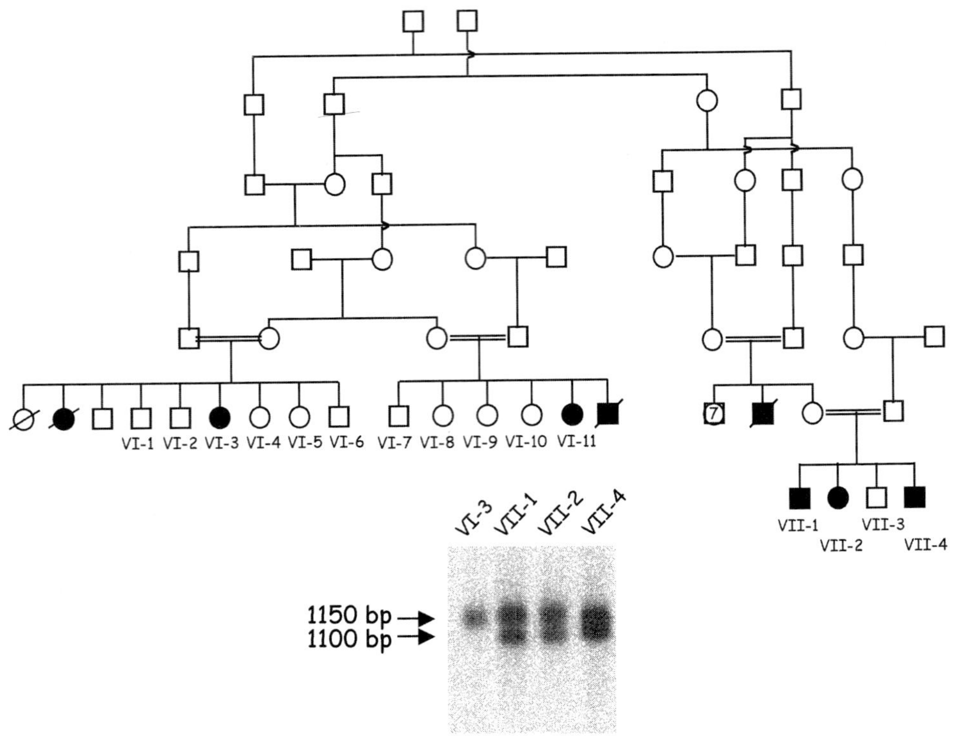

FIGURE 48-7 Pedigree of Swiss family with EPM1. Affected individuals are shown with black symbols. The autoradiogram on the bottom depicts the sizes of the expanded 12-mer repeat in some members of this family. Note the different sizes of CSTB alleles among affected individuals.

patients indicated that the expansion of the dodecamer repeat $(CCCCGCCCCGCG)_n$ which is located about 200 nucleotides upstream of the initiation ATG codon of CSTB is the most common mutation in all populations examined and accounted for all the previously unknown mutations (218 of 236 alleles [27, 32, 60]). Special PCR methods have been developed to accurately measure the size of the repeat expansion. In the series of patients of Lalioti *et al.* [32], the number of the dodecamer repeats found in affected individuals ranged from 30 to 75. Heterozygous parents with one normal and one expanded CSTB allele did not show any particular abnormality and their history of seizures was not different from that of the general population. Within the level of detection of the expansion of the dodecamer repeat, no instance of intergenerational—parent to child—changes in the expansion size were detected [32, 60]. In addition, no anticipation on the severity or age of onset of the disease was observed, mainly because EPM1 is a recessive phenotype and all patients are usually in one generation.

However, the expanded alleles are unstable and the initial (*de novo*) expansions have occurred many times. Analysis of DNA polymorphism haplotypes near the dodecamer repeat in different populations revealed many different haplotypes associated with the expanded alleles [27]. This suggests that the dodecamer expansion occurred independently in different ethnic groups. This is further evidenced as the number of the dodecamer repeat expansion varies between distant affected relatives (Fig. 48-7) in which the haplotypes of affected alleles are the same (Lalioti *et al.,* unpublished). Moreover, the number of the expanded repeats varies in patients from the same ethnic group with the same haplotype [27, 60]. This strongly indicates that the expanded dodecamer repeat is unstable and changes of repeat number occur frequently. In the Finnish population, a putative founding haplotype was detected in 75 of 87 chromosomes studied with the expansion [60]. This striking linkage disequilibrium suggests a single origin of the expansion mutation in this population. It was estimated that the date of the original expansion was at least 2000 years old or about 60–70 generations [17, 59].

Besides the normal alleles with 2 or 3 copies of the dodecamer repeat and the EPM1 alleles with 30–75 copies of the repeat, intermediate alleles of 12–17 copies of the repeat have been observed [32]. These were alleles with 12 and 13 repeat copies in the parents of two CEPH collection pedigrees (102 and 104) from Venezuela. Further genotyping in these two families revealed that the alleles with 12 or 13 copies were often transmitted unstably from parents to offspring (Fig. 48-8). From 21 transmissions of these alleles, 6 repeat expansions (29%) to alleles with 13, 14, 15, or 17 repeats were observed. The instability of these intermediate (perhaps premutation) alleles of the dodecamer repeat in this family is 29×10^{-2}, whereas the mutation rate of some so-called hypermutable polymorphisms is 0.6×10^{-2} [54]. The analysis of haplotypes (all within a genetic distance of less than 2 cM from the repeat) revealed that all six expansions were paternally transmitted [32]. The largest observed expansions in a given meiosis was four repeats (from 13 to 17). Haplotype data in the CEPH families showed that there were no contractions of the premutation alleles. Individuals homozygous for the 12-repeat allele or compound heterozygous for 12 and 17 repeats had no clinical phenotype of EPM1 and therefore these alleles are considered noncausative of EPM1. The clinical significance of intermediate alleles remains unknown and large population studies are needed to establish the probability of expansion to the disease range.

The expansion of the dodecamer (CCCCGCCCGCG)n repeat in EPM1 represents the first case of instability of a repeat unit other than trinucleotides in association with a human disorder. The 12-mer contains a shorter repeat of CCCCG within it, but the expansion involves the entire 12-mer and not the shorter pentamer.

VIII. EXPRESSION OF CSTB GENE WITH THE 12-MER EXPANSION

The CTSB gene is ubiquitously expressed. Northern blot analysis using CSTB cDNA or genomic probes reveals a single mRNA species of about 0.8 kb. Pennacchio *et al.* [44] demonstrated reduced amounts of the CSTB mRNA in lymphoblastoid cell lines from patients with EPM1. Lafrenière *et al.* [27] also demonstrated reduced amounts of the CSTB mRNA in lymphoblastoid cell lines with expansion of the 12-mer repeat. Both of these studies utilized Northern blot analysis. Lalioti *et al.* [32] used RNase protection assays in total RNA from blood leucocytes and/or lymphoblastoid cell lines

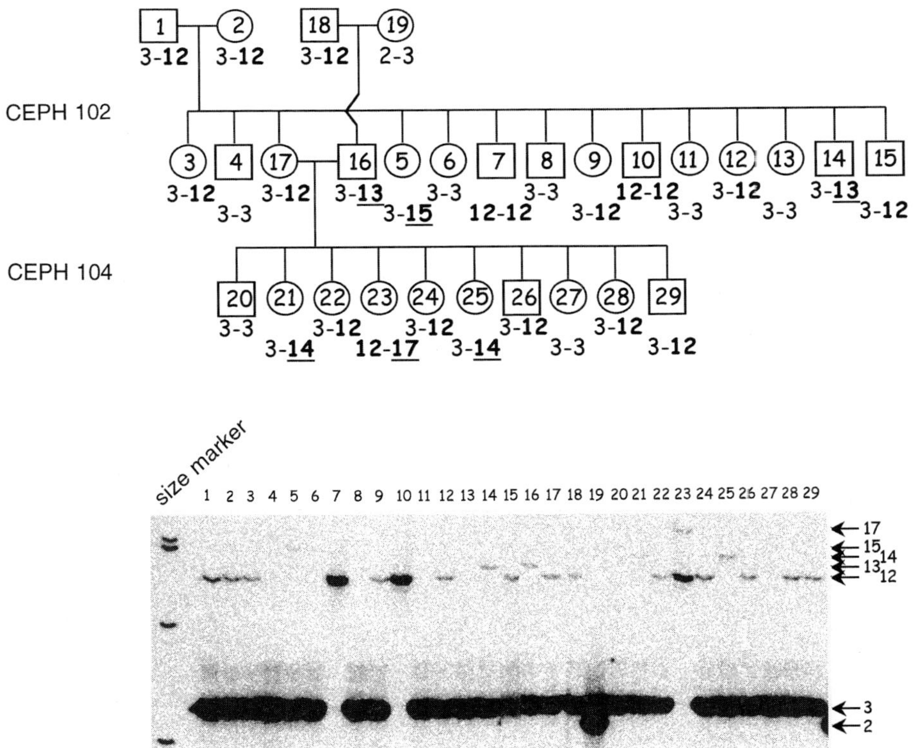

FIGURE 48-8 CEPH pedigrees 102 and 104 showing the number of the 12-mer repeats in the CSTB alleles. For example, 3–12 refers to a genotype with 3 and 12 repeats. *De novo* alleles not present in the parents are underlined. Nonpaternity has been excluded by the use of thousands of polymorphic markers in these families. The autoradiogram on the bottom shows the genotypes of the 12-mer polymorphic repeat in families 102 and 104. The numbers of repeats are indicated on the right. The lane numbering corresponds to the numbers within the pedigree symbols.

from patients, heterozygotes, and normal controls. They observed a marked reduction of the CSTB mRNA in blood leucocytes from patients with 12-mer expansion compared to normals. In contrast, in fibroblasts and lymphoblastoid cell lines, the amount of CSTB mRNA was either normal or only slightly reduced (Fig. 48-9). Thus, the reduction of CSTB gene expression of the expanded CSTB alelles appears to be cell specific [32]. This may account for the selective cell death in the brains of patients with EPM1. DNA methylation differences were not observed for a number of *Hpa*II/*Msp*I sites within the CSTB gene (shown in Fig. 48-2) between normal and EPM1 samples [32]. The low level of mRNA in blood leucocytes compared with the normal level in lymphoblastoid cell lines suggests that the repeat expansion modulates the expression of the CSTB gene in some cell types. This reduction could be due to a simple increase of the distance between DNA binding sites of cell-specific transcription factors and the transcription initiation sites. Another possibility is that the 12-mer is the binding site for transcriptional repressors and the availability of many copies of this sequence results in multiplicity of the negative regulatory effects.

The 12-mer repeat is not within the primary transcript of the CSTB gene, unlike in all the other disorders due to repeat expansions [4]. RNase protection was used to determine the transcription start site of CSTB, and two such sites were observed [32] (Fig. 48-10). The main transcription start site appears to coincide with the expressed sequence tag closest to the 5′ end. Furthermore, 5′ RACE did not detect any additional sequences further upstream of the sites shown in Fig. 48-10. Both potential transcription start sites are on the 3′ side of the 12-mer repeat (about 66 and 77 nucleotides down-

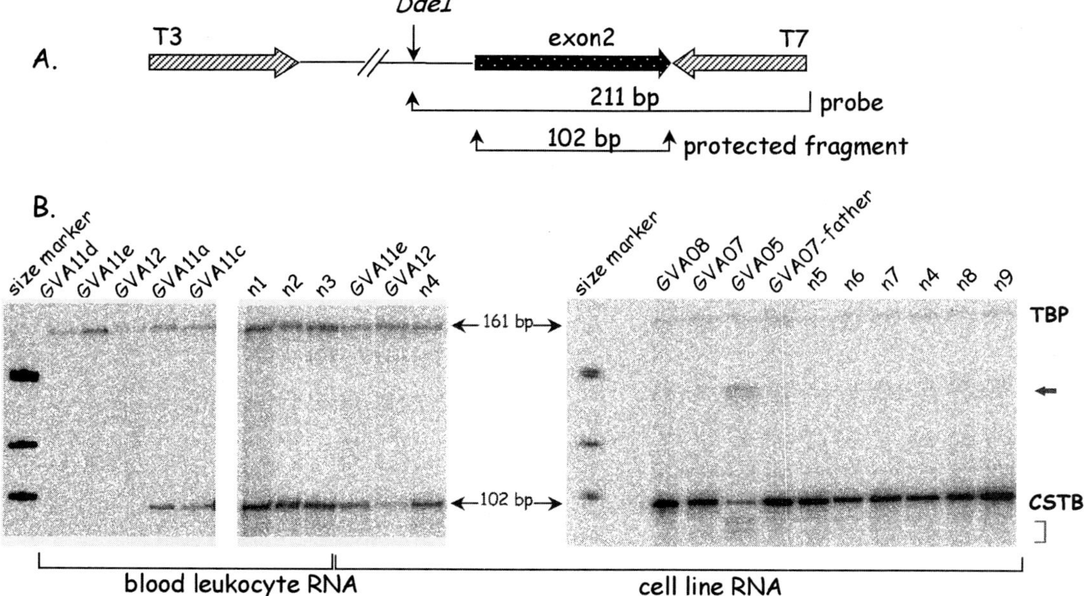

FIGURE 48-9 RNase protection assays of exon 2 of the CSTB gene. (A) Schematic representation of a plasmid containing exons 1 and 2 of CSTB. The plasmid was linearized with *DdeI* and a riboprobe was produced with T7 RNA polymerase. The expected protected fragment is 102 bp. The control probe is a fragment of the TATA-binding protein (TBP) cDNA. The expected control TBP protected fragment is 161 bp. (B) Autoradiogram of RNase protection showing the expression of CSTB relative to TBP in blood leukocytes, lymphoblastoid cell lines, and cultured fibroblasts (GVA08, homozygous for the expansion); n, normals. The arrow and bracket on the right depict the alternatively spliced protected fragments from patient GVA05 who is a compound heterozygote for an expansion and an AG to AC mutation of IVS1 acceptor splice site. These protected products are likely to originate from the utilization of cryptic splice sites in the allele that harbors the splicing mutation. The sizes of the protected fragments are indicated. GVA11d, e, a, and c are four siblings from family GVA11; d and e are homozygous for the repeat expansion, whereas a and c are unaffected. In blood leukocytes, there is a marked reduction of CSTB RNA in patients; in contrast, in most of the cell lines from patients, its expression is not different from normal. Note that in the blood sample of GVA11e there is almost undetectable CSTB RNA, whereas in his cell line this RNA is within the normal limits. Patient GVA12 shows more CSTB RNA in the LBL than in blood leukocytes; however, the expression in LBL is lower than normal.

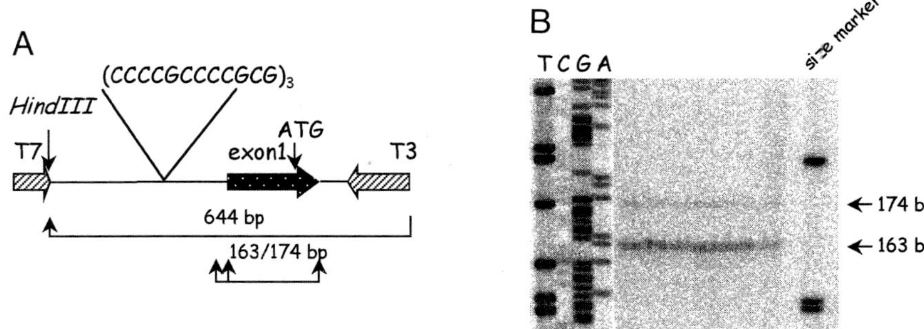

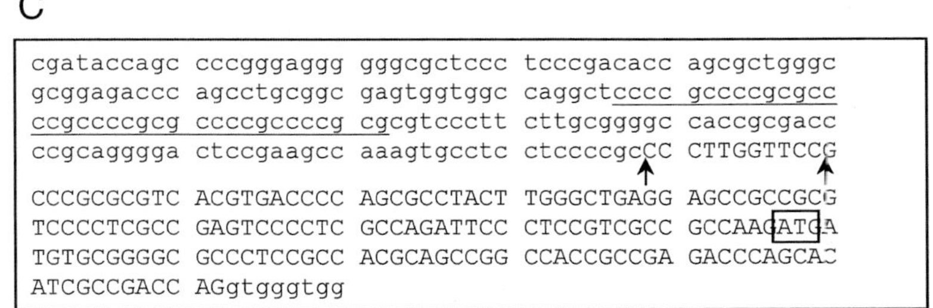

FIGURE 48-10 RNase protection assays of Exon 1 and 5′ flanking region of the CSTB gene. (A) Schematic representation of a plasmid containing exon 1 and about 400 bp 5′ to the ATG initiation codon. The plasmid was linearized with *Hin*dIII, and riboprobe was produced with T3 polymerase. (B) Autoradiogram showing two protected fragments of 163 and 174 nucleotides that include exon 1. The predicted transcription initiation sites are distal to the dodecamer repeat. (C) Partial nucleotide sequence of the CSTB genomic region that includes the 12-mer repeat (underlined), the two predicted transcription initiation sites (arrows), and the translation initiation codon (boxed). Upper case letters denote exon 1 and lower case letters the 5′ flanking region.

stream) and thus the repeat is not included in these transcripts. It is of interest that the 12-mer repeat sequence may contain Sp1 sites.

IX. EXPANSION SIZE AND SEVERITY OF THE PHENOTYPE

The accurate measurement of the size of the 12-mer repeat expansion permitted the study of the potential correlation of the age of onset and the size of the repeat. Data from 28 patients were collected and analyzed. It appears that there is no correlation between the size of the repeat of the age of onset of the disease (Lalioti *et al.*, unpublished). This is in agreement with the observation that siblings that share the same repeat expansion may have different ages of onset. It is possible that once the 12-mer repeat expansion extends beyond a critical threshold, it reduces CSTB expression in certain cells with pathological consequences. Therefore the age of onset of EPM1 no longer depends on the size of the alleles after this threshold is passed by both alleles of the patients. Additional, modifier genes may influence the severity and evolution of the phenotypes.

The pathophysiology of the disease remains a mystery. How does the reduced amount of CSTB in some cells result in the phenotype? As described above, CSTB is an ubiquitously expressed, cytosolic protein that functions as an inhibitor of cysteine proteases, some of which may be involved in apoptosis. The specific cell destruction in the brains of patients with EPM1 may be due to the lack of inhibition of cell-specific proteases and accumulation of degradation products.

X. THE FUTURE

There are many unanswered questions regarding the pathophysiology of the disease. What is the phenotype of a complete absence of cystatin B as, for example, in mice with targeted description of the CSTB gene? What is the mechanism of reduced gene expression due to 12-mer expansion? What is the mechanism of expansion? Which exactly are the cysteine proteases that are inhibited by CSTB in the brain cells that are damaged in EPM1? Is there a function of CSTB other than as protease inhibitor? Could we interfere and correct the defect in cells? The answers to these and other questions will provide the necessary knowledge for more intelligent and effective treatment of EPM1.

Acknowledgments

The author thanks M. Lalioti, H. S. Scott, M. A. Morris, and other members of the laboratory for their contributions to the discovery of the 12-mer repeat expansions, A. Malafosse for his enthusiastic collaboration and expert clinical collections, A. Bottani for the care of the Swiss family, and the laboratories of R. Meyers, D. Cox, G. Rouleau, and A. de la Chapelle for communicating results prior to publication and exciting scientific competition. The research in the author's laboratory is supported by funds from the National Swiss Science Foundation, The European Union/Swiss OFES, and the University and Cantonal Hospital of Geneva. The author and his laboratory colleagues also thank the patients and their families for their collaboration and donation of their precious tissues.

References

1. Abrahamson, M., Barrett, A. J., Salvesen, G., and Grubb, A. (1986). Isolation of six cysteine proteinase inhibitors from human urine: their physicochemical and enzyme kinetic properties and concentrations in biological fluids. *J. Biol. Chem.* **261,** 11282–11289.
2. Abrahamson, M., Ritonja, A., Brown, M. A., Grubb, A., Machleidt, W., and Barrett, A. J. (1987). Identification of the probable inhibitory reactive sites of the cysteine proteinase inhibitors human cystatin C and chicken cystatin. *J. Biol. Chem.* **262,** 9688–9694.
3. Ammann, F., Schweingruber, R., and Paro, M. (1978). A family with progressive myoclonus epilepsy. *Schweiz Arch. Neurol. Neurochir. Psychi.* **123,** 3–15.
4. Ashley, C. T., and Warren, S. T. (1995). Trinucleotide repeat expansion and human disease. *Annu. Rev. Genet.* **29,** 703–728.
5. Barrett, A. J. (1981). Cystatin, the egg white inhibitor of cystein proteinases. *Methods Enzymol.* **80,** 771–778.
6. Berkovic, S. F., Anderman, F., Carpenter, S., and Wolfe, L. S. (1986). Progressive myoclonus epilepsies: specific causes and diagnosis. *N. Engl. J. Med.* **315,** 296–305.
7. Berkovic, S. F., So, N. K., and Andermann, F. (1991). Progressive myoclonus epilepsies: clinical and neurophysiological diagnosis. *J. Clin. Neurophysiol.* **8,** 261–274.
8. Berkovic, S. F., Cochius, J., Andermann, E., and Andermann, F. (1993). Progressive myoclonus epilepsies: clinical and genetic aspects. *Epilepsia* **34**(Suppl. 3), S19–S30.
9. Bode, W., Engh, R., Musil, D., Thiele, U., Huber, R., Karshikov, A., Brzin, J., Kos, J., and Turk, V. (1988). The 2.0 A X-ray crystal structure of chicken egg white cystatin and its possible mode of interaction with cysteine proteinases. *EMBO J.* **7,** 2593–2599.
10. Carpenter, S., and Karpati, G. (1981). Sweat gland duct cells in Lafora disease: diagnosis by skin biopsy. *Neurology* **31,** 1564–1568.
11. Cheng, J. F., and Zhu, Y. (1994). Reverse transcription-polymerase chain reaction detection of transcribed sequences on human chromosome 21. *Genomics* **20,** 184–190.
12. Cheng, J. F., Boyartchuk, V., and Zhu, Y. (1994). Isolation and mapping of human chromosome 21 cDNA: progress in constructing a chromosome 21 expression map. *Genomics* **23,** 75–84.
13. Chen, H., Chrast, R., Rossier, C., Morris, M. A., Lalioti, M. D., and Antonarakis, S. E. (1996). Cloning of 559 potential exons of genes of human chromosome 21 by exon trapping. *Genome Res.* **6,** 747–760.
14. Chumakov, I., Rigault, P., Guillou, S., Ougen, P., Billaut, A., Guasconi, G., Gervy, P., LeGall, I., Soularue, P., Grinas, L., Bougueleret, L., Bellanne-Chantelot, C., Lacroix, B., Barillot, E., Gesnouin, P., Pook, S., Vaysseix, G., Frelat, G., Schmitz, A., Sambucy, J. L., Bosch, A., Estivill, X., Weissenbach, J., Vignal, A., Riethman, H., Cox, D., Patterson, D., Gardiner, K., Hattori, M., Sakaki, Y., Ichikawa, H., Ohki, M., LePaslier, D., Heilig, R., Antonarakis, S. E., and Cohen, D. (1992). Continuum of overlapping clones spanning the entire human chromosome 21q. *Nature* **359,** 380–387.
15. Cochius, J. I., Figlewicz, D. A., Kalviainen, R., Nousiainen, U., Farrell, K., Patry, G., Soderfeldt, B., Frydman, M., Lerman, P., and Andermann, F. (1993). Unverricht-Lundborg disease: absence of nonallelic genetic heterogeneity. *Ann. Neurol.* **34,** 739–741.
16. Cochius, J. I., Carpenter, S., Andermann, E., Rouleau, G., Nousiainen, U., Kalviainen, R., Farrell, K., and Andermann, F. (1994). Sweat gland vacuoles in Unverricht-Lundborg disease: a clue to diagnosis? *Neurology* **44,** 2372–2375.
17. de la Chapelle, A. (1993). Disease gene mapping in isolated human populations: the example of Finland. *J. Med. Genet.* **30,** 857–865.
18. Donis-Keller, H., Green, P., Helms, C., Cartinhour, S., Weiffenbach, B., Stephens, K., Keith, T. P., Bowden, D. W., Smith, D. R., Lander, E. S., Botstein, D., Akots, G., Rediker, K. S., Gravius, T., Brow, V. A., Rising, M. B., Parker, C., Powers, J. A., Watt, D. E., Kauffman, E. R., Bricker, A., Phipps, P., Muuler-Kahle, H., Fulton, T. R., Ng, S., Schumm, J. V., Braman, J. C., Knowlton, R. G., Barker, D. F., Crooks, S. M., Lincoln, S. E., Daly, M. J., and Abrahamson, J. (1987). A genetic linkage map of the human genome. *Cell* **51,** 319–337.
19. Genton, P., Michelucci, R., Tassinari, C. A., and Roger, J. (1990). The Ramsay Hunt syndrome revisited: Mediterranean myoclonus versus mitochondrial encephalomyopathy with ragged-red fibers and Baltic myoclonus. *Acta Neurol. Scand.* **81,** 8–15.
20. Haltia, M., Kristensson, K., and Sourander, A. (1969). Neuropathological studies in three Scandinavian cases of progressive myoclonus epilepsy. *Acta Neurol. Scand.* **45,** 63–77.
21. Koskiniemi, M., Donner, M., Majuri, H., Haltia, M., and Norio, R. (1974). Progressive myoclonus epilepsy. A clinical and histopathological study. *Acta Neurol. Scand.* **50,** 307–332.
22. Koskiniemi, M., Toivakka, E., and Donner, M. (1974). Progressive myoclonus epilepsy. Electroencephalographical findings. *Acta Neurol. Scand.* **50,** 333–359.
23. Koskiniemi, M. (1975). Findings in routine laboratory examination in progressive myoclonus epilepsy. *Acta Neurol. Scand.* **51,** 12–20.
24. Koskiniemi, M., and Palo, J. (1978). Urinary excretion of indican in progressive myoclonus epilepsy without Lafora bodies. The effect of sodium valproate. *J. Neurol. Sci.* **39,** 235–239.
25. Labauge, P., Ouazzani, R., M'Rabet, A., Grid, D., Genton, P., Dravet, C., Chkili, T., Beck, C., Buresi, C., Baldy-Moulinier, M., and Malafosse, A. (1997). Allelic heterogeneity of Mediterranean myoclonus and the cystatin B gene. *Ann. Neurol.* **41,** 686–689.
26. Lafreniere, R. G., de Jong, P. J., and Rouleau, G. A. (1995). A 405-kb cosmid contig and *Hind*III restriction map of the progressive myoclonus epilepsy type 1 (EPM1) candidate region in 21q22.3. *Genomics* **29,** 288–290.
27. Lafreniere, R. G., Rochefort, D. L., Chretien, N., Rommens, J. M., Cochius, J. I., Kalviainen, R., Nousiainen, U., Patry, G., Farrell, K.,

Soderfeldt, B., Federico, A., Hale, B. R., Cossio, O. H., Sorensen, T., Pouliot, M. A., Kmiec, T., Uldall, P., Janszky, J., Pranzatelli, M. R., Andermann, F., Andermann, E., and Rouleau, G. A. (1997). Unstable insertion in the 5′ flanking region of the cystatin B gene is the most common mutation in progressive myoclonus epilepsy type 1, EPM1. *Nature Genet.* **15,** 298–302.

28. Lalioti, M. D., Bottani, A., Morris, M. A., and Antonarakis, S. E. (1996). The Swiss type of progressive myoclonus epilepsy (EPM1) maps to 21q22.3. *Med. Genet.* **2,** 249. [Abstract]
29. Lalioti, M. D., Gos, A., Green, M. R., Rossier, C., Morris, M. A., and Antonarakis, S. E. (1996). The gene for human U2 snRNP auxiliary factor small 35-kDa subunit (U2AF1) maps to the progressive myoclonus epilepsy (EPM1) critical region on chromosome 21q22.3. *Genomics* **33,** 298–300.
30. Lalioti, M. D., Chen, H., Rossier, C., Shafaatian, R., Reid, J. D., and Antonarakis, S. E. (1996). Cloning the cDNA of human PWP2, which encodes a protein with WD repeats and maps to 21q22.3. *Genomics* **35,** 321–327.
31. Lalioti, M. D., Mirotsou, M., Buresi, C., Peitsch, M. C., Rossier, C., Ouazzani, R., Baldy-Moulinier, M., Bottani, A., Malafosse, A., and Antonarakis, S. E. (1997). Identification of mutations in cystatin B, the gene responsible for the Unverricht-Lundborg type of progressive myoclonus epilepsy (EPM1). *Am. J. Hum. Genet.* **60,** 342–351.
32. Lalioti, M. D., Scott, H. S., Buresi, C., Rossier, C., Bottani, A., Morris, M. A., Malafosse, A., and Antonarakis, S. E. (1997). Dodecamer repeat expansion in cystatin B gene in progressive myoclonus epilepsy. *Nature* **386,** 847–851.
33. Lalioti, M. D., Scott, H. S., and Antonarakis, S. E. (1997). What is expanded in progressive myoclonus epilepsy? *Nature Genet.* **17,** 17.
34. Lehesjoki, A. E., Koskiniemi, M., Sistonen, P., Miao, J., Hastbacka, J., Norio, R., and de la Chapelle, A. (1991). Localization of a gene for progressive myoclonus epilepsy to chromosome 21q22. *Proc. Natl. Acad. Sci. USA* **88,** 3696–3699.
35. Lehesjoki, A. E., Koskiniemi, M., Pandolfo, M., Antonelli, A., Kyllerman, M., Wahlstrom, J., Nergardh, A., Burmeister, M., Sistonen, P., and Norio, R. (1992). Linkage studies in progressive myoclonus epilepsy: Unverricht-Lundborg and Lafora's diseases. *Neurology* **42,** 1545–1550.
36. Lehesjoki, A. E., Eldridge, R., Eldridge, J., Wilder, B. J., and de la Chapelle, A. (1993). Progressive myoclonus epilepsy of Unverricht-Lundborg type: a clinical and molecular genetic study of a family from the United States with four affected sibs. *Neurology* **43,** 2384–2386.
37. Lehesjoki, A. E., Koskiniemi, M., Norio, R., Tirrito, S., Sistonen, P., Lander, E., and de la Chapelle, A. (1993). Localization of the EPM1 gene for progressive myoclonus epilepsy on chromosome 21: linkage disequilibrium allows high resolution mapping. *Hum. Mol. Genet.* **2,** 1229–1234.
38. Lundborg, H. (1903). "Die progressive Myoclonus-Epilepsie (Unverricht's Myoclonie). pp. 1–207. Almqvist and Wiksell, Uppsala.
39. Machleidt, W., Thiele, U., Laber, B., Assfalg-Machleidt, I., Esterl, A., Wiegand, G., Kos, J., Turk, V., and Bode, W. (1989). Mechanism of inhibition of papain by chicken egg white cystatin. Inhibition constants of N-terminally truncated forms and cyanogen bromide fragments of the inhibitor. *FEBS Lett.* **243,** 234–238.
40. Malafosse, A., Lehesjoki, A. E., Genton, P., Labauge, P., Durand, G., Tassinari, C. A., Dravet, C., Michelucci, R., and de la Chapelle, A. (1992). Identical genetic locus for Baltic and Mediterranean myoclonus. *Lancet* **339,** 1080–1081.
41. McInnis, M. G., Chakravarti, A., Blaschak, J., Petersen, M. B., Sharma, V., Avramopoulos, D., Blouin, J. L., Konig, U., Brahe, C., Matise, T. C., Chakravarti, A., and Antonarakis, S. E. (1993). A linkage map of human chromosome 21:43 PCR markers at average intervals of 2.5 cM. *Genomics* **16,** 562–571.
42. Meldium, B. S., and Bruton, C. J. (1992). Epilepsy. *In* "Greenfield's Neuropathology"(J. H. Adams and L. W. Duchen, Eds.) 5th ed., pp. 1246–1283. E. Arnold, London.
43. Norio, R., and Koskiniemi, M. (1979). Progressive myoclonus epilepsy: genetic and nosological aspects with special reference to 107 Finnish patients. *Clin. Genet.* **15,** 382–398.
44. Pennacchio, L. A., Lehesjoki, A. E., Stone, N. E., Willour, V. L., Virtaneva, K., Miao, J., D'Amato, E., Ramirez, L., Faham, M., Koskiniemi, M., Warrington, J. A., Norio, R., de la Chapelle, A., Cox, D. R., and Myers, R. M. (1996). Mutations in the gene encoding cystatin B in progressive myoclonus epilepsy. *Science* **271,** 1731–1734.
45. Pennacchio, L. A., and Myers, R. M. (1997). Isolation and characterization of the mouse cystatin B gene. *Genome Res.* **6,** 1103–1109.
46. Rawlings, N. D., and Barrett, A. J. (1990). Evolution of proteins of the cystatin superfamily. *J. Mol. Evol.* **30,** 60–71.
47. Ritonja, A., Machleidt, W., and Barrett, A. J. (1985). Amino acid sequence of the intracellular cysteine proteinase inhibitor cystatin B from human liver. *Biochem. Biophys. Res. Commun.* **131,** 1187–1192.
48. Sato, N., Ishidoh, K., Uchiyama, Y., and Kominami, E. (1990). Molecular cloning and sequencing of cDNA for rat cystatin beta. *Nucleic Acids Res.* **18,** 6698.
49. Schnabel, R., and Gootz, M. (1971). On the substructure of myoclonus bodies in progressive myoclonus epilepsy. *Acta Neuropathol.* **18,** 17–33.
50. Scott, H. S., Chen, H., Rossier, C., Lalioti, M. D., and Antonarakis, S. E. (1997). Isolation of a human gene (HES1) with homology to an Escherichia coli and a zebrafish protein that maps to chromosome 21q22.3. *Hum. Genet.* **99,** 616–623.
51. Serratosa, J. M., Delgado-Escueta, A. V., Posada, I., Shih, S., Drury, I., Berciano, J., Zabal, J. A., Antunez, M. C., and Sparkes, R. S. (1995). The gene for progressive myoclonus epilepsy of the Lafora type maps to chromosome 6q. *Hum. Mol. Genet.* **9,** 1657–1663.
52. Stone, N. E., Fan, J. B., Willour, V., Pennacchio, L. A., Warrington, J. A., Hu, A., de la Chapelle, A., Lehesjoki, A. E., Cox, D. R., and Myers, R. M. (1996). Construction of a 750-kb bacterial clone contig and restriction map in the region of human chromosome 21 containing the progressive myoclonus epilepsy gene. *Genome Res.* **6,** 218–225.
53. Stubbs, M. T., Laber, B., Bode, W., Huber, R., Jerala, R., Lenarcic, B., and Turk, V. (1990). The refined 2.4 A X-ray crystal structure of recombinant human stefin B in complex with the cysteine proteinase papain: a novel type of proteinase inhibitor interaction. *EMBO J.* **9,** 1939–1947.
54. Talbot, C. C., Jr., Avramopoulos, D., Gerken, S., Chakravarti, A., Armour, J. A., Matsunami, N. White, R., and Antonarakis, S. E. (1995). The tetranucleotide repeat polymorphism D21S1245 demonstrates hypermutability in germline and somatic cells. *Hum. Mol. Genet.* **4,** 1193–1199.
55. Thiele, U., Assfalg-Machleidt, I., Machleidt, W., and Auerswald, E. A. (1990). N-terminal variants of recombinant stefin B: effect on affinity for papain and cathepsin B. *Biol. Chem. Hoppe Seyler* **371** (Suppl), 125–136.
56. Turk, V., and Bode, W. (1991). The cystatins: protein inhibitors of cysteine proteinases. *FEBS lett.* **285,** 213–219.
57. Turk, B., Krizaj, I., and Turk, V. (1992). Isolation and characterization of bovine stefin B. *Biol. Chem. Hoppe Seyler* **373,** 441–446.
58. Unverricht, H. (1891). "Die Myoclonie," pp. 1–128. Franz Deuticke, Leipzig.
59. Virtaneva, K., Miao, J., Traskelin, A. L., Stone, N., Warrington, J. A., Weissenbach, J., Myers, R. M., Cox, D. R., Sistonen, P., de la

Chapelle, A., and Lehesjoki, A. E. (1996). Progressive myoclonus epilepsy EPM1 locus maps to a 175-kb interval in distal 21q. *Am. J. Hum. Genet.* **58,** 1247–1253.

60. Virtaneva, K., D'Amato, E., Miao, J., Koskiniemi, M., Norio, R., Avanzini, G., Franceschetti, S., Michelucci, R., Tassinari, C. A., Omer, S., Pennacchio, L. A., Myers, R. M., Dieguez-Lucena, J. L., Krahe, R., de la Chapelle, A., and Lehesjoki, A. E. (1997). Unstable minisatellite expansion causing recessively inherited myoclonus epilepsy, EPM1. *Nature Genet.* **15,** 393–396.
61. Yamakawa, K., Mitchell, S., Hubert, R., Chen, X. N., Colbern, S., Huo, Y. K., Gadomski, C., Kim, U. J., and Korenberg, J. R. (1995). Isolation and characterization of a candidate gene for progressive myoclonus epilepsy on 21q22.3. *Hum. Mol. Genet.* **4,** 709–716.
62. Yamakawa, K., Gao, D. Q., and Korenberg, J. R. (1996). A periodic tryptophan protein 2 gene homologue (PWP2H) in the candidate region of progressive myoclonus epilepsy on 21q22.3. *Cytogenet. Cell Genet.* **74,** 140–145.
63. Yaspo, M. L., Gellen, L., Mott, R., Korn, B., Nizetic, D., Poustka, A. M., and Lehrach, H. (1995). Model for a transcript map of human chromosome 21: isolation of new coding sequences from exon and enriched cDNA libraries. *Hum. Mol. Genet.* **4,** 1291–1304.

CHAPTER 49

Microsatellite Instability in Colon Cancers

JULIE M. CUNNINGHAM, LISA A. BOARDMAN, LAWRENCE J. BURGART, AND STEPHEN N. THIBODEAU

Department of Laboratory Medicine and Pathology, Mayo Clinic, Rochester, Minnesota 55905

I. INTRODUCTION

Our understanding of the molecular pathogenesis of many human cancers has improved significantly in the last several decades. In general, malignant transformation is accompanied by the accumulation in the last several decades. In general, malignant transformation is accompanied by the accumulation of multiple genetic alterations, including the activation of dominantly acting proto-oncogenes and the inactivation of tumor suppressor genes (TSG) (reviewed in [1, 2]). In addition, Nowell *et al.* [3] suggested some time ago that a more generalized genomic instability would contribute to the accumulation of multiple genetic defects, and that an increased rate of acquired mutation would accelerate the development of cancer. Based on the probability of spontaneous mutations, Loeb [4, 5] proposed that a "mutator phenotype" would be required for the multistep process of carcinogenesis.

Maintenance of fidelity during the processes of DNA replication, DNA repair, cell division, cell differentiation, and other cellular functions is a hallmark of a normally functioning cell. Cheng and Loeb [4] have postulated that mutations or exogenous interferences within these critical pathways could lead to genomic instability, manifested as an increase in the frequency of mutations, karyotypic alterations, and rearrangements. Karyotypic instability, characterized by chromosomal losses/gains and structural rearrangements, has long been recognized as a form of genomic instability in solid tumors. Because there is a characteristic progression from the premalignant state to carcinoma, and because there is an abundance of tissues for study, colorectal cancer has provided a singularly useful model for the study of the molecular genetic changes that accompany the transformation from a normal to fully malignant state. In colon cancer, examples of genes implicated in the process of tumorigenesis include the dominantly acting proto-oncogene K-*ras* and the TSGs *APC, p53,* and *DCC* on chromosomes 5, 17, and 18, respectively (reviewed in [6]). Although significant progress has been made in this particular area, a host of genes are involved in the process of carcinogenesis for colon cancer, many of which have yet to be discovered. However, our current understanding of those genes described and the roles they play in both normal and malignant cells has provided a solid framework upon which to expand.

Recently, recognition of alternate pathways for the development of colorectal cancer has begun to emerge. As often happens in science, the convergence of several key areas of research led to this discovery. These include: the observation of microsatellite instability (MIN) in colorectal tumors, the finding of linkage in hereditary nonpolyposis colorectal cancer (HNPCC) along with the observation of microsatellite instability in tumors from these patients, and finally, the elucidation of the mechanisms of DNA repair in bacterial and yeast cells. Work in these areas culminated in the identification of a class of "mutator genes," first predicted by Loeb, which are now known to be involved in DNA mismatch repair.

II. CONVERGENCE OF SCIENCE

A. Microsatellite Instability

Microsatellites are tandem repeats of simple sequences that occur abundantly and are randomly interspersed throughout the human genome. They consist of 10–50 copies of 1- to 6-base-pair motifs, and are characterized by a high degree of polymorphism. Despite the variability observed among individuals, microsatellites are replicated faithfully at each cell division with a new mutation rate of approximately 10^{-3}–10^{-5} [7]. Initially described by Weber and May [8], microsatellites have become the most useful class of DNA polymorphisms for linkage analysis [9] and have been extremely valuable in analyzing tumor DNA for loss of heterozygosity (LOH). It is estimated that there are on the order of 50,000–100,000 (CA)*n* elements scattered throughout the human genome.

In May 1993, three simultaneous reports described a novel type of genomic instability, which has been called microsatellite instability (MIN) or replication error (RER), in a subset of sporadic colorectal cancers and in tumors from patients with HNPCC [10–12]. This type of instability is characterized by alterations in the size (expansion or contraction) of microsatellite DNA within the tumor DNA. Examples of instability at the p53 locus in colon cancer are shown in Fig. 49-1. Interestingly, the alterations observed within the microsatellites appeared to be nonspecific, that is, alterations at many

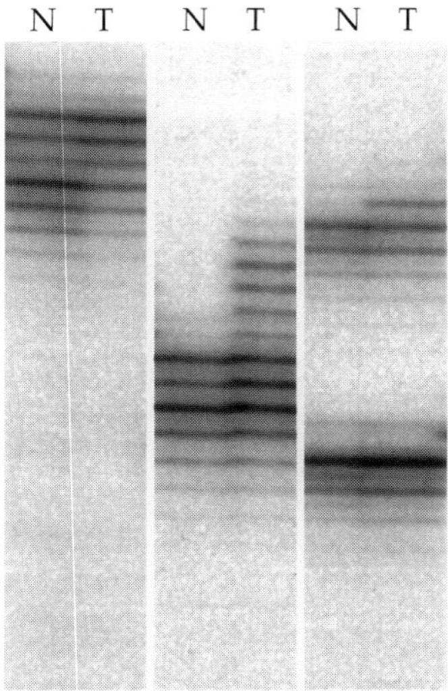

FIGURE 49-1 Examples of LOH and microsatellite instability patterns 1 and 2 at the *p53* locus in three colorectal cancers. N, normal DNA; T, tumor DNA.

loci were observed, irrespective of chromosome or chromosomal locations. Furthermore, the presence of microsatellite instability was associated with the presence of tumor in the proximal colon, improved patient survival, and was inversely correlated with LOH for chromosomes 5q, 17p, and 18q, suggesting that these tumors arose through a mechanism other than that involving LOH [10].

The report by Ionov *et al.* [11] described the finding of somatic deletions in poly (dA/T) sequences and other simple repeats in 12% of colorectal carcinomas. They estimated that these tumors could carry more than 100,000 such mutations and concluded that a previously undescribed form of carcinogenesis might result from a mutation in a DNA replication factor.

Although these findings suggested the presence of a "mutator phenotype" predicted by Loeb, the genes and pathways responsible for such a phenomenon were unknown. The downstream effects of such an abnormality were also unknown. However, based on work with the inherited disorders characterized by trinucleotide repeat expansion, such as Fragile X syndrome and myotonic dystrophy, it was postulated that the instability of microsatellites associated with certain critical (and potentially novel) genes might play a role in tumor initiation and progression.

B. The Genetic Basis of Hereditary Nonpolyposis Colorectal Cancer

One-fifth of adenomatous polyps and colorectal cancers are thought to occur in familial aggregations [13,

14], and a positive family history of colorectal cancer poses an increased risk factor for the disease. The initial observation for the presence of familial aggregation came earlier this century when Aldred S. Warthin undertook a study after learning of a high frequency of cancer in the family of a seamstress, who expressed a fear of suffering the same fate. Following the review of pathology records from 1895 to 1917, Warthin identified familial clustering of colorectal cancers in a number of patients [15]. The family of the seamstress (designated Family "G" in [16]), the prototype cancer family, also had an excess of uterine and gastric cancers, and one branch of the family had a high frequency of hematological malignancies [16]. After more than 50 years, Lynch described the "cancer family syndrome" [16] in which colorectal cancer is the most predominant malignancy. The syndrome was termed "hereditary nonpolyposis colorectal cancer" to distinguish it from the multiple adenomas of familial adenomatous polyposis. The extracolonic manifestations include cancers of the endometrium, stomach, small intestine, urinary tract, and ovary. The terms Lynch syndromes I and II were proposed to describe the absence (Lynch I) or presence (Lynch II) of extracolonic malignancies [17].

Unlike Familial Adenomatous Polyposis, there are no convenient biologic markers to aid in the diagnosis of this disorder. In this particular case, the diagnosis is almost entirely dependent on the family history. As a result criteria to formally identify hereditary nonpolyposis colorectal cancer (HNPCC) families were developed by the International Collaboration Group on Hereditary Non-Polyposis Colorectal Cancer in 1991 [18, 19]. Termed the "Amsterdam criteria," they are: histologically verified colorectal cancer in three or more relatives, two of whom are first-degree relatives; colorectal cancer involving at least two generations; and one or more colorectal carcinoma cases diagnosed before the age of 50 years. Although the Amsterdam criteria helped to create a more homogeneous population for research, it also underestimated the true numbers that might have such a diagnosis. Factors contributing to missed diagnosis include the failure of Amsterdam criteria to take into account extracolonic tumors, poor documentation of diagnosis or causes of death in some families, and an increasing number of small sized families [20].

Evidence for an underlying genetic factor in colorectal cancer came from segregation studies of 11 HNPCC families, which favored an autosomal-dominant mode of inheritance [21]. Analysis of the pattern of transmission of colorectal tumors in multiple studies confirmed that an automsomal-dominant transmission of a single gene could result in susceptibility to both cancer and polyps [13, 22–24].

Cytogenetic studies provided few clues to the location of the gene(s) for HNPCC. After excluding candidate regions on chromosomes 18q and 5q [25, 26], both implicated in the development of sporadic colorectal cancer, a genome-wide search by linkage analysis was undertaken. This later approach resulted in the localization of an HNPCC susceptibility locus on chromosome 2p [27]. Close linkage of disease to microsatellite markers on chromosome 2p (2p16-p15) was found in two large HNPCC kindreds having colorectal cancer with and without endometrial cancer. Of an additional 14 smaller kindreds, 11 displayed varying degrees of linkage to this area, while in three families linkage was formally excluded [12]. In November 1993, linkage to a second locus on chromosome 3 was reported [28]. Linkage studies, therefore, revealed at least two separate loci associated with susceptibility to HNPCC, one localized to chromosome 2 and the other to chromosome 3. However, some families with HNPCC failed to demonstrate linkage to either of the two chromosomal regions [29], suggesting that additional loci were involved.

Suspecting that the susceptibility loci for HNPCC would be a TSG, Aaltonen *et al.* [12] examined tumors for LOH within the chromosome 2-linked region. Although LOH was not detected, tumors did exhibit the phenomenon of microsatellite instability, which these authors called replication error or RER. These data suggested that the genetic susceptibility loci for HNPCC might also be responsible for the genetic instability in a subset of sporadic colorectal cancer, and that the underlying genetic bases of these observations were related.

C. DNA Mismatch Repair

1. Bacterial DNA Repair Systems

The *Escherichia coli* methyl-directed DNA mismatch repair system, MutHLS, is responsible for the repair of most nucleotide mispairs in *E. coli* [30, 31]. Mismatch repair is the recognition and repair of nucleotides that are unpaired or paired with a noncomplementary nucleotide [30]. Repair appears to occur by an excision/resynthesis mechanism that replaces one of the DNA strands containing an incorrectly paired nucleotide. In *E. coli,* this repair system requires several *mut* gene products; MutH, MutL, MutS, and MutU along with single-stranded DNA binding protein, exonuclease I, exonuclease VII, RecJ exonuclease, DNA polymerase III holoenzyme, and DNA ligase (reviewed in [30]). MutS recognizes and binds to mispaired bases. MutL interacts with the MutS–DNA complex, activating MutH, an inefficient endonuclease, which makes a single strand break at hemimethylated GATC sites in DNA on the newly replicated, unmethylated strand. MutH can be replaced by single strand breaks in DNA [32, 33]. Excision requires UvrD (helicase II) and one of the single strand DNA exonucleases (exo 1, exo VII, or RecJ), depending on whether the strand break is 5′ or 3′ to the mispair. Once excision is complete, resynthesis is mediated by DNA polymerase III holoenzyme, single strand binding protein, and DNA ligase.

2. DNA Mismatch Repair in Yeast

Six yeast homologues of the *E. coli* MutS gene (*MSH1-6*) as well as two MutL homologues (*MLH1* and *PMS1*) have been identified in *Saccharomyces cerevisiae* [34–36]. The two MutL homologues are required in yeast to produce the function of the prokaryotic MutL protein, indicating the complexity of mismatch repair in eukaryotes [45]. Studies in yeast have concluded that each of the MutS homologues has a highly restricted function [47]. *PMS1, MLH1,* and *MSH2* are important for correction of mistakes arising during nuclear DNA replication and recombination, while others are critical components in mitochondria (*MSH1*) and recombination pathways (*MSH3* and *MSH6*) [35, 42–46, 48]. The MSH2 protein specifically binds to mismatched DNA base pairs and insertions [49], and mismatch-containing complexes involving MSH2, PMS1, and MLH1 have been identified in yeast, suggesting that, as in *E. coli,* formation of a complex between the yeast MSH2(MutS) and a mispaired base could result in recruitment of PMS1 and MLH1 (MutL) [43, 63]. The yeast MutS homologues MSH6 and MSH3 increase the specificity of MSH2 for single nucleotide G/T and insertion/deletion loop mismatches, respectively [39, 50, 51]. However, there is some overlap in the specificity of the complexes, implying a functional redundancy. While mutations in either *MSH6* or *MSH3* result in relatively mild effects [46], yeast cells harboring null mutations in both genes exhibited microsatellite instability similar to that seen in the cells with a mutant *MSH2* gene [50].

3. Mutator Phenotype

In both bacteria and yeast cells, mutations in the DNA mismatch repair genes result in strong mutator phenotypes, as reflected by increased mutation rates and increased postmeiotic segregation of heterozygous markers. *Escherichia coli* strains with *mutS* and *mutL* mutations, which lead to defective mismatch repair, had 13-fold elevated frequencies of repeat tract instability [52]. Additionally, Strand *et al.* [48] showed that tracts of simple repetitive DNA in yeast were destabilized in yeast cells bearing mutations in any of three genes involved in DNA mismatch repair. Based on these observations, Strand *et al.* [48] proposed that the microsatellite instability observed in some colorectal tumor DNA reflected mutations in genes involved in DNA mismatch repair.

III. MISMATCH REPAIR GENES AND DISEASE/CANCER

A. Cloning of the Human DNA Mismatch Repair Genes

As a result of the series of observations just described, it becomes evident that those genes involved in DNA mismatch repair were ideal candidate genes for the HNPCC susceptibility loci, as well as explaining the phenomenon of tumor microsatellite instability. Utilizing a candidate gene approach, the human homologues of the bacterial and yeast mismatch repair genes were very quickly cloned and localized, and subsequently examined for their involvement in HNPCC [53–56]. Fishel *et al.* [53] and Leach *et al.* [54] were the first to clone the *hMSH2* gene and demonstrate the presence of germline mutations in at least some families with HNPCC.

Several genes have now been identified that encode proteins homologous to components of the bacterial MutHLS system. As in yeast, there are multiple human homologues of the prokaryotic *MutL* (*hMLH1, hPMS1, hPMS2,* and *hPMS3–8*) and *MutS* genes (*hMSH2, hMSH3,* and *hMSH6* (*GTBP*)), which reside on chromosomes 3p21.3, 2q31-33, 7p22, 7q11/7q22, 2p22-p21, 5q11-12, and 2p16, respectively (reviewed in [40]). Germline mutations within both *hMLH1* and *hMSH2* have now been found in a large number of HNPCC kindreds, strengthening the case for the importance of mismatch repair gene in the etiology of familial colorectal cancer [56, 57].

In human cells, DNA mismatches appear to be recognized by heterodimers hMutSα and hMutSβ, comprising hMSH2 with *GTBP*(*hMSH6*) or *hMSH3,* respectively [58–61]. The functional redundancy of hMSH3 and hMSH6 protein may explain why mutations in *hMSH2* are the more prevalent. Heterodimers of hPMS2 and hMLH1, hMUTLα, have also been described [62]. Mutations in any one of the heterodimer components may lead to defective mismatch repair.

It is clear that hMSH2 protein plays a pivotal role in the recognition of base mismatches and certain types of DNA damage. hMSH3, hMSH6, and possibly other interacting molecules can modify the association and specificity of the recognition process, although the precise roles of the different protein homologues, as well as their alternate transcripts, are not completely understood. Alternate transcripts of both *hMLH1* and *hMSH2* have been reported, and appear to be present in both unaffected and affected individuals [63–65]. The biological significance of these variants has yet to be elucidated. Results from one study suggest that at least one variant may be associated with development of cancer in at-risk patients. Individuals with cancer or high-grade dysplasia in the setting of chronic ulcerative colitis (CUC) have a significantly higher frequency of an *hMSH2* variant than either CUC patients without cancer or HGD, or normal blood bank controls [66]. The authors concluded that this variant may be associated with the progression from HGD to cancer in patients with CUC; the variant alone does not confer a phenotypic effect (cancer) unless a second event (CUC) takes place.

3. Inactivation of DNA Mismatch Repair System

1. Gene Mutation

Of the human mismatch repair genes identified to date, the majority of investigations have focused on the mutation analyses of *hMSH2* and *hMLH1*. For HNPCC, current data suggest that approximately 30% of families have germline mutations in *hMSH2* while another 30% have mutations in *hMLH1* [57, 67, 68]. Most *hMSH2* mutations consist of frameshift or nonsense changes, which result in truncated protein products that are presumably functionally defective [67]. Mutations in *hMLH1,* on the other hand, include both frameshift and missense changes. Splice-site defects are common in *hMLH1,* but less frequent in *hMSH2*. These generally result in exon-skipping or production of aberrant transcripts when cryptic splice sites are activated. The *hMSH2* mutations are spread throughout the coding sequence and appear to be randomly distributed, and not necessarily associated with the highly conserved amino acid domains, although there may be some clustering in exon 12 [57]. Wijnen *et al.* [69] reported a clustering of *hMLH1* mutations at exons 15–16, comprising 50% of mutations, the remainder scattered throughout the gene.

Germline mutations in *hMSH2* and *hMLH1* have also been detected in kindreds with the Muir–Torre syndrome, a disorder characterized by the development of both sebaceous and visceral tumors, the latter most commonly being colorectal [70–72]. Mutations in *hMSH2* have also been found in some cases of Turcot's syndrome, a clinical disorder of an individual with a primary brain tumor and multiple colorectal adenomas [73]. Other patients with Turcot's syndrome, however, have germline mutations in the *APC* gene [73]. Although *hMSH2* and *hMLH1* have been analyzed in a large number of patients, analyses of the remaining genes involved in DNA mismatch repair have not been as extensive. Mutations in *hPMS1* and *hPMS2* have been occasionally observed in HNPCC families [74] and in patients with Turcot's syndrome [73]. The limited availability of mutation information for *hPMS1* and *hPMS2,* however, makes it difficult to determine the role of these two genes in HNPCC.

While the role of defective mismatch repair in some families with HNPCC has been well established, the contribution of defects in this pathway for sporadic cancers are not as clear cut. Between 12 and 28% of sporadic colorectal carcinomas exhibit widespread microsatellite instability [10–12, 67, 75, 76]. Current data suggest that mutations in the mismatch repair genes, either germline or somatic, may be infrequent in these cancers [67, 75, 77–81]. Somatic mutations in *hMLH1* have been reported in 10 to 15% of MIN+ colorectal tumors [77, 78]. Mutations in *hMSH2* have been reported to be infrequent in some studies [77, 79] and present in up to 30% of sporadic MIN+ colon tumors in another study [75]. The presence of somatic mutations in *hMSH3* and *hMSH6* has also been infrequent. In one study, a frameshift mutation in *hMSH3* was found in one of sixteen endometrial carcinomas [82], while in another, a mutation in *hMSH6* was reported in a tumor cell line established from a colorectal cancer patient, but the mutation was not detectable in the primary tumor itself [83].

Germline mutations, therefore, occur almost exclusively in HNPCC patients (or other related familial cancer syndromes), while the majority of those mutations found in sporadic colorectal cancer have proven to be somatic. However, the frequency of somatic mutations observed for either *hMSH2* or *hMLH1* in MIN+ sporadic colon cancer is lower than anticipated, suggesting the use of inadequate detection methods in some studies, the involvement of other mismatch repair genes, or other types of genetic alteration.

2. Abnormal Methylation

Altered DNA methylation is one means by which expression and function of a gene may be altered in the absence of a specific gene mutation. The ability of methylation to block transcriptional activation is well documented, particularly within CpG-rich promoters. There is an inverse correlation between the level of gene expression and the extent of DNA methylation at gene promoter regions [84]. Low levels of DNA methylation tend to be seen at the promoter regions of actively transcribed genes, while heavily methylated 5′ regions are found in transcriptionally silent genes [85]. Methylation is thought to play a pivotal role in genomic imprinting and in X-chromosome inactivation, and there is a growing appreciation of a role in the carcinogenic process [86]. Almost all reported cellular oncogenes and tumor suppressor genes have constitutively nonmethylated CpG-rich regions at their 5′ ends in normal cells [84], and at least in some instances, methylation of these CpG islands has been associated with defective expression in the absence of gene mutations (reviewed in [87, 88]. Conversely, hypomethylation of proto-oncogenes has also been described [88].

The possibility that the mistmatch repair genes may be subject to altered expression in the absence of gene mutation was suggested by a study of the relationship between expression of either hMLH1 or hMSH2 proteins and detection of mutations within the respective genes [89]. As noted previously, a number of MIN+ sporadic tumors have no detectable mutations in either *hMSH2* or *hMLH1*. These tumors may have mutations in other mismatch repair genes, may have mutations not detected by the assays employed, or may have altered methylation of promoter sequences for the mismatch repair genes. An examination of the expression of mismatch repair proteins immunohistochemically demonstrated that nearly all MIN+ tumors lacked expression of either hMLH1 or hMSH2, suggesting that other mismatch repair genes are not involved [89]. In another study, Kane *et al.* [90] demonstrated that methylation of the *hMLH1* promoter correlated with a lack of expression in a few sporadic colorectal carcinomas and human tumor cell lines. Four colorectal tumors and two tumor cell lines (SW480 and AN3CA) did not express *hMLH1* and did not harbor mutations within the coding region. Cumulatively, these data suggest that DNA methylation may be another mode of inactivation of the mismatch repair genes, although more extensive studies are required to determine whether or not this is a significant mechanism of mismatch repair protein inactivation in colorectal cancer.

C. Downstream Effects of Defective Mismatch Repair

The finding that a subset of colorectal tumors are characterized by defective mismatch repair lead to the realization that there may be at least two distinct pathways responsible for the process of tumorigenesis in this tumor system. The first pathway is characterized by gross chromosomal changes, resulting in chromosomal rearrangements and aneuploidy, LOH, and the frequent involvement of TSGs such as *APC* and *p53* [1, 2, 6]. The second pathway is defined by the presence of defective mismatch repair, tumor microsatellite instability, and tumors that are near diploid. For those tumors characterized by defective mismatch repair, one of the issues has been to define which target genes would be affected downstream and what role they would play in the neoplastic process. Overall, defective mismatch repair has been shown to increase the overall mutation rate of a cell, and cell lines with defects in mismatch repair have been found to have substantially elevated rates of spontaneous mutation [91]. Examination of the hypoxanthine guanine phosphoribosyl transferase (hrpt) gene has been used as an indicator of overall mutation frequency and has revealed substantially elevated rates of mutation in cells with defective mismatch repair genes [92–95]. The nature of the mutations however, may depend upon which of the mismatch repair genes is defective. Cells with *hMSH2* or *hMLH1* mismatch repair defects had relatively high rates of frameshift mutations, while high frequencies of single base substitutions were more common in cells with altered *GTBP* genes [93, 95].

Thus, given the nature of the "mutator" genes, several possibilities exist for potential downstream consequences. One possibility is that defective mismatch repair increases the cellular mutation rate, resulting in mutations in those genes previously described for colorectal cancer, such as K-*ras, APC,* and *p53*. When examined, the frequency of involvement for K-*ras* appeared to be similar between MIN+ and MIN− tumors in one study [96], but others have reported less frequent alteration in MIN+ tumors [11, 97]. However, several studies have suggested that altered *p53* is observed less frequently in MIN+ tumors [96–99]. When such mutations do occur in MIN+ tumors, there is evidence to suggest that the mutations are of a distinct nature. Additionally, utilizing a mouse model harboring germline mutations in Apc (Min mouse), Reitmar *et al.* [100] have demonstrated that the mechanism of inactivation of the wild-type Apc allele is dependent on the MSH2 status. In MSH2-deficient crosses, inactivation of the wild-type Apc allele occurs in the absence of LOH, suggesting that somatic mutations are responsible for the tumorigenic process. In another study, Lazar *et al.* [101] described six different *APC* mutations, as well as four different *p53* mutations, in a single patient with a germline *hMSH2* mutation, while another report found an excess of *APC* frameshift mutations in MIN+ cases as compared to MIN− tumors [102]. However, no functional assays have been performed to determine whether these mutations are functionally important.

Another possibility is that certain genes may be more susceptible to mutations incurred by defective mismatch repair, especially those possessing repetitive tracts of DNA. Although the function of repetitive units in many cases is still unknown, instability of microsatellites (primarily in the form of repeat expansion) has been shown to be involved in human disease. Heritable unstable DNA elements have been identified as the basis of disease for a number of inherited disorders, such as Fragile X syndrome, spinal and bulbar muscular atrophy, myotonic dystrophy, and Huntington's disease (see other chapters in this book). A computer search for microsatellites associated with genes shows the presence of a large number of such loci. Thus, alterations within repetitive units of DNA could alter the function of a wide spectrum of genes.

In support of this hypothesis, a repetitive tract within the transforming growth factor beta type II receptor (TGFβRII), which plays an important role in negative growth regulation, has been shown to be frequently altered in tumors with defective mismatch repair [103]. This and subsequent studies have shown that such mutations occur in 80–90% of colorectal cancers with microsatellite instability [104–106]. A similar frequency was observed in MIN+ gastric cancers, but the alterations were infrequent in MIN+ endometrial cancers [107]. Importantly, a decrease in the expression of the TGF-βRII mRNA was shown in the mutation positive cases, implying a functional consequence of the mutations detected [103]. These inactivating mutations likely cripple the TGF-β pathway, resulting in loss of growth regulation, and provide a link between DNA repair defects and a specific pathway of tumor progression. Other genes that have demonstrated instability within repetitive units include *E2F-4* [108], the *insulin-like growth factor II receptor* (*IGFIIR*) [109, 110], and the *BAX* [111] genes. Unlike the *TGFβIIR* gene, however, the functional significance of microsatellite instability in these genes is at this point unknown.

IV. TUMOR MICROSATELLITE INSTABILITY

A. Tumor Phenotype

In spite of the studies published to date, there is still confusion as to how one defines tumor microsatellite instability. That is, how many markers should be utilized, which markers should be utilized, and how many markers must display instability before a tumor is defined as being MIN+ (RER+). As a result, there is nonuniformity in study design, and considerable variability in the frequency of microsatellite instability reported for the different tumor types.

Microsatellite instability, as it relates to neoplastic disorders, is a phenomenon widespread throughout the human genome. It involves either the insertion or the deletion of one or more copies of the repetitive unit (e.g., mono- di-, trinucleotide repeat) within the microsatellite being examined. In PCR-based assays, instability can be detected when comparing DNA derived from tumor to that derived from normal tissue. It can be seen either as a novel discrete band in tumor DNA not observed in the corresponding normal DNA (which we have called pattern 2) or as a marked alteration in repeat length, often heterogeneous in nature and appearing as a ladder, which we have called pattern 1 (Fig. 49-1). The pattern of instability and the frequency that a given microsatellite will display instability appears to be marker dependent. For example, some microsatellites are characterized by a high level of instability (i.e., frequently altered in MIN+ tumors) and show predominantly pattern 1 changes, while others are characterized by a low level of instability and show predominantly pattern 2 changes. Others, yet, rarely demonstrate instability, even in those tumors known to be DNA mismatch repair deficient (unpublished observations). The likelihood that a microsatellite will be susceptible to instability may be related to the inherent mutation rate at that locus. Weber and Wong [7] have shown marked variability in the new mutation rates of di-, tri-, and tetranucleotide repeats in CEPH families. We have also observed that those microsatellites that are not polymorphic do not show instability. Given this information, selection of markers will certainly influence the frequency of instability detected for any given tumor system.

The number of markers demonstrating MIN varies for any given tumor. The data so far suggest the presence of several different tumor phenotypes. One of these is characterized by a low (but higher than normal) new mutation rate at the specific microsatellite being examined. A study of 196 non-HNPCC related tumors demonstrated that the incidence of such mutations at specific short tandem repeat loci may range from 0.5% (dinucleotide repeat) to 0.5–2.6% (tri- and tetranucleotide repeats) [112]. In some tumor systems, the frequency of MIN can be quite low. For example, in one study on prostate cancer, 135 microsatellite loci were examined in 55 prostate adenocarcinomas [113]. Overall, only 22 alterations were observed among 5803 genotypes. For any given tumor, therefore, the number of microsatellites exhibiting such alterations will be limited.

Another phenotype is characterized by a higher new mutation rate. An example of this second tumor phenotype may be the microsatellite instability observed in small cell lung carcinoma [114]. A third phenotype is characterized by marked alterations (expansion or contraction) at multiple microsatellite loci. Colon cancer may in fact reflect the presence of these later two tumor phenotypes. Our data [10, 118] suggest that about 15% of colorectal cancer show instability (often as discrete bands) at only a minority of loci (≤30% of markers tested). We have referred to these tumors at MIN+A. Another 15% of tumors, which we have called MIN+B, show marked alterations at a majority of microsatellites examined (>30% of markers tested). Other studies have also shown a bimodal distribution within MIN+ tumors, with cutoffs at 30–50% [76, 115–117]. Data from Thibodeau *et al.* [118], and data of Dietmar *et al.* [115], suggest that the MIN+B phenotype is the result of defective mismatch repair and that MIN+B show distinct clinical and pathologic associations [76, 117, 119], while this is not the case for the MIN+A tumors. For nearly all clinical and pathological parameters examined, the MIN+A tumors resemble the MIN− cases. When assessing tumors for microsatellite instability, therefore, it may be important to take into consideration the frequency of altered microsatellites per tumor. Tumors characterized by different frequencies may have different underlying mechanisms responsible for the instability.

B. Clinical and Pathological Aspects of MIN+ Colorectal Cancer

For both sporadic and HNPCC colon cancers, tumors associated with significant microsatellite instability (defined as MIN+B) have demonstrated certain clinical and histopathological characteristics. These tumors are primarily located proximal to the splenic flexure, are less likely to exhibit LOH, and tend to be near diploid [10, 12, 76, 81, 119, 120]. Left-sided colorectal tumors with microsatellite instability, however, do not necessarily have extracellular mucinous production, nor less frequent p53 expression [121]. Of interest is the observation that, in general, mucinous tumors have a low frequency of *p53* mutations [122], suggesting that such tumors arise independently of alterations in *p53,* consistent with the alternate "mutator phenotype" pathway.

Grossly, colorectal tumors with widespread microsatellite instability (i.e., MIN+B) show a marked preponderance of exophytic growth and tend to be large with mucoid glistening [119]. Histopathologically, these tumors tend to be poorly differentiated with a solid, uniquely cribiformed or signet ring growth pattern. Desmoplasia is diminished in these tumors. Tumor-infiltrating lymphocytes are typically prominent while extracellular mucin overproduction or a Crohn's-like inflammatory infiltrate is variably present. These histologic features are most prominent in right-sided cancer [81, 119, 120, 123, 124]. In fact, proximal to the splenic flexure, histopathology alone has an 84% positive predictive value for detecting colon cancers with significant MIN(MIN+B) [123].

In addition to the correlations just noted, the presence of tumor microsatellite instability in colon cancer may also have prognostic significance. HNPCC patients show improved survival compared to sporadic colorectal cancer as a whole [125], and in several reports, patients with MIN+ colorectal tumors have shown an improved overall survival [10, 76, 81]. HNPCC patients with germline *hMLH1* mutations were shown to have a better survival than those without such mutations, supporting the idea that tumors characterized by defective mismatch repair behave more indolently [126]. In this study, the survival rates of 175 HNPCC patients were compared with those of 14,000 patients with sporadic colorectal cancer diagnosed at <65 years of age. Of the HNPCC patients, 120 were from families segregating a germline *hMLH1* mutation. The overall 5-year survival has 65% for patients with HNPCC compared to 44% for patients with sporadic cancer. At this stage, it will be important to carry out further analyses to definitively determine the prognostic significance of tumor MIN.

C. MIN in Other Tumors

Microsatellite instability is not unique to tumors of the colon. The presence of MIN has been reported in a variety of malignancies, including cancer of the endometrium, stomach, pancreatobiliary system, ovary, small intestine, and renal pelvis and ureters, each of which are part of the HNPCC tumor spectrum (Table 49-1). As is the case for colon cancer, however, MIN in other neoplasms is thought to be nonspecific and secondary to possible common pathways, such as defective mismatch repair. The underlying molecular basis for the MIN in many of these other tumor systems, however, has not been clearly established. In addition, because of differences in the criteria utilized to define MIN, differences in marker selection, and differences in the total number of markers utilized, a comparison of the various studies is quite difficult.

In sporadic gastric adenocarcinomas, the frequency of microsatellite instability (using only widespread MIN as the criteria) has been reported to vary between 0 in young patients [127] to 4–39% [128–135, 152, 185]. MIN+ tumors may be associated with a distinct histopathological phenotype [128, 129, 132–135], are localized to the distal stomach [129, 133] and show improved survival [131, 135]. No instability was observed in EBV-related gastric cancers [130], and although not conclusive given the low number of tested loci, it suggests that EBV-related gastric cancer may arise through a pathway other than DNA mismatch repair defect.

Multiple studies of endometrial cancers from patients with no family history of HNPCC have reported rates of microsatellite instability from 9 to 22% (Table 49-1) [82, 136–140, 186]. Limited clinical information is available from these studies, and only one reported an association of microsatellite instability with advanced grade and poor outcome [137]. Mutations in *hMSH2* or *hMLH1* are infrequent in these tumors, even when they

TABLE 49-1 Microsatellite Instability in Sporadic Tumors

Tumor type	Number of tumors	Number of markers	% of tumors with instability[a]	Comments[b]	Reference
Colorectal[d]	90	4	167		10
	137	10	12		11
	46	7	13	≥2 loci+	12
	18	4	11		147
	49	7	16	≥2 loci+	151
	243	7	17	≥2 loci, ↑ survival, distinct phenotype	76/140
	49	4	24	+ TGFβIIR mutation	152
	21	32	19	Cell lines, ↓ p53 mutation	96
	80	8	6	≥2 loci, no correlation with survival or phenotype	153
	69	5	5.8	All proximal, all + TGFβIIR mutation	154
	43	31	12	>20 loci	115
	54	10	24		156
Gastric	41	4	9.8	+TGβIIR mutation	152
	61	6	39	Distinct phenotype	128
	40	3	15	Distinct phenotype	129
	10	12	0	All <40 yrs age	127
	22	10	0	Negative correlation with p53	157
	46	8	4.3		185
	76	2	33	Associated with ↑ stage, LOH, no MIN in EBV	130
	24	9	8.3	Associated with LOH, p53, APC mutations	131
	52	5	23		132
	34	12	29	Distinct phenotype, ↑ survival	133
	40	7	30	Associated with advanced stage	134
Endometrium	18	7	22	≥2 loci +	140
	36	71	17	Multiple loci +	82
	28	8	21		136
	109	8	9	Associated with ↑ grade, poor outcome	137
	77	5	13	No clinical data	138
	65	8	18.5	IHC hMHLH1 & hMSH2, mutation in 2	139
	30	7	23	> 3/7 loci +	186
Ovarian	78	10	17	Correlates with p53	142
	20	12	10[c]		112
	90	3	12	Mutation infrequent	143
	57	15 (all on 17p)	1.8	"Early" ovarian cancer	141
	19	4	16		147
Pancreas	9	4	67[c]		147
	44	46	61[c]		148
	8	12	12.5		149
Prostate	21	36	0		158
	57	18	44	≥2 loci +	159
	48	17	14.6		160
	30	9	10		161
	66	8	12	≥2 loci	162
	55	155	0		113
Breast	37	11	30	≥2 loci +, ↓ survival, advanced stage	163
	26	7	15	≥3/7 LOCI+, no LOH in MIN+,	150
	42	7	2.3		164
	100	9	1	≥2 loci +	165
	84	7	0	≥2 loci +	140
	26	4	4[c]		147
	104	12	10.6[c]		112
Cervical	13	4	8[c]		147

(*continues*)

TABLE 49-1 (*continued*)

Tumor type	Number of tumors	Number of markers	% of tumors with instability[a]	Comments[b]	Reference
Not Spec	87	7	0		140
Lung, small cell	33	34	24	>10% loci +, associated with LOH, poor prognosis	114
Non-small cell	137	5	5.1		166
	108	6	0		167
	64	8	3.1	≥3 loci +	168
	38	10	15.8	≥3 loci +	169
Glioma	24 GBM	2	21	Only 1 with MIN at both loci; 5/6 with +TGFβIIR mutation had MIN at 1 loci	170
	11 AA		3.4		
	5 Astrol		0		
	54	12	0		112
Bladder	73	1	1.3		171
	200	7	1.5	≥2/7 loci +	172
Leukemia	30	7	0–25[c]	Only in CML in blast crisis	173
	19	5	53	≥2 loci, only in blast crisis	174
Renal	36	7	11.1	≥3/7 loci +. associated with LOH	175
Testis	86	7	0		140
Soft tissue sarcoma	18	4	0		112
SCC head and neck	91	19	7	2/6 with p53 mutations	176
	46	6	21	≥2 loci +	177
	30	14	7	≥2 loci +, correlated with LN metastasis	178
	48	3	23[c]	≥1 loci +	184
BCC	16	4	37.5	≥3/4 loci +, no LOH in MIN+,	183
Hepatocellular	17	4	0		152
	29	4	3[c]		147
Esophageal	28	5	0	Barrett's metaplasia	179
	36		0	Barrett's associated adenocarcinoma	179
	42		0	SCC	179
	20	39	0	SCC	180
	26		0	Barrett's associated adenocarcinoma	180
	18	17	0	SCC	181
	17	128	0		182

[a]Generally ≥30% tested loci with MIN.
[b]IHC, immunohistochemistry; LOH, loss of heterozygosity; MIN, microsatellite instability; LN, lymph node; SCC, squamous cell carcinoma; SCCHN, head and neck SCC; BCC, basal cell carcinoma.
[c]Fraction of loci with MIN unstated.
[d]Representative studies are shown; not intended as a comprehensive list.

exhibit microsatellite instability [136, 139]. Furthermore, no change in protein expression for either hMSH2 or hMLH1 was seen in MIN+, mutation-negative tumors. With few exceptions, these observations suggest that both *hMSH2* and *hMLH1* genes are rarely involved in the instability observed in sporadic endometrial cancers. Risinger *et al.* [82] described the finding of an *hMSH3* mutation in one of twelve endometrial cancers with microsatellite instability, suggesting a role for other mismatch repair genes. The extent to which this applies to other sporadic endometrial tumors has yet to be resolved, as is the involvement of other DNA mismatch repair genes.

The frequency of microsatellite instability reported for sporadic ovarian carcinomas ranges from 1.8% [141] to 17% [112, 142–145] (Table 49-1). Orth *et al.* [146] found a missense *hMSH2* gene mutation in a serous cystadenocarcinoma of the ovary, while Arzimanglou *et al.* [143] found only one somatic mutation, but no germline mutations, in 11 ovarian tumors with microsatellite instability. One report noted the association of microsatellite instability with mutations within the p53 gene. Overall, these data indicate that microsatellite instability is relatively infrequent in sporadic ovarian cancer. Moreover, when instability is observed, neither *hMLH1* nor *hMSH2* mutations are apparent.

Although there are relatively few studies of microsatellite instability in sporadic pancreatic cancer, two studies suggest high rates of instability (61 and 67%) [147, 148], but give insufficient information to determine whether the instability was widespread. One other report notes a frequency of 12% of microsatellite instability in at least two of twelve markers tested [149].

Microsatellite instability has also been reported in a variety of cancers that are not generally part of the tumor spectrum in HNPCC (Table 49-1). There is considerable variability in the frequencies detected among, and within, the different tumors. For example, in breast cancer, microsatellite is not present [140], is present in 11% of tumors at only a single locus [112], or is present at more than one of three loci in 3 of 20 tumors analyzed [150]. As stated earlier, such inconsistency may result from a lack of uniformity of study parameters among the various reports.

V. SUMMARY

In a remarkably short period of time, the genetic basis of one form of hereditary colon cancer (HNPCC) as well as a subset of sporadic colon cancer has been defined. Emanating from work performed in several different areas, a number of genes involved in DNA mismatch repair have been localized, cloned, and subsequently shown to be responsible for a substantial fraction of families with HNPCC. Additionally, some of these same genes have also been shown to be involved in sporadic colon cancer and responsible for the phenomenon of tumor microsatellite instability. These observations have now extended beyond colon cancer into a large number of other tumor systems, suggesting that DNA mismatch repair may be defective in a significant fraction of human malignancies.

Although tremendous progress has been made, many questions remain to be answered. For HNPCC, some of these include defining genotype–phenotype correlations, identifying other susceptibility loci, and identifying those genes that modify the phenotypes for HNPCC (i.e., differences between Lynch I, Lynch II, and Muir–Torre syndrome). For MIN+ sporadic colon cancer, it will be important to further define the role of defective mismatch repair in MIN+A and MIN+B tumors, the mechanism of gene inactivation, and, importantly, the downstream consequences of defective mismatch repair. For other tumor systems, it will be important to further define the cause of microsatellite instability and the role of defective mismatch repair.

References

1. Bishop, J. M. (1991). Molecular themes in oncogenesis. *Cell* **64,** 235–248.
2. Marshall, C. J. (1991). Tumor suppressor genes. *Cell* **64,** 313–326.
3. Nowell, P. C. (1976). The clonal evolution of cancer. *Science* **194,** 23–28.
4. Cheng, K. C., and Loeb, L. A. (1993). Genomic instability and tumor progression: mechanistic considerations. *Adv. Cancer Res.* **60,** 121–156.
5. Loeb, L. A. (1991). Mutator phenotype may be required for multistage carcinogenesis. *Cancer Res.* **51,** 3075–3079.
6. Fearon, E. R., and Jones, P. A. (1992). Progressing towards a molecular description of colorectal cancer development. *FASEB J.* **6,** 2783–2790.
7. Weber, J. L., and Wong, C. (1993). Mutation of human short tandem repeats. *Hum. Mol. Genet.* **2,** 1123–1128.
8. Weber, J. L., and May, P. E. (1989). Abundant class of human DNA polymorphisms which can be typed using the polymerase chain reaction. *Am. J. Hum. Genet.* **44,** 388–396.
9. Weissenbach, J., Gyapay, G., Dib, C., Vignal, A., Morissette, J., Millasseau, P., Vaysseix, G., and Lathrop, M. (1992). A second-generation linkage map of the human genome. *Nature* **359,** 794–801.
10. Thibodeau, S. N., Bren, G., and Schaid, D. J. (1993). Microsatellite instability in cancer of the proximal colon. *Science* **260,** 816–819.
11. Ionov, Y., Peinado, M. A., Malkhosyan, S., Shibata, D., and Perucho, M. (1993). Ubiquitous somatic mutations in simple repeated sequences reveal a new mechanism for colorectal carcinogenesis. *Nature* **363,** 558–561.
12. Aaltonen, L. A., Peltomäki, P., Leach, F. S., Sistonen, P., Pylkkänen, L., Mecklin, J.-P., Jarvinen, H., Powell, S. M., Jen, J., Hamilton, S. R., Petersen, G. M., Kinzler, K. W., Vogelstein, B., and de la Chapelle, A. (1993). Clues to the pathogenesis of familial colorectal cancer. *Science* **260,** 812–816.
13. Cannon-Albright, L. A., Skolnick, M. H., Bishop, D. T., Lee, R. G., and Burt, T. W. (1988). Common inheritance of susceptibility to colonic adenomatous polyps and associated colorectal cancers. *N. Engl. J. Med.* **319,** 533–537.
14. St. John, D. J., McDermott, F. T., Hopper, J. L., Debney, E. A., Johnson, W. R., and Hughes, E. S. (1993). Cancer risk in relatives of patients with common colorectal cancer. *Ann. Int. Med.* **118,** 13–18.
15. Warthin, A. S. (1913). Heredity with reference to carcinoma. *Arch. Int. Med.* **12,** 546–555.
16. Lynch, H. T., and Krush, A. J. (1971). Cancer family G revisited: 1895–1970. *Cancer* **27,** 1505–1511.
17. Lynch, H. T., Shaw, M. W., Magnuson, C. W., Larsen, A. L., and Krush, A. J. (1966) Hereditary factors in cancer. Study of two large midwestern kindreds. *Arch. Int. Med.* **117,** 206–212.
18. Boland, C. R., and Troncale, F. J. (1984). Familial colonic cancer without antecedent polyposis. *Ann. Int. Med.* **100,** 700–701.
19. Vasen, H. F., Mecklin, J. P., Khan, P. M., and Lynch, H. T. (1991). The International Collaboration Group on Hereditary Non-Polyposis Colorectal Cancer (ICG-HNPCC). *Dis. Colon Rectum* **34,** 424–425.
20. Marra, G., and Boland, C. R. (1995). Hereditary nonpolyposis colorectal cancer: the syndrome, the genes, and historical perspectives. *J. Nat. Cancer Inst.* **87,** 1114–1125.

21. Bailey-Wilson, J. E., Elston, R. C., Schuelke, G. S., Kimberling, W., Albano, W., Lynch, J. F., and Lynch, G. T. (1986). Segregation analysis of hereditary nonpolyposis colorectal cancer. *Genet. Epidemiol.* **3,** 27–38.
22. Burt, R. W., Bishiop, T. D., Cannon, L. A., Dowdle, M. A., Lee, R. G., and Skolnick, M. H. (1985). Dominant inheritance of adenomatous colonic polyps and colorectal cancer. *N. Engl. J. Med.* **312,** 1540–1544.
23. Ponz de Leon, M., Sassatelli, R., Benatti, P., and Roncucci, L. (1993). Identification of hereditary nonpolyposis colorectal cancer in the general population. *Cancer* **71,** 3493–3501.
24. Scapoli, C., Ponz de Leon, M., Sassatelli, R., Benatti, P., Roncucci, L., Collins, A., Morton, N. E., and Barrai, I. (1994). Genetic epidemiology of hereditary non-polyposis colorectal syndromes in Modena, Italy: Results of a complex segregation analysis. *Ann. Hum. Genet.* **58,** 275–295.
25. Peltomäki, P., Sistonen, P., Mecklin, J. P., Pylkkänen, L., Jarvinen, H., Simons, J. W., Cho, K. R., Vogelstein, B., and de la Chapelle, A. (1991). Evidence supporting exclusion of the DCC gene and a portion of chromosome 18q as the locus for susceptibility to hereditary nonpolyposis colorectal carcinoma in five kindreds. *Cancer Res.* **51,** 4135–4140.
26. Peltomäki, P., Sistonen, P., Mecklin, J. P., Pylkkänen, L., Aaltonen, L., Nordling, S., Kere, J., Jarvinen, H., and Hamilton, S. R. (1992). Evidence that the MCC-APC gene region in 5q21 is not the site for hereditary nonpolyposis colorectal carcinoma. *Cancer Res.* **52,** 4530–4533.
27. Peltomäki, P., Aaltonen, L. A., Sistonen, P., Pylkkänen, L., Mecklin, J. P., Jarvinen, H., Green, J. S., Jass, J. R., Weber, J. L., Leach, R. S., Petersen, G. M., Hamilton, S. R., de la Chapelle, A., and Vogelstein, B. (1993). Genetic mapping of a locus predisposing to human colorectal cancer. *Science* **260,** 810–812.
28. Lindblom, A., Tannergård, P., Werelius, B., and Nordenskjöld, M. (1993). Genetic mapping of a second locus predisposing to hereditary colorectal cancer. *Nature Genet.* **5,** 279–282.
29. Nyström-Lahti, M., Parsons, R., Sistonen, P., Pylkkänan, L., Aaltonen, L. A., Leach, F. S., Hamilton, S. R., Watson, P., Bronson, E., Fusaro, R., Cavalieri, J., Lynch, J., Lanspa, S., Smyrk, T., Lynch, P., Drouhard, T., Kinzler, K. W., Vogelstein, B., Lynch, H. T., de la Chapelle, A., and Peltomäki, P. (1994). Mismatch repair genes on chromosomes 2p and 3p account for a major share of hereditary nonpolyposis colorectal cancer families evaluable by linkage. *Am. J. Hum. Genet.* **55,** 659–665.
30. Modrich, P. (1991). Mechanisms and biological effects of mismatch repair. *Ann. Rev. Genet.* **25,** 229–253.
31. Radman, M. (1988). Mismatch repair and genetic recombination. *In* "Genetic Recombination" (R. Kucherlapati, and G. R. Smith, Eds.), pp. 169–192. Am. Soc. Microbiol., Washington, DC.
32. Lahue, R. S., Au, K. G., and Modrich, P. (1989). DNA mismatch correction in a defined system. *Science* **245,** 160–164.
33. Längle-Rouault, F., Maenhaut, M. G., and Radman, M. (1987). GATC sequences, DNA nicks and the MutH function in *Escherichia coli* mismatch repair. *EMBO J.* **6,** 1121–1127.
34. New, L., Lui, K., and Crouse, G. F. (1993). The yeast gene *MSH3* defines a new class of eukaryotic MutS homologues. *Mol. Gen. Genet.* **239,** 97–108.
35. Reenan, R. A. G., and Kolodner, R. D. (1992). Characterization of insertion mutations in *Saccharomyces cerevisiae MSH1* and *MSH2* genes: evidence for separate mitochondrial and nuclear functions. *Genetics* **132,** 975–985.
36. Reenan, R. A. G., and Kolodner, R. D. (1992). Isolation and characterization of two *Saccharomyces cerevisiae* genes encoding homologues of the bacterial HexA and MutS mismatch repair proteins. *Genetics* **132,** 963–973.
37. Ross-McDonald, P., and Roeder, G. S. (1994). Mutation of a meiosis-specific MutS homolog decreases crossing over but not mismatch correction. *Cell* **79,** 1069–1080.
38. Hollingsworth, N. M., Ponte, L., and Halsey, C. (1995). MSH5, A novel MutS homolog, facilitates meiotic reciprocal recombination between homologs in *Saccharomyces cerevisiae* but not mismatch repair. *Genes Dev.* **9,** 1728–1739.
39. Marsischky, G. T., Filosi, N., Kane, M. F., and Kolodner, R. (1996). Redundancy of *Saccharomyces cerevisiae* MSH3 and MSH6 in MSH2-dependent mismatch repair. *Genes Dev.* **10,** 407–420.
40. Fishel, R., and Wilson, T. (1997). MutS homologs in mammalian cells. *Curr. Opin. Genet. Dev.* **7,** 105–113.
41. Kramer, W., Kramer, B., Williamson, M. S., and Fogel, S. (1989). Cloning and nucleotide sequence of DNA mismatch repair gene *PMS1* from *Saccharomyces cerevisiae:* homology of PMS1 to prokaryotic MutL and Hex B. *J. Bacteriol.* **171,** 5339–5346.
42. Prolla, T. A., Christie, D.-M., and Liskay, R. M. (1994). A requirement in yeast DNA mismatch repair for *MLH1* and *PMS1,* two homologues of bacterial *Mut L* gene. *Mol. Cell. Biol.* **14,** 407–415.
43. Prolla, T. A., Pang, Q., Alani, E., Kolodner, R. D., and Liskay, R. M. (1994). Interactions between the MSH2, MLH1 and PMS1 proteins during the initiation of DNA mismatch repair. *Science* **264,** 1091–1093.
44. Kramer, B., Kramer, W., Williamson, M. S., and Fogel, S. (1989). Heteroduplex DNA correction in *Saccharomyces cerevisiae* is mismatch specific and requires functional *PMS* genes. *Mol. Cell. Biol.* **9,** 4432–4440.
45. Prolla, T. A., Christie, D.-M., and Liskay, R. M. (1994). Dual requirement in yeast DNA mismatch repair for *MLH1* and *PMS1,* two homologues of the bacterial *mutL* gene. *Mol. Cell. Biol.* **14,** 407–415.
46. Strand, M., Earley, M. C., Crouse, G. F., and Petes, T. D. (1995). Mutations in the *MSH3* gene preferentially lead to deletions within tracts of simple repetitive DNA in *Saccharomyces cerevisiae. Proc. Natl. Acad. Sci. USA* **92,** 10418–10421.
47. Fishel, R., and Kolodner, R. D. (1995). Identification of mismatch repair genes and their role in the development of cancer. *Curr. Opin. Genet. Dev.* **5,** 382–395.
48. Strand, M., Prolla, T. A., Liskay, R. M., and Petes, T. D. (1993). Destabilization of tracts of simple repetitive DNA in yeast by mutations affecting DNA mismatch repair. *Nature* **365,** 274–276. [erratum **368,** 569]
49. Alani, E., Chi, N.-W., and Kolodner, R. (1995). The *Saccharomyces cerevisiae* Msh2 protein specifically binds to duplex oligonucleotides containing mismatched DNA base pairs and insertions. *Genes Dev.* **9,** 923–947.
50. Johnson, R. E., Kovali, G. K., Prakash, L., and Prakash, S. (1996). Requirement of the yeast *MSH3* and *MSH6* genes for *MSH2*-dependent genomic stability. *J. Biol. Chem.* **271,** 7285–7288.
51. Habraken, Y., Sung, P., Prakash, L., and Prakash, S. (1996). Binding of insertion/deletion DNA mismatches by the heterodimer of yeast mismatch repair proteins MSH2 and MSH3. *Curr. Biol.* **6,** 1185–1187.
52. Levinson, G., and Gutman, G. A. (1987). High frequencies of short frameshifts in poly-CA/TG tandem repeats borne by bacteriophage M13 in *Escherichia coli* K-12. *Nucleic Acids Res.* **15,** 5323–5338.
53. Fishel, R., Lescoe, M. K., Rao, M. R. S., Copeland, N. G., Jenkins, N. A., Garber, J., Kane, M., and Kolodner, R. (1993). The human mutator gene homolog MSH2 and its association with hereditary nonpolyposis colon cancer. *Cell* **75,** 1027–1038.

54. Leach, F. S., Nicolaides, N. C., Papadopoulos, N., Liu, B., Jen, J., Parsons, R., Peltomäki, P., Sistonen, P., Aaltonen, L. A., Nyström-Lahti, M., Guan, X. Y., Zhang, J., Meltzer, P. S., Yu, J-W. Kao, R-T., Chen, D. J., Cerosaletti, D. M., Fournier, R. E. K., Todd, S., Lewis, T., Leach, R. J., Naylor, S. L., Weissenbach, J., Mecklin, J.-P., Jarvinen, H., Petersen, G. M., Hamilton, S. R., Green, J., Jass, J., Watson, P., Lynch, H. T., Trent, J. M., de la Chapelle, A., Kinzler, K. W., and Vogelstein, B. (1993). Mutations of a MutS homolog in hereditary nonpolyposis colorectal cancer. *Cell* **75,** 1215–1225.
55. Bronner, C. E., Baker, S. M., Morrison, P. T., Warren, G., Smith, L. G., Lescoe, M. K., Kane, M., Earabino, C., Lipford, J., Lindblom, A., Tannergård, P., Bollag, R. J., Godwin, A. R., Ward, D. C., Nordenskjöld, M., Fishel, R., Kolodner, R., and Liskay, R. M. (1994). Mutation in the DNA mismatch repair gene homologue hMLH1 is associated with hereditary non-polyposis colon cancer. *Nature* **368,** 258–261.
56. Papadopoulos, N., Nicolaides, N. N., Wei, Y.-F., Ruben, S. M., Carter, K. C., Rosen, C. A., Haseltine, W. A., Fleischmann, R. D., Fraser, C. M., Adams, M. D., Venter, J. C., Hamilton, S. E., Petersen, G. M., Watson, P., Lynch, H. T., Peltomäki, P., Mecklin, J.-P., de la Chapelle, A., Kinzler, K. W., and Vogelstein, B. (1994). Mutation of a MutL homolog in hereditary colon cancer. *Science* **263,** 1625–1629.
57. Peltomäki, P., and Vasen, H. F. A. (1997). Mutations predisposing to hereditary nonpolyposis colorectal cancer: database and results of a collaborative study. *Gastroenterology* **113,** 1146–1158.
58. Drummond, J. T., Li, G-M., Longley, M. J., and Modrich, P. (1995). Isolation of an hMSH2-p160 heterodimer that restores DNA mismatch repair to tumor cells. *Science* **268,** 1909–1912.
59. Palombo, F., Gallinari, P., Iaccarino, I., Lettieri, T., Hughes, M., D'Arrigo, A., Truong, O., Hsuan, J. J., and Jiricny, J. (1995). GTBP, a 160-kilodalton protein essential for mismatch-binding activity in human cells. *Science* **268,** 1912–1914.
60. Palombo, F., Iaccarino, I., Nakajima, E., Ikejima, M., Shimada, T., and Jiricny, J. (1996). hMutSb, a heterodimer of hMSH2 and hMSH3, binds to insertion/deletion loops in DNA. *Curr. Biol.* **6,** 1181–1184.
61. Acharya, S., Wilson, T., Gradia, S., Kane, M. F., Guerrette, S., Marsischky, G. T., Kolodner, R., and Fishel, R. (1996). hMSH2 forms specific mispair binding complexes with hMSH3 and hMSH6. *Proc. Nat. Acad. Sci. USA* **93,** 13629–13634.
62. Li, G. M., and Modrich, P. (1995). Restoration of mismatch repair to nuclear extracts of H6 colorectal tumor cells by a heterodimer of human MutL homologues. *Proc. Natl. Acad. Sci. USA* **92,** 1950–1954.
63. Charbonnier, F., Martin, C., Scotte, M., Sibert, L., Moreau, V., and Frebourg, T. (1995). Alternative splicing of *MLH1* messenger RNA in human normal cells. *Cancer Res.* **55,** 1839–1841.
64. Xia, L., Shen, W., Ritacca, F., Mitri, A., Madlensky, L., Berk, T., Cohen, Z., Gallinger, S., and Bapat, B. (1996). Truncated *hMSH2* transcript occurs as a common variant in the population: implication for genetic diagnosis. *Cancer Res.* **56,** 2289–2292.
65. Mori, Y., Shiwaku, H., Fukushige, S., Wakatsuki, S., Sato, M., Nukiwa, T., and Horii, A. (1997). Alternative splicing of *hMSH2* in normal tissues. *Hum. Genet.* **99,** 590–595.
66. Brentnall, T. A., Rubin, C. Y., Crispin, D. A., Emond, M., and Rubin, C. E. (1995). A germline substitution in the human *hMSH2* gene is associated with high-grade dysplasia and cancer in ulcerative colitis. *Gastroenterology* **109,** 151–155.
67. Liu, B., Nicolaides, N. C., Markowitz, S., Willson, J. K., Parsons, R. E., Jen, J., Papadopolous, N., Peltomäki, P., de la Chapelle, A., Hamilton, S. R., Kinzler, K. W., and Vogelstein, B. (1995). Mismatch repair gene defects in sporadic colorectal cancers with microsatellite instability. *Nature Genet.* **9,** 48–55.
68. Peltomäki, P., and de la Chapelle, A. (1997). Mutations predisposing to hereditary nonpolyposis colorectal cancer. *Adv. Cancer Res.* **71,** 93–119.
69. Wijnen, J., Khan, P. M., Vasen, H., Menko, F., van der Klift, H., van den Broek, M., van Leeuwen-Cornelisse, I., Nagengast, F., Meijers-Heijboer, E. J., Lindhout, D., Griffioen, G., Cats, A., Kleibeuker, J., Varesco, L., Bertario, L., Bisgaard, M-L., Mohr, J., Kolodner, R., and Fodde, R. (1996). Majority of *hMLH1* mutations responsible for hereditary nonpolyposis colorectal cancer cluster at the exonic region 15-16. *Am. J. Hum. Genet.* **58,** 300–307.
70. Kruse, R., Lamberts, C., Wang, Y., Ruelfs, C., Bruns, A., Esche, C., Lehmann, P., Ruzicka, T., Rutten, A., Friedl, W., and Propping, P. (1996). Is the mismatch repair deficient type of Muir-Torre syndrome confined to mutations in the hMSH2 gene? *Hum. Genet.* **98,** 737–750.
71. Kolodner, R. D., Hall, N. R., Lipford, J., Kane, M. F., Rao, M. R., Morrison, P., Wirth, L., Finan, P. J., Burn, J., Chapman, P., Earabino, C., Merchant, E., and Bishop, D. T. (1994). Structure of the human MSH2 locus and analysis of two Muir-Torre kindreds for msh2 mutations. *Genomics* **24,** 516–526.
72. Hall, N. R., Murday, V. A., Chapman, P., Williams, M. A., Burn, J., Finan, P. J., and Bishop, D. T. (1994). Genetic linkage in Muir-Torre syndrome to the same chromosomal site as cancer family syndrome. *Eur. J. Cancer* **30A,** 180–182.
73. Hamilton, S. R., Liu, B., Parsons, R. E., Papadopolous, N., Jen, J., Powell, S. M., Krush, A. J., Berk, T., Cohen, Z., Tetu, B., Burger, P. C., Wood, P. A., Taqi, F., Booker, S. V., Petersen, G. M., Offerhaus, G. J. A., Tersmette, A. C., Giardiello, F. M., Vogelstein, B., and Kinzler, K. W. (1995). The molecular basis of Turcot's syndrome. *N. Engl. J. Med.* **3332,** 839–847.
74. Nicolaides, N. C., Papadopoulos, N., Liu, B., Wei, Y. F., Carter, K. C., Ruben, S. M., Rosen, C. A., Haseltine, W. A., Fleischmann, R. D., Fraser, C. M., Adams, M. D., Venter, J. C., Dunlop, M. G., Hamilton, S. R., Petersen, G. M., de la Chapelle, A., Vogelstein, B., and Kinzler, K. W. (1994). Mutations of two *PMS* homologues in hereditary nonpolyposis colon cancer. *Nature* **371,** 75–80.
75. Børresen, A.-L, Lothe, R. A., Meling, G. I., Lystad, S., Morrison, P., Lipford, J., Kane, M. F., Rognum, T. O., and Kolodner, R. D. (1995). Somatic mutations in the *hMSH2* gene in microsatellite unstable colorectal carcinomas. *Hum. Mol. Genet.* **4,** 2065–2072.
76. Lothe, R. A., Peltomäki, P., Meling, G. I., Aaltonen, L. A., Nyström-Lahti, M., Pylkkänen, L., Heimdal, K., Andersen, T. I., Møller, P., Rognum, T. O., Fosså, S. D., Haldorsen, T., Langmark, F., Brøgger, A., de la Chapelle, A., and Børresen, A.-L. (1993). Genomic instability in colorectal cancer: relationship to clinicopathological variables and family history. *Cancer Res.* **53,** 5849–5852.
77. Möslein, G., Tester, D. J., Lindor, N. M., Honchel, R., Cunningham, J. M., French, A. J., Halling, K. C., Schwab, M., Goretzki, P., and Thibodeau, S. N. (1996). Microsatellite instability and mutation analysis of *hMSH2* and *hMLH1* in patients with sporadic, familial and hereditary colorectal cancer. *Hum. Mol. Genet.* **5,** 1245–1252.
78. Wu, Y., Nyström-Lahti, M., Osinga, J., Looman, M. W. G., Peltomäki, P., Aaltonen, L. A., Hofstra, R. M. W., and Buys, C. H. C. M. (1997). *MSH2* and *MLH1* mutations in sporadic replication error-positive colorectal carcinoma as assessed by two-dimensional DNA electrophoresis. *Genes Chromosomes Cancer* **18,** 269–278.

79. Herfarth, K. K., Kodner, I. J., Whelan, A. J., Ivanovich, J. L., Bracamontes, J. R., Wells, S. A., and Goodfellow, P. J. (1997). Mutations in MLH1 are more frequent than in MSH2 sporadic colorectal cancers with microsatellite instability. *Genes Chromosomes Cancer* **18,** 42–49.
80. Beck, N. E., Tomlinson, I. P. M., Homfray, T., Hodgson, S. V., Harocopos, C. J., and Bodner, W. F. (1996). Genetic testing is important in families with a history suggestive of hereditary nonpolyposis colorectal cancer even if the Amsterdam criteria are not fulfilled. *Br. J. Surg.* **84,** 233–237.
81. Bubb, V. J., Curtis, L. J., Cunningham, C., Dunlop, M. G., Carothers, A. D., Morris, R. G., White, S., Bird, C. C., and Wyllie, A. H. (1996). Microsatellite instability and the role of hMSH2 in sporadic colorectal cancer. *Oncogene* **12,** 2641–2649.
82. Risinger, J. I., Umar, A., Boyd, J., Berchuck, A., Kunkel, T. A., and Barrett, J. C. (1996). Mutation of *MSH3* in endometrial cancer and evidence for its functional role in heteroduplex repair. *Nature Genet.* **14,** 102–105.
83. Papadopoulos, N., Nicolaides, N. N., Liu, B., Parsons, R., Lengauer, C., Palombo, F., D'Arrigo, A., Markowitz, S., Willson, J. K., Kinzler, K. W., Jiricny, J., and Vogelstein, B. (1995). Mutations of *GTBP* in genetically unstable cells. *Science* **268,** 1915–1917.
84. Hsieh, C. L. (1994). Dependence of transcriptional repression on CpG methylation density. *Mol. Cell. Biol.* **14,** 5487–5494.
85. Muiznieks, I., and Doerfler, W. (1994). The impact of 5′-CG-3′ methylation on the activity of different eukaryotic promoters: a comparative study. *FEBS Lett.* **344,** 251–254.
86. Jones, P. A., and Gonzaglo, M. L. (1997). Altered DNA methylation and genome instability: a new pathway to cancer? *Proc. Natl. Acad. Sci. USA* **94,** 2103–2105.
87. Bird, A. P. (1996). The relationship of DNA methylation to cancer. *Cancer Surveys* **28,** 87–101.
88. Laird, P. W., and Jaenisch, R. (1996). The role of DNA methylation in cancer genetics and epigenetics. *Ann. Rev. Genet.* **30,** 441–464.
89. Thibodeau, S. N., French, A. J., Roche, P. C., Cunningham, J. M., Tester, D. J., Lindor, N. M., Möslein, G., Baker, S. M., Liskay, R. M., Burgart, L. J., Honchel, R., and Halling, K. C. (1996). Altered expression of hMSH2 and hMLH1 in tumors with microsatellite instability and genetic alterations in mismatch repair genes. *Cancer Res.* **56,** 4836–4840.
90. Kane, M. F., Loda, M., Gaida, G. M., Lipman, J., Mishra, R., Goldman, H., Jessup, J. M., and Kolodner, R. (1997). Methylation of the *hMLH1* promoter correlates with lack of expression of hMLH1 in sporadic colon tumors and mismatch repair-defective human tumor cell lines. *Cancer Res.* **57,** 808–811.
91. Parsons, R., Li, G.-M., Longley, M. J., Fang, W. H., Papadopoulos, N., Jen, J., de la Chapelle, A., Kinzler, D. W., Vogelstein, B., and Modrich, P. (1993). Hypermutability and mismatch repair deficiency in RER^+ tumor cells. *Cell* **75,** 1227–1236.
92. Eshleman, J. R., Lang, E. Z., Bowerfind, G. K., Parsons, R., Vogelstein, B., Willson, J. K. V., Veigl, M. L., Sedwick, W. D., and Markowitz, S. D. (1995). Increased mutation rate at the hprt locus accompanies microsatellite instability in colon cancer. *Oncogene* **10,** 33–37.
93. Malkhosyan, S., McCarty, A., Sawai, H., and Perucho, M. K. (1996). Differences in the spectrum of spontaneous mutations in the hprt gene between tumor cells of the microsatellite mutator phenotype. *Mutat. Res.* **316,** 249–259.
94. Eshleman, J. R., Markowitz, S. D., Donover, P. S., Lang, E. Z., Lutterbaugh, J. D., Li, G.-M., Longley, M., Modrich, P., Veigl, M. L., and Sedwich, W. D. (1996). Diverse hypermutability of multiple expressed sequence motifs present in a cancer with microsatellite instability. *Oncogene* **12,** 1425–1432.
95. Bhattacharyya, N. P., Ganesh, A., Phear, G., Richards, B., Skandalis, A., and Meuth, M. (1995). Molecular analysis of mutations in mutator colorectal carcinoma cell lines. *Hum. Mol. Genet.* **4,** 2057–2064.
96. Thibodeau, S. N., Donovan, K. A., Tester, D. J., Bren, G. D., Halling, K. C., Cunningham, J., Morris, A. V., and Roche, P. C. (1993). Mutator gene effect in colorectal cancer. *Am. J. Hum. Genet.* **53** (Suppl), 1.
97. Konishi, M., Kikuchi-Yanoshita, R., Tanaka, K., Muraoka, M., Onda, A., Okumura, Y., Kishi, N., Iwama, T., Mori, T., Koike, M., Ushio, K., Chiba, M., Nomizu, S., Konishi, F., Utsunomiya, J., and Miyaki, M. (1996). Molecular nature of colon tumors in hereditary nonpolyposis colon cancer, familial polyposis, and sporadic colon cancer. *Gastroenterology* **111,** 307–317.
98. Cottu, P. H., Muzeau, F., Estreicher, A., Flejou, J. F., Thomas, G., and Hamelin, R. (1996). Inverse correlation between RER+ status and p53 mutation in colorectal cancer cell lines. *Oncogene* **13,** 2727–2730.
99. Muta, H., Noguchi, M., Perucho, M., Ushio, K., Sugihara, K., Ochiai, A., Nawaka, H., and Hirohashi, S. (1996). Clinical implications of microsatellite instability in colorectal cancer. *Cancer* **77,** 265–270.
100. Reitmair, A. H., Cai, J.-C., Bjerknes, M., Redston, M., Cheng, H., Pind, M. T. L., Hay, K., Mitri, A., Bapar, B. V., Mak, T. W., and Gallinger, S. (1996). MSH2 deficiency contributes to accelerated APC-mediated intestinal tumorigenesis. *Cancer Res.* **56,** 2922–2926.
101. Lazar, V., Grandjouan, S., Bognel, C., Couturier, D., Rougier, P., Bellet, D., and Bressac-de Paillerets, B. (1994). Accumulation of multiple mutations in tumor suppressor genes during colorectal tumorigenesis in HNPCC patients. *Hum. Mol. Genet.* **3,** 2257–2260.
102. Huang, J., Papadopoulos, N., McKinley, A. J., Farrington, S. M., Curtis, L. J., Wyllie, A. H., Zheng, S., Willson, J. K. V., Markowitz, S. D., Morins, P., Kinzler, K. W., Vogelstein, B., and Dunlop, M. G. (1996). APC mutations in colorectal tumors with mismatch repair deficiency. *Proc. Natl. Acad. Sci. USA* **93,** 9049–9054.
103. Markowitz, S., Wang, J., Myeroff, L., Parsons, R., Sun, L., Lutterbaugh, J., Fan, R. S., Zborowska, E., Kinzler, K. W., Vogelstein, B., Brattain, M., and Willson, J. K. V. (1995). Inactivation of the type II TGF-β receptor in colon cancer cells with microsatellite instability. *Science* **268,** 1336–1338.
104. Parsons, R., Myeroff, L. L., Liu, B., Willson, J. K. V., Markowitz, S. D., Kinzler, K. W., and Vogelstein, B. (1995). Microsatellite instability and mutations of the transforming growth factor β receptor gene in colorectal cancer. *Cancer Res.* **55,** 5548–5550.
105. Akiyama, Y., Iwanaga, R., Saitoh, K., Shiba, K., Ushio, K., Ikeda, E., Iwama, T., Nomizu, T., and Yuasa, Y. (1997). Transforming growth factor β receptor gene mutations in adenomas from hereditary nonpolyposis colorectal cancer. *Gastroenterology* **112,** 33–39.
106. Souza, R. F., Lei, J., Yin, J., Appel, R., Zou, T.-T., Zhou, X., Wang, S., Rhyu, M.-G., Cymes, K., Chan, O., Park, W.-S., Krasna, M. J., Greenwald, B. D., Cottrell, J., Abraham, J. M., Simms, L., Leggett, B., Young, J., Harpaz, N., and Meltzer, S. J. (1997). A transforming growth factor β1 receptor type II mutation in ulcerative colitis-associated neoplasms. *Gastroenterology* **112,** 40–45.
107. Myeroff, L. L., Parsons, R., Kim, S.-J., Hedrick, L., Cho, K. R., Orth, K., Mathis, M., Kinzler, K. W., Lutterbaugh, J., Park, K., Bang, Y.-J., Lee, H. Y., Park, J.-G., Lynch, H. T., Roberts, A. B.,

Vogelstein, B., and Markowitz, S. D. (1995). A transforming growth factor β receptor type II mutation common in colon and gastric but rare in endometrial cancers with microsatellite instability. *Cancer Res.* **55,** 5545–5547.

108. Yoshitaka, T., Matsubara, N., Ikeda, M., Tanino, M., Hanafusa, H., Tanaka, N., and Shimizu, K. (1996). Mutations of E2F-4 trinucleotide repeats in colorectal cancer with microsatellite instability. *Biochem. Biophys. Res. Commun.* **227,** 553–557.
109. Ouyang, H., Shiwaku, H. O., Hagiwara, H., Miura, K., Abe, T., Kato, Y., Ohtani, H., Shiiba, K., Souza, R. F., Meltzer, S. J., and Horii, A. (1997). The insulin-like growth factor II receptor gene is mutated in genetically unstable cancers of the endometrium, stomach, and colorectum. *Cancer Res.* **57,** 1851–1854.
110. Souza, R. F., Appel, R., Yin, J., Wang, S., Smolinski, K. N., Abraham, J. M., Zou, T.-T., Lei, J., Cottrell, J., Cymes, K., Biden, K., Simms, L., Leggett, B., Lynch, P. M., Frazier, M., Powell, S. M., Harpaz, N., Sugimura, H., Young J., and Meltzer, S. J. (1996). Microsatellite instability in the insulin-like growth factor receptor II gene in gastrointestinal tumors. *Nature Genet.* **14,** 255–257.
111. Rampino, N., Yamamoto, H., Ionov, Y., Li, Y., Sawai, H., Reed, J. C., and Perucho, M. (1997). Somatic frameshift mutations in the BAX gene in colon cancers of the microsatellite mutator phenotype. *Science* **275,** 967–969.
112. Wooster, R., Cleton-Jansen, A.-M., Collins, N., Mangion, J., Cornelis, R. S., Cooper, C. S., Gusterson, B. A., Ponder, B. A. J., von Deimling, A., Wiestler, O. D., Cornelisse, C. J., Devilee, P., and Stratton, M. R. (1994). Instability of short tandem repeats (microsatellites) in human cancers. *Nature Genet.* **6,** 152–156.
113. Cunningham, J. M., Shan, A., Wick, M. J., McDonnell, S. K., Schaid, D. J., Tester, D. J., Qian, J., Takahashi, S., Jenkins, R. B., Bostwick, D. G., and Thibodeau, S. N. (1996). Allelic imbalance and microsatellite instability in prostatic adenocarcinoma. *Cancer Res.* **56,** 4475–4482.
114. Merlo, A., Mabry, M., Gabrielson, E., Vollmer, R., Baylin, S. B., and Sidransky, D. (1994). Frequent microsatellite instability in primary small cell lung cancer. *Cancer Res.* **54,** 2098–2101.
115. Dietmaier, W., Wallinger, S., Bocker, S., Kullmann, F., Fishel, R. and Rüschoff, J. (1997). Diagnostic microsatellite instability: definition and correlation with mismatch repair expression. *Cancer Res.* **57,** 749–756.
116. Cawkwell, L., Li, D., Lewis, F. A., Marin, I., Dixon, M. F., and Quirke, P. (1995). Microsatellite instability in colorectal cancer: improved assessment using fluorescent polymerase chain reaction. *Gastroenterology* **109,** 465–471.
117. Samowitz, W. S., Slattery, M. L., and Kerber, R. A. (1995). Microsatellite instability in human colorectal cancer is not a useful indicator of familial colorectal cancer. *Gastroenterology* **109,** 1765–1771.
118. Thibodeau, S. N., French, A. J., Cunningham, J. M., Tester, D., Moon-Tasson, L., Burgart, L. J., Roche, P. C., McDonnell, S. K., Schaid, D. J., and O'Connell, M. J. (1998). Microsatellite instability in colorectal cancer: different mutator phenotypes and the principle involvement of *hMLH1. Cancer Res.* [In press]
119. Kim, H., Jen, J., Vogelstein, B., and Hamilton, S. R. (1994). Clinical and pathological characteristics of sporadic colorectal carcinomas with DNA replication errors in microsatellite sequences. *Am. J. Pathol.* **145,** 148–156.
120. Risio, M., Reato, G., di Celle, P. F., Fizzotti, M., Rossini, F. P., and Foa, R. (1996). Microsatellite instability is associated with the histological features of the tumor in nonfamilial colorectal cancer. *Cancer Res.* **56,** 5470–5474.
121. Ilyas, M., Tomlinson, I. P. M., Novelli, M. R., Hanby, A., Bodmer, W. R., and Talbot, I. C. (1996). Clinico-pathological features and p53 expression in left-sided sporadic colorectal cancers with and without microsatellite instability. *J. Pathol.* **179,** 370–375.
122. Hanski, C., Tiecke, F., Hummel, M., Hanski, M.-L., Schmitt-Gräff, A., Stein, H., and Riecken, E.-O. (1996). Low frequency of p53 gene mutation and protein expression in mucinous colorectal carcinomas. *Cancer Lett.* **103,** 163–170.
123. Krishna, M., Burgart, L. J., French, A. J., Moon-Tasson, L. L., McDonnell, S. K., Schaid, D. J., and Thibodeau, S. N. (1997). Predictive value of histopathology for mutator phenotype (microsatellite instability) in colorectal carcinoma. *Gastroenterology Suppl.* **112,** 4595.
124. Rüschoff, J., Dietmaier, W., Lüttges, J., Seitz, G., Bocker, T., Zirngibl, H., Schlegel, J., Schackert, H. K., Jauch, K. W., and Hotstaedter, F. (1997). Poorly differentiated colonic adenocarcinoma, medullary type. Clinical, phenotypic, and molecular characteristics. *Am. J. Pathol.* **150,** 1815–1825.
125. Lynch, H. T., Alban, O. W.,Recabaren, J. A., Lynch, P. M., and Lynch, J. F. (1981). Prolonged survival as a component of hereditary breast and nonpolyposis colorectal cancer. *Med. Hypoth.* **7,** 1202–1209.
126. Sankila, R., Aaltonen, L. A., Järvinen, H. J., and Mecklin, J. P. (1996). Better survival rates in patients with MLH1-associated hereditary colorectal cancer. *Gastroenterology* **110,** 682–687.
127. Hayden, J. D., Cawkwell, L., Sue-Ling, H., Johnston, D., Dixon, M. F., Quirke, P., and Martin, I. G. (1997). Assessment of microsatellite alterations in young patients with gastric adenocarcinoma. *Cancer* **79,** 684–687.
128. Dos Santos, N. R., Seruca, R., Constância, M., Seixas, M., and Sobrino-Simões, M. (1996). Microsatellite instability at multiple loci in gastric carcinoma: clinicopathological implications and prognosis. *Gastroenterology* **110,** 38–44.
129. Strickler, J. G., Zheng, J., Shu, Q., Burgart, L. J., Alberts, S. R., and Shibata, D. (1994). p53 mutations and microsatellite instability in sporadic gastric cancer: when guardians fail. *Cancer Res.* **54,** 4750–4755.
130. Chong, J., Fukayama, M., Hayashi, Y., Takizawa, T., Koike, M., Konishi, M., Kikuchi-Yanoshita, R., and Miyaki, M. (1994). Microsatellite instability in the progression of gastric carcinoma. *Cancer Res.* **54,** 4595–4597.
131. Semba, S., Yokozaki, H., Yamamoto, S., Yasui, W., and Tahara, E. (1996). Microsatellite instability in precancerous lesions and adenocarcinomas of the stomach. *Cancer* **77,** 1620–1627.
132. Rhyu, M., Park, W., and Meltzer, S. J. (1994). Microsatellite instability occurs frequently in human gastric carcinoma. *Oncogene* **9,** 29–32.
133. Seruca, R., Santos, N. R., David, L., Constancia, M., Barroca, H., Carneiro, F., Seixas, M., Peltomäki, P., Lothe, R., and Sobrinho-Simoes, M. (1994). Sporadic gastric cancers with microsatellite instability display a particular clinicopathologic profile. *Int. J. Cancer* **64,** 32–36.
134. Wu, M., Sheu, J., Shun, C., Lee, W., Wang, J., Wang, T., Cheng, A., and Lin, J. (1997). Infrequent hMSH2 mutations in sporadic gastric adenocarcinoma with microsatellite instability. *Cancer Lett.* **112,** 161–166.
135. Battista, P., Palmirotta, R., Vitullo, P., Veri, M. C., Colalongo, C., Rigoli, L., Fedele, F., Caruso, R., Inferrera, C., Romano, F., Mariani-Costantini, R., Frati, L., and Cama, A. (1997). Microsatellite instability in early gastric cancer. *Int. J. Cancer* **10,** 65–70.
136. Lim, P. C., Tester, D., Cliby, W., Ziesmer, S. C., Roche, P. C., Hartmann, L., Thibodeau, S. N., Podratz, K. C., and Jenkins, R. B. (1996). Absence of mutations in DNA mismatch repair genes in sporadic endometrial tumors with microsatellite instability. *Clin. Cancer Res.* **2,** 1907–1911.

137. Caduff, R. F., Johnston, C. M., Svoboda-Newman, S. M., Poy, E. L., Merajver, S. D., and Frank, T. S. (1996). Clinical and pathological significance of microsatellite instability in sporadic endometrial carcinoma. *Am. J. Pathol.* **14,** 1671–1678.
138. Kobayashi, K., Matsushima, M., Sumiko, K., Saito, H., Sagae, S., Kudo, R., and Nakumura, Y. (1996). Mutational analysis of mismatch repair genes, hMLH1 and hMSH2, in sporadic endometrial carcinomas with microsatellite instability. *Jpn. J. Cancer Res.* **87,** 141–145.
139. Katabuchi, H., van Rees, B., Lambers, A. R., Ronnett, B. M., Blazes, M. S., Leach, F. S., Cho, K. R., and Hedrick, L. (1995). Mutations in DNA mismatch repair genes are not responsible for microsatellite instability in most sporadic endometrial carcinomas. *Cancer Res.* **55,** 5556–5560.
140. Peltomäki, P., Lothe, R. A., Aaltonen, L. A., Pylkkänen, L., Nyström-Lahti, M., Seruca, R., David, L., Holm, R., Ryberg, D., Haugen, A., Brogger, A., Børresen, A. L., and de la Chapelle, A. (1993). Microsatellite instability is associated with tumors that characterize the hereditary non-polyposis colorectal carcinoma syndrome. *Cancer Res.* **53,** 5833–5855.
141. Phillips, N. J., Zeigler, M. R., Radford, D. M., Fair, K. L., Steinbrueck, T., Xynos, F. P., and Donis-Keller, H. (1996). Allelic deletion on chromosome 17p13.3 in early ovarian cancer. *Cancer Res.* **56,** 606–611.
142. Sood, A. K., Skilling, J. S., and Buller, R. E. (1997). Ovarian cancer genomic instability correlates with p53 frameshift mutations. *Cancer Res.* **57,** 1047–1049.
143. Arzimanoglou, I. I., Lallas, T., Osborne, M., Barber, H., and Gilbert, F. (1996). Microsatellite instability differences between familial and sporadic ovarian cancers. *Carcinogenesis* **17,** 1799–1804.
144. Fujita, M., Enomoto, T., Yoshino, K., Nomura, T., Buzard, G. S., Inoue, M., and Okudaira, Y. (1995). MIcrosatellite instability and alterations in the *hMSH2* gene in human ovarian cancer. *Int. J. Cancer* (*Pred. Oncol.*) **64,** 361–366.
145. King, B. L., Carcangiu, M.-L., Carter, D., Kiechle, M., Pfisterer, J., Pfleiderer, A., and Kacinski, B. M. (1995). Microsatellite instability in ovarian neoplasms. *Br. J. Cancer* **72,** 376–382.
146. Orth, K., Hung, J., Gazdar, A., Bowcock, A., Mathis, J. M., and Sambrook, J. (1994). Genetic instability in human ovarian cancer cell lines. *Proc. Natl. Acad. Sci. USA* **91,** 9495–9499.
147. Han, H. J., Yanagisawa, A., Kato, Y., Park, J.-G., and Nakamura, Y. (1993). Genetic instability in pancreatic and poorly differentiated type of gastric cancer. *Cancer Res.* **53,** 5087–5089.
148. Kimura, M., Abe, T., Sunamura, M., Matsuno, S., and Horii, A. (1996). Detailed deletion mapping on chromosome arm 12q in human pancreatic adenocarcinoma: identification of a 1-cM region of common allelic loss. *Genes Chromosom. Cancer* **17,** 88–93.
149. Brentnall, T. A., Chen, R., Lee, J. G., Kimmey, M. B., Bronner, M. P., Haggit, R. C., Kowdley, K. V., Hecker, L. M., and Byrd, D. R. (1995). Microsatellite instability and K-*ras* mutations associated with pancreatic carcinoma and pancreatitis. *Cancer Res.* **55,** 4264–4267.
150. Yee, C. J., Roodi, N., Verrier, C. S., and Parl, F. F. (1994). Microsatellite instability and loss of heterozygosity in breast cancer. *Cancer Res.* **54,** 1641–1644.
151. Aaltonen, L. A., Peltomäki, P., Mecklin, J.-P., Järvinen, H., Jass, J. R., Green, J. S., Lynch, H. T., Watson, P., Tallqvist, G., Juhola, M., Sistonen, P., Hamilton, S. R., Kinzler, K. W., Vogelstein, B., and de la Chapelle, A. (1994) Replication errors in benign and malignant tumors from hereditary nonpolyposis colorectal cancer patients. *Cancer Res.* **54,** 1645–1648.
152. Togo, S., Toda, N., Kanai, F., Kato, N., Shiratori, Y., Kishi, K., Imazeki, F., Makuuchi, M., and Omata, M., (1996). Transforming growth factor β Type II receptor gene mutation common in sporadic cecum cancer with microsatellite instability. *Cancer Res.* **56,** 5620–5623.
153. Ishimaru, G., Adachi, J., Shiseki, M., Yamaguchi, M., Muto, T., and Yokota, J. (1995). Microsatellite instability in primary and metastatic colorectal cancers. *Int. J. Cancer* **64,** 153–157.
154. Akiyama, Y., Iwanaga, R., Ishikawa, T., Sakamoto, K., Nishi, N., Nihei, Z., Iwama, T., Saitoh, K., and Yuasa, Y. (1996). Mutations of the transforming growth factor-β type II receptor gene are strongly related to sporadic proximal colon carcinomas with microsatellite instability. *Cancer* **78,** 2478–2484.
155. Samowitz, W. S., and Slattery, M. L. (1997). Microsatellite instability in colorectal adenomas. *Gastroenterology* **112,** 1515–1519.
156. Nakashima, H., Mori, M., Mimori, K., Inoue, H., Baba, K., Shibuta, K., Kusumoto, H., Haraguchi, M., Ueo, H., and Akiyoshi, T. (1997). Microsatellite instability in Japanese colorectal carcinoma. *Oncol. Rep.* **4,** 387–389.
157. Mironov, N. M., Aguelon, M. A., Potapova, G. I., Omori, Y., Gorbunov, O. V., Klimenkov, A. A., and Yamasaki, H. (1994). Alteration of (CA)n DNA repeats and tumor suppressor genes in human gastric cancer. *Cancer Res.* **54,** 41–44.
158. Watanabe, M., Imai, H., Shiraishi, T., Shimazaki, J., Kotake, T., and Yatani, R. (1995). Microsatellite instability in human prostate cancer. *Br. J. Cancer* **72,** 562–564.
159. Gao, X., Wu, N., Grignon, D., Zacharek, A., Liu, H., Salkowski, A., Li, G., Sakr, W., Sarkar, F., Porter, A., Chen, Y., and Honn, K. (1994). High frequency of mutator phenotype in human prostatic adenocarcinoma. *Oncogene* **9,** 2999–3003.
160. Suzuki, H., Komiya, A., Aida, S., Akimoto S., Shiraishi T., Yatani, R., Igarashi, T., and Shimazaki, J. (1995). Microsatellite instability and other molecular abnormalities in human prostate cancer. *Jpn. J. Cancer Res.* **86,** 956–961.
161. Kagan, J., Pisters, L. L., Troncoso, P., Joe, Y., Parat, J. M., Babaian, R. J., and Von Eschenbach, A. C. (1994). Genetic instability in microsatellite sequences in prostate cancer. *Int. J. Oncol.* **5,** 921–924.
162. Egawa, S., Uchida, T., Suyama, K., Wang, C., Ohori, M., Irie, S., Iwamura, M., and Koshiba, K. (1995). Genomic instability of microsatellite repeats in prostate cancer: relationship to clinicopathological variables. *Cancer Res.* **55,** 2418–2421.
163. Paulson, T. G., Wright, F. A., Parker, B. A., Russack, V., and Wahl, G. M. (1996). Microsatellite instability correlates with reduced survival and poor disease prognosis in breast cancer. *Cancer Res.* **56,** 4021–4026.
164. Sourvinos, G., Kiaris, H., Tsikkinis, A., Vassilaros, S., and Spandidos, D. A. (1997). Microsatellite instability and loss of heterozygosity in primary breast tumours. *Tumor Biol.* **18,** 157–166.
165. Toyama, T., Iwase, H., Iwata, H., Hara, Y., Omoto, Y., Suchi, M., Kato, T., Nakamura, T., and Kobayashi, S. (1996). Microsatellite instability in in situ and invasive sporadic breast cancers of Japanese women. *Cancer Lett.* **108,** 205–209.
166. Ryberg, D., Lindstedt, B., Zienolddiny, S., and Haugen, A. (1995). A hereditary genetic marker closely associated with microsatellite instability in lung cancer. *Cancer Res.* **55,** 3996–3999.
167. Fong, K. M., Zimmerman, P. V., and Smith, P. J. (1995). Microsatellite instability and other molecular abnormalities in non-small cell lung cancer. *Cancer Res.* **55,** 28–30.
168. Pifarré, A., Rosell, R., Monzó, M., De Anta, J. M., Moreno, I., Sánchez, J. J., Ariza, A., Mate, J. L., Martínez, E., and Sánchez, M. (1997). Prognostic values of replication errors on chromosomes 2p and 3p in non-small-cell lung cancer. *Br. J. Cancer* **75,** 184–189.

169. Shridhar, V., Siegfried, J., Hunt, J., del Mar Alonso, M., and Smith, D. I. (1994). Genetic instability of microsatellite sequences in many non-small cell lung carcinomas. *Cancer Res.* **54,** 2084–2087.

170. Izumoto, S., Arita, N., Ohnishi, T., Hiraga, S., Taki, T., Tomita, N., Ohue, M., and Hayakawa, T. (1997). Microsatellite instability and mutated type II transforming growth factor-β receptor gene in gliomas. *Cancer Lett.* **112,** 251–256.

171. Orlow, I., Lianes, P., Lacombe, L., Dalbagni, G., Reuter, V. E., and Cordon-Cardo, C. (1994). Chromosome 9 allelic losses and microsatellite alterations in human bladder tumors. *Cancer Res.* **54,** 2848–2851.

172. Gonzalez-Zulueta, M., Ruppert, J. M., Tokino, K., Tsai, Y. C., Spruck, C. H., Miyao, N., Nichols, P. W., Hermann, G. G., Horn, T., Steven, K., Summerhayes, I. C., Sidransky, D., and Jones, P. A. (1993). Microsatellite instability in bladder cancer. *Cancer Res.* **53,** 5620–5623.

173. Ohyashiki, J. H., Ohyashiki, K., Aizawa, S. Kawakubo, K., Shimamoto, T., Iwaman, H., Hayashi, S., and Toyama, K. (1996). Replication error in hematological neoplasias: genomic instability in progression of disease is different among different types of leukemia. *Clin. Cancer Res.* **2,** 1583–1589.

174. Wada, C., Shionoya, S., Fujino, Y., Tokuhiro, H., Akahoshi, T., Uchida, T., and Ohtani, H. (1994). Genomic instability of microsatellite repeats and its association with the chronic myelogenous leukemia. *Blood* **12,** 3449–3456.

175. Uchida, T., Wada, C., Wang, C., Egawa, S., Ohtani, H., and Koshiba, K. (1994). Genomic instability of microsatellite repeats and mutations of H-, K-, and N-ras, and p53 genes in renal cell carcinoma. *Cancer Res.* **54,** 3682–3685.

176. Ishwad, C. S., Ferrell, R. E., Rossie, K. M., Appel, B. N., Johnson, J. T. Myers, E. N., Law, J. C. Srivastava, S., and Gollin, S. M. (1995). Microsatellite instability in oral cancer. *Int. J. Cancer* **64,** 332–335.

177. Rowley, H., Jones, A., Spandidos, D., and Field, J. (1996). Definition of a tumor suppressor gene locus on the short arm of chromosome 3 in squamous cell carcinoma of the head and neck by means of microsatellite markers. *Arch. Otolaryngol. Head Neck Surg.* **122,** 497–501.

178. Ogawara, K., Uzawa, K., Nakanishi, H., Yokoe, H., Wang, Z., Tanzawa, H., and Sato, K. (1997). Frequent microsatellite instability in oral cancer. *Oncol. Rep.* **4,** 161–165.

179. Meltzer, S. J., Yin, J., Manin, B., Rhyu, M., Cottrell, J., Hudson, E., Redd, J. L., Krasna, M. J., Abraham, J. M., and Reid, B. J. (1994). Microsatellite instability occurs frequently and in both diploid and aneuploid cell populations of Barrett's associated esophageal adenocarcinomas. *Cancer Res.* **54,** 3379–3382.

180. Muzeau, F., Fléjou, J-F., Belghiti, J., Thomas, G., and Hamelin, R. (1997). Infrequent microsatellite instability in oesophageal cancers. *Br. J. Cancer* **75,** 1336–1339.

181. Mironov, N. M., Aguelon A. M., Hollams, E., Lozano, J.-C., and Yamasaki, H. (1995). Microsatellite alterations in human and rat esophageal tumors at selective loci. *Mol. Carcinogen.* **13,** 1–5.

182. Gleeson, C. M., Sloan, J. M., McGuigan, J. A., Ritchie, A. J., Weber, J. L., and Russell, S. E. H. (1996). Ubiquitous somatic alterations at microsatellite alleles occur infrequently in Barrett's associated esophageal adenocarcinoma. *Cancer Res.* **56,** 259–263.

183. D'Errico, M., Calcagnile, A. S., Corona, R., Fucci, M., Annessi, G., Baliva, G., Tosti, M. E., Pasquini, P., and Dogliotti, E. (1997). p53 mutations and chromosome instability in basal cell carcinomas developed at an early or late age. *Cancer Res.* **5,** 747–752.

184. Louhelainen, J., Szyfter, K., Szyfter, W., and Hemminki, K. (1997). Loss of heterozygosity and microsatellite instability in larynx cancer. *Int. J. Oncol.* **10,** 247–252.

185. Keller, G., Rotter, M., Vogelsang, H., Bischoff, P., Becker, K.-F., Siewert, J. R., and Höfler, H. (1995). Microsatellite instability in adenocarcinomas of the upper gastrointestinal tract. *Am. J. Pathol.* **147,** 593–600.

186. Burks, R. T., Kessis, T. D., Cho, K., and Hedrick, L. (1994). Microsatellite instability in endometrial cancer. *Oncogene* **9,** 1163–1166.

Part XVIII

Overview and Summary

CHAPTER 50

Concluding Remarks

ROBERT D. WELLS Department of Biochemistry and Biophysics, Institute of Biosciences and Technology, Texas A&M University, Texas Medical Center, Houston, Texas 77030

STEPHEN T. WARREN Department of Biochemistry, Emory University, Howard Hughes Medical Institute, Atlanta, Georgia 30322

This book is a comprehensive review of the clinical observations, human genetics, biochemistry, molecular biology, and biophysics of human hereditary neurological diseases associated with the expansion of simple repeat sequences. Superbly qualified experts have thoroughly reviewed the subjects on a range of diseases and their underlying mechanisms. Hence, the Editors will only try to identify some of the key outstanding problems that remain in this field.

First, a detailed understanding of the molecular and cellular biological mechanisms involved in the genetic instabilities (expansions and deletions) of the triplet repeat sequences is fundamental for our comprehension of the disease etiologies. Several chapters (33–37) deal with these topics in detail and we may predict that excellent progress will be made in our understanding in the relatively near future. Second, the mechanisms of oligoglutamine expansions which affect protein conformation, enzymatic activities, and protein–protein associations represent an area that will require substantially more attention in the future. In fact, it may be notable that this book does not contain a chapter on the biophysical consequences of oligoglutamine expansion. Third, a detailed understanding of the mechanism of pathogenesis for each of the diseases as a consequence of DNA and/or protein expansions is an area that will require substantial additional work in the future. The excellent chapter (47) by Ross *et al.* lays the groundwork for future investigations.

Fourth, all of these developments in the area of molecular mechanisms will be of benefit for the development of suitable diagnoses and therapeutics which will be applied clinically. In certain cases, gene therapy approaches or other intervention strategies may be useful.

A book of this type raises many more questions than it provides answers. Some of the questions that can be raised include the following:

· Why are the repeat units multiples of three (i.e., triplet repeats) rather than repeats of 4-mers, 5-mers, etc.? It may be noted that myoclonus epilepsy is related to an expansion of a repeating 12-base-pair sequence (see Chapter 48).

· For diseases involving coding repeats, why is the expanded sequence an oligoglutamine rather than an oligoserine or oligoalanine, which would be caused by a shift in the reading frame? It may be noted that synpolydactyly is associated with an expansion of an alanine tract.

· Why are these triplet repeat diseases neurological in nature?

· Why are only humans afflicted by TRS diseases? To date, no animal diseases have been attributed to TRS expansions.

· Why is there an effect of sex on the transmission of some of these diseases (see Section V, Myotonic Dystrophy)?

· Why are only 3 of the 10 possible triplet repeat sequences identified with these diseases? Might at least some of the other 7 TRS be shown in the future to be involved with other diseases?

· How many more diseases exist that are associated with expansions of simple repeat sequences? Note the chapters (Section XI) on candidate disorders revealing anticipation. How many psychiatric syndromes will be associated with triplet repeat expansions in the future?

· Why are different TRS associated with different diseases [i.e., GAA · TTC for FRDA and CTG · CAG for several other diseases (HD, DM, etc.)]. A corollary question is why does one TRS (CTG · CAG) have different mechanisms of action (glutamine expansion for Kennedy's disease, etc., but altered gene expression for DM)?

· What is the role of the flexible and writhed DNA structure for CTG · CAG and CGG · CCG for the associated diseases; likewise, what role does a triplex or other alternate conformation play in the etiology of FRDA?

· Are the TRS replicated and transcribed by the currently known polymerases or might other enzymes be involved?

· When during the cellular life-cycle does the TRS expansion occur?

There are perhaps a few major issues that, when answered, will illuminate a great deal about the basis for dynamic mutations. The first is precisely the mechanism by which certain alleles (i.e., those of affected families) undergo extremely frequent change of repeat length while more common normal alleles are quite stable. It is uncertain if this expansion occurs during scheduled replication, such as during S-phase, or is the result of a gene conversion or repair pathway. Whenever expansion occurs, certain *cis*-acting signals must distinguish an unstable allele from a stable allele. While length and purity of the repeat certainly play a major role, there may well be other *cis* elements which could influence stability. For example, studies in nonmammalian models have clearly implicated the origin of replication as critical as have studies in fragile X syndrome. Could there be normal variation in the position or strength of replication origins flanking repeat loci which could modify instability?

Besides *cis*-acting elements, there is also evidence for *trans*-acting influences. For example, instability in human cells appears limited to germline (often spermatogenesis) and/or early development. Are there proteins or other factors which are uniquely present or absent during these critical phases of instability? Moreover, population variation in these factors or other factors, such as those involved in mismatch repair, could lead to individuals who exhibit higher instability rates at multiple loci or at least at those loci with already primed alleles. Much of the uncertainty about both *cis*- and *trans*-acting elements is the result of our incomplete knowledge of the subtleties of DNA replication in mammalian cells. While further research into basic aspects of mammalian DNA metabolism will certainly enhance our abilities to understand repeat instability, it may also be true that further inquiry into repeat instability will reveal novel insights into this fundamental biological process itself.

Another surely exciting topic for future studies will be the mechanism by which expanded polyglutamine tracts confer novel attributes to certain proteins. Intriguing recent data have implicated nuclear localization along with faulty degradation. The accumulated polyglutamine protein or processed peptide appears to form a relatively stable aggregate in cells destined for degeneration. The similarity with other neurodegenerative processes involved in Alzheimer's disease and the prion diseases is startling and may reveal a general process of neurodegeneration. Since the faulty degradation of expanded polyglutamine proteins appears to be cell-type-specific, where certain cells apparently are able to process the protein normally, there may be cell-type-specific activities which can promote or prevent accumulation of the abnormal protein. Although still much in the future, this scenario has obvious therapeutic implications if one were able to inhibit the activity responsible for leading to accumulation and subsequent cell death.

Many additional questions may also be identified. However, we hope that this book will serve as a stimulus for attracting the interest of other talented and dedicated researchers who may make seminal discoveries on these disorders in the future. This can only be of benefit to science as well as to the patients and families who must endure the devastating consequences of these diseases.

Index

A

D

G

H

I

K

L

M

N

O

P

R

S

T

U

Y

ISBN 0-12-742935-2